UNIVERSITY OF MAINE

RAYMOND H. FOGLER LIBRARY

Proceedings of the
XXVI International Conference on
High Energy Physics

August 6 – 12, 1992 – Dallas, Texas

Proceedings of the XXVI International Conference on High Energy Physics
Volume I

August 6 – 12, 1992 – Dallas, Texas

Editor
James R. Sanford

Sponsored by the Superconducting Super Collider Laboratory
Universities Research Association, Inc.

AIP CONFERENCE PROCEEDINGS NO. 272

Authorization to photocopy items for internal or personal use, beyond the free copying permitted under the 1978 U.S. Copyright Law (see statement below), is granted by the American Institute of Physics for users registered with the Copyright Clearance Center (CCC) Transactional Reporting Service, provided that the base fee of $2.00 per copy is paid directly to CCC, 27 Congress St., Salem, MA 01970. For those organizations that have been granted a photocopy license by CCC, a separate system of payment has been arranged. The fee code for users of the Transactional Reporting Service is: 0094-243X/87 $2.00.

© 1993 American Institute of Physics.

Individual readers of this volume and nonprofit libraries, acting for them, are permitted to make fair use of the material in it, such as copying an article for use in teaching or research. Permission is granted to quote from this volume in scientific work with the customary acknowledgment of the source. To reprint a figure, table, or other excerpt requires the consent of one of the original authors and notification to AIP. Republication or systematic or multiple reproduction of any material in this volume is permitted only under license from AIP. Address inquiries to Series Editor, AIP Conference Proceedings, AIP, 335 East 45th Street, New York, NY 10017-3483.

L.C. Catalog Card No. 93-70412
ISBN   1-56396-125-3 (Vol. I)
          1-56396-126-1 (Vol. II)
          1-56396-127-X (Set)
DOE CONF-920837

Printed in the United States of America.

Table of Contents

VOLUME I

Foreword	xxiii
Conference Organization	xxv
Opening Session	xxvii

PLENARY SESSIONS

Weak Decays, Rare Decays, Mixing, and CP Violation 3
    P. S. Drell
Hadron Spectroscopy and Structure 33
    N. Isgur
Precision Tests of the Electroweak Interaction 56
    L. Rolandi
Tests of QCD 81
    S. Bethke
Hadron and Photon Production of Heavy Quarks 114
    J.N. Butler
Neutrino Mass and Mixing, and Non-Accelerator Experiments 140
    R.G.H. Robertson
Cosmology and Astrophysics 1992 158
    L.M. Krauss
Beyond the Standard Model
    M. Peskin (Paper was not received.)
Nonperturbative Methods 185
    H. Leutwyler
Ultra-Relativistic Heavy-Ion Collisions: Searching for the
Quark-Gluon Plasma 212
    J. Schukraft
Progress in Quantum Field Theory and String Theory 229
    L. Alvarez-Gaumé
Lattice Gauge Theories 241
    R. Petronzio
Recent Developments in Detector Technology 253
    T. Kondo
Status of HERA and the First Results 278
    B.H. Wiik

Future Hadron Collider: The SSC 306
    R.F. Schwitters
Future Hadron Collider: The LHC 321
    C. Rubbia
Electron-Positron Colliders and Other Accelerator Technologies 334
    R.H. Siemann
Conference Summary 346
    S. Weinberg

PARALLEL SESSIONS

1. Weak Decays

The b-Quark Semileptonic Branching Ratio at LEP 367
    R. Clare
Inclusive Measurements of the B-Hadron Lifetime at LEP 373
    M. Pohl
B^0 and B^+ Lifetime Measurements at LEP 379
    M. Feindt
Measurement of Lifetimes of Charged and Neutral Beauty Hadrons from
Fermilab E653 383
    N.R. Stanton
Semileptonic Charmed Meson Decays 388
    G. Bellini
Inclusive Photon Energy Spectrum in B Decays 393
    C. Greub
Selected ARGUS Results on B Meson Decays 397
    Y. Zaitsev
Inclusive Decays of Beauty (& Charm): QCD vs.
Phenomenological Models 402
    I.I. Bigi
Heavy Quark Effective Theory: Applications to Weak Decays 408
    B. Grinstein
Effective Hamiltonian for $\Delta S = 1$ Non-Leptonic Decays Beyond
Leading Logarithms 414
    M. Jamin
Measurements of the Kaon Content in Tau Decays 418
    M.T. Ronan
Measurement of the Mass of the τ Lepton 422
    F.C. Porter
New Results on τ Lepton from CLEO 427
    K.K. Gan
Precise Determination of α_S from τ Decays 435
    E. Braaten
Topological and Hadronic Decays of the Tau 439
    J.M. Roney
Proposed Explanation of Tau Lepton Decay Puzzle: Discrepancy Between
the Measured and the Theoretical Tau Lifetimes 444
    C.K. Jung

2. CP Violation, $B\bar{B}$ Mixing, and Rare Decays

A Measurement of B^0-$\overline{B^0}$ Mixing in Hadronic Z^0 Decays 453
   V. Gibson
ALEPH Results on $B^0\bar{B}^0$ Mixing 458
   J.-P. Lees
A Measurement of B^0-$\bar{B}^0$ Mixing in Z Decays with the L3 Detector 463
   G.J. Bobbink
A Measurement of $B\bar{B}$ Mixing with the DELPHI Detector at LEP 468
   S. Franco
Results on $B\bar{B}$ Mixing and Rare B Decays from CLEO 472
   H. Kroha
Physics at DAΦNE 480
   J. Lee-Franzini
B^0-$\bar{B}^0$ Mixings and Rare B-Decays 484
   A. Ali
Rare Kaon and $b \to s\gamma$ Decays in the Two-Higgs-Doublet Model 492
   C.Q. Geng
Theory of CP-Violation in $K \to 3\pi$ Decays 497
   D. Ebert
Status of Rare K Decays 500
   J.L. Ritchie
Recent Theoretical Development on Direct CP Violation ε'/ε 506
   Y.-L. Wu
Test of CP Violation Using K^0-$\overline{K^0}$ Interferometry 510
   C. Guyot
A Measurement of the Decay $K_L \to \pi^0\gamma\gamma$ by the NA31 Experiment 516
   L. Fayard
Results and Plans for the Fermilab K^0 Decay Program 520
   Y.W. Wah
CP Violation in the Decay $K_L \to \pi^+\pi^-e^+e^-$ 526
   L.M. Sehgal
Theoretical Aspects of $K \to \pi e^+e^-$ and $K \to \pi\gamma\gamma$ 528
   T. Morozumi
Search for Muon Number Violating Decays 532
   H.K. Walter
Measurement of Direct CP-Violation with the Experiments NA31 and
NA48 at CERN 537
   B. Renk

3. Light Quark and Gluonia Spectroscopy

Introductory Remarks: Light Quark and Gluonia Spectroscopy 543
   F.E. Close
A Study of the $E/f_1(1420)$ and of the $f_1(1285)$ in Central Production 544
   M.F. Votruba
New Light Mesons from the Lear Crystal Barrel 548
   C. Amsler

Neutral Light Quark Spectroscopy in Antiproton-Proton Annihilation
at $3000 < \sqrt{s} < 6000$ MeV 552
    G.A. Smith
Results on $f_0(975)$ and $\xi(2.2)$ from BES 556
    Z. Zhipeng
How to Determine the Substructure of the Scalar Mesons at
Around 1 GeV in Mass 562
    F.E. Close
New Two-Photon Results from ARGUS 568
    G. Kernel
Study of the Wave with $J^{PC} = 1^{-+}$ in the Partial-Wave Analysis
of $\eta'\pi^-$, $\eta\pi^-$, $f_1\pi^-$ and $\rho^0\pi^-$ Systems Produced in π^-N Interactions
at $p_{\pi^-} = 37$ GeV/c 572
    Yu.P. Gouz
Spectroscopy of the D-Wave $q\bar{q}$ System; Evidence for Two $J^P = 2^-$ Strange
Meson States Decaying to $K^-\omega$ 577
    B.N. Ratcliff
Another Dirac Oscillator 582
    L.M. Jones
Evidence for the Box Anomaly in η and η' Decays 585
    M. Feindt
Production of Light Quark Resonances in Z^0 Decays 589
    M. Dracos

4. Heavy Quark and Quarkonium States

Evidence for Λ_b Production at LEP 597
    X.C. Lou
Production of $B \to J/\Psi$ in Z° Hadronic Decays 602
    A.M. Segar
Effective Heavy Quark Theory: Matching HQET and Quark Models 608
    F.E. Close
Leading and Subleading Logarithmic QCD Corrections to Bilinear
Heavy Quark Currents 610
    T. Mannel
Constraints on Heavy Meson Form Factors 614
    J. Taron
Implications of Heavy Quark Symmetry and Chiral Dynamics 618
    H.-Y. Cheng
Rigorous QCD Analysis of P-Wave Charmonium Decays 623
    E. Braaten
Recent Results from CLEO on Charm and Bottom Decays 627
    V. Jain
The DØ Muon System and Early Results on its Performance 633
    D. Hedin

5. Electroweak Interactions

LEP Energy Calibration and M_Z 639
   G. Quast
Lepton Forward-Backward Asymmetries 645
   R. Pain
Measurements at LEP of the Forward-Backward Asymmetries of Quarks 651
   T. Wyatt
Measurement of τ Polarisation in Z^0 Decays 657
   K.S. Kumar
Lifetime and Leptonic Branching Ratios of the Tau Lepton 664
   M. McCubbin
A Novel Method to Measure $\Gamma(Z^0 \to b\bar{b})/\Gamma(Z^0 \to \text{hadrons})$ 670
   C. Moisan
Determination of the Number of Light Neutrino Families from
$e^+e^- \to \nu\bar{\nu}\gamma$ 675
   S. Gentile
Electroweak Results from LEP 681
   R. Tanaka
Precise Measurement of the Luminosity at LEP 687
   H. Meinhard
Monte Carlo Program BHLUMI 2.01 for *Bhabha* Scattering at
Low Angles with Yennie-Frautschi-Suura Exponentiation 692
   B.F.L. Ward
ZFITTER: An Analytical Program for Fermion-Pair Production 698
   T. Riemann
First W Decays Observed With the DØ Detector 703
   B. Gobbi
First Measurement of the Left-Right Cross Section Asymmetry in
Z Boson Production at $E_{CM} = 91.55$ GeV 708
   P.C. Rowson
Precision Measurement of the Total Cross Section and Charge Asymmetry
at TOPAZ for $e^+e^- \to \mu^+\mu^-$ and $e^+e^- \to \tau^+\tau^-$ 716
   D.S. Koltick
Precision Measurement of $\sin^2\theta_w$ from νFe Scattering at the Tevatron 721
   T. Bolton
Neutral Current Coupling Constants from $\nu_\mu e^-$ and $\bar{\nu}_\mu e^-$ Scattering 727
   A. Staude
Results from the KARMEN Neutrino Experiment 732
   J.A. Edgington

6. Structure Functions and Spin Physics

NMC Results on Structure Functions 739
   E. Rondio
Preliminary QCD Analysis of Nucleon Structure Functions F_2^p and F_2^d
from NMC 745
   M. Virchaux

Structure Functions, Global QCD Analysis, and Parton Distributions 751
    W.-K. Tung
Measurement of the Cross-Section Ratio σ_n/σ_p in Inelastic Muon-Nucleon
Scattering at Very Low x and Q^2 760
    V. Papavassiliou
Electroproduction Structure Function F_2 in the Low Q^2, Low X Region 765
    B. Badelek
Spin and Strangeness—Open Issues 770
    R.D. Carlitz
Polarized Photon or Proton Primakoff Effect 778
    J. Bernabéu
Deuteronlike Meson-Meson Bound States 784
    N.A. Törnqvist
Constituent Quarks as Skyrmions in Color Space 788
    M. Karliner
Neutrino Production of Dimuons at the Fermilab Tevatron 795
    M.H. Shaevitz
Strangeness Production in Neutrino Interactions 802
    M. Kalelkar
Nuclear Parton Distributions and Extraction of Neutron
Structure Functions 806
    M. Strikman
Measurements of $\gamma\gamma$ Collisions with the OPAL Detector at LEP and
Comparison with QCD Models 812
    J.G. Layter

7. Jets, Fragmentation, Tests of QCD

Large p_T Production of Direct Photons and π^0 Mesons at 500 GeV/c 819
    M. Zieliński
Heavy Flavor Production in π^-—A Collisions at 530 GeV/c 824
    R. Jesik
Polarization Density Matrix at $O(\alpha_s^2)$ of High Q_T $W'S$ in
Hadronic Collisions 829
    E. Mirkes
QCD Tests with CDF 833
    B. Flaugher
First QCD Results from the D0 Detector 842
    H. Weerts
Recent Theoretical Developments for Hadron Collisions 846
    R.K. Ellis
Constraints on the Gluon Density from Bottom Quark and
Prompt Photon Production 853
    R. Meng
Heavy Quarks and QCD in e^+e^- Collisions 859
    T. Behnke
Measurement of the Strong Coupling Constant α_s at LEP 865
    K. Mönig

Multi-Jet Production Rates in Deep-Inelastic Muon-Proton Scattering 872
   C.W. Salgado
Recent Progress in QCD Jet Physics in e^+e^- Collisions 878
   B.R. Webber
QCD Tests with Final State Photons in Hadronic Events at LEP 885
   G. Gratta
QCD Studies of Hadronic Decays of Z^0 Bosons by SLD 892
   D. Muller
Tests of Color Coherence in e^+e^- Collisions 899
   R. Settles
Theoretical Status of α_s Determination 905
   C.J. Maxwell

8. Soft Hadronic Phenomena

Intermittency and Correlations 913
   E.A. De Wolf
Intermittency and Particle Correlations: Theory 919
   I. Sarcevic
A Search for the Signature of a Deconfined Quark-Gluon Phase of
Strongly Interacting Matter in $\bar{p}$-p Interactions at $\sqrt{s} = 1.8$ TeV 926
   L.J. Gutay
Inclusive Particle Production in e^+e^- 931
   M.N.K. Focacci
Resonance Production in e^+e^- Collisions 935
   G.D. Lafferty
Reaction Rates in a Heat Bath 940
   M. Jacob
Diffraction at Collider Energies 943
   L.L. Frankfurt
The Physics of Leading Particles 947
   W.D. Walker
Measurement of the Proton Electromagnetic Form Factors in the
Time-Like Region at 8.9 to 13.0 GeV^2 951
   D. Bettoni
Recent Results on Bose-Einstein Correlations 955
   B. De Lotto
Multiplicity Distributions in e^+e^- and Hadron-Hadron Collisions 960
   R.D. St. Denis

9. Heavy Ion Interactions

Theoretical Overview: Light Ion Lessons, Heavy Ion Hopes 969
   S. Gavin
Thermalization in Ultra-Relativistic Heavy Ion Collisions
at RHIC and LHC 977
   K. Geiger
Strange Anti-Baryons—QGP Versus HG 983
   J. Rafelski

Final State J/ψ Suppression in Nuclear Collisions 991
   H. Satz
Strangeness Signals in Heavy Ion Collisions 1000
   L.P. Remsberg
Antibaryon Production in Relativistic Heavy Ion Collisions 1006
   B.S. Kumar
Dimuon Production in p-W and S-W Collisions 1013
   U. Goerlach
Single Particle Spectra and Two Particle Correlation from NA44,
"The Focusing Spectrometer" at the CERN-SPS 1019
   A. Franz

10. Photon and Hadron Production of Heavy Quarks

A Technique for Observing the Top Quark and Measuring its Mass
at the Tevatron 1027
   G.R. Goldstein
Threshold Effects on Top Production in $\gamma\gamma$ Interactions 1031
   M.C. Gonzalez-Garcia
Systematics of Charm Production in Hadronic Collisions 1035
   P. Hoyer
Photoproduction of Charm Mesons 1038
   R. Gardner
Rare Decay Modes of the D^0, D^+, and D_S Charmed Mesons 1042
   N.M. Cason
Charm Meson Production in 600 GeV/c Pion-Emulsion Interactions 1046
   D.M. Potter
Preliminary Results from Fermilab E789 1050
   J.C. Peng
Feynman-x and Transverse Momentum Dependence of $D^\pm$ and D^0,
$\bar{D}^0$ Production in 250 GeV π^--Nucleon Interactions 1054
   P.E. Karchin
Fermilab E791 1058
   L.M. Cremaldi
Hadroproduction of χ_c States in 530 GeV/c π^- Interactions
with Nuclear Targets 1062
   A. Zieminski
First Observation of Decay Asymmetry and Some Decay Modes of Ξ_c
Produced by Neutrons at Serpukhov Accelerator 1066
   V.D. Kekelidze
Results on Charm Hadroproduction from CERN Experiment WA82 1070
   F. Antinori
Recent ARGUS Results on Charmed Baryon Physics 1076
   J. Stiewe
High Energy Photoproduction of Charm Baryons 1082
   H.W.K. Cheung
Inclusive J/ψ, $\psi(2S)$ and b-Quark Production in $\bar{p}p$ Collisions
at $\sqrt{s} = 1.8$ TeV 1086
   V. Papadimitriou

11. Neutrino Masses and Mixing

Solar Neutrinos Observed by GALLEX at Gran Sasso 1093
   D. Vignaud
Latest Results from the Soviet-American Gallium Experiment 1101
   V.N. Gavrin
What Do Solar Models Tell Us About Solar Neutrino Experiments? 1111
   J.N. Bahcall
MSW Implications of Solar Neutrino Experiments 1117
   S.P. Rosen
A Massive Neutrino in Nuclear Beta Decay? 1123
   E.B. Norman
0.073% (95% C.L.) Upper Limit on 17 keV Neutrino Admixture 1128
   T. Ohshima
An Experiment to Search for a 17 keV Neutrino 1136
   R. Shrock
Recent Double Beta Decay Results 1141
   A. Piepke
The Double Beta Decay Spectra of ^{82}Se, ^{100}Mo, and ^{150}Nd 1148
   M.A. Nelson
What Neutrinoless Double Beta Decay Would Tell Us
About Neutrino Mass 1153
   B. Kayser
Search for Isosinglet Neutral Heavy Lepton with the L3 Detector at LEP 1160
   S. Shevchenko
Hints for Neutrino Masses: A Theoretical Overview 1165
   J.W.F. Valle
Baryogenesis and Neutrino Masses 1171
   R.D. Peccei
Stringy Origin of Neutrino Masses Within the Minimal Supersymmetric
Standard Model 1178
   M. Cvetič

VOLUME II

12. Particle Astrophysics and Non-Accelerator Experiments

Recent Results from Kamiokande on Solar and Atmospheric Neutrinos 1187
   T. Kajita
Axial Strangeness Current Effects in Solar Neutrino Detection 1193
   J. Bernabéu
Recent Results from the CYGNUS Experiment and Plans for the
MILAGRO Experiment 1197
   D.A. Williams
A Search for Astrophysical Point Sources of 100 TeV Gamma Rays
by the UMC Collaboration 1203
   T.A. McKay

Search for UHE γ Sources with the HEGRA Detector 1208
    M. Merck
Detection of TeV Photons from the Active Galaxy Markarian 421 1214
    C.W. Akerlof
Detection of Very High Energy Gamma Rays from the CRAB Source 1218
    P. Baillon
A MACRO Sampler 1222
    M.J. Longo
Strange Quark Matter Search Using the MACRO Detector 1228
    G. Liu
Results from the Soudan 2 Detector 1232
    J.L. Thron
Underground Muons from Point Sources 1238
    M.L. Marshak
Status of the LVD Experiment at Gran Sasso Laboratory 1242
    I.A. Pless
Status of the Lake Baikal Neutrino Detector 1246
    R. Wischnewski
AMANDA South Pole Neutrino Detector 1250
    S. Barwick
A Search for Neutrinoless Double Beta Decay of ^{130}Te with a
Low Temperature Calorimeter 1254
    A. Giuliani
A Pilot Dark Matter Particle Search Experiment 1260
    B.L. Dougherty
Coherent Production/Detection of Light Scalars/Pseudoscalars and
QED Vacuum Polarization 1266
    Y.K. Semertzidis

13. Search for New Particles

CDF Top Search 1273
    J. Huth
Search for Exotic Particles at CDF 1279
    M.S. Gold
Search for New Particles 1285
    H.B. Prosper
Search for New Particles at the $\bar{p}p$ CERN Collider 1290
    M. Nessi
A Study of Double Vertex Events in the Neutrino-Nucleon Interactions 1295
    P. de Barbaro
Search for Leptoquarks at ZEUS 1301
    I. Gialas
Searches for Compositeness at e^+e^- Colliders 1304
    L.F. Thompson
Searches for SUSY Particles at LEP 1309
    G. Wormser
Rare Decays of the Z at LEP 1315
    T. Azemoon

Searches for Standard Model Neutral Higgs Boson at LEP 1321
  T. Mori
A Search for Magnetic Monopoles at LEP 1326
  K. Kinoshita

14. Beyond the Standard Model

Technicolor Phenomenology and Models 1333
  T. Appelquist
Walking Gauge Dynamics and Realistic Technicolor 1342
  R. Sundrum
Complementarity of Resonant and Nonresonant Strong WW Scattering
at SSC and LHC 1347
  M.S. Chanowitz
Observing Electroweak Symmetry Breaking at the SSC 1355
  J.A. Bagger
The Triviality Bound on the Higgs Mass; Its Value and What It Means 1360
  H. Neuberger
Complete One-Loop Analysis of the Minimal Supersymmetric
Higgs Sector 1368
  S. Pokorski
Search for SUSY Higgs Bosons at Future Hadron Colliders 1374
  Z. Kunszt
Higgs Boson Bounds in Non-Minimal Supersymmetric Standard Models 1381
  M. Quirós
String Theory: Lessons for Low Energy Physics? 1386
  M. Dine
Proton Decay and Cosmology Strongly Constrain the Minimal SU(5)
Supergravity Model 1395
  J.L. Lopez
Loops, Cutoffs and Anomalous Gauge Boson Couplings 1401
  D. London
Signatures for Heavy Z' Bosons at Hadron Colliders 1409
  M. Cvetič
A Predictive Ansatz for Fermion Masses 1414
  S. Raby
Top Quark Condensate Models of Electroweak Symmetry Breaking 1424
  W.A. Bardeen
Trinification and the Strong P Problem 1432
  E.D. Carlson
The μ Problem and the Invisible Axion 1436
  J.E. Kim

15. Developments in Field Theory and String Theory

Surface Tension in Hot QCD, Effective Potential and Hot Zero Modes 1443
  C.P. Korthals Altes
Quantum Corrections to Deep Bags 1448
  S.G. Naculich

Three Dimensional Quantum Chromodynamics 1453
   S.G. Rajeev
High Temperature Partition Function of the Rigid String 1457
   Z. Yang
A More Effective Potential 1461
   K. Cahill
Anyons on a Torus 1466
   Y. Hosotani
Supersymmetric Chern-Simons Vortex Systems and Fermion Zero Modes 1470
   B.-H. Lee
Fractional Charge in Perspective 1475
   A.S. Goldhaber
Stability of Vacua and Domain Walls in Supergravity and
Superstring Theory 1479
   M. Cvetič
Non-Perturbative Interactions in String Theory 1485
   B.A. Ovrut
Some Results for Strings at $D > 1$ 1491
   L. Alvarez-Gaumé
The W-String Spectrum 1495
   K.S. Stelle
Duality Symmetries in Calabi-Yau Compactifications 1501
   A. Font
Ramond-Ramond Bosons in Superstring Theory 1505
   L. Dolan
Renormalization Group Coefficients for the Un-Truncated Derivative
Expansion in Effective Field Theory 1510
   V. Bhansali

16. Lattice Gauge Theory

Recent Results from the Columbia Machine 1517
   N.H. Christ
Determining α_s Using Lattice Gauge Theory 1523
   A.X. El-Khadra
The Search for Chiral Fermions from the Lattice: A Status Report 1529
   D.N. Petcher
The Nature of the Continuum Limit in Strongly Coupled Quenched QED 1539
   M.-P. Lombardo
Topological Structure of Lattice QCD Near the Chiral Phase Transition 1545
   M. Müller-Preussker
Properties of Gluon Plasma Bubbles 1549
   M.C. Ogilvie

17. Non-Perturbative Methods

Chiral Structure of the Nucleon 1555
   U.-G. Meißner

Quark Masses, Kaon Masses and $\eta \to 3\pi$ 1562
    J.F. Donoghue
The Low-Energy Effective Action of QCD and
Nambu-Jona-Lasinio Models 1566
    E. de Rafael
Anomalous Gauge-Boson Couplings at Hadron Supercolliders 1572
    G. Valencia
On the Profiles of Jets Initiated by Light and Heavy Quarks 1578
    V.A. Khoze
Baryogenesis at the Electroweak Phase Transition 1583
    D.E. Brahm

18. Astrophysics and Cosmology

COBE DMR Observations of Early Universe Physics 1591
    G.F. Smoot
Inflation, Large-Scale Structure, and COBE 1602
    J.A. Frieman
N-Body Simulations of Cold Dark Matter 1608
    J.M. Gelb
Electroweak Strings: A Brief Overview 1614
    T. Vachaspati
Baryogenesis at the Electroweak Phase Transition: An Overview of
Recent Results 1619
    R.G. Leigh
γ-Ray Bursts and Neutron Star Mergers—Possibly the Strongest
Explosions in the Universe 1626
    T. Piran
Neutrino Physics and Astrophysics 1634
    K. Enqvist
On Axion Driven Baryogenesis 1640
    E.I. Guendelman
Neutralinos as Dark Matter in the Minimal Supergravity Model 1643
    S. Pokorski
Hadronic Instabilities in Very Intense Magnetic Fields 1649
    M. Bander

19. Experimental Techniques

The ICARUS Experiment: Status and Program 1657
    A. Bettini
The First Results from the CRID Detector at SLD 1664
    J. Va'vra
The RICH Counter in the CERN Hyperon Beam Experiment 1671
    H.-W. Siebert
Beam Tests of the ZEUS Barrel Calorimeter 1675
    H.J. Kim
The Accordion Liquid-Argon Project at CERN 1679
    M. Nessi

Tests of the DØ Calorimeter Response in 2–150 GeV Beams 1685
   K. De
Preliminary Results of a Noble Liquid Test Electromagnetic Calorimeter
for GEM 1690
   D. Lissauer
The Limited Streamer Tube System of H1 1694
   J. Tutas
Operation of Thin Effective Substrate Microstrip Gas
Avalanche Chambers 1700
   J.B. Dainton
Proportional Mode Operation of a Gas Pixel Device 1705
   R. Lander
GaAs Solid State Detectors for High Energy Physics 1709
   C. Buttar
The DELPHI Microvertex Detector 1714
   M. Caccia
Tracking and Radiation Tests of Silicon Microstrip Detectors 1721
   P. Skubic
A Silicon Vertex Detector for CLEO 1727
   D. Cinabro
D0 Triggering and Data Acquisition 1732
   B. Gibbard
The Optical Trigger for Beauty Research, Recent Developments
and Perspectives 1738
   Y. Giomataris
Implementation of a 66 MHz Analog Memory as a Front End
for LHC Detectors 1743
   R. Bonino
High-Performance Computing and Distributed Systems 1748
   S.C. Loken
A New Approach to Distributed Computing in High Energy Physics 1753
   P. Avery
Object Oriented Approach to B Reconstruction 1758
   N. Katayama

20. Physics Simulation Methods

Theoretical Basis for QCD Monte Carlo Simulations 1765
   G. Marchesini
Comparisons of Properties of Hadronic Z^0 Decays with
QCD Shower Models 1771
   W. Zuener
TAUOLA Monte Carlo for τ Decays 1777
   Z. Wąs
Physics Simulations at HERA 1781
   H. Jung
The Fast H1 Detector Monte Carlo 1787
   M. Kuhlen

Fourier Parametrization of the Multi-Dimensional
Monte-Carlo Efficiency 1791
    S.A. Sadovsky
CDF and Physics Simulation 1795
    H. Grassmann
The DØ Monte Carlo 1800
    J. Womersley
Physics Simulations for SSC and LHC 1806
    F.E. Paige
Simulations of Ultrarelativistic Heavy Ion Collisions 1812
    X.-N. Wang
Soft Physics Simulations 1820
    J. Ranft

21. New and Planned Detectors and Their Physics Potential

The GEM Experiment at the SSC 1829
    B.C. Barish
The ATLAS Detector for the LHC 1837
    M.A. Parker
CMS: A General Purpose Detector at the LHC 1842
    T.S. Virdee
The H1 Detector at HERA 1849
    F.W. Brasse
The ZEUS Detector—Status August 1992 1856
    A.C. Caldwell
Design and Performance of the SLD Vertex Detector, A 120 MPixel
Tracking System 1862
    C.J.S. Damerell
Operating Results from the DELPHI Ring Imaging Cherenkov (RICH) 1867
    P. Baillon
Preliminary Results from the CMD-2 Detector 1876
    B.I. Khazin
The Status and Plan of Upgrades for BES 1881
    L. Jin
KLOE, A General Purpose Detector for DAΦNE 1885
    P. Franzini
The B Factory Detector for PEP-II: A Status Report 1889
    B. Ratcliff
Status of the Tau-Charm Factory Project and Aspects of the
Detector Design 1897
    R.H. Schindler
The DØ Upgrade Program and Its Physics Potential 1902
    M. Rijssenbeek
Operation of the CDF Silicon Vertex Detector with Colliding Beams
at Fermilab 1908
    F. Bedeschi

22. Status of Existing Accelerators and Future Plans

Status of the Tevatron Collider: Performance and Upgrades 1917
    V.K. Bharadwaj
Long-Term Stability in Proton Storage Rings 1923
    R. Talman
TRISTAN Performance and Plans 1929
    K. Satoh
PEP II: SLAC-Based Asymmetric B Factory 1935
    T. Fieguth
The Status of DAΦNE, The Frascati Φ-Factory 1941
    G. Vignola
JINR Storage Accelerator Complex: C-TAU Factory 1945
    A.N. Sissakian
A New Concept for an Asymmetric Φ Factory to Test CPT and
Study K_S^0 Mesons 1960
    D.B. Cline
LEP Performance and Plans 1966
    S. Myers
SLC Performance and Plans 1971
    N. Phinney
Beamstrahlung in High-Energy e^+e^- Linear Collider QED Background 1979
    K. Kurek

23. Accelerator Technology

Superconducting Accelerator Magnets: A Review of Their Design
and Training 1985
    R.B. Palmer
Superconducting Cavities for Particle Accelerators 1992
    H. Padamsee
Large Scale Cryogenics for Particle Accelerators 2000
    Ph. Lebrun
R&D Status of Linear Collider Technology at KEK 2008
    J. Urakawa
B Factory Technical Challenges 2016
    D.H. Rice
CESR/CLEO Interaction Region Upgrade 2022
    S.D. Henderson
An Approximate Invariant Using Lie Algebra 2026
    T. Garavaglia
Laser Extraction of a 400-MeV H$^-$ Beam 2031
    C. Johnstone

24. First Results from HERA

First Results from the ZEUS Experiment 2037
    B. Löhr
First Results from the H1 Experiment at HERA 2048
    F. Eisele

Conference Photographs	2067
Conference Program	2077
List of Contributed Papers	2105
Conference Participants	2129
Author Index	2167

Foreword

The 26th International Conference on High Energy Physics (ICHEP 92) was held in Dallas, Texas from August 6–12, 1992. Hosted by the SSC Laboratory, the conference was organized under the auspices of the International Union of Pure and Applied Physics with the support of the U.S. Department of Energy, the National Science Foundation, the Texas National Research Laboratory Commission, Universities Research Association, Southern Methodist University, and the University of Texas at Arlington, as well as private institutions and individuals. Invitations were extended to physicists from throughout the world, carrying on the traditions of the "Rochester Conferences" that began in 1950. Thirteen-hundred delegates from 47 countries attended the conference.

A meeting of the scope and size of a Rochester Conference requires intensive efforts from many skilled and talented people. In the case of the 1992 conference, the host site was changed due to unusual world events, and the demands on the conference planners became particularly intense. Therefore, all the individuals who worked on the conference deserve special credit and thanks for the long and strenuous hours they put into making ICHEP 92 a success.

The International Advisory Committee provided overall guidance to the conference organizers. ICHEP 92 was managed by the Local Organizing Committee, backed up by the Conference Chairman and Committee Chairs. Jim Sanford and Catherine Burns, ably assisted by Karen Earley, played central roles in planning and organizing the effort, bearing the brunt of the work and responsibility, and hence deserving much of the credit. Southern Methodist University (SMU) staff members under the direction of Julie Wiksten were very responsive to the needs of the conference. The SMU Physics Department, chaired by Vigdor Teplitz, also provided substantial support to the Local Organizing Committee throughout the planning process. In total, more than 100 "volunteers" from SSC Laboratory and SMU devoted long hours to planning, hosting, and doing the behind-the-scenes work during the conference.

Staging the conference would have been impossible without financial assistance from a variety of agencies, foundations, corporations, and individuals. Their support, combined with the fundraising efforts of the Host Committee, enabled the conference to sponsor a record number of physicists and provide a warm welcome to all attendees.

Another valuable effort came from the Program Committee, who prepared an interesting and relevant physics program. Despite severe time constraints, they were able to develop the topics, select and work with session conveners and speakers, and define the structure of the final program.

From the inception of the conference, the Editorial Committee has been involved in planning for and actively capturing the content of the conference presented in these proceedings. Special credit for this activity should be given to Valerie Kelly, who played a crucial role as the technical coordinator, and to Ed Valauskas, who managed the massive task of running the conference library.

Roy F. Schwitters
Conference Chairman

Conference Organization

Chairman
R. Schwitters

Deputy Chairmen
F. Gilman and R. Kasper

Host Committee

Chair, E. Nye
J. Adams
S. Bartlett
L. Beecherl, Jr.
R. Bolen
G. Bush
D. Cook, III
J. Crawford
J. Evans
K. Granger
C. Hunt
L. Jackson
J. Johnson
J. Junkins
T. Luce, III
M. Meyerson
J. Musolino
V. Prothro
K. Pye
P. Redington
H. Robinson
A. Strauss
L. Temerline
T. Vandergriff

International Advisory Committee

A. Astbury	Canada
N. Cabibbo	Italy
M. Davier	France
C. Escobar	Brazil
S. Fang	P. R. China
C. Jarlskog	Sweden
T. D. Lee	USA
C. Llewellyn-Smith	UK
A. Logunov	Russia
S. Loken	USA
P. Malhotra*	India
L. Okun	Russia
J. Peoples	USA
K. K. Phua	Singapore
B. Richter	USA
C. Rubbia	Switzerland
P. Söding	Germany
H. Sugawara	Japan
Y. Yamaguchi	Japan
A. Zepeda	Mexico
A. Zichichi	Switzerland

* deceased

Program Committee

Chair, C. Baltay	Yale University
J. Ballam	SLAC
L. Bjorken	SLAC
P. Darriulat	CERN
G. Feldman	Harvard University
J. Friedman	MIT
H. Georgi	Harvard University
M. Jacob	CERN
S. Meshkov	SSC Laboratory

Local Organizing Committee

Chair, J. Sanford	SSC Laboratory
K. Anderson	SSC Laboratory
C. Burns	SSC Laboratory
E. Duek	SSC Laboratory
K. Earley	SSC Laboratory
P. Hale	SSC Laboratory
P. Rosen	University of Texas at Arlington
J. Siegrist	SSC Laboratory
R. Stroynowski	Southern Methodist University
V. Teplitz	Southern Methodist University

Editorial Committee

Chair, J. Sanford	SSC Laboratory
P. Dahl	SSC Laboratory
V. Kelly	SSC Laboratory
D. Matthews	Matthews Associates
F. Olness	Southern Methodist University
R. Rooney	SSC Laboratory
E. Valauskas	SSC Laboratory

Opening Session

McFarlin Auditorium, SMU

Opening Remarks

Roy F. Schwitters
Director, Superconducting Super Collider Laboratory
Conference Chairman

I would like to read a letter that came in this morning from the White House:

> I am delighted to send greetings to all those who are gathered in Dallas for the 26th International Conference on High Energy Physics. Special greeting to our visitors from abroad.
>
> History shows us that advances in basic research have had major beneficial impacts on the lives of our citizens. From developing new medical technologies to evolving communications technology, high energy physics plays an increasingly important role in society. Such scientific progress translates into economic growth and opportunities for our people while the knowledge that humankind benefits gives added impetus to solving the mysteries of our universe.
>
> It is particularly fitting that this landmark conference be held in Texas, the planned home of the Superconducting Super Collider, one of our major scientific initiatives. I am committed to ensuring congressional funding so that the SSC begins its operations in 1999 and to ensuring that strong investment in basic research remains a high priority in this country. Through the international science community's tradition of fruitful collaboration, we will continue to reap many rewards from high energy physics. I commend all who have dedicated themselves to exploring the vast potential of this exciting field, and I look forward with you to the enormous benefit of the SSC to your research.
>
> Keep up the terrific work. Barbara joins me in sending best wishes for a successful conference and for every future success.
>
> *George Bush*

We are very pleased to be able to host this conference. Highly unforeseen world events made it impossible to schedule the conference as originally planned. Therefore, we were asked a little over a year ago whether it would be possible to re-schedule the meeting and conduct it here in Dallas. In responding to that request, we were concerned about our ability to host a first-class conference here, especially given all the other activities that are taking place at the SSC Laboratory. So we had to answer several important questions before we could agree to play host.

First of all was the question of venue. Where could we hold a proper scientific meeting, and would adequate space be available? We quickly discovered that there would be no problem with hotels: given that this is August in Dallas, there would be plenty of hotel space. And, of course, we were delighted with the wonderful facilities available here at SMU, which is a fine place for such a meeting. So the venue question was answered.

The next question was that of organization. As you know, our Laboratory staff is preoccupied with the construction of accelerators and detectors and all that entails. Consequently, we have had to call on many people throughout the community—within our Laboratory, throughout the United States, and some of our foreign colleagues, as well as physicists from other institutions here in the Dallas/Fort Worth area. In particular, I want to thank Vigdor Teplitz and Peter Rosen, who have also recently moved to this area to help their institutions build their strength in physics. They have helped us a great deal with the conference organization. The meeting simply wouldn't have been possible, however, without the extraordinary efforts and abilities of Jim Sanford of the SSC Laboratory and his dedicated staff, who have put it all together. Jim has been a hero.

The next critical question was that of the program. It would have been impossible for the Laboratory to do all that was needed to secure the speakers and determine the scientific content. Consequently, we asked that an external program committee be formed that would, on very short notice, assemble this exciting program. Charlie Baltay agreed to chair the committee, and we very much appreciate his efforts and those of his colleagues. We've also had great support from our funding agencies, from the state of Texas, and from various companies, all of which you'll find acknowledged in the conference pamphlet. However, there is one new source of support, one new committee, that I do want to mention in particular, because it is typical of the great support that our new Laboratory enjoys here in Texas. I'm referring to a group of prominent citizens who agreed to form what we've called a "host" committee to help us raise funds and organize some of the extracurricular activities for the conference. Erle Nye, a prominent local businessman, has chaired this committee, and many people from the area have helped out. A significant aspect of the fund raising is that it has allowed us to support the travel and expenses of 120 conferees from various parts of the world who wouldn't otherwise have been able to attend. We are very pleased that they can be here, and we thank the citizens of our "host" committee for making it possible.

Now, with these questions successfully met, we can turn our attention to the conference ahead.

My first experience with the Rochester conference was with the one held 20 years ago in Batavia, Illinois, the 16th International Conference. Then, as today, we were visiting a brand new laboratory. At that time, we were just entering the new world of physics represented by the standard model, which had just appeared on the scene and was to take increasing hold during the next few years. The international conferences since then have recorded the triumph of the standard model. Today, a generation of physicists later, we have a beautiful picture of the subatomic world. But we are trying now to look beyond the standard model, and that is part of the excitement we hope to experience in the sessions ahead of us.

Today we also have a new world of people to deal with: a world, fortunately, where the tensions and threat of nuclear war are greatly reduced. However, this new world faces tremendous economic difficulties and challenges, and it is clear that maintaining the role of high energy physics and getting support for the instruments and tools that physicists need have become increasingly difficult. We witnessed that in June with the disappointing vote against the SSC in the House of Representatives. But then the response of the full international community to that event has been most heartening. The very morning after the House vote, we had letters from so many of you, including CERN and Fermilab, strong statements of support from throughout the world. Since June many physicists have traveled to Washington at their own expense to talk to their representatives and senators. The letters that have poured in to the Laboratory have been very inspiring to all of us who are understandably concerned about the events. The combined effect of everyone's letters and efforts became apparent last week when we received such a powerful vote of confidence from the Senate. Our funding and support for the next fiscal year now look promising, and we have every reason to view the future with optimism. The cohesiveness of the full international community was a big factor in making this possible, and in the corridors and side meetings during the conference ahead we can continue to build the strength of our relationships. And those relationships are vital, for surely the future will be even more difficult than the past, given the new world that we will have to face.

Thank you.

Welcoming Address

Tadao Fujii
Chairman, C11 Commission,
International Union of Pure and Applied Physics

On behalf of the IUPAP C11 Commission on Particles and Fields, it is my great pleasure and honor to welcome you to the 26th International Conference on High-Energy Physics. As the conference chairman mentioned in his opening remarks, we are most fortunate to be able to open this conference today, here in Dallas, after having heard the good news about the important step that was taken by the U.S. Senate to extend the life of the SSC project. I'm certain that this good news will stir lively discussions about the progress and future prospects of our fields in the coming six days of sessions. This conference is also somewhat exceptional in that its site had to be suddenly changed last year from Moscow to Dallas. I want to express our deep gratitude to Professor Roy Schwitters and his colleagues, who took over the hard task of efficiently organizing the conference in such a short time.

In closing, I would like to mention the role of the International Committee for Future Accelerators. Clearly, one of the most important factors in our future success in exploring the deep structure of matter, as well as the origin of the universe, is world-wide international collaboration. In the course of our most recent meeting, in January of this year, ICFA recommended that the design and the use of future large high energy facilities, including the appropriate R&D, should include international participation from the start to ensure that the full intellectual capabilities of the international community are utilized. Following this recommendation, ICFA appointed a subcommittee under the chairmanship of Professor Soergel of DESY to review existing models of international collaboration and prepare models for the future. We sincerely hope that with help from all of you here, the ICFA recommendation will be realized in the near future.

Thank you.

Welcoming Address

Kay Granger
Mayor of Fort Worth

It is my honor to welcome you this morning to this prestigious conference on this prestigious campus. You will be welcomed this morning by two mayors: by myself, mayor of Fort Worth, and by Steve Bartlett, the mayor of Dallas. But we think of ourselves as representing many mayors from many cities and towns that make up the North Texas metroplex. We all share the honor of hosting you here.

I look forward to welcoming you again on Sunday when you are all invited to visit the city of Fort Worth. During that afternoon, you can take a short break from science and visit what we call the "Cowboys and Culture" of North Texas. You will be greeted in downtown Fort Worth in one of our notable parks, the Water Gardens, which has been featured in several motion pictures. We will go from there to what we call our Cultural District. There, in addition to a world renowned zoo and the botanic gardens, you can visit four nationally known museums: the Kimbell Art Museum, the Amon Carter Museum of Western Art, the Museum of Science and History, and the Modern Art Museum. We will also take a brief trip away from what you're doing—working with the future and discovery—back into the past, to the cowboys of Texas in our historic stockyards area.

So it is my great pleasure to be with you this morning, and I hope you will accept our invitation to Fort Worth.

Thank you very much.

Welcoming Address

Steve Bartlett
Mayor of Dallas

We in Texas and we in Dallas believe in science and technology and in the future, and you will find that Texas congressmen are always at the forefront of voting for both the Superconducting Super Collider and for science and research and technology as a whole. Mayor Granger and I between us host sometimes as many as a hundred trade shows, conferences, conventions, and other kinds of seminars in our respective cities every week. Obviously, we have to carefully choose which conferences to attend. It is a mark of the esteem in which we hold you, and the esteem in which we hold science and research generally, that both of us are here today. We welcome you as our honored guests. As scientists you are very much a part of our lives in the Dallas/Fort Worth area and Dallas in particular. You honor us by being here.

Dallas is the eighth largest city in the nation, and a city of considerable strength. We represent the core of the Dallas/Fort Worth area, which is the fourth largest market in the entire country, and the third largest concentration of Fortune 500 corporate headquarters. Within three years, the DFW airport will be the busiest airport in the nation in terms of scheduled takeoffs and landings. On the international scene, Dallas is headquarters for the ninth largest concentration of multinational corporations. We're a center for wholesale trade, for aviation, and for finance. Most significantly, in terms of this conference, the Dallas/Fort Worth area has emerged as a center of applied research in high energy physics. Indeed, it is perhaps the most significant center in terms of the concentration of technology and industry based on technology and, specifically, high energy physics, whether it is Texas Instruments, General Dynamics, LTV, Hughes Training, Fujitsu, Northern Telecom, or Rockwell.

Thus your conference is very timely in the sense that the Dallas/Fort Worth area has always been politically active in trying to ensure that the Superconducting Super Collider will in fact be built. The vote last week in the U.S. Senate confirms our commitment. You are welcome here now and you will be welcomed back early and often. We look forward to the turn of the century when you will be conducting high-energy experiments at the 54-mile Superconducting Super Collider ring just south of us in Ellis County, Texas.

The conference chairman drew attention to a new world; it is a world that we understand here in Dallas. There was a controversial book written not long ago entitled "The End of History." If the end of history means an end of history marked by wars of conquest, then the beginning of history will be marked by competition and cooperation based on scientific endeavor. It will be a world of ideas, a world of science in which we compete with our own minds to discover the secrets of the universe, and to use those secrets in improving the quality of life and the standard of living for generations to come around the world. You then are a part of the new world, of a beginning that will carry into the next century. That is why this conference has such significance for us in Dallas, in the United States, and in the world. The work you are engaged in here will have enormous positive repercussions for the quality of life for humankind for generations to come.

Mayor Granger and I are proud that you would choose to carry on into the brave new world in the Dallas/Fort Worth area. Thank you.

Welcoming Address

Kenneth Pye
President, Southern Methodist University

We are delighted that you are here. I only regret that it was not after the beginning of the school year, for students from some 60 nations study on our campus, and you would have had an opportunity to meet some of your countrymen who are now at home enjoying a summer break.

Although it is impossible for anyone to attend all the sessions, I know you will attend as many as you can, so your schedule is packed. However, if you should have a stray moment, in which you would simply like to meditate on a subject other than physics, I encourage you to wander around our campus. Immediately opposite this hall is the Fondren Library West. On its second floor is the DeGolyer Library, one of the premier libraries of Western material in the United States. If you go out the front door of this building and turn south, the third building on your right is the Meadows School of the Arts. That building is primarily devoted to concert halls, a fine arts library, and theaters, including a new Shakespearean theater to be dedicated next month. Immediately upon entering that building you encounter on your right a small museum of Spanish art. This collection is second only to that of the Metropolitan in the United States. Adjacent to the Meadows School is the theology complex. The first building in that complex is Ridwell Library. Recently renovated and one of the finest libraries in the country, Ridwell maintains collections of art throughout the year for the purpose of broadening the understanding of our student body. Currently, there is an exhibit of Our Lady of Guadalupe, patron saint of Mexico.

Please feel free to ask anyone you meet on the campus any question you may have. I can assure you that everyone is committed to trying to help you and you will be answered with courtesy if not necessarily accuracy. We all want you to leave this university having thoroughly enjoyed your experience. I know you will enjoy the camaraderie of people engaged in the same disciplines, sharing the insights and information that will lead you to the breakthroughs of tomorrow.

Thank you again for being with us. It is a great privilege for us to serve as the site for your conference.

Keynote Address

William Happer
Director, Office of Energy Research
U.S. Department of Energy

Introductory remarks by Dr. Roy Schwitters

Dr. William Happer was nominated by President Bush on May 17, 1991, to be Director of the Office of Energy Research in the U.S. Department of Energy. He was confirmed by the U.S. Senate on August 2, 1991, and sworn into office later that week. As director, Dr. Happer manages the Office of Energy Research, one of the largest sponsors of basic research in the federal government. The OER annual budget is approximately 3 billion dollars, which pays for DOE's programs and basic energy sciences, high energy physics, nuclear physics, the SSC, biological and environmental research, university science education, fusion energy, and scientific computing. In addition to being Director of the Office of Energy Research, Will Happer is also DOE's Science and Technology advisor. In this position, he advises the Secretary of Energy on science and technology issues that cut across DOE programs. Dr. Happer is also responsible for the management of the Department's five multi-program and ten single-program non-weapons laboratories, and the development of department-wide policy for both weapons and non-weapons programs.

Dr. Happer received his B.S. degree in physics from the University of North Carolina in 1960 and his Ph.D. in physics from Princeton University in 1964. He took a position as research associate at Columbia Radiation Laboratory in Columbia University. While at Columbia he served as instructor, assistant professor, and professor in the Department of Physics. He was also co-director of the Columbia Radiation Laboratory from 1971 to 1976 and its director from 1976 to 1979. From 1980 until assuming his current position, Will Happer was professor in the Department of Physics at Princeton University.

Let me say, on behalf of all of us who work at the SSC, that it is truly a pleasure to work with a scientist of Will Happer's caliber and a person of his good sense and good humor. We are pleased and honored that he can be here this morning to address us.

Dr. Happer's Address:

Roy is no more relieved than I am to be here under the conditions that have developed. A friend in Washington tells me that they used to say at the National Security Council that the United States government faces three superpowers in the world: the Soviet Union, China, and the high-energy physics community. As you know, the events in the Senate a few days ago have shown that only one of those powers is left.

I too have a letter I would like to read to you. It is from Admiral James Watkins, the Secretary of Energy and a member of President Bush's cabinet:

> To the delegates of the 26th International Conference on High-Energy Physics.
>
> It is a pleasure to welcome the delegates to the 26th International Conference on High Energy Physics. The Rochester Conference has become through the years since its inception in 1950 one of the preeminent scientific meetings in the world. The conference demonstrates graphically the truly international character of science in which the spirit of world-wide cooperation and intellectual competition drives us to greater and more fundamental understanding of the world about us. It is fitting that you will be at the site for the next great accelerator laboratory, the Superconducting Super Collider. I am pleased that the Senate voted earlier this week by an overwhelming margin to continue funding for the SSC in fiscal year 1993. The SSC represents an important investment in the scientific and technological future of this nation and the world. The President and I are committed to completing the SSC. It is our intention that in less than a decade the SSC will join the world's other great high-energy physics laboratories in providing scientific results which will be reported at future Rochester conferences. I wish all of you the best in your conference.
>
> Sincerely,
>
> James D. Watkins
> Admiral, United States Navy, Retired

So I can assure you that there is tremendous support for the high energy physics community in Washington. My boss, the Admiral, was described somewhat sarcastically in the senate debates last week as the cheerleader for the SSC. I can tell you that Admiral Watkins and the rest of us at the Department of Energy are proud to be cheerleaders for high-energy physics and for all of science. The SSC has very strong support in the international community as well. I think Dr. Schwitters said it right, that the very fine letters and expressions of support that poured in from around the world were a big help in our campaign to reverse the House vote. The reversal of fortunes that you have seen in the last few weeks did not occur by accident, but were the result of very hard work by the supporters of the SSC from all segments of United States society. The SSC certainly got excellent support from the scientific community. But you might be surprised at the outpouring of support that came from common citizens, from some segments of the Congress, and from across the country. As many of you know, President Bush himself paused from some of the heavy burdens of the Presidency to rally support for the SSC. Governor Clinton, the Democratic nominee for President, has also announced his support for the SSC. And there was bipartisan support for the SSC in the overwhelmingly favorable Senate vote. This coalition will continue to work to maintain its strength and support for the SSC in preparation for the conference between the House and the Senate, where we expect to prevail.

In the campaigns to consolidate support for the SSC, we have been forced to explain to many of our fellow citizens the importance of high-energy physics—why it's necessary to build such immensely expensive accelerators. If you follow the debate, you will recognize two major themes in our answers: science for the sake of knowledge, and science that can improve the practical well-being of mankind. For many of us these spiritual and material reasons are not so different from each other, since many of us here believe in the old wisdom that man does not live by bread alone. And I need hardly remind those of you assembled here today of the great traditions of science that lie behind us: the development of mathematics and geometry by the ancient civilizations of Greece, Egypt, and the Far East. Few of us know today very much about the cruel wars, the politics, and the natural disasters of those far off times. But all of us know the Pythagorean theorem—as beautiful today as it was in the days of Euclid. We can trace our scientific tradition through Gallileo's studies of ballistics and falling bodies, through the sublime insights of Isaac Newton's mechanics, through the elucidation of electromagnetism by Oersted, Faraday, and a host of other heroes, through the discovery of radioactivity and the nucleus less than 100 years ago, and finally to the field of high-energy physics, which we will be discussing at this conference. As we look back on this wonderful tradition of science, we have to remember that high energy physics is but one of many scientific tribes that have descended from some ancient scientific father Abraham.

We must deal with other equally admirable tribes in the Department of Energy. The breathtaking discoveries of molecular biology, plate tectonics and geosciences, materials science, space exploration, and many other fields are no less impressive than the triumphs of the standard model. There are more research opportunities than we can afford to pay for in these many fields of science, so partitioning funds to the highest priority project in each area has become increasingly difficult for us. I'll revisit that in a minute, but first let me return to trying to make the case for science and the political arena.

The spiritual values that I mentioned carry much less weight than the promise of material well being that will result from new knowledge. Many scientists take pride in the purity of their work. They feel demeaned when they are asked, "What good is it?" I think this attitude is quite appropriate for an independently wealthy scientist, as many scientists were in previous centuries. I have known one or two like that even in my career. One was the English spectroscopist, D.H. Jackson, who was one of the few people I've met who was wealthy enough to support his own research. Once he wanted to look at the spectrum of some odd rare earth isotope, but the research board refused to give him funds to procure it. So he simply sold one of his race horses and used the proceeds to buy the isotope himself. I remember him showing the spectrum of this isotope in a conference as he spoke about the spectrum of "my" isotope with very, very heavy emphasis on the word "my." Old Jackson had the utmost contempt for whether anyone thought his work was useful or not. But as is often the case, his spectrum played an important role, an unexpected role, later on in the development of certain lasers that we use today. It is hard to make this argument for accidental discovery, accidental benefit to mankind, but of course all of us here know that it is true.

In the debates last week Senator Bryan said, "We must evaluate our research investments on the basis of whether they help our nation in terms of our ability to compete internationally." And much was made of that. Much nonsense was spouted by both sides during the debate, both in the Senate and the House. But I think it's clear that high energy physics *has* had a substantial payoff in the well-being of society. If you look around at the Dallas airport, you'll see klystrons in all the radar sets controlling the incoming air traffic. Of course, these were developed originally for high energy physics at Stanford. Electron synchrotrons—which have sprouted all over the world for studies in materials science—photolithography, and biologic research were originally developed for high energy physics. Of course, we expect the industrialization of superconducting technology to take place right here. During the construction of the SSC, we will see big spinoffs for the commercial world.

There is an even more important contribution that high energy physics has made to the rest of mankind, especially to the scientific community, and that is its demonstration that physicists can work together cooperatively to achieve important scientific goals. Long before scientists in other fields, experimental high energy physicists learned that it was not possible to make an impact without teamwork on an ever more massive scale. The days when one heroic scientist such as Curie or Rutherford or Chadwick, with perhaps an assistant or two, could revolutionize high energy physics were pretty much gone by the middle of this century. So high energy physicists have had to learn to work together to build ever more powerful accelerators and detectors. It is significant that this spirit of cooperation in high energy physics was international from the start. I would bet that half of you here today have spent many years doing important work in countries other than the one you were born in, or the one to which you owe your allegiance now.

As the scientific facilities needed to make progress get larger and larger, they are becoming increasingly difficult for any single country to afford—and that includes the United States. I think we have pretty much reached our limit with the SSC. President Bush has invited all nations to join the United States as partners in building the SSC and its detectors. In our zeal to get going with the exciting science that can be addressed by the SSC, we in the United States probably got started before we had had adequate discussions with potential international partners in this important endeavor. In retrospect, that may have been a mistake. But we are trying hard to correct that mistake and to transform the SSC into a truly international project. When I think about this, I'm reminded of a story about Sam Walton. Some of our international friends perhaps don't know Sam Walton, but he was the richest man in America when he died a few months ago, and he gained his wealth by bringing inexpensive, high-quality goods to small towns across America. He was a genius at doing this, and someone once asked him, "How is it possible, Mr. Walton, that things always work out so well in your business?" Mr. Walton replied, "Good decisions." The questioner said, "Well, that sounds right, Mr. Walton, but how do you make these good decisions?" Mr. Walton said, "Experience." The questioner was still puzzled. "But, Mr. Walton, how do you get experience?" And Mr. Walton said, "Bad decisions." I think, like Mr. Walton, we are learning how to make good decisions on mega-projects for the future. I hope it's not too late to turn the SSC into a good decision.

One area where we think we are doing it right from the start with a scientific mega-project is that of magnetically confined fusion. Many of you know that the break-even point is very close now from magnetically confined fusion. Just last fall deuterium and tritium were loaded for the first time in the Joint European Tokamak or JET. In the first few shots, they were able to get some 2 megawatts of fusion power in some 2 seconds of plasma burning. That's a lot of collisions per second, even for the Super Collider. The luminosity is not so high, but the collision rate is high. It's about the same power as a small jet engine on one of those commuter jets that you see on the runways around here. Within a year we hope to load deuterium and tritium fuel into the Princeton Tokamak where we should generate some 20 megawatts of power for some seconds. Now these two machines are being run by several hundred megawatts of external power, so they are not really breaking even yet. They are putting out a lot of power, but it is only a small percent of what is going in. These are fairly simple machines in principle. The energy is confined in the machine by diffusion, so the confinement time goes up approximately as the square of the linear dimensions.

All knowledgeable people agree that if you make a Tokamak somewhat larger—and I hesitate here to say what "somewhat" means, but it is actually the size of a six-story building somewhat bigger than this auditorium—then you will reach break-even with more power coming out than going in. By the time you look at what goes into such a Tokomak—enormous superconducting coils, vacuum vessels, heat exchangers—the cost is very similar to the cost of the SSC. Nobody is quite sure, but the best estimate is somewhere between $5 and $10 billion United States dollars. This is something that no country is willing to undertake on its own. Therefore we are trying hard to make an international Tokamak that will be a joint effort among Japan, the United States, the European Community, and Russia. The machine has actually been given a name with a catchy acronym: the International Thermonuclear Experimental Reactor, or ITER. Some thought went into the selection of these words and the acronym, because ITER is actually the ancient European verb "to go." It is the same "IDTI" that is still used by our Russian and other Slavic friends here today as their verb "to go." And it's a word that we see in the English-speaking world in such words as itinerary, exit, transit, and many others that come to us from Latin. So let us hope that ITER really will go.

ITER is a slow-moving project compared to what we are used to, including even the SSC. It has a number of very deliberate stages, one of which, the conceptual design, has recently been completed. We are now proceeding into an engineering design where detailed engineering drawings are being assembled in three different centers, one in Japan, one in the United States, and one in Europe. And even now we're moving on with discussions of how to pick a site for ITER. It will, of course, be a difficult thing to choose a site that will leave all four parties feeling fairly treated—feeling that their own national future programs will not suffer if they are not chosen as the site, and that the host country is somehow bearing its fair share of the cost. For example, it is clear that many benefits come to the host country by way of a big new machine. And so perhaps the host country should be expected to pay a larger fraction of the cost of construction. So all these issues are under discussion now. Some of the issues are perhaps already tacitly assumed in the high energy physics community, but in this case we are proceeding very formally, very deliberately, and unfortunately, very slowly. On the other hand, many of my friends in the Office of Management and Budget are quite pleased that things are going slowly. From their point of view, the pace is fine.

The pressures to internationalize scientific mega-projects will, I think, continue and get even stronger as the years go by. The reasons are, first, that machines are getting more and more expensive, and, second, that citizens of all countries are demanding more and more from their governments in terms of social programs. And there simply isn't as much money to go around as there used to be. I think that will certainly be true for the field of high energy physics as well. Those of you assembled here today are the leaders of the international high energy physics community, and you will have to participate in the difficult and time-consuming preparation for new projects in high energy physics in the future. This is not my field, but it seems likely to an outside observer that at some point someone will have to do something about electron physics again. These are very clean and beautiful experiments. It would be nice to move forward once more on that frontier, but it seems likely that the next electron physics machine, perhaps a linear collider, will cost as much as or more than the SSC. It will also have to be built internationally right from the start, perhaps on the ITER model. One good thing about the ITER model is that, as we go forward, treaties are actually made. Ministers get together and sign documents at the highest levels. This makes it more difficult to back out of such a project once it's entered into. It is a benefit you get from slowing down the process. So we hope that we can work out—by some combination of what we learn from constructing the SSC, ITER, and other machines across the world—a way to make science, basic science, a truly international effort in the future, even more so than it has been in the past.

I know that you have not come here, many of you from around the world, to listen to a Washington bureaucrat drone on and on about policy. That is something I personally cannot stand either. Sometimes it begins to drive me crazy, and I have to go and hide in a room and read about physics for an hour or two until I feel better. So I will stop right now.

I thank you all and wish you the most productive, exciting conference that has ever occurred in the Rochester series. You can count on us, the Department of Energy and the federal government, to help as much as we can.

Plenary Sessions

Umphrey Lee Center, SMU

WEAK DECAYS, RARE DECAYS, MIXING, AND CP VIOLATION*

Persis S. Drell and J. Ritchie Patterson
Newman Laboratory of Nuclear Studies
Cornell University
Ithaca, NY 14853-5001

Abstract

We review the new results presented at the XXVI International Conference on High Energy Physics on the topics of weak decays, rare decays, mixing, and CP violation.

INTRODUCTION

Two hundred and twelve abstracts were submitted to this conference on the topics of weak decays, rare decays, mixing, and CP violation, and they were covered by 9 hours of talks in the parallel sessions. We cannot hope to review all the results presented and therefore we will apologize at the start. There are many new results that we will either treat in a cursory fashion, or will fail to mention all together.

The field of weak decays, rare decays, mixing and CP violation is dominated by what we have come to refer to as the Standard Model. In the first half of this review, we will concentrate on the new results from charm and beauty experiments that are helping to define the Standard Model by measuring some of the free parameters of the theory. In the second half, we will concentrate on experiments from τ and K experiments that are trying to probe the validity of the model with tests of its predictive power. Table I lists the experiments we will be discussing and the approximate data samples available to them.

*Work supported by the National Science Foundation.

Table 1: Experiments and Data Samples

Experiment	Data
ARGUS	200,000 $b\bar{b}$
	260,000 $\tau^+\tau^-$
CLEO	935,000 $b\bar{b}$
	1,300,000 $\tau^+\tau^-$
LEP	100,000 $b\bar{b}$
	21,000 $\tau^+\tau^-$
FNAL E799	20×10^9 K_L^0
CERN NA31	2.6×10^9 K_L^0
FNAL E653	8×10^6 triggers
FNAL E687	5×10^8 triggers

DEFINING THE STANDARD MODEL

We will begin by discussing some experiments that can be used to define some of the parameters of the Standard Model. In this category we will discuss results from b and charm physics. The parameters of the Standard Model that must be defined by experiment are the masses of the matter fields (a new measurement of the τ mass is discussed by Luigi Rolandi[1] and a summary of the new limits on neutrino masses is given by R. G. H. Robertson [2]), and the elements of the Cabibbo-Kobayashi-Maskawa (CKM) matrix

[3] that rotates the flavor eigenstates of the strong interaction into the eigenstates of the weak interaction. The B meson is a wonderful laboratory for measuring the elements of CKM matrix, because 4 of the 5 matrix elements involving the third generation are in principle accessible in B decays. We will review the new measurements of the B meson semi-leptonic branching ratios and lifetimes that will allow us to extract updated values for $|V_{ub}|$ and $|V_{cb}|$. We will also discuss measurements of $B\bar{B}$ mixing that give us information on $|V_{td}|$ and $|V_{ts}|$.

There is another less obvious way in which experiments on charm are helping to define the Standard Model. The Standard Model tells us about quarks. In experiments, we deal with hadrons and we must rely on some input from theory to connect the two. The semi-leptonic charm decays are providing a testing ground for much of the machinery we rely on to extract CKM matrix elements in the B system.

b Hadron Lifetimes

Many new results were presented at this conference on b quark and b hadron lifetimes. The lifetime of the B provides the crucial connection between the decay rate to a final state which can be calculated and depends on CKM matrix elements and a branching ratio which is experimentally measured:

$$\frac{Br(B \to X\ell\nu)}{\tau_b} = \Gamma(B \to X\ell\nu). \quad (1)$$

Inclusive b Lifetimes

The so called "inclusive" or "average" b lifetime, τ_b, was first measured in the early 1980's, and at this conference new measurements from LEP of an impressive 5% precision were presented. The LEP experiments choose a b hadron sample by selecting hadronic events with a hard, high p_t lepton. Typically, $p > 3-4$ GeV/c and $p_t > 1$ GeV/c are re-

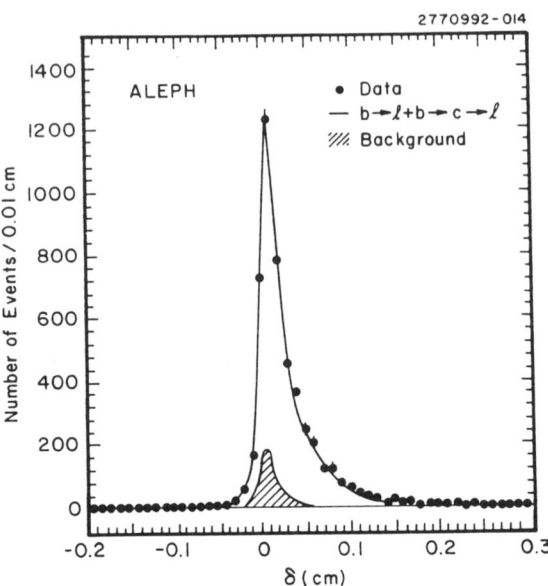

Figure 1. The lepton impact parameter distribution from ALEPH.

quired for the leptons. The b hadron lifetime is extracted from a fit to the impact parameter distribution of the lepton tracks.

Figure 1 shows the lepton impact parameter distribution from ALEPH [4]. The new measurements of the inclusive b lifetime are listed in Table 2, along with the Particle Data Group 1990 world average which does not contain the LEP measurements. The LEP measurements are all high compared to the 1990 world average and of significantly better precision. We will therefore define a LEP average b lifetime using the 4 latest inclusive b lifetime measurements, and assuming a correlated systematic error of 0.04 ps which is dominated by the uncertainty in models of hadronic decay and the fragmentation process.

What does the "average" b lifetime mean? Well, we don't know really and that is a problem. It is the average lifetime of the hadrons produced at, for example, LEP that contain a b quark. It is the lifetime averaged over all species of b hadron: $B_s, \Lambda_b, \overline{B^0}, B^-, B_c...$, with each species weighted by how much of it is pro-

Table 2: Inclusive b Lifetimes

Experiment	τ_b (ps)
ALEPH[4]	1.49 ± 0.07
DELPHI[5]	1.38 ± 0.05
L3 [6]	1.36 ± 0.07
OPAL[7]	1.37 ± 0.09
LEP Average	1.40 ± 0.045
1990 PDG[8]	1.18 ± 0.11

duced in the fragmentation process. What we often need to extract $|V_{cb}|$, particularly from low energy data where only B^- and $\overline{B^0}$ are produced, are the so called exclusive lifetimes for the different species of b hadrons individually. Such measurements were also reported this year.

Exclusive b lifetimes

ALEPH and DELPHI at LEP reported exclusive b hadron lifetimes for $B^-, \overline{B^0}, \Lambda_b$ and B_s. ALEPH and DELPHI tag samples of b hadrons using a lepton that is charge correlated with a hadron in the same jet, compatible with coming from the b semi-leptonic decay. For example, they select samples of B_s decays by requiring a hard, high p_t lepton in the same jet with a D_s reconstructed in $\phi\pi$ or K^*K. The signal from DELPHI demonstrating the existence of the B_s is shown in Figure 2. A clear D_s peak is evident when the D_s and lepton have the opposite sign as expected from B decay (see Figure 3(a)), but no D_s signal is seen when the D_s and lepton have the same sign. A $D_s^+l^-$ correlation is not an unambiguous signal for B_s production. An ordinary B decay can produce such correlated D_s–lepton pairs with either the decay $B \to D_s D$ followed by the semi-leptonic decay of the D (Figure 3(b)) or a $B \to D$ decay accompanied by a pair of $s\bar{s}$ quarks pro-

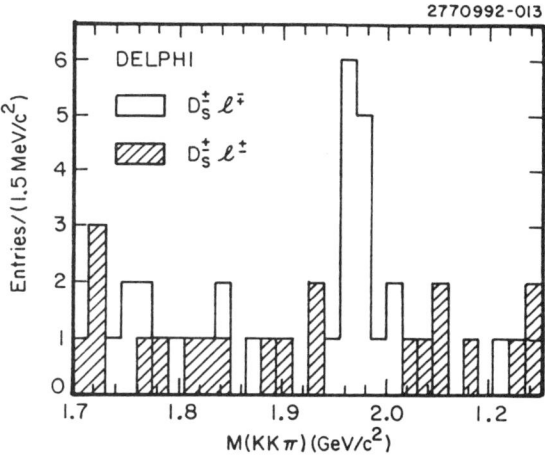

Figure 2. The invariant mass of $KK\pi$ combinations for $D_s^+ l^-$ (solid histogram) and $D_s^+ l^+$ (hatched histogram) combinations from DELPHI.

duced from the vacuum (Figure 3(c)). Both experiments use cuts on the D_s momentum, the lepton momentum, and the invariant mass of the D_s–lepton pair to suppress these backgrounds.

Very similar methods are used to tag samples of Λ_b, $\overline{B^0}$ and B^- using hard leptons charge correlated with a hadron (Λ, D^{*+} or D^0) in the same jet. The hardest samples to tag cleanly are $\overline{B^0}$ and B^-, because there is not a unique final state hadron that tags the charge of the parent B. The $D^{*+}l^-$ sample is dominated by $\overline{B^0}$ decays. Both $\overline{B^0}$ and B^- contribute to the $D^0 l^-$ sample. However, $B \to D^{**}l\nu$ decays feed into both $D^{*+}l^-$ and $D^0 l^-$ final states, and the uncertainty in the branching ratio for that decay dominates the systematic error on the exclusive $\overline{B^0}$ and B^- lifetime measurements.

Once the samples have been isolated, the lifetimes of the various species of b hadrons are measured by forming a vertex using the daughter hadron and the lepton, or by using the lepton impact parameter distribution in the case of the Λ_b. The experiments take advantage of their high resolution silicon vertex detectors

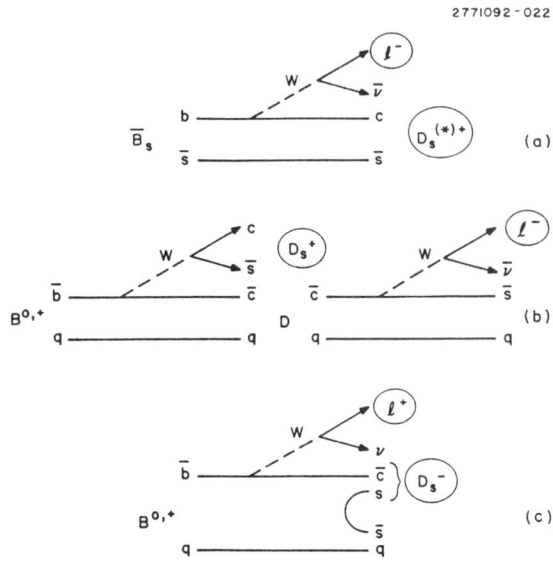

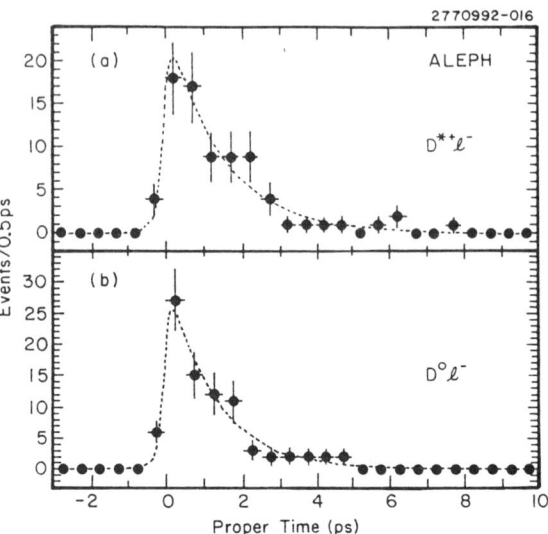

Figure 3. (a)Diagram for $B_s \to D_s \ell \nu$ signal events; (b),(c)The dominant backgrounds from ordinary B decays to the B_s signal.

Figure 4. The proper time distribution for the (a)$D^{*+}l^-$ and (b)$D^0 l^-$ candidates from ALEPH.

which reconstruct the decay vertices with resolutions of a few hundred microns, while the decay lengths being measured are several millimeters. Figure 4 shows the proper time distribution for the $D^{*+}l^-$ and D^0l^- events from ALEPH.

Table 3 shows all of the exclusive lifetimes for b hadrons[9], along with the LEP inclusive b lifetime. It appears that, in contrast to the charm system where the lifetimes of charmed hadrons differ by factors of two from each other, the b hadron lifetimes are rather similar and consistent with the inclusive average b lifetime.

From Table 3, we see that the LEP lifetime results are somewhat in disagreement with reports of a large lifetime difference between charged and neutral b hadrons from a sample of 18 decays measured by the E653 experiment[10] where b hadrons are produced by pions on a nuclear emulsion target. They do not separate their sample into B^- and $\overline{B^0}$ but rather into generic charged and neutral b hadron samples depending on the number of charged tracks in the final state. They find a much longer lifetime for the charged b hadrons.

We can take the average of the ALEPH and DELPHI exclusive B^- and $\overline{B^0}$ lifetime, and form the lifetime ratio, $\tau^-/\tau^0|_{\text{LEP}} = 0.92\pm0.21$ ps. In doing this we have assumed a 0.1 ps systematic error due to the uncertain D^{**} contamination that is common to both experiments and is anticorrelated between τ^- and τ^0. This lifetime ratio can be compared with a new value presented by CLEO that is derived from the semi-leptonic branching fractions of the B^- and $\overline{B^0}$ mesons. CLEO has reconstructed 600 B^- mesons and 490 $\overline{B^0}$ mesons in a variety of hadronic decay modes [11]. They count the leptons in these reconstructed events and get a measurement of the $\overline{B^0}$ and B^- semi-leptonic branching ratios. Under the assumption that $\Gamma(\overline{B^0} \to X\ell^- \overline{\nu}) = \Gamma(B^- \to X\ell^- \overline{\nu})$, they extract $\tau^-/\tau^0|_{\text{CLEO}} = 1.11 \pm 0.30$. Combining this with the LEP result yields

$$\tau^-/\tau^0 = 0.98 \pm 0.17. \qquad (2)$$

Table 3: Exclusive b hadron lifetimes, and LEP average inclusive b lifetime for comparison.

Hadron	τ (ps)
B^- [9]	1.34 ± 0.21
$\overline{B^0}$ [9]	1.46 ± 0.19
B_s [9]	1.05 ± 0.32
Λ_b [9]	0.98 ± 0.23
Neutral b hadrons [10]	$0.81^{+0.36}_{-0.22}$
Charged b hadrons [10]	$3.84^{+3.06}_{-1.37}$
$<b>$ (LEP)	1.40 ± 0.045

Table 4: b semi-leptonic branching fractions

Experiment	$Br(b \to X\ell\nu)$
1992 PDG($\Upsilon(4S)$)	0.107 ± 0.005
L3	0.119 ± 0.006
ALEPH	0.110 ± 0.006
DELPHI	0.100 ± 0.006
OPAL	0.107 ± 0.007

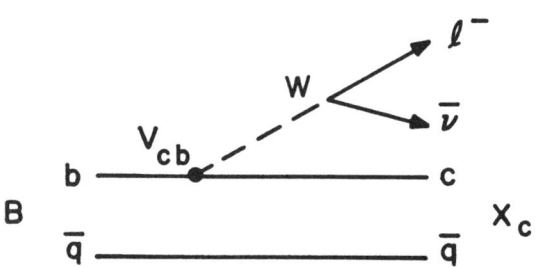

Figure 5. Schematic sketch of a semi-leptonic $b \to c$ decay.

Determination of $|V_{cb}|$

We now want to discuss measurements of $|V_{cb}|$ using the semi-leptonic decay of B mesons. We can schematically draw the semi-leptonic decay of a b quark as the b quark emitting a virtual W and turning into a c quark with coupling strength V_{cb} as shown in Figure 5. The branching ratio, either the inclusive semi-leptonic branching ratio or an exclusive branching ratio to a particular final state like $D^*\ell\nu$, can be related to the decay rate using the appropriate measured B lifetime, and hence $|V_{cb}|$ can be extracted.

Inclusive Determination of $|V_{cb}|$

There are two basic approaches that are used to determine $|V_{cb}|$. One is to use inclusive $B \to X\ell\nu$ decays. This has the advantage of high statistics and all of the LEP experiments reported new measurements of the inclusive b semi-leptonic branching ratio averaged over all species of B-hadron produced in the decay of the Z. The results[12], shown in Table 4, are in very good agreement with each other and with the B semi-leptonic branching ratio measured at low energies by CLEO, ARGUS, and others[13]. It is interesting to note that experiments continue to find a semi-leptonic branching ratio for the b quark that is lower than the 12-13% that can be gracefully accommodated by typical pure spectator models[14]. Recent work that includes nonperturbative corrections to the simple spectator picture are more consistent with these measurements [15].

The LEP experiments can take their measured inclusive semi-leptonic branching ratio and their measured inclusive b lifetime and, under the assumption that the semi-leptonic partial width is the same for all species of b-hadron produced, use the relation:

$$B_{sl}/\tau = \Gamma_{sl} = K_c|V_{cb}|^2 \qquad (3)$$

and extract $|V_{cb}|$. The low energy measurements extract $|V_{cb}|$ in much the same way, except that they must either use an exclusive lifetime, or there is an implicit assumption that all b hadron lifetimes are the same if the inclusive b lifetime is used.

The extraction of $|V_{cb}|$ from the inclusive branching ratios and lifetime is limited by theoretical uncertainties in the proportionality factor K_c in Equation 3 which depend on the different decay models that are used to calculate it. In Table 5, we list the values for $|V_{cb}|$ extracted from LEP and $\Upsilon(4S)$ data, where we have used for K_c, the value obtained from two different models that were fit to the CLEO inclusive lepton spectrum [16]. Averaging the results of the two models quoted in Table 5 and increasing the systematic error to cover the one standard deviation limits for each model we extract:

$$|V_{cb}| = 0.043 \pm 0.008. \quad (4)$$

The error on K_c totally dominates the error on $|V_{cb}|$. It is difficult to reduce the systematic error associated with K_c because in the inclusive method there are very few experimental checks that one can make to verify the theoretical models. The only experimental distribution is the lepton momentum spectrum and the fit to its shape is often used to fine tune the parameters of the models so it does not really provide a test of them. These weaknesses of the inclusive approach can be avoided to some extent by using exclusive decay channels such as $B \to D^* \ell \nu$ or $B \to D \ell \nu$ to measure $|V_{cb}|$.

Exclusive Determination of $|V_{cb}|$

A new measurement of the exclusive semileptonic branching ratio for the decay $\overline{B^0} \to D^{*+} \ell^- \overline{\nu}$ was reported by ARGUS and used to extract $|V_{cb}|$. To measure the branching ratio for this decay, ARGUS selects a sample of events with a D^{*+} and a lepton in them and then calculates the invariant mass of the missing neutrino, assuming the D^{*+} and the lepton came from a $\overline{B^0} \to D^{*+} \ell^- \overline{\nu}$ decay. The neutrino mass squared can be expressed in terms of the energy and momentum of the found D^{*+}, the lepton, and the parent B, as

$$M_\nu^2 = (E_B - (E_{D^*} + E_\ell))^2 - (\vec{P_B} - (\vec{P_{D^*}} + \vec{P_\ell}))^2. \quad (5)$$

For experiments such as ARGUS operating at the $\Upsilon(4S)$ resonance, which decays to a pair of B mesons with no additional particles, $E_B = E_{beam}$, the electron beam energy. The magnitude of $\vec{P_B}$ is known but not its direction; however, it is small (approximately 340 MeV) and is set to zero in Equation 5. This approximation results in some smearing, but when M_ν^2 is plotted for the $D^* l$ events a clear signal is seen at zero M_ν^2 as shown in Figure 6, corresponding to true $\overline{B^0} \to D^{*+} \ell^- \overline{\nu}$ decays. The new ARGUS exclusive branching ratio is $Br(\overline{B^0} \to D^{*+} \ell^- \overline{\nu}) = 0.052 \pm 0.005 \pm 0.006$ [19]. (Note: caution should be used in averaging this with previous CLEO and ARGUS measurements because the D^* branching ratios have changed substantially[20], and the experiments are not consistent in the D branching ratios that they use to extract a result.) $|V_{cb}|$ can be extracted from the measured branching ratio by using a theoretical model.

Typically in order to extract $|V_{cb}|$, models will make an educated guess for wave functions for the initial and final state particles and calculate the matrix elements for the decay or the form factors. For educated guess for wave functions you can read systematic error on $|V_{cb}|$!! The problem is that experimentally one is measuring $K|V_{cb}|^2$, and $|V_{cb}|$ is extracted by using various theoretical models for the form factors to calculate K. The only way to get an error on the proportionality factor, K, is to look at the spread between the available models which is not at all satisfactory. They could all be wrong! The value of $|V_{cb}|$ extracted from the ARGUS $\overline{B^0} \to D^{*+} \ell^- \overline{\nu}$ branching ratio is $|V_{cb}| = 0.040 \pm 0.002 \pm 0.004$ where the systematic error reflects the spread in the theoretical

Table 5: $|V_{cb}|$ from the inclusive semi-leptonic branching ratio.

Model	K_c (ps^{-1})	$|V_{cb}|$(LEP)	$|V_{cb}|$ ($\Upsilon(4S)$)
ACCMM[17]	37 ± 7	0.046 ± 0.005	0.045 ± 0.0045
ISGW[18]	50 ± 10	0.040 ± 0.005	0.039 ± 0.004

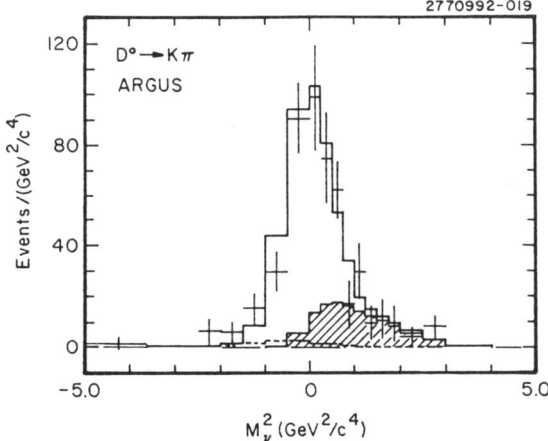

Figure 6. The distribution of M_ν^2 from ARGUS showing the signal for the decay $\overline{B^0} \to D^{*+}\ell^-\overline{\nu}$.

estimates for K from the available models of the exclusive decay.

Heavy Quark Effective Theory

A new theoretical approach to the problem of calculating the form factors for exclusive semi-leptonic heavy quark decays has attracted a lot of attention recently [21]. The basic idea is to notice that a B meson or a charm meson (any light quark bound to a very heavy quark) looks a lot like the hydrogen atom, a light e^- bound to a heavy proton. This is called Heavy Quark Effective Theory or HQET. What does this buy in the experimental extraction of CKM matrix elements?

• Recall that the electron wave function in the hydrogen atom is independent of the mass of the proton (up to hyperfine splittings which are a correction of order m_e/m_p). We might guess then that the light quark part of the meson wave function should be independent of the mass of the heavy quark up to hyperfine corrections of order the meson binding energy, Λ_{QCD}, over the heavy quark mass: Λ_{QCD}/m_Q.

• The heavy quark should behave like a free particle.

• The implication of these two preceding statements is that the meson wave function factors and so, therefore, does the matrix element. For example, when one calculates the hadronic part of the amplitude for a $B \to D^{(*)}\ell\nu$ decay, there is a heavy quark piece describing the decay of a free b quark to a free c quark which is calculable, and there is a light quark overlap integral describing the probability for the light quark cloud in the initial state to turn into the light quark cloud of the final state and this depends on the velocity of incoming and outgoing mesons. This light quark overlap integral is not calculable from first principles, but....

• The light quark overlap integral is universal. It is the same for all heavy pseudoscalar or vector meson to heavy pseudoscalar or vector meson transitions. The overlap integral is called the Isgur-Wise function, $\xi(v \cdot v')$, and it depends on the four velocities v and v' of the incoming and outgoing mesons. All four form factors for $B \to D^{(*)}\ell\nu$ decay can be written as known quantities times ξ!

It is a help that there is now only one unknown in the problem that needs to be modeled, but experimentally this is still unsatisfactory since we know nothing about ξ. What

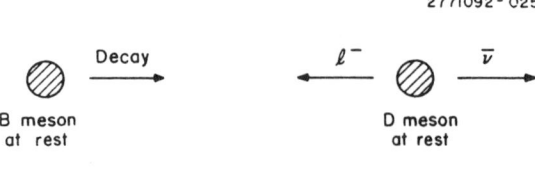

Figure 7. The kinematic configuration of the $B \to D^* \ell \nu$ decay at $q^2 = q^2_{max}$ or $v \cdot v' = 1$.

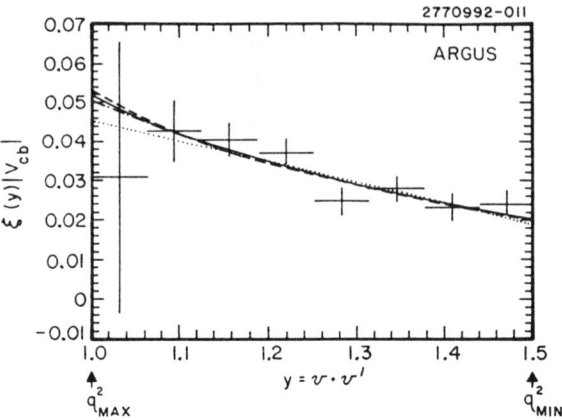

Figure 8. The plot of the differential decay rate vs $y = v \cdot v'$ from ARGUS. The left side of the plot with $y = 1$ corresponds to maximum q^2.

makes HQET attractive is that at zero recoil, when the initial and final state mesons are at rest and the leptons are emitted back to back as shown in Figure 7, $\xi(v \cdot v' = 1) = 1$. At this point, the form factors describing the overlap of the initial and final light quark wave functions are absolutely normalized. This absolute normalization is the result of the fact that at zero recoil, since the light quark wavefunction is independent of the mass of the heavy quark, the light quark wavefunction does not know a c quark has replaced a b quark. There is no velocity mismatch, and the overlap is perfect. At this magic kinematic point, $|V_{cb}|$ can be measured independent of any unknown form factor. Perhaps a simpler way of saying this is that is one trades statistics in data to measure the decay rate in a corner of phase space where the form factor is thought to be well known.

HQET is only an approximation ($M_Q \to \infty$) and there are corrections to it. For $b \to c$ transitions one might expect first order corrections as big as $(\Lambda_{QCD}/M_c) \sim 10 - 20\%$. By a stroke of good fortune, the point of zero recoil that is used to determine $|V_{cb}|$ is protected from first order corrections in (Λ_{QCD}/M_Q), and one need only to worry about corrections of order $(\Lambda_{QCD}/M_Q)^2$ [22].

ARGUS extracts $|V_{cb}|$ from the decay $\overline{B^0} \to D^{*+} \ell^- \overline{\nu}$, taking advantage of HQET, by plotting the differential branching ratio as a function of $y = v \cdot v'$. HQET says that at zero recoil which corresponds to q^2_{max} or $y = 1$ which is the left side of Figure 8, the form factor—the light quark overlap integral between initial and final states—is identically 1, and therefore, when properly normalized by known factors, the left intercept of the plot yields $|V_{cb}|$. To determine the intercept the data are extrapolated from $q^2 < q^2_{max}$. A variety of guesses can be used to estimate how the form factor falls with q^2, and there is a systematic error on the $|V_{cb}|$ as a result. ARGUS finds:

$$|V_{cb}| = 0.050 \pm 0.008 \pm 0.007. \qquad (6)$$

$|V_{cb}|$ extracted using this method agrees with the value of $|V_{cb}|$ derived from the inclusive analyses. The promise here is that there will be improvements to both the systematic and statistical error with more data.

It is extremely important that HQET and other models for calculating form factors be tested. Our extraction of $|V_{cb}|$ depends on them!! There have been several tests of the models from CLEO and ARGUS [19] [23], but there are many more and more precise tests of the models using the exclusive decays that need to be done. Another place that the models can be tested is in the charm semi-leptonic decays.

Semi-leptonic Charm Decay

In B semi-leptonic decays there is no way, apart from the magic zero recoil point (if we believe HQET) of measuring anything except the product of $|V_{cb}|$ times a form factor. In charm decay, one measures $|V_{cs}|$ or $|V_{cd}|$ times a form factor, but one can use the fact that the CKM matrix elements are known independently so the form factors themselves can be measured. This then allows us to test the calculations of the form factors or hadronic matrix elements directly.

The other advantage in the charm decays is that there have been a series of experiments, starting with E691 [24], with sufficient statistics and low background that have fit the differential decay distribution for exclusive semi-leptonic D decays instead of relying on integrated quantities, and they have extracted form factors from data which is a much more powerful technique, and ultimately what one wants to do in the B system.

The fixed target charm experiments use precision silicon vertex detectors to isolate $D^+ \to \overline{K}^{*0} \ell^+ \nu$ decays. They require a good 3-prong secondary vertex in the event with a lepton. The secondary vertex must be detached from the primary by $14 - 20\sigma$. After these cuts, they are left with clean samples of several hundred $D^+ \to \overline{K}^{*0} \ell^+ \nu$ decays with very little background as shown in Figure 9. They then do a multi-dimensional fit to the kinematic quantities specifying the final state and extract the three independent form factors that describe the pseudoscalar to vector decay. The ratios of the form factors can be extracted from the fit. If one then adds in the measured decay rate, $\Gamma(D \to K^* \ell \nu)$, the three individual form factors can be extracted. The form factors, the decay rates and the K^* polarization have all been measured, and it has been found that the ratios of the form factors and their shapes, as measured by the fit to the differential distribution, are in good agreement with most models. However, it has long been known that the measured branching ratio for the $D \to K^* \ell \nu$ transition is smaller than expected from naive quark models, and new results from E687 and E653 confirm this as is shown in Table 6 [25]. (Note that the MARKIII values [26] for the $D^+ \to K^- \pi^+ \pi^+$ branching ratios have been used for the normalizing modes where appropriate.)

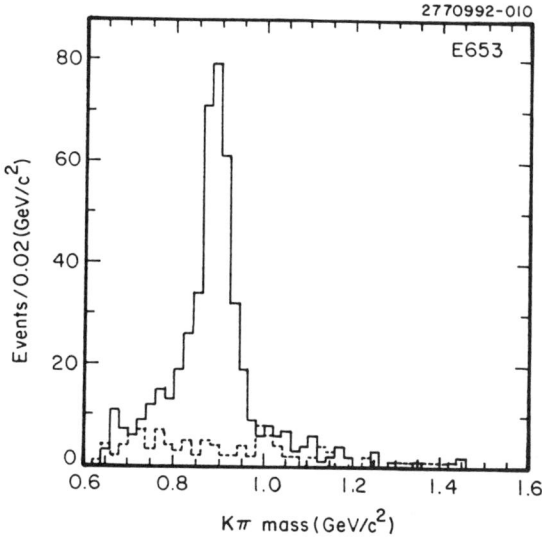

Figure 9. The $\overline{K}^{*0}$ signal from the decay $D^+ \to \overline{K}^{*0} \ell^+ \nu$ from E653.

It is not clear where the models are failing, and whether the agreement with some of the more recent calculations is just fortuitous. We probably cannot use HQET in the $D \to K^*$ transitions because it is not clear the s quark is heavy enough to be considered a heavy quark. Here $(\Lambda_{QCD}/M_s) \sim 50\%$ or larger. Certainly, the experimental machinery developed in charm semi-leptonic decays will need to be applied to the B decays if we ever want to trust any form factor model to help extract a precision measurement of $|V_{cb}|$.

12 Weak Decays, Rare Decays, Mixing, and CP Violation

Table 6: Experimental Results and Theoretical Predictions for $\Gamma(D \to K^*\ell\nu)$

Experiment	$\Gamma(D^+ \to \overline{K^{*0}}\ell^+\nu)$ $(10^{10} s^{-1})$
E687	$5.72 \pm 0.60 \pm 1.1$
E653	3.36 ± 0.90
PDG	3.83 ± 0.54
Theory	$\Gamma(D \to K^*\ell\nu)$ $(10^{10} s^{-1})$
Quark Models	9.5–9.8
QCD Sum Rules	
AOS	7.9 ± 3.9
BBD	4.0 ± 1.6
Lattice	5.2 ± 1.9

Determination of $|V_{ub}|$

We now want to discuss the new measurements of $|V_{ub}|$ that were presented at this conference. The most important thing about V_{ub} is that it be non-zero. This is because if any one of the elements of the 3×3 CKM quark mixing matrix is zero, then it cannot describe the observed CP violation in neutral K decays. That $V_{ub} \neq 0$ seems reasonably well established. The size of $|V_{ub}|$, however, is quite uncertain, both because the experiments do not agree, and the models used to extract $|V_{ub}|$ from the data do not agree.

The measurements of $|V_{ub}|$ come from AR-GUS and CLEO using semi-leptonic B decays. The basic process is the same as for the semi-leptonic $b \to c$ decay except now the b quark turns into a u quark with coupling strength V_{ub} as shown in Figure 10. All the theoretical uncertainties that plagued the extraction of $|V_{cb}|$ are even more of a problem for $|V_{ub}|$. Just as in the measurements of $|V_{cb}|$, $b \to u$ transitions can be searched for inclusively or exclusively.

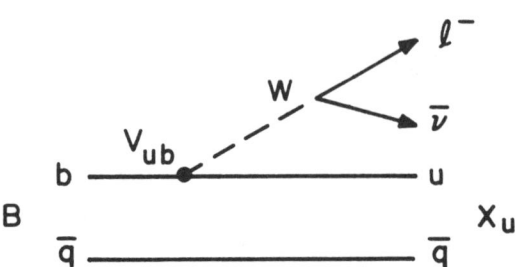

Figure 10. Schematic sketch of a semi-leptonic $b \to u$ decay.

Inclusive Determination of $|V_{ub}|$

The inclusive searches for $b \to u$ transitions look in the single lepton spectrum for leptons from B decay that are kinematically incompatible with coming from a B meson. The b quark will preferentially turn into a c quark when it weakly decays, but charmed quarks are heavy. The lightest mass particle containing a charmed quark is a D meson which has a mass of 1.87 GeV, while in a charmless decay, the final state hadronic mass can be as light as the mass of a pion. This difference in final state hadronic mass is reflected in the momentum of the lepton from the decaying B: the endpoint of the spectrum for leptons from $b \to u$ decays extends about 300 MeV past the endpoint of the spectrum for leptons from $b \to c$ decays. The strategy is therefore to search in the inclusive single lepton spectrum for leptons from B decay with a momentum that is past the endpoint for $b \to c$ decays.

The dominant background to this analysis comes from $e^+e^- \to q\bar{q}$ events from the continuum underlying the $\Upsilon(4S)$, and cuts on event shape are used to suppress those events. The background that remains after the event shape cuts is subtracted using data taken below the $\Upsilon(4S)$ resonance. The data from CLEO[27] after continuum subtraction are shown in Figure 11 and one sees an excess of leptons in

Table 7: Inclusive $|V_{ub}/V_{cb}|$

Model	ARGUS '90	CLEO '90	CLEO '92
ISGW[18]	0.20 ± 0.02	0.15 ± 0.02	0.095 ± 0.027
WBS[30]	0.13 ± 0.02	0.11 ± 0.02	0.065 ± 0.019
KS[31]	0.11 ± 0.01	0.09 ± 0.01	0.053 ± 0.015
ACCMM[17]	0.11 ± 0.01	0.09 ± 0.01	0.062 ± 0.018

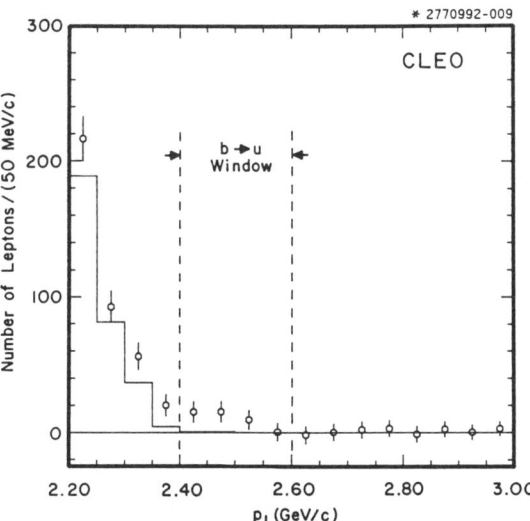

Figure 11. The CLEO lepton endpoint spectrum. The points are the continuum subtracted data. The histogram is the contribution from $b \to c$ decays calculated in the Monte Carlo.

the $b \to u$ region past the endpoint for $b \to c$ decays, but no excess past the $b \to u$ endpoint indicating the continuum has been properly subtracted.

From the lepton excess at the endpoint of the spectrum one needs to use a model to extrapolate the full spectrum and extract $|V_{ub}|$. The extrapolation is large and is very sensitive both to the details of how the models populate the Dalitz plot and which hadronic states dominate right at the endpoint, as well as the wave functions that the models use to calculate the form factors. Different models give very different results. Table 7 lists the new preliminary results for $|V_{ub}/V_{cb}|$ from the inclusive lepton spectrum reported from CLEO using the new CLEOII detector, along with results using similar methods that were previously published by ARGUS[28] and CLEO[29]. As Table 7 illustrates, the use of different models adds a considerable spread to the values of $|V_{ub}|$ extracted from the data. However, for any given model, the new CLEO result suggests a lower value of $|V_{ub}|$ than previously published. It is worth noting that the experiments actually measure $|V_{ub}|^2$. A difference of a factor of 2 in $|V_{ub}|$ between experiments is a factor of 4 in the observed rate!!

Exclusive Search for $b \to u$ Transitions

We will now discuss the searches for exclusive $b \to u$ decays. Both CLEO and ARGUS have searched for exclusive charmless decays and they disagree on the result.

A $b \to u$ transition can make any one of many mesons in the final state: $\pi, \eta, \rho, \omega, ...$ etc. The exclusive searches have so far concentrated on the channels $B \to \rho \ell \nu$ and $B \to \omega \ell \nu$. The analyses make a minimum lepton momentum cut and a variety of kinematic cuts to require consistency with a missing massless neutrino in the event. They also require that the rest of the event (all tracks and showers apart from the lepton and the particles that make candidate ρ or ω) be consistent with coming from the other B in the event. The cuts select events kinematically consistent with a

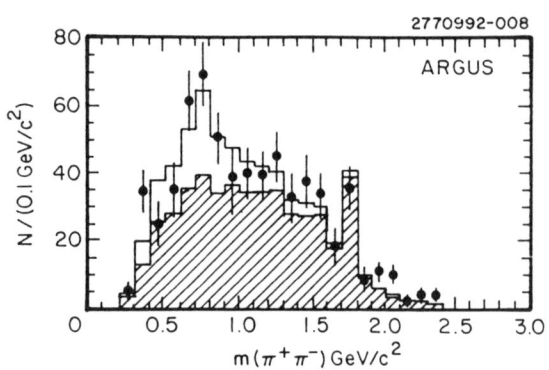
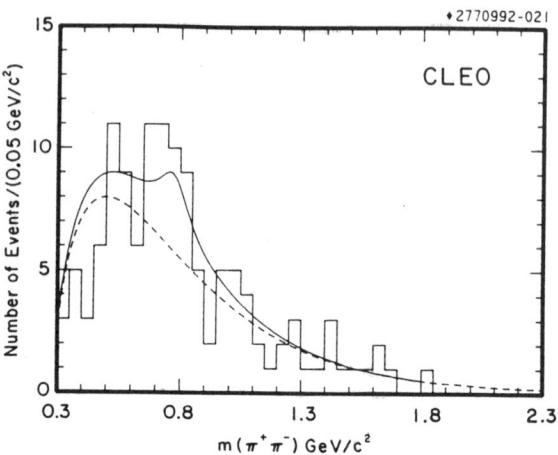

Figure 12. The ARGUS $\pi^+\pi^-$ invariant mass plot illustrating their $B^- \to \rho^0 \ell^- \bar{\nu}$ signal. The data are shown as filled points, the hatched histogram is the contribution from $b \to c$ events as calculated from their Monte Carlo, and the open histogram is the fit to the data.

Figure 13. The CLEO $\pi^+\pi^-$ invariant mass plot for their $B^- \to \rho^0 \ell^- \bar{\nu}$ candidate events in the lepton momentum range $2.3 < p_l < 2.6$ GeV/c. The dashed line indicates the level of contribution from the continuum underlying the $\Upsilon(4S)$. There is no evidence of a signal.

$B \to \rho\ell\nu$ or $\omega\ell\nu$ decay. They then look in the $\pi\pi$ or 3π invariant mass plot for a ρ or an ω which cannot be attributed to $b \to c$ decays or other background.

ARGUS sees a signal in the channel $B^- \to \rho^0 \ell^- \bar{\nu}$ [32]. Figure 12 shows the ARGUS $\pi^+\pi^-$ invariant mass plot. The contribution from $b \to c$ decays, calculated from their Monte Carlo, is shown, and there is a significant excess attributed to $b \to u$ transitions. They use the Monte Carlo to extract how much of the excess is due to $B^- \to \rho^0 \ell^- \bar{\nu}$ versus other $b \to u$ channels. They find

$$Br(B^- \to \rho^0 \ell^- \bar{\nu}) = (1.03 \pm 0.36 \pm 0.25) \times 10^{-3}. \quad (7)$$

CLEO has also looked for exclusive semileptonic $b \to u$ decays and sees nothing[33]. They cut harder in requiring kinematic consistency with a $B^- \to \rho^0 \ell^- \bar{\nu}$ decay, and suppress more background. They see no evidence of a ρ^0 signal in their $\pi^+\pi^-$ invariant mass plot as shown in Figure 13. CLEO also looks for evidence of exclusive semi-leptonic $b \to u$ decays in the modes $B^- \to \omega\ell^-\bar{\nu}$, and $\overline{B^0} \to \rho^+\ell^-\bar{\nu}$. The most convincing lack of a signal is in the $\omega\ell\nu$ channel where, unlike the case for $\rho\ell\nu$, the background does not peak in the signal region. There is no evidence for an ω peak in the 3π invariant mass plot of Figure 14 and there is no evidence of a ρ^+ signal in the $\pi^+\pi^0$ mass plot shown in Figure 15. The CLEO upper limits for the three modes can be combined under the assumptions that

$$\begin{aligned} Br(\overline{B^0} \to \rho^+ \ell^- \bar{\nu}) &= 2Br(B^- \to \rho^0 \ell^- \bar{\nu}) \\ &= 2Br(B^- \to \omega^0 \ell^- \bar{\nu}). \end{aligned} \quad (8)$$

The result from CLEO:

$$Br(B \to V_0 \ell^- \bar{\nu}) < 2 - 3 \times 10^{-4} \quad (9)$$

at the 90% confidence level, where V_0 is a neutral vector meson, contradicts the ARGUS result. (The spread in the CLEO upper limit reflects the different models used to calculate it).

What does this mean for $|V_{ub}|$? Tables 7 and 8 summarize the values for $|V_{ub}/V_{cb}|$ from

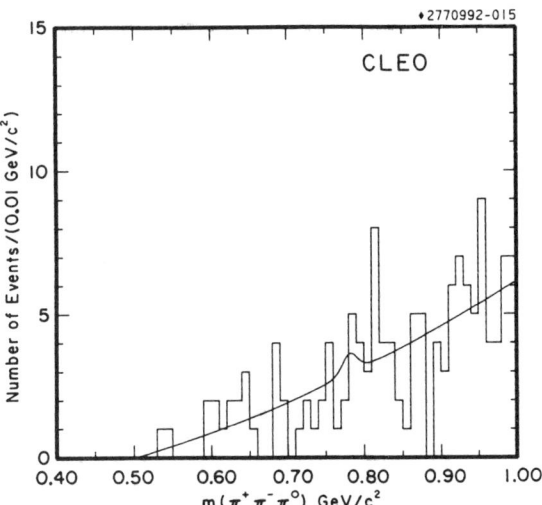

Figure 14. The CLEO $\pi^+\pi^-\pi^0$ invariant mass plot for $B^- \to \omega \ell^- \bar{\nu}$ candidate events in the lepton momentum range $2.3 < p_l < 2.6$ Gev/c

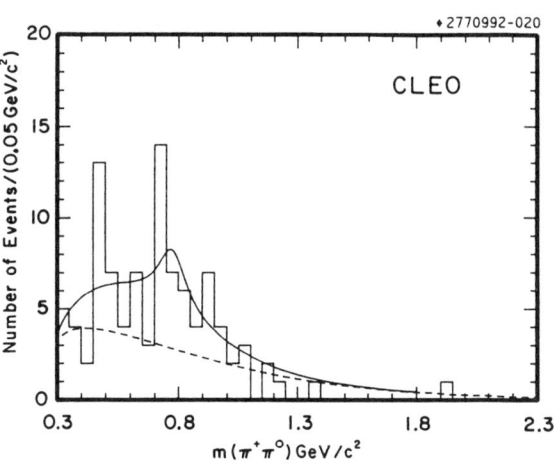

Figure 15. The CLEO $\pi^+\pi^0$ invariant mass plot for $\overline{B^0} \to \rho^+\ell^-\bar{\nu}$ candidate events in the lepton momentum range $2.3 < p_l < 2.6$ Gev/c. The dashed line indicates the level of contribution from the continuum underlying the $\Upsilon(4S)$.

the CLEO and ARGUS inclusive and exclusive analyses, for several of the most popular models. (Note that all the values of $|V_{ub}/V_{cb}|$ in Table 8 have been normalized to the somewhat arbitrary choice of $|V_{cb}| = 0.043$.) If we take a single form factor model, there is a definite lack of consensus on the size of $|V_{ub}/V_{cb}|$. CLEO says $|V_{ub}/V_{cb}|$ is small in the range of 0.1 or less. ARGUS favors a larger value around 0.15 or higher. Furthermore if we now add the theoretical spread, one can choose a value for $|V_{ub}/V_{cb}|$ anywhere from 0.3 down to 0.05.

Table 8: Exclusive $|V_{ub}/V_{cb}|$

Model	ARGUS	CLEO
ISGW[18]	0.30 ± 0.07	< 0.14
WBS[30]	0.16 ± 0.04	< 0.10
KS[31]	0.14 ± 0.03	< 0.08

Is there any hope? The experimental situation will presumably improve as more data and more analyses are done. The prospects for theoretical progress are not as good. There is no magic kinematic point where the form factors are absolutely normalized for $b \to u$ transitions as is the case for $b \to c$ transitions, since the light spectator quark wave function will undergo enormous perturbation as the heavy b quark turns into a light u quark. The only hope HQET can offer is that form factors for $B \to \rho l \nu$, $\pi l \nu$ transitions will be very similar to those for $D \to \rho l \nu$, $\pi l \nu$, $D \to K^* l \nu$ transitions. The argument is basically that the light quark wave function in the initial state doesn't know the difference between being bound to a b or a c quark, so that in the same q^2 range, the form factors for the B and D decays to the same final state, should be the same. Since $|V_{cd}|$ is known from neutrino production of charm, much as was argued in the $b \to c$ case, form factors can be measured and models tested in the charm case and then extended to the b case[21]. It seems to be the only hope for reducing the theoretical uncertainty on $|V_{ub}|$ as measured in semi-leptonic decays!

$B^0\overline{B^0}$ Mixing

We now want to turn to new results in measurements of $B^0\overline{B^0}$ mixing. This is included as an experiment that defines the Standard Model although the very existence of $B^0\overline{B^0}$ mixing is an important test of the model.

The Standard Model suggests that mixing should occur in the $B_d^0\overline{B_d^0}$ and in the $B_s\overline{B_s}$ systems from short distance contributions to the $B-\overline{B}$ mass splitting, ΔM_B that are dominated by box diagrams involving virtual t quarks as shown in Figure 16. The value of $\Delta M/\Gamma$ is given by:

$$\frac{\Delta M}{\Gamma} = \tau_B \frac{G_F^2 M_B}{6\pi^2} B_B f_B^2 |V_{tb}V_{td(s)}^*|^2 M_W^2 S(x_t)\eta \quad (10)$$

where B_B comes from the hadronic matrix element and is thought to be close to unity, f_B is the B meson decay constant, $S(x_t)$ is a known function of $x_t = M_t^2/M_W^2$, and η represents the short distance QCD corrections to the box diagram[34]. In principle we can use B_d or B_s mixing to extract the CKM parameters $|V_{td}|$ and $|V_{ts}|$. In practice, this is hard. There are many quantities needed to extract the matrix elements that we do not know such as the top mass, M_t, and the B meson decay constant, f_B, so the uncertainties on $|V_{td}|$ and $|V_{ts}|$ are large.

Mixing of neutral B mesons has been measured at LEP, the $\Upsilon(4S)$ machines and the $p\bar{p}$ machines. The basic methods are very similar. All experiments take advantage of the fact that b quarks are produced in pairs. The most popular analysis method selects events where both b's decay semi-leptonically and the sign of the lepton is then used to tag the flavor of the parent b. In the absence of backgrounds, if there were no mixing, then only opposite sign

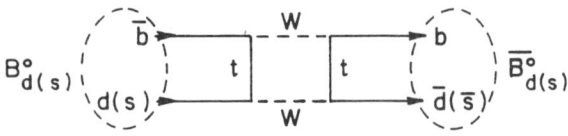

Figure 16. A box diagram that contributes to $B^0\overline{B^0}$ mixing.

dileptons would be produced. Mixing results in same sign lepton pairs.

Of course none of the experiments are background free—in fact the backgrounds are large. A particularly insidious background comes from the B mesons themselves. An event where one B decays semi-leptonically and the other B decays to a charmed hadron which decays semi-leptonically will give like sign dileptons even in the absence of mixing.

All experiments take advantage of the large b quark mass and use high momentum and high p_t leptons to suppress backgrounds to their dilepton sample. For example, L3 selects events with 2 hard leptons in opposite jets and extracts the mixing parameter, χ, using a likelihood fit to the p and p_t spectra of the leptons. χ is just a measure of the branching ratio for a B to mix before it decays semi-leptonically over the branching ratio for the B to decay semi-leptonically and is related to $\frac{\Delta M}{\Gamma}$. For example:

$$\chi_d = \frac{\text{Prob}(\overline{B_d} \to B_d \to \ell)}{\text{Prob}(B_d \to \ell)} \quad (11)$$

$$= \frac{(\Delta M_d/\Gamma_d)^2}{2 + 2(\Delta M_d/\Gamma_d)^2}. \quad (12)$$

At LEP energies one doesn't know which b hadron is the source of the leptons. Both B_s and B_d mesons are expected to exhibit mixing, and the value of χ measured is the average of B_d and B_s mixing, weighted by the amount of B_d and B_s produced in the fragmentation, parametrized as f_d and f_s respectively. CLEO and ARGUS also measure mixing but of B_d mesons only. Low energy results give χ_d. High energy experiments give $f_d\chi_d + f_s\chi_s$ if one assumes that the lifetimes of the different species of b hadrons are the same.

$$\chi(\text{LEP}) = f_d\chi_d + f_s\chi_s \quad (13)$$
$$\chi(\Upsilon(4S)) = f_d\chi_d \quad (14)$$

Table 9 lists the mixing results from the LEP experiments [35], and the results from ARGUS and CLEO[36]. The LEP results show the statistical and systematic error from each experiment. The systematic errors are strongly correlated, and do not include a systematic error due to the uncertainty in the differences of the lifetimes for the different b hadrons. The first error in the ARGUS and CLEO results is the combined experimental statistical plus systematic error, and the second error in each case comes from the uncertainty in the ratio of the charged to neutral B meson lifetime ($\tau^+/\tau^0 = 0.98 \pm 0.17$ from Equation 2) and the error in the production of charged and neutral B mesons at the $\Upsilon(4S)$ ($f^+/f^0 = 1.00 \pm 0.05$)[37]. (We have scaled the results for CLEO and ARGUS to be consistent with our quoted value for τ^+/τ^0).

We see that χ is well measured now both at LEP and by CLEO and ARGUS. The low energy experiments measure χ_d. The results from LEP and the $\Upsilon(4S)$ machines can be combined using values for f_d and f_s, in an attempt to extract χ_s. There are problems, however, with turning these measurements into CKM matrix elements.

• To get $|V_{td}|$ from low energy (CLEO and ARGUS) experiments, one needs to know Bf_B^2, and M_t. M_t is unknown and the theoretical estimates for f_B from lattice QCD or the quark model differ by factors of 2[38]. As a result, despite the precision of the experiments, the range on $|V_{td}|$ is quite large:

$$0.005 < |V_{td}| < 0.018. \quad (15)$$

• To get $|V_{ts}|$, we have the problem of not knowing f_s. This year, crude measurements of f_s were presented for the first time from LEP using the production rate of B_s and b-baryons. Using the experimental values: $f_s = 0.20 \pm 0.07$ and $f_d = 0.35 \pm 0.05$[39], we can combine the LEP and CLEO/ARGUS measurements and we see they are consistent with large B_s mixing:

$$\chi_s = 0.43 \pm 0.17. \quad (16)$$

However, even with more precision on f_s, there is a problem with extracting $|V_{ts}|$. As $\Delta M/\Gamma$ gets large, χ_s saturates at a value of .5 as can be seen from Equation 12, and it becomes very insensitive to the value of $|V_{ts}|$. Using the value of χ_s given in Equation 16, we get

$$(\Delta M/\Gamma)_s = 2.5^{+\infty}_{-2.5} \quad (17)$$

and we see that $|V_{ts}|$ is unconstrained by this measurement.

The moral is that the experimental phenomenon of $B\bar{B}$ mixing is well established and reasonably precisely measured, but the interpretation of the measurements is limited.

TESTING THE STANDARD MODEL

We now want to discuss new experimental results that are allowing us to test the Standard Model picture of weak decays. So far we have concentrated on experiments whose goal has been to measure parameters that define the Standard Model. The rest of the review will be of new results that are probing the validity of the Standard Model. The experiments we will cover are

• New branching fraction measurements in tau decay and the one prong problem
• Tests of $V - A$ structure in τ-decay
• Search for forbidden decays

Table 9: Measurements of the mixing parameter χ from dilepton events.

Experiment	$\chi_{LEP} = f_d \chi_d + f_s \chi_s$	χ_d
DELPHI	$0.121 \pm 0.042 \pm 0.017$	
OPAL	$0.125 \pm 0.017 \pm 0.015$	
ALEPH	$0.137 \pm 0.015 \pm 0.007$	
L3	$0.121 \pm 0.017 \pm 0.006$	
CLEO		$0.153 \pm 0.023 \pm 0.027$
ARGUS		$0.175 \pm 0.041 \pm 0.030$

- Search for higher order processes in the Standard Model

In many ways, the ultimate test of our understanding of the CKM sector of the Standard Model is to understand CP violation. We will end with a report on progress towards this goal.

Tau Decays

Let us start with taus as the first place to look at tests of the Standard Model.

$\tau \to e \nu \bar{\nu}$

The decay, $\tau \to e \nu \bar{\nu}$, should be precisely described by the Standard Model: both the absolute rate and the momentum and angular distributions of the final state leptons are predicted with an expected accuracy of a fraction of a percent. ALEPH[40], ARGUS[41], CLEO[42], DELPHI[43], OPAL[44] and TPC[45] have reported new measurements of the branching fraction $\tau \to e \nu \bar{\nu}$. The new CLEO measurement of the τ electronic branching ratio (B_e) uses events in which both taus have decayed to electrons. The number of two electron events, when corrected for backgrounds, efficiencies and normalized to the tau production cross section and integrated luminosity gives a measurement of B_e^2.

This is not an obvious way to measure B_e

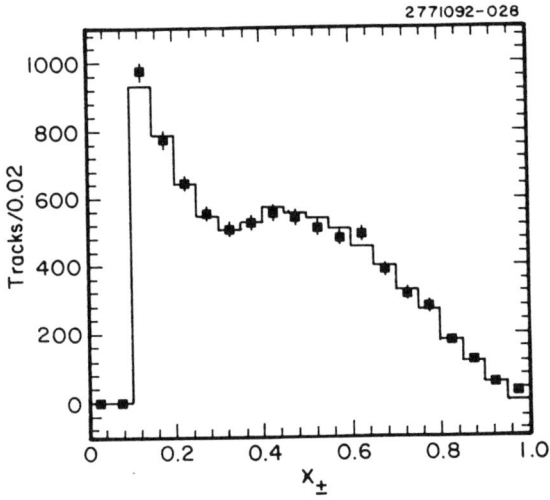

Figure 17. The momentum distribution ($x_\pm = p_{e^\pm}/E_{beam}$) of electrons used in the CLEO measurement of the branching ratio $\tau \to e \nu \bar{\nu}$. Data are filled points and the histogram is Monte Carlo.

due to formidable backgrounds from radiative Bhabhas and two photon events. The momentum spectrum for the di-electron sample is shown in Figure 17 with the Monte Carlo simulated spectrum overlaid. The somewhat unusual shape of the spectrum is the result of the hard cuts requiring that candidate events have missing transverse momentum pointing into the barrel of the detector, consistent with four escaping neutrinos, in order to suppress the Bhabha and two photon backgrounds.

The final result, $B_e = 0.1741 \pm 0.0015 \pm 0.0023$ is consistent with but lower than the

Table 10: New Tau Decay Branching Ratios

Tau Decay Mode[a]	Branching Fractions (%)	
$e\nu\bar{\nu}$	$17.76 \pm 0.33 \pm 0.27$	ALEPH (91) [40]
	$17.3 \pm 0.4 \pm 0.5$	ARGUS [41]
	$17.41 \pm 0.15 \pm 0.23$	CLEO (91) [42]
	$17.62 \pm 0.47 \pm 0.53$	DELPHI (91) [43]
	17.5 ± 0.4	OPAL (90 & 91) [44]
	$18.3 \pm 0.9 \pm 1.3$	TPC/Two-Gamma [45]
$\mu\nu\bar{\nu}$	$17.64 \pm 0.32 \pm 0.20$	ALEPH (91) [46]
	$17.2 \pm 0.4 \pm 0.5$	ARGUS [41]
	$17.73 \pm 0.42 \pm 0.45$	DELPHI (91) [43]
	16.8 ± 0.4	OPAL (90 & 91) [44]
	$17.4 \pm 0.9 \pm 1.1$	TPC/Two-Gamma [45]
$h^-\bar{\nu}$ [b]	$11.7 \pm 0.6 \pm 0.8$	ARGUS [41]
	$12.2 \pm 0.3 \pm 0.4$	OPAL (90 & 91) [47]
	$11.9 \pm 0.7 \pm 0.7$	DELPHI (90) [47]
$h^-\pi^0\bar{\nu}$	$22.6 \pm 0.4 \pm 0.9$ [c]	ARGUS [55]
	$24.83 \pm 0.15 \pm 0.53$ [d]	CLEO [48]
	$22.4 \pm 0.8 \pm 1.3$ [c]	DELPHI (90) [47]
	$23.8 \pm 0.6 \pm 0.7$ [c]	OPAL (90 & 91) [47]
$h^-\pi^0\pi^0\bar{\nu}$ [e]	$8.64 \pm 0.17 \pm 0.44$	CLEO [49]
$h^-3\pi^0\bar{\nu}$	$1.04 \pm 0.07 \pm 0.10$	CLEO [49]
$h^-4\pi^0\bar{\nu}$	$0.15 \pm 0.05 \pm 0.05$	CLEO [49]
$h^-\pi^0\eta\bar{\nu}$	$0.17 \pm 0.02 \pm 0.02$	CLEO [50]
$\pi^-\pi^-\pi^+\bar{\nu}$ [f]	$6.80 \pm 0.1 \pm 0.4$	ARGUS [51]
B_1	$86.6 \pm 0.3 \pm 0.8$	ARGUS [41]
	$84.08 \pm 0.59 \pm 0.45$ [g]	DELPHI (91) [47]
	$84.48 \pm 0.27 \pm 0.23$ [g]	OPAL (90 & 91) [52]
B_5	$0.31 \pm 0.11 \pm 0.07$	DELPHI (91) [47]
	$0.26 \pm 0.06 \pm 0.05$	OPAL (90 & 91) [52]

(a) h represents the sum of π and K
(b) $K^* \to K_L\pi^-$ excluded
(c) $K^* \to K^-\pi^0$ excluded
(d) $K^* \to K^-\pi^0$ included
(e) $K^* \to K_S(\to \pi^0\pi^0)\pi^-$ included
(f) $K^* \to K_S(\to \pi^+\pi^-)\pi^-$ excluded
(g) $K^* \to K_S(\to \pi^+\pi^-)\pi^-$ included

world average and is of comparable precision. The results are summarized in Table 10, along with the new results for other tau decay modes.

Tests of τ Couplings

As is discussed by Luigi Rolandi in his contributions to these proceedings [1], a precision measurement of the tau leptonic branching fraction can be combined with measurements of the tau mass and lifetime to measure the coupling between the τ and the W. The leptonic decays of the tau can also be used to probe the structure of the W−lepton coupling. The distribution of the final charged lepton in angle and momentum contains information about the form of the weak current and allows one to experimentally address to what extent the weak current in tau decay is restricted to a $V - A$ form. For example, if one assumes there are no scalar and tensor currents, then the parameter ρ which characterizes the shape of the lepton momentum spectrum measures the relative amount of $V - A$ versus $V + A$ interaction. The parameter ξ modifies the final state lepton momentum spectrum for different angles θ between the tau spin direction and the lepton momentum [53]. ARGUS has analyzed a sample of $e - \mu$ events and finds $\rho = 0.78 \pm 0.05$ for e^-, 0.76 ± 0.06 for μ^- [54], and $|\xi| = 0.90 \pm 0.13 \pm 0.13$ [55], consistent with the expected values of 0.75 and 1, respectively, for a $V - A$ interaction.

The τ One-Prong Problem

The electronic branching ratio of the tau is only one piece of a nagging problem in tau decay that is a test of the Standard Model only in the broadest sense of the words. For seven years now there has been a concern that there is a deficit of tau decays [56]. The basic question is does the sum of the exclusive branching ratios of the tau equal 100%? In 1985, when many of the exclusive decays had not yet been measured, but the inclusive decays to 1 and 3 charged tracks had, it was noted that with theoretical constraints on the unmeasured modes, there seemed to be a significant deficit (7%) of decays, particularly to final states with 1 charged track. This has since been referred to as the 1-prong problem. There has been speculation about unobserved decay modes causing the deficit but mostly people have hoped that new and more precise measurements would make the problem go away.

Over the last 2 years, all of the LEP experiments as well as ARGUS, CLEO, and TPC have contributed to producing new and precise measurements of almost all of the decay branching ratios of the tau. All new measurements released for the principle modes since the last PDG compilation appear in Table 10. There are two important points concerning these measurements. (1) New measurements of every branching fraction of the tau except B_5 have been reported within the last year or two that are comparable in precision to previous world averages. (2) In EVERY case, for these new experimental measurements, the systematic error is comparable to or greater than the statistical error.

Let us now address the one prong problem or deficit of tau decays. The one prong problem could not arise if the measurements were done by isolating a sample of tau pairs and classifying each of the final states. But in general this is not how it has been done — mostly because for low energy (non-LEP) experiments, it is very difficult to isolate a background-free sample of tau pairs. Instead, exclusive branching fractions are typically measured by identifying a sample of tau decays to a particular final state, and dividing the efficiency-corrected number of events by the number of taus produced, as determined from the integrated luminosity of the experiment. Now, the LEP experiments CAN isolate background-free samples of tau pairs but the exclusive branching fractions that we will talk about from LEP are calculated using the traditional technique described above. The sum of exclusive branching fractions measured in this traditional way could be smaller than the inclusive branching fractions if there were decay modes that no one had explicitly searched for.

Let us see how the new branching fractions affect the 1-prong deficit. The standard procedure is to take all measurements and average them weighted by their combined statistical and systematic error. The entries in the first column of Table 11 shows the 1992 PDG world average [13] values combined with the new measurements from L3, OPAL, DELPHI, ALEPH, ARGUS, CLEO, and TPC which were either published in the past year and not included in the 1992 PDG averages, or submitted to this conference. We see that the sum of all exclusive modes does not equal 100%. It misses by about 6.6σ which is a healthy effect and a bit disturbing. Furthermore, if we compare the world average inclusive one prong rate to the sum of the one prong modes, we see that it is predominantly in the tau decays to one charged particle where we seem to be falling short.

However, the averaging procedure just illustrated is not necessarily a good one. Reason 1 is that we now in many cases have a single measurement that is as precise as the world average which is a combination of many experiments, each with significantly larger errors. Reason 2 is that most of the experimental measurements are now dominated by systematic errors. At the level of precision of these new measurements there may be correlated errors between experiments. Reason 3 is that there is often considerable ambiguity in how experiments have treated the decay $\tau \to K^*\nu$ which contributes to both 3 and 1 prong decays. We have tried treat decays $\tau \to K^*\nu^-$ consistently when combining numbers from different experiments, and have followed the PDG convention[13]. For all these reasons, we consider the one prong problem from a different perspective, by taking, instead of world averages, the best single measurement (smallest fractional error) of each branching ratio and seeing if a deficit still exists when these measurements are summed.

This next column illustrates this procedure. CLEO has the smallest errors on B_e and the multi π^0 decay branching ratios, ALEPH has the smallest errors on the three prong branching ratios and B_μ, and OPAL has the smallest errors on $\tau \to \pi/K\nu$ and the inclusive one and three prong branching ratios, B_1 and B_3. (We have corrected the OPAL and ALEPH numbers to conform to the PDG K^* convention). We see that the exclusive 1-prong branching ratios fall only 1.8σ short of B_1. To make sure that eliminating the one prong problem has not created a three prong problem we check and see the three prong branching ratios again fall only 0.7σ short of B_3. Most importantly, we see that the total sum over all the measured exclusive modes is $97.2 \pm 1.4\%$ which is just 2σ short of 100%, and is hard to get terribly excited about.

Now, there can easily be objections to what we have just done. The obvious objection to this procedure is that ignores lots of excellent experimental data and by increasing the error bars, one makes the effect look less significant. Perhaps a better approach is to take the two or three best experiments of comparable precision and average them. The problem is that it is difficult to make an unbiased selection, and one is still faced with the problem of correlated systematic errors. The conclusion: there does not seem to be strong evidence for a deficit in tau decays. There is clear need for still more work. For example it is interesting to note that one of the reasons that

Table 11: Summary of Tau Decay Branching Ratios(%).

Decay Mode†	New World Average	"Best" Measurement	
$e\nu\bar{\nu}$	17.61 ± 0.16	17.41 ± 0.28	CLEO
$\mu\nu\bar{\nu}$	17.40 ± 0.19	17.54 ± 0.32	ALEPH
$h^-\bar{\nu}$ (incl. $K^* \to K_L\pi^-$)	12.48 ± 0.28	12.66 ± 0.5	OPAL
$h^-\pi^0\bar{\nu}$ (incl. $K^* \to K^-\pi^0$)	23.99 ± 0.34	24.83 ± 0.55	CLEO
$h^-\pi^0\pi^0\bar{\nu}$ (incl. $K^* \to K_S(\to \pi^0\pi^0)\pi^-$)	8.8 ± 0.4	8.64 ± 0.47	CLEO
$h^-3\pi^0\bar{\nu}$	1.22 ± 0.14‡	1.04 ± 0.12	CLEO
$h^-4\pi^0\bar{\nu}$		0.15 ± 0.07	CLEO
$\pi^-\pi^0\eta\bar{\nu}$ (excl. $\eta \to 3\pi^0, \pi^+\pi^-\pi^0$)	0.07 ± 0.01	0.07 ± 0.01	CLEO
Σ 1-prong	81.6 ± 0.7	82.3 ± 1.0	
B_1	85.4 ± 0.2	84.17 ± 0.35	OPAL
$3h^-\bar{\nu}$ (incl. $K^* \to K_S(\to \pi^+\pi^-)\pi^-$)	7.82 ± 0.25	9.80 ± 0.72	ALEPH
$3h^- \geq 1\pi^0\bar{\nu}$ (incl. $K^* \to K_S(\to \pi^+\pi^-)\pi^-$)	5.2 ± 0.4	4.95 ± 0.68	ALEPH
Σ 3-prong	13.0 ± 0.5	14.8 ± 1.0	
B_3	14.5 ± 0.2	15.57 ± 0.34	OPAL
B_5	0.14 ± 0.03	0.10 ± 0.03	HRS
TOTAL (Σ 1-prong + Σ 3-prong + B5)	94.7 ± 0.8	97.2 ± 1.4	

†h represents the sum of π and K.
‡Includes both $\tau^- \to h^-3\pi^0\bar{\nu}$ and $\tau^- \to h^-4\pi^0\bar{\nu}$

the discrepancy was reduced in the 1-prongs was that the OPAL topological branching ratios have very small errors, and find B_1 significantly smaller than the world average and B_3 significantly larger. The other place where the single measurements differ significantly from the world average is the CLEO $\tau \to h^-\pi^0\nu$ and the ALEPH branching ratio for $\tau \to 3h^-\nu$. This last case illustrates the weakness of this method. There is another measurement of comparable precision from ARGUS that differs from the ALEPH number by more than 2σ once the treatment of the K^* and K modes is made consistent between the two. If we use the ARGUS number, the shortfall from 100% become 4.2σ and indicates the beginnings of a 3-prong problem! Future measurements need to concentrate on topological branching fractions and 3-prongs to see if the problem clears up or goes away.

The conclusion that there is no deficit of τ decays is further supported by an analysis from ALEPH where all the tau branching ratios are measured simultaneously [57]. A clean $\tau^+\tau^-$ sample is selected and then all events are distributed among known generic states. The efficiency for each mode as well as the migration matrix which calculates contributions from each decay mode to all other decay modes are calculated. They find that undetected modes can account for less than 2.1% of the total—again very consistent with all decays being accounted for, although some of the branching ratios that they get for the individual modes are quite different from those listed in Table 11.

Rare Decays

We now leave the tau problem and go on to the subject of rare decays. Rare decays can test the Standard Model in two distinct ways.

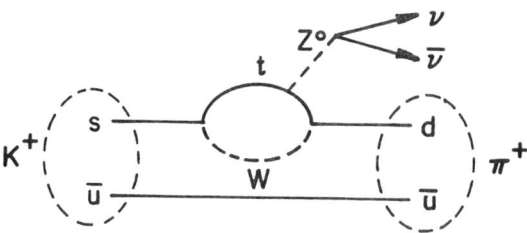

Figure 18. Penguin diagram of the decay $K^+ \to \pi^+\nu\bar{\nu}$

Forbidden Decays

There are many decays that are strictly forbidden in the Standard Model, i.e. $Z^0 \to e^-\mu^+$. Since lepton flavor conservation is added pretty much by hand and its origins are not understood, it is important to put as stringent limits on it as possible because many extensions of the Standard Model allow for the lepton flavor violation at some level[58]. The higher the experimental sensitivity, the greater the restrictions on the mass at which new physics can occur. The most sensitive experiments are now probing mass scales of 200 TeV/c^2[59]. Many new limits were placed on lepton number violating and other forbidden decays of the μ, π, K, τ and Z. These are summarized in Table 12.

Higher Order Decays

There is a second class of rare decays that is allowed in the Standard Model but highly suppressed because these decays can only occur via higher order (non-tree level) processes. Most of the activity here has been in kaon experiments.

One type of higher order process that is particularly interesting is penguin decay. Penguin decays are one loop processes that result in an effective FCNC interaction. An example

Table 12: New Rare Decay Limits or Branching Fractions

Decay Mode	Limit	Confidence Level	Experiment
$K_L \to \mu e$	$< 3.3 \times 10^{-11}$	90%	BNL E791 [59]
$K^+ \to \pi^+ \mu^+ e^-$	$< 2.1 \times 10^{-10}$	90%	BNL E777 [59]
$K^+ \to \pi^+ \nu \bar{\nu}$	$< 5 \times 10^{-9}$	90%	BNL E787 [59]
$K_L \to \mu^+ \mu^-$	$(7.0 \pm 0.9) \times 10^{-9}$		BNL E791 [59]
$K_L \to e^+ e^-$	$< 4.1 \times 10^{-11}$	90%	BNL E791 [59]
$\mu^+ + N \to e^+ + N$	$< 4.4 \times 10^{-12}$	90%	PSI [60]
$Z^0 \to \mu^+ e^-$	$< 1.0 \times 10^{-5}$	95%	ALEPH [61]
$Z^0 \to \tau^+ e^-$	$< 8.0 \times 10^{-5}$	95%	ALEPH [61]
$Z^0 \to \tau^+ \mu^-$	$< 5.5 \times 10^{-5}$	95%	ALEPH [61]
$\tau^+ \to e^+ \pi^0$	$< 2.0 \times 10^{-4}$	90%	ALEPH [62]
$\tau^+ \to \mu^+ \pi^0$	$< 1.2 \times 10^{-4}$	90%	ALEPH [62]
$\tau^+ \to e^+ \gamma$	$< 1.7 \times 10^{-4}$	90%	ALEPH [62]
$\tau^+ \to \mu^+ \gamma$	$< 1.0 \times 10^{-4}$	90%	ALEPH [62]
$\tau^+ \to \mu^+ \gamma$	$< 4.2 \times 10^{-6}$	90%	CLEO [63]

is the decay $K^+ \to \pi^+ \nu \bar{\nu}$ which is an electro-weak penguin, as shown in Figure 18. (Penguin refers to the generic type of one loop process and the electro-weak label refers to the particle radiated in the loop.) Penguins have attracted particular interest among the various suppressed modes because they offer windows on new physics. For example, the presence of an unexpected heavy particle in the loop could significantly alter the rate for such decays. Furthermore, the rates for these decays are sensitive to $|V_{td}|$ and the top quark mass, M_t, and some of the calculations are thought to be reliable enough that someday measurements of these decay rates may help to pin down $|V_{td}|$ once M_t is known[34]. Another motivation to study penguins is that the interference of a penguin amplitude with a Cabibbo suppressed amplitude is expected to result in CP violation [64]. However, for the moment, we are confined to using experimental limits on these decays to test our ability to calculate short distance behavior.

A new limit on $\Gamma(K^+ \to \pi^+ \nu \bar{\nu})$ was presented by BNL E787 [59]. This penguin is particularly attractive because it can be reliably calculated, and the branching ratio is thought to be of order 10^{-10}, which is large enough that there is a reasonable chance to measure it. The experimental signal for the decay is a charged pion from a K^+ decay and nothing else in the event. The main backgrounds, $K^+ \to \mu^+ \nu$ and $K^+ \to \pi^+ \pi^0$, produce monochromatic muons and pions because they are two-body final states, so they can be suppressed by excluding pions with momenta in the contaminated ranges. The new limit of $Br(K^+ \to \pi^+ \nu \bar{\nu}) < 5 \times 10^{-9}$ at the 90% confidence level is approaching the necessary sensitivity to observe this decay.

An outstanding discrepancy in another suppressed decay, $\pi^0 \to e^+ e^-$, that hinted at a violation of the Standard Model, was resolved this year. This is a fourth order electromagnetic decay, and unitarity constraints require that its branching fraction be less than

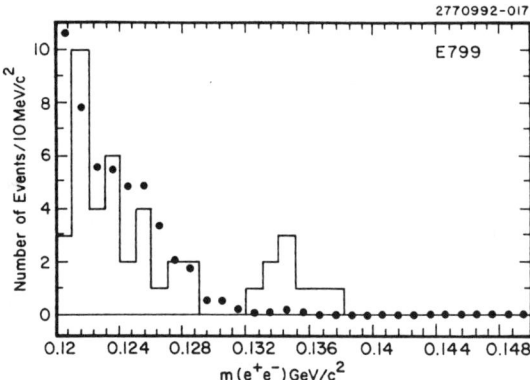

Figure 19. The e^+e^- invariant mass distribution observed by FNAL E799 showing the $\pi^0 \to e^+e^-$ signal. The histogram is the data, and the points are $\pi^0 \to e^+e^-\gamma$ Monte Carlo.

5×10^{-8}. However, past experiments have reported branching fractions[65] higher than this, with an average value of $(18^{+7}_{-6}) \times 10^{-8}$. At this conference BNL E851 and FNAL E799 reported new observations of this mode. To obtain a large event sample, E799 looks at $K_L \to 3\pi^0$ decays where two of the π^0's are reconstructed in their two photon decay mode. After cuts to suppress conversions, shower overlaps, and events with escaping electrons or photons, they obtain the e^+e^- invariant mass distribution shown in Figure 19. One can see that the $e^+e^-\gamma$ events which populate the low mass region contribute little background in the region of the π^0 mass. The new branching fractions are $(6.1 \pm 2.8) \times 10^{-8}$ [59] and $(6.9 \pm 1.8) \times 10^{-8}$ [66] from E851 and E799 respectively, both of which are lower than the previous results and compatible with the unitarity bound.

CP Violation

The final topic we want to discuss is the progress reported in attempts to understand the phenomenon of CP violation. The imaginary phase allowed in the 3×3 CKM matrix is the leading contender as the source of the observed CP violation in the K meson system. If that is indeed its origin, we can make several unambiguous predictions.

(1) ε'/ε should be nonzero barring unfortunate cancellations.

(2) CP violation should occur in decays such as $K_L \to \pi^0 e^+ e^-$

(3) CP violation should occur in the B meson system at a level that may someday be observable.

ε'/ε

The most significant progress in the search for direct CP violation, or any source of CP violation outside of that observed in the neutral K system, has been in the measurements of ε'/ε. There are two experiments that measure ε'/ε: NA31 at CERN and E731 at Fermilab. Neither experiment had new results to report at this conference. The current results reported by the two experiments are[67]:
NA31: $\text{Re}(\varepsilon'/\varepsilon) = (2.3 \pm 0.7) \times 10^{-3}$
E731: $\text{Re}(\varepsilon'/\varepsilon) = (0.60 \pm 0.69) \times 10^{-3}$.
Theoretical predictions for ε'/ε are consistent with both experiments, but the important question remains unresolved: we still do not know if direct CP violation occurs in K decay. Both collaborations expect updated results soon, and both are planning next generation experiments.

$K_L \to \pi^0 e^+ e^-$

CP violation in the decay $K_L \to \pi^0 e^+ e^-$ can occur in two ways. The K_L consists mostly of a CP odd state K_2 with a little of the CP even state K_1 mixed in:

$$K_L \sim K_2 + \varepsilon K_1, \qquad |\varepsilon| \sim 2.26 \times 10^{-3}. \quad (18)$$

The decay channel of interest is the "direct" decay of the CP odd K_2 piece to the CP even state π^0 plus an off shell photon which makes an e^+e^- pair; however, the K_1 component that arises from the well established CP violation through mixing also contributes to the same final state.

The attraction of this decay mode is that the amplitude of both the direct ($K_2 \to \pi^0 e^+ e^-$) and indirect ($K_1 \to \pi^0 e^+ e^-$) transitions are comparable. A disadvantage of this way of searching for direct CP violation is that the expected branching ratio is small (on the order of 10^{-11}), and there is a CP conserving contribution to the decay rate from

$$K_2 \to \pi^0 \gamma\gamma \to \pi^0 e^+ e^- \qquad (19)$$

which could have a branching ratio as high as 10^{-11}. Measurements of $K_L \to \pi^0 \gamma\gamma$ are crucial engineering studies towards the goal of measuring direct CP in the decay $K_L \to \pi^0 e^+ e^-$.

A new measurement of the branching ratio for $K_L \to \pi^0 \gamma\gamma$ was presented by NA31 [68]. The signature for this decay is four photon clusters, with two of them consistent with coming from a π^0, and no charged particles in the detector. The most serious potential background is from $K_L \to 3\pi^0$ which is removed with kinematic cuts. They find $Br(K_L \to \pi^0 \gamma\gamma) = (1.7 \pm 0.2 \pm 0.2) \times 10^{-6}$. They also find that the $\gamma\gamma$ pairs have high invariant mass, consistent with the predictions of chiral perturbation theory. Together these results indicate that the CP conserving contribution will be small compared to direct CP violating contributions in $K_L \to \pi^0 e^+ e^-$ decay.

CP Violation in the B System

CP violation in the B system is still in the engineering stage as well. The decay mode $\overline{B^0} \to \psi K_S$, a favorite mode for looking for CP violation, is well established and there was a new measurement of the branching ratio from CLEO of $Br(\overline{B^0} \to \psi K_S) = (0.102 \pm 0.043 \pm 0.017)\%$[69]. There was also a report from CLEO of a new upper limit on the decay branching ratio $\overline{B^0} \to \pi^+\pi^-$[70], another favorite mode for seeing CP violation in the B system. The upper limit is $Br(\overline{B^0} \to \pi^+\pi^-) = 4.8 \times 10^{-5}$ at the 90% confidence level, which can be compared with the theoretical prediction of 2×10^{-5} for $|V_{ub}/V_{cb}| = 0.1$[71]. The new value of $|V_{ub}/V_{cb}|$ from CLEO is bad news for this mode. There is clearly a long way to go before we see CP violation in the B system!

We hope ε'/ε, a measurement of $K_L \to \pi^0 e^+ e^-$, or direct observation of CP violation the B meson system will either confirm or refute the hypothesis that CP violation has its origins in the fact that there are three generations and that the eigenstates of the strong interactions are diagonal on a different basis than the eigenstates of the weak interactions. Perhaps future experiments can convince us that while we may not understand the why of CP violation, we at least know phenomenologically where it comes from.

Non-Standard Model CP Violation

The origins of CP violation are not firmly established, and it is a good idea to look for evidence of CP violation in sectors apart from decays of heavy flavor mesons where it is expected. One way to search for CP violation in the lepton sector has been to look for an electric dipole moment of the electron by searching for evidence of linear Stark shifts in heavy atoms. No such linear Stark shifts have been seen and the measured electron electric dipole moment is $(-2.7 \pm 9.3) \times 10^{-27} e\text{-cm}$[72].

Another search for CP violation beyond the Standard Model was reported by OPAL[73]. They search for CP violation in $Z^0 \to \tau^+\tau^-$ events where each tau decays to one charged particle (either lepton or hadron). They study the distribution of a CP-odd tensor observable and see no evidence for CP violation. They use their data to put a limit on the weak dipole moment of the tau (CP violation at the $Z\tau\tau$ vertex) of $|d_\tau| \leq 7.0 \times 10^{-17}$ e-cm at 95% confidence level.

Conclusions

In the future, we speculate that the situation we have presented in this review will be somewhat reversed. Tau physics, instead of testing the Standard Model will be used to define the low mass hadronic sector of the model. Already, experiments are attempting to use the ratio of hadronic to leptonic branching ratios of the tau to measure α_s, the strong coupling constant[74], and ARGUS is using the 3π decays of the tau to define the properties of the a_1 resonance[51]. Similarly, CP violation in the B system, penguin decays and ε'/ε, if they become established, will move from the realm of testing the Standard Model to defining it with precision measurements of all the elements of the CKM matrix. Of course, what we, at least, hope for, is that somewhere along the way something will break down and we will at last find a chink in the armor of this all too successful model.

ACKNOWLEDGEMENTS

It is a pleasure to acknowledge the assistance we received from many collaborators and colleagues in preparing this review. We would like to express particular thanks to David Perticone, Henning Schroeder, and Vivek Sharma for the material they provided, to Ed Thorndike for his input on the structure and substance of the review, and to Peter Lepage for educating us about HQET.

REFERENCES

1. L. Rolandi, "Precision Tests of the Electroweak Interaction," these proceedings.
2. R. G. H. Robertson, "Neutrino Masses, Mixing, and Non-Accelerator Experiments," these proceedings.
3. M. Kobayashi and T. Maskawa, "CP Violation in the Renormalizable Theory of Weak Interaction," *Prog. Theor. Phys.* 49, pp. 653-657, (1973).
4. ALEPH Collaboration, "Updated Measurement of the Average b Hadron Lifetime," submitted to XXVI International Conference on High Energy Physics, Dallas, TX, Aug 6-13, 1992.
5. DELPHI Collaboration, "Refined Measurement of the Average Lifetime of B Hadrons using high p_t Muons," submitted to XXVI International Conference on High Energy Physics, Dallas, TX, Aug 6-13, 1992.
6. L3 Collaboration, B. Adeva *et al.*, "Measurement of the Lifetime of B-Hadrons and a Determination of $|V_{cb}|$," *Phys. Lett.* B270, pp. 111-122, (1991).
7. OPAL Collaboration, P. D. Acton *et al.*, "Measurement of the average B hadron lifetime in Z^0 decays," *Phys. Lett.* B274, pp. 513-525, (1992).
8. Particle Data Group, "Review of Particle Properties," *Phys. Lett.* B239, (1990).
9. ALEPH Collaboration, "Measurement of the B^0 and B^+ Meson Lifetimes," "A Measurement of the B_s Lifetime," "A Measurement of the b Baryon Lifetime," submitted to XXVI International Conference on High Energy Physics, Dallas, TX, Aug 6-13, 1992; DELPHI Collaboration, "A Measurement of B Meson Production and Lifetime Using $D\ell$ Events in Z^0 Decays," "B_s^0 tagging at LEP energies using D_s and ϕ mesons," "Measurement of Λ_b Production and Lifetime in Z^0 Hadronic Decays," submitted to XXVI International Conference on High Energy Physics, Dallas, TX, Aug 6-13, 1992. The ALEPH and

DELPHI measurements have been averaged. We have assumed a correlated systematic error of 0.10 ps in the case of the B^- and $\overline{B^0}$ lifetimes. The other errors are predominantly statistical.

10. E653 Collaboration, "Measurement of the Lifetimes of Charged and Neutral Beauty Hadrons,"submitted to XXVI International Conference on High Energy Physics, Dallas, TX, Aug 6-13, 1992.

11. CLEO Collaboration, "Measurement of the $B^+ : B^0$ lifetime ratio,"submitted to XXVI International Conference on High Energy Physics, Dallas, TX, Aug 6-13, 1992.

12. Robert Clare, "The b-quark Semi-leptonic Branching Ratio at LEP, " talk presented at the XXVI International Conference on High Energy Physics, Dallas, TX, Aug 6-13, 1992.

13. Particle Data Group, "Review of Particle Properties," *Phys. Rev.* D45, (1992).

14. M. A. Shifman and M. B. Voloshin, "Hierarchy of Lifetimes of Charmed and Beautiful Hadrons," *Sov. Phys. JETP* 64, pp. 698-728, (1986); M. A. Shifman, "Theory of Weak Interactions: Recent Developments and Problems," *Int. J. Mod. Phys*, A3, pp. 2769-2826, (1988).

15. G. Altarelli and S. Petrarca, "Inclusive Beauty Decays and the Spectator Model,"*Phys. Lett.* B261, pp. 303-310, (1991); I.I. Bigi, N.G. Uraltsev, and A.I. Vainshtein, "Nonperturbative Corrections to Inclusive Beauty and Charm Decays: QCD versus Phenomenological Models," Fermilab Preprint, Fermilab-Pub-92/158-T.

16. CLEO Collaboration, S. Henderson *et al.*, "Measurements of semi-leptonic branching fractions of B mesons at the $\Upsilon(4S)$ resonance,"*Phys. Rev.* D45, pp. 2212-2231, (1992).

17. G. Altarelli *et al.*, "Leptonic Decay of Heavy Flavors: A Theoretical Update," *Nucl. Phys.* B208, pg. 365, (1982).

18. N. Isgur *et al.*, "Semi-leptonic B and D decays in the Quark Model,"*Phys. Rev.* D39, pg. 799, (1989).

19. ARGUS Collaboration, "Semi-leptonic Decays of B to Charged D^*,"submitted to XXVI International Conference on High Energy Physics, Dallas, TX, Aug 6-13, 1992.

20. CLEO Collaboration, F. Butler *et al.*, "Measurement of the $D^*(2010)$ branching fractions," *Phys. Rev. Lett.* 69, pp. 2041 - 2045, (1992).

21. N. Isgur and M. Wise, "Weak decays of heavy mesons in the static quark approximation,"*Phys. Lett.* B232, pg.113, (1989); N. Isgur and M. Wise, "Weak transition form-factors between heavy mesons,"*Phys. Lett.* B237, pg.527, (1990).

22. M. E. Luke, "Effects of subleading operators in the heavy quark effective theory,"*Phys. Lett.* B252, pp. 447-455, (1990); C. G. Boyd and D. E. Brahm, "Vanishing of 1/M corrections at threshold," *Phys. Lett.* B257, pp. 393-398, (1991).

23. CLEO Collaboration, "Lepton Asymmetry Measurements in $B \to D^*\ell\nu_\ell$ and Implications for $V - A$ and the Form Factors," Cornell Preprint, CLNS 92/1156.

24. E691 Collaboration, J. C. Anjos *et al.*, "Measurement of the Form Factors in the Decay $D^+ \to \overline{K}^{*0}e^+\nu_e$,"*Phys. Rev. Lett.* 65, pp. 2630-2633, (1990); E691 Collaboration, J. C. Anjos *et al.*, "A Study of the Decay $D^+ \to \overline{K}^0 e^+\nu_e$," *Phys. Rev. Lett.* 67, pp. 1507-1510, (1991).

25. E653 Collaboration, "Measurement of the Branching Ratio for $D^+ \to \overline{K}^*(892)^0 \mu^+ \nu$" and references therein, submitted to XXVI International Conference on High Energy Physics, Dallas, TX, Aug 6-13, 1992; G. Bellini, "Semi-leptonic Charmed Meson Decays: E687," talk presented at the XXVI International Conference on High Energy Physics, Dallas, TX, Aug 6-13, 1992.

26. MARKIII Collaboration, J. Adler *et al.*, "Re-analysis of Charmed D Meson Branching Fractions," *Phys. Rev. Lett.* 60, pp. 89-92, (1988).

27. Hubert Kroha, "$B^0 \overline{B^0}$ Mixing and Rare Decays from CLEO," talk presented at the XXVI International Conference on High Energy Physics, Dallas, TX, Aug 6-13, 1992.

28. ARGUS Collaboration, H. Albrecht *et al.*, "Observation of Semi-leptonic Charmless B Meson Decays," *Phys. Lett* B234, pp. 409-416, (1990); ARGUS Collaboration, H. Albrecht *et al.*, "Reconstruction of Semi-leptonic $b \to u$ Decays," *Phys. Lett* B255, pp. 297-304, (1991).

29. CLEO Collaboration, R. Fulton *et al.*, "Observation of B-Meson Semi-leptonic Decays to Noncharmed Final States," *Phys. Rev. Lett.* 1, pp. 16 - 20, (1990).

30. M. Wirbel, B. Stech, and M. Bauer, "Exclusive Semi-leptonic Decays of Heavy Mesons," *Zeit. für Phys.* C29, pg. 637, (1985).

31. J. G. Körner and G. A. Schüler, "Exclusive Semi-leptonic Decays of Bottom Mesons in the Spectator Quark Model," *Zeit. für Phys.* C38, pg. 511, (1988).

32. ARGUS Collaboration, "Search for the Decay $B^+ \to \rho^0 \ell^+ \nu$," submitted to XXVI International Conference on High Energy Physics, Dallas, TX, Aug 6-13, 1992.

33. CLEO Collaboration, "Search for Exclusive Charmless Semi-leptonic Decays of the B Meson," submitted to XXVI International Conference on High Energy Physics, Dallas, TX, Aug 6-13, 1992.

34. Andrzej J. Buras and Michaela K. Harlander, "A Top Quark Story: Quark Mixing, CP Violation and Rare Decays in the Standard Model," MPI-PAE/PTh 1/92.

35. L3 Collaboration, "An Improved Measurement of $B^0 \overline{B^0}$ Mixing in Z^0 Decays," submitted to XXVI International Conference on High Energy Physics, Dallas, TX, Aug 6-13, 1992; DELPHI Collaboration, "A Study of $B^0 - \overline{B^0}$ Oscillations using Dileptons from Semi-Leptonic Decay of b Quarks Produced from Z^0," submitted to XXVI International Conference on High Energy Physics, Dallas, TX, Aug 6-13, 1992; OPAL Collaboration, "Updated OPAL Measurement of $B^0 - \overline{B^0}$ Mixing in Hadronic Z^0 Decays," submitted to XXVI International Conference on High Energy Physics, Dallas, TX, Aug 6-13, 1992; ALEPH Collaboration, "Heavy Flavour Physics with Leptons," submitted to XXVI International Conference on High Energy Physics, Dallas, TX, Aug 6-13, 1992.

36. ARGUS Collaboration, "A New Determination of the $B^0 \overline{B^0}$ Oscillation Strength," Desy Preprint DESY 92-050; CLEO Collaboration, "Measurement of $B^0 \overline{B^0}$ mixing," submitted to XXVI International Conference on High Energy Physics, Dallas, TX, Aug 6-13, 1992.

37. G. P. Lepage, "Coulomb corrections for $\Upsilon_{4S} \to B\overline{B}$," *Phys. Rev.* D42, pp. 3251 - 3254, (1990).

38. C. Alexandrou *et al.*, "Bounds on f_B from Lattice QCD," *Nucl. Phys* B374, pp. 263-276, (1992).

39. OPAL Collaboration, "Evidence for b-flavoured Baryon Production in Z^0 Decays at LEP," Cern Preprint CERN-PPE/92-34; DELPHI Collaboration, "B_s^0 Tagging at LEP energies using D_s and ϕ Mesons," submitted to XXVI International Conference on High Energy Physics, Dallas, TX, Aug 6-13, 1992; Vivek Sharma, "Evidence for B_s and Lifetime of B_s and Λ_b," talk presented at the XXVI International Conference on High Energy Physics, Dallas, TX, Aug 6-13, 1992; F. Simonetto, "Mixing from DELPHI," talk presented at the XXVI International Conference on High Energy Physics, Dallas, TX, Aug 6-13, 1992.

40. ALEPH Collaboration, "A Measurement of $\tau \to e\nu_e\nu_\tau$ Branching Ratio with 1991 Data," submitted to XXVI International Conference on High Energy Physics, Dallas, TX, Aug 6-13, 1992.

41. ARGUS Collaboration, H. Albrecht et al., "Measurement of exclusive one-prong and inclusive three-prong branching ratios of the tau lepton," *Zeit. für Phys.* C53, 367 (1992).

42. CLEO Collaboration, "Measurement of the Tau Lepton Electronic Branching Fraction," Cornell preprint CLNS 92/1163.

43. DELPHI Collaboration, "A Study of the Decays of Tau Leptons Produced on the Z Resonance at LEP," CERN preprint CERN-PPE/92-60.

44. M. McCubbin, "Lifetime and Leptonic Branching ratios of the tau lepton," talk presented at submitted to XXVI International Conference on High Energy Physics, Dallas, TX, Aug 6-13, 1992.

45. TPC/Two-Gamma Collaboration, "Measurements of the Kaon Content in Tau Decays," Lawrence Berkeley Laboratory preprint LBL-32377.

46. ALEPH Collaboration, "A Measurement of $\tau \to \mu\nu_e\nu_\tau$ Branching Ratio with 1991 Data," submitted to XXVI International Conference on High Energy Physics, Dallas, TX, Aug 6-13, 1992.

47. J. M. Roney, "Tau topological and hadronic branching ratios at LEP," talk presented at submitted to XXVI International Conference on High Energy Physics, Dallas, TX, Aug 6-13, 1992.

48. CLEO Collaboration, "A measurement of the Branching Fraction $B(\tau^\pm \to h^\pm \pi^0 \nu_\tau)$," submitted to XXVI International Conference on High Energy Physics, Dallas, TX, Aug 6-13, 1992.

49. CLEO Collaboration, "Tau decays with one charged particle plus multiple π^0's," Cornell preprint CLNS 92/1165.

50. CLEO Collaboration, "Measurement of τ decays involving η mesons," Cornell preprint CLNS 92/1159.

51. ARGUS Collaboration, "Analysis of the Decay $\tau^- \to \pi^+\pi^-\pi^-\nu_\tau$ and Determination of the $a_1(1260)$ resonance parameters," submitted to XXVI International Conference on High Energy Physics, Dallas, TX, Aug 6-13, 1992.

52. OPAL Collaboration, "Measurement of the τ Topological Branching Ratios at LEP," CERN preprint CERN-PPE/92-66.

53. Wulf Fetscher, "Leptonic τ decays: How to determine the Lorentz structure of the charged leptonic weak interaction by experiment," *Phys. Rev.* D42, pg. 1544, (1990), and references therein.

54. ARGUS Collaboration, "New Determination of Michel Parameters in Leptonic τ Decays," submitted to XXVI International Conference on High Energy Physics, Dallas, TX, Aug 6-13, 1992.

55. ARGUS Collaboration, A. Golutvin, "τ physics at ARGUS," talk presented at submitted to XXVI International Conference on High Energy Physics, Dallas, TX, Aug 6-13, 1992.

56. Tran N. Truong, "Hadronic τ decay, pion radiative decay, and pion polarizability," *Phys. Rev.* D30, pg. 1509, (1984); F. J. Gilman and S. H. Rhie, "Calculation of Exclusive Decay Modes of the Tau," *Phys. Rev.* D 31, pg. 1066, (1985).

57. ALEPH Collaboration, D. Decamp *et al.*, "Measurement of Tau Branching Ratios," *Zeit. für Phys.* C54, pp. 211-228, (1992).

58. Bruce A. Campbell, "Supersymmetry and neutral-flavor nonconservation," *Phys. Rev.* D28, pg. 209, (1983); E. Eichten *et al.*, "Signatures for Technicolor," *Phys. Rev.* D34, pg. 1547, (1986); Paul Langacker, S. Uma Sankar and K. Schilcher, "$K_L \to \mu e$ in $SU(2)_L \times U(1)$ and $SU(2)_L \times SU(2)_R \times U(1)$ models with large neutrino masses," *Phys. Rev.* D38, pg. 2841, (1988), and references therein.

59. Jack Ritchie, "Rare K decays," talk presented at the XXVI International Conference on High Energy Physics, Dallas, TX, Aug 6-13, 1992.

60. H. Walter, "Search for muon number violating decays," talk presented at the XXVI International Conference on High Energy Physics, Dallas, TX, Aug 6-13, 1992.

61. ALEPH Collaboration, "Search for rare decays of the Z boson," submitted to XXVI International Conference on High Energy Physics, Dallas, TX, Aug 6-13, 1992.

62. ALEPH Collaboration, "Search for lepton flavour violation in tau decays," submitted to XXVI International Conference on High Energy Physics, Dallas, TX, Aug 6-13, 1992.

63. CLEO Collaboration, "A Search for $\tau^- \to \gamma\mu^-$: a test of lepton number conservation," submitted to XXVI International Conference on High Energy Physics, Dallas, TX, Aug 6-13, 1992.

64. F.J. Gilman and M.B. Wise, "The $\Delta I = 1/2$ rule and violation of CP in the six quark model," *Phys. Lett.* B83, pg. 83, (1979).

65. J. S. Frank *et al.*, "A measurement of the branching ratio for the rare decay $\pi^0 \to e^+e^-$," *Phys. Rev.* D28, pg. 423, (1983); J. Fischer *et al.*, "Observation of the $\pi^0 \to e^+e^-$ decay," *Phys. Lett.* B73, pg. 364, (1978).

66. FNAL E799 Collaboration, Yau Wah, " CP, CPT Violation and rare decays of the kaon at Fermilab," talk presented at the XXVI International Conference on High Energy Physics, Dallas, TX, Aug 6-13, 1992.

67. B Winstein, "The E731 Search for Direct CP Violation in $K \to 2\pi$," Proceedings of the Joint International Lepton-Photon Symposium and Europhysics Conference on High Energy Physics, July 25 - August 1, 1991, Geneva, Switzerland, (World Scientific); Giles Barr, "New Results on CP Violation from the NA31 Experiment at CERN," Proceedings of the Joint International Lepton-Photon Symposium and Europhysics Conference on High Energy Physics, July 25 - August 1, 1991, Geneva, Switzerland, (World Scientific).

68. NA31 Collaboration, G. D. Barr *et al.*, "A measurement of the decay $K_L \to \pi^0\gamma\gamma$," *Phys. Lett.* B284, pg. 440, (1992).

69. CLEO Collaboration, "Decay rates and polarization in $B \to \psi$ decay," submitted to XXVI International Conference on

High Energy Physics, Dallas, TX, Aug 6-13, 1992.

70. CLEO Collaboration, "Search for two-body charmless hadronic decays of B mesons," submitted to XXVI International Conference on High Energy Physics, Dallas, TX, Aug 6-13, 1992.

71. M. Bauer, B. Stech and M. Wirbel, "Exclusive Nonleptonic Decays of D, D_s, and B Mesons," *Zeit. für Phys.* C34, pg. 103, (1987).

72. K. Abdullah *et al.*, "New Experimental Limit on the Electron Electric Dipole Moment," *Phys. Rev. Lett.* 65, pp. 2347-2350, (1992).

73. OPAL Collaboration, P.D. Acton *et al.*, "Test of CP-invariance in $e^+e^- \to Z^0 \to \tau^+\tau^-$ and a limit on the weak dipole moment of the τ lepton," *Phys Lett.* B281, pp. 405-415, (1992).

74. E. Braaten, S. Narison and A. Pich, "QCD Analysis of the Tau Hadronic Width," *Nucl. Phys.* B37, pp. 581-612, (1992).

DISCUSSION

A. Sirlin, New York University, USA

Mine is not a question, but an observation concerning another very important weak decay, namely neutron beta decay. There is presently a problem, and I would hope that the experimentalists would follow it up. Namely, there is a highly precise measurement of the neutron beta decay asymmetry from which you can extract G_A/G_V, if you combine this with the lifetime measured by a Russian group. If you combine this with the lifetime of the neutron, you'll get a value for G_V which disagrees with the one obtained from nuclear beta decay. Now, if you take this value of G_V and you try to test for unitarity of the CKM matrix, you overshoot unitarity by about 2σ, in disagreement with the standard model. Furthermore, you cannot even describe this in the $SU(2)_L \times SU(2)_R$ cross section because you will need a mixing parameter between W-left and W-right that is in disagreement with other experiments. So it is very important that somebody repeat the experiment of the neutron asymmetry in neutron beta decay.

M. Danilov, ITEP/SSC, USA

The previous CLEO measurement of $V_{ub} V_{cb}$ in the lepton momentum interval 2.4–2.6 GeV/c was larger than the ARGUS result. The new one is considerably smaller. Are these differences due to statistical fluctuation, or do you think there is a different reason?

Drell

There are more restrictive cuts in the new CLEO analysis, very similar to the cuts in the ARGUS analysis. CLEO II is also much more hermetic than CLEO I.

Danilov

In the previous publication CLEO I analyzed two regions: 2.4 to 2.6 GeV/c and 2.2 to 2.6 GeV/c, and in the region which CLEO II now analyzes the branching ratio was considerably larger than in this table, which ranged from 2.2 to 2.6 GeV/c.

Drell

What I quote is for CLEO II in the lepton momentum interval 2.4 to 2.6 GeV/c. The CLEO I result is averaged over the lepton momentum interval 2.2 to 2.6 GeV/c.

B. Grinstein, SSC Laboratory, USA

I still fail to understand the essence of the problem with the branching fractions in the Tau decay system.

Drell

A discussion is included in the text.

HADRON SPECTROSCOPY AND STRUCTURE*

Nathan Isgur
Continuous Electron Beam Accelerator Facility
12000 Jefferson Avenue
Newport News, VA 23606

Abstract

In this talk I review and comment upon recent developments in hadron spectroscopy and structure. The talk is organized into three main sections dealing with heavy quarkonia ($Q\bar{Q}$), hadrons containing a single heavy quark ($Q\bar{q}$ and Qqq), and hadrons containing only light quarks and glue, although I will emphasize a surprising unity of the phenomena characterizing these systems.

INTRODUCTION

To an audience like this it seems appropriate to begin by addressing the question of whether hadronic physics is still part of high energy physics and, whether it is or not, why you might still be interested in strong interaction physics.

To the first question I would give a qualified "yes" since I recognize in this audience—which must somehow define high energy physics—a substantial number of people whose work is dominantly in this area. My qualification is based on the fact that a growing fraction of the nuclear physics community is now devoted to the solution of the same basic problems.

Even if you are not a member of the strong interaction community of physicists, it seems to me that there are several reasons why you might still wish to follow this subfield:

(1) Our inability to understand from first principles strongly interacting matter, be it hadronic or nuclear, is a basic unsolved problem in describing the world around us. Of course, QCD solves this problem *in principle*, but that is of rather limited consolation to anyone who wants to know "how it works". I for one find this as unsatisfying as being told that QED *in principle* explains all of atomic, molecular, and biological science!

(2) Even if you are not bothered by leaving all of this science behind, you might worry that beyond the Standard Model is an underlying spontaneously broken strongly interacting gauge theory. Such a theory seems likely to be much more complex than QCD, so that it might be sensible to understand the basic physics of a theory we can study in detail experimentally before speculating about the next one.

(3) As a practical matter, it is often necessary to understand strong interaction effects before extracting information from experimental data relevant to other parts of the Standard Model. For example, a heavy quark weakly decays while embedded in a

---

*research supported in part by the U.S. Department of Energy under contract DE-AC05-84ER40150.

hadron, so that the extraction of the Cabibbo-Kobayaski-Maskawa matrix elements depends on being able to relate quark-level amplitudes to hadronic ones.

(4) Finally, and to my taste most importantly, there is every reason to believe that there is still a lot of good physics waiting to be done in the strong interactions. It is of course possible to take the point of view that between QCD and strong interaction phenomena one will encounter nothing but pure complexity, but I believe this is unlikely. Consider, for example, the success of the constituent quark model: surely this is a signal of underlying simplicity waiting to be understood. This is only one of a long list of critical issues waiting to be addressed, including the origin of confinement and chiral symmetry breaking (lattice gauge theory has already taught us a lot—but not yet enough—about these phenomena), the puzzling absence to date of the gluonic degrees of freedom in low energy spectroscopy, and the origin of the Okubo-Zweig-Iizuka rule.

$Q\bar{Q}$: HEAVY QUARKONIA

The discovery of the charmonium system in November 1974 was a watershed in the history of the Standard Model. It simultaneously contributed to electroweak theory by supplying the missing quark of the Glashow-Illiopoulos-Maiani mechanism and to the acceptance of both QCD and the quark model. Since then both the charmonium and Υ families have continued to be very fruitful testing grounds for the Standard Model; the last two years have been particularly productive in this regard.

$Q\bar{Q}$ Theory

One of the most important recent advances in the theoretical treatment of heavy quarkonia has been the development by Lepage and Thacker[1] and collaborators of *nonrelativistic QCD* (NRQCD). Nonrelativistic QCD is *not* a nonrelativistic potential model, but rather a systematic and in principle exact expansion of QCD field theory in inverse powers of the heavy quark mass m_Q. NRQCD is thus an effective field theory which is *equivalent* to QCD. Its great advantage is that it removes m_Q as a scale in the treatment of heavy quarkonia, leaving only $m_Q \alpha_s$ (the "Bohr radius") and $b^{\frac{1}{2}}$ (where b is the QCD string tension) as important scales, thereby enormously expediting the treatment of heavy quarkonia using lattice techniques. The advent of NRQCD marks the end of the era of the hegemony of potential models for $Q\bar{Q}$ systems, and the beginning of *rigorous*, precision tests of QCD based on these systems. An example of this is the addition of α_s determinations from the Υ spectrum to those available from other methods[2].

Another recent advance focuses on a long-standing problem in calculating the total widths of P-wave quarkonia. For example, from diagrams like that of Figure 1, one can calculate that[3]

$$\Gamma(Q\bar{Q}\ ^3P_1) = \frac{|R'(0)|^2}{M_Q^4}\alpha_s^2 \left\{ 0 + \frac{\alpha_s}{\pi}\left[\frac{8}{3}\log\frac{M_a}{\mu} + \text{constant}\right] + \ldots \right\} \quad (1)$$

where μ is an infrared cutoff. In this formula $R'(0)$ is the derivative of the radial $Q\bar{Q}$ wavefunction at the origin (which enters because $R(0) = 0$ for a P-wave state, so that the $Q\bar{Q}$ annihilation is forced to live off the virtual intermediate heavy quark propagator) and the zero inside the bracket is to emphasize that this state has no α_s^2 term (in contradistinction to its $^3P_2$ and $^3P_0$ partners). It has usually been assumed that μ is an unknown nonperturbative parameter which spoils the predictability of the $^3P_1$ and $^1P_1$ hadronic widths. (The $^3P_2$ and $^3P_0$ widths also suffer from such infrared effects at order α_s^3, but these "corrections" have usually been ignored compared to

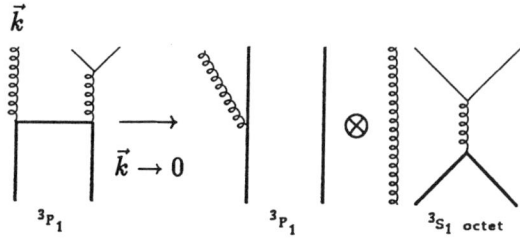

Figure 1. A graph contributing to the total hadronic width of a $^3P_1$ quarkonium state and the factorizing part of this graph which is infrared singular.

the "leading" α_s^2 terms.) Bodwin, Braaten, and Lepage presented to this conference[4] a simple, predictive, and rigorous remedy for this situation. They showed that the infrared divergent part of Fig. 1 factorizes into the product of the amplitude for the process in which the Q or $\bar{Q}$ emits a soft gluon, leaving the $Q\bar{Q}$ system in a $^3S_1$ color octet state, and the amplitude for this state to annihilate through a single hard gluon into a light $q\bar{q}$ pair. The first amplitude is, they show, part of a Fock-space expansion of the state of the system, i.e., the "$^3P_1$ quarkonia" have both a wavefunction for being $Q\bar{Q}$ in a $^3P_1$ state, and a wavefunction for being $gQ\bar{Q}$ with overall $J^{PC} = 1^{++}$ but with the $Q\bar{Q}$ subsystem in a color octet $^3S_1$ state. They go on to show that all of the nonperturbative information required (to this order in α_s) resides in $R'(0)$ and in a $gQ\bar{Q}$ "wavefunction at the origin" which absorbs the μ dependence of eq. (1).

In this very limited review of recent theoretical developments in heavy quarkonia, I will close with two advances in understanding the nonperturbative *gluonic* dynamics of QCD in the presence of static quark sources. The ground state of QCD in the presence of a static $Q\bar{Q}$ pair separated by a distance r is, in the approximation in which all dynamical quarks are ignored, the static $Q\bar{Q}$ potential which we associate with nonrelativistic potential models. Lattice QCD calculations of this gluonic ground state energy $V_0(r)$ are, indeed, in reasonable correspondence with the phenomenological potentials needed in such models. One can also consider the excited states of the glue in the presence of the static $Q\bar{Q}$ sources[5]. In the adiabatic approximation, these excited potentials $V_i(r)$ ($i > 0$) will lead to entirely new spectroscopies in which the quarks and glue are simultaneously excited: the quarkonium hybrids[*1]. Perantonis and Michael[7] have recently been able to obtain clear signals on the lattice for such excited gluonic states and to measure their energies $V_i(r)$ and quantum numbers (see Fig. 2). They find that the first excited potential is doubly degenerate with the quantum numbers and energy of a *transverse phonon* as expected in a flux tube (or string) model[5]. It follows that exotic quarkonia with $J^{PC} = 0^{+-}$, 1^{-+}, and 2^{+-} should exist at modest excitation energies as predicted by Ref. 5.

In an independent development, Olson, Olsson, and Williams[8] presented to this conference analytic evidence for the relativistic flux tube model. They extend the earlier results of Eichten and Feinberg[9], Gromes[10], and Buchmuller[11] on the Wilson loop reduction of QCD. These earlier results had shown that a Lorentz scalar confinement potential matches the spin-dependence of QCD; they extend the comparison to the *spin-independent* v^2/c^2 corrections, finding that a scalar confinement potential fails to give these latter corrections, but that a relativistic flux tube model matches QCD to this order. The main effect is identi-

---

[*1] Hybrids were introduced in the context of the bag model: see Ref. 6 for some early papers.

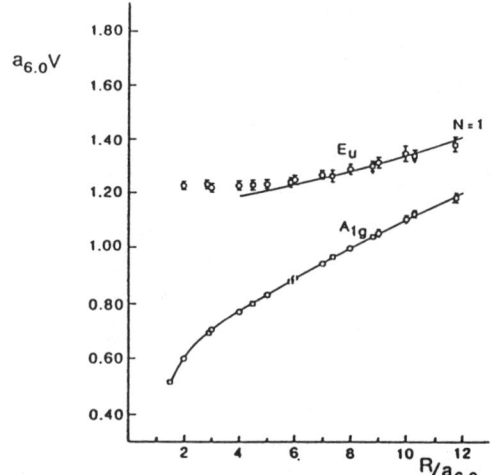

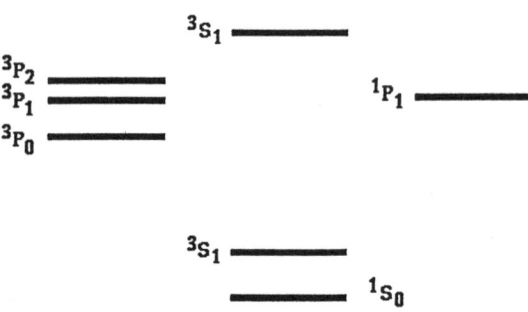

Figure 2. The ground and first excited gluonic energies in the presence of static triplet-antitriplet color sources at separation r (adapted from Ref. 7); the points are lattice data, while the curves are the adiabatic energies of the Nambu-Goto relativistic string. Note that $V_1(r) - V_0(r) = \pi/r$ for large r as expected for a transverse phonon.

fied as being the flux tube moment-of-intertia!

These two results, showing that at low energy the gluonic fields in a static $Q\bar{Q}$ system are string-like, indicate that heavy quarkonium hybrids may be as well described by potential models as ordinary heavy quarkonia. A search for the J^{PC} exotic hybrids of this type certainly now seems warranted. The more speculative extension of these ideas to light quark hybrids[5] may also now be considered to be on a better foundation. I will discuss these states below in the section on Light Quarks and Glue.

$Q\bar{Q}$ Experiment

The headline story of this section is certainly the report given to this conference of the discovery[12] of the long-missing $^1P_1$ charmonium by Fermilab's low energy $\bar{p}$-gas jet collider. They found the h_c at a mass of $3525.2 \pm 0.15 \pm 0.20$ MeV with a width $\Gamma <$ 1.1 MeV (at 90% c.l.). This mass is very close to the center-of-gravity (c.o.g.) of the ^3P_J

Figure 3. The charmonium $^1P_1$ state completes the measurement of the lowest S- and P-wave levels of the positronium-like spectrum started in 1974.

states at 3525.3 ± 0.10 MeV, where it is naively expected. Halzen, Olson, Ollson, and Strong reported[13] to this conference that the one-loop perturbative correction to the $^1P_1 - {}^3P_{\text{c.o.g.}}$ is $+0.7 \pm 0.2$ MeV as observed. However, I am surprised by the smallness of the observed splitting, because I find it difficult to understand why the non-perturbative couplings of the P-waves to virtual decay channels couldn't produce relative shifts amongst these states of order 10 MeV. It would be interesting to understand whether the decoupling from these channels is really as complete as it would appear to be from this result (even though the continuum is only a few hundred MeV's away) or if this degeneracy is mainly accidental. (Altshuler and Silverman reported[14] finding other sources of splittings of this general size.) In any event, the discovery of the $^1P_1$ now completes in a significant way the picture started in the November Revolution of 1974 (see Fig. 3).

While this measurement stole the headlines, other significant results were presented at this conference, including another E760 result[12] giving $\Gamma(\chi_{c_2} \to \gamma\gamma) = 0.34 \pm 0.11$ KeV (compared to 0.8 ± 0.4 KeV expected) and an ARGUS measurement[15] of $\Gamma(\eta_c \to \gamma\gamma) = 12.2 \pm 3.0$ keV (compared to 7 keV expected).

Yet another $\gamma\gamma$ state, the η'_c, remains obscure. Ref. 14 argues that the 90 MeV splitting of the current candidate from the ψ' is difficult to understand; clarifying the existence of this state should clearly be one of the important tasks on E760's agenda.

$Q\bar{q}$ and Qqq: HEAVY-LIGHT SYSTEMS

There are very few cases in which it is possible using analytic methods to make systematic predictions based on QCD in the low-energy, nonperturbative regime. Indeed, the theory has proved so intractable to analytic methods that all such predictions are based not on dynamical calculations, but rather on some symmetry of QCD. Isospin symmetry was the first such symmetry discovered, and we now understand that this approximate symmetry arises because the light quark mass difference $m_d - m_u$ is much smaller than the masses associated with confinement, which are set by the QCD scale Λ_{QCD}. Predictions based on isospin symmetry would, in a world with only strong interactions, be exact in the limit $m_d - m_u \to 0$; corrections to this limit can be studied systematically in an expansion in the small parameters $(m_d - m_u)/\Lambda_{QCD}$ and the electromagnetic fine structure constant α. $SU(3)$ flavor symmetry is similar, but the corrections are larger since $(m_s - m_d)/\Lambda_{QCD}$ is not small. Chiral symmetry $SU(2)_L \times SU(2)_R$ arises in QCD because *both* m_d and m_u are small compared to Λ_{QCD}; it is associated with the separate conservation of vector and axial vector currents. Although spontaneously broken in nature, the existence of this underlying symmetry allows the systematic expansion of chiral perturbation theory in which many low-energy properties of QCD are related to a few reduced matrix elements. If the strange quark mass is also treated as small compared with the QCD scale, then the chiral symmetry group becomes $SU(3)_L \times SU(3)_R$.

Over the last few years there has been progress in understanding systems containing a single heavy quark[16-26] (*i.e.*, a quark with mass m_Q much greater than the scale Λ_{QCD} of the strong interactions). It is now appreciated that there is a new symmetry of QCD, similar to isospin or chiral symmetry, in operation in such systems[21]. This symmetry arises because once a quark becomes sufficiently heavy, its mass becomes irrelevant to the nonperturbative dynamics of the light degrees of freedom of QCD. Consider, as an extreme example, two *very* heavy quarks of masses one and ten kilograms. Although these quarks will live in the usual hadronic "brown muck" of light quarks and glue, they will hardly notice it: their motion will fluctuate only slightly about that of a free heavy quark. Given that such quarks therefore define with great precision their own center-of-mass, we can study hadronic systems built on them in the frame where they act as static sources of color localized at the origin. The equations of QCD in the neighborhood of such an isolated heavy quark are therefore those of the light quark and gluonic degrees of freedom subject to the boundary condition that there is a static triplet source of color-electric field at the origin (*i.e.*, the heavy quark can be treated as a Wilson line). Since this boundary condition is the same for both of our hypothetical heavy quarks (in the static approximation which is essentially perfect given their masses), the solutions for the states of the light degrees of freedom in their presence will be the same (see Fig. 4). Thus *the light degrees of freedom will be symmetric under an isospin-like rotation of the heavy quark flavors into one another* even though the heavy quark masses are *not* almost equal. In particular, the heavy meson and baryon excitation spectra built on any heavy quark will be the same, as will be all amplitudes for the scattering of light hadrons off any state built on the heavy quark.

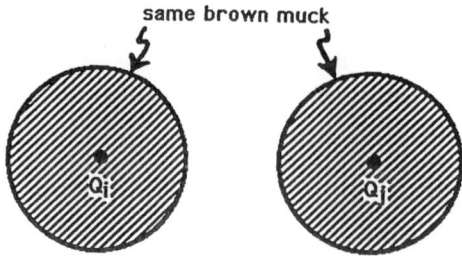

Figure 4. Q_i and Q_j are surrounded by identical brown muck

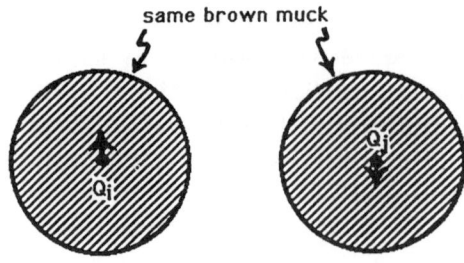

Figure 5. ... even if the spin of Q_j is flipped

The preceeding comments ignored the spin of the heavy quark. This is appropriate in QCD since the spin of a heavy quark decouples from the gluonic field[18]: all heavy quarks look like scalar heavy quarks to the light degrees of freedom. Since the flavor *and* spin of the heavy quark are irrelevant, the static heavy quark symmetry is actually $SU(2N_h)$, where N_h is the number of heavy quarks (see Fig. 5). (The full symmetry group is actually much larger since heavy quarks moving with different velocities cannot be scattered into each other by the strong interactions.) At the spectroscopic level this additional symmetry means that each spectral level built on a heavy quark (unless it happens to have spin zero in its light degrees of freedom) will be a degenerate doublet in total spin.

Heavy quark flavor symmetry is thus analogous to the fact that different isotopes of a given element have the same chemistry: their electronic structure is almost identical because they have the same nuclear charge. The spin symmetry is in turn analogous to the near degeneracy of hyperfine levels in atoms: the electronic structure of the states of a hyperfine multiplet are almost the same because nuclear magnetic moments are small.

In the situation described above where the light degrees of freedom (*i.e.*, light quarks

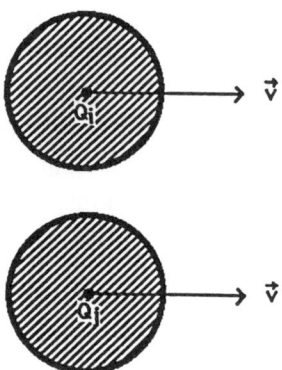

Figure 6. $Q_i(\vec{v})$ is related by the symmetry to $Q_j(\vec{v})$

and antiquarks and the gluons) typically have four-momenta small compared with the heavy quark mass, it is appropriate to go over to an effective theory where the heavy quark mass goes to infinity, with its four-velocity fixed[21,23] (see Fig. 6). The symmetry is based on heavy quarks with fixed *velocity* because the mass and therefore momentum of the heavy quark are irrelevant: the important variable is the motion of the imperturbable center to which the light quarks and gluons must respond dynamically.

The $SU(2N_h)$ spin-flavor symmetry of the heavy quark effective theory is not manifest in the full theory of QCD; it only becomes apparent in the effective theory where the heavy quark masses are taken to infinity. This situation is familiar from our experience with the light quark flavor symmetries of QCD mentioned above. The strong interactions of light quarks q (with masses m_q that are much less than the QCD scale) are greatly simplified by going over to an effective theory where the light quark masses are taken to zero. For N light quarks this effective theory has an $SU(N)_L \times SU(N)_R$ chiral symmetry that is spontaneously broken to the vector $SU(N)_V$ subgroup. Again, the symmetry is not immediately apparent in the full theory of QCD. However, as long as the quark masses are small compared with the QCD scale, they have only a small impact on strong interaction dynamics. Thus the effective theory, where the light quark masses are set to zero, is a good approximation to QCD. The heavy quark flavor-spin symmetry endows us with predictive power much in the same way that light quark chiral symmetry does. For light quark chiral symmetry, it is possible to treat the small quark masses as perturbations and consider the corrections of order m_q/Λ_{QCD} to predictions based on the effective theory where $m_q \to 0$. Similarly, for heavy quark spin-flavor symmetry, it is possible to treat as perturbations the Λ_{QCD}/m_Q corrections to the predictions based on the effective heavy quark theory where $m_Q \to \infty$.

The relationship between operators involving the heavy quarks (e.g., $\bar{Q}\gamma_\mu q$) in the full theory of QCD and operators in the effective theory where the heavy quark masses go to infinity involves some interesting applications of perturbative QCD. Contributions to matrix elements of these operators from loop graphs with virtual momenta comparable to or greater than the heavy quark mass are clearly not correctly reproduced by the effective theory. However, because of asymptotic freedom these differences can be handled by perturbative QCD. In what follows I will suppress the multiplicative matching factors required to connect the low energy effective theory which we are discussing with full QCD; see Ref. 26 for a discussion of these effects and for further references.

Most of the physics underlying heavy quark symmetry has been understood for a long time and has, to some extent, been incorporated into phenomenological models used to predict properties of hadrons containing a single heavy quark.[27,28] What is new is that we now understand that this physics arises from symmetries of an effective theory that is a systematic limit of QCD. Consequently, model-independent predictions are now possible, including important predictions for semileptonic B-meson decay form factors. These are expected to play a vital role in the accurate determination of the values of the Cabibbo–Kobayashi–Maskawa matrix elements V_{cb} and V_{ub} from experimental data. Before looking at these predictions, it is useful to set the stage by considering the implications of heavy quark symmetry for heavy-light spectroscopy[29].

In the limit $m_Q \to \infty$, the spin of the heavy quark $\vec{S}_Q$ and the spin of the light degrees of freedom (i.e., the angular momentum of the light degrees of freedom in the heavy quark's rest frame)

$$\vec{S}_\ell = \vec{S} - \vec{S}_Q , \qquad (2)$$

are separately conserved by the strong interactions (here $\vec{S}$ is the angular momentum of both the heavy quark and the light degrees of freedom in the heavy quark's rest frame, i.e., the total spin). Therefore, in this limit, s_Q, m_Q, s_ℓ, and m_ℓ are good quantum numbers. Since the dynamics are completely independent of the mass and spin of the heavy quark Q it is convenient to classify states containing a single heavy quark by s_ℓ. Then associated with each such state for the light degrees of freedom will

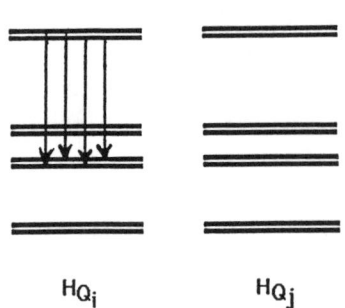

Figure 7. the spectra and transitions of the hadrons built on Q_i and Q_j; note that once the ground states are lined up, the full spectrum of other states built around each heavy quark will match up, i.e., mesons, baryons, continua, and kitchen sinks are all included in this spectral diagram.

be a degenerate doublet of hadrons with total spins (formed from combining the spin of the heavy quark $s_Q = 1/2$ with the spin of the light degrees of freedom s_ℓ)

$$s_\pm = s_\ell \pm 1/2 , \qquad (3)$$

(unless $s_\ell = 0$, in which case a single $s = 1/2$ state is obtained). The flavor symmetry ensures that the spectrum is identical for each flavor Q up to an overall constant mass shift associated with the mass of the heavy quark. Of course, states are also labeled by their parity π (which is the same as the parity of the light degrees of freedom, π_ℓ, since the heavy quark has positive parity) and by other "radial" quantum numbers (see Fig. 7).

To get a better picture of how this works, let's consider the mesons with $Q\bar{q}$ flavor quantum numbers. (Note that although we use the *language* of the constituent quark model, our conclusions will be completely general.) It is reasonable to assume that the ground state mesons with these flavor quantum numbers have $s_\ell = 1/2$ and negative parity, forming a doublet consisting of a spin zero state ($s_- = 0$) which we denote by P_Q and a spin one state ($s_+ = 1$) which we denote by P_Q^*. In the case $Q = c$, these are the D and D^* mesons, and in the case $Q = b$, these are the $\bar{B}$ and $\bar{B}^*$ mesons. In terms of the spin of the heavy quark and the spin of the light degrees of freedom, the states (at rest) are

$$|P_Q\rangle = \frac{1}{\sqrt{2}}[|\uparrow\downarrow\rangle - |\downarrow\uparrow\rangle] , \qquad (4)$$

and

$$|P_Q^*\rangle = \frac{1}{\sqrt{2}}[|\uparrow\downarrow\rangle + |\downarrow\uparrow\rangle] , \qquad (5)$$

where the state $|P_Q^*\rangle$ in eq. (5) has zero component of total spin along the quantization axis $\hat{z}$. In eqs. (4) and (5) the first arrow in a ket refers to the spin of the heavy quark along the z-axis, while the second arrow in a ket refers to that of the light degrees of freedom. Acting with the z-component of the heavy quark spin then gives

$$S_Q^z|P_Q\rangle = \frac{1}{2}|P_Q^*\rangle . \qquad (6)$$

Since $\vec{S}_Q$ commutes with the Hamiltonian, the P_Q and P_Q^* states are degenerate in mass.

Since the heavy quark flavor symmetry applies to the complete set of n-point functions of the theory, not only mass splittings, but also all strong decay amplitudes arising from the emission of light quanta like $\pi, \eta, \rho, \pi\pi$, etc., are independent of heavy quark flavor. For a given heavy quark flavor the spin symmetry ensures that two states with spins $s_\pm$ must have the same total widths. This equality between total widths typically arises in a nontrivial way. The two states of a given multiplet can decay to both states of every available multiplet with distinct partial widths whose sum must be identical. The spin symmetry determines the ratios of these partial widths (see Fig. 7).

The heavy quark symmetry cannot, of course, tell us anything about the spectroscopy of the light degrees of freedom. It can only predict relationships between heavy quark systems involving given states of these degrees of freedom. As we have mentioned for mesons with $Q\bar{q}$ flavor quantum numbers, both the constituent quark model and experiment suggest that the ground states have $s_\ell^{\pi_\ell} = 1/2^-$ giving the $s_-^\pi = 0^-$ and $s_+^\pi = 1^-$ states P_Q and P_Q^*. The constituent quark model also suggests that the lowest lying excited states are likely to be those which correspond to giving the spin 1/2 constituent antiquark a unit of orbital angular momentum resulting in $s_\ell^{\pi_\ell} = 1/2^+$ and $3/2^+$ multiplets. It is easily shown that the $s_+^\pi = 2^+$ state of the $s_\ell^{\pi_\ell} = 3/2^+$ multiplet has decay amplitudes in the proportions $\sqrt{(2/5)} : \sqrt{(3/5)}$ to the states $[P_Q\pi]_{L=2}$ and $[P_Q^*\pi]_{L=2}$ respectively. Its multiplet partner, with $s_-^\pi = 1^+$, decays at the same total rate exclusively to $[P_Q^*\pi]_{L=2}$. Note that the $s_-^\pi = 1^+$ state does not decay to $[P_Q^*\pi]_{L=0}$ even though this is an allowed channel. Similarly, the $s_+^\pi = 1^+$ state of the $s_\ell^{\pi_\ell} = 1/2^+$ multiplet decays exclusively to $[P_Q^*\pi]_{L=0}$, and does not decay to $[P_Q^*\pi]_{L=2}$. Its $s_-^\pi = 0^+$ state decays to $[P_Q\pi]_{L=0}$ with the same total rate*². These predictions are compatible with existing experimental information on mesons containing a charm quark (see below).

Let's now consider some matrix elements of operators in the effective heavy quark theory. The matrix elements we focus on are those that are likely to play an important role in determining the Cabibbo–Kobayashi–Maskawa matrix elements V_{cb} and V_{ub}. From the basic physics of heavy quark symmetry described above, and the cartoons given in Figures 8 and 9, it is not difficult to see that all of the form factors operative in $\bar{B} \to De\bar{\nu}_e$ and $\bar{B} \to D^*e\bar{\nu}_e$ are related. With an unconven-

---

*²These results were first noted by Rosner[30] who obtained them by taking the $m_Q \to \infty$ limit of a quark model calculation. Heavy quark symmetry allows us to see that they are model independent consequences of QCD in that limit.

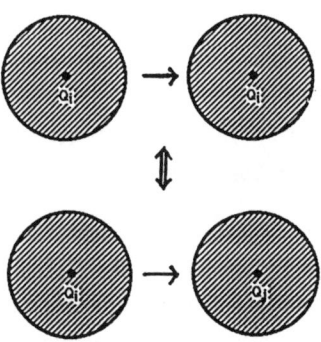

Figure 8. $Q_i \to Q_j$ is related to the elastic transition $Q_i \to Q_i$ by the symmetry

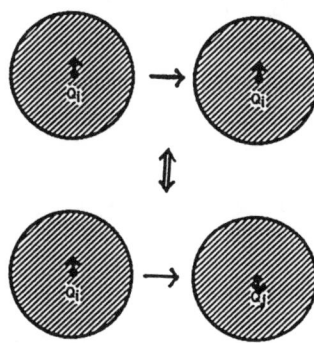

Figure 9. ... even if a spin flips

tional choice of normalizations and form factors which are suitable for a situation where four-velocities and not four-momenta are the relevant variables, namely *³ $\langle P(v')|P(v)\rangle = 2\gamma\delta^3(v'-v)$, etc. and

$$\langle P_{Q_j}(v')|\bar{Q}^{(j)}\gamma_\mu Q^{(i)}|P_{Q_i}(v)\rangle = \tilde{f}_+(v+v')_\mu + \tilde{f}_-(v-v')_\mu . \quad (7)$$

for $\bar{B} \to D$ and

$$\langle P_{Q_j}^*(v',\varepsilon)|\bar{Q}^{(j)}\gamma_\mu\gamma_5 Q^{(i)}|P_{Q_i}(v)\rangle = \tilde{f}\varepsilon_\mu^* + (\varepsilon^* \cdot v)[\tilde{a}_+(v+v')_\mu + \tilde{a}_-(v-v')_\mu] , \quad (8)$$

---

*³Here $v^\mu = p^\mu/m$ is the four-velocity, and $\gamma = v^0$.

and

$$\langle P^*_{Q_j}(v',\varepsilon)|\bar{Q}^{(j)}\gamma_\mu Q^{(i)}|P_{Q_i}(v)\rangle = i\tilde{g}\varepsilon_{\mu\nu\lambda\sigma}\varepsilon^{*\nu}v'^\lambda v^\sigma . \qquad (9)$$

for $\bar{B} \to D^*$, one can show that in fact the form factors characterizing the $P_{Q_i} \to P_{Q_j}$ and $P_{Q_i} \to P^*_{Q_j}$ matrix elements of the vector and axial vector currents are expressible in terms of a single universal function $\xi(w)$, where $w = v \cdot v'$, that is normalized to unity at zero recoil[*4]. Explicitly[21],

$$\tilde{f}_+ = \xi , \quad \tilde{f}_- = 0 , \qquad (10)$$

$$\tilde{f} = (1+w)\xi , \qquad (11)$$

$$(\tilde{a}_+ - \tilde{a}_-) = -\xi , \quad (\tilde{a}_+ + \tilde{a}_-) = 0 , \qquad (12)$$

and

$$\tilde{g} = \xi , \qquad (13)$$

with

$$\xi(1) = 1 . \qquad (14)$$

The function ξ is truly universal: it doesn't depend on the heavy quark's mass or spin, nor does it depend on the current which causes the $Q_i \to Q_j$ transition. The same function would even play a role in the physics of hadrons containing other heavy color triplet particles. Many extensions of the standard model (e.g., supersymmetry and technicolor) contain such heavy spin zero color triplets.

The manipulations leading to eqs. (10) to (14) bear a striking resemblance to the method for deriving the predictions of light quark $SU(3)_V$ symmetry for form factors in the $K \to \pi$ matrix element of the current $\bar{s}\gamma_\mu d$. However, we see from eqs. (7) and (10) that

---
[*4]This variable is called w after the French name for this letter.

heavy quark symmetry does not predict the conventionally defined f_- form factor to be zero and it provides a normalization, not at $q^2 \equiv (p'-p)^2 = 0$, but rather at the maximum value $q^2_{max} = (m_{P_i} - m_{P_j})^2$, corresponding to $w = 1$, where both the initial and final hadrons have the same four-velocity. We call this kinematic point zero recoil, since in the rest frame of the initial hadron the final hadron is also at rest.

Transition matrix elements involving the ground state (isospin-zero) baryons with $Q_i ud$ flavor quantum numbers are even easier to deduce than those involving the ground state mesons. These baryons are denoted by Λ_{Q_i} and we assume that they have $s^{\pi_\ell}_\ell = 0^+$ (this is suggested by the constituent quark model and in the case $Q = c$ is required by experiment). It is straightforward to prove that[31,32]

$$\langle \Lambda_{Q_j}(v',s')|\bar{Q}^{(j)}\Gamma Q^{(i)}|\Lambda_{Q_i}(v,s)\rangle = \eta\bar{u}(v',s')\Gamma u(v,s) , \qquad (15)$$

where η is a universal function of w independent of the heavy quark masses. The heavy quark flavor symmetry implies once again that at zero recoil

$$\eta(1) = 1 . \qquad (16)$$

Apart from the overall factor of η, eq. (15) shows that the hadronic matrix element is like a heavy quark matrix element. This occurs because in a Λ_Q state the spin of the hadron is carried by the heavy quark.

Transition matrix elements between heavy and light states can also be considered. Here, the heavy quark flavor symmetry can be used to relate matrix elements involving different heavy quarks. For example[17,20], with our unconventional normalizations and as usual ignoring QCD matching factors,

$$\langle 0|\bar{q}\gamma_\mu\gamma_5 Q^{(i)}|P_{Q_i}(v)\rangle = \langle 0|\bar{q}\gamma_\mu\gamma_5 Q^{(j)}|P_{Q_j}(v)\rangle . \tag{17}$$

Similar relations hold to light final states like π and ρ provided, in the rest frame of the heavy quark, the final states have the same four-momenta[21]. As a result, heavy quark methods open an interesting avenue for determining the magnitude of the V_{ub} element of the Cabibbo–Kobayashi–Maskawa matrix. Ordinary isospin symmetry plus heavy quark symmetry implies that, for example,

$$\langle \rho(k,\epsilon)|\bar{u}\gamma_\mu(1-\gamma_5)b|\bar{B}(v)\rangle$$
$$= \left(\frac{m_B}{m_D}\right)^{1/2}\left[\frac{\alpha_s(m_b)}{\alpha_s(m_c)}\right]^{-6/25}$$
$$\langle \rho(k,\epsilon)|\bar{d}\gamma_\mu(1-\gamma_5)c|D(v)\rangle , \tag{18}$$

where, in a departure from the simplified discussion of this review, in this case the matching factor $\left[\frac{\alpha_s(m_b)}{\alpha_s(m_c)}\right]^{-6/25}$ has been included explicitly. Eq. (18) is valid in the rest frame of the $\bar{B}$ and D for momenta k not too large compared with the heavy quark masses.[*5] Since in the Cabibbo-suppressed semileptonic decay $D \to \rho\bar{e}\nu_e$ the weak mixing angles are known, the right side of eq. (18) can be determined experimentally. With this information, experimental data on $\bar{B} \to \rho\bar{e}\nu_e$ will allow a determination of $|V_{ub}|$ (see Fig. 10). If one uses light quark $SU(3)_V$ flavor symmetry instead of isospin, then the Cabibbo allowed semileptonic decay $D \to K^*\bar{e}\nu_e$ can be used. The form factors for this decay have already been determined experimentally[34]. Of course this strategy can be used for any convenient light hadronic final state.

It is possible to systematically improve order by order in $\alpha_s(m_c)$ and $\alpha_s(m_b)$ the matching between operators in the full theory of QCD and operators in the effective theory[35,36].

[*5]Model calculations suggest that eq. (18) holds even for k comparable with the heavy charm quark mass[33].

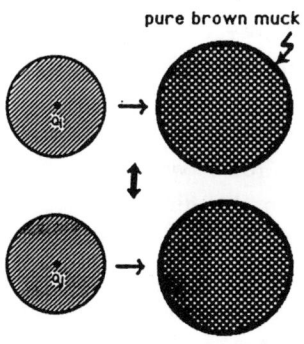

Figure 10. heavy-to-light transitions are also related by the symmetry

This gives calculable perturbative corrections and does not cause any loss of predictive power. Neubert[37] has recently examined the accuracy of the heavy quark symmetry predictions for the decays $\bar{B} \to De\bar{\nu}_e$ and $\bar{B} \to D^*e\bar{\nu}_e$, including both perturbative and $1/m_c$ corrections. As a result of Luke's theorem[38], which protects the zero recoil point from any $1/m_c$ corrections in leading order, he finds that the differential decay rates for these decays near zero recoil receive overall corrections of only $+7\%$ and -1%, respectively. There are therefore good reasons to believe that the determination of V_{cb} is now mainly an experimental problem.

Additional effects of $1/m_Q$ corrections have been studied in many models. Grinstein and Mende[39] and Burkhardt and Swanson[40] have shown how the symmetry is approached in $1+1$-dimensional QCD in the $N_c \to \infty$ limit; these issues have also been addressed in quark models too numerous to mention. The universal function $\xi(w)$ has also been calculated in many different schemes; at this conference we saw new results by Pirjol, Schilcher, and Wu[41] and de Rafael and Taron[42] to join earlier calculations by Radyushkin[43], Neubert[44], Bagan, Ball, Braun, and Dosch[45], and others[46].

Several groups have also realized that there

is a natural way to marry heavy quark and chiral symmetry. This marriage was first arranged by M.B. Wise[47], T.-M. Yan et al.[48], and G. Burdman and J.F. Donoghue[49], and at this conference we heard of further developments in this direction from Cheng and Goity[50]. We also heard at this conference about a number of other interesting developments of the ideas and applications of heavy quark symmetry.

To me, one of the most exciting recent developments pertains to an ancient problem in particle phenomenology: can we understand the nonleptonic weak decays of hadrons? I had just finished some work on this subject at the time of the 1975 Palermo conference when Lipkin warned about the dangers of "Drinking Nonleptonic". He recalled how from time to time over the last 40 years there have been hallucinatory periods from "drinking nonleptonic", i.e., spells of believing that the mystery of the $\Delta I = \frac{1}{2}$ rule and other puzzling features of these decays were on the verge of being solved, but that one always woke up from these giddy periods with a headache. We may still not fully understand the $\Delta I = \frac{1}{2}$ rule, but there is a good chance that we can now at least drink a little nonleptonic of the beauty flavor.

The new development is an extension of heavy quark effective theory by Dugan and Grinstein[51] to prove factorization *for a special class of weak hadronic decays*. Short distance QCD corrections create a low energy effective weak hadronic interaction

$$H_w^{eff} = \frac{G_F}{\sqrt{2}} V_{ud}^* V_{cb} [c_1 h_1 + c_8 h_8] \ , \quad (19)$$

where $h_1 = \bar{c}\gamma^\mu(1-\gamma_5)b \ \bar{d}\gamma_\mu(1-\gamma_5)u$ and $h_8 = \bar{c}\gamma^\mu(1-\gamma_5)\frac{\lambda^i}{\sqrt{2}}b \ \bar{d}\gamma_\mu(1-\gamma_5)\frac{\lambda^i}{\sqrt{2}}u$. If m_c/m_b is fixed, and $m_b \to \infty$, as in the heavy quark limit, then the c quark will recoil with fixed *velocity* against the residual hadronic system. Dugan and Grinstein show that in this limit the decay will be dominated by the c_1 term of H_w^{eff} so long as the $d\bar{u}$ system is at low mass. With this restriction to a collinear $d\bar{u}$ system, they show that the fast moving light quarks decouple from soft gluon exchanges; the octet light pairs therefore do not hadronize into a light $d\bar{u}$ meson, while the singlet pairs factorize from the $b \to c$ process. This leaves the latter process controlled by the same matrix elements as occur in semileptonic $b \to c$ decays, i.e. matrix elements that are governed by heavy quark symmetry. Dugan-Grinstein factorization therefore offers a real chance at a systematic expansion of this class of nonleptonic decays in powers of $1/m_q$ and α_s.

Heavy-light experiment

Given the venue of this meeting and the recent drama which unfolded in Congress, it is clear that this section should begin with the announcement that *ssc is confirmed*. Both ARGUS[52] and FNAL E687[53] reported at this conference the observation of the doubly strange Ω_c at masses of $2716 \pm 5 \pm 5$ MeV and $2707 \pm 2 \pm 5$ MeV respectively. These observations are consistent with both theoretical expectations and with an earlier claim[54] by the CERN hyperon beam experiment WA-62.

We also heard at this conference that ARGUS[55] confirms the $D_{s1}(2536)^+$ in both the $D^{*0}K^+$ and $D^{*+}K_S^0$ channels. We now know that excited D_s spectroscopy matches excited D spectroscopy, and that both are consistent with heavy quark symmetry (see Fig. 11): they display the expected doublet pattern appropriate to an $s_\ell^{\pi_\ell} = \frac{3}{2}^+$ multiplet and the narrow D_{s1} ($\Gamma_{D_{s1}} < 3.9$ MeV) and D_1 required by the symmetry, which predicts that the kinematically allowed D^*K and $D^*\pi$ S-wave decays will be suppressed[29].

The LEP experiments are all beginning to try to exploit the fact that the Z is a good flavor factory, and we heard many re-

Figure 11. The observed $s_\ell^{\pi_\ell} = \frac{3}{2}^+$ spectra of excited D_s and D mesons, displaying the expected doublet structure. The reduced widths (i.e., widths with phase spaced removed) of the multiplet partners are also nearly equal as expected.

ports of progress begin made in studying the B and Λ_b in a high energy environment[56]. Especially impressive was the strong evidence presented[57] by OPAL for the B_s. (They even report one candidate for the decay $B_s \to \psi\phi$ with a mass of 5.36 GeV.) We can expect a steady improvement in the masses, lifetimes, and branching ratios of all of these states from these experiments; they have an especially important role to play in Λ_b and B_s physics where we may be relying on high energy colliders for the bulk of our knowledge. There are many important issues to be resolved here. As a very first step, it is desirable to simply accurately measure the mass of the Λ_b (heavy quark symmetry predicts that $\Lambda_b = \left[\frac{B+3B^*}{4}\right] + \left[\Lambda_c - \left(\frac{D+3D^*}{4}\right)\right] + 0\left(\frac{\Lambda_{QCD}^2}{D}\right)$); eventually, the Λ_b semileptonic decays could prove to be the most accurate means of determining the CKM angles V_{cb} and V_{ub}.

CLEO[58,59] also reported a series of new measurements of isospin-breaking mass differences in the D, D^*, and Σ_c multiplets:

$$D^{*+} - D^{*0} = 3.32 \pm 0.08 \pm 0.05 \text{ MeV} \quad (20)$$
$$D^+ - D^0 = 4.80 \pm 0.10 \pm 0.06 \text{ MeV} \quad (21)$$
$$\Sigma_c^+ - \Sigma_c^0 = 1.1 \pm 0.6 \pm 0.5 \text{ MeV} \quad (22)$$
$$\Sigma_c^{++} - \Sigma_c^+ = -0.2 \pm 0.6 \pm 0.5 \text{ MeV} \quad (23)$$

The first two splittings follow the general pattern of dominance of such splittings by the $d - u$ mass difference. However, there is a surprise in these splittings, in that heavy quark symmetry predicts that they would be equal in the limit $m_c \to \infty$, and (especially given that $D_s^* - D_s \simeq D^* - D$) the 1.5 MeV mass difference between them seems difficult to explain as a $1/m_c$ effect[60]. The second pair of splittings seems superficially within expectations, but I suspect there is a problem here: from them one can form the $\Delta I = 2$ combination $\Sigma_c^{++} + \Sigma_c^0 - 2\Sigma_c^+ = -1.3 \pm 0.7 \pm 0.6$ MeV which is pure electromagnetic and should turn out to be roughly $+1.3$ MeV.

Finally, many new tests of nonleptonic factorization were presented to the conference. There were, first of all, many reported *failures* of factorization for charm decays where Dugan-Grinstein (DG) factorization should not apply but where Bauer-Stech-Wirbel (BSW) factorization[61] has enjoyed considerable success. BSW factorization hypothesizes that a much broader applicability of factorization to *all* amplitudes, not just collinear ones as in DG factorization, can be obtained if the coefficients c_1 and c_8 of eq. (19) are treated as free parameters. BSW also assume that it is possible to reexpress H_w^{eff} in terms of two related hadronic operators with coefficients a_1 and a_2. In fact, the BSW model remains quite successful in those circumstances where only the c_1 term contributes. However, in other circumstances there now seem to be problems. For example, ARGUS[62] has measured

$$\frac{\Gamma(\Lambda_c^+ \to \Sigma^0 \pi^+)}{\Gamma(\Lambda_c^+ \to \Lambda \pi^+)} = \frac{1.8 \pm 0.5 \pm 0.3}{2.2 \pm 0.3 \pm 0.4} \quad , \quad (24)$$

while factorization predicts *zero* since the $I = 0$ $c \to s$ current can't induce a $\Delta I = 1$ $\Lambda_c^+ \to \Sigma^0$ transition. ARGUS also reports[62]

$$\frac{\Gamma(D^0 \to \rho^0 \bar{K}^0)}{\Gamma(D^0 \to K^{*-}\pi^+)} = \frac{0.24 \pm 0.03 \pm 0.03}{0.70 \pm 0.04 \pm 0.04}, \quad (25)$$

much larger than expected from BSW factorization since the process in the numerator proceeds through c_2 alone.

While BSW factorization is experiencing some difficulties, its less ambitious but more rigorous cousin DG factorization has experienced some successes. For example, CLEO has reported[59] an important set of new tests in B decay of DG factorization and heavy quark symmetry. Highlights of their results are given in Table 1. It is clearly too early to drink too heavily of nonleptonic, but there are reasons to remain optimistic.

LIGHT QUARKS AND GLUE

Finally we turn to systems of "pure brown muck". There is a recent revival of interest in these oldest hadronic systems, engendered by both new experimental and theoretical developments. It is an old refrain, but given its importance in structuring the world we live in, our ignorance of the nature of ordinary strongly interacting matter is an embarrassment, and it remains a challenge to close this glaring gap between our knowledge of the basic principles of nuclear and hadronic matter—QCD—and the phenomena which characterize it.

The theory of pure brown muck

One of the most productive theoretical tools in studying light quarks and glue remains the lattice. For example, within the last few years some basic characteristics of the quenched glueball spectrum have become established[63]:

(1) the lightest glueball has vacuum quantum numbers ($J^{PC} = 0^{++}$) and a mass of ~ 1.5 GeV.

(2) the next lightest state is probably a $J^{PC} = 2^{++}$ state with mass ~ 2.2 GeV.

(3) the spectrum really begins above 2 GeV, but in the 2–3 GeV region there is so far no evidence for a J^{PC} exotic state.

These facts indicate that glueball hunting may be a sticky proposition. The 0^{++} meson sector between 1 and 2 GeV is the most poorly understood sector of hadron spectroscopy, and is *the* place that a malicious intelligence would hide a glueball. In addition, looking for a glueball with nonexotic quantum numbers above 2 GeV will be extremely difficult: the meson spectrum in this region will be densely populated with ordinary quarkonium states.

When combined with the recent evidence from the lattice discussed in the $Q\bar{Q}$ section for the flux tube picture of nonperturbative gluon dynamics, the bleak prospects for glueball hunting suggest that the most productive strategy for discovering a gluonic spectroscopy is to concentrate on the J^{PC} exotic hybrids predicted[5] (basically by extrapolation from the $Q\bar{Q}$ case) to be found in the vicinity of 1.8 GeV. Several searches are now underway for such states, which can be expected to be more readily produced than the $Q\bar{Q}$ hybrids.

In addition to this and other lattice activity, there have been continuing theoretical efforts of a more analytical nature. An important focus of activity has been to try to understand the nature of the constituent quark, and amongst the many papers on this subject, those of Manohar and Georgi[64], Kaplan[65], Weinberg[66], and Peris[66] received considerable attention. One of the central goals of this activity has been to attempt to "chiralize" the quark model. Karliner described at this

Table 1. Tests of DG factorization and heavy quark symmetry in B decays; the last two columns indicate which principles are being tested.

channel	expt[59]	theory	factorization?	h.q.s.?
$\bar{B}^0 \to D^{*+}\pi^-$	$1.28 \pm 0.19 \pm 0.30$	1.23 ± 0.17	✓	
$\bar{B}^0 \to D^{*+}\rho^-$	$3.17 \pm 0.43 \pm 0.34$	3.26 ± 0.46	✓	
$\frac{\Gamma_L}{\Gamma_T}(\bar{B}^0 \to D^{*+}\rho^-)$	$90 \pm 7 \pm 5\%$	88%	✓	
$\frac{\bar{B}^0 \to D^+\pi^-}{\bar{B}^0 \to D^{*+}\pi^-}$	$0.96 \pm 0.19 \pm 0.25$	1	✓	✓
$\frac{\bar{B}^0 \to D^+\rho^-}{\bar{B}^0 \to D^{*+}\rho^-}$	$0.97 \pm 0.19 \pm 0.25$	1	✓	✓

meeting[67] a derivation in 1 + 1-dimensional QCD of Kaplan's *ansatz* in which the constituent quark is the "skyrmion" of the colored chiral Lagrangian after bosonization.

The experimental status of conventional brown muck

I am myself particularly fond of the much more mundane suggestion made by Bjorken: a constituent quark is what you get when you take a B meson and remove the heavy quark. To organize *conventional* light quark spectroscopy for you, let me adopt a version of this definition. First, in Figure 12, let's examine the onset of heavy quark symmetry; alternatively, this can be viewed as a demonstration that a constituent quark retains some of the characteristics of a heavy quark. We can see from this diagram that even the $\rho - \pi$ splitting roughly follows the $1/m_Q$ scaling law for the splitting of the $s_\ell^{\pi_\ell} = \frac{1}{2}^-$ multiplet. Note, moreover, that the center-of-gravity of the $s_\ell^{\pi_\ell} = \frac{3}{2}^+$ excited multiplet hardly moves with respect to that of the ground state multiplet, and that the $1/m_Q$ splitting within this excited multiplet remains small.

Figure 13 gives another look at the evolution of hadronic spectra within the (approximately) equal mass quarkonia from the Υ to the π. Once again we see the remarkable similarity of all of these spectra, suggesting that there are no dramatic changes in the underlying physics from the $b\bar{b}$ to the $u\bar{d}$ system, even though m_u, $m_d \ll \Lambda_{QCD}$. Somehow, a constituent light quark seems to behave much like a heavy quark[68].

There were several interesting additions to our knowledge of the spectra of Figure 13 made at this conference. I have already highlighted the discovery[12] of the $c\bar{c}$ $^1P_1$ state $h_c(3526)$. We also heard here that the Beijing e^+e^- machine has confirmed[69] the narrow $\xi(2200)$, which has an economical interpretation as one of the missing[68] F-wave $s\bar{s}$ states, namely $^3F_4$ or $^3F_2$. The LASS experiment, which has been enormously successful in clearing up many questions in strange particle spectroscopy, provided strong evidence[70] in the $s\bar{u}$ sector for both $^3D_2$ and $^1D_2$, the $K_2(1775)$ and $K_2(1820)$; this completes the discovery of expected[68] D-wave $S = -1$ quarkonia. In a related measurement, ARGUS[15] has provided a window on the nature of the $^1D_2$ state $\pi_2(1660)$ by measuring its $\gamma\gamma$ width to be 0.25 ± 0.10 keV. This width is much more understandable from quark potential models[71]

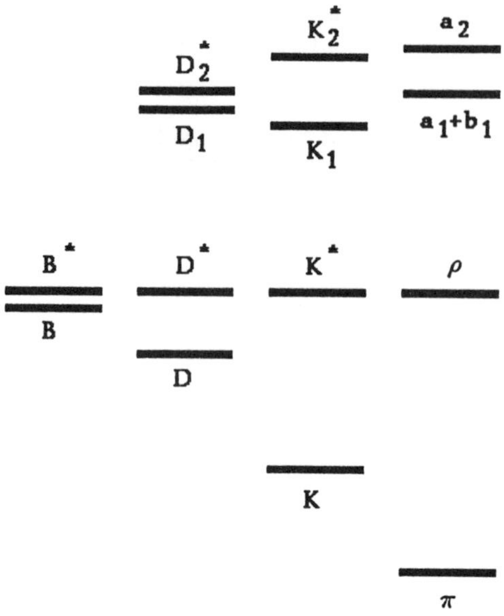

48 Hadron Spectroscopy and Structure

Figure 12. The onset of heavy quark symmetry

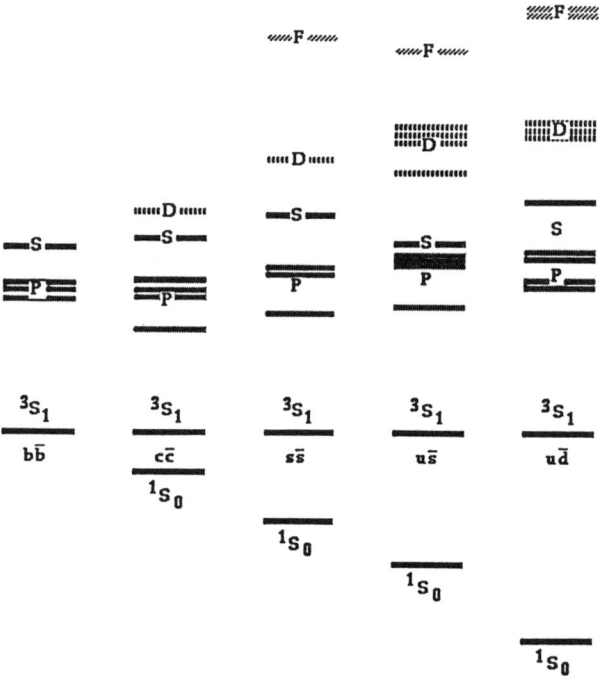

Figure 13. The experimental status of conventional quarkonia

than earlier reported results.

Many other new results in conventional spectroscopy were reported to this conference, but I will mention just one more which brings to a conclusion a long-standing uncertainty about the properties of the a_1 meson. The ARGUS collaboration has reported[72] an analysis of their new sample of $\tau \to \nu_\tau 3\pi$ data which pin down the elusive parameters of this state to be $m = 1211 \pm 7$ MeV and $\Gamma = 446 \pm 21$ MeV, with a D/S ratio for the dominant $\rho\pi$ decay of -0.11 ± 0.02. The mass once again leads to the conclusion that spin-orbit forces are suprisingly small; the width and D/S ratio are in good agreement with those expected in the flux tube model[73].

The experimental status of unconventional brown muck

The experts will have noticed that I have omitted a number of states from my picture in order to emphasize the simplicity of the patterns. There is of course a danger that in doing so I will have conveyed a misleading impression!

There are several classes of states which I have intentionally omitted. First, I have omitted states for which there is very weak evidence and no motivation; I assume that I will be excused for this. At the other extreme, I have omitted some states which are unambiguous, but which I believe are best interpreted within an entirely different framework. Let me begin with some *gedanken* states of this type: $\bar{p}p$ Coulombic bound states exist, but we should clearly omit them from the "elementary" meson spectrum. There are a number of states which are widely (but by no means universally) believed to be of this type—I'll call them "molecular states" since the hadrons themselves are already "quark atoms"—which I have excluded. Foremost amongst the candidates for such states are the old and well-established $J^{PC} = 0^{++}$ states $a_0(980)$ and $f_0(975)$ [formerly the S^* and δ]. Jaffe first suggested[74] that these states were bagged $qq\bar{q}\bar{q}$ states. Recent opinion has shifted to the closely related "$K\bar{K}$ molecule" picture[75] advocated by Weinstein *et al.*, in which these states have much more in common with the deuteron than the proton, and are as appropriately distinguished from "true" mesons as the deuteron is distinguished from a "true" $B = 2$ baryon. There are two other interesting candidates for such states which have been excluded from Figure 13. One is the narrow $f_1(1420)$ which was long interpreted as a member of the $^3P_1$ multiplet, but which now might be interpreted[76,77] as a (virtual) bound state of $K^*\bar{K}$. New experimental results on this state were presented[78] to this conference confirming its 1^{++} assignment and thus the need for a non-quarkonium interpretation of its nature. The other is the newly established $f_2(1520)$ state seen in $\bar{p}p$ collisions, for which convincing evidence was reported at this conference by Amsler[79] and Smith[80]. This state, first reported by the Asterix collaboration and called the AX(1560), has been interpreted by some[81] as a $\bar{p}p$ bound state, and by others[77,82] as a coupled channel bound state of the $\rho\rho - \omega\omega$ system. In either case, it would be appropriately excluded from the "true mesons."

These molecular states have their own interest, of course. Originally, $qq\bar{q}\bar{q}$ states were sought as a new type of hadron. (A "true" multiquark hadron would be a tightly bound system characterized by having all interquark distances comparable). Today I believe their main interest is as a testing ground in novel situations for our very rudimentary understanding of interhadronic forces. One of the goals of strong interaction physics is to understand such forces, since they are ultimately responsible for the existence and properties of nuclei. Of course, we all know that the nucleon-nucleon force can be very well parameterized in terms of meson-exchange-generated forces. However, this parameterization is unlikely to lead to a viable underlying explanation for this and other interhadronic forces. The forces between composite systems can arise not only from particle exchange, but also from constituent exchange. (In fact, in a heavy quark world meson exchange would be negligible compared to quark exchange forces[83].) Studying the forces in exotic hadron-hadron systems can help to delineate the role of these mechanisms (and others) in the mundane nucleon-nucleon case. These two competing mechanisms have been studied by a number of groups. For example, Barnes and Swanson[77,82] systematized the search for attractive meson-meson systems in a quark exchange picture (see also Refs. 74, 75, and 82). Tornquist reported to this conference[82] his interesting results on those channels which have an attractive pion-exchange-induced potential. It is interesting that all groups find attraction in the $\rho\rho - \omega\omega$ system that could be associated with the $f_2(1520)$.

Looking at Figure 13, some may wonder if there is life in spectroscopy beyond the quark model. In two senses, I would argue that there must be. In the first place, no matter how successful the quark model might be, its origin within QCD is not understood. So at the very least there lies ahead the vital task of explaining what the quark model is. This is the task being addressed by work like that I mentioned earlier on the nature of the constituent quark. However, there is more than just theoretical mopping up to be done: we *know* that glueballs and hybrids must exist as degrees of freedom of QCD so there *must* be life beyond the simple pattern of this spectrum. In fact, the absence of evidence for such states in the low energy spectrum seems likely to be a central clue to the existence of the quark model. Moreover, it may well be that as we discover these states experimentally, we will also see the quark model picture break down in such a way that it will be revealed to us what it was in the first place.

I should remark that I may be overstating the present situation. I have already mentioned how the search for glueballs in the 0^{++} sector is hampered by quarkonium chaos there. (For a good discussion, see the review by Close at this conference[84].) However, the GAMS collaboration has reported[85] an interesting candidate state at 1590 MeV which we heard more about at this conference[79]. Given the coexistence in this region of broad $^3P_0$ states, the probable existence of a "$K\bar{K}$ molecule" with these quantum numbers, and the possibility of maximal OZI-violation in this channel[86], much work will be needed to confirm the existence of a glueball in this mass region. Another contender for a glueball used to be the $\theta(1720)$ seen in ψ radiative decays. With the recent evidence that its quantum numbers are really 0^{++} and not 2^{++}, it is no longer such a serious candidate: it could now easily be the first radial excitation of the quarkonium scalar mesons[68]. For either set of quantum numbers, it could also be a $K^*\bar{K}^* - \omega\phi$ molecule[87]. A more interesting and persistent candidate has been proposed in the tensor sector in the 2.0–2.5 GeV mass range by Lindenbaum *et al.*[88]. This group has found a series of $\phi\phi$ resonances produced by $\pi\pi_{\text{virtual}}$ scattering which would be difficult to accommodate as quarkonia. Given the OZI-violating nature of their production, there is also a *prima facie* case for associating these states with glue. Further study of these states is certainly warranted.

As I have already argued, it may be easier ultimately to establish the excitation of the gluonic degree of freedom by finding J^{PC} exotic hybrid mesons (or baryons). Some candidates have been reported to this conference[89], but so far there are no clear resonant signals. On the other hand, the lattice adiabatic potentials described in the $Q\bar{Q}$ section, plus the assumption of smooth evolution of quarkonium spectroscopy into the light quark sector, suggest where to look.

OUTLOOK

It is plausible that hadron spectroscopy, after several years of slow progress, is poised to make significant advances:

- heavy-light systems provide us with the "hydrogen atom" of brown muck, and in so doing severely constrain model building; at the same time our theoretical and experimental knowledge of such heavy-light states is increasing rapidly,

- There are reasons to believe that hybrids, and possibly glueballs, will be discovered in the next generation of hadron spectroscopy experiments; by watching the quark model fail, we should be able to understand better why it works so well, and, more generally,

- new accelerators and detectors, wielded by both high energy and nuclear physicists, will shed new light on this now old but still important problem.

REFERENCES

1. B.A. Thacker and G.P. Lepage, Phys. Rev. **D43**, 196 (1991).
2. See plenary talk PL-12 by R. Petronzio.
3. R. Barbieri, R. Gatto, and E. Remiddi, Phys. Lett. **61B**, 465 (1976); R. Barbieri, M. Caffo, and E. Remiddi, Nucl. Phys. **B162**, 220 (1980).
4. G. Bodwin, E. Braaten, and G.P. Lepage, "Rigorous QCD Predictions for Decays of P-Wave Charmonium", session PA-4B.
5. N. Isgur and J. Paton, Phys. Rev. **D31**, 2910 (1985).
6. T. Barnes, Z. Phys. **C10**, 275 (1981); T. Barnes, F.E. Close, and S. Monaghan, Nucl. Phys. **B198**, 380 (1982); M. Chanowitz and S. Sharpe, Nucl. Phys. **B222**, 211 (1983).
7. S. Perantonis and C. Michael, Nucl. Phys. **B347**, 854 (1990).
8. C. Olson, M.G. Olsson, and K. Williams, Phys. Rev. **D45**, 4307 (1992) and "QCD, Relativistic Flux Tubes, and Potential Models", paper 344.
9. E. Eichten and G. Feinberg, Phys. Rev. **D23**, 2724 (1981).
10. D. Gromes, Z. Phys. **C22**, 265 (1984); **C26**, 401 (1984).
11. W. Buchmüller, Phys. Lett. **112B**, 479 (1982).
12. J. Rosen for the Fermilab E760 collaboration, "Charmonium Results from E760", session PA-4B.
13. F. Halzen, C. Olson, M.G. Olsson, and M.L. Strong, "Mass of the Singlet P State", paper 419.
14. E. Altshuler and D. Silverman, "Predictability of Charmonium Levels for a Range of Fits", paper 428.
15. G. Kernel for the ARGUS collaboration, "Resonance Production by Two Photons", session PA-3B.
16. S. Nussinov and W. Wetzel, Phys. Rev. **D36**, 130 (1987); M.B. Voloshin and M.A. Shifman, Sov. J. Nucl. Phys. **47**, 511 (1988).
17. M.B. Voloshin and M.A. Shifman, Sov. J. Nucl. Phys. **45**, 292 (1987).
18. G.P. Lepage and B.A. Thacker, Nucl. Phys. **B4** (Proc. Suppl.) 199 (1988).
19. E. Eichten, Nucl. Phys. **B4** (Proc. Suppl.) 170 (1988).

20. H.D. Politzer and M.B. Wise, Phys. Lett. **B206**, 681 (1988); **B208**, 504 (1988).

21. N. Isgur and M.B. Wise, Phys. Lett. **B232**, 113 (1989); Phys. Lett. **B237**, 527 (1990).

22. E. Eichten and B. Hill, Phys. Lett. **B234**, 511 (1990).

23. H. Georgi, Phys. Lett. **B240**, 447 (1990).

24. B. Grinstein, Nucl. Phys. **B240**, 447 (1990); **B339**, 253 (1990).

25. J.D. Bjorken, in *Proceedings of the 4th Rencontres de Physique de la Vallee d'Aoste*, La Thuile, Italy, 1990, ed. M. Greco (Editions Frontieres, Gif-sur-Yvette, France, 1990).

26. N. Isgur and M.B. Wise, "Heavy Quark Symmetry", in *B Decays*, ed. S. Stone (World Scientific, Singapore, 1992), p. 158.

27. M. Suzuki, Nucl. Phys. **B258**, 553 (1985); B. Grinstein, M.B. Wise, and N. Isgur, Phys. Rev. Lett. **56**, 298 (1986); T. Altomari and L. Wolfenstein, Phys. Rev. Lett. **58**, 1583 (1987); N. Isgur, D. Scora, B. Grinstein, and M.B. Wise, Phys. Rev. **D39**, 799 (1989).

28. E.V. Shuryak, Phys. Lett. **93B**, 134 (1980); Nucl. Phys. **B198**, 83, (1982); Nucl. Phys. **B328**, 85 (1989).

29. N. Isgur and M.B. Wise, Phys. Rev. Lett. **66**, 1130 (1991).

30. J. Rosner, Comm. Nucl. Part. Phys. **16**, 109 (1986).

31. N. Isgur and M.B. Wise, Nucl. Phys. **B348**, 278 (1991).

32. H. Georgi, Nucl. Phys. **B348**, 293 (1991).

33. N. Isgur, Phys. Rev. **D43**, 810 (1991).

34. J.C. Anjos *et al.*, Phys. Rev. Lett. **62**, 722 (1989); **62**, 1587 (1989); Z. Bai *et al.*, SLAC-PUB-5341 (1990); D. Potter for the Fermilab E653 collaboration, "Charm Meson Production in 600 GeV/c π-Emulsion Interactions", session PA-10A; J. Spengler for the ARGUS collaboration, "New Results in Υ and Charm Physics from ARGUS", session PA-4B; G. Bellini for the Fermilab E678 collaboration "Semileptonic Charm Decays", session PA-1A.

35. A.F. Falk, H. Georgi, B. Grinstein, and M.B. Wise, Nucl. Phys. **B343**, 1 (1990); A. Falk and B. Grinstein, Phys. Lett. **B247**, 406 (1990); X. Ji and M. Musolf, Phys. Lett. **B257**, 409 (1991); G.P. Korchemsky and A.V. Radyushkin, Nucl. Phys. **B283**, 342 (1987); Phys. Lett. **B279**, 359 (1992); M. Neubert, Nucl. Phys. **B371**, 149 (1992).

36. See also B. Grinstein, "Heavy Quark Effective Theory: Application to Weak Decays", session PA-1B.

37. M. Neubert, SLAC preprint SLAC-PUB-5842, 1992; see also T. Mannel, "Leading and Subleading Logarithmic QCD Corrections to Bilinear Heavy Quark Currents", session PA-4A.

38. M. Luke, Phys. Lett. **B252**, 447 (1990).

39. B. Grinstein and P.F. Mende, "Heavy Mesons in Two Dimensions", paper 614.

40. M. Burkhardt and E.S. Swanson, to appear in Phys. Rev. **D**, (1992).

41. Y.L. Wu, K. Schilcher, and D. Pirjol, "QCD Sum Rule Calculation of the $B \to D\ell\nu$ Form Factor", paper 434.

42. J. Taron, "Constraints on Heavy Meson Form Factors", session PA-4A.

43. A.V. Radyushkin, Phys. Lett. **B271**, 218 (1991).

44. M. Neubert, Phys. Rev. **D45**, 2451 (1992).

45. E. Bagan, P. Ball, V.M. Braun, and H.G. Dosch, Phys. Lett. **B278**, 452 (1992).

46. N. Isgur, D. Scora, B. Grinstein, and M.B. Wise, Phys. Rev. **D39**, 799 (1989).

47. M.B. Wise, Phys. Rev. **D45**, R2188 (1992).

48. T.-M. Yan *et al.*, Phys. Rev. **D46**, 1148 (1992).

49. G. Burdman and J.F. Donoghue, Phys. Lett. **B280**, 287 (1992).

50. H. Cheng, "Heavy Quark Symmetry and Chiral Dynamics", session PA-4B; J. Goity, "Chiral Perturbation Theory for SU(3) Breaking in Heavy Mesons", submitted paper.

51. M.J. Dugan and B. Grinstein, Phys. Lett. **B255**, 583 (1991).

52. J. Stiewe for the ARGUS collaboration, "Production and Decays from ARGUS", session PA-10B.

53. H. Cheung for the Fermilab experiment E687 collaboration, "Charm Baryon Production", session PA-10B.

54. A. Biagi *et al.*, Z. Phys. **C28**, 175 (1985).

55. J. Spengler for the ARGUS collaboration, "New Results in Υ and Charm Physics from ARGUS", session PA-4B.

56. See plenary talks PL-1 by P. Drell and PL-3 by J. Butler.

57. The OPAL collaboration, "Evidence for the Existence of the Strange b-Flavoured Meson B_s in Z^0 Decays", submitted paper.

58. The CLEO collaboration, "Measurement of the $D^{*+} - D^+$ and $D^{*0} - D^0$ Mass Differences", paper 900; "Observation of the Charmed Baryon Σ_c^+", paper 901.

59. V. Jain for the CLEO collaboration, "Results on Charm and Bottom", session PA-4B.

60. J. Goity, private communication.

61. M. Bauer, B. Stech, and W. Wirbel, Z. Phys. **C34**, 103 (1987); M. Wirbel, Prog. in Nucl. Part. Phys. **22**, 33 (1988).

62. The ARGUS collaboration, "A Measurement of Asymmetry in the Decay $\Lambda_c^+ \to \Lambda\pi$", paper 278; "A Partial Wave Analysis of $D^0 \to K_S^0 \pi^+ \pi^-$", paper 105.

63. See, for example, C. Michael and M. Teper, Nucl. Phys. **B314**, 347 (1989).

64. A. Manohar and H. Georgi, Nucl. Phys. **B234**, 189 (1984); H. Georgi and A. Manohar, Nucl. Phys. **B310**, 527 (1988).

65. D.B. Kaplan, Phys. Lett. **B235**, 163 (1990); Nucl. Phys. **B351**, 137 (1991).

66. S. Weinberg, Phys. Rev. Lett. **65**, 1181 (1990); **67**, 3473 (1991); S. Peris, Phys. Rev. **D46**, 1202 (1992).

67. M. Karliner, "Quark Solitons as Constituents of Hadrons", session PA-6B.

68. For an explicit model of this type, see S. Godfrey and N. Isgur, Phys. Rev. **D32**, 189 (1985).

69. Zheng Zhipeng, "Results on $f_0(975)$ and $\xi(2.2)$ from BES", session PA-3B.

70. B. Ratcliff for the LASS collaboration, "The $K_2(1770)$ Region: Evidence for Two $J^P = 2^-$ Strange Mesons Decaying to $K\omega$, session PA-3B.

71. E.S. Ackleh and T. Barnes, Phys. Rev. **D45**, 232 (1992).

72. The ARGUS collaboration, "Analysis of the Hadronic Final State of the Decay $\tau \to \pi^- \pi^- \pi^+ \nu_\tau$", paper 252.

73. R. Kokoski and N. Isgur, Phys. Rev. **D35**, 907 (1987); N. Isgur, C. Morningstar, and C. Reader, Phys. Rev. **D39**, 1357 (1989).

74. R.L. Jaffe, Phys. Rev. **D15**, 267, 281 (1977); **D17**, 1444 (1978); R.L. Jaffe and K. Johnson, Phys. Lett. **60B**, 2105 (1976).

75. J. Weinstein and N. Isgur, Phys. Rev. Lett. **48**, 659 (1982); Phys. Rev. **D27**, 588 (1983); **D41**, 2236 (1990).

76. D.O. Caldwell, Mod. Phys. Lett. **A2**, 771 (1987).

77. E.S. Swanson, "Intermeson Potentials from the Constituent Quark Model", to appear in Ann. Phys. (NY), (1992).

78. F. Votruba for the CERN WA76 collaboration, "A Study of the $E/f_1(1420)$ and of the $f_1(1285)$ in Central Production", session PA-3A.

79. C. Amsler for the Crystal Barrel collaboration, "New Light Mesons from the LEAR Crystal Barrel", session PA-3A.

80. G. Smith for the Fermilab E760 collaboration, "Neutral Light Quark Spectroscopy in $\bar{p}p$ Annihilation at $3000 < \sqrt{s} < 6000$ MeV", session PA-3A.

81. C.B. Dover, "Quasinuclear $N\bar{N}$ States", in proceedings of *Intersections between Particle and Nuclear Physics*, AIP Conf. Proc. 243, ed. W.T.H. van Oers (AIP, New York, 1992), p. 381.

82. Yu. S. Kalashnikova, "Four Quark Interpretation of AX", in proceedings of *Hadron '91*, ed. S. Oneda, and D.C. Peaslee (World Scientific, Singapore, 1992), p. 777.; N. Tornquist, Phys. Rev. Lett. **67**, 556 (1991) and paper 702; T. Barnes and E.S. Swanson, Phys. Rev. **D46**, 131 (1992).

83. N. Isgur, "Quark Exchange Forces from a Heavy Quark Perspective", in proceedings of *From Fundamental Fields to Nuclear Phenomena*, ed. J.A. McNeil and C.E. Price (World Scientific, Singapore, 1991), p. 46.

84. F.E. Close, "Scalars and Light Quark Mesons", session PA-3B.

85. D. Alde *et al.*, Phys. Lett. **B201**, 160 (1988).

86. P. Geiger and N. Isgur, "When Can Hadronic Loops Scuttle the OZI Rule?", CEBAF preprint CEBAF-TH-92-24.

87. K. Dooley, E.S. Swanson, and T. Barnes, Phys. Lett. **B275**, 478 (1992).

88. S.J. Lindenbaum, "The Status of Investigations of 0^{++} and 2^{++} Glueball States and New Searches for Exotic Hybrid and Glueball States", paper 249, sheduled for session PA-3A.

89. Y. Guoz for the VES collaboration, "Study of the Wave with $J^{PC} = 1^{-+}$ ... in π^--Nucleon Interactions at $p_\pi = 37$ GeV/c", session PA-3B.

DISCUSSION

M. Gill, Lawrence Berkeley Laboratory, USA

Given that the mass difference between the light and heavy quarks in heavy quarkonium is greater than for the proton and the electron in the hydrogen atom, is there any hope of seeing the analog of hyperfine splitting (the breaking of double nuclear spin degeneracy) in meson spectroscopy?

Isgur

Yes. In fact, we have the opposite problem. The hyperfine interaction breaks the symmetry, and we must analyze the effects this will have on the theory's predictive ability.

A. N. Kamal, University of Alberta, Canada

The ratio of $D^0 \to \pi^0\pi^0$ and $D^0 \to \pi^+\pi^-$ rates, which experimentally appears to be close to 1 (CLEO result), is understood through the final state interactions—the same interactions which are responsible for the resolution of the $D^0 \to K\bar{K}$ and $\pi^+\pi^-$ problem.

Isgur

Correct. However, final state interaction calculations require information on scattering. This is available for $\pi\pi$ states in the D-mass region and so can be used to correct the factorization approximation. However, in general this information is not available in all channels, and in any event such corrections being large violates the basic premises of factorization.

V. Teplitz, Southern Methodist University, USA

Perhaps the answer was given and I missed it, but in the heavy quark effective theory, don't you also expect to have a lot of electromagnetic mass differences like D^* mass differences, B mass differences, B^* and B all being equal?

Isgur

What you expect is that the isospin splitting will be independent of the mass of the heavy quark. So the statement is that the $D^*(+) - D^*(0)$ should be equal to the $D(+) - D(0)$. The numbers were roughly that; in one case, we had 3.5 MeV, and in the other case, 4.5 MeV. So it's not perfect, but the deviation from equality is what you might expect from the residual D^*-D splitting of the doublet. That's about 140 MeV, and that could induce a residual isospin splitting in the difference of differences of about 1 MeV. There is also a prediction that if one measures the analogous isospin splittings in the B-system, that the difference of differences will have dropped by a factor of 3. So there is a very concrete prediction. But it's not true that the splitting should be the same in the B and the D systems, because the charges of the b and c-quarks are different.

PRECISION TESTS OF THE ELECTROWEAK INTERACTION

Luigi ROLANDI
Particle Physics Experiments Division
CERN
CH-1211 Geneva 23

Abstract

Recent measurements of various charged current and neutral current processes are presented and compared with the prediction of the Standard Model of the Electroweak Interaction. No discrepancy was found in the experiments' precision which is often higher than 0.5%. A Standard Model fit to the data allows an indirect determination of the top mass: $m_{top} = 145 \pm 25\ GeV$.

INTRODUCTION

The Electroweak Standard Model is a renormalizable theory and any experimental quantity can be predicted theoretically in terms of a finite number of parameters. These parameters are known with some precision from *a priori* determinations. Any additional measurement can be compared with the prediction. If these agree, the measurement can be used to further constrain the input parameters. Non-agreement indicates inconsistency of the theoretical framework and new physics.

This review summarizes the results of some precise experiments that test the Standard Model in the charged current sector and in the neutral current sector. The various experimental results are presented, compared to the prediction and finally used to constrain the unknown parameters of the model.

Many of the results summarized in this review have been submitted to the Conference as *preliminary*. They may change in the near future when the analyses will be eventually completed.

Charged current lepton universality

Lepton universality predicts that the coupling of the charged current to the three know leptonic doublets are equal. This hypothesis is verified experimentally comparing the decay width of two reactions that are identical on all aspects except the leptonic flavour involved.

Deviation from universality could point, for example, to the existence of a fourth generation neutrino of large mass ($> 46 GeV$) which mixes with one of the known neutrinos[1].

e-µ universality

The measurement of the branching ratio R of the rare pion decay into positron and neutrino ($\pi^+ \to e^+ \nu$) normalized to the normal pion decay $\pi^+ \to \mu^+ \nu$ can be compared with the theoretical calculation R_{TH} thus allowing a test of e-µ universality.

$$R = R_{TH} \left(\frac{g_e}{g_\mu}\right)^2 \quad (1)$$

where g is the coupling constant of the lepton to the charged current. The most recent prediction for R_{TH}[2] is $R_{TH} = (1.2345 \pm 0.0010) \times 10^{-4}$ where the uncertainty arises from uncalculated but bounded pion structure effects.

Two experiments, one performed at TRIUMF[3] and the other at PSI[4], have recently published new precise measurements of R. Positive pions are stopped in an active target and the prompt electrons from the $\pi \to e\nu$ rear decay are separated from the much larger background $\pi \to \mu \to e$ absorbing them in a calorimeter and measuring their energy with high resolution. This information is combined with the pulse height measured in the target, which is different in the two reactions due to the additional energy loss of the muon.

The two experiments obtain consistent results and have similar precision:

$$R = 1.2265 \pm 0.0034 \pm 0.0044 \times 10^{-4} [3]$$

$$R = 1.235 \pm 0.003 \pm 0.004 \times 10^{-4} [4]$$

where the first errors are statistical and the second systematic. They are mainly independent and the two results can be averaged and compared with the theoretical prediction using eqn.1 giving:

$$\frac{g_e}{g_\mu} = 0.9987 \pm 0.0019$$

in agreement with the hypothesis of e-μ universality.

$\mu - \tau$ universality

A test of the μ-τ universality is performed comparing the decay widths of the muon and the tau into electron and two neutrinos. Their ratio is predicted to scale with the 5th power of the mass of the parent lepton:

$$\left(\frac{g_\tau}{g_\mu}\right)^2 = Br(\tau \to e\nu\bar{\nu})\frac{\tau_\mu}{\tau_\tau}\left(\frac{m_\tau}{m_\mu}\right)^5 \quad (2)$$

where m is the mass of the lepton, τ its lifetime and g its coupling constant to the charged current. Radiative corrections[5] are at the level of 0.04% and can be neglected at the present level of precision.

The mass and the lifetime of the muon are known with very small relative errors compared to the corresponding tau quantities and the error on the ratio of the coupling constants is given by:

$$\Delta\left(\frac{g_\tau}{g_\mu}\right) = \frac{1}{2}\left[\frac{\Delta\tau_\tau}{\tau_\tau} \oplus \frac{\Delta Br}{Br} \oplus 5\frac{\Delta m_\tau}{m_\tau}\right] \quad (3)$$

Recent measurements of the mass, lifetime and electronic branching ratio of the τ allow a more precise determination of the ratio of the coupling constants. Using previous measurements this ratio was determined to be consistent with unity at the 2.3 σ level[6].

τ mass measurement

The mass of the τ has been measured at BES in a scan near the $\tau^+\tau^-$ production threshold[7]. Candidate events are identified by requiring $e - \mu$ final state. A likelihood function is used to estimate the τ mass incorporating the $\tau^+\tau^-$ cross section near threshold, the number of candidate events per scan point and the luminosity integrated at each scan point. Fig.1 shows the convergence of the predicted mass with each consecutive scan point, the collected luminosities and the number of candidate events. Using 7 candidate events and the information that no event is found at the other scan points, the value of the mass is fitted from the energy dependence of the cross section

$$m_\tau = 1776.9^{+0.4}_{-0.5} \pm 0.2 \; MeV.$$

The most important systematic error is due to the uncertainty on the selection efficiency. The energy calibration of BES is determined by several scans of the J/Ψ and $\Psi(2S)$ peaks and contributes to the systematics on m_τ with an error of ± 0.09 MeV.

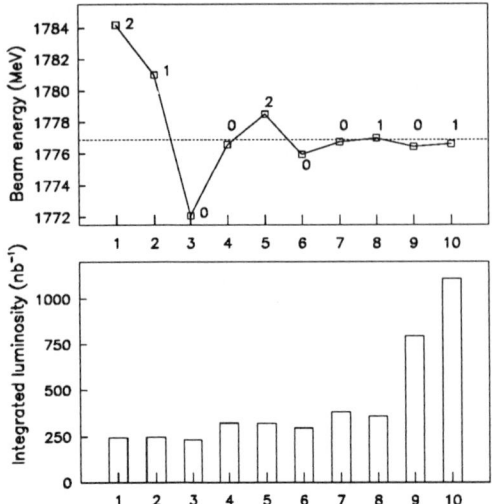

Figure 1. Convergence of the τ mass with the scan points The number of candidate events at each scan point is also indicated.

This result is 7.2 MeV below the previous world average[8] and reduces the error on the τ mass by a factor 6.

Two other independent measurements of the tau mass have been recently performed by Argus[9] ($m_\tau = 1776.3 \pm 2.4 \pm 1.4 \ MeV$) and Cleo[10] ($m_\tau = 1777.6 \pm 0.9 \pm 1.6 \ MeV$) using clever kinematical constraints.

τ lifetime measurement

The lifetime of the τ has been measured with improved precision at LEP using the silicon microvertex detectors installed recently in ALEPH, DELPHI and OPAL. Typical resolutions on the impact parameter are in the range $10 - 20 \mu m$.

The lifetime is measured with various techniques similar to the previous published analysis [11, 12, 13, 14] and also with new methods [15]. In the three-prong decays the secondary vertex is reconstructed allowing a direct measurement of the decay length, while in the one-prong decay the impact parameter technique is used. In the events where both τ particles decay into one prong the correlations among the impact parameters and the azimuthal angles are used in some analysis[11, 15] to increase the sensitivity.

Fig. 2 shows the new results of the LEP experiments [15] compared with the average of the other experiments [8, 16]. The precision on the lifetime has been improved by a factor 2 with respect to the previous world average[8].

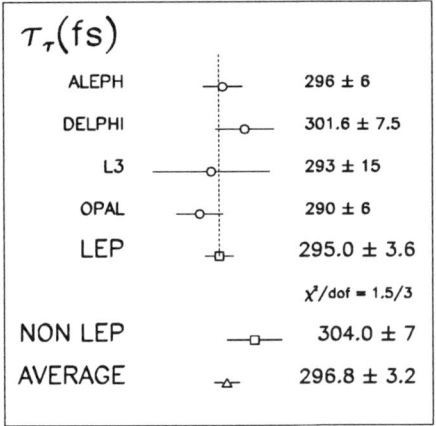

Figure 2. Tau lifetime.

Determination of the Br $\tau \to e\nu\bar{\nu}$

Using its large sample of data, CLEO has measured the electronic branching ratio B_e of the τ [10] selecting events in which both taus decay into electron and normalizing them to the number of τ pairs expected from the luminosity measurement. Their result is

$$B_e = 17.42 \pm 0.15 \pm 0.23\%$$

and is as accurate as the previous world average[8].

New measurements have also been reported from LEP[15] on the electronic and muonic branching ratio of the τ:

$$B_e = 17.68 \pm 0.17 \pm 0.18\%$$
$$B_\mu = 17.33 \pm 0.17 \pm 0.15\%$$

Assuming $e - \mu$ universality, the measured B_μ can be used to improve the accuracy on B_e. $\bar{B}_e$ is obtained as the weighted average between B_e and $B_\mu/0.9728$[1]:

$$\bar{B}_e = 17.76 \pm 0.17\% \quad \text{LEP}$$
$$\bar{B}_e = 17.67 \pm 0.18\% \quad \text{NON-LEP}$$
$$\bar{B}_e = 17.71 \pm 0.12\% \quad \text{AVERAGE}$$

where the NON-LEP average includes the CLEO measurement and previous experiments[8]. The new measurements improve the error on the electronic branching ratio by a factor 1.6 with respect to the previous world average[8].

$\mu - \tau$ universality, results

The measurement of τ mass, lifetime and electronic branching ratio can be now used to perform a test of $\mu - \tau$ universality using eqn. 2.

$$\frac{g_\tau}{g_\mu} = \begin{array}{ll} 0.992 \pm 0.007 & \text{LEP} \quad (1.1\ \sigma) \\ 0.974 \pm 0.012 & \text{NON LEP} \quad (2.2\ \sigma) \\ 0.987 \pm 0.006 & \text{ALL DATA} \quad (2.2\ \sigma) \end{array}$$

The measurement of the mass is so precise that it does not contribute to the error (cfr. eqn.3) on the ratio of the coupling constant. The lifetime and the electronic branching ratio contribute with almost equal weight.

The situation is illustrated in fig. 3, where the measured electronic branching ratio is plotted together with the value derived from the lifetime assuming lepton universality and using eqn.2 with the BES value of the τ mass.

The direct measurements of the branching ratio are very consistent. The LEP value derived from the lifetime is also very consistent with the direct measurement. The NON LEP determination of the lifetime produces a branching ratio that is 2.2 σ higher than the direct measurement, but only 1.1 σ higher

---
[1]$e - \mu$ universality predicts $B_\mu/B_e = 0.9728$ due to the larger μ mass.

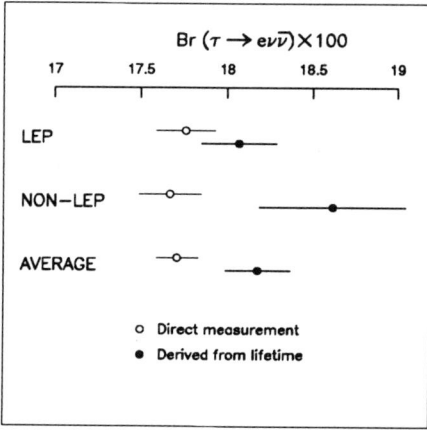

Figure 3. Br $(\tau \to e\nu\bar{\nu})$. The direct measurements are compared with those derived from the lifetime, assuming $\mu - \tau$ universality.

than the LEP value derived from the lifetime. When all data are combined, the agreement between the direct measurement of the branching ratio and the derived one remains at 2.2 σ level.

In conclusion, the new measurements of the τ parameters allow a test of the $\mu - \tau$ universality with an improvement in precision by more than a factor 2 compared to previous measurements[6]. The ratio of the coupling constants is consistent with 1 at the 2.2 σ level.

Neutrino electron scattering

New results on the scattering $\nu e \to \nu e$ have been presented by the CHARM II Collaboration[17].

The signature of the process is a single forward scattered electron producing an electromagnetic shower in the calorimeter. The variable $E_e \theta_e^2$ - product of the energy and square of the scattering angle - is kinematically constrained to values smaller than 1 MeV and is used to separate the signal from the background, mainly due to the reaction $\nu A \to \nu \pi_0 A$.

This scattering is a purely leptonic reaction and its interpretation has little theoretical uncertainties. The cross section is small ($\sigma_{tot}^{\nu e} \sim 10^{-4} \sigma_{tot}^{inclusive}$) resulting in a small number of selected events. In total, about 3000 neutrino electron scattering and 3000 antineutrino electron scattering events have been selected. $\sin^2 \theta$ can be extracted directly from the ratio of the two cross sections:

$$\sin^2 \theta_{\nu e} = 0.232 \pm 0.006 \pm 0.007.$$

From a comparison between the data and the predicted event distribution $d\sigma/dy$ [18], and assuming $g_a^\nu = 1/2$, the vector and axial-vector coupling constants of the electron to the neutral current are measured at $Q^2 \sim 10^{-2} GeV^2$:

$$g_v^e = -0.025 \pm 0.014 \pm 0.014$$
$$g_a^e = -0.503 \pm 0.007 \pm 0.016$$

This results can be directly compared (cfr. fig 4) with the more precise measurement of the same quantities done al LEP at $Q^2 \sim M_z^2$ using the forward backward asymmetry of the electron and the forward backward asymmetry of the tau polarization. Inside the Standard Model the differences of the two couplings at the two scales are negligible due to a partial cancellation between the changes caused by γ-Z transition and the effect of the neutrino charge radius [19]. The agreement is remarkable.

Neutrino nucleon scattering

A new measurement of R, the neutral to charged current interaction rate of neutrino on isoscalar target, has been presented by the CCFR Collaboration[20].

$$R = \frac{\sigma_{NC}^\nu}{\sigma_{CC}^\nu}$$

R is an indirect measurement[21] of the ratio between the W and the Z mass with very

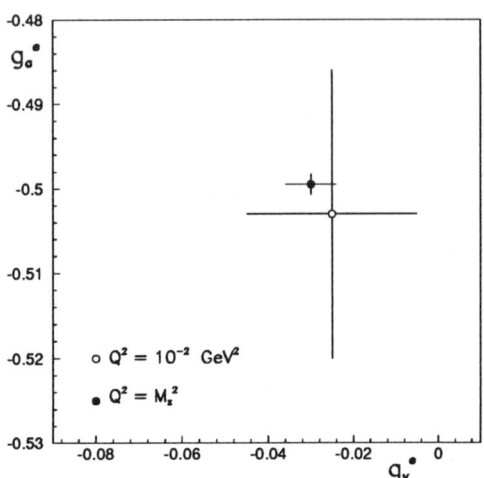

Figure 4. Comparison of results from neutrino electron scattering and LEP.

little dependence on other unknown parameters of the theory

$$R = \left(\frac{M_w}{M_z}\right)^4 \frac{1}{2} \frac{1 - 2\sin^2\theta_w + \frac{10}{9}\sin^4\theta_w(1+r)}{1 - 2\sin^2\theta_w + \sin^4\theta_w} \quad (4)$$

since by numerical accident the dependence of (4) on $\sin^2\theta_w$ is very weak. In this formula r is the ratio between anti-neutrino and neutrino induced charged currents ($r = 0.38 \pm 0.01$[22, 23]).

The CCFR collaboration has selected about 140.000 neutral current and 270.000 charged current events using the Fermilab QUAD-TRIPLET neutrino beam. The separation of the events into the two classes is done using the *event length* measured in the experiment. The main systematic error comes from the model needed for the subtraction of the short charged current events that are misclassified.

The mass ratio is obtained comparing the measured distributions with the result of a complete Monte Carlo simulation:

$$1 - \frac{M_w^2}{M_z^2} = 0.2242 \pm 0.0029 \pm 0.033 \pm 0.0047$$

where the first error is statistical, the second

is due to the experimental systematics and the third to the model used for the interpretation.

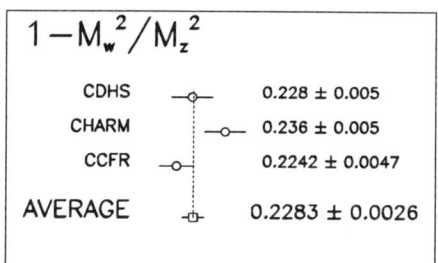

Figure 5. Mass ratio measured by the neutrino nucleon scattering experiments. Only the independent errors are shown and a common systematic error of ±0.0047 has to be added to the average. The values of the mass ratio are quoted at different top masses resulting in possible shifts of the central values of 0.002.

In fig. 5 this result is compared with the measurement done at CERN by CDHS and CHARM[22, 23]. Only the experimental error is shown (including statistics and experimental systematics) since the error due to the model is dominated by the uncertainty on the effective mass of the charm and is very correlated among the three experiments. The correlation is large but not complete since, due to the harder spectrum of the incoming neutrinos, the relative importance of the different terms contributing to the model systematic error is different for CCFR with respect to the two CERN experiments. The combined result of the three experiments is:

$$1 - \frac{M_w^2}{M_z^2} = 0.2283 \pm 0.0026 \pm 0.0045$$

and an indirect determination of the W mass

$$M_w = 80.10 \pm 0.27 \; GeV$$

is obtained using LEP value $M_z = 91.187 \pm 0.007$ GeV. This result is consistent with the direct measurement of the W mass obtained by combining the results of UA2[24] and CDF[25] ($M_w = 80.22 \pm 0.26$ GeV[8]) and has the same precision.

Z lineshape and forward backward asymmetries

The Z parameters have been measured with high precision at LEP using the data collected in years 1990 and 1991 when various scans of the Z peak were performed. The integrated luminosity delivered by LEP to each experiment was about 25 pb^{-1}. The precision on the mass and on the width of the Z depends crucially on the absolute energy calibration of the machine.

The energy calibration of LEP

The absolute energy scale of LEP has been calibrated in 1991[26] using the measurements done with a controlled spin-depolarizing resonance on the vertically polarized beam[27] as main tool. Transverse polarization has been observed using a laser polarimeter during eight dedicated experiments with a level between 5% and 16% under special conditions, distinct from those of the physics runs and at a single nominal energy setting.

A frequency-controlled radial RF magnetic field makes the electron spin precess away from the vertical axis. A depolarizing resonance occurs when the radial magnetic field oscillates at the spin precession frequency $\omega_{dep} = 2\pi \nu_s f_{rev}$. The spin tune ν_s is related to the beam energy via

$$\begin{aligned} E_{beam}(GeV) &= \tfrac{m_e c^2}{a_e} \nu_s \\ &= 0.4406486 \left(N_s + \tfrac{f_{dep}^{res}}{f_{rev}} \right) \end{aligned} \quad (5)$$

where a_e is the gyromagnetic anomaly of the electron, and m_e is the electron mass. The numerical coefficient in eqn. 5 is known with a precision of 2×10^{-7}. The integer part of the spin tune at the Z energy is $N_s = 103$. The intrinsic accuracy of the method is very high and allows a calibration of the average beam energy with an error of ~ 1 MeV. Comparing the spread of six different measurements performed at different times results in an error of

±3.7 MeV on the absolute calibration of the center of mass energy of LEP corresponding to a relative accuracy of 4×10^{-5}.

Other effects have to be taken into account when this calibration is transported to the standard conditions of the physics runs, resulting in a relative accuracy of 5.7×10^{-5}. The main additional source of error is the change with the temperature of the integrated bending magnetic field seen by the electrons.

The absolute calibration is done at a single nominal energy setting (~ 2 GeV above the Z mass). The error on the calibration at the other energy settings relative to this special setting depends on the accuracy of the relation between energy and dipole currents, which is determined by magnetic measurements. This error limits the relative accuracy on the measurement of the Z width to 1.5×10^{-3}.

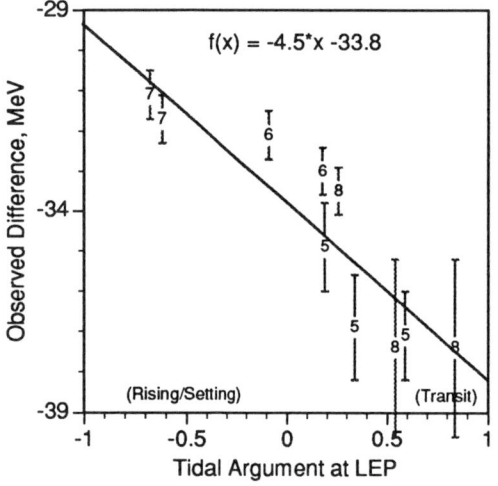

Figure 6. Results of the calibrations done with the resonant depolarization technique as a function of the predicted (normalized) tidal force of the moon and the sun. The numbers at the centre of the error bars indicate the numbering of the calibration experiments.

The error on the Z parameters induced by the machine calibration is very relevant: the error on the mass is dominated by the calibration error, while the error on the width due to calibration uncertainties is as large as the statistical one. It is important to understand how these errors can be reduced in future measurements. The error on the Z width can be improved by measuring the local energy scale of LEP performing an absolute calibration at more than one energy setting.

The error on the Z mass can be improved if the cause for the spread of the beam energies measured with the depolarization technique at different times in ideally identical conditions is understood. A possible explanation is that the radius of LEP changes by a small relative amount (3×10^{-8}) due to the tidal forces inducing a change of the center of mass energy of ~ 10 MeV[28]. The results of the various calibrations are shown in fig. 6 as a function of the tidal force normalized between -1 and +1. The data support this hypothesis which however has to be confirmed by a dedicated experiment. If the data are corrected according to the observed correlation, the spread in the calibrations of ±3 MeV is reduced to ±1 MeV.

Z lineshape

The cross sections $e^+e^- \to f\bar{f}$ for hadronic($q\bar{q}$) and leptonic final ($\ell^+\ell^-$) states are measured at various energies around $\sqrt{s} \sim M_z$. The triggers are highly redundant and their efficiency is essentially 100% for all channels resulting in negligible systematic errors. The event selections are similar to those performed with 1990 data [29, 30, 31, 32] and better performances of the detectors allow a reduction in the systematic errors. The number of selected events and the systematic errors on the event selections are shown in Table 1. The uncertainty in the theoretical cross section of the Bhabha scattering used for the luminosity measurement is reduced to the level of 0.3% by using the BHLUMI[33] computation of the visible cross section which is order α plus Lead-

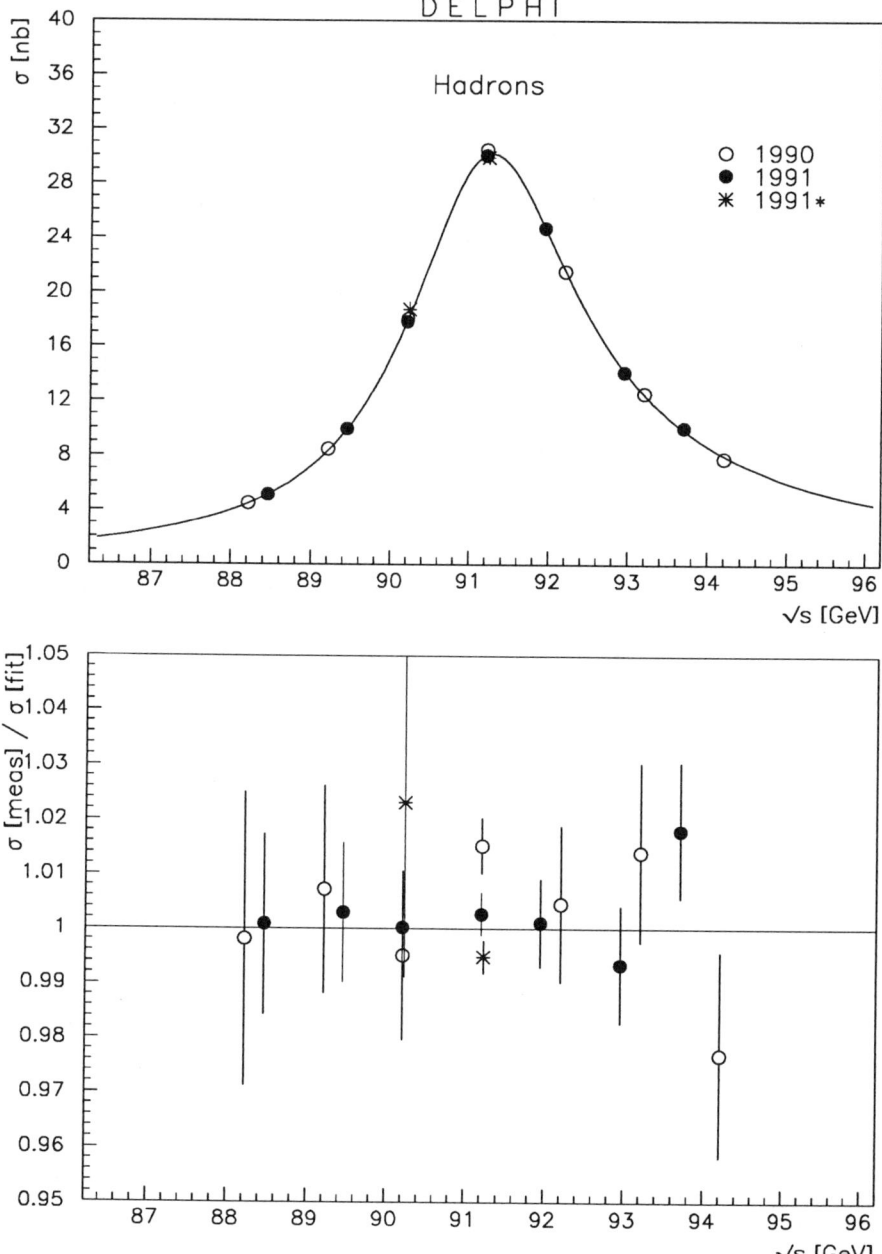

Figure 7. Hadronic cross sections measured by DELPHI. The curve is the result of the five-parameters fit described in the text

Table 1. Number of selected events and systematic errors of the event selections. The data sample corresponds to an integrated luminosity of ~ $25pb^{-1}$ luminosity collected in 1990 and 1991 by each experiment.

		ALEPH	DELPHI	L3	OPAL
Number of events	$q\bar{q}$	452K	368K	425K	454K
	$\ell^+\ell^-$	53K	35K	41K	54K
systematic error	$q\bar{q}$	0.2%	0.3%	0.3%	0.2%
	e^+e^-	0.4%	0.5%	0.6%	0.4%
	$\mu^+\mu^-$	0.5%	0.4%	1%	0.3%
	$\tau^+\tau^-$	0.5%	0.9%	1%	0.8%
experimental systematic error on luminosity		0.45%	0.55%	0.85%	0.7%*

* Opal luminosity systematic error refers to the analysis of 1990 data only

ing Logs resummed. It is complemented with corrections taking into account the high-order terms of the $\gamma - Z$ interference.

The measured cross sections σ^M are fitted to the expression

$$\sigma^M_{f\bar{f}}(s) = \int H(s,s')\sigma^0_{f\bar{f}}(s')ds' \qquad (6)$$

where H is the *radiator function* that takes into account the large (~ 30%) effects of the initial state radiation. σ^0 is the model independent formula of the cross section:

$$\sigma^0_{f\bar{f}}(s) = \sigma^{pole}_{f\bar{f}} \frac{s\Gamma_z^2}{(s-M_z^2)^2 + s^2\Gamma_z^2/M_z^2}$$
$$+\gamma_{exchange} + Interference$$
$$\sigma^{pole}_{f\bar{f}} = \frac{12\pi\Gamma_e\Gamma_f}{M_z^2\Gamma_z^2}$$
(7)

Its first term contains the Z exchange and defines the mass M_z and the width Γ_z of the Z boson. The photon exchange and the interference terms are small and the Standard Model value is assumed within the fits without any loss in generality at the present level of precision.

Lepton forward-backward asymmetries

The forward backward asymmetry $A^M_{FB}(s)$ is determined through a fit of the angular distribution of the cross section measured at at each center of mass energy:

$$\frac{d\sigma(s)}{d\cos\theta} \propto 1 + \cos^2\theta + \frac{8}{3}A^M_{FB}(s)\cos\theta \qquad (8)$$

The measured asymmetry $A^M_{FB}(s)$ is fitted to the expression

$$A^M_{FB}(s) = \frac{\int H'(s,s')\sigma^0_{FB}(s')ds'}{\int H(s,s')\sigma^0(s')ds'}$$

where

$$\sigma^0_{FB} = \int_0^1 \frac{d\sigma^0}{d\cos\theta}d\cos\theta - \int_{-1}^0 \frac{d\sigma^0}{d\cos\theta}d\cos\theta$$

The forward backward cross section σ^0_{FB} depends on the *bare* asymmetry A^0_{FB}:

$$\sigma^0_{FB}(M_z^2) = \sigma^0(M_z^2)(1-\epsilon)(A^0_{FB} + \delta)$$

where the small term ϵ (~ 0.01) is due to the photon exchange and δ (~ 0.002) is the contribution of the imaginary part of the propagator corrections.

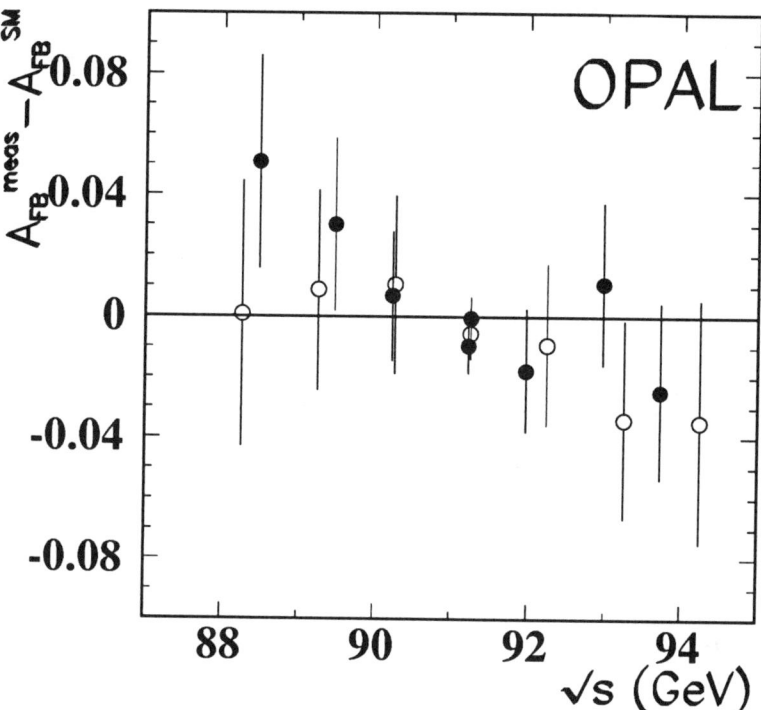

Figure 8. Energy dependence of the lepton forward-backward asymmetry as measured by the OPAL collaboration. The difference between the measured value and the SM prediction for $m_{top} = 150 GeV$ and $m_{higgs} = 300 GeV$ is shown.

This *bare* asymmetry A^0_{FB} can be directly interpreted in terms of the ratio between the vector (g_v) and the axial vector (g_a) coupling constants of the neutral current to the fermions:

$$(A^0_{FB})_x = \frac{3}{4} A^0_e A^0_x$$

$$A^0_x = \frac{2 g^x_v g^x_a}{(g^x_v)^2 + (g^x_a)^2} \quad (9)$$

FIT results

Two fitting programs are used to extract the lineshape parameters and the *bare* asymmetries: ZFITTER[35] and MIZA[36]. They give the same result in 2% of the statistical errors of the fitted parameters[37].

Radiator function H contains the contributions of multiple photon emission and lepton pair production. The error produced on the predicted cross sections by its theoretical uncertainties ($\Delta\sigma/\sigma \sim 4 \times 10^{-4}$[38]) is negligible at the present level of precision.

e^+e^- final state contains additional contributions to the cross section from the t-channel diagrams and from the s-t-channel interference. They are taken into account using program ALIBABA[34] resulting in a systematic error smaller then 0.5%.

There are 9 independent parameters to be fitted: the mass and the width of the Z, the hadronic peak cross section and, for each lepton species, the ratio R between the hadronic and the leptonic partial widths and the *bare*

Table 2. Z lineshape and asymmetry parameters. Lepton universality is not assumed.

Parameter	Average Value	χ^2
$M_Z(GeV)$	91.187 ± 0.007	2.3
$\Gamma_Z(GeV)$	2.492 ± 0.007	1.5
$\sigma_{q\bar{q}}^{pole}(nb)$	41.16 ± 0.18	5.3
R_e	20.92 ± 0.12	3.4
R_μ	20.79 ± 0.10	1.2
R_τ	20.84 ± 0.13	2.5
$(A_{FB}^0)_e$	0.0103 ± 0.0058	1.5
$(A_{FB}^0)_\mu$	0.0120 ± 0.0041	4.7
$(A_{FB}^0)_\tau$	0.0246 ± 0.0048	1.7

Table 3. Z lineshape and asymmetry parameters. Lepton universality is assumed.

Parameter	Average Value	χ^2
$M_z(GeV)$	91.187 ± 0.007	2.3
$\Gamma_z(GeV)$	2.492 ± 0.007	1.5
$\sigma_{q\bar{q}}^{pole}(nb)$	41.16 ± 0.18	5.3
R_ℓ	20.85 ± 0.07	0.5
$(A_{FB}^0)_\ell$	0.0154 ± 0.0027	5.5

forward-backward asymmetry. The number of parameters is reduced to 5 when lepton universality is assumed.

The four LEP experiments give consistent results for all fitted quantities. They have been combined to produce the best average values for the Z lineshape parameters, taking into account errors that are correlated among the experiments. The combination can be done following the procedure described in [39] by taking a simple weighted mean of the fit variables and using the correlation matrix of any of the experiments. The average lineshape and asymmetry parameters from the 9-parameters fit and from the 5-parameters fit are given in Tables 2 and 3. Figure 9 shows the results for the Z mass. Figure 10 shows the comparison of the results obtained by each experiment

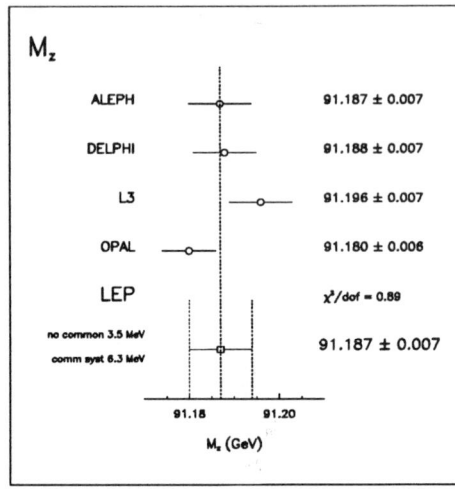

Figure 9. Results for the Z mass.

Table 4. Values of some parameters derived from the result of the 5-parameter fit

Parameter	Average Value
$\Gamma_l(MeV)$	83.33 ± 0.30
$\Gamma_{had}(MeV)$	1737.1 ± 6.7
$(g_v^\ell)^2$	$(1.29 \pm 0.23) \times 10^{-3}$
$(g_a^\ell)^2$	0.2494 ± 0.0009
$\Gamma_{inv}(MeV)$	504.6 ± 5.8

with the standard model prediction. The error on the Z mass is dominated by the uncertainty on the energy calibration of LEP. The error on the width is still statistically limited but the systematic error due to the LEP energy calibration is not negligible. The error on the hadronic peak cross section is dominated in each experiment by the systematic error on the luminosity measurement. Part of this error (0.3%) is due to the theory describing the Bhabha scattering and is completely correlated among the four measurements.

The values of other Z parameters and their errors can be derived from the fitted ones. Table 4 shows the values of some derived parameters using the result of the 5-parameters fit.

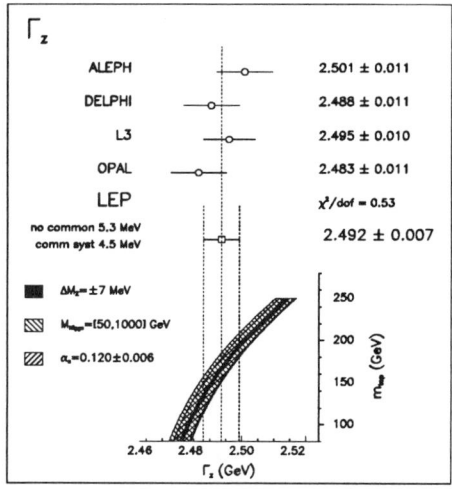
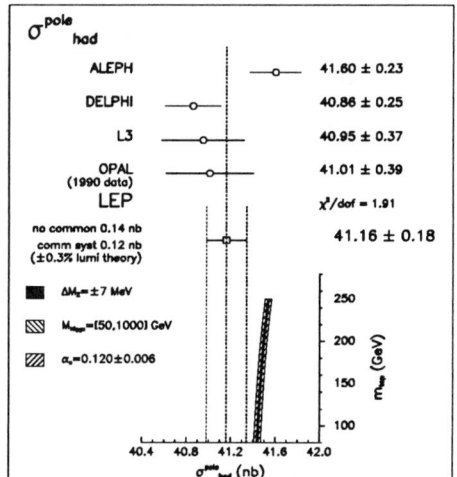
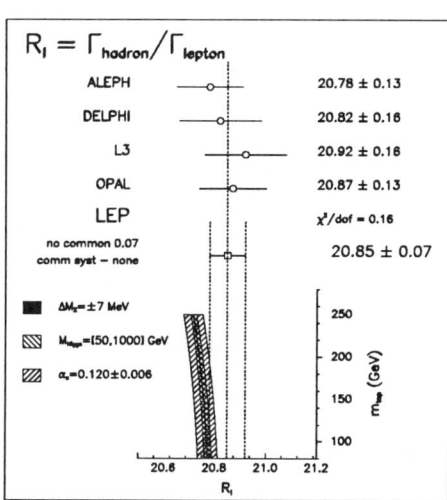
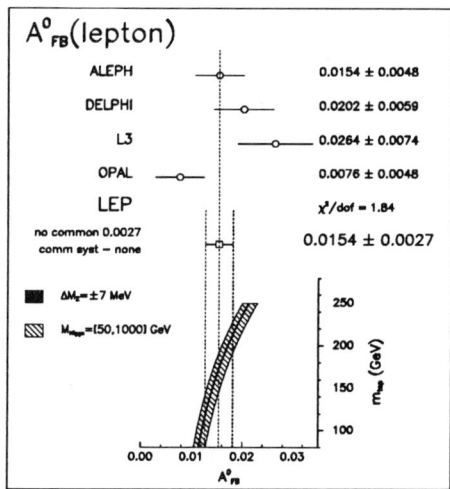

Figure 10 Fit results and Standard Model prediction as a function of m_{top} for the ranges of m_{higgs} and α_s indicated.

Tau polarization asymmetry

The angular dependence of the final state longitudinal polarization of the τ^- produced in the reaction $e^+e^- \to \tau^+\tau^-$ is:

$$P_\tau(\cos\theta) = \frac{d\sigma_R(\cos\theta) - d\sigma_L(\cos\theta)}{d\sigma_R(\cos\theta) + d\sigma_L(\cos\theta)} \quad (10)$$

$$= -\frac{A_\tau + A_e \frac{2\cos\theta}{1+\cos^2\theta}}{1 + A_\tau A_e \frac{2\cos\theta}{1+\cos^2\theta}}$$

where σ_R and σ_L are the cross sections for the production of right-handed and left-handed τ^- and θ is the angle between the e^- and the τ^- directions in the center of mass system. At $\sqrt{s} \simeq M_z$, A_τ (A_e) depends on the ratio of the coupling constants of the $\tau(e)$ to the neutral current.

When averaged on all production angles P_τ is a measurement of A_τ. Also A_e can be extracted when the angular dependence is studied.

The τ polarization is measured by fitting the momentum distribution of its decay product. Particle identification is applied to select the τ decay channel. For each channel the measured spectra are fitted with a linear combination of the Montecarlo predicted spectra for the two different helicities, including background and full detector simulation. V-A is assumed in the charged current τ decay to predict the momentum spectra.

The tau polarization averaged over production angles has been measured at LEP by the four collaborations using the τ^- decays into $e^-\nu\bar{\nu}$, $\mu^-\nu\bar{\nu}$, $\pi^-\nu$ and $\rho^-\nu$. The analysis technique [40] has been improved with respect to the already published result on 1990 data [41, 42, 43], thus reducing systematic errors.

The polarizations obtained by each collaboration in each decay channel are consistent, their averages are shown in fig. 11. Sensitivity is smaller for leptonic decay modes because part of the information is lost in the three body decay with undetectable neutrinos.

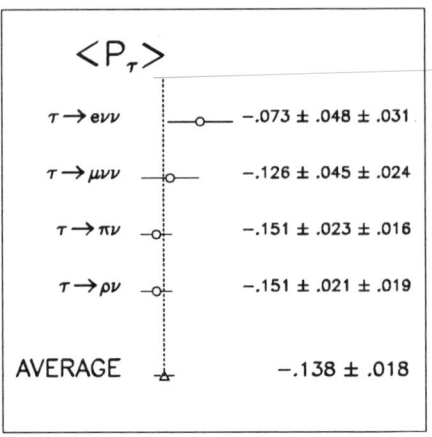

Figure 11. Tau polarization measured using 4 different decay channels. The data shown are averages of the results obtained by the LEP collaborations. The first error is statistical and the second systematic.

The ALEPH collaboration has also studied the angular dependence of the polarization. The polarization averaged over the four different channel is calculated in 9 different $\cos\theta$ bins (cfr. fig. 12). A_e is measured from a fit of the angular dependence (cfr. eqn. 10):

$$A_e = 0.120 \pm 0.029 \pm 0.010.$$

Some corrections have to be made to convert the measured asymmetry A_x to the *bare* asymmetry A_x^0 (cfr. eqn. 9):

$$\begin{aligned}A_\tau^0 = A_\tau \quad &*(1-0.002) &&\sqrt{(s)} \neq M_z \\ &*(1+0.017) &&\text{QED initial state} \\ &*(1+0.005) &&\gamma_{exch} + Interf.\end{aligned}$$

The effect of the final state radiation is taken into account in the fit procedure. The same correction factors apply to A_e.

These correction factors are small. They shift the central value of the measured asymmetry by a small amount compared to the error:

$$A_\tau^0 = 0.140 \pm 0.018$$
$$A_e^0 = 0.122 \pm 0.031$$

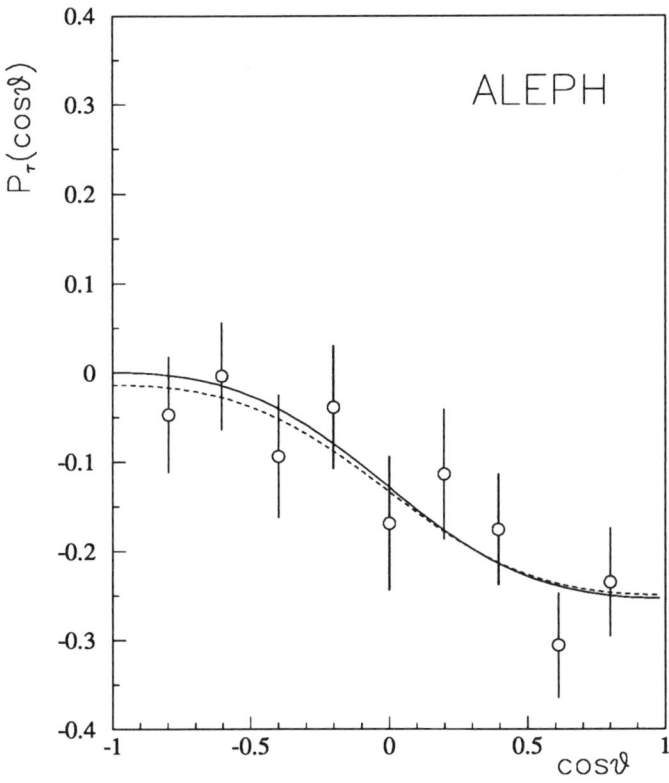

Figure 12. Angular dependence of the τ polarization measured by the ALEPH collaboration. The full curve is the fit of eqn.10 assuming lepton universality. The dashed line is the fit without the universality assumption.

Neutral current lepton universality

The vector and axial-vector coupling constants of the charged leptons to the neutral current can be calculated by combining the measured width of the Z into lepton pairs with the asymmetries. The asymmetries measure the ratio between the coupling constants (cfr eqn. 9) while the partial widths measure the sum of their squares:

$$\Gamma_\ell = \frac{G_F M_z^3}{6\sqrt{2}\pi}\left((g_a^\ell)^2 + (g_v^\ell)^2\right)(1 + \frac{3}{4}\frac{\alpha}{\pi}).$$

The tau polarization is linear in the ratio of the coupling constants thus allowing the determination of the relative sign. The absolute sign is derived from neutrino-electron scattering experiments[18].

Figure 13 shows a comparison among the coupling constants of the three leptons. This comparison is only qualitative since there are correlations between the errors on the coupling constants of different leptons which cannot be shown in the plot.

The coupling constants are equal inside the errors, as is expected from universality. They also agree with the standard model prediction.

The error of g_v^e and g_v^τ is smaller than the error of g_v^μ because of the tau polarization constraint.

A quantitative test of the universality can be performed by comparing ratios R, measured for the three lepton species in the 9-

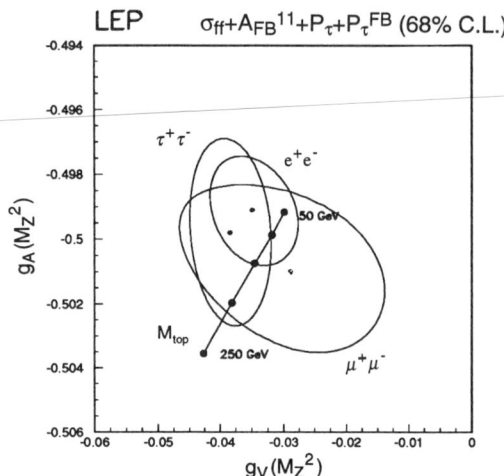

Figure 13. Comparison among the vector and axial vector coupling constants of the neutral current to the leptons.

parameter fit (cfr. tab. 2):

$$\frac{R_e}{R_\mu} = \frac{(g_a^e)^2 + (g_v^e)^2}{(g_a^\mu)^2 + (g_v^\mu)^2} \simeq \left(\frac{g_a^e}{g_a^\mu}\right)^2.$$

This ratio tests the equality of the axial-vector couplings since $g_v \ll g_a$.

$$\frac{g_a^e}{g_a^\mu} = 1.003 \pm 0.003$$

$$\frac{g_a^\tau}{g_a^\mu} = 1.001 \pm 0.004$$

$b\bar{b}$ forward backward asymmetry

The forward backward asymmetry of the b quark produced in the Z decays at LEP has been measured by tagging the $b\bar{b}$ production with the presence of a prompt lepton with high p_T with respect to the parent jet direction. The thrust axis is taken to define the quark direction and the sign of the charge Q_ℓ of the lepton to distinguish b from $\bar{b}$.

The raw asymmetry A_M is measured by fitting the angular distribution

$$\frac{d\sigma}{d\cos\theta_b} \propto 1 + \cos^2\theta_b + \frac{8}{3}A_M\cos\theta_b$$

$$\cos\theta_b = -Q_\ell \cos\theta_{THRUST}$$

Due to the mixing in the $B^0 - \bar{B}^0$ system, the observed b-quark asymmetry is smaller than the actual asymmetry by a factor (1−2χ_B), where χ_B is the probability that a hadron containing b-quark has oscillated into a hadron containing $\bar{b}$-quark at the time of its decay.

The high p_T lepton sample contains also non-prompt leptons from other sources: cascade decay of the b ($b \to c \to \ell$, $b \to \bar{c} \to \ell$), charm decay ($c \to \ell$) and background. The relation between the true (A_{FB}^b) asymmetry and the measured (A_M) one is:

$$A_M = (1 - 2\chi_B)A_{FB}^b [\eta_{b\to\ell} - \eta_{b\to c\to\ell} + \eta_{b\to\bar{c}\to\ell}]$$
$$- A_{FB}^c \eta_{c\to\ell} + A_{BKG}\eta_{BKG}. \quad (11)$$

The fractions η of each source of leptons are determined via Monte Carlo (cfr fig. 15). The sensitivity of the measurement depends crucially on the purity of the sample ($\eta_{b\to\ell}$).

Fig. 14 shows the new results presented by the LEP collaborations[44] which improve the analysis already performed with lower statistics[45, 46, 47, 48].

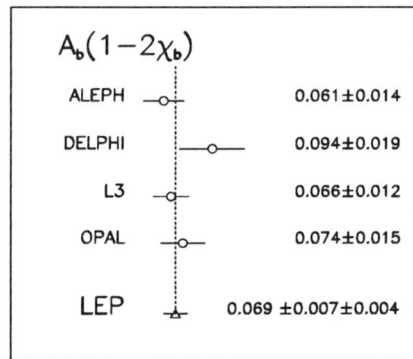

Figure 14. Results for the $b\bar{b}$ asymmetry.

The average shown in fig 14 is done by taking into account common systematic errors. They are due to assumptions on the B decay modes and on fragmentations functions common to the four experiments[44].

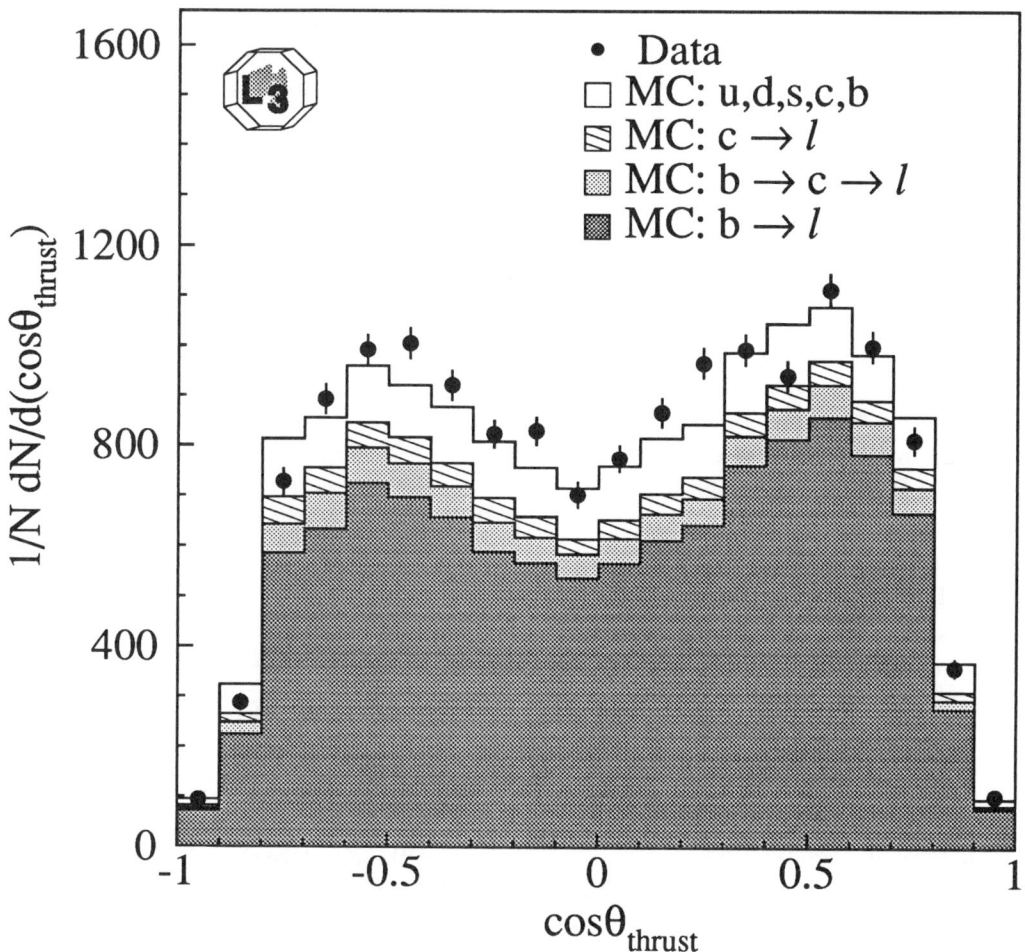

Figure 15. Angular distribution of the thrust axis for events containing a high p_T lepton (>1 Gev) measured by the L3 collaboration. The contributions of the various sources are indicated.

Asymmetry A_{FB}^b is derived using the Lep average value of $\chi_B = 0.126 \pm 0.012$:

$$A_{FB}^b = 0.093 \pm 0.010 \pm 0.006$$

where the first error contains the statistical effects and the uncorrelated systematics and the second contains the correlated systematics including the error on χ_B.

Some corrections have to be made to convert asymmetry A_{FB}^b to the *bare* asymmetry $(A_{FB}^0)_b$ (cfr. eqn. 9):

$$(A_{FB}^0)_b = A_{FB}^b \quad \begin{array}{ll} *(1-0.007) & \sqrt{(s)} \neq M_z \\ *(1+0.040) & \text{QED initial state} \\ *(1+0.023) & \text{QCD final state} \\ *(1+0.0002) & \text{QED final state} \\ *(1-0.003) & \gamma_{exch} + Interf. \end{array}$$

They shift the central value of the measured quantity by $\approx 50\%$ of its error:

$$(A_{FB}^0)_b = \frac{3}{4} A_e^0 A_b^0 = 0.0098 \pm 0.0012.$$

Left-right asymmetry

The annihilation cross-section in e^+e^- collision depends on the helicities of the electron and of the positrons. The left-right asymmetry is defined as:

$$A_{LR} = \frac{\sigma_L - \sigma_R}{\sigma_L + \sigma_R}$$

where σ_L and σ_R are the cross sections for left-handed and right-handed electrons. It is directly related to the electron asymmetry A_e (eqn.9): $A_{LR} = A_e$.

The left right asymmetry has been measured at SLC by the SLD collaboration[49] using a polarized electron source. It is a very robust measurement since systematic errors due to acceptance and luminosity determination are cancelled in the cross section ratio.

With a statistics of 10^3 Z decays and a polarization of 21%, SLD has measured:

$$A_{LR} = A_e = 0.02 \pm 0.07.$$

This result is not yet competitive with LEP, but the foreseen increase in the statistics (10^4 Z decays) and expecially in the polarization (75%)[49] of the electron beam will allow SLD to perform a precise measurement of the electron asymmetry.

The effective sinus: $\sin^2 \theta_W^{eff}$

The effective sinus is defined in terms of the ratio between the vector and axial vector couplings of the lepton to the neutral current:

$$\frac{g_v^\ell}{g_a^\ell} = 1 - 4\sin^2 \theta_W^{eff} \qquad (12)$$

Assuming lepton universality, the measurements of the leptonic forward backward asymmetry and of A_e^0 and A_τ^0 from the τ polarization can be directly converted into $\sin^2 \theta_W^{eff}$ using eqn. 9 and 12.

The $b\bar{b}$ forward backward asymmetry can also be converted into $\sin^2 \theta_W^{eff}$ assuming the Standard Model to evaluate A_b^0. This is a quite safe assumption since in the Standard Model A_b is almost independent from $\sin^2 \theta_W^{eff}$ [50].

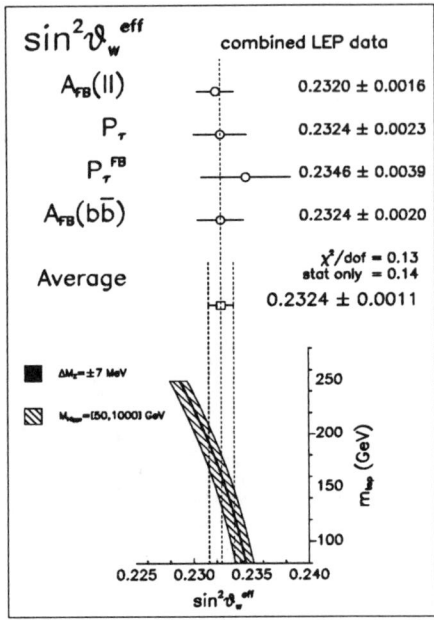

Figure 16. Comparison among different determinations of the effective sinus.

Fig. 16 shows the comparison among the different determinations of $\sin^2 \theta_W^{eff}$. They are all very consistent among each other and also with the Standard Model prediction. The average is:

$$\sin^2 \theta_W^{eff} = 0.2324 \pm 0.0011.$$

Using eqn. 12 this result can be combined with the value of g_a^ℓ from the lineshape fit (cfr. tab. 4) to obtain a more precise evaluation of g_v^ℓ:

$$g_a^\ell = -0.4994 \pm 0.0009$$
$$g_v^\ell = -0.0351 \pm 0.0022$$

The various measurements shown so far (cfr figs. 10 and 16) are all consistent with the Standard Model for a given range of the top mass and each of them can be used to constrain its value.

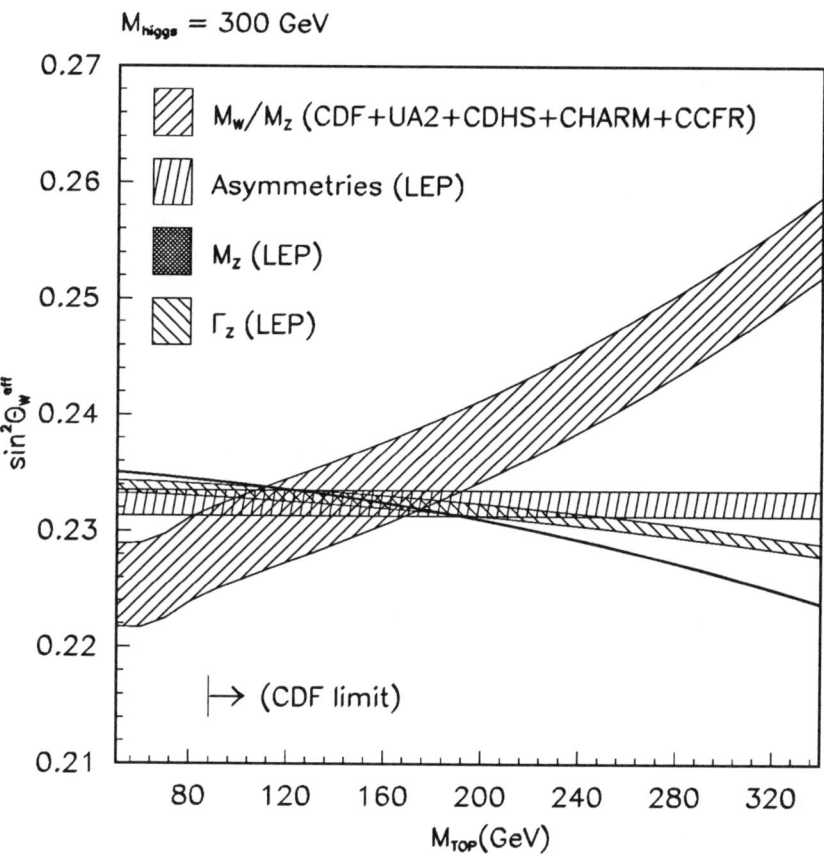

Figure 17. Constraints $\sin^2 \theta_W^{eff}$ versus m_{top} from different measurements corresponding to 1 σ limits. $m_{higgs} = 300$ GeV is assumed.

Fig. 17 shows the consistency among the different constraints: for any value of the top mass and for a fixed value of the Higgs mass (which here is assumed to be 300 GeV) each experimental measurement can be used to predict $\sin^2 \theta_W^{eff}$ within the framework of the Standard Model. The error on the prediction reflects the experimental error on the chosen quantity since all other input parameters have negligible relative error (α, G_F) or are assumed (m_{top}, m_{higgs}). The prediction with its error can be plotted as a function of m_{top} resulting in an allowed (68%) region in the plane $\sin^2 \theta_W^{eff}$ vs m_{top}.

Four different bands are shown in fig 17. They are calculated using the measurements of M_z, Γ_z [2], the asymmetries and mass ratio M_w/M_z. The band produced by the asymmetries is horizontal since they are a direct measurement of $\sin^2 \theta_W^{eff}$. The mass ratio value is obtained by combining the results from the neutrino nucleon scattering with those from

---
[2] The strong coupling constant α_s is needed to predict $\sin^2 \theta_W^{eff}$ from Γ_z. The value $\alpha_s = 0.120 \pm 0.006$ from LEP event shape is used[51] and the uncertainty induced on the prediction by its error is included in the width of the band.

the direct measurement of the W mass and the LEP value of M_z. The correlation between the constraint produced by the mass ratio and the very thigh one produced by M_z is negligible since the relative error on M_z is very small compared to those on the other quantities.

All data are consistent with a top mass within the range 120-180 GeV.

Minimal Standard Model fits

The top mass can be determined by fitting the Standard Model to the various electroweak quantities that are sensitive to it through radiative corrections. In this fit the unknown mass of the Higgs boson is fixed to 300 GeV and the change in the results induced by a change in the Higgs mass from 60 to 1000 GeV is quoted as an additional error.

Two different data sets are used. The first set (A) contains the LEP measurements : M_z, Γ_z, $\sin^2 \theta_W^{eff}$ from the asymmetries, $\sigma_{q\bar{q}}^{pole}$ and R_ℓ. The last two are not sensitive to the top mass and are used to constrain the strong coupling constant α_s. The second set (B) contains the Lep measurements, the mass ratio M_w/M_z from neutrino-nucleon scattering, M_w from UA2[24] and CDF[25], g_a^e and g_v^e from the neutrino electron scattering and Q_w measured in the parity violation experiment in Cesium[52].

Two different fits are performed. In the first fit the value of α_s is further constrained using the measurements done at LEP from the hadronic event shape variables[51] ($\alpha_s = 0.120 \pm 0.006$). In the second this constraint is released. The results of the four fits are shown in table 5.

The χ^2 of the fits is very good, as is anticipated by the consistency of each measurement with the standard model for the same top mass (cfr. fig 17).

The result of fit 1 is equivalent to an indirect determination of the W mass ($M_w = $

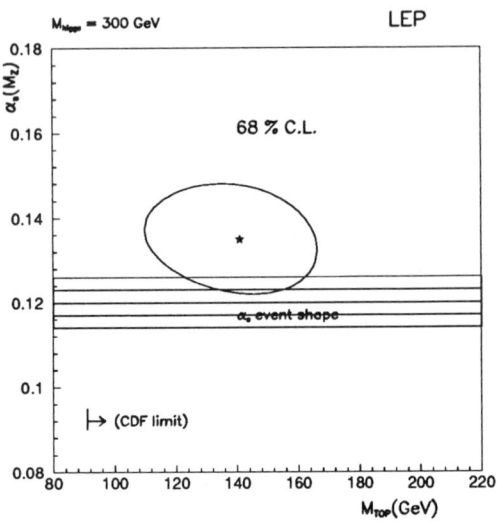

Figure 18. α_s vs m_{top} 68 % C.L. contour as obtained by fit n.4. The comparison with the α_s value obtained from event shape variables is also shown.

$80.17 \pm 0.14 \pm 0.02$ GeV) which is more precise than the direct measurements.

The electroweak data prefer a value of α_s slightly higher than the one derived from the event shape variables, but still consistent with it (cfr. fig. 18). The statistical error on m_{top} is smaller than the uncertainty due to the Higgs mass.

Limits on the Higgs mass

The limit on the Higgs mass obtained by combining the results of the direct search from the four experiments is[53]:

$$m_{higgs} > 60 \; GeV.$$

The effect of the Higgs mass on the radiative correction is small, logarithmic and correlated to the larger effect of the top mass. For this reasons, the electroweak data at the present level of precision cannot constrain the Higgs mass. This is shown in fig 19 where the χ^2 of fit 2 is plotted as a function of the top mass for three different values of the Higgs

Table 5. Results of the standard model fit to top mass and α_s. The first error is statistical, the second is due to the uncertainty on m_{higgs}.

Fit	Data Set	m_{top}(GeV)	α_s	$\chi^2/d.o.f.$
1	A	147^{+23+19}_{-27-22}	constrained	4.6/6
2	B	145^{+17+17}_{-19-19}	constrained	7.9/14
3	A	139^{+24+19}_{-30-22}	$0.135 \pm 0.009 \pm 0.002$	2.6/5
4	B	141^{+17+17}_{-19-19}	$0.135 \pm 0.009 \pm 0.002$	6/13

mass. The data - but in a way that is not significant - prefer lower values of the Higgs mass.

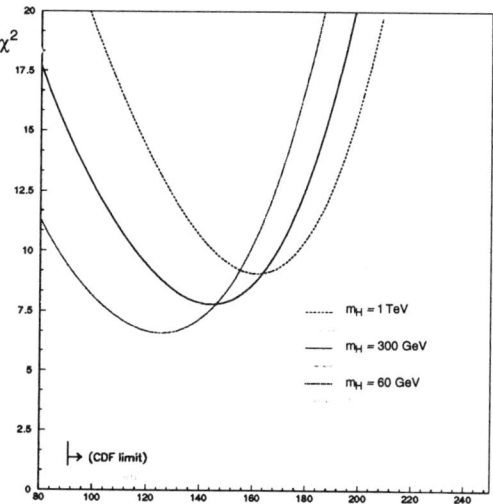

Figure 19. χ^2 of fit 2 as function of m_{top} for three different values of m_{higgs}.

The situation would change if the top were eventually discovered at the Fermilab collider and its mass measured [3] with an error smaller than the statistical error obtained on m_{top} in the Standard Model fit. This measurement could be used to further constrain the fit thus increasing the sensitivity to the Higgs mass.

This sensitivity is illustrated in fig. 20,

---

[3] If $m_{top} \simeq 140$ GeV, CDF should select after cuts 10 $e-\mu$ events plus 80 $\ell-jet$ events from $t\bar{t}$ production with 100 pb^{-1} of integrated luminosity[54].

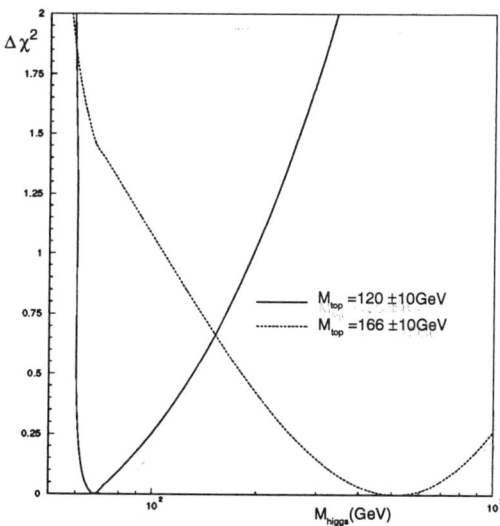

Figure 20. χ^2 variation of fit 2 as function of m_{higgs} for two values of the top mass

where the χ^2 variation of fit 2 is shown as a function of the Higgs mass when a constraint on m_{top} is added to the input data.

The sharp rise of the χ^2 for $m_{higgs} < 60$ GeV is due to the limit set by the direct search. The error on m_{higgs} increases with m_{higgs} since its effect on the radiative corrections is logarithmic. Moreover the top and Higgs masses are positively correlated in the fit: a light top mass implies a light Higgs mass.

If the top mass is light ($\simeq 120$ GeV) the fit prefers a light Higgs mass and could exclude to 1σ Higgs masses larger than 200-300 GeV. If the top mass is heavy ($\simeq 160$ GeV) the

76 Precision Tests of the Electroweak Interaction

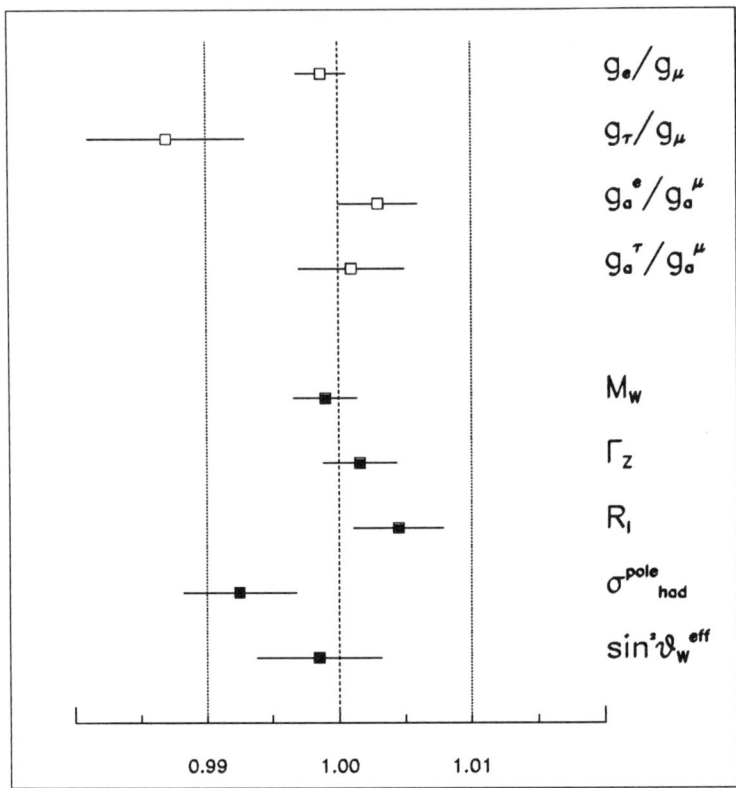

Figure 21. The first four quantities are the ratios of the coupling constants of the leptons which are expected to be equal in the Standard Model. The last five are the ratios between measurements and predictions of the Minimal Standard Model calculated assuming the measured value of the Z mass, $m_{top} = 150$ GeV, $m_{higgs} = 300$ GeV and $\alpha_s = 0.12$. Only experimental errors are shown.

Higgs mass is also heavy and its error would be larger: in this scenario no upper limit on the Higgs mass could be derived from the fit of the electroweak data with the present level of precision even if the top were discovered and its mass measured.

Conclusions

The precise results obtained by the many experiments testing the Minimal Standard Model in various sectors are all in agreement. This is illustrated in fig. 21, where the ratio between measurements and predictions is shown for various observables. The conclusion is therefore that all data can still be consistently interpreted within the electroweak theory in its minimal form. This is a big success of the theory since the experimental precision on many observables is higher than 0.5%.

Table 6. Present and future errors on some electroweak observables.

	Now	End '93	
ΔM_W(MeV)	190	100	CDF+D0
$\Delta \Gamma_z$(MeV)	7	3	LEP
$\Delta \sin^2 \theta_W^{eff}$	0.0011	0.0004	LEP+SLC

The slight discrepancies shown in fig 21 are

not statistically significant and call for more data to further improve the accuracy of the *Precision Tests of the Electroweak interaction*.

LEP, SLC and the Fermilab collider are now running and the various experiments are increasing their statistics. In two years the experimental precision on some crucial quantities (as those shown in fig 17) is expected to improve (cfr. table 6) thus putting the electroweak theory to a stronger test. If CDF and D0 find the top and measure its mass with an error $\simeq 5$ GeV, there is a possibility to be sensitive to the mass of the Higgs boson.

Acknowledgements

It is a pleasure to thank M. Martinez for the enlightening discussions and for the programs for the Standard Model fits. I am grateful to A. Blondel, J. Lefrançois, R. Tanaka and F. Ragusa for the invaluable help given in preparing this review and to many other colleagues for useful discussions and for the help in providing me with the documentation, in particular: G. Altarelli, J-E. Augustin, S. Bethke, T. Bolton, R. Clare, P. Drell, J. Ellis, F. Fidecaro, M. Franklin, S. Ganguli, S. Gentile, K. Kumar, M. McCubbin, H. Meinhard, J. Mildenberger, C. Moisian, T. Mori, G. Myatt, A. Olshevski, J. Panmann, M. Pohl, G. Quast, M. Roney, P. Rowson, M. Sasaki, D. Schaile, V. Sharma, W. Smith, B. Spaan, D. Treille, W. Toki and T. Wyatt.

REFERENCES

1. W.J. Marciano, τ decay puzzle, *Phys. Rev.* D45, R721, (1992).

2. W.J. Marciano, private communication (1990)

3. D.I. Britton *et al*, Measurement of the $\pi^+ \to e^+\nu$ Branching ratio, *Phys. Rew. Lett.* 68, 3000, (1992).

4. G. Czapek *et al*, Branching ratio of the Rare Pion Decay into positron and neutrino, *LHEP-Preprint* BUHE-92-1(1992).

5. W. Marciano and A. Sirlin, Electroweak radiative corrections to τ decay, *Phys. Rev. Lett.* 61, 1815, (1988).

6. M. Danilov, Heavy Flavour Physics, in *Proceedings of the HEP91 Conference*, 1991, p. 333.

7. J.Z. Bai *et al*, Measurement of the mass of the τ lepton, SLAC-PUB 5870 and BEPC-EP-92-01, (1992), and F. Porter, Talk given at this conference in PA-1.

8. Particle Data Group, K. Hikasa *et al*, Review of particle properties, *Phys. Rev. D*, 45, 1 (1992).

9. ARGUS Collaboration, A measurement of the tau mass, *DESY 92-86*, (1992) and A. Golutvin, Talk given at this conference in PA-1.

10. CLEO Collaboration, K. Gan, Tau physics at CLEO, Talk given at this conference in PA-1.

11. ALEPH Collaboration, D. Decamp *et al*, Measurement of the τ lifetime, *Phys. Lett.* B297, 411, (1992).

12. DELPHI Collaboration, P. Abreu *et al*, A measurement of the lifetime of the τ lepton, *Phys. Lett.* B267, 422, (1991).

13. L3 Collaboration, B. Adeva *et al*, Decay properties of the τ lepton measured at the Z resonance, *Phys. Lett.* B265, 451, (1991).

14. OPAL Collaboration, P.D. Acton *et al*, Measurement of the τ lepton lifetime, *Phys. Lett.* B273, 355, (1991).

15. M. McCubbin, Lifetime and branching ratio of the τ, Talk given at this Conference in PA-5.

16. R. Stroynowski, Tau decays at CLEO, in *Proceedings of the HEP91 Conference*,

1991, p. 562.

17. CHARM II Collaboration, A.Staude, NC coupling constants from neutrino electron scattering, talk given at this conference in PA-5.

18. CHARM II Collaboration, P. Vilain et al, Neutral current coupling constants from neutrino and antineutrino electron scattering, *Phys. Lett.* B281, 159, (1992). A.Staude talk given at this conference in PA-5.

19. V.A. Novikov, L.B. Okun, M.I. Vysotsky, Electroweak radiative corrections and the top mass, CERN TH-6053/91. (1991)

20. CCFR Collaboration, T.Bolton, Results from the CCFR experiment, talk given at this conference in PA-5.

21. R.G. Stuart, The $\bar{M}S$ and the $on-shell$ schemes in the analysis of neutrino scattering experiments, *Z. Phys.* C34, 445, 1987.

22. CDHS Collaboration, A. Blondel et al, Electroweak parameters from a high statistics neutrino nucleon scattering experiment, *Z. Phys. C*, 45, 361, (1990).

23. CHARM Collaboration, J.V.Allaby et al, A precise determination of the electroweak mixing angle from semileptonic neutrino scattering, *Z. Phys. C* 36, 611, (1987).

24. UA2 Collaboration, J. Alitti et al, An improved determination of the ratio of W and Z masses at the CERN $p\bar{p}$ collider, *Phys. Lett.* B276, 354, (1992).

25. CDF Collaboration, F. Abe et al, Measurement of the W boson mass in 1.8 Tev $p\bar{p}$ collisions, *Phys. Rew.* D43, 2070, (1991).

26. L. Arnaudon et al, The energy calibration of LEP in 1991, CERN-PPE/92-125 (1992) and G. Quast talk given at this conference in PA5.

27. L. Arnaudon et al, Measurement of LEP beam energy by resonant spin depolarization, CERN-PPE/92-49 (1992).

28. A. Hofmann, Effects modifying the c.m. energy in LEP in *Proceedings of the second workshop on LEP performance*, CERN SL/92-12, (1992).

29. ALEPH Collaboration, D. Decamp et al, Improved measurement of electroweak parameters from Z decays into fermion pairs, *Z. Phys.* C53, 1,(1992).

30. DELPHI Collaboration, P. Aarnio et al, Determination of the Z resonance parameters and couplings from its hadronic decays, *Nucl. Phys.* B367, 511,(1991).

31. L3 Collaboration, B. Adeva et al, Measurement of electroweak parameters from hadronic and leptonic decays of the Z, *Z. Phys.* C51, 179,(1991).

32. OPAL Collaboration, G. Alexander et al, Measurement of the Z lineshape parameters and the electroweak couplings of the charged leptons, *Z. Phys.* C52, 175,(1991).

33. S. Jadach et al, Monte Carlo program BHLUMI 2.01 for Bhabha scattering at low angle with Yenny-Frautshi-Suuna exponentiation, CERN-TH 6230/91, (1991).

34. W. Beenakker et al, Large angle Bhabha scattering, *Nucl. Phys.* B349, 323, (1991).

35. DUBNA-ZEUTHEN radiative correction group, D. Bardin et al, CERN-TH 6443/92, (1992) and references therein.

36. M. Martinez et al, Model independent fitting of the Z lineshape, *Z. Phys.* C49, 645, (1991)

37. T. Riemann, ZFITTER: An analytical program for $e^+e^- \to f\bar{f}$. Talk given at this conference in PA-5.

38. S. Jadach et al, Light pair corrections to the Z lineshape parameters, *Phys. Lett.* B280, 129, (1992).

39. The LEP Collaborations, Electroweak parameters of the Z resonance and the Standard Model, *Phys. Lett.* B199, 224, (1992).

40. K. Kumar, τ polarization, Talk given at this Conference in PA-5.

41. ALEPH Collaboration, D. Decamp *et al*, Measurement of the polarization of the τ lepton produced in Z decays, *Phys. Lett.* B265, 430, (1991).

42. DELPHI Collaboration, P. Abreu *et al*, A study of the decays of the τ leptons produced on the Z resonance at LEP, CERN-PPE 92/60, (1992).

43. OPAL Collaboration, G. Alexander *et al*, Measurement of the branching ratios and τ polarization from τ decays at LEP, *Phys. Lett.* B266, 201, (1991).

44. T. Wyatt, Forward-backward asymmetries, quarks. Talk given at this Conference in PA-5.

45. ALEPH Collaboration, D. Decamp *et al*, Measurement of the forward backward asymmetry in $Z \to b\bar{b}$ and $Z \to c\bar{c}$, *Phys. Lett.* B263, 325, (1991).

46. DELPHI Collaboration, P. Abreu *et al*, A measurement of the $b\bar{b}$ forward backward asymmetry using the semileptonic decay into muons, *Phys. Lett.* B276, 536, (1992).

47. L3 Collaboration, B. Adeva *et al*, A measurement of the $Z \to b\bar{b}$ forward backward asymmetry, *Phys. Lett.* B252, 713, (1991).

48. OPAL Collaboration, M.Z. Akrawy *et al*, A study of heavy flavour production using muons in hadronic Z decays, *Phys. Lett.* B263, 311, (1991).

49. P. Rowson, Polarized electron beam results from SLC, Talk given at this Conference in PA-5.

50. M. Bohm and W. Hollik, Forward-backward asymmetries, in *Z physics at LEP 1*, CERN 89-08.

51. S. Bethke, Test of QCD, Talk given at this Conference in PL-4.

52. M.C. Noecker *et al*, Precision Measurements of Parity Nonconservation in Atomic Cesium: A low-energy test of the electroweak theory, *Phys. Rev. Lett.* 61, 310, (1988).

53. T. Mori, Searches for the neutral Higgs boson at LEP, Talk given at this Conference in PA-13.

54. J. Huth, Top quark search from CDF, Talk given at this Conference in PA-13.

DISCUSSION

V. Khoze, University of Durham, U.K.

What kind of polarimeter has been used for the energy calibration at LEP?

Rolandi

Laser polarimeter based on the compton scattering.

M. Gill, Lawrence Berkeley Laboratory, USA

How does the Tidal Effect change the radius of LEP, and what will be the ultimate error on this quantity?

Rolandi

The whole ring will expand and contract in accordance with the "rubber ball" model of the earth. The level will be as was shown, or about 6 MeV.

J. Haissinski, LAL Orsay, France

To make a full use of a very precise measurement of $\sin^2 \theta_W$ one needs to know with an increased accuracy the running of α_s between $Q^2 \simeq 0$ and $Q^2 \simeq M_z^2$. This requires better data on the cross section $e^+e^- \to hadrons$ at relatively low energies. It would be nice if such measurements could be performed at BEPS in Beijing.

M. Samuel, Oklahoma State University, USA

I would just like to add that an accurate measurement of the total cross-section for e^+e^- to hadrons at low energies is also very crucial for improving the accuracy of the hadronic contribution to the anomalous magnetic moment of the muon. This could be performed at BES, prior to the new experiment on the topic at Brookhaven.

Rolandi

I agree.

TESTS OF QCD

Siegfried Bethke
Physikalisches Institut, University of Heidelberg
Philosophenweg 12
D-6900 Heidelberg, Germany

Abstract

Recent theoretical developments and experimental results on tests of QCD in deep inelastic scattering, in e^+e^- annihilation and in hadron collision processes are reviewed. The topics covered are structure functions, jet physics, prompt photon production, soft gluon coherence effects and a summary of measurements of α_s.

INTRODUCTION

The strong interaction, which is one of the four basic forces between elementary particles, is commonly described by the non-Abelian gauge theory of Quantum Chromodynamics (QCD)[1]. At this conference, recent experimental and theoretical progress concerning hadronic physics and tests of QCD was presented in 3 parallel sessions with more than 40 talks; about 165 contributed papers were submitted to the conference in this field. Due to the limited amount of time and space which is available for this review, not all of these contributions can be adequately summarized, discussed and referenced. Preference will therefore be given to an overview of tests of perturbative QCD, for which both significant data and reliable theoretical predictions exist.

The topics thus covered in detail are listed in Table 1. Each of these topics will be discussed in one of the following sections, whereby the results from different processes like deep inelastic lepton-nucleon scattering (DIS), e^+e^- annihilation and $p\bar{p}$ collisions, which are illustrated in Fig. 1, will be treated

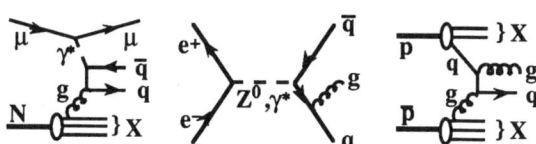

Fig. 1. Examples of diagrams leading to hadronic final states in the processes $\mu N \to \mu q\bar{q}X$, $e^+e^- \to q\bar{q}g$ and $p\bar{p} \to qgX$.

Table 1. Topics presented in this review, and contributing processes.

Topics	DIS	e^+e^-	$p\bar{p}$
Structure Functions	X		X
Jet Physics	X	X	X
Prompt Photons	X	X	X
Soft Gluon Coherence		X	X
Determination of α_s	X	X	X

in parallel. This strategy conforms to the prediction that QCD can describe, within the theoretical uncertainties of the respective calculations, the hadronic aspects of all of these reactions, in the range of center of mass energies and momentum transfers from a few GeV up to several hundreds of GeV, which is experi-

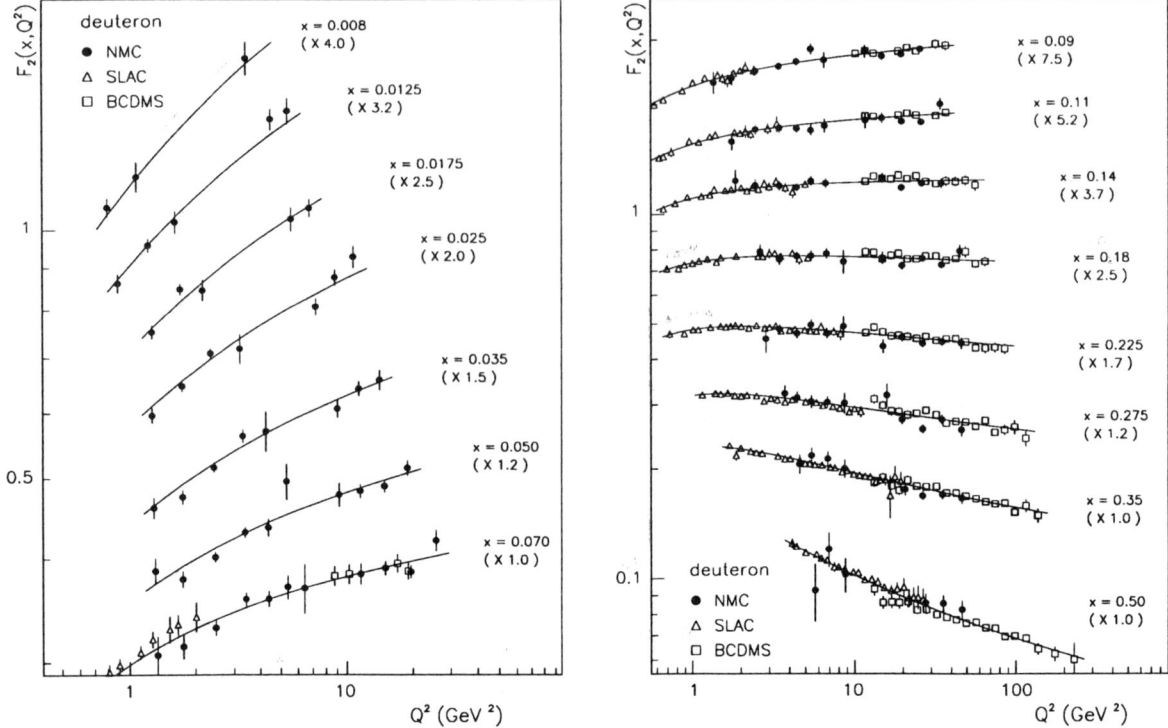

Fig. 2. $F_2(x, Q^2)$ measured from NMC on a D_2 target, compared with corresponding results from SLAC and BCDMS.

mentally available today. Verifications of this prediction are not only vital in testing the internal consistency of the theory, but are also important in understanding the significance of QCD predictions for future experiments.

NUCLEON STRUCTURE FUNCTIONS

Cross sections of physical processes in lepton-nucleon scattering and in hadron-hadron collisions depend on the quark- and gluon-densities in the nucleon. While QCD cannot predict the functional form of these densities, their energy evolution can be precisely calculated. QCD thus predicts, departing from the naive quark-parton model, scaling violations in physical cross sections, which are associated with the radiation of gluons. Such cross sections are usually parametrized by a set of structure functions F_i ($i = 1, 2, 3$). These functions depend on Q^2, which is the quadratic momentum transfer in the scattering process, and on the scaling variable x, which is the relative momentum fraction of the nucleon carried by the struck parton.

New results presented at this conference come from the NMC [2,3] and from the CCFR [4,5] experiments. The high statistics and the large kinematic range of these new data resolve the controversy which previously existed between the BCDMS and EMC as well as between the CDHSW and CCFR data: the measured structure functions $F_2(x, Q^2)$ from NMC, BCDMS, SLAC and CCFR are in good agreement with each other, while those from CDHSW and from EMC are disfavoured. This can be seen in Fig. 2, where F_2 from NMC is compared with the results from BCDMS and SLAC [6], and from Fig. 3 which compares the results from CCFR with those of BCDMS, SLAC, EMC and CDHSW. Note that in these figures the errors of the data are only statisti-

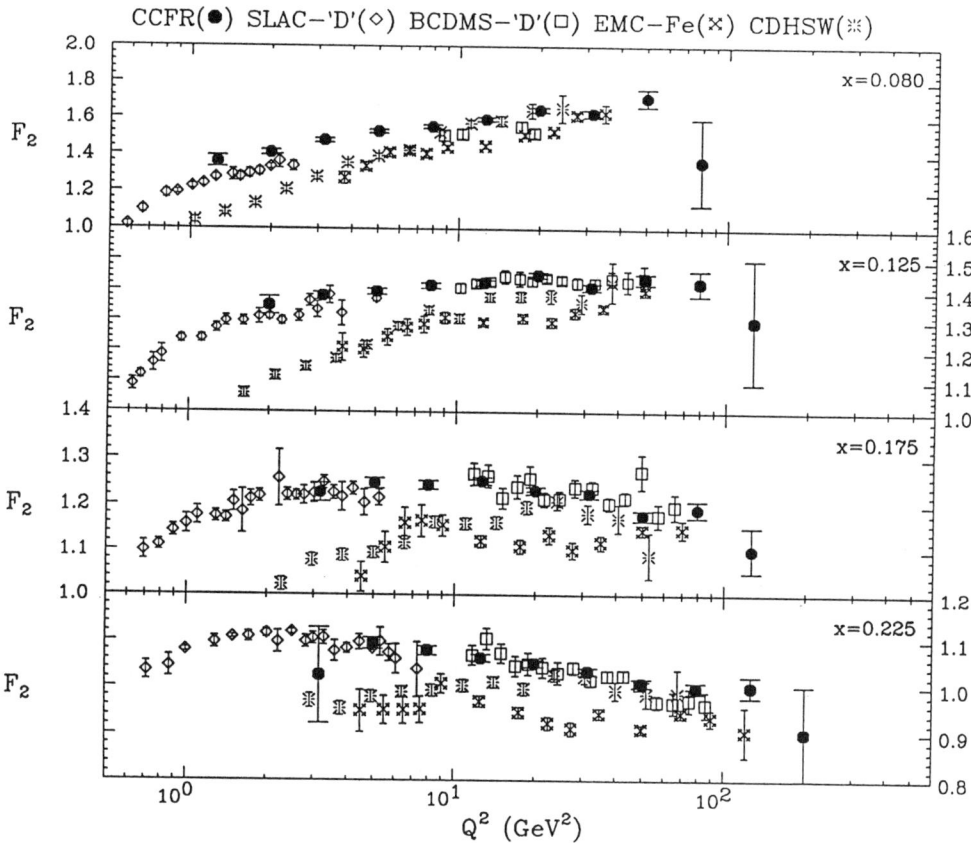

Fig. 3. $F_2(x, Q^2)$ measured from CCFR, compared with results from other experiments.

cal; it was pointed out that the CDHSW data, which seem to show a large deviation from the others, are largely consistent within their systematic errors.

The NMC data extend the measurement of F_2 down to $x = 0.008$, which is about ten times smaller than what was previously available. Previous parametrizations of the parton densities, denoted KMRS [7] and MT [8], turned out to be unable to describe these new data in the region $x < 0.1$, where almost no precise data were available before. This situation, very recently, caused several groups of theorists to re-fit the parton distributions using the new data from CCFR and from NMC, as well as data from prompt photon production and from lepton-pair production (Drell-Yan) [9,10].

The impact of the new NMC data on the new parton density parametrization by Martin, Roberts and Stirling (MRS) [9], compared to the older KMRS extrapolation, is demonstrated in Fig. 4. In addition to an increase of the sea-quark contribution at small x, the authors report some evidence for SU(2) symmetry breaking in the light sea-quark contribution, i.e. a ratio of $\bar{u}/\bar{d} \neq 1$. For $x > 0.1$, the new parametrization caused little change [9]. Similar results were reported from the CTEQ group at this conference [10].

Further new information on the densities of partons, especially of gluons at small x, comes from measurements of the $b\bar{b}$ cross sections at the hadron colliders. In next-to-leading order ($\mathcal{O}(\alpha_s^3)$) QCD, gluon-gluon interactions are the dominant source of bottom-quark production at the present collider energies. In Fig. 5, the $b\bar{b}$ cross sections as measured by UA1 [11] at $\sqrt{s} = 630$ GeV and by CDF [12]

84 Tests of QCD

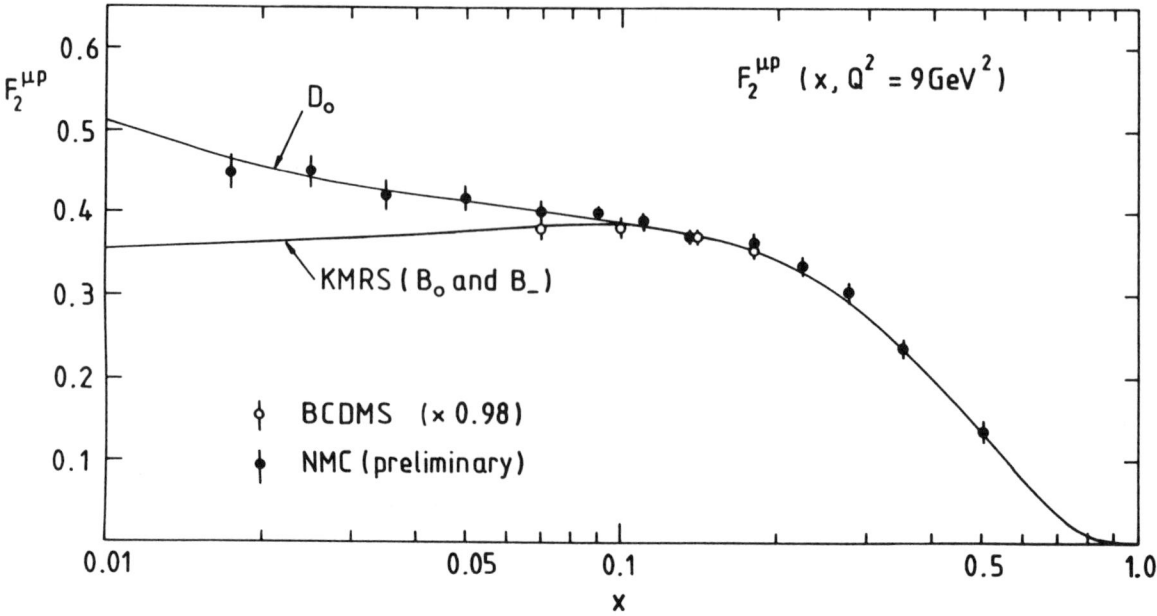

Fig. 4. $F_2(x, Q^2 = 9\,\text{GeV}^2)$ as measured from NMC and BCDMS, compared with the extrapolation of the earlier KMRS and with the new MRS (labelled D_0) parametrization of parton densities.

at $\sqrt{s} = 1.8$ TeV are shown as a function of p_T^{min}, which is the smallest transverse momentum of any of the b-quark jets in one event. The data are compared with $\mathcal{O}(\alpha_s^3)$ QCD calculations [13], which include an earlier parametrization of parton densities [8]. These calculations provide a good description of the UA1 data, but are systematically low if compared to CDF.

Motivated by this difference, a new parametrization of the gluon density was determined [14] such that both data sets can be described simultaneously. The old and new parametrizations of the gluon density $xG(x, Q^2)$, for different values of $\mu = \sqrt{p_{T,min}^2 + M_b^2}$, are displayed in Fig. 6; the new parametrization implies an increased gluon contribution in the region $0.01 < x < 0.1$ and a smaller contribution at $x > 0.1$. The new parametrization was determined in fits to the collider data shown in Fig. 5, to the newest data from deep inelastic lepton-nucleon scattering, and to fixed target prompt photon production data; it then provides a good description of the $b\bar{b}$ cross sections both at $\sqrt{s} = 630$ GeV and 1.8 TeV, as well as of the other data used in the fit [14].

Summarizing the new developments in nuclear structure functions, fits to the DIS data mainly resulted in a change of the sea-quark contributions at small x (the gluon distribution was not changed in these analyses), while $b\bar{b}$ production at hadron colliders seems to indicate the need for significantly different gluon densities. The distributions of sea-quarks and gluons must be largely correlated through the process $g \to q\bar{q}$. An overall analysis of *all* available data, namely the new NMC and CCFR data from DIS, $b\bar{b}$ cross sections from hadron colliders, prompt photon production at fixed target and at hadron collider experiments, and Z^0 and W production and inclusive jet production at hadron colliders, is therefore mandatory.

Parton densities at small x are an important input for theoretical predictions of cross sections at future colliders like the LHC and the SSC, which will operate at $\sqrt{s}$ of 17 TeV and 40 TeV, respectively. At those energies, the uncertainties of the parton densities at

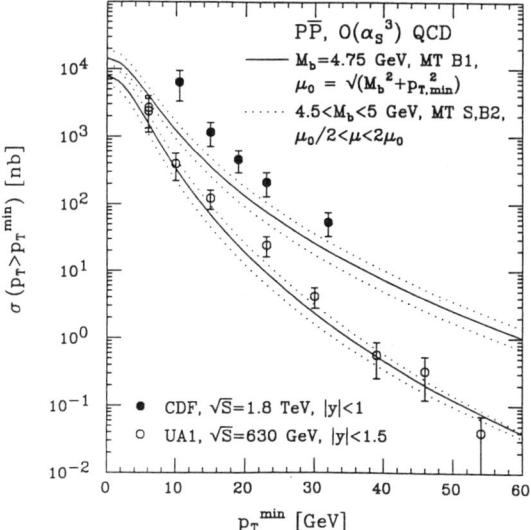

Fig. 5. The integral p_T distribution as a function of p_T^{min} for $p\bar{p} \to bX$, together with predictions from $\mathcal{O}(\alpha_s^3)$ QCD plus earlier parametrizations of parton distributions.

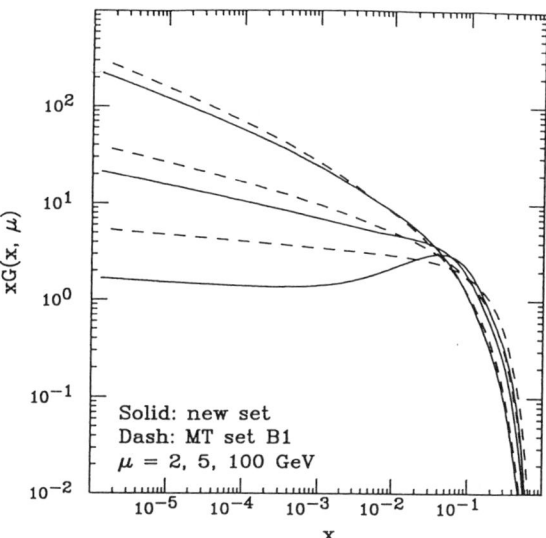

Fig. 6. Distributions of old (dashed) and new (solid) gluon densities.

small and at very small x are the dominant sources of errors. New measurements of structure functions at very small x ($x < 0.01$) will be available soon from experiments at the e-p collider HERA at DESY/Hamburg. HERA started its operation only recently; first and preliminary data have already been presented at this conference [15].

The small-x region is also a challenge for theoretical investigations, since it is the transition region between perturbative and nonperturbative QCD, with the specific feature that α_s is still small in this kinematic range. The current situation of perturbative and of nonperturbative QCD, in the two-dimensional phase space of $1/x$ and Q^2, is summarized schematically in Fig. 7. Some names of theorists who actively work in these fields, as well as the kinematic regions where HERA and LHC are expected to contribute, are also indicated. The theory of the nonperturbative and of the transition regions is still in progress and developing; a review of the present status was given by J. Bartels [16] at this conference.

JET PHYSICS

Collimated jets of hadrons are classically observed in e^+e^- annihilations and at $p\bar{p}$ colliders, at center of mass energies W of the hadronic system which are large enough that the p_T broadening of jet-fragmentation (hadronization) is significantly smaller than the average jet energy, E_{jet}. Typically, this is the case if $2E_{jet} \approx W > 20$ GeV. Evidence for visible jet structures in DIS processes, using a high energy μ beam ($E_\mu = 490$ GeV) at Fermilab, was reported by E665 at this conference [17]. Hadron jets are believed to maintain the kinematic properties of the underlying quarks and gluons; they are therefore ideal tools to test the basic features and predictions of perturbative QCD.

Quantitative studies of jets require an exact definition of resolvable jets. This definition must be applicable to experimental analyses and also to theoretical calculations in order to make a meaningful comparison between the two. The jet algorithms used in e^+e^- and in hadron collision processes are, unfortunately, quite different, such that a direct comparison of jets from these different reactions are not

86 Tests of QCD

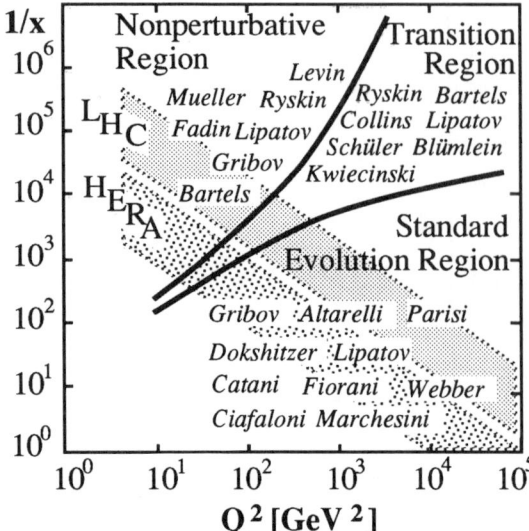

Fig. 7. Schematic diagram of the standard (perturbative QCD) evolution region, of the nonperturbative and the transition region in the $1/x - Q^2$ phase space.

yet possible.

Jets in e^+e^- Annihilation and in DIS

In e^+e^- annihilation, resolvable jets are most commonly defined by an algorithm which was introduced by the JADE collaboration [18]: the scaled pair mass of any two resolvable jets i and j, $y_{ij} = M_{ij}^2/E_{vis}^2$, is required to exceed a threshold value y_{cut}, where E_{vis} is the total visible energy of the event. Starting with all the particles measured in a hadronic event, the pair of particles with the smallest y_{ij} is replaced by (or "recombined" into) a pseudoparticle (or "cluster") k with four-momentum $p_k = p_i + p_j$, as long as $y_{ij} < y_{cut}$. The procedure is repeated until all y_{ij} are larger than y_{cut}, and the remaining clusters of particles are called jets.

The original JADE jet finder, with the metric $M_{i,j}^2 = 2E_iE_j(1-\cos\theta_{i,j})$ which assures small hadronization corrections to jet production rates, is nowadays being replaced by the "Durham" jet finding scheme with $M_{i,j}^2 = 2\min(E_i^2, E_j^2)(1-\cos\theta_{i,j})$ [19–21]. The Durham jet algorithm has certain features which make it very attractive both from experimental and theoretical points of view: its hadronization correction is even smaller than that of the JADE algorithm, it avoids nonintuitive jet recombinations, it allows the resummation of leading and next-to-leading logarithms to all orders, and it also can be applied in DIS and in hadron collisions. In the latter cases, however, a special treatment of the target jets ("X" in Fig. 1) is necessary [22]; see also the presentation of B.R. Webber at this conference [23]. Next-to-leading order (i.e. $\mathcal{O}(\alpha_s^2)$) QCD predictions exist for both the JADE and the Durham algorithm [21], and resummed calculations are available for the Durham scheme [19]

The relative production rates of n-jet events ($n = 2,3,4,...$), $R_n = \sigma_{n-jet}/\sigma_{tot}$, naturally depend on the resolution parameter y_{cut} (for larger y_{cut}, fewer multijet events are resolved) and on the coupling constant α_s. In $\mathcal{O}(\alpha_s^2)$ QCD, for instance, the 3-jet event production rate is given by ($y \equiv y_{cut}$)

$$R_2(y,\mu) = C_1(y)\alpha_s(\mu) + C_2(y, x_\mu)\alpha_s^2(\mu), \quad (1)$$

where μ is the renormalization scale at which α_s is calculated. The coefficients C_1 and C_2 are given by the QCD calculations; they depend on the kinematic definition of resolvable jets (and thus on y_{cut}), and also on the QCD group constants like the number of quark flavors, N_F, and the number of colors, N_C. The next-to-leading order coefficient C_2 also explicitly depends on the renormalization scale factor $x_\mu = \mu/\sqrt{s}$.

Measurements of n-jet productions rates using the JADE algorithm and as a function of the resolution parameter y_{cut}, are available from almost all experiments in e^+e^- annihilation, in the c.m. energy range $\sqrt{s} = 20$ to 93 GeV. At this conference, the SLD collaboration presented their new data [24], based on about 6,000 hadronic Z^0 decays obtained with partly polarized electron beams at the SLAC Linear Collider, and thus joined the scene of QCD studies at the Z^0 resonance which so far had been the exclusive domain of the

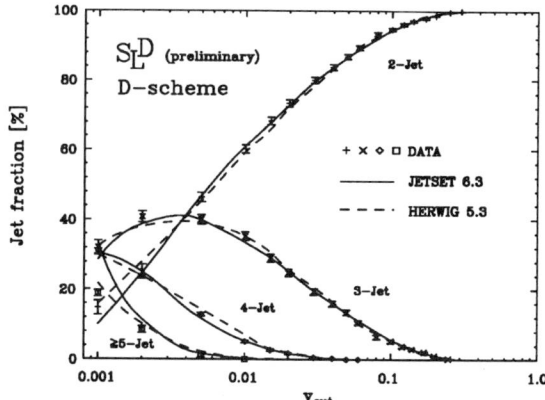

Fig. 8. Production rates of 2-, 3-, 4- and ≥5-jet events in hadronic Z^0 decays, defined with the Durham jet algorithm.

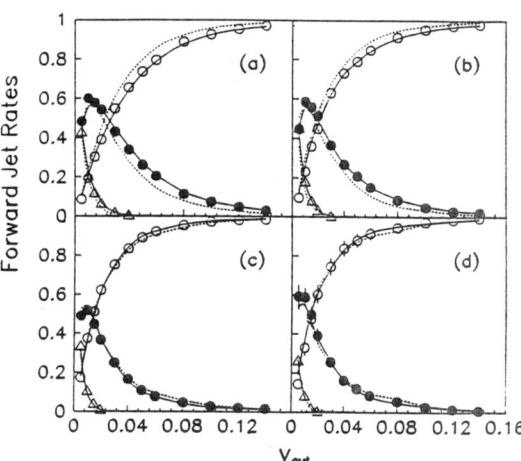

Fig. 9. Forward production rates of 1-, 2- and ≥3-jet events in deep inelastic muon scattering, defined with the JADE jet algorithm for events in the hadronic c.m. energy range $W = 13$ GeV to 18 GeV (a), 18 to 23 GeV (b), 23 to 28 GeV (c) and 28 to 33 GeV (d). The data of E665 are compared with JETSET matrix element (full line) and QCD shower (dashed line) model predictions.

four experiments at LEP. The jet production rates measured by SLD, using the Durham jet finder and as a function of y_{cut}, are presented in Fig. 8. Also shown are the jet rates obtained from the QCD shower plus hadronization models JETSET [25] and HERWIG [26], which describe the data in detail. Similar results on Durham scheme jet rates are available from OPAL [27], based on 127,000 events.

In deep inelastic muon-nucleon scattering, the E665 collaboration analyzed jet production rates [17] in a manner similar to that of e^+e^- experiments. In Fig. 9, the 1-, 2- and ≥3-jet event production rates using the JADE jet algorithm are plotted as a function of y_{cut}. Only those hadrons whose momentum vectors point into the forward hemisphere, which in the hadronic c.m. system is defined by the direction of the outgoing muon, are taken into account, in order to eliminate the (trivial) contribution of the target jet. It is interesting to note that the range of hadronic c.m. energies $13 < W < 33$ GeV, which is available in this experiment, is close to the c.m. energies at which PETRA and PEP operated, $14 < \sqrt{s} < 46$ GeV, and where the first studies of jet rates in e^+e^- annihilation were done [18]. The data of E665 are well described by QCD model calculations, and more quantitative QCD studies, such as determinations of α_s, are expected to come in the near future.

Many detailed studies of jet production rates and of jet dynamics (such as angular correlations and energy distributions) are available from e^+e^- annihilation. In particluar, the data from LEP provide significant tests of the predictions of perturbative QCD, due to high statistics, the high center of mass energy and reduced hadronization effects, and the clean and background free hadronic event samples. Updates of two studies which demonstrate the non-Abelian nature of QCD, i.e. the existence of the gluon self coupling and evidence for asymptotic freedom, will be presented in the following; a more comprehensive review of jet physics from Z^0 decays can be found elsewhere [28].

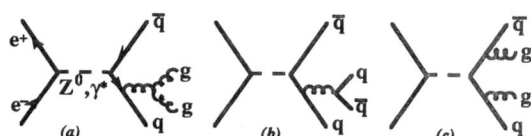

Fig. 10. Generic diagrams leading to 4-jet final states in e^+e^- annihilation, in leading order QCD.

Table 2. Group constants of QCD and of the Abelian vector theory.

	QCD	Abelian
C_F	4/3	1
N_C	3	0
T_F	1/2	3

Table 3. Fit results of group constants N_C/C_F and T_F/C_F.

Experiment	N_C/C_F	T_F/C_F
DELPHI	2.07 ± 0.31	0.34 ± 0.14
ALEPH	2.24 ± 0.40	0.58 ± 0.29

In leading order QCD, 4-jet final states in e^+e^- annihilations are due to double gluon bremsstrahlung, to gluon-quark splitting and to the triple gluon vertex (TGV), as displayed in Fig. 10. The dominant contribution to 4-jet final states is predicted [29] to come from the TGV. In an Abelian model, where the TGV does not exist and which can be constructed by simply replacing the group constants of QCD, C_F, N_C and T_F, with those of U(1) [30] (see Table 2), this part of the cross section is basically taken over by an increased production rate of double gluon bremsstrahlung and 4-quark final states (Fig. 10b,c).

DELPHI [31] and ALEPH [32] study various kinematic distributions of 4-jet events, namely a two-dimensional angular correlation and the 5-dimensional jet pair-mass distribution defined by the momentum vectors of the jets. They determine the group constants N_C/C_F and T_F/C_F in a fit to the $\mathcal{O}(\alpha_s^2)$ matrix element [33]. In these calculations, 4-jet production is described by the sum of five partial cross sections which are proportional to different combinations of the group constants C_F, N_C and T_F. The TGV contribution is proportional to N_C, the numbers of colors, while the process of $g \to q\bar{q}$ is basically proportional to $T_F N_F$, where N_F is the number of quark flavors.

The results of both DELPHI and ALEPH are summarized in Table 3. They are in good agreement with the expectations of QCD, while Abelian models are significantly ruled out (c.f. Table 2). In particular, the non-zero result for N_C/C_F is interpreted as direct evidence for the TGV. In Figure 11, the result of ALEPH is displayed in the two-dimensional N_C/C_F - T_F/C_F plane.

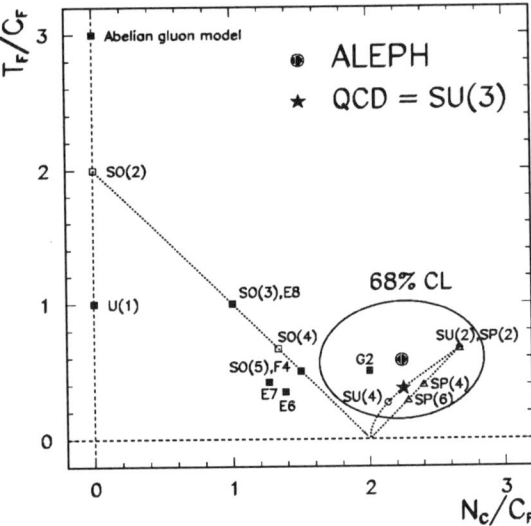

Fig. 11. Fit result of the group constants N_C/C_F and T_F/C_F from a study of 4-jet events by ALEPH, compared with the expectations for several gauge theories.

An intuitive test of asymptotic freedom, and thus of the running coupling constant α_s, can be performed by analyzing the energy dependence of the 3-jet event production rate R_3 [34]. For fixed values of y_{cut}, the energy dependence of R_3 is determined only by the energy dependence of α_s (c.f. Eq. 1). For the JADE algorithm, hadronization effects turn out to be small and (almost) energy independent in regions where $\sqrt{y_{\text{cut}}} E_{cm} > 7$ GeV. In detail, hadronization corrections for jet rates at $y_{\text{cut}} = 0.08$, where the most experimental results have been published, are about 6% with an energy dependent variation of $\pm 2\%$ in the c.m. energy range between 25 GeV and

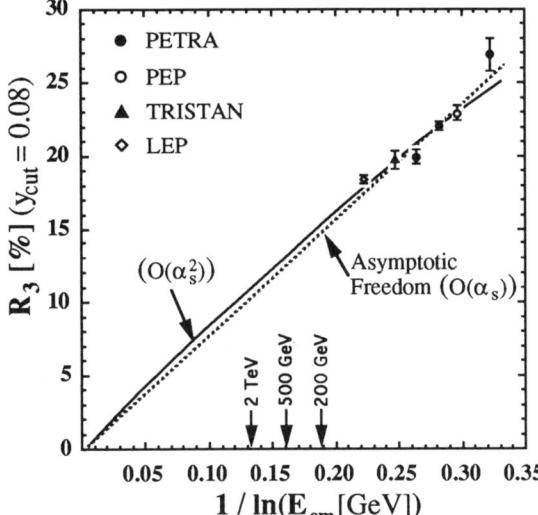

Fig. 12. Three-jet event production rates R_3 ($y_{\text{cut}} = 0.08$) as a function of $1/\ln(E_{cm})$, compared to the prediction of asymptotic freedom.

100 GeV [28], while α_s is expected to change by about 30%. Measurements of R_3, without explicit corrections for hadronization effects, therefore provide a clean test of the running of α_s.

The 3-jet event production rates, as measured by many experiments with the JADE algorithm for $y_{\text{cut}} = 0.08$, in the c.m. energy range between 22 and 93 GeV, are presented in Fig. 12 as a function of $1/\ln(E_{cm})$. The data from different experiments at similar energies are combined [28]. The dashed line is a fit to the leading order QCD prediction, namely $R_3 \propto \alpha_s \propto \frac{1}{\ln E_{cm}}$. The corresponding prediction in $\mathcal{O}(\alpha_s^2)$ is also shown, indicating that higher order terms affect the energy dependence of R_3 only slightly. At infinite energy ($1/\ln(E_{cm}) \to 0$), asymptotic freedom predicts that R_3 and α_s should vanish; an expectation which is in good agreement with the data. The data exclude the possibility of an energy independent coupling with a significance of more than 7 standard deviations [28].

Jets in Hadron Collisions

In hadron collisions, jets are commonly defined by a cone algorithm. Energy clusters measured in segmented electromagnetic and hadronic calorimeters are added to form a single jet if they lie in a cone of radius $R = \sqrt{\Delta\eta^2 + \Delta\phi^2}$, where $\eta = -\ln(\tan\frac{1}{2}\theta)$ is the pseudorapidity, θ is the polar and ϕ is the azimuthal angle of the cluster with respect to the beam axis. Next-to-leading order (i.e. $\mathcal{O}(\alpha_s^3)$ in this case) calculations exist for the inclusive jet production cross section using this type of jet algorithm [35,36] and, more recently, also for the two-jet inclusive cross section [37]; see also the presentation of R.K. Ellis at this conference [38].

The inclusive jet cross section in $p\bar{p}$ collisions at $\sqrt{s} = 1.8$ TeV, measured by CDF [39] as a function of the transverse jet energy E_t, extend over 7 orders of magnitude and were demonstrated to be well described by the $\mathcal{O}(\alpha_s^3)$ QCD predictions. Based on this comparison, a limit on Λ_c, a term characterizing the presence of a possible quark substructure, is set at 1.4 TeV, with 95% confidence level. The cone size dependence of the cross section, for $0.4 \leq R \leq 1.0$, was demonstrated to be in reasonable agreement with the QCD prediction [39]. This is a remarkable success of QCD if one considers that in $\mathcal{O}(\alpha_s^3)$ only one further gluon, in addition to the two energetic jets from the initial scattering process (see e.g. Fig. 1), is generated, which can give rise to a non-vanishing cone size dependence. These results have already been presented at the Geneva conference in 1991 [40,41] and thus shall not be discussed further here.

More recent analyses on jet production at collider energies were contributed to this conference by UA2 [42], CDF [43] and by the D0 collaboration [44], which only recently went into the beam of the Fermilab Tevatron collider. New analyses which involve significant tests of QCD through the comparison with $\mathcal{O}(\alpha_s^3)$ QCD calculations are the study of two-jet angular distributions and a test of scaling violations, which are both based on the CDF data.

The angular distribution of parton-parton

scattering, measured by the angle $\theta^{\#}$ between the beam- and the parton-direction in the rest frame of the jets, depends on the spin of the exchanged particle. This is demonstrated in Fig. 13, where the distribution of $\lambda = \cos\theta^{\#}/\sin^2\theta^{\#}$, normalized by the expectation value at $\lambda = 0$, is displayed for pure spin-1 (i.e. gluon) exchange, for pure spin-$\frac{1}{2}$ and for spin-0 exchange. Also shown is the Born level prediction which corresponds to a mixture of spin-1 and spin-$\frac{1}{2}$ exchange as expected for jet pair masses M_{JJ} of 300 GeV in $p\bar{p}$ collisions at $\sqrt{s} = 1.8$ TeV [37]. The observable λ was chosen such that (leading order) QCD predicts a flat distribution.

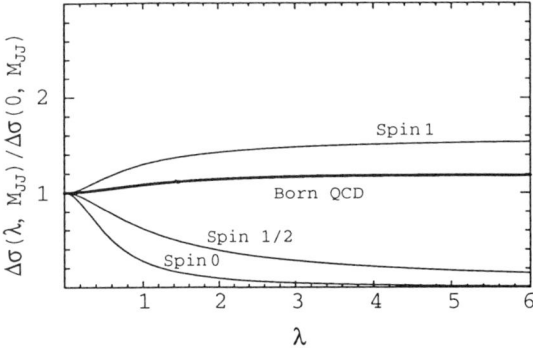

Fig. 13. Angular two-jet distribution as a function of λ, for three choices of exchange quanta and for QCD at Born level.

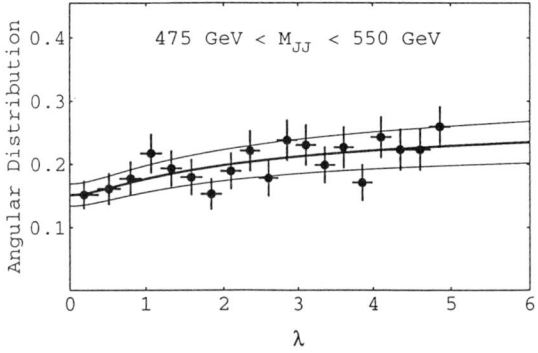

Fig. 14. Angular two-jet distribution as measured by CDF, compared with the $\mathcal{O}(\alpha_s^3)$ QCD predictions of Ellis, Kunszt and Soper.

The two-jet angular distribution, as measured by CDF [45] for jet-pair masses of 475 GeV $\leq M_{jj} \leq$ 550 GeV, is shown in Fig. 14, together with the corresponding $\mathcal{O}(\alpha_s^3)$ QCD predictions of Ellis, Kunszt and Soper [37]. The error bands represent the estimated theoretical uncertainty, while the error bars of the data include the statistical errors only. By comparing with Fig. 13, it is clear that the data distinguish between QCD, i.e. mainly spin-1 exchange, and models with spin-0 or spin-$\frac{1}{2}$ exchange. This result opens an interesting possibility to prove the existence of the gluon self coupling, the TGV, at the hadron collider: if one could show unambiguously that the majority of two-jet events are pairs of gluon-jets, this – in addition to the predominant spin-1 exchange presented above – would directly imply the existence of the TGV.

A test of QCD scaling violations in inclusive jet cross sections could be performed, in principle, by analyzing the cross sections measured at various hadron collider energies. Such measurements are available from experiments at the ISR ($\sqrt{s} = 63$ GeV) and at the Sp$\bar{p}$S (630 GeV), both of which are hadron colliders at CERN, and at the Tevatron at Fermilab (1800 GeV). However, a quantitative study of scaling violations from these data is not possible since the experiments use different jet algorithms, for some of which the corresponding next-to-leading order QCD calculations are not available.

The CDF collaboration has, in addition to their main data sample at $\sqrt{s} = 1800$ GeV, a small amount of data from a run at $\sqrt{s} = 546$ GeV. As a contribution to this conference, CDF thus presented a study of the ratio R of inclusive jet cross sections measured at $\sqrt{s} = 546$ and 1800 GeV, as a function of the scaled variable $x_T = 2E_T/\sqrt{s}$ [43]. Several experimental uncertainties cancel in this ratio, such that the overall systematic uncertainty of this measurement is 10 to 15% (compared to 20 to 30% for jet cross sections at one energy). The theoretical expectation for this ratio is more complicated than in the case of e^+e^- annihilations, where jet cross sections only depend on the energy dependence of α_s: at hadron colliders, the ratio $R = \sigma_{jet}(546 \text{ GeV})/\sigma_{jet}(1800 \text{ GeV})$ also depends on the energy evolution of the parton distributions with $\mu \sim E_T$. The net effect of both the running α_s and of the parton distribution, which have different slopes as a function of x_T, is an almost constant value of $R \sim 1.9$ in the region of $x_T \geq 0.1$, which is experimentally accessible.

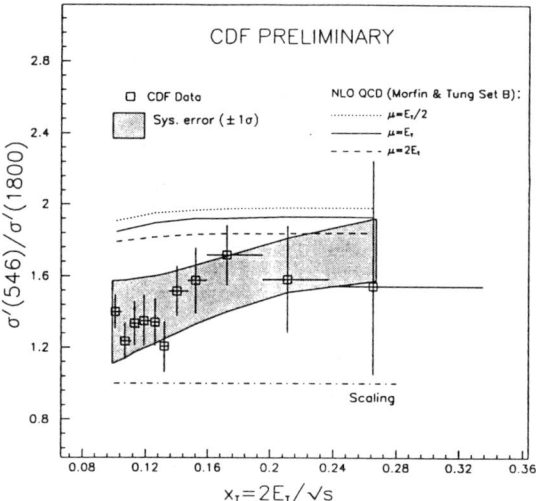

Fig. 15. The ratio of scaled jet cross sections, compared with the $\mathcal{O}(\alpha_s^3)$ QCD predictions for three choices of renormalization scales $\mu = E_T/2$, E_T and $2E_T$. The errors of the data are statistical, and the band indicates the additional experimental systematic uncertainties.

In Fig. 15, the measured ratio of the jet cross sections is compared with the $\mathcal{O}(\alpha_s^3)$ QCD predictions [35], based on the parton distributions of Morfin and Tung [8]. The theoretical uncertainties of the $\mathcal{O}(\alpha_s^3)$ calculations are estimated by variations of the renormalization scale μ between $E_T/2$ and $2E_T$. Also shown is the expectation for the naive parton model in which exact scaling, and thus a ratio of $R \equiv 1$, would hold. The data are clearly not compatible with scaling, but are also not in good agreement with the predicted QCD scaling violations. The reasons for this disagreement are not yet understood: variations of the renormalization scale, the set of structure functions and the cone size for jet reconstruction were studied but could not explain the difference between data and QCD [43]. The new parametrizations of parton distributions, which were discussed at this conference [9,10,14], still remain to be investigated in this analysis, and undiscovered experimental biases can also not be ruled out completely at this stage.

PROMPT PHOTON PRODUCTION

Prompt photons, which are produced in the elementary scattering processes as shown schematically in Fig. 16, provide unique tests of the parton densities and of the details of quark-gluon cascades, since photons are not affected by the color force field or by hadronization after they have been produced. Studies of prompt photon production are an experimental challenge: the cross sections for photon production are typically three to four orders of magnitude smaller than for jet production, and the detection of photons in the hadronic environment requires the application of strict photon isolation cuts in order to reduce the enormous background from decays of π^0 mesons and other particles. The hard selection cuts required in experimental analyses make detailed comparisons with theoretical calculations difficult.

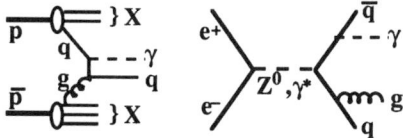

Fig. 16. Examples for prompt photon (γ) production in hadron collisions and in e^+e^- annihilation.

Prompt Photons in Hadron Collisions

The E706 collaboration presented cross sections of prompt photons, measured from reactions of 500 GeV proton and pion beams on hydrogen, copper and beryllium targets, as a function of photon transverse momenta in the range 4 GeV $< p_t <$ 8 GeV [46]. Theoretical calculations in the next-to-leading logarithmic approximation of QCD (NLLA) [47] are consistent with the measured cross section, which decreases, in this kinematic region, by four orders of magnitude.

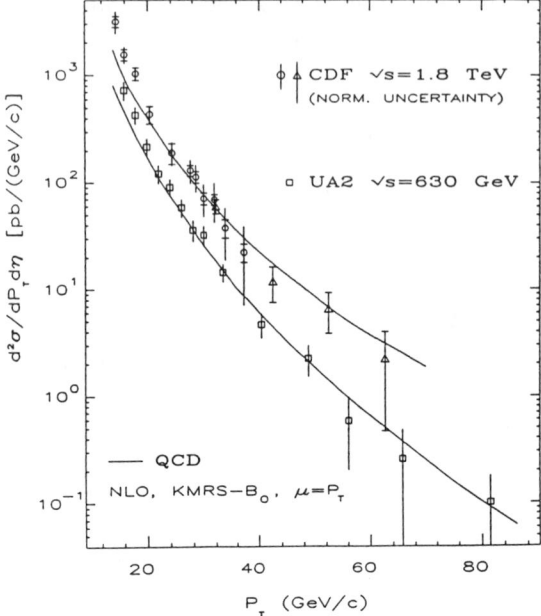

Fig. 17. Isolated prompt photon cross sections from UA2 and from CDF, compared with NLLA QCD predictions.

The prompt photon cross sections from hadron colliders were recently updated by UA2 [48] and CDF [49] and were also presented at this conference [42,43]. The measured cross sections of both experiments are displayed in Fig. 17, as a function of the transverse momentum p_T of the photons (the data of UA2 in this combined figure [49] are not the most recent ones, however). The data are compared with NLLA QCD predictions [50], using the KMRS structure functions [7]. Both data sets agree qualitatively with the calculations but have a steeper slope at low p_T. A possible cause of the difference between data and prediction is the bremsstrahlung process in which a final state quark radiates a photon, and which is expected to contribute mainly at low p_T [51].

Both UA2 [48,42] and CDF [43] also contributed measurements of double prompt photon cross sections, which are suppressed by

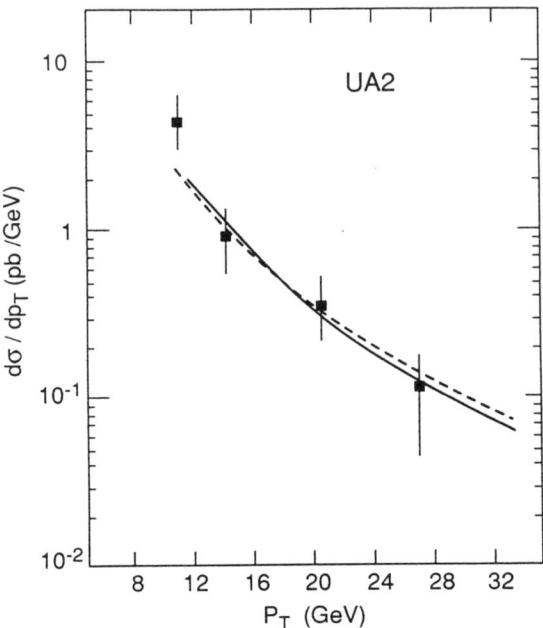

Fig. 18. Cross section for double prompt photon production as measured by UA2, compared to the QCD predictions (see text).

about 6 orders of magnitude compared to the ordinary inclusive hadronic jet cross section. The results from UA2 are displayed in Fig. 18, together with an analytic calculation [52] (dashed line) and an $\mathcal{O}(\alpha_s)$ QCD Monte Carlo prediction [53] (full line). The agreement of the calculations with the data is good, apart from the first bin in p_T, which again may be due to the uncertainties in the contribution of bremsstrahlung diagrams.

The cross section for double prompt photon production measured by CDF [43] turned out to be a factor of 2 to 5 larger than the QCD predictions. The reason for this behaviour is not known so far. Note that the effective range of $x \sim p_T/\sqrt{s}$ to which this measurement corresponds is smaller than in the case of the UA2 data. A change of the gluon distribution at small x could thus be a reasonable explanation for the different behaviour of the UA2 and the CDF data. If the disagreement cannot be explained by an undiscovered experimental bias, this result implies, for example, a much larger background to the process $H \to \gamma\gamma$ in searches

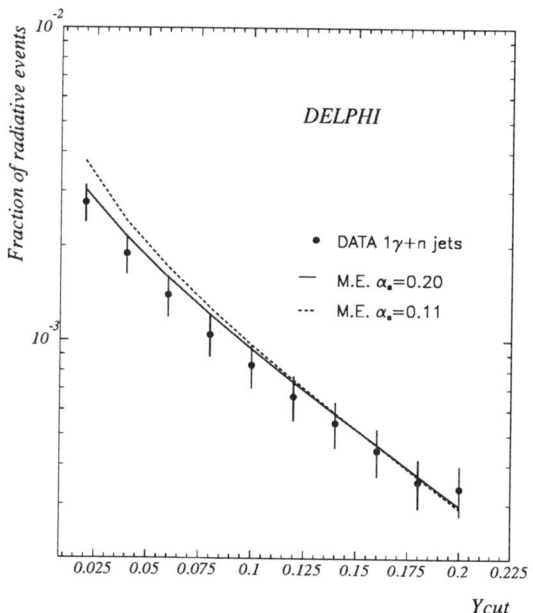 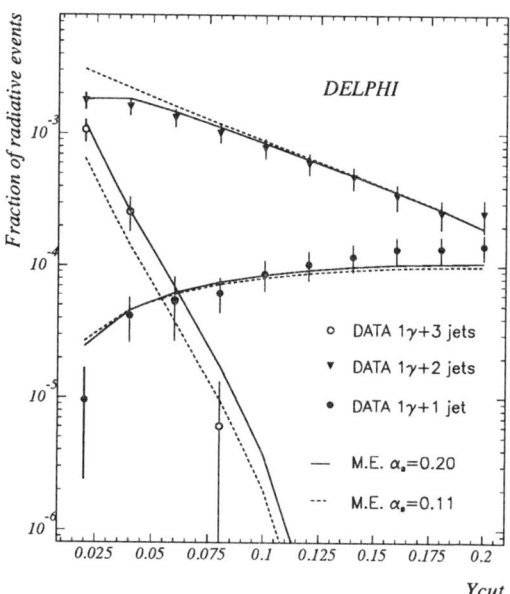

Fig. 19. Fraction of resolved radiative photon events in the hadronic event sample of e⁺e⁻ annihilations at $E_{cm} = 91.2$ GeV: total inclusive rate (left) and fraction of photon events with one, two or three hadron jets (right), as a function of y_{cut}. Data are compared with analytic $\mathcal{O}(\alpha\alpha_s)$ calculations, for different values of α_s.

for the Higgs boson at future collider experiments at LHC and at the SSC than anticipated so far.

Prompt Photons in e⁺e⁻ Annihilations

Significant studies of final state photon radiation in e⁺e⁻ annihilation were first performed at LEP [54], where the background of photons radiated from the initial electron and positron is much reduced due to the resonant cross section at the Z^0 pole. The cross sections for isolated, high energetic photons are studied in terms of the same resolution parameter y_{cut} which is also used, within the JADE jet algorithm, to define hadronic jets. Within each hadronic event which has an isolated, energetic photon, jets are reconstructed from the hadronic system (i.e. not considering the photon) for a given value of y_{cut}. In a second step, only those events are accepted for which the scaled pair masses $y_{\gamma j}$ between the photon (γ) and any jet j are larger than that same value of y_{cut}. The total inclusive production rate of isolated photons and the production rates of events with a photon plus 1, 2 and 3 or more hadronic jets are then studied, as a function of y_{cut}.

At this conference, all four LEP experiments have contributed updates to their analyses of final state photons [55]. Recent improvements in this field come from the theoretical side: there are now three independent calculations of prompt photon production in e⁺e⁻ annihilation, in leading order QCD (i.e. in $\mathcal{O}(\alpha\alpha_s)$), which are available in the form of Monte Carlo generators [56-58]. In addition, three QCD shower and hadronization models, which are based on (next-to-)leading logarithmic approximations of QCD to all orders, also include final state photon radiation: JETSET 7.3 [25], HERWIG 5.4 [26] and ARIADNE 3.2 [59].

In Fig. 19, the fraction of resolved radiative events in the total hadronic event sample and the fractions of events with a photon plus one, two or three hadronic jets are shown, as a function of y_{cut}, as measured by DELPHI [60]. The data, which are corrected for the additional energy- and angular isolation requirements for prompt photons, which are necessary to suppress the background of photons from particle decays, are compared to the analytic predictions of Kramer and Lampe [56]. Reasonable agreement is found in both the overall rate of radiative events and their hadronic jet structure. Similar results are also available from the other LEP experiments [61-63]. QCD shower models also reproduce the measured photon cross sections, whereby ARIADNE seems to be preferred by the data.

At present, the uncertainties in matching the experimental definitions of isolated photons with those used in theoretical calculations limit the precision with which these results can be interpreted. Intense discussions of these issues are under way, such that more quantitative studies can be expected in the near future.

STUDIES OF SOFT GLUON COHERENCE

QCD calculations in fixed order perturbation theory can only describe reactions with up to three (for hadron collisions in $\mathcal{O}(\alpha_s^3)$) or four (for e^+e^- annihilations in $\mathcal{O}(\alpha_s^2)$) quarks and gluons in the final state. Leading logarithmic approximations (LLA), in contrast, allow the calculation and prediction of the dynamics of multiparton final states, leading to a parton shower picture, as shown schematically in Fig. 20. Such calculations can be performed analytically [64,65] or by means of Monte Carlo implementations [25,26,59].

A basic ingredient of LLA calculations is the quantum mechanical effect of soft gluon coherence, which leads to phenomena such as [64] the suppression of low-momentum gluons, the angular ordering of parton emission and the depletion of gluons in certain kinematic regions, which is also called the "string effect". The link between these calculations

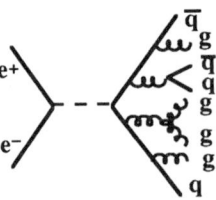

Fig. 20. Development of a parton shower in e^+e^- annihilation.

and the experimental world of hadrons can be made by using Monte Carlo models of hadronization, or by relying on the principle of local parton hadron duality (LPHD) [66], which predicts that the analytical calculations directly provide reliable predictions for hadron spectra, too.

One of the key analyses which studied soft gluon coherence at LEP was the measurement of particle spectra as a function of $\xi_p = \ln(1/x_p)$, where x_p is the particle momentum normalized to the beam energy. These spectra [67,68] are close to the shape of a gaussian and can in fact be well described by the analytic calculations [64], which only depend on Λ_{LLA} (which is related but not identical to $\Lambda_{\overline{MS}}$) and on an overall normalization factor. In particular, the energy dependence of the peak position of these spectra is found to be in good agreement with the predictions which include soft gluon coherence effects [67,68].

More recent calculations provide predictions, in NLLA of perturbative QCD, for two-particle momentum correlations [69],

$$\mathcal{R} = \frac{D_{(2)}(\xi_1, \xi_2)}{D_{(1)}(\xi_1) D_{(1)}(\xi_2)},$$

where $D_{(1)}$ and $D_{(2)}$ are the one- and the two-dimensional distributions of particles in terms of ξ, respectively. In this case, the predictions depend only on the QCD parameter Λ_{LLA}, since normalization factors cancel out in $\mathcal{R}$. OPAL has recently studied these two-particle momentum correlations [70].

An example of the results is shown in Fig. 21, where the measured distribution of $\mathcal{R}$ for particle pairs with $|\xi_1 - \xi_2| < 0.1$ is shown

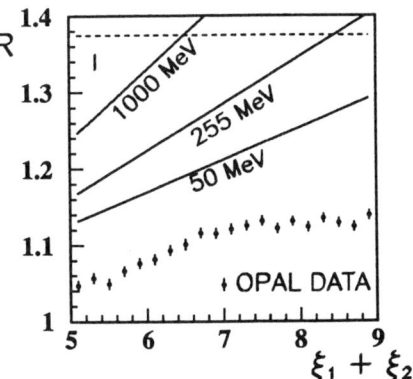

Fig. 21 Two-particle momentum correlations as measured by OPAL, for particle pairs with $|\xi_1 - \xi_2| < 0.1$ and as a function of $\xi_1 + \xi_2$. The data (dots) are compared with analytic NLLA calculations using different values of Λ_{LLA}.

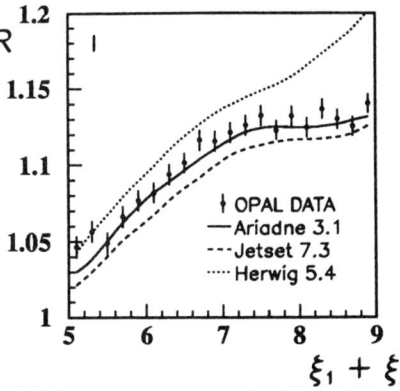

Fig. 22 The same data as in Fig. 21, but compared with several QCD shower plus hadronization models.

as a function of $\xi_1 + \xi_2$. The analytic predictions provide only a poor description of the data, for any reasonable value of the QCD parameter Λ_{LLA}. This is in contrast to the one-dimensional particle distributions, as a function of ξ, which can be well described by the respective calculations, as discussed in the previous paragraph. Higher order corrections to the analytic calculations or, equivalently, hadronization effects, are likely to explain the disagreement between data and theory seen in Fig. 21 [23]. This is further supported by the fact that QCD shower models, which also contain phenomenological parametrizations of the hadronization process, are able to provide a reasonable description of the measured correlations; see Fig. 22. LPHD, in its strict sense, is thus not always a practical assumption.

A similar conclusion was drawn from a study of the second factorial moment of the charged particle multiplicity distribution at LEP [71-73], where the data also could not be described by analytic QCD calculations in NLLA, while the average charged particle multiplicity, as a function of $\sqrt{s}$, was well reproduced. Note that the QCD prediction for the average multiplicity has two free parameters, Λ_{LLA} and an overall normalization factor, while the second factorial moment only depends on Λ_{LLA}, similar to the one-dimensional ξ distribution and for $\mathcal{R}(\xi_1, \xi_2)$, respectivly, as discussed above. Apparently, additional free parameters in LLA calculations, which do not have a particular physical meaning, are able to absorb some of the missing higher order contributions in fits to experimental data.

Another observable which was introduced recently is the ratio of sub-jet multiplicities of 2- and 3-jet events in e^+e^- hadronic final states [74]. It was motivated by a previous experimental result which revealed that the ratio of the average particle multiplicities of gluon and quark jets with similar jet energies is 1.06 ± 0.03 at LEP [75], while the naive QCD expectation predicts, for asymptotic jet energies, a ratio of $N_C/C_F \equiv 9/4$. No theoretical prediction exists for the ratio of hadrons in quark and gluon jets; however, the sub-jet multiplicities of 2- and 3-jet events have been calculated in NLLA of QCD [74]. The idea is to define 3- and 2-jet events with the Durham jet finder at a fixed value of $y_{\text{cut}} \equiv y_1$, and then to study the ratio of the average jet multiplicities of these two event classes, M_3/M_2, for decreasing jet resolution parameters $y_0 < y_1$. In the limit of $y_0 \to 0$, one expects that

$$\frac{M_3}{M_2} \to \frac{2C_F + N_C}{2C_F} = \frac{17}{8},$$

where N_C and C_F are the QCD group constants which relate the relative strength of the gluon-gluon and the quark-gluon coupling, respectively; see Fig. 22 for a graphical explanation of the expectation value of 17/8.

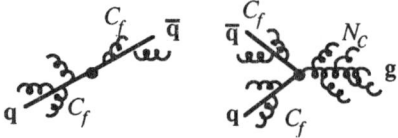

Fig. 22. Subjet multiplicities and relative strength of gluon radiation in 2- and 3-jet final states of e^+e^- annihilation.

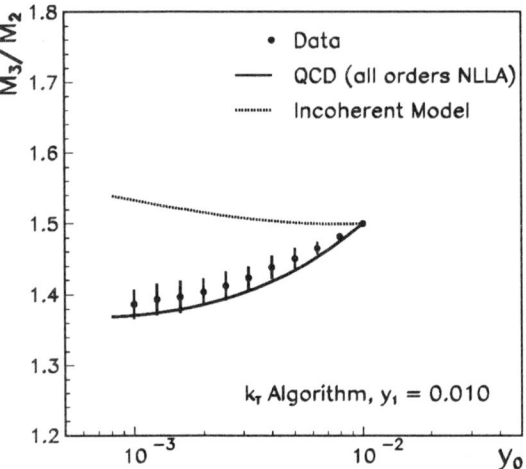

Fig. 23. Ratio of subjet multiplicities for 3- and 2-jet events defined at $y_{\text{cut}} \equiv y_1 = 0.01$; data are from L3 (preliminary).

ALEPH and L3 have contributed preliminary results on sub-jet multiplicities to this conference [72,73]. In Fig. 23, the ratio M_3/M_2 as studied by L3 is plotted as a function of y_0, for $y_1 = 0.01$. The data are compared with the QCD predictions in NLLA [74] and with a model without gluon coherence. The data, starting with the trivial value of $M_3/M_2 = 1.5$ for $y_0 = y_1$, show a decrease of this ratio for decreasing y_0, which is in good agreement with the QCD predictions. The incoherent model predicts that the ratio rises monotonically, in accordance with the naive expectation based on the higher color charge of gluon jets ($N_C > C_F$), but is in apparent disagreement with the data. In the QCD calculation, the decrease of M_3/M_2 is basically due to gluon coherence which partly cancels the effects of the increased color charge of the gluon [74]. This finding can qualitatively explain the small difference between the particle multiplicities of quark- and gluon-jets which was mentioned above.

Many more studies concerning tests of color coherence effects in the intrajet and interjet regions are available from the LEP experiments [72,73,76], such as analyses of azimuthal jet correlations and string effect studies. First steps towards investigating color coherence effects in hadron collisions are reported by CDF [43], where evidence is found for regions of enhanced and depleted gluon radiation in events containing two high and one lower energetic jets. In all the studies presented so far, analytic QCD calculations and/or QCD plus hadronization models, which include the effects of soft gluon coherence, provide a good description of the data, while those models without coherence effects are more or less ruled out.

Coherence effects, however, can often be reproduced by phenomenological string- or cluster-hadronization models [25,26], without explicit inclusion of color coherence. This underlines the close connection between hadronization and the (N)LLA of perturbative QCD, the latter of which is likely to shed more light onto the nonperturbative process of hadron production in the near future. This field is currently developing rapidly.

DETERMINATIONS OF α_s

Determinations of the coupling constant α_s have always been one of the key analyses in hadronic final states of highly energetic particle reactions. Significant progress has recently been achieved through the availability of precise results from e^+e^- annihilation, from deep inelastic lepton-nucleon scattering and from heavy quarkonia decays, leading to evidence of the "running" of α_s in the energy range of 1.78 GeV ($\equiv M_\tau$) to 91.2 GeV ($\equiv M_{Z^0}$) [77]. An overview of the different processes and observ-

Table 4. Processes and Observables from which significant determinations of α_s are derived.

Process	Observable	Theory	Caveats
e^+e^-	hadronic event shapes, jet production rates, energy correlations	NLO and re-summed NLO	hadronization corrections
	$R_Z = \frac{\Gamma(Z^0 \to \text{hadrons})}{\Gamma(Z^0 \to \text{leptons})}$	NNLO	small QCD corrections
	$R_\tau = \frac{Br(\tau \to \text{hadrons})}{Br(\tau \to e\nu)}$	NNLO	nonperturbative corrections
	scaling violations in $\frac{d\sigma}{dx}$ spectra	NLO	only through MC models
	$\frac{\Gamma(\Upsilon \to ggg)}{\Gamma(\Upsilon \to \mu^+\mu^-)}$; ....; J/Ψ; ...	NLO	relativistic corrections
DIS	$\frac{d \ln F_2(x, Q^2)}{d \ln Q^2}$	NLO	higher twist; $g(x, Q^2)$
	$\frac{d \ln F_3(x, Q^2)}{d \ln Q^2}$	NLO	higher twist
$p\bar{p}$	$p\bar{p} \to W + \text{jets}$	NLO	statistics; k-factors
	$p\bar{p} \to b\bar{b}X$	NLO	statistics; exp. systematics
$c\bar{c}$ states	mass difference of 1s and 1p charmonium states	lattice gauge theory	quenched approximation

ables from which significant α_s determinations are available, as well as the order of QCD calculations and the main obstacles for the different analyses, is given in Table 4. The most recent results were reported in several contributions to the parallel sessions of this conference [3,4,78].

The following analyses and results on α_s always adhere to the $\overline{MS}$ renormalization scheme [79]. The dependence of α_s on the renormalization scale μ is parametrized in terms of the QCD scale parameter $\Lambda_{\overline{MS}}$, in next-to-leading (NLO) order perturbation theory, as

$$\alpha_s(\mu) = \frac{12\pi}{\beta_0 \ln(\mu^2/\Lambda^2_{\overline{MS}})}$$
$$\times \left(1 - 6\frac{\beta_1}{\beta_0^2} \frac{\ln(\ln(\mu^2/\Lambda^2_{\overline{MS}}))}{\ln(\mu^2/\Lambda^2_{\overline{MS}})}\right) \quad (2)$$

with

$$\beta_0 = 33 - 2N_F,$$
$$\beta_1 = 153 - 19N_F;$$

where N_F is the number of quark flavors produced in the reaction. In studies of observables for which the next-to-next-to-leading order (NNLO) calculations are available, the corresponding expression for α_s in third order perturbation theory is utilized [80].

The limiting uncertainty in (almost) all α_s determinations, in addition to the "typical" obstacles which are listed in Table 4, are the unknown higher order contributions to the calculations in (N)NLO. These are usually parametrized in terms of the dependence of α_s on the choice of the renormalization scale μ: while in infinite order perturbation theory this dependence cancels out, the next-to-leading and higher order QCD coefficients for each observable, as given e.g. in Eq. 1, explicitly depend on the choice of μ. In most experimental determinations of α_s, variations of μ, like for instance $E_T/2 \leq \mu \leq 2E_T$ in hadron collisions or $\mu_{fit} \leq \mu \leq E_{cm}$ in e$^+$e$^-$ annihilation, are therefore studied to quantify the theoretical scale uncertainty on final results like $\alpha_s(M_{Z^0})$. Experimental fits of both $\Lambda_{\overline{MS}}$ and an "optimal" value of μ_{fit} are sometimes possible; however, μ_{fit} is normally used only to define the range of μ which is included in the estimate of theoretical uncertainties [28].

The treatment of the renomalization scale (and, in addition, of the factorization scale in hadronic reactions) when quoting central values of α_s and their theoretical uncertainty, is not at all uniquely defined, and is in fact controversial [81]. Several theoretical proposals [82-84] for fixing the choice of μ exist. They usually result in values of μ, depending on the observable under study, which are smaller than the physical energy scale of the hard scattering process, like $\mu << E_{cm}$ for jet rates and event shapes in e$^+$e$^-$ annihilation.

Several contributions to this conference make exclusive use of one or the other of these possible scale fixing schemes, for instance S. Sanghera [85] who proposes the removal of the scale uncertainty by relying on the experimental best fit values of μ; or C. Maxwell and J. Barclay [86] who analyse data using the scheme of Grunberg [82] and thereby test non-asymptotic contributions to the $\mathcal{O}(\alpha_s^2)$ predictions for some observables. A. Kataev [87] contributed an extension of the scale fixing method of Brodsky, Lepage and Mackenzie [84] to the NNLO predictions for the hadronic partial width of Z^0 and τ decays, R_Z and R_τ, and L. Surguladze and M. Samuel [88] fix the scale for R_Z and R_τ such that the schemes proposed in refs. 82-84 are satisfied.

Most of the experimental results on α_s which will be discussed in the following contain a certain range of renormalization scales to define the theoretical uncertainties, rather than relying on one of the scale fixing schemes mentioned above. In many cases, the chosen ranges include those schemes as well as physical scales like $\mu = E_{cm}$, such that the estimates of scale uncertainties can be regarded as being rather conservative.

α_s from e$^+$e$^-$ Annihilations

A recent new development in α_s determinations from e$^+$e$^-$ annihilations is to study many different observables in one analysis, using the same data set and identical experimental methods [89,27]. This procedure provides the possibility of verifying the size of the theoretical uncertainty which is necessary to describe all data distributions with one universal value of α_s. NLO QCD predictions (i.e. in $\mathcal{O}(\alpha_s^2)$) are available for about 15 different hadronic event shape, jet production and energy correlation observables [90]. Most of these observables are correlated with each other, but they all have different NLO coefficients and thus, presumably, different (unknown) higher order contributions.

The results of a recent OPAL study of 13 different observables [27] are shown in Fig. 24. The data, like the event shape distributions of Thrust (T), oblateness (O) and heavy jet masses (M_H), jet production rates using different jet recombination schemes and the asymmetry of two-fold energy correlations (AEEC), are corrected for detector acceptance and for hadronization effects, using a number of different QCD plus hadronization models. The coupling constant $\alpha_s(M_{Z^0})$ is then extracted from fits of the analytic $\mathcal{O}(\alpha_s^2)$ QCD predictions to the data, using, in a first step, a renormalization scale of $\mu = E_{cm}$. This leads to the values of $\alpha_s(M_{Z^0})$ plotted in Fig. 24a, where only statistical and experimental, but no theoretical, uncertainties are considered. Within the experimental errors of typically 1% to 3% in α_s, the results from different observables are not compatible with each other; they disagree by as much as eight standard deviations. The need to include theoretical uncertainties

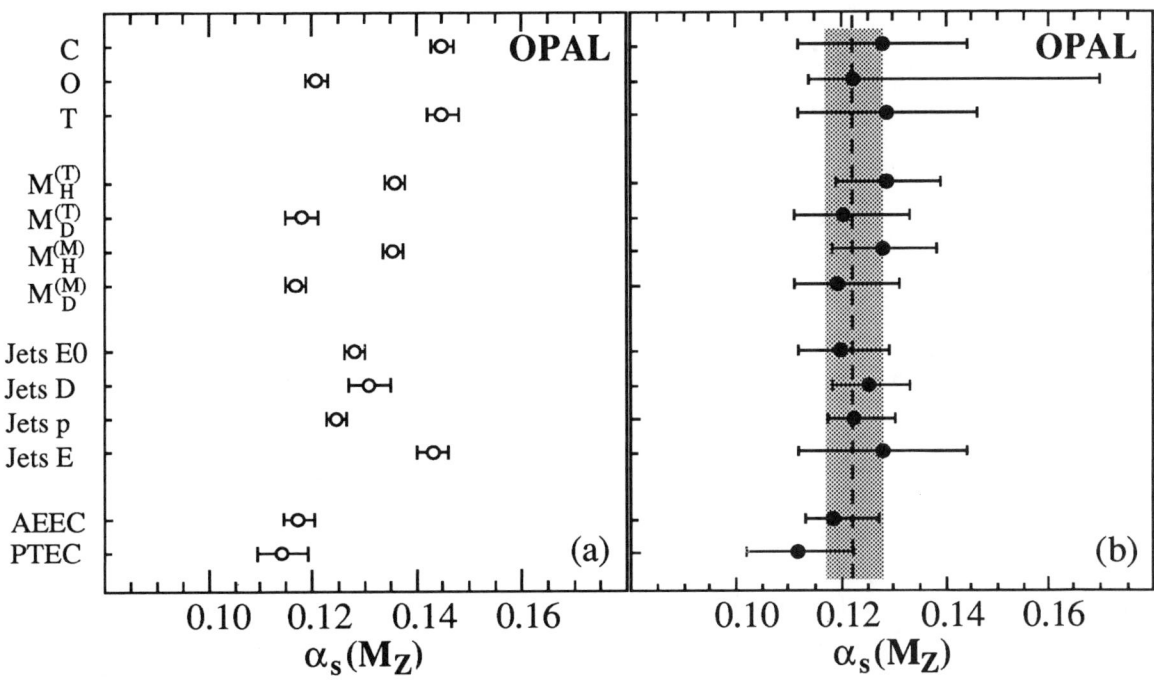

Fig. 24. Measurements of $\alpha_s(M_{Z^0})$ from different observables, including the experimental uncertainties and performing the QCD fits for a fixed renormalization scale $\mu = E_{cm}$; no theoretical uncertainties are taken into account (a). The final values of $\alpha_s(M_{Z^0})$ from the same analysis, now including also theoretical uncertainties (b).

is thus obvious, if QCD is required to give a consistent picture.

In the next step of the analysis, the renormalization scale dependence is studied for each observable, as demonstrated by the corresponding analysis of DELPHI [89] shown in Fig. 25. The functional form of this dependence is different for each observable, typically leading to a decrease of $\alpha_s(M_{Z^0})$ with decreasing renormalization scale, and a minimum at scales as small as a few GeV. The methods for determination of the values of $\alpha_s(M_{Z^0})$ and their theoretical uncertainty are different, in general, for different experiments.

In the OPAL analysis, the central value of $\alpha_s(M_{Z^0})$ for each observable, is defined as the arithmetic mean of the result for $x_\mu \equiv \mu/E_{cm} = 1$ and for x_μ from the best fit; the difference between these two choices is then assigned to be the scale uncertainty. Including hadronization uncertainties and uncertainties due to the minimum parton virtuality to which the data are corrected, leads to the results of $\alpha_s(M_{Z^0})$ shown in Fig. 24b. Comparing this with Fig. 24a, where theoretical uncertainties are not included, one realizes that the overall uncertainties have increased by about an order of magnitude, and that the results from different observables are now in good agreement with each other. Within the theoretical uncertainties assigned in this analysis, the different observables thus lead to a consistent picture with $\alpha_s(M_{Z^0}) = 0.122^{+0.006}_{-0.005}$, where the error is almost entirely theoretical [27].

Since each experiment has chosen its own method of determining and quoting the central value of $\alpha_s(M_{Z^0})$ and its systematic uncertainty, it is mandatory to apply a consistent definition of these quantities before an overall average value of $\alpha_s(M_{Z^0})$ can be quoted. This was recently done in a comprehensive analysis of the published $\alpha_s(M_{Z^0})$ results [91], ba-

100 Tests of QCD

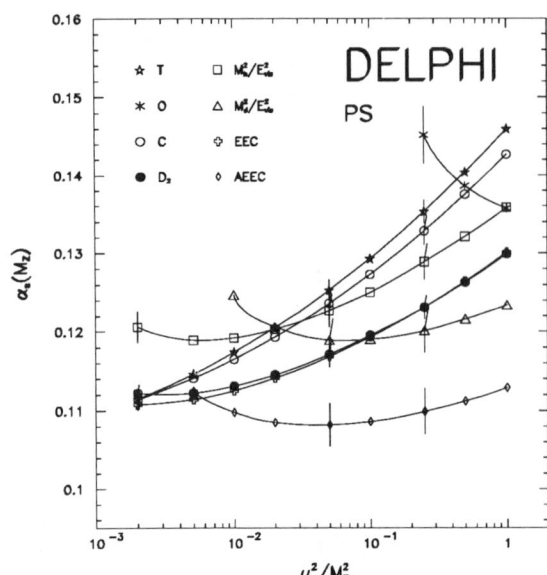

Fig. 25. Dependence of $\alpha_s(M_{Z^0})$ on the renormalization scale factor $x_\mu^2 = \mu^2/E_{cm}^2$, for different observables in $\mathcal{O}(\alpha_s^2)$.

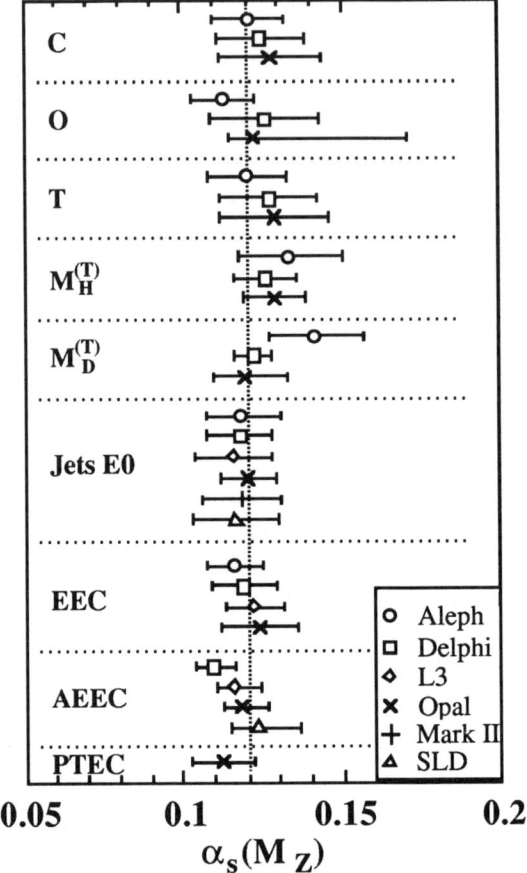

Fig. 26. Compilation of measurements of $\alpha_s(M_{Z^0})$ from event shapes, jet rates and energy correlations, in $\mathcal{O}(\alpha_s^2)$, at LEP and SLC. The errors contain the experimental and theoretical uncertainties, added in quadrature.

sically applying the method which was used by OPAL [27], as described briefly above. Following this method, an update of the results of $\alpha_s(M_{Z^0})$, from all experiments at LEP and SLC, is presented in Fig. 26. Good agreement is found between the experiments, but note that the errors are largely dominated by theoretical uncertainties, which are common to all the experiments. The overall combined result is $\alpha_s(M_{Z^0}) = 0.120 \pm 0.006$.

A new determination of α_s from scaling violations in the inclusive momentum distributions of hadrons, $\frac{1}{\sigma}\frac{d\sigma}{dx}$ where $x = p_{hadron}/E_{beam}$, in the e^+e^- c.m. energy region from 14 GeV to 91.2 GeV, was contributed to this conference by the DELPHI collaboration [92]. DELPHI obtains $\alpha_s(M_{Z^0}) = 0.119 \pm 0.006$, where the error includes both experimental and theoretical uncertainties. This result is derived from a comparison with the prediction of a Monte Carlo implementation of the $\mathcal{O}(\alpha_s^2)$ QCD matrix element combined with a hadronization model, rather than from an analytic QCD calculation.

Recently, calculations have been carried out which contain, in addition to the complete $\mathcal{O}(\alpha_s^2)$ prediction, the resummation of terms of the form $\alpha_s^n \ln^m O$ with $m \geq n$, which yields the so-called next-to-leading logarithmic approximation (NLLA) to all orders. Resummed calculations are available for observables O such as Thrust, heavy jet mass, jet rates in the Durham scheme and energy correlations [19,93-95]; for jet rates, only the leading logs are fully resummed so far. These calculations are expected to provide more accurate predictions of the distributions, especially at high thrust or low jet masses, and they should also have a much reduced dependence on the renormalization scale if compared to the $\mathcal{O}(\alpha_s^2)$ calculations alone. Both these expectations were confirmed in several recent experimental studies [27,96-98].

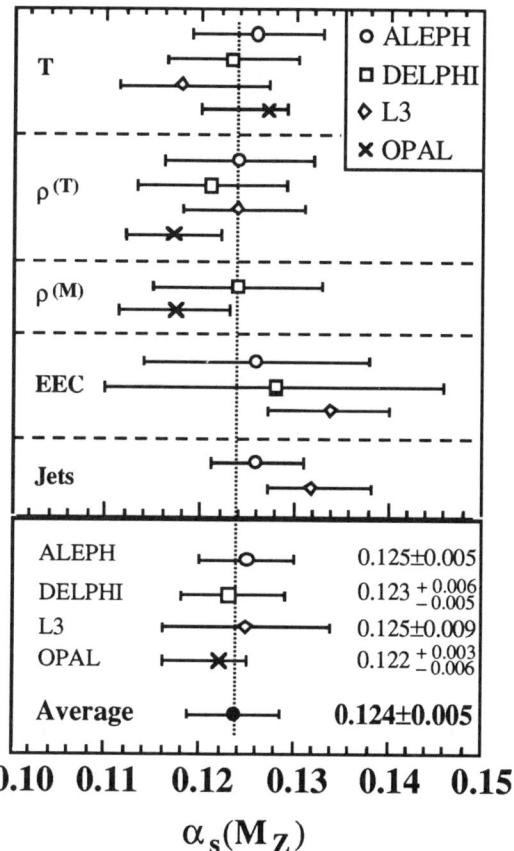

Fig. 27. Summary of measurements of $\alpha_s(M_{Z^0})$ from LEP, using resummed $\mathcal{O}(\alpha_s^2)$ QCD calculations.

A summary of $\alpha_s(M_{Z^0})$ from resummed calculations is given in Fig. 27. In contrast to the results in $\mathcal{O}(\alpha_s^2)$ QCD (c.f. Fig. 26), the central values of $\alpha_s(M_{Z^0})$ are always given for $\mu = E_{cm}$, since small renormalization scales are no longer preferred in resummed calculations. In fact, best fit results are obtained for renormalization scales much closer to $\mu = E_{cm}$ than in the case of $\mathcal{O}(\alpha_s^2)$ QCD. The errors presented in Fig. 27 include experimental and theoretical uncertainties, added in quadrature. Combining these results provides an average value of $\alpha_s(M_{Z^0}) = 0.124 \pm 0.005$, which is in good agreement with the final value from analyses in $\mathcal{O}(\alpha_s^2)$ alone.

An attractive way to determine $\alpha_s(M_{Z^0})$ is a precise measurement of the ratio R_Z of the hadronic and leptonic partial widths of the Z^0,

$$R_Z \equiv \left(\frac{\Gamma_{had}}{\Gamma_{lept}}\right)_{exp} = \left(\frac{\Gamma_{had}}{\Gamma_{lept}}\right)_0 (1 + \delta_{QCD}),$$

since R is not affected by hadronization effects and because the QCD correction δ_{QCD} has been calculated to complete third order ($O(\alpha_s^3)$) perturbation theory [99]. Including quark mass corrections, δ_{QCD} is of the form [100]

$$\delta_{QCD} = 1.05\left(\frac{\alpha_s}{\pi}\right) + 0.9\left(\frac{\alpha_s}{\pi}\right)^2 - 13\left(\frac{\alpha_s}{\pi}\right)^3.$$

The expectation for $(\Gamma_{had}/\Gamma_{lept})_0$, without QCD corrections, is 19.97 with only a small uncertainty due to the unknown masses of the top quark and of the Higgs particle, M_t and M_H.

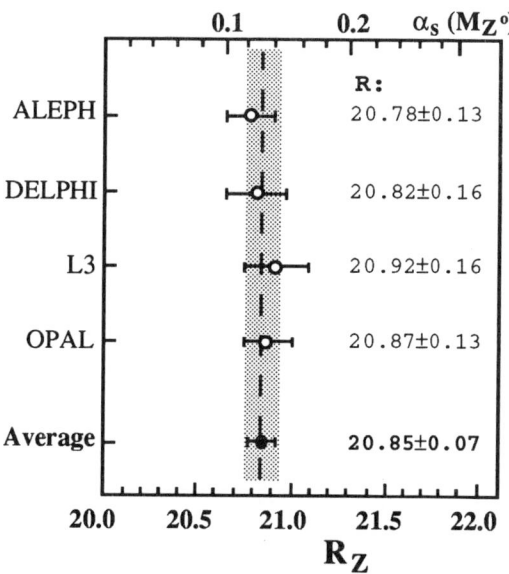

Fig. 28. Compilation of measurements of R_Z from LEP.

The average value of R_Z, summarized in Fig. 28 from the measurements of the four LEP experiments, is $R = 20.85 \pm 0.07$ [101].

This is based on a total of 1.7×10^6 hadronic and 183,000 leptonic Z^0 decays. From this result one infers $\alpha_s(M_{Z^0}) = 0.130 \pm 0.011 \pm 0.004 \pm 0.002$, where the first error is statistical, the second comes from uncertainties in M_t and M_H, and the third error is due to uncertainties of the mass of the b-quark [78]. From a combined fit of M_t and $\alpha_s(M_{Z^0})$, using all the LEP data on the hadronic and leptonic Z^0 line shape and measurements of lepton asymmetries, one obtains $M_t = 139^{+24+19}_{-30-22}$ GeV and $\alpha_s(M_{Z^0}) = 0.135 \pm 0.009 \pm 0.002$, where the first error is experimental, and the second is due to M_H [101]. The renormalization scale dependence of α_s is about ± 0.003, and is thus negligible compared to the current size of the statistical error.

The ratio R_τ of the hadronic and electronic branching fractions of the τ lepton,

$$R_\tau = \frac{B(\tau \to \text{hadrons} + \nu_\tau)}{B(\tau \to e\bar{\nu}_e\nu_\tau)} \equiv \frac{1 - B_e - B_\mu}{B_e},$$

which can be reliably determined by measurements of the electronic and muonic branching fractions B_e and B_μ, is theoretically expected to be given by [102]

$$R_\tau = 3.058(1.001 + \delta_{pert} + \delta_{nonpert}) .$$

Here, δ_{pert} and $\delta_{nonpert}$ are perturbative and non-perturbative QCD corrections; δ_{pert} was calculated to complete $\mathcal{O}(\alpha_s^3)$ and is of similar structure to the one for R_Z [103,102]. The non-perturbative correction was estimated to be $\delta_{nonpert} = -0.007 \pm 0.004$ [102].

The experimental values of R_τ from LEP are summarized in Fig. 29. These values are calculated from the summary of electroweak results given at this conference [101], and lepton universality is assumed by including a phase-space factor of 0.9728 for B_μ. The average value is $R_\tau = 3.64 \pm 0.08$, which, according to the predictions given above, leads to $\alpha_s(M_\tau) = 0.33 \pm 0.04$ in $\mathcal{O}(\alpha_s^3)$ for three quark flavors, $N_F = 3$. This result, which also contains a renormalization scale uncertainty of

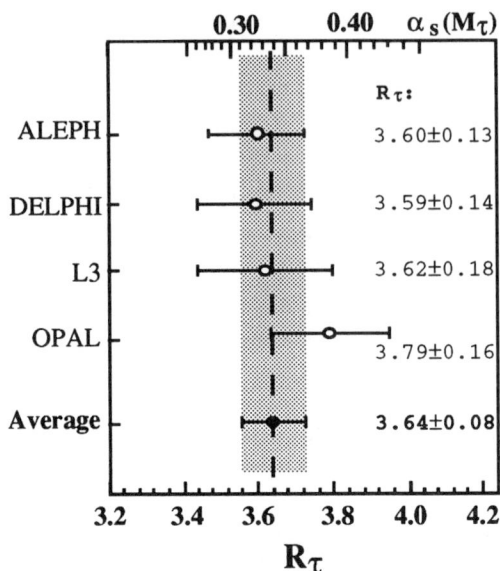

Fig. 29. Compilation of measurements of R_τ from LEP.

± 0.03, added in quadrature, is significantly larger than the value of $\alpha_s(M_{Z^0})$ from event shapes and jet rates measured in Z^0 decay, as is expected for an energy dependent α_s.

Recently, a revised $\mathcal{O}(\alpha_s^3)$ prediction of δ_{pert} became available, which also includes the resummation of leading logarithmic terms [104]. Furthermore, a new method of analysis was proposed, in which $\delta_{nonpert}$ can be simultaneously determined from the data [105], in addition to $\alpha_s(M_\tau)$, instead of relying on the estimates of $\delta_{nonpert}$ mentioned above. This method requires the measurement of weighted integrals of the hadronic invariant mass spectrum of τ decays. The ALEPH collaboration has contributed an analysis to this conference, which is based on this new method and on the improved QCD predictions. Together with the combined LEP average of R_τ as given above and including the renormalization scale uncertainty, one obtains [106] $\alpha_s(M_\tau) = 0.36 \pm 0.04$. This is in good agreement with the result (0.33) given above and therefore indicates that nonperturbative corrections to R_τ are indeed very small.

When extrapolating this value of α_s from

$\mu = M_\tau$ to $\mu = M_{Z^0}$, $\Lambda_{\overline{MS}}$ is adjusted on crossing a quark threshold, such that α_s is a continuous function of μ [80]. This results in $\alpha_s(M_{Z^0}) = 0.121 \pm 0.005$ from τ decays at LEP, where the relative size of the error decreases because of the logarithmic dependence of α_s on μ.

Further results from e^+e^- annihilation, which were contributed to this conference, shall be mentioned briefly: the AMY collaboration at TRISTAN reports a precise determination of α_s from jet production rates [107], which gives $\alpha_s(58 \text{ GeV}) = 0.130 \pm 0.008$. This corresponds to $\alpha_s(M_{Z^0}) = 0.122 \pm 0.007$, in good agreement with the measurements at LEP and SLC. New determinations of the flavor dependence of α_s are available from L3 [108] and from DELPHI [109] at LEP, who determine the ratio $\alpha_s^b/\alpha_s^{udsc}$ from selecting bottom quark events by their semileptonic decays. The combined average of this ratio is [110] $\alpha_s^b/\alpha_s^{udsc} = 1.01 \pm 0.03 \pm 0.03$, which is in good agreement with the flavor independence of α_s predicted by QCD.

α_s from Deep Inelastic Scattering

Scaling violations of structure functions as measured in deep inelastic lepton-nucleon scattering processes (c.f. Figs. 2 and 3) are a powerful tool to determine α_s. QCD relates the logarithmic slopes of structure functions, $d \log F(x, Q^2)/d \log Q^2$, to $\alpha_s(Q)$. The large variety of beam types and target materials, the different structure functions F_2 and F_3, and the large range of x-bins which are available from many experiments, provide the possibility of significant tests of QCD, since one single value of $\Lambda_{\overline{MS}}$ must be able to describe all these different, independent data.

An example of the logarithmic slopes of $F_3(x, Q^2)$, as recently measured by CCFR [4] in ν-Fe scattering, is given in Fig. 30. Note that F_3 is a non-singlet structure function, for which, in contrast to the structure function F_2, the gluon distribution $G(x, Q^2)$ and its re-

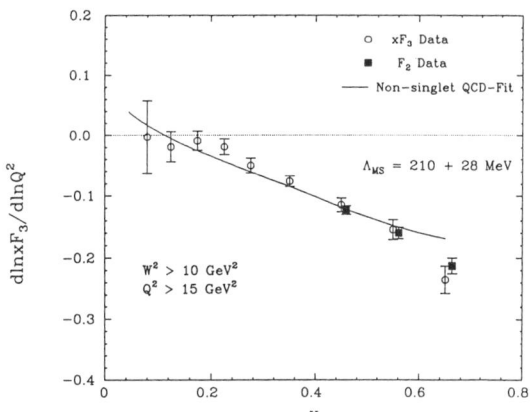

Fig. 30. Logarithmic slopes of $F_3(x, Q^2)$ as measured by CCFR.

Table 5. Measurements of α_s from DIS, converted to $\alpha_s(M_{Z^0})$.

Exp.	beam	targets	$\alpha_s(M_{Z^0})$
BCDMS	μ	C	0.110 ± 0.005
"+SLAC	μ, e	H, D	0.113 ± 0.003
EMC	μ	H	0.108 ± 0.009
NMC	μ	D	0.117 ± 0.005
CCFR	ν	Fe	0.111 ± 0.003
CHARM	ν	CaCO$_3$	$0.115 ^{+0.009}_{-0.012}$

spective uncertainty do not contribute. The data are compared with a fit to the QCD prediction for these slopes, which describe the experimental measurements in the entire x-range. For the precise determination of $\Lambda_{\overline{MS}}$, CCFR substitutes F_2 for xF_3 at large values of x (see Fig. 30), where antiquarks and gluons do not contribute any more. The result is $\Lambda_{\overline{MS}} = (210 \pm 28 \pm 41)$ MeV, for $N_F = 4$ quark flavors, where the first error is statistical and the second systematic.

Similar analyses from other experiments, in general of singlet structure functions F_2, were reported at this conference [3,4] and were recently reviewed in detail e.g. by G. Altarelli [81]. The results, converted to the corresponding values of $\alpha_s(M_{Z^0})$, are summarized in Table 5, where the errors are purely experimental. They are in good agreement with each other, and average to the final value of $\alpha_s(M_{Z^0}) = 0.112 \pm 0.002$. Note that, in principle, the logarithmic slope for each bin of x and for any analysis of structure functions provides a measurement of $\Lambda_{\overline{MS}}$ on its own; the good agreement of the experimental results presented in Table 5, over regions of x, therefore constitutes an important consistency check of QCD.

The scale dependence of α_s in deep inelastic scattering was estimated to be $\Delta \alpha_s(M_{Z^0}) = \pm 0.004$ [6,111]; together with an uncertainty of ± 0.001 in the extrapolation from the average Q^2 of the measurements to the energy scale M_{Z^0}, added in quadrature with the experimental error, the overall final value of $\alpha_s(M_{Z^0})$ from DIS is thus $\alpha_s(M_{Z^0}) = 0.112 \pm 0.005$.

α_s from Hadron Collisions

Only few measurements of α_s from $p\bar{p}$ collider experiments are available, and their precision is significantly lower than the results from e^+e^- annihilation and from DIS experiments. In general, precision measurements at hadron colliders are much more difficult than in reactions which involve leptons, since the quantum numbers and the energy of the initial state of the hard scattering process are unknown. The underlying event caused by the target remnants of the scattered protons and the dependence of the cross sections from the proton structure functions render both the measurement and the theoretical prediction of suitable processes rather complicated.

From a measurement of the ratio of cross sections for W plus one jet and for W without any jet production, the UA2 collaboration determined [112] $\alpha_s(M_W) = 0.123 \pm 0.018 \pm 0.017$. The first error is statistical and the second systematic, mainly from uncertainties in the treatment of the underlying event and from hadronization. Adding them in quadrature, this result corresponds to $\alpha_s(M_{Z^0}) = 0.121 \pm 0.026$. A similar measurement was recently reported from the UA1 experiment [113], which gave $\alpha_s(M_W) = 0.127 \pm 0.026 \pm 0.034$.

The UA1 collaboration determined α_s [114] from their measurement of the $b\bar{b}$ cross section [11] (see Fig. 5) and a fit to the $\mathcal{O}(\alpha_s^3)$ QCD calculations of Nason, Dawson and Ellis [115]. The result is $\alpha_s(20 \text{ GeV}) = 0.138^{+0.028}_{-0.019}$, which can be transformed to $\alpha_s(M_{Z^0}) = 0.109^{+0.016}_{-0.012}$. The error is almost equally due to experimental as well as theoretical, i.e. renormalization scale, uncertainties.

α_s from Heavy Quarkonia Decays

A comprehensive analysis of all available data on quarkonium branching ratios from experiments at SPEAR, CESR, DORIS and VEPP was recently presented by M. Kobel [116]. Values of α_s are determined, in NLO QCD, from ratios R of combined branching ratios like

$$R^{\Upsilon}_{\mu} = \frac{\Gamma(\Upsilon \to ggg)}{\Gamma(\Upsilon \to \mu\mu)},$$

and the corresponding ratios for the J/Ψ. In the nonrelativistic approximation, the dependence on the quarkonium wave function cancels in such ratios. Branching ratios involving the three-gluon decay are especially sensitive to α_s, since in leading order QCD they are proportional to α_s^3.

Kobel finds disagreement between $\Lambda_{\overline{MS}}$ values obtained from charmonium compared to those from bottonium. This disagreement seems to be due mainly to nonrelativistic effects; $<v^2/c^2>$ for the J/Ψ is 0.23, which is considerably larger than for the Υ (≈ 0.08). In the combined fits of Υ and J/Ψ data, relativistic corrections of the form $(1 + D <v^2/c^2>)$ are therefore introduced, with D being a free parameter in the fit. With this additional parameter, all the ratios Γ_{ggg}/Γ_{ll} can be consistently fitted with one single value of $\Lambda_{\overline{MS}}$. The final result is $\alpha_s(M_{Z^0}) = 0.113 \pm 0.001^{+0.007}_{-0.005}$, where the first error is experimental and the second is an estimate of the uncertainties due to the unknown higher order contributions. The latter was determined from generous assumptions of the size of the NNLO coefficients, rather than from variations of the renormalization scale in NLO [116].

α_s from the Mass Splitting of Charmonium States

The splitting of the spin averaged masses of the 1S and the 1P charmonium states was recently calculated using lattice gauge theory (LGT) techniques [117]. LGT provides a nonperturbative means of determining α_s from low energy quantities. The calculations were done ignoring sea-quarks (i.e. the "quenched approximation"); uncertainties arising from this omission and also from the finiteness of the lattice spacing were analyzed and included in the systematic error of α_s.

The bare lattice coupling g_0^2 is determined from a comparison of the experimentally measured mass splitting, $M_{h_c} - (3M_{J/\Psi} + M_{\eta_c})/4 = 458.6 \pm 0.4$ MeV [117], with the result of the calculation; the lattice coupling may then be converted into the $\overline{MS}$ definition of α_s using known perturbative techniques. The largest source of uncertainty arises from the conversion from the zero light quark running coupling constant of the lattice calculation to the four quark running coupling of the real world. The final result is quoted to be $\alpha_s(5~\text{GeV}) = 0.174 \pm 0.012$. Converted to the energy scale of M_{Z^0}, this corresponds to $\alpha_s(M_{Z^0}) = 0.105 \pm 0.004$.

Summary of α_s Measurements

A summary of all the measurements of α_s described in the previous paragraphs is given in Table 6. The table also contains results of previous studies which were not explicitly repeated here, like a determination of α_s from the world average of R_τ [118] (which also includes some preliminary data from LEP), from the total hadronic cross section [119], σ_{had}, and from hadronic event shapes in e^+e^- annihilation [120] below the Z^0 resonance. The values of α_s are given at typical energy scales Q where the actual measurements were done, and are also converted to the energy scale M_{Z^0} by solving analytically the continuity condition [80] of α_s when crossing a quark threshold. The errors given in columns 3 and 4 contain experimental and theoretical errors, added in quadrature; column 5 presents a breakdown of the total error into these two classes of uncertainties. The last column indicates the degree of QCD perturbation theory used to determine α_s, where "LGT" means lattice gauge theory and "resum." stands for resummed $\mathcal{O}(\alpha_s^2)$ calculations.

The values of $\alpha_s(M_{Z^0})$ are compared in Fig. 31. In general, the measurements agree quite well with each other, within their overall errors which are mainly of theoretical nature in most cases. The value of $\alpha_s(M_{Z^0})$ from charmonium mass splitting plus LGT calculation is the smallest, and it also has the smallest error assigned. This particular result is the only one which seems to be incompatible with some of the other α_s determinations, such as the ones from R_τ and from hadronic event shapes at LEP. At this time, this is not considered to be a major problem, since the method of lattice gauge calculations, and in particular the estimate of its uncertainties, is a rather new development in the field of α_s determinations, the reliability of which will have to be verified in similar calculations and applications in the future.

Calculating the weighted average of $\alpha_s(M_{Z^0})$, leaving out the result from charmonium mass splitting, leads to $\alpha_s(M_{Z^0}) = 0.118$. None of the results which are included in this average deviates from this value by more than the size of its assigned error. It is thus concluded that the results are in good agreement with each other, within the limits of their theoretical uncertainties. This is an important consistency check of perturbative QCD which predicts that $\alpha_s(M_{Z^0})$, or α_s expressed at any common energy scale, is a universal number

Table 6. Summary of measurements of α_s.

Process	Q [GeV]	$\alpha_s(Q)$	$\alpha_s(M_{Z^0})$	$\Delta\alpha_s(M_{Z^0})$ exp.	$\Delta\alpha_s(M_{Z^0})$ theor.	Theory
R_τ [LEP]	1.78	0.360 ± 0.040	0.121 ± 0.005	0.003	0.004	NNLO
R_τ [world]	1.78	0.32 ± 0.04	$0.118 ^{+0.004}_{-0.006}$	–	–	NNLO
DIS [ν]	5.0	$0.193 ^{+0.019}_{-0.018}$	0.111 ± 0.006	0.004	0.004	NLO
DIS [μ]	7.1	0.180 ± 0.014	0.113 ± 0.005	0.003	0.004	NLO
$c\bar{c}$ mass splitting	5.0	0.174 ± 0.012	0.105 ± 0.004	0.000	0.004	LGT
$J/\Psi + \Upsilon$ decays	10.0	$0.167 ^{+0.015}_{-0.011}$	$0.113 ^{+0.007}_{-0.005}$	0.001	$^{+0.007}_{-0.005}$	NLO
e^+e^- [σ_{had}]	34.0	0.157 ± 0.018	0.131 ± 0.012	–	–	NNLO
e^+e^- [ev. shapes]	35.0	0.14 ± 0.02	0.119 ± 0.014	–	–	NLO
e^+e^- [ev. shapes]	58.0	0.130 ± 0.008	0.122 ± 0.007	0.003	0.007	NLO
$p\bar{p} \to b\bar{b}X$	20.0	$0.138 ^{+0.028}_{-0.019}$	$0.109 ^{+0.016}_{-0.012}$	$^{+0.012}_{-0.007}$	$^{+0.011}_{-0.010}$	NLO
$p\bar{p} \to W$ jets	80.6	0.123 ± 0.025	0.121 ± 0.024	0.017	0.016	NLO
e^+e^- [scal. viol.]	91.2	0.119 ± 0.006	0.119 ± 0.006	0.001	0.006	NLO
$\Gamma(Z^0 \to$ had.)	91.2	0.130 ± 0.012	0.130 ± 0.012	0.011	0.004	NNLO
Z^0 [ev. shapes]	91.2	0.120 ± 0.006	0.120 ± 0.006	0.001	0.006	NLO
Z^0 [ev. shapes]	91.2	0.124 ± 0.005	0.124 ± 0.005	0.001	0.005	resum.

for all processes. It may be regarded a major success that the result from nonperturbative lattice gauge calculations agrees with the "world average" of $\alpha_s(M_{Z^0}) = 0.118$ to within 10%, although the current error estimate for the quenched approximation suggests a smaller uncertainty at this time.

The agreement of all the results of $\alpha_s(M_{Z^0})$ also imply that there is significant evidence for the running of α_s, since the measurements were done in a wide range of energy scales Q. This evidence can be further quantified in a plot of α_s as a function of Q, from the numbers summarized in Table 6, as shown in Fig. 32. In this figure, the relative size of the assigned uncertainties can also be compared between the measurements, unaffected by the evolution to the scale M_{Z^0} which reduces the relative size of errors. The data are compared with the QCD predictions of a running coupling constant, calculated in $\mathcal{O}(\alpha_s^3)$ for different values of $\Lambda_{\overline{MS}}$, as indicated for $N_F = 5$ quark flavors in Fig. 32. The energy dependence of α_s is distinct, and is in very good agreement with the QCD prediction. Small systematic trends appear to be visible, however: the lower energy results from DIS and from quarkonia decays (but not those from R_τ) seem to prefer

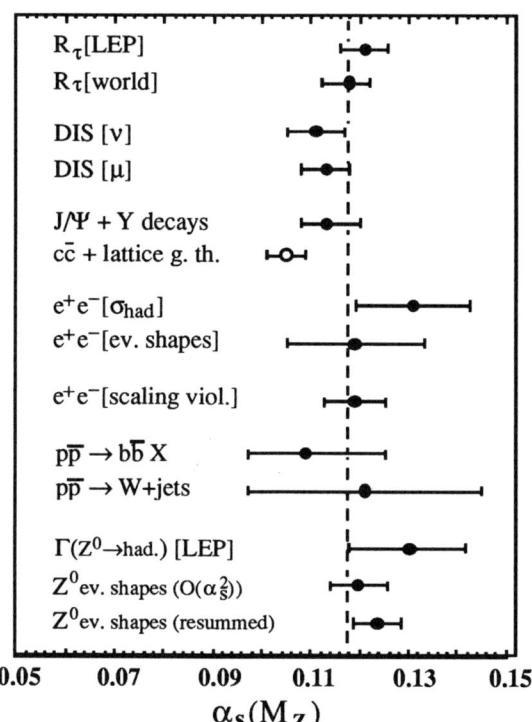

Fig. 31. Summary of measurements of $\alpha_s(M_{Z^0})$.

slightly smaller values of $\Lambda_{\overline{MS}}$ than those from e^+e^- annihilation, as can be seen in Fig. 31, too.

The data, taking an optimistic point of view, can all be described well by QCD with 150 MeV $< \Lambda_{\overline{MS}}^{(5)} <$ 250 MeV, which corresponds to an error in $\alpha_s(M_{Z^0})$ of ± 0.004. With a more pessimistic attitude, however, one can also argue that the range in $\Lambda_{\overline{MS}}$ which is necessary to describe the data is 100 MeV $< \Lambda_{\overline{MS}}^{(5)} <$ 350 MeV, which then corresponds to an error in $\alpha_s(M_{Z^0})$ of ± 0.011. A stringent "determination" of the error of the world average of $\alpha_s(M_{Z^0})$ is not possible, since the uncertainties are mainly of theoretical nature which prevents a classical error analysis. The final world average is thus quoted to be

$$\boxed{\alpha_s(M_{Z^0}) = 0.118 \pm 0.007}\,,$$

where the error corresponds to the average between the optimistic and the pessimistic view

given above. Exactly the same result, however with a different attitude of interpreting the theoretical uncertainties, was recently presented by G. Altarelli [81]. This world average value of $\alpha_s(M_{Z^0})$ now seems to be very stable and solid. Due to the remarkably good agreement between the results quoted from many different processes and observables, there is no apparent reason to argue for an uncertainty which should be much larger than the one given above.

SUMMARY

The large number of significant and precise tests of QCD which were presented at this conference, 20 years after the theory was first formulated, set another milestone in exploring and testing the standard model of the elementary particles and forces.

New, precise measurements of nucleon structure functions became available, with valuable information down to $x \sim 0.008$, extending the x-range towards smaller values by one order of magnitude. The high statistics and the large kinematic range of the new data resolved the controversy between previous measurements. These data have also caused readjustments of the parametrizations of quark distributions, especially at small x. Measurements of $b\bar{b}$ cross sections at hadron colliders indicate that the gluon distributions may also need to be changed. After more than 20 years of experimental and theoretical work on scaling violations in DIS, the field is still developing towards a more consistent and precise picture. An overall and new determination of the parton distributions, using all the available data on structure functions, is mandatory and should soon replace the old parametrizations which are used to calculate hadronic processes for present and for future experiments.

The physics of hadron jets gave further, precise insights into the dynamics and quantum numbers of parton scattering processes. In e^+e^- annihilations at LEP, studies of 4-

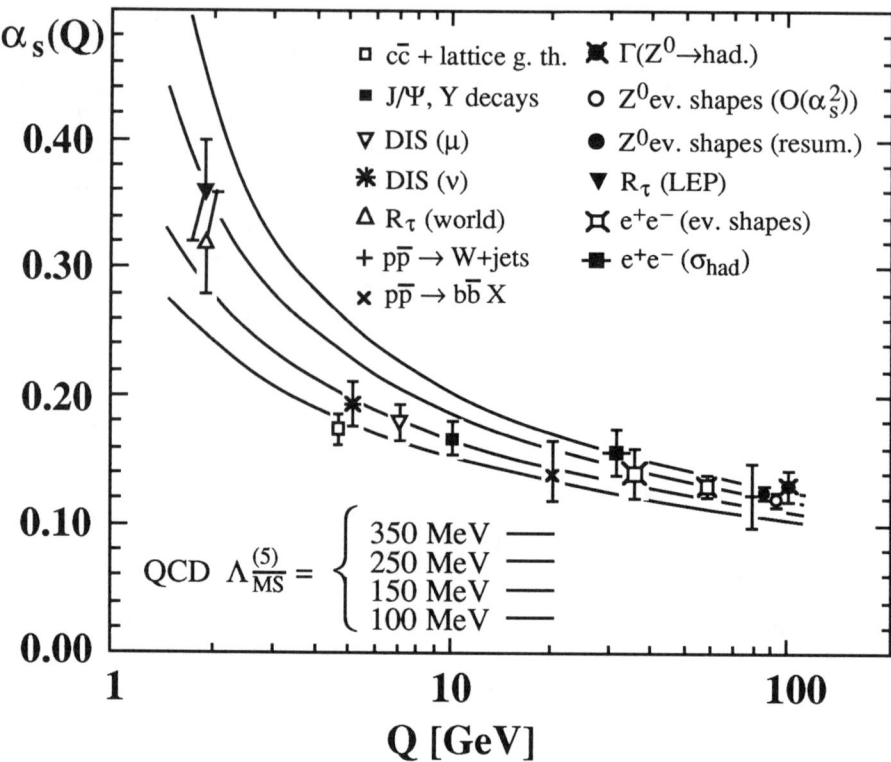

Fig. 32. Summary of $\alpha_s(Q)$.

jet final states gave evidence for the gluon self coupling through experimental fits of the QCD group constants, N_C and T_F. Evidence of asymptotic freedom is demonstrated by the energy dependence of 3-jet event production rates. While jet physics in e^+e^- annihilations has been an effective tool for significant tests of QCD for many years, yet more efficient jet algorithms have been recently introduced and are being utilized in both experimental and theoretical investigations. Observation and first studies of multijet event production in muon-nucleon scattering processes were reported at this conference. In hadron collisions, recent QCD calculations in NLO provided the possibility of detailed studies of the shape of jets and of angular correlations between jets. Significant scaling violations of inclusive jet cross sections at the highest collider energies are observed, but not at exactly the same level as presently expected.

<u>Prompt photon production</u> evolved as a powerful tool to test the dynamics of quarks and gluons in hard scattering processes, since photons are unaffected by the color force field or by hadronization after they have been produced. The experimental challenge to identify prompt photons in dense hadronic environments was accepted by experimentalists in all fields of high energy physics, leading to rather involved measurements, such as cross sections for double prompt photon production in hadron collisions, or photon plus multijet event production rates in e^+e^- annihilation. The data are in reasonable agreement with the theoretical expectations. At this time, the precision of these studies is often limited by the uncertainties in matching the experimental and theoretical definitions of prompt photons.

The boundary between perturbative QCD and the nonperturbative region associated with hadronization is actively studied at LEP. QCD predictions in (next-to-)leading-log approximations, together with the assumption of Local Parton Hadron Duality, are often able to describe successfully momentum spectra and correlations of hadrons; in some cases, however, the data indicate the need for including yet higher order QCD contributions, which can be substantial. The interplay between soft gluon coherence effects and classical hadronization mechanisms like the string effect is under active study, and cannot yet be completely disentangled. First studies of gluon coherence effects are also reported from hadron collisions at large energies.

Determinations of the coupling constant α_s have reached a level of precision which is limited, in almost all cases, by the theoretical uncertainty of the unknown higher order contributions. Calculations are available, in general, in next-to-leading order; only the total hadronic cross section in e^+e^- annihilation, the hadronic decay width of the Z^0 and the ratio of the hadronic over the leptonic τ branching ratios are known to NNLO; a complete NNLO calculation of the Gross-Llewellyn-Smith sum rule in DIS became available after the conference [121]. Resummation of leading and next-to-leading logarithms, in addition to the full NLO, is a new and promising development for the reduction of higher order uncertainties, which is available for some observables in e^+e^- annihilation. The many new precision measurements of α_s from DIS, from heavy quarkonia decays and from e^+e^- annihilation at the Z^0 pole provide significant evidence for the running of α_s, just as predicted by QCD. All results are consistent with a combined world average value of $\alpha_s(M_{Z^0}) = 0.118 \pm 0.007$.

Overall, the final picture of tests of QCD appears to be in a very good state by now, with precision at the level of 6%. Although it seems to be hard to improve the overall accuracy in the near future, many detailed studies and tests are still to be done, especially in the fields of soft hadron production and rare processes like prompt photon production and gluon radiation in tagged heavy quark events. Hera, with its large potential for studies of jet production and of structure functions in wide kinematic ranges of x and Q^2, just started operation and will soon provide significant, new insights into the world of quarks and gluons. Improved QCD calculations in NNLO for some observables in e^+e^- annihilation and in DIS are on the horizon, which may help to understand and to reduce the theoretical uncertainties which currently limit, for instance, the uncertainty of the value of $\alpha_s(M_{Z^0})$ to about 6%. More accurate and interesting results on tests of QCD are therefore expected for the near and not-too-far future.

APOLOGIES

Many studies and presentations on hadronic physics could not be included in this review: for instance new results on intermittency [122,123], on inclusive particle production [124,125], on Bose-Einstein correlations [126], on elastic and diffractive scattering [127], more on QCD studies with heavy quarks at LEP and SLC [110,65] and on theoretical developments in hadron collisions [38]. It was also not possible to include a complete list of references, especially to older publications; the reader is referred to the bibliographies of the references given here.

ACKNOWLEDGEMENTS

I am grateful to the many colleagues who provided me with detailed informations about their results. In particular, I wish to thank F. Diberder, A.X. El-Khadra, S. Ellis, B. Flaugher, J. Huth, M. Kobel, A. Kronfeld, C. Maxwell, S. Mishra, K. Mönig, L. Rolandi, C. Salgado, W.J. Stirling, M. Turner and B.R. Webber for their generous help and for many enlightning discussions.

REFERENCES

1. H. Fritzsch, M. Gell-Mann, 16th Intern. Conf. on High Energy Physics, Chicago-Batavia (1972); H. Fritzsch, M. Gell-Mann and H. Leutwyler, Phys. Lett. B47 (1973) 365; D.J. Gross, F. Wilczek, Phys. Rev. Lett. 30 (1973) 1343; H.D. Politzer, Phys. Rev. Lett. 30 (1973) 1346.
2. E. Rondio (NMC Collab.), these proc.
3. M. Virchaux (NMC Collab.), these proc.
4. S. Mishra (CCFR Collab.), these proc.; Nevis preprints 1459 – 1461 (1992).
5. M. Shaevitz (CCFR Collab.), these proc.
6. M. Virchaux and A. Milsztajn, Phys. Lett. B274 (1992) 221.
7. J. Kwiecinski, A.D. Martin, R.G. Roberts and W.J. Stirling, Phys. Rev. D42 (1990) 3645.
8. J. Morfin and W.K. Tung, Z. Phys. C52 (1991) 13.
9. A.D. Martin, W.J. Stirling and R.G. Roberts, RAL-92-021.
10. W.K. Tung (CTEQ-group), these proc.
11. UA1 Collab., C. Albajar et al., Phys. Lett. B256 (1991) 121.
12. V. Papadimitriou (CDF Collab.), these proceedings.
13. P. Nason, S. Dawson and R.K. Ellis, Nucl. Phys. B303 (1988) 607.
14. R. Meng, these proc.; E.L. Berger, R. Meng and J. Qiu, ANL-HEP-CP-92-79.
15. B. Wijk, plenary talk at this conference.
16. J. Bartels, these proceedings; DESY 92-114 (1992).
17. C. Salgado (E665 Collab.), these proceedings; M.R. Adams et al. (E665 Collab.), Phys. Rev. Lett. 69 (1992)1026.
18. W. Bartel et al. (JADE Collab.), Z. Phys. C33 (1986), 23.
19. S. Catani et al., Phys. Lett. B269 (1991) 432; S. Catani, Proc. 17th Workshop on the INFN Eloisatron Project, Erice, Italy, June 1991.
20. N. Brown, W.J. Stirling, Z. Phys. C53 (1992) 629.
21. S. Bethke, Z. Kunszt, D.E. Soper, W.J. Stirling, Nucl. Phys. B370 (1992) 310.
22. S. Catani, Yu.L. Dokshitzer and B.R. Webber, CERN-TH.6473/92.
23. B.R. Webber, these proceedings.
24. D. Muller (SLD Collab.), these proc.
25. T. Sjöstrand, Comp. Phys. Comm. 39 (1986) 347; T. Sjöstrand and M. Bengtsson, Comp. Phys. Comm. 43 (1987) 367.
26. G. Marchesini and B.R. Webber, Nucl. Phys. B310 (1988) 461; G. Marchesini et al., Comp. Phys. Comm. 67 (1992) 465.
27. OPAL Collaboration, P.D. Acton et al., Z. Phys. C55 (1992) 1.
28. S. Bethke, proc. of the "QCD – 20 Years Later" Conference held at Aachen, Germany, June 1992; HD-PY 92-12.
29. S. Bethke, A. Ricker, P.M. Zerwas, Z. Phys. C49 (1991) 59.
30. G. Kramer, Springer Tracts in Modern Physics, Vol. 102 (1984) 39.
31. DELPHI Collaboration, P. Abreu et al., Phys. Lett. B255 (1991) 466.; updated paper contributed to this conference.
32. ALEPH Collaboration, D. Decamp et al., Phys. Lett. B284 (1992) 151.
33. R.K. Ellis, D.A. Ross and A.E. Terrano, Nucl. Phys, B178 (1981) 421.
34. JADE Collaboration., S. Bethke et al., Phys. Lett. B213 (1988), 235.
35. S. Ellis, Z. Kunszt and D. Soper, Phys. Rev. Lett. 64 (1990) 2121.
36. G. Aversa et al., Z. Phys. C46 (1990) 253.
37. S. Ellis, Z. Kunszt and D. Soper, Phys. Rev. Lett. 69 (1992) 1496.
38. R.K. Ellis, these proceedings.
39. F. Abe et al., CDF Collab., Phys. Rev. Lett. 68 (1992) 1104.
40. R. Plunkett (CDF Collab.), proc. Intern. Lepton-Photon Symp. and Europhysics Conf. on High Energy Physics, Geneva 1991.
41. S. Ellis, proc. Intern. Lepton-Photon Symp. and Europhysics Conf. on High Energy Physics, Geneva 1991.

42. P. Lubrano (UA2 Collab.), these proc.
43. B. Flaugher (CFD Collab.), these proc.
44. H. Weerts (D0 Collab.), these proc.
45. F. Abe et al., CDF Collab., Fermilab-Pub-92/182-E.
46. M. Zielinski, these proceedings. G. Alverson et al., E706 Collaboration, Phys. Rev. Lett. 68 (1992) 2584.
47. P. Aurenche et al., Nucl. Phys. B286 (1987) 509.
48. J. Alitti et al., UA2 Collab., Phys. Lett. B288 (1992) 386.
49. F. Abe et al., CDF Collab., Phys. Rev. Lett. 68 (1992) 2734.
50. H. Baer, J. Ohnemus and J.F. Owens, Phys. Lett. B234 (1990) 127.
51. P. Aurenche et al., LPTHE-Orsay 92/30.
52. P. Aurenche et al., Z. Phys. C29 (1985) 459.
53. B. Bailey, J. Ohnemus and J.F. Owens, DTP/92/18.
54. M. Z. Akrawy et al., OPAL Collab., Phys. Lett. B246 (1990) 285.
55. G. Gratta (L3 Collab.), these proc.
56. G. Kramer and B. Lampe, Phys. Lett. B269 (1991) 401; G. Kramer and H. Spiesberger, DESY 92-022.
57. N. Glover and J. Stirling, DTP 92-52.
58. Z. Kunszt and Z. Trocsanyi, ETH-TH/92-26.
59. L. Lönnblad, proc. of the Workshop on Photon Radiation from Quarks, Annecy, Dec. 1991; CERN-92-04.
60. DELPHI Collab., paper contributed to this conference.
61. P.D. Acton et al., OPAL Collab., Z. Phys. C54 (1992) 193.
62. ALEPH collab., paper contributed to this conference.
63. L3 Collab., paper contributed to this conference.
64. Yu.L. Dokshitzer, V.A. Khoze, A.H. Mueller and S.I. Troyan, *Basics of Perturbative QCD*, Edition Frontieres, France, 1991.
65. V.A. Khoze, these proceedings.
66. D. Amati and G. Veneziano, Phys. Lett. B83 (1979) 87; Ya.I. Azimov et al., Z. Phys. C27 (1985) 65; Z. Phys. C31 (1986) 213.
67. M.Z. Akrawy et al., OPAL Collab., Phys. Lett. B247 (1990) 617.
68. B. Adeva et al., L3 Collab., Phys. Lett. B259 (1991) 199.
69. C.P. Fong and B.R. Webber, Nucl. Phys. B355 (1991) 54.
70. P. Acton et al., OPAL Collab., Phys. Lett. B287 (1992) 401.
71. D. Decamp et al., ALEPH Collab., Phys. Lett. B273 (1991) 181.
72. R. Settles, these proceedings.
73. R. St. Denis, these proceedings.
74. S. Catani, B.R. Webber, Yu.L. Dokshitzer and F. Fiorani, CERN-TH.6419/92.
75. G. Alexander et al., OPAL Collab., Phys. Lett. B265 (1991) 462; J.W. Gary, OPAL Physics Note PN063 (1992), contributed paper to this conference.
76. W. Zeuner, these proceedings.
77. S. Bethke and S. Catani, Proc. of the XXVIIth Rencontre de Moriond, Les Arcs, France, 1992; CERN-TH.6484/92.
78. K. Mönig, these proceedings.
79. W.A. Bardeen, A.J. Buras, D.W. Duke, and T Muta, Phys. Rev. D18 (1978) 3998; A. Buras, Rev. Mod. Phys. 52 (1980) 199.
80. W. J. Marciano, Phys. Rev. D29 (1984) 580.
81. G. Altarelli, proc. of the "QCD – 20 Years Later" Conference held at Aachen, Germany, June 1992; CERN-TH.6623/92.
82. G. Grunberg, Phys. Lett. B95 (1980) 70.
83. P.M. Stevenson, Phys. Rev. D23 (1981) 2916.
84. S.J. Brodsky, G.P. Lepage and P.B. Mackenzie, Phys. Rev. D28 (1983) 228.
85. S. Sanghera, paper contributed to this conference.

86. C.J. Maxwell, these proceedings; D.T. Barclay and C.J. Maxwell, DTP-92/26.
87. A.L. Kataev, CERN-TH.6485/92.
88. L.R. Surguladze and M.A. Samuel, paper contributed to this conference.
89. DELPHI Collaboration, P. Abreu et al., Z. Phys. C54 (1992) 55.
90. Z. Kunszt and P. Nason [conv.] in "Z Physics at LEP 1" (eds. G. Altarelli, R. Kleiss and C. Verzegnassi), CERN 89-08 (1989).
91. S. Bethke, J.E. Pilcher, HD-PY 92/06 and EFI 92-14 (1992), to appear in Ann. Rev. Nucl. Part. Sci. 42 (1992).
92. W. de Boer and T. Kußmaul (DELPHI Collab.), paper contributed to this conference; IEKP-KA/91-11.
93. S. Catani, L. Trentadue, G. Turnock, B.R. Webber, Phys. Lett. B263 (1991) 491.
94. S. Catani, G. Turnock, B.R. Webber, Phys. Lett. B272 (1991) 368.
95. G. Turnock, Cavendish-HEP-92-3.
96. L3 Collab., B. Adeva et al., Phys. Lett. B284 (1992) 471.
97. ALEPH Collab., D. Decamp et al., Phys. Lett. B284(1992) 163.
98. DELPHI Collab., paper contributed to this conference.
99. S.G. Gorishny, A.L. Kataev and S.A. Larin, Phys. Lett. B259 (1991) 144; L.R. Surguladze and M.A. Samuel, Phys. Rev. Lett. 66 (1991) 560.
100. T. Hebbeker, Aachen report PITHA 91/08 (1991) (revised version).
101. L. Rolandi, plenary talk given at this conference.
102. E. Braaten, S. Narison and A. Pich, Nucl. Phys. B373 (1992) 581.
103. E. Braaten, Phys. Rev. Lett. 60 (1988) 1606.
104. F. Le Diberder and A. Pich, Phys. Lett. B286 (1992) 147.
105. F. Le Diberder and A. Pich, Phys. Lett. B289 (1992) 165.
106. F. Le Diberder, private communication.
107. Y.K. Li, AMY Collab., paper contributed to this conference.
108. B. Adeva et al., L3 Collab., Phys. Lett. B271 (1991) 461.
109. DELPHI Collab., paper contributed to this conference.
110. T. Behnke, these proceedings
111. A.D. Martin, R.G Roberts and W.J. Stirling, Phys. Lett. B266 (1991) 273.
112. J. Alitti et al., UA2 Collab., Phys. Lett. B263 (1991) 563.
113. M. Lindgren et al., Phys. Rev. D45 (1992) 3038.
114. A. Geiser (UA1 Collab.), PITHA 92/19 (1992).
115. P. Nason, S. Dawson and R.K. Ellis, Nucl. Phys. B327 (1989) 49.
116. M. Kobel, Proc. of the XXVII Rencontres de Moriond, Les Arcs, France, 1992.
117. A.X. El-Khadra, G. Hockney, A.S. Kronfeld and P.B. Mackenzie, Phys. Rev. Lett. 69 (1992) 729.
118. A. Pich, proc. of the XXVII Rencontres de Moriond, Les Arcs, France, March 1992; CERN-TH.6489/92.
119. R. Marshall, Z. Phys. C43 (1989) 595; G. D'Agostini, W. de Boer and G. Grindhammer, Phys. Lett. B229 (1989) 160.
120. S. Bethke, preprint LBL-28112 (1989).
121. A.L. Kataev, CERN-TH.6604/92.
122. E. de Wolf, these proceedings.
123. I. Sarcevic, these proceedings.
124. M. Kienzle - Foccaci, these proceedings.
125. G. Lafferty, these proceedings.
126. B. de Lotto, these proceedings.
127. L. Frankfourt, these proceedings.

DISCUSSION

P. Schlein, University of California at Los Angeles, USA

The study of deep inelastic scattering (Jet production) in p$\bar{\text{p}}$ interactions, tagged by final state protons with Feynman $x_p > 0.9$, allows a determination of the partonic structure of the proton's soft residue with $(1 - x_p) < 0.1$, which is dominated by the Pomeron. Results from our CERN experiment UA8 on 2-jet production in such events with momentum transfer to the proton in the range 1–2 GeV^2, lead to a rather hard structure of the Pomeron, like $x(1 - x)$. But there also appears to be a δ-function component to the structure. In $\sim 30\%$ of the events, the entire momentum of the Pomeron enters into the hard scattering; this may be the first evidence for the perturbative Pomeron, as was discussed by Frankfurt in a "soft physics" parallel session. These results are being submitted to *Physics Letters*.

G. Bellettini, Pisa, Italy

You presented conclusive evidence for asymptotic freedom based on the dependence of three-jet rate on energy. On the other hand, you had previously shown that this rate depends strongly on the choice of γ cut, the jet-jet separation parameter. How can you reconcile these two apparently conflicting statements?

Bethke

The experimental results shown in that plot were all obtained using the same consistent jet definition, namely the JADE jet finder, using the same value of γ cut = 0.08, which I did not point out explicitly in my talk. For a constant value of γ cut, the theoretical expectation is that the energy dependence comes only from α_s.

Bellettini

Yes, I guessed the plot was made at fixed γ cut. However, in this case would it not have been better to show a family of curves at fixed values of γ cut, possibly all showing consistent asymptotic freedom?

Bethke

These data exist. It's usually not shown, because many of the experiments at lower energy just published data at γ cut = 0.08, which was the canonical value people have agreed to. The plot looks similar at other values of γ cut, but you don't have as many data points. [In fact, a plot similar to what you asked for can be found in Ref. 120, Fig. 8.]

HADRON AND PHOTON PRODUCTION OF HEAVY QUARKS *

J.N. Butler
Fermi National Accelerator Laboratory
Box 500,
Batavia, IL 60510, USA

Abstract

Recent results on the production dynamics of bottom from the Collider Detector Facility, E653, and E672 at Fermilab are reviewed and compared to previous results and theory. Agreement with theory is reasonable but the possible origins of the differences that do exist may lie in higher order corrections or possibly the shape of the gluon distribution. New results on the A-dependence of charm are summarized. Recent results are converging towards an α parameter of 1, independent of x_F. Results on inclusive charm production from Fermilab E769 and E653 and CERN WA82 are reviewed. Recent results on $D - \bar{D}$ correlations in photoproduction from E687 at Fermilab are reported. Many new results on charmed baryon states from CLEO, E687, ARGUS, and Serpukhov BIS-2 are reviewed. This is an area where rapid progress in now occuring. Results on the D^* branching fractions, D_s branching fractions, doubly Cabibbo suppressed decays, and the ratio $\frac{D^\circ \to KK}{D^\circ \to \pi\pi}$ are covered.

INTRODUCTION

Hadron and photon beams are copious sources of charmed and bottom quarks. Technological developments, especially silicon vertex detectors, have made it possible to extract clean signals from the rather high background of normal quark production so that hadron and photon interactions now provide quality information on heavy quarks. There are two areas of study: production dynamics, where measurements of the production characteristics are used to test QCD predictions; and spectroscopy, where new states are sought and decay properties are studied. In this paper, recent results in both areas are reviewed. In the first part, results on bottom hadron production and charm hadron and photon production are presented. In the second part, new results on charm spectroscopy and hadronic weak decays[1] are described along side recent results from e^+e^- collisions. The number of new results in the last year is so large that it is not possible to cover them all in a short review. I apologize, in advance, for the many interesting measurements which I was forced to omit.

B CROSS SECTIONS FROM CDF

The CDF collaboration at Fermilab has recently presented new results on the production of particles containing b-quarks in collisions at

---

*Work supported by the U.S. Department of Energy.

$\sqrt{s} = 1800$ GeV. These results[2], taken together with the results from the UA1 collaboration at CERN[3], provide a good picture of the cross section and p_t dependence of B-production in the central rapidity region.

The cross sections are derived from a variety of 'indirect' signals based on observation of leptons from B-decays and in the CDF case also from completely reconstructed B-mesons. Table 1 lists the various signals used by the two collaborations. Figure 1 shows the signal for fully reconstructed B-mesons observed through the decay $B^+ \to \psi K^+$ and $B^0 \to \psi K^*$ in CDF. Figure 2 shows the cross sections above a value of p_{tmin} vs p_{tmin} from UA1 and from CDF. Here, p_{tmin} is the integrated cross section of all b-quarks (or B-mesons) that are responsible for 90% of the observed signal. QCD calculations[4] based on the α_s^3 calculations of Nason, Dawson, and Ellis[5] are also shown in figure 2. The dashed curves are intended to provide an envelope of 'theoretical uncertainty' introduced by the arbitrariness in the the choice of the renormalization/factorization scale, uncertainty in the size of the QCD Λ parameter, and some uncertainty in the choice of b-quark mass.

It can be seen that the agreement between the theory and the UA1 data is quite good. The CDF data lie about a factor of 2 above the QCD calculation. The agreement is somewhat worse than this at low p_t, where, however, the experimental uncertainties are also largest.

Figure 3 shows preliminary results for the J/ψ and ψ' cross sections from CDF[6]. Shown also are curves for J/ψ and ψ' production from the decay of B-mesons (using the results of Nason, Dawson, and Ellis for the cross section) and, in the case of the ψ a model due to Glover[7] of 'direct' ψ production. Again, there is a disagreement at more than the factor of two level.

While this may be considered a 'satisfactory' level of agreement given the the experimental situation and the arbitrariness and uncertainty in some of the ingredients of the theory, it is still worthwhile to ask what the source of the difference might be. Possibilities include the contributions of even higher order diagrams, the choice of parameters such as μ, Λ, or M_b, or the uncertainties in such basic 'experimental inputs' as the gluon distribution.

The contributions of α_s^4 and higher order processes are clearly indicated by the sensitivity of the cross sections to the choice of the value of μ[4].

It has recently been emphasized that b-production at collider energies provides a good measure of the gluon distribution[4] because it is dominated by gluon fusion mechanisms. The B's observed by CDF are produced by gluons of x between 0.01 and 0.05. The gluon distribution at these x values is not directly measured and is not well constrained by deep inelastic lepton scattering. There has been a recent attempt to modify the gluon distribution to produce better agreement with the CDF data while preserving the agreement with data from all other processes. The result is a new 'improved' gluon distribution which differs the commonly accepted ones. This new distribution, divided by the more standard Morfin-Tung non-singular set B1, is shown in figure 4. In the modified distribution, the gluon density is raised by roughly $\sqrt{2}$ is the region near x $\sim$ 0.01 and it is decreased at x > 0.1 to maintain agreeement with sum rules. The 'modified' distribution also improves the agreement between CDF's measurement of direct photon production and the NLO QCD calculation but does not entirely remove the the discrepancy. More work is needed to understand what the source of the larger than expected cross section might be.

It is also expected that the experimental data will improve soon. The data presented here are from the 1987/88 run of the Fermilab Tevatron Collider and represent an integrated

Table 1. Summary of Datasets for B Cross Sections at Hadron Colliders.

UA1 $\sqrt{s} = 630\,\text{GeV}$:	$|y| < 2.0$
Single μ + jet	$p_t^\mu > 10\,\text{GeV}/c$
	$E_t^{jet} > 10\,\text{GeV}/c$
High Mass $\mu^+\mu^-$	$p_t^\mu > 3\,\text{GeV}/c$
	$6\,\text{GeV}/c^2 < M_{\mu\mu}$
	$M_{\mu\mu} < 35\,\text{GeV}/c^2$
Low mass $\mu^+\mu^-$	$p_t^\mu > 3\,\text{GeV}/c$
	$0.2\,\text{GeV}/c^2 < M_{\mu\mu}$
	$M_{\mu\mu} < 6\,\text{GeV}/c^2$
Inclusive J/ψ	$p_t > 5\,\text{GeV}/c$
CDF $\sqrt{s} = 1800\,\text{GeV}$:	$|y| < 1.0$
Single $e^\pm$	$p_t > 7\,\text{GeV}/c$
	$p_t > 12\,\text{GeV}/c$
Prompt $e^- + D^0$ + c.c.	$p_t(e) > 12\,\text{GeV}/c$
($D^0 \to K^-\pi^+$ + c.c.)	
$B \to \psi K^\pm$	$p_t(\mu) > 3.0\,\text{GeV}/c$
$B \to \psi' X$	$p_t(\psi) > 6.0\,\text{GeV}/c$
$B \to \psi X$	$p_t(\psi) > 6.0\,\text{GeV}/c$

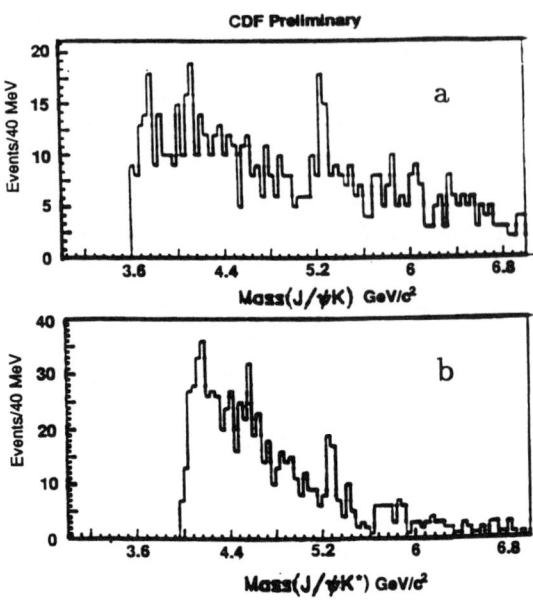

Figure 1. Mass plot for a) $\psi K^\pm$ and b) ψK^{*0} showing a signal for B mesons from CDF.

luminosity of 4.5 pb^{-1}. A new collider run is just beginning at Fermilab. The luminosity goal is 100 pb^{-1}. CDF has geared up to take advantage of this for top AND b-physics by adding a high precision silicon vertex detector (SVX), extending its muon coverage, and making numerous other improvements. Expected yields[8] are shown in table 2. The D0 detector, which is now in full operation, can also contribute to our knowledge of b-production through channels containing leptons.

B PHYSICS RESULTS AT FERMILAB FIXED TARGET ENERGIES

E653

A new measurement of the b-quark cross section for 600 GeV/c π^- has been reported

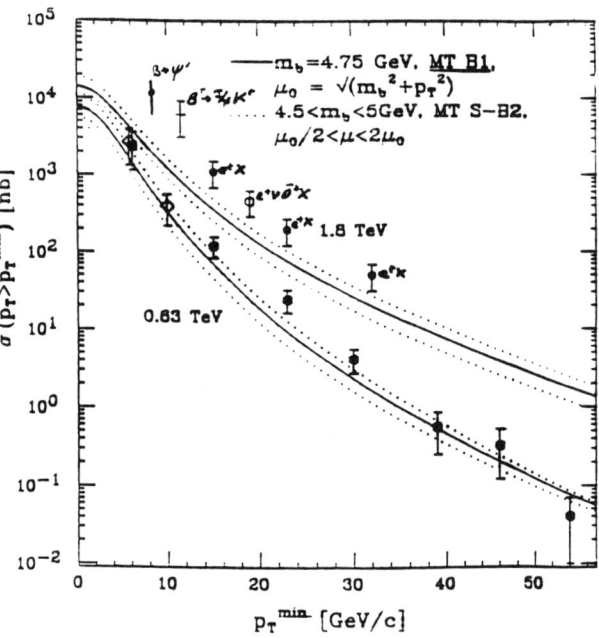

Figure 2. B cross sections integrated from p_{tmin} vs p_{tmin} from UA1 and CDF with NDE calculation superimposed.

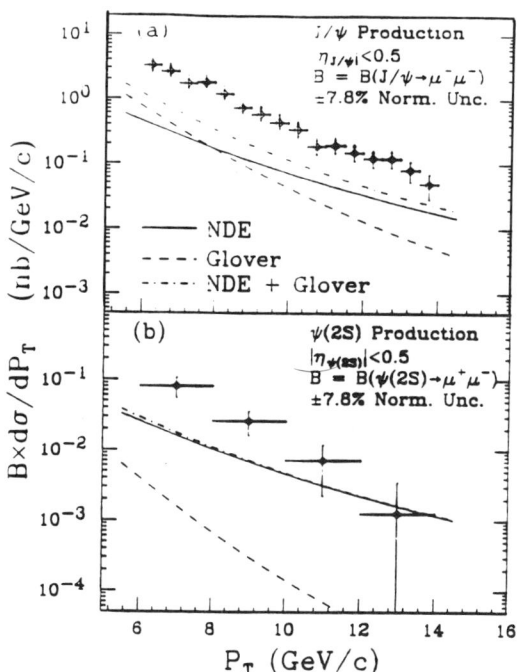

Figure 3. Cross sections from CDF for J/ψ and ψ' production vs. p_t.

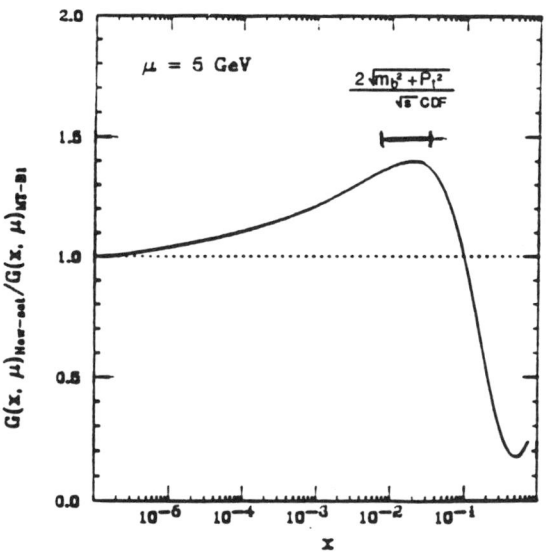

Figure 4. 'Modified' Gluon distribution.

Table 2. Expected rates for various signals in CDF for 100 pb^{-1}.

state	estimated yield
$B \to e\nu X, \mu\nu X$	2.5×10^6
$B \to l\nu D^0(K\pi)$	15,000
$B_s \to l\nu D_s(\phi\pi)$	2,500
$B \to \psi(\mu\mu)X$	3×10^5
$B \to \psi K^\pm$	6,000
$B \to \psi K^*$	5,000
$B \to \psi K^0_s$	2,000
$B \to \psi\phi$	1,000
plus many others	

by Fermilab experiment 653[9]. Nine $b - \bar{b}$ pair events were observed in a hybrid emulsion spectrometer using a muon trigger. The p_t of the trigger muon was required to be greater than 1.5 GeV/c. From these 18 examples of b decays, the x_F and p_t dependence and the total cross section have been extracted.

x_F and p_t dependence

The data are fitted to the form:

$$\frac{d^2\sigma}{dx_F dp_t^2} \propto (1 - |x_F - x_0|)^n \times e^{-bp_t^2} . \quad (1)$$

The result of the fit is:

$$n = 4.55^{+2.85}_{-2.05}(stat) \pm 0.75\,(sys) \quad (2)$$

with the parameter x_0 fixed at 0.075. The fitted value for b is:

$$b = 0.095^{+0.04}_{-0.03}\,(stat\ only) . \quad (3)$$

Within the very large errors, n is not much different than it is for charm. On the other hand, the p_t^2 distribution is much stiffer than it is for charm, as expected. This agrees with the QCD expectation that $b \sim \frac{1}{M_Q^2}$, where M_Q is the heavy quark mass.

Total $b - \bar{b}$ cross section

The total $b-\bar{b}$ cross section (i.e. the 'quark-pair' cross section) is

$$\sigma_{b-\bar{b}} = 33 \pm 11(stat) \pm 7(sys)\,nb/nucleon \quad (4)$$

integrated over all x_F and assuming a linear A-dependence. The result is in good agreement with QCD predictions. The behavior of the cross section with energy is seen to be 'reasonably well' described by the theory over more than two orders of magnitude in the cross section!

Another fact learned from these 9 events is that the two B-particles tend to be produced back-to-back in a plane perpendicular to the beam direction as expected for parton production.

E672

Another Fixed Target experiment, Fermilab E672, has also observed a signal for B-meson production by 530 GeV/c π^-[10]. This experiment triggers on dimuons detected in a toroidal spectrometer downstream of Fermilab E706, a prompt photon experiment. The E706 vertex detector then picks out J/ψ decays downstream of the primary production vertex. The experiment observes a signal for the decay

$$\begin{array}{rcl} B^0 & \to & \psi K^{*0} \\ B^+ & \to & \psi K^+ \end{array} \quad (5)$$

with the ψ subsequently decaying into $\mu^+\mu^-$ and the K^{*0} decaying into $K\pi$. The signal is shown in figure 5. The cross section is still being worked out.

A-DEPENDENCE OF CHARM PRODUCTION

Many Fixed Target charm (and bottom) experiments use solid targets because they must keep the distance from the interaction point to their vertex detector small to achieve

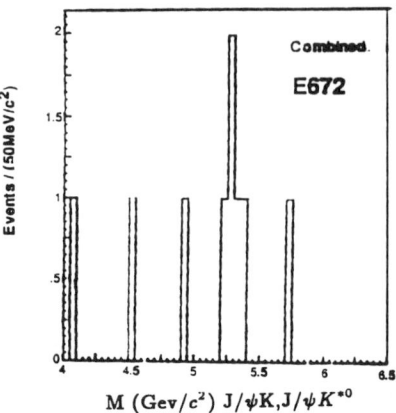

Figure 5. Invariant mass plot for $B^0 \to \psi K^{*0}$ and $B^+ \to \psi K^+$.

good vertex resolution. The ability to compare experiments using different target materials depends on our knowledge of the systematics of the A-dependence of heavy quark production. More fundamentally, in QCD we expect that the production of the heavy quarks takes place inside the nucleus over such a short time scale that it should depend on the parton structure of single nucleons, whereas the formation of the final state hadrons should occur well outside the nucleus[11]. The compact quark structure which propagates from the production point through the nucleus should interact very weakly. The A-dependence of the total heavy quark production cross section, normally described by

$$\sigma = \sigma_0 A^\alpha, \quad (6)$$

should be characterized by $\alpha = 1$. If it were found that $\alpha \ll 1.0$, then the applicability of hard scattering calculations MIGHT be called into question or, at the minimum, modifications would have to be made to account for the presence of the nuclear medium.

A summary of measurements of the A-dependence of charm is given in figure 6[12]. The early measurements in the table appear below the dashed line. They were obtained by a variety of techniques, and cover different regions of x_F. Some use indirect signals, such as prompt muons or neutrinos. Others are plagued by low statistics. They are not in complete agreement. It was suggested that α might vary from a value of about 1.0 at $x_F \approx 0$ to a value of about 0.75 at higher x_F. The decrease of α with increasing x_F is exhibited by lighter particles[13].

New data from Fermilab experiment 769[12] and recent results from CERN WA82[14] appear at the top of the table. They both have high resolution vertex detectors and achieve good yields of fully reconstructed charm particles. E769 measures a value of $\alpha \approx 1.0$ over the x_F range from 0.0 to 0.5 and a p_t range out to 3.0. Their results are shown in figure 7a and b for $D^\pm$ and D^0 production, respectively, on 4 different materials. The large number of fully reconstructed charm decays and the fact that the experiment was configured so that many systematic errors cancelled make this a particularly powerful measurement. Figure 7 c and d show the x_F dependence of α and the exponential dependence of the p_t^2 distributions respectively. WA82 gets a result which is about about 1.3 standard deviations away from 1.0.

The A-dependence of J/ψ and ψ' production by protons at 800 GeV/c has been measured very accurately by the Fermilab E772 experiment[15]:

$$\alpha = 0.920 \pm 0.008 \ . \qquad (7)$$

This number is very similar to the A-dependence of J/ψ photoproduction [16]:

$$\alpha = 0.94 \pm 0.02 \pm 0.03 \text{ (incoherent signal)} \ . \qquad (8)$$

It is reasonable to believe that the A-dependence for open and bound charm ought to be *similar*. The experiments are certainly not inconsistent with this belief. At present, there is no real information on the x_F depen-

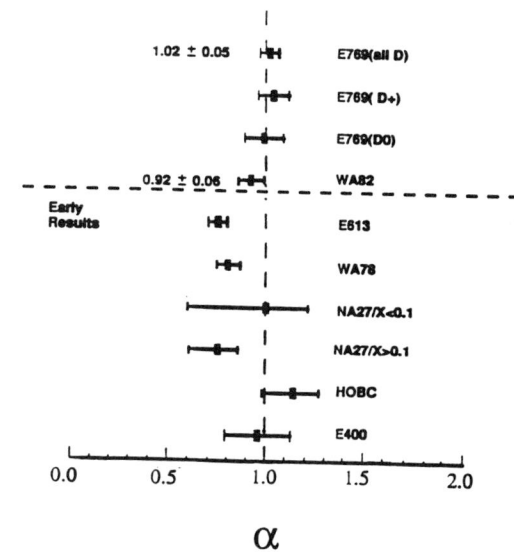

Figure 6. α parameters of charm production.

dence of charm at $x_F \geq 0.5$ although the photoproduction results for J/ψ might be suggestive. It is not known why more recent results using fully reconstructed final states seem to differ systematically from results using prompt muons or neutrinos. Even if there is a slight departure of α from unity, there are several possible explanations which could account for it[17].

INCLUSIVE PRODUCTION OF CHARM

The development of the silicon microstrip detector has finally made it possible to study the hadron and photon production of charmed particles using large numbers of fully reconstructed charm decays. However, only a handful of experiments have been run using this technology so while information on charm production is steadily improving and the quality of the data is much better than it was only a few years ago, the data are less complete than we might wish. Below, we discuss the

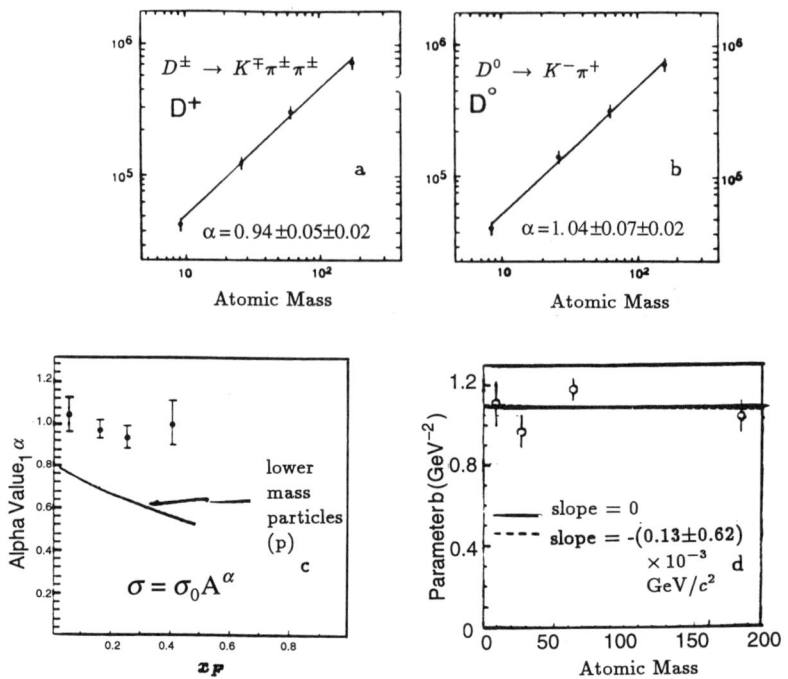

Figure 7. A-dependence of charm meson production by 250 GeV/c π^- a) $\sigma_{tot}(D^+)$ vs A; b) $\sigma_{tot}(D^0)$ vs A; c) α vs x_F; and d) b vs A.

most 'general' features of charm production: the total cross section, the x_F dependence, the p_t dependence, and the possible existence of various special patterns in the inclusive charm production.

TOTAL CHARM CROSS SECTION

Experiments typically measure only a limited number of decay modes of the most abundantly produced charmed particles: $D^{\pm}$, D^0 and $\bar{D}^0$, D_s, and perhaps Λ_c. They also cover only a limited kinematic region. They are, therefore, dependent on the accuracy of branching fractions which so far have only been measured 'adequately' in e^+e^- colliders to extract total cross sections for the observed charmed particles. These must then be extrapolated to other kinematic regions in order to compare them to other experiments and to compare them with models which predict the total $c-\bar{c}$ cross section. Some kinematic regions have not been studied at all in the latest round of experiments. In particular, the region at high x_F, which was explored to some extent by very early experiments at the ISR, has not been revisited. If there is anomalous behavior in this region as suggested by some experiments and some theoretical models, the extrapolations using conventional Leading Order(LO) and Next to Leading Order(NLO) calculations given below will not tell the whole story.

Table 3 lists four measurements of the 'total' $c-\bar{c}$ cross section produced by incident protons[12]. The results are plotted in figure 8 and compared with the work of Altarelli[18], based on an α_s^3 QCD calculation. It can be seen that the experimental results are in good agreement with the calculation both as far as the rate of the rise with energy and the absolute scale. It is significant that agreement is achieved with a very 'reasonable' value for the charmed quark mass of 1.5 GeV/c^2 in the

Table 3. $\sigma_{c-\bar{c}}$ produced by incident protons vs incident proton energy.

exp.	energy (GeV)	cross section μb
NA32	200	$0.90 \pm 0.4 \pm 0.08$
NA27	400	18.8 ± 2.1
E743	800	30^{+6}_{-5}
E653	800	$38 \pm 3 \pm 14$

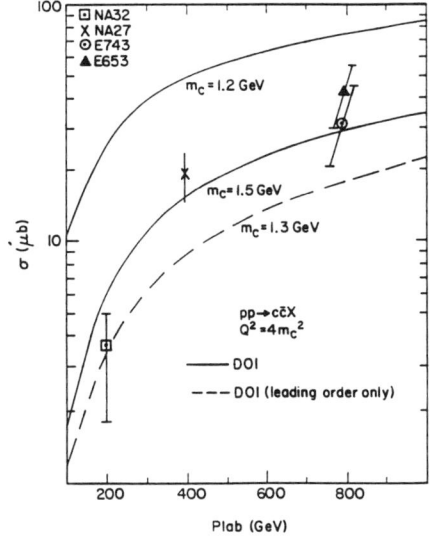

Figure 8. Data on $\sigma_{c-\bar{c}}$ vs energy for charm production by incident protons. The curves represent the calculations of Altarelli et al., for Leading and Next-to-Leading Order QCD.

NLO calculation. The extra α_s^3 diagrams add a large factor to the cross section. If one includes only LO graphs, one needs to use a very low value of the charmed quark mass to push the calculated cross section up high enough to agree with the data.

x_F DEPENDENCE OF CHARM PRODUCTION

The x_F dependence of charm particle production is usually parametrized as:

$$\frac{d\sigma}{dx_F} \propto (1-x_F)^n \qquad (9)$$

Figure 9 shows the values of n measured for the x_F distribution by various experiments[12] for incident pions. The power law dependence centers around 3.5 for pions but may be higher (steeper x_F distribution) for protons. There is little information for incident kaons. Figure 10 shows a comparison between a measured x_F distribution from E769 and the LO QCD prediction for **the charmed quark**[19]. While the agreement is good, this is deceptive because hadronization of the charmed quark is ignored in the calculation. If hadronization is taken into account, the data and LO calculation are in disagreement.

Figure 11 shows a similar result for the x_F distribution from WA82[14]. Along with the data (open circles) are plotted the results of a NLO calculation for bare charmed quarks (i.e. no hadronization) (triangles) and results obtained using Pythia and full string fragmentation of the charmed quarks (filled circles). The PYTHIA calculation actually manages to raise the cross section at high x_F. What appears to happen is that quarks produced at low values of x_F are actually 'dragged' to higher x_F by other partons in the collision– presumably those associated with the beam jet. In the model shown, this effect 'overcompensates' and produces a flatter distribution than is observed. Tuning of parameters can, no doubt, produce good agreement with the data.

It is natural to wonder whether the n value depends on the type of charmed particle being observed. Information on this point is not easy to come by since D_s mesons and charmed baryons are much harder to observe in these experiments than the D-mesons. Interest has focussed on whether there is a so-called leading particle effect. The alleged 'effect'[20] is that charmed particles which contain a light quark that can be supplied by a valence quark from the incident particle should have a stiffer x_F distribution than ones that have no quark in common with the incident particle's valence quarks. This means that in π^- produc-

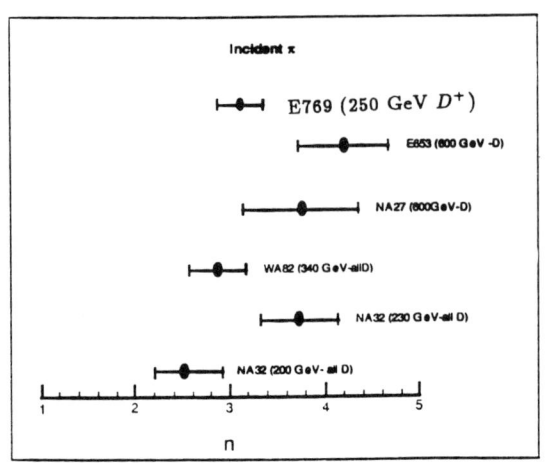

Figure 9. Power law dependence of x_F distribution for various charm final states for several incident particles and energies.

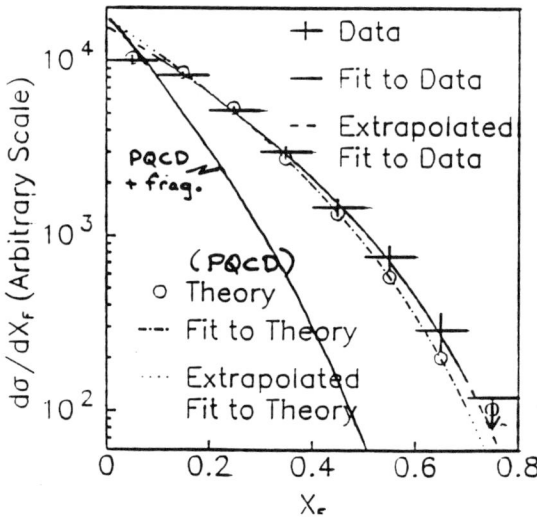

Figure 10. Comparison of measured x_F distribution from E769 and LO QCD calculation (for charmed quarks).

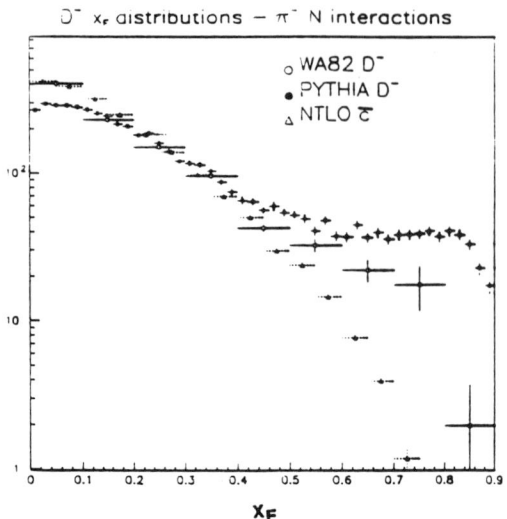

Figure 11. Comparison of measured x_F distribution from WA82, NLO QCD (for charmed quarks), and PYTHIA simulation with string fragmentation.

tion, D^0 and D^-, which are 'leading' should have a higher production cross section at high x_F than $\bar{D}^0$ and D^+, which are non-leading. Originally, these effects were claimed to be quite large and dramatic. The higher statistics experiments do seem to support a SMALL leading particle effect. This is summarized in fig. 12[21].

p_t DEPENDENCE

The p_t dependence of charm production has been measured by many experiments[12] and there is reasonable consistency between the results. The distributions are usually fitted with the function:

$$\frac{d\sigma}{dp_t^2} \propto \exp^{-bp_t^2} . \qquad (10)$$

The results are plotted in figure 13. Experiments are in good agreement and converge to a value of $\approx 1.0\,\text{GeV}/c^2$. The p_t behavior seems

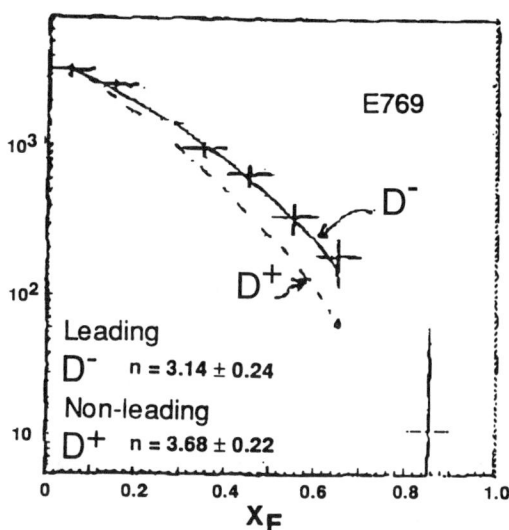

Figure 12. Power law dependence for x_F for leading and non-leading charmed mesons.

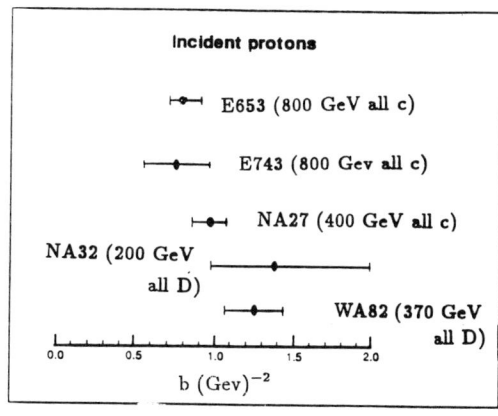

Figure 13. b-parameter of p_t^2 dependence measured in various experiments.

to be independent of x_F, incident particle type and energy, charm type observed, etc. The average value of p_t^2 is close to the expectations of LO QCD.

E769, which has larger samples of fully reconstructed charmed mesons than previous hadroproduction experiments, has examined the p_t distribution at values of p_t above 1 GeV/c. They find[21] that the cross section is no longer well described by the equation above but is better fit by a simple exponential in p_t:

$$\frac{d\sigma}{dp_t^2} \propto \exp^{-b'p_t} \qquad (11)$$

with b' having a value of $2.76 \pm 0.06\,\text{GeV}^{-1}$.

FULLY RECONSTRUCTED $D - \bar{D}$ PAIRS

Correlations between the two charmed(or bottom) particles in events can give additional information on the production mechanism for heavy quarks. However, very little data on such correlations exists. Most of the experiments do not reconstruct completely both charmed particles. This introduces uncertainties in the longitudinal momentum of at least one member of the pair. Therefore, the experiments present data only on angular correlations in the plane perpendicular to the beam direction. Little information is available on the longitudinal or transverse momentum of the pairs or of rapidity correlations between the two charmed particles.

E687, a photoproduction experiment at Fermilab, now has sample of more than 150 events in which both the charmed meson and its anti-charmed partner have been fully reconstructed[22]. This sample represents about 50% of the full dataset accumulated in the 1987,88,90, and 91 runs at Fermilab. Only three decay modes (and their charge conjugates) are used so far in the analysis:

$$\begin{aligned} D^0 &\to K^-\pi^+ \\ D^0 &\to K^-\pi^+\pi^-\pi^+ \\ D^+ &\to K^-\pi^+\pi^+. \end{aligned} \qquad (12)$$

Since the NLO calculation of the cross section is required to get good agreement with the measured total charm photoproduction cross section[23], it makes sense to compare the charm-pair correlations with the NLO ($\alpha_s^2\alpha_{em}$)

result. Unfortunately, very little theoretical work has been done on heavy quark correlations at next-to-leading order except at collider energies[24]. For that reason, we present comparisons between data and the LUND Monte Carlo[25].

Figure 14 shows the three anti-charmed particles when there is a charmed particle observed in the event. The quantity plotted is the 'normalized mass':

$$M_{norm} = \frac{(Mobs - Mpdg)}{\sigma_{Mobs}} \quad (13)$$

where M_{obs} is the mass of the particle, $\sigma_{M_{obs}}$ is its measurement uncertainty, and M_{pdg} is the mass according to the Particle Data Group listings (1992)[26]. Use of this quantity removes the differences in resolution and the slight offset of central mass values among the three states. Figure 15 is a scatterplot of M_{norm} for the charm vs the anti-charmed candidate. A clear signal for double charm production stands out over a small background. The background is subtracted out in every physics distribution which is presented by means of a special weighting technique.

Correlations in azimuthal angle and transverse momentum

In the parton picture of strong interactions, the two heavy quarks should be produced roughly back-to-back in a plane perpendicular to the partons' collision axis. Since the partons themselves carry little intrinsic transverse momentum and the hadronization process occurs mainly along the line of separation of the two heavy quarks, the heavy quark bearing hadrons should be emitted at 180° with respect to each other in a plane perpendicular to the beam. In E687, the acceptance in coplanarity angle is flat and the resolution is about 15 mrad (much less than the size of a bin). Figure 16 shows this azimuthal angle distribution for the two charmed mesons. There

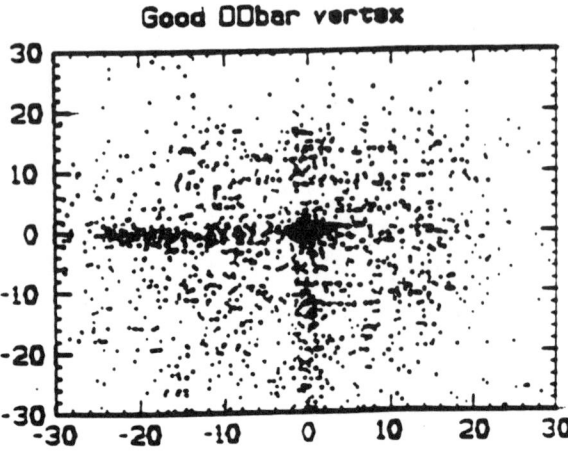

Figure 15. Scatter plot of normalized mass for charmed mesons vs normalized mass for anti-charmed mesons.

is a clear peaking at 180°. The data is well fit by the LUND Monte Carlo using a point-like photon although the azimuthal peaking is more pronounced in the Monte Carlo than in the data.

Because the momentum vectors of both charmed particles are known, it is possible to explore the transverse momentum balance of the two charmed mesons in hopes of learning something about the underlying collision. Figure 17 shows the distribution of the quantity

$$p_{dif}^{vec} = |\vec{p}_t(D) - \vec{p}_t(\bar{D})| \quad (14)$$

for the data and the Monte Carlo. This quantity depends on the dressing of the hadrons and so does not directly measure the magnitude of the mean transverse momentum of the two charmed quarks.

The quantity

$$p_{dif}^{mag} = ||p_t(D)| - |p_t(\bar{D})|| \quad (15)$$

measures the transverse momentum imbalance of the charmed pair. This quantity is related

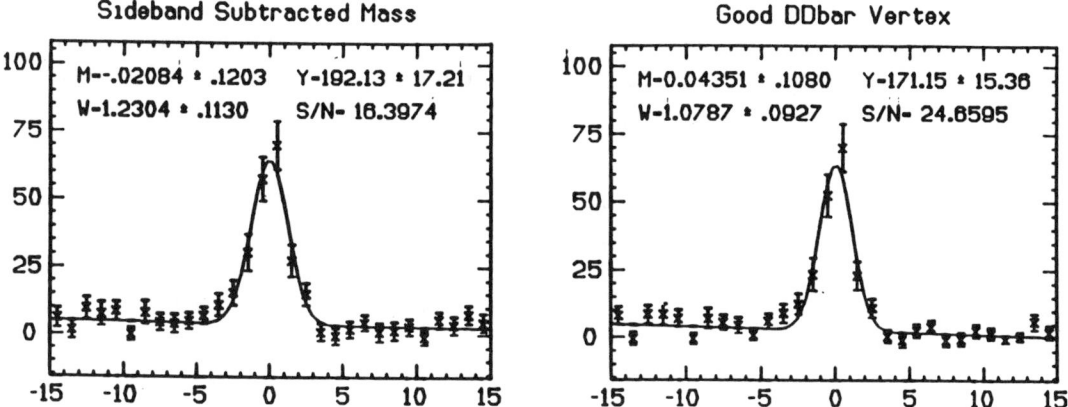

Figure 14. Normalized mass for anti-charmed mesons selected on a) normalized mass of charmed mesons and b) same distribution but with requirement that D and $\bar{D}$ track vectors intersect. Both distributions have background subtracted using sidebands.

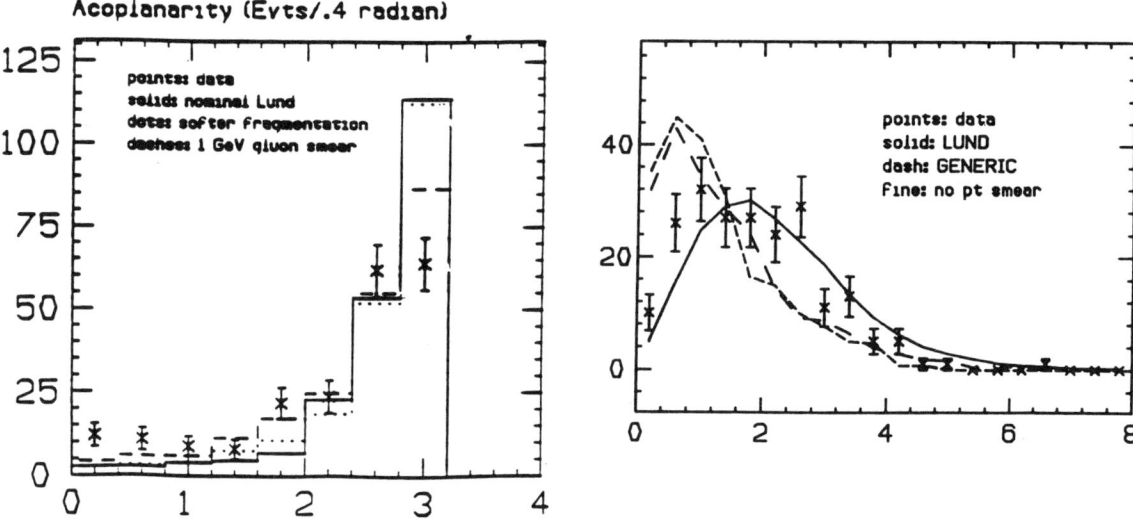

Figure 16. Distribution of relative azimuthal angle between charmed and anti-charmed meson.

Figure 17. Vector p_t difference of charmed and anti-charmed mesons.

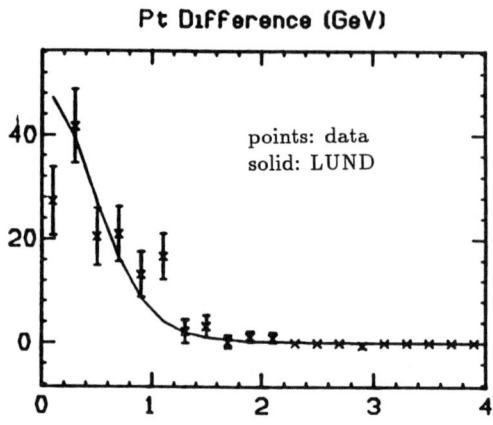

Figure 18. p_{dif}^{mag} of charmed and anti-charmed mesons.

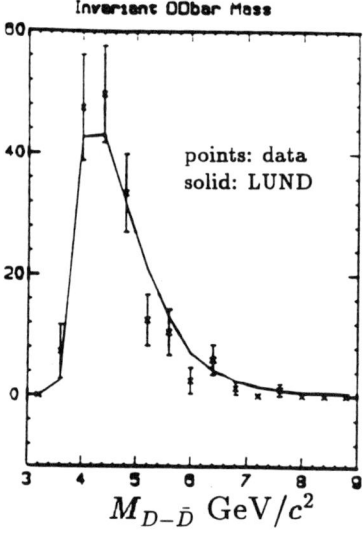

Figure 19. Invariant mass of the charm meson pair.

to the transverse momentum of the colliding partons but is also modified by the hadronization process. It is shown in figure 18.

Invariant Mass of the $D - \bar{D}$ pair

In the photon-gluon fusion process, the invariant mass, $M_{c-\bar{c}}$ of the produced heavy quark pair is related to the Feynman-x, x_g, of the gluon and the center of mass energy squared, s, as:

$$M^2_{c-\bar{c}} = s \times x_g \,. \quad (16)$$

For typical charmed-pair masses of order 5 GeV/c^2, gluons of $x_g \approx 0.05$ contribute. The invariant mass distribution of the charm pair is, of course, modified by the hadronization process. Figure 19 shows the invariant mass distribution of the charmed meson pair. It agrees very well with the Monte Carlo predictions.

Rapidity Correlations

There are expected to be rapidity correlations in heavy quark production. Typically, the two quarks are expected to be produced fairly close together in rapidity. Figure 20 shows the rapidity difference between the charmed and anti-charmed meson. This distribution is also influenced by the hadronization. Again, the agreement with the Monte Carlo is quite good.

Future Prospects for Correlation Studies

E687 will double its sample of fully reconstructed pairs of this kind. In addition, the use of various constraints should also permit the use of incompletely reconstructed charmed mesons coming from D^*'s opposite a fully reconstructed meson. Eventually, this study can be carried out with approximately 2000 events. Inclusive distributions will contain several tens of thousands of events. Correlations can re-

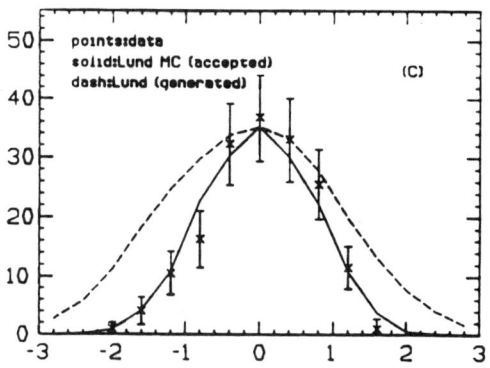

Figure 20. Rapidity difference between charmed and anti-charmed mesons.

veal many aspects of the underlying parton collision. Several other charm experiments at Fixed Target energies may also have data samples like this. In addition, HERA is likely to produce results on inclusive charm photoproduction and possibly reconstructed pairs. It would be very useful to have NLO calculations of all these distributions for photoproduction at these energies and a clear understanding of how to extract information on parton distributions from this type of data. Work along these lines has been started for photoproduction at HERA and is described in a paper by Ali et al.[27], submitted to this conference.

RECENT PROGRESS IN CHARMED BARYON PHYSICS

Knowledge of charmed baryons has lagged behind that of D-mesons for several reasons:

- For charmed baryons, there is no experimental situation in e^+e^- collisions which is as favorable as the $\psi(3770)$ resonance is for charmed mesons;

- The cross sections for charmed baryons are generally smaller than for D-mesons;

Table 4. Relative information on charmed states.

Charm state	Pages in PDG92
D^0, D^+	21
D_s	5
Charmonium	19
Charmed Baryons (all)	5

- The lifetimes of charmed baryons are shorter than those of D-mesons making the use of secondary vertex reconstruction to isolate the charm decay more difficult and less efficient; and

- Many of the decay modes involve neutrons, hyperons, and states that contain neutrals, which are hard to reconstruct.

One striking way to demonstrate this is to simply compare the number of pages of information on charmed mesons in the 1992 Particle Data Group listing[26] for D-mesons to the number for charmed baryons. This comparison is shown in table 4. The situation is even worse than indicated by this comparison because the quality of the results for D-mesons is typically much higher than for charmed baryons. This is illustrated by comparing the errors on the lifetime measurements for the charmed mesons, D^0 and D^+, and the Λ_c. The errors differ by nearly a factor of 5! Other members of the charm baryon family have been barely estabished or not observed at all.

Recently, new data have become available from three experiments which have detectors that can overcome some of the problems noted above and which have completed high luminosity runs. These experiments are the two e^+e^- expereiments, CLEO and ARGUS, and Fermilab photoproduction experiment E687. We review these data below.

Λ_c absolute branching fraction

The Λ_c decay mode which is easiest to see is $pK\pi$. It is a frequent practice to normalize branching fractions of other decay modes to this one. At least at the moment, only e^+e^- facilities have been able to make an attempt at determining the absolute branching fraction for $\Lambda_c \to pK\pi$. The most recent result is from ARGUS[28]. They use the inclusive p and Λ yield on the $\Upsilon(4S)$ to extract the 'inclusive branching fraction', $B_b = BR(B \to \text{baryons})$ with the result:

$$B_b = 6.8 \pm 0.5 \pm 0.3\% . \quad (17)$$

CLEO[29] gets a similar result by a somewhat different method:

$$B_b = 6.4 \pm 0.8 \pm 0.8\% . \quad (18)$$

This baryon yield is entirely attributed to the $\Upsilon(4S)$ decaying to B-mesons which then decay to Λ_c's. Using the measured rate on the $\Upsilon(4S)$ for Λ_c production,

$$BR(\Lambda_c+X)\times BR(\Lambda_c \to pK\pi) = (0.23\pm0.05)\% , \quad (19)$$

one arrives at a final result of

$$BR(\Lambda_c \to pK\pi) = (4.0 \pm 0.3 \pm 0.8)\% . \quad (20)$$

Status of charm-strange baryon spectroscopy

Ω_c

Until recently, very little information existed on the baryon with one charm and two strange quarks. The WA62 Collaboration[30] at CERN reported 3 events in the state $\Xi^-K^-\pi^+\pi^+$ at a mass of $2740 \pm 20\,\text{MeV}/c^2$. Recently, two experiments have reported observations of this state[31],[32].

i. <u>ARGUS observation of $\Omega_c \to \Xi K\pi\pi$</u>

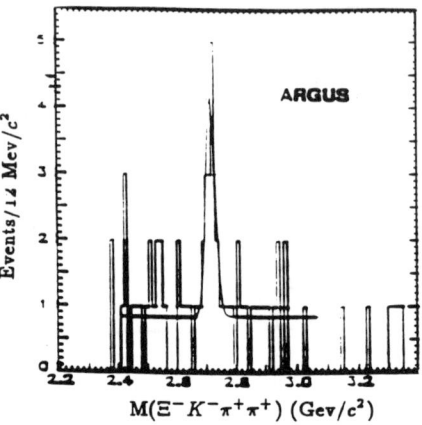

Figure 21. Invariant mass distribution for $\Omega_c \to \Xi K\pi\pi$.

Figure 21 shows the invariant mass plot for $\Xi^-K^-\pi^+\pi^+$. The mass value of the peak is $2719.0 \pm 7.0\,\text{MeV}/c^2$ with a width of $16.6 \pm 6.3\,\text{MeV}/c^2$, consistent with the experimental mass resolution.

ARGUS also reported at this conference the observation of a signal of 5 events in the channel $\Omega^-\pi^+\pi^-\pi^+$ with a mass of 2713 ± 7 MeV/c^2 and now reports and average mass of $2715.5\pm5\pm5$ MeV/c^2[33].

ii. <u>E687 observation of $\Omega_c \to \Omega\pi$</u>

Fermilab Experiment 687 has reported a signal for the decay mode

$$\Omega_c^0 \to \Omega^-\pi^+ . \quad (21)$$

This signal is shown in figure 22. The mass value is $2707.0\pm2.2\pm5.0\,\text{MeV}/c^2$. The width is $7.0 \pm 0.7\,\text{MeV}/c^2$, which is consistent with the spectrometer resolution. This is in reasonable agreement with the ARGUS value of the mass.

Ξ_c

The Ξ_c contains one charm quark, one strange quark, and either a u quark for charge

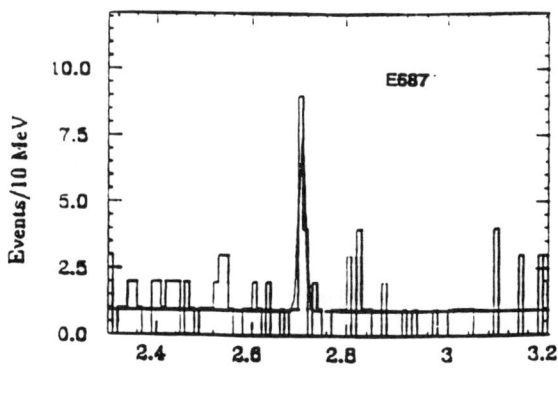

Figure 22. Invariant mass distribution for $\Omega_c \to \Omega^-\pi^+$.

Table 6. Summary of Decay Modes for Λ_c from CLEO

decay mode	CLEO 1.5	CLEO II
$pK^-\pi^+$	1.0	
pK^0	$0.44 \pm 0.07 \pm 0.05$	
$pK^0\pi^+\pi^-$	$0.43 \pm 0.12 \pm 0.04$	
$\Lambda^0\pi^+$	$0.18 \pm 0.03 \pm 0.03$	
$\Sigma^0\pi^+$		0.17 ± 0.3
$\Lambda^0\pi^+\pi^0$		0.41 ± 0.06
$\Sigma^0\pi^+\pi^0$		0.21 ± 0.05
$\Lambda^0\pi^+\pi^-\pi^+$	$0.65 \pm 0.11 \pm 0.12$	
$\Sigma^0\pi^+\pi^-\pi^+$		0.08 ± 0.02
$\Xi^-K^+\pi^+$	$0.15 \pm 0.04 \pm 0.03$	
$\Xi^0 K^+$		$0.066 \pm 0.013 \pm 0.008$
$\Sigma^+K^+K^-$		$0.092 \pm 0.015 \pm 0.011$

Table 5. Summary of Ξ_c States.

experiment	decay mode	mass MeV/c^2	no. of events
Ξ_c^0:			
CLEO	$\Xi^-\pi^+\pi^0$	2467.5 ± 2.5	37 ± 9
CLEO	$\Omega^- K^+$	2470.6 ± 1.5	7.7 ± 3.1
E687	$\Xi^-\pi^+$	2472.1 ± 5.2	20.5 ± 9.2
Ξ_c^+:			
CLEO	$\Xi^0\pi^+\pi^0$	2468.8 ± 4.1	31 ± 9
E687	$\Xi^-\pi^+\pi^+$	2469.1 ± 2.6	23.9 ± 5.7
E687	$\Omega^-K^+\pi^+$	2468.8 ± 5.8	11.9 ± 5.9
E687	$\Lambda^0 K^-\pi^+\pi^+$	2463.5 ± 4.7	12.2 ± 4.7

+1 or a d quark for charge 0. CLEO and E687 have each made new observations of the Ξ_c^0 and the Ξ_c^+ states. Table 5 shows the various states that have been detected and are now being studied. The mass values are consistent within the errors between the two experiments. The observed widths are consistent with resolution in both experiments. Mass plots are shown for CLEO in figure 23 and for E687 in figure 24. E687 will soon have a preliminary value of the Ξ_c lifetime.

Recent results on Λ_c and Σ_c properties

Λ_c decay modes

CLEO has reported on a large number of Λ_c decay modes. Many of these involve photons or π^0's in the final state and are now observable using CLEO's cesium iodide electromagnetic calorimeter. E687 has also observed Λ_c's decaying into several modes.

Table 6 shows a compilation of recent CLEO results on Λ_c decays. All numbers for branching fractions are relative to the branching fraction to $pK\pi$. If we use the number for the absolute branching fraction to $pK\pi$ from ARGUS, we find that about 15% of the total decays of the Λ_c are now accounted for.

Note that decays to Λ^0 are favored over states with Σ^0's and that this effect grows as the associated multiplicity grows. This may be explained if the decays occur predominantly through Y*'s.

The decay mode $\Xi^0 K^+$, shown in figure 25, and the decay $\Xi^0 K^+ K^-$ are possible examples of W-exchange. Charmed baryon decays are thought to be good places to look for W-

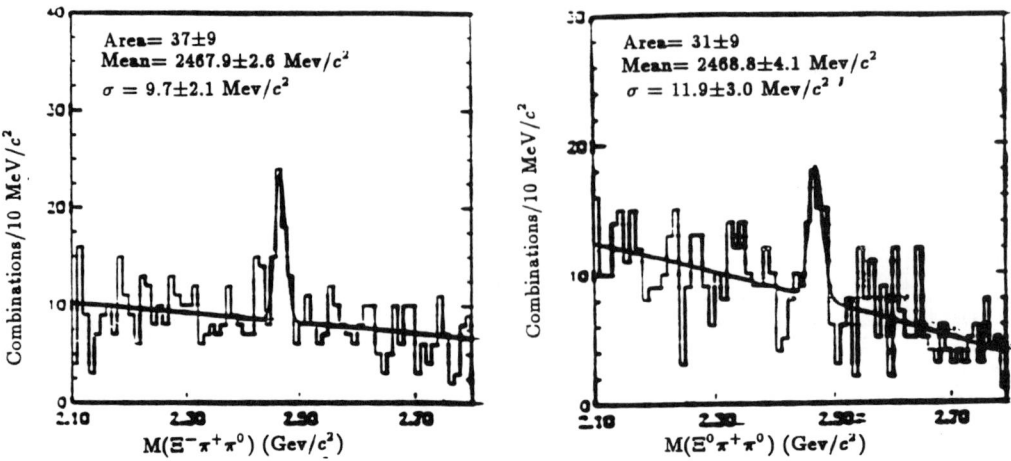

Figure 23. Invariant mass plots for Ξ_c states observed in CLEO.

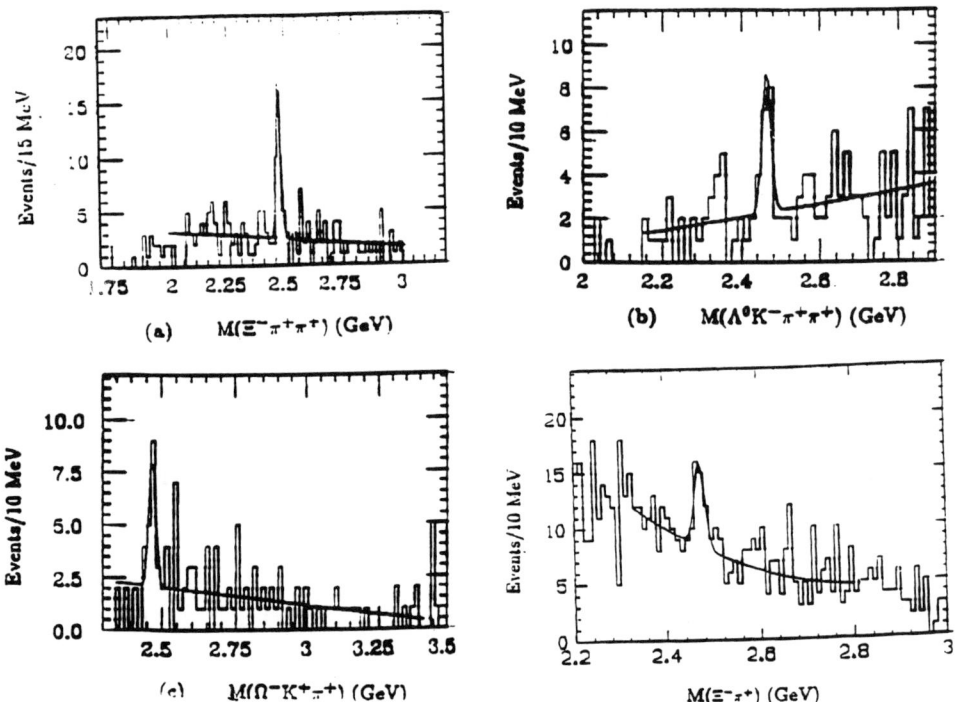

Figure 24. Invariant mass plots for Ξ_c states observed in E687.

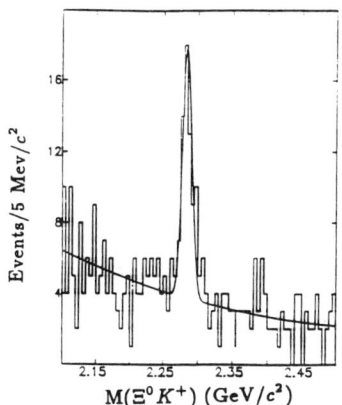

Figure 25. Invariant mass plot for $\Xi^0 K^+$ showing a Λ_c signal from CLEO.

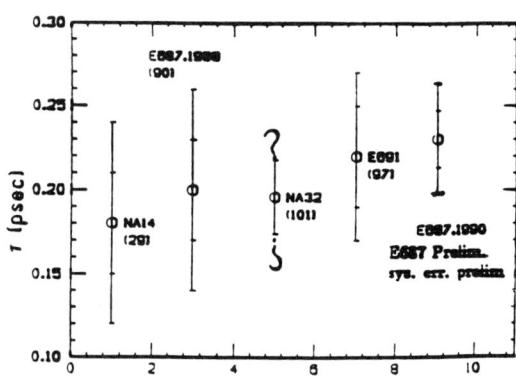

Figure 27. Results on Λ_c lifetime including E687 preliminary result.

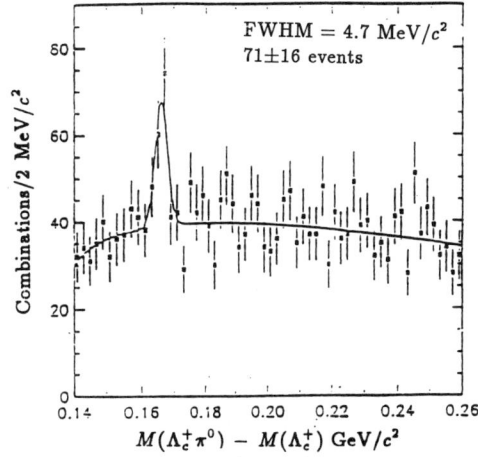

Figure 28. $\Delta m(\Sigma_c^+ - \Lambda_c^+)$ CLEO II.

exchange contributions since they are neither color nor helicity suppressed. However, it is true that final state interactions from non-exchange decays can also produce these final states.

Figure 26 shows preliminary signals from E687 for two Cabibbo suppressed decay modes of the $\Lambda_c - pK^-K^+$ and $p\pi^+\pi^-\pi^+\pi^-$ and 2 other modes–$pK^-\pi^+\pi^0$ and $\Xi^-K^+\pi^+$, which are hard to detect. Branching fractions are being worked out and will be available soon.

Λ_c *lifetime*

E687 has a new PRELIMINARY result for the Λ_c lifetime. The result is derived only from the $pK\pi$ decay mode and reflects about 70% of the data collected in 1990/91. The value shown here will be indicative of the final statistics that may be expected. The systematic errors are expected to improve with further study. The result is:

$$\tau_{\Lambda_c} = 0.230 \pm 0.017 \pm 0.019\,ps \ . \qquad (22)$$

Figure 27 summarizes the measurements of the Λ_c lifetime.

Σ_c *mass differences*

The three members of the charmed baryon isotriplet, the Σ_c's, have also been observed recently by CLEO. In particular, CLEO has a good measurement of the mass of the Σ_c^+. This requires the detection of a cascade π^0 and has been very hard to observe. Their mass difference distribution is shown in figure 28. Their numbers are summarized in table 7.

Recent results from Serpukhov

The EXCHARM collaboration at Serpukhov has reported new results on Ξ_c decays[34]. Charmed baryons are produced near threshold by neutrons of energies between 20 and 60 Gev and are observed in

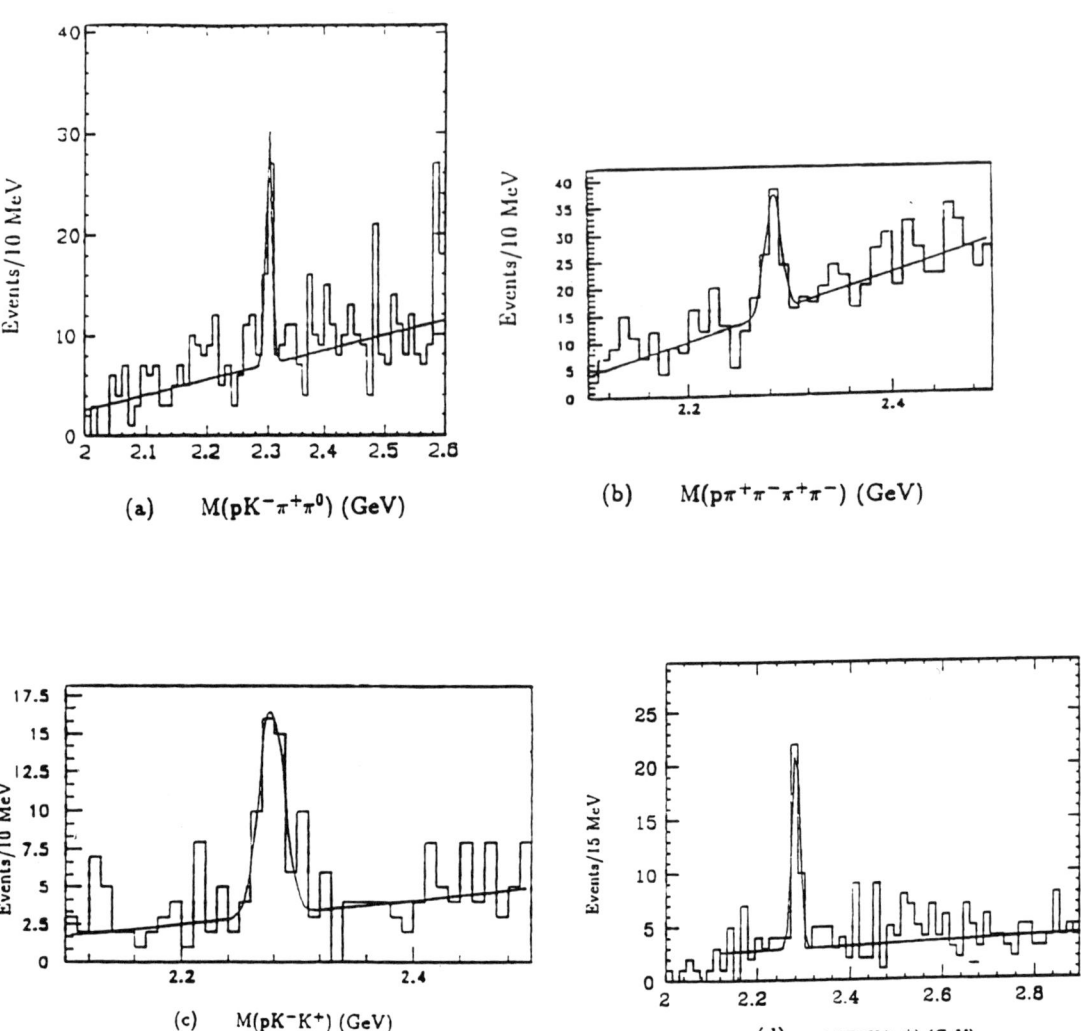

Figure 26. Invariant mass plots for Λ_c states observed in E687.

Table 7. CLEO results for mass differences between Σ_c states and the Λ_c.

difference	CLEO II MeV/c^2
$\Delta m(\Sigma_c^{++} - \Lambda_c)$	167.3±0.5±0.4
$\Delta m(\Sigma_c^0 - \Lambda_c)$	167.3±0.6±0.4
$\Delta m(\Sigma_c^+ - \Lambda_c)$	166.0±0.5±0.2
$\Delta m(\Sigma_c^{++} - \Sigma_c^0)$	0.0±0.8±0.4
$\Delta m(\Sigma_c^+ - \Sigma_c^0)$	-1.3±0.7±0.4

Table 8. EXCHARM results on Ξ_c branching fractions

decay mode	relative branching fraction
Ξ_c^+ :	
$\Lambda^0 K^- \pi^+ \pi^+$	1.0
$p\bar{K}^0 K^- \pi^+$	1.98± 0.76
including :	
$p\bar{K}^0 \bar{K}^{*0}$	0.44±0.23
$\Lambda^* \bar{K}^0 \pi^+$	1.45±0.62
$\Lambda \bar{K}^{*0} \pi^+$	0.43±0.15
Ξ^0 :	
$\Lambda^0 K^- \pi^+$	1.0
$\Lambda^0 \bar{K}^{*0}$	>0.86

the BIS-2 spectrometer. The results support a 'diffractive' production mechanism where a charmed baryon and an anti-charmed meson are produced in the reaction. The Ξ_c^0 was observed in the final state $\Lambda^0 K^- \pi^+$ and the Ξ_c^+ was observed in the states $\Lambda^0 K^- \pi^+ \pi^+$ and $pK_s^0 K^- \pi^+$. Branching fraction results are shown in table 8. Results were also presented on the up-down asymmetry of the K^- in the $pK_s^0 K^- \pi^+$ decay. From the observed asymmetry, the polarization of the Ξ_c^+, produced near threshold, is shown to be greater than 10% at the 90% confidence level.

Outlook for Charm Baryon Physics

There has been a major leap in our knowledge of charm baryon decays and spectroscopy in the last year. Issues like the lifetime hierarchy, the mass spectrum, and the detailed pattern of hadronic decays should now be reexamined theoretically in the light of the new data and the data that is likely to come as analysis continues on existing datasets, experiments which already have data begin reporting results, and new experiments start up. **THIS IS A FIELD WHOSE TIME HAS COME.**

SELECTED ISSUES IN D^0 AND D^+ DECAYS

There are too many new results on D-meson properties to discuss them all here. Below, a few highlights are presented.

D^* branching fractions

Although the spin one charmed mesons– the D^*'s– have been known for more than 15 years, their branching fractions have been recently remeasured in CLEO II[35] with some surprising results. The most significant one is that the 'anomalously high' branching fraction for $D^{*+} \to D^+ \gamma$, reported by MARK III, is now contested by CLEO II. They report a very small branching fraction for this decay– one much more in line with theoretical expectations of 1-2%[36]. Of course, this causes a large increase in the branching fractions for the other two prominent decay modes of the D^{*+}. However, CLEO II also reports very different branching fractions for D^{*0} decays. Table 9 compares recent PRELIMINARY results from CLEO to the values in PDG92[26].

Since many experimental observations involve D^{*+} decays, one does have to keep this situation in mind and many results may need to be revised if these new branching fractions do indeed turn out to be correct.

$D^0 \to \pi^+ \pi^-$ and $D^0 \to K^+ K^-$

Table 9. Branching fractions for D^* mesons.

decay mode	PDG92	CLEO II(prelim)
D^{*+}:		
$D^0\pi^+$	0.54 ± 0.05	$0.68 \pm 0.014 \pm 0.024$
$D^+\pi^0$	0.271 ± 0.028	$0.31 \pm 0.004 \pm 0.016$
$D^+\gamma$	$0.17 \pm 0.05 \pm 0.05$	$0.01 \pm 0.013 \pm 0.005$
D^{*0}:		
$D^0\pi^0$	0.55 ± 0.06	$0.64 \pm 0.024 \pm 0.045$
$D^0\gamma$	0.45 ± 0.06	$0.36 \pm 0.024 \pm 0.045$

Table 10. Results on ratio of K^+K^- to to $\pi^+\pi^-$ branching fractions of D^0.

experiment	$\frac{\Gamma(K^+K^-)}{\Gamma(\pi^+\pi^-)}$
MARK II	3.5 ± 1.8
MARK III	3.7 ± 1.4
ARGUS	2.5 ± 0.7
CLEO	$2.35 \pm 0.37 \pm 0.28$
E691	$1.95 \pm 0.34 \pm 0.22$
WA82	$2.23 \pm 0.81 \pm 0.46$

The ratio of these two decays is a calibration of the degree of flavor symmetry breaking. The current results are shown in table 10. The more recent experiments seem to favor a value close to 2, lower than indicated by the early experiments. For a long time, it was thought that this ratio presented a problem for models of hadronic decays of charmed particles such as that of Bauer, Setch, and Wirbel[37]. However, this ratio can now be reasonably well accounted for by such models provided final state interactions are taken into account[38].

Double Cabibbo Suppressed decays of the D

Double Cabibbo suppressed decays (DCSD) are expected to occur at the level of $tan^4\theta_c$ of Cabibbo favored decays. The sensitivity of current experiments is just at threshold for observing these decays, whose branching fractions into specific modes can be enhanced or suppressed by hadronic effects. E691[39] reports a signal for $D^+ \to \phi K^+$ and WA82 reports a signal in $D^+ \to K^+K^-K^+$. The E691 result gives

$$\frac{\Gamma(D^+ \to \phi K^+)}{\Gamma(D^+ \to K^-\pi^+\pi^+)} = 0.52^{+0.29}_{-0.24} \pm 0.08\% \ . \quad (23)$$

CLEO, which also looked for this decay mode, reports only a 90% confidence level upper limit of 0.57%[35]. The two results are just barely compatible. It is worth noting that, in addition to being Doubly Cabibbo Suppressed, these decay modes also must procede through either an annihilation diagram or a rescattering process.

ARGUS[41] has looked for the DCSD decay $D^0 \to K^+\pi^-$ and E687[42] has looked for $D^+ \to K^+\pi^-\pi^-$. Neither group sees evidence for these decays. Their respective 90% confidence level limits are 0.9% and 0.7%.

$D^0 \to \bar{K}^{*0}\eta$ branching fraction

The $\bar{K}^0\phi$ decay mode of the D^0 was put forth as an example of an exchange diagram[43]. The branching fraction, measured to be about 0.8%[44] is considered to be higher than expected for such a process. It was pointed out by Donoghue[45] that this final state could be fed by final state interactions through the normal spectator decay $D^0 \to \bar{K}^{*0}\eta$. CLEO has now quoted a **preliminary** result for this process[35]:

$$Br(D^0 \to \bar{K}^{*0}\eta) = 1.7 \pm 0.3 \pm 0.4\% \ . \quad (24)$$

This branching fraction is probably big enough to explain the large $\bar{K}^0\phi$ branching fraction. This illustrates clearly the difficulty of extracting quark level information about weak decays in the presence of strong interactions.

Table 11. D_s states from CLEO II containing η's.

decay mode	$\dfrac{BR}{BR(\phi\pi^+)}$
$\eta\pi^+$	$0.54\pm0.09\pm0.06$
$\eta'\pi^+$	$1.20\pm0.15\pm0.11$
$\eta\rho^+$	$2.86\pm0.38\pm^{+0.36}_{-0.38}$
$\eta'\rho^+$	$3.44\pm0.62\pm^{+0.44}_{-0.46}$
$\phi\rho^+$	$1.86\pm0.26\pm^{+0.29}_{-0.40}$

SELECTED ISSUES IN D_s DECAYS

The CLEO detector, exploiting the γ and π^0 reconstruction capabilities of its cesium iodide detector, have observed a large number of D_s decays modes involving η mesons, η' mesons, and charged ρ mesons[46]. Note that about 79±26% of all D_s decay modes are now accounted for. Table 11 shows the branching fractions for some of the decay modes which CLEO has observed.

Comparison of hadronic D_s decays with theory and the $\eta - \eta'$ puzzle

The most widely accepted theoretical framework for explaining two body hadronic decays of mesons has been the Bauer, Stech, Wirbel(BSW) model. The recent data on D_s decays provide a new arena in which this model may be tested. This model has been characterized as being 'moderately successful' in describing hadronic charm decays. Improvements to the model which take into account in detail final state interactions have fixed some of the early disagreements with data as discussed above. One particularly distressing area of disagreement is in the relative branching fractions of the D_s into states with η's and η' mesons. These decays are expected to be dominated by the spectator diagram. Both the η and η' contain a mixture of strange and non-strange quarks. Of the two, the η contains a larger 'strange' component and is lighter in mass leaving a larger phase space for decay. Thus, the η should be always favored over the η'. For the two modes observed, the opposite is true. This indicates that some other diagram must contribute to this process.

One explanation, due to Lipkin[47], is that the annihilation diagram interferes with the spectator diagram constructively in the case of the η' decay and destructively in the case of the η decay. An annihilation amplitude of about 20% of the spectator amplitude would account for the observed effect without causing problems in any other measured property of D_s decays. Such a contribution is not inconsistent with measurements of the annihilation amplitude[48].

SUMMARY AND CONCLUSIONS

- B-physics is already happening at hadron colliders and the data from the new run at Fermilab are eagerly awaited. The initial experience with a microvertex detector in hadron collider should indicate what problems need to be solved before one can begin to exploit the potential of such machines as B-factories. Much better measurements of dynamical distributions will be available and theorists should start to review the status of QCD calculations and all required experimental inputs such as gluon distributions in preparation.

- Charmed Baryon physics has made a big leap forward and merits the renewed attention of theorists.

- The uncertainty and confusion with respect to branching fractions continues. Those who are in a position to help remedy the problem should do so. Uncertainty in branching fractions are often the dominant uncertainty in measurements.

- A comprehensive review of the status of theoretical models of hadronic decays should probably be undertaken in light of all the new data on charm and bottom decays. Inclusion of final state interactions seems definitely to help remove some of the known disagreements between the models and the data.

Many new results have appeared in the last year. Many of the experiments are still in the process of analyzing their data. Several experiments are just beginning to get results, including E791 , E771, and E789 at Fermilab. A new collider run has just started at the Fermilab Tevatron. Preparations are underway for a new round of experiments. CLEO will once again upgrade its detector by adding a microverrtex detector. We should soon begin to get results from HERA. Experiments aimed specifically at charmed-strange stated by using hyperon beams are underway at CERN (WA89) and are being prepared at Fermilab (E781). All-in-all we face the prospect of even more data in the next few years and hopefully many problems will be resolved.

Finally, I have to conclude by saying that we all hope that the new collider run will reveal the illusive top quark. The mass limit set by CDF[49] is now 91 GeV/c^2 and the constraints imposed by the Standard Model ElectroWeak analysis of LEP data point to a mass of around 150 GeV/c^2. Figure 29[50] shows the mass reach that will be achieved in a collider run on integrated luminosity of 100 pb^{-1}. we may hope that the next time this conference convenes the large amount of data that we expect for charm and bottom will be presented in a talk on 'medium heavy' quarks– a talk which will be preceded by one on the properties of a really heavy quark.

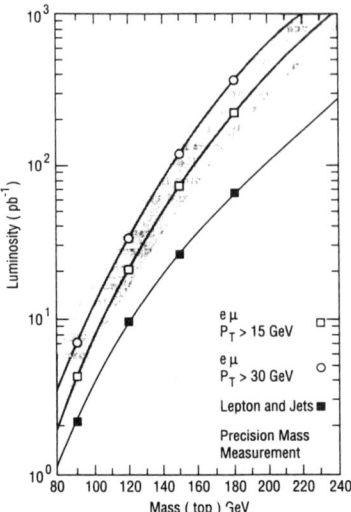

Figure 29. Mass sensitivity of top quark search in next Tevatron run.

ACKNOWLEDGEMENTS

Many people helped in the preparation of this report. I'd like to thank all my colleagues on E687. I drew specifically on papers and talks by J. Wiss, J.P. Cumalat, P.D. Sheldon, L. Moroni, N. Cason, H. Cheung, and R. Gardner. I'd like to thank Keith Ellis, Ed Berger, Sandip Pakvasa, and A. Kamal for useful discussions of theory. John Skarha, Paul Tipton, and G.P. Yeh helped me understand CDF data. J. Bartelt, Ich Jo Kim, V. Jain, and Ed Thorndike helped me with CLEO data. Ken Edwards provided me with a summary of recent ARGUS results. I drew heavily on a recent review and discussions with Jeff Appel. P. Karchin and F. Antinori explained their charm hadroproduction results to me. I'd like to thank P. Drell and S. Bethke for their cooperation and advice in choosing the subject matter so there was little repetition in the various talks. Finally, I want to thank Pat Mascione for all her help, Phil Stebbings for his work on the drawings, and my family for their patience.

REFERENCES

1. Results on charm semileptonic decays are discussed in the article by P. Drell in these proceedings.
2. CDF Collaboration, F. Abe et al., *Phys. Rev. Lett.*, 68, 3403(1992); M. J. Shochet, Fermilab-Conf-91/341-E(1991), and A Measurement of the B-meson and b-Quark Cross Sections at $\sqrt{s} = 1.8$ TeV Using the Decay $B^{\pm} \to J/\psi K^{\pm}$, CDF collaboration; A. Barbaro-Galtieri, Heavy Flavor Physics at Hadron Colliders, Proceedings of the 3rd Topical Seminar on Heavy Flavor Physics, San Miniato, Italy, 1991.
3. UA1 Collaboration, C. Albajar et al., *Phys. Lett.*, B256, 12(1991).
4. E. L. Berger, R. Meng, W. Tung, Implications of the Bottom Quark Cross Section Data at Hadron Colliders, ANL-HEP-PR-92-32, April 17, 1992.
5. P. Nason, S. Dawson, and R.K. Ellis, *Nucl. Phys.*, B203, 607(1988); B327, 49(1989).
6. See V. Papadimitriou, Proceedings, this conference.
7. E. W. N. Glover et al., *Z. Phys.*, 38C, 473 (1988).
8. Private conversations, J. Skarha, P. Tipton, and G. P. Yeh of the CDF collaboration.
9. D. Potter, Charmed meson production in 600 GeV/c π-emulsion interactions, presented at this conference.
10. R. Jesik et al., A Search for Beauty-Meson Production in π^{-}-Nucleon Collisions, paper presented at this conference.
11. For example, F.E. Low, *Phys. Rev. D*, 12, 163(1975); S. Nussinov, *Phys. Rev. Lett*, 34, 1286 (1975) or J. F. Gunion and D. Soper, *Phys. Rev. D*, 15, 2167(1977).
12. J. A. Appel, Hadroproduction of Charm Particles, Fermilab Pub-92/49, Feb. 1992, to appear in Annual Review of Nulcear and Particle Science, 1992.
13. D. S. Barton et al., Experimental study of the A dependence of inclusive hadron fragmentation, *Phys. Rev. D*, 27, 2580(1983).
14. F. Antinori, Results on Charm Hadroproduction from CERN Experiment WA82, presented at this conference.
15. D.M. Alde et al., A Dependence of J/ψ and ψ' Production at 800 GeV/c, *Phys. Rev. Lett.*, 66, 133(1991).
16. M. D. Sokoloff et al., Experimental Study of the A Dependence of J/ψ Photoproduction, *Phys. Rev. Lett.*, 57, 3003(1986).
17. see, for example, S.J. Brodsky and A.H. Mueller, *Phys. Lett.*, B206, 685 (1988).
18. G. Altarelli et al., *Nucl. Phys.*, B308, 724 (1988).
19. P. Karchin, paper presented at this confernence.
20. Aguilar-Benitez et al., *Phys. Lett. B*, 161, 400 (1985).
21. G.A. Alves et al., Feynman-x and Transverse Momentum Dependence of $D^{\pm}$ and $D^0, \bar{D}^0$ Production in 250 GeV π^{-}-Nucleon Interactions, FERMI-PUB-92/208-E, submitted to Phys. Rev. Lett.
22. R. Gardner, Photoproduction of Charmed Mesons, this conference.
23. R.K. Ellis and P. Nason, QCD Radiative Corrections to the Photoproduction of Heavy Quarks, *Nucl. Phys. B*, 312 (1989) 551.
24. M. Mangano et al., *Nucl. Phys.*, B373, 295 (1992).
25. T. Sjostrand, "The Lund Monte Carlo for Jet Fragmentation," *Computer Phys. Comm.*, 39, 347 (1986); and H. U.

Bengtsson, "The LUND Monte Carlo for Hadronic Processes", *Computer Phys. Comm.*, 46 43 (1987). The Monte Carlo used was PYTHIA 5.6 and JETSET 7.3.

26. The Particle Data Group, Review of Particle Properties, *Phys. Rev. D.*, 45, (1992).

27. A. Ali, Heavy Quark Physics at HERA, to be published in the Proceedings of the 1992 Zeuthen Workshop on Elementary Particle Theory: Deep Inelastic Scattering, Teupitz/Brandenburg, April, 1992.

28. ARGUS Collaboration, H. Albrecht et al., Measurement of Inclusive Baryon Production in B Meson Decay, DESY 92-074, May 1992.

29. G. Crawford et al., *Phys. Rev.*, D45 (1992) 752.

30. S. Biagi et al., *Z.Phys.*, C28 175 (1985).

31. ARGUS Collaboration, H. Albrecht et al., First Evidence for the Production of the Charmed, Doubly Strange Baryon, Ω_c, in e^+e^- Annihilation, DESY 92-052, March 1992.

32. H. Cheung, High Energy Photoproduction of Charmed Baryons, this conference.

33. See report by H. Schroder, this conference.

34. See report by V. D. Kekelidze, this conference.

35. the CLEO collaboration, F. Butler et al., CLNS 92/1143, CLEO 92-03, July 1992.

36. G.A. Miller and Paul Singer, *Phys. Rev.*, D37, 2564 and references 1-6 therein; E. Angelos and G. P. Lepage, *Phys. Rev.*, D45, 3021 (1992).

37. M. Bauer, B. Stech, M. Wirbel, *Z. Phys.*, C- Particles and Fields, 34 (1987).

38. A. Czarnecki, A.N. Kamal, Qiping Xu, *Z. Phys.*, C- Particles and Fields, 54, 411-417 (1992).

39. J. Anjos et al., FERMILAB-Pub-91/331; CBPF-NF-036-91.

40. Shown at Physics in Collision, invited talk by Paul Sheldon, Boulder, Colorado, June 1992.

41. H. Albrecht et al., A Search for $D^0 \to K^+\pi^-$, DESY-92-056.

42. see N. Cason, Rare Decay Modes of the D^0, D^+, and D_s Charmed Mesons, in these proceedings.

43. I.I.Y. Bigi and M. Fukugita, *Phys. Lett.*, 91B 121(1980).

44. Frabetti et al., *Phys. Lett.*, B286, 198(1992); Albrecht et al., *Z. Phys.*, C33, 359(198); Bebek et al., *Phys. Rev. Lett.*, 56, 1893 (19); Barlag et al., Phys. Lett., 232B, 561 (19).

45. J. Donoghue, *Phys. Rev.* D33, 1516(1986).

46. CLEO collaboration, P. Avery et al., *Phys. Rev. Lett.*, 68, 1279 (1992); J. Alexander et al., *Phys. Rev. Lett.*, 68, 1275 (1992).

47. H.J. Lipkin, *Phys. Lett.*, B283 421 (1992).

48. J. C. Anjos et al., *Phys. Rev. Lett.*, 62, 125(1989); N. M. Cason, contributed paper, this conference.

49. CDF Collaboration, F. Abe et al., *Phys. Rev. Lett.*, 68, 447 (1992).

50. D. Green and H. Lubatti (ed), *Physics at Fermilab in the 1990's*, p198, World Scientific, Singapore, 1990.

DISCUSSION

B. Ward, University of Tennessee, USA

This is a comment about $\Gamma(D \to K^+K^-)/\Gamma(D \to \pi^+\pi^-)$. Three different calculations by Prof. Kamal's group, by Prof. Chou and Dr. Cheng and by me, using coupled channel, quark-diagrammatic and Lepage-Brodsky perturbative QCD methods, respectively, all gave ~ 2 for this ratio. Thus the ratio is no longer a mystery.

Butler

I agree. I was only aware of but one of those calculations. The issue now is to get better data to further challenge the calculations. A number as high as 2.5 is still not ruled out experimentally.

M. Strikman, Pennsylvania State University, USA

The data on leading charm production indicate an enhancement of the cross section, presumably due to the coalescence with spectator partons. What is known about the ratios of $D^+/D^0/D^-$, $B^\pm/B^0$ at large x_F? Is there any evidence for asymmetry?

Butler

This is a slide that I thought I wouldn't have time to show. But there is something missing here which I'll tell you about in a moment. This is the demonstration of the leading particle effect which does survive these higher experimental statistics. Now this is the D−, which is the leading particle. This is the D+, and they are over-plotted on the same graph. And both W82 and E769 basically give this kind of result. So there's a gap of about ~ 0.5–1.0 in the exponents of the x_F distribution (although with admittedly large errors). Now, this is always described as a SMALL leading particle effect even though it amounts to a factor of 2 in the cross section and x of 0.6. And the reason why it's described as small is just a matter of history. Originally NA27 saw a difference of 6 units in the end parameter. So after that, 0.5 is considered to be small but probably significant. If you're asking whether there are results for other kinds of projectiles, so that you can make more careful tests of this, the answer is "yes". Certainly 769 has data on π^+ and on protons. And if you're asking whether there is an overall asymmetry in these distributions between $D+$ and $D-$, the answer is "yes, there's more". I think there's more $D-$ and that asymmetry emerges as well at low x, and I'm not sure I know what that means. But there is an asymmetry of something like 20 or 30%.

NEUTRINO MASS AND MIXING, AND NON-ACCELERATOR EXPERIMENTS

R.G.H. Robertson
Physics Division
Los Alamos National Laboratory
Los Alamos, NM 87545, U.S.A.

Abstract

We review the current status of experimental knowledge about neutrinos derived from kinematic mass measurements, neutrino oscillation searches at reactors and accelerators, solar neutrinos, atmospheric neutrinos, and single and double beta decay. The solar neutrino results yield fairly strong and consistent indications that neutrino oscillations are occurring. Other evidence for new physics is less consistent and convincing.

INTRODUCTION

Non-accelerator experiments have always played an important role in the physics of elementary particles, particularly the neutrino. Much of our early knowledge of the underlying symmetries that now are embodied in the Standard Model of Glashow, Salam, and Weinberg came from beta decay. The neutrinos continue to intrigue us because they seem to occupy an anomalous position in the Standard Model (lacking right-handed fields and, therefore, mass). Perhaps therein lies the road to the territory that we are convinced must lie beyond the Standard Model.

We focus our attention largely on experimental issues, and subdivide the review into sections that deal with kinematic mass measurements, neutrino oscillation searches at reactors and accelerators, solar neutrinos, atmospheric neutrinos, double beta decay, and 17-keV neutrinos. Proton decay is touched on only briefly. Topics such as theoretical issues, cosmological neutrinos, dark matter, gravitation, and non-accelerator measurements of the Weinberg angle are covered by others at this conference (Krauss, Rolandi).

KINEMATIC MASS MEASUREMENTS

Figure 1 shows the experimental direct upper limits on the masses of the three flavors of neutrino as a function of time. With the exception of the ITEP result[1], no indication of non-zero mass has surfaced.

Tau Neutrino

The high mass of the τ makes a precision determination of the mass of ν_τ difficult, but also permits decays to multihadron final states. The Argus collaboration at DESY has observed[2] 12 decays to 5-pion final states, and a single such event with a mass close to that of the τ is sufficient to constrain severely the mass of ν_τ. An upper limit of 35 MeV at 95% confidence level (CL) has been established in this way. A further 8 decays have been seen, as reported at this conference,[3] but none is very close to the endpoint. However, three new measurements[4] of the τ mass have now

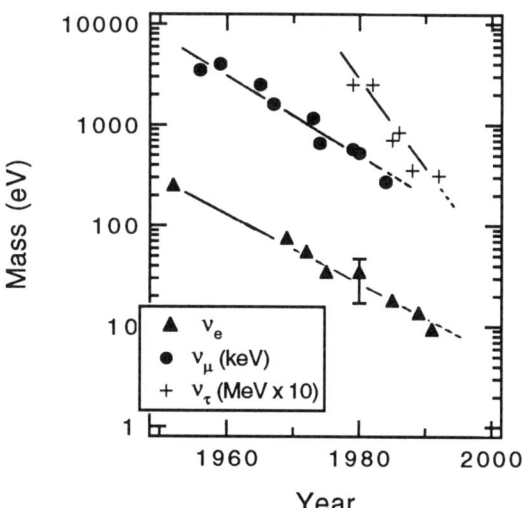

Figure 1. Experimental upper limits on neutrino mass.

Table 1. Mass of τ

Collaboration	Mass, MeV
Particle Data Group	$1784.1^{+2.7}_{-3.6}$
Beijing Spectrometer	$1776.9 \pm 0.4 \pm 0.3$
Argus	$1776.3 \pm 2.4 \pm 1.4$
CLEO II	$1777.6 \pm 0.9 \pm 1.5$

been reported, and these do result in a downward revision of the mass of ν_τ to 31 MeV (95% CL). The mass measurements are summarized in Table I.

The precision is still statistics-limited, and with better resolution and higher statistics, one can look forward to sensitivity in the vicinity of 10 MeV. The only known direct approach to obtaining the mass of ν_τ (and ν_μ) at the level allowed by cosmology (tens of eV for stable neutrinos) is by observing a neutrino burst from a supernova.

Mu Neutrino

The tightest limit on the mass of ν_μ comes from measurements of the μ momentum following the decay of stopped pions. The most recent determination of the mass of the π^+ by Jeckelmann et al. coupled with the muon data of Abela et al. give

$$m_{\nu_\mu} \leq -0.097(72) \text{ MeV}^2,$$

which, with the Bayesian procedure described by the Particle Data Group,[5] yields an upper limit of 270 keV at 90% confidence on the mass of ν_μ.

A new round of experiments at the Paul Scherrer Institute has now reached such a high level of precision that a serious problem in this approach has been discovered. Table 2 gives the recent history of these measurements.

As indicated, the central value for m_ν^2 is over 5σ negative. The mass of the π^+ enters into the calculation, and has been deduced from pionic X-ray spectra. The precision is limited principally by theoretical uncertainties such as electron screening[10] and strong interaction effects (e.g. absorption from 3d state).

Stopped π decay is too subject to theoretical uncertainty to yield m_ν. Instead, it is perhaps the best way to determine m_π. What *can* we use for m_ν? Anderhub et al.[11] used a magnetic 'racetrack' and π decay in flight to obtain,

$$m_\nu^2 = -0.14(20) \text{ MeV}^2$$

$$m_\nu \leq 500 \text{ keV } (90\% \text{CL})$$

The method is (by design) relatively insensitive to m_π and m_μ, and, until some further progress is made on m_π, provides the best direct limit on the mass of m_ν.

Electron Neutrino

All modern determinations of the ν_e mass are searches for a distortion of the beta spectrum of tritium near the 18.6-keV endpoint. There are now 5 recent experiments, all of comparable precision, and all in good agreement (Table 3).

Table 2. Data on $\pi^+ \to \mu^+ + \nu_\mu$ at rest

Collaboration	Ref.	m_μ, MeV	p_μ, MeV/c	m_π, MeV	m_ν^2, MeV^2
Abela et al. 84	6	105.65932(29)	29.79139(83)	139.56761(77)	-0.163(80)
Jeckelmann et al. 88	7			139.56871(53)	-0.097(72)
PDG 88[a]	5	105.658387(34)		139.56737(33)	
Daum et al. 91	8		29.79206(68)	139.56996(67)	
Frosch et al. 92	9		29.79144(20)		-0.127(25)

[a] Electron mass down 8 ppm

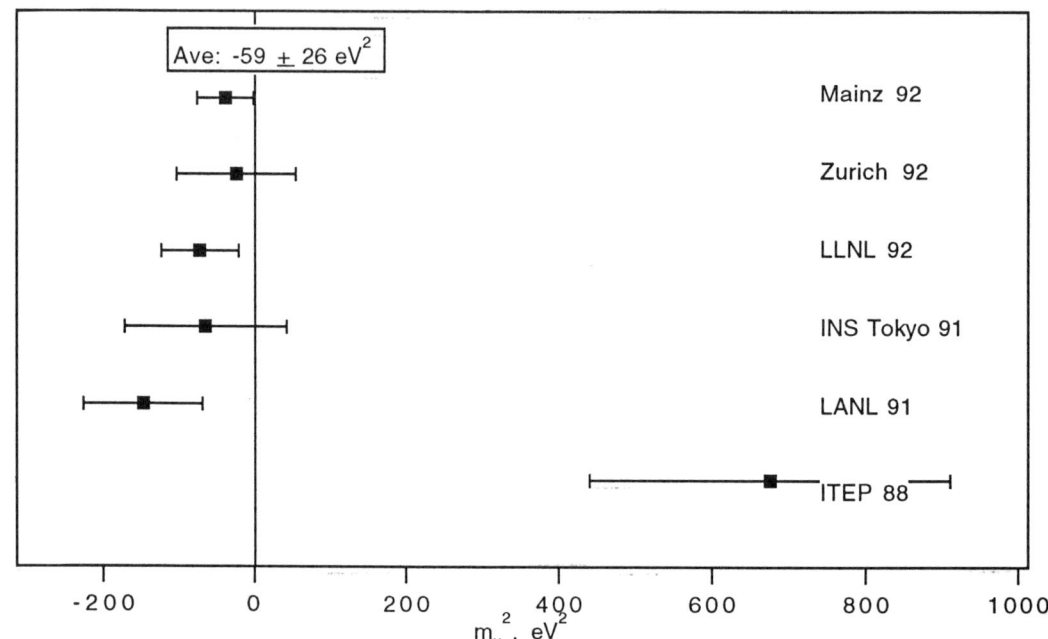

Figure 2. Recent determinations of the mass of $\bar\nu_e$ from tritium beta decay.

As both the Table and Figure 2 make particularly clear, the five experiments contradict the ITEP claim of a non-zero neutrino mass, and it may now be safely concluded that that result is, for some reason, not correct.

However, it is striking that each of the five experiments yields a negative value for m_ν^2. A weighted average of the 5 gives

$$m_\nu^2 = -59 \pm 26 \text{ eV}^2,$$

which is 2.3 σ below 0, and represents only a 1.2% statistical chance (if m_ν is actually 0). If one proceeds nevertheless to find a Bayesian upper limit, it is 5 eV at 95% CL, but it must be emphasized that, with such a low probability, the limit is unsound.

The negative central value may be due to:

1. A statistical fluke.

2. A systematic error in the experiment. Some experimental systematic effect may be influencing one or more of the experiments. (Note that it need not be present in all of them – simply moving one upwards by one to two standard deviations would result in a statisti-

Table 3. Mass of $\bar{\nu}_e$

Institution	Ref.	Mass-squared MeV^2	Limit[a] eV
Los Alamos	79	$-147 \pm 68 \pm 41$	9.3
Livermore	76	$-75 \pm 41 \pm 30$	8.0
INS Tokyo	80	$-65 \pm 85 \pm 65$	13
U. Zürich	81	$-24 \pm 48 \pm 61$	11.5
U. Mainz	82	$-39 \pm 34 \pm 15$	7.2

[a] 95% confidence level

cally unremarkable distribution.) However, the experiments do pass a necessary (although not sufficient) test for being systematic-free,[12] namely that the derived value for the ^{3}H-^{3}He mass difference agrees with independent determinations.

3. A systematic error in the theory. The negative result for m_ν^2 is a symptom of 'too many' counts near the endpoint. That could indicate the existence of an effect outside the conventional atomic and weak-interaction theories, and a number of exotic hypotheses have come to mind. They include a breakdown in the atomic physics calculations of the final-state spectra (particularly unlikely for the case of T$_2$), capture of relic neutrinos (there do not seem to be enough by at least 8 orders of magnitude), tachyonic neutrinos (theoretically unpalatable in concert with bradyon emission), and inner bremsstrahlung of a scalar or pseudoscalar particle that interacts only with neutrinos (e.g. a Majoron: this possibility is intriguing, but appears not to produce the required effect).

A 2.3-standard-deviation effect is not sufficiently large to demand recourse to new physics, and for the moment the phenomenon remains unexplained. Further experimental work is needed to resolve the issue.

NEUTRINO OSCILLATIONS

We are entirely accustomed to the concept that mixing occurs in the quark sector; that is, the weak flavor eigenstates are not the same as the mass (strong-interaction) eigenstates, but are related by a unitary transformation U. It is very likely that, if neutrinos have mass, the same kind of mixing occurs. To search experimentally for such effects (forbidden in the Standard Model), one takes advantage of the fact that neutrino sources and detectors are 'flavor filters.' The probability that a neutrino mixture prepared in flavor λ is detectable in flavor λ' is

$$P_{\lambda\lambda'} = \sum_{kk'} U_{\lambda k} U_{\lambda k'} U_{\lambda' k} U_{\lambda' k'} cos(2.54 \Delta m_{k'k}^2 \frac{L}{E}),$$

where

$$|\nu_\lambda> = \sum_k U_{\lambda k} |\nu_k>,$$
$$\lambda = e, \mu, \tau$$
$$k = 1, 2, 3.$$

The source-detector distance L is in m, the energy E in MeV and Δm^2 in eV^2.

Experiments with $\lambda = \lambda'$ are termed 'disappearance' and those for which $\lambda \neq \lambda'$ are 'appearance.' Owing to the need to determine the neutrino flux precisely, disappearance experiments tend to be limited in sensitivity to values of $P_{\lambda\lambda} \leq 0.95$.

For convenience, the experimental limits are usually presented in the context of two-flavor mixing only, in which case the mixing matrix contains a single undetermined parameter, e.g. an angle, θ. This parameter and the unknown Δm^2 then define a two-dimensional space in which experimental bounds can be placed, but it should be remembered that nature may be more complicated.

Many groups have initially claimed evidence for oscillations, only to discover more mundane explanations for the effects seen. At present, there is no single experiment with reactor or accelerator neutrinos that indicates the presence of oscillations. (We will discuss atmospheric and solar neutrinos below.) An excellent, and still up-to-date, summary has been given by Boehm.[13] We supplement Boehm's review as follows:

1. Conforto[14] has drawn attention to a striking oscillatory behavior in the neutrino data from four high-energy fixed-target experiments. The results can mostly easily be interpreted as ν_e disappearance to a species other than ν_μ, with oscillation parameters:

$$\Delta m^2 = 377 \text{ eV}^2,$$
$$\sin^2 2\theta = 0.48 \pm 0.10 \pm 0.05.$$

These parameters are in conflict with the ν_e disappearance data from the Gösgen Reactor,[15]

$$\text{for } \Delta m^2 \geq 5 \text{ eV}^2,$$
$$\sin^2 2\theta \leq 0.22 \text{ (90\%CL)},$$

as well as with the results of Fermilab experiment E531,[15] which yields,

$$\text{for } \Delta m^2 \geq 100 \text{ eV}^2,$$
$$\sin^2 2\theta \leq 0.18 \text{ (90\%CL)},$$

2. *A propos* of Conforto's observation, the Los Alamos tritium data is being analyzed for evidence of admixture of a neutrino of mass ~ 20 eV with the electron neutrino. Preliminary results disfavor the parameter set found by Conforto at about the 3σ level.

3. The CERN SPS Proposal P254 for direct searches for the interactions of τ neutrinos has been approved. The two experiments, 'Chorus' and 'Nomad' are expected to have sensitivities to $\nu_\mu \to \nu_\tau$ of order 3-4 x 10^{-4} in $\sin^2 2\theta$ for $\Delta m^2 \geq 50$ eV^2, and to $\nu_e \to \nu_\tau$ at the level of 2 x 10^{-2}.

4. New reactor-based experiments are being considered, in addition to the one[13] under construction at the San Onofre complex. A new reactor complex at Chooz in northern France may be the site of a 12-tonne, 1-km experiment,[16] and the fortunate location of the Morton salt mine 12 km from the Point Perry reactors in Ohio may be exploited.[17]

ATMOSPHERIC NEUTRINOS

The interaction of cosmic rays with the upper atmosphere produces showers of hadrons, mostly pions and kaons. The decay sequence

$$\pi^+ \to \mu^+ + \nu_\mu,$$
$$\mu^+ \to e^+ + \nu_e + \overline{\nu}_\mu$$

together with the charge-conjugate reactions, leads to the naive expectation that the flux of μ-flavor neutrinos should be twice the flux of electron-flavor neutrinos. That raises the possibility of a neutrino-oscillation search based on a measurement of the ratio. With baselines ranging from 10 to 10,000 km and energies of 100 to 1000 MeV, a region of parameter space inaccessible to other techniques can be explored. More detailed flux calculations[18-23] take into account the effects of particles ranging out before decay, particles penetrating the earth's crust, directional correlations induced by kinematics and polarization, geomagnetic effects in the primary cosmic-ray spectrum, and nuclear absorption of pions. While the absolute fluxes are uncertain at the level of perhaps a factor of 2, the flavor ratio is considered accurate to about 5% above 50 MeV.

Both the large water Čerenkov detectors, Kamiokande[24] and IMB,[25] find substantial departures from the expected flux ratio in the visible-energy range 100-1000 MeV. The Fréjus[83] and NUSEX[84] experiments, with more limited statistics, find no evidence for the effect, but are not seriously in disagreement either. The results are expressed as the experimental flavor ratio divided by the ratio expected from Monte Carlo simulations based on the theory of neutrino production and absorption, and detector characteristics. They are summarized in Table 4.

Table 4. Atmospheric neutrino flavor ratios.

Collaboration	$\frac{\nu_\mu}{\nu_e}$ Data/$\frac{\nu_\mu}{\nu_e}$ M.C.
Kamiokande	$0.60^{+0.07}_{-0.06} \pm 0.05$
IMB-3	$0.54 \pm 0.05 \pm 0.12$
Fréjus	$1.06 \pm 0.18 \pm 0.15$
	$0.87 \pm 0.16 \pm 0.08^a$
NUSEX	$0.99^{+0.35}_{-0.25} \pm ?$
IMB-3	$1.01 \pm 0.03 \pm 0.11^b$

[a]Fully contained events
[b]Stopping/through muons

If this anomalous ratio is due to $\nu_\mu - \nu_e$ oscillations, there is further information to be obtained on that possibility from upward-going muons. Such muons can only come from neutrino interactions in the rock surrounding the detector and the detector itself – cosmic ray muons themselves cannot penetrate that far. Owing to the larger fiducial mass of the 'target', such interactions are due to higher-energy neutrinos that have travelled greater distances and therefore explore much the same region of oscillation space. Upward going muons that stop in the detector are produced mainly by 3- to 30-GeV ν_μ, while those that pass through are from 30- to 300-GeV ν_μ. From upward going muon rates, the IMB-3 collaboration has been able to rule out a significant part of the parameter space (last line of Table 4), particularly the regions containing the best-fit values. Figure 3 summarizes the data.

In order to conclude that neutrino oscillations are responsible for the anomaly, it is necessary to rule out more mundane explanations. Perhaps there are errors in the calculated flux ratio, or in the calculated cross sections for the interactions of ν_μ and ν_e with ^{16}O. Some of the cross section is contributed by free protons, for which the cross sections are well known,[19] but the main contribution is believed to be quasi-elastic interactions with neutrons and protons in oxygen. The rates are calculated in the framework of the Fermi Gas Model (FGM)[26] with "nuclear corrections" that, in essence, use non-interacting shell-model wavefunctions instead of plane waves. There is relatively little experimental information about these ingredients, but we note that recent work at LAMPF by Koetke et al.[27] on the $^{12}C(\nu_\mu, \mu^-)X$ reaction with neutrinos up to 300 MeV indicate poor agreement with the FGM. Figure 4 shows the results.

We remark that the momentum transferred to recoiling nucleons can be quite small, and that the FGM may be a particularly bad approximation in the part of the spectrum that is nearly elastic. The mass of the μ could then play an important role in suppressing the (ν_μ, μ) cross section.

Yet another explanation has recently been advanced by Mann et al.[28] They observe that it is not clear whether the atmospheric neutrino data indicate a deficiency of ν_μ or an excess of ν_e – the absolute fluxes are not well enough known. Taking the quasi-elastic flux calculated by Bugaev and Naumov,[18] they find good agreement with the μ spectrum, but an excess of electron events that can be attributed to proton decay in the detector: $p \rightarrow e^+\nu\nu$. The partial lifetime of the proton against this mode is found to be 4×10^{31} y. If correct, this

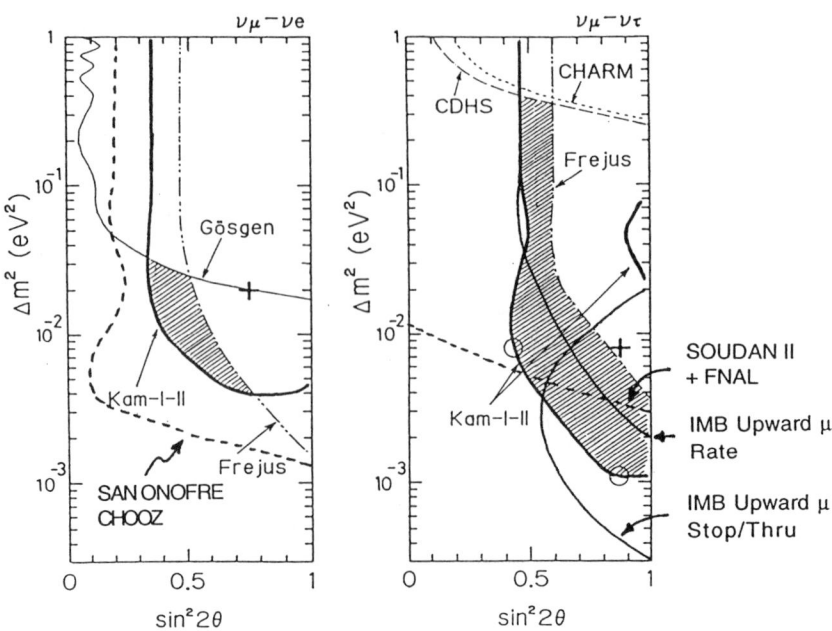

Figure 3. Oscillation parameters in the $\nu_\mu - \nu_e$ and ν_μ disappearance channels. The dotted lines indicate limits to be obtained from future experiments. The crosses are the best fit values from Kamiokande.[24]

would, of course, be a very exciting development.

SOLAR NEUTRINOS

The longstanding solar neutrino problem, which has been extensively reviewed,[29] may be a neutrino-oscillation experiment with a positive result. There are now 4 operating experiments, Homestake, Kamiokande, SAGE and Gallex, and 2 (Sudbury Neutrino Observatory and SuperKamiokande) under construction. Others are in the development phase. The experiments are listed in Table 5. All the operating experiments find lower measured fluxes than are predicted by the "Standard Solar Model" (SSM). It is a commonly held view that the flux of neutrinos from the Sun is hard to calculate – this is not the case. The total luminosity of the Sun is precisely measured, and, if nuclear fusion of H into He provides the energy, then exactly two neutrinos must be produced for each completion of a fusion cycle. Each cycle yields 26 MeV of energy, and (provided ^{8}B is not a major branch, which we know experimentally), at most a few percent of the energy is lost in neutrino emission. Hence the flux can be calculated. What is much more difficult to calculate, of course, is the energy spectrum, because that depends on all the details of nuclear cross-sections, opacities, heavy-element composition, turbulence, etc. The ^{8}B branch, a negligible contributor to energy production, is quite sensitive to the details, whereas the p-p flux is almost independent of them.

There are now two comparably detailed calculations of the solar neutrino spectrum, which are in good agreement when the same input parameters are used. That was not always so, and numerous small corrections and changes have been made to both computations on the way to convergence. The results are summarized in Table 6, with 1σ uncertainties.

The significant differences between the two calculations stem from the inclusion by Bahcall and Pinsonneault of helium diffusion,

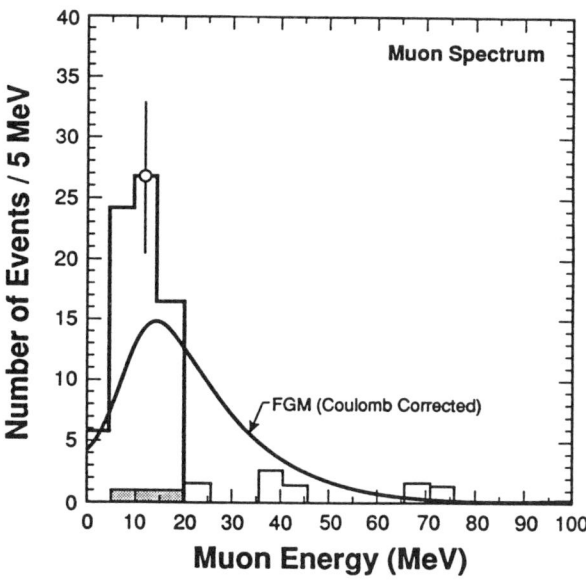

Figure 4. Data on $^{12}C(\nu_\mu, \mu^-)X$ (ref. 27) compared with the FGM

Table 6. Solar neutrino rate calculations.

Target	B-P[a]	T-C[b]
^{37}Cl	8.0(10)	6.4(14) SNU
e^- (H$_2$O)	5.7(1±0.15)	4.3(11) 10^6 cm^{-2}s^{-1}
^{71}Ga	132^{+7}_{-6}	123(7) SNU

[a]Ref. 30 [b]Ref. 31

which increases the ^{8}B flux by 12%, and a different choice of cross section for the ^{7}Be(p,γ)^{8}B reaction. The experimental results are summarized in Table 7.

The SAGE collaboration announced at this meeting[34] results from a second round of data-taking at Baksan. The 1990 and 1991 data sets give, respectively, $20^{+15}_{-20}\pm 32$ and $85^{+22}_{-32}\pm 20$ SNU, and a combined average as given in Table 7. The large systematic error in the first set was assigned to cover the possibility that the counter backgrounds increased with time, as indicated in a few runs. This effect was not seen in the next data set, and so is likely to have been a statistical fluctuation (not improbable with such low statistics). In the second data set, the systematic uncertainty was dominated by radon corrections, and the 'systematic' errors in the two runs are thus largely uncorrelated. The collaboration has applied the Kolmgorov-Cramer-von Mises tests to the data and find that the goodness-of-fit parameter N_w^2 for the combined data is 0.107, which would be exceeded 55% of the time. That indicates that all the data can have come from the same parent distribution. The probability that the early points would be mostly lower than the later ones is not addressed by this statistic, however. The Gallex collaboration does not provide information about goodness-of-fit. Taken at face value, the results of both experiments are in reasonable agreement, with a combined average of 74^{+16}_{-17} SNU. That is definitely not consistent with any SSM, but it falls in the grey area between astrophysical problems and neutrino physics problems. The minimum flux consistent with static hydrogen-burning in the Sun is about 79 SNU.

Table 5. Solar neutrino experiments.

Target	Collaboration	Status	Threshold (MeV) CC	Threshold (MeV) NC	Fid. Mass (tonnes)
^{37}Cl	Homestake	'69-	0.814	-	615 C_2Cl_4
e^-	Kamiokande	'85-	7.5	7.5	680 H_2O
^{71}Ga	SAGE	'90-	0.233	-	30, 57 Ga
^{71}Ga	Gallex	'91-	0.233	-	30 Ga
^{2}H, e^-	SNO	'96-	5.0	2.2	1k D_2O
e^-	SuperKamiokande	'96-	7.5	7.5	22k H_2O
^{11}B, e^-	Borexino	Prop.	0.25	0.25	100 TMB
^{40}Ar, e^-	Icarus	Prop.	2	2	1k LAr
Plus ^{115}In, LHe, ^{205}Tl, Cl, F, I, ^{98}Mo, and others					

Table 7. Solar neutrino rate measurements.

Experiment	Target	Rate	% B-P[a]	% T-C[b]
Homestake[c]	^{37}Cl	2.25±0.24 SNU	28±3	35±4
Kamiokande II & III[d]	e^- (H_2O)	284±29±35 x 10^4 cm^{-2}s^{-1}	50±8	66±11
SAGE I & II[e]	^{71}Ga	58^{+17}_{-24}±14 SNU	44^{+17}_{-21}	47^{+19}_{-23}
Gallex[f]	^{71}Ga	83±19±8 SNU	63±16	67±17

[a]Ref. 30 [b]Ref. 31 [c]Ref. 32 [d]Ref. 33 [e]Ref. 34 [f]Ref. 35

Bludman et al.[36] have made a study of all the extant data except SAGE to learn if they force us to embrace new neutrino physics. They conclude that Mikheyev-Smirnov-Wolfenstein (MSW) matter-enhanced neutrino oscillations explain all the data very well, with the best fit values being

$$\Delta m^2 = (0.3 - 1.2) \times 10^{-5} \text{ eV}^2$$

$$\sin^2 2\theta = (0.4 - 1.5) \times 10^{-2}.$$

A large-angle solution is also permitted but fits less well. [Inclusion of SAGE will drive Δm^2 down and $\sin^2 2\theta$ up by an amount roughly equal to the range.] Petcov and Krastev[37] have also pointed out that, even with only 2 flavors oscillating, a vacuum solution remains viable:

$$\Delta m^2 = (0.5 - 1.1) \times 10^{-10} \text{ eV}^2,$$

$$\sin^2 2\theta \geq 0.75.$$

'Cool-Sun' explanations, in which the core temperature of the Sun is arbitrarily reduced, fail to account for the data: the fundamental conflict is between Homestake and Kamiokande, because they both record mainly the ^{8}B flux. Neutrino oscillations provide a much better account of the data because neutral-current scattering naturally increases the Kamiokande rate relative to the Cl-Ar rate. For the same reason, oscillation into sterile neutrinos is somewhat disfavored. The indications are strong that neutrino oscillations are indeed occurring, but the new generation of experiments – SNO with its explicit neutral-current and spectroscopic capabilities, SuperKamiokande with its large volume, and

Borexino with its low threshold – can provide the definitive proof.

DOUBLE BETA DECAY

Neutrinoless double beta decay enjoys a phase-space advantage over $\beta\beta\nu\nu$ emission (allowed in the Standard Model, but very slow), and searches for it set tight limits on the Majorana masses of electron neutrinos (or of the admixed mass eigenstates that comprise the electron neutrino). The best limits on the effective Majorana mass come from isotopically enriched ^{76}Ge detectors:[38,45]

$$< m_\nu > = \sum_i \xi_i U_{ei}^2 m_i \leq 2 \pm 1 \text{ eV}.$$

The uncertainties are mostly theoretical.

The allowed 2-neutrino mode has now been observed[39,40] in 4 isotopes: ^{76}Ge ^{82}Se, ^{100}Mo, ^{150}Nd, with halflives ranging from 8 to 920 x 10^{18} y (Table 8). Radiochemical experiments[41,42] have also demonstrated the occurrence of double beta decay in ^{130}Te and ^{238}U, although in those cases the mode is unknown. In each of the 4 cases measured spectroscopically, a few extra events have been seen in the region between the energy at which the 2ν distribution has effectively died away and the Q-value. Events are not forbidden there, but their probability is so low ($<10^{-5}$) that they may indicate a departure from the expected physics.

A third possible mode of double beta decay was discovered as a consequence of theories that explain the smallness of neutrino mass in terms of a spontaneously broken B-L symmetry that allows otherwise massless neutrinos to acquire a small Majorana mass.[43,44] A consequence of spontaneous symmetry breaking is the appearance of a massless Goldstone boson, in this case dubbed the 'Majoron.' The Majoron couples only to neutrinos and effectively flips the helicity of the virtual neutrino in neutrinoless double beta decay. The resulting 3-body phase space fits remarkable well with the anomalous 'hard' events observed experimentally. The intensity is 2-3 percent of the 2-neutrino intensity. At this conference Piepke[45] reported on the results from a large isotopic ^{76}Ge detector (Heidelberg-Moscow collaboration). With approximately 4000 2-neutrino decays observed, a small excess (150) is seen in the high-energy region, consistent with the other experiments. However, the shape at lower energies does not seem to fit the 'standard' Majoron hypothesis.

One can also derive limits on Majoron modes from the non-observation of certain decay modes of hadrons,[46] although experiments do not at present have enough sensitivity. Haxton[48] has used the ratio of experimental half-lives[41] for ^{128,130}Te to set an upper limit on the coupling constant $< g_M >$ of 4.2 x 10^{-5}. Finally, Burgess and Cline[47] have shown that 'standard' Majorons do not meet the requirements because the implied Majorana neutrino mass would allow 0ν decays, in contradiction to experiment. If the Majoron carries lepton number -2, then this difficulty can be circumvented. Note that the Burgess-Cline values for the coupling constant differ somewhat from those shown.

Needless to say, if the evidence for anomalous events continues to accumulate to the point where a Majoron of some type is confirmed, it will be a revolutionary development for particle physics. Great progress towards resolving the issue is being made by the experimental groups. The UC Irvine group have increased the magnetic field in their time projection chamber to improve the resolution and the capability for measuring it. Very low background isotopic Ge and Xe detectors are just beginning to accumulate data, with already impressive results.

Table 8. Double-beta-decay experiments (selection).

Isotope	Ref.	Q (MeV) MeV	$T^{2\nu}_{1/2}$ (y) 10^{18} y	$T^{0\nu}_{1/2}$ (y) 10^{23} y	$T^{0\nu M}_{1/2}$ (y) 10^{20} y	$<g_M>$ 10^{-4}
^{82}Se	39	2.995	108^{+26}_{-6}	-	11(4)	2.4(4)
^{100}Mo	39	3.034	$11.6^{+3.4}_{-0.8}$	-	1.5(6)	4.2(4)
^{150}Nd	39	3.367	8	-	0.8(5)	2.1(7)
^{76}Ge	40	2.039	920^{+70}_{-40}	-	200(-)	1.4(-)
^{76}Ge	45	2.039	-	>17 (90%CL)	>400	<1.1
^{130}Te	41	2.533	2670(90)	>0.03 (90%CL)	-	-
^{238}U	42	1.100	2100(600)	-	-	-
$K^+ \to l^+\nu M$	46					<70

THE 17-keV NEUTRINO

The shape of the spectrum in ordinary beta decay has, since the earliest days, been used to set limits on the masses of neutrinos. Recently, many experimental groups have observed 'kinks' in the spectra of ^{3}H, ^{14}C, ^{35}S, ^{45}Ca, ^{63}Ni, and ^{71}Ge at an energy 17 keV below the endpoint, which can be interpreted as an admixture of a 17-keV neutrino with the electron neutrino at an intensity of approximately 1%. The striking uniformity of the results from a wide range of isotopes is illustrated in Fig. 5.

All the positive observations have made use of solid-state detectors. Searches made with magnetic devices have been uniformly negative. The experimental situation is summarized in Table 9. Criticisms have been levelled at both techniques. Bonvicini[49] recently made an exhaustive study of the interplay between statistical and systematic errors in 17-keV experiments and concluded that use of arbitrary shape-correction parameters of Taylor-series form when the true underlying energy dependence of efficiency is unknown can be very dangerous. All magnetic spectrometer experiments have been obliged to do this, owing to the difficulties in determining the response to the necessary accuracy. It was therefore not clear that negative experiments really ruled out the purported effect at the claimed level. Piilonen and Abashian[50] drew attention to small effects due to scattering that were neglected in Si detector experiments and that could possibly induce spurious distortions resembling a massive neutrino.

The balance has now been strongly tipped against a 17-keV neutrino by two new experiments reported at this meeting. A group at the Institute for Nuclear Studies in Tokyo[77] carried out a magnetic-spectrometer study of ^{63}Ni with extremely high statistical precision (2.4 x 10^9 events in the interval 40-60 keV). While arbitrary shape-correction parameters (30 of them) are still required to fit the data (30 independent spectra acquired in three overlapping energy ranges), the overwhelming statistical precision essentially precludes a real 17-keV neutrino distortion from being concealed. The upper limit of 0.073% admixture at 95% CL does not take into account the possible systematic errors in the shape correction, but it is highly unlikely that such a good fit with such high statistics could conspire to conceal a 17-keV neutrino.

An experiment carried out at Argonne National Laboratory by Ahmad et al.[73] makes use

Table 9. Experiments on the 17-keV neutrino.

Collaboration	Ref.	Source	Method	m_ν, keV	$\sin^2 2\theta$
Simpson	51	T in Si	Crystal	17.1(2)	0.03
Haxton	52		Exchange Corrections		
Lindhard & Hansen	53		Screening Corrections		
Simpson (revised)	54,55	T in Si	Crystal	17.1(2)	0.011(3)
Altzitzoglou et al.	56	^{35}S	Magnetic		<0.004 99% CL
Ohi et al.	57	^{35}S	Crystal		<0.0015 90% CL
Apalikov et al.	58	^{35}S	Magnetic		<0.0017 90% CL
Datar et al.	59	^{35}S	Crystal		<0.006 90% CL
Markey & Boehm	60	^{35}S	Magnetic		<0.003 90% CL
Hetherington et al.	61	^{63}Ni	Magnetic		<0.003 90% CL
Hime & Simpson	54	T in Ge	Crystal	16.9(1)	0.011(5)
Simpson & Hime	62	^{35}S	Crystal	16.9(4)	0.0073(9,6)
Hime & Jelley	63	^{35}S	Crystal	17.2(5)	0.0085(6,5)
Sur et al.	64	^{14}C	Crystal	17.1(6)	0.012(3)
Becker et al.	65	^{35}S	Magnetic		<0.006 90% CL
Zlimen et al.	66	^{71}Ge (IB)	Crystal	17.2(12)	0.016(7)
Schonert et al.	67	^{177}Lu	Magnetic		<0.004 68% CL
Hime & Jelley	68	^{63}Ni	Crystal	16.8(4)	0.0099(12,18)
diGregorio et al.	69	^{71}Ge (IB)	Crystal	13.8(18)	0.0080(25)
Bahran & Kalbfleisch	70	T_2 gas	Prop. Ctr.		<0.004 98% CL
Hargrove et al.	71	T_2 gas	Prop. Ctr.		(in progress)
Wark	72	^{35}S	Magnetic		(in progress)
Ahmad et al.	73	^{35}S	Mag.+Cryst.		<0.0025
Simpson	74	^{45}Ca	Crystal	16.1(8)	0.008(?)
Chen et al.	75	^{35}S	Magnetic		
Stoeffl & Decman	76	T_2 gas	Magnetic		(in progress)
Kawakami et al.	77	^{63}Ni	Magnetic		<0.00073 95% CL
Norman et al.	78	^{55}Fe (IB)	Crystal		(no effect seen)

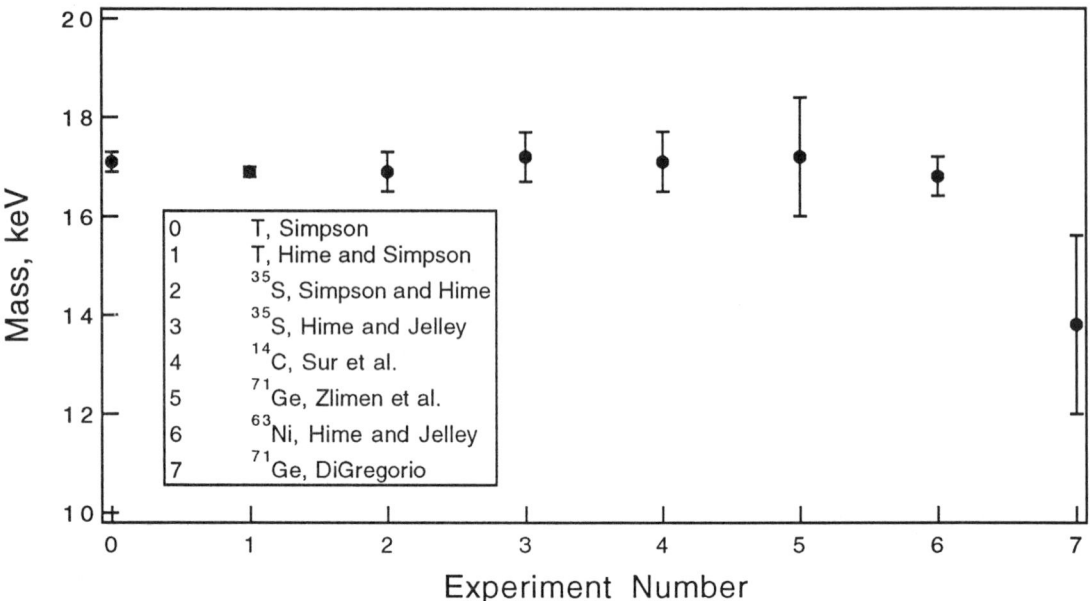

Figure 5. Determinations of the mass of the 17-keV neutrino.

of a thin solid source, a solenoidal magnetic transport, and a silicon detector. With a decreasing field strength along the axis toward the detector, electrons from the source are collimated to a diameter that bears a calculable ratio to the source diameter, without the need for physical collimators that can introduce artifacts into the spectrum. Furthermore, electrons backscattered from the detector are reflected by the magnetic mirror and returned to the detector. Thanks to a geometric efficiency that is essentially 50%, high statistics can be obtained even with very weak (thin) sources, and no arbitrary shape corrections are needed. The ANL group found no evidence for a 17-keV neutrino in ^{35}S decay ($\sin^2\theta < 0.25\%$ at 95% CL). To test the validity of their procedures they added a small amount (1.34%) of ^{14}C activity to the source, and could easily observe the artificial kink it produced. The intensity measured was 1.41(9)%, in excellent agreement with the added amount. This is the more impressive given that the distortion introduced in this way is much 'softer' than the sharp kink produced by a massive neutrino.

The Lawrence Berkeley Laboratory group[78] that has reported evidence for a 17-keV neutrino in ^{14}C continues to see the effect, but has now completed a search in ^{55}Fe internal bremsstrahlung emission without finding a positive signal. While less compelling than the two results presented above, this experiment contradicts earlier indications of the effect in this decay mode.

It must be concluded now that the kink reported in many spectra is *not* an intrinsic feature of the electron spectrum and that there is *no 17-keV neutrino*. The observed effects are due to an as-yet unidentified cause or causes, and it appears possible that a variety of experimental artifacts may have conspired improbably to reproduce it in so many different situations.

SUMMARY

Direct measurements of the masses of the neutrinos have set upper limits on the masses

of
$$\nu_e < 5 \text{ eV},$$
$$\nu_\mu < 500 \text{ keV, and}$$
$$\nu_\tau < 31 \text{ MeV}.$$

In the case of ν_e, the five most recent measurements from tritium beta decay all obtain negative central values for the square of the mass, with a combined significance of 2.3σ. Lacking an explanation for this, the Bayesian upper limit given must be regarded with suspicion. Equally disturbing is the setback in our knowledge of the mass of ν_μ. In the case of stopped π^+ decay, the negative central value for the mass squared is almost certainly connected with theoretical problems in deriving the pion mass.

No terrestrial oscillation experiment now yields evidence for neutrino oscillations, but both the Kamiokande and IMB water Čerenkov detectors show puzzling deficiencies in the muon rates induced by atmospheric neutrinos relative to electron rates. Whether this is due to oscillations, some deficiency in the theoretical interpretation of the production and detection mechanisms, or to some more exotic effect such as proton decay remains an open question. If oscillations are responsible, then parameters in the vicinity of

$$m_x^2 - m_\mu^2 = 10^{-2\pm1} \text{ eV}^2$$
$$\sin^2 2\theta = 0.5$$

are indicated.

The deficiency of solar neutrinos seems more and more likely to require neutrino oscillations for its explanation. Matter-enhanced oscillations à la Mikheyev, Smirnov, and Wolfenstein explain the data from 4 experiments (Homestake, Kamiokande, SAGE, and Gallex) remarkably well, whereas astrophysical interpretations are increasingly strained. Three isolated regions of parameter space are possible:

$$\Delta m^2 = (0.1 - 1.2) \times 10^{-5} \text{ eV}^2$$
$$\sin^2 2\theta = (0.4 - 6) \times 10^{-2},$$

$$\Delta m^2 = (0.1 - 4) \times 10^{-5} \text{ eV}^2$$
$$\sin^2 2\theta = 0.5 - 0.9,$$

$$\Delta m^2 = (0.5 - 1.1) \times 10^{-10} \text{ eV}^2$$
$$\sin^2 2\theta \geq 0.75.$$

Three new experiments, SNO, SuperKamiokande, and Borexino are expected to provide a definitive conclusion about neutrino oscillations in solar neutrinos.

Remarkable experimental progress has been made on the slowest of natural processes, double beta decay. There are now 6 experimental observations of it, 3 spectroscopic, 2 radiochemical, and one seen both ways. Theory generally accounts well for the rates of the allowed 2ν process, increasing confidence in the derived effective Majorana mass, $<2\pm1$ eV. All of the spectroscopic measurements of the electron sum energy spectrum in $\beta\beta\nu\nu$ decay show a small but significant excess of counts just below the Q-value. That may be evidence of a Majoron, but further experimental work is required.

The evidence in favor of a 17-keV neutrino has now been convincingly contradicted by new, highly precise experiments. The observed effects appear to be due to some experimental artifacts not yet fully identified.

Physics seems to be at the threshold of a new discovery. All the evidence points to neutrino mass and oscillations as the explanation of the solar neutrino problem. With such a momentous conclusion at stake, the most careful and detailed experimental work is called for, and physicists are bending with enthusiasm to the task ahead.

REFERENCES

1. S. Boris et al., Pis'ma Zh. Eksp. Teor. Fiz. 45, 267 (1987) [Sov. Phys. JETP Lett. 45, 333 (1987)]; Phys. Rev. Lett 58, 2019 (1987); V.A. Lyubimov, *Proc. Int. Conf. Neutrino '88* (ed. J. Schneps et al., World Scientific, Singapore, 1989), p. 2.

2. A. Albrecht et al., Phys. Lett. B202, 149 (1988).

3. A. Albrecht et al., these proceedings.

4. D. Kreinick, in *Beyond the Standard Model III*, Ottawa, Canada, June 22-24, 1992 (to be published); and private communication.

5. Particle Data Group, Phys. Lett. B170, 1 (1986).

6. R. Abela et al. Phys. Lett. B146, 431 (1984).

7. B. Jeckelmann et al., Phys. Rev. Lett. 56, 1444 (1986).

8. M. Daum et al., Phys. Lett. B265, 425 (1991).

9. R. Frosch, private communication (1992).

10. P. Goudsmit, Paul Scherrer Institute (see ref. 9).

11. H.B. Anderhub et al., Phys. Lett. B114, 76 (1982).

12. R.G.H. Robertson, in *Proc. Franklin Symposium in honor of F. Reines*, Philadelphia, PA, April 29 - May 1, 1992 (to be published).

13. F. Boehm, in *Particles, Strings, and Cosmology (PASCOS-91)*, edited by P. Nath and S. Reucroft, World Scientific, Singapore, 1992, p. 96.

14. G. Conforto, Nuovo Cim. 103A, 751 (1990).

15. N. Ushida et al., Phys. Rev. Lett. 57, 2897 (1986).

16. Y. Declais et al., *Letter of Intent: Search for Neutrino Oscillations at a Distance of 1 km from Two Power Reactors at Chooz*, 1992 (unpublished).

17. R.I. Steinberg et al., *The Perry Experiment: A Long Baseline Reactor Neutrino Oscillation Search*, 1992 (unpublished).

18. E.V. Bugaev and V.A. Naumov, Phys. Lett. B232, 391 (1989).

19. T.K. Gaisser and J.S. O'Connell, Phys. Rev. D 34, 822 (1986).

20. G. Barr, T. K. Gaisser, and T. Stanev, Phys. Rev. D 39, 3532 (1989).

21. T.K. Gaisser, T. Stanev, and G. Barr, Phys. Rev. D 38, 85 (1988).

22. H. Lee and Y. Koh, Nuovo Cim. 105B, 883 (1990).

23. M. Honda et al., Phys. Lett. B248, 193 (1990).

24. K.S. Hirata et al., Phys. Lett. B280, 146 (1992); T. Kajita, these proceedings.

25. IMB Collaboration, Phys. Rev. Lett. 66, 2561 (1991); Phys. Rev. Lett. 69, 1010 (1992).

26. R. Smith and E. Moniz, Nucl. Phys. B43, 605 (1972).

27. D.D. Koetke et al., Los Alamos preprint LA-UR-92-1562 (submitted to Phys. Rev. C) 1992.

28. W.A. Mann, T. Kafka, and W. Leeson, Tufts University preprint TUHEP-92-01 PDK-516.

29. J.N. Bahcall, *Neutrino Astrophysics* Cambridge University Press, Cambridge, UK 1989.

30. J.N. Bahcall and M.H. Pinsonneault, Revs. Mod. Phys. (in press).
31. S. Turck-Chièze, in *Proc. XV Int. Conf. on Neutrino Physics and Astrophysics "Neutrino '92"*, Granada, Spain, June 6-12, 1992 (to be published).
32. K. Lande, in *The Many Aspects of Neutrino Physics*, Fermilab Workshop, Batavia, IL, Nov. 14-17, 1991 (unpublished).
33. K.S. Hirata et al., Phys. Rev. D 44, 2241 (1991); T. Kajita, these proceedings.
34. A.I. Abazov et al., Phys. Rev. Lett. 67, 3332 (1991); A.I. Abazov et al., Nucl. Phys. B (Proc. Suppl.) 19, 84 (1991); V.A. Gavrin, these proceedings.
35. P. Anselmann et al., Phys. Lett. B285, 376 (1992); P. Anselmann et al., Phys. Lett. B285, 390 (1992), D. Vignaud, these proceedings.
36. S.A. Bludman et al., Pennsylvania preprint UPR-0516T (1992).
37. P.I. Krastev and S.T. Petcov, preprint CERN-TH 6539-92 (1992).
38. W.C. Haxton, in *Proc. XV Int. Conf. on Neutrino Physics and Astrophysics "Neutrino '92"*, Granada, Spain, June 6-12, 1992 (to be published).
39. M.K. Moe, M.A. Nelson, M.A. Vient, and S.R. Elliott, in *Proc. XV Int. Conf. on Neutrino Physics and Astrophysics "Neutrino '92"*, Granada, Spain, June 6-12, 1992 (to be published).
40. F.T. Avignone III, et al. Phys. Lett. B256, 559 (1991).
41. T. Bernatowicz et al., submitted to Phys. Rev. Lett. (1992); T. Kirsten, H. Richter, and E. Jessberger, Phys. Rev. Lett. 50, 474 (1983) and Z. Phys. C 16, 189 (1983); A. Alessandrello, Proc. XV Int. Conf. on Neutrino Physics and Astrophysics "Neutrino '92", Granada, Spain, June 6-12, 1992 (to be published).
42. A. L. Turkevich, T.E. Economou, and G.A. Cowan, Phys. Rev. Lett. 67, 3211 (1991).
43. G.B. Gelmini and M. Roncadelli, Phys. Lett. B99, 411 (1981).
44. Y. Chikashige, R.N. Mohapatra, and R.D. Peccei, Phys. Lett. B98, 265 (1981).
45. A. Balysh et al., Phys. Lett. B283, 32 (1992); A. Piepke, these proceedings.
46. V. Barger et al. Phys. Rev. D25, 907 (1982).
47. C.P. Burgess and J. Cline, McGill preprint 92-27; in *"Beyond the Standard Model III"*, Ottawa, Canada, June 22-24, 1992 (to be published).
48. W.C. Haxton, in *Proc. XV Int. Conf. on Neutrino Physics and Astrophysics "Neutrino '92"*, Granada, Spain, June 6-12, 1992 (to be published).
49. G. Bonvicini, CERN preprint CERN EP-92/xx.
50. L. Piilonen and M. Abashian, preprint VPI-IHEP-92/6 (1992).
51. J.J. Simpson, Phys. Rev. Lett. 54, 1891 (1985).
52. W.C. Haxton, Phys. Rev. Lett. 55, 807 (1985).
53. J. Lindhard and P.G. Hansen, Phys. Rev. Lett. 57, 965 (1986); B. Eman and D. Tadic, Phys. Rev. C33, 2128 (1986).
54. A. Hime and J.J. Simpson, Phys. Rev. D39, 1837 (1989).
55. J.J. Simpson, Phys. Lett. B174, 113 (1986).
56. T. Altzitzoglou et al., Phys. Rev. Lett. 55, 799 (1985).

57. T. Ohi et al., Phys. Lett. B160, 322 (1985).

58. A. Apalikov et al., JETP Lett. 42, 289 (1985); V.A. Lyubimov, contr. to *Workshop on the 17 keV Neutrino Question*, Berkeley, CA, December 18-20, 1991 (unpublished).

59. V.M. Datar et al., Nature 318, 547 (1985).

60. J. Markey and F. Boehm, Phys. Rev. C32, 2215 (1985).

61. D.W. Hetherington et al., Phys. Rev. C36, 1504 (1987).

62. J.J. Simpson and A. Hime, Phys. Rev. D39, 1825 (1989).

63. A. Hime and N.A. Jelley, Phys. Lett. B257, 441 (1991).

64. B. Sur et al., Phys. Rev. Lett. 66, 2444 (1991).

65. H. Becker et al., Caltech preprint CALT-63-605 (1991).

66. I. Zlimen et al., Phys. Rev. Lett. 67, 560 (1991).

67. S. Schonert et al., in *Int. Workshop on Electroweak Physics beyond the Standard Model*, Valencia, Spain, October 2-5, 1991.

68. A. Hime and N.A. Jelley, Oxford preprint OUNP 91-21 (1991).

69. D.E. DiGregorio et al., TANDAR preprint LNY584 L-1 SB (1991).

70. M. Bahran and G.R. Kalbfleisch, Oklahoma preprint OKHEP 91-005 (1991).

71. C.K. Hargrove, in *Beyond the Standard Model III*, Ottawa, Canada, June 22-24, 1992 (to be published).

72. D.L. Wark, in *Workshop on the 17 keV Neutrino Question*, Berkeley, CA, December 18-20, 1991 (unpublished).

73. S.J. Freedman, these proceedings.

74. J.J. Simpson, in *Beyond the Standard Model III*, Ottawa, Canada, June 22-24, 1992 (to be published).

75. M. Chen, in *Workshop on the 17 keV Neutrino Question*, Berkeley, CA, December 18-20, 1991 (unpublished).

76. W. Stoeffl and D. Decman, in *The Many Aspects of Neutrino Physics*, Fermilab Workshop, Batavia, IL, Nov. 14-17, 1991 (unpublished).

77. T. Ohshima, these proceedings.

78. E.B. Norman, these proceedings.

79. R.G.H. Robertson et al., Phys. Rev. Lett. 67, 957 (1991).

80. H. Kawakami et al., Phys. Lett. B256, 105 (1991).

81. E. Holzschuh, in *Proc. XV Int. Conf. on Neutrino Physics and Astrophysics "Neutrino '92"*, Granada, Spain, June 6-12, 1992 (to be published).

82. J. Bonn, in *Proc. XV Int. Conf. on Neutrino Physics and Astrophysics "Neutrino '92"*, Granada, Spain, June 6-12, 1992 (to be published).

83. Fréjus collaboration, Phys. Lett. B245, 305 (1991).

84. NUSEX Collaboration, Europhys. Lett. 8, 611 (1989).

DISCUSSION

Mandeep Gill, University of California at Berkeley, USA

Why does COBE suggest a 7 eV neutrino, as you stated in your summary view graph? Was it an estimate of the universe density?

Robertson

It comes from the density fluctuation data and the favored models of structure formation in the universe, which are cold dark matter (e.g., axions, etc.) and hot dark matter (neutrinos); together they suggest a mass value of 7 eV.

COSMOLOGY AND ASTROPHYSICS 1992 *

Lawrence M. Krauss
Center for Theoretical Physics and Dept. of Astronomy
Yale University
Sloane Laboratory, 217 Prospect St.
New Haven CT USA 06511

Abstract

I review recent developments in cosmology and astrophysics relevant to particle physics, focussing on the following questions: What's new in 1992?, What have we learned since the last ICHEP meeting in 1990? and, What are the prospects for the future? Among the topics explicitly discussed are: COBE, Large Scale Structure, and Dark Matter; Big Bang Nucleosynthesis; the Solar Neutrino Problem; and High Energy Gamma Ray Physics.

INTRODUCTION

"It was the best of times. It was the worst of times."
                              Charles Dickens

The subject which I was asked to summarize in one hour spans almost 40 orders of magnitude in energy. Even in the "worst" of times this would be difficult. However, in the times we have been living it is very nearly impossible. Probably no other area in particle physics has produced as many dramatic results in the last two years as the cosmology-astrophysics interface. Thus, deciding to attempt to err on the side of incompleteness rather than coherence (the reader can be the final judge), I have chosen to describe here only that subset of results: (a) which are the most topical for this meeting, and (b) which can be combined together into some pseudo-logical fashion.

Nevertheless, to give some idea of both the range of scales and areas of present interest, I present in figure 1 a graphical view of the field, as a function of energy. This seems to me an appropriate way to organize my comments, and I will generally, although not universally, work in order of ascending energy.

---

* bitnet: Krauss @ Yalehep. Research supported in part by NSF, DOE, and TNRLC

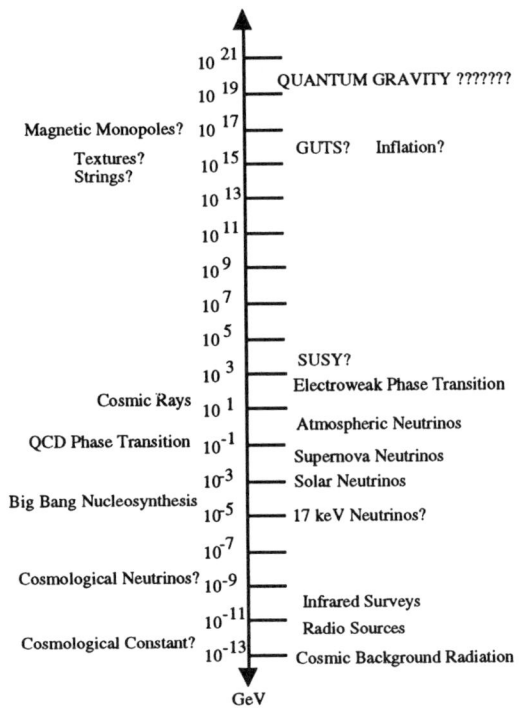

Figure 1. An Overview

There have been active developments in each of the areas listed above. I will not have the time, or the space, to describe all of them, however. Nevertheless, I hope to get to give a flavor of the excitement which has been generated over the past year or two in this field.

1. COBE, LARGE SCALE STRUCTURE, AND DARK MATTER: OR.... THE UNIVERSE STRIKES BACK!

(A) COBE

The popular media reverberated last spring with the news that the Differential Microwave Radiometer (DMR) experiment aboard the COBE (COsmic Background Explorer) satellite had discovered primordial anisotropies in the Cosmic Microwave Background Radiation (CMBR). Referred to variously as the "holy grail", and "the face of God", there was no shortage of hyperbole. Nevertheless, in spite of the signficance attributed to this discovery by the popular press, it *was* in fact extremely significant and, I will claim, of great interest for particle physics as well.

This of course was not the first important observation by the COBE satellite. Recall that in the first 8 minutes of its flight in 1989 COBE measured[1] the spectrum of the CMBR and determined that it was describable by a single black body to better than 1 part in 1000, with a temperature of approx. 2.735 K. This result alone provided strong support for the Big Bang origin of the CMBR, and also ruled out earlier measurements[2] of a high frequency discrepancy with a black body spectrum, which would have required rather exotic sources of energy production during the period 10^5 sec-100,000 years into the Big Bang expansion.

As described by George Smoot at this meeting[3], the DMR experiment aboard COBE provides a *differential* measurement of the CMBR temperature, and not an absolute one. Thus, it has been able to operate long after the liquid He aboard the satellite, necessary for the sensitivity to the absolute temperature, had evaporated. The data on which the observed anisotropy is based is the first two years of COBE data. Another year's data has already been taken and is currently being analyzed.

The DMR experiment employs two independent microwave antennas sampling the sky at an angular separation of 10^0. Three sets of antennae, operating at 90, 53, and 31 GHz respectively were used, and at each frequency two separate channels were available.

In searching for CMBR anisotropies, several larger effects must be first removed. There is a well-known, and well measured, dipole anisotropy in the CMBR signal, at the level of a few parts in a thousand. This is presumably primarily due to the local motion of the satellite with respect to the frame defined by the surface of last scattering of the CMBR. This motion is comprised of a sum of several components: the motion of the satellite around the earth, the motion of the earth around the sun, the sun around our galaxy, the infall of our galaxy to the center of our local group of galaxies, and finally, the large scale drift of our local group of galaxies. The net peculiar motion is on the order of about 600 km/sec, which would be expected to produce a signal of the magnitude observed. Moreover, COBE has now further supported this interpretation by examining the time dependence of dipole signal. The yearly variation of the signal can clearly be seen[3].

Subtracting the measured dipole from the signal, any analysis of the COBE signal must next concern itself with the chief source of background: our galaxy. A great deal of effort has gone into both modelling the galactic signal, and verifying that it does not contaminate the observed residuals[4]. While the galactic signal is least significant in the 90 GHz band, measurements of the rms temperature deviations at 10^0 separations in all three bands do not go to zero as one moves away from the plane of our galaxy, but instead approach a constant value of approximately 30 µK by a galactic latitude of about 25^0. It is this residual signal which is claimed to represent true primordial fluctuations in the CMBR.

With this introduction to the COBE result, I will next address the following 3 issues: (i) What exactly is the COBE result?, (ii) Why is it interesting?, (iii) What does it, and what does it not imply?

(i) The COBE Data:

The first COBE result I have already alluded to. Averaging over the sky at latitudes greater than 30^0 from the galactic plane, COBE reports an rms temperature deviation:

$$\Delta T_{rms} (\theta > 30^0) \approx 30 \,\mu K. \quad (1)$$

Now of course, there is much more information in the CMBR anisotropy than is obtainable from the rms deviation alone. Using a formalism with which particle physicists should be comfortable, it is conventional to define a temperature correlation function, $C(\alpha) \approx <T_1 T_2>$, defined crudely as the average over the sky of the product of temperatures in regions separated by an angle θ. Specifically,

$$C(\theta_{21}{:}\sigma) = \langle \delta T (\hat{x}_1{:}\sigma) \, \delta T (\hat{x}_2{:}\sigma) \rangle_{21} \quad (2)$$

where the average is taken with respect to all positions $\hat{x}_1, \hat{x}_2$ with $\hat{x}_1 \cdot \hat{x}_2 (= \cos \theta_{21})$ fixed, with a smoothing size of σ, the angular response of the detector (a Gaussian FWHM of 7^0).

One can also choose to expand the measured temperature fluctuations across the sky in a multipole expansion:

$$\delta T (\theta, \varphi) = \sum_{l,m} a_{lm} Y_{lm}(\theta, \varphi) \quad (3)$$

If we define the rotationally invariant quantity:

$$a_l^2 = \sum_m |a_{lm}|^2 \quad (4)$$

Then one can define the quadrupole anisotropy:

$$Q = (a_2^2/4\pi)^{1/2} \quad (5)$$

COBE has measured both the full correlation function of the temperature fluctuations, and has also reported a value for the quadrupole anisotropy Q. The measured value is

$$Q = 13 \pm 4 \,\mu K. \quad (6)$$

However, COBE also reports the value for the quadrupole moment determined in a slightly different way. If instead, the correlation function is fit assuming a 'flat', angle independent spectrum (see below), one infers a slightly larger value of the quadrupole:

$$Q_{rms-PS} = 16 \pm 4 \,\mu K. \quad (7)$$

Both values are of use in comparing to theoretical predictions.

While the quadrupole anisotropy is the most significant numerical result quoted by COBE, the correlation function implicitly contains information on all multipoles. Specifically,

$$C(\hat{x}_1 \cdot \hat{x}_2 {:} \sigma) = \frac{1}{4\pi} \sum_{l=0}^{\infty} a_l^2 P_l(\hat{x}_1 \cdot \hat{x}_2) e^{-(l+1/2)^2 \sigma^2} \quad (8)$$

The quadrupole moment dominates this expansion and is thus easiest to extract. Hence the use of this value to characterize the observations.

(ii) Why the interest?

The observed CMBR originated at the epoch when the background matter distribution first became neutral, and decoupled from radiation, at a time of $\approx 3 \times 10^5$ years into the Big Bang explosion. As we look out to high redshifts, and hence to early times, this time defines a "surface", known as the surface of last scattering. Thus, the CMBR provides a redshifted picture of the distribution of radiation at that time. As shown on the schematic picture below, the horizon size at that time would correspond today to an observed angle across the sky of about 1^0.

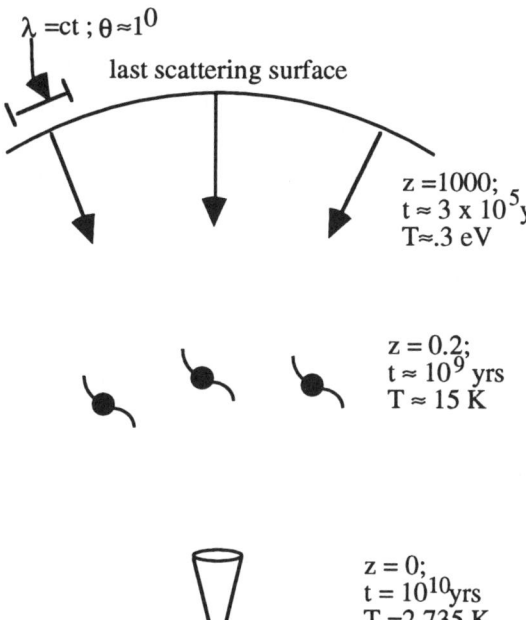

Figure 2. A Schematic View of the Universe

Photons travelling from the last scattering surface to a receiver aboard COBE redshift by a factor of 1000 on average. However, if there are regions of density excess in the dominant energy density at that time, photons leaving such regions at the surface of last scattering will have to "climb up" out of these potential wells, inducing an extra gravitational redshift:

$$\frac{-\Delta v}{v} \approx \frac{1}{3}\Delta\Phi \qquad (9)$$

where Φ is the gravitational potential. Thus, "cool" spots in the CMBR represent density excesses which, if gravity governs the formation of structure, will eventually collapse to form galaxies, or clusters of galaxies. "Hot" spots represent regions of under-density, which will eventually form "voids".

Now, the COBE observations, because of their smoothing scale, are only sensitive to fluctuations in temperature on scales of greater than about 10^0. However, *this is much larger than the scale corresponding the the horizon at the last scattering surface. No causal process since the beginning of the presently observable Big Bang expansion could have moved energy density around on such scales in order to create potential wells!* Thus, if the CMBR has not been further reprocessed since a redshift of $z \approx 1000$, the anisotropies observed by COBE are *primordial*, that is, they must represent initial conditions associated with the Big Bang. Such initial conditions involve the physics appropriate to processes at very high energy, i.e. appropriate to particle physics. That is why the COBE results are both extremely exciting on general grounds, and are also relevant to this meeting.

(iii) What does COBE imply?

(a) Does Cosmic Structure Form due to Gravity?

After the question of what the geometry of the observable universe is, perhaps the central issue in modern cosmology concerns the origin of the observed structures in the universe: galaxies, clusters, and superclusters. The simplest possibility is that such structures formed out of the gravitational collapse of initially small density excesses in the universe. Since gravity is universally attractive, if initially one starts with small fluctuations $\delta\rho/\rho \ll 1$, gravity will cause these to increase. Once they exceed unity, they will separate out from the background expansion and collapse to form bound systems.

After decoupling of matter and radiation at z=1000, matter density perturbations can collapse due to gravity. Simple perturbation theory[5] then implies:

$$\frac{\delta\rho(t_f)}{\rho} = \frac{a(t_f)}{a(t_i)}\frac{\delta\rho(t_i)}{\rho} \qquad (10)$$

as long as the perturbations remain in the linear regime. Here a(t) is the cosmic scale factor at time t. Because the ratio between the cosmic scale factor at decoupling and today is 1000, this implies that in order for fluctuations on the scale of present day galaxies to have grown sufficiently to become O(1) by today, they would have had to have been at least O(10^{-3}) at the time of last scattering. Such density fluctuationas would produce a gravitational redshift in the photon gas:

$$\frac{\delta T}{T} \approx \frac{1}{3}\frac{\delta \rho}{\rho} \geq \frac{1}{3000} \qquad (11)$$

Now, on small angular scales which entered the horizon before decoupling, a variety of dynamical factors due to the matter-radiation coupling must also be considered in order to determine the remnant fluctuations in the CMBR today. For example fluctuations in the baryon fluid, which are strongly coupled to radiation before decoupling, cannot collapse due to the pressure of radiation. Thus, any primordial fluctuations in baryons cannot grow until after decoupling. Hence, the first scale of fluctuations which can grow immediately after coming inside the horizon is that associated with the horizon scale at decoupling, about 2^0 on the sky today.

This argument suggests that if baryons dominated the matter density at the time of decoupling, that their fluctuations should induce an anisotropy at the level given in (11) on scales of about 2^0 in the CMBR, if they are also responsible for galaxy formation. Similarly, this suggests that if the spectrum of primordial fluctuations does not vary much as a function of wavelength, that fluctuations in the CMBR on larger scales, those probed at COBE, should not be much smaller.

The fact that the observed fluctuations in the CMBR are not anywhere near this large is one of the many pieces of indirect evidence that we have that baryonic matter does not dominate the energy density of the universe, at least if galaxy formation is to occur by gravitational collapse. However, before the present COBE observation the upper limits on the anisotropies in the CMBR were so small as to suggest potential problems for this whole idea.

Arguments of the type I have summarized above for baryonic matter[5] suggest that if primordial fluctuations in whatever was the dominant matter in the universe were to grow by gravitational instability to lead to galaxy and cluster formation by the present time, the remnant signature in the CMBR today at COBE scales should exceed $\delta T/T \approx$ few x 10^{-6}. As the previous COBE upper limit was close to this, the whole gravitational instability picture was perilously close to being ruled out. Instead, the observation, by COBE of a $\delta T/T_{quadrupole} \approx 5$ x 10^{-6} is in the range predicted for gravitational instability based models. Thus COBE observations are *consistent* with the standard model of galaxy induced structure formation.

(b) Nature of Primordial Fluctuations

For gaussian initial fluctuations, all information about the primordial spectrum is contained in the power spectrum, P(k), defined in terms of the Fourier transform of the spatial density fluctuation spectrum:

$$P(k) \sim \left| \int \Delta \rho(x) e^{ik \cdot x} dx \right|^2 \qquad (12)$$

The power per logarithmic interval in comoving wavenumber k, or equivalently the rms mass fluctuation on this scale, can then be written:

$$\left(\frac{\delta \rho}{\rho}\right)_k^2 \sim k^3 P(k). \qquad (13)$$

Under the assumption that there is no preferred primordial scale, one assumes P(k) is of a *scale-free* form:

$$P(k) \sim k^n \qquad (14)$$

There are at least *two* conventions for defining the k-dependence of the primordial power spectrum. A particularly natural way for theorists to describe the spectrum is in terms of the amplitude of each mode at the time that mode enters the 'horizon'---the distance over which causal propagation can have taken place since the beginning of the FRW-Big Bang expansion. This is because once the scale of metric perturbations enter the horizon, particle interactions can affect their growth, while before this the evolution of fluctuations on extra-horizon sizes is determined by the equations for the background expansion. Moreover, this is an appropriate scale to discuss the physics responsible for the generation of primordial fluctuations. In inflationary models, for example, the amplitude of primordial fluctuations is fixed as modes are pushed outside the de-Sitter horizon during inflation.

While it is therefore natural to describe the amplitude of fluctuations at horizon crossing, one should remember that describing the spectrum in this way implies that the amplitude of different modes will be specified at different times. Alternatively, one can choose to specify the k-dependence of the amplitude of fluctuations at some distinct, fixed, time. Two commonly chosen examples are: the present time, and the time of decoupling---when the microwave background last scattered.

In advance of any model of primordial fluctuations, several authors (notably Harrison, Zel'dovich, and Peebles and Yu[6]) suggested that the most reasonable ansatz was a *scale-invariant* form. This is a sensible assumption, since it suggests that there is not arbitrarily large power on either small or large scales. In the latter case, this could produce too large an anisotropy in the CMBR, and in the former, it would produce too many small black holes.

If we specify a scale invariant spectrum in its most natural way, as one with constant amplitude at horizon crossing:

$$\left(\frac{\delta\rho}{\rho}\right)_{hor} = const. \qquad (15)$$

then eq. (13), with each mode measured at its horizon crossing, implies that n=-3. However, from the observational viewpoint of COBE, we need to translate this into the k-dependence an observer who is probing fluctuations at a specific time: namely the time of last scattering.

Density fluctuations outside the horizon are not well-defined. That is, they are gauge dependent. Fixing the gauge appropriately, however, one finds that during a matter dominated expansion, density fluctuations outside the horizon can also grow according to (10), in order to keep metric perturbations constant. During a matter dominated expansion $a \sim t^{2/3}$. Thus, an extra-horizon size mode with wavelength λ measured at the time the horizon has size λ_0 will itself cross the horizon at a time $(t/t_0) \approx (\lambda/\lambda_0)^3$. Since during such a time it will grow by a factor $(t/t_0)^{2/3}$, this means that for a perturbation on this scale to have the value given in (15) at the time it crosses the horizon, it must have a value at time t_0 which is $(\lambda_0/\lambda)^2$ smaller. Thus, since $k_0 \sim \lambda_0^{-1}$, one finds

$$\left(\frac{\delta\rho}{\rho}\right)^2_{t_0} = const.^2 \times \left(\frac{k}{k_0}\right)^4 \qquad (16)$$

for modes with $k<k_0$. From (13), we thus find that P(k) for long wavelength modes in a scale invariant spectrum has index n=1 at time t_0.

From the measured angular temperature correlation function, COBE has extracted the index for primordial temperature fluctuations on extra-horizon sized scales at the time of last scattering. The best fit value for the spectral index, including cosmic variance in the predicted spectrum is $n = 1.15^{+0.45}_{-0.65}$, consistent with scale invariance.

(c) Inflation, Dark Matter, and Gravitational Waves:

The COBE results have been claimed to provide support for Inflationary cosmology, with all of its subsequent consequences for dark matter? To what extent is this claim correct? COBE clearly has *not* proved inflation.

However, it is remarkably consistent with all the predictions of inflationary models.

Consider first the nature of the primordial fluctuation spectrum. The COBE observation is consistent with an n≈1 scale invariant spectrum, which is a relatively generic prediction of inflation. However, is it a unique prediction? The answer is both yes and no. As I stated earlier, any sensible model of primordial fluctuations is likely to predict a flat spectrum, and thus models like cosmic strings, Textures, etc., all make this claim. However, there is one fundamental difference between all these models and inflation. In the former cases, one must make the unphysical supposition that all truly primordial fluctuations are zero, and then generate them afterwards. In the latter case any pre-inflationary fluctuation spectrum is *erased* by inflation, leaving behind only that spectrum generated during inflation.

Thus, I think it is fair to say that inflation is the *only* completely comprehensive model of primordial fluctuations. In this sense, this one available model is consistent with the COBE observations. However, this does not preclude that other first principles models may also exist which generate a flat spectrum of primordial fluctuations. We just don't know of any.

In a related vein, the uncertainty in the existing COBE limit is sufficiently large as to accomodate significant deviations from scale invariance. Moreover, it has been increasingly appreciated recently (see below) that inflation allows the possibility that n<1. Thus, if COBE were eventually to be able to prove that n≠1, this might still be compatible with inflationary predictions.

What about the other 'generic' prediction of inflation: that Ω =1, i.e. we live in a flat universe today? COBE cannot probe this directly, but it can indirectly. In the first place, the fact that the overall magnitude of the observed fluctuations, in the range of 10^{-6}, as expected for gravitational clustering in flat cosmologies dominated by exotic dark matter, provides some support for this whole picture. Somewhat more quantitatively, it is well known that the growth of perturbations is more efficient in an Ω =1 flat universe than in open cosmologies with Ω <1. The observed small magnitude of the primordial fluctuations in the CMBR are not small enough to require Ω =1, but it is consistent with this value. In general, like the direct dynamical determinations, they prefer $\Omega \geq 0.2$.

Carrying this line of argument further, one can examine in detail the model predictions for various cosmologies with various different matter contents to see how well the COBE results match onto extrapolations based on observed galaxy clustering and velocities. This is too complex a subject to discuss in detail here, and several talks at this meeting were devoted to the beginnings of such investigations[7,8]. Suffice it to say that when the predictions of the preferred cosmological model, Cold Dark Matter, Ω=1, and a flat spectrum of adiabatic primordial fluctuations, normalized to the galaxy-galaxy two point function at about 8 Mpc, is compared to the COBE observations, the predicted values of $\Delta T/T$ are somewhat smaller than those observed[9]. At the same time, this normalization of the CDM spectrum tends to predict smaller clustering at the large scales probed by the sample provided by the Infrared Astronomical Satellite (IRAS). These problems have lead some to suggest that either: (a) the dark matter distribution is not as biased on small scales as had previously been expected. In this case the CDM predictions match on better to the large scale observations and the COBE data, but appear to predict too much clustering on small scales, (b) the primordial spectrum is "tilted", i.e. it is n<1, giving more power on large scales, (c) Cold Dark Matter must be supplemented by Hot Dark matter, such as one might get from an O(7 eV) neutrino, (d) Cold Dark Matter is dead. I shall come back to this issue a little later.

One factor which was not initially taken into account, but which is now being studied in more detail, is the fact that inflation predicts *two* sources of CMBR anisotropies. In addition to primordial density perturbations, it is predicted that their should also be a primordial spectrum gravitational waves generated during inflation. The heuristic reason for this is that inflation occurs because there is a non-zero vacuum energy density present during the inflationary phase, giving rise to a de Sitter expansion. In such a phase the trace of the energy momentum tensor is non-zero. It is in fact proportional to the cosmological constant Λ, which is in turn proporitional to the vacuum energy density:

$$\langle T^\mu_\mu \rangle \sim \Lambda \sim V \qquad (15)$$

Since gravitational waves couple to the trace of the energy-momentum tensor, they will be generated during any period of expansion when this is non-zero. Their amplitude will be therefore be proportional to V, the vacuum energy density during inflation[10].

Both primordial density perturbations, and gravitational waves will contribute to the observed COBE anisotropy, and since inflation generates both (as will undoubtedly any other scenario responsible for generating the former), it is necessary to consider both sources when attempting to compare inflationary predictions to the COBE observations[11]. This leads to several interesting possibilities. It is possible, at least in principle that the entire COBE quadrupole signal is due to gravitational waves. In this case one gets a constraint on the overall scale of inflation. Because the processes which generate both scalar density perturbations and tensor gravitational waves during inflaton are stochastic, the theoretical predictions themselves involve an inherent uncertainty---the so call "cosmic variance". Because we happen to live in the actual universe, instead of in all possible universes, the value we observe for the magnitude of these fluctuations is just one of a set of possible values with an intrinsic probability distribution. The predictions for each of the 5 components of the quadrupole moment generated by each source of fluctuations are independent, leading to a predicted value distributed as a χ^2 distribution with 5 degrees of freedom. Comparing the observed value, with the predicted value, assuming the entire contribution is due to gravitational waves, one gets a 95% confidence limit[11]:

$$1.5 \times 10^{16} \text{ GeV} < V^{1/4} < 5 \times 10^{16} \text{ GeV} \qquad (16)$$

This value is remarkably close to the predicted scale of grand unification based on an extrapolation of the measured strong, weak, and electromagnetic couplings at LEP[12]. Whether or not this is a coincidence remains to be seen.

In fact, the idea that at least some of the COBE quadrupole is due to gravitational waves is not outrageous. If one extrapolates the favored CDM structure formation model predictions based on observed galaxy-galaxy correlations to the COBE scale on tends to predict a quadrupole which is somewhat smaller than COBE sees, as I earlier remarked, and as I will discuss in somewhat more detail soon.

Of course what one likes, and what one gets may be two different things! Each inflationary model makes predictions not only for the magnitude of scalar and tensor perturbations, but also for their ratio. Thus, aside from the cosmic variance, one is not free to independently vary these two components when comparing predictions and observations.

The inflationary predictions for the total quadrupole anisotropy (Q) and the relative contributions have now been addressed in some detail[13]. It has been pointed out (i.e see Davis et al in ref 13)that not only are the relative contributions of scalar and tensor perturbations to Q coupled, but so is this ratio and the power law dependence, n, of the primordial spectrum itself. In order for gravitational waves to contribute significantly to the observed CMBR quadrupole moment, it appears that n<1 is probably required.

Thus, while COBE has provided a remarkable new observable with the potential to shed a great amount of light on our understanding of the cosmology of the early universe, at present it has not allowed any unambiguous tests of models. In order to unravel the nature of contributions to the quadrupole anisotropy and also determine the nature of the primordial fluctuation spectrum we will have to await the results of a series of new experiments, including South Pole experiments which will anisotropies in the CMBR on very small angular scales. Among other things, the contribution of gravitational waves is expected to be subdominant on such scales even if they were to dominate at large scales. COBE is also continuing to take data for a 3rd year, and the analysis of three year's data will hopefully provide tighter limits on all quantities.

(B) Large Scale Structure

COBE has filled in one piece of a jigsaw puzzle which we hope will eventually lead to a picture of the origin of structure in the universe. Meanwhile, over the last two years work on filling in the other pieces has continued apace.

Most interesting in this regard is new evidence, bolstered by the COBE results described earlier, that gravity is responsible for large scale structure. As early as two years ago, there was still a wide-spread belief that non-gravitational dynamics, à la hydrodynamics, would be required. I think it is fair to say that all the evidence which now exists points firmly towards the idea that gravity is at work.

First and foremost is the fact that the long-awaited approach to homogeneity in the distribution of matter in the universe is now apparent in the largest sky surveys. Prior to these, the overwhelming impression one obtained by looking at the well-known slices of the universe---such as that of the Center for Astrophysics redshift survey with its famous 'stick-man'---was that coherent structures persisted on scales as large as the largest survey. Even surveys on scales as large as ≈ 100 Mpc on a side recently provided evidence for coherent structures on this scale[14].

Such structures are embarassing, in part because of the observed isotropy of the microwave background, and also because these structures should, at a certain point, leave a direct imprint on the CMBR which could have been observable. With this in mind, preliminary results from an ongoing survey are telling. The Las Companas Southern Sky Redshift survey of Oemler et al[15] involves a sparsely sampled set of galaxies in the southern sky with redshifts extending out to 40,000 km/sec (or distances of about 4-800 Mpc, depending upon the value of the Hubble constant). On this scale structures on characteristic scales of O(100) Mpc and smaller are clearly distinguishable, but *no* similar coherent structures are observable on larger scales. The long-awaited approach to homogeneity in the matter distribution may finally have been observed!

Next, additional dynamical support for the role of gravity in structure formation has been one result of an exciting analytical method, called "POTENT", pioneered by Bertshinger, Deckel, and their collaborators[16]. At the same time these studies are providing the first direct evidence that in fact Ω may actual approach unity on large scales.

At the heart of the POTENT analysis is data provided by the IRAS satellite I mentioned earlier. This satellite provide a relatively unbiased all-sky sample of galaxies which have been used to try and extract statistical information about clustering and large scale velocities by a number of groups. In particular, the sample of IRAS galaxies has been systematically probed by other means to measure redshifts. This allows one to extract from the background Hubble expansion the set of peculiar motions of the individual galaxies which may be in response to the local gravitational attraction of regions of overdensity.

One of the problems with measuring redshifts, however, is the fact that one can only extract directly the radial, line-of-sight component of velocities in this manner. At the heart of the POTENT analysis is the recognition that *if* gravity is the cause of peculiar velocities, then one would expect the curl of this velocity field to be smaller than its gradient:

$$|\vec{\nabla} \times \vec{v}\,'| << |\vec{\nabla} \cdot \vec{v}\,'| \qquad (17)$$

In this case the peculiar velocity field can be written in terms of a velocity potential:

$$\vec{v}\,' = -\nabla \varphi \qquad (18)$$

The goal of POTENT is to reconstruct both the complete velocity field and the velocity potential from the observed redshifts of IRAS galaxies. From these, the actual gravitational potentials, and density distributions can be inferred. These latter distributions can then be compared with the actual distribution of luminous objects on the sky. If the agreement is good, it provides strong evidence that in fact the observed structures and associated velocity fields resulted from gravitational collapse.

In practice, since the data are sparse, first the observed radial velocity distribution must be smoothed on the sky. From this, the velocity potential and peculiar velocity field are determined, and the potential and density distribution are determined[16]:

$$\Phi \approx \frac{3}{2}\Omega^{0.4}\varphi \qquad (19)$$

$$\delta \approx -\Omega^{0.6}\vec{\nabla} \cdot \vec{v}\,' \qquad (20)$$

Note that both these quantities depend upon the cosmic density parameter Ω. Thus, the comparison of potent predictions to observations can not only help distinguish the nature of gravitational collapse, but also can help give evidence on the flatness of the universe.

The results of POTENT thus far are quite impressive. The predicted density fields are remarkably similar to the observed distributions of matter. In addition, of more interest, perhaps, is the fact that the value of Ω inferred by this analysis is $\Omega > 0.4$,[17] substantially larger than that inferred by direct dynamical estimates on smaller scales to date.

Having obtained some idea of the underlying density distribution, POTENT can also allow us to probe how gaussian the initial conditions which may have led to this distribution were. Interestingly, POTENT suggests that these initial conditions could only have been gaussian if Ω is large. In particular, assuming gaussian initial conditions gives a bound $\Omega > 0.3$ at the 6-σ level! All previous dynamical estimates on smaller scales suggested that $\Omega < 0.3$, which implies that there may need to be significant dark matter on very large scales in order for these two estimates to be consistent.

While these results on large scale structure are preliminary, they are quite exciting. They not only lend strong support to the conventional notion that gravity causes structure in the universe, but they also support the primary prediction of inflation, namely that $\Omega = 1$. This is the first time in observational cosmology when firm support for both these ideas has been suggested.

(C) Dark Matter

Inextricably tied to both of the issues I have thus far discussed is the question of the nature and distribution of dark matter in the universe. COBE, by discovering evidence for primordial fluctuations, allows us to explore the spectrum of these fluctuations, and in turn allows a comparison not only with theoretical predictions, but also with estimates which are derived from observations of large scale structure. The comparison depends crucially however on the nature of the dominant matter in the universe, which determines how fluctuations grow once they enter the horizon.

The favored generic model of dark matter, based both on particle theory notions, and on large scale structure modelling, is the so-called "Cold Dark Matter" (CDM) model. When combined with the simplest prediction from inflation---an adiabatic n=1 spectrum of primordial fluctuations, the CDM model makes definite predictions about the clustering of matter on all scales once it is fit to the observed clustering at one scale. This is one of its great virtues; it is testable.

In the last two years, a full frontal assault on the CDM model has been launched by observational cosmologists. Recent surveys sensitive to clustering on the scale of ≈ 100 Mpc and smaller have confirmed what appears to be a generic problem for the simplest CDM scenario[18]; too little power is predicted on large scales compared to that observed, and too much power is predicted on small scales, resulting in predicted peculiar velocity fields which are larger than those which are measured.

As a result of these results one now hears the cry: Cold Dark Matter is Dead! shouted from many an ivory tower. I would suggest that, as in the past, reports of the demise of CDM are premature. A number of possibilities remain:

(i) n≠1: I have discussed already the potential benefits of a spectrum which is not precisely scale invariant. In particular, a spectrum with n<1 will result in more large scale power for a fixed power on small scales. Inflation quite plausibly predicts deviations from scale-invariance, especially if gravitational waves are to contribute to the observed COBE signal.

(ii) isocurvature fluctuations: Fluctuations in the density of some material which do not lead to overall metric curvature fluctuations: i.e. they are compensated by corresponding density fluctuations in other species, could easily result in some Cold Dark Matter models[19,20]. Such fluctuations would tend to produce more large scale fluctuations than the corresponding adiabatic ones.

(iii) Cold Plus Hot: In vogue recently has been the idea that Cold Dark Matter is not all there is. Instead, CDM might be supplemented by some material (Hot Dark Matter) which does not cluster so efficiently on small scales. A prime candidate is a light, ≈ 7 eV neutrino, which may be indicated on numerological grounds by present Solar neutrino experiments (discussed later). A 70% CDM, 30% HDM universe appears to have several advantages over a purely CDM universe for structure formation. Perhaps the greatest advantage, in the words of George Smoot is: 'Full Employment for All!'. It is also worth noting that many of the advantages of adding HDM can also be obtained by a non-zero cosmological constant, which may be called for on other grounds[21,22].

It may be that none of these fixes will be required. In the first place, one exciting result of COBE is that it has caused us to rethink the scale where we chose to normalize density fluctuation predictions. Previously, small scale clustering observations had been used. Now, however, it seems to make more sense to normalize them at large scales, where we can observe the primordial fluctuations in the CMBR, instead of fluctuations which have been processed inside the horizon to produce the observed cosmic structures.

If one normalizes the CDM models to the COBE data, one generally can produce the observed--previously coined "excess"-- of power on galaxy cluster-scales. The only remaining problem is the fact that dramatically more clustering is predicted on galactic scales than is observed.

This latter problem brings to the fore a point which I think should be kept in the back of our minds when confronting CDM models with data. All along, and this still persists, the chief problem for CDM scenarios has been from data which is the most suspect. We must remember that we are only just beginning to map large scale structure in the universe. Thus inferences made today based on scant data may be changed tomorrow. Similarly, the present conundrum as I have just described it, is based on some convention for determining how the small scale clustering of dark matter should be reflected in the clustering of luminous galaxies. The fact that the two need not be the same, now known as "biasing", has been an important ingredient in model building, and in computer simulations of CDM-dominated universes. Nevertheless, we really do not know the correspondence. No CDM simulations contain enough physics to unambiguously allow a determinations of which dark matter 'clumps' should be associated with galaxies.

To summarize; CDM models make definite predictions about the clustering of matter in the universe which have trouble confronting the simplest interpretation of the existing data. One of two possibilities ensue: either the models are wrong, or the interpretations are wrong. Nevertheless, in spite of this potential problem, a CDM-dominated $\Omega=1$ universe, with an n=1 spectrum of fluctuations from inflation, comes remarkably close to describing the observed universe. I believe it still remains the best, and simplest testing ground we have. It remains to be seen if it is right.

(D) Conclusions?

The exciting results of observational cosmology obtained over the past two years have led to an embarrassingly consistent picture. Many previously discrepant pieces are beginning to come together. In particular, I believe it is now fair to say that the conventional picture that gravitational collapse is responsible for the observed large scale structure in the universe is now overwhelmingly supported by the evidence at hand. Another fact which several years ago seemed to be unsupported but which now is much more firm, is that the universe is likely to have a critical density of matter. Before the most recent large scale redshift surveys, there was no indication that Ω might be as large as unity. Now, the IRAS analyses uniformly suggest this might be the case. When tied with the COBE observations which support not only the generic predictions of inflation, but also the idea of exotic dark matter, the case for an $\Omega=1$ universe is now much stronger. Finally, we have seen proof that when only one sensible idea exists, it is sometimes correct. COBE has demonstrated that primordial fluctuations are close to being scale invariant. Since theoretically that seemed the only reasonable alternative in advance, this discovery is not so surprising as it is reassuring.

In spite of this growing convergence, however, unsettled problems still remain--which is of course good at the very least because it gives us something to do. The simple Cold Dark Matter dominated universe now has a much greater amount of data to accomodate, and is having a harder time doing it. I expect that we will have to learn a great deal more before this issue is settled. Next, an issue I didn't discuss before. Age estimates, based on models of stellar evolution, *consistently* seem to imply that our galaxy is older than the Universe, if the age of the latter is based on the "Hubble age", determined through the measured expansion. This is more a concern than a problem at present---the uncertainties are still large. But it does serve as a reminder that we may not yet know it all...

What about the future? I think without a doubt the central issue in this whole business is the detection of dark matter! Only then will we have an unambiguous empirical handle on both the early universe, and on the origin of large scale structure. This issue brings up something that is crucial to remember when attempting to utilize cosmology as a probe of particle physics. I think the results we have seen over the past few years very strongly support the notion that arguments from cosmology can be "clean" and even testable. They should be listened to. However, we must never forget the very important difference between an "observation" and "experiment". It is only by combining the latter with the former that we can hope to unambiguously probe the universe. As far as dark matter is concerned, there are ongoing experiments designed to detect various Cold Dark Matter candidates, if they exist, which should be able to report interest limits, if not detections, at the next ICHEP meeting. The possibility that the tau-neutrino might have a cosmologically interesting mass has now taken on renewed importance. Both laboratory experiments, based on ν_μ-ν_τ oscillations, and large underground detectors designed to explore the signal from the next supernova, are interesting and useful tools which may shed light on this important issue.

As far as unravelling the spectrum of primordial density fluctuations, new experiments now underway at the South Pole to measure small scale anisotropies will be crucial, as will experiments able to confirm the COBE results. As far as large scale structure investigations are concerned, besides the new massive redshift surveys which are essential for mapping the universe, there is much work to be done on numerical simulations. As I indicated, in order to understand how to label galaxies in a simulation in which the mass being traced is "dark" we must incorporate the physics relevant to baryonic infall and collapse. Finally, new technologies are allowing an ever increased sensitivity to trace element abundances in stellar systems. This will be essential for probing models of stellar evolution which help us date our galaxy, and also for probing primordial nucleosynthesis, my next topic.

2. BIG BANG NUCLEOSYNTHESIS: OR ....
CAN ONE HAVE TOO MUCH OF A GOOD THING?

Big Bang Nucleosynthesis has been, to date, one of the great success stories of cosmology. Based on the simplest possible model for an expanding universe, and bolstered by well understood physics, unambiguous predictions were made for the abundance of light elements created in the Big Bang expansion. These predictions, which vary by over ten orders of magnitude have been, up to the present time, in remarkable qualitative, and where possible quantitative agreement with "observation". Currently this comparison suggests that the total abundance of baryons in the universe is limited to somewhere between about 1% and 9% of closure density. At the same time it limits the number of relativistic species present during the formation of primordial helium to less than the equivalent of ≈ 3.something light neutrino species, in good agreement with LEP results.

This is not to suggest that controversy does not remain. While the theory of BBN is now quite standard, even allowing for certain uncertainties introduced by possible effects coming from the QCD phase transition, what is by far more uncertain are the measurements and what we can infer from them. I believe that it is fair to say that in spite of several well publicized potential challenges, at this time BBN remains alive and well. Nevertheless, we are at the threshold of making several more precise tests which will in any case allow BBN to be an even stronger probe of early universe cosmology.

The two issues I want to discuss involve the predicted and observed abundances of He and Li in the universe, the most and least abundant of the observed primordial light elements respectively.

The fact that approximately 1/4 of the universe, by weight, is He provided the first definitive success of BBN. Simple arguments, based on the strength of the weak interactions and therefore the abundance of neutrons and protons at the time these interactions freeze out in the early universe immediately pinpointed this as the expected range for primordial He.

This great success has recently become the source of some concern. Observations suggest, for reasons I will shortly outline, that the primordial abundance of ^{4}He is between 22-24% by weight. Nevertheless, utilizing limits obtained by a combination of upper limits on observed D and ^{3}He, one finds that BBN predictions are apparently only consistent with observations if the primordial abundance of ^{4}He is greater than 23.5-23.7%[23,24,25]. (The high value is based on a new investigation of BBN limits incorporating together all theoretical, experimental and observational uncertainties[25]). This is disturbingly close to the claimed upper limit of 24%. Moreover, it is well above the "best fit" value which several authors claim is close to, or even below 23%.

I think that in this case one picture is worth a thousand words.

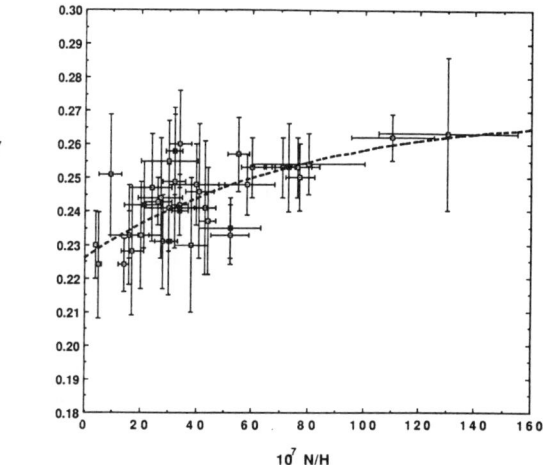

Figure 3: Helium Fraction (Y) versus Nitrogen Abundance (from Olive et al[26])

In order to determine the actual primordial abundance astronomers try to measure the helium content in stars with smaller and smaller abudances of heavy elements, such as oxygen and nitrogen. Such stars are presumably older, because the material in them has been less processed. Based on extrapolating the observed trend in Helium as a function of either oxygen or nitrogen, or some other heavy element, astronomers try and determine the primordial abundance. Above I display one such set of data, and the statistical fit for a relation between helium and nitrogen abundances.

I think it is clear from this picture that while a best fit relation may extrapolate, at low metallicity, to a value near or below 23%, it is also clear that systematic errors are at least as important as statistical ones here. From data like this, it is not clear, to me at least, that a distinction between an upper limit of 24% and 23.7% is meaningful[27]. For example, without a first principles understanding of the helium-nitrogen relation one sensible way to estimate the uncertainty in this relation is to examine the uncertainty on the lowest metallicity point.

While I think the present uncertainties imply that BBN predictions remain safe, these same uncertainties point out more generally the danger in over-interpreting the data. For example, the ^{4}He abundance also is central for the argument which gives an upper limit on the number of neutrinos. Specifically, an upper limit on the sum of primordial D + ^{3}He yields a lower limit on the baryon density of the universe at the time of BBN. Because the predicted ^{4}He abundance rises monotonically with increasing baryon density, putting a lower limit on this latter quantity also puts a lower limit on the predicted ^{4}He, i.e. the value of 23.7% quoted earlier. Now the predicted ^{4}He abundance also increases monotonically with the number of relativistic neutrino species present during BBN. Thus, an observational *lower* bound on ^{4}He puts an upper bound on extra neutrinos. If an upper bound of 24% on ^{4}He is used, a bound of $N_v < 3.4$ has been claimed[24]. However, it is very important to recognize that if one raises the upper bound on ^{4}He to $\approx 24.2\%$, this upper bound on N_ν increases to ≈ 3.6. If one includes additional possible uncertainties in the D+^{3}He bound this number could increase to ≈ 3.7-3.8.

None of these arguments takes away from the power of BBN to limit the number of new particles in nature. However, we have seen, with the 17 keV neutrino, that there may be a world of theoretical difference between 3.4 and 3.6 extra effective species in the radiation gas at $T \approx 1$ MeV. Before hanging one's theory on the hope of being able to distinguish between the BBN predictions for 3.4 and 3.6 species, some appreciation of the uncertainties in the limits on ^{4}He and the other light elements is warranted.

Finally, what if the actual primordial abundance of ^{4}He were less than 23.7%? What might the weak link in BBN then be? My own suspicion is that the D + ^{3}He limit might be revisable upwards. In this case a lower baryon density would be allowed, and thus a lower abundance of ^{4}He. I find this particularly attractive because not only would it allow slightly more ^{7}Li to be produced (see below), but it would also make the BBN predictions for the baryon abundance in the universe closer to the observed abundance of luminous matter. In this case, what you see would be what you get, a possibility I find appealing.

Next to ^{4}He the element which now appears to provide the most sensitive tests of BBN predictions is ^{7}Li. In particular, it is the bound on its primordial abundance which now yields one of the stringent upper bounds on the baryon density today, and hence which gives evidence in favor of exotic dark matter.

There has been some controversy in the past regarding the comparison of BBN predictions with observations. In particular, BBN predicts that as a function of the baryon density the predicted abundance of ^{7}Li reaches a minimum and then rises again. The baryon density which is consistent with observed ^{4}He and D+ ^{3}He abundances falls right near the minimum in the predicted ^{7}Li abundance, near about 10^{-10}.

Looking at very old halo stars, Spite and Spite[28] observed a plateau in the ^{7}Li abundance near a value of 10^{-10}. One can derive upper and lower limits on this abundance which are consistent with the BBN predictions, if one assumes the ^{7}Li abundance in these old objects is indicative of the primordial value, as shown below[29]. (The figure below was used because I have the graphics file on disk. An updated comparison can be obtained in ref 25.)

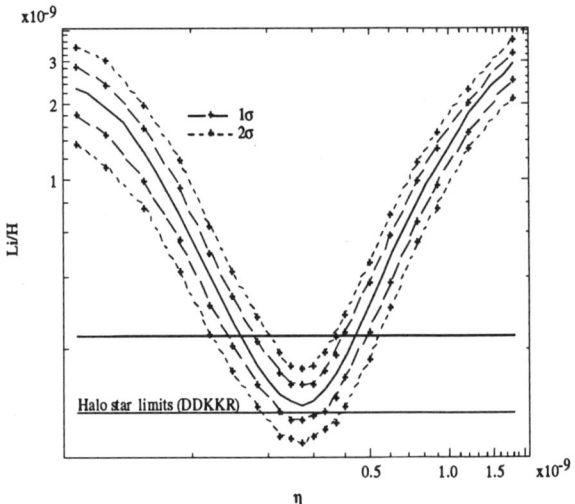

Figure 5: Lithium BBN Predictions and Inferences from Old Halo Stars.

At the same time, there has also been a ^{7}Li plateau observed in another set of much younger stars. The ^{7}Li surface abundance in these stars is an order of magnitude higher than that in the old halo stars. The question then becomes: Which population yields the true primordial value? If the old halo stars do, then Li must be produced in the Galaxy. If the younger stars do, then Li must be destroyed over time in stars. More important, perhaps, if the younger stars do, then it would be very difficult to explain this value in the context of standard BBN for a range of baryon densities consistent with the other elemental abundances.

There have been interesting theoretical arguments in favor of both alternatives. As a result, one would like to have alternate probes of the relevant physics. One of the most interesting of such probes involves another isotope of lithium, ^{6}Li. This isotope is much more fragile than ^{7}Li, and is more easily destroyed in stars. Moreover, BBN estimates suggest that ^{6}Li should be produced primordially at levels far below that of ^{7}Li. Interestingly, however, alternative BBN scenarios often predict a much larger ^{6}Li/^{7}Li ratio than standard BBN.

For this reason, there has been great interest recently when the first-ever measurement of ^{6}Li surface abundance in a stellar system was performed[30]. Observing an old halo star system, an abundance ratio ^{6}Li/^{7}Li ≈ .05 was reported. This is an extremely interesting result, if confirmed. Because ^{6}Li is destroyed much more readily in stars than ^{7}Li, this observation suggests that if this observable remnant ^{6}Li abundance remains in an old halo star, the change in the ^{7}Li abundance, even if the two isotopes were produced initially with *equal* abundance, cannot have been great. This appears to provide strong support for the standard BBN picture.

However, at the same time focus shifts from the question of how much ^{6}Li was destroyed, to the question of how the ^{6}Li abundance got so large in the first place. As I said, standard BBN generically predicts the primordial ^{6}Li/^{7}Li ratio to be small. Is this then a sign of a problem? Again, the answer at this time seems to be: not yet. Arguments regarding Li and Be production in the galaxy by cosmic rays[31] had earlier been used to predict sufficient ^{6}Li production to lead to a potentially observable ^{6}Li/^{7}Li ratio today. But it is not yet clear whether, allowing for stellar destruction of Li, whether as large a ratio as has been observed can be accomodated.

I think observations such as these are leading to the point where precision tests of the BBN picture are possible. At the present time, not only does the overall qualitative agreement of theoretical prediction and observational evidence persist, but so does the quantitative agreement. It now appears that the He abundance alone, in spite of the uncertainties I have concentrated on, provides a constraint on not only the number of neutrino species, but also provides both upper and lower limits on the allowed baryon density in the universe. The situation in the latter case is so tightly constrained that if the actual primordial He abundance could be reliably estimated to less than 5% one could in principle pinpoint a single value of the baryon density which would be consistent with standard BBN. At the same time, the Li situation is evolving quite quickly. It provides an independent probe of both standard and non-standard BBN, and also of models of stellar evolution. Thus the interest in new observations is very great, and should remain a high priority. I have no doubt that new results in this area will in fact occur between now and the next ICHEP meeting.

SOLAR NEUTRINOS OR....
THE SUN ALSO RISES.

Next to COBE, perhaps the most exciting new astrophysical measurements of interest to particle physics have come from two experiments designed to detect the flux of low energy neutrinos from the sun. Solar neutrinos have been observed for more than two decades now, thanks to Ray Davis and colleagues who have used a Cl detector in the Homestake mine. This experiment, primarily sensitive to the high energy B solar neutrinos, has observed on average between 1/4 and 1/3 the expected solar neutrino signal since the 1960's--- establishing the now famous "Solar Neutrino Problem". Since 1989, the Kamiokande large underground water Cerenkov detector, sensitive to only the high energy B flux, has reported a signal which is about 1/2 that predicted by the most often quoted "standard" solar model[32]. This detector has directional information and can therefore unambiguously associate the detected neutrino events with the sun.

Because the high energy B neutrinos are produced in reactions which only contribute marginally to the solar luminosity, there has been great interest in measuring the dominant, low energy pp neutrino flux, with a maximum energy of .42 MeV. In order to be convinced that the observed paucity of events in Homestake and Kamiokande have their origin in new neutrino physics, it has been clear that an experiment sensitive to this dominant flux of solar neutrinos is required. While one could remove the reactions which produce the entire B flux by some as of yet unknown astrophysical mechanism without significantly altering solar energetics, one could not remove the pp flux without turning off the sun.

In the last two years, with an anxious physics community chomping at the bit, the Soviet American Galium Experiment (SAGE) and the European GALLEX experiment have come on line and reported their results. The results have been, at the very least, surprising. I shall not have time to review the experimental details of these efforts (I think Hamish Robertson has done so in his lecture[33]) but will only quote the final results here.

In the table below I list the four operating solar neutrino experiments, their thresholds, and their quoted average rates. To simplify the table I have symmetrized the quoted statistical and systematic errors and added them in quadrature.

Experiment	$E_{threshold}$ (MeV)	Av. Rate/SSM[34]
Homestake	.81	0.28 ± 0.04[35]
KamiokandeI-II	7.5	0.46 ± 0.08[36]
KamiokandeI-III	7.5	0.49 ± 0.08[37]
SAGE (1st yr)	.24	0.15 ± 0.28[38]
SAGE (2 yrs)	.24	0.44 ± 0.21[39]
GALLEX	.24	0.63 ± 0.16[40]

Having the results of these 4 experiments, what have we learned after 25 years?

(1) The sun is probably shining: This is not an entirely facetious remark. As I have indicated, until the most recent set of gallium experiments we had not probed the dominant energy producing reaction in the sun. The first SAGE result was consistent with zero signal, indicating either the sun had turned off, neutrinos were not making it to earth, or the experiment was not detecting the flux. The GALLEX signal, on the other hand, if interpreted as a solar neutrino signal and not a background related signal, implied that there was indeed a non-zero pp flux from the sun. Hence the sun is shining. More recently, after 2 years of running, the SAGE experiment has also quoted a number which is no longer consistent with zero.

In fact, while it is often stated that the pp flux from the sun is model independent, and should account for at least an 80 SNU (0.6 SSM) signal in the gallium detectors, this is not quite correct. As has been stressed, for example by Michel Spiro[41], if we want to espouse what I would call the "agnostic view", which states that we really don't know anything about the sun, even this seemingly innocuous statement about the pp flux is an overstatement. I suspect, based on my discussions with many particle physicists, that many are sympathetic to at least a mild agnosticism, coming from their suspicion of all things astrophysical.

Independent of any solar physics whatsoever, all we really know is the *total* neutrino flux from the sun. This can be immediately calculated from the following facts

(a) the solar luminosity is 10^{33} ergs/sec,

(b) 25 MeV is released per ^{4}He formation

(c) 2 ν's are released per ^{4}He formation.

Combining these together fixes the total solar neutrino flux. If one is willing to ignore all conventional wisdom, one can fix the Be and B neutrino flux by combining the Homestake and Kamiokande results, and therefore fix the expected pp flux to make up the difference. In this case, one *predicts* the gallium signal, assuming all neutrinos which are produced in the sun make it to the earth. The predicted signal is 85 SNU, exactly what GALLEX sees!

One thus might be tempted, and many people have been, to use an argument like this, or else the fact that the GALLEX signal alone is on 2σ away from the standard solar model prediction, to argue that there is in fact no solar neutrino problem. The problem with the above argument is, however, that we *do* know something about the sun! In particular, we know that the Be neutrino flux is much less sensitive to variations in the solar core temperature than is the B flux, and thus it is extremely difficult to imagine varying them in such a way so that the Be flux is more suppressed than the B flux. This is what is required to reconcile the Homestake and Kamiokande results, as the former, which is sensitive to both Be and B reports a smaller signal than the latter which is only sensitive to B neutrinos[42,43]. This leads to my second conclusion:

(2) A neutrino-based solution of the solar neutrino problem remains favored: There are a number of different, but equivalent ways of expressing this fact. They all reduce to the following observation: While the standard solar model prediction is only 2σ away from the GALLEX observation, it is much further away from the other observations. Neutrino-based solutions, however, lie within 1σ of all the observations, for some finite parameter range. Moreover, even any real or imagined astrophysical variations of the standard solar model do not produce agreement of better than $3\text{-}4\sigma$ with all the experiments. One graphical demonstration of this which I find particularly appealing (at least in part because I and my collaborators prepared it[43]) is based on graphing the predictions of various models not in model-parameter space, but rather in experiment vs experiment space. In this case, one can get an immediate visual idea not only of the agreement between observations and theoretical predictions, but also of the "predictivity" of a model. For example, if a model can accomodate all points in this space, the fact that it may agree with a particular set of observations is not that significant. Below I show the predictions of scaling the B, or the B plus Be flux by a variable core temperature, and also the predictions of the favored MSW model for neutrino oscillations, in Homestake-Kamiokande space. In each case, we have allowed Monte-Carlo variations of all relevant solar neutrino parameters within the 1σ uncertainties of the standard solar model[43].

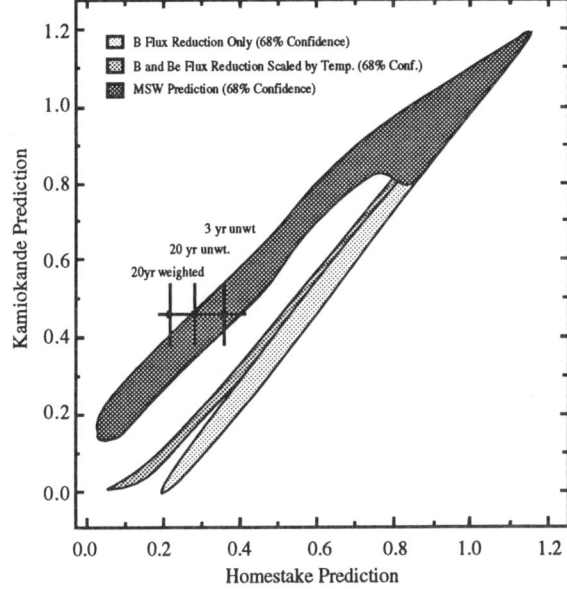

Figure 4: Model Predictions for Rates in Homestake and Kamiokande detectors[43]

As can be seen from this figure, even allowing the solar core temperature to vary arbitrarily does not bring the predictions, including solar model flux uncertainties anywhere near the 20 year averaged data. Indeed, as expected this kind of variation inevitably suggests a greater suppression of the Homestake rate than the Kamiokande rate. The MSW model, for example, on the other hand makes definitive predictions about the relative rates which can be in good agreement with this set of observations.

It is also enlightening to present the same plot for Gallium versus Kamiokande for these same models. I show here the GALLEX and updated SAGE results as well:

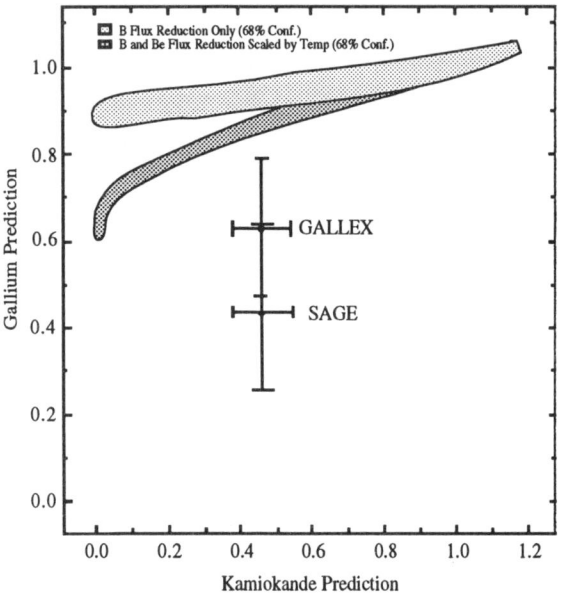

Figure 5: Predictions for Rates in Gallium and Kamiokande Detectors for varying Solar Core Temperatures.

Here the MSW locus of predictions is not shown, because it essentially fills almost all the experimental phase space, i.e. any set of experimental predictions for these experiments can be fit by some MSW parameter. Thus, it is not too surprising that the GALLEX and SAGE data points can be fit. What is worth noting, however, is that while lowering the solar core temperature can lower the gallium prediction to at least the range observed by GALLEX, even here the model fit is not that good, because significantly lower Kamiokande rates would be predicted than are observed. The fit is at best about 2σ in this case.

Thus, as advertised, astrophysical variations do not fit the combined data from the 4 experiments well, while neutrino based solutions, of which the MSW solution has been displayed here, do. It is worth pointing out that the MSW predictions are not unique in this regard. Essentially all neutrino-based models which could be made consistent with the Cl-Kamiokande results, including vacuum oscillations and neutrino magnetic moment based oscillations can be made consistent with all the data.[44] Thus, while a neutrino-based solution of the solar neutrino model remains favored, *which* solution is the right one is not yet unambiguous. This does not stop one from exploring which region of mass-mixing angle space fits the data, and a number of different groups have done just this[45,46,47,48]. Below I show again our own analysis, which includes a global fit to all the current data in Table 1, as well as solar model flux uncertainties, and shows the allowed (90% conf.) vacuum and MSW oscillation regions on the same plot[48]:

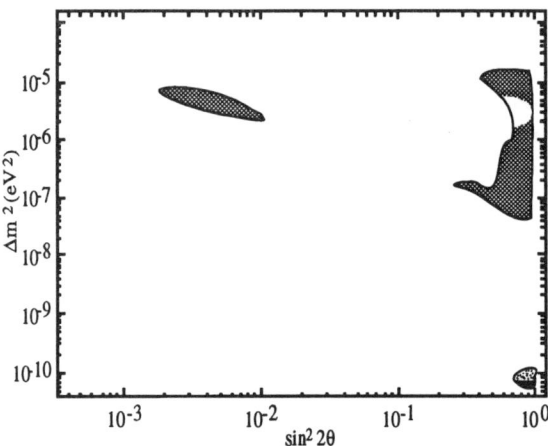

Figure 6: Allowed Range of Neutrino Mass-Mixing for Global Fit to All Experiments[48].

It is even possible to draw (although impossible to read) the allowed region in 3D parameter space in the case that there are oscillations between all of the known species. This general non-uniqueness of the possible solutions sets the stage for my third conclusion:

(3) Gallium experiments have not yet changed the Solar Neutrino Problem.. Specifically, the problem still stands or falls on the Homestake result, even after 25 years: Sad but true. The only range in which gallium could not further enlighten us as to the origin of the solar neutrino problem was in the range observed by GALLEX. Had the number been definitively higher it would have argued strongly in favor of an astrophysical solution. Had it been definitively lower, as SAGE first indicated, it would have argued strongly in favor of a neutrino-based solution. Moreover both the gallium results and the Kamiokande results are not sufficiently far from non-exotic variations in the standard solar model as rule out these possibilities. What was, and still remains damning for these models is the Homestake-Kamiokande combination; specifically the fact that Homestake reports the largest suppression. If the Homestake result were somehow discredited, and there no evidence at this point that it should be, I think much of the basis of the neutrino solution of the solar neutrino problem would evaporate.

(4) What we still need are high statistics counting experiments, and eventually experiments sensitive to the full solar neutrino spectrum. What the current confusion points to, at least in part, is the inadequacy of rare counting experiments. What one would like is several thousand events/year rather than several events per year. This is what SNO, Super-Kamiokande, and to some extent BOREXINO promise. Moreover SNO, with its neutral current capacity, at least in principle, could unambiguously resolve whether neutrino oscillations between the known neutrinos are responsible for the observed signals. Finally, even if this is the case, determining the neutrino mass parameters will still probably require spectral sensitivity to low energy neutrinos[49].

Thus, while this is probably the 10th ICHEP review of the solar neutrino problem over the last 20 years or so, we are unfortunately not yet within clear sight of a solution. Given the probable need for the next generation of detectors, which come on-line around 1996, and given that at least 1 year of running will be required, I predict that the problem will remain with us for at least the 1994 and 1996 meetings. However, the good news is that the answer may appear before the next millenium.

4. COSMIC RAYS AND GAMMA RAYS: OR.....
THAT WHICH IS NOT FORBIDDEN IS REQUIRED.

Cosmic ray physics has sometimes been a murky arena. Yet in the last few years a number of exciting observations have occurred in the field of high energy gamma ray physics which demonstrate exceptionally well the fact that the rationale need not be equated with the mundane. Moreover, while it is less sexy, another important development has taken place. High energy detectors around the world have mastered the art of measuring a Gaussian centered at zero with width 1σ. That is, with increasingly unerring accuracy the new generation of cosmic and gamma ray detectors have reported at this meeting seeing nothing where nothing was expected.[50] I hope this is not too discouraging to the observers, because it means that the field has achieved a level of maturity where both null results and observations are believable and repeatable. I shall concentrate here however on three new and exciting results: the isotropy of Gamma Ray Bursters, and the observation of High Energy gamma rays from the CRAB pulsar and a new more exotic object called Markarian 421, because I believe each of these provides an interesting object lesson for particle physicists.

Before reviewing these areas, however, I want to begin with a more fundamental issue which I have found is unfortunately not often stressed in the experimental reviews I have read and heard. That is: why do cosmic ray physics at all? As far as I can see the central point is the following: Cosmic rays have been observed up to 10^{20} eV. They are isotropic, and the charged particles observed with energies greater than 10^{15} eV are extragalactic. Thus, their origin is of cosmological interest. Moreover, we have no idea at the present time of: (a) the source(s), and (b) the acceleration mechanism which produces such incredibly high energies. It is thus quite likely that when we do learn about these issues, we will learn interesting new physics, or at the very least, discover interesting new objects. Finally, if we are to do astronomy, that is if we want to associate sources with observed signals we need to detect neutral particles such as photons, or neutrinos. It is only in this way that we can be sure that we can point back from the signal to the source.

(a) Gamma Ray Bursts:

I learned from Tsvi Piran's talk at this meeting[51] that gamma ray bursters were discovered by the VELA system of satellites in 1967. These were designed to monitor nuclear weapons testing on earth but could also look outwards for more exotic signals. These objects produce a burst of kev-MeV gamma rays with a duration of between 0.1 and 100 seconds. This time variability suggests that they have a radius less than about 3000 km, making neutron stars prime candidates. The big surprise of the past year or so, however, was the observation by the COMPTON satellite of over 150 bursts during the course of a year with a distribution which was statistically isotropic. This is completely unexpected if the sources of these bursts are in our galaxy. Moreover, the distribution of intensities follows a canonical -3/2 power law which is characteristic of a cosmological distribution rather than a local one.

What is so surprising about the possibility that gamma ray bursters are cosmological in origin is the requirements on their net luminosities in order to produce the observed fluences here on earth, which are in the range of 10^{-5} ergs/cm^2. Assuming isotropic emission one then estimates intrinsic gamma ray luminosities on the order of 10^{51}-10^{53} ergs for these objects for redshifts of order unity.

One explanation for these events, if they are cosmological is particularly intriguing, not merely because of the astrophysical interest, but because it demonstrates that even the most implausible events may happen regularly in the universe. The prime candidate explanation at present appears to involve coalescing neutron starts which should produce huge amounts of energy. What is particularly remarkable about this possibility is that such events are predicted, based on estimates of the neutron star binary abundances in typical galaxies, to occur at a rate of about 10^{-6}/yr per galaxy. However, if one integrates out to a distance of a Hubble length, over which they should be observable, one estimates a predicted event rate of about 1000 events per year. This is more or less the rate which is seen.

I have no idea whether this explanation is correct. If it is, there will no doubt be other interesting signals, including, I would imagine, potential neutrino signals. However, it should remind us that we should not be timid in imagining what new sources are out there.

(b) High Energy Gamma Ray sources and Telescopes:

A new generation of Cerenkov imaging detectors for electromagnetic shower detection[52,53] has brought forward the discovery of a number of interesting sources. What is particularly nice about these detections is that they do not involve statistics. In each case a clear image of the source has been obtained, which can be identified with a known astronomical source. Because this meeting is near the SSC site, I thought it would be particularly appropriate to review the two most interesting sources, both of which have application to SSC physics.

(i) The first source is the Crab pulsar, located at a distance of about 2200 parsecs. This is the highest energy known source of gamma rays in the universe. Gamma rays in the range of 3-15 TeV have been detected at the Whipple observatory[52], and based on the observed flux, and assuming isotropic emission one obtains a continuous luminosity of about 10^{36} ergs, or about 1000 times the solar luminosity, in this energy range. This must be therefore a truly remarkable accelerator!

What makes the Crab even more remarkable is the proximity in energy with that of the SSC center of mass energy. In 1999, when the SSC turns on, it will overtake the Crab and be, barring other discoveries, the highest energy accelerator known not only in Texas, or the US, or the world, but in the Universe! I hope this fact will prove useful in discussions with politicians.

(ii) The second source I want to discuss was also first seen in TeV gamma rays at the Whipple Observatory[52]. It goes by the name of Markarian 421, and is associated with a known optical source at a distance of about 125 Mpc. The gamma ray signal is seen with good (0.1° scale) angular resolution and has been observed continuously over a period of months. Moreover, the image distribution matches exactly the Crab profile. While the maximum energy of the observed signal is not quite as high as the Crab, rather in 1 TeV range, it is not this maximum energy which makes Markarian 421 so remarkable. Using the measured flux in TeV gamma rays one determines an estimated luminosity of $\approx 3 \times 10^{42}$ ergs/sec in this energy range! This is over one billion times the luminosity of the sun, *in TeV gammas!* Put another way, this is equivalent to emitting the energy equivalent of about .001 solar masses per year in TeV gamma rays! Whatever this object is, it is truly remarkable. More relevant to this meeting, whatever it is, it is a very efficient TeV accelerator. It may prove very useful to particle physics to understand how and why.

Going further on these lines, consider the following. Say that Mrk 421 originates with a civilization whose congress was more advanced than ours and approved and funded a working SSC by the present time. Imagine further that the SSC beam was pointing directly at the earth from a distance of 100 Mpc. How would it stand up compared to the flux from Markarian 421? Assuming a beam spread of about 4×10^{-6} radians, which is probably somewhat better than the design specs for the SSC, I derive an observed flux at earth which is approximately 10^{-8} times the measured Mrk flux. Thus Markarian 421 has a luminosity in TeV gamma rays, presumably in every possible direction, of approximately 100 million SSC's!

Moving from these awesome objects, it is perhaps fitting to close this review of cosmic ray physics with an eye to the future. In particular, one's imagination cannot help being captured by the ultimate cosmic ray array, being promoted by Jim Cronin as an internationally funded device. This array would have over 2500 detector stations located in a square grid, about 1.5 km apart, communicating with each other by microwave relays. The beauty of this proposal is that it takes a particle physics approach to the problem of cosmic ray detection. If we are to ever learn about the origin of the highest energy cosmic rays we have learned from experience that we will probably never do so by detecting 1 or 10 events per year. The Array 5000 detector system will have a predicted event rate of particles with energy in excess of 10^{19} ev of 5000 events/year! This is to be compared with the entire integrated world sample of such events which as present is less than 100 events.... In the words of Shelly Glashow, who said when arguing in favor of a now existing accelerator in Geneva: "Do you want to walk, or do you want to fly?"....

5. CONCLUSIONS

Had I more time in this lecture I would certainly have included at least two more topics which I have not been able to treat here. They are: Electroweak Baryogenesis, and The Curse of the Planck Scale. The former has shown that even the standard model can be interesting....in spite of LEP. A great deal of work has been devoted to exploring the possibility that the observed baryon density of the universe, namely all the we can see, was generated dynamically not at the GUT scale, but rather at the electroweak phase transition. In the process, at the very least, new insights into the nature of the electroweak transition have been gleaned. I refer interested readers to a recent review volume[54]. The latter topic relates to the possibility that the Planck scale might not be impotent, as far as low energy physics is concerned. First, it has been argued that the quantum theory of gravity, whatever it may be, can lead to measurable violations of global symmetries at low energies[55]. While this idea has been around for some time[56], it has taken on renewed interest with various cosmological proposals which require unbroken global symmetries below the Planck scale. Second, it has been proposed that physics at the Planck scale might be responsible for ensuring symmetries when nothing else does. In particular, it might solve the strong CP problem[57]. The only problem with these proposals is that we do not have a theory of quantum gravity with which to check them. In particular, for every argument in favor of the first, having to do with gravitational violation of global symmetries, there exists an equally attractive counter-argument. Nevertheless, they do open, at least in principle, the possibility that low energy measurements might tell us something about physics at the Planck scale.

These topics aside, what conclusions can we draw from the developments of the past two years which I have reviewed in depth here? I think first and foremost, we live in interesting times for cosmology:

(i) We now have further proof that the Big Bang *Really* Happened!

(ii) For the first time, the holy trinity of modern cosmology: inflation, a flat universe, and gravitational clustering as the source of large scale structure, may now finally be consistent with observation.

(iii) At the same time, Cold Dark Matter may, or may not, finally be inconsistent with observation.

(iv) Big Bang Nucleosynthesis is still alive and kicking, and at the threshold of several precise tests which can prove it is virus free.

(v) Neutrinos which can prove the sun shines have finally probably been observed. They have not yet solved the Solar Neutrino puzzle, but they do give us cause for hope in 1998. He who waits is rewarded. ..

(vi) Cosmic ray physics has spawned a new generation of detectors. New incredible objects have been discovered, and there is no doubt that more surprises await.

Perhaps the real lesson of these results is that as far as cosmology and astrophysics are concerned, the 1990's could well be the Decade of Discovery. They might do for astophysics what the period 1967-1977 did for particle physics. Of course, this might be a curse instead of a blessing. Nevertheless, it is patently clear that the complementarity between particle physics and cosmology not only persists, it is thriving.

Acknowledgements:

This review would not have been possible without the aid of a number of individuals who helped tutor me in various areas, or who informed me of new, late-breaking results. In particular I want to thank René Ong for his patience and kind explanations of various phenomena in high energy gamma ray physics, which helped clear up many questions I had. I also want to thank Gary Steigman for discussions of Lithium abundances, George Smoot, Gary Hinshaw, Paul Steinhardt and Martin White for discussions (and in the latter case collaborations) on COBE, and also Martin White and Evalyn Gates for their collaboration on issues related to the solar neutrino problem.

REFERENCES

1. J. C. Mather et al, "A Preliminary Measurement of the Cosmic Microwave Background by the Background Explorer (COBE) Satellite," *Ap. J.* 354, pp. L37-L40 (1990).

2. T. Matsumoto et al, "The Submillimeter spectrum of the Cosmic Background Radiation," *Ap. J.* 329, 567-571 (1988).

3. see G. F. Smoot, these proceedings; also G. F. Smoot et al, "COBE Differential Microwave Radiometers: Instrument Design and Implementation," *Ap. J.* 360, 685-695 (1990); G. F. Smoot et al, "Structure in the COBE Differential Microwave Radiometer First-Year Maps", *Ap. J.* 396, L1-L5 (1992).

4. C. L. Bennett et al, "Preliminary Separation of Galactic and Cosmic Microwave Emission for the Cobe Differential Microwave Radiometer", *Ap. J.* 396, L7-L12 (1992).

5. see for example, P.J. E. Peebles, *The Large Scale Structure of the Universe*, Princeton: Princeton U. Press, 1980.

6. E. R. Harrison, "Fluctuations at the Threshold of Classical Cosmology," *Phys. Rev.* D1, 2726-2730 (1970); Ya. B. Zel'dovich, "A Hypothesis, Unifying the Structure and the Entropy of the Universe," *MNRAS* 160, 1P-3P (1972); P.J. E. Peebles and J. T. Yu, "Primeval Adiabatic Perturbation in an Expanding Universe," *Ap. J.* 162, 815 (1970).

7. see J. Frieman,"Inflation", these Proceedings.

8. see J. Gelb, "N-body Simulations of Cold Dark Mattter", these Proceedings.

9. see E. L. Wright et al, "Interpretation of the Cosmic Micreowave Background Radiation Anisotropy Detected by the COBE Differential Microwave Radiometer", *Ap. J.* 396, L13-L18 (1992).

10. A.A. Starobinsky, "Spectrum of Relic Gravitational Radiation and the Early State of the Universe," *JETP Lett* 30, 682-685 (1979).

11. L.M. Krauss and M. White, "Grand Unification, Gravitational Waves, and the Cosmic Microwave Background Anisotropy", *Phys. Rev. Lett.* 69, 869-872 (1992).

12. i.e. see M. Peskin, "Beyond the Standard Model", these proceedings.

13. R. L. Davis et al, "Cosmic Microwave Background Probes Models of Inflation", *Phys. Rev. Lett.* 69, 1856-1859 (1992); D.S. Salopek, "Consequence of the COBE

Satellite for the Inflationary Scenario", DAMTP preprint, submitted to *Phys. Rev. Lett.* ; L. M. Krauss, "COBE, Inflation, and Inflation Scalars", Yale preprint, submitted to *Phys. Rev. Lett.*

14. W. Saunders et al., "The Density Field of the Local Universe," *Nature* 349, 32-38 (1991).

15. A. Oemler, D.L. Tucker, R.P. Kirshner, H. Lin, S.A. Shectman, P.L. Schechter, "The Las Campanas Deep Redshift Survey" to be published in *Observational Cosmology*, A.S.P. Conference Series, 1993.

16. see A. Dekel, "Large Scale Structure: Dynamics" and references therein, to appear in *Proceedings of Rencontres de Blois, Particle Astrophysics,* June 1992, Editions Frontieres, to appear; see also, A. Dekel, E. Bertshinger, A. Yahil, M. Strauss, M. Davis, and J. Huchra, to appear.

17. see A. Yahil, "Extra Galactic Dark Matter" and references therein, to appear in *Proceedings of Rencontres de Blois, Particle Astrophysics,* June 1992, Editions Frontieres, to appear.

18. see J. Gelb, "N-body Simulations of Cold Dark Mattter", these proceedings; also J. Gelb, MIT Ph.D thesis; M.A. Strauss and M. Davis, in *Large Scale Motions in the Universe*, Princeton: Princeton U. Press (1990).

19. D. Seckel and M. S. Turner, ""Isothermal" Density Perturbations in an Axion-dominated Inflationary Universe," *Phys. Rev.* D32, 3178-3183 (1985).

20. L. M. Krauss, "COBE, Inflation, and Inflation Scalars", Yale preprint, submitted to *Phys. Rev. Lett.*

21. see L.M. Krauss, "Non Baryonic Dark Matter" and references therein, to appear in *Proceedings of Rencontres de Blois, Particle Astrophysics,* June 1992, Editions Frontieres, to appear.

22. G. P. E. Efstathiou, W. Sutherland, S.J. Maddox, "The Cosmological Constant and Cold Dark Matter," *Nature* 348, 705-707 (1990).

23. L. M. Krauss and P. Romanelli, "Big Bang Nucleosynthesis: Predictions and Uncertainties", *Ap. J.* 358, 47-59 (1990).

24. T. P. Walker et al, "Primordial Nucleosynthesis Redux", *Ap. J.* 376, 51-69 (1991).

25. M. Smith, L. H. Kawano, and R. A. Malaney, "Experimental, Computational, and Observational Analysis of Primordial Nucleosynthesis", Caltech preprint OAP-716, submitted to *Ap. J.* May 1991.

26. K. A. Olive, G. Steigman and T.P. Walker, "The Upper Bound to the Primordial Abundance of Helium and the Consistency of the Hot Big Bang Model", *Ap. J.* 380, L1-L4 (1991).

27. see ref. 26, and also B.E.J. Pagel, A. Kazlauskas, "Primordial Helium: The Third Decimal Place", Nordita preprint, 92/24A, submitted to MNRAS. 1992.

28. F. Spite and M. Spite, "Lithium Abundance in the Nitrogen-rich Halo Dwarfs," *Astron. Astrophys.* 163, 140-144 (1986).

29. L. M. Krauss and P. Romanelli, "Big Bang Nucleosynthesis: Predictions and Uncertainties", *Ap. J.* 358, 47-59 (1990).

30. V. Smith et al, U. Texas preprint, 1992.

31. G. Steigman and Terry P. Walker, "Production of Li, Be, and B in the Early Galaxy," *Ap.J.* 385, L13-L16 (1992).

32. for a further discussion see J. Bahcall, "Solar Neutrino Predictions", these proceedings.

33. see H. Robertson, "Non-Accelerator Physics", these proceedings.

34. J. Bahcall, *Neutrino Astrophysics*, Cambridge: Cambridge University Press, 1989.

35. see ref. 34; also R. Davis Jr. in *Proceedings of the Seventh Workshop on Grand Unification,* Toyama, Japan 1986, ed. by J. Arafune, Singapore: World Scientific 1986.

36. K. Hirata et al., "Constraints on Neutrino-Oscillation Parameters from the Kamiokande-II Solar-Neutrino Data," *Phys. Rev. Lett.* 65 1301-1304 (1990); "Search for Day-Night and Semiannual Variations in the Solar Neutrino Flux Observed int he Kamiokande-II Detector," *Phys. Rev. Lett.* 66, 9-12 (1991).

37. see T. Kajita, "Solar and Atmospheric Neutrinos", these proceedings.

38. A. I. Abazov et al, "Search for Neutrinos from the Sun Using the Reaction ^{71}Ga(ν_e,e$^-$) ^{71}Ge," *Phys. Rev. Lett.* 67, 3332-3335, (1991).

39. A. Gavrin, "Summary of the Soviet American Gallium Experiment", these proceedings.

40. D Vignaud, "Report on GALLEX", these proceedings.

41. M. Spiro and D. Vignaud, "Solar Model Independent Neutrino Oscillation Signals in the Forthcoming Solar Neutrino Experiments," *Phys. Lett.* B242, 279-284 (1990).

42. J.N. Bahcall and H. A. Bethe, "A Solution of the Solar Neutrino Problem," *Phys. Rev. Lett.* 65, 2233-2235 (1990)

43. see. M. White, L.M. Krauss and E. Gates, "A New Look at the Solar Neutrino Problem", Yale preprint YCTP-P14-92 April 1992, submitted to *Phys. Rev. Lett.*

44. see for example C. S. Lim et al, "Correlation Between Solar Neutrino Flux and Solar Magnetic Activity for Majorana Neutrinos," *Phys. Lett.* B243, 389-395 (1990); P. I. Krastev and S.T. Petcov, "Neutrino Oscillatiosn in Vacuum as a Possible Solution of the Solar Neutrino Problem," *Phys. Lett.* B285, 85-90 (1992); E. Gates, L.M. Krauss, and M. White, "Solar Neutrino Data and Their Implications", *Phys. Rev.* D46, 1263-1273 (1992).

45. X. Shi, D.N. Schramm, and J.N. Bahcall, "Monte Carlo Exploration of Mikheyev-Smirnov-Wolfenstein solutions to the Solar Neutrino Problem", *Phys. Rev. Lett.* 69, 717-720 (1992).

46. S. Bludman et al, Penn preprint, July 1992.

47. P. I. Krastev and S.T. Petcov, "Neutrino Oscillatiosn in Vacuum as a Possible Solution of the Solar Neutrino Problem,", *Phys. Lett.* B285, 85-90 (1992).

48. L.M. Krauss, E. Gates, and M. White, "Solar Neutrino Data, Solar Model Uncertainties, and Neutrino Oscillations", Yale preprint YCTP-P38-92, Sept 1992, submitted to *Phys. Lett. B.*

49. see for example, ref. 43, and J. M. Gelb, W. Kwong and S. P. Rosen, "Implications of new GALLEX results for the MSW solution of the solar neutrino problem", *Phys. Rev. Lett.* 69, 1864-1866 (1992)

50. see D. Williams, "Results from the CYGNUS expt and Plans for MILAGRO", T. McKay, "Search for Astrophysical Gamma Rays above 100 TeV", C. Akerlof, "TeV gamma rays from Markarian 421", P. Baillon, "Detection of Very High Energy Gamma Rays from the Crab Nebula", M. Longo, "Muon and Neutrino Astronomy (MACRO)", these proceedings.

51. see T. Piran, "Gamma Ray Bursters", these proceedings.

52. see C. Akerlof, "TeV gamma rays from Markarian 421", these proceedings.

53. see P. Baillon, "Detection of Very High Energy Gamma Rays from the Crab Nebula"

54. see *Baryon Number Violation at the Electroweak Scale,* ed. L.M. Krauss and S.J. Rey, Singapore: World Scientific, 1992, to appear.

55. eg. see M. Kamionkowski and J. March Russell, "Planck-scale Physics and the Peccei-Quinn Mechanism," *Phys. Lett.* B282, 137-141 (1992); R. Holman et al, "Solutions to the Strong-CP Problem in a World with Gravity," *Phys. Lett.* B282, 132-136 (1992).

56. i.e. see. H. Georgi, L.J. Hall, M. B. Wise, "Grand Unified Models with an Automatic Peccei-Quinn Symmetry, *Nucl. Phys.* B192, 409-416 (1981).

57. eg. see M. Dine et al, "CP and Other Gauge Symmetries in String Theory", *Phys. Rev. Lett.* 69, 2030-2032 (1992); K. Choi, D. Kaplan, and A. Nelson, UCSD preprint PTH 92-11, 1992.

DISCUSSION

D. Brahm, California Institute of Technology, USA

I often hear limits placed on the overall angular velocity of the universe from the lack of a quadrupole anisotropy. Can we turn that around and explain some of the quadrupole anisotropy as evidence of rotation?

Krauss

George Smoot gave an upper limit on that angular velocity in his talk, and it was very small.

NONPERTURBATIVE METHODS*

H. Leutwyler
Institute for Theoretical Physics
University of Bern
Sidlerstr. 5
CH-3012 Bern, Switzerland

Abstract

Recent work on effective field theory is reviewed, focussing on chiral perturbation theory. The principle of the method is outlined. Some of the results obtained with it and some of the problems under investigation are discussed.

INTRODUCTION

Quite a few nonperturbative methods are being used in particle physics. One of these is lattice field theory, covered in a separate review at this meeting[1]. Several others are discussed in papers contributed to this conference: effective Lagrangians, QCD sum rules, high orders of perturbation theory, light cone quantization etc. In particular, nonperturbative methods are required to analyze the low energy structure of QCD, to study baryon number violating processes at the electroweak phase transition and the issue of whether or not these processes might be observable in high energy collisions also calls for theoretical understanding beyond the realm of perturbation theory. As it is impossible for me to review all of this work in the time available, I decided to instead focus on one such method, *effective field theory*. Some of the results obtained with other nonperturbative methods were discussed in one of the parallel sessions[2].

Effective field theory methods can be used in a broad variety of contexts: low energy structure of QCD, Higgs sector of the standard model and its extensions, spin models, solid state physics etc. In fact, one of the remarkable features of effective Lagrangians is their universality. A compound like La_2CuO_4 which develops two-dimensional antiferromagnetic layers and exhibits superconductivity up to rather high temperatures (hence of potential interest for the supercollider, particularly in summer) can be described by an effective Lagrangian[3] which closely resembles the one relevant for QCD! In the following, I will restrict myself to the latter, referring you to the talks of Bagger, Peskin and Valencia[4] for a discussion of effective field theories in the context of electroweak symmetry breaking.

EFFECTIVE LOW ENERGY THEORY OF QCD

At low energies, the behaviour of scattering amplitudes or current matrix elements can be described in terms of a *Taylor series expansion* in powers of the momenta. The electromagnetic form factor of the pion, e.g., may be exanded in powers of the momentum transfer t. In this case, the first two Taylor coefficients are related to the total charge of

---

*Work supported by Schweizerischer Nationalfonds.

the particle and to the mean square radius of the charge distribution, respectively,

$$f_{\pi^+}(t) = 1 + \frac{1}{6} <r^2>_{\pi^+} t + O(t^2). \quad (1)$$

Scattering lengths and effective ranges are analogous low energy constants occurring in the Taylor series expansion of scattering amplitudes.

For the straightforward expansion in powers of the momenta to hold it is essential that the theory does not contain massless particles. The exchange of photons, e.g., gives rise to Coulomb scattering, described by an amplitude of the form $e^2/(p'-p)^2$ which does not admit a Taylor series expansion. Now, QCD does not contain massless particles, but it does contain very light ones: pions. The occurrence of light particles gives rise to singularities in the low energy domain which limit the range of validity of the Taylor series representation. The form factor $f_{\pi^+}(t)$, e.g., contains a cut starting at $t = 4M_\pi^2$, such that the formula (1) provides an adequate representation only for $t \ll 4M_\pi^2$. To extend this representation to larger momenta, one needs to account for the singularities generated by the pions. This can be done, because the reason why M_π is so small is understood: the pions are the Goldstone bosons of a hidden, approximate symmetry[5]. The low energy singularities generated by the remaining members of the pseudoscalar octet ($K^\pm, K^0, \bar{K}^0, \eta$) can be dealt with in the same manner, exploiting the fact that the Hamiltonian of QCD is approximately invariant under chiral $SU(3)_L \times SU(3)_R$. If the three light quark flavours u, d, s, were massless, this symmetry would be an exact one. In reality, chiral symmetry is broken by the quark mass term ocurring in the QCD Hamiltonian

$$H_{\text{QCD}} = H_0 + H_1 \quad (2)$$
$$H_1 = \int d^3x \{m_u \bar{u}u + m_d \bar{d}d + m_s \bar{s}s\}$$

For yet unknown reasons, the masses m_u, m_d, m_s however happen to be small – H_1 can be treated as a perturbation. First order perturbation theory shows that the expansion of the square of the pion mass in powers of m_u, m_d, m_s starts with

$$M_{\pi^+}^2 = (m_u + m_d)B\{1 + O(m_u, m_d, m_s)\} \quad (3)$$

while, for the kaon, the leading term contains the mass of the strange quark,

$$\begin{aligned} M_{K^+}^2 &= (m_u + m_s)B + \ldots \\ M_{K^0}^2 &= (m_d + m_s)B + \ldots \end{aligned} \quad (4)$$

This explains why the pseudoscalar octet contains the eight lightest hadrons and why the mass pattern of this multiplet very strongly breaks eightfold way symmetry: M_π^2, M_K^2 and M_η^2 are proportional to combinations of quark masses, which are small but very different from one another, $m_s \gg m_d > m_u$. For all other multiplets of SU(3), the main contribution to the mass is given by the eigenvalue of H_0 and is of order Λ_{QCD}, while H_1 merely generates a correction which splits the multiplet, the state with the largest matrix element of $\bar{s}s$ ending up at the top.

The effective field theory combines the expansion in powers of momenta with the expansion in powers of m_u, m_d, m_s. The resulting new improved Taylor series, which explicitly accounts for the singularities generated by the Goldstone bosons, is referred to as chiral perturbation theory[6] (χPT). It provides a solid mathematical basis for what used to be called the "PCAC hypothesis".

It does not appear to be possible to account for the singularities generated by the next heavier bound states, the vector mesons, in an equally satisfactory manner. The mass of the ρ-meson is of the order of the scale of QCD and cannot consistently be treated as a small quantity. Although the vector meson dominance hypothesis does lead to valid estimates (an example is given below), a coherent framework which treats these estimates as leading terms of a systematic approximation scheme is not in sight.

The effective low energy theory replaces the quark and gluon fields of QCD by a set of pseudoscalar fields describing the degrees of freedom of the Goldstone bosons π, K, η. It is convenient to collect these fields in a 3×3 matrix $U(x) \in SU(3)$. Accordingly, the Lagrangian of QCD is replaced by an effective Lagrangian which only involves the field $U(x)$ and its derivatives. The most remarkable point here is that this procedure does not mutilate the theory: if the effective Lagrangian is chosen properly, the effective theory is mathematically equivalent to QCD[7].

On the level of the effective Lagrangian, the combined expansion introduced above amounts to an expansion in powers of derivatives and powers of the quark mass matrix

$$\mathcal{M} = \begin{pmatrix} m_u & & \\ & m_d & \\ & & m_d \end{pmatrix} \quad (5)$$

Lorentz invariance and chiral symmetry very strongly constrain the form of the terms occurring in this expansion. Counting $\mathcal{M}$ like two powers of momenta, the expansion starts at $O(p^2)$ and only contains even terms

$$\mathcal{L}_{\text{eff}} = \mathcal{L}_{\text{eff}}^{(2)} + \mathcal{L}_{\text{eff}}^{(4)} + \mathcal{L}_{\text{eff}}^{(6)} + \ldots \quad (6)$$

The leading contribution is of the form

$$\mathcal{L}_{\text{eff}}^{(2)} = \frac{F_\pi^2}{4} \text{tr}\{\partial_\mu U^+ \partial^\mu U\} + \frac{F_\pi^2 B}{2} \text{tr}\{\mathcal{M}(U+U^+)\} \quad (7)$$

and involves two independent coupling constants - the pion decay constant F_π and the constant B occurring in the mass formuale (3), (4). The expression (7) represents a compact summary of the soft pion theorems established in the 1960's: the leading terms in the chiral expansion of the scattering amplitudes and current matrix elements are given by the tree graphs of this Lagrangian.

At order p^4, the effective Lagrangian contains terms with four derivatives such as

$$\mathcal{L}_{\text{eff}}^{(4)} = L_1 [\text{tr}\{\partial_\mu U^+ \partial^\mu U\}]^2 + \ldots \quad (8)$$

as well as terms with one or two powers of $\mathcal{M}$. Altogether, ten coupling constants occur[8], denoted $L_1, \ldots, L_{10}$. Four of these are needed to specify the scattering matrix to first nonleading order. The terms of order $\mathcal{M}^2$ in the meson mass formulae (3), (4) involve another three of these constants. The remaining three couplings concern current matrix elements.

As an illustration, consider again the e.m. form factor $f_{\pi^+}(t)$. To order p^2, the chiral representation reads[9]

$$f_{\pi^+}(t) = 1 + \frac{t}{F_\pi^2}\{2L_9 + 2\phi_\pi(t) + \phi_K(t)\} + O(t^2, t\mathcal{M}) \quad (9)$$

In this example, the leading term (tree graph of $\mathcal{L}_{\text{eff}}^{(2)}$) is trivial. At order p^2, there are two contributions: the term linear in t arises from a tree graph of $\mathcal{L}_{\text{eff}}^{(4)}$ and involves the coupling constant L_9, while the functions $\phi_\pi(t)$ and $\phi_K(t)$ originate in one loop graphs generated by $\mathcal{L}_{\text{eff}}^{(2)}$. The loop integrals contain a logarithmic divergence which is absorbed in a renormalization of L_9 – the net result for $f_{\pi^+}(t)$ is independent of the regularization used. The representation (9) shows how the straightforward Taylor series (1) is modified by the singularites due to $\pi\pi$ and $K\bar{K}$ intermediate states. At the order of the chiral expansion we are considering here, these singularities are described by the one loop integrals $\phi_\pi(t), \phi_K(t)$ which contain cuts starting at $t = 4M_\pi^2$ and $t = 4M_K^2$, respectively. The result (9) also shows that chiral symmetry does not determine the pion charge radius: its magnitude depends on the value of the coupling constant L_9 – the effective Lagrangian is consistent with chiral symmetry for any value of the coupling constants. The symmetry, however, *relates* different observables. The slope of the K_{l_3} form factor $f_+(t)$, e.g., is also fixed by L_9. The experimental value of this slope[10], $\lambda_+ = 0.030$, can therefore be used to first determine the magnitude of L_9 and then to calculate the pion charge radius. This gives $<r^2>_{\pi^+} = 0.42$ fm^2, to be compared with the experimental result, 0.44 fm^2.[11]

In the case of the neutral kaon, the representation analogous to eq. (9) reads

$$f_{K^0}(t) = \frac{t}{F_\pi^2}\{-\phi_\pi(t) + \phi_K(t)\} + O(t^2, t\mathcal{M}). \quad (10)$$

A term of order one does not occur here because the charge vanishes and there is no contribution from $\mathcal{L}_{\text{eff}}^{(4)}$, either. Chiral perturbation theory thus provides a parameter free prediction in terms of the one loop integrals $\phi_\pi(t), \phi_K(t)$. In particular, the slope of the form factor is given by[9]

$$<r^2>_{K^0} = -\frac{1}{16\pi^2 F_\pi^2} \ln \frac{M_K}{M_\pi} = -0.04 \text{fm}^2 \quad (11)$$

to be compared with the experimental value -0.054 ± 0.026 fm^2.[12]

OVERVIEW

In the framework of this short review, the above examples must suffice to illustrate the principle of the method. Many applications have been worked out and many more are under way. Chiral perturbation theory is a very active field of current research (see e.g. ref.[13]) and it is impossible for me to describe this activity in any detail. The present section is to provide a cursory overview and a list of references to the recent literature. A few selected topics of current interest are missing here - they are discussed more thoroughly in the second part of the talk.

$\pi\pi$ Scattering

Pion scattering is the hard core of chirography. It represents one of the rare examples in strong interaction physics where the theoretical predictions[14] are more accurate than the available experimental results. Of particular interest are the S-wave scattering lengths which play a role analogous to the σ-term of πN scattering and which are sensitive to chiral symmetry breaking. As pointed out by Nemenov[15], the decay of $\pi^+\pi^-$-atoms into $\pi^0\pi^0$ offers the opportunity of measuring the S-wave scattering lengths, thereby submitting our understanding of chiral symmetry breaking to a crucial test.

πK Scattering

The corrections of order p^4 to the soft kaon amplitudes are analyzed in ref.[16]. The extension to higher energies, where resonances play an important role, is discussed in ref.[17].

Processes Induced by the Chiral Anomaly

At leading order in the chiral expansion, these processes are described by the Wess-Zumino-Lagrangian[18]. Several authors have analyzed the corrections of order p^6, in particular in view of the decays $\pi^0 \to \gamma\gamma^*, \eta \to \gamma\gamma^*, \eta \to \pi\pi\gamma$ or of the photoproduction reaction $\gamma\pi \to \pi\pi$.[19]

Semileptonic Decays

For nearly all measured semileptonic pion or kaon decays, including radiative transitions, the matrix elements have now been worked out in χPT to order p^4. For a comprehensive discussion, see ref.[20,21]. As pointed out in ref.[22] the available data on the process $K \to \pi\pi e\bar{\nu}$ allow an independent measurement of the coupling constants L_1, L_2 and L_3. The result is in good agreement with the estimates based on the $\pi\pi$ D-waves[8] and confirms the validity of the OZI rule in this case. Many of the τ decay channels can be analyzed by means of χPT, at least in part of phase space[23].

Nonleptonic K Decays

At leading order in the derivative expansion, the effective Lagrangian of the nonleptonic weak interaction involves two independent coupling constants. The $\Delta I = \frac{1}{2}$ rule states that one of these is small compared to the other[24]. Considerable progress in understanding this empirical fact has been achieved in recent years. The rule turns out to be the result of a combination of dynamical effects involving both, hard short distance phenomena and soft final state interactions[25]. Although a calculation of the two effective coupling constants from first principles still involves model dependent steps, the main features of nonleptonic kaon decays are now understood on the basis of the Standard Model.

Recently, the machinery needed for a systematic investigation of the terms of $O(p^4)$ has been set up[26] and used to analyze the decays $K \to 2\pi, 3\pi$ [26,27]. The results are in good agreement with the data which offer several significant tests of the method. As pointed out by D'Ambrosio and Espriu[28] and, independently, by Goity[28], the leading term in the chiral expansion of the amplitude for $K_S \to \gamma\gamma$ originates in a one loop graph and does not involve any unknown parameters (compare the prediction for the charge radius of the neutral kaon discussed above). The prediction is confirmed by NA31[29]. A similar parameter free one loop prediction also holds for the decay $K_L \to \pi^0\gamma\gamma$.[30] The data from both NA31 and E731[31] indicate that this prediction underestimates the rate by a factor of two or three, while the shape of the spectrum is confirmed. The prediction represents the leading term in the chiral expansion; some of the nonleading corrections have already been analyzed[32]. Rescattering effects are typical corrections of this sort and contributions from low lying resonances may also be relevant. The issue is currently under investigation by several groups and related processes such as $K \to \gamma\gamma^*, K \to \pi\gamma\gamma^*, K \to \pi\pi\gamma$ etc. are also under study[33]. For a thorough discussion of the many chiral perturbation theory predictions for K-decays, see ref.[21].

Baryons

The ideas which finally gave birth to χPT originated in an analysis of the πN interaction[34] and many successful predictions involving baryons were established in the sequel. One of the consequences of chiral symmetry, however, has been causing nothing but headaches for many years, viz. the relation which connects the πN scattering amplitude to the σ-term. The culprit was brought to court only rather recently – I will report on the crime below.

Generally, progress in understanding the low energy properties of the baryons is developing at a slow pace. The nonanalytic terms in the chiral expansion of the baryon masses were worked out long ago. They play a significant role in determinations of the ratio $(m_s - \hat{m}) : (m_d - m_u)$ and of the σ-term matrix element from the masses of the baryon octet[35-37]. Another important step was taken in ref.[38] where the πN scattering amplitude was worked out to first nonleading order of χPT. More recently, the interaction of nucleons with low energy photons was also examined beyond leading order. In particular, it was shown that an old low energy "theorem" for π^0 photoproduction receives a correction from infrared singularities inherent in the one loop graphs of χPT[39]. For a discussion of these results and a comparison with the recent accurate photoproduction data, I refer to the report of Meissner[40]. The axial current matrix elements relevant for the semileptonic decays of the hyperons, the magnetic and electric moments and the nonleptonic decays have also been analyzed within χPT – the literature can be traced from ref.[41].

One of the technical complications occurring in these calculations is that, in a relativistic formulation of baryon χPT, the loop contributions renormalize the leading terms of the effective Lagrangian. The complication can be avoided by using a nonrelativistic expansion for the baryon kinematics. Since this, however, merely amounts to a reformulation of the old calculations, the resulting chiral expansions are the same.

The knowledge of the coupling constants occurring in the derivative expansion of the effective baryon Lagrangian is still in a rather primitive state. As this severely limits the predictivity of the theory, it is important to extend the methods used successfully in the mesonic sector (see next section) to the baryon Lagrangian. Jenkins and Manohar[42] analyzed the contributions generated by the first excited state, the decuplet. Since the transitions to this state are enhanced by particularly small energy denominators, it is advantageous to include the decuplet degrees of freedom in the effective Lagrangian ab initio. A detailed discussion of this approach can be found in ref.[43]. Much work still remains to be done here, to provide analogous estimates for the remaining couplings which are related to low energy singularities with other quantum numbers. Baryon χPT is an underdeveloped country.

A very interesting recent development was initiated by Weinberg[44], who showed how to analyze the properties of the nuclear forces on the basis of χPT. The aim is not to improve on the phenomenological models which describe these forces quite successfully, but to understand their main features on the basis of QCD. It is by no means evident why nuclear binding is so weak and even a crude calculation from first principles would represent very significant progress. How does the binding energy depend on the basic parameters of the theory, Λ_{QCD} and $\mathcal{M}$, or on the coupling constants occurring in the derivative expansion of the effective Lagrangian, and what happens, e.g., if the masses of the u- and d-quarks are turned off?

Marriage of Heavy Quark and Chiral Symmetries

If the mass of a constituent is large, the properties of the bound state become independent of both the flavour and the spin of this constituent – *heavy quark symmetry*[45]. Light constituents, on the other hand, are subject to *chiral symmetry*. The marriage of the two symmetries is a hot topic of ongoing research[46].

Temperature

Much like the spontaneous magnetization of a ferromagnet, the quark condensate gradually melts if the system is exposed to a heat bath of increasing temperature. In the massless theory, the condensate is expected to disappear if the temperature reaches a critical value T_c and, for $T > T_c$, chiral symmetry is restored[47].

Chiral perturbation theory has very strong implications for the properties of the low temperature phase. In particular, it allows one to calculate the first few coefficients in the Taylor series expansion of the condensates in powers of T [48]. The result shows that, at low T, the gluon condensate melts much more slowly than the quark condensate. Dispersion and absorption of waves with the quantum numbers of pions or nucleons and other transport phenomena have also been analyzed[49]. Some of these results were briefly discussed in the course of the talk. The material is described in some detail in a recent review[50].

Volume

In particle physics, finite volumes may appear to be beyond reasonable interest. Lattice simulations, however, are necessarily performed on four-dimensional Euclidean boxes and an extrapolation to infinite volume is required to confront the results of these calculations with observation. This can be done in a reliable manner only if the finite size effects are understood. Their magnitude depends on the size L of the box and on the mass gap M occurring in the spectrum of the theory at infinite volume. If the box is large compared to the Compton wavelength of the lightest particle, $ML \gg 1$, then the finite size effects are small corrections of order $\exp(-ML)$. In QCD, the gap is small, $M = M_\pi$, and rather large boxes are needed to eliminate the finite size effects by brute force. In fact, if the quark masses are turned off, the gap disappears and the finite size effects then persist, no matter how large the volume is taken. The phenomenon is related to the well-known fact that spontaneous symmetry breakdown can only take place if the volume is infinite: if the quark masses are turned off at finite volume, chiral symmetry is restored and the quark condensate therefore disappears.

Since the leading finite size effects are generated by the lightest particles, they can be worked out by means of the effective chiral Lagrangian. In particular, the behaviour of the quark condensate in the symmetry restoration region can explicitly be calculated with χPT and several other observables of interest have also been evaluated (for a review, see ref.[50]). The beautiful data obtained for spin models in three and four dimensions[51] confirm the χPT predictions amazingly well. Once lattice simulations of QCD with dynamical quarks will reach sufficiently small quark masses on reasonably large volumes, χPT will likely become a useful tool also for the analysis of these data. Note that quark loops are essential for the chiral properties of the theory – the standard effective Lagrangian analysis is not valid in the quenched approximation. An effective field theory framework for quenched QCD is discussed in ref.[52].

PHYSICS OF THE EFFECTIVE COUPLING CONSTANTS

I now turn to a more detailed discussion of some specific issues arising in χPT. One of the main problems encountered in the effective Lagrangian approach is the occurrence of an entire fauna of effective coupling constants. If these constants are treated as totally arbitrary parameters, the predictive power of the method is equal to zero – as a bare minimum, an estimate of their order of magnitude is needed.

Chiral Scale

Let me first drop the masses of the light quarks and send the heavy ones to infinity. In this limit, QCD is a theoretician's paradise: a theory without adjustable dimensionless parameters. In particular, the effective coupling constants $F_\pi, B, L_1, L_2, \ldots$ are given by pure numbers multiplying powers of Λ_{QCD}. In principle, the numbers are calculable - the available, admittedly crude evaluations of F_π and B on the lattice demonstrate that the calculation is even feasible. As discussed above, the coupling constants $L_1, \ldots, L_{10}$ are renormalized by the logarithmic divergences occurring in the one loop graphs. This property sheds considerable light on the structure of the chiral expansion and provides a rough estimate for the order of magnitude of the effective coupling constants[53]. The point is that the contributions generated by the loop graphs are smaller than the leading (tree graph) contribution only for momenta in the range $|p| \lesssim \Lambda_\chi$, where[54]

$$\Lambda_\chi \equiv 4\pi F_\pi / \sqrt{N_f} \qquad (12)$$

is the scale occurring in the coefficient of the logarithmic divergence (N_f is the number of light quark flavours). This indicates that the derivative expansion is an expansion in powers of $(p/\Lambda_\chi)^2$ with coefficients of order one. The stability argument also applies to the expansion in powers of m_u, m_d and m_s, indicating that the relevant expansion parameter is given by $(M_\pi/\Lambda_\chi)^2$ and $(M_K/\Lambda_\chi)^2$, respectively.

Resonances

A more quantitative picture can be obtained along the following lines. Consider again the e.m. form factor of the pion and compare the chiral representation (10) with the dispersion relation

$$f_{\pi+}(t) = \frac{1}{\pi} \int_{4M_\pi^2}^{\infty} \frac{dt'}{t'-t} \mathrm{Im} f_{\pi+}(t') \qquad (13)$$

In this relation, the contributions ϕ_π, ϕ_K from the one loop graphs of χPT correspond to $\pi\pi$ and $K\bar{K}$ intermediate states. To leading order in the chiral expansion, the corresponding imaginary parts are slowly rising functions of t. The most prominent contribution on the r.h.s., however, stems from the region of the ρ-resonance which nearly saturates the integral: the vector meson dominance formula, $f_{\pi+}(t) = (1 - t/M_\rho^2)^{-1}$, which results if all other contributions are dropped, provides a perfectly decent representation of the form factor for small values of t. In particular,

this formula predicts $<r^2>_{\pi^+} = 0.39$ fm^2, in satisfactory agreement with observation (0.44 fm^2). This implies that the effective coupling constant L_9 is approximately given by[8]

$$L_9 = \frac{F_\pi^2}{2M_\rho^2}. \qquad (14)$$

In the channel under consideration, the pole due to ρ exchange thus represents the dominating low energy singularity – the $\pi\pi$ and $K\bar{K}$ cuts merely generate a small correction. More generally, the validity of the vector meson dominance formula shows that, for the e.m. form factor, the scale of the derivative expansion is set by $M_\rho = 770$ MeV.

Analogous estimates can be given for all effective coupling constants at order p^4, saturating suitable dispersion relations with contributions from resonances[55,56], e.g.

$$L_5 = \frac{F_\pi^2}{4M_S^2} \quad (15a) \qquad L_7 = -\frac{F_\pi^2}{48M_{\eta'}^2} \quad (15b)$$

where $M_S \simeq 980$ MeV and $M_{\eta'} = 958$ MeV are the masses of the scalar octet and pseudoscalar singlet, respectively. In all those cases where direct phenomenological information is available, these estimates do remarkably well. I conclude that the observed low energy structure is dominated by the poles and cuts generated by the lightest particles – hardly a surprise. In some channels, the scale of the chiral expansion is set by M_ρ, in others by the masses of the scalar or pseudoscalar resonances occurring around 1 GeV. This confirms the rough estimate (12). The cuts generated by Goldstone pairs are significant in some cases and are negligible in others, depending on the numerical value of the relevant Clebsch-Gordan coefficient. If this coefficient turns out to be large, the coupling constant in question is sensitive to the renormalization scale used in the loop graphs. The corresponding pole dominance formula is then somewhat fuzzy, because the prediction depends on how the resonance is split from the continuum underneath it.

The quantitative estimates of the chiral scale given above explain why it is justified to treat m_s as a perturbation. At order p^4, the symmetry breaking part of the effective Lagrangian is determined by the coupling constants $L_4, \ldots, L_8$. These constants are immune to the low energy singularities generated by spin 1 resonances, but are affected by the exchange of scalar or pseudoscalar particles. Their magnitude is therefore determined by the scale $M_S \simeq M_{\eta'} \simeq 1$ GeV [see eq. (15)]. Accordingly, the expansion in powers of m_s is controlled by the parameter $(M_K/M_S)^2 \simeq \frac{1}{4}$. The asymmetry in the decay constants, e.g., is given by[56]

$$\frac{F_K - F_\pi}{F_K} = \frac{M_K^2 - M_\pi^2}{M_S^2} + \text{chiral logs}. \quad (16)$$

This shows that the breaking of the chiral and eightfold way symmetries is controlled by the mass ratio of the Goldstone bosons to the non-Goldstone states of spin zero – in χPT, the observation that the Goldstones are the lightest hadrons thus acquires quantitative significance.

Models

There are innumerable attempts at analyzing the low energy structure of QCD on the basis of the QCD Lagrangian and explaining why chiral symmetry breaks down spontaneously: Nambu-Jona-Lasinio model, chiral quark model, stochastic vacuum fields, field strength formulation, dual QCD, etc. The main problem with these models is that they are models. Much of the phenomenological success claimed for some of them originates in the fact that the framework used contains a spontaneously broken symmetry and does not represent a significant test of the specific model. Also, it is known from the work of Ptolemaios et al. that models involving sufficiently many wheels can successfully fit any given set of data. In some of the work, however, a serious attempt is made at establishing an approximation scheme within QCD. The guiding principle is well known from other branches of physics: slow modes enslave fast modes. Only the slow modes require careful treatment in terms of collective variables while the fast ones can be dealt with perturbatively. Even if analytic approximation methods may never reach the numerical accuracy attainable with lattice techniques, it is important to develop such methods, to gain an understanding of the properties of QCD in the infrared regime, in particular of the structure of the ground state. For recent work in this direction, I refer to the report of de Rafael, given in the parallel session devoted to nonperturbative methods[57]. The framework used in that approach contains many of the effective low energy models discussed in the literature as limiting cases and the implications for the magnitude of the effective coupling constants $F_\pi, B, L_1, \ldots$ have explicitly been worked out[57].

As pointed out by Skyrme[58], the effective chiral Lagrangian admits classical solutions with nontrivial topology which can be interpreted as baryonic states. Unfortunately, however, a stable minimum of the classical action only occurs if the terms of order p^4 are in equilibrium with those of order p^2, i.e. at momenta of the order of the chiral scale. This implies that the relevant classical solutions are outside the domain where the quantum fluctuations are under control. The literature[59] offers many variations of the theme - it remains to be seen whether one of these leads to a mathematically coherent formulation of the beautiful original idea.

LIGHT QUARK MASSES

A crude estimate for the order of magnitude of the light quark masses was given[60] shortly after QCD had been identified as a theory of the strong interactions:

$$m_u \simeq 4\text{MeV}, m_d \simeq 6\text{MeV}, m_s \simeq 125-150\text{MeV}. \quad (17)$$

Many papers dealing with the issue have appared since then. Weinberg[61] pointed out that χPT leads to an improved estimate for the ratios $m_u : m_d : m_s$. Using the Dashen theorem[62] to account for e.m. self energies, the lowest order mass formulae given in eqs.(3),(4) imply[61]

$$\begin{aligned} \frac{m_u}{m_d} &= \frac{M_{K^+}^2 - M_{K^0}^2 + 2M_{\pi^0}^2 - M_{\pi^+}^2}{M_{K^0}^2 - M_{K^+}^2 + M_{\pi^+}^2} \\ \frac{m_s}{m_d} &= \frac{M_{K^0}^2 + M_{K^+}^2 - M_{\pi^+}^2}{M_{K^0}^2 - M_{K^+}^2 + M_{\pi^+}^2} \end{aligned} \quad (18)$$

Numerically, this gives

$$\frac{m_u}{m_d} = 0.55 \qquad \frac{m_s}{m_d} = 20.1. \quad (19)$$

The higher order terms in the chiral expansion generate corrections to the mass formulae (18), controlled by the parameter $(M_K/\Lambda_\chi)^2$. Roughly, the numerical result (19) should therefore hold to within 20 or 30%.

The ellipse

The corrections of $O(p^4)$ to the mass formulae (3),(4) were worked out some time ago[8]. In particular, it was shown that these corrections drop out when taking the double ratio

$$Q^2 = \frac{M_K^2}{M_\pi^2} \frac{M_K^2 - M_\pi^2}{M_{K^0}^2 - M_{K^+}^2}. \quad (20)$$

The observed values of the meson masses thus provide a tight constraint on one particular ratio of quark masses,

$$Q^2 = \frac{m_s^2 - \hat{m}^2}{m_d^2 - m_u^2}\{1 + O(\mathcal{M}^2)\} \quad (21)$$

with $\hat{m} = \frac{1}{2}(m_u + m_d)$. The constraint can be visualized by plotting the ratio m_s/m_d versus m_u/m_d [63]. Dropping the corrections of $O(\mathcal{M}^2) = O((M_K/\Lambda_\chi)^4)$, the resulting curve takes the form of an ellipse,

$$\left(\frac{m_u}{m_d}\right)^2 + \frac{1}{Q^2}\left(\frac{m_s}{m_d}\right)^2 = 1 \quad (22)$$

with Q as major semi-axis (the term $(\hat{m}/m_s)^2$ has been dropped as it is numerically very small). The meson masses occurring in the double ratio (20) refer to pure QCD. Using the Dashen theorem to correct for the e.m. self energies, one obtains $Q = 24.1$. For this value of the semi-axis, the ellipse passes through the point specified by Weinberg's mass ratios, eq. (19).

The Dashen theorem is subject to corrections from higher order terms in the chiral expansion. As usual, there are two categories of contributions: loop graphs of order $e^2\mathcal{M}$ and terms of the same order from the derivative expansion of the effective e.m. Lagrangian.

The Clebsch-Gordan coefficients occurring in the loop graphs are known to be large; the corresponding chiral logarithms tend to increase the e.m. contribution to the kaon mass difference[64]. The numerical result depends on the scale used when evaluating the logarithms. In fact, taken by themselves, chiral logs are unsafe at any scale. Their coefficients merely show the size of the cuts due to Goldstone pairs of low energy. In the case of the pion charge radius, e.g., the contribution from the chiral logs is negligibly small compared to the one from the ρ-meson pole, which in the chiral representation (9) manifests itself through the coupling constant L_9.

The information available about the coupling constants occurring in the derivative expansion of the effective e.m. Lagrangian is not satisfactory. Very crude order of magnitude estimates are described in ref.[65]. The leading term in the expansion is known to be adequately accounted for by vector meson exchange graphs[66]. If the quark masses are turned off, the vector meson dominance model[55,66] therefore provides a decent representation of the relevant amplitude. In a recent paper submitted to this conference, Donoghue, Holstein and Wyler[67] assume that this model can be trusted also beyond leading order in the chiral expansion and find that the vector meson singularities generate remarkably large asymmetries of order $e^2 \mathcal{M}$ which increase the value $(K^+ - K^0)_{e.m.} = 1.3$ MeV predicted by the Dashen theorem by about 1 MeV. This enhances the suspicion[64,65] that the Dashen theorem may receive large corrections from higher order terms. To settle the issue, one needs to account for the cuts associated with the chiral logarithms mentioned above as well as for resonances with other quantum numbers. In particular, as discussed in connection with eq.(16), the exchange of scalar particles may give rise to significant contributions.

In the present context, the main point is that even large corrections of this size only lead to a rather modest change in the value of Q, because the mass difference between K^+ and K^0 is predominantly due to $m_d > m_u$: the value $Q = 24.1$ (Dashen) is lowered to $Q = 22.1$ (Donoghue et al.).

The decay $\eta \to 3\pi$ provides an independent measurement of Q: writing the decay rate in the form $\Gamma_{\eta \to \pi^+\pi^-\pi^0} = \Gamma_0/Q^4$, χPT predicts the value of Γ_0 in a parameter free manner[68]. Although the calculation accounts for all corrections of $O(p^4)$, the numerical accuracy is rather modest because, due to strong final state interaction effects, the calculated corrections are quite large. Since the quantity Q enters in the fourth power, the value $\Gamma_{\eta \to \pi^+\pi^-\pi^0} = 283 \pm 28$ eV given by the particle data group[10] still yields a decent measurement[68]: $Q = 20.6 \pm 1.7$. The fact that this result is significantly smaller than the value predicted with the Dashen theorem represents an old puzzle. As noted in refs.[64,65,67], the problem disappears if the e.m. contribution to the kaon mass difference is significantly larger than indicated by the Dashen theorem. In particular, the value $Q = 22.1$ which results from the vector dominance calculation of Donoghue et al. is consistent with the one from η decay.

The theoretical uncertainties in the η decay amplitude could be reduced. The calculation referred to above only accounts for the corrections of order p^4 and includes final state interaction effects[69] and $\eta\eta'$ mixing[70] only to that order. A dispersive analysis along the lines indicated by Khuri and Treiman[71], which uses the χPT predictions only for the subtraction constants would likely lead to a more accurate estimate of the major semi-axis.

The ratio m_u/m_d

Chiral perturbation theory thus fixes one of the two quark mass ratios in terms of the other, to within small uncertainties. The ratios themselves, i.e. the position on the ellipse, are a more subtle issue. Kaplan and Manohar[63] pointed out that the effective Lagrangian contains a hidden symmetry which implies that the corrections to the lowest order result (19) for m_u/m_d cannot be determined on purely phenomenological grounds. They argued that these corrections might be large and that the u-quark might actually be massless. This possibility is of particular interest, because the strong CP problem would then disappear — several authors[72] have given arguments in favour of $m_u = 0$.

Let me show the picture this reasoning leads to. The lowest order mass formulae (3), (4) imply that the ratio m_u/m_d determines the K^0/K^+ mass difference, the scale being

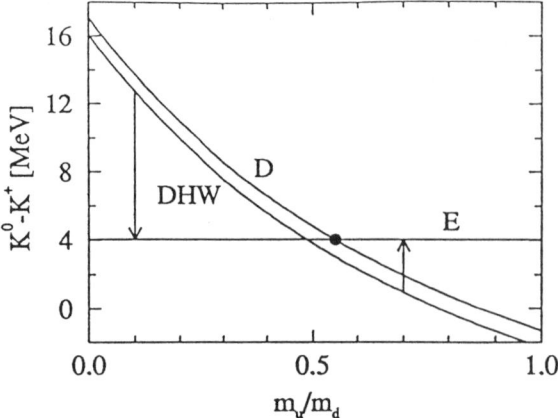

Figure 1. Sensitivity of the first order prediction for the kaon mass difference to m_u/m_d. The two curves differ in the estimate used for the e.m. self energies. E is the experimental value. The dot corresponds to Weinberg's result, eq.(19).

set by M_π:

$$M_{K^0}^2 - M_{K^+}^2 = \frac{m_d - m_u}{m_u + m_d} M_\pi^2 + \ldots \quad (23)$$

The formula holds up to corrections from higher order terms in the chiral expansion and up to e.m. contributions. It is illustrated in Figure 1. The upper curve corresponds to the value $(K^+ - K^0)_{e.m.} = 1.3$ MeV which follows from the Dashen theorem. The correction of Donoghue et al.[67] shifts the result by 1 MeV (lower curve). The horizontal line is the experimental value. Hence the corrections from the higher order terms must generate the contributions shown by the arrows. In particular, if m_u is assumed to vanish, the lowest order mass formula predicts a mass difference which exceeds the observed value by a factor of four. The disaster can only be blamed on the "corrections" from the higher order terms. It is evident that, under such circumstances, it does not make sense to truncate the expansion at first nonleading order and to fool around with the numerics of the effective coupling constants occurring therein. The conclusion to draw from the assumption $m_u = 0$ is that χPT is unable to account for the masses of the Goldstone bosons. The fact that it happens to work remarkably well in other cases must then be accidental. I prefer to conclude that the assumption $m_u = 0$ is not tenable.

In fact, the estimates of the relevant coupling constants given in ref.[55] indicate that the Weinberg ratios only receive small corrections from higher order contributions. The reason why these estimates do not suffer from the ambiguity pointed out by Kaplan and Manohar is the following. The K-M-transformation is a symmetry of the effective Lagrangian, not a symmetry of QCD. In particular, matrix elements of scalar or pseudoscalar operators are not invariant under this transformation[56]. The estimate for L_7, e.g., relies on the sum rules obeyed by the pseudoscalar two-point-functions which do not remain valid if such a transformation is applied. Saturating these sum rules with the lowest lying states, one arrives at the estimate given above, in eq.(15b). In the large N_c limit[8,73], this result is exact. Moreover, the phenomenology of $\eta\eta'$ mixing provides a test of the saturation hypothesis[56]. Using the available information on the mixing angle, one finds that the corrections to the Weinberg ratios, eq.(19), are small, consistent with zero. The value $m_s/\hat{m} = 25.9$, e.g., is confirmed to within an uncertainty of 10 or 15%.

Although the settling of the dust is a slow and somewhat tedious process, large corrections to the Weinberg ratios are firmly ruled out by now. There is some evidence for corrections at the 10% level, such as the change in the value of Q discussed above. Also, there are other sources of information about the ratio $R = (m_s - \hat{m})/(m_d - m_u)$ which compares the breaking of eightfold way symmetry to isospin breaking, but they are subject to systematic uncertainties due to unknown higher order terms. The results for R which follow from the baryon mass pattern and from $\rho\omega$ mixing are perfectly consistent with the Weinberg ratios[36], while ψ' decays favour a smaller value of R [74].

SCALAR FORM FACTOR

A different issue which I wish to discuss in some detail concerns final state interactions. I start with an example which has thoroughly been analyzed by Gasser and Meissner[76]: the scalar form factor,

$$\Gamma(t) \propto \; <\pi(p')\,|\,\bar{u}u+\bar{d}d\,|\,\pi(p)>. \quad (24)$$

Consider the corresponding imaginary part, in the normalization $\Gamma(0) = 1$. χPT leads to an expansion of the form

$$\mathrm{Im}\Gamma(t) = \mathrm{Im}\Gamma_1(t) + \mathrm{Im}\Gamma_2(t) + \ldots \quad (25)$$

where $\mathrm{Im}\Gamma_1$ and $\mathrm{Im}\Gamma_2$ are of order p^2 and p^4, respectively. The leading term is given by ($t \geq 4M_\pi^2$):

$$\mathrm{Im}\Gamma_1(t) = \frac{2t - M_\pi^2}{32\pi F_\pi^2}\left(1 - \frac{4M_\pi^2}{t}\right)^{1/2} \quad (26)$$

and, apart from M_π, only contains the pion decay constant. The corresponding numerical values are shown in Figure 2 (curve 1), taken from ref.[76]. The correction $\mathrm{Im}\Gamma_2(t)$ is a more complicated expression and contains several of the effective coupling constants $L_1, L_2, \ldots$ occurring in $\mathcal{L}_{\mathrm{eff}}^{(4)}$. Fixing these constants phenomenologically, one arrives at the result for $\mathrm{Im}\Gamma_1 + \mathrm{Im}\Gamma_2$ shown in curve 2. Finally, B depicts the result of a dispersive analysis[77], where $\Gamma(t)$ was determined by solving the Muskhelishvili-Omnès integral equations for the $\pi\pi/K\bar{K}$ system. The subtraction constants occurring in that framework are fixed with χPT and the phase shifts are taken from Au et al.[78]

In the low energy domain shown in the Figure, the main features are the following:

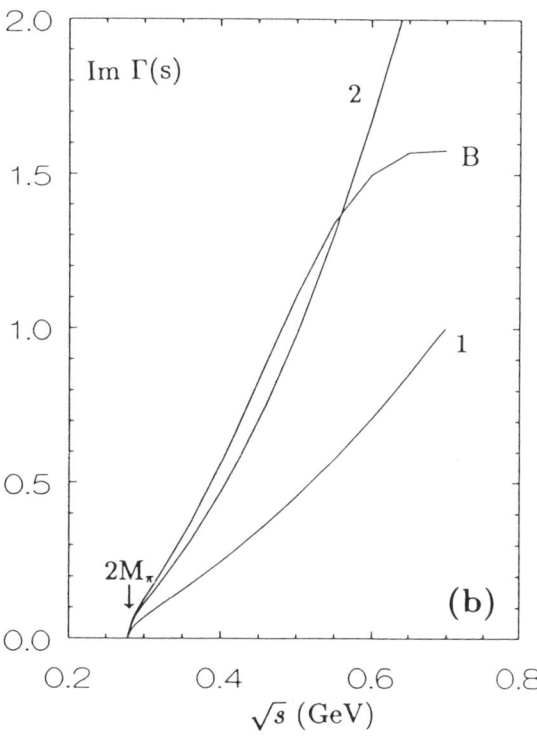

Figure 2. Imaginary part of the scalar pion form factor, taken from Gasser and Meissner[76]. Curves 1 and 2 represent the χPT prediction at first and second order, respectively. B is the result of a dispersive analysis[77]

(i) The final state interaction theorem asserts that, in the elastic region, the imaginary part of the form factor is given by

$$\mathrm{Im}\Gamma(t) = \sin\delta(t)\,|\,\Gamma(t)\,| \quad (27)$$

where $\delta(t)$ is the $\pi\pi$ phase shift for the channel in question (S-wave, $I = 0$).

(ii) The phase shift slowly rises with energy, reaching 90^0 around $\sqrt{t} \simeq 800$ MeV. The chiral representation to first nonleading order describes the phase shift very accurately near threshold. Higher order contributions start playing a significant role only above $\sqrt{t} \simeq 450$ MeV.

(iii) The final state interaction produces a strong cusp at threshold. Between $t = 0$ and $t = 4M_\pi^2$, the form factor therefore rapidly rises, reaching a value around 1.4 at threshold, where it starts picking up an imaginary part. The absolute value continues rising for a while and then starts falling. As a result, the imaginary part exhibits a broad bump resembling a resonance at $\sqrt{t} \simeq 700$ MeV.

The importance of final state interaction effects was repeatedly emphasized by Truong[80]. The example neatly illustrates how the phenomenon manifests itself in χPT. The Figure shows that the leading term in the chiral expansion, ImΓ_1, fails to adequately represent the imaginary part even near threshold. This is understood: the difference between $|\Gamma(t)|$ and 1 counts as a correction of order p^2 and therefore shows up in $\sin\delta\,|\Gamma|$ only at order p^4. At that order, the chiral representation (curve 2) does quite well, up to about $\sqrt{t} \simeq 600$ MeV, simply because it does account for the cusp due to the final state interaction. The Figure shows that, at low energies, the corrections of $O(p^6)$ or higher are remarkably small. For a detailed discussion, see ref.[76]. In this connection, I also mention recent work by Im[81] and by Dobado and Peláez[81], who explore the large N_f limit to all orders of the chiral expansion.

If the expansion is truncated, unitarity is in general only obeyed up to higher order corrections. It may be preferable to work with a representation which explicitly incorporates unitarity and use χPT only for the building blocks of this representation. In the elastic region, this can be done by expressing the partial wave amplitudes in terms of the phase shifts and inserting a suitable parametrization for these. One may, e.g. take the representations of the amplitude which result if the chiral expansion for δ, $tg\delta$ or $cotg\delta$ is truncated at the order at which the χPT calculation has been done. Inverse amplitude representations or suitable Padé approximants also manifestly incorporate unitarity[80]. Like the effective range formula for $cotg\delta$, some of these representations offer the advantage of accomodating resonance poles and can be quite useful as a substitute for a more elaborate dispersive calculation. Manifestly unitary representations of the scattering amplitude necessarily contain higher order terms in the chiral expansion. It is clear, however, that unitarity alone does not fix them: truncating the series for, say, $tg\delta$ and $cotg\delta$, one arrives at two different, unitary amplitudes. Padé approximants provide an intelligent estimate for the order of magnitude of the higher order terms, but there is no reason for the terms generated by this prescription to be correct. In fact, in the case of the form factor discussed above, the leading logarithms are not reproduced correctly[76].

$$\gamma\gamma \to \pi\pi$$

This process is also subject to final state interactions. Since, at low energies, the S-waves dominate, the physics is quite similar to the case of the scalar form factor, except that now, the pions may carry either $I = 0$ or $I = 2$.

Consider first the production of *neutral* pions, $\gamma\gamma \to \pi^0\pi^0$. At leading order in the chiral expansion, this amplitude is given by a one loop graph[82] and only involves the pion decay constant. Numerically, this contribution is shown in Figure 3, taken from a paper by Morgan and Pennington[83] (for related work, see ref.[84]). The data indicated in the Figure are from the Crystal Ball Collaboration[85], scaled by a factor 1.25 to cover full angular range. The shaded area represents the result of a dispersive calculation[83] which uses current algebra to estimate the subtraction constants and is therefore analogous to curve B in Figure 2.

The first order result has been called a "gold-plated prediction of χPT". Now, it is true that the price of gold has come down lately, but there really is no reason for panic.

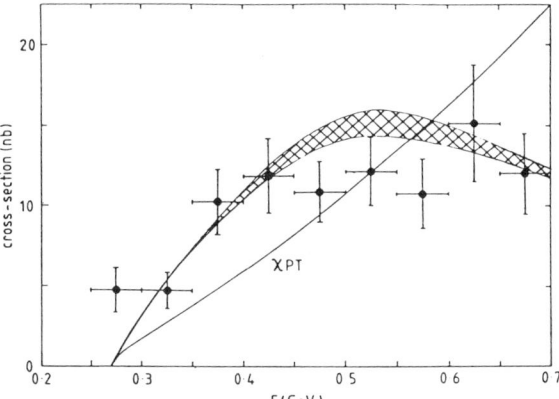

Figure 3. Cross-section for $\gamma\gamma \to \pi^0\pi^0$ as a function of CM energy, taken from Morgan and Pennington[83]. The data are from Crystal Ball[85], scaled by the factor 1.25. The curve represents the first order χPT prediction[82], while the shaded area corresponds to a dispersive evaluation[83].

To estimate the accuracy of a calculation, one needs to understand the physics of those contributions which the calculation neglects. In the present case, the corrections are related to final state interaction effects which, as illustrated by the scalar form factor, can be quite large. Figure 3 indicates that they are sizeable also here, although smaller than in Im$\Gamma(t)$ — the plot shows the square of the amplitude. It is of interest to examine the problem within χPT, calculating the analog of curve 2 in Figure 2. Work in this direction is in progress[86]. The first order prediction[30] for $K \to \pi\gamma\gamma$ is subject to similar corrections (originating in final state interactions in the crossed channel $\gamma\gamma \to K\pi$) but their magnitude yet needs to be worked out[32,33].

In *charged* pion production, $\gamma\gamma \to \pi^+\pi^-$, the situation is rather different, because here, the tree graphs do contribute. The loops only generate corrections which turn out not to be large[82,84]. Indeed, Mark II [87] confirms the prediction of χPT.

The crossed reaction, $\gamma\pi^\pm \to \gamma\pi^\pm$ is also dominated by the Born term (Low theorem). The low energy limit of the remainder is referred to as the *polarizability* of the pion. It is known since a long time[88] that chiral symmetry relates the polarizability to the axial form factor of the decay $\pi \to e\bar{\nu}\gamma$. An old measurement[89], however, disagrees with this prediction. The new experiment planned at Brookhaven should settle the issue.

SIGMA-TERM

As a last item on my list, I briefly discuss the time-honoured low energy theorem[90]

$$F_\pi^2 \bar{D}^+(M_N^2, 2M_\pi^2) = \sigma(2M_\pi^2) \qquad (28)$$

which relates the value of the πN scattering amplitude $D^+(s,t)$ at the Cheng-Dashen point to the matrix element

$$\sigma(t) = \frac{1}{2M_N} < N(p') \mid \hat{m}(\bar{u}u + \bar{d}d) \mid N(p) > \qquad (29)$$

at $t = 2M_\pi^2$. The amplitude $\bar{D}^+(s,t)$ entering in eq.(28) is what remains of $D^+(s,t)$ if the Born term is removed. In fact, the first point to note is that what remains is very little – precision data are needed to reliably determine the l.h.s. in eq.(28). The remarkable work of Höhler and collaborators[91] shows that such a determination is feasible. Relying on the data available ten years ago, these authors obtained $F_\pi^2 \bar{D}^+ \simeq 64$ MeV. This is to be compared with the value $\sigma \simeq 25$ MeV which follows from the lowest order mass formulae for the baryons (assuming that the matrix element $< N \mid \bar{s}s \mid N >$ vanishes and ignoring the difference between $t = 2M_\pi^2$ and $t = 0$, which is of higher order in the chiral expansion). The discrepancy has given rise to rather wild thoughts. In particular, some authors threw the suspicion on the matrix element $< N \mid \bar{s}s \mid N >$, proposing a picture where half of the proton mass would be due to the mass of the strange quark.

Let me go through the various elements of the puzzle, starting with the l.h.s of eq.(28).

We closely examined the evaluation of the scattering amplitude at the Cheng-Dashen point, using dispersion relations[92]. This allowed us, in particular, also to study the sensitivity of the result to the experimental input used in the low energy region which is crucial here. There are several new and precise data sets which were not available ten years ago, but, unfortunately, some of these are mutually inconsistent. This severely limits the accuracy at which the l.h.s. can be determined at this time. New experiments are under way to resolve the inconsistencies. On the basis of the presently available data, we confirm the result of Höhler et al. within errors: we find $F_\pi^2 \bar{D}^+ = 60$ MeV (for details, in particular for a discussion of the various sources of uncertainty, see ref.[93]; roughly, the overall error bar is ± 8 MeV).

Next, consider the equality sign in eq.(28), i.e. the corrections to the low energy theorem. Gasser and collaborators[38] have evaluated the difference between the two sides of the equation within χPT to one loop. Their calculation confirms that the nonanalytic contributions of order $\mathcal{M}^{3/2}$ which occur in the individual terms drop out in the difference. Numerically, what remains from the one loop calculation is less than an MeV. This is the order of magnitude expected for the corrections to the low energy theorem (28).

Next, let us look at the calculation which gives $\sigma \simeq 25$ MeV for the r.h.s. In this calculation, the nonanalytic terms mentioned above do not drop out. Gasser[94] has shown that these contributions generate substantial $SU(3)$-asymmetries in the baryon matrix elements of $\bar{u}u - \bar{d}d$ and $\bar{u}u + \bar{d}d - 2\bar{s}s$. Indeed, these asymmetries are clearly seen when comparing the mass differences generated by $m_d - m_u$ to those due to $m_s - \hat{m}$ in the baryon octet. The nonanalytic contributions increase the prediction for the σ-term by 10 MeV. Unfortunately, the available estimates for the terms of order $\mathcal{M}^2$ are very crude so that the result is subject to theoretical uncertainties, estimated at[94]:

$$\sigma(0) = 35 \pm 5 \text{MeV}. \tag{30}$$

Finally, we need to analyze the difference

$$\sigma(2M_\pi^2) - \sigma(0) = \frac{2M_\pi^2}{\pi} \int_{4M_\pi^2}^{\infty} \frac{dt}{t} \frac{\text{Im}\sigma(t)}{t - 2M_\pi^2}. \tag{31}$$

At leading order in the chiral expansion, the imaginary part of the form factor is given by the tree graphs of χPT. Evaluating the integral, this gives $\sigma(2M_\pi^2) - \sigma(0) = 4.6$ MeV. Does it still come as a surprise that the leading term in the chiral expansion does not provide an adequate representation of the imaginary part? Indeed, in the present case, the final state interaction effects are enormous. The specific reasons are discussed in detail in ref.[95], where we show that the imaginary part can be worked out quite accurately with dispersive techniques. The result,

$$\sigma(2M_\pi^2) - \sigma(0) = 15 \text{MeV} \tag{32}$$

is larger than the leading term in the chiral expansion by a factor of three! This also implies that the slope of the scalar form factor is large: the corresponding mean square radius is twice as large as the mean square radius of the charge distribution.

In summary, I note the following points:

(i) The corrections to the low energy theorem (28) are tiny. Both the analytic continuation required to reach the Cheng-Dashen point and the t-dependence of the form factor are controlled by stable dispersion relations. The low energy theorem can therefore be used to accurately measure the σ-term. In this context, χPT is neither necessary nor sufficient.

(ii) The available data provide an estimate at the 20 % level,

$$\sigma(0) = 45 \pm 8 \text{MeV}. \tag{33}$$

The error bar could be reduced significantly if better data were available.

(iii) In view of the uncertainties, the 10 MeV difference between this result and the prediction (30) is not significant. Wild thoughts can be put to rest. If the difference turns out to be real, it implies that about 130 MeV of the rest energy of the proton stem from the mass of the strange quark. This is quite consistent with what is found in the case of the pion decay constant which decreases by about 9 % if m_s is sent to zero[77].

(iv) Taking the above numbers at face value, the discrepancy between 25 and 60 MeV thus consists of three pieces: 10 MeV from the higher order terms in the chiral expansion of the baryon masses, 10 MeV from the proton matrix element of $\bar{s}s$ and 15 MeV from the t-dependence of the scalar form factor of the proton. The three pieces happen to have the same sign.

CONCLUSION

Effective field theories represent an efficient and mathematically sound method. In particular, the effective low energy theory of QCD – chiral perturbation theory – is a useful tool. The success of χPT originates in two phenomenological observations:

(i) M_π is remarkably small.

(ii) The eightfold way is an approximate symmetry of the strong interactions.

Both of these observations are theoretically understood and are related to the fact that three of the quark masses happen to be small (admittedly, we do not know why they are small). Chiral perturbation theory is a machinery which allows one to investigate the consequences of this fact for the low energy structure of QCD. The method relies on a systematic expansion of the various matrix elements of interest in powers of momenta and quark masses. Plenty of predictions have been worked out with this method. Considerable progress was made in understanding the physics underneath the effective coupling constants occurring in the derivative expansion. Also, there are new results concerning the phenomenology of these couplings [96-99]. Much work, however, yet remains to be done. In particular, the chiral expansion for matrix elements involving baryons or bound states with heavy quarks is awaiting exploration, but there are also many other problems to investigate – just think of $\tau \to 13\pi\nu, 12\pi\nu, \ldots$

Most of the χPT predictions are in good agreement with experiment, but I also discussed several examples where a naive application of lowest order χPT leads to erroneous conclusions. The scalar form factors of the pion and of the proton (σ-term in πN scattering) belong to this category. In these examples, the first nonleading terms occurring in the chiral expansion turn out to be large, for a reason which is understood: final state interaction effects. The low energy behaviour of the $\pi\pi$ phase shifts which determine these effects is perfectly well accounted for by χPT. The example of the scalar pion form factor shows that the final state interaction phenomena clearly manifest themselves also in this framework, but only if the calculation is extended beyond leading order.

Large corrections may also occur in the presence of prominent resonance contributions. The singularities generated by the Δ in πN scattering or by vector mesons in the case of the e.m. self energies are examples of this type. In all cases I know, where there is good evidence for the higher order corrections to be large, the physics of the problem is well understood and can explicitly be worked out.

The ϕ factory DAFNE which is under construction at Frascati will offer plenty of opportunities to test the theory, providing clues for further development.

An application which has barely started

bearing fruit is the analysis of finite size effects in lattice calculations where χPT is likely going to save a considerable amount of time or money.

One of the most remarkable features of effective field theories is their universality. Essentially the same calculations can be and have been successfully applied to other branches of physics, from magnets to the Higgs sector of the Standard Model. The main conclusion to draw, therefore, is that this report on chiral perturbation theory is not going to be the final word on the subject.

Acknowledgement

I thank J. Gasser for support in the exploration of the paper mountain. In this endeavour, some of the local peaks turned out to be rather misleading, because the only papers their authors never fail referring to are those which they have written themselves. This may also apply to the present report, if not as a selection rule then as an approximate conservation law – I apologize for missing references. Finally, I wish to thank H. Bebie, O. Haenni and A. Schenk for their help in the preparation of the manuscript.

REFERENCES

1. R. Petronzio, *Lattice gauge theory*, in these procceddings.
2. V. Zakharov, *High orders of perturbation theory*, in these proceedings;
   D. Brahm, *The electroweak phase transition*, in these proceedings;
   V. Khoze, *B+L violation at high energy*, in these proceedings.
3. P. Hasenfratz and F. Niedermayer, *Phys. Lett.* B268 (1991) 231.
4. J. Bagger, *Observing electroweak symmetry breaking at the SSC*, in these proceedings;
   M. Peskin, *Beyond the Standard Model*, in these proceedings;
   G. Valencia, *Anomalous gauge boson couplings at hadron supercolliders*, in these proceedings.
5. Y. Nambu, *Phys. Rev. Lett.* 4 (1960) 380.
6. For a review, see
   H. Pagels, *Phys. Rep.* 16C (1975) 219;
   H. Georgi, *Weak Interactions and Modern Particle Theory* (Benjamin/Cummings, Menlo Park, 1984);
   J. Donoghue, E. Golowich and B. Holstein, *Dynamics of the Standard Model* (Cambridge Univ. Press, Cambridge, 1992);
   J. Donoghue, *Introduction to nonlinear effective field theory*, in ref.[13];
   H. Leutwyler, *Chiral effective Lagrangians*, Lectures given at Theor. Adv. Study Inst., Boulder, 1991, in *Perspectives in the Standard Model*, eds. R.K. Ellis, C.T. Hill and J.D. Lykken (World Scientific, Singapore, 1992).
7. S. Weinberg, *Physica* A96 (1979) 327.
8. J. Gasser and H. Leutwyler, *Ann. Phys. (N.Y.)* 158 (1984) 142; *Nucl. Phys.* B250 (1985) 465.
9. J. Gasser and H. Leutwyler, *Nucl. Phys.* B250 (1985) 517.
10. Review of Particle Properties, *Phys. Rev.* D45 (1992).
11. S.R. Amendolia et al., *Nucl. Phys.* B277 (1986) 168.
12. W. Molzon et al., *Phys. Rev. Lett.* 41 (1978) 1213.
13. Proc. Workshop on *Effective Field Theories of the Standard Model*, Dobogokö, Hungary, Aug. 1991, ed. U. Meissner (World Scientific, Singapore, 1992).
14. S. Weinberg, *Phys. Rev. Lett.* 17 (1966) 616; J. Gasser and H. Leutwyler, *Phys.*

Lett. 125B (1983) 325.

15. G. Czapek et al., *Letter of intent*, CERN/SPSLC 92-44.

16. V. Bernard, N. Kaiser and U. Meissner, *Nucl. Phys.* B357 (1991) 129; *Phys. Rev.* D43 (1991) 2757.

17. V. Bernard, N. Kaiser and U. Meissner, *Nucl. Phys.* B364 (1991) 283;
    N. Kaiser, $\pi\pi$ and πK *Scattering in* χPT, in ref.[13];
    A. Dobado and J.R. Peláez, preprint Univ. Madrid, FT/UCM/10/92.

18. J. Wess and B. Zumino, *Phys. Lett.* B37 (1971) 95;
    E. Witten, *Nucl. Phys.* B223 (1983) 422.

19. J. Donoghue, B. Holstein and Y. Lin, *Phys. Rev. Lett.* 55 (1985) 2766;
    J. Bijnens, A. Bramon and F. Cornet, *Phys. Rev. Lett.* 61 (1988) 1453; *Z. Phys.* C46 (1990) 599; *Phys. Lett.* B237 (1990) 488;
    J. Donoghue and D. Wyler, *Nucl. Phys.* B316 (1989) 289;
    T.N. Pham, *Phys. Lett.* B246 (1990) 175; *Nucl. Phys. B (Proc. Suppl.)* 23A (1991) 299;
    D. Issler, *Nonrenormalization of the chiral anomaly* and *On the structure of high-order terms in* χPT *with external fields*, preprints SLAC-PUB-4943, 5200 (1990).
    R. Akhoury and A. Alfakih, *Ann. Phys. (N.Y)* 210 (1991) 81;
    J. Bijnens, *Nucl. Phys.* B367 (1991) 709; *The anomalous effective action of QCD*, in ref.[13];
    C.A. Dominguez, *Mod. Phys. Lett.* A2 (1987) 983;
    S. Fajfer, K. Surulitz and R. Oakes, *Phys. Rev.* D44 (1991) 295;
    A. Bramon, E. Pallante and R. Petronzio, *Phys. Lett.* B271 (1991) 237;
    A. Bramon, A. Grau and G. Pancheri, *Phys. Lett.* B277 (1992) 353;
    A. Bramon et al., χPT *for* γPPP *processes*, preprint Univ. Auton. Barcelona, UAB-FT-263-91;
    Ll. Ametller et al., *Phys.Lett.* B276 (1992) 185; *Phys. Rev.* D45 (1992) 986;
    E. Pallante and R. Petronzio, *Anomalous effective Lagrangians and vector resonance models*, preprint Univ. Rome ROM 2F 92/04;
    H. Ito, W. Buck and F. Gross, *The axial anomaly and the dynamical breaking of chiral symmetry in the* $\gamma^* \to \pi^0\gamma$ *reaction*, preprint CEBAF-TH-91-13.

20. J. Bijnens, G. Ecker and J. Gasser, *Semileptonic kaon decays*, in ref.[21] and *Radiative semileptonic decays*, preprint CERN-TH.6625/92.

21. *The DAFNE Physics Handbook*, eds. L. Maiani, G. Pancheri and N. Paver, INFN-Frascati (1992).

22. J. Bijnens, *Nucl. Phys.* B337 (1990) 635;
    C. Riggenbach et al., *Phys. Rev.* D43 (1991) 127.

23. S. Fajfer, K. Surulitz and R. Oakes, $\tau \to \omega\pi\nu$ *decay*, preprint Northwestern Univ. NUHEP-TH-92-9;
    S. Fajfer, Univ. Sarajevo, Abstract submitted to this conference.

24. M. Gell-Mann and A. Pais, in *Proc. 1954 Glasgow Conf. on Nuclear and Meson Physics*, eds. E.H. Bellany and R.G. Moorhouse (Pergamon, London, 1955);
    J. Cronin, *Phys. Rev.* D161 (1967) 1483.

25. W.A. Bardeen, A.J. Buras and J.-M. Gérard, *Nucl. Phys.* B293 (1987) 787; *Phys. Lett.* B192 (1987) 138; *Phys. Lett.* B211 (1988) 343;
    A. Buras, *Nucl. Phys. B (Proc. Suppl.)* 10A (1989) 199;
    B.Y. Blok and M.A. Shifman, *Sov. J. Nucl. Phys.* 45 (1987) 143, 301, 522;

M.A. Shifman, *Nucl. Phys. B (Proc. Suppl.)* 3 (1988) 289;
H.Y. Cheng, *Phys. Rev.* D36 (1987) 2056; *Int. J. Mod. Phys.* A4 (1989) 495;
A. Pich and E. de Rafael, *Phys. Lett.* B158 (1985) 477; *Nucl. Phys.* B358 (1991) 311;
B. Guberina, A. Pich and E. de Rafael, *Phys.Lett.* B163 (1985) 198; *Nucl. Phys.* B277 (1986) 197;
B. Guberina, *Nucl. Phys. B (Proc. Suppl.)* 7A (1989) 213;
B. Stech, *Nucl. Phys. B (Proc. Suppl.)* 1B (1988) 17; *Mod. Phys. Lett.* A6 (1991) 3113;
M. Neubert and B. Stech, *Phys. Rev.* D44 (1991) 775;
J. Goity, *Nucl. Phys. B (Proc. Suppl.)* 7A (1989) 172;
T.N. Pham, in ref.[13];
D. Ebert et al., *Effective chiral Lagrangians for strong, weak and electromagnetic-weak interactions ...*, preprint Zeuthen PHE 91-08.

26. J. Kambor, J. Missimer and D. Wyler, *Nucl. Phys.* B346 (1990) 17; *Phys. Lett.* B261 (1991) 496.

27. J. Donoghue, B. Holstein and G. Valencia, *Phys. Rev.* D36 (1987) 798;
E. Golowich, *Phys. Rev.* D36 (1988) 3516;
J. Donoghue, *Nucl. Phys. B (Proc. Suppl.)* 7A (1989) 59;
S. Fajfer and J.-M. Gérard, *Z. Phys.* C42 (1989) 425;
A.A. Bel'kov et al., *Phys. Lett.* B232 (1989) 118; *Nucl. Phys.* B359 (1991) 322 and *On the origin of the enhancement of CP violating charge asymmetries in $K^+ \to 3\pi$ decays predicted from chiral theory*, preprint DESY 92-106;
H.Y. Cheng, *Phys. Lett.* B238 (1990) 399; *Phys. Rev.* D42 (1990) 72, 3850; D43 (1991) 1579; D44 (1991) 166, 919;
G. Ecker, in *Proc. XXIV Int. Symp. on the Theory of Elementary Particles*, Gosen, Germany (1990);
T. Kurimoto, *Prog. Theor. Phys.* 84 (1990) 658, 675;
T. Morozumi, C.S. Lim and A.I. Sanda, *Phys. Rev. Lett.* 65 (1990) 404;
G. D'Ambrosio, G. Isidori and N. Paver, *Phys. Lett.* B273 (1991) 497;
J. Kambor, $K \to 2\pi, 3\pi$ *decays and the B-parameter in next-to-leading order* χPT, in ref.[13];
M. Neubert, *Z. Phys.* C50 (1991) 243;
G. Ecker, H. Neufeld and A. Pich, *Phys. Lett.* B278 (1992) 337;
J. Kambor et al., *Phys. Rev. Lett.* 68 (1992) 1818;
Y.L. Wu, *Int. J. Mod. Phys.* A7 (1992) 2863;
G. Isidori and A. Pugliese, *Chiral weak Lagrangian for vector mesons and $K \to 3\pi$ decay amplitudes*, preprint Univ. Rome, ROME-849-1991;
G. Isidori, L. Maiani and A. Pugliese, *CP violation in $K \to 3\pi$ decays and lattice QCD B factors*, preprint Univ. Rome, ROME-848-1991;
For a comprehensive review, see
L. Maiani and N. Paver, $K \to 3\pi$ *decays at DAFNE*, in ref.[21].

28. G. D'Ambrosio and D. Espriu, *Phys. Lett.* B175 (1986) 237;
J. Goity, *Z. Phys.* C34 (1987) 341;
F. Buccella, G. D'Ambrosio and M. Miragliuolo, *Nuovo Cim.* 104A (1991) 777;
Z.E.S. Uy, *Phys. Rev.* D43 (1991) 802, 1572;
P. Ko and T.N. Truong, *Phys. Rev.* D43 (1991) 4; D44 (1991) 1616(E).

29. H. Burkhardt et al. (NA31), *Phys. Lett.* B199 (1987) 139.

30. G. Ecker, A. Pich and E. de Rafael, *Phys. Lett.* B189 (1987) 363; B237 (1990) 481;

L. Cappiello and G. D'Ambrosio, *Nuovo Cim.* 99A (1988) 155.

31. G. Barr et al. (NA31), *Phys. Lett.* B242 (1990) 523;
    V. Papadimitriou et al. (E273), *Phys. Rev.* D44 (1991) 573.

32. G. D'Ambrosio, M. Miragliuolo and P. Santorelli, *Radiative non-leptonic kaon decays*, in ref.[21].

33. More than 30 papers dealing with these decays appeared during the last couple of years, some of which were discussed in Parallel Session PA-2. For a comprehensive review, see ref.[32].

34. M. Goldberger and S. Treiman, *Phys. Rev.* 110 (1958) 1178.

35. J. Gasser and A. Zepeda, *Nucl. Phys.* B174 (1980) 445;
    J. Gasser, *Ann. Phys.* 136 (1981) 62.

36. J. Gasser and H. Leutwyler, *Phys. Rep.* C87 (1982) 77;

37. E. Jenkins and A. Manohar, *Phys. Lett.* B281 (1992) 336;
    J. Schechter and A. Subbaraman, *Second order 'flavour' perturbation theory for the baryons*, preprint Syracuse Univ. SU-4228-402;
    B. Schwesinger and H. Weigel, *Nucl. Phys.* A540 (1992) 461.

38. J. Gasser, M. Sainio and A. Švarc, *Nucl. Phys.* B307 (1988) 779.

39. V. Bernard et al., *Phys. Lett.* B268 (1991) 291.

40. U. Meissner, *Chiral structure of the nucleon*, in these proceedings; *Photonucleon processes in χPT*, in ref.[13].

41. A. Krause, *Helv. Phys. Acta* 65 (1990) 3;
    E. Jenkins and A. Manohar, *Phys. Lett.* B259 (1991) 353;
    C. Carone and H. Georgi, *Nucl. Phys.* B375 (1992) 243;
    E. Jenkins, *Nucl. Phys.* B375 (1992) 561;
    V. Bernard et al., *Chiral structure of the nucleon*, preprint Univ. Bern, BUTP-92/15;
    D. Atwood and A. Soni, *χPT constraint on the electric dipole moment of the Λ hyperon*, preprint Brookhaven BML-47557 (1992).
    See also ref.[46].

42. E. Jenkins and A. Manohar, *Phys. Lett.* B255 (1991) 558; B259 (1991) 353;
    E. Jenkins, *Nucl. Phys.* B368 (1992) 190; B375 (1992) 561.

43. E. Jenkins and A. Manohar, *Baryon χPT*, in ref.[13].

44. S. Weinberg, *Phys. Lett.* B251 (1990) 288; *Nucl. Phys.* B363 (1991) 3;
    C. Ordóñez and U. van Kolck, *Chiral Lagrangians and nuclear forces*, preprint Texas Univ. UTTG-01-92;
    T.S. Park, D.P. Min and M. Rho, *χPT and filter phenomena in nuclei*, preprint Seoul National Univ. SNUTP-92-45.
    As shown in a forthcoming paper, the method also applies to π-nucleus reactions (S. Weinberg, priv. commun.)

45. N. Isgur, *Hadron spectroscopy and structure*, in these proceedings.

46. J. Goity, *Phys. Lett.* B249 (1990) 495;
    χPT for $SU(3)$ breaking in heavy meson systems, preprint CEBAF-TH-92-16;
    M.B. Wise, *Phys. Rev.* D45 (1992) 2188;
    B. Grinstein et al., *χPT for f_{D_s}/f_D and B_{B_s}/B_B*, preprint Cal. Inst. Tech., CALT-68-1768;
    G. Burdman and J. Donoghue, *Phys. Lett.* B280 (1992) 287;
    T.-M. Yan et al., *Heavy quark symmetry and chiral dynamics*, preprint National Central Univ. Taiwan, CLNS-92-1138;
    H.-Y. Cheng, *Heavy quark symmetry and*

chiral dynamics, in these proceedings;
P. Cho, χPT *for hadrons containing a heavy quark: the sequel*, preprint Harvard Univ. HUTP-92-A014;
P. Cho and H. Georgi, *Electromagnetic interactions in heavy hadron chiral theory*, preprint Harvard Univ. HUTP-92-A043.

47. Although this picture is widely accepted, it is not firmly established. A different scenario is advocated, e.g., in
L.N. Chang and N.P. Chang, *Phys. Rev.* D45 (1992) 2988.

48. P. Binétruy and M.K. Gaillard, *Phys. Rev.* D32 (1985) 931;
J. Gasser and H. Leutwyler, *Phys. Lett.* B184 (1987) 83;
P. Gerber and H. Leutwyler, *Nucl. Phys.* B321 (1989) 387.

49. J. Goity and H. Leutwyler, *Phys. Lett.* B228 (1989) 517;
H. Leutwyler and A. Smilga, *Nucl. Phys.* B342 (1990) 302;
A. Schenk, *Nucl. Phys.* B363 (1991) 97;
G.M. Welke, R. Venugopalan and M. Prakash, *Phys. Lett.* B245 (1990) 137;
H. Bebie et al., Nucl. Phys. B378 (1992) 95.

50. H. Leutwyler, *Deconfinement and chiral symmetry*, in Proc. Workshop "QCD – 20 Years Later", Aachen, Germany, June 1992, eds. P.M. Zerwas and H.A. Kastrup (World Scientific, Singapore, to be published).

51. A. Hasenfratz et al., *Z. Phys.* C46 (1990) 257; *Nucl. Phys.* B356 (1991) 332;
I. Dimitrović et al., *Nucl. Phys.* B350 (1991) 893; *Phys. Lett.* B268 (1991) 408;
M. Göckeler, K. Jansen and T. Neuhaus, *Phys. Lett.* B273 (1991) 450.

52. S. Sharpe, *Phys. Rev.* D41 (1990) 3233; *Quenched chiral logarithms*, preprint CEBAF-TH-92-12;
C. Bernard and M. Golterman, *Nucl. Phys. B (Proc. Suppl.)*, to appear; χPT *for the quenched approximation of QCD*, preprint Washington Univ. HEP-92-60.

53. H. Georgi, *Weak Interactions and Modern Particle Theory* (Benjamin/Cummings, Menlo Park, 1984);
H. Georgi and A. Manohar, *Nucl. Phys.* B234 (1984) 189.

54. M. Soldate and R. Sundrum, *Nucl. Phys.* B340 (1990) 1;
R.S. Chivukula, M.J. Dugan and M. Golden, *Analyticity, crossing symmetry and the limits of* χPT, preprint Harvard Univ. HUTP-92/A025.

55. G. Ecker et al., *Nucl. Phys.* B321 (1989) 311; *Phys. Lett.* B223 (1989) 425.

56. H. Leutwyler, *Nucl. Phys.* B337 (1990) 108;

57. E. de Rafael, *The low-energy effective action of QCD and Nambu-Jona-Lasinio models*, in these proceedings;
A. Pich, in ref.[13];
An extensive list of references concerning effective low energy models may also be found in ref.[50].

58. T.H.R. Skyrme, *Proc. Roy. Soc. London*, A260 (1961) 127.

59. See, e.g.,
M.M. Islam, *Nonperturbative structure of the nucleon*, preprint Univ. Connecticut (1991); *Z. Phys.* C53 (1992) 253;
J. Ellis, Y. Frishman, A. Hanany and M. Karliner, *Quark solitons as constituents of hadrons*, preprint CERN-TH-6426/92 and the references therein.

60. H. Leutwyler, *Phys. Lett.* B48 (1974) 431; *Nucl. Phys.* B76 (1974) 413;
J. Gasser and H. Leutwyler, *Nucl. Phys.* B94 (1975) 269.

61. S. Weinberg, in *A Festschrift for I.I. Rabi*, ed. L. Motz (New York Acad. Sci, 1977) p. 185.

62. R. Dashen, *Phys. Rev.* 183 (1969) 1245.

63. D. Kaplan and A. Manohar, *Phys. Rev. Lett.* 56 (1986) 2004.

64. P. Langacker and H. Pagels, *Phys. Rev.* D8 (1973) 4620;
K. Maltman and D. Kotchan, *Mod. Phys. Lett.* A5 (1990) 2457.

65. G. Stephenson, K. Maltman and T. Goldman, *Phys. Rev.* D43 (1991) 860.

66. T. Das, G.S. Guralnik, V.S. Mathur, F.E. Low and J.E. Young, *Phys. Rev. Lett.* 18 (1967) 759;
I.S. Gerstein, B.W. Lee, H.T. Nieh and H.J. Schnitzer, *Phys. Rev. Lett.* 19 (1967) 1064;
W.A. Bardeen, J. Bijnens and J.-M. Gérard, *Phys. Rev. Lett.* 62 (1989) 1343.

67. J. Donoghue, B. Holstein and D. Wyler, *Quark masses, $\eta \to 3\pi$ and the kaon mass difference*, preprint Univ. Mass., Amherst (1992).

68. J. Gasser and H. Leutwyler, *Nucl. Phys.* B250 (1985) 539.

69. T.N. Truong, $\eta \to 3\pi, K \to \pi\pi e\nu, \gamma\gamma \to 2\pi^0$ *and S-wave $I = 0$ low energy $\pi\pi$ scattering*, preprint Ecole Polytechnique, Palaiseau, France (1992).

70. R. Akhoury and M. Leurer, *Z. Phys.* C43 (1989) 145;
R. Akhoury and J.M. Frere, *Phys. Lett.* B220 (1989) 258.

71. N. Khuri and S. Treiman, *Phys. Rev.* 119 (1960) 1115;
C. Roiesnel and T.N. Truong, *Nucl. Phys.* B187 (1981) 293.

72. K. Maltman, T. Goldman and G. Stephenson, *Phys. Lett.* B234 (1990) 158;
J. Cline, *Phys. Rev. Lett.* 63 (1989) 1338;
K. Choi, C.W. Kim and W.K. Sze, *Phys. Rev. Lett.* 61 (1988) 794;
K. Choi and C.W. Kim, *Phys. Rev.* D40 (1989) 890;
K. Choi, *How precisely can one determine m_u/m_d* and *Light quark masses and charmonium decays*, preprints U. Cal. San Diego UCSD PTH 91/28 and 92/06.

73. J.-M. Gérard, *Mod. Phys. Lett.* A5 (1990) 391.

74. J. Donoghue and D. Wyler, *Phys. Rev.* D45 (1992) 892 obtain $m_u/m_d \simeq 0.3$ from ψ' decays. This calculation faces the following problems: (i) The chiral expansion of the e.m. contribution[75] to the amplitude $\psi' \to \psi\pi^0$ starts at $O(e^2 p)$ and is *not* suppressed in comparison to the term $O((m_d - m_u)p)$ generated by the quark mass difference. (ii) The pole dominance estimate given for the relevant effective coupling constant only accounts for the η' and neglects scalar resonances which may generate significant flavour asymmetries (compare eq. (16)). (iii) It is not clear that the first term in the multipole expansion, which treats m_c as very heavy, provides an adequate approximation at the accuracy needed here.

75. K. Maltman, *Phys. Rev.* D44 (1991) 751.

76. J. Gasser and U. Meissner, *Nucl. Phys.* B357 (1991) 90; *Phys. Lett.* B 258 (1991) 219.

77. J. Donoghue, J. Gasser and H. Leutwyler, *Nucl. Phys.* B253 (1990) 341.

78. K. Au, D. Morgan and M. Pennington, *Phys. Rev.* D35 (1987) 1633.

79. J. Gasser and U. Meissner, *Phys. Lett.* B258 (1991) 219;
A. Schenk, *Nucl. Phys.* B363 (1991) 97.

80. T.N. Truong, *Phys. Rev.* D61 (1988) 2526; *Phys. Rev. Lett.* 67 (1991) 2260;

A. Dobado, M.J. Herrero and T.N. Truong, *Phys. Lett.* B235 (1990) 134.

81. A. Dobado and J. Peláez, *Phys. Lett.* B286 (1992) 136.

82. J. Bijnens and F. Cornet, *Nucl. Phys.* B296 (1988) 557;
J. Donoghue, B. Holstein and Y.C. Lin, *Phys. Rev.* D37 (1988) 2423.

83. D. Morgan and M. Pennington, *Phys. Lett.* B272 (1991) 134.

84. G. Mennessier and T.N. Truong, *Phys. Lett.* B177 (1986) 195;
A.E. Kaloshin and V.V. Serebryakov, *Z. Phys.* C32 (1986) 279;
D. Morgan and M. Pennington, *Phys. Lett.* B192 (1987) 207; *Z. Phys.* C37 (1988) 431; C48 (1990) 623;
M. Pennington, *Predictions for* $\gamma\gamma \to \pi\pi$; *what Photons at DAFNE will see*, in ref.[21];
P. Ko, *Phys. Rev.* D41 (1990) 1531;
J. Bijnens, S. Dawson and G. Valencia, *Phys. Rev.* D44 (1991) 3555;
S. Bellucci, χPT *predicts* $\pi\pi$ *production in* $\gamma\gamma$ *fusion*, in ref.[21];
C. Im, *Phys. Lett.* B281 (1992) 357;
A. Dobado and J.R. Peláez, *Unitarity and* $\gamma\gamma \to \pi\pi$ *in* χPT, preprint Madrid Univ. FT-UCM-9-92.

85. Crystal Ball Collab., H. Mariske et al., *Phys. Rev.* D41 (1990) 3324.

86. S. Bellucci, J. Gasser and M. Sainio, priv. commun.

87. Mark II Collab., J. Boyer et al., *Phys. Rev.* D42 (1990) 1350.

88. M.V. Terent'ev, *Sov. J. Nucl. Phys.* 16 (1972) 162;
A.I. L'vov and V.A. Petrunkin, in Proc. Workshop "Perspectives in Photon Interactions with Hadrons and Nuclei", eds. M. Schumacher and G. Tamas, *Lecture Notes in Physics*, Vol. 365 (Springer, Berlin, 1990);
B. Holstein, *Comm. Nucl. Part. Phys.* 19 (1990) 239;
D. Babusci et al., *Phys. Lett.* B277 (1992) 158; *Chiral symmetry and pion polarizabilities*, in Proc. Workshop on Physics and Detectors for DAFNE, ed. G. Pancheri, INFN-Frascati (1991); *Vector-meson resonances and pion polarizabilities*, preprint Frascati LNF-92/071 (P); and the references therein.

89. Yu. M. Antipov et al., *Phys. Lett.* B121 (1983) 445; *Z. Phys.* C26 (1985) 495.

90. L. Brown, W. Pardee and R. Peccei, *Phys. Rev.* D4 (1971) 2801.

91. G. Höhler, in *Landolt-Börnstein*, Vol. 9b2, ed. H. Schopper (Springer, Berlin, 1983);
P. Koch, *Z. Phys.* C15 (1982) 161.

92. J. Gasser et al., *Phys. Lett.* B213 (1988) 85;
M. Sainio, *Pion-nucleon update*, in ref.[13];
M.P. Locher, *Nucl. Phys.* A527 (1991) 73.

93. J. Gasser et al., *Phys. Lett.* B253 (1991) 252.

94. J. Gasser, *Ann. Phys. (N.Y.)* 136 (1981) 62; see also ref.[42].

95. J. Gasser et al., *Phys. Lett.* B253 (1991) 260.

96. B. Holstein, *Phys. Lett.* B244 (1990) 83, reanalyzes the e.m. corrections in $\pi \to \mu\bar{\nu}$ with the result $F_\pi = 92.4 \pm 0.3$ MeV.

97. For a new improved determination of L_1, L_2 and L_3, see ref.[22].

98. J. Donoghue and B. Holstein, *Components of a chiral coefficient*, preprint Univ. Mass., Amherst, UMHEP-368 (1992), update the determination of L_{10} from $\pi \to e\bar{\nu}\gamma$. As shown by Dominguez

and Solá[100], the experimental information on the vector and axial vector spectral functions obtained from τ decay[101] allows a test of the low energy theorem for $\pi \to e\bar{\nu}\gamma$. Donoghue and Holstein also update this analysis and conclude that the corresponding value of L_{10} is in good agreement with the one from π decay.

99. The phenomenology of the coupling constants occurring in the effective weak Lagrangian at $O(p^2)$ and $O(p^4)$ is discussed in refs.[26,27]

100. C.A. Dominguez and J. Solá, *Phys. Lett.* B208 (1988) 1311.

101. R.D. Peccei and J. Solá, Nucl. Phys. B281 (1987) 1.

DISCUSSION

H. Rollin, University of Illinois, USA

You mentioned the possibility of examining $\pi^+\pi^-$ atoms. What would one measure, and what would one learn from it?

Leutwyler

One would measure the lifetime of the decay into $\pi^0\pi^0$, thereby learning about the $\pi\pi$ interaction. The decay rate is determined by the S-wave scattering length, which is very sensitive to chiral symmetry breaking. Quite generally, the theoretical predictions concerning symmetry breaking effects are difficult to test experimentally, because they are small. An accurate measurement of the S-wave scattering length would subject the theory to a very sensitive test.

Leo Pillonen, Virginia Polytechnic Institute, USA

Surprisingly, I'm going to try to help chiral perturbation theory. There is a recent interesting paper by Im [and possibly by Peskin] on the large N expansion for $\gamma\gamma \to 2\pi^0$. What they have done is just the bubble summation in the old fashioned way. I have also tried to perform the 1/N correction. In the case of $\gamma\gamma \to 2\pi^0$ this correction turns out to be small. The same thing can be said for the P-wave pion form factors. Doing the bubble summation, which is a way to implement elastic unitarity to all orders, one gets the input for the RMS radius and one gets the width of the ρ-resonance which satisfies the KSFR relation. So I strongly suggest for those involved in χPT to look at the paper that was published recently; it's a good paper and it will implement the idea of elastic unitarity to all orders.

Leutwyler

Yes, thank you for this remark. I am sorry that this and related issues populate some of those transparencies which I did not have enough time to show. In chiral perturbation theory, the limit where the number of flavors is taken large is one way of trying to understand higher order effects. Whenever these effects are large, straightforward χPT is not good enough. The large N limit offers an intelligent extrapolation of the one-loop results to all orders.

B. Grinstein, SSC Laboratory, USA

How can we trust the determination of quark masses at a renormalization scale above the χSB scale (used in connection with GUT's) when the calculation is done at a much smaller scale?

Leutwyler

Mass ratios are independent of scale. In χPT, the normalization used for the quark mass matrix itself does not matter, because only the product of the constant B with $\mathcal{M}$ counts. A change in the normalization of $\mathcal{M}$ merely changes the value of B.

M.M. Islam, University of Connecticut, USA

This is a general question. You have described here an effective field theory. Can you comment on how such a theory can be obtained from QCD?

Leutwyler

For a derivation of the effective field theory, I refer you to Weinberg's 1979 paper (ref. 7). The arguments given there only rely on general properties of QCD at low energies. They imply that the effective field theory describes the low energy structure correctly, in the mathematical sense of the term. The general properties which go into the derivation are generally accepted, but have not rigorously been established for QCD. The question of why the effective coupling constants have the numerical values indicated by phenomenology and how to calculate them on the basis of the QCD Lagrangian is another matter—I tried to explain our current understanding of this question in the course of the talk.

A. Bettini, Padova, Italy

Do you have any predictive power in your theory? For example, if you experiment with a laser beam, what would you expect to find?

Leutwyler

I take it that the question concerns the phase transition to the quark-gluon-plasma. χPT is useful only in the hadronic phase, and only at low temperatures, where the calculated temperature effects are small. The critical temperature is not within the reach of χPT. In connection with heavy ion collisions, the properties of the hadronic phase are important, because the hadrons emitted by the hot spot do not run into the detector straight away. One needs to know the transport properties of the hadronic gas which is formed (such as the mean free path between successive collisions) to extract physics from the observed pion distributions. Here, χPT is of interest.

Bettini

There are physically measurable parameters in question. What temperature, energy, and chemical potential do you need to obtain the transition to a quark-gluon plasma?

Leutwyler

I did not want to give the impression that χPT is the thing to do to analyze the quark-gluon phase. It's the thing to do to analyze the hadronic phase at low energies or temperatures. Invariably, if a droplet of plasma is produced, it will hadronize. If the droplet is big enough, the hadronic gas will reach thermal equilibrium and remain there during cooling, for a while. In the observed pion distributions, one is not seeing the physics of the quark-gluon phase, but the physics of the hadronic phase at the temperature where decoupling occurs. Once the collision is sufficiently energetic, the decoupling temperature will be low enough to be within reach of the chiral expansion.

A. Sirlin, New York University, USA

A theorem exists that relates the slope of the form factor and the charge radius of the π^+, K^0 and K^+, with an error of second order in SU(3), derived from general principles. Is chiral perturbation theory inserted into that theorem nonsense, or approximately correct? If you assume that the form factors are dominated by the vector bosons, it reproduces the Gell-Mann mass formula for the vector bosons.

Leutwyler

I know what you are referring to: the low energy theorem you found yourself. One of the virtues of χPT is that it gives you all of the symmetry relations there are, at a given order in the quark mass expansion. In particular, all of the low energy theorems for the slopes of the vector form factors obtained from current algebra are included as special cases of the chiral representation. We explicitly checked your relation with χPT in 1985 (ref. 9).

ULTRA-RELATIVISTIC HEAVY-ION COLLISIONS: SEARCHING FOR THE QUARK–GLUON PLASMA

Jürgen Schukraft
CERN, Div. PPE
CH-1211 Geneva 23

Abstract

The physics of ultrarelativistic heavy-ion collisions is presented with particular emphasis on the basic concepts used in interpreting the data and some principal problems encountered in this new and rapidly evolving field. A number of selected recent and older results from experiments at CERN and BNL are reviewed. No unambiguous evidence for Quark-Gluon Plasma formation has emerged from the data so far, but some large and significant effects have been observed which clearly show that nucleus-nucleus collisions cannot be described as a straightforward superposition of independent nucleon-nucleon reactions. Results from a number of independent observables (p_t-distributions, strangeness abundances, J/Ψ suppression, particle interferometry) indicate the existence of an extended and strongly interacting system. The first 'exploratory' phase of experimentation seems therefore to confirm that ultrarelativistic heavy-ion collisions are a promising tool to study the properties of bulk hadronic matter under extreme conditions.

INTRODUCTION

The subject of ultra-relativistic heavy-ion collisions

With the advent of ultra-relativistic heavy-ion collisions in the laboratory, at Brookhaven and CERN in 1986, a new interdisciplinary field has emerged from the traditional domains of particle physics and nuclear physics [1–2]. In combining methods and concepts from both areas, the study of heavy-ion reactions at very high energies ($E/m \gg 1$) denotes a new and original approach in investigating the properties of matter and its interactions. In high-energy physics, *interactions* are nowadays derived from *first principles* (gauge theories), and the *matter* concerned consists mostly of *single particles* (hadrons/quarks). In contrast, on nuclear physics scales the strong *interaction* is shielded and can therefore to date only be described in effective or *phenomenological* theories, whereas the *matter* consists of *extended* systems exhibiting collective features. Combining the elementary-interaction aspect of high-energy physics with the macroscopic-matter aspect of nuclear physics, the subject of heavy-ion collisions is the study of *bulk matter* consisting of *strongly interacting* particles (hadrons / quarks). It may therefore be dubbed 'QCD thermodynamics' or 'condensed-matter physics' of elementary particles. The energy scale is given by Λ_{QCD} (the scale parameter of QCD), the pion mass, or the limiting 'Hagedorn temperature', all of which happen to be of the order of 200 MeV. The physics is therefore inherently the physics of 'soft' processes, and the objects under study are the old-fashioned hadrons (π, K, ρ, p, Λ, ...) and light quarks (u, d, s). The language to be used in this field would ideally be the language of thermodynamics, where complex multi-particle states are described in terms of a few macroscopic variables (temperature, density, entropy, etc.).

Figure 1 shows a streamer chamber picture of a heavy-ion reaction (^{32}S on ^{197}Au) at an energy of 200 GeV/A. The resulting 'mess' looks very much like an experimentalist's nightmare: hundreds of final-state particles widely distributed with no discernible structure (besides the purely kinematic concentration at small angles relative to the beam). In more elementary high-energy reactions (pp, e⁺e⁻), a search for simplicity and underlying order usually proceeds in a microscopic direction: from individual *particles* via *jets* to *partons*, described by perturbative *QCD dynamics* in terms of *microscopic variables* (centre-of-mass energy $\sqrt{s}$, momentum transfer, ...). In a similar fashion, we hope to find simplicity in the apparent chaotic final states of A-A reactions along a different direction, applying macroscopic and statistical concepts: from *particles* to statistical *ensembles* described by (non-pertubative) *QCD thermodynamics* in terms of *macroscopic* variables (temperature, pressure, ...).

What makes this field particularly interesting is the prediction of QCD that at high energy densities matter should undergo a phase transition to an entirely new state, the quark-gluon plasma (QGP). At low energy densities, quarks and gluons are bound by the strong force into colourless objects, the hadrons (confinement). In addition, the quarks acquire a large effective mass ($m_u \approx m_d \approx 300$ MeV, $m_s \approx 500$ MeV) via interactions between themselves and the surrounding physical vacuum (broken chiral symmetry). When increasing the energy density by increasing the temperature ('heating') or the matter density ('compressing'), a phase transition might occur towards the QGP, the true perturbative vacuum of QCD, where partons are deconfined and chiral symmetry is approximately restored ($m_u \approx m_d \approx 5$ MeV, $m_s \approx 150$ MeV).

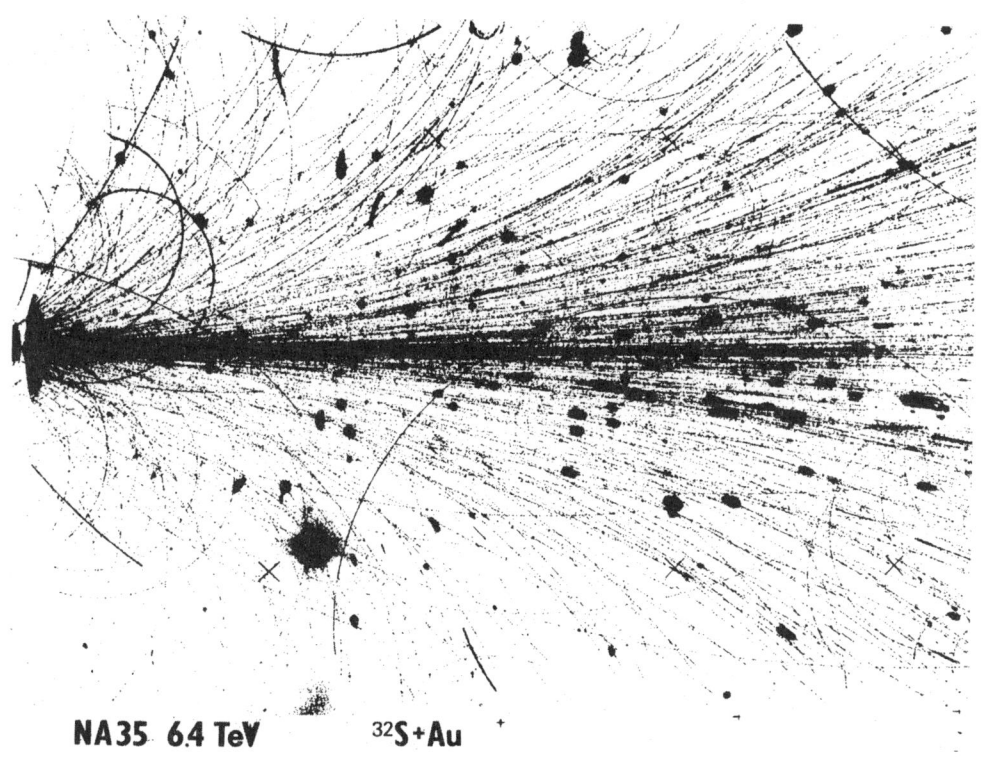

Figure 1. Streamer chamber picture from experiment NA35 of a ^{32}S on ^{197}Au collision at 200 GeV/A.

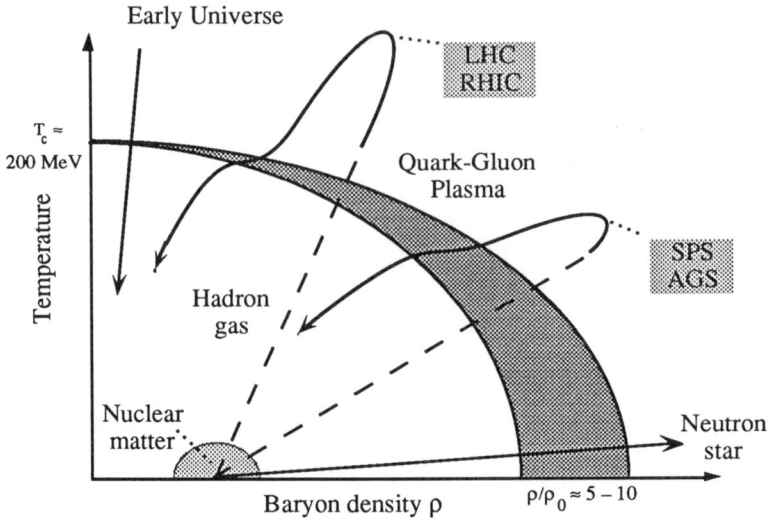

Figure 2. The phase diagram of strongly interacting matter showing the hadronic phase at low temperature and baryon density, the transition region (mixed phase), and the QGP phase. The solid lines illustrate trajectories followed in supernovae explosions, Big Bang evolution, and possibly in heavy-ion reactions at present and future accelerators.

In the context of the Standard Model, the study of this phase diagram of strongly interacting matter (Fig. 2) is not only of interest to study and test QCD on its natural scale (Λ_{QCD}), i.e. in the non-perturbative sector, but it might also shed light on such fundamental questions as the nature of confinement itself and on the process of spontaneous symmetry breaking, which is made responsible for the origin of the 'effective' quark masses (the pion being the assorted Goldstone boson). The early Universe presumably underwent this very phase transition 10^{-5} s after the Big Bang. Critical phenomena that can occur close to a phase boundary, for example long-range density fluctuations (as in condensing water!), might have a bearing on important aspects of cosmology, such as nucleosynthesis, dark matter, and the large-scale structure of the Universe. In astrophysics, the dynamics of supernova explosions and the stability of neutron stars (density $\rho \approx 10\, \rho_{nucleus}$) depends on the compressibility and therefore the equation of state of nuclear matter, and it is even speculated that the core of neutron stars may consist of cold QGP. The study of extreme states of matter created in high-energy nuclear collisions thus provides us with an opportunity of gaining insight into many important aspects in different fields of physics.

Geometry and space–time evolution of heavy-ion collisions

Nuclei are extended objects, and therefore their geometry plays an important role in heavy-ion collisions. Figure 3 shows a sketch of a reaction between asymmetric nuclei A and B; the impact parameter b separates the nucleons into participants, with primary nucleon–nucleon collisions, and spectators, which proceed with little perturbation along the original direction. In the first instances of the reaction, the nucleon–nucleon collisions between the two highly Lorentz-contracted nuclei redistribute a fraction of the original beam energy into other degrees of freedom. After a short time, usually taken to be of the order of 1 fm/c, partons materialize out of the highly-excited QCD field, possibly in the state of an equilibrated QGP. The system expands rapidly, mainly along the longitudinal direction, thereby lowering its temperature, and reaches the critical transition temperature T_c after a few fm/c. Potentially, the matter then spends a long time in the mixed phase, in particular if the transition is of first order. It has to rearrange the many degrees of freedom (partons) of the QGP into the fewer available in the hadron phase, with an associated large release of latent heat. In the last and hadronic phase ('hadron gas' or 'hadron fluid'), the still-interacting system keeps expanding, maybe even in an ordered motion ('collective flow'), to very large dimensions until 'freeze-out' occurs, when interactions cease and the particles stream freely away to be detected in the experiments.

Necessary conditions

In order for heavy-ion collisions to fulfil the promises mentioned above, a number of necessary pre-conditions have to be met and verified by results:

In order to use *macroscopic variables*, the system has to be 'big', i.e. the dimensions need to be much larger than the typical scale of strong interactions ($\gg 1$ fm) and consist of 'many' particles ($\gg 1$).

To use the language of *thermodynamics*, the system has to be in (or near) equilibrium, i.e. its lifetime has to be larger than the relaxation times ($\tau \gg 1$ fm/c). Equilibrium can be reached and maintained throughout the expansion only in a sufficiently interacting system; therefore the number of collisions per particle has to be larger than one.

The *energy densities* ε needed for QGP formation are predicted by QCD to be of the order of 1–3 GeV/fm^3, equivalent to a temperature $T_c \approx 150$–200 MeV or a baryon density $\rho \approx 5$–10 times the nuclear matter density. It has to be verified that these energy densities can indeed be reached in heavy-ion collisions.

Because of the rapid evolution, *experimental observables* will, in general, correspond to an integral over the complete space–time history of the reaction from the first nucleon–nucleon collision until freeze-out, and disentangling the various contributions to a signal from the different phases indeed presents a formidable challenge. Furthermore, a system evolving in equilibrium by definition erases its memory of preceding stages. As in Big Bang cosmology, it is necessary to identify observables that decouple at different times from the expansion and are more sensitive to the early and hot stages of the matter.

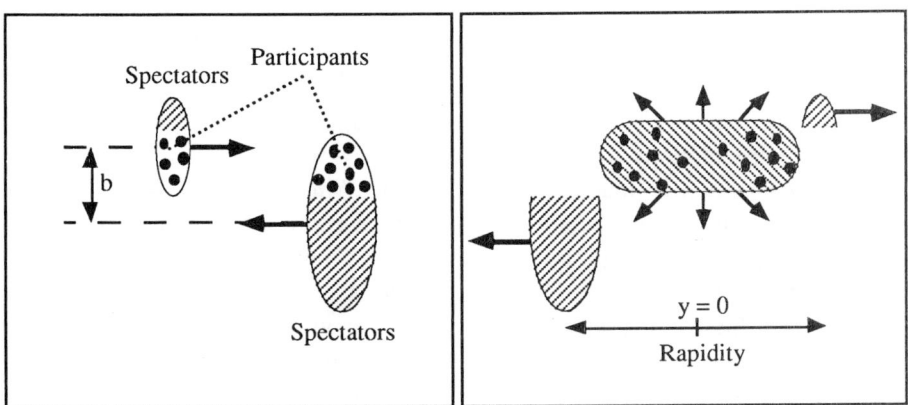

Figure 3. Schematic representation of a heavy-ion collision at impact parameter b, assuming straight-line geometry.

Signals and observables

The various signals that can be observed by experiments and associated (to some extent) with the different stages of the evolution, are summarized below:

INITIAL CONDITIONS: The measurement of global event features is necessary to specify the initial conditions given in a heavy-ion reaction. From baryon distributions, transverse energy- and particle production one can derive in a more or less model dependent way the impact parameter, the initial volume, and the energy density reached on an event-by-event basis.

QUARK-GLUON-PLASMA: Weakly inter-acting probes, which decouple at early times, are the only direct means to gain information on the plasma phase. Direct photons and lepton pairs (virtual photons) are such observables and should emerge as thermal radiation from the heated matter without being altered by final state effects. The strong temperature dependence assures that the thermal radiation is sensitive mainly to the first and hottest stages of the evolution.

Signals originating from hard scattering processes at the very beginning of the reaction are not directly connected to plasma formation; nevertheless they are important as tools to probe the state of the surrounding QCD matter. The suppression ('melting') of heavy quarkonium resonances (J/Ψ, Υ) via Debye screening is characteristic only for the deconfined state of QCD, and the energy loss of jets might be different in a QGP and in hadronic matter ('jet quenching').

In addition, there are speculations about a number of 'exotic' signals, such as massive photons, stable strange matter ('strangelets') or free quarks, which, if observed, might be an unambiguous sign of QGP formation.

PHASE TRANSITION: Strange quarks are abundantly produced in the plasma, but the final number observed in strange hadrons will depend on the details of the hadronization phase transition and to some extend on the following expansion. The presence of the transition could also be signaled by long range fluctuations in multiplicity or by intermittency patterns. The long lifetime associated with a first order phase transition might reflect itself in Bose-Einstein correlations of identical particles (HBT). Ideally, the presence of a phase transition (i.e. constant temperature over a range of energy densities) would reveal itself in a characteristic dependence of the average p_t on energy density.

HADRON GAS: The evolution and cooling of the system is usually described in terms of a hydrodynamic expansion. The collective motion alters the otherwise thermal spectra of produced particles. Thus, the investigation of p_t or m_t spectra could, in principle, yield information on the expansion process. The dynamics of the evolution, in particular the longitudinal expansion of the source, introduce a strong correlation of the position and momentum coordinates of particles, which could, in principle, also be measured via two-particle correlations.

In dense matter, hadronic resonances are predicted to change their characteristics (mass, width, branching ratios) as a consequence of chiral symmetry restoration. If a decay into weakly interacting particles (e.g. lepton pairs) happens inside the medium, we might be able to observe such resonance modifications.

Finally, once the system has reached a certain size and density collisions among the practicles cease and the final hadron distribution freezes-out. The corresponding freeze-out radius is accessible via two-particle correlations.

Experimental facilities

The Alternating Gradient Synchrotron (AGS) at Brookhaven was transformed into a heavy-ion accelerator in 1986; it has been running since then on a regular basis several weeks per year with beams up to ^{28}Si at 14.5 GeV/A (c.m. energy in the nucleon–nucleon system $\sqrt{s} \approx 5$ GeV/A). There are four large experiments and some smaller ones, with a total of 350 users, coming roughly in equal parts from high-energy and nuclear physics.

Earlier this year, beams of Au ions have been succesfully accelerated for the first time in the AGS and delivered to the experiments.

The Super Proton Synchrotron (SPS) at CERN accelerated ^{16}O at 60 ($\sqrt{s} \approx 10$ GeV/A) and 200 GeV/A ($\sqrt{s} \approx 20$ GeV/A) in 1986 and ^{32}S at 200 GeV/A in 1987. After the initial short runs of two weeks each, a new, long-term programme of heavy-ion physics was established at CERN starting in 1990 with several weeks of ^{32}S beams. There are six big electronic detectors and also a number of smaller experiments with a total of 550 users, again half of them coming from nuclear physics.

The early, so-called 'exploratory' phase of heavy-ion collisions (1986–1990) was characterized by the fact that no dedicated machines were used, but rather existing accelerators were upgraded at modest financial expense. Likewise, the experiments made extensive reuse of existing HEP/NP equipment (> 80% at CERN).

RESULTS

Transverse energy E_t and energy density ε

The crucial quantity linking the observed event characteristics with the conjectured formation of a quark–gluon plasma is the energy density ε (energy / volume) achieved in heavy-ion reactions. The corresponding experimental observables are either the number and average momentum of final-state particles created or the transverse energy E_t ($E_t = E \cdot \sin\theta$) emitted perpendicular to the beam direction. As a typical example of the latter, the cross-section $d\sigma/dE_t$ is shown in Fig. 4 for ^{32}S on various targets at 200 GeV/A [3]. The cross-section is integrated in the angular region $96° < \theta_{lab} < 0.5°$, equivalent to essentially 4π in the c.m. system. The absolute amount of E_t released in central collisions of the heaviest systems studied to date (S-U) is indeed remarkable: 450 GeV and about 600 charged particles, exceeding by about two orders of magnitude the values obtained in p–p reactions at the same beam energy.

The shape of the distribution — rapidly falling at low E_t, a rather long and flat region in the middle ('plateau'), followed by an abrupt break and a steep decline ('tail') — is typical for asymmetric A–B reactions (A < B). It is governed by geometry and essentially reflects the number of primary nucleon–nucleon collisions that increases smoothly when going from peripheral ($b \gg 0$, low E_t) to central ($b \approx 0$, high E_t) reactions. Even subtle geometrical effects such as nuclear deformations are borne out by the data. Based on the geometrical picture, the E_t of central collisions between spherical nuclei should grow roughly proportionally with the thickness of the target, i.e. $A_t^{1/3}$. Compared with the only slightly smaller ^{208}Pb target, much larger E_t values are observed for ^{238}U, reflecting the large quadrupole deformation of ^{238}U. High E_t in this case selects reactions where the cigar-shaped U nucleus is preferably aligned with its longer axis parallel to the beam, in this way increasing the effective amount of nuclear matter to the equivalent of a spherical nucleus with mass 400 !

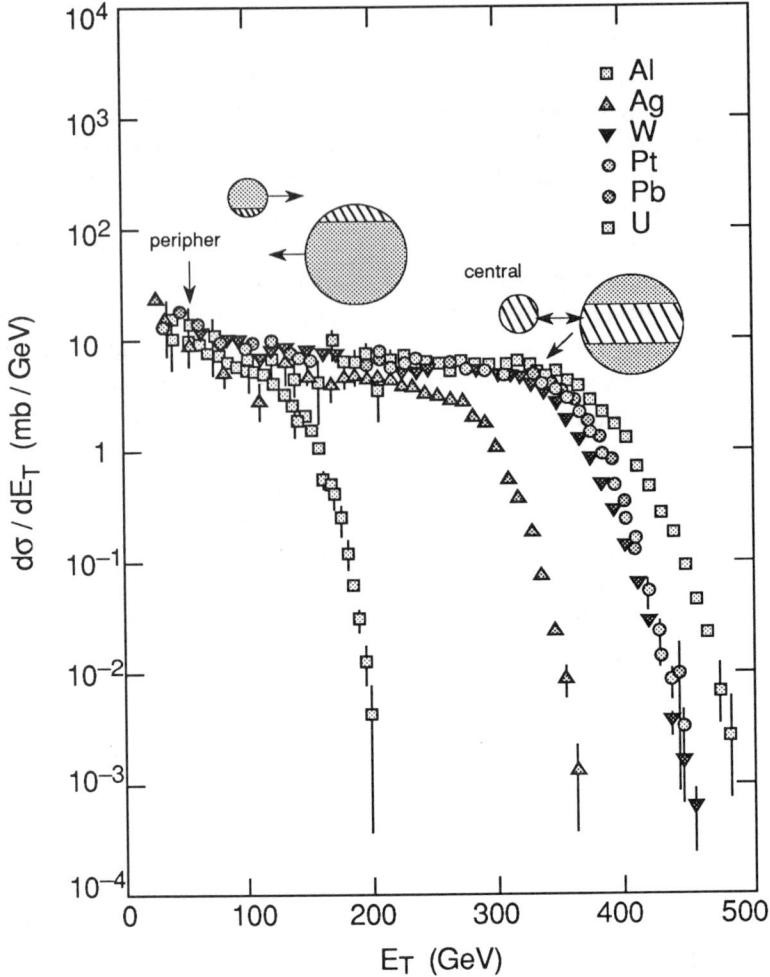

Figure 4. Cross-section for transverse energy production in ^{32}S collisions with a number of different targets at 200 GeV/A. [3]. The inset shows the reaction geometry corresponding roughly to the different regions in E_t.

First preliminary results on transverse energy production with really 'heavy ions' are shown in Fig. 5 from the AGS experiment E877 [4] taken only a few weeks ago (spring '92). Assuming a power law dependence on the atomic mass ($E_t \sim A^\alpha$) when going from Si + Al (broken line) to Au + Au (full line), one finds the exponent α significantly bigger than 1 ($\alpha = 1.11 \pm 0.03$). This indicates a high degree of 'nuclear stopping' and the presence of collective nuclear effects up to and including the heaviest nuclear systems investigated to date.

As no direct observation exists of the initial energy released or of the initial size of the reaction volume, geometrical and kinematical assumptions have to be made in order to convert the measured final energy into an energy density at the very beginning of the reaction. Assuming longitudinal boost invariance ('Bjorken scenario'), the resulting energy densities are estimated to be of the order of 1 GeV/fm^3 at the AGS and 2–4 GeV/fm^3 at CERN, corresponding to temperatures of 150–210 MeV. Despite their somewhat qualitative nature, it is reassuring that these numbers are not only large, about twenty times the energy density of ground-state nuclear matter and still five times larger than the energy density inside a hadron, but indeed of the order of the critical value needed for the phase transition.

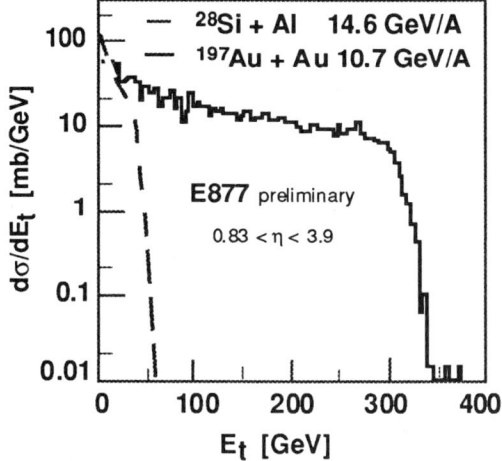

Figure 5. Cross-section for transverse energy production in symmetric collisions of light and heavy systems from experiment E877 at the AGS.

Volumes and lifetimes

The space–time extent of sources emitting radiation or particles can, in principle, be measured via intensity interferometry. Identical particles obey the rules of quantum statistics; the (anti) symmetrization of their wave function leads to correlations in momentum–energy space, which in turn are related, via a Fourier transformation, to the source distribution in space–time. This method, originally introduced by Hanbury-Brown and Twiss (HBT) in 1953 to determine the radii of stars, is now regularly used in particle physics to measure the freeze-out volume (surface of last rescattering). In p–p and e^+e^- reactions, the transverse radii r_t extracted with this method are of the order of 1 fm (size of strings or hadrons). In low-energy nuclear reactions, the sizes correspond precisely to the geometrical interaction region (i.e. to the size of the smaller of the colliding nuclei). The coherence parameter λ which measures the strength of the correlation at zero 4-momentum difference ($\lambda = 1$ for completely chaotic emission) is, however, found in all cases to be significantly smaller than one, a fact that is sometimes attributed to experimental cuts or resonance decays.

The NA35 Collaboration have measured the correlation function with negative pions in a number of nuclear collisions [5]. They find a rather large coherence parameter ($\lambda \approx 1$) and transverse radii between 5 and 6 fm, i.e. values which exceed the projectile size by a factor of about two and indicate a very large and chaotic source which collectively expands from an initial size of 3 fm to a final size of $\approx$ 6 fm. This expansion leads in addition to a lower limit on the lifetime of $\tau > 3$ fm/c. Combining a number of results from p–p and A–B reactions (Fig. 6), the freeze-out radius seems to scale with the particle density like $r_t \approx (dN/dy)^{1/3}$, i.e. it always occurs at roughly the same final matter density.

New high statistics results are available from NA44, an experiment specialized on HBT (see A. Franz in these proceedings). The correlation functions from S + Pb reactions are shown in Fig. 7 for both pions and kaons. A large difference in the width of the correlation function for π's and K's is apparent, which might however partially be due to kinematic effects (Lorentz boost). If the difference persists after the appropriate corrections have been applied (as indicated by a two dimensional analsysis in terms of longitudinal and transverse radii, see A. Franz's contribution), this could signal an earlier freeze-out of the kaons.

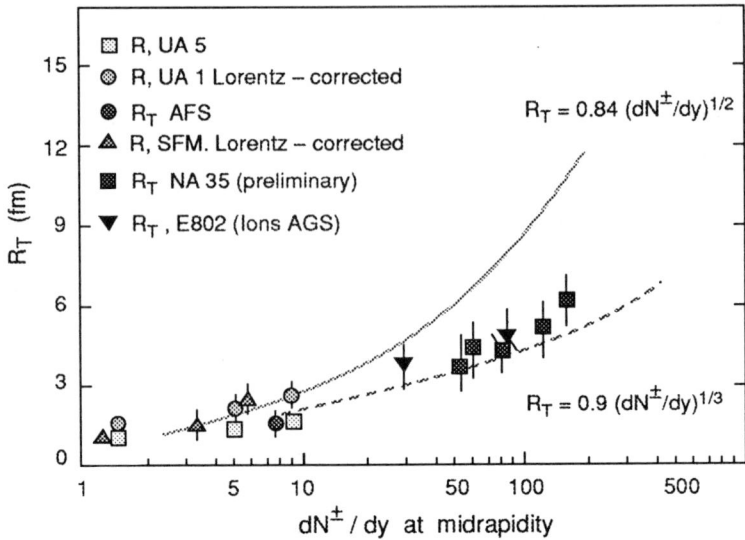

Figure 6. Compilation of transverse source radii obtained at midrapidity in hadron-hadron and nucleus-nucleus reactions as a function of charged particle density [6].

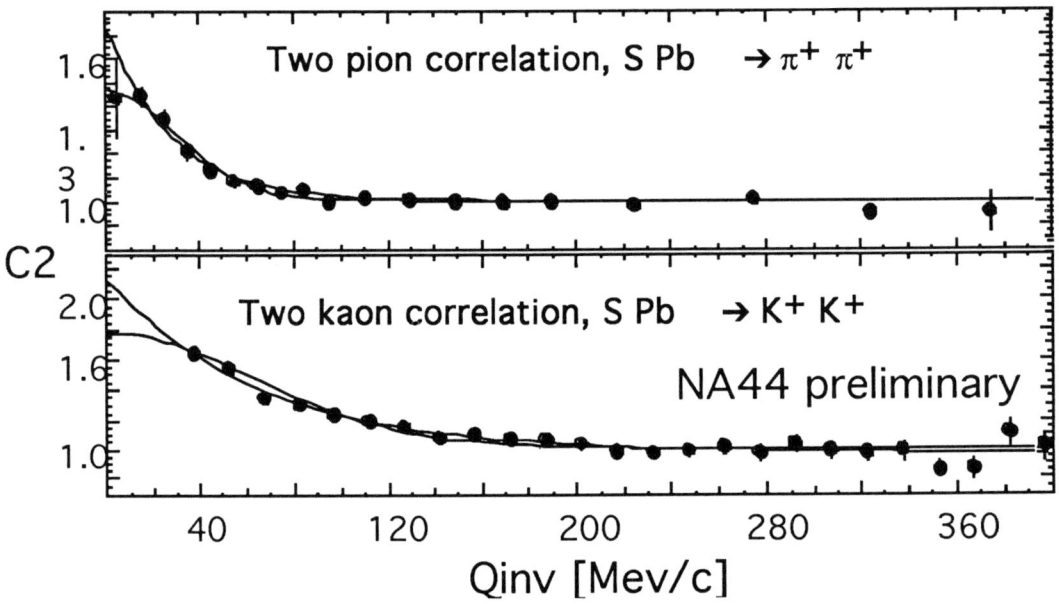

Figure 7. HBT correlation function for pion and kaon pairs in S + PB reactions at 200 GeV/A from NA44 (A. Franz, these proceedings).

Thermal equilibrium

The momentum distribution of particles in thermal equilibrium is given by a Boltzmann law with the average momentum proportional to the temperature. In ultra-relativistic collisions, the longitudinal components (p_l) of the momentum vectors cannot be easily thermalized because of the large asymmetry in the initial state ($p_t = 0$, p_l = beam momentum). Therefore usually only the transverse components (p_t) are considered. They could, in principle, provide a direct measure of the temperature of the system, reveal collective particle flows ('viscosity' of the hadron gas/fluid), and ideally even signal the presence of a phase transition.

The general characteristics of p_t distributions in p–A and A–A reactions can be extracted from Fig. 8, where the cross-section $d\sigma/dp_t^2$ is shown for negative particles (mostly pions) in p–W, O–W and S–W central collisions as measured by NA34 [7]. The distributions differ substantially from the ones measured in p–p (solid line in Fig. 8); they show a strong enhancement both at high p_t (> 1 GeV/c) and at low p_t (< 250 MeV/c). Both the high p_t, usually referred to as 'Cronin' effect, as well as the low p_t rise were actually already observed in the mid 1970s at Fermilab in p–A reactions. The theoretical interest in these deviations has been considerable. The high p_t part can be interpreted in a thermal model with a substantial amount of collective flow; the low p_t part could be the result of some exotic processes, e.g. an approach to pion condensation or the remnant of supercooled QGP droplets. However, more mundane explanations are also possible and even likely, because they naturally describe some systematic trends observed in the data. Resonance decays, in particular of the Δ baryon, could be the source of low p_t pions, and multiple, small-momentum parton scattering inside the nuclei might be responsible for the high p_t enhancement.

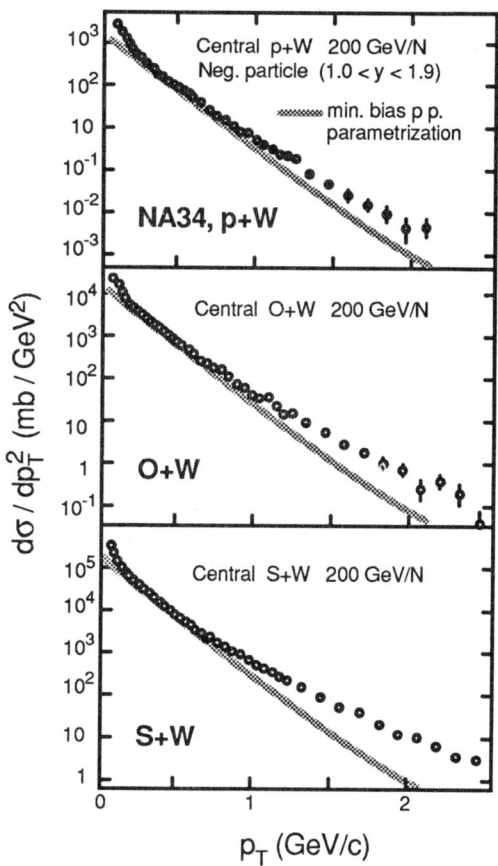

Figure 8. Negative-particle distribution from NA34 for central collisions of p, O, S, (a, b, c) with a W target. For comparison a parametrization of p–p data is shown as a solid line.

Chemical equilibrium

In 'chemical' equilibrium, the abundance of particle species (hadrons/quarks) is again governed by Boltzmann factors, i.e. essentially by the temperature, the respective masses, and a chemical potential. The production of strange quarks is favoured in a QGP because there the mass of the strange quark is reduced to a value comparable to the temperature (chiral symmetry restoration). In addition, the creation of light u and d quarks is hindered by the large number of valence quarks already present in the colliding nuclei (Pauli blocking), leading to a large chemical potential for these quarks that favours s $\bar{s}$ over u $\bar{u}$ and d $\bar{d}$ production.

Results on enhanced strangeness production were already reported shortly after the first data were taken in 1986 [8]. Many experiments have now measured the yield of several particle species containing one or two strange quarks (K^+, K^0, Λ, Ξ and their antiparticles). In most of the cases, their production relative to pions is larger in nucleus-nucleus collisions compared to p-p by a factor of two to five. The K^+/π^+ ratios at the AGS vary smoothly from p+Be, p+Au, Si+Au to Au+Au, increasing from $\approx$ 8% to $\approx$ 25% (see L. Remsberg in these proceedings), whereas K^-/π^- changes only little from 2 to 4 %. These ratios are rapidity dependent and probably reflect a different production mechanism: K^- are more centrally produced, in particular in p-Be collisions, whereas the K^+ behave similar to the protons and shift clearly backwards with increasing target mass. The gradual increase in K^+ production from pp to pA to A-A together with the change in slope of the transverse momentum spectra suggests that rescattering effects of produced particles amongst themselves and with the surrounding cold spectator matter are important. Indeed, microscopical, non-equilibrium Monte Carlo calculations including these effects can reproduce quantitatively the observed results.

Multistrange (anti)baryons are expected to be a particularly sensitive probe of strangeness enhancement from the QGP, because their thermal production depends on a high power of the strangeness density whereas the non-equilibrium production in the hadron gas phase is expected to be suppressed owing to the large mass. The WA85 experiment, optimized to measure strange baryon decays at midrapidity and $p_t > 1$ GeV/c, has found that both Λ and $\bar{\Lambda}$ production increases by $\approx$ 70% when going from p-W to S-W. Whereas the Ξ/Λ ratio is compatible with results from pp and e^+e^- reactions, $\bar{\Xi}/\bar{\Lambda}$ is about five times

greater then the one measured at the ISR, albeit with very large statistical errors. For the reasons mentioned above, a strong overall increase in antibaryon production will be difficult to explain in non-QGP models (see also J. Rafelski in these proceedings).

Like particles with open strangeness, the production of ϕ mesons (s $\bar{s}$) is strongly enhanced in nuclear collisions. Early measurements of NA38 have now been extended to lower p_t by NA34 (see U. Goerlach in these proceedings); they confirm the rise of the $\phi/(\rho+\omega)$ ratio as a function of centrality (see Fig. 9).

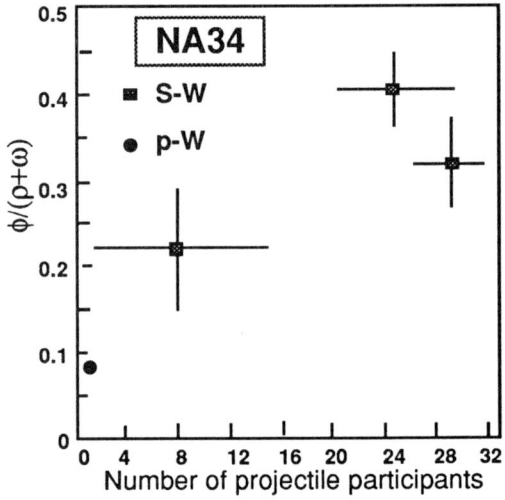

Figure 9. The ratio of ϕ to $\rho+\omega$ production from NA34 for p+W and S+W as a function of the number of projectile participants (essentially proportional to E_t).

The question of 'strangeness as a signal for the QGP' has undergone a rapid development since the first predictions were made. It is now clear that the dynamical evolution through the phase transition and the unavoidable hadron-gas phase can change the strangeness content of matter and readjust the particle ratios to the respective equilibrium values at various stages (which is indeed the very definition of a system evolving in equilibrium!). Strangeness enhancement, in general, is now seen more as a characteristic feature of a system approaching chemical equilibrium rather than as a unique signal for QGP formation. Depending on the assumptions made about the space–time development, the equilibrium values predicted for many strange particles can be very similar in QGP and hadron-gas models. In addition, some of the systematic features of K^+ and Λ production are well described by non-equilibrium rescattering, which is particularly strong in this channel (e.g. $\pi + n \rightarrow K + \Lambda$). On the other hand, no unconventional explanation has so far been put forward for the observed increase of antihyperons ($\bar{\Lambda}, \Xi^-$).

Signals from the QGP

Weakly interacting probes, which decouple at early times, are the only direct means of gaining information on the plasma phase. Direct photons and lepton pairs (virtual photons) are such observables and should emerge as thermal radiation from the heated matter without being altered by final-state effects. Their strong temperature-dependence assures that the thermal radiation is sensitive mainly to the first and hottest stages of the evolution. Unfortunately, they have to be measured in the presence of an immense background of photons and leptons arising from ordinary hadronic decays or hard-QCD processes. So far, this primordial 'background' radiation has not been observed, but the present experimental upper limits ($\gamma/\pi^0 < 10–15\%$) are not yet sensitive enough to measure the predicted yields.

Signals originating from hard-scattering processes at the very beginning of the reaction, such as J/ψ production, are not directly connected to plasma formation; they are nevertheless important as tools to probe the state of the surrounding QCD matter. The original idea [9] that the formation of the J/ψ should be suppressed in a QGP relies on a mechanism analogous to the Debye screening effective during an insulator–conductor Mott transition in QED. The J/ψ is produced during primary nucleon–nucleon collisions mainly via gluon fusion. The confining strong force, which would normally bind the newly created charm quarks within a small, but finite time (formation time) into a J/ψ is, however, screened in the QGP. If the screening radius (Debye radius), which is inversely proportional to the density of colour charges and therefore to the energy density, is smaller than the size of the J/ψ (≈ 0.5 fm), a bound state cannot be formed. The charmed quarks dissolve and separate in space to appear later, after hadronization, as two mesons with open charm. Because of the finite formation time, high-p_t (i.e. 'fast') charm pairs can escape the QGP unaffected, yielding a characteristic, p_t-dependent suppression pattern. This 'melting' process is specific only for the deconfined state of QGP and would therefore qualify as an unambiguous signal; however, the J/ψ is not weakly interacting and final-state rescattering of the J/ψ (absorption) will complicate the situation.

In heavy-ion reactions J/ψ suppression has indeed been observed by the NA38 experiment [10]. Figure 10 shows the dimuon mass spectrum measured in O–U for low-E_t and high-E_t collisions. The two data sets are normalized to the muon pair continuum ('Drell–Yan') outside the resonance region and show a clear suppression of J/ψs by a factor of about two in central collisions compared with peripheral ones. The E_t dependence is shown in more detail for a number of beams and targets in Fig. 11. From the p_t spectra of J/ψs it seems that the suppression is indeed most effective at low transverse momentum. The pattern and the absolute magnitude of the effect are exactly as predicted for a QGP; nevertheless, a number of 'conventional' models are able to describe the data in terms of absorption in a hadron-gas and initial-state parton scattering. In these models, rescattering of the J/ψs in the surrounding medium provides the suppression (e.g. via J/ψ+π −> D+ $\bar{D}$+X), and initial-state parton scattering prior to the gluon fusion shifts the p_t of the surviving J/ψs to larger values (see high-p_t

enhancement of pions earlier). In addition, a change of the nuclear structure functions (gluon shadowing) will lead to a reduced J/ψ production in particular already in p-A reactions (see H. Satz in these proceedings). All these effects, which certainly occur at some level, seem to conspire in such a way as to fake a signal predicted as unambiguous for the QGP. Eventually, however, hadronic and QGP models should be distinguishable because their predictions differ at higher p_t and for heavier charmonium states.

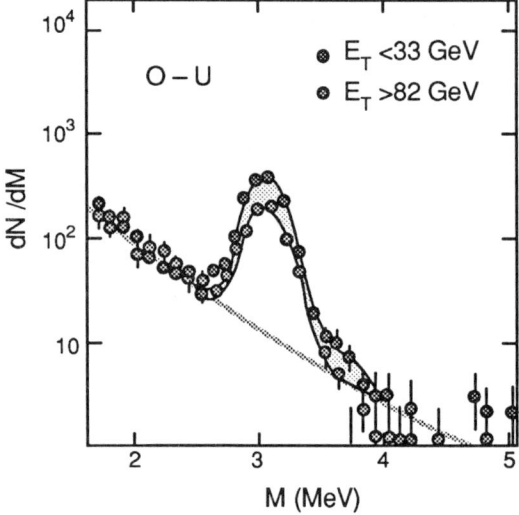

Figure 10. Dimuon mass spectra observed in O-U collisions for low- and high-E_t events. The data are normalized with respect to each other in the continuum region above and below the J/ψ. The solid lines represent fits.

In dense matter, hadronic resonances will change their characteristics (mass, width, branching ratios) as a consequence of chiral symmetry restoration. If a decay into weakly interacting particles (e.g. lepton pairs) happens inside the medium, we might be able to observe such resonance modifications. These measurements are amongst the most difficult to perform, because of large combinatorial backgrounds and the excellent mass resolution required, and only recently have the first experiments able to address some of these points started to take data (see U. Goerlach in these proceedings).

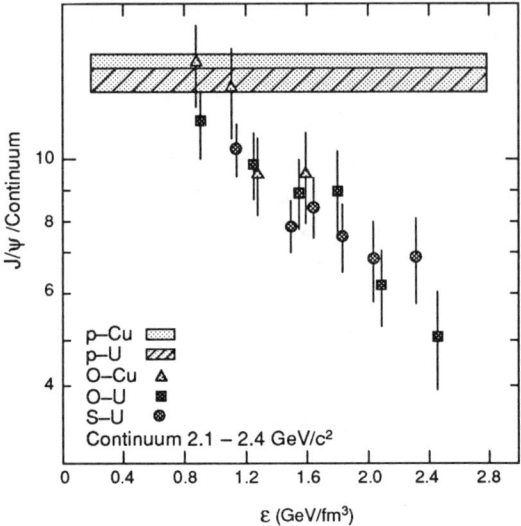

Figure 11. J/Ψ suppresion factor (number of J/Ψ events divided by number of continuum events) as a function of energy density for diffent target-projectile combinations.

In addition, there are speculations about a number of 'exotic' signals, such as massive photons, stable strange matter ('strangelets'), or free quarks, which, if observed, could be a spectacular sign of QGP formation. Free quarks have already been searched for (with negative results), and some strangelet experiments are under consideration at the moment.

CONCLUSIONS AND FUTURE

An impressive amount of data have been collected and analysed in the first five years of accelerator-based ultra-relativistic heavy-ion physics at Brookhaven and CERN. The initial phase, often based on preliminary and sometimes rapidly changing data, was characterized by interpretations ranging from extremely pessimistic ('trivial superposition of p–p collision + geometry') to extremely optimistic ('clear signals of the QGP') views. The field is now becoming mature, the data are consolidating and more-balanced interpretations are emerging. No convincing evidence for the creation of a Quark–Gluon Plasma at present energies and with the (light) heavy ions available to date has been found. Nevertheless, a number of important milestones have been passed that established some necessary pre-requisites vital for the study of dense and strongly interacting matter by means of heavy-ion collisions.

Energy density: The study of global-event features (E_t and multiplicity distributions) has shown that the energy deposited in the reaction volume in the course of a nucleus–nucleus collision is as large as could have optimistically been expected; indeed, in present experiments, the energy density might already be close to or even above the threshold predicted for QGP formation.

Size and lifetime: The transverse size of the reaction zone at freeze-out, as measured by pion interferometry, is increasing by a factor of about two from the initial size; this is possible only in an expanding system with truly collective behaviour of its constituents. The observed large radii therefore constitute the first and unambiguous sign that an extended, strongly interacting system has been created, containing hundreds of particles per unit of rapidity in a final volume approaching 1000 fm^3. These spatial dimensions are certainly large by the standards of particle physics and QCD, and correspondingly macroscopic and statistical concepts should be applicable in the description of ultra-relativistic heavy-ion collisions. In contrast, the lifetime, which is estimated to be only of the order of a few fm/c, is probably at best marginal at present.

Equilibrium: The question of equilibrium and applicability of (QCD) thermodynamics remains unanswered so far. The interpretation of transverse-momentum spectra, which ideally would measure the temperature of a system in thermal equilibrium, and the interpretation of particle ratios, which signal the degree of chemical equilibrium attained in the collision, are complicated by the dynamical evolution from the (possible) QGP state through the phase transition and the unavoidable hadron-gas phase. However, the mere presence of significant differences in the inclusive spectra (both in p_t distributions and particle ratios) between p–p and nucleus–nucleus implies, even in the most conservative (non-equilibrium rescattering) models, at least the presence of extended, dense, and strongly interacting matter. This 'macroscopic' system seems to evolve towards equilibrium distributions by means of rescattering between its constituents. A quantitative analysis of the data, and in particular a search for surviving signals from the QGP, will depend to a large extent on further progress in understanding the hadronization process and the time scales available in the various phases.

Sensitive signals: From the number of experimental observables that are most sensitive to the earliest and hottest stages of the matter only two have been investigated so far: direct photons and J/ψ production. For photons, the experimental accuracy achieved at present is not expected to be sensitive to the thermal radiation of a QGP in the reactions studied so far. On the other hand, J/ψ suppression has been observed with the characteristics predicted for a QGP. For some time, Debye screening in the plasma has been the only explanation that could, with reasonable assumptions, describe the particular E_t- and p_t-dependent suppression pattern. To date, based partially on new results for J/ψ and Drell–Yan production in p–A reactions, it seems likely that a combination of initial-state parton scattering, gluon shadowing, and final-state absorption in the dense reaction zone is sufficient to explain the nucleus–nucleus data.

Taking all evidence together, the first round of 'survey' experiments has shown that an extended and very dense system with collective features has been formed that differs in many aspects from the more elementary hadron–hadron reactions investigated in the past. The first important result is, therefore, that heavy-ion collisions indeed seem to be an appropriate and promising tool for creating and studying the properties of strongly interacting bulk matter. However, it has also become clear that the quest for the QGP will not be a quick or easy task. It requires a systematic and comprehensive search for deviations from theoretical expectations or from smooth extrapolations of existing p–p and p–A data. Only if several such anomalies are found in different observables that cannot be accounted for within a reasonable margin of flexibility, could they provide a basis for a serious claim that the QGP or some other 'new physics' has been discovered. A number of such 'anomalies' have indeed been observed, but conventional theoretical models have been improved from a (partially 'naive') pre-data stage to a level of agreement that does not require radically new physics. This better understanding of conventional physics constitutes the second major achievement attained so far.

We can now consider the 'exploratory' phase in the still very young field of ultra-relativistic heavy-ion collisions to be essentially completed. Falling short of striking discoveries, this phase has nevertheless provided a *principle proof of feasibility* and has even substantiated the expectation that with the next generation of experimentation we shall reach a new and uncharted territory. Approved programmes with real 'heavy' ions foresee ^{197}Au beams at the AGS (11.5 GeV/A) in 1992 and ^{208}Pb beams at the SPS (160 GeV/A) in 1994. Based on our current understanding of the data, these upcoming experiments should lead to matter densities very close to the maximum possible in any laboratory experiment, reaching or even exceeding the ones in the centre of a neutron star. The larger volume, the (slightly) higher energy density, and, most important, the increased lifetime of the reaction zone will all independently help in driving the system further towards equilibrium, whether or not its internal degrees of freedom are of hadronic or of partonic nature.

By the end of the century, a different regime of very high energy density but low baryon density matter will be accessible at new machines. The Relativistic Heavy-ion Collider (RHIC), capable of colliding Au on Au at $\sqrt{s} \approx 200$ GeV/A, is now under construction at BNL and should start operation around 1997. Heavy-ion physics will also play an important rôle in the initial experimental programme of the Large Hadron Collider (LHC) planned at CERN. In particular, the LHC will be the ultimate machine in this field for the foreseeable future. With Pb on Pb at 6.1 TeV/A, corresponding to a total centre-of-mass energy of more than 1200 TeV, we expect particle densities of several thousand per unit of rapidity, a freeze-out volume approaching 100 000 fm^3, and an initial energy density 50 to 100 times larger than the one of normal nuclear matter. Even if, for some reason, no equilibrated QGP were formed at these energies, a hadronic description in terms of individual particles makes no sense either in a system where several dozen hadrons would be piled up on top of each other (>> 10 pions/fm^3). 'New physics' of one kind or another seems therefore bound to appear somewhere along the road towards the Quark-Gluon Plasma.

REFERENCES

1. Recent reviews of the field:

W.M. Geist, *Int. J. Mod. Phys.* A4 (1989) 3717

J. Bartke, *Int. J. Mod. Phys.* A4 (1989) 1319

M.J. Tannenbaum, *Int. J. Mod. Phys.* A4 (1989) 3377

H.R. Schmidt and J. Schukraft, to be publ. in *J. Phys.* G (1992) and preprint CERN-PPE/92 42

H. Gutbrod and H. Stöcker, *Scientific American*, November 1991

J. Gribbin, *New Scientist*, 17 August 1991, 31

2. Original references can be found in the proceedings of the recent Quark-Matter Conferences

'Quark Matter 87', Proc. 6th Int. Conf. on Ultra-Relativistic Nucleus–Nucleus collisions, Nordkirchen, FRG, Aug. 24–28 1987, *Z. Phys* C38 (1988)

'Quark Matter 88', Proc. 7th Int. Conf. on Ultra-Relativistic Nucleus–Nucleus collisions, Lennox, MA, USA, Sept. 26–30 1988, *Nucl. Phys.* A498 (1989)

'Quark Matter 90', Proc. 8th Int. Conf. on Ultra-Relativistic Nucleus–Nucleus collisions, Menton, France, May 7–11 1990, *Nucl. Phys.* A525 (1991)

'Quark Matter 91', Proc. 9th Int. Conf. on Ultra-Relativistic Nucleus–Nucleus collisions, Gatlinburg, TE, USA, Nov. 11–15 1991, *Nucl. Phys.* A544 (1992)

3. T. Åkesson et al, NA34 Collaboration, *Nucl. Phys.* B353 (1991) 1

4. M. Gonin, preprint BNL-47925

5. A. Bamberger et al., NA35 Collaboration, *Phys. Lett.* B203 (1988) 320

6. P. Seyboth et al., NA35 Collaboration, *Nucl. Phys.* A544 (1992) 293c

7. T. Åkesson et al., NA34 Collaboration, *Z. Phys.* C46 (1990) 369

8. Y. Miake et al., E802 Collaboration, *Z. Phys.* C38 (1988) 135

9. T. Matui and H. Satz, *Phys. Lett.* B178 (1986) 416

10. C. Baglin et al., NA38 Collaboration, *Phys. Lett.* B220 (1989) 471, B251 (1990) 465 and 472

DISCUSSION

A. Franz, CERN

The result for the radius parameter extracted from HBT correlation in *SPb* collisions from NA44 shows a large difference between π and K pairs. In particular in the transverse radius component, the radius parameter for kaons is twice the one for pions.

Schukraft

These results are still preliminary and need better understanding of kinematic K effects.

W.D. Walker, Duke University, USA

I wish to point out that there has been an experiment at the Tevatron (E735) which found many of the signals for a quark gluon plasma-strangeness enhancement; lots of baryon-antibaryon production. Our multiplicities are 200 charged particles. The materialization time is the unknown.

Schukraft

The question is one of size. Nuclear-nuclear collisions produce large objects, and thermodynamics may work.

A.M. Baldin, JINR, Dubna

A relativistic invariant approach to the description of hadronic processes has been suggested and is being used in Dubna [A.M. Baldin and L.A. Didenko, *Fortschritte der Physik*, **v38** (1990) p. 261-332]. The asymptotic properties of highly excited nuclear matter have been studied by analyzing the universal characteristic of clusters in three-dimensional rapidity space. The results obtained show that the nuclear collision models applying macroscopic and statistical concepts (energy, density, pressure, etc.) are not realistic.

M. Strikman, Pennsylvania State University, USA

At the parallel session, M. Virchaux reported the results of the global analysis of μN data (SLAC, BLDMS, NMC) which lead to $X\sigma(X, Qo^2 \cong 7\,\text{GeV}^2) \sim (1-x)^{8.9}$. At the same time the models of ψ/J production in pA scattering through gluon fusion requires $X\sigma(XQo^2 \cong 10\,\text{GeV}^2) \sim (1-x)^5$. In view of this inconsistency, naive gluon fusion model of ψ- production is in trouble. Hence to use this model to extract the A-dependence of gluon distribution from the $p + A \to J/\psi + \chi$ data is not justified theoretically.

PROGRESS IN QUANTUM FIELD THEORY AND STRING THEORY

Luis Alvarez-Gaumé
Theoretical Physics Division, CERN
CH - 1211 Geneva 23 Switzerland

Abstract

We review some of the recent developments in Quantum Field Theory and String Theory. A thorough account of these subjects is virtually impossible, and I have therefore chosen a list of topics which should provide a reasonable cross-section.

INTRODUCTION

It is always difficult to address an audience mainly composed of experimental physicists when the subject of discussion is to summarize recent results in the more abstract domains of our field. Although there has been a rather healthy level of activity in the subjects of Quantum Field Theory, String Theory and related topics since the last Rochester Conference (Singapore, 1990), we have not had tantalizing or startling theoretical breakthroughs. Nevertheless, many interesting avenues have been opened. They have not yet been fully explored, and we will need to wait to see how much information can be reaped from them. In a one-hour talk it is virtually impossible to do justice to all interesting papers or research proposals made over the last two years, and a selection has to be made. This is conditioned by the topics presented at the Conference, the expertise and/or prejudices of the rapporteur. Due to these limitations I want to begin by apologizing to all those colleagues who may not find their work among the topics of this lecture.

In Quantum Field Theory, an incomplete list of topics which have been vigorously pursued in the recently includes: i) baryon and lepton number violation in the Standard Model at supercollider energies[1,2] and at high temperatures[3], ii) implications of this process for the generation of baryon asymmetry in the Universe, iii) use of four-dimensional strings to develop new and efficient computational tools[4] in perturbative QCD, iv) the construction of the Master Field in some interesting four-dimensional gauge-invariant theories[5], v) new proposals to study chiral fermions on the lattice[6], vi) low-dimensional Quantum Field Theory and its connections with Condensed Matter Physics, in particular in the study of fractional statistics, the Fractional Quantum Hall Effect and high-temperature superconductors[7], and vii) integrable field theories. The latter is a fast-expanding subject which is developing novel mathematical tools, and new original ways of looking at some simple physical systems. Its main drawback so far is that many of its nice features seem to be inexorably linked to two-dimensional space-time. In the first and second parts of this talk, we will review in some detail parts i), ii) and iii).

In the field of String Theory, an incomplete list of topics investigated in the last few years includes i) soliton solutions in heterotic string theory[8], ii) the computation of some non-perturbative world-sheet effects in string-induced low-energy supergravity effective actions, including mirror symmetries[9], duality[10], threshold effects[11,12] and their implications in space-time supersymmetry breaking; iii) non-critical strings, matrix models of two-dimensional gravity, Liouville theory[13], ground rings[14], etc. and iv) black holes in string theory, both at the fundamental and effective levels[15]. Recent work initiated by Witten[16] proves the existence of two-dimensional black hole solutions in critical string theory. Although these solutions are studied in two dimensions, they keep some of the non-trivial puzzles of four-dimensional solutions. Issues concerning the endpoint of black hole evaporation, loss of quantum coherence and many others, receive a new and refreshing light in this context. Questions of quantum violations of the No-Hair theorems and their consequences have also received a good deal of attention recently[17]. In the last two parts of this talk, I will focus our attention mainly on ii) and iv). Finally, there has also been a lot of activity on the subject of two-dimensional Conformal Field Theory, Topological Field Theory and related areas of Mathematical Physics.

B+L VIOLATION AT HIGH ENERGY

The possibility of B+L violation in the Standard Model was realized in the mid-1970's, when G. 't Hooft[18] presented his resolution of the $U(1)$ problem using instantons. If q represents collectively the quark fields, the baryon current is given by:

$$j_B^\mu = \frac{1}{3} \sum_q \bar{q}\gamma^\mu q = \frac{1}{3} \sum_q (\bar{q}_L \gamma^\mu q_L + \bar{q}_R \gamma^\mu q_R), \quad (1)$$

$L(R)$ stands for the left-(right-)handed part of the quark field. For clarity, we have suppressed color and flavor indices in q. A similar expression can be written for the lepton-number current j_L^μ. Since weak $SU(2)$ only couples to left-handed fields, the conservation of j_B^μ, j_L^μ may be afflicted with quantum anomalies in analogy with the anomaly effect in the explanation of π^0-decay[19]. To test for the conservation of j_B^μ we consider the triangle graph in Fig. 1:

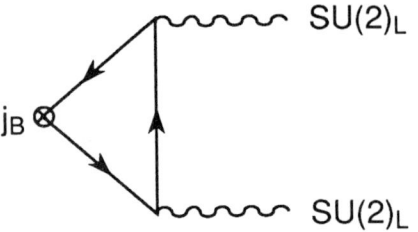

Figure 1. Triangle graph producing an anomaly in the conservation of the baryon-number current.

Since the $SU(2)_L$-currents are coupled to gauge fields, current conservation plus Bose symmetry on the two $SU(2)_L$ vertices completely fix the diagram in Fig. 1. If $\vec{E}^a, \vec{B}^a, a = 1, 2, 3$ are the $SU(2)_L$ electric and magnetic fields, the conservation equation $\partial_\mu j_B^\mu$ is modified to:

$$\partial_\mu j_B^\mu = \frac{\alpha_w}{4\pi} \vec{E}^a \cdot \vec{B}^a \quad (2)$$

The integrated anomaly for three families becomes:

$$\Delta B = \Delta L = -3 \frac{\alpha_w}{4\pi} \int_{-\infty}^{+\infty} dt \int d^3\vec{x} \; \vec{E}^a \cdot \vec{B}^a$$
$$= -3Q \quad (3)$$

The quantity Q is an integer, and it is a number characterizing the topological properties of the gauge field configuration. The argument of the integral in (3) is a total derivative in space-time, and it does not receive contributions to any order of perturbation theory. There are only non-perturbative contributions to B- and L-violating processes in the Standard Model. The configurations responsible for these processes are known as instantons. In a sector of topological charge Q, and in the simplest situation where we ignore contributions from the Higgs field, the gauge field action is bounded below by the instanton action:

$$S \geq S_{int} = \frac{2\pi}{\alpha_w}|Q| \quad (4)$$

This can be interpreted by saying that there is a huge barrier between initial and final configurations differing by B and L quantum numbers. There is a tunnelling suppression in the amplitude of the order of[18] $exp - \frac{2\pi}{\alpha_w} \approx 10^{-78}$. Needless to say, this is unobservable. At extreme conditions, however, such as high temperature or densities[3], (B+L)-violating processes may be unsuppressed. If we plot the potential energy for gauge fields, an oversimplified picture looks like Fig. 2.

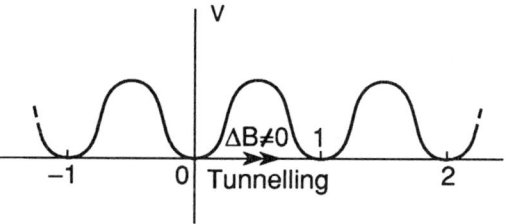

Figure 2. Schematic portrayal of the potential energy for gauge fields.

The minimal height of the barrier between two absolute minima is a saddle-point with a single negative eigenvalue, and it corresponds to a classical finite energy solution to the gauge field-Higgs systems known as the sphaleron[20]. Depending on details which are irrelevant for the present discussion, the sphaleron scale is

$$E_{sp} \sim \frac{\pi M_W}{\alpha_W} \quad (7 - 14 TeV) \qquad (5)$$

where M_W is the W-mass. At finite temperature we can go over the barrier by thermal fluctuations. The Boltzmann factor is small, and the tunnelling contribution is hardly relevant. We can estimate the (B+L)-violation amplitude as[3]:

$$\Gamma[T] \sim e^{-F_{sp}[T]/T} \sim e^{-M_W(T)/T} \qquad (6)$$

where $F_{sp}[T]$ is the sphaleron free energy, and $M_W(T)$ is the temperature-dependent W-mass. Since at high T the low-energy broken symmetries are restored, $M_W(T)$ is small for large T. The sphaleron solution can be used to generate the baryon asymmetry on the Universe.

Although at finite temperature it is reasonable to expect (B+L)-violation, it is more surprising that this phenomenon could also be observed at supercollider energies[21]. The processes envisaged include qq or $q\bar{q}$ collisions with very high multiplicity in the final state. Thus we have in the final state the minimum number of leptons and quarks to saturate the minimal selection rule $\Delta B = \Delta L = -3$, together with a large number of vector mesons and Higgs particles (Fig. 3).

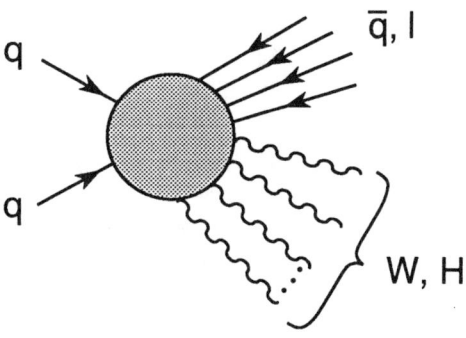

Figure 3. Schematic (B+L)-violating process with high multiplicity in the final state.

The only known way of computing this amplitude is to evaluate the instanton-mediated scattering amplitude and its radiative corrections. The claim is that the inclusion of all high-multiplicity final states leads to an exponentially growing (in energy) contribution which offsets the exponential suppression factor due to instantons. The delicate and controversial issue is how complete this cancellation is. For a process with n_W W's and n_H Higgses, the (B+L)-violating amplitude has the general form

$$A^{\Delta(B+L)}_{n_W, n_H} \sim \frac{(n_W + n_H)!}{M_W^{n_W + n_H}} (\alpha_W)^{(n_W + n_H)/2}$$
$$e^{-2\pi/\alpha_W} \cdot (phase\ space)$$

The total cross-section for (B+L)-violation

can be written as:

$$\sigma_{tot}^{\Delta(B+L)} \sim exp \frac{4\pi}{\alpha_W} F\left(\frac{\sqrt{s}}{E_{sp}}\right) \quad (7)$$

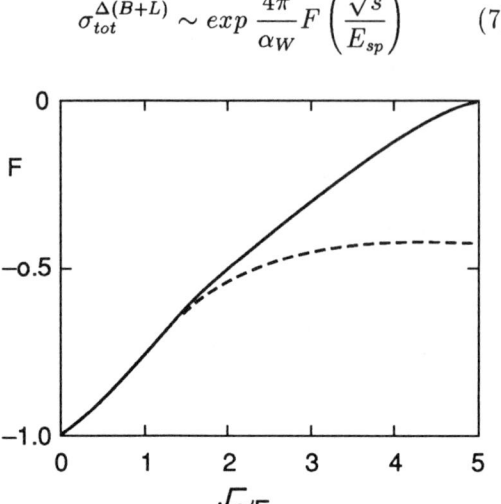

Figure 4. The solid line represents the most optimistic estimate of the functions F in (7). The broken line could represent the expectations of the less enthusiastic.

$\sqrt{s}$ represents the centre-of-mass energy, and E_{sp} is given in (5). The difficult part of the problem is the computation of the function F. In Ref. 1, the computations of many groups are collected, and the first few orders of F are given by:

$$F\left[\frac{\sqrt{s}}{E_{sp}}\right] = -1 + 0.34 \left(\frac{\sqrt{s}}{E_{sp}}\right)^{\frac{4}{3}} - 0.09 \left(\frac{\sqrt{s}}{E_{sp}}\right)^{2}$$
$$+ 0\left((\frac{\sqrt{s}}{E_{sp}})^{\frac{8}{3}}(1 + \ell n \frac{\sqrt{s}}{E_{sp}})\right) \quad (8)$$

There are also problems with violation of s-wave unitarity. Being extremely optimistic, one might be tempted to conclude that F looks qualitatively like the solid line in Fig. 3. A more pessimistic, although probably more realistic assessment of the situation is that after unitarization and higher-order effects are included, one should end up with the broken line in Fig. 3. Hopefully, at the next Rochester Conference we should have a better understanding of this function $F(x)$. More details on this subject can be found in the talks by V. Khoze and V. Zakharov. In the latter, one can find the relation between this problem and the more general issue of higher-order behaviour of perturbation theory, and possible implications for the determination of the energy dependence of α_s in e^+e^- processes. This subject is still quite active and if the most optimistic appraisals turn out to be correct, it will have an important impact on the next generation of hadron colliders.

NEW COMPUTATIONAL RULES IN QCD FROM STRING THEORY

Perturbative theory is one of our basic tools for the understanding of short-distance physics. In QCD specially, perturbative computations are crucial in present and future hadron colliders to understand the Standard Model, and to help discern what lies beyond. It is notorious, however, that traditional Feynman graph computations become rather cumbersome, even for tree-level processes. For example, the tree-level scattering amplitude for two to six gluons involves more than 30 000 graphs. It is obviously prohibitive to evaluate such an amplitude with conventional techniques, and it is quite clear that new computational tools are required. Work by many people[22] has made a number of multijet computations possible. If one wants to include one-loop contributions, the computations are prohibitively difficult. Bern and Kosower[4,23] have proposed a general method to compute one-loop corrections using string theory[24]. They are able to derive a set of simple rules whose result is to effectively combine all diagrams for a given process, perform all cancellations, Lorentz algebra and one-loop momentum integral. The final answer is just the Feynman parameter integrals left at the end of any one-loop computation. The starting point of their method is to consider the n-gluon scattering amplitude in string theory. This amplitude is rather simple to write down, and it automatically sums up all Feynman graphs, organizes them in colour decompositions and leaves us with the Feynman parameter integrals. These simplifications were already known for tree diagrams. What is important in the Bern-Kosower work is that they are also able to derive simple rules for one-loop corrections, and it is not unreasonable to expect that they (or someone else) should be able to extend their rules to higher-loop graphs. In string theories without tachyons, in the limit as the string tension goes to infinity, one is left with the massless excitations of the theory. When we consider loops, all massive states will circulate around the loop, and therefore one has to exercise a certain amount of care to be sure that the only contributions left are those coming from the modes wanted. In particular, the particles in the gravity sector have to be treated separately. At the one-loop level this is not a difficult problem. To control the infinite string tension limit properly, Bern and Kosower choose a consistent four-dimensional heterotic string theory[25]. A flavour of the type of string computations involved is now presented. To compute an n-gluon amplitude at one loop in string theory, we first note that in this case the world-sheet of the string is a torus (Fig. 5).

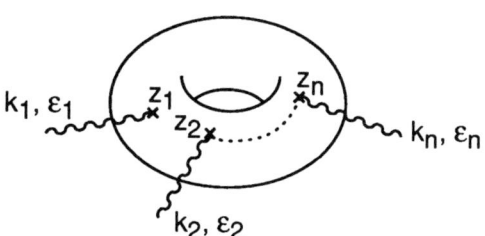

Figure 5. One-loop string diagram needed in the computation of the n-gluon scattering amplitudes.

The torus can be represented as a parallelogram in the complex plane with the opposite side identified as in Fig. 6. The complex parameter τ ($Im\tau > 0$) determines all possible types of one-loop world-sheets contributing to the same process.

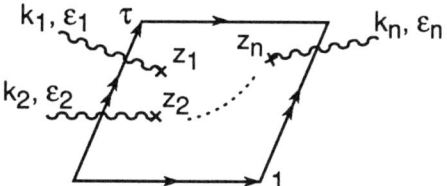

Figure 6. Same diagram as in Fig. 5, but with the torus represented as a parallelogram with opposite ends identified.

The points $z_1, ..., z_n$ on the world-sheet correspond to the insertion of the vertex operators which in string theory represent the emission (or absorption) of a state (in particular a gluon). Each string excitation has an associated vertex operator (for details and references see, for example, ref. 24). We can describe the motion of the string in four-dimensional space-time by a two-dimensional field $X^\mu(z, \bar{z})$, which in the case at hand behaves as a massless free field on the world-sheet. The emission vertex for a given state takes the general form:

$$V^a_{k,\epsilon}(Z, \bar{Z}) = \epsilon(k) \cdot P\left(\frac{\partial X}{\partial Z}, \frac{\partial X}{\partial \bar{Z}}\right) e^{i\lambda k_\mu \cdot X^\mu(Z,\bar{Z})}$$
(9)

where λ is the inverse string tension, $\epsilon(k)$ is the polarization vector of the state emitted, and k its momentum. For gluons, $k^2 = 0, k \cdot \epsilon = 0$. Now, the n-gluon amplitude at the one-loop level is succinctly represented by:

$$A(k_1, \epsilon_1, ..., K_n, \epsilon_n) \sim \int \frac{d^2\tau}{(Im\tau)^2}$$
$$\int d^2 Z_1, ..., d^2 Z_n \cdot \langle V_1(Z_1), ..., V_n(Z_n)\rangle^{\lambda \to 0}_{2D} \quad (10)$$

In (10) the expectation value is computed with respect to a free, two-dimensional massless field theory and it is quite straightforward to write down. The field theory result is obtained as $\lambda \to 0$. In this limit the torus becomes thinner and thinner on one of its dimensions, so that it looks more and more like a particle loop. Technically this is the region $Im\ \tau \to \infty$. As this limit is taken, we must be careful with the contributions coming from the singularities appearing when two vertex operators are on top of each other. Since the operators appearing in (10) belong to a free two-dimensional field theory, these singularities can be evaluated using two-dimensional operator product expansions. When everything is done carefully, one ends up with a rather simple set of rules for writing down the final Feynman parameters' integral. It is important to remark that these rules have also been derived directly from QCD, and that a purely field theoretic formulation using first quantization rules is currently under way[27,28]. It is also clear that when the rules are known, one can devise many independent ways of proving them. However, I find it very difficult to imagine a direct and natural way of arriving at the Bern-Kosower rules without knowing them beforehand. Their original derivation depends crucially on properties of string diagrams which do not have simple or natural counterparts in a purely field-theoretic formulation.

Currently the one-loop computation of $2 \to 3$ gluons amplitude is under way[29], and the results will be reported shortly. It is not hard to guess that many QCD computations will appear in the near future using these techniques. Similarly, work in the direction of extending the Bern-Kosower rules to higher-loop computations is under way[30].

PROGRESS IN FUNDAMENTAL STRINGS

In the subject of fundamental strings, the three topics we will briefly mention comprise: i) progress in the determination of Yukawa couplings and other parameters of the low-energy effective action, ii) black holes and other interesting classical solutions in string theory, and iii) conceptual problems of general relativity as seen from the point of view of string theory.

We do not yet know the basic principles of string theory, although some simple consistency requirements have taken us quite far. Perhaps the conceptually more interesting result in string theory is that it can account for the rather perverse chiral structure of the Standard Model. Furthermore, the gauge groups appearing in string theory, and more concretely in heterotic string theories, are comfortably large to include all known interactions. Gravity appears as an essential part of the theory, and modulo some non-trivial technicalities the theory is well defined in the ultra-violet. The infinities of more standard treatments of quantum gravity do not afflict string scattering amplitudes. Superstring theory is today the only known candidate theory incorporating quantum gravity together with gauge interactions capable of accommodating the Standard Model interactions and with a low-energy spectrum of chiral fermions. From the low-energy point of view (far below the Planck scale), it looks like some supersymmetric extension of the Standard Model. The advantage is that in principle the parameters of the low-energy effective action are calculable once a string compactification (or string vacuum) has been chosen. One of the basic issues not addressed by string theory is the smallness of the cosmological constant Λ. This is quite important in a theory including quantum gravity. If the theory is to make contact with low-energy phenomenology, space-time supersymmetry must be broken, and this usually generates an unacceptably large value for Λ. This argument is, however, based on our use of quantum field theory ideas which may be completely inappropriate in this context. I would even go so far as saying that understanding the smallness of Λ is the single most important problem in trying to make contact between string theory and low-energy physics.

The description of the motion of a string in some given space-time requires a two-dimensional quantum field theory on the world-sheet[24]. These field theories describe the embeddings of the string into space-time.

Different space-times require different field theories. However, not any two-dimensional field theory qualifies. One of the basic requirements is two-dimensional conformal invariance. It can be shown that this condition implies the Einstein equations in string theory. More generally it implies the field equations of the fields representing the string excitations (Einstein equations, Yang-Mills equations, etc.). When considering superstring theory, two-dimensional conformal invariance is replaced by $N = 2$ superconformal invariance if the theory is required to be space-time supersymmetric. Once a given space-time background is chosen, we can proceed to construct the low-energy effective Lagrangian. We are interested mostly in backgrounds of the form $M_4 \times K$, where M_4 is four-dimensional Minkowski space, and K is either a compact space or some suitable conformal field theory. The low-energy Lagrangian will look like any other low-energy $N = 1$ supergravity Lagrangian[31]. Any of them can be expressed in terms of three arbitrary functions: the Kähler potential K which determines the kinetic term of the scalars on the theory, the superpotential W responsible among other things for the Yukawa couplings, and the gauge kinetic function f which determines the gauge coupling constants. The subject of string phenomenology is about the computation of the three function K, W, f. In ordinary model-building, the form of these functions depends very much on how the model is constructed. In string theory these functions are in principle computable directly in terms of the conformal field theory in the given string vacuum solution. The coupling constants of the low-energy Lagrangian are computed in terms of correlation functions of the two-dimensional theory. One of the massless scalars which is ubiquitous in string theory is the dilaton. This is a singlet field whose expectation value is related to the tree-level gauge coupling constants ($\phi = 1/g^2$). One of the specific features of four-dimensional strings is that coupling constants are given in terms of expectation values of fields. They depend on the dilaton field as well as on other scalar fields known as moduli fields. An important problem encountered in the determination of $\langle \phi \rangle$ is that this expectation value is a flat direction in perturbation theory, and therefore non-perturbative effects should generate a non-trivial dilaton potential. As we will see, there are some interesting recent results in this direction. We should also mention that the breaking of space-time supersymmetry is required to have special characteristics in order to obtain reasonably realistic models. The basic scale in the problem is the Planck scale $M_P (\sim 10^{19} GeV)$, and supersymmetry should be broken at scales of the order of $10^{10} - 10^{11} GeV$. This value is not so easy to obtain with current model-building technology. In the resolution of some of the problems listed, some properties of string theory have been applied with encouraging success. One of these properties is known as duality[10]. These transformations are symmetries of string physics which, when interpreted as geometrical operations, relate distinct geometrical configurations. The simplest example of this notion is provided by a string moving on a circle of radius R. Together with the standard states representing string vibrations and translations, we have winding states which wind any number of times around the circle. The end result is that physics is invariant under the transformation $R \to \alpha'/R$, where $\alpha' \sim M_P^{-2}$. This phenomenon has no analogue in conventional field theory. The structure of the symmetry group is only known in a few cases, but it proved to be very useful in constraining the couplings of the low-energy theory[32]. The other symmetry with purely stringy origin is the mirror symmetry[9]. This symmetry is based on the general properties of $N = 2$ superconformal field theories, and again shows that apparently different geometries correspond to just one string model. Using this symmetry, it is possible to compute explicitly some of the Yukawa couplings in the superpotential W. The full physical implications of these symmetries are now being vigorously explored.

In the determination of the gauge coupling constants, we need, as mentioned before, to determine the non-perturbative form of the dilaton effective potential. In quantum field theory non-perturbative effects display a dependence on the coupling constant of the form $exp-1/g^2$. This dependence can be generated by gaugino condensation[33]. The effective potential generated has the form:

$$V_{eff} \sim \sum_a C_a \, exp\left(\frac{24\pi^2}{b_a g_a^2}\right), \qquad (11)$$

where the sum runs over the different factors in the gauge groups responsible for gaugino condensation, b_a are the one-loop coefficients of the beta functions, and $\langle Re \, \phi \rangle$ enters in (11) through the relation between g_a and $\langle Re \, \phi \rangle$. This potential as written has a minimum at $\langle Re \, \phi \rangle = \infty$ corresponding to $g_a = 0$. This changes once loop corrections and threshold effects are included[11]:

$$\frac{1}{g_a^2(p^2)} = \phi \langle Re \, \phi \rangle + b_a \, ln\frac{M_p^2}{p^2} + \Delta_a(T) \quad (12)$$

The $\Delta_a(T)$ are the threshold corrections. The argument T denotes collectively the dependence of the masses of the heavy fields integrated out on light scalar field expectation values. These threshold corrections are crucial for reconciling string model predictions with the current value of the Weinberg angle. The inclusion of (12) into (11) fixes the dilaton expectation value, and this may break space-time supersymmetry[12]. Explicit studies have been carried out so far only in a few models, and hopefully a more general picture will develop in the not very distant future. Many more things could be said about superstring phenomenology, but this seems a good point to stop.

We finally turn to some recent exciting work on string black holes. It is well known that, in General Relativity, collapsing matter under certain mild conditions generates black holes. The singularities in the space-time metric are always supposed to come with a horizon (cosmic censorship hypothesis). A property of classical black holes is to come with no hair[34]. The only parameters which determine a static black hole are its mass, angular momentum and charge. Any other attribute of the matter which collapses to form it disappears. This obviously provides a rather efficient mechanism to violate global symmetries like baryon number. In the mid-Seventies, Hawking[35] made the momentous discovery that black holes radiate thermally when quantum effects of matter fields are taken into account, and he went further to argue that black holes produce an inherent loss of quantum coherence[36]. This has generated a big controversy in the last fifteen years. The computations carried out by Hawking and others involve quantizing fields in the presence of a classical background gravitational field, and the back-reaction is not taken into account, one of the reasons being that, in order to understand it properly, we need a well-defined quantum theory of the gravitational field. The last sentence should, of course, be qualified by string theory, since it is alleged to provide a consistent quantum theory of gravity. Two approaches have been followed in analyzing this problem. On the one hand, one can study some simple toy model of the gravity + dilaton system in two dimensions suggested by the equation of motion of string theory (for details and references, see the recent review in Ref. 15). These simple systems provide a rather precise formulation of Hawking's paradoxes in a situation where we have some control on the approximation used. The second approach began after Witten[16] found an exact string theory in two dimensions based on an $SL(2, \mathbf{R})/U(1)$ coset model which describes a true string black hole in two dimensions. We know from quantum field theory that often in the translation of a problem from four to two dimensions it loses most of its flavour. This does not seem to be the case here. Many of the conceptual issues involved in the four-dimensional problem continue to be non-trivial in the two-dimensional case. Since Witten's solution is perfectly sensible as a conformal field theory (and, more importantly, as a string theory), an important research effort was initiated after Witten's paper to try to understand what answers it gives to black hole paradoxes. Furthermore, recent revisions of the no-hair theorems (see Ref. 17 for details and references) also find a setting in this two-dimensional scenario, in a way possibly connected with the loss of quantum coherence. The authors of Ref. 37, on the basis of the properties of Witten's solution, have proposed solutions to many of the riddles of black hole and string physics. This subject is moving now at full steam, and it seems premature at this stage to review in detail some of the results obtained so far. I just want to leave you with the optimistic impression that some of these hard questions in quantum gravity may find an answer in String Theory.

I hope the reader is convinced after this brief (and very incomplete) survey, that many interesting things are happening in the subjects of Quantum Field Theory and String Theory. Although no startling breakthrough has taken place since the Singapore Conference, it may well be that the seeds for one are contained in some of the topics discussed.

ACKNOWLEDGEMENTS

It is a pleasure to thank the organizers of the XXVI Rochester Conference for their kind invitation to present this survey. I am grateful to E. Alvarez, C. Gomez, L. Ibañez, V. Khoze, W. Lerche, J. Louis, D. Lüst, M. Peskin, E. Rabinovici, A. Schellekens, V. Rubakov and A. Zakharov for many useful discussions.

REFERENCES

1. For details and references, see: A. Ringwald, "Anomalous Baryon Number Violation in High-Energy Scattering", preprint CERN-TH.6479/92.

2. See also the talks presented at this Conference by V. Khoze and A. Zakharov.

3. V. Kuzmin, V. Rubakov and M. Shaposhnikov, "On the Anomalous Electroweak Baryon Number Non-conservation in the Early Universe", *Phys. Lett.* 155B, p. 36, (1985);
P. Arnold and L. McLerran, "Sphalerons, Small Fluctuations and Baryon Number Violation in Electroweak Theory", *Phys. Rev.* D36, p. 581 (1989);
A. Ringwald, "Rate of Anomalous Baryon and Lepton Number Violation at Finite Temperature in Standard Electroweak Theory", *Phys. Lett.* B201, p. 510 (1988), and references therein.

4. Z. Bern and D. Kosower, "Efficient Calculation of One-loop QCD Amplitudes", *Phys. Rev. Lett.* 66, p. 1669 (1991).

5. V.A. Kazakov and A.A. Migdal, "Induced QCD at Large N", preprint PUPT-1322, LPTENS-92-15.

6. D. Kaplan, "A Method for Simulating Chiral Fermions on the Lattice", preprint UCSD-PTH-92-28.

7. See the reprint volume: F. Wilczek, *Fractional Statistics and Anyon Superconductivity*, World Scientific, 1990;
See also E. Fradkin, *Field Theory of Condensed Matter Systems*, Addison Wesley 1991, and references therein.

8. For details and references, see: C. Callan, "Instantons and Solitons in Heterotic String Theory", preprint PUPT-1278, June 1991.

9. W. Lerche, C. Vafa and N. Warner, "Chiral Rings in $N = 2$ Superconformal Theories", *Nucl. Phys.* B324, p. 427 (1989);
L. Dixon, V. Kaplunovsky and J. Louis, "On Effective Field Theories Describing (2,2) Vacua of Heterotic Strings", *Nucl. Phys.* B329, p. 27 (1990);

B.R. Greene and M.R. Plesser, "Duality in Calabi-Yau Moduli Space", *Nucl. Phys.* B338, p. 15 (1990);
P. Candelas, M. Lynker and R. Schimmrigk, "Calabi-Yau Manifolds in Weighted P(4)", *Nucl. Phys.* B341, p. 383 (1990);
B. Greene, *Comm. Math. Phys.* 130, p. 335 (1990);
P. Candelas, X.C. de la Ossa, P.S. Green and L. Parkes, "A Pair of Calabi-Yau Manifolds as an Exactly Soluble Superconformal Theory", *Nucl. Phys.* B359, p. 21 (1991).

10. M.B. Green, J.H. Schwarz and L. Brink, "$N = 1$ Yang-Mills and $N = 8$ Supergravity as Limits of String Theories", *Nucl. Phys.* B198, p. 474 (1982);
K. Kikkawa and M. Yamasaki, "Stability of Self-Consistent Dimensional Reduction", *Phys. Lett.* 144B, p. 365 (1984);
N. Sakai and I. Senda, "Vacuum Energies of String Compactified on Torus", *Prog. Theor. Phys.* 75, p. 692 (1982);
For more details and references on this booming field, see: J.H. Schwarz, "Superstring Compactification and Target Space Duality", preprint CALT-68-1740, May 1991.

11. S. Weinberg, "Effective Gauge Theories", *Phys. Lett.* 91B, p. 51 (1980);
V. Kaplunovsky, "One-loop Threshold Effects in String Unification", *Nucl. Phys.* B307, p. 145 (1988);
L. Dixon, V. Kaplunovsky and J. Louis, "Moduli Dependence of String Loop Corrections to Moduli Dependence", *Nucl. Phys.* B355, p. 649 (1991);
I. Antoniadis, K. Narain and T. Taylor, "Higher Genus String Corrections to Gauge Couplings", *Phys. Lett.* B267, p. 37 (1991).

12. For more details and references, see: J. Louis, "Recent Developments in Superstring Phenomenology", preprint CERN-TH.6492/92.

13. For details and references see, for example: L. Alvarez-Gaumé, "Random Surfaces, Statistical Mechanics and String Theory", *Helv. Phys. Acta* 64, p. 360. 1991;
N. Seiberg, "Notes on Quantum Liouville Theory and Quantum Gravity", Lectures at the 1990 Yukawa Seminar, preprint RU-90-29 (1990).

14. E. Witten, "Ground Ring of Two-Dimensional String Theory", *Nucl. Phys.* B373, p. 187 (1992).

15. A thorough and updated review is: J. Harvey and A. Strominger, "Quantum Aspects of Black Holes", preprint EFI-92-41 (1992).

16. E. Witten, "On String Theory and Black Holes", *Phys. Rev.* D44, p. 314 (1991).

17. A detailed review of this issue with references to the relevant literature is: S. Coleman, J. Preskill and F. Wilczek, *Nucl. Phys.* B378, p. 175 (1992).

18. G. 't Hooft, *Phys. Rev. Lett.* 37, p. 8 (1976); *Phys. Rev.* D14, p. 3432 (1976).

19. For details and references on anomalies, from the Adler-Bell-Jackiw anomaly used here, to gravitational anomalies and other related subjects, see: S. Treiman, R. Jackiw, B. Zumino and E. Witten, *Current Algebra and Anomalies*, Princeton University Press, 1986;
L. Alvarez-Gaumé, "An Introduction to Anomalies", Erice Lectures 1985, eds. G. Velo and A. Wightman, Plenum Press.

20. N. Manton, "Topology in the Weinberg-Salam Model", *Phys. Rev.* D28, p. 2019 (1983);
F. Klinkhamer and N. Manton, "A Saddle-point Solution in the Weinberg-Salam Theory", *Phys. Rev.* D30, p. 2212 (1984).

21. For details and references, see Ref. 1.

22. Details and references can be found in Refs. 4 and 23.

23. Z. Bern and D. Kosower, "The Computation of Loop Amplitudes in Gauge Theories", preprint Fermilab-Pub.-91-111-T (1991); "Practical Applications of Superstrings: New One-Loop Rules for Gauge Theories", preprint PITT-91-03 (1991); "Colour Decomposition of One-Loop Amplitudes in Gauge Theories", *Nucl. Phys.* B362, p. 389 (1991).

24. A comprehensive introduction to string theory is: M.B. Green, J.H. Schwarz and E. Witten, *Superstring Theory*, Vols. I and II, Cambridge University Press, 1987.

25. For details and references on the construction of four-dimensional string theories see, for example, A. Schellekens, *Four-Dimensional String Construction*, North-Holland, 1989.

26. Z. Bern and D.C. Dunbar, "A Mapping Between Feynman and String-Motivated One-Loop Rules in Gauge Theories", preprint PITT-91-17 (1991).

27. M. Strassler, "Field Theory without Feynman Diagrams: One-Loop Effective Actions", preprint SLAC-PUB-5757 (February 1992).

28. The computational rules have inspired the derivation of simplified computational rules starting directly from QCD. One such example was presented at this conference: C.S. Lam, "Navigating Around the Algebraic Jungle of QCD: Efficient Evaluation of Loop Helicity Amplitudes", preprint McGill/92-32.

29. Z. Bern, L. Dixon and D.A. Kosower, work in progress.

30. D.A. Kosower, private communication.

31. E. Cremmer, S. Ferrara, L. Girardello and A. Van Proeyen, "Yang-Mills Theories with Local Supersymmetry: Lagrangian, Transformation Rules and SuperHiggs Effects", *Nucl. Phys.* B212, p. 413 (1983).

32. For reviews and references, see: S. Ferrara and S. Theisen, in *Proceedings of the Hellenic Summer School, 1989*, World Scientific;
D. Lüst, "Duality-Invariant Effective Actions and Automorphic FUnctions for (2,2) String Compactifications", preprint CERN-TH.6143/91.

33. For a review and references see, for example, H.P. Nilles, "Gaugino Condensation and Supersymmetry Breakdown", *Int. J. Mod. Phys.* A5, p. 4199 (1990).

34. On the general subject of black holes and singularity theorems in general relativity, see any of the many excellent books treating the subject. For example, S. Hawking and G.F.R. Ellis, *The Large-Scale Structure of Space-Time*, Cambridge University Press, 1973;
R. Wald, *General Relativity*, Univ. Chicago Press, 1984, and references therein.

35. S. Hawking, "Particle Creation by Black Holes," *Comm. Math. Phys.* 43, p. 199 (1975).

36. S. Hawking, "Breakdown of Predictability in Gravitational Collapse," *Phys. Rev.* D14, p. 2460 (1976).

37. J. Ellis, N. Mavromatos and D. Nanopoulos, "On the W-hair of String Black Holes and the Singularity Problem" preprint CERN-TH.6476/92, and references therein; World-Sheet Duality, Space-Time Foam, and the Quantum Fate of a String Black Hole", preprint CERN-TH.6595/92, and references therein.

DISCUSSION

C.S. Lam, McGill University, Canada

Concerning the 1-loop QCD calculation using the string method of Bern and Kosower, I would like to remark that it is now possible to do it in field theory. This method is as efficient as the string method for 1-loop; it is, however, also applicable to any number of loops as well as processes with external fermions.

V. Khoze, University of Durham, U.K.

I'd like to make a comment concerning the hypothetical observation of the Ringwald-induced instanton phenomena. I'm afraid that even the extreme optimists can't use, for that purpose, e^+e^- collisions; instead they would need e^-e^-, or better $\bar{q}q$ scattering.

Diambrini-Palazzi, University 'La Sapienza

According to recent papers by John Ellis, et al., we should expect that quantum gravity should damp coherent quantum states. Do you agree that this result is consistent with string theory calculations?

Alvarez-Gaumé

In the papers you mention there are suggestions of how string theory would include loss of quantum coherence. However, I am not aware of any detailed analysis proving these suggestions.

Y. Ne'ema, Tel Aviv University

Is the Berne-Kossower technique that you described a revival of the 1969 string, born as an S-matrix amplitude summing (through unitarity) all appropriate Feynman diagrams? Is it a return to the string's origin, using the great progress in 2-dimensional conformal field theory achieved in the 80's?

Alvarez-Gaumé

Yes.

LATTICE GAUGE THEORIES

Roberto Petronzio
Physics Dept. Universita' di Tor Vergata
Via della ricerca scientifica
00133- Roma-Italy

Abstract

Lattice gauge theories are about fifteen years old and I will report on the present status of the field without making the elementary introduction that can be found in the proceedings of the last two conferences. The talk covers briefly the following subjects: the determination of α_S, the status of spectroscopy, heavy quark physics and in particular the calculation of their hadronic weak matrix elements, high temperature QCD, non perturbative Higgs bounds, chiral theories on the lattice and induced theories.

In spite of the very high precision reached by Lep experiments, the uncertainty in the value of the QCD Λ parameter is still rather high[1].
This can be easily understood by considering the well known relation between Λ and the running α_S:

$$\alpha_S(Q^2) = \frac{1}{b\ln(Q^2/\Lambda^2)} \quad (1)$$

from which one can derive how a relative error on α_S at a given scale translates into an error on Λ:

$$\frac{\Delta\Lambda}{\Lambda} = \frac{1}{2b\alpha}\frac{\Delta\alpha}{\alpha} \quad (2)$$

The higher the scale, the lower the value of α_S and the higher the amplification of the error: with the value measured at Lep it amounts to about a factor seven.
It is hard to conceive that Lep will reach a precision better than a few percent and even in this case one would get an uncertainty on Λ of the order of ten percent. The lesson to be drawn is that α_S should be "measured" i.e. related to experimental quantities where it is of order one, i.e. in the non perturbative region. This can only be done within lattice QCD.
At this conference El-Khadra has reported on an estimate of Λ in the $\overline{MS}$ scheme[2]:

$$\Lambda_{\overline{MS}} = 160(_{37},{}^{47})\text{MeV} \quad (3)$$

where the error can be seen to be already very competitive with accelerator estimates.

The determination quoted above proceeds through a few steps. The essential ingredient is a dimensional quantity calculated on the lattice, i.e. expressed in lattice spacing units: in the specific case it is represented by the spin averaged splitting of 1_S and 1_P states of charmonium.

The ratio between the lattice quantity and the physical one provides the value of the lattice spacing in physical units which can be translated into a value for Λ as follows.

The bare coupling constant used in lattice calculations can be seen as the usual running coupling renormalized at the cutoff scale. Equation 1 relates Λ, the cutoff "a" (1/Q) and the bare coupling $\alpha_S(\pi/a)$ and implies that the continuum limit of lattice QCD is at zero bare coupling.

In its simple form the relation holds in leading log accuracy, where the Λ parameter is ill defined. One loop corrections are necessary to relate the determination of Λ in different regularization schemes:

$$\alpha_{scheme1} = \alpha_{scheme2}(1 + C_1 \alpha_{scheme2} + ...) \quad (4)$$

In the particular case of the lattice versus the $\overline{MS}$ scheme these corrections are important and a resummation of the largest terms is needed in order to make the connection between the two schemes perturbative. This can be achieved by following an old suggestion by Parisi, first applied in a study of the scaling properties of the non linear sigma model[3], where the value of the coupling constant is replaced by the perturbative expression of the average plaquette. When α_S tends to zero the replacement is an identity, but for finite values of the coupling it represents a new scheme which reabsorbs the biggest contributions (tadpole digrams) to the difference between lattice and continuum schemes.

Once the value of α_S in the $\overline{MS}$ scheme at the cutoff scale π/a is obtained, by using the parametrization of the particle data group one extracts a value for $\Lambda_{\overline{MS}}$.

The important question is whether the lattice spacing is small enough to make the simulation of discretized QCD a good approximation of the continuum theory. This can be checked through a scaling test which requires that the value obtained remains the same with different lattice spacings. In figure one is shown the dependence upon the lattice spacing of the value of $\Lambda_{\overline{MS}}$ of ref (2).

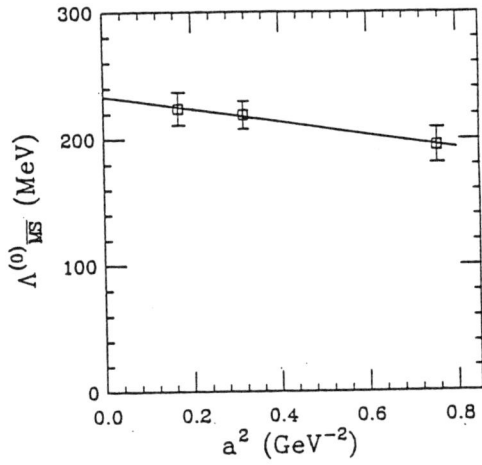

Figure 1. Cutoff dependence of $\Lambda_{\overline{MS}}$ of ref (2).

The fit assumes a quadratic dependence upon the lattice spacing. This is justified by the use of an improved form of the Wilson action (the clover action[4]) which in its standard form would only admit a linear dependence.

In the calculation dynamical fermions were absent and the last step is the connection of the zero flavours Λ_{MS} to the actual one.

This is the model dependent part of the calculation. The idea is to add to the value found for α_s at the scale π/a the difference between the values of α_s with zero flavours and the actual one obtained by evolving the two couplings, normalized to the same value at a typical charmonium momentum of a few hundred MeV, up to the scale π/a.

The systematic errors arising from the phenomenological treatment of the last step should be compared with those of hadronization models in perturbative analysis of hadron jets.

With dynamical fermions correctly incorporated into the calculation, the result would only be affected by systematic errors related to:

i) the determination of the lattice spacing by a physical quantity, i.e. the uncertainty of the lattice calculation. The relative error on "a" turns directly into a relative error on Λ.

ii) the uncertainty in the determination of α_s in a perturbative scheme from the value of α_s in the lattice scheme. From eq.(4), this is of order $\alpha_s^{(L+2)}$ where L is the number of loops included in the calculation: notice that at zero loop one gets an error of order α_s^2 which implies a relative error of order one on Λ. This reflects again the necessity of a one loop calculation to define Λ. Including one loop, the error on Λ is of order α_s and goes to zero with the lattice spacing.

The possibility of minimizing both sources of errors with a single lattice resolution is severely limited by the total size affordable in lattice computations. In order to get a good lattice determination of a non perturbative quantity, without finite volume effects, one needs sizes of a few Fermi and at the same time a lattice resolution where a scaling behaviour is visible. However, asymptotic scaling, i.e. without big next to leading corrections, sets in only at very small distances.

It is very hard to imagine, in the unquenched case, to be able to cover in the near future scale factors bigger than thirty or so. This means going in momentum space from hundred MeV to three GeV at most.

To reach asymptotic scaling, it is necessary to cover a larger scale range by a study of the non perturbative behaviour of the running coupling from large to small distances obtained by matching lattices with different resolutions.

M.Luescher et al.[5] have introduced a definition of a renormalized coupling constant based on the response of the system to variations of an external background chromoelectric field.

This is a physical finite size effect and depends upon the coupling constant renormalized at a scale which is the total lattice volume. The evolution of the coupling constant from large to small distances where the connection with perturbative renormalization schemes can be established is obtained by constructing a step beta function relating the coupling renormalized on a size L to the one on a size 2L. The calculation is repeated with different lattice resolutions, but keeping the total physical size fixed and then extrapolated to zero lattice spacing.

In turn, by changing the coupling constant with a fixed lattice size one can cover several different physical sizes.

Once the shape of the renormalized α_S is available as a function of L in lattice units, the physical units are fixed through the usual procedure of calculating a dimensional quantity. Figure (2) shows the behaviour in physical units of the running alfa for SU(2) pure gauge. A conventional value for the string tension of 425 MeV has been used as input.

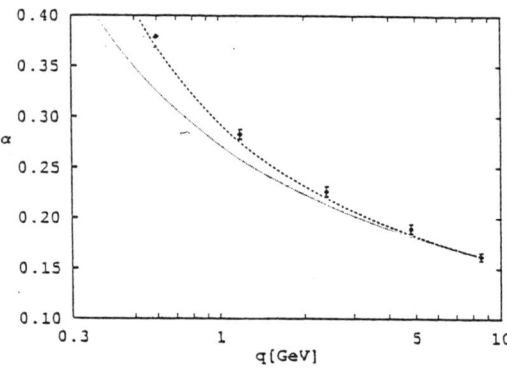

Figure 2. The SU(2) running α_S.

The errors are very small and if such a precision could be reached in the unquenched case, one would have a determination of α_S which would mainly depend upon the accuracy in the calculation of the reference physical quantity.

The question is therefore how far are we from a full QCD calculation with a ten percent error, which would translate directly into a ten percent evaluation of Λ_{MS}.

Quenched spectroscopy calculations have now reached a ten percent precision through various standard improvements:

i)   smaller lattice spacings, corresponding to values for the quantity $6/g^2=\beta$ of 6.2,6.3,;

ii) smaller quark masses, reducing the errors due to the extrapolations required to reach the light physical quark masses;

iii) an improuved Wilson action to obtain a faster convergence to the continuum limit;

iv) optimized choices of the operators whose correlations are used to extract hadron masses, and in particular nonlocal definitions ("walls").

In a standard Edinburgh-(Roma) plot of the ratio of nucleon over proton mass versus the ratio-(squared) of the pion over rho mass, the results presented at the '91 lattice conference by Toussaint[6] reached values around 1.4 with pion's masses roughly equal to half the rho mass. The UKQCD[7] collaboration performed recently an exploratory (i.e. low statistics) analysis of the same quantities shown in figures (3) using both the standard Wilson action and the improuved (clover).

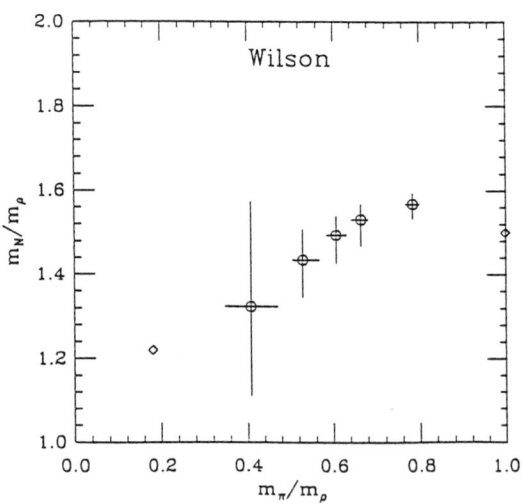

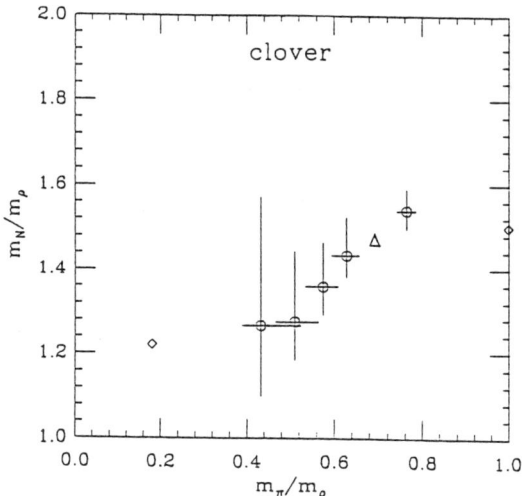

Figure 3. Mass ratio plots for the Wilson and the improuved (clover) action.

The results in the last case give a value for the ratio around 1.3 at a pion mass mass ten percent lighter than the '91 data. On the improved plot I have superimposed the value extrapolated from '88 data (the triangle) by fitting the corrections due to finite lattice spacing which fits with the improved results. Unquenched spectroscopy a year ago had points in the same plot which were similar to the quenched points in '88: in two years from now there will be an increase in computing power (two orders of magnitude) comparable to the increase from '88 to '92. It is conceivable that unquenched simulations will reach the precision of today's quenched ones. For the most recent unquenched results the reader should consult the '92 lattice conference proceedings.

Heavy flavour lattice simulations are a subject where in some cases the experimental data are only poorly measured. In particular, weak hadronic matrix elements are relevant to establish unknown parameters of the electroweak sector. The ε parameter governing CP violation is expressed by:

$$\varepsilon^{K0\text{-}\underline{K0}} = F(\text{mixing angles,CP violating phase, short distance QCD}) \frac{8}{3}f_k^2 M_k^2 B_k \quad (5)$$

in order to know the phase of the K-M matrix, one needs to estimate the quantity:

$$\frac{8}{3}f_k^2 M_k^2 B_k = <K0|O^{\Delta S=2}|\underline{K0}> \quad (6)$$

where $O_{\Delta S=2}$ is the strangeness changing weak Hamiltonian.

The unknown parameter is B_k and there is now a convergence on the results of simulations performed with different formulations of fermions on the lattice: staggered fermions or Wilson fermions.

$$B_{k,\text{staggered}}^{(8)} = 0.72(14)$$

$$B_{k,\text{Wilson}}^{(9)} = 0.85(20) \quad (7)$$

Chiral symmetry implies a matrix element vanishing with the quark mass: while in the staggered fermion case this is automatic, in the Wilson case this is possible only by adding suitable counterterms to the Lagrangian which will restore the correct chiral properties of QCD in the continuum limit, according to a scheme which has been developed over the last few years[9]. The dependence upon the quark mass with the improved Wilson action is shown in figure (4). Within the errors, the results are compatible with a correct chiral behaviour.

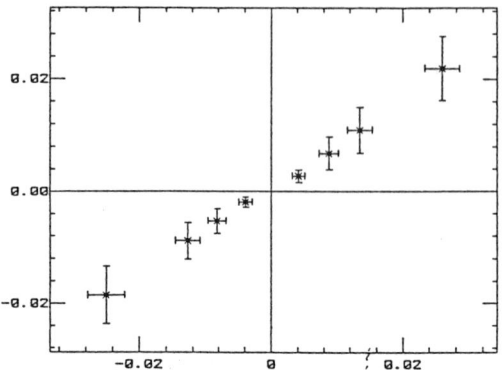

Figure 4. Quark mass dependence of the matrix element of ΔS=2 weak Hamiltonian between kaon states.

For the analogous quantities for B mesons, F_b and B_b, the estimates with both fermion formulations are:

$$f_{B,staggered}=205(40) MeV$$

$$f_{B,Wilson}=180(33) MeV$$

(8)

$$B_{B,staggered}=1.3$$

$$B_{B,Wilson}=1.16(7)$$

A detailed discussion of the subject was presented by V.Soni in the parallel session. I want only to add a comment on these calculations: they are usually performed at quark mass values around the charm mass and then extrapolated to the bottom mass. The infinite mass case is also known where the heavy quark acts only as a static colour source.

The extrapolation is made by joining charm mass data with the infinite mass value with a 1/M type behaviour.

The continuum corrections are of order Λ/M but on the lattice this behaviour is corrected by a function of Λa and of Ma which goes to one when "a" goes to zero. While lattice spacings are small enough to neglect the Λa dependence, this is not the case for the Ma dependence, given that the cutoff is of the order of the charm quark mass. When making fits to the charm data the form of such a dependence should be known and accounted for. This observation applies to another case where one studies mass dependences with mass values of the order of the inverse lattice spacing: the vector-pseudoscalar hyperfine splitting which deviates from the expected behaviour with increasing meson mass. Also the nucleon over rho mass ratio near the infinite quark mass limit could be affected by these corrections.

Pure gauge simulations of QCD at finite temperature have reached a consensus on the order (first) of the deconfining phase transition. When heavy dynamical fermions are introduced the situation cannot change with respect to the quenched case, but for intermediate values of their mass the phase transition disappears. The interesting case is the one of two light quarks where the transition appears again at a temperature where also the chiral restoring phase transition occurs. The ratio between the quark mass and the critical temperature which defines the quarks to be light is around 0.1, which means quark masses below ten-twenty MeV.

The order of the transition is established to be first with four dynamical flavours, while for two it is still under study, with some evidence for a first order type. Present unquenched simulations should be considered rather exploratory. The cutoff is typically only four times the critical temperature and lattice artifacts are expected to severely affect the quality of the results. N.Christ in the parallel session reported on a study of the scaling behaviour of the critical temperature for different numbers of dynamical flavours[10]. The results in fig.(5) show that, while for the pure gauge case an approach to the scaling region can be seen for a cutoff eight or sixteen times larger that the critical temperature, the N_f=2 case is very far from it.

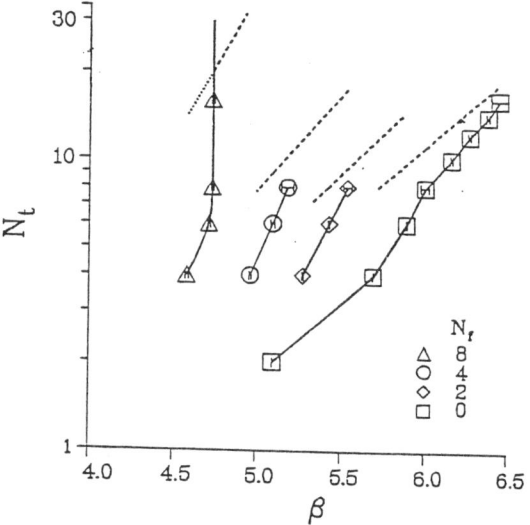

Figure 5. Scaling test of the inverse critical temperature in lattice units (N_t) for different numbers of flavours: the dashed lines are the slopes predicted by asymptotic scaling.

Lattice simulations of the Higgs sector of the standard model have established that the Higgs mass cannot exceed a certain value in the scaling region, i.e. where the ratio of the Higgs mass over the cutoff is less than two.

The original results, obtained with a standard discretization of the Higgs Lagrangian led to numerical values around 650 GeV[11]. Further simulations have been made by adopting discretizations which would minimize Lorentz non invariant operators with respect to the usual hypercubic geometry[12]. A precise definition of the absolute bound is a matter of taste: when the Higgs mass increases, the Landau pole, responsible for the bound, lowers. When it becomes of the order of the Higgs mass (the absolute bound) the theory is by definition a cutoff dependent theory. The original form of the Lagrangian is modified by the presence of non renormalizable operators, remnants of a more complete theory at energy scales beyond the cutoff. They are automatically present in lattice formulations as lattice artifacts which disappear in the continuum limit. For an infrared free theory, a true continuum limit can be reached only for vanishing Higgs mass. Neuberger has reported on studies with different lattice actions[13], i.e. with different sort of non renormalizable operators, trying to establish the heaviest possible Higgs mass compatible with small deviations of the standard predictions in a physical process like Higgs-Higgs scattering. The numerical value of the bound in this case can only be increased to roughly 800 GeV.

If one instead allows big and model dependent cutoff effects, the bound can be significantly increased, like in the example of QCD where the ratio of the mass of the scalar over f_π, which plays the role of the ratio of the Higgs mass over the vacuum expectation value, is around eight, corresponding to a Higgs mass of 1.2 TeV.

The implementation of chiral gauge theories on the lattice has been clarified during the last two years. The difficulties of the lattice approach are related to the Nielsen Ninomiya[14] theorem: under general locality assumptions, on the lattice one has either to break explicitly chirality or to live with more fermion species than naively deducible from the Lagrangian (the fermion doublers). The two possibilities correspond to the theWilson and to the staggered formulation of the fermion action, respectively.

The efforts for defining chiral gauge theories have been concentrated on the Wilson formulation that I will briefly review: the subject has been discussed by Petcher in the parallel session. The way chirality is explicitly broken in the Wilson formulation is by adding a dimension five operator of the form:

$$a\Psi_L \Delta \Psi_R \qquad (9)$$

In chiral gauge theories left and right handed fermions transform differently under the gauge group and the term written above breaks gauge invariance.

The first attempts to solve the problem were inspired from the electroweak standard model. A mass type term for fermions can be made gauge invariant by adding a Higgs field[15]. This can be seen either as a physical Higgs or an unphysical one which should be eliminated by adding suitable counterterms. In the first case it seem impossible to decouple the right handed fields which were introduced to form the Wilson term from the Higgs field. In the second case the class of counterterms needed to decouple it from the spectrum and to recover the correct anomalies should span a whole set of gauge non invariant operators, falsifying the original reason for introducing the Higgs field. This leads essentially to the Rome approach[16], which seems today the only safe treatment of the problem. The idea is to live with a gauge broken theory on the lattice, to add gauge non-invariant counterterms and to tune them to fulfill the BRST identities of the theory. This recipe also correctly allows for topological baryon number violation[17]. The only criticism which may be moved to it is the potential difficulty of controlling satisfactorily the numerical tuning procedure.

It remains the possibility of introducing mirror fermions as physical particles at a neighbour scale: they would solve the problem but would make the chiral theory only a low energy manifestation of a vector-like theory.

The idea of solving the problem by considering the chiral theory as a subset of a vectorlike one is also present in the speculative solution proposed by Kaplan[18]. One constructs a vectorlike theory in four plus one dimensions. The fermion mass has a dependence upon the extra dimension "s" of the form:

$$m_0(s) = \sinh(\mu_0)\theta(s)$$

$$\theta(s) = \begin{cases} -1 & 2 \leq s \leq \frac{L_s}{2} \\ +1 & \frac{L_s}{2}+2 \leq s \leq L_s \\ 0 & s=1, \frac{L_s}{2}+1 \end{cases} \qquad (10)$$

and produces two four dimensional "defects" with associated left and right handed massless particles. When the Wilson term is switched on, a right and a left handed partner on the two defects form together a massive particle which decouples from the low energy spectrum leaving a chiral fermion on a defect and the mirror one on the other.

In a recent paper[19] a simulation of such a model in two plus one dimensions was discussed: the resulting energy-momentum dispersion plot in figure (6), shows that there are chiral particles on the "physical" two dimensional defect.

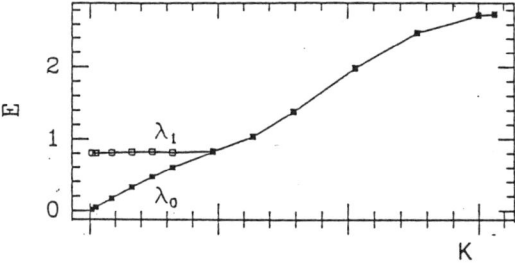

Figure 6. The energy momentum dispersion relation of the two lowest positive eigenvalues of the states confined to the physical two dimensional space.

While fermions can be shown to be "confined" to the defect, i.e. to have wave functions exponentially decreasing away from it, it is not clear to me what will happen to gauge fields and in particular how one can avoid leakages of gauge invariance from the "physical" space.

Theories at the cutoff lattice scale may look very different from the ones at large distances.

For example, they may lack many of the symmetries, Lorentz invariance, chiral symmetry, which will be recovered at large distances in lattice units or, equivalently, at fixed physical distances in the continuum limit. Dynamical effects may change the relevant large distance degrees of freedom of the theory. In the parallel session, M.P.Lombardo reported evidence[20] for the description of strongly interacting non compact QED with a Nambu Jona Lasinio model with an effective four fermion interaction.

The idea of inducing from dynamics a theory different from the one at short distances has been applied to QCD by Migdal and Kazakov[21]. The microscopic degrees of freedom are scalars in the adjoint representation of $SU(N)$ with dynamical gauge invariance: the Lagrangian contains the usual covariant derivatives without an explicit kinetic term for the gauge fields which are a function of the scalars themselves.

The mass of the scalar particles defines an energy scale where the transition bewteen two different types of relevant degrees of freedom is expected to occur. At higher energy scales, the scalars are the propagating particles. At lower energies, the induced theory should be described by the scalar bound states interacting according to gauge invariance, i.e. by the QCD gluons. The identification of the induced theory with QCD is not guaranteed: the theory should have two different correlation lengths, one for the confinement of the scalars and one for the one of the gluons. The main motivation for this approach is the possibility of obtaining an explicit solution of the scalar theory in the large N limit, where many techniques developed for the matrix models can be applied. After integration of dynamical gauge degrees of freedom the effective Lagrangian for the lattice scalar theory is of the form:

$$\sum_x [\ln\phi^2 - 2V(\phi)] + \sum_{<xy>} \ln\left[\frac{\sinh(4\phi(x)\phi(y))}{\phi(x)\phi(y)}\right] \quad (9)$$

where <xy> denotes the sum over nearest neighbours of x.

This differs from an ordinary scalar Lagrangian by the non polynomial terms, without which the theory would have a trivial continuum limit without any interesting dynamics. There are results[22] on numerical simulations of the phase diagram of the theory. It has been found that, in the m_0^2-λ plane (the parameters of the quadratic and of the quartic part respectively of the scalar potential $V(\phi)$) there is a line of a first order phase transition ending with a critical point where a continuum limit does exist, but might be trivial.

Further analytical and numerical analysis of the theory are necessary to establish its connection to QCD.

In summary, quenched QCD is today under control, there are serious attempts to calculate α_s, strange and charm meson quenched weak matrix elements are calculated with typically a 15% accuracy, beauty meson physics suffers from extrapolation biases, high temperature QCD with dynamical fermions has not reached yet a scaling behaviour, the appropriate way of simulating chiral theories on the lattice has been clarified and there is a growing interest in induced theories. The field owes his progress to the developments of algorithms and of computer technology with speeds of a few gigaflops which should increase by the next conference up to a few hundred gigaflops. This should allow a 10-15% determination of unquenched Λ, B physics simulations without systematic errors related to extrapolations, a quantitative study of the plasma phase of QCD with dynamical fermions and exploratory studies of chiral theories.

REFERENCES

1. T.Hebbeker, Phys.Reports 217(1992)70.
2. A.El-Khadra et al., Fermilab pub 91/354-T
3. G.Parisi,proc.High Energy Physics 1980 XX Conf.Madison, ed. L.Durand and L.G.Pondrom (AIP,New York 1981).
4. B.Sheikholeslami and R.Wohlert, Nucl.Phys.B259(1985)572.
5. M.Luescher, R.Sommer, U.Wolff and P.Weisz, preprint CERN-TH-6566 (July 92).
6. D.Toussaint, Nucl.Phys. B(Proc.Suppl.) 26(1992)3.
7. The UKQCD collaboration, Edinburgh preprint 92/506, Southampton preprint SHEP 91/92-15 (April 92).
8. See the review by G.Martinelli in Nucl. Phys.B (Proc.Suppl.) 26(1992)31and the contribution by V.Soni in the lattice parallel session at this conference.
9. G.Martinelli, C.T.Sachrajda, G.Salina and A.Vladikas,Roma La Sapienza preprint 850 (dec.91).
10. F.R.Brown et al., Columbia University preprint, CU-TP-541(July 92).
11. M.Luescher and P.Weisz,Phys.Letters B212(1988)472, J.Kuti, L.Liu and Y.Shen, Phys. Rev.Letters 61(1988)678,

A.Hasenfratz, K.Jansen, J.Jersak, C.B.Lang, T.Neuhaus, H.Yoneyama, Nucl.Phys.B317(1989)81.

12. G.Bhanot, K.Bitar, U.M.Heller, H.Neuberger Nucl.Phys.B353(91)551.
13. H.Neuberger, parallel session at this conference.
14. H.B.Nielsen and M.Ninomiya, Nucl.Phys.B185(81)20.
15. D.Petcher, lattice parallel session at this conference.
16. A.Borrelli, L.Maiani, G.Martinelli, G.C.Rossi and M.Testa, Nucl.Phys. B262(85) 331.
17. T.Banks and A.Dabholkar, Rudgers University preprint RU-92-09.
18. D.B.Kaplan, UCSD preprint PTH/92-16.
19. K.Jensen, UCSD preprint PTH/92-18.
20. A.Kocic, J.B.Kogut, M.P.Lombardo and K.C.Wang, Illinois Univ. Preprint ILL(TH) 92-12.
21. V.A.Kazakov and A.A.Migdal, Princeton Univ. Preprint PUPT-1322 (1992), A.A.Migdal, PUPT -1323(1992).
22. A.Gocksch and Y. Shen, BNL preprint July 1992.

DISCUSSION

M. Karliner, Tel Aviv University

Could you comment on the status of lattice calculations of matrix elements of axial currents, and, in particular, the flavor of singlet axial currents, which is important for the question of proton spin and polarized structure functions?

Petronzio

Calculations of this type have been attempted, in particular connecting this with topological aspects. The results are still not very credible, however.

B. Ward, University of Tennessee, USA

When will we have a direct calculation of B meson parameters, say f_B, at the scale of the B instead of having to extrapolate?

Petronzio

We are presently at 40^4, though, in fact, we need to be at 60^4 lattices, to obtain the needed resolution of $1/(6\ GeV)$ with an acceptable total size.

V. Rubakov, Moscow

Is there any further progress in understanding the field calculations responsible for confinement, chiral symmetry breaking, etc. — that is, of the basic problems that were widely assumed to have been addressed by lattice QCD?

Petronzio

The only contribution I am aware of is one concerning the topological susceptibility: how it would behave at the high-temperature phase transition. It was shown that the fluctuations are drastically reduced beyond the phase transition.

H. Trottier, TRIUMF, Canada

Can you comment on the status of compact versus non-compact (QCD) lattice gauge theory?

Petronzio

The compact theory shows no "peculiarities," and this is the reason for a general shift to the non-compact theory.

RECENT DEVELOPMENTS IN DETECTOR TECHNOLOGY

Takahiko Kondo
Physics Department
KEK, National Laboratory for High Energy Physics
1-1 Oho, Tsukuba-shi, 305, Japan

Abstract

This is a summary of three parallel sessions of "Experimental Techniques including Computing", "Physics Simulation Methods", and "New and Planned Detectors and Their Physics Potential." There were in total 47 talks in these sessions. It is impossible to summarize all of them and also it is not my intention. In this talk, I like to concentrate on a several selected topics.

This report describes a brief summary of (1) new detector plans, (2) close-to-mature technology, (3) computer simulation and (4) new detector technology under intensive R&D, (5) large detectors for SSC and LHC. This summary, however, does not include a serious review on computer simulations due to limited capability of the author.

NEW COLLIDER-DETECTOR MEMBERS

Recently five detector systems became into operation ; CMD-2, SLD, D0, H1 and ZEUS, joining to the collider-detector club. There are also many detector plans in application to the collider-detector club. SDC, GEM, CMS, ATLAS, L3P are for supercolliders. In addition detector plans at lower energy but high intensity colliders are also presented : KLOE, B- and τ-Charm factory detectors.

CMD-2

Fig.1 shows the detector CMD-2 newly upgraded for VEPP-2M, a e^+e^- colliding beam at the Budker Institute at Novosivirsk.[1] During early runs they accumulated 400 nb^{-1} in the CM energy range of 400-1030 MeV. They have scanned the ϕ peak in $K_L K_S$ and $\pi^+\pi^-$ modes. The preliminary results showed Γ_ϕ = 4.71±0.44 MeV for the total width.

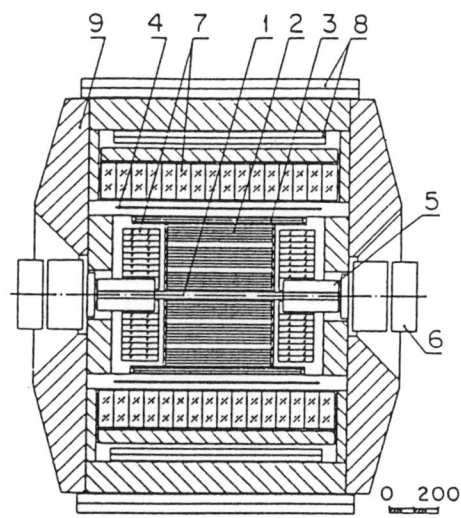

Figure 1. CMD-2 detector at VEPP-2M: 1 vacuum chamber, 2 drift chamber, 3 Z-chamber, 4 superconducting solenoid, 5. compensating solenoid, 6 quadrupoles, 7 calorimeter, 8 muon range chamber, 9 return yokes.[1]

BES

The status of the BES detector at Beijing BEPC was reported.[2] They have accumulated 9 million events around the J/ψ and D_S regions. BES accomplished a precision measurement of τ mass. A quite substantial upgrade is planned for the BEPC operation at higher luminosity.

φ-Factory Detector KLOE

KLOE is a detector proposal for the φ factory DAΦNE under construction in Italy.[3] The major goal of φ factories is to study CP violations in the $K^0\overline{K}^0$ system.[4] In order to reach the level of 10^{-4} in the CP violating parameter Re(ε'/ε), however, the detection efficiency and decay volume definition of $K_L \to \pi^0\pi^0$ must be well controlled, imposing very stringent requirements on EM calorimetry. The proposed EM calorimeter is 0.5 mm thick Pb foils with 1 mm scintillating fibers. The challenge is to reach the energy resolution of $5\%/\sqrt{E}$, the incident position resolution of 1 cm and 100% detection efficiency for >20 MeV photons. These requirements are crucial to reduce the overwhelming background of $K_L \to 3\pi^0$ in the accurate measurement of $2\pi^0$ decays.

B-Factory Detector

A detector plan for PEP-II Asymmetric B factory is presented.[5] In the case of new-generation asymmetric B factories, it is essential to achieve highest possible efficiencies in B-meson reconstruction as well as in longitudinal vertex separation. Among others, π/K separation must be attained up to 4 GeV/c. PEP-II group is pursuing R&D on fast RICH/ CRID, aerogel Cherenkov counters and internal reflection imaging Cherenkov (IRIC) counters. IRIC is a new idea of extracting Cherenkov ring images to outside using total reflection of Cherenkov radiator such as SiO_2 pipes. It is simple and fast. It can be realized with conventional technology. A demonstration on the principle is certainly waited for.

Y. Giomataris brought up an interesting idea for B trigger that might be used for fixed target and even possibly collider experiments.[6] The idea is an optical triggering of off-vertex particles with a curved Cherenkov radiator. A curved radiator with index of refraction of $\sqrt{2}$ can be arranged such that Cherenkov light radiated by particles only from off-vertex points can be trapped in the radiator by total internal reflection. The R&D beam test has confirmed the expected performance. A MC simulation

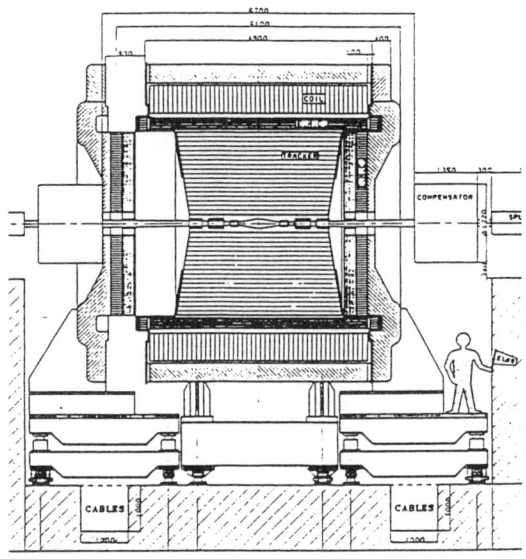

Figure 2: KLOE detector for the φ factory DAΦNE under construction in Italy.[3]

shows that the optical trigger can select b-quark events with an efficiency of 30% and rejection factor of 100 against minimum bias events.

Tau-Charm Factory Detector

Tau-Charm is a proposal of US-Europe collaboration for the τ-charm factory being planned in Spain.[7] It aims at reaching the hadron-lepton separation with hadron contamination at the level of 0.1% around 0.5 GeV/c. This goal is a factor of 40 better than the present MARK III/BES performance. A time-of-flight measurement using scintillating fiber bundles is a unique feature of the proposal. Another key element is longitudinally segmented CsI crystals with photodiode-ASIC readout.

It is quite interesting to point out that the collider accelerators and detectors (φ, τ-C and B factories) being planed are universally aiming at the luminosity of 10^{33} cm^{-2}s^{-1}, though the physics arguments for need of 10^{33} are completely different.

CLOSE-TO-MATURE TECHNOLOGY

There are a few technology that are about finishing the transition from the proof of principles to their large scale application. Below is the description of some of them with lessons we have learned.

Cherenkov Ring Image Counter

Detection of Cherenkov radiation rings is useful in separating K and p's from overwhelming pions in a jet. All three groups, SLD CRID,[8] RICH for hyperon beam[9] and DELPHI RICH[10] demonstrated the practical capability in large-scale detectors. The emission angle of Cherenkov light is accurately measured to identify the charged particles. The drift tubes filled with photo-sensitive drift gas are used to detect Cherenkov light from both gas (C_5F_{12}) and liquid (C_6F_{14}) radiators. The identification of kaons is possible from 2 to 20 GeV. Their experiences tell us, however, that the large scale application is in fact non-trivial. Detection of single photo-electrons as well as handling faint UV light demand ultra-high complexity and cares. For example, gas purity must be kept extremely good and UV-transparent quartz windows must be used to separate the two gas volumes. It needs extreme care to maintain UV sensitive drift gas like C_2H_6+TMAE. The gas causes wire aging rather easily. Furthermore the photon feedback must be adequately suppressed in the vicinity of sense wires. Recognition of a Cherenkov ring is also nontrivial due to

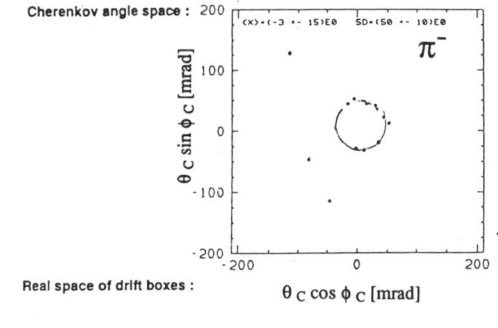

Figure 3: One event display of SLD CRID for a hadronic event with $K_S^0 \rightarrow \pi^+\pi^-$. The displays are in Cherenkov angle space (top) and in real space of drift boxes (bottom). The points in the pictures are photo-electrons.[8]

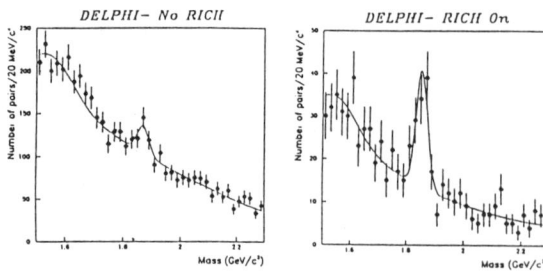

Figure 4: Neutral Kπ mass combination for D^0 candidates without and with RICH in DELPHI detector.[10]

background hits as one can see in Fig. 3, an event sample in SLD CRID. Good news is that all groups observed expected number of photoelectrons per ring (7-8 electrons per ring). DELPHI successfully used RICH signal to identify ϕ, Λ, Λ^C, D$^\pm$ and D^0, an example of which is illustrated in Fig. 4.

Transition Radiation Detector(TRD)

Another technical challenge is a large scale application of transition radiation detection for electron identification in a jet. This is accomplished by VENUS for TRISTAN.[11] The large cylindrical TRD of 3 m in diameter consists of 4 layers of 5 cm thick radiator of polypropylene fibers followed by a x-ray detector with xenon-methane (90/10) gas. As shown in Fig. 5, the TRD introduces a clear separation of electrons from pions for the momentum range of 1 to 30 GeV/c. It identifies non-isolated electrons in jets with pion rejection factor of 0.08 and electron efficiency of 80% with TRD only. The overall pion rejection including lead-glass calorimeter is 0.002 with electron efficiency of 60%. The capability of non-isolated electron identification

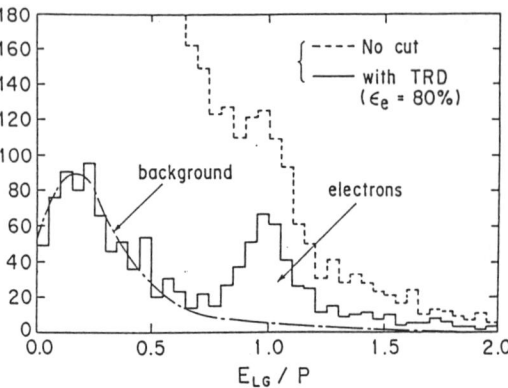

Figure 5: Improvement of e/π separation by the VENUS TRD at TRISTAN. The dashed curve shows electron candidates in hadronic events from e$^+$e$^-$ 58 GeV annihilations and the solid curve is after the trancated-mean pulse height cut of TRD with electron efficiency of 80%.[11]

in hadronic jet events is powerful for b- and c-quark related physics.

Liquid Argon Time Projection Chamber(TPC)

Liquid argon time projection chambers can be useful for future neutrino physics because of many attractive capabilities, such as high density, homogeneity, deadtimelessness, three dimensional reconstruction of minimum ionizing tracks and ionization measurement. A 3-ton prototype ICARUS was developed at CERN.[12] It has 42 cm drift length. A picture of cosmic rays is shown in Fig. 6, in which one sees beautiful tracks of bubble chamber quality. It took more than a decade to successfully develop the LAr TPC since its proposal by C. Rubbia in 1977. The major technical problem is to keep the impurity level down to 0.1 ppb. This is

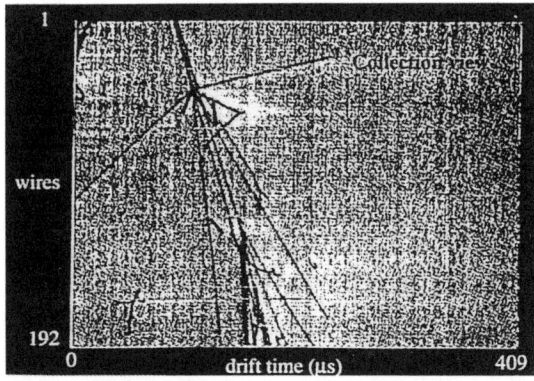

Figure 6: Bubble chamber-like image of cosmic-ray tracks observed in the 3-ton prototype ICARUS liquid argon TPC. [12]

accomplished by recirculation liquefiers with gas-phase purification, and by use of feed-throughs on vetronite support. They obtained S/N ratios of 6 for induced wires and 10 for charge collection wires with a 200 ns sampling time. The detector can also be used as a good and uniform EM calorimeter.

Large-scale Calorimetry : D0, H1 and ZEUS

A number of newly completed large-scale calorimeter systems are reported by D0, H1 and ZEUS groups. Among others, it is noted that the beam test of the D0 LAr/uranium calorimeter demonstrated extremely good linearity down to 2 GeV as shown in Fig. 7.[13] In addition they obtained a good e/π ratio of near 1.02 at high energies after corrections. The energy measurement is inherently degraded by cryostat walls in cryogenic calorimeter. The D0 beam test proved the energy degradation can be compensated to acceptable level using massless gaps and scintillators as compensators. Without these compensators, the D0 calorimeter response drops to about half in the worst case at $\eta=1.15$.

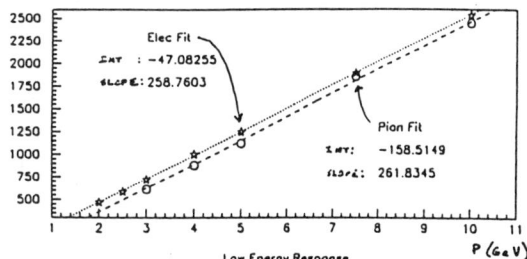
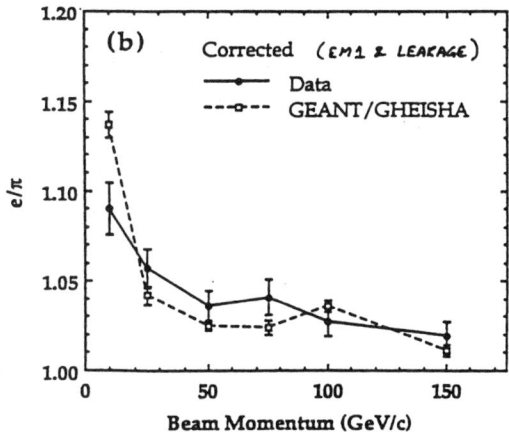

Figure 7. Test beam study of D0 LAr calorimeter showing the excellent linearity down to 2 GeV (top) and e/π ratios (bottom).[13]

The H1 LAr calorimeter with stainless steel absorber adopts a weighting correction method for improving the hadronic energy resolution.[14] This is possible because of its frequent depth segmentation. Their beam test resulted $46.1\%/\sqrt{E} \oplus 2.6\%$ after weighting correction.

ZEUS group found a good use of uranium radioactivity for stability monitoring in their calorimeter. The excellent hermeticity as well as good hadronic energy resolution was achieved. Its test beam resolution is $35\%/\sqrt{E}$ (HAD) with tower-to-tower uniformity is better than 1% and non-linearity of less than 1% in spite of the optical and structural complexity.[15]

CCD Pixel Detector of SLD

The SLD successfully demonstrated a CCD vertex detector working in a severe experimental environment of SLC after 6 years of the extensive R&D program.[16] This is the first time in the history for pixel detectors to be installed and operated. The system consists of 480 CCDs with 120M pixels altogether. Each pixel is a 22 μm square. The system is operated at -80°C to maintain the threshold level at 200 electrons. It is remarkable that the CCD system can run with negligible overlaps even under unusual beam conditions in which the drift chamber HV cannot be turned on. Fig. 8 shows the improvement of vertex finding by the CCD vertex detector. The improvement in r-z plane is even more impressive due to the two dimensional nature of CCDs.

Silicon Microvertex Detectors

Application of silicon microvertex detectors to collider detector comes into fashion. This is due to the technology breakthrough accomplished by the pioneering work of LSI application to silicon microstrip readout by B. Hyams and others at Stanford University.[17]

The DELPHI microvertex detector with 3 layers of single-sided silicon microstrips has been working since 1990.[18] The efficiency of track association in multihadron events turned out to be 96%. It showed the importance of having at least 3 layers in order to reduce ambiguities in track-hit association.

The CDF silicon vertex detector SVX (4 layers, 46 Kch) was recently installed and started to operate in the hadron colliders for the first time in history.[19] The closest layer is only 2.9 cm away from the beam center. The radiation length due to the vertex detector is kept at 2 to 5% within rapidity range of ±4, extremely low for its complexity. It is also noted that they successfully keep the lost channels below 1.5%. The preliminary tracking efficiency is as high as 99% thanks to four-layer design. Fig. 9 is a display of one of two jet events. The fact that the picture is very clean suggests us to throw our preconception that the hadron colliders are dirty and e^+e^- are clean. The reality seems telling that opposite might be happening as one approaches to the beamline as well as to higher energies. Anyhow it is pleasure to watch how big the impact this SVX system gives on t- and b-quark physics.

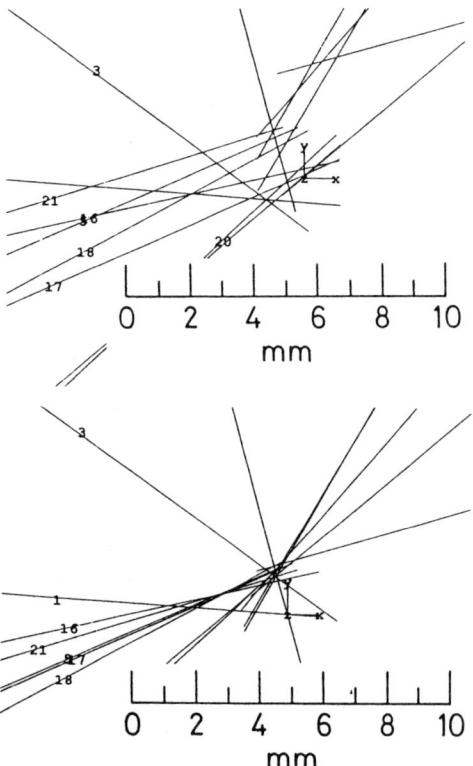

Figure 8. A SLD reconstructed event showing the improvement by CCDs. Tracks without (top) and with (bottom) CCD information.[16]

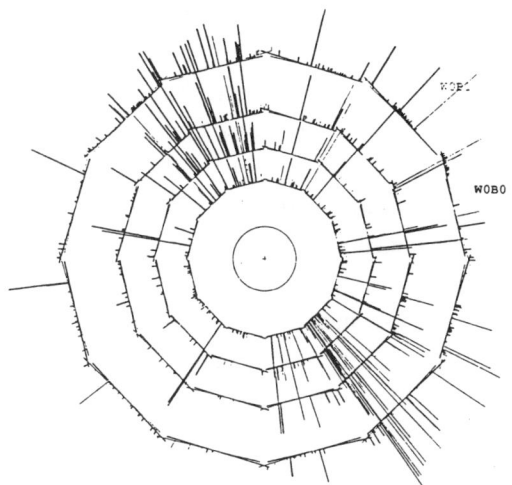

Figure 9. A two-jet event observed in the CDF vertex detector.[19]

CLEO II is planning to install 3 layers of the silicon vertex detector with newly developed double-sided microstrips.[20] The physics gain in this case is large. For example, the effective luminosity for D^+ detection increases by factor 5.3 with improved z vertex resolution of 6 μm from the present 1.2 mm.

COMPUTER SIMULATIONS

Computing technology becomes one of very important elements for high energy physics experiment. There were twelve talks related to the computing technology and simulations. However, the author is not capable to review them due to limited capability. Rather I just list up "key words" that were mentioned in these talks:

- UFMulti and NetQueues[21]
- object oriented programming[22]
- angular ordering[23]
- color coherence[24,29]
- TAUOLA, KORALZ, KORALB[25]
- parton distribution at small x[26]
- fast shower simulation[27,30]
- multi-dimensional MC efficiency[28]
- shower library[30]
- HIJING, jet quenching, parton cascading[32]
- dual parton model[33,32]

NEW DETECTOR TECHNOLOGY UNDER INTENSIVE R&D

We are about entering into new phase of the detector technology. The supercolliders, SSC and LHC, requires new and extraordinary demands on detectors in almost all aspects of the experiments :

<u>Luminosity</u> :

The initial luminosity goals of the SSC and LHC are 10^{33} and 1.6×10^{34} cm^{-2}s^{-1}, respectively, a factor of 100 to 1000 times higher than those of the existing accelerators. As a consequence, the detectors must be radiation tolerant. Event pile-up is a serious problem.

<u>Momentum resolution</u> :

It must be fine enough to reconstruct for example $Z \rightarrow \mu^+\mu^-$ with mass resolution comparable to the natural Z width. Other physics example is to determine charge sign of TeV leptons from W^+W^+ scattering and from new gauge bosons.

<u>Short bunch interval / multiple interactions</u> :

The bunch interval is 15 ns, much shorter than 90 ns at HERA and 3.5 μs at Tevatron. The detector response must be faster or comparable to the bunch interval. Besides, the average number of interactions in one bunch crossing is 1.5 at SSC and 20 at LHC.

<u>Higher multiplicity</u> :

Particle multiplicity of high p_t jets often exceeds 100. Much finer segmentation of detector elements are required. Number of readout channels easily exceeds a few millions.

Missing E_t measurement :

Good hermeticity in calorimetry is required to detect semi-leptonic decays of W and tops and to search super symmetric particles.

Bad S/N ratios :

Interesting events such as Higgs production are in the order of 1 pb while the total inelastic pp cross section is about 100 mb. The S/N ratio is tiny like $1/10^{11}$. Thus we are searching extremely rare events in the ocean of background. Among others $t\bar{t}$ production may be serious sources of backgrounds for various interesting channels.

In the past several years, a large amount of intensive R&D works have been performed in order to meet these severe requirements. The progress has been slow but steady. It is a general feeling that we are able to construct detectors for high-energy supercolliders based on the results of R&D as far as detector technology concerns. Before getting into selected topics of detector R&D, let me summarize the study of radiation levels at supercolliders.

Radiation Environment at LHC/SSC

The radiation level in the interaction regions of hadron colliders has been extensively studied based on the test experiments as well as computer simulations by the SSC-CDG Task Force.[34] The principle contribution of radiation levels comes from the pp collisions. The sources of radiation considered are

1. particles (charged,γ) from pp collisions,
2. EM and hadronic showers in calorimeters,
3. albedo neutrons and γ's from calorimeters.

The energy distribution of albedo neutrons leaking back into tracking cavity peaks at around 1 MeV. It is noted that the effective threshold for damaging silicon devices including microelectronics is 0.16 MeV and therefore the abundant albedo neutrons are quite harmful.[35]

In the estimation of radiation levels, the production of the "minimum bias" events is described by parameterization extrapolated to SSC/LHC energies. Shower developments in calorimeters are simulated by various Monte Carlo programs. The radiation levels thus calculated do not critically depend of the configuration of detectors. As shown In Table 1, however, the neutron flux strongly depends on the absorber materials and the presence of hydrogen atoms in calorimeter.

Table 1. Relative neutron flux.[34]

absorber	detector	relative flux
Uranium	Liq-Ar or Si	1.0
Lead	Liq-Ar or Si	0.5
Uranium	Scintillator	0.3
Lead	Scintillator	0.15

The radiation levels can be parametrized by simple functional forms. Numbers given below are for the standard SSC running condition of $\sqrt{s}$ = 40 TeV, $L = 10^{33}$ cm^{-2}s^{-1} and 10^7 seconds per year. The annual dose Φ_{ch} due to charged secondaries depends only on the transverse radius $r_\perp$ from the beam line as

$$\Phi_{ch} = \frac{40}{r_\perp^2} \text{ MRad / year,}$$

where $r_\perp$ is in cm. Curling tracks enhance the dose by a factor of up to two depending on radius. The annual neutron fluence in the tracking volume is

$$\Phi_n = 1.2 \times 10^{12} (1+\text{reflection}) \text{ n} / \text{cm}^2,$$

in the case of a uranium/scintillator calorimeter with a 2 m inner radius, where "reflection" is mean number of reflections of an albedo neutron by the calorimeter inner wall. The albedo neutrons can be attenuated if one installs hydrogen-rich liners in front of the calorimeter. In the case of the SDC proposal (iron/scintillator), the level is $5(2) \times 10^{11}$ n /cm^2/year without (with) endcap liners at the standard SSC condition.[66]

The radiation dose in the calorimeter peaks strongly at the forward angle. Fig. 10 illustrates a typical distribution of radiation dose.

Table 2 is the summary of the comparison of radiation levels at different colliders with their nominal operating conditions.[36] The comparison takes the energy dependence of secondary production into account. Any SSC detectors aiming at the normal operating condition encounter a factor 2000 more radiation levels than CDF and D0. Detectors at the LHC, if LHC runs at the luminosity of $1.6 \cdot 10^{34}$ cm^{-2}s^{-1} as proposed, must withstand 11 times more dose than those at SSC.

Table 2. Relative radiation levels.[36]

	Tevatron	LHC	SSC
$\sqrt{s}$(TeV)	1.8	16	40
L_{nom}(cm^{-2}s^{-1})	2×10^{30}	1.6×10^{34}	1×10^{33}
$\sigma_{inelastic}$	59 mb	86 mb	100 mb
h*	4.1	6.3	7.5
$<p_\perp>$(GeV/c)	0.46	0.55	0.60
Scale factor	5×10^{-4}	11	1

(* h = charged particles per unit rapidity)

Radiation Damage of Silicon Detectors

Silicon microstrip detector or pixel-type detector is the only plausible technology in the vicinity of the interaction point. The principle reason is the occupancy. There is no other tracking device with granularity as fine as silicon detectors. In addition, silicon carries ideal properties of fast response, high tracking resolution, flexible geometry, compactness, etc. However, there are a few obstacles to be overcome before any large-scale application to supercolliders. They are radiation damage, readout electronics and material thickness.

There have been quite a number of radiation damage studies in the past.[37] These studies more or less came to similar conclusions. The major radiation effects are

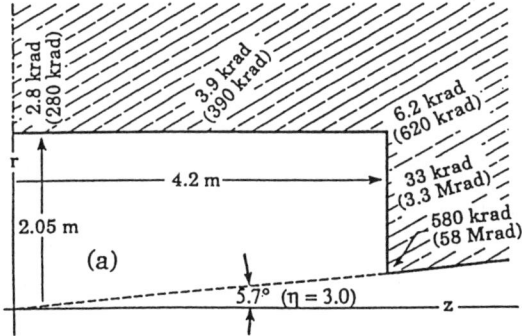

Figure 10. The electromagnetic total dose at the shower maximum in the SDC detector for 1 year at SSC design luminosity and for 10 year operation at L = 10^{34} cm^{-2}s^{-1} (in parenthesis).[66]

- increase of reverse bias current and its room temperature annealing
- bulk type inversion from n into p types
- reduction of charge collection
- charge accumulation in SiO_2 surface

The increase of reverse bias current is caused by defects in the bulk. The leakage current ΔI is proportional to the fluence Φ (particles/cm^2) as

$$\Delta I = \alpha \, \Phi \, V_{bulk}$$

where V_{bulk} is the depleted bulk volume. The damage coefficient α is consistently measured by many groups to be around 5×10^{-17} A·cm. The leakage current decreases at room temperature but a certain fraction of leakage stays for ever.

Especially with neutron (and also proton) radiation, a bulk type inversion from n to p type happens at the fluence of $1-2 \times 10^{13}$ n/cm^2. This was observed first by a strange distortion in position measurement.[38] Later the type inversion was confirmed by the decrease and re-increase of the full depletion voltage as shown in Fig. 11.[39] If the bulk type inversion happens, the p-n junction must move from the front to the back surface. This shift was in fact observed with use of infrared light and alpha sources.[40] The type inversion in n-type bulk is interpreted as an effective creation of acceptor-like defects. Recent study indicates that the situation is more complex. The apparent inversion may be a transition from $p^+/n/n^+$ into $p^+/n/i/p/n^+$.[41] It is fortunate for users that the silicon diode detectors keep working during and after the so-called type inversion.

The effects of increased reverse current can be minimized by employing an AC-coupled readout and shorter shaping time. The maximum dose that silicon device can tolerate is

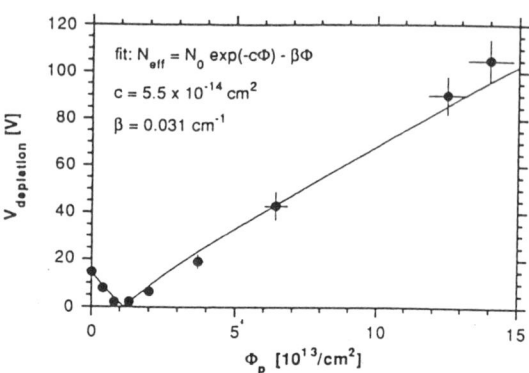

Figure 11. Full depletion voltage of a silicon photodiode as a function of total proton dose.[39]

probably determined by the eventual increase of full depletion voltage beyond 100V or by the reduction of charge collection efficiency. The 2.5% deficit per 10^{13} n/cm^2 was measured.[40] It is now expected that silicon device itself will survive beyond 1 Mrad or several times of 10^{13} n/cm^2.

Double-sided Silicon Microstrip Detector

The material thickness of silicon tracking system must be kept minimum to avoid gamma conversions and to keep high efficiency of electron identification. For this purpose double-sided silicon microstrip detectors (DSSD) have been developed at various places. Among others ALEPH collaboration has installed and operated double-sided detectors successfully. In DSSD, the signal is extracted from the ohmic (opposite to junction) side of silicon detector by keeping the ohmic resistance between strips high. As shown in Fig. 12, insertion of p^+ implantation between n^+ strips of the ohmic side makes a device more radiation tolerant.[42] Oklahoma group performed ^{60}Co source tests of DSSD.[43] The results are encouraging. The bias resistance

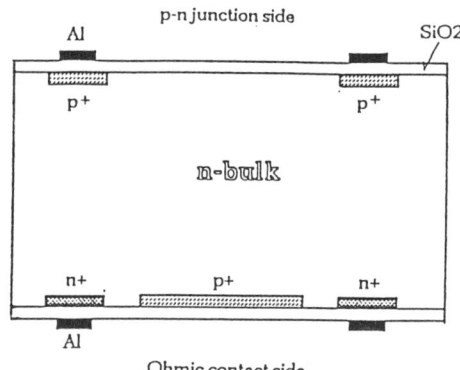

Figure 12. Cross section of radiation-hard double-sided silicon microstrip detector.[66]

showed a variation of a factor 2 at 20~40 Krad. The solution is to use polysilicon resistors as reached by other R&D groups.[44]

GaAs solid state detectors

GaAs is intrinsically radiation harder than silicon. Its electron mobility is 6-times higher than silicon. Therefore GaAs detectors are expected to have a high potentiality for future supercolliders. RD8 group has demonstrated no effects up to 7×10^{13} n/cm^2 of fast neutron irradiation.[45] Several prototype GaAs Schottky barrier detectors, on the other hand, showed only 16% or so for the charge collection efficiency, resulting in the S/N ratio of about 6 for mips. This efficiency was recently improved to 80% by using new commercial substrates or by introducing other types like p-i-n. Thus one of main problems is overcome. Despite of great hope and efforts, the progress seems to be slower than expected.

Microstrip Gas Chambers (MSGC)

This is one of new techniques under intense R&D especially for possible use in LHC/SSC detectors. By means of a photo-lithographic technique, anode and cathode strips are drawn on an insulating substrate with about 200 μm pitch as shown in Fig. 13. This pitch is a factor of ten finer than that of conventional gas wire chambers.

In 1986, A. Oed first introduced the idea of MSGC.[46] Soon after, Pisa group led by R. Bellazzini successfully operated with a gas gain of 10^4 at atmospheric pressure.[47] Successful operation of MSGC in a fixed target experiment with more than 2 months of exposure to a high flux proton beam has been reported also by the same group.[48] Since then a number of groups have studied basic properties of MSGC and gradually it becomes apparent that MSGC carries many promising features.[49] It has been demonstrated that MSGC has

1. proportional gas gain in excess of 10^4,
2. position accuracy of 30 μm for mips,
3. two-track resolution of 250 μm,
4. energy resolution of 11% at 5.9 keV,
5. stability up to 5×10^5 counts/mm^2/s,
6. total charge collection time of ~400 ns,
7. time resolution of ~10 ns.

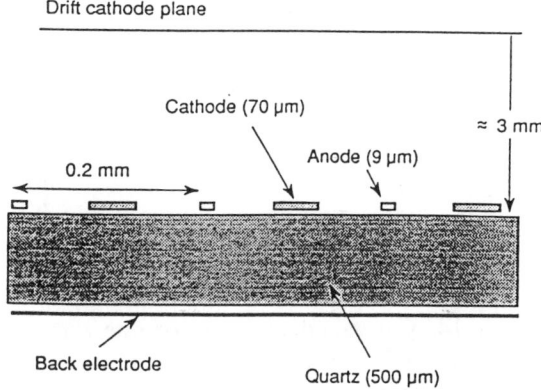

Figure 13. A schematic cross section of MSGC.

The high rate capability as illustrated in Fig. 14 is particularly worth of attention. This superior performance can be inferred from the electric field lines as shown in Fig. 15. Most of positive ions created in the avalanches at the vicinity of the anode travel the short distance to the adjacent cathodes. This point is in sharp contrast to the wire chambers in which positive ions must drift back to the cathodes in the opposite side across the gas ionization space, thus creating a space-charge problem. The combined effects of the finer granularity and shorter pass of positive ions results in two order of magnitude better rate capability than that of wire chambers. Therefore the superiority to proportional wire chambers seems clear in principle.

The advantages of MSGC in comparison with silicon microstrip detectors are (1) signals are typically an order of magnitude larger (10^5 electrons) and (2) it can cover large area rather inexpensively with conventional materials such as plastics while keeping a comparable high position resolution.

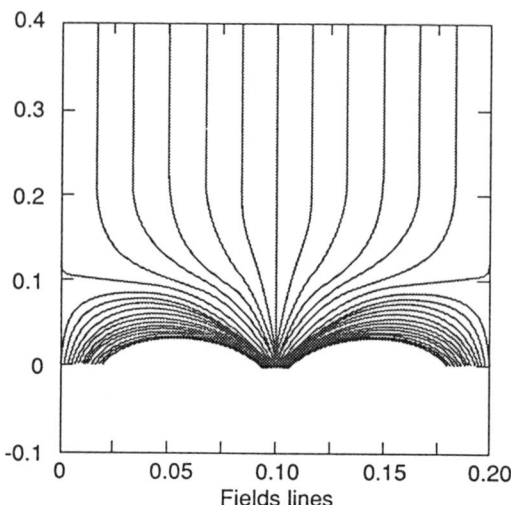

Figure 15. The field-lines of MSGC.[48]

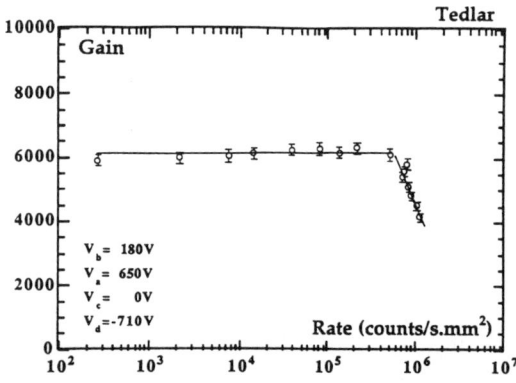

Figure 14. Gas gain dependence on rate with a Tedlar MSGC. The measurement time is several seconds to avoid surface charge-up effects.[50]

There remains, however, a crucial problem to be solved : choice of the substrate material. Many groups observed long-term and time-dependent gain changes. Fig. 16 shows a typical time dependence of relative gain under irradiation for three different substrate materials.[50] From the figure it is apparent that the gain stability strongly depends on the conductivity of the substrates. Most likely this effect is due to charge up of feedback ions on the surface of the insulating material between anode and cathode.

If above conjecture is true, there are two ways to provide the surface conductivity ($10^{11} \sim 10^{14}$ Ω/square) to suppress the charge up ; one to use of a material with medium conductivity and the other to enhance the surface conductivity of a high resistive material by for example ion implantation, chemical treatment or deposition of a thin semiconducting layer. In the former case, the leakage current may be the limitation, while in the latter case

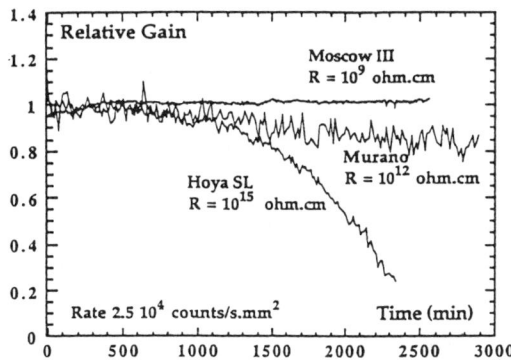

Figure 16. Gain variation as a function of time. R is the bulk conductivity of the substrate.[50]

the stability of surface conductivity might be a problem. So far tried are ion implanted quartz, semiconductive glasses, resistive plastics, and silicon with ion implanted oxide layer.

Another technically cumbersome problem is the protection against discharge. Thin metal strips are easily disrupted by a local discharge. For instance, heavily ionizing recoil protons in the gas generated by MeV neutrons that are abundant in supercollider detectors could initiate a discharge with permanent damages on strips. Two developments are made in this respect ; one is to lower the operating voltage substantially while keeping gas gain same by thinning the substrate to 2 μm and applying voltage in the back-electrode.[51] The other is to mask the edge of anode strips by insulating materials.[52]

Once these fundamental issues are solved, a variety of applications of MSGC can be envisaged. Especially two dimensional or pixel like readout is plausible using back electrode effectively. Application to large scale system for LHC/SSC is highly desirable because of intrinsic high rate capability. It is, however, a long way to proceed because of inadequate experiences as well as various engineering and mass production techniques to be developed.

Fast Gas with CF_4

High rate capability of drift chambers is pursued with gas containing CF_4 after the pioneering work by J. Fisher.[53] The electron drift velocity of CF_4 and hydrocarbon mixture can be 10 cm/μsec at relatively low E/P like 1 V/cm/torr with velocity saturation. It is a factor of two faster than that of conventional drift gas. For high rate use, however, two more factors must be considered ; the wire aging and the space charge effect due to positive ions. Gas mixture of CF_4/iC_4H_{10} showed no apparent wire aging up to 8 C/cm, an order of magnitude better than the conventional gas.[54] CF_4 seems to improve the aging performance drastically. A systematic study of drift properties has indicated that lighter hydrocarbons are better with CF_4 for electron drift velocity.[55] Besides it showed that the positive ion mobility becomes substantially slower as the higher mixing fraction of CF_4. Three fold mixtures of Ar, CF_4 and hydrocarbon are also found to be attractive.

Visible Light Photon Counters (VLPC)

Plastic scintillating fibers of about mm in diameter retain attractive features for fast tracking device with fine granularity. To accomplish a high tracking efficiency, it is essential to use a photodetector with very high quantum efficiency. A new device called VLPC has been developed as a variant of the solid state photomultipliers (SSPM) by the Rockwell International Science Center.[56] VLPC realizes quantum efficiency of 85% with electron gain of 10^4 at λ = 565 nm. Its application, however, is not trivial. The device must be operated at temperature of 6-7K in a liquid helium cryostat.

Accordion LAr/LKr calorimeter

LAr provides many advantages ; readout uniformity, gain stability, easy calibration, flexible granularity, radiation hardness, and insensitivity to magnetic field. But how to make the readout fast enough? How to make it hermetic? Fe/LAr or Pb/LAr calorimeters give moderately good hadronic energy resolution because of non-unity e/h ratios.

Slow charge collection can be improved by pulse differentiation and by locating amplifiers in liquid to minimize the readout inductance.[57]

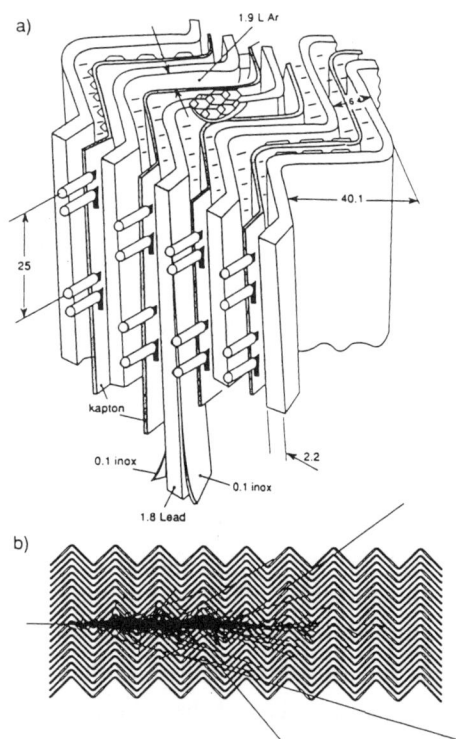

Figure 17. Liquid argon calorimeter with accordion geometry and development of 40 GeV shower in the calorimeter (Monte Carlo simulation).[58]

More aggressively, an accordion-shape electrode was proposed by RD3 collaboration.[58] The idea is to change the geometry of charge collection electrodes into transmission-type as illustrated in Fig. 17. Both absorber plates as well as the readout electrodes are folded into a zig-zag shape in the direction of shower development. Fast front-end electronics are directly mounted in front and at the back of the accordion electrodes. The transverse modulation of electron pulse height arising from accordion geometry is measured to be within ±1% with good agreement with GEANT simulation. The energy resolution with 40 ns peaking time is

$$\frac{\sigma}{E} = \frac{9.6 \pm 0.3\%}{\sqrt{E}} \oplus 0.3 \pm 0.1\% \oplus \frac{0.326 \pm 0.015}{E}$$

where E is in GeV.[59] The last term comes from electronic noise of 70 MeV/cell. A small prototype LAr preshower module was also tested to asses the realistic configuration of ATLAS for LHC. The LAr preshower is useful to measure the position of e and μ's with the resolution of 2 mm/$\sqrt{E}$. The combination of preshower and accordion calorimeter can determine the direction of e and γ's with angular resolution of better than 5 mrad at 40 GeV.

GEM group for SSC experiments tried to improve the energy resolution of EM accordion calorimeter with liquid krypton.[60] Fig. 18 shows the beautiful multiple trace for 110 events of LKr signal with 20 GeV electrons. Although the result is preliminary, they have observed the energy resolution of

$$\frac{\sigma}{E} = \frac{6.3 \sim 6.9\%}{\sqrt{E}} \oplus \frac{0.085}{E},$$

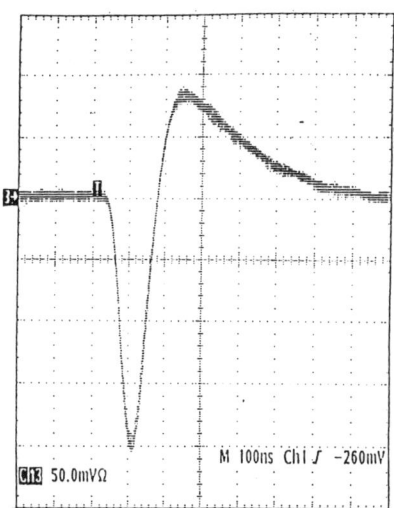

Figure 18. A multiple scope trace of signals for 110 events with 20 GeV electrons observed in the prototype LKr accordion calorimeter. [60]

better than the LAr case of 7.5-$7.8\%/\sqrt{E}$.

Crystal Electromagnetic Calorimetry

Another and quite different approach to attain good energy resolution for e and γ is to use homogeneous calorimeter. Recently scintillating crystals like BaF_2 and CeF_3 have drawn great attention since they are fast and radiation-hard. GEM proposes BaF_2 as an alternative option while CMS and L3P plan to use CeF_3 as a main option for EM calorimetry. Cerium Fluoride CeF_3 found by D. F. Anderson[61] carries many attractive properties for high rate application. It has two scintillation decay constants of 2 and 31 ns at 310 and 340 nm respectively. The slow component is 20 times faster than that of BaF_2. The crystal is tested to be radiation tolerant to about 10 Mrad.[62] It is also noted that a doped crystal $CeF_3(Ba\ 0.67\%)$ is tested to be good up to 100 Mrad.[63]

In this direction, some systematic studies of dense materials for EM calorimetry are progressing in Russia. For example, a new Cherenkov radiator $BaYb_2F_8$, fluoride glass scintillators (FG-1 to FG-5) are being studied at the Lebedev Physical Institute.[64]

VLSI Application

Without extensive use of VLSI, there is no way to handle millions of readout channels from detector elements. Especially candidates of inner trackers such as silicon, MSGC and straw detectors must be supported by electronics attached on the detector.

For example, RD2 group is developing a 66 MHz analog pipeline as a front end for LHC detectors.[65] 66 MHz is needed if one likes to record data synchronous to every collisions. The pipeline structure with depth of at least 1 µs is required to temporarily store all the data until the first level trigger is formed. After the trigger formation, the pipeline data are extracted out with sparse data scanning, multiplexing and analog to digital conversions.

More aggressive strategy is to digitize right at the front end. Although the front end becomes complex in this scheme, subsequent data storage, multiplexing, and transmission becomes much simpler and reliable. Most of the SDC electronics, for example, are planned in this direction.[66] The silicon detector is read-out by ASD's and digital buffers mounted right on the silicon detector. Calorimeter PMT's are read out by current splitters and FADC's followed by digital buffers.

One of noticeable progresses is a TMC (time memory cell) that will be used for straw chamber readout.[67] Its principle is simple. As shown in Fig. 19, the write clock propagates

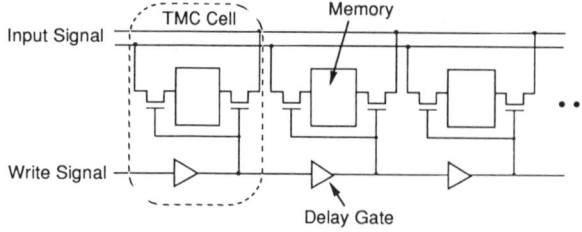

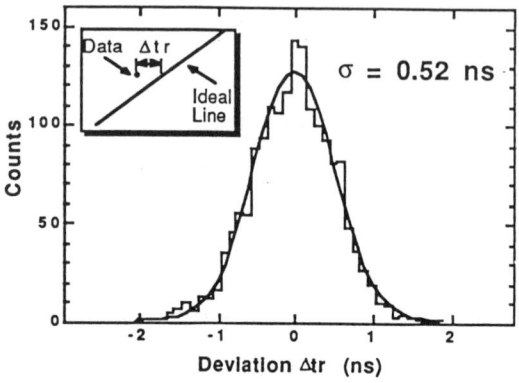

Figure 19. TMC circuit for time measurement: its principle (top) and the accuracy of time measurement with prototype chip TMC1004 with 1 μs pipeline depth.[67]

through 1 ns delay circuits to consecutive memory units. Thus discriminated chamber pulses are directly mapped into memories with 1 ns steps. No deadtimes are introduced and its linearity is excellent. Use of 0.8 μm rule CMOS technology makes the power consumption extremely low like 7 mW/ch.

Networks and Event Building

Rapid progresses in computing power and communication will soon make it possible for scientists to have uniform and unimpeded access to powerful computing systems and data resources regardless of their geographic location. Using a 45 Mbits/s cross-country network, LBL and Pittsburgh Super-computer Center begun a pilot project on interactive communication of 3-dimensional imaging pictures like human brains.[68]

The other challenge is a Gigabit data acquisition system for SSC/LHC. As an example, SDC proposes to employ a scalable parallel event builder. This is a new concept to high energy experiments. Data pieces from detector elements are transmitted in parallel from underground hall to the surface. The event builder converts segmented parallel data into one serial event lump. Since typical event size is 1 Mbytes and the maximum trigger rate is 10 KHz, the event builder must handle up to 10 Gbyte/sec, an aggressive rate in today's standard. The heart of the event builder is a switching network with high bandwidth barrel shifters. Fig. 20 shows the principle of event building schematically. A test bed is constructed based on 16 channel Fibre Channel data switch at LBL,[68] while a prototype 4 GHz switching module is tested at KEK as shown in Fig. 20.[69]

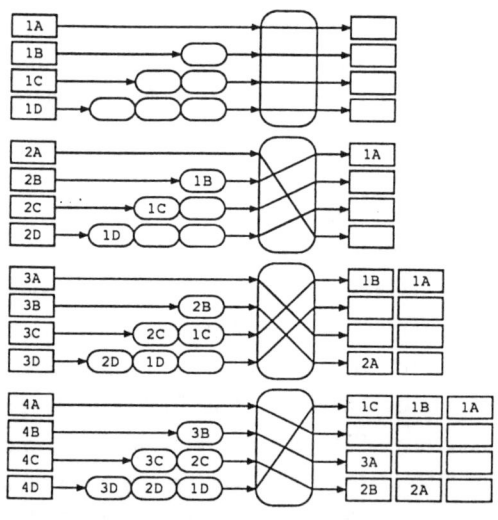

Figure 20. Principle of the event building based on barrel shift switching network. The parallel data are assembled into sequence packets via simple operation of high speed switches.[69]

LARGE DETECTORS FOR SSC AND LHC

In 1987, US DOE started funding for generic detector R&D toward SSC experiments which eventually grew into subsystem R&D for large detector. In late 1990, three letters of intent for large SSC detectors were submitted. At present, two large detector plans SDC (Solenoidal Detector Collaboration) and GEM(Gammas, Electrons and Muons) are getting into the stage of serious engineering design.

A detector R&D committee was formed at CERN to organize the European efforts in 1990. In the general meeting on LHC at Evian March 1992, four expressions of interest for large detectors, three for B-physics and three for heavy ions physics were presented. Letters of intent for large LHC detectors are being prepared ; CMS (Compact Muon Solenoid), ATLAS and L3P.

It is of no use to describe these proposals here, since the designs are not finalized and some early-stage proposals are confusingly full of options. Besides the design details change frequently anyhow. Table 3 lists up the major concepts of these detectors.

Table 3. Design concepts

detector	principal design concepts
SDC	general, large tracking volume
GEM	precision μ, good e and γ
CMS	optimize to μ detection with 4T field
ATLAS	general, precision EM calorimeter
L3P	place all detectors far from I.P.

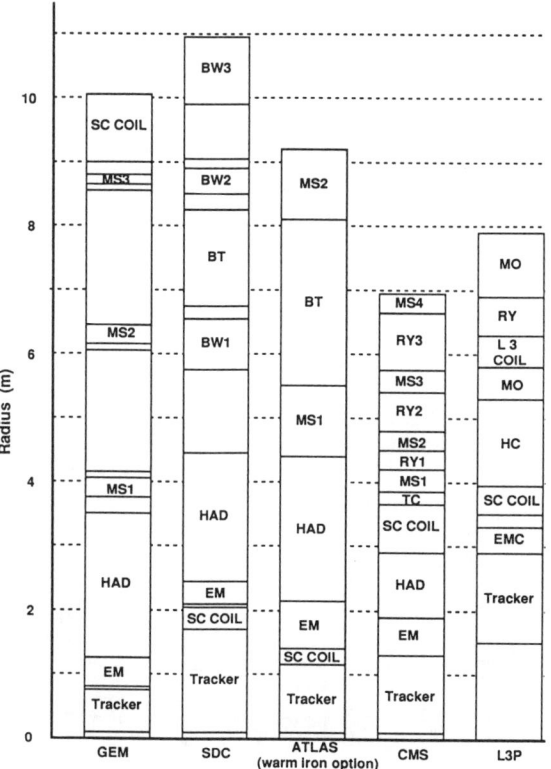

Figure 21. Comparison of proposed detectors for SSC and LHC in the cross section at 90°.

It is worth pointing that, for tracking, all of detector proposals employ the solenoidal field generated by superconducting magnets. Except GEM, all detector challenge to the central magnetic field of 2 Tesla or above. The high field is certainly better not only to improve the momentum resolution but also to confine the abundant soft charged particles into small radii from the beam line.

Somewhat interesting features in these detector proposals to be noted out are;

<u>GEM</u>: all the detectors are inside a large solenoid with no magnetic return yokes. The endcap has a B-field shaper to enhance the momentum resolution in the forward region.

<u>ATLAS</u>: The solenoid and LAr calorimeter share the common cryostat to minimize the

material in front of the EM calorimeter. The preshower detector effectively fills and equalizes the materials in front of EM calorimeter.

<u>CMS</u>: The overall detector size is determined deliverability of the solenoid to CERN.

It is noted that there are differences in approach to the detection of $H \to \gamma\gamma$. This decay mode recently attracts our attention because of super symmetric GUTS model, indicating rather low mass Higgs particles:

GEM : direct H detection with high resolution EM calorimetry with vertex tagging,

SDC : use of associated H production WH, t$\bar{t}$ H $\to \gamma\gamma$ lepton X,

CMS , ATLAS : direct H detection with high resolution EM calorimetry with γ angle measurement,

L3P : direct H detection with high resolution EM calorimetry far away from the IP.

In Fig. 21 the compositions at 90° are plotted with the same scale for comparison. The

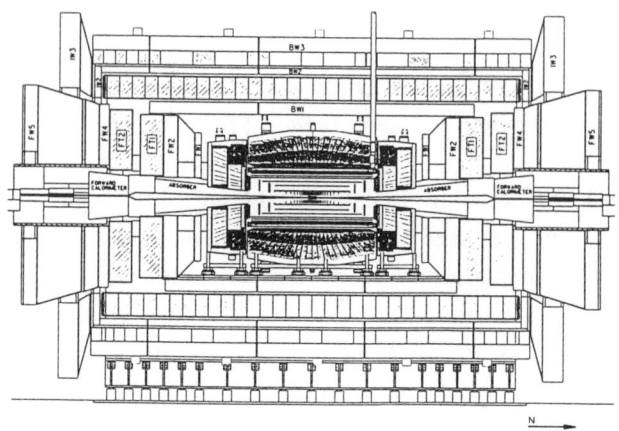

Figure 23. SDC detector for the SSC.[71,66]

schematic pictures of proposed large detectors are shown in Figs. 22-26. For details, the lists of the detector performances, magnets, inner tracking devices, calorimetry, muon detectors as well as number of readout channels are given in Tables 4-8, based on the information available. In the tables, the notation / indicates an alternative option, while [---] indicates a backup or future replacement.

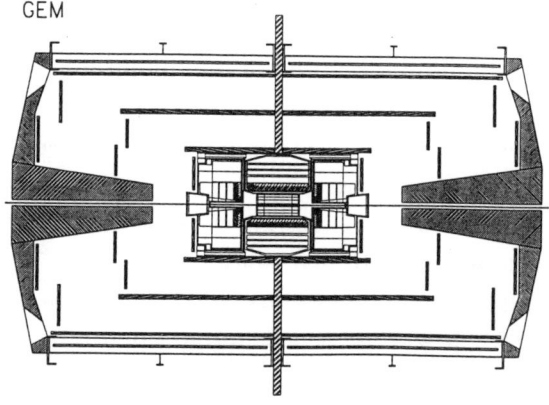

Figure 22. GEM detector for the SSC.[70]

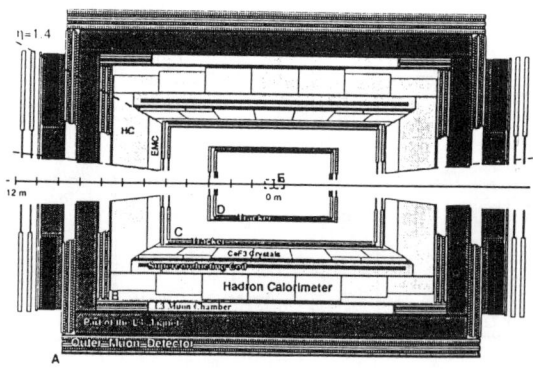

Figure 24. L3P detector for the LHC.[74]

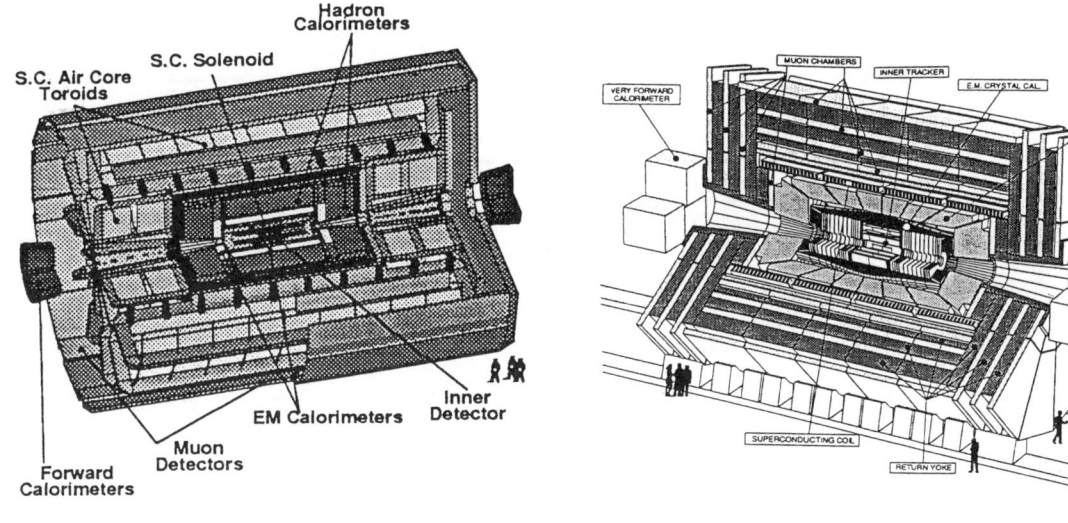

Figure 25. ATLAS detector for the LHC (air toroid option).[72]

Figure 26. CMS detector for the LHC.[73]

Table 4. Expected detector performances

detector	$\Delta p_t/p_t$@90°	$(\Delta p_t/p_t)_\mu$@90° alone	$(\Delta E/E)_{EM}$	$(\Delta E/E)_{HAD}$	$(\Delta E/E)_{FWD}$
SDC	16% @1TeV	25% @1TeV	14%/$\sqrt{E}\oplus$1%	60%/$\sqrt{E}\oplus$4%	100%/$\sqrt{E}\oplus$8%
GEM	1-3×10^{-3}p$_t$ $\oplus$ 2-4%	9% @1TeV	10%/$\sqrt{E}\oplus$0.3% or 2%/$\sqrt{E}\oplus$0.5%	50%/$\sqrt{E}\oplus$2%	10%
CMS	1.2% @100GeV	3% @0.4TeV(cmb)	<3%/$\sqrt{E}\oplus$0.5%	80%/$\sqrt{E}\oplus$<3%	110%/$\sqrt{E}\oplus$9%
ATLAS	5×10^{-4}p$_t\oplus$1%	<20% @1TeV	10%/$\sqrt{E}\oplus$1%	50%/$\sqrt{E}\oplus$3%	100%/$\sqrt{E}\oplus$7%
L3P	0.36% @100GeV	14% @100GeV	0.94%/$\sqrt{E}\oplus$0.14%	100%/$\sqrt{E}\oplus$10%	100%/$\sqrt{E}\oplus$10%

(note : a $\oplus$ b implies $\sqrt{a^2+b^2}$, alone = stand-alone resolution, cmb = combined resolution)

Table 5. Detector magnets

detector	magnetic field	coil radius	stored energy	radiation thickness	magnet for muon
SDC	2.0T	1.8m	143 MJ	1.2 X_0	iron toroids
GEM	0.8T	8.9m	2040 MJ	thick	common solenoid
CMS	4.0T	2.95m	2850 MJ	1.1 λ_I	field in return yokes
ATLAS	2.0T	1.23m	50 MJ	0.92 X_0	air-core / iron toroids
L3P	2.0[3.0]T	3.8m	900[1900]MJ	0.5[0.7] λ_I	field in return yokes

Table 6. Inner tracking detectors

detector	vertex	barrel detector	forward detector
SDC	silicon	straw tube/scintillating fiber	MSGC
GEM	silicon	straw+scintillating fiber/interpolating gas chamber	silicon
CMS	silicon	MSGC/scintillating fiber	MSGC
ATLAS	silicon	silicon + straw-TRD	GaAs, MSGC, straw-TRD
L3P	none	drift tubes+MWPC	MSGC, MWPC

(MSGC : microstrip gas chamber, TRD : transition radiation detector)

Table 7. Calorimetry

detector	EM calorimeter	hadron calorimeter	forward calorimeter
SDC	scintillator-tile fiber	Fe-scintillator-tile fiber	liq-scinti /high pres gas
GEM	Pb-LA/LKr accordion /BaF$_2$	SS-LA-EST / scintillating fiber	LA/liq-scinti/high p gas
CMS	CeF$_3$[Pb-Scinti]	Cu tile-fibers[Cu-Si]	SS-parallel plate cham.
ATLAS	Pb-LA accordion[LA TGT]	Fe-LA-EST /scinti-fiber/scinti-tile	liq-scinti/high pres gas
L3P	CeF$_3$[LKr]	Fe-MWPC	LA+MWPC

(TGT: thin gap turbine structure, EST : electro-static transformer, SS : stainless steel)

Table 8. Muon detection systems

detector	muon tracking detector	trigger detector
SDC	field-shaped drift tube	scintillator
GEM	pressurized drift tube / limited streamer drift tube cathode strip chamber (endcap region)	resistive plate chamber
CMS	drift tube with bunch crossing id capability (DTBX) / wall-less drift chamber(WLDC) / honeycomb strip chamber(HSC) / cathode strip chamber(CSC)	resistive plate chamber / parallel plate chamber
ATLAS	high pressure drift tube / honeycomb strip chamber(HSC) / jet cell chamber(JCC)	chamber stand-alone / resistive plate chamber
L3P	field-shaped drift tube	resistive plate chamber

(/ indicates alternative choices)

Table 9. Number of readout channels

detector	vertex	tracking detector	calorimeters	muon chambers	μ trigger
SDC	5.8M	137K(straw)/473K(scifi) /1.4M(MSGC)	20K(calorimeter), 47K(shower max)	90K	4.5K
GEM	3M	400K(pads)	63K(LA) / 16K(BaF_2),11K(sci)	158K(barrel)+ 252K(endcap)	39K(RPC)
CMS	7-10.5M	10M(MSGC)	243K(CeF_3) 8.8K(scint)	265K(DTBX),242K(CSC)/310K(WLDC)/ 914K(HSC)	140K(RPC) /160K(PPC)
ATLAS	4.2M(Si) 0.8M(GaAs)	2.7M(Si),5M(MSGC), 350K(TRD)	141K(LA), 263K(LA preshower)	190K(JCC)	?
L3P	none	470K(straw),850K(PWC), 172K(MSGC)	200K(EM), 8K(HAD)	70K(drift), 180K(PWC)	200K

These figures and tables tell us that, although the approaches are quite different, the trend is clear. The supercollider detectors become truly bigger in size and weight and complex in detector mechanics and readout electronics. We are entering into a new era.

I would like to conclude this section for the large detector system by quoting a statement by a physicists who experienced an embarrassing long design and construction period of a large detector ;

"A perfect detector like a perfect spouse exists only in dreams; reality imposes some compromises. However, the detector design must be driven by the principal physics goals and not by technological hubris. To overstate the case, one should try to always use obsolete technology.
. . . .
The number of technologies utilized in the detector should be minimized in order to simplify both the operations and the data analysis: not to speak of number of experts needed to operate the detector efficiently......"[68]

SUMMARY AND FUTURE OUTLOOK

New large detector systems CMD-2, SLD, D0, H1, ZEUS started successfully to operate for physics in the past two years. Several proved-in-principle technologies are practically applied to large detectors. These are RICH /CRID, TRD, liquid argon TPC, CCD, and silicon microstrip detectors. The operational experience on these detectors is a good lesson for the future.

The experimental environment for high-luminosity hadron supercolliders is extremely severe. R&D on new technology aiming at these colliders progressed substantially. Efforts on R&D and detector designs are now shifting from the stage of proof-of-principle to the engineering study and financial reality. Large detector plans are rapidly making progress toward new frontier physics at supercolliders of SSC and LHC, as well as at Φ, τ/C, B-factories.

Acknowledgments

The author wishes to express his gratitude to Dr. Makoto Takashima for assisting his preparation of the summary talk at the conference. Thanks are also due to Prof. J. B. Dainton for his kind advice.

REFERENCES

1. B. Khazin, "Preliminary results from the CMD-2 detector," presented to this conference.

2. L. Jin, " Status of BES detector at BEPC," presented to this conference.

3. P. Franzini, "KLOE, A detector for DAΦNE," presented to this conference.

4. M. Fukawa et al., "Physics at a φ-factory," KEK Report 90-12, August 1990, and
   T. Ohshima, "The KEK φ-factory project," KEK Report 91-13, April 1991.

5. B. Ratcliff, "Status of the B factory detector for PEP-II: a progress report," presented to this conference.

6. Y. Giomataris, "The optical trigger for beauty and perspectives," presented to this conference and G. Charpak, Y. Giomataris and L. Lederman, NIM A306(1991)439.

7. R.H. Schindler, "Status of the Tau-Charm project and aspects of the tau-charm detector design," presented to this conference.

8. J. Va'vra, "The first results from CRID," presented to this conference.

9. H.-W. Siebert, "The RICH counter in the CERN hyperon beam experiment," presented to this conference.

10. P. Baillon, "Barrel ring imaging cherenkov detector in DELPHI," presented to this conference.

11. M. Sakuda et al., NIM A311(1992)57, N. Terunuma et al., NIM A323(1992)471 and Y. Fukushima et al., submitted to IEEE 1992 NS Symposium.

12. A. Bettini, "The ICARUS experiment:status & program," presented to this conference and A. Bettini et al., NIM A315(1992)223.

13. K. De, "Test beam studies of the D0 calorimeter with 2-150 GeV beams," presented to this conference.

14. F. Brasse, "Detector H1 for ep collisions," presented to this conference.

15. H.J. Kim, "Beam tests of the ZEUS barrel calorimeter," presented to this conference.

16. C. Damerell, "Design and performance of the SLD vertex detector, a 120 Mpixel tracking system," presented to this conference.

17. B. Hyams et al., NIM 205(1983)99.

18. M. Caccia, "The DELPHI microvertex detector," presented to this conference.

19. F. Bedeschi, "Operation of the CDF silicon vertex detector with colliding beams at Fermilab," presented to this conference.

20. D. Cinabro, "Silicon vertex detector in CLEO II," presented to this conference.

21. P. Avery, "The UFMulti II project," presented to this conference.

22. N. Katayama, "Object oriented approach to B reconstruction," presented to this conference.

23. G. Marchesini, "Perturbative QCD → MC simulation," presented to this conference.

24. W. Zeuner, "Comparisons of properties of hadronic Z0 decays with QCD shower models," presented to this conference.

25. Z. Was, "Monte Carlo programs for t physics," presented to this conference.

26. H. Jung, "Physics simulations at HERA," presented to this conference.

27. H. Kuhlen, "The fast H1 calorimeter simulation," presented to this conference.

28. S. Sadovsky, "Fourier parametrization of the multi-dimensional Monte-Carlo efficiency," presented to this conference.

29. H. Grassman, "CDF and physics simulations," presented to this conference.

30. J. Womersley, "The D0 Monte Carlo," presented to this conference.

31. F. Paige, "Physics simulation for SSC and LHC," presented to this conference.

32. X. Wang, "Heavy ion physics simulations," presented to this conference.

33. J. Ranft, "Soft Physics simulations," presented to this conference.

34. Task Force Report, "Radiation Levels in the SSC Interaction Regions," *SSC-SR-1033*, 1988. Erratum in SSCL-285, 1990.

35. M.G.D. Gilchriese, Editor, "Radiation Effects at the SSC," *SSC-SR-1035*, 1988.

36. D. E. Groom, "Radiation Levels in SSC Detectors," Snowmass 1988, p711.

37. T. Kondo et al., Snowmass 1984, p612,
    T. Ohsugi et al., NIM A265(1988)105,
    M. Hasegawa et al., NIM A277(1989)395.

38. H. Dietl et al., NIM A253(1987)460.

39. M. Nakamura et al., NIM A270(1988)42,
    P. Pitzl et al., NIM A311(1992)98.

40. F. Lemeilleur et al., IEEE NS 39(1991)551.

41. Z. Li and W. Kraner, IEEE NS 39(1991)577.

42. P. Holl et al., IEEE NS 36(1989)251,
    H. Becker et al., IEEE NS 37(1990)101.

43. P. Skubic, "Beam and source tests of double-sided silicon microstrip detectors," presented to this conference.

44. H.J. Ziock et al., IEEE NS 38(1991)269.

45. C. Buttar, "GaAs solid state detectors for high energy physics," presented to this conference.

46. A. Oed, NIM A263 (1988)351.

47. F. Angelini et al., INFN PI/AE 89/2(1989).

48. F. Angelini et al., NIM A315(1992)21.

49. F. Hartjes et al., NIM A289(1990)384.
    X.F. Angelini et al.,IEEENS 37(1990)112,
    X.F. Angelini et al., Nuclear Physics B (Proc. suppl.) 23A(1991)254-260,
    R. Bouclier et al., NIM A315(1992)521,
    M. Geijsberts et al., NIM A313(1992)377.

50. R. Bouclier et al., IEEE NS 39(1992)650 and CERN-PPE/92-53.

51. F. Angelini et al., NIM A314(1992)450 and T. Bowcock, "Operation of thin effective substrate MSGC's," presented to this conference.

52. T. Tanimori et al., "Development of microstrip gas chamber and application to imaging gamma-ray detector," INS-Rep.-939, July 1992.

53. J. Fisher et al., NIM A238(1985)249.

54. R. Openshaw et al., IEEE NS 36(1989)567.

55. T. Yamashita et al., NIM A283(1989)709 and KEK preprint 91-141(1991).

56. M.D. Petroff and M.G. Stapelbrock, IEEE NS 36(1989)158 and M. Atac et al., NIM A320(1992)155.

57. V. Radeka and S. Rescia, NIM A256(1988) 228.

58. RD3 Collaboration, NIM A309(1991)438,
    B. Aubert et al., NIM A315(1992)285.

59. M. Nessi, "The Accordion Liquid-Argon Project at CERN," presented to this conference.

60. D. Lissauer, "The 1992 GEM LAr/Kr accordion calorimeter group," presented to this conference.

61. D.F. Anderson, IEEE NS 36(1989)137 and NIM A287(1990)606, W.M. Mosen and S.E. Derenzo, IEEE NS 36(1989)173.

62. M. Kobayashi et al., NIM A302(1991)443.

63. A.A. Aseev et al., NIM A313(1992)340.

64. A.A. Aseev et al., "New dense, radiation hard materials for electromagnetic calorimetry," contribution to this conference.

65. R. Bonino., "Implementation of a 66MHz analog pipeline as a front end for LHC detectors," presented to this conference.

66. SDC Technical Design Report, SDC-92-201, April 1, 1992.

67. Y. Arai and T. Ohsugi, Snowmass 1986, p455 and Y. Arai et al., IEEE NS 39(1992) 784.

68. S. Loken, "High performance computing and distributed system," presented to this conference.

69. E. Barsotti et al., IEEE NS 37(1990)1216, O. Sasaki et al., submitted IEEE 1992 NS symposium.

70. B. Barish, "GEM" presented in the parallel session of this conference,
GEM Letter of Intent, November 30, 1991

71. T. Kirk, "The Solenoidal Detector Collaboration: General purpose detector & physics goals," presented to this conference.

72. A. Parker, "ASCOT/EAGLE collaboration : General-purpose LHC experiment," presented to this conference,
ATLAS Letter of Intent, CERN/LHCC 92-4, LHCC/12, 1 October, 1992.

73. T. Virdee, "CMS Physics potentials," presented to this conference,
CMS Letter of Intent, CERN/LHCC 92-3, LHCC/11, 1 October, 1992.

74. H. Hofer, "Future of L3 at LHC, L3P," presented to this conference
L3P Letter of Intent, CERN/LHCC 92-5, LHCC/13, 1 October, 1992.

75. M. Derrick, "Calorimetry in ZEUS: Lessons for the future," Snowmass 1988, 791.

Abbreviations used in the references are

NIM : Nuclear Instruments and Methods in Physics Research
IEEE NS : IEEE Transactions on Nuclear Science
Snowmass : Proceedings of the Summer Study organized by APS-DPF.

DISCUSSION

A. Weidberg, Oxford University

A comment on radiation levels at the LHC: a 5 cm polyethylene moderator can reduce the neutron flux by a factor of 10. Tests of irradiated counters show no degradation in performance after a few times 10^{13} n/cm^2. This shows that Si counters can remain operational at the LHC for many years.

Kondo

An SDC study found a reduction of merely a factor of 2, although as noted by Weidberg, detailed studies at CERN showed more reduction factors by moderators, indicating more usefulness of silicon detectors at hadron colliders.

G. Belletini, Sezione Infn di Pisa

You showed an interesting ϕ peak obtained at the Novosibirsk low-energy e^+e^- collider. Do you have a realistic figure for the integrated luminosity in one year?

B. Khazin, Budker Institute of Nuclear Physics

During 1992 the integrated luminosity is expected to be approximately 70 pb^{-1}.

STATUS OF HERA AND THE FIRST RESULTS

B.H.Wiik

II. Institut für Experimentalphysik, Univ. Hamburg, and
Deutsches Elektronen-Synchrotron DESY,
2000 Hamburg 52 - Notkestrasse 85 - Germany

Abstract

The paper is divided into two parts. In the first, the overall layout of HERA and its main parameters are discussed, followed by a brief review of the preaccelerators. The status of the electron and proton rings is followed by the performance of HERA as a collider. The second part opens with a description of the two detectors, H1 and ZEUS, followed by a discussion of experimental conditions. The paper ends with a presentation of the first physics results from both experiments.

1. INTRODUCTION

During the past few weeks the electron-proton collider HERA[1,2] and its large, multipurpose detectors H1[3] and ZEUS[4] have made the transition from a virtual to a real source of data obtained in a new kinematic region.

HERA is the first electron-proton collider ever and it has been constructed within the framework of a novel form of international collaboration where institutions in Canada, China, Czechoslovakia, France, Israel, Italy, Netherlands, Poland, United Kingdom and USA have contributed either components built at home or delegated skilled manpower to work on the project at DESY.

Also the large detectors H1 and ZEUS have been built and will be exploited by an international collaboration. Only some 25% of the 750 physicists presently involved in the programme are from DESY or German universities while the remainder come from 64 institutions in 15 countries.

Thus the construction of HERA and its detectors has been, and the exploitation of the physics will be a truly international effort.

The HERA project was authorized in April 1984 and its construction completed in November 1990 in accordance with the original time schedule. The first electron beam was stored in April 1988 and the first proton beam in August 1991. Electron-proton collisions were observed for the first time in October 1991.

After extensive detector testing with cosmic rays both H1 and ZEUS were installed in the ring during the 1991/92 winter shutdown. The commissioning of HERA resumed in April 1992 and by the end of June luminosity was made available to the experiments.

In the first part of my talk I'll briefly remind you of the HERA layout and parameters. After a short outline of the pre-accelerators, I'll discuss the performance of the HERA electron and proton rings before reviewing its performance as an ep collider.

In the second part I'll first focuss on the detectors, then discuss the experimental conditions and finally present first physics results. Both experiments report[5,6] deep inelastic neutral current data for small values of x, and a measurement of the total photoproduction cross section around 200 GeV in center of mass. Preliminary lower limits on the mass of leptoquarks and on excited states of the electron have also been extracted from the data.

I'll close my talk with a brief summary and a look at the future.

2. HERA

2.1 Overview

HERA is made of two independent accelerators designed to store respectively 820 GeV protons and 30 GeV electrons and to collide the two counterrotating beams head on in four interactions points spaced uniformly around its 6.3 km long circumference.

The main parameters of the HERA rings are listed in Table 1 and the layout of the accelerator complex is shown in Fig. 1.

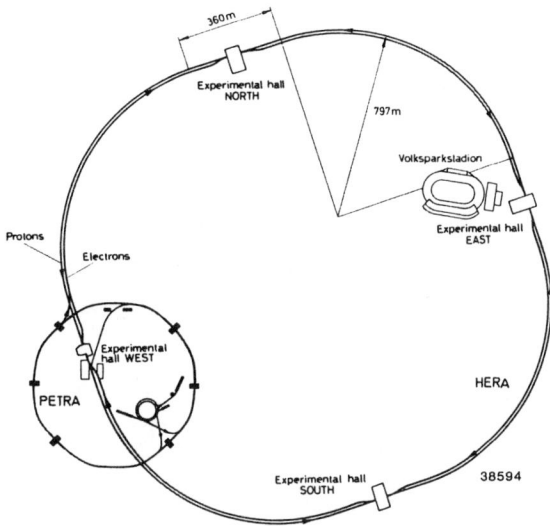

Fig. 1 - The layout of the HERA accelerator complex.

	p-ring	e-ring	units
Nominal energy	820	30	GeV
Polarization time		28	min
Luminosity	1.5×10^{31}		$cm^{-2}s^{-1}$
Space between IR Quad	15		m
Interaction points	4		
Crossing angle	0		mrad
Circumference	6336		
Magnetic field	4.68	0.165	T
Number of particles	2.1	0.8	10^{13}
Number of bunches	210		
Injection energy	40	14	GeV
Filling time	20	15	min
σ_x/σ_y at I.P.	0.29/0.07	0.26/0.07	mm
σ_z at I.P.	110	8.0	mm
Energy loss/turn	6.24×10^{-6}	127	MeV
Circumferential RF voltage	0.2/2.4	260	MV
RF-frequency	52.033/208.13	499.776	MHz
RF-power	1	13.2	MW
Refrigerator	21.0 kW (isothermal at 4.3K) 60 g/s Liq.He 60 kW (40 K - 80 K)		

Table 1 - Main HERA Parameters.

The installed accelerators in the arcs are depicted in Fig. 2. The proton ring is installed 81 cm above the electron ring in the arcs. In the straight sections the proton beam is deflected into the plane of the electron ring. In the region adjacent to the interaction point the protons are guided by normal conducting magnets.

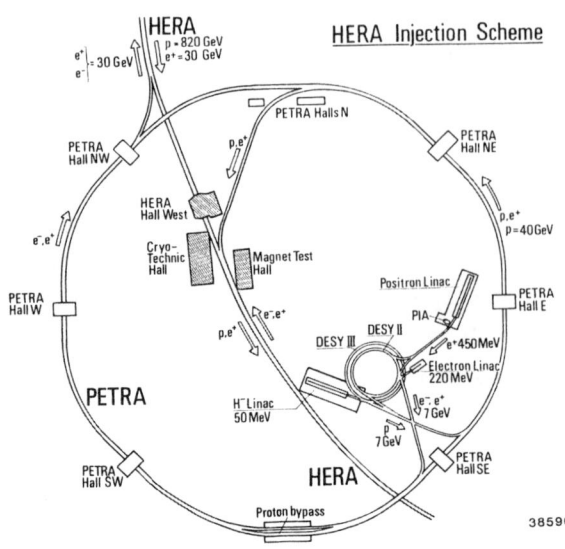

Fig. 3 - The layout of the HERA preaccelerators.

Fig. 2 - The installed HERA electron and proton ring in the arc.

The H1 and ZEUS detectors are installed in straight section North, respectively South. Straight section East has been reserved for new experiments. Straight section West is presently being used as a general utility section incorporating the electron and proton injection systems, the proton beam dump, proton collimators and RF systems.

2.2 *The Preaccelerators*

The layout of the HERA preaccelerators is shown in Fig. 3.

The electron (positron) injection complex is based on available accelerators. Electrons (positrons) from a 500 MeV linear accelerator are injected into a small storage ring, accumulated into a 60 mA single bunch, injected on axis into DESY II, accelerated to 7 GeV and transferred to the modified PETRA II ring. This is repeated at a rate of 12.5 Hz until PETRA II has been filled with 70 bunches each with $0.4 \cdot E11$ electrons (positrons) spaced 28.8 m apart as in HERA, which are then injected into the HERA main ring. This is repeated 2 times until HERA has been filled with 210 bunches. The expected 15 min filling time of the electron ring is dominated by the cycle time of PETRA II. The threshold for multibunch instabilities in PETRA II has been raised from 2.6 mA to the design current of

58 mA, by using a new longitudinal and transverse feedback system.

A whole new chain of preaccelerators has been built in order to inject protons into HERA.

The 50 MeV linear accelerator for negatively charged hydrogen ions is in routine operation and delivers a 6 mA beam to DESY III.

The negatively charged hydrogen ions are stripped upon their injection into DESY III. After a multiturn injection, protons are captured into 11 buckets, spaced 28.8 m apart as in HERA, accelerated to 7.5 GeV, and transferred to PETRA II. Roughly $1.2 \cdot 10^{12}$ protons in 11 bunches reach 7.5 GeV compared to the design intensity of $1.1 \cdot 10^{12}$ protons.

The maximum number of 70 bunches with roughly 40% of the design current has been accummulated in PETRA II and accelerated to the final energy of 40 GeV and transferred to HERA.

2.3 Status of the Electron Ring

While the first run of the electron ring in August 1988 was mainly used to commission technical components, a second run in September 1989 focussed on the performance of the ring. In this run beam currents of 3 mA could be injected, accelerated to 27.5 GeV and stored with a life-time of several hours. The maximum electron energy was limited by the available RF voltage. The commissioning was greatly eased by the reproducibility and stability of the electron ring.

At present the conventional RF system has been augmented by a set of 16 four cell Nb 500 MHz cavities assembled pairwise into 8 cryostats. These cavities have been industrially produced from high purity Niobium (RRR = 300). At the design current of 58 mA the gradient is limited to 2.05 MV/m due to the 100 kW power rating of the input coupler. New couplers rated at 200 kW are now being tested. Without beam all cavities reached 5 MV/m. Initially the design Q-value of $2 \cdot 10^9$ could not be reached in all cases, due to the presence of Nb_xH_y precipitation on the Nb surface.

The beam has been accelerated to 30.3 GeV with both the normal and superconducting RF systems working in parallel. The superconducting system was stable and reliable and it provided roughly 1/4 of the circumferential RF voltage of 160 MV.

The maximum single bunch current was 2.5 mA nearly one order of magnitude above its design value.

The maximum current which can be stored with lifetimes on the order of a few hours is presently limited to roughly 4 mA at 27 GeV. This limit may be due to dust particles trapped in the strong field of the circulating electron beam. A run with a positron beam is planned for the autumn.

The electron beam polarization is determined[7] by measuring the up-down asymmetry of backscattered polarized laser light using a tungsten scintillator sandwich counter.

The observed transverse beam polarization is shown in Fig. 4 as a function of time. The buildup time of 25.8 min is consistent with the measured maximum beam polarization of 58 ± 5%.

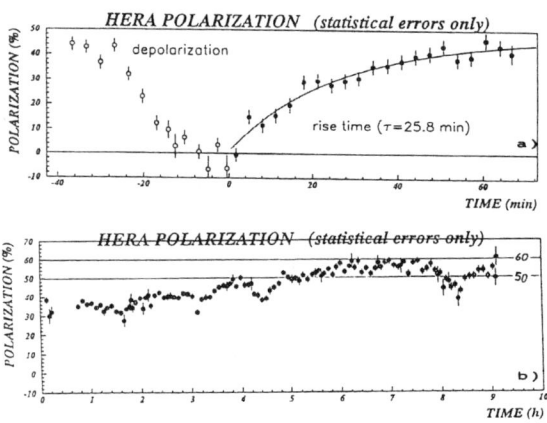

Fig. 4 - a) The transverse electron beam polarization as a function of time. The build up of transverse polarization is clearly seen. b) Beam polarization during a long storage.

A pair of 60 m long magnetic spin rotators designed to turn the vertical spin direction into a state of well defined helicity (or the inverse) have been built and are ready for installation in straight section East.

2.4 Status of the Proton Ring

The cryogenic system and the superconducting magnets are the most challenging components of the HERA proton ring. I will first describe the status and the performance of these components and then discuss the results obtained during the first commissioning runs.

2.4.1 The Refrigeration System

The central refrigerator is located on the DESY site. It is subdivided into three identical plants each providing 6.6 kW isothermally at 4.3 K, 20.4 g liquid helium per second and 20 kW at 40 K to 80 K.

A detailed description[2] of the cryogenic system and its performance can be found elsewhere. Here we only summarize the main results.

The cryogenic plant is very reliable as demonstrated by the ratio of uptime to scheduled time of 0.985. Each of the three coldboxes has now run for about 20000 hrs.

The octant cooldown time of 140 hrs was determined by the condition that mechanical stresses caused by the temperature gradient within a magnet should be limited to 100 MPa which is less than 50% of the critical value. Note that all octants can be cooled down in parallel.

Semiconductor temperature sensors are installed in the single-phase helium volume of each dipole, allowing to monitor the temperature with a precision of 0.02 K. A measurement of the heat load, using these monitors at a known mass flow, yielded a heat load of 5.1 kW at 4.4 K and 28.5 kW at the shield level for the whole ring, including transferlines and feedboxes. These values compare favourably to the proposal values. The heat leak of the magnets alone is 3.6 KW at 4.4 K and 19.6 kW at the shield level, in excellent agreement with heat load measurements on single magnets.

At a current of 5020 A an additional heat load of 21 W at 4.4 K per octant was observed.

The maximum 1-phase helium temperature occurs in the magnets installed immediately downstream of the main current leads. In these magnets the temperature may reach 4.5 K compared to a critical temperature of 6 K at the nominal field of 4.65 T.

The magnet recovery time after a quench is quite short as shown in Fig. 5. After a quench of four magnets at the nominal current of 5000 A it takes 20 min to cool the magnets back down from 20 K to 5 K and another 60 min to reach the operating temperature of 4.4 K.

These quenches were induced by firing heater strips installed adjacent to the dipole coils. Also note that quenches do not propagate into non triggered magnets.

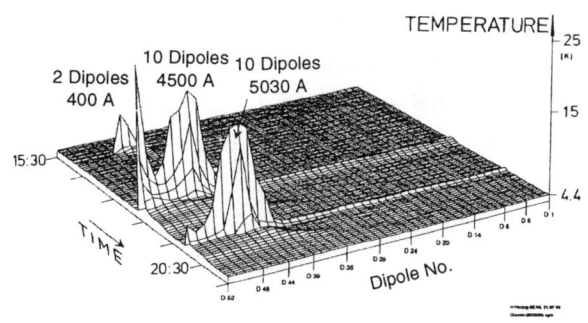

Fig. 5 - The temperature variation along the dipole magnets in an octant is plotted versus time. Groups of magnets are quenched by firing heater strips installed adjacent to the cold coil.

2.4.2 The Superconducting Magnets - A total of 2156 superconducting magnets and correction coils has been installed in the HERA proton ring. In the arcs the superconducting magnets are arranged in 104 cells with an ordering shown in Fig. 6.

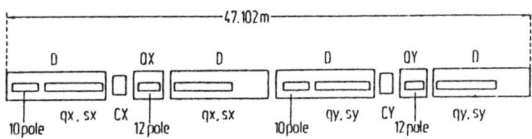

Fig. 6 - A unit cell of the proton ring. D: main dipoles; QX, QY: main quadrupoles, qx, qy: quadrupole correction coils, sx, sy: sextupole correction coils, CX, CY: correction dipoles. In addition, there are 10-pole and 12-pole correction coils.

The design and the performance of the magnets have been reported[2] elsewhere. Here we just summarize the main results. The industrial production of superconducting magnets was a success. During series production, DESY received an average of 8 dipoles and 6 quadrupoles per week, exceeding the contractual rates. Out of 449 dipoles and 246 quadrupoles, only 5 magnets were rejected, four of which had shortened windings and one a bad spot in the superconductor.

All magnets were tested at liquid helium temperature and the results can be summarized as follows:

Nearly 93% of all magnets reached the short sample current at the first or second excitation cycle. None of the magnets quenched below the nominal operating current.

Adjusted to an operating temperature of 4.4 K the average quench current was (6900 ± 130) A for the dipoles and (7840 ± 160) A for the quadrupoles. Only magnets with a quench current above 6500 A were installed in HERA.

The field quality of both the dipole and the quadrupole magnets is better than specified.

The superconducting magnet system has now been in use for more than a year. The magnets in the ring are very reliable and so far faults have not occured in the cold part of the system.

The energy stored in the magnets at 820 GeV exceeds 270 MJ. A sophisticated quench detection and protection system is installed to dump this energy in a safe way if a quench occurs.

We have observed only few beam induced quenches during the first year of operation. This is presumably due to the large safety margin which exist both in field (30%) and in temperature (1.5K).

Also very few spurious quenches have been occurred.

2.4.3 Proton Ring Commissioning - During HERA commissioning PETRA II could deliver a single proton bunch every 5 to 10 minutes with intensities varying between $1 \cdot 10^9$ to $3 \cdot 10^{10}$ protons per bunch.

Also the commissioning of the HERA proton accelerator has been eased by the reliability and stability of the accelerator and by the excellent performance of the beam instrumentation.

The proton beam lifetime at injection is presumably limited by persistent current multipoles in the superconducting magnets.

A lifetime of 10 hrs is reached after a careful correction of these multipoles using correction elements installed directly on the beam pipe as shown in Fig. 6.

In order to determine the required strength of the correction elements at injection and during acceleration, the dipole and sextupole fields are measured continously in cold reference magnets powered in series with the ring magnets.

The acceleration cycle proceeds in two steps. The beam is injected at 40 GeV and then accelerated to 300 GeV. At 300 GeV the optics of the ring is changed from injection to luminosity conditions and then the beam is accelerated to 820 GeV and stored. In the switch from injection to luminosity conditions the maximum value of the β-functions changes from 270 m to 1000 m. Only very small losses occur during the acceleration cycle as shown in Fig. 7.

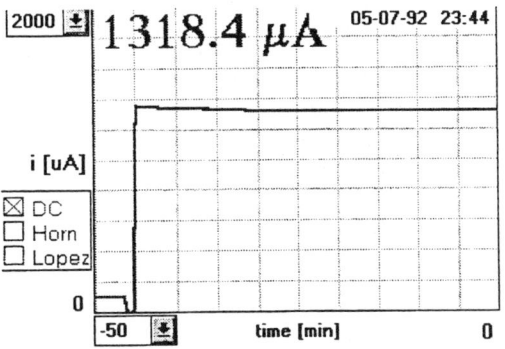

Fig. 7 - The circulating proton current in HERA is plotted as a function of time from injection to storage at 820 GeV.

Because of the long time constant of the main superconducting magnet circuit the instantaneous value of the magnetic field in the dipole reference magnet powered in series with the main circuit is used to derive a reference signal for all other energy dependent systems.

The measured proton beam life time at 820 GeV for 10 bunches each with 20% of its design intensity is several weeks.

2.5 Colliding Beams

The electron and proton ring were operated simultaneously for the first time in October 1991 and the first collisions between a 12 GeV electron bunch and a 480 GeV proton bunch was observed on October 19th.

The luminosity is measured using the reaction $e + p \to e + \gamma + p$ with the electron and the photon detected in coincidence[8,9]. The arrangement of the luminosity monitors is shown in Fig. 8.

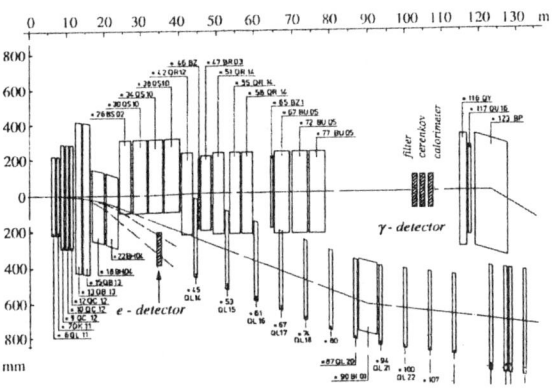

Fig. 8 - The layout of the magnets in the straight section and the luminosity detectors designed to monitor the reaction $e+p \to e+\gamma+p$.

Note the different scale along and transverse to the beam.

The electron-photon coincidence rate measured by the ZEUS luminosity monitor during collisions of 26 GeV electrons and 480 GeV protons is plotted in Fig. 9.

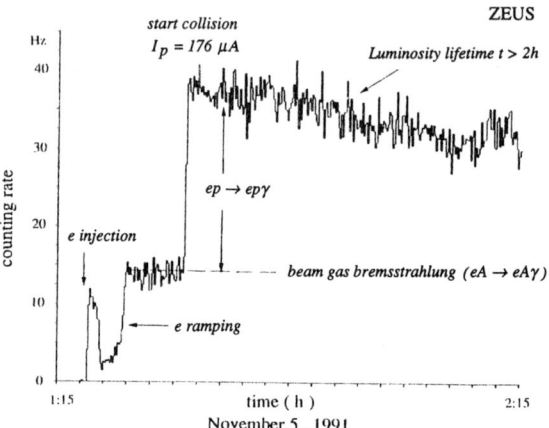

Fig. 9 - The electron-photon coincidence rate as measured with the ZEUS detector as a function of time.

A clear step in the coincidence rate is observed when the timing between the electron and the proton bunch is adjusted for collisions. The observed peak luminosity is about $7 \cdot E27$ cm^{-2}s^{-1} in good agreement with the predicted luminosity.

Recently a maximum luminosity of $1.3 \cdot E29$ cm^{-2}s^{-1} with 9 bunches in each beam has been observed corresponding to 20% of the design luminosity per bunch crossing.

The specific luminosity defined as $L/I_p \cdot I_e$ is a measure of beam overlap and transverse beam dimensions. The specific luminosity has recently been raised to 3.7 cm^{-2}s^{-1}mA^{-2} compared to its design value of 5.4 cm^{-2}s^{-1}mA^{-2}. This demonstrates that the transverse beam dimensions are close to their theoretical values.

The stored proton and electron current versus time for a well centered proton beam colliding with an electron beam of same transverse size is plotted in Fig. 10.

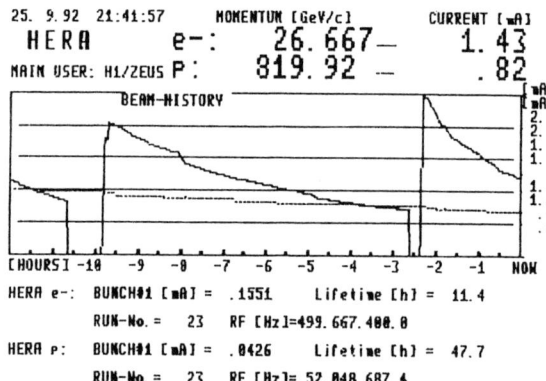

LUMINOSITY RUN

Fig. 10 - The stored proton and electron current as a function of time during luminosity conditions.

Proton beam lifetimes of 50 to 100 hrs are now routinely obtained by matching the size of the electron beam to the size of the proton beam. A proton fill typically lasts 24 to 36 hrs whereas the electron beam is usually dumped and reinjected every 4 to 5 hrs.

The number of colliding bunches stored in each ring has been limited to 10 due to the present limitation on electron beam current. An additional stored electron bunch is used to measure the background in the experiments.

The number of protons per bunch has so far been limited by the proton preinjectors and by losses during beam transfer to 10% to 30% of the design value.

The observed maximum proton beam tune shifts of $\Delta v_{x,p} \simeq 0.001$ and $\Delta v_{y,p} \approx 0.0005$ meet the design values and are so far not a performance limitation. The tune shifts suffered by the electron beam are below their design values.

The measured luminosity and the derived specific luminosity as measured by the ZEUS group as a function of time is plotted in Fig. 11. Whereas the luminosity drops by a factor of 3.5 during the 5 hrs long storage time the specific luminosity remains constant. This demonstrates the stability of the two rings and that the proton emittance remains constant over the fill. It also shows that the proton beam is not strongly influenced by noise in the electron beam - i.e. there is negligible crosstalk between the two beams. This remains true even when the amplitude of the electron feedback system is turned up to its maximum value.

Beam orbits are very reproducible. By following a well defined magnet massage procedure collisions are obtained for the same setting of the correction elements from run to run.

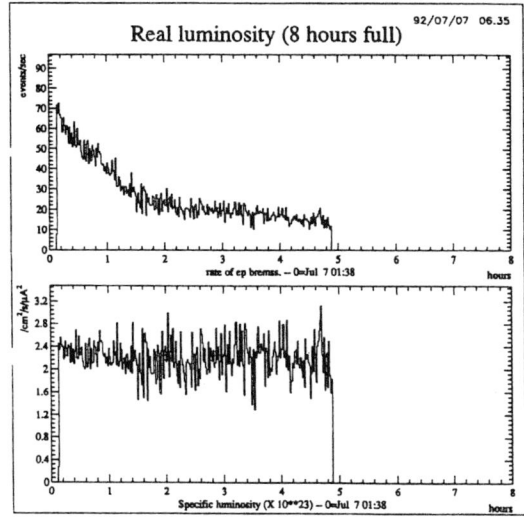

Fig. 11 - The luminosity and specific luminosity as a function of storage time.

3. THE PHYSICS

In a deep inelastic electron-proton event as depicted in Fig. 12 the incoming electron interacts directly with one of the quarks in the proton by means of a spacelike current, charged or neutral. The kinematic variables, Q^2 and v, or the scaled variables x and y, used to describe this process are also defined in Fig. 12.

An electron-proton collision in HERA yields a very simple final state topology. The struck quark will materialize as one or several jets of hadrons whose momentum component transverse to the beam axis is balanced by the transverse momentum of the final state electron (or neutrino). The remains of the proton will appear as a sharply collimated jet of hadrons travelling along the initial proton direction (Fig. 12).

nearly 2 orders of magnitude in Q^2 and v to values of order 30000 GeV^2 and 40000 GeV. Furthermore measurements of the cross section in the scaling region can be extended down to x-values on the order of a few 10^{-5} compared to the present lower limit of roughly 10^{-2}.

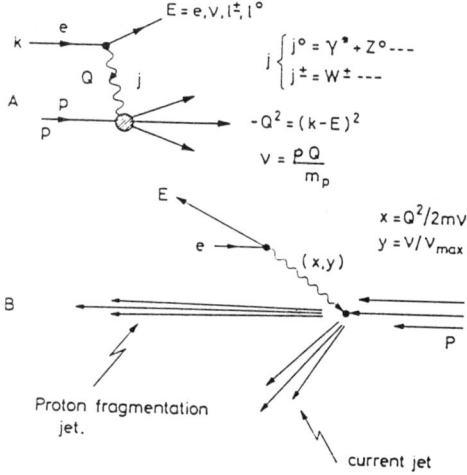

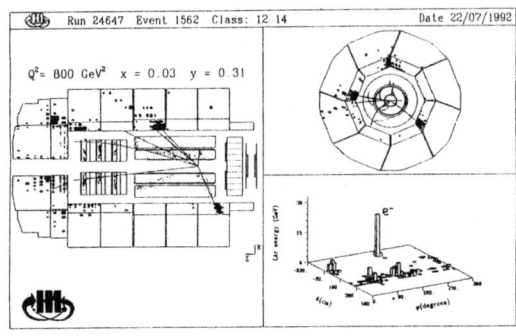

Fig. 13 - Deep inelastic ep event observed by the H1 detector at HERA.

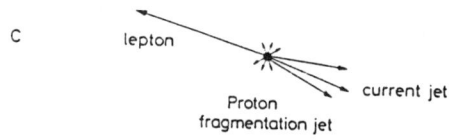

Fig. 12 - The topology of deep inelastic electron-proton interactions at HERA and the definition of kinematic variables.

The predicted deep inelastic event topology is indeed observed at HERA as shown in Fig. 13.

The present maximum values of Q^2 and v of respectively 600 GeV^2 and 400 GeV can be extended at HERA by

A complete discussion of the physics potential of HERA can be found in reference 10.

3.1 The Detectors

A simulated electron-proton event is shown in Fig. 14

The energy of the hadron jets varies between 820 GeV in the forward direction (incident proton direction) to roughly 30 GeV in the backward direction. It is thus clear that a high resolution, finely grained 4π calorimeter is the cornerstone of a general purpose detector at HERA. Since the ratio of

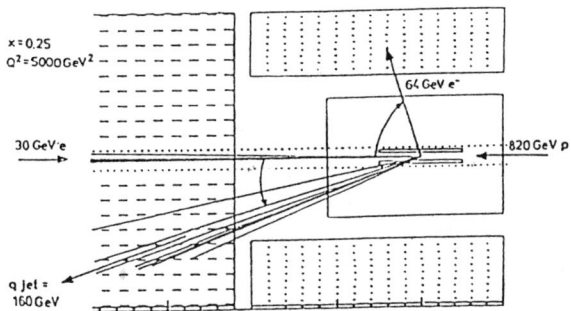

Fig. 14 - The topology of a typical deep inelastic electron-proton event.

neutral to charged pions varies from event to event, it is crucial that the calorimeter has the same response to showering as to non-showering particles.

The H1 Detector

An isometric view of the H1 detector is shown in Fig. 15.

The H1 electromagnetic calorimeter is made of 2.4 mm thick lead plates spaced 2.5 mm apart and the hadronic calorimeter is made of 16 mm thick stainless steel plates with gaps of 2 x 2.5 mm. Both calorimeters are read out using liquid Argon and are installed in a common cryostat.

The liquid Argon read out allows naturally for a fine segmentation both transversely and longitudinally. Transversely, the tower size varies between 3 x 3 and 8 x 8 cm^2 in the electromagnetic calorimeter and between 8 x 8 and 20 x 20 cm^2 in the hadronic calorimeter. Longitudinally, the electromagnetic shower is sampled 3 to 4 times, the hadronic shower 4 to 6 times. There is a total of 45000 independent segments in the detector.

A liquid Argon calorimeter has inherently a different response to showering and non-showering particles.

However, by weighting the measured longitudinal energy loss profile, the H1 group has achieved the same response to electrons and charged pions and hence an improved energy resolution. The energy resolution for electrons and hadrons (after weighting) is respectively:

$$\sigma(e) = \frac{12\%}{\sqrt{E}} \otimes 1\% \quad \sigma(h) = \frac{45\%}{\sqrt{E}} \otimes 1\%$$

The two errors are added in quadrature.

The calorimetry is completed by a forward hadron calorimeter made of copper absorber plates with silicon read-out and a backward electromagnetic lead scintillator sandwich calorimeter.

The calorimeter has been operated since April 1991. The energy calibration has been stable to better than 0.2% and less than 10^{-3} of the read in channels are not operational. Also the relative energy calibration between various units in the central calorimeter is quite good, However, more ep data are needed to reduce the uncertainty on the absolute energy calibration to a value below 5%. The measured energy

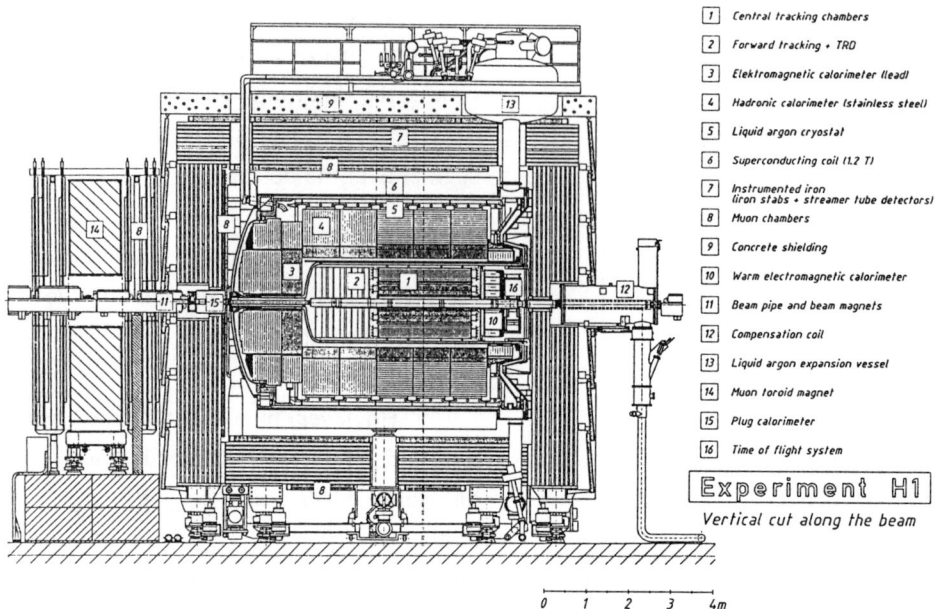

Fig. 15 - An isometric view of the H1 detector.

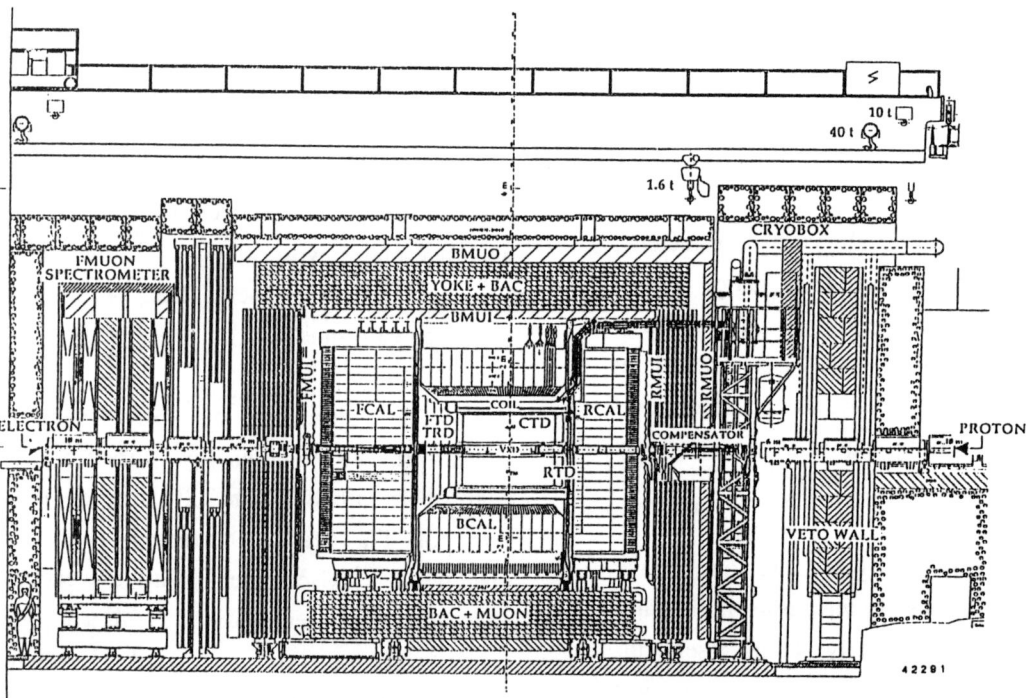

Fig. 16 - Schematic view of the ZEUS detector. The component labels are defined in the text.

resolution of the backward calorimeter is between 2% and 3% for electrons between 10 GeV and 30 GeV. An ultimate resolution of 1% can be reached using ep data.

The liquid Argon read-out can be operated in a magnetic field and the H1 group has chosen to install the calorimeter inside the large superconducting coil, which produces a 1.2 T field over a length of 5.75 m and a diameter of 5.2 m.

In the central region charged particles are being tracked using two jet chambers augmented by two z-drift chambers and two multiwire proportional chambers.

In the forward direction the trajectories are measured using three sets of radial and planar drift chambers. Electrons are identified using transition radiation detectors interleaved with tracking chambers. Trajectories of particles travelling in the backward direction are measured using multiwire proportional chambers. In the forward direction a measured resolution (σ) of 150 μm has been achieved compared to the central chamber values of 190 μm and 350 μm measured in r, Θ, and z respectively.

The calorimeters and the tracking detectors are enclosed and supported by an iron structure of 2200 tons which serves both as the return yoke of the magnet and as a hadronic backing calorimeter and muon--filter. The iron yoke is made of 7.5 cm thick iron plates interleaved with streamer tube chambers.

Muons are identified and measured using three layers of chambers which are installed on both sides and in the middle of the iron structure. In the forward direction muons are detected in a muon spectrometer made of a magnetized iron toroid and six layers of drift chambers. The forward spectrometer covers production angles between 3° and 17° and is designed to detect muons in the range 5 GeV/c to 200 GeV/c.

The ZEUS Detector

A vertical cut of the ZEUS detector along the beam is shown in Fig. 16.

The ZEUS calorimeter is made of 3.3 mm thick 20 cm wide and 4 m long depleted uranium plates, cladded with stainless steel, which are interleaved with 2.6 mm thick scintillator tiles. The tower size, defined by the scintillator tiles, is 5 x 20 cm^2 in the electromagnetic part and 20x20 cm^2 in the hadronic calorimeter. The scintillator tiles are read out using electromagnetic wave length shifters, light guides and photo-multipliers. Longitudinally the calorimeter is segmented into an electromagnetic calorimeter and two hadronic calorimeters. The calorimeter is made of three parts; the central region is surrounded by the barrel (BCAL) calorimeter, the forward (FCAL) and the rear (RCAL) calorimeters which cover the proton, respectively the electron direction. The calorimeters cover in total 99.8% of 4π.

The depleted uranium calorimeter is naturally compensating. The measured ratio of the calorimeter response to showering and nonshowering particles is 1 ± 0.03 over a wide range in energies.

The measured energy resolution of electrons and protons is:

$$\sigma(e) = \frac{17\%}{\sqrt{E}} \otimes 1\%, \quad \sigma(h) = \frac{35\%}{\sqrt{E}} \otimes 1\%.$$

The natural radioactivity of the uranium is used to monitor the calibrations of the calorimeter to better than 0.2%. The noise caused by the uranium decay is on the order of 12-30 MeV depending on cell size. The total noise, integrated over an area of 80 cm x 80 cm which corresponds to the area needed to contain a quark jet, is of order 200 MeV. The calorimeter has a time resolution of roughly 1 ns.

Photomultipliers are sensitive to magnetic fields and the calorimeters are hence installed outside the superconducting coil. The coil, 1.72 m in diameter and 2.8 m long, is 0.9 radiation lengths thick and produces a field of 1.8 T.

The inner tracking detector is located in the magnetic volume. In the central part particles are tracked with a time expansion vertex chamber (VXD) and a vector drift chamber (CTD). The vertex chamber has a measured resolution of 35 μm. The drift-chamber has 9 superlayers each with 8 layers of sense wires yielding a total of 4.608 sense wires. The presently achieved resolution is 100 μm. The forward tracking detector (FDET) contains three planar driftchambers and four transition radiation detectors to identify particles within the jet. Particles emitted along the incident electron direction are measured in the rear tracking detector (RTD) which is made of planar drift chambers. The planar driftchambers have a measured spatial resolution of order 120 μm and the energy loss of a minimum ionizing particle is measured with an uncertainty of 6-7%.

The iron yoke of 2000 tons is segmented into 7.5 cm thick iron plates interleaved with proportional tube chambers (BAC) which serve to measure the energy leaking out of the hadronic calorimeter and to identify muons. Four layers of limited streamer tube chambers are installed on both sides of barrel part (BMUI, BMUO) and the rear part (RMUI, RMUO) of the iron yoke. In the forward direction high energy muons are detected using a torroidal spectrometer equipped with planar drift chambers and limited streamer tube chambers.

Protons scattered under small angles (~ 0.2 mrad) can be measured to within 10 σ of the proton beam size by an elaborate set of 6 Roman pots installed between 23 m and 90 m downstream of the interaction point.

3.2 Experimental Conditions

The HERA design parameters impose rather stringent requirements on the data acquisition system and its ability to recognize and seperate physics events from the much more abundant background events in real time.

3.2.1 Background

The most serious source of background is beam halo protons traversing material upstream and then showering the detector with secondary tracks.

Several methods are used to reduce this source of background to a managable level.

versus the time in the rear calorimeter as measured in the ZEUS detector for events which meet the selection criteria.

A set of precise aperture limiting collimators have been installed in straight section West 1.6 km upstream of the ZEUS detector. A tight collimation reduces the background by one to two orders of magnitude. Both experiments have also installed shielding walls and vetocounters just upstream of the detectors. The remaining events of this type can be removed by a time of flight cut as shown in Fig. 17.

A second class of background events is caused by protons which interact with the residual gas in the beam pipe within the detector.

An example of this class of events observed with the H1 detector is shown in Fig. 18.

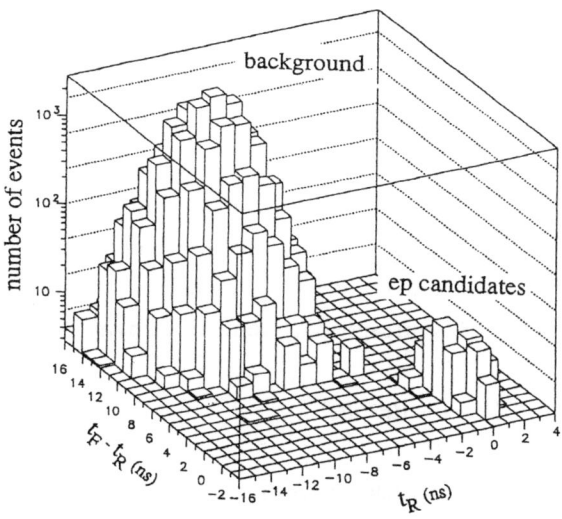

Fig. 17 - Difference in arrival time of signals from the front and the rear calorimeters

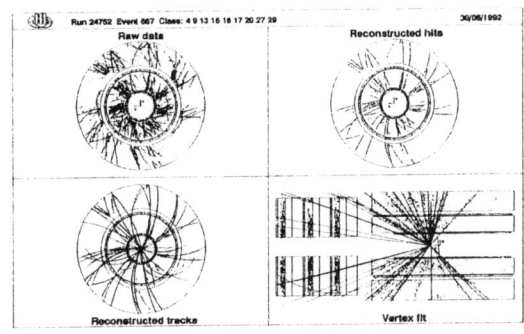

Fig. 18 - Computer reconstructed background event caused by a proton-interacting with the rest gas in the vacuumchamber.

A total of 21 proton tracks was identified using dE/dx information. Obviously this class of events can be reduced by improving the beam pipe vacuum, by a more stringent cut on transverse momentum or by excluding events with many slow protons.

Typically 5% of the design electron current could be stored. At these current levels the synchrotron radiation was effectively masked out and did not cause problems. However, electrons which suffered beam gas interactions or which scattered on the synchrotron masks do lead to a high single rate in the electron luminosity counter. This background was subtracted using the measured pilot bunch rate, properly normalized.

Cosmic rays is another potential source of background events. However, they could be removed using time of flight information from the muon counters.

The z-distribution of neutral current events which meets the selection criteria is shown in Fig. 19.

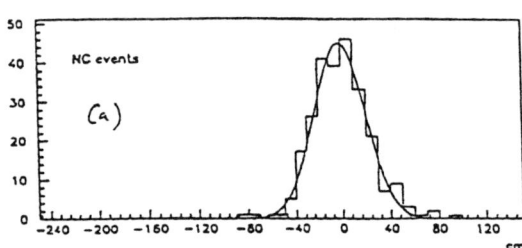

Fig. 19 - The z-distribution of neutral current events which meets the selection criteria. The data were obtained by the ZEUS group.

Note that the event distribution is well fit by a Gaußian distribution with $\sigma_z = 35$ cm in accordance with the observed proton bunch length. This demonstrates that the selection criteria yield a clean sample of deep inelastic neutral current events with a negligible background.

The trigger and data acquisition system must cope with the instantaneous collision rate of roughly 10 MHz and with the large background to signal ratio in a minimum bias trigger configuration. Both experiments select the interesting events using a series of trigger levels. The first level trigger decision needs 2.2 μs (H1) respectively 5 μs (ZEUS). Until then the data from 200000 to 300000 detector elements are stored in dead time less analog or digital pipelines for 23 (H1) or 52 (ZEUS) successive beam crossings. The final event selection is based on a partial event reconstruction using fast processor farms. During the first experimental period both experiments were using rather loose event selection criteria yielding a final event rate of 3-5 Hz corresponding some 600 kbyte/s.

3.3 Physics Results

3.3.1 Deep inelastic neutral current events.

The kinematic region in 1/x and Q^2 available to HERA and to a 600 GeV muon beam incident on a proton at rest is shown in Fig. 20.

The kinematic region can roughly be divided into three areas. A QCD perturbative region located in the lower right hand corner at moderate x and large Q^2, an non perturbative region[11] in the left hand upper corner at small x separated by a transition region defined as the solid curves.

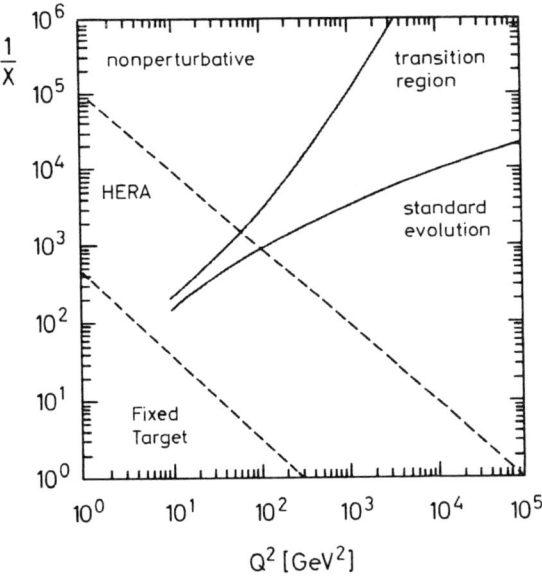

Fig. 20 - The kinematic region in 1/x and Q^2 available to HERA and to a 600 GeV muon beam on a proton at rest. The perturbative and non perturbative regions are separated by a transition region which is also indicated.

In the classic region of QCD the Q^2 evolution of the structure functions can be computed from the Gribov-Lipatov-Altarelli-Parisi equations[12,13]. These equations are defined in terms of the parton splitting functions as indicated in Fig. 21.

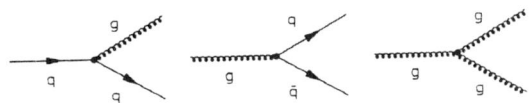

Fig. 21 - The relevant parton splitting diagrams.

At low x-values the structure functions are dominated by the gluon distribution functions xG(x). Extrapolating xG(x), as determined in the perturbative region, towards small x for constant Q^2 leads to a steeply increasing function that violates unitarity.

Since the gluon density increases with 1/x, the gluon-gluon interaction cannot longer be neglected although the gluon-gluon coupling constant $\alpha_s(Q^2)$ is still weak. In this transition region[14] one may be able to use the parton language and the behaviour of the formfactors may be described by adding a recombination term on to the perturbative evolution equations.

A further extrapolation in 1/x yields a very dense partonic system. Although $\alpha_s(Q^2)$ is still small the effective interactions are strong due to the high parton densities. In this region the perturbative approach breaks down and we may not be able to use the parton picture.

The HERA experiments will for the first time be able to explore the transition region and perhaps also the non-perturbative region.

Both experiments report data on ep → e'X. The data are based on an integrated luminosity of 1.5 nb^{-1} (H1) and 3nb^{-1} (ZEUS). Scatter plots of the events in the x,Q^2 plane which meet the selection criteria are shown in Fig. 22.

The kinematic region which has been explored by fixed target experiments is indicated by the hatched area. Well measured events are indicated by the full circles.

The ZEUS collaboration determines the Q^2 and ν of an event by combining the information obtained from a measurement of the final state electron and the hadrons in the current jet, whereas H1 relies on only the measured angle and energy of the scattered electron.

Note that the events are mainly in the backward direction (with respect to the incident proton) and are in range of x-values not accessible to previous experiments.

Computer reconstructed neutral current events are shown in Fig. 15 for a high Q^2 event and in Fig. 23 for a low x event. Note that the events are clean with only a few spurious hits.

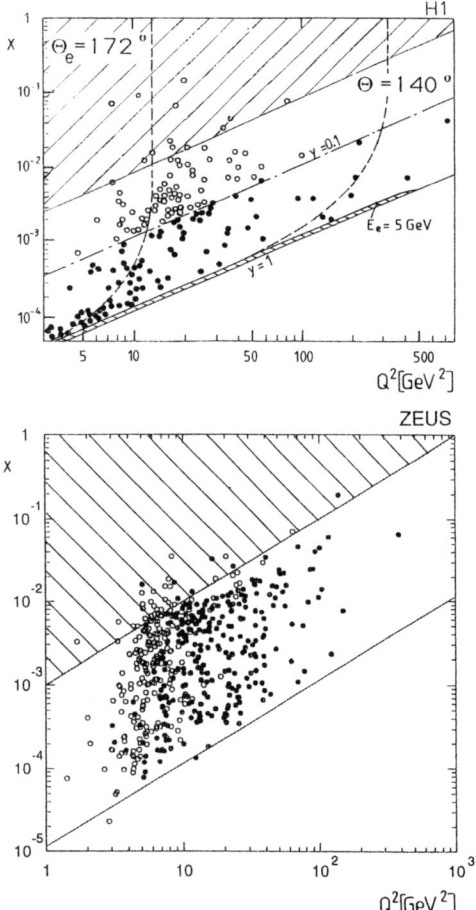

Fig. 22 - A scatter plot of deep inelastic (ep → e'X) events in the x, Q^2 plane measured by the H1 and ZEUS experiments.

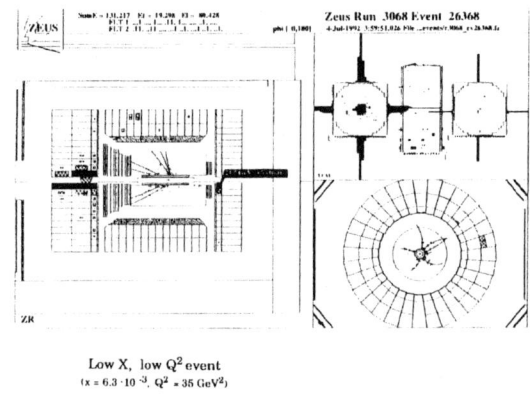

Low X, low Q^2 event
(x = 6.3·10^{-3}, Q^2 = 35 GeV^2)

Fig. 23 - Computer reconstructed neutral current events with Q^2 = 35 GeV^2 and x =

$6.8 \cdot 10^{-3}$.

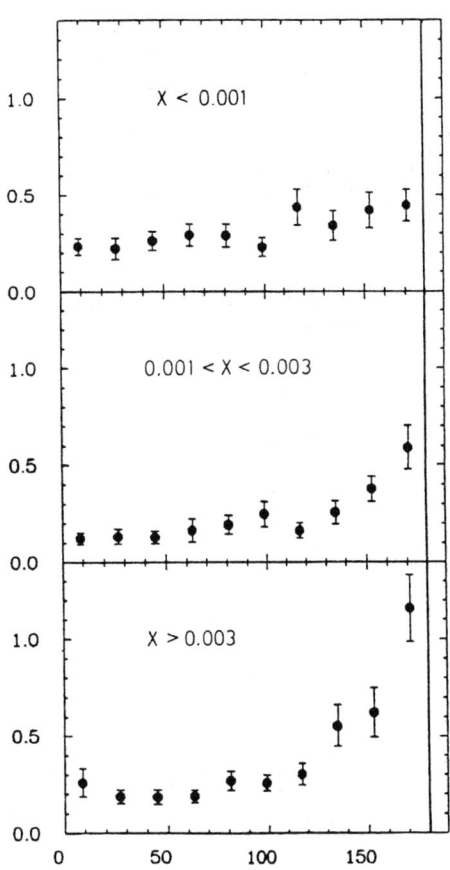

Fig. 24 - Transverse energy flow of hadrons as a function of the opening angle between the electron and the hadrons $\varphi_e - \varphi_{hadron}$ for various cuts in x. The data were obtained by the H1 Collaboration.

A firm prediction of the quark model is that the transverse momentum of the scattered electron should be balanced by the transverse momentum of the hadrons in the current jet. That HERA events have this property is clearly demonstrated in Fig. 24 where the transverse energy flow of the hadrons is plotted as a function of the opening angle $\varphi_e - \varphi_{hadron}$ between the hadron and the electron. This distribution is shown for various cuts in x. There is a clear peaking around $\varphi_e - \varphi_{hadron} \sim 180°$ for events with x > 0.003 consistent with the jet picture.

The Q^2 distributions as measured by the two collaborations, are plotted in Fig. 25a,b and compared to model calculations. Within the rather large error bars the data agree with the predictions.

The scattered electron energy distribution at a fixed angle is sensitive to the gluon x-distribution at low x.

Preliminary energy distributions obtained by the two experiments are shown in Fig. 26 and compared to the theoretical predictions. The prominent peak at the incident electron energy simply reflects the scattering kinematic. The measured electron energy distributions are not yet sensitive to the gluon structure function. However, the quality of the data demonstrates that such detailed comparisons will soon become feasible with improved statistics.

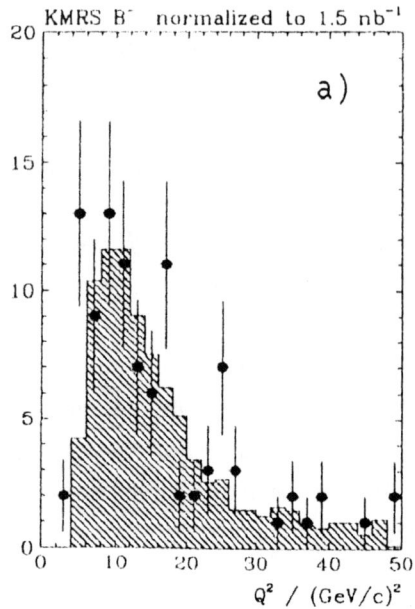

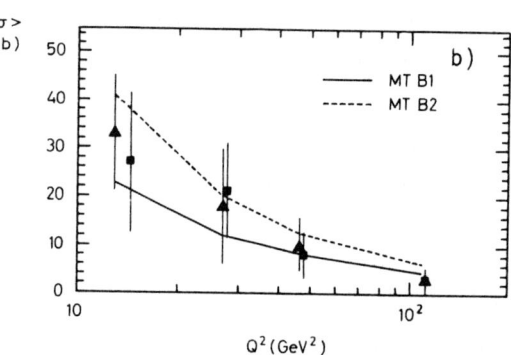

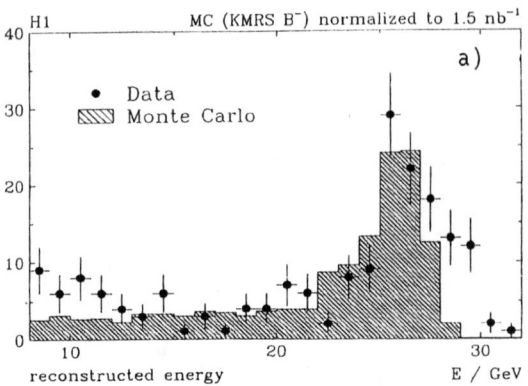

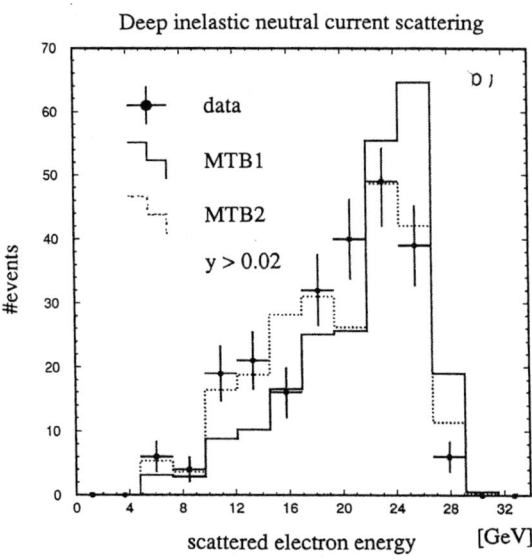

Fig. 25 - The observed Q^2-distribution for events meeting the selection criteria
a) H1-data for values of y from 0.05 to 0.55 compared to an absolutely normalized MC prediction[16].
b) ZEUS-data. The results of two different assumptions in the data analysis are presented and compared to theoretical predictions.

Fig. 26 - The measured energy distribution of the scattered electron in neutral current events: a) Data obtained by the H1 Collaboration and compared to MC predictions b) Data obtained by the ZEUS Collaboration and compared to M.C. predictions.

3.3.2 Photoproduction

The photon has unique properties. On one hand it is a fundamental gauge boson with well defined couplings to basic fermions and other gauge bosons as shown in Fig. 27a. However, a part of the time the photon behaves like a strongly interacting vectorboson (Fig. 27b). For large values of $p_\perp$ the photon interacts via its hadronic constituents, quarks and gluons yielding resolved photon events as shown in Fig. 27c.

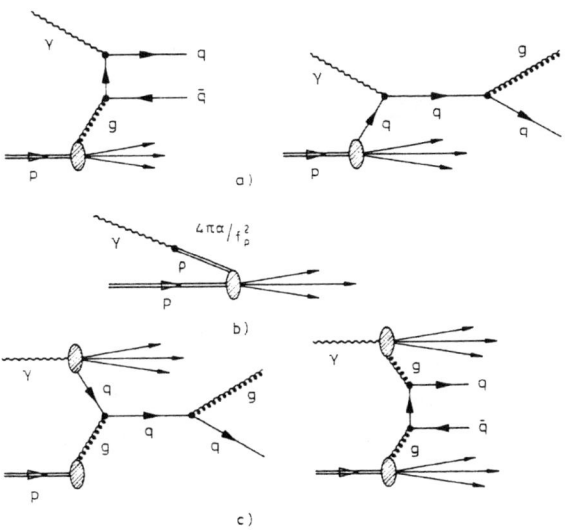

Fig. 27 - Photon - hadron interactions:
a) Pointlike coupling of photons
b) The vectormeson dominance model
c) "Resolved Photon" events induced by the gluon and quark content of the photon

HERA is well suited to study[17] photon induced processes for c.m. energies up to 250 GeV.

For very small values of Q^2 the incident electron beam is equivalent to a well collimated beam of quasi real photons. In this case the electron scattering cross section $\sigma(ep)$ is related to the total photoproduction cross section $\sigma(\gamma p)$ as follows:

$$\frac{d\sigma(ep)}{dy} = \frac{\alpha}{2\pi} \frac{1+(1-y)^2}{y} \sigma(\gamma p) \ln \frac{Q^2_{max}(y)}{Q^2_{min}(y)}$$

The energy and Q^2 of the photon is determined from a measurement of the energy and angle of the scattered electron. Both experiments use the luminosity electron counter for this purpose.

An electron which has emitted a photon in a small cone centered around $0°$ is deflected away form the circulating electron beam in the off axis low β quadrupoles and in the downstream magnets, exits the accelerator vacuum through a thin window, and impinges on the luminosity electron counter.

At ZEUS electrons with scattering angles between $0°$ and 6 mrad and energies E' in the range 0.2E < E' < 0.9E are accepted. This corresponds to a range in Q^2 of the quasireal photon from 10^{-7} GeV^2 to $2 \cdot 10^{-2}$. The electrons with energies between 11 GeV and 16 GeV, corresponding to γp center of mass energies between 186 GeV and 220 GeV are used in the analysis.

The H1 tagger accepts electrons scattered at angles below 5 mrad and energies in the range of 6 GeV to 19 GeV corresponding to a Q^2 range between

10^{-7} GeV^2 and 0.015 GeV^2 and a mean c.m. energy of 200 GeV.

The H1 data were collected concurrent with the deep inelastic data and are based on an integrated luminosity of 1 nb^{-1}.

The bulk of the ZEUS data results from a dedicated photoproduction run with an integrated luminosity of 0.23 nb^{-1}.

The data analysis and the event selection criteria are discussed in detail in reference 5 and 6.

Both experiments report data on the total photoproduction cross section. H1 quotes at a c.m. energy of 200 GeV:

$\sigma_T(\gamma p) = (150 \pm 15(\text{stat.}) \pm 19(\text{syst.}))\mu b$

ZEUS reports at a c.m. energy of 210 GeV:

$\sigma_T(\gamma p) = (154 \pm 16(\text{stat.}) \pm 32(\text{syst.}))\mu b$

The systematic error results from the errors in the measurement of the photon flux and the ep luminosity and in the detection efficiency, all added in quadrature.

The detection efficiency depends on the relative abundance of elastic and diffractive events to low transverse energy inelastic events and hard processes. The total photoproduction cross section has been evaluated for various assumptions on the ratio of these processes consistent with the observed final state topology.

The measured cross sections are plotted in Fig. 28 versus energy together with data at lower energies and various predictions.

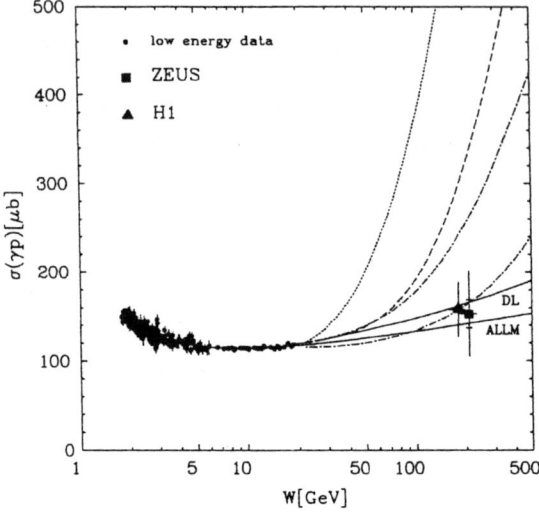

Fig. 28 - The total photoproduction cross sections measured by H1 and ZEUS are plotted versus energy together with low energy data and compared to theoretical predictions.

The data are in good agreement with predictions based on Regge models[18,19] and show an increase with energy similar to that observed in the $\bar{p}$-p total cross section.

Minijet models[17] which simply add the perturbative (QCD) and non-perturbative (soft) contributions predict a steeply rising cross section with energy for reasonable values of $\hat{p}_\perp^{min}$, the lowest momentum below which perturbative QCD cannot be used. These models are in disagreement with the data. Including[20] multiple scattering corrections brings the minijet model into agreement with the observed data.

Hard photon processes which lead to large $p_\perp$ jet events can be induced either by direct photon processes (Fig. 27a) or by resolved photon processes (Fig. 27c). Direct photon processes will yield 2 jet final states in lowest order, whereas resolved photon processes will populate the 3 jet final state.

Large $p_\perp$ photoproduction events at HERA energies display a clear jet structure as seen in Fig. 29.

Based on a jet cone algorithm the H1 group has searched for 2 or more jets final states. They are left with 34 events of which 7 have a tagged electron.

The transverse energy distribution of these events is plotted in Fig. 30 and compared to predictions based on a complete Monte Carlo simulation and on a simulation of the direct process only.

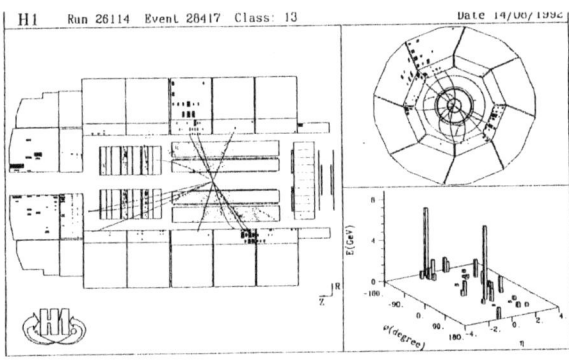

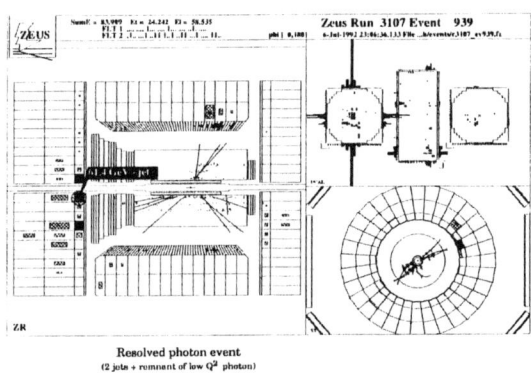

Fig. 29 - Photoproduction events observed by the H1 and the ZEUS detector. Both events show a clear jet structure.

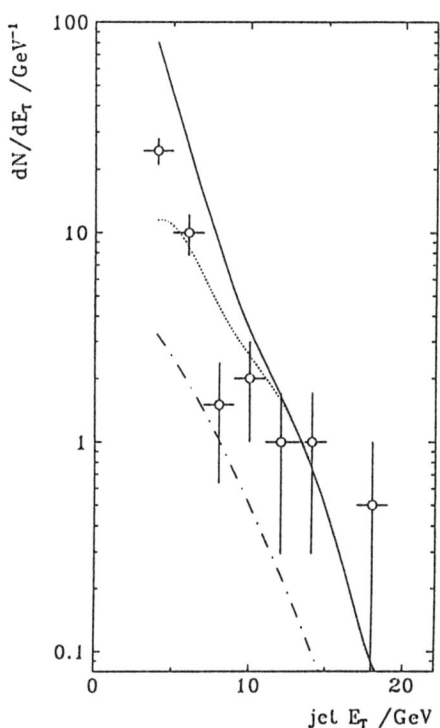

Fig. 30 -The jet transverse energy distribution measured by H1. The solid, respectively the dotted curve are the absolute prediction of the PYTHIA MC programme for $p_\perp^{min} = 1$ GeV, respectively $p_\perp^{min} = 5$ GeV. The prediction for the direct process with only $p_\perp^{min} = 1$ GeV is shown as the dash-dotted line.

The comparison of the data with the MC results suggests that the bulk of the observed photoproduced jet events are due to the "resolved photon" process.

3.3.3 A Search for Leptoquarks

Both experiments report limits on the search for leptoquarks[21], massive particles with both electron and baryon quantum numbers which occur naturally in composite models of leptons and quarks.

HERA as s-channel resonances as indicated in Fig. 31 and their favoured decay mode is into an electron and a quark

The x-distribution of neutral current events obtained by ZEUS which meets the leptoquark selection criteria is plotted in Fig. 31. One event was found for $x > 0.008$ compared to the expected background of 4.2 events.

From these data the ZEUS group derives the following 90% confidence limits:

$g_\ell = 0.3$ $\qquad g_r = 0$
$29.5 \text{ GeV} < M_{LQ} < 47 \text{ GeV}$

$g_\ell = 0.0$ $\qquad g_r = 0.3$
$29.5 \text{ GeV} < M_{LQ} < 67 \text{ GeV}$

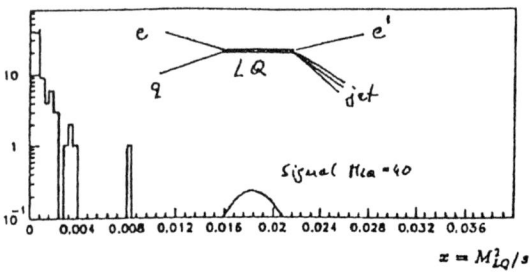

Fig. 31 - The ZEUS x-distribution of neutral current events which meets the leptoquark selection criteria. The expected signal for a leptoquark at 40 GeV at the 90% confidence level is also shown.

Leptoquarks will be produced in

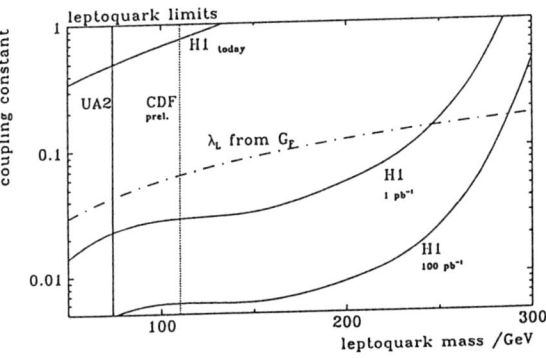

Fig. 32 - Limits on leptoquark mass obtained by the H1 group in ep collision at HERA and by UA2 and CDF in p$\bar{\text{p}}$ collisions.

The mass limits set by the H1 group on the existence of leptoquarks as a function of the coupling strength is shown in Fig. 32 and compared to mass limits obtained in $p\bar{p}$ collisions by UA2 and CDF.

The masslimits which can be reached at HERA for an integrated luminosity of 1 pb^{-1} respectively 100 pb^{-1} are also shown.

4. SUMMARY and OUTLOOK

During the past months HERA and its general purpose detectors H1 and ZEUS have been successfully transformed from a potential into a real source of physics.

The observed ep luminosity per crossing has reached 20% of its design value and we have so far not discovered any fundamental reason why HERA should not be able to meet its design value. The next step will be to increase the beam currents and the number of bunches stored in the two rings from its present value of 10 towards the design value of 210.

Both experiments have been able to extend the measurement of the deep inelastic neutral current cross section by a factor of 100 in x down to x values of order 10^{-4}. Although statistics do not yet allow precise statements on the behaviour of a weakly coupled, densely packed partonic system it is clear that the experiments in this region can be carried out and that data will be forthcoming in the near future.

The experiments have also demonstrated that photoproduction experiments can be carried out at HERA for center of mass energies a factor of 10 above the present fixed target energy range. At these energies the final state jet structure for either direct or "resolved photon" events should become clearly visible.

The experiments have also shown that H1 and ZEUS will soon be able to carry out significant searches for new objects like leptoquarks or excited electrons.

After a brief shutdown the HERA accelerator has resumed operation with the time split between physics runs and accelerator development. The increase in peak luminosity by a factor of two and the fact that many operational aspects have now become routine increased the integrated luminosity available to the experiments by a factor of 5 in just 10 days of running compared to the total integrated luminosity obtained during the month of July.

In addition to the general facilities H1 and ZEUS HERA also offers the intriguing possibility of carrying out high luminosity, high duty cycle fixed target experiments by using an internal target in the two rings.

The proposed HERMES[22] experiment plans to use the longitudinally polarized electron beam on polarized p,d and He^3 gas jet targets to investigate the proton and neutron spin structure.

Another group[23] has proposed to install thin wire targets positioned in the tails of the proton distribution. With this arrangement and the full circulating proton current they estimate a production rate of 60 $b\bar{b}$ pairs every second. Such a rate would be sufficient to measure the CP-violating parametes in the $J/\psi\,\bar{K}^o$ channel.

Both experiments are now under active considerations and - if approved - they would be a new and fascinating aspect of the HERA experimental programme.

References
1. HERA, a proposal for a large electron-proton colliding beam facility at DESY, DESY HERA 81-10
2.. D.Degèle, Proceedings of the Third Europ.Acc.Conf., Berlin, April 1992
   F.Willeke, Proceedings of the XII Int. Acc.Conf., Hamburg, July 1992
   and references therein
3. F.Brasse - for the H1 Collaboration - contribution to this conference
4. D.Caldwell - for the H1 Collaboration - contribution to this conference
5. B.Löhr - for the ZEUS Collaboration - contribution to this conference
6 F.Eisele - for the H1 Collaboration - contribution to this conference
7. HERA Polarimeter Group, contribution to HEACC'92, Hamburg, July 1992 and internal report
8. D.Kisielewska et al., DESY HERA 85-25
9. S.V.Levonian et al., DESY H1-TR113, 1987
10. Physics at HERA, Proceedings of the Workshop, Hamburg, October 29-30, 1991 - edited by W.Buchmüller and G.Ingelmann
11. For a discussion on the low x-region including references see: J.Bartels and J.Feltesse in Ref. 10
12. V.N.Gribov and L.N. Lipatov, Sov. Journ. Nucl.Phys.15, 438 and 675 (1972)
13. G.Altarelli and G.Parisi, Nucl.Phys. 126, 297 (1977)
14. L.V.Gribov, E.M.Levin, and M.G.Ryskin, Phys.Rep. 1001, 1982
15. J.Kwiecisnki, A.D.Martin, R.G.Roberts and W.J.Stirling, Phys.Rev. D42, 3645, 1990
16. J.G.Morfin and W.K.Tung, Z.Phys. C52, 13, 1992
17. For a discussion on low Q^2 Physics at HERA including references see G.A.Schuler in Ref. 10
18. A.Donnachie and P.V.Landshoff, Nucl.Phys. B231, 189, 1984
19. H.Abramowicz, E.M.Levin, A.Levy and U.Maor, Phys.Lett. B269, 465, 1991
20. J.R.Forshaw and J.K.Storrow, Phys.Lett. B268, 116, 1991
21. W.Buchmüller, R.Rückl, D.Wyler, Phys.Lett. B191, 492, 1987
22. The HERMES Collaboration, K.Coulter et al., DESY/PRC/90/1 (1990)
23. Private Communication W.Hoffmann, W.Schmidt-Parzefall.

DISCUSSION

G. Barbiellini, Sezione Infn di Trieste

In your very complete presentation you have not mentioned that all this important work that has brought HERA into production has been realized with a fixed number of people, and the capability of this near-thousand people is quite amazing.

M. Peskin, SLAC, USA

In the measurement of the photoproduction cross section, the electron is identified in the luminosity monitor; thus, there is a minimum Q^2 for this event sample. What is it?

Wiik

The final state electron is tagged between 0° and 5 to 6 mrad corresponding to values of Q^2 between 10^{-7} GeV^2 to $2 \cdot 10^2$ GeV^2. This should be a good approximation to $Q^2 = 0$.

FUTURE HADRON COLLIDER: THE SSC

Roy F. Schwitters
Superconducting Super Collider Laboratory*
2550 Beckleymeade Avenue
Dallas, Texas 75237, USA

Abstract

The design of the SSC is briefly reviewed, including its key machine parameters. The scientific objectives are twofold: a) investigation of high-mass, low-rate, rare phenomena beyond the standard model; and b) investigation of processes within the domain of the standard model. Machine luminosity, a key parameter, is a function of beam brightness and current, and it must be preserved through the injector chain. Features of the various injectors are discussed. The superconducting magnet system is reviewed in terms of model magnet performance, including the highly successful ASST. Various magnet design modifications are noted, reflecting minor changes in the collider arcs and improved installation procedures. The paper concludes with construction scenarios and priority issues for ensuring the earliest collider commissioning.

The SSC is now under construction just south of this campus in Ellis County, Texas. As you know, it is a proton-proton collider enclosed in a race-track-shaped underground tunnel (see Figure 1). The machine consists basically of two arcs housing the two proton rings. The rings are built one above the other in two arcs of bending magnets and focusing magnets. The straight section on the west side of the ring provides the various devices needed to inject the beams and, when required, to dump them. Most important are the interaction regions where the beams will be collided and the experiments mounted. Because the rings are mounted one on top of the other, the beams cross vertically such that there are two collision points on each side of the machine. In addition, on the east side there is an extra utility straight section, where it may be possible someday to extract a beam from the main collider rings, or do other kinds of specialized experiments with internal gas jets and the like. The main campus area, which encloses the buildings for the staff, the injector accelerators, and other operations, is in a large parcel of land on the west side of the machine. On the east side is a smaller campus where, for geological reasons, we will site the very large detectors, which are essentially under construction now. Aside from small service areas around the arcs, where there are refrigerators and power supplies and other facilities, the tunnel of the machine goes underground without disturbing existing farms and countryside. The basic design of the SSC is described in a supplementary design report.

Table 1 lists some of the key parameters of the SSC. The high energy of 20 TeV

---
*Operated by the Universities Research Association, Inc., for the U.S. Department of Energy under Contract No. DE-AC35-89ER40486.

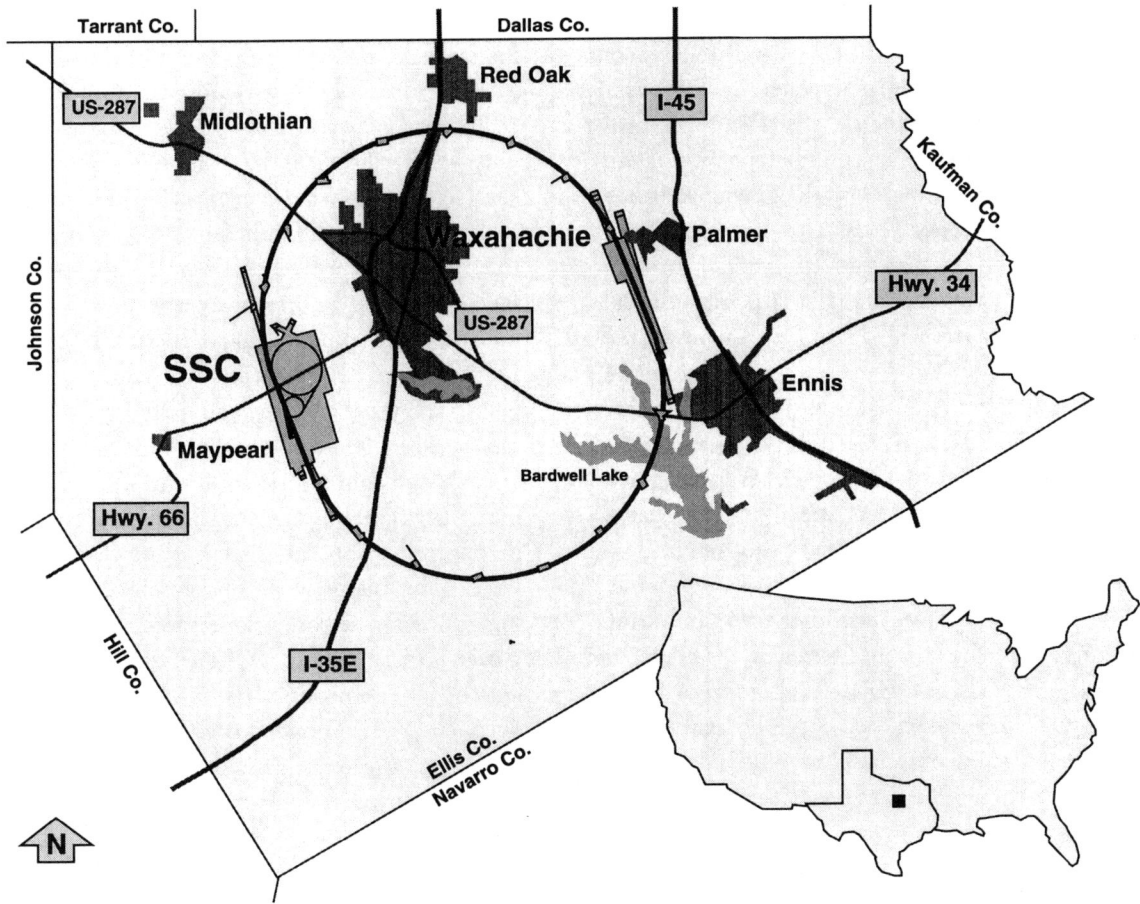

Figure 1. The SSC locale.

Table 1. SSC parameters.

Proton Energy	20 TeV
Circumference of rings	87 km
Protons per r.f. bunch	0.75×10^{10}
Bunch spacing	5 meters
Number of bunches/ring	17,424
Total particle energy/ring	418 megajoules
Emittance (RMS)	1π millimeter-milliradian
Interaction region focal spot size RMS radius, ($\beta^* - 0.5$ m)	5 micrometers
Proton-proton collision rate	60 MHz
Luminosity	1×10^{33} cm^{-2}sec^{-1}
Synchrotron radiation power	8.75 kilowatts/ring

per ring in each of the proton machines is, of course, unique. The beam intensity of roughly 10^{10} protons per bunch is comparable to that in current state-of-the-art proton accelerators.[1] The bunch spacing is an important experimental quality because it has a great impact on how the detectors and associated electronics are designed. In this machine the bunch spacing is 16 nanoseconds, or about 5 meters in distance. An awesome number is the stored energy in the beam of each ring, nearly half a gigajoule. A major engineering aspect of designing such an accelerator is to handle that stored energy properly and safely to prevent it from damaging parts of the ring or the detectors. Another important parameter is the emittance of the beams. The SSC design relies on emittances somewhat smaller than the current figures at accelerators like the Tevatron or HERA. The design luminosity of 10^{33} was chosen after extensive discussions throughout the community on a balance of issues related to expected production rates for physics processes of interest, detector construction, ease of triggering, backgrounds, radiation damage, and other factors to determine a prudent value for the luminosity for launching a major facility like the SSC. The talk of Takahiko Kondo[2] covers many of the issues. Synchrotron radiation begins to be a serious issue in a machine like the SSC. In particular, we are designing for a nominal load of slightly less than 10 kilowatts per ring, which must be absorbed by the cryosystem of the accelerator.

The scientific targets of the SSC have been discussed extensively for years, and I need only touch briefly on them here. The principal motivation for building the SSC is to discover phenomena that will give insight into physics beyond the standard model. The strategy chosen for doing that is to elucidate the nature of electroweak symmetry breaking, which is really an attempt to understand the detailed structure and behavior of the Higgs mechanism as it pertains to the standard model. The goal is not simply the discovery of another particle, because we already know that the Higgs must exist: the W and Z exist and are made of the Higgs field, whatever it is. So the target is really the more difficult one of understanding the full structure of the symmetry-breaking mechanism. In addition, everyone hopes, and many people expect, that there must be new physics beyond the standard model. Various ideas, while not yet convincing, have been discussed at length over the years.

In addition to the high-mass, generally low-rate, rare physics that we can anticipate in exploring for the Higgs, important studies can be made with super colliders involving processes within the context of the standard model. There will be very high rates for top quark production that should permit detailed studies once the top quark is found. Similarly, the exciting questions surrounding B-quark physics need more attention at high energy colliders because, again, of the copious production cross-section for B-quarks. If we give the same kind of attention to the detectors for these facilities that we have given those in our proposed electron-positron factories, we should also be able to contribute substantially to a better understanding of standard model processes. It has been pointed out recently[3] that there will probably be interesting and perhaps exciting low-Q^2 physics involving the Pomeron and the structure of the vacuum.

With the SSC we are trying to design a balanced and diverse experimental program that can address all these topics. But highest priority will go to understanding the nature of symmetry breaking and to learning new physics beyond the standard model. These questions have been studied and Monte Carloed to death by any number of detector proposals: the crucial parameter of the SSC is its high beam energy chosen so that we can find a definitive answer to the question of symmetry breaking within a reasonable period of time. As was discussed throughout this conference, possible masses for the Higgs and relatives of the Higgs will probably span a range from current limits up to the 1 TeV scale (see Figure 2). It is important to note that we now have a solid basis for belief that this full range of possibilities will be addressed by the SSC and fairly soon, too.

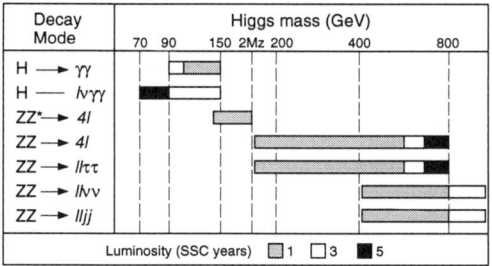

Figure 2. Typical detection limits for the Higgs.

We started our Laboratory about three years ago, taking up residence and rental office space on the southern edge of Dallas (see Figure 3). About two years ago, the first parcel of land was acquired by the state of Texas and turned over to the federal government. This is the land now designated as the N-15 site. About a year ago, we took over a major new building that we call the Central Facility, where now roughly half of the staff resides, in particular most of the people working on the technical design of the accelerators and related systems. Currently, we have a staff of about 2000 distributed among the various facilities.

How well have our engineering designs and technical developments achieved the goals set out for the SSC? The nominal design has a luminosity of 10^{33} cm^{-2}s^{-1} at a beam energy of 20 TeV. As pointed out by Bob Siemann,[4] the expected luminosity can be described as the product of two important parameters. One is the beam brightness, which is the number of particles per bunch, per unit invariant transverse phase space of the bunch; the other is the

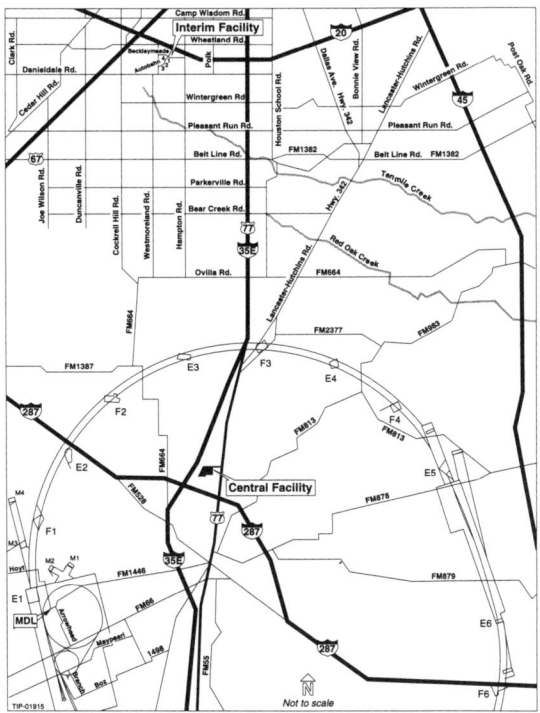

Figure 3. SSC sites.

total current in the ring. Ultimately, the luminosity will be limited by each of these parameters: the brightness and the total current. The nominal luminosity is, we believe, well within the limits that are possible in these accelerators; higher luminosities are ultimately limited by various effects (see Figure 4). In particular, we feel that the brightness figure will actually be limited by the chain of injectors that provide beams to the collider rings. Therefore beam emittance or brightness is something that we have to reflect throughout our designs. The total current for a fixed brightness will be limited at high energies by synchrotron radiation and at low energies by beam-beam phenomena. The latter is essentially the problem of one particle seeing the long line of charge of the other beam as it crosses the interaction region. We believe that it is reasonable to expect substantial increases above the nominal luminosity in the future.

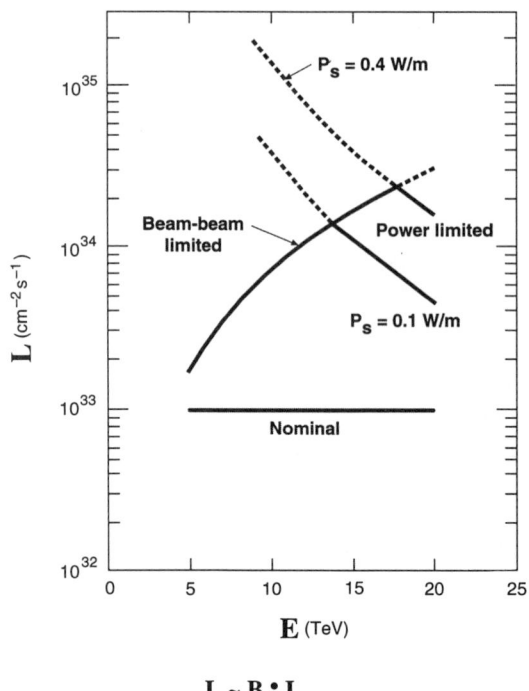

$$L \sim B \cdot I$$
$$B = \text{Brightness } (N_B/\varepsilon_N)$$
$$I = \text{Total current}$$

Figure 4. SSC luminosity potential.

Let me now discuss the brightness issue as it relates to the SSC's hierarchy of injectors (see Figure 5). We start with a linear accelerator; then we feed a low-energy booster, a medium-energy booster, and a high-energy booster. The brightness must be maintained from the beginning. The Linac itself, and its related instruments, are now under construction (see Figure 6). A small forest can be seen just off the right edge of the photo where the campus buildings will eventually be located. The Linac is actually a series of different accelerators (see Figure 7). It starts with an ion source, which has been under operation for well over a year. The ion source has achieved the emittance goals necessary for the full design luminosity. The next stage is an RFQ, which is essentially complete and is undergoing initial performance tests. We have ordered the drift-tube linac, and we are working with our colleagues at the electron-positron facility at the high energy physics laboratory in Beijing, who are building with us the coupled-cavity Linac. The most critical bottleneck in the ultimate brightness of the SSC occurs at the next stage in the transfer from the Linac to the low-energy booster. We have chosen the Linac energy to be 600 MeV. However, the tunnel will be long enough to allow us to increase that energy to 1 GeV if we need to. A change from 600 MeV to 1 GeV has the potential of raising the brightness of the beam by as much as a factor of 3.

The low-energy booster is a demanding machine technically (see Figure 8). It is a 10 Hertz, rapid cycling, proton synchrotron, with a large swing of proton velocity and, hence, frequency. This booster is being built in collaboration with the Budker Institute at Novosibirsk where there is outstanding expertise in this class of machines. The Russians

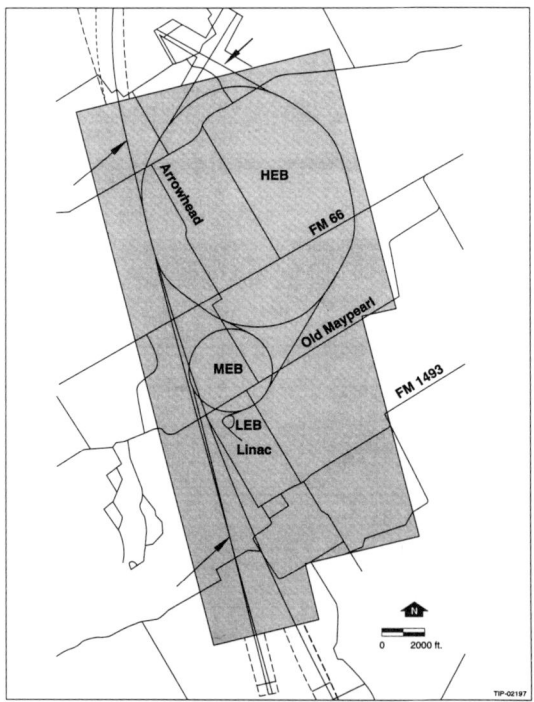

Figure 5. The SSC injectors.

Figure 6. The Linac construction site.

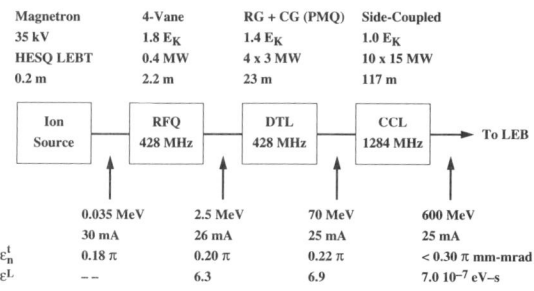

Figure 7. Linac design.

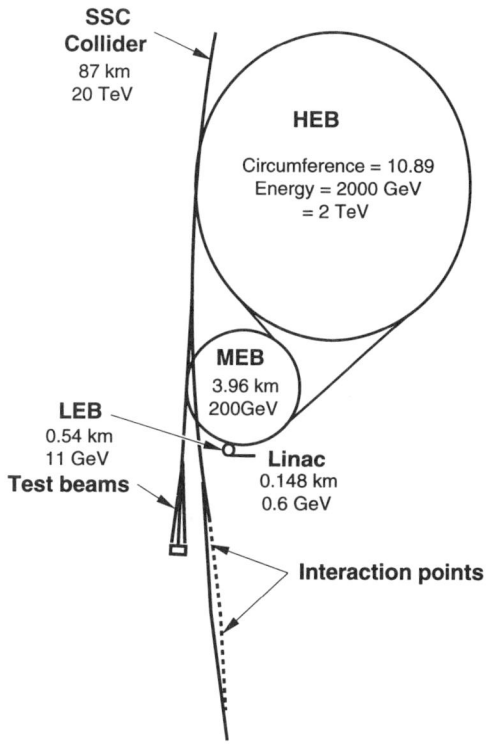

Figure 8. Injector stages.

are providing critical help. The next machine in the series is a relatively conventional proton synchrotron accelerator, much like the Fermilab main-ring injector or, indeed, the main ring itself or the SPS at CERN. We are currently collaborating with Fermilab on the design of the magnets for the medium-energy booster.

The high-energy booster is a rapid cycling, bi-polar 2 TeV synchrotron. It will become the second highest energy accelerator in the world. One of the key challenges of this machine is its bi-polar nature, which is required to inject protons into the two proton collider rings in opposite directions. From the outset we will design a bi-polar cycle in that machine so that it never has a preferred direction. It will inject into one ring and into the other ring, and keep cycling in this way. A critical aspect of maintaining the emittance and beam brightness is in the various transfer lines indicated in Figure 8. In these efforts we are being assisted with key optical components by physicists from India and China and elsewhere.

The rapid cycling nature of the high-energy booster as shown in Figure 9 also puts demands on the superconducting mag-

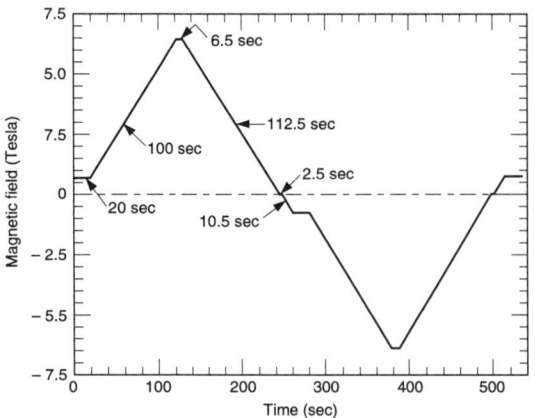

Figure 9. HEB acceleration cycle.

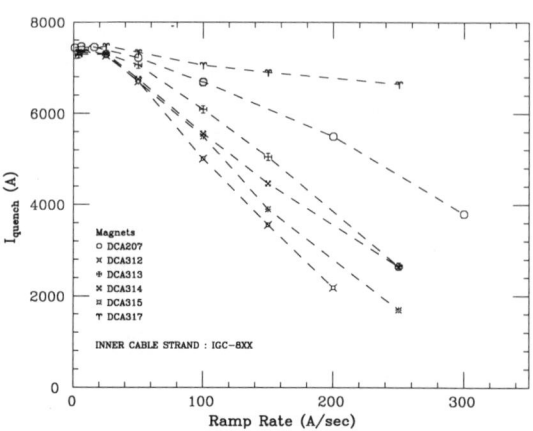

Figure 10. Quench current vs. ramp rate.

nets, which are quite similar in concept to the collider magnets. Some recent results are important and interesting. Figure 10 shows tests of the quench current capabilities of our dipole magnets as a function of ramp rate. The nominal ramp rate for the high-energy booster is 70 amps/second, and there is a wide spread in the currents at which these magnets quench. Some of them quench at relatively low currents at a high ramp rate, and we are actively investigating to understand why. The collider itself ramps at a very low rate, so the ramp rate dependence is really not an issue for the main collider. The quench properties are believed to arise from eddy current heating in the magnet cable during the ramp. In addition to quench properties, there are also effects on the quality of the field associated with the high ramp rate. We have developed over the past few months a detailed model of this phenomenon (see Figure 11). The model describes eddy-current effects by the linkage of flux through connected turns of different wires in a cable; different strands in the cable form a loop around which an EMF can be generated and hence currents can flow. The circulating currents will both heat up the copper matrix of the wire slightly and disturb the quality of the magnetic field. The model has been run on the computer, and it explains rather sat-

isfactorily the phenomena we are observing in terms of the interstrand resistance of the wires as they are pushed into one another during the fabrication of the cable. Figures 12 and 13 show some of the multipoles of the magnets, the non-uniform field components that we observe as a function of current. The important thing to notice is that, in addition to the intrinsic, persistent current phenomena that one sees in the superconducting magnets having to do with the filaments inside the wires, one sees this eddy current effect in the cable. We are now trying to understand this as it relates to the quality control and manufacturability of the cables. A class of magnets exists with resistances high enough that this is not a problem, and we are trying now to control the production of the cable so that it always provides satisfactory magnets. We expect there to be a straightforward engineering solution to the ramp-rate issue.

We have made some minor changes in the final design of the main collider arcs. The lattice was modified by removal of 124 dipoles to produce space in the arcs for utility feeds that match the location of surface facilities, and the magnet interconnect space was increased from 65 to 82.5 cm. The consequences of these changes are shown in Table 2. The peak magnetic field has actually been raised

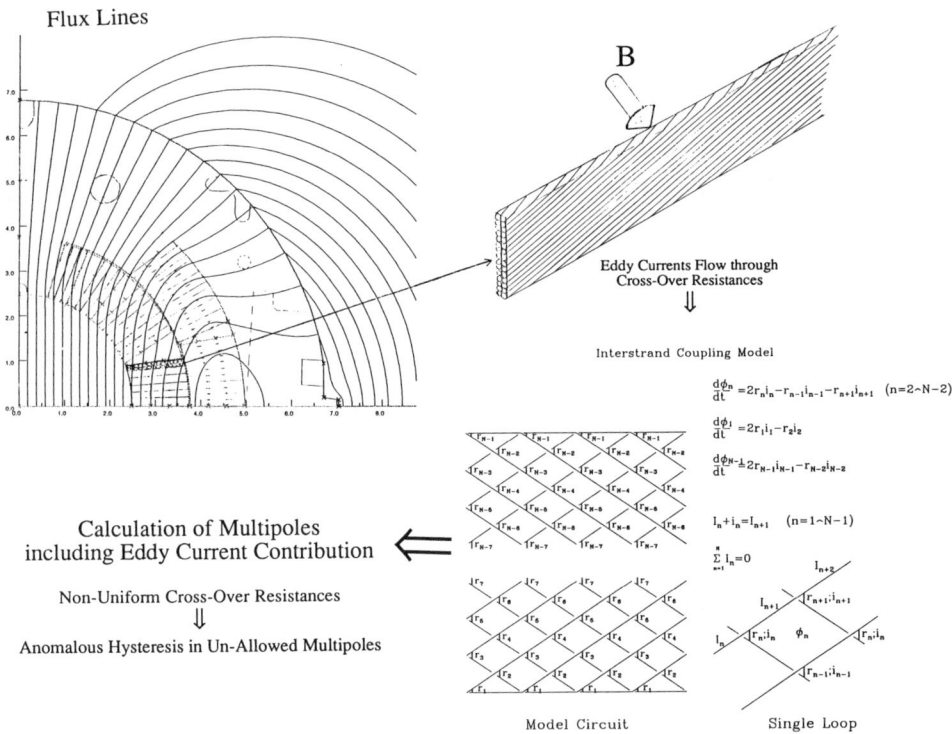

Figure 11. Model of eddy-current effects.

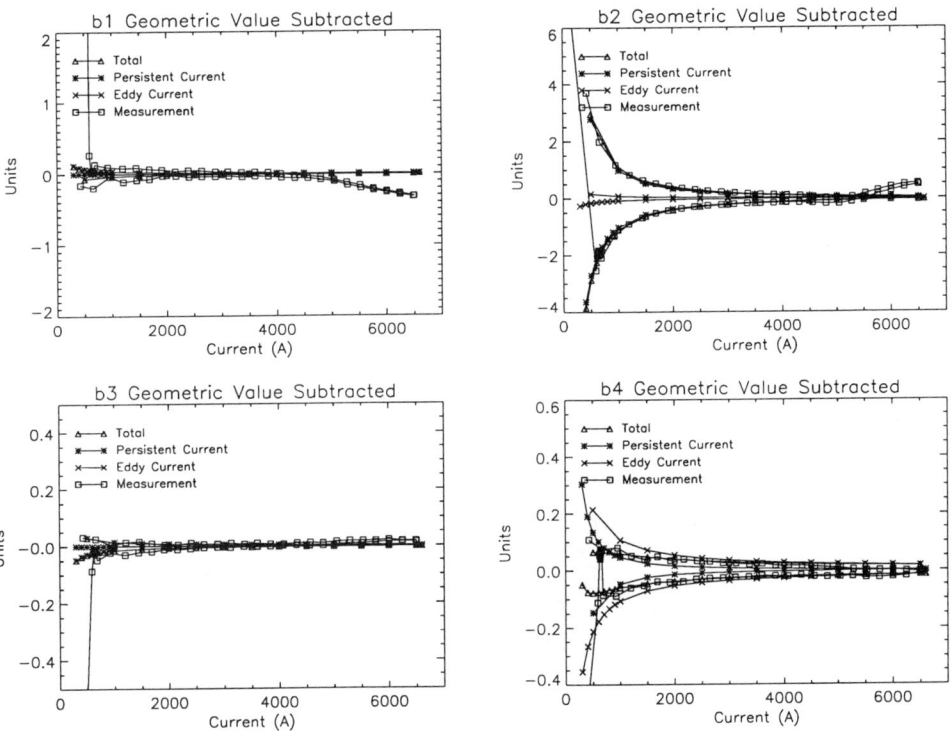

Figure 12. Dipole magnet multipoles (start).

Future Hadron Collider: The SSC

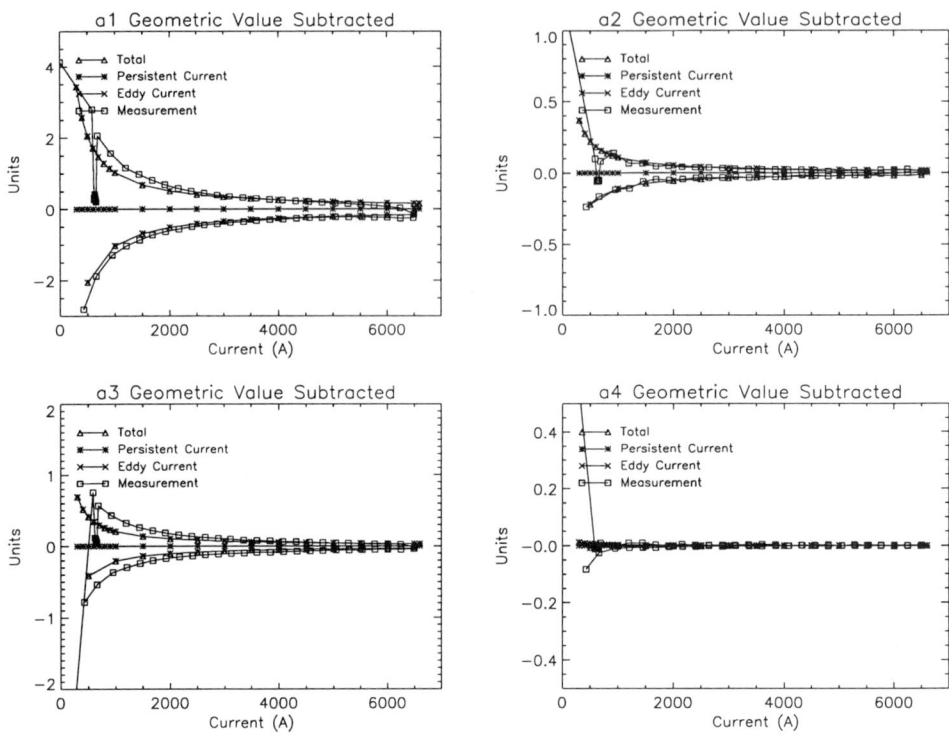

Figure 13. Dipole magnet multipoles (end).

Table 2. Collider arc lattice.

Quantity	SCDR Lattice	Current Lattice
15 m CDM field integral @ 20 TeV	100.008 T-m	101.363 T-m
13 m CDM field integral @ 20 TeV	83.424 T-m	84.469 T-m
15 m CDM magnetic length	15.165 m	14.99 m
CDM field @ 20 TeV	6.6 T	6.762 T*
CDM margin	> 10%	> 10%
Quadrupole integrated gradient @ 20 TeV	1069 T	1069 T
CQM magnetic length	5.2 m	5.025 m
Quadrupole gradient @ 20 TeV	205 T/m	212.7 T/m
Collider operating temperature	4.35°K	4.25°K
Collider operating current @ 20 TeV	6500 A	6668 A

* Increased CDM saturation requires an increase in the quadrupole corrector strength for tracking during acceleration.

slightly to provide more room in the lattice for other components. We have lowered the temperature slightly to keep the operating margin for the magnets; the operating current will then be somewhat over 6600 amps. The precise geometry of the collider and injectors is fixed so that construction of the tunnel can proceed. Figure 14 gives an overview of the tunnel design. In addition to the main tunnel, which is about 14 feet in diameter, there are a number of shafts from the surface down to the tunnel that are used to provide access for utilities or personnel; the oval shafts are used for magnet installation. At present we have completed a triplet of shafts, and we have under contract the four sections of tunnel indicated by the shadings in Figure 14. The contracts cover essentially half of the collider tunnel, and we will be getting under way with tunnel boring machines in September of this year.

Figure 15. Inside magnet delivery shaft.

Figure 16. N-15 construction site.

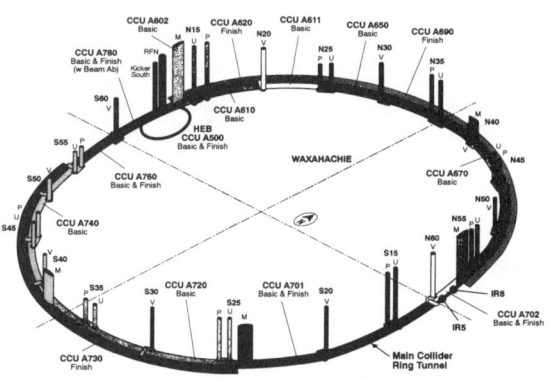

Figure 14. Tunnel design overview.

Figure 15 is a photograph taken down inside the large magnet delivery shaft where they are preparing the way for the large tunnel boring machine. Figure 16 shows the surface area of the magnet delivery shaft. Under construction here is the utility tunnel where the cryogens and power supplies, which will be housed in the buildings shown, will feed this region of the ring. The N-15 service area shown in Figure 16 is typical of those that will be located at intervals around the ring. Twelve of them will be built for the full complex of the high-energy booster and the collider.

The most critical technical components in the collider are the 50-mm dipole magnets. We have been working with Fermilab, Brookhaven National Laboratory, Lawrence Berkeley Laboratory, Saclay, and KEK on the development of the magnets needed for this facility. Elegant engineering is being focused on the ends of the magnets to improve our ability to install them and service them in place. These magnets exhibit excellent mechanical integrity. Figure 17 displays quench curves for several magnets, indicating a healthy operating margin between typical quench currents and the current required for 20 TeV operation with a virtual absence of training quenches. We have just about completed the preliminary development cycle for the collider dipoles. Eighteen of the 50 mm magnets have been built, 12 of them by industry working at Fermilab and Brookhaven. Three more dipoles will be produced in this preliminary phase; two of them will be built at our Magnet Development Laboratory (see Figure 18). We built the MDL on our campus site, and it is already an active laboratory. Soon, magnets will be tested in the Magnet Test Laboratory and constructed at the N-15 site. The MDL also has the capability for mass producing dipoles (see Figure 19). We will probably focus, however, on constructing various special magnets that are needed in relatively small quantities; most of the magnets of standard design will be built by industry.

We are now in the midst of a very critical undertaking called the Accelerator System String Test, which is a full system test of one half-cell of the machine. The test consists of

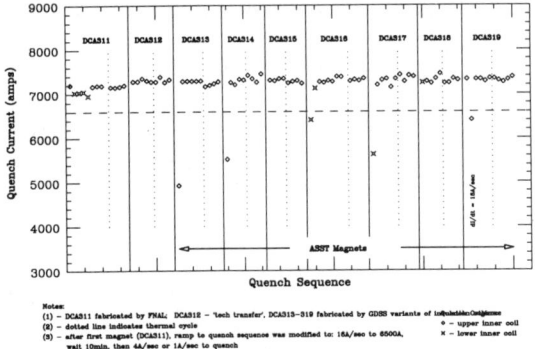

Figure 17. 4.35K quench performance.

Figure 18. The magnet development laboratory.

Figure 19. MDL interior.

five dipole magnets, a quadrupole magnet, and a "spool piece" where all the plumbing and correction coils are found. Also installed at the N-15 site, the test has been cooled down and is now being operated (see Figure 20). It has a full control room where we can begin to test some of the concepts of our control system (see Figure 21). We began to cool down the string at the end of June; the waves of reduced temperature propagating through the string of magnets can be seen in Figure 22.

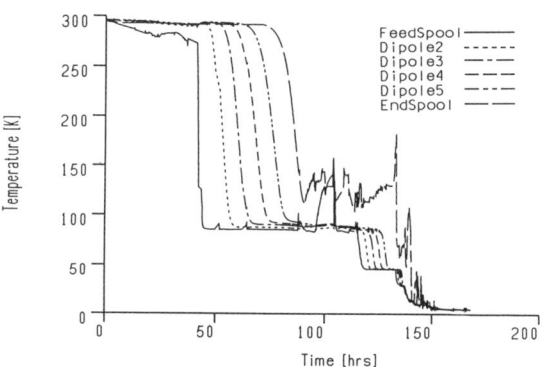

Figure 22. ASST cool down.

Figure 20. The ASST.

Just two years ago, literally to the day, before this string of magnets became superconducting, the land for the N-15 site was purchased by the State of Texas. It was poor, unimproved land, but today there are 100 meters of superconducting magnets there and a lot of related instrumentation. Typical voltages, pressures, and currents during quenches are shown in Figures 23, 24, and 25. In early August we began increasing the current, and we achieved operation at 4000 amps, roughly two thirds of that required for the full 20 TeV operations. (Author s note: The magnet string was successfully ramped to 6.6 T, the nominal SSC operating field, on August 14, and held at that level for some minutes before being lowered to zero.)

Figure 21. ASST control room.

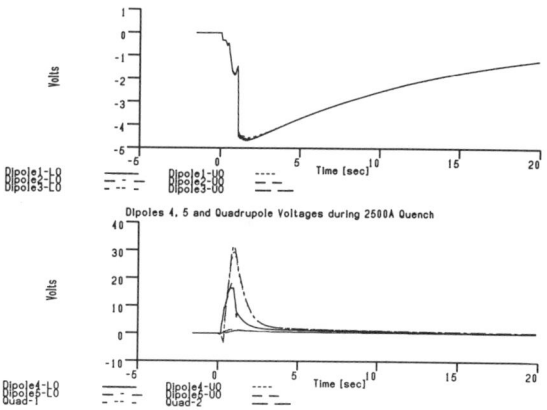

Figure 23. ASST quench characteristics (start).

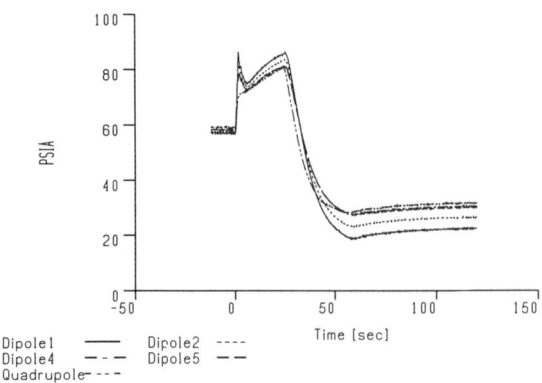

Figure 24. ASST quench characteristics (cont'd).

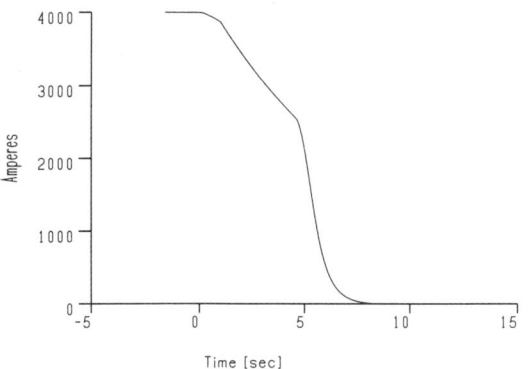

Figure 25. ASST quench characteristics (end).

A significant problem presented by the very high luminosity performance of machines such as the SSC will be the vacuum in the beam tube because of the desorption of gas molecules by synchrotron radiation from the beams. To deal with this vacuum issue, an extensive R&D program is already under way. It is a collaborative, world-wide effort to study problems of photo-desorption of gas off the cold surfaces inside the superconducting magnets. An important decision facing us within the next few months will be whether to put a special liner in the beam tube to intercept the synchrotron radiation.

Regarding schedule, we are already beginning to build the Linac. Depending largely on the total funding that is voted by Congress, we are placing the highest priority on maintaining the collider schedule. Other desirable systems, such as our test beam facility, will have to be delayed, however. We feel it will still be possible, but *difficult*, to complete the machine in the spring of 1999 and to begin commissioning during the summer of that year so that the physics could begin in the fall (see Figure 26). Depending on funding in subsequent years, this schedule may be delayed.

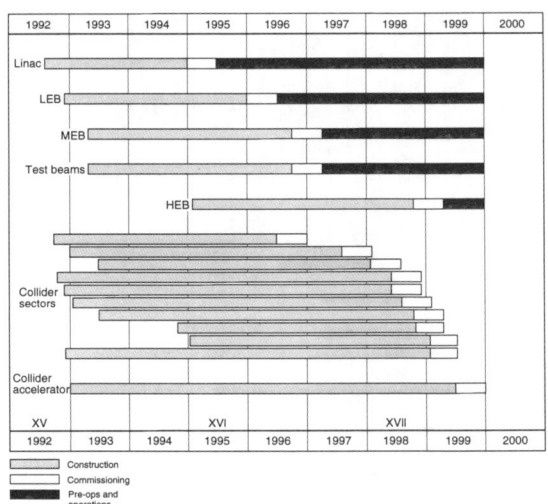

Figure 26. SSC schedule to FY2000.

The SSC's initial scientific program was covered well in the presentation by Takahiko Kondo.[5] We began the process of defining the SSC's initial scientific program with the receipt of Expressions of Interest in June of 1990. To date we have received 21 Expressions of Interest and they run the full gamut from huge detector collaborations down to one-person, one-page proposals. We are now in the process of formal reviews that will move forward to the selection for construction of two large detectors. These are huge international efforts representing roughly half of the U.S. experimental high energy physics community and a comparable, maybe even larger, number of foreign participants. We and our advisory committees feel that it is important to reserve some capital funds for the support of smaller experiments that can address other aspects of SSC physics. We are now in the process of hosting workshops and will be calling for new proposals for smaller experiments sometime within the next two years. The worldwide effort in detector R&D over the last three to five years has been outstanding, giving us confidence that the very large and smaller experiments can be designed to operate at the 10^{33} level of luminosity and perhaps higher.

In conclusion, the scientific opportunity at the SSC will be unparalleled. The machine represents a 20-fold increase in energy beyond what is available today, and it will be able to explore physics beyond the standard model. We are making every effort at our Laboratory to preserve the possibility of a diversity of experimental areas. As of today, much of the Laboratory staff has been assembled, and they are a smoothly working team of the highest caliber. Substantial construction is under way; the string test is in progress (see the author's note above), and two large detector collaborations are moving ahead. We look forward to the view beyond the standard model that the SSC will give us beginning at the turn of the century.

REFERENCES

1. B. Wiik, "Status of HERA and the First Results," presented at the XXVI International Conference on High Energy Physics, Dallas, Texas, August 12, 1992.

2. T. Kondo, "Recent Developments in Detector Technology," presented at the XXVI International Conference on High Energy Physics, Dallas, Texas, August 11, 1992.

3. B. Bjorken "Expression of Interest," "A Full-Acceptance Detector for SSC Physics at Low and Intermediate Mass Scales," SLAC-PUB-5545, May 1991.

4. R. Siemann, "Future Electron-Positron Colliders and other Accelerator Technologies," presented at the XXVI International Conference on High Energy Physics, Dallas, Texas, August 11, 1992.

5. T. Kondo, op. cit.

DISCUSSION

Gil Gilchriese, Lawrence Berkeley Laboratory, USA

While the SSC is not out of danger yet, what would you as the director of the SSC suggest to us, the future users of the machine, to tell our Congressmen to convince them that this is a necessary machine?

Schwitters

I think the best way to get the message to Congressmen is to convince the people who vote for them how important the SSC is. I believe we owe it to society to do our best in explaining the values of high energy physics to our fellow citizens, since they are paying the bill. You should go out and talk to the local Lions club, Cub Scouts, and similar groups. Such an educational effort is a huge task, but it is important and needs doing without delay.

C. Rubbia, CERN

You have shown a table indicating that by the end of the century the machine will be finished as a construction project. Do you have a funding profile associated with it? If you want to spend 8 billion dollars in eight years, it takes at least 1 billion dollars every single year.

Schwitters

The plan for the upcoming fiscal year called for $650 million in federal funding. As you know, the Senate voted for $550 million. We are, of course, analyzing the impact of such a reduction on the schedule. Priority is going to the collider in order to maintain that schedule. It may make it more difficult for us to provide test beams, say, as early as we desire. The shortfall in funds will also lead to inefficiency and increased costs through inflation that will increase the overall construction cost. We are not yet projecting a delay in the completion date for the collider, but continued shortfalls in funding will certainly delay things. The biggest danger we face, in my view, is chronic funding reductions leading to major slippage in schedule. We are doing everything we can to maintain the schedule.

Georgio Belletini, Sezione Infn di Pisa

Recently, a study was made by the accelerator group of the SSC indicating that if you take the start of the machine for physics to be when the luminosity is 10^{30}, it would take 3 calendar years to reach 10^{33} with a factor of ten increase per year. How technically sound is this study, and what are the real limitations?

Schwitters

You are referring to a preliminary "what if" study. What if the luminosity goes up in steps like this? That study represents a set of judgments by physicists and accelerator experts, an attempt to plan a rational early program for the SSC. I would describe it as the initial response, and now we must review the plan with all the various parties involved—detectors and accelerators—and refine it accordingly.

K. Cahill, University of New Mexico, USA

It may be possible to improve injection into the low-energy booster by having an H^- physics facility there, and it might only cost a half million dollars. It would also blunt criticism from the atomic physics community. Is there any movement in that direction?

Schwitters

We received a proposal to do just that. The idea is to study the H^- system as an example of a three-body quantum-mechanical system. There are some interesting resonances and questions that one can study with relativistic H^- beams where the Doppler shift can help. We also plan to have a special channel to divert protons or H^- ions from the Linac to be used in a proton therapy facility. Here one can use laser-stripping to make a fail-safe mechanism to bring the protons out in a safe manner.

FUTURE HADRON COLLIDER : THE LHC

Carlo Rubbia
CERN
CH - 1211 Geneva 23
Switzerland

Abstract

The present strategy for the future of European high energy physics should allow Europe to retain its leading role well into next century by making optimal use of present facilities and within the present level of funding. It concentrates available resources on those areas in which Europe has both experience and an international reputation, and which we believe are most likely to address some of the key open questions on the subject. Within such a strategy, the LHC project is the most natural and cost effective way to address the several fundamental issues which are being formulated with increasing clarity by the exceptional quality of data collected at LEP. LHC is now being actively prepared by an intensive R&D phase, which is expected to lead to the final proposal by end 1993.

STRATEGY FOR THE FUTURE OF PARTICLE PHYSICS IN EUROPE

The international exploitation of LEP, HERA and Gran Sasso are the foundations of the European programme for the 1990's. Our investment of the previous decade will pay off in a series of detailed experimental investigations of the Standard Model and beyond.

However, it is unrealistic to expect all the open questions to be answered by the present programme : indeed some of them cannot even be addressed at LEP or HERA, where collision energies can only reach the 100 to 200 GeV region. These machines will have exhausted their potential for discovery around the end of the 1990's, and, since time scales to design, build and commission new accelerators and detectors are of the order of a decade, we must plan now for their successors.

In order to ensure that Europe can continue to make important contributions in particle physics in the late 1990's and beyond, we must aim at establishing a programme that makes the most out of "pooled" financial resources which at best will remain constant in real value. This implies focussing upon a limited number of central issues. Our immediate priority is to ensure that we exploit fully our experiments at LEP, HERA and Gran Sasso. However, we need to prepare initiatives which will at first complement LEP and HERA and later on take over from them.

To achieve these goals will require flexibility. The nature of particle physics is such that we must be able to change direction in the light of new discoveries or new opportunities that may emerge, despite a heavy infrastructure. We also need to support a few smaller scale activities that impinge strongly on these main objectives : fixed target experiments and a number of "factories" which are now under construction or under consideration, for instance a Φ factory at Frascati[1], a charm-τ factory in Spain[2] and beauty factories in countries such as the United States[3], Japan[4] and Russia[5].

In addition, we must encourage the development of new techniques if we are to stay at the forefront of research in particle physics and continue to provide the new instruments and facilities that are so valuable to other areas of science.

Finally these goals are to be achieved within a collaborative effort on the full European scale. Today about 80% of the European physicists make use of CERN as their primary laboratory and both HERA and Gran Sasso are widely internationalized. Opening to the Central and Eastern European Countries has necessitated some additional temporary sacrifices but has also brought enormous new potentialities. Germany is unified, Poland, Czechoslovakia and Hungary are now CERN Member States and Russia and Israel are CERN-Observers. We are seeking to extend and to strengthen the involvement of the non-European partners through cooperation agreements etc.

I shall now outline, for the late 1990's, a programme which evolves naturally from our present programme, with new activities replacing current activities once machines have been fully exploited or when the research itself forces a change of emphasis.

Our main physics strategy is to target on at least four of the unresolved issues :

− to complete the precision tests of the Standard Model and understand the origin of symmetry breaking. Either the Higgs mechanism is confirmed or some entirely new phenomenology is revealed;

− to understand the three (by now firmly established) generations of quarks and leptons. The third generation (τ, ν_τ, top, b) is still poorly known. The asymmetries of weak interactions, and especially CP violation need to be clarified possibly within the CKM matrix texture;

− to discover the nature of the dark matter in the Universe : either massive neutrinos (hot dark matter) or SUSY particles (cold dark matter) are strong candidates. It is likely that both are needed to explain recent developments in Cosmology;

− to search for quark-gluon plasma and possible phase transitions in hot and highly compressed nuclear matter as predicted by QCD.

The programme we are concentrating upon is modest when compared to all the a priori conceivable possibilities. It concentrates available resources on those areas in which Europe has both experience and an international reputation, and which we believe are most likely to address some of the key open questions on the subject.

STATUS OF THE STANDARD MODEL

The Standard Model of electroweak interactions has so far been tested at LEP to a precision of 1% or better : no deviations have been observed as yet. With over 3 million Z^0s collected by ALEPH, DELPHI, L3 and OPAL, LEP has entered the high precision era and the Z^0 mass determination is now dominated by small systematic uncertainties (m_Z = 91.187 ± 0.007 GeV)[6]. The LEP absolute energy scale is calibrated using resonant depolarization which allows a current precision of ~ 7×10^{-5}. This extreme precision can be illustrated by the

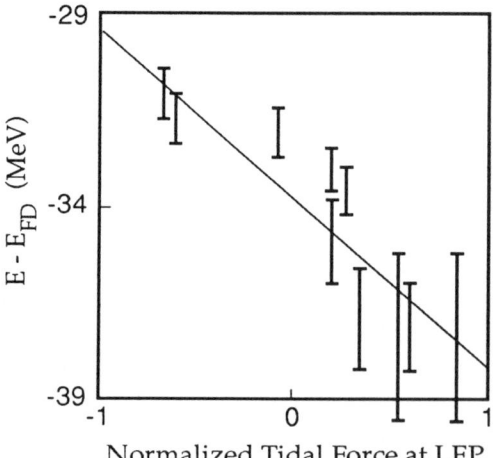

Figure 1. Difference between the energy obtained from resonant depolarization (E) and from magnetic field measurement (E_{FD}) versus the expected moon tidal force at LEP normalized to the range -1 to +1 for a complete moon cycle.

sensitivity to the moon tidal movements which only deform the LEP ring diameter by about 300 μm but produce detectable systematic energy shifts (Figure 1). Once these tidal movements have been taken into account and with a better control of temperature effects one should improve the present precision by a factor of 2~3.

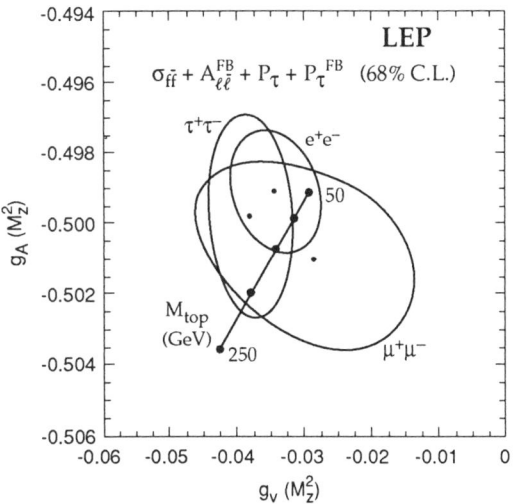

Figure 2. Determination of the axial-vector (g_A) and vector (g_V) weak couplings for the three families of charged leptons produced in Z^0 decays at LEP using the production cross-section, the forward-backward charge asymmetry, the τ polarization and the forward-backward asymmetry of τ polarization.

One can judge the impressive progress from LEP in Standard Model tests by considering for example the determination of the neutral current weak couplings (Figure 2) where lepton universality has been verified down to 1% precision for the axial-vector coupling[6]. The direct determination of the effective mixing parameter, $\sin^2\theta_W^{eff}$ from the measurements of forward-backward asymmetries of lepton and b-quark pairs and of τ polarization has now reached a high precision of 5×10^{-3} (Figure 3).

With forthcoming data from LEP and LEP200 we are aiming at tests of the Standard Model with an accuracy of 0.1% or better to provide crucial clues of what may lie ahead and

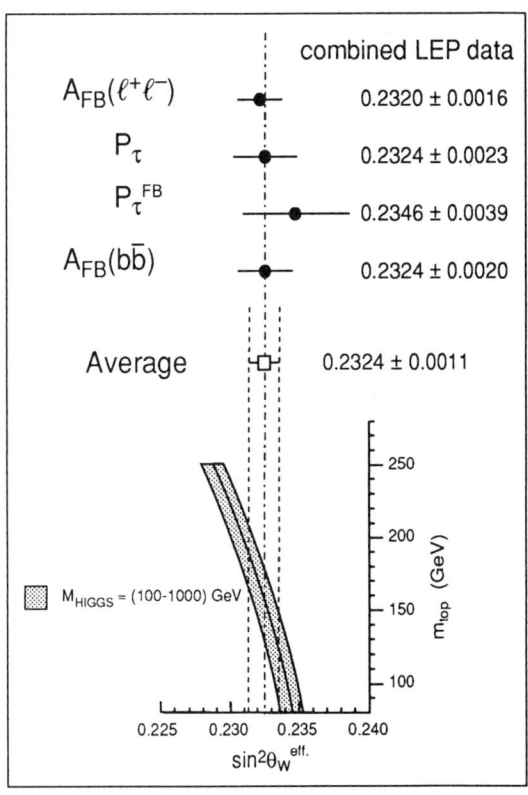

Figure 3. Determination of the effective weak mixing parameter $\sin^2\theta_W^{eff}$ at LEP from lepton pair and $b\bar{b}$ forward-backward asymmetries, τ polarization and forward asymmetry of τ polarization. A comparison is made with the prediction of the Standard Model as a function of the top quark mass and for extreme values of the Higgs mass.

obtain important guide-lines to plan for the "next step". For instance many observables[7] at LEP show sensitivity to the top quark mass (Figure 4) and to the Higgs boson mass and thus provide predictions for these masses within the Standard Model. Using the relation between the Weinberg mixing parameter and the top quark mass in the Standard Model[8] (Figure 5) a W mass accuracy of 50 MeV expected with the next LEP energy upgrade, scheduled to start by 1994 (LEP200) will determine $\sin^2\theta_W$ with an error of only 10^{-3} and therefore constrain the top mass to 10 ÷ 20 GeV which is comparable to the expected experimental accuracy for the direct top observation by hadronic colliders and set rather stringent limits to the possible Higgs mass.

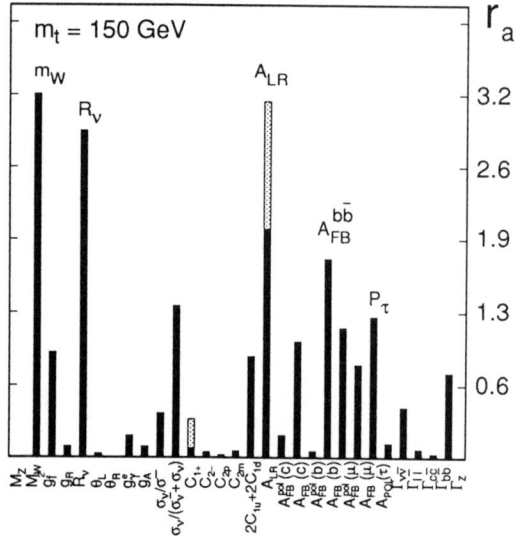

Figure 4. Effect of varying the top quark mass from 100 GeV to 150 GeV on a selected sample of variables. r_a is a measure of the sensitivity of a given variable. It is defined as the ratio of the predicted change to the expected experimental error (see ref. 7).

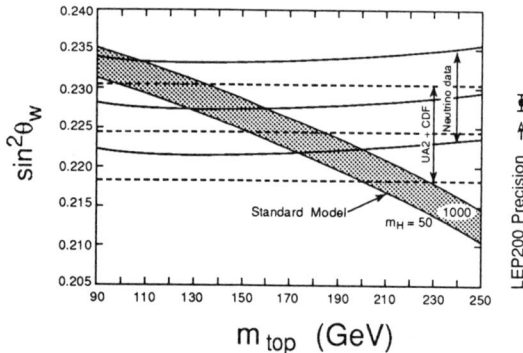

Figure 5. $\sin^2\theta_W = 1 - m^2_W/m^2_Z$ versus the top quark mass. The gray area shows the range of predictions from the Standard Model for extreme values of the Higgs mass. The measurements shown are from the CDF/UA2 determination of m_W/m_Z and from neutrino-nucleus scattering data (ref. 8).

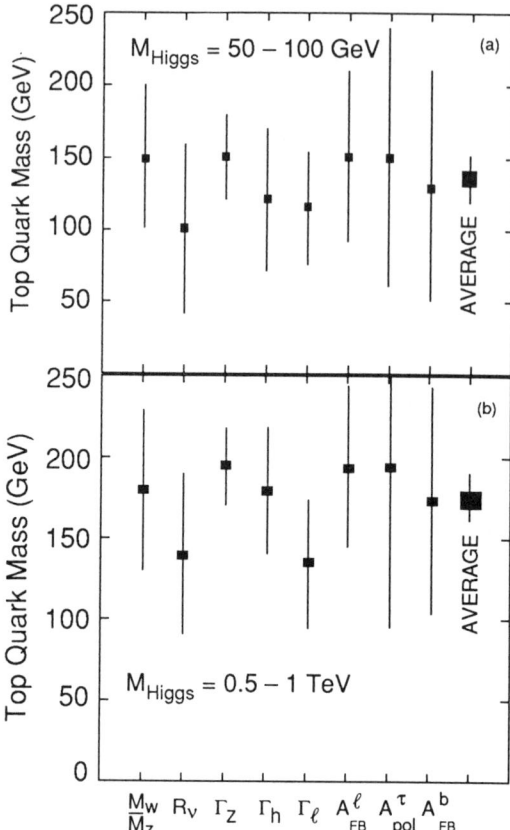

Figure 6. Top quark mass values corresponding to various measurements of Standard Model parameters and for two different Higgs mass ranges : (a)$M_{Higgs} \in$ [50, 100] GeV. The corresponding average top quark mass is 135 ± 17 GeV, (b) $M_{Higgs} \in$ [0.5 - 1] TeV, the corresponding average top quark mass is 177 ± 15 GeV. The combined value is : $m_{top} = 155 \pm 30$ GeV.

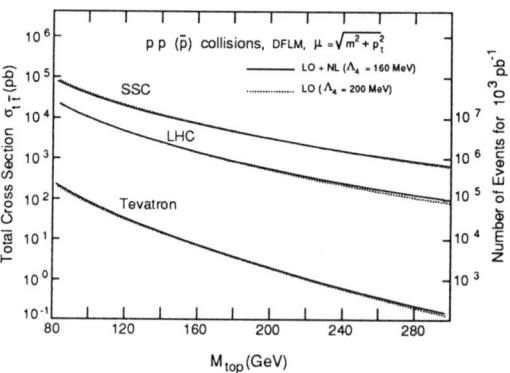

Figure 7. Top quark production cross-sections at hadron colliders (Tevatron, LHC, SSC) as a function of the top quark mass.

So far we have found that :

– the top quark mass is large, $m_{top} = 155 \pm 30$ GeV[8] (Figure 6). Searches at Fermilab become difficult because of the small production cross-section (Figure 7) and clearly LHC will be needed to provide a definite exploration of top quark phenomenology because of the factor 1000 increase in cross-section and because of the much higher luminosity (factor 100);

– the Higgs may well be in the range explorable by LEP200 whose mass reach extends to about 100 GeV. Present data indicate a very low preferred Higgs mass[9] (Figure 8). The direct searches at LEP set a lower mass limit of 60 GeV (95% C.L.)[6] (Figure 9) for the Standard Model Higgs;

– if the Higgs exists and is not found at LEP it will in any case be within the range explorable by LHC since the present upper limit on its mass at 90% C.L. is 1 TeV;

– the SUSY particle mass threshold[10] (Figure 10) is also likely to lie within the explorable range of LHC and at the same time the Minimal SUSY's Higgses could be within the range of LEP200;

– many Technicolor models have been already excluded by LEP[8], but the idea is still alive;

– proton decay and GUT are still "on" but at larger unification masses (10^{16} GeV)[10] which will require new generation experiments (Gran Sasso).

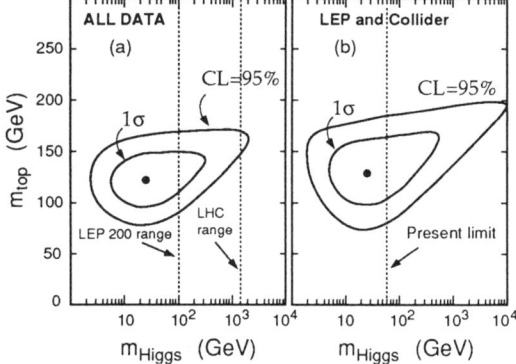

Figure 8. The $\Delta\chi^2 = 1$ and 2.7 contours in the M_{Higgs}, M_{top} plane for (a) the full electroweak dataset, (b) LEP and collider data alone (see ref. 9).

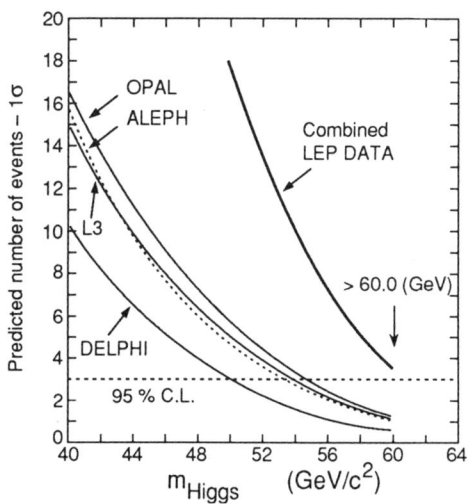

Figure 9. Limits on the Higgs mass from direct searches at LEP. Individual experiment limits and combined limit are shown as function of the Higgs mass.

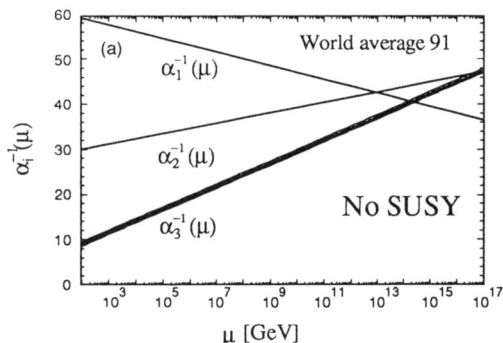

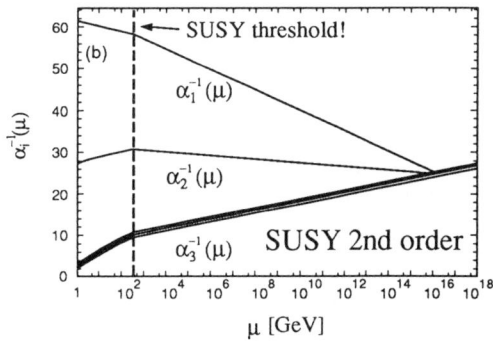

Figure 10. Extrapolation of strong, weak and electromagnetic couplings to the unification scale (a) within the Standard Model, (b) within the minimal supersymmetric model.

In conclusion, LEP physics is a necessary pre-requisite for further studies at the TeV constituent energy range and it must be vigorously pursued as a fundamental preparatory step. With the LEP machine performing better than ever we can only be confident that in the future we shall be able to target with sufficient accuracy where the LHC should find such a new physics. In turn this leads to what one could call a "no-lose" condition for the LHC : either these new particles are found or there is even a bigger surprise signalling a breakdown of the Minimal Standard Model. In this sense LEP and LHC are two closely connected programmes within a common strategy aiming at elucidating a phenomenology which CERN has initiated with the discovery of the Z and W bosons about ten years ago.

THE LHC PROJECT

In order to experimentally investigate the fundamental questions discussed above, whatever mechanism is involved, we need to perform experiments at constituent energies in the 1-2 TeV region. At present, such high energies are accessible only at machines that collide protons, where experiments observe the interactions of their quark and gluon constituents. To reach such an energy domain, a proton beam with energy in excess of 6 TeV is needed. Since we know from dimensional arguments that cross-sections decrease as E^{-2}, we must at the same time master luminosities which are largely in excess of what is customary today.

We will be able to achieve proton beam energies of more than 6 TeV, at unprecedented luminosities well in excess of 10^{33} cm^{-2} s^{-1}, with the proposed Large Hadron Collider (LHC)[11] (Table 1).

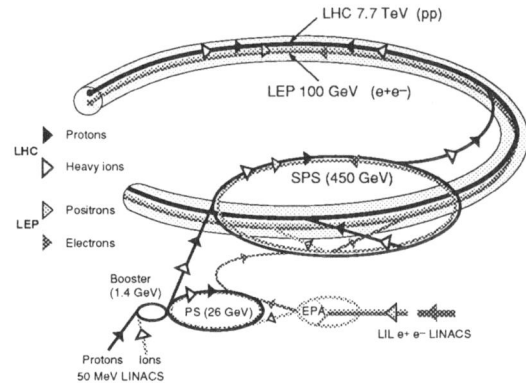

Figure 11. Schematic layout of the LHC injection complex.

Table 1. LHC Main Parameters

		pp	e-p	Pb-ions
Max c.m. energy	(TeV for B=9.5T)	15.4	1.36	1262
Luminosity	(cm^{-2}s^{-1})	1.6 10^{34}	2.8 10^{32}	1.8 10^{27}
Number of bunches		4725	508	800
Bunch spacing	(m/ns)	4.5 / 15	49.4 / 164.7	31.5 / 105
Particles/bunch		10^{11}	9.2 10^{10} 3.0 10^{11}	6.2 10^7
Particles/beam		4.7 10^{14}	4.7 10^{13} 1.5 10^{14}	5.0 10^{10}
Number of experiments		3	1	2
β at interaction point	(m (βx, βy))	0.5	0.85, 0.26 32.7, 3.05	0.5
r.m.s. radius at int. pnt.	(μm (x, y))	15	120, 37	12.4
r.m.s. collision length	(cm)	5.3	3.8	5.3
Crossing angle	(μrad)	200	0	200

As is now traditional at CERN, this new machine would exploit existing facilities. The LHC be built inside the LEP tunnel, and would use the older Proton Synchrotron and Super Proton Synchrotron as injectors (Figure 11), just as LEP does. Successful tests of the Linac/Booster complex (nominal intensity and emittance) have already been carried out, and tests are under way in the SPS to form tightly spaced bunches (10ns) (Figure 12). A 25 ns-spaced-bunch scheme is considered to possibly further increase the luminosity while maintaining the same total circulating current.

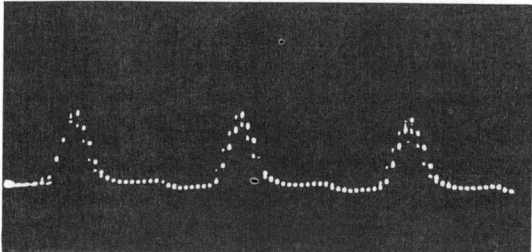

Figure 12. LHC beam in the SPS. The oscilloscope shows 10 ns spaced bunches which have been accelerated to 300 GeV.

To achieve high energies, it is proposed that LHC uses long, high-field superconducting magnets (Figure 13) based on an innovative "two-in-one" design.

This results in substantial savings in cost and construction time. The radius of the LEP tunnel and the maximum field that is likely obtainable in the superconducting magnets, make it possible to reach a maximum proton beam energy in the LHC of around 8 TeV, which is well into the region of interest. The radius of the LEP tunnel was indeed optimized with the LHC in mind. An intensive R&D programme is being carried out. A test of a 10 metre long twin dipole magnet with two sets of HERA coils (Figure 14) has been very successful and has demonstrated the inherent soundness of the "2 in 1" design. The innovating magnet cooling technique uses superfluid helium at 1.9K and is based on that already being installed for the superconducting accelerating cavities of the LEP-200 project.

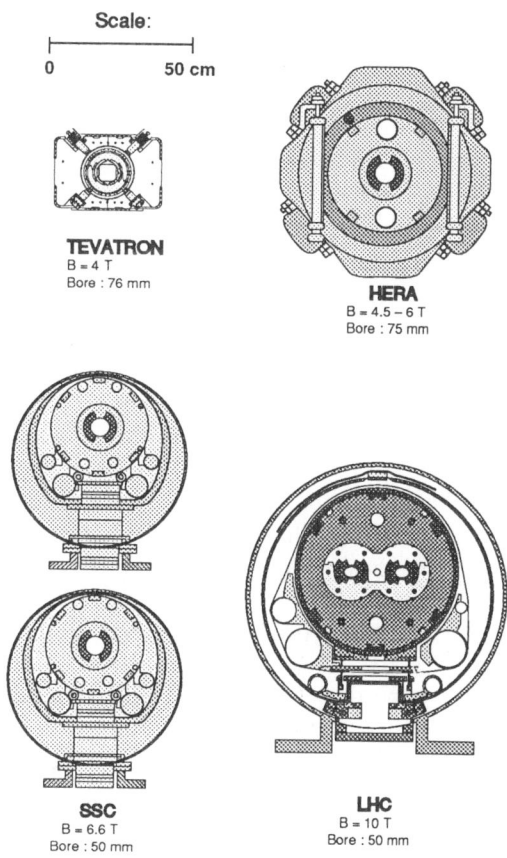

Figure 13. Comparison of the sizes of superconducting dipole magnets for the Fermilab Tevatron, HERA, SSC and LHC.

Figure 14. Photograph of a 10 metre long twin aperture magnet prototype with HERA coils.

The cryogenic scheme was successfully tested in a realistic model showing that it performs with a large safety margin. A full magnet string test is in preparation for 1993.

In a possible plan, the LHC would be ready for experiments in 1999, and from that date on would become a central part of CERN's programme. To obtain a reasonable rate for the interesting 1 TeV collisions at the LHC, the number of protons that collide together will have to be very large, resulting in challenging problems for detector technology.

Development work on detectors and electronics has already begun to face these challenges, and physicists are prominent in study groups set up to examine the design of experiments at the LHC. However, to be ready in 1999, this programme of research and development will have to be substantially increased in the coming years.

THE LHC EXPERIMENTAL PROGRAMME

The experimental programme is the object of a careful strategy in close cooperation with the SPC and ECFA. In October 1990, the Aachen workshop had confirmed the physics objectives and the experimental feasibility[12].

A strong detector R&D programme, coordinated by the Detector Research and Development Committee (DRDC) was designed to stimulate the necessary R&D, harnessing instrumentation activities for LHC and building the strength of the LHC experimental community. More than 1000 people are involved. In March 1992 the Evian Workshop[13] set the stage for detector collaborations when four proton-proton and several heavy ion, B physics and other fixed target expressions of interest were presented.

Following the suggestion of an "adiabatic" approach to the approval of at most two detectors, the four proto-collaborations have started a merging process. ASCOT and EAGLE have merged [at this stage they are considering two options (Figure 15 a&b)]; CMS (Figure 16) and L3P (Figure 17) are in the discussion process.

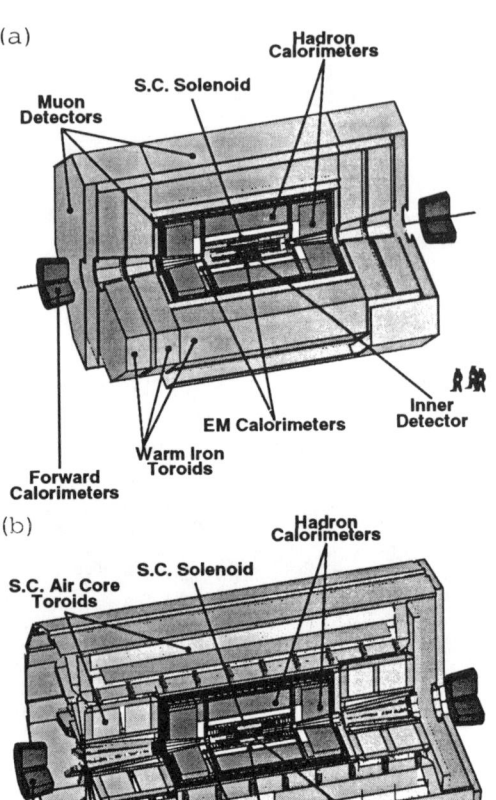

Figure 15. Schematics of the ASCOT-EAGLE detector for LHC : (a) Warm iron toroids option, (b) Superconducting air core toroids option.

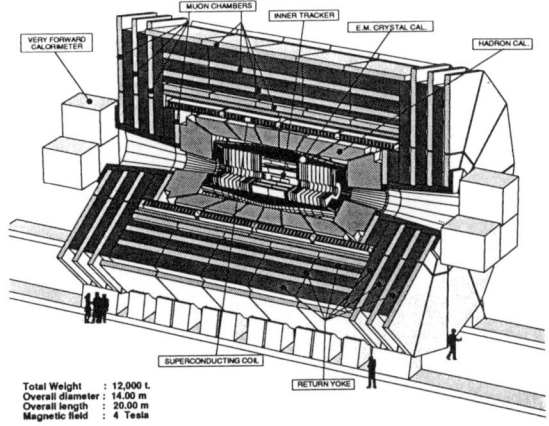

Figure 16. Schematics of the CMS detector for LHC.

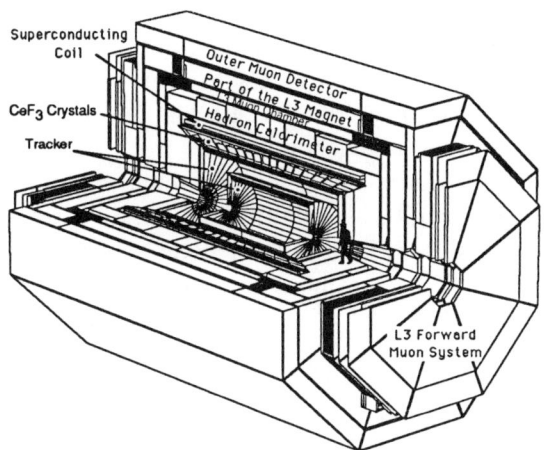

Figure 17. Schematics of the L3P detector for LHC.

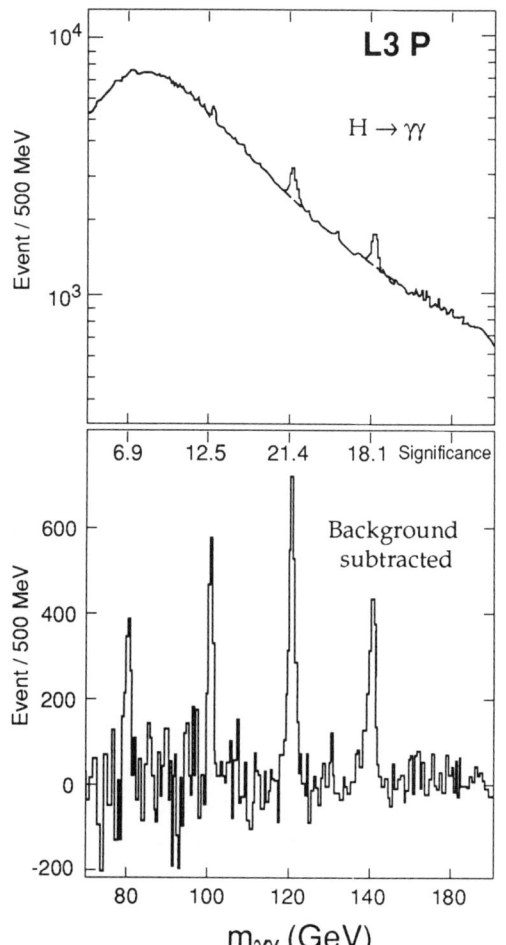

Figure 18. Simulation of the Higgs signal in the two photon channel in the crystal calorimeter of L3P.

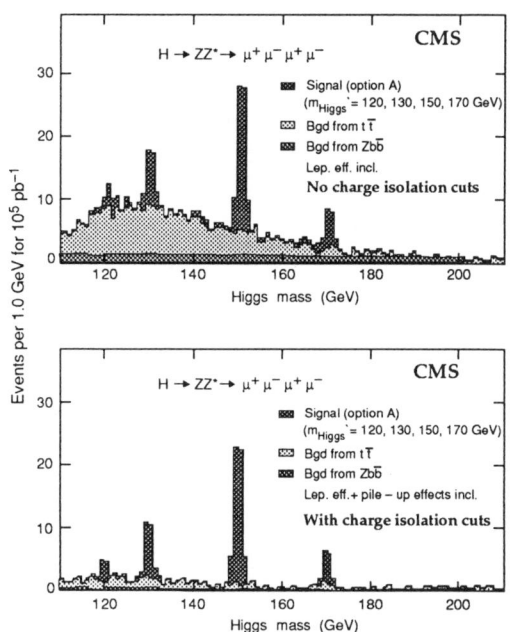

Figure 19. Simulation of the Higgs signal in the four muon channels in the CMS detector.

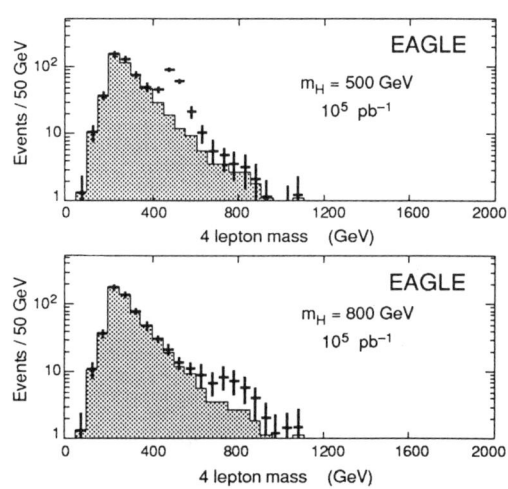

Figure 20. Simulation of the Higgs signal in the four-charged lepton channels (eeee, ee$\mu\mu$, $\mu\mu\mu\mu$) in the EAGLE detector.

It is particularly instructive to observe some of the important physics processes through the "eyes" of the detectors being planned. Detailed detector simulations have confirmed that the experimental environment at LHC would allow a search for the Standard Model Higgs in the whole relevant mass range, illustrated here by the signal in the two-photon channel (Figure 18) for masses from 80 to 150 GeV, in the 4-lepton channel (Figure 19 and Figure 20) for masses up to 800 GeV. Other channels (W pairs) can be used to cover safely the 1 TeV mass region beyond which the Higgs would cease to appeal as a clear resonance.

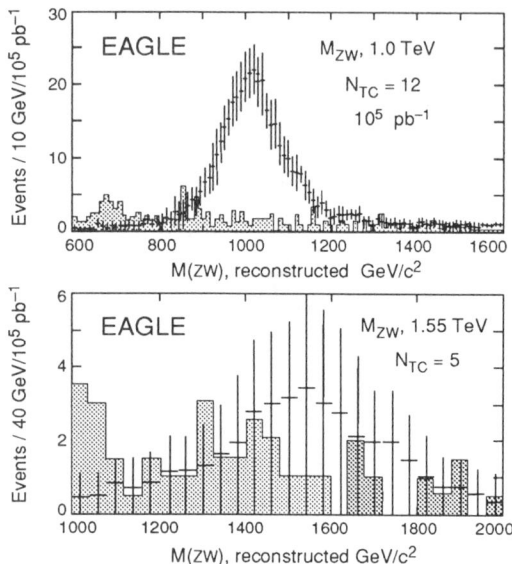

Figure 21. Simulation in the EAGLE detector of $W_L Z_L$ resonances.

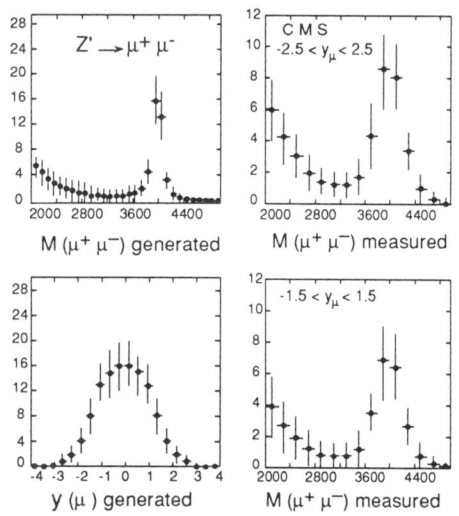

Figure 22. Simulation in the CMS detector of $Z' \to \mu^+\mu^-$ for $m_{Z'} = 4$ TeV in ALR model and 10^5 pb^{-1}.

The study of many other possible physics processes, for instance $W_L Z_L$ (Figure 21) and $Z\gamma$ resonances, heavy Z's (Figure 22) have confirmed that foreseen detectors can cope with the highest luminosity at LHC.

Letters of intent are to be submitted by 1st of October 1992. The LHC Committee is set up with the idea of an "adiabatic" approach to approved programmes with LHCC using the expertise of the DRDC. A conclusion[14] from the Evian Workshop was that, given the limited financial resources, there was a need to optimize detector cost versus detector performance and to consider an evolutionary rather than "all the way" approach for detector construction. In addition, before embarking in the mass production of the detector elements, the equivalent of the so-called "full string test" for the machine is necessary, with the corresponding need of a large amount of test beam and R&D.

B physics at the LHC

Both collider and fixed target modes are considered. Recently a first test using a bent Silicon crystal to extract 120 GeV protons from the SPS was successful (Figure 23) opening the possibility of using the technique at LHC. The

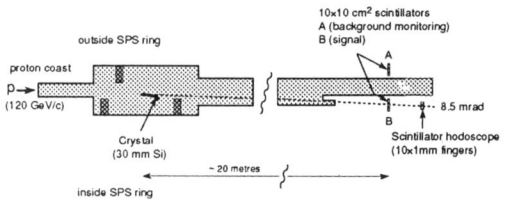

Figure 23. Layout of the test of crystal extraction of beams at the SPS.

$b\bar{b}$ rate at LHC will be very large both in the collider mode where the cross-section is of the order of 1 to 5×10^5 nb (Figure 24) and in the fixed target mode where it is about 500 nb. With a $b\bar{b}$ sample of 10^{13} events in collider mode and 10^{10} to 10^{11} events in the fixed target mode B physics at LHC promises to be exciting. Critical tests of the Standard Model[15] will be performed in the study of $B^0 \bar{B}^0$ oscillations, in the search for rare decay modes and in CP violation. The B sector constitutes a whole field by itself with a very broad physics potential.

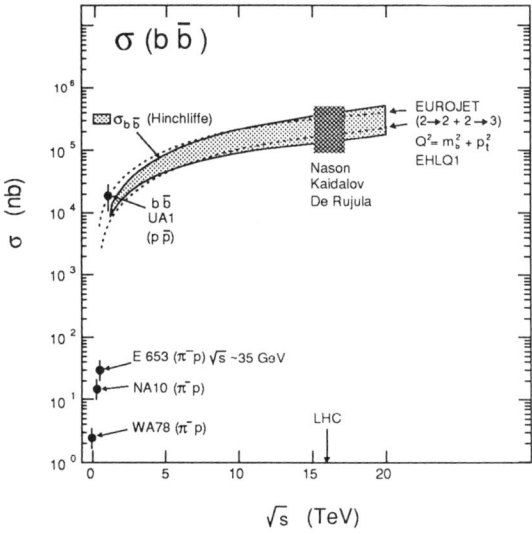

Figure 24. Production cross-section for Beauty at hadron colliders as a function of the centre-of-mass energy.

OTHER PHYSICS POSSIBILITIES AT LHC

The facilities available on the CERN site will allow for other exciting possibilities for the LHC beyond its role as a proton-proton collider. This makes the LHC an even more attractive option for European particle physics beyond the year 2000.

Heavy ion collisions

One possibility actively explored is to collide beams of lead ions at the enormous energies of 800 TeV in the LHC (Table 1). Heavy ion beams are already accelerated in the SPS and it would be only natural to store them in the LHC. Such collisions can produce unprecedented energy densities and volumes, and could therefore reveal fundamental insights into the structure of QCD and perhaps the quark-gluon plasma, a new phase of matter[17] predicted by QCD. Such a collider would be a natural development from the present heavy-ion programme at CERN, in which sulphur ion beams are already produced with an energy of 6.4 TeV per sulphur nucleus, and lead beams are under preparation.

Electron-proton collisions

A further possibility is electron-proton physics at roughly four times the energy available at HERA. This could be done by colliding electron or positron beams of 70 GeV energy from LEP with a 7 TeV proton beam from the LHC (Table 1). Such a possibility is unique to the LHC complex since CERN will have both accelerators on its site. It would allow for instance studies of quark structure down to sizes of around 10^{-19} m[16], a resolution significantly better than that of HERA. The discovery of quark substructure could explain the existence of the three generations of quarks and leptons that are known to exist.

CONCLUSION

Final approval of the Collider and the Detectors requires a decision of the CERN Council, which will start its deliberation procedure at the end of 1993, namely when the R&D phase and the definition of the experimental programme will be completed.

The combination of the various existing and planned accelerators will then give Europe an ensemble of facilities unrivalled anywhere in the world, thus continuing the strong European tradition in high energy physics.

In summary, the most important new initiative for European particle physics over the next decade will lie in the field of proton-proton collisions, some heavy ion collisions being also studied (lead in the CERN SPS). The LHC offers a relatively cheap way to access this crucial area of physics in Europe and has unique capabilities for further development into an electron-proton or heavy-ion collider. It thus has strong support of the European community of physicists and of the Member State Governments.

The SSC, on the other hand, will allow somewhat easier experimental conditions with its exploration of proton-proton collisions at higher energies, but at somewhat smaller luminosities. In broad terms however, the discovery potentials of the two machines appear quite comparable[12].

For both machines we emphasize the vital importance of timely and innovative detector development programmes, a domain where Europe and the United States could develop a strong collaboration, in order to exploit fully the exceptional physics potential of these new accelerators.

REFERENCES

1. *Workshop for Physics and Detectors for DAΦNE the Frascati Φ Factory*, Organized by INFN, April 9-12, 1991; Servizio Documentazione dei Laboratori Nazionali di Frascati, P.O. Box, 13 - I-00044 Frascati, Italy.

2. *Proc. Meeting on the Tau-Charm Factory Detector and Machine*, Sevilla, 1991 (Report TCF-91-3, Sevilla, 1991).

3. "An asymmetric B factory based on PEP, conceptual design", LBL PUB-5303, SLAC-372. "Conceptual design for a B factory based on CESR", CLNS 91–1050.

4. "Accelerator Design of the KEK B-Factory", National Laboratory for High Energy Physics, Tsukuba; Kurokawa S.; B-Factory Accelerator Task Force, Tsukuba, March 1991, p. 169.

5. "INP-Novosibirsk B-Factory", Zhoelents A.A., Novosibirsk, Int. Nucl. Physics, April 1990, p. 6.

6. L. Rolandi, "Precision Tests of the Electroweak Interaction", *Proceedings of the Dallas ICHEP 92*, this conference.

7. P. Langacker et al., "High-precision electroweak experiments : A global search for new physics beyond the Standard Model", *Rev. Mod. Phys.*, Vol. 64, No. 1, January 1992, p. 87-192.

8. G. Altarelli, "Precision Electroweak Data and Constraints on new Physics", *Proceedings of XXVIIth Rencontres de Moriond on Electroweak Interactions and Unified Theories*, Les Arcs, Savoie, France, 15-22 March 1992.

9. J. Ellis et al., "Updated prediction of the top quark mass and implications of its possible confirmation", BARI-TH. 111/92, CERN-TH. 6568/92.

10. U. Amaldi et al., *Phys. Lett.* B260 (1991) 447; W. de Boer, *Proc. Joint Int. Lepton-Photon Symposium and Europhysics Conference on High Energy Physics*, Geneva, 1991, eds. S. Hegarty, et al. (World Scientific, Singapore, 1992), vol. I, pp. 690–693.

11. *Design Study of the Large Hadron Collider (LHC)*, CERN 91–03, 2 May 1991.

12. *Proc. Large Hadron Collider Workshop*, Aachen, Germany, 4-9 October 1990, (CERN 90–10, Geneva, 1990).

13. *Proc. Towards the LHC Experimental Programme Workshop*, 5-8 March 1992, Evian-les-Bains, France.

14. C. Rubbia, Closing Remarks, *Proc. Towards the LHC Experimental Programme Workshop*, 5-8 March 1992, Evian-les-Bains, France.

15. For a review of B physics at LHC see for instance D. Denegri, "Standard Model Physics at the LHC (pp Collisions)", *Proc. Large Hadron Collider Workshop*, Aachen 1990, (CERN 90–10, Geneva, 1990), vol. I, p. 70-80.

16. For a review of e-p physics possibilities at LHC see for instance G. Wolf, *Proc. Towards the LHC Experimental Programme Workshop*, 5-8 March 1992, Evian-les-Bains, France, p. 121-136.

17. For a review of heavy ion physics possibilities at LHC see for instance U. Heinz, *Proc. Towards the LHC Experimental Programme Workshop*, 5-8 March 1992, Evian-les-Bains, France, p. 95-117.

DISCUSSION

B. Nicolescu, Université de Paris-Sud

Is there a realistic possibility of maintaining both proton and antiproton options in the same time frame with the LHC?

Rubbia

As you know, CERN has been a pioneer in antiproton accumulation and still maintains an accumulator system for its low-energy antiproton program. CERN's proton-antiproton collider $S\bar{p}PS$ has been decommissioned, with Fermilab's TeV $\bar{p}p$ collider in continued operation. While concentrating instead on fixed target physics, CERN has retained its antiproton source in conjunction with LEAR. Whether to terminate the latter program, or expand it in the form of SUPERLEAR, will be discussed at the September meeting in Cogne.

M. Peskin, SLAC, USA

It's true that the design luminosity of the SSC is more conservative than that of the LHC; but I thought that the ultimate luminosity of proton-proton colliders is really limited by the detectors, essentially by the resources that the laboratory is able to devote to the detector program. Do you agree with this statement, or could you comment on it?

Rubbia

Certainly. First of all, there are two limits on luminosity. One of them is, naturally, the detector. The other is the onset of synchrotron radiation, which is a very serious vacuum limitation for a machine operating in the 2-4 Kelvin region. The higher the energy, the more synchrotron radiation. Nevertheless, it will be highly advisable to fully exploit the luminosity of the LHC and the SSC — that is, a luminosity of 10^{34} cm^{-2} sec^{-1}. Clearly the luminosity is as important as energy. The maximum energy will depend on available resources. However, the luminosity barrier will be just as important and difficult to handle, the reason being basically that the cross section goes as $1/(energy)^2$. In a machine of extremely high energy, the luminosity, more than the energy, would be the real challenge.

W. Panofsky, SLAC, USA

Dr. Peskin's question was simply in regard to the limiting factor: machine or detector?

Rubbia

I was quite surprised by the relaxed views expressed at Evian-les-Bains concerning the question of coping with a luminosity of 10^{34}. It is not an easy goal, certainly a challenging one, and important to pursue. Let me remind you, however, that not all physics will require 10^{34}; for instance, for top physics at large statistics, 10^{33} will suffice. But striving for a detector with the ultimate capability of 10^{34} is highly desirable for exploiting the high-energy potential of both SSC and LHC, and in this regard I was encouraged by recent developments in detector technology outlined by Dr. Kondo in his presentation.

ELECTRON-POSITRON COLLIDERS AND OTHER ACCELERATOR TECHNOLOGIES

Robert H. Siemann[*]
Stanford Linear Accelerator Center
Stanford, CA 94309

Abstract

The accelerator physics and technology of hadron colliders, heavy quark factories, and linear colliders are reviewed. The status and performance of major high energy accelerators are summarized.

INTRODUCTION

Accelerators and accelerator technology are among the most important factors determining the frontier of particle physics. This paper is a summary of a talk presented at the 1992 International Conference on High Energy Physics. It was a peculiar talk for two reasons. First, the SSC and LHC are the major high energy physics construction projects at the present time, and HERA has just been completed and is starting to operate. Each of these had a dedicated talk, and, therefore, these projects, but not associated accelerator physics and technology, were outside the scope of the talk.

Second, the talk was presented to an audience of particle physicists and does not have the technical detail that you would find at an accelerator conference. The hope is give an overview of the status of major accelerators and a perspective of accelerators and accelerator science to physicists whose work depends critically on that science but who are not directly involved in it.

---

[*] Work supported by the Department of Energy, contract DE-AC03-76SF00515.

HADRON COLLIDERS

The TEVATRON [1]

During the previous TEVATRON collider run in 1988 - 1989 the typical peak luminosity was $1.6 \times 10^{30} \mathrm{cm}^{-2}\mathrm{s}^{-1}$ and the total integrated luminosity was 9.6 pb^{-1}. The long term goal is to reach $5 \times 10^{31} \mathrm{cm}^{-2}\mathrm{s}^{-1}$ with a concomitant increase in integrated luminosity. The luminosity,

$$\mathcal{L} = \frac{1}{4\pi} \frac{N_1 N_2 f_c}{B^2 \sigma_x \sigma_y}, \qquad (1)$$

can be rewritten in different ways to reflect different performance limits. (In this equation N_1 and N_2 are the total number of particles in each beam, f_c is the collision frequency, B is the number of bunches, and σ_x and σ_y are the rms transverse sizes.) The TEVATRON limitations are the number of $\bar{p}$'s and the phase space density, $N/(B\varepsilon)$ where ε is the invariant emittance, of the proton beam. The former comes from the difficulty of producing $\bar{p}$'s and the latter from the tune spread of the proton beam at low energies due to space charge and the beam-beam effects experienced by the $\bar{p}$'s during collisions. Taking account of these limits the luminosity becomes

$$\mathcal{L} = N_{\bar{p}} \frac{N_p}{B\varepsilon_p} \frac{\gamma(f_c/B)}{\pi\beta^*(1+\varepsilon_{\bar{p}}/\varepsilon_p)} . \qquad (2)$$

γ is the beam energy in units of mc^2 and β^* is the "beta function" that measures the depth-of-field of the focus at the collision point. Note that f_c/B is independent of the number of bunches.

The elements for reaching $5 \times 10^{31} cm^{-2} s^{-1}$ are: a) $\bar{p}$ source improvements, b) stronger focusing at the interaction points to reduce β^*, c) electrostatic separators to avoid unwanted collisions and thereby relax one of the limits on $N_p/B\varepsilon_p$ by reducing the beam-beam effect on the $\bar{p}$'s, d) a linac energy increase that reduces the space charge tune spread and relaxes the other limit on $N_p/B\varepsilon_p$, and e) the Main Injector that will allow further intensity increases. The first three of these are completed, and the expected benefits realized. The commissioning of and operating with electrostatic separators is especially important because it is key for realizing the benefits from the linac upgrade and the Main Injector. The linac upgrade is almost complete with commissioning scheduled to begin in the winter of 1993, and Main Injector beam is scheduled for early in 1997.

With these improvements the TEVATRON will be remain a centerpiece of high energy physics activity for the foreseeable future.

UNK [2]

The UNK project would be a 3.0 × 0.4 TeV proton-proton collider with a design luminosity of $10^{33} cm^{-2} s^{-1}$. The tunnel has been excavated and 70% of the components for the (warm) 0.4 TeV ring are completed. Twenty-five full scale superconducting magnets for the 3 TeV ring have been tested successfully, and preparations are in progress for a preproduction run of 100 magnets. Because of the financial problems in the former Soviet Union, completing UNK by 1997-1998 as an international project is being considered.

Accelerator Physics and Technology

Talks in the parallel sessions covered some of the most important accelerator physics and technology of hadron colliders. The following are short summaries of those talks.

Superconducting Magnets [3,4]

There are many interrelated aspects of superconducting magnets - mechanical and quench properties, field quality, cost optimization, etc, and there are detailed articles in accelerator conferences and excellent reviews [5,6] on the subject. New information about mechanical and quench properties is coming from the LHC and SSC, and talks at this conference concentrated on these.

A magnet will quench, become a normal- rather than a super-conductor, at some current. Ideally that current is the "short sample" limit where the current density exceeds the critical current density. Below the short sample limit: i) quenches are caused by a sudden motion of a conductor that raises the temperature of a section of magnet above the critical temperature, ii) once a conductor has moved, it does not return to its original position, and, as a result, iii) magnets "train" - successive quenches tend to occur at higher currents. The mechanics of the magnet determine these training and quench properties. A stiff coil support won't allow conductor motion. The support should include constraints on the ends of the coil, and its stiffness can be increased by tightly clamping the magnet yoke around the collared coil assembly. This is being done in the SSC magnets which reach the short sample limit, about 10% above the design current corresponding to a field of 6.6 T, with one or two quenches at most. The same design philosophy has been used at LBL to build a magnet operating at 1.8 K that reached 10.06 T after five training quenches. The LHC magnetic field will be somewhere between 8.5 and 10 T. One meter long model magnets have reached over 9 T but have had substantial training. Most of the quenches are at the coil ends.

Quenches are caused by losses also. For example, the Rutherford cable used in all superconducting magnets is braided with strands crossing over each other forming loops. These crossovers have resistance; the value depends on the surface coating of the strands. When the magnet current is ramped, EMF's induced in these loops can cause currents and sufficient heating to quench the magnets if the crossovers are not resistive enough. The quench current of some prototype SSC dipoles is more than 10% below the DC value at the ramp rate of the High Energy Booster. Increasing the crossover resistance should solve this particular problem.

The engineering practices to produce good superconducting magnets are understood. Some of the fundamentals of quenches, the effects of cooling, conductor size, inductance, etc, are not. They will need to be understood better as magnet performance is pushed.

Cryogenics [7]

Superconductivity is the technology that determines the energy frontier of particle physics. As a result, cryogenics is becoming increasingly important, and large cryogenic plants are part of most accelerator laboratories. There are 10 - 20 kW of cooling at 4.5 K and several kilometers of cold accelerator at DESY and Fermilab. The capacity at CERN will rise from 19 kW today to roughly 70 kW for LEP200 and 150 kW in the LHC era. The 200 kW capacity of the SSC will come from 10 refrigerators with 20 kW capacity and multiple cooling loops in a continuation of the trend towards efficient, compact cryogenic plants.

Cryogenic engineering must be integrated into a collider design. Some examples follow. Static heat loads, resistive dissipation in superconducting devices, and the impact of beam induced losses (from synchrotron radiation, for example) must be minimized. This affects parameters as basic as whether the magnet body is warm or cold and the size of the magnet bore. Heat loads have large variations in instantaneous power and duration. Quenches make extreme, localized demands for tens of seconds while cooldowns can last several days and extend over a substantial fraction of the accelerator. Energy ramps and steady operation have other load profiles. The magnets and cryogenic system must accommodate this variety. Single-phase and two-phase He have advantages and disadvantages as cooling fluids and can be used in different applications or parts of the cooling loop. For high magnetic fields He II operating below 2 K is feasible although the technology of cold compressors needs further development. The LHC and SSC have included cryogenics as a basic element in the optimization of cost and performance.

Long Term Stability [8]

Single particle motion in nonlinear magnetic fields is the most important accelerator physics issue in large hadron colliders. Nonlinear fields are produced by persistent currents and conductor placement errors in superconducting dipoles and quadrupoles, and nonlinear magnets are used to correct chromatic aberrations. These fields can cause unstable motion and particle losses and have the greatest effect at the injection energy where the momentum is lowest and the ratio of persistent currents to transport currents is largest. Extrapolation from the Tevatron to larger rings is uncertain - the peak energy of the Tevatron is only six times the injection energy, the nonlinearities are small due to the large bore of the Tevatron magnets, and the LHC and SSC have much larger circumferences. The recent success commissioning the HERA proton ring, where the energy ratio is twenty, removes one of those uncertainties and gives confidence in many aspects of stability analysis.

Nonlinear motion is understood in simple situations, but storage rings are complex and have many weak nonlinearities. Computer simulations are the preferred analysis technique. There are several time scales of interest. Stability for a few turns is needed to correct the first turn orbit and establish a circulating beam. Routine accelerator adjustments require stability for roughly 10^4 turns, and filling both rings takes about 10^8 turns. Substantial losses during that time could quench magnets. Recent studies have concentrated on the long time scale. Brute force simulations that follow particles through individual magnets for a million or more turns are possible thanks to increasingly powerful, economic computing. "Survival plots" show the aperture and losses and lead to the conclusion that beam particles are stable for long times.

The remaining question is whether the simulation models contain sufficient physics. At large amplitudes motion is irregular and not deterministic. Stochastic motion and diffusion give a better description. Diffusion has been measured in experiments at the SPS and Tevatron, but these measurements are not understood quantitatively. There are no predictions of the diffusion constants and their dependence on other parameters. The conclusions about long term stability will remain uncertain until these measurements have been explained.

Injectors [9]

The LHC and SSC will achieve high luminosities by a combination of a large number of bunches and beams with high phase space densities. Not surprisingly, this combination determines many of the parameters and specifications of the injector complex and leads to some of the most interesting accelerator physics of large colliders. Counting the linac, its ion source and preaccelerator as one, there are five accelerators in each of these complexes. For example, for the LHC those are the linac, the PS Booster, the PS, the SPS and the LHC itself.

Having a large number of bunches leads to strong coupled bunch instabilities and beam loading. Some of the instabilities are naturally damped by Landau damping; those that aren't must be cured with feedback. Beam loading shifts the optimum RF cavity frequency by amounts comparable to the revolution frequency in some of the accelerators and special RF feedback that can compensate for beam loading is necessary. This has implications i) for all the other RF controls, ii) on the installed RF power, and iii) even the conventional construction by fixing the location of the RF power sources.

The maximum phase space density is determined by the ion source, and, at best, it can be preserved through the acceleration cycle. While both designs allow for some reduction, the densities are higher than those achieved at the SPS and Tevatron. Space charge effects in the low energy boosters, the SSC LEB and the CERN PS, are a central issue and one of the dominant causes of density reduction. Space charge introduces a tune spread that is proportional to the density and $1/\beta\gamma^2$ (β and γ are the usual relativistic quantities: $\beta = v/c$, $\gamma = (1 - \beta^2)^{-1/2}$). The design tune spreads, ΔQ_{SC}, are 0.35 and 0.20 in the SSC LEB and CERN PS, respectively. For comparison, at $\Delta Q_{SC} = 0.35$ in the FNAL Booster there is roughly a factor of two density reduction. Such density reduction is caused by nonlinear resonances within the tune range of particles in the beam. These resonances must be compensated with magnetic nonlinearities. Experiments at the PS that included resonance compensation show that the design phase space density for the LHC can be reached, and LEB simulations lead to the same conclusion for the SSC. In addition, flexibility is being incorporated into the SSC design - the linac housing will be long enough for a future energy upgrade.

With thought about these types of issue and careful design, the LHC and SSC beams should be intense and dense enough to give the design luminosities.

E⁺E⁻ HEAVY QUARK FACTORIES

There is a maximum charge density at the collision point caused by beam produced electromagnetic fields. This limit, called the beam-beam limit, is parametrized by ξ, the beam-beam tune shift. The physics of the beam-beam interaction is not well-understood, but there is extensive experience indicating that designing with $\xi \leq 0.05$ is prudent for e^+e^- storage rings. Equation (1) for the luminosity can be written

$$\pounds = \frac{N(f_c/B)\xi(1+\sigma_y/\sigma_x)\gamma}{2r_e\beta_y^*} \quad (3)$$

when the beam-beam limit is taken into account. This equation can be evaluated for either beam provided the parameters of that beam are used throughout, and the horizontal and vertical tune shifts have been assumed to be equal. Luminosity is increased by raising N, the total number of particles in the beam, reducing β_y^*, and increasing, to the extent possible, the beam-beam limit.

CESR [10]

A series of upgrades including multiple bunches, low β^* optics, and operating with only one interaction point have increased the integrated luminosity roughly an order of magnitude every six years. The present record performance is: $\pounds(peak) = 2.5 \times 10^{32} cm^{-2} s^{-1}$, an integrated luminosity of 1.2 fb^{-1} in 1991, and 1.6 fb^{-1} expected in 1992 (an average of $5 \times 10^{31} cm^{-2} s^{-1}$ for an entire year!).

Upgrades are continuing. New RF cavities and electrostatic separators will remove the present beam current limitations, and an integrated luminosity in the range 2.5 to 3.0 fb^{-1}/year is expected in 1993. Starting in the beginning of 1994 the plan, called "CESR Phase II", is to make further modifications to raise currents and to use a small crossing angle to give a projected increase in the number of bunches to twenty-seven and the luminosity to $6 \times 10^{32} cm^{-2} s^{-1}$. Beyond that, superconducting cavities could be used to raise the luminosity to the 10^{33} range, and, if the past is a guide, the experience gained in the next few years will give additional ideas for raising the luminosity. That experience with high currents and luminosity will be valuable for future heavy quark factories also.

New Projects

There are new projects focused on each of the heavy quarks: Φ-factories at Novosibirsk and Frascati, τC-factories at the Joint Institute for Nuclear Research and in Spain, and B-factories at Novosibirsk, KEK, Cornell, and SLAC. These projects have excellent, well-documented particle physics justifications, and there are advanced, detailed proposals. The Frascati Φ-factory is funded and under construction, but the other projects are waiting for approval or are caught up in the financial uncertainties of the former Soviet Union.

The general consensus on the way to reach high luminosity is:
1. Two rings and a large number of bunches. The luminosity depends on the total number of particles and not on the number of bunches. A large number of bunches reduces single bunch stability problems at the expense of multiple bunch instabilities, but the latter are easier to cure with feedback and RF cavity design. The interaction region tends to be simpler with a large number of bunches because they can be physically smaller without having beam-beam problems. However, the bunches must be separated close to the interaction point, and that adds complications.
2. Flat beams, $\sigma_y \ll \sigma_x$, because that simplifies the interaction region optics and has lower experimental backgrounds.
3. $\beta_y^* \leq 1 - 3$ cm limited by the length of the bunch and momentum dependence of the interaction region focusing if it is made stronger to reduce β_y^* further.

Only the Novosibirsk Φ-factory departs substantively from this consensus. The total currents are huge, 1 - 2 A as compared to ~100 mA in CESR, and heavy demands are placed on the RF and vacuum systems.

Having made these basic choices, work is concentrated on refinements and developing critical components. With the large number of projects and extensive activity it is possible to give only a flavor of the progress. Individual contributions to the conference should be referred to for details [11-15].

The experimental detector and storage ring interaction region are intimately related. They are competing for the same space and influence each other strongly. Experimental backgrounds are particularly important, and there have been extensive simulations of the backgrounds from beam particles degraded by gas scattering and from synchrotron radiation. These simulation have led to major design choices such as the use of flat beams and determine some of the accelerator parameters. They lead to the conclusion that the backgrounds can be reduced sufficiently, also. One simulation has been checked at Cornell [16]. Currents in tracking chambers and trigger rates were compared with Monte Carlo results and found to be in excellent agreement. This shows that backgrounds can be understood, provides a "standard" simulation, and builds confidence in the conclusion of acceptable background levels.

Integration of accelerator and detector components (focusing quadrupoles, compensating solenoids, vacuum pumps, synchrotron radiation masks, vertex detectors, thin beam pipes, and all of the associated fluids and cables) is a major engineering job. B- and τC-factories have to accommodate a small radius interaction region beam pipe for vertex detectors, but the Φ-factories have a new and unusual problem. A large radius pipe is needed for a K^0 decay volume. The focusing quadrupoles must have a small radius for adequate strength, and the transition to the decay region creates an RF cavity-like structure that can extract energy from the beam and lead to heating and interesting beam dynamics.

Beam bunches interact through electromagnetic fields with decay times greater than the time between bunches. Multiple bunch instabilities can be prevented by reducing the Q's, the quality factors, of unwanted resonances. When the Q's cannot be reduced sufficiently, instabilities must be controlled with feedback. High Q resonances can be avoided in many structures, but that is impossible in an RF cavity. However, the Q's for all but the accelerating mode (the fundamental mode) can be reduced by the following. First, only one or two RF cells are combined into a single cavity. That avoids complicated spatial structures of the unwanted higher modes. It has the added benefit of minimizing the fundamental mode power passing through each RF window separating the cavity from the power source. Second, waveguides that are cut-off for the accelerating mode, but not for the higher frequency modes, are placed appropriately to couple strongly to the higher modes and remove energy from them. It has been demonstrated that the Q's of higher modes of room temperature cavities can be reduced sufficiently, to 100 or less, without affecting the fundamental mode.

The same ideas, one or two cells per cavity and strong coupling to higher modes, work for superconducting cavities also. Superconducting cavities have high accelerating gradients, and that has the advantage of reducing the number of cavities. Those gradients can be reached even with a large opening for the beam. With a large opening there are only a few high Q resonant modes and the beam pipe itself becomes the higher mode waveguide. RF absorbers placed in a room temperature section of the beam pipe absorb higher mode energy. The potential advantages of superconducting RF are balanced by uncertainties associated with transferring large RF power to the beams. This is a new mode of operation that will be tested at Cornell in late 1993.

Typical bunch spacings are 1 - 3 m, and feedback systems must detect and correct longitudinal and transverse errors of such closely spaced bunches. This requires i) high frequency detectors, ii) high frequency, high power kickers, and iii) signal processing that can keep track of a large number of bunches. Digital Signal Processing has the bandwidth and data handling capabilities, and it forms the basis of the SLAC/LBL/LLNL feedback system. This feedback has attracted interest at the Advanced Light Source, and a prototype is under construction for testing there.

Conclusions

There is general agreement on the route to high luminosity, and critical elements are being designed, engineered, and tested. The examples above were chosen to give a flavour of these developments. These colliders should reach the design luminosity, and heavy quark factories are sure to be a part of the future of high energy physics.

HIGH ENERGY E⁺E⁻ COLLIDERS

TRISTAN [12].

TRISTAN is running at E_{CM} = 58 GeV, and the luminosity has exceeded 1 pb⁻¹/day. The goal is to accumulate 300 pb⁻¹ total luminosity. It is anticipated that this will be achieved in 1994. Extensive experience has been gained with superconducting RF in a working accelerator. The 48 m of superconducting cavities has operated for several years with an average gradient of 4.7 MV/m.

LEP [17]

A new magnetic configuration compatible with i) $E_{CM} = m_Z$, ii) E_{CM} > W-pair threshold, iii) transverse polarization, and iv) electrostatically separated orbits was commissioned at the beginning of the 1992 LEP run. This was a major change, but the initial difficulties have been solved. At the time of this conference LEP was running with $\mathcal{L} \simeq 10^{31}$cm⁻²s⁻¹ and had an integrated luminosity, over 8 pb⁻¹, exceeding that of the 1991 run at the comparable time.

The beam-beam tune shift and single bunch current are above their design values, and the luminosity and total current are approaching theirs. The total current limit is thought to be caused by the beam-beam interaction at the injection energy where the beams are electrostatically separated but there are residual effects. With the solution of this problem LEP should exceed all of its design goals.

The new magnetic configuration is the foundation for future LEP improvements. The luminosity can be doubled by increasing the number of bunches from four to eight with separated orbits. This mode of operation has been tested successfully, and it should become routine by the end of September of this year. The energy calibration can be improved by resonant depolarization in the same configuration used for collisions.

The new magnet configuration is part of LEP2, the energy increase to above the W-pair threshold. Superconducting RF is the heart of LEP2. At 90 GeV/beam the energy loss per turn is roughly 2.2 GeV. This energy loss must be made up by the RF system. The LEP2 RF system will have 192 superconducting cavities with a gradient of about 5 MV/m together with the 120 existing copper cavities. About 15% of the superconducting cavities have been delivered and accepted, and the full complement is expected in 1994. The physical plant and utilities must be increased substantially also - new RF power sources, cryogenics, water cooling, power supplies, etc. This work is on target for running in 1994 at E_{CM} = 180 GeV.

SLC [18]

1992 has been a critical year for the SLC and for future linear colliders. Despite its peculiarities the SLC is the prototype linear collider, and the feasibility of performing high energy physics experiments at linear colliders must be judged based on the SLC experience. Luminosity, uptime, backgrounds, and polarization are all important and must be achieved simultaneously. The 1992 SLC run has shown that this is possible with the best evidence being the SLD results presented at this conference.

The present SLC performance is: $\mathcal{L} \sim 2 \times 10^{29}$cm⁻²s⁻¹ with peaks up to 2.5×10^{29}cm⁻²s⁻¹, uptime ~ 60 - 70%, sustained running with backgrounds sufficiently low for SLD data taking, and a polarization 22 - 24% at the interaction point. A large number of techniques had to be developed to reach this level of performance. This work is documented in the accelerator conference proceedings and has been summarized by J. Seeman [19]. The recent progress on polarization and feedback are highlighted here.

Producing polarization at the interaction point requires sophisticated physics and technology throughout the accelerator, but the most difficult aspect has proven to be the polarized source itself. There state-of-the-art vacuum, high voltage, and materials engineering combine with incompletely understood physics of quantum efficiency, polarization, and intensity saturation to determine performance. One source meeting the SLC specifications has been operating routinely and reliably since April 1992. The cathode is bulk GaAs with 27% polarization. Depolarization during acceleration, damping, and transport reduce this to 22 - 24% at the interaction point. There are a number of options for increasing the polarization by modifying the cathode: cooling of bulk GaAs to 150 K, changing the cathode to AlGaAs, using thin GaAs and a different wavelength excitation, and using a strained GaAs crystal [20]. The first three are short term (intended for the 1993 run) and promise polarization at the cathode in the 40 - 45% range. Some are being tested now. An operational source using a strained GaAs crystal is still under development and is at least a year away. It would have 85% polarization at the cathode.

Beam properties (orbits for example) vary from pulse-to-pulse and drift. Pulse-to-pulse jitter must be reduced to acceptable levels by stabilizing accelerator components or special beam dynamics like BNS damping [21]. Feedback to control drifts has always been part of the SLC, but recently there has been an innovation of using database driven feedback system based on digital control theory [22]. These feedbacks sample the beam at 20 Hz, correct noise below 2 Hz, and respond to step changes in 0.2 sec. These loops are adaptive: they recognize and compensate for changes in the accelerator, and they are cascaded: they communicate with each other to insure that only one loop corrects a particular error. New loops are implemented with database modifications and modest hardware installation. Feedback has become the preferred solution to many problems, and 22 loops, about four times the number initially planned, are active currently. They are a major contributor to the increased luminosity, uptime and efficiency of the SLC, and this development is important for future linear colliders.

Continuing work of the type illustrated will increase the polarization and luminosity of the SLC, and it will lead to many SLD results at future conferences. The experience gained will be invaluable for the future through the understanding of the practicalities of operating a linear collider for particle physics.

Future Linear Colliders

Some of the principal parameters for linear colliders with E_{CM} = 500 GeV from the 1992 ECFA workshop on e^+e^- linear colliders (LC92) held a week before Dallas are given in Table 1 [23]. There is striking diversity in these parameters representing different judgements about various factors.

The ease of improving technology - The DESY/Darmstadt and lowest frequency JLC are the most conservative in this regard. They take advantage of over forty years of experience with S-band (~3 GHz) RF. Some new components, high power klystrons in particular, are needed. The NLC and the other versions of the JLC extend the basis of present day linacs - high peak power klystrons and modulators - to higher frequencies. Klystrons and accelerator structures must be developed for those frequencies. TESLA relies on substantial improvements in the cost and accelerating gradient of superconducting RF, but it has the advantage of a substantially larger collision spot. VLEPP requires a number of innovations in alignment, jitter reduction, collimation, and beam dynamics to obtain high luminosity with a single intense bunch. CLIC uses a high RF frequency that leads to stringent alignment and

Table 1. Selected Preliminary Linear Collider Parameters for E_{CM} = 500 GeV* [23]

	RF Freq (GHz)	Gradient (MeV/m)	Rep Rate (Hz)	Bunches/ RF Pulse	σ_x/σ_y (nm)	P_B (MW)	Υ	£ (10^{33}cm^{-2}s^{-1})
TESLA	1.3	25	10	800	640/100	16.5	0.035	8
DESY/Darmstadt	3.0	21	50	172	400/32	7.5	0.058	4
NLC (SLAC)	11.4	50	180	90	300/3	4.2	0.10	9
JLC (KEK)	2.8	22	50	55	300/3	1.6	0.20	4
	5.7	40	100	90	260/3	3.6	0.20	7
	11.4	40	150	90	260/3	3.8	0.15	6
VLEPP	14.0	108	300	1	2000/4	2.4	0.06	12
CLIC (CERN)	30.0	80	1700	1-4	90/8	.4-1.6	0.15	1-3

* σ_x and σ_y are the horizontal and vertical beam sizes, P_B is the power of one beam, and Υ is given by eq. (4).
TESLA uses superconducting RF; all others use room temperature RF.

fabrication specifications and significant beam dynamics problems. An elegant "two-beam" RF power source replaces conventional klystrons and requires substantial development.

Costs - New technologies promise significant, but uncertain, cost reductions. Older technologies have better established costs, but these tend to be high and must be lowered through engineering and mass production. The experience of the SSC, an accelerator based on mature technology and a detailed design, teaches us that present day linear collider cost estimates should not be taken seriously.

Extension to higher energies - For room temperature RF, high gradients and high RF frequency tend to be better for high energies. They reduce the accelerator length and improve the energy efficiency, respectively. The NLC, high frequency JLC's, and VLEPP are optimized for E_{CM} = 500 GeV - 1 TeV. The S-band colliders would be straining to go beyond 500 GeV, and the CLIC parameters are those of a multi-TeV collider scaled down to E_{CM} = 500 GeV for purposes of comparison. The energy reach of TESLA depends on how close the fundamental gradient limit of ~50 MeV/m in Nb can be approached.

Experimental backgrounds and energy spectrum - A large number of photons are radiated in the intense electromagnetic fields at the collision point. These "beamstrahlung" photons degrade the energy spectrum of the primary e^+e^- collisions and produce experimental backgrounds directly and through e^+e^- pairs and hadrons from $\gamma\gamma$ collisions. These effects are parametrized by $\Upsilon = \gamma B/B_C$ where B is the effective electromagnetic field strength and $B_C = 4.4 \times 10^{13}$ G is the Schwinger critical field. In terms of beam properties an approximate expression for Υ is

$$\Upsilon = \frac{r_e^2}{\alpha} \frac{\gamma n}{\sigma_L (\sigma_x + \sigma_y)} \quad (4)$$

where n is the number of particles per bunch, σ_L is the bunch length, and r_e and α are the classical radius of the electron and the fine structure constant, respectively. All designs use flat beams to minimize beamstrahlung. Most of the particles produced in electromagnetic processes are in the forward direction and should not cause backgrounds with a properly designed interaction region. Relatively rare e^+e^- pairs with $p_{TRANSVERSE} \geq 20$ MeV/c are an important background, and there was speculation that the "minijet", $\gamma\gamma \to$ hadrons, cross section was large [24]. Preliminary HERA measurements of the total γp cross section indicate that this is not the case [25,26].

Discussions have concentrated on backgrounds originating from collisions themselves. Beam halos are likely to be as serious and providing proper collimation will have a strong influence as designs evolve.

Tolerances and beam power - While there are considerable differences in estimates of the tolerable amount of beamstrahlung, T will be determined by experimental backgrounds. The luminosity can be written in terms of T

$$\mathcal{L} = \frac{1}{4\pi} \frac{\alpha}{r_e^2 mc^2} \frac{T\sigma_L}{\gamma^2} \frac{P_B}{\sigma_y} . \quad (5)$$

Disruption, focusing during the collision, has been neglected. The *free* parameters are σ_L, σ_y, and P_B. Since a bunch can occupy only a small fraction of an RF wavelength, σ_L is one of the many factors affecting the choice of RF frequency. That leaves a tradeoff between beam power and spot size. Effectively, this is a tradeoff between tolerances and beam power.

Bright beams and a final focus system with low order aberrations corrected are required for small spots. Beam brightness depends on the damping rings producing the beams and beam generated fields experienced during acceleration. Damping ring and main accelerator tolerances become more stringent as the brightness increases. The main accelerator tolerances depend strongly on the RF frequency with low frequency favored. That adds tolerances to considerations when selecting the RF frequency. Correcting optical aberrations in the final focus leads to tight alignment, stability, and field quality tolerances there, also. Increasing the beam power relaxes tolerances, but there are limits to beam power. There must be efficient transfer of energy to the beam, and beam handling, collimation and accelerator protection, become difficult as power increases.

Table 1 and the paragraphs following it show the interconnected considerations entering into the *most basic* parameters and the diversity of opinion about the best design. How will the choices be narrowed and the best approach decided? Four important factors will be continued experience with collisions at the SLC, SLC beam dynamics experiments, prototypes of critical components, and prototypes of systems.

Experiments studying beam generated fields, acceleration of flat beams and multiple bunches, advanced orbit correction techniques, and electrodynamics in the strong field regime of beamstrahlung are planned at the SLC. A new facility, the Final Focus Test Beam, is nearing completion. It is a combination of a beam dynamics experiment and a system prototype. Optics with the appropriate corrections will be tested in this prototype final focus. Demagnifications even greater than those planned in many of the designs should be reached.

All of the designs have critical components. Highlights of progress on some of these follows. Multiple bunch acceleration depends on the decoherence or the damping of beam generated fields leading to intrabunch interactions. Decoherence is accomplished by spreading the resonant frequencies of unwanted modes while damping requires reduction of their Q's. Both methods work, and the issue is reducing complexity to lower manufacturing costs. A radial transmission line with an RF choke for the fundamental mode allows damping of higher modes without affecting the fundamental and has the advantage of being a rotationally symmetric structure that could be turned on a lathe [27]. A model cavity has been constructed and shown to have the predicted RF properties.

The gradient of superconducting cavities is limited by field emission. Field emitted electrons extract energy from the cavity fields and generate heat when they strike the wall. Processing the cavities with short burst of high power RF burns up the field emitters which are usually imperfections introduced during manufacturing. Cavities processed in this way have reached gradients in the range 17 - 27 MV/m [28].

X-band (11.4 GHz) colliders need klystrons capable of generating 50 MW of RF power for 1 μs. Klystrons with this power combined with RF pulse compression give a gradient of about 50 MeV/m with a klystron for every 3-4 m of accelerator. A klystron with that performance has been tested successfully at

SLAC [29]. While the efficiency must be improved, it is the first demonstration of an X-band RF power source.

Prototype systems to study both beam dynamics and system engineering are planned or under construction. These include:

1. A 500 MeV TESLA prototype to be constructed at DESY. The aims are to demonstrate a gradient of 15 MV/m, to meet cost goals, and to test a high gradient superconducting linac with beam.

2. A 450 MeV prototype of the DESY/Darmstadt collider that will test long pulse, high power, multiple bunch operation of an S-band linac.

3. The Accelerator Test Facility at KEK combines a 1.5 GeV, S-band linac with a prototype damping ring. The damping ring will produce beams with brightness, single bunch charge, and bunch train structure covering many of the colliders in Table 1. New levels of tolerances, control of beam generated fields, extraction kicker stability, etc will be reached in accomplishing this.

4. Interaction region optics will be studied at the Final Focus Test Beam discussed above.

5. The NLC Test Accelerator planned at SLAC will be a 540 MeV X-band linac based on prototype klystrons and accelerator structures. The primary goal is to construct and reliably operate an engineered section of the NLC linac. Beams dynamics in an X-band linac will be studied also.

6. One of the most challenging aspects of CLIC is the two-beam RF power source. A low energy beam with bunchlets spaced at $\lambda = 1$ cm must be created, accelerated, and energy extracted from it. A beam with this structure will be generated by an RF gun at the CLIC Test Facility. The beam will then be accelerated and energy extraction demonstrated.

These prototypes will narrow the options in Table 1. The widely accepted goal of proponents is a proposal for a high energy linear collider by the mid 1990's.

REFERENCES

1. V. K. Bharadwaj, ICHEP session PA-22.
2. L. P. Soloviev, ICHEP session PA-23.
3. A. Devred, ICHEP session PA-22.
4. R. Palmer, ICHEP session PA-23.
5. R. Palmer and A. V. Tollestrup, *Ann Rev Nucl Part Sci* **34**, 247 (1984).
6. K.-H Meß and P. Schmüser, *CERN Accel School - Supercond in Part Accel*, 87 (1989).
7. Ph. Lebrun, ICHEP session PA-23.
8. R. Talman, ICHEP session PA-22.
9. W. T. Weng, ICHEP session PA-22.
10. D. L. Hartill, ICHEP session PA-22.
11. D. Rice, ICHEP session PA-23.
12. K. Satoh, ICHEP session PA-22.
13. T. Fieguth, ICHEP session PA-22.
14. G. Vignola, ICHEP session PA-22.
15. A. Sissakian, ICHEP session PA-22.
16. S. Henderson, ICHEP session PA-23.
17. S. Myers, ICHEP session PA-22.
18. N. Phinney, ICHEP session PA-22.
19. J. T. Seeman, *Ann Rev Nucl Part Sci* **41**, 389 (1991).
20. T. Maruyama *et al*, *Phys Rev Lett* **66**, 2376 (1991).
21. V. E. Balakin, A. V. Novokhatsky and V. P. Smirnov, *Proc of 12th Int Conf on High-Energy Accel*, 119 (1983).
22. F. Rouse *et al*, SLAC-PUB-5681 (1991).
23. G. Loew, Summary Talk 1992 ECFA Workshop on e+e- Linear Colliders.
24. M. Drees and R. M. Godbole, *Phys Rev Lett* **67**, 1189(1991).
25. B. Löhr, ICHEP session PA-24.
26. F. Eisele, ICHEP session PA-24.
27. T. Shintake, KEK Preprint 92-51 (July, 1992).
28. H. Padamsee, ICHEP session PA-23.
29. G. Caryotakis, 1992 ECFA Workshop on e+e- Linear Colliders.

DISCUSSION

A.M. Baldin, JINR, Dubna

For your information, an accelerator for relativistic heavy nuclei, the Nuclotron, has been under construction during 1987–1991 at Dubna. It utilizes 200 superferric dipoles, has a repetition rate of 0.5–1.0 p.p.s., and beam energy ranging from 100 MeV to 6 GeV per nucleon (Xe, Pb, etc). Maximum proton energy is 12 GeV. One general-purpose detector will be utilized with an internal target. The first experiments are scheduled for early next year; they will study asymptotic behavior of multinucleon interactions, subthresholds and cumulative particle production.

CONFERENCE SUMMARY *

Steven Weinberg
Theory Group, Physics Department
University of Texas
Austin, Texas, 78712

Roy Schwitters has thanked the staff of this conference for their good work. As this is the last talk, I would like to add my own thanks to the staff, and also on behalf of all the participants to thank Roy himself and Vic Teplitz and the other physicists at the Super Collider and at SMU and the other Dallas area universities for their part in making this such an exceptionally well run conference.

This is actually the second time that I have been called on to give the summary talk at one of the Rochester Conferences. The previous time was at Berkeley in 1986, the latest time the conference was held in the United States. You might think that after that experience I would know better than to take on such a difficult task, but in fact nothing could be further from the truth. This is actually an extremely easy job. Everyone knows that it is impossible to review every topic that has been discussed at such a conference, so no one expects it. In fact I found last time at Berkeley that people (at least most people) even forgave me for not mentioning their own work.

It *is* possible to give a brief "coarse-grained" summary of the whole conference. This also is very easy, because it doesn't vary from one conference to another. Here it is:

1. The Standard Model agrees with all data, but has many holes and loose ends.

2. We must still:
   - find the top quark,
   - identify the mechanism (or mechanisms) for CP nonconservation,
   - solve quantum chromodynamics.

3. We don't understand:
   - why the parameters of the standard model take the values we observe,
   - why there are three generations of quarks and leptons,
   - why $SU(3) \otimes SU(2) \otimes U(1)$?

4. Most of these questions revolve around the central unresolved issue concerning the Standard Model: how is $SU(2) \otimes U(1)$ broken? There are two possibilities: The Goldstone bosons which provide the longitudinal parts of the W and the Z are either elementary, or they are not.
   - If the Goldstone bosons are part of a multiplet of elementary scalars, then in order to explain why they are so light compared to a really fundamental scale we probably need to assume supersymmetry.
   - If the Goldstone bosons are composites, then definitely there must be new extra-strong forces, like the

---

*Supported in part by the Robert A. Welch Foundation and NSF Grant 9009850. I wish to thank John Bahcall, Michael Dine, Willy Fischler, Howard Georgi, Nathan Isgur, Vadim Kaplunovsky, Michael Peskin, and Joe Polchinski for helpful conversations regarding topics covered in this report.

technicolor forces reviewed here by Appelquist.

(Peskin in his talk referred to these alternatives as the weak-interaction route and the strong-interaction route.)

5. In order to resolve the issue of electroweak symmetry breaking and break out of the present impasse, we need the

- SSC
- LHC.

* * *

In the remaining 57 minutes of my talk I would like to discuss a few specific topics. These are chosen, not necessarily because they represent the most important things discussed at this conference, but mostly because they are matters that I found interesting and about which I had something that I wanted to say. In 1986 the two topics I chose to discuss were solar neutrino oscillations and string theory. (By the way, Robertson[1] said in his talk the other day that only recently have particle physicists regarded the solar neutrino problem as part of particle physics, which is true only if "recently" extends back at least six years.) These would be a pretty good choice of a pair of special topics for my talk today (which shows how slowly our field is moving), and I will come back to them, but there are a few other topics I want to take up here also. At the end of this discussion of special topics I will say a little about what I think lies ahead for particle physics.

Heavy Quark Symmetry

[1] I will follow the practice here for the most part of just quoting talks given at this conference, chiefly the rapporteur talks. References to the physics literature can be found in the written versions of these talks.

My first topic is heavy quark symmetry. This was discussed by Drell and Isgur in their rapporteur's talks; summaries were given in parallel sessions by Close and Grinstein; and the subject kept coming up in many of the talks in parallel sessions I attended. It is really one of the prettiest developments in the theory of strong interactions in a long time, and one in that in a remarkable way has almost immediately been taken over by experimentalists as part of their standard bag of tricks. Heavy quark symmetry has been explained many times in this conference in a hand-waving way in terms of analogies with atomic physics. Here I would like to offer an explanation (most of which can be found in the papers of the experts), that you can give with your hands strapped to your sides, in terms of the Feynman diagrams of quantum chromodynamics.

Consider a heavy quark line going through a Feynman diagram. Suppose that after emitting momentum to and absorbing momentum from the gluons, it has acquired a four-momentum $mv + k$, where m and v are the mass and initial four-velocity ($v \equiv [\gamma \vec{v}, \gamma]$) of the heavy quark. The components of the four-momentum transfer are limited by an amount of the order of the QCD scale factor Λ, so for $m \gg \Lambda$, the quark propagator may be approximated by

$$\frac{-i(m\slashed{v} + \slashed{k}) + m}{(mv + k)^2 + m^2} \simeq \frac{-i\slashed{v} + 1}{2v \cdot k}. \quad (1)$$

[I am using the usual slash notation, $\slashed{a} \equiv \gamma^\mu a_\mu$, with a metric that has diagonal components $-1, +1, +1, +1$.] This immediately reveals the flavor degeneracy of these theories; as long as we express everything in terms of velocities rather than four-momenta, nothing in matrix elements depends on the heavy quark mass. With N heavy types of heavy quarks, the bound states of one or more heavy quarks plus any number of light quarks and gluons would form $SU(N)$ multiplets of hadrons with equal *binding* energies.

To go further, suppose gluon lines with polarization indices μ, ν, etc. and color indices a, b, etc., are emitted by a heavy quark line before it finally leaves the diagram with spin z-component σ and four-velocity v. By moving all factors $(-i\slashed{v} + 1)/2$ to the end of the line, we easily see that the contribution of the quark-gluon vertices and quark propagators is

$$\bar{u}(\sigma, v) \left(\frac{-i\slashed{v}+1}{2}\right) it_a \gamma^\mu \left(\frac{-i\slashed{v}+1}{2}\right)$$
$$\times it_b \gamma^\nu \left(\frac{-i\slashed{v}+1}{2}\right) \cdots \quad (2)$$
$$= \bar{u}(\sigma, v) \left(\frac{-i\slashed{v}+1}{2}\right) t_a t_b \cdots v^\mu v^\nu \cdots .$$

In other words, nothing depends on the spin z-components of the heavy quarks, aside from kinematic final (and initial) state factors like $\bar{u}(\sigma, v)(-i\slashed{v}+1)/2$, which do not vary from diagram to diagram for any given process. (Despite appearances, this result does not depend on the fact that the quarks have spin $\frac{1}{2}$.) In particular, the positions of the poles in a Feynman diagram will not depend on the spin z-components of any incoming or outgoing heavy quarks, so the bound states of heavy and light quarks will exhibit a spin degeneracy as well as a flavor degeneracy: all of the bound states that are related by rotating only heavy quark spins (as for instance the lowest 0^- and 1^- bound states of a heavy quark and a light antiquark) will have the same mass.

One can also apply this diagrammatic analysis to processes in which a weak interaction induces a transition between hadrons containing different species of heavy quarks. According to the foregoing analysis, the matrix element of a current $\bar{\psi}_{f'} \Gamma \psi_f$ (with Γ an arbitrary 4×4 matrix) between initial and final bound states α and α', consisting of any number of light quarks and/or antiquarks plus one heavy quark respectively of flavor f and f' and four-velocity v and v', must take the form:

$$\langle f', v', \alpha' | \left(\bar{\psi}_{f'} \Gamma \psi_f\right) | f, v, \alpha \rangle$$
$$= \sum_{\sigma', \sigma} C_{\alpha \to \alpha'}(v', \sigma', v, \sigma) \quad (3)$$
$$\times \left[\bar{u}(\sigma', v') \left(\frac{-i\slashed{v}'+1}{2}\right) \Gamma \left(\frac{-i\slashed{v}+1}{2}\right) u(\sigma, v)\right],$$

where $C_{\alpha \to \alpha'}$ is an unknown but flavor-independent function of initial and final heavy quark spin z-components and velocities, that arises from the convolution of the wave functions for states α and α' with the sum of Feynman diagrams in which the heavy quark line is replaced by a product of factors $v^\mu t_a$ for each interaction of gluons with the heavy quark. (These Feynman diagrams could actually be summed explicitly for Abelian gluons, but as far as is known not for real SU(3) gluons.) We can usefully rewrite this formula as a trace:

$$\langle f', v', \alpha' | \left(\bar{\psi}_{f'} \Gamma \psi_f\right) | f, v, \alpha \rangle \quad (4)$$
$$= \text{Tr} \left\{ M_{\alpha \to \alpha'}(v', v) \left(\frac{-i\slashed{v}'+1}{2}\right) \Gamma \left(\frac{-i\slashed{v}+1}{2}\right) \right\},$$

where M is the 4×4 matrix:

$$M_{\alpha \to \alpha'}(v', v) \quad (5)$$
$$\equiv \sum_{\sigma, \sigma'} C_{\alpha \to \alpha'}(v', \sigma', v, \sigma) u(\sigma, v) \bar{u}(\sigma', v').$$

Though M is unknown, its matrix structure is fixed by Lorentz invariance. For instance, for a transition between the lowest 0^- mesons with flavor f and f', the above matrix element must have the same Lorentz transformation properties as the matrix Γ, so $M_{0^- \to 0^-}(v', v)$ must be a scalar. Factors of $\slashed{v}$ and $\slashed{v}'$ do not matter because when multiplied into $(-i\slashed{v}+1)$ or $(-i\slashed{v}'+1)$ they merely yield factors of $+i$. Thus M must here be proportional to the unit matrix:

$$M_{0^- \to 0^-}(v', v) = \frac{1}{2} \xi(-v \cdot v') \mathbf{1}, \quad (6)$$

with a coefficient ξ that depends on the only scalar variable, $v \cdot v'$. This is the celebrated

Isgur-Wise function. If you like analogies, then think of $\xi(-v \cdot v')$ as analogous to the well-known Fermi function $F(Z, W)$, which gives the effect of final state Coulomb interactions in nuclear beta decay. The fact that $\xi(-v \cdot v')$ depends neither on heavy quark flavor nor on the matrix Γ is just like the fact that $F(Z, W)$ does not depend on the nature of the nuclei participating in the beta transition [aside from the energy W, which is analogous to the variable $v \cdot v'$, and the atomic number Z, which is analogous to the fixed color triplet assignment of quarks], or on the matrices [S, V, T, A, or P] appearing in the beta decay Hamiltonian.

It is a challenge to quantum chromodynamics to calculate the Isgur-Wise function, and much effort has been put into this problem, but there is one point where the value of ξ can be obtained without effort. In the special case where $\Gamma = i\gamma^\mu$, $f = f'$, and $v = v'$, the matrix element (4) is given by the conservation of heavy quark flavor as just v^μ [in much the same way that the matrix element for O^{14} beta decay is fixed by the conservation of the isospin vector current], so in this case $\xi = 1$ at the point $v \cdot v' = -1$. But ξ is a universal function, independent of Γ and heavy quark flavors, so in general

$$\xi(1) = 1. \qquad (7)$$

This result is used in determinations of CKM matrix elements like V_{cb} by extrapolating data on weak processes such as $B \to D + \ell + \nu$ to the point $v \cdot v' = -1$.

The same Isgur-Wise function enters into the matrix elements for transitions involving the 1^- mesons that are degenerate with the lightest 0^- mesons. Whatever the wave function for the lowest 0^- mesons, the heavy quark spin degeneracy tells us that the wave function for the degenerate 1^- meson [in its rest frame] with $J_z = 0$ is given by inserting an extra factor $(-1)^{2\sigma}$. We must therefore take $M_{0^- \to 1^-}$ as a scalar matrix function of the final vector meson polarization e^μ that in the rest frame of the final heavy quark for vector meson polarization vector $e^0 = e^1 = e^2 = 0$, $e^3 = 1$ gives a factor $\frac{1}{2}(-)^{2\sigma'}\xi$ when acting to the left on $\bar{u}(v', \sigma')$. Aside from inconsequential terms involving $\not{v}$ and/or $\not{v}'$, the unique matrix satisfying these requirements is

$$M_{0^- \to 1^-, e}(v', v) = \frac{i}{2}\xi(-v \cdot v')\gamma_5 \not{e}. \qquad (8)$$

This can be used in Eq. (4) to calculate the matrix elements for processes like $B \to D^*(2010) + \ell + \nu$. Similar formulas have been derived for $1^- \to 0^-$ and $1^- \to 1^-$ transitions, and for processes involving other hadrons containing one or more heavy quarks.

High Precision Electroweak Physics

Due largely to the great recent success of LEP, electroweak physics has become closer and closer in spirit to quantum electrodynamics, as a branch of physics where high precision is expected. As Rubbia said this morning, we now have three parameters in electroweak physics that we know with quite high precision: the Fermi coupling constant G_F of beta decay (known from muon decay); the fine structure constant α; and the mass of the Z, measured at LEP. These three constants (along with masses and mixing angles for whatever quarks or leptons are involved in a given process) are all you need to calculate any desired matrix elements in the electroweak theory in tree approximation. Beyond the tree approximation one needs to know other parameters, like the top quark and the Higgs mass, but since these enter in loops observable quantities are less sensitive to them. As discussed by Rolandi, with our knowledge of G_F, α, and m_Z we are now able to use the electroweak theory to do high precision calculations of other measured quantities, such as the W mass, the Z leptonic width Γ_ℓ, and $[\sin^2\theta]_Z$ (measured from the forward-backward asymmetry in $e^+ + e^- \to \ell^+ + \ell^-$ at the Z peak), aside from a weak dependence on the top and the Higgs mass. These predictions agree with existing data for a top quark mass between 130 and 170 GeV (to 68 percent confidence) and a Higgs mass which is essentially unconstrained, allowed to be anywhere from 50 Gev to 1 TeV.

Satisfactory as this situation is, we can be pretty sure that it will "soon" (i. e., within a few years) be radically improved. We can count on a further improvement in the accuracy with which m_W, Γ_ℓ, and $[\sin^2 \theta]_Z$ are known, especially through more accurate measurements of m_W at LEP2 and the Tevatron Collider, and perhaps also through an improvement in the measurement of $[\sin^2 \theta]_Z$ at the SLD. But the real change will come when the top quark is discovered and its mass is measured, presumably within a few years at the Tevatron Collider. At that point we shall find ourselves in the position of having a critical test of the simplest (one scalar doublet) version of the electroweak theory, and if that test is failed, of being able to say something about what new physics must be added to this model.

In the last few years it has become customary to parameterize the new physics that may enter in the electroweak theory in terms of what are called oblique radiative corrections. Parameters like the Z mass, the Z leptonic width, etc., would be affected by the top and Higgs mass, as well as by most kinds of new physics that could be added to the minimal electroweak theory, mostly through the 2×2 matrix vacuum polarization of the Z and γ. (For instance, Higgs scalars do not interact very much with u or d quarks or electrons or muons because these quarks and leptons are so light.) Further, as Peskin explained, because experiments are still mostly done at energies that are rather small compared to what we think are the energy scales of new physics, the vacuum polarization can be parameterized by the coefficients of just the first few terms in a Taylor expansion. The most important of these coefficients are called S and T. High precision measurements of m_W, Γ_ℓ, and $[\sin^2 \theta]_Z$ define narrow strips of allowed values in the $S - T$ plane, all running with various slopes from $S < 0$, $T < 0$ to $S > 0$, $T > 0$. For a given top quark mass (measured, say, to an accuracy of 10 GeV) and a Higgs mass between 50 GeV and 1 TeV, the minimal standard model defines a short wedge in the $S-T$ plane, running roughly transverse to these strips. [These strips and wedges are shown in Figure 1. The strips in this figure are centered on values of S and T derived from the current experimental values of m_W, Γ_ℓ, and $[\sin^2 \theta]_Z$.]

There are two possibilities for what will then be found.

- If the wedge intersects all the strips then, depending on where along the wedge the intersection is, we will have a good rough estimate of the Higgs mass in the minimal standard model. (It is often said that observables are only very weakly dependent on the Higgs mass in the minimal standard model, but that is true only if the top quark mass is not known; predictions for a low Higgs mass and a low top quark mass resemble those for a high Higgs mass and a high top quark mass.) If the experimentally allowed ranges of m_W, Γ_ℓ, and $[\sin^2 \theta]_Z$ continue to be centered on their present values, then this intersection would indicate a relatively high Higgs mass for $m_t = 160 \pm 10$ GeV, and a relatively low Higgs mass for $m_t = 120 \pm 10$ GeV.

- On the other hand, if the wedge doesn't intersect the strips, we will have a good clue as to what new physics must be added to the minimal electroweak theory. For instance, technicolor theories

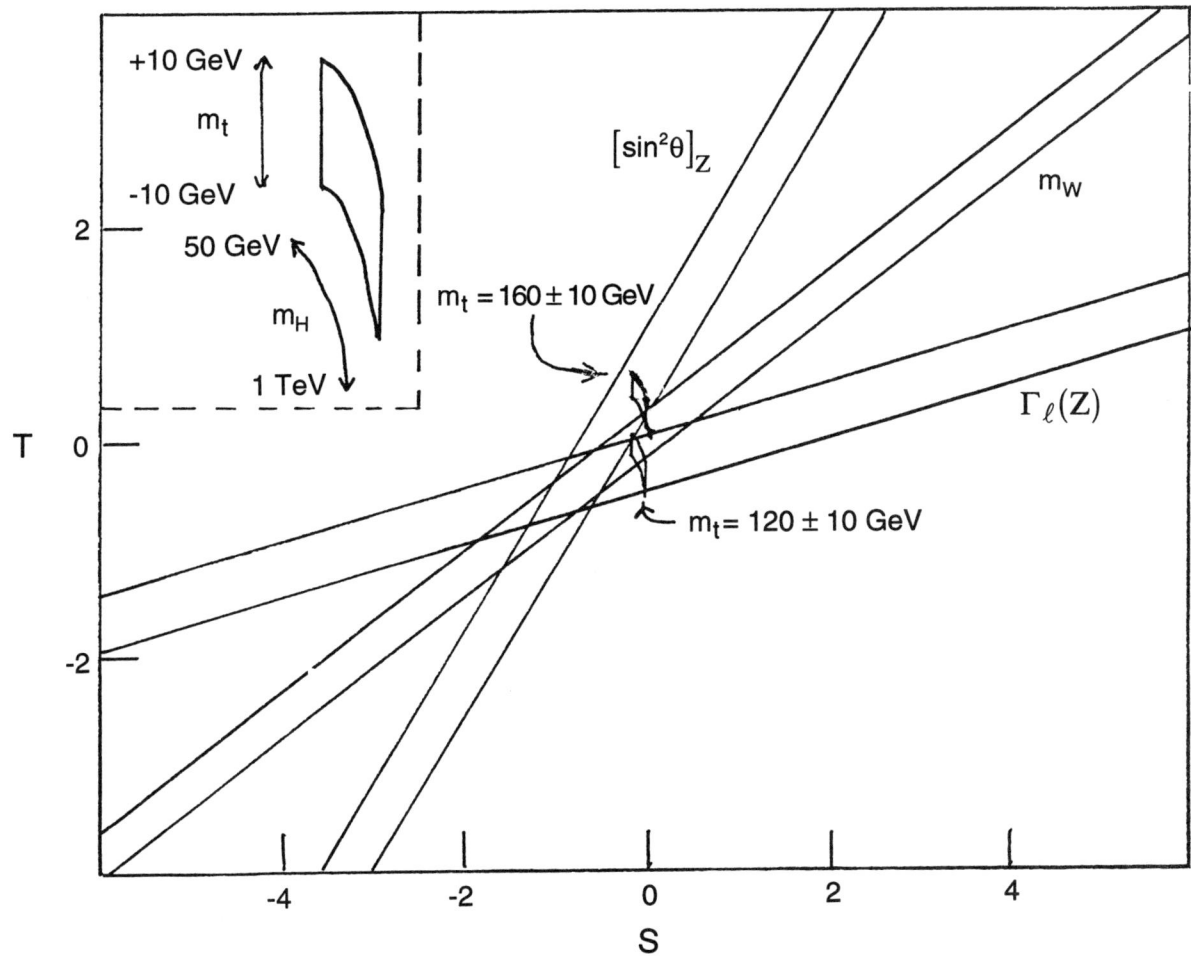

Figure 1. A Future Conference Viewgraph. The strips show the values of the 'oblique' radiative correction parameters S and T that would be experimentally allowed if future experiments were to give a W mass $m_W = 80.15$ GeV $\pm$ 100 MeV; a Z leptonic width $\Gamma_\ell(Z) = 83.33 \pm 0.2$ MeV; or an effective electroweak mixing angle $[\sin^2 \theta]_Z = 0.2320 \pm 0.001$. The wedges show the values of S and T that would be theoretically expected if the top quark were found with a mass of 120 ± 10 GeV or 160 ± 10 GeV, under the assumption that the Higgs mass in the minimal standard model is between 50 GeV and 1 TeV. The blown-up wedge in the insert indicates the values of m_{top} and m_{Higgs} corresponding to different points in the wedge. [This figure is based on one kindly supplied by M. Peskin.]

tend to move the wedges to larger values of S.

All this goes to underline the extreme importance of finding the top quark.

Effective Field Theories

My third topic is not new, but has lately become part of the common language of elementary particle physics. I think that I have heard the words "effective field theory" or "effective Lagrangian" a hundred times in as many different contexts at this meeting. Leutwyler went into effective field theories in some detail, and Peskin applied them in discussing oblique radiative corrections. I remember that after Leutwyler's talk, there was a question from the audience asking how we know that it is legitimate to use these effective field theories in describing the real world, and then someone else came up to me in the lobby and asked the same question. So although this topic is not by any means new, I would like to take a few minutes to explain what we are doing when we use an effective field theory

For this purpose I would like to use an example that is not usually discussed in terms of effective field theories, although I think it's the first example of an effective field theory in the literature. It goes back to the 1930's, when theorists like Heisenberg were calculating the scattering of light by light as a somewhat academic application of the new quantum electrodynamics. In 1936 H.Euler showed that the results for photon-photon scattering amplitudes at photon energies $\omega \ll m_e$, which are obtained in quantum electrodynamics from what would today be called an electron loop graph, could be summarized as a lowest-order perturbation theory result obtained from the Lagrangian density:

$$\mathcal{L}_{\text{eff}} = \frac{1}{2}\left(\vec{E}^2 - \vec{B}^2\right)$$

$$+ \frac{e^4}{360\pi^2 m_e^4}\left[\left(\vec{E}^2 - \vec{B}^2\right)^2 + 7\left(\vec{E}\cdot\vec{B}\right)^2\right] \quad (9)$$

In modern terms, we would say that in order to study processes at energies much less than m_e, we 'integrate out' the electron, replacing the Lagrangian of quantum electrodynamics with the effective Lagrangian (9).

This historical example provides lessons that help us to understand modern effective field theory:

1. We do not really need perturbation theory or even quantum electrodynamics to understand the general form of Eq. (9). Gauge and Lorentz invariance tell us that the most general possible Lagrangian for the electromagnetic field is of the form

$$\mathcal{L}_{\text{eff}} = \frac{1}{2}\left(\vec{E}^2 - \vec{B}^2\right)$$
$$+ c_1\left(\vec{E}^2 - \vec{B}^2\right)^2 + c_2\left(\vec{E}\cdot\vec{B}\right)^2$$
$$+ c_3\left(\vec{E}\cdot\vec{B}\right)\Box\left(\vec{E}\cdot\vec{B}\right) + \cdots, (10)$$

where the ellipsis "$\cdots$" denotes other terms with derivatives and/or more electromagnetic field factors, and c_1, c_2, c_3, etc., are unknown constants. [The factor in the first term is made to be equal to $\frac{1}{2}$ by a canonical normalization of the fields.] Dimensional analysis tells us that c_1 and c_2 have dimensionality $[mass]^{-4}$, while c_3 and all other coefficients have dimensionalities $[mass]^{-n}$, with $n \geq 6$. If (10) is obtained by integrating out 'heavy' charged particles (like the electron) with some typical mass M, then c_1 and c_2 will be proportional to M^{-4}, while c_3 and all other coefficients will be proportional to M^{-6} or higher powers of $1/M$. If we calculate photon scattering amplitudes at photon energies $\omega \ll M$, then the result will be dominated by the Born approximation contribution of just the c_1 and c_2 terms in (10), because all other interactions and higher order graphs will be suppressed by two or more powers of ω/M. [Of course one must go back to quantum electrodynamics to derive the specific values of the coefficients in Eq. (9), though factors like e^4 and $1/8\pi^2$ can be obtained by simply counting vertices and loops in the graphs from which the effective Lagrangian is calculated.]

2. The effective Lagrangian (10) is more than an *aide memoire* for photon-photon scattering amplitudes. It is not hard to show that the dominant terms in the amplitudes for other low-energy photon reactions like $\gamma + \gamma \to \gamma + \gamma + \gamma + \gamma$ are given by using the c_1 and c_2 terms in (10) in the tree approximation. Perhaps more surprisingly, we can regard this effective Lagrangian as the basis of a legitimate quantum field theory, calculating corrections to any process of higher order in ω/M by including photon loop as well as tree diagrams generated by (10). This effective field theory is of course non-renormalizable, but the divergences in these loop graphs are cancelled by renormalization of the coefficients in (10). Non-renormalizable theories are just as renormalizable as renormalizable theories; the only difference is that we have to deal with an infinite number of interaction types. You might suppose that the intrusion of unknown constants from the higher terms in (10) means that this effective theory has lost all predictive power, but that's not true at all. For example, the one-loop contribution to the scattering of light by light gives a term of order $(\omega/M)^8 \log \omega$ with a coefficient given by a known quadratic expression in c_1 and c_2, as well as polynomial terms involving the coefficients of higher terms in (10).

3. The use of perturbation theory with the Lagrangian (10) does not depend on the fact that the fine structure constant is small (though that helps); perturbation theory could be used even if the underlying theory, quantum electrodynamics, were a strongly interacting theory. This is because each loop in the effective field theory introduces additional factors of ω/M into the amplitude for any given process; perturbation theory here is an expansion in powers of ω/M, as well as $e^2/8\pi^2$. This of course is why Euler was able to calculate the leading terms in the photon-photon scattering amplitude by using (9) in lowest order. If $e^2/8\pi^2$ were of order unity we would not be able to derive Eq. (10) from an underlying theory, and we would not be able to calculate values for the constants c_1, c_2, etc., but we could still do perturbation theory with (10) as our effective Lagrangian.

4. The crucial feature of the effective Lagrangian (10) that allows us to use it to generate an expansion for amplitudes in powers of energy is that it involves only non-renormalizable interactions, with coupling constants that necessarily have the dimensions of negative powers of mass. In pure electrodynamics this is a consequence of gauge invariance; renormalizable interactions would be quartic polynomials in the vector potential A^μ without derivatives, and these would not be gauge invariant.

The last remark suggests that we should expect to be able to use effective Lagrangians to do calculations at low energies whenever there are symmetries (like gauge invariance in the above example) that rule out renormalizable interactions. There are other such symmetries. One is general covariance in the theory of gravitation (with no cosmological constant.) Another well-known example is the broken chiral $SU(2) \otimes SU(2)$ symmetry of quantum chromodynamics with massless u and d quarks. This requires that the pion, the Goldstone boson of this broken symmetry, enters into the Lagrangian only in nonrenormalizable couplings involving derivatives of pion fields, such as the leading term in the purely pionic Lagrangian:

$$\mathcal{L}_{\text{eff}} = -\frac{\partial_\mu \vec{\pi} \cdot \partial^\mu \vec{\pi}}{(1 + \vec{\pi}^2/F_\pi^2)^2}. \qquad (11)$$

Effective Lagrangians were introduced into modern particle physics in the 1960s in this context (though we did not know we were doing quantum chromodynamics then.) One recent development in this area is the extension of the effective Lagrangian method to not only the interactions of pions with each other and with single nucleons, but also to the interactions of pions and several nucleons - in other words to the problems of nuclear force and pion scattering on nuclei. The results obtained can be summarized by saying that the chiral Lagrangian approach turns out to justify approximations (such as assuming the dominance of two-body interactions) that have been used for many years by nuclear physicists, though not knowing about the chiral Lagrangian approach it was not quite fair of them to use these approximations.

The heavy quark symmetries discussed earlier have been incorporated into an Effective Heavy Quark Field Theory, and this has been combined with the soft pion theory based on broken chiral symmetries. Effective Lagrangian techniques are also being used today to study not only the ordinary strong interactions, but also technicolor, a hypothetical extra strong force that has been considered as a possible source of the breaking of the $SU(2) \otimes U(1)$ electroweak symmetry. It is easy to invent technicolor models that have an accidental $SU(2) \otimes SU(2)$ symmetry containing the $SU(2) \otimes U(1)$ symmetry of the electroweak interactions, and it's natural to assume that it is spontaneously broken to a sort of isospin symmetry called custodial $SU(2)$, which preserves the usual relation between W and Z masses. This symmetry breakdown of course entails Goldstone bosons that provide the longitudinal parts of the W and the Z, and in most cases additional Goldstone bosons called techni-pions. The effective Lagrangian of these Goldstone bosons is very much like Eq. (11), with a structure dictated by the broken and unbroken symmetries and by the nature of the degrees of freedom. As discussed here by Chanowitz, effective Lagrangians of this sort are actively being used to study the possibilities of finding signs of technicolor at accelerators like the SSC.

This brings us back to the question that was asked after Leutwyler's talk: how do we know that it is legitimate to use these effective Lagrangians? The derivation of a Lagrangian like (11) is not just a matter of 'integrating out' heavy degrees of freedom in quantum chromodynamics, in the way that Euler integrated out the electron to obtain (9), because unlike the electron in QED, the pion does not appear as a field in the QCD Lagrangian. True, the breaking of the $SU(2) \otimes SU(2)$ symmetry implies that there must be a pion particle, and dictates the structure of any field theory that describes this pion, but how do we know that this composite pion can be described by a field theory at all? The answer seems to be that any quantum theory that satisfies Lorentz invariance plus a technical requirement called cluster decomposition plus unitarity will always at sufficiently low energy look like a quantum field theory. Quantum field theory is the only way, we believe, of reconciling these fundamental requirements.

[The clause "at sufficiently low energy" is inserted so that this statement will apply also to string theories, which contain infinite numbers of increasingly heavy particle types, and therefore do not look like quantum field theories at energies comparable to the string mass scale.] Furthermore, the theory of low energy pions must be a quantum field theory described by a broken $SU(2) \otimes SU(2)$ symmetry. The use of Eq. (11) is justified not because we derive it from a field theory like QCD, but because it is the most general chiral invariant theory of pions, aside from terms with more derivatives whose effects are suppressed at energies E much less than a QCD scale of order m_ρ by two or more factors of E/m_ρ.

The same thought occurs to us with somewhat chilling overtones with regard to the Standard Model. We believe that there is some unknown physics at really high energies, roughly of the order of the Planck mass, with unbroken symmetries like $SU(3) \otimes SU(2) \otimes U(1)$ (and perhaps supersymmetry) that keep the particles of the standard model massless before the breaking of $SU(2) \otimes U(1)$ (and supersymmetry). The most general theory we could expect to find at low energies is simply the most general quantum field theory satisfying these symmetries. This is of course the standard model, supplemented with non-renormalizable terms whose effects at energy E are suppressed by powers of E/m_{Planck}. The success of the standard model tells us nothing about whether the underlying theory that describes physics at the Planck scale is a quantum field theory. In a sense this represents the revenge of S-matrix theory, because we now believe that the field theories of which we are so proud, quantum electrodynamics, quantum chromodynamics, even general relativity for that matter, are not what they are because they are truly fundamental field theories, but simply because they are the only way of satisfying the requirements of symmetries and S-matrix theory. Quantum field theory as we use it now is nothing but S-matrix theory made practical. Our best candidate for an underlying theory at the Planck scale is in fact not a quantum field theory in the ordinary sense, which brings me to my next topic.

Superstring Theory

Superstring theories have been studied for over a decade as candidates for a fundamental theory in particle physics. The implications of superstring theories were discussed in the rapporteur talks by Alvarez Gaumé and Peskin, and reviewed here in detail in a parallel session talk by Dine; in this summary I will just make some general remarks about the current state of the theory, with emphasis on recent work on coupling constant unification in these theories.

One change in the last few years has been a movement away from thinking of superstrings as existing in ten or twenty-six dimensions, of which all but the four dimensions of ordinary spacetime somehow become compactified, and toward formulating superstring theories from the beginning in four spacetime dimensions. Not that the previous theories are wrong — it's just that if one starts with a ten dimensional superstring theory and then supposes that six of these dimensions become compactified, what you get is the same as if you had started with a four dimensional superstring theory of an appropriate type. So the four-dimensional approach gives a more general approach to theories that have a chance to describe nature.

A string as it moves through space sweeps out a two dimensional surface, so the study of strings is the study of quantum field theories in two dimensions, with the spacetime coordinates along with other degrees of freedom appearing as fields in these theories. The two surface coordinates in these field theories may be represented as a single complex variable z, but with the complex plane given an arbitrary topology, one that becomes more and more complicated as we go to higher and higher order in perturbation theory.

The action for this two-dimensional theory is a functional of the four spacetime coordinates $x^\mu(z, z^*)$, plus [in the popular 'heterotic' superstring theories] an equal number of spinor coordinates $\psi^\mu(z)$, plus a number of other field variables that are needed to satisfy certain constraints. These constraints arise ultimately from the fact that the spacetime metric has $g_{00} = -1$. This would destroy the positive definiteness of the quantum field theory of the coordinates $x^\mu(z, z^*)$ if it were not for a symmetry known as conformal invariance, which allows us to transform away vibrations of the string into the timelike direction. But conformal invariance in a theory with just four $x^\mu(z, z^*)$'s or four $x^\mu(z, z^*)$'s and four $\psi^\mu(z)$'s is violated by quantum mechanical anomalies similar to the triangle anomaly in QCD, and these anomalies must be cancelled by the other fields added to the action.

Each possible equilibrium state of ordinary fields in four dimensions corresponds to a different two-dimensional conformal field theory, and the vacuum expectation values of the four-dimensional fields correspond to parameters in the conformally invariant action. For instance, there is a term in the string action that (after a suitable choice of coordinates in the complex plane) is of the form:

$$I_{\text{quad}}[x] = -\frac{1}{2} \iint dz\, dz^*\, g_{\mu\nu}(x(z, z^*))$$
$$\times \frac{\partial x^\mu(z, z^*)}{\partial z} \frac{\partial x^\nu(z, z^*)}{\partial z^*}, \quad (12)$$

where $g_{\mu\nu}(x)$ is the gravitational field, which is required by conformal invariance to satisfy the Einstein field equations.

There are some useful general results that apply to this whole class of theories in the tree approximation, where the complex plane is not given any topological complications. One result is a formula for the string mass, a parameter related to the string tension and also to the slope α' of the Regge trajectories on which the excited states of the string lie, and which can most simply be regarded as the mass of the first excited state of a vibrating string. It is given in terms of the string coupling constant (which determines the magnitude of higher order effects) and the Planck mass by

$$M_{\text{string}} \equiv \frac{2}{\sqrt{\alpha'}} = \frac{g_{\text{string}} M_{\text{Planck}}}{\sqrt{8\pi}}. \quad (13)$$

Also, if physics at energy scales below M_{string} is described by an $SU(3) \otimes SU(2) \otimes U(1)$ gauge field theory, then the gauge coupling constants at a renormalization scale μ near M_{string} approach values satisfying the relation:

$$k_{SU(3)}\, g^2_{SU(3)} = k_{SU(2)}\, g^2_{SU(2)} = k_{U(1)}\, g^2_{U(1)} = g^2_{\text{string}}. \quad (14)$$

The constants k are known in the trade as the 'levels' of the Kac-Moody algebra from which the gauge symmetries are derived, but the important thing is that $k_{SU(3)}$ and $k_{SU(2)}$ are in general integers, in most conformal field theories just $+1$. Also, $k_{U(1)}$ takes discrete values (values that are not changed by continuous changes in the conformal field theory), and in a large class of interesting models takes the familiar value of 5/3. In other words the relations among gauge couplings, that until recently have been understood in terms of the 'grand unification' of $SU(3) \otimes SU(2) \otimes U(1)$ in some simple group, occur naturally in string theory without any need for grand unification. In fact, string theory gives a better explanation of these coupling constant relations than grand unification because in grand unified theories the presence of color triplet partners of the Higgs doublet makes it particularly difficult to understand the disparity between the electroweak scale and the grand unification scale, while such GUT partners are absent in string theories with $k_{SU(3)} = k_{SU(2)} = 1$.

Eq. (14) applies to gauge couplings defined at a renormalization scale that is roughly of the order of the string mass scale (13). For more precise information about the value of the scale (or scales) where these relations apply, one must look to a one loop calculation. Such calculations were described here briefly by Alvarez Gaumé, and would have been described by Peskin if time had allowed. Including one-loop corrections, the relation (14) is replaced by

$$\frac{16\pi^2}{g_a^2(\mu)} = k_a \frac{16\pi^2}{g_{\text{string}}^2} + b_a \ln\left(\frac{M_{\text{string}}^2}{\mu^2}\right) + \Delta_a , \quad (15)$$

where b_a are the familiar constants appearing in the one-loop renormalization group equations $[\mu d g_a / d\mu = b_a g_a^3 / 16\pi^2]$ and Δ_a are the so-called threshold terms. In string theory as in field theory, the threshold terms are given by a sum over all species n of heavy particles that are integrated out to obtain the effective "low-energy" gauge theory:

$$\Delta_a = \sum_n c_{na} \ln\left(\frac{M_n^2}{M_{\text{string}}^2}\right), \quad (16)$$

with M_n equal (up to factors of order unity) to the heavy particle masses, and c_{na} constants that depend on the details of the underlying theory. The c_{na} are of order unity, and we expect the M_n to be of the order of M_{string}, so the threshold corrections Δ_n may be expected to be of order unity, which is why we expect Eq. (14) to apply at a renormalization scale μ of order M_{string}.

To make a more precise estimate, it is convenient to rewrite Eq. (15) as:

$$\frac{16\pi^2}{g_a^2(\mu)} = k_a \frac{16\pi^2}{g_{\text{string}}^2} + b_a \ln\left(\frac{M^2}{\mu^2}\right) + \tilde{\Delta}_a , \quad (17)$$

where

$$\tilde{\Delta}_a \equiv \sum_n c_{na} \ln\left(\frac{M_n^2}{M_{\text{string}}^2}\right) + b_a \ln\left(\frac{M_{\text{string}}^2}{M^2}\right), \quad (18)$$

and M is an arbitrary mass. The point is to try to choose M to minimize the typical values of $\tilde{\Delta}_a$, in which case according to (17) M is the renormalization scale at which the couplings most closely satisfy (14). This has been done by a number of authors; what seems to be the best estimate is:

$$M = \frac{e^{(1-\gamma)/2} 3^{-3/4}}{\sqrt{2\pi}} M_{\text{string}} = 0.216 \, M_{\text{string}} . \quad (19)$$

Extrapolating experimental values of the couplings to high energy in the minimal supersymmetric standard model, the couplings $g_{SU(3)}^2/4\pi$, $g_{SU(2)}^2/4\pi$, and $\frac{5}{3} g_{U(1)}^2/4\pi$ are found to converge to a common value $1/(26.3 \pm 2.1)$ at a renormalization scale about equal to 2×10^{16} GeV. [It has recently become possible to be much more precise about this, largely because experimental and theoretical advances have produced a great improvement in our knowledge of $g_{SU(3)}^2$, reported in a plenary session here by Bethke.] With $g_{\text{string}}^2/4\pi$ taken equal to $1/26.3$, Eq. (13) gives $M_{\text{string}} = 1.7 \times 10^{18}$ GeV, and Eq. (19) then gives the renormalization scale for coupling constant unification as $M = 3.6 \times 10^{17}$ GeV. So in other words there is about a factor of twenty discrepancy between the unification scale expected in string theory, proportional to the Planck mass, and the value 2×10^{16} GeV inferred from 'low' energy experiments. In speaking of this as a discrepancy, I can't help feeling a sense of unreality. Here we are, talking about energy scales about 10^{13} to 10^{14} times larger than the largest energies that we can produce in our accelerators, and worried about a discrepancy of a factor of 20!

Of course, one possible resolution of this discrepancy is that there may be more to physics at energies below the string scale than the minimal supersymmetric standard model, and that the couplings have therefore not been extrapolated correctly to very high energies. It is not easy to see what could be added to the minimal supersymmetric standard model that would preserve the *natural* convergence of all three couplings to a common value, while moving the energy at which the convergence occurs to higher values. Alternatively, the resolution of this discrepancy may be that the threshold corrections are larger than we had thought. From (16), we see that this could happen if some of the particles that were 'integrated out' in deriving the $SU(3) \otimes SU(2) \otimes U(1)$ standard model are actually somewhat lighter than M_{string}. Such would be the case if physics immediately below the string scale were described by a Kaluza-Klein theory, with some extra spatial dimensions forming a compact manifold with dimensions somewhat larger than $1/M_{\text{string}}$, or by a four-dimensional grand unified theory, with a simple gauge group broken spontaneously to $SU(3) \otimes SU(2) \otimes U(1)$ at an energy somewhat smaller than the string scale. But it now seems that if Kaluza-Klein or grand unified theories are to have any application to the real world, it will only be in a narrow energy range, roughly from 4×10^{17} GeV down to 2×10^{16} GeV.

These threshold corrections are important in other ways. It has been known for some years that parameters of the conformal field theory such as the dilaton field and modular fields can be fixed only by a non-perturbative dependence of the vacuum energy on these parameters, because in the range of these parameters where perturbation theory is valid, one can show that the vacuum energy has no local minimum. [The dilaton field is particularly important, because it is directly related to the gauge couplings.] We know that the vacuum energy depends non-perturbatively on the couplings $g_a^2(M)$; for instance the coupling $g_a^2(\mu)$ of QCD or some 'hidden sector' gauge field becomes strong at a renormalization scale μ_a of order

$$\mu_a \approx M \exp\left(-\frac{8\pi^2 k_a}{g_a^2(M)}\right), \qquad (20)$$

and the vacuum energy density contains terms of order μ_a^4. But (15) shows that the couplings $g_a^2(M)$ depend on the threshold correction terms Δ_a, which depend in a calculable way on parameters like the dilaton and modular parameter fields, giving the vacuum energy the desired non-perturbative dependence on these parameters.

This analysis has been explored by many authors in the last few years, and there are now many candidates for conformal field theories in which by 'discrete fine tuning' (that is, by choosing specific models out of a list of thousands, but without carefully adjusting continuous parameters) you can make the string coupling come out to have the 'observed' value, $g_{\text{string}}^2/4\pi = 1/26.3$, and you can also arrange to have supersymmetry broken at a scale that would give the gravitino a mass of order 1 TeV, needed to produce electroweak symmetry breaking with the observed strength. This seems to me to represent real progress in the interaction between superstring theory and physics.

Speaking of progress, there has been progress of another sort lately. It has actually become easier to follow the superstring literature, because many superstring theorists are now working on quantum gravity or two-dimensional statistical mechanics, and therefore there is less to read that is relevant to particle physics.

Despite all this progress, superstring theory faces a number of important and difficult problems. One of them is to identify correctly the source of non-perturbative effects. The non-perturbative effects described above are found by studying the effective 'low' energy quantum field theory derived from string theory rather from string theory itself, but we do not know if these the only important non-perturbative effects in string theories. For some years it has been widely assumed that the way to get at other, really stringy, nonperturbative effects is through what's called string field theory. String field theory allows for the creation and annihilation of strings in much the same way that ordinary field theory describes the creation and annihilation of particles. I have never been enthusiastic about string field theory, because it seemed to me to take a beautiful new formalism and make it ugly by trying to make it look like the old formalism of quantum field theory, but who knows?

I would like to offer a modest suggestion as to where we might look for specifically stringy non-perturbative effects without developing a string field theory. Let's recall what Feynman diagrams signal the advent of non-perturbative effects in ordinary quantum field theory. Suppose you want to calculate some process like a vacuum polarization in a pure non-Abelian gauge field theory, at a momentum p that is very small compared to the string scale M_{string}. Instead of introducing running coupling constants, suppose we perversely continue to expand in the coupling defined at M_{string} [essentially g_{string}], as if Gell-Mann and Low had never been born. We would then find a breakdown of perturbation theory, because small factors of $g_a(M_{\text{string}})^2/8\pi^2$ would be accompanied with large factors of $\ln(p/M_{\text{string}})$. These large logarithms come from graphs in which internal massless gauge boson lines approach the mass shell. Now, in string theory the Riemann surfaces that correspond to these diagrams have handles that are pulled out to long thin tubes. But just as in field theory the large logarithms mean that we have to deal with loops within loops within etc., in string theory at low momentum we have to deal with thin handles attached to thin handles attached to etc. The Riemann surfaces from which non-perturbative effects arise, though infinitely complicated, may like a fractal surface have a self-similarity property, of looking qualitatively the same at all scales.

I have no idea how to deal with fractal Riemann surfaces, but it may be simpler than the task of dealing with the corresponding non-perturbative effects in quantum field theory, because as Alvarez Gaumé emphasized, in some respects string theory is much simpler than quantum field theory. Gauge invariance appears more naturally, and there are fewer diagrams. This aspect of string theory has been exploited as a calculational device in quantum chromodynamics, and may perhaps allow us to understand stringy non-perturbative effects that go beyond anything we have been able to understand in quantum field theory.

Perhaps the most fundamental problem facing string theory is that we still do not know even in principle what it is that chooses the correct string theory, corresponding to the correct vacuum. The wrong answer is that the correct vacuum is the one with lowest vacuum energy, because we already know plenty of superstring theories that have negative vacuum energy. There is another possible answer supplied by quantum cosmology. In recent years the study of wormhole effects has suggested that the universe is not in a state in which all coupling constants have definite values, but rather in a quantum mechanical superposition of such states. Some relations among coupling constants are fixed by fundamental principles (such as the fact that the electric charges of the electron and positron are equal and opposite) and would be the same in all terms in this superposition, but any coupling constant that *can* vary continuously from one theory to another would have a continuous range of values in the different terms in the state vector of the universe. There may not be any such free parameters, in which case the following discussion is irrelevant, but our failure to formulate any principle for choosing the vacuum state in string theories suggests that the vacuum energy may be a free parameter that varies continuously from term to term in the state vector of the universe. When the universe becomes large, as it is now, it begins to look like an incoherent mixture of these terms, with various probabilities. In this case, it is only common sense that scientists who worry about the vacuum energy would have to find a value in the fairly narrow range in which life could arise; all the other terms in the wave function are there, but there are no scientists to observe them. The vacuum energy acts in cosmology like Einstein's cosmological constant. It can't be too positive because then galaxies would never have formed, and it can't be too negative because then the universe wouldn't live long enough for life to evolve. On this basis we can understand in a natural way why the vacuum energy is relatively small - some hundred and twenty orders of magnitude smaller than we would guess (a Planck mass per cubic Planck length) from purely dimensional considerations. But these considerations do not tell us that the vacuum energy (or equivalently, the cosmological constant) is zero, or even that it is astronomically negligible, but only that it is less than about a hundred times the present mass density of the universe. This is an interesting bound, because both the cosmological missing mass problem and the cosmic age problem mentioned by Krauss (the fact that age of globular clusters seems to be larger than some estimates of the age of the universe) would be alleviated by a positive cosmological constant corresponding to a vacuum energy roughly ten times the present mass density of the universe. It will be interesting to see whether this is the case.

* * *

This concludes my survey of special topics. I will now stick my neck way out, and try to guess what lies ahead for particle physics.

- All the present experimental challenges to the standard model will disappear. This judgment is based in part on having lived through it all before. Experiments certainly from time to time have required the incorporation of new features in the standard model, such as the tau lepton and the bottom quark. But where new experimental results have proved indigestible, irreconcilable with the general framework of the standard model, they have always gone away. Among these supposed difficulties were the high-y anomaly, parity conserving neutral currents, anomalous trimuon events, and second-class currents. (Some of us suffer just hearing this list.) The two outstanding present experimental difficulties with the standard model are the tau branching ratios discussed by Drell and the seventeen kilovolt neutrino discussed by Robertson. Drell's talk hints that the tau branching ratio problem is beginning to go away (although it certainly hasn't gone away yet), and Robertson came out pretty strongly against the reality of the seventeen kilovolt neutrino. So it takes no great courage to predict that these anomalies will go away as well.

- My second guess (these are all just guesses) is that the electroweak symmetry will turn out to be broken by the vacuum expectation values of elementary scalars that appear in the effective Lagrangian at accessible energies, like the scalar doublet in the original electroweak theory, and that the hierarchy problem will be solved by supersymmetry. I say this for a number of reasons. As Peskin discussed, the technicolor idea requires awkward extensions to produce the quark and lepton masses without introducing new difficulties like flavor changing neutral currents. Also, as Peskin and Rubbia remarked, some simple technicolor theories are already excluded by the high precision electroweak data discussed above. Another reason for this guess is that I find the convergence of the $SU(3) \otimes SU(2) \otimes U(1)$ couplings in the supersymmetric standard model very impressive, and this convergence is easily lost if you mess up the model by adding things like technicolor. My last reason has to do with the solar neutrino problem, and is explained below.

- I would guess that the solar neutrino deficit is real and is in fact explained by the MSW effect. This is in part because of the present state of the neutrino experiments. As discussed by Krauss and in a parallel session by Bahcall, it is not possible by adjusting the temperature at the center of the sun to make the standard solar model fit both the chlorine and Kamiokande data, but calculations based on the MSW effect can fit everything, including the data from SAGE and GALLEX. Furthermore, from a theorist's point of view the neutrino mass-square difference Δm_ν^2 and mixing angle θ_ν that are needed to fit this data are very plausible: θ_ν is like a typical small mixing angle in the CKM matrix, and $\sqrt{\Delta m_\nu^2}$ is a few millivolts, which is just what would be expected in the simplest extensions of the standard model. As is always the case for effective field theories, the Lagrangian of the standard model must be supplemented with non-renormalizable terms that are suppressed by powers of some large mass M, such as 10^{16} GeV. The least suppressed non-renormalizable term is a quartic term of dimension five, involving two factors of both the Higgs doublet and the lepton doublets:

$$\mathcal{L}_5 = \frac{g_{ij}^2}{M} \left[\begin{pmatrix} \phi^0 \\ \phi^+ \end{pmatrix} \cdot \begin{pmatrix} \nu_i \\ \ell_i \end{pmatrix} \right]$$
$$\times \left[\begin{pmatrix} \phi^0 \\ \phi^+ \end{pmatrix} \cdot \begin{pmatrix} \nu_j \\ \ell_j \end{pmatrix} \right], \quad (21)$$

that after electroweak symmetry breaking yields a Majorana neutrino mass matrix:

$$m_{ij}^2 = g_{ij} \langle \phi^0 \rangle_{\text{vac}}^2 / M . \quad (22)$$

If we take the couplings g_{ij} to be of the order of the product of the Yukawa couplings of the i'th and j'th lepton doublets to the scalar doublets, then the largest neutrino mass is of the order of m_{top}^2/M, which for $M \approx 10^{16}$ GeV is indeed a few millivolts. But this attractive picture of the origin of the neutrino masses needed in the MSW effect is only possible if the scalars are elementary. If in a technicolor theory you try to construct lepton number violating non-renormalizable interactions, you must replace the scalar doublet with some sort of bilinear function of techniquark fields, but then (21) is replaced with with an operator of very high dimension, which is strongly suppressed by many factors of $1/M$. Of course, a neutrino mass of a few millivolts might arise from all sorts of possible new physics, like lepton symmetry breaking at the technicolor scale, but there would be no particular reason to expect millivolt neutrino masses. This gives a special importance to studies of solar neutrinos and neutrino oscillations, but the pace of these experiments is unfortunately very slow, a little like real time studies of continental drift. In the next few years we can look forward to SAGE and GALLEX being calibrated with artificial megacurie neutrino sources, to super Kamiokande coming into being, and to the start of the

SNO experiment in Canada. All of these are important, but I want to emphasize that the SNO experiment offers the possibility of measuring a process $\nu + d \rightarrow \nu + p + n$ that arises solely from weak neutral currents and should therefore be unaffected by the MSW effect. If the MSW effect is indeed solely responsible for the observed neutrino deficit, then this neutral current process should be observed at precisely the rate predicted by the standard solar model, which may be our best way of reassuring ourselves that the sun is really well understood.

- We are going to find a great deal of new physics at accessible accelerator energies. With supersymmetry invoked to solve the hierarchy problem, we expect not only sparticles, but also flavor changing neutral current processes that as Peskin emphasized are endemic in supersymmetry theories. Also endemic in supersymmetry theories are CP violations that go beyond the CKM matrix, and for this reason it may be that the next exciting thing to come along will be the discovery of a neutron or atomic or electron electric dipole moment. These electric dipole moments were just briefly mentioned at this conference, but they seem to me to offer one of the most exciting possibilities for progress in particle physics. Experiments here as in solar neutrino physics move very slowly, but I should mention that there has been a lot of progress lately in calculating the electric dipole moment of atoms in various models, with results that are encouraging for future experiments.

- The correct theory underlying the standard model is probably a superstring theory. So far, our best proof consists of asking what else it could be. Superstring theories may be confirmed (and here I'm saying something that Peskin especially wanted me to say) by predictions for the coefficients of soft supersymmetry breaking terms in the supersymmetric standard model. In particular, it has been recently realized that in superstring theories it's typical that the lightest superparticles are gauginos rather than squarks or sleptons. This brings me to my final rash remark:

- Photinos are the cold dark matter needed in galaxy clusters.

* * *

In sticking my neck way out and acting as if we really are beginning to see the final answer I am going against the conventions of conference summarycraft. Michael Peskin showed a cartoon that expresses a more common view: a complacent physicist working away at a computer terminal, oblivious to monsters lurking just behind a wall. It is usual to end a conference summary with a remark that of course we expect that we will discover entirely new physics and that we are very far from anything like a final answer. But just because this is the conventional view and has generally been true in the past does not mean that it is true now. Michael mentioned a geographical metaphor: Columbus expecting to sail straight to the Indies, and not realizing there was something equally interesting in the way, namely America. [By the way, that geographical metaphor was used at a breakfast meeting in 1987 in convincing the Secretary of Energy to support the SSC.] But maybe a different geographical metaphor is more to the point. Imagine polar explorers sitting in the Travellers Club in the late nineteenth century, saying over their port, "You know, no matter how far north one goes, there's always plenty of sea and ice left further north. No matter how far north we go, we shall never get to the North Pole." Well, eventually explorers did get to the North Pole. We too may eventually get to our destination, to a final theory, and possibly sooner rather than later.

DISCUSSION

B. Ward, University of Tennessee, USA

You spoke of a gloomy picture for quantum field theory, with the view that superstring theory represents the ultimate theory. But, as you indicated, superstring theory is based on 2-dimensional conformal field theories, and uses field theory in a robust manner.

Weinberg

I was referring to a quantum field theoretic description of nature in four dimensions. As you know, superstring theory is very different from four-dimensional quantum field theory.

Weak Decays

Parallel Session 1

Perkins Chapel, SMU

Conveners: A. Pich (CERN)
W. Toki (Colorado State)

THE b-QUARK SEMILEPTONIC BRANCHING RATIO AT LEP

Robert Clare
Laboratory for Nuclear Science
Massachusetts Institute of Technology
Cambridge, MA 02139 USA

Abstract

The four LEP experiments have determined the semileptonic branching ratio of b-quarks. The measurements have been performed using hadronic events containing one or more electrons or muons. The average of all four experiments is Br(b $\to \ell\nu$X) = 0.110 ± 0.006 (ℓ = e or μ). The error is dominated by uncertainties in the modeling of b-hadron decays. The result is consistent with measurements at the $\Upsilon(4S)$. In addition, the ALEPH collaboration has made the first determination of Br(b $\to \tau\nu$X) = $(4.20\,^{+0.72}_{-0.68} \pm 0.46)\%$.

INTRODUCTION

In the Standard Model, the semileptonic decay of a b quark is dominated by spectator diagrams. The absolute decay rate for each lepton flavor is given by:

$$\Gamma(b \to \ell\nu X) = \frac{\mathrm{Br}(b \to \ell\nu X)}{\tau_b}$$
$$= \frac{G_F^2 m_b^5}{192\pi^3}(f_c|V_{cb}|^2 + f_u|V_{ub}|^2), \quad (1)$$

where V_{cb} and V_{ub} are elements of the Cabbibo-Kobayashi-Maskawa matrix, and f_c and f_u contain phase space factors and QCD corrections. Thus, measurements of the branching ratio and lifetime can be used to determine these CKM matrix elements.

I report here on measurements of the semileptonic branching ratio of b-quarks performed at LEP. The results from L3, ALEPH and DELPHI are preliminary and include data from the 1991 run. The OPAL results, based on 1990 data, are published. In addition, I report on the first measurement of the b $\to \tau\nu$X branching ratio performed by ALEPH.

EXPERIMENTAL PROCEDURE

Since b-quarks are heavy, the b-hadrons formed in the fragmentation process are very energetic. At LEP energies, the typical b-hadron has about 70% of the beam energy. Thus, leptons from the decays of b-hadrons are also energetic. Another result of the high mass of the b-quark is that the leptons have a high transverse momentum, p_T, with respect to the b-hadron direction. High p and p_T leptons therefore preferentially come from b-hadron decay. The experiments estimate the direction of the hadron using the direction of the jet nearest to the lepton. A typical p_T spectrum in Figure 1 shows that at high p_T the signal process dominates.

All of the experiments use essentially the same techniques to identify leptons in a hadronic environment, although the details vary according to the different detector designs. Muons are identified by looking for tracks that penetrate the typically 6-7 nuclear interaction lengths of the hadron calorimeter. All experiments have outer muon chambers which serve either as complete spec-

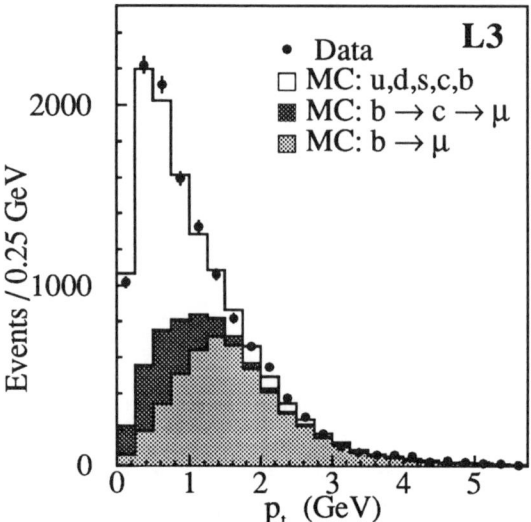

Figure 1. The p_T distribution for muons in the L3 detector. The dots are data. The histograms show the contributions of the signal and the various backgrounds.

trometers or simple taggers. Electrons are identified by clusters in the electromagnetic calorimeter with the correct shower shape, that match with tracks in the central tracking chamber. Some of the experiments also use dE/dx measurements in the tracking chamber. Typically, the identified leptons are required to have at least 3 GeV to reduce backgrounds.

Simple counting can be used to determine the branching ratio. In the absence of backgrounds, the number of events containing a single lepton is:

$$N_\ell \approx 2 N_{\text{had}} \frac{\Gamma_{b\bar{b}}}{\Gamma_{\text{had}}} B \epsilon (1 - B\epsilon), \quad (2)$$

where N_{had} is the total number of hadronic events, B is the semileptonic branching ratio, ϵ is the efficiency for finding the lepton, and $\Gamma_{b\bar{b}}$ and Γ_{had} are the partial widths of the Z^0 into $b\bar{b}$ pairs and hadrons, respectively. The factor 2 comes from the fact that either the b-quark or the $\bar{b}$-quark can decay. The efficiency includes the experimental efficiencies (e.g., lepton identification and geometrical acceptance) as well as kinematic efficiencies (requiring high p, p_T leptons). One can use the Standard Model prediction for $\Gamma_{b\bar{b}}/\Gamma_{\text{had}}$ and solve for the branching ratio.

Another technique is to count the number of events containing two leptons. Again, neglecting backgrounds, this is given by:

$$N_{\ell\ell} \approx N_{\text{had}} \frac{\Gamma_{b\bar{b}}}{\Gamma_{\text{had}}} (B\epsilon)^2. \quad (3)$$

One can combine the single- and di-lepton results to obtain:

$$B \approx \frac{1}{\epsilon} \cdot \frac{2 \frac{N_{\ell\ell}}{N_\ell}}{1 + 2 \frac{N_{\ell\ell}}{N_\ell}}, \quad (4)$$

which is relatively independent of $\Gamma_{b\bar{b}}/\Gamma_{\text{had}}$.

Of course, life is not so simple. There are backgrounds which must be dealt with, such as the cascade decay $b \to c \to \ell X$, the τ decay $b \to \tau \to \ell X$, prompt charm $c \to \ell X$, and fake leptons (punchthrough, K decays and misidentified hadrons). The most serious are the fake leptons and the cascade decay, $b \to c \to \ell X$. The latter one is especially bad, as the branching ratio is not very well known. Because these backgrounds do not have the same spectrum in p and p_T, many of the experiments choose to perform fits in p and p_T space to determine the fractions of signal and background as a function of momentum.

Another serious problem is the modeling of the decays of b-hadrons. The two most popular models are the free-parton model of Altarelli et al.[1] (ACCMM), and the exclusive decay channel model of Isgur et al.[2] (ISGW). These models predict slightly different shapes of the lepton spectrum. For the exclusive models, the amount of high mass 'D' states ("D**") also affects the spectrum. These effects affect the lepton finding efficiency, and thus the branching ratio. All of the LEP experiments use the JETSET[3] Monte Carlo, modified to match the spectrum of CLEO[4] and ARGUS[5] data at the $\Upsilon(4S)$. An example of a study by the L3 Collaboration is shown in Figure 2 where the CLEO/ARGUS data are compared to JETSET predictions without and with the

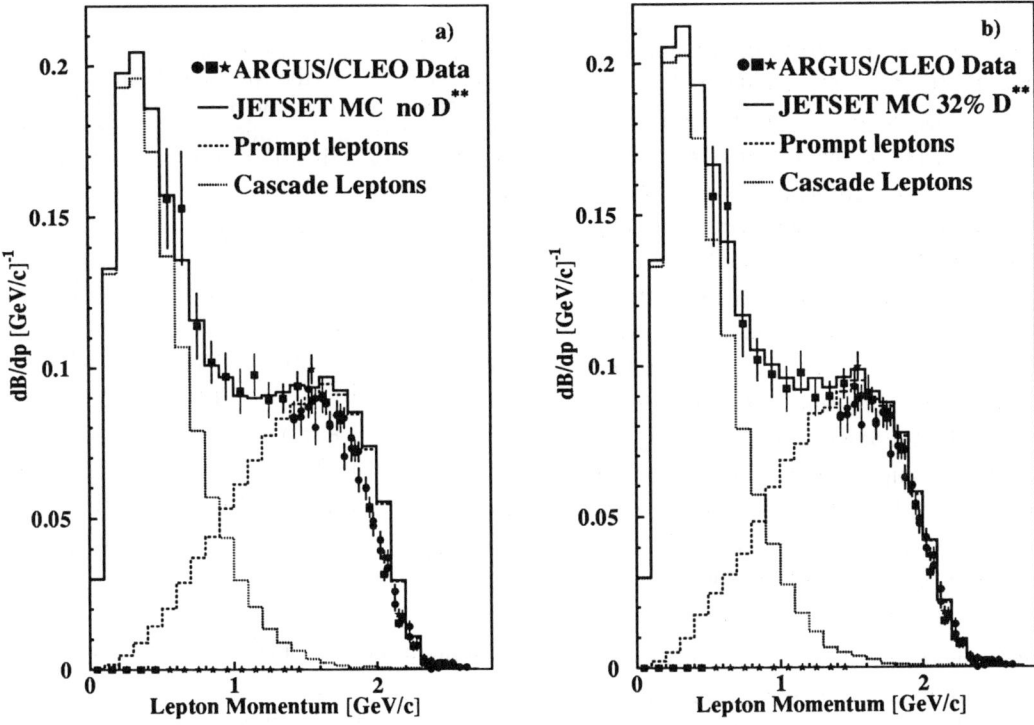

Figure 2. A comparison of the JETSET Monte Carlo to the CLEO/ARGUS $\Upsilon(4S)$ data, without D^{**} (a) and with 32% D^{**} (b). Both cases are with $Br(b \to \ell\nu X) = 0.112$.

addition of D^{**}'s. There is a clear softening of the spectrum with D^{**}'s.

RESULTS

L3[6] have determined the branching ratio using two methods. In both cases they use 1990 and 1991 data, electrons and muons. The first is a fit to the p and p_T spectra assuming the Standard Model value for $\Gamma_{b\bar{b}}/\Gamma_{had}$ (0.217). The preliminary result is $Br(b \to \ell\nu X) = 0.119 \pm 0.001$ (stat) ± 0.006 (sys). The second method uses the ratio of dilepton to single lepton events, resulting in $Br(b \to \ell\nu X) = 0.118 \pm 0.005$ (stat) ± 0.006 (sys). The breakdown of the systematic error is given in Table 1. The dominant errors are the detector understanding (efficiency) and decay model

Table 1. The contributions to the systematic error for the L3 results. (Preliminary).

Contribution	Variation	SM fit	Ratio fit
Efficiency		0.0040	0.0040
$\Gamma_{b\bar{b}}$	± 10 MeV	–	0.0006
$Br(c \to \ell)$	± 0.012	0.0023	0.0005
$\Gamma_{c\bar{c}}$	± 10 MeV	0.0002	0.0001
background	$\pm 15\%$	0.0024	0.0019
b fragment.	$\langle x_E \rangle \pm 0.01$	0.0007	0.0011
c fragment.	$\langle x_E \rangle \pm 0.04$	0.0005	0.0000
fitting bias		0.0010	–
D^{**}	.15 – .32	0.0044	0.0045

understanding.

ALEPH[7] have performed a multi-parameter fit to their 1990 and 1991 single and dilepton data. For the single lepton data they use p and p_T, whereas for the dilepton data they use $p_{T1}p_1 + p_{T2}p_2$ and $\min(p_{T1}, p_{T2})$. They fit simultaneously to seven parameters, including Br(b $\to \ell\nu$X) and $\Gamma_{b\bar{b}}/\Gamma_{had}$. The preliminary results are: Br(b $\to \ell\nu$X) = 0.110 ± 0.004 ± 0.004 and $\Gamma_{b\bar{b}}/\Gamma_{had}$ = 0.211 ± 0.007 ± 0.008, where the first error is statistical and the second systematic. In this fit, the branching ratio is essentially determined from the dileptons, whereas $\Gamma_{b\bar{b}}/\Gamma_{had}$ is determined by the single leptons. There is a 94% correlation between these results due to the relationship given by Eqn. 2. The systematic error contains a contribution of 0.0033 from efficiencies, and 0.0033 from the decay modeling.

DELPHI[8] have determined the branching ratio using a fit to the p and p_T spectra. The result combining published 1990 electron and muon and preliminary 1991 muon data is Br(b $\to \ell\nu$X) = 0.100 ± 0.006, where they have combined the statistical and systematic error (taking into account correlations between the 1990 and 1991 results). The dominant systematic errors are efficiency and the b $\to$ c $\to \ell$X branching ratio. They have also determined the branching ratio using the ratio of dilepton to single lepton events. The preliminary value (using 1990 and 1991 electrons and muons) is Br(b $\to \ell\nu$X) = 0.096 ± 0.008 ± 0.010. The dominant systematic error is from the efficiency.

OPAL[9] have performed a p, p_T fit to their 1990 muon sample resulting in Br(b $\to \ell\nu$X) = 0.104 ± 0.003 ± 0.006. For this result, they have not quoted a systematic error from the decay model. They have also counted high p and p_T electrons from their 1990 data to obtain Br(b $\to \ell\nu$X) = 0.110 ± 0.004 ± 0.009. The systematic error in this analysis includes a contribution of 0.005 from the decay model uncertainty.

The results are summarized in Table 2. The dominant experimental systematic error is the de-

Table 2. Summary of the LEP results on Br(b $\to \ell\nu$X). In the Method column, 'SM' is a result assuming the Standard Model for $\Gamma_{b\bar{b}}/\Gamma_{had}$, 'Ratio' is from the ratio of dileptons to single leptons, and 'Global' is the result of a multi-parameter fit.

Experiment	Method	Result
L3	SM	0.119 ± 0.001 ± 0.006
	Ratio	0.118 ± 0.005 ± 0.006
ALEPH	Global	0.110 ± 0.004 ± 0.004
DELPHI	SM	0.100 ± 0.006
	Ratio	0.096 ± 0.008 ± 0.010
OPAL	SM e	0.110 ± 0.004 ± 0.009
	SM μ	0.104 ± 0.003 ± 0.006
Average		0.110 ± 0.006

termination of the lepton identification efficiencies. For those experiments that have quoted an error due to the decay models, this is the largest error, and is clearly in common with all experiments. To evaluate an average from these results, I have first removed the decay model error, and then averaged the results using the PDG method.[10] The current LEP average is thus

$$\text{Br}(b \to \ell\nu X) = 0.110 \pm 0.003 \pm 0.005, \quad (5)$$

where the first error is the statistical plus experimental systematic error, and the second is the decay model uncertainty. This value is in excellent agreement with the measurements performed at the $\Upsilon(4S)$, where the average Br(b$\to$eX) is 0.107 ± 0.005. Both at LEP and at the $\Upsilon(4S)$ the measurements are somewhat low compared to the theoretical expectations of 0.12–0.15, depending on the model.

$$\text{Br}(b \to \tau\nu X)$$

Because of the mass of the τ lepton, the b $\to \tau\nu$X branching ratio is expected to be about 0.03 in the Standard Model. There has been some recent interest in two Higgs doublet models that in certain circumstances predict substantially higher b → τνX branching ratios.[11] At the same time, the electron and muon branching ratios would be reduced, bringing the theory in line with experiment.

ALEPH has reported a preliminary measurement of the b → τνX branching ratio.[12] Their analysis relies on the fact that in the decay of a b-hadron to a τ, with the subsequent decay of the τ, there are 2 neutrinos in the final state, and thus missing energy.

They first select well contained hadronic events. These events are divided into half-events by a plane perpendicular to the thrust axis. They then reject half-events containing identified electrons or muons, to reject other sources of missing energy such as b → ℓνX. In order to enrich the sample with $b\bar{b}$ events, they require that the event be inconsistent with all tracks coming from the primary vertex. They use their silicon microvertex detector for this.

The missing energy distribution for half-events passing the above selection is shown in Figure 3. From a fit to this distribution ALEPH determined Br(b → τνX) = $(4.20^{+0.72}_{-0.68}\pm 0.46)$% (preliminary), in good agreement with the Standard Model expectation.

CONCLUSIONS

The four LEP experiments have measured the b-quark semileptonic branching ratio to be

$$\text{Br}(b \to \ell\nu X) = 0.110 \pm 0.006. \quad (6)$$

The measurement is now systematics dominated: detector efficiencies typically contribute 2-3% error, while the uncertainties in the decay models contribute 3-4%.

The branching ratio b → τνX has been measured for the first time. The preliminary result is

$$\text{Br}(b \to \tau\nu X) = (4.20^{+0.72}_{-0.68}\pm 0.46)\%, \quad (7)$$

in good agreement with the Standard Model expectation of approximately 3%. This measure-

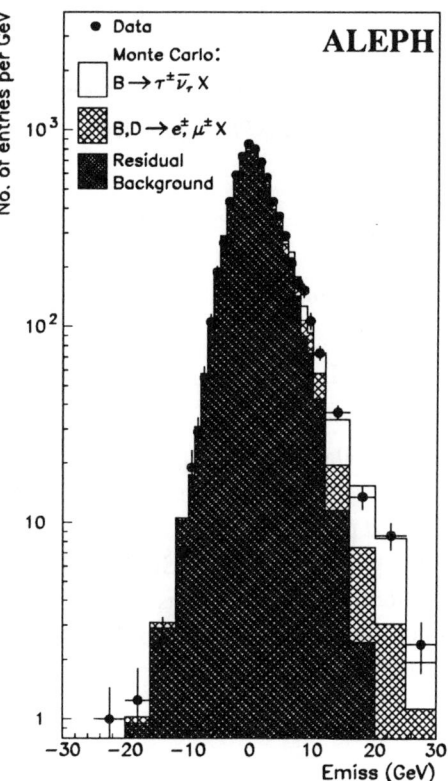

Figure 3. The missing energy distribution for half-events in ALEPH.

ment rules out some exotic two Higgs doublet models.

I would like to thank my colleagues in L3 for their help, as well as Drs. P. Dornan and V. Sharma from ALEPH, Drs. W. Venus and J.-E. Augustin from DELPHI and Drs. P. Wells and J. Kroll from OPAL for providing me preliminary data and for very useful discussions. I acknowledge support from DOE contract DE-AC02-76ER03069.

REFERENCES

1. G. Altarelli, et al., *Nucl. Phys.* **B207**, (1982) 365.

2. N. Isgur, et al., *Phys. Rev.* **D39**, (1989) 799.

3. T. Sjöstrand and M. Bengtsson, *Comput. Phys. Commun.* **43**, (1987) 367;

T. Sjöstrand in "Z Physics at LEPI", CERN-89-08, ed. G. Altarelli, *et al.*, (CERN, Geneva, 1989), volume 3, pp. 143–340.

4. S. Henderson, *et al.*, *Phys. Rev.* **D45**, (1992) 2212.

5. H. Albrecht, *et al.*, *Phys. Lett.* **B249**, (1990) 359.

6. L3 Collaboration, private communication.

7. ALEPH Collaboration, "Heavy Flavour Physics with Leptons", contributed paper to this conference.

8. DELPHI Collaboration, "$\Gamma_{b\bar{b}}$ and $\langle x_E \rangle_B$ using the Z^0 semileptonic decay into muons", contributed paper; DELPHI Collaboration, "A Study of B^0-$\bar{B}^0$ Oscillations using Dileptons from Semi-Leptonic Decay of b Quarks Produced from Z^0", contributed paper.

9. OPAL Collaboration, M.Z. Akrawy, *et al.*, *Phys. Lett.* **B263**, (1991) 311; OPAL Collaboration, P.D. Acton, *et al.*, "A Measurement of Electron Production in Hadronic Z^0 Decays and a Determination of $\Gamma(Z^0 \to b\bar{b})$," CERN-PPE/92-38, submitted to Z. Phys. C.

10. K. Hikasa, *et al.*, The Particle Data Group, *Phys. Rev.* **D45**, (1992) 1.

11. B. Grządkowski and W.-S. Hou, *Phys. Lett.* **B283**, (1992) 427.

12. ALEPH Collaboration, "Measurement of the $B \to \tau^{\pm} \bar{\nu}_\tau X$ Branching Ratio", contributed paper.

INCLUSIVE MEASUREMENTS OF THE B-HADRON LIFETIME AT LEP

Martin Pohl

Inst. für Hochenergiephysik

ETH Hönggerberg

Zürich

Switzerland

Abstract

I review recent, preliminary results of the lifetime of B-hadrons from LEP, obtained using an impact parameter measurement on inclusive B-hadron decays. The average lifetime of B-hadrons produced in Z-decays is found to be 1.40 ± 0.04 ps, which is high compared to the present world average. Interpreting the result within a spectator model, the Cabibbo-Kobayashi-Maskawa matrix element $|V_{cb}|$ is found to be 0.043 ± 0.005.

INTRODUCTION

In the standard electroweak model,[1] the charged current interaction is universal. The partial decay width for b-quarks into leptons $\Gamma(b \to l\nu X)$ is thus the same as that for purely leptonic decays, apart from phase-space factors, the additional complication of Cabibbo-Kobayashi-Maskawa (CKM) matrix elements and QCD corrections. The partial decay width is measured by the ratio of the leptonic branching fraction[2] $\mathrm{Br}(B \to l\nu X)$ of the average B-hadron produced to the lifetime τ_B of that average particle. One can expect that the more inclusive these two measurements are, the more appropriate a spectator model[3] interpretation will be to extract a value of the CKM matrix elements governing b-decays.

Lifetime measurements at LEP are very competitive due to the fact that B-samples are pure and good resolution vertex detectors are used. Pure samples of B-hadron production from Z-decay can be readily obtained using high momentum leptons, with a high transverse momentum with respect to the accompanying hadron jet.[2] Tab. 1 shows the sample composition for the LEP experiments that made new, preliminary data available for this conference. With an efficiency of typically 20%, sample purities of order 80 to 90% are thus obtained.

LIFETIME MEASUREMENT

For the lifetime measurements presented at this conference, L3 and OPAL use only gaseous vertex detectors, while ALEPH and DELPHI also have silicon microstrip devices. The lifetime is measured using the impact parameter of lepton tracks, a quantity defined as the distance of closest approach to the e^+e^- interaction point, signed positive, if the impact lies in the flight direction of the B-hadron, negative otherwise. Since the b-fragmentation is hard, few high momentum tracks originate from other particles than the B-hadron decay products. It is thus often not possible to reconstruct the e^+e^- vertex for an event. Therefore, the average position of the luminous region in a contiguous sample of events is generally used to approximate this vertex. This introduces an additional measurement error due to the size of the luminous region (in LEP: $\simeq 150 \mu m$ horizontally, $\leq 20 \mu m$ vertically).

Table 1. Composition of the LEP b-samples (in %), determined from Monte Carlo (preliminary).

Source	L3	ALEPH	DELPHI
$b \to \mu$	72.5	79.5	47.1
$b \to c/\tau \to \mu$	10.8	8.0	14.1
$c \to \mu$	6.7	4.5	11.9
decay/fake μ	10.0	9.0	27.0
$b \to e$	82.3	87.7	
$b \to c/\tau \to e$	7.7	7.5	
$c \to e$	2.5	3.5	
decay/fake e	7.5	1.3	

For their 1991 data, L3 uses the track measurement in the central tracking chamber (TEC), with a resolution on the impact parameter of about 110 μm. They accept as b-candidates inclusive muons with momentum $p_\mu > 4$ GeV and electrons with energy $E_e > 3$ GeV and transverse momenta with respect to the nearest jet of $1 < p_\perp < 6$ GeV. The resulting sample composition is shown in tab. 1. Fig. 1 shows L3's preliminary impact parameter distributions for $B \to e \nu X$ (2336 candidates) and $B \to \mu \nu X$ (3705 candidates). The distributions are clearly shifted and skewed towards positive values due to the finite B-lifetime. The lifetime fit starts from a set of primordial distributions, one for each event source as listed in tab. 1, obtained from Monte Carlo studies assuming a perfect track detector. These are convoluted with a resolution function determined from suitably defined data samples, taking into account the beam spot size. The final fit is then obtained by a weighted average of these distributions, with weights derived from the predicted sample composition. The resulting fit, as well as its components from the main sources of the sample, is also shown in fig. 1. L3's preliminary result is $\tau_B^e = 1.355 \pm 0.054$ ps for decays into electrons, and $\tau_B^\mu = 1.362 \pm 0.053$ ps for decays into muons. The errors quoted are statistical. A breakdown of systematic errors is shown in tab. 2. The dominant contributions come from the understanding of the primordial distributions and the resolution function of the vertex detector. From both decay channels, L3's preliminary determination of the B-lifetime from 1991 data gives $\tau_B = 1.358 \pm 0.038$ (stat) ± 0.061 (syst) ps, where the common systematic error takes into account correlations between the two decay channels.

Table 2. Breakdown of systematic errors (ps) for the preliminary L3 B-hadron lifetime measurement.

Source	$\mu^\pm$	$e^\pm$
Primordial distribution	0.050	0.050
Sample composition	0.013	0.017
Resolution function	0.050	0.040
c lifetime	0.010	0.010
b fragmentation	0.010	0.010
Hadronic bg shape	0.009	0.009
Total	0.073	0.068

ALEPH's preliminary measurement submitted to this conference[4] uses a two layer, double sided silicon detector with an impact resolution of about 25 μm (plus multiple scattering). Their sample of high momentum ($p > 3$ GeV), high $p_\perp$ (> 1 GeV) inclusive leptons contains 3103 muon candidates and 1806 electron candidates. The sample composition is summarized in tab. 1; the purity is slightly higher than L3's. Fig. 2 shows the impact parameter distributions for both electrons and muons, together with the lifetime fit and the background contributions. The result for electrons is $\tau_B^e = 1.40 \pm 0.05$ ps; for muons it is

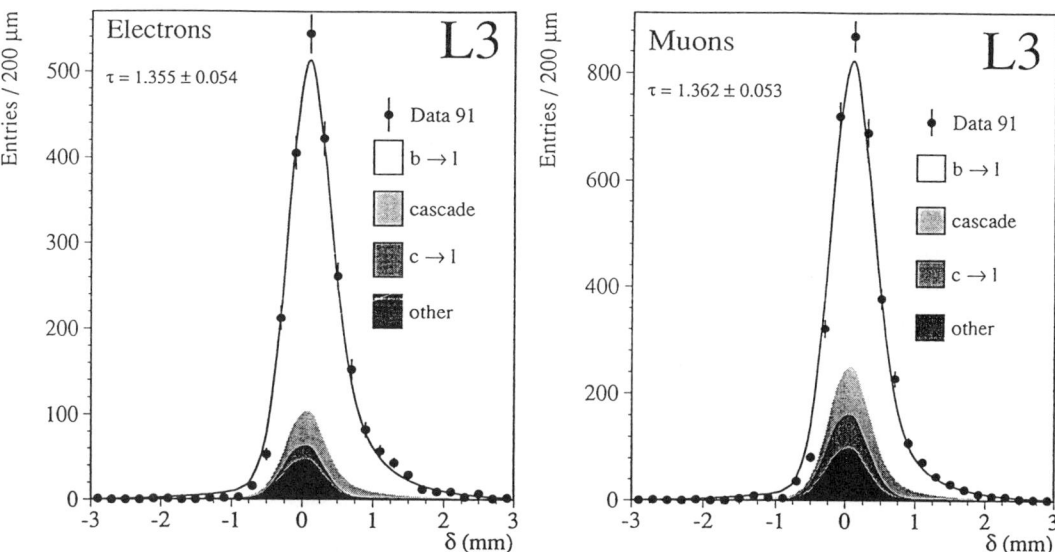

Figure 1. Impact parameter distribution from L3 for a) $B \to e\nu X$ and b) $B \to \mu\nu X$ candidates. Dots are data, full line is the fit result described in the text, shaded areas are the contributions from background sources.

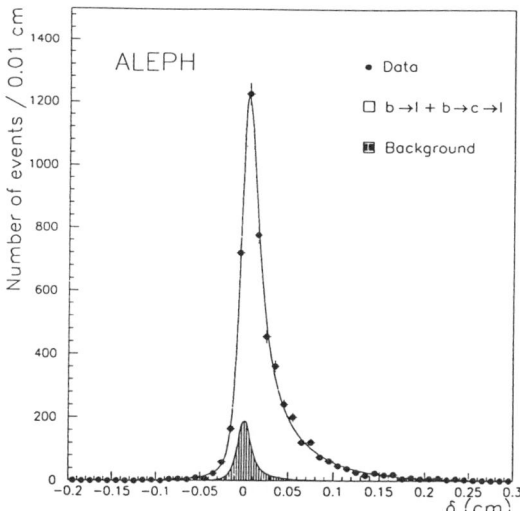

Figure 2. The impact parameter distribution for $b \to \mu$ and $b \to e$ candidates from preliminary 1991 ALEPH data. The curve shows the best fit, the shaded area is the total background contribution.

$\tau_B^\mu = 1.55 \pm 0.04$ ps. Systematic errors are summarized in tab. 3. Since they find no explanation for the apparent difference observed when comparing the electron and muon result, they choose to include a systematic error covering a possible systematic bias on one of them. The combined preliminary result from ALEPH is thus $\tau_B = 1.49 \pm 0.03$ (stat) ± 0.06 (syst) ps, for their 1991 data alone.

The DELPHI detector during 1991 running had a three layer, single sided silicon vertex detector, with an impact parameter resolution of $\sqrt{24^2 + 69^2/p_\perp^2}$ μm; the second term is due to multiple scattering. DELPHI submitted a preliminary lifetime measurement to this conference,[5] based on 3815 candidates for $B \to \mu\nu X$ from 1991, with a somewhat less strict sample definition. Consequently, as shown in tab. 1, their sample purity is lower than for the other LEP experiments. Fig. 3 shows the impact parameter distribution for this sample, together with the lifetime fit and background contributions. The resulting lifetime is $\tau_B = 1.36 \pm 0.05$ (stat) ± 0.05 (syst) ps, the

Table 3. Breakdown of systematic errors (ps) for the preliminary ALEPH B-hadron lifetime measurement.

Source	Syst. Error
Primordial distribution	0.015
Decay and frag. model	0.035
Decay background	0.015
Resolution function	0.015
Sample composition	0.020
c lifetime	0.005
Bremsstrahlung	0.010
Muon-electron discrepancy	0.030
Total	0.058

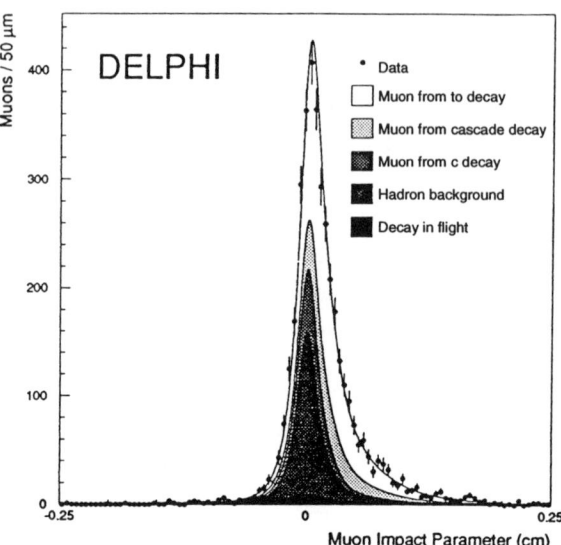

Figure 3. The impact parameter distribution for $b \to \mu$ candidates from preliminary 1991 DELPHI data. The curve shows the best fit, the shaded area is the total background contribution.

sources of systematic error considered are summarized in tab. 4. In addition to this "classical" approach, DELPHI also contributed a new result based on the impact of all tracks in a b candidate event.[6] This measurement is made possible by the fact that in a b event most tracks come from the B-hadron decay and its descendents. The preliminary result is $\tau_B = 1.41 \pm 0.04$ (stat) ± 0.06 (syst) ps.

Table 4. Breakdown of systematic errors (ps) for the preliminary DELPHI B-hadron lifetime measurement.

Source	Syst. Error
Primordial distribution	0.03
Sample composition	0.02
Resolution function	0.02
Decay background	0.02
c lifetime	0.02
Hadronic background	0.01
Fragmentation	0.01
Total	0.05

Fig. 4 summarizes the above mentioned preliminary results together with a published one from the OPAL collaboration.[7] The joint LEP result for the average lifetime of B-hadrons comes out to be $\tau_B = 1.40 \pm 0.04$ ps, when applying the Particle Data Group's method for averaging.[8] This lifetime is higher, though not in significant disagreement with the previous world average[8] of 1.29 ± 0.05 ps.

DETERMINATION OF $|V_{cb}|$

Given this lifetime measurement and the inclusive semileptonic branching ratio from LEP,[2] a spectator model allows one to derive a relation[9] between the CKM matrix elements $|V_{cb}|$ and $|V_{ub}|$ that govern the decay of b-quarks. Neglecting the mass of the u-quark, this relation depends on the following extra parameters:

- the mass of the b-quark, which I assume to be 4.95 ± 0.30 GeV,[10] as evaluated in a spectator model;[3]

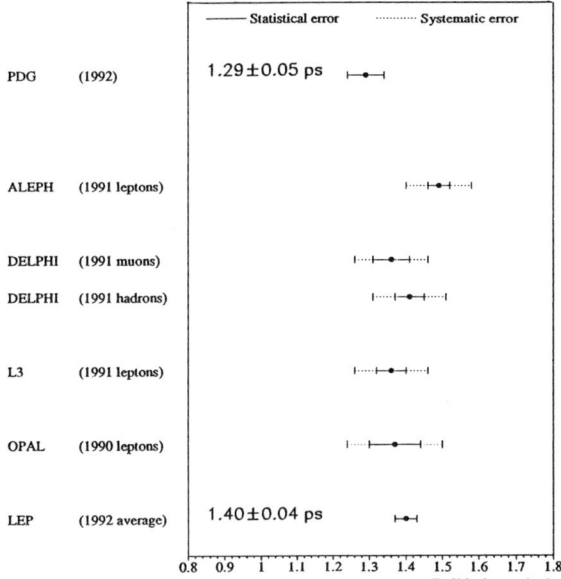

Figure 4. Summary of the b lifetime results from LEP, compared to the previous world average from the PDG. Numbers from ALEPH, DELPHI and L3 are preliminary.

- the mass difference between b- and c-quark, measured to be 3.30 ± 0.02 GeV;[10]

- the value of the strong coupling constant at the relevant scale, $\alpha_s(m_b^2) = 0.20 \pm 0.03$, extrapolated from the measured value $\alpha_s(M_Z^2) = 0.115 \pm 0.009$.[11]

Fig. 5 shows the curve in the $|V_{cb}|$ vs. $|V_{ub}|$ plane that corresponds to the measured LEP B branching fraction and lifetime. Because of a partial compensation of space space factors, the result depends more on the mass difference between b- and c-quark than on the absolute value of the b-mass, at least for small values of $|V_{ub}|$. A small ratio $|V_{ub}|/|V_{cb}| = 0.15 \pm 0.10$ is compatible with measurements on the $\Upsilon(4S)$.[12] Using this value, one finds $|V_{cb}| = 0.043 \pm 0.005$.

ACKNOWLEDGEMENT

I would like to thank my colleagues from the LEP collaborations, especially Drs. J.-E. Augustin, P. Dornan, D. McNally, L. Taylor, W. Venus and P. Wells, for providing me with preliminary data and for useful discussions. This work was partially supported by Schweizerischer Nationalfonds zur Förderung der wissenschaftlichen Forschung.

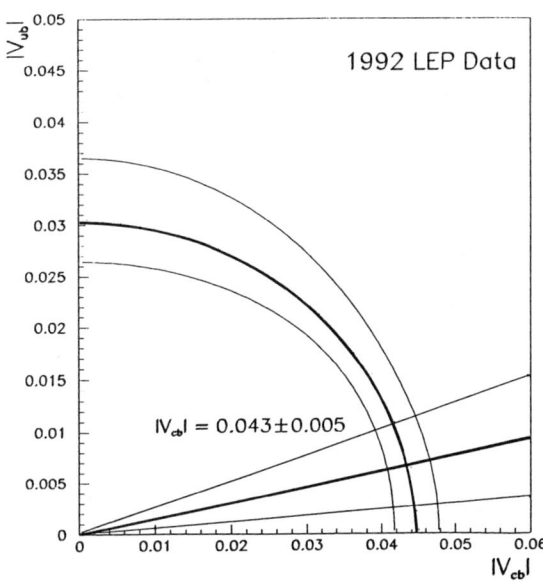

Figure 5. Contour plot of $|V_{ub}|$ vs. $|V_{cb}|$. The curved thick line comes from the LEP measurements of the B-hadron lifetime and semiletonic branching ratio. The thick straight line comes from the ARGUS/CLEO measurement of $|V_{ub}|/|V_{cb}|$. The thin lines correspond to one standard deviation errors.

REFERENCES

1. S.L. Glashow, Nucl. Phys. **22** (1961) 579;
   S. Weinberg, Phys. Rev. Lett. **19** (1967) 1264;
   A. Salam, "Elementary Particle Theory", Ed. N. Svartholm, Stockholm, (1968) 367.

2. R. Clare, "The b-Quark Semileptonic

Branching Ratio at LEP", these proceedings.

3. G. Altarelli *et al.*, Nucl. Phys. **B208** (1982) 365.

4. ALEPH Collaboration, "Updated Measurement of the Average B Hadron Lifetime", contributed paper to this conference.

5. DELPHI Collaboration, "Refined Measurement of the Average Lifetime of B Hadrons Using High p_t Muons", contributed paper to this conference.

6. DELPHI Collaboration, "Inclusive Measurement of the Average Lifetime of B-Hadrons Produced at the Z Peak", contributed paper to this conference.

7. OPAL Collaboration, P.D. Acton *et al.*, Phys Lett. **B274** (1992) 513.

8. The Particle Data Group, K. Hikasa *et al.*, Phys Rev. **D45** (1992) 1.

9. N. Cabibbo and L. Maiani, Phys Lett. **B19** (1978) 109; M. Suzuki, Nucl Phys. **B145** (1978) 420; A. Ali and E. Pietarinen, Nucl. Phys. **B145** (1979) 519.

10. ARGUS Collaboration, H. Albrecht *et al.*, Phys. Lett. **B249** (1990) 359; J.C. Gabriel, Ph.D. Thesis, Univ. of Heidelberg, IHEP-HD/89-1 (1989), unpublished.

11. L3 Collaboration, B. Adeva *et al.*, Phys. Lett. **B248** (1990) 464; ibidem **B257** (1991) 469.

12. CLEO Collaboration, R. Fulton *et al.*, Phys. Rev. Lett. **64** (1990) 16; ARGUS Collaboration, H. Albrecht *et al.*, Phys. Lett. **B255** (1991) 297.

B^0 AND B^+ LIFETIME MEASUREMENTS AT LEP

Michael Feindt
PPE Division, CERN, CH-1211 Geneva, Switzerland
representing the DELPHI Collaboration

Abstract

Using either $D^{(*)}$ lepton correlations or secondary vertex charge measurements, b-quark jets are separated into B^+ and $\bar{B}^0$ enriched samples. Preliminary results on lifetimes of these separated b-samples are presented, exploiting the good vertex resolution of the DELPHI and ALEPH silicon microvertex detectors. Within errors, the B^+ and $\bar{B}^0$ lifetimes appear to be equal.

MOTIVATION

By now, very accurate measurements of the "mean" b lifetime are available. Since a b-quark has enough time to form hadrons before it decays weakly, a "b" might actually be a $\bar{B}^0$, a B^- or B^0_s meson, or a Λ_b baryon. A priori it is not clear whether all of these lifetimes are equal. In case there are large differences between the lifetimes of different b hadron species, the concept of a "mean" b lifetime appears senseless, since the decay length distribution will no longer be described by a single exponential.

In the charmed sector, big lifetime differences appear: $\tau(D^+)/\tau(D^0) = 2.5 \pm 0.1$. These can be understood at least qualitatively in terms of quark interference effects in non-leptonic decay modes. For many important decay modes, e.g. $D \to K\pi$, there are two diagrams leading to the same final state (see fig.1): a "specator" diagram as well as an "internal conversion" diagram. Simple quark line diagrams show that in the case of the D^+ both diagrams lead to the same final state ($\bar{K}^0\pi^+$), whereas in case of the D^0 they lead to the two final states $K^+\pi^-$ and $\bar{K}^0\pi^0$. The main cause of the lifetime difference is a destructive interference between both graphs in D^+ decay. No such interference is possible for D^0 decays, the final states actually being distinguishable.

For B decays, one usually expects the spectator diagrams to dominate by far over the internal conversion diagram, mainly due to the larger b-quark mass (see e.g. the discussions about Heavy Quark Effective Theory [1] and Effective Heavy Quark Theory [2]). Thus, interference effects are believed to be much smaller than in the charm sector and a lifetime ratio around 1 is expected.

This paper reports about direct lifetime determinations on enriched B^0 and B^+ samples, on the basis of decay length measurements using the DELPHI and ALEPH detectors at LEP. Key devices for such analyses are

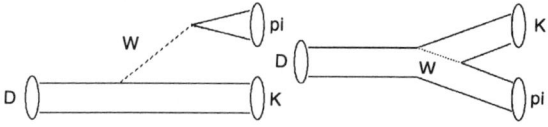

Figure 1. Spectator and internal conversion diagrams in nonleptonic heavy quark decays

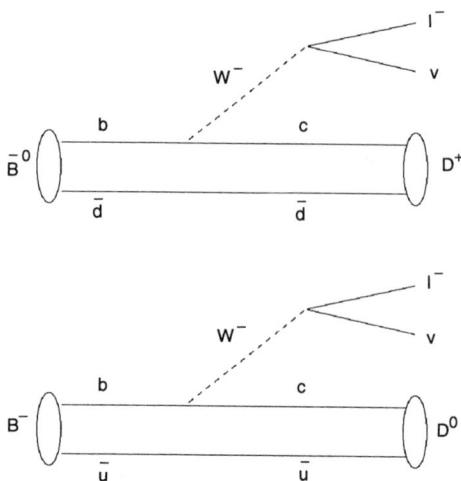

Figure 2. B^0 and B^+ semileptonic decay quark flow lines

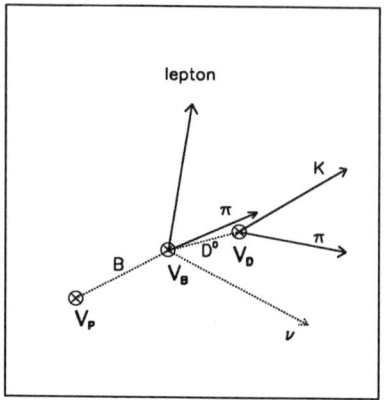

Figure 3. $B \to D^{(*)}l\nu$ vertex geometry

the silicon microvertex- detectors, which allow for track-vertex extrapolation accuracies in the order of $100\mu m$. Details of the analyses presented here may be found in [1–3].

$D^{(*)}$ LEPTON CORRELATIONS

Fig. 2 shows the quark line diagrams for semileptonic B decays into $D^{(*)}$ (which stands for both, D and D^* mesons). It can be seen that neutral B mesons always decay into charged $D^{(*)}$, whereas charged B mesons decay only into neutral $D^{(*)}$ mesons. Furthermore, the weak decay properties of the D mesons ensure that charged kaons from D decays must have the same charge as the lepton of the $B \to Dl\nu$ transition.

Fig. 3 shows the vertex geometry: First $K\pi$ vertices V_D are searched, the D^0 candidates are extrapolated back to search for a vertex V_B with an identified lepton. The B decay length is the distance between the primary vertex and V_B.

Fig. 4 shows the evidence for charmed mesons in multihadronic events with an identified lepton in the same hemisphere: 92 $D^0 \to K^-\pi^+$ candidates, 35 $D^+ \to K^-\pi^+\pi^-$ and together 61 $D^{*+} \to \pi^+ D^0$ candidates (30 in $D^0 \to K\pi$, 31 in $D^0 \to K3\pi$) are found by DELPHI. ALEPH obtained very similar event samples (71 $D^0 \to K\pi$, 71 $D^{*+} \to \pi^+ D^0$). The lifetimes are determined using an event-by-event maximum likelihood fit. The results are (in picoseconds):

reaction	DELPHI	ALEPH
$\bar{B} \to D^0 l^- X$	$1.27^{+0.22}_{-0.18} \pm 0.15$	$1.32^{+0.19}_{-0.18}$
$\bar{B} \to D^+ l^- X$	$1.18^{+0.39}_{-0.27} \pm 0.15$	---
$\bar{B} \to D^{*+} l^- X$	$1.19^{+0.25}_{-0.19} \pm 0.15$	$1.62^{+0.24}_{-0.31}$

As a cross check both collaborations also reconstructed the D lifetimes from the $V_B - V_D$ distance and found them to be compatible with the world averages.

The caveat of this method lies in the fact that $D^{(*)0}$ might denote next to D^0, D^{*0} also higher spin D resonances (D^{**}) or non-resonant $D^{(*)} + n \cdot \pi$. Formerly usually neglected, recent measurements by ARGUS and CLEO II indicate that these contributions might be as large as 40% of the semileptonic branching ratio. Decaying strongly, isospin considerations show that the D^{**} charge is mixed into $D^{(*)}$ and pion charges with some uncertainties, such diffusing the simple B-D charge correlation. First direct evidence for higher D resonance production was presented at this conference [6]. One has to unfold the charge mixing due to this effect using a model for D^{**} production and using measured branching ratios.

To fit B^0 and B^+ lifetimes, simultaneous fits to the D^0 and $D^{(*)+}$ decays are performed, the corresponding sample compositions taken from a model calculation taking into account charge mixing due to D^{**} production and tracking inefficiencies. DELPHI and ALEPH use slightly different assumptions in this model, which leads to a more diagonal matrix in the latter analysis. Also, ALEPH includes the production *rates* in the fit, since $N(B^0)/N(B^+)$ also slightly depends on τ^+/τ^0. The final results are displayed in fig. 6.

SECONDARY VERTEX CHARGE

In this method, pioneered by the DELPHI Collaboration [5], one tries to separate all charged particles of a jet uniquely into two vertices; one primary vertex compatible with the beam spot, and a secondary vertex from the B decay. The finite path length of the D mesons can be neglected compared to the much larger b decay length. All possible

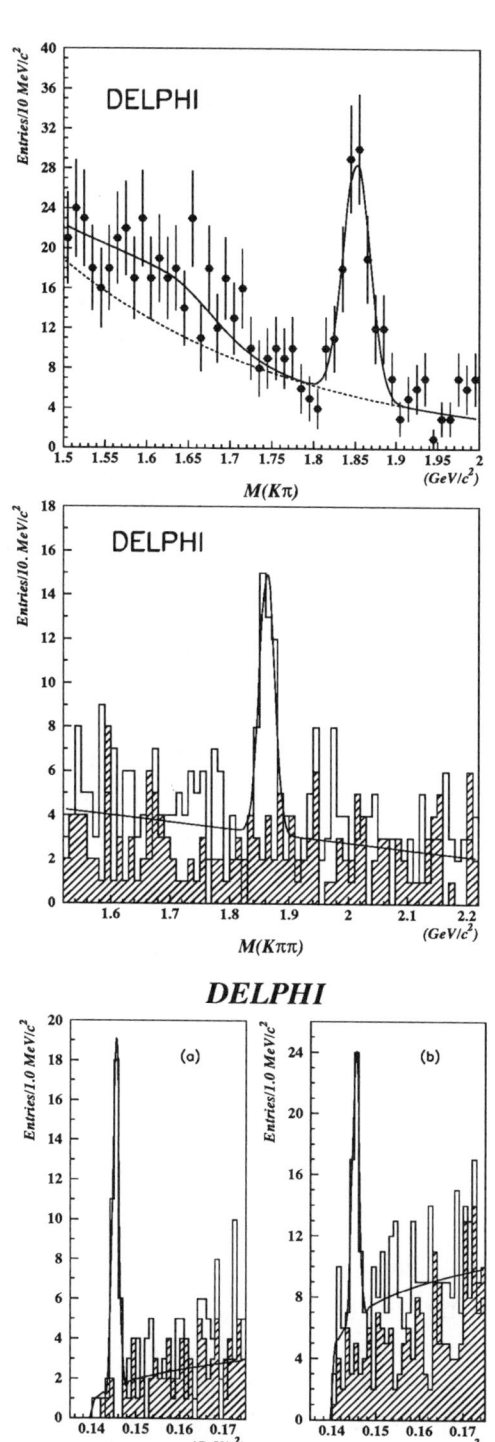

Figure 4. D^0 signal in $K\pi$, D^+ signal in $K\pi\pi$, and D^* signals from Δm plots using $D^0 \to K\pi$ and $D^0 \to K3\pi$

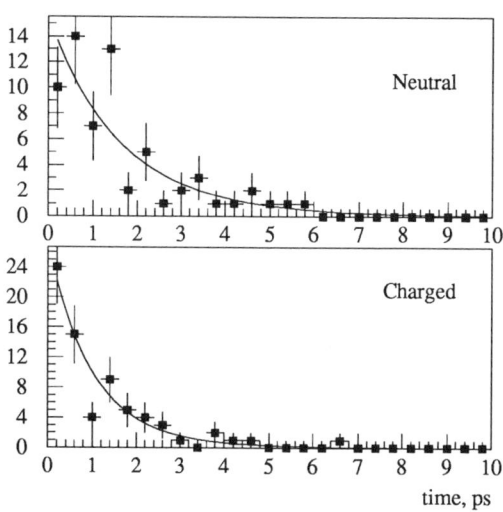

Figure 5. Proper time distributions of charged and neutral secondary vertices

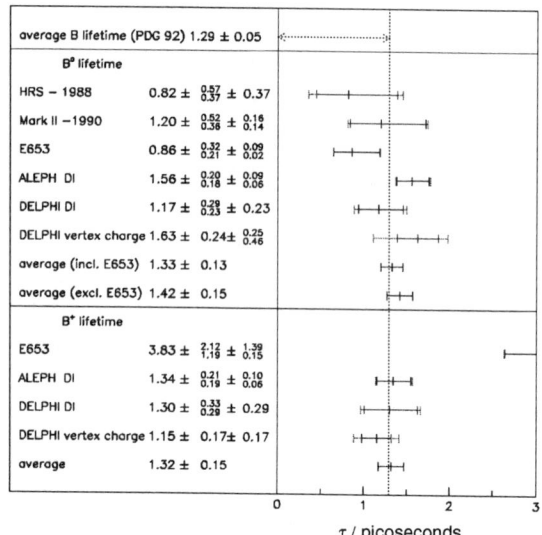

Figure 6. Results on charged and neutral B-lifetimes

permutations of tracks are tried, and an event is kept only if both the primary and the secondary vertex fits have χ^2 probabilities in excess of 10%, and there is one and only one such track combination in the jet. The charge measurement is about 80% correct, with misidentification rates of 10% for $\Delta Q = \pm 1$ and $\approx 1\%$ for $\Delta Q = \pm 2$, as determined from the rate of doubly charged vertices and cross-checked with Monte Carlo. The results of a maximum likelihood fit are: $\tau_{neutral} = 1.81 \pm 0.29\,ps$ and $\tau_{charged} = 0.93 \pm 0.21\,ps$ (see fig.5). These results have been obtained by actually measuring the *excess decay length* beyond the minimum visible decay length which is determined event-by-event from the data. One such is less dependent on Monte Carlo simulations and achieves smaller systematic errors. The results have been checked to be consistent with those obtained using a usual fit method with acceptance correction calculated from Monte Carlo.

Neutral b vertices may stem from $\bar{B}^0$, B_s^0 and Λ_b hadrons, whereas charged b vertices mainly stem from B^+ mesons (and perhaps a very small contribution from Σ_b^+ baryons). For the extraction of B^+ and B^0 lifetimes one has to unfold charge mismeasurements and, mainly for the B^0 lifetime, the influence of B_s^0 and Λ_b decays, for which one has to model production fractions and lifetimes.

All the results are collected in fig. 6, along with former measurements.

CONCLUSIONS

Within errors, the B^+ and B^0 lifetimes measured in e^+e^- reactions are equal to the average b-lifetime. The results from E653 reported at this conference [7] are however only hardly compatible with the measurements presented here. Disregarding that, one can conclude that for B-mesons there is not such an effect as observed in the charm sector.

ACKNOWLEDGEMENTS

I thank D. Bloch, G. Maehlum, W. Murray, S. Schael and G. Wormser (DELPHI Coll.) and J. Walsh (ALEPH Coll.) for explanations, discussions and plots.

REFERENCES

1. P. Drell, B. Grinstein, N. Isgur, and others, this conference.
2. F. Close, this conference.
3. DELPHI Coll., *A measurement of B Meson Production and Lifetime Using Dl⁻ Events in Z^0 Decays*, contributed paper.
4. ALEPH Coll., *Measurement of the B^0 and B^+ Meson Lifetimes*, Internal ALEPH note, unpublished.
5. DELPHI Coll., *A measurement of the mean lifetimes of charged and neutral B mesons*, contributed paper.
6. ARGUS Coll., *Investigation of the Decays $\bar{B}^0 \to D^{*+}l^-\bar{\nu}$ and $\bar{B} \to D^{**}l^-\bar{\nu}$*, contributed paper.
7. E653 Coll., N. Stanton, this conference.

MEASUREMENT OF LIFETIMES OF CHARGED AND NEUTRAL BEAUTY HADRONS FROM FERMILAB E653*

N.R. Stanton
Department of Physics
The Ohio State University
174 W. 18th Avenue
Columbus, OH 43214

Abstract

We report on 9 $b\bar{b}$ pair events produced by a 600 GeV/c π^- beam and detected in the hybrid emulsion spectrometer of Fermilab experiment E653. The measured lifetimes for samples of 12 neutral and 6 charged beauty hadrons are $\tau_{b^0} = 0.81^{+0.34+0.08}_{-0.22-0.02}$ ps, and $\tau_{b^\pm} = 3.84^{+2.73+0.80}_{-1.36-0.16}$ ps.

Fermilab E653 is a hybrid emulsion experiment studying decays of heavy quarks produced by 600 GeV π^-. The apparatus has been described in detail elsewhere[1]. The electronic spectrometer featured an 18-plane silicon microstrip (SMD) vertex detector, a large aperture dipole magnet, 55 drift chamber planes, and a liquid argon calorimeter. This was followed by a second spectrometer for muon analysis, comprised of range steel and 12 drift chamber planes on either side of an iron toroid. The trigger required a beam particle to interact in the target and a muon to penetrate 3900 g/cm^2 of steel. This muon trigger provided enriched samples of beauty and of semimuonic decays of charm.

The experiment used an active target of nuclear emulsion 1.5 cm thick in which the primary interaction and short-lived decays were observed. The high-resolution decay volume was extended by use of two emulsion precision decay analyzers: a thin plate separated from the bulk target by 1.0 cm of low-density foam, and a moving emulsion tape 1.2 cm farther downstream. These decay analyzers, which consisted of thin layers of emulsion on both sides of plastic sheets, were used as precision verniers on spectrometer tracks and gave a fivefold resolution improvement on reconstructed vertices outside the emulsion target. The fiducial decay region, including the emulsion components and the first 6 SMD's, was 12 cm long. In a typical $b\bar{b}$ pair[1] event with 4 decays, half of the decays are inside the bulk emulsion target, and half are outside.

A total of 8.2×10^6 events, corresponding to 2.5×10^8 interactions, were recorded on tape. Reconstructed spectrometer events with a muon transverse momentum $p_{T\mu} > 1.5$ GeV/c were scanned in the emulsion. A scan requiring $p_{T\mu} > 0.8$ GeV/c, now underway, should double the beauty sample. The primary interaction vertex was located for 99.8% of the 6542 events within the emulsion fiducial volume. To eliminate the unwanted events

---

*Work supported in part by the Japan Society for Promotion of Science; the Japan-US Cooperative Research Program for HEP; the Ministry of Education, Science and Culture of Japan; the Korea Science and Engineering Foundation; the Basic Science Research Institute Program, Ministry of Education, Republic of Korea; the US Department of Energy; and the US National Science Foundation.

[1]Since the beauty hadrons observed in this experiment are a mixture of species, we denote a generic beauty meson or hadron by b.

in which the tagged muon came from the primary interaction (via a π or K decay, e.g.), the slopes of the trigger muon, as measured in the spectrometer, were compared to the emulsion slopes at the primary vertex. An event was rejected if the angle difference between the muon and any primary track was less than 2 mrad. To search for decays in the 359 events in which the muon did not come from the primary vertex, tracks from the primary vertex were followed down in the emulsion, vertices outside the emulsion were reconstructed with an interactive program, and unmatched spectrometer tracks were scanned back into the emulsion. A detailed Monte Carlo simulation of both the spectrometer and of the emulsion procedure was used to determine detection efficiencies and resolutions.

To be classified as a $b\bar{b}$ pair, an event must pass three levels of criteria. First, there must be kinematic evidence for multiple heavy vertices, including at least one too heavy for charm. Then, the event must be topologically consistent with $b\bar{b}$ production with subsequent decays to charm; for example, 3 charged decays from the primary is an incorrect topology. Finally, the decays must be kinematically consistent with beauty and charm after daughter decays are assigned. In addition to the 9 $b\bar{b}$ pairs, the 359 events in which the muon did not come from the primary vertex include 175 events consistent with charm, which are very useful for checking systematics.

E653 b and c decays generally have unseen neutrals, so momenta must be estimated. This is done by matching the characteristics (topology, prong count, momenta, ...) of each data event to a large sample of Monte Carlo decays. The momentum distribution of the matched Monte Carlo events is taken to be the probability distribution for the momentum of the data event. A second estimator, similar the one we have used for charm semileptonic decays[2, 3] was used as a check, and for high-statistics simulations.

There are 12 neutral and 6 charged decays

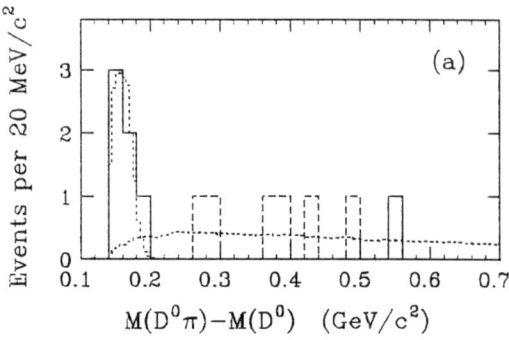

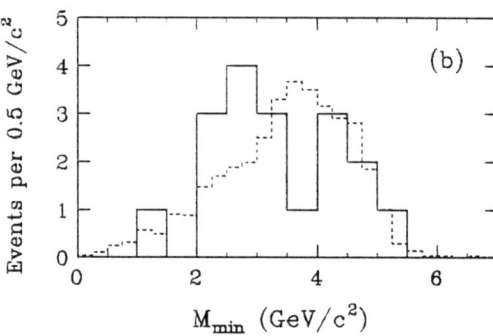

Figure 1. (a) Histograms of the mass difference $M(D^0\pi^\pm) - M(D^0)$ using D^0 and $\pi^\pm$ from beauty decays: right sign pions from data (solid); wrong sign pions from data (long dashes); and Monte Carlo simulation of right sign pions from beauty decays which include $D^{*\pm}$ decays among their decay modes (short dashes).
(b) Histogram of M_{min} for the 18 b and $\bar{b}$ candidates (solid), compared with Monte Carlo simulation (short dashes). The quantity M_{min} is the smallest mass of the parent particle that will allow momentum to be conserved at the decay vertex; it is given by $M_{min} = \sqrt{M_{vis}^2 + p_\perp^2} + p_\perp$, where M_{vis} is the visible invariant mass in the decay, including the measured tracks from the cascade charm.

in the final $b\bar{b}$ sample. All but 3 of the 9 events have 4 decay vertices. The semileptonic decays all have low multiplicity and a muon with high decay $p_\perp$. There is good evidence for 5 D^*'s among the 8 D^0's from b's (Fig. 1a). The distribution in the quantity M_{min} (essentially the transverse mass) of the beauty decays (Fig. 1b) is well-described by the simulation; M_{min}'s are almost all above the charm mass, and they cut off just above the beauty mass, as expected.

The level of background among the $b\bar{b}$ events was estimated by looking at the number and characteristics of data events which passed looser criteria than the final sample, and by studying a Monte Carlo simulation to understand the effectiveness of the tighter criteria in rejecting such backgrounds. It was found that when the topology and detailed kinematics requirements are removed, only one additional data event survives, indicating that the pool of possible background events is small. Then, from a Monte Carlo study of the feedthrough of simulated charm pair events into the $b\bar{b}$ sample (which occurred due to interactions, strange particle decays, and reconstruction errors) the background rejection from the topology and detailed kinematics criteria was found to be a factor of 6. The estimated background is thus about 1/6 event.

The production properties of the $b\bar{b}$ events are summarized in Figs. 2-4. The pair cross section, assuming linear A dependence, is $33 \pm 11 \pm 7$ nb/nucleon, about 700 times smaller than that for charm at this energy[3]. The inclusive x_F distribution (Fig. 2) is described by $d\sigma/dx_F = (1-|x_F-x_0|)^n$, with $n = 5.0^{+2.7+0.0}_{-2.0-0.9}$, similar to charm at this energy, and a positive offset (as expected in π^- production) of $x_0 = 0.06 \pm 0.06$. On the other hand, the inclusive p_T^2 distribution (Fig. 3) is much stiffer than that for charm, as expected; it is described by $d\sigma/dp_T^2 = \exp{-b p_T^2}$, with $b = 0.13^{+0.05}_{-0.04}$. Finally, the distribution in azimuthal angle difference $\Delta\phi$ between b and $\bar{b}$ (Fig. 4) shows that b and $\bar{b}$ tend strongly to be produced back-to-back, more so than for charm pairs[4].

The 175 charm events found during the beauty scan provide good checks on our understanding of systematics. The fraction of charm with found partners, 122/175, agrees quite well with the predictions of the simulation (soft charm, and all-neutral D^0 decays are missed, e.g.). The fractions of found charged

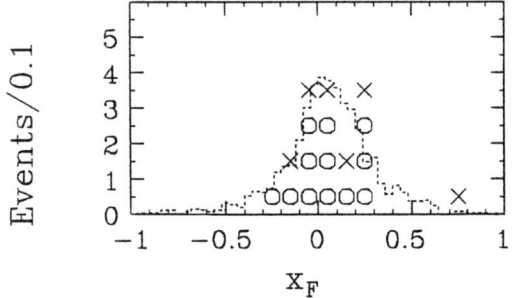

Figure 2. Histogram of Feynman x for the 12 neutral b or $\bar{b}$ decays (circles), and for the 6 $b^{\pm}$ (crosses). The dashed histogram is a Monte Carlo simulation with $n = 5.0$, $x_0 = 0.06$.

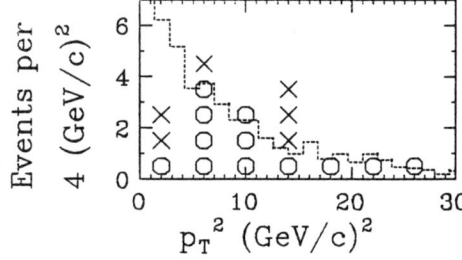

Figure 3. Histogram of p_T^2 for the 12 neutral b or $\bar{b}$ decays (circles), and for the 6 $b^{\pm}$ (crosses). The dashed histogram is a Monte Carlo simulation with $b=0.13$.

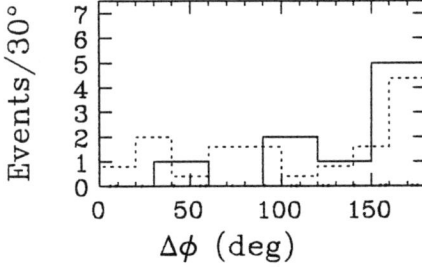

Figure 4. Azimuthal angle difference $\Delta\phi$ between members of $b\bar{b}$ pairs (solid histogram). The dashed histogram is a measured[4] $\Delta\phi$ distribution for charm pairs produced by 800 GeV protons.

and neutral charm decays inside and outside the emulsion are also in satisfactory agreement with simulations. Finally, charm lifetimes from this sample were measured, using the same techniques as for beauty. The D^0 lifetime from 103 such decays is within 5% of the world average, while the charged lifetime from 101 charm semimuonic decays is 24% below the world average for D^+. However, if the charged sample contains 30% D_s decays, consistent with present data[5], there is no discrepancy. These charm lifetime results demonstrate the validity of the momentum estimator and the absence of significant scanning bias for decays.

Fig. 5 shows a distribution of proper decay times from the 18 beauty decays in the completed high-p_T scan sample, plus 6 more "unofficial" decays found so far in the medium-p_T scan now in progress. The curve is the result of the simulation, assuming 1.3 ps lifetime. It is clear that there is an excess of long-lived charged b's. There are fewer charged b decays than neutrals, and the observed charged decays tend to be long. Are short charged decays being misssed? We have studied this issue very carefully with simulations, and can find no such effect. On the contrary, short charged decays are more robust than neutral ones, since they can be found by either followdown or scanback. Finally, there is no such problem in the data with short charged charm.

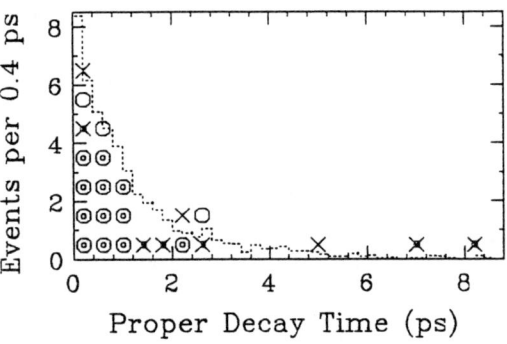

Figure 5. Histogram of estimated proper decay time for neutral (O's) and charged (X's) b candidates. The 18 decays from the completed high-p_T scan are indicated by a central dot to distinguish them from the "unofficial" 3 neutral and 3 charged decays found so far in the medium-p_T scan. The histogram is a Monte Carlo simulation of accepted events, generated with a lifetime of 1.3 ps.

Lifetimes are determined from maximum likelihood fits (see fig. 6), which include effects of detection efficiency and uncertainties in estimated momenta. We have investigated systematic errors by trying both momentum estimators, two models (optimistic and pessimistic) of detection efficiency, and extremes of production properties; all systematic errors from these sources are small compared to statistical errors. We find

$$\tau_{b^0} = 0.81^{+0.34+0.08}_{-0.22-0.02} \text{ ps } (12 \ b^0), \text{ and}$$
$$\tau_{b^\pm} = 3.84^{+2.73+0.80}_{-1.36-0.16} \text{ ps } (6 \ b^\pm).$$

If the 6 decays found so far in the medium-p_T scan are included, the lifetime results change by less than 0.1σ.

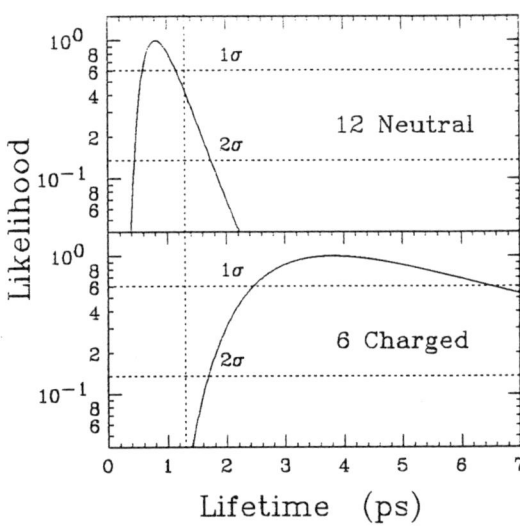

Figure 6. Plots of likelihood vs. lifetime for the 12 neutral and 6 charged decays. The values of log likelihood corresponding to one and two standard deviations are shown as horizontal lines. The vertical line at 1.3 ps indicates the present world average.

Neutral b's produced in hadron collisions at $\sqrt{s} = 33.6$ GeV may well contain B_s mesons and Λ_b baryons in addition to B_d mesons. The observed excess of b^0 over $b^\pm$ in this experiment, while not statistically compelling, may reflect this larger number of neutral states.

Our neutral lifetime result must therefore be regarded as a composite lifetime of an unknown mix of B_d, B_s, and Λ_b. It is 1.3 standard deviations below the present world average[6] (dominated by b's from Z^0 decay) of 1.29 ± 0.05 ps. The charged b's, on the other hand, should be dominated by B_u mesons if the Σ_b baryons decay strongly as expected, and the hadronic production of B_c and Ξ_b are small compared to B_u. As seen in Fig. 6, the charged lifetime from this experiment is more than two standard deviations above the world average for all b. This result is strongly influenced by the two long decays seen in fig. 5, neither of which can be dismissed as background.

Spectator decay models predict equal semileptonic widths for b^0 and $b^\pm$ even if the lifetimes are different due to hadronic effects. Thus if the $b^\pm$ lifetime is significantly longer than that of the b^0, one would expect its semileptonic branching ratio to be correspondingly larger also. However, both ARGUS[7] and CLEO[8] have measured the ratio of B_u^+ and B_d^0 semileptonic branching ratios at the Υ_{4s}, and find them to be equal to 20-25% if B_u^+ and B_d^0 are produced in equal numbers at the Υ_{4s}. Data from this experiment are also not consistent with a larger $b\pm$ semimuonic decay rate: 7 of 12 neutral decays, and 1 of 6 charged decays, were semimuonic[2]. It would be very interesting if the lifetimes for $b^\pm$ and b^0 were indeed significantly different as suggested by this experiment, while the semileptonic branching ratios were equal.

REFERENCES

1. K. Kodama et al., "Hybrid Emulsion Spectrometer for the Detection of Hadronically Produced Heavy Flavor States," *Nucl. Instr. and Meth.* A289, pp. 146-167, (1990).

2. K. Kodama et al., "Measurement of the Relative Branching Fraction $\Gamma(D^0 \to K\mu\nu)/\Gamma(D^0 \to \mu X)$", *Phys. Rev. Lett.* 14, pp. 1819-1822, (1991); K. Kodama et al., "Measurement of the form factor ratios in the decay $D^+ \to \bar{K}^*(892)^0\mu^+\nu$", *Phys. Lett.* B 274, pp. 246-252, (1992); K. Kodama et al., "Measurement of the branching ratio for $D^+ \to \bar{K}^*(892)^0\mu^+\nu$", *Phys. Lett.* B 286, pp. 187-194, (1992).

3. K. Kodama et al., "Charm Meson Production in 600 GeV/c π^- emulsion interactions," *Phys. Lett.* B284, pp. 461-470, (1992).

4. K. Kodama et al., "Charm pair correlations in 800 GeV/c proton-emulsion interactions," *Phys. Lett.* B263, pp. 579-583, (1991).

5. D. Potter, "Charm Meson Production in 600 GeV/c π^- Emulsion Interactions," presented at XXVI International Conference on High Energy Physics, Dallas, Texas, Aug. 6-12, 1992.

6. K. Hikasa et al. (Particle Data Group), "Review of Particle Properties", *Phys. Rev.* D 45, pp. S1 ff., (1992).

7. H. Albrecht et al., "Measurement of the lifetime ratio $\tau(B^+)/\tau(B^0)$," *Phys. Lett.* B232, pp. 554-560, (1989).

8. R. Fulton et al., "Exclusive and inclusive semileptonic decays of B mesons to D mesons," *Phys. Rev.* D43, pp. 651-663, (1991).

---

[2]The trigger enriches $b^\pm$ and b^0 semimuonic decays equally, and the detection efficiencies are nearly equal.

SEMILEPTONIC CHARMED MESON DECAYS

G. Bellini
Dept. of Physics, Milan University and I.N.F.N.
Via Celoria, 16
Milano, Italy, 20133

On behalf of the E687 Collaboration[*]

[*]G. Alimonti[8], D. Alliata[8], V. Arena[13], G. Bellini[8], S. Bianco[5], J.M. Bishop[12], G.R. Blackett[16], G. Boca[13], C.W. Bogart[3], D. Bucholz[11], J.K. Busenitz[12], J.N. Butler[4], B.Caccianiga[8], N.M. Cason[12], C.Castoldi[13], B.G. Cheon[7], H.W.K. Cheung[3], S. Cihangir[4], L. Cinquini[8], D. Claes[11], P. Coteus[3], R. Culbertson[6], S. Culy[3], J.P.Cumalat[3], J.D.Cunningham[12], C.Dallapiccola[3], K.Danyo[16], F.Davenport[9], R.Diaferia[13], M. diCorato[8], M. Enorini[5], F.L.Fabbri[5], J.F.Filasetta[10], P.L. Frabetti[1], I.Gaines[4], P.H.Garbincius[4], R.W.Gardner[6], L.Garren[4], M.Giammarchi[8], G.Gianini[13], J.F. Ginkel[3], B. Gobbi[11], S. Gourlay[4], R. Greene[6], S.V. Greene[3], G.P. Grim[2], T.Handler[16], D.J.Harding[4], D. Hazan[8], P.Inzani[12], G. Jaross[6], W.E.Johns[3], J.S. Kang[7], P. Kasper[4], C.J. Kennedy[12], G.N. Kim[12], K.Y. Kim[7], A.Kreymer[4], P.Lebrun[4], F.Leveraro[8], T.F.Lin[12], K.Lingel[6], A. Lopez[14], S. Malvezzi[8], E.J. Mannel[12], H. Mendez[4], D.Menasce[8], E. Meroni[8], L. Moroni[8], R.J. Mountain[12], M.S. Nehring[3], B.O'Reilly[11], V.S. Paolone[2], S. Park[8], D. Pedrini[8], L.Perasso[8], M. Pisharody[16], D.L. Puseljic[12], S.P. Ratti[13], C.Riccardi[13], R.C. Ruchti[12], A. Sala[8], S. Sala[8], S. Sarwar[5], P.D. Sheldon[17], W.D. Shephard[12], S. Shukla[4], A. Spallone[5], J.A. Swiatek[12], D. Torretta[8], M.Vittone[4], P.Vitulo[13], J.R. Wilson[15], J.Wiss[6], Z.Y.Wu[12], P.M.Yager[2], R.Yoshida[8], A.Zallo[5], M.E. Zanabria[12]

[1]INFN and Univ., Bologna, [2]California Univ., [3]Colorado Univ., [4]Fermilab, [5]INFN-Frascati, [6]Illinois Univ., [7]Korea Univ., [8]INFN and Univ., Milano, [9]North Carolina Univ., [10]Northern Kentucky Univ., [11]Northwestern Univ., [12]Notre Dame Univ., [13]INFN and Univ., Pavia, [14]Puerto Rico Univ., [15]South Carolina Univ., [16]Tennessee Univ., [17]Vanderbilt Univ.

Abstract

Results on the semileptonic channels: $D^0 \to K^+ \mu^- \nu$ +c.c. and $D^+ \to K^{*0} \mu^+ \nu$ +c.c., photoproduced in the E687 experiment at Fermilab, are presented. A preliminary measurement of $\dfrac{\Gamma(D^0 \to K^{*-} \mu \nu)}{\Gamma(D^+ \to K^{*0} \mu \nu)}$ is also discussed.

The Fermilab E687 experiment collected ~600 millions of triggers in three different runs: '87/'88, '90, '91, corresponding to ~10^5 reconstructed golden modes of charmed mesons. The E687 spectrometer, described previously[1], was installed in the wide band area with a photon beam having ~220 GeV mean energy and ~350 GeV top energy.

We report here the results obtained on semileptonic decay modes $D^0 \to K^+ \mu^- \nu$ + c.c. and $D^+ \to K^{*0} \mu^+ \nu$ + c.c., analyzed over ~1/10 of the total statistics ('87/'88 sample).

The $D^0 \to K^- \mu^+ \nu$+c.c. has been analyzed using the sequence $D^{*+} \to D^0 \pi^+$ and c.c.. The procedure goes through the following steps: i) we select events with two reconstructed vertices: the first one (the secondary vertex) is a two prongs corresponding to an identified charged K and an identified high energy muon ($P_\mu > 10$ GeV/c); the second vertex (the primary), upstream to the previous one, must be at least a three prongs with an identified charged pion. A soft cut on the secondary vertex confidence level (>0.05) is applied and a detachment $L/\sigma_L > 3$ is required, where L is the separation between the vertices and σ_L is its error; ii) the straight line between the primary and secondary vertex is assumed as D^0 flight direction and the $M_{K\mu\nu}$ is constrained to the M_D ($M_\nu = 0$). Then P_L^ν is calculated in a frame where $P_L^{K\mu} = 0$ (L≡D^0 flight direction). $E_{K\mu\nu} < 350$ GeV is required and, in case of ambiguity, the solution with the lowest D^* mass is chosen; iii) the selection is done on the D^* charged mass; the D^* decay pion candidate is required to have $P_{tot} \le 15.5$ GeV/c and $P_t \le 0.2$ GeV/c; iv) the right sign is required: $Q_k + Q_\pi = 0$.

In fig. 1a the invariant mass distribution of the right sign combinations $K^- \mu^+ \nu \pi^+$ and c.c. is displayed for events with $L/\sigma_L > 5$. The dashed area is the wrong sign mass distribution ($K^+ \mu^+ \nu \pi^+$, $K^+ \mu^+ \nu \pi^-$, $K^+ \mu^- \nu \pi^+$, and c.c.); the same cuts are applied to the wrong sign events and the normalization is done in the range 2.02-2.4 GeV/c^2.

The total number of events in a mass interval ±7.5 MeV around 2012 Mev as central value is 109±18 events, once subtracted the wrong sign combinations. This number, corrected with the M.C., is not affected by the detachment cut (see fig. 1b, where the data are well reproduced by the M.C.).

The B.R. is measured with respect of $D^0 \to K^- \pi^+$, as reference channel. To select the ($K^- \pi^+$) channel, D^* tagged events from the same data sample used for the ($K^- \pi^+$) channel are selected and the same cuts are applied. The background subtraction consists of 9 events for the (k$\mu\nu$) yield and 2 events for (Kπ); they are mainly due to channels with π^0, π misidentified as μ or K, π o K decay to $\mu\nu$. The measured B.R. is:

$$\frac{BF(D^0 \to K^- \mu^+ \nu + c.c.)}{BF(D^0 \to K^- \pi^+ + c.c.)} = 1.05 \pm 0.2 \pm 0.2.$$

This result is in good agreement with the same (Keν) and (K$\mu\nu$) B.R.'s, measured by other experiments.

The sources of systematic errors are: cut choices, uncertainties in the B.F.'s and in the efficiency, wrong sign subtraction.

The channel $D^+ \to \bar{K}^{*0} \mu^+ \nu$+c.c. $\to$ ($K^- \pi^+$)$\mu^+ \nu$+c.c. is studied selecting the events having a vertex with an identified $K^\mp$, $\pi^\pm$, $\mu^\pm$ ($p_\mu > 10$ GeV/c), and a primary vertex, upstream to the previous one. Cuts are applied to the secondary vertex, asking that no further tracks point to it (CL>0.01), and to the detachment ($L/\sigma_L > 20$). The right sign is required ($Q_K + Q_\mu = 0$), while the constraint that the primary vertex should be within a kinematical cone, which takes into account the charged K, π, μ and the unseen ν (cone cut) seems not effective in our case. Finally the selection is carried out on the K^*. In fig. 2a, b, c the $K^{\mp}\pi^{\pm}$ invariant mass distributions for three different values of the detachment cut (>5, >15, >20) are shown; the wrong sign combinations are already subtracted. The distributions are fitted with the matrix elements, where a K^* Breit Wigner is folded in, time the phase space; the background is well reproduced with the function $x^{a_1} e^{(-a_2 x)}$ where $x=M(K\pi)-[M(K)-M(\pi)]$.

In figs. 2d, e, f, the K^* yields as function of the detachment cut, the secondary vertex confidence level and the isolation cut, are plotted and compared with the MC simulations. The agreement is always good, which means that we have not D^o or D_s contamination, no random background such as a $K\pi$ vertex with a background muon, no background from higher multiplicity decay modes. The total yield is obtained applying a $K\pi$ mass cut: 0.8-1.0 GeV/c^2.

The BF is measured with respect of the reference channel $D^+\to K^-\pi^+\pi^+$:

$$\frac{BF(D^o\to K^{*o}\mu^+\nu)}{BF(D^+\to K^-\pi^+\pi^+)} = 0.67\pm 0.07^{+0.07}_{-0.1}$$

(the $K^{*o}\mu^+\nu$ yield is of course multiplied by 3/2 to account for undetected K^{*o} decays). The sources of systematic error are: the detachment cut, the hadron calorimeter efficiency, the B.F.'s of the background channels, the nuclear scattering in the target.

Our result is in agreement with the measurements obtained by other experiments.

From the fit of the matrix element, modified by the efficiency, we can obtain form factors and Γ_L/Γ_t, the ratio between the longitudinal and transverse components. The form factors, especially the R_v ratio, depends strongly on the systematic errors, while Γ_L/Γ_t is quite insensitive. Then R_v and R_2 need probably a more detailed analysis of the systematic errors. We report here only our measurement of the polarization: $\Gamma_L/\Gamma_t \equiv 1.3\pm 0.3$. This result is not affected by the detachment cut, as it can be deduced from fig. 3. It agrees with a moderate polarization of K^*, as expected from the quark models and the QCD sum rules.

Using the PDG BF's for the reference channels, the E687 D^o and D^+ lifetimes and assuming (following the isospin invariance) $\Gamma(D^o\to K^{*-}\mu^+\nu) = \Gamma(D^+\to K^{*o}\mu^+\nu)$, we obtain:

$$\frac{\Gamma(D^o\to K^{*-}\mu^+\nu)}{\Gamma(D^o\to K^-\mu^+\nu)} = 0.53\pm 0.13^{+0.12}_{-0.13}$$

This result agrees with the measurements of the previous experiments and confirms the disagreement with the quark models.

All the results presented here can be largely improved analyzing all the sample collected, which means to improve the statistics of a factor ~10.

A preliminary measurement of the ratio $\dfrac{\Gamma(D^o\to K^{*-}\mu^+\nu)}{\Gamma(D^+\to K^{*o}\mu^+\nu)}$ has been carried out

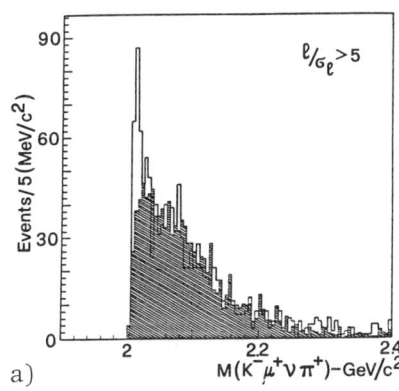

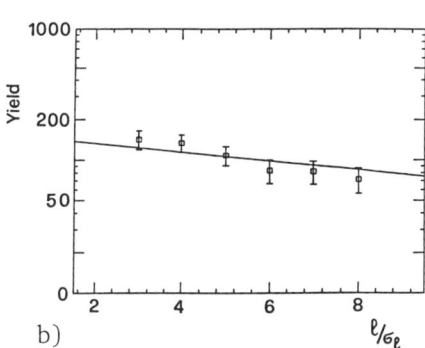

Fig. 1 - a): $K^-\mu^+\nu\pi^+$ and c.c. invariant mass distribution ($L/\sigma_L > 5$); dashed area: wrong sign.
b): D^* yield vs L/σ_L, compared with the M.C. simulations (full line).

Fig. 2 - a,b,c): $(K\pi)^0$ invariant mass distributions (MeV/c^2) for various L/σ_L cuts: >5(a), >15(b), >20(c); they are compared with the matrix element and the background (full line-see text).
d,e,f): K^* yield vs the detachment cut (d), the secondary vertex confidence level (e), the isolation cut (f). The distributions are compared with the M.C. calculations (dashed line).

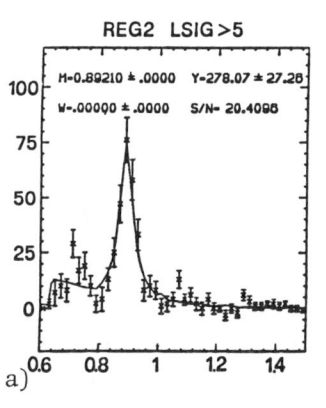
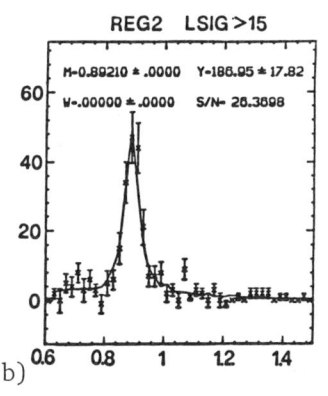
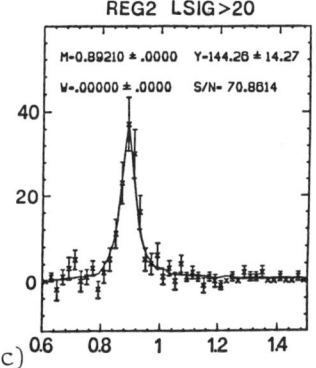

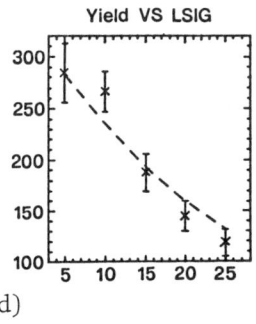
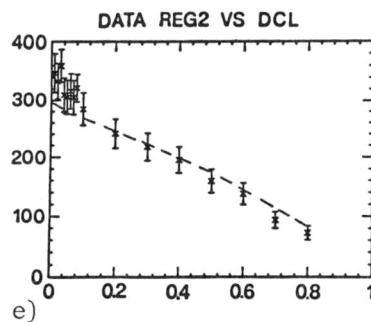
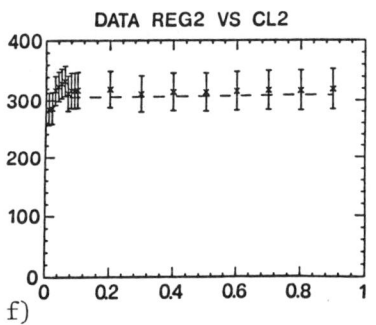

on ~40% of the total data sample to check the isospin invariance. This ratio is obtained by $\dfrac{\text{Yield}(K^{*-}\mu^+\nu)}{\text{Yield}(K^{*0}\mu^+\nu)} \cdot \dfrac{D^+}{D^0} \cdot \dfrac{\tau^+}{\tau^0}$.

The D^+/D^0 and τ^+/τ^0 measurements obtained by E687 have been used: 0.42 ± 0.05 and 2.52 ± 0.17, respectively. The yields are corrected with the efficiencies ($\varepsilon^+/\varepsilon^0=1.43$) and with the B.R.'s: $K^{*0}\to K\pi$, $K^{*-}\to K^0\pi$, $K^0\to K^0_s$, $K^0_s\to\pi\pi$.

The analysis of the $D^+\to(K^-\pi^+)\mu^+\nu$ channel has been done exactly in the same way as described in the previous paragraphs. A total yield of 374 raw decays has been selected.

The channel $D^0\to(K^0_s\pi^-)\mu^+\nu$ was treated following the same procedure used for the $(K\pi)^0\mu\nu$ channel, but in the $(K_s\pi)^-\mu^-\nu$ channel the secondary is only a two prong vertex instead of a three prongs as in the previous case. Then the background is higher. This effect is evident if we compare the $(K^0_s\pi)$ invariant mass distributions of fig. 4 with the equivalent $(K\pi)$ distributions of fig. 2. The same formulas are used for the fitting (full line); in this case the K^* sample has been defined taking into account the area below the Breit Wigner. The selected yield consists of 142 raw decays. This yield shows a dependence on the secondary vertex confidence level, on the isolation cut and on the detachment cut, which is well reproduced by the M.C.. This means that π and μ really come together and that we are really observing a D^0.

The background due to the $(K^{*-}\pi^0\mu^+\nu)$

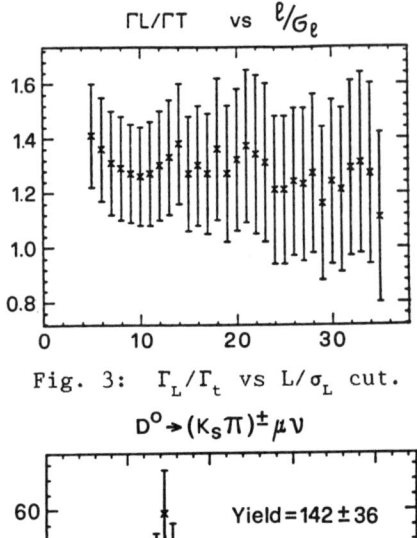

Fig. 3: Γ_L/Γ_t vs L/σ_L cut.

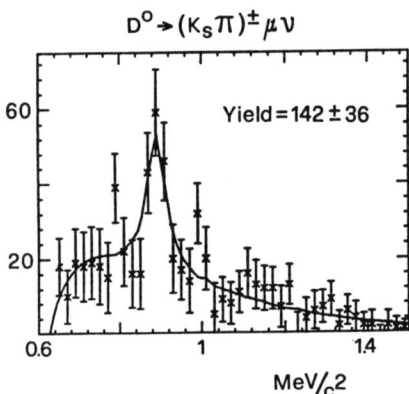

Fig. 4: $(K^0_s\pi)^-$ invariant mass distribution ($L/\sigma_L>5$).

and $(K^{*-}\pi^+\pi^0)$ channels have been subtracted. The final result is:

$$\dfrac{\Gamma(D^0\to K^{*-}\mu^+\nu)}{\Gamma(D^+\to K^{*0}\mu^+\nu)} = 1.60\pm0.43^{+0.2}_{-0.5}$$

The systematic error is mainly due to the uncertainties in the background channel subtraction. This measurement can be improved extending our analysis to the full data sample (~2 time larger).

REFERENCE

1. P.L. Frabetti et al., N.I.M. A320 (1992) 519.

INCLUSIVE PHOTON ENERGY SPECTRUM IN B DECAYS[*]

C. Greub
Institute for Theoretical Physics
University of Zürich
Schönberggasse 9
CH-8001 Zürich, Switzerland

Abstract

We present a theoretical estimate for the inclusive photon energy spectrum in direct decays of B-mesons taking into account both the charged current (CC) and flavour changing neutral current (FCNC) processes. It is shown that the various components in the inclusive spectrum can in principle be disentangled. In particular, the high energy part of the photon energy spectrum is dominated by the electromagnetic penguins. Its measurement could provide the first direct determination of the CKM matrix element $|V_{ts}|$. Furthermore the theoretical uncertainties in the FCNC processes are analyzed.

INTRODUCTION

The inclusive photon energy spectrum in $B \to X\gamma$ is obtained by taking into account both the charged current (CC) transitions $B \to (X_c, X_u) + \gamma$ and the flavour changing neutral current (FCNC) processes $B \to (X_s, X_d) + \gamma$, induced by the electromagnetic penguins (here the subscript q on X_q denotes the quark flavour in the transition $b \to q$). The CC processes are sensitive to the Cabibbo-Kobayashi-Maskawa (CKM) matrix elements $|V_{cb}|$ and $|V_{ub}|$, respectively. On the other hand, the FCNC processes depend on the top quark mass m_t, because in the penguin loops the top quark is involved as internal fermion line. Furthermore, the branching ratio $BR(B \to X_s\gamma)$ and $BR(B \to X_d\gamma)$ depend on $|V_{ts}|$ and $|V_{td}|$, respectively. It turns out, that in these FCNC processes there are rather large theoretical uncertainties. As we will discuss later in some detail, the most important one is the ambiguity in the choice of the renormalization scale μ at which the Wilson coefficients are evaluated. As the m_t dependence of the branching ratio is relatively small ($O(25\%)$ for 100 GeV $\leq m_t \leq 180$ GeV), a measurement of these FCNC processes can hardly be used to get information on the value of m_t. However, as we will show, such measurements could be used to improve the present constraints on $|V_{ts}|$ and $|V_{td}|$.

First we briefly discuss the CC contribution to $B \to X\gamma$. Then we consider the FCNC processes. Finally, we put together the CC and FCNC contributions, leading to the inclusive photon energy spectrum in $B \to X\gamma$.

CC PROCESSES: $B \to (X_c, X_u) + \gamma$

When calculating the CC processes $B \to X_c\gamma$ and $B \to X_u\gamma$ we take into account only spectator diagrams. More precisely, the $B \to X_c\gamma$ process is modelled after the par-

---
[*]Work supported in part by Schweizerischer Nationalfonds.

tonic processes

$$b \to c q \bar{q}' \gamma \quad \text{and} \quad (1)$$

$$b \to c \ell \bar{\nu}_\ell \gamma \quad , \quad (2)$$

where q, q' are quarks and ℓ is a (negatively) charged lepton. Processes (1) and (2) are referred to as the non-leptonic and the semileptonic decays, respectively. The $B \to X_u \gamma$ process is treated similarly. The formulae for all these different channels can be found in ref. 1. We did not take into account the so-called W-exchange two-body decays, since they are generally considered negligible in B decays[2]. Experimentally the most convincing 'proof' of this statement is the (near) equality of the charged and neutral B-meson lifetimes, gotten indirectly through the ARGUS and CLEO measurements of semileptonic branching ratios[3]:

$$\frac{\tau_{B^\pm}}{\tau_{B^0}} = \frac{BR(B^\pm \to D^{(*)0} \ell^\pm \nu)}{BR(B^0 \to D^{(*)-} \ell^+ \nu)} = 0.96 \pm 0.14 \quad (3)$$

Furthermore we have left out the QCD corrections to the decays in eqs. (1) and (2), which are known to be small from analogous studies in non-radiative B-decays. A list of literature concerning these QCD corrections is given in ref. 1.

To take into account the motion of the b quark in the B meson we used the Altarelli et al. model, where the b-quark momentum $\vec{p}$ has a Gaussian distribution[4]:

$$\phi(|\vec{p}|) = \frac{4}{\sqrt{\pi} p_F^3} \exp(-\vec{p}^2/p_F^2) \quad (4)$$

The average Fermi-momentum p_F is obtained from a fit of the ARGUS and CLEO inclusive lepton energy spectrum[5]: $p_F = 0.30 \pm 0.09$ GeV.

FCNC PROCESSES: $B \to (X_s, X_d) + \gamma$

Due to the limited space I concentrate on the case $B \to X_s \gamma$. The reaction $B \to X_d \gamma$ is discussed in ref. 6. On the partonic level we take into account the processes $b \to s\gamma$ and $b \to s\gamma g$. The 3-body decay $b \to s\gamma g$ leads to a non-trivial photon spectrum already at the partonic level[7]. As in the CC processes the partonic results are then folded with the wave function in eq. (4). For details see ref. 8.

Contrary to the CC processes, the short distance QCD corrections enhance the $BR(b \to s\gamma)$ in a significant way and therefore at least the leading logarithmic corrections have to be resummed. Technically this is most conveniently done in the framework of an effective Hamiltonian where the heavy particles (W boson and top-quark) are integrated out. A collection of references concerning these points is given in ref. 1. To leading order in the weak mixing angles, the effective Hamiltonian reads:

$$H_{eff} = -\frac{4G_F}{\sqrt{2}} |V_{ts}|^2 \sum_{i=1}^{8} C_i(\mu) O_i(\mu) \quad . \quad (5)$$

The $O_i(\mu)$ are operators of dimension 6 and the $C_i(\mu)$ are the corresponding Wilson coefficients at the renormalization scale μ. As discussed e.g. in ref. 7 it turns out that only the operators O_2 and O_7 are numerically relevant in our application:

$$O_2 = \bar{c}_\alpha \gamma^\mu L b_\alpha \, \bar{s}_\beta \gamma_\mu L c_\beta \quad (6)$$

$$O_7 = \frac{e}{16\pi^2} \bar{s}_\alpha \sigma_{\mu\nu} (m_b R + m_s L) b_\alpha F^{\mu\nu} \quad . \quad (7)$$

Here $e, F_{\mu\nu}, L$ and R denote the QED coupling constant, the electromagnetic field-strength, the left- and right-handed projection operator, respectively. The detailed form of the Wilson coefficients $C_i(\mu)$ is not repeated here. We only give the generic form:

$$C_i(\mu) = f_i[\eta, C_1(m_W), ..., C_8(m_W)] \quad , \quad (8)$$

with $\eta = \alpha_s(\mu)/\alpha_s(m_W)$. For $\alpha_s(\mu)$ we take the value corresponding to the 2-loop β-function, $N_f = 5$ and $\Lambda_{\overline{MS}}^{(5)} = 225$ MeV[9]. For b

decays the relevant scale μ is somewhere near m_b, but its precise value is not fixed, however. In order to estimate this renormalization scale uncertainty we vary the μ between $(m_b/2)$ and $(2 \cdot m_b)$. The resulting branching ratio is

$$BR(B \to X_s\gamma) = (2-5) \times 10^{-4} \quad , \quad (9)$$

where the rather large uncertainty is mainly due to the μ ambiguity and only to a small extent due to the m_t-variation, as illustrated in figure 2 of ref. 1.

In figure 1 we plot the branching ratio $BR(B \to X_s\gamma)$ as a function of $r \doteq |V_{ts}|^2/|V_{cb}|^2$, for 3 scenarios of μ. Unitarity of the CKM matrix requires $1/4 \leq r \leq 4$ (see ref. 10). In order to illustrate the figure we may <u>assume</u> a <u>hypothetical</u> measurement of $BR(B \to X_s\gamma) = (4 \pm 1) \times 10^{-4}$. This would imply $0.80 \leq |V_{ts}|/|V_{cb}| \leq 1.60$, i.e., a determination of $|V_{ts}|$ to better than a factor of 2.

PHOTON ENERGY SPECTRUM IN $B \to X\gamma$

Putting together the CC and FCNC processes leads to the inclusive photon energy spectrum. In figure 2 the individual contributions are plotted. The spectrum can be divided into 3 regions: [Region 1: $E_\gamma \leq 1.8$ GeV, region 2: 1.8 GeV $\leq E_\gamma \leq 2.1$ GeV, region 3: $E_\gamma \geq 2.1$ GeV]. In region 1 it is dominated by the CC process $B \to X_c\gamma$. In region 2 CC and FCNC processes are comparable. Region 3 is dominated by $B \to X_s\gamma$. Furthermore, in this region the process $B \to X_d\gamma$ is more significant than the CC processes which fall off rapidly. The high frequency part of the photon energy spectrum in $B \to X\gamma$ is therefore the regime of FCNC processes. If strangeness is measurable, the two FCNC components $B \to X_s\gamma$ and $B \to X_d\gamma$ can be disentangled and therefore information on both, V_{ts} and V_{td}, can be obtained.

REFERENCES

1. A. Ali and C. Greub, "Prompt Photon Energy Spectra in B-Decays and Determination of the CKM Matrix Elements," *DESY 92-089 preprint, Phys. Lett B. in press*, (1992).

2. G. Altarelli and S. Petrarca, " Inclusive Beauty Decays and the Spectator Model ," *Phys.Lett.* B261 , pp. 303–310, (1991).

3. H. Schröder, in "QCD–20 Years Later, Aachen, June 9–13, 1992".

4. G. Altarelli et al., "Leptonic Decay of Heavy Flavours: A Theoretical Update," *Nucl. Phys.* B208, pp. 365–380, (1982).

5. R. Fulton et al. (CLEO), "Observation of B-meson Semileptonic Decays to Non-charmed Final States ," *Phys. Rev. Lett.* 64, pp. 16–20, (1990).
H. Albrecht et al. (ARGUS)," Observation of Semileptonic Charmless B-meson Decays ," *Phys. Lett.* B234, pp. 409–416, (1990).
H. Albrecht et al. (ARGUS)," Study of Inclusive Semileptonic B-meson Decays ," *Phys. Lett.* B249, pp. 359–365, (1990).

6. A. Ali and C. Greub, "Rare Decays $B \to X_d\gamma$ in the Standard Model, " *DESY 92-048 preprint, Phys. Lett. B in press*, (1992).

7. A. Ali and C. Greub, "Inclusive Photon Energy Spectrum in Rare B-Decays," *Z. Phys.* C49, pp. 431–438, (1991).

8. A. Ali and C. Greub, "A Profile of the Final States in $B \to X_s\gamma$ and an Estimate of the Branching Ratio $BR(B \to K^*\gamma)$," *Phys. Lett.* B259, pp. 182–190, (1991).

9. G. Altarelli, in "QCD–20 Years Later, Aachen, June 9–13, 1992".

10. M. Aguilar-Benitez et al., "Particle Data Group," *Phys. Lett.* B239, (1990).

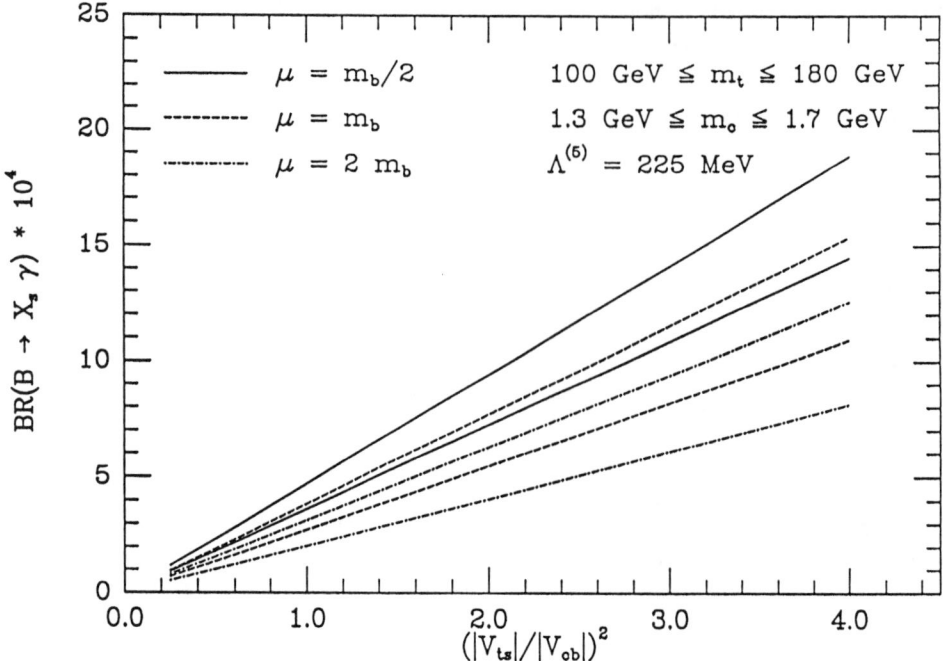

Figure 1. $BR(B \to X_s\gamma)$ for 3 different scenarios of μ.

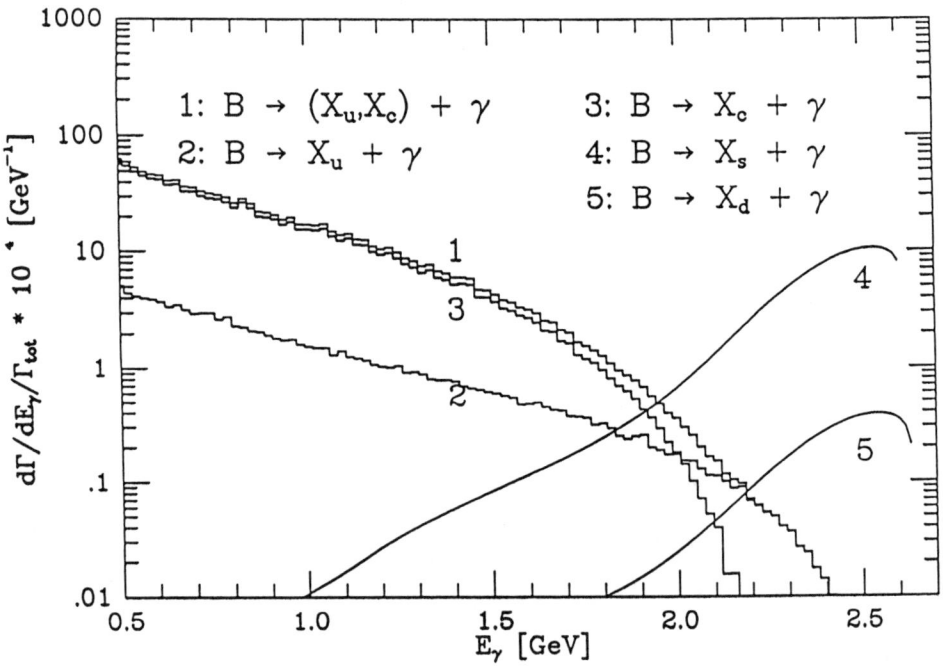

Figure 2. CC and FCNC components contributing to $B \to X\,\gamma$.

SELECTED ARGUS RESULTS ON B MESON DECAYS

Yuri Zaitsev

(ARGUS Collaboration)

Institute of Theoretical and Experimental Physics

117259 Moscow, Russia

Abstract

Semileptonic B meson decays have been studied using the ARGUS detector at DORIS II. The branching ratio for the decay $\overline{B}^0 \to D^{*+}l^-\overline{\nu}$ has been measured. A significant rate for the decay $\overline{B} \to D^{**}l^-\overline{\nu}$ has been observed. From an angular analysis of the cascade $\overline{B}^0 \to D^{*+}(\to D^0\pi^+)l^+\overline{\nu}$, the forward-backward asymmetry A_{FB} and the D^{*+} polarization parameter α have been determined. The decay $\overline{B}^0 \to D^{*+}l^-\overline{\nu}$ have been also measured with partial reconstruction of the D^{*+} meson, and for the first time the inclusive primary electron spectrum in the whole momentum interval has been analyzed.

The study of semileptonic B meson decays can give important information on the CKM matrix elements.

Here we present some recent ARGUS results on semileptonic B meson decays. The data have been taken at the DESY e^+e^- storage ring DORIS II. At the energy of the $\Upsilon(4S)$ an integrated luminisity of 233 pb^{-1} has been collected. A detailed description of the ARGUS detector and its particle identification capabilities can be found in ref.[1]

STUDY OF THE DECAYS $\overline{B}^0 \to D^{*+}l^-\overline{\nu}$ AND $\overline{B} \to D^{**}l^-\overline{\nu}$

To study exclusive semileptonic decays of B meson we selected events containing at least one electron or muon (l^-) * with momentum greater then 1 GeV/c. D^* mesons are reconstructed in the two decay chains $D^{*+} \to D^0\pi^+$ with $D^0 \to K^-\pi^+$ and $D^0 \to K^-\pi^+\pi^+\pi^-$. We include only $(D^0\pi^+)$ combinations from events containing l^- where the measured D^0 mass lies within ± 60 MeV/c^2 of the nominal D^0 mass. The number of $(D^{*+}l^-)$ pairs is determined by fitting a Gaussian for the signal above a background.

We reconstruct the decay $\overline{B}^0 \to D^{*+}l^-\overline{\nu}$ using a missing mass technique developed in ref.[2]. Because B mesons are produced nearly at rest at the $\Upsilon(4S)$ resonance, their momentum can be considered small enough to set $p_B \approx 0$. Using $E_B = E_{beam}$, the recoil-mass squared can be written

$$M_{rec}^2 = (E_{beam} - E_{D^*} - E_{l^-})^2 - (\vec{p}_{D^*} + \vec{p}_{l^-})^2.$$

Fig.1 show the M_{rec}^2 distributions for the two decay chains $D^{*+} \to D^0\pi^+$, followed by $D^0 \to K^-\pi^+$ and $D^0 \to K^-\pi^+\pi^+\pi^-$. In both cases a clear signal at $M_{rec}^2 = 0$ is observed with minor backgrounds from uncorrelated $D^{*+}l^-$ pairs and continuum. For $M_{rec}^2 > 0$ one observes a shoulder which is attributed to the process $\overline{B} \to D^{**}l^-\overline{\nu}$.

---
*References to a specific charge state are to be taken to imply also the charge-conjugate state.

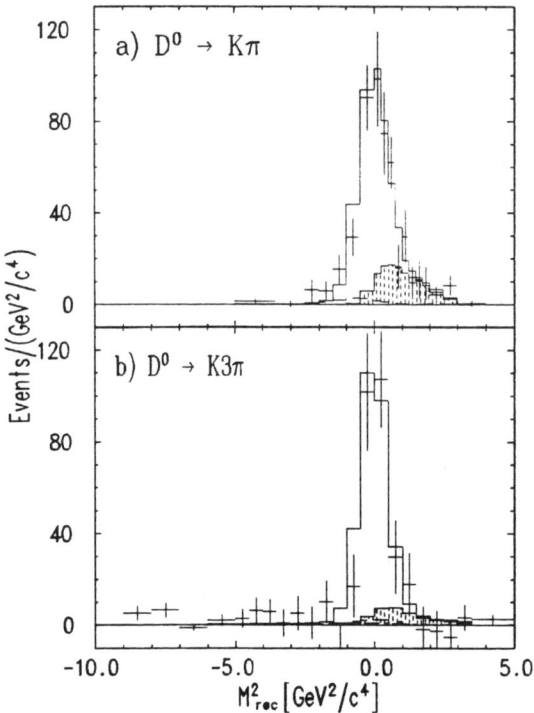

Figure 1. Measured M_{rec}^2 distribution (points with error bars) for the two D^0 decay channels. Solid line shows result of fit. The blank and shaded areas correspond to the rates for decays to D^{*+} and D^{**}. The continuum contribution is shown as dashed line.

Shown in Fig.2 are the invariant mass spectra $M(D^{*+}\pi^-)$ for both $M_{rec}^2 > 0$ and $M_{rec}^2 < 0$. The distributions are simultaneously fit with the same background shape. For the sample with $M_{rec}^2 > 0$, the fit function includes in addition two Breit-Wigner resonances corresponding to P-wave states with masses of 2420 and 2460 MeV/c^2 and full width of 20 MeV/c^2. From the fit, a signal of 30 ± 10 D^{**0} events is obtained for the sum of contributions from both resonances.

From the $63 \pm 15 \pm 6$ $(D^{*+}l^-)$ combinations obtained in the M_{rec}^2 fit, we estimate that 35 ± 9 D^{**0} mesons should be reconstructed in this analysis, in good agreement with the fit result. From the ratio $N(D^{**}l^-)/N(D^{*+}l^-) = 0.27 \pm 0.08 \pm$

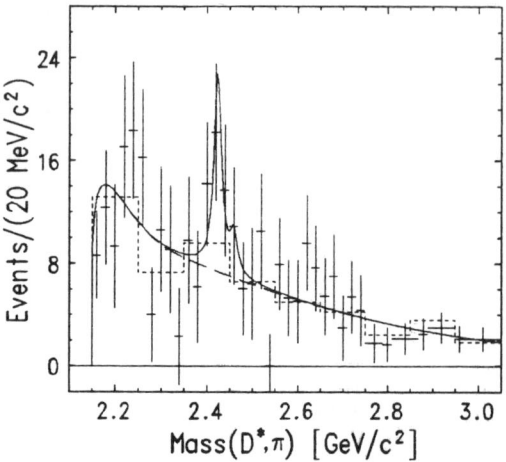

Figure 2. Measured distribution of the $(D^{*+}\pi^-)$ mass (points with error bars) for $M_{rec}^2 > 0$. The dotted histogram – for $M_{rec}^2 < 0$. Solid line is the sum of the background function (dashed line) and two Breit-Wigner curves described in the text.

0.03 and the branching ratio $BR(\overline{B}^0 \to D^{*+}l^-\overline{\nu})$ obtained above, we estimate the $BR(\overline{B}^0 \to D^{**+}l^-\overline{\nu})$ to be $(0.2 \pm 1.0 \pm 0.7)\%$, using the ISGW model[4], and $(3.5 \pm 0.8 \pm 0.6)\%$, using model[5]. This result implies a significant contribution from D^{**} mesons in semileptonic B decays.

Using formalism developed in ref.[6] the Lorentz structure of the decay $\overline{B}^0 \to D^{*+}l^-\overline{\nu}$ has been studied by extracting the differential decay widths as a function of $\cos\theta$, $\cos\theta^*$, q^2 and M_{rec}^2, where θ is the angle between W^- and l^- in the $(l^-\overline{\nu})$ rest frame and θ^* is the angle between D^* and D in the D^* rest frame. The distributions were produced by requiring that the momenta and energies of the D^{*+} and the l^- be consistent with the presumed decay of a $\overline{B}^0$ meson.

The simultanious fit of these four distributions yields:

$A_{FB} = \frac{3}{4} \cdot \frac{\Gamma^- - \Gamma^+}{\Gamma} = 0.20 \pm 0.08 \pm 0.06$,

$\alpha = 2 \cdot \frac{\Gamma^L}{\Gamma^T} - 1 = 1.1 \pm 0.4 \pm 0.2$.

The value for A_{FB} is in agreement with

Table 1. Results of fitting the y-distribution with various parametrizations of the Isgur-Wise function.

		$\xi(y)$	$|V_{cb}| \times 10^3$	ρ	χ^2/df
	A	$1 - \rho^2(y-1)$	$45 \pm 5 \pm 3$	$1.08 \pm 0.11 \pm 0.03$	5.1/6
	B	$\frac{2}{y+1} exp[-(2\rho^2 - 1)\frac{y-1}{y+1}]$	$53 \pm 8 \pm 3$	$1.52 \pm 0.21 \pm 0.10$	4.3/6
	C	$[2/(y+1)]^{2\rho^2}$	$51 \pm 8 \pm 3$	$1.45 \pm 0.19 \pm 0.09$	4.3/6
	D	$exp[-\rho^2(y-1)]$	$50 \pm 8 \pm 2$	$1.37 \pm 0.19 \pm 0.08$	4.4/6

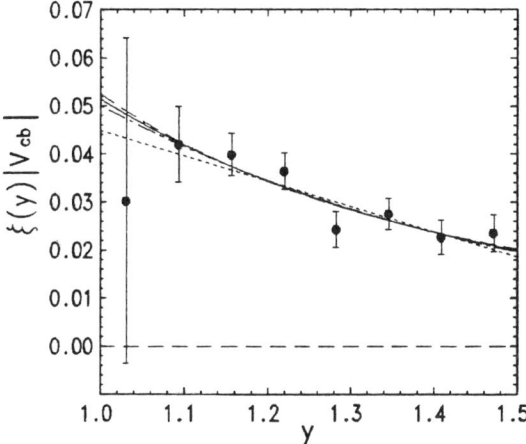

Figure 3. Measured distribution ΔBR $(\overline{B}^0 \to D^{*+}l^-\overline{\nu})/\Delta y$ transformed to correspond to $|V_{cb}| \cdot \xi(y)$. The lines correspond to the fits with each of the four expressions for the Isgur-Wise function (see Tab.1). The dotted line corresponds to the function A.

predictions from the various models showing that the $b \to c$ transitions are left chiral.

Following ref.[7], the model dependence in the $|V_{cb}|$ determination can be considerably reduced by a fit to the y spectrum, where

$$y = (m_B^2 + m_{D^*}^2 - q^2)/(2 \cdot m_B m_{D^*}).$$

The measured $|V_{cb}| \cdot \xi(y)$ distribution is shown in Figure 3 together with the fits with the functions for $\xi(y)$ given in Table 1. The table also includes the fit results for $|V_{cb}|$ and the "charge radius" ρ. The values of $|V_{cb}|$ are determined by the intersection of the fit functions $\xi(y)$ with the ordinate since $\xi(1) = 1$.

The values of $|V_{cb}|$ shown in Table 1 vary from 0.045 to 0.053 depending on the analytical form chosen for the Isgur-Wise function.

STUDY OF THE $\overline{B}^0 \to D^{*+}l^-\overline{\nu}$ DECAY WITH PARTIAL D^{*+} RECONSTRUCTION

Since the energy released in the decay $D^{*+} \to D^0\pi^+$ is small, the directions of the pion is close to that of the D^* and the magnitudes of their momenta are strongly correlated. Using information about the soft pion alone, it is possible to calculate the momentum of the D^* without reconstructing the D^0. From a Monte Carlo simulation it was found, that assuming the direction of the D^* coincides with that of the pion, the momentum of the D^{*+} can be calculated using the formula $p_{D^*} = \alpha p_\pi + \beta$, with parameters $\alpha = 8.23$ and $\beta = 0.41$ GeV/c.

The momentum of the lepton was required to be more than 1.4 GeV/c. The recoil mass distribution after subtraction of the continuum and misidentified hadrons is shown in Fig.4. A clear peak near zero recoil mass in the right-sign combinations corresponds to the signal.

After wrong-sign combination subtraction, the recoil mass distribution shown in Fig.5 was fit with two functions corresponding to the decays $\overline{B}^0 \to D^{*+}l^-\overline{\nu}$ and $\overline{B} \to D^{**}l^-\overline{\nu}$. The fit found the number of

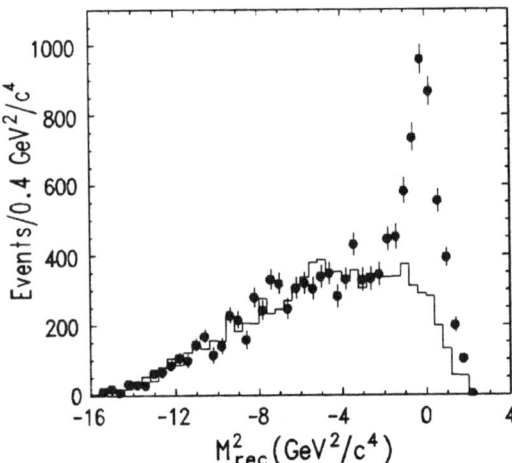

Figure 4. M^2_{rec} spectrum for $l^+\pi^-$ (points) and $l^+\pi^+$ (histogram) combinations. Continuum and fake leptons are subtracted.

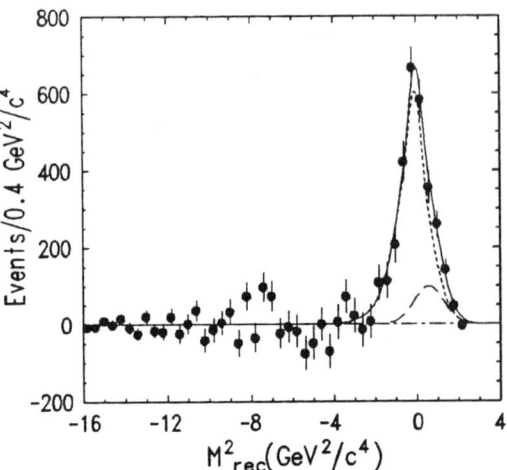

Figure 5. Background subtracted M^2_{rec} spectrum. The curves show the result of the fit described in the text.

the events from the studied decay to be equal to $N(\overline{B}^0 \to D^{*+}l^-\overline{\nu}) = 2693 \pm 183 \pm 105$. Using the CLEO measurement[8] of $BR(D^{*+} \to D^0\pi^+) = (66 \pm 3)\%$ we obtain $BR(\overline{B}^0 \to D^{*+}l^-\overline{\nu}) = (4.6 \pm 0.3 \pm 0.4)\%$. This result is in good agreement with above result and previous measurements[9].

Using the sample of partially reconstructed B^0, an inclusive branching ratio for neutral B mesons can be determined. We selected two samples: a) events containing at least one $l^+\pi^-$ pair with $|M^2_{rec}| < 1$ GeV$^2/c^4$ and b) events containing exactly two leptons each having momentum $p_l > 1.4$ GeV/c and satisfying the constraint in a). Using the acceptance obtained from a Monte Carlo simulation and the semileptonic branching ratios from ref.[3] we get: $BR(\overline{B} \to Xl^-\overline{\nu}) = (9.0 \pm 1.1 \pm 0.8)\%$. Combining this result with our measurement from the exclusive study of $\overline{B}^0 \to D^{*+}l^-\overline{\nu}$, a mean value of $\tau(B^+)/\tau(B^0) = 1.14 \pm 0.18$ is obtained.

STUDY OF LOW MOMENTUM PART OF THE ELECTRON SPECTRUM

Theoretical predictions of the inclusive semileptonic branching ratio for B meson decays based on spectator model are usually above 12%. Experimental values obtained using only the hard part of lepton spectrum and model-dependent extrapolations to low momenta are smaller. To suppress theoretical uncertainties, a measurement of the primary lepton spectrum in almost the whole momentum region was performed.

In the ARGUS detector electrons are well identified over nearly the entire momentum range (> 0.4 GeV/c). The primary electron spectrum can be obtained by subtracting the secondary electron contribution. In order to suppress this contribution we tagged the flavour of one B meson by the sign of the fast lepton ($p_{tag} > 1.4$ GeV/c). Secondary electrons from the untagged B meson have opposite sign to that of primary ones and contribute to the studied spectrum only in the case of $B^0\overline{B}^0$ mixing. The secondary electrons from the tagged B meson are correlated in angle with the tagging leptons and can be considerably suppressed with the cut $\cos\theta_{tag+e^-} > 0$.

The background due to the secondary

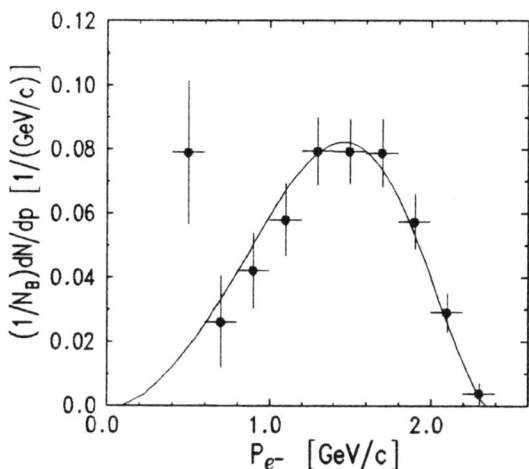

Figure 6. Acceptance corrected momentum spectrum of primary electrons for $\overline{B} \to Xl^-\overline{\nu}$ decays. The full line is the normalized spectrum obtained using IGSW model[4].

electrons originating from the decays of D_s, ψ and τ were estimated by Monte Carlo simulation. Contributions from continuum, hadron misidentification, and electron from converted photons were subtacted from the data.

The remaining background from secondary electrons was also extracted from the data. Electrons tagged with a like-sign fast lepton are mainly secondary from the untagged B meson with a small primary lepton contribution due to $B^0\overline{B}^0$ mixing. The electron spectrum obtained with like-sign tagging was scaled appropriately to take into account $B^0\overline{B}^0$ mixing and a contribution from secondary leptons above 1.4 GeV/c, and then subtracted from the electron spectrum obtained with unlike-sign tagging. The contribution from secondary electrons, produced in the cascade decays of the tagged B meson, was determined by using the angular correlation between the tagging fast lepton and the opposite-sign electron.

The acceptance corrected electron spectrum obtained after all background subtraction is shown in Figure 6. Only statistical errors are shown. Systematic errors increase going to low momenta due to the subtracting procedure. To extract the branching ratio from this spectrum only minimal extrapolation to low momenta is needed. Normalizing the theoretical spectrum from the ISGW model[4] to the data in the region $p_e > 0.4$ we find $BR(\overline{B} \to Xl^-\overline{\nu}) = (10.0 \pm 0.6 \pm 0.6)\%$. Summing up the data using the same ratio for the unmeasured interval $(1.6 \pm 0.1)\%$ we find $BR(\overline{B} \to Xl^-\overline{\nu}) = (11.0 \pm 0.7 \pm 0.7)\%$.

These values agree with the result of our previous analysis[2], where only the hard part of the lepton spectrum was studied. Therefore the "lepton deficit" can not be explained by unexpectedly large fraction of soft leptons.

ACKNOWLEDGEMENTS

I wish to thank the DESY directorate for their financial support which gave me the opportunity to attend this conference.

REFERENCES

1. H.Albrecht et al. (ARGUS), Nucl.Instr. and Methods **A275**(1989)1.

2. H.Albrecht et al. (ARGUS), Phys. Let. **B197** (1987) 452.

3. The Particle Data Group, Phys. Rev. **D45**, No.11 (1992) 1.

4. N.Isgur et al., Phys.Rev. **D39** (1989) 799.

5. S.Balk et al., Preprint MZ-TH-92-22(1992).

6. J.G.Koerner, G.A.Schuler, Phys. Lett. **B226** (1989) 185.

7. M.Neubert, Phys. Lett. **B264** (1991) 455.

8. R.Poling (CLEO), Proc. of LP-EHEP Conf. (1991), Geneva, v.1, p.546.

9. M.Danilov. Proc. of Joint LP-EHEP Conf. (1991), Geneva, v.2, p.331.

INCLUSIVE DECAYS OF BEAUTY (& CHARM): QCD vs. PHENOMENOLOGICAL MODELS *

I.I.Bigi

Dept.of Physics, University of Notre Dame du Lac, Notre Dame, IN 46556
e-mail address: BIGI@UNDHEP, UNDHEP::BIGI

Abstract

A selfconsistent method is presented for treating nonperturbative effects in inclusive nonleptonic and semileptonic decays of heavy flavour hadrons. These effects give rise to powerlike corrections $\propto 1/m_Q^n$, $n \geq 2$ with m_Q denoting the heavy quark mass. The leading corrections to the quasifree 'spectator ansatz' occur for n=2; among other things they *reduce* the semileptonic branching ratio by no more than 10 % in beauty decays and by a factor of roughly two in charm decays. Phenomenological models of heavy flavour decays are unable to mimic these leading corrections by a specific choice of quark masses or by invoking Fermi motion.

GOALS

A detailled and quantitative analysis of the weak decays of heavy flavour states is obviously of fundamental interest. Yet before this unique opportunity can be fully exploited we have to overcome theoretical problems as well as experimental challenges since the decays of beauty and charm states are affected by the strong interactions in an essential way. Those actually enter on two different levels: (i) They modify the weak forces already on the quark level; those effects can be incorporated *perturbatively* through ultraviolet renormalization. (ii) Boundstate dynamics affect the behaviour of the initial state and hadronization that of the final state. Yet those phenomena *cannot* be treated in a *perturbative* framework. It has been customary to employ phenomenological models instead; their connections to QCD are however uncertain. The goal of our analysis is to improve on this unsatisfactory situation by developing a procedure that

- allows to calculate *inclusive* decay rates in terms of the fundamental parameters, namely the quark masses, KM parameters and Λ_{QCD};

- yields a *systematic* classification of the uncertainties;

- contains a prescription for *successive* refinements and

- possesses an *intrinsic* connection to QCD!

I will sketch here only the outlines of our analysis; details can be found in [1, 2]. Two caveats are in order: It should be understood that the

---

*Work done in collaboration with N.G.Uraltsev and A.I.Vainshtein and supported by the NSF under grant PHY92-13313

numbers given below are presented for illustration rather than as the final predictions; those will be obtained in due course. Also our analysis per se does not apply to *exclusive* decays.

GENERAL PROCEDURE

We have adopted the general method outlined in ref.[3]: First one considers the transition operator $\hat{T}(Q \to Q)$ that describes the forward scattering of the heavy quark Q to *second order in the weak Lagrangian.* The *inclusive* decay width for a heavy flavour hadron H_Q is then obtained from the absorptive part of the matrix element of this transition operator taken between the state H_Q. At both steps we employ an expansion in inverse powers of the heavy quark mass m_Q as the central element of our analysis:

(A) The transition operator $\hat{T}$ involves an integration over quark fields in the intermediate state; therefore $\hat{T}$ is nonlocal. Yet it can be expanded into a series of local operators $O^{(i)}$ of increasing dimension with coefficients c_i that contain decreasing powers of m_Q:

$$Im\, \hat{T}(Q \to Q) \simeq \sum_i c_i O^{(i)} \qquad (1)$$

(B) The matrix elements $< H_Q|O^{(i)}|H_Q >$ are evaluated again from a $1/m_Q$ expansion.

Operator Expansion

The contribution from the operator of lowest dimension in eq.(1) will dominate the decay width for $m_Q \to \infty$. The lowest dimensional operator that can appear in this expansion is obtained from integrating out the quarks q_1, $\bar{q}_2$ and q_3 completely from $Q \to q_1\bar{q}_2q_3 \to Q$; this generates the operator $\bar{Q}Q$ with dimension *three*:

$$O^{(3)} = \bar{Q}Q$$

$$c_3 = \frac{G_F^2}{192\pi^3} \cdot N_C \cdot |KM|^2 \cdot m_Q^5 \qquad (2)$$

where KM denotes the appropriate KM factors and N_C the number of colours; the expression given for c_3 thus describes nonleptonic decays. For simplicity we have set the final state quark masses to zero.

It is easily inferred from the m_Q dependance in eq.(2) that taking the matrix element of the operator $O^{(3)}$ between states H_Q will (among other things) generate the *universal* "spectator" contribution to the H_Q decay widths. This will be shown more explicitely later on.

Operators of higher dimensions are readily obtained by cutting intermediate quark lines in $\hat{T}(Q \to q_1\bar{q}_2q_3 \to Q)$:

$$O^{(6)} = (\bar{Q}\Gamma q)(\bar{q}\Gamma Q)$$

$$c_6 = \frac{G_F^2}{12\pi} \cdot |KM|^2 \cdot m_Q^2 \qquad (3)$$

with Γ denoting some combination of γ matrices. It is easy to see that these terms represent "Weak Annihilation/W exchange" (hereafter referred to as WA) and the so-called "Pauli Interference"(=PI)[4]. Furthermore

$$\frac{c_6\, O^{(6)}}{c_3\, O^{(3)}} \propto (1/m_Q)^3 \qquad (4)$$

i.e. these non-spectator contributions which generate differences in the decay widths for the various hadrons H_Q fade away rapidly for m_Q increasing.

It had been claimed[5] that the $1/m_Q$ scaling laws expressed in eqs.(3,4) are invalidated in the presence of gluons and that the WA contribution is greatly enhanced. However a careful analysis of the analytical properties of $Im\, T(Q \to Q)$ shows[1] such claims to be incorrect. The $1/m_Q$ classification stated above thus holds also in the presence of gluons.

So far we have identified dimension *three* and *six* operators. This raises the question whether dimension *four* and *five* operators can enter as well.

A priori there are two candidates for a dimension *four* operator, namely $\partial_\mu(\bar{Q}\gamma_\mu Q)$ and $\bar{Q}\not{D}Q$ with $\mathcal{D}_\mu$ denoting the covariant derivative. Yet the first operator represents a total derivative and is thus irrelevant; the second operator on the other hand can be replaced by $m_Q\bar{Q}Q$ via the equations of motion and thus merely renormalizes the value of m_Q in eq.(2).

There exists one relevant dimension *five* operator, namely the chromomagnetic dipole operator $\bar{Q}i\sigma_{\mu\nu}G_{\mu\nu}Q$ with $G_{\mu\nu}$ denoting the gluon field strength tensor. Accordingly we have:

$$Im\,\hat{T}(Q \to Q) = c_3\bar{Q}Q + c_5\bar{Q}i\sigma\cdot GQ + \\ + c_6(\bar{Q}\Gamma q)(\bar{q}\Gamma Q) + \mathcal{O}(m_Q) \quad (5)$$

Matrix Elements

The relevant matrix elements of the various local operators appearing in eq.(5) can again be evaluated from a heavy quark expansion; in this case it actually amounts to a nonrelativistic expansion:

$$\bar{Q}Q = v_\mu \bar{Q}\gamma_\mu Q - \tfrac{1}{4}m_Q^2 \bar{Q}i\sigma G Q - \\ -\tfrac{1}{2}m_Q^2 \bar{Q}(D^2 - (v_\mu D_\mu)^2)Q + O(1/m_Q^3) \quad (6)$$

with v_μ denoting the relativistic velocity vector. For the same reasons as stated above there is no dimension *four* operator in eq.(6); on the other hand there are now two dimension *five* operators.

Four observations about eq.(6) should be noted at this point:

(i) The first operator on the right hand side of eq.(6) is the generator of the heavy flavour charge Q; its matrix element taken between hadrons H_Q therefore yields unity thus reproducing the "spectator" result.

(ii) Yet even the quasifree operator $\bar{Q}Q$ transcends the spectator ansatz since it contains higher dimensional operators that yield non-perturbative corrections.

(iii) The matrix elements of the second operator are easily obtained:

$$< \Lambda_Q|\bar{Q}i\sigma\cdot GQ|\Lambda_Q > = 0$$

$$< P_Q|\bar{Q}i\sigma\cdot GQ|P_Q > = \\ \tfrac{3}{2}(M^2(V_Q) - M^2(P_Q))\cdot < P_Q|\bar{Q}Q|P_Q > \quad (7)$$

where Λ_Q, P_Q and V_Q denote the lowest baryon, pseudoscalar and vector meson state respectively; their masses are $M(\Lambda_Q)$, $M(P_Q)$ and $M(V_Q)$.

(iv) The third operator can be interpreted as the kinetic energy of the heavy quark Q in the presence of the gluon background field. In all likelihood it has different matrix elements for baryons and mesons. A comparison of the masses of charm and beauty mesons and baryons yields one linear combination, namely

$$(3M_D + M_{D^*} - 4M_{\Lambda_c}) - \\ (3M_B + M_{B^*} - 4M_{\Lambda_b}) \simeq \tfrac{2(M_B-M_D)}{M_D M_B} \cdot \\ (\langle\text{baryon}|\bar{Q}\vec{D}^2Q|\text{baryon}\rangle - \\ \langle\text{meson}|\bar{Q}\vec{D}^2Q|\text{meson}\rangle) \quad (8)$$

We expect that further theoretical analysis will lead to a complete determination of these matrix elements in the future. Yet some nontrivial predictions can be made already now, as will be shown below.

Qualitative Conclusions

The results presented so far can be summarized in a qualitative way as follows: the *inclusive* decay widths of Q *hadrons* can be related to the decay widths of Q *quarks* via an $1/m_Q$ expansion

$$\Gamma(H_Q) = \Gamma(Q) \cdot \\ \cdot [1 + A_2/m_Q^2 + A_3/m_Q^3 + \mathcal{O}(1/m_Q^4)] \quad (9)$$

which contains three lessons:

- The spectator ansatz holds up to terms of order $1/m_Q^2$.

- The first non-spectator effects enter on the $1/m_Q^2$ level, yet they are SU(2) *invariant*; i.e. they yield different lifetimes for baryons on one hand and mesons on the other (see eq.(7)), but not for B_u vs. B_d mesons (or D^+ vs. D^o mesons). E.g. for beauty decays:

$$A_2(\Lambda_b) \neq A_2(B_u) = A_2(B_d) \quad (10)$$

- The conventional non-spectator effects – WA, PI – enter on the $1/m_Q^3$ level:

$$A_3(B^-) < 0,\ 0 < A_3(B_d) < A(\Lambda_b) \quad (11)$$

QUANTITATIVE PHENOMENOLOGY

Two examples are given here to show that specific predictions can be given, what kind of considerations are involved in obtaining them and how they can be made more precise. We will phrase our discussion in terms of beauty decays and subsequently comment on charm decays.

Semileptonic Branching Ratios

A relatively straightforward calculation yields for the semileptonic and nonleptonic transition operator in the external gluon field:

$$\hat{\Gamma}_{SL} = F z_0 (\bar{b} b - \frac{1}{m_b^2} \bar{b} i \sigma \cdot G b) \quad (12)$$

$$\hat{\Gamma}_{NL} = F N_C [A_0 z_0 \cdot (\bar{b} b - \frac{1}{m_b^2} \bar{b} i \sigma \cdot G b) - A_2 z_2 \cdot \frac{4}{m_b^2} \bar{b} i \sigma \cdot G b] \quad (13)$$

$$F = \frac{G_F^2 m_b^5}{192 \pi^3} |V(cb)|^2$$

where we have used the following notation:
(i) z_0, z_2 represent phase space factors that reflect the sizeable mass of the c quark:

$$z_0(x) = 1 - 8x + 8x^3 - x^4 - 12x^2 \log x \quad (14)$$

$$z_2(x) = (1-x)^3;\ x = m_c^2/m_b^2. \quad (15)$$

Obviously $z_0(0) = z_2(0) = 1$ holds.

(ii) A_0, A_2 denote colour factors and QCD radiative corrections lumped into quantities $c_\pm$:

$$A_0 = \frac{1}{2}(c_+^2 + c_-^2) + \frac{c_+^2 - c_-^2}{2N_C} + \mathcal{O}(\alpha_s(m_b^2)) \quad (16)$$

$$A_2 = \frac{c_+^2 - c_-^2}{2N_C} + \mathcal{O}(\alpha_s(m_b^2)) \quad (17)$$

One reads off from eqs. (12,13) that *non-perturbative* corrections to the *semileptonic branching ratio* arise on the $1/m_b^2$ level. More specifically

$$\frac{\delta BR_{SL}(B)}{BR_{SL}(B)} \simeq 6 \cdot \frac{M^2(B^*) - M^2(B)}{m_b^2} \frac{A_2 z_2}{A_0 z_0} \cdot BR_{NL} \quad (18)$$

Since $c_+ < c_-$ one has $A_2 < 0$ and thus also $\delta BR_{SL}(B) < 0$, i.e. *the semileptonic branching ratio is reduced by these nonperturbative corrections!* Numerically one obtains $A_2/A_0 \sim -(0.15 - 0.20)$ and thus $\delta BR_{SL}(B)/BR_{SL}(B) \sim -(0.03 - 0.04)$ if one calculates the QCD radiative corrections on the *leading log* level. *Subleading* QCD corrections might enhance A_2/A_0 by a factor of two to three and thus

$$\frac{|\delta BR_{SL}(B)|}{BR_{SL}(B)} \leq 0.1 \quad (19)$$

i.e. the semileptonic branching ratio is reduced by up to 10 %, say from 12-13% to 11-11.5 %. It should be noted that this reduction *cannot* be mimicked in phenomenological models! Furthermore, it requires "merely" a *perturbative* computation of the next-to-leading log QCD corrections to sharpen the numerical prediction.

Extrapolating this result down to charm decays one realizes that the corresponding reduction there must be considerably larger since $6(M^2(D^*) - M^2(D))/m_c^2 \simeq 2$ instead of

$6(M^2(B^*) - M^2(B))/m_b^2 \simeq 0.15!$ One then finds
$$\frac{\delta BR_{SL}(D)}{BR_{SL}(D)} \sim -\mathcal{O}(50\%) \quad (20)$$

This means that the "normal" semileptonic branching ratio of D mesons is roughly 7-10 % – and thus close to the observed value of D^0 mesons rather than D^+ mesons! – rather than the parton level prediction of $\sim 14\%$. We will come back to this result in the next subsection.

The situation is more involved for the semileptonic decays of baryons: for $BR_{SL}(\Lambda_b)$ is not affected by $\bar{b}i\sigma \cdot Gb$ (see eq.(7)) and thus does *not* get reduced on the $1/m_b^2$ level:

$$BR_{SL}(B) < BR_{SL}(\Lambda_b) + \mathcal{O}(1/m_b^3) \quad (19)$$

Yet on the $1/m_b^3$ level WA enters which will *decrease* $BR_{SL}(\Lambda_b)$. Scaling this down to charm decays one indeed expects $BR_{SL}(\Lambda_c) < BR_{SL}(D)$ as observed.

Lifetime Ratios

A detailled analysis of the analytical structure of $\hat{T}(Q \to Q)$ and of the QCD radiative corrections leads to the prediction [1]

$$\frac{\tau(B^-)}{\tau(B_d)} \simeq 1.08 \ [1.02] \quad (21)$$

for $f_B = 250 \ MeV \ [130 \ MeV]$; i.e. we expect lifetime differences of only a few %! Furthermore those are mainly due to PI with WA being quite irrelevant.

The largest difference between the lifetimes of B_u, B_d and B_s mesons is then expected to arise in a very intriguing fashion: due to $B_s - \bar{B}_s$ mixing there are two neutral strange B meson states with a slightly different mass and somewhat different decay width

$$\frac{\Delta\Gamma(B_s - \bar{B}_s)}{\bar{\Gamma}(B_s)} \simeq 0.23 \ [0.08] \quad (22)$$

for $f_B = 250 \ [150] \ MeV$!

For the lifetimes of beauty *baryons* we expect roughly

$$\frac{\tau(B_d)}{\tau(\Lambda_b)} \sim 1.1 - 1.15$$

There are actually different dynamical aspects involved in the inclusive decays of baryons versus those of mesons since the operator $\bar{b}i\sigma \cdot Gb$ certainly (see eq.(7)) and the operator $\bar{b}\vec{D}^2 b$ presumably possess different matrix elements for meson and baryon states. This is expressed in the following relation:

$$\frac{\Gamma(B) - \Gamma(\Lambda_b)}{\bar{\Gamma}} \simeq -\frac{15}{8}\frac{M^2(B^*) - M^2(B)}{m_b^2}$$
$$-\frac{m_c}{m_b}\frac{1}{4(M_B - M_D)}[(3M_D + M_{D^*} - 4M_{\Lambda_c}) -$$
$$(3M_B + M_{B^*} - 4M_{\Lambda_b})]$$
$$-6\frac{M^2(B^*) - M^2(B)}{m_b^2} \cdot \frac{A_2 z_2}{A_0 z_0} \quad (22)$$

where we have used eqs.(7,8). In addition WA and PI will enhance $\Gamma(\Lambda_b)$ further on the $1/m_b^3$ level.

Extrapolating these findings down to charm decays we find

$$\frac{\tau(D^+)}{\tau(D^0)} \sim 2 \quad (23)$$

which is mostly due to PI with WA contributing only ~ 10 - 20 %. Thus it is D^0 decays that proceed largely "normal" whereas the D^+ lifetime is substantially prolonged. This is in nice agreement with our findings on the semileptonic branching ratios given in the previous subsection!

SUMMARY AND OUTLOOK

We have outlined here a general method that allows to calculate the *inclusive* weak decay rates for heavy flavour hadrons. It is based on a systematic classification and treatment of various contributions in a $1/m_Q$ expansion as given in eq.(9).

Qualitatively one finds on quite general grounds for beauty decays:

$$\tau(\Lambda_b) = \tau(B_d) = \tau(B_u) + \mathcal{O}(1/m_b^2)$$

$$\tau(\Lambda_b) < \tau(B_d) = \tau(B_u) + \mathcal{O}(1/m_b^3)$$

$$\tau(\Lambda_b) < \tau(B_d) < \tau(B_u) + \mathcal{O}(1/m_b^4)$$

i.e.

- There are no corrections of order $1/m_Q$ to the quasifree spectator ansatz.

- The leading nonperturbative corrections arise then on the $1/m_Q^2$ level. They are $SU(3)_{FL}$ invariant, i.e. affect the decays of heavy flavour *mesons* in a uniform way independant of the flavour of the light antiquarks. They enhance the nonleptonic decay width in mesons and lead to a corresponding *reduction* in the semileptonic branching *ratio*.

- These corrections which gauge the approach towards hadron-parton duality cannot be mimicked in phenomenological models.

- The nonperturbative corrections arising on the $1/m_Q^3$ level are *not* $SU(2)_I$ or $SU(3)_{FL}$ invariant. Thus they generate differences in the lifetimes and semileptonic branching ratios among all heavy flavour hadrons H_Q.

- These preasymptotic effects are obviously much larger in charm than in beauty decays.

- We have identified cases where various contributions that are non-leading in $1/N_C$ tend to cancel each other.

A semi-quantitative analysis yields the following numbers:

- $\tau(B^-)$ and $\tau(B_d)$ differ by a few % only.

- $\tau(\Lambda_b)$ and $\tau(B_d)$ could differ by 10-15 %.

- One obtains a factor of two for $\tau(D^+)/\tau(D^0)$.

- $BR_{SL}(B)$ is lowered by up to 10 % or so, whereas $BR_{SL}(D)$ gets reduced by $\mathcal{O}(50\%)$!

It should be understood that some of these numerical predictions are tentative not as a matter of principle, but because they involve *perturbative* corrections that have not been computed yet (but will).

There are some obvious future refinements of this approach that are being tackled: (i) $SU(3)_{FL}$ breaking: we expects such effects to be no larger than a few %, rather than the "canonical" 20 %. (ii) As already indicated at various places the weak decays of heavy flavour *baryons* provide us with new probes of the underlying dynamics. (iii) Beyond total transition rates, one can analyze in such a framework also *distributions*, like the lepton spectra in inclusive semileptonic decays.

REFERENCES

1. I.I.Bigi, N.G.Uraltsev, *Phys.Lett.* **280B** (1992) 120

2. I.I.Bigi, N.G.Uraltsev, A.I.Vainshtein, preprint FERMILAB-PUB-92/158-T, 1992, accept. f.publ. in Phys.Lett.B.

3. M.A.Shifman, M.B.Voloshin, *Sov. Journ. Nucl. Phys.* 41 (1985) 120

4. B.Guberina et al., *Phys.Lett.* B89 (1979) 111

5. M.Bander, D.Silverman, A.Soni, *Phys. Rev. Lett.* 44 (1980) 7, 962(E)

HEAVY QUARK EFFECTIVE THEORY: APPLICATIONS TO WEAK DECAYS*

Benjamín Grinstein
Physics Research Division
Superconducting Super Collider Laboratory
2550 Beckleymeade Ave., MS-2001
Dallas, Texas 75237

Abstract

After a brief historical introduction to the subject, I discuss some of the most important recent developments in the field.

INTRODUCTION

Brief History

Interpretation of measurements of weak decays of heavy mesons is often marred by our inability to calculate hadronic matrix elements. Fortunately, in the limit in which the bottom and charm quarks are considered much heavier than the hadronic scale an approximate symmetry of QCD arises that allows many definite predictions. This symmetry is good only for hadrons with a single heavy quark, but that is fine since these hadrons are the ones of interest for studies of weak decays.

While the present burst of interest in the field stems from the seminal work of Isgur and Wise[1], the main ingredients of their work were already available. Even before the discovery of open charm, *i.e.*, of D mesons, De Rújula, Georgi and Glashow[2] considered a model which incorporated the observation that the spin of a heavy quark decouples in a heavy-light system; see below. They estimated the mass difference of D^* and D mesons to be mush less than that of the corresponding light quark ρ and π mesons, and of the order of the pion mass. Eichten & Feinberg[3] wereled to consider a large mass $1/m$ expansion in their investigations of spin dependent forces in heavy $q\bar{q}$ bound states. Almost a decade later Eichten[4] and Lepage & Thacker[5] suggetsed applying the same techniques to heavy–light systems, thus opening the way towards an effective theory with explicit spin symmetries. Fueled by experimental progress in the measurement of semileptonic decays of B mesons several theoretical calculations of the rates for these decays appeared in the mid '80s[6]–[9]. One of them, by Isgur, Wise and I, pointed out that the matrix element for $\bar{B} \to De\nu$ was most reliably calculated at the kinematic point where the D meson does not recoil in the restframe of the B meson[7]. Nussinov & Wetzel[8] noted this independently. Moreover, they gave a physical explanation that was close in spirit to the modern argument in terms of a HQET, namely, that the state of light degrees of freedom in heavy light systems is independent of the mass of the heavy quark. They had discovered a new flavor symmetry of these heavy-light systems. Unfortunately, their work was, until recently, largely ignored. Influential to Isgur and Wise was the work of Shifman & Voloshin[10], who proposed a theoretical limit

---

*Research funded in part by the Alfred P. Sloan Foundation and by the Department of Energy under contract DE–AC35–89ER40486.

($m_c \to m_b$), now called the "Shifman-Voloshin limit", at which the matrix element for $\bar{B} \to De\nu$ is exactly calculable. The last element used by Isgur and Wise comes from the work of Shifman & Voloshin[11] and of Politzer & Wise[12], who extracted QCD calculable violations to the flavor symmetry predictions in the form of logarithms of the ratio of heavy quark masses. Moreover, Politzer & Wise[13] then attempted to reproduce these calculations working directly within the context of an effective theory.

Progress was made rapidly after publication of the Isgur & Wise papers. On the formal side, a "Heavy Quark Effective Field Theory" (HQET) was finally established[14]. The HQET allows amplitudes to be separated into a perturbatively calculable short distance factor that depends on the heavy mass and a non-perturbative long distance part. The latter is independent of the heavy mass and displays explicitly the heavy quark symmetries. The short distance factor for a heavy to heavy (as in bottom to charm) transition was computed in the leading logarithmic approximation first[15] and shortly thereafter in next to leading log (which was first done *correctly* in ref. [16]). Interest grew rapidly as it became clear that this treatement constitutes a *systemattic* expansion in inverse powers of the heavy mass. The effects of a finite charm-to-bottom ratio, m_c/m_b, were calculated[17] and the general formalism for $1/M$ corrections was developped[18].

Hand-waving

The central idea of the HQET is so simple, it can be described without reference to a single equation. And it should prove useful to refer back to the simple intuitive notion, to be presented below, wherever the formalism and corresponding equations become abstruse.

The HQET is useful when dealing with hadrons composed of one heavy quark and any number of light quarks. In what follows we shall (imprecisely) refer to these as 'heavy hadrons'.

The successes of the constituent quark model is indicative of the fact that, inside hadrons, strongly bound quarks exchange momentum of magnitude a few hundred MeV. We can think of the typical amount Λ by which the quarks are off-shell in the nucleon as $\Lambda \approx m_p/3 \approx 330$MeV. In a heavy hadron the same intuition can be imported, and again the light quark(s) is(are) very far off-shell, by an amount of order Λ. But, if the mass M_Q of the heavy quark Q is large, $M_Q \gg \Lambda$, then, in fact, this quark is almost on–shell. Moreover, interactions with the light quark(s) typically change the momentum of Q by $\sim \Lambda$, but change the *velocity* of Q by a negligible amount, of the order of $\Lambda/M_Q \ll 1$. It therefore makes sense to think of Q as moving with constant velocity, and this velocity is, of course, the velocity of the heavy hadron.

In the rest frame of the heavy hadron, the heavy quark is practically at rest. The heavy quark effectively acts as a static source of gluons. It is characterized by its flavor and color–$SU(3)$ quantum numbers, but not by its mass. In fact, since spin–flip interactions with Q are of the type of magnetic moment transitions, and these involve an explicit factor of g_s/M_Q, where g_s is the strong interactions coupling constant, the spin quantum number itself decouples in the large M_Q case. Therefore, *the properties of heavy hadrons are independent of the spin and mass of the heavy source of color*.

The HQET is nothing more than a method for giving these observations a formal basis. It is useful because it gives a procedure for making explicit calculations. But more importantly, it turns the statement 'M_Q is large' into a *systematic* perturbative expansion in powers of Λ/M_Q. Each order in this expansion involves QCD to all orders in the strong coupling, g_s. Also, the statement of mass and spin independence of properties of heavy hadrons appears in the HQET as approximate internal symmetries of the Lagrangian.

Before closing this section, we point out that these statements apply just as well to a very familiar and quite different system: the atom. The rôle of the heavy quark is played by the nucleus, and that of the light degrees of freedom by the electrons (and the electromagnetic field). That different isotopes have the same chemical properties simply reflects the independence of the atomic wavefunction on the nuclear mass. Atoms with nuclear spin s are $2s + 1$ degenerate; this degeneracy is broken when the finite nuclear mass is accounted for, and the resulting hyperfine splitting is small because the nuclear mass is so much larger than the binding energy (playing the rôle of Λ). It is not surprising that, using M_Q independence, the properties of B and D mesons are related, and using spin independence, those of B and B^* mesons are related, too.

APPLICATIONS

Many applications of the heavy quark expansion, have been developped; it would be impossible to make justice to them in a six page summary. A partial list:

- $B \to D\ell\nu$ and $B \to D^*\ell\nu$
- $B \to \rho\ell\nu$ from $D \to K^*\ell\nu$[19]
- Heavy Hadron Spectrum[20]
- Strong decays of heavy hadrons[20]
- Inclusive semileptonic decay rates[21]
- Semileptonic decays sum rules[21, 22]
- Factorization in $B \to D\pi$[23]
- Semileptonic heavy baryon decays[24]

- Rates for rare B decays[19, 25]
- $e^+e^- \to$ pair of heavy hadrons[17, 26]
- $\Lambda_b \to \Lambda_c D_s$ vs. $\Lambda_b \to \Lambda_c D_s^*$[27]

In what follows I will describe briefly some of the most important recently discovered applications.

Hefty-Handy Lagrangian

Phenomenological lagrangians for heavy mesons[28] and for heavy baryons[29] that incorporate both chiral symmetry and heavy quark symmetries have been proposed. They allow a study of the semileptonic decays $\bar{B} \to DXe\bar{\nu}$ and $\bar{B} \to D^*Xe\bar{\nu}$ where X is a low momentum state of one or more pions. For some quantities, one may also compute the leading chiral symmetry breaking effect due to a large strange mass[30]; if g stands for the D^*-D-π coupling, then $f_{D_S}/f_D - 1 = 0.064(1 + 3g^2)$ and $B_{B_S}/B_B - 1 = 0.052(1 + 3g^2)$. The experimental limit on $\Gamma(D^* \to D\pi)$ constrains g^2 to be less than 0.4[31]. Applications of these new heavy-chiral (or 'hefty-handy') lagrangians are being developed as we write[32].

The bound that is no longer

In my talk I described a bound on the Isgur-Wise function obtained by Taron and de Rafael[33] and I discussed both its phenomenological and its theoretical implications. Because I devoted so much of my talk to it, I think it is appropriate to direct the reader to four recent papers that find an error in the derivation of this bound[34]

Heavy Baryon Form Factor Relations

Six form factors encode the semileptonic decay amplitude $\Lambda_b \to \Lambda_c e\bar{\nu}$. The transition lends itself particularly well to HQET analysis because it is tightly constrained by the heavy quark spin symmetry[24]. Like their mesonic counterparts, the six form factors that parametrize this baryonic process are predicted at leading order in terms of a single Isgur-Wise function. In contrast with their mesonic counterparts, one can prove that, to all orders in the strong coupling constant, five relations among these six form factors remain after $O(1/m_c)$ corrections are included[35]. Defining the form factors through

$$\langle \Lambda_c(v',s')|V^\mu|\Lambda_b(v,s)\rangle = \bar{u}(v',s')[F_1(v\cdot v')\gamma^\mu$$
$$+F_2(v\cdot v')v^\mu + F_3(v\cdot v')v'^\mu]u(v,s)$$
$$\langle \Lambda_c(v',s')|A^\mu|\Lambda_b(v,s)\rangle = \bar{u}(v',s')[G_1(v\cdot v')\gamma^\mu$$
$$+G_2(v\cdot v')v^\mu + G_3(v\cdot v')v'^\mu]\gamma^5 u(v,s), \quad (1)$$

the relations between form factors are, at order $O(\alpha_s(m_c)/m_c)$,

$$\frac{F_1}{G_1} = 1 + \left[\frac{\bar{\Lambda}}{2m_c} + \frac{\bar{\Lambda}}{2m_b}\right]\frac{2}{(v\cdot v'+1)} +$$
$$\frac{4}{3}\frac{\alpha_s(m_c)}{\pi}r + \frac{4}{3}\frac{\alpha_s(m_c)}{\pi}\frac{\bar{\Lambda}}{2m_c}\frac{2(1+r-v\cdot v'r)}{(v\cdot v'+1)}$$
$$\frac{F_2}{G_1} = \frac{G_2}{G_1} = -\frac{\bar{\Lambda}}{2m_c}\frac{2}{(v\cdot v'+1)}$$
$$-\frac{4}{3}\frac{\alpha_s(m_c)}{\pi}r - \frac{4}{3}\frac{\alpha_s(m_c)}{\pi}\frac{\bar{\Lambda}}{2m_c}\frac{2(1+r-v\cdot v'r)}{(v\cdot v'+1)}$$
$$\frac{F_3}{G_1} = -\frac{G_3}{G_1} = -\frac{\bar{\Lambda}}{2m_b}\frac{2}{(v\cdot v'+1)} \quad (2)$$

where

$$r = \frac{\log(v\cdot v' + \sqrt{v\cdot v'^2 - 1})}{\sqrt{v\cdot v'^2 - 1}}. \quad (3)$$

and $\bar{\Lambda}$ is an undetermined constant with unit mass dimensions, expected to be of order of the hadronic scale, $\bar{\Lambda} \sim 500\text{MeV}$. These form factor relations provide a valuable means for assessing the uncertainty in future measurements of the mixing angle $|V_{cb}|$ from semileptonic Λ_b decay.

Measuring $|V_{cb}|$?

The most notorious application of the HQET is to the experimental determination of $|V_{cb}|$. The semileptonic decay rate of the B-meson is dominated by the exclusive decays into D and D^* mesons. The form factors for these are predicted by HQET at the zero recoil point $\vec{p}_D = 0$ in the B-rest-frame, or $q^2 = q^2_{\max} = (M_B - M_D)^2$. Unfortunatelly, the rate vanishes at this kinematic end-point, and an extrapolation is required for the extraction of $|V_{cb}|$. The problem is more severe for the D than for the D^* since the former is helicity supressed at $q^2_{\max}$. The nature of the extrapolation will become irrelevant as much more data becomes available. But for now we must rely on models of hadrons to guide us in constructing reasonable extrapolations. The models available — and corresponding calculations— have not changed from before the advent of the HQET, but one can incorporate the requirement that form factors agree with the HQET prediction at $q^2_{\max}$. The figure below shows my crude attempt at displaying available CLEO and ARGUS data in unit $v\cdot v'$ bins[36]. $\xi(v\cdot v')$ is the Isgur-Wise function, satisfying $\xi(1) = 1$.

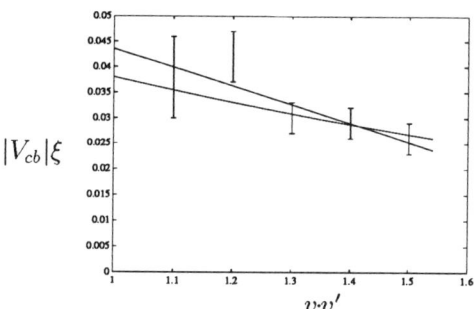

The straight line is a best linear fit to the data $(-0.036 v\cdot v' + 0.080)$. The second curve is a fit to constant $\times \exp(0.7(v\cdot v'-1))$ as suggested by the ISGW model[7, 9].

Considerably more important is the question of whether the incalculable corrections to the HQET predictions might be upsettingly large. It has been established that the HQET predictions of form factors at $q^2_{\max}$ are free of corrections of order $1/M$[35, 37]. Can we expect the $O(1/M^2)$ corrections to be small? The question becomes more pressing in light of results from Monte-Carlo simulations of quenched lattice QCD that indicate that the $1/M$ corrections to the D-meson decay constant, f_D, are of the order of 65%[38].

In an attempt to explore this issue two groups[39, 40] have compared HQET predictions to the soluble 'tHooft model (2-dim planar QCD). Both groups find that the pseudoscalar decay constant is afflicted by large $1/M$ corrections and that the form factor $F(q^2)$ defined in

$$\langle B(p')|\bar{b}\gamma_\mu b|B(p)\rangle = (p+p')_\mu F(q^2) \quad (4)$$

converges rapidly to the Isgur-Wise function. An interesting controversy exists, though. Ref. [39] finds that the matrix element $\langle D(p')|\bar{c}\gamma_\mu b|B(p)\rangle$ at the end-point, $q^2 = q^2_{max}$, is predicted by the HQET to an accuracy of a few percent, while ref. [40] predicts large corrections to the HQET predictions for the form factor f_- of this matrix element. These seemingly contradictory results could be reconciled if it is found that the f_+ form factor itself receives large corrections, just in such a way as to make the whole matrix element at the end-point only slightly corrected (an accurate cancellation of the corrections to f_+ and f_- would be required). More likely is that at least one of refs. [39] and [40] is incorrect.

In either case the general conclusion seems to be that the HQET may converge far faster for some quantities (like the form factor $F(q^2)$ above) than for others (like f_D).

REFERENCES

1. N. Isgur and M.B. Wise, Phys. Lett. **B232**:113(1989); *idem* **B237**:527(1990).

2. A. De Rújula, H. Georgi, S. L. Glashow, Phys. Rev. **D12**:147(1975)

3. E. Eichten and F. L. Feinberg, Phys. Rev. Lett. **43**:1205(1979); *idem,* Phys. Rev. **D23**:2724(1981)

4. E. Eichten, Nucl. Phys. B(Proc. Suppl.)4:170(1988).

5. G. Lepage and B.A. Thacker, Nucl. Phys. B(Proc. Suppl.)4:199(1988)

6. G. Altarelli, et al, Nucl. Phys. **B208**:365(1982); M. Suzuki, Nucl. Phys. **B258**:553(1985); M. Bauer, B. Stech, M. Wirbel, Z. Phys. **C29** 637 (1985); J. G. Korner, and G. A. Schuler, Z. Phys. **C38** 511 (1988); erratum, ibid, **C41** 690 (1989)

7. B. Grinstein, et al, Phys. Rev. Lett. **56**:298(1986)

8. S. Nussinov and W. Wetzel, Phys. Rev. **D36**:130(1987)

9. T. Altomari, and L. Wolfenstein, Phys. Rev. Lett. **58**:1583(1987); Isgur, N., et al., Phys. Rev. **D39**:799(1989)

10. M.B. Voloshin and M.A. Shifman, Sov. J. Nucl. Phys. **47**:511(1988)

11. M.B. Voloshin and M.A. Shifman, Sov. J. Nucl. Phys. **45**:292(1987)

12. H.D. Politzer and M.B. Wise, Phys. Lett. **B206**:681(1988)

13. H.D. Politzer and M.B. Wise, Phys. Lett. **B208**:504(1988)

14. E. Eichten and B. Hill, Phys. Lett. **B234**:511(1990); B. Grinstein, Nucl. Phys. **B339**:253(1990); H. Georgi, Phys. Lett. **B240**:447(1990)

15. A. Falk, et al, Nucl. Phys. **B343**:1(1990)

16. G. P. Korchemskii and A.V. Radyushkin, Phys. Lett. **B279**:359(1992)

17. A.F. Falk and B. Grinstein, Phys. Lett. **B249**:314(1990); A. Falk and B. Grinstein, Phys. Lett. **B247**:406(1990); M. Golden and B. Hill, Phys. Lett. **B254**:225(1991)

18. E. Eichten and B. Hill, Phys. Lett. **B243**:427(1990); A. Falk, B. Grinstein and M. Luke, Nucl. Phys. **B357**:185(1991)

19. N. Isgur and M.B. Wise, Phys. Rev. **D42**:2388(1990)

20. N. Isgur and M. B. Wise, Phys. Rev. Lett. **66**:1132(1991); U. Aglietti, Phys. Lett. **B281**:341(1992)

21. J. Chay, H. Georgi and B. Grinstein, Phys. Lett. **B247**:399(1990); J.D. Bjorken, talk presented at Les Recontré de Physique de la Vallee d'Acoste, La Thuile, Italy, March 18-24, 1990, SLAC-PUB-5278 (1990) unpublished; J.D. Bjorken, I. Dunietz and J. Taron, Nucl. Phys. **B371**:111(1992)

22. N. Isgur and M. B. Wise, Phys. Rev. **D43**:819(1991); A. F. Falk, SLAC–PUB–5689, November 1991

23. M. J. Dugan and B. Grinstein, Phys. Lett. **B255**:583(1991); H. D. Politzer, M. B. Wise, Phys. Lett. **B257**:399(1991); C. Reader, N. Isgur, CEBAF-TH-91-23, unpublished

24. N. Isgur and M.B. Wise, Nucl. Phys. **B348**:276(1991); H. Georgi, Nucl. Phys. **B348**:293(1991); T. Mannel, W. Roberts and Z. Ryzak, Nucl. Phys. **B355**:38(1991); Phys. Lett. **B255**:593(1991)

25. G. Burdman and J. Donoghue, Phys. Lett. **B270**:55(1991)

26. T. Mannel and Z. Ryzak, Phys. Lett. **B247**:412(1990)

27. B. Grinstein, et al, Nucl. Phys. **B363**:19(1991)

28. G. Burdman, J. F. Donoghue, Phys. Lett. **B280**:287(1992); M. B. Wise, Phys. Rev. **D45**:2188(1992); T.-M. Yan, H.-Y. Cheng, C.-Y. Cheung, G.-L. Lin, Y. C. Lin, H.-L. Yu, Phys. Rev. **D46**:1148(1992)

29. H.-Y. Cheng, C.-Y. Cheung, G.-L. Lin, Y. C. Lin, T.-M. Yan, H.-L. Yu, IP-ASTP-07-92, unpublished

30. B. Grinstein, et al, CALT–68–1768, SSCL–Preprint–25, UCSD/PTH 92–05.

31. The ACCMOR collaboration (S. Barlag et al), Phys. Lett. **B278**:480(1992)

32. R. Casalbuoni et al, UGVA-DPT-1992-11-790, unpublished; G. Burdman, UMHEP-382; J. F. Amundson et al, UCSD-PTH-92-31; J. Schechter and A. Subbaraman, SU-4240-519; P. Cho, HUTP-92-A039; J. L. Rosner and M. B. Wise, CALT-68-1807; J.L. Goity, CEBAF-TH-92-16; U. Kilian, J. G. Korner, D. Pirjol Phys. Lett. **B288**:360(1992); L. Wolfenstein, Phys. Lett. **B291**:177(1992)

33. E. de Rafael and J. Taron, Phys. Lett. **B282**:215(1992)

34. J.G. Korner, D. Pirjol, C. Dominguez, MZ-TH-92-48; B. Grinstein, P. F. Mende, SSCL-Preprint-167, BROWN HET-882; A. F. Falk, M. Luke and M. B. Wise, SLAC-PUB-5956; C. E. Carlson et al, CEBAF-TH-92-22, WM-92-113

35. P. Cho and B. Grinstein, Phys. Lett. **B285**:153(1992)

36. S. Stone, *Semileptonic B Decays — Experimental*, in *B Decays*, S. Stone, ed., World Scientific, Singapore, 1992

37. M.E. Luke, Phys. Lett. **B252**:447(1990); H. Georgi, B. Grinstein and M. B. Wise, Phys. Lett. **252B**:456(1990)

38. P. Boucaud et al, Phys. Lett. **B220**:219(1989); C.R. Allton et al, Nucl. Phys. **B349**:598(1991)

39. B. Grinstein and P. F. Mende, Phys. Rev. Lett. **69**:1018(1992)

40. M. Burkhardt and E. Swanson, MIT Report No. CTP-2096

EFFECTIVE HAMILTONIAN FOR $\Delta S = 1$ NON-LEPTONIC DECAYS BEYOND LEADING LOGARITHMS

Matthias JAMIN
Theoretical Physics Division, CERN
CH 1211 Geneva 23, Switzerland

Abstract

An overview on the status of the weak effective hamiltonian for $\Delta S = 1$ decays at the next–to–leading order in perturbation theory is presented. Some theoretical complications which arise at this level of accuracy are discussed. Selected preliminary results, relevant for the $\Delta I = 1/2$ rule and the ratio ε'/ε are given.

INTRODUCTION

In this talk we shall summarize briefly the results of a next–to–leading order calculation of the weak effective hamiltonian for $\Delta S = 1$ non–leptonic decays, performed during the last years. Complete results as well as a discussion on the details of the calculation can be found in refs. [1–5].

The two–loop anomalous dimensions $\mathcal{O}(\alpha_s^2)$ of current–current operators have been calculated in ref. [1], and the extension of this calculation to include QCD as well as electroweak penguin operators is presented in refs. [2] and [3] respectively. The corresponding 10×10 two–loop anomalous dimension matrix $\mathcal{O}(\alpha\alpha_s)$ which is necessary for a consistent treatment of electroweak penguins beyond the leading logarithmic approximation is given in ref. [4], and the effective hamiltonian, incorporating all results of refs. [1–4], is presented in ref. [5].

The resulting hamiltonian is used to study the impact of our next–to–leading order calculation on the ratio ε'/ε, putting special emphasize on the relevance of electroweak penguin operators.

THE EFFECTIVE HAMILTONIAN

The effective hamiltonian relevant for $\Delta S = 1$ decays can be expressed in the form

$$\mathcal{H}_{eff} = \frac{G_F}{\sqrt{2}} \sum_{i=1}^{10} Q_i(\mu) \, C_i(\mu). \quad (1)$$

The set of operators consists of current–current operators $Q_{1,2}$, QCD penguin operators Q_{3-6}, and electroweak penguin operators Q_{7-10}. The explicit form of the operators can be found in ref. [3].

In order to be sensitive to the dependences on the renormalization scheme it is necessary to calculate the anomalous dimensions together with the Wilson coefficient functions of the operators Q_i to the next–to–leading order. In particular, this then allows to

- determine the renormalization scale μ, below which perturbation theory becomes doubtful,

- consistently use $\Lambda_{\overline{MS}}$ in weak decays, and

- trace the scheme dependences stemming from the regularization procedure, the treatment of γ_5 in $D \neq 4$ dimensions, and the form of the operators Q_i.

From the phenomenological point of view, we will be able to see if

- next-to-leading order corrections enhance or suppress the Wilson coefficients $C_i(\mu)$, and

- whether electroweak penguins compensate for the effect of QCD penguins.

The last question is important in the analysis of the ratio ε'/ε, parameterizing direct CP violation in $K \to \pi\pi$ decays.

We will not repeat here the lengthy expressions for the anomalous dimension matrices and the Wilson coefficient functions which can be found in the afore mentioned references [1–5], but rather discuss the main conclusions which can be drawn from the calculation. Let us split the discussion into two parts, first elaborating on some technical subtleties which arise at the next-to-leading level, and second presenting preliminary results of the numerical analysis of the coefficient functions, relevant for the calculation of ε'/ε.

Important technical remarks

Since in practice the coefficient functions are evaluated in the full theory including all quarks and vector bosons at a scale $\mathcal{O}(M_W)$, and the matrix elements of operators are calculated at a scale $\mathcal{O}(1\,GeV)$, the two scales have to be matched using the renormalization group equation. Therefore in vector notation the effective hamiltonian takes the form

$$\mathcal{H}_{eff} = \frac{G_F}{\sqrt{2}} \vec{Q}^T(\mu)\, \hat{U}(\mu, M_W)\, \vec{C}(M_W) \quad (2)$$

where $\hat{U}(\mu, M_W)$ is the evolution matrix between the scales M_W and μ. The coefficients $\vec{C}(M_W)$ are calculable in perturbation theory, $\hat{U}(\mu, M_W)$ comprises the anomalous dimension matrix of $\vec{Q}$, and the matrix elements of the operators $\vec{Q}$ between physical states, which appear in a typical decay, have to be calculated by means of non-perturbative methods, such as, lattice, $1/N$-expansion, chiral perturbation theory, or QCD sum rules.

An important fact to notice is that all three parts in general depend on the renormalization scheme and this dependence is cancelled at the upper end of the evolution in the product $\hat{U} \cdot \vec{C}$ and at the lower end in the product $\vec{Q}^T \cdot \hat{U}$, leaving $\mathcal{H}_{eff}$ scheme independent as it should be. If the coefficient functions as well as the anomalous dimensions are calculated at the next-to-leading order, the cancellation at the upper end of the evolution can be made explicit, however at the present state of calculating hadronic matrix elements, this is not the case for the lower end of the evolution. This remaining scheme dependence can not be cancelled, unless the hadronic matrix elements are calculated in exactly the same scheme as the coefficient functions and the anomalous dimensions, i.e., matching has to be performed not only in the scales, but also in the schemes.

In ref. [2], a rotation to both operators and coefficient functions was suggested, in order to make both quantities scheme independent. Nevertheless, we should remark that this rotation is not unique and would still require the evaluation of matrix elements in the rotated basis. Similar comments apply to the calculation of the B-parameter, relevant for $K^0 - \bar{K}^0$ and $B^0 - \bar{B}^0$ mixing, although in this case, the scheme dependence is less severe [6].

In our calculation of the weak effective hamiltonian we explicitly used two different schemes, namely minimal subtraction with a usual anticommuting γ_5 and the 't Hooft–Veltman prescription with a non-anticommuting γ_5. This allowed us to check that the cancellation of the scheme dependence at the upper end of the evolution does

Table 1. Coefficient functions relevant for the $\Delta I = 1/2$ rule and ε'/ε

	m_t	l.o.	Anatomy	NDR	SI
z_1	150	-0.714	-0.702	-0.485	-0.903
z_2	150	1.408	1.396	1.262	1.554
y_6	150	-0.108	-0.105	-0.098	-0.129
y_8/α	100	-0.029	-0.034	-0.060	-0.045
y_8/α	150	-0.029	0.109	0.125	0.110
y_8/α	200	-0.029	0.215	0.215	0.200

indeed occur, and the residual scheme dependence of the coefficient functions appears only at the lower end of the evolution.

Selected results

In table 1, we present selected preliminary results, relevant for the $\Delta I = 1/2$ rule and ε'/ε, in order to illustrate some of the effects of the next–to–leading order calculation. The coefficient functions are defined as

$$C_i(\mu) = z_i(\mu) + \tau\, y_i(\mu)\,;\quad \tau = -\frac{V_{td}V_{ts}^*}{V_{ud}V_{us}^*}\,,\quad (3)$$

with V_{ij} being elements of the Cabbibo–Kobayashi–Maskawa matrix. We give their values for the leading order, for the treatment according to ref. [7], taking into account top quark effects in the initial condition at the scale M_W, for the next–to–leading order in the NDR scheme, and the scheme independent values for the rotated basis as described in ref. [2]. The coefficient functions are evaluated at $\mu = 1\,GeV$ with $\Lambda_{\overline{MS}}^{(4)} = 300\,MeV$. For the coefficient y_8, the top mass dependence is explicated because of the strong dependence, whereas the other coefficients only show a very weak dependence on m_t.

The following observations can be made on the basis of table 1 and the coefficient functions not shown explicitly:

- Already at $\mu = 1\,GeV$ can the next–to–leading corrections be as large as 20%–30%, indicating that the perturbative evolution should be stopped at $1\,GeV$, and a perturbative evaluation of the coefficients below this scale can not be trusted,

- All four coefficients show a strong dependence on the renormalization scheme. In the case of z_1, z_2, and y_6, this dependence is such that the absolute values of the coefficients are reduced with respect to the leading order in the NDR scheme and are enhanced for the scheme invariant combination. This clearly demonstrates the urgent need for matrix elements, also showing scheme dependence.

- For y_8, including the m_t dependence in the initial condition incorporates the main effect of the next–to–leading order. Nevertheless, for a consistent analysis all next–to–leading contributions have to be included.

- The ratio y_8/y_6, which reflects the importance of electroweak penguins for ε'/ε, is reduced if we consider the scheme independent coefficients. This would be analogous to an observation made in ref. [8] on the basis of the matrix elements in the $1/N$–expansion, namely shifting a possible cancellation between QCD and electroweak penguins to higher values of $m_t \approx 270\,GeV$, as compared to $m_t \approx 220\,GeV$, found in ref. [7].

Conclusions

The next–to–leading order corrections to the $\Delta S = 1$ effective hamiltonian have been calculated in the presence of electroweak penguin operators.

It is found that the perturbative evolution of the coefficient functions should be stopped at $\mu \approx 1\,GeV$, in order to avoid large corrections. At the lower end of the evolutiuon, the resulting coefficient functions depend in general on the renormalization scheme and the treatment of γ_5. This dependence can only be cancelled if the corresponding matrix elements of operators are calculated in the same scheme.

It is evident that the poor knowledge of hadronic matrix elements still is the main obstacle, before considerable progress in the theory of weak decays can be made.

Acknowledgement

The author would like to thank the organizers of the conference for the pleasant atmosphere, and his collaborators for most enjoyable discussions.

REFERENCES

1. A.J. Buras and P.H. Weisz, "QCD Non-leading Corrections to Weak Decays in Dimensional Regularization and 't Hooft–Veltman Schemes", *Nucl. Phys.* B333, pp. 66–99, (1990).

2. A.J. Buras, M. Jamin, M.E. Lautenbacher and P.H. Weisz, "Effective Hamiltonians for $\Delta S = 1$ and $\Delta B = 1$ Non-Leptonic Decays beyond the Leading Logarithmic Approximation", *Nucl. Phys.* B370, pp. 69–104, (1992), Addendum *ibid.* B375, pp. 501–502, (1992).

3. A.J. Buras, M. Jamin and M.E. Lautenbacher, "Two–Loop Anomalous Dimension Matrix for $\Delta S = 1$ Weak Non-Leptonic Decays I: $\mathcal{O}(\alpha_s^2)$", *Technical University Munich preprint* TUM-T31-18/92, (1992).

4. A.J. Buras, M. Jamin and M.E. Lautenbacher, "Two–Loop Anomalous Dimension Matrix for $\Delta S = 1$ Weak Non-Leptonic Decays II: $\mathcal{O}(\alpha\alpha_s)$", *Technical University Munich preprint* TUM-T31-30/92, (1992).

5. A.J. Buras, M. Jamin and M.E. Lautenbacher, "Effective Hamiltonians for $\Delta S = 1$ and $\Delta B = 1$ Non-Leptonic Decays beyond Leading Logarithms in the Presence of Electroweak Penguins", *Technical University Munich preprint*, in preparation.

6. A.J. Buras, M. Jamin and P.H. Weisz, "Leading and Next-To-Leading QCD Corrections to ε–parameter and $B^0 - \bar{B}^0$ Mixing in the Presence of a Heavy Top Quark", *Nucl. Phys.* B347, pp. 491–536, (1990).

7. G. Buchalla, A.J. Buras and M.K. Harlander, "The Anatomy of ε'/ε in the Standard Model", *Nucl. Phys.* B337, pp. 313–362, (1990).

8. J. Heinrich, E.A. Paschos, J.M. Schwarz and Y.L. Wu, "Accuracy of the Predictions for Direct CP Violation", *Phys. Lett.* B279, pp. 140–144, (1992).

MEASUREMENTS OF THE KAON CONTENT IN TAU DECAYS

Michael T. Ronan*
Physics Division
Lawrence Berkeley Laboratory
Berkeley, California, USA, 94720

Abstract

Results on measurements of the kaon content in one-prong and three-prong τ decays are presented for data taken by the TPC/2γ detector at PEP. Using a self-consistent procedure to measure exclusive and inclusive decays, the one-prong analysis extends previous work to kaon decay modes. Three-prong results on $K\pi\pi$, $KK\pi$ and KKK decay modes provide improved branching ratios and a first look at strange axial-vector couplings in τ decays.

INTRODUCTION

The τ one-prong decay problem[1] resides in the discrepancy between the measured topological one-prong branching ratio (85-86%) and the sum of all measured and predicted one-prong exclusive decay modes (~80%). The branching ratios obtained from simultaneous measurements of both inclusive and exclusive τ decays[2], confirm earlier measurements[3] of a larger than expected one-prong plus multiple neutral decay mode, and report a larger three "pion" branching ratio. The latter contributes by both decreasing the topological one-prong branching ratio and by helping to explain the above excess of multiple neutral decays. The balance of the resulting difference divides into several modes at the fraction of a per cent level, well within the systematic error limits of the combined results from earlier experiments.

Using the excellent particle identification capabilities of the PEP-4/9 Time Projection Chamber (TPC), we are in the process of measuring the one-prong decays in a similar manner and extending the analysis to modes including kaons. In this report, I present preliminary one-prong branching ratios from our simultaneous measurements including kaon categories, and an unpublished result[4] on the kaon content in three-prong decays. Finally, I'll comment on possible strange axial-vector resonant contributions to the measured $\tau \to \nu_\tau K\pi\pi$ decays.

DATA SAMPLES

Data were recorded using the TPC/2γ Detector Facility at the SLAC e^+e^- storage ring PEP during 1982-1983 and 1984-1986 with low-field (4 kG) and high-field (13.2 kG) magnets, respectively. The detector and trigger systems are described in detail elsewhere[5].

In our event selection process, we make topological and kinematic cuts to obtain a clean separation into standard Tau 1+1 and 1+3 samples, where the numbers refer to the number of charged particles from the decays of

---

*Representing the TPC/Two-Gamma Collaboration

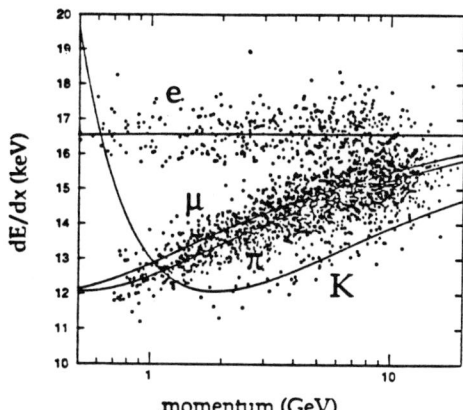

Figure 1. dE/dx ionization loss vs. momentum for 1-prong Tau decays.

the produced tau pair. To reject Bhabha, multihadron and 2-photon backgrounds, we make additional requirements on the scalar momentum sum and on the invariant mass of the 3 non-isolated tracks. In the final analysis we remove any remaining radiative ee or $\mu\mu$ events.

ONE-PRONG TAU DECAYS

One prong τ decays are analyzed in both the Tau 1+1 and 1+3 samples. The dE/dx particle identification for the resulting 2569 one-prong decay candidates observed in our 70 pb^{-1} high-field sample is shown in Fig. 1. We use our electromagnetic calorimetry, muon detection system and excellent TPC dE/dx particle identification to classify tracks into the four possible species: e,μ,π,K.

Electrons tracks are identified as being associated with calorimeter energy deposits greater than 60% of the momentum measured in the TPC. We determine that true electrons satisfy this requirement with 99% efficiency, while pions satisfy this criteria less than 2% of the time for momenta greater than 2 GeV. The additional constraints imposed by our dE/dx measurements reduces the overall pion misidentification probability to well less than 1%.

Muons with momenta greater than 2 GeV are detected in the barrel region as having at least 3 out of 4 drift layers hit. The muon identification efficiency is determined to be 99% and the pion misidentification due to decay or punch through to be less than 2-5% (larger at higher momenta).

Pions and kaons are separated using the measured dE/dx for each track. The expected ionization loss and resolution are used to calculate the efficiency and misidentification on a track-by-track basis. Our classification requirements and overall three sigma separation give average efficiencies of 98% and 85% for pions and kaons, respectively, and a probability for a pion to be misidentified as a kaon of less than a few per cent.

To extract branching ratios, we form a four-component column vector M from the number of one-prong tracks which are classified as e,μ,π or K. The actual number, N, of one-prong τ decays in each mode is related by

$$M = QN \qquad (1)$$

where Q is the track misidentification matrix. As a function of the momentum of each track, we evaluate the misidentification probabilities and invert $Q(p)$ to obtain a track weight, $\mathbf{w} = Q^{-1}(p)\,M$. Summing over all events we obtain values for N_i, the total number of actual τ decays into the ith mode.

The weighted momentum spectrum for inclusive $\pi^{\pm}$ is displayed in Fig. 2a. We find excellent agreement between our measurements and Monte Carlo predictions based on the KORAL-B tau generator[6] using previously measured decay branching ratios including π^0's. In Fig. 2b we find our weighted K momentum spectrum is consistent with present modelling of $\tau \to \nu_\tau K$ and $\tau \to \nu_\tau K\pi^0$.

The exclusive one-prong branching fractions are determined relative to our measured one-prong topological branching ratio to cancel luminosity uncertainties. We obtain the

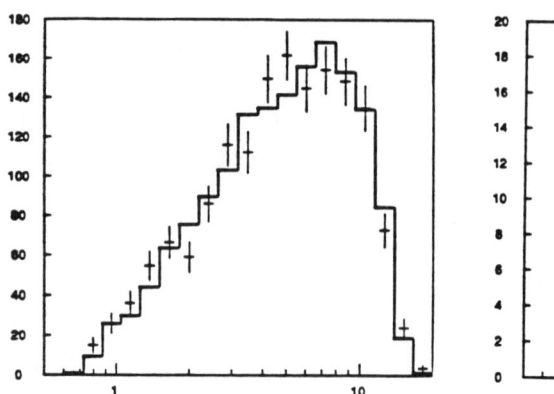
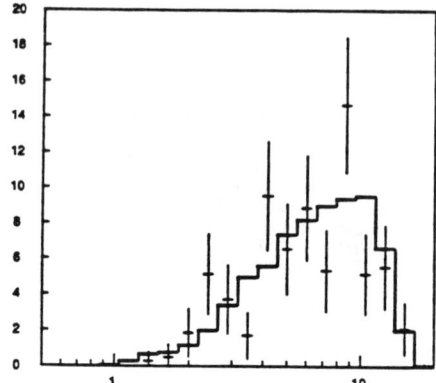

Figure 2. Weighted number of (a) pion and (b) kaon events vs. $\log_{10}$ of momentum in GeV/c for 1-prong Tau decays.

following preliminary results

$$BR(\tau \to \nu_\tau e \nu_e) = 18.3 \pm 1.6 \%$$
$$BR(\tau \to \nu_\tau \mu \nu_\mu) = 17.4 \pm 1.4 \%$$
$$BR(\tau \to \nu_\tau \pi + \geq 0 \; neutrals) = 48.0 \pm 1.6 \%$$
$$BR(\tau \to \nu_\tau K + \geq 0 \; neutrals) = 1.6 \pm 0.4 \%,$$

where the errors are statistical and systematic[7].

THREE-PRONG TAU DECAYS

In hadronic decays of the τ, each good track is either a charged pion or kaon. The dE/dx particle identification for three-prong decays, with electrons from photon conversion removed, is shown in Fig. 3.

Based on individual track assignments similar to those described above, each event is counted in an eight-component vector, M, representing the different possible permutations for the decay modes: $\nu_\tau \pi^- \pi^+ \pi^-$, $\nu_\tau K^- \pi^+ \pi^-$, $\nu_\tau K^- K^+ \pi^-$ or $\nu_\tau K^- K^+ K^-$, where in each case additional neutrals could be present. The true population of the different modes, N, is related by the equation

$$M = PN \quad (2)$$

where P is the <u>event</u> misclassification matrix. This six-dimensional tensor can be expressed as an outer product of track misidentification

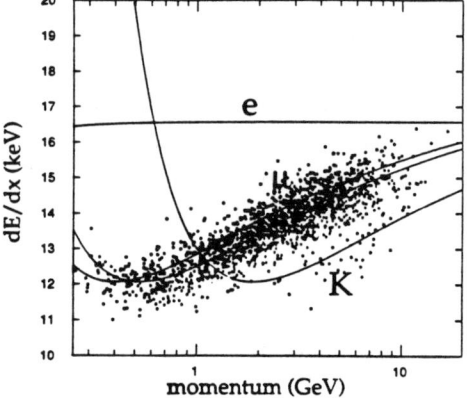

Figure 3. dE/dx ionization loss vs. momentum for 3-prong Tau decays.

matrices, Q:

$$P^a \equiv Q^a(p_1) \otimes Q^a(p_2) \otimes Q^a(p_3). \quad (3)$$

where a is an event label and p_i are the measured parameters for the ith track in the event. Inverting the Q matrices for each track we calculate the inverse of the misclassification matrix to determine an estimator of the true identity of each event, $\mathbf{w} = (P^a)^{-1} M$, which is used in obtaining weighted distributions and which upon summing gives the total numbers of events in each mode.

To provide a check of the event counting, we also perform an extended maximum likelihood fit of the track dE/dx distributions in each event to different decay mode hypotheses. We obtain the following number of events in each decay mode: 470.3 ± 22.0 $\pi\pi\pi$, $23.6\,^{+6.6}_{-5.7}$ $K\pi\pi$, $4.3\,^{+2.7}_{-1.9}$ $KK\pi$, 0.0 ± 0.5 KKK events.

Backgrounds are estimated to be less than 0.5 and 0.3 events in the $K\pi\pi$ and $KK\pi$ channels, respectively, and less than 4 events in the remaining 3π mode. Our estimated systematic error of 20% in these three-prong measurements is mainly due to uncertainties in dE/dx parameterization and in estimation of event selection efficiencies.

To cancel luminosity uncertainties, we determine the exclusive three-prong branching fractions relative to our measured three-prong topological branching ratio obtaining

$$BR(\tau \to \nu_\tau \pi\pi\pi) = 13.7 \pm 0.7 \%$$
$$BR(\tau \to \nu_\tau K\pi\pi) = 0.78\,^{+0.22}_{-0.19} \%$$
$$BR(\tau \to \nu_\tau KK\pi) = 0.18\,^{+0.11}_{-0.08} \%$$
$$BR(\tau \to \nu_\tau KKK) < 0.26 \% \quad (95\% \text{ CL})$$

Our result for the $\nu_\tau KK\pi$ decay mode of the τ agrees with a previous result from DELCO[8], while our measured $BR(\tau \to \nu_\tau K\pi\pi)$ is notably higher than their published result of $0.22\,^{+0.16}_{-0.13}\%$.

In Fig. 4 we plot the weighted $K\pi\pi$ invariant mass spectrum. We see no evidence for a substantial contribution from the lower-mass, strange axial vector meson $K_1(1270)$. However, we do observe a slight excess of events at a mass corresponding to the $K_1(1400)$; analysis of this aspect of our data is continuing.

CONCLUSIONS

In a self-consistent τ decay analysis: For one-prong τ decays, we measure lepton branching ratios which are in good agreement with previous measurements and individual inclusive π and K branching ratios of $B_\pi = 48.0 \pm 1.6\%$ and $B_K = 1.6 \pm 0.4\%$. For the three-prong decays we obtain improved measurements of $B_{K\pi\pi(\pi^0)} = 0.78 \pm 0.2\%$, and $B_{KK\pi(\pi^0)} = 0.18 \pm 0.1\%$. Our analysis of the

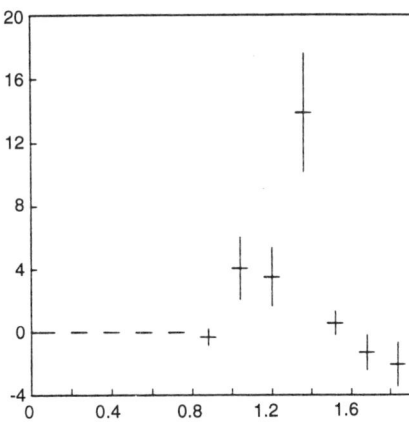

Figure 4. Number of weighted events vs. $K\pi\pi$ invariant mass in GeV/c^2.

neutral component and of the resonant contributions to strange three-prong decays is continuing.

I would like to thank the organizers for an enjoyable, well-organized conference.

REFERENCES

1. F.J. Gilman and S.H. Rhie, Phys. Rev. **D31**, 1066 (1985).
2. D. Decamp et al. (ALEPH Collaboration), Z. Phys. C. **54**, 211 (1992).
3. For a recent review, see K.K. Gan, Proceedings of the Workshop on Tau Lepton Physics, Edited by M. Davier and B. Jean-Marie, Orsay, France, Sept. 24-27, 1990.
4. J.J. Eastman, PhD thesis, University of California, Berkeley, 1990. LBL-30035.
5. H. Aihara et al., LBL-23737 revised, Lawrence Berkeley Laboratory, 1992.
6. Koral-B, Vers. 2.1, S. Jadach and Z. Was.
7. Branching ratios obtained here are statistically independent of our previous measurements, H. Aihara et al. (TPC/2γ Collaboration), Phys. Rev. **D35**, 1553 (1987).
8. G.B. Mills et al. (DELCO Collaboration), Phys. Rev. Lett. **54**, 624 (1985).

MEASUREMENT OF THE MASS OF THE τ LEPTON*

F. C. Porter, representing the BES Collaboration[1]
Lauritsen Laboratory of High Energy Physics
California Institute of Technology
Pasadena, CA 91125

Abstract

The mass of the τ lepton has been measured at the Beijing Electron Positron Collider using the Beijing Spectrometer. A search near threshold for $e^+e^- \to \tau^+\tau^-$ was performed, requiring that one τ decay via $\tau \to e\nu\bar\nu$, and the other via $\tau \to \mu\nu\bar\nu$. The mass value, obtained from a fit to the energy dependence of the $\tau^+\tau^-$ cross section, is $m_\tau = 1776.9^{+0.4}_{-0.5} \pm 0.2$ MeV.

For a conventional charged lepton l, the electronic branching ratio B^e_l, lifetime τ_l, mass m_l, and weak coupling strength $G_{l \to e\nu\bar\nu}$, are related by

$$\frac{B^e_l}{\tau_l} = \frac{G^2_{l \to e\nu\bar\nu}}{192\pi^3} m_l^5, \qquad (1)$$

up to small radiative and electroweak corrections. Equation (1) then implies the following relationship among the above parameters for the τ and μ leptons:

$$\left(\frac{G_{\tau \to e\nu\bar\nu}}{G_{\mu \to e\nu\bar\nu}}\right)^2 = \left(\frac{m_\mu}{m_\tau}\right)^5 \frac{B^e_\tau}{B^e_\mu} \frac{\tau_\mu}{\tau_\tau}. \qquad (2)$$

Particle Data Group[2] (PDG) averages for the above quantities yield $(G_{\tau \to e\nu\bar\nu}/G_{\mu \to e\nu\bar\nu})^2 = 0.941 \pm 0.025$, implying a 2.4 standard deviation disagreement with lepton universality.[3]

*This work was supported in part by the National Natural Science Foundation of China under Contract No. 19290400, by the U. S. Department of Energy under Contract Nos. DE-FG03-92ER40701, DE-FG03-91ER40679, DE-FG02-91ER40676, DE-AC02-76ER03069, DE-AC35-89ER40486, DE-AC03-76SF00515, and DE-FG05-92ER40736, by the Texas National Research Laboratory Commission (Rocky Mountain Consortium for High Energy Physics) under Contract Nos. RGFY91B5 and RGFY92B5, and by the National Science Foundation.

A measurement of the $e^+e^- \to \tau^+\tau^-$ production cross section in the region most sensitive to the τ mass—a few MeV around threshold—provides the means to measure the τ mass with greatly improved precision. This paper presents such a measurement using the Beijing Spectrometer (BES) at the Beijing Electron Positron Collider (BEPC). The $\tau^+\tau^-$ events are identified using the $e\mu$ topology, which provides the best combination of high detection efficiency and low background; the mass value is obtained from a fit to the energy dependence of the cross section. The measurement is independent of the ν_τ mass.

The BEPC[4] operates in the 3 to 5 GeV center-of-mass energy range. Near $\tau^+\tau^-$ threshold, the peak luminosity is 5×10^{30} cm^{-2}s^{-1}, and the spread in the center-of-mass energy of the collider is ≈ 1.4 MeV. The absolute energy scale and energy spread are determined by interpolation between the results of repeated scans of the J/ψ and $\psi(2S)$ resonances.

The BES[4] is a solenoidal detector with a 0.4 T magnetic field. Charged track reconstruction is performed by means of a drift chamber with a solid angle coverage of 85% of 4π. The momentum resolution is $\sigma_p/p = 0.021\sqrt{1+p^2}$ (p in GeV/c). Measurements of

© 1993 American Institute of Physics

dE/dx with resolution 8.5% aid particle identification. Scintillation counters measure the time-of-flight of charged particles over 76% of 4π with a Bhabha resolution of 330 ps. A twelve-radiation-length Pb/gas electromagnetic calorimeter covering 80% of 4π achieves energy resolution $\sigma_E/E = 0.25/\sqrt{E(\text{GeV})}$, and spatial resolution $\sigma_\phi = 4.5$ mrad, $\sigma_z = 2$ cm. A three-layer iron flux return instrumented for muon identification yields spatial resolutions $\sigma_z = 5$ cm, $\sigma_{r\phi} = 3$ cm over 68% of 4π for muons with momentum greater than 550 MeV/c.

The event selection for $e\mu$ candidates requires: (1) exactly two oppositely-charged tracks, with momentum between 350 MeV/c and the maximum for an electron from τ decay, (2) each track's point of closest approach to the intersection point to satisfy $|x| < 1.5$ cm, $|y| < 1.5$ cm and $|z| < 15$ cm, (3) $2.5° < \theta_{\text{acol}} < 177.5°$, $\theta_{\text{acop}} > 10°$, and $(\theta_{\text{acol}} + \theta_{\text{acop}}) > 50°$ (θ_{acop} is the angle between the planes spanned by the beam direction and the momentum vector of e and μ respectively; θ_{acol} is the angle between the momentum vectors of e and μ.), (4) no isolated photons (having a calorimeter energy > 60 MeV and separated from the nearest charged track by $> 12°$), (5) one track identified as a muon in the muon-counter, with calorimeter energy < 500 MeV, and the other track identified as an electron using a combination of calorimeter, dE/dx and time-of-flight information.

Monte Carlo simulations yield a detection efficiency of $\approx 14\%$ for these selection criteria. The background is estimated by applying the same requirements to five million events taken at the J/ψ energy; seven events meet these criteria, corresponding to a background of 0.12 events in the entire $\tau^+\tau^-$ sample.

The likelihood function used to estimate the τ mass incorporates the $\tau^+\tau^-$ cross section near threshold. Including the center-of-mass energy spread Δ, initial state radiation $F(x, W)$,[5] and vacuum polarization corrections $\Pi(W)$,[6] the cross section is

$$\sigma(W, m_\tau) = \frac{1}{\sqrt{2\pi}\Delta} \int_0^\infty dW' e^{\frac{-(W-W')^2}{2\Delta^2}}$$
$$\int_0^{1-\frac{4m_\tau^2}{W'^2}} dx F(x, W') \sigma_1(W'\sqrt{1-x}, m_\tau), \quad (3)$$

where σ_1 is

$$\sigma_1(W, m_\tau) = \frac{4\pi\alpha^2}{3W^2} \frac{\beta(3-\beta^2)}{2} \frac{F_c(\beta)F_r(\beta)}{[1-\Pi(W)]^2}, \quad (4)$$

W is the center-of-mass energy, and $\beta = \sqrt{1-\left(\frac{2m_\tau}{W}\right)^2}$. The Coulomb interaction and final state radiation corrections are described by the functions $F_c(\beta)$ and $F_r(\beta)$.[7]

The likelihood function is a product of Poisson distributions, one for each center-of-mass energy. At each point, the number of expected $e\mu$ events $\langle N \rangle$, is given by:

$$\langle N \rangle = [\epsilon B \sigma(W, m_\tau) + \sigma_B]\mathcal{L}. \quad (5)$$

Here ϵ is the detection efficiency, B is the product branching fraction for $\tau^+\tau^-$ to $e\mu$, $\mathcal{L}$ is the integrated luminosity, and σ_B is the effective background cross section estimated from the J/ψ data sample ($\sigma_B = 0.024$ pb).

Since the range of center-of-mass energies where the $\tau^+\tau^-$ cross section is most sensitive to the τ mass is of the order of the beam energy spread around $\tau^+\tau^-$ threshold, it is important to devise a running strategy to maximize the integrated luminosity in this region. The beam energy is set initially assuming the world average for the τ mass, in this case the PDG value 1784.1 MeV. Then, after each 250–400 nb^{-1} of integrated luminosity, a new estimate of the mass is made using all the data accumulated to that point; in this way a new prediction of the most sensitive energy at which to run is obtained. The energy is changed to this new value if the difference is more than the BEPC step size ($\Delta W \approx 0.4$ MeV). Following this strategy, an integrated luminosity of ≈ 4.3 pb^{-1} has been accumulated at ten energies within a range of 24 MeV.

The sequence of energies is shown in Figure 1; the corresponding data are summarized in Table 1. The ten-step search yielded seven $e\mu$ events. The eleventh and twelfth points in Table 1, taken well above threshold where the cross section is a slowly varying function of the τ mass, provide an improved estimate of the absolute $\tau^+\tau^-$ cross section.

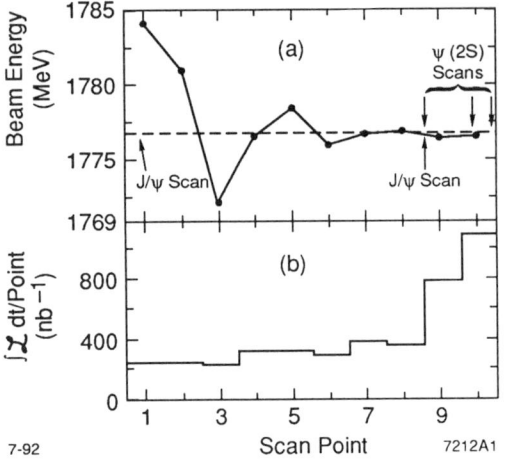

Figure 1. (a) The convergence of the predicted mass with each consecutive scan point; (b) The integrated luminosity accumulated at each point.

In order to account for uncertainties in the efficiency ϵ, the branching fraction product and the luminosity, ϵ is treated as a free parameter in a two-dimensional maximum-likelihood fit for m_τ and ϵ to the data of Table 1. The estimates obtained are $m_\tau = 1776.9$ MeV and $\epsilon = 14.1\%$. The uncertainty in ϵ is equivalent to the uncertainty in the absolute normalization, and is treated as a source of systematic error. The statistical error in m_τ, $^{+0.4}_{-0.5}$ MeV, is determined from the one-parameter likelihood function with ϵ fixed to 14.1% (Figure 2c). The efficiency-corrected cross section data as a function of corrected beam energy and the curve which results from the likelihood fit are shown in Figure 2ab.

Table 1. A chronological summary of the $\tau^+\tau^-$ data.

Scan Point	W/2 (MeV)	Δ (MeV)	$\mathcal{L}$ (nb^{-1})	N ($e\mu$ events)
1	1784.19	1.34	245.8	2
2	1780.99	1.33	248.9	1
3	1772.09	1.36	232.8	0
4	1776.57	1.37	323.0	0
5	1778.49	1.44	322.5	2
6	1775.95	1.43	296.9	0
7	1776.75	1.47	384.0	0
8	1776.98	1.47	360.8	1
9	1776.45	1.44	794.1	0
10	1776.62	1.40	1109.1	1
11	1799.51	1.44	499.7	5
12	1789.55	1.43	250.0	2

Four independent sources of systematic error are considered: uncertainties in the product $\epsilon B \mathcal{L}$, in the absolute beam energy scale, in the beam energy spread, and in the background.

The systematic uncertainty in $\epsilon B \mathcal{L}$ is determined by fixing m_τ at its best estimate value and finding the values of ϵ corresponding to $\pm 1\sigma$ variations in the likelihood function; these efficiencies are 18.3% and 10.6%. Fixing the efficiency to each of these values in turn and fitting for m_τ yields changes in the predicted mass of $\Delta m_\tau = ^{+0.16}_{-0.20}$ MeV. The energy scale is determined from several scans of the J/ψ and $\psi(2S)$ performed during the search (see Figure 1). Assuming a linear relation between measured energy W_M, and the corrected value W, the latter is given by :

$$W = T_\psi + (W_M - M_\psi)\left(\frac{T_{\psi'} - T_\psi}{M_{\psi'} - M_\psi}\right) \quad (6)$$

in the notation of Table 2. At $\tau^+\tau^-$ threshold the resulting mass scale correction is $W - W_M = -0.74$ MeV. The reproducibility of the fits to these scans, together with the other uncertainties listed in Table 2, yields a systematic uncertainty of $\Delta m_\tau = \pm 0.09$ MeV.

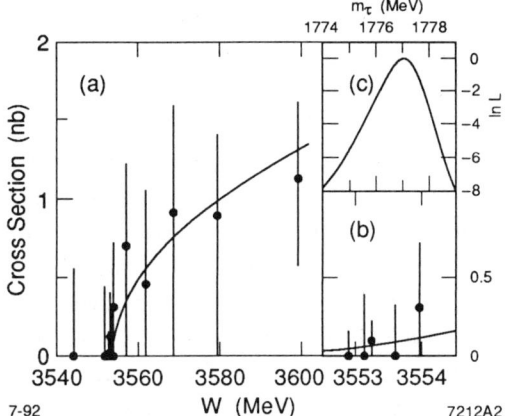

Figure 2. (a) The center-of-mass energy dependence of the $\tau^+\tau^-$ cross section resulting from the likelihood fit (curve), compared to the efficiency-corrected data. The error bar on each data point is computed by integrating the Poisson likelihood function to obtain the interval containing 68% of the area. It should be emphasized that the curve does not result from a direct fit to these data points. (b) An expanded version of (a), in the immediate vicinity of $\tau^+\tau^-$ threshold. (c) The dependence of the logarithm of the likelihood function on m_τ, with efficiency fixed at 14.1%.

Fits to the two resonances are also used to measure the beam energy spread and its variation with center-of-mass energy and beam current. The uncertainty in center-of-mass energy spread is ± 0.08 MeV, yielding a systematic error $\Delta m_\tau = \pm 0.02$ MeV.

Table 2. Contributions to the uncertainty in the energy scale.

Quantity	Error (MeV)
W_M - BEPC center-of-mass energy	0.10
M_ψ - BEPC value for J/ψ mass	0.18
$M_{\psi'}$ - BEPC value for $\psi(2S)$ mass	0.15
T_ψ - PDG value[2] for J/ψ mass	0.09
$T_{\psi'}$ - PDG value[2] for $\psi(2S)$ mass	0.10

The final source of systematic error, uncertainty in the background, is estimated from the 1σ Poisson errors on the seven J/ψ background events and the uncertainty in the hadronic cross section at $\tau^+\tau^-$ threshold. The resulting uncertainty is $\Delta m_\tau = \pm 0.01$ MeV.

These independent systematic errors are added in quadrature to yield a total systematic error of $\Delta m_\tau = ^{+0.18}_{-0.22}$ MeV.

In conclusion, using a maximum likelihood fit to $\tau^+\tau^-$ cross section data near threshold, the mass of the τ lepton has been measured as $m_\tau = 1776.9^{+0.4}_{-0.5} \pm 0.2$ MeV, where the first error is statistical and the second systematic. This result is 7.2 MeV below the PDG average[2] and has significantly smaller errors. Inserting this new value in equation 2, the coupling strength ratio becomes 0.960 ± 0.024 so that the deviation from lepton universality is reduced from 2.4 to 1.7 standard deviations (see Figure 3). Other new results

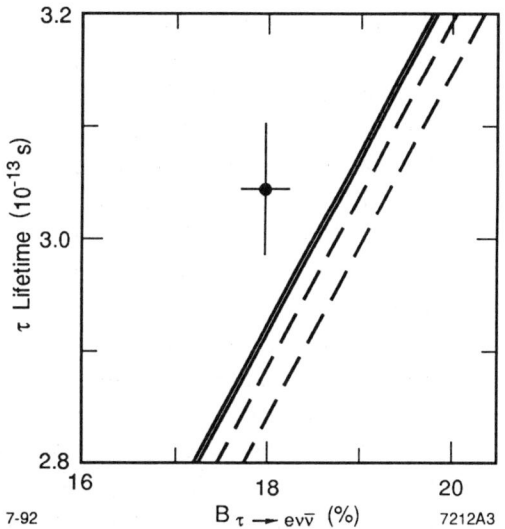

Figure 3. The variation of τ_τ with B_τ^e, given by Eqn. 1 under the assumption of lepton universality; the $\pm 1\sigma$ bands obtained using m_τ from this experiment (solid lines) and using the PDG value (dashed lines) are shown in comparison to the point corresponding to the PDG values (1σ error bars).

on the τ parameters from Argus, CLEO, and LEP have been presented at this conference which modify this comparison – see the summary by Rolandi[8] for the situation including all the new results. It may be noted that this new result for m_τ yields a reduction in the upper limit on m_{ν_τ}, as well as a small reduction in the measured τ lifetime.

REFERENCES

1. The BES collaboration consists of J. Bai, O. Bardon, R. Becker-Szendy, T. Burnett, J. Campbell, S. M. Chen, S. J. Chen, Y. Chen, Z. Cheng, J. Coller, R. Cowan, H. Cui, X. Cui, H. Ding, Z. Du, W. Dunwoodie, C. Fang, M. Fero, M. Gao, S. Gao, W. Gao, Y. Gao, J. Gu, S. Gu, W. Gu, Y. N. Guo, Y. Y. Guo, Y. Han, M. Hatanaka, J. He, D. Hitlin, G. Hu, T. Hu, D. Huang, Y. Huang, J. Izen, Q. Jia, C. Jiang, Z. Jiang, A. Johnson, L. Jones, M. Kelsey, Y. Lai, P. Lang, A. Lankford, F. Li, J. Li, P. Li, Q. Li, R. Li, W. Li, W. D. Li, W. G. Li, Y. Li, S. Lin, H. Liu, Q. Liu, R. Liu, Y. Liu, B. Lowery, J. G. Lu, D. H. Ma, E. C. Ma, J. M. Ma, M. Mandelkern, H. Marsiske, H. Mao, Z. Mao, X. Meng, H. Ni, L. Pan, J. Panetta, F. Porter, E. Prabhakar, N. Qi, Y. Que, J. Quigley, G. Rong, B. Schmid, J. Schultz, J. Shank, Y. Shao, D. Shen, H. Sheng, H. Shi, A. Smith, E. Soderstrom, X. Song, D. Stoker, H. Sun, J. Synodinos, W. Toki, G. Tong, E. Torrence, L. Wang, M. Wang, P. Wang, P. L. Wang, T. Wang, Y. Wang, J. Whitaker, R. Wilson, W. Wisniewski, X. Wu, D. Xi, X. Xia, P. Xie, X. Xie, R. Xu, Z. Xu, S. Xue, R. Yamamoto, J. Yan, W. Yan, C. M. Yang, C. Y. Yang, H. Yao, M. Ye, S. Ye, Z. Yu, B. Zhang, C. Zhang, D. Zhang, H. L. Zhang, H. Y. Zhang, J. Zhang, L. S. Zhang, S. Q. Zhang, Y. Zhang, D. Zhao, M. Zhao, P. Zhao, W. Zhao, J. Zheng, L. Zheng, Z. Zheng, G. Zhou, H. Zhou, L. Zhou, L. Zhou, X. Zhou, Y. Zhou, Q. Zhu, Y. C. Zhu, Y. S. Zhu, and G. Zioulas from the following institutions: Institute of High Energy Physics, Beijing; Boston University; California Institute of Technology; University of California, Irvine; Colorado State University, Fort Collins; Massachusetts Institute of Technology; Stanford Linear Accelerator Center; Superconducting Super Collider Laboratory; University of Texas at Dallas, Richardson; University of Washington, Seattle. This result has been submitted to *Phys. Rev. Lett.*, preprint SLAC-PUB-5870, BEPC-EP-92-01, July 1992.

2. K. Hikasa et al., "Review of Partical Properties," *Phys. Rev.* D45, Part II, (1992).

3. W. J. Marciano, "The Tau Decay Puzzle," *Phys. Rev.* D45, 721, (1992). The measurements quoted in this reference have been superseded by Reference 2.

4. M. H. Ye and Z. P. Zheng, in *Proceedings of the 1989 International Symposium on Lepton and Photon Interactions at High Energies*, Stanford, 1989, 122.

5. É. A. Kuraev and V. S. Fadin, "On Radiative Corrections to e^+e^- Single Photon Annihilation at Large Energy," *Yad. Fiz.* 41, 733, (1985).

6. F. A. Berends and G. J. Komen, "Radiative Corrections to Bhabha Scattering and Mu Pair Production From the Hadronic Vacuum Polarization," *Phys. Lett.* 63B, 432, (1976).

7. M. B. Voloshin, "Topics in Tau Physics at a Tau Charm Factory," Univ. of Minnesota preprint TPI-MINN-89-33-T, Nov. 1989.

8. L. Rolandi, in these proceedings.

NEW RESULTS ON τ LEPTON FROM CLEO

K.K. Gan*
Department of Physics
The Ohio State University
174 W. 18th Avenue
Columbus, OH 43210

Abstract

Several properties of the τ lepton have been measured precisely in e^+e^- annihilation by the CLEO II experiment at the Cornell Electron Storage Ring (CESR). This includes the τ mass and the branching ratios into electron and one charged particle plus multiple π^0's. The implication of these results on the 1-prong decay problem will be discussed.

INTRODUCTION

The τ lepton has been the subject of intense investigation in recent year due to the perplexing problem in the decay into one-charged-particle final states[1]. In the comparison of the measured inclusive and the sum of the measured exclusive decay branching ratios, a few percent of the decays is unaccounted for. In addition, there is a discrepancy in the lifetime that may be related to the problem. The measured lifetime is somewhat longer than that expected from the measured leptonic branching ratio assuming lepton universality[2]. Since the lifetime depends on the τ mass to the fifth power, a more precise measurement of the mass may elucidate the problem. With a large data sample and an electromagnetic calorimeter with excellent energy resolution, the CLEO II detector is in an unique position to investigate the problems. In this paper, we present new and precise measurements of the τ mass and the branching ratios into electron and one charged particle plus multiple π^0's.

DETECTOR

The CLEO II detector is a general purpose spectrometer with excellent charged particle and shower energy detection[3]. Charged particles are measured with three cylindrical drift chambers between 5 cm and 95 cm from the e^+e^- interaction point, with a total of 67 layers. This is surrounded by a scintillation time-of-flight system and a CsI(Tl) calorimeter with 7800 crystals. These detector systems are installed inside a 1.5 T superconducting solenoidal magnet, surrounded by a proportional tube muon chamber with iron absorbers. The momentum resolution is 1.2% for a 5 GeV charged particle and the energy resolution is 1.5% for a 5 GeV shower.

The data sample was collected around the Υ energy region at the Cornell Electron Storage Ring (CESR). The total integrated luminosity of the data used in the analyses varies from 0.76 to 1.43 fb^{-1}, corresponding to 0.6-1.3×10^6 $\tau^+\tau^-$ produced.

---
*Representing the CLEO II Collaboration

τ MASS

The τ mass is extracted by measuring the minimum kinematically allowed τ mass in $\tau^+\tau^-$ candidate events. For a given τ mass, the τ direction must lie on a cone around the observed decay product if the energy of the τ is known. In a $\tau^+\tau^-$ event, ignoring the initial and final state radiations, the two cones (one inverted) must intersect at two lines. If a smaller tau mass is assumed, the cone angle decreases; eventually, the two cones just touch. This corresponds to the minimum kinematically allowed τ mass.

In order to reconstruct the events, both τ's are required to decay hadronically: one charged particle[4] with up to two π^0's. The π^0's are assigned to the nearest charged particle according to the angle. Events with both τ's decay into one charged particle with no π^0 are rejected to suppress two-photon background. If one of the charged particles is identified as an electron or muon, it is also rejected. A total of ~29,000 events passed all cuts.

The minimum mass distribution for the data is shown in Fig. 1. The distribution drops off sharply at the τ mass as expected. The present of events above the τ mass is mainly due to initial and final state radiations, missed π^0, misidentification of π as K etc. The mass spectra from the data and Monte Carlo are fitted with an arctangent curve near the edge with a polynomial background. The result of the fit is shown in Fig. 1. Also shown is the fit to a Monte Carlo sample generated with M_τ = 1784.1 MeV/c^2. The shape of the Monte Carlo spectrum is very similar to the data. The result of the fit to the data is $M_\tau = 1776.6 \pm 0.9$ MeV/c^2.

The absolute energy and momentum scales of the detector are calibrated using $K_S \to \pi^+\pi^-$, $D^0 \to K^-\pi^+$, $D^+ \to K^-\pi^+\pi^+$, $\Lambda \to p^+\pi^-$, $\psi \to \mu^+\mu^-$, and $\pi^0 \to \gamma\gamma$. These and other major source of systematic errors

Table 1. Systematic errors in M_τ.

	Δ	ΔM_τ(MeV/c^2)
Energy	±0.30%	±1.2
Momentum	±0.10%	±0.8
Beam Energy	±0.03%	±0.1
Cuts and Fit	–	±0.5
M_{ν_τ}	< 35 MeV/c^2	+0.9

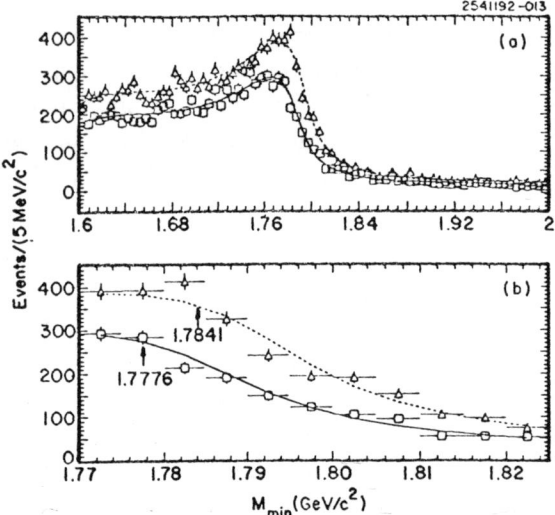

Figure 1. (a) The minimum mass spectra of the data and Monte Carlo (△). The lines show the fit discussed in the text. (b) an expanded view of (a).

are summarized in Table 1. Also shown is the effect of a ν_τ of 35 MeV/c^2. Combining these systematic errors in quadrature yields $M_\tau = 1776.6 \pm 0.9 \pm 1.5^{+0.9}_{-0.0}$ MeV/c^2. The result is significantly smaller the world average[2] but consistent with the recent results from BES[5] and ARGUS[6].

Electron Branching Ratio

The branching ratio for $\tau^- \to e^- \bar\nu_e \nu_\tau$ is measured[7] using e vs. e events,

$$B_e^2 = \frac{N_{ee}}{\epsilon \sigma L},$$

where N_{ee} is the number of e vs. e events observed after correcting for the background, ϵ is the detection efficiency, σ is the cross section, and L is luminosity. This measurement has the advantage that the systematic errors on the acceptance, cross section, and luminosity are halved. The analysis takes advantage of the large data set and the hermetic CsI calorimeter and uses stringent selection criteria to suppress Bhabha and two-photon backgrounds. The suppression is achieved by requiring the acoplanarity angle between the two charged tracks to be in the range, $0.15 < \xi < 1.50$. The Bhabha contamination is further eliminated by rejecting any event that contains a photon with energy greater than ~ 500 MeV. The two-photon background is further reduced by requiring the transverse momentum of events to be greater than 22% of the beam energy and the momentum of both tracks must be greater than 10% of the beam energy. The momentum of both tracks and the missing momentum of the events must point into the central region of the detector. The electron candidate is defined as a particle with shower energy to momentum ratio in the range, $0.85 < E/P < 1.10$, and specific ionization no more than two standard deviations below that expected for an electron.

A total of 3970 events satisfies the selection criteria. The detection efficiency includes three components: acceptance, $\epsilon_a = (11.17 \pm 0.07 \pm 0.15)\%$, electron identification efficiency, $\epsilon_e = (95.76 \pm 0.10 \pm 0.32)\%$, and trigger efficiency, $\epsilon_t = (99.00 \pm 0.13 \pm 0.22)\%$, where the first error is statistical and the second systematic. The two latter efficiencies are very high. The major lost of the acceptance is due to the requirement on the transverse momentum of the event and the momentum on the individual track. The systematic error in the acceptance includes two components, a relative $\pm 1.0\%$ uncertainty in detector modeling and a $\pm 0.8\%$ uncertainty for simulation of tau decay. Decay radiation causes a relative efficiency reduction of $\sim 10\%$, mostly due to the softening of the electron momentum spectrum. Both the electron identification and trigger efficiencies have been measured from the data using radiative Bhabhas.

The backgrounds in the events are small and have been calculated using Monte Carlo. The backgrounds from the two-photon processes, $e^+e^-e^+e^-$, and $e^+e^-\tau^+\tau^-$, are estimated to be $f_{eeee} = (0.62 \pm 0.16 \pm 0.31)\%$ and $f_{ee\tau\tau} = (0.38 \pm 0.09 \pm 0.19)\%$. The migration background from other τ decay is $f_{\tau\tau} = (0.63 \pm 0.15 \pm 0.32)\%$. The Bhabha contamination is estimated to be $f_{ee} = (0.0^{+0.3}_{-0.0})\%$.

The luminosity has been measured using two independent processes, $e^+e^- \to e^+e^-$ and $e^+e^- \to \gamma\gamma$, in which the cross sections can be precisely calculated. From a comparison of two luminosity measurements, the error on the average luminosity is estimated to be $\pm 1.5\%$.

After correcting for the detection efficiency and backgrounds, the result on the electron branching ratio is $B_e = (17.49 \pm 0.14 \pm 0.22)\%$. The systematic error also includes the uncertainty in the τ production cross section, estimated to $\sim \pm 1\%$ (relative). The precision of this measurement is comparable with the world average[2] of all other experiments combined but the measured value is somewhat lower the world average. The branching ratio is significantly lower than the prediction from lepton universality, $B_e = (18.68 \pm 0.37)\%$, based on the current world average of the lifetime[2] and τ mass[8].

The excellent calorimeter also allows for the search for decay radiation. Figure 2 shows the photon energy spectrum observed in the data. The Monte Carlo calculation without the decay radiation severely underestimates the number of photons observed. The spectrum is well reproduced by the Monte Carlo with decay radiation. Decay radiation accounts for $\sim 42\%$ of the photons with energy

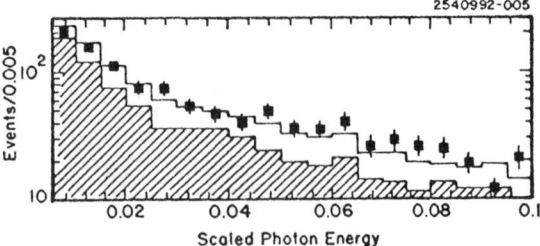

Figure 2. The energy spectrum of the highest energy photon in e vs. e events, normalized to the beam energy. The upper histogram shows the Monte Carlo prediction with decay radiation. The hashed histogram shows the prediction without the radiation.

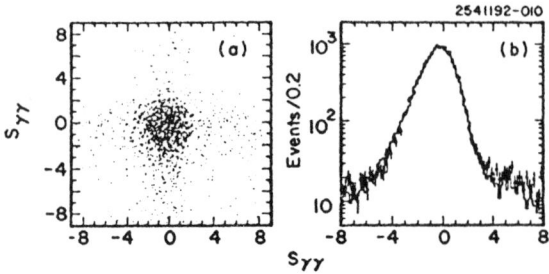

Figure 3. (a) The two-photon invariant mass distribution, in unit of $S_{\gamma\gamma} = (m_{\gamma\gamma} - m_{\pi^0})/\sigma_{\gamma\gamma}$, of one photon pair vs. the other in the data. (b) The two-photon invariant mass distribution for all photon pairs. The histogram shows the Monte Carlo prediction.

> 100 MeV. Others sources for the photons are bremsstrahlung ($\sim 41\%$) in the detector material, initial/final state radiation ($\sim 15\%$), and other τ decay background ($\sim 2\%$). The excess of events without decay radiation is $238 \pm 26 \pm 19$. This is the first direct observation of decay radiation in τ decay.

$h^-\pi^0$ Branching Ratio

The branching ratio for $\tau^- \to h^-\pi^0\nu_\tau$ is measured using $h^+\pi^0$ vs. $h^-\pi^0$ events. As in the case of the B_e measurement, many of the systematic errors are halved. Each event is required to be in the central region (barrel) of the detector, $|\cos\theta| < 0.71$. The acolinearity angle of the two charged tracks must be less than 90°. To suppress two-photon and Bhabha backgrounds, the momentum of each charged track must be greater than 10% of the beam energy and the momentum of at least one track must be less than 85% of the beam energy. The total visible energy of the events must be greater than 40% of the center-of-mass energy and the transverse momentum of the events must be greater than 200 MeV/c.

The π^0 candidates are reconstructed from the photons in the barrel. The invariant mass of each π^0 candidate must be within four standard deviations of the nominal mass, $|S_{\gamma\gamma} = (m_{\gamma\gamma} - m_{\pi^0})/\sigma_{\gamma\gamma}| < 4$. The photon energy is required to be greater than 40 MeV and the π^0 momentum is required to be greater than 10% of the beam energy. Events with unused photons are rejected: barrel photons with energy greater than 100 MeV or endcap photons ($0.71 < |\cos\theta| < 0.95$) with energy greater than 200 MeV.

A total of 6835 events pass the selection criteria. The two-photon invariant mass distribution, in unit of $S_{\gamma\gamma}$, of the events is shown in Fig. 3. A clear enhancement at $\pi^0 - \pi^0$ mass is observed. The signal is extracted by performing a two-dimensional fit, allowing for $\pi^0\gamma\gamma$ and $\gamma\gamma\gamma\gamma$ backgrounds.

The detection efficiency is estimated to be $\epsilon = (6.92 \pm 0.08)\%$ using Monte Carlo. A small correction has been applied to account for the imperfection in the Monte Carlo simulation. The correction is extracted from the data and includes tracking, trigger, and unused photon veto efficiencies. The systematic error in the detection efficiency has been investigated by comparing kinematic distributions of the data with the Monte Carlo, and by varying the cuts used to select the events. From these studies, the systematic error is estimated to be ±3% (relative).

The dominant background is the migration background from other τ decays. The major source of the migration background is $\tau^- \to \pi^- 2\pi^0 \nu_\tau$. The Monte Carlo estimates that the total migration background to be $f_{\tau\tau} = (14.0 \pm 0.3 \pm 0.5)\%$. The systematic error is due to the uncertainty in the branching ratios for the migration background. An addition error of ±1.0% is assigned to $f_{\tau\tau}$ to account for the uncertainty in the relative detection efficiencies for the various decay modes. The hadronic background is estimated to be $f_{qq} = (0.25 \pm 0.14)\%$ and the background from $e^+e^- \to e^+e^-\tau^+\tau^-$ is estimated to be $f_{ee\tau\tau} = (0.11 \pm 0.05)\%$.

After correcting for the detection efficiency and background contamination, the result on the branching ratio is $B_{h\pi^0} = (24.83 \pm 0.15 \pm 0.53)\%$. The systematic error includes the relative uncertainty of ±1.5% in the luminosity and ±1.6% in the τ production cross section. The measured branching ratio is consistent with the world average[2], but with better precision. After correcting[2] for $\tau^- \to K^{*-}\nu_\tau \to K^-\pi^0\nu_\tau$, the branching ratio for $\tau^- \to \pi^-\pi^0\nu_\tau$ is $B_{\pi\pi^0} = (24.35 \pm 0.55)\%$. Normalizing to the world average value for B_e (including the CLEO result presented in this paper) yields $B_{\pi\pi^0}/B_e = 1.38 \pm 0.03$. This is in excellent agreement with the prediction from CVC[9], $B_{\pi\pi^0}/B_e = 1.33 \pm 0.07$, after correcting for the new τ mass[8].

$h^- n\pi^0$ Branching Ratios

The branching ratios for $\tau^- \to h^- n\pi^0\nu_\tau$, for $n = 2$, 3, and 4, have been measured by normalizing to the branching ratio for $\tau^- \to h^-\pi^0\nu_\tau$. This has the advantage that many systematic errors cancel in the ratios. The decay candidates are selected using two different tags: the e and μ tags and the 3-prong tag. Since some of the systematic errors associated with the different tags are very different, this allows a powerful cross check of the measurements. The lepton tags have negligible hadronic background in comparison with the 3-prong tag. However, 3-prong tag has a robust trigger due to the higher multiplicity.

The candidates are required to have 1-1 or 1-3 topology. The 1-3 topology is defined by the plane perpendicular to the highest momentum track. The acolinearity angle of the two tracks in 1-1 events must be less than 90^0 in order to suppress two-photon background. The total energy of the events must be greater than $\sim 25\%$ of the beam energy and the total shower energy must be less than $\sim 80\%$ of the beam energy.

The electron tag is identified by comparing the shower energy deposition in the calorimeter with the momentum measured in the drift chambers and from specific ionization in the drift chambers. The minimum electron momentum allowed is 0.5 GeV/c. The muon tag is identified by requiring the track to traverse at least three absorption length of iron. The minimum muon momentum requirement is 1.0 GeV/c. QED backgrounds in the leptonic tags are suppressed by the requiring the net missing momentum to point into the detector and the transverse momentum of the events to be greater than 200 MeV/c. In the 3-prong tag, the invariant mass of all the detected particles in each hemisphere must be less than 1.7 GeV/c^2. The hadronic events are further reduced by the requirement on the missing mass of the events, $0.5 < [(E_{cm} - E_{vis})^2 - P_{vis}^2]^{\frac{1}{2}} < 7.5$ GeV/c^2. There should be no more than two photons in the 3-prong hemisphere. Radiative Bhabha events with converted photons are rejecting by allowing no more than one electron in each event.

The π^0 candidates are reconstructed from the photons in the barrel region ($|\cos\theta| < 0.80$) of the calorimeter. The minimum photon energy allowed is 60 MeV. For the lepton tags, photons within 20^0 of the lepton direction are ignored. In addition, one of the photons in each π^0 candidate must have an energy greater than 80 MeV and the opening angle of the two photons must be less than 135^0. The momentum vector of each π^0 candidate must lie within 90^0 of the charged track. The energy of any unused photon is required to be less than 100 MeV. In order to increase the detection efficiency for $\tau^- \to h^- 4\pi^0 \nu_\tau$, the minimum photon energy requirement is lowered from 60 to 30 MeV for this decay. For the 3-prong tag, the 1-prong hemisphere is required to contain an even number of photons and events with unused photons are rejected (60 MeV in the barrel and 100 MeV in the endcap, $0.80 < |\cos\theta| < 0.95$).

The $n\pi^0$ candidate is selected based on the combination with lowest reduced chi-square:

$$\chi^2 = \frac{1}{n} \sum_{i=1}^{n} \frac{(m_{\gamma\gamma}^i - m_{\pi^0})^2}{\sigma_i^2} ,$$

where $m_{\gamma\gamma}^i$ is the mass of the ith π^0 candidate. The resolution for the mass, σ_i, is typically between 6 to 8 MeV/c^2. The two-photon invariant mass spectrum for $\tau^- \to h^- \pi^0 \nu_\tau$ is shown in Fig. 4(a), for the lepton tags. A clear π^0 signal with very little background is observed. A four standard deviation cut on the mass is used to defined the normalization sample. In Fig. 4(b-d), the χ^2 for the other three π^0 multiplicities are shown, also with a lepton tag. An enhancement at zero χ^2 is observed for all three photon multiplicities. The number of events for each tag are summarized in Table 2. The invariant mass spectra of the $\pi^- n\pi^0$ candidates from all tags are shown in Fig. 5. There are few events with mass above the τ mass.

The backgrounds are small and have been calculated using both Monte Carlo and data. The background ($f_{\gamma\gamma}$) from two unrelated photons forming a π^0 candidate is estimated from

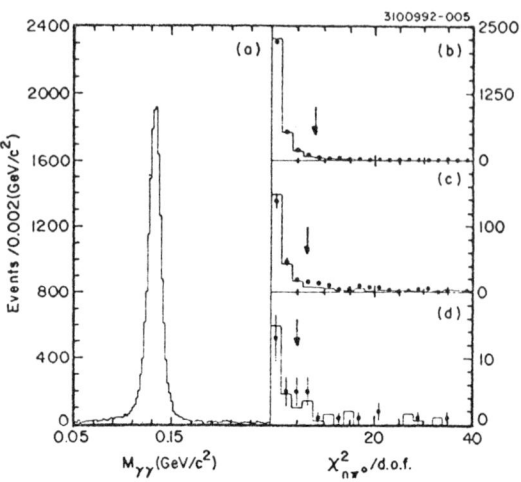

Figure 4. (a) The two-photon mass spectrum of the $1\pi^0$ candidates for lepton tags. (b-d) The χ^2 distribution for the 2, 3, and 4 π^0 candidates for lepton tags. The histograms show the Monte Carlo expectations.

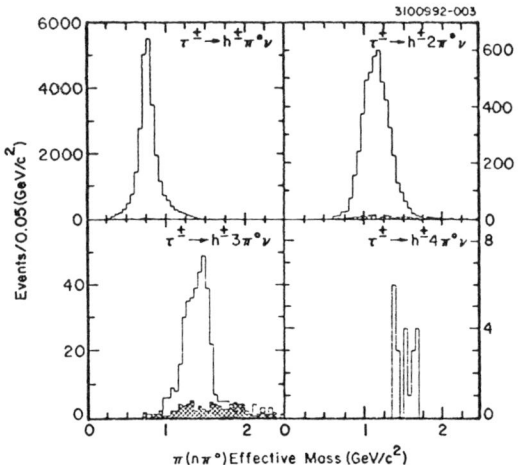

Figure 5. (a) The $\pi^- n\pi^0$ invariant mass spectrum for the 1, 2, 3, and 4 π^0 candidates. The hatched histograms shows the hadronic background expectation. Only the lepton tag is shown in (d).

Table 2. Numbers of events, backgrounds, detection efficiencies, and branching ratios for the $n\pi^0$ candidates in the various tags.

Mode (X)	Tag	Events	Backgrounds (%) $f_{\gamma\gamma}$	$f_{q\bar{q}}$	f_τ	Relative Eff.(%)	$\dfrac{B(\tau^- \to X)}{B(\tau^- \to h^-\pi^0\nu_\tau)}$
$h^-\pi^0\nu_\tau$	e	8935	2.2 ± 0.1	< 0.1	4.3 ± 0.1	100	1
	μ	7470	2.3 ± 0.1	< 0.1	4.6 ± 0.2	100	1
	$3h$	8511	2.0 ± 0.1	2.6 ± 0.1	3.4 ± 0.2	100	1
$h^-2\pi^0\nu_\tau$	e	1639	4.1 ± 0.3	< 0.4	3.6 ± 0.2	53.8 ± 0.8	0.336 ± 0.011
	μ	1434	3.7 ± 0.3	< 1.0	4.4 ± 0.3	53.3 ± 0.8	0.356 ± 0.012
	$3h$	1561	2.9 ± 0.2	5.1 ± 0.4	3.0 ± 0.2	51.2 ± 0.7	0.347 ± 0.011
$h^-3\pi^0\nu_\tau$	e	111	11 ± 2	< 6	12.4 ± 1.7	25.4 ± 0.9	0.041 ± 0.004
	μ	100	11 ± 2	< 6	11.4 ± 1.7	27.0 ± 1.0	0.042 ± 0.005
	$3h$	95	14 ± 2	22 ± 3	6.6 ± 1.2	19.5 ± 0.8	0.039 ± 0.007
$h^-4\pi^0\nu_\tau$	e	9	14 ± 6	< 10	5.7 ± 3.2	16.8 ± 1.1	0.005 ± 0.002
	μ	12	15 ± 6	< 10	4.3 ± 2.3	18.9 ± 1.3	0.007 ± 0.003
	$3h$	4	45 ± 27	< 45	6 ± 6	6.3 ± 0.9	$< 0.012 (90\% CL)$

the tail of the χ^2 distributions. The hadronic background ($f_{q\bar{q}}$) in the lepton tag is calculated using Monte Carlo and in the 3-prong tag, the background is estimated from the data. The migration background from other π^0 multiplicities is estimated using Monte Carlo.

The dominant source of systematic error is the $n\pi^0$ finding efficiency. This has been estimated by varying the cuts on photon energy, angle, and multiplicity and by comparing the branching ratios with the results obtained in a semi-inclusive analysis of the energy spectrum of the unused photons. The signal extraction technique is investigated by comparing the results with n-dimensional side-band subtraction technique. The hadronic background estimate is studied using hadronic events of 3-3 topology from the data.

After correcting for the backgrounds and detection efficiencies, the final results are $B_{h2\pi^0}/B_{h\pi^0} = 0.345 \pm 0.006 \pm 0.016$, $B_{h3\pi^0}/B_{h\pi^0} = 0.041 \pm 0.003 \pm 0.005$, and $B_{h4\pi^0}/B_{h\pi^0} = 0.006 \pm 0.002 \pm 0.002$, where the first error is statistical and the second systematic.

Using the world average of $B_{h\pi^0} = (23.8 \pm 0.8)\%$, the absolute branching ratios are

$$B_{h2\pi^0} = (8.21 \pm 0.15 \pm 0.38 \pm 0.28)\%$$
$$B_{h3\pi^0} = (0.98 \pm 0.07 \pm 0.12 \pm 0.03)\%$$
$$B_{h4\pi^0} = (0.15 \pm 0.04 \pm 0.05 \pm 0.01)\%$$

These results are more precise than, but consistent with, the current world averages. The results on $B_{h3\pi^0}$ and $B_{h4\pi^0}$ represent the first measurement of these decays by *exclusive* reconstruction. These two measurements are consistent with the theoretical expectation from CVC and isospin[1].

CONCLUSIONS

In conclusion, the τ mass and the branching ratios into electron and one charged particle plus multiple π^0's have been measured precisely by the CLEO II experiment. The results are consistent with the world averages and the precision of the branching ratio measurements is comparable or better than the world averages of all other experiments. The measured branching ratios into e, $h2\pi^0$, and $h3\pi^0$ are significantly below the results from ALEPH[11] while the branching ratio for $h\pi^0$ is in good agreement with ALEPH. These precise measurements do not support the hypothesis that the 1-prong problem is due to the underestimates of exclusive branching ratios.

Acknowledgement

The author wishes to thank the OJI program of DOE and SSC Fellowship of TNRLC for their supports.

REFERENCES

1. F. J. Gilman and S. H. Rhie, Phys. Rev. **D31**, 1066 (1985).

2. Particle Data Group, K. Hikasa et al., Phys. Rev. **D45**, Part II, 1 (1992).

3. Y. Kubota et al., Nucl. Instrum. Methods, **A320**, 66 (1992).

4. Through out this paper, all charged particles are assumed to be pions. The charge conjugate state is implied.

5. J. Bai et al., SLAC-PUB-5870, 1992 (submitted to Phys. Rev. Lett.)

6. H. Albrecht et al., Phys. Lett. **292B**, 221 (1992).

7. D. S. Akerib et al., "Measurement of the Tau Lepton Electronic Branching Fraction", CLNS 92/1163, 1992 (submitted to Phys. Rev. Lett.)

8. The τ mass is computed using the results presented in this paper and in Refs. [2], [5], and [6].

9. Y. S. Tsai, Phys. Rev. **D4**, 2821 (1971); W. J. Marciano, Phys. Rev. **D45**, 721 (1992); J. H. Kuhn and A. Santamaria, Z. Phys. **C48**, 445 (1990).

10. M. Procario et al., "Tau Decays with One Charged Particle Plus Multiple π^0's", CLNS 92/1165, 1992 (submitted to Phys. Rev. Lett.)

11. D. Decamp et al., Z. Phys. **C54**, 221 (1992).

PRECISE DETERMINATION OF α_s FROM τ DECAYS*

Eric Braaten
Department of Physics and Astronomy
Northwestern University
Evanston, Illinois 60208 USA

Abstract

The hadronic decay rate of the τ lepton provides a low energy determination of the strong coupling constant that is competitive in precision with any other method. Using the electronic branching fraction of the τ as input, we obtain $\alpha_s(M_\tau) = 0.33 \pm 0.03$. Evolving up to the mass of the Z using the renormalization group, we obtain $\alpha_s(M_Z) = 0.119 \pm 0.004$, in agreement with precise determinations from LEP.

One of the most remarkable aspects of the physics of the τ lepton is that its mass is almost ideally situated to allow a precise low energy determination of the strong coupling constant. For a precise determination of α_s, you need a perturbative signal that stands out above the noise of nonperturbative QCD. It is rather surprising that the hadronic decay rate of the τ provides such a signal, in spite of the fact that its exclusive decay modes are clearly dominated by resonances. A thorough theoretical analysis of the τ hadronic width has recently been carried out by Braaten, Narison, and Pich.[1] The bottom line of this analysis is that τ decay provides a determination of α_s that is absolutely competitive with any others, including the precise high energy determinations coming from LEP.

It is convenient to analyze the hadronic decay rate of the τ in terms of the ratio R_τ obtained by normalizing it to the electronic decay rate:

$$R_\tau = \frac{\Gamma(\tau^- \to \nu_\tau + \text{hadrons})}{\Gamma(\tau^- \to \nu_\tau + e^- \bar{\nu}_e)}. \quad (1)$$

*Invited talk presented at the XXVI International Conference on High Energy Physics, Dallas, Texas (August 6-12, 1992).

This is the analog of the famous ratio R for e^+e^- annihilation into hadrons. A naive prediction for R_τ can be obtained by realizing that the fundamental process underlying the decay of the τ into hadrons is the emission of a virtual W boson which decays into the quark pair $d\bar{u}$ or $s\bar{u}$. The coupling of the W^- to $e^-\bar{\nu}_e$ is identical to its coupling to the quark pair $(\cos\theta_c d + \sin\theta_c s)\bar{u}$, where θ_c is the Cabbibo mixing angle. Thus the ratio R_τ simply counts the number of colors of quarks, which is 3.

Of course there are corrections to the naive prediction $R_\tau = 3$ both from the electroweak interactions and from QCD. The complete prediction for R_τ can be written

$$R_\tau = 3 S_{EW} (1 + \delta_{EW} + \delta_{QCD}). \quad (2)$$

The electroweak corrections include a short distance enhancement factor $S_{EW} = 1.024$ calculated by Marciano and Sirlin[2] and a tiny residual correction of order α calculated by Braaten and Li[3]: $\delta_{EW} = 0.001$. The QCD corrections were first analyzed systematically by Schilcher and Tran[4] in 1984 and independently by Braaten and by Narison and Pich[5] in 1988. The fractional correction δ_{QCD} can be separated into a perturbative contribution δ_{PQCD} and a nonperturbative correction δ_{NPQCD}. The perturbative correction can be expanded in powers of the running coupling constant $\alpha_s(M_\tau)$, where M_τ is the mass of the τ. The first 3 terms in this expansion are known:

$$\delta_{\text{PQCD}} = \frac{\alpha_s}{\pi} + 5.2\left(\frac{\alpha_s}{\pi}\right)^2 + 26.4\left(\frac{\alpha_s}{\pi}\right)^3 \pm 130\left(\frac{\alpha_s}{\pi}\right)^4. \quad (3)$$

The fact that this correction can be expressed as a series in $\alpha_s(M_\tau)$ is not at all obvious. Although the first term α_s/π was written down by Lam and Yan[6] in 1977, the coefficient of $(\alpha_s/\pi)^2$ was not written down explicitly[7] until 1989. The coefficient of $(\alpha_s/\pi)^3$ was calculated by Gorishny, Kataev and Larin and by Samuel and Surguladze in 1991.[8] This term is important because $\alpha_s(M_\tau)$ is not small enough for the α_s^3 correction to be completely negligible. The fourth term in (3) is our conservative estimate[1] of the error due to truncating the perturbation expansion at order α_s^3. It is intended to take into account not only the uncalculated term of order α_s^4 but also reasonable variations in the scale μ of the expansion parameter $\alpha_s(\mu)$.

The ratio R_τ can be used to determine α_s only if the perturbative QCD correction (3) is not overwhelmed by nonperturbative corrections. The mass of the τ is small enough that nonperturbative effects could be large. These effects can be analyzed systematically using the QCD sum rules developed by Shifman, Vainshtein, and Zakharov.[9] The corrections take the form of an expansion in $1/M_\tau^2$, with coefficients that depend logarithmically on M_τ. One finds remarkable suppressions[4,5] among the nonperturbative corrections that are special to the total hadronic decay rate. The $1/M_\tau^2$ corrections are suppressed by 2 powers of light quark masses. The $1/M_\tau^4$ corrections are suppressed either by a light quark mass or by 2 powers of $\alpha_s(M_\tau)$. Thus the dominant nonperturbative corrections fall like $1/M_\tau^6$. Even these are fortuitously small due to cancellations between the vector and axial vector contributions. Our best estimate of these corrections is[1]

$$\delta_{\text{NPQCD}} = -(0.007 \pm 0.004)\left(\frac{1.8 \text{ GeV}}{M_\tau}\right)^6. \quad (4)$$

Thus the nonperturbative QCD corrections are less than 1%.

There are two independent ways of determining the ratio R_τ experimentally. It can be written

$$R_\tau = \frac{\Gamma(\tau) - \Gamma(\tau \to \nu e \bar{\nu}) - \Gamma(\tau \to \nu \mu \bar{\nu})}{\Gamma(\tau \to \nu e \bar{\nu})}. \quad (5)$$

Using the universality of e and μ couplings, this can be expressed as a function of the electronic branching fraction B_e of the τ:

$$R_\tau = \frac{1}{B_e} - 1.973. \quad (6)$$

Alternatively, using the universality of μ and τ couplings as well, R_τ can be expressed in terms of the masses and lifetimes of the μ and τ:

$$R_\tau = \frac{\tau_\mu}{\tau_\tau}\left(\frac{M_\mu}{M_\tau}\right)^5 - 1.973. \quad (7)$$

Unfortunately, there is at present a discrepancy known as the τ lifetime problem[10] between the measurement of B_e and the measurements of M_τ and τ_τ. The value of B_e given in the 1992 Review of Particle Properties[11] gives the ratio $R_\tau = 3.604 \pm 0.081$, which translates into a QCD coupling constant of

$$\alpha_s(M_\tau) = 0.33 \pm 0.03. \quad (8)$$

The 1992 values of M_τ and τ_τ give significantly different results: $R_\tau = 3.28 \pm 0.11$ and $\alpha_s(M_\tau) = 0.20 \pm 0.06$. However a recent precise measurement[9] of the mass of the τ has revealed that the

1992 value was too large by more than 2 standard deviations. Using the new value of the mass, the result is $R_\tau = 3.38 \pm 0.11$, which gives a coupling constant

$$\alpha_s(M_\tau) = 0.24 \pm 0.05, \qquad (9)$$

which differs from (8) by two standard deviations. Unless there really are deviations from the universality of the couplings of the e, μ, and τ, the discrepancy between the measurements of B_e and of M_τ and τ_τ will eventually be resolved.

The electronic branching fraction of the τ provides a determination of $\alpha_s(M_\tau)$ to 10% as given in (8). This does not sound very impressive compared to recent determinations of $\alpha_s(M_Z)$ from LEP that are at the 5% level. However to compare our low energy determination of α_s with the high energy determinations from LEP, we must use the renormalization group to evolve the running coupling constant $\alpha_s(\mu)$ from M_τ up to M_Z. Using the coupling constant (8) obtained from B_e, we find

$$\alpha_s(M_Z) = 0.119 \pm 0.004. \qquad (10)$$

The renormalization group evolution has decreased the coupling constant by almost a factor of 3. More importantly it has a focusing effect on the error bar, which scales roughly as α_s^2. A 10% determination of α_s at the scale M_τ translates into a determination of $\alpha_s(M_Z)$ to better than 4%. The result (10) is in remarkable agreement with the best determinations from LEP. If instead we use for $\alpha_s(M_Z)$ the value (9) which is obtained from the mass and lifetime of the τ, we find

$$\alpha_s(M_Z) = 0.106 \pm 0.009. \qquad (11)$$

The resolution of the discrepancy between (10) and (11) must await the resolution of the τ lifetime problem.

There are several directions in which the theoretical calculation of R_τ could be improved. Data on τ decays can be used to determine more precisely the hadronic matrix elements that appear in the nonperturbative corrections.[13] While R_τ is remarkably insensitive to these matrix elements, this is due to cancellations that are unique to R_τ. The suppression of $1/M_\tau^4$ corrections does not occur for the average value of the hadronic invariant mass, which is sensitive to the gluon condensate. The suppression of $1/M_\tau^6$ corrections does not arise for the difference between the vector and axial vector contributions to R_τ, which is sensitive to the 4-quark condensates. Thus τ decay data can be used to determine these matrix elements. More precise determinations would lead to a sharper estimate of the nonperturbative correction δ_{NPQCD}.

The largest source of theoretical uncertainty comes from higher order perturbative corrections. Our estimate of $\pm 130 \, (\alpha_s/\pi)^4$ for the uncertainty was intended to be conservative. A careful analysis of the renormalization scheme dependence of the perturbative correction could result in a smaller error.[14] It might be possible to decrease the perturbative error further by summing exactly certain large perturbative corrections that come from analytic continuation.[15]

In conclusion, the hadronic decay rate of the τ provides a precise low energy determination of α_s at the scale M_τ. Combined with precision measurements at the scale M_Z from LEP, it provides dramatic quantitative evidence of the running of the strong coupling constant. Due to the remarkable supression of nonperturbative effects and the availability of perturbative calculations to order α_s^3, the theoretical uncertainty in R_τ has been reduced to the 1% level. On the experimental frontier, the most important issue is the τ lifetime problem. Unless there are violations of e-μ-τ universality at a remarkable level in τ decays, this problem will eventually be resolved.

An important byproduct of the higher statistics measurements that may be needed to resolve this problem would be an even more precise determination of α_s from τ decays.

This work was supported in part by the U.S. Department of Energy under Grant DE-FG02-91-ER40684.

REFERENCES

1. E. Braaten, S. Narison, and A. Pich, *Nucl. Phys.* B373, 581 (1992).

2. W. Marciano and A. Sirlin, *Phys. Rev. Lett.* 61, 1815 (1988).

3. E. Braaten and C.S. Li, *Phys. Rev.* D42, 3888 (1990).

4. K. Schilcher and M.D. Tran, *Phys. Rev.* D29, 570 (1970).

5. E. Braaten, *Phys. Rev. Lett.* 60, 1606 (1988); S.Narison and A. Pich, *Phys. Lett.* B211, 183 (1988).

6. C.S. Lam and T.M. Yan, *Phys. Rev.* D16, 703 (1977).

7. E. Braaten, *Phys. Rev.* D39, 1458 (1989).

8. S.G. Gorishny, A.L. Kataev, and S.A. Larin, *Phys. Lett.* B259, 144 (1991); M. A. Samuel and L.R. Surguladze, *Phys. Rev.* D44, 1602 (1991).

9. M.A. Shifman, A.L. Vainshtein and V.I. Zakharov, *Nucl. Phys.* B147, 385 (1979); B147, 448 (1979); B147, 519 (1979).

10. W.J. Marciano, *Phys. Rev.* D45, R721 (1992).

11. Particle Data Group (H. Hikasa et al.), Phys. Rev. D45, Part 2 (1992).

12. BES Collaboration (J.Z. Bai et al.), SLAC preprint SLAC-PUB-5870 (July, 1992).

13. C.A. Dominguez and J. Sola, *Z. Phys.* C40, 63 (1988); V. Giménez, J.A. Peñarrocha and J. Bordes, *Phys. Lett.* B223, 245 (1989); F. Le Diberder and A. Pich, CERN preprint CERN-TH.6422/92 (May 1992).

14. C.J. Maxwell and J.A. Nichols, *Phys. Lett* B236, 63 (1990); M. Haruyama, *Prog. Theor. Phys.* 83, 841 (1990); *Phys. Rev.* D45, 930 (1992); M. Luo and W.J. Marciano, Brookhaven preprint BNL-47187 (February, 1992).

15. A.A. Pivovarov, INR preprint 731/91 (September 1991); F. LeDiberder and A. Pich, CERN preprint TH. 6421/92.

TOPOLOGICAL AND HADRONIC DECAYS OF THE TAU

J. Michael Roney
Institute of Particle Physics
University of Victoria
Victoria, B.C.
V8W 3P6 Canada

Abstract

A summary of recent measurements of the branching ratio of the tau lepton to one, three and five charged particles plus any number of neutral particles performed by the four LEP experiments is presented. Results of exclusive hadronic branching ratios performed by DELPHI, OPAL and ALEPH are summarised. Limits on "missing decay modes" have been set with the ALEPH data and this is compared to the limits on such modes that can set combining recent results from other experiments. The results quoted in this paper should be considered preliminary unless stated otherwise.

INTRODUCTION

The combined measurements of τ decays published prior to 1990 have suggested that there is a discrepancy between the inclusive branching ratio of decay modes containing one charged particle and any number of neutral particles (known as "1-prongs") and the sum of the 1-prong exclusive branching ratios where theoretical constraints were used to limit poorly measured channels[1]. However, evidence against such a "missing decay mode" is provided in analyses of all known exclusive decay modes performed by CELLO[2] and more recently by ALEPH[3]. A description of the state of this controversy in early 1992 may be found in reference[4] in which the sum of the exclusive branching ratios averaged over all experiments except for CELLO and ALEPH is quoted as being $90.6 \pm 1.6\%$ whereas it is quoted to be $100.1 \pm 1.8\%$ averaging only the CELLO and ALEPH results.

Interestingly enough, discrepancies also exist between between the measurements of the inclusive topological branching ratios. For example, the HRS collaboration measures an inclusive 1-prong branching ratio of $86.4 \pm 0.3 \pm 0.3\%$ [5] while the CELLO collaboration reports a value of $84.9 \pm 0.4 \pm 0.3\%$[2].

New measurements of hadronic and topological branching ratios using the relatively low background sample of tau pairs available in the e^+e^- annihilations at LEP ($\sqrt{s} \sim 91$ GeV) have become available since reference [4] was published. All four LEP experiments (L3, DELPHI, OPAL and ALEPH) have measured the topological branching ratios. Currently, the world's most precise measurements have been made by OPAL and those measurements will be briefly described. The measurements by DELPHI, OPAL and ALEPH of the branching ratios of $\tau \to \rho\nu$ and $\tau \to \pi\nu$ will be summarised. The ALEPH measurement of quasi-exclusive branching ratios including multi-π^0 modes will be discussed. The paper will close by giving a status of the "missing decay mode problem" as of August 1992. This

Table 1. Topological Branching Ratios.

	1-prong (%)	5-prong (%)
OPAL	84.48±0.27 ± 0.23	0.26±0.06 ± 0.05
L3	85.6±0.6 ± 0.3	< 0.34 at 95%CL
DELPHI	84.08±0.59 ± 0.45	0.31±0.11 ± 0.07
ALEPH	$85.45^{+0.69}_{-0.73} \pm 0.48$	$0.10^{+0.05}_{-0.04} \pm 0.03$
LEP Average (χ^2/d.o.f.=3.6/3)	84.71±0.27	0.17±0.04

will include new results on the leptonic decay modes from the LEP experiments [6] and new CLEO II results[7] both of which have been presented at this conference.

TOPOLOGICAL BRANCHING RATIOS

The objective of these analyses is to determine B_1, B_3 and B_5: the proportion of tau decays to one, three and five charged particle plus any number of neutral particles. The OPAL[8], L3[9] and ALEPH[3] results have been published whereas the DELPHI results are preliminary.

The OPAL Analysis

The analysis begins by selecting a clean sample of 11262 tau pair events in the barrel region of the OPAL detector. The selection efficiency within this $|\cos\theta| < 0.7$ fiducial region is 92% with a background of 1.9 ± 0.7%. This selection has a relative bias of 0.995 ± 0.001 for events in which both taus decay to a 1-prong (1-1 topology) which arises from cuts designed to remove dimuon backgrounds. A compensating relative bias of 1.015 ± 0.004 is accounted for in the 3-3 topology. No significant bias is present for the other topologies considered. Since the tau pairs are produced from the decay of the Z^0, a cross-check on this selection is provided by comparing the number of selected events after background subtraction (11048) to that expected from the Standard Model (11037).

The analysis proceeds by identifying electron tracks coming from photon conversions and removing them. Two techniques are employed to find conversion electrons, one relying solely on dE/dx whilst the other relies on a reconstruction of the converted photon. The combined efficiency is 89±8%. The error is obtained from a comparison between the data and Monte Carlo of the conversion finding efficiency as determined in control samples of conversions in Bhabha and dimuon events. A track is also removed if it has fewer than 50 out of a possible 159 hits in the OPAL jet chamber or if the momentum is less than 250 MeV. Comparisons between the data and Monte Carlo distributions of the number of hits on the tracks are used to assign the systematic errors associated with this requirement. The remaining tracks in each tau "jet" are counted and if, after subtracting, the number is *even* and if at least one electron track had been found, then the count is increased by one. This simultaneously accounts for the $\tau \to e\nu\bar{\nu}$ decays and reduces the small errors introduced by mis-identifying charged pions as electrons. As only the 1-1, 1-3, 1-5 and 3-3 topologies are considered, only events having a 1-1, 1-2, 1-3, 1-4, 1-5, 2-2, 2-3 or 3-3 topologies are accepted. This removes 71 events of which 58 are expected from the multihadronic background.

A χ^2 fit for B_1, B_3 and B_5 under the constraint $B_1+B_3+B_5=1$ is performed. This constraint removes systematic errors arising from the integrated luminosity and overall tau selection. The χ^2 is:

$$\chi^2 = \sum_{ij} \frac{(n_{ij} - n_{ij}^{bkg} - \sum_{lk} \varepsilon_{kl \to ij} f_{kl} N_{kl})^2}{\sigma_{n_{ij}}^2 + \sigma_{n_{ij}^{bkg}}^2 + \sigma_{\varepsilon_{kl \to ij}}^2 + \sigma_{f_{kl}}^2}$$

where n_{ij} is the number of events with topology i-j, n_{ij}^{bkg} the expected number of background events in the i-j topology, $\varepsilon_{kl \to ij}$ is the migration matrix obtained from the Monte Carlo (where the K_s^0 are all assigned to be 1-prongs in line with the convention described in reference[10]) and f_{kl} is the bias factor from the tau pair selection. $N_{kl} = (2 - \delta_{kl}) B_k B_l N_{\tau\tau}$ where $N_{\tau\tau}$ is the number of tau pairs.

The values for B_1, B_3 and B_5 are $84.48 \pm 0.27 \pm 0.23\%$, $15.26 \pm 0.26 \pm 0.22\%$ and $0.26 \pm 0.06 \pm 0.05\%$, respectively. With a large (-97%) correlation between B_1 and B_3 and a small ($< 20\%$) correlation between B_5 and the other branching ratios. The dominant systematics arise from the non-tau background, track reconstruction, the photon conversion and, in the case of B_1 and B_3, the tau pair selection.

The LEP Results

The values of the 1-prong branching ratios measured by all four LEP experiments are shown in table 1 along with the 5-prong branching ratio measurements. The 1-prong measurements are in good agreement with each other but the average, $84.71 \pm 0.27\%$, is significantly lower than the 1990 PDG world average of $86.4 \pm 0.3\%$ (using the K_s^0 assignment convention of reference [10]). The 5-prong measurements are also consistent with each other and have an average of $0.17 \pm 0.04\%$ which is in good agreement with the previous world average of $0.11 \pm 0.03\%$. A new world average of $0.13 \pm 0.02\%$ is then obtained.

Table 2. Hadronic Branching Ratios.

	$\pi(K)\nu$ (%)	$\rho\nu$ (%)
OPAL	$12.2 \pm 0.3 \pm 0.4$	$23.8 \pm 0.6 \pm 0.7$
ALEPH	12.6 ± 0.6	$24.6 \pm 0.6 \pm 0.9$
DELPHI	$11.9 \pm 0.7 \pm 0.7$	$22.4 \pm 0.8 \pm 1.3$
LEP	12.3 ± 0.4	23.7 ± 0.6
(χ^2/d.o.f.)	0.41/2	1.96/2

EXCLUSIVE HADRONIC DECAYS AT LEP

ALEPH, DELPHI and OPAL have all measured the branching ratios of $\tau \to \pi(K)\nu$ and $\tau \to \rho\nu$. The results are presented in table 2. The ALEPH[3] and DELPHI[11] numbers have been published, whereas the OPAL numbers are preliminary numbers made available for this conference. The LEP average branching ratio to $\pi(K)\nu$ is $12.3 \pm 0.4\%$ and is consistent with the world average value of $11.7 \pm 0.5\%$. A simple weighted average of these results gives a new world average of $12.1 \pm 0.3\%$. For the $\rho\nu$ decay the branching ratio averaged over the three experiments is $23.7 \pm 0.6\%$ and compares well with the previous world averge of $22.2 \pm 1.0\%$[1].

ALEPH tau branching ratio analysis

ALEPH has published an analysis[3] in which the branching ratios of the tau to all known modes are measured. The identification of the leptonic and hadronic modes without π^0s or photons is performed in a conventional manner. In the identification of all other modes (i.e. those containing more than one hadron) two approaches have been taken.

The first is a "quasi-exclusive analysis" which does not explictly require the reconstruction of all π^0's in the tau jet. In that analysis the tau jets identified as *(i)* h$^-$+(π^0,γ), *(ii)* h$^-$+($2\pi^0,\pi^0\gamma,2\gamma$), *(iii)*

$h^-+(3\pi^0 \geq 0\gamma, 2\pi^0 \geq 1\gamma, \pi^0 \geq 2\gamma, \geq 3\gamma)$, *(iv)* 3h and *(v)* 3h + 1 (π^0,γ), 5h + $\geq 0(\pi^0,\gamma)$ are selected and counted. A migration matrix is then used to unfold the branching ratios to the following decay modes: *(i)* $h^-+\pi^0$, *(ii)* $h^-+2\pi^0$, *(iii)* $h^-+\geq 3\pi^0$, *(iv)* 3h and *(v)* 3h+$\geq 1\pi^0$, 5h+$\geq 0\pi^0$.

In the second, "exclusive analysis", the selection is based only on the number of reconstructed π^0s in the tau jet and are classified and counted as *(i)* $h^-+\pi^0$, *(ii)* $h^-+2\pi^0$, *(iii)* $h^-+(3\pi^0 \geq 0\gamma)$ *(iv)* 3h or *(v)* 3h+$\geq 1\pi^0$, 5h+$\geq 0\pi^0$. The branching ratios are extracted for the same decay modes as in the quasi-exclusive analysis. The values extracted in that manner are listed in table 3 along with the values from reference[1] and the preliminary measurements from CLEO II as presented at this conference[7].

The sum of the *quasi-exclusive* branching ratio determinations for modes containing at least one π^0 minus the sum of the *exclusive* branching ratios of the same modes is 0.22±1.6% (this incudes a 0.8% common systematic error associated with π^0 reconstruction). Since the quasi-exclusive analysis allows for the inclusion of new decay modes (i.e. unexpected modes with photons) then this difference has been used to place a limit of 3.4% at 95%CL on the branching ratio of new modes containing π^0's or photons [3].

From table 3 one notes that there is reasonable agreement between the ALEPH measurement and the 1990 world average for most modes. However, the 3h mode as determined by ALEPH is more than two standard deviations higher than the 1990 world average. It may also be noted that nearly all decay modes are measured by ALEPH to have a higher branching ratio than the 1990 world averages. ALEPH's measurements are also consistent with the new CLEO II measurements, but again they are higher.

THE "MISSING DECAY MODE PROBLEM"

As stated in the last section, ALEPH has placed a limit of 3.4% at the 95% CL on new decay modes of the tau that contain π^0s. It is of some interest then to consider the status of the "missing decay mode problem" using the most recent data, that is data for the most part available since 1991. An average of the electronic branching ratios reported at this conference from CLEO II[7], OPAL and DELPHI[6] together with the results from L3[9] and CLEO I[12] yield and average of 17.59±0.21%. The value of the muonic branching ratio obtained from recent OPAL, DELPHI and L3 measurements is 17.14±0.33%[6]. The OPAL and DELPHI measurements of the $\pi(K)$ branching ratio listed above average to 12.14±0.45% whilst the branching ratio to ρ as measured by CLEO II[7], OPAL, DELPHI and ARGUS[13] average to 23.57±0.47%. Including the multi-π^0 modes measured by CLEO II as listed in table 3 (9.61±0.44%), the K^* (1.43±0.17%[14]) and the $\omega\pi^- \geq 0\pi^0$ modes (0.15±0.03% [15]), the sum of the exclusive 1-prong branching ratios is 81.6±0.9%. This is to be compared with the average 1-prong inclusive branching ratio of 84.6±0.3%. The difference of 3.0±0.9 is to be compared to the value of 5.8±1.4% quoted in reference [1] and to the 3.4% that is excluded by the ALEPH analysis. The recent non-ALEPH measurements are consistent with the ALEPH limit on new decay modes and suggest a reduced magnitude as well as a reduced significance of the discrepancy.

As much of this discrepancy has been reduced by the decrease in the 1-prong topological branching ratio, one might expect to see a significant discrepancy in the 3-prong branching ratio arise. This appears to be anticipated by the larger ALEPH measurements of the exclusive 3-prong modes. Given the sizable inconsistencies between the different measurements of these modes[4], it is apparent that

Table 3. Branching Ratios of Multi-π^0 decay modes.

Decay Mode	ALEPH quasi-exclusive (%)	ALEPH exclusive (%)	PDG 1990 (%)	CLEO II (%)
$h+\pi^0$	25.0±0.6	25.3±0.9	22.8±1.6	24.8±0.6
$h^-+2\pi^0$	10.5±0.7	9.1±1.6	7.5±0.9	8.4±0.4
$h^-+\geq 3\pi^0$	1.5±0.4	2.4±0.8	3.0±2.7	1.2±0.1
$3h+\geq 1\pi^0, 5h+\geq 0\pi^0$	5.1±0.3	5.2±0.4	4.7±1.0	
Total(including correlations)	42.1±0.6	41.9±1.2	38±3.4	
3h	9.5±0.7		6.7±0.6	

improved measurements are called for.

SUMMARY

Recent measurements of the tau topological branching ratios at LEP are consistent with each other but give a *signifiantly lower* 1-prong branching ratio as compared with the pre-LEP world average. The difference is 1.7±0.4%. The $\pi(K)$ and ρ branching ratios measured at LEP are consistent with the previous world average but are higher by 2%. Recent measurements of the various exclusive decay modes of the tau indicate that any discrepency between the inclusive and sum of exclusive 1-prong branching ratios is no more than about 3%.

REFERENCES

1. Particle Data Group, "Review of Particle Properties", Phys. Lett. B239, (1990).
2. CELLO Collaboration, H. J. Behrend *et al.*, Phys. Lett. B222, p. 163, (1989).
   CELLO Collaboration, H. J. Behrend *et al.*, Z.Phys. C46 p. 537, (1990).
3. ALEPH Collaboration, D. Decamp *et al.*, CERN-PPE/91-186 (1991).
4. Particle Data Group, "Review of Particle Properties", Phys. Rev. D45, (1990).
5. HRS Collaboration, S. Abachi *et al.*, Phys. Rev. D40 p. 902, (1989).
6. M. McCubbin, "Lifetime and Leptonic B.R.s of the Tau", this conference.
7. K. K. Gan, "Tau Physics at CLEO", this conference.
8. OPAL Collaboration, P. Acton *et al.*, CERN-PPE/92-66 (1992).
9. L3 Collaboration, B. Adeva *et al.*, Phys. Lett. B265, p. 451, (1991).
10. B. C. Barish and R. Stroynowski, Phys. Rep. 157 p. 1, (1988).
11. DELPHI Collaboration, P. Abreu *et al.*, CERN-PPE/92-60 (1992).
12. CLEO Collaboration, R. Ammar *et al.*, Phys. Rev. D45, p. 3976, (1992).
13. A. Golutvin, "Heavy Flavour physics at ARGUS", this conference.
14. CLEO Collaboration, M. Goldberg *et al.*, Phys. Lett. B251, p. 223, (1990).
15. ARGUS Collaboration, H. Albrecht *et al.*, Z.Phys. C41, p. 405, (1988).

PROPOSED EXPLANATION OF TAU LEPTON DECAY PUZZLE: DISCREPANCY BETWEEN THE MEASURED AND THE THEORETICAL TAU LIFETIMES*

Chang Kee Jung
Department of Physics
The State University of New York at Stony Brook
Stony Brook, New York 11794-3800

Abstract

I propose an explanation of the current discrepancy between the measured world average value and the theoretical value of the τ lepton lifetime. I argue that there exist common systematic errors in many of the experiments that use 3-prong τ decays for the lifetime measurement. These systematic effects always shift the measurement towards longer lifetimes, and are caused by the small opening angle of 3-prong τ decays and limited tracking chamber ability to resolve nearby hits. The theoretical τ lifetime agrees well with the measured world average value from the 1-prong τ decays.

The discrepancy between the measured world average value of the τ lepton lifetime and the theoretical value obtained from the current world average values of the τ mass and leptonic branching ratios is one of the few remaining puzzles in the framework of the Standard Model. In this model, since the τ is a sequential lepton with known universal charged current couplings, its lifetime τ_τ is calculable, and directly related to the μ lifetime τ_μ. Neglecting the electron and neutrino masses, the lowest order theoretical prediction for τ_τ in the Standard Model is:[1]

$$\tau_\tau = \tau_\mu (m_\mu/m_\tau)^5 B_l/f,$$

where $m_{\mu,\tau}$ are the masses of the μ and τ respectively; B_l is the τ leptonic branching ratio $B(\tau^- \to l^- \nu_\tau \bar{\nu}_l;\ l = e$ or $\mu)$; and f is a phase-space suppression factor which is unity for $l = e$ and 0.973 for $l = \mu$. When the 1990 Particle Data Group world average value[2] of m_τ and B_e are used, the predicted value of τ_τ becomes $2.83 \pm 0.07 \times 10^{-13} s$ which is about 1.9σ away from the 1990 world average value for the measured τ_τ of $3.03 \pm 0.08 \times 10^{-13} s$. To express the discrepancy in a more convenient way, let's define R_g as the ratio of the two weak coupling constants g_τ^w and g_l^w squared which is a measure of lepton universality, namely:

$$R_g \equiv \left(\frac{g_\tau^w}{g_l^w}\right)^2 = \left(\frac{m_\mu}{m_\tau}\right)^5 \frac{\tau_\mu}{\tau_\tau} \frac{B_l}{f}$$

where g_e^w and g_μ^w are assumed to be identical, and are determined from μ decays. The discrepancy is then expressed as R_g(from B_e) = 0.935 ± 0.034. By combining the results from B_e and B_μ, one obtains R_g(1990 Ave.)= 0.950 ± 0.031 which is about 1.6 σ away from unity. The discrepancy is not overwhelming. However, its persistence and the unusually large number of consistent experimental measurements demand a careful analysis of the situation, especially when new measurements of τ_τ and B_l from LEP and CLEO experiments[3] further confirm the discrepancy, R_g(1992 Ave.)= 0.948 ± 0.023, contrary to gen-

---

*This work was supported in part by Department of Energy contract DE-FG02-92ER40697.

eral expectations. In fact, the significance of the discrepancy has increased to 2.3σ with the addition of new measurements. The discrepancy suggests that the SM, or one or more of the experimental measurements of τ_τ, m_τ and B_l is wrong.

If all the experimental measurements are correct, then a fundamental change is necessary in the SM to accommodate the observed discrepancy. On the other hand, if the SM is correct, one or more of the experimental measurements must be wrong. Indeed, newly reported results[4] on the m_τ indicates that the previous world average value was off by about 7 MeV/c^2. However, such a change in m_τ reduces the significance of the discrepancy only slightly because the new measurements have significantly smaller errors than the previous measurements. Using the new average value of m_τ, 1777.0 ± 0.5 MeV/c^2, one obtains $R_g(1992) = 0.964 \pm 0.021$ which is 1.7σ away from unity. Thus, the discrepancy still persists.

In this Letter, I argue that the systematic bias in τ_τ measurements is the root of the lifetime discrepancy puzzle. By revisiting a previous study,[5] I show that there is a systematic bias towards longer lifetimes in the τ_τ measurement from 3-prong decays which is probably common to most of the experiments.

In general, the τ_τ measurements can be classified into two categories: measurements from 1-prong τ decays and measurements from 3-prong τ decays. In principle, the τ_τ values measured from 1-prong and 3-prong decays should be equal. The topological complication of 3-prong τ decays, however, introduces a unique systematic error to the τ_τ measurements that does not exist in 1-prong decays. The systematic error results from nearby hits generated in the tracking detectors by the three charged particles in a τ decay.

The most obvious bias comes from "hit sharing" among the three reconstructed tracks. A "shared hit" is defined as a recorded drift time in drift chambers which is used in the reconstruction of two or more tracks. The hit sharing problem is particularly serious for the τ decays at high energy colliders, since the decay particles have small opening angles due to the large Lorentz boost. The cylindrical drift chambers most commonly used at e^+e^- collider experiments measure the projected angle of tracks in the plane perpendicular to the beam axis, increasing the probability of hit sharing. (Most previous experiments have used some type of high resolution gas drift chambers to measure the τ lifetime. The use of silicon type vertex detectors or similarly fine grained high resolution vertex detectors for the τ lifetime measurement has been so far minimal.) Since most of the experiments use single-hit electronics readout systems for the drift time measurements, if two or more charged particles pass through the same drift cell, only the information on the shortest drift time can be recorded, resulting in a source of bias in tracking. Because of the limited detector resolution there is no *a priori* way to determine to which track a shared hit should be assigned. Thus, many track fitters simply try all possible combinations to minimize the χ^2 of the fit. Usually, if the χ^2 contribution of a hit is too large, the hit is dropped from the fit. Consequently if two tracks are separated by more than the chamber resolution limit, the confusion is, in principle, resolved and the hit is assigned to the track to which it makes a smaller χ^2 contribution; otherwise the hit is assigned to both tracks, which I define as a shared hit. It should be noted that even when a distinct hit for each track is detected, incorrect assignments can still occur. For experiments that used multihit electronics readout systems, the same problem exists within the two track resolution limit.

Including a misassigned hit in track fitting introduces a distortion in the track trajectory. Ironically, the problem is especially severe in the vertex chambers which are constructed mainly for the purpose of measuring lifetimes and are located very near the beam interaction points. Due to obvious geometrical reasons, hits are shared more often in the vertex layers, and because of its good resolution and proximity to the beam axis, a misassigned hit in the vertex layers can cause a large shift in a track trajectory when extrapolated to the beam interaction region. It is straightforward to show that when hit sharing occurs between the tracks from a decaying particle whose lifetime we wish to measure, it always results in a longer measured decay length and thus long lifetime measurements regardless of the type of lifetime measurement method used. To simplify the problem, consider a case when two nearby reconstructed tracks share a hit. Since the hit is the shorter of the two hits generated by the particles, the track with a misassigned hit will be pulled toward the other, resulting in an apparent longer decay length.[6] In general, hit sharing among the three tracks from τ decay usually results in a longer lifetime.

I examined this hit sharing effect in detail by using the 3-prong τ decays and the $D^0(D^0 \to K\pi)$ decay event samples in the data obtained by the HRS experiment at PEP operated in the e^+e^- center-of-mass energy (e_{cm}) at 29 GeV. A detailed description of the HRS detector can be found elsewhere.[7]

The 3-prong τ decay events used for the study were selected with a 1 vs 3-prong topological cut. Thus, if any hit sharing occurred, it would be among the 3-prong τ decay tracks. In Fig. 1, the mean decay length of the τ candidates measured with the decay length method is plotted against the number of shared hits, counting all combinations between any two of the three tracks in the vertex detector layers. The plot shows clearly an increase in decay length as the number of shared

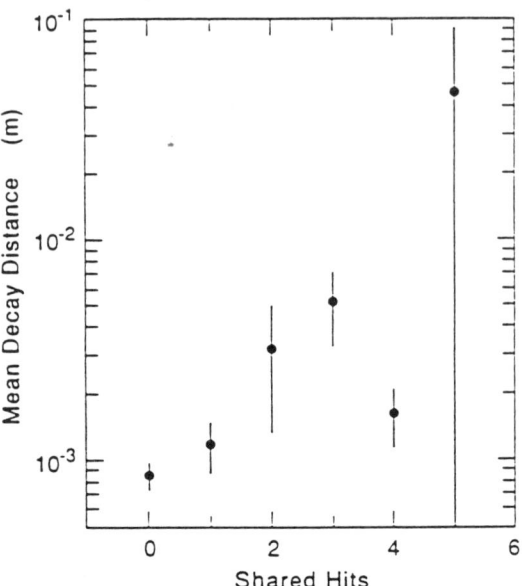

Fig. 1. Number of shared hits vs τ mean decay length for 3-prong τ decays: The error bars reflect possible systematic errors associated with track reconstruction as well as the statistical errors. The typical event by event error in the measured decay length is about 1 mm.

hits increases. It also shows that the error becomes large as shared hits increase. It reflects lack of statistics for those bins but also the fact that as the number of shared hits increase the fluctuation in the measured decay length increases.

With the $D^0 \to K\pi$ events, I examined the effect of the hit sharing between a daughter track of the D^0 and a nearby spectator track that does not come from the D^0 decay. In this case, we expect the daughter tracks would be pulled to either direction, resulting in random fluctuations in the measured decay lengths. When the mean decay length of the D^0 is plotted against the number of shared hits between a daughter and a spectator track, no particular trend for longer or shorter measured mean decay length is observed.

A similar study of measured τ_τ decay

lengths was done by using the data obtained by the Mark II experiment at PEP.[8] The Mark II pattern recognition program does not allow hits to be shared between tracks, *i.e.*, a hit is assigned to only the track to which it contributes the smallest χ^2. Nonetheless, a similar bias can be observed by plotting the mean decay length as a function of the angle ϕ_{ij} between tracks projected in the plane perpendicular to the beam axis as shown in Figure 10 of reference 8. The distribution shows that the decay length is constant (about 600 μm) within errors for ϕ_{ij} greater than 0.1 radian, while it is greater than 1 mm for ϕ_{ij} less than 0.1 radian.

The amount of "raw bias" caused by these effects is large. However, it can be substantially reduced after cuts on such as track quality, decay length and decay length error, and after error scaling in lifetime fitting, even if such procedures are not designed specifically to remove the events with these effects. The net amount of remaining bias depends on the details of the experimental apparatus and the analysis method used. For some measurements, the remaining bias may be negligible. For others it may not.

In principle, one could correct for this effect. The perfect Monte Carlo simulation of each experiment should naturally reflect these effects. But too optimistic and simplified detector models, especially that does not simulate correctly the tails of the hits and interference among the multihits can result in underestimation of the effects. Thus, the best way to completely avoid the bias is to identify and remove the offending events directly from the final data sample. This, of course, reduces the available statistical power of each experiment. Traditionally, when disagreement between the Monte Carlo simulation and the data is observed, one increases the estimated measurement uncertainty to account for unknown systematic errors. Therefore, usually the bias is, in a sense, correctly accounted for in the form of systematic errors for each experiment. The problem arises when the weighted world average is made by combining each measurement assuming independent systematic errors, since the bias due to the hit sharing is common to most of the measurements from 3-prong decays and has the same sign.

In light of the above argument and assuming no specific corrections are made to the τ_τ measurements for the hit sharing effect, the following three general predictions can be made. The first is that the world average τ_τ value measured from 3-prong τ decays should be longer than the average value measured from 1-prong τ decays. The second is that for comparable detector resolution the τ_τ value measured at LEP energies should be longer than the value measured at PEP and PETRA energies, since at LEP the taus are produced with higher Lorentz boost resulting in smaller opening angles and thus more hit sharing. The third is that if a finely grained high resolution tracker, such as a silicon vertex detector, with enough layers to provide a significant weight to track fitting is used, a shorter value for τ_τ will be observed.

Table 1. All statistically independent published τ lifetime measurements from 1-prong τ decays.

Experiment	Measurement	Combined error
MAC 1985	$3.15 \pm 0.36 \pm 0.40$	0.54
MAC 1987	$2.97 \pm 0.26 \pm 0.14$	0.30
JADE 1989	$2.89 \pm 0.33 \pm 0.26$	0.42
DELPHI 1991	$3.21 \pm 0.36 \pm 0.16$	0.39
OPAL 1991	$2.93 \pm 0.13 \pm 0.13$	0.18
L3 1991	$3.18 \pm 0.28 \pm 0.37$	0.46
ALEPH 1992	2.90 ± 0.16	0.16

Table 2. All statistically independent published τ lifetime measurements from 3-prong τ decays.

Experiment	Measurement	Combined error
Mark II 1982	4.60 ± 1.90	1.90
MAC 1982	4.90 ± 2.00	2.00
CELLO 1983	$4.70 ^{+3.90}_{-2.90}$	$^{+3.90}_{-2.90}$
Mark II 1987	$2.88 \pm 0.16 \pm 0.17$	0.23
MAC 1987	$3.16 \pm 0.26 \pm 0.10$	0.28
HRS 1987	$2.99 \pm 0.15 \pm 0.10$	0.18
CLEO 1987	$3.25 \pm 0.14 \pm 0.18$	0.23
ARGUS 1987	$2.95 \pm 0.14 \pm 0.11$	0.18
TASSO 1988	$3.06 \pm 0.20 \pm 0.14$	0.24
JADE 1989	$3.09 ^{+0.35}_{-0.34} \pm 0.11$	0.37
DELPHI 1991	$3.10 \pm 0.31 \pm 0.09$	0.32
OPAL 1991	$3.27 \pm 0.17 \pm 0.11$	0.20
L3 1991	$3.02 \pm 0.36 \pm 0.21$	0.42
CLEO 1991	$3.10 \pm 0.15 \pm 0.07$	0.17
ALEPH 1992	$2.94 \pm 0.25 \pm 0.11$	0.27

In Tables 1 and 2, all published 1-prong and 3-prong τ_τ measurements are compiled separately. As can be seen, the lifetimes measured from 3-prong τ decays dominate the current world average. In Table 3, various averaged values of the τ_τ have been calculated using all published measurements. Although the statistical significance is not compelling, one can see some degree of confirmation of the above predictions, except the third, for which available measurements to date are few: The world average value of τ_τ has not changed with inclusion of new results from the LEP and CLEO experiments but the error has become smaller; the average value of τ_τ measured from 3-prong decays is longer than the aver-

Table 3. Various averages of the τ_τ measurements: The average values are obtained by using all statistically independent measurements. The measurements by the Mark II (1982), MAC (1982) and CELLO (1983) experiments are excluded from the relevant averages, following the PDG recipe.

Average	Ave. measured lifetime
World 1990	3.03 ± 0.08
World 1992	3.03 ± 0.06
1-prong 1990	2.98 ± 0.22
1-prong 1992	2.96 ± 0.10
3-prong 1990	3.03 ± 0.09
3-prong 1992	3.07 ± 0.07
3-prong PEP/PETRA	3.01 ± 0.11
3-prong LEP	3.13 ± 0.14

Table 4. Various R_G values obtained from averages of τ_τ measurements. For 1990 averages, the 1990 PDG values of B_l were used in the calculation.

Average	$R_G(m_\tau=1784.1)$	$R_G(m_\tau=1777.0)$
World 1990	0.956 ± 0.031	0.966 ± 0.030
World 1992	0.948 ± 0.023	0.964 ± 0.021
3-prong 1992	0.935 ± 0.025	0.954 ± 0.024
1-prong 1992	0.970 ± 0.035	0.990 ± 0.035

age value measured from 1-prong decays; and the average 3-prong τ_τ measured at LEP experiments is longer than the average 3-prong τ_τ (1990) measured at PEP and PETRA experiments. In Table 4, various values of R_G are summarized for the m_τ values of the 1990 average (PDG) and the new 1992 average, respectively. As can be seen, the 1-prong averages agree much better with the lepton universality for both m_τ values than the 3-prong averages.

In conclusion, I argue that the cause of the tau decay puzzle, the discrepancy between the measured and the theoretical τ lifetime, comes from the experimental bias in the τ lifetime measurements of 3-prong τ decays. The world average value of the measured tau lifetime is biased towards a longer lifetime due to the biased measurements from 3-prong τ decays. Presently, measurements from 3-prong τ decays dominate the world average value for the τ lifetime; while the theoretical τ lifetime agrees well with the measured world average value from the 1-prong τ decays. The current R_G value obtained using only the 1-prong decays is 0.990 ± 0.035, in excellent agreement with lepton universality.

Without properly accounting for the bias in the 3-prong τ lifetime measurements, I predict, the lifetime discrepancy will not be resolved in the near future even with new precise measurements on the τ mass, the leptonic branching ratio or the lifetime.

I strongly recommend that experimentalists who still have access to data should reconsider earlier analyses keeping in mind the points made in this paper. I also suggest that the particle data group should separate the lifetime measurements into two groups: 1-prong measurements and 3-prong measurements.

I thank K. Hayes, W. Marciano, and K. Riles for their helpful discussions. I also thank the former HRS collaboration for providing data for this analysis.

REFERENCES

1. Y. S. Tsai, *Phys. Rev.* **D4** 2821 (1971). The radiative corrections to the τ_τ and τ_μ calculations are small (~0.4%) and their effects nearly cancel in the equation.

2. Particle Data Group, M. Aguilar-Benitez et al., *Phys. Lett.* **B239** (1990).

3. ALEPH collaboration, D. Decamp et al., *Phys. Lett.* **B279**, 411 (1992); DELPHI collaboration, P. Abreu et al., *Phys. Lett.* **B267**, 422 (1991); OPAL collaboration, P. D. Acton et al., *Phys. Lett.* **B273**, 355 (1991); L3 collaboration, B. Adeva et al., *Phys. Lett.* **B265**, 451 (1991); CLEO collaboration, G. Crawford et al., contributed paper to the LP-HEP conference at Geneva, Switzerland, July 1991.

4. BES collaboration (presented by N. Qi, PRINT-92-0147 Beijing) and ARGUS collaboration (presented by H. Kolanoski) at the APS meeting at Washington, D.C., Apr. 1992.

5. C. K. Jung, *Measurement of the $F^\pm$ meson lifetime*, Ph. D. Thesis, Indiana University (1986).

6. If electronics picks out longer drift distance, the same effect occurs for cases when two nearby particles pass through the same side of a sense wire, while the opposite effect occurs for cases when the particles pass through the opposite side of a sense wire.

7. D. Bender et al., *Phys. Rev.* **D30**, 515 (1984); P. Baringer, C. K. Jung, H. O. Ogren and D. Rust, *Nucl. Instr. Meth.* **A254**, 542 (1987).

8. K. Hayes, in *Proceedings of the Tau-Charm Factory Workshop*, SLAC-Report-343, p. 90 (1989).

CP Violation, B$\bar{\text{B}}$ Mixing, and Rare Decays

Parallel Session 2

Conference Reception, Union Station, Dallas

Conveners: *A. Sanda (Nagoya)*
M. Zeller (Yale)

A MEASUREMENT OF B^0-$\overline{B^0}$ MIXING IN HADRONIC Z^0 DECAYS

THE OPAL COLLABORATION
Presented by
V. Gibson
Cavendish Laboratory
University of Cambridge
Cambridge, UK, CB3 0HE

Abstract

A measurement of B^0-$\overline{B^0}$ mixing from a sample of $\sim 500\,000$ hadronic Z^0 decays recorded with the OPAL detector at LEP is presented. A signal for mixing is observed and an average B^0-$\overline{B^0}$ mixing parameter

$$\chi = 0.125^{+0.017}_{-0.016} \pm 0.015$$

is extracted from the excess of like sign lepton pairs in events containing two lepton candidates.

INTRODUCTION

In the neutral B meson system, the probability that a B^0 meson will transform into its antiparticle can be written as

$$\chi = \frac{N\left(B^0 \to \overline{B^0}\right)}{N\left(B^0 \to \overline{B^0}\right) + N\left(B^0 \to B^0\right)} \quad (1)$$

In Z^0 decays, B^0_s and B^0_d mesons are produced in addition to charged B mesons and b-flavoured baryons such that an average B^0-$\overline{B^0}$ mixing can be defined as

$$\chi = f_d \chi_d + f_s \chi_s \quad (2)$$

where $f_{d,s}$ is the fraction of b quarks that form $B^0_{d,s}$ mesons which mix with a probability $\chi_{d,s}$.

A signature for b quark production is via the semileptonic decay of b-flavoured hadrons. The leptons (either electrons or muons) are produced with a large momentum, p, and large transverse momentum, p_T, with respect to the line of flight of the decaying b hadron and the charge of the lepton is used to tag the charge of the b quark. The signal for B^0-$\overline{B^0}$ mixing is therefore an excess of like sign, high p and high p_T lepton pairs on opposite sides of the event. The main sources of background to the mixing signal arise from

- Primary $b \to \ell$ decays with secondary $b \to c \to \ell$ decays on the opposite side of the event,

- Leptons that do not originate from heavy quarks such as misidentified hadrons, decays of pions and kaons and electrons from photon conversions.

In a previous publication [1], the OPAL collaboration reported a measurement of B^0-$\overline{B^0}$ mixing using data accumulated during 1990. In this analysis, a sample of $\sim 500\,000$ hadronic Z^0 decays from the 1990 and 1991 data are used corresponding to a threefold increase in statistics. In addition, improvements in the algorithms used to identify leptons from semileptonic decays of b hadrons and electrons from photon conversions reduce the experimental systematic errors to a negligible level compared to the uncertainty on the semileptonic branching ratio Br($b \to c \to \ell$).

LEPTON IDENTIFICATION

The data were recorded with the OPAL detector [2] at the CERN e^+e^- collider LEP. Muons are identified using a positional match between the central detector track extrapolated to the muon track segment taking into account energy loss, multiple scattering and measurement errors. Muon candidates are also required to satisfy dE/dx measurements designed to reject kaons and protons. Muon identification in the hadron calorimeter is only used for tracks outside the muon chamber acceptance. The polar angle coverage for muons is $|\cos\theta| < 0.97$. The muon identification efficiency from Monte Carlo simulation is $\sim 80\%$ for $p > 6\,\text{GeV}/c$ and shows good agreement when compared to the measured identification efficiencies for muons from $Z^0 \to \mu^+\mu^-$ at $p \sim 45\,\text{GeV}/c$ and the two-photon process $e^+e^- \to e^+e^-\mu^+\mu^-$ at $p \sim$ 2-6 GeV/c.

Electron identification is based on energy loss measured in the jet chamber, the amplitude in the presampler, the distribution of energy in the electromagnetic cluster associated to a track and the comparison of the track momentum with the electromagnetic cluster energy [3]. The polar angle coverage for electrons is $|\cos\theta| < 0.7$ and $0.83 < |\cos\theta| < 0.9$. The electron identification efficiency is measured from the data by exploiting the independence of the energy loss and calorimeter selection criteria and is typically 50-60% for $p_T > 1\,\text{GeV}/c$.

Lepton candidates are required to have $p > 2\,\text{GeV}/c$ and to satisfy track quality cuts. For each event, the charged tracks and the electromagnetic energy clusters that are unassociated to tracks are grouped into jets according to the JADE scaled invariant jet finding algorithm [4]. The E0 recombination scheme is used with the particle pair effective invariant mass, $x_{min} > (7\,\text{GeV}/c^2)^2$ [5]. Lepton candidates are included in the jet axis calculation and their p_T is calculated with respect to the axis of the jet to which they are associated.

NON-PROMPT BACKGROUNDS

The background to prompt leptons in hadronic Z^0 decays arise from leptons which do not originate from heavy quarks. For muon candidates, these include decays in flight of pions and kaons, hadrons whose interaction products penetrate the detector material, hadrons which do not interact strongly in the material and hadrons which are incorrectly associated with muon detector track segments. The probability for a hadron to fake a prompt muon is $\sim 0.8\%$ per track estimated using Monte Carlo and pions from $K_s \to \pi^+\pi^-$ and $\tau \to 3\pi$ decays.

The probability that a hadron track is identified as an electron is measured from the data using energy loss and calorimeter selection criteria and is typically 0.05% and 0.1% per track for $p < 6\,\text{GeV}/c$ and $p > 6\,\text{GeV}/c$ respectively. Electrons from photon conversions are identified and rejected using a conversion finding algorithm with an efficiency of 84%.

RESULTS

The signature for B^0-$\overline{B^0}$ mixing is an excess of like sign, high p and high p_T lepton pairs on opposite sides of the event in hadronic Z^0 decays. Combining p and p_T into a single variable, p_{comb},

$$p_{comb} = \sqrt{\left(\frac{p}{10}\right)^2 + p_T^2} \quad (3)$$

provides a good separation between primary $b \to \ell$ decays and backgrounds from secondary $b \to c \to \ell$ decays and hadron misidentification. A good discrimination variable is the minimum p_{comb} of the two leptons, p_{comb}^{min}. Requiring that the angle between the two leptons is $> 60°$ results in a data sample of 1914 $\mu\mu$, 1315 μe and 298 ee events. Lepton pair candidates with small opening angles provide an important check on the secondary $b \to c \to \ell$ and non-prompt backgrounds.

The distributions of p_{comb}^{min} are shown in Figure 1 and correspond to 493 000 hadronic Z^0 decays. Monte Carlo events and non-prompt background estimates are used to predict the relative abundance of dilepton candidates from each of the production mechanisms. The JETSET 7.3 Monte Carlo program [6] was used to generate $Z^0 \to b\bar{b}$ and $Z^0 \to c\bar{c}$ events and the fragmentation was performed using the Peterson parameterization [7]. The partial widths $\Gamma(Z^0 \to b\bar{b})/\Gamma(Z^0 \to \text{hadrons}) = 0.217$ and $\Gamma(Z^0 \to c\bar{c})/\Gamma(Z^0 \to \text{hadrons}) = 0.171$ and the semileptonic branching ratios $Br(b \to \ell) = 11.2\%$, $Br(b \to c \to \ell) = 9.8\%$ and $Br(c \to \ell) = 8.0\%$ are used to absolutely normalise the predicted distributions. The $b \to c \to \ell$ branching ratio is calculated using 10.2% for B_d^0 and $B^\pm$ decays from a range of values obtained by CLEO [8] and the JETSET $b \to c \to \ell$ branching ratios for the other b hadrons. The lepton momentum spectra for primary $b \to \ell$ decays is corrected for a 15% $B \to D^{**}\ell\nu$ component.

Counting Method

In this method, events are required to satisfy $p_{comb}^{min} > 1.2\,\text{GeV}/c$. In this region, 89% of the $Z^0 \to b\bar{b}$ events are predicted to contain two prompt leptons and 74% contain two primary semileptonic decays of b-flavoured hadrons. The mixing parameter, χ, is determined from the ratio, R, of the excess of like sign lepton pair events over all lepton pair events. This ratio is shown in Figure 2 together with the predictions for no mixing ($\chi = 0$) and also for $\chi = 0.1$ and 0.2. The number of events, R and the measured value of χ, which when applied to the Monte Carlo

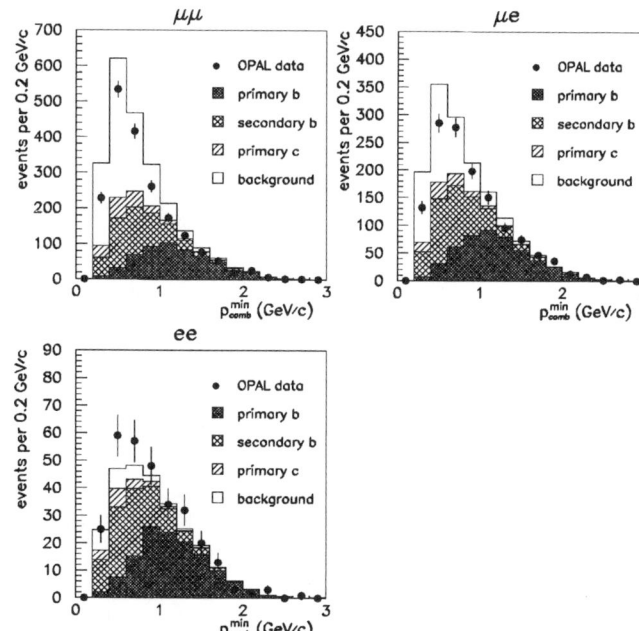

Figure 1. Distributions of p_{comb}^{min} for lepton pairs separated by greater than $60°$.

reproduces the measurement of R, are given in Table 1. The result $\chi = 0.137^{+0.020}_{-0.019}$ (stat) is obtained from the sum of the $\mu\mu$, μe and ee channels.

Fitting Method

A maximum likelihood fit is performed to the distributions of R versus p_{comb}^{min} for the $\mu\mu$, μe and ee data simultaneously. The fit result is $\chi = 0.125^{+0.017}_{-0.016}$ (stat) and has a fit quality chi-squared of 19.2 for 22 degrees of freedom. The value of χ obtained in this method is used in the final result.

SYSTEMATIC ERRORS

The same sources of systematic errors are considered for both methods and are listed in Table 2 together with their effect on the determined value of χ.

The largest systematic in both cases arises from the uncertainty on $Br(b \to c \to \ell)$ due to

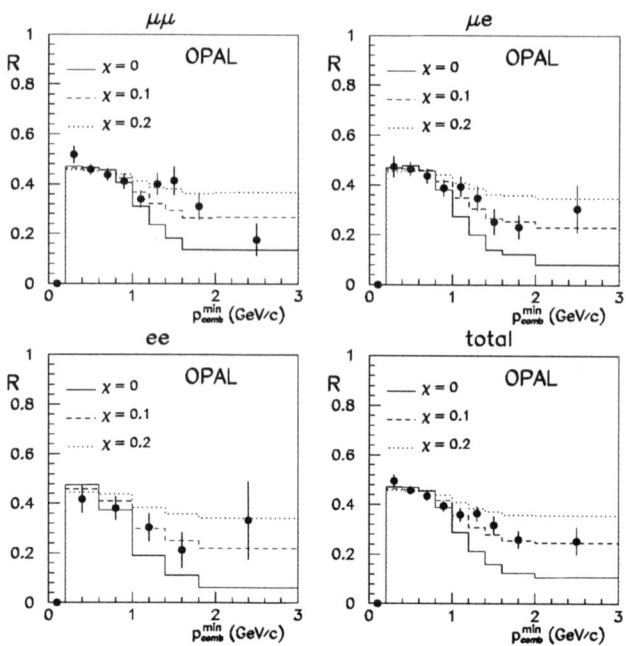

Figure 2. The fraction, R, of the number of large angle like sign lepton pair events versus P_{comb}^{min}.

DISCUSSION AND CONCLUSION

In conclusion, OPAL has observed a signal for B^0-$\overline{B}^0$ mixing using 3527 lepton pair events from $\sim 500\,000$ hadronic Z^0 decays. The average mixing parameter

$$\chi = 0.125^{+0.017}_{-0.016}(\text{stat}) \pm 0.015(\text{syst}) \qquad (4)$$

where the largest systematic error is due to the uncertainty on $\text{Br}(b \to c \to \ell)$. A combined analysis of the OPAL single lepton and lepton pair data will reduce the systematic error on χ. Combining the OPAL measurement with the value of $\chi_d = 0.167 \pm 0.019 \pm 0.018$ measured by CLEO [11], a constraint can be placed on the relation between χ_s and the fraction of b quarks that form B_s^0 mesons, f_s. Assuming that equal fractions of B_d and B_u mesons are produced and that the fraction of b-flavoured baryons is $(9 \pm 4.5)\%$, the constraint is shown in Figure 3. The result is consistent with full B_s^0 mixing ($\chi_s = 0.5$) when f_s is near the predicted value of 12%.

variations introduced by the modelling of the semileptonic decay spectrum and the fraction of B_s^0 and b-flavoured baryons produced in Z^0 decays. The $\text{Br}(b \to \ell)$ is assigned an uncertainty to reflect the range and errors of measurements at LEP and lower energies [8, 9, 10]. and the fraction of D** produced in primary $b \to \ell$ decays is varied between 0 and 32%. $\Gamma(Z^0 \to b\bar{b})/\Gamma(Z^0 \to \text{hadrons})$ and $\text{Br}(c \to \ell)$ are varied by the precision to which they are currently measured. The sensitivity of χ to the b quark fragmentation is assessed by varying $\langle x_E \rangle_b$, the mean energy of the primary b hadrons divided by the beam energy.

Systematic errors due to experimental uncertainties such as non-prompt backgrounds and efficiencies are small compared to the uncertainty on $\text{Br}(b \to c \to \ell)$.

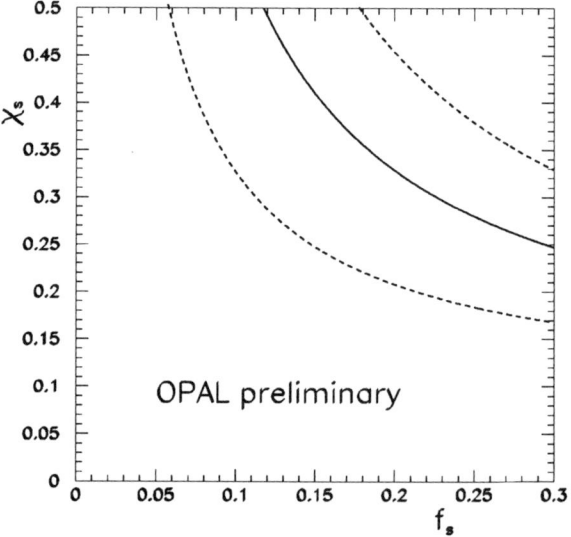

Figure 3. The B_s^0 mixing probability, χ_s, versus the fraction of b quarks that form B_s^0 mesons, f_s. The dashed lines indicate the 1σ errors.

Table 1. Numbers of events with lepton pairs satisfying $p_{comb}^{min} > 1.2$ GeV/c.

	Number of events	Number with like sign leptons	Ratio R	Monte Carlo R			Measured χ
				$\chi = 0$	$\chi = 0.1$	$\chi = 0.2$	
$\mu\mu$	302	109	0.361 ± 0.028	0.186	0.293	0.376	$0.179^{+0.039}_{-0.035}$
$e\mu$	273	79	0.289 ± 0.027	0.154	0.275	0.369	$0.113^{+0.027}_{-0.025}$
ee	75	19	0.253 ± 0.050	0.114	0.252	0.359	$0.101^{+0.043}_{-0.039}$
Total	650	207	0.318 ± 0.018	0.166	0.282	0.372	$0.137^{+0.020}_{-0.019}$

Table 2. Systematic errors on the determination of χ.

Source	Variation Considered	Effect on χ	
		Cut method	Fitting method
$Br(b \to c \to \ell)$	$\pm 16\%$	± 0.009	$+0.011 / -0.010$
$Br(b \to \ell)$	$\pm 10\%$	$+0.006 / -0.008$	$+0.007 / -0.009$
D^{**} fraction	0–32%	± 0.004	± 0.002
$\Gamma(Z^0 \to b\bar{b})/\Gamma(Z^0 \to \text{hadrons})$	$\pm 10\%$	± 0.001	± 0.001
Fragmentation, $\langle x_E \rangle_b$	0.70–0.74	± 0.002	± 0.003
$Br(c \to \ell)$	$\pm 13\%$	± 0.0006	± 0.0002
Muon background	$\pm 30\%$	± 0.003	± 0.003
Electron misidentification	$\pm 25\%$	± 0.0002	± 0.0005
Conversion background	$\pm 50\%$	± 0.0006	± 0.0008
Muon misassociation	$\pm 100\%$	± 0.0002	± 0.0002
Electron efficiency	Constant 0.5	± 0.003	± 0.003
Muon efficiency	$\pm 5\%$ overall change	± 0.0006	± 0.001
	$\pm 25\%$ for $|\cos\theta| > 0.9$	< 0.0001	± 0.0003
Monte Carlo statistics		± 0.004	± 0.004
TOTAL (added in quadrature)		$+0.013 / -0.014$	± 0.015

REFERENCES

1. OPAL Collaboration, P.D.Acton et al., Phys. Lett. **B 276** (1992) 379.
2. OPAL Collaboration, K.Ahmet et al., Nucl. Instr. and Meth. **A 305** (1991) 275.
3. OPAL Collaboration, P.D.Acton et al., Z. Phys. **C 55** (1992) 191.
4. W. Bartel et al., Z. Phys. **C 33** (1986) 23; S. Bethke et al., Phys. Lett. **B 213** (1988) 235.
5. OPAL Collaboration, M. Z. Akrawy et al., Z. Phys. **C 49** (1991) 375.
6. T.Sjöstrand, Comp. Phys. Comm. **39** (1986) 347; M. Bengtsson and T. Sjöstrand, Comp. Phys. Comm. **43** (1987) 367; M. Bengtsson and T. Sjöstrand, Nucl. Phys. **B 289** (1987) 810; OPAL Collaboration, M. Akrawy et al., Z. Phys. **C 47** (1990) 505.
7. C.Peterson et al., Phys. Rev. **D 27** (1983) 105.
8. CLEO Collaboration, S.Henderson et al., Phys. Rev. **D 45** (1992) 2212.
9. J.J. Hernandez et al., Particle Data Book: Phys. Lett. **B 239** (1990) 1.
10. L3 Collaboration, B.Adeva et al., Phys. Lett. **B 261** (1991) 177.
11. CLEO Collaboration, reported by K.Lingel, proceedings of Les Rencontres de Moriond, France, 1992.

ALEPH RESULTS ON $B^0 \bar{B}^0$ MIXING

Jean-Pierre LEES
LAPP
Chemin de Bellevue, B.P. 110
74941 ANNECY LE VIEUX CEDEX, FRANCE

Abstract

We report on a preliminary measurement of the $B^0 \bar{B}^0$ mixing parameter χ using data recorded in 1990 and 1991 with the ALEPH detector. From a global fit to the lepton and dilepton spectra one gets $\chi = 13.7 \pm 1.5 \pm 0.7\%$. After folding in the value obtained from the 1990 jet charge analysis the combined value is $\chi = 13.4 \pm 1.3 \pm 0.8\%$

INTRODUCTION

Since its first evidence by UA1 in 1987[1], $B^0 \bar{B}^0$ mixing has been widely studied, both at experiments running on Υ_{4S}, where only B_d^0 are produced, and at higher energy e^+e^- or hadronic colliders, where both B_d^0 and B_s^0 are produced. In the Standard Model, mixing of B_d^0 and B_s^0 mesons occurs through the box diagram and depends on the mass and weak couplings of the top quark. It is characterized by the ratio x_q of the $B_q^0 \bar{B}_q^0$ mass difference over width (1).

$$x_q = \frac{\Delta M_q}{\Gamma_q} \propto M_t^2 \mid V_{tq} V_{tb}^* \mid^2 \quad (1)$$

Present experimental information on the mixing probabilities is based on time integrated measurements. For that one defines the parameters χ_d and χ_s, which give the probability that a B_d^0 or a B_s^0 has oscillated to its antiparticle (2).

$$\chi_q = Prob(B_q^0 \to \bar{B}_q^0) \simeq \frac{1}{2} \frac{x_q^2}{1+x_q^2} \quad (2)$$

At LEP, all kinds of B hadrons are produced, and one defines an effective mixing parameter χ, which is just the probability that a b quark has hadronised to a B^0 meson which has oscillated (3).

$$\begin{aligned}\chi &= P(b \to B_d^0 \to \bar{B}_d^0) + P(b \to B_s^0 \to \bar{B}_s^0) \\ &= f_d \chi_d + f_s \chi_s \quad (3)\end{aligned}$$

where f_d, f_s are the fraction of B_d^0 and B_s^0 produced. Values commonly assumed are $f_d \sim 40\%$ and $f_s \sim 12\%$.

In this talk, we will report preliminary results obtained by the ALEPH collaboration using about ~ 450000 Z^0 hadronic decays collected in 1990 and 1991. Two independant measurements have been performed, one using dileptons (1990+1991 data), and the other one using a lepton-jet charge correlation method (1990 data only).

LEPTON IDENTIFICATION

The ALEPH detector has been described in detail elsewhere.[2] The lepton identification will only be briefly reminded here (for details, see references [3,4,5]). Only leptons of momenta $P \geq 3$ GeV are considered.

Electrons are identified in the ALEPH detector by matching a charged track measured in the TPC and ITC with an energy deposit consistent with being an electron in the ECAL. The ECAL energy deposits in the four towers around the extrapolation of each charged track is compared with that expected for an electron of the measured momentum. The average depth of the energy deposition in the ECAL is also measured and required to have a value compatible with the value expected for an electron. At least 50 TPC ionisation samples are required, and an electron candidate is rejected if the dE/dx is more than 2.5 standard deviations below the expected value. Photon conversions and Dalitz pairs are removed using a pair finding algorithm. Efficiencies and backgrounds are estimated from the data, as described in ref.[3,5]. Typical efficiencies vary between 60% (low P_t electrons) and 75% (high P_t); π misidentification probability vary between .03 and .08 %, depending on P and P_t.

Muons are Identified by matching a charged track to a pattern of hits in the HCAL; the TPC track has also to extrapolate within 4σ of a hit in the muon chambers. The efficiency and backgrounds to muon identification are estimated in the data by the techniques of ref.[3,4,5]. A mean efficiency of 76% is obtained; π misidentification probability vary between .1 and .6%, depending on P and P_t.

DATA SAMPLE

Leptons are selected from a sample of 154967 events $Z^0 \to Q\bar{Q}$ recorded in 1990 and 291857 events $Z^0 \to Q\bar{Q}$ recorded in 1991. A total of 25578 electron and 35774 muon candidates is obtained. The number of lepton pairs, divided into opposit hemisphere and same hemisphere pairs, is given in tables 1 and 2. A total of 4340 pairs (opposit hemisphere) and 2116 pairs (same hemisphere) is observed.

Table 1. Number of opposit side pairs:

type	++,--	+-	All
ee	313	447	760
$\mu\mu$	635	863	1498
$e\mu$	835	1247	2082
All	1783	2557	4340

Table 2. Number of same side pairs:

type	++,--	+-	All
ee	49	353	402
$\mu\mu$	111	641	752
$e\mu$	163	799	962
All	323	1793	2116

JETS AND PT DEFINITION

Jets are formed using all charged tracks with at least 4 hits in the TPC, neutral electromagnetic and and neutral hadronic energy clusters. The scaled invariant mass clustering algorithm is used to form jets. The jet resolution parameter has been chosen as described in ref.[4] ($y_{cut} = M_{12}^2/E_{vis}^2 = 0.0044$). The P_t of the lepton is then defined as the transverse momentum carried by the lepton with respect to the closest jet, where the lepton momentum is included in the jet momentum vector.

DILEPTONS ANALYSIS

In the dileptons method, one observes $B^0\bar{B}^0$ mixing using a sample of events where both B hadrons decay semileptonically either to muons or electrons. If no mixing occurs, the leptons should have opposit sign, and likesign events are therefore a signature of mixing. Provided equality of semileptonic branching ratios for B_d^0 and B_s^0, the fraction of same

Table 3. Contributions to opposit hemisphere dileptons and mixing dependance of the number of same sign dileptons

process	weight	++,--
$b \to l, b \to l$	$\Gamma_{b\bar{b}} Br^2_{b \to l}$	$2\chi(1-\chi)$
$b \to c \to l, b \to l$	$\Gamma_{b\bar{b}} Br_{b \to l} Br_{b \to c \to l}$	$\chi^2 + (1-\chi)^2$
$b \to \bar{c} \to l, b \to l$	$\Gamma_{b\bar{b}} Br_{b \to l} Br_{b \to \bar{c} \to l}$	$2\chi(1-\chi)$
$b \to c \to l, b \to c \to l$	$\Gamma_{b\bar{b}} Br^2_{b \to c \to l}$	$2\chi(1-\chi)$
$b \to l, b \to bkg$	$\Gamma_{b\bar{b}} Br_{b \to l} Br_{b \to back}$	$\chi_{bk}(1-\chi) + \chi(1-\chi_{bk})$
$b \to c \to l, b \to bkg$	$\Gamma_{b\bar{b}} Br_{b \to c \to l} Br_{b \to back}$	$\chi_{bk}\chi + (1-\chi)(1-\chi_{bk})$
$c \to l, c \to l$	$\Gamma_{c\bar{c}} Br^2_{c \to l}$	0
others	fixed	50%

with $\chi_{bk} = (1-\chi)(1-f) + \chi f$, f being the fraction of same sign in pb-bk evts.

sign dileptons in those events is given by

$$\frac{N_{++,--}}{N_{++,--} + N_{+-}} = 2\chi(1-\chi) \quad (4)$$

with $\chi = f_d \chi_d + f_s \chi_s$

All the different contributions to opposit hemisphere dileptons are listed in table 3, together with the fraction of events going into same signe dileptons. Let's call pb leptons from direct b semi-leptonic decays, sb leptons from cascade decays $b \to c \to l$, pc leptons from direct charm decays and bk all others leptons (background). Dileptons from mixing ($pb - pb$) have a contribution proportionnal to $\Gamma_{b\bar{b}} Br^2_{b \to l}$, while dileptons from cascade decays ($pb - sb$) have a contribution proportionnal to $\Gamma_{b\bar{b}} Br_{b \to l} Br_{b \to c \to l}$. These two processes are the two main contributions to same sign dileptons. In the global fit described below, $\Gamma_{b\bar{b}}$, $Br_{b \to l}$ and $Br_{b \to c \to l}$ are determined from our data, together with mixing parameter χ.

The global fit

This method takes advantage of the redundancy of the data to measure simultaneously $\Gamma_{b\bar{b}}$, $\Gamma_{c\bar{c}}$, the b and c fragmentation, the b and c semi leptonic branching ratios and χ. These parameters are extracted through a binned maximum likelihood fit of single leptons, same and opposit side dileptons spectra.

• single lepton events: they are fitted in the (P, P_t) plane. this part of the fit give constraints on the products $\Gamma_{b\bar{b}} Br_{b \to l}$, $\Gamma_{c\bar{c}} Br_{c \to l}$, and the b and c fragmentation parameters $<x_b>$, $<x_c>$

• Opposit side dileptons: they are fitted in the $(P_\otimes, P_{t\ min})$ plane, with

$$\begin{aligned} P_\otimes &= P_{t1} P_{l2} + P_{t2} P_{l1} \\ P_{t\ min} &= Min(P_{t1}, P_{t2}) \end{aligned} \quad (5)$$

This combination of leptons longitudinal and transverse momenta is shown to be the most discriminant combination of leptons momenta to separate the $pb - pb$ and $pb - sb$ contributions. This part of the fit gives constraints on $\Gamma_{b\bar{b}} Br^2_{b \to l}$, $\Gamma_{b\bar{b}} Br_{b \to l} Br_{b \to c \to l}$ and, to less extent, on $\Gamma_{c\bar{c}} Br^2_{c \to l}$. In addition, splitting into two subsamples of same and opposit charge dileptons allows to measure mixing parameter χ.

• Same side dileptons: this sample is strongly dominated by the $pb - sb$ contribution and gives measurement of $Br_{b \to c \to l}$ in b semi leptonic decays. It is also fitted in the $(P_\otimes, P_{t\ min})$ plane.

Fit inputs

The following assumptions are done when doing the global fit:
- Lepton spectra from b are described by an Altarelli distribution[6], tuned to reproduce latest CLEO data[7].
- Charm spectra in B hadronic decays are correctly modeled by the Monte Carlo (Jetset 7.2 parton shower[8])
- Branching ratio $Br_{b \to c \to l}$ factorize as $Br_{b \to c} Br_{c \to l}$ i.e. one assumes that the charmed hadron composition is the same in charm decays, b hadronic and b semi leptonic decays.
- The branching ratio $Br_{b \to \bar{c}}$ is equal to 14%
- The fraction of same sign events in the $pb - bk$ contribution is equal to 46% (MC prediction)

The mixing, branching ratios and partial width dependance of the different contributions is taken into account as described in table 3.

Systematics and final result

The different systematic errors taken into account are lepton efficiencies and background uncertainities, 10% variation on $Br_{b \to \bar{c}}$ and 10% variation on $Br_{b \to c}$. This last value, quite large, is chosen to take into account the approximation done in factorising $Br_{b \to c \to l}$ as $Br_{b \to c} Br_{c \to l}$. The resulting value obtained for χ is:

$$\chi = .137 \pm .015_{stat} \pm .007_{syst}$$

JET CHARGE ANALYSIS

The jet charge analysis will only be very briefly reminded here, since it's only from 1990 data and has already been published[9]. The basic idea here is to tag the charge of one b looking to a high P, P_t lepton on one side, and to construct a momentum weighted jet charge

Table 4. Detail of systematics taken into account and their effect on χ

source	variation	$\Delta\chi$
e efficiency	$\pm 3\%$	0.05%
e misid	$\pm 20\%$	0.02%
$\gamma \to e^+e^-$, Dalitz	$\pm 20\%$	0.1%
μ effic	$\pm 3\%$	0.08%
punchthrough	$\pm 30\%$	0.07%
decays	$\pm 10\%$	0.18%
$b \to W \to \bar{c}$	$\pm 10\%$	0.1%
$b \to c$	$\pm 10\%$	0.05%
bkg charge asymetry	$\pm 4\%$	0.09%
MC statistics		0.6%

from all jet fragments i in the thrust hemisphere opposit to the lepton. The jet charge is defined by

$$Q_J = \frac{\sum_i |p_{L,i}|^\kappa q_i}{\sum_i |p_{L,i}|^\kappa} \quad (6)$$

and one defines the lepton signed jet charge as

$$Q_{lJ} = -q_l \times Q_J \quad (7)$$

Mixing dependance of jet charge is described by

$$\begin{aligned}\langle Q^{\bar{b}}_{qJ}\rangle &= (1-f_0)\langle Q_{B^+}\rangle \\ &+ (f_0 - \chi)\langle Q_{B^0}\rangle + \chi\langle \bar{Q}_{B^0}\rangle\end{aligned} \quad (8)$$

with $f_0 = f_d + f_s$ fraction of B^0 mesons, $\langle Q_{B^+}\rangle$ mean jet charge for B^+ and B baryons jets, $\langle Q_{B^0}\rangle$ mean jet charge for B^0 which have not mixed and $\langle \bar{Q}_{B^0}\rangle$ mean jet charge for B^0 which have mixed. Taking into account mixing on lepton side, the mixing dependance of lepton signed jet charge is given by:

$$\begin{aligned}\langle Q_{lJ}\rangle^b &= (1-\chi) \cdot \langle Q^{\bar{b}}_{qJ}\rangle + \chi \cdot (-\langle Q^{\bar{b}}_{qJ}\rangle) \\ &= (1-2\chi) \cdot \langle Q^{\bar{b}}_{qJ}\rangle\end{aligned} \quad (9)$$

a weighting factor $\kappa = 1$ is shown to give the best sensitivity to mixing. A mean jet charge

$$\langle Q_{lJ}\rangle^b = 0.0863 \pm 0.0069_{stat} \pm 0.0018_{syst}$$

Table 5. Jet charge systematics

Error source	$\Delta\chi|_{\chi_s=const.}$
$\langle Q^b_{lJ}\rangle$ (stat)	+0.018 -0.018
$\langle Q^b_{lJ}\rangle$ (sys, flavour comp.)	+0.004 -0.004
fragm. and decay model	+0.016 -0.020
Monte carlo statistics	+0.003 -0.003
κ dependance	+0.022 -0.016

is measured in our data. Extracting from it the mixing parameter relies on detailed montecarlo simulations of jet charge distributions, which are somewhat sensitive to assumptions on fragmentation and decay processes. Due to a smaller mean charge of B^0_s jets, the effective mixing parameter measured is given by

$$\chi_{Jet} = f_d\chi_d + 0.72 f_s\chi_s \qquad (10)$$

The value of χ_{Jet} extracted from our data (see [9] for details) is:

$$\chi_{Jet} = .113 \pm .018_{stat} \pm 0.027_{syst}$$

The systematic error, detailed in table 5, is dominated by uncertainities in the fragmentation and decay mechanism of b and c quarks; it also allows for the dependance of the result on the charge weighting scheme and for variations in the flavour composition of the lepton sample.

COMBINED RESULT

Combining Jet charge and dileptons result, one gets:

$$\chi = 0.134 \pm .013_{stat} \pm 0.008_{syst}$$

The error on this result is still dominated by the statistical error, and we can expect a significant improvement of this result in the next years.

REFERENCES

1. C.Albajar et al, *Phys.Lett.* B186, 27. (1987)
2. ALEPH Collab., D.Decamp et al, *Nucl. Instrum. Methods* A 294, 121. (1990)
3. ALEPH Collab., D.Decamp et al, *Phys. Lett* B 244, 551. (1990)
4. ALEPH Collab., D.Decamp et al, *Phys. Lett* B 263, 325. (1991)
5. D.Cinabro, *Ph.D. Thesis* . University of Wisconsin-Madison (1991)
6. G.Altarelli et al, *Nucl. Phys.* B 208, 365. (1982)
7. CLEO Collab., S.Henderson et al, CLNS 91/1101, CLEO 91-7
8. T.Sjöstrand and M.Bengtsson, *(Comput.Phys.Commun.)* 46, 43. (1987)
9. ALEPH Collab., D.Decamp et al, *Phys. Lett* B 284, 177. (1992)

A MEASUREMENT OF B^0-$\bar{B}^0$ MIXING IN Z DECAYS WITH THE L3 DETECTOR

Gerard J. Bobbink
CERN
1211 Geneva 23 Switzerland
and
NIKHEF-H
Amsterdam, The Netherlands
(representing the L3 collaboration)

Abstract

A determination of the B^0-$\bar{B}^0$ mixing parameter χ_B in Z decays has been made based on 410,000 hadronic decays of the Z accumulated in 1990 and 1991 with the L3 detector at LEP. From the analysis of the dilepton events, muons and electrons, we obtain: $\chi_B = 0.121 \pm 0.017$ (stat) ± 0.006 (sys).

INTRODUCTION

In the Standard Model the transformation of a B_d^0 or B_s^0 meson into its antiparticle proceeds via a weak flavor-changing "box diagram", dominated by virtual top quark exchange. The rate of mixing depends on the Cabibbo-Kobayashi-Maskawa matrix elements, V_{td} and V_{ts}, and the top quark mass. The b-hadron semi-leptonic decay modes allow b-hadrons to be tagged since the lepton generally has a high momentum p, due to the hard fragmentation, and a large transverse momentum p_t with respect to the b-quark direction, due to the high b-quark mass. Thus, a distinctive experimental signature of B^0-$\bar{B}^0$ mixing is the observation of like sign dileptons from the decays $B^0 \to \ell^+$ and $\bar{B}^0 \to B^0 \to \ell^+$. The amount of mixing may be expressed as

$$\chi_B = \frac{Br(b \to \bar{B}^0 \to B^0 \to \ell^+ X)}{Br(b \to \text{b-hadron} \to \ell^\pm X)}$$

assuming equal semi-leptonic branching ratios for all hadrons containing a b-quark. Measurements of χ_B at the Z^0 resonance are sensitive to both B_d^0 and B_s^0 mixing, i.e. $\chi_B = f_d \chi_d + f_s \chi_s$, where χ_d and χ_s are the mixing parameters and f_d and f_s are the production fractions of B_d^0 and B_s^0 mesons. In e^+e^- colliders at the $\Upsilon(4S)$ no B_s^0 is produced, thus allowing a direct measurement of χ_d.

EXPERIMENTAL PROCEDURE

The L3 detector, the trigger requirements and the selection criteria for hadronic events containing electrons and muons have been described earlier.[1,2] Muons are identified and measured in the muon chamber system. We require that a muon track consists of track segments in at least two of the three layers of muon chambers, and that the muon track points to the intersection region. Electrons are identified using the BGO and hadron calorimeters, as well as the central tracking chamber (TEC). For this analysis, we have only considered electrons in the barrel region ($|\cos\theta| < 0.69$). For electron candidates we require a cluster in the BGO that is consistent with the shape of an electromagnetic shower, as determined from test beam studies. The cluster centre-of-gravity has to match a TEC track within 5 mrad in the $r\phi$ plane (see figure 1). To avoid mis-matches no other tracks should be present within 5 mrad. A transverse energy-momentum match of $|1/E_t - 1/p_t| < 4\sigma$ is made and to reject misidentified hadrons, we require that the energy deposited in the hadron calorimeter in a cone of half angle 7° behind the electromagnetic cluster is less than 3 GeV.

The momentum of muon candidates is re-

A Measurement of B^0-$\bar{B}^0$ Mixing in Z Decays

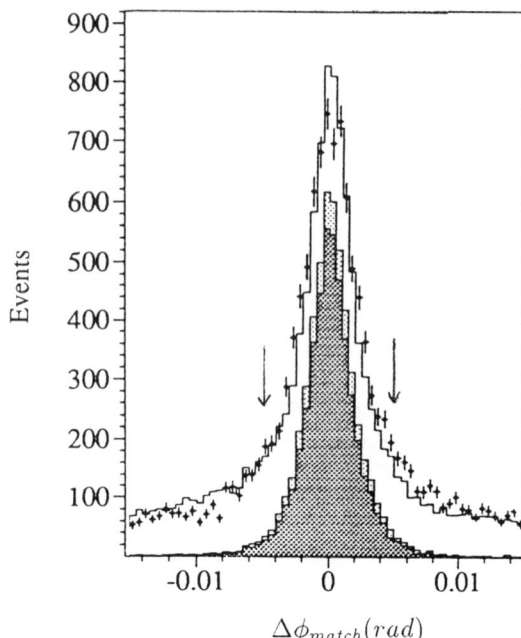

Figure 1. The difference in ϕ between the BGO cluster centre-of-gravity and the TEC track. The points are the data and the histogram is the Monte Carlo simulation. The shaded area represents $b \to e$.

quired to be at least 4 GeV, while the electron energy is required to be at least 3 GeV. From a sample of $Z \to \tau^+\tau^-$ events we have determined the charge confusion to be $0.2 \pm 0.2\%$ for muons and $0.8 \pm 0.3\%$ for electrons.

DI-LEPTON SAMPLE

The signature of B^0-$\bar{B}^0$ mixing is hadronic events with two leptons of the same charge on opposite sides of the event. The angle between the two leptons is required to be larger than 60° to ensure that both leptons are from different b-hadron decays. The transverse momentum of the leptons is measured with respect to the closest jet, where the jet axis has been determined excluding the lepton from the jet. In our sample there are 1303 inclusive dilepton events; in 540 of these, both leptons have $p_t > 1$ GeV. We have also observed 91 events with three inclusive leptons. They were considered in this analysis by using the two leptons with largest transverse momentum with respect to the nearest jet axis.

The number of events and their distribution in various categories is shown in table 1. From

charges	$\mu\mu$	ee	eμ	all
$\ell^+\ell^+$ all p_t	167	17	98	282
$\ell^+\ell^+$ $p_t > 1$ GeV	40	14	32	86
$\ell^-\ell^-$ all p_t	110	20	84	214
$\ell^-\ell^-$ $p_t > 1$ GeV	30	12	31	73
$\ell^+\ell^-$ all p_t	458	65	284	807
$\ell^+\ell^-$ $p_t > 1$ GeV	165	51	165	381
$\ell\ell$ all p_t	735	102	466	1303
$\ell\ell$ $p_t > 1$ GeV	235	77	228	540

Table 1. The dilepton events in the data.

Monte Carlo simulation of $Z \to b\bar{b}$ events, using the LUND parton shower program JETSET 7.3[3] and full detector simulation we expect that the event sample of table 1 consists mainly of events with two prompt B decays. The estimated fractions from various sources are listed in table 2 for $p_t > 1$ GeV.

	Category	$\mu\mu$	ee	eμ
1	$b \to \ell, b \to \ell$	72.6	79.8	80.9
2	$b \to c \to \ell, b \to c \to \ell$	0.5	0.0	0.2
3	$b \to \ell, b \to c \to \ell$	16.1	11.2	11.6
4	$b \to \ell, b \to X$	7.2	8.2	5.2
5	$b \to c \to \ell, b \to X$	1.0	0.7	1.0
6	$b \to X, b \to X$	0.5	0.0	0.4
7	$c \to \ell, c \to \ell$	0.8	0.0	0.2
8	X, X	1.3	0.0	0.4

Table 2. Monte Carlo estimates of the fractions, F_i, (in %) of various event categories for $p_t > 1$ GeV. X indicates a misidentified hadron or leptons from light hadron decays. The $b \to \ell$ fraction includes $b \to \tau \to \ell$ and $b \to \bar{c} \to \ell$ decays.

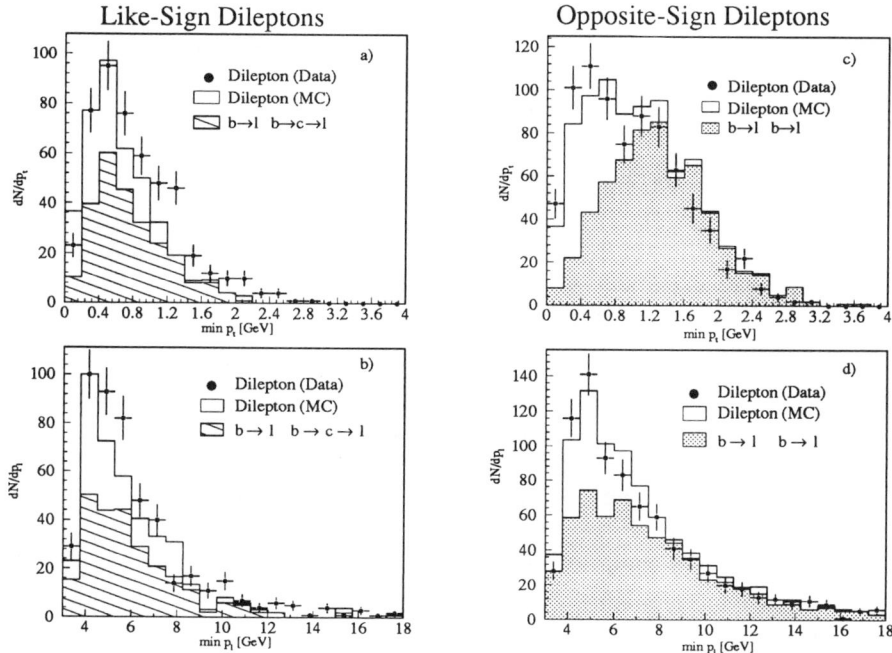

Figure 2. The minimum p_t and p_l for like sign (a and b) and opposite sign (c and d) dilepton events compared to the Monte Carlo expectations with no mixing. The excess of data events in a and b, and the shortage of events in c and d shows the qualitative effect of mixing.

RESULTS

Three methods have been used to measure the mixing parameter $\chi_{_B}$. The first is based on counting the number of high p_t dilepton events with the same charge, while the second and third use different fitting procedures: a 4-dimensional fit to the p and p_t spectra of the dileptons, and a factorized two dimensional fit to the p_l and p_t distributions. The first fit uses the full information of the event, but requires large Monte Carlo statistics to accurately determine the probability functions. The latter fit has the advantage that single lepton events can be used to determine the probability functions, so fewer Monte Carlo events are needed.

COUNTING METHOD

This method is based on the ratio $N^{\pm\pm}/N$ of the number of dilepton events with same charge over all dilepton events requiring $p_t > 1$ GeV for each lepton. This ratio can be expressed as function of $\chi_{_B}$ and the relative fractions F_i given in table 2. For $\mu\mu$, ee and $e\mu$ events we obtain with this method:

$$\chi_{_B} = 0.089 \pm 0.032 \quad (\mu\mu)$$
$$\chi_{_B} = 0.162 \pm 0.056 \quad (ee)$$
$$\chi_{_B} = 0.103 \pm 0.026 \quad (e\mu),$$

where the errors are statistical only. Using a weighted average of the $\mu\mu$, ee, and $e\mu$ results we find: $\chi_{_B} = 0.104 \pm 0.019 \, (stat) \pm 0.013 \, (sys)$, where the systematic error is estimated by varying parameters by their measured or estimated errors.

THE 4-DIMENSIONAL FITTING METHOD

This fit has been described in.[1] An unbinned maximum likelihood fit is performed in 4 dimensions, p and p_t of both leptons. The probability of a data event to come from various sources is determined by the number and type of Monte Carlo events found in a box having the same average value of $(p_1, p_{t1}, p_2, p_{t2})$ as that data event. From these probabilities, a likelihood function is formed, which is then maximized.

This fit and also the factorized fit described below, are less sensitive to changes in branching ratios and background than the counting method. Because this fit samples Monte Carlo events in four dimensions, a large number of simulated events is needed. Production of Monte Carlo events with B mesons decaying leptonically reduced the error due to Monte Carlo statistics but this is still the dominant source of systematic error in this method. From this fit we determine: $\chi_{\scriptscriptstyle B} = 0.124^{+0.018}_{-0.016}$ (stat) ± 0.010 (sys).

THE FACTORIZED FIT METHOD

In this method[2] probability functions are assumed to factorize, and are therefore evaluated independently (using the single lepton data and Monte Carlo) for each lepton as a function of p_l and p_t, where p_l is the lepton momentum along the jet axis. Shown in figure 2 are the p_l and p_t distributions for like sign and opposite sign dileptons.

From a maximum likelihood fit we obtain our final measurement of $\chi_{\scriptscriptstyle B}$:

$$\chi_{\scriptscriptstyle B} = 0.121 \pm 0.017 \pm 0.006,$$

where the systematic error have been estimated by varying parameters by their measured or estimated errors. This determination is in agreement with the determinations from counting and the 4-dimensional fit, as well as with our previous measurement[1] and measurements from other LEP experiments reported at this conference.[4-6]

Separate fits give $\chi_{\scriptscriptstyle B} = 0.088 \pm 0.024$ for $\mu\mu$, 0.158 ± 0.050 for ee, and 0.140 ± 0.028 for $e\mu$ events.

DISCUSSION

To obtain a value of χ_s, a maximum likelihood fit to the data including the results obtained for χ_d[7-9] has been performed using the relation $\chi_{\scriptscriptstyle B} = f_d\chi_d + f_s\chi_s$. The B^0_d and B^0_s fractions, f_d and f_s, are inferred from measurements of the relative production rates of kaons and pions. We have assumed $f_d = 0.40$ and $f_s = 0.12$. These values correspond to a strange quark suppression factor $\gamma_s = f_s/f_d = 0.3$ consistent with measurements at LEP[10] and lower energy e^+e^- colliders.[11,12] The physical constraint, $0 < \chi_d, \chi_s < 0.5$, was not imposed in the fit which yields $\chi_s = 0.46 \pm 0.21$, consistent with maximal mixing in the B^0-$\bar{B}^0$ system. Imposing the physical constraint $0 < \chi_d, \chi_s < 0.5$ gives the one-dimensional limit at the 90% confidence level of $\chi_s > 0.16$.

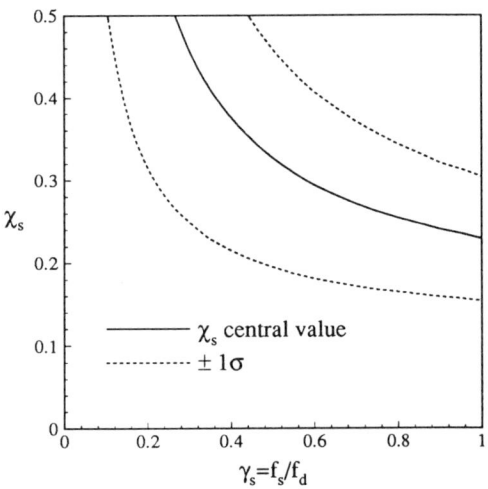

Figure 3. χ_s as a function of $\gamma_s = f_s/f_d$. The one σ errors include a 50% uncertainty on the value of f_B, the fraction of b-baryons produced.

The value of X_s is sensitive to the relative production fractions of different b-hadrons. The dependence of X_s on γ_s is shown in figure 3, up to the SU(3) flavor symmetry limit $\gamma_s = 1$. A b-baryon fraction of $f_B = 0.08$ was assumed. The 1σ errors include a 50% uncertainty on the value of f_B. The effect of the uncertainty is a factor 5 smaller than the statistical errors. The value of X_s is consistent with maximal mixing for any reasonable choice of f_d, f_s and f_B.

CONCLUSIONS

The L3 collaboration has measured mixing in the B^0-$\bar{B}^0$ system using inclusive dilepton events from approximately 410,000 hadronic Z decays, and determined $X_B = 0.121 \pm 0.017 \pm 0.006$. This result is consistent with maximal mixing in the $B_s^0 \bar{B}_s^0$ system: $X_s > 0.16$ at the 90% confidence level.

REFERENCES

1. L3 Collab., B. Adeva *et al.*, *Phys. Lett.* **B 252**, (1990) 703.
2. L3 Collab., B. Adeva *et al.*, *Phys. Lett.* **B 288**, (1992) 395.
3. T. Sjöstrand and M. Bengtsson, *Comput. Phys. Commun.* **43**, (1987) 367; T. Sjöstrand in "Z Physics at LEPI", CERN-89-08, ed. G. Altarelli, *et al.*, (CERN, Geneva, 1989), volume 3, pp. 143–340.
4. V. Gibson (OPAL), these proceedings.
5. J.-P. Lees (ALEPH), these proceedings.
6. F. Simonetto (DELPHI), these proceedings.
7. ARGUS Collab., H. Albrecht *et al.*, *Phys. Lett.* **B 192**, (1987) 245.
8. CLEO Collab., M. Artuso *et al.*, *Phys. Rev. Lett.* **62**, (1989) 2233.
9. ARGUS Collab., H. Albrecht DESY preprint DESY-92-050,1992.
10. OPAL Collab., G. Alexander *et al.*, *Phys. Lett.* **B 264**, (1991) 467.
11. JADE Collab., W. Bartel *et al.*, *Z. Phys.* **C 20**, (1983) 187.
12. TASSO Collab., M. Althoff *et al.*, *Z. Phys.* **C 27**, (1985) 27.

A MEASUREMENT OF $B\overline{B}$ MIXING WITH THE DELPHI DETECTOR AT LEP

Simonetto Franco
INFN and Dipartimento di Fisica dell' Universita di Padova
via Marzolo 8
35100 Padova Italia

Representing The DELPHI Collaboration

Abstract

The data collected during 1990 and 1991 LEP runs with the DELPHI detector have been analyzed to measure the probability χ for a B meson to undergo mixing before decay. A study based on events containing two identified leptons lying on opposite detector sides yielded $\chi = 0.121 \pm^{0.044}_{0.040}(stat.) \pm 0.017(syst.)$; a comparison between the charge of a high p_t muon and the one of the jet recoiling in the opposite direction gave $\chi = 0.075 \pm 0.053(stat. \oplus syst.)$; the combined result is $\chi = 0.101 \pm 0.035(stat. \oplus syst.)$. No signal of CP violation has been observed.

INTRODUCTION

The amount of Beauty flavour mixing at LEP can be measured by comparing at the decay time the charge of the two b quark produced in the electron positron annihilation. The charge of the b quark can be tagged either by means of the charge of the lepton produced in b semileptonic decay (lepton tag), or through the weighted mean of the charges of all the particles contained in the jet produced by b fragmentation and decay (jet tag).

The DELPHI collaboration has measured the mixing parameter χ, defined as

$$\chi = f_d \chi_d + f_s \chi_s \qquad (1)$$

(where $\chi_{d(s)}$ is the amount of mixing of $B_{d(s)}$ meson and $f_{d(s)}$ its production probability), with the data collected during 1990 and 1991 runs of LEP (about 300000 Z^0 hadronic decays have been selected for the following studies) both through a double lepton tag [1] and through a comparison between the charge of the lepton and the one of the opposite recoiling jet [2]. The results of the two measurements are summarized in the following sections.

DILEPTON ANALYSIS

A detailed description of the DELPHI detector can be found in [3]. References [1, 4] contain an exahustive description of lepton tagging. High p_t muons are identified inside muon chamber acceptance ($|(cos\theta)| < 0.865$) with efficiency $\epsilon_\mu = 0.78 \pm 0.02$. The acceptance for electron identification is defined by the HPC fiducial volume ($|(cos\theta)| < 0.65$); the efficiency at high p_t is $\epsilon_e = 0.67 \pm 0.05$. Events have been selected by requiring two leptons lying in opposite hemispheres. This lead to 656 dimuon events, 260 dielectrons and 749 events with a muon and an electron; in 723 cases the two leptons had the same charge.

The quantity

$$R = \frac{N(l^+l^+) + N(l^-l^-)}{N(l^\pm l^\pm)}, \qquad (2)$$

ratio of the number of same sign dileptons to the overall sample, is linked to χ by the relation:

$$R = 2\chi(1-\chi) \quad (3)$$

when both leptons are coming from direct b decay. Background in the same charge sample can originate when at least one hadron in the event passes the lepton selection, or when a lepton from direct b decay is coupled with a cascade lepton in the opposite hemisphere. Semileptonic decays of primary c quarks can only produce opposite charge dileptons. The Monte Carlo R distribution is then fitted to the experimental one, leaving χ as free parameter, in bins of the variable

$$P_{dil} = \sqrt{(\frac{\vec{p}_1 \times \vec{p}_2}{16})^2 + (p_t^{min})^2}. \quad (4)$$

Due to both hard b fragmentation and high b mass, bigger values of P_{dil} correspond to greater b signal to background ratio.

Fig.1 plots R as a function of P_{dil} for data (dots), for a Monte Carlo sample with no mixing (dashed line), and for the fitted Monte Carlo distribution (continuos line).

The fit result is determined by the data with $P_{dil} > 1.5$ and gives :

$$\chi = 0.121 \pm^{0.044}_{0.040} \pm 0.017 \quad (5)$$

The systematic error is dominated by uncertainties on the b semileptonic branching ratio (conservatively varied by ± 10 percent around its expected value) and on the hardness of its fragmentation (the amount of beam energy taken on average by a B hadron, x_E, has been varied in the range $0.68 \div 0.74$).

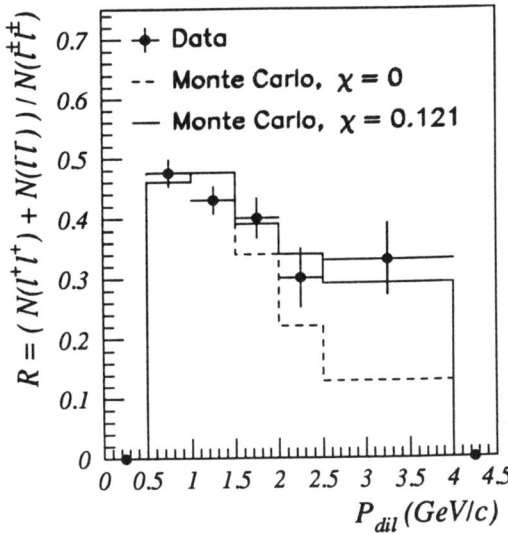

Figure 1. R versus P_{dil}.

LEPTON VS JET CHARGE ANALYSIS

15629 events have been selected containing at least one muon with $p_t > 0.5$ GeV/c. The muon charge is compared to the one of the opposite side recoiling jet, defined as

$$Q^{jet} = \frac{\sum_j (\vec{p}_j \cdot \vec{e}_s)^\kappa q_j}{\sum_j (\vec{p}_j \cdot \vec{e}_s)^\kappa} \quad (6)$$

where $\vec{p}_j$ is the jth particle momentum vector, q_j its charge, $\vec{e}_s$ the sphericity principal axis versor. To increase the sensitivity of the method to the mixing, all the tracks with momentum $p > 200 MeV/c$ are used in the analysis. Monte Carlo is used to determine the correlation between the original b quark charge and jet's one, showing that maximum sensitivity to mixing is obtained for the exponent $\kappa = 1$ in eq. 6. Due to mixing, the average jet charge can be written as

$$<Q^{jet}_\chi> = <Q^{jet}_{\chi=0}> (1-2\chi)$$
$$(1 - 2(f_d\chi_d\Delta_d + f_s\chi_s\Delta_s)). \quad (7)$$

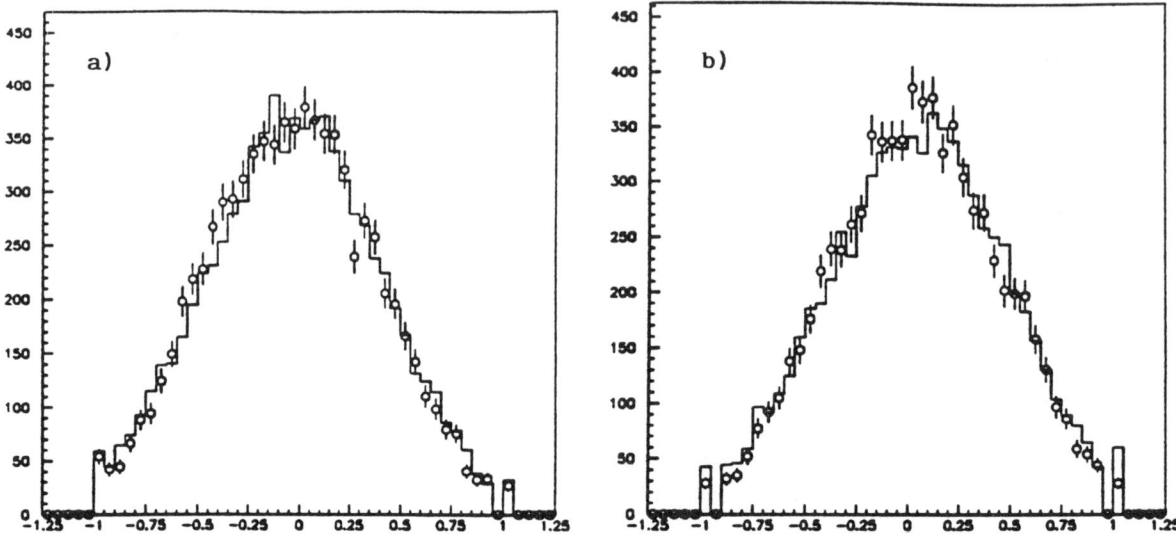

Figure 2. Distribution of the charge of the jet opposite to the positive muon (a), and to the negative one (b). The open circles represent the data, the solid line the fit result.

The first term on right hand side is the average jet charge with no mixing, the second and third reflect the effect of mixing on muon and on jet side respectively. The quantity

$$\Delta_q = \frac{1}{2} \frac{<Q_{B_q}> - <Q_{\overline{B}_q}>}{<Q^{jet}_{\chi=0}>}, (q=d,s) \quad (8)$$

is different for B_d and B_s mesons, as a consequence of loss of the soft π^- coming from the D^{*-} produced in B_d decay (D_s^{*-} can only decay through a photon).

Up to a very good approximation, Monte Carlo shows that equation (7) simplifies by defining the quantity

$$\chi_q = f_d \chi_d + f_{cor} f_s \chi_s \quad (9)$$

different from χ as defined in (1) for the insertion of the factor f_{cor}, expressing the method different sensitivity to B_d and B_s mixing, determined from Monte Carlo to be $f_{cor} = 0.72 \pm 0.03$.

Fit Procedure

Simulated jet charge distribution has been fitted to the experimental one separately for μ^+ and μ^- in different bins of the muon p_t, with the two free parameters

$$\chi_q = \frac{\chi_q^{\mu^+} + \chi_q^{\mu^-}}{2} \quad (10)$$

$$\Delta \chi_q = \frac{\chi_q^{\mu^+} - \chi_q^{\mu^-}}{2}. \quad (11)$$

Values of $\Delta \chi_q$ different from 0 would be a sign of CP violation in the B system. The fitted values are:

$$\chi_q = 0.070 \pm 0.028 \pm 0.024$$
$$\Delta \chi_q = 0.001 \pm 0.028 \pm 0.023 \quad (12)$$

The agreement between data and Monte Carlo can be inspected in fig. 2 where the experimental jet charge distribution is compared to the fitted one both for μ^+ (2.a) and μ^- (2.b). Among the other sources of systematic error inspected (fit method, relative amount of background contamination, μ identification efficiency, $B_d \to D^{*-} X$ branching ratio) uncertainties on the jet charge induced by QCD processes modelling are dominating [2].

COMBINED RESULT

A simple algebraic relation holds between χ and χ_q (see eq. (1),(9)):

$$\chi = \frac{\chi_q}{f_{cor}} + f_d \chi_d (1 - \frac{1}{f_{cor}}) \quad (13)$$

By inserting in (13) the value

$$\chi_d = 0.167 \pm 0.042 \quad (14)$$

as measured by ARGUS and CLEO [5, 6], and the f_d value which can be inferred by DELPHI measurements of B_d, B_s, Λ_b production rates [7, 8, 9], and by the relation $2f_d + f_s + f_{Barions} = 1$,

$$f_d = 0.35 \pm 0.05 \quad (15)$$

one finds, out of the lepton jet analysis

$$\chi = 0.075 \pm 0.053 \quad (16)$$

where the statistical error takes into account the small statistical correlation with the dilepton sample, and has then been quadratically summed to the systematic one. Combining this value with the dilepton study result, the mixing parameter is measured to be

$$\chi = 0.101 \pm 0.035 \quad (17)$$

in good agreement with the Standard Model expectation.

CONCLUSION

The Beauty flavour mixing parameter χ has been measured by the DELPHI collaboration by means of two methods affected by different systematic uncertainties and employing two data sets largely uncorrelated. The combined result

$$\chi = 0.101 \pm 0.035 \quad (18)$$

stands in good agreement with the Standard Model prediction. While room is open for a wide improvement of the dilepton result, in the lepton-jet correlation analysis uncertainties from soft QCD processes are already dominating the systematic error, which is almost of the same size of the statistical one, showing that such a method is near to the end of its possibilities.

REFERENCES

1. U.Amaldi, DELPHI collaboration, "A Study of $B^0 \overline{B^0}$ Oscillations Using Dileptons from Semileptonic Decay of B Hadrons Produced from Z^0"[1].

2. U.Amaldi, DELPHI collaboration, "A Measurement of The $B - \overline{B}$ Mixing Parameter Using Muons and The Charge of The Jet"[1].

3. P.Aarnio et al., DELPHI collaboration, "The DELPHI detector at LEP" *Nucl. Instrum. Methods* A 303 (1991) 233.

4. P.Abreu et al., DELPHI collaboration, "Measurement of the partial width of the Z^0 into $b\bar{b}$ Final States using their Semi-Leptonic decays" CERN-PPE/92-79,1992, submitted to *Zeit. Phys. C*

5. H.Albrecht et al., *Phys. Lett.* B192 (1987) 245.

6. M.Artuso et al., *Phys. Rev. Lett.* 62 (1989) 2233.

7. U.Amaldi, DELPHI collaboration, "A measurement of B mesons production and lifetime using D-lepton events"[1].

8. U.Amaldi, DELPHI collaboration, "B_s^0 tagging at LEP energies using D_s and Φ mesons"[1].

9. U.Amaldi, DELPHI collaboration, Measurement of Λ_b production production and lifetime in Z^0 hadronic decays"[1].

---

[1] submitted to this conference

RESULTS ON $B\overline{B}$ MIXING AND RARE B DECAYS FROM CLEO

Hubert Kroha
Department of Physics and Astronomy
University of Rochester
Rochester, NY 14627
CLEO Collaboration

Abstract

New (and preliminary) results on $B^0\overline{B^0}$ mixing, on inclusive and exclusive semileptonic $b \to u$ transitions, and on charmless hadronic B decays from almost $1 fb^{-1}$ of $\Upsilon(4S)$ data collected with the CLEO II detector are presented. Limits on CP violation from like-sign dilepton events and on the branching ratio for $B^+ \to \tau^+\nu$ are given.

DETECTOR AND DATA SAMPLE

The CLEO II detector[1] at the CESR storage ring has accumulated an integrated luminosity of $0.95\ fb^{-1}$ at the $\Upsilon(4S)$ resonance and of $0.45\ fb^{-1}$ in the continuum below. The data sample corresponds to 973000 $B\overline{B}$ events. Important features of the CLEO II detector are the good momentum resolution of $\Delta p/p \approx 1.3\%$ at $p = 5$ GeV in a 1.5 T magnetic field, particle identification by dE/dx information from 51 independent measurements in the main drift chamber (2σ K/π separation for $p < 0.7$ GeV and $p > 2$ GeV), and excellent position and energy resolution in the electromagnetic calorimeter consisting of 7800 CsI crystals ($\Delta E/E \approx 1.5\text{--}3.8\%$ for photon energies in the range from 5 GeV down to 100 MeV).

Electrons are identified by a combination of E/p, shower shape, and dE/dx information with a typical efficiency and faking probability of $\varepsilon_e \approx 90\%$ and $f_e \approx 0.2\%$, respectively. Muon candidates are required to penetrate at least 5 interaction lengths of iron ($\varepsilon_\mu \approx 90\%$, $f_\mu \approx 1\%$).

$B^0\overline{B^0}$ MIXING

The mixing probability is defined as the ratio

$$\chi_d = \frac{\Gamma(B^0 \to \overline{B^0} \to \ell\nu X)}{\Gamma(B^0 \to \overline{B^0} \to \ell\nu X) + \Gamma(B^0 \to \ell\nu X)}$$
$$= \Lambda_0^{-1} \frac{N_B^{\pm\pm}}{N_B^{\pm\pm} + N_B^{+-}} \quad (1)$$

where the second part of the equation is specific for $B\overline{B}$ production from $\Upsilon(4S)$. $N_B^{\pm\pm}$ and N_B^{+-} are the numbers of like- and opposite-sign dilepton pairs from B semileptonic decays. The parameter

$$\Lambda_0 = \frac{f_0 BR_0^2}{f_0 BR_0^2 + f_+ BR_+^2} \quad (2)$$

is the fraction of dileptons from $B^0\overline{B^0}$ events with $f_0 = 1 - f_+ = BR(\Upsilon(4S) \to B^0\overline{B^0})$ and $BR_{0\,(+)} = BR(B^{0\,(+)} \to \ell^+\nu X)$. In the following, we use $\frac{f_0}{f_+} = 1.00 \pm 0.05$ from the measured B^0–B^+ mass difference and radiative corrections[2] and $\frac{BR_+}{BR_0} = \frac{\tau(B^+)}{\tau(B^0)} = \frac{f_0}{f_+}(1.00 \pm 0.14)$ from measurements of $BR(B \to D^{(*)}\ell\nu)$ [3].

The Dilepton Method

From equation (1), the mixing parameter χ_d is determined by counting like- and opposite-sign dielectron, dimuon, and electron-muon events. Background contributions arise from fake leptons, leptons from $B \to \psi^{(\prime)} X$, $\psi^{(\prime)} \to \ell^+ \ell^-$, continuum leptons, and from secondary charm decays which contribute to like-sign dileptons. Backgrounds are suppressed by cuts in the dilepton opening angle ($-0.8 \leq \cos\theta_{\ell\ell} \leq 0.9$) and in the lepton momentum ($1.5 \leq p_\ell \leq 2.4$ GeV). The lower p_ℓ cut strongly reduces the most severe background from charm decays which is the dominant source of systematic error. This cut has been optimized by Monte Carlo (MC) simulation requiring $\Delta\chi^2_{\text{stat}} + \Delta\chi^2_{\text{syst}}$ to be minimum.

The remaining charm lepton background can be determined from data by fitting the inclusive lepton spectrum with $b \to c\ell\nu$ and $b \to c \to s\ell\nu$ contributions. The charm lepton spectrum is parametrized in a model-independent way using the DELCO[4] $D^{0,+} \to \ell\nu X$ spectrum convoluted with the CLEO[5] D^0 and D^+ spectra. Several models are used to describe $b \to c\ell\nu$ decays. With these inputs, the charm background fraction can be determined within $\pm 30\%$.

After background subtraction, a clear mixing signal is observed: $N^{\pm\pm}_{\ell\ell} = 185 \pm 19 \pm 23$ and $N^{+-}_{\ell\ell} = 2169 \pm 51 \pm 15$. The result for the mixing parameter from dileptons is

$$\chi_d = 0.157 \pm 0.016 \text{ (stat)} \pm 0.018 \text{ (syst)} \\ \pm 0.022 \, (\Lambda_0). \quad (3)$$

The new mixing measurement is not limited by statistics. Figure 1 demonstrates the stability of the measurement for varying minimum momentum cuts. The combination of the first two errors is a factor of 2 smaller than in previous measurements[6] (see Figure 2).

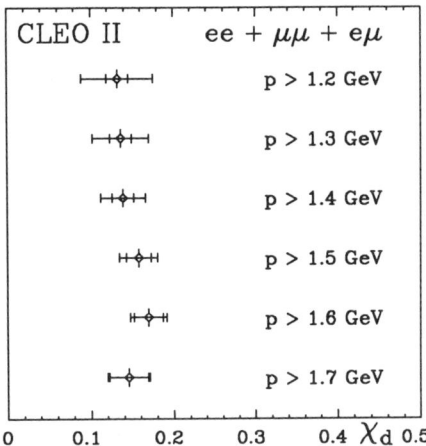

Figure 1. Results for χ_d for different lower p_ℓ cuts. The first thick error bar is statistical, the second is the combined statistical and systematic error.

The last error is due to the uncertainty in Λ_0 which is common to all dilepton mixing measurements at the $\Upsilon(4S)$ and will decrease with improved knowledge of $\tau(B^+)/\tau(B^0)$. In order to reduce the sensitivity to Λ_0, an attempt has been made to efficiently tag $B^0\bar{B}^0$ events.

Partial B^0 Reconstruction

B^0 mesons can be tagged by partial reconstruction in the decay chain $\bar{B}^0_d \to D^{*+}\ell^-\bar{\nu}$, $D^{*+} \to D^0 \pi^+_s$, $D^0 \to K\pi$, ... The standard technique uses a cut in the missing mass squared

$$M^2_\nu = [E_b - (E_{D^*} + E_\ell)]^2 - |\vec{p}_{D^*} + \vec{p}_\ell|^2. \quad (4)$$

This method, however, is limited in statistics. The tagging efficiency can be increased significantly by tagging the D^{*+} decay with the soft π^+_s instead of reconstructing the D^0. Energy and direction of the D^* are estimated by $E_{D^*} \approx \frac{E_{\pi_s}}{E^{\text{cms}}_{\pi_s}} M_{D^*}$ and $\vec{p}_{D^*} \parallel \vec{p}_{\pi_s}$, respectively. The resulting missing mass resolution $\Delta M^2_\nu = 0.9$ GeV^2, is only about twice the resolution for conventional partial reconstruction. For right-sign $\pi^\pm_s \ell^\mp$ combinations with

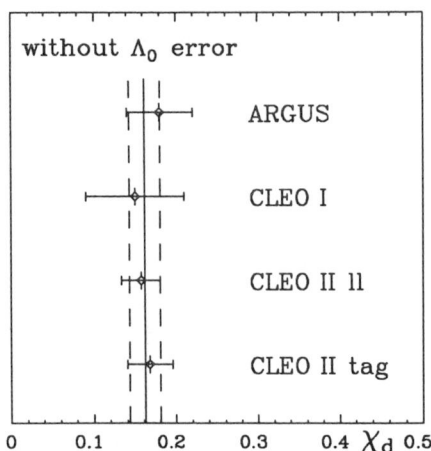

Figure 2. Measurements of the mixing probability χ_d at the $\Upsilon(4S)$.

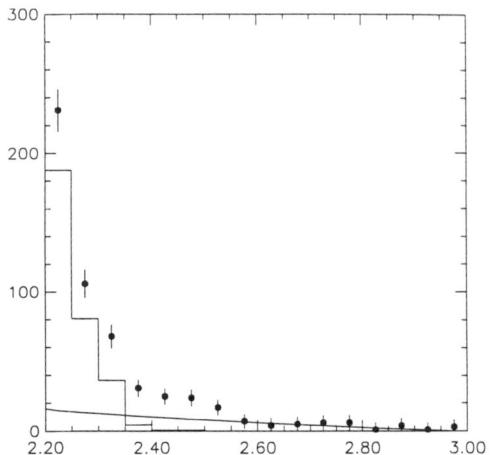

Figure 3. Endpoint region of the inclusive electron and muon spectrum with continuum expectation (smooth curve) and $b \to c\ell\nu$ background (histogram).

$1.4 \leq p_\ell \leq 2.4$ GeV the tag region is defined as $M_\nu^2 > -1.5$ GeV^2.

The spectrum of the second lepton ($p_e > 0.6$ GeV, $p_\mu > 1.4$ GeV) in the tagged events is fitted separately for like- and opposite-sign dilepton events in order to determine the $b \to c\ell\nu$ contributions: $N_{\ell\ell}^{\pm\pm} = 210 \pm 27$ (stat) and $N_{\ell\ell}^{+-} = 1213 \pm 54$ (stat). The purity of the tagged dilepton sample is about 60% corresponding to a $B^0\overline{B^0}$ fraction of 75%. Remaining backgrounds from fake leptons, random tags, and $B \to D^*\pi\ell\nu$ contain B^0 and B^+ decays accounting for a residual Λ_0 dependence which is about a factor of 2 smaller than for untagged dilepton events.
The partial reconstruction method gives

$$\chi_d = 0.167 \pm 0.023 \text{ (stat)} \pm 0.016 \text{ (syst)} \pm 0.011 \, (\Lambda_0) \quad (5)$$

in good agreement with the generic dilepton result.

Note that the two complementary measurements are highly correlated. They are compared to previous measurements[6] in Figure 2. The new world average from dilepton events, $\chi_d = 0.162 \pm 0.019 \pm 0.024 \, (\Lambda_0)$, is dominated by the CLEO II measurement in eq. (3). Combining the average $\Upsilon(4S)$ result with measurements of the mixing parameter $\chi_\ell = f_d\chi_d + f_s\chi_s$ at higher energies[7] allows to extract χ_s depending on the produced fractions $f_{d,s}$ of B_d^0 and B_s^0 mesons, e.g. $\chi_s = 0.47 \pm 0.11$ for $f_d = 0.375$, $f_s = 0.150$ or $\chi_s = 0.48 \pm 0.13$ for $f_d = 0.430$, $f_s = 0.130$.

CP Violating Asymmetry

The CP symmetry is violated in $B^0\overline{B^0}$ mixing if $\Gamma(B^0 \to \overline{B^0} \to \ell^-\bar{\nu}X) \neq \Gamma(\overline{B^0} \to B^0 \to \ell^+\nu X)$ or $\chi \neq \overline{\chi}$. A measure is the CP violating asymmetry in like-sign dilepton events:

$$a_{\text{CP}} = \frac{N_{\ell\ell}^{++} - N_{\ell\ell}^{--}}{N_{\ell\ell}^{++} + N_{\ell\ell}^{--}} = \frac{\chi - \overline{\chi}}{\chi + \overline{\chi} - 2\chi\overline{\chi}}. \quad (6)$$

After the same backround corrections as for the mixing measurement ($\langle\chi\rangle = \frac{1}{2}(\chi + \overline{\chi})$), $N_{\ell\ell}^{++} = 95.9 \pm 13.2 \pm 11.7$ and $N_{\ell\ell}^{--} = 96.8 \pm 13.7 \pm 11.7$ signal events are observed. The asymmetry parameter has the value $a_{\text{CP}} = -(0.0058 \pm 0.099 \pm 0.038)$ corresponding to a 90% C.L. upper limit of $|a_{\text{CP}}| < 0.14$. Note that there is no Λ_0 dependence since only $B_d^0\overline{B_d^0}$ events are involved. The standard

Table 1. Lepton yields in on- and off-resonance data (scale factor 2.19), backgrounds and signal in the endpoint region $2.4 \leq p_\ell \leq 2.6$.

N_{ON}	73
N_{OFF} (fit)	$14.2 \pm 1.6 \pm 1.3$
Yield $\Upsilon(4S)$	$41.8 \pm 9.2 \pm 2.9$
Fakes	1.7 ± 0.9
Leptons $B\bar{B}$	$40.1 \pm 9.2 \pm 3.0$
$b \to c\ell\nu$	$1.3 \pm 1.1 \pm 1.3$
$b \to \psi^{(\prime)} X$	1.0 ± 1.0
$b \to u\ell\nu$	$37.8 \pm 9.3 \pm 3.3$

model prediction is

$$a_{CP} \approx 4\pi \frac{m_c^2}{m_t^2} \frac{\sin\theta_3 \sin\delta}{\sin\theta_2} \approx \mathcal{O}(10^{-3}). \quad (7)$$

CHARMLESS SEMILEPTONIC B DECAYS

Inclusive Search for $b \to u$ Decays

The endpoint region (2.4–2.6 GeV) of the inclusive lepton spectrum above the kinematic limit (2.46 GeV) for $b \to c\ell\nu$ is used to search for $b \to u\ell\nu$ transitions. The most severe background from continuum events is strongly supressed by cuts in the Fox-Wolfram ratio $R_2 < 0.2$, in the missing momentum $|\vec{p}_{miss}| > 1$ GeV, and in the angles of $\vec{p}_{miss}$ relative to the lepton momentum and to the beam axis, exploiting the hermetic detector and the good calorimetry. The cuts are optimized with MC S^2/N. They result in a continuum suppression of 1/66 for a $b \to u\ell\nu$ efficiency of about 40%. Including the lepton momentum cut, the total model-dependent efficiency is about 21%.

The continuum spectrum from off-resonance data after cuts is fit with smooth functions over extended momentum ranges in order to estimate the remaining continuum background in the endpoint region (Figure 3). $b \to c\ell\nu$ background in the endpoint region is very small (see Figure 3) due to the good momentum resolution. Table 1 summarizes the observed yields and background contributions. A clear excess of events over the expected background is observed which is interpreted as a $b \to u$ signal.

The fractional branching ratio in the endpoint region is $\Delta B_{ub}(2.4, 2.6) = (0.48 \pm 0.12 \pm 0.25) \cdot 10^{-4}$. The systematic error is still relatively large since several weak decay models and various $b \to u$ final states are still under investigation for evaluating the efficiency. The $b \to u$ signal observed with the CLEO II detector at high luminosity is considerably smaller than previous results[8] from CLEO I and ARGUS (see the results for $|V_{ub}|/|V_{cb}|$ in Table 3 extracted from the endpoint measurements by model-dependent extrapolation to the full momentum range). CLEO I[8] obtained $\Delta B_{ub}(2.4, 2.6) = (1.8 \pm 0.4 \pm 0.3) \cdot 10^{-4}$.

Search for Exclusive $b \to u$ Decays

In an independent analysis, exclusive semileptonic B decays into charmless vector-meson final states have been searched for, in particular $B^+ \to \omega \ell^+ \nu$, $B^+ \to \rho^0 \ell^+ \nu$, and $B^0 \to \rho^- \ell^+ \nu$ where the ω, ρ^0 and ρ^- mesons are reconstructed in the channels $\pi^+\pi^-\pi^0$, $\pi^+\pi^-$ and $\pi^-\pi^0$, respectively. From general principles, one expects the lepton spectrum for vector meson (V) final states to peak near the kinematical endpoint. Candidate events are divided into a "high" ($p_\ell > 2.3$ GeV) and a "low" ($2.0 < p_\ell < 2.3$ GeV) momentum bin which together cover about 50–75% of the $V\ell\nu$ lepton spectrum. Background sources are from continuum (high), from $b \to c$ decays (low), and combinatoric background from the

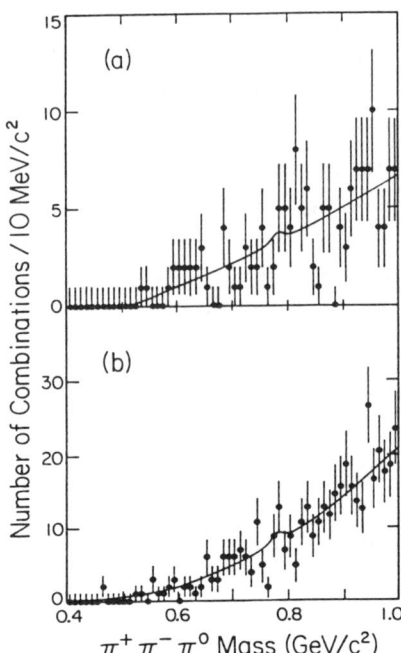

Figure 4. Reconstructed ω mass distributions with fit for $B \to \omega \ell \nu$ with (a) $p_\ell > 2.3$ GeV and (b) $2.0 < p_\ell < 2.3$ GeV.

other B decay in $V\ell\nu$ events.

Background suppression requirements are S^2/N optimized in the MC simulation separately for the high and low p_ℓ bins. They exploit the semileptonic decay hypothesis (neutrino recoil against $V\ell$), consistency of the rest of the event with a B decay, relations between $V\ell$ and the rest of the event, global event shape (R_2), and properties of the $\omega \to \pi^+\pi^-\pi^0$ Dalitz plot. With efficiencies between 2 and 9% depending on models and final states, a background rejection of $\mathcal{O}(10^{-3})$ is achieved.

At the present stage, the analysis consistently aims at conservative upper limits for all channels; no background subtraction is performed. Since the vector-meson content in $b \to c(u)$ (and continuum) events is poorly known, absolute MC predictions for background to the interesting channels are very unreliable. Instead we look for ρ and ω mass peaks above a smooth background. The narrow ω resonance ($\Gamma = 8$ MeV) is insensitive to the background shape and can be fitted with a Breit-Wigner signal curve plus polynomial background (see Figure 4). For the broad ρ resonances ($\Gamma = 152$ MeV), the background contribution is split up into $\rho\ell\nu$ combinatorical bg. from MC simulation and a non-$\rho\ell\nu$ bg. parametrization.

For the high p_ℓ sample, for instance, the following results for potential signal events and 90% C.L. upper limits on branching ratios in the ISGW model are obtained: $B \to \omega \ell \nu$: $N_{\text{fit}} = 1.2 \pm 4.7$ (stat.), $BR < 2.1 \cdot 10^{-4}$; $B \to \rho^0 \ell \nu$: $N_{\text{fit}} = 18 \pm 13$ (stat., fit), $BR < 2.6 \cdot 10^{-4}$; $B \to \rho^+ \ell \nu$: $N_{\text{fit}} = 30 \pm 12$ (stat., fit), $BR < 5.8 \cdot 10^{-4}$. According to isospin symmetry and quark model considerations, the exclusive channels are related by $\Gamma(\overline{B^0} \to \rho^+ \ell^- \bar{\nu}) = 2\Gamma(B^- \to \rho^0 \ell^- \bar{\nu}) \approx 2\Gamma(B^- \to \omega \ell^- \bar{\nu}) =: 2\Gamma(B^- \to V_u^0 \ell^- \bar{\nu})$. Limits for the average branching ratio into a neutral vector meson V_u^0 are summarized in Table 2.

The combined CLEO II limit of $2.1 \cdot 10^{-4}$ (90% C.L.) on the branching ratio for $B^- \to V_u^0 \ell^- \bar{\nu}$ is in contradiction to the ARGUS[9] claim of a signal in the $\rho^0 \ell \nu$ channel with $BR(B^- \to \rho^0 \ell^- \bar{\nu}) = (10.3 \pm 3.6 \pm 2.5) \cdot 10^{-4}$ which would correspond to rather large values of $|V_{ub}|/|V_{cb}|$ on the order of 0.15–0.30 (both results quoted are for the ISGW model). Results for $|V_{ub}|/|V_{cb}|$ from inclusive and exclusive measurements are summarized in Table 3. Both the inclusive and exclusive CLEO II results favour lower values of $|V_{ub}|/|V_{cb}|$ than previous measurements[8].

CHARMLESS HADRONIC B DECAYS

A search has been performed for charmless hadronic B decays, in particular for the decays $B^0 \to \pi^+\pi^-$, $\pi^-\rho^+$, $K^+\pi^-$, and $K^+\rho^-$. The first two are dominated by the spectator

Table 2. Combined $\rho(\omega)\ell\nu$ results and 90% C.L. limits for the decay $B^- \to V_u^0 \ell^- \bar{\nu}$. The second limits are from the inclusive lepton spectrum in the range $2.3 < p_\ell < 2.6$ GeV assuming that $B^- \to \rho(\omega)\ell\nu$ is the only contribution.

Model	$BR(B^- \to V_u^0 \ell^- \bar{\nu})/10^{-4}$ not background subtr.	BR limit/ 10^{-4} exclusive	BR limit/ 10^{-4} inclusive	p_ℓ range
ISGW	$1.36 \pm 0.50 \pm 0.22$	< 2.1	< 1.6	
WSB	$2.15 \pm 0.81 \pm 0.36$	< 3.2	< 2.4	$p_\ell > 2.3$ GeV
KS	$1.65 \pm 0.70 \pm 0.28$	< 2.6	< 1.8	
ISGW	$1.74 \pm 1.72 \pm 0.29$	< 4.1		
WSB	$2.02 \pm 1.95 \pm 0.34$	< 4.7		$2.0 < p_\ell < 2.3$ GeV
KS	$2.06 \pm 1.67 \pm 0.35$	< 4.4		
ISGW	$1.39 \pm 0.42 \pm 0.23$	< 2.1		
WSB	$2.13 \pm 0.75 \pm 0.36$	< 3.2		$p_\ell > 2.0$ GeV
KS	$1.71 \pm 0.65 \pm 0.30$	< 2.6		

Table 3. Results for $|V_{ub}|/|V_{cb}|$ from inclusive and exclusive measurements.

Model	ARGUS incl.	CLEO I incl.	CLEO II incl.	CLEO II excl. 90% C.L.
ISGW	0.18 ± 0.02	0.15 ± 0.02	0.09 ± 0.03	< 0.15
WSB	0.12 ± 0.02	0.11 ± 0.02	0.06 ± 0.02	< 0.11
KS	0.09 ± 0.01	0.09 ± 0.01	0.05 ± 0.02	< 0.08

process. The $\pi^+\pi^-$ final state is a CP eigenstate and interesting for measuring CP violation in mixing at a future B factory. For the last two decays hadronic penguin processes are significant and can interfere with the spectator diagram to induce CP violation.

Upper limits on the branching ratios for these decays (see Table 4) have been improved by almost a factor of 2 compared to the previous best results by CLEO[10] and ARGUS[11]. For $\pi\pi$ and $K\pi$ the limits begin to approach the theoretical predictions[12].

SEARCH FOR $B^+ \to \tau^+\nu$

The branching ratio for the rare decay $B^+ \to \tau^+\nu$ is directly proportional to the CKM matrix element $|V_{ub}|^2$ but experimentally very challenging. The standard model prediction is about $1 \cdot 10^{-4}$. No previously published limits are available. The method employed in this measurement is to tag B^+B^- events by fully reconstructing B^- mesons (716 ± 123 events in the final states $\overline{D^{0(*)}}\pi^-(\rho^-)$, $\psi K(\pi)$, $\chi_c K(\pi)$) and to search for $B^+ \to \tau^+\nu$ in the rest of the event using $\tau \to \ell\nu\nu$, $\rho\nu$, $\pi\nu$, $K\nu$. Zero events have been observed with an expected background of 0.2. This corresponds to a 90% C.L. upper limit of $BR(B^+ \to \tau^+\nu) < 1.3\%$.

CONCLUSIONS

CLEO II has accumulated almost $1 fb^{-1}$

Table 4. 90% C.L. upper limits on selected charmless hadronic B decays.

Channel	N_4S^{obs}	N_{bg}	ε [%]	BR upper limit	previous limit
$B^0 \to \pi^+\pi^-$	4	1.2 ± 0.2	19 ± 3	$4.8 \cdot 10^{-5}$	$9.0 \cdot 10^{-5}$ (Ref. 10)
$B^0 \to K^+\pi^-$	5	1.4 ± 0.2	19 ± 3	$5.6 \cdot 10^{-5}$	$9.0 \cdot 10^{-5}$ (Ref. 10)
$B^0 \to \pi^-\rho^+$	14	6.2 ± 0.7	8.5 ± 1.3	$2.9 \cdot 10^{-4}$	$5.2 \cdot 10^{-4}$ (Ref. 11)
$B^0 \to K^+\rho^-$	5	1.2 ± 0.2	8.5 ± 1.3	$1.1 \cdot 10^{-4}$	none

of $\Upsilon(4S)$ data. Two complementary $B_d^0\overline{B_d^0}$ mixing measurements have been performed which improve the world average error on the mixing probability χ_d by a factor of 2. The first limit on the CP violating asymmetry in mixed like-sign dilepton events has been presented. Evidence for $b \to u\ell\nu$ transitions has been found from the inclusive lepton spectrum endpoint which are, however, observed at a lower level than in previous measurements. Conservative upper limits have been derived for exclusive charmless B decays $B \to \rho(\omega)\ell\nu$ which also support lower values of $|V_{ub}|/|V_{cb}|$ and which are in disagreement with an ARGUS claim of a signal. Limits on charmless hadronic B decays are approaching the theoretical predictions. The first limit on the decay $B^+ \to \tau^+\nu$ has been obtained.

REFERENCES

1. Y. Kubota et al., submitted to *Nucl. Instrum. Methods.*

2. G.P. Lepage, *Phys. Rev.* D42, 3251 (1990); N. Byers and E. Eichten, *Phys. Rev.* D42, 3885 (1990).

3. H. Albrecht et al., *Phys. Lett.* B232, 554 (1989); R. Fulton et al., *Phys. Rev.* D43, 641 (1991); S. Henderson et al., *Phys. Rev.* D45, 2212 (1992); M. Danilov, *Proc. LP-HEP, Geneva* 1991.

4. R.M. Baltrusaitis et al., *Phys. Rev. Lett.* 54, 1976 (1985).

5. D. Bortoletto et al., *Phys. Rev.* D45, 21 (1992).

6. H. Albrecht et al., *Phys. Lett.* B192, 245 (1987); M. Schäfer (ARGUS), *Proc. AIP Conf. 196, Cornell* (1989); M. Artuso et al., *Phys. Rev. Lett.* 62, 2233 (1989).

7. C. Albajar et al., *Phys. Lett.* B262, 171 (1991); F. Abe et al., *Phys. Rev. Lett.* 67, 3351 (1991); H.R. Band et al., *Phys. Lett.* B200, 221 (1988); A.J. Weir et al., *Phys. Lett.* B240, 289 (1990); D. Decamp et al., *Phys. Lett.* B258, 236 (1991); CERN-PPE/92 (1992); P. Abreu et al. (DELPHI Collab.), contributed paper to *XXVI ICHEP, Dallas* 1992; B. Adeva et al., *Phys. Lett.* B288, 395 (1992); P.D. Acton et al. (OPAL Collab.), *Phys. Lett.* B276, 379 (1992); contributed paper to *XXVI ICHEP, Dallas* 1992.

8. R. Fulton et al., *Phys. Rev. Lett.* 64, 16 (1990); H. Albrecht et al., *Phys. Lett.* B255, 297 (1991).

9. M. Paulini (ARGUS), *Proc. LP-HEP, Geneva* (1991); H. Albrecht et al., contributed paper to *XXVI ICHEP, Dallas* 1992.

10. D. Bortoletto et al., *Phys. Rev. Lett.* 62, 2436 (1989); P. Avery et al., *Phys. Lett.* B223, 470 (1989).

11. H. Albrecht et al., *Phys. Lett.* B241, 278 (1990).

12. M.B. Gavela et al., *Phys. Lett.* B154, 425 (1985); M. Bauer, B. Stech, M. Wirbel, *Z. Phys.* C34, 103 (1987); L.-L. Chau and H.-Y. Cheng, *Phys. Rev. Lett.* 59, 958 (1987); L.-L. Chau et al., *Phys. Rev.* D43, 2176 (1991).

PHYSICS AT DAΦNE

Juliet Lee-Franzini
Laboratori Nazionali di Frascati dell'INFN, I-00044, Frascati, Italy
SUNY at Stony Brook, Stony Brook, New York 11794

Abstract

Experimental measurements which can be done at DAΦNE, the ϕ-factory under construction at the Laboratori Nazionali di Frascati dell'INFN, are briefly discussed.

INTRODUCTION

The Frascati ϕ–factory, DAΦNE, project has been described by G. Vignola.[1] DAΦNE is a "factory" of neutral K's which are in a well prepared quantum state, of charged K pairs, as well as of ρ's, η's and the rarer η''s. The high luminosity of DAΦNE will also allow measurements of rare ϕ radiative decays. Another feature at DAΦNE is that, because the ϕ's decay at rest, neutral kaons are produced in collinear pairs, with momenta of 110 MeV/c. The observation of one K guarantees the existence of the other, with determined direction and identity, i.e., K's can be "tagged". By using the KLOE detector proposed for DAΦNE,[2] which has a K_S decay fiducial volume of radius $\sim$8 cm, and K_L decay fiducial volume of radius $\sim$150 cm, a very large numbers of tagged K mesons can be collected per year. This availability of tagged kaons is central to DAΦNE's ability for performing experiments not possible elsewhere, and enables KLOE to be essentially a "self-calibrating" detector, and have an absolute normalization of the K_S, K_L fluxes.

CP AND CPT AT DAΦNE

The KLOE program at DAΦNE is aimed at measuring CP and possible CPT violation in the $K^0, \overline{K}^0 - K_S, K_L$ system, with a sensitivity comparable to the next generation fixed target experiments ($\sim 10^{-4}$), by using both pure K_S, K_L beams obtained from tagging, and can in addition take advantage of the the coherent K_S, K_L states produced which allows performance of *kaon-quantum-interferometry* such that we can measure most parameters of the neutral kaons such as $\Re(\epsilon'/\epsilon)$, $\Im(\epsilon'/\epsilon)$, Δm, $|\eta_{\pi\pi}|$, $\phi_{\pi\pi}$, with improved accuracy. Diagonalizing the K^0-$\overline{K}^0$ mass matrix with non-diagonal elements from $K^0 \rightleftharpoons \overline{K}^0$, we find the physical states K_S and K_L, which we can write (adopting the notation of Ref. 3), without assuming CPT, as

$$|K_S\rangle \propto (1+\epsilon_K+\delta_K)|K^0\rangle + (1-\epsilon_K-\delta_K)|\overline{K}^0\rangle$$

$$|K_L\rangle \propto (1+\epsilon_K-\delta_K)|K^0\rangle - (1-\epsilon_K+\delta_K)|\overline{K}^0\rangle$$

where the (small) complex ϵ_K and δ_K characterize respectively the CP violating and CPT violating parameters in the effective Hamiltonian. Defining the usual amplitude ratios and epsilon parameters:[4]

$$\frac{\langle\pi^+\pi^-|K_L\rangle}{\langle\pi^+\pi^-|K_S\rangle} = \eta_{+-} = \epsilon + \epsilon'$$

$$\frac{\langle\pi^0\pi^0|K_L\rangle}{\langle\pi^0\pi^0|K_S\rangle} = \eta_{00} = \epsilon - 2\epsilon'$$

where η's and ϵ's are all complex, experimental observation of $\epsilon' \neq 0$ would be proof that CP is violated in the decay amplitude.

The standard model, in the context of the CKM quark mixing mechanism, predicts $\Re(\epsilon'/\epsilon) \sim 10^{-3}$, independent of scale factors with smaller values possible, especially for $M_{\rm top} \sim 150$ GeV.[5,6] Actually, if $\Re(\epsilon'/\epsilon)$ were to be identically zero, it could be a signal of physics beyond the SM, but instead of reaching that kind of precision, it is better to search for other, additional, independent ways to test CP–violation. The relationships between $\eta_\pm$, η_{00} and ϵ, ϵ', when one allows CPT violation, remain as above, since it depends only on isospin decomposition, *albeit* both ϵ, ϵ' each acquire terms which violate CP and CPT separately.[3]

At DAΦNE $K^0\bar{K}^0$ are produced in a C-odd state:

$$|i\rangle \propto |K^0, \mathbf{p}\rangle|\bar{K}^0, -\mathbf{p}\rangle - |\bar{K}^0, \mathbf{p}\rangle|K^0, -\mathbf{p}\rangle$$
$$\propto |K_S, -\mathbf{p}\rangle|K_L, \mathbf{p}\rangle - |K_S, \mathbf{p}\rangle|K_L, -\mathbf{p}\rangle.$$

In vacuum the KK pair remains a pure K_S, K_L state.

Defining $\eta_i = \langle f_i|K_L\rangle/\langle f_i|K_S\rangle$, $\Delta t = t_1 - t_2$, $t = t_1 + t_2$, $\Delta\mathcal{M} = \mathcal{M}_L - \mathcal{M}_S$ (where $\mathcal{M}_{S,L} = M_{S,L} - i\Gamma_{S,L}/2$ are the complex masses of K_S and K_L), and $\mathcal{M} = \mathcal{M}_L + \mathcal{M}_S$, the amplitude for decay to states f_1 at time t_1 and f_2 at time t_2, without identification of K_S or K_L is: $\langle f_1, t_1, \mathbf{p}; f_2, t_2, -\mathbf{p}|i\rangle \propto \left(\langle f_1|K_S\rangle\langle f_2|K_S\rangle e^{-i\mathcal{M}t/2}\right) \times \left(\eta_1 e^{i\Delta\mathcal{M}\Delta t/2} - \eta_2 e^{-i\Delta\mathcal{M}\Delta t/2}\right)$.

KAON-INTERFEROMETRY

The decay intensity $I(f_1, f_2, \Delta t = t_1 - t_2)$ to final states f_1 and f_2 is obtained from the above equation by integrating over all t_1, t_2, with Δt constant. For $\Delta t > 0$:

$$I(f_1, f_2; \Delta t) = \frac{1}{2}\int_{\Delta t}^{\infty}|A(f_1, t_1; f_2, t_2)|^2 dt =$$
$$\frac{1}{2\Gamma}|\langle f_1|K_S\rangle\langle f_2|K_S\rangle|^2$$
$$\times \left(|\eta_1|^2 e^{-\Gamma_L \Delta t} + |\eta_2|^2 e^{-\Gamma_S \Delta t}\right.$$
$$\left. - 2|\eta_1||\eta_2|e^{-\Gamma\Delta t/2}\cos(\Delta m\Delta t + \phi_1 - \phi_2)\right),$$

with $\eta_i = A(K_L \to f_i)/A(K_S \to f_i) = |\eta_i|e^{i\phi_i}$, exhibiting interference terms sensitive to phase differences.

The idea of studying correlation in decays of prepared $K^0\bar{K}^0$ pairs, while having a continuous history for over 25 years, received new impetus when the reality of new ϕ–factories inspired a whole group of new studies with increasing relaxing of symmetry violations, for example from ref. 3, which relaxes CPT conservation while retaining the $\Delta S = \Delta Q$ rule in semileptonic decays, to ref. 6 where the last assumption is also relaxed, even though this rule, in the standard model (SM), is violated to only 10^{-7} (in amplitude).[7]

We can perform "kaon-interferometry" by using the decay intensity of the previous equation with appropriate choices of the final states f_1, f_2. For example: 1). $f_1 = f_2$: we can measure Γ_S, Γ_L and Δm, since all the phases disappear. Rates can be measured to $\times 10$ improvement in accuracy and Δm to $\times 2$. 2), $f_1 \neq f_2$: a). with $f_1 = \pi^+\pi^-$, $f_2 = \pi^0\pi^0$, we can measure $\Re(\epsilon'/\epsilon)$, and $\Im(\epsilon'/\epsilon)$. The former by concentrating on large time differences, the latter for $|\Delta t| \leq 5\tau_s$. b). with $f_1 = \pi^+\ell^-\nu$ and $f_2 = \pi^-\ell^+\nu$, we can measure the CPT-violation parameter δ_K, the real part by concentrating on large time difference regions; and the imaginary part for $|\Delta t| \leq 10\tau_s$. c). If $f_1 = 2\pi$, $f_2 = K_{\ell 3}$, this leads to measurements of CP and CPT violation parameters at large time differences, since we measure the asymmetry in K_L semileptonic decays. At small time differences, we obtain Δm, $|\eta_{\pi\pi}|$ and $\phi_{\pi\pi}$. 3). By choosing $f_1 = \pi\pi$, and $f_2 =$ all other decay channels of the other K, the "inclusive method," we improve statistics relative to choosing a single semileptonic channel, and thus obtain the best measurements of the magnitudes of η_{+-} and η_{00} as well as $\Re\epsilon_K + i\Im\delta_K$. 4). By choosing $f_1 = \pi^\pm\ell\nu$, and $f_2 =$ all other decay channels of the partner K one can obtain the best measure of the $\langle K_S|K_L\rangle$ overlap, *i.e.*, determine both $\Re\epsilon_K$ and $\Im\delta_K$. 5). With K_S's, using K_L's for tagging, we can measure the asymmetry in the semileptonic decays of the K_S. The difference between this asymmetry, and that in the semileptonic decays of the K_L's, measures the real part of the CPT-violation parameter δ_K. 6). By measuring the difference in rates into two positive same sign leptons of the K pair from that into two negative same sign leptons we perform the so-called "Kabir" test,[8] which if $\Delta S = \Delta Q$ is assumed, allows us to probe the K^0-$\bar{K}^0$ mass difference to 1×10^{-18}.[3]

We can also use the classical method of the double ratio $\mathcal{R}^\pm/\mathcal{R}^0 = 1 + 6 \times \Re(\epsilon'/\epsilon)$, using tagging to select pure K_S and K_L beams. Other ways of measuring $\Re(\epsilon'/\epsilon)$ from selected final states also exist.[9] While all the methods are not statistically independent, they provide very useful checks because of the very different dependence on all systematic effects while achieving comparable statistical accuracy.[2] In short, at DAΦNE we can improve the present knowledge of all sixteen observables in the K_S, K_L system by up to one or two orders of magnitudes, and, by measuring the real part of δ_K, allow us to separate all the individual CPT violating parameters.[3] If the $\Delta S = \Delta Q$ rule is relaxed, so that 4 new parameters are added to describe the neutral kaon system, only one other experimental input (involving tagging strangeness without use of semileptonic decays of K's) is needed to completely disentangle all eight CP and CPT violating parameters.[6]

OTHER CP VIOLATIONS AT DAΦNE

Mass Matrix CP Violation

So far the only CP violation has been observed in the K_L system. Observation of $K_S \to 3\pi^0$ would constitute a new proof of CP violation. At present the upper limit on the BR is 3.7×10^{-5}. For $\epsilon'_{000}=0$, one expects BR($K_S \to 3\pi^0$) $= 2 \times 10^{-9}$ because we can tag the K_S's by using K_L's, KLOE can collect ~30 events in one year, with zero background. Similarly, we can observe the difference in rates between ($K_S \to \pi^\pm \ell^\mp \nu$), which is expected to be $\sim 16 \times 10^{-4}$ in one year's run at DAΦNE, using known lifetimes and the KLOE geometry. KLOE can measure it to 4×10^{-4}.

Direct CP Violation

Evidence for direct CP violation can be also be obtained from the decays of charged kaons which are copiously produced at DAΦNE. CPT invariance requires that the total rates for $K^\pm \to 3\pi$ be identical while CP requires equality of the partial rates for $K^\pm \to \pi^\pm\pi^+\pi^-$ ($\tau^\pm$) and for $K^\pm \to \pi^\pm\pi^0\pi^0$ ($\tau'^\pm$). The present rate asymmetry is $A = (0.7 \pm 1.2) \times 10^{-3}$ and $A' = (0 \pm 6) \times 10^{-3}$. The possibility of tagging at DAΦNE, permits in fact an exact cancellation of efficiencies in this type of measurements and sensitivities of few$\times 10^{-5}$ can be achieved.

We can also observe differences in the Dalitz plot distributions for K^+ and K^- decays in both the τ and τ' modes. The Dalitz plot population has a slope in the odd pion energy distribution, g. CP invariance must satisfy $g(\tau^+) = g(\tau^-)$ and $g(\tau'^+) = g(\tau'^-)$. Present experimental limits on the asymmetry $A = (g^+ - g^-)/(g^+ + g^-)$ range from few$\times 10^{-2}$ to few$\times 10^{-3}$, while KLOE could reach sensitivities of $\sim 10^{-4}$.

$K^\pm$ radiative decays: differences in rates in the radiative two pion decays of $K^\pm$, $K^\pm \to \pi^\pm\pi^0\gamma$, are also proof of direct CP violation. The present experimental limit for $\Delta\Gamma/2\Gamma$ is $<5\times 10^{-2}$ and the theoretical prediction is $\leq 10^{-3}\times\sin\delta$, where δ is the CKM phase angle. The KLOE sensitivity is $\sim 1.4\times 10^{-3}$.

CHIRAL PERTURBATION THEORY

In the last decade chiral perturbation theory (CHPT) has been extended to the next order terms in the chiral expansion ($\mathcal{O}(m^4)$, $\mathcal{O}(p^4)$, $\mathcal{O}(m^2p^2)$). This extension introduces new parameters which must be determined experimentally. Many new amplitudes can then be predicted.[10]

At lowest order the CHPT relation predicts the slope of the scalar form factor. There is at present disagreement between K^+ data and K_L results and different experiments are mutually incompatible. The world averages are $\lambda_0 = 0.004 \pm 0.007$, for K^+ and 0.025 ± 0.006 for K_L, both in disagreement with the CHPT prediction, 0.017±0.004. KLOE can measure λ_0 for K_L to an accuracy of 1.4×10^{-5}. Similar accuracy are obtained for $K^{\pm}$ and for λ_+.

There is only one measurement of the relevant $K_{\ell 4}$ form factors. KLOE can improve vastly on this topic. These decays also provide another opportunity for the determination of the $\pi\pi$ phase shifts.

Apart from the radiative term, the amplitudes for $K_{\ell 2,\gamma}$, $K_{\ell 2,e^+e^-}$ and $K_{\ell 3,\gamma}$ depend on the K charge radius and three other parameters about which conditions are obtained from pion physics. Additional constraints and checks would follow from the processes listed.

The rate for $K^{\pm} \to \pi^{\pm}\gamma\gamma$ and the $\gamma\gamma$ distributions are uniquely predicted by the chiral lagrangian approach. Dalitz type decays of K mesons and two photon production of pions are also of great interest. Both can be studied with KLOE.

ϕ RADIATIVE DECAYS

We still do not understand the whole scalar meson sector.[11] What is the $f_0(975)$? How much glue is there in the light pseudoscalar mesons? We can measure $\phi \to f_0\gamma$ to great accuracy, in particular determine the sign of $\phi f_0 \gamma$ coupling.[12] We can measure the branching ratio of $\phi \to \eta'\gamma$ to $\sim 1.2\times10^{-6}$ in three months of running, which will answer the second question.[13]

RARE K_S DECAYS

By tagging one can effectively have a beam of K_S, with no background. The pure, tagged DAΦNE K_S beams, with up to 10^{10} kaons per year, will dramatically improve knowledge of K_S branching ratios (most are not measured yet). Searches for $K_S \to \pi^0\nu\bar{\nu}$, $e^+e^-\gamma$, $\mu^+\mu^-\gamma$, $\pi^0 e^+e^-$, $\pi^0\mu^+\mu^-$ etc., down to 10^{-8} or better are possible.

ACKNOWLEDGEMENTS

I wish to thank all the participants of the DAΦNE Theory Workshop, and refer to the upcoming *DAΦNE Physics Handbook* for all the omissions resulting from space constrictions here. I thank my KLOE colleagues and wish to acknowledge both Paolo and Paula J. Franzini for help in preparing this paper.

REFERENCES

1. G. Vignola, these Proceedings.
2. P. Franzini, these Proceedings.
3. C. Buchanan et al., Phys. Rev. **D45**, 4088 (1992).
4. T. T. Wu and C. N. Yang Phys. Rev. Lett. **13**, 380 (1964).
5. For a review of the evolution in the calculation of ϵ'/ϵ, see P. J. Franzini, *Les Rencontres de Physique de la Valleé d'Aoste*, La Thuile, Italy, March 3-9 1991, M. Greco Ed., p. 257.
6. For a comprehensive discussion and references, L. Maiani, "CP and CPT Violation in Neutral Kaon Decays", to appear in the *DAΦNE Physics Handbook*, ed. L. Maiani et al., LNF, Frascati.
7. C. O. Dib and B. Guberina, Phys. Lett. **255**, 113 (1991).
8. P. K. Kabir, Phys. Rev. **D2**, 540 (1970).
9. J. Bernabeu, F. J. Botella and J. Roldán, Phys. Lett. **211B**, 226 (1980).
10. For all references on CHPT, please see *DAΦNE Physics Handbook*, ed. L. Maiani et al., LNF, Frascati.
11. F. Close, these Proceedings.
12. Paula J. Franzini, Won Kim and J. Lee-Franzini, Phys. Lett. **B287** 259, (1992).
13. J. Lee-Franzini, Won Kim, Paula J. Franzini, contribution # 109 to this Conference.

$B^0 - \bar{B}^0$ MIXINGS AND RARE B-DECAYS

Ahmed Ali
Deutsches Elektronen Synchrotron DESY
Notkestr. 85
D-2000 Hamburg 52, FRG

Abstract

I present an update on the Cabibbo-Kobayashi-Maskawa (CKM) matrix elements taking into account the current experimental and theoretical information on weak decays. The resulting fit is combined with the measured value of the B_d^0-$\overline{B_d^0}$ mixing ratio x_d and estimates of the pseudoscalar coupling constants to determine the allowed range of the B_s^0-$\overline{B_s^0}$ mixing ratio x_s in the Standard Model (SM). For the central values of the parameters used we find $8 \leq x_s \leq 24$. Flavour changing neutral current (FCNC) B-decays are briefly reviewed; in particular the role of such decays in determining the CKM matrix elements V_{td} and V_{ts} is emphasized on the example of radiative B-decays, $B \to (X_d + \gamma)$ and $B \to (X_s + \gamma)$.

INTRODUCTION AND OVERVIEW

In SM, mass mixings and FCNC decays have something in common, namely they are allowed only as higher order (loop) processes. For B-hadrons these transitions are dominated by the (virtual) top quark contribution. Hence, their measurements provide valuable constraints on the top quark mass m_t and its CKM couplings V_{td}, V_{ts} and V_{tb}. This information is complementary to the one from the electroweak radiative corrections to the neutral current (NC) and charged current (CC) processes which constrain m_t and (to a smaller extent) the Higgs boson mass through loop effects.

This contribution gives an update on the CKM matrix elements using the fits in ref. 1 and the modifications discussed below. The aim is to test both the consistency of the CKM framework and get constrained estimates for a number of FCNC transitions in B physics. To illustrate this, I review some recent theoretical work, carried out in the SM context, on B^0-$\overline{B^0}$ mixings[1] and the FCNC rare B-decays $B \to X_s + \gamma$ [2] and $B \to X_d + \gamma$ [3]. The former will manifest themselves through exclusive decays such as $B \to K^* + \gamma$ (and higher resonances[4]) and the latter through the CKM-suppressed decay $B \to \rho + \gamma$ (and higher resonances). Also, FCNC B-decays involving dileptons are potentially very interesting [5,6,7]. While m_t will hopefully soon be measured directly, very probably FCNC transitions will be the only source of information for the CKM-matrix elements involving the top quark. In view of this, I emphasize here the quantitative role that such measurements will play in determining the elements of the CKM matrix, V_{ti}.

AN UPDATE ON THE CKM MATRIX

In the Wolfenstein parametrization[8], the CKM-matrix is characterized by the four parameters, λ, A, ρ, η, with η being a complex phase. The present status of these parameters is summarized below. The essential update compared to the values given in the PDG compilation[9] and the work reported in ref. 1 lies in the use of a significantly lower value, $V_{ub}/V_{cb} = 0.085 \pm 0.015$, which is an estimate of the world average of this ratio using the Altarelli et al. model[10]. The various input quantities used for the CKM-matrix fits are discussed below.

First of all, $|V_{us}|$ has been extracted with good accuracy from $K \to \pi e \nu$ and hyperon decays to be [8]

$$|V_{us}| = \lambda = 0.2205 \pm 0.0018 . \quad (1)$$

This agrees quite well with the determination of $V_{ud} \simeq 1 - \frac{1}{2}\lambda^2$ from β-decay[8]:

$$|V_{ud}| = 0.9744 \pm 0.0010 . \quad (2)$$

The parameter A is related to the CKM matrix element V_{cb}, which can be obtained from semileptonic B-decays. Its value has been determined from the inclusive semileptonic decay measurements, which give $|V_{cb}| = 0.042 \pm 0.007$, where the quoted error takes into account the uncertainty due to the semileptonic decay model [11]. Alternatively, and theoretically more reliably, one could use the heavy quark symmetry, relating the *heavy* $\to$ *heavy* hadron transitions, to determine the form factors in the semileptonic decays $B(v) \to (D, D^*)(v') \, l\nu$ (here v and v' are the four-velocities as indicated). Using the heavy quark symmetry one can show that all hadronic form factors describing these decays can be expressed in terms of a single function, the Isgur-Wise function [12], $\xi(v \cdot v')$, having the normalization $\xi(v \cdot v' = 1) = 1$. To extract $|V_{cb}|$, one has to extrapolate the data to the point $y = v \cdot v' = 1$, for which one needs an Ansatz for the Isgur-Wise function $\xi(y)$. Using a linear parametrization for $\xi(y)$ (which is consistent with data), the present measurements of the B-lifetime, $\tau_B = (1.35 \pm 0.05) \times 10^{-12} s$, and the semileptonic branching ratios $BR(B \to D^* l\nu) = (5.1 \pm 0.9)\%$, the ARGUS analysis yields $|V_{cb}| = 0.043 \pm 0.007$ and $|V_{cb}| = 0.045 \pm 0.005$, for the neutral and charged D^* state, respectively[13]. Using other parametrizations, the extracted values for $|V_{cb}|$ are significantly different. For the CKM fits I shall use a value

$$|V_{cb}| = 0.044 \pm 0.006 \quad (3)$$

giving the CKM parameter A

$$A = 0.90 \pm 0.12 . \quad (4)$$

The other two CKM parameters ρ and η are constrained by the measurements of $|V_{ub}/V_{cb}|$, $|\epsilon|$ (the CP-violating parameter in the kaon system), x_d (B_d^0-$\overline{B_d^0}$ mixing) and (in principle) ϵ'/ϵ ($\Delta S = 1$ CP-violation in the kaon system). I shall not use the constraint from the ϵ'/ϵ-analysis but discuss the rest in turn, presenting a fit in which the allowed region of ρ and η and the CKM-unitarity triangle are shown.

So far $|V_{ub}/V_{cb}|$ has been obtained by analyzing the endpoint of the inclusive lepton spectrum in semileptonic B decays. The resulting values are in general model dependent. Using the Altarelli et al. model [10], the earlier ARGUS and CLEO results were compatible with the range $\left|\frac{V_{ub}}{V_{cb}}\right| = 0.12 \pm 0.02$. The new analysis presented by the CLEO collaboration at this conference, however, gives a significantly lower value for this ratio, with the CLEO II measurements being compatible with $|V_{ub}/V_{cb}| \simeq 0.07$ in the same model[11,14].

With the experimental situation in flux, probably it is best to average the old and new CLEO and ARGUS results, yielding

$$\left|\frac{V_{ub}}{V_{cb}}\right| = 0.085 \pm 0.015 . \quad (5)$$

in the Altarelli et al. model [10]. This gives

$$\sqrt{\rho^2 + \eta^2} = 0.39 \pm 0.07 . \quad (6)$$

which is significantly lower than the value $\sqrt{\rho^2 + \eta^2} = 0.63 \pm 0.23$, used earlier [1].

The experimental value of $|\epsilon|$ is

$$|\epsilon| = (2.26 \pm 0.02) \times 10^{-3} . \quad (7)$$

Theoretically, $|\epsilon|$ is essentially proportional to the imaginary part of the box diagram for K^0-$\overline{K^0}$ mixing, and is given by

$$|\epsilon| = \frac{G_F^2 f_K^2 M_K M_W^2}{6\sqrt{2}\pi^2 \Delta M_K} B_K \left(A^2 \lambda^6 \eta\right)$$
$$(y_c \{\eta_{ct} f_3(y_c, y_t) - \eta_{cc}\}$$
$$+ \eta_{tt} y_t f_2(y_t) A^2 \lambda^4 (1-\rho)). \quad (8)$$

Here, the η_i are QCD correction factors, $\eta_{cc} \simeq 0.82$, $\eta_{tt} \simeq 0.62$, $\eta_{ct} \simeq 0.35$ for $\Lambda_{QCD} = 200$ MeV $y_i \equiv m_i^2/M_W^2$, and the functions f_2 and f_3 are given in ref. 1.

One of the unknowns in Eq. (8) is the top quark mass. I shall use a value $100\ GeV \leq m_t \leq 180\ GeV$, which is compatible with the present bounds from electroweak radiative corrections and direct top quark searches. The final parameter in the expression for $|\epsilon|$ is B_K, which represents our ignorance of the matrix element $\langle K^0|(\bar{d}\gamma^\mu(1-\gamma_5)s)^2|\overline{K^0}\rangle$. The evaluation of this matrix element has been the subject of much work. In what follows I shall take

$$B_K = 0.8 \pm 0.2 . \quad (9)$$

Turning now to B_d^0-$\overline{B_d^0}$ mixing, the quantitity x_d, defined below, has been measured to be

$$x_d = 0.67 \pm 0.10 . \quad (10)$$

In the SM, x_d is calculated from the B_d^0-$\overline{B_d^0}$ box diagram, which is dominated by t-quark exchange:

$$x_d \equiv \frac{(\Delta M)_{B_d}}{\Gamma_{B_d}} = \tau_B \frac{G_F^2}{6\pi^2} M_W^2 M_B \left(f_{B_d}^2 B_{B_d}\right)$$
$$\eta_B y_t f_2(y_t)|V_{td}^* V_{tb}|^2 , \quad (11)$$

where $(\Delta M)_{B_d}$ and Γ_{B_d} are, respectively, the mass difference and the average width of the two mass eigenstates, and $|V_{td}^* V_{tb}|^2 = A^2 \lambda^6 \left[(1-\rho)^2 + \eta^2\right]$. Here, η_B is the QCD correction, which has been analyzed in great detail in ref. 15, including the effects of a heavy t-quark. They find that η_B depends sensitively on the definition of the t-quark mass, and that, strictly speaking, only the product $\eta_B(y_t)f_2(y_t)$ is free of this dependence. I will use the next-to-leading order QCD-improved value, $\eta_B = 0.55$. For the B system, the hadronic uncertainty is given by $f_{B_d}^2 B_{B_d}$, analogous to B_K in the kaon system, except that in this case, also f_{B_d} is not measured. And, just like B_K, the evaluation of $f_{B_d}^2 B_{B_d}$ has been the subject of much work. For the CKM fits two ranges for $f_{B_d}^2 B_{B_d}$ are used:

$$(I): \quad f_{B_d}\sqrt{B_{B_d}} = 135 \pm 25\ \text{MeV},$$
$$(II): \quad f_{B_d}\sqrt{B_{B_d}} = 200 \pm 20\ \text{MeV} \quad (12)$$

which represent, respectively, the estimates of the older vintage and more modern Lattice-QCD results [16].

The information regarding the allowed region in ρ-η space can be displayed quite elegantly using the so-called unitarity triangle. Because the CKM matrix is unitary, one has the following relation:

$$V_{ud}V_{ub}^* + V_{cd}V_{cb}^* + V_{td}V_{tb}^* = 0 . \quad (13)$$

This can be (approximately) recast as

$$\frac{V_{ub}^*}{\lambda V_{cb}} + \frac{V_{td}}{\lambda V_{cb}} = 1 , \quad (14)$$

that is, a triangle relation in the complex plane (in ρ-η space).

In order to find the allowed unitarity triangles, the computer program MINUIT has been used to fit the CKM parameters A, ρ and η to the experimental values of $|V_{cb}|$, $|V_{ub}/V_{cb}|$, $|\epsilon|$ and x_d in ref. 1. This fit is updated in this contribution and the results are shown in Fig. 1 and 2 for the coupling constant choice (I) and (II), respectively. Note that the graph for $f_{B_d}\sqrt{B_{B_d}} = 135 \pm 25$ MeV and $m_t = 100$ GeV is a bad fit of the data ($\chi^2/d.o.f. = 2.17$). In all these graphs, the solid line has $\chi^2 = \chi^2_{min} + 1$. For comparison, we include the dashed line, which is the 90% c.l. region ($\chi^2 = \chi^2_{min} + 4.6$). The "best values" of the parameters (ρ, η), together with their ($\chi^2/d.o.f.$), are given in Table 1.

THE UNITARITY TRIANGLE AND x_s

We now discuss the estimates for $x_s \equiv \frac{(\Delta M)_{B_s}}{\Gamma_{B_s}}$. Mixing in the B_s^0-$\overline{B_s^0}$ system follows quite closely that of the B_d^0-$\overline{B_d^0}$ system. The B_s^0-$\overline{B_s^0}$ box diagram is again dominated by t-quark exchange, and the mixing parameter x_s is given by a formula analogous to that of Eq. (11):

$$x_s = \tau_{B_s} \frac{G_F^2}{6\pi^2} M_W^2 M_{B_s} \left(f_{B_s}^2 B_{B_s}\right) \eta_{B_s} y_t f_2(y_t) |V_{ts}^* V_{tb}|^2 \quad (15)$$

Assuming $V_{cb} = V_{ts}$, it follows that one of the sides of the unitarity triangle, $|V_{td}/\lambda V_{cb}|$ can be obtained from the ratio of x_d and x_s:

$$\frac{x_d}{x_s} = \frac{\tau_{B_d} \eta_{B_d} M_{B_d} \left(f_{B_d}^2 B_{B_d}\right)}{\tau_{B_s} \eta_{B_s} M_{B_s} \left(f_{B_s}^2 B_{B_s}\right)} \left|\frac{V_{td}}{V_{ts}}\right|^2. \quad (16)$$

Conversely, the ratio on the l.h.s. can be predicted from the unitarity triangle constraints, as all dependence on the t-quark mass drops out, and we are left with the square of the ratio of CKM matrix elements, multiplied by a

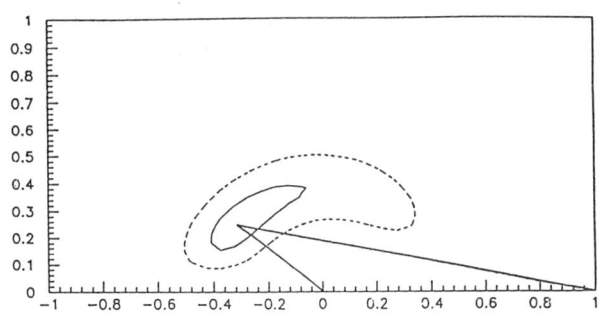

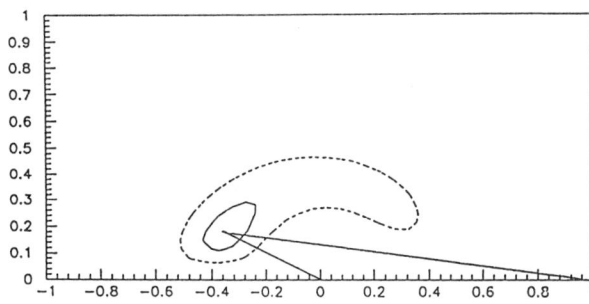

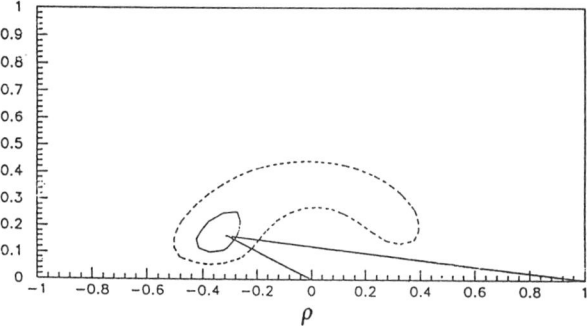

Figure 1. Allowed region in ρ-η space for different values of the standard model parameters. Figs. (a)-(c) have $f_{B_d}\sqrt{B_{B_d}} = 135 \pm 25$ MeV, with $m_t = 100$, 140 and 180 GeV, respectively. The solid line represents the region with $\chi^2 = \chi^2_{min} + 1$; the dashed line denotes the 90% c.l. region. The triangles show the best fit (updated from ref. (1)).

factor which reflects $SU(3)_{\text{flavour}}$ breaking effects. Whether or not x_s can be used to help constrain the unitarity triangle will depend crucially on the theoretical status of the ratio $f_{B_d}^2 B_{B_d}/f_{B_s}^2 B_{B_s}$. I assume that this ratio can be reliably calculated and use the recent Lattice-QCD update[16]

$$\frac{f_{B_s}^2 B_{B_s}}{f_{B_d}^2 B_{B_d}} = 1.19 \pm 0.10. \qquad (17)$$

Setting $\frac{\tau_{B_d} \eta_{B_d} M_{B_d}}{\tau_{B_s} \eta_{B_s} M_{B_s}} = 1$, which should hold to a good accuracy, gives

$$\frac{x_s}{x_d} = 1.19 \pm 0.10 \times \left|\frac{V_{ts}}{V_{td}}\right|^2. \qquad (18)$$

The "best values" for the CKM ratio $|V_{td}/V_{ts}|$ following our fits are given in Table 1. It is interesting to note that this ratio is remarkably stable for the low-f_{B_d} solution (I), but varies considerably for the high-f_{B_d} solution (II), as one varies m_t. Using the central value for x_d in Eq. (18) together with the "best fit" for the CKM triangle gives:

$$8.0 \leq x_s \leq 24.0 . \qquad (19)$$

Thus, to measure x_s, one will have to undertake time-dependent experiments. They are discussed in ref. 1. In summary, a measurement of B_s^0-$\overline{B_s^0}$ mixing will allow a more accurate measurement of $|V_{td}/\lambda V cb|$, and will be an important further test of the unitarity triangle.

FCNC RARE B-DECAYS

The interest in the study of FCNC B-decays is reflected in the large number of inclusive and exclusive decays that have been investigated theoretically [2-7]. In particular, their measurements will provide a normalization of the magnetic moment operator matrix elements, which are dominated by

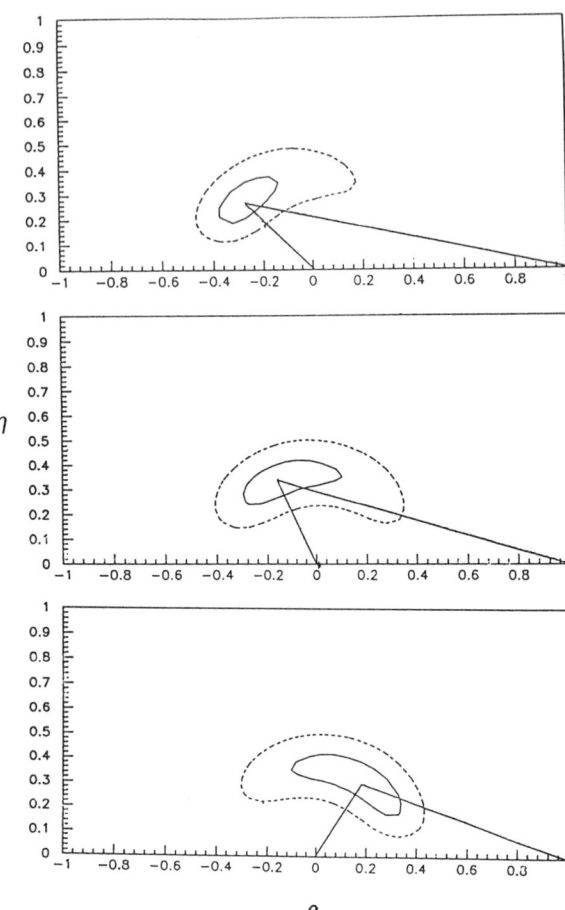

Figure 2. Allowed region in ρ-η space for different values of the standard model parameters. Figs. (a)-(c) have $f_{B_d}\sqrt{B_{B_d}} = 200 \pm 20$ MeV, with $m_t = 100, 140$ and 180 GeV, respectively. The solid line represents the region with $\chi^2 = \chi^2_{min} + 1$; the dashed line denotes the 90% c.l. region. The triangles show the best fit (updated from ref. (1)).

Table 1. The "best values" of the CKM parameters (ρ, η) and the ratio $|V_{td}/V_{ts}|^2$, obtained by a minimum χ^2 fit of the experimental data discussed in the text. Values of m_t and the coupling constant $f_{B_d}\sqrt{B_{B_d}}$ are stated. The resulting minimum χ^2 values from the MINUIT fits are also given (updated from ref. 1.)

m_t (GeV)	$f_{B_d}\sqrt{B_{B_d}}$ (MeV)	(ρ, η)	$|V_{td}/V_{ts}|^2$	χ^2_{min}
100	135 ± 25	$(-0.31, 0.25)$	0.086	2.17
140	135 ± 25	$(-0.35, 0.18)$	0.090	0.46
180	135 ± 25	$(-0.35, 0.16)$	0.089	0.0045
100	200 ± 20	$(-0.28, 0.27)$	0.083	0.25
140	200 ± 20	$(-0.16, 0.33)$	0.071	0.43
180	200 ± 20	$(0.19, 0.31)$	0.036	0.90

the penguin diagrams. For lack of space, I concentrate here on the radiative transitions $b \to s + (g) + \gamma$ and $b \to d + (g) + \gamma$ (where g stands for gluon). The CKM-suppressed rare B-decays of the type $b \to d + (g) + \gamma$ are of particular interest in determining the rather crucial matrix element V_{td}. A good prototype of such reactions is the decay $B \to \rho + \gamma$, which can be used with its CKM-allowed counterpart $B \to K^* + \gamma$ to determine the ratio V_{td}/V_{ts}, as argued in ref. 3 and below.

FCNC Decays $B \to X_s + \gamma$ and $B \to X_d + \gamma$

The framework that has been employed in the calculations of the FCNC B-decays is that of an effective theory with five quarks, obtained by integrating out the heavier degrees of freedom (top quark and $W^\pm$ bosons). To leading order in the small (weak)-mixing angles, a complete set of dimension-6 operators relevant for the processes $b \to s + \gamma$ and $b \to s + \gamma + g$ is contained in the effective Hamiltonian

$$H_{eff}(b \to s\gamma) = -\frac{4G_F}{\sqrt{2}} \lambda_t \sum_{j=1}^{8} C_j(\mu) \hat{O}_j(\mu) \, , \quad (20)$$

where G_F is the Fermi coupling constant, $C_j(\mu)$ are the Wilson coefficients evaluated at the scale μ, and $\lambda_t = V_{tb}V_{ts}^*$. The definitions of the various operators, matching conditions $C_i(m_W)$, and the leading log pertubative QCD corrections giving $C_i(\mu)$ with $\mu \ll m_W$ can be seen in refs. 2-7, where also references to the original work can be found. It is known from earlier studies that the dominant contribution to $b \to s + \gamma$ and $b \to s + g + \gamma$ are due to the four-fermion operator $\hat{O}_2$ and the QED magnetic moment operator $\hat{O}_7$ which are defined as:

$$\hat{O}_2 = (\bar{c}_{L\alpha}\gamma^\mu b_{L\alpha})(\bar{s}_{L\beta}\gamma_\mu c_{L\beta})$$
$$\hat{O}_7 = (e/16\pi^2) \bar{s}_\alpha \sigma^{\mu\nu} (m_b R + m_s L) b_\alpha F_{\mu\nu}$$

$$L = \frac{1-\gamma_5}{2} \, ; \quad R = \frac{1+\gamma_5}{2}$$

Here e denotes the QED coupling constant.

The inclusive rates, photon energy and hadron invariant mass spectra for the decays $b \to s + \gamma$ and $b \to s + \gamma + g$ have been calculated in refs. 2, from where the details can be seen. Here, I quote the inclusive branching ratio for these decays:

$$BR(B \to X_s + \gamma) = (2-5) \times 10^{-4} \quad (21)$$

The central value, $BR(B \to X_s + \gamma) = 3.5 \times 10^{-4}$ (corresponding to $\mu = 5.0\ GeV$ and $m_t = 140\ GeV$), is within a factor 3 away from the present experimental bound on this branching ratio; hence a measurement in this channel is in the wings!

As already stated, the overriding theoretical interest in the study of electromagnetic penguins in B-decays lies in the direct measurement of the CKM matrix elements $|V_{ts}|$ and $|V_{td}|$. The dependence of the branching ratio $BR(B \to X_s + \gamma)$ on the CKM ratio $(|V_{ts}|/|V_{cb}|)^2$ is being discussed by Greub in these proceedings [17]. Note that, despite considerable uncertainties in theoretical estimates, a measurement of the inclusive branching ratio for $B \to X_s + \gamma$ will provide the first direct measurement of the CKM matrix element $|V_{ts}|$.

The modifications that have to be incorporated for the CKM-suppressed radiative rare decays $b \to d+\gamma$ and $b \to d+\gamma+g$ have been discussed in ref. 3. We recall here that for the decays $b \to s + \gamma(+g)$ the effective Hamiltonian was written in the approximation where the CKM factor $\lambda_u = 0$, which is reasonable since $\lambda_u \ll \lambda_c, \lambda_t$ ($\lambda_i \equiv V_{ib}V_{is}^*$). For the decays $b \to d+\gamma$ and $b \to d+\gamma+g$ the CKM factors λ_i are replaced by $\xi_i \equiv V_{ib}V_{id}^*$. Now, ξ_u, ξ_c and ξ_t are all of the same order of magnitude ($A\lambda^3$) and therefore the corresponding approximation $\xi_u = 0$ cannot be made any longer. The differences due to the inclusion of ξ_u to describe the decays $b \to d+\gamma$ and $b \to d+g+\gamma$ can be most easily built in the effective Hamiltonian framework by modifying the operators $\hat{O}_1$ and $\hat{O}_2$, encountered in the decay $b \to s + \gamma$. The dimension-6 operator basis again reads:

$$H_{eff}(b \to d\gamma) = -\frac{4G_F}{\sqrt{2}} \xi_t \sum_{j=1}^{8} C_j(\mu) O_j(\mu) \quad (22)$$

The operators O_1 and O_2 are defined as:

$$O_1 = -\frac{\xi_c}{\xi_t}(\bar{c}_{L\beta}\gamma^\mu b_{L\alpha})(\bar{d}_{L\alpha}\gamma_\mu c_{L\beta})$$
$$\quad - \frac{\xi_u}{\xi_t}(\bar{u}_{L\beta}\gamma^\mu b_{L\alpha})(\bar{d}_{L\alpha}\gamma_\mu u_{L\beta})$$
$$O_2 = -\frac{\xi_c}{\xi_t}(\bar{c}_{L\alpha}\gamma^\mu b_{L\alpha})(\bar{d}_{L\beta}\gamma_\mu c_{L\beta})$$
$$\quad - \frac{\xi_u}{\xi_t}(\bar{u}_{L\alpha}\gamma^\mu b_{L\alpha})(\bar{d}_{L\beta}\gamma_\mu u_{L\beta}) \quad (23)$$

With the operators defined in this basis, the matching conditions, $C_i(m_W)$, and the corresponding Wilson coefficients at the scale μ, $C_i(\mu)$, are precisely the same for the decays $b \to d + \gamma$ and $b \to s + \gamma$.

The decay rate for $B \to X_d + \gamma$ depends on the CKM parameters ρ and η which enter through the CKM factors ξ_t and ξ_c, as already discussed. In contrast to the $B \to X_s + \gamma$ case, the CKM-parametric dependence does not factorize in $B \to X_d + \gamma$, and to get the inclusive branching ratio one has to vary the CKM parameters ρ and η as a function of m_t over the presently allowed range. An estimate of the inclusive branching ratio $BR(B \to X_d+\gamma)$, modelled after the partonic decays $b \to d + \gamma$ and $b \to d + g + \gamma$ (i.e. only the so-called short distance contribution) has been obtained in this way in ref. 3. This analysis is updated here in view of the foregoing discussion in the previous section. The resulting branching ratio as a function of m_t is shown in Fig. 3, yielding

$$BR(B \to X_d + \gamma) = (0.8 - 3.0) \times 10^{-5} \quad (24)$$

for the top quark mass in the range $100\ GeV \le m_t \le 200\ GeV$. The branching ratios $BR(B \to \rho + \gamma)$, as well as the relative rate $BR(B \to \rho + \gamma)/BR(B \to K^* + \gamma)$, using vector meson dominance for the hadronic mass below 1 GeV, have also been estimated in ref. 3. This branching ratio factorizes to

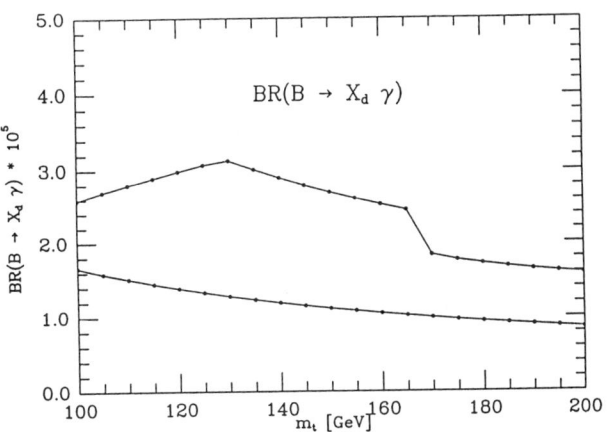

Figure 3. Upper and lower bounds on the branching ratio $BR(B \to X_d + \gamma)$ as function of m_t, obtained by varying over the allowed range in the (ρ, η) plane resulting from Fig. 2 (updated from ref. 3).

a very good approximation, and the updated estimate is:

$$BR(B \to \rho + \gamma) = (3.5 - 8.0) \times 10^{-6} \times \left(\frac{|V_{td}|^2}{1.37 \times 10^{-4}} \right) \quad (25)$$

A much firm prediction is however obtained on the relative branching ratios for the decays $B \to \rho + \gamma$ and $B \to K^* + \gamma$ [3]:

$$\frac{\Gamma(B \to \rho + \gamma)}{\Gamma(B \to K^* + \gamma)} = (0.1) \times \left(\frac{|V_{td}|^2}{1.37 \times 10^{-4}} \right) \quad (26)$$

where I have used the "best fit" CKM parameters" for $m_t = 140$ GeV and the choice (II) for the reference value of $|V_{td}|$. The role of FCNC B-decays in determining the matrix elements V_{ti} is abundantly clear from the analysis discussed here.

I would like to thank Christoph Greub and David London with whom the work reported here was carried out for updating some of the results shown. I also thank Sheldon Stone for a discussion.

REFERENCES

1. A. Ali and D. London, DESY Report 92-075 (1992); to appear in J. Phys. G: Nucl. & Part. Physics, UK.

2. A. Ali and C. Greub, Z. Phys. C49, 431, 1991; Phys. Lett. B259, 182, 1991; ibid B293, 236, 1992.

3. A. Ali and C. Greub, Phys. Lett. B287, 191, 1992.

4. A. Ali, T. Ohl, and T. Mannel, DESY Report 92-113, 1992; to appear in Phys. Lett. B.

5. A. Ali and T. Mannel, Phys. Lett. B264, 447, 1991.

6. A. Ali, T. Mannel and T. Morozumi, Phys. Lett. B273, 505, 1991.

7. A. Ali, DESY Report 92-058 (1992), and in ICTP Series in Theor. Phys.-Vol. 8, World scientific, 1992.

8. L. Wolfenstein, Phys. Rev. Lett. 51, 1945, , 1983.

9. K. Hikasa et al. (Particle Data group), Phys. Rev. D45, No. 11, 1992.

10. G. Altarelli et al., Nucl. Phys. B208, 365, 1982.

11. P. Drell, these proceedings.

12. N. Isgur and M. Wise, Phys. Lett. B232, 113, 1989; ibid B237, 527, 1990.

13. M. Danilov, ECFA B-Workshop, DESY-Hamburg, Oct. 29-30, 1992.

14. S. Stone (Private communication).

15. A.J. Buras, M. Jamin and P.H. Weisz, Nucl. Phys. B347, 491, 1990.

16. O. Pene, ECFA B-Workshop, DESY-Hamburg, Oct. 29-30, 1992.

17. C. Greub, these proceedings.

RARE KAON AND $b \to s\gamma$ DECAYS IN THE TWO-HIGGS-DOUBLET MODEL

C. Q. Geng*
Laboratoire de Physique Nucléaire, Université de Montréal
Montréal, PQ, Canada, H3C 3J7

Abstract

We study rare kaon and $b \to s\gamma$ decays in the two-Higgs-doublet model. We find that the branching ratios of the CP violating rare decays $K_L \to \pi^0 e^+ e^-$ and $K_L \to \pi^0 \nu \bar{\nu}$ can be enhanced relative to the standard model while that of the CP conserving ones $K_L \to \mu\bar{\mu}$ and $K^+ \to \pi^+ \nu \bar{\nu}$ are insensitive to the Higgs effect. We show that the experimental limit on $b \to s\gamma$ could provide the strongest constraint on the Higgs model.

The two-Higgs-doublet model is the simplest possible extension to the standard model (SM) yet it exhibits some of the characteristics of a more complicated scalar structure typical of most beyond the standard models. Here we explore the the phenomenological implications of the model in rare kaon and $b \to s\gamma$ decays.

We consider a $SU(2) \times U(1)$ invariant Lagrangian with two Higgs doublets where one doublet ϕ_u couples to up-type quarks and the other, ϕ_d, to down-type quarks. The only non-standard feature of this Lagrangian is the appearance of extra terms arising from the new scalar fields. We consider only the effect of the charged Higgs. To describe this charged scalar, two parameters need to be introduced. Here we take m_H, the mass of the charged Higgs, and $\xi \equiv v_d/v_u$, the ratio of the vacuum expectation values of ϕ_d and ϕ_u. In the two-Higgs-doublet model, the charged Higgs and quark couplings involve the same Cabibbo-Kobayashi-Maskawa (CKM) matrix as the charged current in the standard model.

To find the allowed values for A, ρ, and η, the parameters of the CKM matrix in the Wolfenstein parametrization, we perform a χ^2 fit to four measurements: ϵ, the CP violation parameter in $K \to \pi\pi$, x_d, the $B_d^0 - \bar{B}_d^0$ mixing, and the ratios $|V_{cb}/V_{us}^2|$ and $|V_{ub}/V_{cb}|$, which are given by[1]

$$
\begin{aligned}
|V_{cb}| &: A = 0.91 \pm 0.14 \\
|V_{ub}/V_{cb}| &: \sqrt{\rho^2 + \eta^2} = 0.50 \pm 0.18 \\
|\epsilon| &: A^2\eta\{0.8 + 1.43 A^2(1-\rho)[B(x_t) \\
&\quad + \xi^4 B_{HH}(x_t, y_t) + \xi^2 B_{WH}(x_t, y_t)]\} \\
&= (0.525 \pm 0.006)\frac{1}{B_K} \\
B_d^0 - \bar{B}_d^0 &: A^2[(1-\rho)^2 + \eta^2][B(x_t) + \\
&\quad \xi^4 B'_{HH}(x_t, y_t) + \xi^2 B'_{WH}(x_t, y_t)] \\
&= (3.1 \pm 0.6)\left(\frac{130\ MeV}{f_B\sqrt{B_B}}\right)^2 \quad (1)
\end{aligned}
$$

where $x_t \equiv m_t^2/M_W^2$, $y_t \equiv m_t^2/m_H^2$, and the various functions can be found in Ref. 2. In Eq. (1), we take $f_B = 220 \pm 30$ MeV, favored by the recent lattice QCD calculations and $B_K = 0.8 \pm 0.2$ and $B_B = 1$, all uncertainties in B_B being hidden in f_B.

To show the effect of the charged Higgs on the CKM matrix elements, we choose $m_t =$

---
*Address after Oct. 1, 1992: Dept. of Phys., Iowa State Univ., Ames, IA 50011

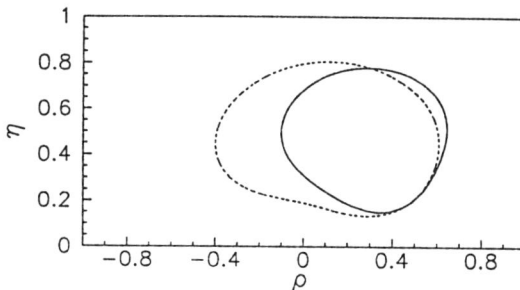

Figure 1. Allowed range for ρ and η in the standard (solid) and two-Higgs-doublet (dashed) models at 90% C.L.

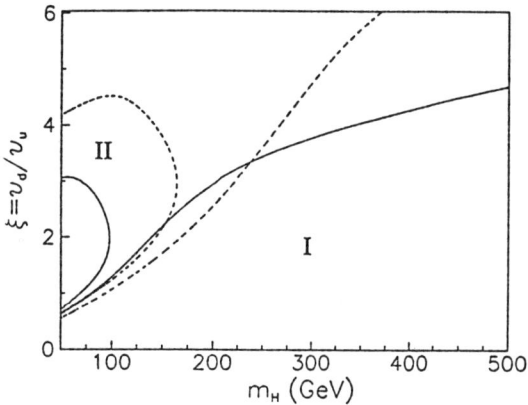

Figure 2. 90% C.L. allowed values for ξ vs m_H with $m_t = 120$ (dotted) and 200 GeV (solid).

120 GeV, $m_H = 150$ GeV, $\xi = 1$ and keep A as a free parameter and we compare, in Figure 1, the results of the fits to the measurements in (1) for the standard and two-Higgs-doublet models. These curves correspond, in the $\rho - \eta$ plane, to $\chi^2 = \chi^2_{min} + \delta$ with $\delta = 4.61$ (90% C.L.). We note that in both models very good fits to the data are found. This figure is typical in the sense that the model with a not too heavy charged Higgs and large f_B, prefers a positive value for ρ. We also see that η has approximately the same ranges as that in the standard model.

The measurements in (1) also constrain the charged Higgs parameter ξ. In Figure 2 we show the 90% C.L. upper limits for ξ in the $m_H - \xi$ plane for $m_t = 120$ and 200 GeV. We note that these limits are consistent with that derived by the unitarity considerations.[2]

Rare Kaon Decays

With the allowed range for the parameters in the two-Higgs-doublet model, we can make predictions for rare kaon decays. We will concentrate on the short distance contributions to the rare decays: $K_L \to \mu\bar{\mu}$, $K^+ \to \pi^+\nu\bar{\nu}$, $K_L \to \pi^0 e^+ e^-$, and $K_L \to \pi^0 \nu\bar{\nu}$ where those to the first and last two modes are CP conserving and violating, respectively. The branching ratios are given by

$$Br(K_L \to \mu\bar{\mu})_{SD} = 4.06 \times 10^{-10} A^4 (1-\rho)^2$$
$$[C_\mu(x_t) + \xi^2 C^{CH}_\mu(x_t, y_t)]^2$$
$$Br(K^+ \to \pi^+ \nu\bar{\nu}) = 10^{-6} \{C_\nu(x_c) + 3.3 \times 10^{-3}$$
$$A^2(1-\rho)[C_\nu(x_t) + \xi^2 C^{CH}_\nu(x_t, y_t)]\}^2$$
$$+1.08^{-11} A^4 \eta^2 \{C_\nu(x_t) + \xi^2 C^{CH}_\nu(x_t, y_t)\}^2$$
$$Br(K_L \to \pi^0 e^+ e^-)_{dir} = 2.6 \times 10^{-14} A^4 \eta^2$$
$$\{[C_V(x_t) + \xi^2 C^{CH}_V(x_t, y_t)]^2$$
$$+[C_A(x_t) + \xi^2 C^{CH}_A(x_t, y_t)]^2\}$$
$$Br(K_L \to \pi^0 \nu\bar{\nu})_{dir} = 4.61 \times 10^{-11} A^4 \eta^2$$
$$[C_\nu(x_t) + \xi^2 C^{CH}_\nu(x_t, y_t)]^2 \quad (2)$$

where $C_\mu(x)$, $C_\nu(x)$, $C_V(x)$, and $C_A(x)$ are defined in Ref. 3 and

$$C^{CH}(x,y) \equiv C^{CH}_\mu(x,y) = C^{CH}_\nu(x,y)$$
$$= \frac{xy(1-y+\ln y)}{4(y-1)}$$
$$C^{CH}_V(x,y) = \frac{2y}{27(y-1)^3}[47y^2 - 79y + 38 +$$
$$\frac{-18y^3 + 36y - 24}{y-1}\ln y]$$
$$+\frac{4(-1+4\sin^2\theta_W)}{\sin^2\theta_W}C^{CH}(x,y)$$

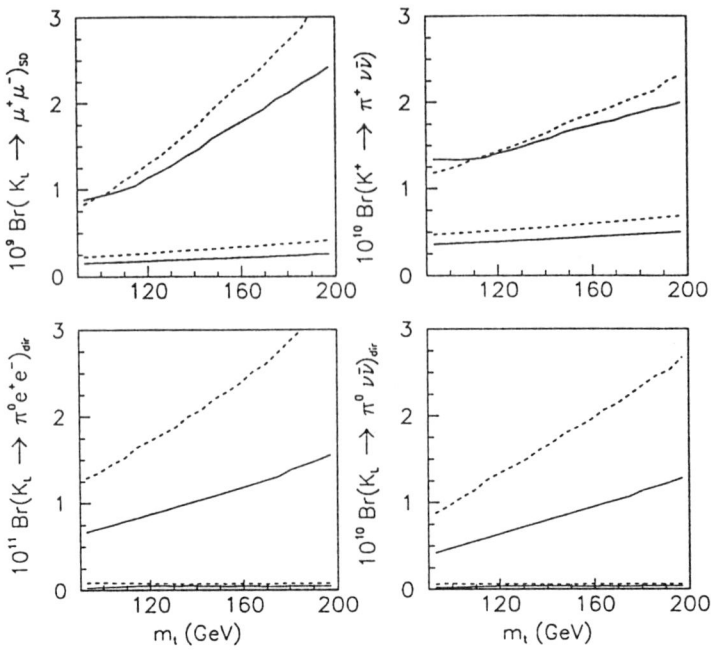

Figure 3. Branching ratios in the standard (solid) and Higgs (dashed) models; $\xi = 1$ and $m_H = 150\ GeV$.

$$C_A^{CH}(x,y) = \frac{4}{\sin^2 \theta_W} C^{CH}(x,y) \quad (3)$$

The calculation of the allowed range for the branching ratios in (2) is performed with the use of a χ^2 fit to the four measurements in (1). Our strategy is to fix m_t (m_t, m_H and ξ) to different values and let the remaining variables A, ρ, η, m_H and ξ (A, ρ and η) vary in the whole range corresponding to a 90% C.L., i.e. $\chi^2 = \chi^2_{min} + 9.24$ (6.25). In this allowed parameter space we look for the upper and lower bounds for each branching ratio. In Figure 3 we plot the branching ratios of the four rare kaon decays in the standard (solid) and two-Higgs-doublet (dashed) models with $\xi = 1$ and $m_H = 150$ GeV as a function of the top quark mass. We see that both upper and lower bounds for the CP conserving modes in the two-Higgs-doublet models are close to the standard ones. For the CP violating decays, the lower bounds are basically the same as the standard ones but the upper bounds are significantly enhanced due to the charged Higgs. For certain values of ξ and m_H the enhancement can be even larger. To see this, in Figure 4, we show the contours corresponding to the upper bounds of $Br(K_L \to \pi^0 e^+ e^-)_{dir} = 1.2$, 1.6 and 2.2×10^{-11} with $m_t = 120$ GeV in the m_H–ξ plane at 90% C.L. The contours for the decay $K_L \to \pi^0 \nu \bar{\nu}$ are similar to the charged lepton mode, but the branching ratios are about a factor of 8 larger.

$b \to s\gamma$

The standard model prediction as well as the impact of QCD correction on the decay have been presented by Ali and Greub.[4] Here we study this decay in the two-Higgs-doublet model. The decay branching ratio is given

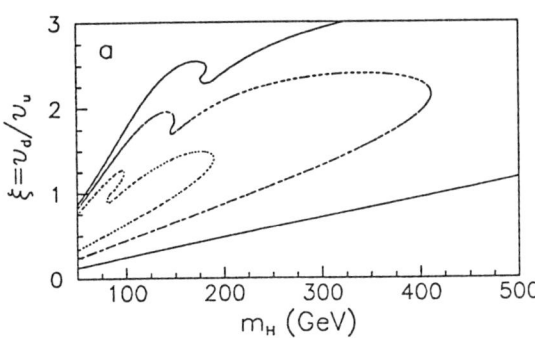

Figure 4. Contours for the upper bounds of Br $(K_L \to \pi^0 e^+e^-)_{dir}$ =1.2 (solid), 1.6 (dashed) and 2.2×10^{-11} (dotted) with $m_t = 120\ GeV$ in the m_H–ξ plane.

by[5]
$$Br(b \to s\gamma) = \frac{\Gamma(b \to s\gamma)}{\Gamma(b \to ce\bar\nu)} Br(b \to ce\bar\nu)$$
$$= \frac{2\alpha}{3\pi} \frac{|C_7(m_b)|^2}{f(m_c/m_b)\eta_0} Br(b \to ce\bar\nu) \quad (4)$$

with $C_7(m_b) =$
$$P_0 + P_{D'}[D'(x_t) + D'_{CH}(y_t) + \frac{1}{3}\xi^2 D'(y_t)]$$
$$+ P_{E'}[E'(x_t) + E'_{CH}(y_t) + \frac{1}{3}\xi^2 E'(y_t)] \quad (5)$$

where P_0, $P_{D'}$, $P_{E'}$, $D'(x)$ and $E'(x)$ are defined in Ref. 6 and
$$D'_{CH}(y) = y\left[\frac{\frac{5}{6}y - \frac{1}{2}}{(1-y)^2} + \frac{y - \frac{2}{3}}{(1-y)^3}\ln y\right]$$
$$E'_{CH}(y) = y\left[\frac{\frac{1}{2}y - \frac{3}{2}}{(1-y)^2} - \frac{1}{(1-y)^3}\ln y\right] \quad (6)$$

We note that the charged Higgs contributions in (5) have the same sign as the standard ones and thus the branching ratio of $b \to s\gamma$ is expected to be always enhanced in the two-Higgs-doublet model.

A limit on the decay can be obtained from the limit $Br(B \to K^*\gamma) < 0.92 \times 10^{-4}$ due to CLEO[7] and theoretical calculations

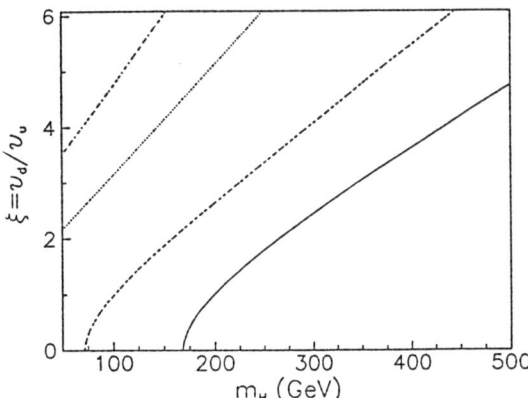

Figure 5. Contours of $Br(b \to s\gamma)$ =0.7 (solid), 1.0 (dashed), 2.3 (dotted), and 5.0×10^{-3} (dash-dotted) with $m_t = 120\ GeV$ in the m_H–ξ plane.

on the ratio of exclusive to inclusive rates $R_{K^*} = \frac{\Gamma(B \to K^*\gamma)}{\Gamma(b \to s\gamma)}$, which range from 4% to 40% in the literature. We thus have the most conservative bound

$$Br(b \to s\gamma) < 2.3 \times 10^{-3}. \quad (7)$$

In Figure 5, we plot the contours of $b \to s\gamma =$ 0.7, 1.0, 2.3 and 5.0×10^{-3} in the $\xi - M_H$ plane for $m_t = 120\ GeV$. We note that the standard model prediction on the decay branching ratio for $m_t = 120\ GeV$ is about 4×10^{-4}. We see that most parts of the allowed region (II) in Figure 2 are ruled out by the bound in (7). Clearly, further measurements on $B \to K^*\gamma$ and better theoretical estimations on R_{K^*} will severely constrain the two-Higgs-doublet model. This has strong impacts on the rare kaon and other rare B decays.[5]

In summary, we have shown that CP violating signals in rare kaon decays $K_L \to \pi^0 e^+e^-$ and $K_L \to \pi^0 \nu\bar\nu$ could be enhanced by the charged Higgs. For a large range of values of ξ and m_H, the two processes could be accessible to future experiments. The CP conserving rare kaon decays on the others hand are insensitive to the presence of the charged Higgs. For $b \to s\gamma$, the branching ratio is significantly enhanced due to the charged Higgs. Experimental limit on the decay could provide the strongest constraint on the two-Higgs-doublet model.

REFERENCES

1. G. Bélanger, C. Q. Geng and P. Turcotte, *Phys. Rev.* **D46** (1992) 2950.

2. C. Q. Geng and J. N. Ng, *Phys. Rev.* **D38** (1988) 2858; A. J. Buras, et al., *Nucl. Phys.* **B337** (1990) 284; V. Barger, J. L. Hewett and R. J. N. Phillips, *Phys. Rev.* **D41** (1990) 3421.

3. G. Bélanger and C. Q. Geng, *Phys. Rev.* **D43** (1991) 140; C. Q. Geng and P. Turcotte, *Phys. Lett.* **278B** (1992) 330.

4. A. Ali and C. Greub, *these Proceedings*.

5. G. Bélanger, C. Q. Geng and P. Turcotte, in preparation.

6. A. J. Buras and M. K. Harlander, *MPI-PAE/Pth 1/92* and references therein.

7. E. H. Thorndike (CLEO Collaboration), Contributed paper # 531, *LP-HEP91* Conference, Geneva, Switzerland (1991).

THEORY OF CP–VIOLATION IN $K \to 3\pi$ DECAYS

A.A.Bel'kov[1], G.Bohm[2], D.Ebert[3], A.V.Lanyov[1], A.Schaale[2]

[1] Particle Physics Laboratory, Joint Institute for Nuclear Research
Head Post Office, P.O.Box 79, 101000 Moscow, Russia

[2] DESY-Zeuthen, Platanenallee 6, O-1615 Zeuthen, Germany

[3] Institut für Elementarteilchenphysik, Humboldt-Universität
Invalidenstraße 110, O-1040 Berlin, Germany

Abstract

We present an analysis of the enhancement of CP-violating charge asymmetries in $K^\pm \to 3\pi$ decays. Calculations of decay amplitudes are performed on the basis of bosonized strong and weak Lagrangians derived from QCD-motivated quark Lagrangians.

Recently, much interest is devoted to the question of a possible enhancement of direct CP-violation effects in $K^\pm \to 3\pi$ decays first proposed in [1]. In this paper from chiral Lagrangians with fourth-order derivative terms, including meson loop rescattering effects, the CP-violating charge asymmetry of the Dalitz plot slope parameter, $\Delta g(K^\pm \to 3\pi)$, was estimated to be of the order of 10^{-3}. On the other hand, in several recent papers [2, 3, 4] estimates have been given which are 1-2 orders of magnitude smaller. By this reason and due to the fact that this problem is surely of great importance for the choice of the future experimental program at ϕ- and K-factories we have reanalyzed this question within our approach taking into account additional effects like fourth-order s-quark mass terms, (π^0, η, η')-mixing, and electromagnetic penguins. For simplicity, we will consider only the above mentioned charge asymmetry of the slope parameter

$$\Delta g(K^\pm \to 3\pi) = \frac{g(K^+) - g(K^-)}{g(K^+) + g(K^-)}.$$

The enhancement effects for other asymmetries, e.g. of branching ratios, are of the same origin.

The starting point of our analysis is the effective Lagrangian describing nonleptonic weak interactions with strangeness change $|\Delta S| = 1$ which is given on the quark level by:

$$\mathcal{L}_w^{nl} = \tilde{G} \sum_{i=1}^{8} c_i \mathcal{O}_i \,.$$

Here $\tilde{G} = \sqrt{2}\, G_F \sin\theta_C \cos\theta_C$ is the weak coupling constant; c_i are Wilson coefficient functions and $\mathcal{O}_i$ are four-quark operators consisting of products of left- and/or right-handed quark currents. The bosonized version of the effective weak Lagrangian can be expressed as a sum of products of $(V \mp A)$ and $(S \mp P)$ meson currents $J^a_{L/R\mu}, J^a_{L/R}$ with slightly modified coefficients ξ_i,

$$\xi_1 = c_1\left(1 - \frac{1}{N_c}\right),\quad \xi_{2,3,4} = c_{2,3,4}\left(1 + \frac{1}{N_c}\right),$$

$$\xi_{5,8} = c_{5,8} + \frac{1}{2N_c} c_{6,7},\quad \xi_{6,7} = c_{6,7} - \frac{2}{N_c} c_{5,8},$$

where the additional terms with color factors $1/N_c$ originate from the Fierz-transformed contribution to the nonleptonic weak effective chiral Lagrangian [5]. The meson currents get contributions from both the p^2- and p^4-terms of the strong interaction Lagrangian.

The corresponding contributions to the p^2- and p^4-parts of the effective Lagrangian of strong interaction are

$$\mathcal{L}^{(p^2)} = \frac{F_0^2}{4} \operatorname{tr}\left(\partial_\mu U \partial^\mu U^+\right)$$
$$+ \frac{F_0^2}{4} \operatorname{tr}\left[M\left(1 + \frac{F_0^2}{6m\langle\bar{q}q\rangle}\partial^2\right)U + \text{h.c.}\right]$$

$$\mathcal{L}^{(p^4)} = \frac{N_c}{32\pi^2}\operatorname{tr}\Big[\tfrac{1}{6}\left(\partial_\mu U \partial_\nu U^+\right)^2$$
$$- \frac{mF_0^2}{\langle\bar{q}q\rangle}\left(1 - \frac{4\pi^2 F_0^2}{m^2 N_c}\right)\partial_\mu U \partial^\mu U^+$$
$$\times \left(MU + U^+ M\right)\Big]$$

where $M = \operatorname{diag}(\chi_u^2, \chi_d^2, \chi_s^2)$ and $\chi_i^2 = -2m_i^0 F_0^{-2}\langle\bar{q}q\rangle$. Here F_0, m_i^0 and m are the (bare) meson decay constant and current and constituent quark masses, respectively.

The $K \to 3\pi$ decay amplitudes are parametrized as follows

$$T_{K^+\to\pi^+\pi^+\pi^-} = 2(\mathcal{A}_{11} + \mathcal{A}_{13})$$
$$- Y(\mathcal{B}_{11} + \mathcal{B}_{13} - \mathcal{B}_{23})$$
$$+ O(X^2, Y^2)$$

with an analogous expression for $T_{K^+\to\pi^0\pi^0\pi^+}$. Here $X = (s_2 - s_1)/m_\pi^2$, $Y = (s_3 - s_0)/m_\pi^2$ are the Dalitz variables and $s_i = (k - p_i)^2$, $s_0 = m_K^2/3 + m_\pi^2$; k, p_i are four-momenta of the kaon and ith pion ($i = 3$ belongs to the odd pion). The Dalitz-plot distribution can be written as a power series expansion of the amplitude squared, $|T|^2$, in terms of the kinematical variables Y and X

$$|T|^2 \propto 1 + gY + hY^2 + kX^2 + \dots$$

where g, h and k are the slope parameters. The isotopic amplitudes $\mathcal{A}_{IJ}$, $\mathcal{B}_{IJ}$ of $K \to 3\pi$ decays have two indices: I, the isospin of the final state, and J, the doubled value of isospin change between the initial and final states. The exact calculations of $\Delta g(K^\pm \to 3\pi)$ have been done using the complex quantities $\mathcal{A}_{IJ}$, $\mathcal{B}_{IJ}$ directly, without introducing the strong phases of final state interactions explicitly.

Let us consider the contributions of the four-quark operators $\mathcal{O}_i$ to the isotopic amplitudes A_2, A_0 and $\mathcal{A}_{IJ}$, $\mathcal{B}_{IJ}$ of $K \to 2\pi$ and $K \to 3\pi$ decays

$$A_2 = -i\sum_{i=1}^{8}\xi_i \frac{\sqrt{3}}{2}\tilde{G}F_0(m_K^2 - m_\pi^2)A_2^{(i)},$$

$$A_0 = -i\sum_{i=1}^{8}\xi_i \sqrt{\frac{3}{8}}\tilde{G}F_0(m_K^2 - m_\pi^2)A_0^{(i)};$$

$$\mathcal{A}_{IJ} = -\sum_{i=1}^{8}\xi_i \left(\tilde{G}\frac{m_K^2 - m_\pi^2}{12}\right)A_{IJ}^{(i)},$$

$$\mathcal{B}_{IJ} = -\sum_{i=1}^{8}\xi_i \left(\tilde{G}\frac{m_\pi^2}{4}\right)B_{IJ}^{(i)}.$$

Using only the currents $J_{L\mu}^{a\,(p^2)}$, $J_L^{a\,(p^2)}$ and the p^2-order part of the strong Lagrangian $\mathcal{L}^{(p^2)}$ it is possible to reproduce the following relations between $K \to 2\pi$, $K \to 3\pi$ isotopic amplitudes in the spirit of the "soft pion" limit:

$$A_{11}^{(i)} = B_{11}^{(i)} = A_0^{(i)},$$
$$A_{13}^{(i)} = A_2^{(i)} \quad (i = 1, 2, \dots, 6),$$

where for nonvanishing amplitudes we have

$$A_0^{(1)} = -A_0^{(2,3)} = -1 = -A_2^{(4)},$$
$$A_0^{(5)} = 4R; \quad R = \frac{\langle\bar{q}q\rangle}{mF_0^2}$$

It is well known that the fourth-order terms of the chiral Lagrangian $\mathcal{L}^{(p^4)}$ and their contributions to the currents $J_{L\mu}^{a\,(p^4)}$ and $J_L^{a\,(p^4)}$ lead to a modification of soft-pion relations for isotopic $K \to 2\pi$ and $K \to 3\pi$ amplitudes.

In particular, there are additional contributions due to of $J_L^{a\,(p^4)}$ to $\mathcal{A}_{11}^{(i)}$ and $\mathcal{B}_{11}^{(i)}$ ($i = 1, 2, 3, 5$):

$$\Delta A_{11}^{(1)} = -\Delta A_{11}^{(2,3)} = -\frac{m_K^2 - 3m_\pi^2}{12\,F_0^2\pi^2},$$

$$\Delta A_{11}^{(5)} = -4R\frac{m_K^2 - 3m_\pi^2}{12\,F_0^2\pi^2};$$

$$\Delta\mathcal{B}_{11}^{(1)} = -\Delta\mathcal{B}_{11}^{(2,3)} = \frac{m_K^2 + 3m_\pi^2}{12\,F_0^2\pi^2},$$
$$\Delta\mathcal{B}_{11}^{(5)} = 4R\frac{m_K^2 + 3m_\pi^2}{12\,F_0^2\pi^2}.$$

One can see that the contribution of the Lagrangian $\mathcal{L}^{(p^2)}$ and the corresponding currents $J_{L\mu}^{a\,(p^2)}$ and $J_L^{a\,(p^2)}$ to the amplitudes with $|\Delta I| = 1/2$ is proportional to $(-\xi_1 + \xi_2 + \xi_3 + 4R\xi_5)$. At the same time the contribution associated to the p^4 corrections is proportional to $(-\xi_1 + \xi_2 + \xi_3 - 4R\xi_5)$. So, ξ_5 cannot be absorbed by a redefinition of the parameters ξ_i. Due to this reason it is possible to separate penguin and nonpenguin contributions in a data fit for $K \to 2\pi$ and $K \to 3\pi$ decays. This different behaviour between penguin and nonpenguin contributions, arising on the fourth-order level, leads to a nonzero value of the interference term of two $\Delta I = 1/2$ amplitudes which becomes the main source of the enhancement of the CP-violating charge asymmetry of the slope parameters $\Delta g(K^\pm \to 3\pi)$ discussed in this talk.

Besides p^4-interactions, also (π^0, η, η')-mixing and one-loop corrections corresponding to meson rescattering modify the soft-pion relations for the isotopic $K \to 2\pi$ and $K \to 3\pi$ amplitudes. At the low-energy scale $\mu = 4\pi F_0 \approx 1$ GeV, the Wilson coefficients $c_i(\mu)$ get probably corrections $O(1/N_c, \mu)$ which cannot be calculated exactly until now. Therefore, in our approach the coefficients c_i, resp. ξ_i, have been treated as phenomenological parameters determined by experiment from the simultaneous analysis of $K \to 2\pi$, $K \to 3\pi$ decays[5].

If we still neglect the contribution of electromagnetic penguin operators, the imaginary part of the coefficient c_5, responsible for the direct CP-violation, can be fixed by the value of $|\varepsilon'/\varepsilon|$, which is considered as experimental input from $K^0 \to 2\pi$ decays, as:

$$|\operatorname{Im} c_5|^{exp} = 0.053^{+0.015}_{-0.011} |\varepsilon'/\varepsilon|.$$

This leads to the following estimates for the charge asymmetries of the slope parameters

$$|\Delta g(K^\pm \to \pi^\pm\pi^\pm\pi^\mp)| = 0.23^{+0.05}_{-0.08} |\varepsilon'/\varepsilon|$$

$$|\Delta g(K^\pm \to \pi^0\pi^0\pi^\pm)| = 0.19^{+0.03}_{-0.08} |\varepsilon'/\varepsilon|.$$

The difference of the first of these values with respect to the one given earlier [1] (by less than a factor 2) represents the model-dependence of this approach. The second value, $\Delta g(K^\pm \to \pi^0\pi^0\pi^\pm)$, is further diminished by the effect of (π^0, η, η')-mixing, formerly not taken into account in this channel. For a final judgement on the predicted Δg-values, one should, however, also take into account the effect of electromagnetic penguins.

Finally, we tried to estimate the effect of electroweak penguin operators $\mathcal{O}_{7,8}$ on the charge asymmetry $\Delta g(K^\pm \to 3\pi)$. As a result we found that taking into account the electroweak penguin does not only not suppress the effect of direct CP-violation in $K \to 3\pi$ decays, but may lead to an additional enhancement in comparison with that seen in $K \to 2\pi$ decays. Of cource, this makes the experimental investigation of $K \to 3\pi$ decays with high statistics even more interesting.

REFERENCES

1. A.A.Bel'kov, G.Bohm, D.Ebert and A.V.Lanyov, Phys. Lett. B232 (1989) 118.

2. H.-Y.Cheng, Phys. Rev. D44 (1991) 919.

3. G.D'Ambrosio, G.Isidori and N.Paver, Phys. Lett. B273 (1991) 497.

4. G.Isidori, L.Maiani and A.Pugliese, Preprint INFN n.848, Roma 1991.

5. A.A.Bel'kov, G.Bohm, D.Ebert, A.V.Lanyov, A.Schaale, DESY-Prepr. 92-106, July 1992.

STATUS OF RARE K DECAYS

Jack L. Ritchie
Department of Physics
University of Texas at Austin
Austin, Texas 78712

Abstract

Recent experiments have achieved unprecedented sensitivites for rare decay modes of the K mesons. Results from experiments at BNL and KEK are described. These experiments search for lepton flavor violation in the decays $K_L^\circ \to \mu e$ and $K^+ \to \pi^+ \mu e$ and highly suppressed Standard Model decays, such as $K_L^\circ \to \mu^+ \mu^-$ and $K^+ \to \pi^+ \nu \bar{\nu}$. New results on rare π° decays obtained in some of these experiments are also presented.

INTRODUCTION

My charge from the conveners was to report on "all of rare K decays excluding FNAL and CERN." The FNAL and CERN experiments were covered by Wah and Iconomidou-Fayard, respectively, in the same session. What is left is the very active program of experiments at BNL and KEK. At BNL, four experiments have produced results; three of these groups are working on major upgrades for the future. One group at KEK has recently completed an experiment and another group is setting up a new experiment. Given the severe time (and space) constraint, it is not possible to cover these experiments in much depth and a rather restricted selection of topics for emphasis is necessary. I regret having to gloss over much important work. I will address three general topics: (1) lepton flavor violation, (2) highly suppressed Standard Model decays, and (3) new results on rare π° decays obtained as a biproduct of these experiments. CP violation and in particular the mode $K_L^\circ \to \pi^\circ e^+ e^-$ will not be discussed; however, I note for completeness that the lowest limit on this mode as of the time of this conference is 5.5×10^{-9} from BNL E845.[1] (All limits in this paper are quoted for the 90% confidence level.)

LEPTON FLAVOR VIOLATION

The conservation of separate lepton number (e.g., muon number) is not understood at a fundamental level, in contrast for example to charge conservation, which is understood as a consequence of the gauge invariance of the electromagnetic interaction. Indeed, most of the widely discussed Standard Model extensions (e.g., extended technicolor, SUSY, right-left symmetry, multiple Higgs particles, etc.) provide mechanisms by which lepton flavor is not conserved and processes such as $K_L^\circ \to \mu e$ and $K^+ \to \pi^+ \mu e$ can occur. Searches for these and other lepton flavor violating decays provide a means of probing for such mechanisms at very high virtual mass scales. For example, for the $K_L^o \to \mu e$ decay, a branching ratio sensitivity of 10^{-11} probes a scale up to about 100 TeV.

The lowest currently published limit on $K_L^o \to \mu e$ is 9.4×10^{-10} from the KEK E137 experiment.[2] A lower sensitivity has been achieved in BNL E791. That experiment is shown in Figure 1. The experiment ran at the BNL AGS, where up to 5×10^{12} protons at 24 GeV were targeted on a one interaction length copper target, followed by sweeping magnets and collimators, to form a neutral beam. The detector was a two arm spectrometer, with two consecutive dipole magnets to assist with the rejection of backgrounds associated with pion decays in the spectrometer. Key features of the detector are precise tracking and kinematic reconstruction ($\sigma_{mass} \simeq 1.4$ MeV), redundant particle identification for both electrons and muons, and a high rate data acquisition system.

Principal backgrounds are due to $K_L^o \to \pi e \nu$, where either the pion decays or both the pion and electron are wrongly identified. The E791 experiment has demonstrated background rejection down to at least the 10^{-11} level. Figure 2 shows a scatter plot of reconstructed mass for candidate μe pairs versus the square of the reconstructed transverse momentum p_T^2 (p_T is defined as the component of the vector sum of muon and electron momenta perpendicular to the direction of the parent K_L^o). E791 ran at the AGS during 1988, 1989, and 1990. Figure 2 is the result of a combined analysis of 1989 and 1990 data sets, from which the preliminary limit on $K_L^o \to \mu e$ is 3.9×10^{-11}. When combined with the result[2] from running during 1988, the final E791 limit is 3.3×10^{-11}.

The KEK E137 experiment differed in details, but was very similar in philosophy. Neither experiment was background limited.

The $K_L^o \to \mu e$ decay would be sensitive to a new axial-vector or pseudoscalar inter-

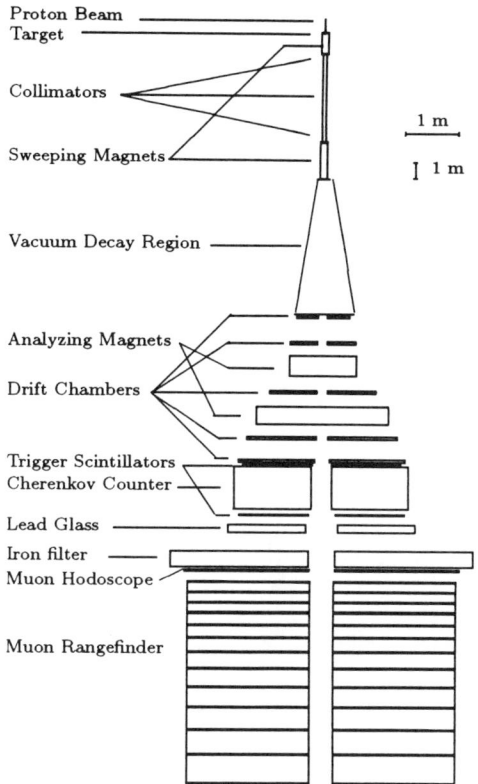

Figure 1. The BNL E791 neutral beam and detector.

action. The $K^+ \to \pi^+ \mu e$ decay is complimentary, since it would be sensitive to vector or scalar interactions. The recently completed BNL E777 has set an upper limit[3] of 2.1×10^{-10} on the mode $K^+ \to \pi^+ \mu^+ e^-$. The best limit on $K^+ \to \pi^+ \mu^- e^+$ is 6.9×10^{-9} from an older experiment.[4]

HIGHLY SUPPRESSED DECAYS

Some highly suppressed Standard Model K decays provide a means of probing short-distance physics and are sensitive to Standard Model parameters such as the top quark mass and the CKM angles. Others are dominated by long-distance physics. Recent work toward understanding the latter cagegory depends on chiral perturbation theory. Quite

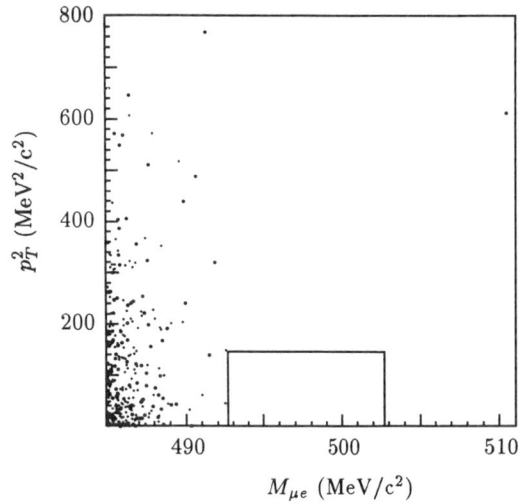

Figure 2. $K_L^o \to \mu e$ candidates from BNL E791. No events appear in the signal region.

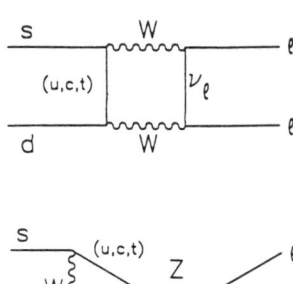

Figure 3. Diagrams for short-distance contributions to $K^+ \to \pi^+ \nu \overline{\nu}$ and $K_L^o \to \mu^+ \mu^-$.

often, it is necessary to understand the long-distance processes in order to isolate the short-distance effects of interest. Here, emphasis will be given to the decays $K^+ \to \pi^+ \nu \overline{\nu}$ and $K_L^o \to \mu^+ \mu^-$, both of which are sensitive to the top quark mass and the V_{td} element of the CKM matrix through the diagrams shown in Figure 3. Regrettably, space does not permit discussion of decays in the second category, such as $K^+ \to \pi^+ e^+ e^-$, $K_L^o \to e^+ e^- \gamma$, and $K_L^o \to e^+ e^- e^+ e^-$, for which recent results exist.

$K^+ \to \pi^+ \nu \overline{\nu}$

This process is dominated by the short-distance graphs of Figure 3. Long-distance effects should not be important, so a measurement of the branching ratio will provide a direct constraint on m_t and V_{td}. The branching ratio is expected to be close to 10^{-10} and so far the process has not been observed. If the decay is observed significantly above the Standard Model level, it would be an indication of new physics.

BNL E787 focuses on this mode. The experiment is performed in a "stopping" K^+ beam, so that the K^+'s decay at rest. The detector, shown in Figure 4, consists of an active scintillating fiber target, where the K^+'s are stopped. The target is surrounded by a cylindrical drift chamber, which in turn is surrounded by a plastic scintillator range stack (where pions stop and subsequently decay). Finally, a lead-scintillator photon veto system surrounds the range stack. A field of 1 Tesla is applied by a solenoidal magnet. The detector is virtually hermetic, with 4π coverage for vetoing photons.

The major backgrounds for $K^+ \to \pi^+ \nu \overline{\nu}$ are the familiar decays $K^+ \to \pi^+ \pi^o$ and $K^+ \to \mu^+ \nu$. Both of these decays create a positively charged particle with a specific momentum, while the $K^+ \to \pi^+ \nu \overline{\nu}$ creates a π^+ with a momentum distributed from 0 to 227 MeV, so that some rejection comes from simply avoiding the $K_{\pi 2}$ and $K_{\mu 2}$ peaks. Additional rejection of the $K^+ \to \pi^+ \pi^o$ comes from vetoing the photons from the π^o, which is done so efficiently that only about 2 in 10^6 π^o's are missed. Additional rejection against $K^+ \to \mu^+ \nu$ is obtained by observing the full π^+ decay chain ($\pi \to \mu \to e$). This is accomplished by digitizing the photomultiplier signals from the range stack with 500 MHz 8-bit

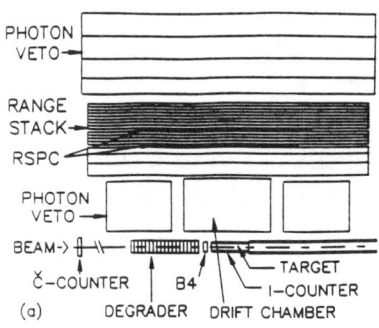

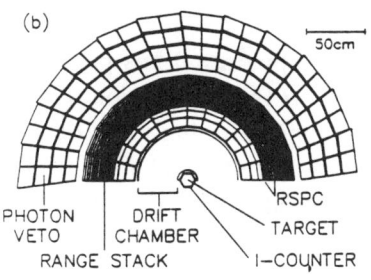

Figure 4. Upper half of the BNL E787 detector, showing (a) side and (b) end views.

transcient digitizers over a period of 10 μsec for each event.

The current upper limit from E787 on $K^+ \to \pi^+ \nu \bar{\nu}$ is 5×10^{-9}. Data on tape should allow this to be improved by roughly a factor of two. This limit is from the region between the $K_{\pi 2}$ and $K_{\mu 2}$ peaks. The group has also demonstrated a capability to adequately reject backgrounds below the $K_{\pi 2}$. From a small sample of engineering data, they have established an independent limit from that region of 2×10^{-8}. In future running the sensitivity of the two regions should be comparable. After the decay is observed, measurements in both regions will make it possible to constrain the form of the decay matrix element.

$K_L^0 \to \mu^+ \mu^-$

This decay is dominated by the long-distance process $K_L^0 \to \gamma\gamma \to \mu^+ \mu^-$, which accounts for the so-called unitarity bound.[6] Using the measured $K_L^0 \to \gamma\gamma$ rate, the two-photon contribution to the branching ratio of $K_L^0 \to \mu^+ \mu^-$ is $(6.8 \pm 0.3) \times 10^{-9}$. Extracting short-distance physics depends on understanding the long-distance contributions, both the unitarity contribution and also contributions from virtual intermediate two-photon states. This subject has been the focus of recent theoretical work[7], but has not progressed to the point where measurements of $K_L^0 \to \mu^+ \mu^-$ can be interpreted with confidence. Nonetheless, this mode has the advantage of having been observed with good statistics.

The KEK E137 experiment observed 179 events and the BNL E791 experiment over 700. This is to be compared with a prior world's sample of 33 events from five experiments. The KEK E137 branching ratio[8] is $(7.9 \pm 0.7) \times 10^{-9}$. The result from the BNL experiment is $(7.0 \pm 0.5) \times 10^{-9}$. This is still preliminary, although only the error bar is likely to change. Figure 5 shows the $\mu\mu$ mass peak for both of these experiments.

$K_L^0 \to e^+ e^-$

The Standard Model contributions to this decay are helicity suppressed. The unitarity bound is 3×10^{-12}. Above this level, the mode is sensitive to a new pseudoscalar strangeness changing neutral current interaction. The lowest published[2] limit at present is 16×10^{-11} from the KEK E137 experiment. A new limit from the BNL E791 experiment, for the full three year data set, is 4.1×10^{-11}.

RARE π^0 DECAYS

The decay $K^+ \to \pi^+ \pi^0$ provides a means for rare K^+ decay experiments also to search for rare π^0 decays. Two new results are described below.

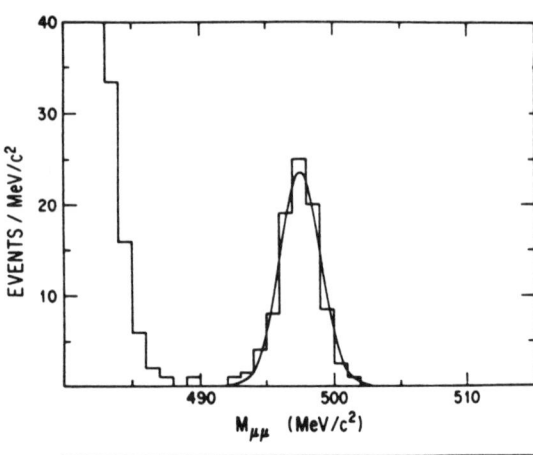

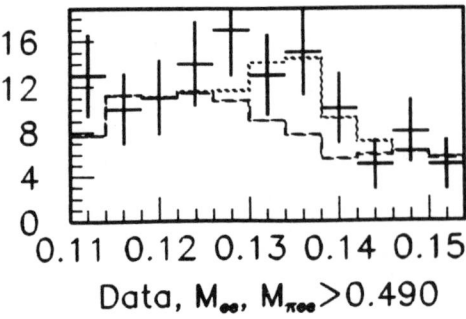

Figure 6. Reconstructed e^+e^- pair mass for events with $M_{\pi ee} > 490$ MeV from BNL E851.

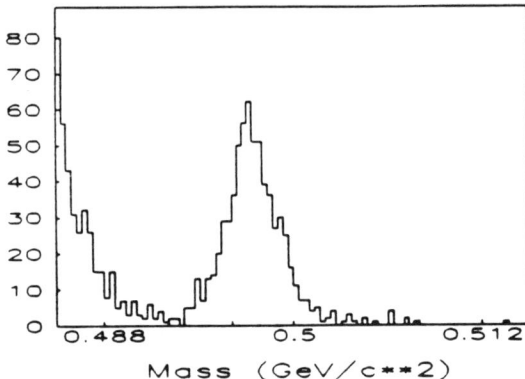

Figure 5. $K_L^0 \to \mu^+\mu^-$ mass peak from KEK E137 (upper) and BNL E791 (lower).

$\pi^0 \to e^+e^-$

The unitarity bound for this decay is 4.75×10^{-8}. A CERN PS experiment[9] reported $B(\pi^0 \to e^+e^-) = 22^{+24}_{-11} \times 10^{-8}$ from tagged π^0's in a K^+ experiment. A LAMPF experiment[10] reported $B(\pi^0 \to e^+e^-) = (17 \pm 6 \pm 3) \times 10^{-8}$, from π^0's produced via $\pi^- p \to n\pi^0$. Both experimental results were substantially above the unitarity prediction. Subsequent theoretical attempts[11] to account for this failed, leading to skepticism regarding these experiments. Also, the SINDRUM experiment[12] reported an upper limit of 13×10^{-8}.

BNL E851 (an extension of BNL E777, using the same apparatus) has produced a new result based on the observation of this decay. The backgrounds are single and double Dalitz decays ($\pi^0 \to e^+e^-\gamma$ and $\pi^0 \to e^+e^-e^+e^-$) were the photon or an e^+e^- pair is missed. These background sources are suppressed by requiring the reconstructed $\pi^+e^+e^-$ mass to be close to the K^+ mass. Figure 6 shows reconstructed e^+e^- mass for events with $\pi^+e^+e^-$ mass greater than 490 MeV. The data show a clear excess centered on the π^0 mass. The dashed curve in Figure 6 is Monte Carlo with no $\pi^0 \to e^+e^-$ contribution and the dotted curve is Monte Carlo with the $\pi^0 \to e^+e^-$ branching ratio at the value favored by the data. The BNL E851 result is $B(\pi^0 \to e^+e^-) = (6.0 \pm 1.8) \times 10^{-8}$. This value is consistent with the unitarity prediction. (See also the talk of Wah, these proceedings, for FNAL results on this mode.)

$\pi^0 \to \gamma X$

The BNL E787 experiment can also use tagged π^0's from K^+ decays to search for exotic decays. No prior limit exists for the decay $\pi^0 \to \gamma X$, where X is a long-lived neutral non-interacting vector particle. Such a particle can arise in Standard Model extensions with an additional $U(1)$ interaction, for example, or

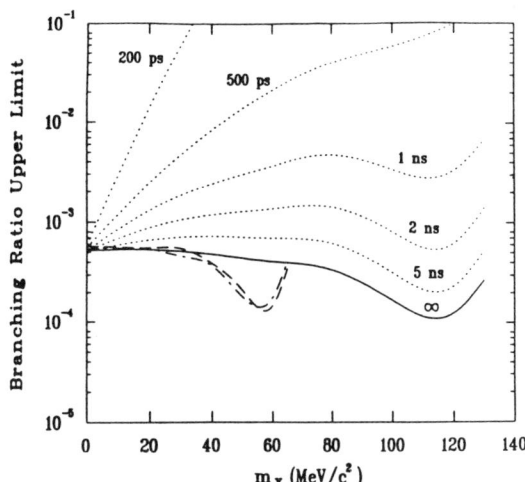

Figure 7. $B(\pi^0 \to \gamma X)$ limits versus X mass, for different X lifetimes, from BNL E787.

in other possible scenarios. The experiment is sensitive to an X with mass below the π^0 mass. The experiment would also be sensitive to three body decays $\pi^0 \to \gamma X X'$, where the X and X' could be neutrinos or supersymmetric particles. Figure 7 shows the experimental branching ratio limit as a function of the X mass. Different limits are indicated for different lifetime assumptions. The dot-dashed and dashed curves correspond to the three body scenario with a phase-space spectrum and a Dalitz-like spectrum, respectively.

FUTURE PROSPECTS

Work is now underway at both BNL and KEK to mount new rare K decay experiments. At BNL where three groups are still active, the thrust is to exploit the four-fold increase in available flux promised by the new Booster and to build on the experiences of the recently completed experiments. BNL E871 and E865 should reach the 10^{-12} level for the lepton flavor violating decays $K_L^0 \to \mu e$ and $K^+ \to \pi^+ \mu^+ e^-$, respectively, and achieve similar sensitivities to other accessible modes. BNL E787 expects to reach an ultimate sensitivity, through beam and detector upgrades, below 10^{-10} and therefore observe $K^+ \to \pi^+ \nu \bar{\nu}$. At KEK, preparations are underway for E162, which will search for $K_L^0 \to \pi^0 e^+ e^-$ close to the 10^{-10} level.

REFERENCES

1. K. E. Ohl et al., *Phys. Rev. Lett.* 64, pp. 2755–2758, (1990).

2. T. Akagi et al., *Phys. Rev. Lett.* 67, pp. 2614–2617, (1991).

3. C. Mathiazhagan et al., *Phys. Rev. Lett.* 63, pp. 2181–2184, (1989).

4. A. M. Lee et al., *Phys. Rev. Lett.* 64, pp. 165–168, (1990).

5. A. M. Diamant-Berger et al., *Phys. Lett.* 62B, pp. 285–289, (1976).

6. L. M. Sehgal, *Phys. Rev.* 183, pp. 1511–1513, (1969).

7. L. Bergström, E. Massó and P. Singer, *Phys. Lett.* 249B, pp. 141-144, (1990); P. Ko, *Phys. Rev.*, D45, 174–177, (1992).

8. T. Akagi et al., *Phys. Rev. Lett.* 67, pp. 2618–2621, (1991).

9. J. Fisher et al., *Phys. Lett.* 73B, pp. 364–368, (1978).

10. J. S. Frank et al., *Phys. Rev.* D28, pp. 423–435, (1983).

11. L. Bergström, *Z. Phys.* C14, pp. 129–134, (1982); L. Bergström, E. Massó, Ll. Ametller, and A. Bramon, *Phys. Lett.* 126B, pp. 117-121, (1983).

12. C. Niebuhr et al., *Phys. Rev.* D40, pp. 2796–2802, (1989).

RECENT THEORETICAL DEVELOPMENT ON DIRECT CP VIOLATION ε'/ε

Yue-Liang Wu
Institut für Physik
Unversität Mainz
Staudinger weg 7
6500 Mainz, Germany

Abstract

In this paper I review recent theoretical development on the direct CP violation parameter ε'/ε in the standard model and the accuracy of predictions for it.

Direct CP violation is measured by the ratio ε'/ε, which has been studied extensively. The motivation to study the direct CP violation is known to understand the fundamental question: is the CP violation observed in 1964[1] in K^0 decay due to a fifth force —the 'superweak' interaction of Wolfenstein model[2], or a complex phase in the quark mixing matrix of the Cabbibo–Kobayashi–Maskawa scheme[3] in the standard model? Since the theoretical predictions for ε'/ε in these two models are different: $\varepsilon'/\varepsilon = 0$ in the superweak model and $\varepsilon'/\varepsilon = O(10^{-3})$[4-6] in the CKM model.

The recent experimental situation is not yet conclusive for these two models, the NA31[7] reported

$$\frac{\varepsilon'}{\varepsilon} = (2.3 \pm 0.7) \times 10^{-3} \quad (1)$$

and the E731[8] reported the value

$$\frac{\varepsilon'}{\varepsilon} = (0.60 \pm 0.69) \times 10^{-3}. \quad (2)$$

As the calculation of ε'/ε in CKM model is not naive as in the superweak model, it has been studied more than ten years. I will briefly review in this talk the recent theoretical development for ε'/ε. In the standard notation

$$\frac{\varepsilon'}{\varepsilon} = 1.25 \times 10^{-3} (\frac{Im\xi_t}{10^{-4}})$$

$$(\frac{C_6(\mu) <Q_6>_0}{0.05})(\frac{1-\Omega}{0.07}). \quad (3)$$

Here (i) $Im\xi_t = |V_{cb}||V_{ub}|sin\delta$ is a parameter of CKM matrix. (ii) $C_6 <Q_6>_0$ arises from the strong penguin. and (iii) $\Omega = \Omega_{EWP} + \Omega_{\eta-\eta'} + \Omega'$, with $\Omega_{EWP} = (C_8 <Q_8>_2 + C_7 <Q_7>_2)/(\omega C_6 <Q_6>_0)$ being the relative contributions of electroweak penguins and $\Omega_{\eta-\eta'}$ the π^0-η-η' mixing. The Wilson coefficient functions C_i are obtained by integrating the renormalization group equations with both QCD and QED corrections present. It has been studied by many groups. The calculations[9-11] up to 1989 were only limited to the case when top quark mass m_t is much smaller than W-boson mass m_W ($m_t \ll m_W$). While the experimental data in 1989 indicated that the top quark mass may be heavy. In fact, in 1984 it was also pointed out by Chou et al.[12] that top quark mass might be larger than W-boson mass ($m_t > m_W$). It is therefore necessary to evaluate the coefficient functions C_i in the case that $m_t > m_W$, which were performed in 1989 by several groups[13-17]. The most interesting observations for $m_t > m_W$ are that the coefficient function C_6 has only a very weak dependence on top quark mass, the coefficient functions C_7 and C_8 have, however, a very strong dependence on top quark mass m_t and are increased very rapidly as top quark mass becomes larger than the W-boson mass, where the renormalization group effects play a crucial rule. Consequently when using the hadronic matrix elements computed with chiral Lagrangians[18] at the tree level, the contributions of electroweak penguin and strong penguin are comparable and important cancellations occur between these two terms when top quark mass is large. Because of this case that one year ago it was thought that for a heavy top quark the ratio ε'/ε could become very small[14-17] and thus imitate the superweak theory[2] for $m_t \sim 210$ GeV.

The estimates for the matrix elements of the operators Q_6, Q_7 and Q_8 are, however, incomplete because they do not reproduce the physical values for the isospin amplitudes A_0 and A_2. For this reason the authors[4-6] improved the estimates for the hadronic matrix elements by including the one-loop corrections in chiral perturbation theory. The numerical calculations found that[4-6] for $\mu = 1$ GeV and $\Lambda_4 = 300$MeV

$$<Q_6>_0 = 2.8 <Q_6>_0^{LO}$$
$$= 1.44(\frac{125 MeV}{m_s})^2 GeV^3 \quad (4)$$
$$<Q_8>_2 = 1.47 <Q_8>_2^{LO}$$
$$= 0.94(\frac{125 MeV}{m_s})^2 GeV^3 \quad (5)$$
$$\Omega_{EWP} = 0.52 \Omega_{EWP}^{LO} \quad (6)$$

which revealed that the importance of the electroweak penguin relative to the gluon penguin is diminished and reduced by a fact 1.9. Therefore the large cancellation between electroweak penguin and strong penguin does not occur any more even when the mass of the top quark is very large. In addition the next-to-leading order corrections to the Wilson coefficient functions were also found[19] to diminish the Ω_{EWP} by $(20 \sim 30)\%$.

The accuracy of the improved estimates was discussed in detail in the most recent paper[6]. One of the criteria is the scale μ − *dependence*. The quantities which enter physical processes are $C_i(\mu) < Q_i >_I$ and, in principle, should be independent of μ. For that a useful approach was developed by Bardeen, Buras and Gerard[20], who suggested to calculate the chiral loops with a finite cutoff M_{cut} and regard M_{cut} to be a physical quantity which is not removed by counter terms, but instead is identified with the renormalization scale μ of the short distance contribution $C_i(\mu)$. It was noted that for $\Lambda_4 = 200$MeV and when μ changes from 0.6 GeV to 1.0 GeV the variation for $C_6 <Q_6>$ is about 18%. These should be compared with a variation of 100% for $<Q_6>$ alone. The improvement for $<Q_8>_2$ is even more impressive, since the variation of $C_8(\mu) <Q_8>_2$ is 1% while $<Q_8>_2$ varies 65%.

Additional support for chiral calculations comes from the fact that they reproduce the transition amplitudes A_0 and A_2. Similar calculations were published earlier[20] and included the one-loop corrections for $<Q_1>_I$ and $<Q_2>_I$. When the new values for $<Q_6>$ were included, a better agreement for A_0 was found[5,6], see table 1.

Moreover, most of the models which favor to the $\Delta I = 1/2$ amplitude A_0 also obtained big values of $<Q_6>_0$. For example, $<Q_6>_0= 1.46(\frac{125 MeV}{m_s})^2 GeV^3$ in a linear σ chiral dynamic model[21] and $<Q_6>_0 = 1.40 GeV^3$ in the bag model[22], comparing these values with eq.(4), a remarkable agreement occur among these models.

The estimates of $Im \xi_t$ depend on elements of the CKM matrix and CP phase δ. The matrix elements $|V_{cb}|$ and $|V_{ub}|$ are determined directly from semileptonic B-meson decays[23,24]

$$|V_{cb}| = 0.041 \pm 0.003, \quad \frac{|V_{ub}|}{|V_{cb}|} = 0.12 \pm 0.03. \quad (7)$$

The estimate of the CP phase is quite standard from fitting the ϵ_k parameter for K^0 mesons and $B^0 - \bar{B}^0$ mixing. Choosing for the

Table 1. Predictions for the amplitude A_0 in unit of 10^{-7} GeV. In parentheses are the earlier values of BBG[20]

m_s(MeV)	100	125	150	175	200
A_0^{th}/A_0^{exp}	1.18	0.96	0.84	0.76	0.72
	(0.78)	(0.67)	(0.63)	(0.61)	(0.60)

additional parameters[25]

$$B_k = 0.75 \pm 0.20 \quad (8)$$

$$\sqrt{\eta_B^{QCD}}\sqrt{B_B}f_B = 120 \pm 18 MeV \quad (9)$$

where the stable values of f_B calculated from QCD are sensitive to the choice of the pole mass of the bottom quark[26]. With these input parameters the range of $Im\xi_t$ is given by

$$0.58 \times 10^{-4}\varepsilon_t \leq Im\xi_t \leq 2.85 \times 10^{-4}\chi_t$$
$$\varepsilon_t \simeq x_t^{1.42}(1 + 0.1 ln x_t) \quad (10)$$
$$\chi_t \simeq x_t^{0.8}\sqrt{1 + 0.754(1-x_t)^{1.6}}$$

with $x_t = 140 GeV/m_t$.

After a brief review for the theoretical development of the calculations for each factors involved in the computation of ε'/ε, it is now in the position to make the conclusions for the ratio ε'/ε. For $100 GeV \leq m_t \leq 250 GeV$

i) The superweak behavor of ε'/ε in the CKM scheme of the standard model is excluded definitely since

$$0.35 \leq 1 - \Omega \leq 0.70 \quad (11)$$

ii) ε'/ε is enhanced by a factor 3.5 due to next-to- leading order contributions

$$C_6 <Q_6>_0 = 3.5 C_6 <Q_6>_0 |_{LO} \quad (12)$$

iii) The big uncertainties still mainly come from the range of $Im\xi_t$.

$$0.24 \times 10^{-4} \leq Im\xi_t \leq 2.6 \times 10^{-4} \quad (13)$$

iv) The allowed values of ε'/ε

$$x_s^2 0.46 \times 10^{-3} \leq \frac{\varepsilon'}{\varepsilon} \leq 8.7 \times 10^{-3} x_s^2 \quad (14)$$

with $x_s = 150 MeV/m_s$

v) Simultaneously the $\Delta I = 1/2$ amplitudes are closer to the data and the dependence of the amplitudes on μ is also improved.

vi) Taking the plausible central values of input parameters, a consistent prediction comes to

$$m_t \sim 140 GeV, \quad \delta \sim 150^0 \quad (15)$$

and

$$\frac{\varepsilon'}{\varepsilon} = 2.6 \times 10^{-3}(\frac{150 MeV}{m_s})^2. \quad (16)$$

It is therefore optimistic to say that the measurement of the ratio ε'/ε is sufficient to distinguish the superweak model and CKM model. Let us wait for the improved new experiments NA48, KTeV, DAPHNE and CP LEAR.

ACKNOWLEDGMENT

I wish to thank E.A. Paschos for many enjoyable collaborations and useful suggestions. Many thanks go to K.C. Chou and T.D. Lee for their encouragement and W.A. Bardeen and J.M. Gerard for their useful discussions.

REFERENCES

1. J. Christenson, J. Cronin, V. Fitch and R. Turlay, Phys. Rev. Lett. 13 (1964) 138.

2. L. Wolfenstein, Phys. Rev. Lett. 13 (1964) 562.

3. N. Cabbibo, Phys. Rev. Lett. 10 (1963) 531; M. Kobayashi and T. Maskawa, Progr. Theor. Phys. 49 (1973) 652.

4. J. Fröhlich, J. Heinrich, E.A. Paschos and J.M. Schwartz, Dortmund preprint DO-TH 02/91 (1991).

5. Y.L. Wu, Intern. Journ. of Mod. Phys. A7 (1992) 2863.

6. J. Heinrich, E.A. Paschos, J.H. Schwartz and Y.L. Wu, Phys. Lett. B279 (1992) 140.

7. K. Kleinknecht, Comments Nucl. Phys. 20 (1992) 281; G.D. Barr, talk at Lepton-Photon Conf. Geneva, August 1991.

8. B. Winstein, talk at Lepton-Photon Conf. Geneva, August 1991.

9. A.I. Vainshtein, V.I. Zakharov and M.A. Shifman, JETP 45 (1977) 670.

10. F.J. Gilman and M.B. Wise, Phys. Rev. D20 (1979) 2392; D27 (1983) 1128; B. Gubrina and R.P. Peccei, Nucl. Phys. B163 (1980) 289.

11. J. Bijnens and M.B. Wise, Phys. Lett. B137 (1984) 245; M. Lusignoli, Nucl. Phys. B325 (1989) 33.

12. K.C. Chou, Y.L. Wu and Y.B. Xie, *Minimum top quark mass and the CP violation parameter ε'/ε*, AS-ITP-84-005 (1984), Beijing report; Chinese Phys. Lett. 1 (1984) 2; K.C. Chou, Proceedings of the Europhysics Topical Conf. on Flavor Mixing in Weak Interaction, held March 1984, at Erice Sicily Italy. Edited by L.-L. Chau.

13. E.A. Paschos, T. Schneider and Y.L. Wu, Nucl. Phys. B332 (1990) 285.

14. J.M. Flynn and L. Randall, Phys. Lett. B224 (1989) 221; Erratum B235 (1990) 412.

15. G. Buchalla, A.J. Buras and M.K. Harlander, Nucl. Phys. B337 (1990) 313.

16. E.A. Paschos, T. Schneider and Y.L. Wu, in: Proc. third Hellenic school on Elementary Particle Physics (Sept. 1989) (World Scientific, Singapore, 1990) pp. 46-73.

17. E.A. Paschos and Y.L. Wu, Mod. Phys. Lett. A6 (1991) 93; in: Proc. of the 25th Intern. Conf. on High Energy Physics, (August 1990) eds. K.K. Phua and Y. Yamaguchi.

18. J. Gasser and H. Leutwyler, Nucl. Phys. B250 (1985) 465,517; W.A. Bardeen, A.J. Buras and J.-M. Gerard, Nucl. Phys. B293 (1987) 787; R.S. Chivukula, J.M. Flynn and H. Georgi, Phys. Lett. B171 (1986) 453; A. Pich, B. Guberina and E. de Rafael, Nucl. Phys. B277 (1986) 197; H.Y. Cheng, Phys. Rev. D36 (1987) 2056.

19. A.J. Buras, M. Jamin, M.E. Lautenbacher and P.H. Weise, Nucl. Phys. B370 (1992) 69.

20. W.A. Bardeen, A.J. Buras and J.-M. Gerard, Nucl. Phys. B293 (1987) 787; Phys. Lett. B192 (1987) 138; Phys. Lett. B211 (1988) 343.

21. T. Morozumi, C.S. Lim and A.I. Sanda, Phys. Rev. Lett. 65 (1990) 404.

22. J.F. Donoghue et al., Phys. Rev. D21 (1980) 186; D23 (1981) 1213; F.J. Gilman and J.S. Hagelin, Phys. Lett. B133 (1983) 443.

23. K. Berkelman and S.L. Stone, Annu. Rev. Nucl. Part. Sci. 41 (1991) 1; J.G. Körner, K. Schilcher, M. Wirbel and Y.L. Wu, Z. Phys. C48 (1990) 663.

24. R. Drell, talk in this proceeding.

25. For a review see, for example, E.A. Paschos and U. Türke, Phys. Rep. 178 (1989) 147 and references therein.

26. For the most recent paper see, for example, Y.L. Wu, Mainz Report, MZ-TH/92-13, 1992, and references therein.

TEST OF CP VIOLATION USING K^0-$\overline{K^0}$ INTERFEROMETRY

The CPLEAR Collaboration

R. Adler[2], T. Alhalel[11], A. Angelopoulos[1], A. Apostolakis[1], E. Aslanides[4,11], G. Backenstoss[2],
C.P. Bee[4,9], J. Bennet[9], V. Bertin[11], J.K. Bienlein[17,a], P. Bloch[4], Ch. Bula[13], P. Carlson[15],
J. Carvalho[5], E. Cawley[9], S. Charalambous[16], M. Chardalas[16], G. Chardin[14], S. Dedoussis[16],
M. Dejardin[4,14], J. Derre[14], M. Dodgson[9,b], J.C. Dousse[7], J. Duclos[14,c], A. Ealet[11], B. Eckart[2],
C. Eleftheriadis[16], I. Evangelou[16], L. Faravel[7], P. Fassnacht[11], J.L. Faure[14], C. Felder[2],
R. Ferreira-Marques[5], W. Fetscher[17], M. Fidecaro[4], A. Filipčič[10], D. Francis[3], J. Fry[9],
C. Fuglesang[15,d], E. Gabathuler[9], R. Gamet[9], D. Garreta[4,14], T. Geralis[13], H.J. Gerber[17], A. Go[3],
P. Gumplinger[17,e], C. Guyot[14], P.F. Harrison[9,f], P.J. Hayman[9], W.G. Heyes[4,h], R.W. Hollander[6],
K. Jansson[15], H.J. Johner[7], K. Jon-And[15], A. Kerek[15], J. Kern[7], P.R. Kettle[13],
C. Kochowski[14], P. Kokkas[8,h], R. Kreuger[6], T. Lawry[3,i], R. Le Gac[12], A. Liolios[16], E. Machado[5],
P. Maley[9], I. Mandić[10], N. Manthos[8], G. Marel[14], M. Mikuž[10], J. Miller[3], F. Montanet[11],
T. Nakada[13], A. Onofre[5], B. Pagels[2], P. Pavlopoulos[2], F. Pelucchi[11], J. Pinto da Cunha[5],
A. Policarpo[5], G. Polivka[2], H. Postma[6], R. Rickenbach[2], B.L. Roberts[3], E. Rozaki[1],
T. Ruf[17], L. Sacks[9], L. Sakeliou[1], P. Sanders[9], C. Santoni[2], K. Sarigiannis[1], M. Schäfer[17],
L. Schaller[7], A. Schopper[4], P. Schune[14], A. Soares[14], S. Szilagyi[15,l], L. Tauscher[2],
C. Thibault[12], F. Touchard[12], C. Touramanis[9], F. Triantis[8], D.A. Tröster[2,j], E. Van Beveren[5],
M. Van den Putte[6], C.W.E. Van Eijk[6], G. Varner[3], S. Vlachos[9], O. Wigger[13],
P. Weber[17], C. Witzig[17,m], O. Wolter[17], C. Yeche[14], D. Zavrtanik[10] and D. Zimmerman[3].

presented by Claude Guyot
DAPNIA/SPP, C.E. Saclay, F-91191, Gif-sur-Yvette CEDEX, France

Abstract

The CPLEAR experiment at CERN is designed to study CP(T) violation in the neutral kaon system through the observation of particle-antiparticle decay asymmetries. Preliminary results of the measurements of the CP violating parameters η_{+-} and ϕ_{+-} in the decay $K^0 \to \pi^+\pi^-$ are reported. Furthermore, the analysis of semileptonic K^0 decays enables the determination of the mass difference Δm between K_L and K_S states and sets a limit on an asymmetry related to a T violating process.

---

[1]University of Athens, Greece, [2]University of Basel, Switzerland, [3]University of Boston, USA, [4]CERN, Geneva, Switzerland, [5]LIP and University of Coimbra, Portugal, [6]Tech. University of Delft, The Netherlands, [7]University of Fribourg, Switzerland, [8]University of Ioannina, Greece, [9]University of Liverpool, UK, [10]J. Stefan Institute and Department of Physics, University of Ljubljana, Slovenia, [11]CPPM, IN2P3-CNRS et Universite d'Aix-Marseille II, Marseille, France, [12]CSNSM, CNRS-IN2P3, Orsay, France, [13]Paul-Scherrer-Institute (PSI), Villigen, Switzerland, [14]DAPNIA/SPP, CE Saclay, France, [15]MSI, Stockholm, Sweden, [16]University of Thessaloniki, Greece, [17]ETH-IMP, Zürich, Switzerland.
[a]now at DESY, Hambourg, Germany, [b]now at ECP/SA, CERN, Switzerland, [c]now at NA48, CERN, Switzerland, [d]now at ESA/EAC, Köln, Germany, [e]now at TRIUMF, Vancouver, Canada, [f]now at QMW, University of London, England, [g]now at CEBAF, Newport News, Virginia, USA, [h]now at SUeR, Fribourg University, Switzerland, [i]now at University

INTRODUCTION

So far, CP violation has only been observed in decays of K_L mesons into two pions final states ($\pi^+\pi^-, \pi^0\pi^0$) and in the charge asymmetry of semileptonic K_L^0 decays. Rather than studying the physical states K_S and K_L, the CPLEAR experiment is unique in making direct use of the flavor eigenstates K^0 and $\overline{K^0}$ [1]. The method used to extract CP violation parameters is based on the measurement of time dependent asymmetries $A_f(t)$ between the decay rates of initially pure K^0 and $\overline{K^0}$ states into a CP eigen state f. The decay rates as a function of the kaon proper time for initially pure K^0 and $\overline{K^0}$ are given by:

$$R(t) \propto e^{-\Gamma_S t} + |\eta_f|^2 e^{-\Gamma_L t}$$
$$\pm e^{-\frac{1}{2}(\Gamma_S + \Gamma_L)t} \cos(\triangle m\, t - \phi_f) \quad (1)$$

where Γ_S and Γ_L are the decay widths of K_S and K_L. The parameter η_f is defined as the amplitude ratio

$$\eta_f = \frac{A(K_L \to f)}{A(K_S \to f)} = |\eta_f| e^{i\phi_f}. \quad (2)$$

For the CP even final states $\pi^+\pi^-$ and $\pi^0\pi^0$, η_f is related to the CP violation parameters in the mass matrix and in the decay amplitude, ϵ and ϵ', as follows:

$$\eta_{+-} = \epsilon + \epsilon', \qquad \eta_{00} = \epsilon - 2\epsilon'. \quad (3)$$

Since $\triangle m \approx \Gamma_S/2$, it is possible to observe the oscillation due to the interference before the K_S component has died away. Therefore, both the magnitude and the phase of the ratio η_f can be obtained from the asymmetry function in the interference region.

For the $\pi^+\pi^-$ final state the decay asymmetry A_f, defined as:

$$A_f(t) = \frac{\overline{R}(t) - R(t)}{\overline{R}(t) + R(t)}, \quad (4)$$

---
of Virginia, USA, ^{j}now at SBS, Basle, Switzerland, ^{k}now at Brookhaven National Laboratory, NY, USA, ^{l}Observer, Atomki, Hungary.

is related to the parameters $|\eta_{+-}|$ and ϕ_{+-} through the equation:

$$A_{+-}(t) = 2Re(\epsilon)$$
$$-\frac{2|\eta_{+-}|e^{\frac{1}{2}(\Gamma_S-\Gamma_L)t}\cos(\triangle m\, t - \phi_{+-})}{1 + |\eta_{+-}|^2 e^{(\Gamma_S-\Gamma_L)t}}. \quad (5)$$

THE CPLEAR EXPERIMENT

At LEAR, the K^0 and $\overline{K^0}$ mesons are simultaneously and symmetrically produced in proton-antiproton annihilations at rest through the reactions

$$p\bar{p} \to K^-\pi^+ K^0, \qquad Br = 2 \times 10^{-3}$$
$$p\bar{p} \to K^+\pi^- \overline{K^0}, \qquad Br = 2 \times 10^{-3}. \quad (6)$$

Flavor conservation of the strong interaction allows strangeness tagging of the neutral kaons by the sign of the accompanying $K^\pm$ mesons. The simultaneous and symmetric production of K^0 and $\overline{K^0}$ together with a symmetrical detection of their decay states have the considerable advantage of minimizing systematic effects. Also, this approach gives access to the analysis of the $K^0/\overline{K^0}$ asymmetry for various decay channels.

A cross-section of the CPLEAR detector is shown in Fig.1, displaying a typical event of the type $p\bar{p} \to K^+\pi^-\overline{K^0}$ and the subsequent decay of the $\overline{K^0}$ into $\pi^+\pi^-$.

This cylindrical detector is 3.6 m long with a radius of 1 m, placed inside a 0.44 T solenoidal magnetic field. The antiprotons are stopped and annihilate inside a spherical gas hydrogen target at 16 atm pressure. Charged particle tracking is performed with two multiwire Proportional Chambers, followed by six Drift Chambers and two layers of Streamer Tubes. Charged kaons and pions are identified using the Particle Identification Detector (PID) [2], consisting of a Scintillator-Čerenkov-Scintillator sandwich

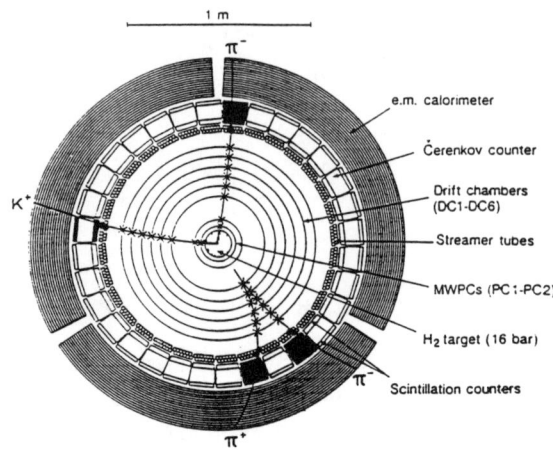

Figure 1. *The CPLEAR detector*

(SCS). The threshold for producing light in the Čerenkov counter is 300 MeV/c for pions and 700 MeV/c for kaons. Therefore $K^{\pm}$ mesons produced in the reactions (6) with momenta less than 700 MeV/c are required to have a $S\overline{C}S$ pattern in the PID, which has a rejection efficiency for pions of more than 99%.

A high granularity electromagnetic calorimeter [4] (5.8 X_0), made of 18 layers of Pb converters and 4x4.5 mm^2 streamer tubes, is used for photon detection and electron identification. Its good spatial resolution (5 mm for a single photon impact point) is well suited for reconstructing neutral final states.

The small production fraction of the desired *golden channels* (6) requires a high collision rate of $\approx$ 1 MHz. Consequently, in order to provide an efficient online event selection a fast and sophisticated trigger system has been developed [3]. The different trigger levels include $K^{\pm}$ identification, track parametrisation, kinematic cuts at the primary vertex and parametrisation of clusters in the electromagnetic calorimeter.

FIRST DATA AND RESULTS

The first results on the analysis of the two charged pions decay channel, based on data taken in 1990, have already been published [5]. The results presented at this conference are based on the data taken in 1990 and 1991 with only part of the trigger processors installed. Two exclusive trigger types were used:

- A *short distance* trigger which demands a kaon candidate in the Čerenkov counter, a transverse momentum cut on this candidate and a configuration of hits in the chambers compatible with a two or four tracks topology, at least two of them being primary tracks (i.e. with a hit in the proportional chambers). This trigger was used to study the K/π separation, $K^0/\overline{K^0}$ biases and normalisation (e.g. induced by the different strong interactions of K^+ and K^-) and the decays involving π^0s (two and three pions final states).

- A *long distance* trigger which requires at least four tracks candidates, only two of them being primary tracks. It was used to look at decays into $\pi^+\pi^-$ in the interference region (proper time larger than about 5 τ_S) where the asymmetry is maximum.

Two charged pions final states

In the offline analysis, the event selection required a four tracks topology with a kaon candidate of momentum larger than 350 MeV. The remaining events were then passed through kinematical and geometrical constrained fits to minimize the residual background from three pions and semileptonic final states. In addition, the constraints improve the resolution on the measured K^0 decay length (0.1 τ_S). About 4 10^5 events from the long distance trigger and 2 10^5 events from the short distance trigger were selected for this analysis. The acceptance corrected decay distribution for the sum of K^0 and $\overline{K^0}$ mesons is displayed in Fig.2. For proper times $t \lesssim 10\tau_S$ it is clearly dominated by two pions decays whereas at large lifetime a significant fraction of background remains, consisting mainly of semileptonic K_L decays. From a fit to the data, leaving the K_S lifetime and the amount of background K_L decays as free parameters, the value for the K_S lifetime was found to be

$$\tau = (0.98 \pm 0.01)\tau_S \ . \qquad (7)$$

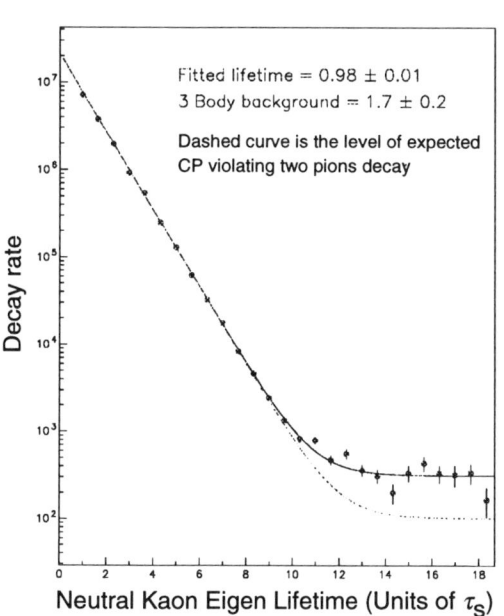

Figure 2. *The decay rate for K^0 and $\overline{K^0}$*

Due to different interactions of K^+ and K^- in the detector material the total number of tagged K^0 and $\overline{K^0}$ has to be normalized before extracting the asymmetry. The normalisation $\alpha = N(\overline{K^0})/N(K^0)$ was determined from the data at short decay times where the interference is small compared to α. Within the present statistical errors, the ratio turns out to be independent of the neutral kaon momentum with a value of 1.14 ± 0.01. With the normalized rates, the asymmetry $A_{+-}(t)$ displayed in Fig.3 was obtained. No acceptance correction has been applied since this cancels out in the ratio Eq.(4). The dashed curve is a fit to the data using Eq.(5) with η_{+-} and ϕ_{+-} as free parameters. The data agree well with the fit for $t < 10\tau_S$, whereas at larger lifetimes the magnitude of the asymmetry is reduced by background contamination from K_L semileptonic decays as shown by the fit which includes the contribution from this background. The values of the fitted parameters are:

$$\begin{aligned} |\eta_{+-}| &= (2.30 \pm 0.10 \pm 0.04) \times 10^{-3} \\ \phi_{+-} &= 44.9° \pm 2.5° \pm 0.5° \ , \end{aligned} \qquad (8)$$

where the first error is statistical and includes the error on the normalisation and the second is systematic. The uncertainty from the $K^0/\overline{K^0}$ normalisation comes mainly from the limited statistics of events at small proper time due to the short running time with the short distance trigger. It affects significantly only the measurement of ϕ_{+-}. Without this contribution, the statistical error on ϕ_{+-} would be 1.8°. The main sources of systematical uncertainties arise from the correction for the regeneration effects (-0.08 10^{-3} on η_{+-} and +1° on ϕ_{+-}) and from the magnitude of the three bodies background. The uncertainty of $\pm 0.8°$ on ϕ_{+-} induced by the limited knowledge of Δm [6] has not been included.

Semileptonic decays

The identification of electrons originating from semileptonic decays proceeds via the analysis of the scintillators and Čerenkov signals for low momentum tracks (below 250 MeV/c) or via the shower topology in the

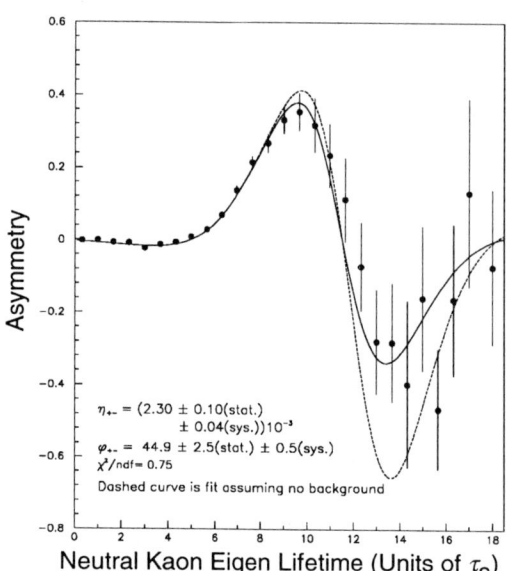

Figure 3. *The decay asymmetry* $A_{+-}(t)$

asymmetry $A_T(t)$ defined as

$$A_T(t) = \frac{\overline{N}^+ - N^-}{\overline{N}^+ + N^-} \quad (9)$$

would reflect a difference in the probabilities $P(K^0 \to \overline{K^0})$ and $P(\overline{K^0} \to K^0)$, thus indicating a departure from T invariance. Assuming CPT invariance, one expects $A_T(t) \simeq 4Re(\epsilon)$. Again assuming the $\Delta S = \Delta Q$ rule, the oscillation frequency Δm can be determined from the asymmetry $A_{\Delta m}(t)$ independently of the CP violation parameters:

$$\begin{aligned} A_{\Delta m}(t) &= \frac{N^+ + \overline{N}^- - (\overline{N}^+ + N^-)}{N^+ + \overline{N}^- + \overline{N}^+ + N^-} \\ &= \frac{2\cos(\Delta m\, t)\, e^{-\frac{1}{2}(\Gamma_S + \Gamma_L)t}}{e^{-\Gamma_S t} + e^{-\Gamma_L t}} .(10) \end{aligned}$$

The determination of T violation depends on the efficiency ratio $\eta = \eta(e^-\pi^+)/\eta(e^+\pi^-)$ as well as on the normalisation α already described in the $K^0 \to \pi^+\pi^-$ analysis. The value for η obtained from the data is 1.036 ± 0.024. The asymmetry distributions are shown in Fig.4 and Fig.5. A fit to these distributions yields the following results:

calorimeter at higher momenta. In the following pre-analysis based on long distance triggers, electron candidates were identified with the PID only thus reducing the statistics to about 50% of the total sample. After applying constrained fits to reduce the background from two and three pions decays, $4\ 10^4$ events were left in the final data sample. The four possible semileptonic decays of neutral kaons are given below with the corresponding numbers of events:

$$\begin{aligned} N^- &: K^0 \to \pi^- e^+ \nu_l, \ \overline{N}^+ : \overline{K^0} \to \pi^+ e^- \overline{\nu}_l \\ N^+ &: K^0 \to \pi^+ e^- \overline{\nu}_l, \ \overline{N}^- : \overline{K^0} \to \pi^- e^+ \nu_l . \end{aligned}$$

While the two processes in the first row originate from $\Delta S = \Delta Q$ transitions, the others correspond to $\Delta S = -\Delta Q$ which is highly suppressed in the Standard Model. These decays can however proceed from the particle/antiparticle oscillation. Consequently, assuming $\Delta S = \Delta Q$, a non-zero value of the

$$\begin{aligned} A_T(t) &= (8.5 \pm 7.6 \pm 15.5) \times 10^{-3} \\ \Delta m &= (0.523 \pm 0.018 \\ &\quad +0.007/-0.001) \times 10^{10} \hbar s^{-1} \end{aligned}$$

where the first error is statistical and the second systematic. The dominant contributions to the systematic uncertainties come from the normalisation and efficiency corrections (for A_T) and the level of residual background contamination (for A_T and Δm). The values of Δm and $Re(\epsilon)$ (deduced from A_T assuming CPT invariance) are in agreement with the world averages [6].

CONCLUSIONS

These preliminary results have demonstrated the validity of the CPLEAR approach

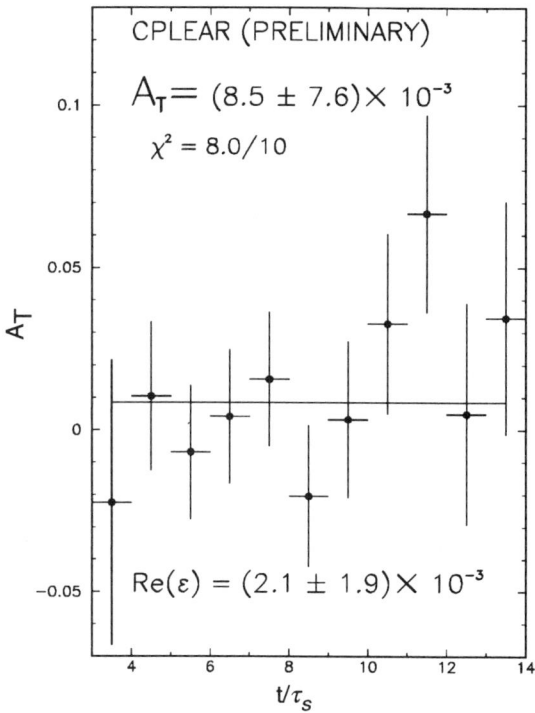

Figure 4. *The asymmetry $A_T(t)$*

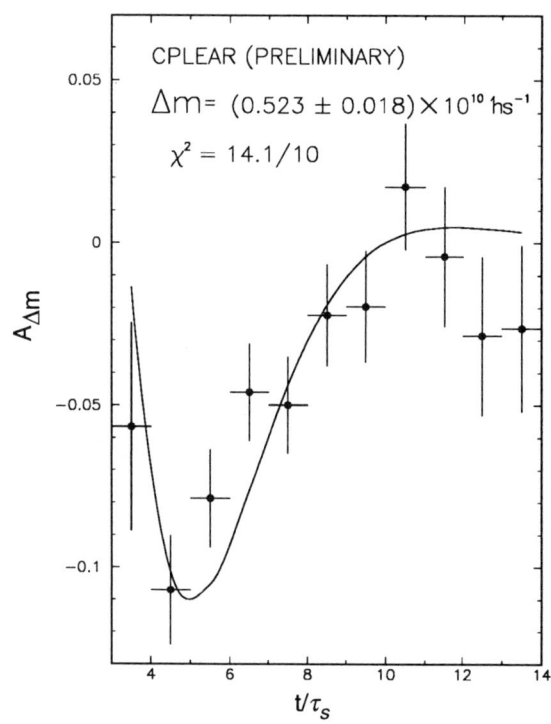

Figure 5. *The asymmetry $A_{\Delta m}(t)$*

based on the comparison of the decay properties of particles and antiparticles. The measurements of the CP violating parameters η_{+-} and ϕ_{+-} are in agreement with the world average values [6]. The analysis of semileptonic K^0 decays will soon give access to the first direct evidence of T violation. The completion of the full trigger system for the data taking periods starting in 1992 will yield a major improvement on the significance of these measurements together with new results on the three pions decay channel and a test of the $\Delta S = \Delta Q$ rule.

REFERENCES

1. L. Adiels et al., Proposal CERN PSCC/85-6 PSCC/P82 (1985).

2. A. Angelopoulos et al., (CPLEAR Collaboration), Nucl. Instr. and Meth. **A311** (1992), 78.

3. D. Tröster et al., (CPLEAR Collaboration), Nucl. Instr. and Meth. **A279** (1989), 285.

4. R. Adler et al., (CPLEAR Collaboration), Preprint CERN PPE/92-68 (1992), to appear in Nucl. Instr. and Meth.

5. R. Adler et al., (CPLEAR Collaboration), Phys. Lett. **B286** (1992) 180.

6. Particle Data Group, Phys. Rev. **D45** (1992), 1.

A MEASUREMENT OF THE DECAY $K_L \to \pi^0\gamma\gamma$ BY THE NA31 EXPERIMENT

Lydia Fayard

*Laboratoire de l'Accélérateur Linéaire, IN2P3 - CNRS et Université de Paris-Sud,
91405 Orsay Cedex, France.*

NA31 is the CERN-Mainz-Orsay-Pisa-Siegen dedicated experiment for a precise measurement of ε'/ε. It turns out that, thanks to the high accumulated statistics over 3 different data taking periods, a search for rare Kaon decays has been favoured. Several results on measurements or limits have been published[1], in particular the first observation of $K_L \to \pi^0\gamma\gamma$ decay[2]. We report here the final analysis of $K_L \to \pi^0\gamma\gamma$ using the whole available statistics collected with the NA31 detector in 1986, 1988 and 1989 run periods.

The Interest of $K_L \to \pi^0\gamma\gamma$.

First of all, the decay $K_L \to \pi^0\gamma\gamma$ may be used to test chiral perturbation theory and some semi-empirical models describing the Kaon system[3]. These models predict for the lowest order a branching ratio around 0.7×10^{-6}, a 2γ final State with $J = 0$ and a $\gamma\gamma$ invariant mass spectrum extending from $2m_{\pi^0}$ to the kinematical limit $M_K - m_{\pi^0}$. Higher order chiral expansion or (and) other calculations using vector (or scalar) meson intermediate states[4], give a branching ratio between 0.7×10^{-6} and 3×10^{-6} depending on the parameter a_V which characterizes the $J = 2$ 2γ final state importance. Moreover, a precise measurement of the $K_L \to \pi^0\gamma\gamma$ branching ratio, may be used to estimate the CP-conserving contribution to the decay $K_L \to \pi^0 e^+ e^-$ via two photon intermediate state[5]. The $J = 0$ amplitude is strongly helicity suppressed. This is not the case for the $J = 2$ state. If both contributions are small enough the decay $K_L \to \pi^0 e^+ e^-$ is dominated by CP violating contributions and then, the experimental study of that mechanism would be easier.

Experimental set-up and data selection.

The NA31 beam and detector have already been described in detail elsewhere[6]. We only summarize here the relevant parts for this analysis.
- A Liquid Argon Calorimeter (LAC) using x and y readout, measures photons with $7.5\%/\sqrt{E}$ energy resolution and $\pm .5$ mm position resolution. Clusters are resolved in that device from a distance of 1.5cm in projection.
- Four ring-shaped anticounters veto events with photons escaping the calorimeter acceptance.
- Two wire Chambers spaced by 25 meters are placed in front of LAC.
- An Hadronic calorimeter behind the LAC is used as veto for hadronic events.

The initial data sample consists of 5×10^8 events. For $K_L \to \pi^0\gamma\gamma$ we apply the following requirements:

* They have only 4 reconstructed electromagnetic showers each of them with energy between 5 and 100 GeV, at a distance from the beam axis greater than 16cm and 2.5cm far away from the calorimeter borderlines. All showers are

separated at least by 5cm in each projection.

* Total energy of 4 showers is between 60 and 170 GeV.
* No spacepoints in the upstream wire chamber and no more than 1 in the downstream one, to allow for backsplach from the calorimeter.
* Energy barycentre in LAC within 5cm from the beam axis. This cut rejects events with large missing transverse energy.

Assuming Kaon Mass for the four showers, we reconstruct then the longitudinal position of the vertex Z_K. In the case of events with undetected particles, Z_K is shifted from the true position towards the calorimeter. We therefore require Z_K to lie in the first 20 meters downstream of the final collimator to reduce the $3\pi^0$ background.

The four showers are grouped into 2 pairs. Events having 2 $\gamma\gamma$ combinations with both invariant masses compatible with m_{π^0} are rejected to reduce the $K_L \to 2\pi^0$ component.

The $\pi^0\gamma\gamma$ candidates are required to have one pair $\gamma_1\gamma_2$ with invariant mass M_{12} compatible with π^0 mass (between 125 and 145 Mev). The invariant mass of the other pair $\gamma_3\gamma_4$ is called M_{34}. 5889 $K_L \to \pi^0\gamma\gamma$ candidates survive the preceeding selection, the majority of them being background coming from $K_L \to 3\pi^0$ decays. A Monte Carlo simulation of a large number of $K_L \to 3\pi^0$ decays shows that the surviving the cuts described aboved sample, has the following characteristics:

* The more important part of background comes from decays upstream the final collimator.
* Almost 60% of these events results from events with 2 photons missing the detector. In that case the shift of the reconstructed Z_K is greater than 7.5 m.
* The remaining 40% of the sample corresponds to events with one missing photon and 2 overlapping showers in the calorimeter. The Z_K shift is also greater than 7.5 m. That kind of events can be partially identified since the width of the reconstructed shower is in general broader than for isolated photons. We reject then events if the rms of the shower energy distribution is greater than 1.1 cm.

At that level, the remaining sample is still dominated by $K_L \to 3\pi^0$ decays, which can be separated into 2 types:

- Events with one π^0 within acceptance and correctly measured.
- Events with 2 π^0's within acceptance having two of their 4 photons overlapping.

In the first case, we can estimate the true K_L vertex assuming π^0 mass and using energies and positions of its 2 photons. For each pair of showers, we compute a decay vertex assuming the invariant mass to be the nominal π^0 mass. The maximum value of all 2γ combinations (but less than Z_K -7.5m), will be kept as estimator of the true K_L decay vertex. We call it Z_π.

For the second case, we attempt by splitting the energy of the shower to reconstruct the common decay vertex of two π^0's. Among the twelve combinations (4 photons taken by 3) we keep the one giving the closest to Z_k-7.5m decay vertex. This estimator is called $Z_{\pi\pi}$. Finally we use as unique estimator Z_{max}, the maximum of Z_π and $Z_{\pi\pi}$. The event is then rejected if Z_{max} is greater than -15m for Z_K> 10m or greater than (-Z_K+5m) for Z_K less than 10m. These values are defined by optimization resulting from Monte Carlo studies. The efficiency of that cut against $K_L \to 3\pi^0$ remaining background, is 99.6% while 54% of the $\pi^0\gamma\gamma$ signal is lost.

94 candidates remain after all cuts. Fig 1 shows a significant peak in the M_{12} distribution around the π^0 mass. We define a signal region between 132.5 and 137.5 MeV which corresponds to 1.5 times the expected resolution. 63 events are in this region. The residual $2\pi^0$ and $3\pi^0$ background may be estimated by linear interpolation in the signal region from side bands (125-130 MeV and 140-145 MeV). Estimating in that way an expected background of 6. ± 1.7 events, we conclude that we have observed 57 ± 8.1 $K_L \to \pi^0\gamma\gamma$ events within our cuts.

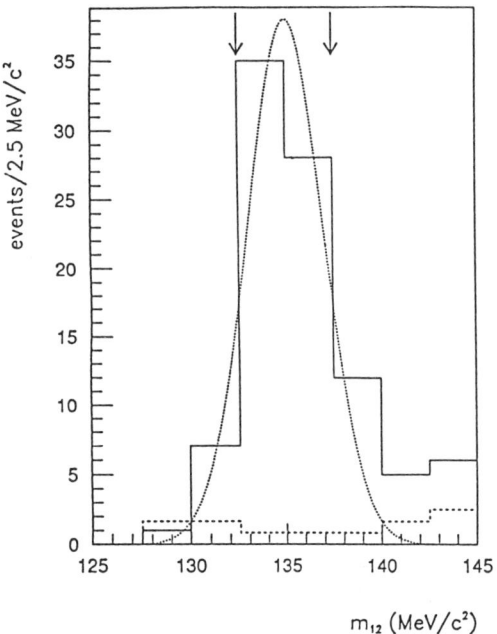

Figure 1 : Invariant mass M_{12} for $K_L \to \pi^0\gamma\gamma$ candidates (solid) and for expected background from $K_L \to 2\pi^0$ and $K_L \to 3\pi^0$. The Gaussian represents the expected resolution.

Results.

To measure the branching ratio we normalize the 57 events to the $K_L \to 2\pi^0$ sample ($4.24 \times 10^{+4}$ events). The ratio of acceptances for $\pi^0\gamma\gamma$ and $2\pi^0$ within our cuts is 0.72 ± .02. Using the $K_L \to 2\pi^0$ branching ratio from ref. 7 we obtain:

$$\Gamma(K_L \to \pi^0\gamma\gamma)/\Gamma(K_L \to \text{all}) = (1.7 \pm .2 \pm .2) \times 10^{-6}$$

the systematic error takes into account background estimate uncertainties, energy scale and acceptance variation with the parameter a_V.

In order to investigate the decay mechanism of the observed decay, we study the distribution of the M_{34} invariant mass of the photon pair. The data (fig 2) shows a clear accumulation of events in high M_{34} values (above 240 MeV/c²) as predicted by chiral models or models involving two photons in J = 0 final state. In the region below 240 MeV we only have 3 events when we expect 1.5. We conclude that in that region, there is no evidence of signal.

We performed then a maximum likelihood fit of a_V to the data set, as a function of the two Dalitz

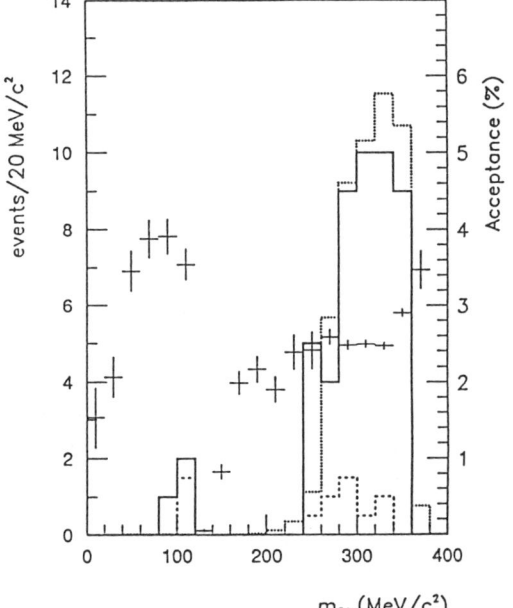

Figure 2 : Invariant mass M_{34} for $K_L \to \pi^0\gamma\gamma$ candidates with M_{12} in the signal region (solid), in the side bands properly normalised (dashed) and for $K_L \to \pi^0\gamma\gamma$ events simulated with $a_v = 0$. The acceptance is given by crosses.

variables M_{34} and $Y=|E_3-E_4|/M_K$. The result is :

$$a_V = -.05^{+.14}_{-.17}$$

which can be translated into a 90% confidence level limit of

$$-.32 < a_V < .19$$

The observed rate of $\pi^0\gamma\gamma$ combined with this limit, gives that, the CP-conserving contribution to the decay $K_L \to \pi^0 e^+ e^-$ should be smaller than 4.5×10^{-13} at 90% C.L.

Conclusions

We have measured the branching ratio of the decay $K_L \to \pi^0\gamma\gamma$ in agreement with expectations from chiral perturbation theory. The invariant mass distribution of 2 photons favours models with dominant $J = 0$ two photon system. The smallness of a_V implies that the CP-conserving contribution to $K_L \to \pi^0 e^+ e^-$ is small compared to the CP-violating contributions, and then gives confidence for a future measurement of CP violation in that particular decay mode.

Acknowledgements

I would like to thank Laurent Serin for his kind help during the preparation of this talk.

REFERENCES

[1] G. Barr et al.: Phys. Lett. B214(1988)303
G. Barr et al.: Phys. Lett. B235(1990)356
G. Barr et al.: Phys. Lett. B259(1991)389

[2] G. Barr et al.: Phys. Lett. B242(1990)523

[3] G. Ecker, A. Pich, E. de Rafael : Phys. Lett. B189(1987)363 L. M. Sehgal: Phys. Rev D6(1972)367 P. Ko and J. L. Rosner: Phys.Rev. D40(1989)377

[4] G. W. Intemann: Phys. Rev. D13(1976)653. L. M. Sehgal: Phys. Rev. D38(1988)808. T. Morozumi and H. Iwasaki: Prog. Theor. Phys. 82(1989)371. J. Flynn and L. Randall: Phys. Lett. B216(1989)221. G. Ecker, A.Pich and E. de Rafael : Phys. Lett. B237(1990) 481. L. M. Sehgal: Phys. Rev. D41 (1990) 161. J. Bignens, S. Dawson, G. Valencia: Phys. Rev. D44 (1991) 3555.

[5] J. F. Donoghue, B. R. Holstein and G. Valencia : Phys. Rev. D35(1987)2769 C. O. Dib, I. Dunietz, F. J. Gilman: Phys. Rev. D39(1989)2639. A. Barker et al.: Phys. Rev. D61(1990)3546. K. E. Olh et al.: Phys. Rev. Lett. 64 (1990) 2755.

[6] H. Burkhardt et al.: Phys. Lett B206(1988)169 H. Burkhardt et al.: Phys. lett. B199 (1987)139. H. Burkhardt et al.: Nucl. Inst. and Meth. A268(1988)116.

[7] Particle Data Group G. Yost et al .: Phys. Lett. B239(1990)1.

RESULTS AND PLANS FOR THE FERMILAB K^0 DECAY PROGRAM*

Yau W. Wah

Department of Physics and the Enrico Fermi Institute
The University of Chicago
5640 S. Ellis Avenue
Chicago, Illinois 60637

Abstract

We present preliminary branching ratio results on $K_L \to \pi^+\pi^-\pi^0\gamma$, and $\pi^0 \to e^+e^-$. We also observed twenty examples of $K_L \to \pi^+\pi^-e^+e^-$. These results come from data collected in the Fermilab fixed target experiment E799 Phase I test run which finished data taking in January 1992.

INTRODUCTION

Results presented here come from a series of Fermilab kaon experiments (E-731, E-773, E-799 Phase I) which were designed to measure direct CP violating and CPT violating parameters and search for CP violating rare kaon decays. We report a preliminary branching ratio determination of $K_L \to \pi^+\pi^-\pi^0 \gamma$ from the E-731 data; a prelimenary mesurement of the branching ratio of $\pi^0 \to e^+e^-$ and a report of first observation of twenty $K_L \to \pi^+\pi^-e^+e^-$ events from the E-799 Phase I data. We also briefly report on the status of the E-731 ε'/ε measurement and the future Fermilab neutral kaon program (KTeV).

---

* A collaboration of U. of Chicago, Elmburst, Fermilab, U. of Illinois-Urbana, UCLA, Rutgers, and U. of Colorado-Boulder. The author wishes to acknowledge the support of an OJI grant from the US Department of Energy.

$K_L \to \pi^+\pi^-\pi^0 \gamma$ branching ratio determination

$K_L \to \pi^+\pi^-\pi^0 \gamma$ has both inner bremsstrahlung and direct emission decay amplitudes[1,2,3] and has never been observed before. Figure 1 shows the reconstructed transverse momentum square vs. invariant mass for $\pi^+\pi^-\pi^0 \gamma$ final state. Figure 2 shows the vertical projection of Figure 1 with P_t^2 less than 100 MeV^2. There are 444 events total within ± 6.4 MeV of the nominal kaon mass with an estimated background of 15 events. Figure 3 shows the distribution of the gamma center of mass energy. Normalizing to 1% of the ~1.8 million $K_L \to \pi^+\pi^-\pi^0$ events simultaneously collected within a fiducial decay length of 37 meter and energy of kaon between 20 and 220 GeV, BR($K_L \to \pi^+\pi^-\pi^-\gamma$, $E\gamma > 20$MeV) / BR($K_L \to \pi^+\pi^-\pi^0$)=(2.04± 0.15 ± 0.08)x10^{-4} where the first error is statistical and the second error is systematics. With higher statistics, it is possible to observe the direct emission amplitude which could be as big as 10% of the inner-brem. Results from a higher statistics sample (factor of 2) with different E_γ threshold will be available soon.

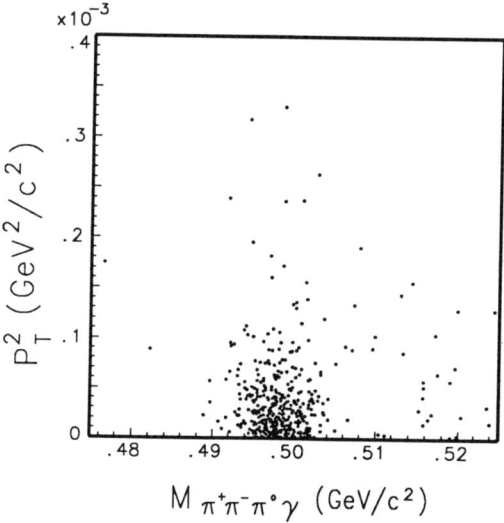

Figure 1. Scatterplot of square of the kaon transverse momentum vs $\pi^+\pi^-\pi^0\gamma$ invariant mass for the $K_L \to \pi^+\pi^-\pi^0\gamma$ data sample.

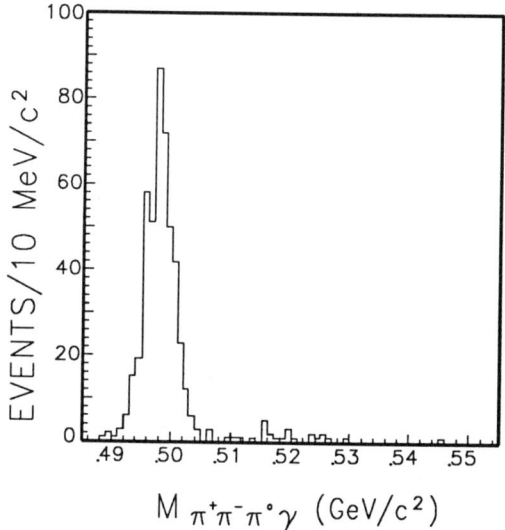

Figure 2. Distribution of $\pi^+\pi^-\pi^0\gamma$ invariant mass with $P_t^2 < 100 \, \text{MeV}^2$ for the $K_L \to \pi^+\pi^-\pi^0\gamma$ sample.

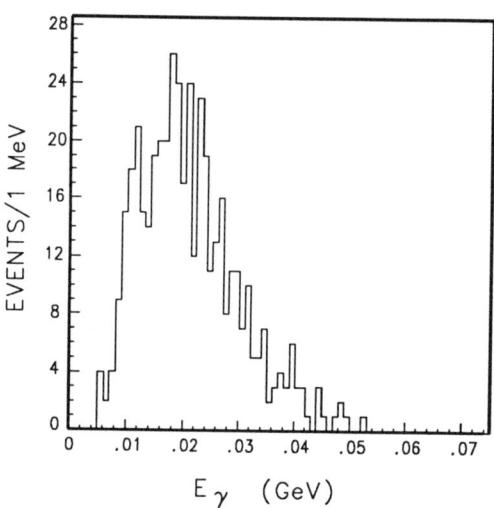

Figure 3. Distribution of energy of γ in the kaon center of mass frame for the $K_L \to \pi^+\pi^-\pi^0\gamma$ sample after all cuts.

First $K_L \to \pi^+\pi^-e^+e^-$ observation

Theories[4,5] suggest that this decay mode could be interesting to study CP violation. Experimentally, the background is $K_L \to \pi^+\pi^-\pi^0$ with π^0 Dalitz decay. Figure 4 shows the $\pi^+\pi^-e^+e^-\gamma$ invariant mass distribution with the γ detected. Figure 5 shows the $\pi^+\pi^-e^+e^-$ invariant mass distribution the γ undetected. To suppress the background, kinematics cuts consistent with the π^0 Dalitz were applied. Figure 6 shows the final $\pi\pi ee$ invariant mass distribution with the "right" signs shown at the top, and the "wrong" signs shown at the bottom. This is the first observation of $K_L \to \pi^+\pi^-e^+e^-$ and result on the branching ratio will be available soon.

Measurement of $\pi^0 \to e^+e^-$ branching ratio

A new technique using "tagged" π^0 from the copious $K_L \to 3\pi^0$ decay was used to measure the $\pi^0 \to e^+e^-$ branching ratio which has a contentious history [6,7,8]. Figure 7 shows the e^+e^- invariant mass distribution for the tagged π^0 Dalitz decay with the Dalitz γ detected. Figure 8 shows a signal of 9 $\pi^0 \to e^+e^-$ events with no Dalitz γ detected. The dots are the Monte Carlo simulation of the Dalitz background and is absolutely normalized with nine times higher statistics. Other backgrounds with two single Dalitz, double Dalitz, and external conversion are negligible. Systematic studies included momentum scale and shifts, calorimeter energy smearing and shifts. Taking into account of 100% uncertainty of background normalization, the subtraction is 0.86±1.5 events. Net signal is 8.1±3.3 events. The signal acceptance is 2.24% and we determined BR $(\pi^0 \to e^+e^-) = (6.9 \pm 2.8) \times 10^{-8}$.

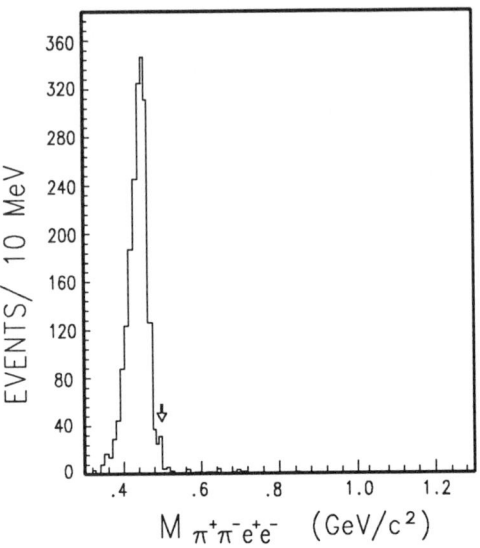

Figure 5. Distribution of $\pi^+ \pi^- e^+ e^-$ invariant mass for the four track sample with no gamma observed in the final state.

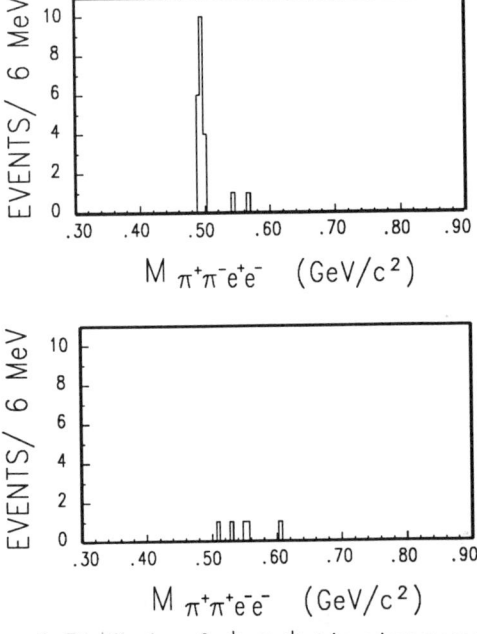

Figure 6. Distribution of $\pi^+ \pi^- e^+ e^-$ invariant mass for the four track sample. The top figure shows 20 signal events around the kaon mass. The bottom figure shows the mass distribution of the "opposite sign tracks" sample.

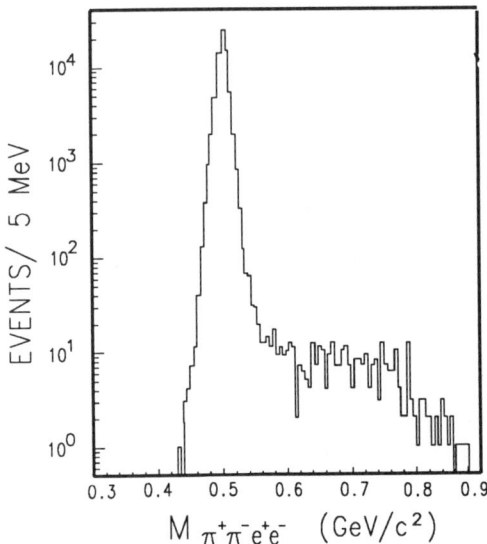

Figure 4. Distribution of $\pi^+ \pi^- e^+ e^- \gamma$ i invariant mass for $K_L \to \pi^+ \pi^- \pi^0$ with π^0 Dalitz decay, all final state particles were measured.

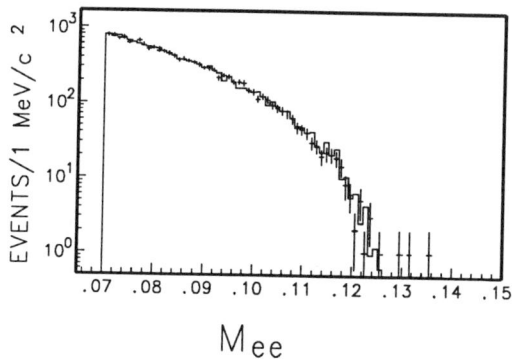

Figure 7. Distribution of e^+e^- invariant mass for the π^0 Dalitz decay sample. Solid line is Monte Carlo simulation.

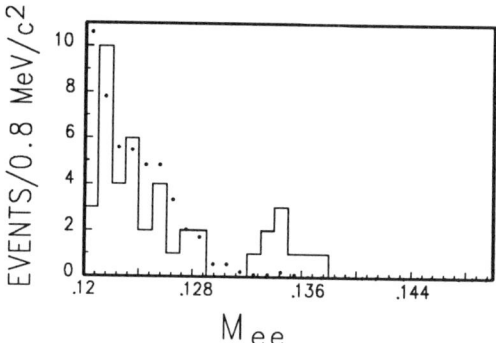

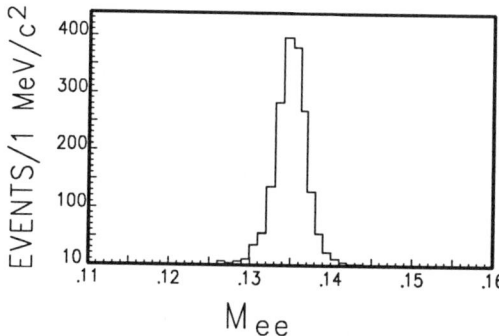

Figure 8. Distribution of e^+e^- invariant mass after all cuts. Dot in the top figure is Monte Carlo simulation of background. Bottom figure shows Monte Carlo simulation of signal unnormalised.

The history of the $\pi^0 \rightarrow e^+e^-$ branching ratio measurements is shown in Figure 9. This result demonstrated the decay in flight technique is viable to make a clean measurement and that there seems not to be as much "room" for large additional contributions beyond the QED "unitarity" limit.

Other results

The preliminary result [9] for the FULL E731 data set is $(\varepsilon'/\varepsilon)_{2\pi} = (6.0 \pm 5.8$ (stat.) ± 3.2 (syst) $) \times 10^{-4}$ that combine to: $(\varepsilon'/\varepsilon)_{2\pi} = (6.0 \pm 6.9) \times 10^{-4}$. The result is still consistent with zero. Our group has also made very precise determinations of other parameters of the neutral kaon system with the same 2π data samples, using exactly the same techniques of background subtraction and acceptance corrections. The reported results, where both statistical and systematic errors are included, are: $\tau_S = (0.8912 \pm 0.0013) \times 10^{-10}$ sec ;

$\Delta\phi = (0.5339 \pm 0.0020) \times 10^{10}$ sec^{-1};

$\phi_{+-} = (43.2 \pm 1.6)^0$;

$\Delta\phi = \phi_{+-} - \phi_{00} = (-0.6 \pm 1.6)^0$

These results are either comparable to or exceed in precision the best previous determinations. In particular, the value for ϕ_{+-} is in good agreement (as expected from CPT symmetry) with the expected "superweak phase", $43.5^0 \pm 0.14^0$. The result for τ_S is three times better than the best previous determination, and the value for $\Delta\phi$ is a factor of two better than earlier results.

Furthermore, analysis is going on with an additional sample of $K_L \rightarrow 2\pi^0$ events (about a factor of two increase) so that the final statistical error of $(\varepsilon'/\varepsilon)_{2\pi}$ is expected to be

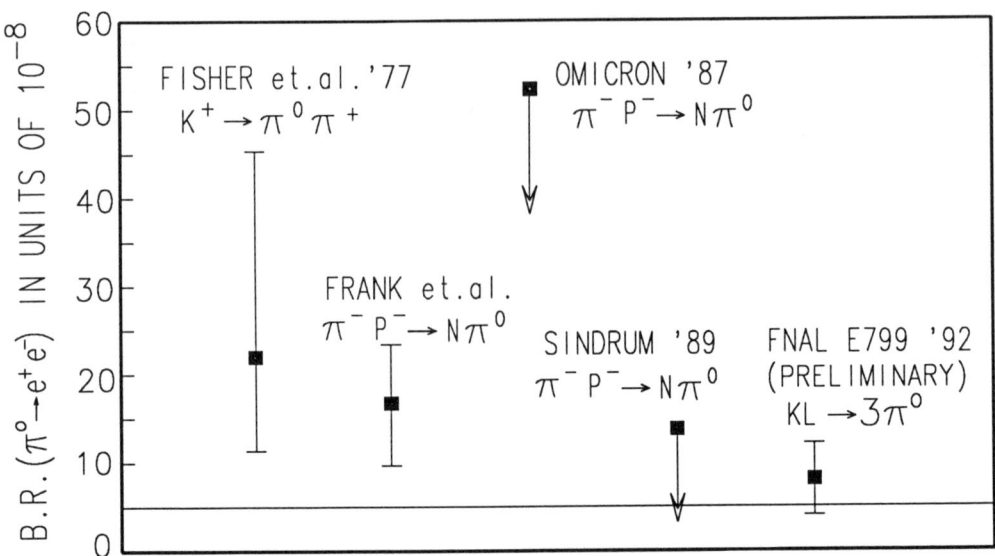

Figure 9 History of the $\pi^0 \to e^+e^-$ measurements

$\pm 5.2 \times 10^{-4}$. A result [10] of BR($K_L \to \pi^0 \nu \bar{\nu}$) < 2.2 $\times 10^{-4}$ (90% CL) is also published from E-731 data. This mode is most interesting that it is mediated by "pure" direct CP violating amplitude.

SUMMARY AND OUTLOOK

Preliminary results are (i) first measurement of BR ($K_L \to \pi^+\pi^-\pi^-\gamma$, $E\gamma > 20$ MeV) / BR ($K_L \to \pi^+\pi^-\pi^0$) = (2.04 ± 0.15 ± 0.08) $\times 10^{-4}$ where the first error is statistical and the second error is systematic ; (ii) first observation of $K_L \to \pi^+\pi^- e^+e^-$; and (iii) BR ($\pi^0 \to e^+e^-$) = (6.9 ± 2.8)$\times 10^{-8}$, this result comes from a by far cleanest sample of 9 events with 0.9 event background and suggests that sizable additional contributions to the QED unitarity decay amplitude seems unlikely.

Fermilab KTeV (Kaon at Tevatron) program which is experiment E-799 Phase II and E832, will measure $(\varepsilon'/\varepsilon)_{2\pi}$ to an accuracy of 1×10^{-4} and will search for the CP violating rare decay modes $K_L \to \pi^0 l^+ l^-$ ($l = e, \mu, \nu$) with single event sensitivity of $\sim 10^{-12}$.

REFERENCES

1. R.Ferrari and M.Rosa-Clot, "Bremsstrahlung in $K_L \to 3\pi$ Decays", *IL Nuovo Cimento* Vol. LVI A, 582 (1968).
2. A. Neveu and J. Scherk, "Electromagnetic Corrections to $K_L \to 3\pi$ Rates", *Phys. Lett.* 27B, 384(1968).
3. S. Fajfer, K.Suruliz & R. J.Oakes, "Effective low-energy, large-N Lagrangian calculation of the $K_L \to \pi\pi\pi\gamma$ decay modes", *Phys. Rev.* D42, 3875 (1990).

4. L. M. Sehgal and M. Wanniger, "CP violation in the decay $K_L \to \pi^+\pi^-e^+e^-$ ", *Phys. Rev.* D46, 1035(1992).

5. D. P. Majumdar and J. Smith, " Current Algebra, Field-Current Identity, the $K_2 K_1$ Electromagnetic Transition, and the Decays $K_L \to \pi^+\pi^-e^+e^-$", *Phys. Rev.* Vol.187, 2039 (1969).

6. J. Fischer et.al., "Observation of the $\pi^0 \to e^+e^-$ Decay", *Phys. Lett.* 73B, 364 (1978).

7. J. S. Frank et. al., "Measurement of the branching ratio for the rare decay $\pi^0 \to e^+e^-$ ", *Phys. Rev.* D28, 423(1983).

8. C. Niebuhr et. al., "Search for the decay $\pi^0 \to e^+e^-$ ", *Phys. Rev.* D40, 2796 (1989).

9. E731 Collab., presented by B. Winstein in : *Proc. Joint Internat. Lepton-Photon Symp. and Europhys. Conf. on High Energy Physics* (Geneva, 1991); eds. S. Hegarty, K. Potter, and E. Quercigh.

10. G. Graham et. al., "A Search for the decay $K_L \to \pi^0 \nu \nu$ "; Accepted by Phys. Lett. B.

CP VIOLATION IN THE DECAY $K_L \to \pi^+\pi^-e^+e^-$

L. M. Sehgal
Institut für Theoretische Physik (E)
Rheinisch–Westfälische Technische Hochschule
D–5100 Aachen, Germany

Abstract

The decay $K_L \to \pi^+\pi^-e^+e^-$ is analysed in a model containing (i) a CP-conserving amplitude associated with the $M1$ transition in $K_L \to \pi^+\pi^-\gamma$, (ii) an indirect CP-violating amplitude related to the bremsstrahlung component of $K_L \to \pi^+\pi^-\gamma$, and (iii) a direct CP-violating term related to the short distance interaction $s\bar{d} \to e^+e^-$. Interference of the first two components produces a large CP-violating asymmetry ($\sim 14\%$) in the distribution of the angle Φ between the e^+e^- and $\pi^+\pi^-$ planes. The full angular distribution contains two further CP-violating observables. Effects of direct CP violation are estimated to be small.

This is a report on an analysis of the reaction $K_L \to \pi^+\pi^-e^+e^-$ that has just been published [1], and an extension of this work that is in progress [2].

The decay $K_L \to \pi^+\pi^-e^+e^-$ can be envisaged, in the first instance, as a conversion process related to the radiative decay $K_L \to \pi^+\pi^-\gamma$. The latter is empirically known [3] to contain two components: a bremsstrahlung piece related to the CP-violating decay $K_L \to \pi^+\pi^-$, and a CP-conserving magnetic dipole component. Interference of these terms produces a CP-violating circular polarization of the photon in $K_L \to \pi^+\pi^-\gamma$. The conversion process $K_L \to \pi^+\pi^-e^+e^-$ may be viewed as a means of probing this polarization, by studying the correlation of the e^+e^- plane relative to the $\pi^+\pi^-$ plane.

The reaction $K_L \to \pi^+\pi^-e^+e^-$ can also proceed via non–radiative mechanisms. For example, the charge radius of the K^0 gives a CP-conserving contribution with the $\pi^+\pi^-$ in an s-wave [4]. In addition one expects CP violating contributions associated with the short–distance interaction $s\bar{d} \to e^+e^-$.

In this general scenario, the decay amplitude of $K_L \to \pi^+(p_+)\pi^-(p_-)e^+(k_+)e^-(k_-)$ can be parametrized as

$$A(K_L \to \pi^+\pi^-e^+e^-)$$
$$= e\,|f_s|\left[g_{Br}\left(\frac{p_+^\mu}{p_+\cdot k} - \frac{p_-^\mu}{p_-\cdot k}\right)\right.$$
$$\left. + g_{M1}\epsilon_{\mu\nu\rho\sigma}k^\nu p_+^\rho p_-^\sigma + \frac{g_P}{m_K^2}\frac{k^2}{s_\pi - m_K^2}(p_+ - p_-)^\mu\right]$$
$$\cdot\frac{e}{k^2}\,\bar{u}(k_-)\gamma_\mu v(k_+)$$
$$+\text{short–distance contributions} \qquad (1)$$

with $k = k_+ + k_-$ and $s_\pi = (p_+ + p_-)^2$. Here f_S is the amplitude of $K_S \to \pi^+\pi^-$, and the parameters g_{Br}, g_{M1} and g_P are given by $g_{Br} = \eta_{+-}e^{i\delta_0(m_K^2)}$, $g_{M1} = i\,(0.76)e^{i\delta_1(s_\pi)}$, $g_P = -\frac{1}{3}<R^2>_{K^0}m_K^2 e^{i\delta_0(s_\pi)}$, $\delta_{0,1}$ being the s- and p-wave $\pi\pi$ phase shifts.

In Ref. [1], we calculated the decay rate of $K_L \to \pi^+\pi^-e^+e^-$ and the spectrum of e^+e^- and $\pi^+\pi^-$ masses, neglecting short–distance effects. The branching ratio was determined to be $3.1 \cdot 10^{-7}$. A significant CP-violating asymmetry was found in the Φ-distribution of the process, Φ being the angle between the e^+e^- and $\pi^+\pi^-$ planes:

$$A = \left(\int_0^{\pi/2} d\Phi - \int_{\frac{\pi}{2}}^{\pi} d\Phi\right) \frac{d\Gamma}{d\Phi} \bigg/ \left(\int_0^{\pi} d\Phi \frac{d\Gamma}{d\Phi}\right)$$
$$= 15\% \cos(\Phi_{+-} + \delta_0(m_K^2) - \bar{\delta}_1 - \frac{\pi}{2})$$
$$\approx 14\% \qquad (2)$$

where $\bar{\delta}_1$ denotes an average p-wave phase ($\sim 10^0$). *

We have extended the above analysis in two directions [2].

(a) The angular distribution of the process is found to have the structure

$$\frac{d\Gamma}{d\cos\Theta_l d\Phi} = K_1 + K_2 \cos 2\Theta_l$$
$$+ K_3 \sin^2\Theta_l \cos 2\Phi + K_4 \sin 2\Theta_l \cos 2\Phi$$
$$+ K_5 \sin\Theta_l \cos\Phi + K_6 \cos\Theta_l$$
$$+ K_7 \sin\Theta_l \cos\Phi + K_8 \sin 2\Theta_l \sin\Phi$$
$$+ K_9 \sin^2\Theta_l \sin 2\Phi \qquad (3)$$

Here Θ_l is the angle between the e^- momentum vector (in the e^+e^- c.m. frame) and the dilepton momentum in the K^0 rest frame. CPT invariance ensures $K_6 = 0$. CP violation manifests itself in the coefficients K_4, K_7 and K_9. Integrating over $\cos\Theta_l$, the only CP-violating term that survives is K_9. This is the asymmetry calculated in Ref. [1].

(b) Effects of direct CP violation have been calculated using the short-distance Lagrangian [5]

$$\mathcal{L}_{eff} = \frac{G_F}{\sqrt{2}} \alpha \sum_q V_{qs}^* V_{qd} \qquad (4)$$
$$\cdot [F_V(x_q)Q_V + F_A(x_q)Q_A + F_M(x_q)Q_M]$$

with

$$Q_V = [\bar{s}\gamma_\mu(1-\gamma_5)d][\bar{e}\gamma_\mu e]$$
$$Q_A = [\bar{s}\gamma_\mu(1-\gamma_5)d][\bar{e}\gamma_\mu\gamma_5 e]$$
$$Q_M = [\bar{s}(i\, m_s \sigma_{\mu\nu} k^\nu (1-\gamma_5)$$
$$+ i\, m_d \sigma_{\mu\nu} k^\nu (1+\gamma_5))d] \cdot \frac{1}{k^2}\bar{e}\gamma_\mu e \qquad (5)$$

Here F_V, F_A, F_M are known functions of $x_q = m_q^2/m_W^2$. The hadronic matrix element $< \pi^+\pi^-|\bar{s}\gamma_\mu(1-\gamma_5)d|K_L >$ can be related to the matrix element $< \pi^+\pi^-|\bar{s}\gamma_\mu(1-\gamma_5)u|K^+ >$ measured in K_{l4} decay. The term Q_A, containing an axial vector electron current, contributes uniquely to K_7. This coefficient is thus a specific measure of direct CP violation.

Our preliminary estimates suggest that the direct CP-violating contributions are very small. The CP-violating coefficients in Eq. (3), normalized to K_1, are $K_4/K_1 = -5 \cdot 10^{-3}$, $K_7/K_1 = -2 \cdot 10^{-6}$, $K_9/K_1 = -0.26$. By comparison, the CP conserving terms are $K_2/K_1 = +0.25$, $K_3/K_1 = +0.15$, $K_5/K_1 = -1.7 \cdot 10^{-2}$, $K_8/K_1 = -2 \cdot 10^{-3}$. The major CP-violating effect is the asymmetry related to the coefficient K_9, which was calculated in Ref. [1]. This is dominated by indirect CP violation. Details of these results will be published shortly [2].

References

[1] L. M. Sehgal and M. Wanninger, Phys. Rev. D **46**, 1035 (1992); and Erratum.

[2] P. Heiliger and L.M. Sehgal, in preparation.

[3] E 731 Collaboration, E. Ramberg, Fermilab Report No. Fermilab-Conf-91/258, 1991.

[4] D. P. Majumdar and J. Smith, Phys. Rev. **187**, 2039 (1969).

[5] C. O. Dib, I. Dunietz and F. S. Gilman, Phys. Rev. D **39**, 2639 (1989).

---

*In Ref. [1], the factor "i" in g_{M1} was omitted, resulting in a cosine factor $\cos(\Phi_{+-} + \delta_0(m_K^2) - \bar{\delta}_1)$, and a correspondingly lower asymmetry $\approx 4\%$. See Erratum in Ref. [1].

THEORETICAL ASPECTS OF $K \to \pi e^+ e^-$ AND $K \to \pi \gamma \gamma$

T. Morozumi
Department of Physics
University of Toronto
Toronto, Ontario, Canada M4V 1A7

T. Kurimoto
College of General Education
Osaka University
Toyonaka
Osaka, Japan, 560

A. I. Sanda
Deptartment of Physics
Nagoya University
Chikusaku
Nagoya, Japan, 464

Abstract

We propose a lagrangian which may describe low energy physics of QCD (≤ 1 GeV). The lagrangian which incorporates SU(3) symmetry breaking and low lying mesons (spin ≤ 1) may give us new aspects for K meson physics. $K \to \pi e \bar{\nu}$, $K^+ \to \pi^+ e^+ e^-$ and $K_L \to \pi^0 \gamma \gamma$ are discussed by using the lagrangian and preliminary results are presented.

INTRODUCTION

Recently, it has been shown that the chiral lagrangian including vector, axial, and scalar mesons can describe low energy physics of QCD (≤ 1 GeV). This has been done by calculating the coefficients of the higher order derivative terms of non-linear σ model.[1] By incorporating those mesons in addition to pseudo-Nambu Goldstone bosons (π and K mesons), we can go beyond chiral perturbation. We have some advantage by doing this: (1) Because chiral perturbation relies on momentum expansion, we have to stay within the energy scale where the momentum expansion converges. Including the resonances explicitly, we may go further up in energy. All orders of momentum expansion are included in the scheme while chiral perturbation requires resummation of a finite series of the terms when infinite series could be important. (2) In K, η, and η' decays, those higher order derivative terms and / or effects of resonances may not be neglected, because the energy scale is 500 MeV, almost as high as the vector meson mass. Sometimes, the resonances contribute in the internal line of Feynman diagrams and if the mass of the resonances is near the external momentum scale, the resonances do not decouple and the contribution can be large. (3) The physical quantities such as charge radius of pions can be written in terms of vector mesons masses and couplings.

CHIRAL LAGRANGIAN WITH SPIN 0 AND SPIN 1 FIELDS

Keeping these things in mind, we propose a chiral lagrangian with spin 0 (scalar, pseudoscalar) and spin 1(vector, and axial vector) mesons.[2][3][4][5]

$$\begin{aligned}L =~& \frac{f^2}{4}tr(DUDU^\dagger) + \frac{1}{2}Tr(F_A F_A + F_V F_V) \\ &+ Tr(D\Sigma D\Sigma) - Ms_{ij}^2 \Sigma_{ij}\Sigma_{ji} \\ &+ \frac{F_\pi^2}{4} rTr(m^0)(U+U^\dagger) \\ &- M_A^2 Tr(A^2) + 2\frac{i}{g'} M_A^2 Tr(A\alpha_\perp) \\ &- M_V^2 Tr(V - \frac{i}{g}\alpha_\parallel)^2 \\ &+ 4\sqrt{2} C_d Tr(\Sigma \alpha_\perp \alpha_\perp) \\ &+ \sqrt{2} C_m Tr(\Sigma(\Omega m^0 \Omega + \Omega^\dagger m^0 \Omega^\dagger) r). \end{aligned}\quad(1)$$

where

$$\begin{aligned}\alpha_\perp &= \frac{\Omega D_R \Omega^\dagger - \Omega^\dagger D_L \Omega}{2i}, \\ \alpha_\parallel &= \frac{\Omega D_R \Omega^\dagger + \Omega^\dagger D_L \Omega}{2i}, \\ \Omega &= e^{i\frac{\pi}{F_\pi}}, U = \Omega^2, \\ m^0{}_{ij} &= m^0{}_i \delta_{i,j}, D_{R(L)} = d + A_{R(L)},\end{aligned}\quad(2)$$

where $m^0{}_i (i=1,2,3)$ are current quark masses for u, d, and s quarks. $A_R(A_L)$ are external gauge fields. Σ is scalar nonet and V and A are vector and axial vector nonet respectively. We include explicit SU(3) breaking effect into the scalar meson mass matrix. Covariant derivative and field strengthes are defined in the following.

$$\begin{aligned}D\Sigma &= d\Sigma + g[V,\Sigma], \\ F_{V\mu\nu} &= d_\mu V_\nu - d_\nu V_\mu + g[V_\mu, V_\nu] \\ F_{A\mu\nu} &= d_\mu A_\nu - d_\nu A_\mu + g[V_\mu, A_\nu] \\ &\quad - g[V_\nu, A_\mu]\end{aligned}\quad(3)$$

In order to calculate radiative rare decays and Non-leptonic weak decays, we use QCD corrected weak lagrangian derived by Gilman and Wise, combined with the hadronic currents of the lagrangian in Eq.(1).

$$L_W = \frac{G_F}{\sqrt{2}} s_1 c_1 c_3 \sum_i R_i(\mu) Q_i(\mu) \quad(4)$$

where Q_i is a set of four quark operators. R_i are Wilson coefficients evaluated at a certain scale $\mu \simeq 1\text{GeV}$. Q_i can be written in terms of quark bilinears. Assuming that these bilinears can be written in terms of hadronic Noether currents derived by the lagrangian in Eq.(1), we can calculate hadronic matrix elements of the weak lagrangian in Eq.(4).

CHARGE RADIUS, $\frac{F_K}{F_\pi}$, K_{e3} WEAK FORM FACTOR AND $K^+ \to \pi^+ e^+ e^-$

By using the lagrangian in Eq.(1), it is straightforward to calculate charge radius of pions.

$$<r_\pi^2> = 3\frac{1}{g^2 F_\pi^2}. \quad(5)$$

g can be determined by both pion charge radius and $\rho \to \pi\pi$ decay. By using g determined from $\rho \to \pi\pi$, we predict the charge radius.

$$\begin{aligned}<r_\pi^2>_{theory} &= 0.46(fm^2) \\ <r_\pi^2>_{experiment} &= 0.439 \pm 0.008 \end{aligned}\quad(6)$$

The charge radius of kaons will be given in ref.(5). SU(3) breaking of the decay constants of pions and kaons comes from the tadpole diagram of scalar mesons. (The term proportional to C_m in Eq.(1).) In the limit of $m^0{}_u = m^0{}_d = 0$, the SU(3) breaking of the decay constants is given by the following formula:

$$\frac{F_K}{F_\pi} - 1 = \frac{4 C_d C_m M_K^2}{M s_{33}^2 F_\pi^2} \quad(7)$$

where Ms_{33} is a $S\bar{S}$ component of scalar meson's mass matrix. Vector form factors of K_{e3} decay and $K^+ \to \pi^+ e^+ e^-$ are given by the following formulae:

$$Amp.(K^+ \to \pi^+ e^+ e^-) = \frac{G_F}{\sqrt{2}} s_1 c_1 c_3 \frac{\alpha}{\pi} F_{ee}(q^2)$$

$$(p_K + p_\pi)^\mu \bar{e}\gamma_\mu e \quad (8)$$

$$Amp.(K^+ \to \pi^0 e^+ \nu_e) = \frac{G_F}{\sqrt{2}} s_1 c_1 c_3 F_+(q^2)$$
$$(p_K + p_\pi)^\mu \bar{e}\gamma_\mu(1-\gamma_5)\nu_e \quad (9)$$

Both of the vector form factors are expanded by momentum transfer:

$$F_{ee(+)}(q^2) = 1 + \lambda_{ee(+)} q^2/m_\pi^2 \quad (10)$$

The expressions for $\lambda_{ee(+)}$ will be given in ref.(5) and will be compared with experimental results.[6]

$K_L \to \pi^0 \gamma\gamma$ AND $K_L \to \pi^0 e^+ e^-$

The measuerment of this mode is useful to estimate CP conserving contribution of $K_L \to \pi^0 e^+ e^-$. Also this mode can be a test of chiral lagrangian From theoretical side, two different mechanisms have been proposed for this decay. One is 1 loop diagrams of pions: The contribution includes the diagram $K_L \to \pi^0(\pi^{+*}\pi^{-*}) \to \pi^0 \gamma\gamma$. (We use $*$ to denote off-shell particles.) This contribution has a typical differential decay rate ; higher rate is predicted above the 2π threshold : $4m_\pi \leq m_{2\gamma} \leq m_K - m_\pi$. Because in this energy region, the outgoing π^0 is soft, then $J_{2\gamma} = 0$ is dominant for 2γ angular momentum. Thus it leads a chirality suppression for CP conserving amplitude of $K_L \to \pi^0 e^+ e^-$. Experiments support this spectrum.[7][8] Other contribution comes from the diagram such as $K_L \to \pi^{0*}(\eta^*, \eta'^*) \to \rho^*(\omega^*)\gamma_1 \to \pi^0 \gamma_1 \gamma_2$. This contribution has the higher rate for lower invariant mass of 2γ. If this contribution is large, then CP conserving contribution may not be negligible in $K_L \to \pi^0 e^+ e^-$ since $J_{2\gamma} = 2$ is dominant and chirality suppression does not occur. The experiments [7][8] suggest that this contribution in $K_L \to \pi^0 \gamma\gamma$ is lower than some of the theoretical calculations.[9] Here we focus on the second contribution and discuss what

is the problem of the previous estimate. The lagrangan in Eq.(1) requires that these vector contribution are negligible. There are two main uncertainties in the previous estimates. (a) The $K_L - \pi^0$ weak transition amplitude has some model dependence. In the previous calculation, this quantity is estimated by using the lowest order weak chiral lagrangian. $L^{CPTH}{}_W = g_8 F_\pi^2 Tr(\lambda_6 dU dU^\dagger)$. The coupling constant is fixed so that it reproduces $K \to (2\pi)_{I=0}$ amplitude. However this determination may overestimate $K_L \pi$ transition amplitude. (See below.) We will show this explicitly by using a model which dynamically enhances $\Delta I = 1/2$ amplitude. The dynamics is related to a scalar resonance thus it includes all order of the momentum expansion implicitly. In chiral perturbation language, if the higher order derivative terms are as important as the lowest order term, the fitting with the lowest order chiral lagrangian may not be appropriate. (b) We have to take into account the weak transitions not only in the initial state but in the final state and the intermediate state. In addition to the diagrams considered in the previous anlysis [9], the following contribution appears: (A) $K_L \to K^* \gamma_1 \to \rho(\omega)\gamma_1 \to \gamma_1, \gamma_2, \pi^0$. (B) $K_L \to K^* \gamma_1 \to K^0, \gamma_1, \gamma_2, \to \gamma_1, \gamma_2 \pi^0$ In the former contribution, we need $K^* - \rho(\omega)$ transition, amplitude which can be calculated by using our lagrangian. These diagrams tend to cancel each other.

Now we go back to the estimation of $K - \pi$ transition amplitude. By using the lagrangian of the lowest order chiral perturbation the $K\pi$ transition amplitude normalized by the experimental value of $A_0(K \to (\pi\pi)_{I=0})$ is given by:

$$\frac{<\pi^0|H_W|K_0>}{A_0} = 5.6 \times 10^{-2} (GeV) \quad (11)$$

On the other hand our lagrangian can reproduce A_0 amplitude by using QCD coefficients ($R_1 = -0.87, R_2 = 1.5, R_6 = -0.1$), $m_0^s = 120 MeV$ and by assuming the existence of $I = 0$ scalar meson around 700 MeV. In our chiral lagrangian, about 70% of the amplitude A_0 is explained by the penguin amplitude, which is enhanced by the momentum dependence of the $I = 0$ scalar meson propagator and strong interaction between $I = 0$ scalar meson and $(2\pi)_{I=0}$ final states.[10] Such enhancement mechanism is particularly important for $K \to 2\pi$ decays and there is no such a dynamical enhancement in $K\pi$ transition amplitude. Preliminary study shows that our model predicts the order of 0.1 smaller value for the same rato defined in Eq.(11). This excercise shows that the $K\pi$ amplitude was overestimated in the previous analysis.

CONCLUSIONS

We propose a chiral lagrangian with spin 0 and 1 mesons and give a few results like charge radius of pion, SU(3) breaking of decay constants. The final results will be given in ref.(5). The lagrangian which we propose here can be used both for semileptonic and non-leptonic K decays and may lead to an unified treatment for K decays in general.

REFERENCES

1. G. Ecker, J. Gasser, A. Pich and E.De Rafael, The role of resonances in chiral perturbation theory, *Nucl Phys.*B321 pp311-342 (1989).

2. S. Weinberg, Non-linear realizationi of chiral symmetry, *Phys. Rev.*166,pp1568-1577 (1967).

3. M. Bando, T. Kugo, S. Uehara, K. Yamawaki, T. Yanagida, Is ρ meson dynamical gauge boson of hidden local symmetry ?,*Phys. Rev. Lett* 54 pp1215-1218 (1985).

4. D. Ebert and H. Reihardt, Effective Chiral Hadron Lagrangian with Anomalies and Skyrme terms from quark flavour dynamics, *Nucl Phys*B271 pp188-216 (1986).

5. T. Kurimoto, T. Morozumi, and A. I. Sanda. In preparation.

6. C. Alliegro et. al. Study of the decay $K^+ \to \pi^+ e^+ e^-$. *Phys. Rev. Lett.* 68, pp278-281.

7. G. D. Barr et.al. Measurement of the decay $K_L \to \pi^0\gamma\gamma$. *Phys.Lett B* 284 pp 440 (1992).

8. V. Papadimitriou et.al. A measurement of the branching ratio of the decay $K_L \to \pi^0\gamma\gamma$.*Phys. Rev. D* 44 pp573-576 (1991).

9. L. M. Sehgal, CP violation in$K_L \to \pi^0 e^+ e^-$: Interference of one photon contribution and two photon exchange *Phys. Rev. D38*, pp808-813 (1989), T. Morozumi and H. Iwasaki,*Prog. Theor. Phys.* pp371-379 (1989). CP conserving contrinution in the decays $K_L \to \pi^0\gamma\gamma$ and $K_L \to \pi^0 e^+ e^-$.

10. E. P. Shabalin, $K_s \to 2\pi$ decays in a theory with an effctive chiral lagrangian, *Sov. J. Nucl. Phys.* 48 pp172 (1989), T. Morozumi, C. S. Lim and A. I. Sanda, Chiral Weak Dynamics, *Phys. Rev. Lett* 65 pp404-407 (1990).

SEARCH FOR MUON NUMBER VIOLATING DECAYS

H.K. Walter *
Paul Scherrer Institute
CH-5232 Villigen PSI
Switzerland

Abstract

The SINDRUM II experiment at PSI searches for the coherent conversion of a muon into an electron in titanium. The theoretical motivation, the SINDRUM II detector and the results of the first data-taking period in 89 are presented. In total 3.2 million events have been analyzed and no candidate for the process $\mu^- Ti \to e^- Ti$ has been found. An upper limit of $B_{\mu e} < 4.4 \cdot 10^{-12}$ (90% C.L.) is obtained for the branching ratio. The plans to lower the sensitivity by two orders of magnitude are discussed. A muonium-antimuonium conversion experiment is being set up at PSI using the refurbished SINDRUM I detector. The goal of the experiment is to improve the sensitivity for the effective coupling constant $G_{M\bar{M}}$ by a factor of ~100.

INTRODUCTION

The standard model of particle physics, although most successful and uncontradicted by experiment, is believed to be an effective low energy approximation of a more fundamental theory unbroken at higher energies. The driving force for this belief is the quest for simplicity and a minimum of ad hoc assumptions, i.e. mainly aesthetical. The most unsatisfactory aspects of the standard model is the lack of understanding of the multitude of interactions and flavors and of mass generation. In particular the origin of a triplet of families is mysterious and experiments establishing connections between these families are of fundamental importance. Neutrino masses and mixings and lepton flavor changing effective neutral currents belong to this category. The processes $\mu^+ \to e^+\gamma$, $\mu^+ \to e^+e^+e^-$, $\mu^- + (A,Z) \to e^- + (A,Z)$, $K_L \to \mu e$ and $K^+ \to \pi^+ \mu e$ are very sensitive probes since high intensity beams and unique signatures allow to measure very small branching ratios, which in turn probe very high energy scales for new physics.

MUON ELECTRON CONVERSION

Neutrinoless muon-electron conversion is especially sensitive to new physics since the constituents of the nucleus (A,Z), say Ti, add coherently to the conversion rate. The aim of our new experiment at PSI is to search for $\mu - e$ conversion in Ti with a sensitivity which is 2 - 3 orders of magnitude better than achieved previously[1].

---

*Representing the SINDRUM I and II collaborations (ETH Zürich, Univ. Zürich, Paul Scherrer Institute Villigen, RWTH Aachen, Univ. Heidelberg, Univ. Swierk, JINR Dubna, Univ. Tbilisi, Yale Univ.)

Backgrounds and beam

The experimental signature is a single monoenergetic electron emitted from the ground state of the muonic Ti-atom which has a lifetime of 329 ns[2]. The energy of the electron is $E_{e^-} = m_\mu c^2 - B_\mu - R_{nucl.}$, where m_μ is the muon mass, B_μ the binding energy of the muonic atom ground state and $R_{nucl.}$ the recoil energy of the nucleus; it equals $E_{e^-} = 104.27$ MeV. The sensitivity may be limited by potential backgrounds which are:

Beam muons: i) Muon decay from the atomic ground state (MIO), which has an endpoint energy of the signal itself. ii) Radiative capture of the muon followed by internal or external production of an asymmetric e^+e^- pair, of which the positron remains undetected. The endpoint energy depends on the initial and final state mass difference ($\Delta E = 3.99$ MeV for Ti). Both decays can be suppressed by good energy resolution, the former can be used as an independent normalization.

Beam pions: Radiative pion capture with asymmetric pair production. This background must be suppressed by a combination of low beam contamination and a prompt veto, where the former must be more and more effective, as larger intensities and smaller branching ratios are attempted. With 200 μA proton beam intensity and at a muon momentum of 100 MeV/c a μ^- flux of $\sim 10^7$ s^{-1} with a π/μ contamination of $\sim 10^{-7}$ could be achieved at the μE1 channel of PSI, of which $\sim$25% was stopped in a 140 mm $\phi \times$ 300 mm long Ti target with effective density of 0.103 g/cm^3. A prompt beam veto further reduced the pion contamination.

Beam electrons: Beam electrons and electrons from muon decay in flight scattered off the target with a momentum around 100 MeV/c could fake a $\mu - e$ conversion signal. The choice of a small beam momentum, a prompt beam veto, pulse height selection in the beam counter,

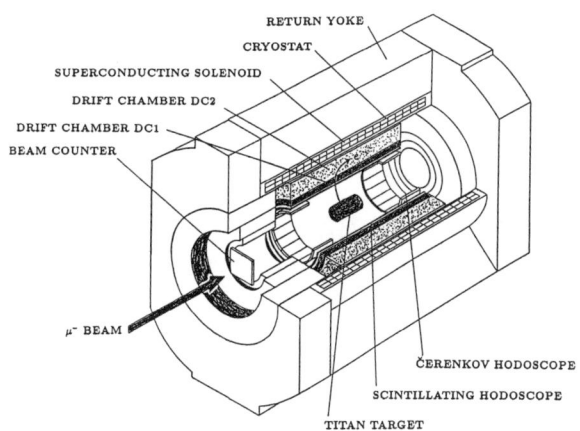

Figure 1. Cut-away view of the SINDRUM II detector (not to scale).

and a veto on a RF-correlated phase segment can suppress these electrons.

Cosmic rays: A large variety of background types is induced by cosmic rays. Electrons of $\sim$100 MeV can be produced by showers in the yoke iron, knock-out of electrons or pair production from the target or other detector material by charged cosmic particles or photons or neutrons. Active and passive shielding as well as event topologies showing associated track activity must be used to suppress this background.

The detector SINDRUM II

The setup of the spectrometer is shown in Fig. 1. A detailed description of the apparatus and its performance can be found in ref. 1.

The performance was tested using cosmic ray muons and positrons from the decay $\pi^+ \to e^+\nu_e$. A momentum resolution of $\sigma = 0.6\%$ was measured not including target absorption, which increases it to 1% for the $\mu - e$ conversion signal. The spectrometer was brought into operation a few months before the 1990-91 shutdown of the PSI accelerator. A data taking period of one month mainly in the 50 MHz mode followed and during 2.15×10^6 s about 3.2×10^6 events were written on tape. Test and calibration measurements completed this data taking period. In the following the data evaluation and results of a preliminary analysis are discussed.

Analysis of the 1989 run

First the helical trajectory of the particle which triggered the detector readout is reconstructed and translated into the momentum vector at the track origin which is defined as the point of closest approach to the detector axis. In a second step the presence of additional tracks is checked which is mainly used to recognize cosmic ray background. In a third step the information from the beam counter is analyzed to identify scattered beam electrons and electrons from pion capture.

Two thirds of the reconstructed events are scattered beam electrons which have been strongly suppressed by rejecting prompt events. The final analysis is based on the momentum spectra of different classes of events (Fig. 2). In the distribution of the delayed electrons shown in Fig. 2a no event is found in the region 98-105 MeV/c where 86% of the hypothetical conversion events are expected.

The 90% C.L. upper limit for the branching ratio thus is:

$$B_{\mu e} < \frac{2.3}{N_{stop} f_{cap} A_g^{\mu e} \times A_t^{\mu e} \times \varepsilon_{tot}^{\mu e}} \quad (1)$$

whereas the number of MIO events detected is:

$$N^{MIO} = N_{stop}(1 - f_{cap}) A_g^{MIO} A_t^{MIO} \varepsilon_{tot}^{MIO} \quad (2)$$

where N_{stop} = number of muons stopped in the target during the experiment, f_{cap} = fraction of muons captured by the Ti nucleus ($85.3 \pm 0.3\%$) [2], A_g = geometrical acceptance of spectrometer, A_t = acceptance of the online trigger conditions on the hodoscopes and chambers, ε_{tot}

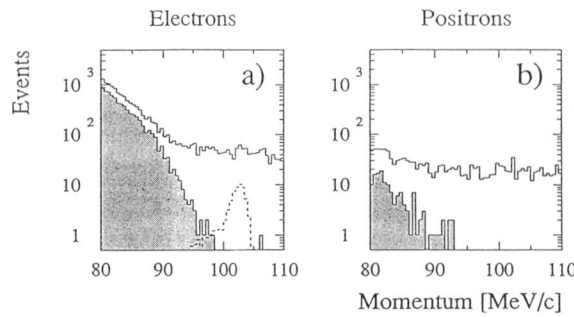

Figure 2. Momentum distributions from delayed electrons (a) and positrons (b) which originate in the target with (gray) and without (white) rejection of cosmic ray background. The dashed histogram in part a is the distribution for conversion events generated by a Monte Carlo simulation under the assumption of a branching ratio of 10^{-10}.

= product of various efficiencies, like intrinsic and reconstruction efficiencies, cuts, losses, etc. Table 1 gives the corresponding quantities which have been obtained by measurement and/or Monte Carlo simulations. The number of muon stops was evaluated using the beam counter signals and the stop fraction from measurement and Monte Carlo simulations of the beam to be $(4.24 \pm 0.35) \times 10^{12}$. This number was cross-checked using ~ 16000 measured MIO events and the theoretical branching ratio as $(4.20 \pm 0.80) \times 10^{12}$. The value $(4.23 \pm 0.32) \times 10^{12}$ was used and the corresponding upper limit on the branching ratio is:

$$B_{\mu e} < 4.4 \times 10^{-12} \text{ (90\% C.L.)}.$$

As shown by a Monte Carlo simulation the measured spectrum can be described both in shape and in rate by normal muon decay in orbit. It has been demonstrated that in the final analysis of this first run the sensitivity can be improved by about 20%.

The processes contributing background in the momentum region of interest are in order of their rate: cosmic ray events, scattered beam electrons, scattered electrons from muon decay in flight and pion induced events. One type of cosmic ray events could only be rejected at the cost of a reduction in acceptance. These are asymmetric e^+e^- pairs produced in the target by isolated high energy photons.

The events show pronounced peaks at angles which are not shielded by either concrete area walls or the return yoke and could be rejected by 16% cuts in the acceptance. In future experiments additional shielding will eliminate this problem. A rejection improvement by a factor of three of asymmetric pair production background has been achieved by introducing a second hodoscope at a radius of ~130 mm with which associated positrons can be vetoed.

In 1992 data have been taken with a Pb target and in 1993 the Ti-experiment will be continued. Various rate and efficiency improvements will allow a 10 times lower sensitivity. In addition, studies for a pion to muon convertor (PMC) at the new high intensity beam line πE5 have been carried out, which would allow to lower the final sensitivity by yet another order of magnitude.

Muonium-Antimuonium Conversion

The spontaneous conversion of muonium ($M = \mu^+e^-$) into antimuonium ($\overline{M} = \mu^-e^+$) violates lepton generation number by two units. Although not provided in the standard model, it can be expected in many extensions to it. For historic reasons the strength of the new interaction is described by a coupling constant $G_{M\overline{M}}$ from an effective $V - A$ type Hamiltonian. It is connected to the probability $P_{\overline{M}}$ of observing $\overline{M}$ atom decay from a system which started as pure M through

$$P_{\overline{M}} = (2.57 \times 10^{-5})(G_{M\overline{M}}/G_F)^2 , \quad (3)$$

where G_F is the Fermi coupling constant of the weak interaction.

The $M - \overline{M}$-conversion is of particular interest for left-right symmetric models and supersymmetric theories. A coupling strength of $\leq 10^{-5}$ has been estimated for models in which $M - \overline{M}$-oscillation is mediated by massive Majorana neutrinos[3]. Within generic SUSY models a value of $G_{M\overline{M}} \approx 10^{-2} G_F$ can be expected for slepton masses of the order of 100 GeV[4].

The recent development of intense sources of thermal M in vacuum has enabled and stimulated very sensitive searches for spontaneous $M - \overline{M}$-conversion[5]. Positive muons at subsurface momentum are stopped in a SiO_2 powder target and form M by charge exchange. The atoms diffuse to the surface and leave the target with Maxwell-Boltzmann velocity distribution at thermal ($T \approx 300\,K$) energies into the surrounding vacuum.

Muonium atoms decaying at rest in vacuum liberate fast positrons with a well known energy spectrum (Michel spectrum) which has a maximum at 52.83 MeV. The electron in the atomic shell of the M atom is left behind. These electrons have a distribution of kinetic energies which averages to 13.5 eV with a rapidly falling tail towards higher energies. In the case of a decaying antimuonium atom there is a fast (Michel) electron and a slow positron from the atomic shell. Both particles can be uniquely identified and they can be observed in coincidence.

The goal of the new experiment[6] presently under way at PSI (Fig. 3) is a sensitivity for the coupling constant of $G_{M\overline{M}} < 10^{-3}$. Compared to a recent LAMPF experiment[5] this is an improvement by a factor of $2.5 \cdot 10^4$ in rate, which mainly is achieved by a large solid angle for the electron from μ^--decay.

The refurbished SINDRUM I detector will be used as high energy spectrometer operated at 0.1 T magnetic field. The field strength is a compromise between good e^- identification and

Table 1. Acceptances and efficiencies
for $\mu - e$ conversion and MIO

	$\mu - e$	MIO
A_g	0.500	22×10^{-8}
A_t	0.754	0.269
$A = A_g \times A_t$	0.376 ± 0.022	$(5.9 \pm 0.9) \times 10^{-8}$
ε_{tot}	0.386 ± 0.032	0.441 ± 0.048
$A \times \varepsilon$	0.145 ± 0.015	$(2.60 \pm 0.49) \times 10^{-8}$

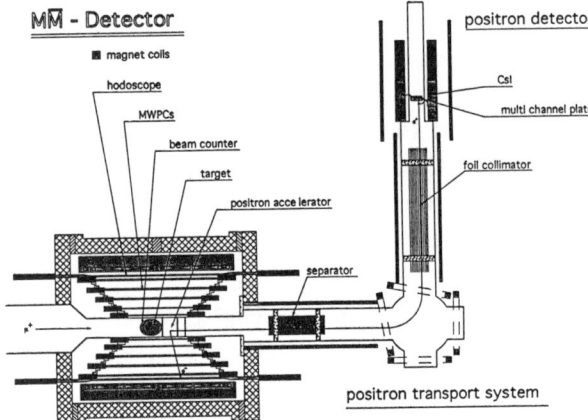

Figure 3. Setup of the new muonium-antimuonium experiment currently under way at PSI. The refurbished SINDRUM I detector is used to identify the fast Michel particles. Atomic positrons are accelerated to approximately 10 keV and guided in a magnetic field of 0.1 T to a position sensitive multichannel plate detector. 511 keV γ's from positron annihilation are detected in a set of twelve CsI crystals surrounding the MCP.

moderate suppression of the $M - \overline{M}$-conversion probability by the degeneration-lifting magnetic splitting. This detector has an acceptance for the Michel particle of about 60%. The complete apparatus is shown in Fig. 3 and described in ref. 1.

The experiment has been started in the πE3 beam area which delivers 1×10^6 μ^+/sec with 100 μA proton current at 21 MeV/c. After the realization of the high proton current plans at PSI one can expect at least 5×10^6 μ^+/sec. Within 10^7 sec of data taking one expects 0.13 background events with a 5 nsec coincidence window. Thus, the experiment will be sensitive to spontaneous $M - \overline{M}$-conversion at a level which is interesting for comparison with the model of Herczeg and Mohapatra[4].

REFERENCES

1. H.K. Walter and K. Jungmann, to be published in Z. Phys.

2. T. Suzuki et al., Phys. Rev. C35 (1987) 2212.

3. A. Halprin, Phys. Rev. Lett. 48 (1982) 1313.

4. P. Herczeg and R. Mohapatra, to be published in Phys. Rev. Lett. (LAMPF Preprint LA-UR 92-2089).

5. B.E. Matthias et al., Phys. Rev. Lett. 66 (1991) 2716.

6. PSI Exp. R-89-06.1, W. Bertl and K. Jungmann, spokesmen.

MEASUREMENT OF DIRECT CP – VIOLATION WITH THE EXPERIMENTS NA31 AND NA48 AT CERN

Burkhard Renk
CERN, Edinburgh, Mainz, Orsay, Pisa and Siegen Collaboration
Cambridge, CERN, Edinburgh, Ferrara, Mainz, Perugia, Pisa, Saclay,
Siegen, Torino and Vienna Collaboration
Universität Mainz
D 6500 Mainz

ABSTRACT

The NA31 experiment has measured the CP violation parameter ε'/ε. The result of data collected in 1988 is $\text{Re}(\varepsilon'/\varepsilon) = (1.7 \pm 1.0) \times 10^{-3}$. A preliminary result of data collected in 1989 is $\text{Re}(\varepsilon'/\varepsilon) = (2.1 \pm 0.9) \times 10^{-3}$. Combining these two results with the original result from the 1986 data set we obtain $\text{Re}(\varepsilon'/\varepsilon) = (2.3 \pm 0.7) \times 10^{-3}$, which is a more than three standard deviation evidence for direct CP violation. A new experiment NA48 is under construction which aims for a significant reduction of the statistical and the systematical errors in order to reach a combined error not exceeding 2×10^{-4}.

CP VIOLATION IN K_L DECAYS

Direct CP violation leads to different decay characteristics for K_L and K_S mesons and therefore to a deviation of the double ratio

$$R = \frac{\Gamma(K_L \to \pi^0\pi^0)}{\Gamma(K_L \to \pi^+\pi^-)} \times \frac{\Gamma(K_S \to \pi^+\pi^-)}{\Gamma(K_S \to \pi^0\pi^0)}$$

from unity. The parameter $\text{Re}(\varepsilon'/\varepsilon)$ is related to R as follows:

$$\text{Re}(\varepsilon'/\varepsilon) = (1 - R)/6.$$

The theoretical expectation [1] for $\text{Re}(\varepsilon'/\varepsilon)$ depends on the mass of the top and the strange quark and lies in the range $(0.5-3.0) \times 10^{-3}$, much smaller or even negative values are not in agreement with today's understanding of the standard model explanation of the origin of CP violation.

PREVIOUS RESULTS ON ε'/ε

The NA31 collaboration at CERN [2] and the E731 collaboration at FNAL [3] have published results of

$$\text{Re}(\varepsilon'/\varepsilon) = (3.3 \pm 1.1) \times 10^{-3}$$

and

$$\text{Re}(\varepsilon'/\varepsilon) = (-0.4 \pm 1.4 \pm 0.6) \times 10^{-3}$$

respectively. The NA31 result is based on a 27% sample of their total data, while the FNAL result is based on 20%. A result of NA31 on another 28% sample [4] was presented in 1991:

$$\text{Re}(\varepsilon'/\varepsilon) = (1.7 \pm 1.0) \times 10^{-3}$$

The E731 experiment [5] has given a preliminary result on their full data sample:

$$\text{Re}(\varepsilon'/\varepsilon) = (0.6 \pm 0.69) \times 10^{-3}.$$

MEASUREMENT OF Re(ε'/ε) BY THE NA31 EXPERIMENT

The experimental method of the NA31 experiment at CERN is alternate running with K_L and K_S beams with simultaneous recording of the two decay modes. The K_L mesons are produced at a target 245 m upstream of the detector, while the K_S beam is produced at a movable target inside the decay region of the K_L. The beam and the detector have been described in detail elsewhere [6].

The event samples collected during the 1988 and 1989 runs are shown in table 1, together they make 73% of the total NA31 data sample.

Event statistics:	1988	1989
$K_L \to \pi^0\pi^0$	110K	180K
$K_L \to \pi^+\pi^-$	290K	470K
$K_S \to \pi^0\pi^0$	560K	630K
$K_S \to \pi^+\pi^-$	1380K	1530K

Table 1: Event samples collected by the NA31 experiment.

While the background contribution in the K_S decay modes is very small, it needs to be determined carefully for the K_L sample. Table 2 summarizes the background contributions.

The uncertainty of the background subtraction is one of the major contributions to the systematic error. The dominant contributions are listed in table 3. The numbers in the table only apply to the 1988 data sample.

Mode	Source	Contribution
$K_S \to \pi^+\pi^-$	$\Lambda \to p\pi^-$	< 0.01%
	$n+A \to p\pi^-$	0.06%
$K_S \to \pi^0\pi^0$		< 0.1%
$K_L \to \pi^+\pi^-$	$K_L \to \pi e \nu$	0.49%
	$K_L \to \pi \mu \nu$	0.25%
	$K_L \to \pi\pi\pi^0$ and $K_L \to \pi\pi\gamma$	0.04%
	$n+A \to p\pi^-$	0.12%
	TOTAL	0.9% ± 0.15%
$K_L \to \pi^0\pi^0$	$K_L \to \pi^0\pi^0\pi^0$	
	1988	3.2% ± 0.16%
	1989	2.6% ± 0.17%

Table 2: Summary of the background contributions for the NA31 analysis.

$K_L \to \pi^+\pi^-$ background	0.15%
$K_L \to \pi^0\pi^0$ background	0.17%
Energy calibration and stability	0.14%
Inefficiencies	0.23%
Monte - Carlo	0.10%
Accidental correction	0.20%
Total systematic error	0.4%
Statistical error	
1988	0.4%
1989	0.3%

Table 3: Dominant uncertainties on the double ratio R for the NA31 analysis

The preliminary result achieved for the 1989 data sample is

$$R = 0.988 \pm 0.003 \pm 0.004$$

and

$$\text{Re}(\varepsilon'/\varepsilon) = (2.1 \pm 0.9) \times 10^{-3}.$$

Averaging the three NA31 results, taking into account common systematic errors, we obtain

$$Re(\varepsilon'/\varepsilon) = (2.3 \pm 0.7) \times 10^{-3},$$

which is more than a three standard deviation evidence for direct CP violation.
The world average (CERN and FNAL data) is

$$Re(\varepsilon'/\varepsilon) = (1.5 \pm 0.5) \times 10^{-3},$$

again three standard deviations away from zero.

NA48: A NEW EXPERIMENT FOR DIRECT CP VIOLATION

The search for direct CP violation is the only way to clarify the origin of CP violation, therefore a substantial improvement of the measurements is needed. Since for the CP experiments the statistical and systematical errors are of equal size, a new generation of experiments is needed with much bigger event samples and reduced systematic uncertainties.

The primary goal of the experiment NA48 at CERN [7] is to measure the CP violation parameter $Re(\varepsilon'/\varepsilon)$ with an accuracy of 2×10^{-4}. In order to gather 10 times more events than NA31, the experiment is sited in the CERN high intensity area and uses faster detector components and a much faster readout. In a running time of two to three years, 3×10^6 events of the type $K_L \rightarrow \pi^0 \pi^0$ can be accumulated, which are required for a result with 1×10^{-3} statistical uncertainty on the measurement of the double ratio R and 1.7×10^{-4} in ε'/ε.

The experiment is designed to take K_L and K_S events in both decay modes simultaneously with two nearly collinear K_L and K_S beams produced concurrently. A small fraction of the protons which penetrate the K_L target 240 m upstream of the detector is transferred via channeling in a bent cristal to a K_S target directly at the beginning of the decay region. K_L and K_S events are distinguished by tagging the protons producing the K_S component. The tagging counter needs a time resolution better than 500 ps, in the test of a prototype a resolution of 130 ps has been achieved.

Using this method, the accidental corrections and inefficiencies are identical for both event types and cancel for the double ratio. Also the energy scales are the same for K_L and K_S events. Thus, three of the five major systematic uncertainties of NA31 are reduced substantially in the new experiment.

Charged mode decays ($K_{L,S} \rightarrow \pi^+\pi^-$) are measured in a magnetic spectrometer with a central dipole magnet and two sets of large high precision drift chambers on each side. Each single plane is designed for a resolution better than 100μm, resulting in an average mass resolution for the Kaon mass of 3 MeV/c^2.

Neutral mode decays ($K_{L,S} \rightarrow \pi^0\pi^0$) are recorded in a homogeneous liquid krypton calorimeter designed for high rate capability, excellent energy and space resolution and sub-nanosecond time resolution. First prototype runs have been done in 1991 showing a time resolution of 600ps and a spatial resolution of 1 mm for electrons.

Using these two detector components, a 0.1% background contribution to both decay modes is expected. The improved mass resolution reduces the backgrounds and background uncertainties, the remaining two major contributions to the systematic error, by an order of magnitude. Thus a systematic uncertainty not exceeding 1×10^{-3} can be be achieved for the measurement of the double ratio R.

Using the higher event rate together with the excellent resolution, the experiment can also be used for the analysis of rare kaon decays. The sensitivity for rare decays which can be achieved using the data samples for the measurement of ε'/ε is 10^{-9}.

SUMMARY

The NA31 collaboration has reported a preliminary value for the CP violation parameter $\mathrm{Re}(\varepsilon'/\varepsilon)$:

$$\mathrm{Re}(\varepsilon'/\varepsilon) = (2.3 \pm 0.7) \times 10^{-3}.$$

A new experiment NA48 is under construction which aims for a measurement with an accuracy of

$$\Delta\mathrm{Re}(\varepsilon'/\varepsilon) = 0.2 \times 10^{-3}.$$

A first run of this experiment is expected in 1994.

REFERENCES

[1] Y.L. Wu, in Proceedings of this conference,.
[2] H. Burkhardt *et.al.*, Phys. Lett. B 206 (1988) 169.
[3] J. R. Patterson *et.al.*, Phys. Rev. Lett. 64 (1990) 1491.
[4] G.Barr, in Proceedings of Lepton-Photon Symposium and Europhysics Conference on High Energy Physics, Geneva, 1991, 179.
[5] B. Winstein, in Proceedings of Lepton - Photon Symposium and Europhysics Conference on High Energy Physics, Geneva, 1991, 186.
[6] H. Burkhardt *et. al.*, Nuclear Instruments and Methods A268 (1988) 116.
[7] The experiment has been approved in 1991.

The NA48 collaboration:

Univ. Cambridge: R.S. DeWolf, P.A. Elcombe, F.J. Munday, M.A. Parker, T.O. White. *CERN:* G.D. Barr, P. Buchholz, D. Cundy, N. Doble, L. Gatignon, A. Gonidec, P. Grafström, G. Kessler, A. Norton, D. Schinzel, H. Taureg, H. Wahl. *Univ. Edinburgh:* N. McKay, K.J. Peach, E. Veitch, L.L.J. Vick, A. Walker. *Univ. Ferrara:* D. Bettoni, R. Calabrese, P. Dalpiaz, J. Duclos, E. Luppi, M. Martini, F. Petrucci, F.Rossi, M. Savrie. *Univ. Mainz:* T. Beier, H. Blümer, K. Kleinknecht, F. Leber, P. Mayer, B. Renk, H.Rohrer, J. Staeck, A. Wagner, O. Zeitnitz. *Univ. Perugia:* M. Calvetti, P. Cenci, P. Lariccia, P. Lubrano, F. Tondini. *Pisa:* L. Bertanza, A. Bigi, P. Calafiura, R. Casali, R. Carosi, M.C. Carrozza, C. Cerri, R. Fantechi, I. Manelli, V. Marzulli, A. Nappi, G. Pierazzini. *CEN Saclay:* J. Alliti, J. Cheze, M. De Beer, P. Debu, B. Peyaud, B. Vallage, J. Zsembery. *Univ. Siegen:* M. Holder, A. Kreutz, M. Rost, W. Weis, R. Werthenbach. *Univ. Torino:* C. Biino, A. Ceccucci, R. Cester, F. Marchetto, E. Menichetti, A. Migliori, R. Mussa, S. Palestini, N. Pastrone, L. Pesando, M. Sozzi. *Akad.d.Wiss.Vienna:* E. Griesmayer, M. Markytan, G. Neuhofer, M. Pernicka, F. Szonsco, A. Taurok, C.E. Wolf.

Light Quark and Gluonia Spectroscopy

Parallel Session 3

Hughes-Trigg Student Center, SMU

Conveners: F. Close (Rutherford)
A. Kirk (CERN)

INTRODUCTORY REMARKS: LIGHT QUARK AND GLUONIA SPECTROSCOPY

F E Close
Rutherford Appleton Laboratory,
Chilton, Didcot, Oxon,
OX11 0QX, England

Spectroscopy is the field where the dynamics of non-perturbative QCD are revealed in most detail; the challenge is to decode the message. This area is where the non-perturbative interactions of gluons may be revealed and the strategic aim is to determine whether glueballs or other excited gluonic states, "hybrids", exist. Although lattice simulations and QCD inspired models differ in details, they tend to agree in their prediction that the lightest glueball is a scalar, $J^{PC} = 0^{++}$, and that it probably has mass between 1 and 2 GeV. It is therefore particularly exciting that, at last, there appear to be several scalar mesons being identified in experiments in this mass range. Indeed in this session we will hear rather convincing evidence for the existence of meson states "beyond" the $Q\bar{Q}$ model. If we are convinced that they are genuine, the next stage will be to devise tests that can help to determine their dynamical structure — gluonic, multiquark, meson molecules etc.

To help orient the presentations, consider first the charmonium spectroscopy spanning 3-4 GeV. $1S, 1P, 2S$ and ^3D, states are rather clear, in particular the $^3P_{0,1,2}$ lie roughly midway between $1S$ and $2S$ or ^3D, as is natural in potential models. Since 1988 there has been the possibility that a similar completion of the $1P$ levels for the light quarks is now being achieved in the 1-2 GeV region. The LASS and GAMS data (some of which will be mentioned by Ratcliff later) enabled me to summarise in ref 1 a possible set of assignments that I have updated in light of the 1992 PDG, ref 2

$^1P_1$	$b_1(1235)$	$h_1(1190)$	$h_1(1380)$	$K(1270)$
$^3P_2$	$a_2(1320)$	$f_2(1270)$	$f_2(1510)$	$K(1430)$
$^3P_1$	$a_1(1260)$	$f_1(1285)$	$f_1(1510)$	$K(1400)$
$^3P_0$	$a_0(1320)$	$f_0(1400)$	$f_0(1525)$	$K(1430)$

There is now rather good evidence for an additional 1^{++} state $f_1(1420)$ and Votruba will give news about this state. Support for the $f_0(1400)$ as the (dominantly) $^3P_0$ $q\bar{q}$ state comes from Crystal Ball and CELLO on $\gamma\gamma$ who find a substantial coupling in accord with predictions (3) for the $^3P_0$ state. Thus the $^3P_0$ scalars appear to be complete and so one must determine whether other scalar mesons, such as $f_0(975), a_0(980), f_0(1590)$ and $f_0(1720)$ are real and, if so, what their dynamical structure is. We will hear news about them during the experimental presentations. I shall then propose a strategy to aid identifying the dynamical structure of the $f_0(975)$ and $a_0(980)$ in particular.

REFERENCES

1. F.E. Close, p 432 *Proc. of Storrs APS Meeting*, World Scientific 1988.
2. Particle Data Group, Phys. Rev. **D45**, 1 (1992).
3. T. Barnes, F. Close and Z. Li, Phys. Rev. **D43**, 2161 (1991).

A STUDY OF THE E/f_1(1420) AND OF THE f_1(1285) IN CENTRAL PRODUCTION

WA76 Collaboration
Athens-Bari-Birmingham-CERN-Collège de France

Presented by M.F. Votruba, Birmingham, U.K.

T.A. Armstrong[4a], M. Benayoun[5], W. Beusch[4], I.J. Bloodworth[3], J.N. Carney[3],
C.J. Dodenhoff[3], C. Evangelista[2], B.R. French[4], B. Ghidini[2], M. Girone[2],
A. Jacholkowski[4], J. Kahane[5], J.B. Kinson[3], A. Kirk[4], K. Knudson[4], V. Lenti[2],
Ph. Leruste[5], A. Malamant[5], J.L. Narjoux[5], F. Navach[2], A. Palano[2],
N. Redaelli[4b], L. Rossi[4c], M. Sené[5], R. Sené[5], M. Stassinaki[1],
G. Vassiliadis[1], O. Villalobos Baillie[3], M.F. Votruba[3] and G. Zito[2]

1) Athens University, Nuclear Physics Department, Athens, Greece
2) Dipartimento di Fisica dell'Università and Sezione INFN, Bari, Italy
3) University of Birmingham, Physics Department, Birmingham, U.K.
4) CERN, European Organization for Nuclear Research, Geneva, Switzerland
5) Collège de France, Paris, France
a) Present address: Pennsylvania State University, University Park, USA
b) Present address: INFN and Dipartimento di Fisica, Milan, Italy
c) Present address: INFN and Dipartimento di Fisica, Genoa, Italy

Abstract

In central production the WA76 experiment, at the CERN Omega Spectrometer, observes the f_1(1285) decaying to $K_s^0 K^{\pm} \pi^{\mp}$, $\eta \pi^+ \pi^-$, $\pi^+ \pi^- \pi^+ \pi^-$ and $\rho^0 \gamma$.

The E/f_1(1420) is only observed to decay to $K^* \bar{K}$. A spin-parity analysis of the $K_s^0 K^{\pm} \pi^{\mp}$ system combined with the observation of E/f_1(1420)$\rightarrow K_s^0 K_s^0 \pi^0$ determines the quantum numbers of the E/f_1(1420) to be $I^G(J^{PC}) = 0^+(1^{++})$.

The evidence for non-$q\bar{q}$ mesons, such as glueballs, hybrids or multiquark states, can only come from the comparison of light mesons produced by several dynamical sources. One of these is the Double Pomeron Exchange process. The WA76 experiment, at the CERN Omega Spectrometer, with its double-particle exchange trigger at incident momenta of 85 GeV/c and 300 GeV/c, provides a wealth of data on central production ($|x_F| < 0.15$) of mesons decaying to $\pi\pi, K\bar{K}, K\bar{K}\pi, \eta\pi\pi, 4\pi, \rho\gamma$ and other exclusive channels. For a recent comprehensive review see ref.[1]

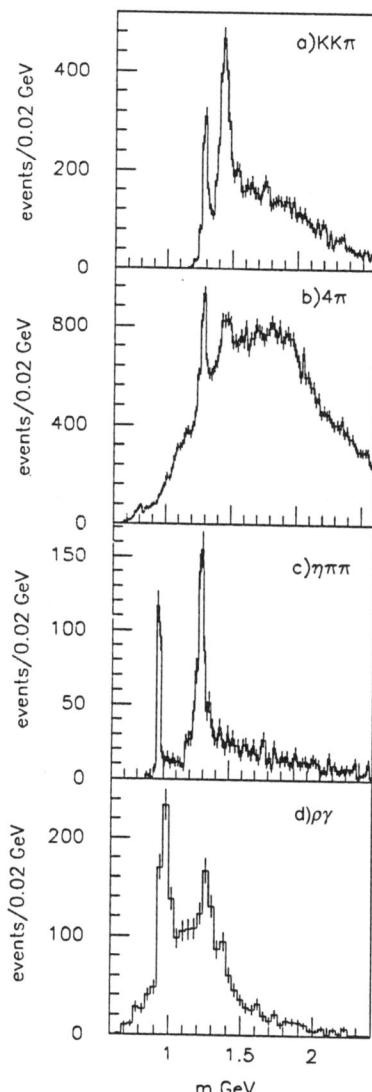

Figure 1. The effective mass spectra for reactions (1) to (5).

In this paper we concentrate on the 1.4 GeV mass region, studying the $E/f_1(1420)$ meson and comparing it with the well known $f_1(1285)$ state. Energy-momentum balancing events were isolated for the following reactions:

$$pp \to p_f(K_s^0 K^{\pm} \pi^{\mp}) p_s \qquad (1)$$

$$\pi^+ p \to \pi^+ (K_s^0 K^{\pm} \pi^{\mp}) p_s, \text{ (at 85 GeV/c)} \qquad (2)$$

$$pp \to p_f(2\pi^+ 2\pi^-) p_s \qquad (3)$$

$$pp \to p_f(\pi^+ \pi^- \eta) p_s \qquad (4)$$

$$pp \to p_f(\pi^+ \pi^- \gamma) p_s \qquad (5)$$

$$pp \to p_f(K_s^0 K_s^0 \pi^0) p_s \qquad (6)$$

at 300 GeV/c, where the subscripts f, s indicate the fastest and the slowest particle in the laboratory, respectively.

In fig.1(a) the combined $K_s^0 K^{\pm} \pi^{\mp}$ mass spectrum from reactions (1) and (2) shows prominent signals of $f_1(1285)$ and $E/f_1(1420)$. Fig. 1(b) shows the 4π mass spectrum for the reaction (3). The $f_1(1285)$ is seen here together with an enhancement at 1450 MeV. This enhancement is not due to the 4π decay of the $f_1(1420)$; it is a new state, we have called it the X(1450) meson having mass 1449 ± 4 MeV, $\Gamma = 78 \pm 18$ MeV, decaying via $a_1(1260)\pi$, I = 0, wave. We determined its quantum numbers as $J^{PC} = 2^{++}$ or 1^{-+}, see ref.2. Fig. 1(c) shows the $\eta \pi^+ \pi^-$ mass spectrum for

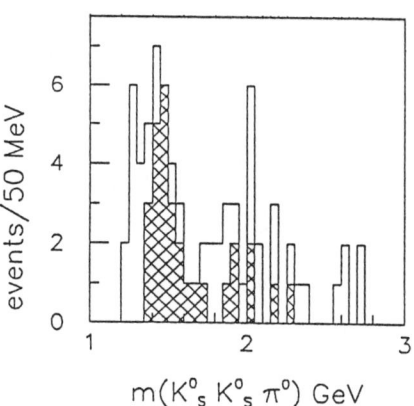

Figure 2. The effective mass spectrum for reaction (6).

reaction (4). A strong signal of $f_1(1285)$ is present; there is no signal in the $f_1(1420)$ mass region. Fig. 1(d) shows the $\rho^0\gamma$ effective mass distribution for reaction (5) where signals for the $\eta'(958)$ and the $f_1(1285)$ are visible. Finally, fig. 2 shows evidence for a $K^*\overline{K}$ signal (shaded) of the E/$f_1(1420)$ in the $K_S^0 K_S^0 \pi^0$ mass spectrum from reaction (6); for more detail see ref.[3]. Thus we see that the $f_1(1285)$ decays into $K_S^0 K^\pm \pi^\mp$, $\eta \pi^+ \pi^-$, $2\pi^+ 2\pi^-$ and $\rho^0 \gamma$ channels. The E/$f_1(1420)$ is only seen to decay into $K_S^0 K^\pm \pi^\mp$ and $K_S^0 K_S^0 \pi^0$ states.

In the spin-parity analyses we use the Dalitz plot analysis with Zemach tensors and standard isobar model assumptions. In the low mass region of the $\eta\pi\pi$ system we have considered the following intermediate states: $\delta/a_0(980)\pi$, $\varepsilon/f_0(1400)\eta$ and $\rho(770)\eta$, with waves up to spin 1. Interference was allowed between waves having the same spin-parity. The results of the max. likelihood fits are shown in fig. 3. The $J^{PC} = 1^{++}\delta/a_0\pi$ is the dominant wave and peaks at the $f_1(1285)$ mass. There is no evidence for E/$f_1(1420)$ production; this agrees with our analysis of the centrally produced $K_S^0 K^\pm \pi^\mp$ system, which gives a contribution consistent with zero in this mass region. The $J^{PC} = 0^{-+}$ waves are small and do not show resonant behaviour.

In the low mass region of the $K_S^0 K^\pm \pi^\mp$ system we have considered the $\delta/a_0(980)\pi$ and $K^*\overline{K}$ intermediate states with $J^{PC} = 0^{-+}$, 1^{++}, 1^{-+}, 1^{+-}. The results of the max. likelihood fits are shown in fig. 4. The $J^{PG} = 1^{++}$ wave dominates and shows a clear resonant signal at the E/$f_1(1420)$ mass. The fitted curve shown in fig. 4 corresponds to m = 1430 ± 4 MeV and Γ = 58 ± 10 MeV. There is a small $J^{PG} = 0^{-+}$ $K^*\overline{K}$ contribution. A fit with a simple B.-W. gives m = 1425 ± 13 MeV, Γ = 71 ± 31 MeV. There is a relatively small 1^{+-} contribution which is non-resonant. The absence of any $\delta/a_0(980)\pi$ contribution in the 1.4 GeV mass region agrees well with the observed absence of a signal in the $\eta\pi\pi$ system.

In the $f_1(1285)$ mass region of the $K_S^0 K^\pm \pi^\mp$ system the fit (not shown) gives a dominant wave $1^{++}\delta/a_0(980)\pi$ at (70 ± 6)%, with 30% phase space background. We thus confirm production in the central region of the E/$f_1(1420)$ with the only observed decay mode being $K^*\overline{K}$ and determine its quantum numbers to be $J^{PG} = 1^{++}$. This result, combined with the positive C-parity assignment based on the evidence of E/$f_1(1420)$ signal in the $K_S^0 K_S^0 \pi^0$

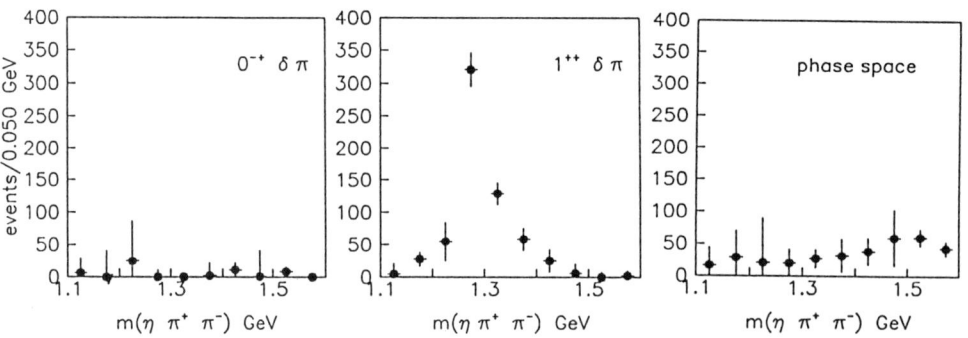

Figure 3. Dalitz plot analysis of the $\eta\pi^+\pi^-$ system.

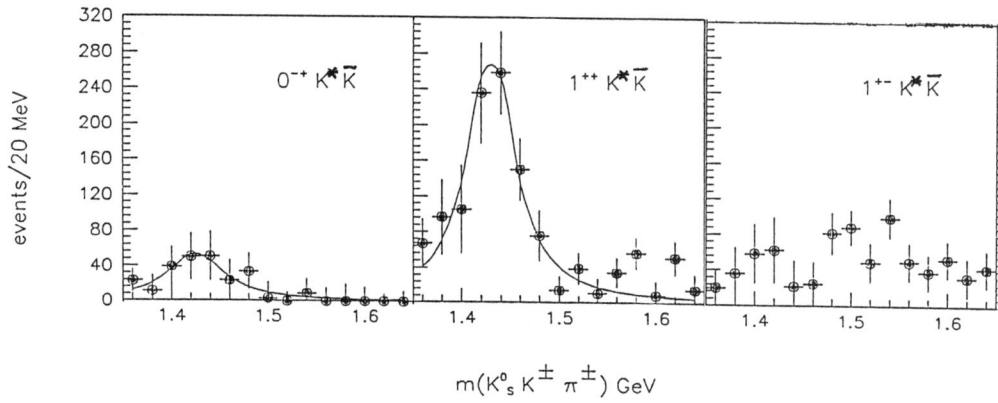

Figure 4. Dalitz plot analysis of the $K_s^0 K^\pm \pi^\mp$ system.

channel, allows us, using the relation $G = C(-1)^I$ to determine the isospin to be zero. Therefore the quantum numbers of $E/f_1(1420)$ are $I^G(J^{PC}) = 0^+(1^{++})$. We also set[3] the upper limits for the

B.R.$(E/f_1(1420) \to \eta\pi\pi) < 0.1$ (95%c.l.) and

B.R.$(E/f_1(1420) \to \rho^0\gamma) < 0.08$ (95%c.l.).

We note that the cross-section for the centrally ($|x_F| < 0.15$) produced $f_1(1285)$ is found to decrease with the incident momentum while the cross-section for the $E/f_1(1420)$ appears to remain constant,

85 GeV/c: $\sigma(f_1(1285)) = 230 \pm 70$ nb,
  $\sigma(f_1(1420)) = 400 \pm 120$ nb,
300 GeV/c: $\sigma(f_1(1285)) = 135 \pm 40$ nb,
  $\sigma(f_1(1420)) = 400 \pm 110$ nb.

In conclusion, we have studied the centrally produced $f_1(1285)$ meson with $J^{PC} = 1^{++}$, decaying mainly through the $\delta/a_0(980)\pi$ into $K_s^0 K^\pm \pi^\mp$, $\rho^0\gamma$ and $2\pi^+ 2\pi^-$ channels. A spin-parity analysis of the $E/f_1(1420)$ state gives $I^G(J^{PC}) = 0^+(1^{++})$; this meson is seen only in the $K^*\bar{K}$ channel in the $K_s^0 K^\pm \pi^\mp$ and $K_s^0 K_s^0 \pi^0$ final state. The $E/f_1(1420)$ appears to be an extra resonance which does not easily fit into quark model[1,3]. It is interesting to understand what it really is: a hybrid meson, a $K^*\bar{K}$ molecule or a multiquark state.

REFERENCES

1. A. Palano, Resonance Production in Central Hadronic Collisions, invited talk to Photon-Photon'92, IXth Int. Workshop, University of California, San Diego in La Jolla, March 1992; CERN Prepr.PPE 92-93, May 1992.

2. T.A. Armstrong et al., Evidence for new states produced in Central region in the reaction pp $\to$ p$_f(2\pi^+ 2\pi^-)$p$_s$ at 300 GeV/c, Phys. Lett. B228 pp.536-540, (1989).

3. T.A. Armstrong et al., Further study of the $E/f_1(1420)$ meson in Central Production, CERN Prepr. PPE 92-94, May 1992; to be publ. in Z.Phys.C.

NEW LIGHT MESONS FROM THE LEAR CRYSTAL BARREL

C. Amsler
Physik-Institut
Universität Zürich
Schönberggasse 9
CH-8001 Zürich, Switzerland
(On behalf of the Crystal Barrel Collaboration)

Abstract

Two new isoscalar mesons are observed in $\bar{p}p$ annihilation at rest at LEAR: a tensor meson decaying to $\pi^0\pi^0$ with mass 1515 MeV in the annihilation channel $\bar{p}p \to 3\pi^0$ and a scalar meson decaying to $\eta\eta$ with mass 1560 MeV in the channel $\eta\eta\pi^0$. A search for other scalars in the channel $\omega\pi^0\pi^0$ is reported.

INTRODUCTION

The Crystal Barrel collaboration is searching for new light mesons with masses below 2 GeV by studying low energy $\bar{p}p$ annihilation at LEAR at rest and in flight up to a $\bar{p}$ momentum of 2 GeV/c. Apart from detecting charged annihilation products (pions and kaons), the apparatus is capable of detecting and reconstructing final states with a high γ multiplicity from π^0, η, η' and ω decays. This paper reports on recent results for $\bar{p}p$ annihilation at rest into final states involving only photons.

Antiprotons with a momentum of 200 MeV/c are stopped in a 4 cm long liquid hydrogen target. A silicon counter in front of the target monitors the incident $\bar{p}$ flux (typically 10^4 $\bar{p}$/s). The target is surrounded by two cylindrical proportional wire chambers, a jet driftchamber and a barrel shaped calorimeter consisting of 1380 CsI(Tl) crystals with photodiode readout. The assembly is located in a solenoidal magnet providing a homogeneous field of 1.5 T parallel to the incident $\bar{p}$ beam. A more detailed description of the apparatus can be found in ref.[1]. The data presented here were collected by vetoing charged final states and are subsamples of the 20 million zero prong annihilations that were collected so far, corresponding to $\sim 5 \times 10^8$ annihilations at rest.

A TENSOR MESON AT 1515 MEV

Following the request from the conveners, I shall briefly summarize our results on the reaction $\bar{p}p \to 3\pi^0$ published last year [2]. The $3\pi^0$ Dalitz plot can be described by contributions from $f_2(1270)$ and $f_2(1810)$ decaying to $2\pi^0$, by the S-wave $2\pi^0$ final state interaction [3] and by a 2^{++} isoscalar resonance $f_2(1520)$ with mass 1515 and width 120 MeV, decaying to $\pi^0\pi^0$. The new state $f_2(1520)$ is produced by annihilation from the $\bar{p}p$ atomic orbitals with roughly equal contributions from $^1S_0$, $^3P_1$ and $^3P_2$. Figure 1 shows the $2\pi^0$-invariant mass distribution. Assuming a branching ratio of 0.76 ± 0.23 % for the $3\pi^0$ final state [4], one finds a branching ratio for $\bar{p}p \to f_2(1520)\pi^0 \to 3\pi^0$ of (0.20 ± 0.07) %.

This state has been confirmed in the same channel but at higher energies where more phase space is available [5]. Formerly called AX, it had been observed earlier by the As-

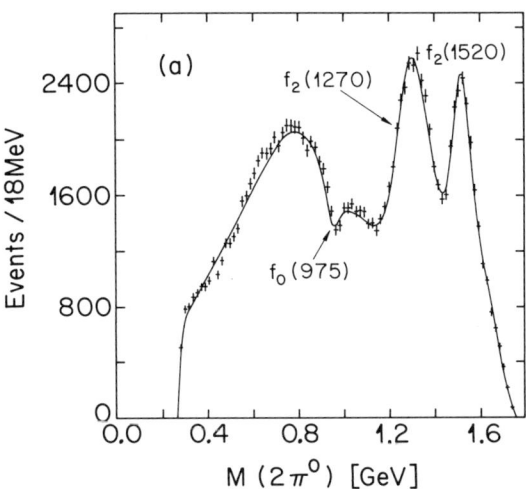

Figure 1. $2\pi^0$-invariant mass distribution showing the new $f_2(1520)$ meson. The full curve gives the result of the Dalitz plot fit.

terix collaboration at LEAR in the final state $\pi^+\pi^-\pi^0$ in hydrogen gas [6]. Its mass was slightly higher (1565 MeV). However, the analysis of the charged channel performed by Asterix is limited by the complexity of the interfering $\rho^+\rho^-$ bands under AX. The production and decay rates are consistent for both experiments. AX has also been confirmed by the Obelix experiment in $\bar{n}p \to \pi^-\pi^+\pi^+$ [7].

The meson $f_2(1520)$ is not $f_2'(1525)$ since the latter is mainly decays to $\overline{K}K$ while no obvious signal is observed in the channel $K\overline{K}\pi^0$. The 2^{++} $\bar{q}q$ multiplet is already complete. The first isoscalar radial excitation is predicted around 1820 MeV [8] for which the meson $f_2(1810)$ introduced above is a candidate. Hence $f_2(1520)$ is probably exotic. Its most likely nature is a multiquark meson or a deeply bound $\overline{N}N$ state [9]. The ratio of AX$\to \eta\eta$ over AX$\to \pi^0\pi^0$ decay rates, obtained from the analyses of the $3\pi^0$ and $2\eta\pi^0$ channels (next section) is about 10%, in fair agreement with predictions from the $\overline{N}N$ model [9].

A NEW SCALAR MESON AT 1560 MEV

The channel $\bar{p}p \to \pi^0\eta\eta \to 6\gamma$ has been studied at rest with a sample of 1.1×10^7 zero prong events [10]. A 7C kinematic fit leads to 22492 $\pi^0\eta\eta$ events. The Dalitz plot and the invariant mass projections are shown in Fig. 2. The horizontal and vertical bands in the Dalitz plot are due to $a_0(980) \to \eta\pi^0$ (Fig. 2a). One also observes two diagonal bands which correspond to states decaying into $\eta\eta$. Both states are clearly seen in the $\eta\eta$ mass projection (Fig. 2b). They are also observed when one η decays to $3\pi^0$ (10γ final state, Fig. 2c). An amplitude analysis of the Dalitz plot distribution has been performed [10]. The peak at 1400 MeV is identified with $f_0(1400) \to \eta\eta$ and the peak around 1560 MeV with a new scalar $X(1560) \to \eta\eta$ with mass 1560 ± 25 MeV and width 245 ± 50 MeV. Taking into account all η decay modes, we estimate that the fraction of all annihilations into $X(1560) \to \pi^0\eta\eta$ is about 0.1%.

$X(1560)$ may be identical to $f_0(1590)$ reported by the GAMS collaboration [11]. We have however searched for the $\eta\eta'$ decay mode of this state by analyzing the final state $\eta'\eta\pi^0(\eta' \to \eta\pi^0\pi^0)$, leading to 10$\gamma$ and set an upper limit for the ratio $\eta\eta'/\eta\eta$ of 0.23 [12], in contrast with GAMS which reported 2.7 ± 0.8. No signal for $\pi^0\pi^0$ or $K\overline{K}$ decay has been reported so far, but an analysis of the channel $K^+K^-\pi^0$ and a reanalysis of our $3\pi^0$ data allowing for a $2\pi^0$ scalar contribution in this mass region are in progress.

Two isospin zero $\bar{q}q$ scalar mesons are already known ($f_0(1400)$ and $f_0(1525)$) and the first radial excitation is expected (and already reported) around 1720 MeV. This suggests that $X(1560)$ is exotic. It could correspond to the predicted 0^{++} ground state glueball predicted in this mass region by lattice gauge theories.

550 New Light Mesons from the Lear Crystal Barrel

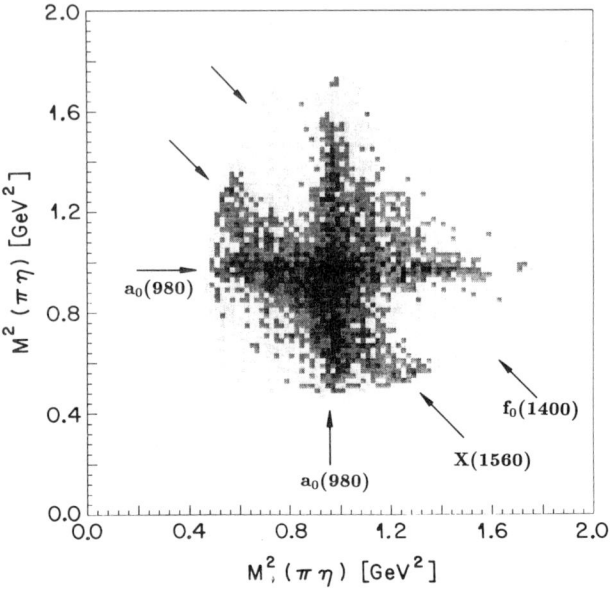
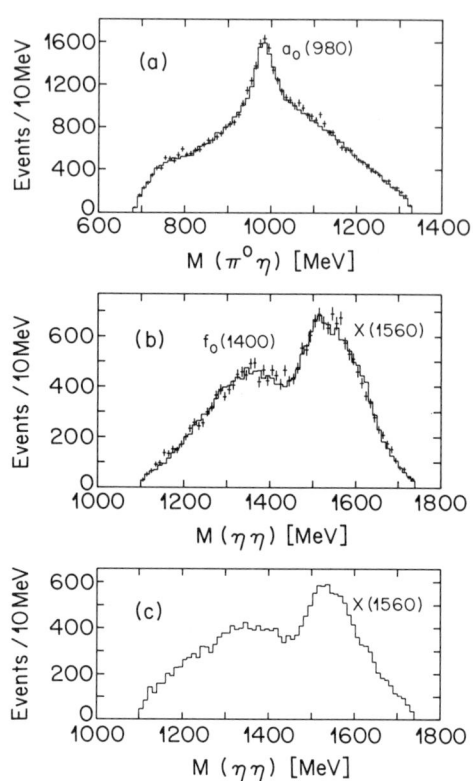

Figure 2. Dalitz plot and invariant mass projections for the channel $\bar{p}p \to \pi^0\eta\eta \to 6\gamma$ showing a new meson X(1560) decaying to $\eta\eta$; (a) $\pi^0\eta$-invariant mass distribution; (b) $\eta\eta$ mass distribution; (c) $\eta\eta$ mass distribution when one η decays to $3\pi^0$ (10γ final state).

SEARCH FOR OTHER SCALARS

A natural parity band of isoscalar 0^{++}, 1^{--} and 2^{++} $\overline{N}N$ bound states is predicted by potential models of the $\overline{N}N$ interaction. As mentioned above, $f_2(1520)$ is a candidate for the 2^{++} state. In this case, its 0^{++} partner would be expected around 1.1 GeV [9]. Bubble chamber experiments have indeed reported evidence for such a 0^{++} meson at 1.1 GeV decaying to $\pi^+\pi^-$ in the channels $\bar{p}p \to \pi^+\pi^-\omega$ [13] and $\bar{p}n \to \rho^-\pi^+\pi^-$ [14].

We have analyzed the channel $\bar{p}p \to \pi^0\pi^0\omega$ ($\omega \to \pi^0\gamma$) leading to 7 photons. Compared to $\pi^+\pi^-\omega$, the analysis of $\pi^0\pi^0\omega$ is greatly simplified by the absence of the strong $\rho\omega$ contribution. Figure 3 shows the $\pi^0\gamma$-mass distribution from a sample of 2.5 million zero prong events. The $\omega\pi^0\pi^0$ Dalitz plot and the corresponding projections are shown in Fig. 4 (18'951 events).

A reasonable Dalitz plot fit is obtained

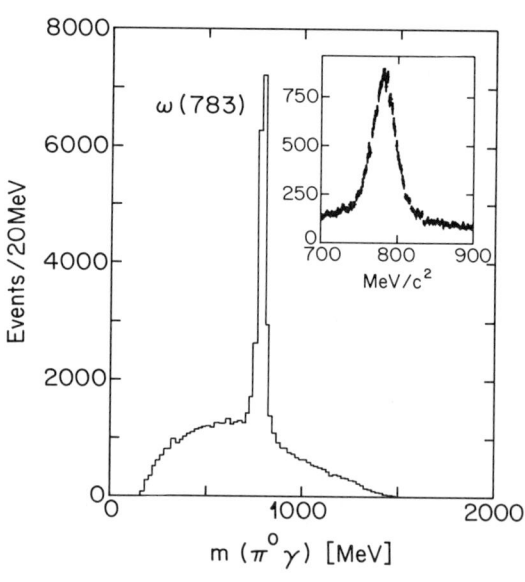

Figure 3. $\pi^0\gamma$-invariant mass distribution (3 entries/event) for $\bar{p}p$ annihilation into $3\pi^0\gamma$.

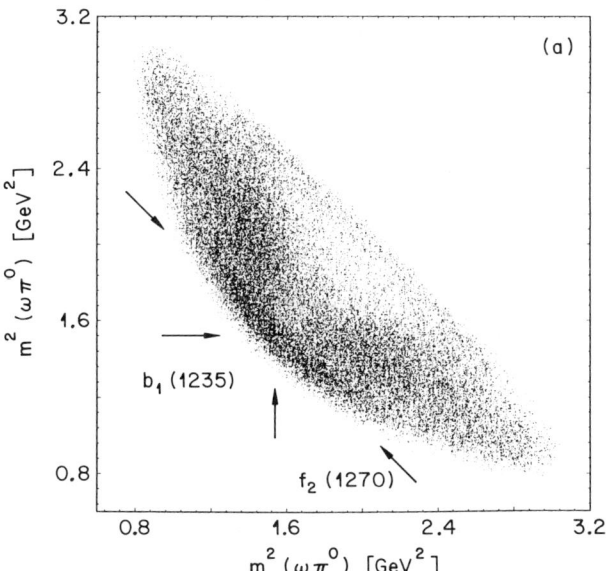
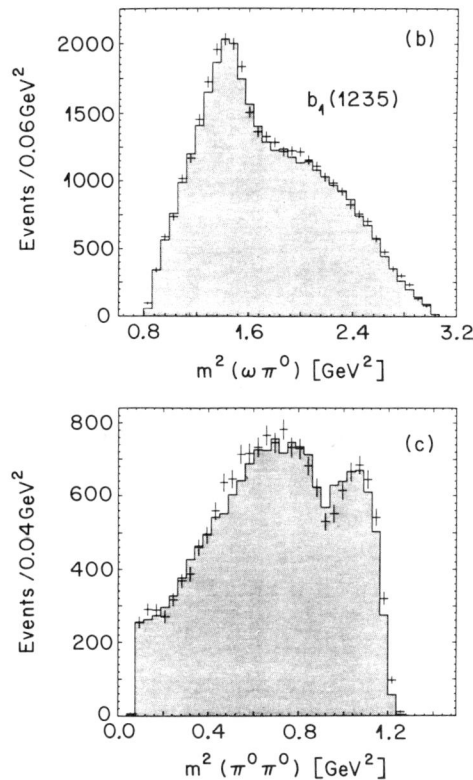

Figure 4. Dalitz plot and invariant mass projections for the channel $\bar{p}p \to \pi^0\pi^0\omega$. The data are shown by crosses, the best fit by histograms.

with contributions from the intermediate resonance $b_1(1235) \to \pi^0\omega$ which is clearly observed in the $\pi^0\omega$ mass distribution, from $f_2(1270) \to \pi^0\pi^0$ and with the $\pi^0\pi^0$ S-wave amplitude [3]. A peak in the $\pi^0\pi^0$ mass plot is indeed observed around 1.1 GeV. The amplitude analysis shows, however, that this peak is 2^{++} and can be ascribed to the tail of the broad $f_2(1270)$ meson produced in $\bar{p}p \to f_2(1270)\omega$. No scalar meson around 1100 MeV is required. Additional resonances decaying to $\omega\pi^0$, in particular radial excitations of the ρ or the exotic $C(1480)$ [15] are not required.

REFERENCES

1. E. Aker et al., NIM A321, 69, (1992)
2. E. Aker et al., PL B260, 249, (1991)
3. K.L Au et al., PR D35, 1633, (1987)
4. S. Devons et al., PL B47, 271, (1973)
5. G.A. Smith, this parallel session
6. B. May et al., PL B225, 450, (1989), Z. Phys. C46, 191, 203, (1990)
7. A. Adamo et al., PL B287, 368, (1992)
8. S. Godfrey and N. Isgur, PR D32, 189, (1985)
9. C.B. Dover et al., PR C43, 379, (1991)
10. C. Amsler et al., PL B291, 347 (1992)
11. D. Alde et al., PL B201, 160, (1988)
12. I. Augustin et al., Second Biennial Conf. on Low-Energy Antiproton Physics, Courmayeur, Italy, September 14–19, 1992
13. R. Bizzarri et al., NP B14, 169, (1969)
14. I. Daftari et al., PRL 58, 859, (1987)
15. S. Bityukov et al., PL B188, 383, (1987)

NEUTRAL LIGHT QUARK SPECTROSCOPY IN ANTIPROTON-PROTON ANNIHILATION AT 3000 < $\sqrt{s}$ < 6000 MEV

Gerald A. Smith
Department of Physics
Penn State University
303 Osmond Lab
University Park, PA 16802

For the
E760 (Fermilab) Collaboration

Abstract

Six photon final states, produced in proton antiproton annihilations at $\sqrt{s}$ = 2980 and 3097 MeV, have been studied in Fermilab experiment E760. A meson which decays into $2\pi^0$ with mass 1510±7 MeV and width 104±20 MeV has been observed. We identify it with the $f_2(1520)$ which has been reported recently by the Crystal Barrel Collaboration at LEAR in $\bar{p}p$ annihilations at rest.

INTRODUCTION

Included among investigations of charmonium states produced in antiproton-proton annihilations in the Fermilab Antiproton Accumulator Ring (Experiment E-760), we have studied multiphoton final states which constitute background to all-neutral charmonium decays. These investigations exploit the unique characteristics of the experiment, including the point-like interaction of the beam and gas-jet target and the fine granularity of the lead-glass calorimeter. In combination, these offer the possibility of measuring photon energies and directions with the same precision as charged particles. The E-760 detector has been described in detail in previous publications.[1,2]

The data reported in this paper were taken in conjunction with scans at the J/ψ and η_c charmonium resonances. Since we experienced some problems decelerating a large stack of antiprotons to the formation energies of these resonances, which are below the Accumulator transition energy, we performed these scans with the end-tail of a stack, typically with < 2×10^{11} circulating antiprotons. For each scan we accumulated at these energies between 200 nb^{-1} and 500 nb^{-1} of luminosity, integrated over 30 to 50 hours of data-taking.

Each event was kinematically fitted to a 6 photon hypothesis with the constraints of total four-momentum conservation. Two photon invariant masses were then calculated for all 15 pairs. Each two-photon pair with an invariant mass in the range 70-200 MeV (449-649 MeV) was identified as coming from a π^0 (η) decay into two photons. Events were then kinematically fitted to:

$$\bar{p} + p \rightarrow \pi^0 + \pi^0 + \pi^0 \quad (1)$$

$$\bar{p} + p \rightarrow \pi^0 + \pi^0 + \eta \quad (2)$$

$$\bar{p} + p \rightarrow \pi^0 + \eta + \eta \quad (3)$$

$$\bar{p} + p \rightarrow \eta + \eta + \eta \quad (4)$$

Reactions were selected if the confidence level exceeded 10%, the value above which the probability distributions are flat. In cases when there was more than one possible assignment, the hypothesis with the highest probability was chosen. It was found that 9.2% of events fitting reaction (1) were ambiguous with one of reactions (2-4). An additional 0.9% were ambiguous among other π^0 assignments internal to reaction (1). A total of 1.25×10^6 (0.25×10^6) examples of reaction (1) have been identified at 2980 (3097) MeV.

RESULTS

In Figure 1 we present Dalitz plots for reaction (1) at (a) 2980 and (b) 3097 MeV. The number of entries has been reduced to avoid saturation effects which limit the eye's ability to discern structures. One can see very clearly the presence of the $f_2(1270)$ and narrow bands centered at 1510 MeV on both plots. We emphasize that the 1510 MeV bands are seen in both regions of overlap and non-overlap with the $f_2(1270)$. Therefore, it is not an interference effect. There appears to be an intensification of events near 4.0 GeV^2 in the $f_2(1270)$ bands, which may be interference between the $f_2(1270)$ and another state near 2 GeV.

We show in Figure 2 uncorrected (shaded) and Monte Carlo acceptance corrected (all) mass projections at (a) 2980 and (b) 3097 MeV. The acceptance was calculated by generating events in a four dimensional space including the mass, production angle and two decay angles of the di-pion system. Again, $f_2(1270)$ and 1510 MeV enhancements are clearly seen. Fits to the mass region 1.1-1.6 GeV in the corrected data using two Breit-Wigner resonant structures and a polynomial background give a mass and width for the $f_2(1270)$ (Table 1) which are in good agreement with established values.[3] A variety of background fits have been attempted. As expected, the χ^2 decreases as the complexity of the background is allowed to increase to third order (dashed curves). However, we find that only the widths are weakly (± 20%) affected by this procedure. A large, broad enhancement is also seen centered near 1950 MeV with a width of approximately 200 MeV. It is possible this is (a) the $f_4(2050)$,[3] or (b) a reflection of the $f_2(1270)$, or (c) a new object, or (d) some combination of the above. This mass region will be the subject of further study.

In a recent experiment at LEAR,[4] the Crystal Barrel (CB) Collaboration reported the existence of the $f_2(1520)$,[3] a $2\pi^0$ meson state with M = 1515±10 MeV, Γ = 120±10 MeV, and J^{PC} = 2^{++}, in reaction (1) at rest ($\sqrt{s}$ = 1876 MeV). Because of the excellent agreement of our measurement of the mass and width of the 1510 MeV object (Table 1), we identify it with the $f_2(1520)$. In our Dalitz plot (Figure 1) and mass projections (Figure 2) we observe the $f_2(1270)$ in agreement with the CB, but

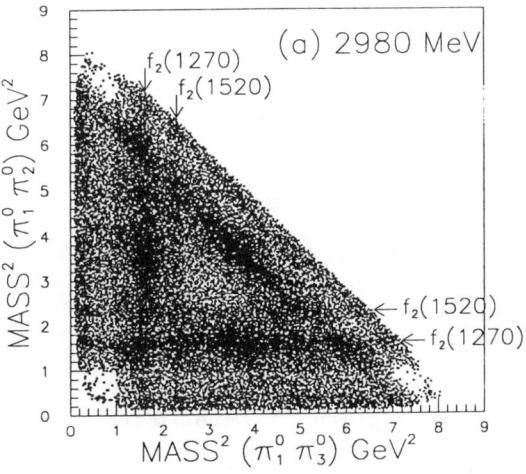 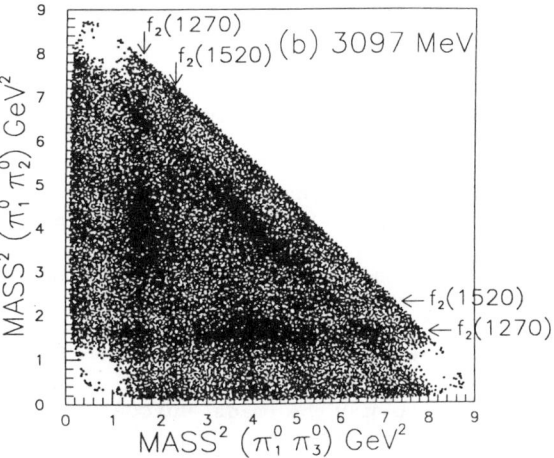

Fig. 1. Dalitz plots for reaction (1) at (a) 2980 and (b) 3097 MeV c.m. energy.

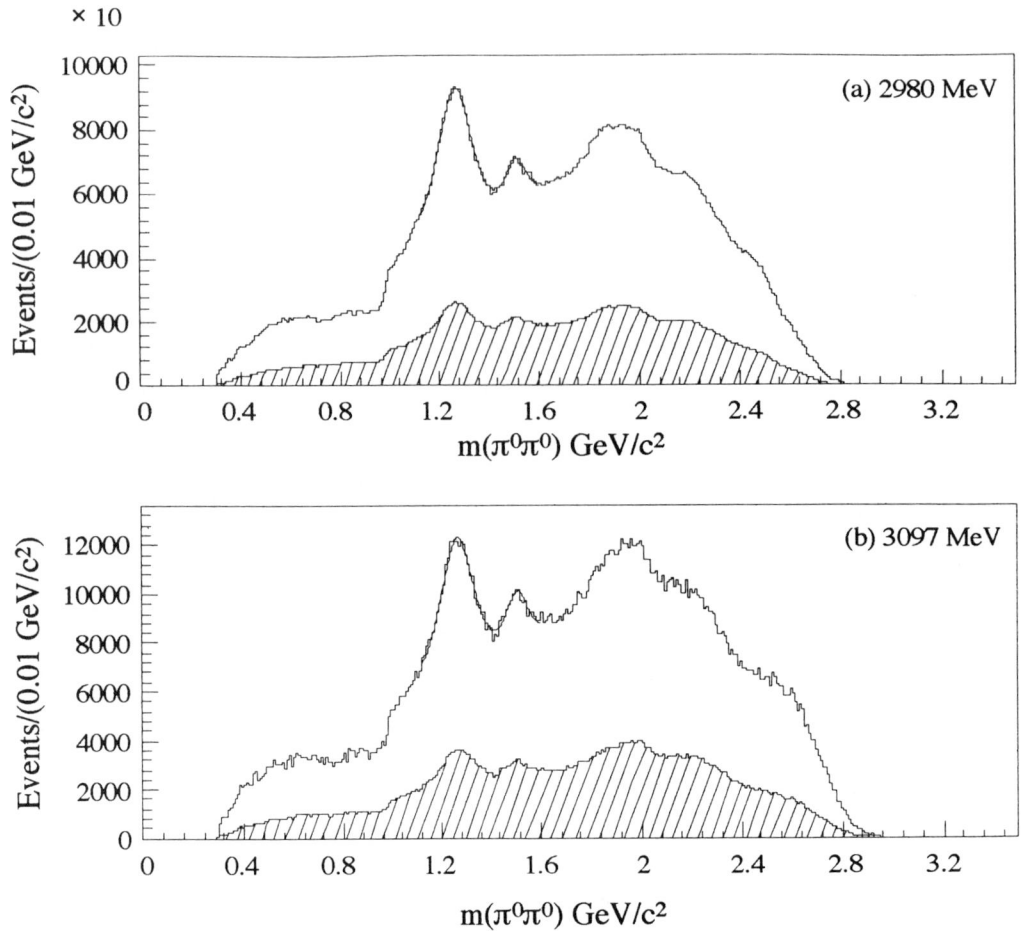

Fig. 2. Uncorrected (shaded) and Monte Carlo acceptance corrected (all) dipion mass projections for reaction (1) at (a) 2980 and (b) 3097 MeV c.m. energy.

we do not find evidence for the broad structure at M = 1790±50 MeV reported by the CB.

Possible earlier evidence for the $f_2(1520)$ comes from Gray et al.,[5] Bridges et al.[6] and May et al.,[7] in at-rest annihilations. Our observation of the $f_2(1520)$ is consistent with the argument that it may be made uniquely in annihilation reactions. Dover et al.[8] have recently offered a theoretical interpretation of these results in terms of a bound state model of the antiproton-proton system.

Analysis of reactions (2-4) is presently in progress to determine the extent of production of the $f_2(1520)$ in these channels.

We gratefully acknowledge the technical support from our collaborating institutions and the outstanding contribution of the Fermilab Accelerator Division Antiproton Department. This work was funded by the U.S. Department of Energy, the National Science Foundation, and the Italian Istituto Nazionale di Fisica Nucleare.

Table 1: Integrated luminosities, cross sections, and fitted masses and widths.

$\sqrt{s}$	Int.L(nb^{-1})	$\sigma(\mu b)$	State	Mass(MeV)	Width(MeV)
(a) 2980 MeV	1760	5.0±2.5	$f_2(1270)$	1267±7	194±20
			$f_2(1520)$	1512±10	102±30
(b) 3097 MeV	460	3.2±1.6	$f_2(1270)$	1265±7	203±20
			$f_2(1520)$	1507±10	106±30

References
[1] T.A. Armstrong et al., Nucl. Phys. B **373**, 35 (1992).
[2] T.A. Armstrong et al., Phys. Rev. Lett. **68**, 1468 (1992).
[3] Review of Particle Properties, Phys. Rev. D **45**, 1 (1992).
[4] E. Aker et al., Phys. Lett. B **260**, 249 (1991).
[5] L. Gray et al., Phys. Rev. D **27**, 307 (1983).
[6] D. Bridges et al., Phys. Rev. Lett. **57**, 1534 (1986).
[7] B. May et al., Z. Phys. C **46**, 191 (1990); 203 (1990).
[8] C.B. Dover et al., Phys. Rev. C **43**, 379 (1991).

RESULTS ON $f_0(975)$ AND $\xi(2.2)$ FROM BES

BES collaboration
Zheng Zhipeng
Institute of High Energy Physics,
Academia Sinica,
P.O.Box 918,Beijing,10039,PRC

ABSTRACT

Study on $J/\psi \to \phi f_0(975)$ gives its branching ratio, the helicity amplitude ratio, and the mass and width of $f_0(975)$. For $\xi(2.2)$, the mass, width and branching ratio were measured in $J/\psi \to \gamma K^+K^-, \gamma K_s^0 K_s^0$ and $\gamma\eta\eta$.

Introduction

From the beginning of 1990 to May 1991, BES(Beijing Spectrometer) was running in three separated periods, during which $9 \times 10^6 J/\psi$ events were collected. A part of these data were reconstructed, calibrated and analyzed, and some results on following respects were obtained:

1) the measurement of J/ψ widths.

2) the nature of $f_2(1270)$.

3) the observation of $\theta/f_2(1720)$ and $\iota/\eta(1440)$.

4) the study of $J/\psi \to \gamma\rho^0\rho^0$.

After the J/ψ running, from Nov. 1991 to Feb. 1992, the τ mass measurement was successfully conducted. Then the data of D_s are collected since the Spring of 1992.

Among all above, the preliminary results on $f_0(975)$ and $\xi(2.2)$ are given in this paper.

Part I. $f_0(975)$

The scalar mesons are less understood due to their quite unconventional experimental properties[1]. The problem is focused on the $f_0(975)$ and $a_0(980)$. $f_0(975)$ (or named s(975),S*) is relatived to multiquark state, $K\bar{K}$ molecule, and glueball[2]. Heavy effort has been put on this state.

We studied $f_0(975)$ through the following decay channel of J/ψ :

$$J/\psi \to \phi f_0, \phi \to K^+K^-, f_0 \to \pi^+\pi^-$$

1) The candidate events to the final state $K^+K^-\pi^+\pi^-$ must satisfy the following criteria :

—— four charged tracks with zero total charge.

—— at least 2 kaons are determined by TOF in 3σ.

—— 4-constraint kinematic fit, $\chi^2 < 36$.

—— ϕ signal is selected by $|M_{K^+K^-} - 1.02| < 0.015$ GeV.

The mass distribution of the surviving events is shown in fig.1. They are from $7 \times 10^6 J/\psi$ events.

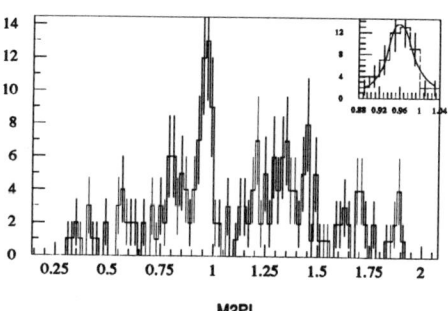

Fig.1 $M_{\pi\pi}$ Recoiling against ϕ in $J/\psi \to \phi\pi\pi$

2) Mass Distribution Analysis —— Coupled Channel Fit

We fit the $f_0(975)$ signal by using Maximum Likelihood Method with the following coupled-channel formular, and considering the $K\bar{K}$ threshold[3].

$$\frac{d\sigma}{dm} \sim \left| \frac{M_R \Gamma_\pi}{M_R^2 - m^2 - iM_R(\Gamma_\pi + \Gamma_K)} \right|^2$$

where,

$$\Gamma_\pi = g_\pi q_\pi = g_\pi \sqrt{\frac{m^2}{4} - M_\pi^2}$$

$$\Gamma_K = g_K q_K$$
$$= \begin{cases} g_K \sqrt{\frac{m^2}{4} - M_K^2} & above\ threshold \\ ig_K \sqrt{M_K^2 - \frac{m^2}{4}} & below\ threshold \end{cases}$$

g_π, g_K are the coupling constants.

The mass resolution in this region is 7 MeV. The fitting is not sensitive to g_K, so we fixed it at 0.15. the result is:

$$M_R = 0.956 \pm 0.005\ GeV$$
$$g_\pi = 0.060 \pm 0.026$$

Which gives $g_K/g_\pi = 2.5$, showing that $f_0(975)$ couples more strongly to $K\bar{K}$.

The corresponding pole position is $(966 \pm 5) - i(15 \pm 7)\ MeV$. And the branching ratio is

$$Br(J/\psi \to \phi f_0(975)) = (2.9 \pm 0.4 \pm 0.7) \times 10^{-4}$$

3) Some Comparision.

The pole position and $Br(J/\psi \to \phi f_0(975))$ from the relative experiments and the value listed in PDG are compared in Table 1.

Table 1: The comparison of $f_0(975)$ result

	Pole Position $(m - i\Gamma/2)MeV$	Br $(J/\psi \to \phi f_0(975))$ $\times 10^{-4}$
P.D.G.	$(975.6 \pm 3.1) - i(16.8 \pm 2.8)$	(3.2 ± 0.5)
MARKII	$(974 \pm 4) - i(14 \pm 5)$	(2.6 ± 0.6)
MARIII		$(3.4 \pm 0.5 \pm 0.8)$
DM2		$(3.6 \pm 0.6), 1986$ $(4.6 \pm 0.4 \pm 0.8),$ 1988
BES	$(966 \pm 5) - i(15 \pm 7)$	$(2.9 \pm 0.4 \pm 0.7)$

4) Angular Distribution Fit

The helicity formulism of the process $J/\psi \to VX, V \to P_1^+ P_2^-, X \to P_3^+ P_4^-$ is[5]:

$$W(\theta_v \theta_1 \phi_1 \theta_m \phi_m) \sim$$
$$\sum I_{\lambda_J, \lambda'_J}(\theta_v) \cdot A_{\lambda_v, \lambda_x} \cdot A_{\lambda_{v'}, \lambda_{x'}}$$
$$\cdot D^{1*}_{\lambda_v, 0}(\theta_1, \phi_1, 0) \cdot D^1_{\lambda_{v'}, 0}(\theta_1, \phi_1, 0)$$
$$\cdot D^{s*}_{-\lambda_x, 0}(\theta_m, \phi_m, 0) \cdot D^s_{-\lambda'_x, 0}(\theta_m, \phi_m, 0)$$

Where, S is the spin value of X. A_{λ_v, λ_x} is the helicity amplitude of the process $e^+e^- \to J/\psi \to VX$. θ_v is the angle between e^+ direction and V direction in the J/ψ rest frame, (θ_1, ϕ_1) is the polar and azimath angle of P_1^+

direction in V rest frame. And (θ_m, ϕ_m) the polar and azimath angle of P_3^+ direction in X rest frame. Define the following helicity amplitude ratios:

$$X = \frac{A_{11}}{A_{10}}, \quad Y = \frac{A_{12}}{A_{10}},$$

$$Z_1 = \frac{A_{00}}{A_{10}}, \quad Z_2 = \frac{A_{01}}{A_{10}}$$

Where, the conservation of parity and time inverse invariant of the process are used, the maximum number of the independent amplitudes in this kind of process is only 5, and they are relatively real.

The $cos\theta_m$ distribution (fig.2) is flat. the acceptance efficiency is also flat.

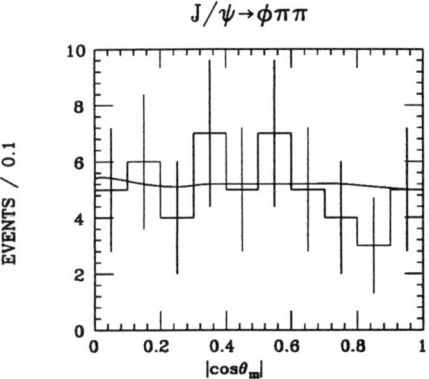

Fig.2. $cos\theta_m$ Distribution

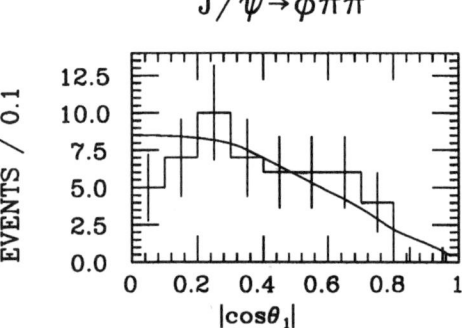

Fig.3 $cos\theta_1$ Distribution

This shows $f_0(975)$ is a scalar.

$$J^{PC}(f_0(975)) = 0^{++}$$

So, W is independent on $cos\theta_m$. It can be written as following,

$$W \sim (1+cos^2\theta_v)sin^2\theta_1 - sin^2\theta_v sin^2\theta_1 cos(2\phi_1)$$
$$+ 2sin^2\theta_v \cdot Z_1^2 \cdot cos^2\theta_1$$
$$- 2Z_1 sin(2\theta_v) sin\theta_1 cos\theta_1 cos\phi_1$$

Taking the acceptance efficiency of all the angular distributions into consideration, Maximum Likelihood Fitting gives:

$$Z_1 = 0.0 \pm 0.2$$

The fit result projected on the angle θ_1 is demonstrated in Fig. 3.

Part II. $\xi(2.2)$

In 1983, MARK-III oberserved $\xi(2.2)$ at 2.2 GeV in the process of $J/\psi \to \gamma K^+ K^-$ and $J/\psi \to \gamma K_s^0 K_s^0$ [4]. However DM2 said they didn't see the narrow resonance at this region from these two process [5]. As to the fix target experiments, some of them support the existence of $\xi(2.2)$, while others do not. At BES, three decay channels were studied for finding the signal of $\xi(2.2)$. They are:

$$J/\psi \to \gamma K^+ K^-$$
$$J/\psi \to \gamma K_s^0 K_s^0$$

and,

$$J/\psi \to \gamma \eta \eta$$

For each channels, the selection criteria and preliminary results are as following:

(I) $J/\psi \to \gamma \xi, \xi \to K^+ K^-$

(I.a) event selection

——— two tracks with opposite charge and 1 upto 4 shower clusters.

—— Good $K^\pm$ selection. At least one charge track determined by TOF to be a K rather than a π. The collinear angle must be greater than 5.7° or the energy for the cluster less than 1 GeV. $\mid U \mid = \mid E_{missing} - P_{missing} \mid \leq 0.25 \text{GeV}$. And $\delta_T = T^+ - T^- - (t_{K^+} + t_{K^-} t_{\pi^+} t_{\pi^-})/2 > -0.06$ nanosecond.

—— Good γ selection. $E_\gamma > 100$ MeV, $P_{t_0}^2 = 4P_{miss}^2 Sin^2\theta_0/2 < 0.0035 GeV^2$, the difference between the direction determinted by shower and direction from the interaction point to the first hit point in the shower $\delta_c < 0.08$ and $cos\theta_{\gamma K^\pm} < 0.98$.

—— 4-C kinematical fitting requires $\chi^2_{\gamma K^+ K^-} < 30$ and $\chi^2_{\gamma \pi^+ \pi^-} > \chi^2_{\gamma K^+ K^-}$.

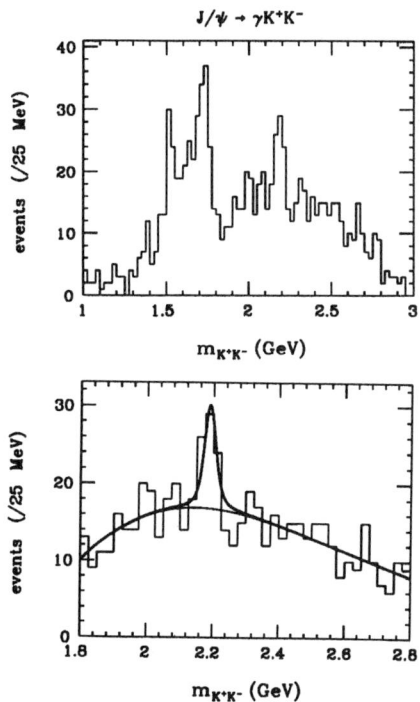

Fig.4 $J/\psi \to \gamma\xi, \xi \to K^+K^-$

(I.b) The preliminary result based on $2.6 \times 10^6 J/\psi$ events (Fig. 4):

$$m = (2190 \pm 10 \pm 17)MeV$$

$$\Gamma = (22 \pm 26)MeV$$
$$B_r(J/\psi \to \gamma\xi, \xi \to K^+K^-)$$
$$= (7.6 \pm 4.1 \pm 2.4) \times 10^{-5}$$

the statistical significance is $d \sim 4.0\sigma$.

(II) $J/\psi \to \gamma\xi, \xi \to K_s^0 K_s^0, K_s^0 \to \pi^+\pi^-$

(II.a) the selection criteria :

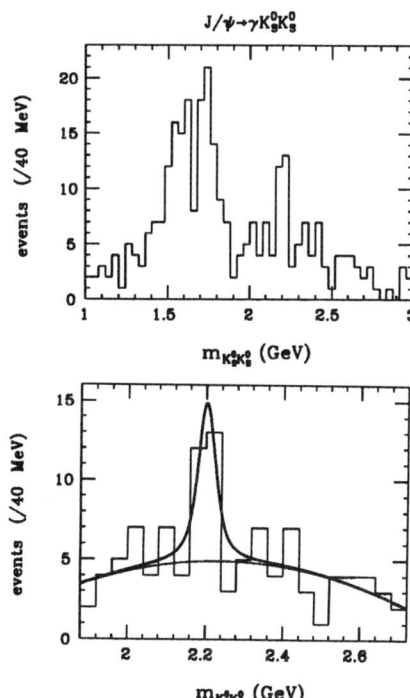

Fig.5 $J/\psi \to \gamma\xi, \xi \to K_s^0 K_s^0$

—— 4 charged tracks with total charge zero and at least 1 shower clusters (deposited energy over 100 MeV).

——K_s^0 selection. two K_s^0 are required which are selected by : $\mid M_{\pi^+\pi^-} - M_{K_s^0} \mid < 50 MeV$ and the distance of K_s^0 vertex to the origin in xy plane $R_{xy}(1)$ or $R_{xy}(2) \geq 1cm$, R_{xy} 25cm, $\delta Z_{K_s^0} \leq 8cm$, and $\mid cos\theta \mid = \mid \vec{P}_{K_s^0} \cdot \vec{R}_{xy} \mid / \mid \vec{P}_{K_s^0} \mid \cdot \mid \vec{R}_{xy} \mid < 0.5$ for each K_s^0.

—— after dE/dX correcting and swimming the track parameters to K_s^0 point, 4-C kinematical fit requires $\chi^2_{\gamma\pi^+\pi^-\pi^+\pi^-} < 50$.

— cut after 4-C fit , $P_{t_0}^2 \leq 0.005 GeV^2$,
$|U| \leq 0.4 GeV$ and $\delta M^2 = (M_{vertex1} - M_{K_s^0})^2 + (M_{vertex2} - M_{K_s^0})^2 < 0.001 GeV^2$.

(II.b) The preliminary result based on $7.1 \times 10^6 J/\psi$ events (Fig . 5) :

$$m = (2202 \pm 14 \pm 13) MeV$$

$$\Gamma = (38 \pm 25) MeV$$

$$B_r(J/\psi \to \gamma\xi, \xi \to K_s^0 K_s^0)$$
$$= (2.7 \pm 4.1 \pm 0.9) \times 10^{-5}$$

the statistical significance is $d \sim 3.6\sigma$.

(III) $J/\psi \to \gamma\xi, \xi \to \eta\eta, \eta \to \gamma\gamma$

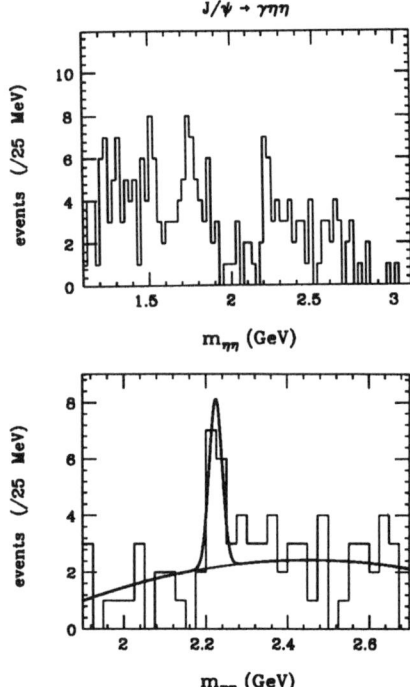

Fig.6 $J/\psi \to \gamma\xi, \xi \to \eta\eta$

(III.a) the selection criteria :

— Good γ selection : $E_\gamma > 80 MeV$, $\delta_c < 0.4$ and $\theta_{\gamma\gamma} > 8°$.

— $J/\psi \to 5\gamma$ Event Selection .

$N_{good \, \gamma} \geq 5, 2GeV < E_{total} < 4GeV$,

$P_{total} < 1.1 GeV$ and 4-C fit to $J/\psi \to 5\gamma$, $\chi^2 < 50$.

— $J/\psi \to \gamma\eta\eta$ Event Selection . $|M_{\gamma_i\gamma_j} - M_{\pi^0}| > 70 MeV$ for all pairs , $(M_{\eta_1\gamma_2} - M_{\eta^0})^2 + (M_{\gamma_3\gamma_4} - M_{\eta^0})^2 < 12.5 MeV^2$, and 6-C fit to $J/\psi \to \gamma\eta\eta$, $\chi^2 < 20$ and $|cos\theta^*_{\gamma\eta_i}| < 0.9$.

(III.b) The preliminary result based on $5.7 \times 10^6 J/\psi$ events(Fig . 6) :

$$m = (2224 \pm 8 \pm 17) MeV$$

$$\Gamma = (16 \pm 9) MeV$$

$B_r(J/\psi \to \xi, \xi \to \gamma\eta\eta) = (11.4 \pm 5.2 \pm 3.4) \times 10^{-5}$

the statistical significance is $d \sim 3.5\sigma$.

Table 2 and Table 3 show the results on $\xi(2.2)$ from BES and from MARK-III . For further confirming and understanding of the $\xi(2.2)$ nature,

1) We need to improve the data calibration , and to increase the data statistics.

2) We also need to improve the π/K separation .

3) we should take notice to the processes of $J/\psi \to \gamma\phi w$, and $J/\psi \to \gamma\eta\eta'$.

4) And the J^{PC} of $\xi(2.2)$ would also be measured on the base of a large statistics.

SUMMARY

For $f_0(975)$, the pole position was determined as $(966 \pm 5) - i(15 \pm 7) MeV$, and the branching ratio is measured to be $Br(J/\psi \to \phi f_0(975)) = (2.9 \pm 0.4 \pm 0.7) \times 10^{-4}$. The helicity amplitude ratio $Z_1 = A_{00}/A_{10} = 0.0 \pm 0.20$ was obtained . As for $\xi(2.2)$, first, a distinct and narraw resonance at the energy region of

2.2 GeV was observed by analyzing the processes of $J/\psi \to \gamma K^+ K^-$, $J/\psi \to \gamma K_s^0 K_s^0$ and $J/\psi \to \gamma \eta \eta$. Secondly, the corresponding mass, width and branching ratio of them were given. The masses obtained from three channels are slightly defferent, but consistant within the error bar.

Table 2: $\xi(2.2)$ results on BES

	Mass	Width	Br ($\times 10^{-5}$)	d
K^+K^-	$2190 \pm 10 \pm 17$	22 ± 26	$7.6 \pm 4.1 \pm 2.4$	4.0σ
$K_s^0 K_s^0$	$2202 \pm 14 \pm 13$	38 ± 25	$2.7 \pm 1.1 \pm 0.9$	3.6σ
$\eta\eta$	$2224 \pm 8 \pm 17$	16 ± 9	$11.4 \pm 5.2 \pm 3.4$	3.5σ

Table 3: $\xi(2.2)$ results on MARK-III

	Mass (MeV)	Width (MeV)	Br ($\times 10^{-5}$)	d
K^+K^-	$2230 \pm 6 \pm 14$	$26^{+20}_{-16} \pm 17$	$4.2^{+1.7}_{-1.4} \pm 0.8$	4.5σ
$K_s^0 K_s^0$	$2232 \pm 7 \pm 7$	$18^{+23}_{-15} \pm 10$	$3.2^{+1.6}_{-1.3} \pm 0.7$	3.6σ
$\eta\eta$	-	-	< 7	-

References

[1] Nils A. Törnqvist, Phys. Rev. Letts. 49(1982)624

[2] The 23rd International Conference on hadron spectroscopy, Hadron '89.

R. Jaffe, Phys. Rev. D15(1977)267

J. Wenstein and N. Isgur, Phys. Rev. D27(1983)558

G. Mennessier et. al., Gluonium and the 0^{++} spectrum, Phys. Letts. 158B(1985)153

[3] S.M. Flatte Phys. Lett. 63B(1976)224

[4] K. F. Einsweiler, Ph. D. Thesis, SLAC-Report-278(1984)

[5] J. Augustin et al. Contrib. to Int. Symp. on Lepton-Photon Interact. at High Energies, Leipzeig, 1984;

Orsay Report, No. LAL-84/30, Oct. 1984

HOW TO DETERMINE THE SUBSTRUCTURE OF THE SCALAR MESONS AT AROUND 1 GEV IN MASS

F E Close
Rutherford Appleton Laboratory,
Chilton, Didcot, Oxon,
OX11 OQX, England

Abstract

The scalar mesons $f_0(975)$ and $a_0(980)$ can be produced in ϕ radiative decays and in $\gamma\gamma$ collisions as well as in $p\bar{p} \to \pi\pi\eta, \pi\eta\eta$, or $\pi\pi\pi$. The ratio of the f_0 and a_0 production rates may determine the substructure of these mesons.

INTRODUCTION

20 years after the discovery of QCD and 30 years after the invention of the quark model we still do not know whether glueballs or hybrid states exist. This is an unsatisfactory gap in our knowledge. There are as many predictions for the spectrum of glueballs as there are theorists who have worked on the problem; however they agree on one thing — the lightest glueball is predicted to be a scalar, $J^{PC} = 0^{++}$.

In order to identify this, and other gluonic states, it is a good strategy to understand better the spectroscopy of the conventional $q\bar{q}$ hadrons, in particular those with $c = +$. To do this, $\gamma\gamma$ physics is particularly useful and recently we have developed tests that will help to identify $q\bar{q}$ and $q\bar{q}g$ states[1,2,3]. This has already had positive results in that it has enabled two groups[4] to isolate the candidate $f_0(1400)$ as a $q\bar{q}$ state coupling to $\gamma\gamma$ with canonical strength.

There are also two, nearly degenerate, scalars $f_0(975)$ and $a_0(980)$ with $I = 0, 1$ respectively whose internal structure has long been controversial: they may be $q\bar{q}, K\bar{K}$ molecules[5], $qq\bar{q}\bar{q}$[6,7] or excitations of negative energy quarks[8]. In ref. 9 we summarised reasons why the $^3P_0(q\bar{q})$ interpretation of these states is unlikely; however a unitarised version of the quark model (ref. 10) in which scalar mesons are strongly mixed with the meson-meson continuum avoids some of the $q\bar{q}$ problems. I will show how radiative decays of the ϕ meson (at DAΦNE or VEPP) and gamma-gamma physics can help to discriminate among the various hypotheses. In particular the ratio of branching ratios

$$\frac{\Gamma(\phi \to \gamma a_0)}{\Gamma(\phi \to \gamma f_0)} \quad ; \quad \frac{\Gamma(\gamma\gamma \to a_0)}{\Gamma(\gamma\gamma \to f_0)}$$

can provide rather sharp tests of their substructure.

ARE f_0, a_0 VACUUM FRIENDLY?

Gribov has suggested that the $a_0(980)$, $f_0(975)$ may be $q\bar{q}$ excitations of negative energy quark states that arise in his model of confinement. If this is so, then these mesons should be prominent in processes where the "vacuum" is excited. Examples are ψ decays, e^+e^- annihilation in the region away from the leading $q, \bar{q}$ jets, the central region in hadron scattering and $p\bar{p}$ annihilation.

Qualitatively this may be in accord with data. The $f_0(975)$ was seen as a bump in $\psi \to \phi f_0 \to \phi \pi \pi$ and at this conference we have heard that BEPC confirm this[11]. AFS and WA76 at CERN saw a sharp drop[12] in the central $\pi\pi$ spectrum associated with the $f_0(975)$. Now we learn that the $f_0(975)$ is being seen[13] in Z^0 decays by DELPHI; it will be interesting to see if it is dominantly being produced away from the leading jets. Finally the $p\bar{p}$ at LEAR is particularly promising[14]. Here $p\bar{p} \to \pi\pi\eta$ produces both the a_0 and the f_0 as $p\bar{p} \to \pi a_0$ or $f_0 \eta$. The channels $\pi\eta\eta$ and $\pi\pi\pi$ enable $a_0 \eta$ and πf_0 production rates to be measured. Hence we may hope for absolute rates and also ratios

$$p\bar{p} \to \pi a_0 : \pi f_0 : \eta a_0 : \eta f_0$$

that will teach us about the nature of these states. There is also preliminary report of $a_0(980)$ as a shoulder in the $\pi\eta\eta$ spectrum at Fermilab[15].

The cleanest probes of substructure are likely to be in $\gamma\gamma \to S$ and $\phi \to \gamma S$ ($S = a_0$ or f_0). In a recent paper[9] we pointed out a simple (and to us unexpected) experimental test which sharply distinguishes among these alternative explanations. We showed that the rates for $\phi \to f_0 (975) \gamma \to \pi\pi\gamma$ and $\phi \to a_0 (980) \gamma \to \pi\eta\gamma$ in the quarkonium, glueball, and $K\bar{K}$ molecule interpretations differ significantly. The absolute branching ratios are predicted to lie in the range 10^{-4} to 10^{-6} depending upon the structure of these scalar mesons (ref. 9). Furthermore, the ratio of branching ratios

$$\frac{\phi \to a_0(980)\gamma}{\phi \to f_0(975)\gamma}$$

also may prove to be an important datum in that it can have a model-dependent value anywhere from zero to infinity (see Table 2)! This should be measured at VEPP and DAFNE (ref. 16).

RATIOS OF BRANCHING RATIOS AS TESTS OF SUBSTRUCTURE

The ratio of branching ratios is a particularly interesting measurement. The ratio of $\Gamma(\phi \to \gamma a_0)/\Gamma(\phi \to \gamma f_0)$ is approximately zero if they are quarkonia (the f_0 being $s\bar{s}$ and the a_0 being OZI decoupled), it is approximately unity if they are $K\bar{K}$ systems, while for $q^2\bar{q}^2$ the ratio is sensitively dependent on the internal structure of the states. This sensitivity in $qq\bar{q}\bar{q}$ arises because $\phi \to S\gamma$ is an $E1$ transition whose matrix element, being proportional to $\Sigma e_i \vec{r}_i$, probes the electric charges of the constituents weighted by their vector distance from the overall centre of mass of the system. Thus, although the absolute transition rate for $S = qq\bar{q}\bar{q}$ depends on unknown dynamics, the ratio of a_0 to f_0 production will be sensitive to the internal spatial structure of the scalar mesons through the relative phases in $I = 0$ and 1 wavefunctions and the relative spatial distributions of quarks and antiquarks.

For example, suppose that the state's constituents are distributed about the centre of mass with the structure $(q\bar{s})(\bar{q}s)$, where q denotes u or d and (ab) represents a spherically symmetric cluster. Then

$$\left\{ \begin{array}{c} f_0 \\ a_0 \end{array} \right\} = \frac{1}{\sqrt{2}}[(u\bar{s})(\bar{u}s) \pm (d\bar{s})(\bar{d}s)] \quad (1)$$

and the $E1$ matrix element will be

$$M \sim [(e_u + e_{\bar{s}}) \pm (e_d + e_{\bar{s}})] = e_{K^+} \pm e_{K^0}$$

and hence the ratio $\Gamma(\phi \to \gamma f_0)/\Gamma(\phi \to \gamma a_0)$ will be unity. The quarks are distributed *as if* in a $K\bar{K}$ molecular system (which is a specific example of this configuration) and only the absolute branching ratio will distinguish $q^2\bar{q}^2$ from $K\bar{K}$.

If the distribution is $(q\bar{q})(s\bar{s})$ then the matrix element

$$M \sim [(e_q + e_{\bar{q}}) - (e_s + e_{\bar{s}})] = 0.$$

Here the quark distributions mimic $\pi^0 \eta$ (in the a_0) or $\eta\eta$ (in the f_0). In this case the absolute branching ratios will be suppressed. Most interesting is the case where $S = D\bar{D}$, where D denotes a diquark, i.e. where

$$\left\{ \begin{array}{c} f_0 \\ a_0 \end{array} \right\} = \frac{1}{\sqrt{2}}[(us)(\bar{u}\bar{s}) \pm (ds)(\bar{d}\bar{s})] \quad (2)$$

in which case

$$M \sim [(e_u + e_s) \pm (e_d + e_s)]$$

so that

$$\frac{\Gamma(\phi \to \gamma a_0)}{\Gamma(\phi \to \gamma f_0)} = (\frac{1+2}{1-2})^2 = 9.$$

The absolute rate in this case depends on an unknown overlap between $K\bar{K}$ and the diquark structure; nonetheless the dominance of a_0 over f_0 would be rather distinctive. For convenience these possibilities are summarised in Table 2.

It is even possible that finer details of a diquark substructure may be resolved. As the strange quark is more massive than the nonstrange, one may anticipate that

$$<r>_s \, < \, <r>_{u,d}$$

Thus if

$$\frac{<r>_s}{<r_q>} = x$$

then the ratio

$$\frac{\Gamma(\phi \to \gamma a_0)}{\Gamma(\phi \to \gamma f_0)} = (\frac{3}{1+2x})^2$$

for a $K\bar{K}$-like "bag" and for a $D\bar{D}$-like "bag" one has

$$\frac{\Gamma(\phi \to \gamma a_0)}{\Gamma(\phi \to \gamma f_0)} = (\frac{3}{1-2x})^2.$$

Starting from the DD configuration there is a continuous transition as x journeys from 1 (the "symmetry limit") to -1 (which corresponds to a KK-like structure). It is when $x = \frac{1}{2}$ that the "distorted diquark" phase kills the f_0 production. If $x \to 0$ one has the "transition" configuration where $s\bar{s}$ are at the origin and thereby "frozen out" of the $E1$ transition. This gives a ratio of 9 (which will of course also be the ratio for a $u\bar{u} \pm d\bar{d}$ picture, as in Gribov's theory, Ref 8). I summarise these results in diagrammatic form in table 3.

SCALAR MESONS IN $\gamma\gamma$ PHYSICS

The ϕ radiative is sensitive to the constituent charges (via the γ) and the $s\bar{s}$ content (via the ϕ); in $\gamma\gamma$ coupling one has a complementary probe of the internal charge structure of the produced mesons. The combination of $\phi \to \gamma S$ and $\gamma\gamma \to S$ may definitively establish the nature of the a_0 and f_0 states.

Unlike the ϕ radiative decays, there are data on $\gamma\gamma$ couplings to the scalars. The data were [17]

$$\Gamma(\gamma\gamma \to a_0) = 0.19^{+0.17}_{-0.14} \, kev$$
$$\Gamma(\gamma\gamma \to f_0) = 0.30 \pm 0.10 \, kev \quad (3)$$

and tantalisingly similar to Barnes' [18] calculation of the width for $K\bar{K}$ molecules (0.2 kev), and an order of magnitude smaller than expected for a $q\bar{q}$ $^3P_0$ [1-3] (though there is a possible loophole in the latter case since in the $m_q \to 0$ limit the $^3P_0 \gamma\gamma$ width is predicted to vanish). If the $K\bar{K}$ loop dominates the interaction ($\gamma\gamma \to K\bar{K} \to S$) then the above widths are very natural [19] but possible contributions from $\pi\pi$ in the case of the f_0 confuses this simple result. The modelling of the $\gamma\gamma$ width needs more detailed study. However a recent analysis by Morgan and Pennington[20] has modified these numbers. The current world average[21] is

$$\Gamma(\gamma\gamma \to a_0)B(a_0 \to \eta\pi) = 0.24 \pm 0.08 \, kev$$
$$\Gamma(\gamma\gamma \to f_0) = 0.56 \pm 0.11 \, kev \quad (4)$$

and the absolute magnitude of a_0, and the ratio of the two, depends on the poorly determined

Table 1. ϕ photodecays to quarkonia

quarkonium	formula	ϕ branching ratio
$f_0 = \sqrt{\frac{1}{2}}(u\bar{u} + d\bar{d})\ ^3P_0$	$0^{a)}$	$\lesssim 10^{-6}$
$f_0 = s\bar{s}\ ^3P_0$	$\frac{4\alpha|d_{f_0\phi}|^2\omega^3}{243}$	$\simeq 1 \times 10^{-5}$
$a_0 = \sqrt{\frac{1}{2}}(u\bar{u} - d\bar{d})\ ^3P_0$	$0^{b)}$	$\lesssim 10^{-6}$

a) proceeds through $\omega - \phi$ and $f_0 - f_0'$ mixing
b) proceeds through $\omega - \phi$ mixing only

Table 2. Some qualitative implications of $\frac{\Gamma(\phi \to a_0 \gamma)}{\Gamma(\phi \to f_0 \gamma)}$

scalar meson constitution	$\frac{\Gamma(\phi \to a_0 \gamma)}{\Gamma(\phi \to f_0 \gamma)}$	absolute branching ratios	comments
$K\bar{K}$ molecule	$1^{(a)}$	$a_0 \simeq f_0 \simeq 4 \times 10^{-5}$	$K\bar{K}$ dominates loop diagrams see ref. 9
$q^2\bar{q}^2$: $\ \ K\bar{K}$-like "bag"	$1^{(b)}$		
$\quad\quad\quad D\bar{D}$-like "bag"	$9^{(c)}$	rates probably	see section 5
$\quad\quad\quad (n\bar{n})(s\bar{s})$-like "bag"	-	$< 10^{-6}$, see section 5	of ref. 9
$(q\bar{q})^3P_0$: $\ f_0 = n\bar{n}$	-		
$\quad\quad\quad\quad\quad f_0 = s\bar{s}$	$\simeq 0$	see Table 1	see Table 1
f_0 glueball, a_0 quarkonium		$\leq 10^{-6}$	see text

a) neglecting $I = 0, 1$ mixing effects.
b) if $\frac{\langle r_s \rangle}{\langle r_q \rangle} \equiv x$, then this ratio is $(\frac{3}{1+2x})^2$.
c) if $\frac{\langle r_s \rangle}{\langle r_q \rangle} \equiv x$, then this ratio is $(\frac{3}{1-2x})^2$.

$B(a_0 \to \eta\pi)$. This has often been approximated by unity (as in ref. 18 and the numbers cited above in eq. 3) whereas the value could easily be below 50% (ref. 21). Thus one cannot eliminate equality for the two $\gamma\gamma$ widths. Given that $B(a_0 \to \eta\pi) \leq 1$ we can at least limit the ratio

$$\frac{\Gamma(\gamma\gamma \to f_0)}{\Gamma(\gamma\gamma \to a_0)} \leq \frac{21 \pm 4}{9 \pm 3} \quad (5)$$

The ratio of $\gamma\gamma$ ratios is a direct probe of the substructure.

In the case of a KK molecule, dominated by the KK loop, the ratio will be unity. Contrast this with a simple $q\bar{q}$ picture. If

$$a_0 = (u\bar{u} - d\bar{d})/\sqrt{2}$$
$$f_0 = (u\bar{u} + d\bar{d})/\sqrt{2}$$

then

$$\frac{\Gamma(\gamma\gamma \to f_0)}{\Gamma(\gamma\gamma \to a_0)} = \frac{25}{9}$$

due to the amplitude being proportional to the squared charges of the constituents. This would apply to Gribov's picture where the f_0, a_0 are excitations of negative energy light flavours. Within the large errors this ratio too is still allowed though hardly favoured. Hence improvement in

Table 3. Strange quarks ♠ and antiquarks ♡, and nonstrange quarks ● and antiquarks ○ form a $qs\bar{q}\bar{s}$ state. The spatial configuration relative to the c.m. is indicated in column 1 and a qualitative description of the state is in column 2 (D refers to "diquark"). The ratio of ratios is in column 3.

Spatial Configuration	Description	$\Gamma(\phi \to \gamma a_0)/\Gamma(\phi \to \gamma f_0)$
● ○ ♠ ♡	$D\bar{D}$	9
● ○ ♠ ♡	distorted $D\bar{D}$	∞
♡ ● ○ ♠	frozen ss	9
● ○ ♡ ♠	distorted $K\bar{K}$	2
● ○ ♡ ♠	$K\bar{K}$	1

the $\gamma\gamma$ widths (and in particular the $\eta\pi$ branching ratio of the a_0) could be rather crucial.

If the a_0, f_0 are compact four quark states, the ratio of widths is less well defined and depends on the nature of the virtual intermediate state between the two photons. A coherent production involving only the ground state (Born) meson is one extreme; a sum over a complete set of intermediate states, which effectively may be modelled by incoherent Compton scattering is another extreme.

For the coherent case one has for $\Gamma(\gamma\gamma \to f_0)/\Gamma(\gamma\gamma \to a_0)$

$$K\bar{K} \quad 1$$
$$D\bar{D} \quad \frac{25}{9}$$

For the incoherent case one would have $\frac{25}{9}$ for all cases in the limit where the $s\bar{s}$ production is frozen out; if $s\bar{s}$ is produced by gluons with the same strength as the non-strange, the ratio rises to $\frac{49}{9}$. This latter value is probably ruled out empirically.

Finally one has the possibility that each photon produces a single $q\bar{q}$. In this case the matrix element is proportional to $(e_u \pm e_d)e_s$ and hence the f_0/a_0 ratio will be $\frac{1}{9}$. This is ruled out unless the $B(a_0 \to \eta\pi)$ is very small and the $\Gamma(\gamma\gamma \to a_0)$ an unlikely 5 keV. Thus in these simple examples alone the ratio can vary from $\frac{1}{9}$ to $\frac{49}{9}$. More realistic modelling seems to be needed.

In any event, the combination of $\gamma\gamma \to S$ and $\phi \to \gamma S$ data for both a_0 and f_0 can make important contributions to clarifying the structure of these states.

SUMMARY

There appears to be too many scalar mesons in the 1-2 GeV region for the $q\bar{q}$ quark model to accommodate. This may be a clue to the excitation of glue and existence of glueballs. However it may be that radial excitations that are expected to lie above 2 GeV in the quark model are, for some reason, displaced to low masses. To eliminate this we need a study of $\bar{p}p$ above 2 GeV (Fermilab, LEAR in flight) and an intensive study of $\gamma\gamma$ to clarify the $C = +q\bar{q}$ spectroscopy by exploiting the known $\gamma\gamma$ selection rules (ref. 3).

In addition we need clarification on the spin of $\theta(1720)$ — is it really a scalar meson? — and establishing decay channels of $f_0(1590)$ other than $\eta\eta$. We need confirmation that there are indeed $f_0 f_0'$ lurking beneath the $f_2(1270)$ and $f_2(1510)$. And we can look forward to rates and ratios for $a_0(980)$ and $f_0(975)$ production.

The immediate strategy for scalar glueball hunting is
1) Clarify what $a_0(980)$ and $f_0(975)$ are
2) Confirm that all the 1-2 GeV states are real

3) Show that the "ordinary" spectrum continues beyond 2-3 GeV, in particular with $\bar{p}p$ and $\gamma\gamma$ production experiments. This may show the existence of scalars "beyond quarks".

REFERENCES

1. T. Barnes, F.E. Close and Z.P. Li, Phys. Rev. **D43**, 2161 (1991).
2. F.E. Close and Z.P. Li, Z. Phys. **C54**, 147 (1992).
3. E. Ackleh, T. Barnes and F.E. Close, Phys. Rev. **D**, (in press).
4. Crystal Ball, J. Bienlein et al, DESY 91-145, *Proc 9th Photon Photon Workshop*, La Jolla 1992; CELLO, J. Harjes, *Proc Hadron 91*.
5. J. Weinstein and N. Isgur, Phys. Rev. Lett. **48**, 659 (1982), Phys. Rev. **D27**, 588 (1983), Phys. Rev. **D41**, 2236 (1990).
6. R. Jaffe, Phys. Rev. **D15**, 267, 281 (1977); Phys. Rev. **D17**, 1444 (1978).
7. R. Jaffe and F. Low, Phys. Rev. **D19**, 2105 (1979).
8. V. Gribov, private communication.
9. F.E. Close, N. Isgur and S. Kumano, Nucl. Phys. B (in press), RAL-92-026, CEBAF 92-13, IUNTC 92-16.
10. N.A. Tornquist, Phys. Rev. Lett. **49**, 624 (1982).
11. Z. Zheng (these proceedings).
12. AFS Collaboration
    WA76
13. DELPHI Collaboration, submitted to Phys. Letters.
14. C. Amsler, Crystal Barrel, these proceedings.
15. G. Smith, E760, these proceedings.
16. VEPP: J. Thompson. DAFNE: J. Franzini; submission to this session. See also P. Franzini et al, Phys. Lett. **B287**, 259 (1992).
17. D. Antreasyan et al., Phys. Rev. **D33**, 1847 (1986); Mark II, Phys. Rev. **D42**, 1350 (1990); Crystal Ball, Phys. Rev. **D41**, 3324 (1990).
18. T. Barnes, Phys. Lett. **B165**, 434 (1985).
19. N. Achasov and G. Shestakov, Z. Phys. **C41**, 309 (1988); T. Truong, *Proc of Hadron 89*.
20. D. Morgan and M. Pennington, Phys. Lett. **B258**, 444 (1991).
21. Particle Data, Group, Phys. Rev. **D45**, S1 (1992).

NEW TWO–PHOTON RESULTS FROM ARGUS*

G. Kernel
Institut J. Stefan and Oddelek za fiziko
Univerza v Ljubljani
Jamova 39
SI-61111 Ljubljana, Slovenia

Abstract

This paper presents data on two–photon production of a pair of vector mesons, and on resonance production of three pions. Vector pair production was studied in three channels: $\gamma\gamma \to \omega\rho$, $\gamma\gamma \to \rho\phi$ and $\gamma\gamma \to \omega\phi$. For the first channel also a partial–wave analysis was performed of the reaction $\gamma\gamma \to \pi^+\pi^+\pi^-\pi^-\pi^0$. The reaction $\gamma\gamma \to \rho\phi$ was observed for the first time. There is also evidence for the reaction $\gamma\gamma \to \omega\phi$. The $\gamma\gamma \to \pi^+\pi^-\pi^0$ reaction is dominated by the formation of a_2 in its helicity 2 state. A partial–wave analysis has revealed the $\pi_2(1670)$ state with a two–photon partial width of (0.25 ± 0.15) keV which is significantly lower than indicated by previous experiments. There is an indication for an additional peak at 1.8 GeV with a prefered spin–parity assignment 2^-.

INTRODUCTION

The integrated cross sections for two–photon production of a pair of vector mesons vary typically by two orders of magnitude from channel to channel. The reaction $\gamma\gamma \to \rho^0\rho^0$ plays a dominant role with its large cross section, while $\rho^0\phi$, $\omega\phi$ and $\phi\phi$ channels have so small cross sections that have not yet been detected. In this paper we present the results of a partial–wave analysis of the reaction $\gamma\gamma \to \rho^0\omega$, and for the first time an observation of $\gamma\gamma \to \rho^0\phi$ and an indication for $\gamma\gamma \to \omega\phi$.

Gamma–gamma interactions represent an important tool for studying meson resonances since two–photon partial widths $\Gamma_{\gamma\gamma}$ should be closely related to their quark and gluon content. In addition, only resonances with certain quantum numbers are populated, and determination of spins and parities follows from a partial–wave analysis. Here we present results of a study of resonances that appear in the reaction $\gamma\gamma \to \pi^+\pi^-\pi^0$.

The data used for the present study were taken at an average energy of 10.2 GeV and correspond to an integrated luminosity of 456 pb^{-1}. For properties and performance of the ARGUS detector we refer to ref. [1] and for the method of analysis to ref. [2]. To ensure that only photons that are almost real contribute to the analyzed data a typical cut of $\sum_i \vec{p}_{T,i} < 110$ MeV/c was applied to the sum of transversal momenta. In this way also a large proportion of non–gamma–gamma events is removed from the data.

---

*ARGUS Collaboration: DESY, Dortmund, Erlangen–Nürnberg, Hamburg, Heidelberg, Karlsruhe, Ljubljana, Lund, Montreal, Moscow, Ottawa, Toronto.

TWO–PHOTON PRODUCTION OF VECTOR PAIRS

The analysis of $\gamma\gamma \to \pi^+\pi^+\pi^-\pi^-\pi^0$ shows that only about 12% of the cross section is due to the reaction $\gamma\gamma \to \rho^0\omega$. The cross section for $\gamma\gamma \to \rho^0\omega$ is displayed in figure 1a. Preliminary analysis indicates a substantial contribution of $J^P = 2^+$ waves.

Figure 1b shows the invariant–mass spectrum for $\gamma\gamma \to \rho^0\phi$. It is estimated that the contribution to the cross section of processes other than $\gamma\gamma \to \rho^0\phi$ does not exceed 22%.

The results of the analysis for $\gamma\gamma \to \omega\phi$ are seen in figure 1c. There seems to be some evidence for an $\omega\phi$ signal in the invariant–mass region between 1.9 and 2.3 GeV/c² with a cross section of about (0.72 ± 0.38) nb. For the rest of the mass region only an upper limit can be set. The above figure could also be translated into an upper limit of 1.6 nb at 95% confidence level.

RESONANCES INDUCED BY THE REACTION $\gamma\gamma \to \pi^+\pi^-\pi^0$

It is known that the process $\gamma\gamma \to \pi^+\pi^-\pi^0$ proceeds mainly through the resonace a_2 in a dominant helicity 2 amplitude. It is tempting to see whether also some other resonances could be observed. In particular, a structure was found recently[3,4] at invariant masses above a_2 which was identified as $\pi_2(1670)$. It is felt that a partial–wave analysis is necessary in order to establish the formation of resonances in this process and to measure their widths. The present study was focused to these subjects.

A sample of 3684 events satisfied the selection criteria for $\gamma\gamma \to \pi^+\pi^-\pi^0$. The background was determined from the experimental data on contributing processes. The largest

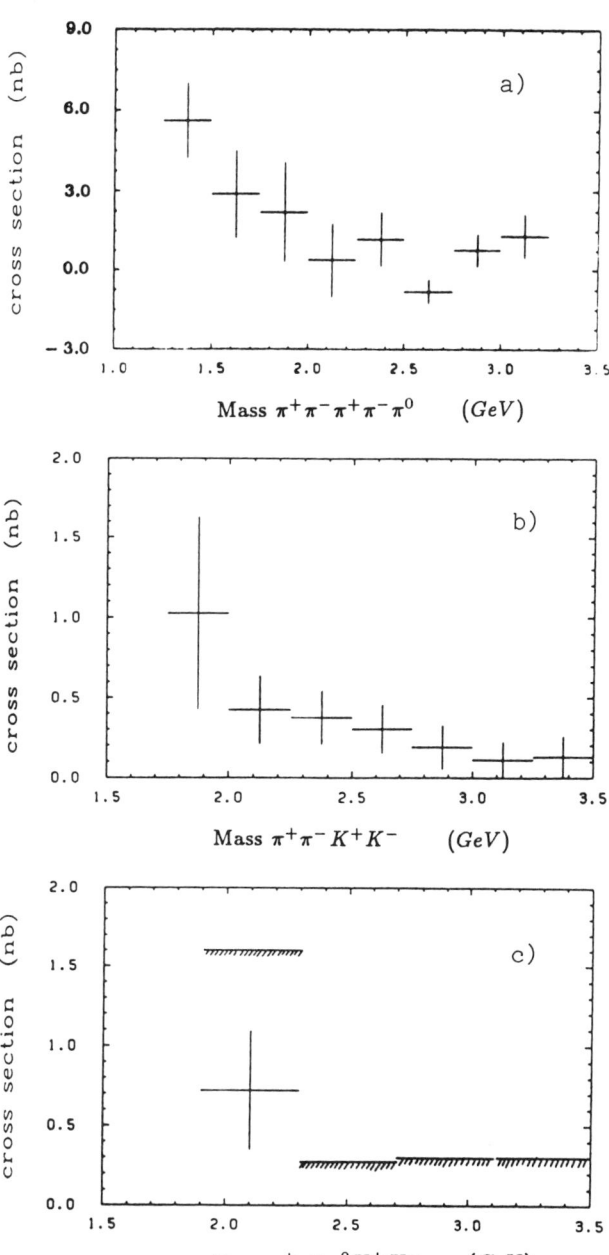

Figure 1. Cross sections for production of vector pairs by two photons: $\gamma\gamma \to \rho^0\omega$ (a), $\gamma\gamma \to \rho^0\phi$ (b) and $\gamma\gamma \to \omega\phi$ (c). In the later case upper limits (95% conf. level) for the cross section are also shown.

Table 1. Partial Waves Taken Into Account in the Analysis of $\gamma\gamma \to \pi^+\pi^-\pi^0$.

(Spin^{Parity}, Helicity)	Final State
$(2^+, \pm 2)$	$\rho^\pm \pi^\mp$
$(2^+, 0)$	$\rho^\pm \pi^\mp$
$(2^-, 0)$	$\rho^\pm \pi^\mp$
$(2^-, 0)$	$f_2 \pi^0$
$(0^-, 0)$	$\rho^\pm \pi^\mp$
$(0^-, 0)$	$f_2 \pi^0$
Isotropic	$\pi^+\pi^-\pi^0$

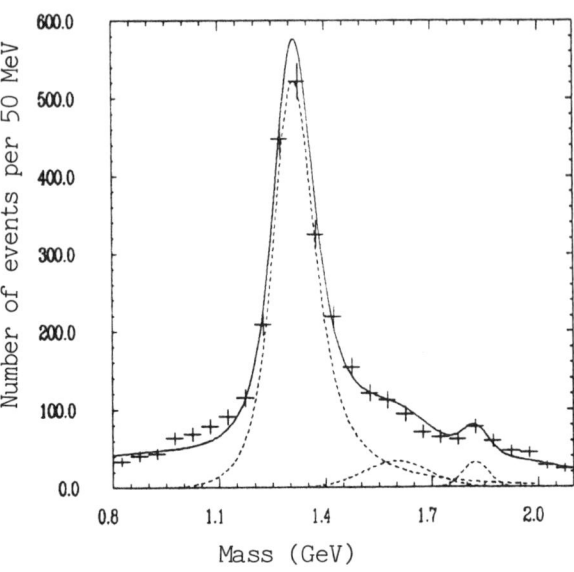

Figure 2. Measured $\pi^+\pi^-\pi^0$ mass spectrum obtained after event selection. the full line represents the sum of the contributions as determined by the partial-wave analysis: $a_2(1320)$, $\pi_2(1670)$ and $X(\approx 1800)$ (dashed lines) and a smooth spectrum of nonresonant 3π production.

contribution to the background comes from $\gamma\gamma \to \pi^+\pi^-\pi^0\pi^0$ where one π^0 was not detected. It amounts to 200 events as obtained from data of ref.[5]. Other sources contribute less than 10 events and can be neglected.

The experimental data were analysed using the maximum likelihood method in a similar way as it was done in the case of $\gamma\gamma \to \pi^+\pi^-\pi^+\pi^-$ [2]. The partial waves that were taken into account are listed in table 1.

Note that only $(2^+, \pm 2)$ and $(2^+, 0)$ are involved in $\gamma\gamma \to a_2$. A small contribution from $f_0(975)\pi^0$ was also detected but was neglected in the final analysis. The analysis was devided into two parts according to two regions of the three-pion invariant mass: 0.8 – 1.5 GeV/c^2 and 1.35 – 2.1 GeV/c^2. As expected, the low-mass region is dominated by $(2^+, \pm 2)$ from a_2. There are small contributions of $(0,0) \to \rho^\pm\pi^\mp$ [presumably $\pi(1300)$] and the "isotropic" channel. Using the PDG data[6] for the branching ratio $a_2 \to 3\pi$ a value of (0.90 ± 0.15) keV was obtained for the radiative partial width $\Gamma_{\gamma\gamma}(a_2)$. The contribution of $\Gamma_{\gamma\gamma}(J_z = 0)$ to the a_2 two-photon width is about 1%. For the product $\Gamma_{\gamma\gamma}(\pi_{1300}) \cdot$ Br$(\pi_{1300} \to \rho^\pm\pi^\mp)$ we found an upper limit of 0.1 keV (90% conf. level). The method was tested with events that were Monte-Carlo generated, and it was found that input data were consistently reproduced by the analysis.

In the region $m(3\pi) > 1.35$ GeV/c^2 the two-pion invariant-mass spectrum shows two peaks identified as f_0 and f_2. This is a clear indication that this region cannot be solely attributed to a_2 which cannot decay into $f_0\pi^0$ or $f_2\pi^0$. In the partial-wave analysis use was made of the known a_2 resonance parameters in order to maximize the accuracy. In the maximum-likelihood analysis we have, therefore, fixed the well established contribution of

a_2 and varied the amplitudes for

$$(2^-, 0) \to \left\{ \begin{array}{c} f_2 \pi^0 \\ \rho^\pm \pi^\mp \end{array} \right\},$$

$$(0^-, 0) \to \left\{ \begin{array}{c} f_2 \pi^0 \\ \rho^\pm \pi^\mp \end{array} \right\},$$

and isotropic distribution.

Figure 2 shows the results of the partial–wave analysis.

SUMMARY

In the vector pair production by two photons a partial wave analysis was performed of the reaction $\gamma\gamma \to \rho^0 \omega$. It was found that the $\gamma\gamma \to \rho^0 \omega$ contribution to the total $\gamma\gamma \to \pi^+ \pi^+ \pi^- \pi^- \pi^0$ cross section amounts to 12%. A signal is seen for the first time in the $\gamma\gamma \to \rho^0 \phi$ reaction. There is evidence for $\omega\phi$ production in two–photon interactions. No signal was observed for the reaction $\gamma\gamma \to \phi\phi$. A value $\Gamma_{\gamma\gamma} = (0.90\pm0.15)$ keV and a helicity ratio $\frac{\Gamma_{\gamma\gamma}(J_z = \pm 2)}{\Gamma_{\gamma\gamma}} = 0.99\pm0.05$ were deduced for a_2. An upper limit $\Gamma_{\gamma\gamma} \cdot \mathrm{Br}(\rho^\pm \pi^\mp) < 0.1$ keV was obtained for the resonance $\pi(1300)$. A peak at about 1620 MeV/c^2 in the three–pion invariant mass was identified as π_2. There is an indication of a state at 1800 MeV/c^2 with a prefered spin–parity assignemnt $J^P = 2^-$.

The presented results should be considered as preliminary since more detail simulation is needed to achieve the final accuracy.

REFERENCES

1. ARGUS Collaboration, H. Albrecht et al., "ARGUS: A Universal Detector at DORIS II," *Nucl. Instr. and Meth.* A275, pp. 1-48, (1989).

2. ARGUS Collaboration, H. Albrecht et al., "Observation of spin–parity 2$^+$ dominance in the reaction $\gamma\gamma \to \rho^0 \rho^0$ near threshold," *Z. Phys. C - Particles and Fields* 50, pp. 1-10, (1991).

3. CELLO Collaboration, H.-J. Behrend et al., "$a_2(1320)$ and $\pi_2(1670)$ formation in the reaction $\gamma\gamma \to \pi^+ \pi^- \pi^0$," *Z. Phys. C - Particles and Fields* 46, pp. 583-591, (1990).

4. Crystal Ball Collaboration, D. Antreasyan et al., "First observation of $\gamma\gamma \to \pi_2 \to \pi^0 \pi^0 \pi^0$," *Z. Phys. C - Particles and Fields* 48, pp. 561-566, (1990).

5. ARGUS Collaboration, H. Albrecht et al., "A Measurement of $\gamma\gamma \to \rho^+ \rho^-$," *Phys. Lett.* B217, pp. 205-219, (1989).

6. Particle Data Group, "Review of Particle Properties," *Phys. Rev.* D45, No. 11, (1992).

STUDY OF THE WAVE WITH $J^{PC} = 1^{-+}$ IN THE PARTIAL-WAVE ANALYSIS OF $\eta'\pi^-$, $\eta\pi^-$, $f_1\pi^-$ AND $\rho^0\pi^-$ SYSTEMS PRODUCED IN $\pi^- N$ INTERACTIONS AT $p_{\pi^-} = 37$ GeV/c.

VES Collaboration

Yu.P. Gouz, E.B. Berdnikov, S.I. Bityukov, G.V. Borisov, R.I. Dzhelyadin,
Yu.M. Ivanyushenkov, A.N. Karyukhin, I.A. Kachaev, Yu.A. Khokhlov,
G.A. Klyuchnikov, V.F. Konstantinov, M.E. Kostrikov, V.V. Kostyukhin, A.A. Kriushin,
V.V. Lapin, V.D. Matveyev, V.F. Obraztsov, A.P. Ostankov, D.I. Ryabchikov,
V.K. Semenov, E.A. Starchenko, N.K. Vishnevsky, E.V. Vlasov, A.M. Zaitsev,
Institute for High Energy Physics, 142284, Protvino, Moscow region, Russia,
G.M. Beladidze, T.A. Lomtadze, E.G. Tskhadadze,
Institute of Physics, 380077, Tbilisi, Georgia

Abstract

The wave with exotic quantum numbers $J^{PC} = 1^{-+}$ is studied in $\eta'\pi^-$, $\eta\pi^-$, $f_1\pi^-$ and $\rho^0\pi^-$ systems produced in $\pi^- N$ interactions at $p_{\pi^-} = 37$ GeV/c in the high-statistics experiment at IHEP VES setup. The magnitude of $J^{PC} = 1^{-+}$ wave in $\eta'\pi^-$ system significantly exceeds that in $\eta\pi^-$ system. Resonance-like structure with mass $M = 1.62 \pm 0.02$ GeV and width $\Gamma = 0.24 \pm 0.05$ GeV is observed in $\rho^0\pi^-$ system.

INTRODUCTION

The $J^{PC} = 1^{-+}$ wave has been studied using the data of high-statistics experiment at the VES setup of IHEP 70-GeV proton synchrotron at the incident beam momentum of $p_{\pi^-} = 37$ GeV/c. The setup is a large aperture magnetic spectrometer which includes proportional and drift chambers and a lead-glass electromagnetic calorimeter. The description of the VES setup can be found in [1, 2]. Statistics available for physical analysis is $\sim 8 \cdot 10^7$ reconstructed events in two data samples ($2 \cdot 10^7 + 6 \cdot 10^7$).

EXPERIMENTAL RESULTS

$\eta'\pi^-$ system

The $\eta'\pi^-$ system has been studied in the reaction

$$\pi^- N \to \pi^- \eta' N \quad (1)$$
$$\hookrightarrow \pi^+\pi^-\eta$$
$$\hookrightarrow \gamma\gamma$$

Effective mass distributions of $\gamma\gamma$, $\pi^+\pi^-\eta$, $\eta'\pi^-$ and t-distribution are shown in Fig.1 a), b), c), d), respectively. Events from two side bands near η'-meson peak have been used for the background subtraction. Total of ~ 4200 "$\eta'\pi$" events and ~ 1100 "background" events are selected for the analysis.

The partial-wave analysis of $\eta'\pi^-$ system has been done in 50-MeV bins on $\eta'\pi^-$ mass in $0 < -t' < 1$ GeV^2 interval. The waves with $J^{PC} M\eta = 0^{++}0-, 1^{-+}0-, 1^{-+}1-, 2^{++}0-, 2^{++}1-, 1^{-+}1+, 2^{++}1+$, further noticed as S0, P0, P-, D0, D-, P+, D+, are used in the fit. The following results are obtained:

- Contributions of all waves with unnatu-

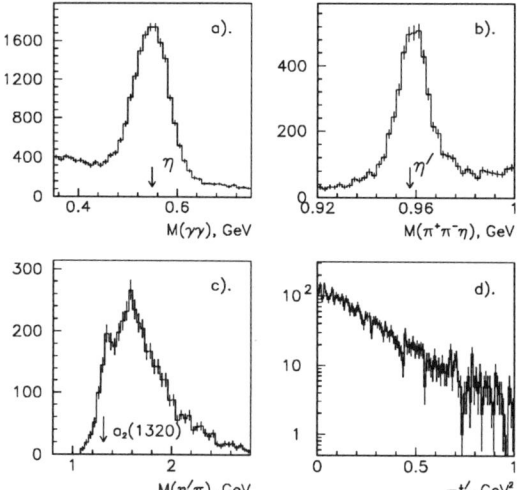

Figure 1. Effective mass distributions of $\gamma\gamma$ (a), $\pi^+\pi^-\eta$ (b), $\eta'\pi^-$ (c); t-distribution for $\eta'\pi^-$ events (d).

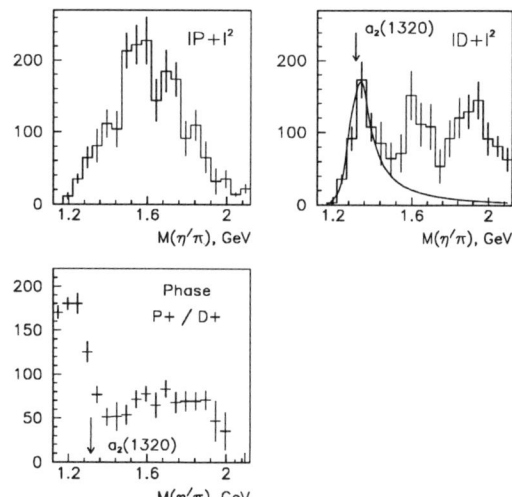

Figure 2. P+ and D+ waves and their relative phase in $\eta'\pi^-$ system.

ral parity exchange ($\eta = -1$) is compatible with zero;

- P+ wave is dominant; it is peaking at $M \sim 1.6$ GeV;

- In D+ wave the $a_2(1320)$ meson is seen, with some additional structure(s) at $M_{\eta'\pi^-} > 1.5$ GeV.

P+ and D+ waves and their relative phase are shown in Fig.2. Discrete ambiguities of the solution have been studied using the method given in [3]. All eight independent solutions are equal within the error limits.

$\eta\pi^-$ *system*

The $\eta\pi^-$ system has been studied in the reaction

$$\pi^- N \to \pi^- \eta N \quad (2)$$
$$\hookrightarrow \pi^+\pi^-\pi^0$$
$$\hookrightarrow \gamma\gamma$$

Effective mass distributions of $\gamma\gamma$, $\pi^+\pi^-\pi^0$, $\eta\pi^-$ and t-distribution are shown in Fig.3. Total of ~ 27000 "$\eta\pi$" events and ~ 5500 "background" events are selected for the analysis.

The partial-wave analysis of $\eta\pi^-$ system has been done in 40-MeV bins on $\eta\pi^-$ mass in $0 < -t' < 1$ GeV^2 interval using the same set of waves as that for $\eta'\pi^-$ system. In $\eta\pi^-$ system D+ wave dominates, while all other waves are negligibly small. In particular, 1^{-+} $\hat{\rho}^-(1405)$-meson and 0^{++} $a_0^-(1320)$-meson (see [6, 7]) are not observed.

The magnitudes $|T|^2$ (the wave intensities divided by the phase space and standard Blatt-Weisskopf factors) for P+ wave in $\eta'\pi^-$ and $\eta\pi^-$ systems are shown in Fig.5. The magnitude for $\eta'\pi^-$ system significantly exceeds that for $\eta\pi^-$ system. This feature had been predicted ([4], see also [5] and references therein) for 1^{-+} hybrids ($q\tilde{q}g$).

$f_1(1285)\pi^-$ *system*

The $f_1(1285)\pi^-$ system has been studied in the reaction

$$\pi^- N \to \pi^- f_1 N \quad (3)$$
$$\hookrightarrow \pi^+\pi^-\eta$$
$$\hookrightarrow \gamma\gamma$$

574 Study of the Wave with $J^{PC}=1^{-+}$

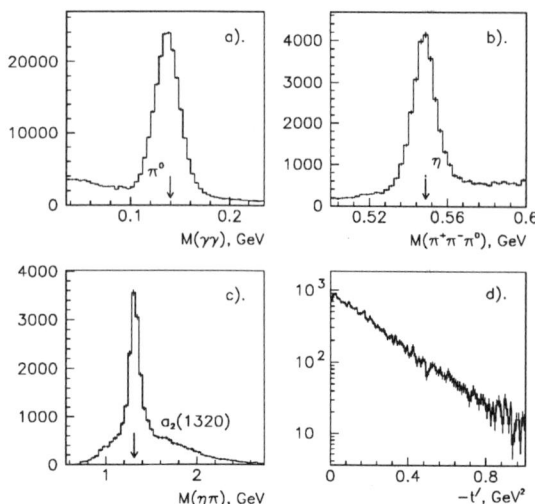

Figure 3. Effective mass distributions of $\gamma\gamma$ (a), $\pi^+\pi^-\pi^0$ (b), $\eta\pi^-$ (c); t-distribution for $\eta\pi^-$ events (d).

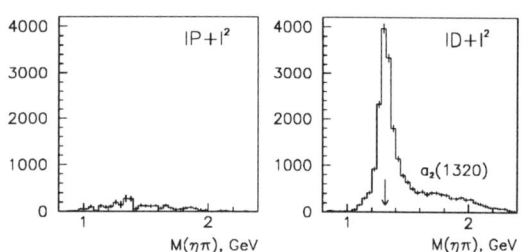

Figure 4. P+ and D+ waves in $\eta\pi^-$ system.

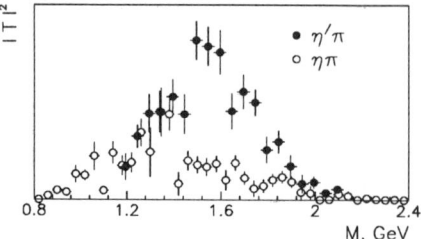

Figure 5. Magnitudes of P+ wave in $\eta'\pi^-$ and $\eta\pi^-$ systems.

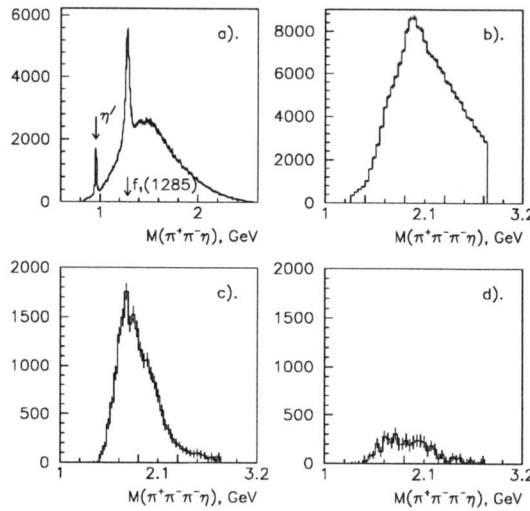

Figure 6. The effective mass spectrum of $\pi^+\pi^-\eta$ subsystem (a); total $\pi^+\pi^-\pi^-\eta$ statistics (acceptance corrected) (b); results of PWA for the $f_1\pi^-$ waves with $J^{PC}M\eta = 1^{++}0+$ (c) and $1^{-+}1+$ (d).

Data are obtained from the partial-wave analysis of $\pi^+\pi^-\pi^-\eta$ system, because $f_1(1285)$ is not a narrow resonance. Total of $1.02 \cdot 10^5$ $\pi^+\pi^-\pi^-\eta$ events has been used in the analysis. The effective mass spectrum of $\pi^+\pi^-\eta$ subsystem is shown in fig.7 a), the peak of $f_1(1285)$ meson is clearly seen.

The partial-wave analysis of $\pi^+\pi^-\pi^-\eta$ system has been done in 40-MeV bins on $\pi^+\pi^-\pi^-\eta$ mass in $0 < -t' < 1$ GeV^2 interval. Amplitudes were calculated using Zemach tensor formalism; fitting procedure of Illinois PWA program used. The $\eta'\pi^-$ events are excluded from this analysis because of difficulties for the analysis caused by the narrowness of η'-

meson. The set of 17 nonzero waves including 1^{-+} ($f_1\pi$) has been defined iteratively. The main wave is $J^{PC}M\eta = 1^{++}0 + (f_1\pi^-)$. The 1^{-+} wave is seen but small. The results on 1^{++} and 1^{-+} $f_1\pi^-$ waves are shown in fig.7 c), d).

$\rho^0\pi^-$ system

The partial-wave analysis of $\pi^+\pi^-\pi^-$ system has been done using the version of Illinois PWA program. Number of events selected for the analysis is $\sim 2.8 \cdot 10^6$, which is about $\frac{1}{7}$

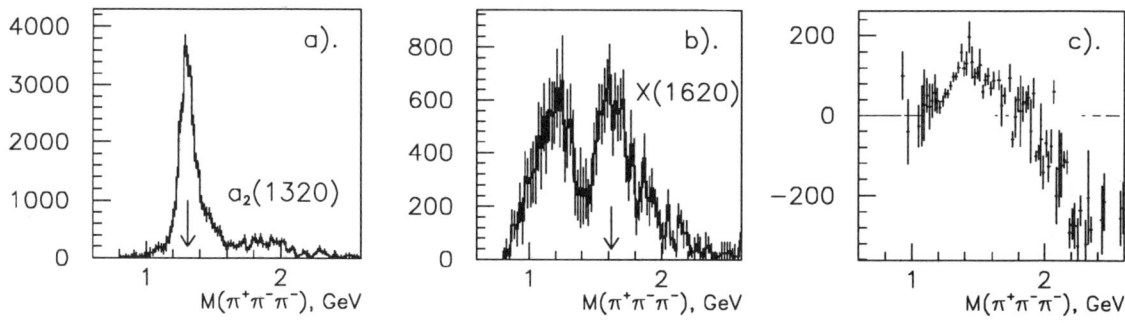

Figure 7. The $\rho^0\pi^-$ waves with $J^{PC}LM\eta = 2^{++}D1-$ (a) and $1^{-+}P1+$ (b); their relative phase (c).

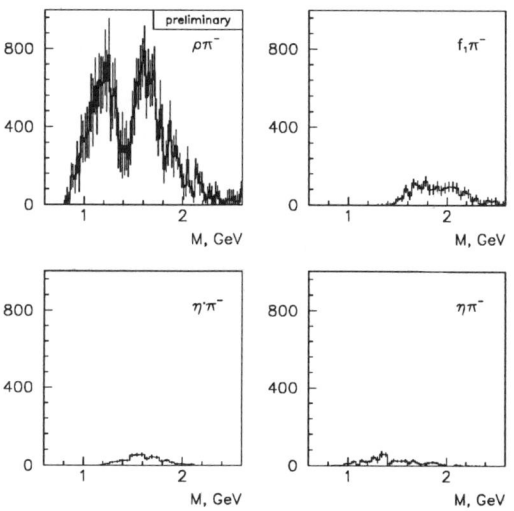

Figure 8. Relative 1^{-+} wave intensities in $\rho^0\pi^-$, $f_1\pi^-$, $\eta'\pi^-$ and $\eta\pi^-$ systems, arbitrary units.

of the total $\pi^+\pi^-\pi^-$ statistics. The PWA has been done independently in four t' intervals:

$0 \ \ \ < -t' < 0.06$ GeV^2
$0.06 < -t' < 0.2 \ \ $ GeV^2
$0.2 \ < -t' < 0.6 \ \ $ GeV^2
$0.6 \ < -t' < 1 \ \ \ \ $ GeV^2.

Isobars in $\pi^+\pi^-$ subsystem are ϵ ($\pi^+\pi^-$ S-wave), ρ^0 and f_2 in standard parametrization. Number of waves included in the analysis is 30. All well-established (as, for example, $\pi_2^-(1670)$, $\pi^-(1300)$, $a_2^-(1320)$) and some not well-established (as $\pi^-(1770)$) are clearly seen, with proper phase variation between them. The waves with $J^{PC}LM\eta = 2^{++}D1+$ and $1^{-+}P1+$ in the $\rho^0\pi^-$ system at the second t'-interval and their relative phase are shown in Fig.7. It is seen that 1^{-+} wave is *not small* ($\sim \frac{1}{6}$ of $a_2^-(1320)$); it has not only a threshold bump, as could be expected, but in addition it shows a second peak at higher masses. It should be noticed as a preliminary result that this second peak can be interpreted as a resonance (there is a proper phase variation with respect to 2^{++} wave); its parameters are: $M = 1.62 \pm 0.02$ GeV, $\Gamma = 0.24 \pm 0.05$ GeV. It is worth mentioning that the peak in $\eta'\pi^-$ P+ wave could have the same origin; the behaviour of the relative phase between P+ and D+ waves cannot be easily interpreted because the D+ wave is not flat enough in this mass region (see Fig.2).

Relative intensities of 1^{-+} wave in $\rho^0\pi^-$, $f_1\pi^-$, $\eta'\pi^-$ and $\eta\pi^-$ systems are shown in fig.8.

CONCLUSIONS

The following results are obtained studying the production of $J^{PC} = 1^{-+}$ wave at the beam momentum of 37 GeV/c:

1. The wave with $J^{PC} = 1^{-+}$ dominates in the $\eta'\pi^-$ final state. It peaks at $M_{\eta'\pi^-} \sim 1.6$ GeV/c;

2. In the $\eta\pi^-$ system the $J^{PC} = 2^{++}$ is dominant. The contribution of all other waves (including $J^{PC} = 1^{-+}$) is small.

3. $|T(1^{-+} \to \eta'\pi^-)|^2 > |T(1^{-+} \to \eta\pi^-)|^2$ at $M_{\eta'\pi^-(\eta\pi^-)} > 1.4$ GeV.

4. In the $f_1(1285)\pi^-$ final state $J^{PC} = 1^{++}$ wave dominates. The 1^{-+} wave is small but seen.

5. The $J^{PC} = 1^{-+}$ is observed in the partial-wave analysis of the reaction $\pi^- N \to \pi^+\pi^-\pi^- N$. The structure in $J^{PC} = 1^{-+}$ ($\rho^0\pi^-$) wave is observed which could be interpreted as a resonance with $M = 1.62 \pm 0.02$ GeV and $\Gamma = 0.24 \pm 0.05$ GeV (the resonance interpretation must be considered as preliminary).

REFERENCES

1. S. I. Bityukov et al, "Observation of resonance with mass $m = 1814$ MeV decaying into $\pi^-\eta\eta$," Phys. Lett. B 268, pp. 137–141, (1991).

2. A. Zaitsev, "Recent results from VES detector at IHEP," presented at the IVth International Conference on Hadron Spectroscopy, Univ. of Maryland, College Park, USA, August 12-16, 1991.

3. S. Sadovsky, "On the ambiguities in the partial wave analysis of $\pi^- p \to \eta\pi^0 n$ reaction," Preprint IHEP 91-75, Protvino, 1991.

4. F. E. Close and H. J. Lipkin, "New experimental evidence for four-quark exotics. The Serpukhov $\phi\pi$ resonance and GAMS $\eta\pi$ enhancement," Phys. Lett. B 196, pp. 245–250, (1987).

5. F. Iddir et al, "$q\tilde{q}g$ hybrid and $q q \tilde{q} \tilde{q}$ diquonium interpretation of the GAMS 1^{-+} resonance," Phys. Lett. B 205, pp. 564–568, (1988).

6. M. Boutemeur and M. Poulet, "A new scalar meson decaying into $\eta\pi^0$ and results of $\eta'\pi^0$ analysis at 100 GeV/c," in *Proceedings of the IIIrd International Conference on Hadron Spectroscopy*, 1989, pp 119–125.

7. D. Alde et al, "Evidence for a 1^{-+} exotic meson," Phys. Lett. B 205, pp. 397–400, (1988).

SPECTROSCOPY OF THE D–WAVE $q\bar{q}$ SYSTEM; EVIDENCE FOR TWO $J^P = 2^-$ STRANGE MESON STATES DECAYING TO $K^-\omega$*

B.N. RATCLIFF,[1] D. ASTON,[1] N. AWAJI,[2] T. BIENZ,[1] F. BIRD,[1] J. D'AMORE,[3]
W. DUNWOODIE,[1] R. ENDORF,[3] K. FUJII,[2] H. HAYASHII,[2] S. IWATA,[2]
W.B. JOHNSON,[1] R. KAJIKAWA,[2] P. KUNZ,[1] Y. KWON,[1] D.W.G.S. LEITH,[1]
L. LEVINSON,[1] T. MATSUI,[2] B.T. MEADOWS,[3] A. MIYAMOTO,[2] M. NUSSBAUM,[3]
H. OZAKI,[2] C.O. PAK,[2] P. RENSING,[1] D. SCHULTZ,[1] S. SHAPIRO,[1]
T. SHIMOMURA,[2] P. K. SINERVO,[1] A. SUGIYAMA,[2] S. SUZUKI,[2]
T. TAUCHI,[2] N. TOGE,[1] K. UKAI,[4] A. WAITE,[1] S. WILLIAMS[1]

[1]Stanford Linear Accelerator Center, Stanford University, CA 94309, USA
[2]Department of Physics, Nagoya University, Furo-cho, Chikusa-ku, Nagoya 464, JAPAN
[3]Department of Physics, University of Cincinnati, Cincinnati, OH 45221, USA
[4]Institute for Nuclear Study, University of Tokyo, 3-2-1 Midori-cho, Tanashi-shi, Tokyo 188, JAPAN

ABSTRACT

Evidence is presented for two $J^P = 2^-$ strange mesons; one at ~ 1.77 and the other at ~ 1.82 GeV/c^2. These states have been observed in a partial wave analysis of the $K^-\omega$ system in the reaction $K^-p \to K^-\pi^+\pi^-\pi^0 p$ where the strange mesons decay into $K^-\omega$ and the ω then decays to $\pi^+\pi^-\pi^0$. The data set contains $\sim 10^5$ $K^-\omega p$ events at 11 GeV/c taken with the LASS spectrometer at SLAC.

INTRODUCTION

Even though evidence for the quark model is very strong, the correct $q\bar{q}$ quark model assignments of all the known mesons are far from clear. There are a number of states that are experimentally "missing," even for the low spin multiplets, and there is considerable controversy in the assignment of several of the light non-strange mesons. This is particularly true for the 0^{++} multiplet where there are "too many" candidate states. Table 1 reproduces the suggested assignments of the Particle Data Group (PDG).[1] The most obvious "hole" is in the D-wave 2^- sector, where there is no specific $q\bar{q}$ combination with good candidates for both singlet and triplet 2^- states. This is true even for the strange meson spectrum, which is the best understood of any $q\bar{q}$ system. Figure 1 shows the strange meson spectrum from the LASS/SLAC Kp program before the analysis presented here today. Even though the number of strange states observed is quite large, with orbitally excited states up to 5^- and with a significant number of triplet and radially excited candidates, the expected level structure is only complete for the $L = 0$ and $L = 1$ ground states. Completing the D-wave ($L = 2$) singlet and triplet levels would sharpen comparison of the experimental data with the models considerably,[2] particularly for the spin-dependent forces.

In this paper, we present evidence for two strange 2^- states in the $K_2(1770)$ region. These results are taken from a high-statistics study of the $K\omega$ system produced in the reaction

$$K^-p \to K^-\pi^+\pi^-\pi^0 p \quad (1)$$

at 11 GeV/c. The data were obtained with the Large Aperture Superconducting Solenoid

---

* Work supported by Department of Energy contract DE–AC03–76SF00515; the National Science Foundation under grant Nos. PHY82-09144, PHY85-13808, and the Japan-US Cooperative Research Project on High-Energy Physics.

Table I. From PDG Phys. Rev. D III. 69 1 June (1992)									
$N^{2S+1}L_J$	J^{PC}	$u\bar{d}, u\bar{u}, d\bar{d}$ $I=1$	$u\bar{u}, d\bar{d}, s\bar{s}$ $I=0$	$c\bar{c}$ $I=0$	$b\bar{b}$ $I=0$	$\bar{s}u, \bar{s}d$ $I=1/2$	$c\bar{u}, c\bar{d}$ $I=1/2$	$c\bar{s}$ $I+0$	$\bar{b}u, \bar{b}d$ $I+1/2$
1^1S_0	0^{-+}	π	η, η'	η_c		K	D	D_s	B
1^3S_1	1^{--}	ρ	ω, ϕ	$J/\psi(1S)$	$\Upsilon(1S)$	$K^*(892)$	$D^*(2010)$	$D_s(2110)$	$B^\beta(5330)$
1^1P_1	1^{+-}	$b_1(1235)$	$h_1(1170), h_1(1380)$			K_{1B}†	$D_1(2420)$	$D_{s1}(2536)$	
1^3P_0	0^{++}	$a_0(980)$	$f_0(1400), f_0(975)$	$\chi_{c0}(1P)$	$\chi_{b0}(1P)$	$K_0^*(1430)$			
1^3P_1	1^{++}	$a_1(1260)$	$f_1(1285), f_1(1510)$	$\chi_{c1}(1P)$	$\chi_{b1}(1P)$	K_{1A}†			
1^3P_2	2^{++}	$a_2(1320)$	$f_2(1270), f_2'(1525)$	$\chi_{c2}(1P)$	$\chi_{b2}(1P)$	$K_2^*(1430)$	$D_2^*(2460)$		
1^1D_2	2^{-+}	$\pi_2(1670)$							
1^3D_1	1^{--}	$\rho(1700)$	$\omega(1600)$	$\psi(3770)$		$K^*(1680)$			
1^3D_2	2^{--}					$K_2(1770)$			
1^3D_3	3^{--}	$\rho_3(1690)$	$\omega_3(1670), \phi_3(1850)$			$K_3^*(1780)$			
1^3F_4	4^{++}	$a_4(2040)$	$f_4(2050), f_4(2220)$			$K_4^*(2045)$			
2^1S_0	0^{-+}	$\pi(1300)$	$\eta(1295)$	$\eta_c(2S)$		$K(1460)$			
2^3S_1	1^{--}	$\rho(1450)$	$\omega(1390), \phi(1680)$	$\psi(2S)$	$\Upsilon(2S)$	$K^*(1410)$			
2^3P_2	2^{++}		$f_2(1810), f_2(2010)$		$\chi_{b2}(2P)$	$K_2^*(1980)$			
3^1S_0	$0^{\pm}$	$\pi(1770)$	$\eta(1760)$			$K(1830)$			
† The K_{1A} and K_{1B} are nearly 45° mixed states of the $K_1(1270)$ and $K_1(1400)$.									

Table 1. The $q\bar{q}$ quark model assignments suggested by the PDG[1] for most of the experimentally known low spin meson systems.

(LASS) spectrometer at SLAC, which is described in detail elsewhere.[3] Since the LASS spectrometer was not equipped with a photon detector, the π^0 in the final state is not seen directly, but is reconstructed in the missing π^0 channel. The $K^-\omega p$ sample of $\sim 10^5$ events obtained in this experiment is at least 25 times larger than that obtained in any previous experiment.

DATA AND ANALYSIS

The $\pi^+\pi^-\pi^0$ mass spectrum of Figure 2 shows a clear ω signal, with signal to background ratio about one to one in the signal region (0.72–0.84 GeV/c^2). It is clear from the Dalitz plot (not shown) that the high-mass $K\omega$ region overlaps with substantial production of several baryon resonances, so events with $M_{p\omega} < 2.28$ or $M_{pK} < 2.0$ GeV/c^2 are eliminated. The $K^-\pi^+\pi^-\pi^0$ effective mass distribution in the ω region [Fig. 3(a)] shows peaks in the $K^-\omega$ threshold region and in the region around 1.75 GeV/c^2. The shaded histogram shows the events with the baryon resonance region removed. Most of the high mass $K^-\omega$ events lie in the overlap region and are removed by this cut.

The analysis, more details of which can be found elsewhere,[4] is performed using joint decay spherical-harmonic moments in the $K^-\omega$ Gottfried-Jackson frame and the ω rest frame, using the normal to the decay plane as the analyzer. The $\pi^+\pi^-\pi^0$ mass spectrum shown in Fig. 2 contains a significant background under the ω peak region (6.72 to 0.84 GeV/c^2). Each moment is background subtracted using the ω sideband regions indicated (0.64 to 0.70 GeV/c^2 and 0.86 to 0.92 GeV/c^2) and acceptance corrected, after which the partial

The background and acceptance corrected $K^-\omega$ mass distribution obtained is shown in Fig. 3(b). The main features are similar to those observed for the uncorrected data. There is a strong peak at threshold, and a large bump in the 1.7 to 1.8 GeV/c^2 region, with some evidence for a smaller structure around 1.5 GeV/c^2.

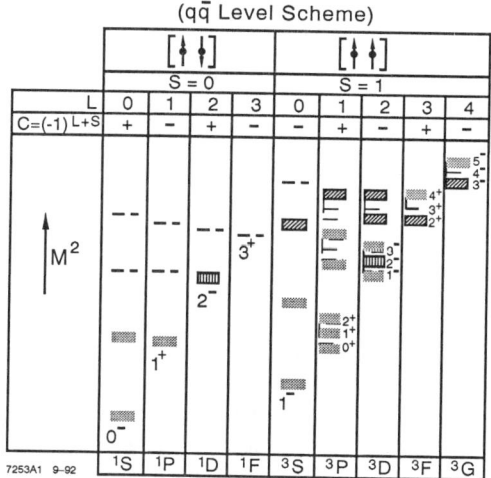

Figure 1. Level diagram (Grotrian plot) for the strange meson states known before this analysis. All of these states have been observed in the LASS/SLAC group B program. The dashed lines indicate the lowest lying states expected in the quark model. The mass levels are illustrative only. The states indicated by cross-hatching are clearly observed, and have generally been confirmed; in a few cases there are possible classification ambiguities. Only one strange 2^- meson has been confirmed. It could be either the S=0 or the S=1 state, or perhaps a mixture, as shown by the vertically lined boxes. The states indicated by diagonal lines are more speculative and require confirmation.

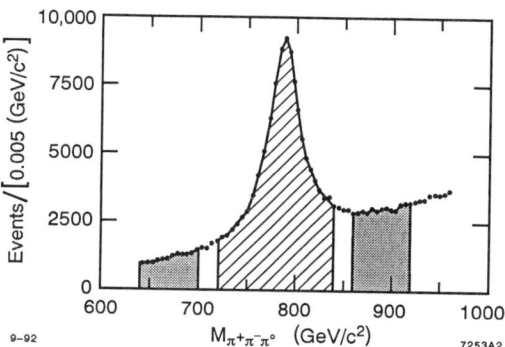

Figure 2. The $\pi^+\pi^-\pi^0$ invariant mass distribution; the signal region is diagonally lined while the background region is shaded.

THE PARTIAL WAVE STRUCTURE

The low mass $K^-\omega$ region is dominated by 1^+ waves (not shown), while the mass bump around 1.75 GeV/c^2 is dominantly 2^-. Figure 4 shows the incoherently summed intensity of all the 2^- waves. There is a large and rather broad bump centered around 1.75 GeV/c^2. The much smaller $K^-\omega$ decays of the leading $K_2^*(1430)$ (not shown) and the $K_3^*(1780)$ are also observed with branching ratios which are consistent with predictions from SU(3). Figure 5 shows the real and imaginary parts of the PWA amplitudes for the $J^P = 2^-$ and 3^- waves that are significant in the 1.75 GeV/c^2 region. In addition to the bump in the 3^- amplitude corresponding to $K_3^*(1780)$ production, there is a substantial and rather complicated structure in the different 2^- amplitudes.

However, this structure can be explained in a straightforward manner as follows: First, the 3^- amplitudes are fit to a single Breit-Wigner (B-W) resonance model to define a phase reference in the 1750 MeV/c^2 mass region. Then the 2^- and 3^- amplitudes are fit simultaneously with their relative phases and magnitudes as free parameters. Two different models are compared: (1) that the 2^- waves observed in the 1750 mass region come from a

580 Spectroscopy of the D-Wave $q\bar{q}$ System

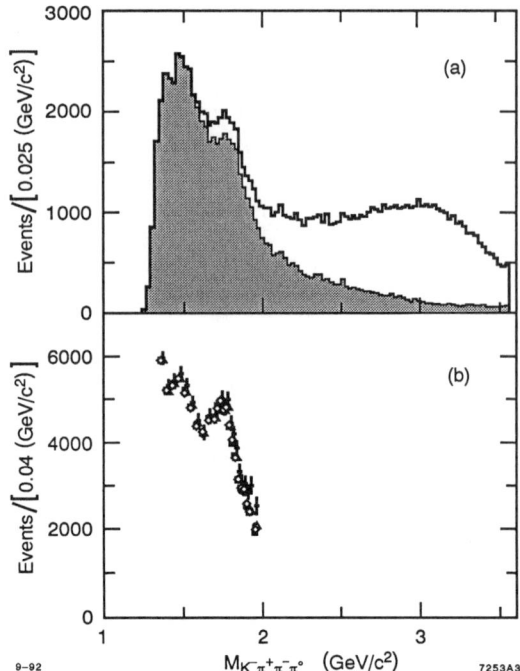

Figure 3. The $K - \pi^+\pi^-\pi^0$ invariant mass distribution for events with $0.1 < |t'| < 2.0$ (GeV/c)^2; (a) the unshaded curve contains all events that satisfy $0.72 < M_{3\pi} < 0.84$ GeV/c^2 while the shaded portion contains events with $M_{p\omega} > 2.28$ and $M_{pK} > 2.0$ GeV/c^2; (b) the background-subtracted and acceptance-corrected mass distribution; the points with error bars are the measured values and the other points are the values obtained in the PWA fit discussed in the text.

single resonance, and (2) that they come from two resonances. The dotted curves in Fig. 5 show the fit results of hypothesis (1). The fitted mass and width of the 2^- resonance are 1728 ± 7 MeV/c^2 and 221 ± 22 MeV/c^2, respectively. The χ^2 is 128.9 for 116 degrees of freedom. The one-resonance fit does not reproduce the 2^-1^+F wave at all well. Moreover,

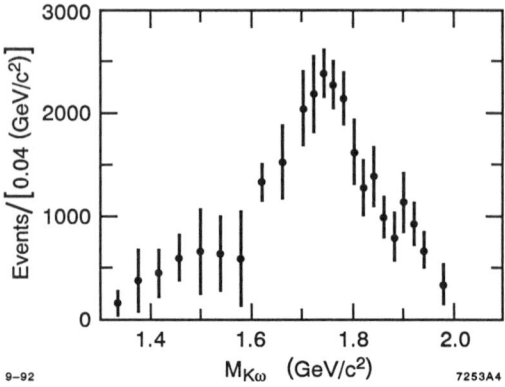

Figure 4. The summed intensities of the 2^- waves.

the dip at ~ 1.84 GeV/c^2 in the $Re(2^-0^+P)$ and the tail of $Re(2^-0^+F)$ are not well represented by the fit.

On the other hand, the fit results of hypothesis (2) represented by the solid curves in Fig. 5 reproduce all of the amplitudes very well and provide a significantly better fit to the data with a χ^2 of 70.6 for 110 degrees of freedom. The fitted masses of the two resonances are 1773 ± 8 and 1816 ± 13 MeV/c^2 and the fitted widths are 186 ± 14 and 276 ± 35 MeV/c^2, respectively. In this model, the 2^-1^+F wave is almost entirely the higher mass resonance, but the other amplitudes contain rather large contributions from both resonances, and overall, each resonance is observed with nearly equal strength into P and F waves. The χ^2 difference is almost 60 units between the one and two resonance models, which is a nominal "Gaussian" significance level for the second resonance of more than 7σ. In fact, since the PWA typically has a number of nearby solutions, the error bars shown and used in the fit tend to be overestimated compared to Normal errors. Thus, the absolute values of the χ^2 for these fits, the difference between the models, and the significance of the second resonance in the fit, will all tend to be underestimated. Clearly,

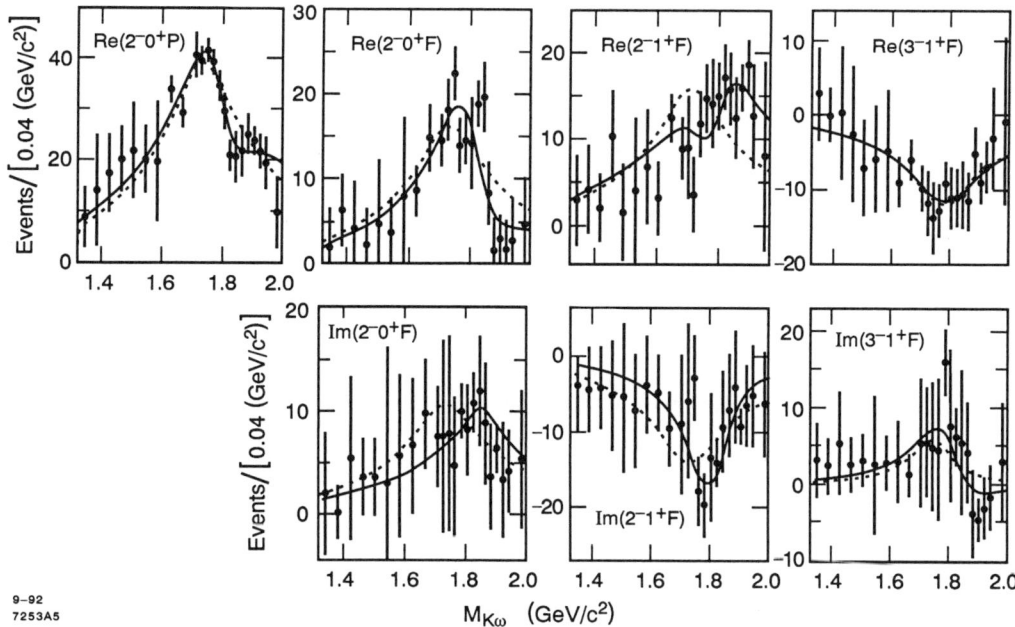

Figure 5. The real and imaginary parts of the $K^-\omega$ 2^- and 3^- amplitudes; the lines show the results of the fits described in the text.

the data strongly prefer the model with two 2^- resonances.

SUMMARY

A partial wave analysis of a high-statistics sample of $\sim 10^5 K^-\omega p$ events provides very good evidence for two 2^- strange resonances with masses around 1773 and 1816 MeV/c^2. This is the only $q\bar{q}$ D-wave spectrum with good candidates for all ground state singlet and triplet levels. The singlet/triplet classification of these states is unclear since strange mesons are not eigenstates of charge conjugation C, and the experimental states can therefore be mixed (as are the $^1P_1$ and $^3P_1$ strange states). It is interesting that Godfrey and Isgur[2] predict masses of 1780 and 1810 MeV/c^2 for the unmixed $^1D_2$ and $^3D_2$ states respectively, remarkably close to the experimental values. Kokowski and Isgur[5] also predict that the pure states are essentially decoupled, with the lower mass state decaying mostly to P-wave and the higher mass state to F-wave. Experimentally, production can complicate matters, but the 2^-1^+F wave is explained as mostly the higher mass state, as predicted. On the other hand, both resonances contribute significantly to the other P and F wave amplitudes and, overall, the resonances appear with roughly equal total strengths in the P and F wave amplitudes.

REFERENCES

1. K. Hikasa et al., Phys. Rev. D45 (1992).
2. See, for example, S. Godfrey and N. Isgur, Phys. Rev. D32(1985) 189.
3. D. Aston et al., SLAC-REP-298 (1986).
4. D. Aston et al., SLAC-PUB-5634 (1991); also see, Ph.D. thesis of Y. Kwon (to be published).
5. R. Kokowski and N. Isgur, Phys. Rev. D35(1987) 907.

ANOTHER DIRAC OSCILLATOR*

Lorella M. Jones
Physics Department
University of Illinois at Urbana-Champaign
1110 W. Green Street
Urbana, IL 61801-3080 (USA)

Abstract

A novel form for the "Dirac" harmonic oscillator is discussed.

Both nuclear and particle physics have situations which can be approximated by studying a spin $\frac{1}{2}$ particle trapped in an infinitely deep potential well. In dealing with such cases, it is always useful to know the results of similar exactly soluble problems. For this reason, there has been some activity in the literature dealing with equations for the Dirac oscillator.[1,2,3,4]

Today I want to show you a different approach to creation of a Dirac equation for the harmonic oscillator. Beginning with the free particle equation

$$(\gamma^\mu \partial_\mu)\Psi = (\text{const})\Psi \tag{1}$$

we see that the "mathematically obvious" oscillator equation should take the form

$$(\gamma^\mu \mathcal{O}_\mu)\Psi = (\text{const})\Psi \tag{2}$$

where $\mathcal{O}_\mu$ is a 4-vector *composed entirely of space-time operators*, and such that the square of $\mathcal{O}_\mu$ is a standard harmonic oscillator.

For simplicity, normalize the 1-d harmonic oscillator to have equation

$$\left(-\frac{\partial^2}{\partial x^2} + x^2\right)\Psi = E\Psi \tag{3}$$

*This work is supported in part by the U.S. Department of Energy under contract DOEACO 276ER011955 Task P and grant DOEFG0291ER40677 Task P

with eigenvalues $E = 2m + 1$. Now consider the Hermitian operator

$$\mathcal{D} = -i\left(\frac{\partial}{\partial x} - Qx\right) \tag{4}$$

where Q is the inversion operator ($Qx = -xQ$, $Qt = -tQ$, etc.). From the property

$$\mathcal{D}^2 = -\frac{\partial^2}{\partial x^2} + x^2 - Q = H - Q \tag{5}$$

we see that $\mathcal{D}$ is basically the "square root" of the normal harmonic oscillator. Sample eigenvalues and eigenstates of $\mathcal{D}$ are

$$\begin{aligned} \lambda &= 0 & & e^{-x^2/2} \\ \lambda &= +2 & & (1 - 2x^2 + 2ix)e^{-x^2/2} \\ \lambda &= -2 & & (1 - 2x^2 - 2ix)e^{-x^2/2} \end{aligned} \tag{6}$$

A 4-vector of operators $\mathcal{O}_\mu$ based on $\mathcal{D}$ is easily constructed.

$$\begin{aligned} \mathcal{O}_x &= -i\left(\frac{\partial}{\partial x} - Qx\right); \\ \mathcal{O}_z &= -i\left(\frac{\partial}{\partial z} - Qz\right); \\ \mathcal{O}_y &= -i\left(\frac{\partial}{\partial y} - Qy\right); \\ \mathcal{O}_t &= i\left(\frac{\partial}{\partial t} + Qt\right) \end{aligned} \tag{7}$$

We then are led directly to the Dirac oscillator equation

$$(\gamma_o \mathcal{O}_t - \gamma_x \mathcal{O}_x - \gamma_y \mathcal{O}_y - \gamma_z \mathcal{O}_z)\Psi = W\Psi = \lambda\Psi \tag{8}$$

The equation has some non-zero eigenvalues of the form $\lambda^2 = -4(N+1)$. This is exactly what we expect for a harmonic oscillator. (The minus sign comes from our "time minus space" metric). See my preprint[5] and the Appendix below for the detailed form of the eigenstates.

There are also some zero eigenvalue solutions. The eigenfunctions corresponding to these have a sort of "stringlike" shape: if α is an arbitrary 2-component spinor[6] and f and g are arbitrary functions, these $\lambda = 0$ solutions take forms like

$$\begin{pmatrix} \alpha \\ -\sigma_z\alpha \end{pmatrix} f(t+z)e^{(t^2-x^2-y^2-z^2)/2}$$
$$\begin{pmatrix} \alpha \\ \sigma_z\alpha \end{pmatrix} g(t-z)e^{(t^2-x^2-y^2-z^2)/2} \quad \text{and} \quad (9)$$

Another interesting feature of the equation is that the square of the operator, W^2 has a "spin-orbit"–like piece of the form

$$\gamma_x\gamma_y(\mathcal{O}_x\mathcal{O}_y - \mathcal{O}_y\mathcal{O}_x) = \gamma_x\gamma_y 2Q\left(x\frac{\partial}{\partial y} - y\frac{\partial}{\partial x}\right) \quad (10)$$

Comparison of our equation with the Moshinsky-Szczepaniak Dirac oscillator as written in Refs. 1 & 2 shows that the matrix structure is different. Some features are similar, however: The squares of their eigenvalues are linear in an integer, and they do have a spin-orbit coupling.[3]

The scalar relativistic oscillator was studied by Kim & Noz[7]. Their equation might be cast in the form

$$\left(\mathcal{O}_t^2 - \mathcal{O}_x^2 - \mathcal{O}_y^2 - \mathcal{O}_z^2\right)\Psi = (\text{const})\Psi \quad (11)$$

which is the obvious "Klein-Gordon" equation corresponding to our "Dirac" equation.

As yet, none of these approaches has been used extensively in practical applications. Hence, it is not clear whether any of them are correct. The derivation given here does, however, teach us some interesting and possibly more generally applicable things: (i) The inversion (parity) operator can be used in an equation; (ii) There is a Hermitian operator which is basically the square root of the harmonic oscillator, and (iii) The solutions displayed in Eq. 9 have the feature that "positive energy" and "negative energy" solutions seem to correspond to right-moving and left-moving "strings".

APPENDIX

To find the non-zero eigenvalue states with $\lambda^2 = -4(N+1)$, use states of the form

$$\begin{pmatrix} \chi_u \\ \chi_\ell \end{pmatrix} \quad \text{with} \quad \chi_\ell = \pm Q\chi_u \quad (12)$$

(since we observe that, if Ψ is a solution to Equation (8), so is $Q\gamma_5\Psi$).

The solutions with $\chi_\ell = +Q\chi_u$, using variable $\xi = t^2 - r^2$, take the form

$$e^{-\xi}\chi_u = h_1(\xi) + (t - \boldsymbol{\sigma}\cdot\mathbf{r})h_2(\xi) \quad (13)$$

Whereas those for which $\chi_\ell = -Q\chi_u$ take the form

$$e^{-\xi}\chi_u = h_1(\xi) + (t + \boldsymbol{\sigma}\cdot\mathbf{r})h_2(\xi) \quad (14)$$

The auxiliary functions h_1 and h_2 for both cases obey the coupled differential equations

$$\begin{aligned} 2i\frac{\partial h_1}{\partial \xi} &= \lambda h_2 \\ i\left(4h_2 + 2\xi h_2 + 2\xi\frac{\partial h_2}{\partial \xi}\right) &= \lambda h_1 \end{aligned} \quad (15)$$

and the eigenvalues $\lambda^2 = -4(N+1)$ are determined by requiring that h_1 and h_2 be represented by finite series.

REFERENCES

1. M. Moshinsky and A. Szczepaniak, "The Dirac Oscillator," *J. Phys. A.* 22, pp L817–L819 (1989)

2. M. Moreno and A. Zentella, "Covariance, CPT and the Foldy-Wouthuysen Transformation for the Dirac Oscillator," *J. Phys. A.* 22, pp L821–L825 (1989)

3. J. Benitez, R. P. Martinez y Romero, H. N. Nunez-Yepez and A. L. Salas-Brito, "Solution and Hidden Supersymmetry of a Dirac Oscillator," *Phys. Rev. Letters* 64, pp 1643–1645 (1990)

4. O. Castenos, A. Frank, R. Lopez and L. F. Urrutia, "Soluble Extensions of the Dirac Oscillator with Exact and Broken Supersymmetry," *Phys. Rev. D* 43, pp 544–547 (1991)

5. L. M. Jones, "Another Dirac Oscillator," Preprint ILL-(TH)-91-24 (submitted to this conference)

6. I use the Bjorken and Drell gamma matrices. J. D. Bjorken and S. D. Drell, *Relativistic Quantum Mechanics*, New York, McGraw Hill (1964), p 282

7. Y. S. Kim and M. E. Noz, *Theory and Applications of the Poincaré Group*, Boston, Reidel (1986) p 111 et. seq.

EVIDENCE FOR THE BOX ANOMALY IN η AND η' DECAYS

Michael Feindt
PPE Division, CERN, CH-1211 Geneva, Switzerland
representing the authors of ref. [1]

Abstract

Refitting all available data on η and η' decays into $\pi^+\pi^-\gamma$ we find an important non-resonant contribution below the dominating ρ-resonance. We show that the shape and size of the continuum is consistent with predictions if we identify it as stemming from the box anomaly.

INTRODUCTION AND OUTLINE

Lots of experiments having reconstructed η' decays into $\pi^+\pi^-\gamma$ have noticed that the invariant $\pi^+\pi^-$ mass spectrum is not well described by the expectation of a single ρ resonance. We therefore reviewed and refitted all available data [1]. In order to get a reliable ρ resonance parametrization we also refitted $e^+e^- \to \pi^+\pi^-$ cross sections. Surprisingly, these are not well fitted by a single ρ with standard PDG values, which leaves some freedom in the analysis; and we adopt two extreme models.

We find that η and η' decays are much better described when a continuum term is introduced. A possible explanation is that it might stem from the box anomaly. We evaluate the experimental findings in the framework of Chanowitz equations [2] in terms of the pseudoscalar nonet decay constants f_π, f_8 and f_1 as well as the octet-singlet mixing angle θ_{PS}. We find parameters in the expected range. Internal consistency is achieved in both our ρ-models, however one solution needs integrally charged quarks.

TRIANGLE AND BOX ANOMALIES

The well known triangle anomaly shown in fig. 1a is responsible for the $\gamma\gamma$ decays of the π^0, η and η'. The matrix element reads

$$M(X \to \gamma\gamma) = B_X(k_1 \cdot k_2)\ \epsilon_{\mu\nu\rho\sigma}k_1^\mu k_2^\nu \varepsilon_1^\rho \varepsilon_2^\sigma$$

In this expression k_i and ε_i^δ ($i=1,2$) are the 4-momenta and polarization vectors of the two photons. At low energy, i.e. at the unphysical point where all 4–momenta vanish, the following relations can be established by e.g. PCAC and the Wess Zumino Lagrangian [2]:

$$B_\pi(0) = \frac{\alpha_{em}}{\pi\sqrt{3}}\frac{\sqrt{3}}{f_\pi}$$

$$B_\eta(0) = -\frac{\alpha_{em}}{\pi\sqrt{3}}\left[\frac{\cos\theta_{PS}}{f_8} - 2\sqrt{2}\,\xi\frac{\sin\theta_{PS}}{f_1}\right]$$

$$B_{\eta'}(0) = -\frac{\alpha_{em}}{\pi\sqrt{3}}\left[\frac{\sin\theta_{PS}}{f_8} + 2\sqrt{2}\,\xi\frac{\cos\theta_{PS}}{f_1}\right]$$

Here ξ is a discrete quantity which is equal to 1 for standard fractionally charged quarks and some special "designer" integral charge quark models (ICQM), but equal to 2 for all common formulations of ICQM. The couplings $B(k_1\cdot k_2)$ are believed to exhibit only very weak energy dependence, and in the following we consider them to be energy independent.

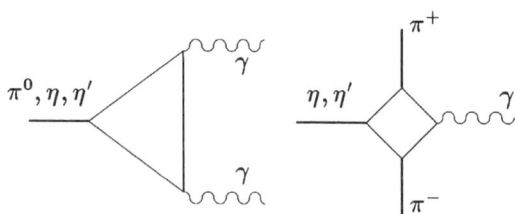

Figure 1. a) Triangle anomaly; b) box anomaly

Using the measured $\gamma\gamma$ widths of the π^0, η and η' [3], the coupling constants can be deduced from

$$\Gamma(X \to \gamma\gamma) = \frac{m_X^3}{64\pi}|B_X(0)|^2 \; , \; X = \pi^0, \eta, \eta'$$

to be $B(\pi^0) = [24.5 \pm 0.3]\,TeV^{-1}$, $B(\eta) = [-23.7 \pm 3.8]\,TeV^{-1}$ and $B(\eta') = [-32.1 \pm 0.9]\,TeV^{-1}$. Note the good accuracy of the prediction in case of the π^0: the prediction using $f_\pi = 93.15\,MeV$ [4] from $\pi^\pm \to \mu^\pm \nu$ is $B(\pi^0) = 24.9\,TeV^{-1}$.

Although much less known, the same theories predict another anomaly occuring in processes involving three pseudoscalar particles: the so-called box anomaly depicted in fig 1b. The corresponding matrix element can be written

$$M(X \to \pi\pi\gamma) = E_X(p_+k, p_-k)\,\epsilon_{\mu\nu\rho\sigma}\varepsilon^\mu k^\nu p_+^\rho p_-^\sigma$$

where $p_\pm$ are the 4–momenta of the outgoing $\pi^\pm$, k and ε are respectively the photon 4–momentum and its polarization vector. At low energies, the functions E_X are given by [2]:

$$E_\eta(0) = -\frac{e}{4\pi^2\sqrt{3}}\frac{1}{f_\pi^2}\left[\frac{\cos\theta_{PS}}{f_8} - \sqrt{2}\frac{\sin\theta_{PS}}{f_1}\right]$$

$$E_{\eta'}(0) = -\frac{e}{4\pi^2\sqrt{3}}\frac{1}{f_\pi^2}\left[\frac{\sin\theta_{PS}}{f_8} + \sqrt{2}\frac{\cos\theta_{PS}}{f_1}\right]$$

Whereas the connection of the triangle anomaly constants to measurable decay rates is obvious, the occurance of the ρ-resonance in the $\eta/\eta' \to \pi^+\pi^-\gamma$ decays obscures a direct connection of E_X and decay rates.

OUR MODEL FOR $\eta/\eta' \to \pi^+\pi^-\gamma$

Knowing that a pure ρ resonance does not describe well experimental data we are led to the following model (see fig. 2): We assume that the radiative decays $\eta/\eta' \to \rho^0\gamma$ and the non-resonant production due to the anomaly are actually two distinct processes, and the overall amplitude is the coherent sum of both contributions. Furthermore, the anomaly strengths B_X and E_X are assumed to be energy independent. Finally, for quantitative fits to data we need a model for the ρ-resonance. We use a standard relativistic Breit-Wigner amplitude whose width is parametrized as

$$\Gamma_\rho(m) = \Gamma_0\left[\frac{q_\pi(m)}{q_\pi(m_\rho)}\right]^3\left(\frac{m_\rho}{m}\right)^\lambda$$

where the q^3 dependence is due to the p–wave matrix element and phase space, and any deviation of λ from 1 signals deviations from elementary pointlike couplings or the influence of left-hand singularities. This choice of parametrization is in accord with analyticity, for details see [1].

FITTING $e^+e^- \to \pi^+\pi^-$ IN THE ρ REGION

To get reliable parameters m_ρ, $\Gamma_\rho(m_\rho)$ and λ for the η/η' fits, we refitted the most recent $e^+e^- \to \pi^+\pi^-$ data, using ρ and ω Breit-Wigner amplitudes and a free $\rho - \omega$-mixing angle φ. We chose this process since it is a clean, pure QED initial state which is believed to be free of uncertainties due to fi-

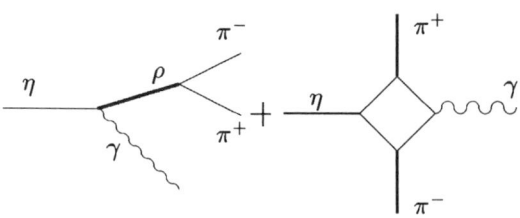

Figure 2. Model amplitude for $\eta/\eta' \to \pi^+\pi^-\gamma$

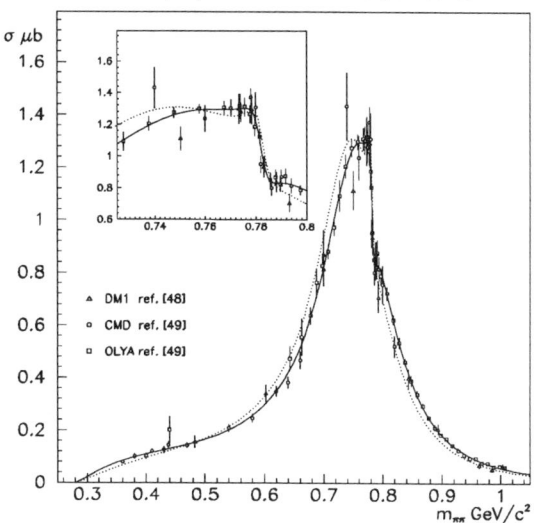

Figure 3. Model M1 (solid line) and PDG value prediction (dotted line) compared to measured cross sections $\sigma(e^+e^- \to \pi^+\pi^-)$

nal state interactions, interferences, production mechanisms etc.. To our surprise, the standard PDG ρ parameters do not give a good description (see dotted curve in fig 3; $\chi^2/ndf = 498/80$). We achieve a better fit by leaving m_ρ and Γ_ρ free (model M2, $\chi^2/ndf = 90/78$), but here we need a ρ mass almost degenerate to the ω mass, more than 20σ away from the PDG value ($m_\rho = 780.8 \pm 0.5\,MeV$, $\Gamma = 153 \pm 2\,MeV$, $\lambda = 0.66 \pm 0.05$). Another possibility is to allow for a small continuum term in the amplitude, and to fix the mass to the PDG value (model M1). Here we achieve a perfect description of the data with $\chi^2/ndf = 61/77$, $\Gamma = 142 \pm 2\,MeV$, $\lambda = 1.75 \pm 0.08$ and $\phi = (111.4 \pm 3.2)°$, see solid curve in fig. 3. Note that the continuum cross section needed is tiny ($0.006\,\mu b$ at the ρ mass); it is the interference term which is changing the details in the peak region. It is not clear whether such a term is theoretically justified, and for our argumentation of the continuum in the η/η' decays stemming from the box anomaly the need to introduce a

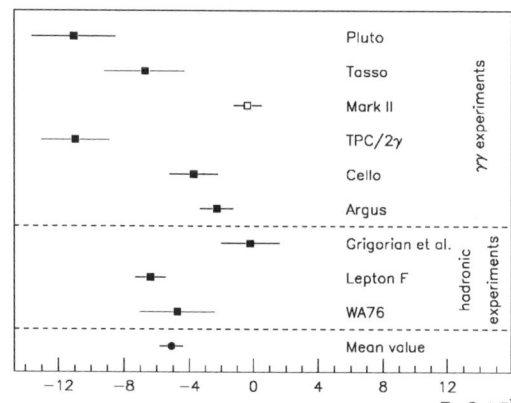

Figure 4. Results of fits for $E_{\eta'}$, using ρ parametrization M1

non-explained continuum into $e^+e^- \to \pi^+\pi^-$ is not very satisfactory.

FITS TO η AND $\eta' \to \gamma\pi^+\pi^-$

Using the two models M1 and M2 outlined above, we fitted all available $\pi^+\pi^-$ mass spectra. Some of the spectra were available only without acceptance correction; we tried to reconstruct corrected spectra wherever this was possible from formulae and curves given in the publications, for details see [1]. For the η decay, we used the data of Layter et al. [5], and for the η' data from PLUTO, TASSO, TPC/2γ, CELLO, ARGUS, Grigorian et al., Lepton F and WA 76 (for references see [1]). Although there is some scattering in these data, the mean values extracted for $E_{\eta'}$ using both models are clearly incompatible with zero in both ρ parametrization models. The results are:

$$\begin{array}{ll} model\,M1 & model\,M2 \\ E_\eta = -5.0 \pm 1.5 & E_\eta = -4.5 \pm 0.7 \\ E_{\eta'} = -5.1 \pm 0.7 & E_{\eta'} = -2.1 \pm 0.5 \end{array}$$

We recently learned that there are additional large clean η' samples from DM2 and Mark III in the reaction $J/\psi \to \gamma\gamma\pi^+\pi^-$ whose $\pi^+\pi^-$ mass spectra should be analyzed. Also the data now taken at the BES experiment can improve the experimental situation soon.

QUANTITATIVE ANALYSIS

We now insert the experimental values into the four equations for B_η, $B_{\eta'}$, E_η and $E_{\eta'}$. First we treat ξ as a free parameter, with the result $\xi = 1.2 \pm 0.2$ in model M1 and $\xi = 2.7 \pm 0.7$. Since ξ can only take the values 1 or 2, we see that we achieve internal consistency in our model if either model M1 (i.e. a small continuum exists in $e^+e^- \to \pi^+\pi^-$) is correct and $\xi = 1$, i.e. quarks are fractionally charged; or model M2 (i.e. the ρ mass is almost as high as the ω mass) and $\xi = 2$, i.e. quarks are integrally charged. In both these cases, the results for the pseudoscalar nonet parameters are:

quantity	model M1	model M2
f_π/f_1	$0.96^{+0.03}_{-0.05}$	0.48 ± 0.02
f_π/f_8	$0.46^{+0.43}_{-0.38}$	0.58 ± 0.16
$\theta_{PS}[°]$	$-26.7^{+9.6}_{-8.4}$	-24.1 ± 4.2
ξ	1	2
χ^2/ndf	0.54/1	0.03/1
χ^2/ndf alternate ξ	12/1	42/1

Thus, the parameters determined in both models are consistent with the expectations from e.g. mass formulae, chiral perturbation theory etc. (see [1]). From our model, it is presently impossible to distinguish between these solutions. In our view the unexpectedly large freedom in the ρ parameter determination and the apparent discrepancy between the precise $e^+e^- \to \pi^+\pi^-$ data and the PDG parameters without some additional contribution is an interesting question itself and needs further investigation.

SUMMARY

We have shown that a small continuum in addition to the dominating ρ resonance considerably improves fits to $\pi^+\pi^-$ mass spectra in $\eta/\eta' \to \pi^+\pi^-\gamma$ decays. We interpret this continuum as stemming from the box anomaly and evaluate the pseudoscalar coupling constants in the framework of Chanowitz equations. Using two possible ρ parametrizations, we get two solutions; one with $\xi = 1$ (conventional QCD) and another one with $\xi = 2$ (integer charged quarks). Both these solutions are self-consistent (4 equations, 3 parameters), and the parameters f_1, f_8 and θ_{PS} are in the range expected from other determinations.

ACKNOWLEDGEMENTS

I want to thank the organizers of the Hadron Spectroscopy Session of the Conference for the "last second" possibility to present the results. In addition, my warmest thanks go to my collaborators in this work, M. Benayoun, Ph. Leruste, J.L. Narjoux, K. Šafařík (College de France), A. Kirk (CERN) and M. Girone (INFN Bari).

REFERENCES

1. M. Benayoun, M.Feindt, M. Girone, A. Kirk, Ph. Leruste, J.L. Narjoux, K. Šafařík, College de France Preprint LPC 92-27, submitted to Nucl. Phys. B

2. M.S. Chanowitz, Phys. Rev. Lett. **35** (1975) 977, Phys. Rev. Lett. **44** (1980) 59; Procs. VIth Int. Workshop on Photon–Photon Collisions, Lake Tahoe 1984

3. S. Kawabata, Procs. Joint Int. Lepton-Photon Symposium and Europhysics Conference on High Energy Physics, Geneva 1991

4. Particle Data Group, Phys. Lett. B **239** (1990) 1

5. J.G. Layter et al., Phys. Rev. D7 (1973) 2565

PRODUCTION OF LIGHT QUARK RESONANCES IN Z⁰ DECAYS

Marcos DRACOS
CERN–CRN Strasbourg
(representing DELPHI collaboration)

Abstract

A study of inclusive ϱ°, $K^{*0}(892)$, $f_0(975)$ and $f_2(1270)$ production in hadronic Z^0 decays is presented. The obtained results are in a good agreement with the predictions of the JETSET 7.3 PS and HERWIG 5.4 models exept the $f_2(1270)$ production which is overestimated by HERWIG. A $p\bar{p}$ candidate resonance at $m_{p\bar{p}}$=2.17 GeV is presented and the ratio Ω/Ξ is found to be 27% $\pm$ 10%.

INTRODUCTION

Using 200000 hadronic Z^0 decays collected in 1991 at LEP with the DELPHI detector [1] an analysis of light resonances without using particle identification is presented.

A well known difficulty in resonance studies arises when particle identification is not available and each particle is assigned a pion or kaon mass depending on the invariant-mass distribution under study ($\pi\pi$ or $K\pi$). This leads not only to an increased combinatorial background but also to the problem of "reflections" where the resonance signals in a particular particle combination (e.g. $K\pi$) distort the invariant-mass spectrum of other combinations (e.g. $\pi\pi$).

Unfortunatly only 60000 events dispose particle identification provided by the DELPHI RICH so far. These events have been used to explore the $p\bar{p}$ invariant mass spectrum in searching of exotic resonances (i.e baryonium - four quark bound state). The same events have been used to reconstruct the heavy strange baryons through the decays $\Xi \to \Lambda^0 \pi$ and $\Omega \to \Lambda^0 K$. The results given for these two latter studies are preliminary and need more statistics to be confirmed.

LIGHT RESONANCES

The fitting procedure applied to the invariant-mass distributions to extract resonance cross sections and, in particular, the treatment of reflections used in this study are very similar to those described in [2].

The $\pi^+\pi^-$ and $K^\pm\pi^\mp$ invariant-mass distributions in the total available range of x_p ($= p/p_{beam}$), as well as in each x_p-interval are fitted with the expressions:

$$d\sigma/dM_{\pi\pi} = \beta_{K^\circ}BW_{K^\circ} + BG_{\pi\pi}(\alpha_{\pi\pi} + \quad (1)$$
$$\beta_{\varrho^\circ}BW_{\varrho^\circ} + \beta_{f_o}BW_{f_o} + \beta_{f_2}BW_{f_2})$$

$$d\sigma/dM_{K\pi} = BG_{K\pi}(\alpha_{K\pi} + \beta_{K^*}BW_{K^*} + \\ \beta_{K_2^*}BW_{K_2^*}), \quad (2)$$

where α and β are fitted parameters and BW is a Breit-Wigner function with free parameters the central mass M_o and total width Γ supposed to be a sum of the natural width Γ_o and the experimental resolution width Γ_R (for a justification, see for example [3]). The function used for the background BG is:

$$BG = (M - M_{th})^{\gamma_1} exp(\gamma_2 M + \gamma_3 M^2 + \gamma_4 M^3 + \gamma_5 M^4) \qquad (3)$$

M_{th} is the threshold mass for the relevant mass combination and γ_i are free parameters [1].

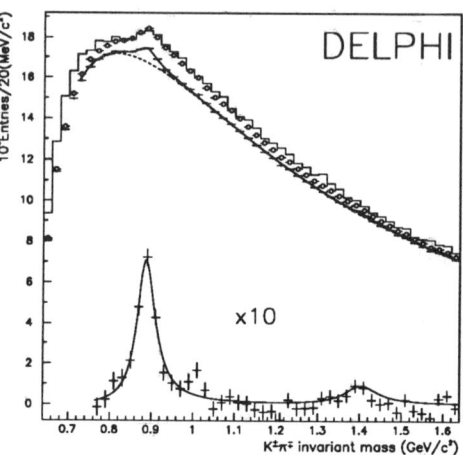

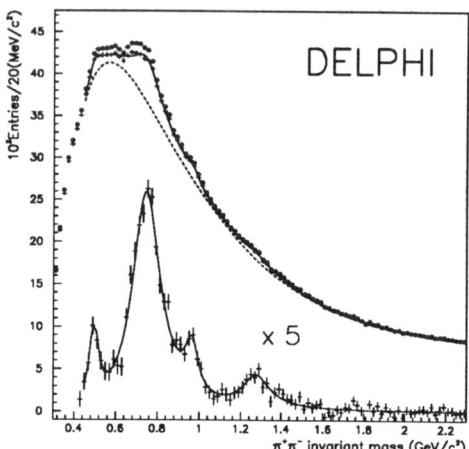

Figure 1. The raw $\pi^+\pi^-$ invariant-mass spectrum in the range $x_p > 0.1$ (open dots) and the one corrected for reflections (crosses). The solid curve is the result of the fit to expression (1). The dashed curve shows the estimate of the background. The lower plot is the corrected data after background subtraction with the curve showing the resonance contributions.

The reflections have been treated in the following way: Each $\pi^+\pi^-$ combination is weighted by the product of $W_{K^+\pi^-}$ and $W_{\pi^+K^-}$ which are the probabilities that the combination does not belong to the $K^{*0}(892)$ or $K^{*0}(1430)$ and to the $\overline{K}^{*0}(892)$ or $\overline{K}^{*0}(1430)$ respectively (similar for each $K^+\pi^-$). The weights are defined as:

$$W_{\pi\pi} = \frac{\alpha_{\pi\pi} BG_{\pi\pi}}{d\sigma/dM_{\pi\pi}}, \qquad (4)$$

$$W_{K\pi} = \frac{\alpha_{K\pi}}{\alpha_{K\pi} + \beta_{K^*} BW_{K^*} + \beta_{K_2^*} BW_{K_2^*}}. \qquad (5)$$

---

[1] Other forms of background parameterization have also been tried and give the same results within errors (for more details see [4])

Figure 2. The raw $K^{\pm}\pi^{\mp}$ invariant-mass spectrum in the range $x_p > 0.05$ (see fig.1 for explanations). The histogram (solid line) shows the distribution obtained from Monte-Carlo data (using JETSET).

This procedure has been checked with events generated by JETSET 7.3 PS and HERWIG 5.4. Consistent results are obtained.

The raw $\pi^+\pi^-$ and $K^{\pm}\pi^{\mp}$ invariant mass distributions and those corrected for reflections are shown in figures 1 and 2 respectively. The invariant mass distributions exhibit clear ϱ^0, $f_0(975)$ and $f_2(1270)$ signals, the peak related to the $K_S^0 \to \pi^+\pi^-$ decay (fig.1), $K^{*0}(892)$ and $K^{*0}(1430)$ signals (fig.2). However, at the present level of statistics the $K^{*0}(1430)$ cross section cannot be reliably determined.

The DELPHI simulation program is used to correct the extracted cross sections for geometrical acceptance, kinematical cuts, particle interactions within the detector material and other detector imperfections.

The invariant mass distributions from the simulated events are in good agreement with the data (fig.2).

The measured average resonance multipli-

Table 1. Average resonance multiplicities per hadronic event in indicated x_p-ranges in comparison with the JETSET 7.3 PS and HERWIG 5.4 predictions.

Resonance	x_p-range	Experiment	JETSET	HERWIG
ϱ^0	$x_p > 0.1$	0.51 ±0.05	0.54	0.51
	$x_p > 0.05$	0.83 ±0.07	0.90	0.85
	$x_p > 0$	1.43 ±0.12	1.55	1.40
$K^{*0}(892)$	$x_p > 0.1$	0.36 ±0.09	0.45	0.34
	$x_p > 0.05$	0.64 ±0.12	0.68	0.53
	$x_p > 0$	0.97 ±0.18	1.04	0.84
$f_0(975)$	$x_p > 0.1$	0.042±0.026	-	-
	$x_p > 0.05$	0.10 ±0.03	-	-
$f_2(1270)$	$x_p > 0.1$	0.11 ±0.04	-	0.24

cities per hadronic event in the x_p-ranges (table 1) are obtained by normalisation of the determined resonance cross sections to the total hadronic cross section σ_{hadr}. For the vector mesons, they are in a good agreement with the JETSET 7.3 PS and HERWIG 5.4 predictions.

The measured ratio:

$$f_2(1270)/\varrho^0 = 0.22 \pm 0.08 \qquad (6)$$

for $x_p > 0.1$ is in agreement with the tensor-to-vector meson ratios in the full x_p-range measured in hadronic reactions: $K_2^{*+}(1430)/K^{*+}(892) = 0.25 \pm 0.04$ [5], $K_2^{*0}(1430)/K^{*0}(892) = 0.23 \pm 0.08$, $f_2(1270)/\varrho^0 = 0.26 \pm 0.05$ [6], 0.30 ± 0.06 [2] and 0.24 ± 0.03 [7]. The ratio:

$$f_2(1270)/f_0(975) = 3^{+7}_{-1} \qquad (7)$$

for $x_p > 0.1$ can be compared with the corresponding values of 2 ± 1 in e^+e^- annihilation at 29 GeV [8] and 4.1 ± 1.5 in pp-interactions at $\sqrt{s} = 27.5$ GeV [7]. It is of interest that all these results are consistent, within large errors, with the simple spin statistics prediction of 5 for the ratio of the tensor-to-scalar mesons.

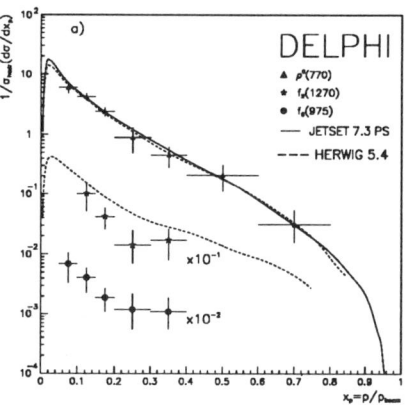

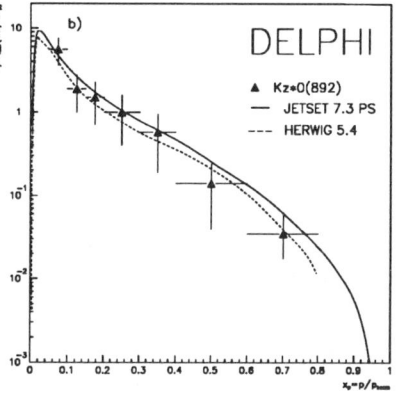

Figure 3. $1/\sigma_{hadr} \cdot d\sigma/dx_p$ for a) ρ^0, $f_2(1270)$ and $f_0(975)$ and b) for $K^{*0}(892)$.

The differential cross sections, $1/\sigma_{hadr} \cdot d\sigma/dx_p$, for the ϱ^0, $f_2(1270)$, $f_o(975)$ and $K^{*0}(892)$ are shown in fig.3, together with the JETSET and HERWIG predictions for vector mesons and the HERWIG prediction for the $f_2(1270)$. The $f_o(975)$ production is not included neither in JETSET nor in HERWIG. The agreement between data and models for vector mesons is quite impressive. The $f_2(1270)$ production is overestimated by HERWIG.

p$\bar{\text{p}}$ INVARIANT MASS

For this analysis secondary vertices are reconstructed using particles identified as protons (anti-protons) by the RICH (covering the polar angles between 42^0 and 148^0). In figure 4 the mean Cerenkov angle measured by the RICH is plotted versus the momentum of the particle. The formation of clear bands corresponding to the pion, kaon and proton signature demonstrates the detector identification capability. The distance between the primary and secondary vertex (in the Rϕ-plane) is required to be less than 1 mm[9]. All tracks suspected to be affected by any detector problem have been rejected.

The invariant mass distribution of these candidates is shown in figure 5. An enhancement of entries appears arround 2.17 GeV/c^2. Possible kinematical reflections on both data and Monte–Carlo have been exclusively reconstructed. The pion rejection factor has been found to be very high. The enhancement can not be attributed to any reflection of other resonances. (for example from $\Lambda^0 \to$ pπ where the

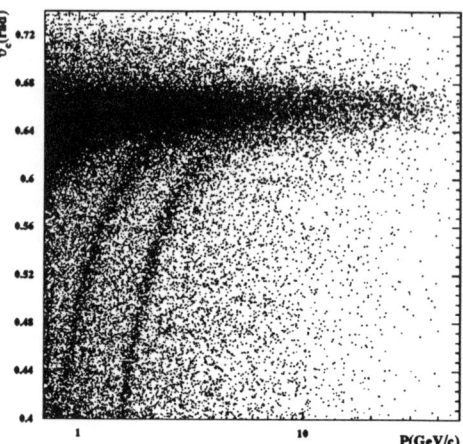

Figure 4. Mean Cerenkov angle versus the particle momentum

pion is wrongly identified as proton – all these cases have been reconstructed and removed if a peak was seen arround the mass of the considered resonance).

The grey distribution of figure 5 is obtained by the wrong sign combinations (pp+$\bar{\text{p}}\bar{\text{p}}$) and it has been fitted by a polynomial distribution. By scaling this curve in order to have the same number of entries in the p$\bar{\text{p}}$ and pp+$\bar{\text{p}}\bar{\text{p}}$ distribution an estimation of the background level is obtained. The excess arround 2.17 GeV/c^2 is significant by 2.5 standard deviations.

Assuming that this peak comes from a uu($\bar{\text{u}}\bar{\text{u}}$) resonance[10], we speculate that its formation is easier in more than two-jet events simply because the vacuum excitation is more important due to gluon emission than in two-jet events. Figure 6 shows the $m_{\text{p}\bar{\text{p}}}$ distribution for more than two-jet events and for two-jet events (grey distribution). Indeed the enhancement arround 2.17 GeV/c^2 comes for events having more than two jets. For events having only two jets the distribution is compatible with uniform background.

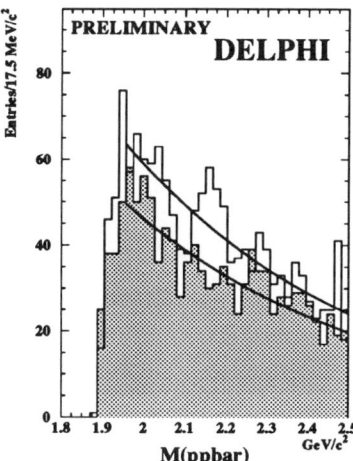

Figure 5. $m_{p\bar{p}}$ distribution (wrong sign combinations in grey)

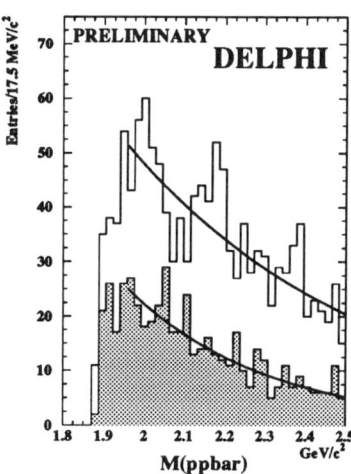

Figure 6. $m_{p\bar{p}}$ for $Njets \geq 3$ and $Njets = 2$ (grey)

Ω/Ξ RATIO

Another resonance production which has to do with the vacuum excitation is the $\Omega^-(\bar{\Omega})$ as Ω is composed by three s-quarks (not so easy to bring three s-quarks together). The production of $\Xi \to \Lambda\pi$ and $\Omega \to \Lambda K$ (with $\Lambda \to p\pi$) has been studied by using the RICH to identify protons and kaons in order to remove reflections of the one particle into the other one[11]. The ratio Ω/Ξ is found to be $27\% \pm 10\%$ which is significantly higher than what JETSET gives (3%).

REFERENCES

1. DELPHI collab., Nucl.Inst.and Meth. A303 (1991) 233.
2. NA22-EHS Collab., N.M.Agababyan et al., Z.Phys. C41 (1989) 539.
3. Mirabelle Collab., P.Granet et al., Nucl. Phys. B140 (1978) 389.
4. DELPHI Collab., "Measurement of inclusive production of Light Meson Resonances in Hadronic Decays of the Z^{0}", submitted to this Conference.
5. Mirabelle Collab., I.V.Ajinenko et al., Z.Phys. C24 (1984) 103.
6. Mirabelle Collab., P.V.Chliapnikov et al., Nucl. Phys. B176 (1980) 303; Z.Phys. C12 (1982) 113.
7. NA27 (LEBC-EHS) Collab., M.Aguilar-Benitez et al., Z.Phys. C50 (1991) 405.
8. HRS Collab., S.Abachi et al., Phys. Rev. Lett. 16 (1986) 1990; Phys. Lett. B199 (1987) 151.
9. M.Dracos, S.Tzamarias DELPHI 92–115 PHYS 221 (1992).
10. Chan Hong-Mo, RAL-89-072 (1989).
11. M.Dracos, S.Tzamarias, J.Werner DELPHI 92–33 RICH 49 (1992).

Heavy Quark and Quarkonium States

Parallel Session 4

Reunion Tower, Dallas

Conveners: B. Grinstein (SSCL)
M. Tuts (Columbia)

EVIDENCE FOR Λ_b PRODUCTION AT LEP

XinChou Lou
Department of Physics, Indiana University
Bloomington, Indiana 47405, USA

(Representing the OPAL Collaboration)

Abstract

Evidence is observed for the production of b-flavoured baryons at LEP via the study of correlations between the Λ baryons and prompt leptons from b hadron decays. Signals for the Λ_b are also seen in its semileptonic decays $\Lambda_b \to \Lambda_c^+ \ell^- \overline{\nu}_\ell X$, where the charmed baryon Λ_c^+ is reconstructed in the $pK^-\pi^+$ mode. Based on these results, the existence of b-flavoured baryons is established.

INTRODUCTION

The b-flavoured baryons have been predicted by the quark model [1]. However experimental information about their existence and their properties are very limited [2, 3]. Decays of the Z^0 boson provide a copious source of b hadrons, making it possible to search for and to study the b baryons at LEP. Evidence for b baryons from $\Lambda\ell$ correlations and observations of the decay $\Lambda_b \to \Lambda_c^+\ell^-\overline{\nu}_\ell X$ from LEP experiments are presented in this paper. A search for the decay $\Lambda \to J/\psi\Lambda$, which was reported by the UA1 experiment, is also described.

Λ-LEPTON CORRELATIONS

Semileptonic decays of b baryons are expected to produce $\Lambda\ell^-$ and $\overline{\Lambda}\ell^+$ pairs but not $\Lambda\ell^+$ and $\overline{\Lambda}\ell^-$ pairs, where $\ell = e$ or $\ell = \mu$ and stems from direct decays of the b quark. (Throughout this paper the anti-particle state is implied whenever an particle name is mentioned.) Due to the large mass of the b quark, these leptons have large momenta and transverse momenta (p_T) relative to the b hadron direction. The Λ baryons resulting from b baryon decays tend to carry large fraction of the b baryon momenta. At LEP the b and $\overline{b}$ quarks are well separated topologically in Z^0 decays. In the lab frame the $\Lambda\ell^-$ pairs from b baryons decays are well contained in a small cone and can not be confused with Λ and ℓ^- from decays of different b hadrons in the same event.

Thus an effective tag for b baryons at LEP is via the correlation between energetic Λ baryons and energetic and high p_T e or μ, where the $\Lambda\ell^-$ pair is contained in a common jet.

ALEPH [4] and OPAL [5] have published evidence for b baryons based on their observation of excess $\Lambda\ell^-$ pairs. A preliminary study by DELPHI [6] of their data also indicates the production of b baryons in Z^0 decays. The $p\pi^+$ invariant mass distributions for $p\pi^+\ell^-$ and $p\pi^+\ell^+$ combinations are shown in Figures 1, 2 and 3 as obtained by ALEPH, OPAL and DELPHI, respectively. Clear ex-

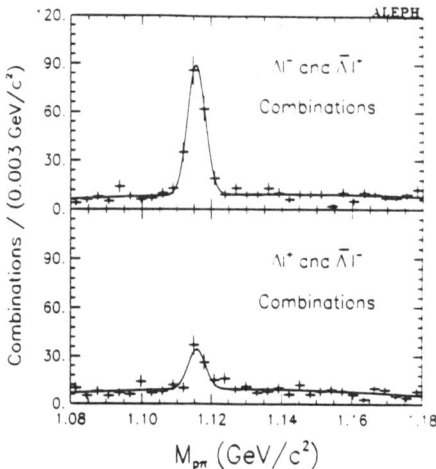

Figure 1. The $p\pi^-$ invariant mass distributions from ALEPH for a) $p\pi^-\ell^-$ and b) $p\pi^-\ell^+$ combinations.

cess in the $p\pi^+\ell^-$ combination is observed in these distributions.

Detailed study of sources of $\Lambda\ell^\pm$ events are carried out by these experiments. In addition to b baryon decays, $\Lambda\ell^\pm$ events can come from semileptonic baryonic decays of B mesons ($\Lambda\ell^-$), semileptonic decays of the Λ_c^+ ($\Lambda\ell^+$), Λ baryons combined with fake leptons ($\Lambda\ell^\pm$) and association of Λ and anti-baryons which is misidentified as a lepton ($\Lambda\ell^-$). Contributions from these sources to the excess $\Lambda\ell^-$ events are found to be very small and can not account for the observed excess. Therefore the excess $\Lambda\ell^-$ events are interpreted as coming from semileptonic decays of b baryons.

The numerical results are summarized in Table 1. The quantity $f(b \to \Lambda_b) \cdot B(\Lambda_b \to \Lambda\ell^-\bar{\nu}_\ell X)$ measured by these experiments are consistent, where $f(b \to \Lambda_b)$ is the b baryon production rate per b quark in Z^0 decays. It is noticeable that the systematic errors are larger than or comparable with the statistical errors; along with large uncertainties on $B(\Lambda_b \to \Lambda\ell^- X)$, it is very difficult to extract $f(b \to \Lambda_b)$ reliably.

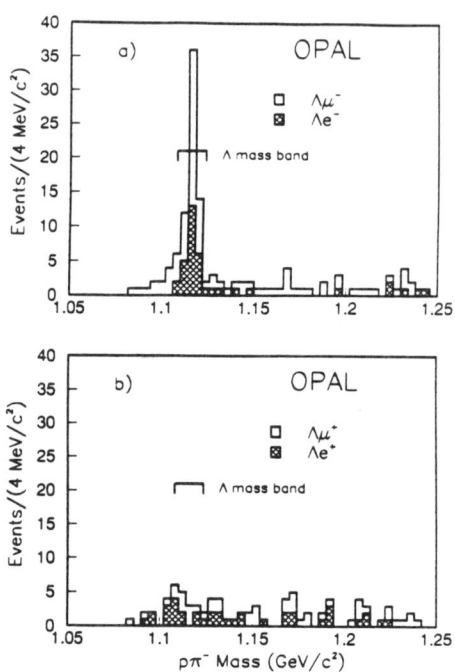

Figure 2. The $p\pi^-$ invariant mass distributions from OPAL for a) $p\pi^-\ell^-$ and b) $p\pi^-\ell^+$ combinations.

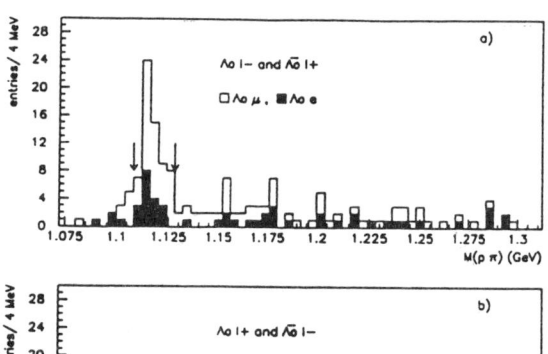

Figure 3. The $p\pi^-$ invariant mass distributions from DELPHI for a) $p\pi^-\ell^-$ and b) $p\pi^-\ell^+$ combinations.

Table 1. Results on Λ-lepton Correlations in Z^0 Decays at LEP

	ALEPH	DELPHI	OPAL
Data Sample Used (No. of Z^0 bosons)	451 000	329 000	458 583
Excess $\Lambda\ell^-$ Events	117.0±18.0	30.0±10.0	55.0±9.0
Expected Background	$\sim^{+6.7}_{-(4.2\pm5.6)}$	$^{+1.1}_{-0.4}$	$^{+3.1}_{-0.3}$
b Baryon Events	$121.2\pm18.0^{+21.9}_{-22.9}$	$30.0\pm10.0^{+0.4}_{-1.1}$	$55.0\pm9.0^{+0.3}_{-3.1}$
Product B R*($\times 10^{-3}$)	$3.8\pm0.6^{+0.9}_{-1.1}$	$3.8\pm1.2\pm0.8$	$2.9\pm0.5\pm0.7$

* Product B R = $f(b \to \Lambda_b) \cdot B(\Lambda_b \to \Lambda\ell^- X)$

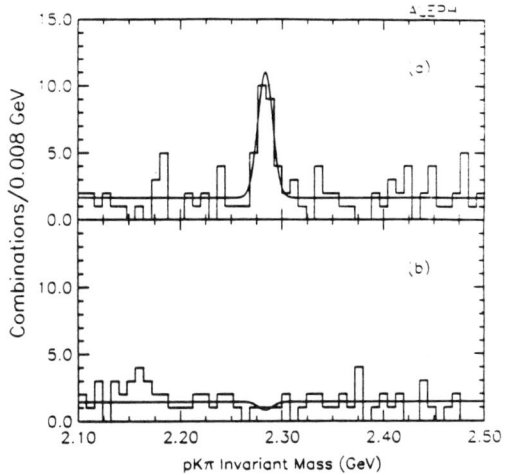

Figure 4. $pK^+\pi^-$ invariant mass distributions measured by ALEPH.

OBSERVATION OF $\Lambda_b \to \Lambda_c^+ \ell^- \overline{\nu}_\ell X$

The decay $\Lambda_b \to \Lambda_c^+ \ell^- \overline{\nu}_\ell X$ is expected to dominate the semileptonic Λ_b decays [7, 8]. Semileptonic decays of other b-flavoured baryons are not expected to form the final state $\Lambda_c^+ \ell^- \overline{\nu}_\ell X$. For example, the Ξ_b and Ω_b baryons decay preferentially into $\Xi_c \ell^- \overline{\nu}_\ell X$ and $\Omega_c \ell^- \overline{\nu}_\ell X$ [7, 9], respectively, with subsequent weak decays of the Ξ_c and Ω_c; the Σ_b baryons are expected to be heavy enough [10] to decay strongly or radiatively into the Λ_b. A $\Lambda_c^+ \ell^- \overline{\nu}_\ell X$ signal will thus unambiguously identify the Λ_b.

ALEPH [11] and OPAL have observed signals for $\Lambda_b \to \Lambda_c^+ \ell^- \overline{\nu}_\ell X$ in Z^0 decays. The charmed baryon Λ_c^+ is reconstructed in its

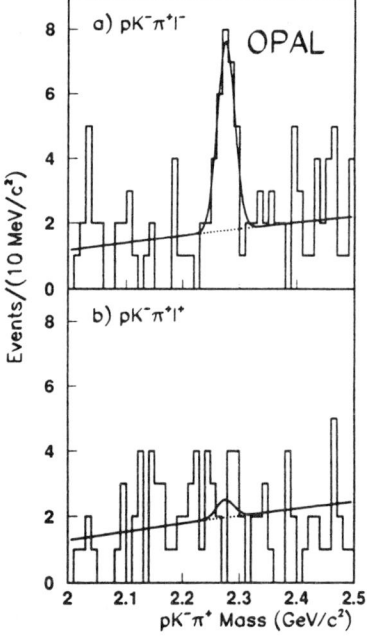

Figure 5. $pK^+\pi^-$ invariant mass distributions measured by OPAL.

Table 2. Results on $\Lambda_b \to \Lambda_c^+ \ell^- \overline{\nu}_\ell X$ at LEP

	ALEPH	OPAL
Background	0.4±0.3	0.5±0.6
Λ_b Events	21.0±5.0	22.0±5.6±0.9
Product B R * ($\times 10^{-2}$)	1.5±0.35±0.45	1.5±0.35±0.26

* See text for definition.

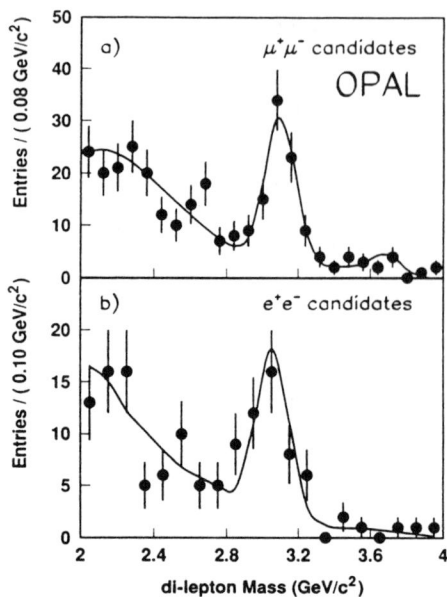

Figure 6. Di-lepton invariant mass distributions as observed by OPAL.

$pK^-\pi^+$ mode. Due to large boost of the Λ_b produced in Z^0 decays, the Λ_c^+ particles exhibit a hard momentum distribution and their decay vertices are displaced on the average ~ 2 mm away from the primary interaction point. Thus both experiments select high momentum $pK^-\pi^+$ combinations and require that they form a secondary vertex which has a positive decay length with respect to the average beam spot. In addition, the $pK^-\pi^+\ell^-$ system is demanded to have large invariant mass (>3.5 GeV/c^2) in order to suppress combinatorial background. The $pK^-\pi^+$ invariant mass distributions for the $pK^-\pi^+\ell^-$ and $pK^-\pi^+\ell^+$ combinations are shown in Figures 4 and 5, as selected by ALEPH and OPAL, respectively. Λ_c^+ signals are observed in the $pK^-\pi^+\ell^-$ combinations. The data are consistent with no Λ_c^+ events in the $pK^-\pi^+\ell^+$ combinations, which can not result in from the Λ_b as it does not decays into $\Lambda_c^+\ell^+X$.

The final results are summarized in Table 2. The product branching ratio $f(b \to \Lambda_b) \cdot B(\Lambda \to \Lambda_c^+\ell^-\overline{\nu}_\ell X)$ determined by both experiments are in good agreement, where the branching ratio $B(\Lambda_c^+ \to pK\pi^+)$ has been assumed to be $(4.3\pm1.0)\%$ [12]. Adding the errors in quadrature, the average product branching ratio is found to be $(1.5\pm0.3\pm0.3)\%$. Assuming $B(\Lambda_b \to \Lambda_c^+\ell^-\overline{\nu}_\ell X) = 11\%$, the product branching ratio can be translated into $f(b \to \Lambda_b) = (13.6\pm2.7\pm2.7)\%$.

SEARCH FOR $\Lambda_b \to J/\psi \Lambda$

The UA1 experiment [3] has reported an observation of $\Lambda_b \to J/\psi\Lambda$ with a branching ratio of $(1.8\pm1.0)\%$. The decay $\Lambda_b \to J/\psi\Lambda$ provides a clean channel to search for the Λ_b at LEP. An observation of a Λ_b signal at LEP would provide an independent confirmation of the UA1 result.

Production of J/ψ has been observed in Z^0 decays at LEP [13]. As an example, the J/ψ signals from OPAL are shown in Figure 6. Approximately 120 J/ψ mesons are reconstructed above an estimated background of 22 events. Using the same data sample as in the $\Lambda\ell$ correlation study, ALEPH and OPAL combine J/ψ mesons with Λ baryons to form $J/\psi\Lambda$ invariant mass distribution. No $J/\psi\Lambda$ candidates are found with an invariant mass greater than 5.2 GeV/c^2 after selections. More details of their analyses are given in Table 3. Assuming the UA1 branching ratio for $\Lambda_b \to J/\psi\Lambda$, ALEPH and OPAL expect to see ~ 3.8 and 2.0 ± 1.6 $\Lambda_b \to J/\psi\Lambda$ events, respectively.

Table 3. Results on $\Lambda_b \to J/\psi\Lambda$ Search at LEP

	ALEPH	OPAL
$\Lambda_b \to J/\psi\Lambda$ Candidates	0	0
$f(b \to \Lambda_b)$ (%) Assumed	9.0	7.2±1.2±3.4
Expected Events	3.8	2.0±1.6

ALEPH assumes $f(b \to \Lambda_b)=9\%$ to derive an 90% confidence level (CL) upper limit of $B(\Lambda_b \to J/\psi\Lambda) <1.1\%$. Based on 0 candidate for $\Lambda_b \to J/\psi\Lambda$, OPAL concludes that $f(b \to \Lambda_b) \cdot B(\Lambda_b \to J/\psi\Lambda) <0.23\%$ at 90% CL.

SUMMARY

Evidence for b-flavoured baryons from excess $\Lambda\ell^-$ events in Z^0 decays are observed by ALEPH, DELPHI and OPAL experiments at LEP. Observations of $\Lambda_b \to \Lambda_c^+\ell^-\bar{\nu}_\ell X$ by ALEPH and OPAL experiments confirm the results based on $\Lambda\ell$ correlations. The data indicate a Λ_b production rate of $f(b \to \Lambda_b) =(13.6\pm2.7\pm2.7)\%$ per b quark in Z^0 decays. These results provide firm experimental bases for the existence of the b-flavoured baryons. Searches for $\Lambda_b \to J/\psi\Lambda$ have been performed by ALEPH and OPAL. With data samples used, they are not able to confirm the UA1 observation.

ACKNOWLEDGEMENT

I thank my colleagues on OPAL, V. Sharma (ALEPH) and W. Venus (DELPHI) for providing me with results on this subject from their respective experiments.

REFERENCES

1. See, for example, S. Capstick and N. Isgur Phys. Rev. **D 34** (1986) 2809;
   N. Isgur and M. B. Wise, Phys. Rev. Lett. **66** (1991) 1130.
2. Particle Data Group, J. J. Hernandez *et al.*, Phys. Lett. **B 239** (1990) 1.
3. UA1 Collaboration, C. Albajar *et al.*, Phys. Lett. **B 273** (1991) 540.
4. ALEPH Collab., D. Decamp *et al.*, Phys. Lett. **B 278** (1992) 209;
   ALEPH Collab., D. Buskulic *et al.*, CERN-PPE/92-138, submitted to Phys. Lett. **B**.
5. OPAL Collab., P. D. Acton *et al.*, Phys. Lett. **B 281** (1992) 394.
6. DELPHI Collab., P. Abreu *et al.*, *Measurement of Λ_b production and lifetime in Z^0 hadronic decays*, submitted to this conference.
7. S. Rudaz and M. B. Voloshin, Phys. Lett. **B 252** (1990) 443.
8. T. Mannel and G. A. Schuler, DESY 91-095, Sept. 6, 1991.
9. N. Isgur and M. B. Wise, Nucl. Phys. **B 348** (1991) 276.
10. W. Kwong and J. L. Rosner, Phys. Rev. **D 44** (1991) 212.
11. ALEPH Collab., D. Buskulic *et al.*, CERN-PPE/92-73, submitted to Phys. Lett. **B**.
12. CLEO Collab., G. Crawford *et al.*, Phys. Rev. **D 45** (1992) 752.
13. OPAL Collab., G. Alexander *et al.*, Phys. Lett. **B 266** (1991) 485;
    L3 Collab., B. Adeva *et al.*, CERN-PPE/92-99, submitted to Phys. Lett. **B**.

PRODUCTION OF $B \to J/\Psi$ IN Z° HADRONIC DECAYS

Report for the LEP Collaborations — Aleph, L3 and Delphi

A.M. Segar
(Oxford-Delphi)

Abstract

New results for J/Ψ production in Z° decays from Aleph, L3 and Delphi are presented. An average for the Branching ratio $Br(Z^\circ \to J/\Psi + X)$ of $(4.1 \pm 0.4) \times 10^{-3}$ is obtained, and an average lifetime for the B sample generated in Z° decays is 1.34 ± 0.14 ps. A total of 11 exclusive B decays to J/Ψ have been observed.

The process of interest is $Z^\circ \to BX$; $B \to J/\Psi + Y$ as illustrated by the diagram (or its conjugate):

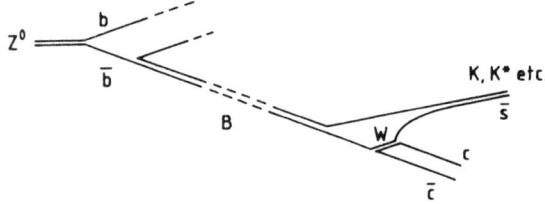

The J/Ψ is identified in decays $J/\Psi \to e^+e^-$, $J/\Psi \to \mu^+\mu^-$ which at the same time allow location of the secondary B decay vertex. Present data is limited to about 400k Z°, or 40 to 90 J/Ψ events per experiment.

The Opal experiment reported the first observation of the process [1]. New, preliminary results from Aleph, L3 and Delphi are reported here[6].

1. EVENT SELECTION AND EFFICIENCY

Each experiment applied cuts to the charged tracks and to the lepton pair. Delphi data includes the silicon strip vertex detector, two layers in '90 and three in '91. Aleph included Si data, with both $r\phi$ and z coordinates, in their '91 data.

The lepton identification method, electron or muon, differs among the experiments and for details reference must be made to the individual experimental papers.

With the above selection, and its muon id, Delphi obtains a $J/\Psi \to \mu^+\mu^-$ reconstruction efficiency by Monte Carlo of $34.4 \pm 1.5\%$ for '90 data, and $32.5 \pm 1.5\%$ for '91. Aleph gives $22.1 \pm 0.8\%$ for $J/\Psi \to e^+e^-$ and $41.0 \pm 0.8\%$ for $J/\Psi \to \mu^+\mu^-$. L3 find $12 \pm 1\%$ for e^+e^- and $26 \pm 1\%$ for $\mu^+\mu^-$.

2. MASS PEAKS AND BACKGROUNDS

With events selected as above, mass peaks are observed at the J/Ψ, examples being the μ pairs from Delphi Fig.1a, the μ pairs from Aleph Fig.1b, and the e pairs from L3 Fig.1c.

With a suitable parametrisation for the background, the peaks can be fitted to gaussians of central value consistent with the J/Ψ mass and with width reflecting the experimental resolution e.g. Delphi averaging '90 and '91 gives

$$M_{J/\Psi} = 3.089 \pm 0.011 \text{ GeV}/c^2$$

$$\sigma_{J/\Psi} = 0.051 \pm 0.010 \text{ GeV}/c^2$$

and Aleph gives
$$M_{J/\Psi} \quad 3.094 \pm 0.003 \text{ GeV}/c^2$$
$$\sigma_{J/\Psi} = 0.028 \pm 0.004 \text{ GeV}/c^2$$

The background to the J/Ψ can be studied by plotting same sign pairs or μe pairs in the data, and by the use of the JETSET 7.3 hadronisation in a Monte Carlo. Aleph used the latter with proportions of 40% B_d, 40% B_u, 11% B_s and 9% b baryons.

The background is largely combinatorial, that is random combinations of pairs of tracks one or both of which may be misidentified leptons. At the same time the two lepton tag is a selector of $b\bar{b}$, so there are important physics backgrounds from b's:

(i) $b \to c$ "cascade" decays illustrated by

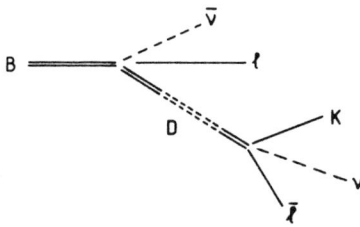

where the $\ell\bar{\ell}$ might be reconstructed as a false secondary vertex.

(ii) Direct J/Ψ production, mainly in the process

$$Z^\circ \to q\bar{q}g^* \to gg\Psi \quad \text{but also} \quad Z^\circ \to c\bar{c}\Psi$$

The contribution of these processes has been estimated theoretically[2,3] and it is found to be of order 3% and 0.3% respectively of the J/Ψ events in the peak. The massive gluon yields J/Ψ of low momenta, see Fig.2. Fits by Aleph (Fig.2a), L3 (Fig.2b) and Delphi are consistent with a small contribution from the $gg\Psi$ process. The theoretical estimate is only

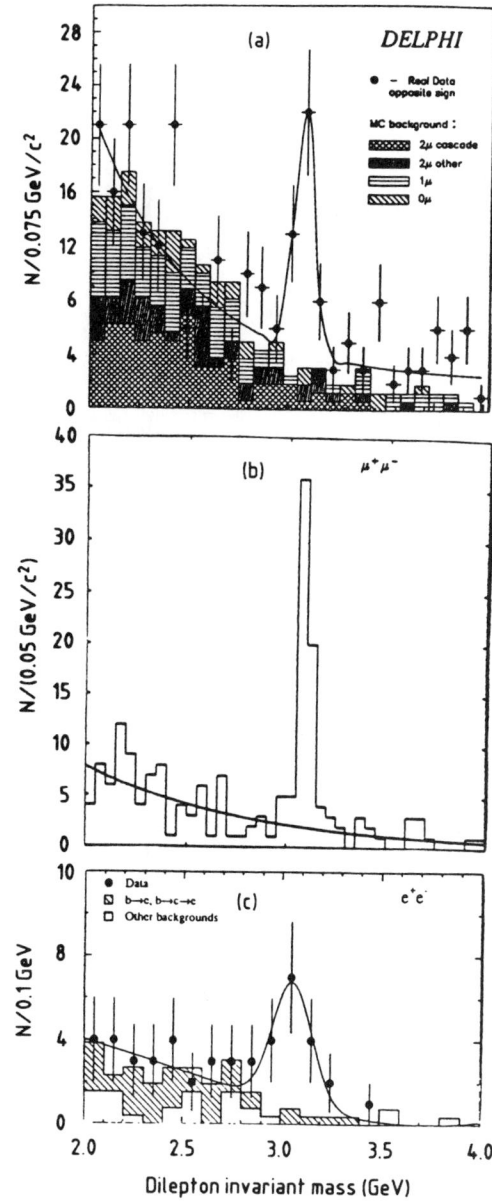

Figure 1. Distributions of dilepton invariant mass for a) μ pairs from Delphi, b) μ pairs from Aleph and c) e pairs from L3.
Fitted curves with J/Ψ peak plus background are shown on a) and c), where the histograms of the backgrounds estimated by Monte Carlo are also drawn.

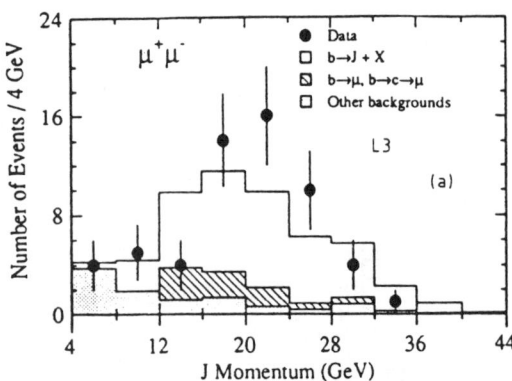

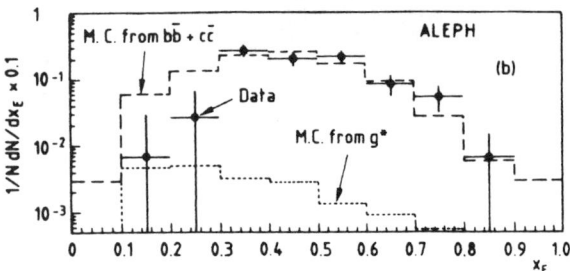

Figure 2. Distributions in a) J/Ψ momentum for L3, and b) $x_E = E_{J/\Psi}/E_{beam}$ for Aleph. The data points are acceptance corrected, and non-J/Ψ background has been subtracted. Normalised histograms of Monte-Carlo expectations are shown, and in a) the cascade decay and other J/Ψ background components are indicated in the shaded histograms, whilst in b) the g^* component is represented by the the dotted histogram.

The relatively low figure for Aleph is due to their energy flow algorithm — the total energy measured in the hemisphere containing the jet is required to be > 0.85 (beam energy). This cut reduces their background by approximately a factor of two, removing effectively those events containing unseen neutrinos.

3. BRANCHING RATIOS

The Z^0 branching ratios for the J/Ψ can be obtained from the efficiency and background studies outlined above. The results are summarised in Table 1, and for all four experiments combined correspond to 60 e^+e^- and 175 $\mu^+\mu^-$ events. If we assume that errors may be combined and are independent between experiments, an average of the experimental results gives

$$\text{Br}(Z^0 \to J/\Psi + X) = (4.1 \pm 0.4) \times 10^{-3}$$

In the table, the branching ratio for $B \to J/\Psi + Y$ is also given, the result being closely consistent with that from Argus and Cleo[4,5] where the content of B species differs, the agreement lending support to the heavy quark effective field theory.

in leading order, with scale assumed to be m_{g^*} so is uncertain by a factor of order 2. L3 use the angular distribution between the jet thrust axis and the J/Ψ to obtain an upper limit to the $gg\Psi$ process

$$\text{Br}(Z^0 \to q\bar{q}g^* \to X\Psi)$$
$$< 7 \times 10^{-4} \text{ at 90\% confidence}$$
$$\text{or } < 18\% \text{ of } Z^0 \to X\Psi$$

The backgrounds under the peaks vary among the experiments

Opal 32% Aleph 15% L3 26% Delphi 35%

4. EXCLUSIVE EVENTS

Both Aleph and Delphi have made use of their vertex detectors to obtain a sample of exclusive events. Fig.3 is an example of such a Delphi event, where $J/\Psi \to \mu^+\mu^-$ is combined with a $K^{*0} \to K^+\pi^-$ decay.

There are 11 exclusive events in all, summarised in Table 2.

The exclusive channels hold promise for the future when perhaps ten times the number of Z^0 should become available.

5. B LIFETIME

Each of the experiments has given an analysis of the lifetime figure for the B sample

Table 1. Summary of Branching Ratios (Br)

Expt.	Br$(Z° \rightarrow J/\Psi + X)$	Br $(B \rightarrow J/\Psi + Y)$ for B's from $Z°$.
Opal 163k hadronic Z 13 e^+e^- 28$\mu^+\mu^-$	$(5.3 \pm 0.9 \pm 0.5) \times 10^{-3}$ stat. syst.	[adjusted for Mark III '92 Br.]
Aleph 441k 32 e^+e^- 60 $\mu^+\mu^-$	$(3.81 \pm 0.41 \pm 0.26) \times 10^{-3}$	$(1.21 \pm 0.13 \pm 0.08) \times 10^{-2}$
L3 410k 15 e^+e^- 43 $\mu^+\mu^-$	$(4.1 \pm 0.7 \pm 0.3) \times 10^{-3}$	$(1.3 \pm 0.2 \pm 0.2) \times 10^{-2}$
Delphi 358k 44 $\mu^+\mu^-$	$(4.15 \pm 0.75 \pm 0.53) \times 10^{-3}$	$(1.33 \pm 0.25 \pm 0.18) \times 10^{-2}$
Totals 60 d^+e^- 175 $\mu^+\mu^-$	Average $(4.1 \pm 0.4) \times 10^{-3}$	For B's from $\Upsilon(4S)$; $(1.31 \pm 0.21) \times 10^{-2}$

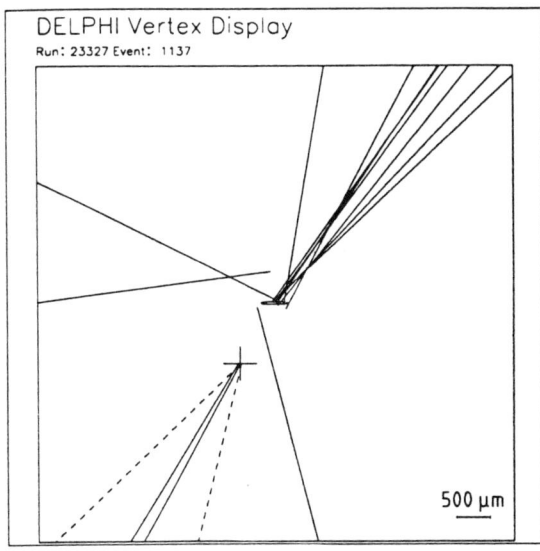

Figure 3. An exclusive $B \rightarrow J/\Psi$ decay observed by Delphi. The $r\phi$ plane is displayed, in the vertex region for a candidate for the decay chain

$B° \rightarrow J/\Psi K^{*°}$, $J/\Psi \rightarrow \mu^+\mu^-$, $K^{*°} \rightarrow K^+\pi^-$.

Dashed lines indicate the muons reconstructed at the secondary vertex.

studied in $Z°$ decays. The analysis is complicated by the fact that there are backgrounds at zero lifetime ($Z° \rightarrow q\bar{q}(gg\Psi)$) and backgrounds of longer effective lifetime than the b ($Z° \rightarrow bX \rightarrow cY$ cascade decays). The Figs.4 a, b, and c show the distance or the time distributions obtained, and the results for the lifetime are given in Table 3. Assuming that the errors can be combined we get 1.34 ± 0.14 ps as the average lifetime.

Within the fairly large errors, this lifetime agrees with those measured for different B samples, again lending some support to the heavy quark effective field theory, excluding say b baryon lifetimes different by factors of two from those of B mesons.

Table 2. Summary of exclusive J/Ψ events.

Aleph		
$B^{\pm} \to J/\Psi K^{\pm}$	$B^{\circ} \to J/\Psi K^{\circ}_S$	$B^{\circ} \to J/\Psi K^{*\circ}$
5 events	1 event	1 event
$\text{Br}(B \to J/\Psi K)$		
$(0.22 \pm 0.10 \pm 0.02) \times 10^{-2}$	[Argus and Cleo[4,5] 0.08×10^{-2}]	
Delphi		
$B^- \to J/\Psi K^-$	$B^{\circ} \to J/\Psi K^+ \pi^-$	$B^- \to J/\Psi K^- \pi^+ \pi^-$
1 event	2 events	1 event

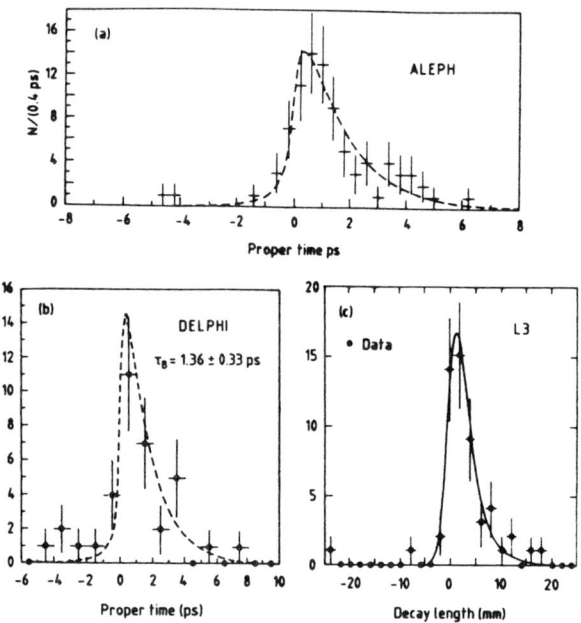

Figure 4. a) Distribution in proper time of events in the J/Ψ peak from Aleph. The curve is the result of an unbinned maximum likelihood fit to this distribution and simultaneously to the out-of-peak background, including 2% of prompt J/Ψ.
b) Delphi proper time distribution, with a similar maximum likelihood fit.
c) L3 data, and a maximum likelihood fit to the decay length distribution.

REFERENCES

1. Opal Collaboration. "Observation of J/Ψ production in multi-hadronic Z° decays", *Phys. Lett.* **B266**, pp.485-496, (1991).

2. V. Barger, K. Cheung, W-Y. Keung, "Z-boson decays to heavy quarkonium", *Phys. Rev.* **D41**, pp.1541-1546, (1990).

3. K. Hagiwara, A.D. Martin, W.J. Stirling, "J/Ψ production from gluon jets at LEP", *Phys. Lett.* **B267**, pp.527-531, (1991).

4. Argus Collaboration. "Exclusive hadronic decays of B mesons", *Z. Phys.* **C48**, pp.543-551, (1990).

5. Cleo Collaboration. "Study of the decay $B \to \Psi X$", *Phys. Rev.* **D34**, pp.3279-3285, (1986).

   Cleo Collaboration. "Inclusive and exclusive decays of B mesons to final states including charm and charmonium mesons", *Phys. Rev.* **D45**, pp.21-35, (1991).

6. Papers submitted to the ICHEP Dallas '92 Conference:

   Aleph Collaboration. "Measurements of B lifetime and branching fractions with

Table 3. Summary of B lifetime measurements

Opal	$1.32^{+0.31}_{-0.25} \pm 0.15$ ps [see[1] for study of systematics]				
L3	$1.34 \pm 0.20 \pm 0.10$ ps				
Systematic error estimation:					
p_B	τ_{backgd}	$backgd$ fraction	$\Delta\ell$	$t-d$ in TEC	
0.02	0.04	0.06	0.05	0.05	$= \pm 0.10$ps total
Aleph	$1.35^{+0.19}_{-0.17} \pm 0.05$ ps				60% with VDET hits
Systematic error estimation:					
p_B	Δp_B	$\Delta\ell$	beam spot	$backgd$ fraction	prompt J/Ψ $backgd$
0.02	0.01	0.01	0.01	0.03	$^{+0.03}_{-0.02}$
MC check	$= \pm 0.05$ ps total				
0.02					
Delphi	$1.36 \pm 0.33 \pm 0.12$ ps				'91 data only, with 3 layer
Systematic error estimation:					
p_B	$backgd$	prompt J/Ψ $backgd$		MC check	μVTX
0.02	$^{+0.05}_{-0.11}$	$^{+0.09}_{-0.05}$		0.02	$= \pm 0.12$ ps total
	[Δp_B, $\Delta\ell$, beam spot contributions included in the stat. error]				

the J/Ψ tag in Z° decays". (Also submitted to *Phys. Lett. B*)

L3 Collaboration. "Inclusive J production in Z° decays". (Also submitted to *Phys. Lett. B*). Preliminary B lifetime analysis.

Delphi Collaboration. "A study of $B \to J/\Psi$ production in Z° hadronic decays with Delphi".

EFFECTIVE HEAVY QUARK THEORY: MATCHING HQET AND QUARK MODELS

F E Close
Rutherford Appleton Laboratory,
Chilton, Didcot, Oxon,
OX11 OQX, England

Heavy quark effective theory (HQET) deals with the limit $m_Q \to \infty$ where the velocity v_μ is fixed. Quark models by contrast tend to deal with $m \to \infty$ and an expansion in v/c. We[1] have been investigating how to match the HQET and the quark model for composite states containing a heavy quark thereby making an "effective" heavy quark theory, EHQT. This turns out to be non-trivial, raises interesting questions and may give some pedagogic insights. Ultimately we hope that it may enable sensible estimates to be made about the size of $1/M(1/M^2)$ effects and the hadron dependence in ratios of heavy- light decays such as $B \to M\ell\nu/D \to M\ell\nu$ where $M = K, K^*, \rho$ etc.

Consider first the $m_Q \to \infty$ limit where $p_\mu = m_Q v_\mu$. The effective spinor h_v then satisfies the Dirac equation $\slashed{v} h = h$ and so

$$h_v = \sqrt{\frac{1+v_0}{2}} \begin{pmatrix} 1 \\ \frac{\vec{\sigma} \cdot \vec{v}}{1+v_0} \end{pmatrix} \quad (1)$$

So far so good. But now consider a meson $Q\bar{q}$ where the heavy flavour undergoes weak decay. The heavy quark and the meson velocity change from v to v'. However, for this to be an exclusive process, the light $\bar{q}$ velocity must equal that of the meson: how does the light $\bar{q}$ spectator adjust from v to v'?

This puzzle has been noticed in different ways in the literature[2], Korner et al[3] suggest that the $\bar{q}$ suffers a dynamical velocity kick, and hence they generate the universal Isgur-Wise form factor $F(v.v')$ from this light quark

interaction. Later in EHQT, we shall see how the velocity change occurs and shall question the above form factor argument.

At $0(1/M)$ we write

$$p_Q^\mu = m_Q v^\mu + K^\mu \quad (2)$$

and the full spinor becomes

$$\psi \equiv (1 + \frac{\slashed{K}}{2m}) \begin{pmatrix} 1 \\ \frac{\vec{\sigma} \cdot \vec{v}}{1+v_0} \end{pmatrix} \quad (3)$$

where the $\slashed{K}$ acts on the left. The essence of making an "effective heavy quark theory" is to make a particular choice for K. In this idealised example of free non-interacting quarks where QCD has been switched off, this is

$$K^\mu \equiv k^\mu + \frac{\vec{v} \cdot \vec{k}}{1+v_0} v^\mu \quad (4)$$

(In the context of mass reparametrisation this corresponds to $k.v = -k_0$ i.e. $\delta m = k_0 \neq 0$. Note also that if $k = 0(\Lambda_{QCD})$, then $K = 0(v_0\Lambda)$). Substitute (4) into (3) and one finds

$$\psi \simeq \begin{pmatrix} 1 \\ \frac{\vec{\sigma} \cdot \vec{v}}{1+v_0} \end{pmatrix} + \begin{pmatrix} \frac{\vec{\sigma} \cdot \vec{v}}{1+v_0} \\ 1 \end{pmatrix} \frac{\vec{\sigma} \cdot \vec{k}}{2m_Q}. \quad (5)$$

where now $\vec{\sigma}.\vec{k}$ appears **on the right**.

A PHYSICAL PICTURE

Consider a "meson" at rest ($\vec{v}_M = 0$) where its overall spin is well defined, consisting of "brown muck" and heavy quark with three momenta $\pm\vec{k}$. The spinor of the heavy quark in absence of interaction is

$$Q(\vec{v}=0;\vec{k}) = \begin{pmatrix} 1 \\ \frac{\vec{\sigma}\cdot\vec{k}}{2m_Q} \end{pmatrix} \quad (6)$$

Now boost to velocity $\vec{v}$ and one obtains

$$Q(\vec{v},\vec{k}) \simeq \begin{pmatrix} 1 + \frac{\vec{\sigma}\cdot\vec{v}\vec{\sigma}\cdot\vec{k}}{2m_Q(1+v_0)} \\ \frac{\vec{\sigma}\cdot\vec{v}}{1+v_0} + \frac{\vec{\sigma}\cdot\vec{k}}{2m_Q} \end{pmatrix} \quad (7)$$

The $\sigma.v.\sigma.k$ term is a Wigner rotation arising from two boosts (k followed by v) and is crucial for consistency with fundamental low energy theorems in Compton scattering[1,4]. When dealing with heavy quark in a heavy composite hadron it may be dangerous to expand matrix elements based on the $\begin{pmatrix} 1 \\ \frac{\sigma \cdot p}{E+m} \end{pmatrix}$ structure present in eq (7) as this is exact only if $\vec{v} \times \vec{k} = 0$. Notice that the physical spinor eq (7), corresponding to a free quark with relative momentum $\vec{k}$ boosted to overall velocity $\vec{v}$, is identical to the heavy quark expansion at eq (5); in particular note the physical significance of $\sigma.k$ appearing on the right. Note that this suggests that in covariant trace formalism one should write $(1 + \rlap{/}{v})(1 + \frac{\rlap{/}{k}}{2m})\Gamma$ rather than $(1+\frac{\rlap{/}{k}}{2m})(1+\rlap{/}{v})\Gamma$ if wishing to match onto dynamical models (where $\Gamma = \gamma_5, \rlap{/}{\epsilon}$ for pseudoscalar or vector meson). This matching has arisen as a result of a particular choice of variables, eq (4), or mass parametrisation. These ideas can be generalised to include QCD interactions on the heavy quark

ELECTROMAGNETIC INTERACTION

Consider a "meson" consisting of a heavy quark, charge e mass m and a quasi-free electrically neutral spectator system. The (k independent piece of the) spatial vector current using the spinor eq (7) (M = hadron mass)

$$\vec{J} = \frac{1}{2}(\vec{v}+\vec{v}') - \frac{iM}{2m_Q}\vec{\sigma}_Q \times (\vec{v}'+\vec{v})$$

$$-\frac{i}{8}(\vec{v}_0' - \vec{v}_0)(\frac{2M}{m_Q}\vec{\sigma}_Q - \vec{\sigma}_T) \times (\vec{v} + \vec{v}') \quad (8)$$

The appearance of σ_T rather than σ_Q in the final term arises from the careful treatment, in particular including the spin rotation of the spectator system transforming from v to v'.

By working to $0(v,k)$ one sees that the **momentum** of the spectator system is unaltered. There is a transfer from relative (k) to overall (v) induced by the recoil of the system, with

$$K - K' = (M - m_Q)(v' - v) \quad (9)$$

The probability to make this transfer depends upon the relative momentum distribution in the hadron which depends on the binding dynamics. It is thus that the effective form factor $F(v.v')$ obtains. In the limit where $m_Q \to \infty$ and QCD is turned off, $F(v.v') = \delta_{vv'}$ and the model is unrealistic. For sensible dynamics to obtain it appears that binding and $0(1/M)$ effects must be included.

REFERENCES

1. F.E. Close and Z.P. Li, RAL-92-022, *Physics Letters* (in press);
   F.E. Close and A. Wambach, work in progress.

2. For example, M.J. Dugan, M. Golden and B. Grinstein, *Phys. Letters* **B**, (1992).

3. J. Korner, Nucl Phys. **B**(*Proc. Sup*) 21, 366 (1991);
   F. Hussein, J. Korner and G. Thompson, *Ann. Phys.* **206**, 334 (1991).

4. S. Brodsky and J. Primack, ' *Ann. Phys.* **52**, 315 (1969).

LEADING AND SUBLEADING LOGARITHMIC QCD CORRECTIONS TO BILINEAR HEAVY QUARK CURRENTS

Thomas Mannel Wolfgang Kilian Panagiotis Manakos

Institut für Kernphysik
Technische Hochschule Darmstadt
Schlossgartenstr. 9, D–6100 Darmstadt, Germany

Abstract

We present a complete calculation of the two loop anomalous dimension and the corresponding Wilson coefficient for bilinear currents of heavy quarks.

INTRODUCTION

The understanding of systems involving heavy quarks has considerably improved by employing the so called heavy quark limit [1]. In this limit, where the masses of the heavy quarks are sent to infinity, new symmetries arise, which allow to relate form factors of heavy meson transition amplitudes in a completely model independent way.

The corrections to this limit fall into two classes: The recoil corrections of the order $1/m_H$ and the QCD radiative corrections of the order $\alpha_s(m_H)$ where m_H is the mass of the heavy quark. The latter type of corrections may be calculated systematically by using the Feynman rules of HQET as formulated e.g. in [2]. Using the renormalization group of HQET one may sum logarithms of the type $\ln m_H$ by calculating the anomalous dimensions of the operators of interest perturbatively.

We present a two loop calculation of the anomalous dimension of heavy quark bilinear currents of the form $\bar{h}_{v'} \Gamma h_v$ where Γ is some arbitrary combination of Dirac matrices. As it is the case for the one loop result the anomalous dimension of this current depends on the velocities of the heavy mesons; it is real for t channel processes like heavy quark decays or scattering processes off heavy quarks, but it acquires an imaginary part in the s channel, i.e. for processes like heavy meson pair creation in e^+e^- annihilation. This is due to the fact that the imaginary part of the matrix element in the s channel as calculated in the full theory contains a logarithmic dependence on the heavy mass m_H. After the renormalization group resummation of these logarithms this imaginary part turns into a phase of the matrix element which is the nonabelian analogue of the Coulomb phase of QED.

The anomalous dimension for the t channel was already discussed by Korchemsky and Radyushkin in [3]. Our result for the t channel agrees with the one given in [3]. An independent calculation of the t channel anomalous dimension was performed by Ji in [4] which is not in agreement with our result.

t CHANNEL ANOMALOUS DIMENSION

At the two loop level one has to evaluate nine diagrams which are depicted in fig.1. The incoming heavy quark is moving with a velocity v_1 and has a residual momentum p_1, i.e. it has the (offshell) momentum $P = m_1 v_1 + p_1$. Similarly, the outgoing heavy quark has the momentum $P' = m_2 v_2 + p_2$. The integrals cal-

Fig.1: Two Loop Diagrams

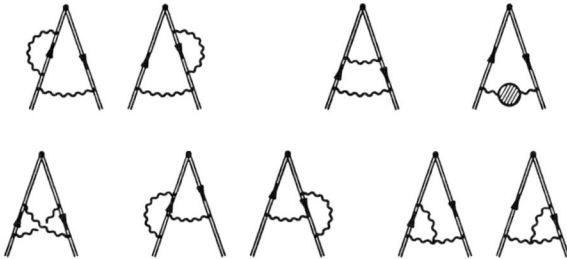

culated below depend only on $v_1 \cdot v_2$, $\omega = v_1 \cdot p_1$ and $\xi\omega = v_2 \cdot p_2$. We introduce the hyperbolic angle variable θ with

$$v_1 \cdot v_2 = \operatorname{ch}\theta \qquad \sqrt{(v_1 \cdot v_2)^2 - 1} = \operatorname{sh}\theta. \quad (1)$$

The calculation is performed in dimensional regularization where the dimension $4-\epsilon$. Due to infrared divergencies, which occur in intermediate steps of the calculation, an infrared regularization is needed. We have chosen to regularize the infrared singularities by keeping the external momenta off shell, i.e. by assigning residual momenta p_1 and p_2 to the external quark lines. The individual diagrams may depend only on the ratio ξ of the two scalar products $v_1 \cdot p_1$ and $v_2 \cdot p_2$. Although the individual contributions of the diagrams in fig.1 depend on ξ, the final result for the anomalous dimension is independent of the particular IR regularization scheme. A different scheme for IR regularization chosen in [3], where a cut off in the Feynman parameter space has been used, yields the same result.

The anomalous dimensions in HQET depend in general on the velocities which becomes here a dependence on the hyperbolic angle θ. Performing the calculation as described in [5] we obtain for the t channel anomalous dimension

$$\gamma_1 = C_F(1 - \theta \operatorname{cth}\theta) \quad (2)$$
$$\gamma_2 = C_F C_A \left((\theta \operatorname{cth}\theta)^2 \left[-s_1(\theta) + \frac{1}{2}s_2(\theta) \right] + \frac{1}{2} \right.$$

$$+ (\theta \operatorname{cth}\theta - 1)\left[\frac{\pi^2}{24} - \frac{67}{36} \right]$$
$$+ (\theta \operatorname{cth}\theta)\left[(s_1(\theta) + s_4(\theta) - s_5(\theta)) \right] - 1 \Big)$$
$$+ 5\frac{n_f T_F C_F}{9}(\theta \operatorname{cth}\theta - 1) \quad (3)$$

where the correct normalization $\gamma_i(\theta = 0) = 0$ ($i = 1, 2$) is obtained by taking into account the self energy contributions. Here we have introduced the auxiliary functions

$$s_1(\theta) = \frac{1}{\theta}\int_0^\theta d\psi\, \psi \operatorname{cth}\psi \quad (4)$$
$$s_2(\theta) = \frac{2}{\theta^2}\int_0^\theta d\psi\, \psi^2 \operatorname{cth}\psi \quad (5)$$
$$s_4(\theta) = \frac{1}{\theta}\int_0^\theta d\psi\, \frac{\psi \operatorname{cth}\psi}{1 - \left(\frac{\operatorname{sh}\psi}{\operatorname{sh}\theta}\right)^2} \ln\frac{\operatorname{sh}\theta}{\operatorname{sh}\psi} \quad (6)$$
$$s_5(\theta) = \frac{1}{\theta}\int_0^\theta d\psi\, \frac{1}{1 - \left(\frac{\operatorname{sh}\psi}{\operatorname{sh}\theta}\right)^2} \ln\frac{\operatorname{sh}\theta}{\operatorname{sh}\psi} \quad (7)$$

Note that our result agrees with the one obtained by Korchemsky and Radyushkin [3].

It is well known that the two loop anomalous dimension vanishes for the abelian case with no light flavors. This case is obtained by setting $C_A = n_f = 0$ which is a simple check of our result.

s CHANNEL ANOMALOUS DIMENSION

The result (3) is valid in the region relevant for weak decays and scattering processes

$$-\infty < q^2 < (m_1 - m_2)^2, \quad (8)$$
$$q^2 = m_1^2 + m_2^2 - 2m_1 m_2 v_1 \cdot v_2$$

which we call the t channel region. However, for the discussion of e.g. pair creation of heavy mesons in e^+e^- annihilation one may also be interested in the s channel region

$$(m_1 + m_2)^2 < q^2 < \infty \quad (9)$$

which corresponds to the replacement $v_2 \to -v_2$ or, equivalently, $v_1 \cdot v_2 \to -v_1 \cdot v_2$.

For the one loop result the continuation is trivial and the result is

$$\gamma_1(\theta - i\pi) = \gamma_1(\theta) + C_F i\pi \coth\theta \qquad (10)$$

Note that due to the analytic continuation an imaginary part appears, while the real part remains unchanged in one loop order.

The analytic continuation of the two loop result (3) is more involved, since the integrals along the real axis expressing the functions $s_i(\theta)$ become contour integrals in the complex θ plane. The integrands of the functions s_i have singularities on the imaginary axis at $\theta = ki\pi$ for integer $k \neq 0$ and a logarithmic branching point at $\theta = 0$. With a suitable choice of the integration contour [5] we obtain

$$\gamma_2(\theta - i\pi) = \gamma_2(\theta)$$
$$+ C_F C_A \frac{\pi^2}{2} \left(\theta \coth^2\theta - \coth\theta + 1\right)$$
$$- \theta \coth\theta \, s_6(\theta) - \ln \operatorname{ch}\theta)$$
$$+ i\pi C_F C_A \left(\theta \coth^2\theta (-s_1(\theta) + \ln(2\operatorname{sh}\theta))\right.$$
$$+ \theta \coth\theta (-s_7(\theta) + s_8(\theta))$$
$$+ \coth\theta \left[\frac{1}{4}L_2(\operatorname{th}^2\theta) + \frac{1}{2}\ln^2 \operatorname{ch}\theta\right.$$
$$\left.\left. - \ln \operatorname{sh}\theta \ln \operatorname{ch}\theta - \ln(2\operatorname{sh}\theta) + \frac{67}{36}\right]\right)$$
$$- i\pi \frac{5n_f T_F C_F}{9} \coth\theta. \qquad (11)$$

where we introduced three new functions

$$s_6(\theta) = \frac{2}{\pi\theta}\int_0^{\pi/2} d\psi \frac{\psi \cot\psi}{1 + \left(\frac{\sin\psi}{\operatorname{sh}\theta}\right)^2}$$

$$s_7(\theta) = \frac{2}{\pi\theta}\int_0^{\pi/2} d\psi \frac{\psi \cot\psi}{1 + \left(\frac{\sin\psi}{\operatorname{sh}\theta}\right)^2} \ln\frac{\operatorname{sh}\theta}{\sin\psi}$$

$$s_8(\theta) = \frac{2}{\pi\theta}\int_0^{\pi/2} d\psi \frac{1}{1 + \frac{\sin^2\psi}{\operatorname{sh}^2\theta}} \ln\frac{\operatorname{sh}\theta}{\sin\psi}$$

and L_2 is the dilogarithmic function

$$L_2(z) = -\int_0^z dt \frac{\ln(1-t)}{t}.$$

Note that here unlike the one loop case the real part of the anomalous dimension is different in the s and t channel.

WILSON COEFFICIENTS

The relevant quantities entering observables are the Wilson coefficients $C(\mu, \theta)$ defined by

$$C(\mu_1, \theta) \langle H(v_1)|\bar{h}_{v_1}\Gamma h_{v_2}|H(v_2)\rangle_{\mu_1} \quad (12)$$
$$= C(\mu_2, \theta) \langle H(v_1)|\bar{h}_{v_1}\Gamma h_{v_2}|H(v_2)\rangle_{\mu_2}$$

where $H(v_i)$ denotes the heavy mesons and Γ is an arbitrary combination of Dirac matrices. The subscripts μ_i denote the renormalization points of the matrix elements. The Wilson coefficients obey the usual renormalization group equation which is solved using the two loop anomalous dimensions calculated above. One obtains for C for the running between m_c and μ is up to two loops

$$C(\mu, \theta) = C(m_c, \theta) \left(\frac{\alpha_s(m_c)}{\alpha_s(\mu)}\right)^{-\frac{\gamma_1(\theta)}{2\beta_1}} \quad (13)$$
$$\left(\frac{1 + \frac{\beta_2}{\beta_1}\frac{\alpha_s(m_c)}{\pi}}{1 + \frac{\beta_2}{\beta_1}\frac{\alpha_s(\mu)}{\pi}}\right)^{-\frac{\gamma_2(\theta)}{2\beta_2} + \frac{\gamma_1(\theta)}{2\beta_1}}$$

where the two loop approximation has to be used for $\alpha_s(\mu)$. The one loop result is obtained by omitting the last factor and inserting the one loop approximation for α_s. The initial value $C(m_c, \theta)$ is given by the appropriate matching at the scale $\mu = m_c$.

In order to get a quantitative idea of the results we have evaluated the integrals s_i numerically. The results are displayed in figures 2 and 3.

In fig.2 we present the results for the Wilson coefficient for $m_c = 1.8$ GeV at the renormalization point $\mu = 0.6$ GeV as a function of $v_1 \cdot v_2$. The choice $\mu = 0.6$ GeV for figs.2 and 3 is certainly at the lower limit of possible choices, since the terms neglected are of the order $\alpha_s^2(\alpha_s \ln(m_c/\mu))^n$ for $n \geq 0$ and perturbation theory breaks down.

The dotted line is the one loop result which we have plotted in units of $C(m_c)$ which is

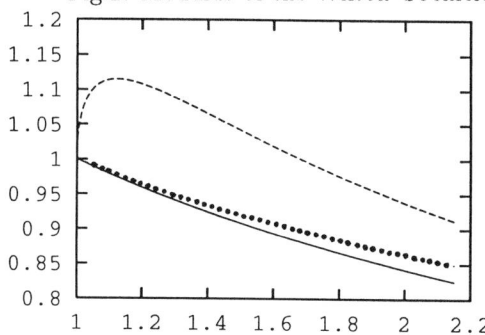

Fig.2: Modulus of the Wilson Coefficient.

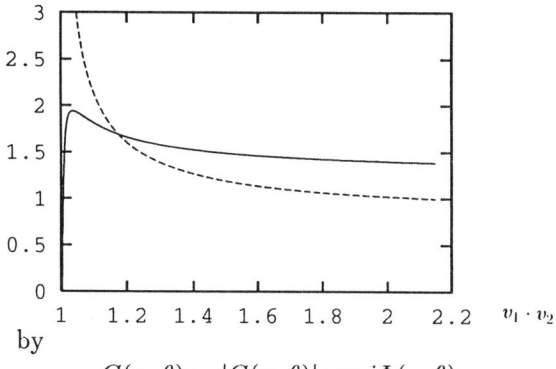

Fig.3: s channel Phase of the Wilson Coefficient.

the Wilson coefficient obtained from the leading log scaling above m_c and tree level matching at m_c. The strong coupling α_s is inserted in one loop approximation with a value of $\Lambda_{QCD}^{(3)} = 0.3$ GeV. At threshold the anomalous dimension vanishes and thus the value of the Wilson coefficient for $v_1 \cdot v_2 = 1$ is given by $C(m_c)$. In the range relevant for weak $b \to c$ decays the corrections due to one loop scaling are of $\mathcal{O}(10\%)$. At one loop level the absolute value of the Wilson coefficient is the same both in the s and t channel.

The solid and the dashed line in fig.2 is our two loop result for the Wilson coefficient for $\mu = 0.6$ GeV, again in units of $C(m_c)$. In order to be consistent $C(m_c)$ has to include the two loop scaling above m_c and the one loop matching, which consists of a θ dependent $\mathcal{O}(\alpha_s)$ correction and an additional operator [5]. The strong coupling α_s has been inserted in two loop approximation. The solid line is the t channel result. Compared to the leading order result there is a small correction of less than 3%. The Wilson coefficient in the s channel is represented by the dashed line in fig.2. Due to the different real part of the anomalous dimension in the s and t channels the two loop Wilson coefficient in the s channel has a different shape; close to threshold one has a positive contribution of about 12% whereas the corrections become negative for values $v_1 \cdot v_2 > 1.7$.

However, in the s channel the Wilson coefficient picks up an additional phase Φ defined by

$$C(\mu, \theta) = |C(\mu, \theta)| \exp i\Phi(\mu, \theta).$$

The phase $\Phi(\mu, \theta)$ is plotted in fig.3 where the dashed line is the one loop result while the solid line represents the two loop result. Both the one and two loop result for the phase show a $1/\theta$ behaviour near threshold.

REFERENCES

1. N. Isgur and M. Wise, Phys. Lett. **B232** (1989) 113; Phys. Lett. **B237** (1990) 527. H. Politzer and M. Wise, Phys. Lett. **B206** (1988) 681; Phys. Lett. **B208** (1988) 504. M. Voloshin and M. Shifman, Sov. J. Nucl. Phys. **45** (1987) 292; Sov. J. Nucl. Phys. **47** (1988) 511. E. Eichten and B. Hill, Phys. Lett. **B234** (1990) 511. B. Grinstein, Nucl. Phys. **B339** (1990) 253. H. Georgi, Phys. Lett. **B240** (1990) 447. T. Mannel, W. Roberts and Z. Ryzak, Nucl. Phys. **B 368** (1992) 204. A. Falk, H. Georgi, B. Grinstein and M. Wise, Nucl. Phys. **B343** (1990) 1.

2. see the last paper of ref.[1].

3. G. Korchemsky and A. Radyushkin, Nucl. Phys. **B 283** (1987) 342; Preprint CPT-91/P.2629 , UPRF-91-316 (1991).

4. X. Ji, Phys. Lett. **B 264** (1991) 193.

5. W. Kilian, P. Manakos and T. Mannel, Darmstadt Preprint IKDA 92/9.

CONSTRAINTS ON HEAVY MESON FORM FACTORS [*]

Josep Taron
CPT, CNRS-Luminy, Case 907
F-13288 Marseille Cedex 9
France

Abstract

In the limit $m_c, m_b \to \infty$ (m_c/m_b fixed) the transition amplitudes $<D^*|V-A|B>$ and $<D|V-A|B>$ feature just one unknown, the Isgur-Wise form factor. Assuming that it has the usual analyticity properties of form factors in Quantum Field Theory, unitarity puts tight bounds on its allowed values. These bounds uncover a value of $|V_{cb}|$ smaller than previous fits of ARGUS and CLEO data, done with parametrizations of the Isgur-Wise form factor which do not verify the bounds.

The study of semileptonic decays of hadrons containing one heavy quark was one of the main thrusts [1] for the discovery of approximate new symmetries [2] in hadron physics, which introduce great simplifications in the description of such decays. These new *spin* and *flavour* symmetries are exact in the limit of infinitely heavy b and c quarks. The low lying pseudoscalar and vector mesons become degenerate in mass (e.g., $m_{B^*} = m_B = m_b$); also, the matrix elements of the transitions

$$B \to D e \bar\nu \quad B \to D^* e \bar\nu \qquad (1)$$

are fixed up to a common unknown form factor $\xi(v.v')$, the Isgur-Wise function, which plays a central rôle in the heavy quark description of the decays in (1). The shape of this function is not fixed by symmetry requirements alone; they impose, however, that $\xi(v.v')$ should be a function of only the product of the B and D (D^*) meson four velocities, v and v', respectively, and that the normalization at the kinematic point of zero recoil $v.v' = 1$ should be fixed to unity,

$$\xi(v.v' = 1) = 1. \qquad (2)$$

$\xi(v.v')$ is a universal function since it does not depend on the heavy flavour of the mesons involved in (1).

The Isgur-Wise function may thus be regarded as the limit of the form factor $F(q^2)$ of the matrix element ($q = p - p'$)

$$<B(p')|\bar b \gamma_\mu b|B(p)> = (p+p')_\mu F(q^2), \quad (3)$$

when $m_b \to \infty$. The normalization (2) expresses the fact that the current $\bar b \gamma_\mu b$ is conserved and $F(q^2 = 0) = 1$ is the b-charge of the B meson. The analytic properties of $\xi(v.v')$ are thus those of $F(q^2)$ in the $m_b \to \infty$ limit: $\xi(v.v')$ is an analytic function in the cut complex $v.v'$ plane, with a cut running from $-\infty$ to -1.

Further detailed knowledge of $\xi(v.v')$ would require a full QCD calculation in the nonperturbative regime, which is beyond reach. Nevertheless, upper and lower bounds

---

[*]Work supported by *EEC Science Twinning Grant* SCI-000337, and *CYCIT* AEN-90-0033 (Spain).

Table 1. Numerical values for the upper and lower bounds on $\xi(v.v')$ (see Fig. 1).

$v.v'$	$f_-(v.v')$	$\sqrt{\frac{2}{1+v.v'}}$
1.0	1.000	1.000
1.1	0.957	0.976
1.2	0.918	0.953
1.3	0.881	0.933
1.4	0.848	0.913
1.5	0.817	0.894
1.6	0.789	0.877

on $\xi(v.v')$ can be derived using unitarity and analyticity arguments, which constrain it rather tightly over the range $1 \leq v.v' \leq \frac{1}{2}(\frac{m_B}{m_{D^*}} + \frac{m_{D^*}}{m_B}) \simeq 1.6$, relevant for the decays in (1).

The upper bound

The upper bound was already first derived by Bjorken [3] in the context of inclusive semileptonic charmed B decays,

$$\xi(v.v') \leq \sqrt{\frac{2}{1+v.v'}}. \qquad (4)$$

The amount of deviation from $\sqrt{\frac{2}{1+v.v'}}$ reflects the room for extra light hadrons (π's,...) in the final state. $\sqrt{\frac{2}{1+v.v'}}$ itself is the form factor in a model of free noninteracting quarks, first introduced by Suzuki [4]; here the current $\bar{c}\gamma_\mu(1-\gamma_5)b$ turns a free b quark into a free c quark, but it can certainly not excite any light mode.

The lower bound

The lower bound follows from general considerations of unitarity, analyticity, the property of asymtotic freedom of QCD, as well as the simplifying features that arise in the infinite heavy quark mass limit: namely that all semileptonic transitions in (1) share the same unknown form factor, and that this form factor takes the value unity at zero recoil.

The derivation of the lower bound proceeds in four steps [5]. First, one evaluates the correlator of two currents $\bar{b}\gamma_\mu b$ at momentum transfer $q^2 = 0$: in the large b-mass limit the calculation can be done with perturbative QCD because the resonance region and the threshold $q^2 = 4m_B^2$ are pushed far away from $q^2 = 0$. The correlator reduces to one-loop of quarks [6]: higher order corrections in $\alpha_s(m_b)$ are vanishingly small due to asymptotic freedom and non-perturbative power corrections à la SVZ [7] become also negligible, suppressed by inverse powers of m_b. Secondly, recall that the spectral function of the two current correlator gets positive contributions from all possible hadronic states with the quantum numbers of the current $\bar{b}\gamma_\mu b$ and evaluate the (additive) contribution of the states $|\bar{B}B>$, $|\bar{B}^*B>$, $|\bar{B}B^*>$, $|\bar{B}^*B^*>$ to the spectral function. In a third step one relates the two previous results through a dispersion relation, which reads

$$1 \geq \frac{5}{6}\int_{-\infty}^{-1} d\omega \frac{(2-\omega)(-1-\omega)^{3/2}}{(1-\omega)^{7/2}} |\xi(\omega)|^2, \qquad (5)$$

where $\omega = v.v'$. Notice that it is only in the large b-mass limit that the four states are related to each other by the spin-flavour symmetry, and that all of them contribute proportionally to the same unknown factor $|\xi(\omega)|^2$. The constraint (5) on an integral over the cut $\omega \leq -1$ of $|\xi(\omega)|^2$ multiplied by some phase space weight function finally results in upper and lower bounds for $\xi(v.v')$ as a bonus of analyticity. The analytic structure of $\xi(v.v')$ implies, for $v.v' \geq 1$:

$$f_-(v.v') \leq \xi(v.v') \leq f_+(v.v'), \qquad (6)$$

where

$$f_\pm(v.v') = \frac{\xi(1)}{(1+z)^2(1+(2\sqrt{6}-5)z)} \times$$

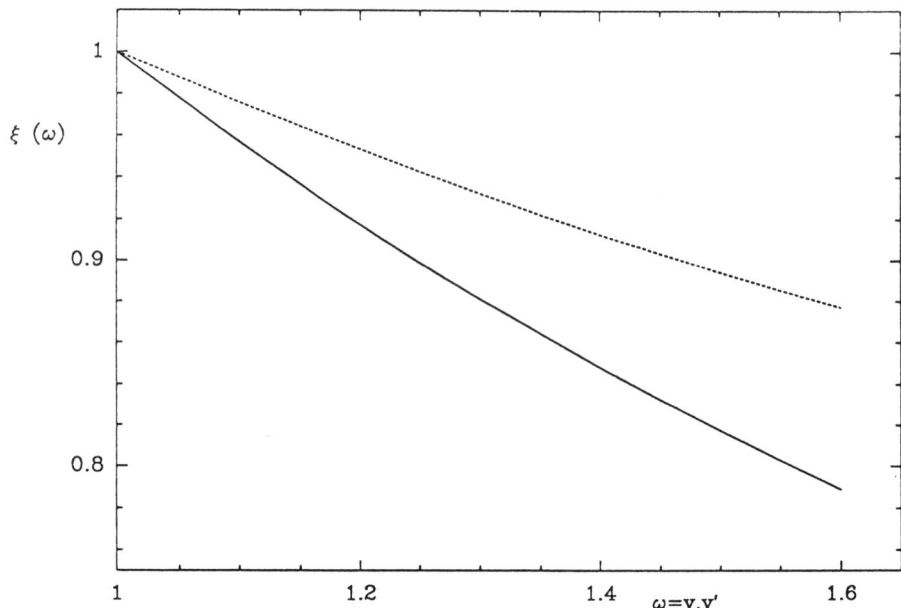

Figure 1. Plot of the lower and upper bounds: $f_-(v \cdot v')$ (solid line) and $\sqrt{\frac{2}{1+v \cdot v'}}$ (dashed line), respectively.

$$\left[\sqrt{1-z} \pm \frac{z}{\sqrt{1+z}} \sqrt{\frac{1}{(C\xi(1))^2} - 1} \right],$$

$z = \frac{\sqrt{\frac{1+v \cdot v'}{2}} - 1}{\sqrt{\frac{1+v \cdot v'}{2}} + 1}$, and $C = \frac{\sqrt{5\pi}}{8}\left(\frac{1}{\sqrt{2}} + \frac{1}{\sqrt{3}}\right) \simeq 0.636$. We set $\xi(1) = 1$, according to (2). The upper bound turns out to be greater than Bjorken's upper bound (4). The lower bound, though, is nontrivial. Both upper and lower bounds are shown in Fig. 1.

The value of $|V_{cb}|$

What can be learnt about $|V_{cb}|$, from these bounds? The bounds point towards a smaller value of $|V_{cb}|$ than the ones currently quoted in the literature (for a review, see Persis Drell's talk at this Conference). The reason for that is the flatness of the bounds at $v \cdot v' = 1$. Indeed, the prediction

$$-0.45 \leq \xi'(1) \leq -0.25 \qquad (7)$$

is, at any rate, significantly smaller (in magnitude) than the value $\xi'(1) \sim -1$ obtained by several parametrizations of $\xi(v \cdot v')$ [8] upon fitting the available ARGUS [9] and CLEO [10] data. However, data are still poor around $v \cdot v' = 1$ to allow for a significant determination of the derivative $\xi'(1)$.

There are several attempts to compute $\xi(v \cdot v')$ using QCD sum rules [11,12], which seem to converge to functions closer to the window allowed by the bounds in (4) and (6). The most recent calculation [12] seems to yield $\xi'(1) = -0.54$ and the discrepancy is at the 20% level only. (This is just the scale invariant part of the result in ref. 12. The disagreement claimed by the authors between their result and the bounds is based on an inappropriate comparison: in ref. 12 the bounds are compared to an "Isgur-Wise" function that depends on an arbitrary scale whereas $F(q^2)$ in (3) is a physical, observable quantity and so its $m_b \to \infty$ limit, $\xi(v \cdot v')$, should also be. The "Isgur-Wise" function in ref. 12 is not normalized to unity at $v \cdot v' = 1$ either).

Disregarding the finite mass corrections, the infinitely heavy quark mass expressions yield ($\tau_B = 1.40 \pm 0.045$ ps, *LEP* average, as quoted in Persis Drell's talk at this Conference)

$$|V_{cb}|_{\infty\ mass\ approximation} = 0.030 \ . \quad (8)$$

This fit has a statistical error of ±0.002 and another error of ±0.002 due to having fit the data with a strip of allowed values rather than with a function. Finite mass corrections are expected to raise the central value in (8) by an amount not larger than 20% [13].

Aknowledgements This work was done in a most fruitful and enjoyable collaboration with Eduardo de Rafael. We aknowledge M. Danilov, K. Reim and Y. Zaitzev, and D. Bortoletto and S. Stone for provinding the tables of numerical data from ARGUS and CLEO $B \to D^* e \bar{\nu}$ experiments for us.

REFERENCES

1. M.A. Shifman and M.B. Voloshin, "On production of D and D^* Mesons in B meson decays", *Sov. J. Nucl. Phys.* 45, p. 292 (1987).

2. N. Isgur and M.B. Wise, "Weak decays of heavy mesons in the static quark approximation", *Phys. Lett.* B 232, p. 113 (1989); "Weak transition form-factors between heavy mesons", *Phys. Lett.* B 237, p. 527 (1990).

3. J.D. Bjorken, "New symmetries in heavy flavour physics", SLAC-PUB-5278 (1990).

4. M. Suzuki, "Spectator theory of final state spins in semileptonic decays of heavy flavoured mesons", *Nucl. Phys.* B 258, p. 553 (1985).

5. E. de Rafael and J. Taron, "Constraints on heavy meson form factors", *Phys. Lett.* B 282, p. 215 (1992).

6. V.A. Novikov, L.B. Okun, M.A. Shifman, A.I. Vainshtein, M.B. Voloshin, V.I. Zakharov, "Charmonium and gluons", *Phys. Repts.* 41 C, p. 1 (1978).

7. M.A. Shifman, A.I. Vainshtein, V.I. Zakharov, *Nucl. Phys.* B 147, pp. 385, 448, 519 (1979).

8. J.L. Rosner, "Determination of pseudoscalar charmed-meson decay constants from B-meson decays", *Phys. Rev.* D 42, p. 3732 (1990). T. Mannel, W. Roberts and Z. Ryzak, "Testing the heavy quark effective theory in $\bar{B}^0 \to D^{(*)+} e^- \bar{\nu}$", *Phys. Lett.* B 254, p. 254 (1991). M. Neubert, "Model-independent extraction of $|V_{cb}|$ from semileptonic decays", *Phys. Lett.* B 264, p. 455 (1991). G. Burdman, "On the extraction of $|V_{cb}|$ from semileptonic B decays", *Phys. Lett.* B 284, p. 133 (1992).

9. ARGUS Collaboration (H. Albrecht, et al.), "Measurement of the decay $B^- \to D^{*0} l^- \bar{\nu}$", *Phys. Lett.* B 275, p. 195 (1992). Data on the decay $\bar{B}^0 \to D^{*+} l^- \bar{\nu}$, from ARGUS contributed paper to these Proceedings.

10. D. Bortoletto and S. Stone, "Factorization test using $\bar{B}^0 \to D^{*+}\pi^-$ and an estimate of f_D, using $B \to DD_s^-$", *Phys. Rev. Lett.* 65, p. 2951 (1990).

11. A.V. Radyushkin, *Phys. Lett.* B 271, p. 218 (1991). M. Neubert, *Phys. Rev.* D 45, p. 2451 (1992).

12. E. Bagan, P. Ball and P. Gosdzinsky, Heidelberg pre-print HD-THEP-92-40 (1992).

13. E. de Rafael and J. Taron, "Determination of $|V_{cb}|$ in the large heavy quark mass limit", CPT-CNRS(Luminy) pre-print CPT-92/P.2797 (1992).

IMPLICATIONS OF HEAVY QUARK SYMMETRY AND CHIRAL DYNAMICS *

Hai-Yang Cheng
Institute of Physics
Academia Sinica
Taipei, Taiwan 11529
Republic of China

Abstract

The low-energy strong, electromagnetic and heavy-flavor-conserving nonleptonic weak interactions of heavy hadrons with light pseudoscalar mesons are studied in a formalism which incorporates both heavy quark symmetry and chiral symmetry. Some implications are discussed, including the strong decays of D^* and Σ_c, the electromagnetic decays of D^*, B^*, Σ_c, Ξ'_c, the semileptonic weak decays $B \to D(D^*)\pi\ell\bar{\nu}$, and the nonleptonic weak decays $\Xi_c \to \Lambda_c\pi$, $\Omega_c \to \Xi'_c\pi$.

INTRODUCTION

The objective of this work[1] is to exploit various low-energy interactions of the heavy mesons and heavy baryons with the Goldstone bosons π, K and η. For this purpose, we presented an effective Lagrangian formalism which combines heavy quark symmetry[2] and chiral symmetry. Many interesting applications are discussed in this talk.

DYNAMICS OF HEAVY HADRONS

A heavy meson contains a heavy quark and a light antiquark. The ground states comprise the usual 1^- and 0^- mesons, which will be denoted by P^* and P, respectively. It is a simple matter to write down the chiral Lagrangian involving P and P^* and their couplings to the Goldstone bosons.[1,3,4] Because of space limit, we will not write down the explicit expressions for the chiral Lagrangians, which can be found in ref.[1a]. This meson Lagrangian contains two independent coupling constants f_Q and g_Q. Now the heavy-quark flavor symmetry gives the dependence of f_Q as $f_Q = \sqrt{M_P M_{P^*}} f$, while spin symmetry relates g_Q to f_Q via $g_Q = \frac{1}{2}f$. The nonrelativistic quark model has a simple prediction for the value of f. It turns out that

$$f = -2 \quad \text{for } g_A^q = 0.75, \qquad (1)$$

where g_A^q is the axial-vector coupling constant of the light quark. The value of g_A^q is chosen in such a way that the correct value of $g_A^{\text{nucleon}} = 1.25$ is reproduced.

The heavy baryon of interest is that constructed from a heavy quark and two light quarks. The two light quarks form a symmetric **6** or an antisymmetric antitriplet $\bar{3}$ in flavor SU(3) space. The most general chiral-invariant Lagrangian in the baryon sector[1,5] has six coupling constants $g_1, g_2, ..., g_6$ and its explicit form is given in ref.[1a]. As the meson case, heavy-quark spin symmetry reduces the six coupling constants to two independent ones, say g_1 and g_2. Again, the quark model

---

*This work was supported in part by the National Science Council of the Republic of China.

Table 1. The predicted branching ratios of the D^* mesons.

decay mode	Γ(keV)	Br(%)	CLEO II (%)
$D^{*+} \to D^0 \pi^+$	95	67.3	$68.1 \pm 1.0 \pm 1.3$
$D^{*+} \to D^+ \pi^0$	44	31.2	$30.8 \pm 0.4 \pm 0.8$
$D^{*+} \to D^+ \gamma$	2	1.5	$1.1 \pm 1.4 \pm 1.6$
$D^{*+} \to$ all	141		
$D^{*0} \to D^0 \pi^0$	68	66.7	$63.6 \pm 2.3 \pm 3.3$
$D^{*0} \to D^0 \gamma$	34	33.3	$36.4 \pm 2.3 \pm 3.3$
$D^{*0} \to$ all	102		

has a simple prediction for these two couplings

$$g_1 = 0.25, \quad g_2 = -0.75\sqrt{\frac{2}{3}}. \quad (2)$$

APPLICATIONS

As a first application, we consider the strong decays of D^*. The results are depicted in Table 1. For completeness we have also included the radiative decay $D^* \to D\gamma$ (see below). It is evident that the agreement between theory and the very recent experimental measurement of CLEO II [6] is excellent. This means that the dynamics of the D^* decay is completely determined by chiral symmetry and the quark model. It follows from Table 1 that the total widths of the D^* are

$$\Gamma(D^{*+}) = 141 \, \text{keV}, \quad \Gamma(D^{*0}) = 102 \, \text{keV}. \quad (3)$$

Note that our prediction of $\Gamma(D^{*+})$ is very close to the upper limit 131 keV (90% CL) published by the ACCMOR Collaboration[7]. Experimentalists are urged to perform more precision measurement of $\Gamma(D^*)$.

For the heavy baryons, the strong decay $\Sigma_c \to \Lambda_c \pi$ has been observed experimentally, though its actual rate has not been measured. We find

$$\Gamma(\Sigma_c^+ \to \Lambda_c^+ \pi^0) = 2.45 \, \text{MeV} \quad (4)$$

for $g_A^q = 0.75$. Of course, the test of heavy-quark symmetry must await the experimental discovery of, for example, $\Sigma_b \to \Lambda_b \pi$, $\Sigma_c^* \to \Lambda_c \pi$ in the future.

Needless to say, the semileptonic decay $B \to D(D^*)\pi \ell \bar{\nu}$ provides a nice test on the synthesis of heavy quark and chiral symmetries. Of course, the applicability of chiral Lagrangians requires that the emitted pion be soft. We have worked out the spectrum of the electron, π, D and D^*. Figure 1 shows the electron spectrum of the non-resonant decay $B^0 \to D^+ \pi^0 e^- \bar{\nu}_e$ with a cutoff of 200MeV for the pion momentum. (Resonance here means that the invariant mass of D and π is within 3.3 MeV of the D^* mass.) For the purpose of illustration we have adopted the prarmetrization[8]

$$\xi(v \cdot v') = 1 - \rho^2(v \cdot v' - 1) + c(v \cdot v' - 1)^2 \quad (5)$$

with $\rho \sim 1.08$ and $c \sim 0.62$ for the Isgur-Wise function. The corresponding branching ratios are listed in Table 2.

Table 2. Branching ratios of $B \to D(D^*)\pi e \bar{\nu}_e$

p_π cutoff	100 MeV	200 MeV
$Br(B^0 \to D^+\pi^0 e^-\bar{\nu}_e)_{\text{NR}}$	4×10^{-4}	7×10^{-4}
$Br(B^0 \to D^{*+}\pi^0 e^-\bar{\nu}_e)$	7×10^{-6}	4×10^{-5}

Experimentally, the non-resonant decay

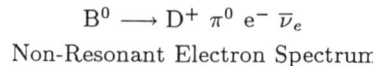
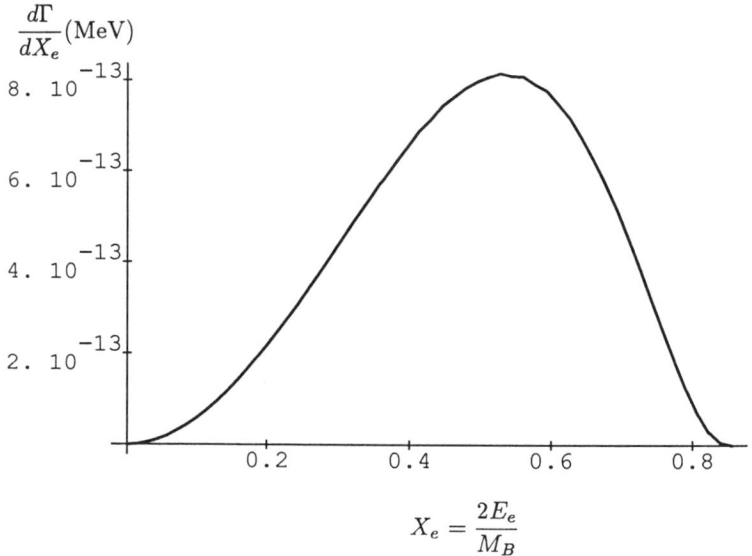

Figure 1. The non-resonant electron spectrum as a function of X_e.

$B^0 \to D^+ \pi^0 e^- \bar{\nu}_e$ with a soft pion emission should be observable in the near future.

HEAVY-FLAVOR-CONSERVING NONLEPTONIC WEAK DECAYS

We have applied heavy quark symmetries to heavy-flavor-conserving nonleptonic decays.[1b] Some examples are $\Xi_Q \to \Lambda_Q \pi$ and $\Omega_Q \to \Xi_Q \pi$. The idea is simple: In these decays only the two light quarks inside the heavy baryon will participate in weak interactions, whereas the heavy quark behaves as a "spectator". The effective $\Delta S = 1$ weak chiral Lagrangian is given by

$$\mathcal{L}^{\Delta S=1} = h_1 \text{tr}(\bar{B}_{\bar{3}} \xi^\dagger \lambda_6 \xi B_{\bar{3}})$$
$$+ h_2 \text{tr}(\bar{B}_6 \xi^\dagger \lambda_6 \xi B_6)$$
$$+ h_3 \text{tr}(\bar{B}_6 \xi^\dagger \lambda_6 \xi B_{\bar{3}}) + h.c.$$
$$+ h_4 \text{tr}(\bar{B}_6^{*\mu} \xi^\dagger \lambda_6 \xi B_{6\mu}^*), \quad (6)$$

where $B_{\bar{3}}$ and B_6 denote an antitriplet and sextet heavy baryon, respectively, and B_6^* refers to a spin $\frac{3}{2}$ heavy baryon field. It can be shown that $h_3 = 0$ and $h_2 = -h_4$ in the heavy quark limit. Thus, the heavy quark symmetry predicts two independent coupling constants that describe the transition between two $B_{\bar{3}}$ and the transitions between two B_6 or $B_{6\mu}^*$.

However, in the MIT bag model and the diquark model we have $h_2 = h_4 = 0$; that is, only $B_{\bar{3}} - B_{\bar{3}}$ transitions have nonzero amplitudes. The coupling constant h_1 is calculated in both models. The antitriplet transitions, such as $\Xi_c \to \Lambda_c \pi$, are found to have the decay rates

$$\Gamma(\Xi_c^0 \to \Lambda_c^+ \pi^-) = 1.7 \times 10^{-15} \text{ GeV},$$
$$\Gamma(\Xi_c^+ \to \Lambda_c^+ \pi^0) = 1.0 \times 10^{-15} \text{ GeV}. \quad (7)$$

The $B_6 - B_6$ transition $\Omega_c^0 \to \Xi_c'^+ \pi^-$, which vanishes in the chiral limit, receives a finite factorizable contribution as a result of symmetry breaking effects. Its decay rate is estimated to be

$$\Gamma(\Omega_c^0 \to \Xi_c'^+ \pi^-) = 4.3 \times 10^{-17}\, \text{GeV}. \quad (8)$$

Using the the theoretical values of the charmed-baryon lifetimes[9]

$$\begin{align} \tau(\Xi_c^0) &= 1.5 \times 10^{-13} s, \\ \tau(\Xi_c^+) &= 3.3 \times 10^{-13} s, \\ \tau(\Omega_c^0) &= 1.3 \times 10^{-13} s, \end{align} \quad (9)$$

we at last get the branching ratios

$$\begin{align} Br(\Xi_c^0 \to \Lambda_c^+ \pi^-) &= 3.8 \times 10^{-4}, \\ Br(\Xi_c^+ \to \Lambda_c^+ \pi^0) &= 5.0 \times 10^{-4}, \\ Br(\Omega_c^0 \to \Xi_c'^+ \pi^-) &= 0.9 \times 10^{-5}. \end{align} \quad (10)$$

We urge the experimentalists to check carefully our predictions: (1) the decay rates of allowed $B_{\bar{3}} - B_{\bar{3}}$ transitions; (2) the absence of $B_{\bar{3}} - B_6$ and $B_6^* - B_6$ transitions in the limit of heavy quark symmetry; (3) the weaker predictions by the MIT bag model and the diquark model that in the symmetry limit $B_6 - B_6$ and $B_6^* - B_6^*$ nonleptonic weak transitions should not occur. These transitions are, therefore, suppressed relative to the allowed $B_{\bar{3}} - B_{\bar{3}}$ transitions. We already see that a transition of this kind $\Omega_c^0 \to \Xi_c'^+ \pi^-$ can proceed via factorizable processes, but with a branching ratio smaller by one order of magnitude.

RADIATIVE DECAYS

We have aslo studied the radiative decays of heavy mesons and heavy baryons[1c]. The chiral Lagrangians for the electromagnetic interactions of heavy hadrons consist of two pieces: one from gauging electromagnetically the strong-interaction chiral Lagrangian, and the other from the anomalous magnetic moment interactions of the heavy baryons and mesons. Due to the heavy quark spin symmetry, the latter contains only one independent coupling constant in the meson sector and two in the baryon sector. These coupling constants only depend on the light quarks and can be calculated in the nonrelativistic quark model.

Applications of our formulism to the radiative decays $D^* \to D\gamma$ were already given in Table 1. Predictions for the electromagnetic decays of D_s^*, B^*, Σ_c and Ξ_c' are[1c]

$$\begin{align} \Gamma(D_s^{*+} \to D_s^+ \gamma) &= 2.4\, \text{keV}, \\ \Gamma(B_u^{*+} \to B_u^+ \gamma) &= 0.84\, \text{keV}, \\ \Gamma(B_d^{*0} \to B_d^0 \gamma) &= 0.28\, \text{keV}, \\ \Gamma(\Sigma_c^+ \to \Lambda_c^+ \gamma) &= 93\, \text{keV}, \\ \Gamma(\Xi_c'^+ \to \Xi_c^+ \gamma) &= 16\, \text{keV}, \\ \Gamma(\Xi_c'^0 \to \Xi_c^0 \gamma) &= 0.3\, \text{keV}. \end{align} \quad (11)$$

This together with the partial rate of $\Sigma_c^+ \to \Lambda_c^+ \pi^0$ given by Eq.(4) yields the total decay width of Σ_c^+

$$\Gamma_{tot}(\Sigma_c^+) = 2.54\, \text{MeV}, \quad (12)$$

and the branching ratio of $\Sigma_c^+ \to \Lambda_c^+ \gamma$

$$Br(\Sigma_c^+ \to \Lambda_c^+ \gamma) = 3.8\%. \quad (13)$$

Acknowledgments

I wish to thank my collaborators C.Y. Cheung, G.L. Lin, Y.C. Lin, T.M. Yan, and H.L Yu for a most enjoyable collaboration and to Bill Dimm and G.L. Lin for providing me the figure.

REFERENCES

1. T.M. Yan, H.Y. Cheng, C.Y. Cheung, G.L. Lin, Y.C. Lin and H.L. Yu, (a) *Phys. Rev.* D46, 1148 (1992); (b) CLNS 92/1153; (c) CLNS 92/1158.
2. N. Isgur and M. Wise, *Phys. Lett.* B232 113 (1989); *ibid* B237, 527 (1990).
3. M. Wise, *Phys. Rev.* D45, R2188 (1992).
4. G. Burdman and J. Donoghue, *Phys. Lett.* B280, 287 (1992).
5. P. Cho, *Phys. Lett.* B285, 145 (1992).
6. F. Butler et al., CLNS 92/1143 (1992).
7. S. Barlag et al., *Phys. Lett.* B278, 480 (1992).
8. G. Burdman, *Phys. Lett.* B284, 133 (1992).
9. H.Y. Cheng, *Phys. Lett.* B289, 455 (1992).

RIGOROUS QCD ANALYSIS OF P-WAVE CHARMONIUM DECAYS

Eric Braaten*
Department of Physics and Astronomy
Northwestern University
Evanston, Illinois 60208 USA

Geoffrey T. Bodwin
High Energy Physics Division
Argonne National Laboratory
Argonne, Illinois 60439 USA

G. Peter Lepage
Newman Laboratory of Nuclear Studies
Cornell University
Ithaca, New York 14853 USA

Abstract

A new factorization theorem for the decay rates of P-wave charmonium states is presented. Infrared divergences encountered in previous perturbative calculations of P-wave decays are avoided by taking into account the annihilation of the charmed quark and antiquark in a color-octet S-wave state. Using recent precision measurements of the total widths of the χ_{c1} and χ_{c2} states as inputs, we predict all the remaining light-hadronic and electromagnetic decay rates of the P-wave charmonium states.

The discovery of charmonium in November 1974 launched a revolution in particle physics. Within weeks, calculations of the decay rates of the S-wave charmonium states were submitted to Physical Review Letters.[1] The decays were assumed to occur through a short distance process in which a charmed quark and antiquark annihilate from a color-singlet S-wave state. The decay rates to leading order in α_s could then be obtained from the corresponding positronium decay rates simply by changing the coupling constant and inserting a color factor. It was also assumed implicitly in these calculations that the infrared divergences that inevitably arise at higher orders in α_s could all be factored into a single nonperturbative parameter, the non-relativistic wavefunction at the origin. The assumption has never been put on a rigorous theoretical basis, but it is supported by explicit calculations[2] of the S-wave decay rates to next-to-leading order in α_s.

Given the success of the S-wave calculations, one might expect that a similar approach could be used to calculate the annihilation rates of P-wave charmonium states. The natural assumption is that these decay rates should also be calculable using perturbative QCD in terms of a single nonperturbative parameter, the derivative of the radial wavefunction at the origin $R'(0)$. Unfortunately, this assumption is wrong. Explicit calculations by Barbieri and various collaborators[3] reveal that there are infrared divergences at order α_s^3 which cannot be factored into $R'(0)$. The results of Barbieri et al. for the decay rate into light hadrons of the $^3P_0$, $^3P_1$, $^3P_2$, and $^1P_1$ states are

---
* Invited talk presented at the XXVI International Conference on High Energy Physics, Dallas, Texas (August 6-12, 1992).

$$\Gamma(\chi_{c0}) = \Gamma_0 \alpha_s^2 \left[6 + \frac{\alpha_s}{\pi} \left(\frac{8}{3} \log \frac{M_c}{\mu} + 52.6 \right) \right], \quad (1)$$

$$\Gamma(\chi_{c1}) = \Gamma_0 \alpha_s^2 \left[0 + \frac{\alpha_s}{\pi} \left(\frac{8}{3} \log \frac{M_c}{\mu} - 1.3 \right) \right], \quad (2)$$

$$\Gamma(\chi_{c2}) = \Gamma_0 \alpha_s^2 \left[\frac{8}{5} + \frac{\alpha_s}{\pi} \left(\frac{8}{3} \log \frac{M_c}{\mu} - 7.7 \right) \right], \quad (3)$$

$$\Gamma(h_c) = \Gamma_0 \alpha_s^2 \left[0 + \frac{\alpha_s}{\pi} \left(\frac{20}{9} \log \frac{M_c}{\mu} - 0.1 \right) \right], \quad (4)$$

where $\Gamma_0 = |R'(0)|^2/M_c^4$, M_c is the mass of the charmed quark, and μ is an infrared cutoff. All four decay rates have a logarithmic infrared divergence at order α_s^3 which cannot be factored into $R'(0)$. The presence of infrared divergences has not prevented some phenomenologists from using these calculations. The infrared cutoff μ has typically been replaced by the binding energy or by the inverse of the mean radius of the bound state in a potential model calculation. These are however just arbitrary prescriptions. A perturbative QCD calculation with infrared divergences is meaningless, because higher order corrections are not suppressed by the running coupling constant $\alpha_s(M_c)$.

A rigorous QCD calculation of the decay rate requires the clean separation of long distance effects, which are inherently nonperturbative, from short distance effects, which can be calculated in perturbation theory. The infrared divergences in (1)-(4) indicate that there are nonperturbative contributions beyond those represented by $R'(0)$. That the coefficient of the logarithm in the α_s^3 term is identical for χ_{c0}, χ_{c1}, and χ_{c2} is an important clue to the nonperturbative mechanism. It implies that the divergence is independent of the total angular momentum state of the $c\bar{c}$ pair. An analysis of the decay rates for the χ_{cJ} states reveals that the divergence comes from the subprocess $c\bar{c} \rightarrow g q \bar{q}$, where the initial $c\bar{c}$ pair is in a color-singlet ^3P_J state, and from the region of phase space where the momentum of the final state gluon approaches 0. The divergent part of the matrix element can be factored into the amplitude for the $c\bar{c}$ to radiate the soft gluon, making a transition to a color-octet $^3S_1$ state, and the amplitude for the subsequent annihilation from the color-octet S-wave state via the process $c\bar{c} \rightarrow q\bar{q}$. Similarly, the divergent part of the matrix element for decay of the h_c can be factored into the amplitude for $c\bar{c}$ to radiate a soft gluon, making a transition from the color-singlet $^1P_1$ state to a color-octet $^1S_0$ state, and the amplitude for the subsequent annihilation from the color-octet S-wave state via the process $c\bar{c} \rightarrow gg$. This analysis reveals that there are two distinct short distance mechanisms that contribute to these decays. A P-wave state can decay not only by the annihilation of $c\bar{c}$ in a color-singlet P-wave state, but also by the annihilation of $c\bar{c}$ in a color-octet S-wave state. The first mechanism involves the dominant $c\bar{c}$ component of the bound state wavefunction, but the c and $\bar{c}$ must overcome the P-wave angular momentum barrier in order to annihilate. The second mechanism involves the small $c\bar{c}g$ component of the bound state wavefunction. The suppression from the wavefunction is compensated by the fact that annihilation from the S-wave state is not hindered by any angular momentum barrier.

The results of this analysis can be summarized by rigorous factorization formulas for the inclusive annihilation rates of the P-wave charmonium states,[4] which are correct to leading order in v^2, where v is the typical velocity of the charm quark in the charmonium state, and to all orders in $\alpha_s(M_c)$:

$$\Gamma(\chi_{cJ} \to X) = H_1 \hat{\Gamma}_1\left(c\bar{c}(^3P_J) \to X\right) \quad (5)$$
$$+ H_8 \hat{\Gamma}_8\left(c\bar{c}(^3S_1) \to X\right),$$
$$\Gamma(h_c \to X) = H_1 \hat{\Gamma}_1\left(c\bar{c}(^1P_1) \to X\right) \quad (6)$$
$$+ H_8 \hat{\Gamma}_8\left(c\bar{c}(^1S_0) \to X\right).$$

The factors $\hat{\Gamma}_1$ and $\hat{\Gamma}_8$ are color-singlet and color-octet subprocess rates, respectively, and can be calculated as perturbation expansions in $\alpha_s(M_c)$. At leading order in $\alpha_s(M_c)$, they can be extracted from the calculations of Barbieri et al.[3] The nonvanishing subprocess rates are

$$\hat{\Gamma}_1\left(c\bar{c}(^3P_0) \to gg\right) = \frac{4\pi}{3}\alpha_s(M_c)^2, \quad (7)$$

$$\hat{\Gamma}_1\left(c\bar{c}(^3P_2) \to gg\right) = \frac{16\pi}{45}\alpha_s(M_c)^2, \quad (8)$$

$$\hat{\Gamma}_8\left(c\bar{c}(^3S_1) \to q\bar{q}\right) = \pi\alpha_s(M_c)^2, \quad (9)$$

$$\hat{\Gamma}_8\left(c\bar{c}(^1S_0) \to gg\right) = \frac{5\pi}{6}\alpha_s(M_c)^2. \quad (10)$$

The factors H_1 and H_8 in (5) and (6) are nonperturbative hadronic matrix elements. The matrix element H_1 is determined by the radial wavefunction $R(r)$ for the $c\bar{c}$ component of the P-wave state:

$$H_1 = \frac{9}{2\pi}\frac{|R'(0)|^2}{M_c^4}. \quad (11)$$

The matrix element H_8 depends on the wavefunction for the $c\bar{c}g$ component of the bound state. While the matrix elements H_1 and H_8 can in principle be calculated using lattice QCD simulations, they can also be determined phenomenologically from decays of the χ_{c1} and χ_{c2} states:[4]

$$H_1 = \frac{45}{16\pi}\frac{\Gamma(\chi_{c2} \to \ell.h.) - \Gamma(\chi_{c1} \to \ell.h.)}{\alpha_s(M_c)^2}, \quad (12)$$

$$H_8 = \frac{1}{\pi}\frac{\Gamma(\chi_{c1} \to \ell.h.)}{\alpha_s(M_c)^2}, \quad (13)$$

where $\Gamma(\chi_{cJ} \to \ell.h.)$ is the decay rate of the χ_{cJ} into light hadrons. The E760 collaboration at Fermilab has recently made precise measurements of the widths of the χ_{c1} and χ_{c2}.[5] Combining them with previous measurements of the radiative branching fractions of the χ_{c1} and χ_{c2}[6], we determine their decay rates into light hadrons. From (12) and (13), we then obtain $H_1 \simeq 15$ MeV and $H_8 \simeq 3$ MeV.

Having determined H_1 and H_8, we are able to predict all the other electromagnetic and inclusive hadronic decay rates of the P-wave states.[4] The $^1P_1$ state h_c was recently discovered by the E760 collaboration.[7] Our predictions for its total width and the branching fractions for its radiative decay into η_c and its decay into a hard photon plus light hadrons are

$$\Gamma(h_c) = (1.0 \pm 0.2) \text{ MeV}, \quad (14)$$

$$B(h_c \to \eta_c + \gamma) = (50 \pm 10)\%, \quad (15)$$

$$B(h_c \to \gamma + \text{light hadrons}) = (2 \pm 1)\%. \quad (16)$$

Our prediction (14) for its total width is consistent with the upper bound obtained by the E760 collaboration.[7] Our predictions for the total width of the χ_{c0} and the branching fractions for its radiative decay into ψ and its electromagnetic decay into two photons are

$$\Gamma(\chi_{c0}) = (5 \pm 2) \text{ MeV}, \quad (17)$$

$$B(\chi_{c0} \to \psi + \gamma) = (2 \pm 1)\%, \quad (18)$$

$$B(\chi_{c0} \to \gamma\gamma) = (7 \pm 4) \times 10^{-4}. \quad (19)$$

Our prediction for the branching fraction for the electromagnetic decay of the χ_{c2} into two photons is

$$B(\chi_{c2} \to \gamma\gamma) = (4 \pm 2) \times 10^{-4}. \quad (20)$$

This differs significantly from previous measurements[6], but is in good agreement with the recent measurement of the E760 collaboration.[7] Our predictions (17) and (18) for the total width of the χ_{c0} and its radiative branching fraction differ significantly from previous measurements. We expect that more accurate measurements of the decays of the χ_{c0} will also bring them into agreement with the QCD predictions.

Our analysis of the decays of the P-wave charmonium states is based on leading-order QCD calculations. The precision of the predictions could be significantly improved by calculating all of the subprocess rates to next-to-leading order in α_s. The methods we have used to solve the problem of infrared divergences in P-wave charmonium decay rates can also be applied to P-wave quarkonium production processes. A leading-order analysis of P-wave charmonium production in B meson decays has been carried out.[8] A similar analysis is required to obtain the correct QCD predictions for photoproduction, leptoproduction, hadroproduction, and collider production of P-wave quarkonium states. The solution to the infrared problem of P-wave quarkonium decays has extended the domain of applicability of perturbative QCD to a whole new class of hadronic process.

This work was supported in part by the U.S. Department of Energy, Division of High Energy Physics, under Contract W-31-109-ENG-38 and under Grant DE-FG02-91-ER40684, and by the National Science Foundation.

REFERENCES

1. T. Appelquist and H.D. Politzer, *Phys. Rev. Lett.* 34, 43 (1975); A. de Rujula and S.L. Glashow, *Phys. Rev. Lett.* 34, 46 (1975).

2. R. Barbieri, G. Curci, E. d'Emilio, and E. Remiddi, *Nucl. Phys.* B154, 535 (1979); P. Mackenzie and G. Lepage, *Phys. Rev. Lett.* 47, 1244 (1981).

3. R. Barbieri, R. Gatto, and R. Kögerler, *Phys. Lett.* 60B, 183 (1976); R. Barbieri, R. Gatto, and E. Remiddi, *Phys. Lett.* 61B, 465 (1976); R. Barbieri, M. Caffo, and E. Remiddi, *Nucl. Phys.* B162, 220 (1980); R. Barbieri, M. Caffo, R. Gatto, and E. Remiddi, *Phys. Lett.* 95B, 93 (1980); *Nucl. Phys.* B192, 61 (1981).

4. G.T. Bodwin, E. Braaten and G.P. Lepage, *Phys. Rev.* D46, R1914 (1992).

5. E760 collaboration (T.A. Armstrong et al.), *Nucl. Phys.* B373, 35 (1992); *Phys. Rev. Lett.* 68, 1468 (1992).

6. Particle Data Group (J.J. Hernandez et al.), *Phys. Lett.* B239, 1 (1990).

7. J. Rosen, "Charmonium Results from E760," in Proceedings of the XXVI International Conference on High Energy Physics, Dallas (August 6-12, 1992) ed. J.R. Sanford (American Institute of Physics, New York, 1993).

8. G.T. Bodwin, E. Braaten, T.C. Yuan and G.P. Lepage, Argonne preprint ANL-HEP-TH-92-63 (August, 1992).

RECENT RESULTS FROM CLEO ON CHARM AND BOTTOM DECAYS

Vivek Jain
Department Of Physics
Vanderbilt University
Nashville, TN 37235, USA

Abstract

New results on Charm and Bottom decays from the CLEO experiment are presented. The data used in the analysis were collected at the Cornell Electron Storage Ring between 1990 and 1992.

INTRODUCTION

In this paper, I present recent results from the CLEO experiment on Charm and Bottom decays. The results on Bottom decays are based on 920 pb^{-1} of data collected on the $\Upsilon(4S)$ and 420 pb^{-1} collected on the continuum, 60 MeV below $\Upsilon(4S)$. Charm results include another 210 pb^{-1} of data collected on the $\Upsilon(3S)$ and on the continuum above the $\Upsilon(4S)$, for a total of 1550 pb^{-1}.

THEORETICAL OVERVIEW

Several models[1,2] have been proposed to explain heavy quark hadronic decays. They have enjoyed some success in the charm sector, especially in explaining the lifetime difference between the D^+ and D^0. The simplest is based on the factorization approach[1], in which a decay amplitude is expressed as the product of two independent currents, and each of these currents is written in terms of form factors. These models work well for "external" W spectator decays, but not so reliably for "internal" W spectator decays. In addition, none of these models can reliably predict the contribution of "W exchange" diagrams or account for the effects of Final State Interactions (FSI). The factorization approach is expected to be more reliable for B decays, since the energy release is larger, and the decay particles move away from each other before they can interact with each other and give rise to final state interactions.

CHARM DECAYS

(1) $D^0 \to \overline{K}^0 \pi^0$

This decay mode is an example of an "internal W" spectator diagram. From naive colour counting, the ratio $\frac{B(D^0 \to \overline{K}^0 \pi^0)}{B(D^0 \to K^- \pi^+)}$ was expected[3] to be 1/25 - 1/18; experimentally this ratio is approximately 0.6 ± 0.15.[4] The enhancement is thought to be due to FSI (via "$K\pi$" scattering). In Figure 1, we present data from CLEO, where the D^0 momentum is greater than $0.5E_{beam}$ ($X = P/E_b > 0.5$). The fit yields 1942 ± 64 events. Normalizing to $D^0 \to \overline{K}^0 \pi^+ \pi^-$, and using the Mark III branching ratio (= $6.4 \pm 1.1\%$) for this mode, we obtain $\mathcal{B}(D^0 \to \overline{K}^0 \pi^0) = 2.8 \pm 0.1 \pm 0.6\%$. This agrees well with previous measurements[4], but disagrees with the new E691 result.[5]

(2) $D^0 \to \overline{K}^{*0} \eta$

This decay is particularly interesting for

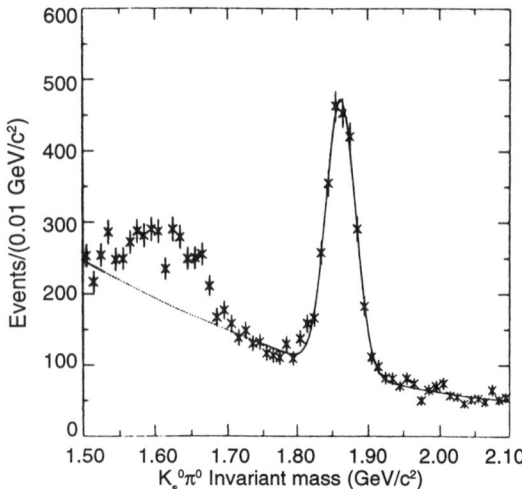

Figure 1. Invariant mass for $\bar{K}^0\pi^0$ mass combinations.

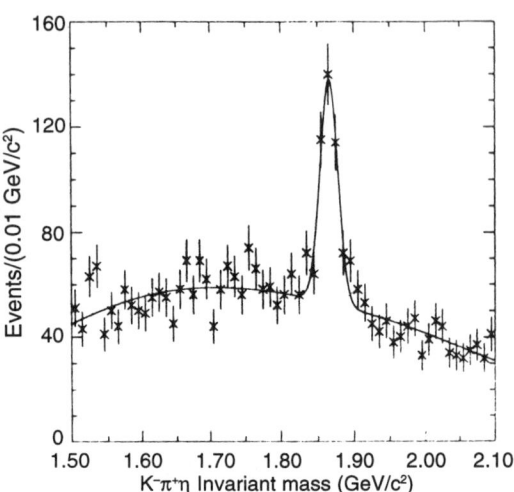

Figure 2a. Invariant mass for $K^-\pi^+\eta$ mass combinations.

studying FSI. When the decay $D^0 \to \bar{K}^0\phi$ was first observed it was thought to be evidence for W-exchange type diagrams. However, the measured[4] branching ratio ($\approx 0.8\%$) was too large to be explained by lowest order diagrams. An alternative explanation was proposed by Donoghue[6], who pointed out that the decay could proceed through the intermediate state $\bar{K}^{*0}\eta$ (internal W spectator diagram), which would undergo strong interactions to produce $\bar{K}^0\phi$. According to this model, $\mathcal{B}(D^0 \to \bar{K}^{*0}\eta)$ has to be at least 2%, in order to explain the measured branching ratio for $D^0 \to \bar{K}^0\phi$. Factorization models[1], on the other hand, predicted $\mathcal{B}(D^0 \to \bar{K}^{*0}\eta)$ to be 0.3%.

In Figure 2a, we present the invariant mass of $K^-\pi^+\eta$ combinations, for $X_{D^0} > 0.5$. The $\bar{K}^{*0}$ is reconstructed in the $K^-\pi^+$ mode, and the η in the $\gamma\gamma$ mode.

In Figure 2b we show the $K^-\pi^+$ mass combinations for events within the D peak. A fit to the $K^-\pi^+$ mass peak yields 226 ± 28 events. We subtract the contribution from the D sideband region (12 ± 14 events), to obtain a net yield of 214 ± 31 events. Normalizing to the Mark III branching ratio[7] for $D^0 \to K^-\pi^+\pi^0$ ($= 13.1 \pm 1.8\%$) we obtain $\mathcal{B}(D^0 \to \bar{K}^{*0}\eta) = 1.8 \pm 0.3 \pm 0.4\%$. The previous results[4] were either upper limits or had larger errors. Our result agrees with the explanation based on FSI.

(3) $D^+ \to \phi K^+$

This mode is an example of a doubly Cabbibo suppressed decay (DCSD). It can proceed either via a W-exchange diagram or through FSI, where the intermediate state $K^{*+}\eta$ rescatters to produce ϕK^+, in the same manner as in $D^0 \to \bar{K}^0\phi$. Recently, E691[8] has measured this branching ratio to be $4.0^{+2.2}_{-1.8} \pm 0.6 \times 10^{-4}$. This rate seems too large to be due to the DCSD W-exchange diagram (one expects a large rate like this to be due to a spectator diagram and not W-exchange[8]). However, the diagram based on FSI could enhance the expected rate.

In Figure 3, we present the invariant mass of ϕK^+ combinations for $X_{D^+} > 0.5$. The fit yields 24 ± 16 events, implying a 90% C.L. upper limit of 45 events. Normalizing to $D^+ \to \phi\pi^+$, where we observe 617 ± 51 events, and

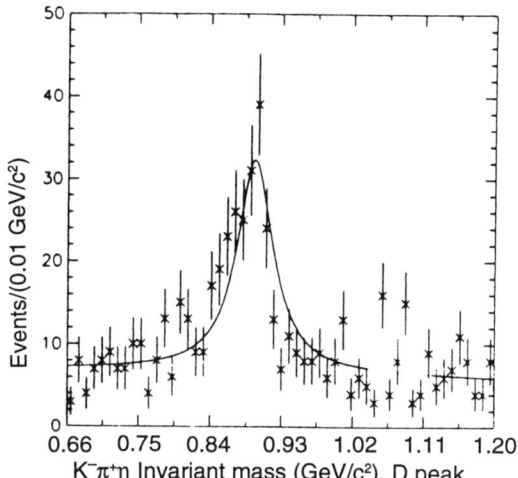

Figure 2b. Invariant mass for $K^-\pi^+$ mass combinations falling within the D^0 peak.

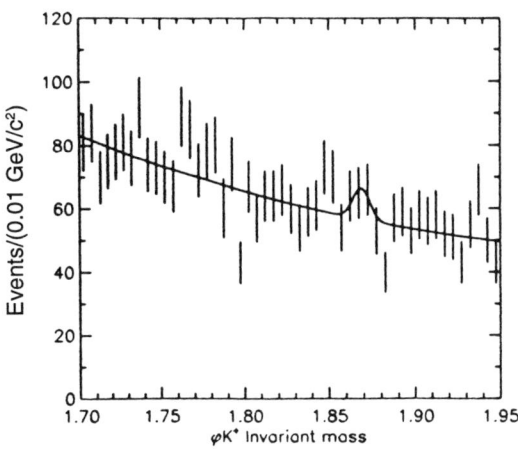

Figure 3. Invariant mass for ϕK^+ combinations.

using the world average[4] for $\mathcal{B}(D^+ \to \phi\pi^+) = 0.6 \pm 0.08\%$, we obtain $\mathcal{B}(D^+ \to \phi K^+) < 5.2 \times 10^{-4}$ (90% C.L. upper limit). This result agrees with the E691 result.[8] We will be able to shed more light on this mode with additional data.

(4) W-exchange decays of the Λ_c

The world average[4] for Λ_c lifetime, $1.91^{+0.15}_{-0.12} \times 10^{-13}$ s, is considerably shorter than charm meson lifetimes. This suggests that there may be non-spectator decays for the Λ_c; W-exchange decays are a possibility since they are not helicity suppressed as in the case of charm mesons.

We have searched for Λ_c decay modes which could conceivably come from W-exchange diagrams, and present results on two such modes, $\Lambda_c \to \Sigma^+ K^+ K^-$ and $\Lambda_c \to \Xi^0 K^+$. In Figure 4, we show $\Sigma^+ K^+ K^-$ mass combinations, where Σ^+ is detected in the $p\pi^0$ mode. The fit yields 49 ± 9 events for $X_{\Lambda_c} > 0.36$. Normalizing to $\Lambda_c \to pK^-\pi^+$ yields, $\frac{\mathcal{B}(\Lambda_c \to \Sigma^+ K^- K^+)}{\mathcal{B}(\Lambda_c \to pK^-\pi^+)} = 0.076 \pm 0.014 \pm 0.011$.

In Figure 5, we present evidence for $\Lambda_c \to \Xi^0 K^+$ decays, where Ξ^0 is detected in the $\Lambda\pi^0$ final state. The fit yields 48 ± 8 events for $X_{\Lambda_c} > 0.36$. Again, normalizing to $\Lambda_c \to pK^-\pi^+$, we obtain, $\frac{\mathcal{B}(\Lambda_c \to \Xi^0 K^+)}{\mathcal{B}(\Lambda_c \to pK^-\pi^+)} = 0.077 \pm 0.014 \pm 0.012$.

These two decay modes combine to be about 15% of $\Lambda_c \to pK^-\pi^+$, implying that W-exchange may play a large role in charm baryon decays. These results agree with recent theoretical calculations[9].

BOTTOM DECAYS

(1) Tests of Factorization

As mentioned above, factorization is expected to be more reliable for B decays. If factorization holds, the ratio of hadronic decays can be related to semileptonic decays. For instance[10], $R_\pi = \frac{\Gamma(\overline{B}^0 \to D^{*+}\pi^-)}{\frac{d\Gamma}{dQ^2}(B \to D^*l\nu)|_{Q^2 = m_\pi^2}} = 6\pi^2|V_{ud}|^2 f_\pi^2 |a_1|^2$, where Q^2 is the 4-momentum transferred from the B meson to the D^* meson, a_1 accounts for perturbative and non-perturbative QCD effects, and f_π is the pion decay constant. Similarly, one can test this equality for $Q^2 = m_\rho^2, m_{a_1}^2, m_{D^*}^2$, etc.

In Figures 6 and 7, we present the beam constrained mass distributions for $\overline{B}^0 \to$

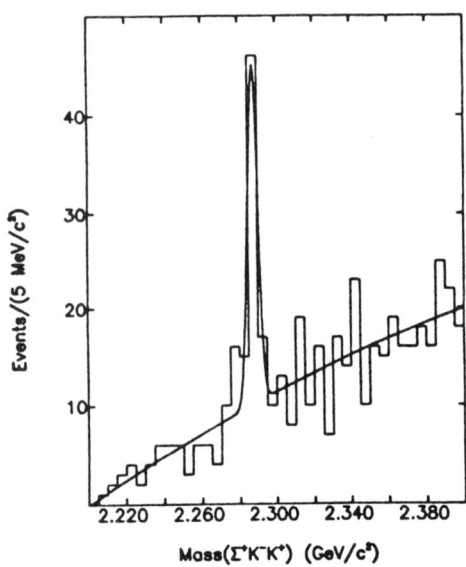

Figure 4. Invariant mass for $\Sigma^+ K^- K^+$ combinations.

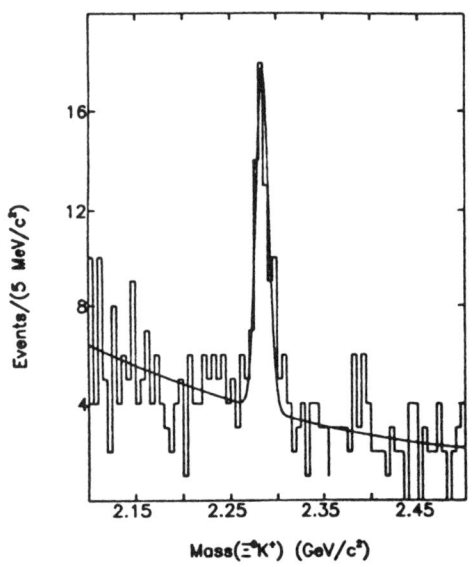

Figure 5. Invariant mass for $\Xi^0 K^+$ combinations.

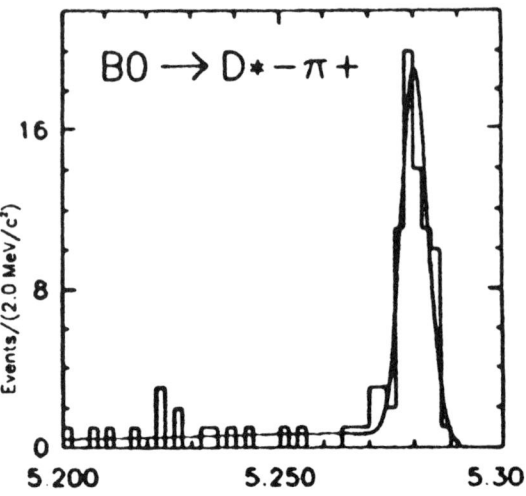

Figure 6. Beam constrained mass for $D^{*+}\pi^-$ combinations.

$D^{*+}\pi^-$ and $\overline{B}^0 \to D^{*+}\rho^-$ decays, respectively. D^{*+}'s are detected in the $D^0\pi^+$ mode, and D^0's in the $K^-\pi^+$, $K^-\pi^+\pi^0$, $K^-\pi^+\pi^-\pi^+$ final states. The fit yields 66.3 ± 9.0 events for $\overline{B}^0 \to D^{*+}\pi^-$, and 80 ± 10.5 events for $\overline{B}^0 \to D^{*+}\rho^-$. We searched for non-resonant $\pi\pi$ production in the latter case, and set an upper limit of 6 events (90% C.L.). We choose to ignore this contribution in the following considerations (the $B \to D^{**}\pi$ contribution is also small and is ignored). Using "world averages" for $a_1, V_{ud}, f_\pi, f_\rho$, and theoretical models to interpolate $\frac{d\Gamma}{dQ^2}(B \to D^*l\nu)|_{Q^2=m_\pi^2}$ to $Q^2 = m_\pi^2, m_\rho^2$, we obtain $R_\pi(\exp) = 1.28 \pm 0.19 \pm 0.30 \text{GeV}^2$, whereas we expect $R_\pi(\text{theo}) = 1.23 \pm 0.17$ GeV^2. Similarly, we obtain $R_\rho(\exp) = 3.17 \pm 0.43 \pm 0.74 \text{GeV}^2$, whereas theoretically we expect this ratio to be $R_\rho(\text{theo}) = 3.26 \pm 0.46 \text{GeV}^2$. This test indicates that factorization does hold for B decays.

(2) Measurement of $\frac{\tau_{B^+}}{\tau_{B^0}}$. A study of $\frac{\tau_{D^+}}{\tau_{D^0}}, \frac{\tau_{K^+}}{\tau_{K^0}}$ led to a better understanding of hadronic K, D decays, especially effects of non-spectator diagrams, interference between in-

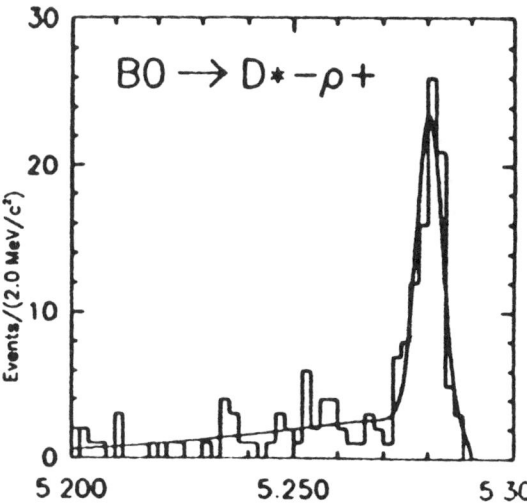

Figure 7. Beam constrained mass for $D^{*+}\rho^-$ combinations.

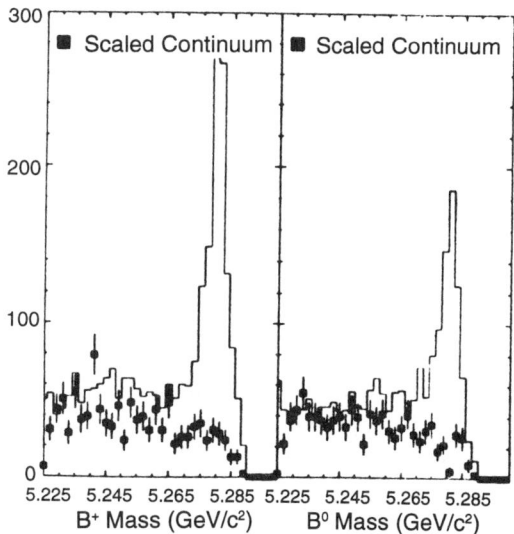

Figure 8. Beam constrained mass for B^+ and B^0 candidates.

ternal and external W diagrams, etc. Since, we expect these effects to be small for B mesons, this ratio should be close to 1. In the $\Upsilon(4S)$ region, $\frac{\tau_{B^+}}{\tau_{B^0}}$ is measured by using the relationship between lifetimes and semileptonic branching ratios, i.e., $\frac{\tau_{B^+}}{\tau_{B^0}} = \frac{\mathcal{B}(B^+ \to D^{(*)}l\nu)}{\mathcal{B}(B^0 \to D^{(*)}l\nu)}$ (assuming that $\Gamma_{SL}^+ = \Gamma_{SL}^0$). Usually, the individual semileptonic branching ratios are measured, and their ratio gives the ratio of lifetimes. This measurement, however, depends on the relative production cross-section of B^0 to B^+ ($\frac{f_{00}}{f_\pm}$), which is not very well known.

CLEO has recently measured $\frac{\tau_{B^+}}{\tau_{B^0}}$ using a sample of fully reconstructed B^+ and B^0 mesons. Here, $\mathcal{B}(B \to Xl\nu) = \frac{\text{No. of leptons}}{\text{No. of tagged B's}}$. Thus, this measurement is independent of the B^0 and B^+ production cross-sections. The B mesons are reconstructed using the CABS approach[11] in the $D\pi, D^*\pi, D\rho, D^*\rho, Da_1, D^*a_1, \psi K(K^*)$ modes. In Figure 8, we show the beam constrained mass for B^+ candidates and B^0 candidates. The solid line is the ON $\Upsilon(4S)$ data, and the dark squares represent the scaled continuum.

In Table 1, we give details of this analysis. The unobserved lepton spectrum is corrected by the ISGW model which assumes no $B \to D^{**}l\nu$ decays, and also by the ISGW* model which assumes 26% of B semileptonic decays are $B \to D^{**}l\nu$. Taking the ratio of semileptonic branching ratios corrected by the ISGW model, we get $\frac{\tau_{B^+}}{\tau_{B^0}} = 1.11 \pm 0.29 \pm 0.08$ (using the ISGW* model, yields a very similar result). This result is consistent with previous results from CLEO and ARGUS: $\frac{\tau_{B^+}}{\tau_{B^0}} = 0.93 \pm 0.16 \times \frac{f_{00}}{f_\pm}$, where $\frac{f_\pm}{f_{00}}$ is expected to be within 1.0-1.2.

CONCLUSIONS

The combination of the CLEO detector which has excellent tracking and photon detection capabilities, and the Cornell Electron Storage Ring facility which provides vast amounts of luminosity (≈ 1000 pb^{-1} /year on the $\Upsilon(4S)$ + continuum below the $\Upsilon(4S)$), promises a rich program of Charm and Bottom studies.

Table 1. Tagged B^0, B^+ sample

	No. of tagged B's	No. of leptons ISGW/ISGW*	Br. ratio ISGW
B^+	602 ± 58	72.1 ± 11.1 / 77.1 ± 11.8	$12.0 \pm 2.0 \pm 1.3\%$
B^0	493 ± 44	44.7 ± 7.8 / 47.8 ± 8.4	$10.8 \pm 2.1 \pm 1.4\%$

ISGW model has 0% $B \to D^{**} l\nu$, whereas ISGW* has 26% $(D^{**}l\nu)/(Xl\nu)$.

ACKNOWLEDGEMENTS

We gratefully acknowledge the effort of the CESR staff in providing us with excellent luminosity and running conditions. This work was supported by the National Science Foundation and the U.S. Dept. of Energy. I would like to thank my colleagues for helping me with details of their analysis; Simon Patton and Jon Urheim for helping me with LaTeX.

REFERENCES

1. M. Bauer et al., Z. Phys. C 34, 103(1987).

2. A. N. Kamal et al., Phys. Rev. D 35, 3515 (1987); L.L. Chau et al., Phys. Rev. D 36, 137 (1987); B.Yu Blok et al., Sov. J. Nucl. Phys., 45, 522 (1987); A.J. Buras et al., Nucl. Phys. B268, 16, (1986).

3. I.I. Bigi, Heavy Quark Physics, AIP Conference Proceedings 196, eds. P.S. Drell and D.L. Rubin, (1989).

4. K. Hikasa et al., Phys. Rev. D 45, 1 (1992).

5. J.C. Anjos et al., Phys. Rev. D 46, R1 (1992).

6. J.F. Donoghue, Phys. Rev. D 33, 1516 (1986).

7. J. Adler et al., Phys. Rev. Lett. 60, 89 (1988).

8. J.C. Anjos et al., FERMILAB Pub 91/331.

9. J. Korner et al., DESY Preprint 92-049.

10. J.D. Bjorken, Conf. on Weak Decays of Heavy Quarks, Institute for Theoretical Physics, May 21, 1990.

11. N. Katayama, Proceedings of the Int. Conf. on Computing in High Energy Physics, Eds. Y. Watase, F. Abe, pp439-444, 1991.

12. I.I. Bigi and B. Stech, Proc. of Workshop on High Sensitivity Beauty Physics at Fermilab, Eds. A.J. Slaughter et al., pp239, 1988.

THE DØ MUON SYSTEM AND EARLY RESULTS ON ITS PERFORMANCE*

David Hedin for the DØ Collaboration[1]
Department of Physics
Northern Illinois University
DeKalb, IL, 60115

The DØ detector is a large, general-purpose detector designed to take full advantage of the 2 TeV energy of the Fermilab collider. The design of the experiment emphasizes accurate identification, complete angular acceptance, and precise measurement of the decay products of W and Z bosons: charged leptons (both electrons and muons), quarks and gluons, which emerge as collimated jets of particles, and noninteracting particles, such as neutrinos. The primary physics goals of DØ include searching for new phenomena, such as the top quark or particles outside the standard model, and high-precision studies of the W and Z bosons. In addition, the excellent muon identification will allow the study of b-quark production and decay. This report will describe DØ's muon system, give preliminary measurements of chamber and trigger rates, and discuss muon identification.

The DØ detector is shown in Figure 1. It consists of three major hardware systems: calorimetry, muon detection, and central tracking, which together allow fairly complete characterization of most proton-antiproton collision events.[2] The central tracking system consists of four drift-chamber systems (vertex, central, and two end systems) and transition radiation detectors for electron identification. Surrounding the central tracking system are three uranium/liquid-argon calorimeters. The uranium is a dense medium, allowing containment of high-energy hadron showers in a relatively short depth, as well as equal response to electrons and hadrons, while the liquid-argon ionization medium gives ease of calibration, stability, radiation hardness, and the ability to build in fine segmentation in all three coordinates. Finally, surrounding the calorimeters is the muon system,[3] consisting of five iron toroids and three layers of proportional drift tube chambers.

The muon toroids are used to measure the muon momentum and absorb all remnant portions of hadron showers. The central toroid is 1.09 m thick while the ends are each 1.52 m thick. The toroids are operated with an average field of 1.9 T. The momentum resolution is dominated by multiple scattering with a typical value of 20%. The combined calorimeter plus toroid thickness varies from about 14 λ in the central region to 19 λ in the end regions. This thickness reduces backgrounds from hadronic punchthroughs to a negligible level,[4] but there is still a significant singles rate in the muon chambers from hadron-induced spray.

Three layers of drift chambers, one between the calorimeter and toroid and two outside the toroid, are used to measure muon trajectories. The wide-angle muon system consists of 164 chambers, using 10 cm cells, which cover the angular region greater than about 10 degrees. These chambers combine drift time measurement with time division and vernier

---

*This work was supported in part by the U.S. Department of Energy and the National Science Foundation.

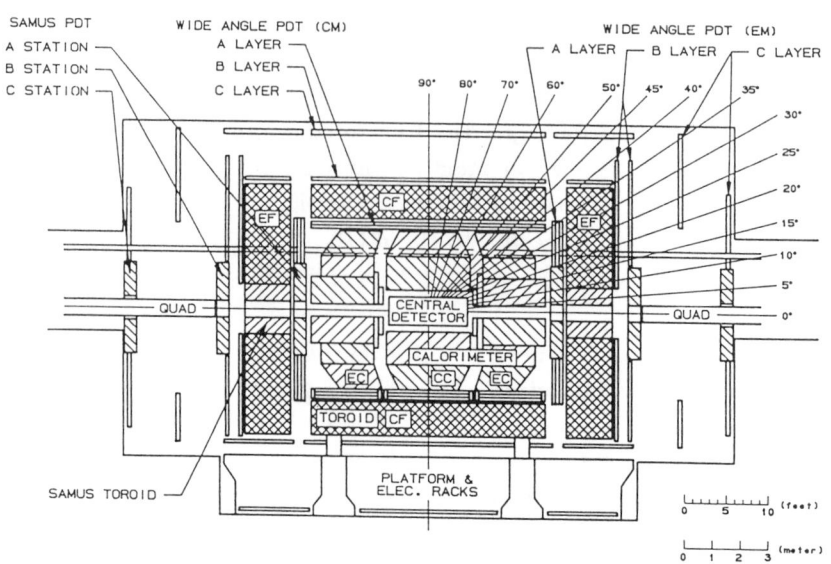

Figure 1. Elevation view of the DØ detector.

pads to obtain 3D points. The innermost layer has four measurement planes while the outer two have three each so that most muons are measured with ten 3D points. In the small-angle region between 5 and 20 degrees, six modules of 3.0 cm drift cells are used with each module having six planes in an XX,YY,UU configuration. The smaller cell size is needed in this region to reduce cell occupancy and to increase the p_t threshold of the trigger. Figure 2 gives the muon geometric acceptance as a function of η with the requirement that the muon goes through at least two of the three layers. The muon acceptance is greater than 70% for all $|\eta| < 3.2$. Also shown is the acceptance requiring all three layers to be hit. Gaps in the central region coverage are due to support and service structures.

There are three stages to the DØ muon trigger. The first two (Level 1 and 1.5) use hardware logic and fast processors to make trigger decisions in about 3.5 and 20 μsec respectively.[5] The drift time and time division information is ignored at these stages and the hit cells are used as simple hodoscopic elements of 1.5 to 6 cell widths in Level 1 and

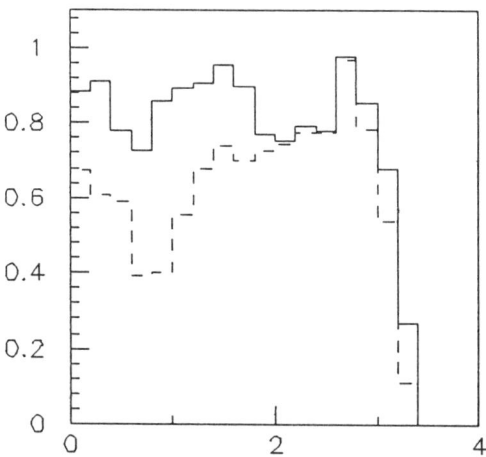

Figure 2. Muon geometric acceptance versus η for all reconstructed tracks (solid) and for 3-layer tracks (dashed).

one-half cell width in Level 1.5. The muon Level 2 trigger uses a VAX processor farm and the digitized data to reconstruct muon candidates. The Level 2 trigger is a subset of the offline reconstruction software and is designed to take about 250 ms per event. The muon-related portion of the data acquisition bandwidth will be about 400 Hz passing Level 1, 80 Hz passing Level 1.5, and 0.3 Hz passing Level 2 and then written to tape.

Level 1 triggers have two primary sources: cosmic rays in the central region and hadron-induced spray in the forward region. A set of scintillation counters covering the central ceiling is used to reduce cosmic ray triggers. Hadron-induced triggers are often due to uncorrelated hits. The average event has about 75 hits in the wide-angle chambers with most of these hits at small angles. To deal with this, smaller effective cell size is used at Level 1 at lower angles and all three layers are required to produce a valid trigger. We are currently implementing the muon trigger. As of the writing of this report we have measured a Level 1 single muon trigger rate of 10 μb and 75 μb for the regions $|\eta| < 1.0$ and $|\eta| < 1.7$ respectively. The dimuon trigger rate for $|\eta| < 1.7$ is 4 μb. The Level 1.5 trigger is observed to have a further rejection factor of 2-5 and we are in the process of optimizing its use.

Muon identification utilizes information from the muon system itself, the central tracking, and the calorimeter. The primary backgrounds are from cosmic rays, which are out of time with the beam crossing and do not intercept the primary vertex, and hadron-induced spray (often uncorrelated chamber hits) which produces poorly fit tracks which also tend to miss the vertex. Good tracks in the muon chambers are defined by using appropriate track quality cuts (for example, on the χ^2 of the fit and the number of hits on the track) and that there are no muon chambers along the track without hits. The hit density is largest at small angles and our multi-layer chamber coverage is also best in this region (with some muons hitting 24 wire chamber planes). Using only muon chamber information, we can also require that the track projects to the vertex; cuts on this can be used in the Level 2 trigger prior to analyzing the central tracking data.

The DØ calorimeter is sensitive to minimum ionizing energy depositions, and the presence of such energy along the muon track helps to eliminate non-muon backgrounds. Muon chamber tracks are also matched with the central detector tracks. In addition to properly projecting to the muon chamber hits, the matched central track can be required to have a good fit and small impact parameter. This aids in eliminating non-muon and cosmic backgrounds, and can also tag π/K decays. Finally, timing information in the central tracking, the muon chambers, and the scintillation counters on the central top can be used to remove any remaining cosmic ray muons.

Figure 3 shows an event display of a dimuon event a jet near each muon. The tracks reconstructed in the muon chambers project to the vertex, to minimum ionizing energy in the calorimeter, and to tracks in the central tracking system. The muon system is relatively clean both near the jets and in the angular regions closer to the beam.

After the initial three months of data collection at the Fermilab $p\bar{p}$ collider, we have found that the muon system has performed to its design expectations. The chambers are able to operate in the collision hall without any abnormal failures. The chamber rates are reasonable, agreeing within a factor of ± 2 with Monte Carlo predictions, as are the trigger rates. Triggering on muons at low angles and low p_t is still a difficult problem, and we are in the process of optimizing the trigger settings. We are also still working on our initial calibration of electronic channels and alignment from survey values so that physics processes

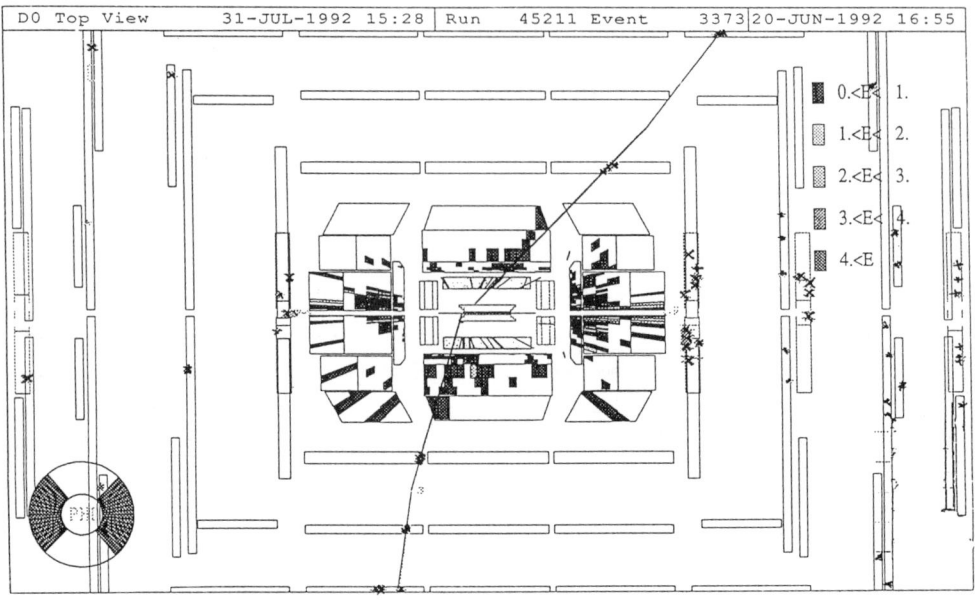

Figure 3a. Top view of a dimuon candidate event

can be searched for and studied at the earliest possible date.

REFERENCES

1. Uniandes, Arizona, Brookhaven, Brown, California-Riverside, CBPF, CINVESTAV, Columbia, Delhi, Fermilab, Florida State, Hawaii, Illinois-Chicago, Indiana, Iowa State, LBL, Maryland, Michigan, Michigan State, Moscow State, NYU, Northeastern, Northern Illinois, Northwestern, Notre Dame, Panjab, IHEP-Protvino, Purdue, Rice, Rochester, CEN-Saclay, Stony Brook, SSCL, Tata Institute, Texas-Arlington, Texas A & M.

2. *DØ Design Report*, Fermilab (1984).

3. C. Brown, et al., *NIM*, A279, 331 (1989); J.M. Butler, et al., *NIM*, A290, 122 (1990).

4. D. Green, et al., *NIM*, A244, 356 (1985).

5. M. Fortner, et al., *IEEE Transactions on Nuclear Science*, 38, 480 (1991).

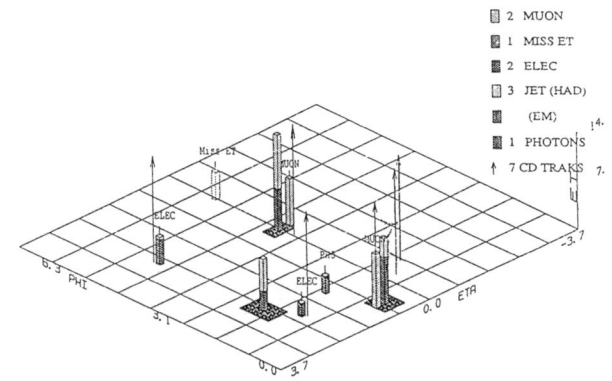

Figure 3b. Lego plot (in η and ϕ) of a dimuon candidate event.

Electroweak Interactions
Parallel Session 5

Downtown Dallas

Conveners: F. Linde (CERN)
W. Marciano (BNL)

LEP ENERGY CALIBRATION AND M_Z

Günter Quast
Deutsches Elektronen-Synchrotron (DESY)
Notkestr. 85
2000 Hamburg 52, Germany

Abstract

This report describes the calibration of the beam energy of the Large Electron-Positron Collider (LEP) at CERN for the 1991 running period. An intensive program of investigations and the utilization of resonant de-polarization of transversely polarized electron beams led to improved measurements of the mass, M_Z, and the width, Γ_Z, of the Z^0 boson. The *preliminary* results are: $M_Z = (91.187 \pm 0.0035 \pm 0.0063)$ GeV and $\Gamma_Z = (2.492 \pm 0.0053 \pm 0.0045)$ GeV, where the first error is statistical and the second arises from the uncertainties of the energy calibration.

INTRODUCTION

The large electron-positron collider, LEP, at CERN measures the Z^0 parameters with high accuracy. LEP started providing luminosity to the four experiments ALEPH, DELPHI, L3 and OPAL in summer 1989. By the end of 1991, about 2×10^6 Z^0 decays have been recorded; about $\frac{2}{3}$ of the luminosity was taken at an energy close to the peak of the Z^0 resonance and the rest was taken in "scanning" mode at six different energies off-peak in order to determine the central position and the width of the resonance from the energy dependence of the e^+e^- annihilation cross section. The error on the mass, M_Z, is affected by the precision to which the absolute energy scale of LEP is known, and the width, Γ_Z, is affected by the uncertainty on the distance of the energy points.

Figure 1 shows the dependence of the uncertainty on the top quark mass on the accuracy to which M_Z is known using recent LEP measurements as reported in by L. Rolandi at this conference. The quality of the measurements requires a precision on M_Z better than about 50 MeV in order not to be limited by the knowledge of M_Z. If, in the future, M_{top} is measured directly and the precision on other quantities improves, Standard Model tests will require a much better knowledge of M_Z. LEP is the unique place to precisely measure this fundamental quantity of nature, and the main emphasis in the following chapters is on the dominant systematic error of this measurement, arising from the uncertainties of the energy of the electron and positron beams in LEP [1].

ENERGY CALIBRATION IN 1991

The momentum of the beam particles at central orbit is proportional to the magnetic field in the bending dipoles integrated over the path of the particles. In total, there are 3368 main ring dipoles made of concrete-steel cores,

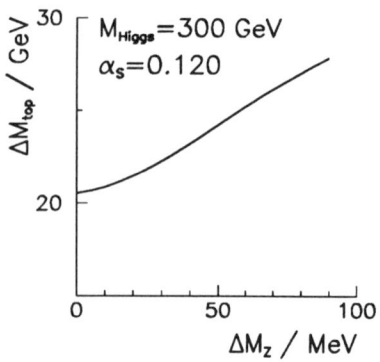

Figure 1. ΔM_{top} as a function of ΔM_Z.

the properties of which change gradually with time ("magnet aging"). Several tools are available in order to measure the magnetic field as seen by the circulating beams:

1. The Field Display
measures by means of a rotating coil the magnetic field in a reference magnet, with an all-iron core, which is powered in series with the main ring magnets. These measurements are frequently available and are used as a reference value for the beam energy for each fill of the machine, to which corrections are applied based on other calibration techniques.

2. The Flux Loop
consists of a closed electrical loop embracing all dipoles; the integrated induced voltage when changing the magnet current from 0 to its nominal value is a measure of the magnetic field. Flux loop calibrations are performed regularly and are used to monitor the time dependence of the magnetic field in the bending dipoles.

3. Proton Calibration Runs
were performed several times by injecting protons at an energy of 20 GeV into LEP. Since protons are not ultra-relativistic at that energy, their speed and hence their momentum can be determined from the frequency of the RF voltage applied to the acceleration cavities when the proton beam is at central orbit. This method measures the beam energy to high accuracy at 20 GeV ($\sim 10 \times 10^{-5}$) however, flux loop measurements are then needed to extrapolate to 45.6 GeV, leading to a substantial degradation of the precision ($\sim 20 \times 10^{-4}$).

4. Calibration by Resonant De-polarization
is based on a measurement of the frequency with which the spin of transversely polarized electrons precesses under the influence of the bending field. The precision of each measurement is very high ($<2\times 10^{-5}$), but the flux loop is still needed for tracking in time. This method is the most direct measure of the beam energy close to physics conditions and is the main tool for the 1991 energy calibration.

Calibration by Resonant De-polarization

Transverse polarization was first observed in LEP in Aug. 1990 [2] by means of a Compton-scattering laser polarimeter, after careful beam tuning and removal of depolarizing sources like the experimental solenoids. Towards the end of the 1991 running period, polarized electron beams with a polarization level of 10-20% at an energy of ~ 45.6 GeV were exposed to a weak frequency-controlled radial magnetic field produced by a kicker magnet. Under such conditions, a de-polarizing resonance occurs if the resulting spin kick away from the vertical axis is in phase with the spin precession, i.e. $\omega_{dep} = \omega_s \equiv 2\pi \nu_s f_{rev}$, where ν_s is the spin tune (number of spin precessions per turn) and f_{rev} is the revolution frequency of the beams. The spin tune depends on the beam energy,

$$\nu = \frac{(g_e - 2)}{2} \frac{E_{beam}}{m_e c^2} \equiv \frac{E_{beam}}{0.4406486(1)}$$

and is 105.53 at an energy of 46.5 GeV.

Since the de-polarizing field is encountered only once per turn, the method is sensitive only to the fractional spin tune, $\delta_s = \nu - \text{int}(\nu)$, and a resonance occurs also at the "mirror frequency" $1 - \delta_s$. However, the integer part

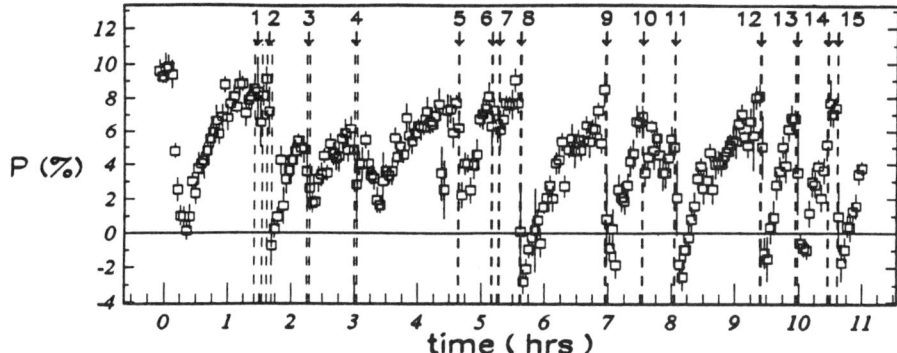

Figure 2. Polarization signal obtained 16 Sept. 1991

is well known from other calibration methods, since a tune change of one corresponds to 440 MeV change of energy, and the mirror ambiguity can be resolved by slightly altering the energy. Four measurements of the beam energy were successfully performed in 1991 [3]. Figure 2 shows an example; successive sweeps of the frequency of the kicker magnets were performed, numbered from 1 to 15 with the dashed lines indicating their duration. When the resonance frequency was crossed, the beams de-polarized, as first seen in sweep number 2. The resonance-crossing was typically located with a precision corresponding to ~1 MeV in beam energy. However, the observed scatter of all the measurements taken during the scan is larger (3.7 MeV), indicating that the beam energy was not always exactly the same. The advantage of this calibration method is that the effects of constant magnetic fields (the earth's field and remanent fields) are included as well as contributions from correction dipoles and from the quadrupoles and sextupoles in cases where the beams are not at central orbit.

Time Dependence of the Beam Energy

Flux loop calibrations were performed regularly during the 1991 running period in order to monitor the stability of the bending magnets. A correlation was discovered with the magnet temperature, measured on 8 reference magnets. Laboratory tests showed that this effect is really due to a change of the magnetic field and not merely to a temperature dependence of the measuring technique. From these tests and from fits to flux loop calibrations taken at different temperatures a correction of $(1.\pm 0.25)\times 10^{-4}/\,°C$ was evaluated. The temperature spread during physics running was about 1 °C.

Figure 3 shows the corrections to the energy from the field display obtained from flux loop measurements corrected to the same temperature. A constant time dependence before and after the step around day 690 was assumed, as indicatded by the two lines. The step of ~30 MeV in centre-of-mass energy corresponds to a change to the cooling system in August 1991, which might have led to a change of the magnet temperatures relative to the measured temperature. The 1991 energy scan started just after the step, one calibration with 20 GeV protons was done before and two were performed thereafter. The precision of the proton calibrations was not high enough to conclusively confirm that the step corresponds to a genuine change in beam energy. The fills before the scan were calibrated by taking the average of the proton calibration

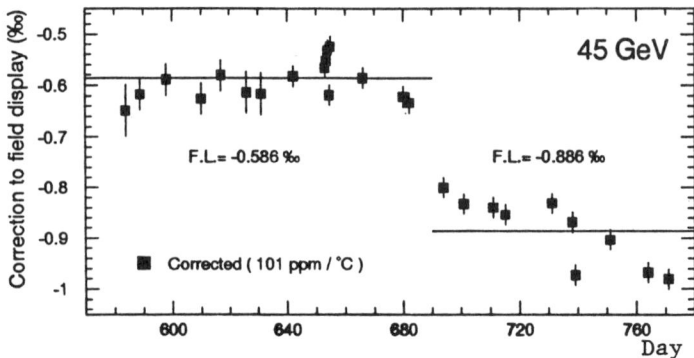

Figure 3. Flux Loop calibrations in 1991

and the extrapolation of the polarization results by means of the flux loop, where half the size of the step was taken as real; the combined error is 18 MeV.

Figure 3 also shows a possible slope in the behavior of the flux loop during the energy scan, which could not be excluded from the polarization data and which was therefore taken into account as a systematic error of ±2 MeV.

Misaligned RF Cavities

An alignment error of the radio-frequency cavities causes differences of the centre-of-mass energies between the four experiments. Copper cavities are installed around the L3 and OPAL interaction regions, which are operated at two frequencies such that the energy oscillates between the acceleration cavities and storage cavities coupled with them. The cavities are aligned for the lower frequency and thus are too far apart for the other one; therefore beam particles on the incoming side of the interaction region arrive too early at the cavities and gain an energy which is ∼8 MeV higher than for ideal alignment, whereas they gain to little on the other side. The result is an offset of the centre-of-mass energies in the L3 and OPAL experiments of about 15 MeV; the precise value depends on the power distribution on the RF cavities. When cavities around an interaction point are powered unequally, *e.g.* due to cavity trips, energy changes of a few MeV occur at all four interaction points. The uncertainties on the energy offsets are small, ∼2 MeV for an individual fill, since the effect is mostly due to the well-known geometrical alignment of the copper cavities and changes only little when operational parameters such as the synchrotron tune or the RF power distribution vary. There is almost no dependence of the energy offset on the centre-of-mass energy and therefore Γ_Z remains unaffected.

Tidal Effects

A possible explanation for most of the scatter of the beam energy observed from the depolarization calibration are tidal forces exerted by the sun and the moon which distort the spheroidal shape of the earth. These lead to changes of the radius, R, of LEP, which translate into changes of the energy, since the circumference of the closed path taken by the particles, determined by the RF frequency, does not change. The relative change of momentum can be calculated using

$$\Delta p/p = (1/\alpha_c)\Delta R/R,$$

where $\alpha_c = 3.87 \times 10^{-4}$ is the momentum compaction factor for the 1991 optics. A change of the 4.24 km LEP radius by 0.15 mm implies a change of the beam energy of ∼4 MeV. Figure 4 shows the differences in beam energy

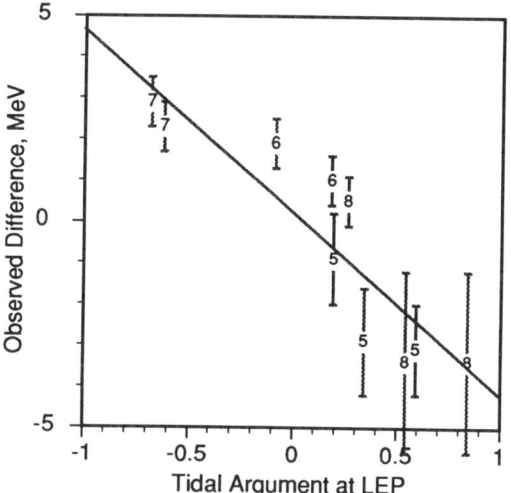

Figure 4. Beam energy from de-polarization

as a function of the tidal force normalized between -1 and +1. The numbers, 5-8, indicate different fills during which the measurements were made, and the error bars give the range in energy within which de-polarization occurred. A clear correlation is indeed seen, however, more measurements are needed where all other parameters affecting the energy, mainly the temperature and corrector dipole settings, are well controlled. At present, the scatter is used as an error estimate.

Calibration Results

The centre-of-mass energy in physics runs during the energy scan was obtained from the field display values corrected for the average deviation from the field display as obtained from the measurements with resonant de-polarization. The correction to the field display value of the centre-of-mass energy for runs at 93 GeV is (-61.0±5.3) MeV at 93 GeV. The error is composed as follows:
- the spread of the de-polarization results divided by the square root of the number of measurements (±3.7 MeV)
- the effects of the temperature difference between polarization runs and physics runs (±3 MeV)
- the possible slope in the flux loop calibration (±2 MeV)
- an error on the RF corrections due to uncertainties of the average operational parameters of the machine (±1 MeV)

Fills at energies other than 93 GeV have a contribution from an observed non-linearity in the excitation curve of the dipoles leading to a correction of (2.0 ± 1.5) MeV$\times(93-E_{cm}/\text{GeV})$. In addition, there is a possible random energy-point-to-energy-point error arising from systematically different settings of machine parameters at different energies, estimated to be $\Delta E^{set}/E = 3\times 10^{-5}$.

Since only five fills were taken at off-peak points in 1991, it is important to consider fill-to-fill fluctuations arising from the non-reproducibility of the beam energy. The spread of the polarization data $(\sim 8\times 10^{-5})$, and energy changes due to dipole temperature variations $(\pm 3\times 10^{-5})$ and RF instabilities $(\pm 2\times 10^{-5})$ lead to an estimate for the fill-to-fill reproducibility of the energy of $\Delta E^{rep}/E = 10^{-4}$; this error is reduced according to the number of fills per energy point.

In summary, the energy error of each scan point at mean energy E_i with n_i fills contributing to it is described by:

$$\frac{\Delta E_i}{E_i} = \left(\frac{\Delta E}{E}\right)^{abs} \oplus \left(\frac{\Delta E}{E}\right)^{set} \oplus \frac{1}{\sqrt{n_i}}\left(\frac{\Delta E}{E}\right)^{rep}$$
$$\oplus \frac{(93-E_i/GeV)}{E_i}\Delta E^{non-lin}$$

where

$(\Delta E/E)^{abs} = 5.7\times 10^{-5}$
$(\Delta E/E)^{set} = 3\times 10^{-5}$
$(\Delta E/E)^{rep} = 10\times 10^{-5}$
$\Delta E^{non-lin} = 1.5$ MeV

MEASUREMENT OF M_Z AND Γ_Z

M_Z and Γ_Z are extracted from combined fits to the cross sections for $e^+e^- \to hadrons$ and $e^+e^- \to leptons$ measured at different centre-of-mass energies (7 points in 1990, one point in 1991 before and 7 points during the energy scan). The fits assume a Breit-Wigner line shape with a width depending on the squared centre-of-mass energy, which is convoluted with a radiator function in order to take into account photonic corrections [4]. The errors from the energy calibration were taken into account in the fitting procedures by constructing the error correlation matrix between all scan points in 1990 and 1991. The knowledge of the absolute energy scale of the 1990 data was much less precise ($\sim$ 20 MeV) and correlated with that of 1991, and, therefore, this information was not used in the determination of M_Z; the 1990 data do, however, contribute to the measurement of Γ_Z, since point-to-point errors were similar.

Figure 5 shows the results for M_Z. The individual experimental results are quoted without the common error arising from the energy calibration. The combined result is
$M_Z = (91.187 \pm 0.0035_{stat.} \pm 0063_{LEP})$ GeV. The dominant uncertainty comes from the LEP energy calibration. In contrast, Γ_Z is still dominated by the statistical error:
$\Gamma_Z = (2.492 \pm 0.0053_{stat.} \pm 0.0045_{LEP})$ GeV.

FUTURE IMPROVEMENTS

It is hoped that polarization will be achievable at energy points other than 93 GeV. This would allow the point-to-point error to be reduced and therefore lead to a reduced systematic error on Γ_Z, well matching the expected increase in statistics. More frequent polarization measurements under well controlled machine conditions will allow an improved understanding of the effects of magnet temperatures, the flux loop drift, corrector settings and tidal forces. A realignment of the copper cavities could eliminate the dependence of the beam energy on the radio-frequency system. All together, a precision of a few MeV on the absolute and relative energy scale of LEP does not seem beyond reach.

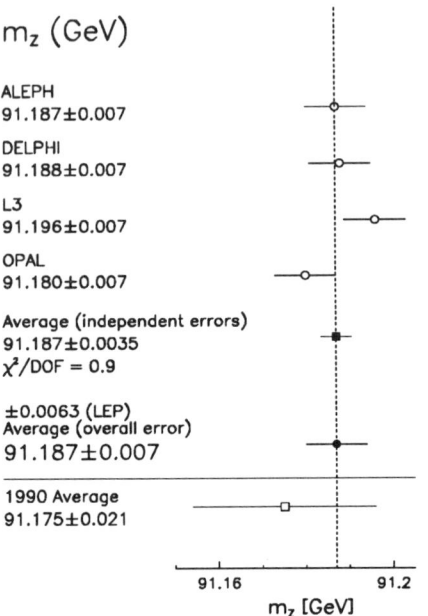

Figure 5. Measurements of the Z^0 mass.

REFERENCES

1. L. Arnaudon et al., *The Energy Calibration of LEP in 1991*. CERN-PPE/92-125 and CERN-SL/92-37(DI), July 1992

2. L. Knudsen et al., *First Observation of transverse Beam Polarization in LEP*, Phys. Lett. B270 (1991)

3. L. Arnaudon al., *Measurement of LEP Beam Energy by Resonant Spin Depolarization*, CERN-PPE/92-49, subm. to Phys. Lett. B

4. F.A. Berends et al., *Z Physics at LEP1*, CERN 89-08, ed. G. Altarelli et al., Vol. 1 (1989), p89

LEPTON FORWARD-BACKWARD ASYMMETRIES

R. Pain
DELPHI Collaboration
LPNHE-Paris
4, Place Jussieu
Paris, France, 75005

Abstract

Results of Forward-Backward Aysmmetries with Leptons measured at Z^0 energies are presented. Details of the analysis by the DELPHI Collaboration are given together with the most recent values of the peak Aymmetries for electrons, muons and taus obtained by ALEPH, DELPHI, L3 and OPAL Collaborations at LEP.

INTRODUCTION

In view of recent improvements in the determination of the beam energies at LEP[1], The four LEP experiments were performing combined lineshape and asymmetry analysis of their 1990 and 1991 data samples. Complete results of these analysis have been presented at this conference[2].

In this report I will concentrate on the analysis of the Forward-Backward Leptonic asymmetries measured by the DELPHI Collaboration. I will describe the selection of electrons, muons and taus final states together with the flavor independent analysis. All numbers given in this paper are preliminary.

In 1990 and 1991, data taking was organized such as about 2/3 of the data was collected at a centre-of-mass energy, $\sqrt{s}$, equal to the Z^0 mass. The remaining data was collected during several "energy scan" at about seven different center-of-mass energies on the Z^0 resonance peak. During these two years, DELPHI collected about 450,000 Z^0 events.

Results of the Forward-Backward leptonic asymmetries measured with the DELPHI dectector[3] during 1990 data taking have already been published[4]. In 1991, the statistics was more than twice the 1990 one with an improved detector (in particular, the use of a silicon microvertex[5]) and better understanding of the detector which make the overall uncertainty on the measured leptonic asymmetries about a factor 2 smaller.

The Forward-Backward asymmetry, A_{FB}, can be estimated by two different methods :

1. from the ratio

$$A_{FB} = \frac{\sigma_F - \sigma_B}{\sigma_F + \sigma_B} \quad (1)$$

("counting method"), where σ_F and σ_B are the cross-sections in forward and backward hemispheres respectively after extrapolation to the full angular acceptance.

2. from a likelihood fit of the differential cross-section

$$\frac{d\sigma}{d\cos\theta} = 1 + cos^2\theta + \frac{8}{3}A_{FB}cos\theta \quad (2)$$

where the last term gives the Forward-Backward asymmetry.

Although the likelihood method gives in principle better precision, different systematics contribute to these measurements and both methods were used for muon and tau final states. For electrons, the t-channel (and interference channel) contribution has to be subtracted and only the counting method was applied.

SELECTION OF ELECTRONS

The electron Forward-Backward asymmetry $A^e_{FB}(s)$ was measured in the barrel region only because of large t-channel contribution. Two independent analysis were performed. The first one, similar to the 1990 one[4], requires two back to back energy clusters in the Barrel electromagnetic detector (HPC) together with less than 4 charged tracks in the barrel tracking system which is based on the vertex detector (VD), the inner detector (ID), the time projection chamber (TPC) and the outer detector (OD).

In a second analysis, two independent selection criteria were applied which make it possible to compute selection efficiencies with high precision. The first set of cuts is based on the VD and the HPC only asking for more than two opposite tracks (and less than 4) in the VD and two electromagnetic clusters in the HPC. The second set of cuts relies on the barrel tracking system without making use of the VD together with energy deposition in the barrel hadron calorimeter (HCAL) compatible with that of electrons and hit distribution in the OD matching the OD electron hypothesis. Events were retained for which the polar angle of both electrons are in the range $44° < \theta < 136°$ and the acollinearity between the two electrons is smaller than $10°$.

The global selection efficiency and trigger efficiency amount to $(97.1 \pm 0.2)\%$ and $(99.6 \pm 0.2)\%$ respectively.

The main background comes from the reaction $e^+e^- \to \tau^+\tau^-$ (1.1% at $\sqrt{s} = M_Z$) and two photon events with electron pair final state (1.5% at $\sqrt{s} = M_Z$). The background coming from hadron final states was found to be negligible.

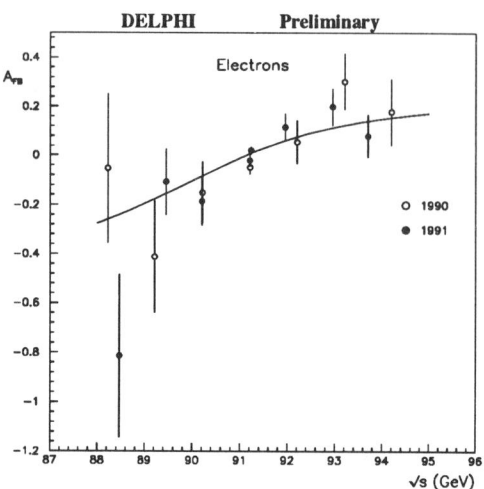

Figure 1. Electron Forward-Backward asymmetry as function of $\sqrt{s}$

As in 1990, the charge was determined from the tracking system and from the difference of the azimuthal position of the impact point on the HPC when the first method failed. The contribution to the systematic uncertainty from charge determination amount to $\delta A_{syst} = \pm 0.002$.

Figure 1 shows the measured values of A^e_{FB} for each energy point in 1990 (empty circle) and in 1991 (full circle) together with the result of the global electroweak fit which will be discussed later.

SELECTION OF MUONS

The analysis procedure for the selection of $e^+e^- \to \mu^+\mu^-$ candidates was very similar to that of 1990[4]. The polar-angle range was further extended to $10° < \theta < 170°$ and extensive use of the vertex detector was made to allow better background rejection. Tracks were retained if their momentum was greater than 15 *GeV* and acolinearity less than 10°.

It was required that each particle was identified as a muon by either the electromagnetic calorimeter or the hadron calorimeter or the muon chambers both in the barrel and forward regions. For the muon chambers, identification was based on the association of the position of the muon chambers hits with those expected from the extrapolation of the tracks. For the hadronic and electromagnetic calorimeters, it was required that the energy deposited was consistent with that expected for a minimum ionising particle.

The identification efficiency of each of the sub-detectors was estimated as a function of θ. The overall muon identification efficiency was found to be 97.4%. The sign of the electric charge was obtained from track momentum determination. The uncertainty from charge misidentification was found to be negligible. The resulting sample contained 11465 events.

The cosmic background was further reduced in 1991 by the use of time of flight detector (TOF) and OD timing informations. the remaining background amount to 0.2%. The main background comes from $\tau^+\tau^-$ final state (2.6%) and was computed using KORALZ Monte-Carlo and subtracted.

The muon Forward backward asymmetry was computed with both counting and likelihood methods. The two methods give compatible results. The overall systematic uncertainty was then estimated to be $\delta A^\mu_{syst} = \pm 0.003$. The values obtained with likelihood method are shown on Figure 2 together with the result of the global electroweak fit (solid line). On Figure 3 the region near $\sqrt{s} = M_Z$ has been enlarged and preliminary value for the ongoing 1992 data taking has been added. Two peak points are reported for 1991 data taking because of different machine condition resulting in different center-of-mass energy.

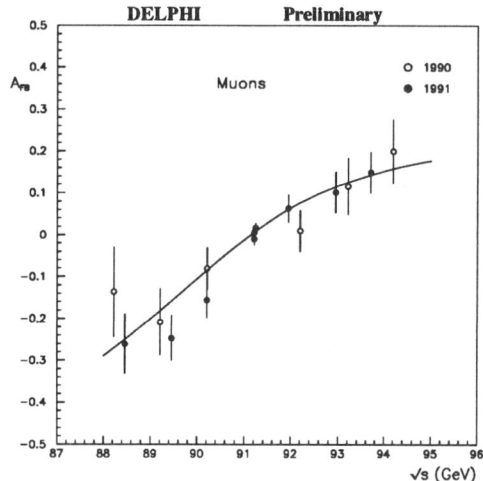

Figure 2. Muon Forward-Backward asymmetry as function of $\sqrt{s}$

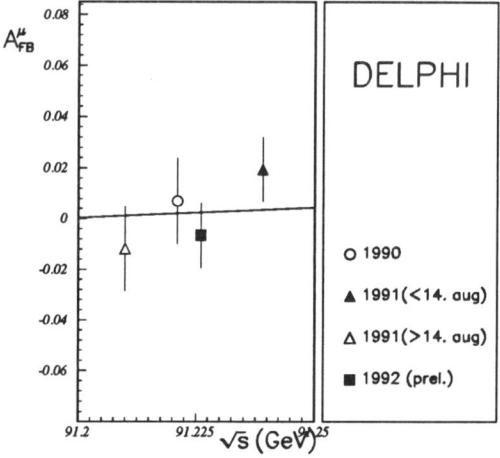

Figure 3. Muon Forward-Backward asymmetry near $\sqrt{s} = M_Z$

SELECTION OF TAUS

The selection of $e^+e^- \to \tau^+\tau^-$ events consisted of a combination of topological cuts based on the charged particle tracking and cuts using electromagnetic calorimetry in order to separate the $\tau^+\tau^-$ signal from the different backgrounds.

The background from hadronic events was minimized by demanding a maximum of 6 charged particles, one of which had to be isolated in angle from all the other charged particles in the event by at least 160° and be in the polar angle range $25° < \theta < 155°$. The rejection of leptonic Z^0 decays was done asking for the "radial" electromagnetic energy, $E_{rad} = \sqrt{E_1^2 + E_2^2}$ to be smaller than E_{beam} for rejecting $e^+e^- \to e^+e^-(\gamma)$ events and and the radial momentum, $P_{rad} = \sqrt{P_1^2 + P_2^2}$ be smaller than P_{Beam} for rejecting $e^+e^- \to \mu^+\mu^-(\gamma)$ events. The remaining resonant background amounts to $(1.9 \pm 0.4)\%$.

In order to minimize the contamination of events from two photon reactions ($e^+e^- \to e^+e^- f^+f^-$), it was required that the total visible energy be greater than $8\ GeV$ and that the missing transverse momentum with respect to the beam direction be greater than $0.4\ GeV$. The remaining cross section for two photon processes was calculated to be (4.2 ± 0.9) pb in the acceptance.

The selection efficiency was determined from simulated data produced with KORALZ Monte-Carlo to be $(81.6 \pm 0.7)\%$ in the acceptance region and the trigger efficiency was found to be $(99.9 \pm 0.1)\%$.

To calculate A_{FB}, one has to be able to reconstruct the $\tau^+\tau^-$ pair production direction and the charges. Thrust axis is used for the direction. For the charge determination, the sign of the total charge in each hemisphere is used. The systematic uncertainty from charge determination is estimated to be $\delta A = 0.001 A$.

Because of its t-channel contribution, the

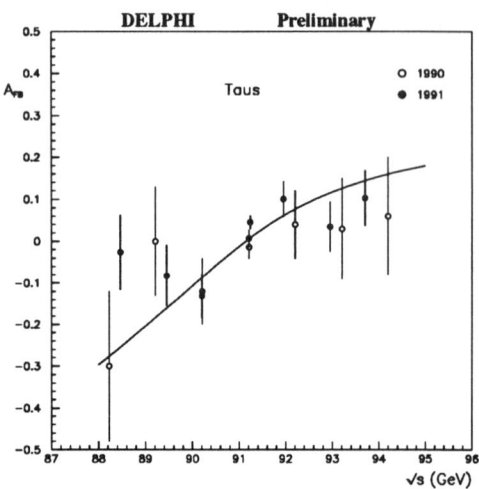

Figure 4. Tau Forward-Backward asymmetry as function of $\sqrt{s}$

high QED asymmetry of the $e^+e^- \to e^+e^-$ background had to be taken into account. It introduced a shift $\Delta A_{FB} \approx 0.9\delta$ (δ is the relative amount of t-channel bhabhas in the sample) which was subtracted.

The tau Forward-Backward asymmetry was computed with both counting and likelihood methods. The two methods give compatible results. The values obtained with the maximum likelihood method are shown on Figure 4 together with the result of the global electroweak fit. The overall systematic error is $\Delta A_{FB} = 0.015 A_{FB} \oplus \frac{0.0013}{\sigma(nb)}$.

FLAVOR INDEPENDENT ANALYSIS

In this analysis, the leptonic decays $Z^0 \to l^+l^-$ where $l = e, \mu, \tau$) were selected without trying to separate the three flavors. This approch allows a very efficient selection of leptonic events since no tight cuts are needed to separate the different leptonic channels. In addition, since leptonic channels are the main background one to another, the total background is smaller than in the flavor dependent analysis.

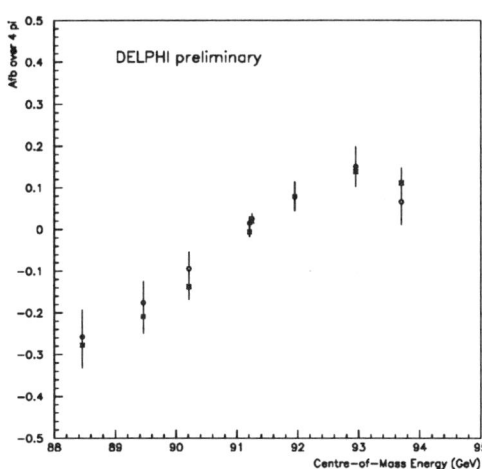

Figure 5. Flavor independent Forward-Backward asymmetry as function of $\sqrt{s}$

The event selection was based on the tracking system only. Its takes advantage of the low multiplicity, back-to-back topology and high visible momentun of the leptonic final states. It was restricted to the barrel region ($43° < \theta < 137°$) and the event acolinearity was required to be less than $20°$.

The remaining hadronic and two-photon backgrounds were computed using Monte-Carlo to be $(0.3 \pm 0.2)\%$ and (7 ± 1) pb respectively.

To avoid systematic errors associated with track superposition and bad charge determination in τ decays, only 1-1 topology events with oppositively-charged particles were retained. A total of 18172 events were selected in the 1991 data sample. The asymmetry was computed by the counting method. In Figure 5, the measured flavor independent asymmetry is plotted as function of the centre-of-mass energy (open circles) together with the average of the 3 independent leptonic asymmetries (stars). The values agree very well.

DETERMINATION OF PEAK ASYMMETRIES

The four LEP experiments agreed on a common set of electroweak parameters to extract from their data and to compare. The Forward-Backward leptonic peak asymmetries, $A_{FB}^{0,l}$ are of these parameters together with the mass of the Z^0, the width, the peak cross-section and the ratios of the leptonic partial width and the hadronic width, R^l. These parameters are obtained from a combined fit to the hadronic cross-section and leptonic cross-sections and asymmetries as function of the center-of-mass energy. The DELPHI Collaboration makes use of a program called ZFITTER[7] to do these fits. At first, a 9 parameters fit is performed to extract the leptonic asymmetries separately for electron, muons and taus. The combined 1990 and 1991 analysis gives :

$$A_{FB}^{0,e} = 0.013 \pm 0.013$$
$$A_{FB}^{0,\mu} = 0.015 \pm 0.008$$
$$A_{FB}^{0,\tau} = 0.033 \pm 0.010$$

The values are positive as expected in the electroweak theory and agree well with the lepton universality hypothesis. Therefore assuming universality, one can performe a 5 parameters fit and extract the leptonic Forward-Backward peak asymmetry, $A_{FB}^{0,l}$ to :

$$A_{FB}^{0,l} = 0.0202 \pm 0.0059$$

which show a 4 standard deviation positive Forward-Backward Leptonic peak asymmetry in agreement with the standard model prediction.

The leptonic Forward-Backward peak asymmetry being defined as :

$$A_{FB}^{0,l} = \frac{3\frac{g_v^2}{g_a^2}}{(1+\frac{g_v^2}{g_a^2})^2} \qquad (3)$$

one can derive a value of the electroweak mixing angle from the asymmetry alone using the relation :

$$\frac{g_v^2}{g_a^2} = (1 - 4\sin^2\theta_W^{eff}(M_Z))^2$$

Table 1. Lepton Peak Asymmetries

	$A_{FB}^{0,e}$	$A_{FB}^{0,\mu}$	$A_{FB}^{0,\tau}$
ALEPH	0.014 ± 0.009	0.007 ± 0.007	0.027 ± 0.008
DELPHI	0.013 ± 0.013	0.015 ± 0.008	0.033 ± 0.010
L3	0.017 ± 0.014	0.031 ± 0.010	0.028 ± 0.016
OPAL	-0.002 ± 0.012	0.005 ± 0.008	0.016 ± 0.008

One obtains for the effective electroweak mixing angle :

$$\sin^2 \theta_W^{eff}(M_Z) = 0.2294 \pm 0.0031$$

which happens to be, according to recent calculations[8], numerically equal to the $\overline{MS}$ scheme value $\sin^2 \theta_W^{\overline{MS}}(M_Z)$. This measurement agrees very well with other LEP and $p\bar{p}$ as well as low energy measurements.

Table 2. Leptonic Peak Asymmetry

Exp.	$A_{FB}^{0,l}$
ALEPH	0.0154 ± 0.0048
DELPHI	0.0202 ± 0.0059
L3	0.0264 ± 0.0074
OPAL	0.0076 ± 0.0048
Average	0.0163 ± 0.0037

CONCLUSION

The LEP Collaborations have performed precise measurements of the Forward Backward leptonic peak asymmetries. The values obtained for electrons, muons and taus final states by the 4 experiments are given in table 1. All measurements are in good agreement and reached similar precision.

A more sensitive measurement is obtained mixing the three leptonic flavors. The values obtained by the 4 LEP Collaborations together with the weighted average are reported in table 2. The four measurements are in very good agreement. The uncertainty being still dominated by statistics, these measurements will profit very soon from the foreseen future increase of LEP luminosity.

REFERENCES

1. G. Quast,"LEP energy calibration; Measurement of M_Z", in the parallel session on Electroweak interactions at this conference.

2. L. Rolandi, "Electroweak results", in plenary session at this conference.

3. P. Aarnio et al., DELPHI Collaboration, "The DELPHI detector at LEP", Nucl. Inst. and Meth. A303 (1991) 233-276.

4. P. Abreu et al., DELPHI Collaboration, "Determination of Z^0 resonance parameters and couplings from its hadronic and leptonic decays", Nucl. Phys. B 367 (1991) 511-574.

5. M. Caccia,"DELPHI microvertex detector" in the parallel session on Experimental techniques at this conference.

6. D. Yu Bardin et al., Nucl. Phys. B 351 (1991) 1.

7. D. Yu Bardin et al., Z. Phys. C44 (1989) 493 and Comp. Phys. Comm. 59 (1990) 303.

8. D. Yu Bardin, private communication.

MEASUREMENTS AT LEP OF THE FORWARD-BACKWARD ASYMMETRIES OF QUARKS

Presented on behalf of the ALEPH, DELPHI, L3 and OPAL collaborations by:

Terry Wyatt
Department of Physics, Schuster Laboratory
The University, Manchester M13 9PL. U.K.

Abstract

The measurements by the four LEP experiments of the forward-backward asymmetries of quarks are reviewed. In combining the measurements I have taken special care in the evaluation of common systematic errors that arise from our imprecise understanding of the production and decay of hadrons containing b and c quarks.

1 INTRODUCTION

The cross section for the production of quark-antiquark pairs in e^+e^- annihilation has the form:

$$\frac{d\sigma}{d(\cos\theta)} \propto 1 + \cos^2\theta + b\cos\theta \quad (1)$$

where θ is the angle between the incoming e^- and the outgoing quark. The number of 'forward' events (N_F) is defined to be the number of events for which $\theta < \frac{\pi}{2}$. Similarly, N_B is the number for which $\frac{\pi}{2} < \theta < \pi$. The term proportional to $\cos\theta$ in equation 1 leads to a 'forward-backward asymmetry' (A_{FB}) given by:

$$A_{FB} = \frac{N_F - N_B}{N_F + N_B} = \frac{3}{8}b \quad (2)$$

The e^+e^- collider LEP has so far been operated at centre of mass energies (E_{cm}) within ± 3GeV of the Z^0 pole. At $E_{cm} = M_Z$ the production of $q\bar{q}$ pairs is dominated by Z^0 exchange and A_{FB} is given in lowest order by:

$$A_{FB} = 3\left(\frac{a_e v_e}{a_e^2 + v_e^2}\right)\left(\frac{a_q v_q}{a_q^2 + v_q^2}\right) \quad (3)$$

where a_e, v_e are the axial and vector couplings of the electron and a_q, v_q are the equivalent couplings for the quark flavour 'q'. The main physics motivation for measuring A_{FB} comes from the fact that it is sensitive to the effects of electroweak radiative corrections of the type shown, for example, in figure 1. The size

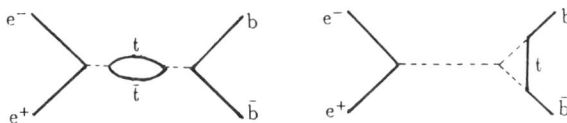

Figure 1: Examples of electroweak radiative corrections to A_{FB}.

of these corrections depends on the unknown mass of the top quark, as is shown below for A_{FB}^b in figure 4 and for A_{FB}^b in figure 6. In order to achieve a sensitivity to M_{top} at the level of ~ 20GeV we need to measure A_{FB}^b to an accuracy of ± 0.005 or better.

In order to measure A_{FB} experimentally the jet originating from the quark must be distinguished from the jet originating from the antiquark. The techniques described in this report are: using the semileptonic decay $b \to l$ to measure A_{FB}^b (section 2) ; using the semileptonic decay $c \to l$ and tagged $D^{*\pm}$ candidates to measure A_{FB}^c (section 3) ; using a momentum weighted jet charge to measure an average A_{FB} without distinguishing between individual quark flavours (section 4). The combined data sample collected by the LEP experiments up to the end of 1991, on which results are presented here, corresponds to about two million multihadronic Z^0 decays.

2 MEASUREMENT OF A_{FB}^b

All four experiments have used the semileptonic decay b → l to measure A_{FB}^b[1],[2]. Because of the hard fragmentation and large mass of B mesons, cuts in momentum (p≳3GeV) and transverse momentum to the nearest jet (p_T≳1GeV) are used to enhance the contribution from b → l. The sign of the lepton candidate enables b and b̄ jets to be distinguished.

In addition to the prompt semileptonic decay b → l a number of other sources contribute to the high p, p_T lepton sample. These are shown schematically in figure 2. The forward-backward asymmetry of the various contributions is shown in the third column of the figure. This leads to an experimentally observed asymmetry (A_{FB}^{raw}) that is given by:

$$A_{FB}^{raw} = (f_{b \to l} - f_{b \to c \to l} + f_{b \to \bar{c} \to l})(1 - 2\chi)A_{FB}^b$$
$$-f_{c \to l}A_{FB}^c + f_{background}A_{FB}^{background} \quad (4)$$

where f_i is the fraction of the high p, p_T lepton sample that originates from source 'i' and $\sum_i f_i = 100\%$. χ is the average probability that a lepton from b → l originated from a meson that had undergone $B\bar{B}^o$ mixing before decaying. It can be seen that uncertainties in the various terms in equation 4 will lead to systematic errors in the estimation of A_{FB}^b from A_{FB}^{raw}. The relevant systematic uncertainties in A_{FB}^b associated with each of the physical contributions to the high p, p_T lepton sample are indicated in the fourth column of figure 2. The contribution of the non-prompt background depends on details of the individual experiments and leads to systematic errors that are presumably uncorrelated among the measurements from the four experiments. The remaining contributions to the fourth column of figure 2 reflect our imprecise understanding of the production and decay of hadrons containing b and c quarks. These uncertainties cause correlated systematic errors in the measurements from the four experiments, which must be taken into account when the results are combined.

The size of the correlated systematic errors will depend on the fractional composition f_i of the high p, p_T lepton samples. These are given for each of the four experiments in table 1. It

	l	$f_{b \to l}$ [%]	$f_{b \to c, \bar{c} \to l}$ [%]	$f_{c \to l}$ [%]	$f_{background}$ [%]
DELPHI	e	53	12	11	24
	μ	63	11	9	17
L3	e	82	5	2	11
	μ	72	7	7	14
ALEPH	e	85	7	4	4
	μ	79	7	4	10
OPAL	e	78	10	6	6
	μ	77	7	4	12
'average'		80	7	4	9

Table 1: Composition of the high p, p_T lepton samples.

can be seen that the high p, p_T lepton samples of ALEPH, L3 and OPAL all have a similar composition. We can therefore expect that the correlated systematic errors on the extracted A_{FB}^b values will be of a similar size for each of these samples. In order to estimate the size of these correlated errors I will consider a lepton sample corresponding to the 'average' composition defined in the last row of table 1. The correlated systematic errors will be considered in two groups: those arising from our imprecise knowledge of the decay of B hadrons and those arising from the contribution of $Z^0 \to c\bar{c}$.

Our knowledge of B hadron semileptonic decay comes largely from measurements at the Υ_{4s}. For example, figure 3 shows the momentum spectrum of prompt leptons from B hadron decay measured by the CLEO experiment[3]. In order to extract the branching ratios B(b → l), B(b → c, c̄ → l) from these data a prediction must be made for the momentum spectrum of leptons from these two processes. Table 2 shows the branching ratios extracted by CLEO[3], using the predictions of three different models of B hadron

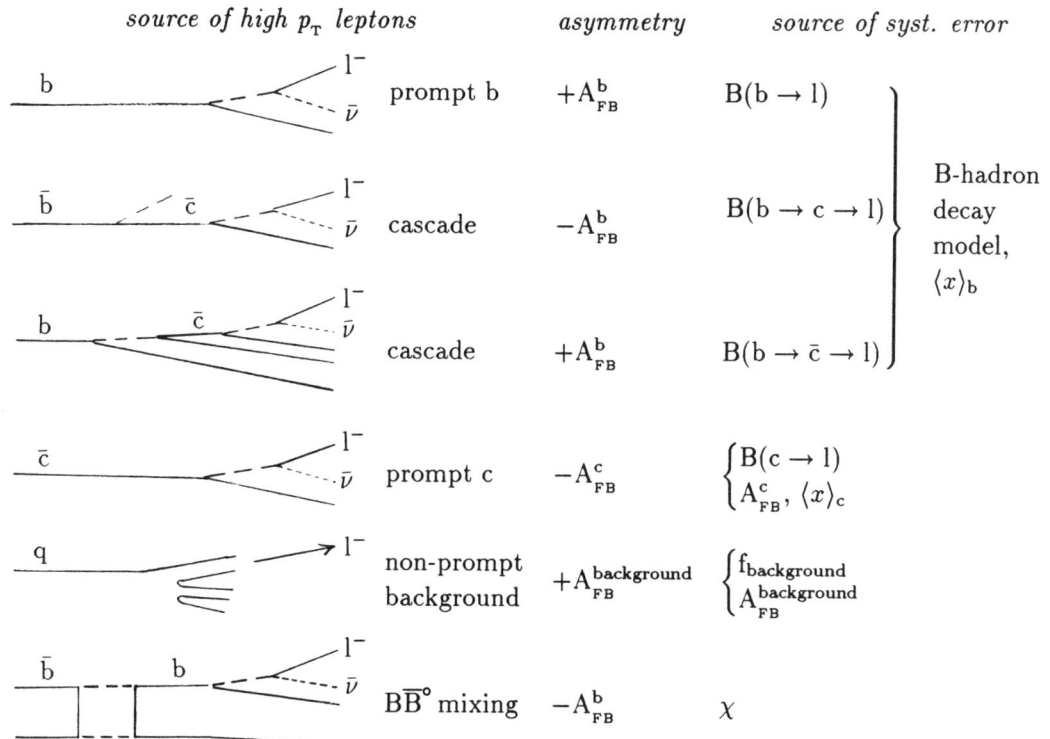

Figure 2: Contributions to the high p, p_T lepton sample.

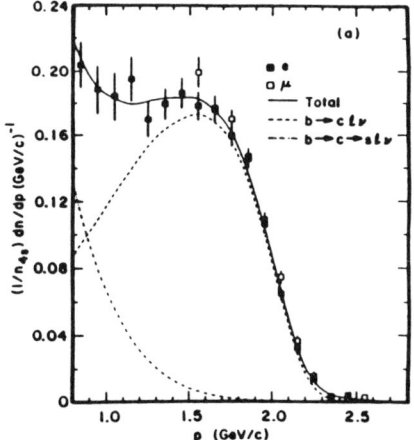

Figure 3: The momentum spectrum of prompt leptons measured by CLEO.

decay. Note that the variations due to the models are large (~10%) and anticorrelated. The differences in the shape of the lepton p_T spectrum at LEP predicted by the three models changes the expected efficiency of the cut in p_T. Together these effects lead to a fractional error in the corrected asymmetry value of: $\Delta A_{FB}^b / A_{FB}^b \approx 3\%$.

branching ratio	B decay model		
[%]	ISGW	ACM	ISGW**
$B(b \to l)$	9.9	10.5	11.2
$B(b \to c, \bar{c} \to l)$	11.3	9.7	9.0

Table 2: B decay branching ratios measured by CLEO.

At LEP $B(b \to c, \bar{c} \to l)$ is expected to be lower than that measured by CLEO. This is because at the Υ_{4s} only B^0 and $B^\pm$ mesons are produced, whereas at LEP we expect the production of also B_s and Λ_b. These are expected to decay predominantly to D_s and Λ_c, respectively, which have semileptonic branching ratios significantly smaller (~5%) than the average of D^0 and $D^\pm$ (~10%). If we assume: $B(b \to c, \bar{c} \to l)_{LEP} \approx (.94 \pm .06) B(b \to c, \bar{c} \to l)_\Upsilon$ and take into account the experimental statistical and systematic errors on the measurements of $B(b \to l)$, $B(b \to c, \bar{c} \to l)$ at the Υ_{4s} this leads to a fractional error in the corrected asymmetry of: $\Delta A_{FB}^b / A_{FB}^b \approx 2\%$.

The contribution to A_{FB}^{raw} in equation 4 due to $Z^0 \to c\bar{c}$ is $-f_{c \to 1} A_{FB}^c$. Assuming: the value $A_{FB}^c = 0.072 \pm 0.027$ as calculated in section 3 of this note; an uncertainty in the c quark fragmentation function corresponding to $0.54 < \langle x \rangle_c < 0.56$; and an uncertainty in the average semileptonic branching ratio $\Delta B(c \to 1)/B(c \to 1) \approx 15\%$, leads to a fractional error: $\Delta A_{FB}^b / A_{FB}^b \approx 3\%$.

Combining the three errors given above gives a total correlated systematic error of $\Delta A_{FB}^b / A_{FB}^b \approx 5\%$. The experimental measurements[1],[2] of A_{FB}^b are presented in the upper part of figure 4. The A_{FB}^b values are

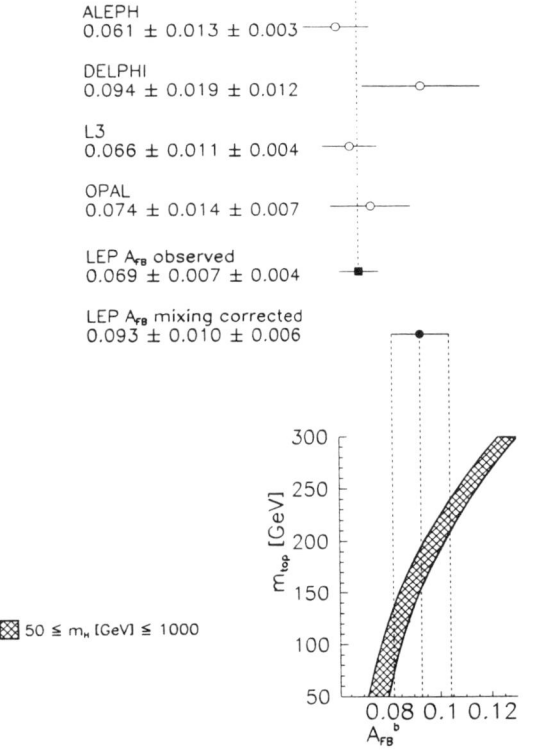

Figure 4: Summary of the results on A_{FB}^b.

not corrected for $B\overline{B}^0$ mixing. The first errors are statistical and the second are the systematic errors quoted by the experiments[1]. Re-

---

[1] The A_{FB}^b value from L3 is obtained from a combined fit to A_{FB}^b and A_{FB}^c. The error in A_{FB}^b due to the uncertainty in A_{FB}^c thus contributes to the statistical

moving from the systematic error of each experiment the contribution they assign to the sources considered above as correlated, gives the following uncorrelated systematic errors:

ALEPH DELPHI L3 OPAL
±0.002 ±0.011 ±0.004 ±0.005

Combining the four measurements gives:
$$A_{FB}^b = 0.069 \pm 0.007 \pm 0.004$$
where the first error is the statistical and uncorrelated systematic and the second is the correlated systematic error corresponding to $\Delta A_{FB}^b / A_{FB}^b \approx 5\%$ as calculated above.

Correcting for $B\overline{B}^0$ mixing using the combined LEP measurement $\chi = 0.126 \pm 0.012$ [4] gives the final result:
$$A_{FB}^b = 0.093 \pm 0.010 \pm 0.006$$
where the uncertainty in the mixing correction has been added to the correlated error.

In the lower part of figure 4 the final result is compared to the prediction of the standard model as a function of M_{top}. The experimental results are not corrected for electroweak and QCD radiative effects. These are, however, taken into account in the standard model prediction, which was obtained with the program ZFITTER [5]. A fit with this program yields the result: $M_{top} = 176^{+49}_{-68}$ GeV, for $M_h = 300$ GeV. Alternatively, the value of A_{FB}^b given above corresponds to: $\sin^2 \theta_W^{eff} = 0.2321 \pm 0.0020$.

3 MEASUREMENT OF A_{FB}^c

Measurement of A_{FB}^c using $c \to 1$

Two experiments have measured A_{FB}^c using the semileptonic decay $c \to 1$ [6]. In general leptons from $c \to 1$ are produced at lower p, p_T than those from B meson decay. The results:

	A_{FB}^c	(stat.)	(syst.)
ALEPH	0.064	±0.039	±0.030
L3	0.083	±0.038	±0.027

have been obtained from a combined fit to the p, p_T spectrum that gives both A_{FB}^b

---

error of this fit. I have removed this contribution since it is considered as a correlated systematic error.

and A_{FB}^c. Because there are sizable backgrounds in the low p, p_T region both from non-prompt sources and from B meson decay, the statistical and systematic errors are rather large. Of the sytematic errors quoted by the experiments I have considered the following two groups to be 100% correlated between the two measurements:

correlated syst. errors	ALEPH	L3
$B(c \to l)$, $B(b \to l)$, $\langle x \rangle_c$, $\langle x \rangle_b$	0.017	0.014
$A_{FB}^{background}$	0.011	0.020

This leaves uncorrelated systematic errors of: ±0.022 for ALEPH and ±0.012 for L3. The combined result is then:

$$A_{FB}^c = 0.074 \pm 0.030 \pm 0.022$$

where the first error is the statistical and uncorrelated systematic and the second is the correlated systematic error.

Measurement of A_{FB}^c using tagged $D^{*\pm}$

Three experiments have used the low Q value in the decay $D^{*\pm} \to D^0 \pi^+$ in order to tag $D^{*\pm}$ for measurments of A_{FB}^c [2]. In addition to the decay $D^0 \to K^- \pi^+$, which leads to a narrow peak in the $K^- \pi^+$ mass spectrum at M_{D^0}, OPAL and DELPHI have used the decay $D^0 \to K^- \pi^+ \pi^0$, where the undetected π^0 leads to a broad peak in the $K^- \pi^+$ mass spectrum below M_{D^0}. The available event samples are summarized in Table 3. The measured values of A_{FB}^c are given below:

	A_{FB}^c	(stat.)	(syst.)	E_{cm} range
ALEPH	0.045	±0.076	±0.004	peak ±1
DELPHI	0.107	±0.075	±0.013	peak only
	0.064	±0.049	±0.024	peak only
OPAL	-0.085	±0.130	±0.032	$\langle E_{cm} \rangle$=89.7
	0.290	±0.100	±0.032	$\langle E_{cm} \rangle$=92.7

and presented as a function of E_{cm} in figure 5. The errors are predominantly statistical and combining the three measurements at the Z^0 peak assuming no correlations yields the result:

$$A_{FB}^c = 0.070 \pm 0.038$$

In figure 6 the two measurements of A_{FB}^c from

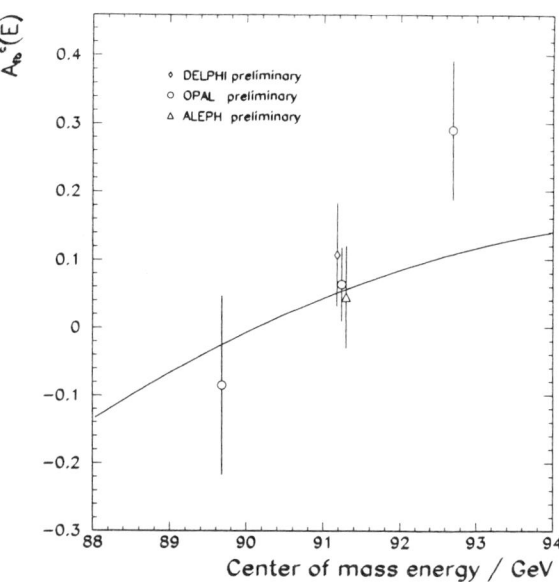

Figure 5: A_{FB}^c from tagged $D^{*\pm}$ versus E_{cm}.

leptons and tagged $D^{*\pm}$s are combined and the result:

$$A_{FB}^c = 0.072 \pm 0.027$$

is compared with the prediction of the standard model, which was obtained as a function of M_{top} with the program ZFITTER [5].

4 MEASUREMENT OF A_{FB} USING JET CHARGES

In quark fragmentation the leading hadrons carry information on the charge of the primary quark and tend to be produced with high momentum. Using this fact three experiments have used a momentum weighted jet charge to distinguish, on a statistical basis, the jets originating from the positively and negatively charged quarks and thus measure A_{FB} [7]. No attempt has been made to distinguish events originating from the different flavours of primary quarks. Since the up-type and down-type quarks are expected to have an asymmetry of the same sign, but they have charges of opposite sign, there is a partial cancelation between the two types of quarks in the average charge asymmetry measured by such methods. Monte Carlo simulations are needed to evaluate the efficiency with which the correct

656 The Forward-Backward Asymmetries of Quarks

	E_{cm} range	D^0 mode	$\langle x \rangle_e$ range	combinatorial background fraction [%]	background subtracted number $D^{*\pm}$	purity $Z^0 \to c\bar{c}$ [%]
OPAL	all E_{cm}	$K^-\pi^+$	> 0.5	13	312	78
		$K^-\pi^+\pi^0$	> 0.5	47	385	78
ALEPH	peak ±1GeV	$K^-\pi^+$	> 0.5	5	296	78
DELPHI	peak only	$K^-\pi^+$	> 0.3	30	188	54
		$K^-\pi^+\pi^0$	> 0.4	40	115	68

Table 3: The tagged $D^{*\pm}$ samples.

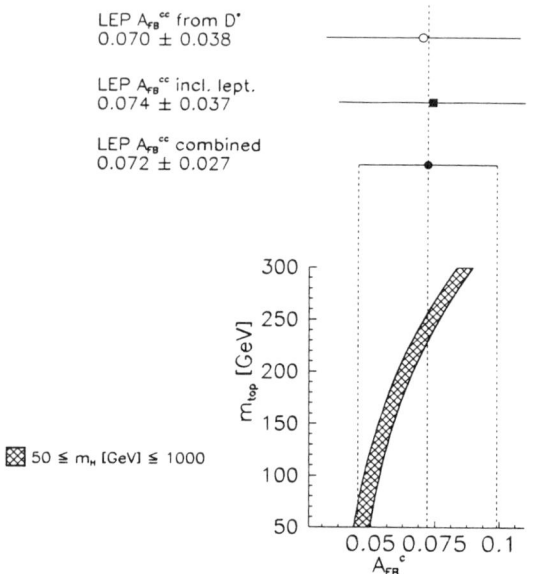

Figure 6: Combined $l^{\pm}$ and $D^{*\pm}$ measurement of A_{FB}^c.

charge assignment is made. The sensitivity of the calculated efficiency to details of the quark fragmentation model is the dominant source of systematic error in these analyses. Unfortunately, not all the experiments quote a value of the average charge asymmetry corrected for the above effects; the only way the results can be directly combined is at the level of the extracted $\sin^2 \theta_W^{eff}$ values given below:

	$\sin^2 \theta_W^{eff}$	(uncorrel.)	(correl.)
OPAL	0.2321	±0.0028	±0.0020
DELPHI	0.2345	±0.0030	±0.0027
ALEPH	0.2295	±0.0029	±0.0040
combined	0.2323	±0.0017	±0.0027

where the first error is the statistical and uncorrelated experimental systematic and the second is the correlated systematic error, which comes mainly from quark fragmentation. From the errors on the combined result given above, it is clear that unless significant progress can be made in controling the fragmentation uncertainties such measurements have reached the limit of their accuracy and are no longer competitive with direct measurement of A_{FB}^b described in section 2.

Acknowledgements: I am very grateful to Bob Clare, Günter Quast, Dorothee Schaile, Vivek Sharma and Pippa Wells for their kind help during the preparation of this work.

REFERENCES

1. L3 Collaboration, CERN-PPE/92-121;
2. ALEPH, DELPHI, OPAL Collaborations, preliminary results submitted to this conference.
3. CLEO Collaboration, CLNS 91/1101.
4. To be published in the proceedings of the XII International Conference on Physics in Collision, Boulder, June 1992.
5. D. Yu. Bardin et al, Berlin-Zeuthen 89-08.
6. ALEPH Collaboration, Phys. Lett. B263 (1991) 325.
7. ALEPH Collaboration, Phys. Lett. B259 (1991) 377, and preliminary results submitted to this conference. DELPHI Collaboration CERN-PPE/91-213. OPAL Collaboration CERN-PPE/92-139.

MEASUREMENT OF τ POLARISATION IN Z^0 DECAYS

Krishna S. Kumar *
Department of Physics,
Harvard University,
Cambridge, MA 02138, USA. †

Abstract

We present measurements of the τ polarisation in $e^+e^- \to \tau^+\tau^-$ from the four LEP experiments. From these measurements, the ratios of the vector and axial vector coupling constants of the electron and the tau lepton to the weak neutral current are obtained to be $g_{Ve}/g_{Ae} = 0.066 \pm 0.015$ and $g_{V\tau}/g_{A\tau} = 0.070 \pm 0.009$ respectively, constituting a sensitive test of e-τ universality in weak neutral current interactions. In the Standard Model, this leads to a value of the weak mixing angle at $Q^2 = m_Z^2$ of $\sin^2\theta_{\text{eff}} = 0.2326 \pm 0.0020$.

INTRODUCTION

The Standard Model has had great success as the effective theory of electroweak interactions at energy scales below 100 GeV. The fine structure constant α, the Fermi coupling constant G_F, and the mass of the Z^0 boson m_Z are each now known with a precision of better than 10^{-4}.[1,2] Using these as input, the model predicts the couplings of the fermions to the vector bosons as functions of the masses of the top quark (m_t) and the Higgs boson (m_H) at a given Q^2. The remarkable success of LEP has allowed precise measurements of neutral current couplings at $Q^2 = m_Z^2$. A comparison of theoretical estimates with these experimental measurements constitutes several stringent tests of the Standard Model, helps constrain m_t and m_H and may provide clues towards new physics at higher energies.

---

*Representing the L3 Collaboration.
†E-Mail: kkumar@cernapo.cern.ch ; Mailing Address: Div. PPE, CERN, CH 1211, Geneva 23, Switzerland.

For unpolarised e^+e^- beams, the polarisation $\mathcal{P}_f$ of final state fermions in $e^+e^- \to Z^0 \to f^+f^-$ is sensitive to the parity-violating components of the weak neutral current interaction. $\mathcal{P}_f$ is the asymmetry in the total production cross-section σ of positive ($h = +1$) and negative ($h = -1$) helicity fermions:

$$\mathcal{P}_f = \frac{\sigma(h=+1) - \sigma(h=-1)}{\sigma(h=+1) + \sigma(h=-1)}. \quad (1)$$

If the weak neutral current contains only vector and axial-vector couplings g_V and g_A, helicity conservation in the massless limit implies $\mathcal{P}_{f^-} = -\mathcal{P}_{f^+} \equiv \mathcal{P}_f$.

Further, due to the parity violation in Z^0 production, $\mathcal{P}_f$ varies with the polar angle θ of f^- with respect to the e^- direction:

$$\mathcal{P}_f(\theta) = -\frac{A_f + A_e\left(\frac{2\cos\theta}{1+\cos^2\theta}\right)}{1 + A_f A_e\left(\frac{2\cos\theta}{1+\cos^2\theta}\right)}. \quad (2)$$

A_e and A_f are functions of g_V and g_A:

$$A_e = \frac{2g_{Ve}/g_{Ae}}{1+(g_{Ve}/g_{Ae})^2}, \quad A_f = \frac{2g_{Vf}/g_{Af}}{1+(g_{Vf}/g_{Af})^2}, \quad (3)$$

thus making it possible to measure them simultaneously and independently. This must be contrasted to the case of the forward backward charge asymmetry (A_{FB}), which is sensitive to the product $A_e A_f$. It is possible to analyze the final state polarisation in the case of τ leptons since it decays close to the interaction region.

The measurement of $\mathcal{P}_\tau$ has many advantages as a precision electroweak measurement. Being the measurement of an asymmetry, it is insensitive to the absolute luminosity. A_e and A_τ are approximately linear in the couplings; thus $\mathcal{P}_\tau(\theta)$ measures the relative sign of g_V and g_A as well as tests e-τ universality in weak neutral current interactions. Nonzero values for A_e and A_τ constitute observations of parity violation in Z^0 production and decay respectively. In the Standard Model, $A_e = A_\tau \simeq 2(1 - 4\sin^2\theta_W)$, demonstrating the large sensitivity of the measurement to the the weak mixing angle. $\mathcal{P}_\tau$ is relatively insensitive to radiative corrections, unlike A_{FB}.

EXPERIMENTAL TECHNIQUE

Due to the short decay length of τ leptons and the parity violating V $-$ A structure of the weak charged current decay, $\mathcal{P}_\tau$ can be deduced from an analysis of the kinematics of τ decays.[3] τ leptons of opposite helicity have different decay angular distributions in the τ rest frame, and thus different energy distributions in the laboratory frame. However, deviations from the V $-$ A structure of the weak charged current would modify these decay distributions. In the following, we assume that any such deviations are negligible. The kinematics of the two body decays $\tau^- \to \pi^-\nu_\tau$, $\rho^-\nu_\tau$ and $a_1^-\nu_\tau$ and the three body decays $\tau^- \to e^-\bar{\nu}_e\nu_\tau$ and $\mu^-\bar{\nu}_\mu\nu_\tau$* have been studied,

*In all cases, charge conjugated decays are also used. In decays involving π^-s the corresponding decays involving K^-s are also used.

Channel	J	Branch. Ratio	Anal. Power	Stat. Weight
$e^-\bar{\nu}_e\nu_\tau$	1/2	0.178	0.37	0.07
$\mu^-\bar{\nu}_\mu\nu_\tau$	1/2	0.175	0.37	0.07
$\pi^-\nu_\tau$	0	0.125	1.00	0.32
$\rho^-\nu_\tau$	1	0.230	0.87	0.50
$a_1^-\nu_\tau$	1	0.070	0.40	0.03

Table 1. Sensitivity to $\mathcal{P}_\tau$ for relevant decay modes.

which together comprise 77% of all τ decays.

For all the decay modes, the differential cross section with respect to the energy of the decay particle can be calculated as a function of the $\mathcal{P}_\tau$.[4] Table 1 summarizes the sensitivity of each decay mode to $\mathcal{P}_\tau$.[5] The analyzing power is maximal for $\tau^- \to \pi^-\nu_\tau$ since π^- is spinless, the decay is two-body and the neutrino is purely left-handed. For the two-body decays into vector particles, the analyzing power is smaller since there are two allowed spins states. The sensitivity is enhanced by analyzing their subsequent decays,[5,6] gaining information about the spin state from the decay angular distribution. After accounting for the various branching fractions, the largest statistical significance comes from $\tau^- \to \rho^-\nu_\tau$. For the three-body decay modes, the analyzing power is significantly smaller.

In each case, $\mathcal{P}_\tau$ is measured by obtaining the linear combination of the $h = +1$ and $h = -1$ monte carlo distributions (i.e. reweighting), of the relevant kinematic variables which best fits the data. The Monte Carlo distributions are obtained after event generation[7] and full detector simulation and after applying the same selection as that for the data. In the fits, $\mathcal{P}_\tau$ and the overall normalization are left as free parameters. The dominant systematic errors arise from distortions of the energy distributions from efficiency corrections, calibration and final state radiation, and from uncertainties in the background subtraction mainly from other τ decays.

DATA ANALYSIS

The significantly higher multiplicity of multihadronic events and the relatively small cross-section of the two photon background due to the Z^0 resonance allows the isolation of a pure sample of leptonic Z^0 decays. To remove decays to dielectrons and dimuons, additional cuts are applied to reject events with two identified electrons and muons. An important consideration in the preselection is to minimize the variation in the selection efficiency as a function of the visible energy, which is sensitive to $\mathcal{P}_\tau$ since the τ^+ and τ^- helicities are fully correlated. The results for each channel from the four experiments are summarized in Figure 1.* In what follows, we

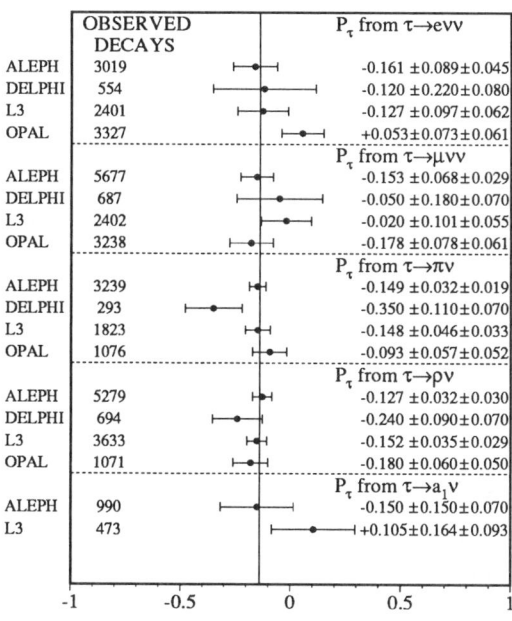

Figure 1. Compilation of all $\mathcal{P}_\tau$ measurements

describe the selection for all the channels using one specific analysis from the 4 experiments as an example.

---

*Aleph and Opal results are preliminary. The Delphi results use 1990 data only

$\tau^- \to e^- \bar{\nu}_e \nu_\tau$

τ decays to electrons are characterized by an isolated track pointing to the center of gravity of a narrow and symmetric shower in the electromagnetic calorimeter, with little activity in the hadron calorimeter. The main backgrounds come from $Z^0 \to e^+e^-(\gamma)$ where one of the electrons has been misidentified, and from $\tau^- \to \rho^- \nu_\tau$ decays with an energetic π^0.

Opal associates tracks with electromagnetic clusters and only considers those that have at most two associated tracks. The energy of the cluster E_{cl} is required to match the track momentum p_{tr}: $0.7 \leq E_{cl}/p_{tr} \leq 2.0$. In addition, the hadronic energy associated with the jet containing the cluster is required to be small and an angular match in the azimuth between the track and the signal in the presampler is required. Dielectron events are rejected by limiting the electromagnetic cluster energy: $\sum E_{cl} \leq 0.8\sqrt{s}$.

The estimated efficiency is 49% with a total background of 5.7%. The electron energy spectrum is corrected for energy dependent inefficiencies, detector resolution and final state radiation and fitted to the analytical formula. The most important systematic error is due to the correction for biases in the preselection.

$\tau^- \to \mu^- \bar{\nu}_\mu \nu_\tau$

τ decays to muons are characterized by an isolated track pointing to a minimum ionising shower in the calorimeters in addition to hits in the muon chambers behind the calorimeters. The main backgrounds come from $Z^0 \to \mu^+\mu^-(\gamma)$ where one of the muons interacts in the calorimeters, and from $\tau^- \to \pi^- \nu_\tau$ where the π^- does not interact in the calorimeters.

In the Delphi analysis, the electromagnetic energy in a cone of 30° around the isolated track is required to be less than 0.3 GeV. In addition, the particle is required to give a signal in the hadron calorimeter consistent with that expected from a minimum ionising particle and hits in the muons chambers or the outermost layer of the hadron calorimeter.

The estimated efficiency is 39% with a total background of 4.4%. The momentum spectrum of the selected events is corrected for the effects of resolution and background. $\mathcal{P}_\tau$ is obtained from the reweighting procedure using monte carlo events of the two helicities. The most important systematic error is from uncertainties in the background subtraction.

$\tau^- \to \pi^- \nu_\tau$

τ decays to pions are characterized by an isolated track pointing to an energetic shower in the calorimeters. The main backgrounds come from τ decays to π^-s with associated π^0s which are misidentified due to their proximity to the π^- shower.

Aleph uses the longitudinal and transverse segmentation of the electromagnetic calorimeter to identify electrons and photons and uses this as a veto for the pion selection. In addition, a pattern recognition algorithm in the hadron calorimeter using spatial information of the hits as well as matching with tracks facilitates separation of pions from muons.

The estimated efficiency is 61% with a background of 6.5%. $\mathcal{P}_\tau$ is obtained from the pion momentum spectrum using the reweighting procedure. The most important systematic error arises from uncertainties in the particle identification procedure. Figure 2 shows the pion momentum spectrum, clearly indicating a negative polarization.

$\tau^- \to \rho^- \nu_\tau$

These decays are characterized by an isolated track pointing to an energetic shower, accompanied by one or two electromagnetic showers close to the π^- shower. The main

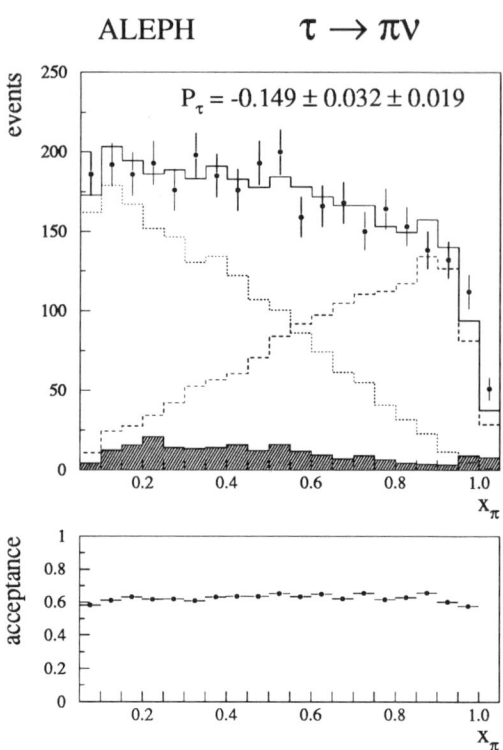

Figure 2. The pion energy distribution as a function of $x_\pi = E_\pi/E_{\text{beam}}$ in the Aleph analysis. The dashed and dotted lines show the contributions from $h = +1$ and -1 respectively. The shaded histogram is the estimate of the background. The lower plot shows the estimated selection efficiency.

background comes from τ decays involving more than one π^0 of which one or more are not reconstructed.

To facilitate discrimination between these decays, L3 uses an algorithm for finding neutral clusters in the electromagnetic calorimeter (BGO) with the emphasis on finding π^0 showers merged with charged particle showers. The contribution to the energy in the BGO from the π^- is estimated using the impact point of the charged track as well as an average shape corresponding to this impact point calculated from a testbeam study. After subtracting this contribution, secondary clusters are reconstructed which are attributed to

π^0s provided they match the expected shower profile. Candidates are required to have exactly one π^0 close to the track. The momentum measured in the tracking chamber is required to match the remaining calorimetric energy within errors. Figure 3 shows the $\pi^-\pi^0$ invariant mass for selected candidates.

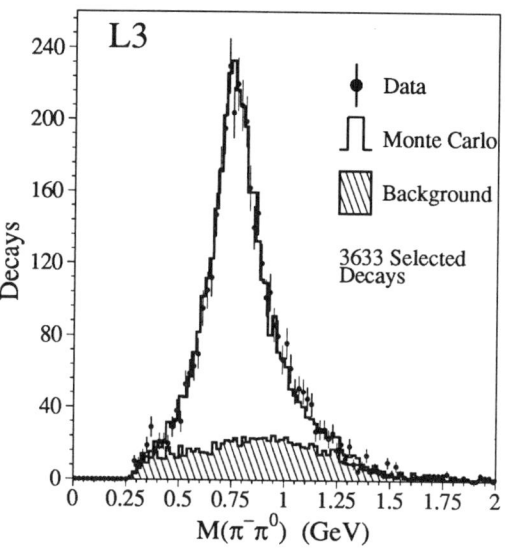

Figure 3. Invariant mass of the $\pi^-\pi^0$ for selected candidates in the L3 analysis.

The selection efficiency is 38% with a background of 17%, mostly from $\tau^- \to a_1^- \nu_\tau$ decays. The information from the ρ^- decay is incorporated by calculating θ^* and ψ^* from the 4-vectors of the π^- and π^0.[5] θ^* is the angle between the τ line of flight and the ρ in the τ rest frame and ψ^* is the angle between the ρ and the π in the ρ rest frame. The reweighting procedure is carried out to the two-dimensional distributions in $\cos\theta^*$ and $\cos\psi^*$. To maximize the analyzing power, the fit is performed separately in 9 different bins in the range $0.35\text{GeV} \le m_\rho \le 1.25\text{GeV}$ and then averaged. The most important systematic error arises from uncertainties in the π^- energy calibration.

LEP AVERAGE FOR $\frac{g_{V\tau}}{g_{A\tau}}$ and $\frac{g_{Ve}}{g_{Ae}}$

For the calculation of the average value of $\mathcal{P}_\tau$, 0.003 has to be subtracted from those measurements which were obtained with the reweighting procedure to account for initial and final state radiation and $\gamma - Z^0$ interference. Given the current statistical errors, no significant common systematic errors are foreseen among the results from the different experiments. The LEP average is

$$\mathcal{P}_\tau(Q^2 = m_Z^2) = -0.140 \pm 0.018 .$$

ALEPH

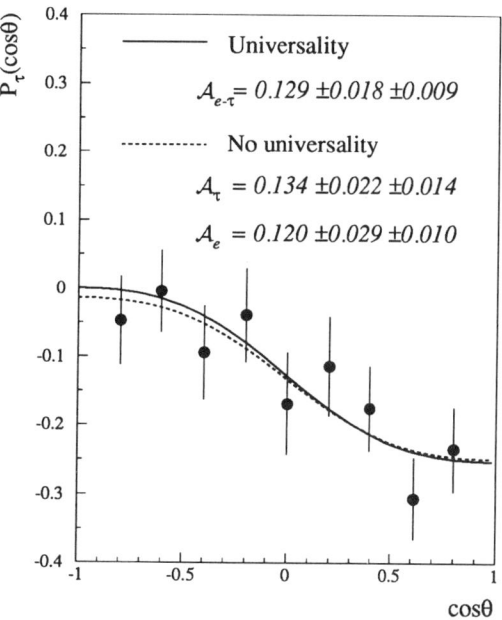

Figure 4. $\mathcal{P}_\tau(\theta)$ vs $\cos\theta$ from Aleph

Aleph has determined A_e and A_τ simultaneously by dividing the data sample in bins of $\cos\theta$, where θ is the polar angle between the τ^- with respect to the e^-, and fitting all the channels in a given bin simultaneously. Figure 4 shows these measurements as well as result of fits using Eqn. 2 with and without the assumption of lepton universality.

With their 1990 data, Opal separated their electron, muon and single pion samples into 2 samples each depending on whether the τ^- was travelling forward or backward. From this, the forward backward polarisation asymmetry was obtained, which provides a measurement of the electron coupling ratio.

The averages from the four experiments for g_V/g_A, obtained using eqn. 3, are summarized in Fig. 5.

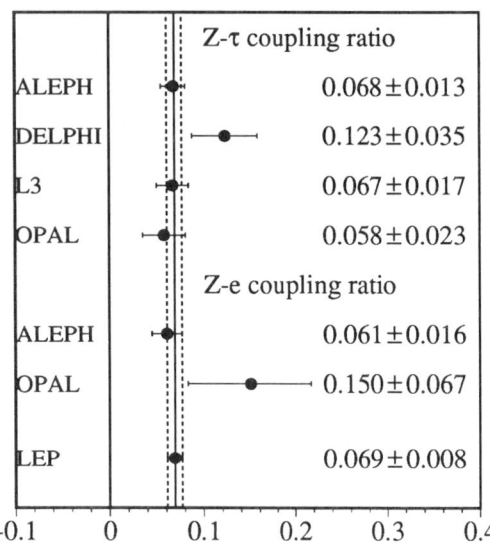

Figure 5. LEP measurements for $g_{V\tau}/g_{A\tau}$ and g_{Ve}/g_{Ae}

CONCLUSIONS

Results

The measurements of $\mathcal{P}_\tau$ presented above represent a summary of recently published measurements,[8] as well as improvements due to increase in statistics. From these, we obtain the LEP averages for the ratio of the vector and axial vector coupling constants effective at $Q^2 = m_Z^2$:

$$\frac{g_{Ve}}{g_{Ae}} = 0.066 \pm 0.015; \quad \frac{g_{V\tau}}{g_{A\tau}} = 0.070 \pm 0.009$$

consistent with lepton universality. From this, a value for the effective weak mixing angle is obtained:

$$\sin^2 \theta_{\text{eff}} = 0.2326 \pm 0.0020 ,$$

consistent with other measurements on the Z^0 peak.[9] It should be emphasized that these measurements are dominated by statistical errors and that there are no significant correlations in the systematic errors among the four experiments.

Outlook

In the near future, we expect measurements of $\mathcal{P}_\tau$ with higher statistics and new measurements of $\mathcal{P}_\tau(\theta)$ from all four experiments. The statistical precision will also be improved by better exploitation of the energy correlation between the τ leptons in the same event and by using the acollinearity angle between them.[10] In the interpretation of these measurements, we have assumed the V − A structure of the weak charged current in τ decays. The current experimental data,[11] though supporting this hypothesis, are not very precise. With improved statistics, LEP can begin to probe the structure of the weak charged current in τ decays by assuming the neutral current couplings as determined from other processes on the Z^0 peak.[12]

I would like to thank my colleagues in L3 for their help, as well as Dr Michael Schmitt from Aleph and Drs. Michael Roney and Makoto Sasaki from Opal for providing me with preliminary data.

REFERENCES

1. K. Hikasa *et al.*, Review of Particle Properties, *Phys. Rev.* **D45**, No. 11, (1992).

2. G. Quast, "LEP Energy Calibration, Measurement of m_Z", these proceedings.

3. Y. S. Tsai, *Phys. Rev.* **D4** (1971) 2821.

4. S. Jadach, Z. Was *et al.* in *Z Physics at LEP1*, CERN Report CERN-89-08, eds G. Altarelli, R. Kleiss and C. Verzegnassi (CERN, Geneva, 1989) Vol. 1, p. 235.

5. A. Rougé, Z. Phys. **C48** (1990) 75.

6. K. Hagiwara, A. D. Martin, D. Zeppenfeld, *Phys. Lett.* **B235** (1990) 198.

7. S. Jadach and Z. Was, *Comput. Phys. Commun.* **35** (1985);
R. Kleiss, "Z Physics at LEP", CERN-8908 (1989), Vol. III, p. 1.

8. Aleph Collab., D. Decamp *et. al.*, Phys. Lett. **B265** (1991) 430;
Opal Collab., G. Alexander *et. al.*, Phys. Lett. **B266** (1991) 201;
Delphi Collab., CERN Preprint, CERN-PPE/92-60, (1992);
L3 Collab., CERN Preprint, CERN-PPE/92-132, (1992).

9. R. Tanaka, "Electroweak Results from LEP", these proceedings.

10. R. Alemany *et. al.*, CERN Preprint, CERN-TH/6191/91, (1991).

11. Argus Collab., H. Albrecht *et al.*, *Phys. Lett.* **B246** (1990) 278;
Argus Collab., H. Albrecht *et al.*, *Phys. Lett.* **B250** (1990) 164.

12. P. Privitera, CERN Preprint, CERN-PPE/92-88, (1992).

LIFETIME AND LEPTONIC BRANCHING RATIOS OF THE TAU LEPTON

Martin McCubbin
Department of Physics
Oliver Lodge Laboratory
University of Liverpool
Liverpool, England

Abstract

This paper describes recent measurements of the tau lifetime and leptonic branching ratios, concentrating on results from DELPHI at LEP but also mentioning results from other experiments, in particular those from other LEP experiments. The DELPHI analysis is based on around 6500 $\tau^+\tau^-$ pairs in the barrel region, taken in 1991. The combined LEP results show that the ratio of the tau to muon Fermi coupling constants is consistant with unity.

INTRODUCTION

The Standard Model gives a prediction for the tau lifetime[1,2]:

$$\tau_\tau = \tau_\mu \left(\frac{G_\mu}{G_\tau}\right)^2 \left(\frac{m_\mu}{m_\tau}\right)^5 BR(\tau \to e\nu\nu). \quad (1)$$

Assuming $G_\tau = G_\mu$ and using the world average values[3] for $BR(\tau \to e\nu\nu)$, τ_μ, m_μ and m_τ, equation (1) gives

$$\tau_\tau = 286 \pm 4 fs,$$

which is 2σ away from the world average tau lifetime[3]

$$\tau_\tau = 305 \pm 6 fs.$$

Further measurements of the tau lifetime are needed to resolve or confirm this discrepancy, or alternatively to test $\tau - \mu$ universality.

THE DELPHI DETECTOR

The DELPHI detector has been described in detail elsewhere[4]. The microvertex detector[5], vital for the lifetime measurement, comprises 3 layers of single-sided silicon detectors arranged cylindrically around the interaction point at radii of 6cm, 9cm and 11cm. The 50μm readout pitch gives an intrinsic r-ϕ resolution of 6μm.

The track extrapolation resolution at the vertex is found to be

$$\sigma = \sqrt{26^2 + \left(\frac{65}{p_t(GeV/c)}\right)^2} \mu m,$$

i.e. approximately 30μm for tau decays.

The centre of the interaction region was determined every 100 $Z^0 \to q\bar{q}$ events, with a precision of 10μm. This was taken as the $\tau^+\tau^-$ production point. The extent of the interaction region is given by $\sigma_x = 145\mu m$ and $\sigma_y = 7\mu m$, where x and y are the horizontal and vertical directions in the plane transverse to the beam, respectively.

TAU LIFETIME

DELPHI uses four methods for measuring the tau lifetime, which will be reviewed in this section. Results from the other LEP experiments are also mentioned.

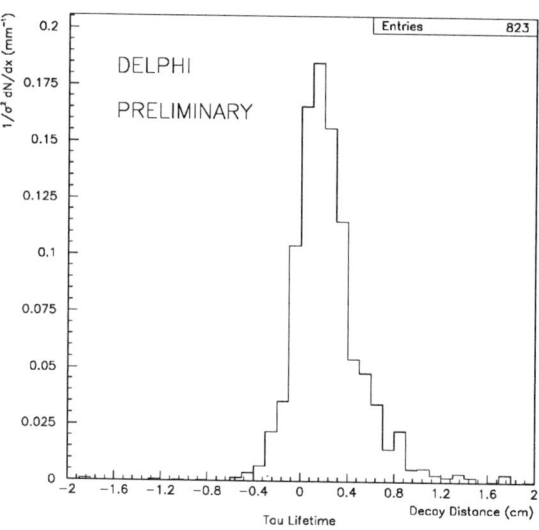

Figure 1. The 3-prong decay distance distribution (DELPHI).

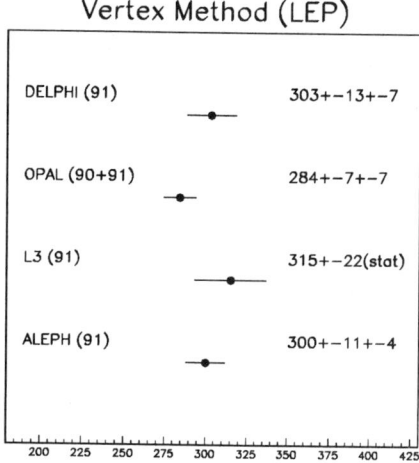

Figure 2. Recent LEP tau lifetime numbers (vertex method).

Vertex Method

This method[6] involves looking at $\tau^+\tau^-$ events which decay into a 1 versus 3 charged-track topology, reconstructing the decay vertex on the 3-prong side, and determining the decay distance from the production point. After event selection, track and vertex quality cuts, 823 secondary vertices are found. The decay distance distribution is shown in figure 1. A maximum likelihood fit to the data gives

$$\tau_\tau = 303 \pm 13(stat) \pm 7(sys) fs .$$

Other recent measurements of the tau lifetime at LEP[9,13,14] using this method are shown in figure 2.

Impact Parameter Method

For this method[6], $\tau^+\tau^-$ events are selected which decay into 1 versus 1 charged-track topologies. The impact parameter[4] for 1-prongs is related to the tau lifetime, and is signed using the charged track in the opposite hemisphere to define the tau direction. After event selection and track quality cuts there are 6113 1-prong tau decays. The signed impact parameter distribution is shown in figure 3. A maximum likelihood fit to the data gives

$$\tau_\tau = 304 \pm 11 \pm 6 fs .$$

Some recent measurements of the tau lifetime at LEP[10,13,14] using this method are shown in figure 4.

Decay Angle Correlation Method

Again, $\tau^+\tau^-$ events decaying into 1 versus 1 charged-track topologies are used. This method[7] relies on the fact that the difference in impact parameters for the two tracks is proportional to the projected acoplanarity, with a constant of proportionality related to the lifetime. In addition to the event selection made for the impact parameter method, cuts are made in order to reject radiative events. After cuts, 2873 events remain in the projected acoplanarity range -0.2 to 0.2 (radians). The mean difference in impact parameter versus the projected acoplanarity is shown in figure

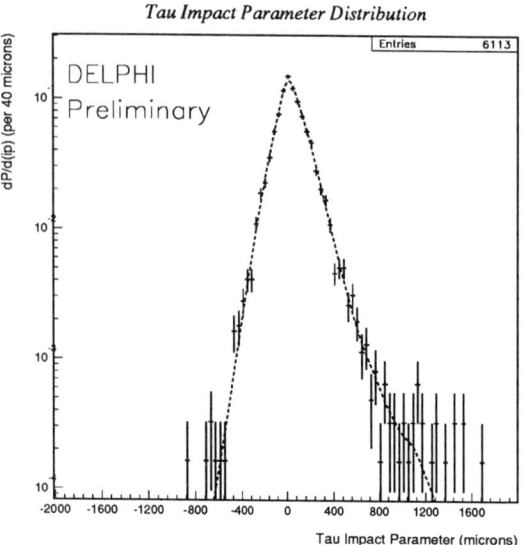

Figure 3. The signed impact parameter distribution with fit (DELPHI).

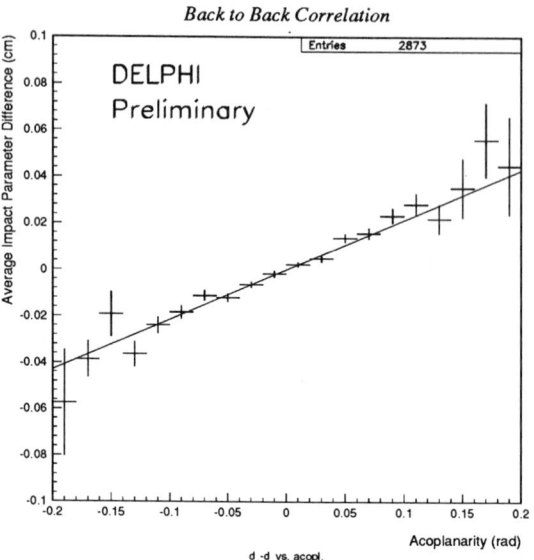

Figure 5. Mean impact parameter difference versus projected acoplanarity with fit (DELPHI).

5. A straight line fit to the data, including the effect of biases due to the remaining radiative events, gives

$$\tau_\tau = 299 \pm 11 \pm 6 fs .$$

ALEPH has carried out a measurement of the tau lifetime using this method for their 1991 data[11], with the result

$$\tau_\tau = 311 \pm 12 \pm 4 fs .$$

Missed Distance Method

Here the sum of the impact parameters for events in which $\tau^+\tau^-$ decays into 1 versus 1 charged-track topologies is used to extract the tau lifetime[8]. The same data sample as in the impact parameter difference method is used, together with extra track quality cuts; this leaves 2353 events. The distribution of impact parameter sum, or missed distance, is shown in figure 6. A log likelihood fit to the data gives

$$\tau_\tau = 301 \pm 9 \pm 6 fs .$$

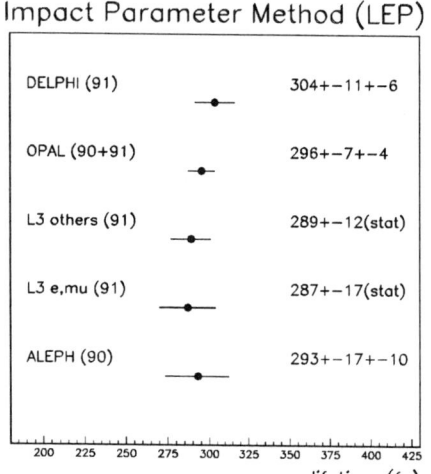

Figure 4. Recent LEP tau lifetime numbers (impact parameter method).

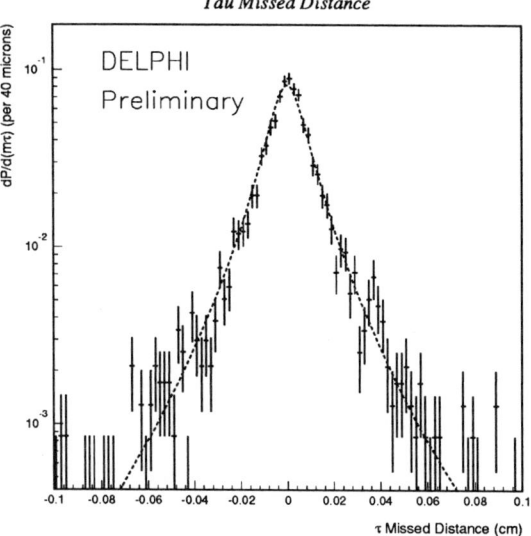

Figure 6. The missed distance distribution with fit (DELPHI).

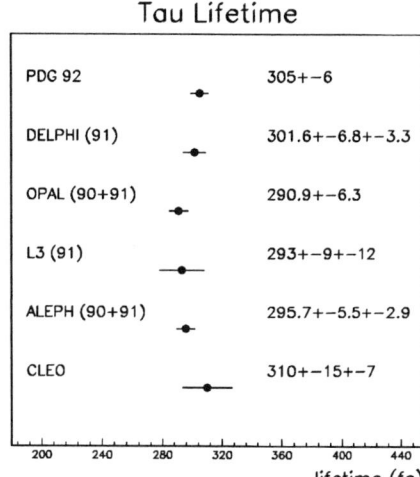

Figure 7. Summary of recent tau lifetime results.

ALEPH has carried out a measurement of the tau lifetime using a similar method on their 1991 data sample[12], and obtains the result

$$\tau_\tau = 287 \pm 8 \pm 5 fs \ .$$

Summary of Tau Lifetime Results

Figure 7 shows a summary of recent measurements of the tau lifetime[6-15]; the numbers quoted for each LEP experiment are derived by combining the results for each method used by that experiment.

TAU LEPTONIC BRANCHING RATIOS

The 1991 DELPHI $\tau^+\tau^-$ event sample for the measurement of $BR(\tau \to e\nu\nu)$ and $BR(\tau \to \mu\nu\nu)$ comprises around 6000 events, from which 1272 $\tau \to e\nu\nu$ decays are selected with an efficiency of 60%, and 1579 $\tau \to \mu\nu\nu$ decays are selected with an efficiency of 72%. The results for the leptonic branching ratios

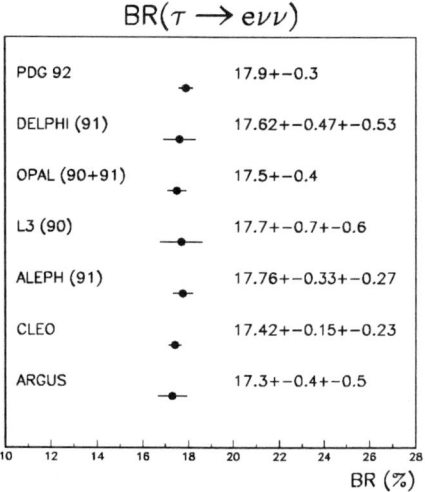

Figure 8. Summary of recent results on the electronic branching ratio.

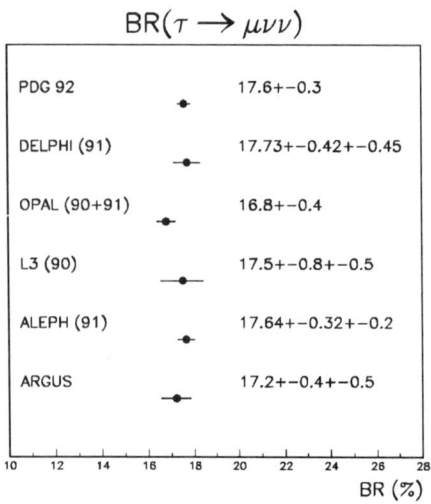

Figure 9. Summary of recent results on the muonic branching ratio.

are

$$BR(\tau \to e\nu\nu) = 17.62 \pm 0.47 \pm 0.53\%,$$

$$BR(\tau \to \mu\nu\nu) = 17.73 \pm 0.42 \pm 0.45\%.$$

Figures 8 and 9 show the comparison of these numbers with other recent leptonic branching ratio results[16-21].

CONCLUSIONS

Averaging the LEP values, not taking into account common systematic errors, yields

$$BR(\tau \to e\nu\nu) = 17.63 \pm 0.26\%,$$

$$BR(\tau \to \mu\nu\nu) = 17.34 \pm 0.24\%,$$

$$\tau_\tau = 295.3 \pm 3.7 fs\ .$$

Using these values, and a value for the tau mass $m_\tau = 1777.1 \pm 0.5$ MeV/c^2, derived by averaging the measurements from three different experiments[20,22,23], equation (1) gives

$$\frac{G_\tau}{G_\mu} = 0.987 \pm 0.010,$$

which is a consistant with unity. Including the ARGUS and CLEO values mentioned in this paper for $BR(\tau \to e\nu\nu)$ and the tau lifetime, we get

$$\frac{G_\tau}{G_\mu} = 0.983 \pm 0.008.$$

REFERENCES

1. Y.S.Tsai, "Decay correlations of heavy leptons in e$^+$e$^- \to$ l$^+$l$^-$", *Phys. Rev.* D4, pp. 2821-2837, (1971).

2. H.B.Thacker and J.J.Sakurai, "Lifetimes and branching ratios of heavy leptons", *Phys. Lett.* 36B, pp. 103-105, (1971).

3. Particle Data Group, M.Aguilar-Benitez et al., *Phys. Rev.* D45, part 2, (1992).

4. DELPHI collaboration, "The DELPHI detector at LEP", *Nucl. Instr. and Meth.* A303, pp. 233-276, (1991).

5. "DELPHI microvertex detector", submitted to this conference and *Nucl. Instr. and Meth.*

6. "A measurement of the tau lifetime", DELPHI collaboration, submitted to this conference.

7. "A measurement of the tau lifetime using the impact parameter difference method", DELPHI internal note 92-131 PHYS 231, (1992).

8. "A measurement of the tau lifetime using the miss distance method", DELPHI internal note 92-132 PHYS 232, (1992).

9. "Tau lifetime with the decay length method using VDET", ALEPH internal note 92-22, (1992).

10. ALEPH collaboration, Decamp et al., "Measurement of the tau lepton lifetime", *Phys. Lett.* 279B, pp. 411-421, (1992).

11. "Measurement of the tau lifetime using the impact parameter difference method: 1991 data with VDET", ALEPH internal note 92-73, (1992).

12. "Using the sum of impact parameters to measure the tau lifetime", ALEPH internal note 92-27, (1992).

13. OPAL collaboration, private communication.

14. L3 collaboration, private communication.

15. Stroynowski, "Tau Decay at CLEO", *Proceedings of the LP-HEP '91*, page 562, (1991).

16. "A measurement of $\tau \to e\nu\nu$ branching ratio with 1991 data", ALEPH internal note 92-94, (1992).

17. "A measurement of $\tau \to \mu\nu\nu$ branching ratio with 1991 data", ALEPH internal note 92-57, (1992).

18. OPAL collaboration, private communication.

19. L3 Collaboration, Adeva et al., "Decay properties of tau leptons measured at the Z^0 resonance", *Phys. Lett.* 265B, pp. 451-461, (1991).

20. "New results on tau lepton from CLEO", submitted to this conference.

21. ARGUS collaboration, Albrecht et al., "Measurement of exclusive one-prong and inclusive three-prong branching ratios of the τ lepton", *Zeit. Phys.* C53, pp. 367-374, (1992).

22. "Tau physics at Argus", submitted to this conference.

23. "Measurement of the mass of the tau lepton", BES collaboration, submitted to this conference.

A NOVEL METHOD TO MEASURE $\Gamma(Z^0 \to b\bar{b})/\Gamma(Z^0 \to \text{hadrons})$

Christian Moisan
Laboratoire de physique nucléaire
Université de Montréal
C.P. 6128, Succ. A, Montréal
Canada H3C 3J7

Sijbrand de Jong
CERN
CH-1211 Genève 23
Switzerland

Abstract

A new method to measure $\Gamma_{b\bar{b}}/\Gamma_{\text{had}}$ at the Z^0 resonance is presented involving the identification of $Z^0 \to b\bar{b}$ events by the semi-leptonic decay of b-flavoured hadrons and their relatively long lifetime. Using several combinations of tagged hemispheres allows the determination of identification efficiencies from the data. The method is therefore insensitive to uncertainties in b quark fragmentation and properties of b-flavoured hadrons such as decay modelling, branching ratios, relative production rates and specific lifetimes. The result is compared to other presently available measurements from LEP.

INTRODUCTION

According to the Minimal Standard Model, the partial width for the decay $Z^0 \to b\bar{b}$ features only a weak dependence on the top quark mass. This near independence results from the partial cancellation of the top quark vacuum polarisation amplitude by final state vertex corrections to the process $Z^0 \to b\bar{b}$. An accurate measurement of $\Gamma(Z^0 \to b\bar{b})$, better than 1%, would provide a stringent constraint on the Standard Model. An observed deviation from the Standard Model prediction would indicate new physics not obscured by M_{top}.

Measurements that have been established to date suffer from systematic uncertainties that limit the effectiveness of such tests.[1,2] These systematic uncertainties arise from relying on the detailed modelling of the properties of b-flavoured and charmed hadrons and of the detector to determine the tagging efficiency and the purity of the tagged sample. In the following we present a novel method that is insensitive to many of these systematic effects and reduces the sensitivity to the other sources of systematic errors.

THE MIXED TAG METHOD

The mixed tag method to measure $\Gamma_{b\bar{b}}/\Gamma_{\text{had}}$ makes use of two techniques to identify $Z^0 \to b\bar{b}$ events that are uncorrelated. The first method uses the fact that b-flavoured hadrons have a sizable semi-leptonic decay rate. Due to the hard fragmentation of b quarks and the mass of b-flavoured hadrons, the decay leptons normally have a large momentum and transverse momentum with respect to the direction of the parent jet. In this analysis only electrons will be considered. The second method is based on the fact that b-flavoured hadrons have a long lifetime compared to many hadrons that do not contain b quarks. Together with the hard fragmentation of b quarks, the result is that b-flavoured hadrons produce decay tracks with sizable impact parameter values with respect to the primary vertex.

Event Samples

The analysis described here considers the 1990 and 1991 hadronic event samples recorded by the OPAL detector at LEP. The 1991 data sample is limited to those runs in which the silicon microvertex detector was operational. Each event is divided into two hemispheres according to the thrust axis. Events are divided into jets by the JADE algorithm.[3]

Electron Tag

The identification of electrons is described elsewhere.[4] The presence of a b-flavoured hadron in a hemisphere is identified by the electron tag if: *at least one electron candidate track is found in the hemisphere with momentum greater than 4.0 GeV/c and transverse momentum greater than 0.8 GeV/c with respect to its parent jet including the electron.*

The Forward Multiplicity Tag

Decays of b-flavoured hadrons are expected to often include several charged tracks with a large positive impact parameter value with respect to the primary vertex, d_0. The probability for the presence of a b-flavoured hadron decay vertex can be parametrised by the *forward multiplicity*,[5] that is the number of tracks with significance above a given threshold. The *significance* is defined as $S = d_0/\sigma(d_0)$. The presence of a b-flavoured hadron in a hemisphere is identified by the forward multiplicity tag if: *at least 2 tracks are found in the hemisphere with significance greater than 2.5 for the 1991 data and significance greater than 1.5 for the 1990 data.*

The Four Independent Observables

Using the electron and forward multiplicity tags described above, four independent observables can be measured.

- f_v the fraction of hemispheres tagged by forward multiplicity,
- f_l the fraction of hemispheres tagged by an electron,
- f_{vv} the fraction of events with a forward multiplicity tag in both hemispheres,
- f_{lv} the fraction of events with an electron tag in one hemisphere and a forward multiplicity tag in the other, where events with an electron tag and a forward multiplicity tag in each hemisphere are counted twice.

There are five unknown parameters describing the rates of these topologies: the efficiency and purity for each of the two tags and the fraction of $Z^0 \to b\bar{b}$ events in the total event sample. This means that one of the five parameters has to be determined in another way. The lepton background was chosen for this purpose. The four observables can be expressed as:

$$f_v = \epsilon_v \frac{\Gamma_{b\bar{b}}}{\Gamma_{had}} + \rho_v^c \frac{\Gamma_{c\bar{c}}}{\Gamma_{had}} + \rho_v^{uds} \frac{\Gamma_{uds}}{\Gamma_{had}}, \quad (1)$$

$$f_l = \epsilon_l \frac{\Gamma_{b\bar{b}}}{\Gamma_{had}} + \Delta_c \frac{\Gamma_{c\bar{c}}}{\Gamma_{had}} + \Delta_{uds} \frac{\Gamma_{uds}}{\Gamma_{had}}, \quad (2)$$

$$f_{vv} = \epsilon_v^2 \frac{\Gamma_{b\bar{b}}}{\Gamma_{had}} + (\rho_v^c)^2 \frac{\Gamma_{c\bar{c}}}{\Gamma_{had}}$$
$$+ (\rho_v^{uds})^2 \frac{\Gamma_{uds}}{\Gamma_{had}}, \quad (3)$$

$$\frac{f_{lv}}{2} = \epsilon_l \epsilon_v \frac{\Gamma_{b\bar{b}}}{\Gamma_{had}} + \rho_v^c \Delta_c \frac{\Gamma_{c\bar{c}}}{\Gamma_{had}}$$
$$+ \rho_v^{uds} \Delta_{uds} \frac{\Gamma_{uds}}{\Gamma_{had}}, \quad (4)$$

with the following constraint for $\Gamma_{uds}/\Gamma_{had}$:

$$\frac{\Gamma_{uds}}{\Gamma_{had}} = 1 - \frac{\Gamma_{b\bar{b}}}{\Gamma_{had}} - \frac{\Gamma_{c\bar{c}}}{\Gamma_{had}}. \quad (5)$$

The introduced symbols are defined as:

ϵ_v the efficiency to select a b quark hemisphere with the forward multiplicity tag,

ϵ_l the efficiency to select a b quark hemisphere with the electron tag,

ρ_v^c the probability to select a c quark hemisphere with the forward multiplicity tag,

ρ_v^{uds} the probability to select an u, d or s quark hemisphere with the forward multiplicity tag,

Δ_c the probability to select a c quark hemispheres by the electron tag, and

Δ_{uds} the probability to select an u, d or s quark hemispheres by the electron tag.

The equations are solved for $\Gamma_{b\bar{b}}/\Gamma_{had}$, ϵ_v, ϵ_l and ρ_v^c.

The parameter ρ_v^{uds} is related to ρ_v^c through

$$\rho_v^{uds} = \kappa \rho_v^c . \qquad (6)$$

The parameter κ is determined from a Monte Carlo study by:

$$\kappa = \frac{N_v^{uds}/N_{tot}^{uds}}{N_v^c/N_{tot}^c} , \qquad (7)$$

where N_v^{uds} and N_{tot}^{uds} are the number of hemispheres tagged by forward multiplicity for u, d and s quarks and total number of hemispheres for these quark species in the Monte Carlo respectively. The numbers N_v^c and N_{tot}^c are defined in analogy for c quarks.

The Z^0 decay width $\Gamma_{c\bar{c}}/\Gamma_{had}$, is obtained from the LEP average:[1]

$$\frac{\Gamma_{c\bar{c}}}{\Gamma_{had}} = 0.171 \pm 0.017 . \qquad (8)$$

The background to the electron tag is divided into five classes:

N_γ number of electrons from photon conversions,

N_{Dalitz} number of electrons from Dalitz decays,

N_{decay} number of electrons from the semi-leptonic decay of hadrons containing only light (u, d and s) quarks,

N_{misid} number of hadrons misidentified as electrons, and

N_c^{prompt} number of electrons from prompt semi-leptonic decay of charmed hadrons.

Using this, the light quark background probability can be written as:

$$\Delta_{uds} = \frac{N_{misid} + N_\gamma + N_{Dalitz} + N_{decay}}{2N_{had}} , \qquad (9)$$

and the charm background probability can be expressed as:

$$\frac{\Gamma_{c\bar{c}}}{\Gamma_{had}}\Delta_c = \frac{N_c^{prompt}}{2N_{had}} + \Delta_{uds}\frac{\Gamma_{c\bar{c}}}{\Gamma_{had}} , \qquad (10)$$

The study of these sources of electron background is detailed elsewhere.[4]

Equations (1) to (4) are solved through a fit to the four free parameters with a χ^2 function. Because of the appearance of quadratic terms for the efficiencies, the equations are only strictly valid when the tagging probabilities are constant over the full range of any given variable. However, these efficiencies are known to vary as a function of the polar angle θ. Therefore, to assure the validity of the equations, they are solved in several sub-ranges of $|\cos\theta_{thrust}|$. The equations were also solved for individual sub-ranges in number of jets, to estimate the effect of hard gluon radiation. Possible biases from the beam spot ellipse were investigated by solving the equations in bins of azimuthal angle, ϕ. A consistency check of the method has also been performed on six Monte Carlo samples generated with different values for $\Gamma_{b\bar{b}}/\Gamma_{had}$, giving excellent agreement.

Table 1. Systematic Errors on $\Gamma_{b\bar{b}}/\Gamma_{had}$

$\Gamma_{c\bar{c}}/\Gamma_{had}$	±0.002
N_c^{prompt}	±0.004
$N_\gamma + N_{Dalitz}$	±0.002
$N_{misid} + N_{decay}$	±0.001
κ	±0.003
Forward Multiplicity Cuts	±0.008
Azimuthal Correlations	±0.002
Jet Multiplicity	±0.002
Thrust Cut	±0.002
Total	±0.010

RESULT

The combined preliminary result for $|\cos\theta_{thrust}| < 0.6$ and thrust value $T > 0.8$ is:

$$\frac{\Gamma_{b\bar{b}}}{\Gamma_{had}} = 0.218 \pm 0.013 \text{ (stat)} \pm 0.010 \text{ (syst)}.$$

The statistical uncertainty on the determination of $\Gamma_{b\bar{b}}/\Gamma_{had}$ is dominated by the statistical errors on the fractions f_l and $f_{l\nu}$.

The systematic uncertainty is broken down in Table 1. It is dominated by the variation of the result when the cuts defining the forward multiplicity tag are changed to scan a wide range of efficiencies and purities. Since the lepton and lifetime tag efficiencies are determined from the data itself, the solution of the equations does not require the input of b quark fragmentation models, b-flavoured hadron semi-leptonic decay rates, or b-flavoured hadron lifetimes. However, the result depends on c quark fragmentation and charmed hadron semi-leptonic decay rates. This causes an important uncertainty in the background determination of the lepton tag, manifesting itself as the second biggest contribution to the systematic uncertainty on $\Gamma_{b\bar{b}}/\Gamma_{had}$.

SUMMARY

A new method to measure $\Gamma_{b\bar{b}}/\Gamma_{had}$ was introduced. The motivation for this method is to reduce the sensitivity of the measurement to a number of unknown quantities, such as the b quark fragmentation process, the b quark semi-leptonic decay rate, modelling of the semi-leptonic decays, b-flavoured hadron lifetimes, and also to reduce the sensitivity to modelling by Monte Carlo simulations. In order to achieve this goal, mutually independent tagging information was used.

Figure 1 summarises the available measurements of $\Gamma(Z^0 \to b\bar{b})$ from the LEP collaborations assuming:[6,7]

$\Gamma(Z^0 \to \text{hadrons}) = (1740 \pm 12)$ MeV,
$BR(b \to l + X) = 0.113 \pm 0.010 \pm 0.006.$

Published results have been discussed in recent conferences.[1,2] The preliminary result submitted by the DELPHI Collaboration to this conference that uses their 1990 and 1991 single lepton sample[8] does not show an improved accuracy on the determination of $\Gamma(Z^0 \to b\bar{b})$, since based on a conventional approach. The preliminary results from neural network analyses[8,9] are also largely limited by the Monte Carlo systematic. The OPAL preliminary result discussed here shows a clear improvement of the systematic uncertainty. Fitting differential distributions of the lepton and di-lepton samples, ALEPH has recently presented a preliminary measurement of $\Gamma_{b\bar{b}}/\Gamma_{had}$,[9] that is also very promising from the point of view of the size of systematic errors.

Averaging all these measurements yields
$\Gamma(Z^0 \to b\bar{b}) = (373 \pm 9)$ MeV,
in excellent agreement with the expectation from the Standard Model.

REFERENCES

1. P. S. Wells, "Partial Widths of the Z^0 to $b\bar{b}$ and $c\bar{c}$ at LEP," in *proceedings of the 4*th

Method to Measure $\Gamma(Z^0 \to b\bar{b})/\Gamma(Z^0 \to \text{hadrons})$

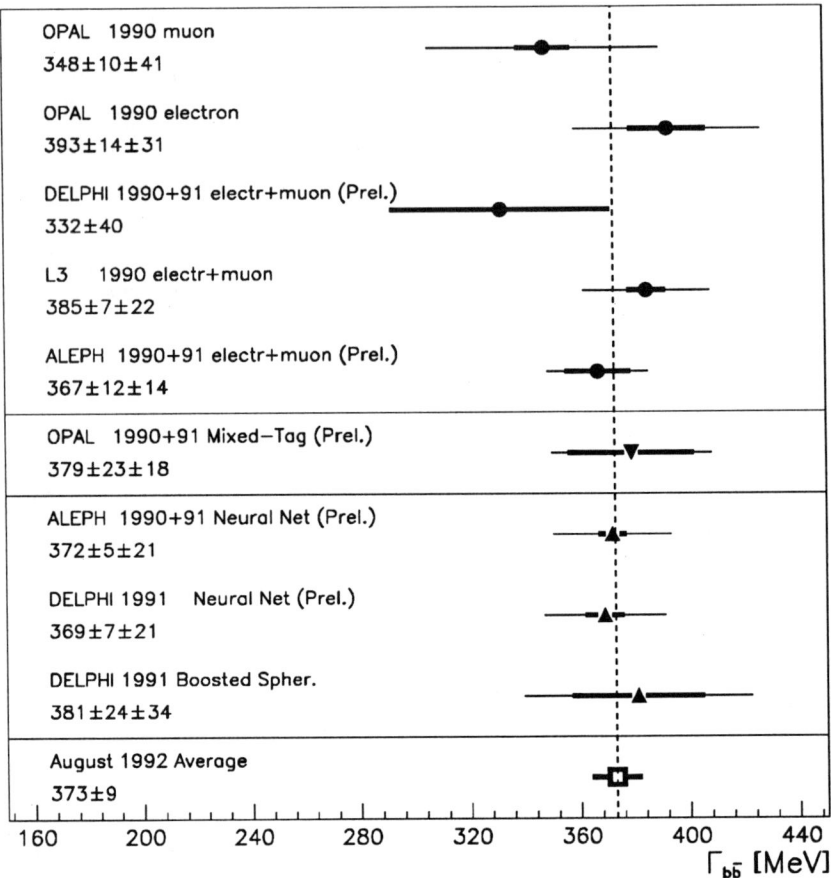

Figure 1. Status of $\Gamma(Z^0 \to b\bar{b})$ from LEP.

International Symposium on Heavy Flavour Physics, 1991, pp. 423–433.

2. J. Kroll, "Heavy Flavour Physics at LEP," presented at the XXVII Rencontres de Moriond, Moriond, France, March 15-22, 1992.

3. M. Z. Akrawy et al., *Z. Phys.* **C49**, pp. 375–384, (1991).

4. P. D. Acton et al., *CERN-PPE/92-38*, (March 6, 1992), submitted to Z. Phys. C.

5. R. Jacobsen et al., *Phys. Rev. Lett.* **67**, pp. 3347–3350, (1991).

6. The LEP Collaborations, *CERN-PPE/91-232*, (December 20, 1991), submitted to Z. Phys. C.

7. B. Adeva et al., *Phys. Lett.* **B261**, pp. 177–187, (1991).

8. P. Abreu et al., "Classification of the Hadronic Decays of the Z^0 into b and c Quark Pairs using a Neural Network" and "$\Gamma_{b\bar{b}}$ and $\langle x_E \rangle_B$ using the Z^0 semi-leptonic decay into muons," contributed papers to this conference.

9. B. Brandl, "Measurement of $\Gamma(Z^0 \to b\bar{b})$," contribution to the XXVII Rencontres de Moriond, Moriond, France, March 15-22, 1992.

DETERMINATION OF THE NUMBER OF LIGHT NEUTRINO FAMILIES FROM $e^+e^- \to \nu\bar{\nu}\gamma$

Simonetta Gentile *
CERN ,
and
Dipartimento di Fisica,
Universitá della Calabria,
87036 Rende (Cosenza),Italia.

Abstract

In this paper the determination of the number of the light neutrino families, N_ν, obtained from the measurement of the cross-section of the radiative process $e^+e^- \to \nu\bar{\nu}\gamma$ at LEP is discussed. The first result, $N_\nu = 3.0 \pm 0.4 \pm 0.2$, (73 events) was obtained from the OPAL Collaboration. A recent measurement, $N_\nu = 3.14 \pm 0.24 \pm 0.12$ (202 events) obtained by L3, is described in some detail.

INTRODUCTION

Only a few years ago the number of fermion generations and (in SU(3)xSU(2)xU(1)) the number of light neutrino types, N_ν, were among the important questions in the Standard Model. The LEP experiments had provided the answer to this question and this can be considered one of the most fundamental results obtained. Each neutrino species (with a mass lighter than $M_Z/2$)contributes to the Z^0 width via the decay $Z^0 \to \nu\bar{\nu}$. The number of neutrino families N_ν can be determined in two ways:

- *Indirect determination*: assuming that all visible Z^0 decays lead to charged leptonic and hadronic final states, this method is based on the measurement of the Z^0 partial width to invisible final states. The invisible width is defined as $\Gamma_{invisible} = \Gamma_Z - \Gamma_{hadrons} - 3\Gamma_{leptons}$, where $\Gamma_Z, \Gamma_{hadrons}, \Gamma_{leptons}$ are respectively the total, hadronic and leptonic Z^0 widths.

*E-Mail: Gentile@vxcern.cern.ch ; Mailing Address: Div. PPE, CERN, CH 1211, Geneva 23, Switzerland.

- *Direct determination* : This method is based on the measurement of the cross section for the process $e^+e^- \to \nu\bar{\nu}\gamma$ and the comparison with its predicted dependence on the number of neutrino types, N_ν.

The theoretical inputs from the Standard Model used in these determinations are different and complementary which results in producing *two independent determinations of N_ν*.

Invisible width

In the Standard Model, each neutrino family contributes with $\Gamma_{\nu\bar{\nu}}$ to $\Gamma_{invisible}$: $\Gamma_{invisible} = N_\nu \Gamma_{\nu\bar{\nu}}$. The calculated value is $\Gamma_{\nu\bar{\nu}} = 166.8 \pm 1.5 \; MeV$ and its ratio to the charged lepton width (particularly insensitive to radiative corrections) is:

$$\frac{\Gamma_{\nu\bar{\nu}}}{\Gamma_{leptons}} = 1.993 \pm 0.004.$$

The uncertainty corresponds to a variation of the top mass between 90 and 250 GeV. The measurement of this ratio obtained from the four LEP collaborations (ALEPH,DELPHI,OPAL,L3)[1] leads to a number of neutrino families:

$$N_\nu = 3.00 \pm 0.05$$

Using this method, the dominant contribution to the error comes from the theoretical uncertainty on the luminosity (± 0.02), but there is also a significant contribution from the systematics in the analysis of hadronic and charged leptonic decay modes.

This result *excludes* a fourth fermion family, associated with a light ($< 45\ GeV$) neutrino.

Radiative method

This paper is mainly devoted to the discussion of the second method : the direct measurement of the cross section of the radiative process

$$e^+e^- \to \nu\bar{\nu}\gamma \ . \tag{1}$$

The signature of such events is a single photon arising from initial state radiation.

The cross section for the process (1) can be written as[2] :

$$\frac{d^2\sigma}{dE_\gamma d\cos\theta_\gamma} = H(E_\gamma, \cos\theta_\gamma, s)\sigma_0(s')$$

where H is a radiator function describing the initial state radiation of a photon of energy E_γ at an angle θ_γ with respect to the beam axis, s is the square of the center of mass energy, and $\sigma_0(s')$ is the cross-section for the process $e^+e^- \to \nu\bar{\nu}$, at the "reduced" center of mass energy given by $s' = s(1 - 2E_\gamma/\sqrt{s})$.
In lowest order and by approximating the W contribution by a four-point interaction, σ_0 is given by :

$$\sigma_0(s) = \frac{G_F^2 s}{12\pi}\left(2 + \frac{N_\nu(g_v^2 + g_a^2)}{[1 - s/M_Z^2]^2 + \Gamma_Z^2/M_Z^2} + \frac{2(g_v + g_a)[1 - s/M_Z^2]}{[1 - s/M_Z^2]^2 + \Gamma_Z^2/M_Z^2}\right) \tag{2}$$

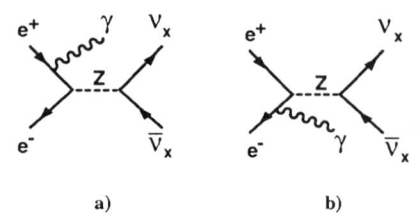

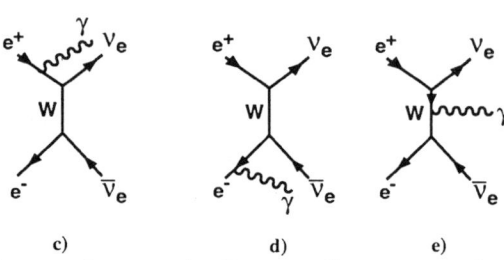

Figure 1. Lowest order Feynman diagrams contributing to $e^+e^- \to \nu\bar{\nu}\gamma$.

The dominant term is proportional to the number of light neutrino families N_ν and represents the square of the amplitude for Z^0 exchange terms (diagrams a),b) in fig.1) summed over all neutrino types. The first term arises from the square of the W exchange (diagrams c),d),e) in fig.1), which contributes only to ν_e production, and the third is due to $W - Z^0$ interference. These terms contribute less than 3 % at collision energies within 3 GeV of the Z^0 resonance as scanned by LEP in the 1990 and 1991 runs. Moreover the contribution of these terms is independent of N_ν because they give rise only to $\nu_e\bar{\nu}_e$ pairs. The most favoured collision energy is well above the Z^0 mass, where the initial photon radiation, which brings the electron-positron center-of-mass energy back to the Z^0 resonance, would be easily observed and allow for a precise measurement of the invisible width.

A still good experimental condition for this measurement is realized at energies at least

3 GeV above the Z^0 mass, where the ratio between the signal and the QED background is favourable. This is clearly demonstrated from the differential cross-section, plotted in fig.2, as a function of photon energies at different collision energies.

However, this condition was not often achieved because the center-of-mass energy during the last two years was mainly at the Z^0 peak where the photons from $e^+e^- \to \nu\bar{\nu}\gamma$ have low energies with a rapidly falling spectrum.

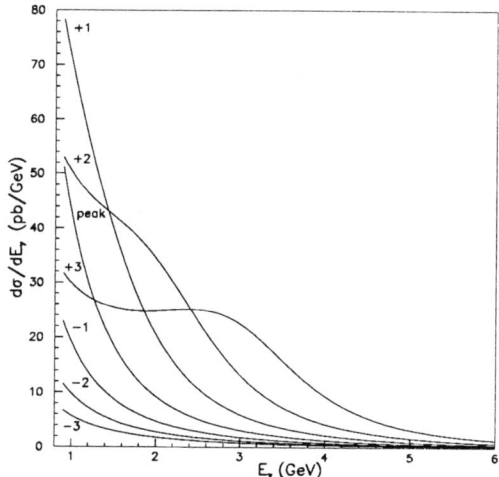

Figure 2. Cross-section for $e^+e^- \to \nu\bar{\nu}\gamma$ as a function of E_γ, for various center-of-mass energies

EXPERIMENTAL TECHNIQUE

In this situation events with low-energy photons, near 1 GeV, have to be used. The measurement becomes more difficult and the detector has to satisfy more stringent experimental conditions:

- An *efficient trigger* for low energy photons (down 1 GeV), corresponding to 1% of the center-of-mass energy, deposited in the electromagnetic calorimeter.

- A *good control of background*. Large backgrounds, as discussed later, originate from processes in which some final state particles escape detection at angles close to the beam or in detector holes.

- A *good knowledge of the electromagnetic scale*. As shown in the previous figure, this condition originates from the rapidly falling photon energy spectrum.

The signal

The selection of $\nu\bar{\nu}\gamma$ candidates requires the identification of events with:

(A) A neutral electromagnetic energy deposit in the electromagnetic calorimeter.

(B) No other activity in the detector.

These very simple criteria imply the inclusion of various backgrounds. We now discuss the main one.

The background

The main sources of background to the single photon events are from QED processes,[2] in which apart from the photon, all final state particles, mainly produced at small polar angle, escape detection.

The dominant background comes from:

- *Radiative Bhabha scattering*:

$$e^+e^- \to e^+e^-\gamma . \qquad (3)$$

at low Q^2, when both electrons escape detection and only the photon is seen in the detector.

The requirement that final state electrons be below the detector angular acceptance puts some limit on the p_t of the photon. But in the range of energies considered, near the Z^0 mass, this eliminates also a part of signal. The p_t cut must be relaxed, allowing a fraction of background events to be included in the sample. Therefore it is necessary to have a good understanding of this process in the analysis [3,5,6].

Other possible sources of background include:

- *Two-photon processes*:

$$e^+e^- \rightarrow e^+e^- X \ . \quad (4)$$

where X is $f_2, \pi^0, \eta, \eta', a_2$, and only one photon is observed from the final state X. This constitutes a potentially large background and the spectra of single photons from these sources are expected to be softer than from radiative Bhabha scattering. This channel needs to be understood, when very low-energy single photons are studied.

Moreover another possible contribution comes from four-fermion final states:

$$e^+e^- \rightarrow e^+e^- l^+ l^- \gamma \ . \quad (5)$$

when all charged particles escape detection (l is for lepton).

- *Annihilation in three photons*:

$$e^+e^- \rightarrow \gamma\gamma\gamma \ . \quad (6)$$

- *Other Z^0 decays*: Background from events in which Z^0 decays into undetected charged particles and one photon (*i.e.* $\mu^+\mu^-\gamma$ where the two muons are unseen).

- *Other backgrounds*: Can arise from cosmic rays, beam halo, beam-wall and beam-gas interactions.

The first result on single photon events was obtained from the OPAL collaboration from an analysis of data collected during 1990. This

Experiment	L_{int} pb^{-1}	N_{Obs}	N_ν
OPAL[3]	5.3	73	$3.0 \pm 0.4 \pm 0.2$
ALEPH[4]	5.29	147	$2.7 \pm 0.3 \pm 0.2$
L3-90[5]	2.9	61	$3.2 \pm 0.6 \pm 0.2$
L3-91[6]	9.6	202	$3.14 \pm 0.24 \pm 0.12$
LEP Coll Invis.width[1]			3.00 ± 0.05

Table 1. Summary of single photon measurements at LEP.

result is based on 73 events, corresponding to an integrated luminosity $L = 5.3\ pb^{-1}$ at e^+e^- c.m. energies between 88 and 94 GeV[3]. From the fit to the analytical formula they obtain a number of light neutrinos: $N_\nu = 3.00 \pm 0.4 \pm 0.2$. This result and similar ones from ALEPH[4] and L3 [5] are summarised in Table 1, toghether with the value obtained from invisible width analysis[1]. A more recent result has been reported by the L3 collaboration, based on data collected during 1991[6]. We discuss it in more detail in the following paragraph.

DATA ANALYSIS

The L3 detector, with a very good electromagnetic calorimeter, is in a favourable position to carry out this measurement. It is designed to measure energy and position of leptons and photons with high precision [7]. The electromagnetic calorimeter has good energy and coordinate resolution for photons pointing to the vertex, good linearity and, very importantly, a negligible uncertainty on the absolute energy scale.

Another important experimental aspect is the trigger. The single photon trigger is entirely based on the electromagnetic calorimeter and requires a neutral electromagnetic deposit greater than 0.9 GeV (1.5 GeV in 1990) and greater than 80 % of the total electromagnetic energy. A dedicated trigger has been implemented on isolated electrons from $e^+e^- \to e^+e^-\gamma$, providing a measurement of single photon trigger efficiency [6]. This efficiency, checked also with a Monte Carlo simulation, ranges from 59% at the Z^0 peak to 71 % at 3 GeV above the peak with a systematic error of 1.3%.

The selection of candidates is based on the requirement of a neutral electromagnetic deposit with energy E_γ greater than 0.9 GeV in the region of the electromagnetic calorimeter barrel from 45° to 135° (condition (A)) and a veto, to suppress QED background, on the presence of other particles, extended to a minimum angle of 1.5° with respect to the beam axis, defined from the luminosity monitor (condition (B)).

After applying the selection cuts, the sample consists of 291 single photon candidates. The trigger, selection and the total efficiencies are reported in fig.3 as a function of center of mass eneries. The energy spectrum of these candidates is shown in fig.4 together with the Monte Carlo prediction of expected signal for $N_\nu=3$ and the backgrounds. Radiative Bhabha scattering background corresponds to 94 events. Smaller backgrounds such as two-photon processes, $\gamma\gamma\gamma$ and $\mu^+\mu^-\gamma$ consist of 15 events. The cosmic contamination is less than 3 events and other backgrounds are negligible. In the energy spectrum of single photon can-

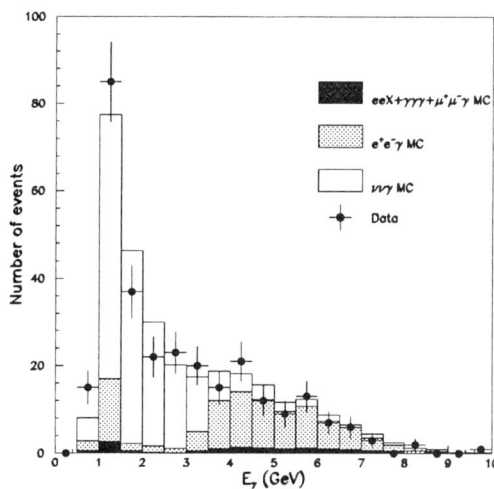

Figure 4. Energy spectrum of single photons and Monte Carlo predictions.

didates (fig.4) the region in the energy range 0.9 $GeV \leq E_\gamma \leq 3.5$ GeV has the best signal-to-background ratio. Applying this cut, 202 events survive. The number of light neutrino families N_ν has been extracted by performing a maximum likelihood fit to the number of candidates. The analytical formula used is an improved Born approximation of equation (2) with an s-dependent Z^0 width to take into account the electroweak corrections. The cross section can be written as:

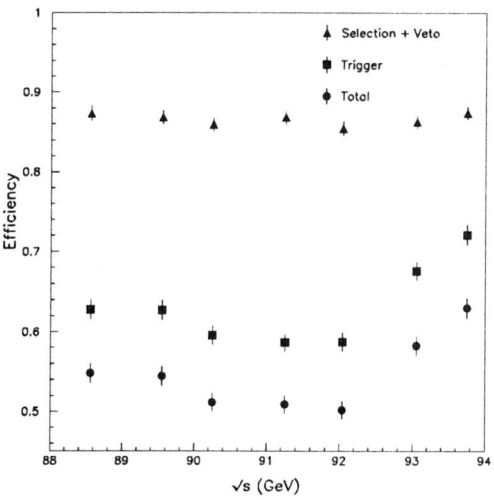

Figure 3. Selection, trigger and total efficiency at different c. m. energies.

$$\sigma_0(s) = \frac{12\pi}{M_Z^2} \frac{s\Gamma_e N_\nu \Gamma_{\nu\bar{\nu}}}{(s-M_Z^2)^2 + s^2\Gamma_Z^2/M_Z^2} + W\ terms$$

where $\Gamma_{\nu\bar{\nu}}$ is the decay width of the Z^0 in a neutrino pair with standard model couplings, M_Z, Γ_Z and Γ_e are our measured values[9] for the Z^0 mass, the total width and the electron partial width respectively. The result of the fit gives $N_\nu = 3.14 \pm 0.24$ (stat).

The various contributions to the systematic error come from: trigger efficiency (± 0.04), luminosity (± 0.03), event selection (± 0.02), background subtraction (± 0.09), cosmic ray contamination (± 0.02), theoretical uncertainty (e.g. top mass) and errors on the measurements of Z^0 parameters (± 0.05). This leads to a total systematic error ± 0.12 and the final value, based on an integrated luminosity of $L = 9.6 \, pb^{-1}$, is:

$$N_\nu = 3.14 \pm 0.24(\text{stat}) \pm 0.12(\text{sys})$$

This corresponds to a Z^0 invisible width of:

$$\Gamma_{invisible} = 524 \pm 40 \text{ (stat)} \pm 20 \text{ (sys) MeV}$$

This improves upon previously published results by L3 and OPAL[3,5]. The corrected cross-section is shown in fig.5 as a function of the c.m. energy along with the expectations from $N_\nu = 2, 3, 4$ and from the best fit.

CONCLUSIONS

The determination of the number of light neutrino families obtained by measuring the single photon cross-section at LEP at energies around the Z^0 pole gives as a most precise result (from L3):

$$N_\nu = 3.14 \pm 0.24(\text{stat}) \pm 0.12(\text{sys}).$$

This constitutes a determination of N_ν which is independent, both from a theoretical and an experimental point of view, from the measurement based on the lineshape determination.

In the near future we expect measurements from all LEP experiments based on data which is now being collected to give a determination of N_ν with higher statistics and better systematics.

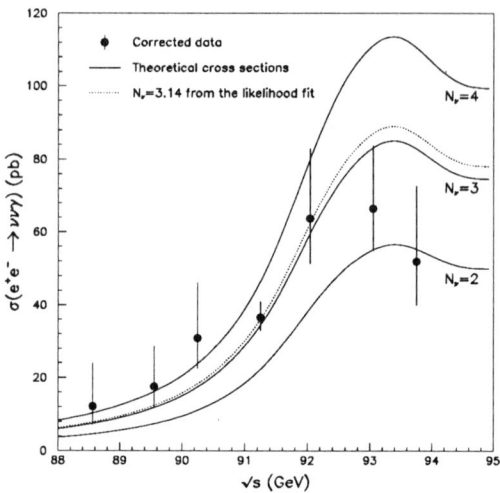

Figure 5. Corrected single photon cross section as function of the c.m. energy, compared with predictions

REFERENCES

1. The LEP Collab., *Phys. Lett.* **B276** (1992) 247.

2. L.Trentadue et al. in *Z Physics at LEP1*, CERN/89-08,Vol.1,p. 129.

3. M.Z. Akrawy et al., OPAL Collab., *Z. Phys.* **C50** (1991) 373.

4. A.Buijs, pres. to Joint Int. Lepton-Photon Symp. and Europhysics Conf. On High Energy Physics,(1991),Geneva (Suisse).

5. B.Adeva et al., L3 Collab., *Phys. Lett.* **B275** (1992) 209.

6. O.Adriani et al., L3 Collab., CERN Preprint, CERN-PPE/92-128, (1992).

7. B.Adeva, et al., L3 Collab *Nucl.Inst. Meth.* **A289** (1990) 35.

8. O.Nicrosini and L. Trentadue, *Nucl.Phys.* **B318** (1989) 1.

9. B.Adeva et al., L3 Collab., *Z. Phys.* **C51** (1991) 179.

ELECTROWEAK RESULTS FROM LEP

Reisaburo TANAKA

Laboratoire de Physique Nucléaire et des Hautes Energies
Ecole Polytechnique, IN2P3-CNRS
Route de Saclay, F-91128 Palaiseau CEDEX, FRANCE

Abstract

With 2 million Z^0 decays, the LEP experiments at CERN have performed high precision tests of the electroweak theory. Preliminary results of 1990 and 1991 data are reported.

INTRODUCTION

Each of the 4 LEP experiments at CERN has accumulated nearly one-half million Z^0 decays, corresponding to an integrated luminosity of about 20 pb^{-1}, by the end of 1991. Since the last publication of LEP data [1], significant improvements have been achieved in both LEP energy measurements with transverse beam polarization [2] and experimental systematic errors [3].

LINESHAPE MEASUREMENT

The Z lineshape can be described as the convolution of a Breit-Wigner resonance cross section $\sigma^0(s)$ with initial-state radiation,

$$\sigma_{f\bar{f}}(s) = \int H(s,s') \sigma^0_{f\bar{f}}(s') ds', \quad (1)$$

where $H(s,s')$ is the initial-state radiator function. The peak cross section is expressed as,

$$\sigma^0_{peak} = \frac{12\pi}{M_Z^2} \frac{\Gamma_e \Gamma_f}{\Gamma_Z^2}. \quad (2)$$

The lineshape of the lepton forward-backward asymmetry can be described in a similar way.

Fitting Method

The cross section measurement and lepton forward-backward asymmetry measurement were fit simultaneously. The fitting programs used are MIZA [4] for ALEPH and ZFITTER [5] for DELPHI, L3 and OPAL. The programs agree well below the experimental accuracy.

The procedure used to extract the LEP averages was as follows:

- Each LEP experiment provides a set of most independent parameters,
  5 parameter ($M_Z, \Gamma_Z, \sigma^0_{had}, R_l, A^{0\,l\bar{l}}_{FB}$)
  9 parameter ($M_Z, \Gamma_Z, \sigma^0_{had}, R_e, R_\mu, R_\tau, A^{0\,e^+e^-}_{FB}, A^{0\,\mu^+\mu^-}_{FB}, A^{0\,\tau^+\tau^-}_{FB}$).

- Assign systematic error common to the LEP experiments,
  $\Delta M_Z = 6.3$ MeV and $\Delta \Gamma_Z = 4.5$ MeV from LEP energy uncertainty,
  $\Delta \sigma^0_{had}/\sigma^0_{had} = 0.3\%$ from luminosity (theory).

- Construct 20×20 or 36×36 covariance matrix taking correlated errors into account.

Table 1. Measured Z parameters by LEP experiments.

		ALEPH	DELPHI	L3	OPAL
M_Z	(GeV)	91.187±0.009	91.188±0.010	91.196±0.010	91.180±0.009
Γ_Z	(GeV)	2.501±0.012	2.488±0.012	2.495±0.011	2.483±0.012
σ_{had}^0	(nb)	41.60±0.27	40.86±0.28	40.95±0.39	41.01±0.41*
R_l		20.78±0.13	20.82±0.16	20.92±0.16	20.87±0.13
$A_{FB}^{0\,l\bar{l}}$	(%)	1.54±0.48	2.02±0.59	2.64±0.74	0.76±0.48
R_e		20.69±0.21	20.79±0.28	21.24±0.24	21.01±0.26
R_μ		20.88±0.20	20.92±0.22	20.72±0.27	20.65±0.18
R_τ		20.77±0.23	20.69±0.30	20.63±0.31	21.16±0.25
$A_{FB}^{0\,e^+e^-}$	(%)	1.40±0.93	1.3±1.3	1.7±1.4	-0.2±1.2
$A_{FB}^{0\,\mu^+\mu^-}$	(%)	0.74±0.72	1.48±0.83	3.1±1.0	0.47±0.76
$A_{FB}^{0\,\tau^+\tau^-}$	(%)	2.69±0.82	3.3±1.0	2.8±1.6	1.65±0.82

* OPAL peak cross section is from 1990 data.

- Minimize $\chi^2 = \Delta^T \text{COV}^{-1} \Delta$ to obtain the best fit values.

The summary of measured Z parameters from each LEP experiment is given in Table 1. Combined results are given in Table 2, where the third column shows the common systematic error among experiments due to the LEP energy uncertainty and the theoretical uncertainty on the luminosity calculation. When combining the LEP data, additional common systematic errors due to t-channel subtraction in the electron channel (0.5% of total t-ch subtraction) and due to the LEP energy uncertainty in the lepton forward-backward asymmetry (0.05%) were found to be negligible compared with experimental error, and thus were ignored.

The overall χ^2 was 17.2/15 dof for the 5 parameter fit (assuming lepton universality) and was 26.2/27 dof for the 9 parameter fit (no lepton universality).

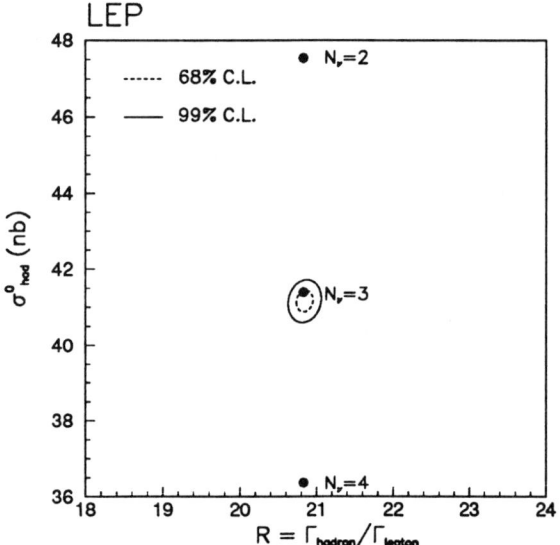

Figure 1. Contour plot of hadronic peak cross section versus R_l for combined LEP data together with the Standard Model prediction.

Extracted Z Parameters

The assumption of three light neutrino

species (Standard Model constraint), as shown in Table 2, results in a better determination of the strong coupling constant $\alpha_s(M_Z^2)$ than using $R_l = \Gamma_{had}/\Gamma_{l\bar{l}}$ [6],

$$\alpha_s(M_Z^2) = 0.130 \pm 0.012. \quad (3)$$

These values are about 1.5 σ higher than those determined with event shape analysis.

The number of light neutrino species N_ν is determined by assuming lepton universality as,

$$N_\nu = \left(\frac{\Gamma_{\nu\bar{\nu}}}{\Gamma_{l\bar{l}}}\right)_{SM}^{-1} \cdot \left\{\sqrt{\frac{12\pi R_l}{M_Z^2 \sigma_{had}^0}} - R_l - 3\right\}, \quad (4)$$

where the Standard Model prediction is $\Gamma_{\nu\bar{\nu}}/\Gamma_{l\bar{l}} = 1.993 \pm 0.004$. The result is consistent with $N_\nu = 3$ as shown in Figure 1.

ASYMMETRY MEASUREMENT

The effective vector and axial-vector couplings are defined via,

$$\begin{cases} g_V(M_Z^2) = g_A(M_Z^2)\left(1 - 4\sin^2\theta_W^{\text{eff}}\right), \\ g_A(M_Z^2) = -\frac{1}{2}\sqrt{\rho^{\text{eff}}}, \end{cases} \quad (5)$$

where the effective electroweak mixing angle $\sin^2\theta_W^{\text{eff}}$ is defined through the vector and axial-vector effective coupling of electron,

$$\sin^2\theta_W^{\text{eff}} = \frac{1}{4}\left(1 - \frac{g_{Ve}}{g_{Ae}}\right). \quad (6)$$

The lepton forward-backward asymmetry, tau polarization, tau polarization forward-backward asymmetry, b-quark charge asymmetry and leptonic partial width can be expressed in terms of effective couplings :

$$\begin{aligned} A_{FB}^{0\,l\bar{l}} &= \frac{3}{4}A_e A_l, \\ P_\tau &= -A_\tau, \\ P_\tau^{FB} &= -\frac{3}{4}A_e, \\ A_{FB}^{0\,b\bar{b}} &= \frac{3}{4}A_e A_b, \\ \Gamma_{l\bar{l}} &= \frac{G_F M_Z^3}{6\pi\sqrt{2}} \cdot \left(g_{Vl}^2 + g_{Al}^2\right) \cdot \left(1 + \frac{3\alpha}{4\pi}\right), \end{aligned} \quad (7)$$

Table 2. Combined Z parameter from 4 LEP experiments.

	LEP	comm.	χ^2
M_Z (GeV)	91.187±0.007	0.0063	2.3
Γ_Z (GeV)	2.492±0.007	0.0045	1.5
σ_{had}^0 (nb)	41.16±0.18	0.12	5.3
R_l	20.85±0.07	—	0.5
$A_{FB}^{0\,l\bar{l}}$ (%)	1.54±0.27	0.05	5.5
$\Gamma_{l\bar{l}}$ (MeV)	83.33±0.30	0.18	
Γ_{had} (MeV)	1737±7	3.8	
Γ_{inv} (MeV)	505±6	3.1	
N_ν	3.04±0.04	0.02	1.2
$\alpha_s(M_Z^2)$	0.135±0.009*	—	
R_e	20.92±0.12	—	3.4
R_μ	20.79±0.10	—	1.2
R_τ	20.84±0.13	—	2.5
$A_{FB}^{0\,e^+e^-}$ (%)	1.03±0.58	0.05	1.5
$A_{FB}^{0\,\mu^+\mu^-}$ (%)	1.20±0.41	0.05	4.7
$A_{FB}^{0\,\tau^+\tau^-}$ (%)	2.46±0.48	0.05	1.7
Γ_e (MeV)	83.19±0.36	0.18	
Γ_μ (MeV)	83.69±0.53	0.18	
Γ_τ (MeV)	83.49±0.63	0.18	

* The determination with Standard Model fit by assuming $N_\nu = 3$.

Table 3. Effective $\sin^2\theta_W^{\text{eff}}$.

	LEP Avr.	$\sin^2\theta_W^{\text{eff}}$
$A_{FB}^{0,l\bar{l}}$	0.0154±0.0027	0.2320±0.0016
P_τ	-0.140±0.018	0.2324±0.0023
P_τ^{FB}	-0.092±0.023*	0.2346±0.0039
$A_{FB}^{0,b\bar{b}}$	0.098±0.011	0.2324±0.0020
Asymmetry Average		0.2324±0.0011
Lineshape $\Gamma_{l\bar{l}}$		0.2334±0.0011

* Only ALEPH data were used.

where $A_f = 2g_{V_f}g_{A_f}/(g_{V_f}^2 + g_{A_f}^2)$.

All the asymmetry data are evaluated at M_Z. Deconvolution of the initial- and final-state QED/QCD radiative correction, subtraction of γZ and pure γ term are performed for all data. The effective electroweak mixing parameter $\sin^2\theta_W^{\text{eff}}$ was extracted from the combined LEP measurement for various asymmetry data as shown in Table 3 together with $\sin^2\theta_W^{\text{eff}}$ from leptonic partial width. The effective electroweak coupling from lineshape has been extracted by using the formula,

$$\Gamma_{l\bar{l}} = (1 + \kappa^{\text{eff}}) \cdot \frac{\alpha(M_Z^2)M_Z}{48\sin^2\theta_W^{\text{eff}}\cos^2\theta_W^{\text{eff}}}$$
$$\times \left[1 + (1 - 4\sin^2\theta_W^{\text{eff}})^2\right]\left(1 + \frac{3\alpha}{4\pi}\right) \quad (8)$$

where $\kappa^{\text{eff}} = 0.0040 \pm 0.0020$ is the effective parameter which absorbs the running Z self-energy. The same error is obtained with all asymmetry data and with leptonic partial width.

Lepton Universality

To test lepton universality in the neutral current, the lineshape, lepton forward-backward asymmetry, tau polarization and tau polarization forward-backward asymme-

Table 4. Effective vector and axial-vector coupling constants.

flavour	$g_V(M_Z^2)$	$g_A(M_Z^2)$
e	-0.035±0.005	-0.4991±0.0011
μ	-0.029±0.011	-0.5010±0.0017
τ	-0.038±0.004	-0.4998±0.0019
l	-0.0352±0.0025	-0.4995±0.0009

try have been combined for all LEP data to determine effective leptonic vector and axial-vector coupling. The result is given in Table 4, and shown in Figure 2. The axial-vector coupling is determined by the leptonic partial width, while asymmetry data measure the ratio $g_V(M_Z^2)/g_A(M_Z^2)$. Basically, the error on electron couplings is $\sqrt{2}$ times smaller than others due to the existence of the eeZ coupling in both the initial and final states. It also benefits from P_τ^{FB} measurement which is the measure of A_e. Significant improvement is achieved for g_{V_τ} determination through the tau polarization measurement.

When converted into the coupling ratio of muons and taus with respect to electrons, one finds,

$$\frac{g_{V\mu}}{g_{Ve}} = 0.83 \, {}^{+0.42}_{-0.31},$$
$$\frac{g_{V\tau}}{g_{Ve}} = 1.10 \, {}^{+0.22}_{-0.18},$$
$$\frac{g_{A\mu}}{g_{Ae}} = 1.0037 \pm 0.0040,$$
$$\frac{g_{A\tau}}{g_{Ae}} = 1.0013 \pm 0.0042. \quad (9)$$

Thus lepton universality in the neutral-current processes holds at the 0.4% level.

STANDARD MODEL FIT

By assuming the Standard Model, one determines the top mass through electroweak

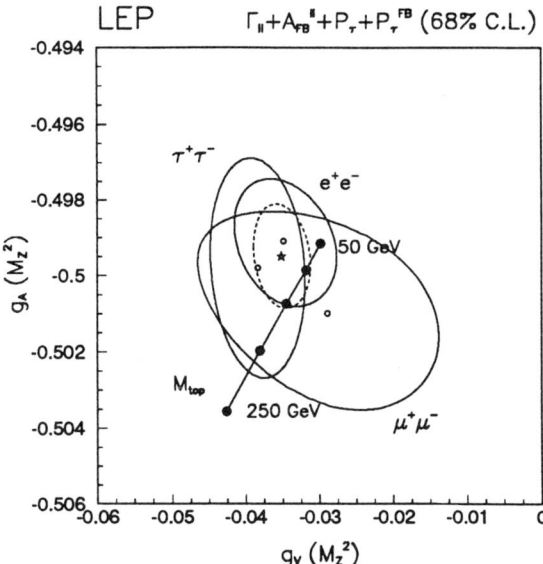

Figure 2. The effective vector and axial-vector coupling. The dashed line is the result with assuming lepton universality. The Standard Model with $M_{Higgs} = 300$ GeV is also shown for various top mass.

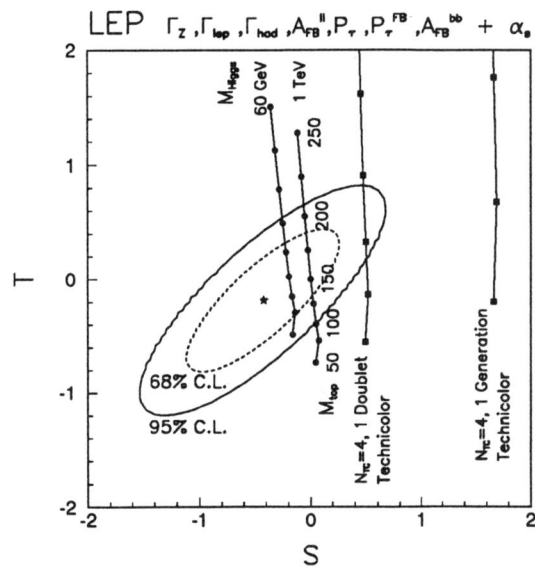

Figure 3. The parameter S and T determined from LEP data alone.

radiative corrections. Since LEP-I data alone cannot determine Higgs mass, one normally fixes the Higgs mass. Additional constraint can be imposed by using the $\alpha_s(M_Z^2)$ measurement from event shape analysis at LEP.

When M_{Higgs} is fixed at 300 GeV, and $\alpha_s(M_Z^2)$ is treated as free parameter, combining all the LEP data gives,

$$M_{top} = 139 {}^{+24}_{-29} {}^{+19(M_H=1TeV)}_{-23(M_H=50GeV)} \text{ GeV}$$
$$\alpha_s(M_Z^2) = 0.135 \pm 0.009 \pm 0.002(M_H) \quad (10)$$

where $\chi^2 = 2.6/5$ dof.

If one performs the constrained fit with $\alpha_s(M_Z^2)$ determined from event shape analysis $\alpha_s(M_Z^2) = 0.120 \pm 0.006$ [6], one finds

$$M_{top} = 148 {}^{+22}_{-26} {}^{+19(M_H=1TeV)}_{-23(M_H=50GeV)} \text{ GeV}$$
$$\alpha_s(M_Z^2) = 0.125 \pm 0.005 \pm 0.002(M_H) \quad (11)$$

where $\chi^2 = 4.8/6$ dof. This fit also yields a determination of $\sin^2 \theta_W^{\text{eff}}$ and M_W,

$$\sin^2 \theta_W^{\text{eff}} = 0.2328 \pm 0.0007$$
$$M_W = 80.17 {}^{+0.14}_{-0.15} \text{ GeV}. \quad (12)$$

Oblique Electroweak Correction

To probe the physics beyond the Standard Model, a fit for the S and T parameters has been performed [7]. With LEP data alone, the parameter U is irrelevant. All the LEP data (alone) were used to fit S, T and $\alpha_s(M_Z^2)$ with the constraint on $\alpha_s(M_Z^2)$ measured with event shape analysis at LEP. From the Z lineshape measurement, information on total and partial widths ($\Gamma_Z, \Gamma_{l\bar{l}}$ and Γ_{had}) was used with a correlation between them taken into account. The reference point was taken at $M_{top} = 150$ GeV and $M_{Higgs} = 1$ TeV. A conservative error was assigned for the strong coupling constant $\alpha_s(M_Z^2) = 0.120 \pm 0.010$. This constrained fit yields,

$$\begin{cases} S & = -0.42 \pm 0.45 \\ T & = -0.19 \pm 0.41 \\ \alpha_s(M_Z^2) & = 0.126 \pm 0.007 \end{cases} \quad (13)$$

with $\chi^2 = 2.3/5$ dof. The result is shown in Figure 3, where a technicolor model is also shown. Both S and T are consistent with zero, contrary to the previous result in 1990 when negative S was preferred [7].

CONCLUSION

High precision tests of the electroweak theory have been performed at LEP with typically 0.4% accuracy. No significant deviation from the Minimal Standard Model was observed. Although LEP-I alone has no sensitivity to the Higgs mass, the top mass has been determined to ± 20% accuracy with LEP data.

As for future measurements, the LEP energy uncertainty could be reduced to the order of 3 MeV on the Z mass and total width if one performs the energy scan again and measures the beam polarization at $M_Z \pm 2$ GeV. With more statistics, each experiment is expected to reduce its systematic error. Asymmetry data are still statistically limited. These measurements can be further improved with the forthcoming data in 1992.

I would like to thank colleagues from the LEP electroweak working group who worked for the results presented in this paper. I had largely appreciated discussions and suggestions from my colleagues on ALEPH, Alain BLONDEL, Manel MARTINEZ and Luigi ROLANDI. I also would like to thank Michael E. PESKIN for helpful discussions.

REFERENCES

1. The LEP Collaborations,
   Phys. Lett. B276, pp. 247, (1992).

2. L. Arnaudon et al.,
   'The Energy Calibration of LEP in 1991',
   CERN-PPE/92-125 (1992).

3. ALEPH, DELPHI, L3 and OPAL Collaboration,
   contribution to ICHEP92, Dallas, Aug. 6-12, 1992.

4. M. Martinez et al.,
   Zeit. für Phys. C49, pp. 645, (1991).

5. D. Bardin et al.,
   Zeit. für Phys. C44, pp. 493, (1989),
   Phys. Lett. B229, pp. 405, (1989),
   Comp. Phys. Comm. 59, pp. 303, (1990),
   Nucl. Phys. B351, pp. 1, (1991)
   CERN-TH 6443/92 (1992).

6. S. Bethke, in these Proceedings ICHEP92, Dallas, Aug. 6-12, 1992.

7. M. E. Peskin and T. Takeuchi,
   Phys. Rev. Lett. 65, pp. 964, (1990),
   Phys. Rev. D46, pp. 381, (1992).

PRECISE MEASUREMENT OF THE LUMINOSITY AT LEP

Helge Meinhard
PPE Division
European Organisation for Nuclear Research (CERN)
CH-1211 Genève 23, Switzerland

Abstract

This paper describes the determination of the luminosity as performed by the various LEP collaborations. The total systematic uncertainty of the luminosity determination has been put to the 0.6 % level, 0.3 % of which is due to the uncertainty of the theoretical prediction of the cross section. Further improvements on the experimental errors are foreseen.

INTRODUCTION

Luminosity is defined as the ratio of the number of events N_{ref} observed for some reference reaction, and the cross section σ_{ref} for the reference reaction:

$$\mathcal{L} = \frac{N_{\text{ref}}}{\sigma_{\text{ref}}} \qquad (1)$$

At electron-positron colliders, Bhabha scattering $e^+e^- \to e^+e^-(\gamma)$ at small angles has always been used as a reference reaction. At small values of the angle, its cross section varies steeply with the scattering angle. Hence, the integrated cross section at small angles is large, but a precise knowledge of the detector acceptance is crucial for a precise determination of $\mathcal{L}$. At small angles, the pure QED contributions to the cross section dominate, the most important non-QED term being the interference between the γ exchange in the t channel and the Z exchange in the s channel.

For electroweak measurements at LEP or SLC, cross sections of the Z decaying into fermion pairs are determined as the number of observed events after corrections, divided by $\mathcal{L}$. The precise measurement of $\mathcal{L}$ is hence as important as the determination of the rate of the process being studied. A 1 % error in $\mathcal{L}$, for example, causes an error of 0.08 on the number of light neutrino species.

EXAMPLE: THE ALEPH $\mathcal{L}$ MEASUREMENT [1]

Apparatus

The ALEPH luminosity calorimeter is a sampling calorimeter comprising 38 layers of lead sheets and proportional wires. The total thickness corresponds to 24.6 radiation lengths. The sensitive region extends from about 45 mrad to 190 mrad. There are two modules of 180^0 azimuthal coverage on each side of the interaction point. The anode wires are read out in common for each plane of a 180^0 module. The cathode planes are segmented into pads of about 3 cm × 3 cm which form towers pointing projectively to the interaction point. For the readout, each tower is segmented into three stories.

The position of boundaries between pads with respect to the common centre of a pair of adjacent modules is known to an accuracy of 125 μm. Showers, which are formed by combining adjacent towers with significant energy deposition, show an energy resolution of 3.5 % and a position resolution of 1.2 mm for a 45

GeV electron.

Until the end of 1991, a nine-layer tracking chamber was mounted in front of the calorimeter which has been used for intensive systematic studies. It was removed to provide space for the new silicon-tungsten calorimeter to be incorporated into ALEPH in September 1992.

Event selection

A fiducial area is defined using the boundaries between adjacent pad towers. It excludes edge towers as well as a region at large scattering angles which is shadowed by insensitive support material, and is hence difficult to simulate. The non-fiducial area is limited by circles and straight lines, and fully comprises the fiducial area.

The standard method, which fixes the absolute scale of the luminosity, requires that the highest energetic cluster on side 1 be inside the fiducial area, using the energy sharing of the cluster between towers on both sides of the boundary to obtain a more precise acceptance definition. On side 2, the centroid of the highest energy cluster is required to be within the non-fiducial area. The roles of side 1 and 2 are interchanged with each ALEPH trigger. Furthermore, the energies of the selected clusters in terms of the beam energy must both exceed 0.44; the sum must be larger than 0.6 times the CMS energy. The acoplanarity of the two selected clusters must not exceed 10^0.

In addition, a high-statistics method has been used which is normalized to the standard method such that the sum over the luminosities for each energy point agree. It omits the fiducial cut of the standard method, making the non-fiducial cut on side 2 the only one on the cluster position. The cuts on energies and acoplanarity remain unchanged. This method provides a 50 % higher statistics as compared to the standard method, and hence reduces the luminosity error on the individual energy points.

Cross section

For the calculation of the Bhabha cross section, the Monte Carlo generator BHLUMI [2] has been used. It is a multiphoton generator which includes leading logarithms to all orders of the electromagnetic coupling constant. The theoretical uncertainty of the QED part is claimed to be 0.25 %. However, all Z contributions are implemented at Born level only. BABAMC [3] has been used to determine a correction due to the first order non-tree level contributions of the γ-Z interference, which amounts to (0.32 ± 0.03) %. Effects of higher order photonic corrections have been estimated using a semianalytical program (ALIBABA). The generated events were then run through a full detector simulation based on GEANT, and processed through the event reconstruction and the luminosity analysis like the real data. One million accepted events were obtained, leading to a reference cross section of (26.35 ± 0.03) nb at a reference CMS energy of 91.2 GeV.

In order to extrapolate the cross section to various CMS energies, an approximation formula has been used which takes into account the trivial QED $1/s$ energy scaling as well as the energy dependence of the Z-γ interference, the Z resonance contributions both in forward and backward direction, and radiative effects. The parameters of the formula have been determined using large numbers of BABAMC generated events at energies scanning the whole energy range used for the data analysis.

Systematic errors

The systematic errors of the ALEPH 1991 luminosity measurement are summarized in table 1. The experimental contributions total to 0.44 % which becomes 0.46 % when the

Table 1. Systematic errors of the ALEPH luminosity measurement, in %

Fiducial side cut	0.32
Alignment, mechanics of one module	0.22
Statistics of simulation	0.14
Non-fiducial side cut	0.13
Energy cut	0.08
Acoplanarity cut	0.05
Background subtraction	0.03
Beam parameters, ext. alignment	0.02
(1990: 0.6)	0.44
Normalization of high stat. method	0.11
Total experimental error	0.46
Error on BHLUMI QED part	0.25
Weak contributions (BABAMC)	0.15
Total theoretical error (1990: 0.3)	0.30
Total systematic error on luminosity	0.54

uncertainty of the normalization of the high-statistics selection method is included. This can be compared to the 0.6 % quoted for the 1990 analysis. Adding the 0.3 % error attributed to the theoretical prediction of the cross section, the total error for the 1991 analysis is 0.54 %.

Upgrades

In September 1992, a new silicon-tungsten calorimeter SICAL will be installed in front of the luminosity calorimeter. Its sensitive range extends from 24.4 to 57.8 mrad, thus making smaller scattering angles with larger Bhabha cross sections visible. The total fiducial cross section of LCAL and SICAL is 4.6 times that of LCAL alone.

The modules cover 180⁰ in azimuth. Two of them are mounted on each side of the interaction point. Each one comprises 12 staggered layers of two tungsten plates, a G10 plate carrying the ceramic support for the silicon crystals, and a thin Aluminum support plane. The total material corresponds to 24 radiation lengths. The silicon pads are divided radially into 16 segments with a pitch of 5.3 mm and azimuthally into 32 segments with an opening angle of 11.25⁰. Adjacent layers are rotated in azimuth with respect of each other; hence, there are no dead zones, in contrast to the luminosity calorimeter.

The resolution for 45 GeV electrons, derived from Monte Carlo studies and test beam measurements, is 3.3 % for the energy, 0.2⁰ for the azimuthal angle, and 250 μm for the radius of the shower centroid at the centre of pads, which becomes as small as 15 μm at pad boundaries. The distance between any pad boundaries within a pair of adjacent modules, which is crucial for a precise $\mathcal{L}$ measurement, is known to 25 μm. The overall installation and support, which is rather uncritical, is precise to 200 μm.

The luminosity determination with SICAL will be done following the same ideas as with LCAL. There will be a fiducial cut using energy fractions across pad boundaries and a non-fiducial cut using shower centroids, the sides on which they are applied alternating with each trigger. Energy and acoplanarity cuts will be the same as with LCAL. A first estimate shows that a systematic error of 0.15 % is feasible.

OVERVIEW OF OTHER LEP EXPERIMENTS

L3[4]

The L3 luminosity monitor consists of one BGO calorimeter on each side of the interaction point, covering scattering angles between 25 and 70 mrad. Each calorimeter comprises 304 BGO crystals, yielding resolutions of 1.5 % for the energy, 0.4 mrad for the polar angle, and 0.5⁰ for the azimuthal angle of a 45 GeV electron. Two samples of Bhabha events are defined, and averaged later, by applying a fiducial cut on either side which requires the shower centroid to be at least one crystal away from the calorimeter edge. In addition, asymmetric energy cuts (0.8 and 0.4 E_{beam}) and an acoplanarity cut (10^0) are applied. The systematic experimental error has been quoted to be 0.7 % for the 1990 analysis and will go down for the 1991 results.

The cross section has been determined to (88.5 ± 0.3) nb, using BABAMC and a full detector simulation. For the 1991 results, BHLUMI will be used which allows the reduction of the theoretical error to 0.3 %.

Before the beginning of the 1993 data taking, a silicon strip tracking device will be installed in front of the calorimeters. It contains two layers with radial and one layer with azimuthal readout, and will improve the knowledge of the calorimeter acceptance to 50 μm, thus allowing for an experimental error of the luminosity determination of below 0.3 %.

OPAL [5]

The OPAL luminosity monitor consists of a lead/scintillator sandwich calorimeter which is azimuthally segmented. The readout is performed both at the inner and outer acceptance edge. Its energy resolution is 2.5 % for a 45 GeV electron. After four (of 24 in total) radiation lengths, it is interspersed with a layer of tube chambers, providing polar and azimuthal angles of the shower centroids. Drift chambers, mounted in front of the calorimeter, measure the impact point of the Bhabha electrons and are used to check the tube chamber position.

The mean value of the polar angles of the scattered electrons is required to be within 58 and 110 mrad, the individual angles being restricted by 48 and 120 mrad. The azimuthal angles must be well apart from the vertical and horizontal planes in order to avoid shadowing by insensitive material. The sum of the energies must exceed 2/3 E_{CMS}, the acoplanarity must not exceed 20^0. The experimental error on $\mathcal{L}$, quoted to be 0.7 % for the 1990 results, is not yet known for the 1991 analysis.

The cross section of 21.9 nb has been obtained from BABAMC so far with a quoted error of 0.4 %. From now on, BHLUMI will be used, allowing for a reduction of this error to 0.3 %. Apart from the standard method, a high-statistics method is used which does not put any requirements on the shower positions. Its cross section is roughly twice of that of the standard method.

In spring 1992, additional drift chambers were installed in front of the calorimeter with their sense wires much closer to the inner acceptance. This lead to an improved understanding of the tube chamber information. In spring 1993, it is planned to install a silicon-tungsten calorimeter which is built very much along the same principles as the ALEPH one, except that it has 18 rather than 12 sampling layers, the first 14 ones with a smaller thickness, and the radial pad size has only about half the ALEPH value, hence providing a better resolution of the radial positions of shower centroids. The fiducial cross section is about 70 nb.

DELPHI [6]

DELPHI measures the luminosity with a lead/scintillator sandwich calorimeter comprising 288 towers on each side of the interaction point. In order to define the acceptance accurately, a precision lead mask with projective edges, hiding the innermost towers and those next to the vertical plane, was installed in front of one calorimeter. On the masked side, it is required that the shower centroid is not in the edge towers. On the non-masked side, less strict acceptance cuts are applied. In addition, there are cuts on acoplanarity and energy as well as a protection against low-energy particles hitting the fibres directly, thus faking a high energy deposition. The experimental error on $\mathcal{L}$, given to be 0.8 % for the 1990 analysis, is claimed to be 0.5 % for the 1991 data, the reduction being mainly due to the installation of a silicon strip tracking device in front of the non-masked calorimeter.

The cross section of (27.12 ± 0.04) nb has been obtained using BABAMC and a full detector simulation for the 1990 analysis with an error of 0.5 %. For the 1991 data, BHLUMI has been used, which decreased the theoretical error to 0.3 %.

As with other LEP experiments, there are two pairs of small silicon-tungsten calorimeters installed beyond the mini-beta quadrupoles around DELPHI, each of which comprises 12 silicon plates and 3 silicon strip planes. The visible cross section is about 500 nb. Since an absolute normalization is very difficult, its data are normalized to the luminosity obtained from the standard monitor.

In September 1992, the lead mask will be replaced by a tungsten mask, improving both the mechanical accuracy and the amount of shadowing material. In 1994, it is planned to replace the calorimeter by a lead/scintillator sandwich with two interspersed silicon strip planes. The readout towers will be projective. Due to its smaller radius, the visible cross section more than doubles.

CONCLUSIONS

Compared with the startup of LEP when luminosity errors of 2 to 5 % were quoted, significant progress has been made both in the understanding and handling of the experimental equipment, which has been improved partly, and in the theoretical predictions of the Bhabha cross section at small scattering angles. Due to this progress, the smallest total systematic error of the luminosity quoted by the LEP experiments is now 0.54 %. This also represents a significant progress compared with the previous generation of electron-positron collider experiments at PETRA and PEP, where this figure was eventually about 2.5 to 3 %. From the planned upgrades of the luminosity monitors at LEP, a further significant improvement can be expected in the near future.

REFERENCES

1. ALEPH Collaboration, "Measurement of the absolute luminosity with the ALEPH detector", *Z. Phys.* C53, 375–390 (1992)

2. S. Jadach, E. Richter-Wąs, B.F.L. Ward, Z. Wąs, "Monte Carlo program BHLUMI 2.01 for Bhabha scattering at low angles with Yennie-Frautschi-Suura exponentiation", *Comp. Phys. Comm.* 70, 305–344 (1992)

3. M. Böhm, A. Denner, W. Hollik, "Radiative corrections to Bhabha scattering at high energies (I). Virtual and soft photon corrections", *Nucl. Phys.* B304, 687–711 (1988)
   F.A. Berends, R. Kleiss, W. Hollik, "Radiative corrections to Bhabha scattering at high energies (II). Hard photon corrections and Monte Carlo treatment", *Nucl. Phys.* B304, 712–748 (1988)

4. L3 Collaboration, "Measurement of electroweak parameters from hadronic and leptonic decays of the Z^0", *Z. Phys.* C51, 179–203 (1991)

5. OPAL collaboration, "Measurement of the Z^0 line shape parameters and the electroweak couplings of charged leptons", *Z. Phys.* C52, 175–207 (1991)

6. DELPHI Collaboration, "Determination of Z^0 resonance parameters and couplings from its hadronic and leptonic decays", *Nucl. Phys.* B367, 511–574 (1991)

MONTE CARLO PROGRAM BHLUMI 2.01 FOR BHABHA SCATTERING AT LOW ANGLES WITH YENNIE-FRAUTSCHI-SUURA EXPONENTIATION*

B.F.L. Ward
Department of Physics and Astronomy
University of Tennessee, Knoxville, Tennessee 37996-1200
and
SLAC, Stanford University, Stanford, California 94309
S. Jadach
CERN, Theory Division, Geneva 23, Switzerland,
and
Institute of Nuclear Physics, Krakow, ul. Kawiory 26a, Poland,
E. Richter-Was
Chair of Computer Science, Jagellonian University, Krakow, ul. Reymonta 4, Poland
Z. Was
Institute of Nuclear Physics, Krakow, ul. Kawiory 26a, Poland

ABSTRACT

We discuss the Monte Carlo program BHLUMI 2.01 for small angle Bhabha scattering with an overall precision of .25%. The QED calculation at this precision level is essential for the luminosity measurement at LEP/SLC experiments. We summarize the theoretical development and implementation of our Yennie-Frautschi-Suura $\mathcal{O}(\alpha)$ exponentiated low angle Bhabha scattering event generator and discuss our outlook for the future.

In what follows, we present the current status of the theoretical development and implementation of the Monte Carlo program BHLUMI 2.01 [1-5], which was recently implemented at LEP/SLC experiments for calculating the QED expectations for the luminosity process $e^+e^- \to e^+e^- + n(\gamma)$ at .25% precision. BHLUMI 2.01 is the end product of a series of calculations [1-4] in which we have made new calculations of the various QED radiative corrections to small angle Bhabha scattering at Z^0 energies. For reference, we note that the respective luminosity process measurement is important for all other experimental measurements in Z^0 physics [6,7]. Indeed, for illustration, we recall that the luminosity is a strongly varying function of time, so that the typical measurement procedure in which all data are averaged over long time intervals means that any un-

---

*Work supported in part by the US DOE, contracts DE-AC03-76SF00515 and DE-FG05-91ER40627, by TNRLC grant RCFY9201 in support of the Superconducting Super Collider Laboratory, and by Polish Ministry of Education grants KBN 203809101 and 223729102.

certainty on the luminosity $\mathcal{L}$ will propagate into all measured observables such as cross sections, asymmetries, etc. For some observables, the error on $\mathcal{L}$ enters directly into the error of their measurement: for the error on N_ν, the number of massless neutrino generations as determined from the Z^0 invisible width, we have $\Delta N_\nu \sim 7\Delta\mathcal{L}/\mathcal{L}$. Hence, it is an important theoretical issue to keep the theoretical uncertainty in the expectation for $\mathcal{L}$ significantly below the attendant experimental uncertainty.

In early 1991, this pure experimental error went down below the level of 1% so that the theoretical uncertainty on the QED radiative corrections to Bhabha scattering at small angles became an essential or even the dominant component of the total error on the luminosity measurement. We emphasize that, in the luminosity measurement, the t-channel dominated QED calculation is regarded as known and calculable from the QED Lagrangian density to arbitrary accuracy. In practice, at a certain precision level, effects due to real and virtual photon radiation enter and such effects have to be calculated to an appropriate precision level. In fact, we must be definitive about what we mean when we speak about the precision level of a theoretical calculation. As defined in Ref. 8, we devide the uncertainty on a theoretical calculation into two components: a physical component which refers to contributions from higher perturbative orders and a technical component which refers to biases due to numerical instabilities, programming bugs, calculational approximations, etc. Both of these contributions to the theoretical precision have to be calculated.

We recall that the radiative corrections to small angle Bhabha scattering were calculated for the PETRA/PEP experiments in Ref. 9 using standard perturbative techniques at $\mathcal{O}(\alpha)$ with respect to the Born approximation. The result of this calculation was encoded in the Monte Carlo (MC) event generator OLDBAB, so that implementation of arbitrary detector cuts and experimental efficiencies in the calculation was readily achieved. During the LEP1 Workshop [10], OLDBAB was extended in Ref. 11 to the complete $\mathcal{O}(\alpha)$ MC event generator BABAMC in which the pure weak corrections to low angle Bhabha scattering at $\mathcal{O}(\alpha)$ were also included. The precision tag of OLDBAB (and BABAMC) was around 1% and was definitely insufficient for the 1991 LEP data.

The task of reducing QED uncertainty in the luminosity measurement in papers [1-3] was divided into three corresponding steps:
- calculation of the technical uncertainty of the older $\mathcal{O}(\alpha)$ calculations,
- new calculation of the higher order effects beyond $\mathcal{O}(\alpha)$ keeping strict control on technical precision,
- determination of the total (technical + physical) precision of the multiphoton $\mathcal{O}(\alpha)$ exponentiated Monte Carlo generator BH-

LUMI.

Note that the event generator BHLUMI existed much earlier [12] but its development was not continued because at the time it was not obvious what kind of experimental error in the luminosity measurement would be reached in LEP experiments.[#1]

In the first step of Ref. 1 an effort was made to get control on the technical precision of the existing $\mathcal{O}(\alpha)$ calculation, as implemented in the OLDBAB Monte Carlo program. In this non-trivial task we have used the following method: We have calculated analytically the integrated $\mathcal{O}(\alpha)$ cross section and several basic inclusive cross sections in the presence of semi realistic cuts allowing for hard bremsstrahlung. The most difficult part of the calculation was the precise analytical integration over the hard photon bremsstrahlung phase space. The agreement of the MC with the analytical results at the unprecedented level of 0.03% (for the total cross section) provided solid proof that we control technical precision of the combined (MC and analytical) calculation at this level for a wide range of the parameters defining the experimental trigger. In the process of the above tests we have made a series of modifications and improvements on the OLDBAB program. The new version we have named OLDBIS to avoid confusion with the original code.

This first step of Ref. 1 was immediately profitable because with the new results (tables in Rref. 1) all LEP experiments (or anybody else) could check whether they are using a correct version of the $\mathcal{O}(\alpha)$ QED calculation for small angle Bhabha down to 0.03% precision. In particular, LEP collaborations could check immediately if the BABAMC program used to analyse 1990/91 data was compatible with these new $\mathcal{O}(\alpha)$ results/tests. On the other hand the results of Ref. 1 could be treated as a *necessary* starting point (baseline calculation) for study of the higher order corrections beyond the $\mathcal{O}(\alpha)$. One should keep in mind that it is impossible to talk about higher order corrections without knowing very precisely $\mathcal{O}(\alpha)$ corrections simply because they are defined as a difference between the given higher calculation and the $\mathcal{O}(\alpha)$ calculation! Note that in Ref. 1 we have found a very useful feature of the QED corrections to small angle Bhabha scattering, i.e. we have shown that the QED interference among emission from the upper electron line and lower positron line can be neglected. This phenomenon is very similar to the smallness of the QED interference among the initial and final state bremsstrahlung at the narrow resonance.

In the next step described in Ref. 2, we took aim at the biggest correction beyond $\mathcal{O}(\alpha)$, i.e., $\mathcal{O}(\alpha^2 L^2)$, the second order

---

[#1] The other reason was that we could not test and improve on BHLUMI before the new higher order calculation of ref. [2] with technical precision below 0.1% existed.

leading-logarithmic (LL) correction where $L = \ln(|t|/m_e^2)$. This effect is for typical experimental cuts about 0.5% and it was calculated using two independent MC calculations. The very small difference between two MC results being 0.03% was taken as the technical precision of the result.[#2] The LL calculation was also implemented in the form of the MC event generator LUMLOG such that arbitrary cuts could be imposed. Note that the MC program LUMLOG calculates the difference $\mathcal{O}(\alpha^2 L^2) - \mathcal{O}(\alpha L)$. In the lower part of Fig. 1 we show this difference as a function of the energy cut, see Refs. 2,3 for more details. This difference was used (for given cuts) to upgrade the existing $\mathcal{O}(\alpha)$ calculation (BABAMC) already used in the LEP 1990/91 data at the level of the total cross section (not for differential distributions). The other smaller perturbative subleading contribution $\mathcal{O}(\alpha^2 L)$ was estimated in several different ways and was found to be at the level of 0.2%. The $\mathcal{O}(\alpha^3 L^3)$ correction was directly calculated. The total QED uncertainty was estimated/calculated at the level of 0.3%. Note that in the above calculations we have used the experimental trigger of the calorimetric type which does not distinguish the electron and photon and measures their total energy. This kind of measurement eliminates corrections due to final state QED bremsstrahlung, in the LL approximation.

The OLDBIS+LUMLOG (or BABAMC

[#2] Note that technical precision was in this case calculated and not merely estimated.

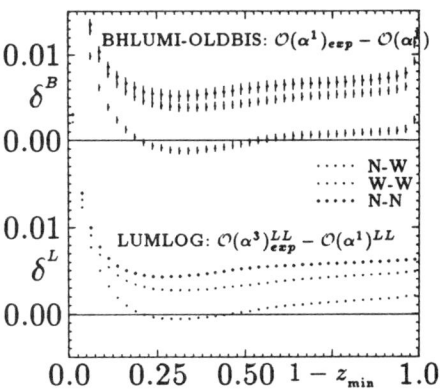

Fig. 1. Second and higher order correction for asymmetric calorimetric trigger Ξ_{NW} in BHLUMI and LUMLOG for varying energy cut z_{min}. Plotted are $\sigma^{B,L} = (\sigma_1^{B,L} - \sigma_2^{B,L})/\sigma_{Born}$, where σ_{Born} is calculated for the Narrow-Narrow angular range. $\sigma_1^B = \sigma\{\mathcal{O}(\alpha^1)exp\}$ is from BHLUMI and unexponentiated $\sigma_2^B = \sigma\{\mathcal{O}(\alpha^1)\}$ is from OLDBIS [8,1]. $\sigma_1^L = \sigma\{\mathcal{O}(\alpha^3)exp\}$ and ($\sigma_2^L = \sigma\{\mathcal{O}(\alpha^1)\}$ are both from LUMLOG [2]. We plot $\delta^{B,L}$ separately. Their difference represents the missing $\mathcal{O}(\sigma^2 L^2)$ bremsstrahlung correction in BHLUMI (together with its technical precision). It is less than 0.15% as quoted in Table 2b of Ref. 2. Vacuum polarization, Z and s channel γ are switched off. MC statistical errors correspond to 2.5 10^7 of weighted events from BHLUMI and 10^8 events from OLDBIS and LUMLOG.

+LUMLOG) solution of Ref. 2 was a temporary solution, not very convenient because it was not implemented in the form of the single MC event generator. It was limited to the total cross section (although for a wide class of cuts) and not really applicable for the differential cross sections. In the third step of Refs. 3,4 we have taken the multiphoton MC generator of Ref. 12 which is based on the Yennie-Frautschi-Suura $\mathcal{O}(\alpha)$ exponentiation and we have proven, using the OLDBIS+LUMLOG calculation, that for semi-realistic experimental cuts close to the typical LEP experimental trigger the total cross section from BHLUMI stays very close, to within 0.15%, from the $\mathcal{O}(\alpha) + \mathcal{O}(\alpha^2 L^2)$ result. We have also improved on many aspects of the BHLUMI program with respect to the original version of Ref. 12 (numerical instabilities due to extreme smallness of the electron mass). The essence of the above test/exercise is well depicted in Fig. 1. The great advantage of BHLUMI, important for the data analysis, is that it simulates multiphoton events in a very realistic way and it is able to calculate reliably many inclusive distributions. This is in fact very helpful in reducing/understanding many subtle apparatus effects.

The new version 2.01 of the BHLUMI program was published in Ref. 4 and is now in use for 1991 LEP data analysis. It also contains as a Monte Carlo sub-generators the programs OLDBIS and LUMLOG such that the patient user is able to reproduce all important tests from Refs. 1,2 and 3 for himself.

Here, we should note that the overall precision of BHLUMI 2.01 was quoted conservatively in Ref. 3 as .25%. One contribution to this uncertainty, that from higher order corrections to the $\gamma - Z^0$ interference effect, we estimated as .03% at $\sqrt{s} = M_{Z^0}$ whereas our more recent $\mathcal{O}(\alpha)$ exponentiated result for this effect is $\sim$.1%, 2 σ higher than the original estimate. Others [6] have recently estimated this effect at $\lesssim$.2%, 5σ higher than the original estimate. We are currently implementing this higher order correction precisely into BHLUMI 2.01 and, in the interim period, we agree with the previous speaker that the overall precision of BHLUMI 2.01 is conservatively .3% if one considers non-QED effects also.

In this short note we concentrate only on the recent work of the present authors on the problem of QED corrections to small angle Bhabha. We refer the reader to Ref. 5 for a complete list of references to recent works in this subject.

REFERENCES

1. S. Jadach, E. Richter-Was, B.F.L. Ward, Z. Was, Phys. Lett. **B253** (1991) 469, (CERN preprint TH-5888).

2. S. Jadach, E. Richter-Was, B.F.L. Ward and Z. Was, Phys. Lett. **B260** (1991) 438 (CERN preprint TH-5995).

3. S. Jadach, E. Richter-Was, B.F.L. Ward and Z. Was, Phys. Lett. **B268** (1991) 253 (CERN preprint TH-6118).

4. S. Jadach, E. Richter-Was, B.F.L. Ward and Z. Was, "Monte Carlo program BHLUMI-2.01 for Bhabha scat-

tering at low angles with Yennie-Frautschi-Suura exponentiation", CERN TH-6230 preprint, Comp. Phys. Commun. (1992) in press.

5. B.F.L. Ward, Proc. of Geneva Lepton Photon Conference, Geneva 1991.

6. H. Meinhard, preceding talk.

7. F. Dydak, in Proc. of Singapore Conference, 1990.

8. Z. Was and S. Jadach. "QED predictions and their systematic errors", Proc. of Amsterdam Conf. on the Monte Carlo Simulation, 1991, CERN preprint TH-6159 (1991).

9. F.A. Berends and R. Kleiss, Nucl. Phys. **B228** (1983) 537.

10. G. Altarelli, R. Kleiss and C. Verzegnassi, Z Phys. at LEP1, CERN-89-08, Vol. 3, 1989.

11. M. Böhm, A.Denner and W. Hollik, Nucl. Phys. **B304** (1988) 687; F.A. Berends, R. Kleiss and W. Hollik, Nucl. Phys. **B304** 712.

12. S. Jadach and B.F.L. Ward, Phys. Rev. **D40** (1989) 3582.

ZF*T*T*ER*

AN ANALYTICAL PROGRAM FOR FERMION-PAIR PRODUCTION

Tord Riemann
DESY – Institut für Hochenergiephysik
Platanenallee 6
Zeuthen, Germany, O–1615

Abstract

I discuss the semi-analytical codes which have been developed for the Z line-shape analysis at LEP I. They are applied for a model-independent and, when using a weak library, a Standard Model interpretation of the data. Some of them are applicable for New Physics searches. The package ZF*T*T*ER* serves as an example, and comparisons of the codes are discussed. The degrees of freedom of the line shape and of asymmetries are made explicit.

INTRODUCTION

Among the most important results from LEP I experiments are the precise values of mass and width of the Z boson, $\delta M_Z/M_Z = 7\,\text{MeV}/91.187\,\text{GeV} = 0.008\%$, $\delta\Gamma_Z/\Gamma_Z = 7\,\text{MeV}/2.492\,\text{GeV} = 0.3\%$ [1]. Although their structure is much richer, the weak loop corrections may be properly described by a single number: the effective weak mixing angle $\sin^2\theta_W^{eff} = 0.2324 \pm 0.0011$. The corresponding on mass shell weak mixing angle is $\sin^2\theta_W \approx 0.226$. Their difference is a measure of the relevance of weak loop corrections.

The consistency of predictions and measurements for the Z line shape, i.e. the total cross section of the reaction

$$e^+e^- \to (\gamma, Z) \to f\bar{f}(n\gamma), \quad (1)$$

heavily relies on the exact treatment of the radiative corrections; for a model-independent analysis, these are the photonic corrections which are potentially large and apparatus dependent. If cross sections are to be interpreted in the Standard Model, a weak library has to be used in addition in order to describe the hard-scattering process.

All the relevant line-shape codes rely on an ansatz of the following type:

$$\sigma_{T,pol}(s) = \int ds'\,\sigma^o_{T,pol}(s')\rho_T(\frac{s'}{s}, \cos\theta_{\max}), \quad (2)$$

$$\sigma_{FB}(s) = \int ds'\,\sigma^o_{FB}(s')\rho_{FB}(\frac{s'}{s}, \cos\theta_{\max}), \quad (3)$$

where the σ^o describe the hard-scattering process and the radiator functions ρ the QED corrections. The latter are result of a three-fold integration and may cover some cuts on the photon phase space and an acceptance cut $\cos\theta_{\max}$. In the simplest case, it is $\rho_T^{hard} = \frac{\alpha}{\pi}(L_e - 1)(1 + s'^2/s^2)/(1 - s'/s)$ [2]. Further, due to the soft photon dominance at LEP I the relation $\rho_T \approx \rho_{FB}$ holds with good accuracy.

At LEP I, the measurable asymmetries are

$$A_{FB} = \frac{\sigma_{FB}}{\sigma_T}, \quad A_{pol} = \frac{\sigma_{pol}}{\sigma_T}. \quad (4)$$

They are important for the determination of additional parameters besides mass and width of the Z in a so-called global fit to (1).

SEMI-ANALYTICAL PROGRAMS

Stimulated by the great success of the LEP I collaborations in their aim to perform measurements with unpreceded accuracy, a variety of semi-analytical programs has been developed and updated by several groups of authors, some of them realizing the model-independent (MI) approach, others using the Standard Model (SM), and some of them both. In an obvious notation, I give here a list which, of course, cannot be exhaustive:

- ALIBABA [3] (σ_T, A_{FB}; SM)
- CALASY [4] (σ_T, A_{FB}; MI, SM)
- BCMS [5] (σ_T; MI, SM)
- CMNPPP [6] (σ_T, A_{FB}; MI, SM)
- MIZA [7] (σ_T, A_{FB}; MI)
- $_{ZF}I^TT_{ER}$ [8, 9, 10, 11] (σ_T, A_{FB}; MI, SM)
- ZSHAPE [12] (σ_T; SM)

Details on the above programs, especially on the treatment of the photonic corrections, may be found in the references.

THE PROGRAM $_{ZF}I^TT_{ER}$

For the line-shape analysis, the DELPHI, L3, and OPAL collaborations use $_{ZF}I^TT_{ER}$, and ALEPH uses MIZA. The $_{ZF}I^TT_{ER}$ is a flexible code with many options. These are described in [8]. In particular, the photonic corrections may be taken into account with three different sets of radiator functions $\rho_{T,FB}$ [9]:

- no cut
- cuts on the fermion scattering angle $\cos\theta$ and on the photon energy (i.e. s')
- cuts on the fermion scattering angle $\cos\theta$, on the acollinearity ξ of the fermions, and on the fermion energy.

They may be combined with several treatments of the hard-scattering process (with different sets of free parameters):

- Standard Model [10, 11] (M_Z, m_t, M_H, α_s)
- Effective Couplings [8] (M_Z, Γ_Z, g_v, g_a)
- Partial Widths [10] ($M_Z, \Gamma_Z, \Gamma_e, \Gamma_f$)
- S-Matrix Approach [13] ($M_Z, \Gamma_Z, R(ZZ), J(\gamma Z)$).

The resulting flexibility, enhanced by some flags in the weak library DIZET, enables the user to simulate closely the experimental set-up. Further, detailed comparisons with other line-shape codes are possible.

EXTENSIONS

Due to the modular structure of $_{ZF}I^TT_{ER}$, extensions for a covering of New Physics are possible with not too much effort. For the example of Z' physics, I refer to ZEFIT [14]. The determination of the parameters $\epsilon_{1,2,3}$, which are related to S, T, U, is possible with ZFEPSLON [15].

COMPARISONS

Comparisons between the different codes are extremely important since some radiative corrections are large (those from photonic bremsstrahlung reaching the order of 30%), while the quantum effects which we are searching for are much smaller (1 - 2.5 per mille). Assuming the validity of the Standard Model, there are different levels of complexity which may be gone through step by step in a comparison. With fixed input (M_Z, m_t, M_H, α_s), one may look at several quantities:

On mass shell mixing angle $\sin^2\theta_W$

Since $\sin^2\theta_W = 1 - M_W^2/M_Z^2$, this mixing angle contains for known M_Z a prediction of the W mass (or of Δr). As an example, we quote from table 11 of [8] the following numbers: $\sin^2 \theta_W = 0.22681, 0.22690, 0.22695$. They are calculated for $M_Z = 91.170$, $m_t = 150$, $M_H = 500$ GeV from three independent programs (Hollik [16], $_{ZF}I^TT_{ER}$, Degrassi et al. [17]). The deviations are of the order of fractions of a per mille (under well-defined, adjusted assumptions).

Total and partial widths Γ_Z, Γ_f

The total width is sum of the partial widths $\Gamma_f \sim \rho_f[g_v^{eff}(f)^2 + g_a^{eff}(f)^2]$. With presently yet sufficient accuracy, a 'universal' effective weak mixing angle may be derived from the effective couplings: $g_v^{eff}/g_a^{eff} = 1 - 4|Q_f|\sin^2\theta_W^{eff}$. For definiteness, this effective weak mixing angle is calculated from Γ_e. Evidently, the on mass shell weak mixing angle and the effective one contain qualitatively different physical information. The relation between them may be described by a weak form factor κ, $\sin^2\theta_W^{eff} = \kappa \sin^2\theta_W$. Again from table 11 of [8]: $\sin^2\theta_W^{eff} = 0.23291, 0.23300, 0.23306$ (from Hollik, $_{ZF}I^TT_{ER}$, Degrassi et al.). These numbers correspond to a mean value of $\kappa \approx 1.0269$.

Improved Born approximation σ^o

The effective (or improved) Born cross sections contain the complete weak corrections. These corrections are covered by four weak form factors ρ, $\kappa_e \approx \kappa_f$, $\kappa_{ef} \approx \kappa_e \kappa_f$. The indicated relations between the κ_i are fulfilled at LEP I energies by their leading contributions (e.g. the leading top quark corrections). In the same approximation, the κ_i of the cross sections agree *at the Z resonance* with those introduced in the Z partial widths. This may be understood intuitively from the relation $\sigma^o(Z,Z) \sim \Gamma_e \Gamma_f/[(s-M_Z^2)^2 + M_Z^2 \Gamma_Z^2]$.

The cross section σ

Finally, the ultimate test of a program concerns the cross-section prediction including the photonic corrections (and perhaps not at the same time the weak loops: in case of programs with a model-independent ansatz).

For a comparison of σ between $_{ZF}I^TT_{ER}$ and ZSHAPE and ALIBABA, I again refer to [8]. The agreement is at the per mille level. The BCMS program and $_{ZF}I^TT_{ER}$ have similar agreement for σ at LEP I energies [18], as do $_{ZF}I^TT_{ER}$ and the CMNPPP program [6] (where in addition σ^o is compared). For $_{ZF}I^TT_{ER}$ and MIZA, a comparative fit has been performed [19]. Some measured cross sections (from preliminary 1991 ALEPH data) have been fitted with the following results from $_{ZF}I^TT_{ER}$, MIZA: $M_Z = 91.19142, 91.19148$ GeV; $\Gamma_Z = 2.5109, 2.5109$ GeV; $\sigma_{had}^o = 41.637, 41.650$ nb; $\Gamma_{had}/\Gamma_l = 20.688, 20.692$; $A_{FB,l}^o = 0.0114, 0.0106$.

Summarizing this section, one may conclude that the different programs agree with each other - within their assumptions on the underlying theory - at a level of one or several per mille. Taking into account the uncertainty of the assumptions, especially on the effects of higher orders, the uncertainty of the predictions will amount to several per mille. They safely allow to determine the Z boson parameters within the errors quoted above. In case of a reduction of the experimental errors by a factor of two, one should re-investigate the potential theoretical errors induced by the lineshape programs.

DEGREES OF FREEDOM

In model-independent fits to the lineshape data, often the following set of variables, or a similar one, is used for a parametrization [20]: $M_Z, \Gamma_Z, \sigma_{had}^{o,peak}, R_{had} = \sigma_{had}^{o,peak}/\sigma_{lept}^{o,peak}, A_{FB}^{lept}$. Here, I will discuss the question of a truly model-independent parametrization. In [5, 13], it has been shown that there exists exactly one, unique cross-section formula which is valid at LEP I with extremely high accuracy (and which can be improved if necessary); I quote it here in the form as derived from an S-matrix ansatz [13]:

$$\sigma^{o,S}(s) = \frac{r_\gamma}{s} + \frac{sR + (s - M_Z^2) J}{(s - M_Z^2)^2 + M_Z^2 \Gamma_Z^2} + \frac{r_0}{M_Z^2} + \cdots \quad (5)$$

Since r_γ is predicted by QED and r_0, being non-resonating and arizing from quantum corrections, is strongly suppressed, the cross section depends on only four real constants: M_Z, Γ_Z, R, J. Here, the R describes the pure Z exchange cross section, and J the γZ interference. One should mention that the definitions of mass and width of the Z boson will slightly deviate from the values which became common in recent years since in (5) a constant width function is introduced [21].

Due to correlations, the knowledge of R, J (or a combined fitting together with M_Z, Γ_Z) is substantial for a precise determination of mass and width of the Z. So, the peak position, whose uncertainty is a measure of the uncertainty of M_Z, depends on J_T/R_T:

$$\delta\sqrt{s_{peak}} = \delta M_Z + (\text{QED corrs.}) + \frac{1}{4}\delta\frac{\Gamma_Z^2}{M_Z}\frac{J_T}{R_T}. \quad (6)$$

Taking the ratio of two cross-section formulas, one can derive a simple, universal formula for asymmetries at LEP I [22]:

$$\left.\begin{array}{c} A_{FB} \\ A_{pol} \end{array}\right\} = A_0 \left[1 + C_{QED}(s)\frac{A_1}{A_0}\left(s - M_Z^2\right)\right]. \quad (7)$$

In leading order, it is

$$A_0 = \frac{R_A}{R_T}, \qquad \frac{A_1}{A_0} = \frac{J_A}{R_A} - \frac{J_T}{R_T}. \quad (8)$$

The smooth function C contains QED corrections, including the radiative tail for energies beyond the resonance. Any asymmetry contains only two unknown real constants R_A, J_A. Here one should have in mind that the M_Z, Γ_Z, R_T, J_T are known from the corresponding line shape. Of course, the discussion applies to the leptonic and hadronic channels separately. This doubles the number of constants to be determined, if no additional model-dependent assumptions are introduced (e.g. the effective, universal weak mixing angle, or a prediction of the J_T from the quark parton model). A first application of the described formalism to L3 data of 1990 and 1991 may be found in [23].

The S-matrix parameters are related to those introduced above; e.g. $A_{FB}^{lept} \approx A_{0,FB}^{lept}$, $\sigma_{had}^{o,peak} \approx r_\gamma/M_Z^2 + R_{T,had}/\Gamma_Z^2$. The exact relations are slightly more involved. Strictly speaking, the so-called model-independent approaches to σ^o with a parametrization in effective couplings or in partial Z widths are (quite efficient) approximations to the unique model-independent cross section [5, 13].

SUMMARY

Basically, there are two languages for a line-shape description:

- Standard Model and extensions

- Model-independent ansatz

The semi-analytical programs realize them both in several ways, and in combination with one or the other approach to the QED corrections. They agree with each other within 1 - 2.5 per mille. The perhaps most flexible (and nevertheless fast) program is $_{ZF}$I$^T T_{ER}$.

The Z line shape allows to determine four physical parameters:

$$M_Z, \Gamma_Z, R_T, J_T.$$

Thus, a serious line-shape fit is based on at least five different beam energies.

Global fits to reaction (1) at LEP I include asymmetries. These depend each on two additional parameters:

$$R_A, J_A.$$

A serious fit needs asymmetry data from at least three different beam energies.

If the LEP collaborations will aim at experimental errors for M_Z and Γ_Z as small as 3 MeV [1], a new round of careful checks and, possibly, improvements of the existing lineshape programs would be recommended.

Acknowledgements

First of all, I would like to thank all my collegues, theorists and experimentalists, who contributed, in one way or the other, to $_{ZF}I^TT_{ER}$. Much of the results covered in this contribution would not have been possible without the long collaboration with Dima Bardin. Thanks also to S. Kirsch and S. Riemann who prepared codes and performed tests, comparisons and fits. Finally, I acknowledge the kind hospitality at the Theory Division of CERN where much work on the Z line shape has been initiated and performed.

REFERENCES

1. L. Rolandi, these proceedings.
2. G. Bonneau, F. Martin, Nucl. Phys. B27 (1971) 381.
3. W. Beenakker, F.A. Berends, S.C. van der Marck, Nucl. Phys. B349 (1991) 323.
4. S. Jadach, Z. Was, in *Z Physics at LEP 1*, CERN 89-08, 1989 (G. Altarelli et al., eds.).
5. A. Borrelli, M. Consoli, L. Maiani, R. Sisto, Nucl. Phys. B333 (1990) 357.
6. M. Cacciari, G. Montagna, O. Nicrosini, G. Passarino, R. Pittau, to appear in Phys. Lett B (1992); F. Piccinini, R. Pittau, INFN Pavia prepr. (1992); and references therein.
7. M. Martinez, L. Garrido, R. Miquel, J. Harton, R. Tanaka, Z. Physik C49 (1991) 645.
8. D. Bardin, M. Bilenky, A. Chizhov, A. Olshevsky, S. Riemann, T. Riemann, M. Sachwitz, A. Sazonov, Yu. Sedykh, I. Sheer, program $_{ZF}I^TT_{ER}$; D. Bardin et al., CERN-TH. 6443/92 (1992).
9. D. Bardin et al., Nucl. Phys. B351 (1991) 1; Phys. Lett. B255 (1991) 290; M. Bilenky, A. Sazonov, Dubna prepr. E2-89-792 (1989).
10. A. Akhundov, D. Bardin, T. Riemann, Nucl. Phys. B276 (1986) 1.
11. D. Bardin, M. Bilenky, G. Mitselmakher, T. Riemann and M. Sachwitz, Z. Physik C44 (1989) 493.
12. F.A. Berends, G.J.H. Burgers, W.L. van Neerven, Nucl. Phys. B297 (1988) 429.
13. A. Leike, T. Riemann, J. Rose, Phys. Lett. B273 (1991) 513.
14. A. Leike, S. Riemann, T. Riemann, CERN-TH. 6545/92 (1992), to appear in Phys. Lett. B.
15. A. Gurtu, unpublished.
16. W. Hollik, Fortran package DELTAR; Fortschr. Physik 38 (1990) 165; prepr. MPI-Ph/92-116(1992).
17. G. Degrassi, S. Fanchiotti, A. Sirlin, Nucl. Phys. B351 (1991) 49.
18. D. Bardin, private information.
19. M. Martinez, private information.
20. The LEP Collaborations, Phys. Lett. B276 (1992) 247.
21. D. Bardin, A. Leike, T. Riemann, M. Sachwitz, Phys. Lett. B206 (1988) 539.
22. T. Riemann, CERN-TH. 6590/92 (1992), to appear in Phys. Lett. B.
23. S. Kirsch, S. Riemann, L3 note # 1233 (1992).

FIRST W DECAYS OBSERVED WITH THE DØ DETECTOR[*]

Bruno Gobbi, for the DØ Collaboration[†]
Department of Physics and Astronomy
Northwestern University
Evanston, Illinois, USA 60208

Abstract

We have observed the first few W decays into electrons at the Tevatron, using the newly commissioned DØ detector. Preliminary results are presented. The number of observed decays as well as the transverse momentum of the electron, the neutrino and their transverse mass distributions, are consistent with expectations.

The DØ project is a second generation detector designed to operate at the Fermilab 2 TeV Tevatron. This project was proposed in 1984 and the first data were recorded this June at the start of the present 1992 Tevatron running period.

---

[*]This work was supported in part by the U.S. Department of Energy.

[†]The D0 collaboration includes: Universidad de los Andes (Colombia), University of Arizona, Brookhaven National Laboratory, Brown University, University of California at Riverside, CBPF (Brazil), CINVESTAV (Mexico), Columbia University, Delhi University (India), Fermilab, Florida State University, University of Hawaii, University of Illinois at Chicago, Indiana University, Iowa State University, Lawrence Berkeley Laboratory, University of Maryland, University of Michigan, Michigan State University, Moscow State University (Russia), New York University, Northeastern University, Northern Illinois University, Northwestern University, University of Notre Dame, Panjab University (India), IHEP-Protvino (Russia), Purdue University, Rice University, University of Rochester, CEN-Saclay (France), SUNY at Stony Brook, Superconducting Supercollider Laboratory, Tata Institute of Fundamental Research (India), University of Texas at Arlington, Texas A & M University.

The detector includes three main components: the Central Tracking system (CD, described in some detail in this paper), the Calorimetry and the Muon system. Figure 1 shows an isometric view of the detector. The Calorimeter is a liquid argon-uranium system with an electromagnetic section followed by two hadronic sections.[1] The information is read out according to a tower geometry. The muon spectrometer includes magnetized

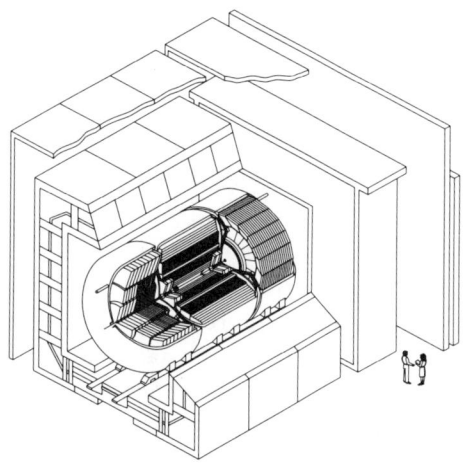

Figure 1. Isometric view of the DØ detector.

toroids and three layers of drift chambers.[2] The total number of electronics channels is 117,000. The main features of the DØ detector are the large solid angle coverage extending down to theta of 2 degrees and the excellent identification of leptons and jets. The detector is compact and does not have a central magnetic field.

All accepted events have a vertex defined by three or more tracks. Electrons are identified first in the electromagnetic section of the calorimeter and then in the transition radiation detector. Also present must be a single minimum ionizing track originating from the interaction vertex and pointing to the electromagnetic shower. Muons are recognized by a track in the muon spectrometer that extrapolates through the central tracking to the vertex. Neutrinos are considered whenever there is missing energy in the calorimeter. Jets are detected in the calorimeter.

THE DØ TRACKING SYSTEM

The DØ Tracking system includes three units of drift chambers.[3] A fourth component, the Transition Radiation Detector (TRD), is also considered to be part of the tracking. Figure 2 shows a cross section of the DØ central tracking. Two of the drift chambers, the Vertex (VTX) and the Central (CDC) are cylindrical units with a jet geometry. They measure tracks in a range of (pseudorapidity) $|\eta| < 1$. The Forward Drift Chambers (FDC), made with Theta- and Phi-modules, extend the coverage to $|\eta| = 3.0$.

The VTX includes three shells made by cylinders of carbon fibers. The inner unit is segmented in 16 cells and the outer two in 32 cells. Cells in the same layer are separated by planes of cathode wires (163 μm gold plated Al). Each cell has 8 radial sense wires (25 μm $NiCoTin$) in the middle, 0.457 cm apart. They are staggered by $\pm 100\,\mu$m.

Pairs of guard wires at ground separate adjacent sense wires. Depending on the VTX-layer, the electrons are drifted over a distance of 1.1 to 1.6 cm. All VTX sense wires are read out at both ends. Charge division is used to measure the coordinate along the beam. The chamber uses a mixture of 95.5% CO_2 and C_2H_6 at atmospheric pressure. The drift velocity is 7.4 μm/ns at the operating voltage of 1.0 kV/cm.

The CDC includes four cylindrical layers of drift chambers.[4] Each chamber is built with cells made with panels of Rohacell foam, wrapped in Kevlar and covered with Kapton. Field shaping electrodes are located on the walls of these cells. Each center plane of a cell has seven sense wires (30 μm gold-plated W) separated by 0.6 cm and staggered by 200 μm. Pairs of guard wires (163 μm gold plated Al) separate adjacent sense wires. The maximum drift distance is 7 cm. Each cell also includes two delay lines (drift velocity 0.24 cm/ns) which measure the track position along the beam direction. Cells from adjacent layers are staggered by a half-cell. The sense wire information is read out at one end of the wire only.

The FDC units, figure 3, include two types of modules. The PHI units, of cylindrical shape, have radial wires. There are 36 cells in each PHI unit. Cells are separated by a divider and have 16 sense wires (30 μm gold plate W), 0.8 cm apart and staggered by 200 μm. One guard wire (163 μm goldplated Al) separates two adjacent sense wires. The THETA units are made with four quadrants of 6 cells each accomodating 8 sense wires, 0.8 cm apart and staggered by 200μm. Two guard wires separate sequential sense wires. Each cell has a rectangular cross section and is made with light materials. The three cells near the beam are half cells. Each cell contains one delay line. CDC and FDC are operated at atmospheric pressure with $Ar/CO_2/CH_4$ (93/3/4%), and

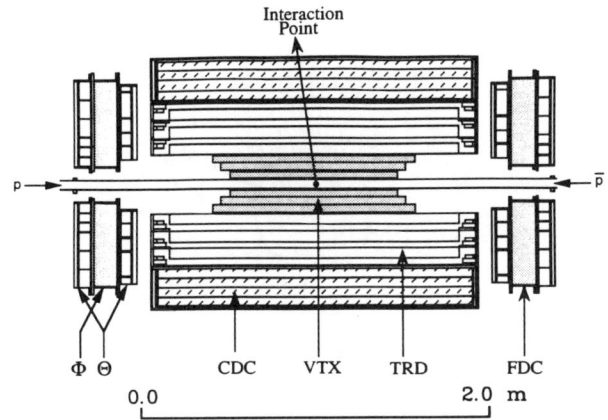

Figure 2. The DØ central tracking system.

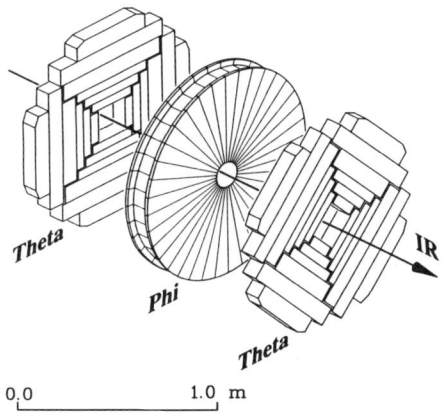

Figure 3. Isometric view of one FDC.

the drift velocity is 35 µm/ns.

The transition radiation detector (TRD)[5] is located between the VTX and the CDC. It consists of three cylindrical modules and its function is to identify electrons in a range of $|\eta| < 1$. The radiator, at small radius, is made of 382 foils of polypropylene, 18 µm thick and 150 µm apart. Outside the radiator there is a shell filled with a mixture of $Xe/CH_4/C_2H_6$ (91/7/2%). A grid of cathode wires at 0 V separates this gas volume into a drift region and, at larger radius, a cylindrical PWC. The 256 wires of the PWC run parallel to the beam direction and measure, in addition to the dE/dx of the tracks, the ionization of the electrons photoproduced by the transition radiation in the Xe of the drift region.

Read-Out Electronics of the CD

All four components of the central tracking are read out by the same electronics. The charge collected by a wire is first amplified then shaped and amplified again. The analog signals are then digitized by a 106 MHz wave form digitizer FADC and zero suppressed. The information of the 5,700 channels is processed in parallel in VME crates. The digitization of the FADC starts synchronously with each beam crossing. The resultant data is stored in a buffer and then either goes through the suppression cycle or is dumped depending on the decision of the level-1 trigger frame. The retained events are transfered to the level-2 microvax farm. Two pieces of information are extracted from the FADCs' data: the position of the hits and the ionization of the tracks. The hit finding algorithm makes use of the pulse shape to determine the track position. The area under each pulse is a measure of the dE/dx. This is used by the TRDs to identify e and by the drift chambers to identify γ that convert into e^+e^-.

Test Beam Results of the CD

All the central detector chambers have been tested in test beams. The position resolution measured with the VTX, in the R-Φ plane, for a single track, is between 40 and 60 µm for drift distances larger than 0.2 mm. This resolution has been determined using the triple time difference technique.[6] The position resolution along the wire obtained from the charge division is 1.5 cm. Both CDC and FDC measure the position of a single track along the drift distance with a sigma of 180 µm. This result is obtained by fitting straight lines

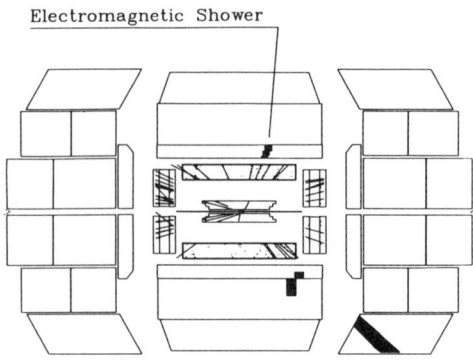

Figure 4. R-z view of a $W \to e$ decay.

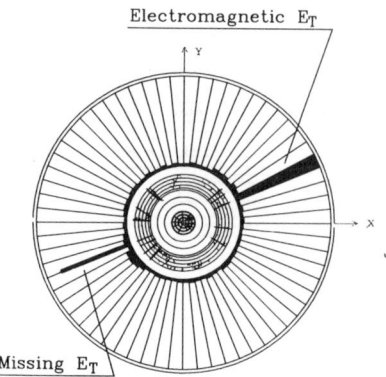

Figure 5. Same event seen along the beam.

through the chamber. This precision remains the same within 10% over the entire drift distance of 7 cm. A single delay line measures the z coordinate in the CDC with a sigma of 3 mm.[7] The two track separation, measured using e^+e^- from converted gammas (1.5 m upstream), is 0.5 cm with 90 % efficiency for identifying the tracks.[8] (The same quantity calculated by displacing individual tracks relative to each other is 0.2 cm). The dE/dx resolution is 13 % when retaining 75 % of the hits in the CDC or FDC. This gives a rejection factor of 50 for 2 mip (e^+e^-) for a 90 % probability of identifying a one mip particle.[9] The efficiency for identifying tracks with the resolution quoted is above 98 %. The test results of the TRD predict that the three modules will have a pion rejection factor of 50 for a 90 % probability of identifying the electron.

DØ TRIGGERING SYSTEM

The triggering system of the DØ detector requires first the coincidence between particles produced at small angle and on both sides of the interaction (level-0). A hardware trigger processor selects events based on the information in the calorimeter as well as in the muon spectrometer (level-1). An additional selection is achieved by analysis done on a farm of microprocessors (level-2). Data can be collected simultaneously for up to 30 different combinations (triggers) of requirements of level-1 and level-2. The events satisfying the required 'trigger' criterias are then stored for further analysis.[10]

The present 'trigger' for $W \to e\nu$ demands first a minimum bias interaction (level-0). Second, it requires from the level-1 an $E_T > 15\,\text{GeV}$ in the electromagnetic section of the calorimeter. Events are anlysed in the level-2 and they are retained whenever clusters of electromagnetic energy with $E_T > 25\,\text{GeV}$ are identified. A test on the shape of the shower is also performed. The offline analysis, in addition to fully reconstructing clusters of electromagnetic energy, compares the longitudinal shape of the shower to that measured in a test beam with an identical calorimeter module. A one mip track must also connect the event's vertex with the centroid of the electromagnetic cluster.

FIRST $W \to e\nu$ DECAYS

In the 1992 Tevatron running period, prior to this conference, about $100\,\text{nb}^{-1}$ have been delivered. Half of this luminosity has been used to debug and calibrate the detector. A fraction of the remaining luminosity has been dedicated to the study of Ws decays.

The present preliminary analysis of this

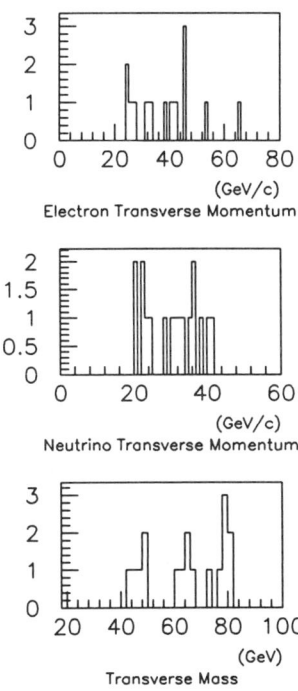

Figure 6. Results.

data has identified 17 candidates of W decaying into $e\nu$ with the p_T of the electron and of the neutrino larger than 20 GeV/c. An analysis of IVBs decaying into muon has produced 10 candidates. Two Z decays into e^+e^- have also been identified. The number of events observed is consistent with our expectations. Figure 4 displays the information recorded in the calorimeter and part of the drift chambers for a W candidate decaying into $e\nu$. The tracks associated with the interaction vertex are visible. Cells of the calorimeter with more than 1 GeV are also shown for the entire azimuthal angle. A cluster of energy deposited in the electromagnetic section of the central calorimeter is clearly shown. In figure 5 the same candidate is seen along the beams' direction. The detector has measured 39.5 GeV of electromagnetic energy in one direction and an almost equally large amount of missing energy in the exact opposite direction.

The transverse momentum of the electron and neutrino are plotted in figure 6. The two electrons with a $p_T > 40\,\mathrm{GeV}$ correspond to events where the electron is produced together with a jet. Figure 5 shows the transverse mass distribution of the 17 candidates. The distributions of these three plots are consistent with what is expected for the decay of a W with a mass of about 80 GeV into $e\nu$.

The goal of DØ towards the study of the IVBs for the 1992 Tevatron running period is to measure the mass of the W with a precision of 160 MeV. This will be achievable with the expected luminosity of 25 pb^{-1}. This measurement together with the prediction of the Standard Model will set new limits on the mass of the top quark.

REFERENCES

1. Harry Weerts, *these Proceedings*

2. David Hedin, *these Proceedings*

3. A.R. Clark et al., "The Central Tracking Detectors for DØ" in *Nucl. Instr. and Methods* A279, pp. 243-248, (1989)

4. T. Behnke, Ph.D. thesis, SUNY, Stony Brook, LI NY 11794-3800, 1989

5. J.F. Detoeuf et al., "The DØ Transition Radiation Detector" in *Nucl. Instr. and Methods* A265, pp. 157-166, (1988)

6. A.R. Clark et al., "DØ Vertex Drift Chamber Construction and Test Results" in *Nucl. Instr. and Methods in Physics Research* A315, pp. 193-196, 1992

7. D. Pizzuto, Ph.D. thesis, SUNY, Stony Brook, LI NY 11794-3800, 1991

8. J. Bantly, Ph.D. thesis, Northwestern University, Evanston Il. 60208 USA, 1991

9. S. Rajagopalan, Ph. D. thesis, Northwestern Univ., Evanston Il. 60208 USA, 1991

10. Bruce Gibbard, *these Proceedings*

FIRST MEASUREMENT OF THE LEFT-RIGHT CROSS SECTION ASYMMETRY IN Z BOSON PRODUCTION AT $E_{CM} = 91.55$ GEV

The SLD Collaboration[°]
presented by P.C. Rowson
Columbia University, Department of Physics, New York, NY 10027
Stanford Linear Accelerator, Stanford University, Stanford, CA 94309

Abstract

The left-right cross section asymmetry for Z boson production in e^+e^- annihilation (A_{LR}) has been measured at $E_{cm} = 91.55$ GeV with the SLD detector at the SLAC Linear Collider (SLC) using a longitudinally polarized electron beam. The electron polarization was continually monitored with a Compton scattering polarimeter, and was typically 22 %. We have accumulated a sample of $\sim 10,200$ Z events. We find that $A_{LR} = 0.100 \pm 0.044 \pm 0.003$ where the first error is statistical and the second is systematic. From this measurement, we determine the weak mixing angle defined at the Z boson pole to be $\sin^2 \theta_W^{\text{lept}} = 0.2378 \pm 0.0056$.

INTRODUCTION

This paper presents the result of a measurement of the left-right cross section asymmetry (A_{LR}) in the production of Z bosons by e^+e^- collisions performed by the SLD Collaboration at the SLAC Linear Collider.[1]

The left-right asymmetry is defined as follows,

$$A_{LR} \equiv \frac{\sigma(e^+e^-_L \to Z) - \sigma(e^+e^-_R \to Z)}{\sigma(e^+e^-_L \to Z) + \sigma(e^+e^-_R \to Z)}, \quad (1)$$

where $\sigma(e^+e^-_L \to Z)$ and $\sigma(e^+e^-_R \to Z)$ are the production cross sections for Z bosons with left-handed and right-handed electrons, respectively. Within the context of the Standard Model, this quantity is a sensitive function of the electroweak mixing parameter $\sin^2 \theta_W^{\text{lept}}$,[2]

$$A_{LR} = \frac{2v_e a_e}{v_e^2 + a_e^2} = \frac{2\left[1 - 4\sin^2\theta_W^{\text{lept}}\right]}{1 + \left[1 - 4\sin^2\theta_W^{\text{lept}}\right]^2}, \quad (2)$$

where v_e and a_e are the vector and axial vector coupling constants of the Z boson to the electron current. Note that A_{LR} is sensitive to the initial state couplings and is insensitive to the final state couplings.[3] The left-right asymmetry has the following properties:[3] it is sensitive to virtual electroweak corrections; it is insensitive to real radiative corrections; it is a weak function of center-of-mass energy, E_{cm}, near the Z pole; and it is expected to be relatively large, in the range 0.10-0.15.

We measured A_{LR} by counting all hadronic decay modes of the Z boson (the sample also contains $\tau^+\tau^-$ final states) for each of the two longitudinal polarization states of the electron beam. The measurement does not require an absolute luminosity measurement or any knowledge of the absolute detector acceptance and efficiency.[4]

THE POLARIZED ELECTRON BEAM AT THE SLC

The SLAC Linear Collider (SLC) was designed to produce, accelerate, and collide a spin-polarized electron beam.[5] A diagram of the SLC is shown in Figure 1. The polarized electron source consists of a GaAs photocathode that is illuminated by a circularly polarized laser beam.[6] The emitted electrons

---

[*]Work supported by Department of Energy contract DE-AC03-76SF00515.
[°]List of authors follows the list of references.

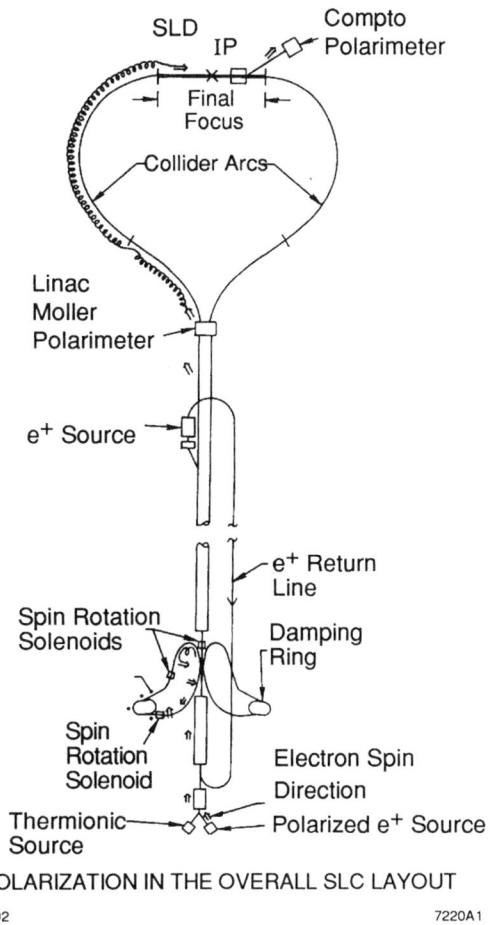

Figure. 1. The SLC. The electron spin direction is indicated by the double-arrow.

are longitudinally polarized, and the electron helicity is changed randomly on a cycle-by-cycle basis (the SLC operates at 120 Hz). The polarization of the emitted electrons is typically 28 %.

A system composed of the dipole magnets in the transfer line from the linac to the damping rings, and a superconducting solenoid magnet, is used to rotate the longitudinal polarization of the beam into the vertical direction to preserve polarization during storage in the damping ring. A system composed of two superconducting solenoids and the dipole magnets in the return line to the linac is used to reorient the polarization vector upon extraction from the damping ring. This system has the ability to provide nearly all polarization orientations in the linac. A fractional polarization loss of 5 % occurs in the damping ring.

The electron pulse is transported through the North Arc and Final Focus systems of the SLC to the interaction point (IP) of the machine. Polarization loss in the arcs due to energy dispersion is expected to be 5–10 fractionally, while the net spin rotation due to the arc system is sensitive to the parameters of the orbit and is measured empirically. The spin rotation system is adjusted to maximize the longitudinal polarization at the IP.

After passing through the interaction point, the longitudinal component of the electron beam polarization is measured with a Compton polarimeter. The Compton polarimeter, which will be described in the next section, measured a typical IP polarization of 22 %.

The electron and positron beams are then transported to the south and north beam dumps, respectively, where precision energy spectrometers[7] are located upstream of the beam dumps and monitor the beam energies continually. The mean electron and positron energies were measured to be 45.71 GeV and 45.84 GeV, respectively. The mean center-of-mass energy was $E_{cm} = 91.55 \pm 0.04$ GeV.

THE POLARIZATION MEASUREMENT

The Compton polarimeter continually monitors the longitudinal polarization of the electron beam after it has passed through the IP and before it is deflected by dipole magnets. Polarimeter data are acquired continually for intervals of 20,000 SLC cycles (∼3 min) and are logged in summary form onto SLD data tapes. A diagram of the polarimeter is shown in Figure 2. The electron beam collides with a circularly polarized photon beam which is produced by a frequency-doubled Nd:YAG laser of wavelength 532 nm. The scattered and unscattered beams remain unseparated until they pass through a pair of dipole magnets.

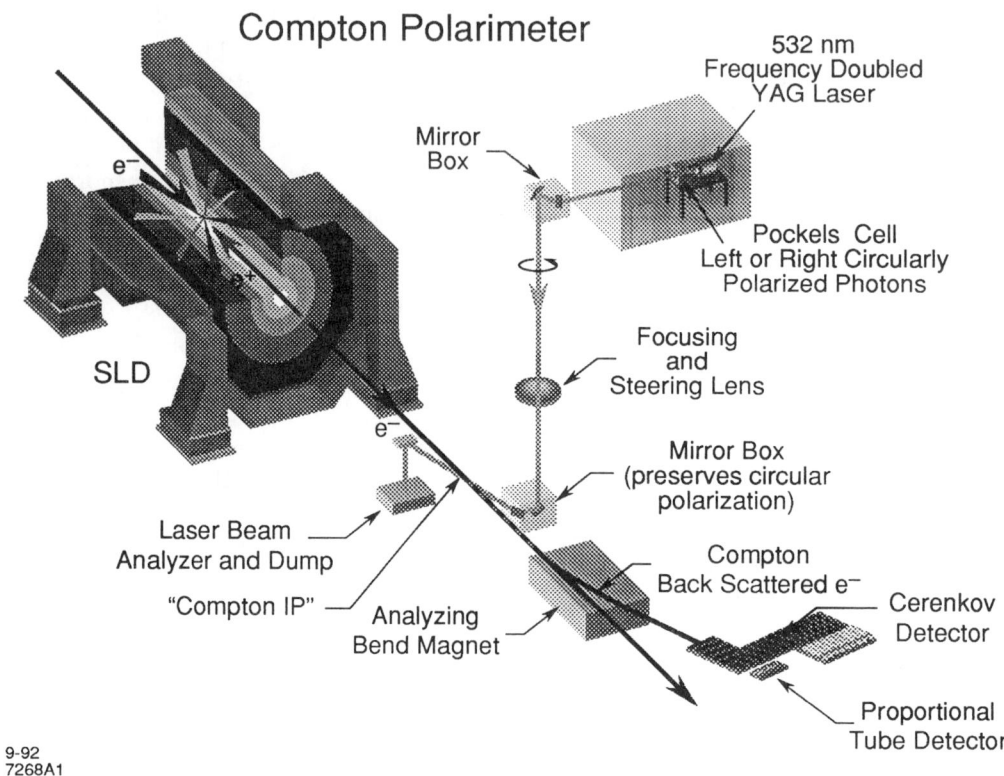

Figure 2. A schematic diagram of the Compton Polarimeter.

The scattered electrons are dispersed horizontally and exit the vacuum system through a thin window. Electrons in the energy interval 15-30 GeV are detected and momentum analyzed by a pair of redundant multichannel detectors (a Cherenkov detector and a proportional tube detector). We measure the counting rates in the detectors for anti-parallel and parallel photon/electron beam helicities; given the laser polarization the asymmetry formed from these rates determines the electron beam polarization.[8] The circular polarization of the laser beam at the Compton IP was measured to be 93±2%. The absolute helicity of the laser polarization was determined from comparison with a calibrated quarter-wave plate.[9] In order to avoid systematic effects, the sign of the circular polarization is changed randomly on sequential laser pulses.

The channel-by-channel polarization asymmetry as measured by the Cherenkov detector is shown in Figure 3. The solid histogram represents the best fit of the data to a convolution of the theoretical asymmetry and a simulated response function of the spectrometer. The errors reflect the systematic uncertainties in the transverse position of the detector and the momentum scale of the spectrometer, which are determined from measurements of the minimum electron energy point and the zero-asymmetry point.

Including effects due to the Compton polarimeter spectrometer and laser systems, we estimate the total relative systematic error on the polarization ($\frac{\delta P}{P}$) to be 3%.

We have performed a number of checks of the polarization measurement. The polarimeter measures the electron scattering rate for two helicity states of electrons and two helicity states of photons. From these rates we form

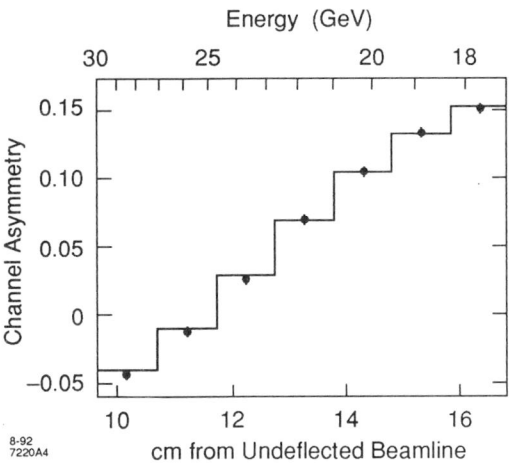

Figure 3. The polarization asymmetry measured by seven channels of the Cerenkov detector. The solid histogram represents the best fit of the data to a convolution of the theoretical asymmetry and a simulated response function of the spectrometer.

two non-zero asymmetries and two null asymmetries, and we verify that the nonzero asymmetries are consistent and that the null asymmetries are consistent with zero.

An additional systematic error would arise if the average beam polarization at the Compton interaction point differed from the luminosity-weighted average beam polarization at the SLC interaction point (the true polarization). We have investigated a number of possible effects, none of which exceeds the level of a relative 0.1%. For example, the SLC collision point and the electron-photon collision point are separated by a lattice of quadrupole magnets. The beam divergence is different at the two points and the beam direction could be different at the two points. We estimate that these effects cause the measured and true polarizations to differ by less than a relative 0.07%. For details on other effects, see references.[10,1]

THE SLD DETECTOR AND EVENT SELECTION

The polarized e^+e^- collisions are measured by the SLD detector which has been described elsewhere.[12] This analysis makes use of the liquid argon (LAC)[13] and warm iron (WIC)[14] calorimeter systems to measure the energies of final state particles, the central drift chamber (CDC) to reconstruct the trajectories of charged particles in a solenoidal magnetic field of 0.6 T, and the luminosity monitoring system (LUM)[15] to measure the rate of small-angle Bhabha scattering events.

The sample of Z decays used here was selected via a calorimetric analysis based largely upon the LAC. The calorimetric analysis must distinguish Z events from several backgrounds that are unique to the operation of a linear collider and differ from those encountered at e^+e^- storage rings. The backgrounds fall into two major categories: those due to low energy electrons and photons that scatter from various beamline elements and apertures, and those due to high energy muons that traverse the detector parallel to the beam axis (due to the low average current in the SLC, backgrounds caused by beam collisions with residual gas in the beamline are negligible). We make use of the fine segmentation and tower geometry of the LAC to suppress both backgrounds. All electromagnetic and hadronic LAC towers used in the analysis are required to satisfy a combination of tower threshold cuts and criteria that select against radially isolated energy deposition in a combined electromagnetic-hadronic tower. All events are required to satisfy a set of global event cuts based on total visible energy and energy balance.

Our Z events are associated with polarization measurements by proximity in time, where we require that all acceptable events must have been recorded by the SLD detector within 1 hour of a polarization measurement.

As we describe in the next section, a control sample is provided by small-angle Bhabha scattering events selected using the LUM system. The accepted Bhabha scattering cross section is approximately twice the total cross section for hadronic Z final states.

The sign of the electron beam helicity is supplied to the SLD data acquisition system via two redundant data paths. The correct synchronization of the helicity signals with triggered and logged events is verified by the following procedure: The positron beam is turned off. The electron source is modified to deliver beam for only one of the two electron helicity states. Data are logged with a low threshold LAC trigger or a random trigger. An offline analysis is then used to verify that radiation is observed in the various detector subsystems only for events of the expected helicity. This test has been performed on seven occasions to date. During the tests, the rate of improperly synchronized pulses was less than 0.05% at 95% confidence.

We estimate that the combined efficiency of the trigger and the Z selection criteria is about 92% for hadronic Z decays. Comparing this selection procedure with one that is based upon tracking information[16] and by applying our selection procedure to Monte Carlo events, we estimate that the residual beam-related background in the Z sample is less than 1%. The contribution of two-photon processes to the Z sample has been estimated by a Monte Carlo simulation to be less than 0.1%. Another component of our sample, tau lepton pairs, constitute an estimated $1.5 \pm 0.5\%$ of the sample. Since tau pair events manifest the correct value of A_{LR}, we do not remove them from the sample. Final state e^+e^- events are removed since the presence of the t-channel photon exchange subprocess dilutes the value of A_{LR}. We apply an e^+e^- identification procedure which searches for large and highly localized energy deposition in the electromagnetic section of the LAC. The residual e^+e^- background in the hadronic Z sample is about 0.5%.

A total of 10,224 Z events and 25,615 small-angle Bhabha events satisfy the selection criteria. We find that 5,226 of the Z events and 12,832 of the small-angle Bhabha events were produced with left-handed electron beam and 4,998 of the Z events and 12,783 of the Bhabha events were produced with right-handed beam.[17]

DETERMINATION OF A_{LR}

The left-right asymmetry is defined in equation (1) in terms of the cross sections for completely polarized electron beams colliding with an unpolarized positron beam. For the case where luminosity, event detection efficiency, electron polarization and backgrounds are helicity-independent (and we will justify these assumptions), the following simple expression holds :

$$A_{LR} = \frac{A_{meas}}{\mathcal{P}} = \frac{1}{\mathcal{P}} \cdot \left(\frac{N_L - N_R}{N_L + N_R} \right), \quad (3)$$

where A_{meas} is the observed asymmetry, $\mathcal{P}$ is the luminosity-weighted, average polarization and N_L and N_R are the total event counts produced by left- and right-handed electron beam respectively.

The helicity dependence of event acceptance is negligibly small.[4] The Compton polarimeter measures the difference between left- and right-handed beam polarization to be less than 5×10^{-3}. There is considerable reason to believe that the left-right SLC luminosity asymmetry is also quite small. The use of a Pockels cell to reverse the source laser helicity provides a very powerful constraint upon possible differences in the left-handed and right-handed electron beams produced by the photoemission gun. In addition, since the couplings of the electron spins to fields of the beam transport system are much weaker than the corresponding couplings of the electron charges, we expect that the beams remain nearly identical as they are damped, accelerated, and transported to the interaction point. Finally, the use of random sign reversal of the electron beam helicity insures that there are no correlations between the beam helicity and periodic variations in the SLC luminosity.

In order to investigate a possible left-right luminosity difference, we have compared a number of electron beam parameters by helicity. We verify that the beam currents, energies, and energy spreads are independent of the beam helicity. The beam position and direction at the end of the linac are also independent of helicity. We verify that the flux of beamstrahlung photons produced by beam-beam interactions is independent of the beam helicity.[18] From these checks we conservatively estimate that the left-right SLC luminosity asymmetry is less than 10^{-3}. We have also checked that the numbers of left-handed and right-handed pulses logged by the SLD data acquisition system are equal to within statistical errors.

Finally, we note that the left-right asymmetry of the small-angle Bhabha scattering cross section is expected to be very small ($\sim 10^{-4}$). Therefore, the numbers of small-angle Bhabha events produced from left-handed and right-handed beams, N_{lumL} and N_{lumR} respectively, measure the relative left-handed and right-handed luminosities. The relative luminosities may be expressed in terms of the luminosity asymmetry, $A_{lum} \equiv (N_{lumL} - N_{lumR})/(N_{lumL} + N_{lumR})$, which we currently measure to be $A_{lum} = 0.002 \pm 0.006$.

The average polarization can be estimated from measurements of the beam polarization that are made when valid Z events are recorded,

$$\mathcal{P} \equiv \frac{1}{N_Z} \sum_{i=1}^{N_Z} \mathcal{P}_i, \qquad (4)$$

where N_Z is the total number of Z events, and $\mathcal{P}_i$ is the polarization that was measured when the ith event was logged. We evaluate the luminosity-weighted polarization to be

$$\mathcal{P} = 22.4 \pm 0.7\%,$$

where the error is dominated by the systematic uncertainty on the polarimeter measure-

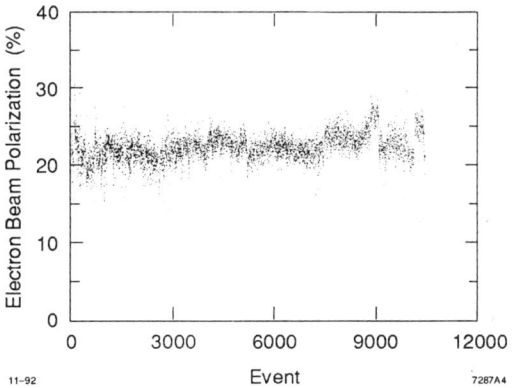

Figure 4. The electron beam polarization as sampled when each Z event was logged.

ments. Using equation (3), we find the left-right asymmetry to be

$$A_{LR} = 0.100 \pm 0.044(\text{stat.}) \pm 0.003(\text{syst.}).$$

The systematic error is dominated by the error on the polarization determination.

CONCLUSIONS AND PLANS

We report a measurement of the left-right asymmetry in the Z boson production cross section at $E_{cm} = 91.55$ GeV. Using a sample of 10,224 hadronic events, we find the left-right asymmetry to be $0.100 \pm 0.044 \pm 0.003$. We calculate the electroweak mixing parameter to be

$$\sin^2 \theta_W^{\text{lept}} = 0.2378 \pm 0.0056(\text{stat.} + \text{sys.}),$$

where we have corrected the result to account for the deviation of the SLC center-of-mass energy from the Z-pole energy and for initial state radiation (the correction from the result given by equation (2) is $+0.0003$)[19]. Our result is consistent with recent results from the LEP experiments.[20]

In the future, we expect to make a number of improvements. We anticipate significant improvements in SLC luminosity, leading to an accumulated data sample of more than 50,000 events during the upcoming 1993 run. We expect to increase the electron polarization to a value above 40%. It may be possible to make use of the recent development of strained lattice cathodes[21] to achieve source polarizations in excess of 80%.

REFERENCES

1. The result given here supersedes the result presented at Dallas on 8-7-92. The new number is the result of improved statistics and the correction of an overall sign error in the Compton polarimeter analysis.
2. We define the parameter $\sin^2\theta_W^{\text{lept}}$ in terms of the ratio of vector and axial vector coupling strengths of the Z to the electron, $\sin^2\theta_W^{\text{lept}} \equiv (1 - v_e/a_e)/4$.
3. Many of the important properties of A_{LR} are discussed in D.C. Kennedy, B.W. Lynn, C.J.-C. Im, and R.G. Stuart *Nuc. Phys.* **B321**, 83 (1989).
4. The product of detector acceptance and efficiency can vary according to decay mode provided the acceptance-efficiency product for each final state is an even function of $\cos\theta$, where θ is the polar angle of the final state fermion with respect to the electron beam direction. This condition is required to cancel terms in the differential cross section that are odd in $\cos\theta$. An even acceptance-efficiency product follows automatically if the efficiency for detecting a fermion at some polar angle θ is equal to the efficiency for detecting an antifermion at the same polar angle. Detectors with axially symmetric solenoidal magnetic fields (*e.g.* the SLD) have this property.
5. D. Blockus *et al.*, Proposal for Polarization at the SLC, April 1986.
6. The SLC polarized electron gun is described in D. Schultz *et al.*, SLAC-PUB-5768, 1992, and the laser system is described in J. Frisch *et al.*, SLAC-PUB-5965, 1992.
7. J. Kent *et al.*, SLAC-PUB-4922, March 1989.
8. See S.B. Gunst and L.A. Page, *Phys. Rev.* **92**, 970 (1953) or H.A. Olsen, *Applications of QED*, Springer Tracts in Modern Physics, Vol. 44, p.83, (1968).
9. The orientations of the fast and slow axes of the quarter-wave plate were checked by comparing the phase difference induced by the plate with the calculable phase shift induced by total internal reflection in a glass prism.
10. See Reference 7 and G.S. Abrams *et al.*, *Phys. Rev. Lett.*, **63**, 2173 (1989).
11. K. Yokoya and P. Chen, SLAC-PUB-4692, September 1988, and P. Chen private communication.
12. The SLD Design Report, SLAC Report 273, 1984.
13. D. Axen *et al.*, The Lead-Liquid Argon Sampling Calorimeter of the SLD Detector, SLAC-PUB-5354, 1992 (to be published in Nuclear Instruments and Methods).
14. A.C. Benvenuti *et al.*, Invited talk at the 2nd International Conference on Calorimetry in High Energy Physics, Capri, Italy, October 1991, SLAC-PUB-5713.
15. S.C. Berridge *et al.*, Proceedings of the 1991 Nuclear Science Symposium, Santa Fe, NM, vol. 1, page 495; SLAC-PUB-5694 (to be published in IEEE Trans. Nucl. Sci.).
16. D. Muller in these proceedings.
17. We use the theoretical sign of the Compton scattering asymmetry and the measured helicity of the Compton laser to infer the correct sign of the beam helicity.
18. The flux of beamstrahlung photons produced at the IP is related to many of the parameters that determine the luminosity but is not a direct measure of the luminosity.
19. Our own calculation agrees with the results given by the EXPOSTAR program described in D.C. Kennedy *et al.*, *Z. fur Phys.*, **C53**, p.617, 1992, and a modified version of the ZSHAPE program described in CERN 89-08, vol. 3, p. 50, 1989.
20. L. Rolandi in these proceedings.
21. T. Maruyama, *et al.*, *Phys. Rev. B*, **46**, 4261 (1992).

LIST OF AUTHORS

K. Abe,[20] I. Abt,[28] P. D. Acton,[3] G. Agnew,[3] C. Alber,[26] D. F. Alzofon,[19] P. Antilogus,[19] C. Arroyo,[5] W. W. Ash,[19] V. Ashford,[19] A. Astbury,[32] D. Aston,[19] Y. Au,[5] D. A. Axen,[24] N. Bacchetta,[9] K. G. Baird,[17] W. Baker,[19] C. Baltay,[36] H. R. Band,[34] G. J. Baranko,[26] O. Bardon,[15] F. Barrera,[19] R. Battiston,[10] D. A. Bauer,[22] A. O. Bazarko,[5] A. Bean,[22] G. Beer,[32] R. J. Belcinski,[29] R. A. Bell,[19] R. Ben-David,[36] A. C. Benvenuti,[7] R. Berger,[19] S. C. Berridge,[31] S. Bethke,[14] M. Biasini,[10] T. Bienz,[19] G. M. Bilei,[10] F. Bird,[19] D. Bisello,[9] G. Blaylock,[23] R. Blumberg,[19] J. R. Bogart,[19] T. Bolton,[5] S. Bougerolle,[24] G. R. Bower,[19] R. F. Boyce,[19] J. Brau,[30] M. Breidenbach,[19] T. E. Browder,[19] W. M. Bugg,[31] B. Burgess,[19] D. Burke,[19] T. H. Burnett,[33] P. N. Burrows,[15] W. Busza,[15] B. L. Byers,[19] A. Calcaterra,[12] D. O. Caldwell,[22] D. Calloway,[19] B. Camanzi,[8] L. Camilleri,[5] M. Carpinelli,[11] J. Carr,[26] S. Cartwright,[29] R. Cassell,[19] R. Castaldi,[11,27] A. Castro,[9] M. Cavalli-Sforza,[23] G. B. Chadwick,[19] O. Chamberlain,[14] D. Chambers,[19] L. Chen,[35] P.E.L. Clarke,[3] R. Claus,[19] J. Clendenin,[19] H. O. Cohn,[31] J. A. Coller,[2] V. Cook,[33] D. Cords,[19] R. Cotton,[3] R. F. Cowan,[15] P. A. Coyle,[23] D. G. Coyne,[23] W. Craddock,[19] H. Cutler,[19] A. D'Oliveira,[25] C.J.S. Damerell,[18] S. Dasu,[19] R. Davis,[19] R. De Sangro,[12] P. De Simone,[12] S. De Simone,[12] T. Dean,[19] F. Dejongh,[4] R. Dell'Orso,[11] A. Disco,[36] R. Dolin,[22] R. W. Downing,[28] Y. C. Du,[31]

R. Dubois,[19] J. E. Duboscq,[22] W. Dunwoodie,[19] D. D. Durrett,[26] G. Eigen,[4] B. I. Eisenstein,[28] R. D. Elia,[19] W. T. Emmet, II,[36] R. L. English,[18] E. Erdos,[26] J. Escalera,[19] C. Fan,[26] M. J. Fero,[15] J. Ferrie,[19] T. Fieguth,[19] J. Flynn,[19] D. A. Forbush,[33] K. M. Fortune,[28] J. D. Fox,[19] M. J. Fox,[19] R. Frey,[30] D. R. Freytag,[19] J. I. Friedman,[15] J. Fujimoto,[13] K. Furuno,[30] M. Gaillard,[19] M. Gallinaro,[12] E. Garwin,[19] T. Gillman,[18] A. Gioumousis,[19] G. Gladding,[28] S. Gonzalez,[15] D. P. Gurd,[21] D. L. Hale,[22] G. M. Haller,[19] G. D. Hallewell,[19] V. Hamilton,[19] M. J. Haney,[28] T. Hansl-Kozanecka,[15] H. Hargis,[31] J. Harrison,[33] E. L. Hart,[31] K. Hasegawa,[20] Y. Hasegawa,[20] S. Hedges,[3] S. S. Hertzbach,[29] M. D. Hildreth,[19] R. C. Hilomen,[19] D. G. Hitlin,[4] T. A. Hodges,[32] J. Hodgson,[19] J. J. Hoeflich,[19] A. Honma,[32] D. Horelick,[19] J. Huber,[30] M. E. Huffer,[19] E. W. Hughes,[19] H. Hwang,[30] E. Hyatt,[5] Y. Iwasaki,[20] J. M. Izen,[28] P. Jacques,[17] C. Jako,[19] A. S. Johnson,[2] J. R. Johnson,[34] R. A. Johnson,[25] S. Jones,[19] T. Junk,[19] S. Kaiser,[19] R. Kajikawa,[16] M. Kalelkar,[17] H. Kang,[19] I. Karliner,[28] H. Kawahara,[19] R. K. Keeler,[32] M. H. Kelsey,[4] H. W. Kendall,[15] D. Kharakh,[19] H. Y. Kim,[33] P. C. Kim,[19] R. King,[19] M. Klein,[4] R. R. Kofler,[29] M. Kowitt,[14] N. M. Krishna,[26] R. S. Kroeger,[31] P. F. Kunz,[19] Y. Kwon,[19] J. F. Labs,[19] R. R. Langstaff,[32] M. Langston,[30] R. Larsen,[19] A. Lath,[15] J. A. Lauber,[26] D.W.G. Leith,[19] L. Lintern,[18] X. Liu,[23] M. Loreti,[9] A. Lu,[22] H. L. Lynch,[19] T. Lyons,[15] J. Ma,[33] W. A. Majid,[28] G. Mancinelli,[10] S. Mánly,[36] D. Mansour,[19] G. Mantovani,[10] T. W. Markiewicz,[19] T. Maruyama,[19] G. R. Mason,[32] H. Masuda,[19] L. Mathys,[22] G. Mazaheri,[19] A. Mazzucato,[9] E. Mazzucato,[8] J. F. McGowan,[28] S. McHugh,[22] A. K. McKemey,[3] B. T. Meadows,[25] D. J. Mellor,[28] R. Messner,[19] A. I. Mincer,[4] P. M. Mockett,[33] K. C. Moffeit,[19] R. J. Morrison,[22] B. Mours,[19] G. Mueller,[19] D. Muller,[19] G. Mundy,[19] T. Nagamine,[19] U. Nauenberg,[26] H. Neal,[19] D. Nelson,[19] V. Nesterov,[19] M. Nordby,[19] M. Nussbaum,[25] A. Nuttall,[19] H. Ogren,[6] J. Olsen,[19] C. Oram,[21] L. S. Osborne,[15] R. Ossa,[19] G. Oxoby,[19] L. Paffrath,[19] A. Palounek,[15] R.S. Panvini,[35] H. Park,[30] M. Pauluzzi,[10] T. J. Pavel,[19] F. Perrier,[19] I. Peruzzi,[12] L. Pescara,[9] D. Peters,[24] H. Petersen,[19] M. Petradza,[19] M. Piccolo,[12] L. Piemontese,[8] E. Pieroni,[11] R. Pitthan,[19] K. T. Pitts,[30] R. J. Plano,[17] P. R. Poffenberger,[32] R. Prepost,[34] C. Y. Prescott,[19] D. Pripstein,[14] G. D. Punkar,[19] G. Putallaz,[19] P. Rankin,[26] B. N. Ratcliff,[19] T. W. Reeves,[35] P. E. Rensing,[19] J. D. Richman,[22] R. Rinta,[19] L. P. Robertson,[32] L. S. Rochester,[19] L. Rosenson,[15] J. E. Rothberg,[33] A. Rothenberg,[19] P. C. Rowson,[5] J. J. Russell,[19] D. Rust,[6] E. Rutz,[25] P. Saez,[19] B. Saitta,[8] A. K. Santha,[25]

A. Santocchia,[10] O. H. Saxton,[19] T. Schalk,[23] P. R. Schenk,[32] R. H. Schindler,[19] U. Schneekloth,[15] M. Schneider,[23] D. Schultz,[19] G. E. Schultz,[26] B. Schumm,[14] A. Seiden,[23] L. Servoli,[10] M. H. Shaevitz,[5] J. T. Shank,[2] G. Shapiro,[14] S. L. Shapiro,[19] H. Shaw,[19] D. J. Sherden,[19] T. Shimomura,[19] A. Shoup,[25] R. L. Shypit,[24] C. Simopoulos,[19] K. Skarpaas,[19] S. R. Smith,[19] A. Snyder,[19] J. A. Snyder,[36] R. Sobie,[24] M. D. Sokoloff,[25] E. N. Spencer,[23] S. St. Lorant,[19] P. Stamer,[17] H. Steiner,[14] R. Steiner,[1] R. J. Stephenson,[18] G. Stewart,[28] P. Stiles,[19] I. E. Stockdale,[25] M. G. Strauss,[29] D. Su,[18] F. Suekane,[20] A. Sugiyama,[16] S. Suzuki,[16] M. Swartz,[19] A. Szumilo,[33] M. Z. Tahar,[2] T. Takahashi,[19] G. J. Tappern,[18] G. Tarnopolsky,[19] F. E. Taylor,[15] M. Tecchio,[9] J. J. Thaler,[28] F. Toevs,[33] N. Toge,[19] M. Turcotte,[32] J. D. Turk,[36] T. Usher,[19] J. Va'Vra,[19] C. Vannini,[11] E. Vella,[33] J. P. Venuti,[35] R. Verdier,[15] P. G. Verdini,[11] B. F. Wadsworth,[15] A. P. Waite,[19] D. Walz,[19] D. Warner,[2] R. Watt,[19] S. J. Watts,[3] T. Weber,[19] A. W. Weidemann,[31] J. S. Whitaker,[2] S. L. White,[31] F. J. Wickens,[18] S. A. Wickert,[22] D. A. Williams,[23] D. C. Williams,[15] R. W. Williams,[33] S. H. Williams,[19] R. J. Wilson,[2] W. J. Wisniewski,[4] M. S. Witherell,[22] M. Woods,[19] G. B. Word,[17] J. Wyss,[9] R. K. Yamamoto,[15] J. M. Yamartino,[15] C. Yee,[19] S. J. Yellin,[22] A. Yim,[19] C. C. Young,[19] K. K. Young,[33] H. Yuta,[20] G. Zapalac,[34] R. W. Zdarko,[19] C. Zeitlin,[30] M. Zolotorev.[14] and P. Zuchelli,[8]

LIST OF INSTITUTIONS

[1] Adelphi University, [2] Boston University, [3] Brunel University, [4] California Institute of Technology, [5] Columbia University, [6] Indiana University, [7] INFN Sezione di Bologna, [8] INFN Sezione di Ferrara and Universitá di Ferrara, [9] INFN Sezione di Padova and Universitá di Padova, [10] INFN Sezione di Perugia and Universitá Perugia, [11] INFN Sezione di Pisa and Universitá di Pisa, [12] INFN Lab. Nazionalli di Frascati, [13] KEK National Laboratory, [14] Lawrence Berkeley Laboratory, University of California, [15] Massachusetts Institute of Technology, [16] Nagoya University, [17] Rutgers University, [18] Rutherford Appleton Laboratory, [19] Stanford Linear Accelerator Center, [20] Tohoku University, [21] TRIUMF [22] University of California, Santa Barbara, [23] University of California, Santa Cruz, [24] University of Cincinnati, [26] University of Colorado, [27] Universitá di Genova, [28] University of Illinois, [29] University of Massachusetts, [30] University of Oregon, [31] University of Tennesee, [32] University of Victoria, TRIUMF, [33] University of Washington, [34] University of Wisconsin, [35] Vanderbilt University, [36] Yale University.

PRECISION MEASUREMENT OF THE TOTAL CROSS SECTION AND CHARGE ASYMMETRY AT TOPAZ FOR $e^+e^- \to \mu^+\mu^-$ AND $e^+e^- \to \tau^+\tau^-$

David S. Koltick
Department of Physics, Purdue University, 1396 Physics Building, West Lafayette, IN 47907

Abstract

The process $e^+e^- \to \mu^+\mu^-$ and $e^+e^- \to \tau^+\tau^-$ have been studied in the energy range $\sqrt{s}$ = 52 - 61.4 GeV, using the TOPAZ detector at TRISTAN. From an integrated luminosity of L = 74.0 pb-1, lowest-order cross sections and forward-backward asymmetries are measured to be:

$$\langle \sigma_{\mu\mu} \rangle = 25.4 \pm 0.9 \pm 1.2\, pb, \langle A_{\mu\mu} \rangle = -32.2 \pm 3.1 \pm 1.1\%,$$

$$\langle \sigma_{\tau\tau} \rangle = 27.1 \pm 1.1 \pm 1.2\, pb, \langle A_{\tau\tau} \rangle = -33.9 \pm 4.9 \pm 1.0\%,$$

at an average energy of $\langle\sqrt{s}\rangle$ = 57.87 GeV. From the measured asymmetry we derived axial vector couplings of $a_e a_\mu = 0.96 \pm 0.09 \pm 0.01$, and $a_e a_\tau = 1.01 \pm 0.14 \pm 0.01$.

INTRODUCTION

A precision measurement of the forward-backward asymmetry [1] of $\mu^+\mu^-$ and $\tau^+\tau^-$ pairs produced in e^+e^- collisions below the $Z°$ pole provides a unique test of Standard Model [2] physics. At TRISTAN the strength of the exchange of the $Z°$ and the photon are of comparable strength. Their relative strengths are shown in Figure 1. At the highest TRISTAN energies the contribution of the $Z°$ to the total hadronic cross section represents ~2 units of R, while the photon contributes ~4 units of R. This is in contrast to LEP where the $Z°$ pole overwhelms the photon to the extend that it can be neglected as if it did not exist. The experiments at TRISTAN can be thought of a modern version of a double slit Young's experiment as shown in Figure 2. The slits S_1 and S_2 are replaced with the $Z°$ and the photon. For interference to occur it must be that the quantum numbers of the underlying quantum states are identical. To the extent that they are not the interference pattern will be changed. Because the two interfering states are of comparable strength a strong interference can take place. The standard Model predicts an asymmetry for leptons of -34% at $\sqrt{s}$ =58 GeV. Because this asymmetry is large, even with typical statistic and systematic limitations of a few percent, this asymmetry can be observed at ~10 standard deviations from QED.

These data then can be used to determine:
1. The weak couplings a_μ and a_τ off the $Z°$ pole.
2. Test the universality of the μ and τ by comparing $\sigma_\mu \sigma_\tau$, $A_\mu A_\tau$, $a_\mu a_\tau$
3. Place limits on the compositness of μ and τ.
4. Place constraints on the properties of $Z°$.
5. Place limits on higher mass neutral bosons.

The theory of electroweak interactions allow the cross section of the production of μ and τ pairs to be written,

$$\frac{d\sigma}{d\Omega} = \frac{\alpha^2}{4s}\left[R(1+\cos^2\theta) + B\cos\theta\right] \quad (1)$$

The interaction strength of the two processes is expressed as a single coupling constant α. R and B depend only on the electric and weak charges of the leptons and the properties of the $Z°$ boson namely its mass and width. A single coupling constant, α, is used to describe the interaction of both particles.

The total cross section is proportional to R,

$$R = Q_e^2 Q_{\mu,\tau}^2 + 2Q_e Q_{\mu,\tau} v_e v_{\mu,\tau} \text{Re}\{x\} \\ + (v_e^2 + a_e^2)(v_{\mu,\tau}^2 + a_{\mu,\tau}^2)|x|^2 \quad (2)$$

© 1993 American Institute of Physics

and the forward backwards asymmetry is proportional to

$$B = 4Q_e Q_{\mu,\tau} a_e a_{\mu,\tau} \mathrm{Re}\{x\} + 8 v_e v_{\mu,\tau} a_e a_{\mu,\tau} |x|^2$$

where $x = \dfrac{1}{16 \sin^2 \theta_W \cos^2 \theta_W} \dfrac{s}{s - M_z^2 + i\Gamma_z M_z}$.

the asymmetry is then

$$A = \frac{3}{8} \frac{B}{R}.$$

DETECTOR AND LUMINOSITY

The data for these measurements were taken by the TOPAZ group at KEK in Japan. Other TRISTAN groups [3],[4] have presented results using integrated luminosities of 17.7 pb^{-1} and 27.3 pb^{-1} respectively. For this analysis, we have a total luminosity of 74.0 pb^{-1}. Data collection is continuing and a final data sample of ~ 300 pb^{-1} is expected.

Detailed descriptions of the TOPAZ detector are available in other literature [5]. Briefly, TOPAZ employs a time projection chamber (TPC) as the primary tracking element in a multipurpose cylindrical detector. A time of flight (TOF) system, together with a trigger chamber (TCH), provide pre-trigger information for charged tracks in the TPC. A superconducting magnetic coil produces a uniform field of 1 T over the inner detector volume.

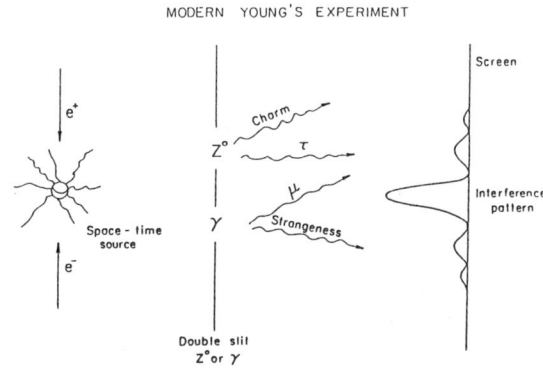

Fig. 2 Modern Young's Experiment

Outside the TPC, electromagnetic calorimetry is achieved with a barrel calorimeter (BCL) composed of lead-glass modules, and an end cap calorimeter (ECL). Three layers of drift chambers, interspersed with iron absorbers, make up a muon detection system (MDC) outside the BCL.

Integrated luminosities are determined by counting Bhabha events in the ECL. The BCL and small angle Bhabha counters near the beam pipe provide a luminosity check at angles above and below the ECL. The luminosity value has a systematic error of 4.0%, common to all energies, which has been determined by comparing the measurement at different azimuthal regions within the ECL. This is the dominant contribution to systematic errors in this analysis. The total systematic error found for the µ-pair analysis is 4.6% and for the τ-pair analysis 4.4%.

µ PAIR ANALYSIS

We selected events with the following properties for our µ pair sample: (1) Two good TPC tracks that have momenta that exceed E$_{beam}$/3. (2) The tracks must have an acollinearity that is less than 10° and be located within the good track-trigger region, 0.10 < |cos(θ)| < 0.75. (3) The sum of all calorimeter energy is less than 5.0 GeV, and the largest single shower is less than 2.0 GeV. (4) TOF hit times of between 5 and 12 ns (measured from the beam crossing) are associated with each track.

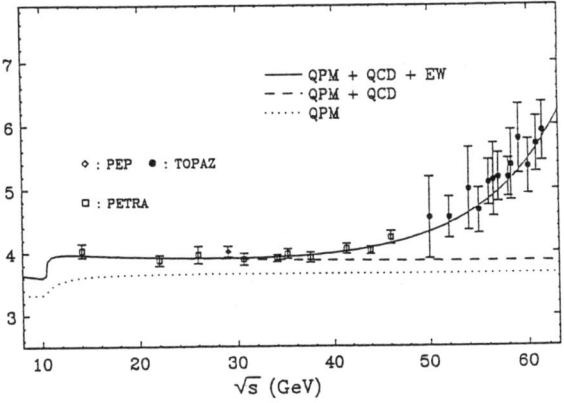

Fig. 1 Hadronic R ratio at TRISTAN

The difference of these two times must be less than 4 ns to exclude cosmic rays.

After selection, the residual backgrounds are 1.1±0.1% from τ pairs, and 0.6 ± 0.3% from ee(μμ) events. cosmic rays are estimated to contribute 0.5 ± 0.3%. The wrong charge assignment probability is about 0.8%.

The cross section was obtained by correcting the number of observed events, $N^{\mu\mu(\tau\tau)}$, for the estimated backbround, $N^{background}$, trigger efficiency, η^{trig}, detection efficiency, η^{eff}, and radiative processes, $(1+\delta)$. This is expressed as

$$\sigma_{\mu\mu(\tau\tau)} = \frac{N^{\mu\mu(\tau\tau)} - N^{background}}{L\eta^{trig}\eta^{eff}(1+\delta)},$$

where L is the integrated Luminosity. The event distribution in cosθ was likewise corrected in individual bins.

The radiative correction was determined by the Monte Carlo program written by J. Fujimoto and Y. Shimizu[6] which includes a full electroweak calculation to $O(\alpha^3)$ by the "on-shell" renormalization scheme. Typical correction factors are $(1+\delta)_{\mu\mu} = 1.68$ and $\eta^{eff}_{\mu\mu} = 0.290$ at $\sqrt{s} = 58$ GeV. The relatively large radiative correction includes photons with energies up to the maximum kinematically allowed value,

$$k_{max}/E_{beam} = (1-u^2), \text{ where } u = M_{\mu(\tau)}/E_{beam}.$$

The detection efficiency accounts for all events distributed over the entire 4π solid angle. Radiative events with an acollinearity >10° and events outside the detector's acceptance account for most of the events removed in the selection. For events within the detector satisfying the acollinearity cut, the efficiency is $\eta^{eff}_{\mu\mu} \sim 85\%$, and the contribution from radiative processes is only 1.7%.

τ PAIR ANALYSIS

We selected τ pairs where at least one τ decays to a single charged particle and the other decays to either 1 or 3 charged particles. Events of a 3-3 topology were discarded because they contain a large hadronic background and represent only a small fraction of the τ sample. The actual number of charged tracks observed can differ from the true topology, due to detector interactions and tracking efficiencies, so events are classified as either 1-1 or 1-n, where n >1.

$\sqrt{s}$ (GeV)	$\eta^{eff}_{\mu\mu}$	$(1+\delta)_{\mu\mu}$	$\sigma_{e^+e^- \to \mu^+\mu^-}$ (pb)	$\eta^{eff}_{\tau\tau}$	$(1+\delta)_{\tau\tau}$	$\sigma_{e^+e^- \to \tau^+\tau^-}$ (pb)	σ_{theory} (pb)
52.0	0.291	1.676	33.5 ± 4.6	0.201	1.377	32.5 ± 5.6	33.0
54.0	0.289	1.676	18.0 ± 8.0	0.196	1.384	23.6 ± 11.8	30.8
55.0	0.289	1.676	27.1 ± 4.5	0.197	1.382	36.3 ± 6.6	29.8
56.5	0.255	1.676	19.9 ± 7.1	0.174	1.379	33.2 ± 11.7	28.5
57.0	0.253	1.677	21.7 ± 3.8	0.173	1.387	27.9 ± 5.5	28.0
58.0	0.287	1.676	25.1 ± 1.0	0.193	1.387	25.9 ± 1.3	27.2
58.7	0.290	1.676	32.2 ± 5.1	0.201	1.377	32.8 ± 6.6	26.7
60.0	0.284	1.672	28.3 ± 4.5	0.193	1.386	33.5 ± 6.3	25.8
60.8	0.285	1.674	28.3 ± 4.2	0.193	1.387	25.1 ± 5.0	25.3
61.4	0.284	1.666	14.5 ± 3.2	0.193	1.385	21.2 ± 5.0	24.9
⟨57.87⟩	–	–	⟨25.4 ± 0.9⟩	–	–	⟨27.1 ± 1.1⟩	⟨27.4⟩

Table 1

Selected events satisfied the following: (1) The number of tracks is less than 9 and at least one jet contains only one track. the acoplanarity of the event is less than 0.015. (3) TOF times for at least two tracks in opposite jets differ by less than 5.0 ns. (4) Visible energy (charged plus neutral) $> \frac{\sqrt{s}}{4}$ and momentum balance along the beam axis, $|p_z| < 0.4$. where the neutral energy is the sum of all calorimeter energy not matched to TPC tracks. (5) The total calorimeter energy is between 2.0 GeV and $0.70\sqrt{s}$, and no individual shower exceeds $0.95 E_{beam}$. (6) At least one jet has no electron tracks. (7) The acollinearity of single tracks (for 1-1 events) or jet axes (for 1-n events) is between 1° and 15°. (8) All single tracks or jet axes are within the detector region $|\cos\theta| < 0.75$.

After slection, remaining backgrounds were estimated to be 1.5±0.2% from hadronic events, 0.8±0.5% from ee(γ), 1.6±0.1% from μμ(γ), 1.2±0.6% and 1.4±0.6% from two photon ee(ee) and ee(ττ) reactions, respectively. Cosmic ray and ee(μμ) backgrounds are negligible. Of 572 τ events selected, 312 were of the 1-1 topology while 260 were of the 1-n topology, consistent with event simulation.

The direction of the τ was taken along the thrust axis, with the sign of the charge determined by that of the single prong decays. the thrust axis is estimated to differ from the true τ direction by less than 4° for 95% of the events.

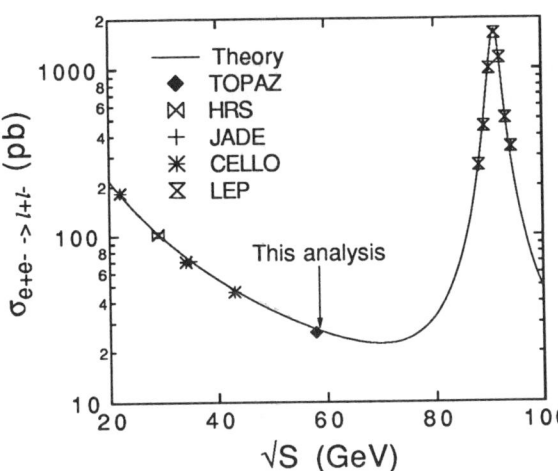

Fig.3 The combined cross sections for $\mu^+\mu^-$ and $\tau^+\tau^-$

The probability of wrong charge assignment is about 3%, which is well accounted for by our detector simulator, and therefore does not contribute a systematic uncertainty. The τ pair analysis proceeds in the same manner as for μ pairs, with the following modifications: The full radiative correction is only $(1+\delta)_{\tau\tau} = 1.39$ with a corresponding total detection efficiency $\eta^{eff}_{\tau\tau} = 0.193$ at $\sqrt{s} = 58$ GeV. In addition, 95% of all τ events satisfy the neutral energy trigger conditions, for which there is no measureable inefficiency. Only 5% of the events rely solely upon the track trigger, whose inefficiency is small. We therefore take $\eta^{trig} \approx 100\%$ with negligible error.

RESULTS

The integrated luminosity and measured cross sections at various values of $\sqrt{s}$ are listed in table 1. Also listed are the detection efficiency and radiative correction factors. The data at all values of $\sqrt{s}$ can be combined (weighted by luminosities) to yield

$$\langle \sigma_{\mu\mu} \rangle = 25.4 \pm 0.9(stat.) \pm 1.2(sys.) \text{pb}$$
$$\langle \sigma_{\tau\tau} \rangle = 27.1 \pm 1.1(stat.) \pm 1.2(sys.) \text{pb}$$

at an average energy of $\langle \sqrt{s} \rangle = 57.87$ GeV. In the ratio of these values, the dominant systematic error from the luminosity measurement cancels; we obtain

$$\frac{\langle \sigma_{\tau\tau} \rangle}{\langle \sigma_{\mu\mu} \rangle} = 1.07 \pm 0.06(stat.) \pm 0.03(sys.),$$

which shows no significant deviation from the prediction of lepton universality.

Shown in Fig. 3 is the combined cross section [7] for μ and τ from this analysis along with those obtained from other energies.

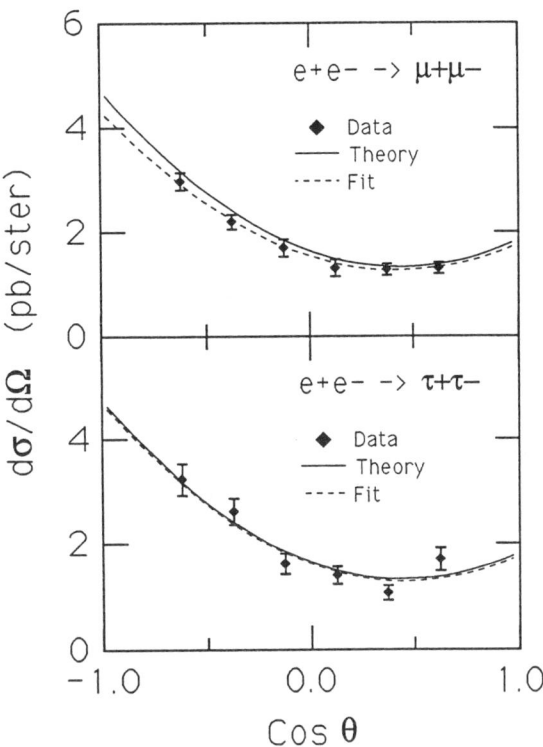

Fig. 4 The angular distribution for μ and τ pairs.

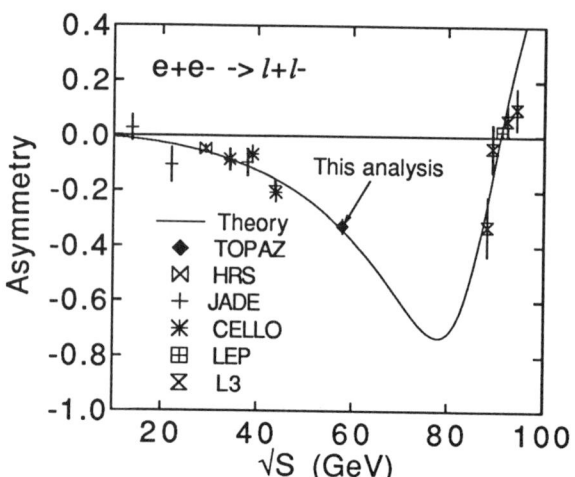

Fig. 5 Combined data for the forward-backward asymmetry.

The forward backward asymmetry at the average energy is obtained by fitting the angular distributions, shown in fig. 4, to the equation

$$(\frac{d\sigma_{\mu\mu(\tau\tau)}}{d\Omega}) = N(1 + \cos^2\theta + \frac{8}{3}A_{\mu\mu(\tau\tau)}\cos\theta),$$

where the parameter N is introduced to account for the overall normalization in the fit. The results are

$$\langle A_{\mu\mu}\rangle = -32.2 \pm 3.1(stat.) \pm 1.1(sys.)\%$$
$$\langle A_{\tau\tau}\rangle = -33.9 \pm 4.9(stat.) \pm 1.0(sys.)\%$$

Fig. 5 shows the combined data for μ and τ pairs as a function of energy [7].

These cross sections and asymmetries are to be compared to Standard Model predictions of

$$\sigma_{theory} = 27.43 pb$$
$$A_{theory} = -33.50\%,$$

also at $\langle\sqrt{s}\rangle$ = 57.87 GeV. For both $e^+e^- \rightarrow \mu^+\mu^-$ and $e^+e^- \rightarrow \tau^+\tau^-$ processes, the results are consistent with the theoretical predication. We calculate the axial vector couplings from the theoretical asymmetry where we assume the vector coupling values are $a_e a_\mu$ = 0.96 ± 0.09 ± 0.01 and $a_e a_\tau$ = 1.01 ± 0.14 ± 0.01 respectively. these values are also consistent with theoretical values of $a_e = a_\mu = a_\tau = -1$, and with measurements at LEP energies [8].

REFERENCES

[1] R. Budny, Phys. Lett. B45(1973)340.

[2] S.L. Glashow, Nucl. Phys.22(1961)579; A. Salam, Phys. Rev. 127(1962)331; S. Weinberg, Phys. Rev. Lett. 19(1967)1264.

[3] [AMY] A. Bacala et al., Phys. Lett B218(1989)112.

[4] [VENUS] K. Abe et al., Z. Phys. C48(1990)13.

[5] (TPC) T. Kamae et al., Nucl. Instr. and Meth. A252(1986)423; A. Shirahashi et al., IEEE Trans. on Nucl. Sci. NS-35(1988)414; (TOF) T. Kishida et al., Nucl. Instr. and Meth. A254(1987)367; (BCL) S. Kawabata et al., Nucl. Instr. and Meth. A270(1988)11; (ECL) K. Fujii et al.., Nucl. Instr. and Meth. A236(1985)55; J. Fujimoto et al., Nucl. Instr. and meth. A256(1987)449.

[6] J. Fujimoto et al., Prog. Theor. Phys. Supplement, 100(1990); M. Igarashi et al., Nucl. Phys. B263(1986)347; K. Aoki et al., Prog. Theor. Phys. Supplement, 73(1982); A. Hioki, Prog. Theor. Phys., 68(1982)2134.

[7] B. Howell, Ph.D. thesis, August 1992, Purdue University.

[8] The LEP Collaborations, Phy. Lett. B276(1992)247.

PRECISION MEASUREMENT OF $\sin^2\theta_w$ FROM νFe SCATTERING AT THE TEVATRON

T. Bolton, C. Arroyo, K.T. Bachmann, A.O. Bazarko, R.E. Blair,
C. Foudas, B.J. King, W.C. Lefmann, W.C. Leung, S.R. Mishra, E. Oltman,
P.Z. Quintas, S.A.Rabinowitz, F. Sciulli, W.G. Seligman, M.H. Shaevitz
Columbia University, New York, NY 10027

F.S. Merritt, M.J. Oreglia, B.A. Schumm
University of Chicago, Chicago, IL 60637

R.H. Bernstein, F. Borcherding, H.E. Fisk, M.J. Lamm,
W. Marsh, K.W.B. Merritt, H. Schellman, D.D. Yovanovitch
Fermilab, Batavia, IL 60510

A. Bodek, H.S. Budd, P. de Barbaro, W.K. Sakumoto
University of Rochester, Rochester, NY 14627

W.H. Smith, T.S. Kinnel, P.H. Sandler
University of Wisconsin, Madison, WI 53706

Abstract

We present the results of a preliminary determination of $\sin^2\theta_W$ from a study of νFe scattering in the Lab E detector at Fermilab. The analysis is based on a sample of 5×10^5 events with a mean neutrino energy of 166 GeV. Our result, $\sin^2\theta_W = 0.2242 \pm 0.0044_{expt.} \pm 0.0047_{model}$, is the highest energy high statistics determination of the weak mixing angle using neutrino data.

INTRODUCTION

The Standard Model of Electroweak Physics requires as input five parameters to describe high energy processes: α, G_F, M_Z, m_t, and m_h. The latter two enter into low energy calculations, such as extracting $\sin^2\theta_W$ from νN scattering, via loop corrections to the W and Z self-energies. The radiative corrections to the effective $\sin^2\theta_W$'s measured in different interactions are quadratic in m_t for sufficiently large top mass, but only logarithmic in m_h. The strong dependence on the top mass, coupled with the LEP measurements of the Z^0 mass to an accuracy of 0.02% allows indirect determinations of m_t via precision measurements of neutral current νN scattering. Existing neutrino measurements of $\sin^2\theta_W$ already constrain the top mass to $m_t < 200$ GeV, a limit quite competitive with those attained by collider experiments. [1] Once the top mass is known, there is the possibility that sufficiently high precision measurements of $\sin^2\theta_W$ in different processes will constrain the Higgs mass or, perhaps through an inconsistency, point to New Physics. Thus, considerable motivation exists to improve upon the determination of $\sin^2\theta_W$ in the neutrino sector. [2]

We report herein on a new preliminary determination of $\sin^2\theta_W$ from Fermilab experiment E770 by the CCFR collaboration.

DATA SAMPLE

The data were taken with the Lab E neutrino detector during the 1988 fixed target run of the Tevatron. The Lab E detector [3] consists of a 680 ton non-magnetized iron calorimeter followed by an iron toroid muon spectrometer. The toroid is not used in this analysis. The calorimeter consists of 84 planes of 3m×3m×10 cm steel absorb er. Scintillation counters of the same transverse dimension sample the energy every 10cm (0.6λ, $5.9X_0$) of steel, permitting a hadron energy resolution of $\frac{\Delta E}{E} = \frac{0.89}{\sqrt{E}}$. The scintillators also determine the event timing and establish the longitudinal vertex position. Transverse vertex position is provided by double planes of drift chambers placed at 20 cm intervals of steel; the x,y vertex resolution, using the chambers, is 5.0 cm.

Three essential quantities enter into the $\sin^2\theta_W$ analysis: the event length, the e vent radial vertex position, and the event energy.

The length is defined as $L = place - cexit + 1$. *Place* is the most upstream of the first pair two consecutive scintillation counters that each record at least 4 minimum ionizing particles (mips) equivalent energy; and *cexit* is the first counter upstream of the first occurrence of a gap of three consecutive counters each with energy less than 0.25 mip downstream of place. Note that *place* = 1 or *cexit* = 1 refers to the most downstream counter. We separate events into three categories based on L and *cexit*: *short* events with $L < 31$ counters (3 m of steel), which are mostly neutral current events; *toroid* events with *cexit* < 3, which are mostly charged current events exiting the calorimeter and entering the toroid; and *intermediate* events with $L > 30$ counters and *cexit* > 3, which are dominantly charged current events in which the muon exits the calorimeter or ranges out. To make this classification sensible, we require *place* to be at least 3.4 m of steel (34 counters) from the downstream end of the calorimeter and 0.6 m of steel (6 counters) from the upstream end.

The event transverse vertex position is obtained from a weighted sum of drift chamber hits from the first two drift chamber planes downstream of *place*. To minimize the number of charged current events with $L < 31$ which exit the sides of the detector and to suppress electron neutrino background, we require the event radius to be less than 76.2 cm.

The event energy is defined as the sum of pulse heights from the first twenty counters in the event, starting with *place*. We require events to have at least 30 GeV of visible energy. This cut insures that the calorimeter is being used well within its linear regime, guarantees 100% trigger efficiency, and strongly suppresses non-deep-inelastic processes and the cosmic ray background.

After all cuts, our event sample contains 1.51×10^5 short events, 2.93×10^5 toroid events, and 0.42×10^5 intermediate length events, with a mean neutrino energy of 166 GeV. This represents the highest energy high statistics neutral current neutrino measurement to date.

ANALYSIS TECHNIQUE

To determine the weak mixing angle, we attempt to reproduce the measured quantity $R_{30} \equiv \frac{N_{short}}{N_{toroid}}$ with a Monte Carlo. The Monte Carlo includes as ingredients the QCD corrected quark-parton model, a parametrization of the quadrupole-triplet neutrino beam, and a detailed description of the Lab E detector. To first, approximation, $R_{30} \simeq R_\nu$, where R_ν is the ratio of neutral current to charged current cross sections. Previous $\sin^2 \theta_W$ determinations consisted of extracting a suitably corrected value of R_ν and then applying the Llewellyn-Smith formalism [4]. We forego that approach here and instead attempt to adjust our Monte Carlo to match our data using $\sin^2 \theta_W$ as the single free parameter. The composition of short length events is approximately 60% ν_μ neutral current, 22% ν_μ charged current, 9% $\bar{\nu}_\mu$ neutral current, 1% $\bar{\nu}_\mu$ charged current, and 8% ν_e charged and neutral current.

The parton distributions in our Monte Carlo are obtained directly from structure function measurements by the same experiment. [5] Corrections must be made for the non-isoscalar target, QED radiative effects, the strange and charm sea of the nucleon, the longitudinal structure function, and charm production. The last is the most important. Charm production is modeled via the slow rescaling formalism in which charm threshold effects are parametrized using an effective charm mass, m_c. From an analysis of 6000 opposite sign dimuon from this experiment [6]: $m_c = 1.34 \pm 0.34$ GeV.

The dominant component of the neutrino flux arising from two body charged pion and kaon decays is directly measured in the experiment. A more serious issue is the electron neutrino flux. Since all ν_e interactions produce short length events, a mis-estimate of this component of the flux translates into a large error on $\sin^2 \theta_W$. There are two components to the ν_e flux. The first, originating from $K^+ \to \pi^0 e^+ \nu_e$ can be tightly constrained from $K_{\mu 2}$ decays, whose contribution from the flux can be separated from $\pi_{\mu 2}$ contribution by exploiting the well known energy-radius correlation. The second component of the ν_e flux is from $K_L^0 \to \pi^- e^+ \nu_e$; these decays contribute $\sim 17\%$ of the ν_e events. This source is important because the primary proton beam is targeted at 0 degrees relative to the Lab E detector to maximize total flux for charged current measurements. The K_L component cannot be constrained using $K_L^0 \to \pi^- \mu^+ \nu_\mu$ decays because the neutral $K_{\mu 3}$ decays are a small fraction of the charged $K_{\mu 2}$ decays. The K_L^0 component must instead be calculated using other experimental data on K_L^0 production. [7]

A check on the ν_e flux calculation can be obtained from the data itself by exploiting the different longitudinal energy distribution of charged current ν_e events compared to ν_μ neutral current events. The quantity η_3 is defined as $\eta_3 = 1 - E_3/E_{tot}$, where E_3 is the sum of the energy in the three most upstream counters and E_{tot} is the total energy. Muon neutrino neutral current events are characterized by a broad distribution in η_3, reflecting the large fluctuations in hadronic shower length. Electron neutrino charged current events, by contrast, have an η_3 distribution that is peaked towards small values of η_3 since a substantial portion of the event energy is carried by an electron. It is possible to obtain the shape of the η_3 distribution for neutral current events empirically from charged current events in which the muon has been removed, and to obtain the electron neutrino shape by convolving hadron showers from charged current events with electron showers from a test beam. Figure 1 shows the fraction of short events which are attributed to charged current ν_e as a function of the hadron energy cut as determined by a flux Monte Carlo and by the η_3 technique. The two independent calculations are consistent to within the errors of $\pm 4.6\%$ for the flux Monte Carlo and $\pm 6.8\%$ for the η_3 fits. Combining the two results, the estimated uncertainty for the ν_e flux is $\pm 3.8\%$. This error is dominated by the 20% uncertainty in K_L^0 production of ν_e.

The critical detector parameters that must be modeled are the calorimeter respo nse to hadrons, muons, and electrons; and the efficiency and noise characteristics of the counters with respect to minimum ionizing particles. The electron, pion, and m uon calorimeter performance has been studied for energies from 8-400 GeV in test bea m runs in 1984, 1987, and 1991; hence accurate parametrizations based on real data are available. Muon energy loss in the 20 counter energy definition region is particularly important as the presence of the muon generates an asymmetry in the charged current vs. neutral current flux. Our muon energy loss parametrization is tuned with a large sample of "straight through" beam muons taken at the same time as the neutrino data. Noise and efficiency are measured counter-by-counter as a function of radial position using beam muons. The detector portion of the Monte Carlo is thus almost completely a parametrizat ion of actual data. A small exception is the correction of the event longitudinal vertex for albedo effects. This correction is based on Geant/Gheisha. [8] However, we have been able to verify the Geant correction, on average, by comparing the event vertex determ ined from tracking information in dimuon events versus that obtained from the calorim eter.

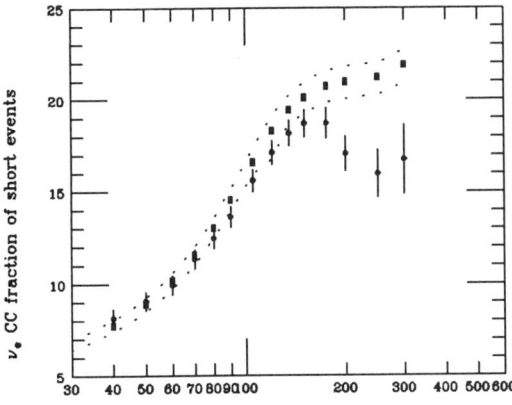

Figure 1. The fraction of short events which are attributed to charged current ν_e interactions. The points are the η_3 determination, while the bands represent the prediction of a beam Monte Carlo.

RESULT

Our Monte Carlo predicts

$$\sin^2\theta_W - 0.230 = -1.525 \cdot (R_{30} - 0.5182)$$

for events satisfying our fiducial and energy cuts. The data give

$$R_{30} = 0.5151 \pm 0.0016_{stat},$$

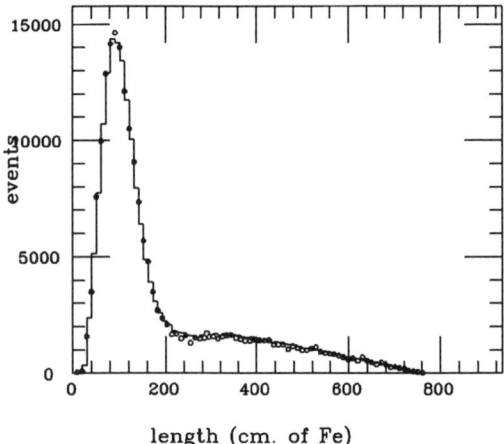

Figure 2. Event length for data (solid histogram) and Monte Carlo (plotted points) with $\sin^2\theta_W{}^{uncorr.} = 0.2338$.

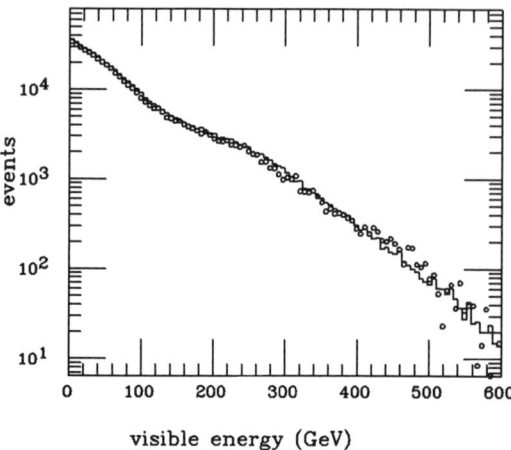

Figure 3. Visible energy distribution for data (solid histogram) and Monte Carlo (plotted points) with $\sin^2\theta_W{}^{uncorr.} = 0.2338$.

implying an uncorrected value :

$$\sin^2\theta_W{}^{uncorr.} = 0.2338 \pm 0.0029_{stat} \quad \text{(preliminary)}.$$

The statistical error includes the contribution of Monte Carlo statistics and will in the near furure be reduced to the data value of ±0.0023. Figures 2 and 3 show the distributions of event length, visible energy, and radius for the data with Monte Carlo predictions overlaid. The agreement between data and Monte Carlo is good.

CORRECTIONS AND SYSTEMATIC ERRORS

Three additional corrections have been made to $\sin^2\theta_W$ to give the final result: 1) A correction of +0.0011 to $\sin^2\theta_W$ due to cosmic rays. This correction has been measured using events taken out of time with the neutrino beam. 2) A correction of -0.0017 to take into account special kinematic properties of charm production associated di-lepton events in the charged current sector which are absent in neutral currents. An example is the presence of missing energy in events contained a charm quark which has decayed semi-leptonically. 3) A correction of -0.0085 to account for radiative corrections not taken into account in our Monte Carlo. Corrections due to muon bremsstrahlung are sensitive to our cuts and are thus explicitly taken into account via the formalism of De Rujula et al. [10] The remaining correction is obtained from Sirlin and Marciano [9]. This radiative correction procedure has been checked against a calculation by Bardin; [11] the two methods agree to 0.001 in $\sin^2\theta_W$.

After applying the corrections, our preliminary value for $\sin^2\theta_W$ is

$$\sin^2\theta_W = 0.2248 \pm 0.0029_{stat} \pm 0.0033_{exp} \pm 0.0047_{model}$$

Systematic errors are broken down in Table 1. The two most important for this experiment are those associated with charm production (±0.0034) and the ν_e flux (±0.0027). Combining the statistical and experimental systematics: $\sin^2\theta_W = 0.2242 \pm 0.0044 \pm 0.0047$. This can be compared with the two highest precision existing measurements, $\sin^2\theta_W = 0.228 \pm 0.005 \pm 0.005$ by CDHS [12] and $\sin^2\theta_W = 0.236 \pm 0.005 \pm 0.005$ by CHARM [13]. Our value is consistent with these two results, though slightly lower.

The largest model error is due to uncertainties in charm production in the charged current sector. While the magnitude of this systematic error is not much less than previous analysis, we feel this contribution is under much better control. The 30 GeV visible energy cut that we impose lessens the influence of the charmed threshold. More important however, we have measured all of the parameters of the slow rescaling model of charm production in the same experiment. Charm production is thus accurately parametrized, independent of the theoretical validity of the charm production model, for our experiment. Final analysis of the dimuon data from this experiment should reduce the charm associated error by 25%. Because charm production is sensitive to details of the model, e.g. the sea quark distributions, and to experimental details like the neutrino energy spectrum, one is cautioned against simply substituting our charm mass and its error into other experimental analyses to obtain "better" values of $\sin^2 \theta_W$. The slow rescaling model parameters used in this analysis accurately parametrize charm production in Fermilab E744/770.

The largest experimental systematic uncertainty is from lack of knowledge of the ν_e flux, which is itself driven by the poorly constrained contribution of K_L^0 decays to the ν_e flux. Significant reduction of this error in this experiment can only be achieved through better understanding of K_L^0 production and decay in the quad-triplet beam; the prospects for this seem poor.

NEXT GENERATION NEUTRINO MEASUREMENTS

A factor of two reduction in the errors on $\sin^2 \theta_W$ in neutrino experiments can be achieved only by substantially reducing the ν_e flux and charm mass errors. Fermilab E815 [15] is an experiment designed to do just that. A new sign-selected wide band beam coupled with the factor of two higher intensity available at the Tevatron since the completion of several successful accelerator upgrades will permit precision cross section measurements for both neutrinos and anti-neutrinos. Sufficient anti-neutrino statistics will be available to measure the quantity [14]

$$R^- = \frac{\sigma_{NC}^\nu - \sigma_{NC}^{\bar\nu}}{\sigma_{CC}^\nu - \sigma_{CC}^{\bar\nu}} \simeq \frac{1}{2} - \sin^2 \theta_W.$$

Since the cross section differences cancel all non-valence quark contributions to R^-, this quantity is almost completely insensitive to charm production parameters. It is estimated that the error due to charm production uncertainty will be of order ± 0.0005 using the R^- formalism.

The new beam will also reduce most of the uncertainty attributed to ν_e. This will be accomplished by targeting the primary proton beam at 6.8 mrad so that the neutral kaons point away from the neutrino detector. Since the ν_e error is dominated by the K_L^0 uncertainty a significant improvement will be achieved.

Experiment E815 should achieve a *total* error on $\sin^2 \theta_W$ of ± 0.003. This factor of two increase in sensitivity will keep the neutrino measurements in step with collider programs in terms of sensitivity to deviations from the standard model.

REFERENCES

1. D.C. Kennedy and P. Langacker, Phys. Rev. **D44** (1991) 1591; P. Langacker and M. Luo, Phys. Rev. **D44** (1991) 817; P. Langacker, M. Luo, and A. Mann, Rev. Mod. Phys. **64** (1992) 87.

Table 1. Summary of Present E770 Errors on Weak Mixing Angle

SOURCE	ERROR
data statistics	0.0023
Monte Carlo statistics	0.0018
TOTAL STATISTICAL	0.0029
$\nu_\mu, \bar\nu_\mu$ flux	0.0010
energy scale, resolution	0.0009
length determination	0.0014
ν_e flux	0.0027
TOTAL EXP. SYSTEMATIC	0.0033
structure functions	0.0003
non-isoscalarity ($\delta \frac{U_V}{D_V} = \pm 10\%$)	0.0017
long. struct. func. ($\delta R_L = \pm 10\%$)	0.0014
charm mass ($\delta m_c = \pm 0.34 GeV$)	0.0034
strange sea ($\delta \kappa = \pm 0.07$)	0.0006
charm sea ($\delta \frac{c}{s} = 15\%$)	0.0019
higher twist	0.0005
radiative corrections	0.0010
TOTAL MODEL	0.0047

2. The weak mixing angle extracted in neutrino scattering from the ratio of the charged current to neutral current cross sections is, due to a near cancellation of two corrections, nearly equal to the on-shell definition: $\sin^2 \theta_W^\nu \simeq \sin^2 \theta_W^S = 1 - M_W^2/M_Z^2$ to within ± 0.002 for top masses less than 230 GeV. W.J. Marciano and A. Sirlin, Phys. Rev. **D26** (1980) 2695; R.G. Stuart, Z. Phys. **C34** (1987) 445.

3. W.K. Sakumoto *et al*, Nucl. Instru. Meth., **A294** (1990) 179.

4. C.H. llewellyn-Smith, Nucl. Phys. **B228** (1983) 205.

5. P.Z. Quintas *et al.*, Nevis Report 1461, Dec. 1991. Submitted to Phys. Rev. Lett.; W.C. Leung *et al*, Nevis Report 1460, Dec. 1991, Submitted to Phys. Rev. Lett.

6. M.H. Shaevitz, Nucl. Phys. **B19** (proc. supp.) (1991) 270; S.A. Rabinowitz *et al.*, submitted to Phys. Rev. Lett

7. P. Skubic *et al.*, Phys. Rev. **D18** (1978) 3115.

8. R. Brun *et al.*, CERN-DD/78/2, 1978.

9. A. Sirlin and W. J. Marciano, Nucl. Phys. **B189** (1982) 442.

10. A. De Rujula, R. Petronzio, and A. Savoy-Navarro, Nucl. Phys. **B154** (1979) 394.

11. D. Yu. Bardin and O. M. Fedorenko, Sov. J. Nucl. Phys. **30**, (1979) 418; and Private Communication.

12. H. Abramowicz *et al.*, Phys. Rev. Lett. **57** (1986) 298

13. J.V. Allaby *et al.*, Z. Phys. **C36** (1987) 611.

14. E.A. Paschos and L. Wolfenstein, Phys. Rev. **D7** (1973) 91.

15. "Precision Measurements of Neutrino Neutral Current Interactions Using a Sign-Selected Beam", Fermilab Proposal P-815, 1991.

NEUTRAL CURRENT COUPLING CONSTANTS FROM $\nu_\mu e^-$ AND $\bar{\nu}_\mu e^-$ SCATTERING

The CHARM II Collaboration

B. Akkus [10], E. Arik [10], R. Beyer [2], F.W. Büsser [3], A. Capone [8], M. Caria [7], A.G. Cocco [7], D. De Pedis [8], E. Di Capua [9], U. Dore [8], B. Eckart [7], A. Ereditato [7], D. Favart [4], G. Fiorillo [7], W. Flegel [2], C. Foos [3], A. Frenkel-Rambaldi [8], L. Gerland [3], P. Gorbunov [5], G. Grégoire [4], E. Grigoriev [5], H. Grote [2], K. Hiller [11], V. Khovansky [5], E. Knoops [4], T. Layda [3], V. Lemaître [4], W. Lippich [6], P.F. Loverre [8], D. Macina [8], A. Maslennikov [5], T. Mouthuy [2], R. Nahnhauer [11], A. Nathaniel [6], F. Niebergall [3], H. Overas [2], V. Palladino [7], J. Panman [2], G. Piredda [8], G. Rädel [3], S. Ricciardi [9], H.E. Roloff [11], A. Rozanov [2], B. Saitta [9], R. Santacesaria [8], M. Serin-Zeyrek [10], R. Sever [10], P. Stähelin [3], A. Staude [6], P. Strolin [7], P. Tolun [10], P. Vilain [1], J. Vogt [6], T. Voss [3], G. Wilquet [1], K. Winter [2], G. Zacek [2] and V. Zacek [2].

1) Inter-University Institute for High Energies (ULB-VUB), Brussels, Belgium.
2) CERN, Geneva, Switzerland.
3) II. Institut für Experimentalphysik, Universität Hamburg, Hamburg, Germany.
4) Université Catholique de Louvain, Louvain-la-Neuve, Belgium.
5) Institute for Theoretical and Experimental Physics, Moscow, Russian Federation.
6) Sektion Physik der Universität München, Munich, Germany.
7) Università e Istituto Nazionale di Fisica Nucleare (INFN), Naples, Italy.
8) Università "La Sapienza" e Istituto Nazionale di Fisica Nucleare (INFN), Rome, Italy.
9) Università di Ferrara e Istituto Nazionale di Fisica Nucleare (INFN), Ferrara, Italy.
10) YEFAM, Turkey.
11) DESY - Institut für Hochenergiephysik, Zeuthen, Germany.

presented by A. Staude

Abstract

We report on determinations of the neutral current coupling constants of the electron from the study of muon neutrino and antineutrino scattering off electrons. The results were obtained with the CHARM II detector which was exposed to the CERN wide band neutrino beam during the years 1987 to 1991. We present the determination of g_V and g_A from the measurement of $\nu_\mu e$ and $\bar{\nu}_\mu e$ cross sections with 80% of the data and a preliminary result for $\sin^2\theta_W$ from the ratio of $\nu_\mu e$ and $\bar{\nu}_\mu e$ cross sections, using the full data sample. We furthermore give results from the comparison of neutrino electron scattering with inverse muon decay and present the first measurement of the y-distributions for $\nu_\mu e$ and $\bar{\nu}_\mu e$ scattering. The agreement between our results and those from the measurements at LEP demonstrates the validity of the Standard Model over a range in momentum transfer which spans six orders of magnitude.

INTRODUCTION

The scattering of muon neutrinos off electrons is well suited for the measurement of the neutral current (NC) couplings g_V and g_A of the electron to the Z^0. The differential cross sections for $\nu_\mu e \to \nu_\mu e$ ($\bar{\nu}_\mu e \to \bar{\nu}_\mu e$) are:

$$\frac{d\sigma}{dy} = \frac{G_F m_e E_\nu}{2\pi}\left\{[g_V + (-) g_A]^2 + [g_V - (+) g_A]^2 (1-y)^2\right\} \quad (1)$$

G_F is the Fermi coupling constant, m_e the electron mass, E_ν the neutrino energy and y the relative energy transfer from the neutrino to the electron. In the Standard Model (SM), g_V and g_A are related to the electroweak mixing angle θ_W and the ratio ρ of the neutral to charged current coupling strength :

$$g_V = \rho\left(-1/2 + 2\sin^2\theta_W\right), \quad g_A = -\rho/2 \qquad (2)$$

In the CHARM II experiment neutrino electron scattering is studied at small momentum transfer ($Q^2 \approx 10^{-2}$ GeV^2). The comparison of the results with the precise measurements at the Z^0 pole with LEP [1] ($Q^2 \approx 10^4$ GeV^2) requires the application of electroweak radiative corrections and hence allows to check these, providing thus a test of the SM.

APPARATUS, SIGNAL AND BACKGROUNDS, NEUTRINO FLUXES

The CHARM II detector [2] consisted of a fine grained target calorimeter composed of glass plates and streamer tubes with 550 tons fiducial mass, followed by a muon spectrometer. For electrons with a recoil energy $E_e = 10$ GeV, typical in the experiment, the energy resolution is 1.4 GeV and the resolution of the angle θ_e between the electron and the neutrino direction is 5.4 mrad in each projection.

In neutrino electron scattering, the variable $E_e\theta_e^2 = 2m_e(1-y)$ is smaller than 1 MeV. The signature of the process is therefore a single forward going electromagnetic shower. Background arises from semileptonic neutrino reactions on nucleons and nuclei which may also give single electromagnetic showers : by single coherent or diffractive π^0 production, quasi-elastic charged current (CC) reactions of electron neutrinos and inclusive reactions with a predominantly electromagnetic final state. The individual backgrounds differ mainly by their E_e distributions. The $E_e\theta_e^2$ distribution for all of them, however, is much wider than for neutrino electron scattering.

The detector was exposed to the CERN wide band ν and $\bar{\nu}$ beams, whose main components are ν_μ resp. $\bar{\nu}_\mu$ with about 10% contamination ("wrong components") of $\bar{\nu}_\mu$, ν_e and $\bar{\nu}_e$ resp. ν_μ, $\bar{\nu}_e$ and ν_e. The abundances of ν_μ and $\bar{\nu}_\mu$ and their energy spectra were determined from quasielastic semileptonic CC events, the corresponding quantities for electron neutrinos by a Monte Carlo calculation. The fluxes were determined from well understood neutrino reactions on nucleons or nuclei. The absolute flux in the ν_μ beam was measured to $\pm 4.7\%$ with inclusive neutrino reactions as well as CC interactions [3]. The flux of the $\bar{\nu}$ beam was calculated using the flux ratio of the two beams which was measured to $\pm 2.2\%$ with four different monitor reactions (inclusive, CC, quasielastic, single π^0 production) and was also determined from the muon fluxes in the muon shield of the beam [4].

DETERMINATION OF g_V AND g_A FROM ABSOLUTE CROSS SECTIONS

The results of the analysis are based on data taken in 1987 - 1990 and have been published recently [3].

We determined g_V and g_A from a fit of a Monte Carlo prediction to the two-dimensional distributions for single electromagnetic showers $N_\nu(E_e, E_e\theta_e^2)$ and $N_{\bar{\nu}}(E_e, E_e\theta_e^2)$, observed in the ν and $\bar{\nu}$ beams. To construct the prediction we estimated the contribution of all beam components to the neutrino electron scattering signal from their respective cross sections, the fluxes, acceptances and resolutions. The couplings g_V and g_A are treated as free parameters. We added the four modeled distributions for the semileptonic backgrounds. Their abundances are additional free parameters. The fit is performed for the kinematical region $0 < E_e\theta_e^2 < 72$ MeV, 3 GeV $< E_e <$ 24 GeV.

The fourfold ambiguity expected for the result of the fit from (1) alone is reduced to a twofold one because $\approx 10\%$ of the events are arising from ν_e or $\bar{\nu}_e$ interactions, to which a CC amplitude contributes as well. One of the two remaining solutions can be excluded by results

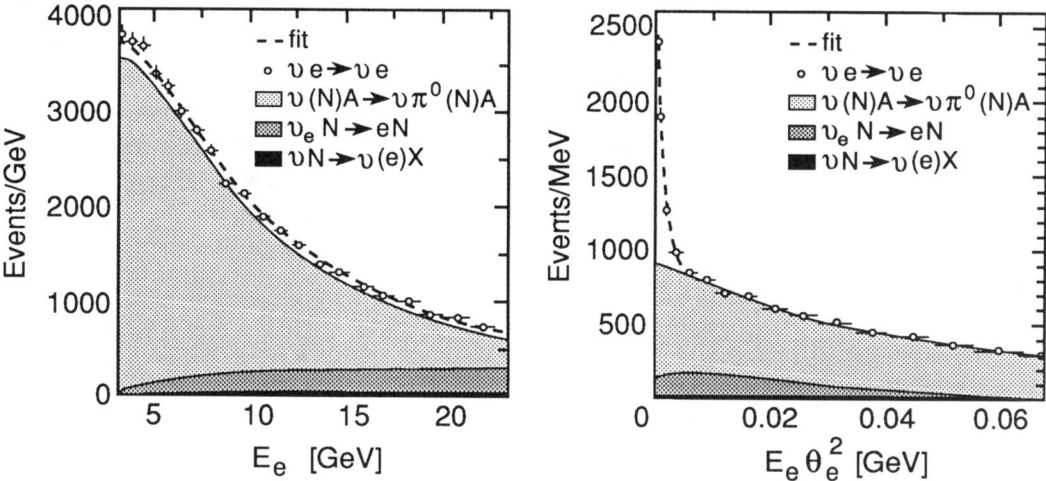

Figure 1. Experimental data for the ν beam (shown in circles) and the result of the best fit. The different background components are cumulative.

from $e^+e^- \to e^+e^-$ experiments [1]. Fig. 1 shows the projections of $N_\nu(E_e, E_e\theta_e^2)$ together with the result of the fit. The best fit gives

$g_V = -0.025 \pm 0.014$ (stat) ± 0.014 (syst),
$g_A = -0.503 \pm 0.007$ (stat) ± 0.016 (syst).

These values for g_V and g_A are effective values for neutrino electron scattering at low Q^2. Only higher order QED corrections [5], which are experiment dependent, have been applied, but no electroweak corrections. Fig. 2 shows our result together with those from previous muon neutrino electron scattering experiments [6, 7] and the latest result from LEP [1].

To compare our result with the result from LEP, electroweak radiative corrections have to be applied. However, since the two major contributions accidentally cancel, these corrections are small (0 to −0.005 for g_V and 0 to 0.004 for g_A [5]) for a wide range of top and Higgs masses and may therefore be neglected.

MEASUREMENT OF $\sin^2\theta_w$ FROM THE RATIO OF $\nu_\mu e$ AND $\bar\nu_\mu e$ CROSS SECTIONS

We present here a preliminary result based on the full data sample, which contains about 3000 neutrino electron events each for the ν and $\bar\nu$ beam. The result of a similar analysis based on one half the statistics (data from 1987 - 1989) has already been published [4].

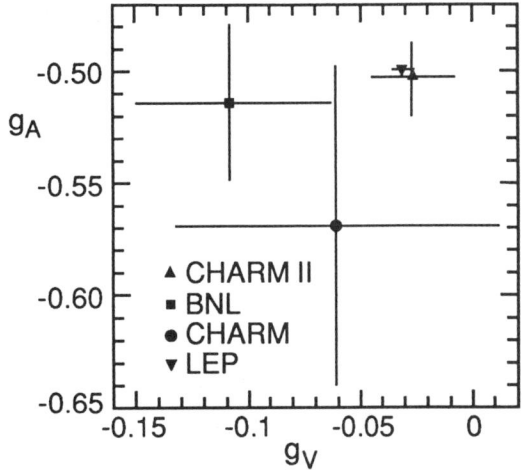

Figure 2. Comparison of our result with results from previous muon neutrino electron scattering experiments [6, 7] and with the latest LEP result [1] in the g_V - g_A plane. The errors are the statistical and systematic errors added in quadrature. For the LEP result, the error of g_A is smaller than the symbol.

The ratio $\sigma(\nu_\mu e) / \sigma(\bar{\nu}_\mu e)$ depends only on g_V / g_A or, alternatively, only on $\sin^2\theta_w$ (see (1) and (2)). Experimentally, the measurement of the cross section ratio is much simpler than the measurement of absolute cross sections. Many uncertainties, e.g. for efficiencies, acceptances and resolutions, cancel to a large extent and only the ratio of the beam fluxes has to be known.

Our analysis method is quite equivalent to the determination of $\sin^2\theta_w$ from a cross section ratio. We performed a fit to N_ν and $N_{\bar\nu}$, as described in the previous chapter, but replaced in the predictions g_V and g_A by $\sin^2\theta_w$ and assumed $\rho = 1$ to estimate the contribution from ν_e and $\bar\nu_e$ to the νe signal. Furthermore, we allow the pre-dictions for N_ν and $N_{\bar\nu}$ to vary with a common scale factor. The result of the best fit is

$$\sin^2\theta_w = 0.232 \pm 0.006 \text{ (stat)} \pm 0.007 \text{ (syst)}.$$

The electroweak corrections, required to compare this result with the effective value of $\sin^2\theta_w$ measured at LEP, are again negligible.

DETERMINATION OF $\sin^2\theta_w$ AND ρ FROM NEUTRINO ELECTRON SCATTERING AND INVERSE MUON DECAY

This analysis has been performed using the data taken 1987 - 1990 and is still preliminary.

The cross section for the inverse muon decay (IMD) $\nu_\mu + e \to \mu^- + \nu_e$, which is a CC reaction, does not depend on g_V or g_A.

IMD events are characterized by a single outgoing muon with small transverse momentum $p_\perp$ ($p_\perp^2$ is kinematically constrained to less than $2E_\mu m_e$). They appear as a peak at $p_\perp^2 = 0$ in the $p_\perp^2$-distribution of quasielastic ν_μ events. Since the $p_\perp$ - shape of the background at small $p_\perp^2$ is significantly influenced by nuclear effects, we used, after some corrections, the $p_\perp^2$ distribution of quasielastic $\bar\nu_\mu$ events for the background subtraction [8]. We obtained ≈ 14000 IMD events for the ν beam and ≈ 4000 for the $\bar\nu$ beam, due to its ν_μ component.

Since the cross section is known, IMD events have been used to determine the ν_μ flux in the ν and the $\bar\nu$ beam. The fluxes of all other beam components were then inferred from the measured beam compositions and energy spectra. We used these fluxes to calculate the predictions for N_ν and $N_{\bar\nu}$ and expressed the free parameters g_V and g_A by $\sin^2\theta_w$ and ρ. Again, we determined the free parameters by a fit and obtained

$$\sin^2\theta_w = 0.239 \pm 0.014,$$
$$\rho = 0.994 \pm 0.042$$

where statistical and systematic errors have been added in quadrature.

Since the absolute fluxes are determined with IMD events, the measurement is equivalent to a NC/CC determination for leptonic reactions.

THE SHAPE OF THE y-DISTRIBUTIONS FOR $\nu_\mu e$ AND $\bar\nu_\mu e$ - SCATTERING

As can be seen from (1), the shapes of the differential cross sections for $\nu_\mu e$ and $\bar\nu_\mu e$ scattering depend only on the ratio g_R^2/g_L^2 of the right- and lefthanded couplings g_R and g_L of the electron to the Z^0, where :

$$g_L = 1/2 (g_V + g_A), \quad g_R = 1/2 (g_V - g_A). \quad (3)$$

We have extracted the y-distributions from N_ν and $N_{\bar\nu}$ in a nearly model independent analysis. Only the knowledge about compositions and energy spectra of the beams is required.

We first constructed the ($E_e, E_e\theta_e^2$) distributions for $\nu_\mu e$ ($\bar\nu_\mu e$) events in the ν ($\bar\nu$) beam subtracting from N_ν and $N_{\bar\nu}$ the background of semileptonic events and the contributions of the wrong beam components to the neutrino electron events. The background of semileptonic reactions was determined by a fit of the modeled background components, with their abundances as free parameters, to N_ν and $N_{\bar\nu}$ in the kinematical region 6 MeV < $E_e\theta_e^2$ < 72 MeV, which contains no νe signal. The fit was extrapolated to the signal region. The contribution

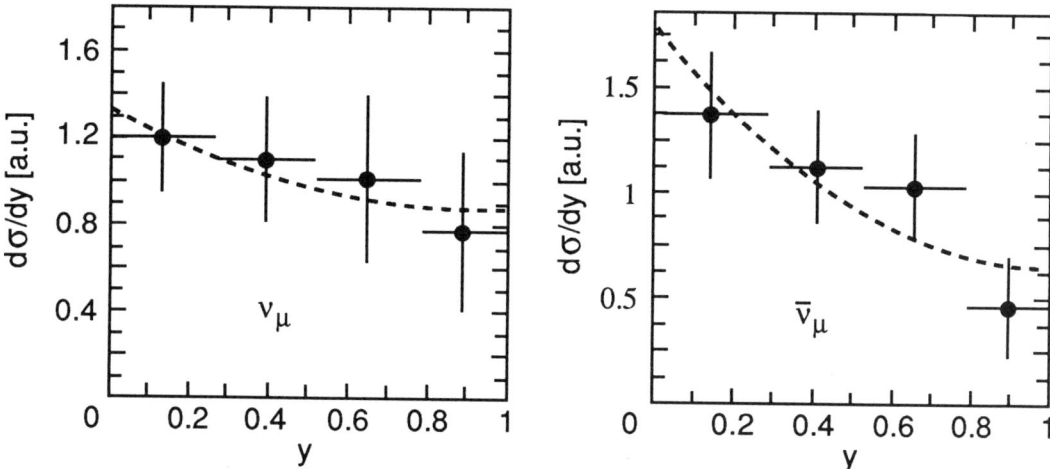

Figure 3. Unfolded y - distributions for $\nu_\mu e$ and $\bar{\nu}_\mu e$ scattering (arbitrary units). The dashed line corresponds to $g_R^2 / g_L^2 = 0.60$.

of νe events from the wrong beam components was calculated using the measured abundances for these components. This introduces a model dependence since the cross sections for the different neutrino species depend differently on $\sin^2\theta_W$ and ρ. We have assumed $\sin^2\theta_W = 0.2337$, as determined at LEP [9], and $\rho = 1$. Due to the small contribution from the wrong beam components to the signal the model dependence is, however, small and, compared to the statistical errors, negligible.

The shapes of the $(E_e, E_e\theta_e^2)$ distributions for $\nu_\mu e$ resp. $\bar{\nu}_\mu e$ events depend on the y-shape of the cross sections as well as on the ν_μ resp. the $\bar{\nu}_\mu$ energy distribution and the experimental resolutions. We determined the shapes of the differential cross sections $d\sigma/dy$ by unfolding the $(E_e, E_e\theta_e^2)$ distributions with a regularized unfolding procedure [10] which assumes nothing but smoothness for the shape of the y-distributions. Fig. 3 shows the results.

The measured y-distributions demonstrate, based on angular momentum conservation alone, the need for a NC coupling to right-handed electrons, as predicted by the SM. The best fit to the distributions gives

$$g_R^2 / g_L^2 = 0.60 \pm 0.19 \text{ (stat)}$$

in agreement with the SM parameters determined in this experiment by the other methods.

REFERENCES

1. L. Rolandi, "Precision tests of the Electroweak Interaction", *these proceedings*.
2. K. de Winter et al., *Nucl. Instrum. Methods* A278 (1989) 670.
3. P. Vilain et al., *Phys. Lett.* B281 (1992) 159.
4. D. Geiregat et al., *Phys. Lett.* B259 (1991) 499.
5. D.Yu. Bardin and V.A. Dokuchaeva, *Nucl. Phys.* B246 (1984) 221; Preprint JINR E2-86-260 (1986).
   D.Yu. Bardin, NUFITTER, a program to calculate electroweak radiative corrections for neutrino electron scattering.
6. J. Dorenbosch et al., *Z. Phys.* C41 (1989) 567
7. L.A. Ahrens et al., *Phys. Rev.* D41 (1990) 3297.
8. D. Geiregat et al., *Phys. Lett.* B247 (1990) 131.
9. The LEP collaborations : ALEPH, DELPHI, L3 and OPAL, *Phys. Lett.* B276 (1992) 247.
10. V. Blobel, Preprint DESY 84/118, Hamburg 1984.

RESULTS FROM THE KARMEN NEUTRINO EXPERIMENT

J.A.Edgington, A.Malik, B.Seligmann
Department of Physics, Queen Mary & Westfield College
Mile End Road, London E1 4NS, U.K.

G.Drexlin, V.Eberhard, K.Eitel, H.Gemmeke, G.Giorginis, W.Grandegger, M.Kleifges,
J.Kleinfeller, R.Maschuw, P.Plischke, J.Rapp, F.Raupp, J.Wochele, J.Wolf, S.Wölfle, B.Zeitnitz
Institut für Kernphysik I, Kernforschungszentrum Karlsruhe
Postfach 3640, W-7500 Karlsruhe, Germany

B.Bodmann, F.Burtak, E.Finckh, A.Glombik, T.Hanika, J.Hößl, W.Kretschmer, R.Meyer, F.Schilling
Physikalisches Institut, Universität Erlangen-Nürnberg
Erwin-Rommel-Strasse 1, W-8520, Erlangen, Germany

A.Dodd
Rutherford Appleton Laboratory
Chilton, Didcot OX11 0QX, U.K.

N.E.Booth
Department of Nuclear Physics, Oxford University
Keble Road, Oxford OX1 3RH, U.K.

Abstract

The KARMEN experimental program at the pulsed beam-stop source ISIS studies the interactions of neutrinos with energies up to 53 MeV in a 56 tonne scintillation calorimeter. Major aims include studies of specific weak couplings, weak nuclear form factors, and ν_μ - ν_e universality, by the measurement of charged current (CC) and neutral current (NC) neutrino interactions with ^{12}C. We present relevant results from the first two years of data taking.

INTRODUCTION

The KARMEN experiment is performed at the neutron spallation facility ISIS at Rutherford Appleton Laboratory. The decay of stopped pions, produced in the U/D_2O target by the 800 MeV pulsed proton beam, yields equal fluxes of ν_μ, ν_e and $\bar{\nu}_\mu$ with energies up to 52.8 MeV according to the sequence $\pi^+ \rightarrow \mu^+ + \nu_\mu$; $\mu^+ \rightarrow e^+ + \nu_e + \bar{\nu}_\mu$. The combination of a narrow proton spill (2 x 130 ns bunches at 50 Hz) with the different π^+, μ^+ lifetimes results in a prompt ν_μ burst within ~0.5 μs of beam-on-target, followed by ν_e, $\bar{\nu}_\mu$ emission, unaccompanied by ν_μ, at later times. The duty factor ~10^{-4} allows highly effective suppression of cosmic background. A high resolution calorimeter[1] containing 56 t of organic liquid scintillator is housed in a 6000 t steel bunker at a mean distance from the production target of 18 m of which 8 m is steel, reducing neutron fluxes by ~10^{-16}. The calorimeter, with >98% active mass, serves as a live target for studying various ν-induced reactions on ^{12}C (reported here) and ^1H.

NEUTRINO REACTIONS ON ^{12}C: THEORY

Figure 1 shows the superallowed transitions linking the ground state of ^{12}C to the $(1^+,1)$ analogue states of the $A = 12$ system. There have been several[2-5] calculations of the transition amplitudes of which those[3-5] that use the elementary particle treatment[6] are model-independent, requiring as input only the four weak nuclear form factors $F_M(q^2)$, $F_A(q^2)$, $F_P(q^2)$ and $F_T(q^2)$. In the absence of second-class currents, matrix elements for the charged (CC) and neutral (NC) current reactions are related, assuming isospin symmetry, by the Wigner-Eckart theorem. The form factors therefore contain all specifically nuclear information.

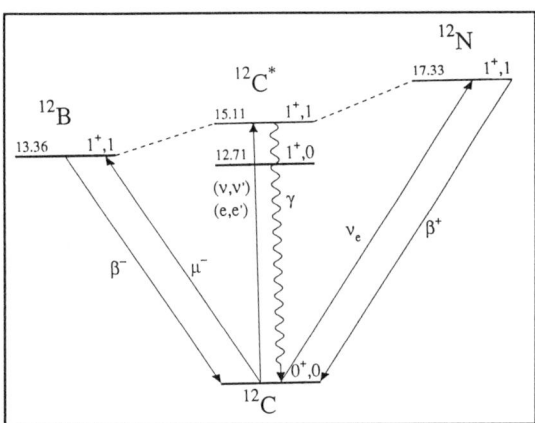

Figure 1. Transitions in the A=12 system.

The value of the weak magnetism form factor F_M at $q^2=0$ is related by the CVC hypothesis to the rate of the M1 γ-decay of the 15.1 MeV state in ^{12}C. The ft-values and angular distributions of β-decays of the analogue states in ^{12}B and ^{12}N allow $F_A(0)$ and $F_T(0)$ to be calculated; the latter is found to be small. Measurements of inelastic electron scattering and of μ⁻ capture on ^{12}C give the q^2 dependence of F_M and F_A respectively, with the result that $F_A(q^2) \propto F_M(q^2)$; different authors[4,5] use similar explicit forms which differ from dipole dependence by <5% in the low q^2 range accessible to KARMEN. The same form is taken for $F_T(q^2)$ and the contribution of the pseudoscalar term F_P is ignored on account of its m_e^2-dependent coefficient. Thus the axial vector form factor $F_A(q^2)$ dominates. The ability of KARMEN to measure CC and NC reactions ^{12}C$(\nu_e,e^-)^{12}$N$_{g.s.}$ and ^{12}C$(\nu,\nu')^{12}$C$^*(1^+,1)$ permits detailed study of ν-nucleus interactions, of the structure of weak currents, ν_μ-ν_e universality, and neutrino oscillations (not discussed here).

FIRST RESULTS FROM KARMEN

The exclusive CC reaction $^{12}C(\nu_e,e^-)^{12}N_{g.s.}$

Detection of this exclusive reaction is based on a spatially correlated delayed coincidence between an electron from the inverse β-decay of ^{12}C during the ν_e time window and a positron from the subsequent ^{12}N decay, which uniquely identifies ν-induced transitions to the ground state of ^{12}N. The data sample used for the present analysis was taken between April 1990 and June 1992 representing 2641 C of protons on target. Software cuts[7] on the time, energy and spatial correlation of the prompt and delayed signals selected 130 coincidence events. Proof that these ν-candidates are indeed due to exclusive CC reactions is given by Figure 2; the measured time and energy distributions of the prompt and delayed signals are in very good agreement with what one expects from the reaction sequence ^{12}C$(\nu_e,e^-)^{12}$N$_{g.s.} \rightarrow ^{12}$C+$e^+$+$\nu_e$. In particular the time distribution of the prompt signal, Figure 2(b), clearly shows these events to have been induced by ν_e from μ⁺-decay. From 123.7 ± 11.4 events left after background subtraction (signal:background ≈ 20:1) the cross section flux-averaged over the energy distribution of the ν_e was deduced to be

$$\langle \sigma_{excl\ CC} \rangle = [8.0 \pm 0.75_{stat} \pm 0.75_{syst}] \times 10^{-42}\ cm^2$$

Calculations[2,4,5] yielding values in the range (8.0-9.4) x 10^{-42} cm², with 10% uncertainties, are in good agreement, as is a lower resolution experiment[8] at LAMPF.

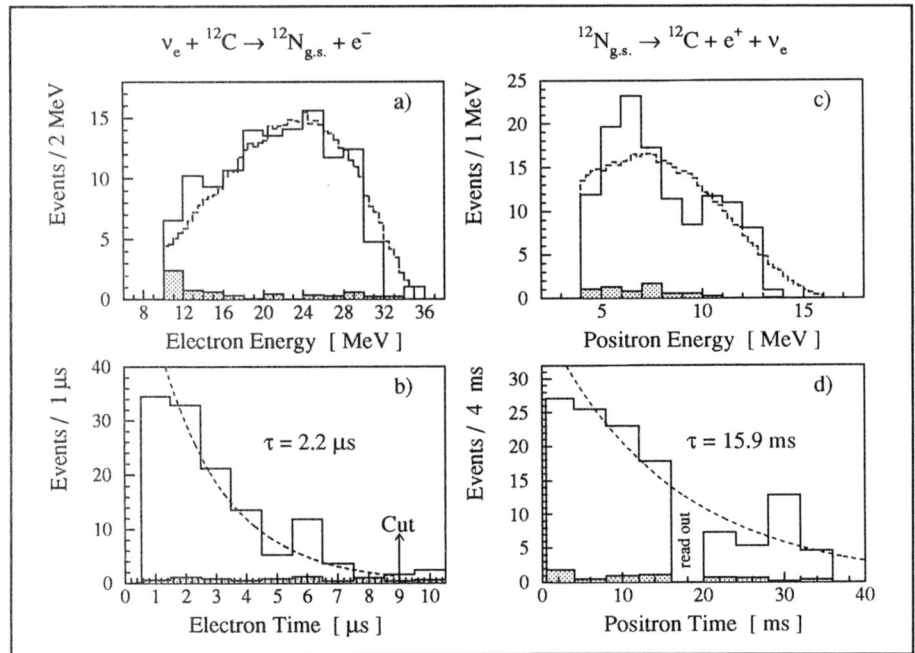

Figure 2. Energy and time distributions of prompt and delayed signals following cuts to select CC reactions on ^{12}C. Energy spectra are compared to Monte Carlo simulations of $^{12}C(\nu_e,e^-)^{12}N_{g.s.}$ (broken lines), time distributions are shown with the decay curves of μ^+ and ^{12}N superimposed. Normalised background with beam off is shown shaded.

The excellent energy resolution of our detector allowed the neutrino energy distribution to be reconstructed from the energy spectrum of prompt electrons, and the CC cross section evaluated in four energy intervals (Figure 3). This first measurement of energy dependence of a ν-nucleus cross section shows clearly the threshold behaviour, while better statistics will allow study of the q^2 dependence of F_A.

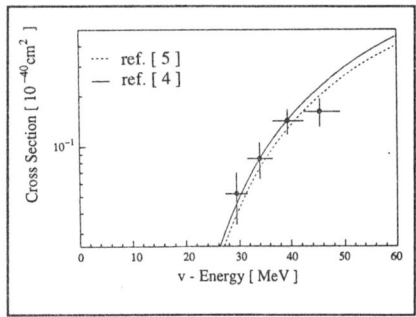

Figure 3. Energy dependence of $^{12}C(\nu_e,e^-)^{12}N_{g.s.}$

The NC reaction $^{12}C(\nu,\nu')^{12}C^*(1^+,1;15.1\ MeV)$

Improved trigger conditions since July 1990 have enabled us to make the first observation of a weak nuclear excitation, namely this NC transition between discrete nuclear states. The signal for this process is detection of a localised scintillation event of 15 MeV visible energy from photons emitted as the (1^+1) analogue state of ^{12}C decays back to the ground state with a 94% γ-decay branching ratio. In order to optimise the signal:background ratio evaluation[9] was restricted to the $\nu_e,\bar\nu_\mu$ time window of 0.5-3.5 μs after beam-on-target. The energy spectrum of events satisfying these criteria is shown, with "beam off" background subtracted, in Figure 4. Above 17 MeV there is a broad distribution of events attributable to inclusive CC reactions. Between 11 and 16 MeV lies a distinct peak which we ascribe to the reaction $^{12}C(\nu,\nu')^{12}C^*(1^+,1)$. The time distribution in this energy interval shows an exponential decrease

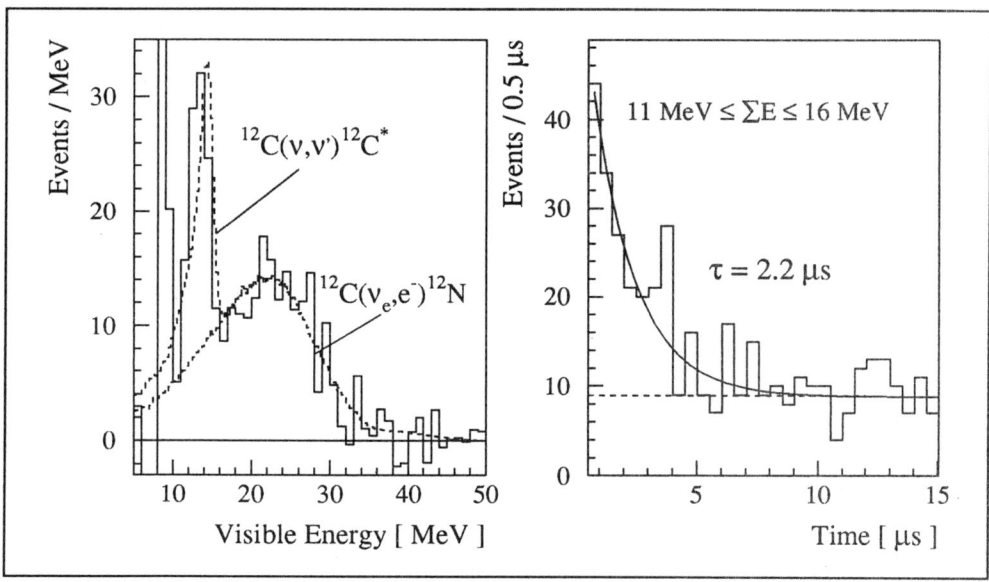

Figure 4. Spectra of single prong events satisfying the criteria described in the text. Left: Visible energy after background subtraction; broken lines show Monte Carlo simulations for NC excitation of ^{12}C, and inclusive CC reactions. Right: Time with respect to beam-on-target, cosmic background (broken line) *not* subtracted.

with a time constant of 2.2 µs indicating that these events are indeed induced by ν_e and $\bar{\nu}_\mu$ from µ$^+$ decay. We find 68 ± 13 such events leading to a flux averaged cross section for the sum of ν_e and $\bar{\nu}_\mu$ transitions of

$$\langle\sigma_{NC}\rangle = [9.5 \pm 1.8_{stat} \pm 1.35_{syst}] \times 10^{-42} \text{ cm}^2$$

again in good agreement with theory[2,3,5].

The inclusive CC reaction $^{12}C(\nu_e,e^-)^{12}N^*$

The broad structure in the energy spectrum of Figure 4 is almost all assigned to inclusive CC reactions on ^{12}C. The time distribution of events above 17 MeV again shows the 2.2 µs decay constant of µ$^+$ decay showing that they too are ν-induced. There are small contributions from ν-e scattering and from the inclusive reaction $^{13}C(\nu_e,e^-)^{13}N^*$. The former is calculable with confidence, the latter is less well understood but scales roughly with the (natural) isotopic composition of the scintillator. Together they amount to only a few percent of the events with visible energy above 17 MeV. The number of transitions to the ^{12}N ground state is directly known from our measurement of the exclusive channel, and the kinematic distribution is well reproduced by a Monte Carlo simulation (*cf* Figure 2(a)). The remaining events are thus ascribed to CC transitions to excited states of ^{12}N, all of which are proton unstable. Their energy spectrum is well fitted by a simulation which takes into account the known level structure of ^{12}N. From 108 ± 18.5 events we determine the flux-averaged cross section for the reaction $^{12}C(\nu_e,e^-)^{12}N^*$ to be

$$\langle\sigma_{CC \text{ exc.N}}\rangle = [12.3 \pm 2.2_{stat} \pm 1.5_{syst}] \times 10^{-42} \text{ cm}^2$$

contrdicting theoretical estimates of 3.7×10^{-42}cm^2 [2] and 6.4×10^{-42}cm^2 [10], and substantially larger than expected from measurements[8] at LAMPF of the exclusive, and total inclusive, cross sections. This discrepancy is unresolved and demonstrates a need, here and in similar situations in ν-astrophysics, for further study of ν-induced transitions to excited nuclear states.

EQUALITY OF $\bar{\nu}_\mu$-ν_e NC COUPLING

Apart from an isospin factor $1/\sqrt{2}$, the matrix elements of the dominant isovector axial vector currents are the same[4,5] for the NC (CC) reactions $^{12}C(\nu,\nu')^{12}C^*(1^+ 1)$ and $^{12}C(\nu_e,e^-)^{12}N_{g.s.}$. As the former is induced by fluxes of $\bar{\nu}_\mu$, ν_e of equal intensity, the ratio

$$R = \sigma_{NC}(\nu_e+\bar{\nu}_\mu)/\sigma_{CC}(\nu_e)$$

is about 1, provided that ν_e and $\bar{\nu}_\mu$ couple in the same way to the Z°. The exact expectation for this ratio, taking into account both the small ν-$\bar{\nu}$ difference in the NC matrix elements and the slightly different energy spectra of ν_e and $\bar{\nu}_\mu$, is[4] $R = 1.08$, whereas from our measurements we get $R = 1.19 \pm 0.26$. Better statistics will soon improve this flux independent test of flavour universality in the neutrino NC coupling.

A NEW LIMIT ON SCALAR COUPLING

Muon decay is described in the standard model by a chiral hamiltonian in which the left-handed vector coupling constant $g^V_{LL}=1$, all other couplings being zero. Measurements on charged leptons in μ^+ decay are consistent with this; in particular the Michel parameter

$$\rho = \tfrac{3}{4}\{1-[|g^V_{LR}|^2+|g^V_{RL}|^2+2|g^T_{LR}|^2+2|g^T_{RL}|^2 + \Re(g^S_{LR}g^{T*}_{LR}+g^S_{RL}g^{T*}_{RL})]\}$$

describing the positron energy spectrum is found to have its V-A value of ¾. Fetscher has recently remarked[11], however, that the good energy resolution of the KARMEN calorimeter allows a direct measurement of the ν_e energy spectrum described by the analogous parameter

$$\omega = 3/16\{|g^S_{LL}|^2+|g^S_{RR}|^2+4|g^V_{LR}|^2+4|g^V_{RL}|^2 + |g^S_{LR}+2g^T_{LR}|^2+|g^S_{RL}+2g^T_{RL}|^2\}$$

and that an experimental upper limit on ω would for the first time set an upper limit on the left-handed scalar coupling constant $|g^S_{LL}|$.

In deriving the energy dependence of the inverse β-decay $^{12}C(\nu_e,e^-)^{12}N_{g.s.}$ (Figure 3) we have implicitly assumed the V-A form for μ decay. However Fetscher shows[11] that a non-zero value of ω has a significant effect only near the maximal ν energy and that the slope of the excitation function is otherwise almost independent of ω, allowing the effect of $\omega \neq 0$ to be disentangled. By parametrising our results he finds $\omega \leq 2.6 \times 10^{-2}$ and derives a new limit of

$$|g^S_{LL}| \leq 0.37 \text{ (68\% c.l.)}.$$

CONCLUSIONS

In its first two years the KARMEN experiment has made a precision measurement of an allowed CC nuclear transition and confirmed the dominance of the axial vector form factor. A measurement of reactions leading to excited states reveals deficiencies in the understanding of ν-induced forbidden transitions. We have made the first observation of a NC transition between nuclear states, and confirmed the assumption of minimal breaking of isospin symmetry. Our data provide a direct, flux-independent test of the universality of ν-Z° couplings, and enable a new upper limit to be set on scalar couplings in the weak interaction.

REFERENCES

1. G.Drexlin et al, Nucl.Inst.Meth. **A289** (1990) 490
2. T.W.Donnelly, Phys.Lett. **B34** (1973) 93; and private communication (1992)
3. J.Bernabéu & P.Pascual, Nucl.Phys. **A324** (1979) 365
4. M.Fukugita, Y.Kohyama and K.Kubodera, Phys.Lett. **B212** (1988) 139
5. S.L.Mintz et al, in Prog.Nucl.Phys. (Elsevier, 1991) p.290
6. C.W.Kim & H.Primakoff, in Mesons in Nuclei I (North-Holland, 1979) p.67
7. B.Bodmann et al, Phys.Lett. **B280** (1992) 198
8. D.A.Krakauer et al, Phys.Rev. **C45** (1992) 2450
9. B.Bodmann et al, Phys.Lett. **B267** (1991) 321
10. E.Kolbe et al, Nucl.Phys. **A540** (1992) 599
11. W.Fetscher, preprint, ETH Zürich (1992)

Structure Functions and Spin Physics

Parallel Session 6

Sanctuario de Guadalupe, Dallas

Conveners: M. Karliner (Tel Aviv)
S. Mishra (Harvard)

NMC RESULTS ON STRUCTURE FUNCTIONS

Ewa Rondio
NMC Collaboration
Institute for Nuclear Studies
Hoza 69
Warsaw, Poland

Abstract

The results of the New Muon Collaboration measurements of deep inelastic muon scattering on hydrogen and deuterium are presented. The data extend the range of previous measurements towards low x. They are in good agreement with the SLAC and BCDMS results and show differences with respect to the EMC ones. At low x these data are not described by existing parton parametrisations. The Gottfried sum calculated from these data agrees with the previously published NMC result and is significantly lower than $1/3$, the value expected for a flavor symmetric sea, and agrees with the previously published NMC result. The difference between the ratio of longitudinally to transversely polarised virtual photon cross sections for the deuteron and proton, $R^d - R^p$, is consistent with zero down to $x = 0.01$.

INTRODUCTION

The nucleon structure function $F_2(x, Q^2)$ reflects the momentum distribution of quarks in the nucleon, an important aspect of its internal structure. Knowledge of the structure function of the proton (F_2^p), and the deuteron (F_2^d) has steadily improved in recent years, due to deep inelastic electron and muon scattering experiments[1-3], but significant discrepancies between some of these results remain.

In deep inelastic scattering of charged leptons the differential cross section for the one-photon exchange can be written as:

$$\frac{d^2\sigma(x, Q^2)}{dx\,dQ^2} = \frac{4\pi\alpha^2}{Q^4} \cdot \frac{F_2(x, Q^2)}{x} \cdot$$
$$\cdot \left\{ 1 - y - \frac{Q^2}{4E^2} + \frac{y^2 + Q^2/E^2}{2(1 + R(x, Q^2))} \right\}, \quad (1)$$

where $-Q^2$ is the four-momentum transfer squared and E is the energy of the incident muon. The two scaling variables x and y are defined as $x = Q^2/2M\nu$ and $y = \nu/E$, where ν is the energy of the virtual photon and M the proton mass. $F_2(x, Q^2)$ is the nucleon structure function and R the ratio of longitudinally to transversely polarised virtual photon absorption cross sections.

In order to determine F_2 from the present data, a parametrisation of R has to be assumed. In this analysis the parametrisation of reference [4] was used.

THE EXPERIMENT

This experiment (NMC–NA37) was performed at the M2 muon beam line of the CERN SPS. The data presented here were taken during 1986 and 1987 at nominal incident muon energies of 90 and 280 GeV. The spectrometer was an upgraded version of the

EMC apparatus. Improvements relevant to this analysis are described in references [5]

The measurement was simultaneous on the proton and deuteron using two similar pairs of 3 m long targets exposed alternately to the beam. In one pair the upstream target was liquid hydrogen and the downstream target liquid deuterium, while in the other pair the order was reversed. The simultaneity of the measurements greatly reduced the uncertainty of the relative normalisation between the proton and deuteron structure functions. This allows a very precise measurement of the structure function ratio F_2^n/F_2^p.

It was found that some of the large drift chambers used to reconstruct the scattered muon tracks suffered inefficiencies due to large event-related backgrounds. As these inefficiencies are not fully understood, these chambers were not used in analysis of F_2 structure functions. The spectrometer's acceptance was then limited by the size of the smaller proportional chambers at the same position. In the analysis of the structure function ratio however all the avaliable data were included.

STRUCTURE FUNCTION ANALYSIS

An iterative method was employed to extract the structure functions. In this method the spectrometer acceptance was determined with a Monte Carlo simulation; each accepted Monte Carlo event was weighted with the inclusive cross section, i.e. the one-photon exchange cross section together with contributions from radiative and other higher order processes. These weights were computed from an initial choice of F_2 and a fixed parametrisation of R [4]. A comparison of the normalised yields of data and accepted Monte Carlo events permitted new values of $F_2(x, Q^2)$ to be determined. Parametrisations of the new F_2 values were used to recompute the one photon cross section and the radiative contributions for use in the subsequent iteration. The procedure was repeated until the values of F_2 changed by less than 0.2% – typically after two or three iterations.

Radiative Corrections

The radiative contributions to the cross section were calculated using the method of Akhundov, Bardin and Shumeiko [6]. This procedure contains the most complete treatment of higher order corrections available. The inputs to the calculation were taken from recent descriptions of available data as discussed in reference [7]. In the kinematic range of the present measurement the largest radiative contributions to the cross section are less than 35%.

The procedure was compared with that of Mo and Tsai [8] including vacuum polarisation by quark and τ loops and electroweak interference terms. The differences between the results from the two schemes were always less than 2%.

Results on F_2

The kinematic region covered by the data is $0.006 \leq x \leq 0.6$ and $0.5 \leq Q^2 \leq 55$ GeV^2.

The measured structure functions for the proton are shown versus Q^2 for each bin in x in figure 1. The data clearly exhibit the scaling violations expected from perturbative QCD. The slopes, $d \ln F_2/d \ln Q^2$, are strongly positive at low x and become negative at larger values of x. The error bars represent the statistical errors. The systematic errors due to the radiative corrections, incident and scattered muon energy calibrations, reconstruction inefficiency, functional form of the parametrisation and the acceptance uncertainty were added in quadrature and are shown as the bands. These bands are plotted relative to the function fitted to the data. The overall

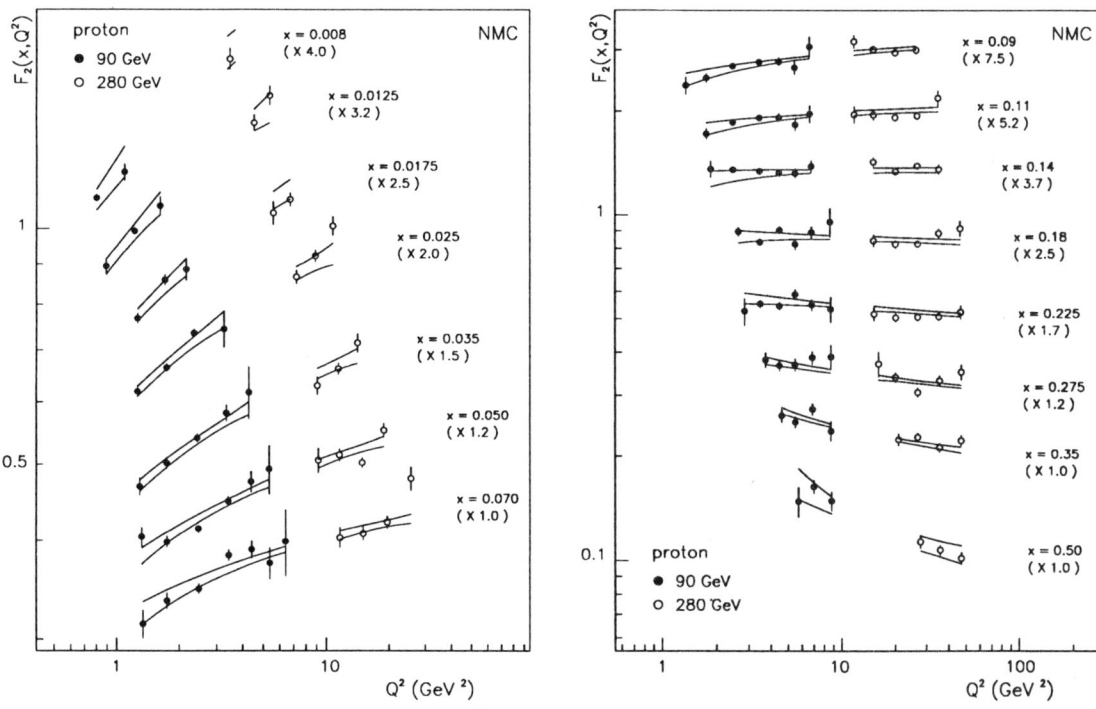

Figure 1. The proton structure function F_2^p. Data in each x bin have been scaled by indicated factor for clarity. The error bars represent statistical errors, the bands the total systematic error.

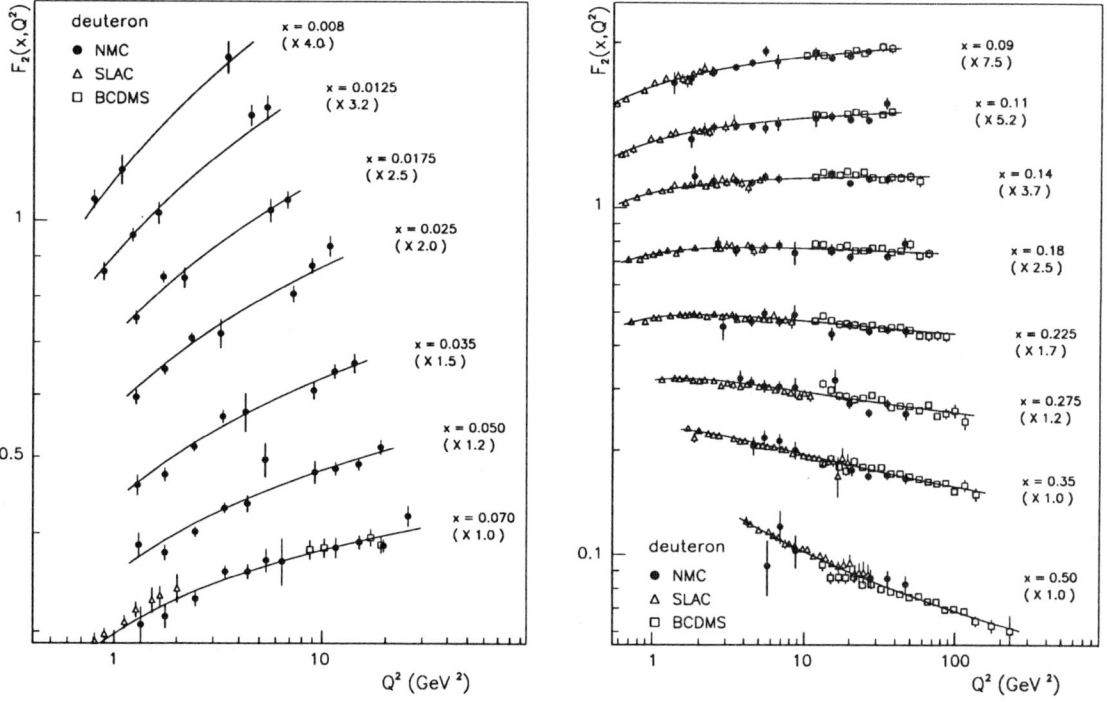

Figure 2. The results of F_2^d from present experiment (filled symbols) compared with those of SLAC (triangles) and BCDMS (squares). The curve is the result of the fit to all three data sets.

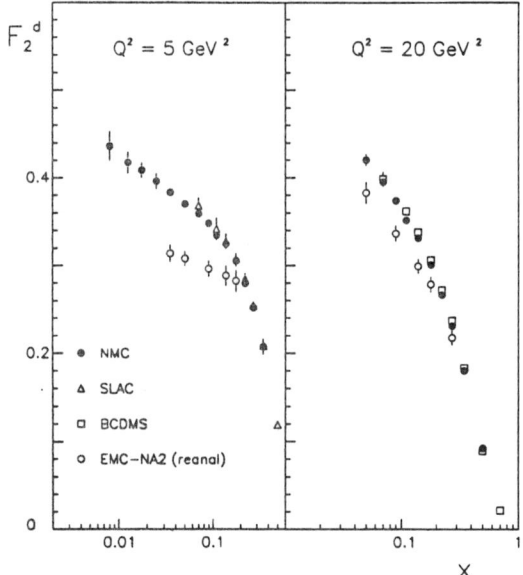

Figure 3. The results of F_2^d from the NMC experiment compared with those of EMC, SLAC and BCDMS as a function of x at Q^2 of 5 and 20 GeV^2.

normalisation uncertainty 1.6%(2.6%) of the 90(280 GeV)) data is not included in the error bands.

Comparison with Other Experiments

The data for the deuteron are shown together with those of SLAC [3] and BCDMS [2] in fig.2. It shows very good agreement between all three experiments. The present data cover part of the Q^2 region of each of the other experiments, and extend to much lower x. The curve plotted in this figure is the result of a fit to the three data sets.

The EMC-NA2 data [1] have recently been re-analysed [9], using the QCD prediction for R in place of the $R = 0$ assumed in the original analysis. These deuteron data are compared with the present results as a function of x for two different values of Q^2 in figure 3. The SLAC and BCDMS data are also shown. Systematic differences with EMC of up to 20%

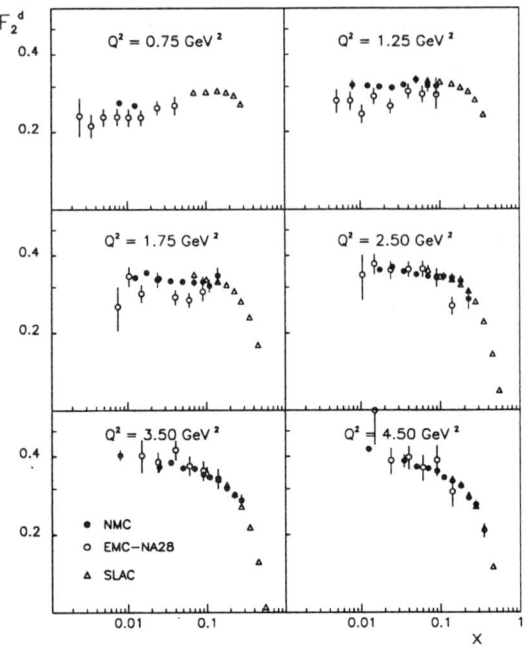

Figure 4. The results on F_2^d from NMC compared with those of EMC-NA28 and SLAC as a function of x at several Q^2 values.

at low x are seen. In the light of our studies of the reconstruction losses in the large drift chambers it seems likely that the discrepancies at low x are due to such inefficiencies affecting the EMC data. A comparison with previously published proton data [1,2,3] leads to similar conclusions.

In figure 4 the deuteron data are plotted versus x for several bins of Q^2 together with the EMC-NA28 low x measurement [10], and the SLAC data [3]. The EMC-NA28 data are in fair agreement with the present results. Of interest in this figure is the clear x-independence of the structure functions for $Q^2 \leq 2.5$ GeV^2 and $x \leq 0.1$ as expected from a simple Regge theory.

The low x behaviour of the structure functions (or the parton distributions) is important in determining the reaction rates to be expected in future experiments at higher energies (LHC, SSC). The comparison of the present results with the structure functions as calculated from recent phenomenological parton distributions [11] shows that none of them can properly describe the new data in the low x region.

GOTTFRIED SUM

In the quark-parton model the Gottfried sum $S_G = \int (F_2^p - F_2^n) dx/x$ can be expressed, under the assumption of isospin symmetry between proton and neutron, in terms of quark charges and their momentum distributions as:

$$S_G = \frac{1}{3} + \frac{2}{3} \int_0^1 (\bar{u}(x) - \bar{d}(x)) dx. \quad (2)$$

The first term gives contribution from the valence quarks, the second term vanishes for a flavor symmetric sea ($\bar{u} = \bar{d}$). The result presented by NMC [12] was determined from the precisely measured structure function ratio, F_2^n/F_2^p, and then avaliable F_2^d. The value obtained in this way is $S_G = 0.240 \pm 0.016$ with systematic and statistical errors added in quadrature. This is significantly below quark-parton model expectations for flavor symmetric sea ($\bar{u} = \bar{d}$).

The contribution from the measured range in x (0.004-0.8) was found to be $0.227 \pm 0.007(stat) \pm 0.014(syst)$. The ratio F_2^n/F_2^p from the present analysis is consistent with that presented in [7], albeit with larger statistical errors (as only a subset of the data is used). If the presently determined F_2^d had been used in the previous analysis, the value of the Gottfried sum in the measured range would be 0.234 ± 0.008, where the error is statistical only. This is in agreement with the published value. Both results for $F_2^p - F_2^n$ and the integral are shown in fig.5. Furthermore, the Gottfried sum obtained directly from F_2^p and F_2^d was found to be consistent with this result (see M.Virchaux contribution to this session).

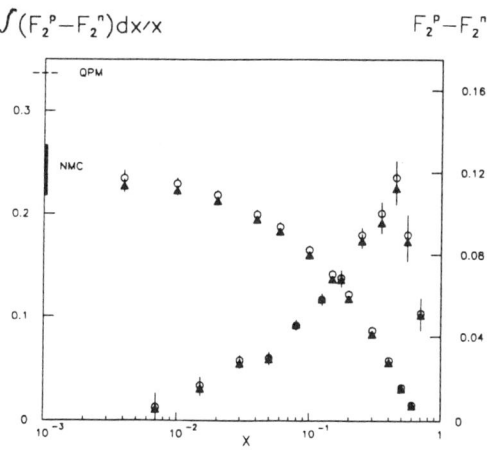

Figure 5. The difference $F_2^p(x) - F_2^n(x)$ (scale on the right) and values of the integral from x to 1 (on the left scale) at $Q^2 = 4$ GeV^2. The triangles are the results from ref.12 obtained with the old F_2^d and the circles with the new F_2^d presented here (ref.5).

$R^d - R^p$ AT LOW X

From the high precision measurement of the cross section ratio σ^d/σ^p at different energies it is possible to estimate whether the function $R(\sigma_L/\sigma_T)$ is different for the deuteron and proton.

The NMC has measured this ratio with high precision at 90 and 280 GeV incident muon energies with negligible errors for the relative normalisation. This ratio can be expressed in terms of structure function ratios, R^d and R^p (independent of the beam energy E_i) as follows:

$$\frac{\sigma^d}{\sigma^p}(E_i) = \frac{F_2^d}{F_2^p} \cdot \frac{1 + R^d}{1 + R^p} \cdot \frac{1 + z_i R^d}{1 + z_i R^p} \quad (3)$$

with all the dependence of E_i in z_i defined by:

$$\frac{1}{z_i} = 1 + \frac{1}{2} \cdot \frac{y_i^2 + Q^2/E_i^2}{1 - y_i - Q^2/4 E_i^2}. \quad (4)$$

The sensitivity to differences in R is much larger than to individual R values. From the fit to cross section ratios σ^d/σ^p, R differences were obtained. The method is discussed in detail in reference [13]. The values

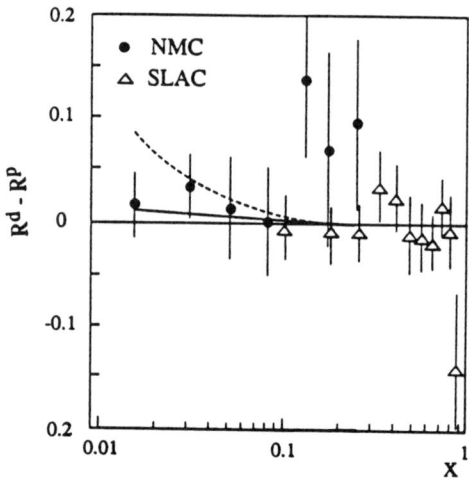

Figure 6. The measured values of $R^d - R^p$ as a function of x. Also shown are the results from SLAC analysis. The line corresponds to QCd prediction and the dashed line to QCD prediction with gluon distribution enlarged by 10%.

for $\Delta R = R^d - R^p$ are shown as a function of x in fig.6 and compared with SLAC results [4] at high x. No significant x dependence is seen and the average value $\Delta R = 0.031 \pm 0.016(stat) \pm 0.011(syst)$ is compatible with zero.

SUMMARY

We have presented new measurements of the proton and deuteron structure functions over a wide kinematic range: $0.006 \leq x \leq 0.6$ and $0.5 \leq Q^2 \leq 55$ GeV^2. The data exhibit logarithmic scaling violations down to small values of x, even at low Q^2. In the range of overlap with the previous SLAC and BCDMS data good agreement is observed between the three experiments. Clear x and Q^2 dependent differences with the EMC-NA2 data are seen. Recent parametrisations of parton distributions fail to describe precisely the structure functions at low x. The Gottfried sum value determined with the new F_2^d is consistent with the previous determination and significantly lower than 1/3. The $R^d - R^p$ is consistent with zero down to $x = 0.01$.

REFERENCES

1. EMC-NA2, J.J.Aubert et al., *Nucl.Phys.* B259 (1985) 1; *Nucl.Phys.*B293(1987)740.
2. BCDMS, A.C. Benvenuti et al., *Phys.Lett.* B223 (1989) 485; *Phys.Lett.* B237 (1990) 592.
3. L.W. Whitlow et al., *Phys.Lett.* B282 (1992) 475.
4. L.W. Whitlow et al., *Phys.Lett.* B250 (1990) 193.
5. NMC, P. Amaudruz et al., preprint CERN PPE /91-124, accepted for publ. in *Phys.Lett.*
6. A.A. Akhundov et al, *Sov.J.Nucl.Phys.* 26 (1977) 660; 44 (1986) 988; JINR-Dubna preprints E2-10147 (1976), E2-10205 (1976), E2-86-104 (1986); D. Bardin and N. Shumeiko, *Sov.J.Nucl.Phys.* 29 (1979) 499.
7. NMC, P. Amaudruz et al., *Nucl.Phys.* B371 (1992) 3.
8. L.W. Mo and Y.S. Tsai, *Rev.Mod.Phys.* 41 (1969) 205; Y.S. Tsai, SLAC-PUB-848 (1971).
9. K. Bazizi and S.J. Wimpenny, preprint UCR/DIS/91-02.
10. EMC-NA28, M. Arneodo et al., *Nucl. Phys.* B333 (1990) 1.
11. H Plothow-Besch, "Parton Density Functions" in *Proc. 3rd Workshop on Detector and Event Simulation in High Energy Physics* 1991; Program PDFLIB in CERN Program Library Pool W5051 (1991).
12. NMC, P. Amaudruz et al., *Phys.Rev.Lett.* 66 (1991) 2712.
13. NMC, P. Amaudruz et al., preprint CERN PPE /91-134, accepted for publ. in *Phys.Lett.*

PRELIMINARY QCD ANALYSIS OF NUCLEON STRUCTURE FUNCTIONS F_2^p AND F_2^d FROM NMC

Marc Virchaux
DAPNIA/SPP, C.E. Saclay
F-91191, Gif-sur-Yvette CEDEX, France,
for the New Muon Collaboration.

Abstract

We present a preliminary next-to-leading order QCD analysis of measurements of the nucleon structure function F_2 by the NMC collaboration. The measured scaling violations of F_2 are in good agreement with the expectations from perturbative QCD. The flavour non-singlet and singlet quark distributions as well as the gluon distribution are extracted with good precision down to low values of x (0.008). These results are compared to low-x extrapolations based on recent parton distribution fits. Also given is a preliminary result on the strong coupling constant α_S.

INTRODUCTION

Structure functions, F_2, measured in deep inelastic scattering are an important source of information on the constituents of nucleons. A measurement of their dependence on the Bjorken scaling variable x and on Q^2 permits the underlying parton distributions to be extracted. These play an essential role in the interpretation of measurements at present and future hadron colliders. The measurement of Bjorken scaling violations in deep inelastic scattering provides a test of QCD and allows for a precise determination of the only free parameter in the theory : the strong coupling constant α_S.

In the QCD parton model, scaling violations of F_2 originate from the Q^2 evolution of quark and gluon distributions. This Q^2 evolution is due to the QCD processes of gluon radiation and $q\bar{q}$ pair creation and is described by the Altarelli-Parisi evolution equations [1]. In addition, non perturbative contributions to the Q^2 dependence of F_2 are due to the interaction of the struck quark with the spectator quarks. These higher twist contributions, which are not described by the QCD evolution equations, behave like a power series in $1/Q^2$ and may thus become important at low values of Q^2 (see e.g. [2]).

We present here a next-to-leading order QCD analysis of preliminary measurements [3] of the nucleon structure function F_2 from deep inelastic muon scattering on hydrogen and deuterium targets by the New Muon Collaboration (NMC, CERN-NA37). The final version of these data were presented in this session together with a comparison with results from previous experiments [4]. The unique feature of the NMC measurements is the high statistics gathered down to very low values of x (0.008).

QCD ANALYSIS : METHOD

We have simultaneously fitted the QCD prediction to the measurements of F_2^p and

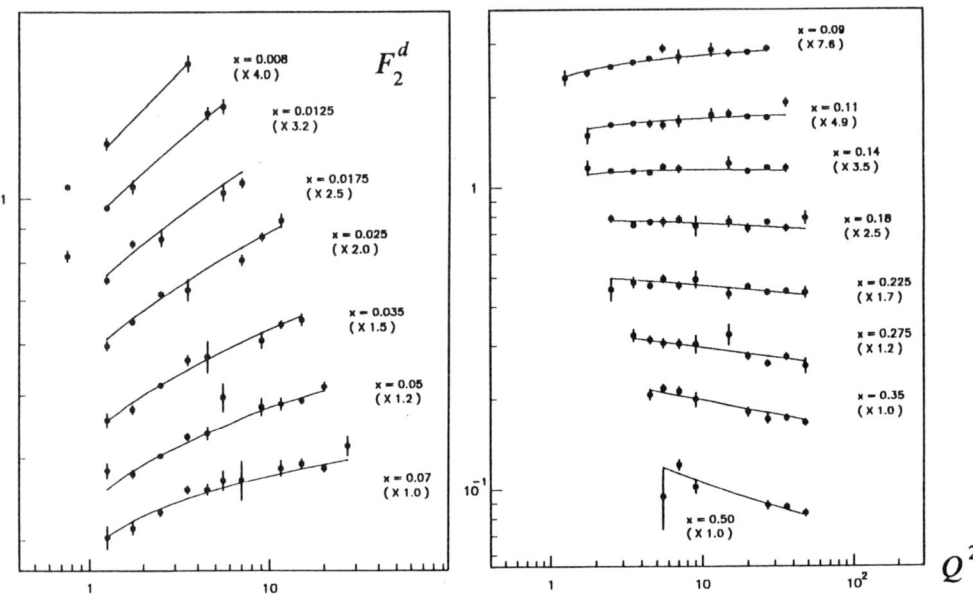

Figure 1. The structure function F_2^d. The curves correspond to the QCD fit to the data described in the text.

F_2^d. The large scaling violations at low x observed in the data are primarily due to gluons. The main aim of the present QCD analysis is then to extract, albeit indirectly through the Altarelli-Parisi equations, the gluon distribution at low x with good accuracy.

For this purpose the structure functions were decomposed into flavour non-singlet and singlet parts. In contrast to the singlet structure function, the Q^2 evolution of the non-singlet structure functions, or quark distributions, do not depend on the gluon. The quark distributions and the gluon distribution were parametrised as functions of x at a fixed value of $Q^2 = 7$ GeV^2. The parametrisation of the gluon distribution was chosen to be $xG(x) = A(1-x)^\eta P(x)$ with $P(x)$ a two parameter function constructed such that it could only be different from unity for $x < 0.1$. The Q^2 evolution of the parton distributions was obtained from a numerical integration of the Altarelli-Parisi equations in next-to-leading order. The calculation was performed in the $\overline{MS}$ renormalisation scheme using a program developed from the one described in [5]. The agreement between the QCD predictions for $F_2^p(x, Q^2)$ and $F_2^d(x, Q^2)$, computed from the quark and gluon distributions, and the data was quantified through a χ^2 evaluation. Optimal values for the 10 parameters describing the x dependence of the input distributions were then obtained from a minimisation of this χ^2 based on MINUIT [6]. In addition a free parameter was introduced to describe the relative normalisation of the 90 and 280 GeV data sets.

To avoid complications due to the correlation at low x between the gluon distribution and the strong coupling constant α_S we used as an input to the fit the result from ref. [2]: $\alpha_S(M_Z^2) = 0.113 \pm 0.003 \pm 0.004$, where the first error is experimental and the second theoretical. In the fit the momentum sum rule was imposed, that is, the singlet quark distribution and the gluon distribution were extrapolated to $x = 0$ and their integrals required to add up to unity. The fit was performed simultane-

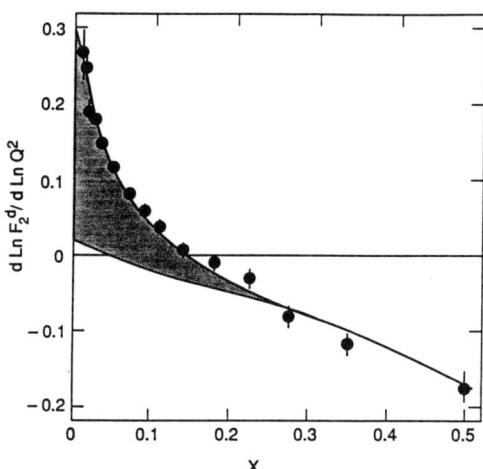

Figure 2. The slope $d\ln F_2^d/d\ln Q^2$ versus x. The upper (lower) curve corresponds to the QCD prediction with (without) the contribution from gluons.

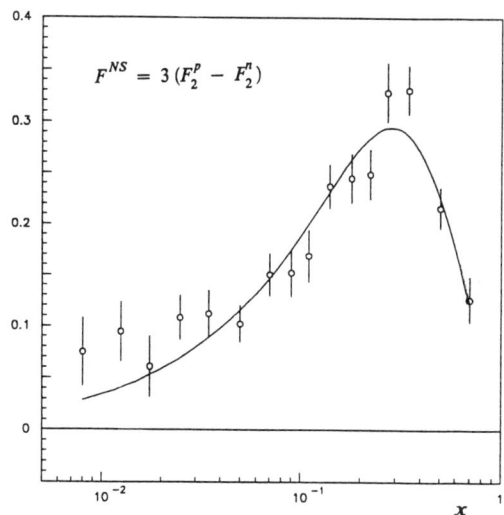

Figure 3. The non-singlet structure function $F_2^p - F_2^n$ at a fixed value of Q^2. The curve corresponds to the fit described in the text.

ously on F_2^p and F_2^d measured at $Q^2 > 1$ GeV^2; this requirement excluded from the analysis four data points at low x.

In QCD analyses of structure functions it is usually assumed that the Q^2 evolution of F_2^d is that of a pure flavour singlet. We have explicitly taken into account the small non-singlet component of F_2^d (proportional to the difference between strange and charmed quark distributions) with an assumed x dependence from the parton distribution set KMRS-B0 of Kwieciński et al. [7]. Its influence on the result of the fit was found to be negligible.

QCD ANALYSIS : RESULTS

The QCD fit provided a good overall description of the data ($\chi^2/df = 315/249$, statistical errors) as shown in figure 1 for F_2^d. The relative normalisation of the 90 and 280 GeV data sets was found to be 1.024 ± 0.004 ; in the fit the 90 GeV data were lowered by 1.2% with a corresponding rise of the 280 GeV data. This fitted value is compatible with the estimated relative normalisation uncertainty

(2.2%).[1] The slopes $d\ln F_2^d/d\ln Q^2$, averaged over Q^2, are plotted versus x in figure 2. The lower curve in figure 2 (which is indicative only) corresponds to the QCD prediction in absence of gluons : the shaded area illustrates the large contribution to the slope from gluons at low x.

The non-singlet structure function $F_2^p - F_2^n$ is shown at a fixed value of $Q^2 = 7$ GeV^2 in figure 3, together with the result from the fit. The integral of the corresponding quark distribution, parametrised as $A\,x^\alpha(1-x)^\beta$, yielded a value for the Gottfried sum : $S_G = 0.242 \pm 0.020$ (stat+syst), in agreement with the result of ref. [8].

The x dependence of the deuteron quark distribution (corresponding to $18/5\,F_2^d$) and of the gluon distribution are shown in figure 4 at a fixed value of $Q^2 = 7$ GeV^2. The shaded bands around the curves indicate the statistical and experimental systematic errors added in quadrature : the deuteron quark (gluon)

---
[1]This preferred relative normalisation shift is reduced (from 2.4% to about 1.1%) when using the final version of the NMC data [4], the QCD analysis of which is under way.

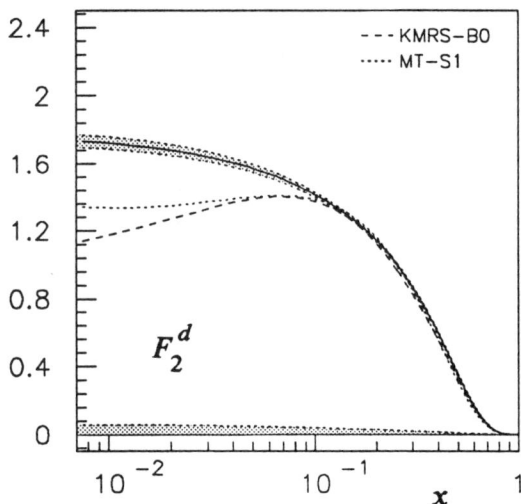
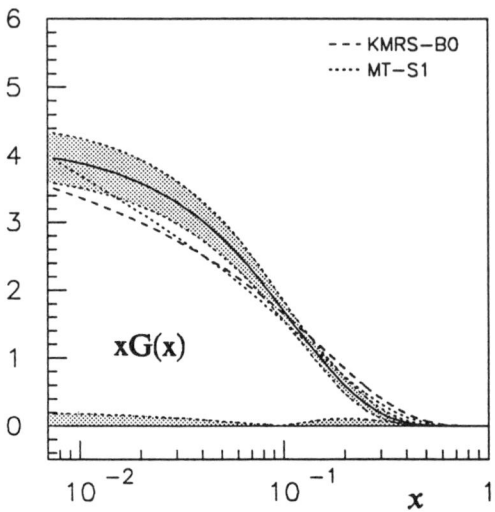

Figure 4. The deuteron quark distribution (left) and the gluon distribution (right) obtained from the QCD analysis described in the text. The combined statistical and systematic errors are indicated by the shaded bands around the curves. The normalisation error is shown by the bands at the bottom. Also shown are the distributions from KMRS-B0 and MT-S1.

distribution is extracted with an uncertainty of about ±3% (±10%) at $x = 0.01$. This does not include the uncertainties due to a 3% normalisation error on F_2 (shown separately by the bands at the bottom of figure 4) nor that due to the error on the input value of α_S (see below). The fractions of the nucleon momentum carried by quarks and gluons at $Q^2 = 7$ GeV^2 were found to be 0.56 and 0.44 respectively. Compared to results from previous structure function data, this determination of the gluon distribution extends to much lower values of x and is more precise. This is illustrated in figure 5 where the present determination is compared to previous ones from the CDHSW measurements [9] and the combined SLAC and BCDMS data [2].

Releasing the momentum sum rule did not significantly change the fitted distributions and yielded for the total momentum carried by quarks and gluons a value of 1.03 ± 0.04 (stat+syst), consistent with unity. The sensitivity of the fitted quark distributions to the value of α_S was found to be negligible whereas raising (lowering) $\alpha_S(M_Z^2)$ by the experimental error of 0.003 lowered (raised) the gluon distribution at low x by about one standard deviation. This influence of α_S on the gluon distribution was expected as, for given F_2 data at low x, the fitted gluon distribution varies in inverse proportion to the assumed value of α_S. So the constraint of the momentum sum rule which can legitimately be imposed in the fit because of the narrowness of the unmeasured region at low x (0. to 0.008) should yield a measurement of α_S. Indeed, leaving α_S a free parameter in the fit, we obtain a (preliminary) value of $\alpha_S = 0.244 \pm 0.023$ (stat+syst) at $Q^2 = 10$ GeV^2. Extrapolation to M_Z^2 gives: $\alpha_S(M_Z^2) = 0.117 \pm 0.005$ (experimental error), consistent with earlier results from deep inelastic scattering [2] and from other processes [10]. In contrast to previous measurements from deep inelastic scattering however, this fairly precise determination of α_S arises mainly through the constraint of the momen-

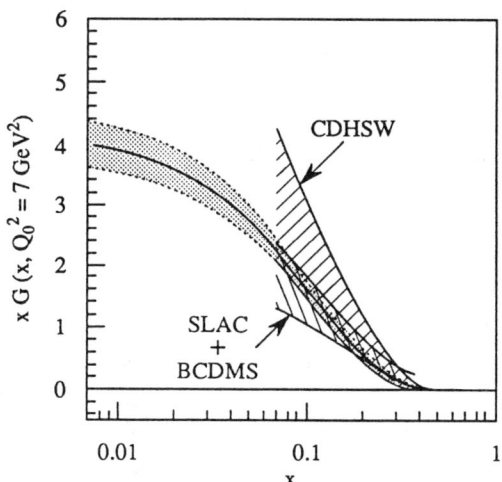

Figure 5. The gluon distribution determined in this analysis compared to previous determinations in deep inelastic scattering [2, 9].

tum sum rule, and not from the measured scaling violations at $x > 0.3$.

To investigate the sensitivity to possible higher twist effects, we fitted the data to the parametrisation

$$F_2(x, Q^2) = F_2^{LT}(x, Q^2) \cdot \left\{1 + C(x)/Q^2\right\}$$

where F_2^{LT} obeys the next-to-leading order QCD evolution equations and $C(x)/Q^2$ is a phenomenological description of the twist-four contribution to the Q^2 dependence of F_2. Above $x = 0.2$, $C(x)$ was taken to be that of ref. [2] (averaged over proton and deuteron) whereas at lower values of x it was assumed to be constant : $C = -0.04$ GeV^2 (see figure 6 left). The fit resulted in a slightly better description of the data ($\chi^2/df = 298/249$) with a gluon distribution lowered by about half a standard deviation at low x (figure 6 right). No significant change was observed in the fitted quark distributions. Leaving α_S a free parameter in the fit, yielded a fitted value $\alpha_S(M_Z^2) = 0.114$. We conclude that the effect of higher twist contributions at low x, if they exist at all, is likely to be within the quoted statistical and systematic errors.

Also shown in figure 4 are the deuteron quark distribution and the gluon distribution from the parton distribution sets of Kwieciński et al. (KMRS-B0) [7] and of Morfin and Tung (MT-S1) [11]. Whereas the gluon distributions obtained by these authors are in reasonable agreement with the present result, large discrepancies of up to 30% are seen for the deuteron quark distribution at $x < 0.1$. In their analyses, the region of low x was not constrained by data. Because sizable discrepancies at low x are observed for essentially all available parton distribution sets [12], it is likely that the low x constraint imposed by the present structure function data will have a significant impact on calculations of hard scattering cross sections at the energies of future colliders.

CONCLUSIONS

Due to their high statistics down to very low x values, the NMC results on F_2^p and F_2^d are the first structure function data to allow a precise determination of the gluon distribution in the nucleon. This extension of the data to low x also makes it legitimate to constrain the QCD fit with the momentum sum rule thus constraining the value of α_S. Finally it is remarkable that the Q^2 evolution of the data can be described by perturbative QCD all the way down to 1 GeV^2.

ACKNOWLEDGEMENTS

It is a pleasure to thank Michiel Botje and Alain Milsztajn for their help in preparing this document.

REFERENCES

1. G. Altarelli and G. Parisi, Nucl. Phys. B126 (1977) 298.
2. M. Virchaux and A. Milsztajn, Phys. Lett. B274 (1992) 221.

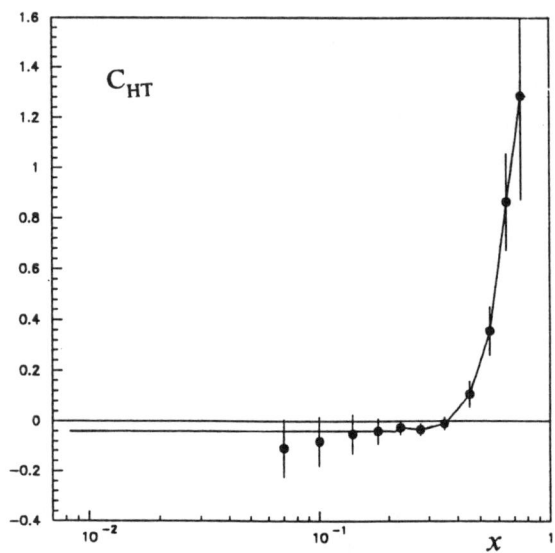 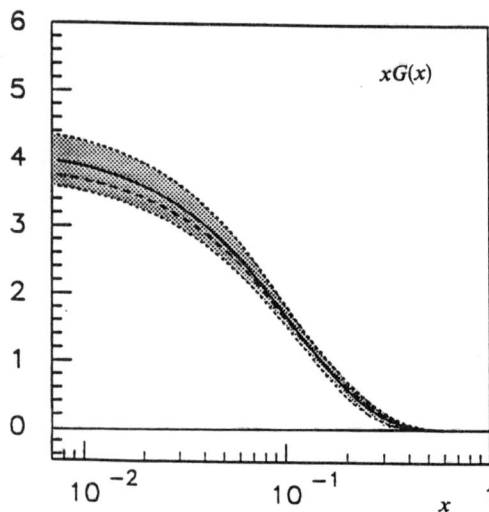

Figure 6. (left) The higher twist coefficient $C(x)$ from ref. [2], averaged over proton and deuteron (solid points). The curve indicates the extrapolation to low values of x used in the fit described in the text. (right) The gluon distribution obtained from the fit without (solid curve) and with (dashed curve) inclusion of higher twist effects.

3. I.G. Bird, in: Proc. LPHEP91 Conf. (Geneva,1991);
   M. van der Heijden, Ph.D Thesis, University of Amsterdam, 1991;
   I.G. Bird, Ph.D Thesis, Free University, Amsterdam, 1992.
4. E. Rondio, these proceedings.
5. A. Ouraou, Ph.D. Thesis, Université de Paris-XI, 1988;
   M. Virchaux, Ph.D. Thesis, Université de Paris-VII, 1988.
6. F. James and M. Roos, MINUIT 89.12j, CERN Program Library long write-up D506 (1989).
7. J. Kwieciński et al., Phys. Rev. D42 (1990) 3645.
8. NMC, P. Amaudruz et al., Phys. Rev. Lett 66 (1991) 2712.
9. CDHSW, P. Berge et al., Z. Phys. C49 (1991) 187.
10. S. Bethke and S. Catani, preprint CERN-TH.6484/92 (May 1992).
11. J.G. Morfin and W.K. Tung, Z. Phys. C52 (1991) 13.
12. Program PDFLIB / CERN Program Library (W5051); H. Plothow-Besch, 'Parton Density Functions', Proc. 3rd Workshop on Detector and Event Simulation in High Energy Physics, Amsterdam, 8-12 April 1991.

STRUCTURE FUNCTIONS, GLOBAL QCD ANALYSIS, AND PARTON DISTRIBUTIONS

Wu-Ki Tung
Michigan State University

Abstract

Recent progress on structure function measurements in muon- and neutrino- Deep Inelastic Scattering by the NMC and CCFR collaborations has significant impact on global QCD analysis and parton distribution determination. This review consists of: (i) a brief survey of existing parton distributions and comparison of their predictions at small-x compared to the new data; (ii) a discussion of issues relevant to new quantitative QCD global analysis; (iii) a summary of results obtained by recent global analysis using the new data; (iv) relation of these results to current collider physics measurements at the Tevatron; and (v) challenges and prospects.

INTRODUCTION

In the current theoretical framework, high energy lepton-hadron and hadron-hadron interaction cross-sections σ, both in Standard Model and in "New Physics" processes, are related to calculable fundamental particle (parton) interaction cross-sections $\hat\sigma$ by the QCD factorization Theorms[1] as a sum of integrals convoluting the later with universal *Parton Distributions* $f_A^a(x,\mu)$. The parton distributions functions can, in principle, be determined from analyzing a set of standard experiments — particularly deep inelastic scattering structure functions. As both theory and experiments have advanced greatly in recent years, it is important that this QCD Parton Model be applied with the requisite care in order to achieve the desired level of accuracy. In particular, it is necessary to recognize that, beyond leading order in QCD, both the "hard cross-section" $\hat\sigma$ and the parton distribution functions $f_A^a(x,\mu)$ are renormalization-scheme and renormalization-scale dependent whereas, of course, the physical cross-sections σ are independent of these theoretical artifices.

QCD global analysis consists of the consistent and systematic comparison of existing reliable QCD calculations with experimental data for as wide a range of processes as possible in order to both test the validity of the perturbative QCD framework and to extract the universal parton distributions. This endeavor depends on continued progress on the theoretical understanding of the high energy processes and on the accurate measurements of the relevant cross-sections. The list of standard processes consists of, in rough order of maturity, deep inelastic scattering (DIS), lepton-pair production (LPP, or "Drell-Yan process"), W- and Z-production, high-p_t direct-photon production, high-p_t jet-production, heavy flavor production, ... etc.

RECENT EXPERIMENTAL RESULTS

Data used in previous global analyses consist of those from the SLAC-MIT, EMC, CDHSW, BCDMS deep inelastic scattering experiments; the E288, E605 lepton-pair production experiments; and the WA70 direct photon production experiment.[2] Important recent experimental results come from the

NMC measurements of F_2^n/F_2^p, $F_2^p - F_2^n$, and $F_2^{p,d}$ (particularly at small x)[3] using a muon beam; and from the CCFR measurement of $F_{2,3}^{Fe}$ using neutrinos and antineutrinos.[4] These experiments, because of their extended kinematic coverage, their accuracy, and their high statistics, have very significant impact on the QCD global analysis – as we will describe in this review.

In the near future, we anticipate new results in DIS from E665 (small x, nuclear targets, ... etc.) and from HERA[5] which will extend structure measurements in both (x, Q) variables by about two orders of magnitudes. A fixed-target direct photon experiment (E706) at Fermilab will soon yield extensive data on this important process which is sensitive to the gluons inside the nucleon. From hadron collider physics, new quantitative measurements of LPP (DY), direct photon, W-, Z-production as well as Jets and W + Jets production in the current run at the Tevatron will propel the phenomenological study of these processes from mere consistency checks on QCD to important complementary sources of information on nucleon parton structure. We shall describe aspects of this emerging part of QCD analysis later in this review.

BRIEF SURVEY OF PARTON DISTRIBUTION SETS

First generation leading order (LO) QCD parton distributions include well-known names such as Field-Feynman, Buras-Gaemers, DO, GHR, EHLQ, ... etc.[8] They relied on experimental data of the early 1980's (EMC83, CDHS83, ...) which are, by now, very much out-dated. First next-to-leading order (NLO) distributions were obtained by the DFLM group (based mainly on the CHARM experiment), and the MRS group (using the more extensive set of data from EMC, BCDMS, CDHSW, E288, ...) in 1988. (N.B. Because the DFLM analysis did not include the muon-scattering data, the associated distributions produce predictions which deviate from the BCDMS data by 4–5 standard deviations in various kinematic regions.) Several special purpose parton distribution analyses have since been carried out focusing on specific issues: e.g. ABFOW (direct photon data), and GRV (dynamically generated sea-quarks).[8] They were not designed for general applications.

Results of recent-past comprehensive analyses (circa 1990), which have provided the main sources of information on current quantitative QCD applications, consist mostly of the HMRS, KMRS distributions,[6] and the MT distributions.[7] Both groups used the (BCDMS, EMC, CDHSW, E605) experiments; and the MRS group also included the WA70 photon and preliminary NMC F_2^n/F_2^p data.

In view of the advent of the significant NMC and CCFR results, it is of interest to ask whether these parton distributions are still adequate compared with the these high statistics data, especially in the newly extended small-x region? Fig.1 shows 3 representative graphs comparing NMC and CCFR small-x data with predictions from two existing MT distribution sets (B1 and B2) as well as a new fit (labeled CTEQ-normal) to be discussed later. We see that the curves labeled MT-B1, which is more representative of most existing parton distribution sets, miss the rising of the structure function F_2 at small-x. It is to be noted, however, that the MT-B2 set (with "singular" extrapolation in the $x \to 0$ limit) does predict structure functions $F_2^{\mu,\nu}$ in general agreement with the new data; although it does not do as well with F_3^ν. A more complete comparison (plots of which are omitted from this written report because of space limitations) confirms these observations. We conclude that new global analysis of available data is indeed necessary.

For quantitative next-to-leading order (NLO) QCD phenomenological study based on data of high accuracy, a meaningful global analysis must take into account a number of theoretical and experimental considerations. In addition to maintain a consistent choice of renormalization-scheme for the parton distributions and the parton (hard) cross-section calculations, it is important to be aware of gluon-quark mixing effects which may upset naive order counting in processes dominated by sea-quarks.[8] Theoretical uncertainty associated with scale-dependence of finite-order perturbative approach must be taken into account. In direct photon production, proper attention is needed to include the proper fragmentation functions (bremsstrahlung contribution) for inclusive cross-sections and to introduce consistent isolation and cone definitions (theoretically and experimentally) for isolated cross-sections. Some of these effects are particularly significant at collider energies.

For the very high-statistics DIS experiments, statistical errors have been reduced to such a low level that systematic errors often dominate the experimental uncertainties. The proper inclusion of systematic errors in the global analysis becomes an important issue. They cannot be routinely neglected, as often done in the past. Another issue is the relative normalization of measured cross-sections from the variety of experiments included in a global analysis. In practice, one finds that for the rather accurate DIS experiments, stable relative normalization factors emerge from most phenomenological studies: to within 1-2%, SLAC, BCDMS, CCFR, NMC: 1.00, 0.975, 0.95, 1.00. On the other hand or DY and Dir γ experiments, the values of the normalization factors vary widely in global fits if they are allowed to float; and their values are correlated with those of the QCD parameters. Additional issues of concern are: (i) uncertainties

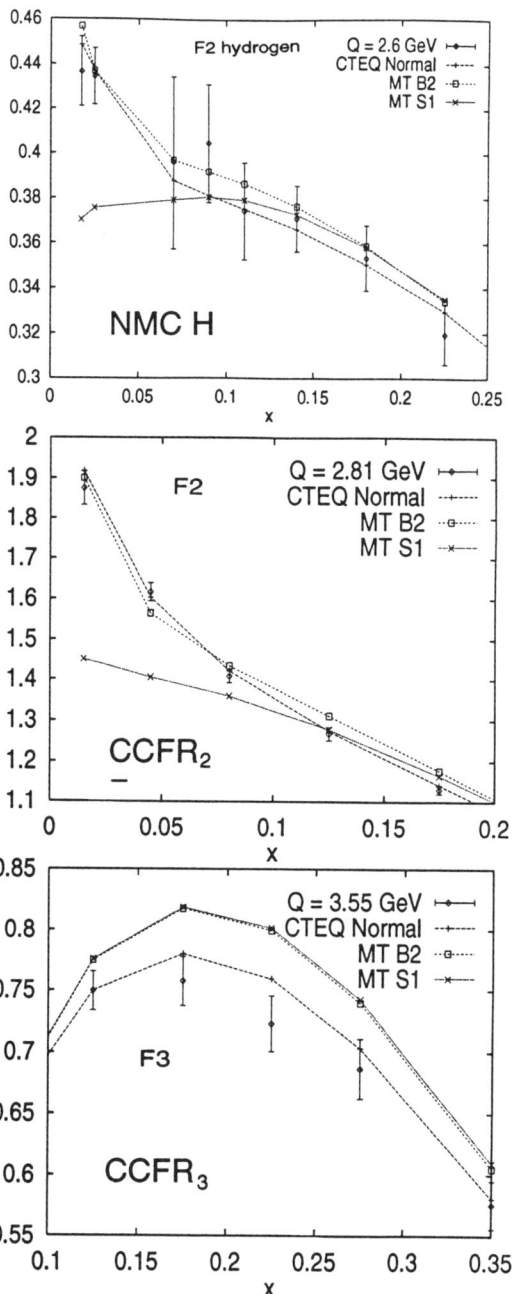

Figure 1. Comparison of new small-x data with predictions of earlier parton distribution sets MT-B1 and B2 as well as with those of a new set CTEQ-normal.

on heavy-target correction applied to neutrino scattering data; (ii) validity of the assumption that deuterium structure functions are the incoherent sum of nucleon structure functions – e.g. small nuclear shadowing effect can lead to significant corrections to the determination of $F_2^p - F_2^n$,[9] and shadowing effects in the small-x region are not fully understood.; (iii) experimental corrections such as *charm- threshold correction* for neutrino charged current cross-section. (This effect should be included in the theory rather than treated as an experimental correction.)

CURRENT GLOBAL ANALYSES AND PARTON DISTRIBUTIONS

New global analyses have been carried out by the MRS group[10] and by the CTEQ collaboration[11] in 1992. The inputs to these analyses are briefly summarized in Table I. Due to limitation of time (and space!) we can only summarize the salient features of these new analyses, drawing mostly from the extensive CTEQ work that the author is familiar with.

The single most important general feature of the new comprehensive analyses is the extra-ordinary quantitative agreement of the NLO-QCD parton framework with the very high statistics DIS experiments (BCDMS, CCFR, and NMC) over the entire kinematic range these experiments cover; and the consistency of this framework with all other available experiments in LPP (DY) and direct photon production as well. Using the least χ^2 fitting method, the overall χ^2 is typically around 700 for around 800 data points in a relatively unconstrained fit $\succeq 20$ QCD and parton distribution (shape) parameters as well as some overall relative normalization factors for the experimental data sets. The χ^2 is higher, but still acceptable, (by 50 - 80) if the shape parameters are reduced to 12 - 15 (according to some

	MRS RAL-92-021, DTP/92/16	CTEQ (Preliminary)
NMC	$F_2^{H,D}$-Preliminary F_2^p/F_2^n, $F_2^p - F_2^n$	$F_2^{H,D}$-Preprint[3] F_2^p/F_2^n, $F_2^p - F_2^n$
CCFR	$F_{2,3}^{Fe}$-Preliminary (x<0.4, Errors enlarged)	$F_{2,3}^{Fe}$-Preprints[4] Submitted to PRL
BCDMS	$F_2^{H,D}$	$F_2^{H,D}$
LPP (DY)	(used for consistency check)	E605: $ds/dM^2 dy$ (120 pts)
Dir. g Prod.	WA70: ds/dp_t^2 (8 pts)	WA70: $ds/dp_t^2 dy$ (40 pts), UA(6), E706 (1988)
Sys. Err.	Included as pt.-to-pt. Err.	Included as pt.-to-pt. Err.
Fitting method	Least Chi-square (Duke-Owens-Roberts)	Least Chi-square (MINUIT)

Figure 2. Summary of new global analyses.

conventional prejudices) and if the freedom of re-normalization of data sets for the DY and direct photon experiments is restricted. In any case, the χ^2 for the individual experiments are all ~ 1 per degree of freedom,[1] underlying the impressive consistency. This is clearly confirmed by inspection of comparison plots of theoretical curves with all available data. An indication of this is give in Fig. 1 where the curves labeled CTEQ-normal represent results from the new analysis; the complete set of comparisons is omitted in this written report. In this general context, we note:

- The small-x ($\sim 0.01 - 0.08$) data from NMC and, also from CCFR, require sub-

---

[1]The one exception is the NMC data set at 280 GeV which typically give χ^2 per degree of freedom of ~ 1.4. However, most of the excess χ^2 come from data points in the $x = 0.225$ bin in which the theoretical values agree extremely well with the more accurate BCDMS data.

stantially increased sea-quark distributions in this region. We found this also produces indirect effect on the shape of the gluon distribution such that the momentum fraction carried by the gluon at $Q = 2\ GeV$ is reduced from the previously accepted value ~ 0.45 by 10% to 0.40 or slightly below. This fraction is partially recovered after evolution to larger Q (see below). (Note that in NLO QCD, the gluon momentum fraction is not exactly tied to the integral of the neutrino structure function F_2.)

- As a free-parameter in the global fit, $\Lambda_{QCD}(5fl)$ is found to be typically $\sim 220 MeV$ in these new fits. Fits with $\Lambda_{QCD}(5fl)$ fixed at $140 MeV$ have substantially higher χ^2, by about 25 for 800 pts. (This lower value of Λ corresponds to $\Lambda_{QCD}(4fl) \sim 220 MeV$ which was preferred by previous fits and $dF(x,Q)/dlnQ$ analyses by the BCDMS and CCFR groups.) The new higher value is closely related to the association of greater freedom to parton distribution shapes required by the new data, cf. below. (Restricting this freedom reproduces the original value for Λ obtained in the MT fits but results in overall χ^2 even higher than mentioned above.) Higher value of Λ leads to somewhat faster QCD evolution at low Q (cf. comment on evolution of gluon fractional momentum above).

- Precise measurements of F_2 for both H and D by NMC strongly suggest the absence of SU(2) flavor symmetry for the sea-quarks, assuming corrections to deuterium structure functions are not significant. (This assumption has been questioned.[9]) Fits with SU(2) symmetric sea enforced have substantially higher χ^2 (about 25 units for 800 pts.). Fits with unequal $\bar{u}$ and $\bar{d}$ distributions suggest, in general, larger $\bar{d}$ then $\bar{u}$.

In addition to understanding the precise reasons for the new features of these findings, there are a few loose ends which need to be considered: (i) Proper treatment of systematic errors as correlated errors and detailed study of these effects have not yet been done by any group. (ii) Charm production contribution to the structure functions, significant in some regions of phase space, have not been carried out consistently. (The gluon-fusion mechanism is as important as the strange-quark scattering mechanism,[12] but has not yet been included in these analysis.) (iii) Should the CDHSW neutrino data also be included in the global analysis? On the one hand, in most regions of the (x,Q) plane where CDHSW data are in agreement with the higher statistics CCFR ones, the inclusion of the former will have no effect on the analysis; on the other hand, in the few regions of disagreement (which affect the lnQ derivatives), it does not make much sense to do a combined fit before the experimental discrepancies are resolved. (Some effort is being made to re-analysis CDHSW data.)

COLLIDER PHYSICS AND PRECISION STUDIES OF THE SM

Especially with the current high luminosity run at the Fermilab Tevatron, many hadron collider processes will become important sources of quantitative information on both standard-model and unconventional physics. Already, preliminary data on W-, Z-production, Lepton-pair production, Direct-photon production, Jet production, and Heavy flavor production are putting the conventional QCD parton model to test. The expected increase in integrated luminosity by an order of magnitude will dramatically increase the impact of these measurements. On the one hand, low mass Drell-Yan cross-sections, small-p_t direct photons and jets, and B-meson production at the Tevatron all probe the parton structure of the nucleon to small-x region unreachable from fixed-target processes (down to, say, 0.005). On the other hand, some of these processes probe the parton structure in different combinations than the available fixed target ones, hence provide complementary information even in regions of overlapping kinematics (such as Drell-Yan in $p\bar{p}$, rather than pp, interaction, and in W-, Z-production).

How do the newly obtained parton distribution functions described in the previous section impact on expectations for collider measurements? Preliminary study done by the CTEQ Collaboration indicate that: (i) the behavior of the new distributions in the small-x region imply increased cross-section for all the above mentioned collider processes. This is gratifying since comparisons of existing parton distribution predictions to preliminary CDF data[13] on these processes have consistently found the latter being below the measurements. Detailed comparisons will be pursued as these data are published and as new data become available. We have already found, by studies using assumed data with errors reduced to anticipated levels, that, for instance, the low-mass lepton-pair measurements should have a significant impact (in the context of a global analysis) in the determination of the small-x behavior of the parton distributions.. Fig. 3 illustrates this point. Shown in the first plot are the CDF data points compared with (i) predictions from MT-B1 and B2 distributions. We see that, as with the NMC and CCFR data, the second set agrees with the new data better; (ii) prediction from the new CTEQ-normal fit (which used the NMC and CCFR, but not the CDF, data). This curve improves the agreement over both MT-B1 and B2; and (iii) finally, results obtained from a dedicated fit which includes the new CDF data with errors reduced by a factor of 10 (which is a little more than the anticipated improvement for the current run) in order to give them useful statistical weight in the global fit. The fact that the new collider data can have a significant impact in QCD global analysis can clearly be seen from the progression of improvements and by comparison of the parton distributions, especially $\bar{u}(x,Q)$ and $g(x,Q)$, determined from these fits. Fig.3b shows the $\bar{u}(x,Q)$ distributions at $Q = 2.0~GeV$ determined by: (a) the CTEQ normal fit which does not use the Drell-Yan data; (b) the dedicated fit as described in (iii) above, and (c) two additional dedicated global fits including simulated collider data with central values one standard deviation higher (lower) than the current values, and with errors reduced as in (iii) above. We see that the values of the Lepton-pair production cross-section at small τ are directly reflected in the $\bar{u}$ behavior. Fig.3c shows the corresponding gluon distribution curves for these four fits. The differences are quite significant, as can easily be seen. We note that all these distributions fit the established DIS data sets and the direct photon data.

Other collider processes are also of interest. Waiting to be studied are the sensitivity of Jet cross-sections, W+jets... etc. to the new parton parameters.

It has been known for some time that certain classes of precision tests of the Standard Model, e.g. the determination of the Weinberg angle in DIS, the main source of uncertainty is from that of parton distributions. With the new parton distributions based on much more accurate data than before, it is important to quantify the remaining uncertainties and study the implications for QCD effects in precision electro-weak physics. This has yet to be done.

CHALLENGES AND PROSPECTS

• Although the *gluon distribution* is much

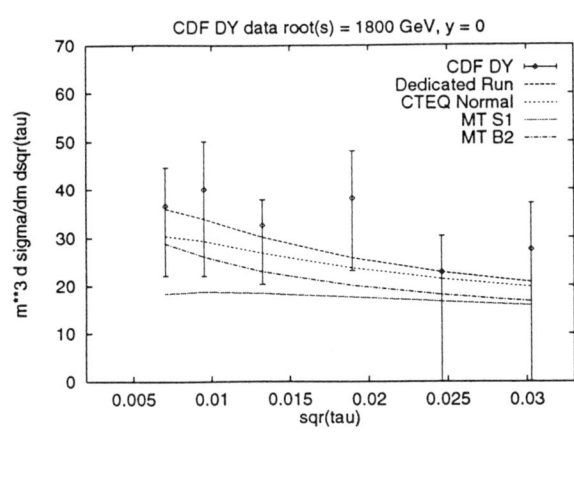
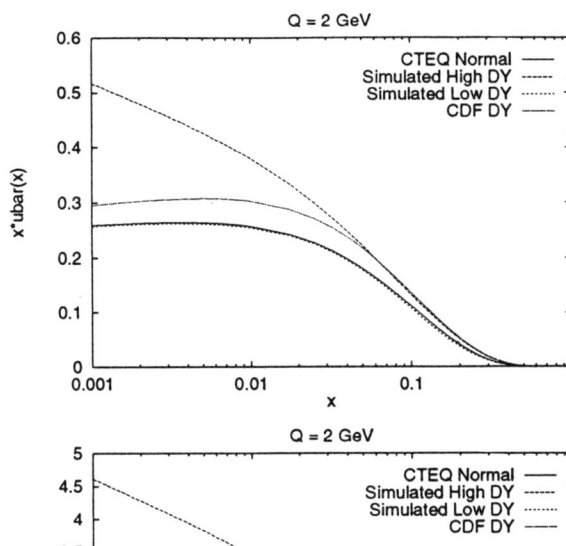
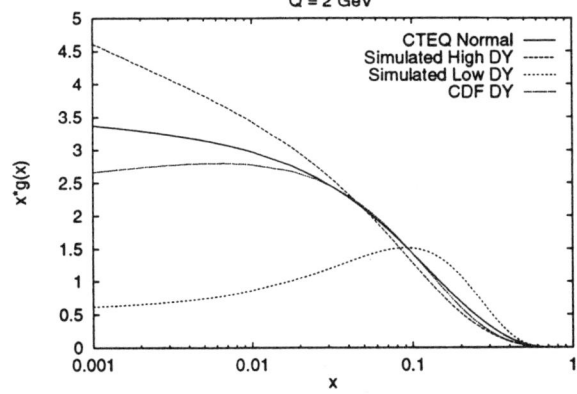

Figure 3. (a) Comparison of CDF lepton-pair production data with old and new parton distribution predictions; (b) $\bar{u}$ distributions determined from 4 fits (cf. text); (c) the gluon distribution from the same fits.

better determined now then most assume (cf. Owens & Tung in [2]); it still needs to be pinned down much more in order to do understand hadron structure and to calculate quantitatively standard model and new physics processes in current and future colliders. New data on Dir. γ, B-production, and Jet cross-sections at the Tevatron, anticipated in the next few years, will contribute substantially in this important task. In addition, the HERA program has important components targeting on the determination of the gluon distribution: measurement of the longitudinal structure function, j/ψ production, and open charm and bottom production are all sensitive to the gluon. True progress on these fronts depends both on the availability of new data and on resolving many theoretical issues involved in quantitative QCD predictions specific to these processes.

- The flavor dependence of the sea quark distributions is still controversial: is it SU(3) symmetric or is the strange quark suppressed with respect to the non-strange ones (i.e. is $\kappa \equiv 2s/(u+d)$ equal to 1 or is it close to 1/2 as existing LO analyses of di-muon charged current DIS dimuon production data seem to indicate? The consistency of the traditional LO analysis has been seriously questioned because of the importance of gluon-sea-quark mixing effects.[12] New analysis of the high statistics CCFR data based on the self-consistent NLO QCD formalism soon will yield more definitive answer to this question.[14]

- In the same vein, is the non-strange sea-quark sector SU(2) symmetric? Although the NMC $(F_2^p - F_2^n)$ data strongly suggest unequal $\bar{u}$ and $\bar{d}$ distributions inside the nucleon, thus the violation of the Gottfried Sum-rule,[15] this issue is far from being settled. As mentioned earlier the extraction of $(F_2^p - F_2^n)$ from the measured F_2^p/F_2^D ratio maybe non-trivial and subject to modifications due to nuclear shadowing.[9] Another possibility is that the SU(2) symmetric sea solution may yet be revived when the global analysis progresses toward incorporating other new processes which are sensitive to the difference between $\bar{u}$ and $\bar{d}$, such as Drell-Yan and W/Z data.

- Small-x behavior of Parton distributions must be understood (theoretically) and determined (experimentally) better. The theoretical issues originate from the mufti-scale nature of the problem which, depending on the relative sizes of the various ratios of the relevant scales, require completely new mathematical techniques to analyze and "resum" the perturbation series. Cf. Desy 90 Workshop on Small-x Physics, and Fermilab 92 Workshop on Diffractive Processes and Small-x. New experimental results from E665, DY, B-production at the Tevatron and, especially HERA will provide testing ground for some of the new ideas on small-x physics and point the way to new thinking on this poorly understood part of QCD.

In conclusion, global QCD analysis and the quantitative determination of parton distributions depend on the systematic application of the perturbative QCD formalism to a wide variety of SM processes in a theoretically and experimentally consistent way; these tasks touch upon all aspects of high energy physics; and they require coordinated efforts by many physicists – both theorists and experimentalists – in order to obtain meaningful quantitative results. It should be noted also that because of the non-trivial nature of the perturbative QCD beyond the leading order, the available parton distributions must be applied to relevant calculations in the proper way (e.g. consistent in the order and in renormalization schemes) by "'users" in order to avoid creating confusion and to ensure progress.

Acknowledgement: The research reported is supported in part by the National Science Foundation and by Texas National Research Laboratory Commission. The global analysis work are done in collaboration with Jim Botts, Jorge Morfin, J.W. Qiu, and H. Weerts. Discussions with other members of the CTEQ Collaboration have been very helpful.

REFERENCES

1. J. Collins and D. Soper in *Perturbative QCD*, Ed. A. Mueller, World Scientific (1989).

2. For detailed references, see various reviews in *Proceedings of the Workshop on Hadron Structure Functions and Parton Distribution Functions*, Ed. Geesaman et.al., World Scientific (1990); also, see J. Owens and W.K. Tung, in *Ann. Rev. Nucl. Sci.* (1992).

3. NMC Collaboration, P.Amaudruz et al, Preprint CERN-PPE/92-124

4. CCFR Collaboration, Nevis Preprints, S.R.Mishra et al., #1459; W.C.Leung et al., #1460; and P.Z.Quintas et al., #1461;

5. *Physics at HERA*, Ed. W. Buchmuller and G. Ingelman, DESY, 1991.

6. P.N. Harriman, A.D. Martin, R.G. Roberts & W.J. Stirling, *Phys. Rev.* **D42**, 798 (1990).

7. Jorge G. Morfin and Wu-Ki Tung, *Z. Phys.* **C52** 13, (1991).

8. See Wu-Ki Tung, "Overview of Structure Functions and Parton Distributions", in Ref.2.

9. M. Strikman, this conference; and private communications.

10. A. Martin, R. Roberts, and J. Stirling, Durham and Rutherford Preprint (DTP/92/16, RAL-92-021)

11. J. Botts, J. Morfin, J.W. Qiu, W.K. Tung, and H. Weerts (in preparation); the CTEQ collaboration (Coordinated Theoretical/Experimental Project on QCD Phenomenology and Tests of the Standard Model) consists of, in addition to the above authors (as members of its global fit subgroup), R. Brock, J. Collins, J. Huston, S. Mishra, F. Olness, J. Owens, J. Pumplin, J. Smith, D. Soper, and G. Sterman.

12. M.G. Aivazis, F. Olness & Wu-Ki Tung, *Phys. Rev. Lett.* **65**, 2339 (1990).

13. CDF direct photon data: *Phys. Rev. Lett.* **68**, 1992; CDF b-quark production data: F. Ukegawa (Ph.D. Thesis), Univ Tsukuba Preprint UTPP-40 (1991); CDF Lepton-pair production data: (in preparation).

14. A. Bazarko and M. Shaevitz, private communications.

15. NMC, *Nucl. Phys.* **B371**,3 (1992).

MEASUREMENT OF THE CROSS-SECTION RATIO σ_n/σ_p IN INELASTIC MUON-NUCLEON SCATTERING AT VERY LOW x AND Q^2

Vassili Papavassiliou*
Physics Division
Argonne National Laboratory
9700 South Cass Avenue
Argonne, Illinois 60439-4828

(Representing the Fermilab E-665 Collaboration)

Abstract

Preliminary results are presented on the measurement of the cross-section ratio σ_n/σ_p in inelastic μN scattering obtained by the E-665 experiment using the Fermilab 490 GeV/c muon beam and liquid H_2 and D_2 targets. The results extend the previously measured x range by two orders of magnitude, down to 2×10^{-5}, at $Q^2 > 10^{-2}\,\text{GeV}^2/c^2$. The ratio is consistent with 1 throughout the new range.

INTRODUCTION

Recently, there has been considerable interest in the very small x ($x < 10^{-2}$) behavior of the nucleon structure functions (where x is the Bjorken scaling variable), soon to be probed at HERA. Theoretical calculations based on Perturbative Quantum Chromodynamics (PQCD) make interesting predictions for this kinematic range.[1,2] In current fixed-target experiments, the low-x region can be accessed only at the expense of low Q^2 ($Q^2 < 1$ GeV$^2/c^2$), where PQCD is not applicable. Here Q^2 is the invariant mass squared of the virtual photon that mediates the interaction. This kinematic range is frequently studied using Regge theory. A recent study[3] gives predictions for the structure functions of the nucleon at low x down to $Q^2 = 0$, extrapolating from data at higher Q^2.

Recent data[4] on the proton and neutron structure functions F_2^p and F_2^n cover the range $x > 0.008$, $Q^2 > 0.75$ GeV$^2/c^2$. The ratio F_2^n/F_2^p has been measured[5] down to $x = 0.002$ and is seen approaching 1 at the lowest x values. Similarly, data[6] on the total real-photon ($Q^2 = 0$) absorption cross sections on H_2 and D_2 show that the proton and neutron cross sections differ by only a few percent in the range $\nu = 4 - 18$ GeV, where ν is the photon energy in the laboratory frame, approaching a common value at large ν.

This paper presents preliminary results on the ratio of inelastic cross sections σ_n/σ_p in muon-nucleon scattering that include, for the first time, the range $x = 2 \times 10^{-5} - 2 \times 10^{-3}$, with $Q^2 > 0.01$ GeV$^2/c^2$ and $\nu > 50$ GeV, obtained by Fermilab experiment E-665. This ratio is equal to the ratio F_2^n/F_2^p, under some assumptions. to be discussed at the end.

---

*Work supported by the U.S. Department of Energy, Nuclear Physics Division, under Contract No. W-31-109-ENG-38.

THE EXPERIMENT

Data were taken in the Fermilab muon beam using the E-665 double-dipole open-geometry spectrometer[7] during the 1987–88 fixed-target run. The relevant data for this analysis were taken on two 1.1 m-long liquid targets, a D_2 target (density 0.16 g/cm^2) and a H_2 target (0.07 g/cm^2), at 490 GeV/c mean beam momentum. The Small-Angle Trigger (SAT)[7] triggered on the absence of a veto signal at the expected position of the unscattered beam, determined by a processor for each incoming beam track, detected in a set of 7 scintillator planes. The acceptance of the trigger extended to scattering angles smaller than 0.1 mrad, inside the phase space of the unscattered beam.

The scattered muon was identified in the off-line analysis by its ability to penetrate a set of steel and concrete absorbers; its trajectory was reconstructed with the help of proportional and drift chambers. An electromagnetic calorimeter was used to discriminate between electrons and hadrons and to detect neutral particles depositing electromagnetic energy.

EVENT SELECTION

Events were selected requiring both the incoming and scattered muon tracks to be fully reconstructed and fitted to a vertex with a χ^2 probability better than 10^{-3} within the fiducial volume of the target. The following kinematic cuts were applied to constrain the data sample away from regions where the resolution in the kinematic variables is poor or the radiative corrections are too large:

$$E_B > 400 \text{ GeV}$$
$$Q^2 > 10^{-2} \text{ GeV}^2/c^2$$
$$x > 2 \times 10^{-5}$$
$$0.1 < y < 0.9$$
$$\nu > 50 \text{ GeV}$$
$$|\phi - \pi| > 0.2 \text{ rad}$$

Here E_B is the beam energy, the photon energy ν is the energy transfer in the laboratory, $y = \nu/E_B$ is the fractional energy transfer and ϕ is the azimuthal angle of the scattered muon around the incoming muon direction. This last cut eliminated events with the outgoing muon in the horizontal (bending) plane, for which the vertex was poorly determined. The resolution in x is approximately constant at $\sim 15\%$ for $x > 0.001$ but this value increases to about 50% at the lowest x.

Rejection of electromagnetic backgrounds

In addition to the above cuts, two independent methods were employed to separate inelastic events from the large electromagnetic backgrounds, not related to the nucleon structure, that dominate at $x < 10^{-3}$. These backgrounds are due to elastic μe scattering in the target and to muon bremsstrahlung. The first type appears as a peak in the x distribution centered at $x = m_e/m_p \approx 5 \times 10^{-4}$, where m_e and m_p are the electron and proton masses, respectively, with a width determined by the experimental resolution. The second background type appears as events with large energy loss (ν) reconstructed with very small scattering angles and therefore at very low x. Figure 1 shows the event distribution in $\log_{10} x$ after the kinematic cuts (solid line).

In the first method, electromagnetic events were rejected by selecting only those events for which the total electromagnetic energy E_{CAL} deposited in the calorimeter was less than 0.4ν, or for which the bremsstrahlung planarity was greater than 10^{-5}. This quantity is defined from the momentum vectors $\mathbf{p}$ and $\mathbf{p}'$ of the incoming and outgoing muons and the vector $\mathbf{k}$ connecting the vertex to the largest-energy cluster on the calorimeter, as

$$\mathcal{P} = \frac{|(\mathbf{p} \times \mathbf{p}') \cdot \mathbf{k}|}{|\mathbf{p}||\mathbf{p}'||\mathbf{k}|}. \quad (1)$$

The resulting distribution is shown in Figure 1

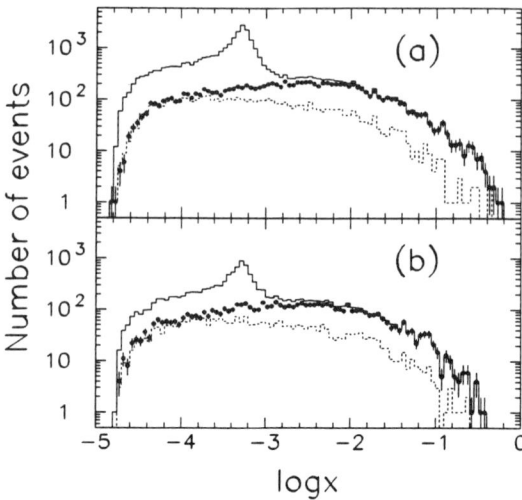

Figure 1. Event distribution vs. $\log_{10} x$ from H_2 (a) and D_2 (b) after kinematic cuts only (solid line), kinematic and calorimeter cuts (full dots) and kinematic cuts plus the hadron requirement (dotted line).

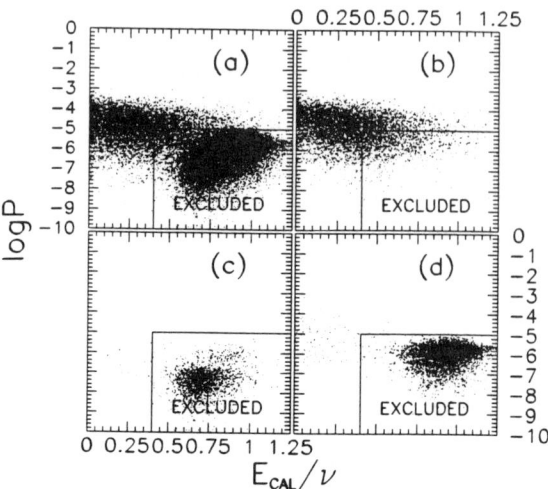

Figure 2. $\log_{10} \mathcal{P}$ vs. E_{CAL}/ν for all events (a) and for three different event types, inelastic, bremsstrahlung and μe (b–d), classified without use of the calorimeter.

as black dots. Figure 2 shows the event distribution from the H_2 target in the $\log_{10} \mathcal{P}$ vs. E_{CAL}/ν plane for all events (a); inelastic events, as determined by the second method below (b); μ-bremsstrahlung events, events with $x < 10^{-4}$ and no additional tracks besides the scattered muon (c); and elastic μe events, identified as events with only one muon and one negative secondary track in the final state (d). The distributions from the D_2 target look very similar. It can be seen that the cuts remove essentially all electromagnetic events of either type, while most inelastic events are retained.

In the second method, inelastic events were explicitly selected by requiring at least two hadrons of the same sign to be associated with the event. Here a "hadron" was defined as any reconstructed track compatible with coming from the event vertex, other than the scattered muon or other muons. The same-sign requirement was imposed to avoid contamination from γ photons converting into e^+e^- pairs. The result is shown in Figure 1 as a dotted line. At low x, similar numbers of events are selected by the two methods. At higher x, however, the number of events with the second method is much smaller, due to smaller ν values and because of smaller acceptance for low-momentum tracks in the forward spectrometer.

RESULTS

The ratio σ_n/σ_p is derived from the event yields from the H_2 and D_2 targets and the target densities and beam fluxes, assuming $\sigma_d = (\sigma_p + \sigma_n)/2$ and correcting only for events originating in the target vessel and a $\sim 5\%$ contamination of the D_2 target with HD molecules. In particular, no corrections were applied for Fermi motion, the effect of which is expected[8] to be negligible at $x < 0.6$, or nuclear shadowing effects in the deuteron. Also, no radiative corrections were performed, other than the removal of the backgrounds described above. The remaining radiative corrections on the ratio are not expected to be large.

Figure 3 shows the ratio σ_n/σ_p as a func-

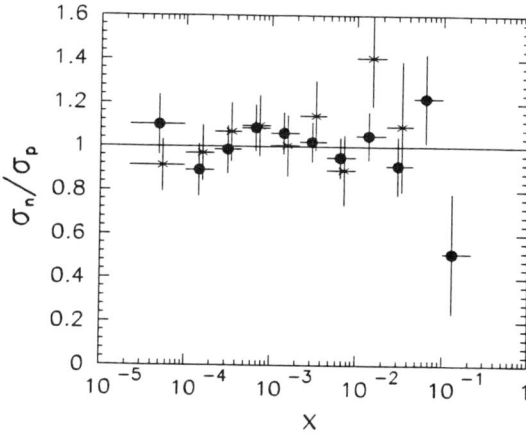

Figure 3. σ_n/σ_p with calorimeter cuts (•) and hadron requirement (×). The vertical error bars show the statistical errors and the horizontal the extent of the x bins. The two sets of points are slightly displaced for clarity.

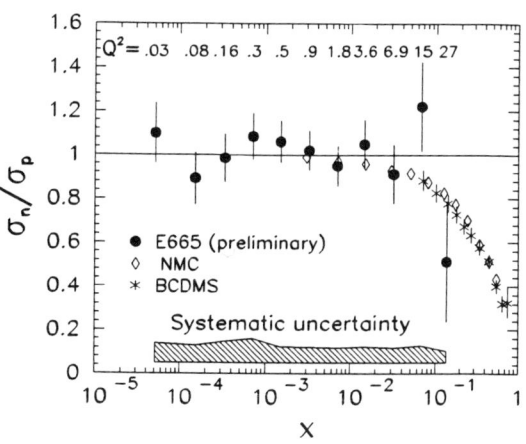

Figure 4. Preliminary results on σ_n/σ_p vs. x (•). The error bars show the statistical errors, while the shaded band shows the estimated systematic uncertainty. Also shown are previous results on F_2^n/F_2^p from NMC[5] (◊) and BCDMS[9] (×).

tion of x, as derived by the two methods, the calorimeter cuts (full dots) and the hadron requirement (exes). The two methods give compatible results, within the statistical uncertainties, giving confidence that the electromagnetic backgrounds have been removed properly. The results from the first method will be used, because of the better statistics, especially at higher x.

Systematic errors include a ±2% uncertainty on the relative beam fluxes and a ±3% uncertainty on the relative acceptance for the two data samples, which were taken at different running periods. These translate into a ±7% total normalization uncertainty on σ_n/σ_p. As a further check on the relative normalization, the cross-section ratio for elastic μe scattering from the D_2 and H_2 targets was measured to be 1.01 ± 0.03, consistent with 1.

The calorimeter was cross-calibrated for the two data-taking periods using electrons from elastic μe events. The uncertainty in the subtraction of the electromagnetic backgrounds due to residual differences in the calorimeter response were included in the systematic error. This uncertainty is sizable only at $x < 10^{-3}$.

Figure 4 shows the ratio obtained with the calorimeter cuts (black dots). The band at the bottom of the plot shows the estimate for the size of the systematic uncertainties. The average Q^2 value in each x bin is shown at the top of the plot. The W of the data (the invariant mass of the hadronic final state) is always greater than 10 GeV, far above the resonance region, with an average varying from 25 GeV at the lowest x to 12 GeV at the highest.

Also shown in Figure 4 are the results for F_2^n/F_2^p from the two highest-statistics previous experiments, NMC[5] (open diamonds) and BCDMS[9] (exes). This structure-function ratio is equal to the cross-section ratio if the ratio R of transverse-to-longitudinal cross sections for the neutron is equal to that for the proton, as suggested by an analysis of data from SLAC[10] at higher x, and assuming that the radiative corrections affect the two targets equally. These corrections are still under investigation.

The effect of nuclear shadowing on deuterium is at present unknown, as no experimental measurements exist. Calculations[11,12] predict an effect of up to 2 − 4% at some values of x and Q^2, but these calculations do not extend to the kinematic range of this experiment. If the effect exists, the true values of σ_n/σ_p are higher than presented here.

DISCUSSION

The new results extend the low-x range of the measurement by two orders of magnitude. The ratio is equal to 1, within errors, throughout this range, consistent with the extrapolation of the NMC measurement. The validity of this extrapolation was crucial for the derivation[5] of the integral of the difference $F_2^p(x) - F_2^n(x)$ and the conclusion that the Gottfried sum rule[13] is violated. The present data bridge the gap between the high-x, high-Q^2 deep inelastic scattering data[5] and those from real photoproduction[6] at lower ν.

The equality of the neutron and proton cross sections for $x \to 0$ is expected from Regge arguments.[1,3] Our result is consistent with this expectation, provided any shadowing effects are small compared to the experimental uncertainties. Alternatively, if this equality is assumed to be true, then the above result suggests the absence of significant nuclear shadowing effects in the deuteron, within the statistical and systematic uncertainties.

REFERENCES

1. B. Badelek et al., "Small x Physics," DESY Report No. DESY 91-124, 1991 (to be published).
2. E. Levin, "Nucleon Structure Functions at Small x," DESY Report No. DESY 91-110, 1991 (to be published).
3. B. Badelek and J. Kwieciński, "Electroproduction Structure Function F_2 in the low Q^2, low x region," Phys. Lett. (to be published).
4. P. Amaudruz et al., "Proton and Deuteron F_2 Structure Functions in Deep Inelastic Muon Scattering," Phys. Lett. (to be published).
5. P. Amaudruz et al., "Gottfried Sum from the Ratio F_2^n/F_2^p," Phys. Rev. Lett. 66, pp. 2712–2715, (1991).
6. D. O. Caldwell et al., "Total Hadronic Photoabsorption Cross Sections on Hydrogen and Complex Nuclei from 4 to 18 GeV," Phys. Rev. D7, pp. 1362–1383 (1973).
7. M. R. Adams et al., "A Spectrometer for Muon Scattering at the Tevatron," Nucl. Instrum. Methods A291, pp. 533–551, (1990).
8. L. L. Frankfurt and M. R. Strickman, "On the Problem of Extracting the Neutron Structure Function from eD Scattering," Phys. Lett. B76, pp. 333–336, (1978).
9. A. C. Benvenuti et al., "A Comparison of the Structure Functions $F_2(x, Q^2)$ of the Proton and the Neutron from Deep Inelastic Muon Scattering at High Q^2," Phys. Lett. B237, pp. 599–604, (1990).
10. L. W. Whitlow et al., "A Precise Extraction of $R = \sigma_L/\sigma_T$ from a Global Analysis of the SLAC Deep Inelastic e-p and e-d Scattering Cross Sections," Phys. Lett. B250, pp. 193–198, (1990).
11. B. Badelek and J. Kwieciński, "Shadowing in Inelastic Lepton-Deuteron Scattering," Nucl. Phys. B370, pp. 278–298, (1992).
12. V. R. Zoller, "Nuclear Shadowing in Deuteron and the Gottfried Sum Rule," Phys. Lett. B279, pp. 145–148, (1992).
13. K. Gottfried, "Sum Rule for High-Energy Electron-Proton Scattering," Phys. Rev. Lett. 18, pp. 1174–1177, (1967).

ELECTROPRODUCTION STRUCTURE FUNCTION F_2 IN THE LOW Q^2, LOW X REGION*

Barbara Badełek
Institute of Experimental Physics
Warsaw University
ul. Hoża 69
00-681 Warsaw, Poland

and

Jan Kwieciński
Department of Theoretical Physics
H. Niewodniczański Institute of Nuclear Physics
ul. Radzikowskiego 152
31-342 Cracow, Poland

Abstract

The nucleon structure function F_2 in the low Q^2, low x region is constructed. The analysis has been motivated by the results of the electroproduction experiments for which measurements at low values of x have been performed at the expense of lowering Q^2 down to 1 GeV^2 or less. Contributions from both the parton model with QCD corrections suitably extended to the low Q^2 region and from the low mass vector mesons were taken into account. The former contribution results from the large Q^2 structure function analysis which includes the recent F_2 measurements by the New Muon Collaboration. Predictions of the model are compared with the results of the electroproduction measurements.

The nucleon structure functions in the low x region ($x < 0.05$ or so) have recently been measured in a very precise way by the NMC[1] and the CCFR Collaboration[2]. Adding the earlier albeit less precise low x data of the EMC–NA28[3] and the higher x measurements of SLAC[4] and BCDMS[5] we obtain a reliable set of data on F_2 over a wide x ($0.006 < x < 0.9$) interval. However the present fixed target measurements correlate low x region with the low Q^2 one, $Q^2 \sim 1 GeV^2$ and less. Due to the conservation of the electromagnetic current, F_2 must vanish as Q^2 goes to zero so it is clear that scaling cannot be a valid concept at $Q^2 << 1 GeV^2$. Theoretical models assuring a smooth transition from the scaling to non-scaling regions and applicable in a wide Q^2 interval, including photoproduction ($Q^2 = 0$), are thus necessary to understand the experimental results. Besides that the models might be of a high practical value for the analysis of the high Q^2 measurements (e.g. in the radiative corrections procedure).

The purpose of this talk is to present the results of a detailed analysis of the electroproduction structure function $F_2(x, Q^2)$ at low x and for arbitrary Q^2 using the generalised vector meson dominance (GVMD) ideas linking it with the QCD structure function analysis at large Q^2. More details can be found in ref.[6].

In GVMD the structure function F_2 is represented by the contribution of the (infinite number of) vector mesons which couple to vir-

---

*Supported in part by the Polish Committee for Scientific Research grant 2 0198 91 01.

tual photon. The contribution of large mass vector mesons is determined by the structure function in the large Q^2 region, F_2^{AS}, which is described by the QCD improved parton model and assumed as being given. In ref.[7] we proposed the unified description of the structure function using these ideas. It was however based on the F_2^{AS} parametrisation which has become outdated[8]. A method presented here is based on the most recent parton distributions[9,10]. Also a simplified form of the asymptotic contribution is used without loosing its dynamical content.

Apart from GVMD there exist several phenomenological parametrisations of F_2 extrapolating it from the scaling region to low values of Q^2, ref.[11-14]. However these parametrisations are not linked with the conventional QCD evolution. Nor is the low Q^2 behaviour related to the explicit vector meson dominance, known to dominate at low Q^2, ref.[15].

The structure functions in the low Q^2 region have also been extensively discussed using the so called dynamically calculated parton distributions where the QCD improved parton model is directly extended to the very low scales Q^2 much below 1 GeV^2, ref.[16]. They have also been discussed within the two gluon exchange model for the virtual photon - nucleon scattering[17].

We begin from the following representation of the structure function $F_2(x, Q^2)$, ref.[7]:

$$F_2[x = Q^2/(s + Q^2 - M^2), Q^2]$$
$$= \frac{Q^2}{4\pi} \sum_v \frac{M_v^4 \sigma_v(s)}{\gamma_v^2 (Q^2 + M_v^2)^2}$$
$$+ Q^2 \int_{Q_0^2}^{\infty} dQ'^2 \frac{\Phi(Q'^2, s)}{(Q'^2 + Q^2)^2}$$
$$\equiv F_2^{(v)}(x, Q^2) + F_2^{(p)}(x, Q^2) \quad (1)$$

The definition of various kinematical variables is standard, $s \equiv W^2$ where W is the invariant mass of the electroproduced hadronic system and M is the nucleon mass. The function $\Phi(Q^2, s)$ is expressed as follows:

$$\Phi(Q^2, s) = -\frac{1}{\pi} Im \int^{-Q^2} \frac{dQ'^2}{Q'^2} F_2^{AS}(x', Q'^2) \quad (2)$$

The asymptotic structure function $F_2^{AS}(x, Q^2)$ is assumed to be given. It may be obtained from the QCD structure function analysis in the large Q^2 region. By construction, $F_2(x, Q^2) \Rightarrow F_2^{AS}(x, Q^2)$ for large Q^2. The second term in (1) can be looked upon as the extrapolation of the (QCD improved) parton model for arbitrary Q^2. The first term corresponds to the low mass vector meson dominance part since the sum extends over the low mass vector mesons. Contribution of vector mesons heavier then Q_0 is included in the integral in (1). Choosing the parameter $Q_0^2 > (M_v^2)_{max}$ where $(M_v)_{max}$ is the mass of the heaviest vector meson included in the sum one explicitly avoids double counting when adding two separate contributions to the structure function. Note that Q_0 should be smaller than the mass of the lightest vector meson not included in the sum. The quantities $\sigma_v(s)$ are the vector meson-nucleon total cross sections determined by the πN and KN total cross sections and γ_v^2 can be related in the standard way to the leptonic width of the vector meson v [7,15]. It should be noted that the representation (1) is written for fixed s and is expected to be valid at $s \gg Q^2$, i.e. at low x but for arbitrary Q^2.

We propose to simplify the representation (1) for the partonic part $F_2^{(p)}(x, Q^2)$ i.e.:

$$F_2^{(p)}(x, Q^2) = \frac{Q^2}{(Q^2 + Q_0^2)} F_2^{AS}(\bar{x}, Q^2 + Q_0^2) \quad (3)$$

where

$$\bar{x} = \frac{Q^2 + Q_0^2}{s + Q^2 - M^2 + Q_0^2} \equiv \frac{Q^2 + Q_0^2}{2M\nu + Q_0^2} \quad (4)$$

Simplified parametrisation (3) connecting $F_2^{(p)}(x, Q^2)$ to F_2^{AS} by an appropriate change of the arguments of the latter posseses all the main properties of the second term in (1). First of all it is evident that $F_2^{(p)}(x,Q^2) \Rightarrow F_2^{AS}(x,Q^2)$ for large Q^2. Moreover the parametrisation of $F_2^{(p)}$ defined by (3) preserves the analytic properties of the second term in (1).

It should be stressed that apart from the parameter Q_0^2 which is constrained by physical requirements described above the representation (3) does not contain any other free parameters except of course those which are implicitly present in the parametrisation of parton distributions defining F_2^{AS}.

The deuteron structure function F_2 calculated from the representation (1) with $F_2^{(p)}(x,Q^2)$ given by the eq. (3) is shown in figs 1 and 2. The structure function $F_2^{AS}(x,Q^2)$ was obtained from the most recent MRS parametrisation of the parton distributions using the D^- fit[9,10]. The MRS analysis incorporates in particular the new, NMC[1] and CCFR Collaboration[2] measurements at $Q^2 > 5 GeV^2$.

The parametric forms of parton distributions in a proton at the reference scale $Q^2 = 4 GeV^2$ were the same as in ref.[9]. The numerical values of various parameters appearing in those forms were obtained from the fit based on the Altarelli–Parisi equations in the leading log Q^2 approximation[10].

In the vector meson contribution, $F_2^{(v)}(x,Q^2)$, we included ρ, ω and ϕ states. All parameters defining $F_2^{(v)}(x,Q^2)$ in eq. (1) including those describing the energy dependence of σ_v are the same as in ref.[7]. Low sensitivity of the results to the value of Q_0^2 is illustrated by small differences obtained when taking $Q_0^2 = 1.2 GeV^2$ and $Q_0^2 = 1.5 GeV^2$.

The following properties of the Q^2 and x dependence of the results should be noted:

1. An "approach to scaling" (i.e. Q^2 dependence of F_2 at Q^2 less than 1 GeV^2 or so) is clearly visible in fig. 1. The change of curvature comes from the factor Q^2 in eq. (1) and the magnitude of this change, particularily at $Q^2 \ll 1\ GeV^2$, is controlled by the vector meson contribution, a non-trivial test of this mechanism. Although the vector meson contribution $F_2^{(v)}$ dominates in the very low Q^2 region ($Q^2 < 1\ GeV^2$) the partonic component $F_2^{(p)}$ still gives a significant contribution (i.e. at least 20% - 30%) there.

2. At $Q^2 = 10 GeV^2$ the structure function F_2 calculated from the model differs from $F_2^{AS}(x,Q^2)$ by less than 3%.

3. The x dependence at fixed Q^2 reflects both the energy dependence of σ_v and the x-dependence of F_2^{AS}. Expectations coming from Regge theory are incorporated in the parametrisations of the $\sigma_v(s)$, ref.[7] and of the input parton distributions at the reference scale $Q^2 = 4\ GeV^2$, refs [9,10,18]. The increase of $F_2(x,Q^2)$ with decreasing x reflects the increase of the total cross sections $\sigma_v(s)$ with increasing s as well as the increase of $F_2^{(p)}(x,Q^2)$ with decreasing x. The increase of $F_2^{(p)}(x,Q^2)$ with decreasing x is weaker at small Q^2 than the similar increase in the large Q^2 region. The s dependence of $\sigma_v(s)$ is also relatively weak in the relevant region of s, ref.[7]. As the result the x dependence of $F_2(x,Q^2)$ is relatively weak for low Q^2, fig.2.

Within errors and keeping in mind the overall normalisation uncertainties of the data sets (up tp 3% for the NMC data[1] and up to 7% for the EMC–NA28[3]) the model reproduces the measurements very accurately. The good agreement of data and model for large Q^2, i.e. in the region where the structure function F_2 is practically equal to F_2^{AS} is natural since the NMC and BCDMS data from the region $Q^2 > 5\ GeV^2$ were used for fitting the MRS parton distributions parametrisations[9,10] from which our asymptotic structure function F_2^{AS} was constructed. The agreement of the model with the data also for moderately large values of

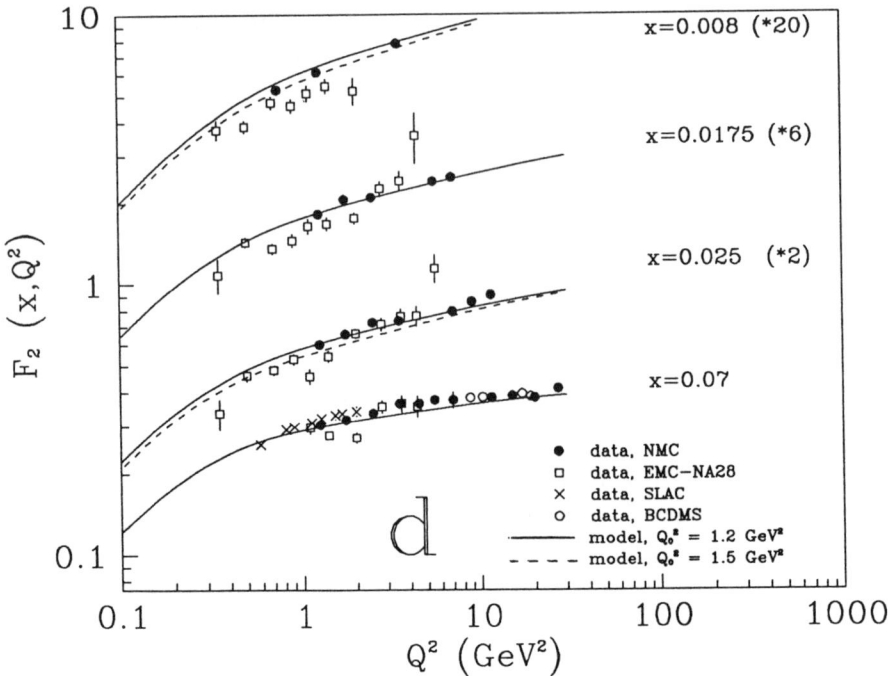

Figure 1. The deuteron structure function F_2 calculated from (1) with $F_2^{(p)}(x,Q^2)$ given by (3) as a function of Q^2 for different intervals of x. For clarity both the data and the theoretical calculations in each x bin have been scaled by the indicated factors. Continuous and dashed lines correspond to $Q_0^2 = 1.2\ GeV^2$ and $Q_0^2 = 1.5\ GeV^2$ respectively. The data come from NMC[1], EMC-NA28[3], SLAC[4] and BCDMS[5]. The error bars represent the statistical and systematic errors added in quadrature.

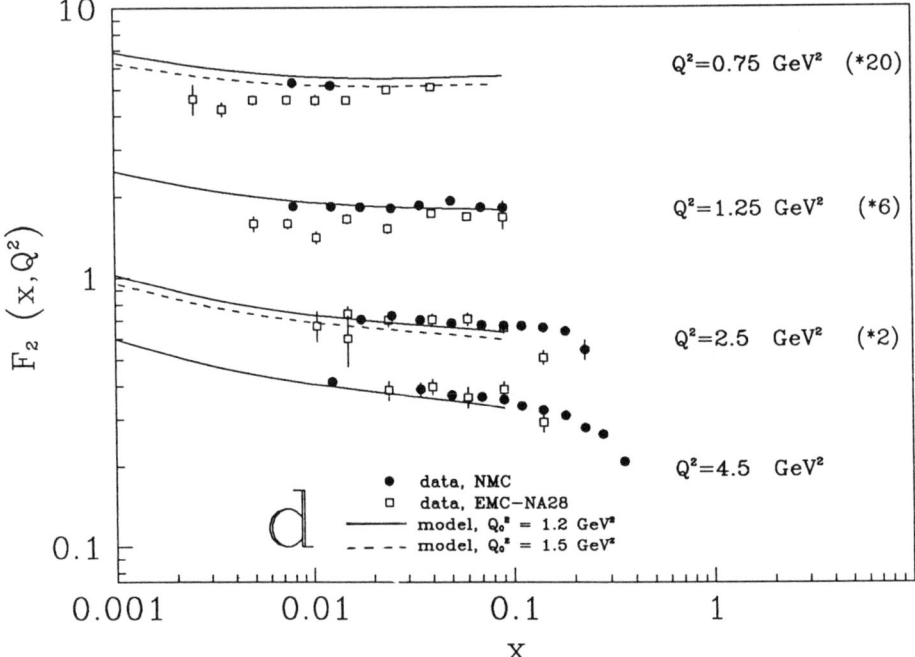

Figure 2. As in fig.1 but the structure functions are shown as functions of x for different intervals of Q^2.

Q^2 (5 $GeV^2 < Q^2 <$ 20 GeV^2 or so) where the data are well described by $F_2^{AS}(x,Q^2)$ alone[9] is less trivial since the vector meson dominance part $F_2^{(v)}$ is still non-negligible there. The new predictions of the model apply to the low Q^2 region ($Q^2 <$ 5 GeV^2) and in particular to the very low Q^2 values ($Q^2 <$ 1 GeV^2) where the vector dominance part starts to dominate. Agreement of our predictions with experimental data which unfortunately are still scarce in this very low Q^2 region is the most important test of our scheme.

We have also examined the $Q^2 = 0$ (photoproduction) predictions of the model. The relevant formula for the total photoproduction cross section reads:

$$\sigma_{\gamma p}(s) = \lim_{Q^2 \Rightarrow 0} 4\pi^2 \alpha_{em} \frac{F_2}{Q^2} \quad (5)$$

where F_2 is given by formulae (1), (3) and (4). The model predicts a correct shape of the $\sigma_{\gamma p}(s)$ but overestimates its magnitude by about 10 − 15%. Inclusion of nondiagonal terms describing transitions like $\rho \Rightarrow \rho'$ brings this number down but at the same time results for F_2 at $Q^2 > 0$ are also affected.

The FORTRAN code computing F_2 for a given x and Q^2 is available on request from: BADELEK@HOZAVX.FUW.EDU.PL or KWIECINSKIJ@VSB01.IFJ.EDU.PL

ACKNOWLEGEMENTS

We are most grateful to Dick Roberts for kindly supplying us with the parametrisation of the LLO MRS input parton distributions.

REFERENCES

1. NMC, P.Amaudruz et al., CERN-PPE/92-124 and to appear in *Phys. Lett.*
2. CCFR, S.R.Mishra et al., *Nevis preprint* 1465, January 1992.
3. EMC–NA28, M. Arneodo et al., *Nucl. Phys.*, B333, pp. 1–47, (1990).
4. SLAC, L.W. Whitlow, Ph.D. Thesis, *SLAC report* 357 (1990); L.W.Whitlow, *Phys. Lett.*, B282, pp. 475–482, (1992).
5. BCDMS, A.C.Benvenuti et al., *Phys. Lett.*, B237, pp. 599–604,(1990).
6. B.Badełek and J.Kwieciński, *Warsaw University preprint* IFD/3/1992 and to appear in *Phys. Lett.*.
7. J.Kwieciński and B.Badełek, *Z. Phys.*, C43, pp. 251–260, (1989).
8. E.Eichten et al., *Rev. Mod. Phys.*, 56, pp. 579– 707, (1984).
9. A.D.Martin, W.J.Stirling and R.G.Roberts, *preprint* RAL-92-021 and DTP/92/16, April 1992.
10. R.G.Roberts, private communication.
11. SLAC E61, S.Stein et al., *Phys. Rev.*, D12, pp. 1884–1919,(1975).
12. A.Donnachie and P.V.Landshoff, *Nucl. Phys.* B244, pp. 322–336, (1984) .
13. NMC, P.Amaudruz et al., *Nucl. Phys.* B371, pp. 3–31, (1992).
14. H.Abramowicz et al., *Phys. Lett.* B269, pp. 465–476, (1991).
15. T.H. Bauer et al., *Rev. Mod. Phys.* 50, pp. 261–436, (1978).
16. M. Glück, E. Reya and A. Vogt, *Z. Phys.* C48, pp. 471–482, (1990); *Z. Phys.* C53, pp. 127–134, (1992).
17. N.N. Nikolaev and B.G. Zakharov, *Z. Phys.* C49, pp. 607–618, (1991).
18. J. Kwieciński, A.D. Martin, R.G. Roberts and W.J. Stirling, *Phys. Rev.* D42, pp. 3645–3659, (1990).

SPIN AND STRANGENESS — OPEN ISSUES[*]

Robert D. Carlitz
Department of Physics & Astronomy
University of Pittsburgh
Pittsburgh, PA 15260

Abstract

This talk provides a brief review of recent developments relating to the spin substructure and strangeness content of the nucleon.

NEW EXPERIMENTS

Long-term Program

The recent interest in the spin substructure and strangeness content of the nucleon has resulted from progress in a long-term program of experiments involving the scattering of polarized leptons from polarized hadrons. The first such experiments[1] (E80 and E130 at SLAC) explored $e - p$ scattering in the Feynman x range $0.1 < x < 0.65$ with momentum transfers Q in the range $1 \text{ Gev}^2 < Q^2 \text{ Gev}^2 < 10$. The EMC group at CERN explored $\mu - p$ scattering at higher energies,[2] extending the x range down to 0.015 and carrying values of Q^2 out to 75 Gev^2.

These experiments measure quantities which were thought to have a simple and straightforward interpretation in terms of the quark model, and the SLAC experiments appeared to be consistent with these expectations. The closer look provided by the EMC experiment cast doubt on this simple picture, and a number of new experiments[3] are now in various stages of completion with an eye toward resolving the new questions raised by the EMC data.

- The SMC experiment at CERN will scatter muons in the energy range of 100−200 Gev from polarized proton and deuterium targets. This allows one to extend the x range as low as 0.003 for $Q^2 = 1$ Gev^2. Furthermore it will provide information on the spin structure of the neutron as well as that of the proton. New target technology allows much more rapid spin reversals than were possible in the EMC experiment, with a resultant reduction in systematic errors.

- E142 at SLAC will scatter electrons from polarized tritium, with x values in the range $0.04 < x < 0.7$ to obtain information on the spin structure functions of the neutron.

- E143 at SLAC will cover the same x range with polarized protons and polarized deuterons in the polarized target. This experiment will allow one to compare the spin structure functions of the proton and neutron and test the Bjorken sum rule.

- HERMES will scatter the internal electron beam at HERA (with an energy of 30 − 35 Gev) from hydrogen, deuterium

---

[*]Supported in part by the National Science Foundation.

and tritium in an internal storage cell. The storage cell provides an atomic target, making it possible to reverse the target spin on a pulse by pulse basis and thus obtain extremely low systematic errors.

- HELP proposes to scatter $50 - 100$ Gev electrons in the internal beam at LEP from a polarized gas jet.

The list of new experiments shows a compressed time frame in two senses. New target technology allows one to reverse the target spins more rapidly and promises significant reductions in the systematic errors that have traditionally plagued such experiments. Furthermore one sees that the relatively slow development of this entire subfield promises to be accelerated significantly in the next few years. There will be accurate data obtained for the spin structure functions of both the proton and the neutron during this time period.

Measured Quantities

The experimental quantity which is extracted in the experiments just discussed is a longitudinal asymmetry

$$A^p = \frac{\sigma(e^\uparrow p^\uparrow) - \sigma(e^\uparrow p^\downarrow)}{\sigma(e^\uparrow p^\uparrow) + \sigma(e^\uparrow p^\downarrow)}, \quad (1)$$

in terms of which one defines a spin structure function

$$g_1^p(x, Q^2) = A^p F_1(x, Q^2). \quad (2)$$

Odd moments of $g_1^p(x, Q^2)$ with respect to x correspond to matrix elements of local operators constructed from the quark and gluon fields. In particular, for the first moment of g_1^p we can write

$$\Gamma_1 = \int dx g_1^p(x, Q^2) = \sum_i e_i^2 \Delta q_i'(Q^2)$$
$$\times \left(1 - \frac{\alpha_S(Q^2)}{\pi} + \mathcal{O}(\alpha_S^2)\right), \quad (3)$$

where

$$2M_p s^\mu \Delta q_i' = <ps|\bar{q}_i \gamma^\mu \gamma_5 q_i|ps>. \quad (4)$$

The quantities $\Delta q_i'$ denote the fraction of the proton's chirality associated with quarks of flavor i. The total chirality associated with quarks in the proton is thus

$$\Sigma' = \sum_i \Delta q_i'. \quad (5)$$

Naively, one would associated a quark's chirality with that quark's spin, but this expectation could be upset by anomalies in the chiral current, and experiment suggests this may be the case.

Simple Expectations

A naive association of chirality and spin would let one write

$$g_1(x, Q^2) = \frac{1}{2} \sum_i e_i^2 [q^\uparrow(x, Q^2) - q^\downarrow(x, Q^2)],$$
$$(6)$$

where $q^\uparrow$ and $q^\downarrow$ refer to the probability of finding quarks with spin up or down in a proton polarized to have its spin up (in its direction of motion).

The nonrelativistic quark model provides an explicit prescription for the spin distribution of quarks within the proton, with the result that

$$\begin{aligned}\Delta u' &= 4/3 \\ \Delta d' &= -1/3 \\ \Delta s' &= 0\end{aligned} \quad (7)$$

These values imply that $\Gamma_1 = 5/18$ and $\Sigma' = 1$. The values of $\Delta u'$ and $\Delta d'$ also lead to a prediction that the ratio of the axial vector and vector couplings of the proton should be $g_A/g_V = 5/3$. The experimental value of this ratio is 25% smaller than this prediction, so a simple model which respects the breaking of SU(6) symmetry might reduce each of the $\Delta q_i'$

by a similar amount. This leads to a broken SU(6) prediction of

$$\Gamma_1 \simeq 0.21,$$
$$\Sigma' \simeq 0.75. \quad (8)$$

A somewhat more sophisticated model was provided by the work[4] of Ellis and Jaffe. They assumed that the strange quark component $\Delta s'$ should vanish and used data from hyperon beta decays to determine $\Delta u'$ and $\Delta d'$ in terms of the couplings F and D of the axial vector current:

$$\Delta u' = 2F \simeq 0.93,$$
$$\Delta d' = F - D \simeq -0.33, \quad (9)$$

which leads to the expectations

$$\Gamma_1 \simeq 0.19,$$
$$\Sigma' \simeq 0.60. \quad (10)$$

Models of SU(6) symmetry breaking more elaborate than the one discussed above lead to a picture equivalent to that of Ellis and Jaffe. One expects that SU(6) breaking should not involve constant ratios in x but should lead to differences in the functional form of the quark distribution functions, so that

$$d(x, Q^2) \ll u(x, Q^2) \quad \text{as} \quad x \to 1$$
$$\Delta q(x, Q^2)/q(x, Q^2) \to 0 \quad \text{as} \quad x \to 0$$

One can incorporate these features in a model which associates all of the proton's spin with its valence quarks, parametrizing the spin distributions and constraining the parametrization by sum rules and the constraints of Regge asymptotic behavior. Such a model[5] provided an adequate description of the data from E80 and E130 but failed badly when confronted with the data of the EMC experiment.

Surprising Results

The data from experiments E80 and E130 did not extend to low enough values of x to test models of Γ_1 very precisely, so prior to the EMC experiment there was no reason to suspect the validity of the Ellis-Jaffe sum rule. The EMC data, combined with data from hyperon beta decays and independent of any assumption about the strange quark content of the proton, provide the following results:

$$\Delta u' = 0.782 \pm 0.032 \pm 0.046,$$
$$\Delta d' = -0.471 \pm 0.032 \pm 0.046, \quad (11)$$
$$\Delta s' = -0.190 \pm 0.032 \pm 0.046.$$

The measured values for $\Delta d'$ and $\Delta s'$ are both larger than what is suggested by the simple models discussed above. Furthermore, the total quark chirality,

$$\Sigma' = 0.120 \pm 0.094 \pm 0.138 \quad (12)$$

is actually consistent with zero, far below its expected value.

CONSEQUENCES OF EMC RESULTS

Polarized Neutrons are Interesting

The results of the EMC experiment, if confirmed by more sensitive experiments now in progress, have important consequences for the spin structure function of the neutron, $g_1^n(x, Q^2)$. Consider the predictions of the previously-discussed models for the ratio Γ_1^n/Γ_1^p:

$$\text{SU(6)} \Rightarrow \Gamma_1^n/\Gamma_1^p = 0$$
$$\text{Ellis/Jaffe} \Rightarrow \Gamma_1^n/\Gamma_1^p \simeq -0.13$$
$$\text{EMC} \Rightarrow \Gamma_1^n/\Gamma_1^p \simeq -0.50$$

One sees that if the EMC result is valid, the neutron's spin structure function should be of the same order of magnitude as that of the proton. This is a surprising result, since the non-relativistic quark model would have g_1^n vanish.

The experiments now under way at CERN and SLAC will answer this question directly, with simultaneous measurements of the spin structure functions of the proton and the neutron. The difference of the first moments of these structure functions, $\Gamma_1^p - \Gamma_1^n$, provides an important test of QCD known as the Bjorken sum rule[6]. Direct evaluation of this sum rule will allow one to distinguish between QCD predictions and the predictions of non-QCD models[7] which have been brought forward as explanations of the EMC data.

The small x behavior of $g_1^n(x, Q^2)$ should be particularly interesting. While $g_1^p(x, Q^2)$ appears to be positive for all x, models for $g_1^n(x, Q^2)$ typically provide a function which is positive near $x = 1$ but which changes sign and is negative near $x = 0$. The negative region of g_1^n dominates the moment integral and gives the expected negative value for Γ_1^n.

Strange Quarks may be Abundant

If $\Delta s'$ is as large as is suggested by the EMC data, one should be able to see similarly large effects for the strange form factor of the nucleon and for the strange quark distribution functions $s(x, Q^2)$ and $\Delta s(x, Q^2)$. Indeed, an analysis[8] of neutrino-proton elastic scattering produces a value of the strange axial vector coupling G_1^s consistent with the EMC analysis. Parity violating effects in the elastic scattering of polarized electrons could provide similar information on the strange vector couplings F_1^s and F_2^s. Measurements of parity violating effects in deep inelastic scattering can also yield information on the strange quark distributions.

Gluons may be Highly Polarized

One of the more intriguing possibilities raised by the EMC experiment is the idea[9-11] that gluons within a polarized proton may themselves to highly polarized. The argument is that, because of the axial anomaly, the quark helicity $\Delta q_i'$ can receive a contribution from polarized gluons, and that one should make a decomposition

$$\Delta q_i' = \Delta q_i - \frac{\alpha_S}{2\pi} \Delta G. \quad (13)$$

Here Δq_i (without a prime) denotes the quark spin fraction and ΔG the gluon spin fraction. One could accomodate the expectations of the nonrelativistic quark model in the EMC data if one admitted the possibility of a large gluon spin fraction,

$$\Delta G \sim 5 - 6. \quad (14)$$

Since the postulated breakup of $\Delta q_i'$ into a quark piece and a gluon piece has no effect on the total quantity measured in the EMC process, it's clear that this one experiment is insufficient to test the hypothesis of a large gluon spin. One needs additional experiments which are directly sensitive to the gluon spin fraction. Several such possibilities have been proposed.

- $e^\uparrow p^\uparrow \to e + 2\ jets + X$.[11] The asymptotic values of Δq_i and $\alpha_S \Delta G$ are both expected to be constants on the basis of the QCD evolution equations. A 2-jet process initiated by polarized quarks will give a cross section proportional to $\alpha_S \Delta q_i$ while a 2-jet process initiated by polarized gluons will give a cross section of order $\alpha_S \Delta G$. For large Q^2 the gluon process should be dominant, and the measured 2-jet cross section can determine the magnitude of ΔG. This argument indicates that ΔG can be isolated from Δq although this specific process will be a very hard one to measure.

- $p^\uparrow p^\uparrow \to \gamma + X$.[12] This process is one of several which involve a polarized beam and target and which are sensitive to the gluon spin distribution within the proton. Two other such processes are listed below.

- $\gamma^\uparrow p^\uparrow \to J/\Psi + X$.

- $p^\uparrow p^\uparrow \to \text{jet} + X$.

- $pp^\uparrow \to \mu^+\mu^- + X$. This process differs from the preceding ones in that it involves a single spin asymmetry and requires only a polarized beam or target. Such processes are easier to study than double spin asymmetries, but the expected asymmetry for the single spin process is smaller by a factor of $\alpha_S(Q^2)$, as we shall discuss later in this talk.

THEORETICAL CHALLENGES

Explain small Σ'

If one accepts the EMC data at face value, then it is striking that Σ' might be so small – even consistent with zero. A number of theoretical explanations of this phenomenon have been suggested. Among them are the following:

- Skyrmions.[13]

- Nonperturbative evolution.[14]

- The axial anomaly and the U(1) problem.[15]

- Lattice computations.[16]

It is probably fair to say that none of these explanations has so far proved to be either conclusive or complete. Certainly there is no argument so solid as to suggest that the results inferred from the EMC data *must* be correct.

Construct New Models

As more data is gathering on detailed properties of the gluon and strange quark spin distributions $\Delta G(x, Q^2)$ and $\Delta s(x, Q^2)$, it is obviously desirable to have models for these quantities. Some steps have been made in this direction in the course of developing proposals

for measuring these quantities, but more work needs to be done in this area.

Explain Anomalies

We have already mentioned that the anomalous divergence of the axial vector current, $\partial_\mu \bar{q} \gamma^\mu \gamma_5 q$, is of some relevance for the quantities Σ', ΔG and Δs. It is intriguing to note that another anomalous divergence, $\partial_\mu T^{\mu\nu}$ is relevant for matrix elements of the scalar operators $\bar{q}_i q_i$. This is another place where the norelativistic quark model seems to fail badly, since the matrix element $<p|\bar{s}s|p>$ seems to be fairly large[17]. Perhaps there is some connection between anomalous currents and failures of the naive quark model. Certainly the naive quark model cannot account for these anomalies, which result from details of the quark's interaction with the gluon field.

Constituent Quarks vs. Current Quarks

A more general context for the type of question that we have been discussing is provided by the idea of consituent quarks – in terms of which the nucleon has a simple structure – and current quarks – in terms of which the current of QCD have a simple structure. For some applications the differences between the two kinds of quarks are insignificant, while for others they appear to be crucial. At present we have no general rule for which quantities should be describable by the non-relativistic quark model and which quantities require a more detailed description even to understand their orders of magnitude.

OTHER STRUCTURE FUNCTIONS

So far in this talk we have concentrated upon those spin structure functions which occur at leading twist in deep inelastic scattering. These structure functions are the easiest spin structure functions to measure and the only ones for which extensive data has been accumulated to date. During the last year there have been important advances in the theoretical understanding of higher twist operators in deep inelastic scattering and leading twist effects in purely hadronic processes. This information will be important for experiments that will be performed in the next few years.

Higher Twist

There has been some level of confusion with regard to higher twist operators and transverse spin distributions in deep inelastic scattering which dates back to Feynman's book[18] introducing the parton model. Moments of the structure function $g_1(x, Q^2)$ correspond to matrix elements of a set of twist 2 operators. There is another structure function accessible in deep inelastic scattering – $g_2(x, Q^2)$ – whose moments correspond to a set of twist 3 operators. The structure function g_2 has no simple interpretation in terms of probabilities; rather it refers to correlations among quarks and gluons within the polarized proton. Since the parton model ignores the interactions of quarks and gluons, it does not provide an adequate framework for discussion of these effects. Recent discussions of the subject[19] have clarified this picture considerably.

An examination of the twist three operators is a means of gaining more detailed information on the substructure of the nucleon. Such information may be a key to the understanding of such issues as the current quark – constituent quark puzzle. Models of these structure functions will be needed to interpret experimental data which should be forthcoming from HERA. Such models must include aspects of the quark-gluon interaction to make any sense. This is possible in the bag model of the nucleon, and should ultimately be possible in lattice calculations as well.

Transverse Spin

Transverse spin asymmetries involve helicity flips. Hence, in deep inelastic scattering processes, transverse spin effects are suppressed by factors of m_q/Q and are impractical to measure. In purely hadronic processes such as $p^\uparrow p^\uparrow$ scattering, the helicity flip of one quark can be compensated by the flip of another quark, and transverse spin effects are observable as leading twist effects. These effects will probably first be measured at the RHIC spin facility. Theorists have an interesting opportunity to construct models for the transverse spin structure functions in advance of these measurements.

There exists a large body of experimental data involving single transverse spin asymmetries. Such processes are suppressed at the leading order in QCD, but significant asymmetries are still observed at moderate values of Q^2. It would be interesting to connect these measurements with measurements of the leading twist effects in $p^\uparrow p^\uparrow$ scattering.

Recently it has been noted[20,21] that leading twist effects involving a single transversely polarized particle in the initial state can be observed if one measures the polarization of particles in the final state. Thus, for example, in the process

$$pp^\uparrow \to \text{jet}^\uparrow + X \qquad (15)$$

one can study the angular distribution of particles within the jet and extract information on the jet's handedness.

Another related approach[22] focusses upon the process

$$ep^\uparrow \to e + h + X \qquad (16)$$

and examines the distribution of a given hadron h about the axis of the recoil jet. This distribution relates directly to the jet's transverse polarization. Both methods promise to provide information on the leading twist transverse spin structure function of the proton.

Single Spin Asymmetries

We have mentioned previously the proposal to examine the gluon spin fraction through measurements of single spin asymmetries in the process

$$pp^\uparrow \to \mu^+\mu^- + X. \quad (17)$$

The spin dependence of this process involves the quantity $\vec{\sigma} \cdot \vec{p}^+ \times \vec{p}^-$. It is necessary to measure two particles in the final state, e.g., p^+ and p^-, to extract a longitudinal spin distribution in this process. Furthermore, since $\vec{\sigma} \cdot \vec{p}^+ \times \vec{p}^-$ is odd under time reversal, the production amplitude must involve a nontrivial phase, which implies that it must arise from one loop (or higher) processes at the parton level. This implies parton level asymmetries of order $\alpha_S/(2\pi)$. Detailed calculations[23] show that this estimate is correct.

To examine ΔG in processes of this type one should select events in which the $\mu^+\mu^-$ pair moves in a direction opposite that of the polarized proton, e.g. forward production from a polarized target. Using detailed models of the contributing structure functions one finds[24] typical expected experimental asymmetries on the order of 1%-2% over a large range of phase space. If gluons within the proton are highly polarized, then ΔG should be responsible for about 1/2 of the observed asymmetry.

REFERENCES

1. M.J. Alguard *et al.*, *Phys. Rev. Lett.* **37**, 1261 (1976); *ibid* **41**, 70 (1978); G. Baum *et al.*, *Phys. Rev. Lett.* **51**, 1135 (1983).

2. EM Collaboration, J. Ashman *et al.*, *Phys. Lett.* **B206**, 364 (1988); *Nucl. Phys.* **B328**, 1 (1989).

3. A recent summary may be found in the proceedings of the SLAC workshop on High Energy Electroproduction and Spin Physics, SLAC-392 (1992).

4. J. Ellis and R.L. Jaffe, *Phys. Rev.* **D9**, 1444 (1974).

5. R.D. Carlitz and J. Kaur, *Phys. Rev. Lett.* **38**, 673 (1977).

6. J.D. Bjorken, *Phys. Rev.* **148**, 1467 (1966).

7. G. Preparata and J. Soffer, *Phys. Rev. Lett.* **61**, 1167 (1988).

8. D. Kaplan and A. Manohar, *Nucl. Phys.* **B310**, 527 (1988).

9. A.V. Efremov, O.V. Teryaev, JINR preprint E2-88-287, unpublished (1988).

10. G. Altarelli and G.G. Ross, *Phys. Lett.* **B212**, 391 (1988).

11. R.D. Carlitz, J.C. Collins and A.M. Mueller, *Phys. Lett.* **B214**, 229 (1988).

12. E. Berger and J.-W. Qiu, *Phys. Rev.* **D40**, 778 (1989).

13. S. Brodsky, J. Ellis and M. Karliner, *Phys. Rev. Lett.* **206B**, 309 (1988); J. Ellis and M. Karliner, *Phys. Lett.* **B213**, 73 (1988).

14. R.L. Jaffe, *Phys. Lett.* **B193**, 101 (1987).

15. G. Veneziano, *Mod. Phys. Lett.* **A4**, 1605 (1989).

16. J. Mandula, *Phys. Rev. Lett.* **65**, 1403 (1990).

17. T.P. Cheng and R. Dashen, *Phys. Rev. Lett.* **26**, 594 (1971).

18. R.P. Feynman, *Photon-Hadron Collisions*. Reading, MA: Benjamin, 1971.

19. J.P. Ralston and D.E. Soper, *Nucl. Phys.* **B152**, 209 (1979); X. Artru and M. Mekhfi, *Z. Phys.* **C 45**, 669 (1990); J.C. Collins in *Proceedings of the Polarized Collider Workshop*, J.C. Collins, S. Heppelmann and R. Robinett, eds. New York: A.I.P,

1991). I. Balitsky and V. Braun, *Nucl. Phys.* B361, 93 (1991); R.L. Jaffe and X.-D. Ji, *Phys. Rev. Lett.* 67, 552 (1991); J.-W. Qiu and G. Sterman, *Nucl. Phys.* B353, 105, 137 (1991).

20. A.V. Efremov, L. Mankiewicz and N.A. Törnqvist, *Phys. Lett.* B284, 394 (1992).

21. R.D. Carlitz, J.C. Collins, S. Heppelmann, R.L. Jaffe and G. Ladinsky, Penn State preprint PSU/TH/101 (1992).

22. J.C. Collins, Penn State preprint PSU/TH/102 (1992).

23. R.D. Carlitz and R.S. Willey, *Phys. Rev.* D45, 2323 (1992).

24. R.D. Carlitz and R.S. Willey, University of Pittsburgh preprint 91-10 (1991).

POLARIZED PHOTON OR PROTON PRIMAKOFF EFFECT[*]

J. Bernabéu and J. Vidal
Departament de Física Teòrica, Universitat de València, e IFIC,
Centre Mixt Univ. Valencia-CSIC, E-46100 Burjassot, Spain

L.N. Epele, H. Fanchiotti, C.A. García Canal and G.A. González Sprinberg
Departamento de Física, Universidad Nacional de La Plata,
C.C. 67, 1900 La Plata, Argentina and CONICET, Argentina

Abstract

A proposal to determine the axial coupling of the proton for the neutral strangeness current is discussed. By means of the $\gamma - Z - \pi^o$ triangle anomaly, the parity violating asymmetries for polarized photon or polarized proton Primakoff effect filter the couplings so as to leave the proton axial coupling only. We calculate the relevant observables induced by the electroweak interference and study the regions of energy and Q^2 of possible experimental interest.

INTRODUCTION

The spin-dependent structure function $g_1^p(x, Q^2)$ of the proton, as determined by the EMC-experiment[1], has given rise to a lively debate[2] about the composition and the spin structure of the proton. Together with previous SLAC data for electron scattering EMC-data led to the first moment

$$\int_0^1 dx\ g_1^p(x,Q^2) = 0.126 \pm 0.018$$
$$< Q^2 >_{EMC} = 11 GeV^2 \quad (1)$$
$$< Q^2 >_{SLAC} = 5 GeV^2$$

By means of the OPE in the limit $Q^2 \to \infty$. Eq.(1) measures the helicity carried by the different quarks (and antiquarks) in a polarized proton, weighted by the appropriate square of charges for each quark. Defining the polarized quark moments Δq, Eq.(1) determines the combination $\frac{4}{9}\Delta u + \frac{1}{9}\Delta d + \frac{1}{9}\Delta s$. Another two combinations, members of the octet under flavour-SU(3), are obtained from the analysis of semileptonic baryon decays: $\Delta u - \Delta d$ and $\Delta u + \Delta d - 2\Delta s$. This analysis gives

$$\Delta u = 0.78 \pm 0.06$$
$$\Delta d = -0.47 \pm 0.06 \quad (2)$$
$$\Delta s = 0.19 \pm 0.06$$

In particular, the flavour singlet part of this first moment is found to be anomalously small

$$\Delta\Sigma \equiv \Delta u + \Delta d + \Delta s = 0.12 \pm 0.17 \quad (3)$$

and it gave rise to the so called "spin crisis".

Another probe of the flavour content of the proton is provided by the weak neutral axial current, for which the operator is

$$J_\lambda^{A,Z} = \bar\Psi_u \gamma_\lambda \gamma_5 \Psi_u - \bar\Psi_d \gamma_\lambda \gamma_5 \Psi_d - \bar\Psi_s \gamma_\lambda \gamma_5 \Psi_s \quad (4)$$

For elastic low Q^2 neutral current processes, the polarized quark moment Δq for definite flavour gives the corresponding coupling constant for axial currents

$$< p|\bar\Psi_q \gamma_\lambda \gamma_5 \Psi_q|p > \stackrel{Q^2 \to 0}{\longrightarrow} (\Delta q)\bar p \gamma_\lambda \gamma_5 p \quad (5)$$

---

[*]This work has been supported in part by CICYT, under Grant AEN 90-0040.

so that the axial coupling of the proton (and neutron) for neutral current interactions is determined as

$$G_A^p = \Delta u - \Delta d - \Delta s = 1.44 \pm 0.006$$
$$G_A^n = -\Delta u + \Delta d - \Delta s = -1.06 \pm 0.06 \quad (6)$$

when the result (2) is used. In terms of nucleonic isospin, there is therefore, in addition to the well known isovector axial coupling g_A, an isoscalar axial coupling f_A, such that

$$g_A = \Delta u - \Delta d = 1.254 \pm 0.006$$
$$f_A = -\Delta s = 0.19 \pm 0.06 \quad (7)$$

A non-vanishing value of the last term has thus been suggested by the EMC-data. The question is: Is it possible to measure G_A^p, or even f_A directly? In section 2 we discuss several proposals studied previously by different authors. In the proposal presented here, the parity violating asymmetries in the Primakoff effect for polarized photons or polarized protons, we shall see that the neutral vector coupling is filtered and only G_A^p is left in the observables. A detailed study is presented in the subsequent Sections.

NEUTRAL CURRENT PHENOMENA

A non-vanishing value for the axial coupling of the strange quark current f_A of the proton has measurable consequences in neutral current phenomena related to their isoscalar axial components. There are different proposals for an experimental observation of this new component, which can be searched for elastic neutrino-proton scattering[3]. An experiment with intermediate energy neutrinos on low-Q^2 $\nu p \to \nu p$ is underway at LAMPF. Its information content is $(G_V^p)^2 + 3(G_A^F)^2$. A general discussion on possible sources of information on f_A is given in Ref.4.

With incident neutrinos, besides the urgent νp scattering, there are suggestions[5] to look for axial transitions in nuclei and thus select the relevant coupling. This program can be accomplished if the nuclear physics ingredients are well under control. Particularly promising[6] is the ^{12}C excitation $\nu\ ^{12}C \to \nu\ ^{12}C^*(1^+, I = 0; 12MeV)$ with intermediate energy neutrinos; this isoscalar level decays by α-emission. The filter of quantum numbers selects the strength given by f_A^2, with a small contamination of the magnetic vector coupling $f_A\ G_M$. The use of reactor neutrinos can be envisaged[4] for a target like ^7Li, looking for the axial transition $\nu\ ^7Li \to \nu\ ^7Li^*\ (0.48MeV)$; this level decays by γ-emission. With isospin 1/2, the cross section is proportional to the combination $(g_A + 0.87 f_A)^2$.

The parity-violating asymmetries in the scattering of polarized electrons on protons and nuclei are induced by the electroweak interference between γ- and Z-exchanges. A number of experiments have been proposed[7] to look for the matrix elements of the axial, as well as the vector, strange quark-current. On general grounds, the term in the P-odd observable proportional to G_A contains, as a factor, the vector coupling of the electron v^e too. This last coupling suffers from the "1/4-effect" and it is very small. Due to this effect, the proposed experiment $\vec{e}\,p \to e\,p$ at CEBAF will look for strangeness components in the weak vector current, which contribute to the asymmetry with the axial coupling a^e of the electron. The strange charge form factor will be looked for in the coherent nuclear scattering $\vec{e}\ ^{12}C \to e\ ^{12}C$, an experiment now underway at BATES.

Similar comments apply to the nuclear-spin-dependent parity violating (NSDPV) effects in atomic physics: the axial hadronic current appears with the vector coupling of the electron, small. In a $Q^2 = 0$ situation, such as that for atoms, there cannot be strange components in the nuclear vector current. although the (BNSDPV)-effect has probably been seen[8] in the hyperfine lines of Cs, its magnitude is controlled in heavy atoms by the nuclear anapole moment and not by neutral currents. Better prospects to look for the neutral axial couplings of protons and neutrons appear from the (NSDPV)-effects in light muonic atoms[9].

The main point of this article is to show that the polarized Primakoff effect is particularly adapted to provide selected information on the axial coupling G_A^p. In precise terms, the parity violating asymmetry for circularly polarized photons gives G_A^p, which interferes with the magnetic structure G_M of the proton, and the corresponding asymmetry for longitudinally polarized protons goes like $G_A^p G_E$, with G_E the electric form factor of the proton.

PRIMAKOFF EFFECT

The Primakoff Effect[10] consists of the coherent photoproduction of π^o by the nuclear Coulomb field. Its origin is mediated by the triangle chiral anomaly[11] for the vector-vector-axial currents. The anomalous axial current, in the chiral limit, has the divergence

$$\partial_\mu A^\mu = \frac{e^2}{16\pi^2} D^{\gamma\gamma} F_{\mu\nu} \tilde{F}^{\mu\nu} \qquad (8)$$

where $F_{\mu\nu}$ is the e.m. tensor and $D^{\gamma\gamma}$ is the $\gamma - \gamma - \pi^o$ anomaly, given by

$$D^{\gamma\gamma} = N_c Tr[\{Q^{em}, Q^{em}\}\frac{\tau_3}{2}] = 1 \qquad (9)$$

For later considerations, we should remember that, even if π^o only involves (u,d)-quark flavours, due to anomaly cancellation in the standard theory one has $3(Q_u^2 - Q_d^2) = Q_e^2$, which is the result (9).

Primakoff effect experiments have been performed in the past[12] in order to measure the π^o-lifetime. The cross section shows the following prime features:
- it is proportional to the decay width of the π^o into two photons, and to the square of the charge of the target.
- it is concentrated in the forward region, with a characteristic sharp peak around $\theta \simeq m_\pi^2/(2E^2)$, where θ is the angle between the incident photon and the outgoing π^o.

In the Primakoff production there exists also a neutral weak current contribution through the $\gamma - Z - \pi^o$ vertex. This contribution will be of the order $\frac{G_F Q^2}{\alpha}$, relative to the e.m. one, so one has to look for parity-violating asymmetries[13] in order to disentangle it. The P-odd observables will be induced by the weak-electromagnetic interference for polarized photons or polarized protons.

Let us consider the triangle loop associated with the $\gamma - Z - \pi^o$ vertex. Due to C-symmetry, the combination of currents (VAA) vanishes [this is manifestly seen by the cancel-

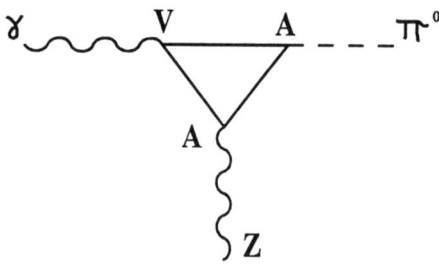

lation of the two running senses in the loop]. These are good news. The only parity violating amplitude is given by the diagram

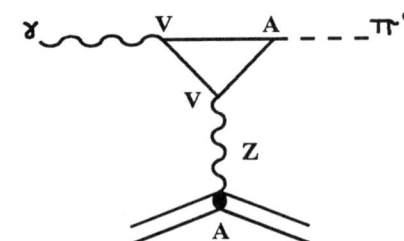

in which we note: 1) it is proportional to the $\gamma - Z - \pi^o$ anomaly $D^{\gamma Z}$; 2) the parity-violating interference automatically selects the weak neutral <u>axial</u> current of the proton, with coupling G_A^p.

As we are in a low Q^2 regime, where PCAC can be used, this method can provide a way to measure the $D^{\gamma Z}$-anomaly. Its value is[14]

$$D^{\gamma Z} = \frac{N_c}{s_w c_w} Tr[\{Q^{em}, V^Z\}\frac{\tau_3}{2}] = \frac{1 - 4s_w^2}{4 s_w c_w} \quad (10)$$

which appears surprisingly small: $D^{\gamma Z} \simeq 0.05 \leftrightarrow D^{\gamma\gamma} = 1$. This will suppress to sought-for interference. Why is there a "1/4-effect" in this case, with only (u,d)- quarks running into the loop? Again, due to anomaly cancellation in the standard theory, one has $3(Q_u v^u - Q_d v^d) = Q_e v^e$! Bad news.

P-ODD ASYMMETRIES

All the observable quantities of interest are obtained from the electromagnetic and weak amplitudes as

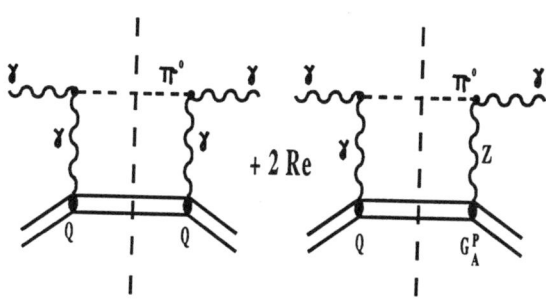

$$L^{\nu\mu}\{W_{\nu\mu}^{c.m.} - \frac{1 - 4s_w^2}{4\pi} \frac{G_F}{\sqrt{2}} \frac{q^2}{\alpha} W_{\nu\mu}^I\}, \quad (11)$$

where the non-baryonic tensor $L^{\nu\mu}$ is common to both terms in Eq.(11), $\gamma - \gamma$ and $\gamma - Z$. In fact, its structure is of the form

$$L^{\nu\mu} = L_S^{\nu\mu} + i L_A^{\nu\mu}(h) \quad (12)$$

where $L_S^{\nu\mu}$ is real, symmetric and independent of the photon helicity h, whereas the second term is imaginary, antisymmetric and linear in h.

The electromagnetic $\gamma - \gamma$ tensor $W_{\nu\mu}^{c.m.}$ has the following structure

$$W_{\nu\mu}^{c.m.} = W_{\nu\mu,S}^{c.m.} + i W_{\nu\mu,A}^{c.m.}(s) \quad (13)$$

where $W_{\nu\mu,S}^{c.m.}$ is real, symmetric and independent of the proton polarization s: it is the analogue of $W_{1,2}$ in deep inelastic scattering and here it is determined by the two Sachs form factors, G_E and G_M, of the proton. The second term, on the contrary, is imaginary, antisymmetric and linear in s: it is the analogue of $G_{1,2}$. By inspection of Eqs. (12) and (13), we see that the contraction $L^{\nu\mu} W_{\nu\mu}^{c.m.}$ cannot induce <u>separate</u> linear terms in h <u>or</u> s. As our aim is the extraction of G_A^p in $W_{\nu\mu}^I$, the recipe here (as opposed to that for Spin Dependent Structure Functions $G_{1,2}$) is: <u>not to take both polarized photon and proton (simultaneously) asymmetry</u>. Then the $\gamma - \gamma$ term is just given by the unpolarized cross section $L_S^{\nu\mu}, W_{\nu\mu,S}^{c.m.}$, the conventional Primakoff effect.

The interference $\gamma - Z$ tensor $W_{\nu\mu}^I$ has the following structure

$$W_{\nu\mu}^I = i W_{\nu\mu,A}^I + W_{\nu\mu}^I(s) \quad (14)$$

where the first term is imaginary, antisymmetric and independent of the proton polarization s: it is the analogue of W_3 and it is given by the axial-magnetic interference $G_A^p G_M$. This is due to the fact that only the transverse components of the proton current interfere if no measurement is made with proton polarization. One thus builds a first parity-violating observable from

$$L_A^{\nu\mu}(h) W_{\nu\mu,A}^I,$$

which corresponds to the symmetry for circularly polarized photons

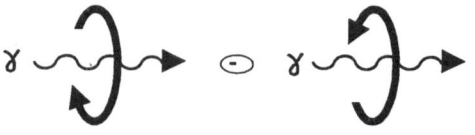

$$A^\gamma = \frac{d\sigma(h=+) - d\sigma(h=-)}{d\sigma(h=+) + d\sigma(h=-)} =$$
$$= \frac{1-4s_w^2}{4\pi} \frac{G_F}{\sqrt{2}} \frac{(-q^2)}{\alpha} \frac{G_A^p G_M}{G_E^2} \frac{q^2-m_\pi^2}{2ME} \quad (15)$$

One recognizes in A^γ all the relevant factors: the relative anomaly $D^{\gamma Z} \leftrightarrow D^{\gamma\gamma}$, the weak-e.m. interference and the specific axial-magnetic interference versus the square of the charge.

The second term of Eq.(14) is real and linear in s: it is given by the axial-electric interference $G_A^p G_E$, which will manifest itself if a measurement is made with proton polarization. One thus builds a second parity-violating observable from

$$L_S^{\nu\mu} W_{\nu\mu}^I(s) ,$$

which corresponds to the asymmetry for longitudinally polarized protons

$$A^p \equiv \frac{d\sigma(s=+) - d\sigma(s=-)}{d\sigma(s=+) + d\sigma(s=-)} =$$
$$= \frac{1-4s_w^2}{4\pi} \frac{G_F}{\sqrt{2}} \frac{(-q^2)}{\alpha} \frac{G_A^p}{G_E} \quad (16)$$

As seen, A^p is a function of Q^2 only, with no explicit dependence on the incident energy E.

CONCLUSIONS

The parity violating asymmetries for polarized photon (15) or polarized proton (16) Primakoff effect filter the couplings of the proton so as to leave the weak neutral axial coupling G_A^p only. These asymmetries, due to the interference between γ- and Z-exchanges, are mediated by the $\gamma - Z - \pi^o$ anomaly, which is here applied in a situation of (PCAC)-regime.

Inspection of Eqs.(15) and (16) tells us that, if the Primakoff process can still be identified by its characteristic $(sin^2\theta/Q^4)$- dependence, one should move to the upper tail of the peak in order to reach $Q^2 \sim 0.2-0.4 GeV^2$: the lower cross section is compensated by higher asymmetries. In Table 1 we give the results for the cross section and the two asymmetries for the incident energies E=1, 4 and 7 GeV, and $Q^2 = 0.2-0.4 GeV^2$, assuming G_A^p as given by Eq.(6).

In the case of circular polarization of photons, the A^γ-asymmetry decreases with increasing energy, at fixed Q^2. For longitudinal polarization of protons, the A^p-asymmetry is independent of energy, at fixed Q^2; higher energies increase the number of events.

E (GeV)	θ (rad)	Q^2 (GeV^2)	$\frac{d\sigma}{d(\cos\theta)}$ (μ barn)	A^γ	A^p
1	0.46	0.2	2.2×10^{-3}	6.6×10^{-7}	-2.0×10^{-6}
1	0.70	0.4	9.8×10^{-4}	3.0×10^{-6}	-4.3×10^{-6}
4	0.11	0.2	3.8×10^{-2}	1.5×10^{-7}	-2.0×10^{-6}
4	0.15	0.4	2.0×10^{-2}	5.4×10^{-7}	-4.3×10^{-6}
7	0.06	0.2	0.12	8×10^{-8}	-2.0×10^{-6}
7	0.09	0.4	5.4×10^{-2}	3.8×10^{-7}	-4.3×10^{-6}

Table 1. Primakoff cross section and P-odd asymmetries

REFERENCES

1. J. Ashman et. al., EMC experiment, *Phys. Lett.* B206 (1988) 364; *Nucl. Phys.* B328 (1989) 1.

2. G. Altarelli, *Lectures at the International School of Subnuclear Physics*, Erice, CERN-TH 5675/90.

3. L. Wolfenstein, *Phys. Rev.*, D19 (1979) 3450;
   D.B. Kaplan and A. Manohar, *Nucl. Phys.*, B310 (1988) 527;
   J. Ellis and M. Karliner, *Phys. Lett.* B213 (1988) 73.

4. J. Bernabeu, *Nucl. Phys.* A518 (1990) 317.

5. T. Suzuki et. al., *Phys. Lett.* B252 (1990) 323;
   T. Suzuki, *Nucl. Phys.* A515 (1990) 609;
   E. M. Henley et al., *Phys. Lett.* B269 (1991) 31.

6. J. Segura, J. Bernabeu and F.J. Botella, in preparation

7. R.D. Mckeown, *Phys. Lett.* B219 (1989) 140; *Nucl. Phys.* A532 (1991) 499.
   D.H. Beck, *Phys. Rev.* D39 (1989) 3248; *Nucl. Phys.* A532 (1991) 507.
   R. Decker et al., *Nucl. Phys.* A512 (1990) 626.

   J. Bernabeu et. al., *Phys. Lett.* B282 (1992) 177.

8. M.C. Noecker, B.P. Masterton and C.E. Wieman, *Phys. Rev. Lett.* 61 (1988) 310.

9. J. Bernabeu, J. Bordes and J. Vidal, *Z.Phys* 41 (1989) 679;
   J. Missimar and L.M. Simons, PR-90-03, PSI (1990);
   J. Bernabeu, FTUV/91-53.

10. H. Primakoff, *Phys. Rev.* 81 (1951) 899;
    A. Halprin et al., *Phys. Rev.* 152 (1966) 1295.

11. S.L. Adler, *Phys. Rev.* 177 (1969) 2426;
    J.S. Bell and R. Jackiw, *Nuovo Cim.* A60 (1969) 47.

12. Browman et al., *Phys. Rev. Lett* 33 (1974) 1400.

13. J. Bernabeu, J. Vidal, L.N. Epele, H. Fanchiotti, C.A. Garcia Canal and G.A. Gonzalez Sprinberg, in preparation.

14. M. Jacob and T.T. Wu, *Phys. Lett.* B232 (1989) 529.

DEUTERONLIKE MESON-MESON BOUND STATES

Nils A. Törnqvist
Research Institute for High Energy Physics
University of Helsinki
Siltavuorenpenger 20
SF-00170 Helsinki, Finland

Abstract

Arguments are summarized for why many of the best non-$q\bar{q}$ candidates may be deuteronlike bound states of two vector mesons or $K\bar{K}^*$. Using pion exchange as a guide for where to look for bound states, only a few deuteronlike meson-meson resonances are expected. Four of these are likely to be the already seen non-$q\bar{q}$ candidates: $f_1(1420)$, $f_2(1520)$, $f_0(1590)$, $f_0(1710)$. In addition, one expects a $K_0^*(\approx 1660)$ composed of $K^*\omega$ and $K^*\rho$. In the charm sector one expects a $D^*\bar{D}^*$ composite $\chi_{c0}(\approx 4000)$ and an axial state $D\bar{D}^*$ composite $\chi_{c1}(\approx 3870)$. The last mentioned one implies a state doubling in the axial charm sector near 3900 MeV, similar to the $f_1(1420) - f_1(1510)$ puzzle in the light quark sector. The second resonance is the expected $c\bar{c}\, 2^3P_1$ $\chi_{c1}(\approx 3950)$. The $\chi_{c1}(\approx 3870)$ is particularily interesting, since it should be narrow and an almost unavoidable prediction of pion exchange. Namely, the exchanged pion can be almost on shell giving a Coulomb-like potential, due to the mass relation $m_{D^*} \approx m_D + m_\pi$.

INTRODUCTION

This talk is a summary of my arguments for why many of our best non-$q\bar{q}$ candidates are likely to be deuteronlike bound states of two vector mesons or of $K\bar{K}^* + c.c.$.

No one doubts that the deuteron exists; So why should we not have deuteronlike meson-meson bound states, where the nucleons are replaced by mesons, in particular by vector mesons? For the deuteron we know that pion exchange is a reasonable first approximation which provides the binding, so it seems natural to ask the questions: When does pion exchange couple and when does it give a strong attraction? Of course, pion exchange need not provide all the binding. In fact, even for the deuteron we know that other effects are also important, and looking at e.g., the Paris NN potential[1] one can estimate that only a little more than 50% of the attractive potential is given by simple one-pion exchange. But, it seems very likely that should meson-meson composites exist which are large like the deuteron, i.e., with a radius of about 2 meson radii (1.5 fm) pion exchange should be an important contribution. Heavier meson exchanges have about 5 times shorter range, and similarily quark interchange forces are expected to be of short range. They would thus be important for small baglike multiquark configurations. The pion is nearly massless on the hadronic scale, as required by chiral symmetry, and clearly provides the main dynamics by which we can separate tightly bound baglike multiquark states from deuteronlike meson-meson composites.

Below we first discuss why a new terminology is needed and then summarize our arguments for why many of the observed non-$q\bar{q}$ candidates may be deuteronlike. At the end the predictions for new states are given.

TERMINOLOGY

Multiquark or meson-meson bound states are often referred to as "*molecules*" for historical reasons. However, many physicists do not find that name very well chosen. Usually we do not use names which already have another meaning. Furthermore, for meson-meson composites the analoguous objects in nuclear-molecular physics would be nuclei, not molecules. The latter are, of course, orders of magnitudes larger objects, and usually much more complex structures than are composites of two mesons. In fact, using the anology within the baryon sector one is often discussing states composed of two mesons, and then the term "deuteronlike" is used. But a "deuteronlike molecule" is certainly nonsense and terminology clashes like this should be avoided in order not to increase the confusion already present through the scarse data at hand.

Hence, I would like to suggest that if one wants to use the term *molecule* it should be used only in the context of a baglike multiquark states. In such *quark molecules* the quarks do not clearly cluster into two or more $q\bar{q}$ pairs or qqq baryons. Thus the term *quark molecule* would also represent multiquark composites with baryon number $\neq 0$ such as 6-quark composites which are not deuteronlike.

Then, in models where one builds the bound states from well established mesons, such that the quarks are clearly separated to $q\bar{q}$ pairs we need a new term. I suggested[2] the acronym *deuson* for "*deu*teronlike me*son* bound states", or generally for composites built from two mesons. These are thus comparatively loosely bound large composites with a radius larger than ordinary mesons by, say, about a factor of two. The term naturally generalizes to *trisons*, *helisons* (from heliumlike) etc., would heavier nucleuslike multi-mesonic systems (*nusons*) be found. In fact, one could speculate that the $X(3100)^3$ could be a *helison* composed mainly of $K^*\omega\omega$ (where the ω's represent also ρ's). One would need a rather large binding of the order of 100 MeV, but heliumlike systems are expected to have large binding, and with 4 constituents there are relatively more potential terms, which should strengthen the binding.

Of course, in some cases it may be difficult to distinguish such *deusons* from *quark molecules*. It is clear, as eloquently discussed by Maltman and Isgur[4], that there is a smooth transition between the two configurations. But, at least for distances of about, say, 1.5 fm it should be simpler, and more economical, to talk of the mesonic degrees of freedom, than of multiquarks clustering into near colorless $q\bar{q}$ pairs. For nuclei it is, of course, much more economical to talk of them in terms of bound multinucleons than in terms of multiquarks, so the same may well be true for multimesons. Hence, it should be useful to have a clear terminology to distinguish which picture one is talking about.

THE NON-$q\bar{q}$ CANDIDATES

Here I summarize the arguments why many of the best non-$q\bar{q}$ candidates are likely to be deuteronlike, i.e., deusons. We shall discuss the $f_0(1710)$ ("the θ"), the $f_0(1590)$ ("the G)", the $f_2(1520)$ ("the Asterix") and the $f_1(1420)$ ("the E"). These are our best non-$q\bar{q}$ candidates below 2 GeV in the sense that they pass the best model independent test we have for finding non-$q\bar{q}$ states, "the state counting test": When all nearby slots for the $q\bar{q}$ states are already occupied by well established resonances, the remaining ones must be non-$q\bar{q}$. The arguments which favor the deuson model for these are listed below in (i)-(iv). For the $f_0(1710)$[5], and for the $f_1(1420)$[6] other model calculations also favor a similar interpretation.

(i) *They all lie near an important threshold.*
The $f_0(1710)$ lies near the $K^*\bar{K}^*$ threshold (1790 MeV). The $f_2(1520)$ and the $f_0(1590)$ (the mass of the latter is found to be smaller at 1550 MeV by the crystal barrel[7]) are close to the $\rho\rho$ and $\omega\omega$ thresholds ($\approx$ 1553 MeV). Finally, the $f_1(1420)$ lies near the $K\bar{K}^*$ threshold (1390 MeV).

(ii) *Their decay modes with large phase space are strongly suppressed.*
For the $f_0(1710)$ the $K\bar{K}$ mode is dominant (38%), while $\eta\eta$ is (18%) and $\pi\pi$ is only (4%). For the $f_0(1590)$ the ratios of the branching ratios are $\eta\eta'/\eta\eta/K\bar{K}/\pi^0\pi^0 \approx$ 2.7/ 1/ < 0.6/ < 0.3. Assuming, as the Particle Data Group[8], that the Asterix $f_2(1520)$ resonance is the same as seen previously by Bridges et al.[9] (both are seen in $N\bar{N}$ annihilations) the $f_2(1520)$ decays predominantly into 4π via $\rho\rho$, but only 5% into $\pi\pi$. Finally, the $f_1(1420)$ is seen essentially only in the $K\bar{K}^*$ mode, whereas $\pi a_0(980)$ is small. For ordinary $q\bar{q}$ mesons, for small flavor singlet glueballs, or for small multiquark states one expects much larger branching ratios to especially the $\pi\pi$ mode. To understand this odd behaviour, a simple solution is to assume a strongly falling form factor in momentum space. This means that in coordinate space we have large objects, larger than the usual size of about 0.7 fm for mesons. For 4-quark systems this leads naturally to configurations where two $q\bar{q}$ pairs are separated relatively far from each others, i.e., to deuteronlike meson-meson bound states or deusons.

(iii) *There is strong attraction from pion exchange for precisely those meson-meson channels noted in (i) whose I^G, J^{PC} are those of the experimental candidates.*
This is my strongest argument favouring the deuson model. If the structures are large in extent as suggested in (ii) one expects the longest range forces to play an important role. This means to a first approximation pion exchange. So one should ask the questions: When does pion exchange couple, when is it attractive, and when is it particularly strong? As a first estimate one can look at the crossing matrix for one-pion exchange, and find numbers measuring the relative strength of one-pion exchange[2]. The remarkable result is that one finds strong attraction for only a few quantum numbers of meson-meson configurations, and that in most cases these quantum numbers coincide with those of the above experimental candidates! In addition, one finds a few new predicted deusons (see below) to be looked for.

(iv) *The total widths of the non-$q\bar{q}$ candidates are not incompatible with the expected width from constituent decay.*
A natural objection to this deuteronlike picture is that if the constituents are unstable ρ's, K^*'s or ω's, the deusons should decay simply through their constituent decay. In particular the ρ has a large width of 150 MeV and therefore, the deusons could become too broad compared to the experimental widths of our candidates. However, it is easy to convince oneself that the naive bound ($\Gamma < \sum \Gamma_{\text{constituent}}$) is satisfied for the mentioned candidates. If the binding energy is large, (as for the $f_0(1710)$, almost 80 MeV), this constituent decay would be suppressed. It would be very interesting to see whether there is a small $K\pi\bar{K}\pi$ decay mode from the tails of the K^*'s in the $f_0(1710)$ as expected in our picture.

PREDICTIONS AND CONCLUSIONS

Does the deuson model predict new states which should have been seen, or could be seen in the future? Within the light meson sector one predicts a $(K^*\omega - K^*\rho)/\sqrt{2}$ scalar I=1/2 deuson near 1660 MeV, i.e., a $K_0^*(\approx 1660)$, which should couple mainly to $K^*\rho$. It would have been difficult to see in previous experiments as its coupling to $K\pi$ or $K\rho$ should be small. In addition, it would be hidden under the broad ($q\bar{s}$ 2^3S_1 or 1^3D_1) vector resonance $K^*(1680)$). It is too heavy to be seen in $N\bar{N}$ at rest, but $N\bar{N}$ in flight can soon be studied at LEAR (and later at SuperLEAR). This should be a good place to look for the $K_0^*(\approx 1660)$. For other quantum numbers in the light meson sector (assuming only 0^{-+} and 1^{--} mesons as constituents) pion exchange is either absent, repulsive or weak. Thus one does not expect deusons which should already have been seen.

However, the best and most dramatic predictions are in the heavy quark sector. One expects a "sister of the $K^*\bar{K}^*$ $f_0(1710)$" or a $D^*\bar{D}^*$ scalar deuson near the threshold 4020 MeV (a $\chi_{c0}(\approx 4000)$) with a width like the $f_0(1710)$ of about 50 MeV to $D\bar{D}$.

Even more interesting is the "sister of the $f_1(1420)$": a $D\bar{D}^*$ axial deuson expected near 3870 MeV, a $\chi_{c1}(\approx 3870)$. As parity forbids $D\bar{D}$ decay it can decay only through charm annihilation and should thus be very narrow. Since we expect another axial meson, the 2^3P_1 $c\bar{c}$ near 3950 MeV ($\chi_{c1}(\approx 3950)$), we should have in the charm sector a repetition of the state doubling puzzle among the light axials ($f_1(1420)$ - $f_1(1510)$). The Fermilab $\bar{p}$ accumulator or SuperLEAR should be able to see these heavy deusons.

The $\chi_{c1}(\approx 3870)$ $D\bar{D}^*$ deuson is particularily difficult to avoid in our scheme, since due to the mass relation $m_{D^*} = m_D + m_\pi$ the exchanged pion can be essentially on shell. Therefore, the potential becomes really very long range, - almost Coulomb-like. Both constituents D and D^* are essentially stable (the latter width is expected to be in the neighborhood of 50 keV). Thus this deuson should be very narrow, apart from the small charm annihilation. As the pion can be almost on shell one could also picture this deuson candidate as being part of the time composed of $D\bar{D}\pi$, - a heavy scalar $D\bar{D}$-core surrounded by a pionic cloud partly in the form of an almost real P-wave pion.

More detailed model calculations are called for and would be highly welcome.

REFERENCES

1. M. Lacombe, B. Loiseau, J.M. Richard, R. Vinh Mau, J. Côté, P. Pirès and R. de Tourreil, *Phys. Rev.* C21, 861 (1980).

2. N.A. Törnqvist, *Phys. Rev. Lett.* 67, 556, (1992);
Proc. of Hadron 91, College Park, Maryland USA, 1991, Eds. S. Oneda, D.C. Peaslee, World Scientific (1992), 795.

3. M. Bourquin et al., Phys. Lett. B172 (1986) 113; A.N. Aleev et al., JINR D1-89-642, Rapid Comm., 19-86.

4. K. Maltman and N. Isgur, *Phys. Rev. Lett.* 50, 1827 (1983); *Phys. Rev.* D29, 952 (1984);
N. Isgur, *Acta Phys. Austriaca* Suppl. XXVII, 177-266 (1985).

5. K. Dooley, E.S. Swanson and T. Barnes, *Phys. Lett.* B275, 478 (1992).

6. R. Longacre, *Phys. Rev.* D42, 874 (1990).

7. C. Amsler et al., (the Crystal Barrel Collaboration), CERN preprint (October 1991), CERN-PPE/91-188, Talk at NAN 91 conference, Moscow, July 8-11, (To appear in proceedings).

8. K. Hikasa et al., (The Particle Data Group), *Phys. Rev.* D45, 1 (1992).

9. D. Bridges et al., *Phys. Rev. Lett.* 56, 211 (1986); 56, 215 (1986);
L. Gray et al., *Phys. Rev.* D27, 307 (1983).

CONSTITUENT QUARKS AS SKYRMIONS IN COLOR SPACE

Marek Karliner *
Raymond and Beverly Sackler Faculty of Exact Sciences
School of Physics and Astronomy
Tel-Aviv University, 69978 Tel-Aviv, Israel

Abstract

We exhibit static solutions of QCD in two dimensions that have the quantum numbers of baryons and mesons, constructed out of quark and anti-quark solitons of the bosonized Lagrangian. In isolation the latter solitons have infinite energy, corresponding to the presence of a string carrying the non-singlet color flux off to spatial infinity. When N_c solitons of this type are combined, a static, finite-energy, color singlet solution is formed, corresponding to a baryon. Similarly, static meson solutions are formed out of a soliton and an anti-soliton of different flavours. The stability of the mesons against annihilation is ensured by flavour conservation. Our results can be viewed as a derivation of the constituent quark model in QCD_2, allowing a detailed study of constituent mass generation and of the heavy quark symmetry.

INTRODUCTION

One of the key outstanding problems in strong interaction physics is the derivation of hadron spectroscopy from QCD, the underlying theory. Quarks were first postulated as constituents of hadrons to describe qualitatively the spectroscopy of mesons and baryons containing the three lightest u, d and s quark flavours. Subsequently, it was realized that the short-distance properties of strongly-interacting matter could be described exactly in terms of current quarks and the asymptotic freedom of QCD. The phenomenological successes of current algebra and chiral symmetry implied that the current light (u, d, s) quarks must be much lighter than the original constituent quarks, and the relation between current and constituent light quarks awaits clarification[1]-[4]. The mystery of the relationship between light current and the apparently heavier constituent quarks is only deepened by the successes of calculations of baryon properties made using the Skyrme model[6], a soliton in the low-energy chiral approximation[7] to QCD in terms of bosonic matrix variables. Constituent quarks do not appear in the Skyrme model, their rôles being usurped by coherent states of current quarks.

In $1 + 1$ dimensions, the spectrum and interactions of mesons in QCD_2 were first discussed in the framework of the large-N_c expansion.[8] For baryons, non-Abelian bosonization methods[9] applied to QCD_2[10] have made it possible to obtain the low-lying spectrum[11]-[15] in the case of an unbroken light flavour symmetry, again without any reference to the idea of constituent quarks. More recently, explicit asymptotic static soliton solutions of the bosonized heavy quark theory have been exhibited.[4] These have the quantum numbers of quarks and an infinite energy

---

*Reporting on joint work with J. Ellis, Y. Frishman and A. Hanany.

associated with a color flux tube of infinite length. There are also qualitative and group-theoretical indications that such a mechanism could be responsible for appearance of constituent quarks in QCD$_4$, but the relevant dynamics is as yet unknown.[2]

Recently we extended[5] this approach by exhibiting static soliton solutions of QCD$_2$ that have the quantum numbers of baryons and mesons. These new solutions are color singlets and have finite energy. The solutions with baryon number zero are bound states of the quark and anti-quark solitons, while those with non-zero baryon number are bound states of N_c quark solitons, corresponding to mesons and baryons, respectively. They provide a theoretical laboratory in which the concept of a constituent quark can be dissected. They also provide insight into the QCD description of heavy-light $Q\bar{q}$ mesons such as the D and B, and baryons with one or two heavy quarks, to which the previous heavy-quark effective potential and light-quark chiral approaches have not been applicable. We show that the D and B mesons are likely to contain OZI-evading densities of quark–anti-quark pairs that are absent in the naïve constituent quark description, and could play observable rôles in their dynamics and decays.

MESONS FROM SOLITONS IN QED$_2$

Some of the interesting nonperturbative phenomena in QCD$_2$ have close analogues in QED$_2$ and are easy to derive, once the bosonized form of the Lagrangian is known. In this section we present a rederivation of the relevant results obtained long ago, via Abelian bosonization, by Coleman[16], and add some new results of our own, namely explicit solutions in the case of broken flavour symmetry. We believe the reader will find this section useful for developing physical intuition for the discussion of QCD$_2$ to follow in the next section.

The Lagrangian of multi-flavour massive QED in two dimensions is

$$\mathcal{L} = \sum_k \bar{\psi}_k(i\slashed{D} - m_k)\psi_k - \frac{1}{4}F_{\mu\nu}F^{\mu\nu} \quad (1)$$

where k is the flavour index. The bosonized version of (1), after F is integrated out, reads[16]

$$\begin{aligned}\mathcal{L} =& \frac{1}{2}\sum_k(\partial_\mu\chi_k)^2 - \frac{e^2}{2\pi}(\sum_k \chi_k)^2 \\ &+ \sum_k m_k^2 \cos\sqrt{4\pi}\chi_k\end{aligned} \quad (2)$$

where $Q = (1/\sqrt{\pi})\sum_k[\chi_k(\infty) - \chi_k(-\infty)]$ is the total electric charge. The equations of motion in the static case read

$$\chi_k'' - 4\alpha(\sum_l \chi_l) - \sqrt{4\pi}m_k^2 \sin\sqrt{4\pi}\chi_k = 0 \quad (3)$$

where $\alpha \equiv e^2/4\pi$. If we take $\chi_k(-\infty) = 0$, in order to obtain finite energy static solutions we must have:

$$\sum_l \chi_l(\infty) = 0; \quad \cos\sqrt{4\pi}\chi_k(\infty) = 1 \quad (4)$$

We see that only states with zero total charge Q are allowed, which is what one expects, since QED$_2$ is confining. From (4) it follows

$$\chi_k(\infty) = \sqrt{\pi}\,n_k, \quad n_k = 0, \pm 1, \pm 2, \ldots \quad (5)$$

For two flavours, from (4) and (5), taking the lowest non-trivial n_k's, namely $n_1 = 1, n_2 = -1$, we obtain

$$\chi_1(\infty) = \sqrt{\pi}; \quad \chi_2(\infty) = -\sqrt{\pi} \quad (6)$$

The b.c. (6) correspond to a meson built out of a soliton and an anti-soliton. Eqs. (3) with b.c. (6) can be solved explicitly. When $m_1 = m_2$ it is easy to see that $\chi_1(x) = -\chi_2(x)$ and the "string tension" term proportional to α in (3) vanishes, leading to two "mirror" decoupled sine-Gordon equations for χ_1, χ_2.

When $m_1 \neq m_2$, a solution can be found numerically. For $m_1 \to \infty$ we can neglect the α term in (3) get a free soliton of mass m_1,

$$\chi_1(x) \xrightarrow[m_1 \to \infty]{} \sqrt{\pi}\,\theta(x) \qquad (7)$$

as follows from the explicit form of the sine-Gordon solution. Then, for the light flavour, $k = 2$,

$$\chi_2'' - 4\alpha\chi_2 - \sqrt{4\pi}\,m_2^2 \sin\sqrt{4\pi}\chi_2 = 4\alpha\sqrt{\pi}\,\theta(x)\,. \qquad (8)$$

Thus the light "anti-quark" field χ_2 feels a point-like "source" term due to the heavy "quark" χ_1. We will now show that a very similar phenomenon occurs in QCD_2.

HADRONIC SOLITONS

Two non-Abelian bosonizations of QCD_2 have been developed, one in terms of $SU(N_c) \times U(N_f)$ bosonic variables[11] where N_c is the number of colors and N_f is the number of flavours, and the other[12] in terms of $U(N_c \times N_f)$ bosonic variables.[1] In the $U(N_c \times N_f)$ scheme there are static solutions that have the quantum numbers of baryons and mesons, constructed out of quark and anti-quark solitons. In isolation the latter solitons have infinite energy, corresponding to the presence of a string carrying the non-singlet color flux off to spatial infinity. When N_c solitons of this type are combined, a static, finite-energy, color singlet solution is formed, corresponding to a baryon. Similarly, static meson solutions are formed out of a soliton and an anti-soliton of different flavours. The stability of the mesons against annihilation is ensured by flavour conservation. In the $SU(N_c) \times U(N_f)$ scheme no such solutions exist[5]. Intuitively this is because in such "product" scheme the $U(N_f)$ part of the field

---
[1]The specific case of $SU(N_c)$, $N_f = 2$ has also been considered in a mixed Abelian – non-Abelian formalism[17].

describes the mesons directly, so in the product scheme mesons are "fundamental" fields, rather than "composites".

After the gauge fields are integrated out, one obtains[12] the effective exact $U(N_c \times N_f)$ flavour symmetric bosonic action $S_{eff}[u]$ where u is an element in $U(N_c \times N_f)$. We shall look for solutions with u in diagonal form[11],

$$u_{\alpha\alpha' j j'} = \delta_{\alpha\alpha'}\,\delta_{jj'}\,e^{-i\sqrt{4\pi}\chi_{\alpha j}} \qquad (9)$$

$\alpha = 1, \ldots, N_c$; $j = 1, \ldots, N_f$. In general, for non-equal masses, in the static case,

$$S_{eff}[u] = -\frac{1}{2}\sum_{\alpha j}\int \left[\chi_{\alpha j}'(x)\right]^2$$
$$-2\alpha_c \sum_{\alpha}\int\left[\sum_l \chi_{\alpha l} - \frac{1}{N_c}\sum_{\beta l}\chi_{\beta l}\right]^2$$
$$+ \sum_{\alpha j}\int m_j^2 \cos\sqrt{4\pi}\chi_{\alpha j} \qquad (10)$$

where $\alpha_c = e_c^2/4\pi$ and m_j stands for the j-th mass. The similarity to (2) is obvious, the main difference being the extra $1/N_c$ piece in the interaction term. Its presence and the minus sign in front of it are due to the non-Abelian nature of the theory. They make it possible for quark solitons to combine not just into mesons but into baryons as well.

In analogy with QED_2, it turns out[5] that in order to have finite energy static solutions and taking $\chi_{\alpha j}(-\infty) = 0$, we must have

$$\frac{1}{\sqrt{\pi}}\chi_{\alpha j} = n_{\alpha j} \quad \text{integers}$$
$$\sum_l \chi_{\alpha l} = \sqrt{\pi}\sum_l n_{\alpha l} = \sqrt{\pi}n \text{ indep. of } \alpha \quad (11)$$

The baryon number of any given flavour l is given by

$$B_l = \sum_{\alpha} n_{\alpha l} \qquad (12)$$

Combining eqs. (11) and (12) we get the total baryon number

$$B = \sum_l B_l = nN_c \qquad (13)$$

which clearly is an integer multiple of N_c.

EXPLICIT SOLUTIONS

For the mesonic solutions, we need $B = 0$ and hence $n = 0$. Let us consider the case of a meson containing a quark of flavour $l = 1$ and an anti-quark of flavour $l = 2$, and no other constituents. Thus, we need

$$\sum_l n_{\alpha l} = 0$$
$$\sum_\alpha n_{\alpha l} = \begin{cases} 1 & l = 1 \\ -1 & l = 2 \\ 0 & l \geq 3 \end{cases} \quad (14)$$

Let us now give an example of a solution in the $U(N_c \times N_f)$ scheme. As expected, such a solution will have components in a non-factorizable part in the color-flavour space.

The asymptotic boundary conditions are:

$$\chi_{\alpha j}(-\infty) = 0; \quad \chi_{\alpha j}(+\infty) = \sqrt{\pi} n_{\alpha j} \quad (15)$$

where the set $\{n_{\alpha j}\}$ must satisfy the constraints (14). A possible solution is

$$n_{11} = 1; \quad n_{12} = -1 \quad (16)$$

with all other $n_{\alpha j}$ being zero. Having specified the asymptotic boundary conditions at $x \to \pm\infty$, we must now see whether a solution exists for all x. In general such solutions can only be found numerically.[2] The case of an exact $SU(N_f)$ symmetry, i.e. equal quark masses, is an exception where an explicit analytical solution can be found[5]:

$$\chi_{11}(x) = -\chi_{12}(x) = \frac{2}{\sqrt{\pi}} \tan^{-1}\left[\exp\left(\sqrt{4\pi} m \, x\right)\right], \quad (17)$$

with all others vanishing identically.

Recently there has been much discussion in the literature of the so-called heavy quark symmetry.[19] Here this symmetry manifests itself in a rather clear way. Consider a $Q\bar{q}$ meson made out of a heavy quark Q and a light anti-quark $\bar{q}$, such as the D- or B-mesons. When m_Q is much larger than the scale of the theory, $m_Q \gg e_c$, its profile tends to a theta function,

$$\chi_{1Q} \xrightarrow[m_Q/e_c \to \infty]{} \sqrt{\pi}\,\theta(x) \quad (18)$$

as in the QED$_2$ case, while the baryonic current of the heavy quark tends to a delta function,

$$J_Q^B = \frac{1}{\sqrt{\pi}} \sum_\alpha \partial_x \chi_{\alpha Q} \xrightarrow[m_Q/e_c \to \infty]{} \delta(x) \quad (19)$$

Thus the heavy quark acts as a static color source, while the profile $\chi_{\alpha q}$ of the light quark becomes independent of m_Q.

The presence of the static color source (19) makes the total energy of the $Q\bar{q}$ system finite. Had there been no other color source, the light flavour profile $\chi_{1\bar{q}}$ on its own would correspond to a configuration with a net baryon number -1, one "unit" of color charge and a finite energy density per unit length, $2\pi\alpha_c(1-1/N_c)^2$, associated with a color flux tube of infinite length and resulting in the total energy being infinite. This is reminiscent of the single quark solution discussed in Ref. [4]. It gives precise meaning to the intuitive concept of quark confinement: isolated quarks have infinite energy because flux conservation forces them to emit a flux tube which has no sink to absorb it.

Multi-quark baryonic solutions can be obtained in a similar way to the meson solutions. For example, taking $N_c = 3$ and $B = 3$ (a 3-quark state[3]) we find $n = 1$ (cf. eq. (13)). One possible solution is

$$n_{11} = n_{21} = n_{31} = 1 \quad (20)$$

Corresponding to a "uuu"-like baryon, i.e. a "Δ^{++}". When quark masses are equal, $m_1 = m_2 = \ldots = m_{N_f}$, this solution coincides with one found earlier in the "product scheme"[11],

---

[2]We have used the subroutine package COLSYS.[18]

[3]In our normalization a single quark carries one unit of baryon number.

in which the color part is "frozen", i.e. the string tension vanishes identically. The vanishing of string tension in (20) is caused by a mechanism similar to the one occurring in the meson with equal quark masses described above, leading to

$$\chi_{11}(x) = \chi_{21}(x) = \chi_{31}(x) = \frac{2}{\sqrt{\pi}} \tan^{-1} e^{\sqrt{4\pi}m\,x} \tag{21}$$

There is an important difference, however, between the meson and the baryon cases. The vanishing of the string tension term in the meson is a manifestation of the fact that the chromoelectric fluxes of the quark and anti-quark cancel each other. This phenomenon has its counterpart in QED_2[16], as discussed previously. The cancellation of fluxes of N_c quarks has no such counterpart and can only occur in a non-Abelian theory. Another difference between the baryonic and mesonic solitons is that the latter do not exist in the "product scheme".

An intrinsically new 3-quark solution occurs when more than one flavour appears in the nontrivial solution, for example,

$$n_{11} = n_{22} = n_{33} = 1 \tag{22}$$

This corresponds to a baryon in which each quark has a different flavour. Such a solution is particularly interesting when quark masses are taken to be non-equal, corresponding to an "uds"-like baryon, or to a baryon in which one quark is much heavier than the QCD scale, such as the Σ_c or Σ_b, again serving as a theoretical laboratory for the study of the heavy quark symmetry discussed earlier.[4]

One can also study baryons containing two heavy quarks. For $N_c = 3$, the light quark distribution in such a baryon is the same as in a $\bar{Q}q$ meson. The physical reason for this is that the two heavy quarks are essentially at rest and act as a static color source. They are color triplets and combine to a color anti-triplet, $\mathbf{3} \otimes \mathbf{3} \to \mathbf{3^*}$, i.e. the effective field seen by the light quark q in a QQq baryon is that of a static anti-quark. This gives precise meaning to the concept of constituent quark in QCD_2. The QQq case is particularly clear, since this type of baryon contains only one light constituent quark, while in Qqq or qqq baryons there are two or three such objects, superimposed nonlinearly.

SEMI-CLASSICAL QUANTIZATION

As has already been mentioned, one of the specific applications which motivated this study was that to mesons containing just one heavy quark Q. In the constituent quark language, these would be described as $Q\bar{q}$ mesons, where $\bar{q}$ represents some light constituent quark (u, d or s). Examples include the D, D_s and B mesons. To describe such mesons within our QCD_2 approach, we need to consider a quark mass matrix with one heavy eigenvalue $M \gg e_c$, and $N_f - 1$ (typically three) light eigenvalues $m \ll e_c$. In such a case, the light quark degrees of freedom should be quantized semi-classically, as was already done for baryons made out of light current quarks.[11] The resulting lump is the best QCD model we can derive for the concept of a light constituent quark. However, clearly it is a coherent state containing in some sense an infinite number of current quarks, at least in the massless limit.

COMMENTS ON D AND B PHYSICS

In the previous sections we have described QCD_2 solitons which could serve as models for $Q\bar{q}$ mesons such as the D or B. In addition to giving some insight into the concept of a constituent quark from the point of QCD, this study may also give some new insights into the dynamics and weak decays of D and

---
[4]For a different approach to heavy-quark baryons in the chiral soliton framework, see Ref. [20].

B mesons.[5] In particular, we would like to comment on the existence and possible phenomenological rôle of non-valence quarks in the D and B meson wave functions.

There are various phenomenological indications that the proton wave function contains a significant density of non-valence $\bar{s}s$ quarks. Moreover, their abundance relative to $\bar{u}u$ and $\bar{d}d$ quarks is qualitatively reproduced by Skyrme model soliton calculations. The relative abundances of $\bar{s}s$, $\bar{u}u$ and $\bar{d}d$ have also been calculated in baryonic QCD_2 solitons,[14] and found to be qualitatively similar to the QCD_4 results.[21]-[23] Here we invert the logic: QCD_2 mesonic solitons contain calculable non-valence quark densities, and we would expect QCD_4 mesonic solitons and hence physical D and B mesons to contain similar non-valence quark densities.

The calculation of the different light quark densities in a QCD_2 mesonic soliton parallels very closely that in a baryonic soliton.[14] A typical result is

$$\langle D|\bar{s}s|D\rangle / \langle D|\bar{u}u|D\rangle = \frac{1}{2} \quad (23)$$

and similarly for B mesons. We might expect the ratio (23) to be qualitatively similar, though possibly smaller by about a factor 2,[14] for the realistic case of QCD_4 with light flavour $SU(3)$ breaking.[21]-[23]

The presence of significant amounts of $\bar{s}s$ quarks in the D and B mesons implies that the naïve OZI rule forbidding disconnected quark diagrams can be evaded.[25] This might also have implications for D and B production dynamics, but here we only emphasize some possible implications for D and B decays.

- 1. Annihilation diagrams: $c\bar{s} \to u\bar{d}$ could be more important for D^0 and D^+ mesons than is normally supposed when

---

[5] $Q\bar{q}$ mesons in QCD_2 have also been recently studied in the large-N_c limit.[24]

only the D_s wave function is assumed to contain strange quarks.

- 2. The final states from D and B decays could contain more strange particles than is normally supposed when $\bar{s}s$ pairs need to pop out of the QCD vacuum or be created at the weak vertex. This could help explain the surprisingly large[26] branching ratios for $D^0 \to \phi K^0$ and $D^+ \to \phi K^+$. This observation could also have implications for attempts to estimate the ratio $|V_{cs}/V_{cd}|$ of Kobayashi-Maskawa matrix elements on the basis of strange final states.[26] It might also have implications for the ratios of $D \to K\bar{K}$ and $\pi\pi$ final states.

- 3. The $\bar{s}s$ pairs could provide an additional source for $B \to \phi + X$. At this time it is premature to compare this with the data.

Detailed investigations of these possibilities should await more realistic calculations in QCD_4, however.

SUMMARY AND OUTLOOK

We have shown that the spectrum of QCD_2 includes finite-energy mesonic and baryonic solitons. These solitons can be regarded as bound states of the infinite-energy single-quark solitons that we found previously. They provide meaning for the previously fuzzy concept of a constituent quark, at least in QCD_2. A particularly interesting application is to the study of mesons and baryons containing both heavy and light quarks. A constituent light quark is seen to be a semi-classical coherent state containing an indefinite number of light $\bar{q}q$ pairs, among which non-valence flavours have as much as one half of the abundance of the valence flavour. This observation could have phenomenological implications for the dynamics and weak decays of D and B mesons.

The next step is to extend the analysis of this paper to the realistic case of four dimensions. This may be possible for the lowest-lying $Q\bar{q}$ mesons if they are describable by spherically symmetric wave functions, which could be analyzed using an effective two-dimensional field theory in (r,t) coordinates. We are now investigating this possibility.

ACKNOWLEDGEMENTS

This research was supported in part by the Einstein Center at the Weizmann Institute and by the Basic Research Foundation administered by the Israel Academy of Sciences and Humanities and by the United States-Israel Binational Science Foundation (BSF), Jerusalem, Israel.

REFERENCES

1. H. Georgi and A. Manohar, *Nucl. Phys.* **B310**(88)527.
2. D. B. Kaplan, *Phys. Lett.* **B235**(90)163; *Nucl. Phys.* **B351**(1991), 137.
3. S. Weinberg, *Phys. Rev. Lett.* **65**(90)1181; U. Ellwanger and B. Stech, *Z. Phys.* **C49**(91)683.
4. J. Ellis, Y. Frishman and M. Karliner, *Phys. Lett.* **B272**(91)333.
5. J. Ellis, Y. Frishman A. Hanany and M. Karliner, *Nucl. Phys.* **B382**(92)189.
6. T.H.R. Skyrme, *Proc. Roy. Soc. London* **A260**(1961)127; E. Witten, *Nucl. Phys.* **B223** (1983)422, *ibid* 433; G. Adkins, C. Nappi and E. Witten, *Nucl. Phys.* **B228**(1983)433; for $N_f=3$ see: E. Guadagnini, *Nucl.Phys.* **236**(84)35; P.O. Mazur, et al., *Phys. Lett.* **147B**(1984), 137.
7. E. Witten in Lewes Workshop Proc.; A. Chodos et al., Eds; World Sci., 1984.
8. G.'t Hooft, *Nucl. Phys.* **B72**(74)461; **B75**(1974),461; see [5] for full set of refs.
9. E. Witten, *Comm. Math.Phys.* **92**(84)455.
10. D. Gonzales and A.N. Redlich, *Nucl. Phys.* **B256**(1985)621.
11. G.D. Date, Y. Frishman and J. Sonnenschein, *Nucl. Phys.* **B283**(1987), 365.
12. Y. Frishman and J. Sonnenschein, *Nucl. Phys.* **B294**(1987), 801.
13. Y. Frishman and W. J. Zakrzewski, *Nucl. Phys.* **B331**(1990)781.
14. Y. Frishman and M. Karliner, *Nucl. Phys.* **B334**(1990), 339.
15. K. Hornbostel, S. J. Brodsky and H.C. Pauli, *Phys. Rev.* **D41**(1990)3814.
16. S. Coleman, *Ann. Phys.* **101**(76)239.
17. D. Gepner, *Nucl. Phys.* **B252**(1985)481.
18. U. Ascher, J. Christiansen and R.D. Russel, *ACM Trans. Math. Soft.* **7**(81)209.
19. For a recent review, see B. Grinstein, SSCL-PREPRINT-17, Dec 1991,
20. M. Rho et al., *Phys. Lett.* **B251**(90)597.
21. J. Donoghue and C. Nappi, *Phys. Lett.* **B168**(1986)105; J. Donoghue, in Proc. of II-nd Int. Conf. on πN Physics.
22. H. Yabu, *Phys. Lett.* **B218**(89)124; D. B. Kaplan and I. Klebanov, *Nucl. Phys.* **B335**(90)45; I. Klebanov in Proc. NATO ASI Cargese, 1989.
23. H. Weigel et al., *Phys. Rev.* **D42**(90)3177.
24. B. Grinstein and P.F. Mende, *Phys. Rev. Lett.* **69**(92)1018.
25. J. Ellis, E. Gabathuler and M. Karliner, *Phys. Lett.* **B217**(88)173; for recent review & data, see R. Decker et al., CERN-PPE 92-010 (1992).
26. See Ref. [5] for references and discussion.

NEUTRINO PRODUCTION OF DIMUONS AT THE FERMILAB TEVATRON

M.H. Shaevitz, C. Arroyo, K.T. Bachmann, A.O. Bazarko, R.E. Blair,
T. Bolton, C. Foudas, B.J. King, W.C. Lefmann, W.C. Leung, S.R. Mishra,
E. Oltman, P.Z. Quintas, S.A.Rabinowitz, F. Sciulli, W.G. Seligman
Columbia University, New York, NY 10027

F.S. Merritt, M.J. Oreglia, B.A. Schumm
University of Chicago, Chicago, IL 60637

R.H. Bernstein, F. Borcherding, H.E. Fisk, M.J. Lamm,
W. Marsh, K.W.B. Merritt, H. Schellman, D.D. Yovanovitch
Fermilab, Batavia, IL 60510

A. Bodek, H.S. Budd, P. de Barbaro, W.K. Sakumoto
University of Rochester, Rochester, NY 14627

W.H. Smith, T.S. Kinnel, P.H. Sandler
University of Wisconsin, Madison, WI 53706

Abstract

Neutrino and antineutrino interactions with two muons in the final state have been studied by the CCFR collaboration in the Fermilab Tevatron neutrino beam. The rate of neutrino- and antineutrino-induced prompt same-sign dimuon production in steel was measured using a sample of 220 $\mu^-\mu^-$ events and 15 $\mu^+\mu^+$ events with $P_\mu > 9$ GeV/c, and energies between 30 GeV and 600 GeV. In addition, a sample of 5044 ν_μ and 1062 $\bar{\nu}_\mu$ induced opposite-sign ($\mu^\mp\mu^\pm$) events was also observed with $P_{\mu 1} \geq 9$ Gev/c, $P_{\mu 2} \geq 5$ GeV/c and $30 \leq E_\nu \leq 600$ GeV. The opposite-sign data support the slow rescaling model of charm production with a value of $m_c = 1.31 \pm 0.24$ GeV/c^2. The CKM matrix element V_{cd}=.209±0.012 and the nucleon strangeness content, η_s=.064$^{+0.0075}_{-0.0065}$ are also extracted from the data. The first measurements of the Q^2 dependence of the strange quark densities, $xs(x)$, are also presented.

INTRODUCTION

We report on an experimental study of neutrino induced interactions with two muons of the same and opposite electric charge in the final state. Events with two muons of the same charge are expected to be mostly due to pion and kaon decay in the hadron showers of charged current events. Opposite sign events result from reactions that produce a charmed quark, followed by the leptonic decay of the charmed particle. Since charmed quarks are predominantly produced by valence down quarks and strange sea quarks, a measurement of the strange content of the nucleon can be extracted from the production rate and kinematics of the opposite charge events.

EXPERIMENT

The neutrinos were provided by the Fermilab Tevatron Quadrupole-Triplet neutrino beam (QTB), which yielded a high intensity and high energy ν_μ and $\bar{\nu}_\mu$ beam. We accumulated 3.7×10^6 total charged-current triggers, with neutrino energies between 30 GeV and 600 GeV, in two data runs (Fermilab experiments E744 and E770). The CCFR detector [1, 2] consisted of a 690 ton iron-target calorimeter instrumented with liquid scintillation counters and drift chambers, followed by a 420 ton iron-toroid muon spectrometer. Single muon and dimuon events were selected from the sample of charged-current triggers if they passed conditions ensuring proper reconstruction in the detector.

To ensure that the muon did not exit the side of the detector before reaching the spectrometer, the angle of the muon at the vertex was required to be less than 250 mr with respect to the incident neutrino direction. In addition, the momentum of the muon was required to be at least 3 GeV/c at the front face of the muon spectrometer and 9 GeV/c at the event vertex, when corrected for energy loss in the target. The time of the track, determined from the drift chambers in the muon spectrometer, was required to be within 36 ns of the time obtained from the calorimeter counters and triggering toroid hodoscopes. About 1.5 million ν_μ and 0.3 million $\overline{\nu_\mu}$ induced charged-current events passed these selection criteria.

A sample of 15000 candidate multimuon events were selected using two independent criteria based on calorimeter counter pulse heights downstream of the end of the hadron shower and indications of two tracks in the calorimeter drift chambers. The efficiency of this initial selection was 99% [3]. Pictures of the candidate events were scanned and about 4% of them were fixed interactively by physicists for errors in track reconstruction. This was a minor effect since 99% of the final sample of same-sign dimuons in E744 were found without interactively refitting the tracks [4].

SAME-SIGN DIMUONS

Sources of prompt same-sign dimuons may include second-order quantum chromodynamic processes such as $c\bar{c}$ gluon bremsstrahlung and $D^0 - \overline{D^0}$ mixing. Non-prompt same-sign dimuons are produced by decaying pions or kaons in the hadron shower of a charged-current event. Since prompt sources cannot be distinguished from non-prompt sources in our apparatus, the non-prompt meson decay background must be subtracted from the observed number of same-sign dimuons to obtain the rate of prompt same-sign dimuon production. The rate of prompt same-sign dimuon production at energies below 200 GeV was measured by previous neutrino experiments to be somewhat higher than expected from theoretical predictions [5, 6, 7, 8, 9, 10, 11]. These measurements also seemed to indicate an increase in the prompt rate with increasing neutrino energy. However, in Fermilab experiment E744, the CCFR collaboration reported results with neutrino energies up to 600 GeV that were consistent with Standard Model predictions and with zero [3]. These measurements did not exhibit a strong energy dependence. The results presented here are based on data from both E744 and E770 with more than double the statistics of E744. In addition, we have made new improved measurements of non-prompt muon-production which lead to a substantial reduction in the systematic errors in the background calculation [12].

Same-sign dimuon events were selected from the candidate multimuon events with two tracks that passed the muon-track cuts described above. To ensure that the muons originate from the same incident neutrino, the transverse separation between the two tracks at the point of closest approach was required to be less than 15 cm. In addition, their time difference as determined from the tracks in the toroid gaps was required to be less than 28 ns. If there was a third track with muon momentum at the vertex greater than 3.1 GeV/c, the event was identified as a trimuon and eliminated from the dimuon sample. The number of trimuons misidentified as dimuons was a background that was calculated from the observed trimuon events and amounted to 8.66 ± 5.54 $\mu^-\mu^-$ events and 1.02 ± 0.65 $\mu^+\mu^+$ events.

Of the 1.8 million charged-current events, there were 220 $\mu^-\mu^-$ and 15 $\mu^+\mu^+$ events with $P_\mu >$ 9 GeV/c. The non-prompt background was broken down into two categories, the primary decay background and the secondary decay background. The primary decay background comes from events in which one of the primary hadrons at the hadron vertex decays to produce the second muon. The secondary decay background comes from events in which a secondary hadron – a hadron produced in the subsequent interactions of the primary hadrons – decays to produce the second muon. The calculation of these backgrounds is taken from experimental data. The neutrino charged-current cross section is taken from the most recent CCFR structure functions [13]. The probability of a primary decay is given by parametrizations of a Monte Carlo calculation that is based on electroweak fragmentation measurements by neutrino bubble chamber experiments and well known interaction and decay probabilities [14]. The probability of a secondary decay is given by parametrizations of a Monte Carlo calculation based on newly measured muon-production rates in hadron-induced showers by the CCFR collaboration [15].

The 10% systematic uncertainty of the shower simulation is determined by its agreement with the test beam measurements used to set the level of muon-production in the simulation and the accuracy of the measurements. The total systematic uncertainty in the secondary decay background is 14.8% for incident neutrinos and 21.2% for incident antineutrinos. This includes the error from the input spectrum of primary hadrons: 10.9% for incident neutrinos and 18.7% for incident antineutrinos. There is an additional 10% error from the shower simulation and a 2% uncertainty from the interaction lengths of hadrons.

There is an additional background due to two charged-current events that come from the same RF bucket, which are produced by a neutrino in the same position within the detector. Such events are called overlays. Most are eliminated with the cuts described above on the time of passage of the muon tracks relative to the trigger time and on the transverse dis-

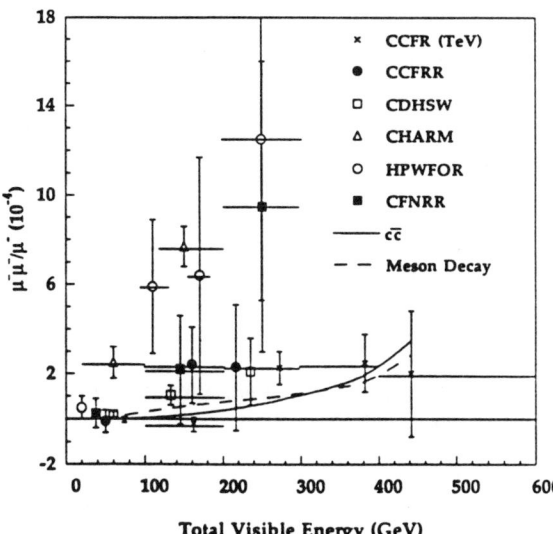

Figure 1. The prompt same-sign dimuon rates relative to the charged-current rate compared to previous experiments. The CCFR TeV points and the dotted line representing the 90% C.L. upper limit are from this experiment. The dashed line represents the energy dependence of the meson-decay background rate, decreased by 0.4 to match the level of the same-sign dimuon data excess. The solid line represents the rate due to $c\bar{c}$ gluon bremsstrahlung increased by a factor of 60 to equal the level of the same sign excess.

tance of closest approach. The overlay background was 1.10 ± 0.44 $\mu^-\mu^-$ events and 0.06 ± 0.03 $\mu^+\mu^+$ events [12].

The meson decay background comprises 94% of the total background, contributing 56.32 ± 8.35 $\mu^-\mu^-$ and 3.75 ± 0.80 $\mu^+\mu^+$ events from secondary decays, and 109.45 ± 11.95 $\mu^-\mu^-$ and 12.04 ± 2.25 $\mu^+\mu^+$ events from primary decays. The trimuon background comprises about 4% of the background, and the overlay background accounts for less than 1% of the background. The 220 $\mu^-\mu^-$ events have a total background from meson decays, misidentified trimuons, and overlays of 175.54 ± 19.35 events, while the 25 $\mu^+\mu^+$ events have a total background of 16.87 ± 3.04 events. This yields an observed prompt excess of 44.46 ± 24.38 $\mu^-\mu^-$ events and 8.14 ± 5.17 $\mu^+\mu^+$ events, where the error is statistical and systematic combined. The shapes of the kinematic distributions for the same-sign dimuon data and the meson-decay background are reasonably consistent[16].

The final prompt rates for visible energies between 30 GeV and 600 GeV and $P_\mu > 9$ GeV/c are $(5.4 \pm 2.3) \times 10^{-5}$ or less than 9.2×10^{-5} at the 90% C.L. per charged-current event for incident ν_μ, and $(5.2 \pm 3.3) \times 10^{-5}$ or less than 10.5×10^{-5} at the 90% C.L. per charged-current event for incident $\overline{\nu_\mu}$. The errors include statistical and systematic uncertainties. The energy dependence of the measured rate is shown in Figure 1.

Figure 1 also shows the rate due to the meson-decay background, which was multiplied by 0.4 to bring it down to the level of the data excess for shape comparison. The shape of the energy dependence for the meson-decay background agrees with the data. To ensure that no other sources of same-sign dimuons contribute to this excess, we calculated the expected rate due to prompt processes predicted by the standard model. For example, the range of rates expected from a $c\bar{c}$ gluon bremsstrahlung calculation – based on the work of Barger et al. [17] and Cudell et al. [18] – is shown in Figure 1. For this calculation, we set the mass of the charm quark parameter, m_c, to 1.3 ± 0.3 GeV/c^2, as recently measured by the CCFR collaboration in opposite-sign dimuon production [19]. The structure functions are the CCFR QTB structure functions [13, 20] and the fragmentation of the c-quark to a D-meson is modeled with the Peterson fragmentation function [21]. The calculated $c\bar{c}$ gluon bremsstrahlung rate of same-sign dimuons with $P_\mu > 9$ GeV/c for the energies of this experiment is $(0.09 \pm 0.39) \times 10^{-5}$ per charged-current event for incident ν_μ or less than 0.7×10^{-5} at the 90% C.L. This calculated $c\bar{c}$ gluon bremsstrahlung rate is too small to be considered an important contribution to the measured rate of prompt same-sign dimuon production.

A comparison of our results with the prompt ν_μ-induced same-sign rates from other experiments is shown in Figure 1. The rates presented in this publication do not show the energy dependence suggested by some of the measurements prior to 1988. However, they agree with results reported previously by the CCFR collaboration [3] in 1988. A more detailed discussion of previous experiments can be found in Reference [12].

(Note that the comparison of rates as a function of energy for different kinds of neutrino beams is uncertain. This is because the visible energy for same-sign events is smaller than the visible energy for charged-current events, and the energy distribution of the neutrino beam differs for QTB, wide-band, and narrow-band beams. This has not been accounted for in the comparison of Figure 1.)

OPPOSITE SIGN DIMUONS

The distinctive opposite sign dimuon signature serves as a unique and highly sensitive probe of the the strange sea content of the nucleon and heavy charm production. The strange quark structure function is of particular theoretical interest in the exploration higher order mass corrections [22], while the threshold behavior associated with the heavy charm mass is critical to the extraction of the weak mixing angle, $sin^2\theta_W$, from neutrino neutral current data.

The heavy charm quark is expected to introduce an energy threshold in the dimuon production rate. This effect is described through the slow rescaling model [23], in which ξ, the momentum fraction carried by the struck quark, is related to the kinematic variable $x = Q^2/2ME_\nu y$ by the expression $\xi = x(1 + m_c^2/Q^2)$. Representing the momentum distribution of the s and d quarks within the nucleon as $s(\xi)$ and $d(\xi)$ the cross section for neutrino production of dimuons may be written:

$$\frac{d^2\sigma(\nu N \to \mu^-\mu^+ X)}{d\xi\, dy} = \frac{G^2 M E_\nu}{\pi} \{\xi d(\xi)|V_{cd}|^2 + 2\xi s(\xi)|V_{cs}|^2\}$$
$$\times [1 - \frac{m_c^2}{2ME_\nu\xi}]D(z)B_c \quad (1)$$

where the function $D(z)$ describes the fragmentation of charm quarks into charmed hadrons and B_c is the semileptonic branching ratio for charmed hadrons. The analogous equation for antineutrinos is found by substituting $d(\xi) \to \bar{d}(\xi)$ and $s(\xi) \to \bar{s}(\xi)$.

Previously published results from E744 [24] described opposite sign dimuon data for $30 \le E_\nu \le 600$ GeV with $P_\mu \ge 9$ Gev/c and $\theta_\mu \le 250$ mrad demanded for both muon tracks. By combining the two samples, requiring $E_{had} \ge 10$ GeV, and lowering the $P_\mu 2$ cut to 5 GeV/c for $E_{had} \le 130$ GeV a sample of 5044 ν_μ 1062 $\bar{\nu}_\mu$ induced $\mu^\mp \mu^\pm$ events are observed, a more than threefold statistical enhancement. Muonic decays of non-prompt π and K mesons comprise the primary dimuon background of $796.5\pm11.5\, \nu_\mu$ and $118.0\pm2.1 \bar{\nu}_\mu$ events to the above sample [16]. This background has been extensively studied and measured using experimental data as described in the section on same-sign dimuons.

In order to extract values for the various physics parameter, the data were compared to simulated events. The single and dimuon events were simulated using Monte Carlo techniques. Quark and antiquark momentum densities were obtained from the CCFR structure functions [13] using a modified Buras-Gaemers parameterization [25]. The strange quark x dependence is assumed to be given by $s(x) \propto (1-x)^\beta$ with the magnitude set by the parameter $\kappa = 2S/(\overline{U}+\overline{D})$ (where $S = \int_0^1 xs(x)dx$, etc.). The normalization is set by the ratio of data to Monte Carlo for the charged-current single muon events.

A multiparameter χ^2 minimization is used to compare the data and Monte Carlo events binned in five E_{vis} ($= E_{\mu_1} + E_{\mu_2} + E_{had}$) bins and ten x_{vis} ($= Q_{vis}^2/2M(E_{had} + E_{\mu_2})$) bins. The Monte Carlo event weights are shifted by varying m_c, β, κ, and B_c, to minimize χ^2, yielding best values for the parameters and their errors.

The largest source of systematic uncertainty is the charm fragmentation, modeled using the Peterson function $D(z) = 1/z(1 - 1/z - \epsilon/(1-z))^2$ [26]. The Monte Carlo is fit to the data for various fixed values of ϵ, and a study of the distribution of $Z_{vis} = E_{\mu_2}/(E_{\mu_2} + E_{had})$ permits a measurement of $\epsilon = .22 \pm .05$. This value is combined with the E531 emulsion result [27] (analyzed for $W^2 > 30$ GeV^2) of $\epsilon = .18\pm.06$ to yield a neutrino average $\epsilon = .20\pm.04$. (This value is consistent with that from the ARGUS [28] and CLEO [29] e^+e^- experiments, which find $\epsilon = .19 \pm .03$ and $.156 \pm .015$ respectively.) The uncertainty in ϵ is included directly in the fitting procedure through an additional term in the overall χ^2.

Other systematic errors are found by varying parameters within their uncertainties. These include: π/K background, the relative P_μ and E_{had} energy scale, dimuon data selection, R_{LONG}, and the u_v/d_v ratio [30]. Assuming the PDG values [14] of $|V_{cd}|^2 = .0484$ and $|V_{cs}|^2 = .9494$, the multiparameter fit yields:

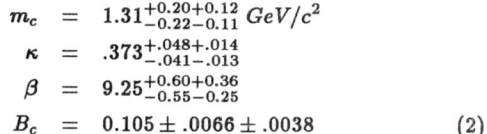

$$m_c = 1.31^{+0.20+0.12}_{-0.22-0.11}\ GeV/c^2$$
$$\kappa = .373^{+.048+.014}_{-.041-.013}$$
$$\beta = 9.25^{+0.60+0.36}_{-0.55-0.25}$$
$$B_c = 0.105 \pm .0066 \pm .0038 \qquad (2)$$

where the first error is statistical and the second is systematic. The χ^2 of 42.5 for 46 degrees of freedom suggests excellent agreement between the data and Monte Carlo.

The difference between the strange sea exponent β and that of the total sea $\alpha = 6.75$ (at $Q^2 = 18.08$), where $x\bar{q}(x) \propto (1-x)^\alpha$, provides a quantitative indication that the strange sea is softer than the $\bar{u}$ and $\bar{d}$ sea.

The value of κ in (2) is lower than previous CCFR results [24],[14]. This is due to an increase in the non-strange sea for the latest measured structure functions [13]. Defining the strange sea content of the nucleon as $\eta_s = 2S/(U+D)$, the measured value of κ and $R = \bar{Q}/Q = .195$ at $Q^2 = 18.08\ GeV^2/c^2$ combine to yield:
$$\eta_s = 0.064^{+.0072}_{-.0062} \pm .0020$$
This result is consistent with η_s from the previous publications.

The ratio of dimuon to single muon production serves as a direct test of the slow rescaling hypothesis. The acceptance corrected rates exhibit an energy dependence characteristic of heavy charm quark production. Once corrected for this threshold with $m_c = 1.31$ the rates flatten out, exhibiting only the sharp, low energy threshold behavior associated with the production of heavy charmed mesons (See Figure 2).

The strange quark momentum distributions $xs(x)$ are found from the observed dimuon event distributions, corrected for acceptance and charm mass effects using the slow rescaling model. This extraction is made using a leading order formalism and, therefore, may need corrections for higher order effects. These strange sea structure functions, shown in Figure 3 demonstrate scaling violations analogous to those seen in non-strange quarks.

If the CKM matrix elements are not assumed then the results of the fits in (2) can be rewritten in terms of the products:
$$|V_{cd}|^2 B_c = 5.09 \pm .32^{+.17}_{-.16} \times 10^{-3}$$
Substitution of the neutrino world average charm branching ratio [19] $B_c = .116 \pm .010$ yields:
$$|V_{cd}| = .209 \pm .011 \pm .0035$$

Figure 2. Opposite-sign dimuon rates versus E_ν for ν_μ (top) and $\bar{\nu}_\mu$ (bottom) data. Rates corrected for acceptance, smearing, and kinematic cuts are indicated by squares. Those corrected for slow rescaling with $m_c = 1.31$ are given by circles. The curves indicate the slow rescaling model prediction with $m_c = 1.31 GeV/c^2$ (dotted) and $m_c = 0.0 GeV/c^2$ (dashed).

CONCLUSIONS

This experiment used a large sample of same-sign dimuon events obtained at the Fermilab Tevatron, increasing the statistical significance of the total same-sign dimuon rate. Furthermore, detailed measurements of muon-production in hadron showers to calculate the secondary component of the meson-decay background reduced the systematic uncertainty in the rate of same-sign production. The rate of prompt same-sign dimuon production with $P_\mu > 9$ GeV/c in ν_μ-N interactions for incident energies between 30 GeV and 600 GeV is $(5.3 \pm 2.4) \times 10^{-5}$ per ν_μ charged-current event or less than 9.2×10^{-5} at the 90% C.L. For incident $\overline{\nu_\mu}$ the rate is $(5.2 \pm 3.3) \times 10^{-5}$ per charged-current event or less than 10.5×10^{-5} at the 90% C.L.

The opposite sign dimuon data support the slow rescaling hypothesis for a value of $m_c = 1.31 \pm .24$. The charm mass error constitutes the single largest source of theoretical uncertainty in precision measurements of the weak mixing angle, $sin^2\theta_W$. The new result will reduce this uncertainty significantly, from .0034 to .0024. The CKM matrix element is found to be $|V_{cd}| = .209 \pm .012$. The nucleon strangeness content is measured to be $\eta_s = .064^{+.0075}_{-.0065}$ and the strange sea is found to be softer than its non-strange counterpart. The measurement of the Q^2 dependence of the strange sea structure function $xs(x)$ may be used to test perturbative-evolution predictions and evaluate flavor asymmetry in the sea.

Recently, there has been much theoretical work [22] to extend the leading-order formalism of neutrino charm production to higher orders. Since a dominant contribution to this process is scattering off strange sea quarks, it is expected that the next-to-leading order terms due to gluon quark-pair splitting will be significant. Our group is now in the process of applying the next-to-leading order calculations to the analysis of the data and expect to have updated results within the next several months.

REFERENCES

1. W.K. Sakumoto et al., Nucl. Instrum. Methods **A294** 179 (1990).
2. F.S. Merritt et al., Nucl. Instrum. Methods **A245** 27 (1986).
3. B.A. Schumm et al., Phys. Rev. Lett. **60** 1618 (1988).
4. B.A. Schumm, Ph.D. Thesis, University of Chicago, 1988.
5. K. Lang et al., Z. Phys. **C33**, 483 (1987).
6. J.G.H. DeGroot et al., Phys. Lett. **86B**, 103 (1979).
7. M. Holder et al., Phys. Lett. **70B**, 396 (1977).
8. H. Burkhardt et al., Z. Phys. **C31**, 39 (1985).
9. M. Jonker et al., Phys. Lett. **107B**, 241 (1981)
10. T. Trinko et al., Phys. Rev. **D23**, 1889 (1981).
11. K. Nishikawa et al., Phys. Rev. Lett. **46**, 1555 (1981).
12. P.H. Sandler, Ph.D. Thesis, University of Wisconsin-Madison, 1992.
13. S.R. Mishra et al., Nevis Preprint #1459, submitted to Phys. Rev. Lett. (1992).
14. Review of Particle Properties, Phys. Lett. **B239** (1990).
15. P.H. Sandler et al., Phys. Rev. D **42**, 761 (1990).
16. P.H. Sandler et al., submitted to Z. Phys. C., 1992.
17. V. Barger, W.Y. Keung, R.J.N. Phillips, Phys. Rev. **D25**, 1803 (1982).
18. J.R. Cudell, F. Halzen, and K. Hikasa, Phys. Lett. **B175**, 227 (1986).
19. M.H. Shaevitz, Nucl. Phys. B (Proc. Suppl.) **19**, 270 (1991); S.A. Rabinowitz et al., Nevis Preprint #1483, submitted to Phys. Rev. Lett. (1992)
20. W.C. Leung, Ph.D. Thesis, Columbia University, 1991.
21. C. Peterson et al., Phys. Rev. **D27**, 105 (1983).
22. M.A. Aivazis, F.I. Olness, and W.K. Tung, Phys. Rev. Lett. **65**, 2339 (1990); V. Barone, et al., Phys. Lett. **B268**, 279 (1991).
23. H. Georgi and H.D. Politzer, Phys. Rev. D **14**, 1829 (1976); R.M. Barnett, Phys. D **14**, 70 (1976).
24. C. Foudas, et al., Phys. Rev. Lett. **64**, 1207 (1990).
25. A.J. Buras, K.J.F. Gaemers, Nucl. Phys. **B132**, 249 (1978). At $Q^2 = 18.08$ the parton densities are given by $xu_v(x) = 2.718x^{0.61}(1-x)^{2.95}$, $xd_v(x) = 0.577xu_v(x)(1-x)$, and $sea(x) = 1.260(1-x)^{6.75}$.
26. C. Peterson, et al., Phys. Rev. D **27**, 105 (1983).
27. N. Ushida, et al., Phys. Lett. **B121**, 292 (1983).
28. H. Albrecht, et al., Phys. Lett. **B150**, 235 (1985).
29. D. Bortoletto, et al., Phys. Rev. **D37**, 1719 (1988).
30. D. Allasia, et al., Phys. Lett. **B249**, 366 (1990).

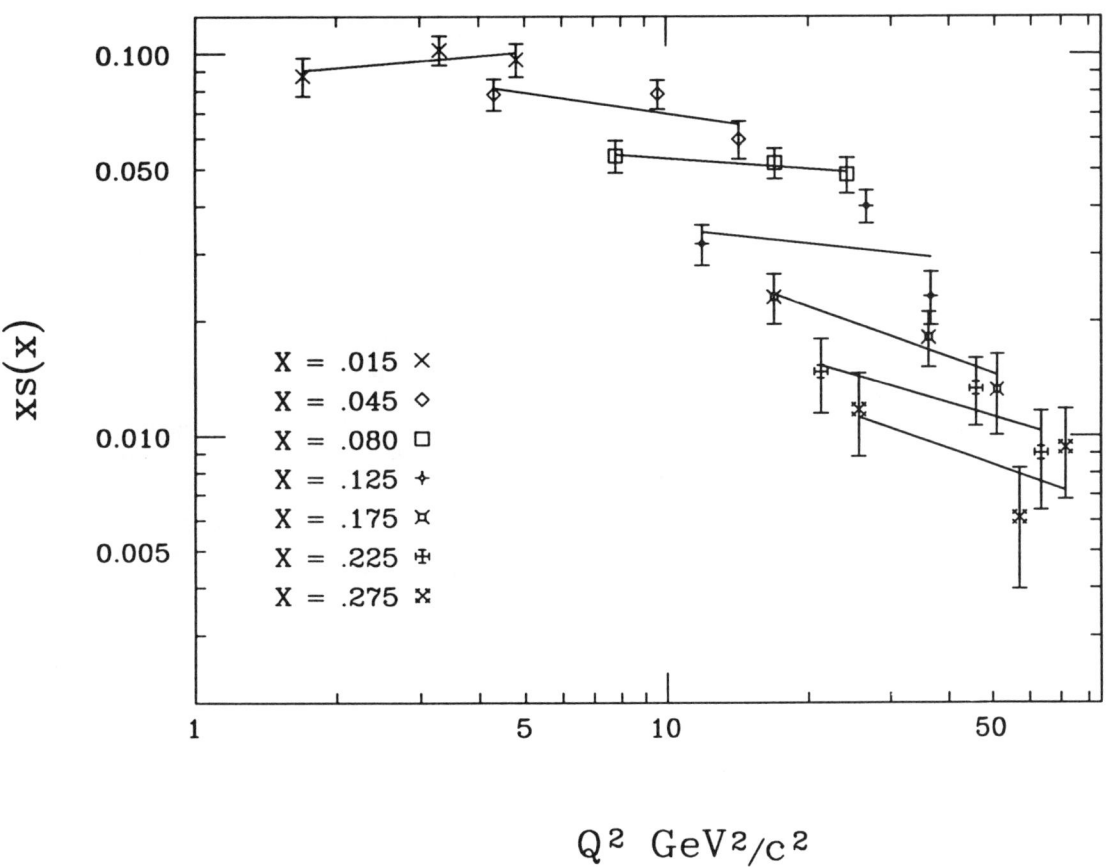

Figure 3. Strange sea structure functions $xs(x)$ versus Q^2 for several values of x. The lines are power-law fits to the data. Errors are statistical. An additional 11% scale error arises due to the uncertainty in κ.

STRANGENESS PRODUCTION IN NEUTRINO INTERACTIONS

M. Kalelkar (for the E632 Collaboration[*])
Department of Physics and Astronomy
Rutgers University
New Brunswick, NJ 08903

[*]E632 Collaboration: Berkeley, Birmingham, Brussels, CERN, Chandigarh, Fermilab, Hawaii, IHEP Serpukhov, Illinois Tech., Imperial Coll., ITEP Moscow, Jammu, Munich (MPI), Moscow State Univ., Oxford, Rutgers, Saclay, Stevens, Tufts

Abstract

A study has been made of K^0, Λ and $\bar{\Lambda}$ particle production in charged-current neutrino interactions at higher energy ($\langle E_\nu \rangle$ = 150 GeV) than any previous study. The experiment was done at Fermilab using the 15-ft. bubble chamber, and the data sample consists of 968 observed neutral strange particles. Production rates per event have been measured to be $(40.8 \pm 4.8)\%$ for K^0, $(12.7 \pm 1.4)\%$ for Λ, and $(1.5 \pm 0.5)\%$ for $\bar{\Lambda}$; they are significantly higher than in lower-energy experiments. The dependence of rates on event variables has been measured, and single-particle distributions obtained as well. Comparisons have been made with previous experiments.

INTRODUCTION

There have been a number of published results[1] on neutral strange particle production in charged current (CC) ν_μ and $\bar{\nu}_\mu$ interactions. However, most of these experiments have been at average neutrino energies of about 50 GeV. We have done a study of neutral strange particle production in an experiment (E632) at the Fermilab Tevatron featuring higher neutrino energies ($\langle E_\nu \rangle$ = 150 GeV) than any previous work.

EXPERIMENTAL PROCEDURE

E632 took data in two runs. The neutrino beam was produced by the quadrupole triplet train, which focused secondaries produced by the interactions of 800 GeV protons from the Tevatron. The detector was the 15-ft. bubble chamber filled with a neon-hydrogen mixture (75% molar neon in the first run, 63% in the second). A fiducial volume of about 15 m³ was employed, in which ν and $\bar{\nu}$ CC events were produced in a ratio of 6 to 1, and with mean energies of 150 GeV and 110 GeV respectively. The chamber was in a magnetic field of 3 T.

Photographs of the chamber made by three cameras were examined, and events on film were measured. All tracks from the primary vertex were measured, as well as associated vees (possible neutral strange particle decays) and γ conversions. All tracks were geometrically reconstructed, and kinematic fitting of vees and γ's was performed. The sample reported here corresponds to 1.66×10^{17} protons on target.

The bubble chamber was equipped with arrays of proportional tubes: an Internal Picket Fence (IPF) to determine the event time, and an External Muon Identifier (EMI) to identify muons.[2] Muon candidates were required to have a momentum exceeding 5 GeV/c, to leave the chamber without interacting, and to have a good 2-plane match in the EMI, corresponding to traversal through 7 to 11 hadronic interaction lengths after the bubble chamber. CC events were required to have a muon satisfying these criteria, and a total hadronic mass over 2 GeV.

Vees and γ's were required to make constrained fits with probability greater than 0.1% to the hypotheses $K_S \rightarrow \pi^+\pi^-$, $\Lambda \rightarrow p\pi^-$, $\bar{\Lambda} \rightarrow \bar{p}\pi^+$ or $\gamma(p) \rightarrow (p)e^+e^-$. Vee vertices were also required to be at least 1 cm from the primary vertex and at least 20 cm from the chamber wall. Fit ambiguities were resolved by comparing the probabilities of the competing fits. An algorithm for resolving the ambiguities was devised in such a manner as to produce an isotropic K_S decay angular distribution, which is the distribution of the cosine of the angle between the K_S line of flight and the direction of a decay pion.[3]

A total of 968 vees comprised the final sample, corresponding to 6263 ν and 1115 $\bar{\nu}$ CC events. Table 1 gives the numbers of K_S, Λ and $\bar{\Lambda}$ from ν and $\bar{\nu}$ CC interactions.

Table 1. Observed numbers of neutral strange particles from ν and $\bar{\nu}$ charged current events.

Vee	From ν CC	From $\bar{\nu}$ CC
K_S	502	93
Λ	285	57
$\bar{\Lambda}$	27	4

PRODUCTION RATES

To determine inclusive production rates, the vees were weighted for random scan loss, geometric detection efficiency, loss due to interaction before decay, loss at small flight times, energetic vee loss, and Λ loss due to low-momentum decay π^-. Average experimental weights were 1.87 for K_S, 1.99 for Λ, and 2.56 for $\bar{\Lambda}$. An additional weight for decay branching ratio was then also applied, and Table 2 gives the fully corrected average multiplicities per CC event. The quoted errors include statistical errors as well as the uncertainties in calculating each of the correction factors, and the uncertainty in the ambiguity resolution algorithm. The ratio of Λ to K^0 production is 0.31±0.05 for ν, and 0.26±0.06 for $\bar{\nu}$.

Table 2. Average neutral strange particle multiplicities per charged current event. The K^0 rate includes $K^0 + \bar{K}^0$.

Vee	ν CC events	$\bar{\nu}$ CC events
K^0	0.408±0.048	0.454±0.078
Λ	0.127±0.014	0.118±0.019
$\bar{\Lambda}$	0.015±0.005	0.010±0.007

DIFFERENTIAL RATES

To study the dependence of neutral strange particle production rates on incident neutrino energy, we have measured for each event all charged tracks and associated neutrals. To correct for missing energy (mostly neutrals that do not materialize within the bubble chamber) we have employed the Bonn method,[4] which is based partly on transverse momentum balance. The average correction to the visible energy is only about 9%, and the resulting energy spectrum of events is in good agreement with a Monte Carlo calculation using the beam profile. For the rest of this section, results are shown only for ν events, because there are too few $\bar{\nu}$ events.

Fig. 1 shows the K^o and Λ production rates as a function of neutrino energy. For comparison, published rates from previous experiments are also shown. For $E_\nu < 100$ GeV, our results are consistent with these older results. Above 100 GeV, however, where there are no previous measurements, we observe a significant increase in neutral strange particle production, especially for the K^o.

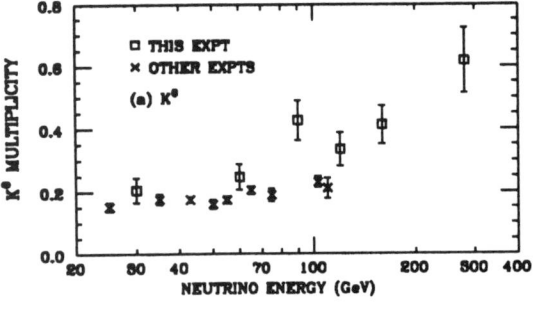

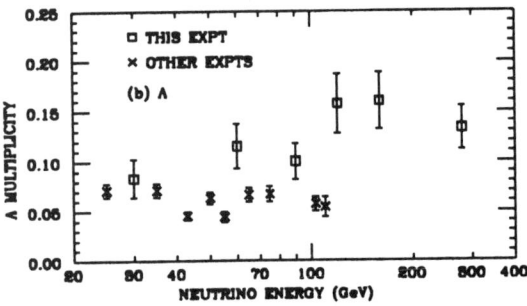

Figure 1. Average multiplicities of (a) K^o and (b) Λ as a function of neutrino energy.

We have also found that the K^o production rate increases sharply with W^2 (the square of the hadronic effective mass), Q^2 (the invariant square of the four-momentum transfer from the neutrino to the muon), and y_B (the fractional energy transfer to the hadronic system). Λ production increases less significantly with these variables. Both K^o and Λ production rates are independent of the Bjorken scaling variable x_B.

To obtain information on production mechanisms, we have examined various single-particle distributions. Fig. 2 shows the K^o and Λ rapidities in the hadronic center of mass, normalized to the number of CC events, so that the ordinate represents the average multiplicity of a particular particle per unit interval of rapidity. There is significant production of K^o at higher values of rapidity than in previous experiments; the asymmetry between the numbers of forward (F) and backward (B) K^o is A = (F-B)/(F+B) = 0.35±0.02, which is appreciably greater than the value of 0.16±0.02 reported by the highest-statistics previous experiment.[5] The Λ is mainly produced in the target fragmentation region, although there is some central production as well. The forward-backward asymmetry is A = -0.47±0.03, compared to -0.71±0.02 for Ref. 5.

Fig. 3 shows the ratio of K^o to π^- normalized rapidity distributions. The π^- is believed to be almost exclusively produced centrally in ν CC interactions, because it does not contain the u quark which is in the current jet. The K^o/π^- ratio in Fig. 3 unambiguously indicates enhanced production of K^o as the rapidity increases. In the very forward direction, about one-quarter of the particles might be kaons.

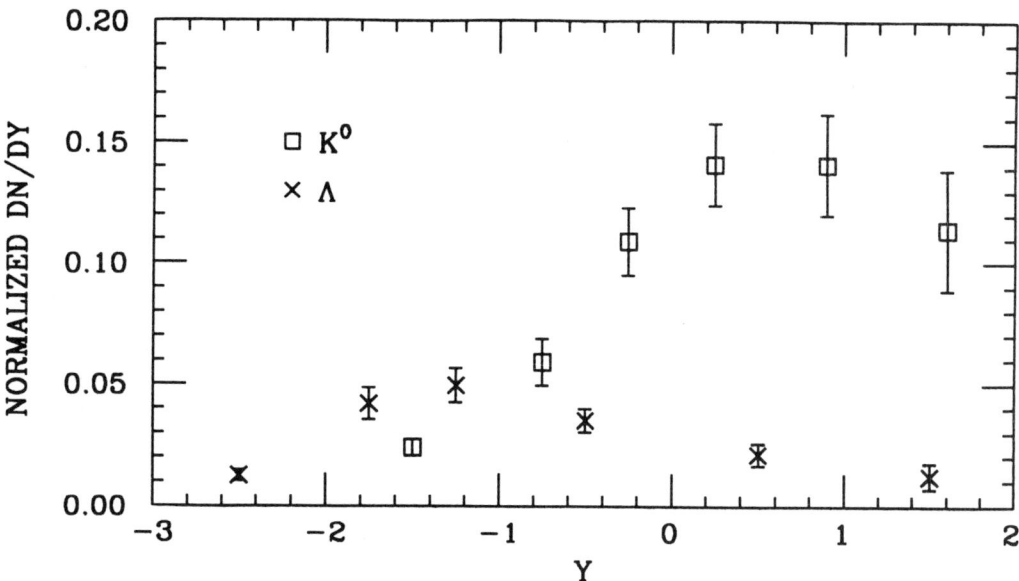

Figure 2. Normalized center-of-mass rapidity distributions for K^0 and Λ.

This work was supported in part by the U.S. Dep't. of Energy and the National Science Foundation, and by the U.K. Science and Energy Research Council.

REFERENCES

1. The most recent paper, which includes a list of all previous ones, is S. Willocq et al., Z. Phys. C 53, 207 (1992).

2. Details about the IPF and EMI are given in V. Jain et al., Phys. Rev. D 41, 2057 (1990).

3. For more details about the event selection criteria, see D. DeProspo, Ph.D. Thesis, Rutgers University, 1991.

4. H.G. Heilmann, Bonn Internal Report No. WA21-int 1, 1978 (unpublished).

5. N.J. Baker et al., Phys. Rev. D 34, 1251 (1986).

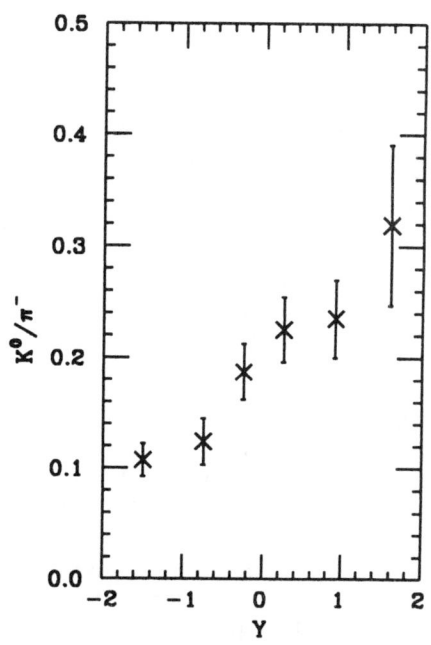

Figure 3. Ratio of K^0 to π^- normalized rapidity distributions.

NUCLEAR PARTON DISTRIBUTIONS AND EXTRACTION OF NEUTRON STRUCTURE FUNCTIONS

Mark Strikman

Pennsylvania State University, State College, PA 16802, U.S.A.

Abstract

Current information on the nuclear modification of parton distributions is summarized. Implications for the current global analyses of the parton distributions include i) significant increase of $d(x)/u(x)$ ratio at x close to 1 making in close to predictions of perturbative QCD, indications of $\bar{d}(x)/\bar{u}(x) > 1$ at x=0.007, suggesting that deviation of the Gottfried sum rule from 1/3 is predominantly due to region of very small x. First deep inelastic data of BCDMS at x>1 are also discussed.

INTRODUCTION

Experimental contributions presented at this conference clearly demonstrate, that after many years of measurements the results of different experiments finally agree with each other on a few percent level. Obviously, this greatly simplifies attempts of determining parton distributions in nucleon based on the global analyses of hard phenomena. However many of the key input data are obtained using deuteron targets and, in the case of neutrino scattering, using iron targets. To ensure consistency of the analyzed data samples the deuteron and nuclear data should be corrected for nuclear distortions of the parton distributions. Introducing such corrections seems to be feasible now, since after several years of dedicated experiments (prompted by the observation of the EMC effect) a coherent picture of parton distributions in nuclei seems to emerge. It appears that different physical phenomena are responsible for the observed "zigzag" x-dependence of $R_A(x,Q^2) = \frac{2}{A}F_{2A}(x,Q^2)/F_{2D}(x,Q^2)$: nuclear shadowing provides natural explanation of the small x depletion; deviations of nucleus wave function from the simple many nucleon picture of nucleus are responsible for the dip at $x \in 0.4-0.7$, while increase of R_A at higher x is related to the structure of the short-range correlations in nuclei (for recent detailed review see[1]). In this talk we will consider three examples where nuclear corrections are definitely important: role of nuclear shadowing in determining F_{2n}/F_{2p} ratio at x→0, determination of the d/u ratio at x≥0.5, and quarks in nuclei at x>1.

Small x physics.

It follows from practically model independent analysis of the representation of $F_{2T}(x,Q^2)$ through the commutator $<T|J_\mu(y)J_\nu(0)|T>$ that in the $x \to 0$, $q_0 \to \infty$ limit in the nucleus rest frame γ^* converts into a hadronic state at the distances
$$y_3 = l \sim 1/2m_N x \gg 2R_A. \quad (1)$$
In QCD due to color screening the effective interaction cross section, σ, depends on the transverse area, S, occupied by the color:
$$\sigma \sim S, \quad \text{for small S} \quad (2)$$
and only slowly increases with S for $S > \pi r_\pi^2$. $\sigma_{\gamma^* A(N)}$ can be calculated by applying the Gribov representation for γ^*A scattering which is modified in QCD as[2,3]
$$\sigma_{\gamma^* A}(Q^2,\nu) = \quad (3)$$
$$= \frac{\alpha}{3\pi} \int_0^\infty \frac{R(M^2)M^2\,dM^2}{(M^2+Q^2)^2} \int d\sigma_{"M^2"A}^{(S)}(Q^2,\nu)$$
for $x \ll 1/4R_A m_N$. Here M is the mass of

the intermediate hadronic state and $R(M^2) = \sigma(e^+e^- \to \text{hadrons})/\sigma(e^+e^- \to \mu^+\mu^-)$. $R(M^2) \approx 3 \Sigma e_q^2$ where sum goes over the flavors which can be produced at given M^2 and it is determined by the quark loop. Contribution of small transverse momenta in the loop corresponds to large color separation: $r_t \sim 1/k_{to}$, and hence to $\sigma_{(q\bar{q})N} \sim \sigma_{\pi N}$, while $q\bar{q}$ pairs with large k_t interact with cross section $\sim \alpha_s^2(k_t^2)/k_t^2$. Neglecting contribution to the shadowing from the $q\bar{q}$ jets with $k_t > k_{to}$ one can calculate nuclear shadowing using eq.2. From comparison with the data for A=1 we find that i) The soft contribution dominates in $F_{2N}(x \sim 10^{-2}, Q^2 \sim \text{few GeV}^2)$, $\sigma(q\bar{q}, k_{qt} < k_{ot}) \approx (1 - 0.5)\sigma_{\pi N}$. ii) Color screening which leads to decrease of the cross section $\propto r_t^2$ and hence to the Bjorken scaling at small x is present at least for $r_t^2 < 0.7 r_\pi^2$. Results of the calculations[1-3] which are rather insensitive to the input parameters can be written at $x < 1/4R_A m_N$ as

$A_{eff}/A = (1-\lambda)^2 \sigma_{tot}(\pi A)/A\sigma_{tot}(\pi N) + \lambda$, (4)

where $\lambda = 0.3 - 0.4$. It is rather straightforward to include effects of the finite longitudinal distances (eq.1) for $x < 0.05-0.1$ [3], at $0.2 > x \geq 0.0.05$ one has to include effects of enhancement of the parton distributions (see next section). Results of calculations agree reasonably well with existing high precision NMC data for μA scattering[4] and the Drell-Yan data for A-dependence of the antiquark distributions[5], see figs.1,2. It rather strongly depends on Q^2 because of scaling violation and absence of shadowing at $x \geq 0.1$. Shadowing is expected to be present in all channels — sea, valence quarks, gluons.

Practically all existing calculations of nuclear shadowing at $x \geq 0.01$ are explicitly or implicitly based on eq.2. Thus, though details of calculations differ, they lead to similar A-dependence of the shadowing.

Therefore if one adjusts parameters of the models to fit the observed shadowing for light nuclei, the predictions of shadowing for the deuteron turn out to be very close. Thus it appears that the shadowing in μD scattering can be estimated with reasonable accuracy (Note that nonlinear effects[6] (parton-parton fusion) which may become significant at smaller x have similar A-dependence and would not affect such extrapolation). We find[7]
$F_{2D}(x,Q^2)/\{F_{2p}(x,Q^2)+F_{2n}(x,Q^2)\} =$ (5)
$1 - (2.3 \pm 0.2)\%$ at $x < 10^{-2}$.
As a result the NMC data[8] at x=0.007 correspond to
$F_{2n}(0.007)/F_{2p}(0.007) = 1.03 \pm 0.015 \pm 0.023$,
while extrapolation from x=0.08 assuming $\bar{u} = \bar{d}$ leads to 0.94. This estimate indicates that $\bar{d} > \bar{u}$ at small x: $\bar{d}(0.007)/\bar{u}(0.007) \sim 1.1$. Assuming that $x\bar{d} - x\bar{u} \propto x^{0.5-0.7}$ we found deviation of the Gottfried integral from 1/3 observed by NMC (which is somewhat bigger than one given in the paper due to nuclear shadowing effects) is saturated by $\int (\bar{d}-\bar{u})dx$ over the $0 < x < 0.02$ region. The very recent Drell-Yan data[9] implicitly support this conclusion. The authors find no indications of $\bar{d} > \bar{u}$ at $x \geq 0.1$ on the level predicted in the models where the Gottfried integral is saturated by the contribution to $\int(\bar{d}-\bar{u})dx$ from the region $x \geq 0.1$. Thus the only way to reconcile these data with the NMC measurements of the Gottfried integral is to assume that large contribution comes from the x<0.05 region not covered by the Drell-Yan experiment.

ENHANCEMENT OF PARTON DISTRIBUTIONS IN THE X ~ 0.1 REGION

Since parton distributions are shadowed at small x the exact QCD sum rule for the baryon charge

$$\int_0^A \frac{1}{A} V_A(\tilde{x}, Q^2) d\tilde{x} - \int_0^1 V_N(x, Q^2) dx = 0, \quad (6)$$

where $\tilde{x} = AQ^2/2m_A q_0 = Am_N/m_A x$ predicts that valence quark distribution is enhanced at higher x. At the same time the exact QCD sum rule for the conservation of the total nucleus momentum implies that gluon and/or sea distribution is enhanced at higher x. Analysis of the data shows that actually the gluons are enhanced though antiquarks are suppressed for all x. This difference of the nuclear effects for valence and for sea quarks, and for gluons makes it impossible to apply universal rescaling of all parton distributions by the same amounts as for F2A ratios.

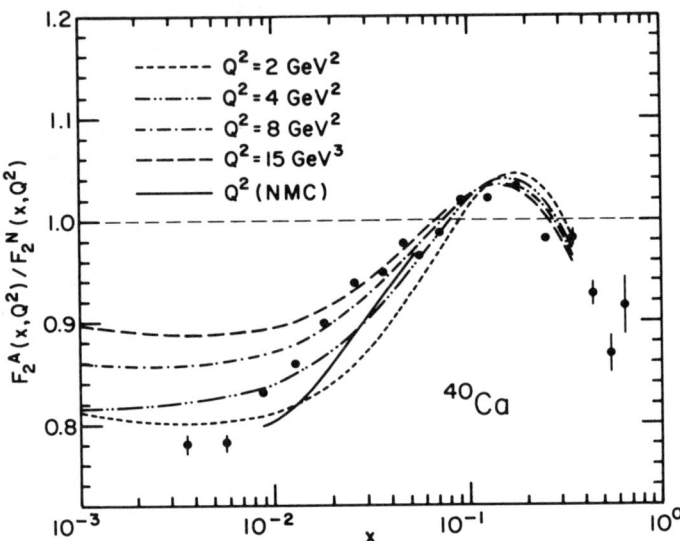

Figure 1. Q^2 - evolution of F_2^{Ca}/F_2^D ratio. NMC data are from [4], curves use [11].

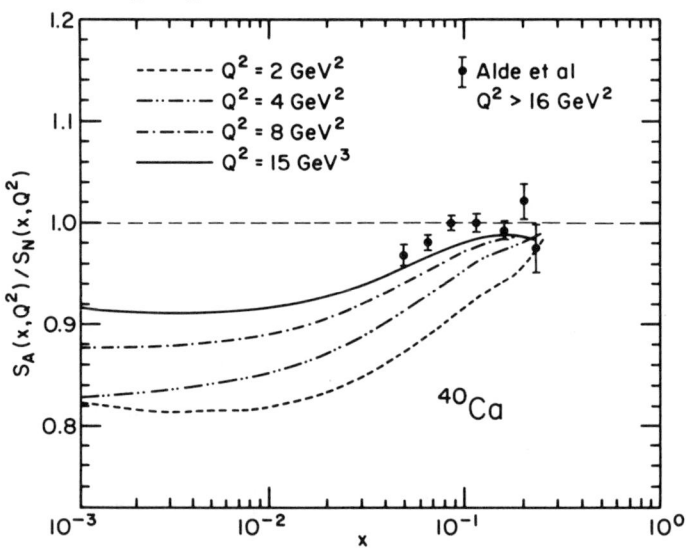

Figure 2. Q^2 - evolution of $\bar{q}_2^{Ca}/\bar{q}_2^D$. The Drell-Yan data are from [5], curves are from [10].

Obviously the shadowing/enhancement pattern noticeably changes the x dependence of the parton distributions in the 0.01<x<0.1 region. For example, in this x-range the ratio F_{3Fe}/F_{3N} decreases as $x^{0.1}$, while it flattens at smaller x. It is necessary to check to what extend this affects the experimental determination of $\int_0^1 F_{3Fe}(x)dx$.

We used the NMC data for Ca for the ratio of the second moments of F_{2Ca} and F_{2D} (which have the smallest relative normalization errors) and the exact momentum sum rule to calculate the ratio of momentum fractions carried by gluons in Ca and in a free nucleon, $\gamma_G(Ca)$[10]. Neglecting the change of the gluon momentum in D and possible change of the strange sea we find using final NMC and SLAC data:

$$\gamma_G(Ca) = (1.4 \pm 0.3 \pm 0.9)\% \quad (7)$$

By including these effects $\gamma_G(Ca)$ would increase by 0.1-0.5%. Since for small x, $G_A(x,Q^2)$ is expected to be shadowed, the enhancement of $G_A(x,Q^2)$ in nuclei is likely to be concentrated at $x \sim 1/(2m_N r_{NN})$, where $r_{NN} \sim 1-2$ fm is characteristic distance in NN interaction, i.e. 0.03<x<0.15 where it can exceed 10(20)% for A=40(A=∞) and $Q^2 \sim Q_0^2 \sim 1-2$ GeV^2.[10] This gluon enhancement contributes significantly to the scaling violation of the antiquark ratio shown in Fig.2. Therefore small shadowing for q_A observed in Ref.5 for x~0.04 and $Q^2 > 16 GeV^2$ corresponds to much larger shadowing for $Q^2 \sim Q_0^2$ and same x.

To summarize, all parton distributions are shadowed at small x, while at larger x only valence quark and gluon distributions are enhanced. It is important to take into account this pattern in the global analyses of the data. It is worth noting also that this pattern of nuclear modification of parton distributions is difficult to understand if intermediate range nuclear forces are dominated by exchanges of ordinary mesons.

d(x)/u(x) AT x≥0.45 AND THE EMC EFFECT.

The EMC effect has firmly established that in the region of x<0.8 the nuclear parton distributions are modified in the opposite direction to one expected from the Fermi motion effects. Consequently the currently used procedure [11] for extraction of the $F_{2n}(x)/F_{2p}(x)$ ratio has to be modified to take into account the EMC effect. It was pointed out in [12] that this can be done in a wide region of x in a practically model independent way. Indeed, since essential longitudinal distances, l, (eq.1) for x > 0.2 are smaller than the average internucleon distances and since nuclei are rather dilute systems, any deviations of $R_A(x,Q^2)$ from 1 for x sufficiently below 1 should be proportional to the mean nuclear density $<\rho_A(r)>$:

$$R_A(x,Q^2) - 1 = f(A)\phi(x,Q^2) \quad (8)$$

where

$$f(A) \sim <\rho_A(r)> - <\rho_D(r)> \quad (9)$$

where $<\rho_A(r)> = \int \rho_A^2(r)d^3r/A$, and $\int \rho_A(r)d^3r = A$. This estimate is valid for most of the models suggested for the explanation of the EMC effect and it reasonably agrees with the data,[13] see comparison in Refs.2,12,14. This factorisation should break down at x≥0.8 where higher order terms become important in the decomposition in particular for the Fermi motion contribution. As a result for 0.2<x<0.8 we can correct for nuclear effects in the deuteron using data on the EMC effect with heavier nuclei as

$$\frac{F_{2D}(x,Q^2)}{F_{2p}(x,Q^2)+F_{2n}(x,Q^2)} - 1 = \quad (10)$$

$$= \frac{<\rho_D>}{<\rho_{Fe}>-<\rho_D>}\left[\frac{F_{2Fe}(x,Q^2)}{F_{2D}(x,Q^2)}-1\right]$$

where $<\rho_D>/\{<\rho_{Fe}>-<\rho_D>\} \approx 0.25$. The SLAC analysis[15] has compared values of $R(x) \equiv F_{2n}/F_{2p}$ obtained using preEMC effect procedure [11] and procedure given by eq.(10). One can see from Fig.3. that results differ qualitatively for x ≥0.5. The new procedure leads to significantly larger value of the ratio, which is consistent with the pQCD estimate of Ref.16, $R(x\to 1)=3/7$. Further experiments are necessary to probe this ratio. The possible options include study of nuclei with similar densities and different n/p ratios, like ^{9}Be, and ^{12}C (S.Rock, private communication), tagged neutron structure function data ($l+D \to l'$ + backward proton +X reaction), the ratio of cross sections of W^+ and W^- production in $p\bar{p}$ collisions at large y.

SUPERFAST QUARKS IN NUCLEI - x>1.

Measurement of quark distributions at x>1 would enable to study directly the A-dependence and the quark structure of the short-range correlations in nuclei, providing a unique information about properties of small superdense nuclear drops, since the final state interaction does not contribute in the leading twist. Measurements in these kinematics are very difficult due to resolution problems and low cross section - F_{2A} small and decreases with x as exp-bx with b~15. First data at really large 200 > Q^2 > 50 GeV^2 were presented in the contributed paper of BCDMS [17], see Fig.4 where data for highest Q^2 range are presented (according to Ref.17 the data in the highest Q-range have the smallest systematic errors. The analysis required a detailed knowledge of the resolution function of the spectrometer. One of the impressive test was the study of the x>1 events for the scattering off hydrogen, which was found to be described by the detector resolution function without adjustable parameters. Our estimates of the cross section based on the current information about magnitude of short-range correlations in nuclei and the color screening model of the EMC effect agree reasonably with these measurements.

The observed cross section is very small and inclusion of the x>1 region in the evolution equations is likely to have very small effect of the current analysis of the scaling violation.

Natural next step would be to perform detailed measurements using SLAC linac for $Q^2 \in$ (10-20) GeV^2 where leading twist contribution starts to dominate, as well as to perform dedicated measurements for $Q^2 \sim$ 100 GeV^2.

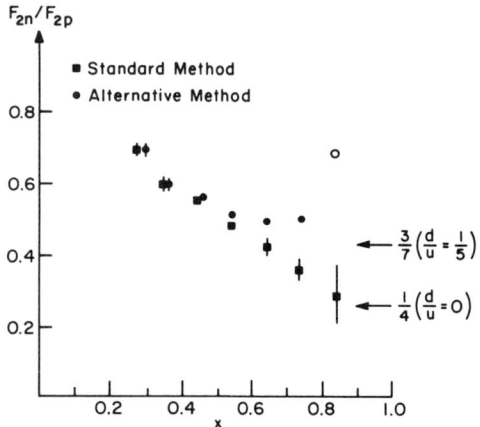

Figure 3. Results of extraction of the F_{2n}/F_{2p} ratio from the F_{2D}/F_{2p} data using "standard method" [11] and Alternative (density method)[12]. Open circle at x=0.85 indicates that the density method is applicable for x<0.8 only.

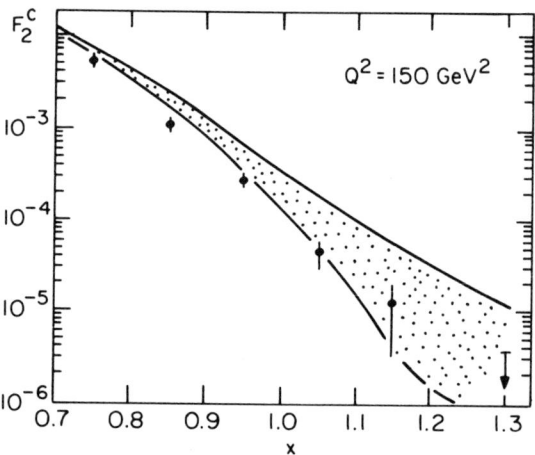

Figure 4. Comparison of the BCDMS data[16] with the model of Ref.3 which takes into account short-range correlations in nuclei and suppression of small size configurations in bound nucleons.

To summarize, inclusion of nuclear effects in the global analyses of parton distribution is clearly necessary. It would lead to a significant modification of these distributions both for large and for small x.

I would like to thank L.Frankfurt and S.Liuti for collaboration on the subjects discussed in the talk.

REFERENCES

1. L.L. Frankfurt & M.I. Strikman, PARTON STRUCTURE OF NUCLEI AS SEEN BY LEPTON PROBES: FROM X $\sim 10^{-3}$ TO X=1 AND BEYOND, in "Modern topics in electron scattering", B. Frois and I. Sick, eds., World Scientific (1991) 762-797.

2. L.L. Frankfurt & M.I.Strikman, HARD NUCLEAR PROCESSES AND MICROSCOPIC NUCLEAR STRUCTURE. PHYS. REP. 160 (1988) 235-427

3. L.L. Frankfurt & M.I. Strikman, SHADOWING AND ENHANCEMENT OF QUARK DISTRIBUTIONS IN NUCLEI AT SMALL x. Nucl. Phys. B316,pp. 340-355, 1989.

4. P.Amaudruz et al., PRECISION MEASUREMENT OF THE STRUCTURE FUNCTION RATIOS F2(HE)/F2(D), F2(C)/F2(D), F2(CA)/F2(D), Z.Phys.C51,387-394, 1991.

5. D.M.Alde et al., NUCLEAR DEPENDENCE OF DIMUONS AT 800-GEV, Phys.Rev.Lett. 64,pp.2479-2482,1990.

6. A.H.Mueller and J.Qiu, GLUON RECOMBINATION AND SHADOWING AT SMALL VALUES OF X. Nucl.Phys.B268, pp.427-452,1986.

7. L.L. Frankfurt and M.I. Strikman, PARTON DISTRIBUTIONS FROM X=10^{-3} TO X=1 AND BEYOND, In proceedings of IV Conference Intersections between particle and nuclear physics 1991, AIP conference proceedings, 243, pp.762-766 (1992).

8. P.Amaudruz et al, THE RATIO F2(N)/F2(P) IN DEEP INELASTIC MUON SCATTERING, Nucl.Phys.B371,pp.3-31,1991.

9. P.L.McGaughey, et al., LIMIT ON THE $\bar{d}/\bar{u}$ ASYMMETRY OF THE NUCLEON SEA FROM DRELL-YAN PRODUCTION, Phys.Rev..Lett, 69,pp.1726-1728.

10. L.L.Frankfurt, M.I.Strikman, and S.Liuti, EVIDENCE FOR ENHANCEMENT OF GLUON AND VALENCE QUARK DISTRIBUTIONS IN NUCLEI FROM HARD LEPTON NUCLEUS PROCESSES,Phys.Rev.Lett., 66, pp1725-1728, (1990)

11. L.L. Frankfurt & M.I. Strikman, HIGH-ENERGY PHENOMENA, SHORT RANGE NUCLEAR STRUCTURE AND QCD. Phys. Rep. 76 pp.215-347,1981.

12. L.L. Frankfurt & M.I. Strikman, POINTLIKE CONFIGURATIONS IN HADRONS AND NUCLEI AND DEEP INELASTIC REACTIONS WITH LEPTONS: EMC AND EMC LIKE EFFECTS.Nucl.Phys.B250 pp.143-176, 1985. Preprint LINP-84-929 (1984) 15p.

13. R.G. Arnold,et al,MEASUREMENTS OF THE A-DEPENDENCE OF DEEP INELASTIC ELECTRON SCATTERING FROM NUCLEI, Phys.Rev.Lett.,52,pp.727-730,1984.

14. R.L.Jaffe,F.E.Close,R.G.Roberts, and G.G.Ross, ON THE NUCLEAR DEPENDENCE OF ELECTROPRODUCTION, Phys.Lett., B134,pp. 449- 453, 1984.

15. L.W. Whitlow et al, PRECISE MEASUREMENTS OF THE PROTON AND DEUTERON STRUCTURE FUNCTIONS FROM A GLOBAL ANALYSIS OF THE SLAC DEEP INELASTIC ELECTRON SCATTERING CROSS-SECTIONS. Phys.Lett.B282,pp.475-482,1992.

16. G.Farrar, D.R.Jackson, PION AND NUCLEON STRUCTURE FUNCTIONS NEAR X=1, Phys.Rev.Lett.35,pp.1416-1419,1975.

16. I.Savin (BCDMS), NUCLEAR STRUCTURE FUNCTIONS IN CARBON NEAR x=1, Contributed paper.

MEASUREMENTS OF γγ COLLISIONS WITH THE OPAL DETECTOR AT LEP AND COMPARISON WITH QCD MODELS*

J.G. Layter
Department of Physics
University of California
Riverside, California USA 92521

Abstract

Samples of muon pairs, tau pairs, and hadrons produced in single tagged two-photon collisions have been obtained with the OPAL detector. The tagging electron is detected in the Forward Detector. A variety of triggers select the remainder of the event. From the muon pair sample the QED structure function F_2 has been extracted. Measured event distributions from the tau pair sample agree with QED calculations. Distributions of Q^2, W, and x obtained from the hadron sample agree well with a Monte Carlo model with contributions from QCD, Vector Meson Dominance, and the production of charmed quarks and tau leptons.

INTRODUCTION

The reaction $e^+e^- \to e^+e^-\mu^+\mu^-$ is a pure QED reaction and is in principle well understood. Consequently it was one of the first two-photon reactions studied in the PEP-PETRA era,[1] and this is the case as well at LEP. On the contrary, two-photon tau pair production was never observed at lower energies since the γγ mass range was too low. This paper describes the first direct observation of tau pairs in two-photon collisions, made at LEP in the OPAL detector. The F_2 structure function of the photon has been measured in the reaction $e^+e^- \to hadrons$ at lower energy colliders[2] for values of Q^2 from 0.1 to over 100 GeV$^2/c^2$. LEP offers the possibility of testing QCD over a wider range of Q^2 if sufficient luminosity becomes available. At the present time sufficient statistics are available in the region $4 < Q^2 < 30$ GeV$^2/c^2$ to provide a meaningful check of earlier results.

MUON PAIR PRODUCTION

Kinematics

The cross section for deep inelastic scattering of an electron (or positron) from a nearly real virtual photon associated with the opposing positron (or electron) can be written in terms of the structure functions F_1 and F_2 as

$$\frac{d\sigma}{dxdy} = \frac{16\pi\alpha^2 EE'}{Q^4}[(1-y)F_2(x,Q^2) + xy^2 F_1(x,Q^2)] \quad (1)$$

The structure functions F_1 and F_2 are functions of the variables x and Q^2, where Q^2 is the momentum transfer squared of the photon, as determined by the measurement of the

---

*Reporting an analysis performed by A. Buijs, C. Sbarra, J.E. Conboy, C.P. Howarth, B.W. Kennedy, M.E. Lehto, and D.J. Miller.

scattered lepton. The kinematic variable x is defined as

$$x = Q^2/(Q^2 + W^2) \qquad (2)$$

where W is the invariant mass of the muon pair. The variable y is defined as

$$y = 1 - \frac{E'}{E} \cos^2(\frac{\theta'}{2}) \qquad (3)$$

For sufficiently small values of y, the two-photon cross section can be written as a function of F_2 only. In contrast to the hadronic structure functions, the leptonic structure functions can in principle by calculated in QED.

Monte Carlo Simulation

Monte Carlo simulations using the Vermaseren generator[3] are compared to data distributions presented here. This generator includes multiperipheral and bremsstrahlung diagrams from among the lowest order diagrams contributing to the process $e^+e^- \to e^+e^-\mu^+\mu^-$. A sample of 15000 events corresponding to an integrated luminosity of 60.3 pb^{-1} was produced with the following cuts imposed: the scattering angle of the electron had to be at least 40 mrad; at least one muon had to have a momentum greater than 900 MeV/c and be in an angular range satisfying $|\cos\theta| < 0.9$.

Event Selection

Events fall into two classes depending on whether the tag is made in the OPAL Forward Detector or in the Barrel or End Cap lead glass arrays. The latter class are the interesting high Q^2 events, with $\theta_{tag} > 200$ mrad and $Q^2 > 100$ GeV$^2/c^2$, but since they are currently small in number they will not be discussed further.

The former class have Q^2 between 4 and 30 GeV$^2/c^2$ with an average value of 8.0 GeV$^2/c^2$.

For these events, the selection requirements separate cleanly into those for the tag and those for the rest of the event. The tagging electron must be in the fiducial volume of the Forward Detector, $47 < \theta_{tag} < 120$ mrad, and its energy must exceed 0.5 E_{beam}. In the remainder of the event, there must be two and only two charged tracks of opposite sign, at least one of which must have a momentum greater than 1 GeV/c and in the polar angle region satisfying $|\cos\theta| < 0.8$. For the other track the requirements are 300 MeV/c and $|\cos\theta| < 0.95$.

The $e^+e^- \to e^+e^-\mu^+\mu^-$ final state is identified by requiring the presence of one identified muon and rejecting the event if there is an identified electron. Muons are identified either by possessing a reconstructed track segment in the muon chambers or by having certain hit patterns in the instrumented iron hadron calorimeter. Electrons are identified by their E/p ratio if their momentum is greater than 2 GeV/c, and by dE/dx for momenta between 0.3 and 2 GeV/c. In neither case can there be a hadron calorimeter signal. After these identification requirements, there are 738 events, to be compared to 725 events in the Monte Carlo sample.

Extraction of F_2

The QED structure function is extracted from the data using the following procedure: *i)* Events are generated according to the theoretical luminosity function, but with a constant $F_2 = 1/\alpha$. The distribution of the production angle of the muon with respect to the photon direction in the $\gamma\gamma$ center of mass is generated according to the QED prediction. *ii)* These events are passed through the detector simulation and the analysis cuts. *iii)* The x-distribution of the data is divided by the corresponding Monte Carlo distribution, thus eliminating the effects of the detector and of

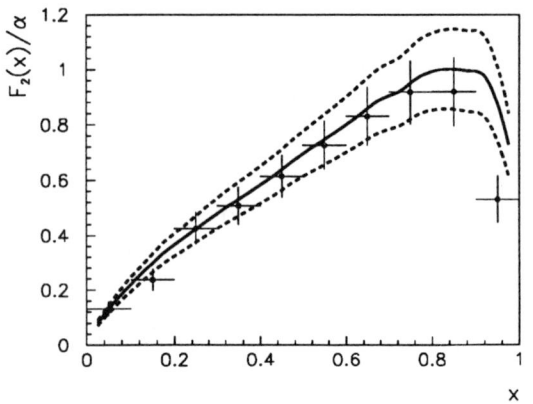

Figure 1. The measured values of the structure function F_2 as a function of x, extracted from the low Q^2 sample. The solid line is the QED expectation; the dashed lines are one-sigma deviations of the systematic error due to the selection efficiency.

the photon flux factor. The resulting F_2 distribution, shown in Figure 1, can be compared directly with the QED calculation.

The measurement of F_2 agrees well with an analytical form of the structure function[4] incorporated in a simple Monte Carlo generator.[5] There is no indication of an excess over the QED expectation. An order of magnitude more statistics is needed before the measurement of F_2 can be extended to the high Q^2 range.

TAU PAIR PRODUCTION

Monte Carlo Simulation

The Monte Carlo simulation again uses the Vermaseren generator with the decay of the tau leptons obtained from routines in the JETSET library. A sample of 5000 events corresponding to an integrated luminosity of 162 pb^{-1} was produced requiring that the scattering angle of the electron had to be at least 40 mrad and at least one charged particle from the tau decay had to have a momentum greater than 900 MeV/c and be in an angular range satisfying $|\cos\theta| < 0.9$.

Event Selection

After certain detector status requirements were satisfied, the data sample corresponded to 17.1 pb^{-1} from 1990 and 1991 running. Events were triggered with either of two independent triggers: either the coincidence of a 15 GeV or greater energy deposition in the forward detector and a charged track in the central detector, or 35 GeV or more in either of the two forward detectors. The overall trigger efficiency was found to be 97.5 ± 0.7%.

In the subsequent selection, the requirements for the tagging electron were the same as described above for the case of the muon pairs. Furthermore, events were required to have two and only two good tracks of opposite charge. Events with photon conversions or with additional tracks were rejected. At least one of the tracks had to have a momentum of at least 1 GeV/c and a polar angle satisfying $|\cos\theta| < 0.75$. For the other track, the corresponding requirements were 300 MeV/c and $|\cos\theta| < 0.95$.

For the identification of the $\tau^+\tau^-$ final state, only those events were considered in which one tau decayed into an electron and the other into a muon or a hadron with possible additional neutral particles. Both electron identification and non-electron (π, μ, K) identification was carried out by cuts on the dE/dx signal in a momentum range between 0.3 and 3.0 GeV/c, for which there was an e-π separation of 5σ. The resulting sample of 15 events is shown in Figure 2.

Backgrounds

Backgrounds from several sources have been considered: i) $\gamma\gamma \to e^+e^-$, estimated by counting tracks in the tail of the dE/dx

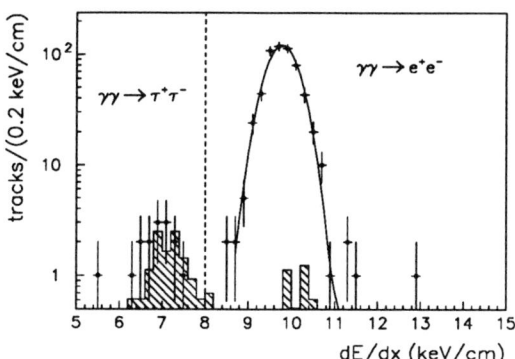

Figure 2. Projection of the energy loss spectrum for tracks in events in which the other track is identified as an electron. The shaded histogram corresponds to a Vermaseren Monte Carlo tau pair simulation normalized to the integrated luminosity of the data sample. The dashed line indicates where the cut was applied

distribution of the data; ii) $\gamma\gamma \to \mu^+\mu^-$ was estimated from the Monte Carlo; iii) events with larger multiplicities or with photon conversions with one or more tracks not reconstructed; iv) $\gamma\gamma \to K^+K^-$ or $\gamma\gamma \to p\bar{p}$, which enter if a kaon or proton lies in the e-K or e-p crossover regions of the dE/dx plot; v) $e^+e^- \to \tau^+\tau^-(\gamma)$ events, A simple addition of these sources leads to an estimated background of 4.8±2.1 events. The $\tau^+\tau^-$ sample is thus 10.2±4.4 events, in good agreement with the Monte Carlo estimate of 12.2 ± 1.1 ± 1.9 events. Here the systematic error is taken to be identical to the estimated systematic error for the muon reaction and includes a conservative estimate of the luminosity. More precise comparisons between the $\gamma\gamma \to \mu^+\mu^-$ and $\gamma\gamma \to \tau^+\tau^-$ reactions await a larger data sample.

HADRON PRODUCTION

The appropriate kinematic variables have been discussed above. In contrast to the leptonic case, the hadrons present both theoretical and experimental problems. The theoretical problem[6] revolves around the separation of the "point-like" part of F_2, calculable in QCD, from the "hadronic" part, for which there is no clearcut calculational prescription. The experimental problem arises from the fact that not all of a hadronic event can be seen because of detector limitations. Hence the visible energy of the event W_{vis} is less than the true energy W_{true}, and x_{vis} is greater than x_{true}. This difficulty can be dealt with by "unfolding"[7] the data to recover the true values, but the efficacy of this step depends on having a good Monte Carlo description of the data.

Event Selection

The hadronic event trigger is essentially the same as that for the tau pairs. The efficiency of this trigger is ~ 93%, and has been shown to be independent of x. The tagging electron must as always be in the fiducial volume of the Forward Detector, and its energy must exceed 0.66 E_{beam}. As an antitag requirement, no cluster exceeding 0.25 E_{beam} can be in the opposite hemisphere.

In the central detector, there must be three or more charged tracks, at least one of which must have a momentum greater than 1 GeV/c. The tracks must satisfy certain quality cuts, and their invariant mass W_{track} must be greater than 2 GeV/c^2. Cuts on transverse momentum in and out of the tag plane are used to remove residual backgrounds. There are 582 events that satisfy these requirements from the combined 1990-1991 data sample. The average Q^2 for the sample is 10.2 GeV$^2/c^2$.

Monte Carlo Simulation

A Monte Carlo program F2GEN has been developed which generates events according to given F_2 structure functions. For the hard contribution we have used the AMY parametrization[8] of F_2 as given by Kapusta's "all order QCD" approach,[9] with a cutoff parameter p_t^0 which represents the lowest allowed transverse momentum of the virtual quark. It is assumed that the contribution with $p_t < p_t^0$ comes entirely from VMD (Vector Meson Dominance) of the hadronic part of the photon, with $F_2/\alpha = 0.2(1-x)$. For VMD, the decay of the $q\bar{q}$ system is assumed to be either pointlike, for a fraction f_{point} of all VMD events, or else peripheral, with the mean transverse momentum of the outgoing quark jet $\langle p_t \rangle = 300$ MeV/c. Additionally $c\bar{c}$ and $\tau^+\tau^-$ events are generated with the Vermaseren Monte Carlo. Fragmentation is effected with the LUND string fragmentation model.

Events from all five samples (QCD, VMD pointlike, VMD peripheral, $c\bar{c}$, and $\tau^+\tau^-$) are generated according to the measured experimental luminosity, processed through standard OPAL simulation and reconstruction programs, and analyzed with the same selection criteria as the real data. The two parameters p_t^0 and f_{point} are then varied to make the combined sample agree with the experimental x_{track} distribution. (At this time only charged tracks are included in the analysis.) Good agreement can be reached with the values $p_t^0 \approx 0.1$ and $f_{point} \approx 0.2$. Comparison of the distribution of W_{track} with the Monte Carlo result with these parameters is shown in Figure 3. Unfolding of the data using this Monte Carlo is currently under way and the structure function over the range $4 < Q^2 < 30$ GeV$^2/c^2$ will be reported shortly.

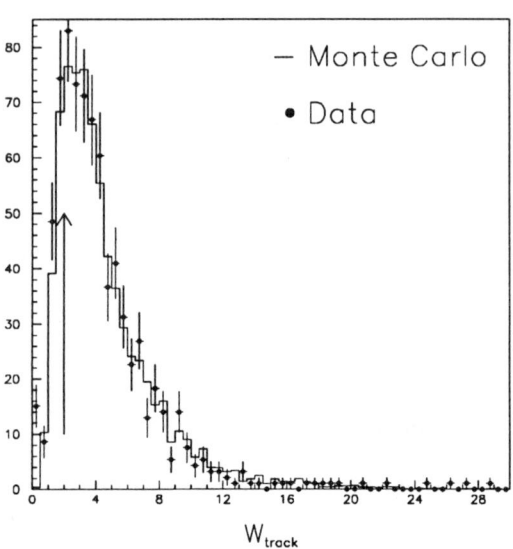

Figure 3. Comparison of data and Monte Carlo distributions of W for tracks. The arrow shows the cut on events with $W < 2$ GeV/c^2.

REFERENCES

1. M.P. Cain, et al., (PEP-9 Collaboration), Phys. Lett. B147 p. 232-6, (1984).
2. Ch. Berger, et al., (PLUTO Collaboration), Z. Phys. C27, p. 249, (1985).
   W. Bartel, et al., (JADE Collaboration), Z. Phys. C30, p. 545, (1986)
   B. Adeva, et al., (MARK-J Collaboration), Phys. Rev. D38, p. 2665, (1988)
   H.-J. Behrend, et al., Z. Phys. C43, p. 1, (1989).
3. J.A.M. Vermaseren, Nucl. Phys. B229, p. 347, (1983).
4. V.M. Budnev, et al., Phys. Rep. 15, p. 181, (1975).
5. W.G.J. Langeveld, Ph.D. Thesis (unpublished), Rijksuniversiteit Utrecht, p. 63-8, (1985).
6. J.H. Field, F. Kapusta, and L. Poggioli, Z. Phys. C36, p. 121-9, (1987).
7. V. Blobel, CERN Report 85-09.
8. T. Sasaki, et al., (AMY Collaboration) Phys. Lett. B252, p. 491-8, (1990).
9. F. Kapusta, Z. Phys. C42, p. 225-9, (1989).

Jets, Fragmentation, Tests of QCD

Parallel Session 7

"An Evening in Ellis County" Picnic

Conveners: J. Huth (Fermilab)
V. Khoze (Durham)

LARGE p_T PRODUCTION OF DIRECT PHOTONS AND π^0 MESONS AT 500 GeV/c *

M. Zieliński
Department of Physics and Astronomy
University of Rochester
Rochester, NY 14627, U.S.A.
Representing the E706 Collaboration.

Abstract

We report results from Fermilab experiment E706 on the production of direct photons and neutral mesons at large transverse momenta.[1] The data are for 500 GeV/c π^- and p beams incident on Be and Cu targets. Cross sections for π^0 and direct-photon production over the kinematic range $3.5 < p_T < 10$ GeV/c and $-0.7 < y < 0.7$ are compared with next-to-leading log QCD calculations using several recent sets of parton distribution functions.

INTRODUCTION

In the past few years, there has been a significant effort to improve the understanding of the parton structure of hadrons.[2] This activity was to much extent stimulated by the availability of several new sets of high-accuracy data on Deep Inelastic Scattering (DIS) of leptons (from BCDMS, EMC and CDHSW groups, and more recently from NMC and CCFR), production of Drell-Yan (DY) pairs (E605) and direct photons (WA70, E706, UA6, UA2, CDF). Several sets of parton distributions, based on comprehensive ("global") analyses of the new data, became available from the MRS et al.,[3] MT[4] and GRV[5] groups. While some of the previous fits (MRS et al., ABFOW/ABFKW[6]) used the direct photon measurements from WA70 (together with DIS and DY data), the forthcoming parametrizations by the CTEQ group[7] will also include the E706 data.

In the framework of global analyses, direct photon production provides strong constraints on the gluon distribution and, in addition, allows tests of QCD phenomenology. It offers more simplicity compared to other large-p_T processes (e.g., jet or inclusive hadron production) and is directly sensitive (unlike DIS) to the gluon distribution. In particular: 1. The number of the contributing diagrams is smaller – to lowest order only the Compton ($qg \to q\gamma$) and the annihilation ($q\bar{q} \to g\gamma$) subprocesses contribute; 2. There is no fragmentation to unfold in the case of the inclusive photon spectrum (except for the bremsstrahlung contribution); 3. A complete next-to-leading log (NLL) QCD calculation is available.[8]

Although more complex, theoretical description of high-p_T production of mesons (π^0 being of interest here) has reached a similar level of maturity: a full NLL calculation of relevant QCD diagrams[9] and a NLL parametrization of π^0 fragmentation functions[10] became available recently, allowing additional tests of the theory.

---

*This work was supported by the DOE, NSF and the UGC of India.

E706

Fermilab experiment E706 is designed to measure production of direct photons and associated particles. The apparatus features a large ($\approx$3m diameter) liquid argon calorimeter with a finely segmented electromagnetic section (EMLAC) and a hadronic part, and a charged particle spectrometer employing silicon microstrip detectors in the target region and multiwire proportional chambers downstream of an analysis magnet. The experiment accumulated $\approx 12\text{pb}^{-1}$ of π^- beam data and $\approx 9\text{pb}^{-1}$ of proton beam data at 500 GeV/c, and $\approx 11\text{pb}^{-1}$ of proton beam data at 800 GeV/c, on a combination of Be, Cu and H_2 targets. The results presented in this paper are obtained from only $\approx 0.5\text{pb}^{-1}$ of π^-Be and $\approx 0.8\text{pb}^{-1}$ of pBe data acquired during the 1987-88 Fermilab fixed target run at 500 GeV/c.

ENERGY SCALE

Because of the steep fall-off of the cross sections with p_T, the inclusive large-p_T measurements are very sensitive to small variations of the absolute energy scale of the detector. Therefore, particular care has been applied in the determination of the energy scale of the EMLAC. For this calibration, we used events in which one of the high-energy π^0 photons converted into an e^+e^- pair. The energy response of the EMLAC to electrons of known momentum was measured (the tracking system was calibrated using $K_S^0 \to \pi^+\pi^-$ and $J/\psi \to \mu^+\mu^-$ decays), and used to determine an empirical correction for energy losses. A corresponding Monte Carlo correction was applied to photons. Using this energy scale, the position of the peak in the $\pi^0 \to \gamma e^+e^-$ mass distribution (Fig. 1) does not exhibit any residual dependence on the γ energy (electrons were measured using the tracking system), and the $\pi^0 \to \gamma\gamma$, $\eta \to \gamma\gamma$ and $\omega \to \pi^0\gamma$ masses are

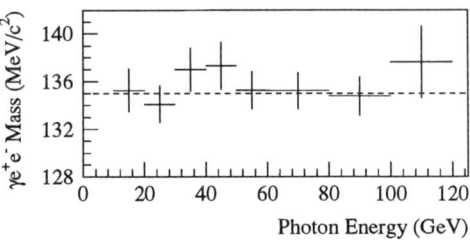

Figure 1. Mean value of the π^0 mass for events in which one of the decay photons converted, determined using the tracking momentum for e^- and e^+ and the EMLAC energy for γ.

all within 1% of their accepted values (Fig. 2, ω data not shown).

DIRECT PHOTON BACKGROUNDS

The major background to the direct photon signal in E706 comes from asymmetric two-photon decays of π^0 mesons, where the smaller photon is not detected; additional background is caused by electromagnetic decays of η, ω, η' and K_S^0 mesons. These backgrounds were subtracted statistically, based on Monte Carlo studies of single photon candidates expected from the measured yields of neutral mesons.

Another source of background, typical to fixed target experiments, is due to muons in the beam halo that occasionally shower in the calorimeter, simulating large-p_T triggers. This background was rejected successfully on-line (using two veto walls upstream of the target) and off-line (using additional veto wall requirements, and timing and direction information for the trigger shower obtained from the EMLAC).

RESULTS

The inclusive π^0 cross section per nu-

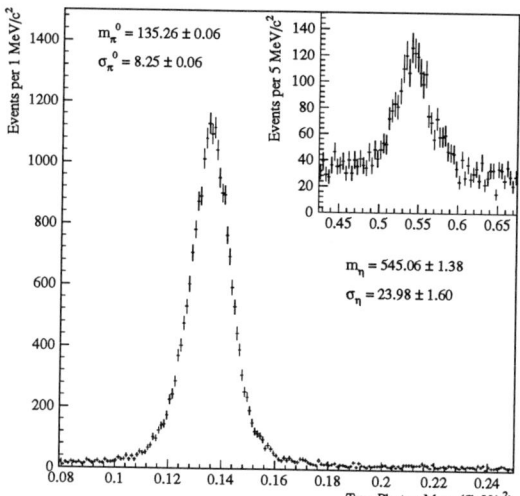

Figure 2. Two photon mass distribution in the π^0 and η (insert) regions, for $p_T > 3.5$ GeV/c and photon energy asymmetry < 0.75.

cleon for pBe and π^-Be is presented in Fig. 3, averaged over the rapidity range $-0.7 < y < 0.7$. The curves represent the NLL QCD calculation of Aversa et al.,[9] using the ABFOW/ABFKW parton distributions and a preliminary NLL π^0 fragmentation parametrization,[10] for two choices of scales $Q^2 = p_T^2$ and $Q^2 = p_T^2/4$. The theoretical calculations (which assume a nucleon target) were approximately corrected for the nuclear dependence of the π^0 production (the factor $A^{\alpha-1} \approx 1.2$ for $A = 9$ was used, measured using our Be and Cu data). The agreement of the measurements with the calculation for $Q^2 = p_T^2/4$ is good, but the large dependence of the predictions on the choice of scale may indicate that even the NLL QCD description may be not fully adequate.

The inclusive direct photon cross section per nucleon for pBe and π^-Be is presented in Fig. 4, averaged over the rapidity range $-0.7 < y < 0.7$. The curves are NLL QCD calculations of Aurenche et al.,[8] using the AB-

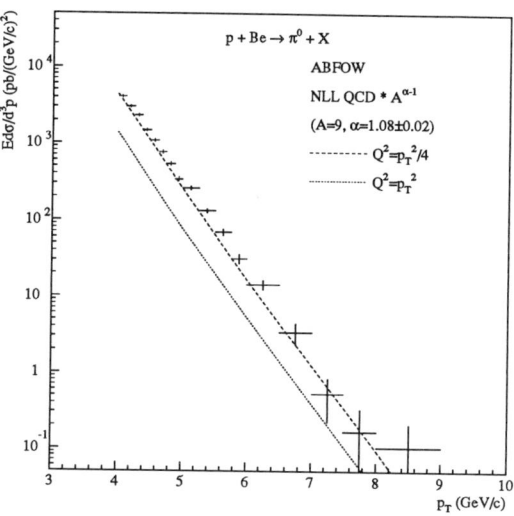

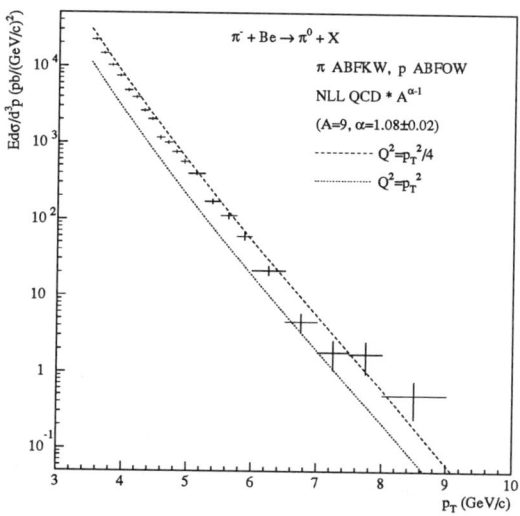

Figure 3. Invariant cross section per nucleon for π^0 production for p and π^- interactions with Be. The curves represent QCD predictions as described in the text.

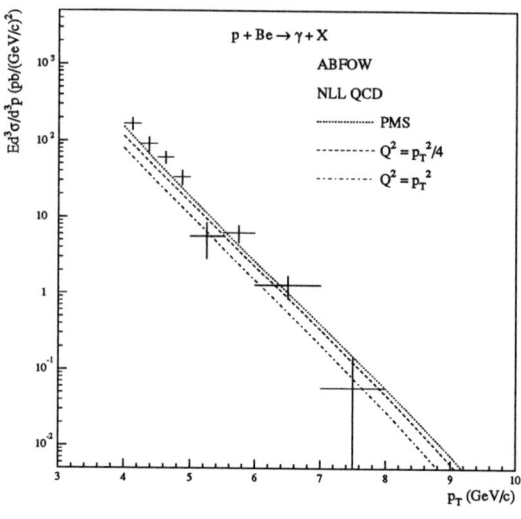
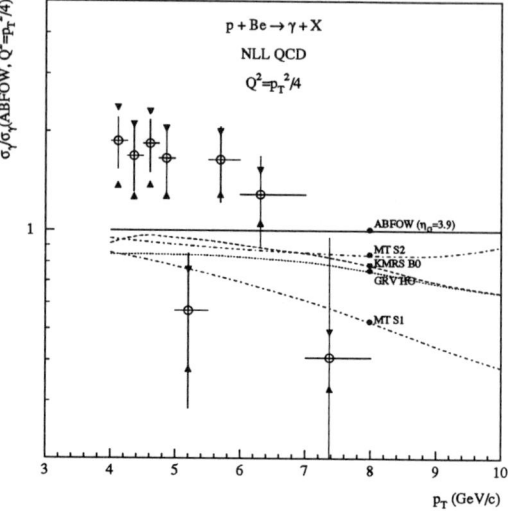
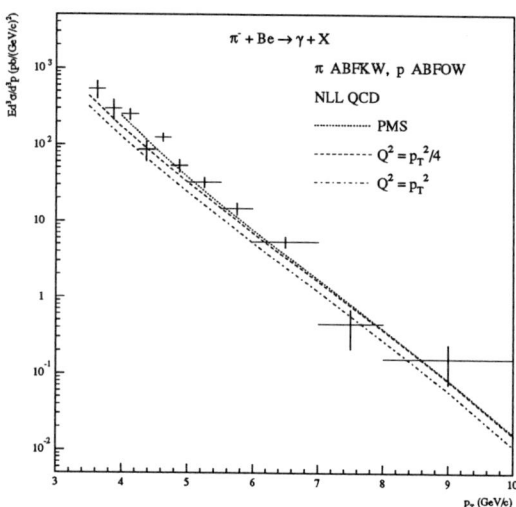
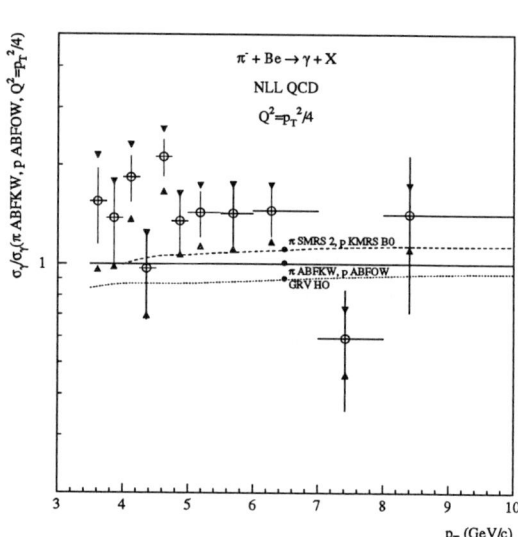

Figure 4. Invariant cross section per nucleon for γ production for p and π^- interactions with Be. The curves represent QCD predictions as described in the text.

Figure 5. Comparison of cross sections for direct-photon production evaluated using various input parton distribution functions for pBe and π^-Be interactions. Vertical bars correspond to statistical and the triangles to systematic uncertainties.

FOW/ABFKW parton distributions for scales $Q^2 = p_T^2$, $Q^2 = p_T^2/4$, and the "optimized" Q^2 obtained using the principle of minimum sensitivity (PMS). (In the photon case, no nuclear dependence correction was applied; the data from the 87-88 run, within the large error bars, are consistent with linear A-dependence of direct photon production.) As before, a strong dependence of the theoretical calculation on the choice of scales is observed. The agreement with data seems best using the PMS determination of scales (PMS tends to yield lowest values for the scales, and hence highest cross section).

In Fig. 5 we present the ratio of data to the calculation using the ABFOW/ABFKW fits as function of p_T, together with calculations using other recent parametrizations of parton distributions from the KMRS/SMRS, MT and GRV groups (the scales are $Q^2 = p_T^2/4$ in all cases). The various calculations are in agreement, typically within 20%, but low relative to data, for this choice of scales. Both experimental and theoretical uncertainties tend to cancel in ratios of cross sections. This is demonstrated in Fig. 6, where the data and theory agree well for $\sigma(\pi^-\text{Be} \to \gamma X)/\sigma(p\text{Be} \to \gamma X)$, and the scale dependence of the prediction is considerably smaller than for individual cross sections.

I thank P. Aurenche, J. Owens, J. Qiu and W.-K. Tung for discussions, and J. Ph. Guillet and M. Werlen for providing the NLL π^0 calculation programs.

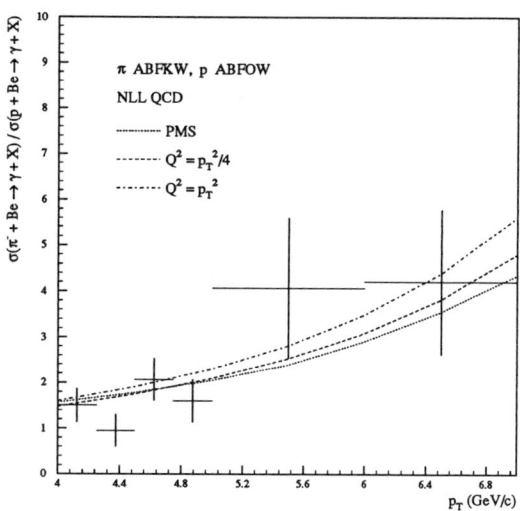

Figure 6. Ratio of π^-Be to pBe cross sections for direct-photon production. Theoretical curves are for ABFOW/ABFKW parton distributions.

REFERENCES

1. G. O. Alverson et al., Phys. Rev. Lett. 68, 2584 (1992), and Phys. Rev. D45, R3899 (1992).
2. A comprehensive collection of fits to the parton distributions has been assembled in CERN's PDFLIB library; the documentation points to original publications.
3. J. Kwiecinski, A. D. Martin, R. G. Roberts and W. J. Stirling Phys. Rev. D42, 3645 (1990); P. J. Sutton et al., Phys. Rev. D45, 2349 (1992).
4. J. G. Morfin and W.-K. Tung, Z. Phys. C52, 13 (1991).
5. M. Glück, E. Reya and A. Vogt, Z. Phys. C53, 127 (1992), ibid. C53, 651 (1992).
6. P. Aurenche et al., Phys. Rev. D39, 3275 (1989), and Phys. Lett. 233B, 517 (1989).
7. W.-K. Tung, in these Proceedings.
8. P. Aurenche et al., Phys. Lett. 140B, 87 (1984), and Nucl. Phys. B286, 509 (1987), ibid. B297, 661 (1988).
9. F. Aversa et al., Nucl. Phys. B327, 105 (1989).
10. J. Guillet, in Proceedings of the XXVIIth Rencontres de Moriond, Les Arcs, Savoie, France, March 1992, and private communication.

HEAVY FLAVOR PRODUCTION IN π^- –A COLLISIONS AT 530 GeV/c*

R. Jesik[3], V. Abramov[2], Yu. Antipov[2], B. Baldin[2], R. Crittenden[4], L. Dauwe[6], C. Davis[5], S. Denisov[2], A. Dyshkant[2], A. Dzierba[4], V. Glebov[2], H. Goldberg[3], A. Gribushin[4], V. Koreshev[2], J. Krider[1], A. Krinitsyn[2], R. Li[4], S. Margulies[3], T. Marshall[4], J. Martin[4], H. Mendez[3], A. Petrukhin[2], V. Sirotenko[2], P. Smith[4], J. Solomon[3], T. Sulanke[4], R. Sulyaev[2], F. Vaca[3], A. Zieminski[4]

E672 Collaboration

[1] Fermi National Accelerator Laboratory, Batavia, Illinois 60510
[2] Institute for High Energy Physics, Serpukhov, Russia
[3] University of Illinois at Chicago, Chicago, IL 60680
[4] Indiana University, Bloomington, IN 47405
[5] University of Louisville, Louisville, KY 40292
[6] University of Michigan at Flint, Flint, MI 48502

S. Blusk[7], C. Bromberg[4], P. Chang[5], B. C. Choudhary[2], W. H. Chung[7], L. de Barbaro[8], W. Dlugosz[5], J. Dunlea[8], E. Engels, Jr.[7], G. Fanourakis[8], G. Ginther[8], K. Hartman[6], J. Huston[4], V. Kapoor[2], C. Lirakis[5], F. Lobkowicz[8], S. Mani[1], J. Mansour[8], A. Maul[4], R. Miller[4], E. Pothier[5], R. Roser[8], P. Shepard[7], D. Skow[3], P. Slattery[8], L. Sorrell[4], N. Varelas[8], D. Weerasundara[7], C. Yosef[4], M. Zielinski[8]

E706 Collaboration

[1] University of California-Davis, Davis, California 95616
[2] University of Delhi, Delhi 11 00 07, India
[3] Fermi National Accelerator Laboratory, Batavia, Illinois 60510
[4] Michigan State University, East Lansing, Michigan 48824
[5] Northeastern University, Boston, Massachusetts 02115
[6] Pennsylvania State University, University Park, Pennsylvania 16802
[7] University of Pittsburgh, Pittsburgh, Pennsylvania 15260
[8] University of Rochester, Rochester, New York 14627

Abstract

We report recent results on a search for beauty-meson production by 530 GeV/c negative pions incident on Cu and Be targets via observing the B decay into J/ψ modes. Our data, taken during the 1990 Fermilab fixed-target run using a high-mass dimuon trigger, yielded 11×10^3 reconstructed $J/\psi \to \mu^+\mu^-$ events. A subset of these events have J/ψs coming from secondary vertices, a characteristic of heavy-quark decay. We also have observed a few candidate events in the exclusive decay modes $B^+ \to J/\psi + K^+$ and $B^0 \to J/\psi + K^{0*}$ (and their charge conjugates), in which the B mass and lifetime are fully reconstructed.

---

*Work supported by DOE, NSF, and IHEP (Serpukhov)

Introduction

Experimental observation of hadronic heavy-quark production yields important information on strong interactions. Fermilab experiment E672 was designed to look at processes that result in muons in the final state: specifically, Drell-Yan and heavy-quark (b,c) production. QCD diagrams up to order α_s^3 for heavy quark production have recently been calculated, and predictions have been made for the beauty-quark total cross section.[1,2] One of this experiment's goals is to measure this cross section.

Since the first observation of a single $B\bar{B}$ pair by WA75[3], experimental data on B-meson hadroproduction has come from four experiments: WA78[4], UA1[5], CDF[6], and E653[7]. All of the experiments observe Bs through decays containing muons. CDF and UA1 have reconstructed invariant-mass signals in exclusive decay modes involving the J/ψ. With our dimuon trigger, we can observe the J/ψ decay modes of the B-meson, and also dimuons from semi-leptonic $B\bar{B}$ pair decays. This paper reports our analysis of $B \rightarrow J/\psi + X$ decays.

Apparatus and Data Taken

Experiment E672 uses an open-geometry detector optimized to study high-mass dimuon production. The apparatus is located in Fermilab Meson-West beam line. It also serves E706, an experiment studying the production of direct photons. The outputs of all detectors are available to both experiments. The E672 muon detector consists of a toroid magnet, two petal-shaped scintillation-counter hodoscopes (daisy counters), and six PWCs with a total of 20 signal planes.[8] The Meson-West apparatus also includes a liquid-argon calorimeter, a forward steel/scintillator calorimeter, and a charged-particle spectrometer consisting of a 16-plane silicon-microstrip vertex detector (SSD), a dipole magnet, 2 strawtube chambers (16 planes), and 4 PWCs (16 planes).[9] The acceptance of the combined detector for muon pairs is non-zero only in the forward direction of the beam-nucleon c.m. frame ($x_F > 0$), peaking at $x_F = 0.30$. The acceptance in p_T for J/ψs is flat up to 3.5 GeV/c.

Dimuon pretriggers were formed by coincidences between the daisy-counter hodoscopes.

A fast (10 μs) processor was used to calculate, and trigger on, the effective mass of track combinations in the muon PWCs. Data was taken during the 1990 Fermilab fixed-target run using a 530 GeV/c π^- beam incident on Cu and Be targets. Approximately 5.6×10^6 dimuon triggers were accumulated during this run, with a corresponding luminosity of approximately 8 pb^{-1}. Figure 1 shows the reconstructed opposite-sign dimuon invariant-mass distribution in the J/ψ range. A two-resonance-plus-background fit to this sample yields $11{,}200 \pm 440$ J/ψs with a mass resolution of 85 MeV/c^2, and 341 ± 89 ψ's.

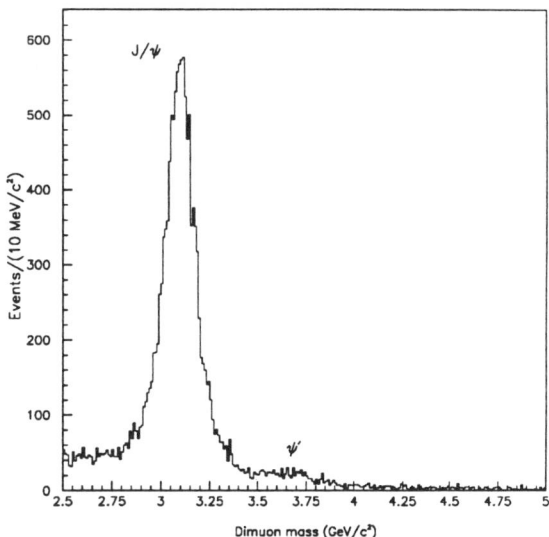

Fig. 1. $\mu^+\mu^-$ invariant mass in the J/ψ range.

J/ψs From Secondary Vertices

$B \rightarrow J/\psi + X$ decays are uniquely tagged by J/ψs emerging from secondary vertices. Although the branching ratio into J/ψ is small (1.12%)[10,11], this decay is advantageous in that the muons provide a clean trigger and the background under the J/ψ is small. Many of the decays into J/ψ have only charged particles in their final states. This offers the additional advantage that the B mass and lifetime can be fully reconstructed, giving a very clean signal. A significant background to secondary-vertex J/ψs from B decays is produced by events in which a high-momentum particle from the primary interaction interacts further downstream in the target and produces a J/ψ (secondary interactions). Events with J/ψs from vertices in regions where only air is present are particularly interesting because they do not suffer significantly from this background.

Of the 11×10^3 J/ψs from the 1990 sample, approximately 11% have more than one reconstructed vertex. In 64% of these events, there was a possibility that both muons came from a secondary vertex. A dimuon-oriented vertex refit was done for these events. The procedure was as follows: Dimuons in the mass range $2.85 \text{ GeV}/c^2 < M_{\mu\mu} < 3.35 \text{ GeV}/c^2$ were refit, constraining the mass to the Particle Data Group (PDG) J/ψ mass value of 3.097 GeV/c^2.[12] A three-dimensional, impact-parameter-minimizing vertex refit was then performed using the 1C-fit J/ψ track parameters. The resulting vertex was then used as a seed for an iterative vertex-fitting procedure that associated other charged tracks to this vertex. The reconstructed J/ψ vertex Z position distribution from the data is shown in Fig. 2a. The target consists of two 0.8 mm thick pieces of Cu followed by two pieces of Be, 3.71 cm and 1.12 cm thick, respectively. These elements are clearly resolved in the reconstructed vertex plot.

The sample was then searched for events in which this dimuon vertex was downstream of the primary vertex, yielding 857 events. In order to clean the sample, several cuts were applied to these events: Only events with at least three fully reconstructed tracks from the primary were kept (631 events passed this cut). Fiducial-volume cuts were made for both primary and secondary vertices (577 events passed this cut). A significance greater than 3 was required for both the longitudinal and transverse separations between the primary and secondary vertices, with significance defined as the separation divided by the combined vertex uncertainty. An absolute longitudinal separation of 3 mm was also required (121 events passed these cuts). Secondary J/ψ vertices with more than four associated hadrons were discarded in order to reduce the background from secondary interactions (73 events passed this cut). The primary- and secondary-vertex Z-position distributions for the seventy-three events passing these cuts are shown in Fig. 2b and 2c, respectively.

The background to this signal from false vertex

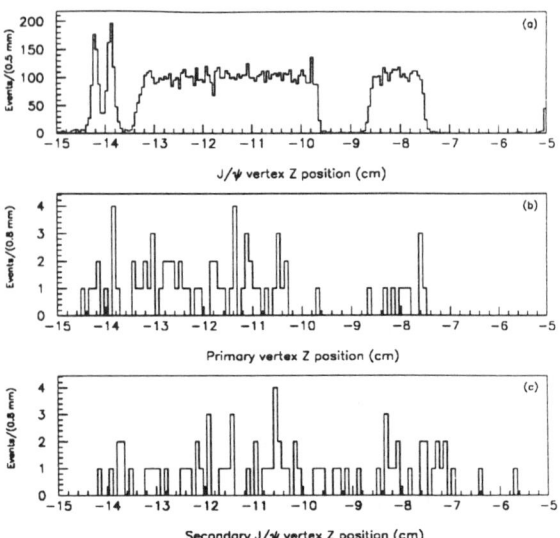

Fig. 2. Vertex Z position distributions: (a) J/ψ vertex position for all events; (b) primary vertex position for the 73 events passing the cuts; (c) secondary J/ψ vertex position.

reconstruction for this sample is estimated to be 4 ± 2 events.[13] Although the requirement of low secondary multiplicities certainly improves the likelihood of the J/ψ coming from a B decay, secondary-interaction events remain in the sample. Preliminary Monte Carlo simulation indicates that we should expect approximately 80 secondary-interaction events to pass the vertex quality cuts in the 11.2K J/ψ sample, with 25 to 40 also passing the multiplicity cut.[14]

There are two reasonably large regions in the fiducial volume of the target-SSD system where only air is present (-9.61 cm < Z < -8.62 cm, and -7.50 cm < Z < -5.55 cm). It is highly unlikely that events with secondaries in these gaps are from secondary interactions. The positions in the Z-Y plane of the J/ψ vertices in these gaps are shown in Fig. 3 (with Z error bars). There are ten events in which the J/ψ vertex is at least three standard deviations away from a region where target material is present. Assuming their vertices are distributed uniformly across the volume, 1 ± 1 of the estimated 4 ± 2 background events from fake vertex reconstruction will occur in the gaps. With these considerations, we report a preliminary signal of 9 ± 3 secondary-vertex J/ψ events from B decays in which the secondary vertex occurs in a region where

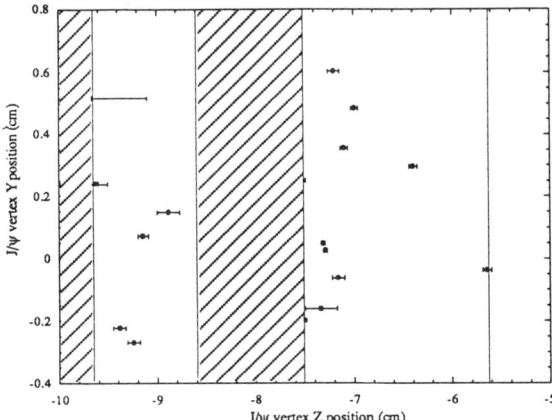

Fig. 3. Secondary J/ψ vertex position in the Y-Z plane showing events with vertices in gaps where no material is present. Hatched areas represent target material; the line at -5.6 cm represents a Mylar-foil window.

only air is present. Extrapolating from the nine gap events, we estimate 26 ± 10 $B \rightarrow J/\psi + X$ events in the entire fiducial volume.[13] This number added to the estimated number of events from secondary interactions (33 ± 7) and false vertex reconstruction (4 ± 2) gives a total of 63 ± 12 events, quite consistent with the 73 ± 9 events found. We are in the process of checking systematic effects, especially Monte Carlo based cuts and assumptions, in order to estimate the B-meson hadroproduction cross-section using this signal.

Exclusive B Decay Modes

The decay modes $B \rightarrow J/\psi + K$ and $B \rightarrow J/\psi + K^*$ have significant channels that contain only charged particles in the final state, allowing full reconstruction of the B mass and momentum. Together with the decay length given by the vertex reconstruction, these quantities fully specify the lifetime of the decaying particle. For the $B^{\pm} \rightarrow J/\psi + K^{\pm}$ decay modes, a three-prong vertex refit was performed on single charged tracks (assumed to be kaons) in combination with the 1C-fit-J/ψ muon tracks. The track's momentum vector was added to that of the J/ψ to form a candidate B momentum vector. This vector was projected back from the secondary to the primary vertex, and a transverse impact parameter (δ_v) was calculated.

The event was considered a B decay candidate if it passed the following cuts: Only hadrons with a transverse impact parameter less than 20 μm with respect to the secondary dimuon vertex were used. A quality cut ($\chi^2 < 4$) was imposed on the three-prong vertex fit. Monte Carlo-based momentum cuts were imposed on the candidate kaon decay track in order to reduce the background of hadrons from the underlying event: the track's p_T with respect to the beam axis was required to be greater than 0.5 GeV/c, and tracks with momentum greater than 150 GeV/c were discarded. All of the cuts of the secondary-vertex J/ψ analysis described previously were also used in this analysis, with a secondary-vertex hadron multiplicity requirement of two tracks or less. The final cut was that the total momentum vector of the secondary-vertex tracks had to point back to the primary vertex such that $\delta_v < 120$ μm. Four events from the secondary vertex J/ψ data sample survived these cuts. Their invariant-mass distribution is shown in Fig. 4a. Two events fall in the expected B-mass region, one from each of the candidate charged decay modes; $B^+ \rightarrow J/\psi + K^+$ and $B^- \rightarrow J/\psi + K^-$.

The search for the exclusive decay mode $B^0 \rightarrow J/\psi + K^{0*}$ (and its charge conjugate) proceeded as follows: K^{0*}'s were observed by their decay into $K^{\pm}\pi^{\mp}$ pairs. A kinematic criterion for

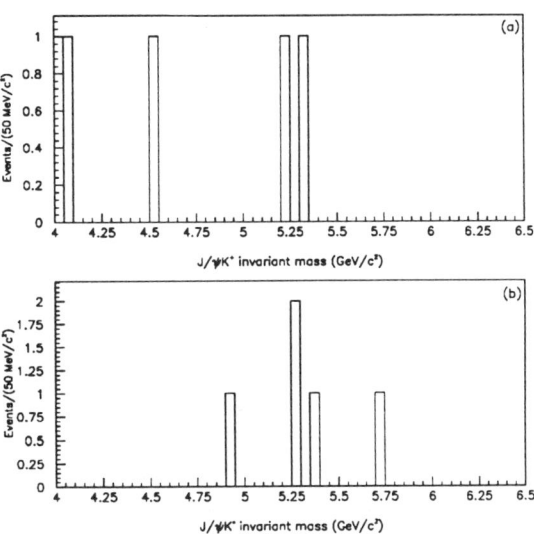

Fig. 4. Reconstructed invariant mass: (a) $J/\psi K^{\pm}$ combinations, (b) $J/\psi K^{0*}$ combinations.

kaon identification was used to reduce combinations: if one of the tracks in a pair had momentum greater than half of the other, it was assigned to be the kaon. Charged track pairs from the secondary-vertex J/ψ sample were then selected in the observed K^{0*} mass region and combined with the J/ψ in the same manner as in the charged-kaon analysis. Only $K^{\pm}\pi^{\mp}$ pairs having resultant p_T greater than 0.5 GeV/c with respect to the beam axis were kept, and the candidate B momentum was required to point back to the primary vertex to within 200 μm. Five events passed these cuts, three of which have reconstructed masses in the expected B-signal range (see Fig. 4b). The secondary vertex for one of the events in the B-mass range occurs in a gap region where only air is present.

The combined $J/\psi K^{\pm}$ and $J/\psi K^{0*}$ invariant-mass distribution is shown in Fig. 5. There is a clear excess of events near the nominal B-meson mass. A background analysis using primary vertex events subject to the same cuts as the B candidate sample shows no evidence for arbitrary enhancement in the B-mass region due to the imposed cuts.[13] The event characteristics, including the decay particle lifetimes, for the five B-decay candidates are shown in Table 1. The lifetimes for these events are consistent with the measured mean B lifetime of 1.29 ps.[12]

Table 1. B-candidate event characteristics, P is the B momentum.

Decay Mode	B Mass (GeV/c²)	P (GeV/c)	Decay Length (cm)	Lifetime (ps)
$B^+ \to J/\psi + K^+$	5.31	281	2.91	1.8
$B^- \to J/\psi + K^-$	5.24	163	0.49	0.5
$\bar{B}^0 \to J/\psi + \bar{K}^{0*}$	5.36	287	0.41	0.3
$B^0 \to J/\psi + K^{0*}$	5.29	210	0.51	0.4
$\bar{B}^0 \to J/\psi + \bar{K}^{0*}$	5.25	198	1.49	1.3

Conclusion

We have found candidates for B-meson decays in modes involving the J/ψ. In seventy-three events, a J/ψ was observed to emerge from a low-multiplicity secondary vertex, a characteristic of b-quark decay. While there is a background to this signal from secondary interactions, in ten of these events the J/ψ vertex occurs in air gap regions where these interactions are highly unlikely. The reconstructions of the exclusive B decays resulting in $J/\psi + K^{\pm}$ and $J/\psi + K^{0*}$ final states, while small in number, provide a clean signal in the invariant-mass distribution. The studies necessary to extract a cross section measurement from these signals are currently in progress.

References

1. P. Nason, S. Dawson, and R. K. Ellis, Nucl. Phys. **B303**, 607 (1988); **B327**, 49 (1989).
2. E. L. Berger, "Heavy Flavor Production", ANL-HEP-CP-88-26, June 1988.
3. WA75 Collab., J. Albanese et al., Phys. Lett. **158B**, 186, (1985)
4. WA78 Collab., M. Catanesi et al., Phys. Lett. **187B**, 431, (1987)
5. UA1 Collab., C. Albajar et al., Phys. Lett. **186B**, 237, (1987); **256B**, 163, (1991); **273B**, 540, (1991)
6. CDF Collab., F. Abe et al., Phys. Rev. Lett. **68**, 3404, (1992)
7. E653 Collab., R. Sidwell, in *The Vancouver Meeting*, eds. David Axen et al., (World Scientific, 1992) p. 516
8. V. Abramov et al., FERMILAB-Pub-91/62-E, Mar., 1991.
9. G. Alverson et al., "Production of Direct Photons and Neutral Mesons at Large Transverse Momenta by π and p Beams at 500 GeV/c", in preparation
10. ARGUS Collab., H. Albrecht et al., Z. Phys. **C48**, 543, (1990)
11. CLEO Collab., D Bortoletto et al., Phys. Rev. **D45**, 145, (1992)
12. Review of Particle Properties, Phys. Rev. **D45**, (1992)
13. R. Jesik, "B-Physics at E672", E672 Note, May 1992
14. V. Sirotenko, "Estimation of Background to B-Signal Due to J/ψ Production in Secondary Interactions", E672 Note, April 1988

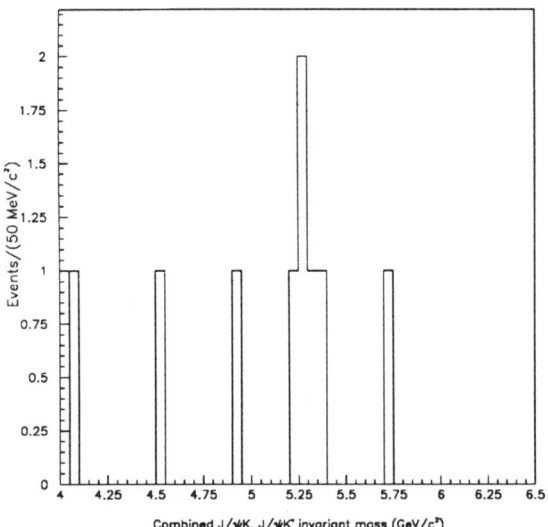

Fig. 5. Combined $J/\psi K^{\pm}$, $J/\psi K^{0*}$ invariant mass.

POLARIZATION DENSITY MATRIX AT $O(\alpha_s^2)$ OF HIGH Q_T $W'S$ IN HADRONIC COLLISIONS *

Erwin Mirkes

Institut für Theoretische Teilchenphysik
Universität Karlsruhe
Kaiserstr. 12, Postfach 6980
7500 Karlsruhe 1, Germany

Abstract

We propose the measurement of the angular decay distribution of leptons from W's produced at high-q_T in hadronic collisions. The angular distribution is characterized by nine angular coefficients A_i corresponding to nine polarization density matrix elements of the W. We show that one can measure six of these angular coefficients without the complete reconstruction of the decay kinematics due to the unobservability of the neutrino. We present $O(\alpha_s^2)$ results of the transverse momentum distribution of these coefficients at zero rapidity of the W using Morfin-Tung parton distribution functions.

The measurement of the angular distribution of leptons from $W's$ provides a detailed test of the production mechanism (Drell-Yan process including QCD corrections) of polarized gauge bosons [1]. From the fact that the gauge boson is produced at high-transverse momentum q_T one can define an event plane spanned by the beam and the gauge boson's momentum direction which provides a reference plane for a detailed study of lepton-hadron correlation effects. Technically the physics of lepton-hadron correlations is described by contraction of the lepton tensor $L_{\mu\nu}$ with the hadron tensor $H^{\mu\nu}$

$$H^{\mu\nu} L_{\mu\nu}$$

where $L_{\mu\nu}$ acts as an analyzer of the polarization of the W. It is thus possible to test the underlying production dynamics encoded in the hadron tensor $H_{\mu\nu}$ in a much more detailed way than is possible by rate measurements alone.

The lowest order QCD processes which contribute to large q_T-W production are:

$$H^{\mu\nu}(\text{born}): \begin{array}{rcl} q + \bar{q} & \to & W + G \\ q + G & \to & W + q \end{array} \quad (1)$$

At $O(\alpha_s^2)$ we have the following tree and loop processes that contribute to $H^{\mu\nu}$

$$H^{\mu\nu}(\text{tree}): \begin{array}{rcl} q + \bar{q} & \to & W + G + G \\ q + \bar{q} & \to & W + q + \bar{q} \\ q + G & \to & W + q + G \\ q + q & \to & W + q + q \\ G + G & \to & W + q + \bar{q} \end{array} \quad (2)$$

$$H^{\mu\nu}(\text{loop}): \begin{array}{rcl} q + \bar{q} & \to & W + G \\ q + G & \to & W + q \end{array} \quad (3)$$

To obtain the q_T distribution of the W one has to integrate over the phase space of the final state partons in (1)-(3) with the W held fixed at a given q_T. The analytical results are too lengthy to be reproduced here. They can be found in [1]. The general structure of the

---

*Acknowledgements: I thank J.G. Körner and G.A. Schuler for their advice in this calculation.

hadronic tensor can be readily exhibited by writing down the most general covariant expansion. One has (q denotes the momentum of the W):

$$\begin{aligned}
H^{pc}_{\mu\nu} &= H_1 \left(g_{\mu\nu} - \frac{q_\mu q_\nu}{q^2} \right) \\
&+ H_2\, \tilde{p}_{1\mu} \tilde{p}_{1\nu} \\
&+ H_3\, \tilde{p}_{2\mu} \tilde{p}_{2\nu} \\
&+ H_4\, (\tilde{p}_{1\mu} \tilde{p}_{2\nu} + \tilde{p}_{1\nu} \tilde{p}_{2\mu}) \\
&+ i H_5\, (\tilde{p}_{1\mu} \tilde{p}_{2\nu} - \tilde{p}_{1\nu} \tilde{p}_{2\mu}) \quad (4)
\end{aligned}$$

$$\begin{aligned}
H^{pv}_{\mu\nu} &= H_6\, i\,\epsilon(\mu\nu p_1 q) \\
&+ H_7\, i\,\epsilon(\mu\nu p_2 q) \\
&+ H_8\, (\tilde{p}_{1\mu}\, \epsilon(\nu p_1 p_2 q) + \mu \leftrightarrow \nu) \\
&+ H_9\, (\tilde{p}_{2\mu}\, \epsilon(\nu p_1 p_2 q) + \mu \leftrightarrow \nu)
\end{aligned}$$

where we have employed current conserved momenta variables $\tilde{p}_{i\mu} = p_{i\mu} - (p_i q)/q^2 q_\mu$. Note that H_5, H_8 and H_9 receive only contributions from the absorptive parts of the one-loop diagrams in (3).

An equivalent representation of the hadron tensor is obtained in the so called helicity basis by calculating the helicity projections

$$H_{mm'} = \epsilon^*_\mu(m) H^{\mu\nu} \epsilon_\nu(m') \quad (5)$$

$(m, m' = +, 0, -)$ where

$$\begin{aligned}
\epsilon_\mu(\pm) &= \tfrac{1}{\sqrt{2}} (0; \pm 1, -i, 0) \\
\epsilon_\mu(0) &= (0; 0, 0, 1)
\end{aligned} \quad (6)$$

are the polarization vectors for the W defined with respect to the Collins-Soper frame (see below): The resulting 3×3 matrix $H_{mm'}$ represents the *polarization density matrix* of the gauge boson which is being analyzed in the subsequent leptonic decay:

$$H_{mm'} = \begin{pmatrix} H_{++} & H_{+0} & H_{+-} \\ H_{0+} & H_{00} & H_{0-} \\ H_{-+} & H_{-0} & H_{--} \end{pmatrix} \quad (7)$$

The most general angular decay distribution of the leptons from a gauge boson produced through the subprocesses in (1)-(3) is given by

$$\frac{d\sigma}{dy\, dq_T^2\, d\cos\theta\, d\phi} = \frac{d\sigma_{U+L}}{dy\, dq_T^2} \frac{3}{8} \frac{1}{2\pi} \Bigg[$$

$$\text{p.c.} \begin{cases} & (1 + \cos^2\theta) \\ + \tfrac{1}{2} A_0 & (1 - 3\cos^2\theta) \\ + A_1 & \sin 2\theta \cos\phi \\ + \tfrac{1}{2} A_2 & \sin^2\theta \cos 2\phi \\ + A_7 & \sin\theta \sin\phi \end{cases} \quad (8)$$

$$\text{p.v.} \begin{cases} + A_3 & \sin\theta \cos\phi \\ + A_4 & \cos\theta \\ + A_5 & \sin^2\theta \sin 2\phi \\ + A_6 & \sin 2\theta \sin\phi) \end{cases} \Bigg]$$

Here θ and ϕ are the polar and azimuthal angles of the leptons (or jets) in the gauge boson rest frame and y denotes the rapidity of the gauge boson. Note that the total (angle integrated) rate σ^{U+L} is factored out from the r.h.s. of eq. (8). The angular coefficients A_i are linearly related to the density matrix elements in (7) (for details see [1]) and to the structure functions H_i in (4).

In general it will not be possible to reconstruct the W rest frame from the measured three-momentum of the charged lepton and the missing transverse momentum approximating the neutrino p_T. However, there is a frame where one can obtain θ and ϕ in (8) uniquely modulo a sign ambiguity in $\cos\theta$ unaffected by the unknown longitudinal momentum of the neutrino. This is the Collins-Soper (CS) frame [2] where the initial $P\bar{P}$-pair lies in the xz-plane and the z-axis bisects the angle between the proton and negative antiproton momenta. The components of the lepton momentum in the CS frame in terms of measurable lab momenta can be found in sec. 2 of [1]. The $\theta \leftrightarrow \pi - \theta$ ambiguity in the polar angle implies that only the angular coefficients A_0, A_2, A_3, A_5 and A_7 in (8) can be determined in a W production experiment.

The angular coefficients A_0, A_1, A_2 and A_7 receive contributions from the parity conserving (p.c.) part of the hadron tensors $H_{\mu\nu}$ i.e. from H_1, H_2, H_3, H_4 and H_5 in (4). The remaining four A_3, A_4, A_5, A_6 are proportional to the parity violating (p.v.) part of $H_{\mu\nu}$. They change sign under parity (P) transformation. The relevant coupling coefficients are [1]:

$$\begin{aligned} A_0, A_1, A_2 &\approx (v_l^2 + a_l^2)(v_q^2 + a_q^2) \\ A_3, A_4 &\approx v_l a_l v_q a_q \\ A_5, A_6 &\approx (v_l^2 + a_l^2) v_q a_q \\ A_7 &\approx v_l a_l (v_q^2 + a_q^2) \end{aligned} \quad (9)$$

Here $v_q(v_l)$ and $a_q(a_l)$ denotes the vector and axial vector coupling of the gauge boson to the quark (lepton) respectively. Note that $A_5 - A_7$ are so called "T-odd" observables. They are linearly related to H_5, H_8 and H_9 in (4) and obtain only contributions from the absorptive part of the one-loop amplitudes in (3). As a result the "T-odd" observables are zero at $O(\alpha_s)$. Note also that new CP-violating interactions beyond the standard model would also give contributions to $A_5 - A_7$ [3].

We will now present some numerical results for the measurable angular coefficients A_0, A_2, A_3, A_5 and A_7 in the Collins-Soper frame for $y = 0$. First numerical results of the y integrated angular coefficients A_0 and A_2 for the dominant $O(\alpha_s^2)$ contributions are given in [4]. To be specific we shall evaluate the q_T distribution for the angular coefficients A_i for W^+ production at $\sqrt{S} = 1.8$ TeV for the contributing parton subprocesses in eq. (1)-(3). We use the parton density parametrization set 3 of Morfin Tung [5].

We identify the scales used in the coupling constant and in the parton distribution function and set them equal to $\mu^2 = (m_W^2 + q_T^2)/2$.

Fig. 1 shows the q_T distribution of A_0, A_2 and A_3 in $O(\alpha_s)$ (dashed) and $O(\alpha_s^2)$ (solid). A_0 and A_2 are increasing functions of q_T reaching about 0.9 at high q_T. These two coefficients are simply related by $A_0 = A_2$ at the Born level [6]. Turning to the NLO corrections to A_0 and A_2 (including all parton subprocesses in (1)-(3)) one finds that there are positive corrections to A_0 whereas the $O(\alpha_s^2)$ contributions leads to negative corrections to A_2. The p.v. coefficient A_3 remains relatively small even at large q_T. This is different if one integrates over the whole y range. In this case A_3 reaches about 20% at $q_T = 100$ GeV [1], whereas A_0 and A_2 are not changed when integrating over the rapidity. Turning now to the $O(\alpha_s^2)$ contributions to the T-odd angular asymmetries A_5 and A_7 shown in fig. 2, one observes that the contributions are relatively small even at large q_T. The largest asymmetry is expected for A_5 where the magnitude of the asymmetry exceeds -0.016 at $q_T=100$ GeV. These angular distributions reverse the sign, when switching from $V - A$ to $V + A$ theory (see eq. (9)).

Note that in the limit $q_T \to 0$ all coefficients A_i vanish, implying that the polar angle distribution reduces to the $(1 + \cos\theta)^2$ prediction of the Drell Yan mechanism. It appears that the $O(\alpha_s^2)$ corrections to the angular coefficients A_i are not large (less than 10 %) when they are normalized to the NLO rate σ^{U+L}. However, from both the CDF and D0 experiments at FERMILAB large numbers of W's are expected over the next years [7]. Therefore it might become possible to test these QCD predictions in detail.

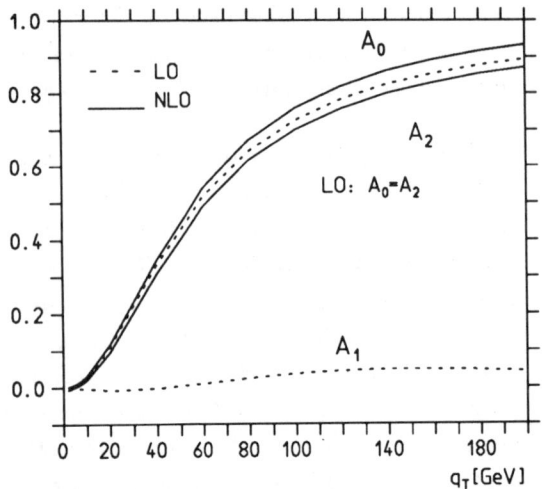

Figure 1
Angular coefficients A_0, A_2 and A_3 at $y = 0$ as a function of q_T for $p\bar{p} \to W^+ + X$ at $\sqrt{S} = 1.8$ TeV. Shown are the ($O(\alpha_s)$) (dashed) and $O(\alpha_s^2)$ corrected (solid) contributions.

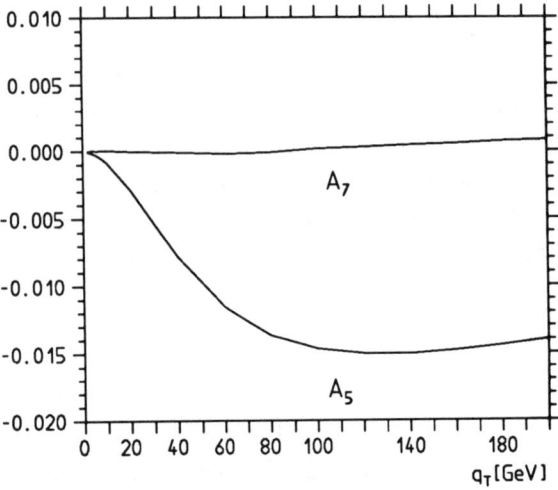

Figure 2
"T-odd" angular coefficients A_5 and A_7 at $y = 0$ as a function of q_T for $p\bar{p} \to W^+ + X$ at $\sqrt{S} = 1.8$ TeV.

REFERENCES

1. E. Mirkes, Karlsruhe preprint TTP92-12, to appear in *Nucl. Phys.* **B** (1992)

2. J.C. Collins and D.E. Soper, *Phys. Rev.* **D 16** (1977) 2219

3. A. Brandenburg, J.P. Ma, R. Münch and O. Nachtmann, *Z. Phys.* **C 51** (1991) 225

4. E. Mirkes, J.G. Körner and G.A. Schuler, *Phys. Lett.* **B 259** (1991) 151.

5. J. Morfin and Wu-Ki Tung, FERMILAB-Pub-90/74 (1990)

6. C. S. Lam and Wu-Ki Tung, Phys. Rev. D18 (1978) 2447; ibid D21 (1980) 2712; Phys. Lett. 80B (1979) 228.

7. H. Grassmann, FERMILAB-Conf-92/105 (1992)

QCD TESTS WITH CDF*

CDF Collaboration[†]
Presented by Brenna Flaugher
Fermi National Accelerator Laboratory
Batavia, Illinois U.S.A. 60510

Abstract

Measurement of scaling violations, the inclusive photon and diphoton cross sections as well as the photon-jet and jet-jet angular distributions are discussed and compared to leading order and next-to-leading order QCD. A study of four-jet events is described, with a limit on the cross section for double parton scattering. The multiplicity of jets in W boson events is compared to theoretical predictions.

INTRODUCTION

This paper presents a summary of the recent QCD results using the CDF detector[1] and data collected during the 1988-1989 running of the Fermilab proton-antiproton collider. In addition to the 4.4 pb^{-1} collected at $\sqrt{s}$ = 1800 GeV, 8.6 nb^{-1} of data were taken at $\sqrt{s}$ = 546 GeV to allow a measurement of the scaling violations predicted by QCD. Results are presented for the ratio of the scaled jet cross sections, and for the photon and diphoton cross sections. Measurement of the jet-jet and photon-jet angular distributions provide tests of QCD at larger pseudorapdities than is typically included the inclusive cross section measurements. Topological variables of four-jet events are compared to QCD predictions and a limit on the double parton scattering cross section is derived. Finally, a study of jet multiplicity in $W \to e+\nu$ and $W \to \mu+\nu$ events is presented.

JET IDENTIFICATION

CDF uses a cone algorithm for jet identification[2], where the radius of the cone is defined as R=$\sqrt{\Delta\eta^2 + \Delta\phi^2}$. The calorimeter energy falling within the cone is summed to give a single four-vector for each jet. Typically, a cone size of R=0.4, 0.7 or 1.0 is used.

Recently, $O(\alpha_s^3)$ calculations[3] for jet production have been performed in which a similar cone algorithm is employed. Cones are drawn around the partons in $\eta-\phi$ space. If two partons happen to fall within the cone they are merged into one "jet". In contrast to leading order calculations, this produces, at the parton level, a dependence of the predicted cross section on the cone size and jet shape. Figure 1 shows the measured cross section as a function of cone size[4].

At the parton level, the jet shape, $\psi(r)$, is

---

*Supported by the U.S. Dept. of Energy, contract number DE-AC02-76CH03000.

[†]The CDF Collaboration: ANL, Brandeis, UCLA, U. Chicago, Duke, FNAL, INFN-Frascati, Harvard, U. Ill., Johns Hopkins, KEK, LBL, MIT, U. Mich. INFN-Padova, U. Penn., INFN-Pisa, Purdue, Rochester, Rockefeller, Rutgers, Texas A&M, Tufts, Tsukuba, U. Wisconsin

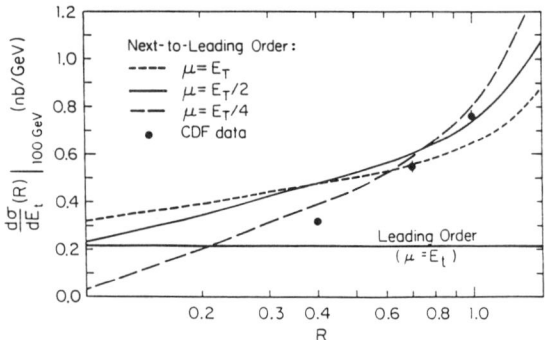

Figure 1. Inclusive Jet Cross section for 100 GeV Jets for cone sizes R = 0.4, 0.7, and 1.0.

defined as the fraction of energy falling within a cone of radius r, normalized to the energy within a cone of R=1.0 and averaged over many events.

Experimentally, $\psi(r)$ is defined as the sum of the P_T of tracks falling within a cone of radius r, around the calorimeter cluster centroid, and normalized to the total track P_T within a cone of radius 1.0. Figure 2 shows the jet shape for 100 GeV jets[5], compared to $O(\alpha_s^3)$ calculations.

INCLUSIVE JET CROSS SECTION

With the data taken at $\sqrt{s} = 546$ and 1800 GeV, CDF is in a unique position to test the predictions of QCD for scaling of the jet cross sections. Previously, it was necessary to compare results from different experiments, and thus the systematic errors would not cancel, or such a large change (more than a factor of 3) in center of mass energy was not possible.

Without QCD evolution, the naive parton model predicts that the scaled jet cross section, $\sigma' = P_T^4(Ed^3\sigma/dp^3)$, will be independent of $\sqrt{s}$ when plotted as a function of the variable $X_T = 2P_T/\sqrt{s}$. Thus, the ratio of the scaled cross sections as measured at $\sqrt{s}$=546

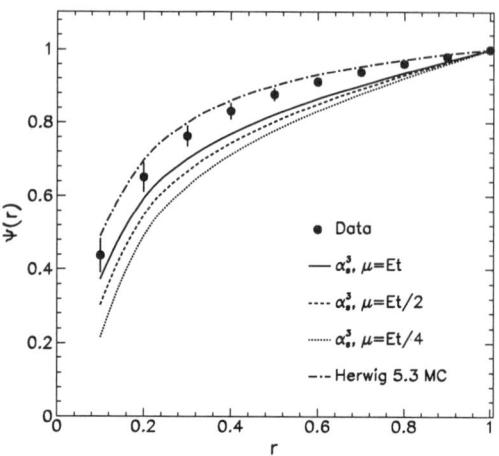

Figure 2. Jet shape for 100 GeV jets compared to next-to-leading order predictions.

and 1800 GeV should be 1. However, when evolution of the structure functions and of α_s are included in the theory, exact scaling is broken, and this ratio is expected to have an average value (at CDF energies) of about 1.8.

The measurement of the inclusive jet cross section at $\sqrt{s}$ =1.8 TeV has been discussed in Reference [4] and is in good agreement with QCD predictions. Similar techniques were used in the analysis of the $\sqrt{s}$=546 GeV data. The systematic errors for the two samples and the ratio are shown in Figure 3. Notice that the systematic error is greatly reduced in the ratio of the cross sections.

Figure 4 shows the ratio of the scaled jet cross sections as measured by CDF. The band indicates the size of $\pm 1\sigma$ of systematic uncertainty. To determine the significance of the result we form an average value for the data and theory over the X_T range 0.101-0.265. We observe a 1.5-2.4 σ difference between QCD and the data, where the range comes from different choices of structure functions and renormaliza-

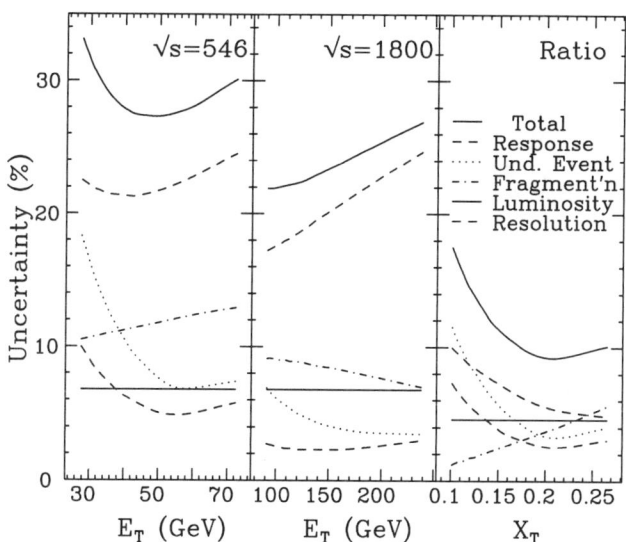

Figure 3. Systematic errors for the scaled jet cross sections at $\sqrt{s}$ = 546 and 1800 GeV.

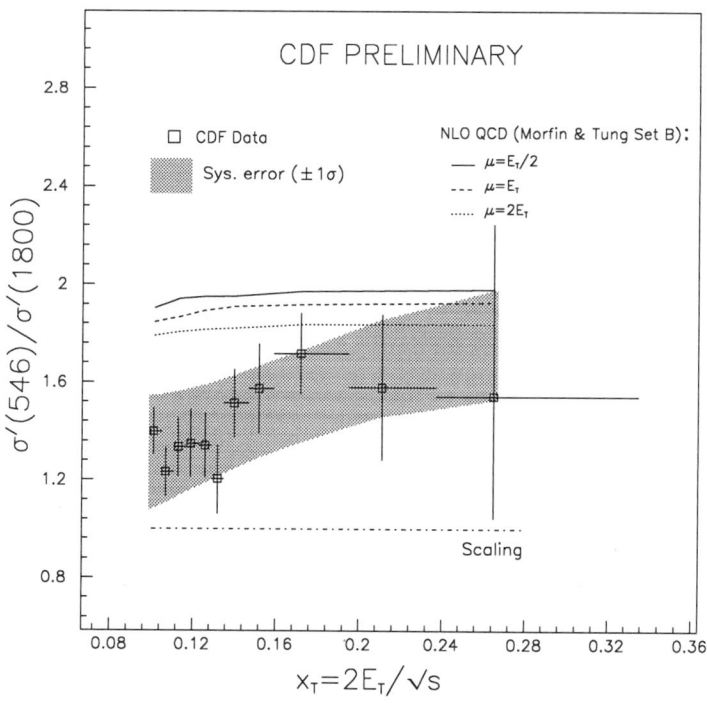

Figure 4. Ratio of the scaled jet cross sections at $\sqrt{s}$ = 546 and 1800 GeV.

tion scale. Scaling is excluded by the data at the 95% confidence level.

INCLUSIVE PHOTON CROSS SECTION

Photons produced directly from the hard collision provide a probe of the gluon structure functions and an energy measurement which is free from the effects of fragmentation. Photon identification at CDF is described in Reference [6]. Separation of photons from the background (mainly π^os) is accomplished through the comparison of the shower profiles from the data with shower profiles from test beam electrons.

Figure 5 shows the photon P_T spectrum as measured by CDF[6]. The theoretical predictions are all next-to-leading order and the renormalization scale is P_T. The data seems to have a steeper slope at low P_T than the theoretical predictions. The effect of higher order terms, bremsstrahlung diagrams and new structure functions[7] are under study. At present, the range in the predictions from different choices of renormalization scale and the disagreement between theory and data in the low P_T region, preclude the separation of the effects of different gluon structure functions.

The production of di-photon events provides another test of QCD and is an important background to Higgs$\to \gamma\gamma$ for the SSC. Events with two photons were identified using similar cuts to the single photon analysis except the E_T cut was 10 GeV on each photon[8]. Figure 6 shows the measured di-photon cross section compared to the theoretical predictions. Each photon is entered in the plot.

ANGULAR DISTRIBUTIONS

QCD predicts that the jet-jet angular distribution is dominated by Rutherford-like t-channel gluon exchange (spin 1). The photon-jet final state is expected to be relatively flat since it is dominated by t-channel quark ex-

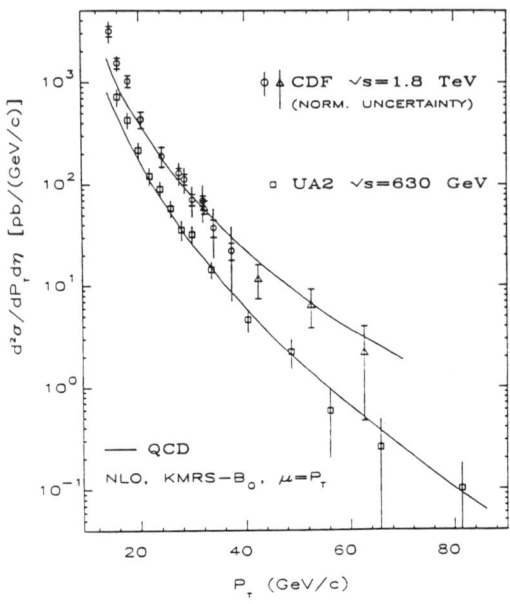

Figure 5. Inclusive Photon Cross section

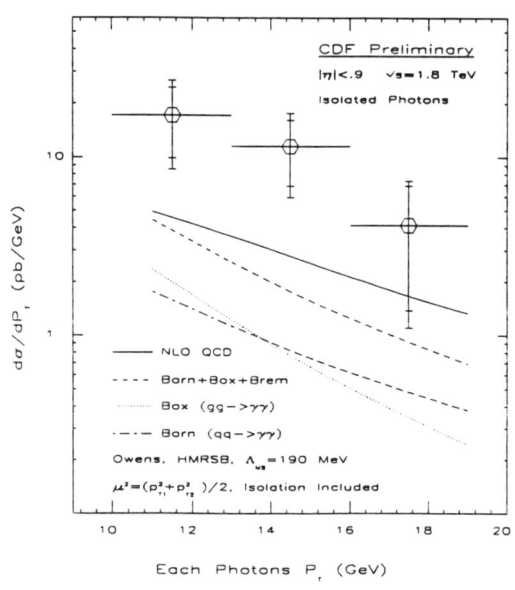

Figure 6. Di-Photon Cross section.

change (spin 1/2). This is directly reflected in the angular distribution.

The cross section for dijet events can be written in terms of the mass, M_{JJ}, the center-of-mass scattering angle, θ^*, and longitudinal boost of the dijet system, $\eta_{boost} = (\eta_1 + \eta_2)/2$, where η_1 and η_2 are the pseudorapidities of the two highest E_T jets[9]. The dijet mass of an event is calculated using the four-vectors of the leading two jets. The scattering angle is related to η_1 and η_2 by the equations $\eta^* = (\eta_1 - \eta_2)/2$ and $\cos\theta^* = \tanh\eta^*$.

Similarly, in photon-jet events, the variables η^* and $\cos\theta^*$ are defined using the pseudorapidities of the photon, η_γ, and the jet, η_{jet}. The jet is defined as the vector sum of the jets which fall in a 120° cone opposite to the photon direction in ϕ. To avoid uncertainties associated with the measurement of the jet energy, the invariant mass of the photon-jet system is calculated using the P_T of the photon and η^*: $M_{\gamma,jet} = 2P_{T,\gamma}\cosh\eta^*$.

In the comparisons to theoretical predictions, the jet-jet angular distribution is plotted in terms of the variable $\chi = e^{2|\eta^*|}$. For t-channel exchange, which dominates at large η^*, the $dN/d\chi$ spectrum is expected to be flat and thus insensitive to smearing effects. In addition, the signal of quark compositeness would show up as a peak at low χ.

Figure 7 shows the data[9] compared to $O(\alpha_s^2)$ and $O(\alpha_s^3)$ calculations[10] for HMRSB structure functions. The data and the theoretical curves are normalized to unit area. Four sets of Morfin-Tung structure functions were tested (S, B1, B2 and E); they gave the same confidence levels to within 2%. From this data a limit on the quark compositeness energy scale of $\Lambda_c >1000$ GeV has been derived.

Figure 8 shows photon-jet angular distribution compared to leading order and next-to-leading order QCD calculations. The dijet angular distribution as previously measured[11] by CDF has also been included since it was measured at low mass ($M_{jj} \geq 148$ GeV) and thus is closer than the current dijet data to the mass used in the photon-jet measurement. Although the statistics are limited, the photon-jet data appears to be flatter than the dijet data as expected from the spin of the propagators.

FOUR-JET EVENTS

QCD predicts that the dominant mechanism for the production of four-jet events is a 2→2 parton interaction plus double gluon bremsstrahlung. An alternative hypothesis is that four-jet events could be produced by double parton interactions in which two partons in each of the incoming hadrons collide and produce two di-jet events. The double parton scattering events are expected to have two sets of well balanced dijets, randomly oriented with respect to one another. In double bremsstrahlung events, the 3rd and 4th highest E_T jets are expected to be found preferentially near the leading two jets. Topological variables are used to compare the data to the theoretical predictions. Figure 9 shows plots of the cosine of the angle between each jet pair, compared to a leading order four-jet QCD Monte Carlo[12]. Clearly the data is well described by the theoretical predictions.

Using variables sensitive to the double-dijet structure of the events, the data is fit to an admixture of QCD double bremsstrahlung and double parton scattering predictions. It was found that the best fit corresponded to a small contribution (5.2%, 2σ from 0) from double parton scattering. From this we derive a limit on the double parton scattering cross section of $\sigma_{DP} <0.12$ μb for partons with $E_T > 18$ GeV at the 95% confidence level. To compare between experiments the limit is usually quoted in terms of the effective cross section $\sigma_{eff} \equiv \sigma_{dijet}^2/2\sigma_{DP}$, where σ_{dijet} is the

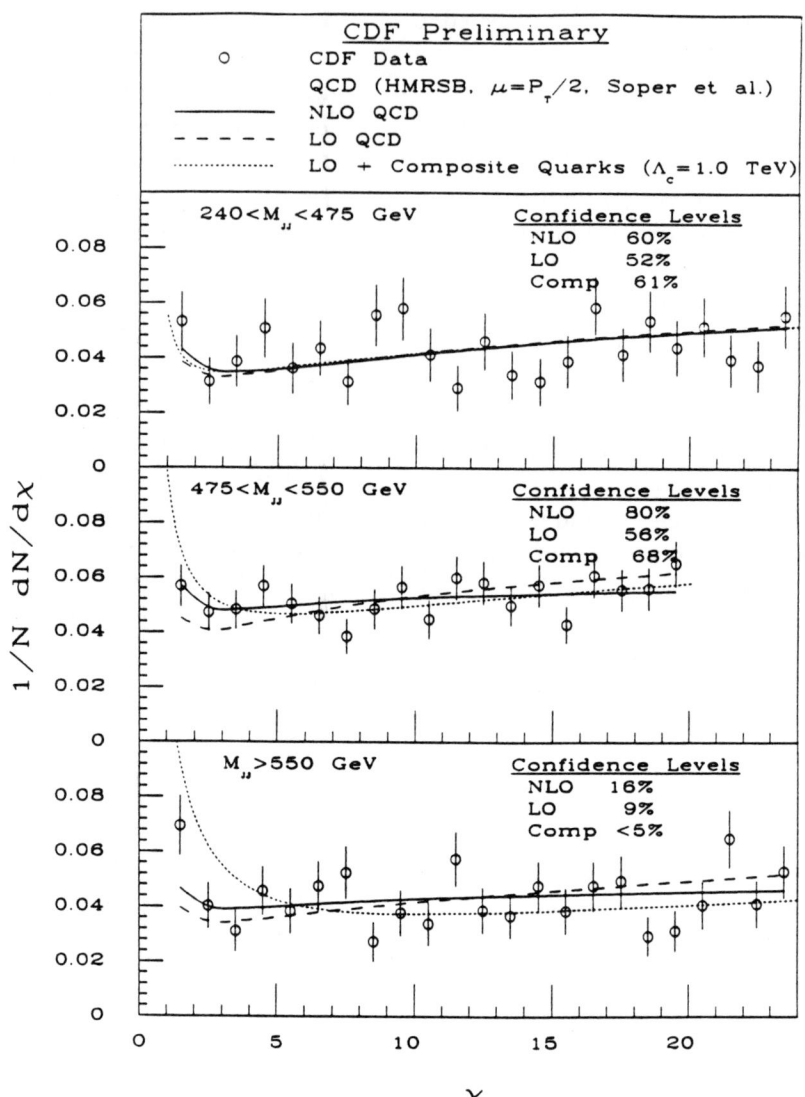

Figure 7. Dijet angular distribution.

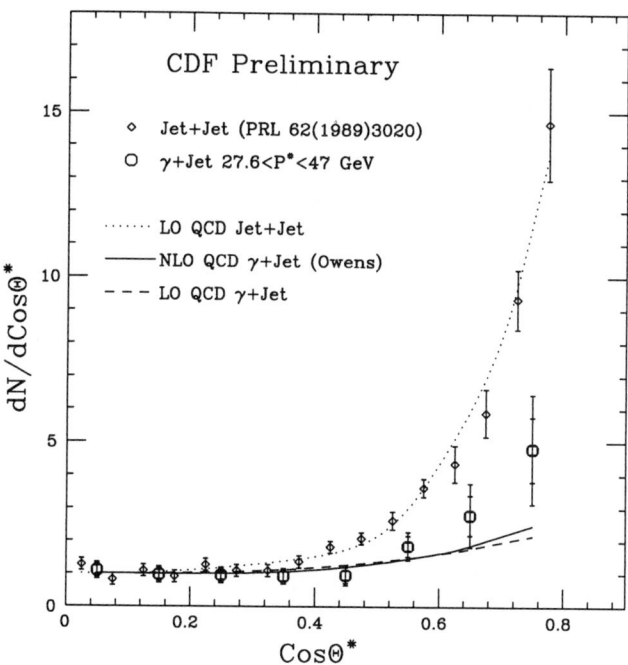

Figure 8. Photon-jet angular distribution.

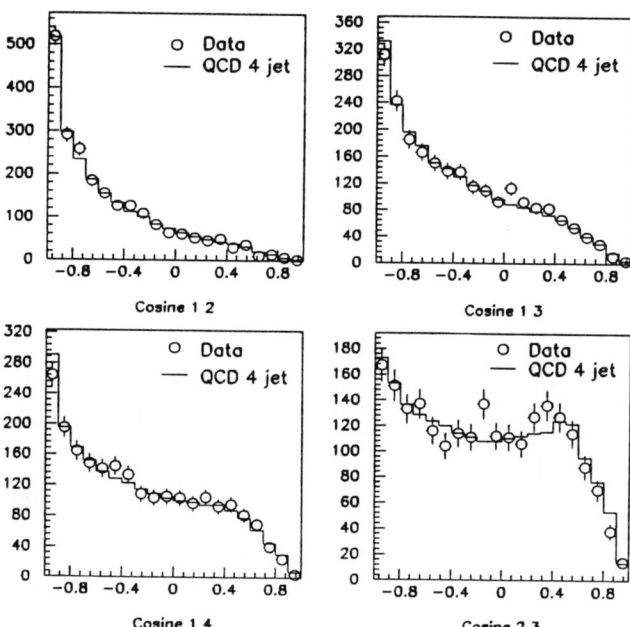

Figure 9. Angular separation between jets in four-jet events compared to QCD.

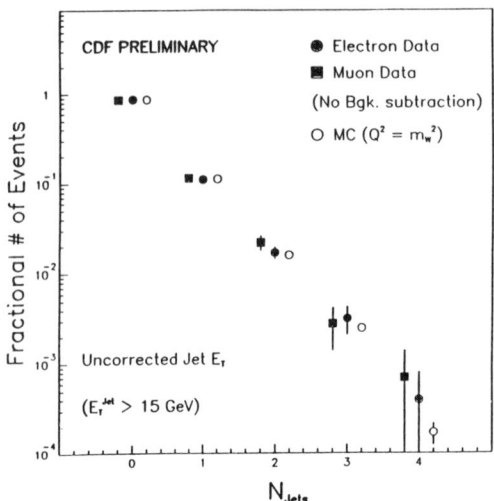

Figure 10. Jet multiplicity in W events for electron and muon W-decays.

dijet cross section. Our limit is $\sigma_{eff} > 3.9$ mb at the 95% confidence level.

JET MULTIPLICITY IN W-BOSON EVENTS

A measurement of the jet multiplicity in high P_T W events provides another test of QCD. Events with a vector boson plus jets are one of the major backgrounds in the search for the top and the Higgs. Events in which the W decays into a electron + neutrino or a muon + neutrino are selected using the standard CDF cuts as described in Reference [13]. The leptons are restricted to the central region ($|\eta| < 1.1$ for electrons and $|\eta| < 0.6$ for muons) while the jets are allowed to be further forward, $|\eta| < 2.4$. A total of 4.05 pb^{-1} of electron data and 3.54 pb^{-1} of muon data was collected. Figure 10 shows the jet mulitiplicity distribution in W events compared to Monte Carlo parton level calculations[14] plus full detector simulation. The theory shows good agreement with the data.

CONCLUSIONS

The inclusive jet cross section has been measured by CDF at two center-of-mass energies: $\sqrt{s} = 546$ GeV and 1800 GeV. The ratio of the scaled jet cross sections is a measure of the scaling violations predicted by QCD. The data has been found to consistant with QCD at the 1.5-2.4σ level and the hypothesis of exact scaling has been ruled out at the 95% confidence level.

Cross sections for direct photons and diphotons have been measured. Some disagreements with the theory have been observed, but the effect of new structure functions, higher order terms and bremsstrahlung are still under study.

The dijet and photon-jet angular distributions have been measured and are well described by the theoretical predictions. A compositeness limit of $\Lambda_c > 1000$ GeV has been derived from the dijet data.

Leading order QCD provides a good description of the four-jet data. Preliminary limits on the double parton scattering cross section of $\sigma_{DP} < 0.12\mu b$ and $\sigma_{eff} > 3.9$ mb have been derived. Finally, the multiplicity of jets in W events has been measured, and is in good agreement with QCD.

REFERENCES

1. F. Abe et al. (CDF Collaboration), *Nucl. Instr. Meth.* A271, pp.387, (1988).

2. F. Abe et al. (CDF Collaboration), *Phys. Rev. D* 45, pp.1448, (1992).

3. S. Ellis, Z. Kunszt, D. Soper, *Phys. Rev. Lett.* 64, pp.2121, (1990).

4. F. Abe et al. (CDF Collaboration), *Phys. Rev. Lett.* 62, pp.613, (1989).

5. F. Abe et al. (CDF Collaboration), *Submitted to Phys. Rev. Lett.*, FERMILAB-PUB-92/167-E.

6. F. Abe et al. (CDF Collaboration), *Phys. Rev. Lett.* 68, pp.2734, (1992).

7. New low x NMC data was released at this conference. See talks by E.Rondio, M. Virchaux, and W.Tung.

8. R. Harris, "QCD at CDF", in proceedings of the XXVII Recontre de Moriond, 1992; FERMILAB-CONF-92/146-E.

9. F. Abe et al. (CDF Collaboration), *Submitted to Phys. Rev. Lett.*, FERMILAB-PUB-92/182-E.

10. S. Ellis, Z. Kunszt, D. Soper, *Phys. Rev. Lett.* 69, pp.1496, (1992).

11. F. Abe et al. (CDF Collaboration), *Phys. Rev. Lett.* 62, pp.3020, (1989).

12. I. Hinchliff, PAPAGENO Monte Carlo program (in preparation).

13. F. Abe et al. (CDF Collaboration), *Phys. Rev. D* 44, pp.29, (199).

14. F.A. Berends, W.T.Giele, H.Kuijf and B. Tausk, FERMILAB-PUB-90/213-T.

FIRST QCD RESULTS FROM THE D0 DETECTOR

H.Weerts[1]**
Dept. of Physics & Astronomy
Michigan State University
East Lansing, Michigan 48824

Abstract

The D0 detector at the Fermilab Tevatron proton-antiproton collider has recently started taking data. In this paper we present first and very preliminary results on measurements with jets in this detector. Also the D0 calorimeter and triggers for jets will be described.

INTRODUCTION

The D0 detector is a general purpose, second generation collider detector designed and built to study proton-antiproton interactions at the Fermilab Tevatron at a centre of mass energy of 1.8TeV. The detector was completed in February of 1992 and observed its first collisions in May 1992. Figure 1 shows an overall view of the detector. The main characteristics are: hermetic and uniform coverage for muons, electrons and jets and no central magnetic field. In this paper we shall describe in some detail the calorimeter of the experiment and we refer to references 1,2,3 and 4 for a description of the other major detector components and the data acquisition system. Results presented here are very preliminary and from data taken during the first two months of operation. This corresponds to an integrated luminosity of approximately 300nb^{-1} delivered, of which 56nb^{-1} were written to tape.

DESCRIPTION OF THE CALORIMETER

The D0 calorimeter uses uranium as an absorber and liquid argon as the charge sensitive medium. Figure 2 shows a cut out view of the calorimeter and the central tracking system. The calorimeter consists of 3 cryostats: the central part covering pseudo rapidity $|\eta|<1.2$ and the forward calorimeters covering $1.2 < |\eta| < 4.0$. Fig. 2 illustrates the hermetic and uniform coverage over the total rapidity and azimuth region. The calorimeter has been in operation since February of this year. The change in the response (monitored by argon gaps instrumented with α and β sources) has been less than 0.2% over the last 5 months and the purity of the argon used is better than 0.7ppm of oxygen equivalent.
Longitudinally the calorimeter starts with an electromagnetic section, subdivided into 4 segments of thickness 2, 2, 7 and 10 radiation lengths. This is followed by a fine hadronic uranium section consisting of 3 depth segments

---

[1]For the D0 collaboration:Universidad de los Andes (Colombia), University of Arizona, Brookhaven National Laboratory, Brown University, University of California at Riverside, Centro Brasiliero de Pesquisas Fisicas(Brazil), CINVESTAV(Mexico), Columbia University, Delhi University(India),Fermi National Accelerator Laboratory, Florida State University, University of Hawaii, University Illinois at Chicago, Indiana University, Iowa State University, Lawrence Berkeley Laboratory, University of Maryland, University of Michigan, Michigan State University, Moscow State University(Russia) ,New York University, Northeastern University, Northern Illinois University, Northwestern University, University of Notre Dame, Panjab University(India), Institute for High Energy Physics at Protvino(Russia), Purdue University, Rice University, University of Rochester, CEN Saclay(France), State University of New York at Stony Brook, Superconducting Supercollider Laboratory, Tata Institute of Fundamental Research(India), University of Texas at Arlington, Texas A&M University

**This work was supported in part by the U.S. Department of Energy and the National Science Foundation.

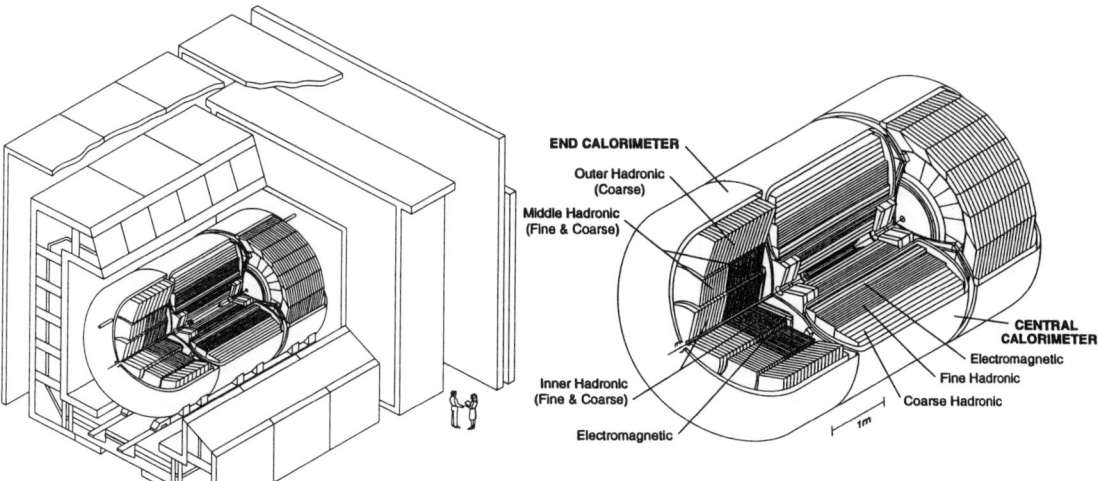

Figure 1. Overview of the D0 detector

Figure 2. The D0 liquid argon calorimeter

each one interaction length (λ) long. The last section is a Cu or Fe coarse section which accounts for 3λ in the central region and up to 5λ in the forward region. Transversely the segmentation in pseudo-rapidity and azimuth(ϕ) is $\Delta\eta \times \Delta\phi = 0.1 \times 0.1$ up to $\eta=2.5$ where it doubles in size in both dimensions. This results in a total of about 50,000 electronics channels being read out. Included in this channel count are special massless gaps and scintillator tiles located between the cryostats. These detectors are used to correct for the amount of dead material in this area [5]. The fine transverse and longitudinal segmentation results in a nearly projective tower geometry which makes identification of jets very straightforward (see Fig. 3). The noise characteristics of the calorimeter are excellent. Typical noise depends on the size of the cell and varies from 15 MeV/cell in the electromagnetic sections to 70 MeV/cell in the fine hadronic sections. This noise is completely dominated by fluctuations in the dark current generated by the uranium radioactivity. The least count for a cell is 3-5MeV/cell depending on the sampling fraction of the cell. Minimum ionizing particles can be easily observed [5]. The lego plot of a three jet event in Fig. 3 shows how quiet the calorimeter is and how clearly jets can be identified.

The energy response of the calorimeter to single electrons and pions has been measured between 2 and 150 GeV in a testbeam[5]. This single particle response has been transferred to D0 and it is this response which is being used for the jet energy scale in this paper. No corrections to the jet energy are applied which arise from nonlinearities

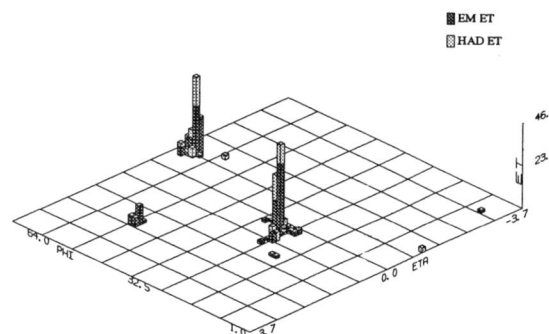

Figure 3. D0 event display of a 3 jet event.

in the single particle response. For this paper jets are defined by a fixed cone size algorithm with a cone $\Delta R = \sqrt{\Delta^2\eta + \Delta^2\phi} = 0.7$.

JET TRIGGERS

At the first level(L1) of the trigger, jets are selected by requiring a jet trigger tower, with transverse size $\Delta\eta \times \Delta\phi = 0.2 \times 0.2$ and summed over electromagnetic and fine hadronic depth segments, to be above a certain transverse energy threshold. A jet trigger in general consists of requiring one or more trigger towers, possibly in selected η,ϕ regions, to be above threshold. Data available for this presentation only used a trigger instrumented for $|\eta| < 1.6$ and the data shown are all with this restriction. Currently the trigger covers $|\eta| < 3.2$ and soon will be extended to $\eta=4$. Figure 4 shows the L1 trigger efficiency as a function of the reconstructed jet transverse energy(E_t) for requiring 1 trigger tower to contain more than 3GeV of transverse energy. On average a trigger tower contains about 1/3 of the total jet energy and because of the width of the distribution this trigger does not become 100% efficient until jet transverse energies of 22GeV. The inclusive jet triggers consist of a set of thresholds, as mentioned above, which allow us to measure the

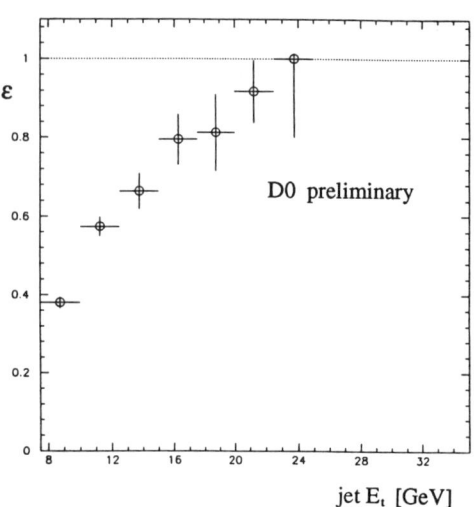

Figure 4. Level-1 trigger efficiency(ϵ) as a function of jet E_t

total jet E_t spectrum.
At the second stage of the trigger, jets are identified and reconstructed using a cone size algorithm very similar to the offline reconstruction. It is here that the final threshold is applied to the total jet E_t as measured inside the cone (for more details see ref. 3).

JET RESULTS

The inclusive jet E_t spectrum using all jet triggers is shown in figure 5. As mentioned above this is the uncorrected jet transverse energy simply using the singe particle response from the testbeam. No corrections are applied for out of cone energy or contributions from the underlying event. The spectrum displays the characteristic steep fall with increasing E_t and extends out to 110GeV for this subset of data, which corresponds to an integrated luminosity of about 2nb^{-1}. Figures 6a and 6b show the azimuthal and pseudorapidity distributions for a subset of jets with jet measured $E_t > 15$GeV. Within errors the distribution in azimuth is flat as would be expected, except for a slight drop at large ϕ which is attributed to a trigger bias in the L1 jet trigger. It also shows no contamination in the area $\phi=1.7$ to 1.9 rad where the main ring accelerator passes through the calorimeter. The pseudorapidity distribution clearly demonstrates the trigger bias due to the fact that the trigger was only instrumented down to $|\eta|=1.6$. The distribution within the covered rapidity region is not completely flat as expected, which is due to

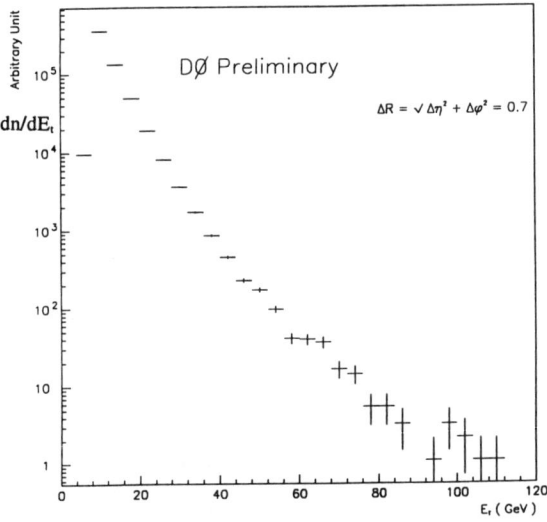

Figure 5. The inclusive jet E_t distribution

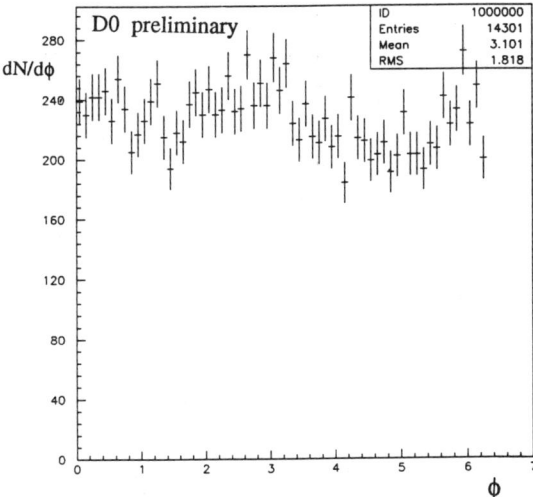
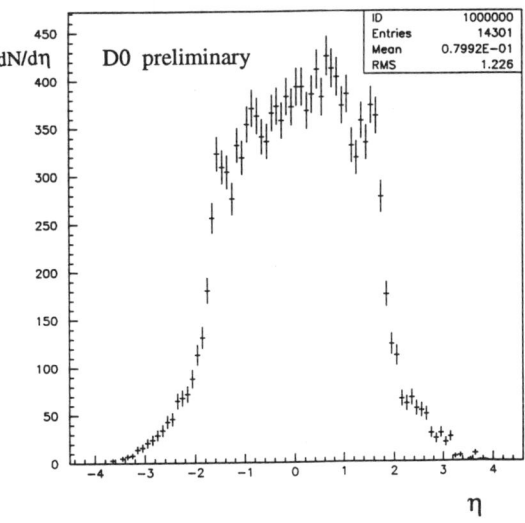

Figure 6a. dN/dφ distribution for jets with $E_t > 15$ GeV.

Figure 6b. dN/dη distribution for jets with $E_t > 15$ GeV

the vertex position being shifted by about 10cm from the nominal center position in the detector. The entries in this distribution for $|\eta|>1.6$ are due to additional jets found in an event if the jet that triggered is in the instrumented region.

SUMMARY

The D0 detector has been taking data since May of this year. Very preliminary results on triggering and reconstruction of jets have been presented. These first results indicate that the detector is working well and we look forward to extending the existing measurements of jet cross sections and angular distributions into rapidity regions not explored before at Tevatron energies. With the fine transverse segmentation it is also intended to study the jet cone size dependence of these quantities.

REFERENCES

1. D.Hedin, "The D0 Muon System and early Results on its Performance.", Paper contributed to this conference.
2. H.Prosper, "Search for New Particles", Paper contributed to this conference.
3. B.Gibbard, "D0 Triggering and Data Acquisition.", Paper contributed to this conference.
4. B.Gobbi, "First W decays observed with the D0 detector.", Paper contributed to this conference.
5. Kaushik De ,"Tests of the D0 Calorimeter Response in 2-150GeV beams", and references therein. Paper contributed to this conference.

RECENT THEORETICAL DEVELOPMENTS FOR HADRON COLLISIONS

R. K. Ellis
Theory Department
Fermilab
P. O. Box 500
Batavia, IL 60510, USA

Abstract

Recent theoretical developments in the description of hadronic collisions at high energy are reviewed.

INCLUSIVE CROSS SECTIONS

The description of inclusive cross-sections in QCD is well understood. The cross section for any hard scattering process may be written schematically as,

$$\sigma = \int f(x_a,\mu) f(x_b,\mu) \hat{\sigma}(x_a, x_b, \alpha_S(\mu), \frac{M^2}{\mu^2}) \quad (1)$$

where f are the parton structure functions and the short distance cross-section $\hat{\sigma}$ is calculable as a perturbation series in α_S.

$$\hat{\sigma} = \hat{\sigma}_0 + \alpha_S \hat{\sigma}_1 + \alpha_S^2 \hat{\sigma}_2 + \ldots \quad (2)$$

The scale M characterises the hardness of the interaction. Physical predictions need parton distribution functions, a choice for the renormalisation and factorisation scale μ and a value for α_S. The experimental status of the measurement of these quantities is reviewed by Bethke in his plenary talk. I will not review the determination of these quantities but will concentrate instead on the progress which has been made in the calculation of short distance cross-sections. Most of the recent progress has been in the description of electro-weak processes at high energy colliders. $O(\alpha_S)$ corrections are now available for most processes of current interest.

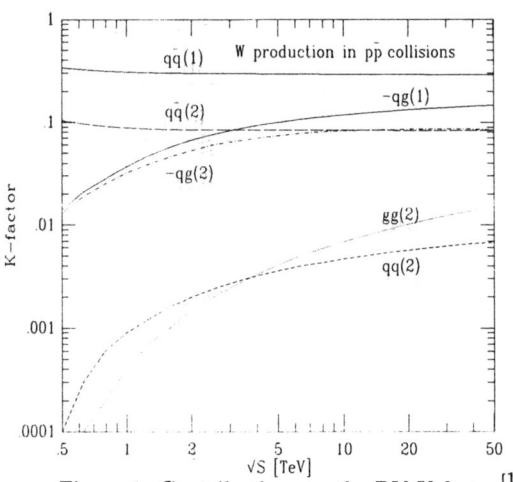

Figure 1. Contributions to the DY K-factor[1]

Drell Yan Process

The Drell Yan process occupies a strategic importance in QCD collisions. It is the simplest process involving two incoming hadrons and has historically been of great theoretical importance in establishing the validity of Eq. (1) in QCD. Lepton pair production occurs in lowest order by $q\bar{q}$ annihilation, in $O(\alpha_S)$ by $q\bar{q}$ and qg processes and in $O(\alpha_S^2)$ by $q\bar{q}, qg, gg$ and qq processes. The calculation of the short distance cross section has been performed through order $O(\alpha_S^2)$ [1]. Experimental testing of the Drell-Yan mechanism is now mainly performed in the production of

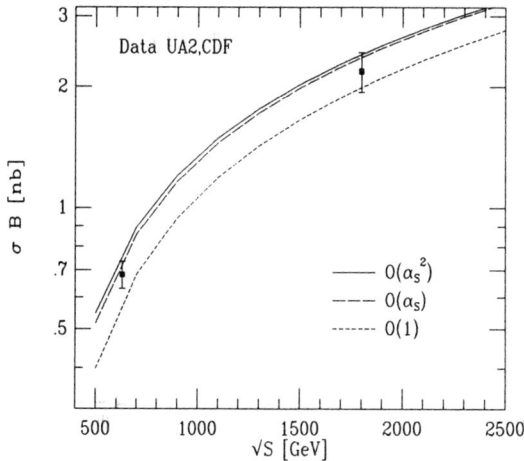

Figure 2. W cross-sections compared with data

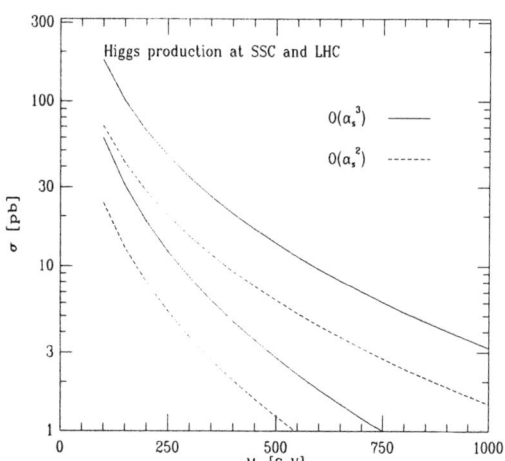

Figure 3. Higgs boson production beyond the leading order.

W and Z bosons. Fig. 1 shows the various contributions to W production in $p\bar{p}$ collisions expressed as a fraction of the lowest order $q\bar{q}$ contribution. The $O(\alpha_S)$ and $O(\alpha_S^2)$ contributions to the qg process, $qg(1)$ and $qg(2)$, are shown reversed in sign. These corrections are calculated in terms of parton distribution functions derived from Deep Inelastic Scattering (DIS). In $O(\alpha_S)$ the net correction is about 30% and is helpful in achieving agreement with the data as shown in Fig. 2. The $O(\alpha_S^2)$ corrections are less than 10%, in part because of a cancellation between $q\bar{q}$ and qg contributions as shown in Fig. 1.

Higgs boson production

The cross-section for the production of the standard model Higgs boson is important to estimate rates at the next generation of pp colliders. Both the gluon fusion processs $gg \to HX$[2],[3]. and the vector boson fusion process $q\bar{q} \to VVX \to HX$ [4] have been calculated beyond the leading order. The gluon fusion calculation has been performed in order α_S^3 in the heavy top quark limit, $m_t \gg M_H$. In this limit the resulting cross-section is independent of the value of the top quark mass. If $m_t = 140$ GeV, the values of the cross section for $M_H > 280$ GeV where the approximation $m_t \gg M_H$ is not valid, illustrate the sensitivity of the cross section to a new heavy flavour. Fig. 3.shows that the radiative corrections are large, so large as to shed doubt on the validity of the perturbation series.

In contrast the vector boson fusion cross sections are found to be increased by less than 12% in next to leading order. The W fusion cross section can be approximately expressed as the product of two deep inelastic structure functions, F_2. Hence the corrections are small when expressed in the DIS scheme[4] . Note however that the vector boson fusion mechanism is less important than the gluon-gluon fusion mechanism, because the top quark is so large. This is true except for very large Higgs masses[3].

Production of W's at large p_t

Remarkably enough the p_t distribution of vector bosons is calculable in QCD for all values of p_t. Fig. 4 shows the result of such a calculation[5] compared with data from CDF. The dash-dotted and dashed lines show respectively the contribution of soft-resummation and the soft resummation including the per-

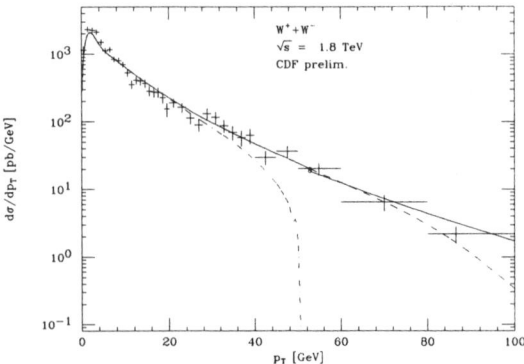

Figure 4. W transverse momentum

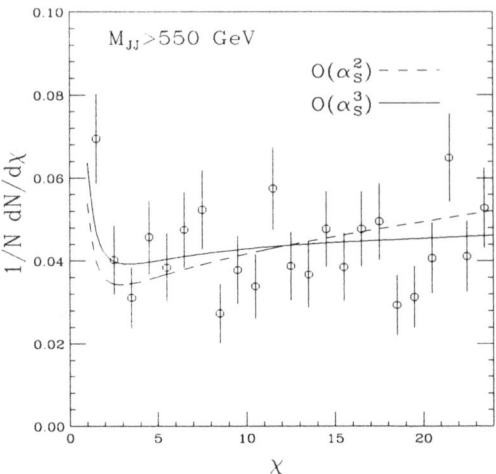

Figure 6. CDF dijet angular dependence[16].

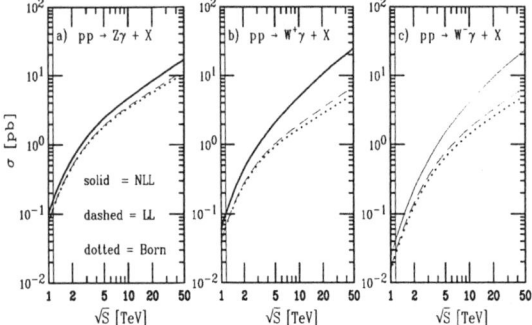

Figure 5. Radiative corrections to $V\gamma$ production[6]

turbative matching.

Vector boson pair production

Cross sections for the production of pairs of electroweak vector bosons are of interest because they probe the three boson coupling, which is an untested sector of the electroweak theory. Fig. 5 gives an example of theoretical results[7],[6] on $V\gamma$ production, where V is a W or a Z. At high energy the radiative corrections are very large, because of the large gluon flux which enters through the qg processes. In addition $\sigma_{W\gamma} < \sigma_{Z\gamma}$ in the Born approximation because of the presence of the radiation zero in the $W\gamma$ process. The radiation zero is not present in the qg process and the correction for the $W\gamma$ process is very large.

Results on corrections to ZZ production are given in refs. [8],[9],[10]. Results on corrections to WZ production are given in refs. [11],[12]. Results on corrections to WW production are given in ref. [13].

Two jet production

Ellis, Kunszt and Soper[15] have calculated QCD jet cross sections beyond the leading order. They have written a program which calculates any 'infrared safe' jet cross section through $O(\alpha_S^3)$. This program uses the $O(\alpha_S^3)$ matrix elements calculated in ref. [14]. As an example, Fig. 6 shows the dijet angular distribution compared to data from the CDF collaboration. The variable χ is defined as $(1 + \cos\theta)/(1 - \cos\theta)$ where θ is the center of mass scattering angle. The scale μ is taken to be the $p_T/2$. The slope in χ of the angular distribution is very sensitive to the choice of scale in the Born approximation. This sensitivity to the choice of scale is much diminished in the $O(\alpha_S^3)$ calculation.

EXCLUSIVE JET CROSS SECTIONS

This section gives details of jet cross sections. I shall start with the progress in tree graph calculations where a produced parton is identified with a jet. A major impetus for these calculations has been the need to calculate the standard model backgrounds to new processes.

Implications for top quark search

One of the most important applications of the multijet calculations is the estimate of the background to the top quark search. A $t\bar{t}$ event produced in a hadronic collision comes mainly from $q\bar{q}$ annihilation if $m_t > 100$ GeV. Note that W gluon fusion does not lead to an observable signal[17] at $\sqrt{(s)} = 1.8$ TeV. A $t\bar{t}$ event gives rise to an observable $W + n$-jet signature ($n \leq 4$) when one of the W's from top decay undergoes a semileptonic decay. The background is due to the production of a W in association with QCD jets. Fig 7 shows the cross-section for both as a function of the top quark mass[18]. Both the signal and the background are calculated at tree graph level. The jets are defined using standard cuts of the CDF collaboration (for details see ref. [18]). The theoretical uncertainty is estimated by varying the renormalisation scale between $m_t/2 < \mu < 2\sqrt{(m_t^2 + p_T^2)}$ for the signal and between $M_W/2 < \mu < 2\sqrt{(M_W^2 + p_T^2)}$ for the background. This plot shows the importance of the four jet channel for top discovery. Note however that these plots do not include the discrimination against the background obtained when a b-quark is identified. With b tagging the $W+4$ jets background is effectively removed.

The top quark cross sections are known beyond the leading order[19]. This allows us to reduce the theoretical error on the top quark cross section shown in Fig. 7. The upper values of the top quark cross sections shown in Fig. 7 are favoured by NLO calculations. Resummation of the dominant terms in $O(\alpha_S^3)$ also tends to increase the cross-section[20].

Jet cross-sections beyond the leading order

Tree level predictions suffer from a num-

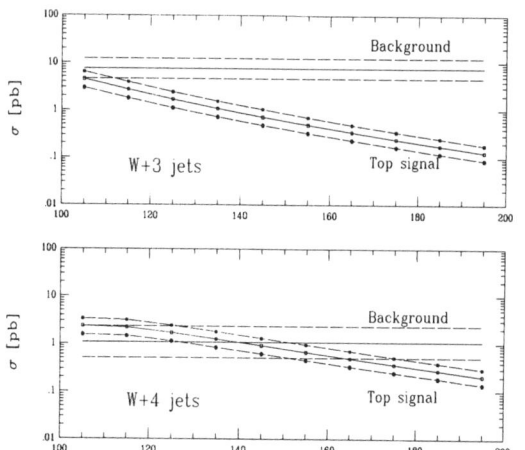

Figure 7. W + jet cross sections from background and signal

ber of deficiencies. First, they have no parton merging cone size dependence. The number of partons in the final state determines the number of jets. Two partons which are produced close together can never be merged to give a single jet. Second they are sensitive to variations of the size of the coupling constant α_S. This is particularly true of cross-sections which begin in order α_S^n for large n. For example the W plus four jet cross section occurs in order α_S^4. A 20% uncertainty in α_S leads to a factor 2 uncertainty in the cross section. Estimates beyond leading order can be done either using a Monte-Carlo program or using fixed order perturbation theory. Both methods have their advantages. Here I describe only advances in fixed order techniques.

In fixed order perturbation theory one can sensibly define a cross section for any 'infrared safe' quantity. This is technically complicated because it requires regulation and cancellation of contributions from real and virtual emission graphs, but presents no conceptual difficulties. Fig. 8 shows an example of such a calculation[21] for a $W + 1$ jet cross-section plotted as a function of E_T. Standard CDF cuts are used. The softening of the jet E_T spectrum beyond the leading order is due to

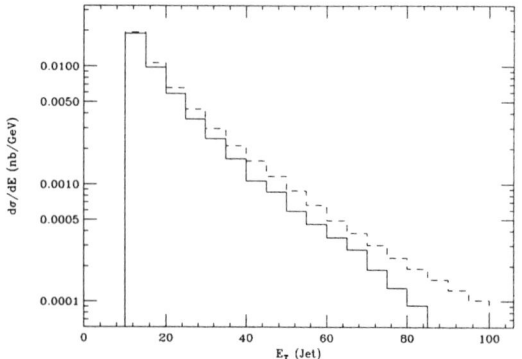

Figure 8. The $W^{\pm}+1$ jet cross section vs. the transverse energy of the jet. The dashed line is leading order, the solid line is next to leading order.

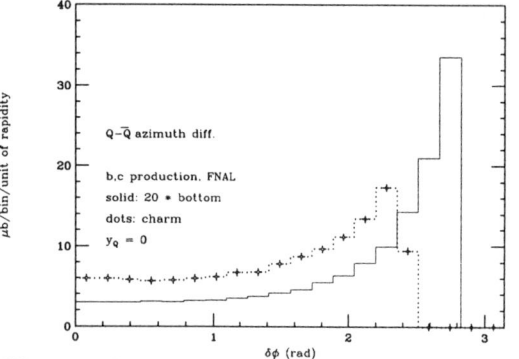

Figure 9. Azimuthal correlation of two heavy quarks

the use of a fixed E_T cut to define the jets. At high tranverse energy the accompanying soft radiation scales with E_T. Thus at higher E_T the accompanying soft radiation is much more likely to promote the event to the two jet sample, leading to a smaller $W+1$ jet sample.

As a second example of fixed order Monte Carlo I show the azimuthal correlations in the transverse plane of charm and bottom quarks produced in $p\bar{p}$ collisions at $\sqrt{S} = 1.8$ TeV. At $O(\alpha_S^3)$ the cross-section in the back-to-back region is negative. This is a signal that fixed order perturbation theory is not sufficient in this region. In Fig.9 the azimuthal correlation is shown[22]. The contribution of bins near the back-to-back region have been added together until they yield a net positive result.

NEW CALCULATIONAL TECHNIQUES

Higher order calculations in QCD become very complex, but the final answers can often be cast in a simpler form than the intermediate expressions. In the past few years a number of methods have been invented to streamline the approach to simple answers. These include,

1. Colour ordered subamplitudes[23].
2. Helicity amplitude techniques[23].
3. String Motivated techniques[24],[25]

I shall spend some time describing the lessons of the string exercises for standard field theory calculations[25]. It turns out that some of the simplicity of the string theory results, emerges naturally using two choices of gauge. I shall describe these two gauges since they are perhaps not well known to this community.

Gervais-Neveu gauge

In a traditional Feynman gauge the Lagrangian is given by

$$L = -\frac{1}{2}\{F^{\mu\nu}F_{\mu\nu}\} - \{(\partial_\mu A^\mu)^2\} + \ldots \quad (3)$$

where A and F are matrices in the fundamental representation of $SU(3)$ and the curly braces indicate that the trace is taken over the colour degrees of freedom. This leads to the traditional three gluon vertex, which written in colour ordered notation is,

$$V^{\epsilon_1\epsilon_2\epsilon_3} = g\sum\{t^{c_1}t^{c_2}t^{c_3}\}[\epsilon_1 \cdot \epsilon_2(2\epsilon_3 \cdot p_2 - \epsilon_3 \cdot p_3) \\ + \text{cyclic}] \quad (4)$$

ϵ_i label the polarisations of the lines which are not necessarily physical, ($\epsilon_i \cdot p_i$ is not necessarily zero). The sum runs over non-cyclic permutations of the labels 1, 2 and 3. In the Gervais-Neveu gauge[26] the gauge fixing term is modified as follows

$$L = -\frac{1}{2}\{F^{\mu\nu}F_{\mu\nu}\} - \{(\partial_\mu A^\mu - igA_\mu A^\mu)^2\} + \ldots \quad (5)$$

The three gluon vertex is now given by

$$V^{\epsilon_1 \epsilon_2 \epsilon_3} = g \sum \{t^{c_1} t^{c_2} t^{c_3}\} [\epsilon_1 \cdot \epsilon_2 (2\epsilon_3 \cdot p_2) + \text{cyclic}] \quad (6)$$

In this gauge the term proportional to $\epsilon_3 \cdot p_3$ is not present, even for an internal line. The colour ordered four gluon vertex is also simplified. The Feynman gauge propagator is unchanged.

Background field method[27]

In this case the gauge fixed Lagrangian is given by,

$$L = -\frac{1}{2}\{F^{\mu\nu}(A+Q) F_{\mu\nu}(A+Q)\} \\ -\{(D(A)_\mu Q^\mu)^2\} + \ldots \quad (7)$$

The gauge invariance with respect to the background field A is never broken. The three gluon vertex for one background field (label 1) interacting with two quantum fields, Q, (labels 2 and 3) is given by,

$$V^{\epsilon_1 \mu_2 \mu_3} = g f^{c_1 c_2 c_3} [g^{\mu_2 \mu_3} \epsilon_1 \cdot (p_2 - p_3) \\ + 2\epsilon_1^{\mu_3} p_1^{\mu_2} - 2\epsilon_1^{\mu_2} p_1^{\mu_3}] \quad (8)$$

Note that the momenta of the quantum fields (the loop momentum) appears only in the first term leading to a considerable simplification of the Lorentz algebra.

Conclusions

The proponents of superstring motivated techniques emphasize that there is more to the methods derived from superstring theory than just an astute choice of gauge in the normal field theory. In particular, the superstring techniques bypass a lot of the Lorentz algebra required in a standard field theory calculation. Note also that the rules derived by Bern and Kosower have been derived without using string theory by Strassler[28]. There are two open questions for the superstring motivated techniques. Can the extensions to fermions be handled gracefully? Can the techniques go beyond the proving stage and yield new results for physics?

REFERENCES

1. W.L. van Neerven and E.B. Zijlstra, "The $O(\alpha_S^2)$ corrected Drell-Yan K factor in the DIS and $\overline{MS}$ scheme", *Nucl. Phys. B* B382, 11, (1992) and references therein.

2. A. Djouadi et al., "Production of Higgs Bosons in proton colliders," *Phys. Lett B* 264, 440–446, (1991).

3. S. Dawson, "Radiative corrections to Higgs boson production," *Nucl. Phys. B* B359, 283, (1991).

4. T. Han, et al., "Structure function approach to vector boson scattering in pp collisions", Fermilab-Pub-92-171-T (Jun 1992).

5. P. B. Arnold and R. P. Kauffman, "W and Z production at next-to-leading order, From large q_t to small", *Nucl.Phys.*B349,381-413, 1991.

6. J. Ohnemus, "Order α_S calculations of hadronic $W^\pm\gamma$ and $Z\gamma$ production", DTP-92-54 (Jul 1992).

7. J. Smith, et al., "QCD corrections to the reaction $p\bar{p} \to W\gamma X$", *Z. Phys.* C44,267 (1989).

8. B. Mele, et al., "QCD radiative corrections to Z boson pair production in hadronic collisions",*Nucl.Phys.*B357,409-438,1991.

9. J. Ohnemus and J.F. Owens, "An Order α_S calculation of hadronic ZZ production", *Phys. Rev.*D43,3626-3639,1991.

10. E.W.N. Glover and J.J. van der Bij, "Z boson pair production via gluon fusion", *Nucl.Phys.* B321, 561 (1989).

11. J. Ohnemus, "An Order α_S calculation of hadronic $W^\pm Z$ production",*Phys. Rev.*D44, 3477-3489,1991.

12. S. Frixione et al., "Strong corrections to WZ production at hadron colliders", GEF-TH-2-1992 (Jan 1992).

13. J. Ohnemus, "An Order α_S calculation of hadronic W^+W^- production", Phys.Rev.D44,1403-1414,1991.

14. R. K. Ellis and J. Sexton, "QCD radiative corrections to parton-parton scattering", Nucl. Phys.269,445,(1986).

15. S. D. Ellis et al., "Two jet production in hadron collisions at order α_S^3 in QCD",Phys.Rev.Lett.69,1496,(1992).

16. F. Abe et al., "The dijet angular distribution in $p\bar{p}$ Collisions at $\sqrt{S} = 1.8$ TeV", Fermilab-Pub-92-182-E (1992).

17. R.K. Ellis and S. Parke, "Top quark production by W gluon fusion", FERMILAB-PUB-92-132-T (May 1992).

18. F.A. Berends et al., "Top search in Multijet Signals", FERMILAB-PUB-92-196-T (Aug 1992).

19. R.K. Ellis, "Rates for top quark production", Phys. Lett. B259,492 (1991) and references therein

20. E.Laenen et al., "All order resummation of soft gluon contributions to heavy quark production in hadron-hadron collisions" Nucl.Phys. B369 543 (1992).

21. W.T. Giele, et al., Next-to-leading order calculations in jet physics. FERMILAB-CONF-92-213-T (Aug 1992).

22. M.L. Mangano, et al., Heavy quark correlations in hadron collisions at next-to- leading order. Nucl.Phys.B373,295-345,1992.

23. See, for example, M.L. Mangano and S.J. Parke, "Multi - parton amplitudes in gauge theories", Phys. Reports. 200, 301 (1991).

24. Z. Bern and D.A. Kosower, "Efficient calculation of one loop QCD amplitudes", Phys. Rev.Lett. 66,1669-1672,1991, Nucl.Phys. B379,451-561,1992.

25. Z. Bern and D.C. Dunbar, "A Mapping between Feynman and string motivated one loop rules in gauge theories", Nucl.Phys. B379,562-601,1992.

26. J.L. Gervais and A. Neveu, "Feynman Rules for Massive Gauge Fields," Nucl. Phys. B 46, 381–401, (1972).

27. See, for example, L.F. Abbott, "The background field method beyond one loop," Nucl. Phys. B 185, 189, (1981).

28. M.J. Strassler, "Field theory without Feynman diagrams: One loop effective actions", SLAC-PUB-5757 (Feb 1992).

CONSTRAINTS ON THE GLUON DENSITY FROM BOTTOM QUARK AND PROMPT PHOTON PRODUCTION *

Edmond L. Berger and Ruibin Meng

High Energy Physics Division, Argonne National Laboratory, Argonne, IL 60439

Jianwei Qiu

Department of Physics and Astronomy, Iowa State University, Ames, IA 50011

Abstract

In next-to-leading order quantum chromodynamics, gluon-gluon interactions dominate the production of bottom quarks at hadron collider energies, and gluon-quark interactions control inclusive prompt photon production at large transverse momentum in pp collisions at fixed-target energies. Using such data, in conjunction with data from deep inelastic lepton scattering, we determine a new gluon density whose shape differs substantially from that derived from previous fits of data. The new set of parton densities provides a good fit to bottom quark, prompt photon, and deep inelastic data, including the most recent NMC and CCFR results.

Quantum chromodynamics is used routinely to compute expected cross sections for the production of heavy quarks and of direct (or prompt) photons in hadron interactions.[1] Calculations of the short distance, hard scattering cross section have been carried out to next-to-leading order in perturbation theory.[2,3] After convolution of the hard scattering cross section with parton probability densities[4] determined from fits to data on other processes, cross sections are derived for the production of heavy quarks[2] and of prompt photons at large transverse momentum[3] in hadron-hadron scattering. Theoretical uncertainties in the final answer arise from various sources. Principal among these is the relatively poorly constrained gluon density $G(x,\mu)$, especially in the relevant region of

fractional momentum x. For heavy quark production, $x \simeq 2M_T/\sqrt{s}$; $M_T^2 = p_T^2 + M_Q^2$, where p_T and M_Q are the transverse momentum and mass of the heavy quark, and $\sqrt{s}$ is the center-of-mass energy of the hadronic collision. For prompt photon production at large transverse momentum, $x \simeq 2p_T/\sqrt{s}$. In $G(x,\mu)$, μ is the factorization scale. The sensitivity of predictions to $G(x,\mu)$ is strong.[5] For example, at $\sqrt{s} = 1.8$ TeV, gluon-gluon subprocesses account for over 90% of the bottom quark production cross section at small values of p_T, and for over 50% even at $p_T = 80$ GeV. Existing gluon distributions fail to reproduce the data[6] at $\sqrt{s} = 1.8$ TeV.

In this paper we use the Fermilab CDF[6] and CERN UA1[7] collider data on bottom quark production, the CERN WA70[8] and Fermilab E706[9] fixed-target prompt photon production data, and data from deep inelastic lepton scattering[10-12] to make a new determination of the gluon density. We remark that we

---

*Work supported by the U.S. Department of Energy, Division of High Energy Physics, Contract W-31-109-ENG-38 and DE-FG02-92ER40730.001.

do not include prompt photon data at collider energies in this work. Such data include important fragmentation contributions and, for experimental reasons, require a photon isolation selection, both of which complicate the analysis.[13] Our analysis is carried out entirely within the usual context of the factorization assumption and employs next-to-leading order hard scattering cross sections.[2,3] Roughly stated, the ranges of x probed by the different experiments are: $0.01 < x < 0.06$ by CDF; $0.03 < x < 0.16$ by UA1; $0.20 < x < 0.65$ by E706; and $0.35 < x < 0.61$ by WA70. The collider bottom quark data and the fixed target prompt photon data therefore provide support for $G(x, \mu)$ in non-overlapping but nearly contiguous ranges of x.

An unguided simultaneous fit to data from bottom quark production, prompt photon production, and deep inelastic lepton scattering would not be successful since there are many more data points from deep inelastic lepton scattering, with much smaller uncertainties. The strategy adopted was an iterative one: starting with an established set of parton densities, we forced the gluon determination by fitting first to the bottom quark and prompt photon cross sections and then, with that gluon density as a starting distribution, refitting the prompt photon, the deep inelastic, and the bottom quark cross sections together. The fits we report here were carried out in the $\overline{MS}$ factorization scheme. The collider bottom quark data points were treated as entirely independent of each other in our fits. We recognize the attendant imprecision since the experimental errors are correlated, having a common overall normalization uncertainty. Thus, our procedure assigns more weight to the collider bottom quark data than they deserve.

To determine our starting distribution, we began with the parton densities of the Morfin-Tung set B1. We retained the form of the va-

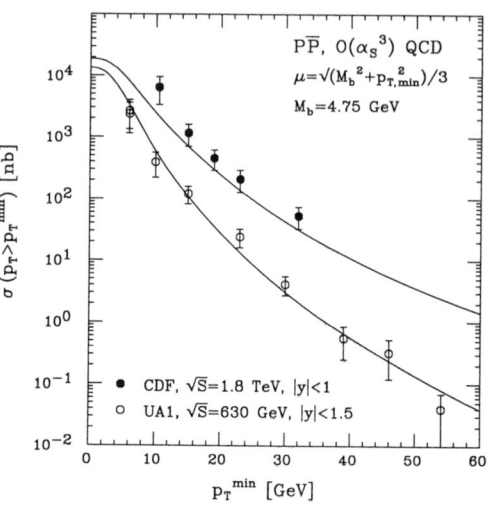

Figure 1. The solid curves show the results of our fit to the CDF and UA1 data. They are obtained from convoluting the $O(\alpha_s^3)$ QCD hard scattering cross section with the new parton densities determined from our combined fit to the deep inelastic scattering, bottom quark, and prompt photon data.

lence quarks and the x dependent shape of sea (quarks and antiquarks). The parametrization of the gluon density was altered to fit the bottom quark and prompt photon cross sections, and the normalization of sea was adjusted to satisfy the momentum sum rule. Evolution of the parton densities was carried out to two-loop level.

Our first conclusion[5] is that it is not possible to fit the CDF data adequately unless μ is allowed to decrease below $\mu/M_T = 0.5$. The CDF data prefer values of μ in the vicinity of $\mu = \frac{1}{3} M_T$. Settling on $\mu = \frac{1}{3} M_T$, we used the gluon and other densities determined from the fit to the CDF and UA1 bottom quark data as starting distributions in the simultaneous fit to the bottom quark data, prompt photon data, and data from deep inelastic lepton scattering,[10-12] along with data from mass-

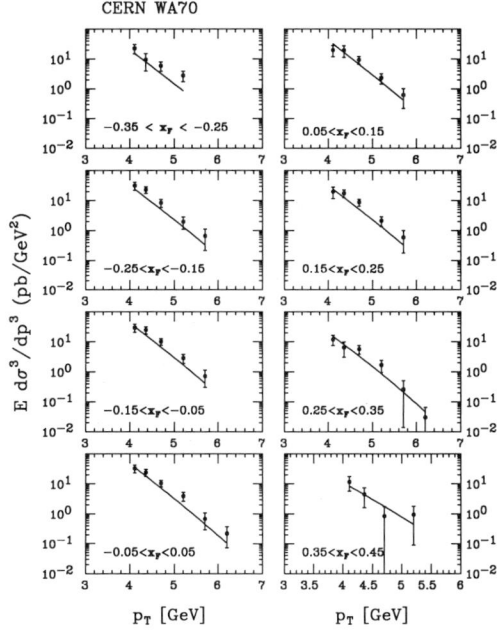

Figure 2. The solid curves show the results of our fit to the CERN WA70 data. They are obtained from convoluting the $O(\alpha_s^2)$ hard scattering cross section with the new parton densities.

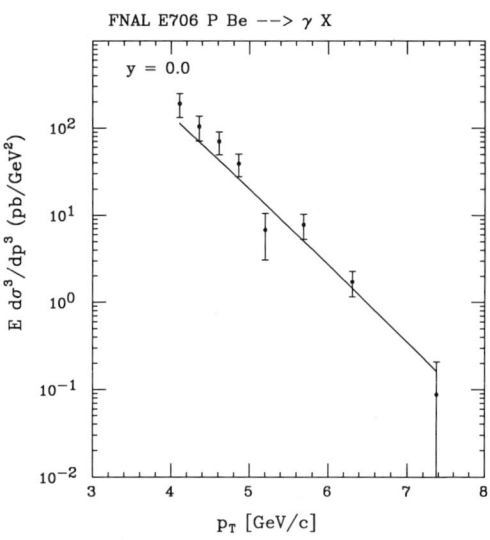

Figure 3. As in Fig. 2, but for the Fermilab E706 data.

ive lepton pair production.[14] We limited our fit to the deep inelastic data having $Q > 3.16$ GeV and hadron energy $W > 4$ GeV. In fitting the deep inelastic data our scale choice remained $\mu = Q$, but, as remarked above, we fixed $\mu = \frac{1}{3} M_T$ in the case of the bottom quark data. For the prompt photon data we fixed $\mu = \frac{1}{2} p_T$.

A satisfactory fit was obtained to the combined data set. Comparisons with the bottom quark data are presented in Fig. 1, comparisons with the CERN WA70 and Fermilab E706 data are presented in Figs. 2 and 3, and two examples of the fit to deep inelastic data are shown in Fig. 4 and Fig. 5. In Table 1 we list the values of χ^2 from the combined fit.

In Fig. 6 we show comparisons of our new gluon density with that from MT set B1 for three values of the scale μ, and in Fig. 7

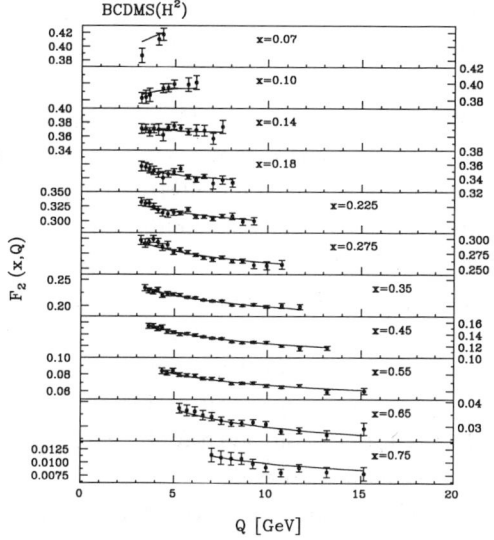

Figure 4. The solid curves show the results of our fit to the BCDMS hydrogen data set, $\mu p \to \mu X$. They are obtained from the new parton densities determined from our combined fit to the deep inelastic scattering, bottom quark, and prompt photon data.

we show the x dependence of the ratio of the new gluon density divided by the density in Morfin-Tung set B1. We note that the bottom quark data require a considerable increase in $xG(x,\mu)$ in the neighborhood of $x = 0.05$ for values of $\mu \simeq 5$ GeV.

The qualitative features of the curve in Fig. 7 are easy to understand. The CDF bottom quark data are sensitive to gluons with values of $x = 2M_T/\sqrt{s}$ in the range of $0.01 < x < 0.06$, and the UA1 bottom quark data to gluons in the range $0.03 < x < 0.16$. The magnitude of the CDF data exceeds predictions based on earlier gluon densities, including set B1 by about a factor of 2. Since gluon-gluon scattering dominates, the data require an increase in the normalization of the gluon density by roughly $\sqrt{2}$ in the relevant range of x. The fixed-target prompt photon data support the magnitude of $xG(x,\mu)$ for $x \geq 0.2$. In order to satisfy the momentum sum rule, the increase of the gluon density at intermediate values of x must be compensated by a decrease elsewhere, resulting in the depletion observed in Fig. 7.

The enhancement of $xG(x,\mu_o)$ near $x = 0.05$, required by the CDF data, also appears to force the new initial distribution to decrease moderately as x decreases below 0.01, in contrast to the MT B1 distribution which increases gently as $x \to 0$. We note, however, that the CDF bottom quark data with $|y| < 1$ do not place any constraint on the behavior of $xG(x,\mu)$ for very small x, i.e., $x < 10^{-3}$, and that QCD evolution rapidly reduces the difference as μ increases.

In this paper, we have focused on the strong constraints on the gluon density provided principally by bottom quark production data at collider energies. We have shown that a new gluon density may be derived from the CDF and UA1 data. While substantially different in character from published gluon densities, the new density is nevertheless compat-

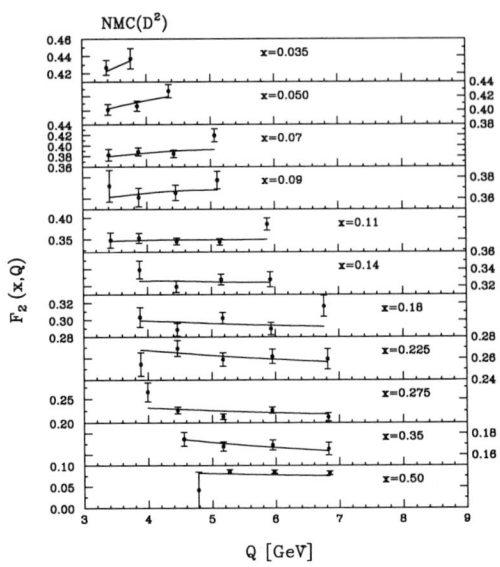

Figure 5. As in Fig. 4, but for the NMC deuteron data set.

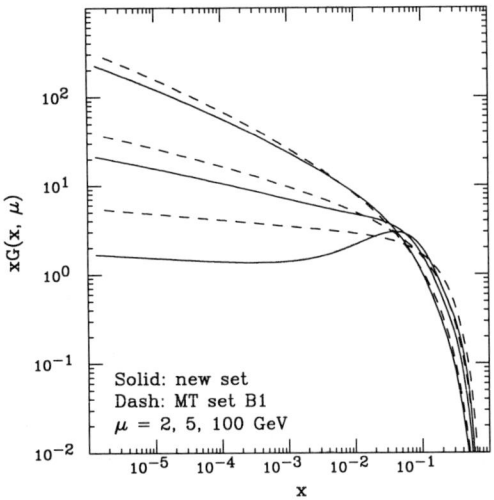

Figure 6. Solid and dashed curves show the behavior of our new gluon density and that from MT set B1 as a function of x, for three values of the factorization scale μ.

Table 1. Values of χ^2 from the combined fit to the prompt photon, bottom quark, deep inelastic lepton scattering and massive lepton pair production data. The top line specifies the data sets, and the second line lists the values of χ^2 divided by the number of data points. The BCDMS data are renormalized by a factor 0.975. The CCFR F_2 data are renormalized by a factor 0.94. The Fermilab E605 data on massive lepton pair production are renormalized by a factor 0.8.

UA1	CDF	WA70	E706	BCDMS		CDHSW		NMC		CCFR		E605
				F_2^H	F_2^D	F_2	xF_3	F_2^H	F_2^D	$0.94\times F_2$	xF_3	$\times 0.8$
0.64	1.15	0.98	1.36	0.68	1.08	0.58	0.51	1.64	1.15	2.34	0.88	0.90

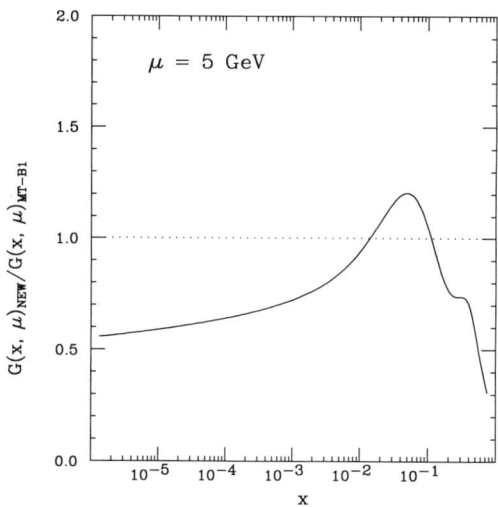

Figure 7. Ratio of the new gluon density to the gluon density in Morfin-Tung set B1. The ratio is evaluated at scale $\mu = 5$ GeV.

ible with the data on fixed target prompt photon production, deep inelastic lepton scattering, and massive lepton pair production that had been used for previous determinations of $xG(x,\mu)$. It has been observed[15] that the new NMC data[11] support an increase in the magnitude of the parton densities in the small x region. This trend is consistent with the effect we have determined by fitting the CDF data.

The longitudinal structure function $F_L(x, Q^2)$ measured in deep inelastic lepton scattering has often been recommended for a determination of $xG(x, Q^2)$. Experiments at HERA should be valuable, especially in the region $0.001 < x < 0.1$ where the new gluon density manifests a significant increase in magnitude.

We regard the work presented here in the spirit of a proof of principle that the CDF data are constraining and can be accommodated, more than in the spirit of a strict determination of $xG(x,\mu)$. Important questions revolve about the role of contributions in order α_s^4 and beyond. A more definitive quantitative fit awaits better understanding of these contributions and the greater statistical precision promised from forthcoming runs at the Tevatron.

REFERENCES

1. For a recent treatment, including references to prior work, consult E. L. Berger and R. Meng, Phys. Rev. **D46**, 169 (1992).

2. P. Nason, S. Dawson, and R. K. Ellis, Nucl. Phys. **B303**, 607 (1988); **B327**, 49 (1989); W. Beenakker, H. Kuijf, W. L. van Neerven, and J. Smith, Phys. Rev. **D40**, 54 (1989); W. Beenakker, W. L. van Neerven, R. Meng, G. Schuler, and J. Smith, Nucl. Phys. **B351**, 507 (1991).

3. E. L. Berger and J. Qiu, to be published; P. Aurenche, R. Baier, and M. Fontannaz, Phys. Rev. **D42**, 1440 (1990) and references therein.

4. J. G. Morfin and W. K. Tung, Z. Phys. **C52**, 13 (1991); J. Kwiecinski, A. D. Martin, R. G. Roberts, and W. J. Stirling, Phys. Rev. **D42**, 3645 (1990); M. Diemoz, F. Ferroni, E. Longo, and G. Martinelli, Z. Phys. **C39**, 21 (1988).

5. E. L. Berger, R. Meng, and W-K. Tung, Phys. Rev. **D46**, 1895 (1992).

6. CDF Collaboration, M. J. Shochet, Fermilab-CONF-91/341-E (1991).

7. UA1 Collaboration, C. Albajar et al., Phys. Lett. **B256**, 121 (1991).

8. WA70 Collaboration, M. Bonesini et al., Z. Phys. **C37**, 535 (1988); **C38**, 371 (1988).

9. E706 Collaboration, G. Alverson et al., Phys. Rev. Lett. **68**, 2584 (1992).

10. BCDMS Collaboration, A. C. Benvenuti et al., Phys. Lett. **B223**, 485 (1989), **B237**, 599 (1990); CDHSW Collaboration, J. P. Berge et al., Zeit. Phys. **C49**, 187 (1990).

11. NMC Collaboration, P. Amaudruz et al., CERN-PPE/92-124, submitted to Phys. Lett.

12. CCFR Collaboration, S. R. Mishra et al., Nevis Preprint 1459, submitted for publication in Phys. Rev. Lett.

13. E. L. Berger and J. Qiu, Phys. Rev. **D44**, 2002 (1991).

14. E605 Collaboration, C. N. Brown et al., Phys. Rev. Lett. **63**, 371 (1988).

15. A. D. Martin, R. G. Roberts, and W. J. Stirling, Rutherford Appleton Laboratory preprint RAL-92-021 (1992).

HEAVY QUARKS and QCD in e^+e^- COLLISIONS

Ties Behnke
II. Institut für Experimentalphysik der Universität Hamburg and DESY
DESY
Notkestrasse 85
D–2000 Hamburg 52
Germany

Abstract

Recent investigations of QCD aspects of heavy quark events in e^+e^- collisions are presented. Using three jet fractions in these events the flavor independence of the coupling constant of QCD, α_s, is verified: $\alpha_s^b/\alpha_s^{udsc} = 1.01 \pm 0.03 \pm 0.03$. The average scaled energy of charmed and bottom hadrons in decays of the Z^0 is determined. Comparing these results with measurements at lower energies clear evidence for scaling violation is established. Studies of the shape of the fragmentation function do not yet allow a significant discrimination between different theoretical models. A study of the differences between quark and gluon jets performed in a model independent way establishes, that gluon jets are softer and broader compared to quark jets in the same environment.

INTRODUCTION

QCD as the theory of the strong interaction has been very successful in describing the features of hadronic events observed in e^+e^- collisions. The large amount of data now available from LEP makes flavor dependent analyses of QCD properties possible. In this review first results from LEP and SLC on a number of such studies are presented, together with results from lower energy colliders.

THE FLAVOR DEPENDENCE OF α_s

In the Standard Model of the strong and elektroweak interactions the strong force is expected to be flavor blind. In particular this implies that the coupling constant of QCD, α_s, does not depend on the flavor of the particles interacting.

At LEP energies α_s has been determined with a precision of better than 6%: $\alpha_s = 0.118 \pm 0.007$ [1]. The error is now dominated by theoretical uncertainties, mostly due to the scale uncertainty. A large number of different measurements exist, but only for the normal mixture of flavors [2].

At LEP L3 and DELPHI have investigated the flavor dependence of α_s [3,4] by measuring the ratio $\alpha_s^b/\alpha_s^{udsc}$. They select bottom events by their semileptonic decays with inclusive prompt electrons or muons in the final state [5,6].

The strong coupling constant α_s is determined from the fraction R_{3jet}^b of 3–jet events observed. This ratio is in first order QCD directly proportional to α_s^b. Both collaborations use the JADE-algorithm [7] to define jets.

The DELPHI collaboration performs a second analysis, in which the momentum spectra of the different decays contributing to the inclusive lepton momentum are used to further

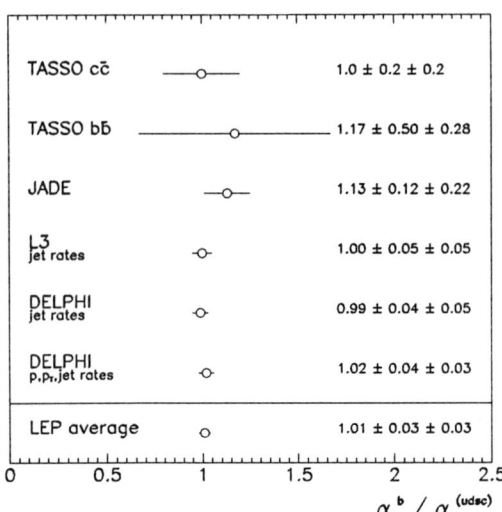

Figure 1. Measured Ratio $\alpha_s^b/\alpha_s^{udsc}$ at Petra and LEP energies

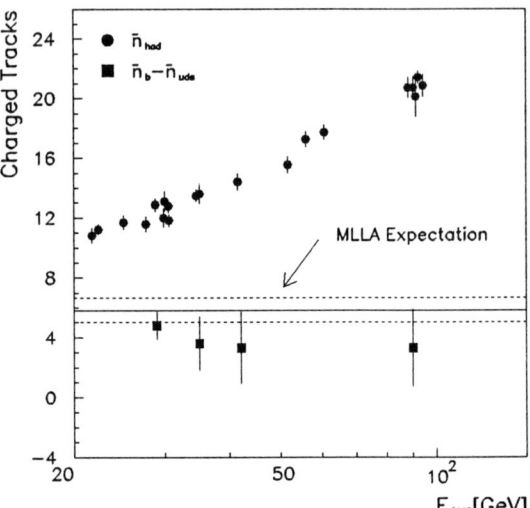

Figure 2. Energy dependence of the total multiplicity (circles) and the multiplicity difference between b and light quarks (squares)

constrain the measurement.

The results of the different methods are shown in Figure 1. Also included are results obtained at PETRA energies [8,9,10]. The L3 measurement and the second DELPHI measurement have been combined to form the LEP average of

$$\frac{\alpha_s^b}{\alpha_s^{udsc}} = 1.01 \pm 0.03 \pm 0.03. \quad (1)$$

This result is in good agreement with the hypothesis of flavor independence of QCD.

MULTIPLICITIES

The MARKII collaboration studies the multiplicity in heavy quarks events to extract information on the differences between heavy and light quark events [11,12]. They analyze the events in terms of the non–leading multiplicity. This quantity is defined as the number of all tracks, which are not decay products of the heavy, leading quark.

B hadrons are identified by their long lifetime. A total of 196 events are selected by requiring that a secondary vertex with $b/\sigma_b > 3.0$ exists, where b is the decay length reconstructed for this vertex, and σ_b the error on this number. Counting only tracks not coming from the decay of the B hadron they find an average non–leading multiplicity in b–events of $\bar{n}_{nl} = 12.0 \pm 1.8 \pm 0.6$.

Recent theoretical calculations in MLLA [13,14] show, that the difference between the light and heavy quark multiplicities is accessible through the non–leading multiplicity, and that it is constant with energy. In Figure 2 the MARKII result and lower energy results are compared with results from these calculations. Within errors the difference $\bar{n}_b - \bar{n}_{uds}$ is constant and agrees reasonably well with the prediction. Clearly the MARKII analysis is very much statistically limited.

FRAGMENTATION FUNCTIONS OF HEAVY QUARKS

At LEP energies hadrons containing heavy quarks are mostly produced at the elektroweak vertex, not in the cascade. They therefore provide a tool to investigate directly the properties of the heavy quarks produced. Contributions from gluon splitting are expected to be

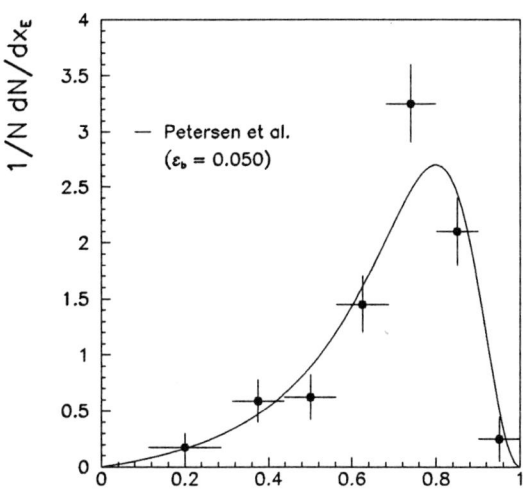

Figure 3. Bottom fragmentation function as determined by the L3 collaboration [6]

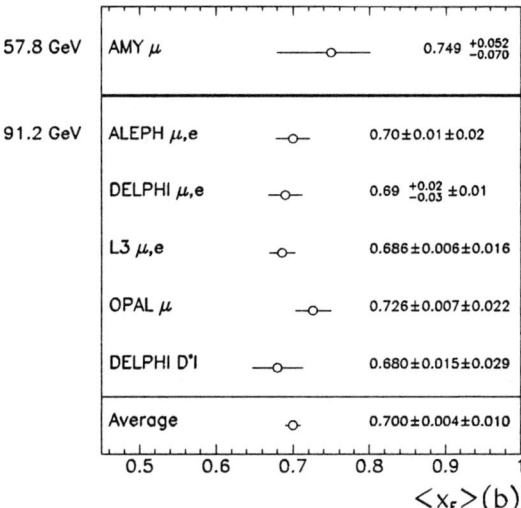

Figure 4. Mean energy $<x>_{b\to l}$ at 57.8 GeV and 91.2 GeV

very small.

Two variables are widely used in the parametrization of the heavy quark fragmentation, x_E and z. The first, x_E, is defined as $x_E = 2 \cdot E_{hadron}/E_{cms}$. It includes the effects of hard gluon radiation in the development of the cascade. The second, z, is more closely related to QCD models. A commonly used definition is $z = (E + p_\parallel)_{hadron}/(E + p_\parallel)_{quark}$ [14]. While x_E is directly experimentally accessible, z can only be calculated in the context of a specific model.

The Fragmentation Function of Bottom Hadrons

The fragmentation function for b–quarks so far has only been determined indirectly from their semileptonic decay into electrons or muons. Bottom events are tagged by an inclusive high momentum lepton. The momentum of the bottom hadron is reconstructed by unfolding the lepton spectrum from the observed one. All four LEP experiments determine the average momentum of bottom hadrons in this way [5,6,15,16].

The L3 and the OPAL experiment study the shape of these spectra in some detail. In Figure 3 the x-distribution as determined by L3 is shown. The OPAL collaboration performs a similar study, using the variable z instead of x. They compare their result with a number of theoretical parametrizations of the b–fragmentation function [17]. Within errors no distinction between them is possible. Both datasets agree with a often used parametrization proposed by Petersen et. al. [18].

The DELPHI collaboration attempts to directly reconstruct B meson decays and measure their energy by combining a sample of D mesons with an inclusive prompt lepton of the correct charge in the same jet [19].

The results of the different analyses are presented in Figure 4. Also shown is a measurement by the AMY collaboration [20] obtained at an energy of 57.8 GeV at the TRISTAN collider. Combining the LEP results leads to a LEP average of

$$x_E(b) = 0.700 \pm 0.004 \pm 0.010 .$$

The Fragmentation Function of Charm Hadrons

A direct measurement of the charm fragmentation function is possible through the investigation of fully reconstructed decays of charmed hadrons. Most accessible is the $D^{*\pm}$ meson and its decay $D^{*\pm} \to D^0 \pi^\pm; D^0 \to K^\mp \pi^\pm$. ALEPH, DELPHI and OPAL have selected samples of D^* mesons in this decay channel [21,22,23].

On the Z^0 peak about 50% of all D^* mesons are produced in $c\bar{c}$ decays, the remainder mostly in $b\bar{b}$ decays. The collaborations use different method to disentangle them. ALEPH and DELPHI try to determine from other variables the source of D^* mesons. ALEPH uses a topological discriminator made up from 9 event shape variables and optimized for b to c separation [24]. DELPHI utilizes the long lifetime of B hadrons to discriminate D^* from b and D^* from c. The OPAL collaboration tries to use as much information as available from lower energy experiments to estimate the contribution of decays from B–mesons.

ALEPH and OPAL repeat the analysis performed with inclusive leptons to extract information on the c–fragmentation in a manner equivalent to what is done in measuring the b–fragmentation function [15,16].

In Figure 5 the results of the measurements are summarized. Also included is a measurement of the AMY collaboration [20]. Combining the measurements using lepton tags at Z^0 energies

$$<x>_{c\to l} = 0.522 \pm 0.010 \pm 0.020$$

is found. From the events tagged with D^* mesons a mean x_E of

$$<x>_{c\to D^*} = 0.494 \pm 0.009 \pm 0.010$$

is determined.

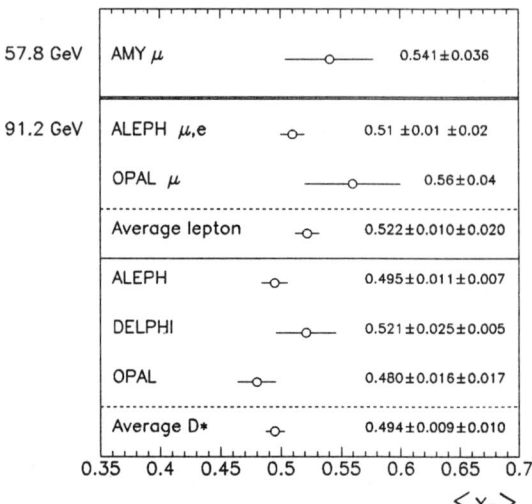

Figure 5. Comparison of different measurements of $<x>_{c\to D^*}$ and $<x>_{c\to l}$ at 57.8 GeV and 91.2 GeV

The Energy Dependence of the Fragmentation Function

In Table 1 the mean energies for charmed and bottom hadrons determined with the D^* tag and the inclusive lepton tag are listed. Values at energies below the LEP energies are taken from a review by Peter Maettig [25]. Also listed are the ratios between the mean energies at different energies. As discussed in a recent paper by Dokshitzer, Khoze and Troyan [14], meaningful predictions are possible for this ratio, since non–perturbative effects tend to cancel.

In Figure 6 the evolution of $<x>_{c\to D^*}$ with center of mass energy is plotted. Superimposed are prediction of the ARIADNE Monte Carlo Model [26] for different values of the QCD scale parameter $\Lambda_{ARIADNE}$. The data exhibit clear evidence for scaling violations.

In Figure 7 the ratio of the the mean energies from D^* and from leptons are plotted and compared with the expectations [13,14]. Within the errors the data correspond to a common value of $\Lambda_{LLA} \approx 426^{+150}_{-140}$ MeV, in good agree-

Table 1. Mean x_E for charm and bottom hadrons measured at different center of mass energies. Also shown are the ratios between different energies as described in the text

E_{cms}	$<x>_{c\to D^*}$	$\frac{<x>_{D^*}(91GeV)}{<x>(E_{cms})}$	$<x>_{b\to l}$	$\frac{<x>_b(91GeV)}{<x>_b(E)}$
10.4	$0.727 \pm 0.010 \pm 0.010$	$0.680 \pm 0.016 \pm 0.017$		
32.0	$0.587 \pm 0.011 \pm 0.010$	$0.842 \pm 0.022 \pm 0.022$	0.789 ± 0.010	0.887 ± 0.021
91.0	$0.494 \pm 0.009 \pm 0.010$		$0.700 \pm 0.004 \pm 0.010$	

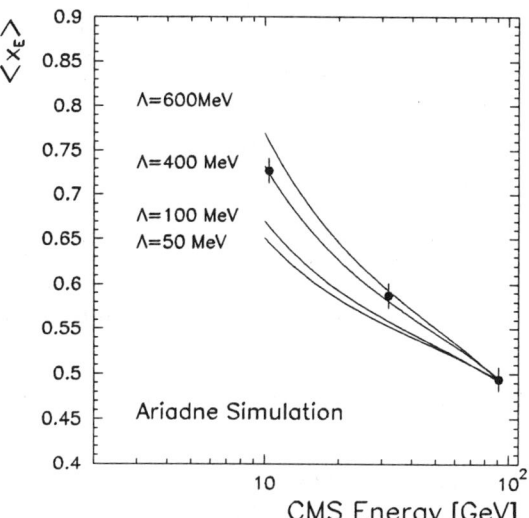

Figure 6. Evolution of $<x>_{c\to D^*}$ with center of mass energy. The curves are from a ARIADNE simulation, normalized to the observed value at 92.4 GeV.

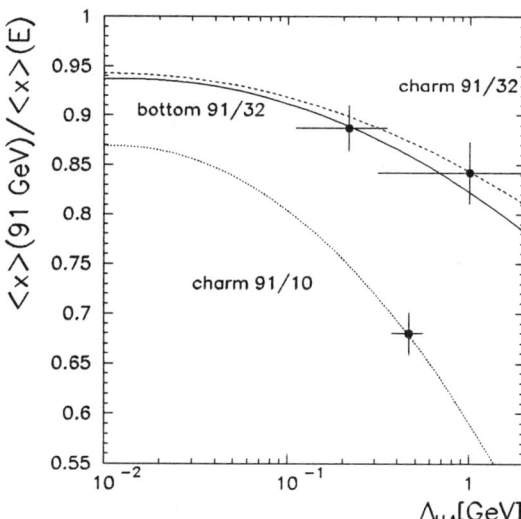

Figure 7. Ratio of $<x>$ at different center of mass energies. The curves are the expectation from MLLA [13,14]. The points are the data as listed in table 1.

ment with other measurements.

QUARK GLUON DIFFERENCES

It has been demonstrated that in a model independent way quark and gluon jets can be separated at LEP energies, using a special symmetric class of three jet events together with the identification of prompt inclusive leptons [27,28,29]. This technique has been used by the OPAL collaboration to study differences between quark and gluon jets.

OPAL finds, that gluon jets are significantly softer and broader than quark jets. The ratio of the mean multiplicity of the two samples is determined to be

$$\frac{<n>_{gluon}}{<n>_{quark}} = 1.06 \pm 0.03 \ . \quad (2)$$

From naive colour counting this ratio is expected to be 9/4. However this naive expectation is not directly applicable for this special class of events. More experimental and theoretical work is needed before a clear conclusion is possible.

CONCLUSION

In this paper results from QCD analyses of events tagged as decays of heavy quarks have been presented. Analyses of the dependence of the strong coupling constant on the flavor support strongly the QCD prediction, that α_s is independent of the flavor:

$$\alpha_s^b/\alpha_s^{udsc} = 1.01 \pm 0.03 \pm 0.03 \; .$$

The fragmentation properties of heavy quarks have been studied using lepton tags and fully reconstructed decays of the D* . Clear evidence for scaling violations are found. At this time no decisive discrimination between different proposed shapes of the fragmentation function is possible.

The study of QCD aspects of heavy quarks is just starting. Together with the rapidly increasing theoretical understanding of this subject precision test should very soon become possible.

ACKNOWLEDGEMENTS

I like to acknowledge the generous help provided by my collegues at the other e^+e^- experiments in gathering the material for this review.

REFERENCES

1. S.Bethke, Plenary talk at this conference.
2. S.Bethke, *Heidelberg HD - PY - 92 - 06* (1992).
3. DELPHI, preliminary, paper submitted to this conference.
4. L3, *Phys. Lett. B 271* (1991) 461.
5. DELPHI, *CERN PPE/92-79* (1992).
6. L3, *Phys. Lett. B 261* (1991) 177.
7. JADE, *Z. Phys C 33* (1986) 23.
   JADE, *Phys. Lett. B 213* (1988) 235.
8. TASSO, *Z. Phys. C 42* (1989) 17.
9. TASSO, *Phys. Lett. B 135* (1984) 317.
10. Günter Eckerlin (JADE) PhD. thesis Heidelberg (1990), unpublished.
11. MARKII, *Phys. Rev D 46* (1992) 453.
    B.A.Schumm et.al., *LBL-32725* (1992).
12. V.Khoze, contribution to this conference, *SLAC PUB 5909* (1992).
13. Y.Dokshitzer et.al. *LU TP 92 - 13* (1992).
14. see e.g. J.Chrin, *Z. Phys. C 36* (1987) 163.
15. ALEPH, preliminary, paper submitted to this conference, (1992).
16. OPAL, *Phys. Lett. B 263* (1991) 311.
17. OPAL, preliminary, *OPAL physics note PN-034* (1991).
18. C.Petersen et. al., *Phys. Rev. D 27* (1983) 105.
19. DELPHI, preliminary, paper submitted to this conference (1992)
20. AMY, preliminary, paper submitted to this conference (1992).
21. ALEPH, preliminary, paper submitted to this conference (1992).
22. DELPHI, preliminary, paper submitted to this conference (1992).
23. OPAL, preliminary, paper submitted to this conference, *OPAL physics note PN-092* (1992).
24. B.Brandl et. al., *PCCF RI 92.02* (1992). P.Henard (ALEPH), 4th International Symposium on Heavy Flavor Physics, ORSAY (1991).
25. P.Maettig, *Bonn HE 91-19* (1991).
26. ARIADNE Monte Carlo, L.Lönnblad, ARIADNE3.2.
27. OPAL, *Phys. Lett. B 265* (1991) 462.
28. OPAL, *Phys. Lett. B 261* (1991) 334.
29. OPAL, preliminary,*OPAL physics note PN-063* (1992).

MEASUREMENT OF THE STRONG COUPLING CONSTANT α_s AT LEP

Klaus Mönig
PPE Division
CERN
CH-1211 Genéve 23

Abstract

A summary on the measurement of the strong coupling constant α_s at the LEP e^+e^- collider from the hadronic width of the Z, from the hadronic branching ratio of the τ and from the shape of the hadronic final states is given.

INTRODUCTION

The theory of strong interactions (QCD) has only one free parameter (e.g. $\Lambda_{\overline{MS}}$) from which the coupling constant at any renormalization scale μ^2 can be computed. Thus the theory is fully determined once α_s is measured at some scale and QCD can be tested by measuring it from different processes or by checking that the energy dependence is as predicted. In addition, a precise knowledge of α_s is desired since unified theories predict a meeting of the three coupling constants of the SM at some energy scale and α_s has by far the largest error at the moment.
The LEP observables which are sensitive to α_s are calculated in perturbation theory. Many event shape variables are known to second order and some of them have been improved recently by next to leading log calculations. The hadronic partial width of the Z and the hadronic branching ratio of the τ are calculated to third order; for the latter also some leading terms have been resummed.
The renormalization scale μ^2 at which a quantity is evaluated in QCD is completely arbitrary and for an exact calculation the result should not depend on this. However, in fixed order, α_s changes when calculated at a different scale and the compensation for each term occurs only at the next order by terms of the type $\ln\frac{\mu^2}{s}$. The compensation is not perfect in $O(\alpha_s^n)$, but the uncertainty is of $O(\alpha_s^{n+1})$, so the apparent scale dependence can be used to estimate higher order terms where the scale is varied around some "natural" scale.

α_s FROM THE HADRONIC WIDTH OF THE Z

The partial width for the Z going into a fermion antifermion pair is given by

$$\Gamma_f = \frac{G_F M_Z^3}{6\pi\sqrt{2}}(g_V^2 + g_A^2)N_c^{(f)}(1 + \delta_{QCD}^{(f)})$$

with:

$$N_c^{(f)} = \begin{cases} 1 & (leptons) \\ 3 & (quarks) \end{cases}$$

where the hadronic width is defined as the sum of the quark partial widths. The QCD correction to the width vanishes for leptons and has the form [1,2]

$$\delta_{QCD} = \frac{\alpha_s}{\pi} + (1.41 - f(m_t))\left(\frac{\alpha_s}{\pi}\right)^2 - 12.8\left(\frac{\alpha_s}{\pi}\right)^3$$

for hadrons neglecting mass effects. The top mass dependent function $f(m_t)$ enters as a correction to the axial coupling at second order

Table 1. Results of the LEP experiments on R_τ

	R_τ	$\alpha_s(m_\tau^2)$	$\alpha_s(M_Z^2)$
ALEPH	3.60 ± 0.13	0.349 ± 0.066	$0.120 \begin{array}{c}+0.006\\-0.008\end{array}$
DELPHI	3.59 ± 0.15	0.319 ± 0.054	0.120 ± 0.007
L3	3.62 ± 0.18	0.36 ± 0.09	$0.122 \begin{array}{c}+0.008\\-0.011\end{array}$
OPAL	3.80 ± 0.16	$0.389 \begin{array}{c}+0.073\\-0.075\end{array}$	$0.123 \begin{array}{c}+0.006\\-0.008\end{array}$

The DELPHI and OPAL numbers are without the resummation

and has a value of about 0.5. The mass dependent corrections to the b partial width have been calculated by Kühn et al. [3] and are included in the result.

The vector- (g_V) and axial vector (g_A) couplings are affected by electroweak radiative corrections, which receive quite sizeable contributions depending on the unknown top and Higgs masses. These dependencies can, however, be eliminated to a large extent if instead of Γ_{had} itself the ratio R of the hadronic to the leptonic partial width is used to measure α_s. From the LEP avarage [4] of $R = 20.85 \pm 0.07$ one obtains

$$\alpha_s(M_Z^2) = 0.130 \pm 0.011 \pm 0.004 \pm 0.002,$$

where the first error is experimental, the second is due to the unknown top and Higgs masses and the third error comes from the uncertainties in the mass dependent QCD corrections. The residual top/Higgs mass error is calculated by taking $m_t = 130 \pm 30\,GeV$ as obtained from a global fit to the electroweak data [4] and varying m_H between 60 and 1000 GeV.

α_s FROM THE HADRONIC BRANCHING RATIO OF THE τ

The ratio of the hadronic to the electronic branching ratio of the τ is sensitive to QCD corrections similar to the corresponding ratio of the Z but at a much lower scale. It is given by [5]

$$R_\tau = 3(|v_{ud}|^2+|v_{us}|^2)(1+\delta_{EW}+\delta_{nonpert}+\delta_{pert})$$

with

$$\delta_{EW} = \frac{5}{12} + 2\ln\frac{M_Z}{m_\tau} \simeq 0.019$$
$$\delta_{nonpert} \simeq -0.01 \pm 0.004$$
$$\delta_{pert} = \frac{\alpha_s}{\pi} + 5.202\left(\frac{\alpha_s}{\pi}\right)^2 + 26.37\left(\frac{\alpha_s}{\pi}\right)^3$$

The v_{ui}'s are the relevant Kobayashi-Maskawa Matrix elements.

For the perturbative QCD prediction the 3rd order formula is given here. Recently a resummation of some leading terms has been calculated [6] which was used by some collaborations. The effect of this is a change in $\alpha_s(M_Z^2)$ of $+0.002$

R_τ can be either calculated from the leptonic branching ratio of the τ as done by all collaborations [7,8,9,10] or from the τ-lifetime, assuming Lepton universality. The L3 number is a combination of both methods [11].

Table 1 gives the results of the 4 LEP collaborations for R_τ, $\alpha_s(m_\tau^2)$ and $\alpha_s(M_Z^2)$.

The nonperturbative corrections can be measured if in addition to R_τ the mass spectrum of the hadronic decay products $\left(\frac{dN}{ds}\right)$ is known. The ALEPH collaboration [8] has measured the moments

$$D_{kl} = \frac{1}{N} \int_0^{m_\tau^2} \left(1 - \frac{s}{m_\tau^2}\right)^k \left(\frac{s}{m_\tau^2}\right)^l \frac{dN}{ds} ds$$

for $k = 1$, $l = 0, 1, 2, 3$.

They perform a fit to these moments plus R_τ where α_s and three parameters for the non perturbative corrections are varied simultaneously. They find $\alpha_s(m_\tau^2) = 0.354 \pm 0.067$ which leads to $\alpha_s(M_Z^2) = 0.120 \pm 0.007 \pm 0.001$, where the second error is due to the uncertainty from the extrapolation of α_s from m_τ to M_Z, in which case the flavour thresholds of c and b quarks have to be taken into account. If the LEP mean value of $R_\tau = 3.645 \pm 0.077$ is used together with the ALEPH D_{kl}'s one obtaines $\alpha_s(m_\tau^2) = 0.360 \pm 0.040$ and $\alpha_s(M_Z^2) = 0.121 \pm 0.004 \pm 0.001$

α_s FROM EVENT SHAPES

Up to now the differential distributions for approximately 15 event shape variables have been calculated to $O(\alpha_s^2)$ and for 4 of them the next to leading log calculations are complete.

Since the Geneva conference DELPHI has finalized their $O(\alpha_s^2)$ analysis with 8 variables where special emphasis is put on a coherent treatment of the parton shower and the matrix element hadronization models [12]. Their final result is $\alpha_s(M_Z^2) = 0.113 \pm 0.007$ where the error is dominated by theoretical uncertainties.

OPAL has published a very complete analysis with 11 variables [7]. In this analysis it is clear that the results from the different quantities do not agree when they are all analysed with a renormalization scale equal to the centre of mass energy and when only the exper-

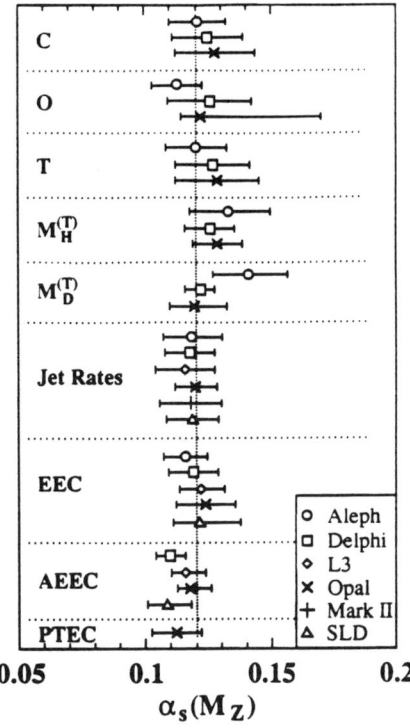

Figure 1. Measurement of the strong coupling constant from event shape variables in 2nd order from the LEP/SLC experiments. The avaraging over the different scales is done according to the OPAL method

imental errors are taken into account. However, if the theoretical error is estimated from the scale dependence and the central value is taken as a mean over the different scales the agreement is good. Fig. 1 shows the results of all LEP/SLC experiments where the averaging for the different scales is done by S. Bethke as closely as possible to the OPAL method [13]. From this $\alpha_s(M_Z^2) = 0.122 ^{+0.006}_{-0.005}$ is derived for OPAL and for all LEP/SLC experiments $\alpha_s(M_Z^2) = 0.120 \pm 0.006$.

Analysis in NLLA

Recently for some event shape variables it could be shown that they exponentiate and the

QCD corrections in next to leading log have been calculated [14]. As not even all first order terms are included in a NLLA calculation the calculation has to be matched with the corresponding fixed order calculation. This procedure is not unique. One way is to do the full matching in the logarithm and then exponentiate the result (lnR-matching), an alternative is to exponentiate every single contribution and then do the matching (R-matching). A third method, that should be theoretically preferred [14], is very close to the lnR-matching and will for that reason not be treated separately here. Up to now the thrust (T), the energy-energy correlation (EEC), the heavy jet mass with respect to the thrust axis $M_H^{(T)}$ and the heavy jet mass obtained from the minimization of the light plus heavy jet mass squared $M_H^{(M)}$ are available in NLLA. For the differential two jet rate (D_2) and the mean number of jets $<N_{jet}>$ in the so called Durham algorithm only the leading logs have been calculated. After summing the lnR terms, the only "natural" scale left is the centre of mass energy. So that scale is used by all collaborations except OPAL, which uses a slightly different scale for their central value.

and the same fit range as for the $O(\alpha_s^2)$ was used. The theoretical uncertainty was estimated from the difference of the results with a renormalization scale of $\mu^2 = s$ and from the scale that gave the best χ^2 in the fit using the same procedure as they applied in the $O(\alpha_s^2)$ analysis. Fig. 2 shows their analysis for $M_H^{(T)}$ with the fit results for the NLLA and the $O(\alpha_s^2)$ calculation. One can clearly observe the large improvement from the NLLA in the two jet region.

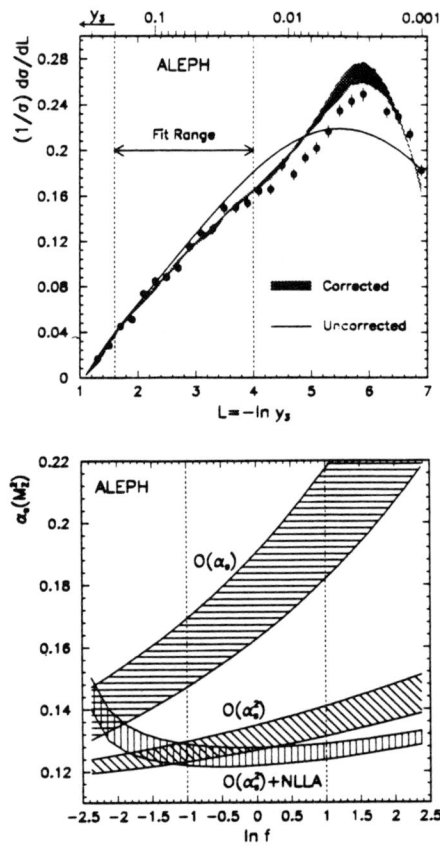

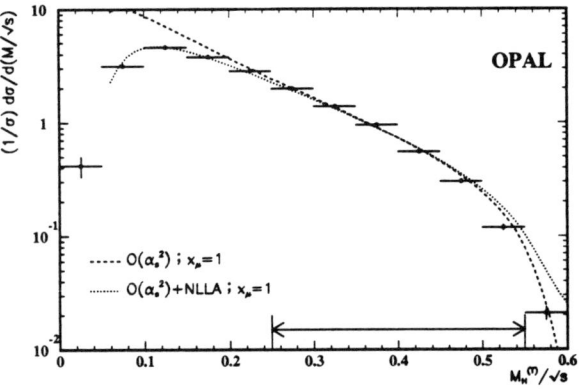

Figure 2. Measurement of the heavy jet mass by the OPAL collaboration

Figure 3. a) Measurement of the differential two jet rate by the ALEPH collaboration, b) scale dependence and matching ambiguity for this quantity

OPAL has analyzed $1-T$, $M_H^{(T)}$ and $M_H^{(M)}$ [7]. The analysis was done in the lnR-matching

ALEPH has analysed $1-T$, $M_H^{(T)}$, EEC and D_2 using an optimized fit range for the

NLLA calculations [15]. They take the mean of the lnR- and the R-matching and estimate the theoretical uncertainty from the matching ambiguity and the scale variation $-1 < \ln \frac{\mu^2}{s} < 1$. As this is a consistent procedure also D_2, where only the leading log is complete, can be used. Fig. 3a) shows the differential two jet rate. The band gives the fit result for different hadronization models, the solid line the prediction, if no hadronization correction is performed. Fig. 3b) shows the scale dependence and the matching ambiguity for fixed first and second order and for the NLLA analysis. One can see that both are significantly reduced for the NLLA case.

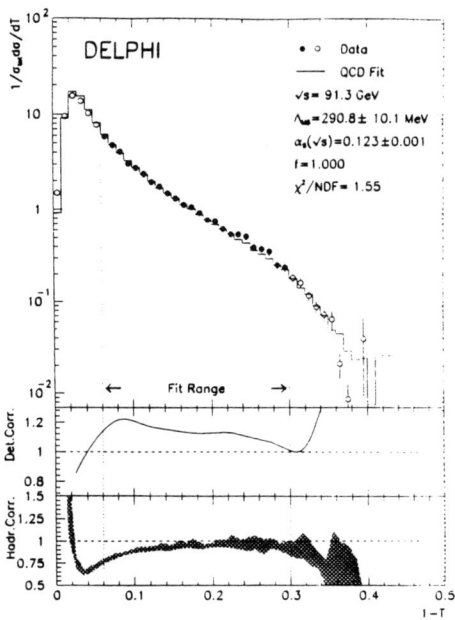

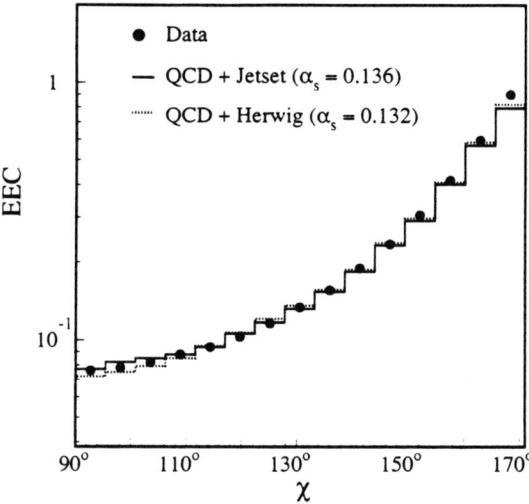

Figure 4. Measurement of the energy energy correlation by the L3 collaboration

L3 has analysed $1 - T$, $M_H^{(T)}$, EEC and $< N_{jet} >$, where the latter is not included in the mean because of the missing next to leading terms [16]. They use the lnR-matching and estimate the theoretical uncertainty from a scale variation between $0.25s$ and $4s$ or alternatively from the spread of the individual quantities. Both estimates agree. Fig. 4 shows their EEC analysis with a fit using either Jetset or Herwig for fragmentation corrections. The difference in the results is a typical hadronization uncertainty.

Figure 5. Measurement of the thrust by the DELPHI collaboration

DELPHI has analyzed $1 - T$, $M_H^{(T)}$, $M_H^{(M)}$, and EEC [17]. For the first three quantities they perform 2 analyses. At low y ($y = 1 - T, M_H^{(T)}, M_H^{(M)}$) the resummed terms dominate completely so that there is no matching ambiguity, the detector and hadronization corrections are, clearly, sizeable. The second analysis, which is used for the central values is done at intermediate y, where the matching ambiguity exists but the corrections are small. The two analyses agree only for the lnR-matching so this is used for the final results. For the EEC the first analysis is not possible. So for the second the mean of the two matchings is taken and the difference is included in the error estimation. Apart from that the theoretical uncertainty is taken from a scale variation between $0.5s$ and $2s$. Fig. 5 shows their $1 - T$ analysis together with the detector and hadronization corrections.

Fig. 6 compares the results of all experiments for different variables and derived

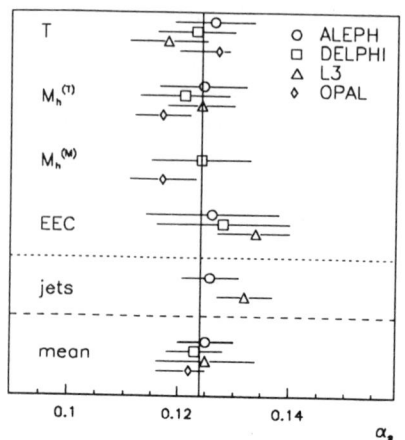

Figure 6. Summary of the NLLA results from LEP

means. From that an overall mean of

$$\alpha_s(M_Z^2) = 0.124 \pm 0.005$$

can be found.

α_s FROM SCALING VIOLATIONS

The DELPHI collaboration has done an analysis of the scaling violations in the scaled momentum distributions $\frac{1}{\sigma}\frac{d\sigma}{dx}$, $x = \frac{2p}{\sqrt{s}}$ using published data from PEP, PETRA, TRISTAN and ALEPH and unpublished data from DELPHI [18].
According to the Altarelli-Parisi equations the scaling violations should depend on α_s. However these equations cannot be used directly since the fragmentation functions of all quarks need to be known separately. This is especially important since the flavour composition changes strongly from the machines at lower energies to LEP. Instead α_s has been determined by fitting the Lund ME Monte-Carlo simultaneously to all data leaving $\Lambda_{\overline{MS}}$, the light quark and the heavy quark fragmentation functions as free parameters. This becomes possible by setting the jet resolution parameter to $y_{cut} = (9.1 GeV)^2/s$. For the fit only data at large Q^2 and large x have been used: $\sqrt{s} > 25\,GeV$, $x > 0.2$. To estimate the experimental systematic error, different combinations of experiments have been fitted and the known systematic errors of the different points were taken into account in the fit. To estimate the theoretical uncertainties the scale factor has been varied between 0.01 and 1, and the fit has been performed in different x ranges and with different y_{cut}. However, since only the difference between distributions at different energies matters the theoretical uncertainties turn out to be small. Fig. 7 shows the

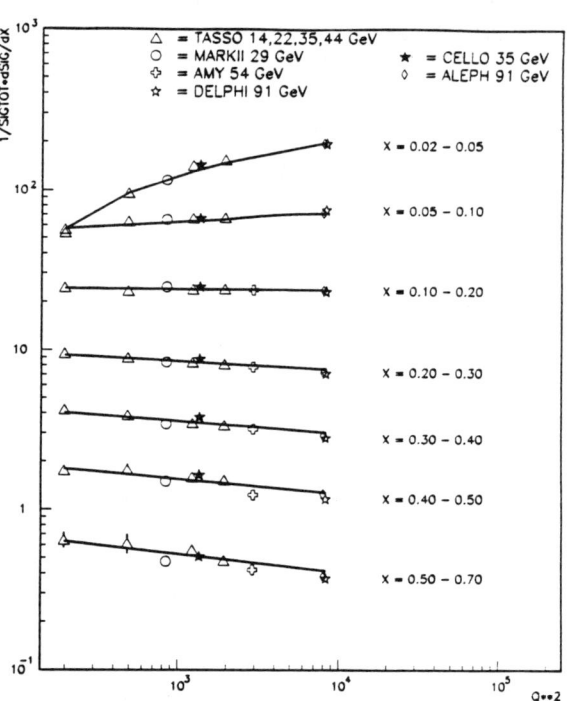

Figure 7. Measurement of the strong coupling constant from scaling violations

measured x-distributions together with the result of the fit. The agreement is good even outside the fit range. As a final result it was found

$$\alpha_s(M_Z^2) = 0.119 \pm 0.006,$$

Table 2. Summary of LEP results on α_s

Process	$\alpha_s(M_Z^2)$
Γ_{had}	0.130 ± 0.012
R_τ	0.121 ± 0.005
event shapes $O(\alpha_s^2)$	0.120 ± 0.006
event shapes NLLA	0.124 ± 0.005

which agrees well with the value obtained from event shapes and with the value from the scaling violations in deep inelastic scattering [19] of $\alpha_s(M_Z^2) = 0.113 \pm 0.005$.

CONCLUSIONS

Table 2 summarizes the LEP mean values on α_s for the different processes.

With R_τ and Γ_{had} LEP has two similar processes at different energy scales. From this it can be shown that α_s really runs. Comparison of the hadronic width with the event shape variables and scaling violations proves that α_s is process independent. As a mean of all LEP measurements one finds

$$\alpha_s(M_Z^2) = 0.123 \pm 0.005.$$

ACKNOWLEDGEMENTS

I would like to thank G. Altarelli, A. Pich, B. Webber and many of my colleagues from the LEP experiments for usefull discussions.

REFERENCES

1. Z. Kunszt and P. Nason, *Z Physics at LEP 1*, CERN 89-08, Vol. 1 373
2. S.G. Gorishny, A.L. Kataev and S.A. Larin, *Phys. Lett.* B259 (1991) 144
   L.R. Surguladze and M.A. Samuel, *Phys. Rev. Lett.* 66, 560 (1991)
3. K.G. Chetyrkin, J.H. Kühn, *Phys. Lett.* B248 (1990) 359
4. L. Rolandi, *Tests of Electroweal Interactions*, proceedings of this conference
5. E. Braaten, S. Narison, A. Pich, *Nucl. Phys.* B373 (1992) 581
6. F. Le Diberder, A. Pich, CERN-TH 6421/92
7. OPAL collaboration, P.D. Acton et al., *Zeit. Phys.* C 55 (1992) 1
8. ALEPH collaboration, note submitted to this conference
9. L3 collaboration, B. Adeva et al., *Phys. Lett.* B265 (1991) 451
10. DELPHI collaboration, private comunication
11. L3 collaboration, private comunication
12. DELPHI collaboration, P. Abreu et al., *Zeit. Phys.* C 54 (1992) 55
13. S. Bethke, J.E. Pilcher, HD-PY 92/06 to appear in Ann. Rev. Nucl. Part. Sci.
14. S. Catani, G. Turnock, B.R. Webber and L. Trentadue, *Phys. Lett.* B263 (1991) 491
    S. Catani, G. Turnock and B.R. Webber, *Phys. Lett.* B272 (1991) 368
    G. Turnock, Cavendish-HEP-92/3 (1992)
    S. Catani et al. CERN TH 6328/91
    S. Catani et al., *Phys. Lett.* B269 (1991) 432
15. ALEPH collaboration, D. Decamp et al., *Phys. Lett.* B284 (1992) 163
16. L3 collaboration, B. Adeva et al., *Phys. Lett.* B284 (1992) 471
17. DELPHI collaboration, P. Abreu et al., Paper submitted to this conference
18. DELPHI collaboration, P. Abreu et al., Paper submitted to this conference
    W. de Boer, T. Kußmaul, IEKP-KA/92-11
19. A.D. Martin, W.J. Stirling, R.G. Roberts, *Phys. Lett.* B266 (1991) 73

MULTI-JET PRODUCTION RATES IN DEEP-INELASTIC MUON-PROTON SCATTERING

Carlos W. Salgado*
Fermi National Accelerator Laboratory
P.O.Box 500
Batavia, IL 60510, U.S.A.

Abstract

Measurements of forward multi-jet production rates in deep-inelastic muon-proton scattering are presented. Data were taken with a 490 GeV muon beam incident on a hydrogen target. Jets were defined using the JADE jet finding algorithm. The measured rates are presented as function of W, the hadronic center-of-mass energy and the jet resolution parameter, y_{cut}, in energies up to W=33 GeV. Good agreement is found in comparisons with predictions of the QCD-inspired Lund Monte Carlo models. Non-perturbative QCD production mechanisms, inside the Lund Model, can not reproduce the results for energies greater than $W \simeq 20$ GeV. Sensitivities of the jet rate measurements to the low x ($x \simeq 0.02$) gluon content of the nucleon and the evolution of α_s are studied.

INTRODUCTION

Deep-inelastic lepton-nucleon scattering (DIS) has been associated with measurements of nucleon structure functions[1] and, through them, the parton distributions and the strong coupling constant, α_s.[2] Measurements of DIS cross-sections represent one of the cleanest ways to obtain the parton distributions and to measure α_s. Studies of differential cross-sections observing the final hadronic system produced in DIS, as for example multi-jet production, can also be used for the determination of these quantities. However, these measurements have not acquired the same degree of precision as the structure functions measurements. Measurements of multi-jet production rates, by the E665 experiment at Fermilab[3], have been possible for the first time due to the highest muon beam energies available at the Tevatron.

First order (α_s) QCD corrections to the DIS Born cross-section include two new final state processes: gluon bremsstrahlung and photon-gluon fusion, shown in figure 1. For a fixed beam energy, the kinematics of the order α_s final state partonic system are defined by five variables. Two variables describe the electroweak vertex kinematics, Q^2: the negative square of the virtual photon four-momentum and $x = Q^2/2Pq$, and three describe the final partonic phase space, $x_p = Q^2/2pq, z = pp_1/pq$ and ϕ the azimuthal angle of the partonic plane relative to the μ scattering plane. The fraction of the nucleon momentum carried by the struck parton, η, is given by $\eta = x/x_p$, since $p = \eta P$. The total hadronic center-of-mass energy is W, related to the previous vari-

*Representing the Fermilab E665 Collaboration

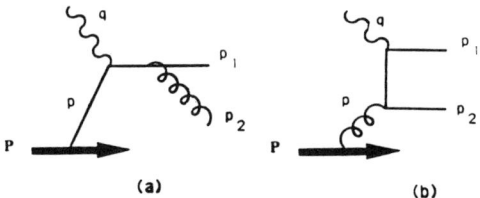

Figure 1. First order α_s corrections to the deep-inelastic muon-proton scattering. (a) Gluon bremmstrahlung (b) photon-gluon fusion.

ables by $W^2 = (P+q)^2 \simeq Q^2(1-x)/x$.

The production rates for these processes have been calculated in perturbative QCD[4]. At order α_s, they are directly proportional to α_s, in contrast to the more complicated total DIS cross-section expressions.

After fragmentation, given a particular jet finding algorithm and enough center-of-mass energy, these events can be observed as three-jet events. They can be described in the virtual photon-proton center-of-mass system as two jets produced by the fragmentation of the final partonic states (mostly going forward with respect to the the direction of the incoming virtual photon) and one backward going jet produced by the target remnants. The usual nomenclature for this configuration is (2+1) jets ((1+1) for the Born term).

DATA

The E665 experiment[5] used a 490 GeV muon beam which struck a 1.15 m long hydrogen target. Charged particles reconstructed in the tracking system and fitted to the primary vertex and neutral particles reconstructed in the electromagnetic calorimeter are used. The tracking system is essentially in the forward direction (data from the streamer chamber detector that surrounded the target region are not included in this analysis). The event sample is defined by appling the following kine-

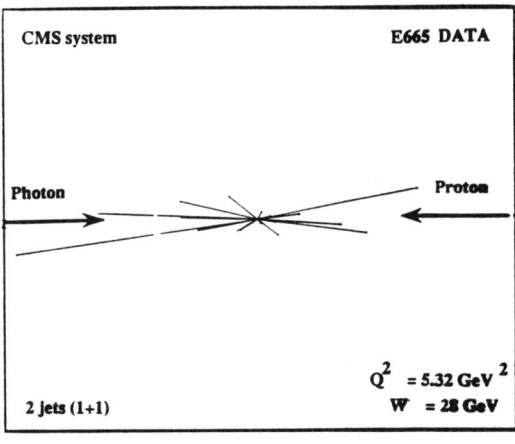

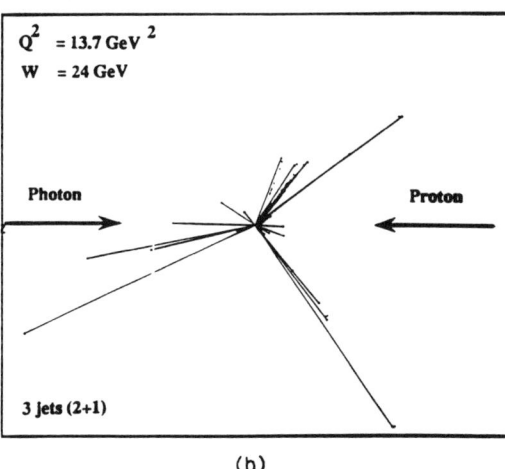

Figure 2. E665 observed events boosted to the virtual photon-proton center-of-mass frame and projected into the hadronic plane: (a) a (1+1)-jets event, (b) a (2+1)-jets event.

matical cuts: $Q^2 > 4.0\ (GeV/c)^2$, $\nu > 40 GeV$, $x > 0.003$ and $0.05 < \nu/E_{Beam} < 0.95$, where ν is the energy of the virtual photon in the lab frame. Events from coherent photon bremsstrahlung were identified as such if $\nu > 200$ GeV and $E_{Cal}/\nu > 0.035$ and removed from the data sample. Only particles going forward in the virtual photon-proton center-of-mass system were considered. More details on the analysis can be found in reference [3].

Figure 2 shows two observed E665 events boosted into the virtual photon-proton center of mass system and projected onto the hadronic plane (the plane containing the virtual photon and in which the sum of the squared transverse momenta of the hadrons is maximized). The line segments are proportional to the particle momenta and neutral and charged particles are shown. Electronic and streamer chamber detector information is included. Figure 1a shows the expected topology of a (1+1) jet event and figure 1b shows a (2+1) jet event, with the backward jet in the direction of the proton and the two forward jets in the direction of the photon. A qutative jet analysis, such as the measurement of n-jet rates, needs a jet "counting" algorithm and corrections for acceptance and detector efficiencies.

JET RATES MEASUREMENTS

To define the number of jets in an event we use the JADE jet finding algorithm[6]. The JADE algorithm was designed to resemble the technique use in perturbative QCD calculations to overcome soft and colinear singularities. In the experimental implementation of the algorithm, a test variable, $y_{ij} = 2E_iE_j(1-cos\theta_{ij})/(\epsilon W)^2$ is calculated for each pair of hadrons, i and j. $E_{i,j}$ are the particle energies, θ_{ij} the angle between them and ϵ an efficiency factor applied to W defining the energy scale. All quantities are calculated in the virtual photon-proton center-of-mass system. The minimum y_{ij} is compared to the jet resolution parameter, y_{cut}. If $y_{ij}^{min} < y_{cut}$ the two particles four-momenta are added to form a new particle k, such that $p_k^\mu = p_i^\mu + p_j^\mu$. The procedure is repeated until y_{ij}^{min} is larger or equal to y_{cut}. The resulting combined particles are called jets. All charged particles are assumed to be pions and all neutral particles to be photons. The forward n-jet rates are the ratios between the number of events with forward n-jets and the total number of events. We found that using $\epsilon = 0.5$ when applying the algorithm to the raw data minimized the detector efficiency and acceptance corrections. The final results were corrected for this choice of scale and are presented for $\epsilon = 1.0$.

A GEANT[7]-based Monte Carlo simulation of our detector was used to correct the data distributions for geometrical acceptance, reconstruction efficiency and resolution. The Lund Monte Carlo (LEPTO 5.2 and JETSET 6.3)[8] was used as the physics generator. Both versions of the Lund generator, matrix elements (ME) and parton showers (PS), were able to reproduce many aspects of the data, including the uncorrected jet rates, with similarly good accuracy. The data distributions were corrected bin by bin using:

$$R_n^{corrected}(y_{cut}, W) = \frac{R_n^{MC\ true}}{R_n^{MC\ recon}} R_n^{data}(y_{cut}, W)$$

where $R_n^{MC\ true}$ was obtained applying the JADE algorithm to the primary forward hadrons generated by the Monte Carlo and $R_n^{MC\ recon}$ was obtained from Monte Carlo generated events subjected to the identical analysis as to the data.

We estimated the systematic uncertainties in the jet rates due to the event and particle se-

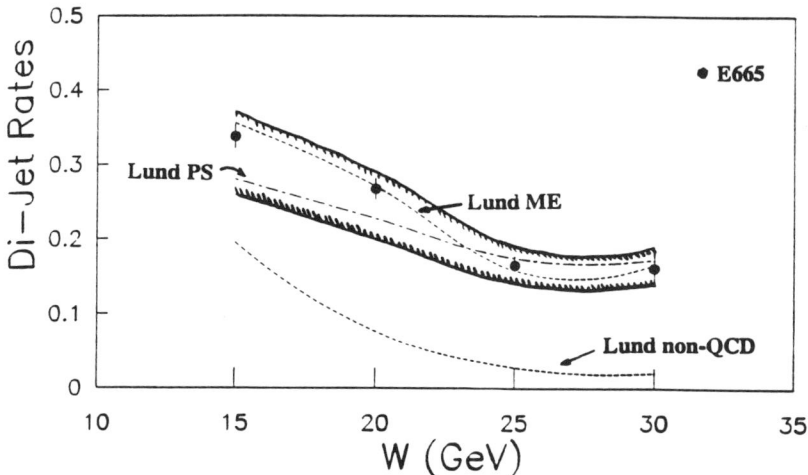

Figure 3. Forward di-jet rates versus W, for $y_{cut} = 0.04$. Different Lund model predictions are also shown.

lection criteria to be less than ±0.01, and that due to our ability to model the acceptance and efficiency of the apparatus to also be less than ±0.01. We varied the initial choice of energy scale in the definition of y_{ij} from values of ϵ between 0.3 and 1, and found that the final corrected jet rates values were very stable, at the ±0.015 level. The most important uncertanty came from the physics generator used within the Monte Carlo. We estimated this uncertainty by comparing the corrected rates using the ME and the PS options of the Lund Model. The differences in the resultant rates were Q^2 dependent but were always less than 0.04. The combined systematic uncertainties were estimated adding the model dependent uncertainties to the sum in quadrature of all the others. This total systematic uncertainty is always less than 0.06.

Figure 3 shows the forward di-jet rates versus W for a value of $y_{cut} = 0.04$. Also shown are the predictions of the Lund ME and PS Monte Carlos. Default values for the parameters in both Lund options and Morfin & Tung-LO parton distributions[9] were used. Good agreement is observed with the QCD based models, but the naive parton model based Lund Monte Carlo ("non-QCD") can not explain the data.

Error bars represent binomial statistical errors and the marked region, the estimated systematic uncertainties.

QCD ANALYSIS

QCD information from the hadronic system might be extracted if hadron-parton duality can be established; in other words, if the hadronic jet rates measure the underlying partonic jet rates predicted by perturbative QCD. Figure 4 shows the corrected forward hadronic jet rates compared to the forward partonic jet rates as simulated by the ME Lund Monte Carlo (obtained by applying the JADE finding algorithm to the partons generated by the Lund Monte Carlo). They agree at the 20% level. The next step will be to find the relation between the forward produced rates and the (2+1) jet rates calculated by perturbative QCD. Different jet finding algorithms might also be explored to minimize these corrections. We are still working to disentangle these effects.

The goal is to obtain information about the gluon distribution function, $G(\eta, Q^2)$ and the strong coupling constant, α_s. This information can be obtained by comparing data and

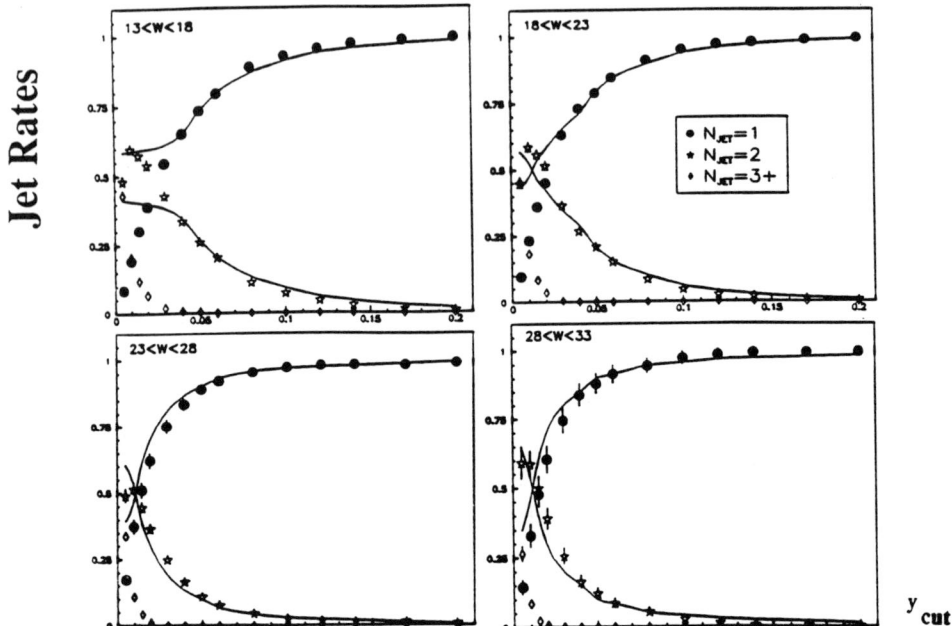

Figure 4. Forward jet rates versus y_{cut} in four W bins. The solid lines are forward partonic jet rates from the Lund ME Monte Carlo simulation.

perturbative QCD predictions. I will present here our first attempts to estimate sensitivity of the jet rates to $G(\eta, Q^2)$ and α_s. To fix the final partonic state phase space, we chose to fix $x = 0.02$ and $y_{cut} = 0.04$ (limits of the phase space integration). If we assume that all parton distributions[9] and the mass of the quarks are known, the only unknown quantity in the (2+1) jet rates is α_s.

Figure 5a shows predictions for the (2+1) jet rates from the ME Lund Monte Carlo model at the partonic level, for fixed (at $Q^2 = 7 GeV^2$) and running α_s. The Lund Monte Carlo predicts, in our energy regime, a maximum difference of rates between α_s fixed and running, of the order of 0.02. As explained before, the uncertainties associated with our present measurements are of the order of 0.06.

The gluon distribution function enters the (2+1) jet rates through an integral of the form, $\int_x^1 G(x/x_p, Q^2) dx_p$. At our values of x the photon-gluon fusion process is expected to have a big contribution to the (2+1) jet cross-section, thus we expect to be sensitive to the form of $G(\eta, Q^2)$.

To study the sensitivity of the jet rates to the gluon distribution, we chose the form, $\eta G(\eta, Q^2) = A_0 \eta^{-A_1}(1-\eta)^{A_2}$ and studied the jet rates sensitivity to changes to the value of A_1. We keep A_2 fixed (from LO Morfin &Tung distributions[9]) and change A_0 to maintain the sum over the gluon momenta constant when changing A_1. Figure 5b shows the partonic (2+1) jet rates obtained from the Lund ME Monte Carlo for three values of A_1; 0.0, 0.4 and 0.7. The absolute difference between the rates is of the order of 0.09. Comparing to our present experimental uncertainty (of the order of 0.06), we expect to be sensitive to values of A_1 of the order of $\Delta A_1 = 0.7$.

CONCLUSIONS

Measurements of forward multi-jet production rates have been made in the range $13 \leq W \leq 33$ GeV using the JADE jet finding algorithm.

Comparisons, at the hadron level, to ME and

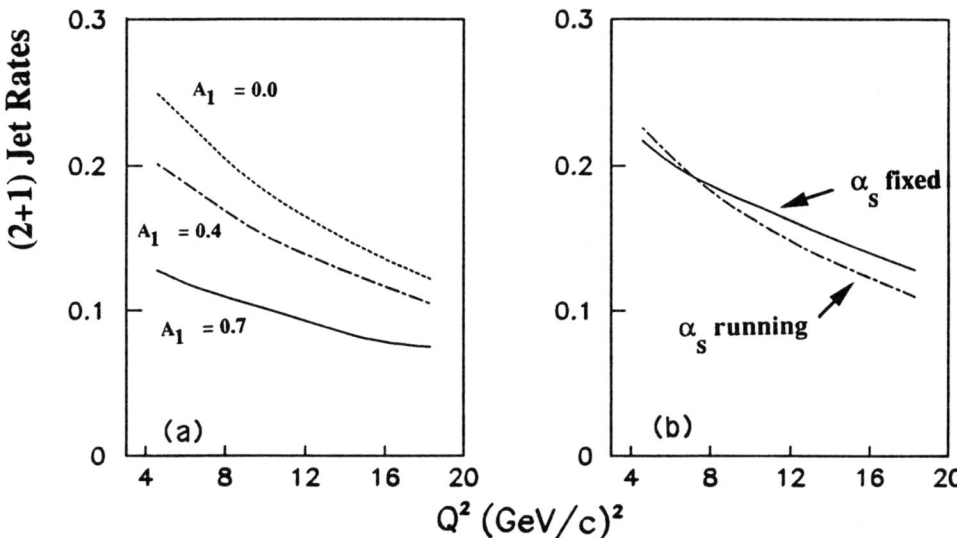

Figure 5. Lund ME Monte Carlo partonic (2+1) jet rates versus Q^2, (a) for three different choices of the gluon distribution, and (b) for α_s fixed (at $Q^2 = 7\,GeV^2$) and running.

PS Lund Model predictions are in good agreement with the data, whereas the predictions of the bare parton model (non-QCD) Lund Monte Carlo are excluded, at least for energies greater than $W \simeq 20 GeV$. A decrease in the resolved di-jet rates is observed as a function of W, in qualitative agreement with perturbative QCD.

Preliminary results from the QCD analysis indicate that the use of the jet rates measurements to determine the running of α_s is not possible, at this time, due to the size of our statistical and systematic uncertainties. On the other hand, these measurements can be used to constrain the gluon distribution at low values of x ($x \simeq 0.02$).

REFERENCES

1. S.R. Mishra and F. Sciulli, Annu. Rev. Nucl. Part. Sci. **39**, 259 (1989).
2. Proceedings of the Workshop on Hadron Structure Functions and Parton Distributions, edited by D. Geesaman, J. Morfin, C. Sazama and W. K. Tung, World Scientific, Singapore, (1990).
3. M.R. Adams, et al., Phys.Rev. Lett. **69**, 1026 (1992).
4. T. Brodkorb, J.G. Körner, E. Mirkes and G.A. Schuler, Z.Phys. C **44**, 415 (1989); D. Graudenz, Phys. Lett. B **256**, 518 (1991).
5. M.R. Adams, et al., Nucl. Inst. and Meth. **A291**, 533 (1990).
6. W. Bartel, et al., Z. Phys. C **33**, 23 (1986).
7. R. Brun, et al., CERN-DD/EE/84-1 (1987).
8. G. Ingelman, in *The Lund Monte Carlo for Lepto-Nucleon Scattering - LEPTO, Version 5.2*, unpublished Program Manual..
9. J.G. Morfin and W.K. Tung, Z. Phys. C **52**, 13 (1991).

RECENT PROGRESS IN QCD JET PHYSICS IN e^+e^- COLLISIONS

B.R. Webber[*]
Theory Division
CERN
CH-1211 Geneva 23
Switzerland

Abstract

Recent progress is reviewed in the following areas of e^+e^- jet physics: exponentiation of large higher-order logarithms and its application to improve determinations of the strong coupling constant from event shape distributions; jet clustering algorithms and studies of jet multiplicities using the new $k_\perp$-algorithm; studies of two-particle correlations in jet fragmentation.

INTRODUCTION

Over the past two years, the great wealth of new data on hadronic final states in Z^0 decay has prompted a strong revival of activity in QCD jet physics. A better understanding of jet properties is important, not only to test QCD itself and to determine its fundamental parameter α_S, but also to be able to use jet signatures for other kinds of physics. In this brief review, I concentrate on some of the recent theoretical progess. Comparisons between these theoretical ideas and experiment are presented in other contributions to this session, and in the plenary talk on QCD.

LARGE LOGARITHMS

Many of the most detailed current tests of QCD depend on comparing e^+e^- annihilation data with predictions based on perturbation theory. For totally inclusive quantities, such as the hadronic cross section, the perturbative prediction is a well-behaved power series in α_S, with coefficients that are simply numbers of order unity. For many quantities of interest, however, the coefficients may contain large logarithmic factors that have to be resummed to all orders before a reliable prediction can be obtained.

Quantities that are still 'sufficiently inclusive' have a single logarithm for each power of α_S, which can be resummed using renormalization group techniques. The scaling violation in the jet fragmentation function $D(x,Q)$ (the distribution in energy fraction x of hadrons in a jet produced at a momentum scale Q) is of this type at values of x that are not close to the kinematic limits. In this case the large logarithm is of the form $\ln(Q_1/Q_2)$ where Q_1 and Q_2 are two widely separated scales at which the fragmentation functions are compared.

In such quantities, logarithms associated with soft gluon emission cancel between real and virtual contributions. When this cancellation is incomplete, for example near the boundary of phase space (i.e. at very small or very large x in the fragmentation function) we

[*]On leave of absence from Cavendish Laboratory, University of Cambridge, UK.

have a 'semi-inclusive' quantity, which is finite in perturbation theory but has up to two large logarithms for each power of α_S.

Representing the large logarithm by L, a semi-inclusive quantity $F(L)$ thus has a perturbative expansion of the general form

$$F(L) \sim 1 +$$
$$+ \alpha_S(A_{12}L^2 + A_{11}L + A_{10} + \ldots)$$
$$+ \alpha_S^2(A_{24}L^4 + A_{23}L^3 + A_{22}L^2 + \ldots) \quad (1)$$
$$+ \alpha_S^3(A_{36}L^6 + A_{35}L^5 + A_{34}L^4 + \ldots)$$

Now suppose we make a change of scale in the definition of α_S:

$$\alpha_S(\mu^2) \to \alpha_S(f\mu^2) = \alpha_S - \beta_0 \alpha_S^2 \ln f + \ldots \quad (2)$$

There must be a corresponding change in the coefficients A_{ij} in order to cancel the scale dependence order by order

$$\begin{aligned} A_{22} &\to A_{22} + A_{12}\beta_0 \ln f + \ldots \\ A_{34} &\to A_{34} + A_{24}\beta_0 \ln f + \ldots \end{aligned} \quad (3)$$

It follows that in order to measure α_S (or equivalently the value of the QCD scale Λ in a particular renormalization scheme) from a semi-inclusive distribution one needs to know the values of the *next-to-next-to-leading* coefficients $A_{n,2n-2}$ to all orders.

EXPONENTIATION

For some quantities there are factorization theorems [1] that imply *exponentiation* of logarithms in the following sense:

$$\begin{aligned} F(L) \sim\ & (1 + C_1\alpha_S + C_2\alpha_S^2 + \ldots) \\ & \times \exp[\alpha_S(G_{12}L^2 + G_{11}L) \\ & + \alpha_S^2(G_{23}L^3 + G_{22}L^2 + G_{21}L) + \ldots] \end{aligned} \quad (4)$$

The corrections to this result vanish as $L \to \infty$. Notice that in the exponent the coefficients are such that $G_{n,m>n+1} = 0$, that is, one finds only $n+1$ instead of $2n$ large logarithms in the expansion of $\ln F$ to order α_S^n. A change of scale in α_S changes the *next-to-leading* coefficients in the exponent, G_{22} etc. Thus when exponentiation can be proven it suffices to know the leading and next-to-leading logarithmic coefficients in the exponent in order to determine α_S.

In QED, factorization of matrix elements for multiple soft photon emission, and exponentiation of the corresponding logarithms, follow from the effective independence of successive emissions. In QCD the gauge boson is itself charged and therefore the situation is more complicated. It turns out, however, that for sufficiently inclusive quantities there is a cancellation between leading non-Abelian logarithms from real and virtual contributions, and remaining next-to-leading logarithms can be resummed by replacing α_S by a running coupling with a suitable argument. In this way one obtains a generalized form of exponentiation for the non-Abelian case [1, 2].

Factorization of matrix elements alone is not sufficient for the exponentiation of large logarithms in the distribution of an observable quantity. Factorization of the phase space in the relevant region is also needed. As discussed below, this can be proven for certain e^+e^- event shape measures near the two-jet configuration, leading to the possibility of measuring α_S from event shape distributions in this semi-inclusive region.

EVENT SHAPES

The fitting of hadronic event shape distributions in e^+e^- annihilation is by now a well established method for α_S determination. A number of infrared-safe event shape variables have been defined and computed in full to second order in α_S [3]. The perturbative predictions can be compared with experimental data after correction for the non-perturbative hadronization process. However, there remain some serious uncertainties about how best to perform such comparisons:

i) *Range to be fitted.* Near the kinematic limits, the perturbative expansions for event shapes involve large double logarithms of the type described above. In comparing data with fixed-order calculations one must therefore use only a restricted region well away from phase space boundaries, whereas most of the data are close to the limit corresponding to a two-jet configuration.

ii) *Hadronization corrections.* The most successful hadronization models are those incorporated in parton-shower event simulation programs [4], which are based on the assumption that hadronization is a local process in phase space, involving a rather large number of final-state partons at a low momentum transfer scale. Such models are not suitable for estimating corrections to fixed-order perturbative calculations, which involve at most four partons in second order.

iii) *Renormalization scale dependence.* The result obtained for α_S depends on the renormalization scale μ used in the perturbative calculation. This dependence represents a residual uncertainty due to orders higher than those computed. Intuitively, one would expect $\mu \sim Q$ to be the natural scale, but in many cases $\mu \ll Q$ appears to give a better fit, with a somewhat smaller value of $\alpha_S(Q^2)$. The strong scale dependence, combined with the preference for such small scales, has been the biggest source of uncertainty in α_S determinations from event shapes at LEP [5].

In a series of recent papers [6]-[10], it was shown that all of the above types of uncertainty can be reduced by resummation of large logarithms in event shape variables that satisfy the conditions for exponentiation: i) Resummation allows analyses to be extended closer to the two-jet kinematic limit, giving better statistical precision; ii) It corresponds closely to what is done, at a less precise level, in parton shower simulations, and so it can be combined more reliably with hadronization corrections estimated from models based on parton showers; iii) It reduces the renormalization scale dependence of the prediction, by taking into account an important class of higher-order corrections. Most importantly, it reproduces, at a scale $\mu \sim Q$, the effect of using a scale $\mu \ll Q$ in the second-order calculation, and thus it removes the motivation for using a small scale.

The event shape variables considered in detail in Refs. [6]-[10] are the thrust, the heavy jet mass, the energy-energy correlation, and two new quantities related to the transverse broadening of jets. In all cases, some transformation of variables is required before factorization occurs. For example, in the thrust distribution double logarithms of $1 - T$ occur in the high thrust region, $T \to 1$. In this region we have

$$T \sim 1 - (k_1^2 + k_2^2)/Q^2 \quad (5)$$

where k_1 and k_2 are the invariant masses of the two final-state jets and Q is the c.m. energy. Thus the cross section for thrust $T > 1 - \tau$ is given at small τ by

$$\sigma(1 - T < \tau) \sim \int dk_1 dk_2 J(Q, k_1) \\ \times J(Q, k_2)\Theta(\tau Q^2 - k_1^2 - k_2^2) \quad (6)$$

where $J(Q, k)$, the jet mass distribution at scale Q, factorizes and exponentiates to next-to-leading order [11, 12]. In order to factorize the phase space, we perform a Laplace transformation to obtain

$$\sigma(1 - T < \tau) \sim \\ \frac{1}{2\pi i} \int \frac{dN}{N} \exp(N\tau) \left[J_N(Q)\right]^2 \quad (7)$$

where

$$J_N(Q) = \int_0^\infty dk J(Q, k) \exp(-Nk^2/Q^2). \quad (8)$$

Thus the Laplace transform of $\sigma(1 - T < \tau)$ factorizes to next-to-leading order. Inverting the transform gives a prediction of the distribution in the high-thrust region, which can be combined with the fixed-order prediction for lower T after subtracting all the terms up to order α_S^2 that have been exponentiated. The resulting prediction [6] is more stable with respect to hadronization corrections and renormalization scale dependence. The corresponding predictions of other event shapes show similar good features [7, 8, 9].

By now, several experimental studies have been performed comparing these predictions with data [13], and more are in progress. Many of the results are reviewed in other contributions to these Proceedings, so only a few general remarks will be made here.

The most important general feature of the comparisons is that they no longer show a preference for small values of the renormalization scale μ: indeed a fit usually cannot be obtained for $\mu \ll Q$. Since the principal large logarithms have been explicitly resummed, there is no dynamical argument for using a scale much different from Q, and most resummed analyses quote a theoretical uncertainty corresponding to a rather small range of scales around $\mu \sim Q$. The resulting values of α_S are slightly larger than those obtained from earlier fixed-order analyses: the small scales preferred by the latter tend to be correlated with a smaller value of α_S.

The resummed predictions of event shape distributions still have their limitations, which should not be overlooked. At very high thrust, where the value of $\alpha_S \ln(1-T)$ is large, the predictions still become unreliable. At present energies the estimated hadronization corrections also become uncomfortably large in the same region. At low thrust, the resummed terms are not enhanced but instead the non-logarithmic contributions become very small. Thus the terms of order α_S^3 and beyond generated by resummation, which are reliable only at high thrust, may significantly distort the prediction in the low-thrust region. This effect may be assessed and controlled by varying the lower limit of the fitted region, and by comparing different schemes for matching the resummed and fixed-order predictions [13].

JET ALGORITHMS

Fixed-order analyses of multijet fractions in e^+e^- annihilation are another popular way of measuring α_S, and so one would like to apply the exponentiation technique to these quantities. It is also vital to understand QCD multijet final states when searching for new physics in hadronic channels. As in the case of event shapes, the QCD matrix elements have the required factorization properties, but factorization of the phase space is also necessary in order for generalized exponentiation to occur.

The most common way of defining jets in e^+e^- collisions has been by a jet clustering algorithm of the JADE type [14]. One defines a resolution parameter y_{cut} and combines the pair of particles (ij) of lowest invariant mass M_{ij} into a cluster if $y_{ij} \equiv M_{ij}^2/Q^2 < y_{\text{cut}}$. This procedure is iterated for the particles and clusters until no more combination is possible, at which stage the remaining clusters are called jets. The n-jet fraction of events is a function of the jet resolution, $R_n(y_{\text{cut}})$, whose perturbative expansion involves double logarithms of y_{cut}.

It turns out that, for the JADE algorithm as outlined above, even the leading logarithms of y_{cut} do not exponentiate, because the phase space defined by the invariant mass cut does not factorize [15]. Neglecting the particle (or cluster) masses, the resolution variable is

$$y_{ij} = 2E_iE_j(1 - \cos\theta_{ij})/Q^2 \ . \qquad (9)$$

The factor of $E_i E_j$ builds in an attractive kinematic correlation between soft gluons, which for example enhances the 3-jet fraction at the expense of those for 2- and 4-jets. As a result, the resummation of logarithms of $y_{\rm cut}$ for the unmodified JADE algorithm appears a hopeless task.

The lack of exponentiation in the JADE algorithm has led to the introduction of similar algorithms with different resolution variables to overcome this problem. The best choice seems to be the $k_\perp$- or 'Durham' algorithm [16]-[19], in which Eq. (9) is replaced by

$$y_{ij} = 2\min(E_i^2, E_j^2)(1 - \cos\theta_{ij})/Q^2 . \quad (10)$$

The decoupling of E_i and E_j leads to exponentiation of leading and next-to-leading logarithms for the 2-jet fraction and to generalized exponentiation for higher multijets [17, 18]. As expected, the hadronization corrections also appear to be under better control [19]. The issue of renormalization scale dependence, which as we discussed concerns next-to-next-to-leading logarithms, is still under investigation.

The $k_\perp$-algorithm can be generalized to lepton-hadron and hadron-hadron collisions in a way that leads to universal factorization of initial-state mass singularities [20].

JET MULTIPLICITIES

The exponentiation properties of the jet fractions defined according to the $k_\perp$-algorithm (10) make it possible to predict the average multiplicity of jets using this algorithm [21], either as a function of c.m. energy Q at fixed resolution $y_{\rm cut}$, or as a function of $y_{\rm cut}$ at fixed Q. The latter predictions are more useful since they can be tested with great precision at $Q = m_Z$. Very good agreement has been found [13], although the central value obtained for $\alpha_S(m_Z)$ is a little higher than in the event shape measurements.

Predictions in which one first selects a 2- or 3-jet sample at a given resolution $y_{\rm cut} = y_1$ and then studies the 'sub-jet' multiplicity in each jet as a function of $y_{\rm cut} < y_1$ have also been made [22]. Since the first two jets are initiated by quarks while the third is due to a gluon, one may hope to learn about the fragmentation of gluon jets from this analysis. The naive expectation that the gluon jet has a higher multiplicity in proportion to its colour charge squared, i.e. $C_A/C_F = 9/4$ times that of a quark jet, is not fulfilled at present energies, due to QCD coherence effects which strongly suppress the fragmentation of the gluon jet in the 3-jet sample. Hence one typically finds that the predicted multiplicity in the 3-jet sample is less than 3/2 times that in the 2-jet events. The experimental results [23] are in agreement with these predictions.

Because of the good theoretical properties of the $k_\perp$-algorithm, much more detailed predictions can be made concerning the distributions of sub-jet within jets. Comparisons with experiment, possibly using events with tagged quark jets, should cast much light on the properties of gluon jets in the near future.

TWO-PARTICLE CORRELATIONS

New information on jet fragmentation in e^+e^- collisions has been provided by measurement of the two-particle momentum distribution [24]. Defining

$$\xi = \ln\left(\frac{1}{x}\right) \quad (11)$$

where $x = 2p/Q$ is the momentum fraction, let $D(\xi)$ and $D(\xi_1, \xi_2)$ represent the single-particle and two-particle inclusive ξ-distributions, respectively. Then the two-particle correlation function,

$$R(\xi_1, \xi_2) = D(\xi_1, \xi_2)/D(\xi_1)D(\xi_2) \quad (12)$$

is predicted to have the form [25]

$$R = c_1 + c_2(\xi_1 + \xi_2) + c_3(\xi_1 - \xi_2)^2 + \ldots \quad (13)$$

where the coefficients c_n have power series expansions in $a \equiv \sqrt{2\beta_0 \alpha_S}$:

$$\begin{aligned} c_1 &= 1.375 - 1.262a + d_1 a^2 + \ldots \\ c_2 &= 0.877 a^3 + d_2 a^4 + \ldots \quad (14) \\ c_3 &= -1.125 a^4 + d_3 a^5 + \ldots \end{aligned}$$

Here d_n represent unknown higher-order coefficients. Table 1 shows the values of the coefficients obtained from experiment [24], together with the predictions (14), taking $a = 0.4$ at $Q = m_Z$. The last column shows the values of the unknown higher-order coefficients d_n that would be required for perfect agreement with experiment. We see that they are all of order unity or smaller, showing that the agreement is as good as should be expected, given the rather large value of the expansion parameter a. This agreement lends support to the idea that these two-particle correlations are coming primarily from the perturbative phase of jet evolution, rather than from non-perturbative processes such as resonance decays.

ACKNOWLEDGEMENTS

It is a pleasure to acknowledge my collaborators on these topics, S. Catani, Yu.L. Dokshitzer, L. Trentadue and G. Turnock, and to thank also S. Bethke, G.D. Cowan, T. Hebbeker and D.R. Ward for valuable conversations and comments.

Table 1. Coefficients in Two-Particle Correlation Function (13).

n	c_n (OPAL)	c_n (QCD)	d_n (fit)
1	0.928(2)	$0.870 + 0.160 d_1$	0.36
2	0.025(3)	$0.056 + 0.026 d_2$	-1.21
3	-0.021(3)	$-0.029 + 0.010 d_3$	0.78

REFERENCES

1. For a review and earlier references, see J.C. Collins and D.E. Soper, Ann. Rev. Nucl. Part. Sci. 37 (1987) 383.

2. G. Sterman, Nucl. Phys. B281 (1987) 310; S. Catani, E. d'Emilio and L. Trentadue, Phys. Lett. 211B (1988) 335; S. Catani and L. Trentadue, Phys. Lett. 217B (1989) 539; Nucl. Phys. B327 (1989) 353; Nucl. Phys. B353 (1991) 183.

3. Z. Kunszt, P. Nason, G. Marchesini and B.R. Webber, in 'Z Physics at LEP 1', CERN 89-08, vol. 1, p. 373.

4. For a review see B.R. Webber, Ann. Rev. Nucl. Part. Sci. 36 (1986) 253.

5. ALEPH Collaboration, D. Decamp et al., Phys. Lett. 234B (1990) 399; Phys. Lett. 255B (1991) 623; CERN preprint PPE/90-196; DELPHI Collaboration, P. Aarnio et al., Phys. Lett. 240B (1990) 271; Phys. Lett. 247B (1990) 167; P. Abreu et al., Zeit. Phys. C54 (1992) 55. L3 Collaboration, B. Adeva et al., Phys. Lett. 237B (1990) 136; Phys. Lett. 248B (1990) 464; OPAL Collaboration, M.Z. Akrawy et al., Phys. Lett. 235B (1990) 389; Phys. Lett. 252B (1990) 159; Zeit. Phys. C47 (1990) 505; Zeit. Phys. C49 (1991) 375; SLD Collaboration, K. Abe et al., Stanford preprints SLAC-PUB-5904, 5905.

6. S. Catani, L. Trentadue, G. Turnock and

B.R. Webber, Phys. Lett. 263B (1991) 491.

7. S. Catani, G. Turnock and B.R. Webber, Phys. Lett. 272B (1991) 368.

8. G. Turnock, Cambridge preprint Cavendish-HEP-91/3.

9. S. Catani, G. Turnock and B.R. Webber, CERN preprint TH.6570/92, to be published in Phys. Lett. B.

10. S. Catani, L. Trentadue, G. Turnock and B.R. Webber, Cambridge preprint Cavendish-HEP-91/11, CERN preprint TH.6640/92.

11. S. Catani and L. Trentadue, Phys. Lett. 217B (1989) 539; Nucl. Phys. B327 (1989) 353; Nucl. Phys. B353 (1991) 183.

12. S. Catani, G. Marchesini and B.R. Webber, Nucl. Phys. B349 (1991) 635.

13. OPAL Collaboration, P.D. Acton et al., Zeit. Phys. C55 (1992) 1; ALEPH Collaboration, D. Decamp et al., Phys. Lett. 284B (1992) 163; L3 Collaboration, O. Adriani et al., Phys. Lett. 284B (1992) 471; H. Furstenau, in Proc. XXVII Recontres de Moriond, *QCD and High Energy Hadronic Interactions*.

14. JADE Collaboration, S. Bethke et al., Phys. Lett. 213B (1988) 235.

15. N. Brown and W.J. Stirling, Phys. Lett. 252B (1990) 657; Zeit. Phys. C53 (1992) 629.

16. Report of Hard QCD Working Group, in Proc. Durham Workshop on Jet Studies at LEP and HERA, J. Phys. G 17 (1991) 1537.

17. S. Catani, Yu.L. Dokshitzer, M. Olsson, G. Turnock and B.R. Webber, Phys. Lett. 269B (1991) 432.

18. S. Catani, in 'QCD at 200 TeV', Proc. 17th INFN Eloisatron Project Workshop, Erice, Italy, June 1991, ed. L. Cifarelli and Yu.L. Dokshitzer (Plenum Press, New York); S. Catani, Yu.L. Dokshitzer, M. Olsson, G. Turnock and B.R. Webber, Cambridge preprint Cavendish-HEP-91/12, in preparation.

19. S. Bethke, Z. Kunszt, D.E. Soper and W.J. Stirling, Nucl. Phys. B370 (1992) 310.

20. S. Catani, Yu.L. Dokshitzer and B.R. Webber, Phys. Lett. 285B (1992) 291; S. Catani, Yu.L. Dokshitzer, M.H. Seymour and B.R. Webber, CERN preprint in preparation.

21. S. Catani, Yu.L. Dokshitzer, F. Fiorani and B.R. Webber, Nucl. Phys. B377 (1992) 445.

22. S. Catani, Yu.L. Dokshitzer, F. Fiorani and B.R. Webber, CERN preprint TH.6419/92, to be published in Nucl. Phys. B.

23. OPAL and ALEPH Collaborations, these Proceedings.

24. OPAL Collaboration, P.D. Acton et al., Phys. Lett. 287B (1992) 401.

25. C.P. Fong and B.R. Webber, Phys. Lett. 241B (1990) 255; Nucl. Phys. B355 (1991) 54.

QCD TESTS WITH FINAL STATE PHOTONS IN HADRONIC EVENTS AT LEP

Presented by Giorgio Gratta
Caltech
for the L3 Collaboration

Abstract

We report on studies of energetic, isolated photons in hadronic events at LEP. Since the main source of these photons is expected to be quark final state radiation, they represent a powerful tool to study the short distance structure of QCD. Furthermore events with quark final state radiation can be used to simulate the kinematics of hard gluon emission and therefore study details of the fragmentation process.

Introduction

The study of energetic, isolated photons in hadronic Z^0 decays at LEP offers an important probe of the short-distance structure of QCD[1]. Although the elementary processes of photon and gluon emission are closely related, photons have the advantage of appearing directly in the final state, whereas gluons undergo a complex evolution into hadrons.

Previous studies of final state radiation at lower energy e^+e^- colliders[2] were limited by a large contamination due to photons radiated from the initial state electrons and positrons. At the Z^0 resonance this background is strongly suppressed, making LEP an ideal laboratory for this study. The primary background to direct photons at LEP is the decay of energetic neutral hadrons into photons which are unresolved in the detector.

We present comparisons of the data collected in the last two years by the four LEP experiments[3] with theoretical models.

Two complementary approaches are currently available for a theoretical description of final state quark radiation produced at the Z^0 resonance: matrix element and leading logarithms (or parton shower) methods. In the matrix element approach, the cross section is expanded systematically in powers of the electromagnetic coupling constant, α, and the running strong coupling constant, α_s [4]. Since the calculations require some cutoffs in order to avoid infrared and collinear divergences some care has to be applied in order to match the experimental cuts to the quantities calculated by the theory. In the leading logarithm approach, the cross section is expanded to all orders in the coupling constants, but only the leading logarithmic terms are used. These logarithmic terms dominate the cross section for collinear and soft radiation, and this is the limit in which this approach is strictly applicable. A smooth extrapolation to hard and isolated radiation is possible, but not necessarily accurate. Several Monte-Carlo programs are available for describing the quark photon and gluon radiation using both the leading logarithm and the matrix element approach. We present results using the programs: ARIADNE[5], HERWIG[6] and JETSET[7]. These programs differ in the variables they use for a leading logarithmic expansion of the cross section, but all perform an ad-hoc matching of the first branch to the appropriate first order matrix element ($\mathcal{O}(\alpha)$ for a $q \to q\gamma$ branch, $\mathcal{O}(\alpha_s)$ for a $q \to qg$ branch). The modelling of the hadronization effects is the same in ARIADNE and JETSET, but different in HERWIG.

Background reduction

Since an hadronic Z^0 decay contains on average $\approx$ 20 photons of non prompt origin, mostly coming from the decays of neutral hadrons like π^0 or η, while prompt photons above few GeV occur only in a few percent of the events one can easily see that background rejection for this physics channel is a hard task. The general strategy adopted by all LEP experiments is to have a first drastic background reduction by requiring the photons to be quite energetic and isolated with respect to the hadronic activity in the event. Although in the detail cuts are different for each of the experiments, mostly depending on the characteristics of the different detectors used, we can summarize this first step of the analysis by saying that events are selected if a neutral cluster of energy greater that 5 (or 7 for some analysis) GeV is observed in the electromagnetic calorimeter, and the shower center of gravity is isolated by more than $\approx 20°$ with respect to the axis of the closest jet. Jets are usually formed using the JADE[8] or Durham[9] algorithms excluding the candidate photon from the event and using recombination parameters corresponding to $y_{cut}(\text{JADE}) \approx 0.05$. These cuts have very strong rejection power on the background that is usually coming from within the jets and little effect on high p_t processes like the prompt photon signal. In this way the backgroud is reduced to the same order of magnitude as the signal.

Further reduction is obtained usually by studying the shape of the photon candidate shower to check its consistency in being produced by a prompt photon. It has to be remarked that this technique, although powerful against a large part of the neutral hadron decay kinematics, becomes quite insensitive in the case of very hard hadrons (above $\approx$ 30 GeV for LEP detectors), and is of no use in the case of very asymmetric decays, where the two photons of the decay have very large angular opening. Nevertheless it is of primary importance to evaluate and subtract the remaining background using variables, like the ones characterizing the shower shapes, that are independent from the physics of the fragmentation processes. In fact our knowledge of fragmentation at large angle from the jet axes is quite crude and to perform harder isolation cuts to reject more background is equivalent to relay on the uncertain Monte Carlo simulation of fragmentation in the jet tail regions.

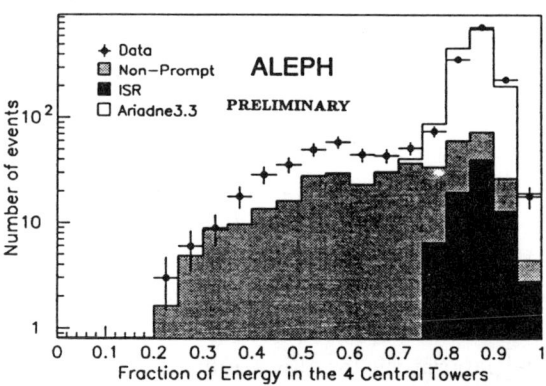

Fig. 1 Shower shape parameter distribution for the ALEPH[10] analysis. Prompt photons populate the peak on the right.

In Fig. 1 we show the distribution of the shower shape parameter used in the ALEPH analysis[10]. Prompt photons populate the region above ≈ 0.75, while the tail at lower values is mostly due to neutral hadron decays. A substantial excess of background in the data respect to the Monte Carlo prediction is clearly visible. In order to verify this effect L3 has reconstructed asimmetrical neutral hadron decays into two photons in a kinematic region similar to the one covered by Fig. 1. The two photon invariant mass distribution (Fig. 2) shows clear peaks in the π^0 and η peaks and confirms the excess of events remarked above.

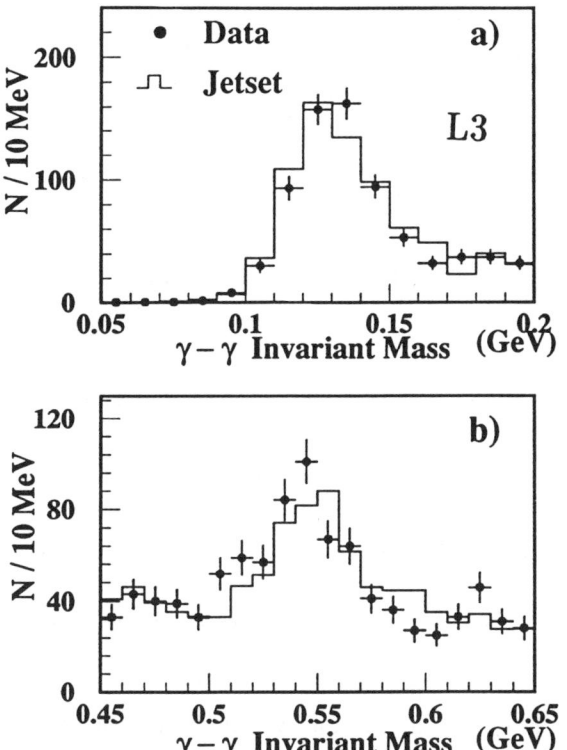

Fig. 2 Invariant mass distribution for isolated cluster pairs for data and JETSET from the L3 analysis. The Monte Carlo has been rescaled by a factor 1.84 in the π^0 region (a), and by a factor 2.11 in the η region (b).

Parallel analyses studying isolated charged tracks generally find the same trend, so that we are led to conclude that Monte Carlo generators underestimate the cross section of isolated, energetic (as defined above) hadrons by a factor 1.5 to 2.5, depending upon the specific channel and cuts used.

This result is then used to rescale the number of background events which is subtracted to obtain a clean sample of prompt photons. Initial state radiation is also subtracted at this stage using the prediction of electroweak Monte Carlo generators assumed to be accurate enough to the purpose. Using data collected in 1990 and 1991 each experiment obtains a sample of the order of 1000 events, the exact number depending on the values of the cuts and on the angular region from where the photons are accepted (some experiments use only photons in the barrel calorimeter region where the initial state radiation contribution is minimal). Efficiencies for selecting photons above energies of order 5 GeV are $\approx 50\%$.

The systematic error on the total number of prompt photon events (and therefore on the normalization of the differential cross sections used in the comparison with the theoretical calculations) is below $\approx 6\%$, with some experiments claiming substantially lower figures depending on their perception of the uncertainty in the background subtraction.

Comparison with QCD models

Results are usually compared with QCD calculations at hadron level or at parton level, and some unfolding of detector and (in the second case) fragmentation effects has to be performed. This procedures usually result in a further contribution to the systematic error. In Fig. 3 we show the comparison of data with three parton shower Monte Carlo calculations for the photon energy, angle to the closest jet and p_t to the event Thrust axis in the L3 analysis. Integrating the distributions we obtain a branching ratio for the production of events with a photons above 5 GeV and isolated by more than 20° with respect to the closest jet axis, relative to the production of hadronic events of:

$$\frac{\Gamma(Z \to \gamma + \text{hads})}{\Gamma(Z \to \text{hads})} =$$

$$5.2 \pm 0.3(\text{stat}) \pm 0.4(\text{syst}) \cdot 10^{-3}$$

to be compared to:

JETSET : $[4.53 \pm 0.04\,(\text{stat})] \cdot 10^{-3}$
HERWIG : $[6.09 \pm 0.04\,(\text{stat})] \cdot 10^{-3}$
ARIADNE: $[6.13 \pm 0.04\,(\text{stat})] \cdot 10^{-3}$.

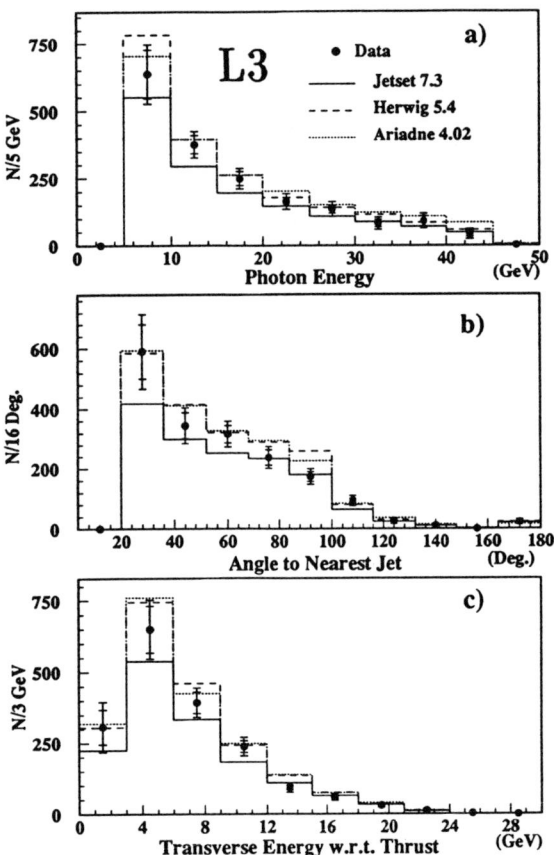

Fig. 3 Distributions of (a) the final state photon energy, (b) the angle between the photons and the nearest jet, and (c) the transverse energy of the photons with respect to the event Thrust axis for the L3 analysis.

Comparisons of data with matrix element calculations have been performed initially using analytical formulae[4,11] and more recently using Monte Carlo generators based on the same calculations[12,13,14]. The comparison of data with two of these generators is shown in Fig. 4 for the ALEPH analysis[10]. In this case the variable chosen for the comparison is the y_{cut} between the photon and the jets in the event. We remark that from a qualitative point of view matrix element calculations always tend to overestimate the experimental cross section.

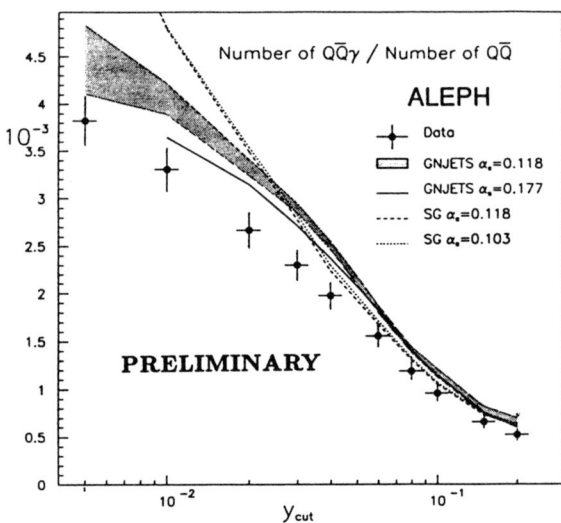

Fig. 4 Prompt photon event rate as a function of y_{cut} compared with the matrix element calculations GNJETS[12] and SG[13] for the ALEPH analysis[10]. The band around the GNJETS prediction corresponds to the range given by the variations of the cutoff parameter.

This is also true for two of the parton showers (ARIADNE and HERWIG) while the other parton shower considered (JETSET) clearly underestimates the measurement. Although this situation is very clear in the case of the parton showers, where all experiments basically agree in methods and results, in the comparison with matrix element calculation there is still confusion especially regarding the way of handling the cutoffs used to avoid singularities in the calculation. For this reason at the present time different experiments still present their data in ways that make precise quantitative comparisons very difficult. However the subject is under intense study both from the theoretical and from the experimental side, so that it is natural to forsee that the last remaining uncertainties in the correct use of matrix element calculations will soon be solved.

The string effect

Events with hard, isolated photons can also be used to explore some properties of fragmentation precesses by simulating the kinematic boost due to hard gluon emission by using non-colored hard photons. More than 10 years ago the observation was made at PEP and PERTA[15] that in three jet events less particles are produced in the region between the two quark jets relative to the other inter-jet regions. This effect was recently confirmed at LEP[16].

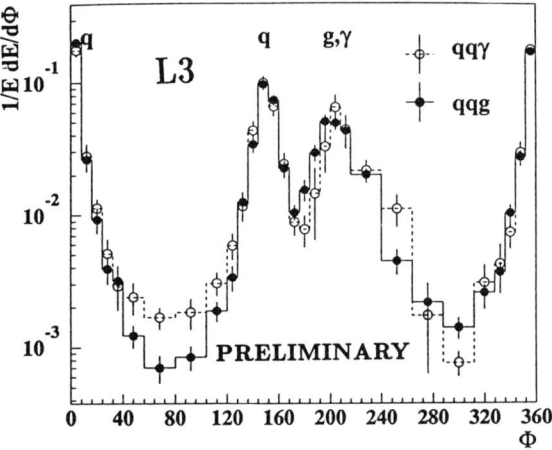

Fig. 5 Comparison of energy flow in events with photon or gluon radiation in the L3 analysis[19]. The *string effect* is visible as a depletion of region around 70° in the case when a gluon is emitted.

This asymmetry in the particle (and energy) flow in three jet events was predicted in the contest of the string fragmentation model[17], and can also be explained by analytical QCD calculations including coherence effects[18]. In Fig. 5 we show the energy flow in the event plane for three jet events and for prompt photon events in the L3 analysis[19]. The energy flow is measured as the energy weighted distribution of the angles of all calorimetric clusters with respect to the axis of the first jet. Jets are simply energy ordered in the $q\bar{q}\gamma$ case while in the $q\bar{q}g$ case an inclusive muon with energy exceeding 4 GeV is used to tag the second quark jet. In this way the probability for the third jet to be associated to the gluon is $\approx 85\%$. In both cases we restrict ourselves to events where the angle Φ between the two quark jets is in the range $145° < \Phi < 155°$. The angle between the first jet and the gluon (or the photon) varies from 90° to 170°. In total 60 $q\bar{q}\gamma$ events and 105 $q\bar{q}g$ events contribute to the plot in Fig. 5. In the region between the two quark jets one can clearly see a depletion in the case of $q\bar{q}g$ events. The ratio of the integrals of the energy flows in the range $40° < \Phi < 104°$ is measured to be 2.1 ± 0.5 where the error includes both statistical and systematic contributions. Both JETSET and HERWIG reproduce the measurements.

Conclusions

After three years of data taking at LEP the world sample of prompt photons in hadronic Z0 decays has reached a size that allows rather stringent tests of the QCD/QED calculations. The availability of these data has led to the development of new analytical calculations and Monte Carlo programs that properly account for the competition between photon and gluon radiation in hadronic final states. At present models are generally in qualitative agreement with data although some differences still exist. In the domain of parton shower simulations it is unclear why different programs based on very similar approximations give quite different results. In the (newer) field of Matrix Element generators on the other hand there is still considereble discussion on the range of validity of different programs in terms of internal variables and experimental cuts. We expect the next several months to be crucial for a better comprehension of Matrix Element calculations, while a new challenge will arise from the completion of the 1992 LEP data taking that should see the doubling of the event samples.

Finally the string effect in fragmentation has been observed at LEP in a model independent way by comparing energy flows in three jet and two jet plus photon events.

References

1. H. Fritzsch, M. Gell-Mann and H. Leytwyler, Phys. Lett. **B 47** (1973) 365; D.J. Gross and F. Wilczek, Phys. Rev. Lett. **30** (1973) 1343; D.J. Gross and F. Wilczek, Phys. Rev. **D 8** (1973) 3633; H.D. Politzer, Phys. Rev. Lett. **30** (1973) 1346; K. Koller, T.F. Walsh and P.M. Zerwas, Z. Phys. **C 2** (1979) 197; E. Laerman et al., Nucl. Phys. **B 207** (1982) 205.

2. MAC Collab,. H.R. Band et al., Phys. Rev. Lett. **54** (1985) 95; TASSO Collab., W. Braunschweig et al., Z. Phys. **C 41** (1988) 385; VENUS Collab., K. Abe et al., Phys. Rev. Lett. **63** (1989) 1776; JADE Collab., D.D. Pitzl et al., Z. Phys. **C 46** (1990) 1.

3. ALEPH Collab., D. Decamp et al., Phys. Lett. **B 264** (1991) 476; DELPHI Collab, P. Abreu et al., Z. Phys. **C 53** (1992) 555. L3 Collab. O. Adriani et al., CERN-PPE/92-131, submitted to Phys. Lett. B; OPAL Collab, M.Z. Akrawy et al., Phys. Lett. **B 246** (1990) 285. OPAL Collab, P.D. Acton et al., Z. Phys. **C 54** (1992) 193.

4. G. Kramer and B. Lampe, Phys. Lett. **B269** (1991) 401.

5. ARIADNE 4.02; L. Lönnblad, DESY Preprint 92-046 (1992), to be published in Comp. Phys. Comm.

6. HERWIG 5.4; G. Marchesini et al., Comp. Phys. Comm. **67** (1992) 465.

7. JETSET 7.3; T. Sjöstrand, Comp. Phys. Comm. **39** (1986) 347; T. Sjöstrand and M. Bengtsson, Comp. Phys. Comm. **43** (1987) 367.

8. JADE Collab, W. Bartel et al., Z. Phys. **C 33** (1986) 23; JADE Collab, S. Bethke et al., Phys. Lett. **B 123** (1988) 235.

9. Yu. Dokshitzer, in "Workshop on Jets at LEP and HERA", Durham, UK, Dec. 1990; N. Brown and J. Stirling, Z. Phys. **C53** (1992) 629; S. Catani et al., Phys. Lett. **B 269** (1991) 432; S. Bethke, Z. Kunszt, D.E. Soper, W.J. Stirling, Nucl. Phys. **B370** (1992) 310.

10. ALEPH Collab., Private Communication.

11. OPAL Collab, P.D. Acton et al., Z. Phys. **C 54** (1992) 193.

12. G. Kramer, H. Spiesberger, DESY 92-022.

13. N.W.N. Glover, W.J. Stirling, DTP 92-52.

14. Z. Kunszt, Z. Trocsanyi, ETH-TH/92-26.

15. JADE Collab., W. Bartel et al., Phys. Lett. **B 101** (1981) 129; JADE Collab., W. Bartel et al., Z. Phys. **C21** (1983) 37; JADE Collab., F. Ould-Saada et al., Z. Phys. **C39** (1988) 1; MARK II Collab., P.D. Sheldon et al., Phys. Rev.

Lett. **57** (1986) 1398; TASSO Collab., M. Althoff *et al.*, Z. Phys. **C29** (1985) 29; TPC Collab., H. Aihara *et al.*, Z. Phys. **C28** (1985) 31; TPC/2γ Collab., H. Aihara *et al.*, Phys. Rev. Lett. **57** (1986) 945.

16. OPAL Collab., M.Z. Akrawy *et al.*, Phys. Lett. **B 261** (1991) 334; L3 Collab., B. Adeva *et al.*, Z. Phys. **C55** (1992) 39.

17. B. Andersson, G. Gustafson and T. Sjöstrand, Phys. Lett. **B 94** (1980) 211; B. Andersson, G. Gustafson and T. Sjöstrand, Z. Phys. **C6** (1980) 235.

18. Y.I. Azimov *et al.*, Phys. Lett. **B 165** (1985) 147; Y.I. Azimov *et al.*, Jad. Fiz. **43** (1986) 149; V.A. Khoze, "Color Coherence Physics at the Z", CERN-TH 5849-90.

19. L3 Internal Note No.1217.

QCD STUDIES OF HADRONIC DECAYS OF Z^0 BOSONS BY SLD

The SLD Collaboration*

presented by

David Muller, SLAC

Abstract

Z^0 bosons have been produced by collisions of longitudinally polarized electrons with unpolarized positrons at the SLAC Linear Collider and their decays have been recorded by the SLD experiment. We present preliminary QCD results based on the first 6000 such decays. We find good agreement between the inclusive properties of these data and the predictions of perturbative QCD plus fragmentation models. The strong coupling, α_s, has been measured by three methods: jet rates yield $\alpha_s(M_Z) = 0.119 \pm 0.002$ (stat.) ± 0.003 (exp. syst.) ± 0.014 (theor.); energy-energy correlations yield $\alpha_s(M_Z) = 0.121 \pm 0.002 \pm 0.004 \pm ^{0.016}_{0.009}$; and the energy-energy correlation asymmetry gives $\alpha_s(M_Z) = 0.108 \pm 0.003 \pm 0.005 \pm ^{0.008}_{0.003}$.

INTRODUCTION

The SLAC Linear Collider (SLC) produces electron-positron annihilation events at the Z^0 resonance which are recorded by the SLC Large Detector (SLD)[1]. The first physics run began in February 1992. SLC performance continued to improve during the run, routinely achieving Z^0 production rates of 10-20 per hour. By the end of August, about 12,000 Z^0s had been accumulated. Approximately 6000 hadronic Z^0 decays were used in the analysis presented here.

A major achievement of the 1992 run was the delivery of an intense beam of longitudinally polarized electrons. Details of the polarization program and a preliminary measurement of the left-right cross section asymmetry were contributed separately to this conference[2]. In this paper we study in detail the structure of hadronic Z^0 decays, compare with the predictions of perturbative QCD plus fragmentation models, and measure the strong coupling, α_s, by three established techniques.

THE SLD AND EVENT SELECTION

The detector is described in detail elsewhere[1]. The micro-vertex and Cherenkov Ring Imaging Detectors were not used in this analysis, but are described in separate contributions to this conference[3].

Charged particles were tracked in the Central Drift Chamber (CDC), which consists of 80 layers of axial or stereo sense wires, contained in a 0.6T axial magnetic field. Particle energies were measured in the Liquid Argon Calorimeter (LAC) and Warm Iron Calorimeter, which are segmented into approximately 40,000 projective towers.

Two triggers were used for hadronic events, one requiring a total LAC energy greater than 8 GeV, the other requiring at least two well-separated tracks in the CDC. Events were then required to pass two loose selections of hadronic events, one based on the topology of energy deposition in the LAC, the other on the number and topology of charged tracks in the CDC.

The analysis presented here used charged tracks measured in the CDC. A set of cuts was applied to select well-measured tracks and events well-contained within the detector acceptance. Tracks were required to have:

- a fit quality of $\sqrt{2\chi^2} - \sqrt{2N_{df}-1} < 15$,

- a closest approach to the beam axis within 10 cm, and within 20 cm along the axis of the nominal interaction point,
- a polar angle, θ, with respect to the beam axis within $|\cos\theta| < 0.8$, and
- a minimum momentum transverse to the beam axis of $p_\perp > 150$ MeV/c.

Events were required to have:
- a minimum of five such tracks,
- no track with measured momentum, $p > 100$ GeV/c,
- a thrust axis with polar angle, θ_T, with respect to the beam within $|\cos\theta_T| < 0.71$, and
- a minimum charged visible energy, $E_{vis} > 0.2 M_Z$, where all tracks were assigned the charged pion mass.

A total of 3837 events survived these cuts. The background is dominated by an estimated contribution of < 0.5% from tau pair events.

HADRONIC EVENT PROPERTIES

We have studied global event variables, including thrust, oblateness, sphericity and aplanarity, as well as inclusive track variables, such as rapidity, momentum, and transverse momentum in and out of the event plane. In addition, we have selected a sample of 3-jet events using a y_{cut} (see below) of 0.02, in order to examine the scaled jet energies and the polar angles of the most energetic jet and the event plane, as well as the Ellis-Karliner angle[4].

For each of these quantities, we compared the distributions from the data with the predictions of two perturbative QCD plus fragmentation Monte Carlo programs, JETSET 6.3[5] and HERWIG 5.3[6]. For JETSET, we used a parameter set tuned by TASSO[7] at $\sqrt{s} = 35$ GeV. For HERWIG, we used the default parameters. For each model, 10,000 events were generated and passed through a detailed simulation of the SLD and the same reconstruction, event selection, and analysis as the data.

For all variables studied, both models give a good description of the data. The distributions of thrust, oblateness, and transverse momentum in and out of the event plane are shown in Fig. 1 as examples. These results confirm predictions[8] of the JETSET simulation made before data at the Z^0 were available, and are in agreement with results from experiments at LEP[9].

JET RATES AND α_S

The measurement of jet production rates provides an intuitive way to determine the strong coupling, α_s, since in first order perturbative QCD the rate of three-jet events is directly proportional to this coupling. Jets are often reconstructed using the "JADE algorithm"[10], in which the lowest mass pair of particles is iteratively clustered together until all $m_{ij}^2 > y_{cut} E_{vis}^2$. The number of clusters remaining is defined to be the jet multiplicity of the event. We have used the E, E0 and p clustering schemes[11], as well as the recently-introduced "Durham" or $k_\perp$ scheme[12].

Jet multiplicity rates were calculated from our data as a function of the resolution parameter, y_{cut}, and from the simulations described above, which were found to reproduce the data. The data were therefore corrected to the parton level using the JETSET simulation, and compared with theoretical calculations. Figure 2 shows the quantity $D_2(y_{cut})$, which is the distribution of the value of y_{cut} for which the event changes from a two-jet event to a three-jet event, for the Durham scheme. Also shown are two fits to the data of a calculation by Kunszt and Nason[13]. The calculation has two parameters, $\Lambda_{\overline{MS}}$, which is related to α_s, and the QCD renormalization scale, μ, the choice of which is not theoretically well-defined. In one fit (dashed line) μ was fixed to the Z^0 mass. In the second (solid line) it was a free parameter. Both fits are able to describe the data, however the $\Lambda_{\overline{MS}}$ values are quite different and the fitted value of μ is very small.

Figure 3 shows the value of $\alpha_s(M_Z)$ calculated from the fitted $\Lambda_{\overline{MS}}$ with fixed μ, as a function of μ for each of the schemes studied. There is substantial variation between the four schemes for any fixed μ, and the schemes show strong and different μ dependences, although low fitted values of μ are obtained in each case. In order to quote a result, we first averaged the

QCD Studies of Hadronic Decays

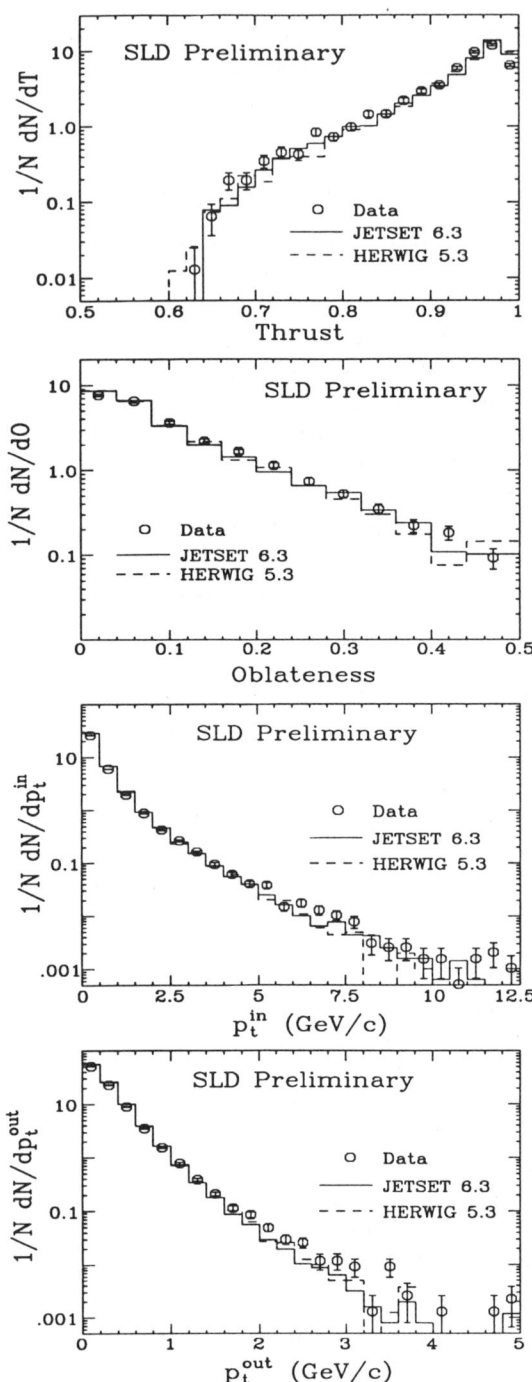

Figure 1. Comparison of (a) thrust, (b) oblateness, (c) p_t^{in} and (d) p_t^{out} distributions from our data (points with error bars) with predictions of the JETSET (solid line) and HERWIG (dashed line) simulations.

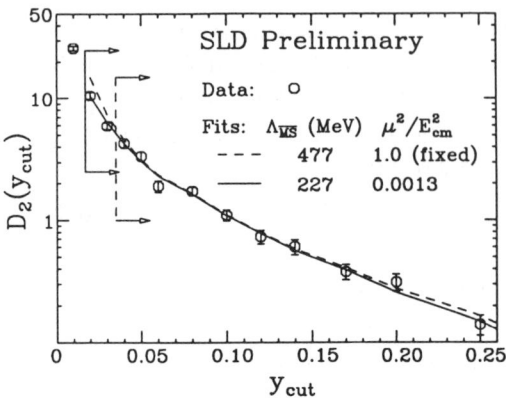

Figure 2. The corrected differential two-jet rate in the Durham scheme. The calculations of Kunszt and Nason have been fitted to the data with the renormalization scale fixed (dashed) and free (solid). The fit ranges[14] are indicated by the arrows.

α_s values from the two fits (μ free and $\mu = M_Z$) for each scheme, then averaged over the four schemes. Our preliminary result is $\alpha_s(M_Z) = 0.119 \pm 0.002 \pm 0.003 \pm 0.014$. The first error is statistical. The second error is experimental systematic, evaluated by varying the analysis cuts and detector simulation. The third error is theoretical and is dominated by the largest observed variation with μ, although it also includes contributions from varying hadronization simulations and the differences between the jet-finding schemes.

ENERGY-ENERGY CORRELATIONS

Another quantity sensitive to the strong coupling is the energy-weighted distribution of opening angles, χ, between particle pairs, or energy-energy correlation[15], EEC(χ) ≡

$$\langle \frac{1}{2\Delta\chi} \int_{\chi - \frac{\Delta\chi}{2}}^{\chi + \frac{\Delta\chi}{2}} \sum_{ij} \frac{E_i E_j}{E_{vis}^2} \delta(\chi' - \chi_{ij}) d\chi' \rangle ,$$

where the average is over all events in the sample. The region around $\chi \sim \pi/2$ is sensitive to hard gluon emission. Since the EEC uses tracks

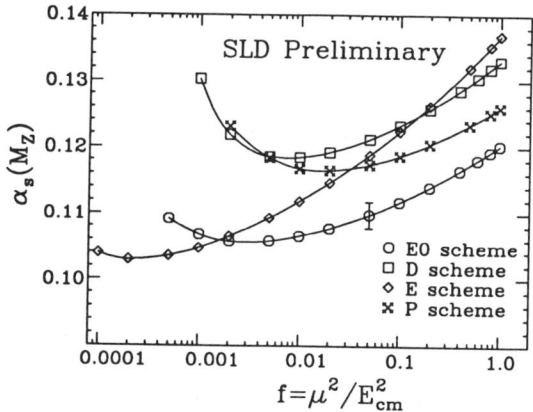

Figure 3. Renormalization scale dependence of the α_s measurement for the four clustering schemes. The size of the statistical error is indicated on one point.

directly, this method is insensitive to ambiguities in jet finding. The asymmetry, $AEEC(\chi) = EEC(\pi-\chi) - EEC(\chi)$, is also sensitive to α_s and is expected to be less sensitive to details of hadronization.

The EEC and AEEC were derived from our data and from the two Monte Carlo simulations. Both simulations reproduced the data, and the data were corrected to the parton level and compared with four theoretical calculations[13,16]. Figure 4 shows the corrected data along with fits to one calculation. Here also, there is considerable ambiguity in the choice of renormalization scale. Figure 5 shows the μ dependence of the fitted $\Lambda_{\overline{MS}}$ value for each calculation. All fits give adequate descriptions of the data. However, there is substantial variation between the calculations, and each calculation shows a strong dependence on the renormalization scale.

For the purpose of quoting a result, we took the fit from Kunszt and Nason at f=0.1 as our central $\Lambda_{\overline{MS}}$ value and calculated α_s. This yields $\alpha_s(M_Z) = 0.121 \pm 0.002 \pm 0.004 \pm^{0.016}_{0.009}$ for the EEC and $\alpha_s(M_Z) = 0.108 \pm 0.003 \pm 0.005 \pm^{0.008}_{0.003}$ for the AEEC. In both cases, the first error is statistical, the second experimental systematic and the third theoretical. The experimental systematic errors were evaluated by varying

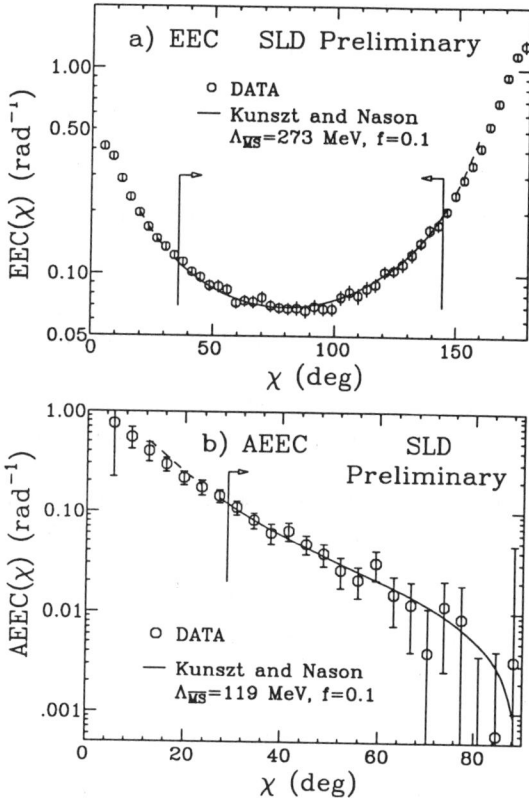

Figure 4. The measured (a) energy-energy correlation and (b) its asymmetry. The solid lines are fits using calculations of Kunszt and Nason over the regions indicated by the arrows.

the analysis cuts and fit ranges. The theoretical error dominates and is due mostly to the renormalization scale dependence, but also takes into account hadronization and differences between the four calculations.

SUMMARY AND CONCLUSIONS

Properties of hadronic decays of Z^0 bosons have been measured by the SLD at SLAC. These properties are reproduced by the perturbative QCD plus fragmentation Monte Carlo programs JETSET and HERWIG.

These events have been used to measure the strong coupling, α_s, by three methods, with the results $\alpha_s(M_Z)$ =

$0.119 \pm 0.002 \pm 0.003 \pm 0.014$ (Jet Rates)

$0.121 \pm 0.002 \pm 0.004 \pm {}^{0.016}_{0.009}$ (EEC)

$0.108 \pm 0.003 \pm 0.005 \pm {}^{0.008}_{0.003}$ (AEEC)

In each case, the first error listed is statistical, the second is from experimental systematics, and the third is our estimate of the theoretical uncertainty. The theoretical errors are dominated by uncertainties in the choice of renormalization scale.

These results are all in agreement with results from experiments at LEP[17] within *experimental* errors. The AEEC gives a smaller value of α_s than the other two methods which is significant if only experimental errors are considered.

ACKNOWLEDGEMENTS

The SLD Collaboration is indebted to all the SLAC staff whose efforts resulted in the successful operation of the SLC which produced the events used in this paper. We wish to thank S. Bethke, G.W. Gary, I. Knowles, G. Kramer, Z. Kunszt, H.-J. Lu, P. Osland, T. Sjöstrand, B. Ward and B. Webber for helpful comments relating to this analysis. This work was supported by Department of Energy contract DE-AC03-76SF00515.

REFERENCES

[1] SLD Design Report, SLAC Report 273 (1984)

[2] See contributions to these proceedings by N. Phinney and P. Rowson.

[3] See contributions to these proceedings by C. Damerell and J. Va'Vra.

[4] J. Ellis and I. Karliner, Nucl. Phys. **B148** (1979) p. 141.

[5] T. Sjöstrand, Comput. Phys. Commun. **39** (1986) p. 347, **43** (1987) p. 367.

[6] G. Marchesini and B.R. Webber, Nucl. Phys. **B310** (1988) p. 461, G. Abbiendi, *et al.*, Cavendish-HEP-90/26 (1990).

[7] TASSO collab., Z. Phys. **C41** (1988) p. 359.

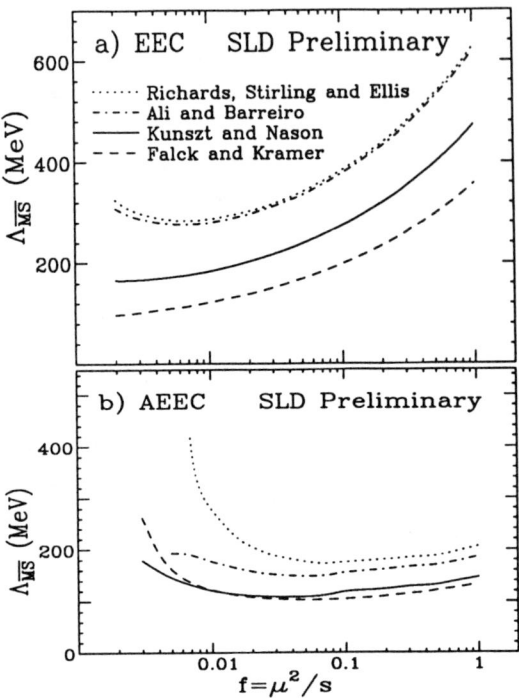

Figure 5. Variation of the fitted $\Lambda_{\overline{MS}}$ with the assumed renormalization scale in (a) EEC and (b) AEEC.

[8] P.N. Burrows, Z. Phys. **C41** (1988) p. 375.

[9] OPAL collab., Z. Phys. **C47** (1990) p. 505.

[10] JADE collab., Z. Phys. **C33** (1986) p. 23.

[11] OPAL collab., Z. Phys. **C49** (1991) p. 375.

[12] N. Brown and W. J. Stirling, Z. Phys. **C53** (1991) p. 692.

[13] Z. Kunszt *et al.*, Z Physics at LEP I, Vol. I, CERN-89-08 (1989) p. 373.

[14] OPAL collab., CERN-PPE 92/18 (1992).

[15] C.L. Basham *et al.*, Phys. Rev. Lett. **41** (1978) p. 1585; Phys Rev. **D17** (1978) p. 2298; Phys. Rev. **D19** (1979) p. 2018.

[16] D.G. Richards, W.J. Sterling and S.D. Ellis, Nucl. Phys. **B229** (1983) p. 317; A. Ali, and F. Barreiro, Nucl. Phys. **B236** (1984) p. 269; N.K. Falck and G. Kramer, Z. Phys. **C42** (1989) p. 459.

[17] See contribution to these proceedings by S. Bethke.

*K. Abe[20], I. Abt[28], P.D. Acton[3], G. Agnew[3], C. Alber[26], D.F. Alzofon[19], P. Antilogus[19], C. Arroyo[5], W.W. Ash[19], V. Ashford[19], A. Astbury[32], D. Aston[19], Y. Au[5], D.A. Axen[24], N. Bacchetta[9], K.G. Baird[17], W. Baker[19], C. Baltay[36], H.R. Band[34], G.J. Baranko[26], O. Bardon[15], F. Barrera[19], R. Battiston[10], D.A. Bauer[22], A.O. Bazarko[5], A. Bean[22], G. Beer[32], R.J. Belcinski[29], R.A. Bell[19], R. Ben-David[36], A.C. Benvenuti[7], R. Berger[19], S.C. Berridge[31], S. Bethke[14], M. Biasini[10], T. Bienz[19], G.M. Bilei[10], F. Bird[19], D. Bisello[9], G. Blaylock[23], R. Blumberg[19], J.R. Bogart[19], T. Bolton[5], S. Bougerolle[24], G.R. Bower[19], R.F. Boyce[19], J. Brau[30], M. Breidenbach[19], T.E. Browder[19], W.M. Bugg[31], B. Burgess[19], D. Burke[19], T.H. Burnett[33], P.N. Burrows[15],W. Busza[15], B.L. Byers[19], A. Calcaterra[12], D.O. Caldwell[22], D. Calloway[19], B. Camanzi[8], L. Camilleri[5], M. Carpinelli[11], J. Carr[26], S. Cartwright[29], R. Cassell[19], R. Castaldi[11,27], A. Castro[9], M. Cavalli-Sforza[23], G.B. Chadwick[19], O. Chamberlain[14], D. Chambers[19], L. Chen[35], P.E.L. Clarke[3], R. Claus[19], J. Clendenin[19], H.O. Cohn[31], J.A. Coller[2], V. Cook[33], D. Cords[19], R. Cotton[3], R.F. Cowan[15], P.A. Coyle[23], D.G. Coyne[23], W. Craddock[19], H. Cutler[19], A. D'Oliveira[25], C.J.S. Damerell[18], S. Dasu[19], R. Davis[19], R. De Sangro[12], P. De Simone[12], S. De Simone[12], T. Dean[19], F. Dejongh[4], R. Dell'Orso[11], A. Disco[36], R. Dolin[22], R.W. Downing[28], Y.C. Du[31], R. Dubois[19], J.E. Duboscq[22], W. Dunwoodie[19], D.D. Durrett[26], G. Eigen[4], B.I. Eisenstein[28], R.D. Elia[19], W.T. Emmet, II[36], R.L. English[18], E. Erdos[26], J. Escalera[19], C. Fan[26], M.J. Fero[15], J. Ferrie[19], T. Fieguth[19], J. Flynn[19], D.A. Forbush[33], K.M. Fortune[28], J.D. Fox[19], M.J. Fox[19], R. Frey[30], D.R. Freytag[19], J.I. Friedman[15], J. Fujimoto[13], K. Furuno[30], M. Gaillard[19], M. Gallinaro[12], E. Garwin[19], T. Gillman[18], A. Gioumousis[19], G. Gladding[28], S. Gonzalez[15], D.P. Gurd[21], D.L. Hale[22], G.M. Haller[19], G.D. Hallewell[19], V. Hamilton[19], M.J. Haney[28], T. Hansl-Kozanecka[15], H. Hargis[31], J. Harrison[33], E.L. Hart[31], K. Hasegawa[20], Y. Hasegawa[20], S. Hedges[3], S.S. Hertzbach[29], M.D. Hildreth[19], R.C. Hilomen[19], D.G. Hitlin[4], T.A. Hodges[32], J. Hodgson[19], J.J. Hoeflich[19], A. Honma[32], D. Horelick[19], J. Huber[30], M.E. Huffer[19], E.W. Hughes[19], H. Hwang[30], E. Hyatt[5], Y. Iwasaki[20], J.M. Izen[28], P. Jacques[17], C. Jako[19], A.S. Johnson[2], J.R. Johnson[34], R.A. Johnson[25], S. Jones[19], T. Junk[19], S. Kaiser[19], R. Kajikawa[16], M. Kalelkar[17], H. Kang[19], I. Karliner[28], H. Kawahara[19], R.K. Keeler[32], M.H. Kelsey[4], H.W. Kendall[15], D. Kharakh[19], H.Y. Kim[33], P.C. Kim[19], R. King[19], M. Klein[4], R.R. Kofler[29], M. Kowitt[14], N.M. Krishna[26], R.S. Kroeger[31], P.F. Kunz[19], Y. Kwon[19], J.F. Labs[19], R.R. Langstaff[32], M. Langston[30], R. Larsen[19], A. Lath[15], J.A. Lauber[26], D.W.G. Leith[19], L. Lintern[18], X. Liu[23], M. Loreti[9], A. Lu[22], H.L. Lynch[19], T. Lyons[15], J. Ma[33], W.A. Majid[28], G. Mancinelli[10], S. Manly[36], D. Mansour[19], G. Mantovani[10], T.W. Markiewicz[19], T. Maruyama[19], G.R. Mason[32], H. Masuda[19], L. Mathys[22], G. Mazaheri[19], A. Mazzucato[9], E. Mazzucato[8], J.F. McGowan[28], S. McHugh[22], A.K. McKemey[3], B.T. Meadows[25], D.J. Mellor[28], R. Messner[19], A.I. Mincer[4], P.M. Mockett[33], K.C. Moffeit[19], R.J. Morrison[22], B. Mours[19], G. Mueller[19], D. Muller[19], G. Mundy[19], T. Nagamine[19], U. Nauenberg[26], H. Neal[19], D. Nelson[19], V. Nesterov[19], M. Nordby[19], M. Nussbaum[25], A. Nuttall[19], H. Ogren[6], J. Olsen[19], C. Oram[21], L.S. Osborne[15], R. Ossa[19], G. Oxoby[19], L. Paffrath[19], A. Palounek[15], R.S. Panvini[35], H. Park[30], M. Pauluzzi[10], T.J. Pavel[19], F. Perrier[19], I. Peruzzi[10,12], L. Pescara[9], D. Peters[24], H. Petersen[19], M. Petradza[19], M. Piccolo[12], L. Piemontese[8], E. Pieroni[11], R. Pitthan[19], K.T. Pitts[30], R.J. Plano[17], P.R. Poffenberger[32], R. Prepost[34], C.Y. Prescott[19], D. Pripstein[14], G.D. Punkar[19], G. Putallaz[19], P. Rankin[26], B.N. Ratcliff[19], T.W. Reeves[35], P.E. Rensing[19], J.D. Richman[22], R. Rinta[19], L.P. Robertson[32], L.S. Rochester[19], L. Rosenson[15], J.E. Rothberg[33], A. Rothenberg[19], P.C. Rowson[5], J.J. Russell[19], D. Rust[6], E. Rutz[25], P. Saez[19], B. Saitta[8], A.K. Santha[25], A. Santocchia[10], O.H. Saxton[19], T. Schalk[23], P.R. Schenk[32], R.H. Schindler[19], U. Schneekloth[15], M. Schneider[23], D. Schultz[19], G.E. Schultz[26], B. Schumm[14], A. Seiden[23], L. Servoli[10], M.H. Shaevitz[5], J.T. Shank[2], G. Shapiro[14], S.L. Shapiro[19], H. Shaw[19], D.J. Sherden[19], T. Shimomura[19], A. Shoup[25], R.L. Shypit[24], C. Simopoulos[19], K. Skarpaas[19], S.R. Smith[19], A. Snyder[19], J.A. Snyder[36], R. Sobie[24], M.D. Sokoloff[25], E.N. Spencer[23], S. St. Lorant[19], P. Stamer[17], H. Steiner[14], R. Steiner[1], R.J. Stephenson[18], G. Stewart[28], P. Stiles[19], I.E. Stockdale[25], M.G. Strauss[29], D. Su[18], F. Suekane[20], A. Sugiyama[16], S. Suzuki[16], M. Swartz[19], A. Szumilo[33], M.Z. Tahar[2], T. Takahashi[19], G.J. Tappern[18], G. Tarnopolsky[19], F.E. Taylor[15], M. Tecchio(9), J.J. Thaler[28], F. Toevs[33], N. Toge[19], M. Turcotte[32], J.D. Turk[36], T. Usher[19], J. Va'Vra[19], C. Vannini[11], E. Vella[33], J.P. Venuti[35], R. Verdier[15], P.G. Verdini[11], B.F. Wadsworth[15], A.P. Waite[19], D. Walz[19], D. Warner[2], R. Watt[19], S.J. Watts[3], T. Weber[19], A.W. Weidemann[31], J.S. Whitaker[2], S.L. White[31], F.J. Wickens[18], S.A. Wickert[22], D.A. Williams[23], D.C. Williams[15], R.W. Williams[33], S.H. Williams[19], R.J. Wilson[2], W.J. Wisniewski[4], M.S. Witherell[22], M. Woods[19], G.B. Word[17], J. Wyss[9], R.K. Yamamoto[15], J.M. Yamartino[15], C. Yee[19], S.J. Yellin[22], A. Yim[19], C.C. Young[19], K.K. Young[33], H. Yuta[20], G. Zapalac[34], R.W. Zdarko[19], C. Zeitlin[30], M. Zolotorev[14], and P. Zuchelli[8].

[1] Adelphi University,
[2] Boston University,
[3] Brunel University,
[4] California Institute of Technology,
[5] Columbia University,
[6] Indiana University,
[7] INFN Sezione di Bologna,
[8] INFN Sezione di Ferrara and Università di Ferrara,
[9] INFN Sezione di Padova and Università di Padova,
[10] INFN Sezione di Perugia and Università Perugia,
[11] INFN Sezione di Pisa and Università di Pisa,
[12] INFN Lab. Nazionalli di Frascati,
[13] KEK National Laboratory,
[14] Lawrence Berkeley Laboratory, University of California,
[15] Massachusetts Institute of Technology,
[16] Nagoya University,
[17] Rutgers University,
[18] Rutherford Appleton Laboratory,
[19] Stanford Linear Accelerator Center,
[20] Tohoku University,
[21] TRIUMF,
[22] University of California, Santa Barbara,
[23] University of California, Santa Cruz,
[24] University of British Columbia,
[25] University of Cincinnati,
[26] University of Colorado,
[27] Università di Genova,
[28] University of Illinois,
[29] University of Massachusetts,
[30] University of Oregon,
[31] University of Tennessee,
[32] University of Victoria,
[33] University of Washington,
[34] University of Wisconsin,
[35] Vanderbilt University,
[36] Yale University.

TESTS OF COLOR COHERENCE IN e^+e^- COLLISIONS

Ronald Settles
Max-Plank-Institut für Physik
Werner-Heisenberg-Institut
Föhringer Ring 6
D-8000 München 40

Abstract

Recent tests of color coherence in e^+e^- experiments are described. The quantum mechanics of color flow leads to interference effects which can be studied in the final state hadronic system. QCD predictions are compared with recent data on particle momentum spectra, two-particle momentum correlations, the string effect, and energy-multiplicity-multiplicity correlations.

INTRODUCTION

This report is a brief review of the subject of "color coherence" in e^+e^- annhiliations and some of the experimental tests, including mainly those contributed to this conference. Color coherence deals with the parton emission in the shower development of high energy jets [1]–[11]. The quantum mechanics of the gluon radiation affects the hadronic final states of e^+e^- annhiliation [12]–[16]. In general there is destructive interference among infrared gluons which has consequences for *intrajet* and *interjet* properties.

This physics is also referred to as "QCD coherence" or "gluon coherence" or "interference". One has three handles with which to study this subject: the theoretical QCD predictions, the Monte Carlo models, and the data. Due to the shortness of this conference report, what follows cannot claim completeness in the discussion of the theory or of the experimental tests. Apologies to all whose contributions to this subject could not be directly referenced here.

Intrajet coherence lead to angular ordering of the parton emission; each successive gluon in the shower development is radiated at a smaller angle than its predecessor. Examples of jet properties which manifest this "QCD coherence of the first kind" are the particle momentum spectra, the two-particle momentum corelations, the particle multiplicity, and the related subjet multiplicity.

The *interjet* effects are sometimes referred to as "QCD coherence of the second kind": the angular structure of particle flows between hard partons lead to the well-known "string effect" seen in the angular distribution of particles between hard partons in three-jet events and to "energy-multiplicity-multiplicity" correlations seen in the azimuthal distribution of particles emitted about the quark direction.

For the phenomena listed above, there are QCD predictions which are based on perturbation-theory approximations of varying levels of sophistication: leading order (LO), next-to-leading order (NLO), double-log approximation (DLA), leading-log approximation (LLA), modified leading-log approximation (MLLA), and so forth [12]–[16]. Since these calculations pertain to the partons and no one yet can calculate the transition of partons into color-neutral hadrons, the question is, can we see these effects of the soft partons in our data? Here the hypothesis of local parton-hadron duality (LPHD) is invoked [12,14], according to which the inclusive hadron distributions are assumed to be proportional to the corresponding parton-level distributions, so that hadronization does not affect the coherence signal to a significant degree. The hope is that deviations from LPHD will give clues to the hadronization regime and the physics of confinement.

In addition to the QCD predictions, which tend to be limited by missing higher orders or not-quite-working LPHD, one can also look for these effects through the eyes of the Monte Carlo generators. Some of the more frequently-used ones on the market are JETSET [18], HERWIG [19], ARIADNE [20], COJETS [21], and NLLjet [22]. With these models, one can study the *parton shower* generation and the *fragmentation*, and thereby try to separate the shower from the fragmentation effects. For the *parton shower*, JETSET, HERWIG, ARIADNE, and NLLjet have coherence included, but each using a different approximation thechnique, COJETS has an incoherent parton shower, while in JETSET it can be switched off in various ways. As for the *fragmentation*, JETSET, ARIADNE, and NLLjet have Lund string fragmentation, HERWIG has cluster fragmentation, and COJETS has independent fragmentation, which again in JETSET can be switched on. Thus many tools exist to help learn about the late stages of the hadronization process.

In the following, the tests of color coherence via particle momentum spectra [1]–[6], two-particle momentum correlations [7],

string effect [8], and energy-multiplicity-multiplicity correlations (EMMC) [6,9,10] will be described briefly, with emphasis on the first and last topics. The topics of multiplicity and sub-jet multiplicity will not be covered here since they are the subject of a separate report at this conference [11]. At the end an attempt will be made to draw conclusions as to how well the QCD perturbation approximations along with LPHD are being confirmed (or not) by the data.

HADRON MOMENTUM SPECTRA

The spectrum of final-state hadrons, expressed in terms of the variable $\xi_p = \ln \frac{1}{x_p}$, $x_p = \frac{2p}{\sqrt{s}}$, is roughly Gaussian in shape (the exact form is predicted in MLLA) with functional dependence $\frac{d\sigma}{\sigma d\xi_p} = N(Y) f(\xi_p, Y, Q_0)$ where $Y = \ln \frac{\sqrt{s}}{2\Lambda_{eff}}$, Λ_{eff} is an effective QCD scale parameter, and Q_0 is the minimum parton-virtuality cutoff. For light hadrons the limiting spectrum with $Q_0 = \Lambda_{eff}$ is predicted, while for heavy hadrons $Q_0 \geq \Lambda_{eff}$. New analyses fo this type have been done by TOPAZ [1] and DELPHI [2].

MLLA predicts the position of the peak ξ_{max}, which should have the numerical value of 3.8 for $\sqrt{s} = M_Z$. Also ξ_{max} shoud decrease with increasing particle mass. The measurements of the LEP collaborations find $\xi_{max} \simeq 3.65 \pm 0.05$. There is a problem here, in that Monte Carlo studies show [6] that the position of the ξ_p distribution is sensitive to the decay of the final state hadrons. However the evolution of ξ_{max} with energy (see below) seems to be reproduced by the prediction, albeit with adjustment of the free parameter Λ_{eff}.

The peak is predicted to grow with energy, as given by $\xi_{max} = \frac{1}{2}Y + B\sqrt{\frac{Y}{16N_C}} + \mathcal{O}(1)$ where $B = (\frac{1}{2}N_C - \frac{2}{3}\frac{N_f}{N_C^2})/(\frac{11}{3}N_C - \frac{2}{3}N_f)$. In

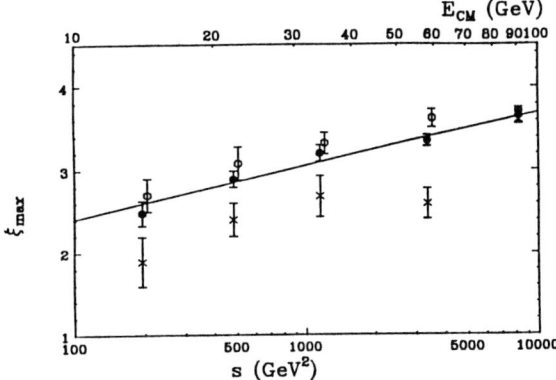

Figure 1. The evolution of the peak of the $\ln\frac{1}{x_p}$ distribution with energy for charged pions (open circles), all charged particles (closed circles), and $K^{\pm}$ (crosses). Points below $\sqrt{s}=40$GeV are from TASSO [3], at 58GeV from TOPAZ [1], and at 91 GeV from OPAL [4], and L3 [5]. A straight line $a\log\sqrt{s} + b$ is fit to the closed circles gives $a=0.548\pm0.052$, whereby 0.5 is expected from MLLA and 1.0 from phase space.

Figure 1 is shown an example compilation [1] of prediction and measurement of the behavior of ξ_{max}. One can mention at this point that the energy dependence at a fixed energy (e.g., $\sqrt{s} = M_Z$) can also be studied using the so-called restricted-cone analysis [6,15], by replacing Y by $\ln\frac{\sqrt{s}\sin\theta}{2\Lambda_{eff}}$, where θ is the opening angle of the cone.

One sees in Fig. 1 that the data show the expected rise in ξ_{max} with energy and a decrease in ξ_{max} with particle mass. The fits of the various spectra to the Gaussian-like curve as predicted should proceed by fitting Λ_{eff} to the light hadrons (pions), giving typically $\Lambda=150$MeV, and then using this value fit to the heavier hadrons with Q_0, which should increase with particle mass. This procedure seems to work more-or-less, but the errors are still a bit large [1,2] for it to be a sensitive test. More data from the LEP experiments will be available soon to enable such tests.

2-PARTICLE MOMENTUM CORREL.

In the framework of LLA, the two-particle momentum correlation function $R(\xi_1, \xi_2)$ has been calculated [23], where $R(\xi_1, \xi_2) = \frac{D^{(2)}(\xi_1,\xi_2)}{D^{(1)}(\xi_1)D^{(1)}(\xi_2)}$, and $D^{(i)}(\xi_i) = \frac{d\sigma}{\sigma d\xi_p}$, $i=1,2$. These correlations have been measured by OPAL [7] for charged particles in the range $2.5 \leq \xi_p \leq 4.5$, for which data agree satisfactorily with the QCD predictions. However, the QCD prediction for $R(\xi_1, \xi_2)$ only reproduces the qualitative trends in the data, but is unable to account for it quantatively.

In such a case, the hope would be that the QCD prediction can be improved, and one must turn to the Monte Carlo generators. Going through the various models [18]–[22], they all seem to describe the data at the hadron level reasonably well, whether or not coherence effects are included at the parton level. This indicates that hadronization plays an important rôle in two-particle momentum correlations. However, ARIADNE does do the best overall job of reproducing the data.

STRING EFFECT

The string effect is the well-known phenomenon measured in three-jet events since the PETRA-PEP days: the particle density in the angular region between quark and antiquark jets is depleted relative to that between quark and gluon jets. It is called the "string effect" because it was predicted by the Lund string fragmentation [18] and reproduces the data better than independent fragmentation schemes. QCD coherence also predicts the existence of this effect, which arises from soft gluon interference in the $q\bar{q}$ interjet region. Thus the two rather different explanations of the data are that Lund says that it happens in the fragmentation, while QCD coherence says it happens in the parton shower.

This effect has been measured by practically all e^+e^- experiments over the last 10 years, and OPAL [8] has recently reexamined it making use of the Z^0 data and presently available Monte Carlo tools [18]–[21]. What is measured is the ratio $\tilde{R} = \frac{N_{31}}{N_{12}}$ (1,2,3 number the q, q, g jets) of the number of particles within the central 50% of the angular interval between quark and gluon jets to that between the two quark jets. The measured value is $\tilde{R} = 1.84 \pm 0.09$. Further results of this thorough study are the following: Models with coherent showers *and* string or cluster fragmentation reproduce the data well. JETSET incoherent shower + string fragmentation is too low, $\tilde{R} = 1.59 \pm 0.01$, so that this is an argument in favor of coherence. However, *all* models with independent fragmentation are even lower, $\tilde{R} \sim 1.2$, which would tend to mean that the fragmentation is more important. In general, both fragmentation and coherence effects are needed to explain the data.

This effect can be studied in other ways. One way is to compare the particle densities in $q\bar{q}g$ and $q\bar{q}\gamma$ events [15,17]. The ratio in the $q\bar{q}$ interjet region for the two reactions is $\frac{N_C^2-2}{2(N_C^2-1)} = \frac{7}{16}$. This can be used to separate the fragmentation and shower effects. Another way employing the same dynamical origin is described in the next section.

EMMC

The energy-multiplicity-multiplicity correlation originates in the emission of two soft gluons $e^+e^- \to q_1 \bar{q}_2 g_3 g_4$. The cross section $\sigma_0 \sigma(q_1\bar{q}_2 g_3 g_4) = \sigma(q_1\bar{q}_2 g_3)\sigma(q_1\bar{q}_2 g_4)C$ where the correlation term C is predicted [24] to be $C = 1 + \frac{N_C^2}{N_C^2-1}(\frac{\cos\phi_{34}}{\cosh\eta_{34}-\cos\phi_{34}})$, where $\eta_j = -\ln\tan\frac{\theta}{2}(j=3,4)$, $\eta_{34} = \eta_3 - \eta_4$, and $\phi_{34} = \phi_3 - \phi_4$. θ and ϕ are the polar and azimuthal angles relative to the $q\bar{q}$ axis, the q and $\bar{q}$ being nearly collinear. Thus the ratio of C at $\phi_{34} = 180°$ to that at 90° is just $\frac{7}{16}$, so that comparing with the above paragraph, this suppression has just the same dynamical origin as the string effect. This is the basis for the proposed EMMC analysis [24].

Such an analysis has been done by ALEPH [6], DELPHI [10], and OPAL [9]. The idea is to take all triplets of particles, one particle gives the reference direction and its energy is put in the bin corresponding to the azimuthal angle between the other two, which should be in some appropriately chosen rapidity interval about the reference particle. The energy weighting seeks out the jet direction, the rapidity interval avoids unwanted effects, e.g., from three-jet events, and the azimuthal binning measures directly the interference effect described above (of course, with proper normalization).

The results are that the global features of the data are described by all Monte Carlo models, and that the ratio of C at $\phi_{34} = 180°$ to that at 90° is not $\frac{7}{16} = 0.44$ as predicted by QCD, but ~ 0.8. This and Monte Carlo studies led DELPHI to conclude that "(it is) difficult to interpret the observation of EMMC correlations as evidence for coherence." The earlier ALEPH analysis concluded that "more studies were necessary."

However, the discrepancy between QCD prediction and the data may be attributed to next-to-leading order effects [25]. And the OPAL analysis went one step further, in particular by subtracting the data from the Monte Carlo models and investigating the models [18]–[21] with and without coherence in detail. The results are seen in Figure 2. The conclusion was that "...correlations at $\phi \geq 90°$ ...clearly favor models which include interference effects in the parton shower development..."

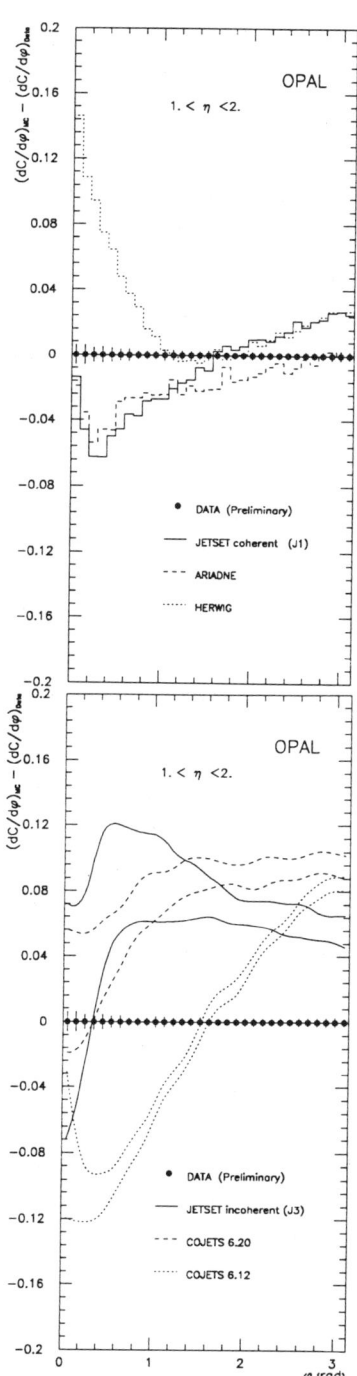

Figure 2. The EMMC comparison between Monte Carlo models and OPAL data [9]. The differences between models and data are plotted. The data were corrected using JETSET. For the incoherent shower models, the bands correspond to a variation of the main parameters by ±1 standard deviation.

Conclusions

- There is an impressive list of QCD predictions using approximations to color coherence, which must be tested in a systematic way by the experiments: particle momentum spectra, two-particle momentum correlations, multiplicity distributions, the string effect, and energy-multiplicity-multiplicity correlations, to name a few.

- The difficulty arises from the fact that quantitative comparisons of QCD predictions with the data tend to suffer due to missing higher orders in the approximations and because LPHD is also only approximate. In general one is relegated to qualitative comparisons.

- It looks promising to continue the approach of making detailed comparisons using the Monte Carlo models and the many options available for including or excluding coherence effects. Delimiting the borderline between soft gluons and fragmentation will be a tough nut to crack, but ultimately must be.

REFERENCES

1. TOPAZ Collab., "Test of QCD Prediction of Inclusive Spectrum in Hadron Jets in e^+e^- Annhiliation," *submitted to this conference* (1992).

2. DELPHI Collab., "Production of $K^0{}_s$ and Λ at small x in the Hadronic Decays of the Z^0," *submitted to this conference*, DELPHI 92-88 PHYS 199 (1992).

3. TASSO Collab., *Z. Phys.* C22, p.307 (1984).

4. OPAL Collab., *Phys.Lett.* B247, p.617 (1990).

5. L3 Collab., *Phys. Lett.* B259, p.199 (1991).
6. ALEPH Collab., "Measurement of Charged Particle Distributions Sensitive to QCD Coherence in Hadronic Z Decays," *submitted to the Lepton-Photon Conference Geneva 1991*, ALEPH Note 91-103 (1991).
7. OPAL Collab., "A study of two-particle momentum correlations in hadronic Z^0 decays," *submitted to this conference*, CERN-PPE/92-89 (1992).
8. OPAL Collab., "Further Studies of the String Effect," *submitted to this conference*, OPAL Internal Physics Note PN049 (1991).
9. OPAL Collab., "QCD coherence studies using two particle azimuthal correlations", *submitted to this conference*, OPAL Internal Physics Note PN081 (1992).
10. DELPHI Collab., "Interjet Correlations in Hadronic Z^0 Decays," *submitted to this conference*, DELPHI 92-87 PHYS 198 (1992).
11. Richard D. St. Denis, "Multiplicity Distributions in e^+e^- and Hadron-Hadron Collisions," *these conference proceedings*.
12. Yu.L. Dokshitzer and S.I. Troyan, *Proc. 19th Winter School of the LNPI*, Vol.I, p.144; LNPI-922 (1984).
13. A.H. Mueller, *Nucl. Phys.* B213, p.85 (1983), (*Nucl. Phys.* B241, p.141 (1984)), *Nucl. Phys.* B228, p.351 (1984).
14. Ya.I. Azimov, *Z. Phys.* C27, p.65 (1985), *Z. Phys.* C31, p.213 (1986).
15. Yu.L. Dokshitzer, V.A. Khoze, S.I. Troyan, and A.H. Mueller, *Rev. Mod. Phys.* 60, p.373 (1988).
16. Yu.L. Dokshitzer, V.A. Khoze, and S.I. Troyan, *Perturbative QCD, ed. A.H. Mueller*, World Scientific Singapore, p.241 (1989).
17. Z. Kunszt, P. Nason, G. Marchesini, and B.R. Webber, CERN 89-08, Vol.1, p.373 (1989).
18. B. Andersson, G. Gustafson, G. Ingelman, and T. Sjöstrand, *Phys. Rep.* 97, p.31 (1983); M. Bengtsson and T. Sjöstrand, *Phys. Lett.* 185B, p.435 (1987).
19. G. Marchesini and B. Webber, Cavendish - HEP - 88/7 (1988); G. Marchesini and B. Webber, *Nucl. Phys.* B310, p.461 (1988); I.Knowles, *Nucl. Phys.* B310, p.571 (1988).
20. L. Lönnblad, DESY 92-046 (1992).
21. R. Ordorico, DFUB 92-6 (1992).
22. K. Kato and T. Munehisa, *Comp. Phys. Comm* 64, p.67 (1991), *Mod. Phys. Lett.* A1, p.345 (1986), *Phys. Rev.* D36, p.61 (1987), *Phys. Rev.* D39, p.156 (1989).
23. C.P. Fong and B. Webber, *Nucl. Phys.* B355, p.54 (1988).
24. Yu.L. Dokshitzer, V.A. Khoze, P. Marchesini, and B.R. Webber, *Phys. Lett.* B245, p.243 (1990).
25. Yu.K. Dokshitzer and P. Marchesini, Lund LU 92-19 (1992) and private communication.

THEORETICAL STATUS OF α_s DETERMINATION

C.J. Maxwell
Centre for Particle Theory
University of Durham
South Road
Durham, U.K.

Abstract

We review the methods used by the LEP collaborations to extract $\alpha_s(M_Z)$ and suggest that they have little theoretical justification. A formalism allowing a sensible estimate of the uncertainties is proposed.

INTRODUCTION

The LEP collaborations all have high statistics data samples which enable them to make accurate measurements of a wide range of e^+e^- QCD observables - jet fractions, thrust distributions, hadronic width of Z^0, energy-energy correlations ... For most of these quantities non-perturbative effects such as hadronization corrections are expected to be small and next-to-leading order (NLO) calculations are available in renormalization group (RG) improved QCD perturbation theory. By comparing such calculations with data one should then be able to extract Λ_{QCD} the fundamental SU(3) Standard Model parameter (equivalently $\alpha_s(M_Z)$ in the $\overline{MS}$ renormalization scheme).

Such α_s determinations are hampered by the fact that truncated perturbative predictions are not unique but depend on the chosen renormalization scale, more generally the renormalization scheme (RS). This RS dependence problem has no universally accepted solution and has formed the basis of an extensive literature.

In this talk we wish to review the way in which this problem has been dealt with in the extractions of $\alpha_s(M_Z)$ by the LEP collaborations[1-4] and to suggest that these determinations have little theoretical justification. We propose a formalism[5] which allows one to learn about the relative size of uncalculated higher-order corrections and to sensibly estimate the theoretical uncertainty in α_s.

RS DEPENDENCE

We begin by briefly reviewing the RS dependence problem. Consider a generic (LEP) observable R expanded in powers of the RG improved coupling

$$a \equiv \frac{\alpha_s}{\pi}$$

$$R = a + r_1 a^2 + r_2 a^3 + \ldots \quad (1)$$

For the e^+e^- R-ratio the next-to-NLO corrections (up to r_2) are available. For all the other LEP observables only NLO corrections up to r_1 are known.

Using the notation of Stevenson[6] the coupling a is determined by the beta-function equation

$$\frac{da}{d\tau} = -a^2(1+ca+c_2 a^2+\ldots) \quad . \quad (2)$$

Here $\tau = b \ln \mu/\tilde{\Lambda}$ with μ the renormalization scale and $\tilde{\Lambda}$ the dimensional transmutation mass scale. b and c are RS invariants. In N_f-flavour QCD with three colours, $b = \frac{1}{6}(33-2N_f)$ and $c = (153-19 N_f)/2(33-2 N_f)$. The remaining coefficients $c_2, c_3, \ldots$ are RS dependent. The ensemble $\{\tau, c_2, c_3, \ldots\}$ may be taken to label the RS. The coupling a and coefficients r_i depend on RS in such a way that when summed to all-orders R is RS-dependent. For instance

$$R_{NLO}(\tau) = a(\tau) + r_1(\tau) a^2(\tau), \quad (3)$$

where $a(\tau)$ is obtained by truncating eq. (2) at NLO and integrating to obtain

$$\tau = \frac{1}{a(\tau)} + c \ln\left[\frac{ca(\tau)}{1+ca(\tau)}\right] \equiv F(a(\tau)), \quad (4)$$

and this transcendental equation can be solved to give $a(\tau) = F^{-1}(\tau)$. Recognizing that $\rho_0 = \tau - r_1(\tau)$ is an RS-invariant one can write explicitly

$$R_{NLO}(\tau) = a(\tau) + a^2(\tau) F(a(\tau)) - \rho_0 a^2(\tau). \quad (5)$$

When plotted versus $a(\tau)$ R_{NLO} has the inverted parabolic shape shown in Fig. 1, provided that $\rho_0 > 0$.

Figure 1. Generic form of a NLO approximant plotted versus $a(\tau)$.

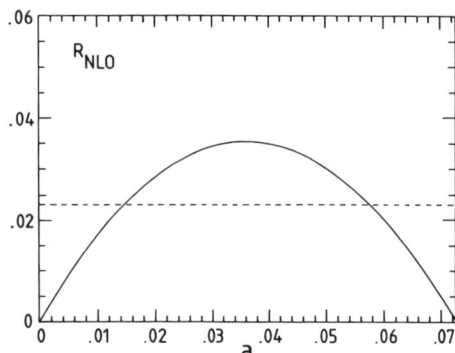

Figure 1. Generic form of a NLO approximant plotted versus $a(\tau)$.

If Q denotes the single massive scale on which the dimensionless R depends (i.e. $\sqrt{s}$ the e^+e^- c.m. energy for the LEP observables) then

$$\rho_0(Q) = b \ln \frac{Q}{\tilde{\Lambda}_{RS}} - r_1^{RS}(\mu=Q) \equiv b \ln \frac{Q}{\bar{\Lambda}}, \quad (6)$$

where $r_1^{RS}(\mu=Q)$ is a Q-independent constant, and $\bar{\Lambda} = \exp(r_1^{RS}/b)\tilde{\Lambda}_{RS}$ is an RS-invariant but observable-dependent constant. Clearly for $Q > \bar{\Lambda}$ we are guaranteed that $\rho_0 > 0$. Changing $\tilde{\Lambda}_{\overline{MS}}$ changes ρ_0 and moves the curve in Fig. 1 up and down. Taking the horizontal dashed line to represent the experimental value of the observable we see that for suitably large $\tilde{\Lambda}_{\overline{MS}}$ the data will intersect the curve for two values of τ. An infinity of $\tau, \tilde{\Lambda}_{MS}$ values will fit the data perfectly.

In order to extract $\tilde{\Lambda}_{\overline{MS}}$ we need to specify τ. This is precisely the RS dependence problem. Various "solutions" of the problem, i.e. motivations of particular choices of τ have been proposed:

(i) PMS or Principal of Minimal Sensitivity.[6]

Choose τ so that $\frac{dR^{NLO}(\tau)}{d\tau} = 0$, i.e. so that one is at a stationary point where the prediction is least sensitive to τ.

(ii) FAC or Fastest Apparent Convergence Criterion.[7,8]

(Effective Charge) Choose $\tau = \rho_0$ so that $R_{NLO}(\rho_0) = a(\rho_0)$ and $r_1(\rho_0) = 0$, so that the NLO coefficient vanishes.

(iii) Physical Scale.

Choose $\mu \simeq Q$ the "physical" scale in the problem, i.e. $\sqrt{s}$ in e^+e^- annihilation. The problem is that $\tau = b \ln \frac{\mu}{\tilde{\Lambda}_{RS}}$ is the relevant variable and one can always change the RS (and hence $\tilde{\Lambda}_{RS}$) and μ, but preserve τ. μ has no physical meaning!

(iv) Fit μ and $\tilde{\Lambda}$ to data.

Over some range of the dependent kinematical variables (thrust, jet resolution cut ...) demand stable best fits for a μ (independent of these variables) and $\tilde{\Lambda}$.

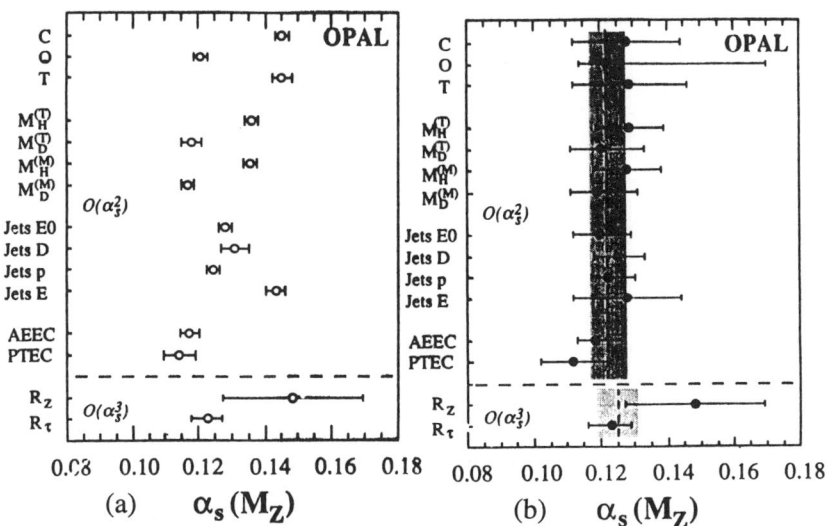

Figure 2. (a) $\alpha_s(M_Z)$ from fitting to OPAL data with $\mu = M_Z$.
(b) Result from fitting as described in text.

The LEP collaborations use a mixture of approaches (iii) and (iv) in extracting $\alpha_s(M_Z)$.[1-4] OPAL for instance[1] extracts α_s from the data fixing $\mu = M_Z$ and then μ and $\tilde{\Lambda}$ are fitted as in (iv) to yield another value of α_s. The quoted central value of $\alpha_s(M_Z)$ is then taken to lie halfway between these extremes and the error bar is given by the interval between them. Bethke[9] has reported at this conference an overall value of $\alpha_s(M_Z)$ obtained by fitting all the LEP experimental data in this way $\alpha_s(M_Z) = .120 \pm .006$. We show in Fig. 2(a) the very scattered values of $\alpha_s(M_Z)$ obtained from the OPAL data[1] assuming $\mu = M_Z$, and Fig. 2(b) shows the greatly enlarged error bars but more consistent central values of $\alpha_s(M_Z)$ obtained using the above procedure. The fitted μ tends to be considerably less than M_Z, and this is also true of the scales obtained from applying PMS or FAC, these two criteria giving almost identical results. Sanghera[10] has also submitted papers reporting similar analyses.

Whilst such fits may be justified as a pragmatic experimental response to the lack of theoretical consensus over the scheme dependence issue we find little else to recommend them.

What is needed is a formalism[5] in which "uncalculated higher-order corrections", which represent the irreducible error in the α_s extraction, have a physical meaning.

EFFECTIVE CHARGE FORMALISM

The complications of the RS dependence problem can be mollified by noting that a <u>unique</u> perturbative result for $\frac{dR(Q)}{d\ln Q}$ in terms of $R(Q)$ exists[5,7,8]

$$\frac{dR}{d\ln Q} = -b\rho(R) = -bR^2(1+cR+\rho_2 R^2+\ldots) \quad (7)$$

This equation is just the beta-function equation in the Effective Charge Scheme ($\equiv$ Fastest Apparent Convergence) where $a = R$ and the beta-function coefficients are $c_k^{EC} = \rho_k$, with the ρ_k Q-independent, RS invariant combinations of the r_k and c_k in any other scheme; $\rho_2 = r_2 + c_2 - r_1 c - r_1^2$, ... Whilst $dR/d\ln Q$ can in principle be directly measured from experiment, colliders are usually designed to make high statistics measurements of R itself at fixed Q, so we need to integrate eq. (7). In QCD $b > 0$, so

Theoretical Status of α_s Determination

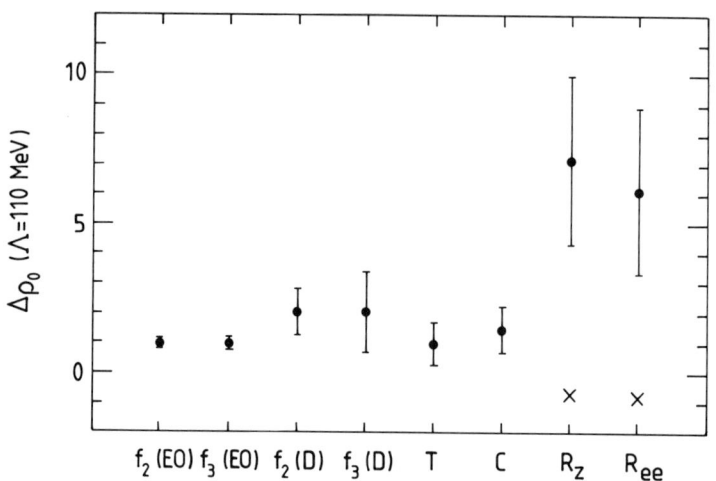

Figure 3. $\Delta\rho_0$ of eq. (10) with $\tilde{\Lambda}_{\overline{MS}}^{(5)} = 110$ MeV, plotted for OPAL data.

assuming that $\rho(R(Q)) > 0$ for $Q > Q_0$, Asymptotic Freedom applies and we are guaranteed that as $Q \to \infty$, $R \to 0$ with

$$F(R) = \rho_0 - \Delta\rho_0 = b\ell n \frac{Q}{\tilde{\Lambda}_{RS}} - r_1^{RS} - \rho_2 R^2 + \ldots \quad (8)$$

where

$$\Delta\rho_0 = \int_0^R dx \left| -\frac{1}{\rho(x)} + \frac{1}{x^2(1+cx)} \right| \quad (9)$$

$\tilde{\Lambda}$ of eq. (6) arises as a constant of integration. As $Q \to \infty$, $\Delta\rho_0 \to 0$ and $F(R) \simeq b\ell n \frac{Q}{\tilde{\Lambda}_{RS}} - r_1^{RS}(Q)$.

So given a NLO calculation $\tilde{\Lambda}_{RS}$ may be extracted using the asymptotic expression. Conversely and usefully we may use the experimental data for $R(R_{exp})$ to see how small $\Delta\rho_0$ is at current Q by forming

$$\Delta\rho_0^{exp}(Q) = b\ell n \frac{Q}{\tilde{\Lambda}_{\overline{MS}}} - r_1^{\overline{MS}}(\mu=Q) - F(R_{exp}(Q)). \quad (10)$$

We can then measure $\Delta\rho_0$ up to $\tilde{\Lambda}_{\overline{MS}}$ which is, however, a universal additive constant so that relative differences can be measured. Choosing $\tilde{\Lambda}_{\overline{MS}}^{(5)} = 110$ MeV to accord with the central value of $\tilde{\Lambda}$ obtained from a recent non-perturbative quenched lattice estimate[11] of the 1P-1S splitting in the charmonium system 110^{+36}_{-31} MeV ($\alpha_s(M_Z) = .105 \pm .004$), and using OPAL data,[1] we obtain the $\Delta\rho_0$ plot of Fig. 3. f_2, f_3 indicate the 2 and 3-jet fractions using the EO and D (Durham) algorithms, T and C are thrust and the shape variable C, respectively, and R_Z denotes the average LEP data for $(\Gamma_{had}(Z)/\Gamma_{lep}(Z))$, and $R_{e^+e^-}$ the R ratio at $\sqrt{s} = 34$ GeV off the Z^0. The crosses denote the absolute next-NLO ($O(a^3)$) prediction for the R-ratios[12] obtained using eq. (9) with $\rho(x)$ truncated.

All observables, apart from R_Z and $R_{e^+e^-}$, are nicely consistent with a horizontal straight line. This strongly suggests that $\Delta\rho_0$ is indeed small at LEP energies for these quantities since it is unlikely that $\Delta\rho_0$ is large and approximately equal for uncorrelated observables. Fitting to $\Delta\rho_0 = 0$ one obtains $\alpha_s(M_Z) = .108 \pm .004$, consistent with the lattice estimate and in good agreement with electroproduction data. There are large errors in the experimental measurements of R_Z and $R_{e^+e^-}$, but it would appear that there may be large sub-asymptotic corrections for these quantities.

$R_{e^+e^-}$ AT LOW ENERGIES

We conclude by briefly mentioning an interesting submitted paper by Mattingly and Stevenson.[13] Letting R denote the QCD correction to the e^+e^- total hadronic cross-section at c.m. energy Q and using the $\overline{MS}$ calculations[12] up to $O(a^3)$ for five quark flavours, eq. (7) becomes

$$\frac{dR}{d\ell nQ} = -3.83(1 + 1.26R - 15.1R^2), \quad (11)$$

so there is an apparent fixed point, $\rho(R^*) = 0$, for $R^* = 0.3$. This is not realised for $N_f = 5$, but evolving R down in energy through flavour thresholds using eq. (11), and changing $c(N_f)$ and $\rho_2(N_f)$ with a suitable continuity condition, R freezes to the $N_f = 2$ fixed point $R^* \simeq 0.4$. One can compare the R(Q) obtained with low energy data, where both are smeared as suggested by Poggio, Quinn and Weinberg, to suppress non-perturbative effects. As shown in Fig. 4 excellent agreement is obtained in the low energy ($Q \lesssim 1$ GeV) region with a $\Delta = 3$ GeV^2 smearing parameter.

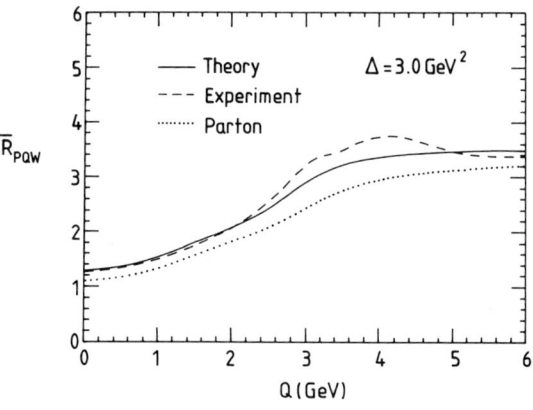

Figure 4. PQW smeared R ($\Delta = 3$ GeV^2) as evolved using eq. (11) (theory) and from data (experiment).

REFERENCES

1. OPAL Collaboration, M.Z. Akrawy et al, CERN preprint PPE/92-18 (1992).

2. DELPHI Collaboration, P. Abreu et al, Phys. Lett. B252, 149 (1990).

3. ALEPH Collaboration, D. Decamp et al, Phys. Lett. B255, 623 (1991).

4. L3 Collaboration, B. Adeva et al, Phys. Lett. B257, 469 (1990).

5. D.T. Barclay and C.J. Maxwell, Durham preprint DTP-92/26 (1992).

6. P.M. Stevenson, Phys. Rev. D23, 2916 (1981).

7. G. Grunberg, Phys. Lett. B95, 70 (1980);
G. Grunberg, Phys. Rev. D29, 2315 (1984).

8. A. Dhar and V. Gupta, Phys. Rev. D29, 2822 (1984).

9. S. Bethke, plenary talk at this conference.

10. S. Sanghera, contributed paper.

11. A.X. El-Khadra, G. Hockney, A.S. Kronfeld and P.B. Mackenzie, FERMILAB-PUB-91/354-T (1991).

12. L.R. Surguladze and M.A. Samuel, Phys. Rev. Lett. 66, 560 (1991);
S.G. Gorishny, A.L. Kataev and S.A. Larin, Phys. Lett. B259, 144 (1991).

13. A. Mattingly and P.M. Stevenson, contributed paper.

Soft Hadronic Phenomena

Parallel Session 8

Stockyards and Coliseum Arena, Fort Worth

***Conveners: K. Goulianos (Rockefeller)
W. Kittel (Nijmegen)***

INTERMITTENCY AND CORRELATIONS

E.A. De Wolf
Univserstaire Instelling Antwerpen
Universiteitsplein 1
Wilrijk, Belgium, B2610

Abstract

This paper presents a selection of new experimental results on particle fluctuations in hadron induced collisions. Three-dimensional analyses and correlations studies in invariant mass are shown to offer interesting perspectives towards better understanding of "intermittency" in particle physics.

INTRODUCTION

The subject of "intermittency" (IM) in particle physics developed from questions on a possibly dynamical origin of rare events, with very large multiplicity fluctuations, observed in several experiments [1]. The "dynamics"—as opposed to "statistics"—interpretation was first explored in [2]. The authors introduced the notion of IM to particle physicists and supplied them with the tools to tackle the problem. Since then IM-studies became almost synonymous with "analyses of (normalized) factorial moments".

The factorial moments $F_q(\delta) = \tilde{F}_q/<n>^q$ ($\tilde{F}_q(\delta) = <n(n-1)\ldots(n-q+1)>$) of the particle multiplicity (n) distribution are studied for a sequence of phase space domains (volume δ) by consecutive subdivision of an initial volume in M^d equal subdomains; $d = 1, 2, 3$ is the dimension of δ. To gain in statistics, F_q's of individual cells are averaged over M^d cells and over events ("horizontal analysis"). In a "vertical analysis" the order of averaging is interchanged.

By "intermittency" in the HEP-sense is meant that F_q is a power-law in δ:

$$F_q \sim (1/\delta)^{\phi_q}, \quad \phi_q > 0. \quad (1)$$

The slope ϕ_q is the "intermittency index" from which the "anomalous dimension" $d_q = \phi_q/(q-1)$ is derived. Validity of (1) in a certain δ-range implies scale-invariance. The moments $\tilde{F}_q$ are equal (for identical particles) to an integral of the inclusive q-particle density ρ_q over a hypercube. Decomposing the ρ_q in connected correlation functions via a cluster expansion, one expresses F_q in terms of factorial cumulants as [3]

$$F_2 = 1 + K_2; \quad F_3 = 1 + 3K_2 + K_3; \quad \text{etc..} \quad (2)$$

Inspired by Multifractal Theory other observables have been advocated and studied in some detail [4, 5]. Here we cannot enter this complementary and interesting subject.

EXPERIMENTAL RESULTS

Brief overview

The status of IM is summarized in several excellent recent reviews [7, 8]. Its is usually stated that "intermittency exists in all types of reactions" (e^+e^-, lh, hh, hA, AA). By this

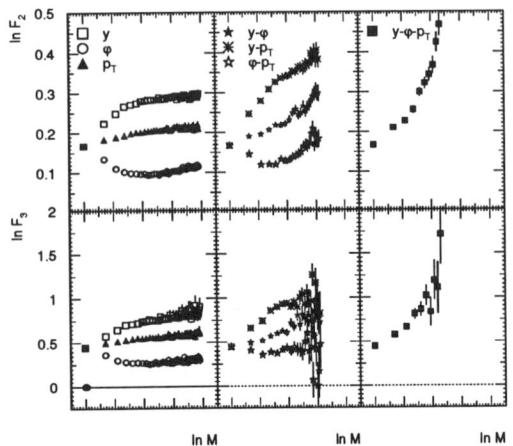

Figure 1. NA22 data for π^+/K^+p collisions at 250 GeV/c showing factorial moments in transformed y, ϕ, $\ln p_T$ in different dimensions [6].

is actually meant that the $F_q(\delta)$'s in 2D and 3D momentum space rise with decreasing δ. The most prominent features of F_q-data found in most experiments are illustrated in Fig. 1 where new NA22 data are displayed [6]. The one-dimensional (1D) F_q's usually flatten off and saturate at large M. The 2D moments follow more closely eqn.(1) However, 3D moments show concave upward bending and increase faster than a power-law.

These facts are now understood to imply that IM "lives in 3D". Projection on lower-dimensional subspaces dilutes the effect [9].

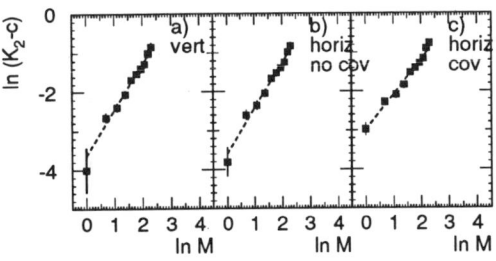

Figure 2. NA22 data on K_2 in three dimensions (transformed) for charged particles; (a) "vertical", (b) "horizontal", (c) "horizontal" with fit (see text) including covariance matrix [6].

The Fiałkowski observation

It is fair to say that, at present, no consistent interpretation has emerged from the many data on F_q's and d_q's, although interesting regularities were noted, in particular the so-called Ochs-Wosiek relation (see [7, 8]). We could mention the dependence of d_q on $n_{(ch)}$ or p_T which is different in hadron collisions and in e^+e^- annihilation [7] and believed to reflect different dynamics.

The situation was much clarified by Fiałkowski [10]. Using 3D data on μp, π/Kp, pA, and AA collisions, he noted that F_2 shows a surprisingly high degree of "universality". Writing $F_2(\delta)$ as

$$F_2(\delta) = A + B\,\delta^{-\phi_2}, \qquad (3)$$

he found ϕ_2 to have a value of $\sim 0.4 - 0.5$ in all the processes considered. The B-values also turn out to be quite similar. These observations contradict the hitherto accepted idea that d_q's and F_q's become smaller the more complex the process due to the superposition of production "sources". The author therefore speculated that IM is a "universal collective" effect.

Relation (3) suggests to plot the cumulant K_2 rather than F_2 itself. Fig. 2 shows new 3D data from NA22. The lines are fits to $K_2(\delta) = c + a'\delta^{-\phi_2}$ with $\phi_2 = 0.39 \pm 0.06$. K_3 and K_4 also grow approximately linearly with $\ln M$ (not shown), with $\phi_3 = 0.60\pm0.05$, $\phi_4 = 1.28\pm 0.24$ (c being fixed at zero). This confirms the conjecture in [10] and suggests validity of the form (3) also for K_3 and K_4.

A possible reason for "universality" becomes clear when one realizes [10] that (for F_2) δ is directly related to the four-momentum difference squared $Q^2 = -(p_1 - p_2)^2$ or the invariant mass (M_{inv}) of the particle pairs. Consequently, (3) implies that the 2-particle correlation function is a power-law in M_{inv} or in Q^2. The latter is singular while the former is

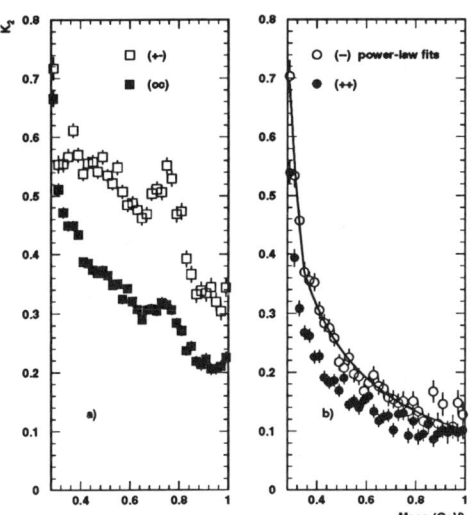

Figure 3. From NA22: $K_2(M_{\text{inv}})$ for cc, $+-$, $--$ and $++$ pairs of tracks with c.m. rapidity $-2 < y < 2$, in $K^+/\pi^+ p$ collisions at 250 GeV/c[11]. The solid line is a power-law fit (see text).

finite (on the positive M_{inv}-axis).

Via a historically tortuous path starting in 1D rapidity-space and ending in 3D transformed space, the data finally seem to tell us that an IM-analysis should be performed in (Lorentz-invariant) multiparticle variables! Interestingly, the conjectured power-law in M_{inv} of the correlation function is precisely the one predicted in (old-fashioned) Dual Mueller-Regge theory. There, power-laws are the rule rather than an exception[1]. Could it be that the Fialkowski universality merely derives from the "universality" of Regge-trajectory parameters and their couplings?

CORRELATIONS IN INVARIANT MASS

The Differential method

Having concluded that IM-data seem to prefer invariant mass variables, we turn to a discussion of this topic. The invariant

---
[1]Whether the Mueller-Regge formalism is valid for very small M_{inv} is quite unclear, however

mass dependence of the 2-particle correlation function was, to our knowledge, first studied by Berger *et al.* [12] more than 15 years ago. Starting from the correlation function $C_2(p_1, p_2) \equiv \rho_2(p_1, p_2) - \rho_1(p_1)\rho_1(p_2)$, the authors defined the function

$$C_2(M_{\text{inv}}) = \rho_2(M_{\text{inv}}) - \rho_1 \otimes \rho_1(M_{\text{inv}}). \quad (4)$$

obtained after integration of C_2 over all variables except M_{inv}. $\rho_2(M_{\text{inv}})$ is the normalized 2-particle invariant mass spectrum. The "background term" $\rho_1 \otimes \rho_1(M_{\text{inv}})$ is the integral of $\rho_1(p_1)\rho_1(p_2)$ with M_{inv} fixed. For the data to be shown, it is obtained from "uncorrelated" ("mixed") events built by random selection from a track-pool. We further utilize the function $K_2(M_{\text{inv}}) = C_2(M_{\text{inv}})/\rho_1 \otimes \rho_1(M_{\text{inv}})$.

The analysis in [12], based on low statistics pp data at 205 GeV/c, demonstrates that $K_2^{+-}(M_{\text{inv}})$ and $K_2^{\pm\pm}(M_{\text{inv}})$ follow an approximate power-law, which they write as

$$K_2(M_{\text{inv}}) = (M_{\text{inv}}^2)^{\alpha_X(0)-1}, \quad (5)$$

The notation reminds the interpretation of (5) in terms of the Mueller-Regge formalism. $\alpha_X(0)$ is the appropriate Regge-intercept; $X = R$ for non-exotic pairs and $X = E$ for exotic ones. The ratio K_2^{--}/K_2^{+-} was further seen to to fall as M_{inv}^{-2}, consistent with $\alpha_R(0) - \alpha_E(0) = 1$. Not relying on Mueller-Regge theory, the authors argued that most of the correlations at small M_{inv} are due to resonance decays into three or more pions [12, 13].

The analysis in [12] gives much new insight in the origin of IM. From the start, one now realizes that different charge-states should be treated separately since the M_{inv} dependence is very different. This fact, obvious in M_{inv} but much less so in rapidity, was not commonly appreciated until recently and important information remained hidden.

The Berger et al. method has now been applied by NA22. Figure 3 presents (preliminary) data on $K_2(M_{\text{inv}})$ [11]. $K_2^{+-}(M_{\text{inv}})$ has a prominent ρ^0 peak but is quite flat near threshold (contamination from Dalitz's is clearly visible in the first bin of subfigure a). $K_2^{--}(M_{\text{inv}})$ falls much faster. A fit of $K_2 \sim (M_{\text{inv}}^2)^{-\beta}$ yields $\beta^{--} = 1.29 \pm 0.04$, $\beta^{++} = 1.46 \pm 0.03$, $\beta^{+-} = 0.17 \pm 0.02$, in agreement with [12] and consistent with the relation $\alpha_R(0) - \alpha_E(0) = 1$. NA22 also finds that cuts on transverse momentum or relative azimuthal angle $\delta\phi$ strongly affect the shape of $K_2^{+-}(M_{\text{inv}})$ but have hardly any effect on $K_2^{--}(M_{\text{inv}})$ for $M_{\text{inv}} < 0.5$ GeV/c^2. This means that $K_2^{--}(M_{\text{inv}})$ is essentially a function only of M_{inv} (or Q^2) and demonstrates again the advantage of M_{inv} compared to other variables[2].

The Correlation Integral method

Inspired by techniques from statistical mechanics and mathematics [14], an elegant method has been proposed [15] to measure factorial moments with higher precision. The Correlation Integral (CI) method has now been utilized by UA1-MB [16], NA22 [6] and DELPHI [17].

To measure F_q one simply counts the number of q-tuples that have a "distance" (say $\delta y'$) smaller than a given distance δy; δy is, as usual, varied over a some range. The number of q-tuples is normalized by the corresponding number obtained from "mixed" events. Various "distance" measures can be used, a now popular choice being Q^2, also familiar from Bose-Einstein (BE) studies. For a detailed account of the method we must refer to [15].

From the definition it is seen that the CI-method (in Q^2) is just a "cumulative" version of the differential method of [12] discussed above.

---

[2]Further interpretation—using a power-law in M_{inv}— of data in d dimensions ($d = 1, 2, 3$), and their inter-relation, is given in [11]

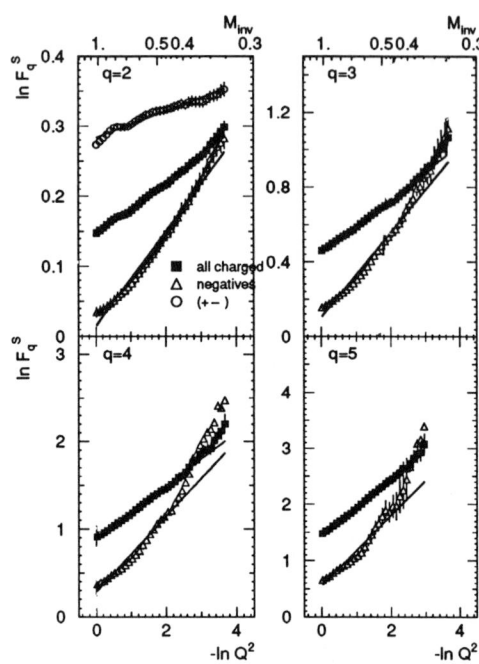

Figure 4. Correlation Integral data from NA22 in Q^2 [6]. Lines represent a power-law.

Figure 4 shows NA22 CI-data for different charge combinations. UA1-MB 2-particle results are presented in Fig. 5 where the "differential" CI $f_2(Q^2)$ is plotted ($F_2 = \int_0^{Q^2} f(Q_1^2) \, dQ_1^2$). From these data important conclusions can be drawn.

i) The moments for like-sign q-tuples increase much faster with decreasing Q^2 than the unlike-sign ones.

ii) The all-charged moments at small Q^2 are dominated by the like-sign ones. Thus, correlations among identical particles are the most likely cause of the IM-effect. This supports the early suggestion [18] that BE-type correlations play a dominant role in IM but does not explain a power-law. A way-out, with possibly far-reaching consequences, is to assume [19] that the size and/or shape of the production source(s) are subject to self-similar fluctuations due to self-criticality of parton showers.

Whether BE-effects are the sole cause of intermittency is not settled though. For ex-

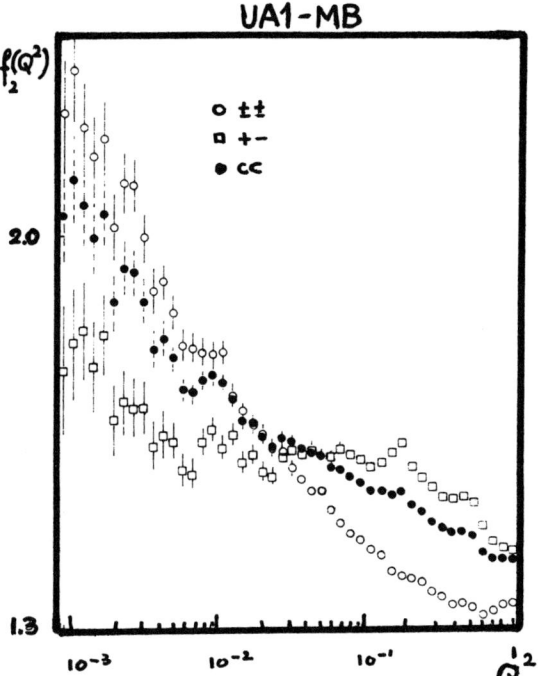

Figure 5. UA1-MB: $f_2(Q^2)$ (see text) for $\bar{p}p$ collisions at $\sqrt{s} = 630$ GeV [16].

ample, Fig. 4 shows that the negative-particle data follow an approximate power-law curve already at large Q^2 where a "static" BE-effect, with a "radius" of ~ 1 fermi, is negligible. Moreover, they start to exceed the curve around $-\ln Q^2 \sim 2$, where a conventional BE-effect begins to produce visible effects. A Mueller-Regge type power-law, together with a standard BE-effect localized near the threshold thus seems a viable—albeit less exciting—alternative to the Białas conjecture. This interpretation finds extra support in a recent successful comparison of FRITIOF+BE with negative particle 3D OAu data [20].

SUMMARY AND OUTLOOK

Although many features of IM-data remain to be clarified, significant progress has been made quite recently.

i) "Intermittency" is a 3D phenomenon.

ii) A proper choice of variables is crucial. Much effort and ingenuity was invested in analyses using single-particle variables (e.g. $y, \phi, \ln p_T$) which now seem far from ideal. This is true in particular for so-called "transformed" variables[3]. As shown here, Lorentz-invariant many-particle variables, such as invariant mass, are eminently suitable. This is hardly a surprising remembering e.g. multiperipheral-type models

iii) Much more can be learned from differential than from highly averaged observables which are insensitive to "details". A clear example are studies of "all-charged" data which have long hidden the importance of the BE-effect.

iv) There is converging evidence that "intermittency" is mostly, if not exclusively, due to strong dynamical correlations in identical particle systems of very small invariant mass. This needs further verification in e^+e^- annihilation. Understanding these correlations involves knowledge of the dynamics of a low-energy strongly interacting hadron system—a "dense pion liquid". Treatment at the amplitude level will automatically include BE interference, but must also incorporate the whole complex of final-state interactions (resonant and non-resonant amplitudes, reflections, etc.); a formidable task. Still, all these effects may conspire and allow for a simple solution in terms of concepts, such as multifractality. This happens e.g. when analysing the level-density in certain complex nuclei [21].

v) One may speculate that a low energy dense pion liquid remembers little, if anything at all, of the collision process that created it. This would explain the Fiałkowski "universality". It needs to be verified in a careful comparison of e^+e^- with hadron data. Preliminary results from DELPHI and UA1-MB

---

[3]How could a theory or model be mathematically formulated in terms of variables whose definition can only be given after the theory is solved?

[17] already show encouragingly little difference in the 2-particle CI's for same-charge pairs at small Q^2. The same holds for opposite charges. From such data one may learn about the relative importance of the perturbative and the hadronization phase (if such a distinction makes at all sense) in hadronic collisions and in e^+e^- annihilation.

One cannot avoid to remark that the recent developments have lead along a path quite distinct from the original one. With the exception of the intriguing "self-similar interaction volume" idea of [19], it is now less obvious that the fractality concepts, eminently useful in studies of non-linear random media, will eventually offer new insight into multiparticle production. This does not mean that the field should be abandoned. On the contrary, further work must clarify, for example, the reasons for the dramatic failures of hadron collision models[4] and should improve hadronization models, also in e^+e^- annihilation. The meaning of empirical regularities (charge-, multiplicity-, p_T-dependence etc.) should be reassessed and the Ochs–Wosiek relation understood. Finally, new tools should be tried in preparation for truly high multiplicity experiments where the real power of the "intermittency idea" may one day be revealed.

REFERENCES

1. (a) T.H. Burnet et al. (JACEE): *Phys. Rev. Lett.* **50** (1983) 2062; (b) G.J. Alner et al. (UA5): *Phys. Rep.* **154** (1987) 247; (c) M. Adamus et al. (NA22): *Phys. Lett.* **B185** (1987) 200; M.I. Adamovich et al. (EMU01): *Phys. Lett.* **B201** (1988) 397.
2. A. Białas, R. Peschanski, *Nucl. Phys.* **B273** (1986) 703; *Nucl. Phys.* **B308** (1988) 857.
3. A.H. Muller, *Phys. Rev.* **D4** (1971) 250.
4. R.C. Hwa, *Phys. Rev.* **D41** (1990) 1456; C.B. Chiu, R.C. Hwa, *Phys. Rev.* **D43** (1991) 100.
5. V. Arena et al. (IHSC), paper 707, this conf.
6. N. Agababyan et al. (NA22), paper 126, this conf.
7. N. Schmitz, "A Review of experimental results on intermittency" in *Proc. XXI Int. Symp. on Multiparticle Dynamics*, Wu Yuanfang and Liu Lianshou eds. (World Scientific, Singapore, 1992) pp. 377–393.
8. A. De Angelis, P. Lipa, W. Ochs, "Fractal structure and intermittency in multiparticle production" in *Proc. Joint Int. Lepton-Photon Symposium and Europhysics Conf. on High Energy Physics*, S. Hegarty, K. Potter, E. Quercigh eds (World Scientific, Singapore, 1992) pp. 377–393.
9. W. Ochs, *Phys. Lett.* **B247** (1990) 101; idem *Z. Phys.* **C50** (1991) 339; A. Białas, J. Seixas, *Phys. Lett.* **B250** (1990) 161.
10. K. Fialkowski, *Phys. Lett.* **B272** (1991) 139; idem "Universal Intermittency Slopes?" in Proc. Ringberg Workshop on Multiparticle Production R.C. Hwa, W. Ochs, N. Schmitz eds. (World Scientific, Singapore 1992) pp. 238–248.
11. E.A. De Wolf, "Understanding Intermittency" in *Proc. XXII Int. Symp. on Multiparticle Dynamics*, Santiago de Compostela, July 1992, to be publ.
12. E.L. Berger, R. Singer, G.H. Thomas, T. Kafka, *Phys. Rev.* **D15** (1977) 206.
13. G.H. Thomas, *Phys. Rev.* **D15** (1977) 2636.
14. J.P. Grassberger, *Phys. Lett.* **97A** (1983) 227; H.G.E. Hentschel, I. Procaccia, *Physica* **8D** (1983) 435.
15. P. Lipa, P. Carruthers, H.C. Eggers, B. Buschbeck, "The correlation integral as probe of multiparticle correlations" Arizona preprint AZPH-TH/91-53.
16. B. Buschbeck et al. (UA1-MB), this conf.
17. F. Mandl (UA1-MB and DELPHI), "Correlation Integral Studies in DELPHI and UA1" in *Proc. XXII Int. Symp. on Multiparticle Dynamics*, Santiago de Compostela, July 1992, to be publ., and private communication.
18. M. Gyulassy in *Festschrift L. Van Hove*, A. Giovannini, W. Kittel eds. (World Scientific, Singapore 1990), p. 479.
19. A. Białas, *Acta Phys. Pol.* **B23** (1992) 561.
20. K. Kadija, P. Seyboth, *Phys. Lett.* **B287** (1992) 363.
21. T.A. Brody, *Rev. Mod. Phys.* **53** (1981) 385.

---

[4]This topic is being actively pursued at present but cannot be covered here

INTERMITTENCY AND PARTICLE CORRELATIONS: THEORY

Ina Sarcevic
Department of Physics
University of Arizona
Tucson, AZ 85721

Abstract

We review recent theoretical developments in understanding the intermittency phenomenon observed in high-energy leptonic, hadronic and nuclear collisions. In particular, we discuss self-similar cascading and QCD parton showers, models with phase transitions, and the three-dimensional statistical field theory of multiparticle density fluctuations.

INTRODUCTION

The first unusually large local fluctuations in rapidity distribution were observed in a high multiplicity cosmic ray event [1], followed by the famous NA22 "spike" event [2]. The analysis proposed to study these fluctuations by measuring the factorial moments, which act as filters for the spike events, showed that these spectacular events were not the result of statistical fluctuations. The observation of a power-law behavior of the factorial moments, i.e. $F_p \sim \delta y^{-\nu_p}$, in a sufficiently large range of rapidity scales, δy, was claimed to be a signal of a dynamical "intermittent" behavior, in analogy with the onset of turbulence in hydrodynamics [3]. In particle physics, this would correspond either to a self-similar cascade process or to the behavior of a statistical system near a critical point. The simplest example of the self-similar cascade is the chain decay of hadronic "clusters", the initial heavy-mass "cluster" decaying into smaller clusters, which in turn decay into still smaller clusters. This leads to a power-law behavior for the normalized multiplicity moments, with exponents related to the (multi-)fractal dimension [4]. On the other hand, if a system undergoing a second-order phase transition is close to its critical point, the correlation functions exhibit power-law singularities. This implies that scaling laws and intermittency exponents are related to anomalous dimensions representing a simple (mono-) fractal [5]. The attractive idea of using intermittency (i.e. multiparticle fluctuations present for a large range of scales) to study the fractal structure of high-energy collisions has inspired extensive experimental and theoretical work [6].

EXPERIMENTAL RESULTS

In the past few years, several experimental groups have investigated intermittency signals by measuring factorial moments defined as [3]

$$F_p(\delta) = \frac{<n(n-1)\ldots(n-p+1)>}{<n>^p} =$$

$$= \frac{1}{M(\delta)^p} \sum_{m=1}^{M} \int_{\Omega_m} \prod_i d^3 x_i \frac{\rho_p(\vec{x}_1 \ldots \vec{x}_p)}{(\bar{\rho}_m)^p} \quad (1)$$

where n is the number of particles in a bin m, M is the total number of bins, δ is the phase space region and ρ_p is p-particle density correlation function. Initially, the analysis was done in rapidity (i.e. $\delta \equiv \delta y$ and $\delta y = Y/M$) and factorial moments were found to increase with decreasing bin size [6]. Many experiments claimed to observe intermittency, i.e. $F_p \sim \delta y^{-\nu_p}$, even though all the data clearly showed a tendency to level off at small δy. It has been shown that all one-dimensional hadronic data can be described by exponential two-particle correlations and the linked-pair ansatz for higher-order correlations, without invoking any singular behavior for the correlations [7]. Similar conclusions were reached for one-dimensional leptonic and nuclear data [8]. Recently, Ochs has pointed out the importance of performing the experimental analysis in three dimensions (rapidity, p_T and azimuthal angle), and has indicated how projection onto two or one dimension can reduce or destroy the "intermittency" signal [9]. In the past year, multidimensional analyses have been performed for e^+e^- collisions (by the CELLO, DELPHI, ALEPH and OPAL Collaborations), for hadronic collisions (by the NA22 and UA1 Collaborations) and for nuclear collisions (by the KLM, EMU01 and NA35 Collaborations) [10]. As predicted by Ochs, all experiments observe a stronger intermittency signal in higher dimension. In e^+e^- collisions, all the multidimensional, as well as one dimensional, data were found to be consistent with the existing parton shower Monte Carlos, indicating that the observed increase of the factorial moments with decreasing phase space size is the consequence of the parton cascade. In the case of hadronic collisions, none of the existing Monte Carlos can account for the observed effect and it seems likely that it is some combination of a perturbative parton cascade with soft-type interactions. Experimental analysis of the semi-hard (i.e. "minijet") events could help unravel this problem. In high-energy heavy-ion collisions, there is no theory which explains the observed rise of the moments and all the existing Monte Carlo programs for heavy-ion collisions fail to reproduce the data. It has been argued that the unusually large fluctuations signal a phase transition from quark-gluon plasma to hadronic matter. So far, no conclusive prediction for the signature of quark matter has been identified and, as a consequence the study of the unusually large density fluctuations has attracted considerable attention, especially in as much as they may reveal the presently unknown dynamics of particle production.

THEORETICAL MODELS

Cascade Models

The simplest multidimensional model that leads to intermittency is the so-called α-model [3]. In this model, the particle density in each sub-interval is a product of random numbers with common distribution so that the cascade is self-similar. The factorial moments are power-laws, while the one-dimensional projections show saturation at the small δy (i.e. the intermittency signal gets lost). This model does not describe the data well and should be used only as a toy-model. In e^+e^- collisions, the dominant mechanism for particle production is a QCD parton cascade, which implicitly violates scaling because of the running coupling constant, the angular cutoff, and the formation of hadronic resonances. It has been shown that statistical models, such as those that employ log-normal and negative binomial distributions, fail to reproduce the data in the small region of phase space, confirming the importance of QCD parton cascading in the underlying dynamics [11]. In the case of high-energy hadronic collisions, one can construct a simple self-similar cascading model for multiparticle production. At very high energies, the phase space available for particle production is large enough to allow a self-similar cascade with many branches to develop. Clearly, this new mechanism for multiparticle production has some threshold energy. For example, at $\sqrt{s} = 20 GeV$ the cascade has only a few branches, since the maximum rapidity available ($Y \equiv \ln s$) is only slightly above the resonance formation threshold. In contrast, at SSC energies, $Y \geq 20$ and a self-similar cascade with many branches can develop. The threshold energy for this self-similar cascade mechanism is of the order of a few hundred GeV. Since we expect that at these energies the application of perturbative QCD is well justified, this self-similar cascade should be related to low-p_T jet production. In support of this conjecture, we mention the recent observations of "minijets", which indicate that the fraction of "semi-hard" events responsible for low-p_T jet production increases very quickly with energy. For example, at $\sqrt{s} \sim 20 - 50 GeV$ it is only a few percent, while at CERN Collider energies, it is about $15 - 17\%$. At SSC energies, one expects that most of the events will be "semi-hard". In the simple self-similar cascade model [4], a collision takes place in several steps. First, a "heavy mass particle" is created (this could be a jet, for example). Then, this particle decays into two lighter particles of mass m_1, which by the conservation of energy is related to the mass m_0 of the initial particle ($m_0^2 = 4(m_1^2 + p_1^2)$). This pattern continues until the initial mass is reduced to the mass of the resonance ($m_{\pi\pi} \sim 0.5 GeV$). Such a one-dimensional self-similar cascade model is found to agree well with the UA5 data on multiplicity moments in different rapidity regions [4] and it would be very interesting to test it at Tevatron and SSC energies. With a better understanding of nonperturbative QCD, one might be able to derive the intermittency exponents, which in this case resemble multifractals. Deriving the probability distribution by solving stochastic evolution equations, in which quarks and gluons branch with Altarelli-Parisi-type probabilities, has also been shown to reproduce much of the data [12]. A model that contains two mechanisms has been proposed, in which parameters of the negative binomial distribution have a physical interpretation [13]. Furthermore, multifractality and phase transitions have been studied in a two-component self-similar model, that purports to identify the signature of the formation of the quark-gluon plasma [14].

Intermittency and Phase Transitions

The study of intermittency in the 2D Ising Model provides a simple illustration of the connection between the critical behavior or scale invariance of the underlying theory and the corresponding intermittency exponents [5]. The intermittency exponents (ν_p) of the block spin moments were found to be equal to $D(p-1)$, where D is related to the critical exponents of the Ising Model. Bialas and Hwa [15] conjectured that data that follow this behavior indicate that the system has undergone a phase transition from a quark-gluon plasma to a hadronic gas. By fitting the one-dimensional factorial moments with straight lines (i.e. by assuming that the moments do not saturate at small δy) and then by determining whether these slopes follow monofractal (signal of quark-gluon plasma) or multifractal (signal of the cascading) behavior, they claimed that the QGP has been observed in S-Au data. These data were later reanalyzed and somewhat different exponents were found, $\nu_p = (p^{1.6} - 1)/(p - 1)$ [9]. Recently, the exact solution for the factorial moments in the 1D Ising Model have been shown to display intermittency and these results have been compared to high-energy nuclear data [16]. The universality of the factorial moments, namely that all the moments, F_p, can be expresses as some power of F_2 ($F_p \sim (F_2)^{c_p}$), was found to be present in all high-energy collisions [17]. The fundamental reason for this behavior is still not understood. Finally, there have been some attempts to interpret hadronic data in terms of a second-order phase transition by evoking a two-component model (one component corresponds to the Feynman-Wilson "gas" and the other to the critical behavior) [18].

Cumulant Expansion and Factorial Cumulants

In order to examine the true higher-order correlations, the trivial contributions from two-particle correlations need to be subtracted. The connection between the factorial moments and the correlations can be seen in Eq. (1). The F_p can be expressed in terms of the bin-averaged cumulant moments:

$$F_2 = 1 + K_2$$

$$F_3 = 1 + 3K_2 + K_3$$

$$F_4 = 1 + 6K_2 + 3(K_2)^2 + 4K_3 + K_4 \quad (2)$$

where

$$K_p(\delta) = \frac{1}{M(\delta)^p} \sum_m \int_{\Omega_m} \prod_i d^3 x_i \; k_p(\vec{x}_1 \ldots \vec{x}_p)$$

and

$$k_2(1,2) = \frac{\rho_2(\vec{x}_1, \vec{x}_2)}{<\rho(\vec{x}_1)><\rho(\vec{x}_2)>} - 1 \quad (3)$$

$$k_3(1,2,3) = \frac{\rho_3(1,2,3)}{<\rho(1)><\rho(2)><\rho(3)>} - \sum_{perm}^{(3)} \frac{\rho_2(1,2)}{<\rho(1)><\rho(2)>} + 2.$$

Clearly, if there are no true, dynamical correlations, the cumulants K_p vanish.

It has been found that K_2 decreases from lighter to heavier projectiles, especially in the case of Sulfur. Furthermore, in hadronic collisions K_3 and K_4 are non-negligible (for example, K_3 contributes up to 20% to F_3 at small δy), while in nucleus-nucleus collisions, at the same energy, these cumulants are compatible with zero [19]. This implies that there are no statistically significant correlations of order higher than two for heavy-ion collisions (i.e. the observed increase of the higher-order factorial moments F_p is entirely due to the dynamical two-particle correlations). This conclusion was found to hold even in a higher-dimensional analysis [20].

It is intuitively clear that rescattering of initially correlated particles by downstream constituents should decorrelate those initial correlations. More quantitative calculations are needed to explain these phenomenological results, in particular for the anticipated rapidity fluctuations at RHIC and LHC energies, in order to see whether suppression of multiplicity cumulants, and the attendant dominance of factorial moments by two-particle cumulants, continues to hold. Even if strong space-time fluctuations should occur, of the sort associated with the transition to a quark-gluon plasma phase, the rapidity moments must obey the identities of Eqs. (2). In this case, however, we expect the higher cumulant moments to suddenly increase, to reflect the presence of the more violent bulk fluctuations that precede hadronization.

Statistical Field Theory of Multiparticle Density Fluctuations

We have seen that particles produced in high-energy heavy-ion collisions exhibit only two-particle correlations, indicating that perhaps higher-order correlations are washed out by rescattering of the initially correlated particles. Presently, there is no theory that describes this phenomena. Recently, a three-dimensional statistical field theory of density fluctuations which has these features has been proposed [21]. This model was formulated in analogy with the Ginzburg-Landau theory of superconductivity [22]. The large number of particles produced in ultrarelativistic heavy-ion collisions justifies the use if a statistical theory of particle production. The formal analogy with the statistical mechanics of a one-dimensional "gas" was first pointed out by Feynman and Wilson [23] and was later further developed by Scalapino and Sugar [24] and many others [25]. The idea is to build a statistical theory of the macroscopic observables by imagining that the microscopic degrees of freedom are integrated out and represented in terms of a few phenomenological parameters and by postulating that this theory will eventually be derived from a more fundamental theory, such as QCD.

While in the G-L theory of superconductivity the field (i.e. the order parameter) represents superconducting pairs, in the particle production problem, the relevant variable is the density fluctuation. The "field" $\phi(\vec{x})$ is a random variable which depends on the rapidity of the particle and its transverse momentum p_t and it is identified with the density fluctuation (specifically, $\phi(\vec{x}) = \frac{\rho(\vec{x})}{<\rho(\vec{x})>} - 1$, so that $<\phi> \equiv 0$).

Even though particles produced in high-energy collisions need not be in thermal equilibrium, one can still introduce a functional of the field ϕ, $F[\phi]$, which plays a role analogous to the free energy in equilibrium statistical mechanics. In principle one should be able to derive this functional from the underlying dynamics.

We start by introducing our functional $F[\phi]$ in the Ginzburg-Landau form

$$F[\phi] = \int_0^Y dy \int_{p_t \leq p_{t,max}} \frac{d^2 p_t}{P^2} [a^2(\partial_y \phi)^2 + a^2(\nabla_{(p_t/P)}\phi)^2 + M^2\phi^2 + V(\phi)],$$

where Y is the rapidity gap between projectile and target, and a and M are phenomenological parameters that depend on control parameters of the considered reaction (such as total energy and mass number(s)). All physical quantities can be obtained in terms of ensemble averages appropriately weighted by $F[\phi]$ with the corresponding "partition function" $Z = \int \mathcal{D}\phi e^{-F[\phi]}$. For example, field correlations, which from our definition of the field are related to particle correlations, are given by

$$< \phi(\vec{x_1})\phi(\vec{x_2})...\phi(\vec{x_p}) > = \frac{1}{Z} \int \mathcal{D}\phi e^{-F[\phi]} \phi(\vec{x_1})\phi(\vec{x_2})...\phi(\vec{x_p}),$$

where $\vec{x} \equiv (y, \vec{z} \equiv \vec{p_t}/P)$.

In particular, using the definition of the field one finds that the field correlations correspond to the following cumulant particle correlations k_p:

$$< \phi(\vec{x_1})\phi(\vec{x_2}) > \equiv k_2(1,2)$$

$$< \phi(\vec{x_1})\phi(\vec{x_2})\phi(\vec{x_3}) > \equiv k_3(1,2,3) \quad (4)$$

$$< \phi(\vec{x_1})\phi(\vec{x_2})\phi(\vec{x_3})\phi(\vec{x_4}) > \equiv k_4(1,2,3,4) + \sum_{perm}^{(3)} k_2(1,2)k_2(3,4),$$

where the k_p's are defined by Eqs. (3).

Clearly, *if* the interaction term in the functional $F[\phi]$ is not present (if $V(\phi) \equiv 0$), all the higher-order odd-power correlations vanish and even-power correlations can be expressed in terms of two-field correlations. This implies that $k_p = 0$ for $p \geq 3$, while the two-particle correlations are given by

$$\langle \phi(\vec{x_1})\phi(\vec{x_2}) \rangle = \frac{\gamma}{2\pi\xi} e^{-|\vec{x_1}-\vec{x_2}|/\xi}/|\vec{x_1}-\vec{x_2}|$$

$$\langle \phi(y_1)\phi(y_2) \rangle = \gamma e^{-|y_1-y_2|/\xi},$$

where $\gamma = 1/4aM$ and $\xi = a/M$ and the second equation applies for the one-dimensional case considered below. Note that the three-dimensional correlation function has a singular, Yukawa-type form.

In the three-dimensional case, K_2 is obtained by numerical integration over the appropriate phase space region. The results are found to be in good agreement with multidimensional data [26]. The three-dimensional cumulant obeys a power-law behavior, i.e. $K_2(\delta) \sim 1/\delta$, as observed experimentally, and all the two-dimensional and one-dimensional projections saturate in the small bins. The results for one-dimensional field theory can be found in Ref. 21. Both coefficients, "mass" M and "kinetic coefficient" a, are found to increase with the complexity of the system. The value of the correlation length ξ usually determines how far the system is from the critical point. When $\xi \to \infty$ (or $M \to 0$), the system goes through the phase transition. The fitted values for ξ ($\xi \sim O(1)$) do not indicate critical behavior for the system at present energies. Approaching a critical point or, more generally, a phase transition will presumably change the behavior of the two-particle as well as the multi-particle correlations. Therefore, the appealing possibility that we may study the phase transition from hadronic matter to a quark-gluon plasma and provide further constraints on theoretical models by measuring three-dimensional density fluctuations at higher energies (e.g. at RHIC and LHC) certainly deserves further investigations.

CONCLUSIONS

Recent measurements of intermittency and particle correlations in leptonic, hadronic and heavy-ion collisions show the presence of unusually large nonstatistical fluctuations in a range of phase space intervals. All e^+e^- data are found to be consistent with the Standard Parton Shower Monte Carlos, confirming the importance of QCD as the underlying dynamics. However, there is no Monte Carlo that describes the hadronic data, where particle correlations (i.e. intermittency) become even stronger at higher energies. An interesting possibility is that the particle production at high energies is govern by the self-similar cascading of the hadronic "clusters" or low-p_T jets, resulting in scaling behavior of the multiplicity moments with multifractal exponents. The phase space available at the SSC and LHC is large enough that one might be able to search for fractal structure in multi-particle production and thereby, gain insight into the scale invariance of "soft" partonic cascades. Since in high-energy heavy-ion collisions, strong collective effects have been observed, the possibility of creating the quark-gluon plasma at RHIC and LHC seems realistic. However, whether and how one can use intermittency/multiparticle correlations to detect this new phase of matter is not clear and deserves further theoretical study, in particular by connecting present "models" to the fundamental theory of the strong interactions, QCD.

REFERENCES

1. T. H. Burnett et al. (JACEE Collaboration), *Phys. Rev. Lett.* **50**, 2062 (1983).
2. M. Adamus et al. (NA22 Collaboration), *Phys. Lett.* **B185**, 200 (1987).
3. A. Białas and R. Peschanski, *Nucl. Phys.* **B273**, 703 (1986); *Nucl. Phys.* **B308**, 857 (1988).
4. I. Sarcevic and H. Satz, *Phys. Lett.* **B233**, 251 (1989).
5. H. Satz, *Nucl. Phys.* **B320**, 613 (1989); J. Wosiek, *Acta Physica Polonica* **B19**, 629 (1990).
6. For a review see *Intermittency in High Energy Collisions*, eds. F. Cooper, R. Hwa, and I. Sarcevic (World Scientific, Singapore, 1991).
7. P. Carruthers and I. Sarcevic, *Phys. Rev. Lett.* **63**, 1562 (1989);

P. Carruthers, H. C. Eggers, and I. Sarcevic, *Phys. Lett.* **B254**, 258 (1991).

8. H. C. Eggers, Ph. D. thesis, University of Arizona, 1991.

9. W. Ochs, *Z. Phys.* **C50**, 379 (1991).

10. For a review of the experimental data see, E. De Wolf, these Proceedings; A. De Angelis, P. Lipa and W. Ochs, in *Proceedings of the LEP-HEP 1991 Conference*, Geneva July 25-August 1, 1991.

11. K. Fialkowski. W. Ochs and I. Sarcevic, *Z. Phys.* **C54**, 621 (1992).

12. A. H. Chan, L. K. Keat and C. K. Chew, *Branching Processes for Intermittency in Multihadron Production*, contribution to the *Soft Hadronic Session*, these *Proceedings*.

13. A. N. Sissakian and G. T. Torosyan, *Two-component Scheme, Clusters in the Central Region and "Intermittent" Behavior of Multiparticle Characteristics*, contribution to the *Soft Hadronic Session*, these *Proceedings*.

14. M. Blazek, *Antiproton-proton Interactions at $\sqrt{s} = 1.8 TeV$, Multifractality and Phase Transitions*, contribution to the *Soft Hadronic Session*, these *Proceedings*.

15. A. Bialas and R. Hwa, *Phys. Lett.* **B253**, 436 (1991).

16. J. Seixas, *Mod. Phys. Lett.* **A6**, 1237 (1991), Ling-Lie Chau and Ding-Wei Huang, *Phys. Lett.* **B283**, 1 (1992).

17. W. Ochs, in Ref. 10 and L. L. Chau and D. W. Huang, *Universality and Intermittency of Factorial Moments in $\bar{p}p$, e^+e^- and Comparisons with Exact Solutions from Ising Model*, contribution to the *Soft Hadronic Session*, these *Proceedings*.

18. N. G. Antoniou et al., *Phys. Rev.* **D29**, 1470 (1984); *Phys. Rev.* **D14**, 3578 (1976);
N. G. Antoniou et al., *Phys. Lett.* **B260**, 199 (1991); **B245**, 619 (1990).

19. P. Carruthers, H. C. Eggers, I. Sarcevic, *Phys. Rev.* **C44**, 1629 (1992).

20. M. I. Adamovich et al. (EMU01 Collaboration), University of Washington preprint, UWSEA-PUB-92-07, submitted to *Phys. Rev. D*; see also preliminary NA35 data in *Proceedings of the Workshop on Multiparticle Production, Fluctuations and Fractal Structure*, eds. R. Hwa, W. Ochs and N. Schmitz (World Scientific, Singapore, 1992), pg. 206.

21. H-Th. Elze and I. Sarcevic, *Phys. Rev. Lett.* **68**, 1988 (1992).

22. V. L. Ginzburg and L. D. Landau, *Zh. Eksp. Teor. Fiz.* **20**, 1064 (1950).

23. R. P. Feynman, unpublished; K. Wilson, Cornell Report No. CLNS-131 (1970), unpublished.

24. D. J. Scalapino and R. L. Sugar, *Phys. Rev.* **D8**, 2284 (1973); J. C. Botke, D. J. Scalapino, and R. L. Sugar, *Phys. Rev.* **D9**, 813 (1974); **D10**, 1604 (1974).

25. P. Carruthers and I. Sarcevic, *Phys. Lett.* **B189**, 442 (1987); I. Dremin and M. T. Nazirov, in *Proceedings of the Workshop on Multiparticle Production, Fluctuations and Fractal Structure*, eds. R. Hwa, W. Ochs and N. Schmitz (World Scientific, Singapore, 1992), pg. 340; R. C. Hwa and M. T. Nazirov, *Phys. Rev. Lett.* **69**, 741 (1992).

26. H. C. Eggers, H-Th. Elze and I. Sarcevic, University of Arizona preprint, AZPH-TH/92-25.

A SEARCH FOR THE SIGNATURE OF A DECONFINED QUARK-GLUON PHASE OF STRONGLY INTERACTING MATTER IN $\bar{p}-p$ INTERACTIONS AT $\sqrt{s}$=1.8 TEV

Laszlo J. Gutay
Department of Physics, Purdue University, W. Lafayette, IN 47907

E-735 Collaboration - T. Alexopoulos, C. Allen, E. W. Anderson, V. Balamurali, S. Banerjee, P. D. Beery, P. Bhat, J. M. Bishop, N. N. Biswas, A. Bujak, D. D. Carmony, T. Carter, Y. Choi, P. Cole, R. De Bonte, V. De Carlo, A. R. Erwin, C. Findeisen, A. T. Goshaw, L. J. Gutay, A. S. Hirsch, C. Hojvat, J. R. Jennings, V. P. Kenney, C. S. Lindsey, C. Loomis, J. M. LoSecco, T. McMahon, A. P. McManus, N. Morgan, K. Nelson, S. H. Oh, N. T. Porile, D. Reeves, A. Rimai, W. J. Robertson, R. P. Scharenberg, S. R. Stampke, B. C. Stringfellow, M. Thompson, F. Turkot, W. D. Walker, C. H. Wang, J. Warchol, D. K. Wesson, and Y. Zhan

Abstract

Multiplicity distributions, particle ratios and the average transverse momenta of particles are presented as function of collider energy at $\sqrt{s}$ = 300, 540, 1000, and 1800 GeV. The dramatic variation of these quantities starting at $\sqrt{s}$ = 300 GeV, suggests the onset of a new particle production mechanism. The $<p_t>$ versus n_c curve is measured as function of n_c at $\sqrt{s}$ =1800 GeV and compared with e^+-e^- data.

The experiment FNAL- E735 was performed at the C0 intersection region. The accelerator operated at $\sqrt{s}$ = 300, 540, 1000 and 1800 GeV energies. High statistics was available only at 1800 GeV. The intersection region was surrounded with an array of 240 scintillation detectors and a cylindrical drift chamber to measure the charged particle multiplicity (n_c) created in the collision. It covered the -3.25 < η < +3.25 pseudorapidity interval and 0 < ϕ < 360° azimuthal angle. For the subset of particles, which pass through the -0.36 < η < 1.0 and 0° < ϕ < 18° solid angle interval, the particle momentum and mass was also measured by a magnetic spectrometer time of flight (TOF) system. The apparatus is described in Reference 1.

We find (not shown) that the KNO scaling breaks down at high multiplicities above $\sqrt{s}$ = 300 GeV. In Figs. 1a, 1b, and 1c the energy dependence of $<p_t>$ and particle ratios are shown. Note the rapid rise above $\sqrt{s}$ = 300 GeV. The heavier the particle, the larger the $<p_t>$ increase. These results suggest that a new type of particle production sets in above $\sqrt{s}$ = 300 GeV. To test one of the possible hypotheses, the onset of a first order hadrons → quark-gluon phase transition, we looked for the known[2] shape of the temperature (T) versus entropy density (S) phase diagram at constant pressure (P). As in earlier thermodynamic studies of hadrons, we assumed that $<p_t>$ is related to temperature and dn_c/dy to entropy density: $<p_t>$ ∝ T, dn_c/dy ∝ S. The resulting $<p_t>$ versus n_c curve is shown in Fig. 2a. Note that the curves for pions, kaons and antiprotons are very different. All show the expected $<p_t>$ rise. However, the expected $<p_t>$ plateau in the mixed phase region holds only for pions but without a second rise. One could speculate that: (a) At higher n_c the pion curve will show a second rise. (b) With higher resolution a small plateau would emerge for both protons and kaons. (c) The difference in the curves arises from the fact that they were produced at different pressures (m ∝ P). It can be shown that transverse flow[3] does not explain the differences between the π, K and $\bar{p}$ results.

The suggestion that the results in Fig. 2a is a reflection of the onset of mini-jets[4] will be discussed next. Since mini-jets are low energy gluon jets, one has to turn to high energy, highly inelastic e^+-e^- collisions to study their properties. In Fig. 2b the p_t versus n_c data is given for our $\bar{p}-p$ data[5] and the corresponding LEP data[6]. A sharp rise in $<p_t>$ is also observed for the e^+-e^- data. Since the JETSET

Monte Carlo program describes the e^+-e^- data in meticulous detail, we studied its predictions[7] for $q-\bar{q}$, $q-\bar{q}-g$, $q-\bar{q}-g-g$ (2 jets, 3 jets, 4 jets) reactions. Authors in Ref. 7 found that for a given number of jets, the $<p_t>$ is almost independent of n_c. However, the average p_t increases as the number of jets increases. (See Fig. 3). As the multiplicity increases, the mixture of events with 3, 4, etc. jets increases and thus $<p_t>$ increases. Although the p_t variable for e^+-e^- collision is defined with respect to the jet axis and the jet angular distributions are not the same in the two reactions, the qualitative picture is clear.

The low value and n_c independent plateau for pions suggest that pions emerge from $q-\bar{q}$ jet fragmentation. The sharply rising $<p_t>$ curve for kaons and antiprotons suggest that their production mechanism is related to multi-gluon production. The large value of $<p_t>$ for baryons and its dramatic energy dependence has been predicted by JETSET. This result by Blocki et al.,[8] from Warsaw University was submitted to this session to review.

In conclusion, we have not found a unique and convinving evidence for QGP formation. We have found evidence for the onset of a new production mechanism, which is most likely gluon radiation. Continued probing of the highly inelastic region of $\bar{p}-p$ interaction is promising provided the details of jet fragmentation into baryons is clearly understood.

REFERENCES

1. T. Alexopoulos et al., Phys. Rev. Lett. **64**, 991 (1990).

2. M. W. Zemansky, "Heat and Thermodynamics", p. 336, McGraw-Hill (1951).

3. P. Levai and B. Muller, Phys. Rev. Lett. **67**, 1519 (1991).

4. X. N. Wang and M. Gyulassy, Phys. Rev. **D44**, 44 (1991).

5. T. Alexopoulos et al., Phys. Rev. Lett. **60**, 1622 (1988).

6. DELPHI Collaboration, Phys. Lett. **B276**, 254 (1992).

7. M. Szczekowski and G. Wilk, Phys. Rev. **D44**, R577 (1991).

8. J. Blocki, M. Szczekowski and G. Wilk, "Multiplicity Dependence of Mean Transverse Momentum for Strange-Particles and Baryons in e^+e^- Annihilations and in Hadron-Hadron Collisions" (submitted to Parallel Session PA-8A of this conference).

928 The Signature of a Deconfined Quark-Gluon Phase

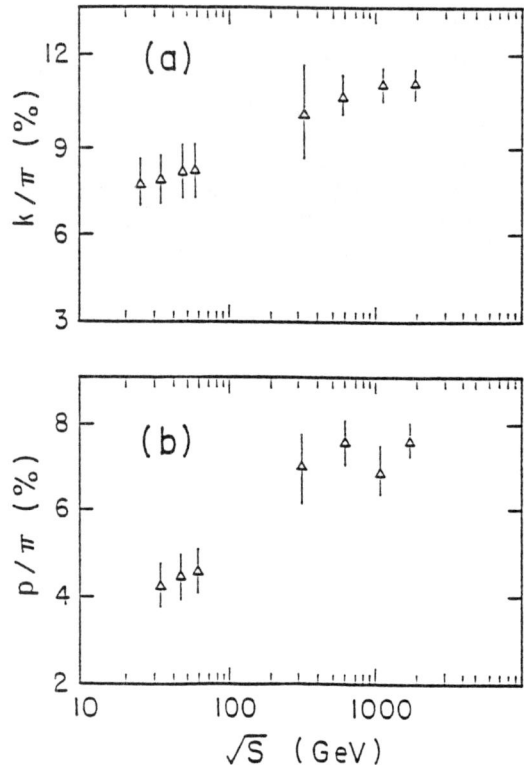

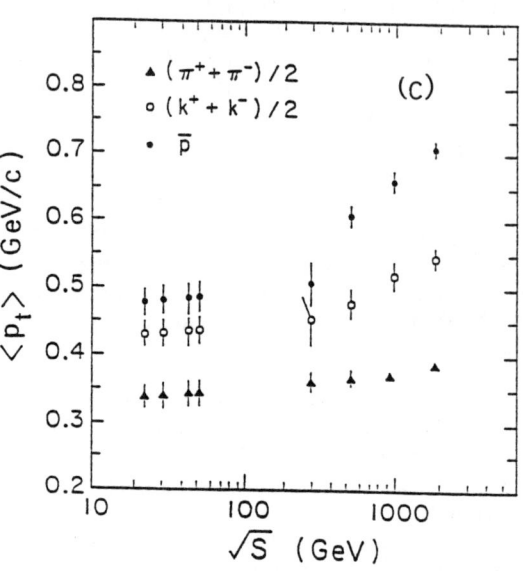

Fig.: 1c The energy dependence of the $<p_t>$ for π, K and $\bar{p}$.

Fig.: 1a The energy dependence of the K/π ratios.

1b The energy dependence of the $\bar{p}/\pi$ ratios.

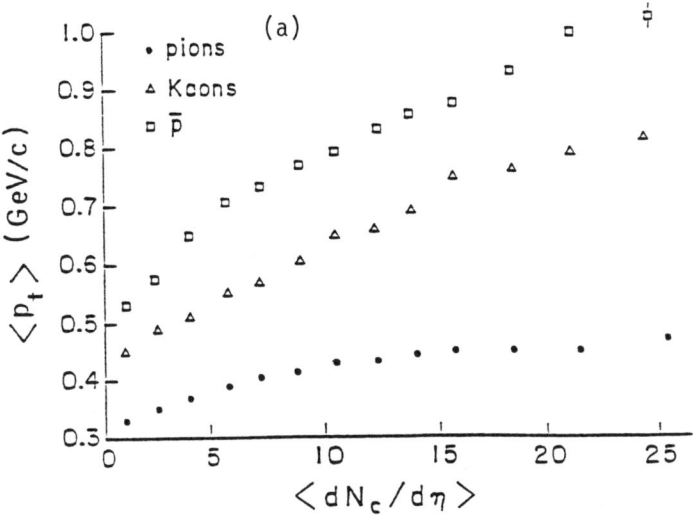

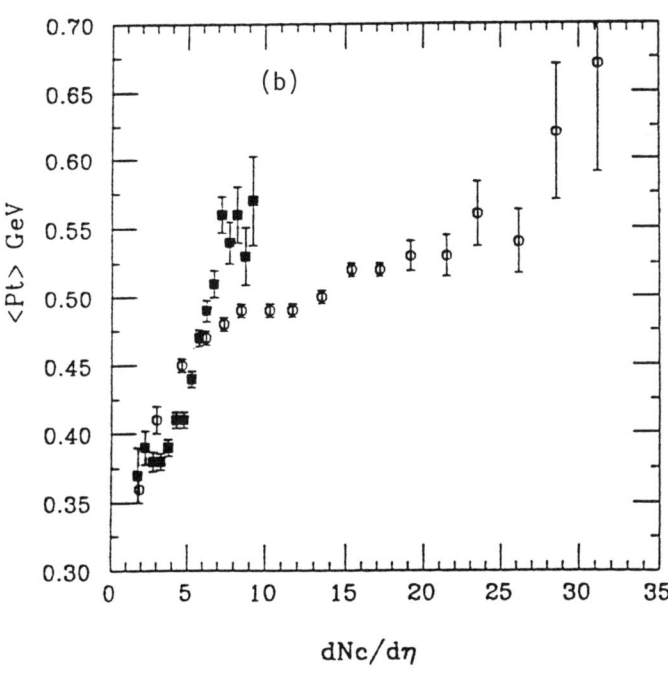

Fig: 2a The <p_t> versus $dN_c/d\eta$ curve for pions, kaons and antiprotons.

2b The <p_t> versus dN_c/η curve for unseparated particles in $\bar{p}$-p and e^+-e^- collisions.

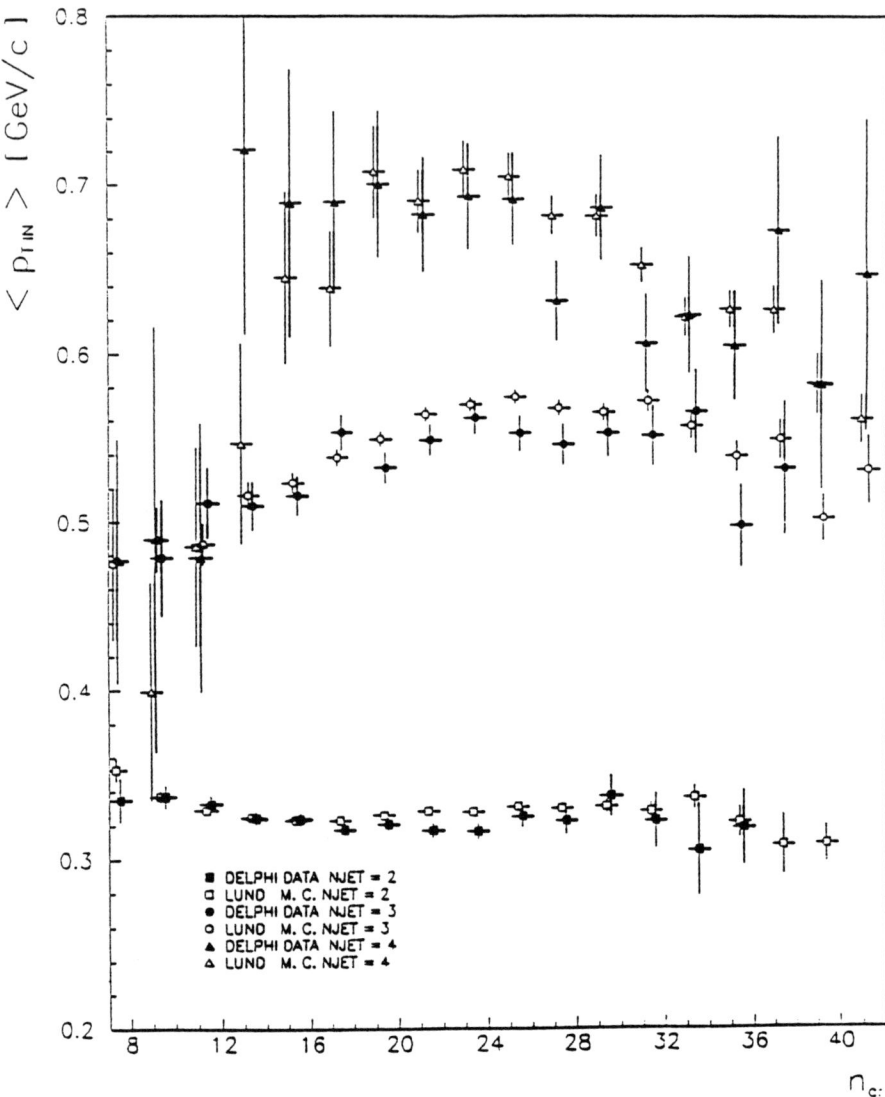

Fig.: 3 Dependence of $\langle p_t \rangle$ as function of n_c for 2 jet, 3 jet and 4 jet events.

INCLUSIVE PARTICLE PRODUCTION IN e^+e^-

Maria Novella Kienzle Focacci
DPNC
Université de Genève
24,quai E.Ansermet
CH-1211 Genève

Abstract

The recent results on the energy spectra of the low lying mesons and baryons, as measured at LEP , are described. The data are presented in the framework of the analytical QCD calculations whith the hypothesis of local parton hadron duality.

INTRODUCTION

The inclusive production of hadrons in the e^+e^- interactions has been systematically studied by the four LEP experiments : Aleph,Delphi,L3,Opal. The study of the low lying ($J^P = 0^-$) mesons and of the ground state baryons , the spin 1/2 octet and the spin 3/2 decuplet, is almost complete. The new data, presented at the conference, will be reviewed and discussed in the framework of analytical QCD calculations.

LOW LYING MESONS

The analysis of the energy spectrum of the charged hadrons at LEP ($\sqrt{s} = 91$ GeV) and its comparison with lower energy e^+e^- data has already been published [1]. The study of identified charged pions and kaons is still in progress, since it requires a good understanding of the dE/dx measurement; while the neutral pion was already studied by L3 [1] and the neutral kaon (K^0_S) by Opal and Delphi[1]. The $J^P = 0^-$ octet is then completed by the analysis of $\eta \rightarrow \gamma\gamma$ presented at this Conference by Aleph [2] and L3 [3].

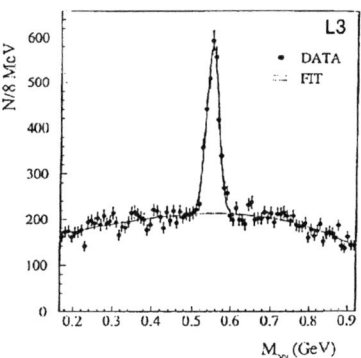

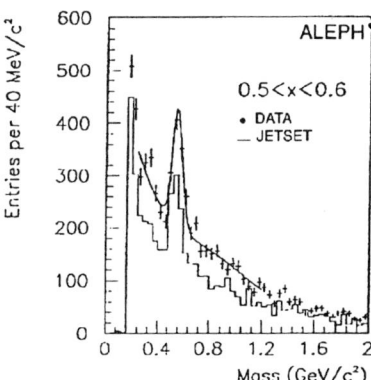

Figure 1. $\gamma\gamma$ effective mass distribution.

© 1993 American Institute of Physics

Although the two experiments cover a different momentum range : $0.035 < x_p < 0.225$ for L3 ($x_p = p_{hadron}/E_{beam}$) and $0.1 < x < 0.9$ for Aleph ($x = E_{hadron}/E_{beam}$) the data are in good agreement and they are well reproduced by the JETSET and HERWIG Monte Carlos.

Aleph also observes a small η' signal in the $\pi^+\pi^-\eta$ effective mass. The experimental production ratio $\eta'/\eta = 0.23 \pm 0.06 \pm 0.05$ for $x > 0.1$ is much smaller than the JETSET prediction (0.82) but consistent with the HERWIG value of 0.28.

THE HYPERONS

The identification of hyperons take advantage of the long flight distances : $c\tau = 7.9, 4.9, 2.46$ cm for Λ, Ξ and Ω respectively. The data selections require a secondary vertex clearly displaced from the primary vertex. Only Delphi uses the microvertex detector to improve the reconstruction efficiency and the angular coverage. The dE/dx measurement, with a resolution better than 5%, is used by Aleph and Opal to highly reduce the combinatorial background. The Λ [4, 5], Ξ [5, 6] and Ω [5, 6] are identified by their decay $p\pi^-$, $\Lambda\pi^-$ and ΛK^- respectively [1] (fig 2 a,b,c). The wrong charge combination provides an excellent method to determine the remaining background under Ξ^- and Ω^-. The $\Sigma(1385)$ and $\Xi^0(1530)$ decay strongly into $\Lambda\pi^-$ and $\Xi^-\pi^+$ respectively. The background under the Ξ^0 may still be subtracted by the wrong combinations (fig 2 d) but the combinatorial background under the Σ is very important (fig 2 e). Table 1 lists the yield per event of the low lying hadrons, the data have been integrated over the full momentum range using JETSET for the extrapolation to the unobserved regions. The weighted average of all experiments is given everywhere the data are well consistent. Only for the Ω^- production Opal and Aleph are somewhat in disagreement so their values are quoted separately.

MLLA MODEL

The hadron production is well described by the QCD perturbative calculations of the Modifided Leading Logarithm Approximation (MLLA [7]) which predicts that the bremsstrahlung like energy spectrum of the soft gluons is reduced, in the low energy side, by colour quantum interference. The analytical calculations can be performed at all orders taking into account single and double logarithms. The connection with the experimental hadron distribution is given by the hypothesis of a Local Parton Hadron Duality (LPHD): the hadron production cross section is assumed to be proportional to the parton distribution at the end of the perturbative cascade. To put in evidence the gluon interference effect the data are displayed as function of the variable $\xi_p = ln(1/x_p)$ and fitted by a Gaussian shape [5] or by an analytical function [3]:

$1/\sigma_h \, d\sigma/d\xi_p = N(\sqrt{s}) \, f(\sqrt{s}, \Lambda_{eff}; \xi_p)$

where only N and Λ_{eff} are free parameters, while the cutoff parameter of the QCD cascade is set to $Q_0 = \Lambda_{eff}$ ("limiting spectrum"). Fig 3 shows the ξ distribution for the $\eta, K^0, \Lambda^0, \Xi^-$.

The position of the maximum of these distributions, ξ^*, characterizing the gluon interference phenomena, is listed in table 2 and compared with the values obtained previously for the π^0. All the maxima are at a similar position except for the pion which has a much larger value of ξ^*.

The value of ξ^* increases with beam energy, fig 4 shows a compilation of π^0 and η data, the logarithmic increase is well reproduced by the MLLA model.

---

[1] In the text only the particle sign and decay is indicated, although the experiments measure equally particles and antiparticles.

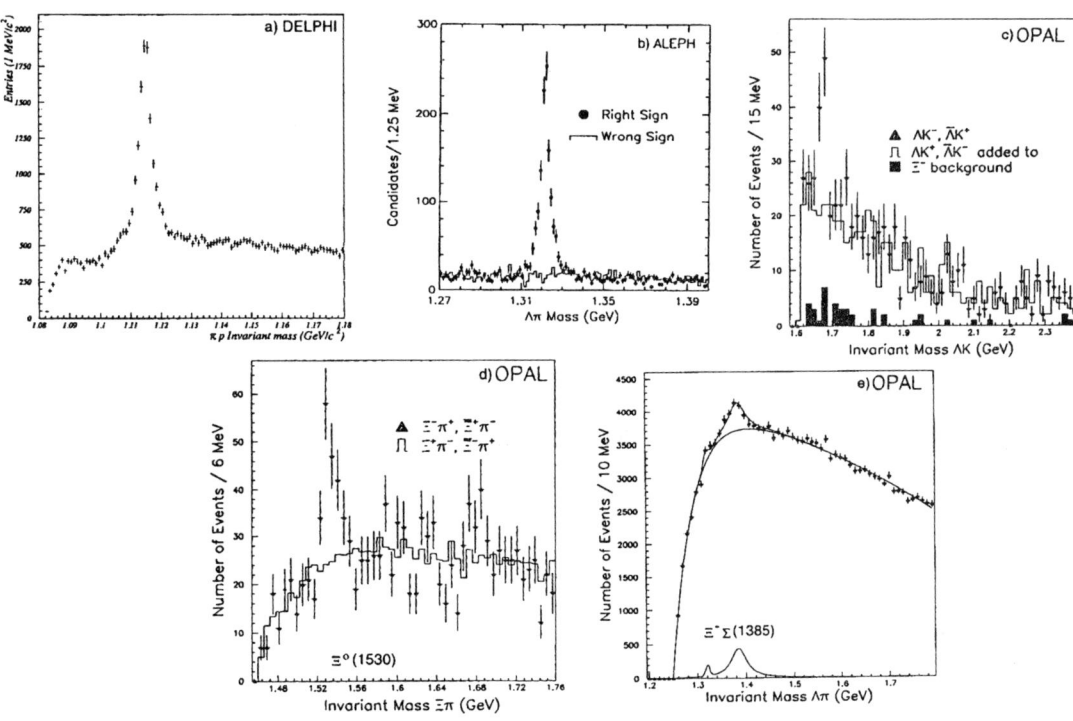

Figure 2. Some effective mass spectra of hyperons

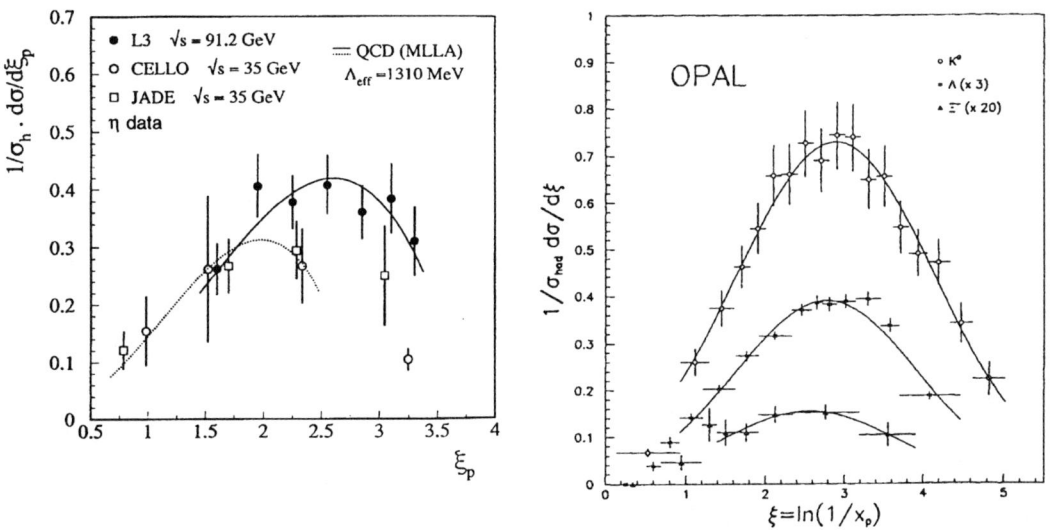

Figure 3. $\xi_p = ln(1/x_p)$ distribution for the $\eta, K^0, \Lambda^0, \Xi^-$

Table 1. Integrated hadron yield

type	mass(Mev)	yield	experiments	ref.
charged		20.9 ± 0.2	all	1
π^0	135.0	9.8 ± 0.7	L3	1
K^0	497.7	2.12 ± 0.06	Opal,Delphi	1
η	548.8	1.06 ± 0.16	Aleph,L3	2/3
η'	958.	0.17 ± 0.06	Aleph	2
Λ	1115.6	0.351 ± 0.019	Delphi,Opal	4/5
Σ^-	1385.	0.0380 ± 0.0062	Opal	5
Ξ^-	1321.	0.0239 ± 0.0014	Aleph,Delphi,Opal	6/4/5
Ξ^0	1530.	0.0063 ± 0.0014	Opal	5
Ω^-	1672.	0.0012 ± 0.0005	Aleph	6
Ω^-	1672.	0.0050 ± 0.0015	Opal	5

Table 2. Maximum value of the ξ distribution

type	value	experiments	ref.
π^0	4.11 ± 0.18	L3	1
K^0	2.91 ± 0.04	Delphi	4
K^0	2.91 ± 0.04	Opal	1
η	2.60 ± 0.15	L3	3
Λ	2.82 ± 0.07	Delphi	4
Λ	2.77 ± 0.05	Opal	5
Ξ^-	2.57 ± 0.11	Opal	5

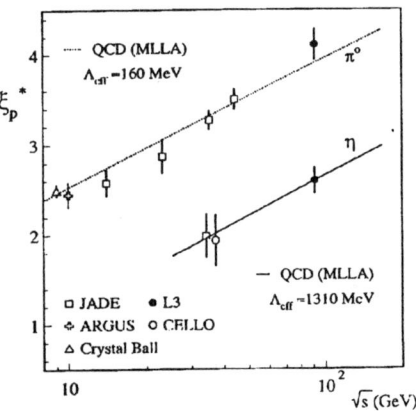

Figure 4. Energy dependence of the peak position ξ^* for π^0 and η production.

REFERENCES

1. All references preceeding this conference may be found in :
   T.Hebbeker,"Tests of Quantum Chromodynamics in Hadronic Decays of Z^0 bosons produced in e^+e^- annihilation," to be published in Physics Reports (1992).

2. The Aleph Collaboration, "Measurement of the Production Rates of η and η' in hadronic Z Decays," CERN-PPE/92-74

3. The L3 Collaboration, *Phys.Lett.* B 286, p. 403, (1992).

4. Delphi Collaboration, "Production of K_S^0 and Λ at small x in the Hadronic Decays of the Z^0," Delphi 92-88 PHYS 199 (30 june 1992)

5. Opal Collaboration, "A Measurement of Strange Baryon Production in Hadronic Z^0 Decays," CERN-PPE/92-118

6. The Aleph Collaboration, "Ξ and Ω Production in Z decays," contributed paper to ICHEP 92 Conference.

7. Yu.L.Dokshitzer,V.A.Khoze,S.I.Troyan, *Int.J.Mod.Phys.* A 7, p. 1875, (1992).

RESONANCE PRODUCTION IN e^+e^- COLLISIONS*

G. D. Lafferty
Department of Physics
University of Manchester
Manchester, GB–M13 9PL

Abstract

Two processes which give rise to light meson resonances in e^+e^- collisions are $\gamma\gamma$ reactions and inclusive production in jets from e^+e^- annihilation. New results are presented from $\gamma\gamma$ collisions at ARGUS and from hadronic Z^0 decays in DELPHI and OPAL.

INTRODUCTION

Studies of meson resonance production in e^+e^- reactions have been pursued for many years in two main areas. In $\gamma\gamma$ reactions, $e^+e^- \to e^+e^- X$, the system X, when produced via two almost real photons, has a restricted set of possible quantum numbers. This enables rather complete analyses of the properties of the final states, and the reactions allow tests of Vector Meson Dominance. However the numbers of events available have been somewhat low and there are many channels still to be studied in detail. The second area, inclusive resonance production in e^+e^- annihilation, enables the testing of QCD-inspired models of parton fragmentation since the resonances arise in hadronization of $q\bar{q}$, $q\bar{q}g$ etc. states from intermediate virtual photons or real Z^0's. Many physics measurements at LEP are now limited by systematic errors arising from incomplete knowledge of the structure of multihadronic events, so that measurements of inclusive particle and resonance rates are vital to improve the physics simulations and reduce these systematic errors.

RESONANCES IN $\gamma\gamma$ REACTIONS

The reaction $\gamma\gamma \to$ two vector mesons

Although an unexpectedly large cross section has been measured for the reaction $\gamma\gamma \to \rho^0\rho^0$, probably due to interference of isospin 0 and isospin 2 states, rates for other vector meson pairs are much lower. ARGUS[1] have now made a partial wave analysis of their $\gamma\gamma \to 5\pi$ data including the $\rho^0\omega$ channel, and have in addition made the first observations of $\rho^0\phi$ and $\omega\phi$ production.

In $\gamma\gamma \to \rho^0\omega$, which accounts for 7.8% of the total 5π cross section, the main contribution is found to come from the $(J^P, J_z) = (2^+, 2)$ partial wave. In $\gamma\gamma \to \pi^+\pi^- K^+K^-$ there is a clear correlation between the ρ^0 signal in $\pi^+\pi^-$ and the ϕ in K^+K^- as is shown in figure 1. The measured cross section for $\rho^0\phi$ is compatible with the previous upper limit, and rises sharply at threshold, falling slowly with

---

*Presented at the XXVI International Conference on High Energy Physics, Dallas, USA, 6–12 August 1992.

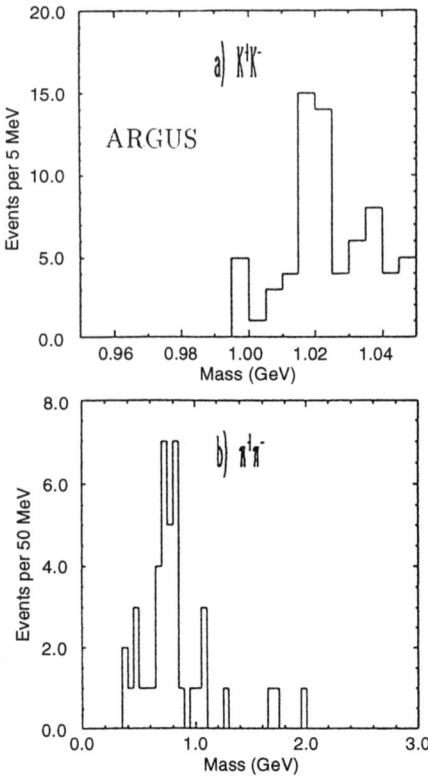

Figure 1. (a) Mass of K^+K^-; and (b) mass of $\pi^+\pi^-$ for K^+K^- mass within 10 MeV of the ϕ mass (ARGUS).

mass. Evidence for $\gamma\gamma \to \omega\phi$ is seen in the $\pi^+\pi^-\pi^0 K^+K^-$ state, with a cross section of 0.72 ± 0.38 nb for the mass range 1.9–2.3 GeV. Above 2.3 GeV a new upper limit of 0.3 nb has been set.

The reaction $\gamma\gamma \to \pi^+\pi^-\pi^0$

The channel $\gamma\gamma \to \pi^+\pi^-\pi^0$ is known to be dominated by a_2 production in the helicity 2 state; additionally there has been evidence for $\pi_2(1670)$. The intensity of the helicity zero contribution to a_2 is of interest since relativistic corrections could allow as much as 6%. The ARGUS data[2] have been analysed using a maximum likelihood partial wave analysis employing a full set of possible contributions. The mass region 0.8–1.5 GeV is dominated by the $(J^P, J_Z) = (2^+, \pm 2)$ $\rho^\pm\pi^\mp$ wave due to a_2, with a $J_Z = 0$ contribution of only about 1%. The partial width was measured to be $\Gamma_{\gamma\gamma}(a_2) = 0.90 \pm 0.15$ keV. For the $\pi(1300)$, the PWA gave an upper limit $\Gamma_{\gamma\gamma}(\pi(1300))\mathrm{Br}(\pi(1300) \to \rho^\pm\pi^\mp) < 0.01$ keV. The data also show a peak at higher mass, identified with the π_2, and evidence for an additional peak at 1.8 GeV which cannot be identified with a known resonance. A similar peak was already seen by ARGUS in $K^+K^-\pi^0$.

RESONANCES IN HADRONIC Z^0 DECAY

Vector Mesons: $\rho(770)$, $K^*(892)$ and $\phi(1020)$

Both DELPHI[3] and OPAL[4] have data on the inclusive production of the $\rho(770)^0$, $K^*(892)^0$ and $\phi(1020)$ vector mesons in Z^0 decay. All three states are identified in the corresponding two-particle mass spectra: $\pi^+\pi^-$, $K^\pm\pi^\mp$ and K^+K^- respectively. OPAL makes use of dE/dx measurements of the charged tracks in order to separate pions from kaons with good efficiency over a large momentum range. Like-charge particle combinations are used as a measure of the backgrounds in the unlike-charge mass spectra which contain the resonances. The result is that the resonances are cleanly identified in the appropriate mass spectra on rather low effective backgrounds. As an illustration, figure 2 shows the OPAL data for $K^\pm\pi^\mp$ (points with error bars) compared to an absolute prediction of the Jetset72 Monte Carlo with its default parameter set. It is clear that the Monte Carlo model overestimates the K^* rate. OPAL then use fits, in bins of the scaled momentum variable x_p, to appropriate contributions as deduced from Monte Carlo simulations in order to measure for the resonances the fragmentation functions, $1/\sigma \cdot d\sigma/dx_p$, and the overall

Table 1. Mean multiplicities of resonances per multihadronic Z^0 decay

Resonance	Multiplicity	Experiment	x_p Range	Jetset	Herwig
$\rho(770)^0$	$1.43 \pm 0.12 \pm 0.22$	DELPHI	$x_p > 0$	1.55	1.40
$K^*(892)^0$	$0.76 \pm 0.07 \pm 0.06$	OPAL	$x_p > 0$	1.06	0.77
	$0.97 \pm 0.18 \pm 0.31$	DELPHI	$x_p > 0$	1.04	0.84
$K^*(892)^\pm$	$0.72 \pm 0.02 \pm 0.08$	OPAL	$x_p > 0$	1.10	0.82
	$1.33 \pm 0.11 \pm 0.24$	DELPHI	$x_p > 0$		
$\phi(1020)$	$0.086 \pm 0.015 \pm 0.010$	OPAL	$x_p > 0$	0.189	0.113
	$0.077 \pm 0.019 \pm 0.033$	DELPHI	$x_p > 0.2$	0.060	0.078
$f_0(975)$	$0.10 \pm 0.03 \pm 0.019$	DELPHI	$x_p > 0.05$		
$f_2(1270)$	$0.11 \pm 0.04 \pm 0.03$	DELPHI	$x_p > 0.1$		0.24

branching ratios to $\pi\pi$. Figure 3 shows the inclusive $\pi^+\pi^-$ mass spectrum with the fitted curves; the lower data are after subtraction of the fitted background and clearly show the resonance contributions from ρ^0, f_0 and f_2 (as well as a peak due to K_S^0). As with the ρ^0, the fitted f_0 mass, at 961 ± 4 MeV, is somewhat below the PDG value. The measured multiplicities are given in table 1.

Both groups report evidence for $K_2^*(1430)$ production in the inclusive $K^\pm\pi^\mp$ spectra (see for example figure 2), but neither has sufficient statistical accuracy in the data analysed so far to present a useful measurement.

Discussion

As mentioned previously, measurements of inclusive resonance yields serve two main functions: as tests of models based on QCD, such as Jetset and Herwig, and as a means of improving the physics simulations of multihadronic Z^0 decays. Both DELPHI and OPAL have compared their vector meson measurements with the predictions of Jetset (with parton showers) and Herwig using parameter sets found to describe well the particle yields at PEP and Petra energies and with modifications to some parameters needed to reproduce global event properties at LEP. Table 1 lists the predictions of the models; the two groups use different program versions but the resonance yields are not strongly affected. The ρ^0 yield is in agreement with both models within the rather large errors, and with the caveat that there still appear to be some poorly understood systematic effects in the resonance line shape in the data. The K^* measurements are in good agreement with Herwig, while the Jetset predictions are considerably higher. Similarly the ϕ rate is in better agreement with Herwig than with Jetset. Thus it appears that, for these mesons at least, the variation of yield with energy is better simulated by Herwig.

The fragmentation functions, $1/\sigma.d\sigma/dx_p$, have also been compared with the behaviours predicted by the models. In general there is good agreement in the shapes although the OPAL data for the K^{*0} and ϕ, figure 4, indicate that the momentum spectra may be too hard in the Monte Carlo simulations. Figure 5 shows the fragmentation functions from DELPHI for ρ^0, f_0 and f_2. The f_0 is simulated by neither program, and the f_2 by default only in Herwig. It can be seen that although the rate for f_2 is high in the simulation, the shape is similar to that for the f_0 and is reproduced

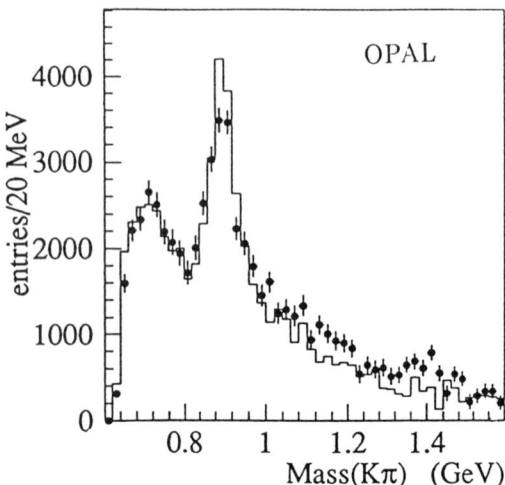

Figure 2. Mass of inclusive $K^{\pm}\pi^{\mp}$ combinations (OPAL).

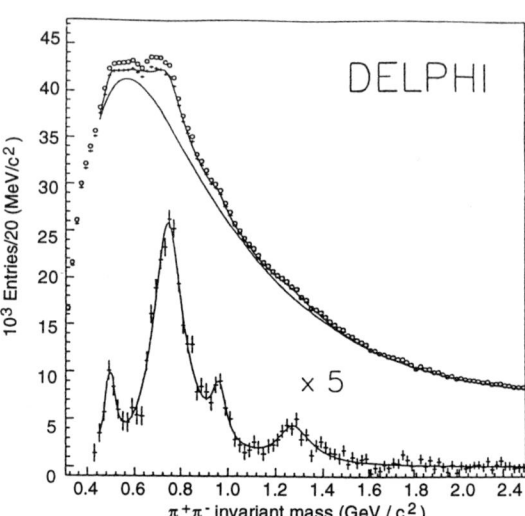

Figure 3. Mass of inclusive $\pi^{\pm}\pi^{\mp}$ combinations (DELPHI) for $x_p > 0.1$. Open dots show raw data while crosses are data corrected for reflections.

mean multiplicities per Z^0 decay.

DELPHI make no use of particle identification but instead form the mass spectra using assumed particle types. The spectra are then fitted, again for ranges of x_p, using analytical functions to represent resonance line shapes and background contributions. Problems due to reflections, arising from lack of particle identification, are treated using a weighting technique and iterating the fits. Because of this lack of particle identification, the resulting measurement errors are considerably larger than those obtained by OPAL.

Both groups report problems with the $\rho(770)^0$. The fitted mass quoted by DELPHI, 757±2 MeV, is well below the particle data group value and is inconsistent with it. Nevertheless DELPHI extract rates for the ρ^0 from their fits. OPAL however point out that the lineshape of the ρ^0 appears to be significantly distorted, especially at low momentum, and that a lack of proper understanding at present prohibits reliable measurement of the intensity. They find that a better representation of the observed line shape is obtained in the Jetset simulations if use is made of the option to generate Bose-Einstein effects. OPAL state that good measurements of the inclusive ρ^0 will have to await better understanding of these effects.

Table 1 gives a summary of the measured vector meson multiplicities per hadronic Z^0 decay (note that the two ϕ measurements are over different x_p ranges). For completeness the table also contains a measurement of $K^*(892)^{\pm}$ previously published by DELPHI[5], and a recent result from OPAL[6] which is significantly lower than the DELPHI result but in good agreement with the OPAL K^{*0} measurement. The $K^{*\pm}$ results are based on analyses of inclusive $K^0_S \pi^{\pm}$ spectra and are independent of the neutral vector meson analyses.

Scalar and tensor mesons: $f_0(975)$, $f_2(1270)$ and $K^*_2(1430)$

In their fits to the $\pi^+\pi^-$ spectra, DELPHI allow, in addition to contributions from backgrounds and ρ^0, terms due to the $f_0(975)$ and $f_2(1270)$ mesons, both of which have large

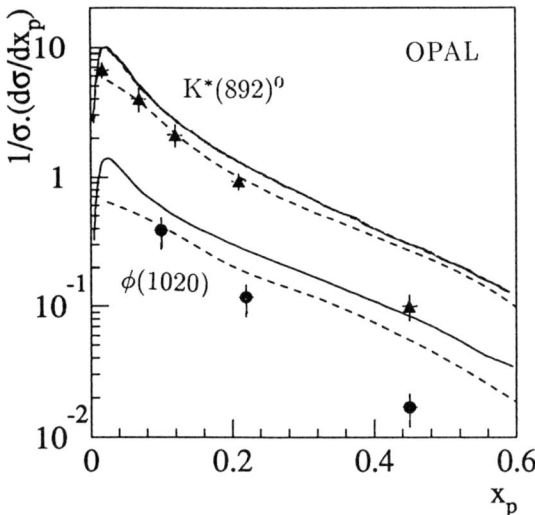

Figure 4. Fragmentation function for K* (triangles) and ϕ (circles) (OPAL). The curves show predictions of Jetset (solid) and Herwig (dashed).

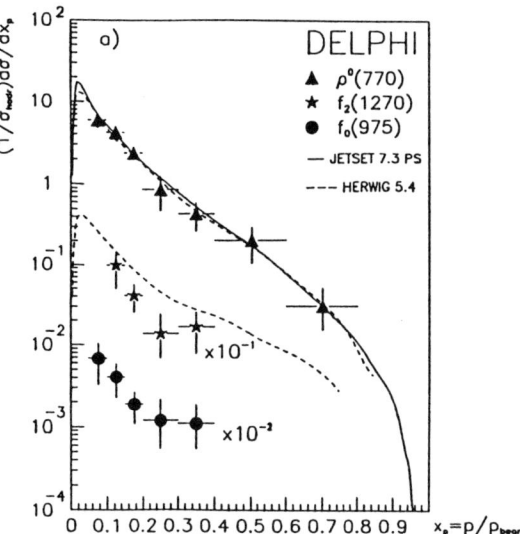

Figure 5. Fragmentation function for ρ^0, f_0 and f_2 (DELPHI). The curves show predictions of Jetset (solid) and Herwig (dashed).

by Herwig. The DELPHI result indicates that f_2 production is an important feature of inclusive mass spectra in Z^0 decay and should be included in Jetset simulations of LEP physics. For work of high precision, the f_0 may also be important.

Studies of meson resonances in Z^0 decays are only just beginning, and significantly more knowledge of rates for identified stable particles and resonances will be needed before a consistent picture emerges. Then it may be possible to gain a much better understanding of the details of quark and gluon hadronization.

REFERENCES

1. The ARGUS Collaboration, H. Albrecht et al., "Production of Two Vector Mesons in Gamma-Gamma Reactions," presented at the XXVI International Conference on High Energy Physics, Dallas, USA, August 6-12, 1992.

2. The ARGUS Collaboration, H. Albrecht et al., "Resonance Production by Two Photons," presented at the XXVI International Conference on High Energy Physics, Dallas, USA, August 6-12, 1992.

3. The DELPHI Collaboration, "Measurement of Inclusive Production of Light Meson Resonances in Hadronic Decays of the Z^0," presented at the XXVI International Conference on High Energy Physics, Dallas, USA, August 6-12, 1992.

4. The OPAL Collaboration, P. D. Acton et al., "Inclusive Neutral Vector Meson Production in Multihadronic Z^0 Decays," presented at the XXVI International Conference on High Energy Physics, Dallas, USA, August 6-12, 1992; CERN-PPE/92-116; to be published in Z.Phys.C.

5. The DELPHI Collaboration, P. Abreu et al., "Production of Strange Particles in the Hadronic Decays of the Z^0," Phys.Lett. 275B, 231(1992).

6. The OPAL Collaboration, "A Study of $K^{*\pm}$ Production in Z^0 Decays," OPAL Physics Note, PN083, August, 1992.

REACTION RATES IN A HEAT BATH

M. Jacob
Theory Division
CERN
CH-1211 Geneva 23

and

P.V. Landshoff
DAMTP, University of Cambridge
Cambridge CB3 9EW, United Kingdom
and Theory Division, CERN

Abstract

The presence of a heat bath of photons gives an extra effective mass to charged particles and changes the effective strength of the interaction responsible for a decay. All effects are calculated and it is shown that, to order αT^2, the decay rate is not modified.

This is a short report on work which is more extensively described in the paper Phys. Lett. B281 (1992) 114 [1].

We consider the decay of a neutral particle into two charged spinless particles in order to study the influence of a heat bath. The heat bath corresponds to a thermal distribution of photons described by a partition function

$$Z = \frac{1}{1 - e^{-\beta\omega}} \quad (1)$$

where $\beta^{-1} = T$.

The masses of all the particles involved are supposed to be larger than T so that only the photons are thermalized in practice.

We study this relatively simple case as illustrative of the effects associated with the extra radiative corrections due to the heat bath which (i) brings a mass shift Δm^2 proportional to αT^2, (ii) provides vertex corrections and (iii) gives corrections corresponding to absorption from the heat bath and induced radiation into it. This altogether results in a change $\Delta\Gamma$ of the reaction rate Γ.

In physics we deal with systems which, for all practical reasons, we can consider as decoupled from the rest of the world. However, we cannot fully dissociate them from some background radiation. We have indeed the cosmic background radiation, as recently studied in detail by COBE. We may also consider with particular interest reaction rates in the Sun, in a quark-gluon plasma, or in the early Universe.

This question is how big these effects are. We show that they globally vanish to order αT^2 whilst, at higher temperature, the mass shift effect becomes dominant.

This is a topical problem[2,3] which we approach with a new method, thus bringing simplicity and clarity to an overall picture not lacking in conflicts[3]. This is clearly a "soft process", due to thermal photons and hence fully calculable.

Our new approach consists of:

(i) The use of a thermal propagator in Feynman graphs;

(ii) the use of "old-fashioned" regularization techniques based on the introduction of spurions at each vertex, with resulting factors of the type $e^{-\eta|t|}$ in the amplitude;

(iii) a computation limited to directly measurable physical quantities, such as a decay rate.

A typical contribution to the decay rate is provided by the Figure. The calculation of this graph, with the dotted line implying particles on the mass shell, is one of the <u>additive</u> contributions to the rate.

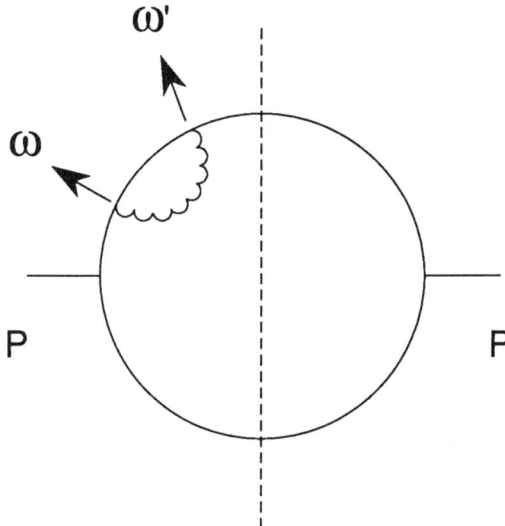

Figure. A contribution to the rate coming from the presence of the heat bath. The photon propagator is limited to D(2). The dashed line indicates the mass shell condition. The spurion energies at the photon vertices are explicitly displayed.

In the photon propagator, one can consider additively the standard contribution and the thermal contribution. The effect of the heat bath is then obtained by considering the contribution of the difference Δ_β between the full propagator and the standard one. We follow standard methods[2] and find

$$\Delta_\beta(k) = 2\pi\, \delta(k^2)\, \frac{1}{e^{\beta|k_0|} - 1} \qquad (2)$$

The rate is calculated as:

$$\Gamma \sim \frac{1}{Z} \mathrm{tr}\, e^{-\beta H_\gamma} a(P)\, MM^+ a^+(P) \qquad (3)$$

where $a^+(P)$ is the creation operator for the initial particle of four-momentum P, H_γ is the photon Hamiltonian and M is the transition amplitude. We limit ourselves to order α.

We regularize the divergences due to the propagator before the loop (Figure) by letting energies (ω) flow at each vertex and by taking at the end only the limit $\omega \to 0$. This old-fashioned method is very well adapted to a heat bath which may not exist asymptotically in space and time.

The resulting lack of energy conservation during the calculation is treated in terms of a power expansion of the overall δ function

$$\delta(E_1 + E_2 + \Delta E - E) = \sum_\nu \frac{(\Delta E)^\nu}{\nu!} \delta^{(\nu)}(E_1 + E_2 - E)$$

(4)

where ΔE is the global spurion energy.

The different contributions to the effect of the heat bath on the decay rate Γ, coming from the graph of the Figure, and the others, can easily be sorted out as:

(i) $\Delta\Gamma_1$, the result of a mass shift $\Delta m^2 \sim \alpha T^2$;

(ii) $\Delta\Gamma_2$, the result of vertex corrections;

(iii) $\Delta\Gamma_3$, the result of absorption from the heat bath and induced emission into it.

All infra-red divergences appearing in $\Delta\Gamma_2$ and $\Delta\Gamma_3$ mutually cancel but, furthermore, all effects cancel to order αT^2 between the three correction terms.

Decay rates are therefore very stable with respect to temperature, <u>despite</u> the induced mass shift. When the temperature increases, whilst αT remains small to justify the perturbative calculation, $\Delta\Gamma_1$ keeps rising like T^2, whilst $\Delta\Gamma_2 + \Delta\Gamma_3$ behaves eventually like T only.

Is this cancellation to order T^2 a general characteristic of all physical quantities such as reaction rates and cross-sections? Other cases are investigated.

REFERENCES

1. M. Jacob and P.V. Landshoff, *Phys. Lett.* B281, 114 (1992).

2. J.I. Kapusta, *Finite Temperature Field Theory,* Cambridge University Press, 1989.

3. A. Niegawa, *Phys. Lett.* B247, 351 (1990);
   P.V. Landshoff and J.I. Kapusta, *J. Phys.* G15, 267 (1989);
   Y. Ueda, *Phys. Rev.* D9, 1383 (1981);
   G. Peressutti and B.S. Skagerstam, *Phys. Lett.* 110B, 406 (1982);
   J.L. Cambri et al., *Nucl. Phys.* B209, 372 (1982);
   M. Le Bellac, *Schladming Lectures*, 1991;
   A. Niemi and G.W. Semenoff, *Nucl. Phys.* B230, 181 (1984);
   A. Niegawa and T. Kakashiba, *Nucl. Phys.* B370, 335 (1992);
   J.F. Donoghue and B.R. Holstein, *Phys. Rev.* D28, 340 (1983); D29, 3004 (1983);
   W. Keil, *Phys. Rev.* D40, 1179 (1989);
   L. Dolan and R. Jackiw, *Phys. Rev.* D9, 3320 (1974).

DIFFRACTION AT COLLIDER ENERGIES

L. L. Frankfurt
Department of Physics, FM-15
University of Washington
Seattle, Washington 98195

Abstract

The aim of this talk is to outline lessons with "soft" hadron physics to explain a) feasibility to observe and to investigate color transparency, color opacity effects at colliders; b) significant probability and specific features of hard diffractive processes; c) feasibility to investigate components of parton wave functions of hadrons with minimal number of constituents. This new physics would be more important with increase of collision energy.

"SOFT" HADRON PHYSICS

i) Pomeron exchange with the trajectory $\alpha(t) = 1.08 + 0.2t$ provides a remarkable first approximation to near forward elastic scattering from ISR to Tevatron (cf. discussion in Ref. 1). Odderon exchange may help[2] to explain anomalously large value of

Re A $p\bar{p}$/ Im A$\bar{p}p/_{t=0}$ observed by UA (4).

ii) Supercritical pomeron ($\alpha(0) > 1$) reasonably describes energy dependence of elastic cross sections of $pp, \bar{p}p$ scattering within forward peak and of total cross sections (cf. Refs. 1,4 and references therein).

iii) Inelastic intermediate states (diffractive dissociation of initial hadrons) should be included into eikonal approximation.[5]

iv) ISR data on single diffraction can be reasonably described by triple pomeron formulae.[6] If so, Reggeon Calculus predicts more narrow distribution in multiplicities for central cluster X than in pp scattering due to smallness of triple pomeron vertex. At the same time UA(1) data for the double diffraction: $\bar{p}p \to X_1 + X + X_2$ found too fast increase of multiplicity with M_X for central cluster X. Presence of trigger for $E_T > 3$ GeV which effectively suppresses the contribution of small masses makes theoretical interpretation difficult.

Observed dependence of total and elastic cross sections on initial energy is a temporary phenomenon. It should be changed to avoid contradiction with unitarity condition that partial waves cannot exceed 1. Really partial waves of $\bar{p}p$ collision with impact parameter $b = 0$ extracted from Tevatron data on total and elastic cross sections are $|f|b = 0)| = 1/2$ and increase with energy.

COLOR COHERENT EFFECTS

For sufficiently large projectile energies the length of spatial transition between different quark-gluon configurations $|n\rangle$ in the wave function of projectile hadron h exceeds r_T - the radius of a target T. Really uncertainty principle shows that:

$1/(En - Eh) \approx$

$$2E_h / \left[\sum_i (m_i^2 + k_{it}^2)/\alpha_i - m_h^2 \right] > r_T \quad (1)$$

In the eikonal approximation Eq. (1) means that due to the large Lorentz factor orbits of constituents in the fast hadron h are frozen during collision and the total cross section σ_{hT} becomes almost incoherent sum of cross sections for the different configurations $|n\rangle$. In pQCD[8] and in nonperturbative regime[9] different configurations have different cross sections depending on its transverse radius. So the fundamental quantity is $P_h(\sigma)$ - the distribution over the value of cross section. Such a quantity formally arises in the formalism of scattering states.[10]

Four moments of $P_h(\sigma) - \langle \sigma^n \rangle \equiv \int P(\sigma) \sigma^n d\sigma$ were extracted from experimental data. $\langle \sigma^0 \rangle$ is conservation of probability; $\langle \sigma^1 \rangle$ is average cross section. $\langle \sigma^{(2)} \rangle$ is extracted from single diffraction in Ref. 11 and from inelastic shadowing correction to σ_{pd}^{tot} in Ref. 12. $\langle \sigma^3 \rangle$ has been extracted from diffractive dissociation of proton of deuteron.[12] The distribution Eq. (12) reproduces all these momenta[12]:

$$P_N(\sigma) = 0.66(\sigma/(\sigma + a)) \exp - \left(\frac{\sigma - \sigma_0}{1.1\sigma_0} \right)^6 \quad (2)$$

where $\sigma_0 = 0.89\langle \sigma \rangle; a = 0.1\sigma_0$. Note that $P_N(\sigma) \sim \sigma$ at $\sigma \to 0$ since a nucleon is a system of 3 quarks.[17]

New Phenomena at Colliders

1. Significant probability of $P_N(\sigma \ll \langle \sigma \rangle)$ leads to numerous color transparency (ct) phenomena and gives possibility to probe QCD in a new way (see review[16]). The most striking example is the significant transparency for spectator nucleons in the central nucleus-nucleus collisions. This physics may reveal itself starting from CERN nuclear beam energies.[17]

2. Analysis performed in Ref. (15) shows that ct phenomena suggested for a nucleon projectile (cf. review[16]) should be much more pronounced for the pion projectile. This is physics for FNAL, KAON.

3. Large ($\sim 1/2$) probability of $P_N(\sigma)$ for $\sigma > \langle \sigma \rangle$ leads to numerous color opacity phenomena:

i) Broadening of large E_T tail seems to be observed by NA34 in heavy ion collisions.[18]

ii) Percolation phase transitions and peculiar fractals are expected in central heavy ion collisions.[19] Use of this phenomenon may help to search for other phase transitions.

HARD DIFFRACTIVE PROCESSES

One of the hard processes relevant for $P(\sigma \to 0)$ is the single diffractive dissociation: $\bar{p} + p_i \to \text{``}M_X\text{''} + p_f$ in the kinematics

$$|M_X^2/s| \ll 1, \left| [(P_i - P_f)^2 = t]/s \right| \ll 1.$$

Requirement of large p_t trigger selects small spatial size configuration in the initial (final) nucleons. Three phenomena work in the same direction. i) radiation is kinematically suppressed ii) for collision of black bodies diffraction is small. So the contribution of configurations with large cross section is suppressed. iii) Analysis of electromagnetic form factors of hadrons in realistic models of hadrons[20] shows that color screening phenomenon may reveal itself for $Q^2 > 1$ - 2 GeV^2. So there is a good chance for the applicability of pQCD in the single diffraction at $|t| \geq 1$ GeV^2. In pQCD for small α_s cross section is given by gluon ladder and has the form:[21]

$$\frac{d\sigma p\bar{p}}{dt\,dM_X^2} \to {}^{\text{``}M_X\text{''}+P} = \alpha_S^2(t) \frac{\left(\frac{S}{M_X^2}\right)^{2w}}{\alpha_S^2 \ell n^3(s/M_X^2)}$$

$$1/M_X^2 \left|\frac{t}{M_X^2}\right| G_N\left(\frac{(t)}{M_X^2}, |t|\right) |F_{|PN}^g(t)|^2 \quad (3)$$

Here $w(t) = 12\alpha_s(t)\ell n2/\pi$ and $F_{|PN}$ is the two gluon form factor of a nucleon. Eq. (4) is a striking prediction of pQCD. Really soft physics predicts that the peak at $x = 1$ and $t = 0$ ($M_X^2/S = 1 - x$) should soften with the increase of t. But Eq. (4) means that the peak at $x \to 1$ should become even more pronounced at larger t.

We want to draw attention that in the eikonal approximation (soft physics) in difference from pQCD the role of multipomeron exchanges increases with increase of t (cf. Ref. 22).

At large t where pQCD dominate one can calculate diffractive production of high p_t jets:

$$p + p \to \{jet_1 + jet_2 + X\} + p$$

Distinctive prediction of pQCD is that gluon structure function of the pomeron at large z ($z = (\vec{p}_{jet_1} + \vec{p}_{jet_2})/(\vec{p}_c - \vec{p}_f)$) behaves as[21,23]

$$z\, G_{|p}(z) \sim \frac{1}{1-z},\, \delta(1-z) \quad (4)$$

This prediction should be confronted with predictions for the nonperturbative pomeron:

$$z\, G_{lp}(z) \sim (1-z)^n \text{ with } n = 1 - 5$$

(cf. Ref. 24 and refs. therein). Thus in pQCD pomeron behaves like point-like particle.

Recent UA(8) data found that in 30% of events dijets carry the whole momentum of the pomeron.[25]

REFERENCES

1. M.M. Block and A. White "4th Blois Workshop on Elastic and Diffractive Scattering", *Nucl. Phys. B.* (Proc. Suppl.) 25, (1992).
2. P. Cauron and B. Nicolescu, "The Importance of the Measurement of ρ for the understanding of the High Energy Hadron-Hadron Scattering," Contributed paper N154.
3. D. Berhard et al., UA4 Collaboration, p. 4, B198, 583 (1987).
4. E. Gottsman et al. "A Quantitative Description of the Pomeron", Contributed paper N48.
5. R. Engel et. al. "Extrapolation of Hadron Cross Sections to Collider Energies within the Two Component Dual Parton Model", Contributed paper N277.
6. K. Coulianos, *Phys Rep.* 11, 169 (1983).
7. D. Joyce, et al. "Double Pomeron Exchange in $\bar{p}p$ Interactions at 0.63 TeV", Contributed paper N229.
8. F.E. Low, *Phys. Rev.* D12, 163, (1975) S. Nussinov, *Phys. Rev. Lett.* 34, 1286 (1975), *Phys. Rev.* D14, 246, (1976).
9. L.L. Frankfurt and M.I. Strikman, *Phys. Rep.* 160, 235 (1988).
10. E.L Feinberg and I.V. Pomeranchuk, *Suppl. Nuovo Cimento.* 111, 652, (1956). M.I. Good and W.A. Walker, *Phys. Rev.* 120, 1857, (1960).
11. H. Miettinen and J. Pumplin, *Phys. Rev.* D18, 1696, (1978), *Phys. Rev. Lett.* 42, 204, (1979).
12. B. Blättel et al., "Cross-Section Fluctuations," Contributed paper, *Nucl. Phys. A.* 544, 479, (1992).
13. C. Bertch et al., *Phys. Rev. Lett.* 47, 297, (1981).

14. L.L. Frankfurt, to be published.
15. B. Blättel, *et al.*, to be published, Contributed papers 650, 682.
16. L. Frankfurt and M. Strikman, *Prog. Part. Nucl. Phys.*, 27, 135, (1991).
17. L. Frankfurt and M. Strikman, *Phys. Rev. Lett.*, 66, 2289, (1991).
18. H. Heiselberg, *et. al.*, *Phys. Rev. Lett.*, 67, 2946, (1991).
19. A. Bulgac and L. Frankfurt, to be published.
20. L. Frankfurt *et al.*, "Comment for Particle and Nuclear Physics", September, 1992.
21. L. Frankfurt and M. Strikman, *Phys. Rev. Lett.*, 63, 1914, (1989).
22. V. Abramovsky and R. Betman, Contributed paper N763.
23. L. Frankfurt, Talk at FAD Meeting, Dallas, March, 1992.
24. G. Ingelman and P. Schlein, *Phys. Lett* 152B, 256, (1985).
25. P. Schlein, Comment after Bartke talk at this Conference.

THE PHYSICS OF LEADING PARTICLES

W.D. WALKER
Physics Department
Duke University
Durham, N.C. 27708-0305

Abstract

We review recent results reported at this conference on leading particles. The results show that the leading particles are the result of fragmentation that occurs at considerable distance from the struck nucleus. We also show that the quarks (diquarks) or partons undergo relatively little absorption if they traverse nuclear matter. Quark and diquark nucleon cross sections are estimated.

We are interested in studying leading particles from hadron-nuclear collisions in order to give some insights into parton-nucleon or quark-nucleon interactions. We present results from 100 GeV $\pi^\pm$, $\bar{p}$, p interactions with Mg, Ag and Au nuclei. We also report results from 800 GeV p interacting with a variety of nuclei. These experiments allow one to use the struck nuclei as a laboratory with dimensions of a 2-10 Fermis. One can think of these results crudely in terms of the additive quark model. The basic idea is that the leading particles are the result of the fragmentation of the spectator quark (or diquark) from the projectile after one of the other constituents has interacted. Thus we have the possiblity of seeing the effects of interactions that occur within a few Fermis of the initial interaction which has disturbed the initial state of the incident particle.

Four years ago Miettenin and Stevenson[1] showed that most leading particles originate from the periphery of struck nuclei. One of their plots is shown in Figure 1. They show a plot of cross section for the production of the leading particle versus the nuclear radius. The linearity of the plot leads one to believe that the number of such particles is proportional to the area of the halo around the edge of the nucleus.

A new result reported at this conference (Fermilab E706)[2] was an experiment which measured relative numbers of π^+, K^+, P at 540 GeV coming from 800 GeV P-nuclear collisions. The particle identification was done on a pulse to pulse basis on a variety of targets. The relative number of π-K-P's is measured with considerable precision and the comparison between different elements should be quite precise also. The results of their experiment are shown in Table I. The remarkable result is that the relative numbers of π, K, P change by a few percentage points as one goes to Be to Pb. The

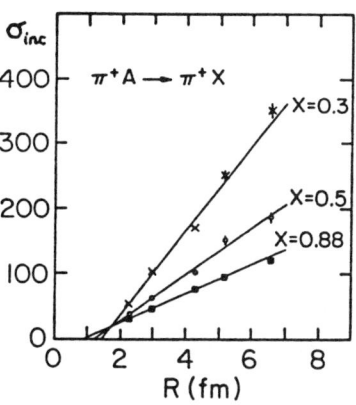

Figure 1. Cross section vs. nuclear radius.

© 1993 American Institute of Physics

Table I. The A Dependence of Leading Particle Production by 800 GeV Protons (530 GeV Secondary Energy).

	$\pi_{fraction}$ %	$K_{fraction}$ %	$P_{fraction}$ %	Secondary Beam Per Primary Proton
Be	5.36	1.03	93.61	$.573 \cdot 10^{-4}$
C	5.43	1.06	93.51	$.536 \cdot 10^{-4}$
Al	5.74	1.09	93.17	$.474 \cdot 10^{-4}$
Cu	5.92	1.10	92.98	$.407 \cdot 10^{-4}$
Sn	6.13	1.07	92.80	$.325 \cdot 10^{-4}$
W	6.16	1.08	92.76	$.317 \cdot 10^{-4}$
Pb	6.15	1.08	92.77	$.329 \cdot 10^{-4}$
Error	±1%	±4%	±0.5%	±0.6%

average multiplicity of pions produced in these collisions will vary by at least a factor of two. We note that the relative percentage of π, K and P vary by less than one percentage point as we go from Be to Pb. This must mean that the distance from the nucleus at which fragmentation occurs must be at least 30-40 Fermis downstream and perhaps several times more than this.

We now turn to the description of the results of a bubble chamber experiment E597.[3] In this experiment we looked at the hadron collision with Ag, Au and Mg nuclei. We noted that if we looked at particles of more than 20 GeV/c (P_{in} = 100 GeV/c) that their average momentum did not seem to change very much as we went from peripheral to central collisions. This means that we have a beam of partons or quarks whose absorption we can readily measure if we can get a measure of the nuclear matter that has been traversed.

One of the great advantages of a bubble chamber is that one can look at many aspects of a given collision. We look at both low energy and high energy products and can readily identify relatively low energy protons. The identification of the low energy protons is the key to getting a handle on the impact parameter of a given collision. A collision producing lots of low energy protons certainly must be a fairly central collision whereas a peripheral collision will produce a small number of protons. We show in Figure 2 the general features of these collisions as a function of the number of protons produced (Np). For small Np the number of particles (mainly π's) produced are the same for all three elements (Δ = Mg, O = Ag, ☐ = Au). As we look at higher Np collisions there are more low energy particles produced for the collisions with the heavier nuclei. The number of leading particles decrease as Np increases. We have made the correlation more quantitative by estimating the distance traversed by the shower particles in the nuclear matter. We estimate the volume of a cone that the nuclear particles sweep out as they emerge from the nucleus. We presume that all the nucleons intercepted by the cone are ejected. The cone size is chosen to reproduce the distribution in number of the observed protons. A characteristic curve is shown in Figure 3 for Ag. A by product of this calculation is the distribution of flight paths of the products in nuclear matter for a given value of Np. This is precisely the

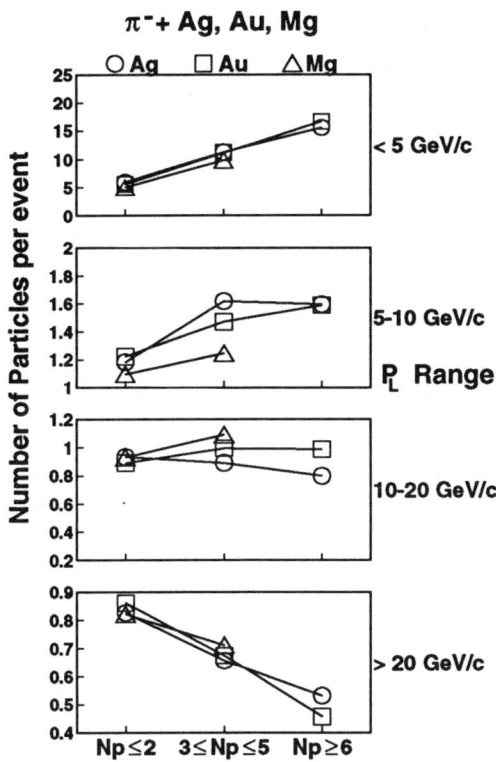

Figure 2. Number of particles produced in different elements(Mg, Ag, Au) vs. the number of visible protons(Np).

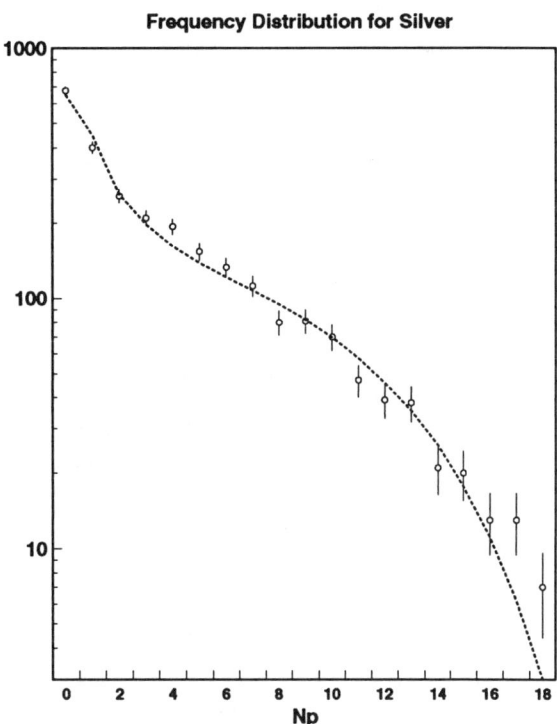

Figure 3. Curve of frequency of occurrence of the number of protons from Ag.

quantity that we need in estimating the leading particle absorption coefficient. We believe that this absorption is directly related to the quark (or diquark) nucleon cross section. For each value of Np we calculate the average (and dispersion) in the areal density in nuclear matter traversed by the products of the initial nuclear interaction. The results of this calculation are plotted against the counting rate for leading particles with $p \geq 20$ GeV/c (O) and $p \geq 30$ GeV/c (●). The results are shown in Figure 4. We can calculate the quark-nucleon or diquark-nucleon cross section from these slopes. The results of these calculations are given in Table II. We note that the cross sections seem to be relatively small (as compared to the π-nucleon cross section) by about a factor of two. There

also seems to be a dependence on X_F for the diquark cross section.

This possibility has been emphasized by Strikman.[4] We believe that if anything our cross sections are probably upper limits of the true cross sections. The fact that the cross sections are small is probably the result of the fact that a sizeable fraction of the π-nucleon cross section is the result of gluon interaction. We know that in the case of the nucleon that

Table II. Cross sections in mb.

	$P \geq 20$ GeV/c	$P \geq 30$ GeV/c
π^+	4.5 ± .9	4.9 ± 1.0
π^-	4.4 ± .5	5.5 ± 1.0
p	5.4 ± 1.8	7.2 ± 2.1
$\bar{p}$	4.1 ± .7	7.1 ± 1.2

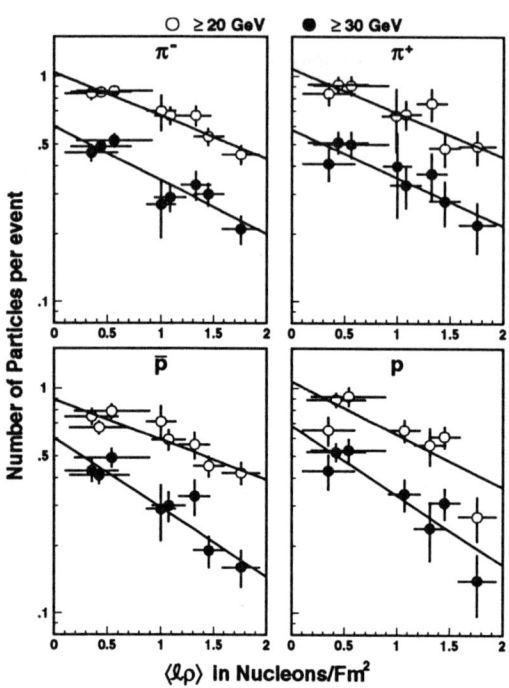

Figure 4. The number of leading particles per event with p ≥ 20 GeV/c (O) and p ≥ 30 GeV/c (●) vs. the estimated flight path in nuclear matter.

1/2 of the momentum is carried by the gluons in the particle. In a paper submitted to this conference by Navarra[5] they have estimated that 1/2 of the cross section in this energy range is due to gluon interaction and the fraction increases with increasing energy.

Strikman[4] has pointed out that the diquark cross section will likely be dependent on the fraction of the momentum carried by the diquark. The diquark tends to dissociate as it progresses through the nuclear matter.

There are other data which seem to show that the quark-nucleon cross section is small. A recent Fermilab fixed target experiment[6] has looked at dijets produced in p-nuclear collisions. Indeed the cross section per nucleon is not very A dependent, however the coplanarity of the dijets with the incident particle indicates that there may be considerable scattering of the jet quarks as they emerge.

As mentioned above we believe that our cross sections are upper bounds for absorption of the quark. If for example some of the glue is stripped out of a π then the q and $\bar{q}$ would continue on through the nucleus. In such a case we would possibly see the fragmentation of both of these products. Consequently we might believe that we see the absorption effects of both products, hence our quoted cross section is too large. Our results differ from earlier results on stopping power[7] in that we look at central collisions specifically.

I acknowledge many useful conversations with Professors B. Mueller, P. Carruthers, M. Strikman and T. Fields in particular.

REFERENCES

1. H. E. Miettenin and P. M. Stevenson, *Phys. Lett. B* **199**, 591, (1987).

2. C. Johnstone *et al*, Report on E706 to Conference (A241).

3. A review of E597 can be found in W. D. Walker and P. C. Bhat, *Hadronic Multiparticle Production*, edited by P. Carruthers, p. 153, (1988).

4. M. Strikman, private communication.

5. F. Navarra, Report to Conference (A198).

6 M. D. Corcoran *et al*, *Phys. Lett. B* **259**, 209, (1991).

7. W. Busza and A. S. Goldhaber, *Phys. Lett.* **139B**, 235, (1984).

MEASUREMENT OF THE PROTON ELECTROMAGNETIC FORM FACTORS IN THE TIME-LIKE REGION AT 8.9 TO 13.0 GeV^2

Diego Bettoni
INFN - Sezione di Ferrara, Italy
for the E760 Collaboration[7]

Abstract

Cross sections for the reaction $p\bar{p} \to e^+e^-$ have been measured at $s = 8.9, 12.4,$ and $13.0 \ GeV^2$. The cross sections have been analyzed to obtain proton electromagnetic form factors in the time-like region.

INTRODUCTION

In this paper we present a new measurement of the cross section for the reaction :

$$p\bar{p} \to e^+e^- \qquad (1)$$

at $\sqrt{s} = 3.0 \ GeV$, $3.5 \ GeV$, and $3.6 \ GeV$. These measurements were made as a part of Fermilab experiment E760, which is dedicated to the study of charmonium by resonant formation in $p\bar{p}$ annihilations [4]. The differential cross section $d\sigma/d(cos\theta^*)$ for process (1) can be expressed in terms of the proton magnetic and electric form factors G_M and G_E[5].

The electromagnetic form factors of the proton have been measured very precisely in the space-like region[1] up to $q^2 = 31 \ (GeV/c)^2$. Precise results for G_M in the time-like region ($s = -q^2 > 0$) exist only for a small interval near threshold, $4m_p^2c^4 \leq s \leq 4.2 \ GeV^2$ [2], whereas for larger momentum transfers only upper limits have been established by earlier $e^+e^- \to p\bar{p}$ and $p\bar{p} \to e^+e^-$ experiments[3].

Experiment E760 has been carried out at the antiproton accumulator of the Fermilab Antiproton Source. The circulating beam of stochastically cooled antiprotons (up to $4 \times 10^{11} \bar{p}$) intersects an internal hydrogen gas jet target to provide instantaneous luminosities up to $9 \times 10^{30} cm^{-2}s^{-1}$, which allow the measurement of cross sections of the order of a few pb, corresponding to the values expected for process (1) in our energy range.

The detector is a non-magnetic spectrometer with full azimuthal (ϕ) coverage and polar angle (θ) acceptance ranging from 2° to 70°. The central detector ($12° < \theta < 70°$) has cylindrical symmetry around the beam axis and consists of : two scintillator hodoscopes (H1 and H2) ; a straw-tube drift chamber, a radial projection chamber and a multiwire proportional chamber which provide the inner tracking ; a multicell threshold Čerenkov counter for electron/pion separation; two planes of limited streamer tubes and a planar multiwire proportional chamber which provide the outer tracking; an electromagnetic calorimeter consisting of 1280 lead-glass blocks pointing to the interaction region, with a 20-fold segmentation in θ and a 64-fold segmentation in ϕ The forward detector, extending the θ acceptance down to 2° consists of a veto hodoscope, a three-layer straw-tube drift chamber and a Pb/scintillator calorimeter. The luminosity monitor is provided by a silicon detector mounted at $\theta = 86.5°$, which

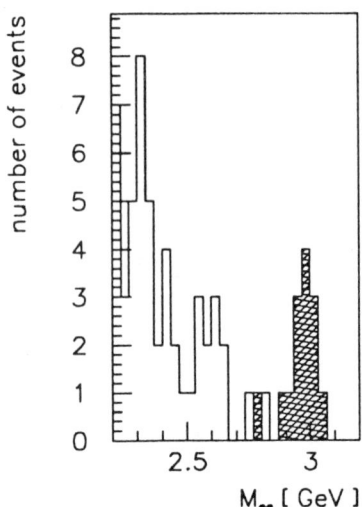

Figure 1. e^+e^- invariant mass distribution.

measures the direction and energy of the recoil proton from forward antiproton elastic scattering.

DATA ANALYSIS

Off-line analysis of the data is based on the identification of two electron tracks collinear in the center of mass, with an invariant mass compatible with the center of mass energy ($\sqrt{s}$) of the $p\bar{p}$ system. The fiducial range of polar angle in which the electrons are accepted is $15° < \theta < 60°$. A preliminary selection requires the two highest energy showers in the central calorimeter to be associated with the electron track candidates. The electron identification is based on the pulse height information from the hodoscope H2 and the Čerenkov counter, dE/dx information from the RPC, and the transverse shape of the energy deposition in the central calorimeter[4]. The e^+e^- invariant mass distribution at this stage of the selection is shown in fig. 1 for $\sqrt{s} = 3.0\ GeV$.

The low invariant mass background, due to residual Dalitz pairs or photon conversions from the vast π^0 component of the $p\bar{p}$ annihilation, disappears when we require only two showers in the electromagnetic calorimeter and impose the requirements of two body kinematics on the electron directions. By requiring $178.3° < \phi_{ee} < 181.7°$ and $177° < \theta^*_{ee} < 183°$ (where ϕ_{ee} is the azimuthal angle between the 2 electrons and θ^*_{ee} is the sum of the center of mass polar angles of the two electrons with respect to the antiproton direction) we select the events shown as the shaded area in fig. 1. These events cluster around the right center of mass energy with a spread compatible with the experimental mass resolution and constitute our final sample. The efficiency of the above event selection, determined by using the e^+e^- events from the J/ψ and ψ' scans, is found to be $\epsilon_{sel} = (0.79 \pm 0.03)$. The overall efficiency, which includes the trigger, is $\epsilon = (0.73 \pm 0.03)$.[4]

Two-pion final states are the main source of background for reaction (1). The charged pion contamination is measured from a sample of resolved $p\bar{p} \to \pi^+\pi^-$ events, collected with a dedicated hadron trigger, from which we measure a combined calorimeter and Čerenkov rejection for $\pi^+\pi^-$ events of 1.6×10^9. This, together with the known $p\bar{p} \to \pi^+\pi^-$ cross section [6], leads to an estimated background cross section $\leq 0.01\ pb$, which is negligible when compared to the measured cross section.

The $p\bar{p} \to \pi^0\pi^0$ annihilation is also a source of possible background, when each of the two π^0's either undergoes a Dalitz decay or has one of the two decay photons converted in the 0.2 mm stainless steel wall of the beam pipe. This background has been estimated from the data collected in the same experiment with an all neutral trigger designed to detect $p\bar{p} \to \gamma\gamma$ events. Taking into account the conversion probability (2%) per photon and the probability for both pairs to be misidentified as single electrons (5%) we estimate a background cross section from photon conversions to be $\leq 0.016\ pb$ at $\sqrt{s} \approx 3.0\ GeV$ and $\leq 0.002\ pb$ at $\sqrt{s} \approx 3.5\ GeV$, which is also negligible. Similar contributions come from one or both π^0's

Table 1. Summary of results for the magnetic form factor of the proton.

s (GeV^2)	$\mathcal{L}$ (pb^{-1})	$N_{e^+e^-}$	σ_{corr} (pb)	$|cos\theta_{cm}|$	$|G_M|$ (a)	$|G_M|$ (b)
8.9	2.8 ± 0.1	14	$6.8^{+2.3}_{-1.8}$	$0 \div 0.45$	$0.033^{+.006}_{-.004}$	$0.039^{+.007}_{-.005}$
12.4	17.7 ± 0.9	11	$0.85^{+.34}_{-.25}$	$0 \div 0.60$	$0.013^{+.003}_{-.002}$	$0.014^{+.003}_{-.002}$
13.0	6.0 ± 0.3	4	$0.91^{+.72}_{-.44}$	$0 \div 0.62$	$0.013^{+.005}_{-.003}$	$0.015^{+.006}_{-.004}$

decaying into a Dalitz pair. In conclusion, we estimate the total background from all sources to be less than 0.3 events among the 29 events observed at the three energies.

The e^+e^- events from the decay in the tails of the J/ψ and ψ' resonances constitute a possible background for the form factor events of reaction (1). This contribution has been calculated to be negligible.

RESULTS

For a given integrated luminosity $\mathcal{L}$, and overall efficiency ϵ, the number of events is:

$$N = \epsilon\, \mathcal{L}\, \sigma_{corr}, \quad (2)$$

with

$$\sigma_{corr} = \int_{-cos\theta^*_{max}}^{cos\theta^*_{max}} d(cos\theta^*) \frac{d\sigma}{d(cos\theta^*)} \quad (3)$$

In Table 1 we present the values of $s = -q^2$, $cos\theta^*$ and σ_{corr}. The cross sections σ_{corr} are obtained from eq. 2. Due to the limited statistics and angular coverage we cannot derive G_E and G_M separately. The values of $|G_M|$ are extracted from σ_{corr} under the two assumptions: (a) $|G_E| = |G_M|$ and (b) neglecting the term containing G_E and are reported in table 1. It is to be noted that the values of $|G_M|$ determined under the two approximtions differ by less than 15%. The values of the proton magnetic form factor, under assumption (a), are shown in fig. 4, along with the earlier results from the literature [2].

The dashed curve represents a $1/q^4$ fit to the data for $-q^2 \geq 5\, GeV^2$, which corresponds to the simplest prediction for the asymptotic behaviour of the form factors based on the quark counting rules. As can be seen the data are well fitted by this functional dependence. It is interesting to note that the numerical values of $|G_M|$ in the time-like region are nearly twice as large as those measured in the corresponding space-like region.

We gratefully acknowledge the technical support from our collaborating institutions and the outstanding contribution of the Fermilab Accelerator Division. This work was supported in part by the U.S. Department of Energy, the U.S. National Science Foundation and the Italian Istituto Nazionale di Fisica Nucleare.

REFERENCES

1. R.G.Arnold et al., *Phys. Rev. Lett.* **57**, 174(1986)
   P.E.Bosted et al., *Phys. Rev. Lett.* **68**, 3841(1992).

2. G.Bardin et al.,*Phys. Lett.* **B255**, 149(1991); ibid. **B257**, 514(1991) and references therein.

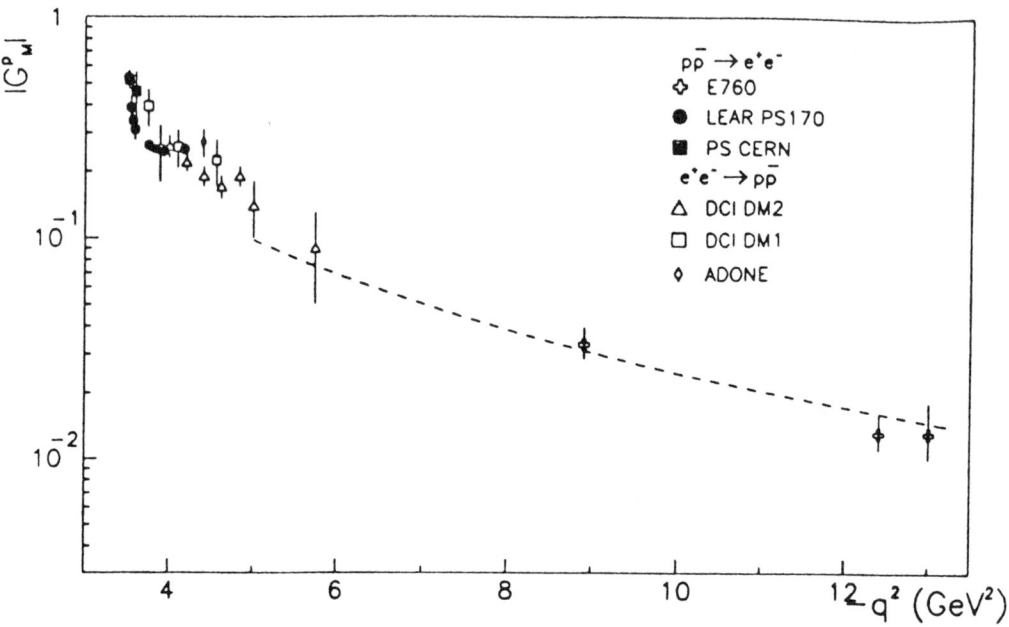

Figure 2. The magnetic form factor $|G_M|$ as a function of $-q^2$.

3. C. Baglin et al., *Phys. Lett.* B163, 400(1985);
   D. L. Hartill et al., *Phys. Rev.* 184, 1415(1969);
   M. Conversi et al., *Nuovo Cimento* 40, 690(1965)

4. T.A. Armstrong et al., FERMILAB-Pub-92/245-E, submitted to *Physical Review D*.
   T.A. Armstrong et al., *Nucl. Phys.* B373, 35(1992).

5. A. Zichichi, S.M. Berman, N. Cabibbo, R. Gatto, *Nuovo Cimento* 24, 170(1962)

6. V. Flaminio et al., *Compilation of Cross Sections - III : p and $\bar{p}$ Induced reactions*, CERN-HERA 84-01 (1984)

7. V. Bharadwaj, M. Church, A. Hahn, S. Hsueh, W. Marsh, J. Peoples Jr., S. Pordes, P. Rapidis, R. Ray, S. Werkema (Fermilab);
   D. Bettoni, R. Calabrese, P. Dalpiaz, P.F.-Dalpiaz, M. Fabbri, A. Gianoli, E. Luppi, M. Martini, F. Petrucci, M. Savrié (Ferrara);
   A. Buzzo, M. Macrí, M. Marinelli, M. Pallavicini, C. Patrignani, M.G. Pia, A. Santroni, A. Scalisi (Genova);
   D. Broemmelsiek, J. Fast, K. Gollwitzer, M. Mandelkern, J. Marques, J. Schultz, A. Smith, M.F. Weber, G. Zioulas (U.C. Irvine);
   D. Dimitroyannis, C.M. Ginsburg, M. Masuzawa, J. Rosen, M. Sarmiento, K.K. Seth, S. Trokenheim, J. Zhao (Northwestern);
   T. Armstrong, A. Hasan, R. Lewis, A.M.-Majewska, J. Reid, G.A. Smith, Y. Zhang (Penn State);
   C. Biino, G. Borreani, A. Ceccucci, R. Cester, R. Dibenedetto, F. Marchetto, E. Menichetti, A. Migliori, R. Mussa, S. Palestini, N. Pastrone, L. Pesando, G. Rinaudo, B. Roccuzzo, M. Sozzi (Torino).

RECENT RESULTS ON BOSE-EINSTEIN CORRELATIONS

B. DE LOTTO
Dipartimento di Fisica dell'Universita' and INFN
Via Fagagna 208
I-33100 Udine, Italy

Abstract

Recent experimental results on Bose-Einstein correlations are reported, with special emphasis on new studies in e^+e^- interactions.

INTRODUCTION

An enhancement in the production of pairs of pions of the same charge and similar momenta produced in high energy collisions was first observed in antiproton annihilations and attributed to Bose-Einstein (BE) statistics appropriate to identical pion pairs[1].

Bose-Einstein correlations between pion pairs can be used to study the space-time structure of the hadronization source[2]. This has been done for hadron-hadron (hh), heavy ion, muon-hadron and e^+e^- collisions (see Ref. 3 for introductions to the subject and reviews, and Ref. 4 for results of e^+e^- experiments below the Z^0 energy). Measurements at the LEP e^+e^- collider have been presented recently[5,6,7].

To study the enhanced probability for the emission of two identical bosons it is useful to define a correlation function R:

$$R(p_1, p_2) = \frac{P(p_1, p_2)}{P(p_1)P(p_2)}$$

where $P(p_1, p_2)$ is the two-particle probability density, subject to Bose-Einstein symmetrization, and $P(p_i)$ is the corresponding single particle quantity for a particle with four-momentum p_i. In practice, $P(p_1)P(p_2)$ is often replaced by a reference 2-particle distribution $P_o(p_1, p_2)$, which, ideally, resembles $P(p_1, p_2)$ in all respects, apart from the lack of Bose-Einstein symmetrization.

If $f(x)$ is the space-time distribution of the source, $R(p_1, p_2)$ takes the form

$$R(p_1, p_2) = 1 + |G[f(x)]|^2,$$

where $G[f(x)] = \int f(x) e^{-i(p_1-p_2)\cdot x} dx$ is the Fourier transform of $f(x)$. Thus by studying the correlations between the momenta of particle pairs one can determine the distribution of their points of origin. Several parametrizations of the correlation function R can be found[3].

A common parametrization for $R(p_1, p_2)$ is the Goldhaber Lorentz invariant expression

$$R(Q) = 1 + \lambda e^{-r^2 Q^2} \qquad (1)$$

where the correlation function R is expressed as a function of the four momentum difference of the pair $Q^2 = -(p_1 - p_2)^2$ and the particle source is assumed to have a Gaussian shape in the rest frame of the pair. The parameter r gives the source size and λ measures the strength of the effect, and can vary from 0 to 1.

The systematics coming from the definition of an appropriate reference sample, from the different parametrizations of the source density, from the models of the emission of the

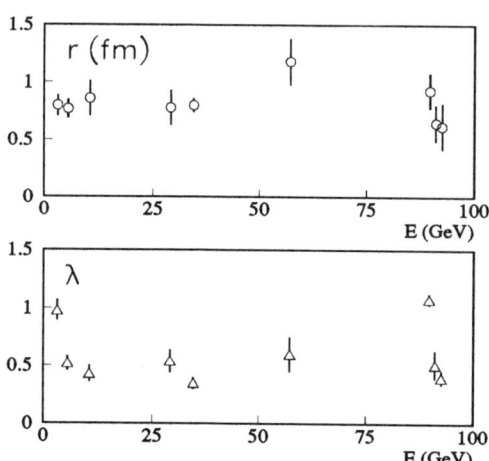

Figure 1. Dependence of the radius of the interaction region (upper) and of the strength parameter (lower) on the center of mass energy, for e^+e^- interactions. Recent results from ALEPH and DELPHI are the rightmost ones.

particles, from the estimate of the effect of the particles not directly emitted from the primary source, and from the effect of final-state interactions[9,10], make it difficult to compare results from different experiments.

I will thus try only to present some highlights from the experimental results after the 1991 Europhysics Conference on High Energy Physics. An exhaustive review of recent results before that conference is presented in Ref. 11.

RESULTS FROM e^+e^- INTERACTIONS

In Fig. 1, a summary of the results on the source radius r (a) and on the strength parameter λ (b) from pion interferometry studies in e^+e^- interactions is presented.

The recent results from ALEPH[6] and DELPHI[7] are quoted with an error dominated by the systematic error induced by the effect of choosing unlike-sign couples or couples from a mixing technique as a reference sample.

Most experiments indicate an independence of the two parameters on the primary energy.

Two preliminary particularly interesting results have been presented in this conference.

$K_S^0 K_S^0$ Interference

Because of the fact that the production rate of strange particles at center of mass energies so far explored is much lower than that of pions, up to now very few studies have been published on correlations effects in kaon pairs, and all come from hadron-hadron interactions[12]. Preliminary results on $K_S^0 K_S^0$ interference in e^+e^- interactions at the Z^0 peak have been presented by OPAL[13], and recently reported by DELPHI[14].

The special interest in this study is due to the fact that K_S^0 pairs should interfere even if they come from a $K^0 \bar{K}^0$ system, namely a non identical boson pair. The reason why pairs of K^0 should display a full Bose-Einstein correlation is detailed in Ref. 15. When a pair of K^0 is produced, there are two possibilities:

a. The couple is made by two K^0, or by two $\bar{K}^0$. In this case, the two kaons are identical bosons, and then subject to Bose-Einstein condensation. This eventuality, according to JETSET 7.3[16] based on Parton Shower, happens at LEP in about 35% of the cases.

b. The couple is made by one K^0 and one $\bar{K}^0$. In this case, the two kaons are different particles. The wavefunction of the $|K^0 \bar{K}^0>$ system can be projected into two eigenvectors of the charge conjugation operator:

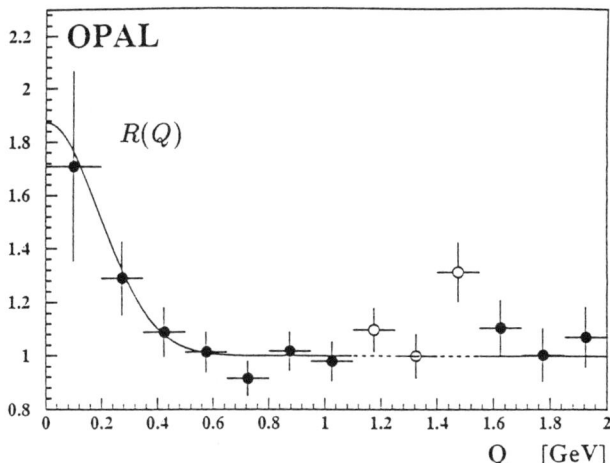

Figure 2. Correlation function of K_S^0 couples, as a function of Q (GeV/c), as measured by the OPAL experiment at LEP.

$$|K^0\bar{K}^0>_{C=\pm 1} = \frac{1}{2}|K^0(p_1)\bar{K}^0(p_2)> \pm \frac{1}{2}|\bar{K}^0(p_1)K^0(p_2)>,$$

where p is the momentum of the kaons in their center of mass.

In the limit of $Q=0$ (i.e., in the region of the Bose-Einstein effect), $p_1 \simeq p_2$, and the probability amplitude for the $C = -1$ state is zero. Thus the state made by a K_S^0 and a K_L^0 is vetoed.

What one experimentally detects is a couple of identical particles, namely a K_S^0 K_S^0 couple. Thus, one expects the Bose-Einstein statistics to hold for this couple.

The results on the correlation function from OPAL are displayed in Fig. 2. A Monte Carlo sample has been taken as a reference distribution. OPAL finds $r = 0.72 \pm 0.17 \pm 0.19$ fm, $\lambda = 1.12 \pm 0.33 \pm 0.29$. DELPHI finds $r = 0.50 \pm 0.14 \pm 0.17$ fm, $\lambda = 0.67 \pm 0.28 \pm 0.12$. The results are consistent with those obtained for pions.

Flavour-Tagged Bose-Einstein Interference

DELPHI uses a Neural Network for studying a possible flavour-dependence of the effect. The events are sliced in probability of coming from the hadronization of a $b\bar{b}$ pair[17].

The strength parameter is consistent with decreasing as the probability of coming from a $b\bar{b}$ pair increases (Fig. 3, left). This can be explained by the fact that, in a $b\bar{b}$ event, many particles come from the decay of a B meson, and not from the primary source. The results of the fit to r and λ are displayed in Fig. 3 (left), where the averages (consistent with Ref. 7) are marked as dashed lines. The radius r of the source of the hadronization does not display a clear dependence on the percentage of b events present in the sample. The average number of particles produced via strong interaction in $b\bar{b}$ events should be smaller than for light quarks. One could thus expect a smaller radius of the hadronization region. This is not observed, although, when one studies the dependence of the radius on the multiplicity, a trend consistent with being proportional to $<n>^{1/3}$ (Fig. 3, right) is seen.

RESULTS FROM hh COLLISIONS

Here the most recent results come from experiments E735[18] and CDF[19] running at the Fermilab proton-antiproton Collider at a center of mass energy of 1.8 TeV. Again the measurements of the radius r and of the strength parameter λ are independent of the center of mass energy and of the type of interaction.

An explicit dependence of the source dimension r as a function of the pion multiplicity is found by E735, as shown in Fig. 4 together with a comparison with results from lower energy experiments. They also find a decreasing of the λ parameter with increasing multiplicity. The same experiment finds that both r and the lifetime decrease with increasing momentum of the pion pair, while the strength of

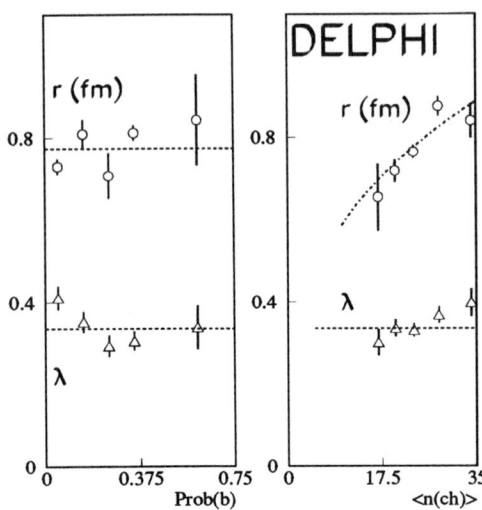

Figure 3. Left: Radius r and strength parameter lambda, as obtained from the fit of the ratio of like- to unlike- pion pairs. versus the probability of belonging to the class of $b\bar{b}$ events. Right: r and lambda, as obtained from the fit of the ratio of like- to unlike- pion pairs, for different multiplicities.

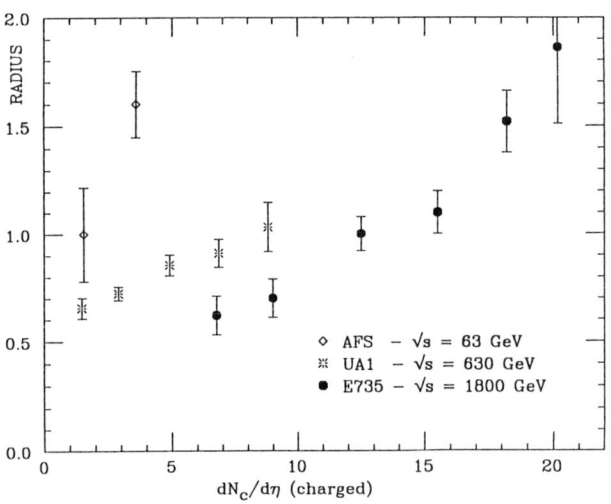

Figure 4. Interaction radius as a function of the charged particle multiplicity per unit of pseudorapidity.

the interference, λ, does not depend measurably on pion pair momentum.

Finally, a study of the production region of identical pions at $38 GeV/c$ π^- − $Nuclei$ interactions with high p_t particles does not see any dependence of the radius r on the atomic number of the target[20].

CONCLUSIONS

After the 1991 Europhysics Conference, new results have been presented on the measurement of the radius of the interaction region and of the strength parameter in high energy e^+e^- and hadron-hadron collisions. The results are consistent with previous measurements, and no indications of energy dependence of the Bose-Einstein effect emerge.

Recent analyses of correlation between K_S^0, and flavour-tagged analyses, promise to open a new chapter in the study of the Bose-Einstein interference.

REFERENCES

1. G. Goldhaber et al., *Phys. Rev.* **120** (1960) 300.

2. G.I. Kopylov and M.I. Podgoretskii, Sov. J. *Nucl. Phys.* **19** (1974) 215; G.I. Kopylov, *Phys. Lett.* **50B** (1974) 472.
G. Cocconi, *Phys. Lett.* **49B** (1974) 459.
M. Deutschmann et al., *Nucl. Phys.* **204B** (1982) 333.
A. Giovannini and G. Veneziano, *Nucl. Phys.* **B130** (1977) 61.
M. G. Bowler, *Z. Phys.* **C29** (1985) 617.

3. B. Lörstad, *Int. J. Mod. Phys.* **A4** (1989) 2861.
W. A. Zajc, in "Hadronic Multiparticle Production", P. Carruthers ed., World Scientific (1988) 235.

4. H. Aihara et al., (TPC-PEP4), *Phys. Rev.* **D31** (1985) 996.
P. Avery et al., (CLEO), *Phys. Rev.* **D32** (1985) 2294.

I. Juricic et al (Mark II), *Phys. Rev* **D39** (1989) 1.

M. Althoff et al. (TASSO), *Z. Phys.* **C30** (1986) 355.

R. C. Walker (AMY), UR-1176 (1990).

5. P.D. Acton et al. (OPAL), *Phys. Lett.* **267B** (1991) 143.

6. D. Decamp et al. (ALEPH), CERN-PPE 91/183.

7. P. Abreu et al. (DELPHI), *Phys. Lett.* **B286** (1992) 201.

8. T. Akesson et al. (AFS), *Z. Phys.* **C36** (1987) 517.

   M. Althoff et al. (TASSO), *Z. Phys.* **C30** (1986) 355.

9. M. Gyulassy, S.K. Kaufmann and L.W. Wilson, *Phys. Rev.* **C20** (1979) 2267.

   M. G. Bowler, *Phys. Lett.* **B275** (1991) 69.

10. M. G. Bowler, *Z. Phys.* **C39** (1988) 81.

11. S. Marcellini, Proceedings of the 1991 Lepton-Photon Symposium (Geneva 1991), S. Hegarty et al. eds., p. 750.

12. D. Bertrand et al., *Nucl. Phys.* **B128** (1977) 365.

    A.M. Cooper et al., *Nucl. Phys.* **B139** (1978) 45.

    T. Akesson et al., *Phys. Lett.* **B155** (1985) 128.

    M. Aguilar-Benitez et al., *Z. Phys.* **C54** (1992) 21.

13. G. Alexander et al. (OPAL), contributed to this conference.

14. A. De Angelis, L. Vitale and J. Pennanen, DELPHI internal note, CERN, August 1992.

15. H. Lipkin, *Phys. Lett.* **B219** (1989) 474; WIS Preprint 92/29, March 1992.

16. T. Sjöstrand, *Comp. Phys. Comm.* **27** (1982) 243, ibid. **28** (1983) 229.

    T. Sjöstrand and M. Bengtsson, *Comp. Phys. Comm.* **43** (1987) 367.

17. A. De Angelis and P. Del Giudice, Udine Report 92/08/AA (September 1992), to be published in the Proceedings of the 2nd Workshop on Neural Networks, Isola d'Elba (Italy), June 1992.

18. T. Alexopoulos et al., contributed to this conference.

19. F. Rimondi, private communication.

20. L. L. Gabunia et al., KFKI-1992-07/A, contributed to this conference.

MULTIPLICITY DISTRIBUTIONS IN e^+e^- AND HADRON-HADRON COLLISIONS

Richard D. St. Denis
Max-Plank-Institute für Physik
Werner-Heisenberg-Institut
Föhringer Ring 6
D-8000 München 40

Abstract

Log-normal, negative binomial, and Krasznovszky-Wagner distributions are shown to fit the e^+e^- and hadron-hadron final state charged multiplicities. KNO scaling holds in e^+e^- from $\sqrt{s} = 5$ to 91 GeV. QCD predictions for average multiplicity behaviour with $\sqrt{s}$ describe the data well although the shape predictions are influenced strongly by higher order effects. QCD calculations of subjet multiplicities are shown within their expected region of validity to agree well with the data.

INTRODUCTION

Results on charged particle multiplicities from LEP and some reexamination of results from the hadron colliders have appeared recently and are summarized here [1]–[8]. To begin, the experimental problem is revisited. Physical motivations which lead to a variety of functions proposed to describe these distributions and their $\sqrt{s}$ dependence are given. QCD predictions for clusters of particles, subjet multiplicities, include coherence effects and offer a recasting of the question of multiplicities in quark and gluon jets in terms of a mass resolution parameter [9]. The results of the present calculations are compared to data. Similar coherence effects in protons, implications for supercollider physics and suggested means for observing these effects are finally discussed.

CHARGED PARTICLE MULTIPLICITIES

THE EXPERIMENTAL ISSUE

The experimental problem for charged particle multiplicities comes from the fact that the observed number of particles must be related to the true number of particles in the event. This relation is calculated with Monte Carlo which convolutes detector response with a particle production model. Uncertainties in this computation dominate the systematic error in the individual probabilities to observe a certain number of particles as well as in the moments of the probability distributions. With more sophisticated techniques the particle production model dependence can be reduced [1]; however, some systematic error on the probability distribution remains, manifesting itself through correlations between nearly equal numbers of particles. These correlations (usually unpublished) make it very difficult and dangerous to interpret fits to the corrected distributions in terms of χ^2/DOF.

SOME DISTRIBUTIONS EXPLORED

One desires to distinguish various physical models based on the functional behaviour of the distributions they predict. Of the large number of functions and models available, the following are reviewed here: the negative binomial, the log-normal, and the Krasznovszky-Wagner (KW) distibutions. In addition, leading order or next-to-leading order QCD predictions are discussed.

- **Negative Binomial Distribution (NBD):** A model leading to the negative binomial distribution is one in which the final state particles come from some number, N_c, of Poisson distributed "clans" of particles, where each clan contributes some coherently distributed number n_c of particles [10]. The probability to observe k particles for an average multiplicity, $\bar{n}$, is related to these clan model parameters by [11] $\bar{N}_c = \bar{n}/\bar{n}_c = k \log(1 + \bar{n}/k)$.

- **Log-normal Distribution (LND):** One model leading to the log-normal distribution is a multivariate branching process [12]. In this model, the probability is described in terms of both the momentum available and the current number of particles. At each branchpoint one envisions a tree having a thickness associated with the amount of energy available and the number of branches is the number of particles. One starts at the trunk with energy at $\sqrt{s}$ and is cut off naturally by the pion mass at the thinnest branch.

- **KW Distribution:** A model leading to the KW distribution [13] envisions multiparticle production through an interaction based on some overlap function which is described by an impact parameter distribution. This function is characterized by three parameters: A, m, and the average multiplicity, $\langle n \rangle$. The parameter A is a scale breaking parameter and should be constant with $\sqrt{s}$ if the process obeys KNO scaling. The parameter m should be fixed for a given kind of physics, e.g. m=1.5 for all e^+e^- physics. This distribution includes many special cases of other physics models depending on parameter settings, i.e. the gamma function and Weibull are among the distributions included. It also includes as a special case a model based on three dimensional strings or colliding projectiles [16].

Distributions Confront Data

The multiplicity distributions have been fitted with the above functions in a number of circumstances. These are described here.

- **Single particle multiplicities:** The NBD and LND fits to ALEPH [1] multiplicities are shown in Fig. 1 together with the differences between data and fits. It can be seen that there is some slight preference for the LND over the NBD.

- **KW fits and $\sqrt{s}$ dependence:** The KW distribution has been fit to e^+e^- data [4] as well as hadron collider data [5]. For the electron machines, the value of A is shown as a function of $\sqrt{s}$ in Fig 2. Its constancy demonstrates the KNO scaling for this physics process. The fits are all reasonable and furthermore, with the value of A, all moments are predicted correctly.

The fits to proton-(anti)proton inelastic and non-diffractive data require different values of m; this implies different physics for all cases. For example, the non-diffractive pp and $p\bar{p}$ have values of m=1.25 and 0.9 respectively. The m=1.25 value does not fit well in the case of non-diffractive $p\bar{p}$. The behaviour of A shows KNO scaling in $p\bar{p}$ inelastic and in ISR data for pp inelastic. The variation of A in $p\bar{p}$ non-diffractive data is consistent with the KNO scale breaking observed in UA5.

- **Fits to multiplicities within jets:** DELPHI [2] has clustered jets using the metric

$$y_{ij} = \frac{2E_i E_j (1 - \cos\theta_{ij})}{E_{vis}^2} \quad (1)$$

to define the smallest resolvable scaled invariant mass one can resolve for clusters of particles. They find that over a range in y_{ij} where the number of clusters approximately matches the number of partons, the multiplicity distribution of the event is fit well by a NBD. Numerically, the fits are good for $y_{ij}=$ 0.005, 0.01, and 0.02, but deteriorate for 0.04. They also note that the ratio of the dispersion in the individual jets, D_i, to average number of particles, $\langle n_i \rangle$, scales with the square root of the number of jets:

$$\frac{D_i}{\langle n_i \rangle} / \frac{D_w}{\langle n_w \rangle} = \sqrt{j}, \quad (2)$$

where w indicates the whole event dispersion or multiplicity.

- **$\sqrt{s}$ dependence of average event multiplicities:** A variety of fits to $\sqrt{s}$ dependence of the average multiplicity have been performed (see Fig. 3 and

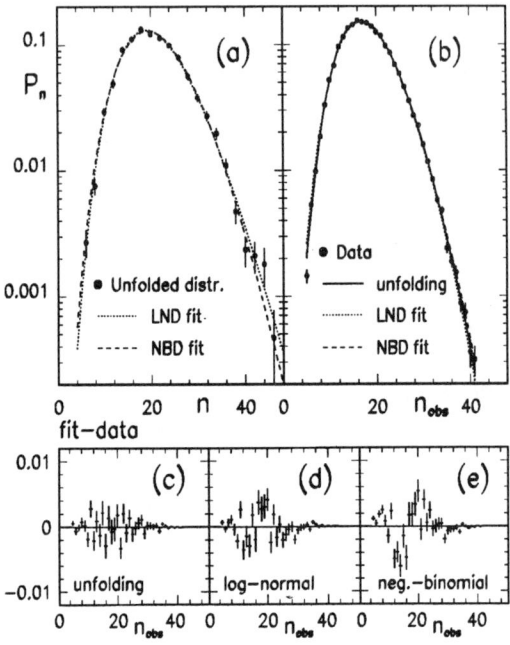

Figure 1. Fits and residuals for Unfolded, LND, and NBD to ALEPH data.

Table 1) [1]. These include a variation on the standard phenomenological ansatz, $\langle n \rangle = a + b \log s + c \log^2 s$, a QCD inspired behaviour, and the LND prediction.

Finally, a fit of the QCD prediction of the $\sqrt{s}$ dependence of the second binomial moment shows that the next-to-leading order fit approaches the data (relative to the leading order) but one still needs higher order and non-perturbative contributions to get good agreement (i.e. as implemented in LUND 7.2 [15].)

Gluon Interference Effects

DELPHI [2] has examined the problem of comparing multiplicities in gluon and quark jets and concludes that there seems to be no

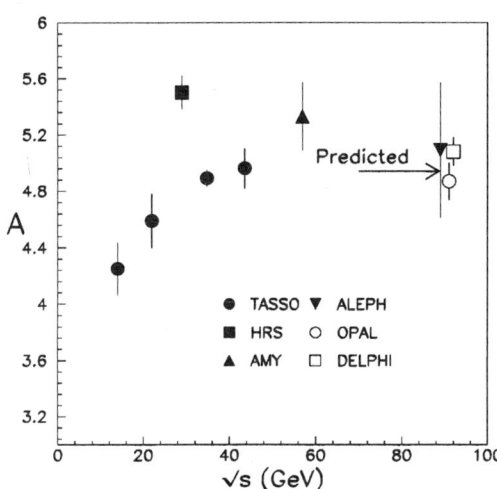

Figure 2. Energy dependences of the A parameter in the KW distribution for e^+e^- data.

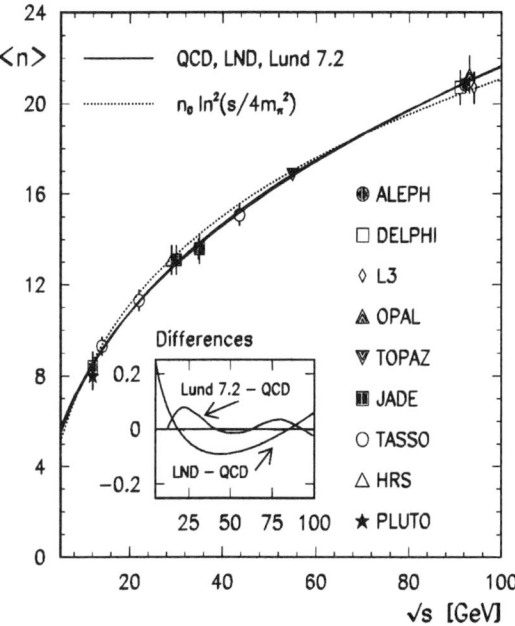

Figure 3. Energy dependence of multiplicites and a variety of fit functions.

Physics	Formula
Ansatz	$\langle n \rangle = n_0 \log^2 s/4m_\pi^2$
QCD	$\langle n \rangle = a\alpha_s^{b_1} \exp \frac{b_2}{\sqrt{\alpha_s}}$
LND	$\langle n(s) \rangle =$ $(1 + \langle n(s_0) \rangle)(s/s_0)^\gamma - 1$

Table 1. Functions describing the energy dependence of multiplicities.

difference within the large uncertainties of the method. This question has been addressed in a more rigorous way by examining the number of jets within jets [6] [7]. The procedure is as follows. One classifies all events by the number of clusters of particles, N, for a given scaled mass resolution value, y_1. For each of the classes, the number of subclusters, M_N, appearing for $y_0 \leq y_1$ is plotted. This is compared to a resummed next-to-leading log QCD prediction [9]. In fact, to address the ratio of three to two jet multiplicities, M_3/M_2 is plotted as shown in Fig. 4 for a relatively small value of y_1. In the limit of large and small y_1 calculations can be performed and the data agree with the calculations in their expected region of validity. For an intermediate value of y_1, the data lie between the calculations. Color counting would indicate the multiplicity ratio in three jet events relative to two jet events to be 17/8. However two effects act to reduce this. First, the momenta of the gluons in the three jet events are smaller thereby reducing the multiplicity. Second, there are important gluon coherence effects that also act to reduce the multiplicity.

These gluon coherence effects can also manifest themselves in the proton structure function [8]. As one goes to small x, the number of gluons predicted by the Altarelli-Parisi equations increases rapidly and at some point the wave functions overlap, leading to inter-

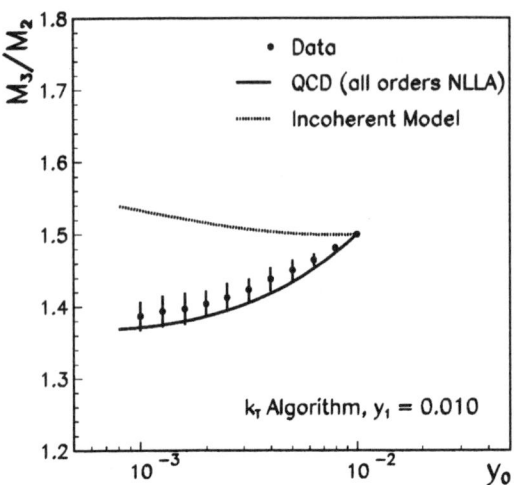

Figure 4. L3 Data for subjet multiplicites.

ference. As a result of this the properties of particle multiplicities should be modified. This overlap should set in around x=0.001 implying that one has to be at a $\sqrt{Q^2}$ of about 20(8)GeV for SSC(LHC) energies. The majority of data that might show this effect would be in the diffractive region. Therefore calculations of multiplicities have been performed using the structure functions that include the coherence effects plus "semi-hard QCD" and supercritical pomeron exchange as described in [8]. According to this prediction, the average number of charged particles can vary from about 100 to 165 depending on the details of the overlap of the gluon wave functions. The shape of the KNO distribution is also sensitive to these details.

Conclusions

- One should beware of the meaning of χ^2/DOF when evaluating fit quality of various ansaetze. Systematic errors and correlations between bins are important.

- From the multiplicity distributions, the NBD may not be the best fit, the LND looks reasonable and KW seems to do well in e^+e^-. The KW distribution gives various parameters that scan a range of models but as a result make it difficult to have a precise physical interpretation in terms of a concrete model.

- The KW distribution fits show that the physics changes for diffractive and non-diffractive pp and $p\bar{p}$ collisions. It provides a convenient way to parameterized deviations from KNO scaling.

- NBD fits work for events with a fixed number of individual jets.

- QCD predicts the $\sqrt{s}$ dependence of the average multiplicity accurately, but only roughly predicts the shape of the distribution.

- The problem concerning the ratio of the number of particles in a quark jet and in a gluon jet has been reformulated in terms of subjet multiplicities. In this form, the data are described well within the expected region of validity of the QCD calculations.

- There may be something to learn about the overlap of gluon wave functions at small x in supercollider minimum bias events.

REFERENCES

1. ALEPH Collab., D. Decamp et. al., "Measurement of the Charged Particle Multiplicity Distribution in Hadronic Z Decays," *Phys. Lett. B* 273, pp. 181–192, (1991).

2. DELPHI Collab., P. Abreu et. al., "Charged Particle Multiplicity Distributions for Fixed Number of Jets in Z^0 Hadronic Decays," *CERN-PPE preprint* 92–64(1992).

3. S. Krasznovszky, "Analysis of the KW Distribution Using a Generalization of the Meijer's G-Function," *KFKI preprint* 1992-13/A.

4. S. Krasznovszky and I. Wagner, "Description of Charged-particle Multiplicity Distributions in e^+e^- Annihilation at TRISTAN and LEP1 Energies from AMY,ALEPH,OPAL and DELPHI Collaborations," *submitted to this conference:P158*.

5. S. Krasznovszky and I. Wagner, "Description of Charged-particle Multiplicity Distributions in $p\bar{p}$ Collisions at 200, 546, and 900 GeV Energies from UA5," *submitted to this conference:A282*.

6. ALEPH Collab., D. Decamp et. al., "Study of Sub-jet Multiplicities in Hadronic Z Decays," *submitted to this conference:A845*.

7. L3 Collab., S. Banerjee, "Average Cluster Multiplicities in 3- and 2-Jet Events at the Z^0 resonance," *submitted to this conference,L3 Internal note* 1171 (May 26, 1992).

8. D. Pertermann, J. Ranft, and F. W. Bopp, "Hadron Production at Supercolliders in the Two Component Dual Parton Model and the Small x Behaviour of the Structure Functions," *University of Leipzig and University of Siegen preprints* UL-92-4, SI-92-2.

9. S. Catani, B.R. Webber, Yu.L. Dokshitzer, and F. Firorani, "Average Multiplicites in Two- and Three-Jet e^+e^- Annihilation Events," *CERN-TH and Lund Univerity preprints* CERN-TH.6419/92, LU TP 92-9.

10. A. Giovannini and L. Van Hove, "Negative Binomial Multiplicity Distributions in High Energy Hadron Collisions," *Z. Phys. C* 30, pp. 391-400, (1986).

11. A. Giovannini and L. Van Hove, "Negative Binomial Properties and Clan Structure in Multiplicity Distributions," *Acta Phys. Pol.* B19, pp. 495-510, (1988).

12. G. Wrochna, "Multiparticle Production as a Bivariate Branching Process," *Warsaw Univerisity Preprint* IFD/5/1990.

13. S. Krasznovszky and I. Wagner, "Description of Charged-particle Multiplicity Distributions in e^+e^- Annihilation at the Available Energies of 14,22,34.8 and 43.6 GeV from TASSO," *Phys. Lett. B* 213, pp. 103–106, (1988).

14. UA5 Collab.,"UA5: A General Study of Proton-antiproton Physics at $\sqrt{s} = 546$ GeV," *Phys. Reports* 154, pp. 247-383.

15. M Bengtson and T. Sjöstrand, "Coherent Parton Showers Versus Matrix Elements - Implications of PETRA/PEP Data," *Phys. Lett. B* 185, pp. 435-440, (1987).

16. Chou Kuang-chao, Liu Lian-sou, and Meng Ta-chung, "Koba-Nielsen-Olesen Scaling and Production Mechanism in High Energy Collisions," *Phys. Rev. D* 28, pp. 1080–1085, (1983).

Heavy Ion Interactions

Parallel Session 9

Meyerson Symphony Center, Dallas

Conveners: U. Heinz (Regensburg)
W. Zajc (Columbia)

THEORETICAL OVERVIEW: LIGHT ION LESSONS, HEAVY ION HOPES*

Sean Gavin
Department of Physics
Brookhaven National Laboratory
Upton, NY 11973

Abstract

Progress in understanding relativistic nuclear collisions is surveyed.

INTRODUCTION

Experiments using light ion beams of atomic masses $A \sim 30$ have been underway since 1986 at the Brookhaven AGS and the CERN SPS at the respective energies $\sqrt{s} \sim 5\ A$ GeV and $20\ A$ GeV. The first truly heavy ion runs with a gold beam began this spring at the AGS. In this talk I will survey our progress towards an understanding of nuclear collision dynamics, focusing on those issues that are relevant to Au+Au at the AGS (see Ref. [1] for a complete overview). In view of what we have already learned from the light ion data, I will argue that the prospects for producing matter at extreme density in these experiments are excellent.

LIGHT ION LESSONS

The aim of the AGS and SPS light ion programs is to ascertain the prospects for studying dense strongly interacting matter in heavy ion collisions. Important questions are:

1. How much energy does the projectile lose, i.e., what is the stopping power of the target [2]?

2. How is this energy partitioned? Does the heavy ion system pass through a state near local thermal equilibrium?

3. How are strangeness and baryon number partitioned among the produced particles? Does the system approach chemical equilibrium?

4. What are the maximum energy and baryon densities? Where and when in the course of the collision are these conditions reached?

5. Are there any precursors of collective phenomena?

The term "collective phenomena" includes the sought-after possibility of quark gluon plasma formation but, more generally, refers to any interesting many body flow or particle production phenomena. Experience at the lower Bevalac energy $\sim 1\ A$ GeV indicates that evidence of collective flow can be obtained only through careful analysis of the heaviest AA collisions [3].

Extreme Models

To illustrate the range of stopping power, equilibration, density, and collectivity, we consider the following extreme scenarios. In

---

*This work was supported in part by contract number DE-AC02-76CH00016 with the U. S. Department of Energy.

the independent fragmentation scenario, one supposes that the nuclear collision is a superposition of independent nucleon-nucleon, NN, subcollisions. Final state scattering of the primary nucleons and produced particles are entirely neglected. This scenario has been investigated using the first Monte Carlo event generators, such as the LUND/Fritiof model and early versions of VENUS [4]. These models yield a minimum of stopping power and allow for no collective behavior.

To characterize the opposite, fully collective extreme, one can imagine that the projectile and target participants stop each other completely to form a spherically symmetric fireball. Following an early thermal equilibrium stage, the fireball expands and rarefies until it's constituents cease to interact — the asymptotic particles are then formed. The evolution of the fireball has been studied extensively in the context of the Landau hydrodynamic formulation [5]. [N.B. Not all Landau models assume the spherical symmetry of a fireball [6].]

Extreme Models vs. Data

Chapman and Gyulassy have pointed out that neither fireball nor independent fragmentation models stand up to a systematic comparison to the AGS light ion data [7]. They emphasized that fireball models give a proton rapidity distribution that disagrees seriously with measurements from BNL experiments E802 and E814, see refs. cited in [8]. The angular distribution of particles $dN/d\Omega$ from a spherically symmetric source is constant about the center of mass, so that the rapidity distribution is roughly

$$dN/dy_{\text{fireball}} = (2\pi dN/d\Omega) \, d\cos\theta_{\text{cm}}/dy$$
$$\propto \{\cosh(y - y_{\text{cm}})\}^{-1/2}, \quad (1)$$

where $y \approx y_{\text{cm}} - \ln(\tan\theta_{\text{cm}})$, plus small corrections that depend on the particle mass. This

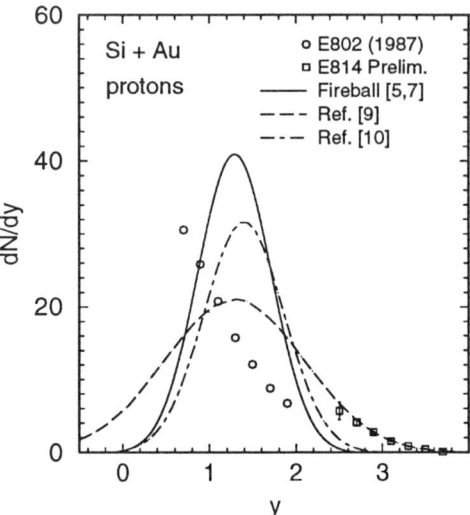

Figure 1. Measured proton rapidity density compared to fireball models from [7].

distribution is peaked near the participant center of mass rapidity, which is $y_{\text{cm}} \sim 1.4$ for central Si+Au (one counts as participants those target and projectile nucleons in the direct path of the collision). Fig. 1 from [7] shows proton data from E802 and E814 compared with rapidity distributions of participant protons calculated for both the spherical fireball (including the mass corrections) and the variant models of ref. [9] and ref. [10]. The results are clearly *above* the data for $1 < y < 2$. Note that this discrepancy cannot be remedied by an appropriate treatment of spectator nucleons. In contrast, the rapidity spectra of secondaries from AGS and SPS are peaked at y_{cm}, although the measured distributions are substantially more broad than the fireball shape [5].

The independent fragmentation picture is also wrong. While LUND model calculations agree with the proton rapidity density, Chapman and Gyulassy find that the pion dN/dy are in serious disagreement with data [7]. The enhancement of strangeness production studied by E802 and E810 is also in conflict with independent fragmentation models. The K^+

production per projectile nucleon in central Si+Au is almost twice that in minimum bias p+Au [11], and roughly four times the independent fragmentation model expectations [12]. Taken together, the various failures of independent fragmentation provide strong indications of the final state interactions omitted in that picture.

At the SPS, J/ψ suppression [13] and ϕ enhancement [14] further support the need to include final state interactions and, moreover, suggest that high densities of secondary particles are achieved [15, 16]. The J/ψ suppression measured for oxygen and sulphur beams on uranium targets at $\sqrt{s} = 20$ A GeV can be attributed to $N + J/\psi$ and meson+$J/\psi \to D\overline{D} + X$ reactions, but not to $N + J/\psi$ interactions alone [15]. The meson densities needed to explain the data exceed ~ 1 fm^{-3} in models, as discussed by Satz in these proceedings.

HADRONIC TRANSPORT MODELS

To describe heavy ion dynamics consistently at AGS-SPS energies, one must address systems that are neither completely independent nor fully collective. Most phenomenological success has been in the context of an hadronic transport formulation. As in independent fragmentation one assumes that hadrons are formed in NN subcollisions, only now one follows the trajectories of the particles and allows them to scatter. Rudimentary collective behavior associated with equilibration and flow can occur if scattering is sufficiently frequent.

In an hadronic formulation, one uses particle data to include various two body scattering processes, such as $\pi N \to \Delta$, $\pi N \to K\Lambda$, and $\pi\pi \to K\overline{K}$. However, cross sections for many reactions, e.g. $\Delta N \to \Delta N$ or $\pi\rho \to \omega\eta$, are not known and must essentially be guessed. In addition, formation times for the production of secondaries must be introduced to simulate uncertainty principle effects. To constrain this vast array of parameters, one can turn to lepton-nucleus and hadron-nucleus data [17, 18]. Workers have found that N, π, and K spectra depend primarily on a tractable subset of these parameters, so that the practical task of understanding heavy ion experiments is not as formidable as it might seem.

The most comprehensive phenomenological models applicable at AGS energies are the Monte Carlo cascade simulations RQMD [19] and ARC [20]. At SPS energies and above, one has a variety of models, such as HIJING discussed by X.-N. Wang in these proceedings (see also the papers by Werner, Amelin et al., Kawrakov et al., and Umar et al. in [1]). In principle, these models should describe all pp, lA, hA, and AB data in their respective energy ranges with a single set of parameters. In addition, hosts of simpler but less comprehensive models have been employed to understand qualitative aspects of specific phenomena, e.g. [25, 26] discussed later.

At the higher RHIC and LHC energies, a large fraction of the particle production in nuclear collisions will result from semihard processes accessible via perturbative QCD methods [21]. To describe the initial stages of such collisions, Geiger has developed the partonic transport model presented in this session. The likelihood that such a QCD-based description can be applicable is one of the strongest motivations for moving to higher energies.

Transport Models vs. Data

So far, the comprehensive cascade simulations are in reasonable agreement with experiment. Regarding the stopping power, proton rapidity distributions from ARC are in excellent accord with results from E814 at high rapidity and E802 at midrapidity [22]. The average rapidity loss per baryon is found to be ~ 1.5 units, as expected from simple multiple scattering extrapolations from pA data. Such a rapidity loss is nearly sufficient to bring projectile participants of initial rapidity $y = 3.4$ to rest in the center of mass on average — there is significant stopping of the projectile. However, the target participants in the asymmetric Si+Au collision are not accelerated into a fireball, cf. Fig. 1.

The measured strangeness enhancement discussed earlier constitutes circumstantial evidence of the scattering of secondaries necessary for thermal and chemical equilibration. The measured increase of the production of K^+ and Λ from p+p to p+Au to Si+Au can be attributed to reactions such as $N\Delta \to K\Lambda N$, $\pi N \to K\Lambda$, and $\pi\pi \to K\overline{K}$. See the literature for detailed comparisons of RQMD [12, 23] and ARC [20, 22] to data.

There is however no direct quantitative indication of the densities, collectivity, or equilibration from the current light ion experiments. Signals of thermal equilibration are particularly subtle, since transverse momentum p_T spectra from $pp \to \pi + X$ already *appear* thermal, $d\sigma/dp_T^2 \sim \exp\{-(p_T^2 + m_\pi^2)^{1/2}/T\}$, where T is a temperature-like parameter. At the SPS, NA35 has measured an enhancement of charged particles relative to this pp form for $p_T < 200$ MeV in central O+Au at midrapidity. Kataja and Ruuskanen pointed out that Bose enhancement can produce this effect, since pion densities at the SPS are likely ~ 1 fm^{-3} [24].

Transport models of nonequilibrium pion evolution indeed find a tendency towards this Bose enhancement, though the light ion systems are not fully equilibrated [25, 26]. Nevertheless, there are several caveats: Theoretically, the full range of the Bose effect is extremely small — the difference between noninteracting and fully thermalized pion spectra in O+Au is at most 20% [25]. Moreover, a nuclear enhancement of baryon resonance production must also contribute to the enhancement, since a larger low p_T shift is measured at target rapidity ([25] and [26] only treat meson resonances). To estimate the significance of these various contributions at the 20% level, it is necessary to include Bose statistics in more comprehensive cascade models that include baryons. Finally, the measurement of this small effect is quite intricate, since one must correct the charged particle spectra for misidentified electrons. SPS data with complete particle identification are forthcoming [27].

As at the SPS, there is no consensus among AGS model builders as to which microscopic ingredients are necessary to explain the data. For example, ARC attributes the measured proton p_T distribution to rescattering of the primary nucleons and baryon resonances [22], while RQMD requires the addition of a strong repulsive mean field interaction for baryons [23].

HEAVY ION HOPES

Comprehensive simulations like ARC and RQMD can be used to explore the spacetime dynamics of nuclear collisions. By following the trajectories of particles in model events, one can extract indications of the energy and baryon densities achieved in real collisions. However, the spacetime picture so obtained is consistent only if the computed final particle spectra agree with all available data.

ARC predictions for central Au+Au at 11.6 A GeV beam energies indicate a gaussian-shaped pion rapidity distribution with a peak value $dN/dy \approx 80$ for π^+ at midrapidity, and a width that is roughly 60% larger than that of a fireball, cf. Eq. (1) [28]. RQMD pion distributions for 10 A GeV Pb+Pb are roughly $\sim 40\%$ higher [29].

Collective Flow and High Densities?

Collective flow has proven useful in probing the dynamics of nuclear collisions [3]. At the highest energies, hydrodynamic calculations suggest that a transverse collective expansion driven by pressure can reflect the equation of state of the underlying fluid, see, e.g. [30]. Despite early optimism, no clear transverse flow effects have been seen in light ion collisions. The most recent intensity interferometry measurements at BNL and CERN for pions and kaons imply transverse radii that are consistent with the projectile rms radii [27, 31].

In contrast to the lighter systems, ARC calculations suggest that transverse expansion can be substantial in Au+Au at the AGS. To study the expansion, the distribution of the transverse radii for particles at the position of their last interaction is constructed. Pang, Schlagel and Kahana find a 60% increase in the average $\langle r_\perp \rangle$ of pions and protons compared to the rms radius of Au. A relatively modest $\sim 30\%$ increase is computed for Si+Au.[40] The impact of such a small Si+Au increase on the interferometry signal is within the uncertainties in the analysis of this signal [32].

To illustrate the conditions that can be obtained in a central Au+Au collision, Pang *et al.* follow the time evolution of the baryon density ρ_b and average transverse momentum $\langle p_T \rangle$ for baryons in a test cell of 1 fm radius in the interior of the collision volume. Fig. 2 shows the $\langle p_T \rangle$-ρ_b trajectory of this cell for Si+Si (squares) and Au+Au (circles). The trajectory is traversed counterclockwise, with the plotting symbols spaced a time interval 1 fm/c apart in the cell's local rest frame. In essence, $\langle p_T \rangle^2/m$ measures the kinetic energy in the transverse degrees of freedom of baryons in the cell, where m is the nucleon mass. If the nuclei simply fly through one another without any interaction, then geometry gives a peak density $\rho_0/\rho_{nm} \sim 2\gamma_{cm} \sim 5.2$ for $\gamma_{cm} = 2.6$ and $\rho_{nm} = 0.16$ fm^{-3} for nuclear matter. For Si+Si (and similarly for Si+Au), the baryons in the cell are compressed beyond ρ_0, but pass through the high density region very rapidly.

The trajectories for Au+Au and Si+Si are qualitatively different. The compression of baryons is greater in the heavier system. More importantly, the matter in the cell remains at the most extreme densities for a much longer time. If I add the dashed line to Fig. 2 as an "imaginary boundary" between the low and high energy density regions, then the Au+Au spends 2-3 times longer in the high density region.

Observe the cusp shape at the highest density in Au+Au. I note that the trajectory at the cusp and afterward scales very nearly as $\langle p_T \rangle \propto \rho_b^{1/3}$, as expected for adiabatic expansion. Si+Si does not exhibit this scaling, except perhaps in the final ~ 2 fm/c. Does this mean that matter in the interior of a Au+Au collision approaches local equilibrium? More detailed model analyses are needed to see if the flow in ARC really is adiabatic and thermalized in this region. However, Sorge *et al.* find similar interior densities and strong indications of thermalization using RQMD [23].

If the cell really were in local equilibrium with its surroundings, then the temperature would be $T \approx 2\langle p_T \rangle^2/\pi m$. In a static, equilibrium system, the dashed curve would roughly correspond to the QCD phase boundary, assuming a first order chiral restoration transition at $T_c = 200$ MeV for $\rho_b = 0$. Of course, this cell is centrally located, and one expects the density to be higher there than in the periphery. Furthermore, the dynamics of the collision might proceed differently, *e.g.*, if quark gluon plasma were to form. Nevertheless, I emphasize these rather conservative models suggest that Au+Au can produce exactly the conditions that we hope for!

Collective Particle Production?

If the conditions in real Au+Au experiments are as dramatic as simulations suggest, then one can expect the occurrence of nov-

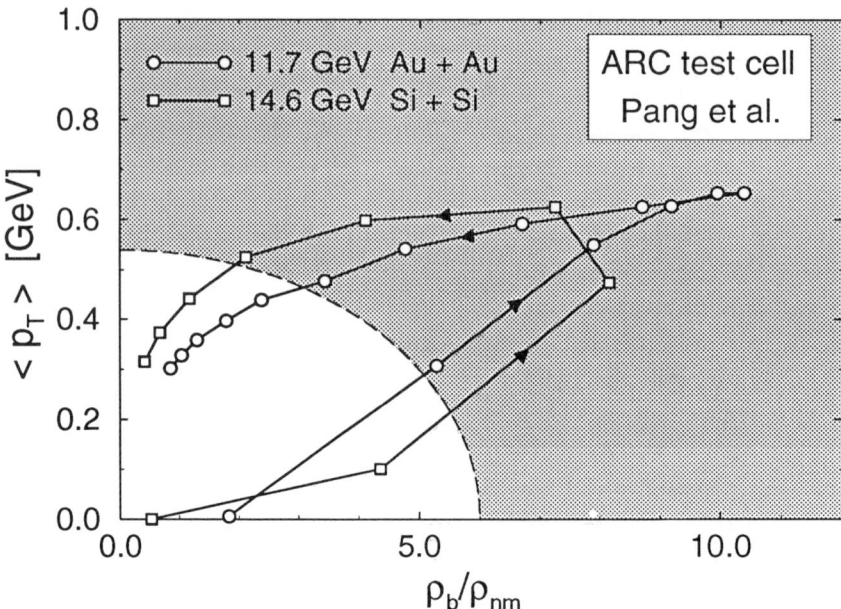

Figure 2. Time evolution of density and $\langle p_T \rangle$ for baryons in a test cell from ARC [28].

el collective phenomena associated with quark gluon plasma formation. Such phenomena may appear as a discrepancy between hadronic transport model predictions and data. The challenge will then be to prove that this discrepancy cannot be accounted for within the uncertainties in the models.

Where do we look for such discrepancies and what can we learn from them? One persistant puzzle concerns the order of the chiral restoration transition in QCD. Wilczek [33] has stressed that recent lattice results for two light flavors [34] are consistent with a second order transition. However, the latest simulations with smaller lattice spacing exhibit behavior tantalizingly close to that of a first order transition [35]. In view of these ambiguous results, let us ask if there are observables in heavy ion collisions that can depend on the nonequilibrium dynamics of the chiral transition.

Many authors have speculated that the chiral transition can enhance baryon-antibaryon pair production. Chiral restoration can contribute to this effect, e.g., by enhancing quark-antiquark pairs due to plasma formation, and by reducing the $\pi\pi \to N\overline{N}$ threshold as the in-medium hadron masses tend to zero. [Color rope formation [36] and minijet production can also enhance baryon pairs, but these mechanisms do not contribute at AGS energies.] Interestingly, DeGrand and Ellis, Heinz and Kowalski have argued by analogy with the cosmological Kibble mechanism that $N\overline{N}$ pairs can be enhanced in the Skyrme model due to the non-trivial topological structure induced by the formation of metastable chiral domains [37]. This mechanism explicitly depends on both the order of the transition and its nonequilibrium nature in AA collisions.

It is extremely interesting that antiproton formation can depend on the order of the chiral transition; other chiral probes are discussed in [38]. One potential problem with the $\overline{p}$ probe is that annihilation may wipe out any initial excess. The mean free path for $\overline{p}$'s is $\approx \{\rho_b \sigma_a\}^{-1} < 0.3$ fm, since the annihilation cross section is $\sigma_a \sim 40 - 100$ mb and the density can exceed $\rho_b \sim 1$ fm^{-3}.

In the absence of any collective production, however, we expect antiproton production to be *suppressed* [39]. Specifically, Gyulassy, Plümer, Venugopalan and myself predicted that annihilation reduces the $\bar{p}/p$ ratio in AA relative to pp collisions.

ARC calculations [40] indeed find that annihilation modifies the dependence of $\bar{p}$ production on A and centrality in accord with the AGS light ion data discussed by Kumar in these proceedings. However, there are uncertainties in the cascade description of $\bar{p}$ dynamics at $\sqrt{s} = 5$ A GeV. The production in NN subcollisions is near threshold and, consequently, is very sensitive to the treatment of rescattering. More significant is the uncertainty in the medium effects in the annihilation. Kahana et al. point out that the large σ_a can be shielded, since the interacting $\bar{p}$ can scatter in the medium before their antiquark content is annihilated.

These large uncertainties leave ample room for enhancement in light ion collisions. To exhibit possible enhancement effects in Au+Au, it is useful to study the centrality dependence of $\bar{p}$ and p production [41]. The more central is the collision, the larger is the baryon density and, correspondingly, the greater is the annihilation. On the other hand, the higher energy density in central collisions can lead to chiral restoration in a larger fraction of the collision volume, which in turn can produce more initial baryon pairs. ARC calculations find that $\bar{p}/p$ at midrapidity can decrease by a factor of ~ 5 as the impact parameter is varied from 10 fm to 0 [40]. It would be extremely interesting if this year's measurements reveal an enhancement relative to these predictions!

I am grateful to Yang Pang, Tom Schlagel, and Sid Kahana for furnishing me with ARC results, and to M. Bloomer, S. Chapman, S. Gottlieb, M. Gyulassy, T. Matsui, R. Pisarski, W. Schaffer, and W. Zajc for stimulating discussions.

REFERENCES

1. *Proc. Quark Matter '91 Conf.*, T.C. Awes, F.E. Obenshain, F. Plasil, M.R. Strayer and C.Y. Wong eds., *Nucl. Phys.* A544 (1992).

2. Here, "stopping power" refers to the momentum distribution of baryons, rather than the spacetime evolution of baryons.

3. H.H. Gutbrod, A.M. Postkanzer and H.G. Ritter, *Rep. Prog. Phys.* 52, 1267 (1989).

4. B. Anderson et al., *Nucl. Phys.* B281, 289 (1987); J.A. Casado et al., *Z. Phys.* C33, 541 (1987); K. Werner, *Phys. Lett.* 208B, 502 (1988).

5. E. Schnedermann, J. Sollfrank, and U. Heinz, TPR-92-29, to appear in *Proc. NATO Advanced Study Institute on Particle Production in Highly Excited Matter*, Il Ciocco, Italy, July 12-24 1992, H.H. Gutbrod ed., Plenum Press, New York 1992.

6. U. Ornik, F.W. Pottag and R.M. Weiner, in *Hadronic Matter in Collision 1988*, P.A. Carruthers and J. Rafelski eds., Singapore, 1989, p. 310; R. Venugopalan and M. Prakash, *Phys. Rev.* C41, 221 (1990); R. Venugopalan, PhD Thesis, Stony Brook (1992), unpublished.

7. S. Chapman and M. Gyulassy, *Phys. Rev. Lett.* 67, 1210 (1991); *Phys. Rev.* C45, 2952 (1992).

8. S. Nagamiya in [1], p. 5c.

9. J. Stachel and P. Braun-Munzinger, *Phys. Lett.* B216, 1 (1989).

10. G. Brown et al., *Phys. Rev.* C43, 1181 (1991); C. M. Ko et al., *Phys. Rev. Lett.* 66, 2577 (1991); 67, 1811 (1991).

11. T. Abbott et al (E802), *Phys. Rev. Lett.* 66, 1567 (1991); K. J. Foley (E810) in [1], p. 335c.

12. R. Mattiello et al., *Phys. Rev. Lett.* **63**, 1459 (1989).

13. C. Baglin et al., *Phys. Lett.* **B270**, 105 (1991), and refs. therein.

14. R. Ferreira (NA38) in [1], p. 497c; M.A. Mazzoni (NA34) in [1], p. 623c.

15. S. Gavin and R. Vogt, *Nucl. Phys.* **B345**, 104 (1990); R. Vogt in [1], p. 615c; S. Gavin, H. Satz, R.L. Thews, and R. Vogt, in progress.

16. P. Koch, U. Heinz, and J. Pisut, *Phys. Lett.* **B243**, 149 (1990); F. Grassi and H. Heiselberg in [1], p. 619c.

17. A. Bialas and M. Gyulassy, Nucl. Phys. **B291**, 793 (1987).

18. M. Gyulassy and M. Plümer, Nucl. Phys. **B346**, 1 (1990).

19. H. Sorge, H. Stöcker, and W. Greiner, *Ann. Phys. (N.Y.)* **192**, 266 (1989).

20. Y. Pang, T.J. Schalgel, and S.H. Kahana, *Phys. Rev. Lett.* **68**, 2743 (1992).

21. K. Kajantie et al. *Phys. Rev. Lett.* **59**, 2517 (1987); J.P. Blaizot and A.H. Mueller, *Nucl. Phys.* **B289**, 847 (1987); K.J. Eskola et al. *Nucl. Phys.* **B323**, 37 (1989).

22. T.J. Schlagel, S.H. Kahana, and Y. Pang, BNL-47608, to appear *Phys. Rev. Lett.* (1992).

23. H. Sorge et al., *Phys. Rev. Lett.* **68**, 286 (1992); A. Jahns et al., *Phys. Rev. Lett.* **68**, 2895 (1992).

24. M. Kataja and P.V. Ruuskanen, *Phys. Lett.* **B243**, 181 (1990); P. Gerber et al, *Phys. Lett.* **B246**, 513 (1990).

25. S. Gavin and P.V. Ruuskanen, *Phys. Lett.* **B262**, 326 (1991); S. Gavin in [1], p. 459c.

26. H.W. Barz, P. Danielewicz, H. Schulz, and G.M. Welke, MSUCL-805 (1992); G.M. Welke and G. Bertsch, *Phys. Rev.* **C45**, 1403 (1992).

27. M. Sarabura (NA44 Collab.), in [1], p. 125c.

28. T.J. Schlagel, Y. Pang, and S.H. Kahana, in preparation.

29. H. Sorge et al., *Phys. Lett.* **B243**, 7 (1990).

30. G. Baym et al., *Nucl. Phys.* **A407**, 541 (1983); M. Kataja et al. *Z. Phys.* **C55**, 153 (1992).

31. W.A. Zajc (E802 Collab.), in [1], p. 237c; P. Seyboth (NA35 Collab.), in [1], p. 293c.

32. M. Gyulassy and S.S. Padula, in [1], p. 537c; *Phys. Lett.* **B217**, 181 (1989).

33. F. Wilczek, IASSNS-HEP-91/65, (1991); R. Pisarski and F. Wilczeck, *Phys. Rev.* **D29**, 338 (1984).

34. F.R. Brown et al., *Phys. Rev. Lett.* **65**, 2491 (1990).

35. C. Bernard et al., IUHET-222, (1992); N.H. Christ in these proceedings.

36. J. Knoll, *Z. Phys.* **C38**, 187 (1988); H. Sorge et al., *Phys. Lett.* **B289**, 6 (1992).

37. T.A. DeGrand, *Phys. Rev.* **D30**, 2001 (1984); J. Ellis et al., *Phys. Lett.* **B233**, 223 (1989).

38. Y. Takahashi and S. Nagamiya, *Nucl. Phys.* **A525**, 623c (1991); J.D. Bjorken, *Int. J. Mod. Phys.* **A7**, 4189 (1992).

39. S. Gavin, M. Gyulassy, M. Plümer and R. Venugopalan, *Phys. Lett.* **B234**, 175 (1990).

40. S.H. Kahana, Y. Pang, T.J. Schlagel and C.B. Dover, in preparation.

41. S. Gavin *Nucl. Phys.* **A525**, 459c (1991).

THERMALIZATION IN ULTRA-RELATIVISTIC HEAVY ION COLLISIONS AT RHIC AND LHC*

Klaus Geiger
School of Physics and Astronomy
University of Minnesota
116 Church St S.E.
Minneapolis, MN 55455, U.S.A.

Abstract

The dynamics of partons in ultra-relativistic ^{197}Au+^{197}Au collisions in the future collider experiments at RHIC and LHC during the first 3 fm/c are simulated in full six-dimensional phase-space within a parton cascade model to compute the entropy production and the space-time dependent energy densities, temperatures, etc., in the central collsion region. The partons' evolution from pre-equilibrium towards the formation of a thermalized quark-gluon plasma is quantitatively analyzed resulting in predictions for the energy densities $\varepsilon \simeq 15$ - 31 GeV fm^{-3} and associated temperatures $T \simeq 295$ - 345 MeV at $\sqrt{s} = 200$ - 6300 A GeV. The multiplicity of final pions from the plasma is estimated from the amount of entropy produced, yielding a huge $dN^{(\pi)}/dy \simeq 1900$ - 3400. An equation of state is extracted together with initial conditions for the further hydrodynamical space-time evolution of the matter.

INTRODUCTION

The production and observation of quark-gluon plasmas (QGP)[1] in ultra-relativistic heavy ion collisions is one of the most ambiteous goals of the experimental program of the BNL 'Relativistic Heavy Ion Collider' (RHIC) and the CERN 'Lepton Hadron Collider' (LHC). In this context the thermalization properties of the quark-gluon matter formed in the central collision region during the first few fm/c need to be investigated in order to gain an understanding of a QGP formation in these reactions. The theoretical approach that I would like to advocate here, is to follow the system of interacting partons in complete phase-space and time, from the moment of nuclear touch, through a pre-equilibrium phase, towards the establishment of a thermal (and chemical) equilibrium. From the coresponding partons' phase-space distributions one can then extract the degree of equilibration and, if the system is sufficiently thermalized, the initial and boundary conditions as well as the equation of state for a relativistic hydrodynamical description of the subsequent expansion of the QGP and its freeze out.

THE PARTON CASCADE MODEL

The model framework for the analysis is provided by the parton cascade model[2,3] which recently has been supplemented by a suitable

---

*This work was supported by the U.S. Department of Energy under Contract No. DOE/DE-FG02-87ER-40328.

phenomenological hadronization scheme[4]. In the parton cascade model ultra-relativistic nuclear collisions are decribed as the time evolution of the partons' phase-space distributions. The space-time development is formulated within renormalization group improved QCD perturbation theory, embedded in the framework of relativistic transport theory. The dynamics of the dissipative processes during the early stage of the nuclear reactions is simulated as the evolution of multiple internetted parton cascades associated with quark and gluon interactions. At the end of the perturbative QCD phase the hadronization is modelled as a recombination of the final state partons to form color singlet clusters, followed by the fragmentation ('decay') of these clusters into observable hadronic states. It was demonstrated in Ref. 4 that the parton cascade model combined with this cluster hadronization scheme sets a consistent framework to simulate and study the time evolution of hadron-hadron and nucleus-nucleus collisions in complete phase-space, from the first instant of collision to the final particle yield.

RESULTS

To study the space-time evolution of the partons and their thermodynamics in ultra-relativistic heavy ion collisions at the collider energies of RHIC and LHC, I performed a series of simulations of central $Au + Au$ collisions with various beam energies $\sqrt{s} = 200$, 1000, 2000, 4000, 6300 A GeV (A=197), using the parton cascade model. The analysis was focussed on the microscopic dynamics of partons in a central phase-space volume with its spatial part taken to be a cylinder of 2 fm in length and 6.7 fm ($= R_{Au}$) in radius, centered at the nuclear center-of-mass. In the following I will refer to this spatial volume as *central cylinder*.

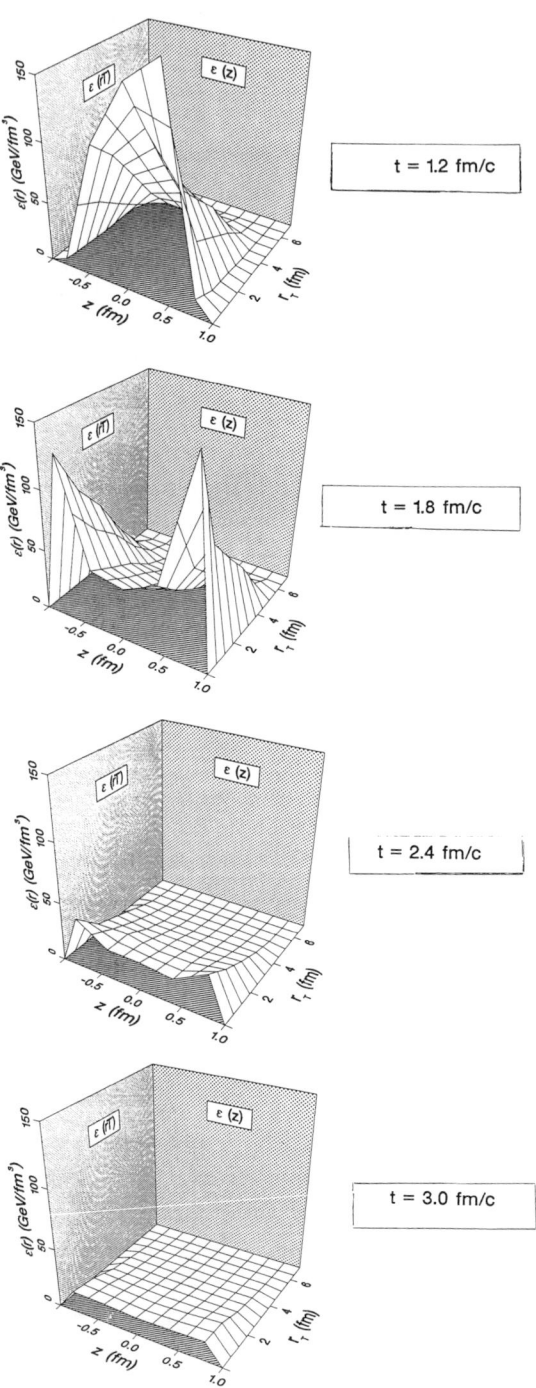

Figure 1. Energy density profiles within the central cylinder at different times during a $Au + Au$ collision with $\sqrt{s} = 200$ A GeV. The first picture is already 0.2 fm/c after the moment of maximum compression.

Energy Densitiy Profiles

In Fig. 1 the time development of the energy density profile in this central cylinder is shown exemplarily for $Au + Au$ at $\sqrt{s} = 200$ A GeV. The maximum energy density in this case is reached at $t \simeq 1$ fm/c around $z = 0$ fm. The first picture is therefore already 0.2 fm/c after the maximum density has been achieved. One observes that the further space-time evolution is at first characterized by a large longitudinal flow of the fast particles which then escape with progressing time from the central cylinder in opposite directions forming two receding fronts. On the other hand, the softer partons expand slowly and eventually establish an isotropic plateau between -1 fm $\leq z \leq$ 1 fm and in transverse direction up to $r_\perp \simeq$ 4 fm. Thus, the partons establish a homogeneous density within a volume of $\simeq 100$ fm^3 at a constant value of $\varepsilon \simeq 15$ GeV fm^{-3} (for $\sqrt{s} = 200$ A GeV).

Time Evolution of Densities

Fig. 2 demonstrates the process of equilibration in terms of the time evolution of the energy, particle and entropy densities, $\varepsilon(t)$, $n(t)$ and $s(t)$, respectively, within a thin disc centered at $z = 0$ fm. This *central slab* was taken to be $-0.2 \leq z \leq +0.2$ fm in longitudinal extent and $0 \leq r_\perp \leq 6.7$ fm in transverse direction, i.e. the middle slice of the aforementioned central cylinder. The densities reach their absolute maximum in the moment of highest compression and decrease fast as the high rapidity particles move apart, so that only the partons with small rapidities remain in the central slab. The final values at $t_f = 3$ fm/c are in the ranges $\varepsilon(t_f) \simeq 15$ - 31 GeV fm^{-3}, $n(t_f) \simeq 18$ - 28 fm^{-3} and $s(t_f) \simeq 68$ - 120 fm^{-3} for $\sqrt{s} = 200$ - 6300 A GeV. Similarly as in the case of the energy density profile in Fig. 1, at $t = t_f$ these values hold not only for the central slab, but are represen-

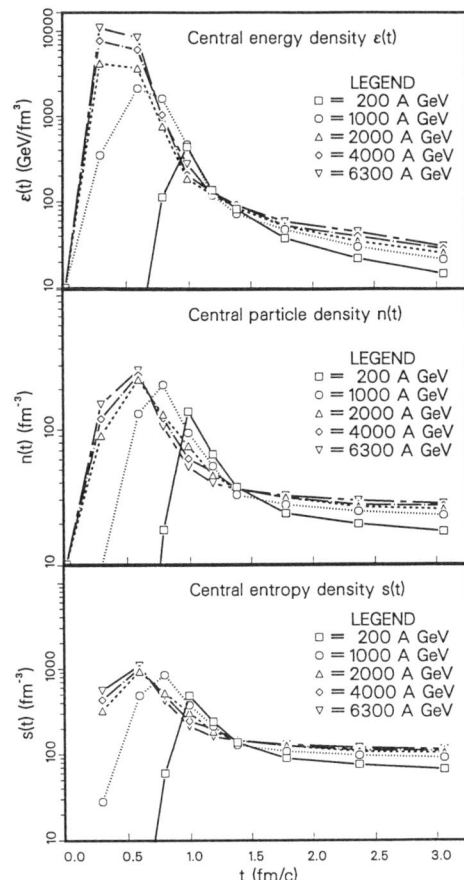

Figure 2. Energy, particle and entropy densities in the central slab (defined in the text) as a function of time in $Au + Au$ collisions at $\sqrt{s} = 200$ - 6300 A GeV.

tative for the densities throughout the central cylinder at least up to and $r_\perp = 4$ fm.

Temperatures in the Central Region

Fig. 3 a) displays the time evolution of the temperatures $T(t) = 4/3\, s(t)/\varepsilon(t)$ associated with the central energy and entropy densities shown in Fig. 2. The Fig. 3 b) illustrates the beam energy dependence of the final temperatures of the parton matter in the central volume at $t_f = 3$ fm/c. They increase $\propto \ln(\sqrt{s})$ from $\simeq 295$ MeV ($\sqrt{s} = 200$ A GeV)

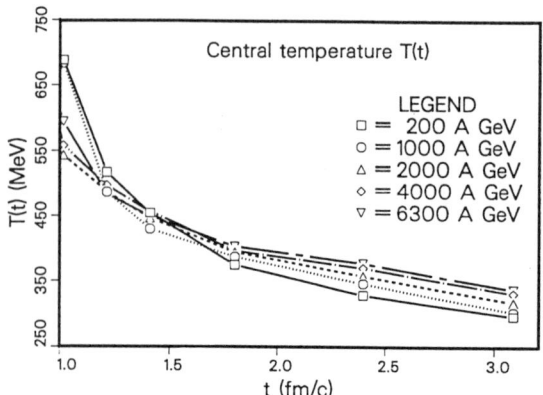

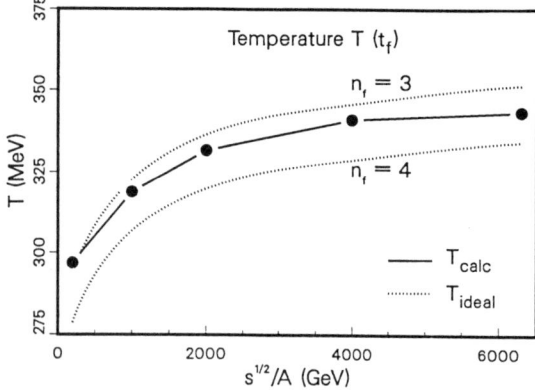

Figure 3. a) Time evolution of central temperatures associated with the densities $e(t)$ and $s(t)$ of Fig. 2. b) Final values of temperatures in the central cylinder at $t_f = 3$ fm/c as a function of beam energy.

to $\simeq 345$ MeV ($\sqrt{s} = 6300$ A GeV). The dotted lines correspond to the temperatures of an ideal gas of gluons and quarks with the same energy and entropy densities and with $n_f = 3$, respectively $n_f = 4$ flavor degrees of freedom. Obviously the distribution of partons calculated with the parton cascade model closely resembles an ideal gas with a slightly increasing effective number of 3 - 4 quark flavors.

Entropy Production

A good indicator for the degree and the time scale of a thermalization is the specific entropy, i.e., the produced entropy per secondary parton, since it necessarily must vanish in space-time when the system reaches an equilibrium state. Fig. 4 a) displays the time development of the specific entropy $(S/N)(t)$ for the various beam energies, wheras Fig. 4 b) shows the corresponding number of partons $N(t)$ present at time t in the central cylinder defined before. The curves show a rapid build-up of S/N and relax approximately exponential to reach their final values between 3.9 and 4.3. Comparing these values with $(S/N)_{ideal} \simeq 4$ for an ideal gas of non-interacting massless quarks and gluons, one sees that the difference between the resulting entropy of the realistic model calculation and the idealized case amounts only to $\simeq \pm 0.2 - 0.3$. Although the model includes massive quarks and accounts for interactions among the partons (which however at 3 fm/c have reduced to relatively infrequent, mostly elastic scatterings), the system of partons looks also from this point of view effectively like an almost ideal gas. The variation with beam energy is due to a weak temperature dependence of the ratio between entropy and the number of partons. Since, as shown in Fig. 3 b), the temperature of the parton matter and the effective flavor degrees of freedom in the central region grow slowly with the beam energy, an increase of the entropy per particle is natural.

Entropy and Pion Multiplicities

From the calculated total produced entropy S on the parton level and the fact that the entropy is conserved, one can estimate the associated multiplicity of pions[5,6] that are produced in at mid-rapidity in central $Au + Au$ collisions[3]. The resulting pion multiplicity per unit rapidity around $y = 0$ $dN^{(\pi)}/dy$ is shown

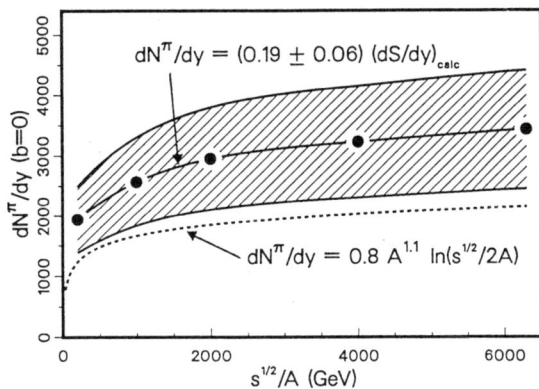

Figure 5. Prediction for the multiplicity of pions produced in the central collision region in central $Au+Au$ collisions as a function of beam energy. The model result is obtained by relating the total produced entropy by the partons to the number of pions per unit rapidity at $y = 0$. The lower curve shows an estimate from an extrapolation of pp and pA data taken from Ref. 7.

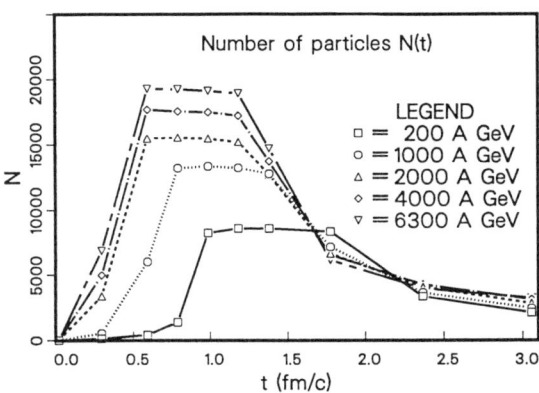

Figure 4. a) Production of entropy per secondary parton as a function of time. b) Corresponding number of partons in the central cylinder that contribute to the total entropy production.

in Fig. 5 as a function of the beam energy. It ranges from $\simeq 1900$ ($\sqrt{s} = 200$ A GeV) to $\simeq 3400$ ($\sqrt{s} = 6300$ A GeV).

Equation of State - Fluid Dynamics

Finally, as explained in Ref. 3, one may extract an effective equation of state of the form $p(T) = a(T)T^4$ that interrelates the pressure p and the temperature T via a function $a(T)$ that contains the space-time history of the parton cascade evolution from the moment of nuclear contact to the point of equilibration at $t = 3$ fm/c. The initial temperature $T_i \equiv T(\tau_i^h, w) = 86.2$ MeV $w^{-0.09} \ln(w)$ and the corresponding $a(T_i) \equiv a(\tau_i^h, w) = 4.95 + 0.3 \left[\ln(w/200)\right]^{0.8}$ can be parametrized in terms of the variable $w = \sqrt{s}/AGeV$ and refer to an energy dependent initial proper time for a hydrodynamical description $\tau_i^h = 2 - 2.6$ fm/c for $\sqrt{s} = 200 - 6300$ A GeV. The associated initial conditions can be similarly expressed through the energy, particle and entropy densities at τ_i^h. The boundary conditions are given by the geometry of the plasma of partons, i.e. by the spatial extension of the central cylinder. These estimates can be used as an input to a hydrodynamical description of the further space-time evolution of the plasma until it freezes out to yield final hadronic states.

SUMMARY

A selection of results of a complete phase-space analysis, obtained on the basis of the parton cascade model was presented to shed some light on the possible space-time development of central Au + Au collisions with $\sqrt{s} = 200$ A GeV - 6300 A GeV during the first 3 fm/c. The thermodynamics of the pre-equilibrium stage and the process of parton thermalization was analyzed in terms of the time evolution of energy, particle and entropy densities of the secondary partons in the central collision region, defined as a cylindrical volume of 2 fm in length and a radius $R_{Au} = 6.7$ fm centered at the nuclear center-of-mass. The results imply that by a time $t_f \simeq 3$ fm/c after the moment of nuclear contact of the collisions very hot and dense plasmas of deconfined quarks and gluons with an effective number of quark flavors $n_f = 3 - 4$ have been formed in the central region. The equilibrated quark-gluon matter extends over a volume of at least 100 fm^3. An extracted equation of state of the form $p(T) = a(T)T^4$ where p and T are the pressure and the temperature of the matter, describes the state of the plasmas in a way that closely resembles an ideal gas of partons. The energy densities $\varepsilon \simeq 15 - 31$ GeV fm^{-3} and temperatures $T \simeq 295 - 345$ MeV of the plasmas at proper times $\tau \simeq 2 - 2.6$ fm/c for $\sqrt{s} = 200 - 6300$ A GeV are very large. Connected with this is an intense entropy production which can be related to the multiplicity of pions produced from the plasmas, yielding $dN^{(\pi)}/dy \simeq 1900 - 3400$ around $y = 0$.

REFERENCES

1. See e.g. *Quark Matter '91, Proceedings of the Ninth International Conference on Ultra-relativistic Nucleus-Nucleus Collisions*, Gatlinburg, TN, U.S.A., 1991 (to appear in *Nucl. Phys.* A).

2. K. Geiger and B. Müller, *Nucl. Phys.* B 369, pp. 600, (1992).

3. K. Geiger, "Thermalization in Ultra-relativistic Nuclear Collisions," Parts 1 and 2, University of Minnesota preprints, 1992 (to appear in *Phys. Rev.* D).

4. K. Geiger, "Particle Production in Ultra-relativistic Nuclear Collisions," University of Minnesota preprint, 1992 (submitted to *Phys. Rev.* D).

5. J. D. Bjorken, *Phys. Rev.* D 27, pp. 143, (1983).

6. R. C. Hwa and K. Kajantie, *Phys. Rev.* D 32, pp. 1109, (1985).

7. H. Satz, "Heavy ion physics at very high energies", talks given at the *CERN Heavy Ion Workshop*, CERN, 1991.

STRANGE ANTI-BARYONS — QGP VERSUS HG

Johann Rafelski*
Department of Physics, University of Arizona
TUCSON, AZ 85718

Jean Letessier and Ahmed Tounsi
Laboratoire de Physique Théorique et Hautes Energies[†]
Université PARIS 7, 2 Place Jussieu, F-75251 CEDEX 05, France

Abstract

We study quark-gluon plasma (QGP) and hadronic gas (HG) models of the central fireball presumed to be the source of abundantly produced strange (anti-)baryons in S → W collisions at 200 GeV A. We consider how multi-strange (anti-)baryon multiplicities depend on strangeness conservation and compare the HG and QGP fireball scenarios. We argue that the total particle multiplicity emerging from the central rapidity region as well as the variation of production rates with changes in the beam energy allows to distinguish between the two reaction scenarios.

INTRODUCTION

Kinetic strange particle production models [1] show that abundant strangeness is suggestive of QGP formation in relativistic nuclear collisions. Even more specific information about the nature of the dense matter can be derived studying strange quark and antiquark clusters, which are more sensitive to the environment from which they emerge [2], here in particular particle density. Therefore the relative production abundances of strange and multi-strange baryons and anti-baryons where studied experimentally [3]. They turned out to be particularly sensitive probes of the thermal conditions of their source [4]. It has already been demonstrated that the observed particle abundances are in agreement with a picture of explosively disintegrating QGP fireball [5].

It can be also argued [6, 7] that these results are compatible with the scenario of an equilibrium HG fireball, though here one lacks an accepted mechanism for strangeness production. But perhaps more importantly, it is hard to imagine a hadronic gas at temperature well above the pion mass, *viz.* $T = 215$ MeV which is the required temperature for the hadronic gas interpretation of the strange antibaryon data to work [7].

The problem we address here is how one can eliminate experimentally the unnatural for the circumstances possibility of a hadronic gas fireball as the underlying strange particle source. We suggested [7] that a simple distinction of these two phases derives from the inherent difference with regard to their entropy content S given a fixed and conserved property, such as baryon number content B which can be determined experimentally. B is seen as being well understood in terms of the nucleon number of the combined system of the projectile nucleus and the target tube of nu-

---

*In part supported by US-DOE grant DE- FG02-92ER40733
[†]Unité associée au CNRS UA 280

clear matter cut out in the collision from the much larger target nucleus. Baryon number of the fireball can decrease only by particle radiation in the final disintegration of the fireball, beyond which we assume that the scattering between the different components have ceased and the relative abundances carry the information about the property of the source.

On the other hand, once the pre-equilibrium reactions have been terminated, and the particle momentum distributions have reached their thermal form, entropy production effectively has ceased, even if a phase transition occurs from a primordial phase to the final HG state. Hence both baryon number and entropy content of the isolated fireball remain constant and their ratio in a theoretical description is rather model independent. Therefore a supplementary measurement, which will permit to define the properties of this source is the multiplicity per participating baryon in the fireball which is directly related to entropy. While the hadronization of the entropy rich QGP fireball is presently not understood, we take advantage here of the fact that in any case a substantially enhanced particle multiplicity must result, as compared to the HG scenario. This can e.g. arise if the QGP fireball were to evaporate emitting hadronic particles sequentially. For $T \sim 215$ MeV [8], HG leads to about 40% of the particle multiplicity expected for QGP scenario [7].

Alternatively, one may be able to distinguish the two cases considering the response of the measured parameters to changes in energy or/and size of the colliding nuclei [9]. We note that the hadronic gas particle abundances are certain functions of the three thermal parameters, T the temperature, μ_B the baryo-chemical potential, and μ_s the strange-chemical potential. The constraint to a fixed strangeness fixes in HG a relation between μ_B and μ_s. Ideally, the number of strange and antistrange quarks are equal, but pre-equilibrium emission can introduce a small asymmetry. The strange-chemical potential μ_s will in case of QGP formation always remain independent of μ_B, and near to the value $\mu_s = 0$, while the saturation of the phase space described by a factor γ (see below) will approach unity for increasing size of the hot fireball. Neither result is generally correct for the case of HG, and values of μ_s and γ have considerable impact on the strange particle abundances.

To be able to forecast both for QGP and HG scenarios the behavior of strange particle abundances as the nuclear collision energy changes we study here how the strange-chemical potential μ_s relates to the baryo-chemical potential μ_B, for several choices of energy density (temperature). We consider three bench mark temperatures: aside of $T \simeq 200$ MeV appropriate for the CERN–SPS energy range we also include in our discussion $T \simeq 150$ appropriate for the lowest conditions with probable QGP formation and $T \simeq 300$ MeV, which we judge appropriate for the BNL–RHIC facility under construction.

Our work [7, 9] is in detail very different from the parallel effort of Cleymans and Satz [6, 10]. We study particle ratios at fixed transverse mass $m_\perp > 1.5$ GeV. We allow for the strangeness phase space to be only partially saturated, and we consider the degree of saturation to be experimentally measurable with the result to be compared to the kinetic theory of strangeness production. We allow the strangeness to be unbalanced (up to 10%). We compare the entropy contents for the different scenarios and confront the findings with the observed particle multiplicity. We consider S–S experimental results not to be in the same class as the S–W results (at 200 GeV A) due to the different stopping, and do not combine the data in our analysis of the experimental results. We do not use kaon data, as kaons, unlike strange anti-baryons can arise from peripheral, spectator related processes and are therefore not necessarily witnesses of the same stages of the collision.

THERMAL FIREBALL MODEL

We assume the formation in the collision of a region of space containing much of the energy and baryon number available, with the hadronic particles sharing the accessible energy — this we call a central (rapidity) fireball. Our discussion is unaffected by the presence of a collective flow, and the thermal parameters of the fireball are deduced from the experimental results. In this analysis we imply that a rapid disintegration of the fireball ensues its initial formation. Therefore it is possible to use a simple average value of the thermal parameters for the entire fireball neglecting the influence of flow on the spectra [11]. Recent studies of the dynamics of QGP to HG transition [12] find that such a scenario is not impossible for the hot and dense fireball we consider here.

For a hadronic fireball created in central S→W collisions one has about 108 baryons in the geometric interaction tube at small impact parameter. In the HG fireball scenario we find that the strange pair abundance is about 0.4 per baryon, given the thermal parameters determined earlier [7], corresponding to about 40 strange particle pairs. In QGP fireball scenario the strangeness pair abundance per baryon at $\gamma = 0.7$ is about one, i.e., there is a strange particle yield enhancement by about factor 2.5 as compared to hadronic gas interactions.

u-d asymmetry

The ratio of the net number of down and up quarks in the fireball

$$R_f = \frac{\langle d - \bar{d} \rangle}{\langle u - \bar{u} \rangle}, \qquad (1)$$

arising in a S→W-tube collisions is $R_f^{S-W} \simeq$ 1.08 and in Pb–Pb collisions it is $R_f^{Pb-Pb} \simeq$ 1.15. Taking this into account we denote: $\mu_q = (\mu_d + \mu_u)/2$ and the asymmetry is related to $\delta\mu = \mu_d - \mu_u$. The value of $\delta\mu$ is at each fixed T given by the value of R_f, in dependence on the assumed structure of the source, i.e., the equation of state. In the region of T, μ_B of interest to us we find that $\delta\mu/\mu_q \sim R_f - 1$; though small, the difference between the chemical potentials of u and d quarks is not negligible.

Partition function

We have distinguished between the u, d quarks in our calculations and had computed the HG partition function distinguishing the flavor content of strange and non-strange hadrons. We have summed explicitly the contributions of hadronic particles with the mesons included up to mass 1690 MeV, nucleons up to 1675 MeV and Δ's up to 1900 MeV. Our procedure will be evident when we discuss the strange particle sector explicitly below. We note that higher hadronic resonances would matter only if their number were divergent as is the case in the Bootstrap approach of Hagedorn [13] and the HG was sufficiently long lived to populate all high mass resonances. Our simple minded approach to describe the HG phase is not sufficiently precise as soon as we ask questions which are dependent either on the ever increasing mass spectrum of particles or on the proper volume occupied by the particles [14]. However, quantities such as condition of zero strangeness, fixed entropy per baryon are independent of the absolute normalization of the volume and of the renormalization introduced by the diverging spectrum and hence can be considered in the approach we take.

In order to simplify the comparison between our work and Ref. [6] that while we always use μ_s the strange (quark) chemical potential, these authors use instead the chemical potential μ_S of the kaons ($q\bar{s}$). The relation is: $\mu_s = \mu_q - \mu_S = \mu_B/3 - \mu_S$. Here we employed the relation between the baryo-chemical potential and quark chemical potential $\mu_B = 3\mu_q$ as follows from the baryon number carried by quarks. In general it is not necessary to introduce different chemical potentials (or fugacities) for the hadronic gas phase, as the chemical potential of each HG species is simply the sum of the potentials of the constituent quarks, viz. for a proton $\mu_p = 2\mu_u + \mu_d$, etc. Frequently, instead of chemical potentials the fugacities $\lambda_i = \exp(\mu_i/T)$ are used.

In the Boltzmann approximation the partition function for the strange particle fraction of the hadronic gas, $\mathcal{Z}_s$ in the notation of Ref. [15] is

$$\ln \mathcal{Z}_s = \frac{VT^3}{2\pi^2} \left\{ (\lambda_s \lambda_q^{-1} + \lambda_s^{-1} \lambda_q) \gamma F_K \right.$$
$$+ (\lambda_s \lambda_q^2 + \lambda_s^{-1} \lambda_q^2) \gamma F_Y$$
$$+ (\lambda_s^2 \lambda_q + \lambda_s^{-2} \lambda_q^{-1}) \gamma^2 F_\Xi$$
$$\left. + (\lambda_s^3 + \lambda_s^{-3}) \gamma^3 F_\Omega \right\} , \quad (2)$$

where the kaon (K), hyperon (Y), cascade (Ξ) and omega (Ω) degrees of freedom in the hadronic gas are included successively. Given the recently measured values of the thermal parameters we are obliged to sum over some more strange hadronic particles than was done in Eq. 4 of Ref. [15]. The phase space factors F_i of the strange particles are:

$$F_K = \sum_j g_{K_j} W(m_{K_j}/T) ,$$
$$F_Y = \sum_j g_{Y_j} W(m_{Y_j}/T) ,$$
$$F_\Xi = \sum_j g_{\Xi_j} W(m_{\Xi_j}/T) ,$$
$$F_\Omega = \sum_j g_{\Omega_j} W(m_{\Omega_j}/T) . \quad (3)$$

We have included kaons up to 1650 MeV, hyperons up to 1750 MeV, cascades up to 1820 MeV and also omegas up to 2250 MeV. We use the notation $W(x) = x^2 K_2(x)$, and K_2 is the modified Bessel function. The factor γ in Eq. 2 allows us to consider the strange particle gas away from absolute chemical equilibrium which corresponds to the value $\gamma = 1$. In general, if there is not sufficient time to make strangeness (but sufficient time to exchange strange quarks between the carriers, which we implicitly assumed above) the partition function applies with $\gamma < 1$. The value of the factor γ is determined by the dynamics of strangeness production. Its measurement is only possible in the comparison of abundances of hadrons comprising different numbers of strange (or antistrange) quarks. The value $\gamma = 0.7 \pm 0.1$ [4] arising from the WA 85 results [3] is suggestive of QGP based strangeness production mechanisms.

Strangeness balance

We consider a HG fireball in which the number of s and $\bar{s}$ quarks is (nearly) equal. The condition that the total strangeness vanishes takes the form

$$0 = \langle s \rangle - \langle \bar{s} \rangle = \lambda_s \frac{\partial}{\partial \lambda_s} \ln \mathcal{Z}_s . \quad (4)$$

This is an implicit equation relating λ_s with λ_q for each given T. A slight generalization is obtained allowing that a small fraction imbalance in strange quark numbers arising from pre-equilibrium emission [9] and effects of up to 10% were considered.

CONSTRAINT BETWEEN μ_s AND μ_B

In order to better understand the numerical results we first study an aspect of the condition (4) analytically. We seek for $\mu_s = 0$, viz. $\lambda_s = 1$, non trivial values of μ_B^0 (different from the trivial solution $\mu_B^0 = 0$) for which strangeness balances out. We find the exact answer to be:

$$\mu_B^0 = 3 \cosh^{-1} \left(\frac{F_K}{2 F_Y} - \gamma \frac{F_\Xi}{F_Y} \right) . \quad (5)$$

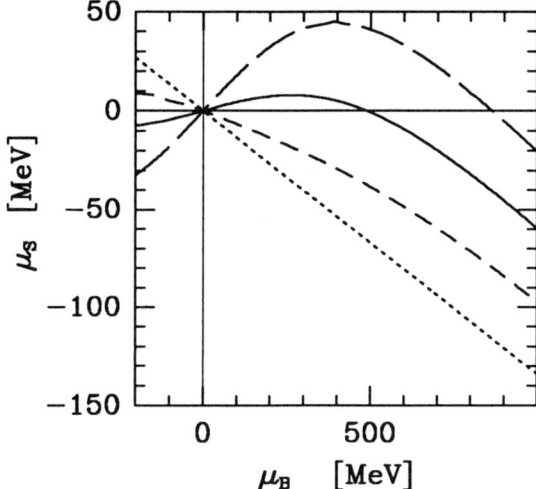

Figure 1. Strange-chemical potential μ_s versus baryochemical potential μ_B for zero strangeness fireball. Long-dashed line corresponds to $T = 150$ MeV, solid line to $T = 200$ MeV, and dashed line to $T = 300$ MeV. The dotted line is the limiting curve for large T.

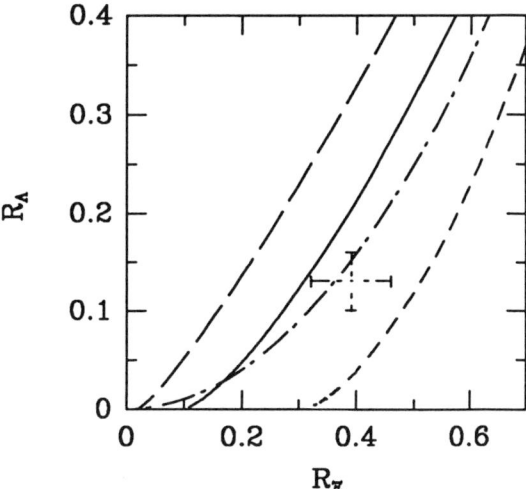

Figure 2. R_Λ versus R_Ξ. Long-dashed line corresponds to $T = 150$ MeV, solid line to $T = 200$ MeV, and dashed line to $T = 300$ MeV in HG. The dashed-dotted line corresponds to QGP.

There is a real solution only when the argument on the right hand side is greater than unity. It turns out that this condition is a sensitive function of the temperature T and of the hadronic resonances included in Eq. 3. For any given spectrum used to compute the phase space factors F_i there is a temperature T_0 beyond which no such solution is possible — this occurs since F_K/F_Y is a monotonically decreasing function of T (see Fig. 1, Ref. [15]).

In consequence of the above observations we expect that there is a domain of temperatures for which even a considerable change of μ_B does not induce a significant change of μ_s. For the complete set of resonances we have included in the phase space factor, this quite peculiar behavior occurs just at the temperature $T \simeq 215$ MeV. In Fig. 1 we present the constraint between μ_s and μ_B at fixed $T = 200$ MeV (solid curves), 150 MeV (long-dashed curves), 300 MeV (short-dashed curves) and 1,000 ($\simeq \infty$) MeV (dotted curves) for $\gamma = 0.7$ (the choice $\gamma = 1$ influences this result in-significantly). The solid lines corresponding to $T = 200$ MeV is indeed leading to quite small values of μ_s for all $\mu_B < 500$ MeV.

Strange baryon ratios

This behavior explains why in the vicinity of $T = 215 \pm 15$ MeV the QGP and HG are leading to the same particle abundances: the constrain to zero strangeness in HG is consistent with μ_s characteristic of QGP phase. We also find as Fig. 1 clearly shows, that this ambiguity does not persist at higher or lower temperature. To make this point more quantitative, we now study strange baryon ratios arising from a thermal fireball *constrained* to vanishing strangeness.

Comparing *spectra* of particles with their antiparticles within overlapping regions of $m_\perp$, the Boltzmann and *all* other statistical factors cancel, and their respective abundances are only functions of fugacities [2]. In Fig. 2 we show for the case of exactly vanishing strangeness the resulting relation of $R_\Xi = \overline{\Xi^-}/\Xi^-$ with $R_\Lambda = \overline{\Lambda}/\Lambda$. In addition to the

HG results for temperatures $T = 200$ MeV (solid line), $T = 150$ MeV (dashed line) and $T = 300$ MeV (dotted line) we show the case $\mu_s = 0$ corresponding to QGP source (dashed-dotted line). The cross corresponds to the result reported by the WA85 experiment [3]. As can be seen, the QGP curve will nearly coincide with the $T = 215$ MeV curve in the HG case, as we noted before [7].

In addition to the baryon ratios one can also consider the ratio of kaons to hyperons, again at fixed $m_\perp$. Because of the experimental procedures used, which rely on the observation of the disintegration of neutral strange particles into two charged products a comparison of the K_s (here s stands for *short*) with the Λ (which includes the Σ^0 abundance) is available. Considering $R_K = K_s/(\Lambda + \Sigma^0)$ as function of R_Λ we find that there is poor sensitivity of the result to the nature of the fireball, R_K is a good measure of the baryo-chemical potential [1] of the source of these particles, which may be in part the fragmentation region of the heavy target nucleus.

Measurement of γ

We have introduced the factor γ which characterizes the approach to saturation density of the strange quark abundance. It can be experimentally measured by determining the product:

$$\frac{\Xi \cdot \overline{\Xi}}{\Lambda \cdot \overline{\Lambda}} = \gamma^2 \ . \qquad (6)$$

It is easy to see that all spectral and chemical factors cancel in this combination of particle abundances, and the only unbalanced factor is the phase space density of strangeness. We refer to reference [9] for further details.

DISCUSSION

In the μ_B–T plane the condition of zero strangeness combined with the condition $\lambda_s = 1$ leads to the curve shown in Fig. 3

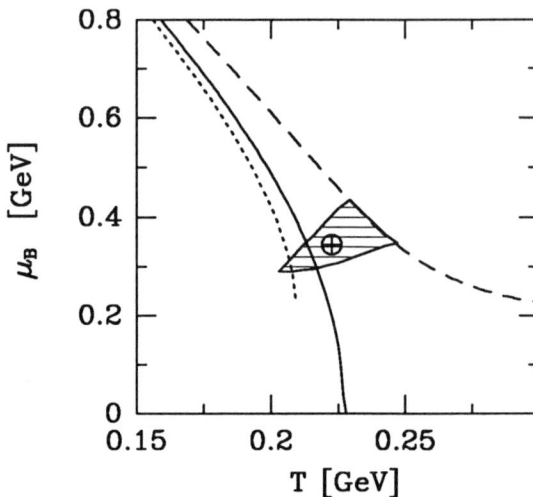

Figure 3. Solid line shows in the μ_B–T plane the condition of zero strangeness in HG fireball assuming the QGP-like condition $\lambda_s = 1$; dashed line $\lambda_s = 0.95$; dotted line $\lambda_s = 1.02$. Hatched is the region compatible with the experimental WA85 data. The $\oplus$ corresponds to $T = 220$ MeV, $\mu_B = 340$ MeV, the central point for QGP fireball.

by the solid line. Dashed line is the case $\lambda_s = 0.95$ and dotted line $\lambda_s = 1.02$. The upper and lower boundary of hatched area arise from $\mu_B/T = 3 \cdot 0.52 \pm 0.01$ and from the constraint obtained from the K^-/Λ ratio reported [3]. We find that if we are willing to accept a hadronic gas [6] at temperature of $T \simeq 200 - 210$ MeV, it could indeed be the source of strange particles — a puzzle in such an interpretation is the condition of $\lambda_s \simeq 1$ which is natural for QGP, and does not have at present any special founding for the HG state.

Particle Multiplicity

The properties of the HG and QGP fireballs are considerably different in particular with regard to the entropy content. Both states are easily distinguishable in the regime of values μ_B, T shown in Fig. 3. We find for the entropy per baryon $\mathcal{S}^{HG}/\mathcal{B} = 21.5 \pm 1.5$. Consequently, the pion multiplicity which can be expected from such a HG fireball is 4 ± 0.5.

This is less than half of the QGP based expectations we found in [7], and clearly the difference is considerable in terms of experimental sensitivity. Checking the theoretical sensitivity we find that the point at which the entropy of HG and QGP coincide *and* strangeness vanishes *and* $\lambda_s \simeq 1$ is at $T \simeq 135$ MeV, $\mu_B \simeq 950$ MeV, quite different from the region of interest here. We note that charged particle multiplicity *above* 600 in the central region has been seen [16] in heavy ion collisions corresponding possibly to a total particle multiplicity of about 1,000, as required in the QGP scenario for the central fireball we described above.

Some of these emulsion particle multiplicity data are shown as function of rapidity in Fig. 4. Here we draw [16] D_Q, the difference in the number of positively and negatively charged particles normalized by their sum. All up to date scanned (15) events of 200 GeV A S-Ag interactions with the "central" trigger being the requirement for the total charged multiplicity > 300. Reaction spectators (target fragments) are not observed in this experiment. In absence of strange particles we find assuming pion symmetry $\pi^+ = \pi^- = \pi^0 = \pi/3$:

$$D_Q \equiv \frac{N^+ - N^-}{N^+ + N^-} \to \frac{\mathcal{B}}{\pi} \frac{1.5}{1 + R_f + 1.5\,\mathcal{B}/\pi} \quad (7)$$

At central rapidity a value of 0.08–0.09 is found. For a hadronic gas with small strange particle component we would have expected based on the entropy argument a value more than twice as large. In a numeric calculation in which we take $\lambda_s = 1 \pm 0.05$ and fix the temperature for each μ_B such that strangeness is conserved, we find:

$$D_Q = \frac{\mu_B}{1.3\text{ GeV}} \quad \text{for } \mu_B < 0.6 \text{ GeV}. \quad (8)$$

This HG result is extremely simple, considering the complexity of the calculation. We thus see that in the HG scenario for the strangeness

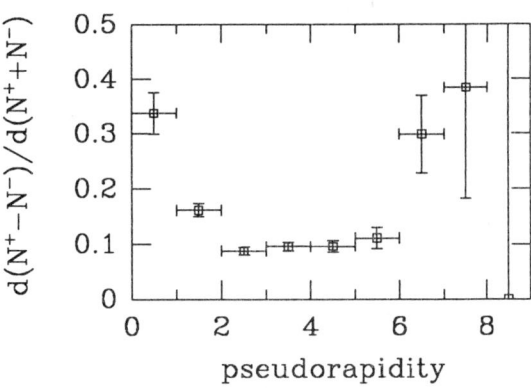

Figure 4. Emulsion data for charged particle multiplicity as function of pseudo-rapidity: difference of positively and negatively charged particles normalized by sum of both polarities. (Courtesy of CERN-EMU 05 collaboration, Y. Takahashi et al. [16]).

source which *has to have* $\mu_B \sim 0.35$ GeV, see Fig. 3, Eq. 8 implies $D_Q \sim 0.27$, which is incompatible with the EMU 05 data [16], as is seen in Fig. 4. Along with this observation that HG is incompatible with he combined EMU 05 and WA85 data goes the fact we discussed here at length that the characteristic property of QGP is the persistence of the fireball parameters $\mu_s = 0$, $\gamma \sim 1$ irrespective of the temperature reached.

Dependence on Temperature

Only for a temperature in the vicinity of 200–215 MeV the HG scenario also leads to the value $\mu_s = 0$, while there is no ready explanation how $\gamma \sim 1$ is reached. We have identified the reason for this coincidence to be the peculiar behavior of μ_s as function of μ_B when the constrain on the total fireball strangeness is imposed. Thus it is quite natural to expect that at the lower energy accessible to CERN–SPS (60 GeV S-beam and about 50 GeV Pb-beam) a lower value of temperature accompanied by higher value of μ_B will be reached. We stress that *if the strange particles emerge from nearly zero strangeness fireball* formed at a given temperature (150, 200, 300 MeV) it is imperative that the observed particle ratios fall on the here presented lines of Fig. 2,

In conclusion we note that the currently available strange particle spectra, though consistent both with QGP and/or HG at $T > 200$ MeV, are strongly favoring the QGP interpretation (because of $\lambda_s = 1$, $\gamma \simeq 1$). This case is considerably strengthened by the consideration of the emulsion multiplicity data. Our discussion therefore strongly suggests that strange anti-baryon abundances be measured along with non-strange particle multiplicities and that it is desirable to widen the study of strange anti-baryon ratios to higher *and* lower nuclear beam energies.

REFERENCES

1. P. Koch, B. Müller and J. Rafelski, *Phys. Rep.* 142, pp. 167–262, (1986).

2. J. Rafelski and M. Danos, *Phys. Lett.* B192, pp 432- 437 (1987); J. Rafelski, *Phys. Rep.* 88, pp. 331–346 (1982).

3. S. Abatzis et al., *Phys. Lett.* B270, pp. 123--127 (1991) and references therein.

4. J. Rafelski, "Strange and hot matter", to appear in *Nucl. Phys.* A, (1992) and references therein.

5. J. Rafelski *Phys. Lett.* B262, pp. 333–340, (1991).

6. J. Cleymans and H. Satz, "Thermal Hadron Production in High Energy heavy Ion Collisions", preprint CERN-TH 6523/92 and BI-TP 92/08.

7. J. Letessier, A. Tounsi and J. Rafelski, "Hot hadronic Matter and Strange Anti-Baryons", Preprint PAR/LPTHE/92-23, and AZPH-TH/92-21 June 1992 *Phys. Lett.* B in press.

8. J. Rafelski, H. Rafelski and M. Danos, "Strange Fireballs", Preprint AZPH-TH/92-7, submitted to *Phys. Lett.* B.

9. J. Letessier, A. Tounsi, U. Heinz and J. Rafelski, "Strangeness Conservation in Hot Fireballs", Preprint PAR/LPTHE/92-27, TPR-92-28 and AZPH-TH/92-23, to be submitted for publication.

10. J. Cleymans "Production of Strange Particles in High Energy Heavy Ion Collisions", preprint BI-TP-92/18, presented at *Quarks-92* Zvenigorod, Russia, May 1992, and references therein.

11. E. Schnedermann, J. Sollfrank and U. Heinz "Fireball Spectra", preprint TPR-92-29, to appear in proceedings of the NATO-ASI *Particle Production in Highly Excited Matter*, Plenum. Pub. Co, H. Gutbrod and J. Rafelski, editors, New York 1993.

12. L. Csernai and J. Kapusta, presentation at the NATO-ASI *Particle Production in Highly Excited Matter*, to appear in proceedings, Plenum. Pub. Co, H. Gutbrod and J. Rafelski, editors, New York 1993.

13. R. Hagedorn, I. Montvay and J. Rafelski, "Thermodynamics of Nuclear Matter from the Statistical Bootstrap Model", Erice proceedings, *Hadronic Matter at Extreme Density*, N. Cabbibo and L. Sertorio, editors, also CERN-TH 2605 (1978).

14. M. Kataja, J. Letessier, P. V. Ruuskanen and A. Tounsi, *Z. Physik* C–Particles and Fields 55, pp. 153–162, (1992).

15. J. Rafelski *Phys. Lett.* B190, pp. 167–172 (1987).

16. Y. Takahashi et al., CERN-EMU 05 collaboration, private communication.

FINAL STATE J/ψ SUPPRESSION IN NUCLEAR COLLISIONS

Sourendu Gupta and Helmut Satz
Theory Division, CERN
CH-1211 Geneva 23, Switzerland

Abstract

The hadroproduction of J/ψ, ψ' and Υ on nuclear targets shows two "initial state" effects: a modification of the structure functions in nuclei leads a reduced production relative to $p-p$ interactions, and parton scattering in nuclear matter broadens the transverse momentum spectrum in comparison to that in $p-p$ collisions. If charmonium or bottonium states are to be used as probes for the nature of dense matter formed in nucleus-nucleus collisions, both these effects must first be taken into account. We do this and discuss the resulting final state features of J/ψ and ψ' production in high energy heavy ion experiments.

A suppression of the J/ψ peak in the dilepton spectrum, relative to the Drell-Yan continuum, was predicted as a signal for quark deconfinement in nuclear collisions [1]. Such a suppression was subsequently observed in O-Cu, O-U and S-U collisions at CERN [2]. However, a similar suppression is also known in hadron-nucleus interactions, which show a reduction of J/ψ, ψ' and Υ production with increasing nuclear mass number A, in contrast to A-independent overall Drell-Yan rates [3][4]. The suppression in hadron-nucleus collisions can hardly be related to quark deconfinement. In fact, it was recently shown to arise as a consequence of modified gluon structure functions in nuclei (gluon shadowing) [5]. Similarly, the transverse momentum spectra for J/ψ and Drell-Yan production in h-A and A-B interactions are broadened by initial state modifications of the parton distributions [6]–[8]. Clearly, all initial state effects must be removed from the data before they can be used to probe quark deconfinement.

The purpose of this paper is two-fold. First, we show that an interesting pattern of J/ψ suppression remains even after the removal of initial-state effects from the nucleus-nucleus collision data: the suppression now appears to end for primordial energy densities ϵ_o below about 1 GeV/fm^3, and it becomes essentially independent of the transverse momentum of the J/ψ. Next we consider the pattern of suppression expected for J/ψ and ψ' under variations of ϵ_o in the absorption and in the deconfinement scenarios; here we draw special attention to features which, we believe, must be further clarified by experiment in order to distinguish between these two possible suppression mechanisms. – We begin by recalling the two independent initial-state effects which must be accounted for, then we apply the corresponding corrections to nucleus-

© 1993 American Institute of Physics

nucleus collision data and discuss their effect on plasma lifetime arguments, and finally we indicate the regions of phase-space where the two mentioned scenarios make distinct predictions.

The inclusive cross section for hadroproduction of charmonium or bottonium is given in the framework of perturbative QCD as a convolution of three factors: the distribution of the partons (quarks or gluons) in the incident hadrons or nuclei, the cross section for the interaction of these partons to produce a $c\bar{c}$ or a $b\bar{b}$ pair, and the binding of these heavy flavour quarks to the appropriate resonance. Denoting these three elements by F, $d\hat{\sigma}$ and H respectively, the hadroproduction cross section can be written as

$$d\sigma(s, M) = \left(\sum_{ij} F_i(x_1, q_1)\, F_j(x_2, q_2)\, d\hat{\sigma}_{ij}\right) H(Q)\, d\Gamma(Q)\,. \quad (1)$$

Here x_1 and x_2 are the fractional logitudinal momenta of the partons from beam and target, respectively, and q_1 and q_2 are the transverse momenta of these partons when they undergo the hard scattering described by the cross section $d\hat{\sigma}$. The CMS energy of the colliding hadrons is $\sqrt{s}$, and $d\Gamma$ is the Lorentz-invariant phase space measure. The $c\bar{c}$ or $b\bar{b}$ pair has four-momentum Q, which we also take to be the four-momentum of the eventual charmonium or bottonium state (of mass M), since the resonance binding is a soft process.

The cross section $d\sigma$ has an expansion in the renormalised coupling α_S. The lowest order term (order α_S^2) in this series can be identified with the cross section for the production of the resonance with $Q_T = q_1 + q_2$ and a given value of $x_F = Q_L/\sqrt{s}$. Since the factorised form (1) demands that the intrinsic transverse momentum distribution of partons is strongly damped, it yields the cross section for low Q_T. Terms of higher order give rise to larger transverse momenta. This perturbation series would have to be resummed to yield the full Q_T spectrum of the resonance. For small Q_T, the resummation would reproduce results of the lowest order term, calculated with the assumption that the partons carry a small intrinsic transverse momentum; for $Q_T \gtrsim M$ one would recover the first perturbative correction. In the intermediate region the resummation would yield a form interpolating between the two.

The inclusive cross section can be computed using parton distributions extracted from other experiments – for example, from deep inelastic scattering or direct photon production. These measured distributions depend only on the longitudinal momenta of incoming partons. To use them, we must define the complete distribution function in the form

$$F_i(x, q) = f_i(x)\, h_i(q), \quad \text{where} \quad \int dq\, h_i(q) = 1. \quad (2)$$

The distribution function $h_i(q)$ describes the transverse momentum spectrum of a parton of species i at the interaction; it contains the intrinsic momentum of the partons in beam or target nucleons, plus whatever modifications these partons have experienced in their traversal of nuclear matter up to the point of hard interaction.

In the rest of this paper we shall restrict ourselves to $Q_T \lesssim M$ and approximate the cross section $d\sigma$ by the order α_S^2 QCD cross section. In this region, the distribution over Q_T can be thought to arise entirely from $h_i(q)$ through the convolution

$$G_{pA}(Q_T) = \int d^2q_1 d^2q_2\, h_p(q_1) h_A(q_2) \delta^2(q_1 + q_2 - Q_T). \quad (3)$$

We can now use the kinematical relations

$$x_1 = \frac{1}{2}\left(\sqrt{4\tau + x_F^2} + x_F\right),$$

$$x_2 = \frac{1}{2}\left(\sqrt{4\tau + x_F^2} - x_F\right),$$

$$x_1 x_2 = \tau = \frac{M^2}{s}. \quad (4)$$

The parton cross section $d\hat{\sigma}$ includes only contributions from gluon fusion and the annihilation of light quark-antiquark pairs; it does not contain any nuclear effects. For x_F close to zero, there is also no nuclear effect on the resonance binding of the $c\bar{c}$ or $b\bar{b}$ system. The parton distribution functions F, however, do depend on the nuclear medium, in the form of gluon shadowing and the EMC effect, as well as through initial state broadening of the parton distributions before the fusion or annihilation process. Our aim is to determine the effect of these initial state phenomena from data for $p-A$ collisions, apply the result to nucleus-nucleus interactions and see what remains of the suppression pattern beyond these initial state features.

We first consider the role of gluon shadowing in $p-A$ collisions. In [5] it was shown that in the kinematic region of interest here, the cross section (1), integrated over Q_T, is strongly dominated by gluon fusion. For the ratio of charmonium or bottonium hadroproduction cross sections on nuclei of mass number A to that on hydrogen, this leads to

$$R_{A/p}(\sqrt{s}, M, x_F) = R_{A/p}(x_2) = \frac{g_A(x_2)}{g_p(x_2)}, \quad (5)$$

where g_A and g_p are the gluon structure functions in the target nucleus and the target proton, respectively. For any given A, the ratio (5), which a priori depends on the three variables $\sqrt{s}$, M and x_F separately, thus becomes a function of the one dimensionless variable x_2 only. It was shown in [5] that for J/ψ, ψ' and Υ production with incident pion and proton beams at $\sqrt{s} \geq 20$ GeV, all data indeed fall on one universal curve in x_2. For $x_2 \lesssim 0.2$, this curve decreases with x_2, as expected if there is shadowing of nuclear gluon densities. The resulting behaviour can be parametrised in the form

$$R_{A/p}(x_2) = 1 + a \ln A \ln x_2, \quad (6)$$

and a fit to the data leads to $a = 0.021 \pm 0.001$. Gluon shadowing thus gives rise to a suppression of charmonium production on nuclear targets, in comparison to the corresponding production on a proton target. Let us see what remains of the experimentally observed suppression pattern in nucleus-nucleus collisions when this is taken into account.

The NA38 experiment at CERN provides the ratio of the J/ψ signal to the Drell-Yan continuum for p-Cu, p-U, O-Cu, O-U and S-U collisions [2][9]. Following the analysis in [5], we calculate the shadowing effects relative to p-p collisions in each of these cases. From

$$R_{(AB)/(pp)} = \frac{g_A(x_1)g_B(x_2)}{g_p(x_1)g_p(x_2)}$$
$$= R_{A/p}(x_1) R_{B/p}(x_2) \quad (7)$$

and Eq.(6), we get for $\sqrt{s} = 20$ GeV the results shown in Table 1. We have assumed $\langle x_F \rangle \simeq 0$, and thus $x_1 = x_2 = M/\sqrt{s}$. At this value of x_1 and x_2, the quark structure function ratios entering the Drell-Yan rates are approximately unity. Hence if we divide the data for the various channels by the corresponding values of Table 1, we obtain the suppression beyond that due to shadowing effects in the initial state. In Fig.1, the resulting final state suppression is shown as a function of the energy density, using the data given in [9]. In particular, we note:

- for the highest energy densities in nuclear collisions, there still is a J/ψ suppression of around 50 % relative to $p-A$ and $p-p$ collisions;

- at the lowest energy densities measured in nuclear collisions, i.e, around 1 GeV/fm^3, there is no suppression relative to $p-A$ and $p-p$ collisions, once shadowing is taken into account.

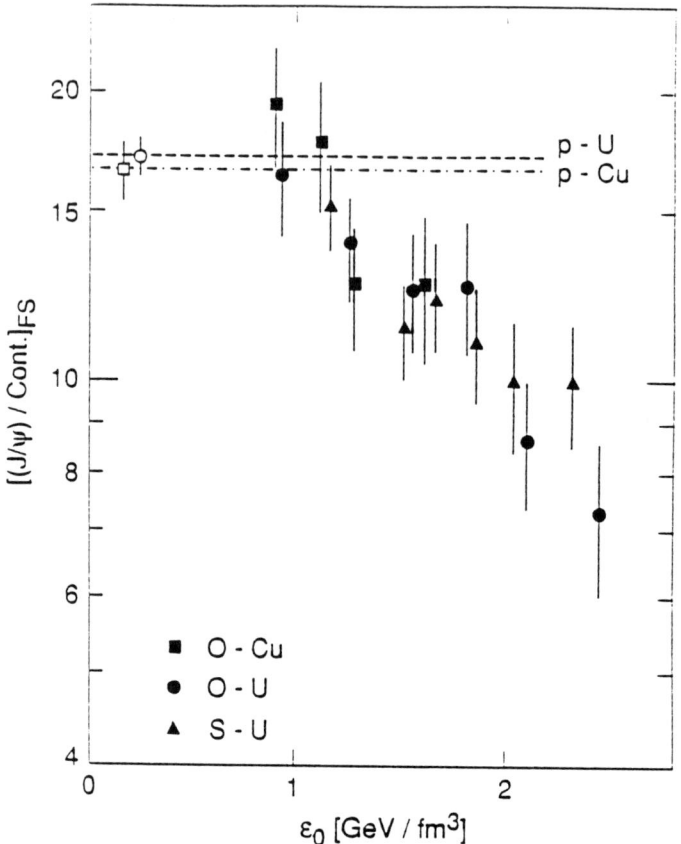

Fig. 1 Final state J/ψ suppression in nucleus-nucleus collisions as a function of the energy density, (a) from [9], (b) from [2].

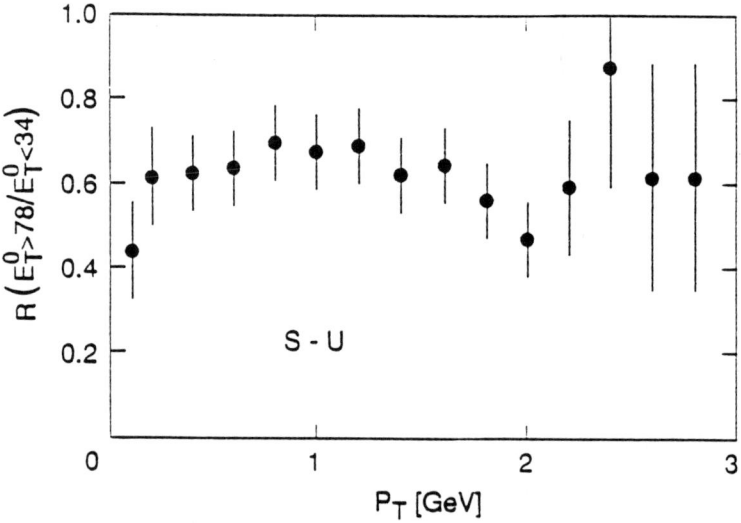

Fig. 2 Final state J/ψ suppression in $S-U$ collisions as a function of the transverse momentum, from [10].

The pattern seen in Fig.1 is thus very suggestive of a threshold for suppression around $\epsilon \simeq 1$ GeV/fm^3, as expected if quark deconfinement is the relevant mechanism.

We now turn to initial state parton scattering and its effect on the transverse momentum behaviour of the suppression [6]–[8]. We can parametrise the partonic transverse momentum distributions $h_i(q)$ of Eq.(2) in the form

$$h(q) = (1/2\pi\langle q^2\rangle)\, e^{-q^2/\langle q^2\rangle}. \qquad (8)$$

The transverse momentum distribution of the charmonium or bottonium states created in p-p or p-A collisions through the fusion of two partons is now easily obtained as

$$G_{pA}(Q_T) = (1/2\pi\langle P_T^2\rangle_{pA})\, e^{-Q_T^2/\langle P_T^2\rangle_{pA}}, \qquad (9)$$

where $\langle P_T^2\rangle_{pA} = \langle q^2\rangle_p + \langle q^2\rangle_A$.

Using these results, we obtain for the differential ratio $R_{A/p}(x_2, Q_T)$ of charmonium or bottonium hadroproduction on nuclei and nucleons the form

$$R_{A/p}(x_2, Q_T) = \frac{g_A(x_2)}{g_p(x_2)}\, H_{A/p}(Q_T), \qquad (10)$$

with

$$H_{A/p}(Q_T) = \{\langle P_T^2\rangle_{pp}/\langle P_T^2\rangle_{pA}\} \\ \exp\{Q_T^2(\langle P_T^2\rangle_{pp}^{-1} - \langle P_T^2\rangle_{pA}^{-1})\}. \qquad (11)$$

Experimentally, it was found that the transverse momentum distribution in $p-A$ collisions is broader than for $p-p$. A measure of this broadening is provided by the quantity

$$\Delta_{pA} = \langle P_T^2\rangle_{pA} - \langle P_T^2\rangle_{pp}. \qquad (12)$$

Data with a 200 GeV proton beam incident on platinum and hydrogen give $\Delta_{pPt} = 0.34\pm0.06$ GeV^2 [3].

This result has been interpreted as multiple initial-state parton scattering [6]–[8]. In this approach it is assumed that in the passage of the target proton through the nucleus, successive interactions broaden the intrinsic transverse momentum distribution of partons. Then the mean squared transverse momentum with which partons enter the hard interaction is proportional to the number of previous interactions and hence to the mean path-length L_A of the projectile in the nucleus. Simple geometric considerations lead to the conclusion that $L_A = (3/4)R_A$, where $R_A = 1.2A^{1/3}$ is the target radius. This gives us

$$\Delta_{pA} \simeq \delta(3/4)R_A, \qquad (13)$$

which becomes exact for large A. Combining data from the NA3 experiment [3] with the more recent NA38 data on Cu and U targets at the same energy, a consistent parametrisation can be obtained with $\delta = 0.065 \pm 0.012$ GeV^2/fm [10].

In nucleus-nucleus collisions, the broadening arises due to the motion of the projectile through the target, as well as the other way around. Hence, for central collisions of two large nuclei A and B

$$\Delta_{AB} = \langle P_T^2\rangle_{AB} - \langle P_T^2\rangle_{pp} \\ \simeq \delta(3/4)(R_A + R_B). \qquad (14)$$

The effective path length L_{AB} depends on the impact parameter b. It has its maximum value, $L_{AB} = (3/4)(R_A + R_B)$, at $b = 0$, and decreases with increasing b, vanishing at $b = R_A + R_B$. This b-dependence translates into a dependence on the transverse hadronic energy E_T [6]–[8][10], since more central collisions lead to higher multiplicities and hence a higher E_T in the final state. The E_T dependence of Δ_{AB} for charmonium spectra from nucleus-nucleus collisions was shown to be fully accounted for by this effect [10]. We thus have to check what final state pattern for the tranverse momentum distribution itself remains when the "beam rotation" due to initial state transverse momentum broadening is included.

From NA38 we have the ratio $R_{H/L}$ of the Q_T distributions at high and at low E_T. To remove the effect of initial state parton scattering, $R_{H/L}$ has to be multiplied by

$$G_L/G_H \equiv \{\langle P_T^2\rangle_H/\langle P_T^2\rangle_L\}$$
$$\exp\{-Q_T^2(\langle P_T^2\rangle_L^{-1} - \langle P_T^2\rangle_H^{-1})\}. \quad (15)$$

The parameters $\langle P_T^2\rangle_H$ and $\langle P_T^2\rangle_L$ entering here can either be determined from geometric considerations relating E_T to impact parameter and path length, or they can be taken from the data. Since the data fit the geometric picture [10], we use them directly. In Fig. 2 we show the result for $S-U$ collisions. As expected, the transverse momentum dependence of the suppression is now essentially removed; we have a constant suppression of around 60 % up to $Q_T \simeq 2-3$ GeV. For transverse momenta beyond this range, neither the factorized form (1) nor the Gaussian distribution in Eq.(8) makes any sense.

Let us summarize: we find that the final state J/ψ suppression in nuclear collisions, obtained after removing the effects of both gluon shadowing and initial state parton scattering, appears to set in rather abruptly at an energy density around 1 GeV/fm^3 and shows little dependence on the transverse momentum of the J/ψ. What consequences does this behaviour have for the attempts to account for the observed suppression in terms of either quark deconfinement by colour screening [1] or absorption in dense hadronic matter [11]–[13]?

Deconfinement sets in at a critical threshold energy density $\epsilon_c \simeq 1$ GeV/fm^3, although the threshold can be softened by quantum mechanical modifications of J/ψ formation [14]. Since sufficiently fast J/ψ's can escape from the deconfining medium in space or in time, deconfinement can also lead to a transverse momentum dependence [15]. The overall suppression observed by NA38 disappears for $Q_T \simeq 2-3$ GeV; if interpreted in terms of an "escape" from the plasma, this leads to extremely short lifetimes for the deconfining medium (less than one fermi between plasma formation and critical point) [16][17]. We have seen, however, that once initial state parton scattering is removed, very little Q_T dependence remains. This in turn removes the argument for very short plasma lifetimes; if there is deconfinement in the primordial final state, the plasma persists for at least some 2-3 fm, in accord with estimates from one-dimensional hydrodynamical expansion. The surviving J/ψ's are in fact not those from $c\bar{c}$ pairs which are fast enough to escape from the medium before they have separated more than the diameter of the physical J/ψ, but rather those which are produced in the cooler edges of the interaction region. In addition, the mixed origin of the observed J/ψ has to be taken into account [18]: about 70% are produced directly as ^1S $c\bar{c}$ states (we denote these states by ψ in what follows), the remainder come from χ_c decay [19]. While the χ_c melts essentially at the deconfinement point ϵ_c, the energy density needed for the melting of the ψ is about twice this value [20]. Hence much of the observed J/ψ suppression would come from a melting of the χ_c in the matter produced in present experiments. Let us see what this means quantitatively.

In the plasma scenario, the survival probability for the J/ψ depends only on the fraction of the interaction region "cool" enough to prevent melting. In a central collision of a small on a large nucleus one finds for nuclei of constant density an energy density profile of the form [21]

$$\epsilon_o(r) \sim \epsilon_o\left(1 - (r/R_B)^2\right)^{2/3}, \quad (16)$$

where R_B is the radius of the (smaller) projectile nucleus and r the radial distance from its centre. It leads to a survival probability

$$S_x(\epsilon_o) = [\Theta(\epsilon_x - \epsilon_o) + \Theta(\epsilon_o - \epsilon_x)(\epsilon_x/\epsilon_o)^{9/4}], \quad (17)$$

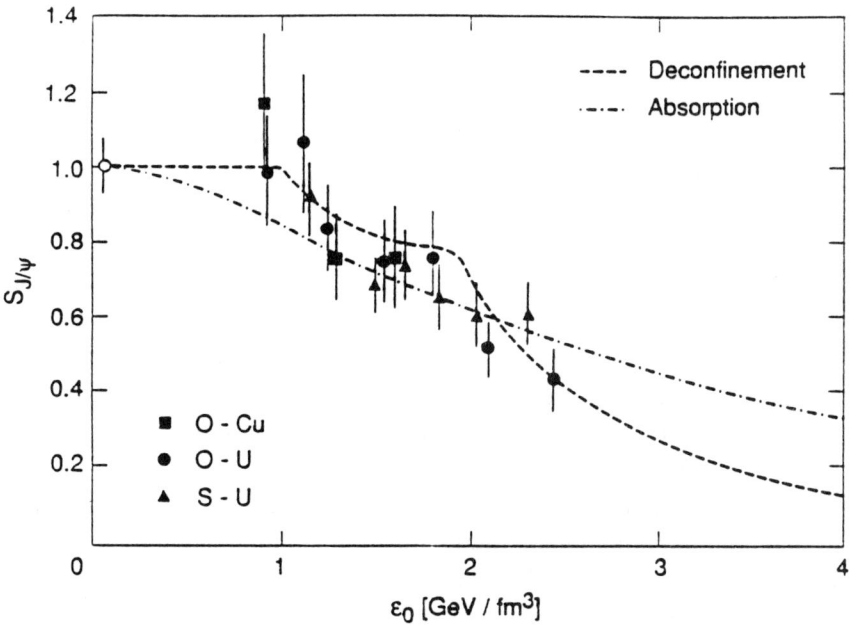

Fig. 3 Final state J/ψ suppression in nucleus-nucleus collisions as a function of the energy density [9], compared to the forms predicted by deconfinement (solid line) and absorption (dashed line). The data are normalised to the $p - U$ value, with errors as indicated on the open circle.

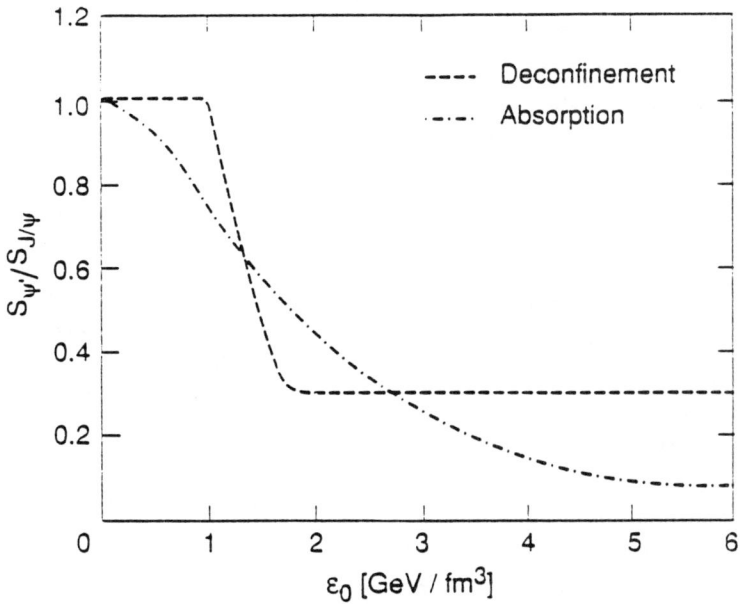

Fig. 4 The ratio of the survival functions $S_{\psi'}/S_{J/\psi}$ as a function of the energy density for deconfinement (dashed line) and absorption (dash-dotted line).

where ϵ_x is the energy density needed to melt state x; with $\epsilon_\chi \simeq \epsilon_c$ and $\epsilon_\psi \simeq 1.9\,\epsilon_c$ [20], we obtain

$$\begin{aligned}S_{J/\psi}(\epsilon_o) &= 0.7[\Theta(1.9\epsilon_c - \epsilon_o) \\ &+ \Theta(\epsilon_o - 1.9\,\epsilon_c)(1.9\,\epsilon_c/\epsilon_o)^{9/4}] \\ &+ 0.3\,[\Theta(\epsilon_c - \epsilon_o) \\ &+ \Theta(\epsilon_o - \epsilon_c)(\epsilon_c/\epsilon_o)^{9/4}]\end{aligned} \quad (18)$$

as the (parameter-free) form of the survival function for the J/ψ. We see in Fig. 3 that it agrees fairly well with the present data. Note that if more than 30% of the observed J/ψ's come from χ_c decay, the survival probability will drop faster. It will, however, retain its bump structure around 2 GeV/fm^3, as long as the measured J/ψ comes from two channels with different dissociation temperatures.

In the range accessible to present experiments, absorption is quite independent of the transverse momentum of the J/ψ, and it can also give a sufficiently strong suppression [11]–[13]. It cannot, however, account for the threshold behaviour which seems to appear when gluon shadowing is removed (see Fig.1). Instead, we obtain from a simple absorption picture an exponentially increasing suppression, with a survival probability of the form

$$S_{J/\psi}(\epsilon_o) \simeq \exp\{-\sigma n(\epsilon_o)\tau_o \ln[n(\epsilon_o)/n_f]\}, \quad (19)$$

where σ is the effective absorption cross section, ϵ_o the initial energy density, n the corresponding density of constituents causing the absorption, n_f the effective freeze-out density, and $\tau_o \simeq 1$ fm the equilibration time. Using geometric estimates for the total cross sections [22], we have $\sigma_t(\psi) \simeq 3$ mb and $\sigma_t(\chi_c) \simeq 7$ mb. Taking into account the χ_c/ψ mixture of the J/ψ, this gives us $\sigma_t(J/\psi) \simeq 4$ mb. To fit the existing J/ψ suppression with the form (19), we assume that absorptive processes provide half of the total cross section, and that the initial density $n(\epsilon_o)$ of the constituents (including resonances) is 2.5 times smaller than that obtained from the observed pions [11]–[13]. The resulting J/ψ suppression is also shown in Fig. 3. Within the experimental errors, it agrees with the data as well as the deconfinement form (18). Clearly better statistics are needed.

Finally we want to point out that a comparison of data for J/ψ and ψ' production can provide a further way to distinguish deconfinement and absorption. If quark deconfinement is the dominant suppression mechanism, then the ratio of ψ' to J/ψ production is

$$S_{\psi'}/S_{J/\psi}(\epsilon_o) = \frac{\Theta(1.9\epsilon_c - \epsilon_o)}{0.3 + 0.7(\epsilon_o/\epsilon_c)^{9/4}} + 0.3\,\Theta(\epsilon_o - 1.9\epsilon_c) \quad (20)$$

where we have used $\epsilon_{\psi'} \simeq \epsilon_c$ [20]. The resulting behaviour is shown in Fig. 4. We note that it is only the mixed χ/ψ origin of the J/ψ which leads to a variation with energy density, and this only for $\epsilon_c \leq \epsilon_o \leq 1.9\epsilon_c$. In the absorption scenario, on the other hand, both the J/ψ signal itself and the ratio of ψ' to J/ψ production should decrease exponentially with increasing energy density,

$$\frac{S_{\psi'}}{S_{J/\psi}}(\epsilon_o) \simeq \exp\{-\Delta\sigma n(\epsilon_o)\tau_o \ln[n(\epsilon_o)/n_f]\}, \quad (21)$$

where $\Delta\sigma = \sigma(\psi') - \sigma(J/\psi)$ denotes the difference between the absorption cross sections of the ψ' and the J/ψ. The corresponding behaviour of the $\psi'/(J/\psi)$ ratio, with the same input parameters as above and $\sigma_t(\psi') \simeq 11$ mb, is included in Fig. 4. We note that absorption and deconfinement lead to quite different behaviour at almost all energy densities. High statistics data thus could do much to clarify the nature of J/ψ and ψ' suppression.

REFERENCES

1. T. Matsui and H. Satz, *Phys. Lett.* B178, 416 (1986).

2. C. Baglin et al., *Phys. Lett.* B220, 471 (1989); *Phys. Lett.* B251, 465 (1990); *Phys. Lett.* B251, 472 (1990); *Phys. Lett.* B255, 459 (1991).

3. J. Badier et al., *Z. Phys.* C20, 101 (1983).

4. D. M. Alde et al., *Phys. Rev. Lett.* 66, 133 (1991) and *Phys. Rev. Lett.* 66, 2285 (1991).

5. S. Gupta and H. Satz, "Gluon Shadowing and Charmonium/Bottonium Production on Nuclear Targets", CERN Preprint CERN-TH.6400/92, February 1992.

6. S. Gavin and M. Gyulassy, *Phys. Lett.* B214, 241 (1988).

7. J.-P. Blaizot and J.-Y. Ollitrault, *Phys. Lett.* B217, 386 (1989).

8. J. Hüfner, Y. Kurihara and H.J. Pirner, *Phys. Lett.* B215, 218 (1988).

9. S. Papillon, Thèse, Université de Paris-Sud, Orsay IPNO-T.91.03, March 1991.

10. C. Baglin et al., *Phys. Lett.* B268, 453 (1991) and *Phys. Lett.* B262, 362 (1991).

11. J. Fťáčnik, P. Lichard and J. Pišút, *Phys. Lett.* B207, 194 (1988);

    J. Fťáčnik et al., *Z. Phys.* C42, 139 (1988).

12. S. Gavin, M. Gyulassy and A. Jackson, *Phys. Lett.* B207, 257 (1988).

13. R. Vogt et al., *Phys. Lett.* B207, 263 (1988).

14. T. Matsui, Ann. Phys.(NY) 196, 182 (1989);

    V. Černý et al., *Z. Phys.* C46, 481 (1990);

    J. Cleymans and R. L. Thews, *Z. Phys.* C45, 391 (1990).

15. F. Karsch and R. Petronzio, *Phys. Lett.* B193, 105 (1987).

16. F. Karsch and R. Petronzio, *Phys. Lett.* B212, 255 (1988).

17. J. Blaizot and J.Y. Ollitrault, *Phys. Lett.* B199, 499 (1987).

18. F. Karsch and R. Petronzio, *Z. Phys.* C37, 627 (1988).

19. Y. Lemoigne et al., *Phys. Lett.* 113B, 509 (1982).

20. F. Karsch, M.T. Mehr and H. Satz, *Z. Phys.* C37, 617 (1988).

21. F. Karsch and H. Satz, *Z. Phys.* C51, 209 (1991).

22. J. Hüfner and B. Povh, *Phys. Rev. Lett.* 58, 1612 (1987).

TABLE 1

Suppression by gluon shadowing in nuclear collisions at $\sqrt{s} = 20$ GeV, for $\langle x_f \rangle \simeq 0$, from [5].

Reaction	$R_{(A-B)/(p-U)}$
p – Cu	0.837
p – U	0.786
O – Cu	0.747
O – U	0.701
S – U	0.679

STRANGENESS SIGNALS IN HEAVY ION COLLISIONS*

Louis P. Remsberg
Chemistry Department
Brookhaven National Laboratory
Upton, NY 11973

Abstract

The experimental data on strange meson and strange baryon production in relativistic heavy ion collisions are reviewed.

INTRODUCTION

Enhanced production of strange particles was suggested early on as a possible signature for the formation of the quark-gluon plasma.[1] This follows from the observation that, while the thresholds for ΛK and KK production are 700 and 1000 MeV, respectively, that for $s\bar{s}$ production is only about 300 MeV. Quark pairs are expected to copiously produced in the plasma, and the high temperature expected in this deconfined state should thus result in a relative enhancement of strange quark pairs in the plasma. This in turn might be expected to yield an enhancement of strange particle production relative to that from ordinary hadronic interactions. However, the fate of these strange quarks as the plasma rehadronizes is not easy to predict, and a calculation which takes into account multiple hadronic interactions during heavy-ion collisions, but include no plasma formation, has predicted modest strangeness enhancements.[2] Thus strange particle production cannot be regarded as an unambiguous signal for quark-gluon plasma formation. Yields of strange anti-baryons remain the most promising an indicator of plasma production.[3] Nevertheless, in the absence of deconfinement, the cross sections for the various strange particles are good diagnostics for the details of the heavy-ion collisions.

The data come from the heavy-ion programs at the BNL AGS, with 14.6 A·GeV/c p, O, and Si beams and (since 1992) a 11.7 A·GeV/c Au beam, and at the CERN SPS with 200 A·GeV/c p, O, and S beams. Most of the data initially are in the form of p_t or m_t spectra. Although much useful information is contained in the slope parameters, which range from ~150 to ~225 MeV, the systematic errors in the determination of these slopes are often large and poorly known, and will not be discussed here.

AGS DATA

Two experiments at the AGS, E802 and E810, have reported data on strange particles. E802 has employed a magnetic spectrometer with excellent particle identification to measure charged-particle momentum spectra. One of it's first findings was a substantial enhancement in the yields of charged kaons.[4] This is best illustrated in Figure 1 which compares charged-particle yields from p+Be, p+Au,[5] and central Si+Au reactions.[6] The latter are divided by 28 to give the yields per interacting projectile nucleon. One sees that the dn/dy values for pions in the vicinity of mid-rapidity are approximately constant, whereas those for K^+ increase from p+Be to p+Au to central Si+Au. The behavior of the values for K^- are similar, but with less statistical significance. The K^+/π^+ ratios for $1.2 \leq y \leq 1.4$ in p+Be, p+Au, and central Si+Au are $(7.8\pm0.4)\%$, $(12.5\pm0.6)\%$, and $(18.2\pm0.9)\%$; the errors are statistical only.

---

*This work has been supported by the U.S. Department of Energy under contract DE-AC02-76CH00016.

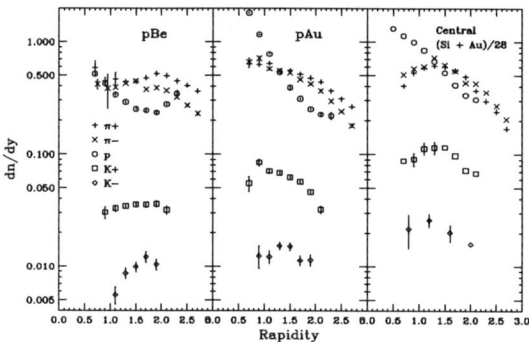

Figure 1. Rapidity distributions dn/dy for pions, kaons, and protons in p+Be, p+Au, and central Si+Au collisions at 14.6 A·GeV/c from E802. The Si+Au data are plotted as (dn/dy)/28.

The p+Be reaction effectively mimics pp (or, more precisely, pd), while p+Au demonstrates the effects of secondary interactions in the Au target. The enhancement in the K^+/π^+ ratio for the heavy-ion projectile is demonstrated in the increase from p+Au to central Si+Au.

The earlier E802 results suggested that the dn/dy distribution for K^+ was shifted to smaller rapidities compared with those for pions. The latest E802 data,[6] with better statistics and increased y and p_t coverage, are included in Figure 1. Although these data are statistically consistent with the earlier data, it is now clear that the K^+ distribution peaks at the same rapidity as do the pions, and has the same shape. The K^- distribution is less well known, but the shape is consistent with that of the pions.

A study of the K^+/π^+ ratio as a function of both collision centrality and target was recently carried out by E802.[7] The collision geometry can be well defined with a zero-degree calorimeter which effectively functions as a beam stop. It measures the total kinetic energy of the projectile spectators after an interaction, the number of which is obtained by dividing by

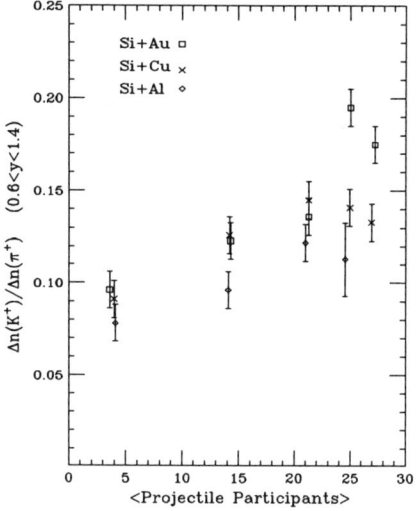

Figure 2. K/π ratios plotted for each target as function of number of projectile participants from E802.

13.5 GeV. The number of projectile participants, a convenient measure of collision centrality, is obtained by subtracting the number of spectators from 28. The K^+/π^+ ratio as a function of the number of participants for Si + Al, Cu, and Au targets is shown in Figure 2. Several points should be noted: a) the values of the ratio for Al are slightly lower than those for Cu and Au, reflecting the incomplete "stopping" in the Al target; b) the values for non-central collisions with Cu and Au are essentially identical; and c) although there is a general increase with increasing centrality, the ratio shows a jump for central Si+Au collisions. The increase in the K^+/π^+ ratio with both increasing target size and centrality, suggests the importance of reinteractions, and that they increase the production of kaons more than that of pions.

The RQMD model[2] explicitly takes reinteractions into account, and a comparison of K/π ratios calculated by RQMD with the E802 experimental values is shown in Figure 3. It appears that the calculation slightly underesti-

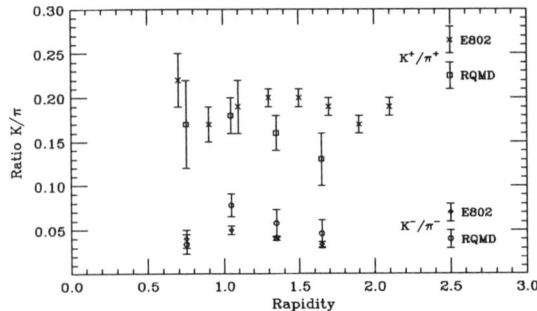

Figure 3. Comparisons of experimental (E802) and calculated (RQMD) K/π ratios for central Si+Au.

mates the K^+/π^+ ratio and slightly overestimates the K^-/π^- ratio, but RQMD actually does a remarkably good job of reproducing these data.

The data from E810 complements that from E802. A TPC (without dE/dx) is used which thus identifies K^0's and Λ's from secondary vertices. Their recent results for Si+Si are shown in Figure 4.[8] These data provide complete coverage in y because the symmetric systems can be reflected about y_{cm}. The reflected data for K_0's are in very good agreement with the E802 data for $(K^+ + K^-)/2$ from Si+Al. Also shown is comparable data from pp interactions multiplied by 28,[9] which indicates an enhancement of a factor of about two for the Si reactions. The Λ rapidity distribution from the Si+Si reaction is strongly peaked at mid-rapidity, in contrast to that from pp collisions, confirming the idea that the enhanced Λ production comes from multiple interactions.

The E810 data for Si+Pb are shown in Figure 5. The K_0 data for Si+Pb is also compared with the $(K^+ + K^-)/2$ data from E802 for Si+Au. The two data sets are consistent although the comparison is not rigorous because of the isospin factor. It is interesting to note that where there is overlap between the E802 data for protons and the E810 data for Λ's ($1.5 \leq y \leq 2.1$) the Λ/p ratio is about 1/4. Comparisons of the E810 data with results from the

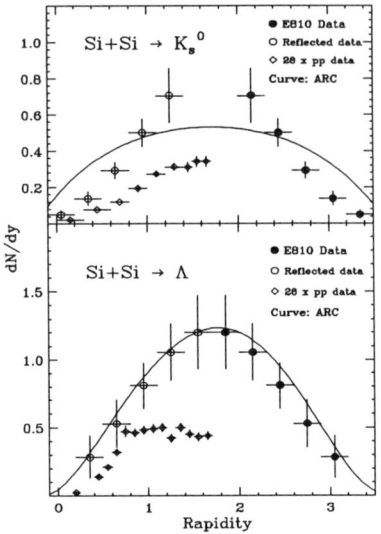

Figure 4. Rapidity distributions for K^0's and Λ's in central Si+Si collisions at 14.6 A·GeV/c from E810.

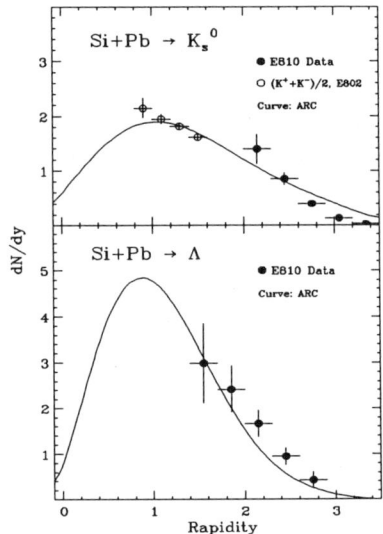

Figure 5. Rapidity distributions for K^0's and Λ's from central Si+Pb collisions at 14.6 A·GeV/c from E810.

ARC code[10] are also included in figures 4 and 5. The agreement is seen to be excellent for the Λ's and fair for the K_0's.

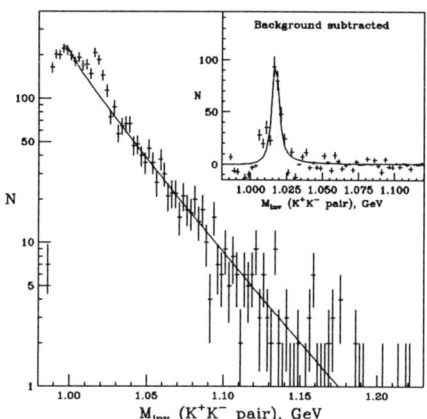

Figure 6. Invariant mass distribution of K^+K^- pairs from central Si+Au collisions at 14.6 A·GeV/c from E859. The inset shows the ϕ-peak with the background subtracted.

E859, the first follow-on experiment to E802 with the addition of a second-level trigger to the E802 spectrometer to select rare particles, is obtaining extensive data on kaon spectra. These are still being analyzed, but one interesting result, shown in Figure 6, is the observation of a substantial ϕ-meson peak in a sample of K^+K^- pairs. The ϕ, which consists of a $s\bar{s}$ pair, has been suggested as an interesting strangeness signal.[11] The mass and width obtained are consistent with the known values and the spectrometer resolution, but a detailed acceptance calculation has yet to be completed so no statement about the yield can be made.

The first results on kaon production with the Au beam were obtained by E866, a follow-on experiment to E802/E859. Because of the high multiplicities resulting from Au-Au collisions, data are so far only available from back angles. The K^+/π^+ ratio resulting from central Au+Au collisions in the rapidity interval $0.55 \leq y \leq 0.85$ is $(25\pm2)\%$, which is substantially higher than that from central Si+Au collisions. A summary of all the E802/E866 K^+/π^+ ratios in this rapidity interval for central collisions of the various projectiles with Au targets is shown in Figure 7. The K^-/π^- ratio for central Au+Au does not

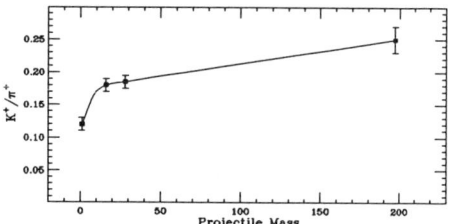

Figure 7. K^+/π^+ ratios in central collisions p+Au, O+Au, Si+Au, and Au+Au in the rapidity interval $0.55 \leq y \leq 0.85$ from E802/E866.

show a similar increase over that from Si+Au. However, this may be a consequence of the relatively higher threshold for K^- production and the lower beam energy (per nucleon.)

SPS DATA

The SPS data primarily come from three experiments, NA35, NA36, and WA85, all of which detect and reconstruct V^0's. NA35 used a streamer chamber and has reported yields of K^0's, Λ's and anti-Λ's from p+S and S+S collisions at 200 A·GeV/c.[12] They observed an enhancement of about two in the multiplicities of all three strange particles relative to those of negative hadrons for central S+S collisions, compared to p+S and peripheral S+S colli-

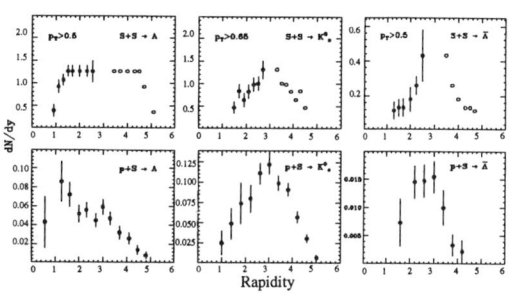

Figure 8. Rapidity distributions of Λ, K^0, and anti-Λ in central S+S and in p+S collisions from NA35. The open circles in the upper row are reflections of the measured data about mid-rapidity.

sions. The rapidity distributions for central S+S (for particles above the indicated p_T cutoff) and p+S collisions are shown in Figure 8. The distributions for K^0's and anti-Λ's are peaked at mid-rapidity, but those for Λ's are different. The Λ-distribution for p+S collisions shows a peak in the target fragmentation region with a gradual decrease going forward, more or less as expected, but that for central S+S collisions exhibits a broad plateau. This presumably indicates a gradual shift in the predominant Λ production mechanism with rapidity.

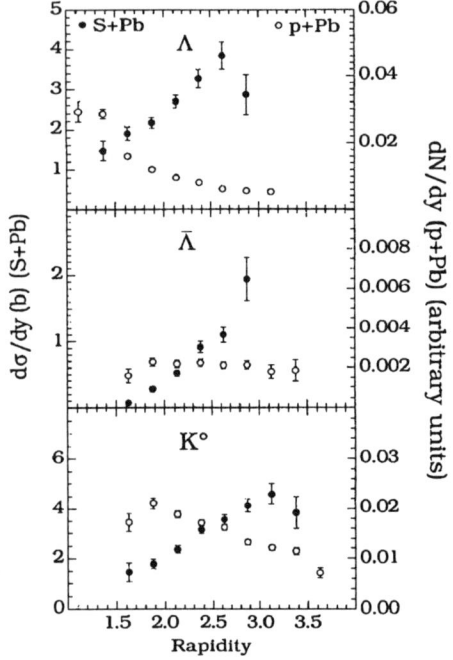

Figure 9. Rapidity distributions of Λ, anti-Λ, and K⁰ in S+Pb and p+Pb collisions at 200 A·GeV/c from NA36.

NA36 employed a TPC to detect V⁰'s.[13] Their recent data on the production of K⁰'s, Λ's, and anti-Λ's from p+Pb and S+Pb collisions at 200 A·GeV/c are presented in Figure 9. These rapidity distributions are qualitatively similar to those of NA35 for the S target with the notable exception of the Λ's resulting from S+Pb collisions. The Λ distribution exhibits a similar increase towards mid-rapidity as do those of the K⁰'s and anti-Λ's. This again implies a significant change in the production mechanism with increasing rapidity. The larger effect with the Pb target is probably due to the increased stopping and resulting higher energy-density than is the case for the S target. These data for S+Pb also show an increase, by factors of 2-3, in the ratios of the multiplicities of the strange particles to those of negative hadrons for non-peripheral collisions relative to the ratios from peripheral collisions and p+Pb.

The WA85 experiment uses the CERN Omega Spectrometer with MWPC's to reconstruct V⁰'s in the limited kinematic range $2.3 < y < 3.0$ and $p_T > 1.0$ GeV/c.[14] However, they are able to reconstruct Ξ^-'s and anti-Ξ^-'s, and thus their Λ and anti-Λ yields can be corrected for contributions from Ξ decay. (All of the measurements discussed here do not distinguish Σ^0 from Λ.) The WA85 results show smaller enhancements in the ratios of K⁰, Λ, and anti-Λ multiplicities to negative-hadron multiplicities from S+Pb collisions relative to those from p+Pb collisions. This may be a consequence of the limited kinematic coverage of this experiment. The most interesting results from WA85 concern Ξ production. The relative yields of the hyperons are given in Table 1.

Table 1. Relative hyperon production ratios for S+W data.

Ratio	$m_T > 1.72$ GeV	$1 < p^T < 2$ GeV/c
anti-Λ/Λ	0.13±0.03	0.13±0.03
anti-Ξ/Ξ	0.39±0.07	0.30±0.07
Ξ/Λ	0.20±0.04	0.11±0.02
anti-Ξ/anti-Λ	0.6 ±0.2	0.33±0.11

The anti-Ξ/anti-Λ ratio is substantially larger for S+W systems than for any other reaction from which it has been measured. The significance of these results are discussed at length by Rafelski.[6]

The two di-muon experiments at CERN, NA38[15] and HELIOS,[16] have reported an enhancement in the $\phi/(\rho+\omega)$ ratio for central

S+W collisions at 200 A·GeV/c relative to p+W or non-central S+W. The available data are still preliminary, but the enhancements reported are consistent with those obtained by the other CERN experiments for neutral strange particles.

CONCLUSIONS

All of the relevant heavy-ion experiments at the AGS and the SPS have reported enhancement factors of ~2-3 in the production of strange particles in central nucleus-nucleus collisions relative to that in peripheral nucleus-nucleus or proton-nucleus collisions. Although these results are consistent with a transient phase transition to a quark-gluon plasma, they can also be explained by multiple hadronic reactions. Thus this strangeness enhancement cannot by itself be taken as an indicator of formation of the quark-gluon plasma. However, the excessive production of doubly-strange antibaryons remains a promising candidate.

REFERENCES

1. J. Rafelski and B. Müller, "Strangeness Production in the Quark-Gluon Plasma," *Phys. Rev. Lett.* 48, pp. 1066-1069, (1982); and P. Koch, B. Müller, and J. Rafelski, "Strangeness in Relativistic Heavy Ion Collisions," *Phys. Rep.* 142, pp. 167-262, (1986).

2. R. Mattiello, H. Sorge, H. Stöcker, and W. Greiner, "K/π Ratios in Relativistic Nuclear Collisions: A Signature for the Quark-Gluon Plasma?," *Phys. Rev. Lett.* 63, pp. 1459-1462 (1989).

3. J. Rafelski, "Strange and Hot Matter," *Nucl. Phys. A* 544, pp. 279c-292c, (1992); J. Rafelski, "Strange Antibaryons: QGP Versus Hadron Gas," *these proceedings*.

4. T. Abbott, *et al.*, "Preliminary spectrometer results 4.from E802," *Z. Phys. C* 38, pp. 135-139, (1988).

5. T. Abbott *et al.*, "Comparison of p + A and Si + Au Collisions at 14.6 GeV/c," *Phys. Rev. Lett.* 66, pp. 1567-1570, (1991).

6. M. Gonin, "Hadron Spectra from Si+A Collisions at 14.6 A·GeV/c," presented at the International Nuclear Physics Conference, Wiesbaden, Germany, July 26 - August 1, 1992

7. T. Abbott, *et al.*, "Centrality Dependence of K^+ and π^+ multiplicities from Si+A collisions at 14.6 A·GeV/c," *Phys. Lett. B*, in press (1992).

8. S.E. Eiseman, *et al.*, "Rapidity Distributions of K^0's and Λ's Produced by 14.6 A·GeV/c Si beams on Si and Pb Targets," *Phys. Lett. B*, in press, (1992).

9. V. Blobel, *et al.*, "Multiplicities, topological cross sections, and single particle inclusive distributions from pp interactions at 12 and 24 GeV/c," *Nucl. Phys. B* 69, p. 454, (1974).

10. Y. Pang, T.J. Schlagel, and S.H. Kahana, "Cascade for Relativistic Nuclear Collisions," *Phys. Rev. Lett.*, 68, pp. 2743-2746, (1992).

11. A. Shor, "ϕ-Meson Production as a Probe of the Quark-Gluon Plasma," *Phys. Rev. Lett.* 54, pp. 1122-1125, (1985).

12. J. Bartke, *et al.*, "Neutral strange particle production in sulphur-sulphur and proton-sulphur collisions at 200 GeV/nucleon," *Z . Phys. C*, pp. 191-200, (1990).

13. E. Anderson, *et al.*, "Results from CERN experiment NA36 of strangeness production," *Nucl. Phys. A* 544, pp. 309c-320c, (19920, and D. Greiner, private communication.

14. S. Abatzis, *et al.*, Strange particle production in sulphur-tungsten interactions at 200 GeV/c per nucleon," *Nucl. Phys. A* 544, pp. 321c-334c, (1992), and F. Navach, private communication.

15. M.C. Abreu, *et al*, "Muon pair production in heavy ion interactions at 200 GeV per nucleon," *Nucl. Phys. A* 544, pp. 209c-222c, (1992).

16. U. Goerlach, "Results of the HELIOS collaboration on low mass dilepton and soft photon production in p-Be, p-W and S-W collisions," *Nucl. Phys. A* 544, pp. 109c-124c, (1992), and U. Goerlach, "New results in di-muon and vector resonanceproduction at 200 GeV/c," *these proceedings*.

ANTIBARYON PRODUCTION IN RELATIVISTIC HEAVY ION COLLISIONS

B. Shiva Kumar
A.W. Wright Nuclear Structure Laboratory,
Yale University, New Haven, CT 06511, U.S.A.

Abstract

Relativistic heavy ion collisions have been used successfully to create states of hot and dense nuclear matter, and maybe even quark matter. Theoretical predictions have suggested that in the event of quark matter formation, the production of anti-baryons in nucleus-nucleus collisions is expected to be enhanced. I will describe measurements of antiproton and antideuteron distributions at AGS energies.

INTRODUCTION

The enhanced production of antibaryons has been predicted to be a signature for quark matter formation and chiral symmetry restoration[1-4]. Several experiments have studied the distributions of antibaryons (strange[5-7] and non-strange[8-10]) as functions of rapidity, y, transverse momentum, p_t, and centrality. The observed abundances of the strange antibaryons are in excess of what is reasonably expected to be produced from a hadron gas. Strangeness enhancement has been discussed in detail by two other papers in this volume[11,12]. I shall therefore devote this paper to a discussion of antiproton ($\bar{p}$) distributions, and further restrict myself to a discussion of data measured at AGS energies ($\approx$15 GeV per nucleon).

In the *stopping* regime, heavy ion interactions result in the formation of a baryon rich collision volume. It has been suggested[13] that antiprotons could be used to probe this volume since they interact very strongly in a baryon rich environment. Also, since antiprotons are produced within this volume, their distributions are expected to be a probe of the collision dynamics, especially at energies close to the $\bar{p}$ production threshold. I shall begin by describing the simpler p-p and p-A collisions. We can then see how the information gleaned from such studies may be used to understand the more complex nucleus-nucleus collisions.

P-P AND P-A MEASUREMENTS

There have been several measurements of $\bar{p}$ production in proton-proton, and proton-nucleus collisions[14-20]. The general features of the data (see table 1) are the following. The $\bar{p}$ rapidity distributions are peaked near y_{NN}, the nucleon-nucleon center of mass rapidity. The yield per event of antiprotons is relatively independent of target. At projectile rapidities, with increasing target mass, the antiproton yield per event decreases, and at lower rapidities, there is a corresponding increase seen. The data imply a small shift to lower rapidities of the antiproton distributions with increasing target mass. The integrated $\bar{p}$ yield per event is relatively independent of target mass. The $\bar{p}$ transverse momentum (p_t) distributions can be described by exponentials in transverse mass ($m_t = \sqrt{m^2 + p_t^2}$) using the functional form

$$E\frac{d^3N}{dp^3} = C \cdot exp[-(m_t - m)/B],$$

Reference	Targets and beam momentum	Kinematics
Allaby et al.	H, Be, B4C, Al, Cu, and Pb, 19.2 GeV/c	Secondary Momenta 4.5 to 16 GeV/c. 12.5 to 70.0 milliradians
Dekkers et al.	H, Be, and Pb targets 19 and 24 GeV/c	Secondary momenta of 1 to 12 GeV/c 0 and 0.1 milliradians
Eichten et al.	H, Be, B4C, Al, Cu, and Pb 24 GeV/c	Secondary momenta of 4.5 to 16 GeV/c 17 to 127 milliradians
Vaisenberg et al	Be, Al, Cu, and Au 10 GeV/c	Secondary momenta 0.57 to 1.85 GeV/c 188 milliradians
Abbott et al	Be, Al, Cu, and Au 14.6 GeV/c	Secondary momenta 0.4 to 3.7 GeV/c 0.09 and 0.24 milliradians

Table 1: A summary of antiproton measurements in p-A collisions at energies of ≈ 15 GeV.

where the inverse slope parameter B is ≈140 MeV at beam energies of 24 GeV, and 93±18 MeV at 14.6 GeV. The data are consistent with $\overline{p}$ production occurring predominantly in first collisions (collisions at the maximum possible $\sqrt{s}$) between projectile and target nucleons. The extent of the shift towards target rapidity of the $\overline{p}$ distributions is a measure of the number of the antiprotons being produced in second and subsequent collisions. This can be estimated to be of the order of 20%. Strong annihilation effects should have manifest themselves in the $\overline{p}$ rapidity distributions moving away from target rapidity. Since the data seem to indicate the opposite effect, this argues in favor of minimal annihilation of the produced antiprotons.

Reference	Targets and beam momentum	Antiproton kinematics
Abbott et al. (E802)	Al and Au 14.6 GeV	Secondary momenta 0.4 to 3.7 GeV/c 0.09 and 0.24 milliradians
Aoki et al. (E858)	Al, Cu, and Au 14.6 GeV	Secondary momenta of 1 to 12 GeV/c 0.0 to 7.0 milliradians
J. Barrette et al. (E814)	Al, Cu, and Pb 14.6 GeV	Secondary momenta of 1 to 5 GeV/c 0.0 to 17.0 milliradians

Table 2: A summary of antiproton measurements in Si-A collisions at energies of ≈ 15 GeV per nucleon.

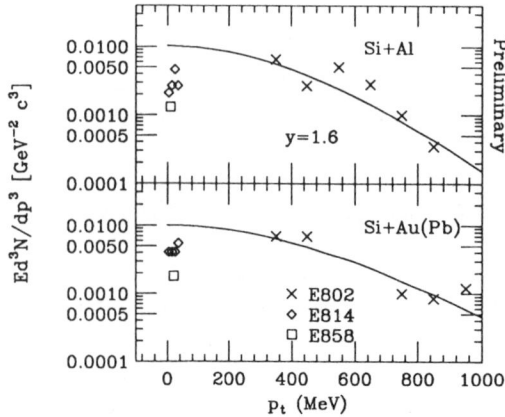

Figure 1: The antiproton invariant multiplicity measured by the E802, E814, and E858 experiments for Si+Al, Au, and Pb interactions. The curves are exponential fits in m_t to the E802 data.

A-A COLLISIONS

Three experiments have measured $\bar{p}$ distributions in nucleus-nucleus collisions[8-10]. They are described in Table 2. Experiment E802 has measured $\bar{p}$ yields at $p_t > 300$ MeV, and the data of E814 and E858 were measured near p_t=0. The nucleus-nucleus collision environment is different from p-p and p-A collisions in that the collisions result in the production of a baryon dense nuclear volume.

The relative yield of antiprotons measured by the E802 collaboration is 0.75(15) for the Al target relative to Au. E858 reports 0.69(8), and 0.82(16) for the Al and Cu targets relative to Au. E814 measures 0.69(8) and 0.94(10) for the Al and Cu targets relative to Pb. These numbers, in particular those of E858 and E814 measured in similar kinematic regions, are in good agreement with each other.

Experiment E802 has made a measurement of the slopes of the p_t distributions. Experiments E814 and E858 do not have sufficient coverage in p_t to make such measurements. The inverse slope parameters of fits to the p_t distributions are determined to be about 141 MeV and are not strongly dependent on target species, or centrality. The slope parameters in A-A collisions are larger, as expected, than those measured for p-A collisions, and for p-p collisions at higher energies. Some of the reasons for this differencee are the mulitple collision contribution to the production of antiprotons, and the elastic scattering of antiprotons on their way out of the nuclear medium. The contribution from fermi-motion in the target should already be evident in the change of the slope parameters in going from p-p to p-A collisions. There is an additional change in slope in going from p-A to A-A collisions due to fermi motion in the projectile. The present statistical uncertainty in the measurements of the p_t distributions precludes a quantitative understanding of the collision dynamics, or how the slope parameters for the antiproton distributions change with projectile-target combinations, and collision centrality. Improved data sets should however become available in the near future. Figure 1 shows the invariant multiplicity measured by E802, E858, and E814, plotted as functions of p_t for y=1.6. The fits to the data are curves of the functional form described above. The parameters chosen for the fits and the data of E802 were obtained from the thesis of J. Costales[21]. The data for E814 are from the thesis of V. Greene[22], and those of E858 are from ref. 9. The uncertainties on measured points are not shown since this plot is intended to demonstrate trends rather than draw conclusions from data as yet unpublished in refereed journals. It is apparent that the curves through the E802 data overpredict the yield at low p_t where the measurements of E814 and E858 have been made. This could indicate the presence of a suppression of antiprotons at small p_t.

In order to examine the measurements of E858 and E814, figure 2 plots the invariant cross section at p_t=0 measured by the three experiments, as functions of rapidity. The data points for E802 are extrapolations to p_t=0 using curves of the functional form described above. Again, the difference between the measurements of E814, E858 and extrapolations to p_t = 0 of measurements of E802 are evident at all rapidities. Further, a discrepancy between the measurements of E814 and

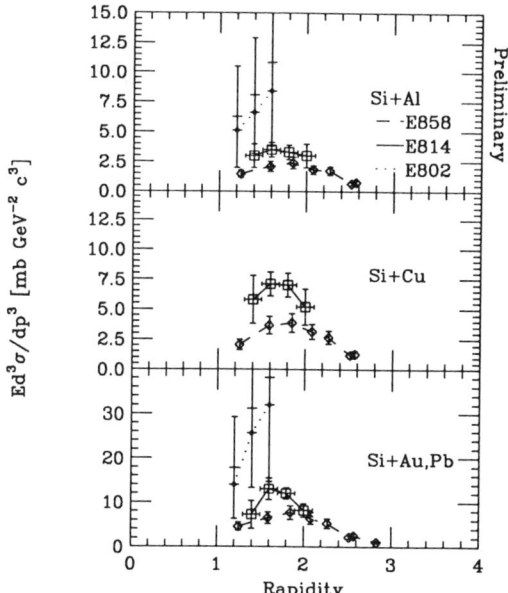

Figure 2: The invariant crosssections at $p_t=0$ as measured by E814 and E858, and obtained from fits to the data of E802.

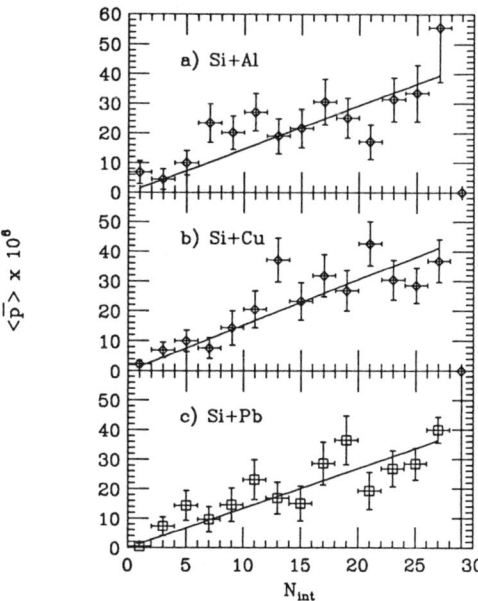

Figure 3: The yield of antiprotons per event measured by E814 in Si+A interactions plotted as a function of the number of nucleons in the projectile that have interacted.

E858 is evident for all three targets. These measurements are made in overlapping kinematic regions, and should therefore agree. The differences are as yet unresolved.

Experiments E802 and E814 have studied the centrality dependence of antiproton yields. While the E802 measurement studies minimum bias and central collisions (defined to be events in the uppermost 7% of the charged particle multiplicity distribution), E814 has measured the antiproton yield as a function of centrality. Figure 3 shows the yields of antiprotons measured by E814 as functions of centrality for Si ions interacting with Al, Cu, and Pb targets. Centrality is determined by the number of nucleons in the ^{28}Si projectile that have interacted. This number is determined by calorimetric measurements of energy deposited in a forward calorimeter. The ratios of yields in central collisions to those in minimum bias collisions is 2.7 and 2.25 for the Al and Pb targets, respectively. This agrees within uncertainties with the ratios 2.2 and 2.4 measured by E802 for Al, and Au targets.

A surprising feature evident in the data is the lack of target dependence. If $\bar{p}$ production increased linearly with the number of projectile nucleons that have interacted, we expect the yield of antiprotons to increase as is observed. However, if there is strong $\bar{p}$ annihilation in the baryon rich collision volume created in nucleus-nucleus collisions, we expect a suppression of the $\bar{p}$ yield that increases with centrality, and target mass. Such effects, if present in the data, are not very strong.

MODEL CALCULATIONS

There are several models that use Monte Carlo techniques to simulate $\bar{p}$ production in nucleus-nucleus collisions[21-29]. Since the center of mass energies in the collisions studied by the E814, E802, and E858 experiments are very close to the threshold energy for $\bar{p}$ production, most of the production of antiprotons occurs, in most calculations, in nucleon-nucleon collisions where both the projectile and target nucleon are interacting for the first time.

The Venus model (version 3.07)[23] predicts a rather strong target and centrality dependence that is not observed in the data. The model underpredicts the yields of experiment E802 for minimum bias collisions for the Al target, and does reasonably well in reproducing the data for central Si+Au interactions[21].

The Fritiof model (Version 1.7 modified for the AGS)[24] when compared to the data of E802 is able to describe the target and centrality dependence of $\bar{p}$ production[21], but overpredicts the observed yields by a factor in excess of about 5. The abundance of antiprotons depends strongly on a diquark suppression factor. This is obtained from e^+e^- string fragmentation data. Further, it has been sugested[26] that an inadequate treatment of the geometry of the color flux tube "strings" formed in nucleus-nucleus interactions results in an excess in the number of produced antiprotons.

The RQMD model[26] uses a string fragmentation procedure similar to that used by Fritiof. The procedure is modified by the implementation of diquark suppression at the ends of flux tubes. This results in a reduction in the abundance of antiprotons produced per nucleon-nucleon collision, giving much better agreement between the model and p-p data. In the nucleus-nucleus collision environment however, RQMD predicts that the initial production of antiprotons is enhanced by collective effects. Baryons are excited to high lying resonances in multiple collision processes, and subsequently decay to antiprotons. In this manner the initial production of antiprotons in Si+Al collisions is a factor of 3-4 higher than first-collision scaling of p-p data, and for Si+Au, it is a factor of 4-5 higher. However, 65% of the produced antiprotons are annihilated in the Si+Al collision environment, and about 85% in the Si+Au environment. The RQMD model, using a formation time of 2.5 fm/c (a time, following an interaction which produces an antiproton, during which the particle is being "formed" and is therefore not allowed to interact with its free-space cross section) is fairly successful in the description of the data of E802. The model has not yet been appied to the data of experiments E858 and E814. Similar collective enhancements in the production of antiprotons, as in RQMD, have been proposed as an explanation for the large sub-threshold $\bar{p}$ production cross sections[27].

Models based on the scaling of first collisions[21,22,25] are also able to provide a reasonable description of the available experimental data, if they use large formation times (>3fm/c). Shorter times result in large losses of the produced antiprotons because the $\bar{p}$ annihilation cross sections are of the order of 100 millibarns for antiprotons moving slowly, 400 MeV on average, relative to the nucleon-nucleon center of mass[29].

The ARC model[28] is also able to provide a rather good description of the E802 data. The model uses a formation time of 1 fm/c, which is closer to what is an accepted for other particle species at AGS energies. The $\bar{p}$ production, as in the simple models described above, happens predominantly in first collisions. However, in this model the large $\bar{p}$ annihilation cross sections are 'screened', and the model therefore predicts that only about 25% of the produced antiprotons are annihilated within the medium. The process of screening can be explained thus. The 100 millibarn $\bar{p}$ annihilation cross section implies a range of a few fermi. In a high nuclear density environment, this distance is larger than the mean distance between particles. The antiproton, therefore, has a large probability for being struck, for example by a pion, before it has a chance to annihilate, reducing this cross section relative to its free-space value. The extent of the reduction is proportional to the density of the system. Such effects are not considered in the RQMD calculation. Preliminary indications are that ARC is able to describe the E814 data, and the $\bar{p}$ measurement of E802 for p-A interactions. These calculations should become available soon.

Does a high nuclear density environment suppress the production of antiprotons? RQMD suggests that it does, and ARC says it does not. While RQMD says the initial production of antiprotons is enhanced in the nucleus-nucleus collision environment, ARC says it is not. There are some observable consequences of these model predictions which have not been tested adequately as yet. One is the ability of the models to reproduce the centrality and target independence seen in the antiproton data of E814. It would be rather fortuitous if, as suggested by RQMD, increased antiproton production was always balanced, as functions of target, and centrality, by increased antiproton absorption. Models that predict (or assume) that the in-medium antiproton annihilation cross sections are large will also predict a shift in the mean of the rapidity of the antiproton distributions in asymmetric target-projectile systems, as functions of centrality. The expected shift away from y_{NN} is evident in the RQMD calculation shown in ref. 26. The model predictions using the shorter formation times are peaked nearly half a unit in rapidity above y_{NN}. The distributions of antiprotons produced in peripheral collisions are expected to be centered close to y_{NN}, and very similar to those in p-p collisions. The ARC calculation does not predict such a rapidity shift. Another aspect of the data that merits further scrutiny is the change in the m_t distributions as functions of target and projectile, and centrality. Large annihilation cross sections tend to reduce the number of scatters that an antiproton can experience on its way out of the collision environment. This and the fact that antiprotons are being produced close to threshold at AGS energies makes the slopes (see equation above) of their distributions in m_t smaller than those for protons. If the in-medium annihilation cross sections are "screened" as suggested by ARC, the calculated slopes should be larger that those predicted by RQMD. Hence antiproton m_t and y distributions measured with smaller statistical and systematic uncertainties are needed in order to clarify the details of the reaction mechanisms leading to the observed antiproton distributions, and to distinguish between the RQMD and ARC model predictions. Some data will become available soon from experiments E866, and E878. A comprehensive measurement is planned by experiment E864.

ANTIDEUTERONS

The only experiment that has thus far been able to measure antideuteron distributions is E858. The experiment reports[9] that the ratio of $\bar{d}/\bar{p}$ invariant cross sections at $y=1.7$ for minimum bias Si+Au interactions is $1 \pm 0.7 \times 10^{-5}$. This is larger than what is expected by direct production (antideutron production thresholds are at higher energies than available in 14.6 GeV n-n collisions), but smaller than predicted by a simple coalescence model. The data are interpreted as implying that antiprotons are coming from a large source even in non-central events, an effect that could be due to either long $\bar{p}$ formation times or freeze-out times. This statement relies on a comparison between the E858 $\bar{d}$ data, and those measured for deuterons by experiment E802 (for $y<1.25$). What is compared is a coalescence factor B, which is the ratio of the invariant cross section of the composite particle divided by the product of the invariant cross sections of its ingredients, evaluated at the appropriate momenta. It bears emphasis that these factors decrease as one goes from $y<1.25$ to $y=1.7$. Hence a better estimate of what the coalescence prediction of antideuteron yields should be can be made using the E814 deuteron data at $y=1.7$, numbers that should become available soon[30]. Also, deuteron and antideuteron distributions from experiment E878 for Si+Au and Au+Au collisions should become available soon. A relativistic coalescence calculation[31], and a calculation using RQMD[32] have predicted ratios of $\bar{d}/\bar{p}$ invariant cross sections at the level seen by experiment E858. However, these predictions are for central collisions. Hopefully, these calculations will be repeated for minimum bias interactions and will enable us to understand whether antinucleus production is at expected levels, or far less.

SUMMARY

Several experiments have measured the distributions of antiprotons and antideuterons at AGS energies. The present quality of the available data clearly indicates the presence of several interesting differences between the p-p and nucleus-nucleus collision environments. Detailed measurements of antiproton distributions should allow us to probe important aspects of the collision dynamics leading to the formation of nuclear matter at high density.

1. T. DeGrand, Phys. Rev. **D30**, 2001 (1984).

2. U. Heinz, P.R. Subramanian, and W. Greiner, Z. Phys. **A318**, 247 (1984).

3. J. Rafelski, and B. Müller, Phys. Rev. Lett. **48**, 1066 (1982).

4. J. Ellis, U. Heinz, and H. Kowalski, Phys. Lett. **B233**, 223 (1989).

5. WA85 Collaboration, S. Abatzis et al., Phys. Lett. **B270**, 123 (1991).

6. NA36 Colaboration, E. Anderson et al., Submitted to Phys. Lett.

7. NA35 Collaboration, J. Bartke et al., Z. Phys. **C48**, 191 (1990).

8. E802 Collaboration, A. Abbott et al., Phys. Lett. **B271**, 447, (1991) and Quark matter 90.

9. E858 Collaboration, A. Aoki et al., Phys. Rev. Lett., in press, and P. Stankus et al., Nucl. Phys. **A544**, 603c (1992).

10. E814 collaboration, J. Barrette et al., preprint Yale 40609-1091. and S.V. Greene et al., Nucl. Phys. **A544**, 599c (1992).

11. J. Rafelski, *ibid*.

12. L. Remsberg, *ibid*.

13. S. Gavin, M. Gyulassy, and M. Plümer, Phys. Lett. **B234**, 175 (1990).

14. D. Dekkers et al., Phys. Rev. **137**, B962 (1965).

15. J.V. Allaby et al., CERN 70-12 (1970).

16. T. Eichten et al., Nucl. Phys. **B44**, 333 (1972).

17. A.M. Rossi et al., Nucl. Phys. **B86**, 403 (1975).

18. U. Amaldi et al., Nucl. Phys. **B86**, 403 (1975).

19. A.O. Vaisenberg et al., JETP Lett, **29**, 660 (1979).

20. E802, T. Abbott et al., preprint, to be published.

21. J. Costales, Ph. D. Thesis, MIT (1990).

22. S. V. Greene, Ph. D. Thesis, Yale (1992).

23. Venus, K. Werner and P. Koch, Phys. Lett. **B242**, 251 (1990).

24. Fritiof, B. Nillson-Almquist and E. Stenlund, Comput. Phys. Commun. **43**, 387 (1987).

25. B. Shiva Kumar, S.V. Greene, and J.T. Mitchell, Yale 40609-1092, unpublished.

26. RQMD, A. Jahns, H. Stocker, W. Greiner, and H. Sorge, Phys. Rev. Lett. **68**, 2595 (1992).

27. G. Batko, W. Cassing, U. Mosel, K. Niita, and Gy. Wolf, Phys. Lett. **B256**, 331 (1991).

28. ARC, Y. Pang and T. Schlagel, private communication.

29. P. Koch and C.B. Dover, Phys. Rev. **C40**, 145 (1989).

30. J.V. Germani, private communication.

31. C.B. Dover et al., Phys. Rev. C44, 1636 (1991).

32. A.S. Botvina, A. Jahns, H. Stocker, and W. Greiner, unpublished.

DIMUON PRODUCTION IN p-W AND S-W COLLISIONS

Ulrich Goerlach
representing the HELIOS-3 Collaboration
DESY
Notkestrasse 85
D 2000 Hamburg 52
Germany

Abstract

The HELIOS-3 experiment at the CERN SPS has measured the dimuon production in 200 GeV/c proton–tungsten and sulfur–tungsten collisions in a wide kinematic region. First results on dimuon pairs in the mass range between production threshold and the J/Ψ are presented, showing a drastic change in the spectral shape of the mass distribution, manifesting itself in a much smaller ratio of the vector resonances ρ/ω and Φ above the continuum in S-W than in p-W collisions.

INTRODUCTION

Thermal electromagnetic radiation emitted during an ultra-relativistic heavy–ion collision is an important experimental signature in the search for the formation of a Quark-Gluon-Plasma (QGP).

Virtual photons radiated by the QGP-phase lead to a dilepton continuum which is characterized by the temperature of the plasma. Also in a hadronic gas or a mixed phase electromagnetic radiation will occur. However, in this case the shape of the dilepton spectrum will be significantly modified by the coupling of the photon to the vector resonances, in particular the ρ.

Experimentally we can divide the dilepton continuum into the so called low–mass region below the ρ resonance and the high mass region above the Φ resonance. In hadronic interactions we know that the high mass region is composed of charm and Drell-Yan production, whereas the low–mass region is dominated by the Dalitz decays of the mesons.

The low–mass region has been studied in proton nucleus collisions for more than a decade. See Ref. [1] and references therein. Only very recently it could be shown by the HELIOS-1 collaboration measuring electron and muon pair production in p-Be interactions that the low mass region is quantitatively reproduced by the sum of Dalitz decays of mesonic resonances. For greater detail see [2, 3]. In the following we will report on the dimuon mass distributions in proton– and sulphur–tungsten interactions.

THE HELIOS-3 EXPERIMENT

HELIOS-3 is a fixed target experiment at the CERN SPS using proton and sulphur beams of 200 GeV/c per nucleon. It is developed from the HELIOS-1 experiment and optimized for heavy ion collisions. As shown in Fig. 1, muons are measured in a magnetic spectrometer based on a 4.1 Tm superconducting dipole magnet, 32 planes of MWPCs and 3 scintillator hodoscopes for muon identification

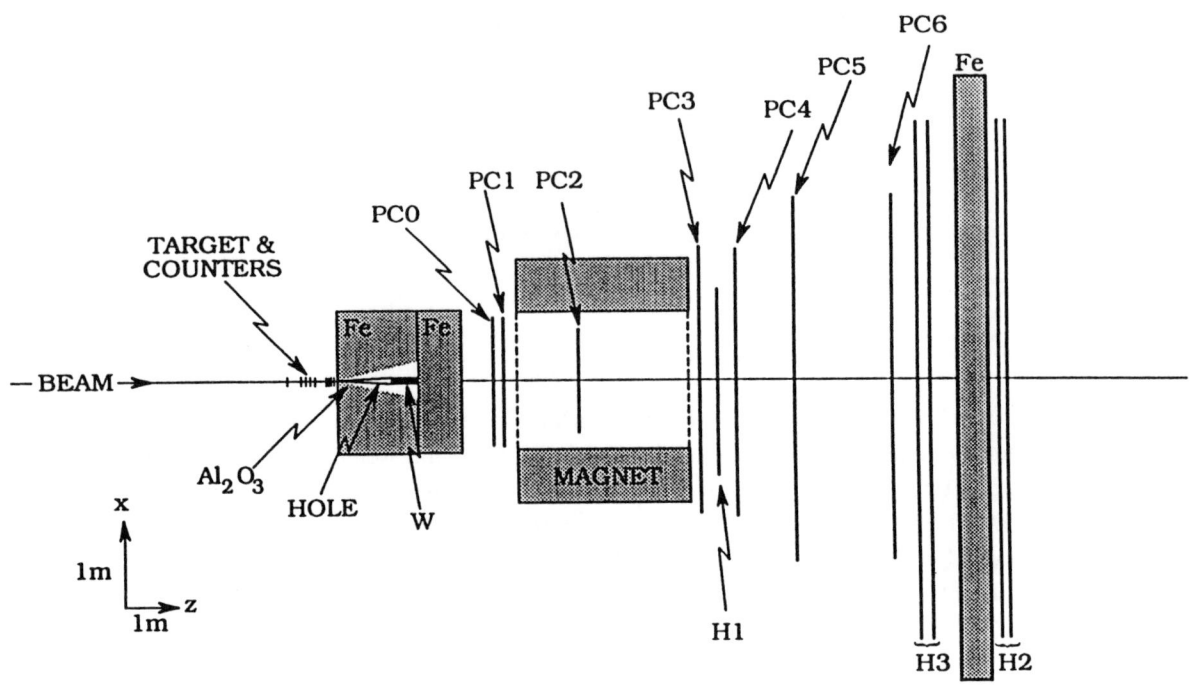

Figure 1. The HELIOS-3 experiment.

and trigger purposes.

In order to minimize the large combinatorial background from decays of $\pi^\pm$ and $K^\pm$ into muons we reduced the distance between the target and the hadron absorber to 25 cm. A special hadron absorber was constructed, consisting of Al_2O_3 in the first 6 interaction length, followed by 100 cm of Fe. With this configuration a mass resolution of 80 MeV at 1 GeV was achieved.

However, the very short target–absorber separation leads to a further experimental problem in the low mass region, namely to distinguish direct muons originating from interactions of the beam in the target from muons produced in interactions of secondaries in the absorber. For leading particles or non-interacting projectile nucleons (spectators) this problem is greatly reduced by a central ±8 mradian hole in the absorber, which stops these particles 1.45 meters away from the target in a tungsten block which is at the end of the hole inside the absorber. The remaining contributions to the dimuon mass spectrum from secondary interactions in the absorber has to be estimated by Monte Carlo.

Two sets of sufficiently segmented silicon ring counters were used to measure the charged particle multiplicity in the event thus providing a centrality-trigger.

DATA REDUCTION AND ANALYSIS

In 1990, with beams of 200 GeV/c per nucleon, we recorded 2 million dimuon triggers in p–W and a total of 12 million in S–W interactions, a large fraction of them being central collisions (9 million).

In the first step of the data reduction cleanup cuts on the beam counters were applied.

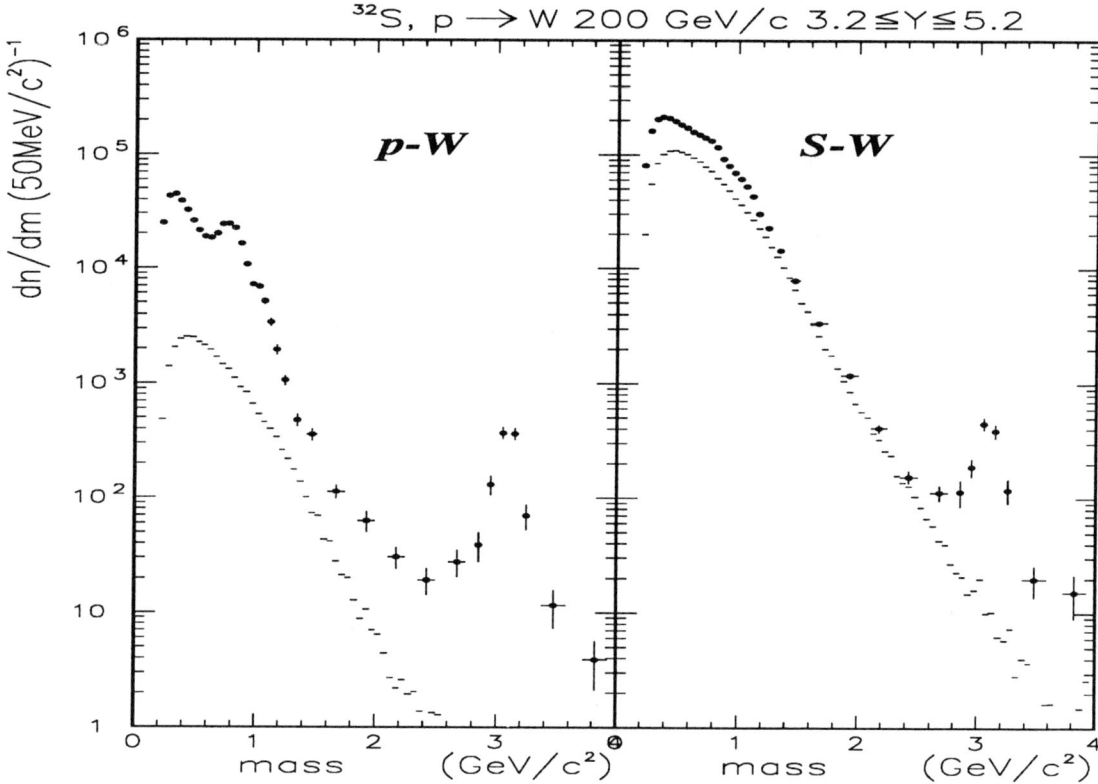

Figure 2. The raw mass spectrum and the combinatorial background in p-W and S-W.

Then, after the off line verification of the trigger conditions, the track reconstruction in the spectrometer was performed. Finally, identified muons were selected and tested for a good vertex on the beam axis at the target position: a three standard deviation(σ) cut on the z-coordinate of the point of closest approach to the beam axis was imposed on the muon candidates. The error(σ) is given by multiple scattering and depends strongly on the angle and the p_T of the muon.

A kinematic range in the M_T–rapidity plane of the reconstructed muon pair was selected in a region with an acceptance of sufficient reliability. The particular choice of this cut is motivated by the properties of the spectrometer: the low momentum acceptance limit for muons is given by the energy loss of the absorber and the strength of the magnetic field. The upper limit is dominated by kinematics. Both expressions depend on the polar angle, or on the rapidity if expressed in terms of M_T. The rapidity interval is basically determined by the angular acceptance of the spectrometer.

The cuts were parameterized as follows, $Y_{\mu\mu}$ and $M_{T,\mu\mu}$ being the rapidity and the transverse mass of the muon pair.

$$\sqrt{(2m_\mu)^2 + \left(\frac{2p_\mu^{min}}{\cosh Y_{\mu\mu}}\right)^2} \leq M_{T,\mu\mu}$$

$$M_{T,\mu\mu} \leq 70\%(kin.lim.)$$

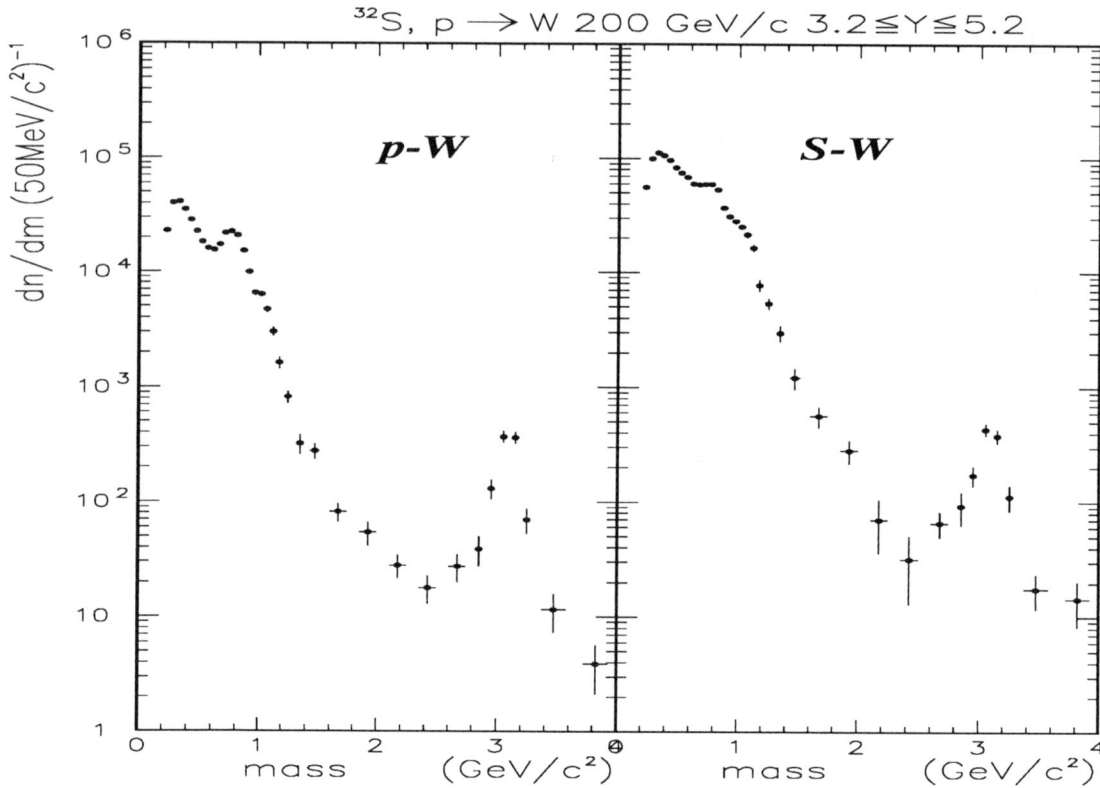

Figure 3. The background subtracted and acceptance corrected dimuon mass distributions for p-W and S-W collisions.

$$3.2 \leq Y_{\mu\mu} \leq 5.2$$
$$p_\mu^{min} = 7.5 \ GeV/c$$

Decays of charged hadrons, in particular pions and kaons, are a major source of muons leading to a combinatorial dimuon background. The size of it can be estimated by the number of measured like sign muon pairs according to:

$$N(\mu^+\mu^-) = 2 \cdot R(N)\sqrt{N(\mu^+\mu^+)N(\mu^-\mu^-)}$$

The factor $R(N)$ depends crucially on the charged multiplicity and the charge asymmetry in the event. For 200 GeV/c collisions we found $R(N) = 1.57 \pm 0.1$ for p–W and $R(N) = 1.10 \pm 0.06$ for central S–W collisions. These values have been established by GEANT Monte Carlo simulations with the event generator VENUS [4]. The errors are given by variations of the Monte Carlo and uncertainties in the normalization procedure.

To obtain the correct shape of the combinatorial background we constructed muon pairs using muons from different events. By this method the spectral shape can be obtained with negligible statistical errors. In Fig. 2 we show the mass spectra of p-W and S-W collisions together with the combinatorial background. It is obvious, that for S-W interaction this subtraction is most critical, especially at higher masses.

As already mentioned above, the short distance between the target and the hadron absorber leads to an additional experimental background from secondaries producing muon pairs by interactions in the dump. This background has to be simulated. In order to calibrate our Monte Carlo simulation of this process in magnitude and spectral distributions, we took data with 25, 50 and 100 GeV/c pions and 200 GeV/c protons impinging directly on the absorber. This background depends on the total multiplicity and the number of beam-spectators in the event. However, in our present calculations, for central collisions its total contribution is very small in the kinematic region selected in this analysis: only about less than 7% of the low mass pairs can be attributed to this process. Above the Φ resonance this contribution is negligible.

RESULTS

In Fig. 3 we present the mass distribution of muon pairs in p-W and central S-W interactions integrated over the described kinematic range after the subtraction of the combinatorial and dump induced background. Within this region the muon pairs have been corrected for the geometrical acceptance and the reconstruction efficiency. The S-W data sample is dominated by small impact parameters containing only about 5% peripheral events.

Besides the strongly increased Φ cross-section (see Ref. [5]), we can notice that the ratio of the dimuon continuum to the vector meson resonances changes drastically when we compare sulphur to proton collisions. We also note the change in slope of the low mass continuum.

If we normalize the two distributions to each other at the ρ/ω mass peak we find that the continuum is about twice as high in S-W collisions. This apparent higher level of the continuum or, alternatively, the reduction of the vector mesons seem to reach out up to 2 GeV in mass. At the J/Ψ we find with this normalization a relative suppression which is compatible with the results of experiment NA38.

We consider the different appearance of the mass distribution as significant and an important result. We like to point out that it is nearly independent of Monte Carlo corrections and does not rely on any model.

Presently, we are comparing our measured distributions with predictions for the lepton yield form Dalitz decays in the low mass region and to Drell-Yan and charm production in the higher mass region. A preliminary result on the low mass region was reported already in [3] showing for p-W an acceptable agreement with the prediction of Dalitz decays. It is perhaps not surprising that a simple extrapolation from p-Be to S-W will fail to describe the S-W data, which are so different in shape. More refined estimates are being worked on.

CONCLUSIONS

Striking differences in the mass distribution of muon pairs in p-W and S-W collisions have been observed, which suggests a different composition of the dimuon continuum. This could point to a dramatic change in the production ratios of vector mesons or possibly to a different or new dimuon production mechanism.

REFERENCES

1. U. Goerlach, "Review on Low Mass Dilepton and Soft Photon Production", presented at the International Workshop on Quark Gluon Plasma Signatures, Strasbourg, (1990), proceedings p. 305.

2. W. Willis, "Soft Electromagnetic Processes in Hadronic Collisions", presented at the Photon Lepton conference, Geneva (1991).

3. U. Goerlach, "Results of the HELIOS collaboration on Low Mass Dilepton and Soft Photon Production in p-Be, p-W and S-w Collisions", Nucl. Phys. A544 (1992) 109-124.

4. K. Werner, Phys. Rev. Lett. 62 (1989) 2460.

5. M.A. Mazzoni,"Measurement of the $\Phi/(\rho+\omega)$ ratio in p–W and S–W interactions", Nucl. Phys. A544 (1992).

SINGLE PARTICLE SPECTRA AND TWO PARTICLE CORRELATION FROM NA44, "THE FOCUSING SPECTROMETER" AT THE CERN-SPS

H.Atherton[2], H.Bøggild[8], J.Boissevain[6], K.Bussmann[2], M.Cherney[4], E.Chesi[2], G.DiTore[2], J.Dodd[3], J.Downing[4], S.Esumi[5], C.W.Fabjan[2], A.Franz[2], K.H.Hansen[8], T.Humanic[9], T.Ikemoto[5], B.Jacak[6], R.Jayanti[9], H.Kalechofsky[9], T.Kobayashi[12], R.Kuatadze[10], Y.Y.Lee[9], M.Leltchouk[3], D.Linker[3], B.Lörstad[7], N.Maeda[5], A.Medvedev[3], Y.Miake[13], A.Miyabayashi[7], M.Murray[6], S.Nagamiya[3], S.Nishimura[5], E.Noteboom[4], S.U.Pandey[9], P.Peters[4], F.Piuz[2], V.Polychronakos[1], M.Potekhin[3], G.Poulard[2], D.Rahm[1], J.M.Rieubland[2], A.Sakaguchi[5], M.Sarabura[6], K.Shigaki[11], J.Simon-Gillo[6], W.Sondheim[6], T.Sugitate[5], J.Sullivan[6], Y.Sumi[5], J.Sunier[6], H.Sletten[2], H.Tam[3], H.vanHecke[6], D.Williams[2], W.Willis[3], T.Zhu[3]

[1]Brookhaven Nat. Lab., [2]CERN, [3]Columbia Univ., [4]Creighton Univ., [5]Hiroshima Univ., [6]Los Alamos Nat. Lab., [7]Lund Univ., [8]Niels Bohr Inst., [9]Pittsburgh Univ., [10]Tbilisi State Univ., [11]Univ.of Tokyo, [12]Tsukuba Nat. Lab., [13]Tsukuba Univ.

Achim Franz
CERN - PPE
CH 1211 Geneva 23
Switzerland

Abstract

We present preliminary results on the M_T spectra of identified pions, kaons, and protons around central rapidity originating from collisions of 450GeV/c incident protons on Be, S and Pb targets as well as 200AGeV/c Sulfur ions on S and Pb targets. Two particle Hanbury-Brown Twiss interferometry results are also given for identified pion and kaon pairs.

APPARATUS

NA44, the "Focusing Spectrometer", is a second generation experiment[1] based on the results from the first round of experiments with ultrarelativistic heavy ions at the AGS and CERN (for a review of the results see e.g. the references 2, 3, and the references therein). NA44 covers the phase space at low p_T and midrapidity achieving a very good particle identification and momentum resolution with special emphasis on small momentum differences for two particle measurements. Fig.1. shows a schematic view of the setup. The incoming particles are tagged by a beam counter (intrinsic time resolution $\sigma \sim 35 ps$)[4]. A scintillation counter behind the target serves as an interaction trigger and to select central events. It is followed by a silicon pad counter which measures the charged particle multiplicity in the range $1.5 < \eta < 3.3$. Two Dipole magnets select particles of a certain nominal momentum ($p_{nom} \pm 20\%$) and charge. Note that only one charge state at a given setting can be measured. Particles are then focused into the acceptance of tracking hodoscopes by three superconducting quadrupoles. The three tracking hodoscopes consist of 50, 60 and 50 vertical scintillator slats, respectively, and achieve a TOF resolution of $\sigma \sim 100 ps$[5]. On the trigger level two threshold Cerenkov counters select the particle type and are also used in the off-line analysis.

All other detectors shown in the drawing are not used in the analysis presented and are not discussed further.

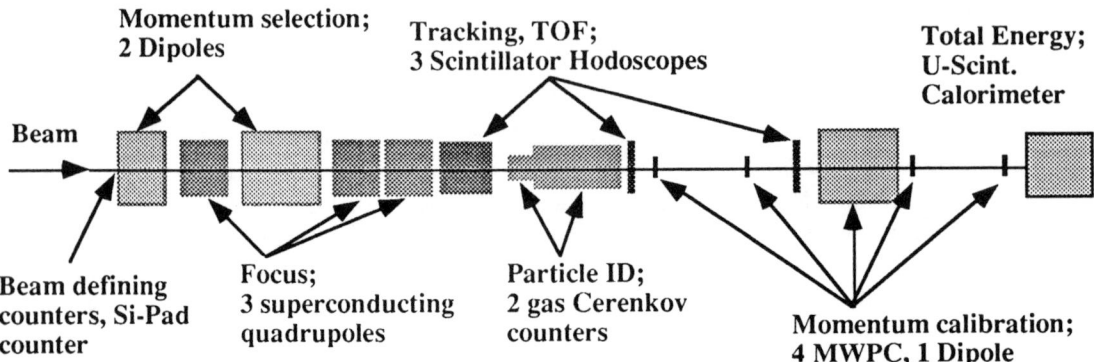

Figure 1. Schematic side view of the NA44 experimental setup.

DATA PRESENTED

In the following, data samples from single and two particle triggers are presented. The single particle data were taken at 8GeV/c nominal momentum where the acceptance lies between $3.0 \leq y \leq 3.5$ for kaons and $2.6 \leq y \leq 2.9$ for protons. For pions it ranges from y=4.8 at the lowest p_T to y=3.2 at the highest. The pair data were taken at 4GeV/c nominal momentum. Here the acceptance for kaons lies at y=2.8 and the pions range from y=3.2 to y=4.0.

SINGLE PARTICLE SPECTRA

The single particle p_T spectra for all systems studied can be well described by an exponential in M_T ($M_T = \sqrt{p_T^2 + M_0^2}$, M_0= rest mass of the particle). Fig. 2 shows as an example of these distributions the relative normalized yield as a function of K_T ($K_T = M_T - M_0$) for pions, kaons and protons produced in 200AGeV/c S-S interactions. Table 1 summarizes the inverse slopes. The contamination of the proton data by kaons is less than 1% and the kaon data sample itself contains less than 5% pion contamination. The acceptance and reconstruction corrections are studied by Monte Carlo but the systematic error estimate is still in a preliminary state, so data shown include statistical errors only.

Several experiments[6,7] have reported an increase in the yield of pions at low p_T, whereas others[8] do not measure such a rise. We find that the ratio of the spectra for pPb/pBe is constant over p_T. The same is true for the yield ratios of SS/pBe or SPb/pBe. It should be noted that our measurements were taken forward of central rapidity compared to the experiments of references 6 and 7.

Table 1. Inverse slopes of M_T distribution in MeV/c^2

Particle	Rapidity	pBe	pPb	SS	SPb
$\pi^{\pm}$	3.2 - 4.0	149 ± 4	138 ± 3	144 ± 1	129 ± 3
K^+	3.0 - 3.5	115 ± 2		176 ± 5	200 ± 7
K^-	3.0 - 3.5	123 ± 3	118 ± 2	171 ± 4	185 ± 6
p^+	2.6 - 2.9	91 ± 4		156 ± 8	176 ± 9
p^-	2.6 - 2.9	97 ± 3	100 ± 4	140 ± 8	118 ± 9

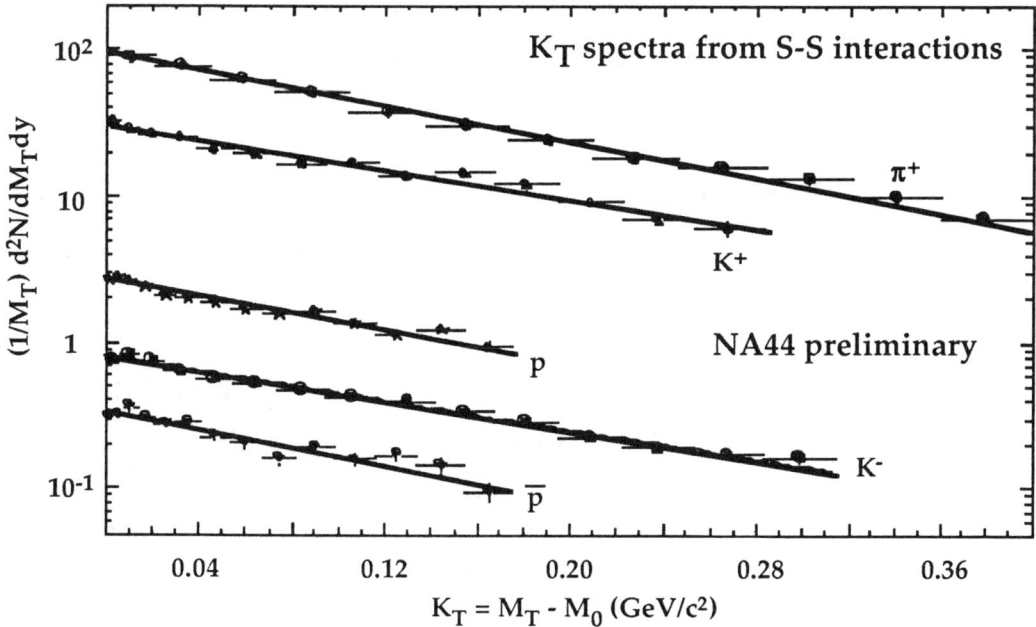

Figure 2. KT spectra for identified pions, kaons and protons. Error bars are statistical only.

TWO PARTICLE CORRELATION

Two particle interferometry, following the work of Hanbury-Brown and Twiss[9] (see also reference 10 and the references therein), can yield spatial and temporal information about the particle emitting source. The multi dimensional correlation function (C2) can be parametrized as a function of the momentum difference (Q) between the two particles with an exponential distribution ($C2=1+\lambda\exp\{-2QR\}$) or by a Gaussian source distribution ($C2=1+\lambda\exp\{-Q^2R^2\}$); in a two dimensional analysis where the momentum difference vector is projected along (Q_L) and perpendicular (Q_T) to the beam direction. Fig.3 shows the correlation function for identified pion and kaon pairs together with an exponential and Gaussian fit. The distribution for the pions includes systematic errors as described by NA44[11], which also includes details about the analysis and the applied corrections. In the kaon sample, the systematic error is not yet well understood, and therefore the two datapoints for the lowest Q_{inv} are left out. For these data points the correction due to momentum resolution and Coulomb correction are largest and are currently under investigation. In Table 2 the fit parameters are summarized. For the one dimensional distribution an exponential fit gives a better χ^2 compared to a Gaussian fit but both are statistically acceptable. It should be noted that our results are similar to measurements at the AGS[12,13] taken at lower beam momentum. The second line in Table 2 are final results[11], all other data are preliminary.

Although R_{inv} is not a measure of the space distribution of the source, as it contains a time component as well, the difference between the pion and kaon distributions is striking. The difference persists in the transverse projection R_T where boost effects are minimal. It may indicate an earlier freeze-out of the kaons compared to the pions.

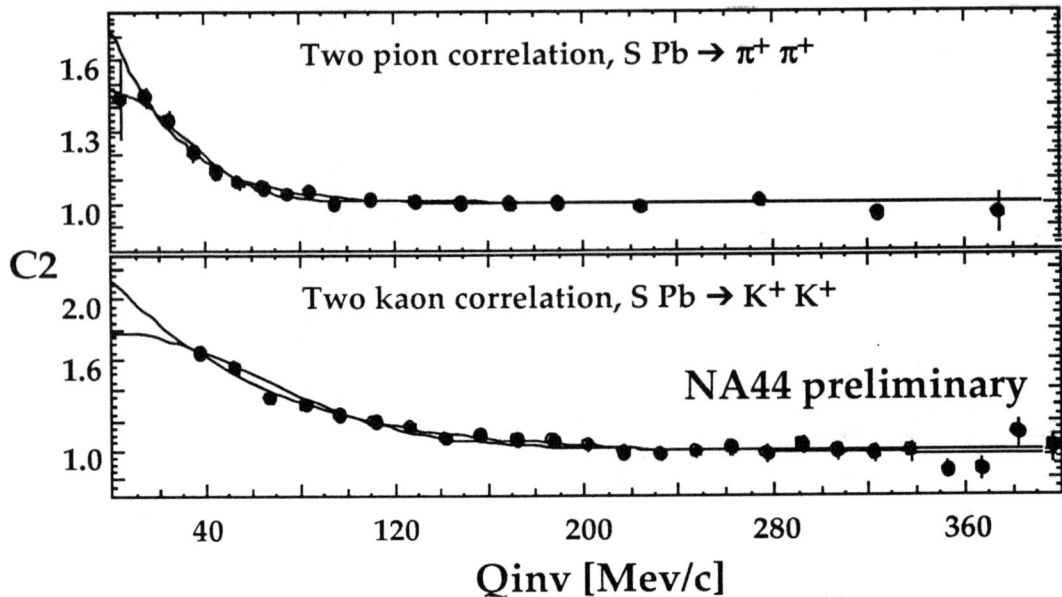

Figure 3. HBT Correlation function (C2) for pion and kaon pairs. The error bars for the pion data include systematic errors, while for the kaons the errors are statistical only

Table 2. Radius and λ parameters for one and two dimensional Gaussian and exponential source distributions

one dimensional		Gaussian source distr		exponential source distr	
System	Pair	λ	R [fm]	λ	R [fm]
p Pb	$\pi^+\pi^+$	0.43 (0.03)	2.00 (0.22)		
S Pb	$\pi^+\pi^+$	0.46 (0.04)	4.50 (0.31)	0.77 (0.08)	3.54 (0.33)
S Pb	K^+K^+	0.67 (0.04)	2.02 (0.11)	1.31 (0.12)	1.79 (0.12)
S Pb	K^-K^-	0.77 (0.08)	2.17 (0.20)	1.59 (0.30)	2.02 (0.27)

two dimensional		Gaussian source distribution		
System	Pair	λ	R_T [fm]	R_L [fm]
S Pb	$\pi^+\pi^+$	0.53 (0.06)	3.89 (0.14)	4.38 (0.21)
S Pb	K^+K^+	0.78 (0.05)	2.16 (0.13)	1.77 (0.18)
S Pb	K^-K^-	0.90 (0.11)	2.15 (0.21)	2.18 (0.33)

REFERENCES

1. H.Bøggild et. al., "Proposal for a Focusing Spectrometer for One and Two Particles", CERN/SPSC/88-37, SPSC/P239

2. Proceedings of the Sixth International Conference on Ultrarelativistic Nucleus-Nucleus Collisions, Quark Matter 1987, Zeitschift für Physik C38, (1988) 1

    Proceedings of the Seventh International Conference on Ultrarelativistic Nucleus-Nucleus Collisions, Quark Matter 1988, Nucl. Phys. A498, (1989) 1

    Proceedings of the Eighth International Conference on Ultrarelativistic Nucleus-Nucleus Collisions, Quark Matter 1991, Nucl. Phys. A544, (1992) 1

3. H.R.Schmidt, J.Schukraft, "The Physics of Ultra-Relativistic Heavy-Ion Collisions", CERN-PPE/92-42, March 1992

4. N.Maeda et.al., "A Gaseous Beam-Counter with Time Resolution of 24ps for Relativistic Nuclear Beams", to be submitted to Nucl. Inst. and Methods

5. T.Kobayashi and T.Sugitate, Nucl. Instr. and Meth. A287, (1990) 389

6. H.Ströbele et. al., NA35 Coll., Z.Phys. C38, (1988) 89.

7. T.Akesson et.al., Helios Coll., Z.Phys.C46 (1990) 361.

8. Y.Takahashi et.al., EMU05 Coll., Nucl. Phys. A525, (1991) 591c.

9. R.Hanbury-Brown and R.Q.Twiss, Nature 178, (1956) 1046.

10. B.Lorstad, Int. J. Mod. Phys. A4, (1988) 2861.

11. H.Bøggild et. al., "Identified pion interferometry in heavy ion collisions at CERN", NA44 Coll., to be submitted to Phys. Lett.

12. T.Abbott et. al., Phys. Rev. Lett. (1992) 1030.

13. R.J.Morse, "Bose-Einstein correlation measurements in 14.6 A GeV/c nucleus nucleus collisions", Ph.D. Thesis, Massachusetts Institute of Technology, 1990.

Photon and Hadron Production of Heavy Quarks

Parallel Session 10

Dallas Hall, SMU

Conveners: J. Cumalat (Colorado)
S. Paul (CERN)

A TECHNIQUE FOR OBSERVING THE TOP QUARK AND MEASURING ITS MASS AT THE TEVATRON

Gary R. Goldstein[*] and K. Sliwa[*]
Department of Physics
Tufts University
Medford, Massachusetts 02155 USA

and

R.H. Dalitz
Department of Theoretical Physics
University of Oxford
1 Keble Road, Oxford OX1 3NP UK

ABSTRACT

A method is introduced for separating top quark production from Standard Model background in the channel in which one top quark decays semi-leptonically and its anti-quark decays hadronically into three jets. The method is applied to simulated CDF data and discriminates top (with $m_t > 120$ GeV) from background.

The mass of the top quark exceeds 91 GeV[1], a bound deduced from the empirical dilepton inclusive total cross-section. Unless its mass is much greater than 200 GeV, it should be produced by the Tevatron collider, but formidable backgrounds have made its identification difficult. However, owing to its large mass, top decay is expected to have some unique features that make the separation from background feasible, as we will show.

Top quarks will decay primarily into a b-quark and a real W boson[2], and for top masses above about 120 GeV, this weak decay will occur more rapidly than hadronization[3]. Hence the final state in top decay will satisfy strong kinematic constraints. A scheme for analysing t-t̄ production in p-p̄ annihilation through **leptonic decay modes** using such constraints was developed[3], applied to the one possible
_____
[*] U.S. Dept.of Energy grant.

candidate event[4] and yielded a mass of about 130 GeV. The W also decays through q-q̄ channels. A larger signal for top pair production is expected for the event configuration

$$p+\bar{p} \to t+\bar{t}+X; \quad t \to l+\nu+b(\text{jet});$$
$$\bar{t} \to q(\text{jet})+\bar{q}(\text{jet})+\bar{b}(\text{jet}) \quad (1)$$

which we study here.

For high mass top quarks the cross section for (1) is expected to stand out over the Standard Model background[5] of W boson and multiple jet production. We will show herein that a pronounced separation of signal from background can be achieved by requiring that individual events satisfy the detailed kinematic criteria of real t-t̄ production. A probabilistic mass distribution can be defined, the "mean probability distribution", that will facilitate that separation.

In an event like (1) the 4-momenta of the three jets associated with the hadronic W-decay will determine a total

4-momentum for the hadronically decaying top; p_t and the mass m_t* will be determined. The lepton and remaining b-jet 4-momentum, along with the mass m_t* just determined, define an ellipse of possible 3-momenta for the other top quark[3]. The hypothesis of limited total transverse momentum for the partons leads to the requirement that the transverse vector $-(p_t)_T$ be near points on the ellipse. The combination of the existence of an ellipse for this m_t* and a near cancellation of transverse momenta constitute our kinematic fitting criteria. Any event with a kinematic configuration satisfying these severely restrictive criteria will be a candidate for top production.

The probability that a candidate event actually corresponds to process (1) depends on the top mass through the product of probabilities for the leptonic decay mode, the hadronic decay mode, the total transverse momentum (for simplicity taken as a step function of width $0.1 m_t*$) all multiplied by the probability for the parton production mechanism favored for the particular kinematics, with the structure functions $F_1(x_1)$ and $F_2(x_2)$ weighted by the q-q̄ or gluon fusion differential cross section.

Ideally, for a perfectly measured event containing a high transverse energy lepton, considerable missing transverse energy (for a neutrino) and at least four jets with clearly identified flavor, the task of checking whether or not the top production hypothesis is likely to be correct is straightforward. The likelihood for that event, with a specific identification of jet flavors, would be achieved by simply evaluating the *a posteriori* probability for the given kinematic configuration and applying Bayesian statistics.

For real data, however, there are two complications that have considerable impact: (a) the lack of flavor identification for the jets, which requires us to consider all possible jet combinations, in accord with the given event; (b) the sizeable uncertainties in jet energy measurements, wich is dealt with as follows.

Each event in any sample of real Tevatron data, or simulated data as we consider here, has a set of "measured" lepton and jet 4-momenta. A combination of four jets is tentatively assigned: one to the semileptonic decaying top and three to the hadronically decaying partner. The three jets, with 4-momenta p_1, p_2, p_3, form the tentative top (or antitop) momentum $t = p_1 + p_2 + p_3$, with top "mass" $m_t*^2 = t_\mu t^\mu$.

The measured value of a jet energy represents one of a continuum of possible "true" values, with frequency of occurence given by an empirical probability distribution[6]. The jet energies can be varied away from those "measured" values and weighted accordingly. As they are varied, two of the jets in the hadronic mode are constrained to have invariant mass equal to $M_W = 80.6$ GeV.

Each subsequent kinematic configuration is submitted to the same fitting procedure, candidate values are obtained and the *a posteriori* probability for that mass m_t* determined. When all the configurations are processed a probability distribution versus top mass for that jet combination is obtained. It is a measure of the probability that that combination fits the top hypothesis for each mass.

All combinations that lead to candidate jet configurations are treated on the same footing. The total probability for the observed jet configuration is then obtained as a function of top mass m_t^* by summing the separate probabilities calculated as above for each acceptable jet combination. Summing all candidate events to obtain a single total or mean probability distribution $P(m_t^*)$ provides the means to separate "top events" from "non-top events".

To test the fitting procedure advocated here we have used simulations of $t\bar{t}$ production and W + multi-jet Standard Model "background" events. For the former

$$p+\bar{p}\to X+t+\bar{t}\to X+W^++W^-+b+\bar{b}$$
$$\to X+l^\pm+4\text{jets} \quad (3)$$

we use two different simulations, a "toy model" purely parton level simulation and a more sophisticated calculation using ISAJET[7] to simulate the jet fragmentation and gluon emission. To include the CDF detector effects in the latter the code QFL is used[8].

We show in Fig.1 the mean probability distribution $P(m_t^*)$ calculated for 1000 randomly chosen toy model events. Gaussian probabilities appropriate to the CDF determination of jet energies have been included and detector cuts have been applied. We note that the peak lies close to the input 140 GeV, being due to "correct" combinations and that the peak dominates the contributions from the widely spread "wrong" combinations.

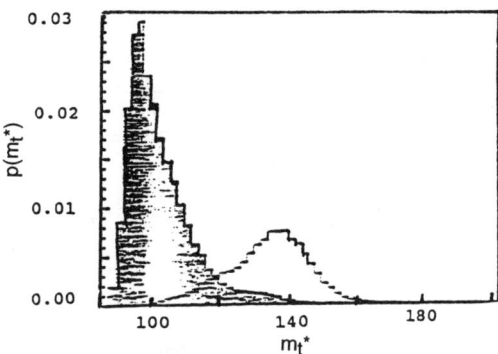

Figure 1. $P(m_t^*)$ for toy $t\bar{t}$ events and W+jets (shaded).

The mean distribution $P(m_t^*)$ deduced from a sample of 500 ISAJET+QFL events, with the same cuts, is shown in Fig.2. It is known that the use of ISAJET here gives the same total cross section for $t\bar{t}$ production (14±2 pb for m_t=140 GeV at the Tevatron energy) as does the QCD calculation of Nason, Dawson and Ellis[9]. (The toy model events have been normalized to that total cross section of 14 pb.) This

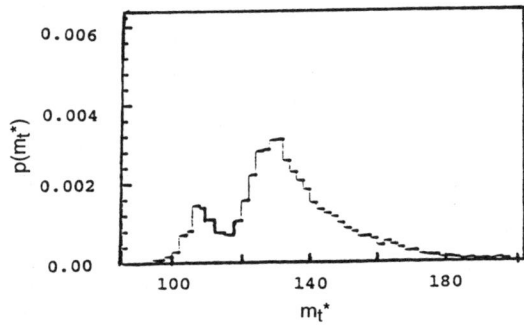

Figure 2. $P(m_t^*)$ for ISAJET $t\bar{t}$ events.

distribution shows a sharp peak at 130 GeV, 10 GeV below the input mass, and its FWHM is about 25 GeV. The shift in the peak relative to the input mass is believed to be due to gluon radiation, taken into account in ISAJET, but is not yet fully understood. The secondary peak arises from the "wrong" combinations of jets.

For the Standard Model background Berends, et.al.[5] have provided full tree-level calculations of the cross sections for the processes

$$p+\bar{p} \to X+W^{\pm}+n(\text{jets}) \to X+l^{\pm}+n(\text{jets}) \quad (4)$$

where n≤4, which they have used to simulate the background provided by (4) to the top-antitop production and decay events (3). We have analyzed this QCD background by applying our $P(m_t^*)$ procedure to a large sample of "W + 3 jets" and "W + 4 jets" generated[10] with the VECBOS Monte Carlo program along with the ISAJET+QFL for jet evolution and detection. The events were subjected to the same analysis as $t\bar{t}$ events. The resulting mean probability distribution as a function of m_t^* and normalized to the luminosity of 4 pb^{-1} (the integrated luminosity of 1988-1989 CDF run) is shown shaded and superimposed on the toy events in Fig.1. There is very little Standard Model background left above m_t=120 GeV.

The completely different forms of the probability distributions for t-$\bar{t}$ simulations versus W+jets demonstrates that the mean probability distribution function for real data will effectively distinguish t-$\bar{t}$ production from Standard Model background[11].

REFERENCES

1. F. Abe, et.al.,(CDF Collab.) Phys. Rev. Lett. **68**, 447(1992); Phys. Rev. **D45**, 3921 (1992).
2. I.Bigi, Phys.Lett. **B175**, 233 (1986).
3. R.H.Dalitz, G.R.Goldstein, Phys.Rev.**D45**, 1531 (1992); Phys. Lett. **B287**, 225 (1992).
4. F. Abe, et.al.,(CDF Collab.) Phys.Rev.Lett. **64**, 147 (1990); K.Sliwa,"Search for top quark at Fermilab Collider", in Proc. 4th Intl. Symposium on Heavy Flavour, Orsay, France, 1991, Editions Frontieres, p.567.
5. F.A.Berends, et.al., Nucl.Phys. **B357**, 32 (1991).
6. F.Abe, et.al.,(CDF Collab.), Phys. Rev. Lett. **68**, 1104 (1992); we appreciate the help of Naor Wainer for providing us with the subroutine that incorporates jet errors.
7. F.Paige and S.Protopopescu, in "Proceedings of the Summer Study on the Physics of the SSC, Snowmass, Colorado, 1986, ed. R.Donaldson and J.Marx (DPF,American Physical Society, New York, 1986), p.320.
8. C.Newman-Holmes, J.Freeman, in Proc. Workshop on Detector Simulation for the SSC, Argonne, 1987, ed. L. Price (ANL-HEP-CP-80-51), pp.190,285.
9. P.Nason,S.Dawson, R.K.Ellis, Nucl.Phys. **B303**, 607 (1988).
10. We appreciate the work of Jose Benlloch in generating the VECBOS samples. Cuts were imposed at the generator level on jets of ΔR>0.6, p_T>10 GeV, $|\eta|$<2.5 and p_T>10 GeV, $|\eta|$<1.2 on leptons.
11. A more thorough account is given in G.R.Goldstein, K.Sliwa, R.H.Dalitz, Tufts U. preprint TH-G-92-01 (1992).

THRESHOLD EFFECTS ON TOP PRODUCTION IN $\gamma\gamma$ INTERACTIONS*

M. C. Gonzalez-Garcia
Physics Department
University of Wisconsin
1150 University Ave.
Madison, WI 53706, USA

Abstract

The exchange of gluons between heavy quarks results in an enhancement of their production near threshold. We study QCD threshold effects in $\gamma\gamma$ collisions. The results are relevant to top production by beamstrahlung and laser back-scattering in future linear collider experiments

The QCD threshold enhancement of top production in e^+e^- and hadron collisions has been profusely studied.[1,2] Here I present the results of the analysis done in Ref.[3] on this effect in $\gamma\gamma$ collisions. We consider two possible sources of photons in e^+e^- linear colliders: beamstrahlung and laser back-scattering.

In future e^+e^- linear colliders, the strong electromagnetic fields associated with the high charge density in the bunches, subject particles to very strong accelerating forces just prior to or during the collision. As a result photons are radiated. This is known as beamstrahlung.[4] The luminosity of photons produced depends on the characteristics of the beams. The desired photon luminosity can, in fact, be achieved by tuning the beam parameters. The aspect ratio, G,[3] provides a good measure of beamstrahlung, with large photon luminosities associated with small values of G. We will focus on the design for the 500 GeV collider NLC[5] ($G \simeq 2.7$). We illustrate how results change for different beam profiles and increased energy. For instance for high photon luminosity we tune to round beams ($G = 1$).

Beamstrahlung photons have a soft spectrum. Hard photons can be obtained by laser back-scattering. Here intense γ beams are generated by backward Compton scattering of soft photons from a laser.[6] The laser frequency is chosen to make impossible that the back-scattered photon interact with the laser and create e^+e^- pairs. In our numerical calculations, we assumed a frequency $\omega_0 \simeq 1.26$ eV for the NLC. For further details on the photon luminosities see Ref.[3] and references therein.

The enhanced two-photon luminosity, whether from beamstrahlung or laser back-scattering origin, is the source of a large number of $q\bar{q}$ pairs[8] via two different mechanism. Quarks can be generated by a direct photon process, where the photons couple directly to charged quarks. Alternatively, photons can interact via their quark and gluon constituents. This is referred to as a "resolved" photon process which is described in terms of the structure function of the partons in the photon.[7]

For each process, the total cross section is obtained by folding the elementary cross section with the photon luminosity.

*Work supported by the U.S. DOE under contract No. DE-AC02-76ER00881, and by the University of Wisconsin Research Committee.

$$\sigma(e^+e^- \rightarrow \gamma\gamma \rightarrow i+j \rightarrow t\bar{t})(s) = \int_{z_{min}}^{z_{max}} dz\, \frac{dL_{ij}}{dz}\hat{\sigma}(i+j \rightarrow t\bar{t})(\hat{s}=z^2 s) \quad (1)$$

$z = \sqrt{\tau} = \sqrt{\hat{s}}/\sqrt{s}$, where $\sqrt{s}$ is the total e^+e^- CM energy and $\sqrt{\hat{s}}$ the $\gamma\gamma$ CM energy, and dL_{ij}/dz stands for the differential luminosity of the partons i and j. The expressions for the luminosities dL_{ij}/dz can be found in Ref.[3]

The exchange of gluons between associatively produced heavy quarks modifies significantly their production cross section near the threshold. Moreover, for a very heavy quark, like the top, non-perturbative QCD effects are small. Therefore the threshold behaviour can be computed perturbatively,[11] and the modifications of the cross section near threshold due to QCD can be calculated in terms of a Coulomb-like interaction between t and $\bar{t}$

$$V_{S,8}(r) = C_{S,8}\frac{\alpha_s}{r} \quad \text{with} \quad C_S = -\frac{4}{3} \quad C_8 = \frac{1}{6} \quad (2)$$

The potential is attractive when $t\bar{t}$ pair is in the colour singlet channel, and it is repulsive in the colour octet state. Since the interaction is attractive in the singlet channel, the formation of bound states by multiple gluon exchanges between the t and the $\bar{t}$ can in principle occur. However, if the top quark is heavier than ~ 140 GeV, the formation time of the bound state is larger than the toponium lifetime and the resonance structure disappears.[12] These interactions nevertheless lead to significant modifications of the cross section near threshold. This mechanism is analogous to the Coulomb rescattering in QED discussed by Sommerfeld[13] and Sakharov.[14]

In the narrow width approximation, we can obtain the QCD effects near the threshold replacing, in the tree-level cross sections, the usual threshold factor $\beta = \sqrt{1 - \frac{4m_t^2}{\hat{s}}}$ by

$$\beta|\Psi_{S,8}(0)|^2 = \beta \frac{C_{S,8}\frac{\pi\alpha_s}{\beta}}{1 - \exp(-C_{S,8}\frac{\pi\alpha_s}{\beta})} \quad (3)$$

$\Psi_{S,8}(0)$ is the wave function at the origin for the colour singlet (S), and octet (8) channels respectively. Near threshold ($\beta \to 0$), the cross section in the colour singlet channel is increased since β is substituted by the non-vanishing factor $4\pi\alpha_s/3$, and the octet channel cross section is exponentialy suppressed.

Equation (3) can be interpreted as the exponentiated version of the first order QCD corrections near the threshold. It does not include the effects of bound states below threshold. These states are confined into a very small energy region and their contribution to the total cross section, which is obtained by integration over all CM energies, is rather small.

The value of the cross section for the different subprocesses is given in Table 1 for $m_{top} = 120$ GeV and $\sqrt{s} = 500$ GeV. For resolved photon processes, we present predictions for two parametrizations of the photon structure functions: the one given by Drees-Grassie (DG)[10] and the one by Levy-Abramowicz-Charchula (LAC3),[9] which are respectively characterized by a soft and a hard gluon distribution. The correction to each subproces is as follows. In the direct $\gamma\gamma$ interactions, the $t\bar{t}$ pair is produced in a singlet state and the correction leads to an increase in the cross section. $t\bar{t}$ produced in $\gamma(g)+\gamma$ collision and $\gamma(q)+\gamma(\bar{q})$ annihilation are in the colour octet channel and the correction is negative and small. In $\gamma(g)+\gamma(g)$ fusion the final state is a mixture of color singlet and octet states in a ratio 2 to 5 given by the color factors. Since the enhancement in the singlet channel is much larger than the suppression in the octet channel the net correction to gg is positive. We also see that the contribution of "resolved" photons to the total $\gamma\gamma$ cross section is small as a result of the suppression of their luminosity at high values of x. Since the direct singlet channel dominates, *the threshold effect results in a significant enhancement of the total cross section*. This enhancement is roughly a factor 2 for beamstrahlung and more than 50% for laser back-scattering.

Table 1. Cross sections for top production at 500 GeV for $m_t = 120$ GeV. For each process the left (right) column is the cross section without (with) the threshold factors. For "resolved" photon processes the upper (lower) number is the cross section with DG (LAC3) parametrization

process	cross section (pb)						
e^+e^-	0.70				0.94		
photon-photon	laser		G=2.7			G=1	
$\gamma + \gamma$	0.74	1.2	9.0×10^{-3}	1.8×10^{-2}		0.18	0.36
$\gamma + \gamma(g)$	4.2×10^{-3}	3.8×10^{-3}	1.2×10^{-5}	1.1×10^{-5}		3.5×10^{-4}	3.1×10^{-4}
	1.7×10^{-2}	1.5×10^{-2}	4.8×10^{-5}	4.4×10^{-5}		1.4×10^{-3}	1.3×10^{-3}
$\gamma(g) + \gamma(g)$	5.4×10^{-7}	7.1×10^{-7}	1.4×10^{-9}	1.8×10^{-9}		3.1×10^{-8}	4.2×10^{-8}
	9.7×10^{-6}	1.2×10^{-5}	2.4×10^{-8}	3.1×10^{-8}		5.5×10^{-7}	7.2×10^{-7}
$\gamma(q) + \gamma(\bar{q})$	2.5×10^{-4}	2.2×10^{-4}	8.8×10^{-7}	7.8×10^{-7}		2.6×10^{-5}	2.3×10^{-5}
	2.8×10^{-4}	2.5×10^{-4}	9.7×10^{-7}	8.7×10^{-7}		2.9×10^{-5}	2.6×10^{-5}

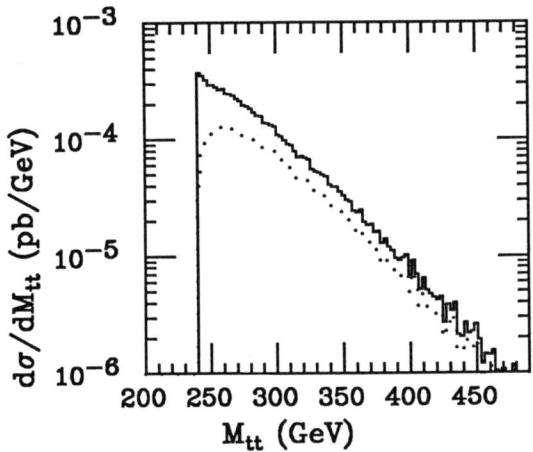

Figure 1. $t\bar{t}$ invariant mass distribution for $m_t = 120$ GeV for beamstrahlung photons with G=2.7. The solid (dotted) lines show the distribution with (without) the threshold factors

In Fig. 1 we show the invariant mass distribution of the $t\bar{t}$ pair. The corrections are larger for small invariant masses when $t\bar{t}$ pair is produced near threshold. This explains why the QCD corrections are larger in $\gamma\gamma$ than in e^+e^-. For the same reason the correction is small for laser back-scattering where the luminosity at low x is suppressed.

The dependence on the top mass and on the collider energy is shown in Fig.2. As expected, the QCD corrections increase slightly with the mass and decrease with the CM energy. Beamstrahlung, for round beams, can give a substantial contribution to $t\bar{t}$ production. The threshold corrections make this contribution even larger. At $\sqrt{s} = 500$ GeV the two photon contributions is at most 10% for the original design. However, for a circular beam more than 50% of the $t\bar{t}$ pairs with $m_{top} < 110$ GeV are produced in two photon collisions. Since $\gamma\gamma$ cross section increases with energy while e^+e^- one decreases, the two photon contributions are much more important at 1 TeV. However, top quarks produced by beamstrahlung photons preferentially populate the low p_T region[3] and so do the prompt leptons from their decay. Therefore their signature suffers from a large background from b and c produced in e^+e^- and in $\gamma\gamma$ processes.

The advantage of photon interactions is more dramatic for laser back-scattering. At $\sqrt{s} = 500$ (1000) GeV, for $m_{top} < 130$ (280) GeV, a "$\gamma\gamma$ collider" can produce more $t\bar{t}$ pairs than the corresponding e^+e^- collider. The background from c and b can be efficiently suppressed because it is concentrated at lower p_T-values than the one from the top signal. Fur-

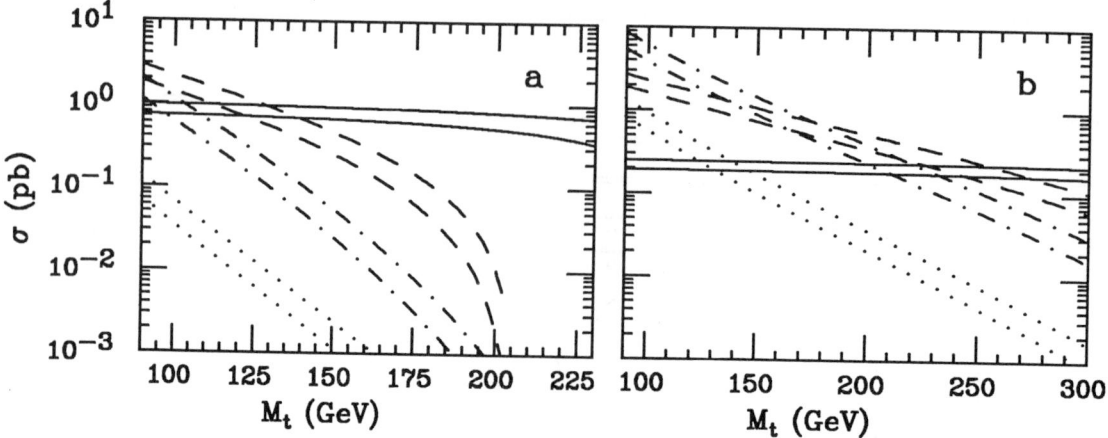

Figure 2. Total $t\bar{t}$ production cross section as a function of m_t for (a) $\sqrt{s} = 500$ GeV and (b) $\sqrt{s} = 1$ TeV. Solid lines are the cross sections for direct e^+e^- production, dashed lines for production with back-scattered laser photons, and dotted (dot-dashed) lines for production with beamstrahlung photons with G=2.7 (G=1). In all cases the upper (lower) lines show the cross section with (without) the threshold factors

thermore, the separation of the signal from the background is easier than in direct e^+e^- production.[3]

REFERENCES

1. C. Berger et al., Phys. Lett. **86B**, 413 (1979); I. S. Güsken, J. H. Kühn and P. M. Zerwas, Phys. Lett. **155B**, 185 (1985); J. Feigenbaum, Phys. Rev. D**43**, 264 (1991); W. Kwong, Phys. Rev. D**43**, 1488 (1991);

2. V. S. Fadin and V. A. Khoze, JETP Lett. **46**, 525 (1987); V. S. Fadin and V. A. Khoze, Sov. J. Nucl. Phys. **48**, 309 (1988). V. Fadin, V. A. Khoze and T. Sjöstrand, Z. Phys. **C48**, 613 (1990).

3. O. Eboli, M.C. Gonzalez-Garcia, F. Halzen and S. Novaes, MAD/PH/701.

4. R. J. Noble, Nucl. Instrum. & Methods **A256**, 427 (1987). R. Blankenbecler and S. D. Drell, Phys. Rev. D**36**, 277 (1987). M. Jacob and T. T. Wu, Phys. Lett. **197B**, 253 (1987).

5. R. B. Palmer, Ann. Rev. Nucl. Part. Sci. **40**, 529 (1990).

6. F. R. Arutyunian and V. A. Tumanian, Phys. Lett. **4**, 176 (1963); R. H. Milburn, Phys. Rev. Lett. **10**, 75 (1963).

7. E. Witten, Ncl. Phys. **B120**, 189 (1977).

8. F. Halzen, C. S. Kim and M. L. Stong, Phys. Lett. **B274**, 489 (1992). M. Drees and R. M. Godbole DESY 90-044 and BU 92/1.

9. H. Abramowicz, K. Charchula and A. Levy, Phys. Lett. **B269**, 458 (1991).

10. M. Drees and K. Grassie, Z Phys. **C28**, 458 (1991).

11. I. Bigi, Y. Dokshitzer, V. Khoze, J. H. Kühn and P. M. Zerwas, Phys. Lett. **B181**, 157 (1986). M. B. Voloshin, Nucl. Phys. **154**, 365 (1979); H. Leutwyler, Phys. Lett. **98**, 447 (1981).

12. K. Fujikawa, Prog. Theor. Phys. **61**, 1186 (1979); I. S. Güsken, J. H. Kühn and P. M. Zerwas, Nucl. Phys.**B262**, 393 (1985);

13. A. Sommerfeld, Atombau und Spektrallinien, Bd. 2, Braunschweig, Vieweg 1939.

14. A. D. Sakharov, JETP **18**, 631 (1948).

SYSTEMATICS OF CHARM PRODUCTION IN HADRONIC COLLISIONS*

Paul Hoyer
Department of Physics
Univerity of Helsinki
Siltavuorenpenger 20 C
SF – 0170 Helsinki, Finland

Abstract

Charmed particles produced in hadron collisions have a harder momentum spectrum than predicted by QCD at leading twist. This and related features of the data can be understood as resulting from coalescence of the charm quark with co-moving light quarks, and a contribution of intrinsic charm at large values of x_F.

Heavy quark production provides an interesting testing ground for Perturbative QCD. The inclusive production cross-section for a hadron H containing a heavy quark Q in a hadronic collision $h + A \to H + X$ has the general form

$$\sigma(hA \to HX) = \sum_{ab} G_{a/h} G_{b/A} \hat{\sigma}(ab \to Q + X) D_{H/Q} \quad (1)$$

Here the G's are structure functions of the colliding hadrons, $\hat{\sigma}$ is the hard subprocess and D describes the heavy quark fragmentation process $Q \to H + X$. The structure and fragmentation functions G, D are *universal*, i.e., they are independent of the underlying hard process (described by $\hat{\sigma}$) and are known from other reactions. The "higher twist" corrections to the expression (1) are suppressed by inverse powers of the mass M of the heavy quark.

Measurements of the corrections to the leading twist QCD prediction (1) for charm production at large Feynman $x_F = 2p_H/\sqrt{s}$ are particularly interesting, since

---

*This report is based on work with R. Vogt and S.J. Brodsky.[1]

(i) In contrast to charm production in e^+e^- annihilations and to (forward) charm photoproduction, the charm quark produced in the fragmentation region of a hadron is accompanied by co-moving light quarks. Particularly at large x_F, the coalescence of charm and light quarks of similar velocities is enhanced.[2]

(ii) At large x_F, the corrections to Eq. (1) are of $\mathcal{O}(1/M^2(1-x_F))$. An analysis[3] of the combined QCD limit $M^2 \to \infty$, $x_F \to 1$, with $M^2(1-x_F)$ fixed, revealed new leading order diagrams. The key characteristic of these new diagrams is that the charm quark pair couples to more than one beam parton. Such "intrinsic charm" contributions can be important in charm production at large x_F. Data on deep inelastic lepton scattering in fact show[4] that for this reaction the higher twist terms *are* enhanced at large x_F.

THE MODEL

We have made a phenomenological study of the importance of the higher twist effects (i) and (ii) in the available data on charm production[1]. The coalescence (i) of light and heavy quarks maintains the velocity of both quarks - hence the momentum of the charmed hadron will be close to that of the charm quark. We describe this by replacing the standard "Peterson" fragmentation function[5], which is known to fit charm production in e^+e^- annihilations, by a "δ-function" fragmentation, for which the charm hadron and quark momenta are equal.

The intrinsic charm contribution (ii) is calculable in QCD (given the multi-parton distributions in the beam hadron). Since this calculation has not yet been attempted, we used a phenomenological form for the intrinsic charm contribution to the production cross-section,

$$\frac{d\sigma_{ic}}{dx_1 \cdots dx_n} = N_n \frac{\delta(1 - \sum x_i)}{(m_h^2 - \sum(\widehat{m}_i^2/x_i))^2} \quad (2)$$

Here $n = 4$ ($n = 5$) for incoming mesons (baryons) h, corresponding to intrinsic $q\bar{q}c\bar{c}$ ($qqqc\bar{c}$) states in the wave function. The normalization N_n was fixed so that the relative contribution of (2) to the total charm cross-section (including the fusion contribution (1)) was the same as that deduced from the A-dependence of J/ψ production[6,7].

The coalescence of charm and light quarks can lead to quantum number correlations ("leading particle effects") between the projectile and charmed hadron. Since the light and heavy quarks must have similar *velocities*, their momentum ratio is $\sim m/M$, where m is the (effective) mass of the light quark. Thus charm quarks produced by the fusion process, which have low x_F, will coalesce mainly with sea quarks, implying negligible leading particle effects. The intrinsic production mechanism, on the other hand, favors[3] coalescence of the charm quark with valence quarks of the projectile. We included this effect in the hadronization of intrinsically produced heavy quarks.

RESULTS

As an illustration of our results, Fig. 1 shows our fit to the recent, high statistics E769 data[8] on $\pi^- A \to D + X$ at 250 GeV/c. We observe:

- The experimentally observed x_F-distribution is harder than the QCD prediction at leading twist (solid line in (a)). The δ-function fragmentation scheme including intrinsic charm (dot-dashed line in (a)) gives good agreement with the data.

- The size of the intrinsic charm contribution (2) is seen from the difference between the dashed and solid lines in (a). Clearly, high statistics and good acceptance at large x_F is needed to study this component of the production cross-section.

- The leading particle effect is expected to be significant only for $x_F \geq 0.5$, as seen by comparing the predictions for the D^- (leading, Fig. 1(b)), and D^+ (non-leading, Fig. 1(c)) momentum distributions. This is consistent with the data.

We emphasize that both higher twist mechanisms (i) and (ii) are enhanced at low transverse momentum, $p_\perp \leq 1$ GeV. Hence one expects $\langle p_\perp \rangle$ to decrease as x_F increases.

The importance of coalescence and intrinsic charm may be enhanced for projectiles containing heavy valence quarks. There is experimental evidence for anomalously large charm production with hyperon beams.[9]

The experimental data on the dependence of the open charm production cross section on the nuclear number A of the target is still inconclusive at large x_F. The intrinsic charm cross-section is expected[3,7,10] to behave as $\sim A^{2/3}$, as compared to A^1 for the leading twist fusion contribution. Such an anomalous A-dependence is clearly seen in the data on J/ψ production[6], showing the importance of higher twist effects in hidden charm production. The high x_F J/ψ cross section is larger than predicted by the fusion process, and its nuclear dependence is consistent with $A^{2/3}$, as expected from intrinsic charm.[7]

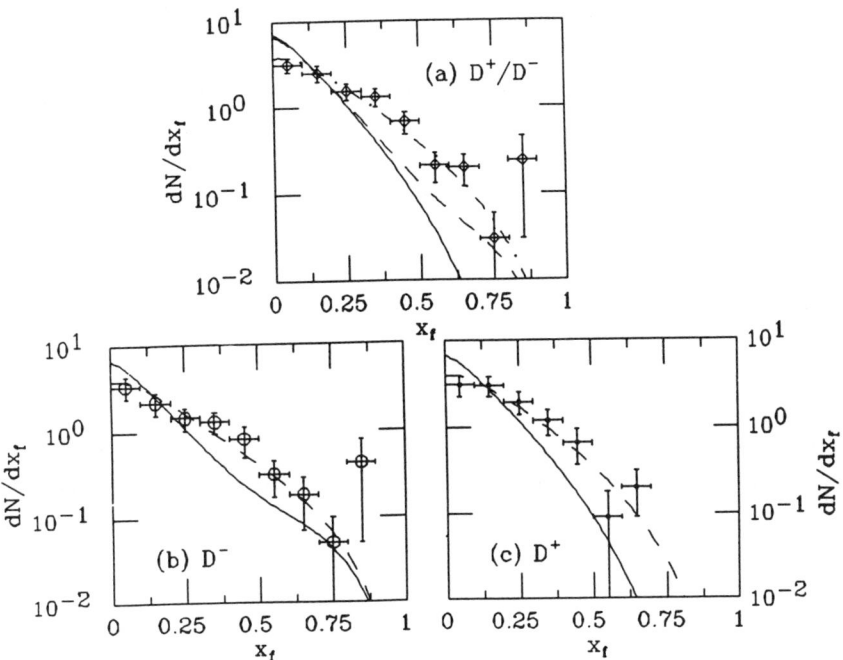

Figure 1. x_F distributions of D mesons produced in $\pi^- A$ collisions at 250 GeV/c (E769 data[8]). (a) The solid line shows the prediction of leading twist QCD, using a Peterson[5] fragmentation scheme for $c \to D + X$ which fits the data on D production in e^+e^- annihilations. The dashed curve shows the effect of adding the intrinsic charm contribution (2) to the fusion cross section (1). The dot-dashed curve results from fusion + intrinsic charm at the quark level, i.e., for δ-function fragmentation. (b) D^- production compared to our model, including both the fusion and intrinsic production mechanisms. The solid curve is obtained with Peterson fragmentation and the dashed curve with δ-function fragmentation. (c) As in (b), for D^+ production.

REFERENCES

1. R. Vogt, S. J. Brodsky and P. Hoyer, SLAC-PUB-5827(1992), to be published in *Nucl. Phys. B*.
2. S. J. Brodsky, J. F. Gunion and D. E. Soper, *Phys. Rev.* **D36**, 2710 (1987).
3. S. J. Brodsky, P. Hoyer, A. H. Mueller and W.-K. Tang, *Nucl. Phys.* **B369**, 519 (1992).
4. M. Virchaux and A. Milsztajn, *Phys. Lett.* **B274**, 221 (1992).
5. C. Peterson, D. Schlatter, I. Schmitt and P. Zerwas, *Phys. Rev.* **D27**, 105 (1983).
6. J. Badier et al., *Z. Phys.* **C20**, 101 (1983).
7. R. Vogt, S. J. Brodsky and P. Hoyer, *Nucl. Phys.* **B360**, 67 (1991).
8. J. A. Appel, Fermilab-Pub.-92/49, to appear in *Ann. Rev. Nucl. Part. Sci.* **42** (1992). See also P. Karchin, these Proceedings.
9. S.F. Biagi, et al., *Z. Phys.* **C28**, 175 (1985).
10. S.J. Brodsky and P. Hoyer, *Phys. Rev. Lett.* **63**, 1566 (1989).

PHOTOPRODUCTION OF CHARM MESONS*

Robert Gardner
for the Fermilab E687 Collaboration[1]
Department of Physics
University of Illinois at Urbana-Champaign
1110 W. Green St.
Urbana, IL 61801, USA

Abstract

Preliminary analysis of data from the Fermilab high energy photoproduction experiment E687 for events containing two fully reconstructed charmed mesons is presented. Experimental results on correlations between the $D\overline{D}$ pairs are compared to predictions by the Lund model.

INTRODUCTION

We have analyzed data collected by the Fermilab high energy photoproduction experiment E687 for events containing two fully reconstructed charmed mesons. Correlations between two fully reconstructed D's can be exploited to test QCD production models[2] by comparing, for example, the transverse acoplanarity angle between the D and $\overline{D}$. In this paper we establish a very clean signal for $D\overline{D}$ pairs and make *preliminary* comparisons of their dynamical properties to predictions by the Lund model[3].

The E687 detector, which is described in detail elsewhere[4], is a large aperture multiparticle spectrometer. A microvertex detector consisting of 12 planes of silicon microstrips arranged in three views provides high resolution tracking allowing the separation of primary and secondary vertices. Deflections of charged particles by two analysis magnets of opposite polarity are measured by five stations of multiwire proportional chambers. Three multicell Čerenkov counters operating in threshold mode are used for particle identification. The photon beam is derived from a 315 GeV/c electron beam with a $\sigma = 13\%$ momentum spread. The electron beam impinges on a 27% radiation length lead foil producing bremsstrahlung photons which are directed to a 4 cm long Be target. The average photon energy for the data sample was 220 GeV.

$D\overline{D}$ EVENT SELECTION

For our event sample we used decay modes of the D mesons which are copiously produced with high acceptance, namely, $D^+ \to K^-\pi^+\pi^+$, $D^0 \to K^-\pi^+$, $D^0 \to K^-\pi^+\pi^+\pi^-$, together with their charged conjugates. The principal cutting tool used to isolate the D signals is the decay flight distance[5] ℓ/σ_ℓ, which ranged between 0 and 7 depending on the candidate topology. The signal is developed by considering unique $D\overline{D}$ candidate combinations and plotting the mass of the D candidate subject to cuts on the $\overline{D}$. Fig. 1(a) shows the D invariant mass distribution re-

---

*Work supported in part by the National Science Foundation, the US Department of Energy, the Italian Istituto Nazionale di Fisica Nucleare and Ministero della Università e della Ricerca Scientifici.

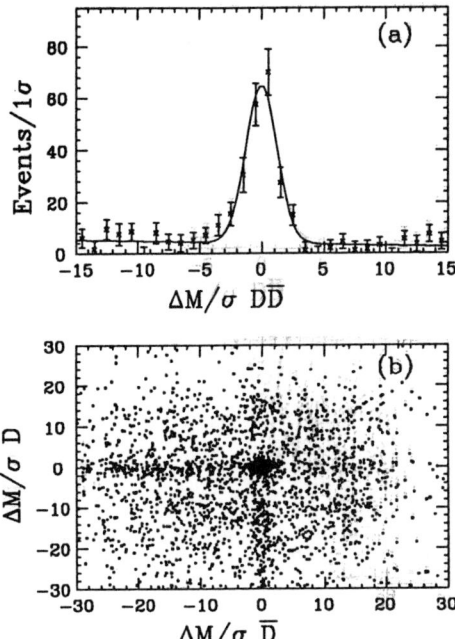

Figure 1. (a) Background subtracted mass distribution (in measurement standard deviations away from nominal value, $\Delta M/\sigma$). (b) Normalized D vs. $\overline{D}$ mass scatter plot.

quiring that the $\overline{D}$ candidate mass was reconstructed within 2σ of the nominal value. Single D and random backgrounds are subtracted using $\overline{D}$ sidebands 4-8σ away from the signal peak. The sideband subtracted signal yield is 192 ± 17 with a very good signal to noise ratio of ≈ 16.

Additional evidence of the $D\overline{D}$ signature, shown in Fig. 1(b), is obtained by making a scatter plot in which the D mass appears on the vertical axis and the $\overline{D}$ mass on the horizontal. Clearly there is a pileup of events in region where the $D\overline{D}$ bands overlap. Events accumulating in bands below the D or $\overline{D}$ signal regions are largely composed of partially reconstructed D's.

D^+ INCLUSIVE DISTRIBUTIONS

Having established our $D\overline{D}$ signal, we next evaluate the inclusive dynamical production properties predicted by the Monte Carlo[6] using a high statistics D^+ sample which is shown if Fig. 2(a). This very clean sample consists of 4400 ± 72 reconstructed $D^{\pm} \to K^{\mp}\pi^{\pm}\pi^{\pm}$ decays having a signal to background ratio of 17. In Figs. 2(b-d) the energy, p_t^2 and x_f distributions are plotted. The solid curves on the figures represent Monte Carlo predictions, which are found to be in excellent agreement with the data. Neither the data nor the Monte Carlo have been corrected for acceptance.

$D\overline{D}$ CORRELATIONS

We turn now to the study of $D\overline{D}$ events. In Fig. 3(a) we plot the p_t difference for the $D\overline{D}$ pair. The distribution exhibits very good agreement with the Monte Carlo. The longitudinal correlation between the charmed mesons is shown by the rapidity difference ΔY. In Fig. 3(b) we plot ΔY for data and the Monte Carlo and again find them in good agreement. We caution that the acceptance in this variable is not flat and the distribution should be viewed accordingly.

Total Mass and Dressing Sensitivity

Fig. 4(a) shows the $M_{D\overline{D}}$ distribution from data (points), and the Monte Carlo (solid). The $M_{D\overline{D}}$ distribution is related to the photon-gluon center of mass energy in the context of the photon gluon fusion model[7].

The extent to which $D\overline{D}$ distributions reflect underlying partonic processes depends in part on how much the dressing processes smear the D momenta as they emerge from the fragmentation. The fraction of energy of the D's with respect to the incident photon is an indication of the dressing severity. Fig. 4(b) shows the fractional distribution, Z, given by

$$Z = \frac{E_D + E_{\overline{D}}}{E_\gamma}. \qquad (1)$$

Both the data and Monte Carlo exhibit rather

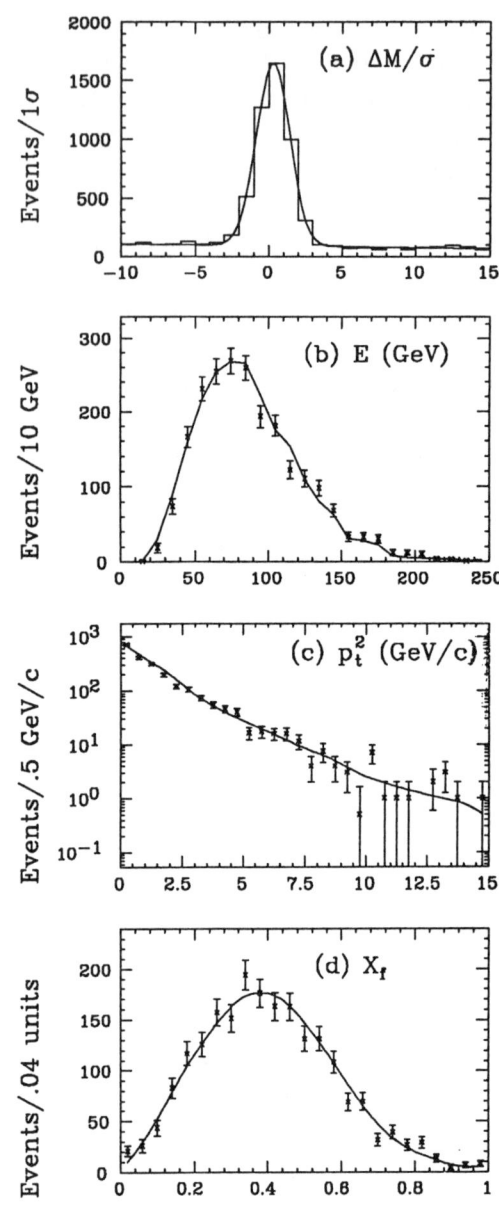

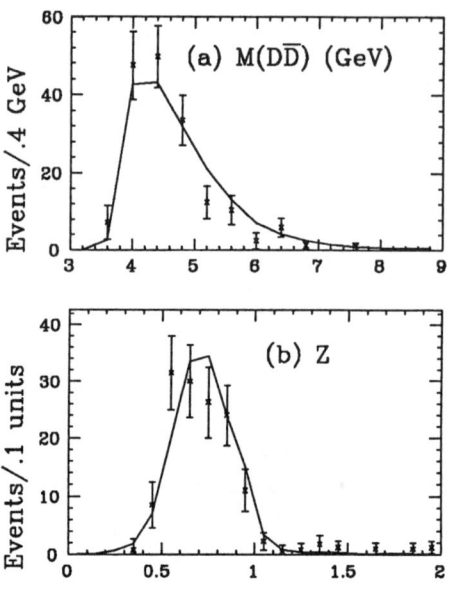

Figure 3. $D\overline{D}$ difference distributions: (a) p_t, (b) ΔY. (Data are represented by points and Monte Carlo by the solid curves.)

broad distributions and imply that on average the accepted D mesons carry approximately 75% of the initial photon energy.

Acoplanarity $\Delta\phi$

The acoplanarity angle $\Delta\phi$, plotted in Fig. 5., is the azimuthal angle between the D and $\overline{D}$ momentum vectors in the plane transverse to the photon direction. The data peaks at π radians corresponding to back to back $D\overline{D}$ production, though tends to be somewhat flatter than the Monte Carlo. This may be due in part to the intrinsic p_t spread of the target gluon, assumed to be Gaussian distributed in p_t with $\sigma = .44$ GeV/c. When σ is increased to 1 GeV/c (dashed histogram of Fig. 5) the Monte Carlo is in better agreement with data. A QCD calculation has been applied for the analogous distribution in hadroproduction[8] though the case for photoproduction has yet to be published.

Figure 2. Inclusive properties of $D^\pm \to K^\mp \pi^\pm \pi^\pm$ decays: (a) normalized invariant mass, (b) energy, (c) p_t^2, (d) x_f. (In Figs.(b-c), data are represented by points and results from simulated Monte Carlo events are indicated by the solid curves.)

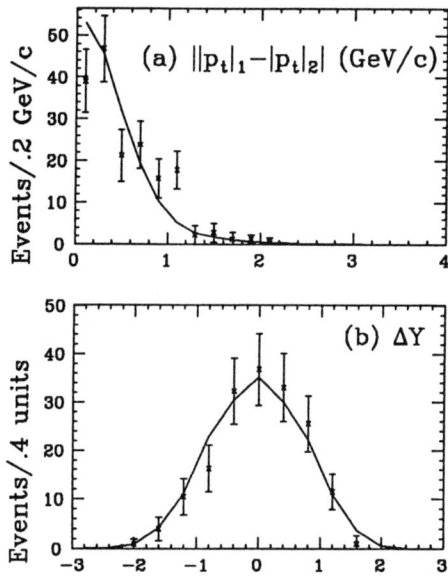

Figure 4. (a) Total mass $M_{D\overline{D}}$, (b) charm energy fraction Z. (Data are represented by points and Monte Carlo by the solid curves.)

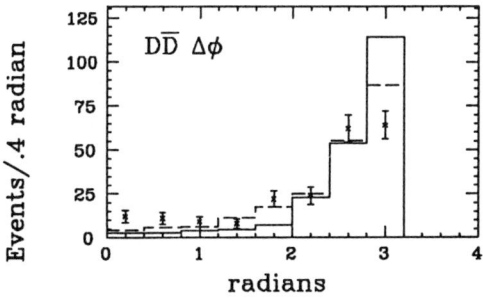

Figure 5. $D\overline{D}$ Acoplanarity. Data are represented by points and Monte Carlo by the solid histogram. The dashed histogram is for the Monte Carlo when the target gluon is given a larger intrinsic p_t spread.

CONCLUSIONS

We have presented very convincing evidence for photoproduced states containing fully reconstructed $D\overline{D}$ pairs, and have examined many of their kinematical and dynamic properties. Our preliminary results indicate good agreement with most predictions from a Monte Carlo based on the Lund model.

REFERENCES

1. Please see G. Bellini, *these proceedings*, for a list of collaborators and institutions in the E687 Collaboration.

2. For example, R.K. Ellis, P. Nason, "QCD Radiative Corrections to the Photoproduction of Heavy Quarks", *Nucl. Phys.* B 312 (1989) 551.

3. T. Sjostrand, "The Lund Monte Carlo for Jet Fragmentation," *Computer Phys. Comm.* 39 (1986) 347. H.-U. Bengtsson, "The Lund Monte Carlo for Hadronic Processes," *Computer Phys. Comm.* 46 (1987) 43.

4. E687, P. L. Frabetti *et al.*, "Description and Performance of the E687 Spectrometer," FERMILAB Pub-90/258E (1990), to be published in *Nucl. Instrum. Methods*.

5. The variable ℓ, with error σ_ℓ, is the signed 3 dimensional separation between vertices.

6. The Monte Carlo consists of the Lund packages[3] PYTHIA 5.6 for the charm photoproduction and JETSET 7.3 for fragmentation, combined with simulation algorithms for the E687 apparatus.[4]

7. L.M. Jones, H.W. Wyld, "Charmed-Particle Production by Photon-Gluon Fusion," *Phys. Rev.* D 17 (1978) 759.

8. M. Mangano, P. Nason and G. Ridolfi, "Heavy-Quark Correlations in Hadron Collisions at Next-to-Leading Order," *Nucl. Phys.* B 373 (1992) 295.

RARE DECAY MODES OF THE D^0, D^+, AND D_S CHARMED MESONS*

N.M. CASON for the E687 COLLABORATION**
UNIVERSITY OF NOTRE DAME
NOTRE DAME, IN 46556

Abstract

Preliminary results on decay modes of charmed particles produced in the high energy photoproduction experiment E687 in the 1990-91 fixed target run at Fermilab are presented. Modes discussed include: the K^-K^+ decay of the D^0; the doubly Cabibbo suppressed $K^+\pi^+\pi^-$ decay of the D^+; the $\pi^-\pi^+\pi^+$ decay mode of the D^+ and the D_s; and $K^-\rho^+$ and $K^*\pi$ substructure in the $K^-\pi^+\pi^0$ decay of the D^0.

We report preliminary results obtained on decay modes of charmed mesons produced in the Fermilab high energy photoproduction experiment E687. The results discussed here are based on the data taken during the 1990-91 fixed target run. The E687 spectrometer has been described previously.[1]

The data samples used for the analyses in this paper come from less than half of the 1990-91 data. Since these analyses are taking place at several collaborating institutions, the samples are not identical for all modes discussed. Each of the analyses is carried out on a subset of the 1990 data sample which is approximately 40% of the total 1990-91 data sample. In some cases, the 1987-88 data are combined with the 1990-91 data. When the text refers to a charm state in this paper, the charge conjugate state is always included unless specifically noted. In addition it should be emphasized that the data analysis is still in its early stages and all results must be taken to be preliminary at this time.

Although this paper summarizes the results of several different analyses each with a different set of data cuts, the procedures used to isolate charm signals have a common thread. In general, the analyses use either a "candidate-driven" algorithm or a "vertex-driven" algorithm to isolate the charmed meson. Charm signals are isolated using vertex quality cuts, vertex separation cuts, vertex isolation (track sharing) cuts, and Cerenkov cuts.

Shown in Fig.1 are the K^-K^+ and $K^-\pi^+$ effective mass distributions for an event sample using

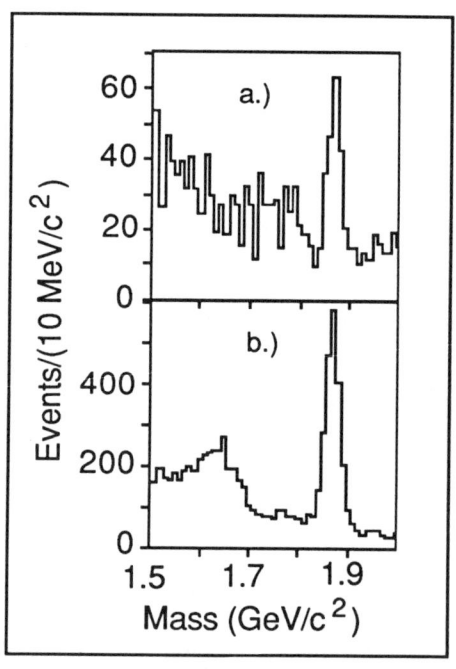

Figure 1. a.) The K^-K^+ effective mass distribution using the vertex-driven algorithm; and b.) the $K^-\pi^+$ effective mass distribution using the vertex-driven algorithm.

---

*Supported by the US DOE and NSF and by INFN
**See G. Bellini in these proceedings for a list of E687 collaborators and institutions.

the vertex-driven algorithm. The data in this figure are selected with: standard Cerenkov cuts; $D/\sigma_D<7$; and $L/\sigma_L >5$. Here D is the transverse miss distance at the primary vertex for the secondary vertex and σ_D is its error; and L is the separation of the vertices with σ_L being its error. (The data sample used was slightly more than 50% of the 1990 subset.) Decays of the D^0 to K^-K^+ and $K^-\pi^+$ are observed. The enhancement in Fig. 1b near 1.65 GeV/c^2 is due primarily to D^0 decays to $K^-\pi^+\pi^0$ where the π^0 is not detected. A preliminary branching ratio of $\Gamma(D^0\to K^-K^+)/\Gamma(D^0\to K^-\pi^+)$ =0.123±0.015 was obtained from fits to the distributions of Fig. 1. This value is consistent with the current world average[2] of 0.113±0.007.

A search for direct CP violation in the decay $D^0\to K^-K^+$ has been made using $D^0(\overline{D^0})$ decays which have been identified by association with a $D^{*+}(D^{*-})$ decay. The association is made using the sign of the pion emitted in the D^* decay. The asymmetry A is defined:

$$A = \frac{|\Gamma(D^0\to K^-K^+)-\Gamma(\overline{D^0}\to K^-K^+)|}{\Gamma(D^0\to K^-K^+)+\Gamma(\overline{D^0}\to K^-K^+)}.$$

The distributions in Fig. 2 show the K^-K^+ distributions for the D^0 and $\overline{D^0}$ samples using the candidate driven algorithm, where the kaons have been identified by the Cerenkov system. There is also a 1% cut on the quality of the decay vertex and a cut requiring $L/\sigma_L > 4$. We observe 61.7±9.0 D^0 decays and 66.0±8.6 $\overline{D^0}$ decays. In order to get a limit on A, a correction for the relative D^{*+} and D^{*-} cross section has been made following the above procedure using the $D^0(\overline{D^0})$ decays to $K\pi$. When the correction factor of 1.030±.058 is applied (to the D^0 mode) an asymmetry of 0.019±0.10 is obtained. Converting this to an upper limit, we obtain $A<0.184$ at the 90% confidence level. The current limit[3] on A is $A<0.45$ (90%).

A measurement (or limit) on the doubly

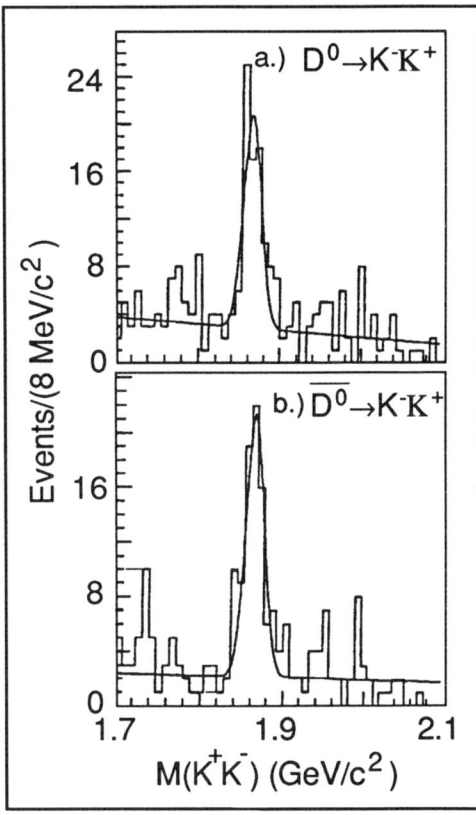

Figure 2. The K^-K^+ effective mass distribution using D^* tagging to obtain separate D^0 and $\overline{D^0}$ signals.

Cabibbo suppressed decay (DCSD) $D^+\to K^+\pi^+\pi^-$ is found by forming the ratio of the signal (or limit on the signal) in this decay mode to that of the allowed decay to $K^-\pi^+\pi^+$. Since these two decays have similar kinematics, it is only necessary to apply the same cuts to the two modes and determine the signal observed in each mode. Shown in Fig. 3 are the mass distributions for: a.) the allowed $K^-\pi^+\pi^+$ mode; and b.) the DCSD $K^+\pi^+\pi^-$ mode. For these distributions, there are kaon Cerenkov requirements; vertex quality and isolation cuts; and a vertex separation cut of $L/\sigma_L > 15$. The signal in the allowed channel in Fig. 3a consists of 2438 ± 52 events with a fitted mass of 1867.8 ± 2 MeV/c^2 and $\sigma = 10.8 \pm 0.2$ MeV/c^2. In the DCSD channel of Fig. 3b, there is no evidence for a signal. Fixing the mass and width expected for a real signal, the fitting procedure yields a signal com-

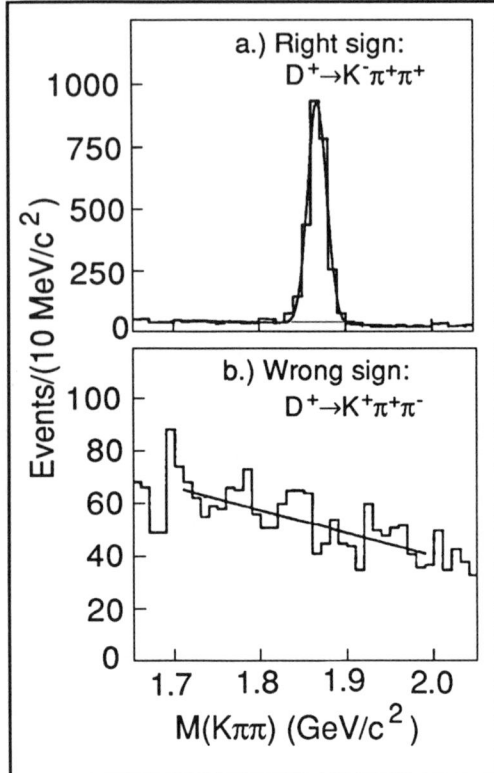

Figure 3. The right-sign and wrong-sign $K\pi\pi$ effective mass distributions used to set a limit on the DCSD mode of the D^+.

patible with zero with an error of ± 15 events. The preliminary 90% confidence level upper limit is thus given as:

$$\mathrm{BR}\left(\frac{D^+ \to K^+\pi^+\pi^-}{D^+ \to K^-\pi^+\pi^+}\right) < 0.01 \text{ (90\% C.L.)}.$$

The signal in Fig. 3a is being investigated for possible contributions from background reflections and the preliminary upper limit quoted here will be improved by these studies as well as by improved statistics.

Shown in Fig. 4a is the $\pi^-\pi^+\pi^+$ effective mass distribution for events found using the candidate-driven algorithm with confidence level cuts and isolation cuts. Other cuts applied include: a momentum cut on the 3π candidate (p > 60 GeV/c);

$L/\sigma_L > 7$; and $\sigma_\ell < 1000\mu$. Clear signals are present at both the D^+ and the D_S mass regions. Indications are that these peaks are $\pi^-\pi^+\pi^+$ decays of these states. Shown for reference in Fig. 4b and 4c are the $K^-\pi^+\pi^+$ and $\phi\pi$ effective mass distributions for events with the same cuts (except for different Cerenkov cuts) as the sample in Fig. 4a. The normalizing modes of the $D^+ \to K^-\pi^+\pi^+$ and the $D_S \to \phi\pi$ are present with very low background. These modes are being used to determine the corresponding branching ratios.

Shown in Fig. 5a is the $K^-\pi^+\pi^0$ effective mass distribution for the sample of events where two photons are detected by the inner electromagnetic calorimeter and which reconstruct to an effective

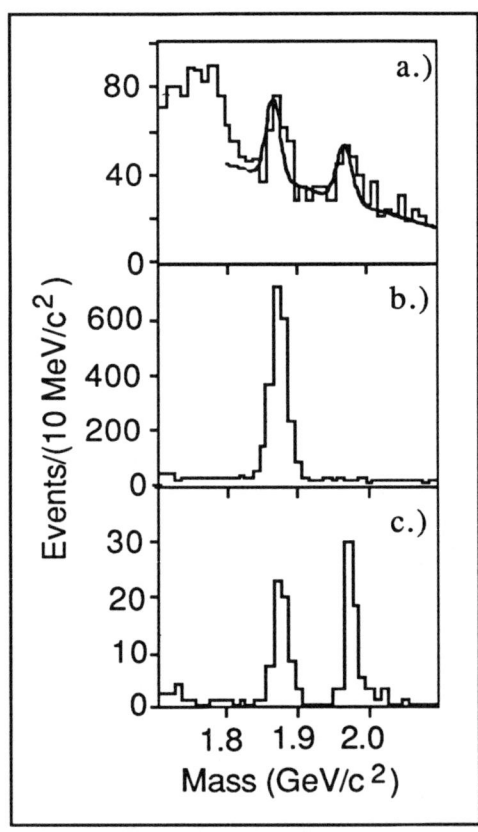

Figure 4. Distributions of the: a.)$\pi^-\pi^+\pi^+$; b.)$K^-\pi^+\pi^+$; and c.)$\phi\pi$ effective masses for the data sample described in the text.

mass $0.09 < M(\gamma\gamma) < .16$ GeV/c^2. Events are required to have a D* tag ($|M(D^*) - M(D)|$ must lie within 2 MeV/c^2 of nominal). Cerenkov and vertex isolation cuts were applied as was a vertex

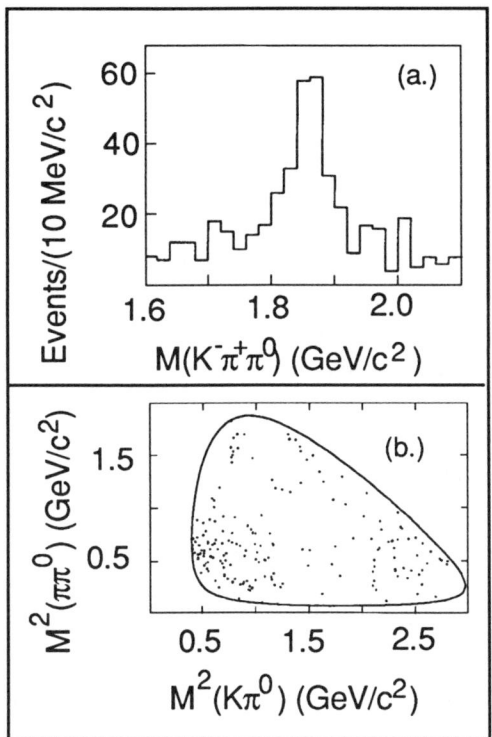

Figure 5 - a.) The $K^-\pi^+\pi^0$ effective mass distribution for the events described in the text; and b.) the Dalitz plot for the events within 2σ of the nominal D^0 mass.

separation cut of $L/\sigma_L > 5$. In Fig. 5b is shown the Dalitz plot for the events in the D^0 mass region. The most prominent feature of the distribution is the clustering of events in the fore-aft decay direction in the region of the $\rho(770)$ meson.

Shown in Fig. 6 are the projections of the Dalitz plot along with preliminary results of an amplitude analysis using the method described in an earlier publication.[4] The fit is a maximum likelihood fit with the coherent sum of amplitudes from the non-resonant, the $K^-\rho^+$, the $K^{*0}\pi^0$, and the $K^{*-}\pi^+$ channels. The fit includes effects due to efficiency, background (as estimated by wrong sign combinations obtained from the data using

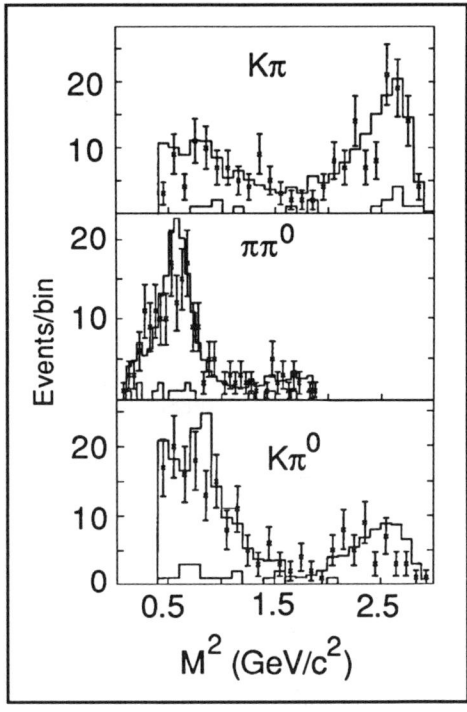

Figure 6 - Dalitz plot projections for the events in Fig. 5b with results of the amplitude analysis shown as the histogram and data shown as the points with error bars. The background is shown as the inner histogram.

the sign of the parent D^* to select the correct charge), and resolution. This preliminary fit indicates that the decay is dominated by the $K^-\rho^+$ mode (branching fraction of $76\pm7\pm9\%$) which is reasonable given the structure of the Dalitz plot. The $K^*\pi$ modes make up most of the rest of the decays with $20\pm5\pm5\%$ $K^{*-}\pi^+$ and $9\pm4\pm1\%$ $K^{*0}\pi^0$. And additional fraction of $5\pm3\pm2\%$ is due to non-resonant decay. (The fractions do not add up to 100% because of interference effects.)

References

1. P.L. Frabetti et al., Nucl. Instr. and Meth. **A320**, 519 (1992).
2. K. Hikasa et al., Phys. Rev. **D45**, S1 (1992).
3. J.C. Anjos et al., Phys. Rev. **D44**, R3371 (1991).
4. P.L. Frabetti et al., Physics Letters **B286**, 195 (1992).

CHARM MESON PRODUCTION IN 600 GEV/C PION-EMULSION INTERACTIONS

D. M. Potter
Physics Department
Carnegie Mellon University
Pittsburgh, PA 15213
(for the E653 collaboration)

Abstract

Results of the second run of experiment E653 at Fermilab are briefly reviewed. Preliminary new results are presented for an upper limit on coherent charm production and an upper limit for BR($D^0 \to \overline{K}^* \pi \mu \nu$). New results for the production and semileptonic decays for D_S mesons are also presented.

INTRODUCTION

Experiment E653 at Fermilab was designed to study production and decay of charm and beauty particles. In the second run, a 600 GeV/c negative pion beam interacted in a 1.5 cm long nuclear emulsion target. Tracks were reconstructed in an 18 plane silicon microstrip vertex detector and momentum analyzed in a magnetic spectrometer with 55 drift chamber planes. The experiment was triggered by a muon, which had passed through an independent downstream magnetic spectrometer. This run recorded 8.2×10^6 triggers, corresponding to 2.5×10^8 interactions.

TWO PRELIMINARY NEW RESULTS

Preliminary new results on three topics will be presented; two of these are briefly summarized below, and the third will be discussed in more detail.

Upper Limit for BR($D^0 \to K^- \pi^+ \pi^- \mu^+ \nu$)

A sample of 4-prong vertices in which one of the tracks was the trigger muon was analyzed to search for the decay mode $D^0 \to K^- \pi^+ \pi^- \mu^+ \nu$. No acceptable candidates were found. From this result, the limit BR($D^0 \to K^- \pi^+ \pi^- \mu^+ \nu$) /BR($D^0 \to K^- \mu^+ \nu$) $<\sim 4\%$ is obtained. The corresponding limit for BR($D^0 \to \overline{K}^* \pi \mu \nu$) is less than 0.2%, which can be compared to the Particle Data Group[1] value of < 1.6%.

Limit on Coherent Charm Production

After the main run of E653 was completed, the experiment was run in a parasitic mode with 800 GeV/c protons to search for coherent production of charm particles. To this end, the emulsion target was replaced with an active silicon wafer target, and the muon trigger was replaced with a multiplicity jump trigger. The latter was designed to trigger on the increase in charge multiplicity resulting from charm decays. This run collected 0.5×10^6 triggers with a trigger rate corresponding to 3% per interaction. No charm signals were seen. The preliminary result, which depends on the mass of the diffracted system, is that the coherent production cross section for charm is less than

25 μb per silicon nucleus. This corresponds to less than 2.5% of the total charm cross section, which has also been measured[2] in E653.

D_S SEMILEPTONIC DECAYS

Motivation and Models

Interest in semileptonic decay modes of the D_S has recently focussed on their role in setting the absolute scale for D_S branching ratios. The technique is to fix the D_S cross section by relating the decay rate for $D_S^+ \to \phi\mu\nu$ to that measured for $D \to \overline{K}^*\mu\nu$ when both modes are observed in the same experiment. It is assumed that the ratio $\Gamma(D_S^+ \to \phi\mu\nu)/\Gamma(D \to \overline{K}^*\mu\nu)$ has less model dependence than the absolute decay widths, which are rather poorly predicted by quark models. This assumption can be generalized. Thus, the measured value of $\Gamma(D \to \overline{K}^*\mu\nu)$ sets the decay width not only for $D_S^+ \to \phi\mu\nu$, but also for $D \to \rho\mu\nu, \omega\mu\nu$, and $D_S^+ \to \overline{K}^{*0}\mu\nu$, and $\Gamma(D \to \overline{K}\mu\nu)$ the decay widths for $D_S^+ \to \eta\mu\nu, \eta'\mu\nu$, $D \to \pi\mu\nu, \eta\mu\nu, \eta'\mu\nu$, and $D_S^+ \to \overline{K}^0\mu\nu$. At at greater level of detail, the form factor ratios measured for $D^+ \to \overline{K}^{*0}\mu\nu$ fix those for $D_S^+ \to \phi\mu\nu$, as well as for the other modes with vector mesons. Consequently, there are many possible tests of the model input used to establish the D_S cross section.

Preliminary Results

In this experiment, we have a preliminary result for the D_S cross section which relies on the model input described above, and additional results which allow the model to be tested. Figure 1 shows the dikaon mass distribution for 3-prong vertices which include the trigger muon. The two hadron tracks in the vertex were assigned the kaon mass, since particle identification was not available. A clear

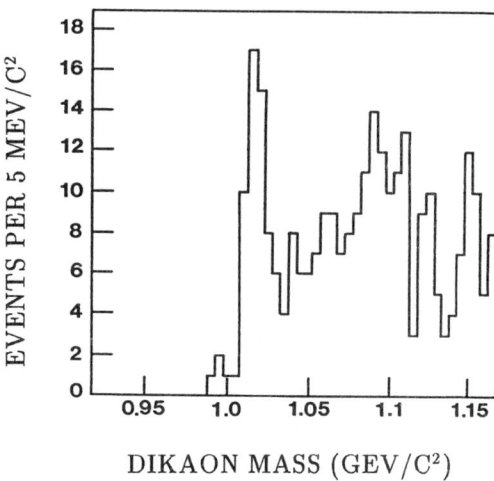

Figure 1. Dikaon mass distribution for a sample of 3-prong vertices which include the trigger muon.

ϕ signal of 36 events on a background of 16 is observed. These events provide a clean sample of $D_S^+ \to \phi\mu\nu$ decays, and a comparison of these to our $D^+ \to \overline{K}^{*0}\mu\nu$ signal yields the following result:

$$\left(\frac{\sigma(D_S^+)}{\sigma(D^+)}\right)_{x_F>0} \cdot \left(\frac{\Gamma(D_S^+ \to \phi\mu\nu)}{\Gamma(D^+ \to \overline{K}^{*0}\mu\nu)}\right) =$$

$$0.66 \pm 0.16 \pm 0.15$$

The ISGW model[3] predicts the ratio of decay rates to be 1.02; with this and our measurement[4] of $\sigma(D^\pm)$, we obtain $\sigma\left(\pi^- N \to D_S^\pm; x_F > 0\right) = 5.6 \pm 1.4 \pm 1.9\,\mu b$/nucleon. Fits of the kinematic distributions to $(1-x_F)^n$ and $e^{-b \cdot p_T^2}$ yielded $n = 5.4^{+1.4}_{-1.2}{}^{+0.8}_{-0.7}$ and $b = 1.09^{+0.25}_{-0.21} \pm 0.03$. These results are similar to our measurements[1] for D^+ and D^0, for which we obtained $n = 4.2$ and $b = 0.76$.

A fit to the angular and q^2 distributions of the decay yields the form factor ratios, $A_2(0)/A_1(0)$ and $V(0)/A_1(0)$, evaluated at $q^2 = 0$. Table 1 presents results for these and also for the ratio of the longitudinal to trans-

Table 1. Form factor ratios and the ratio of transverse to longitudinal decay widths for $D_S^+ \to \phi\mu\nu$ and comparison to those for $D^+ \to \overline{K}^{*0}\mu\nu$ and to various model predictions.

	$A_2(0)/A_1(0)$	$V(0)/A_1(0)$	Γ_L/Γ_T
$D_S^+ \to \phi\mu\nu$			
E653 (this work)	$2.1^{+0.8}_{-0.7}{}^{+0.1}_{-0.2}$	$2.5 \pm 1.1^{+0.3}_{-0.4}$	$0.5 \pm 0.3 \pm 0.1$
ISGW[3]			1.15
BKS[5]	$0.78 \pm 0.08^{+0.17}_{-0.13}$	$2.00 \pm 0.19^{+0.20}_{-0.25}$	
$D^+ \to \overline{K}^{*0}\mu\nu$			
E653[6]	$0.82^{+0.22}_{-0.23} \pm 0.11$	$2.00^{+0.34}_{-0.32} \pm 0.16$	$1.18 \pm 0.18 \pm 0.08$
ISGW[7]	1.0 ± 0.3	1.4 ± 0.4	1.1 ± 0.2
AW/GS[8]	0.8	1.9	1.2
BKS[5]	$0.70 \pm 0.16^{+0.20}_{-0.15}$	$1.99 \pm 0.22^{+0.31}_{-0.35}$	

verse decay widths, Γ_L/Γ_T, which is computed from the measured form factor ratios.

Figure 2 shows the dipion mass distribution for 3-prong vertices which include the trigger muon. Above 0.4 GeV/c^2 the distribution is dominated by $D^+ \to \overline{K}^{*0}\mu\nu$, where the kaon is identified as a pion. Between 0.3 and 0.4 GeV/c^2, there are contributions from $D_S^+ \to \eta\mu\nu$ and/or $D_S^+ \to \eta'\mu\nu$. In this analysis, neither the η nor the η' is fully reconstructed; only the charged particles in their decays are observed. The resulting dipion mass distribution is dominated by the $\pi^+\pi^-\pi^0$ mode of the η, and $\pi^+\pi^-\eta, \eta \to neutrals$ modes of the η'; the dipion mass distributions are virtually identical for these modes. Figure 2 includes a simulated mass distribution for the 2-prong decays of the η and η', where the mix is that corresponding the decay widths predicted[3] for $D_S^+ \to \eta\mu\nu$ and $D_S^+ \to \eta'\mu\nu$; the simulation is in good agreement with the data. Also shown in Fig. 2 is the background obtained from events in which both hadrons have the same charge.

If the two kaons from the decay of a ϕ

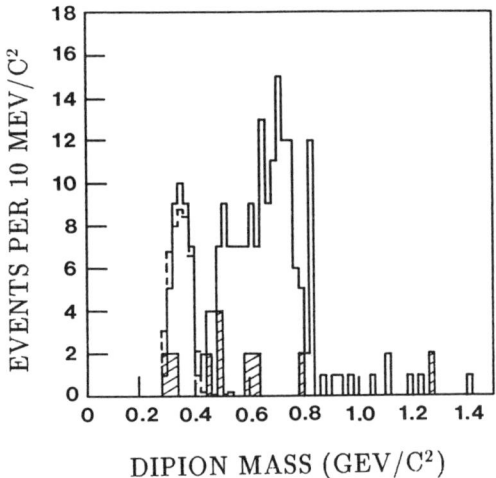

Figure 2. Dipion mass distribution (solid) for a sample of 3-prong vertices which include the trigger muon. Also shown are a simulation (dashed) and the background (hatched), both of which are described in the text.

are assigned the pion mass, the corresponding dipion mass also peaks between 0.3 and 0.4 GeV/c^2, and ϕ's have been excluded in the final results. However, the dikaon mass distribution for the events in Fig. 2 between 0.3 and 0.4 GeV/c^2 is not dominated by the ϕ. We have found no decay modes which could reasonably account for these events, other than $D_S^+ \to \eta\mu\nu$ and $D_S^+ \to \eta'\mu\nu$.

For the η and η' modes to which this experiment is sensitive, the product of efficiency times branching ratio is the same for $D_S^+ \to \eta\mu\nu$ and $D_S^+ \to \eta'\mu\nu$. Since there is no efficient way to distinguish between the two decays, we quote the result:

$$\frac{\Gamma(D_S^+ \to (\eta + \eta')\mu\nu)}{\Gamma(D_S^+ \to \phi\mu\nu)} = 2.4 \pm 1.2$$

The ISGW model[3], as modified according to the discussion in *Motivation and Models*, predicts 0.87 for this ratio.

Discussion

From Table 1 it can be seen that the form factor ratio $A_2(0)/A_1(0)$ for $D_S^+ \to \phi\mu\nu$ is larger than that for $D^+ \to \overline{K}^{*0}\mu\nu$ and also than predicted by several models. Similarly, $\frac{\Gamma(D_S^+ \to (\eta+\eta')\mu\nu)}{\Gamma(D_S^+ \to \phi\mu\nu)}$ is larger than predicted. Although these discrepancies are not large, additional tests of the model input are needed before it can be used with confidence in the determination of the D_S cross section, and hence the normalization of the D_S branching ratios.

REFERENCES

1. K. Hikase, et al., (Particle Data Group), Phys. Rev. **D45**, 1 (1992).

2. K. Kodama, et al., Phys. Lett. **B263**, 573 (1991).

3. D. Scora and N. Isgur, University of Toronto preprint, UTPT-89-29.

4. K. Kodama, et al., Phys. Lett. **B284**, 461 (1992).

5. C. W. Bernard, A. X. El-Khadra, and A. Soni, Phys. Rev. **D45**, 869 (1992).

6. K. Kodama et al., Phys. Lett. **B274**, 246 (1992).

7. N. Isgur, D. Scora, B. Grinstein and M. B. Wise, Phys. Rev. **D39**, 799 (1989).

8. T. Altomari and L. Wolfenstein, Phys. Rev. **D37**, 681 (1988), and F. J. Gilman and R. L. Singleton, Phys. Rev. **D41**, 142 (1990).

PRELIMINARY RESULTS FROM FERMILAB E789

J. C. Peng, J. Boissevain, T. A. Carey, D. M. Jansen, R. Jeppesen, J. S. Kapustinsky
D. W. Lane, M. J. Leitch, J. W. Lillberg, P. L. McGaughey, J. M. Moss
Los Alamos National Laboratory

G. Brown, D. Isenhower, M. Sadler, R. Schnathorst, R. Schwindt
Abilene Christian University

G. Gidal, P. M. Ho, M. S. I. Kowitt, K. B. Luk, D. Pripstein
Lawrence Berkeley Laboratory

L. M. Lederman, M. H. Schub
University of Chicago

C. N. Brown, W. E. Cooper, H. D. Glass, K. N. Gounder, C. S. Mishra
Fermi National Accelerator Laboratory

M. Apolinski, W. Luebke, D. M. Kaplan, V. M. Martin, R. S. Preston,
J. Sa, V. Tanikella
Northern Illinois University

R. L. Childers, C. W. Darden, D. Snodgrass, J. R. Wilson
University of South Carolina

Y. C. Chen*, G. C. Kiang, P. K. Teng
Academia Sinica, Taiwan
National Cheng Kung University, Taiwan*

Abstract

Fermilab experiment 789 studies low-multiplicity decays of neutral D and B mesons in a high-rate fixed-target environment. Preliminary results from the 1991 run are presented.

INTRODUCTION

Experiment 789 at Fermilab was designed to measure low-multiplicity decays of B mesons produced in a high-rate fixed-target environment. The existing E605/E772 spectrometer, which was used in previous experiments to detect hadron and lepton pairs with good mass resolution and high rate capability, was significantly upgraded for E789 (Figure 1). In particular, a silicon microstrip vertex spectrometer and a vertex trigger processor were installed. The main goals of the E789 experiment are 1) to measure the B production cross section at 800 GeV via the detection of inclusive $B \to J/\psi + X$ decays, 2) to search for charmless dihadron decay modes such as $B \to \pi^+\pi^-, K\pi, K^+K^-, \bar{p}p$. The sensitivity for these measurements clearly depends on the rate capability of the spectrometer and on the performance of the silicon microstrip detectors in this high-rate environment (which is similar to those anticipated in future SSC experiments). Preliminary results from analysis of a fraction of the data taken in the 1991 fixed-target run are presented in this paper.

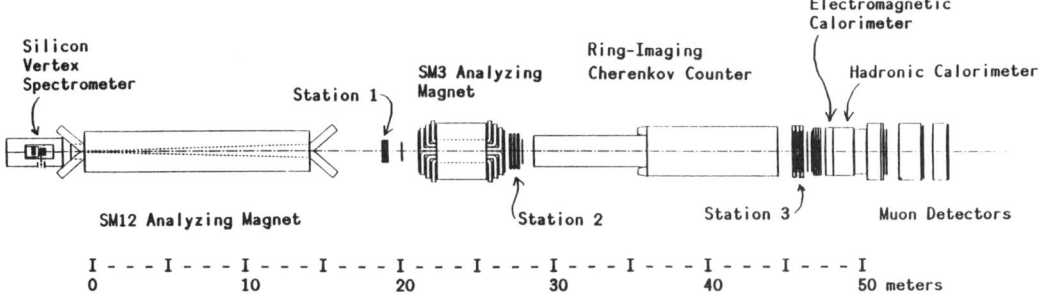

Figure 1. E789 apparatus (plan view).

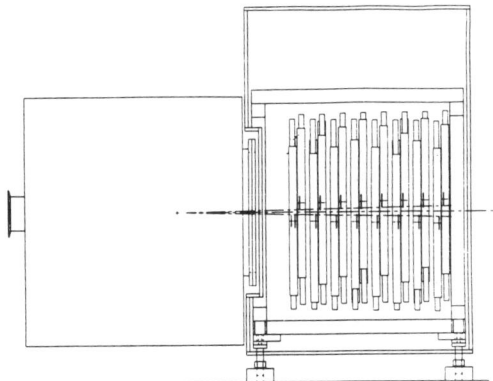

Figure 2. E789 silicon spectrometer.

SUMMARY OF E789 1991 RUN

During the 1991 run, an 800 GeV proton beam was incident on one of several thin wire targets, ranging from 0.1 mm to 0.2 mm high and 0.8 mm to 3 mm thick. Sixteen silicon microstrip detectors (SMD's) were positioned downstream of the target to cover the angular range from 20 mr to 60 mr above and below the beam axis (Figure 2). The silicon detectors, type 'B' from Micron Semiconductor, have 5×5 cm^2 area, 300 μm thickness, and 50 μm pitch. The SMD's were oriented to measure either the Y (vertical) or the U,V coordinates (5° stereo angle). Signals from 8,544 silicon strips were individually read out via Fermilab 128-channel amplifier cards[1] and LBL discriminators[2] synchronized to the accelerator RF. The discriminated signals were then transmitted through $\approx$ 400 ns of multiconductor cable to coincidence registers. A vertex processor[3], which finds tracks in the silicon detectors and selects events with decay vertices downstream of the target, was also implemented and functional. The use of a thin target localizes the primary interaction vertex and greatly simplifies the design of the vertex processor, which only needs to determine the decay vertex.

Two different settings of the spectrometer, which separately optimize the acceptance for charm or beauty decays, were used in the 1991 run. The charm running served to check the performance of the SMD's and the vertex trigger processor. In addition, the nuclear dependence of D meson production was measured on beryllium and gold targets, which should provide information related to the origin of the J/ψ A-dependence observed[4] in E772 at the same beam energy. The vertex

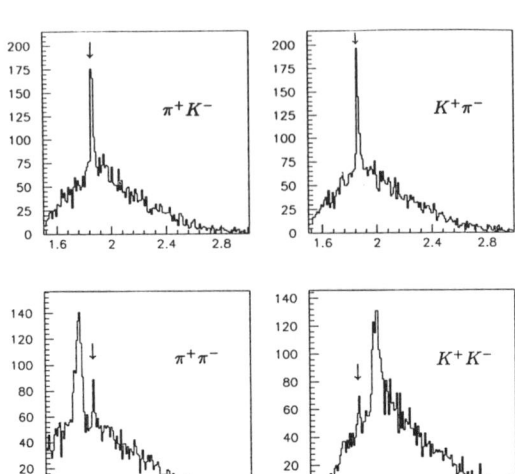

Figure 3. Mass spectra for dihadron events reconstructed with various assumptions for the hadron species. The arrows indicate the D decay peaks.

trigger processor and the upgraded data acquisition system enabled us to take up to 10^{10} protons per beam spill on a 1.5 mm-thick gold target (4 MHz interaction rate).

For the beauty running, the spectrometer was set at a configuration to accept both $B \to J/\psi X$ and $B \to h^+h^-$ decays. We took data at 5×10^{10} protons per pulse on a 3-mm-thick gold target, corresponding to a 50-MHz interaction rate, without using the vertex trigger processor.

PRELIMINARY RESULTS FROM E789

Figure 3 shows the dihadron mass spectra obtained from an analysis of ≈ 10 % of our charm data sample. To effectively reject the dihadron background, we require that both tracks do not point back to the target, and that the proper lifetime is greater than 0.6 ps. The efficiency of each silicon plane is found to be better than 90 % and the impact parameter resolution of ≈ 30 μm gives a Z-vertex resolution of ≈ 1 mm for typical D events. More details on the performance of the SMD's in this high rate experiment have been reported elsewhere[5]. Information from the ring-imaging Cherenkov detector has not yet been used for π/K identification, and the plots in Fig. 3 are obtained by assigning either pion or kaon mass to the hadrons. The $D^0 \to \pi^+ K^-, \bar{D}^0 \to K^+\pi^-$, and $D^0, \bar{D}^0 \to \pi^+\pi^-, K^+K^-$ decays are clearly visible in Fig. 3. Although the D^0 mass resolution shown in Fig. 3 is ≈ 10 MeV, more recent studies show that a 5 MeV mass resolution is obtained with an improved field map of the analyzing magnet.

Figure 4 shows the dimuon mass spectrum from a preliminary analysis of ≈ 15 % of our beauty data sample. Good silicon tracks are required for events shown in Fig. 4, but no vertex cuts are applied. Approximately 15,000 J/ψ and 300 ψ' events are observed. To search for $B \to J/\psi X$ events, the impact parameters of both muon tracks are required to be greater than 150 μm and the decay vertex is required to be at least 7 mm downstream of the target (7 mm $< Z_{vertex} <$ 5 cm). Fig. 5(a) shows that

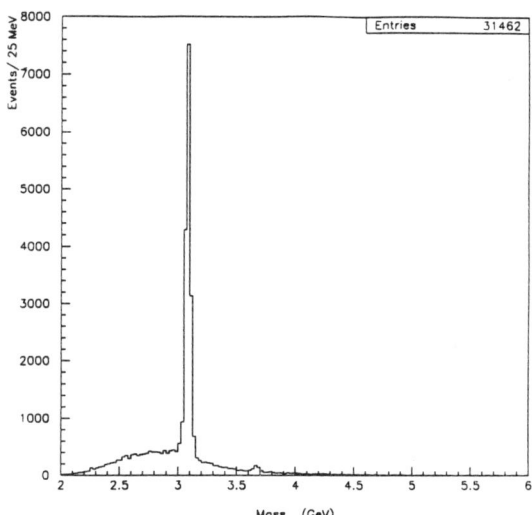

Figure 4. Mass spectrum for dimuon events.

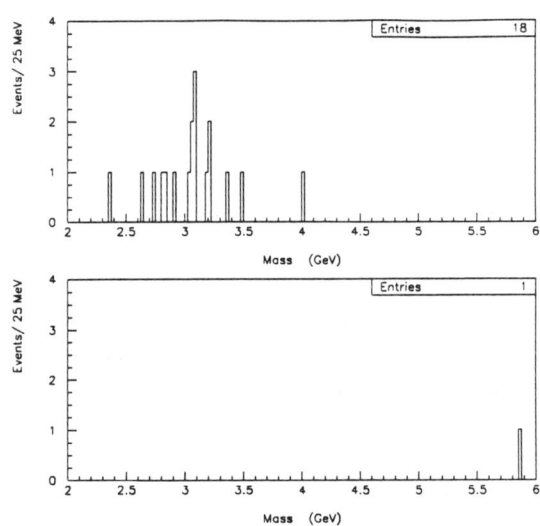

Figure 5. Mass spectra for dimuon events passing a) downstream Z-vertex cuts, b) upstream Z-vertex cut.

a total of eighteen events, six of them in the J/ψ mass region, survive these cuts. These six events are considered candidate events for $B \to J/\psi X$ decays. To estimate the background which might be caused by silicon tracking errors, we also select events with decay vertices upstream of the target, namely -5 cm $< Z_{vertex} <$ - 7 mm. Fig. 5(b) shows that this background is unimportant. More data need to be analysed before a more definitive b signal and b-production cross section can be obtained.

In summary, the E789 experiment explores the feasibility to study b-physics in a high-rate fixed-target environment. Preliminary results show that the silicon vertex spectrometer works well. Six $B \to J/\psi X$ candidate events were observed from an analysis of $\approx$ 15 % of the dimuon data sample.

REFERENCES

1. D. Christian et al., IEEE Trans. Nucl. Sci. NS-36 (1989) 507.
2. B. T. Turko et al., "A Multichannel Discriminator System for Silicon Strip Detector Readout," to appear in IEEE Trans. Nucl. Sci.
3. C. Lee et al., IEEE Trans. Nucl. Sci. 38 (1989) 461.
4. D. M. Alde et al., Phys. Rev. Lett. 66 (1991) 133.
5. J. S. Kapustinsky et al., "Radiation Damage Effects on the Silicon Microstrip Detector in E789," presented at the International Conference on Advanced Technology and Particle Physics, Como, Italy, 1992.

FEYNMAN-x AND TRANSVERSE MOMENTUM DEPENDENCE OF $D^{\pm}$ AND $D^0, \overline{D}^0$ PRODUCTION IN 250 GeV π^--NUCLEON INTERACTIONS*

Paul E. Karchin
Physics Department
Yale University
J.W. Gibbs Laboratory, P.O. Box 6666
New Haven, CT 06511
Representing the Fermilab E769 Collaboration[1]

Abstract

We measure the differential cross section with respect to Feynman-x (x_F) and transverse momentum (P_T) for charm meson production using targets of Be, Al, Cu and W. In the range $0.1 < x_F < 0.7$, $d\sigma/dx_F$ is well fit by the form $(1 - x_F)^n$ with $n = 3.9 \pm 0.2$. The difference between n values for D^- and D^+ is 1.1 ± 0.4. However, we find an asymmetry of 0.18 ± 0.04 favoring the production of D^- compared to D^+. In the lower P_T range, < 2 GeV, $d\sigma/dP_T^2$ is well fit by the form $\exp(-b \times P_T^2)$ with $b = 1.03 \pm 0.04$ GeV^{-2}, while in the higher P_T range, 0.8 to 3.6 GeV, it is well fit by the form $\exp(-b' \times P_T)$ with $b' = 2.76 \pm 0.06$ GeV^{-1}. The shape of the differential cross section has no significant dependence on atomic mass of the target material.

Although charm was co-discovered using hadronic interactions in 1974, the production mechanism is still not well established. In perturbative QCD, charm is hadroproduced by gluon-gluon fusion and quark-anti-quark annihilation. The total and differential cross sections for charm quark production have been calculated using perturbative QCD by Nason, Dawson, and Ellis[2] (NDE) including terms up to next-to-leading order (NLO). While some measurements of total cross section and differential cross section in x_F and P_T are consistent with the NDE predictions, other experiments (listed in ref. 6.) report either a total cross section larger than can be accommodated by NLO QCD and/or show large differences in the production of leading versus non-leading charm. (Leading charm particles have at least one valence quark that was transferred from the projectile.) To explain some of these measurements, Combridge[3] and Brodsky et al.[4] postulated production from the charm component of the parton sea of hadrons.

Fermilab experiment E769 addresses these issues with a high statistics study of exclusively reconstructed charm particles using 250 GeV secondary beams with identified $\pi^{\pm}$, $K^{\pm}$ and p on thin foil targets of Be, Al, Cu and W. The data were collected during the 1987-88 running period. In this letter we report differential cross sections measured using the decays $D^+ \to K^-\pi^+\pi^+$ and $D^0 \to K^-\pi^+$ (and charge conjugates). Results on the total charm cross section and the dependence on atomic mass and incident particle type are in preparation.

The open-geometry spectrometer used in E769 was substantially the same as previously used in Fermilab photoproduction experiment E691[5]. More detailed descriptions of the apparatus are found in ref. 6 and references quoted

---

*This work was supported, in part, under D.O.E. contract DE-AC02-76ER-03075.

therein. For E769, a differential Cerenkov counter (DISC) was installed for beam particle identification. With the DISC set to tag kaons, the beam sample used in this analysis consisted of 95% π^- with a contamination of 3% K^- and 2% $\bar{p}$. The trigger required total transverse energy in the calorimeters > 5.5 GeV. This reduced the event rate by a factor of 3 while maintaining an efficiency of about 75% for charm. About 400 million events were recorded in the entire data set, 150 million of these from π^- interactions, and the remainder from π^+, $K^\pm$, and p interactions.

After reconstruction and analysis cuts, which are described in ref. 6, the resulting invariant mass plots for the two decay channels are shown in Fig. 1. The numbers of signal events, determined from binned maximum likelihood fits, are 700 ± 24 for the charged D and 607 ± 29 for the neutral D. To determine the differential distributions, we fit a mass plot for each bin of x_F and P_T. The acceptance was calculated from a complete Monte Carlo simulation which included the effects of the resolution, geometry and efficiency of all detectors, interactions in the apparatus, and all analysis cuts. The acceptance for charged and neutral D's is similar. As a function of x_F, it varies between 2% and 6% in the range 0.0 to 0.8 with a maximum at 0.25. The acceptance in P_T increases from 4% to 8% over the range 0 to 4 GeV. Systematic errors in the acceptance include the uncertainties in trigger simulation and detector efficiencies. The systematic errors are small compared to statistical errors in the data and thus we quote only one error (statistical) in our results.

Our measured $d\sigma/dx_F$ for charged and neutral particles combined is shown in Fig. 2. We performed a least-squares fit to the data using the functional form $(1-x_F)^n$ which gives a good fit in the range $0.1 \leq x_F \leq 0.7$ with $n = 3.9 \pm 0.2$. As shown in Fig. 2, the form $(1-x_F)^n$ does not give a good fit to the data or to the theoretical prediction for $x_F < 0.1$. The theoretical prediction we show is that of NDE for charm quark production for a quark mass of 1.5 GeV, scaled to the data. A fit to these theory points over the same range as the data yields $n = 4.25$. While the prediction matches the data well, we note that the effect of fragmentation of the quarks into mesons is not taken into account in the calculation of NDE.

Our results for combined charged and neutral D's are compared with those from other experiments in Table I. Our values of n are consistent with those measured in NA32[7] but our value for the difference in n between leading and non-leading charm (0.3 ± 0.4) does not confirm the large difference suggested by the NA27 measurement[8] of 6.1 ± 1.5. To best test for a leading particle effect versus x_F, we analyzed our x_F distributions separately for D^- (leading) and D^+ (non-leading). We chose charged D's because the leading/non-leading character is the same whether the charged D is directly produced or results from the decay of a D^*, while this is not the case for the neutral D's. We measure $n(D^+) - n(D^-) = 1.1 \pm 0.4$.

We further test for a leading particle effect by calculating the asymmetry, $A(x,y) = [\sigma(x) - \sigma(y)]/[\sigma(x) + \sigma(y)]$, where $\sigma(x)$ is the number of events with meson x divided by the acceptance for meson x, as a function of x_F, integrated over $x_F > 0$. We obtain $A(D^-, D^+) = 0.18 \pm 0.04$ and $A(D^0, \overline{D}^0) = -0.06 \pm 0.05$. Thus we see a 4.5σ asymmetry in the charged D channel but no effect in the neutral channel.

Our results for $d\sigma/dP_T^2$ are plotted in Fig. 3. At high P_T (> 2 GeV) there is a significant deviation from the functional form $\exp(-b \times P_T^2)$, an effect not seen in experiments with smaller statistics. To compare our measurement with those from other experiments, we fit our data to the above form in a limited range as shown in Fig. 3 and in Table I.

Table 1. Production parameters for incident π^- and for neutral and charged D mesons, combined, from this experiment compared to other measurements.

Expt.	E769	NA32	NA27
P_{beam} (GeV)	250	230	360
Target(s)	Be, Al, Cu, W	Cu	H
x_F Fit Range	0.1 to 0.7	0.0 to 0.8	0.0 to 0.9
n(all)	3.9±0.2	3.74±0.23±0.37	3.8±0.63
$n(D^-, D^0)$	3.7±0.3	$3.23^{+0.30}_{-0.28}\pm0.32$	$1.8^{+0.6}_{-0.5}$
$n(D^+, \overline{D}^0)$	4.0±0.3	$4.34^{+0.36}_{-0.35}\pm0.43$	$7.9^{+1.6}_{-1.4}$
Fit Range in P_T^2 (GeV^2)	0 to 4	0 to 10	0 to 4.5
b(all)(GeV^{-2})	1.03±0.04	0.83±0.03±0.02	$1.18^{+0.18}_{-0.16}$
$b(D^-, D^0)$	1.07±0.05	0.74±0.04±0.02	
$b(D^+, \overline{D}^0)$	0.99±0.05	0.95±0.05±0.03	

For $P_T > 0.8$ GeV, $d\sigma/dP_T^2$ is well fit by the form $\exp(-b' \times P_T)$. This is shown in Fig. 4 which also shows the prediction for charm quarks using LO QCD. This is also well fit with the same functional form. In addition to the previous requirement, $P_T > 0.8$ GeV, we restrict the fit range to $P_T < 2.4$ GeV so that we can compare b' values for various sub-samples, which were measured with lower statistics. Fitting the theory points in the same range yields $b' = 2.16$ GeV^{-1}. For the full sample we use the larger range $0.8 < P_T < 3.6$ GeV to obtain our best value of $b' = 2.76 \pm 0.06$ GeV^{-1}.

If charm production is due to short distance collisions and fragmentation, then there should be little dependence of the x_F, P_T, or P_T^2 distributions on the atomic mass of the target material. To test this hypothesis, we fit these distributions separately for Be, Al, Cu and W for the values of n, b' and b (reported in ref. 6). We see no significant dependence of these parameters on atomic mass.

REFERENCES

1. Centro Brasileiro de Pesquisas Físicas, (Brazil), Fermilab, Univ. of Mississippi, Northeastern Univ., Univ. of Toronto (Canada), Tufts Univ., Univ. of Wisconsin, Yale Univ.

2. P. Nason, S. Dawson, K. Ellis, *Nucl. Phys.* B327, pp. 49, (1989).

3. B.L. Combridge, *Nucl. Phys.* B151, pp. 429, (1979).

4. S.J. Brodsky et al., *Phys. Lett.* 93B, pp. 451 (1980).

5. E691 Collab., J.R. Raab et al., *Phys. Rev.* D37, pp. 2391 (1988).

6. E769 Collab., G.A. Alves et al., submitted to *Phys. Rev. Lett.*, Fermilab Pub-92/208 (1992).

7. ACCMOR Collaboration, S. Barlag et al., *Z. Phys.* C49, pp. 555 (1991).

8. NA27 LEBC-EHS Collab., M. Aguilar-Benitez et al., *Phys. Lett.* 161B, pp. 400 (1985); 201B, pp. 176 (1988).

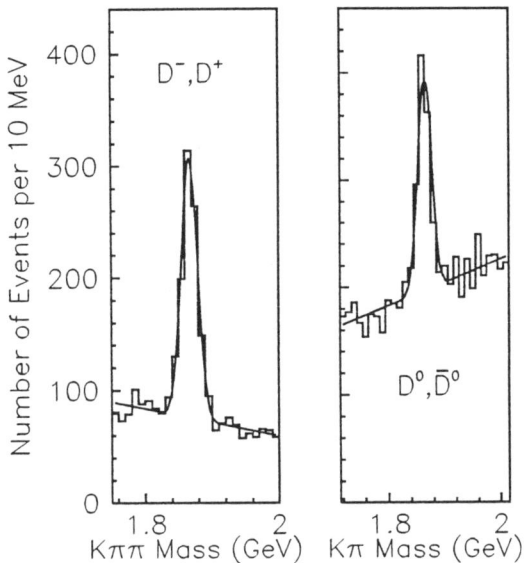

Figure 1. Invariant mass distribution for (left plot) $D^+ \to K^-\pi^+\pi^+$ (+ c.c.) and (right plot) $D^0 \to K^-\pi^+$ (+ c.c.).

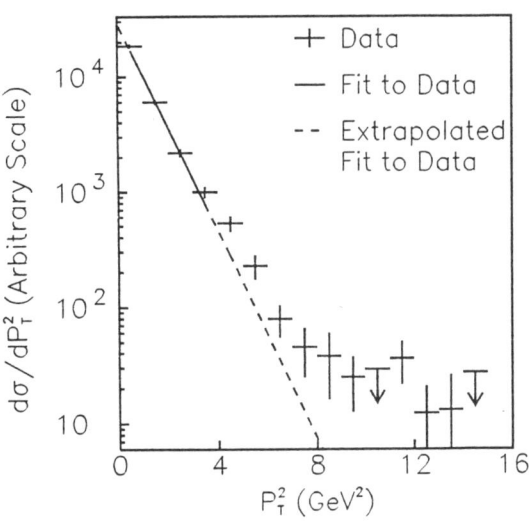

Figure 3. Measured $d\sigma/P_T^2$ versus P_T^2 and a fit over the range 0 to 4 GeV^2 (solid curve), extrapolated to the full range (dashed curve). The arrows indicate 90% C.L. upper limits.

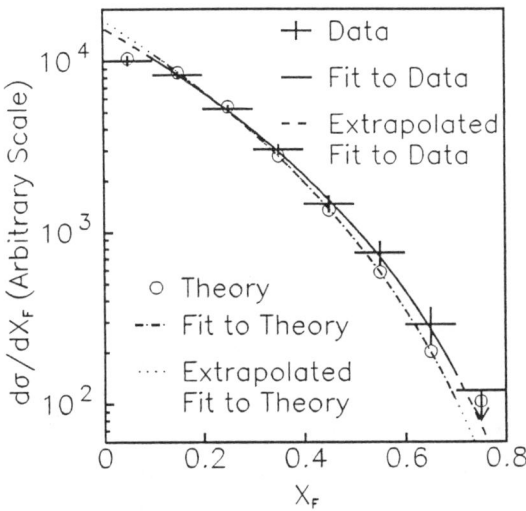

Figure 2. Measured $d\sigma/dx_F$ and comparison with the NLO QCD prediction for quarks[2]. The arrow indicates 90% C.L. upper limit. The solid (dashed-dotted) curve is from a fit to the measurements (theory) over the range 0.1 to 0.7, and the dashed (dotted) curve shows the extrapolation of this fit over the full range.

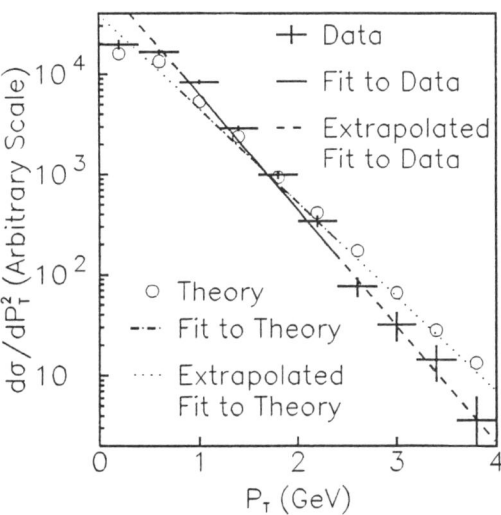

Figure 4. Measured $d\sigma/P_T^2$ versus P_T and comparison with the LO QCD prediction for quarks. The solid (dashed-dotted) curve is a fit to the measurements (theory) over the range 0.8 to 2.4 GeV, and the dashed (dotted) curve shows the extrapolation of this fit over the full range.

FERMILAB E791 *

L.M. Cremaldi,[6] E.M. Aitala,[6] F.M.L. Almeida,[9] S. Amato,[1] J.C. Anjos,[1] J.A. Appel,[4] D. Ashery,[10] J. Astorga,[12] S. Banerjee,[4] S. Beck,[10] I. Bediaga,[1] G. Blaylock,[2] S.B. Bracker,[11] P.R. Burchat,[2] R. Burnstein,[5] T. Carter,[4] I. Costa,[1] K. Denisenko,[4] C. Darling,[14] P. Gagnon,[2] S. Gerzon,[10] K. Gounder,[6] D. Granite,[7] M. Halling,[4] C. James,[4] P.A. Kasper,[5] S. Kwan,[4] J. Lichtenstadt,[10] B. Lundberg,[4] J.R.T. de Mello Neto,[1] R. Milburn,[12] J. M. de Miranda,[1] A. Napier,[12] A. Nguyen,[7] A.B.d'Oliveira,[3] K.C.Peng,[5] M.V. Purohit,[8] B.Quinn,[6] S. Radeztsky,[13] A. Rafatian,[6] A.J. Ramalho,[6] N.W. Reay,[7] K. Reibel,[7] J.J. Reidy,[6] H. Rubin,[5] A. Santha,[3] A.F.S. Santoro,[1] A. Schwartz,[8] M. Sheaff,[13] R.A. Sidwell,[7] H. daSilva Carvalho,[9] J. Slaughter,[14] M.D. Sokoloff,[3] M. Souza,[1] N. Stanton,[7] K. Sugano,[2] D.J. Summers,[6] S. Takach,[14] K. Thorne,[4] A. Tripathi,[7] D. Trumer,[10] S. Watanabe,[13] J. Wiener,[8] N. Witchey,[7] E. Wolin,[14] D. Yi[6]

[1]Centro Brasileiro de Pesquisas Fisicas, Rio de Janeiro
[2]University of California, Santa Cruz, CA 95064
[3]University of Cincinnati, Cincinnati, OH 45221
[4]Fermilab, Batavia, IL 60510
[5]Illinois Institute of Technology, Chicago, IL 60616
[6]University of Mississippi, Oxford, MS 38677
[7]Ohio State University, Columbus, OH 43210
[8]Princeton University, Princeton, NJ 08544
[9]Universidade Federal do Rio de Janeiro
[10]Tel-Aviv University, Tel-Aviv 69978
[11]317 Belsize Drive, Toronto, Ontario M4S1M7
[12]Tufts University, Medford, MA 02155
[13]University of Wisconsin, Madison, WI 53706
[14]Yale University, New Haven, CT 06511

Abstract

Fermilab E791, a very high statistics charm particle experiment, recently completed its data taking at Fermilab's Tagged Photon Laboratory. Over 20 billion events were recorded through a loose transverse energy trigger and written to 8mm tape in the 1991-92 fixed target run at Fermilab. This unprecedented data sample containing charm is being analysed on many-thousand MIP RISC computing farms set up at sites in the collaboration. A glimpse of the data taking and analysis effort is presented. We also show some preliminary results for common charm decay modes. Our present analysis indicates a very rich yield of over 200K reconstructed charm decays.

---

*This work was supported by the U.S. D.O.E. and N.S.F. the U.S.-Israel Binational Science Foundation, and the Brasilian Conselho Nacional de Desenvolvimento Científico e Tecnológico.

INTRODUCTION

E791 is the fourth in a series of charm particle experiments performed at Fermilab's Tagged Photon Lab (TPL) over the past several years. The charm sample is produced through 500 GeV/c $\pi^- N$ interactions in a platinum-diamond target. The data is recorded with a low bias transverse energy trigger (E_T) formed in the hadron and electromagnetic calorimeters. The goal is to reconstruct over 100K charm decays for high statistics studies and a for a close look at rare charm decay physics. As in the past, E791 uses a high precision silicon vertex detector and a large open geometry spectrometer to extract charm decays on low backgrounds.

The success of the E791 hinges on its high data rate capability and offline reconstruction of data. A high rate data acquisition system was built [1] to record a multi-billion event data sample necessary for extracting charm with such high statistics. The challenge ahead is the offline reconstruction of this very large event sample. Computing costs are dropping at a rate of about a factor of two each year, facilitating the timely processing of this large data sample.

SPECTROMETER

The spectrometer features 23 planes of silicon microstrip detectors covering a ± 100 mrad solid angle in the forward direction. The magnetic spectrometer consists of two horizontally bending dipoles with a combined p_T kick of about .5 GeV/c. 35 planes of drift chambers are interleaved throughout for charged particle tracking. Typical momentum resolutions for high mass charm states are in the 8-12 MeV/c^2 range depending on the decay multiplicity and decay energy available.

The target consisted of a series of foils, first .5 mm Pt foil followed by four 1.5 mm C foils, each spaced 1.5 cm apart. This arrangement optimizes the detection of short lived states decaying in the air gaps between foils and provides sufficient interaction rate for the experiment. The target represents about .03 interaction length and produces about a 40 KHz interaction rate during our beam spill.

The beam particles are tracked into the target by a set of 8 upstream MWPC's positioned in the beamline and 6 planes of silicon microstrips (SMDs) located just upstream of the target assembly. The beam tracking allows a precise transverse beam location ($7\mu m$), while the z position of the interaction point can be easily isolated in a target foil. We use this beam constraint in locating the primary vertex interaction point with high efficiency.

Particle identification is provided by two multi-cell threshold Cherenkov counters[2], giving π/K separation in the 6-72 GeV/c momentum range. An electromagnetic and hadron calorimeter[3,4] provide good e/π separation, as well as photon identification. Muons are tagged in a pair of hodoscopes following steele absorber at the downstream end of the spectrometer.

DATA SAMPLE

E791 recorded data with a loose transverse energy trigger (E_T) and beam track requirement. The E_T triggered event rate was about 9000 events/sec. E791's high rate data acquisition system, capable of logging 10 MB/sec., recorded about 4000 events each second during spill and interspill. These events were written to 8mm tape. 24000 data tapes were recorded in the 6 month running period, corresponding to 20 billion events.

COMPUTING

With such a large data sample, computing becomes a major issue. We have so far implemented two large RISC based computing farms at the University of Mississippi and

Ohio State University for reconstruction of E791 data. Each of these farms is equivalent to about 900 and 1500 MIPs respectively, with plans for future expansions.

These startup farms went into action in February 1992, just a month after the fixed target run had ended. At that time, using a preliminary version of our rconstruction code, we extracted our first charm signals.[5,6]

Future farm activity is planned at Fermilab on the large IBM and Silicon Graphics systems set up at their computing center. In addition, the CBPF group is adapting a set of ACPII processors at FNAL for E791 use, in a joint FNAL-CBPF effort. In all E791 will have over 7000 MIPs of dedicated computing contributing to the 1-2 year reconstruction effort.

CHARM PHYSICS

The physics potential from E791 is enormous. It is the highest statistics charm experiment done to date. We will be able to improve the lifetime measurements of D-mesons to an unprecedented accuracy and make substantial improvements to charm baryon lifetime measurements. Our high statistics D-meson samples can be used to search for $D^0 - \overline{D}^0$ mixing, singly and doubly Cabibbo suppressed decays as $D^+ \to K^+\pi^+\pi^-$, and $D^0 \to K^+\pi^-$. We will be able to make major contributions to charm semileptonic decays, especially in the Cabibbo suppressed decays as $D^0 \to \pi^- l^+ \nu$. Searches for flavor changing neutral currents, $D^+ \to \pi^+\mu^+\mu^-$, and for CP violation in D decays, such as in $D^0 \to K^+K^-$, may be made. E791's good efficiency for detecting Λ, Ξ, and K_s vee decays will give us high sensitivity to many rare charm baryon decay channels.

Thus far our efforts have been focused on optimization of reconstruction code and surveys of the such charm decay modes as $D \to K\pi, K2\pi$, and $K3\pi$ decays. For these studies, a vertex separation cut corresponding to 6-8 σ_z is made between primary and secondary charm vertex. We require that the charm decay points back to the primary or there be transverse momentum balance about the D-meson line of flight. Charm signals in these decay modes look very promising and are displayed in Figure 1. We presently estimate a reconstructed charm yield of over 200K events into these modes.

CONCLUSION

E791 has reached a milestone in recording a data set with more than 200K fully reconstructable charm decays. We are improving our code and will begin a physics pass on the data soon. We are looking forward to the start of our charm physics analyses.

REFERENCES

1. S. Amato et al., "The E791 Parallel Architecture Data Acquisition System", to be published in NIM.

2. D. Bartlett et al., Nucl. Inst. Meth. **A260**(1987) 55.

3. J.A. Appel et al., Nucl. Inst. Meth. **A243**(1986) 361.

4. V.K.Bharadwaj et al., Nucl. Inst. Meth. **228**(1985) 283., D.J. Summers, Nucl. Inst. Meth. **228**(1985) 290.

5. D.J. Summers et al., "Charm Physics at Fermilab E791", UMS/HEP/92-020, May 28, 1982. XXVIIth Rencontre de Moriond, Electroweak Interactions and Unified Theories, Les Arcs, Francs, 15-22 March 1992.

6. K. Thorne et al.,"Update on Hadroproduced Charm at TPL", FERMILAB-Conf-92/174. XXVIIth Rencontre de Moriond, QCD and High Energy Hadronic Interactions, Les Arcs, Francs, March 1992.

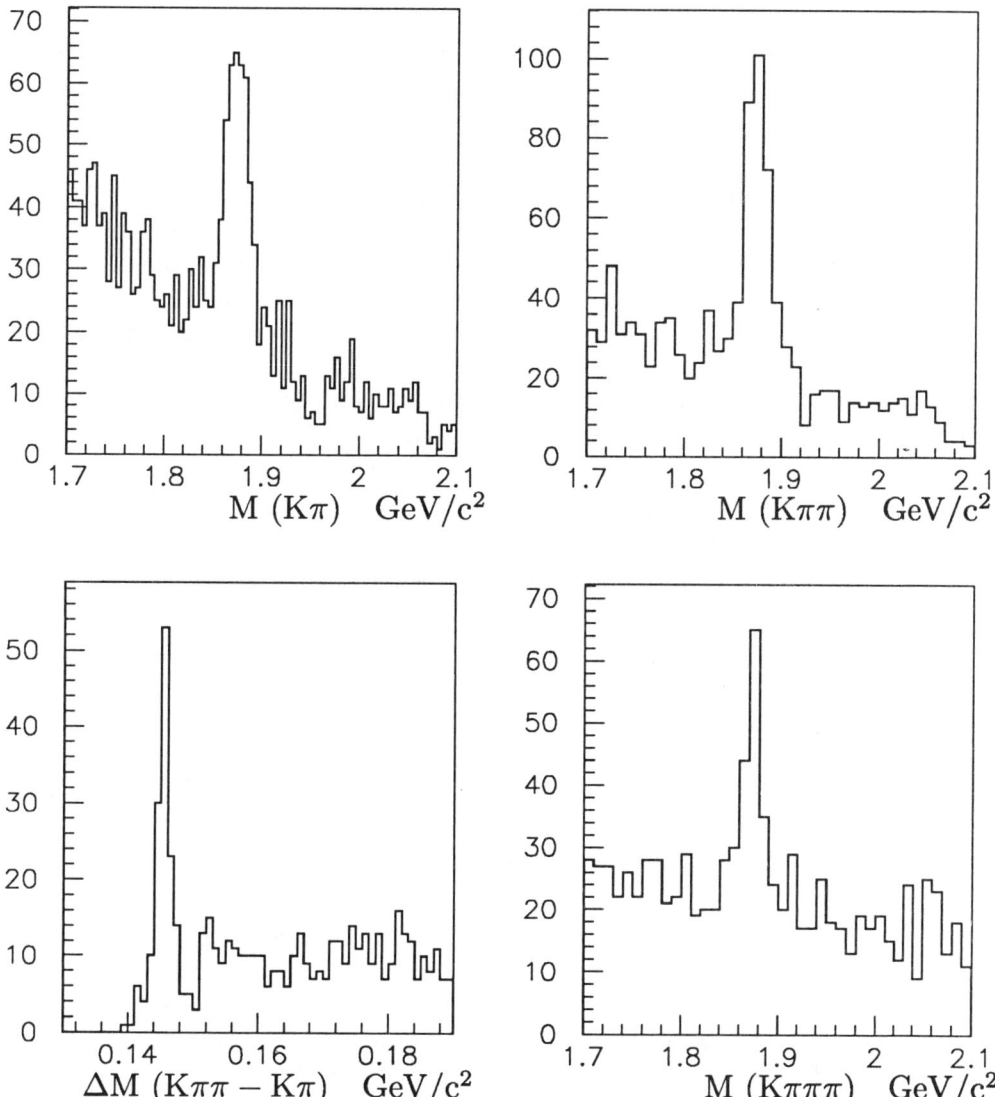

Figure 1. Charm signals extracted from a small fraction of the E791 data set.

HADROPRODUCTION OF χ_c STATES
IN 530 GeV/c π^- INTERACTIONS WITH NUCLEAR TARGETS

Presented by A. Zieminski[2]

R. Li[2], V. Abramov[3], Yu. Antipov[3], B. Baldin[3], R. Crittenden[2], L. Dauwe[6],
C. Davis[5], S. Denisov[3], A. Dyshkant[3], A. Dzierba[2], V. Glebov[3], H. Goldberg[4],
A. Gribushin[2], R. Jesik[4], S. Kartik[2], V. Koreshev[3], J. Krider[1], A. Krinitsyn[3],
S. Margulies[4], T. Marshall[2], J. Martin[2], H. Mendez[4], A. Petrukhin[3], V. Sirotenko[3],
P. Smith[2], J. Solomon[4], T. Sulanke[2], C. Thoma[6], F. Vaca[4], A. Zieminski[2]

E672 Collaboration

[1] *Fermi National Accelerator Laboratory, Batavia, IL 60510*
[2] *Indiana University, Bloomington, IN 47405*
[3] *Institute for High Energy Physics, Serpukhov, USSR*
[4] *University of Illinois at Chicago, Chicago, IL 60680*
[5] *University of Louisville, Louisville, KY 40292*
[6] *University of Michigan at Flint, Flint, MI 48502*

G. Alverson[5], S. Blusk[7], C. Bromberg[4], P. Chang[5], B. C. Choudhary[2], W. H. Chung[7],
L. de Barbaro[8], W. Dlugosz[5], J. Dunlea[8], E. Engels, Jr.[7], G. Fanourakis[8], G. Ginther[8],
K. Hartman[6], J. Huston[4], V. Kapoor[2], C. Lirakis[5], F. Lobkowicz[8], P. Lukens[3],
S. Mani[1], J. Mansour[8], A. Maul[4], R. Miller[4], E. Pothier[5], R. Roser[8],
P. Shepard[7], D. Skow[3], P. Slattery[8], L. Sorrell[4], N. Varelas[8], D. Weerasundara[7],
C. Yosef[4], M. Zielinski[8]

E706 Collaboration

[1] *University of Californi-Davis, Davis, California 95616*
[2] *University of Delhi, Delhi 11 00 07, India*
[3] *Fermi National Accelerator Laboratory, Batavia, Illinois 60510*
[4] *Michigan State University, East Lansing, Michigan 48824*
[5] *Northeastern University, Boston, Massachusetts 02115*
[6] *Pennsylvania State University, University Park, Pennsylvania 16802*
[7] *University of Pittsburgh, Pittsburgh, Pennsylvania 15260*
[8] *University of Rochester, Rochester, New York 14627*

Abstract

We are studying production of χ_c states in 530 GeV/c π^- interactions with several targets. χ_c mesons are observed in the mode ($\chi \to J/\psi + \gamma$). Only photons that converted to e^+e^- pairs are used in the reconstruction of the χ_c mesons. Preliminary analysis shows that the fraction of observed J/ψs coming from χ_c radiative decays is $0.44 \pm 0.09 \pm 0.08$, and that the relative production rate of χ_{c1} to χ_{c2} is 1.3 ± 0.6.

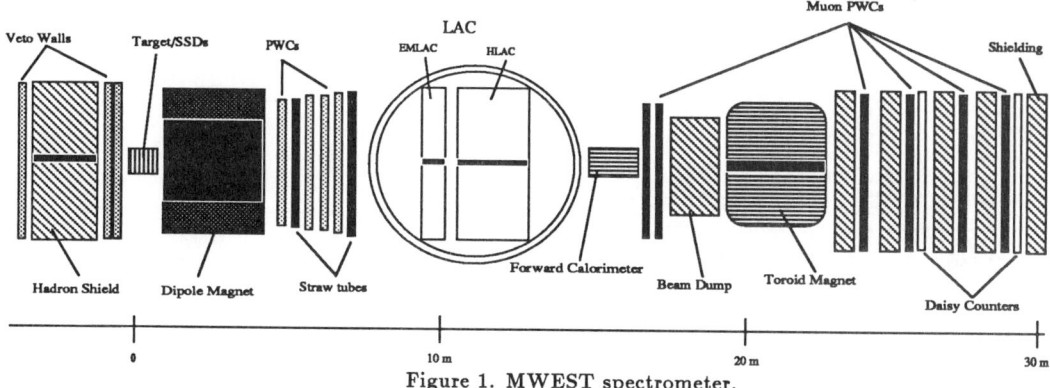

Figure 1. MWEST spectrometer.

INTRODUCTION

Production of the p-wave charmonium states (χ_c) in hadronic collisions provides important information on the QCD subprocesses involved. The p-wave charmonia can couple to two gluons (virtual or on-shell) while the s-wave vector states couplings require at least three gluons. As a result, according to perturbative QCD, the χ_c states are expected to have larger production cross sections than the J/ψ by at least a factor of $1/\alpha_S$ if the contribution from sea-quark fusion is ignored.[1] The χ_{c1} and χ_{c2} are the only p-wave charmonia below the open charm threshold with significant branching ratios for $\chi_c \to J/\psi + \gamma$. Therefore, we expect a large fraction of observed J/ψs to come from radiative χ_{c1} and χ_{c2} decay.[2] Furthermore, the QCD leading diagram of gluon-gluon fusion suppresses production of χ_{c1} relative to χ_{c2} because of angular momentum conservation and the zero mass of the gluon. Thus, measurement of relative production rate of the two states will help to test the picture of perturbative QCD and the detail of possible nonperturbative effects, such as color evaporation[3], in charmonium production.

APPARATUS

The MWEST spectrometer at Fermilab is

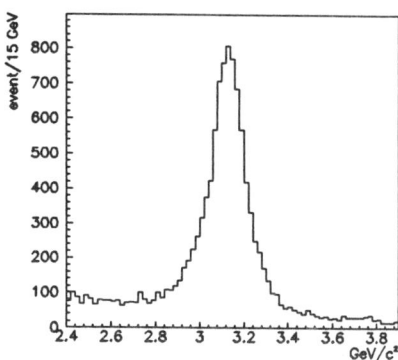

Figure 2. $\mu^+\mu^-$ invariant mass in J/ψ region.

designed to simultaneously study productions of dimuon pairs and high p_T direct photons in interactions of protons and pions incident on several targets. The spectrometer is situated in the Meson West beamline. Figure 1 shows a layout of the spectrometer.[4,5]

During 1990, data was taken with 530 GeV/c π^- beam incident on both Cu and Be targets. We collected 5.6 million triggers. About 80% of the data is used in this analysis. Aproximately 9K J/ψs were reconstructed from this sample. The dimuon mass spectrum in the J/ψ region is shown in Fig. 2. The mass resolution of 85 MeV/c^2 is consistent with the Monte Carlo simulation. The x_F range of the reconstructed J/ψs is between 0.1 and 0.8. At this stage of analysis, the data from the straw drift tubes has not been used.

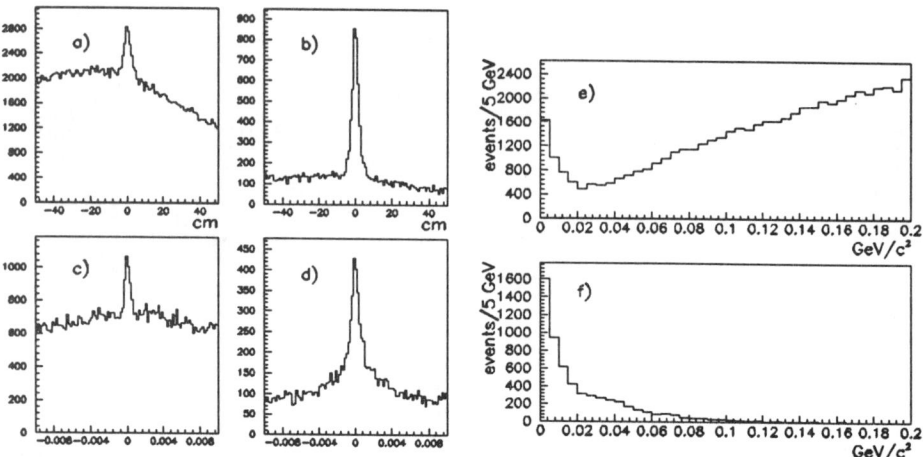

Figure 3. a) z position of the intersection point of two oppositely charged tracks relative to the center of dipole magnet in the x-view without cut; b) same as a) but requiring the difference in the y slopes of the two tracks to be less than 0.003; c) difference in the y slopes without cuts; d) same as c) but requiring the x view intersection point to be at the magnet center within ±10 cm; e) invariant mass of all oppositely-charged track pairs, assuming they are e^+e^-; f) same as e) but with both of the cuts in b) and in d).

PRODUCTION OF χ_c STATES

In order to achieve the mass resolution required for separating the two χ_c states, we looked for J/ψ events with e^+e^- pairs resulting from photon conversions. An e^+e^- pair appears as a single track in the silicon-strip detector before the dipole magnet, and as two distinct tracks in the PWCs immediately after the magnet (With PWCs we always refer to these PWCs unless stated otherwise). The direction of the magnetic field in the dipole is along the y axis defined as perpendicular to the horizontal plane. Therefore, the electron and positron tracks should nearly overlap in the y view, while in the x view the projection of their downstream segments should intersect near the center of the magnet.

Figure 3 shows the distributions of the y-slope differences and the difference in z between the x view crossing point and the center of the magnet for the oppositely-charged track pairs in the PWCs. The obvious correlation between the peaks in the two distributions indicates that the peaks are due to photon conversion pairs. The e^+e^- pairs also contribute to a clean peak near zero mass in the e^+e^- invariant-mass spectrum shown in Fig. 3e, 3f. In our analysis, a photon conversion pair is defined as a pair of oppositely-charged tracks satisfying the following requirements: a) their invariant mass (assuming they are e^+e^-) is less than 30 MeV/c^2; b) the difference in the y slopes of their PWC segments is less than 0.003 ; c) the x intercept of their PWC segments is within ±10 cm of the center of the dipole. The $e^+e^-e^+e^-$ invariant mass distribution shown in Fig. 4a exhibits the π^0 peak. The enhancement near zero is due to multiple combinations of tracks.

J/ψs are defined as the dimuons with invariant mass 2.9 GeV/$c^2 \leq M_{\mu^+\mu^-} \leq$ 3.3 GeV/c^2. With the constraint that the $\mu^+\mu^-$ invariant mass is 3.097, an χ^2-fit is done for each J/ψ to improve the momentum accuracy. In a Monte Carlo study, we calculated the detection efficiencies of J/ψ and the e^+e^- pairs from χ_{c1} state, assuming the initial x_F and p_T distribution for χ_{c1} is the same as we measured for J/ψs. Decay of χ_c was assumed to be isotropic. GEANT3 package was used in the Monte Carlo simulation. All the known

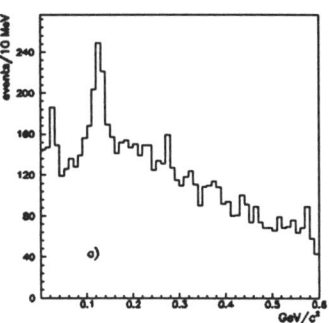 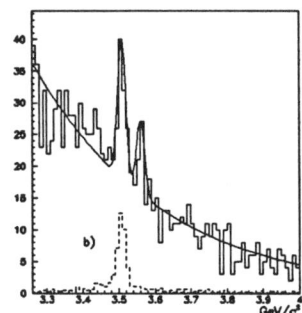 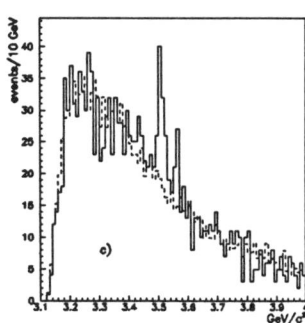

Figure 4. a) $e^+e^-e^+e^-$ mass distribution. b) $e^+e^- J/\psi$ mass distribution with fit. The dashed line is the mass distribution calculated from Monte Carlo χ_{c1} events. c) Calculated background (dashed line) overlapped with the mass distribution.

materials in front of the dipole magnet were included in the simulation in order to get the proper photon conversion probabilities. The ratio of global detection efficiencies for the χ_cs relative to J/ψs was found to be 0.018±0.003. The errors are mainly due to uncertainties in χ_c and J/ψ initial distributions and in the photon reconstruction efficiencies. The Monte Carlo estimated χ_c mass resolution is 11±2 MeV/c^2.

The $e^+e^- J/\psi$ mass distribution is shown in Fig 4b. Two peaks corresponding to χ_{c1} and χ_{c2} are clearly visible. The background was determined by mixing e^+e^- pairs and J/ψs from different events and was normalized to the number of events in the region $M(\gamma J/\psi) \leq 3.4 GeV/c^2$. In figure 4c, the calculated background (dashed line) is shown with the complete mass spectrum. A fit to the background shape plus two Gaussians, with fixed widths of 11 MeV/c^2, gives 51±11 and 19±9 χ_{c1} and χ_{c2} events, respectively. The measured χ_{c1} and χ_{c2} mass was 3.509±0.003 GeV/c^2 and 3.560±0.004 GeV/c^2, respectively. Using the relative acceptance mentioned above and the total number of J/ψs and χ_cs observed, we find that 44±9±8% of the observed J/ψ production cross section are due to χ_c productions and decays. The first uncertainty is statistical and second systematic. Assuming that the acceptance for the two χ_c states is the same and taking into account the different branching ratios[6], we determine the relative production $\sigma(\chi_{c1})/\sigma(\chi_{c2})$ to be 1.3 ± 0.6. The uncertainty here is statistical. These results are consistent with previous measurements made at lower energies.[7]

REFERENCES

1. R. Baier and R. Rückl, Z. Phys. C19, 251 (1989).

2. C. E. Carlson and R. Suaya, Phys. Rev. D18, 760 (1978).

3. M. Glück, et al., Phys. Rev. D17, 2324 (1978).

4. V. Abramov et al., FERMILAB-Pub-91/62-E, Mar, 1991.

5. G. Alverson et al., "Production of Direct Photons and Neutral Mesons at Large Transverse Momenta From π_- and p Beams at 500 GeV/c", in preparation.

6. Review of Particle Properties, Phys. Rev. D45, # 11, 1992.

7. T. B. W. Kirk et al., Phys. Rev. Lett. 42, 619 (1979), Y. Lemoigne et al., Phys. Lett. 113B, 509 (1982), F. Binon et al., Nucl. Phys. B239, 311 (1984), S. R. Hahn et al., Phys. Rev. D30, 671 (1984), D. A. Bauer et al. Phys. Rev. Lett. 54, 753 (1984), E705 Collaboration, FERMILAB-Pub-92/140-E, May, 1992

FIRST OBSERVATION OF DECAY ASYMMETRY AND SOME DECAY MODES OF Ξ_c PRODUCED BY NEUTRONS AT SERPUKHOV ACCELERATOR

presented by V. D. Kekelidze
JINR, Dubna, Russia, 141980

EXCHARM Collaboration: Dubna – Alma-Ata – Bucharest – Moscow – Minsk – Prague – Sofia – Tbilisi

Abstract

Search for decays of charmed strange baryons Ξ_c^o and Ξ_c^+ produced by 40–60 GeV neutrons in the BIS–2 experiment is presented. A decay asymmetry is shown indicating Ξ_c^+ polarization. New decay modes of Ξ_c^o and Ξ_c^+ have been observed, as well as the known one.

The experiment has been performed with the BIS–2 Spectrometer [1] at the 70 GeV Serpukhov Accelerator. The mean energy of the neutron beam is $\sim$ 40 GeV. The advantages of study charm hadroproduction at such low energy are: the low multiplicity of charged particles which leads to the minimum of combinatorial background, and the possibility of such a threshold phenomena as the charmed baryon polarization [2].

The presented results are based on 64 $\cdot 10^6$ accepted neutron nucleus interactions. Ξ_c were searched via their decays into $\Lambda K^- + \pi's$ and $K_S^o p K^- \pi^+$. Around $756 \cdot 10^3$ and $618 \cdot 10^3$ Λ and K_s^o have been accepted, respectively. Two Cherenkov Counters (14– and 7–cells) filled correspondingly with Freon and Air at the normal pressure, were used to identify charged hadrons [3].

21 076, 2 802 and 2 452 combinations of the final states

$$\Lambda + K^- + \pi^+ \quad (1)$$
$$\Lambda + K^- + \pi^+ + \pi^+ \quad (2)$$
$$K_s^o + p + K^- + \pi^+ \quad (3)$$

have been selected, respectively, using Cherenkov Counter signals. The experimental mass resolution of the states (1–3) is around 10 MeV/c^2. The spectrum of the invariant mass of the final state (1) does not contain any significant peak. Whereas, the distribution of the combinations with the $K^-\pi^+$ invariant mass from the $\bar{K}^*$ mass region ($\pm\Gamma$) shows a significant peak (50/61) at the Ξ_c mass region (fig.1). Statistical significance of this peak is > 6 st. dev. This indi-

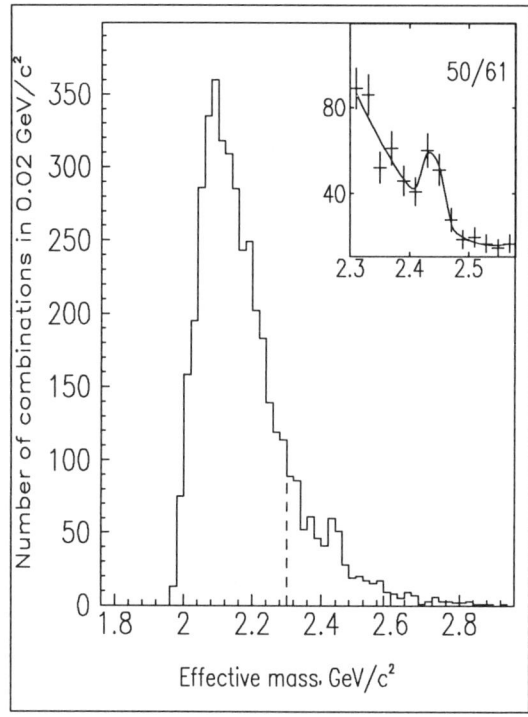

Figure 1: Spectrum of mass $\Lambda K^*(892)^0$

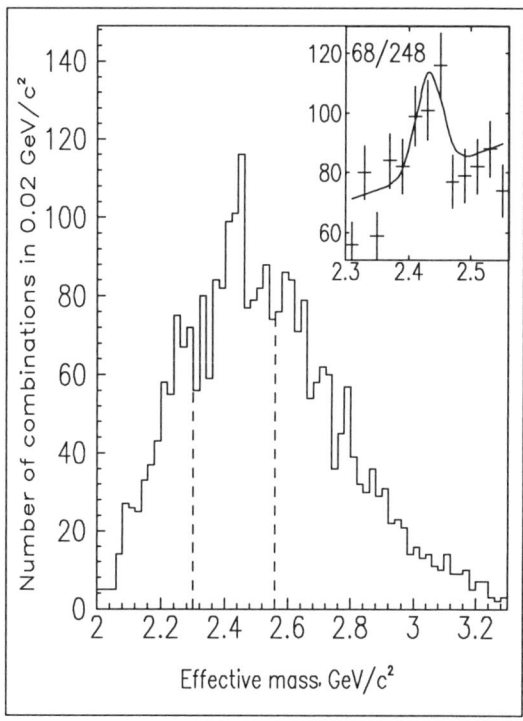

Figure 2: Spectrum of mass $\Lambda K^-\pi^+\pi^+$

cates the cascade decay:

$$\Xi_c^o \rightarrow \Lambda + \bar{K}^*(892)^o. \quad (4)$$

This decay rate in the final state (1) was estimated to be (at 90 % CL):

$$\frac{\Gamma(\Lambda \bar{K}^*(892)^o)}{\Gamma(\Lambda K^-\pi^+)} \geq \mathbf{0.86}. \quad (5)$$

The invariant mass spectra of the final states (2) and (3) are presented in figs. 2 and 3, respectively. Peaks (68/248 and 58/144) are seen (> 4 st. dev. each) at the Ξ_c mass region in the both spectra. These peaks indicate the observation of Ξ_c^+ decays into (2) and (3). Charmed baryon Λ_c^+ decays into $K^o p\pi^+\pi^-$ and $\Lambda(3\pi)^+$

have been seen in the BIS–2 experiment earlier [4]. This allows one to obtain a mass difference between Ξ_c^+ and Λ_c^+ with a minimum experimental bias. The obtained mass difference is:

$$\Delta m = \mathbf{162} \pm 10(st) \pm 15(sys). \quad (6)$$

Cascade decays of Ξ_c^+ into

$$\bar{K}^o + p + \bar{K}^*(892)^o \quad (7)$$
$$\Lambda(1520) + \bar{K}^o + \pi^+ \quad (8)$$
$$\Lambda + \bar{K}^*(892)^o + \pi^+ \quad (9)$$

have been estimated the analysing the appropriate subsystem invariant mass spectra.
The estimations of branching ratios of all the observed decays as well as

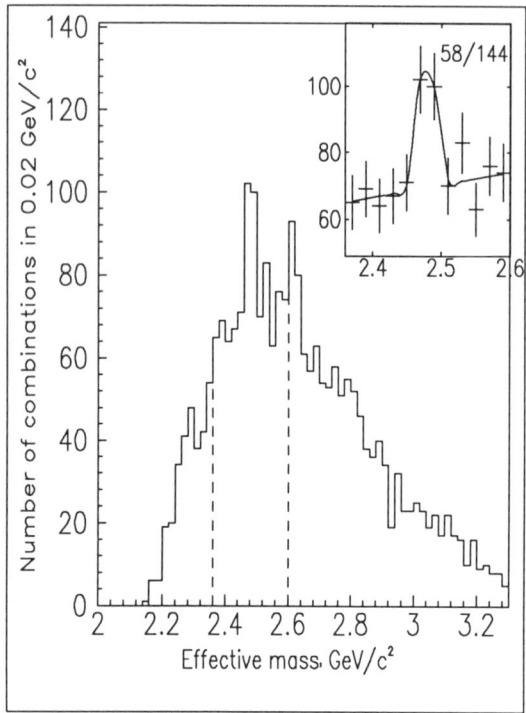

Figure 3: Spectrum of mass $K_s^0 p K^- \pi^+$

the results of the hyperon experiment WA-62 [5] are presented in Table 1.

P-violating asymmetry of the observed decays (1-3) has been studied. For each of the k-th decay product considered as an analyzer particle, the up-down asymmetry:

$$A_k = \frac{N_k(up) - N_k(down)}{N_k(up) + N_k(down)} \quad (10)$$

was calculated. Where $N_k(up)$ and $N_k(down)$ are numbers of the k-th particle emitted up and down off the Ξ_c^+ production plane, whose normal is defined with momenta of neutron and Ξ_c: $\vec{N} = [\vec{n} \times \vec{\Xi}_c]$.

The significant asymmetry has been

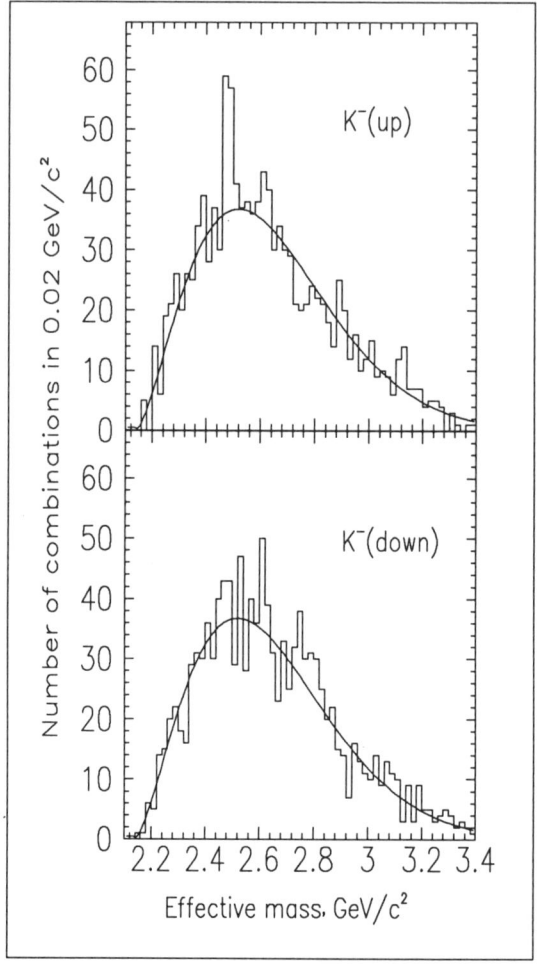

Figure 4: Spectra of mass $K_s^0 p K^- \pi^+$ for K^- emitted "up" and "down".

observed only for the decay (3). Fig. 4 shows such an asymmetry for K^-. The asymmetries of K^- and π^+ emission are +0.52 ± 0.26 and –0.62 ± 0.27, respectively, while the asymmetry of the background K^-'s is around zero (± 0.02). The study of systematics gives a possible error of +0.24/–0.11 caused mainly by the accuracy of the $\vec{n}$ – vector definition (+0.20/–0.09) and estimation of Cherenkov

Table 1: Ratios of the observed decay widths to the known one.

i-th Decay	$\Gamma_i / \Gamma(\Lambda K^-\pi^+\pi^+)$	
	BIS–2	WA–62
$p\ \bar{K}^o\ K^-\ \pi^+$ including:	**1.98 ± 0.76**	< 0.08
$\quad p\ \bar{K}^o\ \bar{K}^*(892)^o$	**0.44 ± 0.23**	
$\quad \Lambda^*(1520)\bar{K}^o\ \pi^+$	**1.45 ± 0.62**	
$\Lambda\ \bar{K}^*(892)^o\pi^+$	**0.43 ± 0.15**	< 0.70

Counter efficiency for different cells (+0.13/−0.06). Thus, the K^- asymmetry measured for the decay (3) is:

$$A_{K^-} = \mathbf{0.52} \pm .26(st)\,^{+.24}_{-.11}(sys). \quad (11)$$

Decay asymmetry should be a consequence of the decaying particle polarization: $\wp = 2 \cdot A/(\sum_i \alpha_i \cdot w_i)$, where α_i (≤ 1) are the decay parameters, and w_i – the weights of the i-th decay submode. Taking into account the obtained asymmetry (11) one could estimate that:

$$\wp \geq \mathbf{0.1} \quad (12)$$

at $\geq 90\%$ CL.
This result shows the Ξ_c^+ polarization and their weak decay observation.

CONCLUSION

New Ξ_c^o and Ξ_c^+ decay modes (4) and (3, 7-9) have been observed. The branching ratio of the observed decay $\Xi_c^+ \to \bar{K}^o p K^- \pi^+$ is larger than it was estimated earlier in the experiment [5].

Ξ_c^+ polarization (12) is shown for the first time. This might be a result of the polarization of c-quark produced in hadron interactions near the threshold.

ACKNOWLEDGMENTS

The authors are greatly obliged to I.A.Savin, A.N.Sissakian and N.E.Tyurin for support of this study.

REFERENCES

1. Aleev A. N. et al., JINR, P1–89–854, Dubna, 1989;
2. Aleev A. N. et al., JINR, D1–84–859, Dubna, 1984;
3. Gus'kov B. N. et al., JINR, P1–86–248, Dubna, 1986;
4. Aleev A.N. et al., Jorn. of Nucl. Phys.,v.37, p.1474 (1983);
5. Biagi S.F. et al., Z. Phys., C28, p.175 (1985).

RESULTS ON CHARM HADROPRODUCTION FROM CERN EXPERIMENT WA82

F. Antinori, D. Barberis, W. Beusch, M. Davenport, J.P. Dufey, B.R. French, K. Harrison,
A. Jacholkowski, A. Kirk, E. Lamanna, J.C. Lassalle, F. Muller[†],
N. Redaelli, C. Roda and M. Weymann

CERN, Geneva, Switzerland.

A. Forino, R. Gessaroli, P. Mazzanti, A. Quareni and F. Viaggi[†]

Dipartimento di Fisica and INFN, Bologna, Italy.

R. Anselmi, V. Casanova, M. Dameri, R. Hurst, P. Novelli,
B. Osculati, L. Rossi and G. Tomasini

Dipartimento di Fisica and INFN, Genova, Italy.

A. Buys, F. Grard and P. Legros

Université de Mons-Hainaut and IISN, Mons, Belgium.

M. Adamovich, Y. Alexandrov, S. Kharlamov, P. Nechaeva and M. Zavertyaev

Lebedev Physical Institute, Moscow, Russia.

Abstract

Experiment WA82 has collected data from 1987 to 1989 with the Ω' spectrometer at the CERN SPS. The aim of WA82 was a high statistics study of charm hadroproduction, using a silicon microstrip vertex detector and an impact parameter trigger. Latest results on the nuclear dependence of charm production and on the x_F distributions of D^+ and D^- mesons are presented and discussed.

1. INTRODUCTION

One of the main open problems in charm physics is the relative importance of perturbative and non-perturbative QCD phenomena in the hadronic production of charm mesons. The nuclear dependence of charm hadroproduction and the longitudinal distributions of charm hadrons are two main benchmarks in this respect: perturbative QCD predicts a linear rise of the charm hadroproduction cross section with the mass number of the target nucleus, and if perturbative QCD alone were at play one would expect the x_F distributions for all charm hadrons to reproduce those of the c and $\bar{c}$ quarks, and therefore to be all very similar.

Experiment WA82 was run at the CERN Ω' spectrometer. The experiment was based on a silicon microstrip telescope, which allowed triggering and off-line detection of secondary vertices. The collected charm sample, consisting of about 3000 fully reconstructed decays, has allowed study of the production and decay mechanisms of charm hadrons. In section 2 some features of the experimental apparatus of WA82 relevant to the paper are briefly reviewed, the analysis chain and the data sample are treated in section 3. Section 4 is devoted to the description of the measurement of the nuclear dependence of charm hadroproduction, while the results on the x_F distributions of $D^\pm$ mesons are discussed in section 5. Section 6 contains the summary and the conclusions.

[†]Deceased

2. EXPERIMENTAL APPARATUS

π^- (340 GeV/c) or p (370 GeV/c) beams collided on a 2 mm thick target, which was divided vertically into two portions of different materials (W/Si or W/Cu depending on the runs) to allow the study of the nuclear dependence of charm production.

The silicon vertex detector[1] was 50 cm long and consisted of fourteen planes with pitches ranging from 10 μm to 50 μm. Eight additional 20 μm pitch planes were used as beam detector. The typical accuracy on the measurement of the position of secondary vertices was 500 μm (longitudinal) and 30 μm (transverse).

The microstrip detectors were sitting inside the Ω' spectrometer, equipped with its standard multi-wire and drift chambers. The magnetic field intensity was 1.8 $Tesla$ at the center of the Ω magnet. The accuracy on the measurement of momenta was $(dp/p)_{MEAS} \simeq 3 \cdot 10^{-4} p(GeV)$, while the multiple scattering contribution was $(dp/p)_{SCAT} \simeq 10^{-3}$. The apparatus also featured a RICH detector and an electromagnetic calorimeter, which have not been used for the analysis presented in this paper.

Three vertex detector planes and one beam plane were also employed in the impact parameter trigger[2]: selecting events with at least one track with an impact parameter to the primary vertex in a range between 100 μm and 1000 μm, this trigger produced an enrichment factor of the charm sample of about 15.

3. DATA ANALYSIS

WA82 took data in 1987, 1988 and 1989. The π^- beam was used in all three years, on a W/Si target in 1987 and 1988 ($1.8 \cdot 10^7$ triggers), and on W/Cu in 1989 ($3 \cdot 10^7$ triggers). In addition, a proton beam run was performed in 1988 (10^7 triggers).

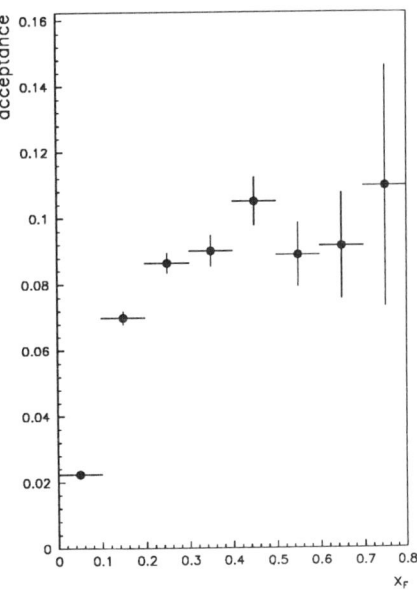

Figure 1. x_F acceptance for $D^+ \to K^- \pi^+ \pi^+$

The full data sample has been used for the analysis presented in this paper. The analysis chain started with a filter program, selecting events with evidence for secondary vertices using only the microstrip information in the non-bending projection. The events thus selected were passed through the full track and vertex space reconstruction program. Some cuts were applied in order to clean up the samples: the primary vertex of the event had to be reconstructed inside the target region; the secondary vertex had to be well separated from the primary vertex (at least 6 σ_{VTX} where $\sigma_{VTX} = \sigma_{PRIM} \oplus \sigma_{SEC}$) and from the target (at least 2 σ_{SEC}), and had to point back to the primary vertex within 60 μm. Vertices with an error on the calculated invariant mass bigger than 30 MeV/c^2 were rejected. The acceptance for D mesons of the apparatus, trigger and analysis chain extends to high x_F (as an example the $D^+ \to K^- \pi^+ \pi^+$ x_F acceptance is shown in figure 1).

Invariant mass peaks were fitted to gaussian shapes, and charm candidates were se-

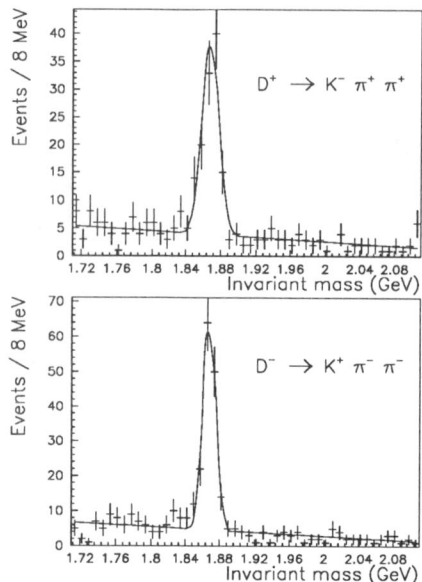

Figure 2. Invariant mass distributions for $D^+ \to K^-\pi^+\pi^+$ and $D^- \to K^+\pi^-\pi^-$

lected by invariant mass cuts of ± 3 fitted standard deviations (typically $\sigma_M = 5 \div 8\ MeV/c^2$) around the fitted central value. The amount of background under the peaks was estimated by straight line fits. The results of two such fits are shown in figure 2. The final charm sample thus obtained consists of some 3000 fully reconstructed charm hadrons.

4. NUCLEAR DEPENDENCE

The cross section for particle production on nuclei is usually parametrized as $\sigma(A) = \sigma_0 A^\alpha$. From a theoretical point of view, α should approach the surface value of 2/3 for long distance phenomena (full screening), and the volume value of 1 for pointlike interactions (in particular $\alpha = 1$ is the prediction for perturbative QCD phenomena).

Experimentally, $\alpha \simeq 3/4$ for the total hadron-nucleus cross section, and for strangeness production.[3,7] $\alpha \simeq 3/4$ has also been measured for charm production by some early beam dump experiments.[4,5]

In WA82 the beam illuminates simultaneously the two halves (W and Si or Cu) of the target, this allows to eliminate several possible sources of systematic uncertainty in the measurement of $\alpha(D)$.[6]

In order to perform the calculation, the number of D mesons renconstructed in events with the primary vertex lying in each half of the target must be corrected for the relative luminosities on the two halves (measured by regular beam trigger sampling) and for the biases due to the trigger, the event reconstruction and the analysis (as determined by a detailed Monte Carlo).

As a control of our procedure we have performed the calculation of α for our K_S^0 sample. We obtained:

$$\alpha(K_S^0) = 0.72 \pm 0.02 \qquad for \quad <x_F> = 0.05$$

in good agreement with published data.[3,7]

The same procedure applied to the charm sample gives

$$\alpha(D) = 0.92 \pm 0.06 \qquad for \quad <x_F> = 0.24$$

The error is dominated by statistics, and the systematic uncertainty arising from Monte Carlo corrections is small and is included in the quoted error. Recent open charm and J/ψ experiments[8] report α values ranging from 0.85 to 1.

There have been some indications that α may decrease with x_F in charm hadroproduction.[8,9,10] Our data, shown in figure 3, do not indicate, within the errors, any decrease of $\alpha(D)$ with x_F.

5. x_F DISTRIBUTIONS FOR D^+ AND D^-

If the longitudinal distributions for charm mesons were to follow those of the produced charm quarks, i.e. if no non-perturbative phenomena were present, the x_F distributions for all D mesons should be similar, and in particular independent of the D meson light quark content.

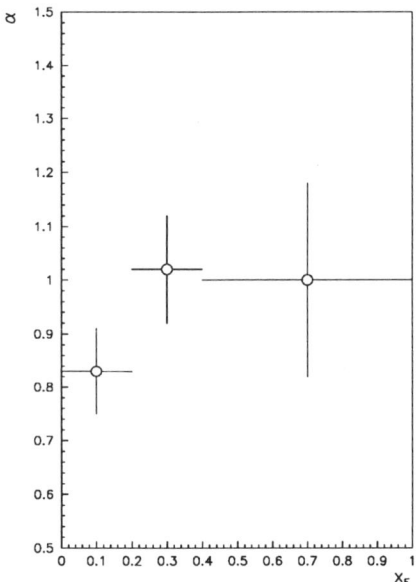

Figure 3. $\alpha(D)$ versus x_F

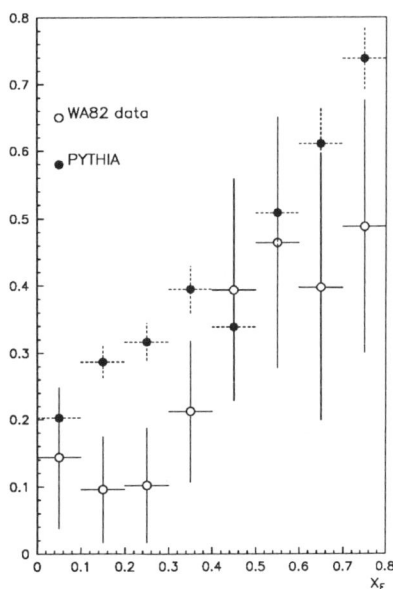

Figure 4. $(D^- - D^+)/(D^- + D^+)$

Early results from NA27[11,12] indicated enhancement of leading charmed mesons (charmed mesons containing a beam valence quark) at high x_F in π^- induced reactions. This "leading particle effect" has been confirmed by a beam dump experiment[13] but not by NA32.[14]

With a π^- ($\bar{u}d$) beam, D^+ ($c\bar{d}$) is non-leading, and D^- ($\bar{c}d$) is leading. The D^0 ($c\bar{u}$) and the $\bar{D}^0$ ($\bar{c}u$) cannot be unambiguously considered as respectively leading and non-leading: they are in the case of direct production, but if they are produced in resonance decay, their leading/non-leading character may be swapped (as for example in the decay $D^{*+}(c\bar{d}) \to D^0(c\bar{u}) + \pi^+(u\bar{d})$). This cannot happen in the $D^\pm$ case, as $D^{*0} \to D^+\pi^-$ decays are forbidden by energy conservation, and the remaining $D^{*+} \to D^+\pi^0$ and $D^{*+} \to D^+\gamma$ decays conserve the "leading-ness" of the charmed meson. For this reason only D^+ and D^- samples have been used in our analysis.

For the x_F distributions study, the D^+ and D^- candidate samples have been selected with $\pm 3\sigma_M$ cuts on the invariant mass distributions. Background samples have been defined as the contents of the two sidebands in the $K^-\pi^+\pi^+$ and $K^+\pi^-\pi^-$ invariant mass distributions in the range 1740 MeV - 1830 MeV and 1900 MeV - 2000 MeV, renormalized to the amount of background under the D^+ and D^- peaks. This background was then subtracted from the D^+ and D^- candidate x_F distributions. The $D^\pm$ sample thus selected consists of 358 ± 12 $D^+ \to K^-\pi^+\pi^+$, 515 ± 9 $D^- \to K^+\pi^-\pi^-$ from π^- interactions, and 97 ± 3 $D^\pm \to K^\mp\pi^\pm\pi^\pm$ from p interactions.

The $(D^- - D^+)/(D^- + D^+)$ distribution (figure 4) shows a clear excess of D^- over D^+, which increases with x_F. A statistical test on the D^- and D^+ distributions has been performed, combining a χ^2 test and a run test. The probability that the D^- and D^+ distributions be two random samplings of the same limit distribution is less than 5%.

After acceptance correction, the integral

excess of D^- over D^+ for positive x_F is

$$\frac{D^-}{D^+} = 1.38 \pm 0.13$$

The PYTHIA[15] event generator qualitatively reproduces this effect as a hadronization effect in the Lund string fragmentation model: at our energies the $\bar{d}$ quark necessary for building a D^+ meson must essentially be produced by string fragmentation. On the contrary a d quark in the forward hemisphere is very often present in the final state as a spectator quark from the beam π^-. Therefore there are additional ways for producing a D^- than a D^+, particularly at high x_F, near to the projectile rapidity, so an excess of D^- over D^+, increasing with increasing x_F is predicted by PYTHIA.

The $(D^- - D^+)/(D^- + D^+)$ distributions from WA82 and from PYTHIA are compared in figure 4. The exact amount of the effect is not reproduced by PYTHIA, which also predicts too large an integral value of the excess of D^- over D^+ (a factor 2, compared with our measured 1.38 ± 0.13).

The D^+ and D^- x_F acceptance corrected distributions measured by WA82 are compared with the PYTHIA predictions in figures 5 and 6. Also shown are the $c/\bar{c}$ distributions from next-to-leading-order (NTLO) QCD calculations[16] (whose shape is compatible with that of the PYTHIA leading-order + parton shower $c/\bar{c}$ ones, not shown). The PYTHIA D^+ and the NTLO c quark x_F distributions are very similar, apart from a small deviation at high x_F. Both are compatible with our data within the experimental accuracy. A much more marked deviation of the PYTHIA charmed hadron distribution from the NTLO charmed quark distribution, due to the mechanism outlined above, is visible in the D^- case. WA82 D^- data lie in between the two.

In the proton beam case, the D^+ and D^- x_F distributions are compatible with each

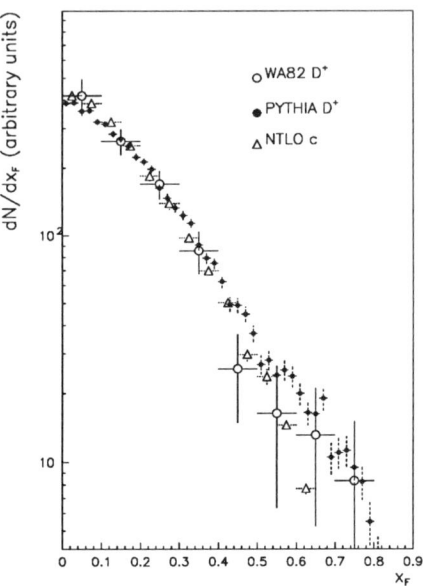

Figure 5. D^+ x_F distributions, π^- interactions

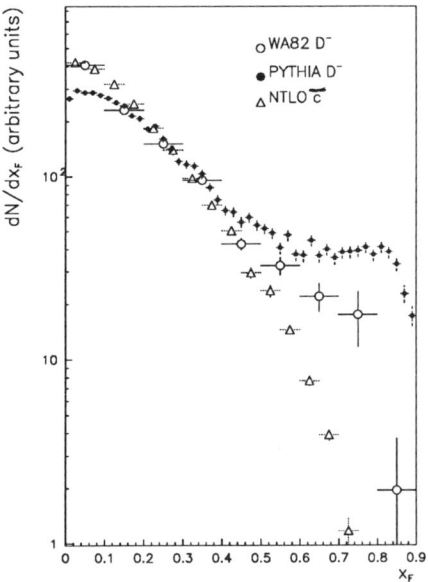

Figure 6. D^- x_F distributions, π^- interactions

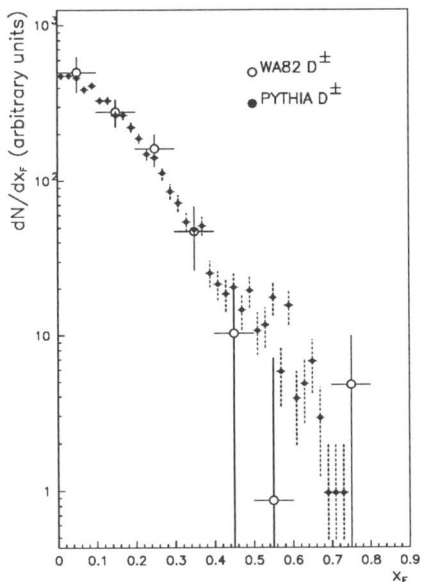

Figure 7. $D^{\pm}$ x_F distributions, p interactions

other within our statistics. The cumulative $D^{\pm}$ x_F distribution is plotted in figure 7 along with the PYTHIA prediction. The two curves are compatible, and steeper than the corresponding pion beam ones, as expected from the difference in the gluon structure functions, which are steeper in protons than in pions.

6. SUMMARY AND CONCLUSIONS

We have studied the nuclear dependence of charm hadroproduction and the longitudinal distributions of D^+ and D^- mesons in $\sqrt{s} \simeq 25$ GeV π^- - nucleus and p - nucleus interactions. We have measured a value of $\alpha(D) = 0.92 \pm 0.06$. There is no indication of a variation of $\alpha(D)$ with x_F in our data. A clear excess of D^- over D^+, increasing towards high x_F, is observed in the π^- beam sample. The measured integral excess for positive x_F is $D^-/D^+ = 1.38\pm 0.13$, and the probability that the two x_F distributions be random samplings of the same limit distribution is less than 5%. Comparison with theory shows that the nuclear dependence of the integral charm production cross section is compatible within 1.5σ with the perturbative QCD prediction, while non-perturbative effects are probably required to explain the relative amounts and the longitudinal distributions of the different charmed mesons.

REFERENCES

1. M.Adamovich et al., *Nucl. Inst. Meth.* A309 (1991), 401.
2. M.Adamovich et al., *IEEE Trans. Nucl. Sci.* 37 (1990), 236.
3. D.S.Barton et al., *Phys. Rev.* D27 (1983), 2580.
4. H.Cobbaert et al., *Zeit. Phys.* C36 (1987), 577.
5. M.E.Duffy et al., *Phys. Rev. Lett.* 55 (1985), 1816.
6. M.Adamovich et al., *Phys. Lett.* B284 (1992), 453.
7. P.Skubic et al., *Phys. Rev.* D18 (1978), 3115.
8. D.M.Alde et al., *Phys. Rev. Lett.* 66 (1991), 133.
9. M.Mac Dermott and S.Reucroft, *Phys. Lett.* B184 (1987), 108.
10. S.Kartik et al., *Phys. Rev.* D41 (1990), 41.
11. M.Aguilar-Benitez et al., *Phys. Lett.* B168 (1986), 170.
12. M.Aguilar-Benitez et al., *Zeit. Phys.* C31 (1986), 491.
13. J.L.Ritchie et al., *Phys. Lett.* B138 (1984), 213.
14. S.Barlag et al., CERN/EP 88-04.
15. H.-U.Bengtsson and T.Sjöstrand, *Comp. Phys. Comm.* 46 (1987), 43.
16. P.Nason, private communications.

RECENT ARGUS RESULTS ON CHARMED BARYON PHYSICS

J. Stiewe
(ARGUS Collaboration)
Institut für Hochenergiephysik der Universität
Schröderstr. 90
D-6900 Heidelberg, Germany

Abstract

We report on the decays $\Lambda_c \to \Xi^- K^+ \pi^+$ and $\Lambda_c \to \Xi^{*0} K^+$, our evidence for the Ω_c through the decay channels $\Omega_c \to \Xi^- K^- \pi^+ \pi^+$ and $\Omega_c \to \Omega^- \pi^+ \pi^+ \pi^-$, observation of semileptonic Ξ_c^0 decays, and the determination of the branching fractions for B meson decays into baryons, and $\Lambda_c^+ \to p K^- \pi^+$.

INTRODUCTION

We present some recent ARGUS results on charmed baryon physics: Observation of the decays $\Lambda_c^+ \to \Xi^- K^+ \pi^+$ and $\Lambda_c^+ \to \Xi^{*0} K^+$, the latter being a candidate for a W - exchange process; first evidence for the production of the charmed, doubly strange baryon Ω_c in $e^+ e^-$ annihilation through its decays $\Omega_c \to \Xi^- K^- \pi^+ \pi^+$ and $\Omega_c \to \Omega^- \pi^+ \pi^+ \pi^-$; observation of semileptonic Ξ_c^0 decays; and finally a determination of the branching ratio for B meson decays into baryons which allows to extract $BR(\Lambda_c^+ \to p K^- \pi^+)$.

The data has been taken at the $e^+ e^-$ storage ring DORIS II of DESY in the energy range of the Υ resonances, and in the neighbouring continuum. Charged hadron and lepton identification was made on the basis of specific energy loss in the drift chamber gas, velocity measurement in the time-of-flight counters, energy deposition and shower profile in the electromagnetic calorimeter, and hits in the muon chambers surrounding the detector. A detailed description of the detector ARGUS and its particle identification capabilities can be found in [1]. *

OBSERVATION OF THE DECAYS $\Lambda_c \to \Xi^- K^+ \pi^+$ and $\Lambda_c \to \Xi^{*0} K^+$

This analysis was based on a data sample of 386 events/pb taken at an average energy of $\sqrt{s} = 10.4 GeV$. Ξ^- hyperons [†] were reconstructed through their decays to $\Lambda \pi^-$, and Λ's were identified in the decay mode $\Lambda \to p \pi^-$. Accepted Λ and Ξ^- hyperons were subjected to a mass constraint fit. $\Xi^- K^+ \pi^+$ combinations surviving a series of cuts against backgrounds were required to have x_p greater than 0.5, where $x_p = p(\Xi^- K^+ \pi^+)/p_{max}$, and $p_{max} = \sqrt{E_{beam}^2 - m^2(\Xi^- K^+ \pi^+)}$. The resulting $\Xi^- K^+ \pi^+$ mass spectrum shows a clear peak at the Λ_c^+ mass.

A fit with a third order polynomial times

---

*This work was supported by the German Bundesministerium für Forschung und Technologie under contract number 055HD21P.
†References to a specific charge state should be taken to imply the charge-conjugate state, too.

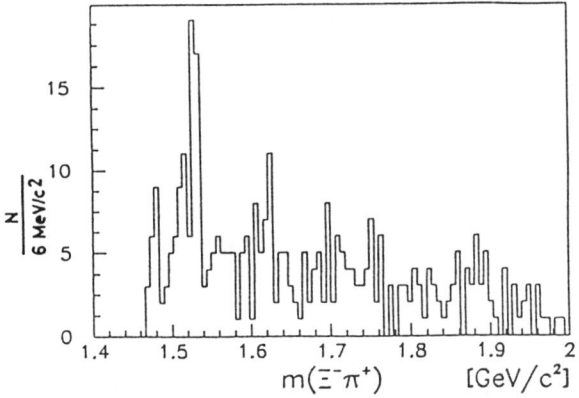 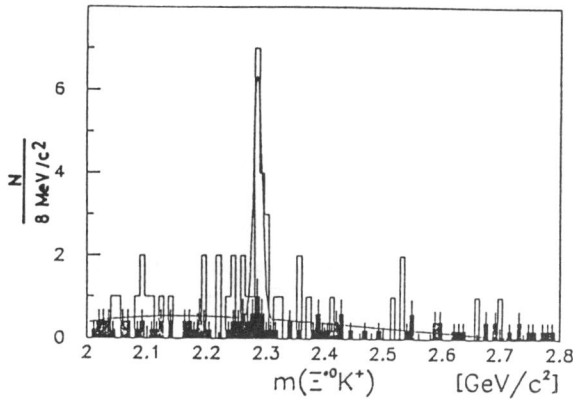

Figure 1. a) Invariant $\Xi^-\pi^+$ mass spectrum from all accepted $\Xi^-K^+\pi^+$ combinations, showing the Ξ^{*0} signal. (b) The $\Xi^-K^+\pi^+$ mass spectrum after a cut around the Ξ^{*0}. The shaded region shows the background from the Ξ^{*0} sidebands.

a square root threshold factor to describe the background, and a Gaussian with width fixed to 7 MeV/c^2 as determined in a Monte Carlo simulation to describe the signal, yields 33.6±6.7 events at a mass of (2284.8 ± 1.8) MeV/c^2, in good agreement with the nominal value [2]. The "wrong sign" (i.e. $M(\Xi^-K^+\pi^-)$) spectrum with the same cuts applied, does not show any enhancement in the signal region. To examine the $\Xi^-\pi^+$ submass, the mass spectrum of all $\Xi^-\pi^+$ combinations surviving the cuts is displayed in Fig. 1a. There is a clear signal at the Ξ^{*0} mass of 1.532 GeV/c^2. To determine how much this contributes to the Λ_c^+ signal, and how much of the Λ_c^+ is non-resonant, the Λ_c^+ signal was fitted with a cut around the Ξ^{*0} mass (Fig. 1b). The Λ_c^+ signal from the Ξ^{*0} sidebands was used to estimate the background to be subtracted. The result is that (11.4 ± 3.9) events can be attributed to the decay $\Lambda_c \to \Xi^{*0}K^+$. With the reconstruction efficiencies obtained from a Monte Carlo and detector simulation procedure, these numbers convert into products of cross section and branching ratio as follows:

$$\sigma(\Lambda_c^+) \cdot BR(\Lambda_c \to \Xi^-K^+\pi^+) = (1.7\pm0.3\pm0.2)pb, \quad (1)$$

and

$$\sigma(\Lambda_c^+) \cdot BR(\Lambda_c \to \Xi^{*0}K^+) = (0.6\pm0.2\pm0.1)pb. \quad (2)$$

This implies that $(35\pm17)\%$ of the decay $\Lambda_c \to \Xi^-K^+\pi^+$ proceeds via a two body intermediate state.

The branching ratio for $\Lambda_c \to \Xi^{*0}K^+$ has been calculated by Körner and Krämer to be 0.5% [3]. Using the ARGUS results

$$BR(\Lambda_c^+ \to pK^-\pi^+) = (4.0\pm0.8\pm0.3)\% \quad (3)$$

[4], see also below, and

$$\sigma(\Lambda_c^+) \cdot BR(\Lambda_c^+ \to pK^-\pi^+) = (12.0\pm1.1\pm1.3)pb \quad (4)$$

[5], our value for $BR(\Lambda_c \to \Xi^{*0}K^+)$ is $(0.2\pm0.1)\%$. In the absence of final state rescattering effects, the decay $\Lambda_c \to \Xi^{*0}K^+$ is a candidate for a W - exchange process which is

assumed to play a dominant role in charmed baryon decays.

EVIDENCE FOR THE PRODUCTION OF THE Ω_c

The only evidence for the existence of the Ω_c so far came from the CERN WA 62 Collaboration who observed a cluster of three events in the invariant $\Xi^- K^- \pi^+ \pi^+$ mass, with mean and spread of (2740 ± 20) MeV/c^2 [6]. We have looked for the Ω_c in the above decay channel [7], and in the channel $\Omega^- \pi^+ \pi^+ \pi^-$. The data sample used for this study corresponded to 389 events/pb taken at the $\Upsilon(4S)$ resonance energy and in the neighbouring continuum.

The hyperons Ξ^- and Ω^- were reconstructed through their decays to $\Lambda \pi^-$ and ΛK^-, respectively. Λ hyperons were identified through their decays to $p\pi^-$ with well identified decay vertices. As the Λ's are themselves daughter particles of long lived parents, their decay vertices were required to have a distance of at least 4 cm (2 cm) from the main vertex. Mass constraint fits have been applied for intermediate states (Λ, Ξ^-, Ω^-).

To suppress background from non - charm fragmentation processes, the scaled momentum $x_p = p/p_{max}$, where p_{max} is the maximum momentum accesible for the particle combination considered, was required to be larger than 0.4 (0.5). As a powerful means to further suppress background, the momentum vectors of all candidate decay particles from an Ω_c decay were required to point into the same "hemisphere", the hemispheres being defined by a plane perpendicular to the event thrust axis which was taken as an approximation to the primary c - quark axis.

The results are displayed in Figs. 2a and 2b. Fig. 2a shows the invariant $\Xi^- K^- \pi^+ \pi^+$ mass spectrum after the cuts, together with the result of fitting a Gaussian plus a flat background shape to the data. A satellite peak below the Ω_c signal could be identified as due to cross talk from the decay $\Xi_c^0 \to \Xi^- \pi^- \pi^+ \pi^+$, with the π^- misidentified as a K^-, and has been removed from the plot. Careful checks have been made, for either decay channel, to ensure that the signals are neither due to reflexions from other charmed baryon decays, nor to accidental kinematical correlations. The two decay channels which lead to the same net final state $\Lambda K^- \pi^- \pi^+ \pi^+$ are orthogonal with the exception of one event that shows up in both signal regions.

The fit yields 9.9 ± 3.6 entries at a mass of (2719 ± 6) MeV/c^2, the width being $\sigma = (13.8 \pm 4.9)$ MeV/c^2, in agreement with that one expected from a Monte Carlo simulation, (13.5 ± 2.6) MeV/c^2. The significance of the signal has been estimated by fitting the distribution by a background shape alone, i.e. assuming that there is no signal at all, and calculating the Poissonian probability that the number of entries in the signal region is a fluctuation of the background. The resulting probability is $2 \cdot 10^{-3}$, corresponding to a Gaussian significance of 3.1 standard deviations.

The $\Omega^- \pi^+ \pi^+ \pi^-$ invariant mass distribution, after cuts, is displayed in Fig. 2b, again with the result of the fit. The Gaussian contains 6.5 ± 3.2 entries at a mass of (2713.0 ± 5.1) MeV/c^2, corresponding to 2.4 Gaussian standard deviations. The second channel, however, confirms the mass determined in the first one, and allows to calculate a weighted mean of

$$M(\Omega_c) = (2716 \pm 4 \pm 2) MeV/c^2.$$

For the ratio of branching ratios we obtain

$$\frac{BR(\Omega_c \to \Omega^- \pi^+ \pi^+ \pi^-)}{BR(\Omega_c \to \Xi^- K^- \pi^+ \pi^+)} = 0.26 \pm 0.16.$$

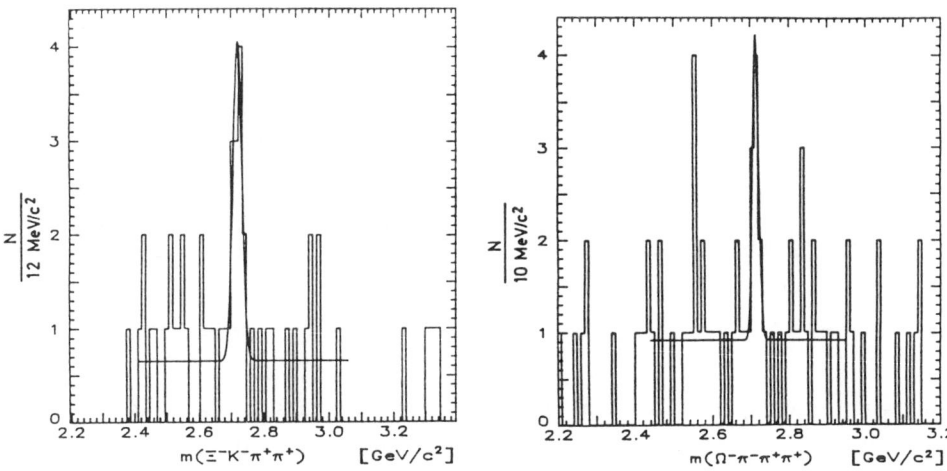

Figure 2. a) Invariant mass spectrum of all accepted $\Xi^- K^- \pi^+ \pi^+$ combinations. (b) Invariant mass spectrum of all accepted $\Omega^- \pi^- \pi^+ \pi^+$ combinations.

OBSERVATION OF Ξ_c^0 SEMILEPTONIC DECAYS

The semileptonic decay $\Xi_c^0 \to \Xi^- l^+ X$ is studied through correlations between Ξ^- hyperons and electrons or muons. The data used for this analysis comprise an integrated luminosity of 493.3 events/pb taken at the $\Upsilon(1S)$, $\Upsilon(2S)$, $\Upsilon(4S)$ resonances, and in the neighbouring continuum. Background from Bhabha, beam - gas and beam - wall events has been removed by appropriate cuts. Possible contributions from B meson decays have been suppressed by cuts in the scaled momentum $x_p (>0.45)$ and the second Fox - Wolfram moment (> 0.35).

The Ξ^- was reconstructed through its decay to $\Lambda \pi^-$, and the Λ through its decay to proton and π^-. Electrons and muons were identified on the basis of a combined likelihood ratio which includes dE/dx, time-of-flight, shower counter, and muon chamber information. Fig. 3a shows the invariant mass distributions for $\Lambda \pi^-$ candidates that have a "right - sign" lepton (electron or muon) in the same event, and a combined ($\Lambda \pi^-$ + lepton) - mass less than that of the Ξ_c^0 baryon [2]. The distribution was fitted with a Gaussian signal and a third order polynomial background with an exponential threshold factor. The fit to the right-sign $\Lambda \pi^-$ mass spectrum yields a signal with a mean mass of 1.32213 ± 0.00063 GeV/c^2, a width $\sigma = 2.1 \pm 0.5$ MeV/c^2 and 23.6 ± 5.9 $\Xi^- l^+$ events. Breaking this down into individual lepton channels, one finds 16.1 ± 4.9 $\Xi^- e^+$ and 7.5 ± 3.3 $\Xi^- \mu^+$ events. Also, using the above mean mass and width, a wrong-sign signal of 3.0 ± 2.5 $\Xi^- l^-$ events is observed.

In order to extract the (Ξ^- + lepton) - invariant mass distribution, the following background sources have been identified and corrected for:

1. Chance coincidence of a real Ξ^- with a real lepton, both from separate sources ("random correlations"). Monte Carlo studies show that this kind of background is negligible for $x_p > 0.45$ since baryon and lepton tend to come from different jets. For the same reason, it tends to populate the mass spectrum well above the signal region.

2. Chance coincidence of a fake Ξ^- with a real

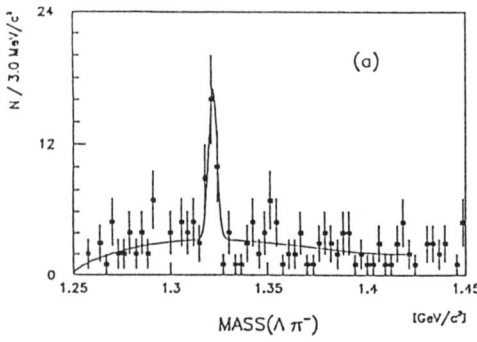 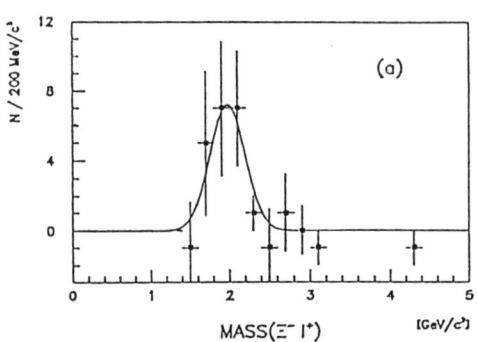

Figure 3. a) Invariant mass spectrum of $\Lambda\pi^-$ combinations for events with a right-sign lepton. b) $\Xi^- l^+$ right-sign invariant mass distribution after background subtraction. The full line represents the corresponding Monte Carlo spectrum.

lepton ("Fake Ξ^-"). This background can be estimated by extrapolating from the Ξ^- sidebands to the Ξ^- signal region.

3. Chance coincidence of a real Ξ^- with a fake lepton. This background is estimated using the known misidentification rate per hadron, and by multiplying it to the hadronic track multiplicity per event, the number of Ξ^- candidates, and the appropriate kinematical efficiencies.

4. Misidentification of pions as leptons in decay modes involving a Ξ^-. These are: $\Xi_c^0 \to \Xi^- \pi^+$, $\Xi_c^0 \to \Xi^- \pi^+ \pi^+ \pi^-$, $\Xi_c^+ \to \Xi^- \pi^+ \pi^+$, and $\Lambda_c \to \Xi^- K^+ \pi^+$. This background can be estimated by multiplying the expected number of continuum generated events of the above four decay channels with the lepton-hadron misidentification rate and the $\Xi^- l^+$ reconstruction efficiency.

After subtraction of the backgrounds discussed above, a signal of 18.1 ± 5.9 events on top of a background of 19.4 ± 1.5 events is left. The $\Xi^- l^+$ right sign invariant mass distribution, after background subtraction, is shown in Fig. 3b together with the Monte Carlo generated curve (full line).

For the sum of the electron and muon channels, this leads to

$$\sigma(e^+e^- \to \Xi_c^0 X) \cdot BR(\Xi_c^0 \to \Xi^-(e^+ + \mu^+)X) = (1.44 \pm 0.47 \pm 0.25) pb. \quad (5)$$

Breaking this down into the individual lepton channels, we get

$$\sigma(e^+e^- \to \Xi_c^0 X) \cdot BR(\Xi_c^0 \to \Xi^- e^+ X)$$
$$= (0.71 \pm 0.28 \pm 0.12) pb, \quad (6)$$

and

$$\sigma(e^+e^- \to \Xi_c^0 X) \cdot BR(\Xi_c^0 \to \Xi^- \mu^+ X)$$
$$= (0.83 \pm 0.50 \pm 0.15) pb. \quad (7)$$

Assuming lepton universality and averaging, we finally get

$$\sigma(e^+e^- \to \Xi_c^0 X) \cdot BR(\Xi_c^0 \to \Xi^- l^+ X)$$
$$= (0.74 \pm 0.24 \pm 0.09) pb. \quad (8)$$

The systematic error receives contributions from uncertainties in the integrated luminosity, track reconstruction and lepton identification efficiencies, variation of fit and background parameters, and parametrization of the x_p-spectrum.

THE BRANCHING RATIO OF B MESONS INTO BARYONS

The branching ratio of B mesons into baryons can be extracted from a set of observations with only few assumptions [4]. Since B meson decays are dominated by $b \to c$ transitions, it is assumed that stable baryons (here protons and Λ's) are always produced either as decay products of a charmed baryon, or in the fragmentation of the remaining B meson decay products. The production of non - charmed baryon antibaryon pairs in the fragmentation is expected to be suppressed.

There are 12 observable quantities: N_p, N_Λ, $N_{p\bar{p}}$, $N_{p\bar{\Lambda}}$, $N_{\Lambda\bar{\Lambda}}$, $N_{l^\pm p\bar{p}}$, N_{l+p}, N_{l-p}, $N_{l+\Lambda}$, $N_{l-\Lambda}$, $N_{l+p\bar{\Lambda}}$, $N_{l-p\bar{\Lambda}}$. Here, N means the number of events observed, as labelled by the subscripts; l is either an electron or muon. These 12 quantities can be expressed in terms of five unknown parameters: The inclusive branching fraction into baryons, B_b, the branching ratio for the decay of charmed baryons into protons and Λ's, θ_p and θ_Λ, and the relative fractions of protons and Λ's produced in the hadronization processes, f_p and f_Λ. After determining the 12 observable numbers of events, the system of 12 equations with five unkowns can be solved by a χ^2 minimization procedure. From this fit, the inclusive branching ratios $B \to baryons$ is determined to be $(6.8 \pm 0.5 \pm 0.3)\%$ which is in good agreement with the number published by the CLEO Collaboratrion [8], $(6.4 \pm 0.8 \pm 0.8)\%$.

This number allows to determine the branching ratio $BR(\Lambda_c \to pK^-\pi^+)$ from the product of branching ratios $BR(B \to \Lambda_c X) \cdot BR(\Lambda_c \to pK^-\pi^+)$ which has been measured to be $(0.28 \pm 0.05)\%$ [9]. With the assumption that B meson decays to Ξ_c and Ω_c can be neglected, one obtains

$$BR(\Lambda_c^+ \to pK^-\pi^+) = (4.0 \pm 0.8 \pm 0.3)\%.$$

REFERENCES

1. H. Albrecht et al. (ARGUS), Nucl. Instr. and Methods **A275** (1989) 1
2. The Particle Data Group, Phys. Rev. **D45**, No. 11 (1992) 1
3. J. G. Körner and M. Krämer, Exclusive Non-Leptonic Charm Baryon Decays, **DESY 92-049**, March 1992
4. H. Albrecht et al. (ARGUS), Measurement of Inclusive Baryon Production in B Meson Decays, **DESY 92-074**, May 1992
5. C. Charlesworth, Ph. D. Thesis, Toronto, 1992
6. S. Biagi et al. (WA62), Z. Phys. **C28** (1985) 175
7. H. Albrecht et al. (ARGUS), Evidence for the Production of the Charmed, Doubly Strange Baryon Ω_c in e^+e^- Annihilation, **DESY 92-052**, March 1992
8. G. Crawford et al. (CLEO), Phys. Rev. **D45** (1992) 752
9. H. Albrecht et al. (ARGUS), Phys. Letters **B207** (1988) 109

HIGH ENERGY PHOTOPRODUCTION OF CHARM BARYONS*

Harry W. K. Cheung
for the Fermilab E687 Collaboration[1]
Dept. of Physics
University of Colorado, Campus Box 390
Boulder CO 80309.

Abstract

Preliminary analysis of Charm Baryons from Fermilab high energy photoproduction experiment E687 is presented. The results include the first observation of $\Omega_c^0 \to \Omega^-\pi^+$ and a preliminary result for the Λ_c^+ lifetime of 0.230±0.017±0.019 ps.

INTRODUCTION

We have analyzed data collected by the Fermilab high energy photoproduction experiment E687 for events containing protons, Λ^0's, Ξ^-'s and Ω^-'s. The entire data sample of E687 is 510 million hadronic triggers; the data presented in this paper is from about 50% of the total data sample.

The E687 detector is described in detail elsewhere[2]. The experiment uses a photon beam of 50–350 GeV (mean energy 220 GeV) on a 4 cm Be target. A microvertex detector consisting of 12 planes of silicon microstrips arranged in three views provides high resolution tracking, allowing the separation of primary and secondary vertices. Deflections of charged particles by two analysis magnets of opposite polarity are measured by five stations of multiwire proportional chambers (PWCs). Three multi-cell Čerenkov counters operating in threshold mode are used for particle identification.

*Work supported in part by the National Science Foundation, the US Department of Energy, the Italian Istituto Nazionale di Fisica Nucleare and Ministero della Università e della Ricerca Scientifica.

EVENT SELECTION

Λ's are reconstructed through the decay $\Lambda \to p\pi^-$, and Ξ^-'s and Ω^-'s are reconstructed through the decay channels $\Xi^- \to \Lambda\pi^-$ and $\Omega^- \to \Lambda K^-$. Only Ξ/Ω decays downstream of the silicon microstrip detectors are used in this analysis as these have a measured microstrip Ξ/Ω track. These decays are reconstructed by intersecting the daughter $\pi(K)$ and the Λ and by requiring that the resultant momentum vector agrees to within 2 milliradians with an unmatched microstrip track created by the charged hyperon prior to decay. Figures 1(a) and (b) show the $\Lambda\pi^-$ and ΛK^- invariant mass plots respectively, (charged conjugate states are implicitly included).

All of the charm signals presented are obtained with a candidate-driven vertex finder using the silicon track information. A description of this technique can be found in reference 3. The distance L between the primary and secondary vertices is calculated and divided by σ_L, the error on that difference. One of the principal cutting tools used to isolate Charm baryon signals is the significance of the sepa-

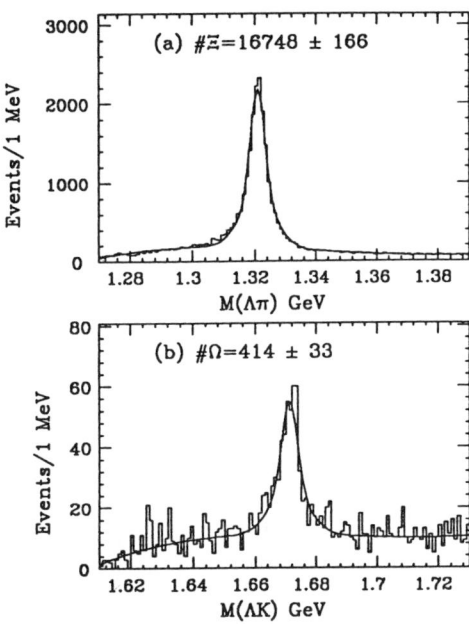

Figure 1. Mass plots for (a) $\Lambda\pi$ and (b) ΛK showing reconstructed Ξ's and Ω's.

ration of the primary and secondary vertices given by L/σ_L.[4]

Other selection critera used in the analysis include particle identification using the Čerenkov counters and various vertexing cuts.

$$\Lambda_c^+$$

We present in figure 2(a) the invariant mass plots for Λ_c^+ decaying to $pK^-\pi^+$ for $L/\sigma_L > 4$. Using the reconstructed $\Lambda_c^+ \to pK^-\pi^+$ (and charged conjugate) decays, we made a measurement of the Λ_c lifetime. The decay length is used and a binned maximum likelihood fit method is used. This method is described in detail in reference 3. The relatively high statistics in this sample allow detailed studies for systematic effects in our lifetime measurement. We found no systematic effects outside the statistical errors. An upper limit on the systematic error has been estimated and is quoted in the following preliminary result for the Λ_c lifetime:

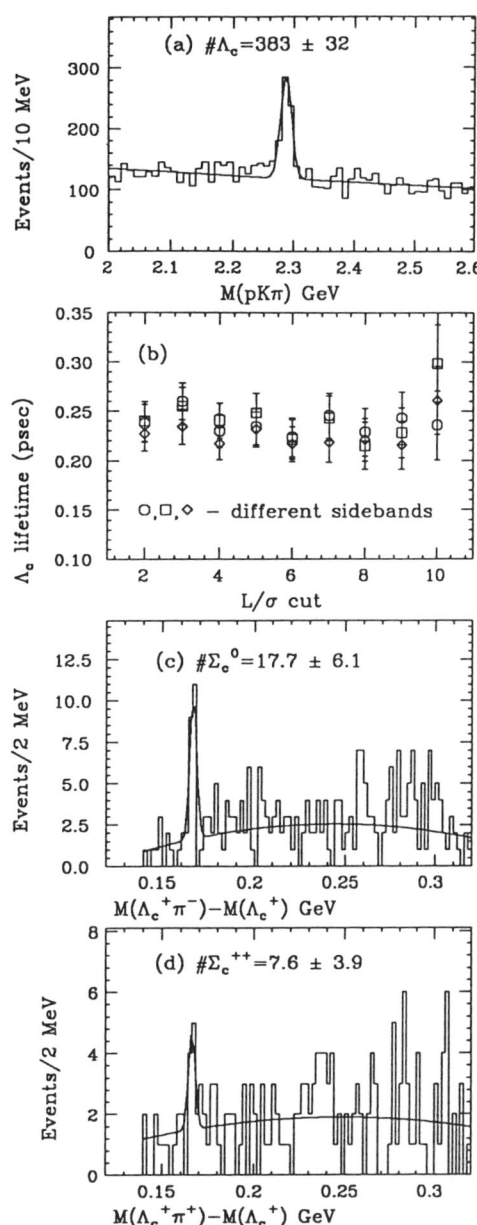

Figure 2. (a) $pK\pi$ mass plot showing reconstructed Λ_c's. (b) Λ_c lifetime vs L/σ_L cut using different sidebands.[3] (c) $\Sigma_c^0 - \Lambda_c^+$ mass difference. (d) $\Sigma_c^{++} - \Lambda_c^+$ mass difference.

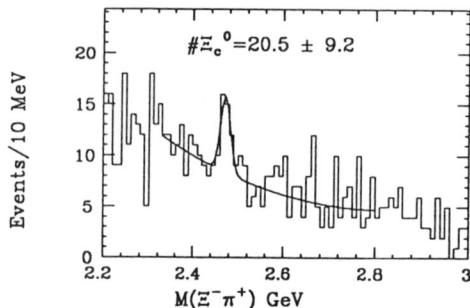

Figure 3. $\Xi^-\pi^+$ mass plot showing reconstructed Ξ_c^0's.

$0.230\pm0.017(stat.)\pm0.019(syst.)$ ps. We show in fig.2(b) the measured Λ_c lifetime using different background sidebands[3] as a function of different L/σ_L cuts. Many other systematic studies were also done, including trying a different likelihood fit method based on event-by-event probabilities,[5] and studying the effect of backgrounds from reflections of D_s^+ and D^+ decays to $K^+K^-\pi^+$.

We present in figure 2(c) a $M(pK^-\pi^+\pi^-)$-$M(pK^-\pi^+)$ mass difference plot showing 17.7 ± 6.1 Σ_c^0's, with a $\Sigma_c^0 - \Lambda_c^+$ mass difference of 166.2 ± 0.6 MeV (only the statistical error is quoted). We only see a weak signal for the $\Sigma_c^{++} \to \Lambda_c^+\pi^+ \to pK^-\pi^+\pi^+$ of 7.6 ± 3.9 events as shown in fig.2(d).

We have also analyzed and found Λ_c^+ decays to the following channels: $pK^-\pi^+\pi^0$, pK^+K^-, $p\pi^+\pi^-\pi^+\pi^-$, $pK_s^0\pi^+\pi^-$, pK_s^0 and $\Xi^-K^+\pi^+$. These were presented at the conference but unfortunately there is insufficient room to show them here.

$$\Xi_c^0$$

We present in fig.3 a mass plot for $\Xi_c^0 \to \Xi^-\pi^+$. The neutral strange charmed baryon has been very elusive thus far, possibly due to a very short lifetime which renders our vertex separation cut (L/σ_L) less effective. We observe 20.5 ± 9.2 events at a mass of 2472.1 ± 5.2 MeV/c^2 and width of 11 ± 5 MeV/c^2, which

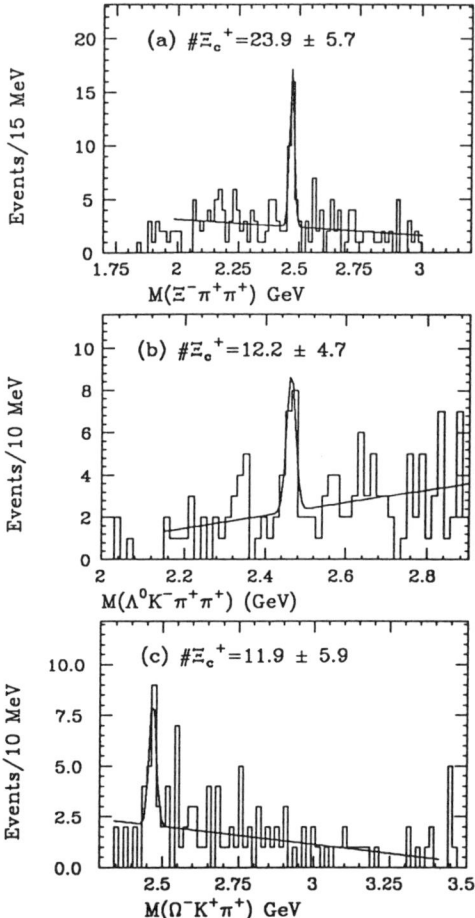

Figure 4. Mass plots for (a) $\Xi^-\pi^+\pi^+$ (b) $\Lambda K^-\pi^+\pi^+$ and (c) $\Omega^-K^+\pi^+$ showing reconstructed Ξ_c^+'s.

are consistent with a Ξ_c^0 of zero width.

$$\Xi_c^+$$

We observe Ξ_c^+ in 3 separate decay modes: 23.9 ± 5.7 events in $\Xi^-\pi^+\pi^+$ (mass of 2469.1 ± 2.6 MeV/c^2), 12.2 ± 4.7 events in $\Lambda^0K^-\pi^+\pi^+$ (mass of 2463.5 ± 4.7 MeV/c^2), and 11.9 ± 5.9 events in $\Omega^-K^+\pi^+$ (mass of 2468.6 ± 5.8 MeV/c^2). All the masses are consistent with the current world average of 2466.4 ± 2.1 MeV/c^2, and the widths are consistent with Monte Carlo widths assuming zero intrinsic widths. A lifetime measurement is forthcom-

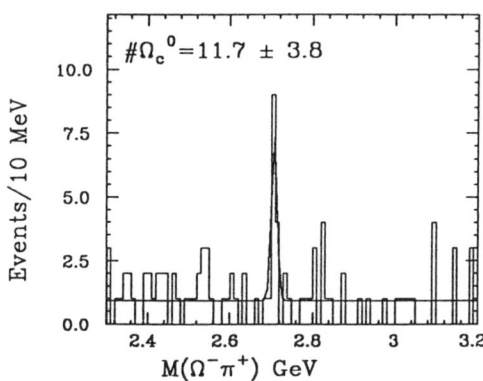

Figure 5. $\Omega^-\pi^+$ mass plot showing reconstructed Ω_c^0's.

ing, and preliminary analysis based on a maximum likelihood fit favors a lifetime longer than that of Λ_c^+. The mass plots are presented in Fig. 4.

Ω_c^0

The first evidence[6] for the Ω_c^0 came from CERN experiment WA-62 in which 3 events in the invariant mass plot $\Xi^-K^-\pi^+\pi^+$ were found to be clustered around 2740 ± 20 MeV/c^2. Recently the ARGUS collaboration has found further evidence[7] for this channel with 12.2 ± 4.5 entries at a mass of 2719.0 ± 7.0 ± 2.5 MeV/c^2.

We have looked at the $\Omega^-\pi^+$ decay channel. It is our expectation that this channel will have little background as the expected signal is very far from the peak of phase space. Figure 5 shows the $\Omega^-\pi^+$ invariant mass distribution. A total of 11.7 ± 3.8 events above background are found in the peak. The yield of events is determined by a maximum likelihood fit to a flat background plus a gaussian. The peak occurs at a mass of 2707.0 ± 2.2 ± 5.0 MeV/c^2 in agreement with the ARGUS mass. The measured width of 7.0 ± 0.7 MeV/c^2 is consistent with a mass of zero width as determined from Monte Carlo simulation. A lifetime analysis of the Ω_c^0 is forthcoming.

CONCLUSIONS

A wealth of Charm baryon decays have been reconstructed in experiment E687. With the full data sample, we expect to obtain precise measurements on the Λ_c^+, Ξ_c^+, Ξ_c^0 and Ω_c^0 lifetimes. Some relative branching ratios of the Λ_c^+ and of the Ξ_c^+ will also be measured.

REFERENCES

1. Please see G. Bellini, *these proceedings*, for a list of collaborators and institutions in the E687 Collaboration.

2. E687, P. L. Frabetti *et al.*, "Description and Performance of the E687 Spectrometer", FERMILAB Pub-90/258E (1990), to be published in Nucl. Instrum. Methods.

3. E687, P. L. Frabetti *et al.*, "A measurement of the D^0 and D^+ lifetimes", Phys. Lett. B263, pp.584–590, (1991).

4. The variable L is the signed 3 dimensional separation between vertices and σ_L is the error on L computed on an event by event basis including effects of multiple coulomb scattering.

5. E687, P. L. Frabetti *et al.*, "Measurement of the Λ_c^+ and D_s^+ lifetimes", Phys. Lett. B251, pp.639–644, (1990).

6. S. Biagi *et al.*, "Properties of the Charmed Strange Baryon A^+ and Evidence for the Charmed Doubly Strange Baryon T^0 at 2.74 GeV/c^2.", Z. Phys. C28, pp.175, (1985).

7. H. Albrecht *et al.*, "Evidence for the Production of Charmed, Doubly Strange Baryon Ω_c in e^+e^- Annihilation", DESY 92-052, March 1992.

INCLUSIVE J/ψ, $\psi(2S)$ AND b-QUARK PRODUCTION IN $\bar{p}p$ COLLISIONS AT $\sqrt{s} = 1.8$ TEV *

CDF Collaboration[†]
Presented by Vaia Papadimitriou
Fermi National Accelerator Laboratory
Batavia, Illinois U.S.A. 60510

Abstract

Inclusive J/ψ and $\psi(2S)$ production has been studied in $\bar{p}p$ collisions at $\sqrt{s} = 1.8$ TeV with the Collider Detector at Fermilab. The products of production cross section times the branching fraction of $J/\psi(\psi(2S))$ to $\mu^+\mu^-$ are reported as functions of the $J/\psi(\psi(2S))$ P_T in the kinematic range $P_T > 6$ GeV/c and $|\eta| \leq 0.5$. The products of the integrated cross section times branching fraction and the b-quark production cross section calculated from these values are also reported.

INTRODUCTION

This paper presents a study of the reactions $\bar{p}p \to J/\psi(\psi(2S))$ X $\to \mu^+\mu^-$X at $\sqrt{s} = 1.8$ TeV using 2.6 ± 0.2 pb^{-1} of data collected during the 1988-1989 running period of the Fermilab $\bar{p}p$ collider. This study is important for the investigation of charmonium production mechanisms in $\bar{p}p$ collisions[1], and it is the first measurement of J/ψ and $\psi(2S)$ cross sections at Tevatron energies. The J/ψ and $\psi(2S)$ signals are also important for the study of the production of b quarks at low P_T;[2,3] this paper reports the inclusive b-quark production cross section.

*Supported by the U.S. Dept. of Energy, contract number DE-AC02-76CH03000.
†The CDF Collaboration: ANL, Brandeis, UCLA, U. Chicago, Duke, FNAL, INFN-Frascati, Harvard, U. Ill., Johns Hopkins, KEK, LBL, MIT, U. Mich. INFN-Padova, U. Penn., INFN-Pisa, Purdue, Rochester, Rockefeller, Rutgers, Texas A&M, Tufts, Tsukuba, U. Wisconsin

J/ψ, $\psi(2S)$ SIGNALS

The components of the Collider Detector at Fermilab (CDF) relevant for this analysis are the central tracking chamber (CTC) which is in a 1.4116–T axial magnetic field and the central muon chambers that provide muon identification in the pseudorapidity region $|\eta^\mu| < 0.61$. The data sample used for the analysis was collected with a multi-level central dimuon trigger. From events passing this trigger, a sample of opposite sign dimuons was selected with the following cuts: $P_T^\mu > 3.0$ GeV/c for each muon; less than a 3σ difference in position between each muon chamber track and its associated, extrapolated CTC track, where σ is the calculated uncertainty due to multiple scattering, energy loss, and measurement uncertainties; a common vertex along the beam axis for the two muons; $|\eta| \leq 0.5$ and $6.0 < P_T < 14.0$ GeV/c for the muon pair. The resulting mass distributions after all cuts are shown in Fig. 1. The J/ψ and $\psi(2S)$ peaks were each fit to a Gaussian line shape plus a linear background. The

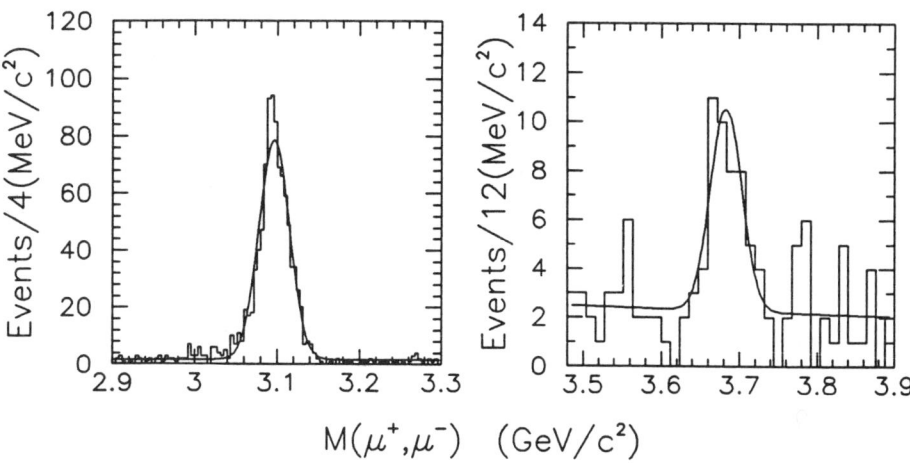

Figure 1. $\mu^+\mu^-$ mass distribution for (a) J/ψ and (b) $\psi(2S)$. The histogram corresponds to the data and the solid curve is a fit to a Gaussian plus a linear background.

number of J/ψ candidates above background within a $\pm 2.5\sigma$ mass signal region, $3.05 < m_{\mu^+\mu^-} < 3.15$ GeV/c^2, is 889 ± 30 and the resulting J/ψ mass is (3.0965 ± 0.0007) GeV/c^2 with $\sigma = (18.5 \pm 0.6)$ MeV/c^2. The number of $\psi(2S)$ candidates above background within a $\pm 2.5\sigma$ mass signal region, $3.63 < m_{\mu^+\mu^-} < 3.73$ GeV/c^2, is 35 ± 8 and the $\psi(2S)$ mass is (3.683 ± 0.005) GeV/c^2 with $\sigma = (20 \pm 4)$ MeV/c^2.

J/ψ, $\psi(2S)$ CROSS SECTIONS

After correcting the J/ψ and $\psi(2S)$ signals for trigger efficiency, kinematic and geometric acceptance and all other experimental efficiencies as a function of P_T, we obtain the J/ψ and $\psi(2S)$ differential cross sections which are displayed in Fig. 2 as functions of P_T. The circles correspond to the data. The vertical error bars are from statistical fluctuations and the P_T-dependent systematic uncertainties added in quadrature. Theoretical predictions for the two types of processes expected to dominate J/ψ and $\psi(2S)$ production are also plotted.

The solid curve in Fig. 2a (2b) is a next-

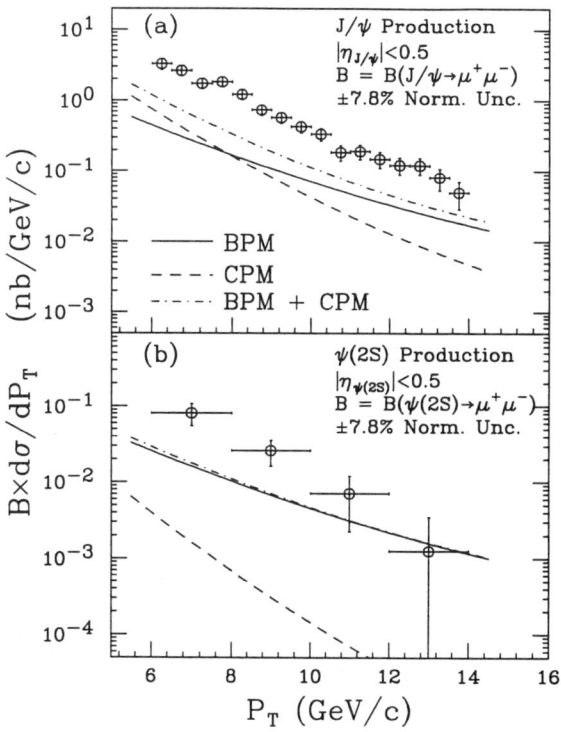

Figure 2. The product $B \times \left(\frac{d\sigma}{dP_T}\right)$ vs. P_T for (a) $J/\psi \to \mu^+\mu^-$ and (b) $\psi(2S) \to \mu^+\mu^-$.

to-leading-order calculation by Nason, Dawson and Ellis (NDE)[4] of the production of b-quarks, leading to B-mesons by modeling the energy sharing between the quark and the meson with the Peterson fragmentation function[5] and using $\epsilon_P = 0.006 \pm 0.002$. The B mesons subsequently decay to J/ψ ($\psi(2S)$) whose momentum spectra are given by ARGUS(CLEO)[6,7]. We refer to this overall calculation as B-production model (BPM). Uncertainties in this prediction arise because the b mass is comparable to the P_T of this experiment and in a range where neglected higher order terms may be significant. Further, the strongest leading-order processes for b production are gluon fusion at low x where the structure functions are not well known.

The dashed curve in Fig. 2a (2b) corresponds to J/ψ's ($\psi(2S)$'s) from direct charmonium production[1,8]. We refer to this overall calculation as the charmonium production model (CPM). The direct $\psi(2S)$ production was found[8] to be very small, ~ 25 times smaller in magnitude than the data.

The sum of these two contributions (BPM and CPM) is also plotted in Fig. 2. In Fig. 2a we fit the theory to the data by summing the two theoretical contributions with independent normalization factors. With no normalization constraints a good fit is obtained with $\sim 69\%$ J/ψ production from CPM and $\sim 31\%$ J/ψ production from BPM. However, a previous CDF study[3] of $B^\pm \to J/\psi K^\pm$ showed that the BPM calculation underestimates the b-quark cross section by a factor of 5.5 ± 2.8 which indicates that $(75^{+25}_{-40})\%$ of our J/ψ's come from B decays. Using this information we find that the 90% C.L. upper limit on the BPM contribution is $\sim 60\%$. If future measurements exceed this value, then one must conclude that not only the normalization of BPM, but also the P_T-dependence of at least one of the models is wrong.

The products of the inclusive production cross section times branching fraction in the kinematic range $P_T > 6$ GeV/c and $|\eta| \le 0.5$ are:

$$\sigma(\bar{p}p \to J/\psi\, X) \times B(J/\psi \to \mu^+\mu^-) =$$

$$6.88 \pm 0.23(stat)\, ^{+0.93}_{-1.08}\, (syst)\ \text{nb and}$$

$$\sigma(\bar{p}p \to \psi(2S)X) \times B(\psi(2S) \to \mu^+\mu^-) =$$

$$0.232 \pm 0.051(stat)\, ^{+0.029}_{-0.032}\, (syst)\ \text{nb},$$

where an extrapolation of the cross sections for values of $P_T > 14$ GeV/c was carried out. The central values of the $J/\psi(\psi(2S))$ production cross sections are given for zero polarization, and the systematic uncertainties due to unknown polarizations are $^{+6.5\,(4.4)\%}_{-11.2(7.3)\%}$.

The other major systematic uncertainty in both the J/ψ and $\psi(2S)$ production cross sections is due to the trigger efficiency and was estimated to be $\pm 9\%$.

b-QUARK CROSS SECTION

In order to determine the b-quark cross section, we use the measurement of the $J/\psi(\psi(2S))$ inclusive production cross sections, the ratio of $J/\psi(\psi(2S))$ to b-quark cross sections as determined by a Monte Carlo technique[9,4,5,6,7], the combined branching ratios[10,11] $B(B \to J/\psi(\psi(2S))\, X) \times B(J/\psi(\psi(2S)) \to \mu^+\mu^-)$, Br_2, and the fraction f_B of $J/\psi(\psi(2S))$'s from B meson decays:

$$\sigma^b_{exp}(P_T^b > P_T^{min}, |y^b| < 1) =$$

$$\frac{Br_1 \times \sigma^c_{exp}(P_T^c > 6\text{GeV/c}, |\eta^c| < 0.5) \times R \times f_B}{2 \times Br_2}, \quad (1)$$

where

$$R = \frac{\sigma^b_{BPM}(P_T^b > P_T^{min}, |y^b| < 1.0)}{\sigma^c_{BPM}(P_T^c > 6\text{GeV/c}, |\eta^c| < 0.5)},$$

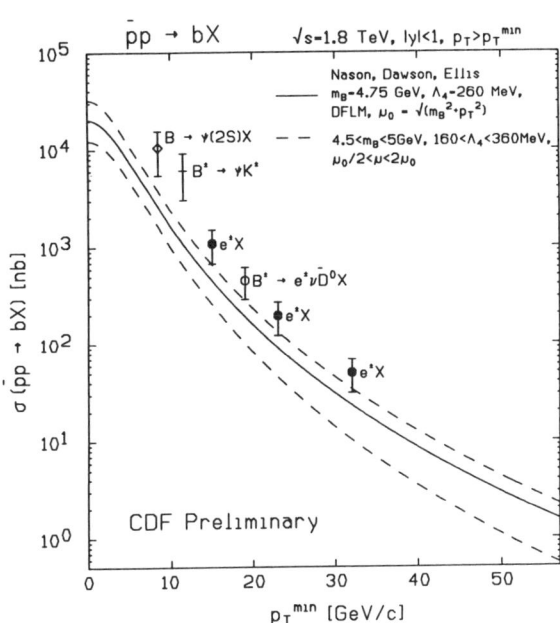

Figure 3. The b-quark production cross section.

$Br_1 = B(J/\psi(\psi(2S)) \to \mu^+\mu^-)$ and the index "c" stands for "$J/\psi(\psi(2S))$". P_T^{min} is determined by the Monte Carlo program and is chosen such that approximately 90% of the produced $J/\psi(\psi(2S))$ have $P_T^b > P_T^{min}$; we have set $P_T^{min} = 8.5$ GeV/c for this analysis. Assuming the fraction f_B to be unity, we get:

$$\sigma^b(P_T^b > 8.5 \text{GeV}/c, |y^b| < 1) = 18.9 \begin{array}{c}+4.7\\-5.0\end{array} \mu b$$

using J/ψ's and

$$\sigma^b(P_T^b > 8.5 \text{GeV}/c, |y^b| < 1) = 10.5 \begin{array}{c}+5.0\\-5.1\end{array} \mu b$$

using $\psi(2S)$'s. The fraction is believed to be close to one for $\psi(2S)$'s[1,8,12] but not for J/ψ's. In Fig. 3 we display the b-quark cross section derived by using $\psi(2S)$'s together with b-quark cross sections derived from other CDF analyses. The b-quark cross section we get using $\psi(2S)$'s is approximately 1.5 standard deviations higher than the theoretical calculation[4], in reasonable agreement with Ref. 3.

REFERENCES

1. E. W. N. Glover, A. D. Martin, W. J. Stirling, "J/ψ Production at Large Transverse Momentum at Hadron Colliders," *Z. Phys. C* 38, pp. 473–478, (1988).

2. C. Albajar et al., "Beauty Production at the CERN $\bar{p}p$ Collider," *Physics Letters B* 256, pp. 121–127, (1991).

3. F. Abe et al., "Measurement of the B-Meson and the b-Quark Cross Sections at $\sqrt{s} = 1.8$ TeV Using the Exclusive Decay $B^\pm \to J/\psi K^\pm$," *Phys. Rev. Lett. B* 68, pp. 3403–3407 (1992).

4. P. Nason, S. Dawson, and R. K. Ellis, "The Total Cross Section for the Production of Heavy Quarks in Hadronic Collisions," *Nucl. Phys. B* 303, pp. 607–633, (1988).

5. C. Peterson et al., "Scaling Violations in Inclusive e^+e^- Annihilation Spectra," *Phys. Rev. D* 27, 105 (1983).

6. H. Schroder, personal communication.

7. W. Chen, Ph.D Thesis, Purdue University, May, 1990.

8. E. W. N. Glover, personal communication.

9. N. Ellis and A. Kernan, "Heavy Quark Production at the CERN $p\bar{p}$ Collider," *Physics Reports* 195, pp. 23–125 (1990).

10. H. Albrecht et al., "B Meson Decays into Charmonium States," *Physics Letters B* 199, pp. 451–456 (1987).

11. M. S. Alam et al., "Study of the Decay $B \to \psi X$," *Phys. Rev. D* 34, pp. 3279–3285 (1986).

12. S. D. Ellis, M. B. Einhorn, C. Quigg, "Comment on Hadronic Production of Psions," *Phys. Rev. Letters* 36, 1263 (1976).

Neutrino Masses and Mixing

Parallel Session 11

Downtown Dallas with "Old Red" Courthouse

Conveners: B. Kayser (NSF)
J. Valle (Valencia)

SOLAR NEUTRINOS OBSERVED BY GALLEX AT GRAN SASSO

D. Vignaud
DAPNIA / Service de Physique des Particules
CE Saclay
F - 91191 Gif-sur-Yvette Cedex

Abstract

The GALLEX experiment, installed in the Gran Sasso Underground Laboratory, has measured the rate of production of ^{71}Ge from ^{71}Ga by solar neutrinos, using 30.3 tons of gallium in the form of gallium chloride solution. 14 measurements carried out in one year of operation give, after corrections for side reactions and other backgrounds, an average value of $(83 \pm 19\,(\text{stat.}) \pm 8\,(\text{syst.}))$ SNU (1σ) due to solar neutrinos. This result, two standard deviations below the predictions of solar model calculations (122-132 SNU), constitutes the first observation of solar pp-neutrinos. The possible deficit is interpreted, together with those of the chlorine and Kamiokande experiments, in terms either of astrophysics or of neutrino oscillations (MSW effect).

1. INTRODUCTION

The main purpose of the GALLEX solar neutrino experiment [1] is the detection of the pp-neutrinos (ν_{pp}) which are produced in the primary proton-proton fusion reactions in the core of the Sun.

The solar ν_e interact with the ^{71}Ga isotope of a gallium target, producing the radioactive ^{71}Ge isotope, whose lifetime is $T_{1/2} = 11.43$ d. The reaction threshold is 233 keV, well below the maximum energy of the ν_{pp} (420 keV). The production rate of ^{71}Ge given by standard solar models (SSM) [2,3,4] is around 122-132 SNU (solar neutrino unit) of which about 55 % are expected to come from ν_{pp}. The other main sources are ν_{Be}, from the ^{7}Be electron capture reaction, (about 25 %) and ν_B, from the ^{8}B decay (about 10 %), the remaining coming from the pep reaction and the CNO cycle.

The experiment is performed by extracting the ^{71}Ge atoms which are produced during an exposure time of several weeks and by then observing their decays. The experimental setup includes the target (100 tons of a liquid GaCl$_3$ solution containing 30.3 tons of gallium), a germanium extraction system and a ^{71}Ge counting system. It is installed 125 km east of Rome, in the Gran Sasso Underground Laboratory, where it is shielded by about 1000 m of rock.

The experimental procedure is described in section 2. The data analysis and results are presented in section 3. Implications of the result in terms of astrophysics or neutrino masses are discussed in section 4. Section 5 briefly describes the next steps which include a calibration with an artificial neutrino source. Conclusions are given in section 6.

More experimental details are given in reference [5] and the implications of the GALLEX determination of the solar neutrino flux are detailed in reference [6].

2. EXPERIMENTAL PROCEDURE

Two tanks (70 m^3 each), called A and B, are available to contain the target solution: a process tank (A) which is equipped with a large tube in order to hold the ^{51}Cr neutrino source and a second tank (B), needed for safety reasons, equipped with a smaller sweeping system and which has been used until now for the first data taking.

A solar neutrino capture rate of 132 SNU corresponds to 1.19 ^{71}Ge atom produced per day and yields only few atoms to count after a three-week exposure. Before each run about 1 mg of non radioactive germanium isotopes (^{72}Ge, ^{74}Ge, ^{76}Ge) or of natural germanium is put in the tank and dissolved in the solution. This carrier has two purposes: to check the germanium extraction efficiency and to be used as the counting gas once transformed into germane. Germanium atoms are incorporated into GeCl$_4$ molecules, which are volatile in presence of HCl, but dissolve in pure water. ^{71}Ge atoms and germanium carrier are extracted from the gallium solution by passing through it about 2000 m^3 of nitrogen during about 20 h (in the B tank). The germanium is then absorbed in the pure water of large absorption columns (about 30 l of water). The germanium extraction efficiency is measured for the germanium carrier by an atomic absorption method and is typically 99%. A series of smaller columns serves to concentrate the germanium in a smaller volume of about 1 l of water. The GeCl$_4$ is then back-extracted in 50 ml of tritium-free water and this final solution is used to produce germane (GeH$_4$). The germane is purified from remaining impurities (air, radon, CO$_2$,...) by gas chromatography before filling the counters. To optimize the counting performance, the gas mixture introduced in the low-background counters at a slight overpressure is germane (30%) and old xenon (70%).

^{71}Ge decays by electron capture and the resulting spectrum of Auger electrons and X-rays absorbed in the counter has two peaks at 1.2 keV and 10.4 keV, corresponding respectively to L and K-capture. The counters are calibrated using the 35 keV X-rays from a cerium source, which produce fluorescent X-rays in the xenon, giving peaks at 1.03, 5.09 and 9.75 keV. Energy resolutions (FWHM) of the counters are typically 26% for the K-peak and 43% for the L-peak.

In order to reach sufficiently low background rates, pulse shape analysis of the proportional counter pulses is done to discriminate between fast pulses originating from point-like ^{71}Ge decays and background from Compton scattering with a spatially extended track. The Camac-based counting system is thus equipped with a fast transient digitizer (sampling every 0.5 ns) to record the shape of each preamplifier output pulse. Good events are selected in the two-dimensional plot risetime versus energy as the fast pulses in the L and in the K regions. The optimal acceptance "windows" are determined with counters spiked with a large amount of ^{71}Ge. The counting efficiencies, including dead volume effects, are determined for each counter and are typically 30% for the L-window and 35% for the K-window.

The counters are placed into a copper box which is inserted into a passive copper shielding or into the well of a NaI(Tl) detector (active shielding). The NaI detector is used either as an anticoincidence detector device or in coincidence mode to distinguish ^{71}Ge from ^{69}Ge decays. This setup is installed in a steel vessel filled with low activity lead and closed by an air-tight box to prevent radon from entering, the radon produced inside being trapped. The complete system, including the associated electronics, is inside a Faraday cage connected to a "microvax" through an optical link.

The counters backgrounds are typically 0.10 cpd (fast L pulses) and 0.05 cpd (fast K pulses) in the active counting positions and 0.07 cpd (fast L+K) in the passive side.

^{68}Ge can be produced in the gallium solution by cosmic ray interactions via (p,2n) or (p,4n) reactions on Ga. It decays via electron capture (100 %), yielding events indistinguishable from ^{71}Ge decays except for the much longer half life (288 days). About 17 million ^{68}Ge atoms were produced in the target before it was brought underground. Normal purging of the solution was not sufficient to remove them. It seems that a small fraction (about 0.1 %) of the ^{68}Ge was attached to trace impurities (e.g. polysilicic acids) and released slowly. Because of the high initial concentrations, this counting level was much higher than the one expected for ^{71}Ge produced by solar neutrinos. Heat treatment and more than 20 desorption runs, performed between July 1990 and April 1991, were necessary to reduce the ^{68}Ge component to an acceptable level (a few atoms released per day of standing time, yielding less than 0.1 count per day).

3. DATA ANALYSIS AND RESULTS

The data taking for solar neutrinos started on 14 May 1991 when the residual ^{68}Ge component has become comparable to the counter background. This first data taking period ("Gallex 1") ended on 29 April 1992. In order to determine the counter background rates properly, it is necessary to count several months, much longer than a few ^{71}Ge lifetimes. The data reported here concern 14 runs and have to be considered as preliminary since the counting time has not elapsed for all runs.

The first phase of data treatment is to eliminate obvious background events. ^{71}Ge decays without the production of γ-rays so the NaI is used as a veto for counters on the active side. Decays of residual ^{222}Rn and its short-lived daughters are mostly eliminated by ignoring all events occuring less than 3 h after or 15 mn before an overflow event (> 16 keV). This efficiently vetoes cascades where at least one of the initial two α's of the radon decay chain occurs in the active volume of the counter. The number of L and K events eliminated by this cut is equivalent to about 20 SNU or about three ^{222}Rn atoms introduced into the counter filling.

The second phase uses the pulse shape. Only fast pulses events which have a risetime between about 20 and about 40 ns are considered. Exact values are derived for each measurement from the calibration with the cerium source. These calibrations are done periodically to check the stability of the counting system.

The energy spectrum for the selected fast events (for all 14 runs) is shown in figure 1 for different counting periods. The L- and K-peaks of ^{71}Ge are clearly seen in the first counting period corresponding to the first mean life of ^{71}Ge . They disappear in the subsequent counting periods. The identical curves in fig. 1a,1b,1c are an handdrawn simulation of the expected background to facilitate cross comparisons. Figure 1d shows the energy spectrum of a ^{71}Ge calibration sample.

To quantify the magnitude of the signal, events that are candidate for ^{71}Ge decays are selected by applying cuts in i) energy (energy acceptance windows extend to 2 FWHM on each side of the peak position), ii) risetime (risetime cuts accept 97.7 % of L-pulses and 95 % of K-pulses). The selected events are used as input for a maximum likelihood analysis [7]. Using the time distribution of the events, this analysis determines the number of events due to a component decaying with the ^{71}Ge lifetime and a number

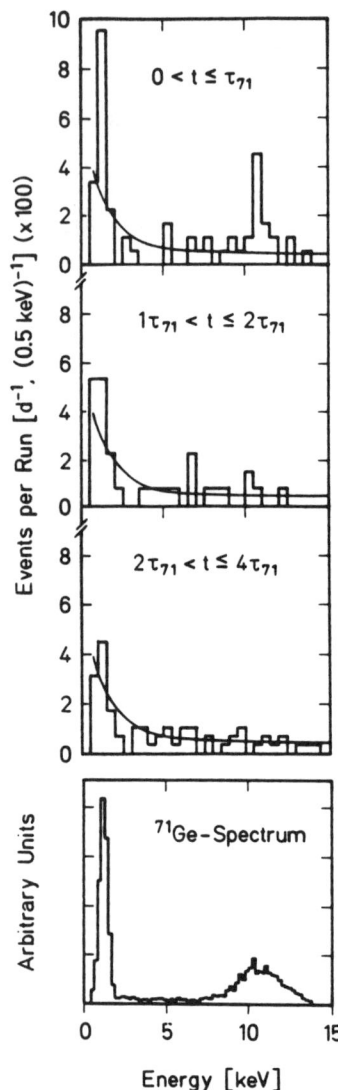

Figure 1. Energy spectra for the fast events of the 14 solar neutrino runs during different counting periods after the end of extraction : a) $(0\tau-1\tau)$ (τ is the ^{71}Ge mean life) ; b) $(1\tau-2\tau)$; c) $(2\tau-4\tau)$; d) measured spectrum from a ^{71}Ge calibration sample. Handdrawn curves below a), b) and c) are identical ; they simulate the expected background and are drawn to facilitate the comparisons.

due to a time-independent background. A known background component decaying with the ^{68}Ge lifetime is also assumed so as to take into account the residual ^{68}Ge whose magnitude is determined independently.

The result for the combined 14 runs gives a mean production rate of $(90 \pm 19) \times 10^{-36}$ ^{71}Ge atoms per target atom per second. Next the known contributions which are not due to solar neutrinos (side reactions and residual backgrounds) have to be subtracted : a) 3.7 ± 1.1 SNU for the cosmic muon induced background, b) 0.15 ± 0.10 SNU for the fast neutron induced background, c) < 0.2 SNU for actinides in the target, d) 2 ± 1 SNU for radon background not suppressed by the cuts described above, e) 1 ± 1 SNU for ^{69}Ge coming from muons or ν_B . This yields a net rate associated with solar neutrinos of $(83 \pm 19\,(\text{stat.}) \pm 8\,(\text{syst.})$ SNU). The contributions to the systematic error come from the side reactions, from the ^{68}Ge correction mentioned above (± 5.5 SNU), from the uncertainties in the chemical yield (± 1.5 SNU), the counting efficiency (± 3.6 SNU) and the risetime cut (± 1.8 SNU). Results for individual runs can be found elsewhere [5].

Analyzing separately the L- (K-) windows, yields a rate of 104 ± 31 SNU (70 ± 23 SNU). These results do not differ significantly from the combined result.

In order to check the reliability of the result, some tests have been performed : i) the overall result is notably stable to changes in the energy and risetime cuts, even to complete opening of the risetime windows. ii) In a maximum likelihood analysis in which the mean life τ of the decaying component is left free, the fit yields $\tau = (13.5 ^{+5.2}_{-3.5})$ days, compatible within statistics with 16.49 d for ^{71}Ge .

Another experiment, SAGE, in Baksan (Caucasus, Russia), uses gallium metal as a target. They published last year [8] a result for the 1990 data (20^{+15}_{-20} (stat.) ± 32 (syst.) SNU), two standard deviations below the GALLEX value and in strong disagreement with SSM. Using new data from 1991 runs, new values have been presented at this Conference [9] (58^{+17}_{-24} (stat.) ± 14 (syst.) SNU) in reasonable agreement with the GALLEX value.

4. INTERPRETATION

As stated in the introduction, the gallium experiments are sensitive to all solar neutrino sources, particularly to neutrinos coming from the primary proton-proton fusion reaction in the Sun. The GALLEX result is two standard deviations below the predictions of standard model calculations [2,3,4] which are between 122 and 132 SNU. Considered on its own, it does not imply any problem with the models.

The radiochemical chlorine experiment, which is sensitive mainly to ν_{Be} and ν_B, has observed since 1968 a neutrino flux three to four times smaller than the predictions of standard solar models [10]. Since 1988, the real-time Kamiokande experiment, which is sensitive mainly to ν_B, has observed about half the expected rate [11].

A simultaneous interpretation of these three experiments clearly require either modifications of solar models in order to predict a lower ν_e flux, or neutrino "oscillations" which imply that solar ν_e are transformed into ν_μ or ν_τ. These two possibilities are now discussed.

A basic difference between the gallium experiments and the chlorine or Kamiokande experiments is that the response of gallium is heavily weighted by the primary pp reaction, the only neutrino producing reaction in the dominant ppI cycle ([2,3,4]). Because of this close link between the dominant ppI cycle and the well measured solar luminosity, the gallium response to the solar neutrino flux is very much less model dependent; this, in turn, implies that it might be possible to produce concordance with the chlorine and Kamiokande results by modifying the relative contributions of the various cycles.

As an example, a cut-back of the highly temperature-dependent ^{8}B branch, a less drastic cut-back of the less sensitive ^{7}Be branch, balanced by an increase in the pp branch, might lead to a successful meeting with the chlorine and Kamiokande results as well as the GALLEX result. A "successful" adjustment would be a reduction of the ^{8}B and CNO neutrino flux to 0.3 of the SSM value (of reference [2]), of the ^{7}Be neutrino flux to 0.35 of the SSM value and an increase of the pp-neutrino flux to 1.05 times the SSM value. This maintains the solar luminosity at its nominal value and it would put Kamiokande at 0.3 of the SSM value, within 2σ of the published measured value; chlorine at 2.6 SNU, within 1σ [10]; and gallium at 97 SNU (of which about 15 SNU only are due to other sources than pp and pep neutrinos), within 1σ of the GALLEX value. GALLEX can truly claim to have finally "seen" primary pp-neutrinos.

To be more quantitative, we have attempted to fit the central temperature of the Sun, T_c, to the observed fluxes by a phenomenological mock-up. It is assumed that the neutrino flux of each branch can be characterized by just a simple temperature dependence $(T_c)^{n_i}$. The exponents n_i are taken from the "1000-model" calculation of reference [2] : $n_{pp} = -1.2$, $n_{Be} = 8$ and $n_B = 18$; the exponents for the CNO neutrinos are arbitrarily chosen as 18, equal to that for the ν_B. Then T_c is varied so as to minimize the χ^2 of the σ-weighted combination of the chlorine, Kamiokande and GALLEX results, relative to the different models of references [2,3,4]. The results are simply summarized: the T_c values range from a lowering of 4.4 to 6.1 % and χ^2 that correspond to confidence levels that are all below 5 %. The main constraint comes about from the chlorine and Kamiokande values, with a GALLEX value that is at the high end of its 1σ range (about 110 SNU). (Similar conclusions were reached recently by Bludman et al.[12]).

A change in central temperature T_c of the order of 5% is a large amount in terms of changes in astrophysical parameters. Furthermore, the fits to helioseismological data can be obtained only by accompanying the temperature decrease with the proper change of chemical composition near the solar center. Moreover, if taken at face values, the confidence levels associated with such a fit (less than 5%) make it a very poor bet. It indicates the extent of the stretch that is needed and shows that a simple temperature variation is only marginally compatible with our understanding of the physical processes which govern stellar evolution.

Solar neutrino oscillations have long ago been suggested as an explanation of the unexpectedly low result of the chlorine experiment. With the discovery of matter enhanced neutrino oscillations (the so-called MSW effect [13]), this hypothesis has even become more popular. If we accept the solar model as given, we can analyze all discrepancies between solar model calculations and the results of the solar neutrino experiments in terms of alteration of the properties of solar neutrinos during their passage from the solar core to the detector.

The two-neutrino version assumes that the important mixing is between the ν_e and one other, usually assumed to be the ν_μ. The parameter space is defined by Δm^2 (difference between the two squared masses), which enters the formula as $\Delta m^2/2E_\nu$, and $\sin^2 2\theta$, where θ is the mixing angle. Because the flavor changing probabilities depend on E_ν and because the various branches of the solar energy cycle differ in neutrino energies, the MSW effect has distinguishable effects, depending on the energy weightings, between the different solar neutrino experiments. The MSW effect allows one, through an appropriate choice of parameters, to suppress any desired region of the ν spectrum.

Each experiment defines in the (Δm^2, $\sin^2 2\theta$) plane a triangular region where the reduction factor experimentally observed for the ν_e flux is in agreement with the MSW calculations. The areas common to the different experiments are good candidates for neutrino oscillation parameters. Such plots have been often shown and a recent one can be found in [6] (see also [14]).

To rely more on the statistical significance of the analysis, the three experimental results have been fitted, using a χ^2 method, to the flux reduction predicted by the MSW mechanism in the (Δm^2, $\sin^2 2\theta$) plane.

Figure 2 shows the two black areas selected at 90% confidence level. One lies at Δm^2 equal to 3×10^{-6} to 10^{-5} eV^2 and $\sin^2 2\theta$ in the 4×10^{-3} to 10^{-2} region. The second is for a wide range of Δm^2 at a $\sin^2 2\theta$ close to the maximum value. On the same plot, the area inside the full line represents the region excluded by the GALLEX result (99% confidence level) almost independently of solar models. Within this zone, a gallium rate of smaller than 40 SNU is predicted by solar models ranging from the standard models to "minimal" models with only ν_{pp}.

Searches for day-night (produced by the effect of the Earth) and semiannual variations of the ν_B flux were carried out by the Kamiokande collaboration [15]. No such short-time variation was observed, within statistical error, which requires the exclusion of the region between the dotted lines in figure 2.

The results shown in figure 2 are not significantly dependent on the choice of either a given standard solar model or of different inputs. Calculations done using the Turck-

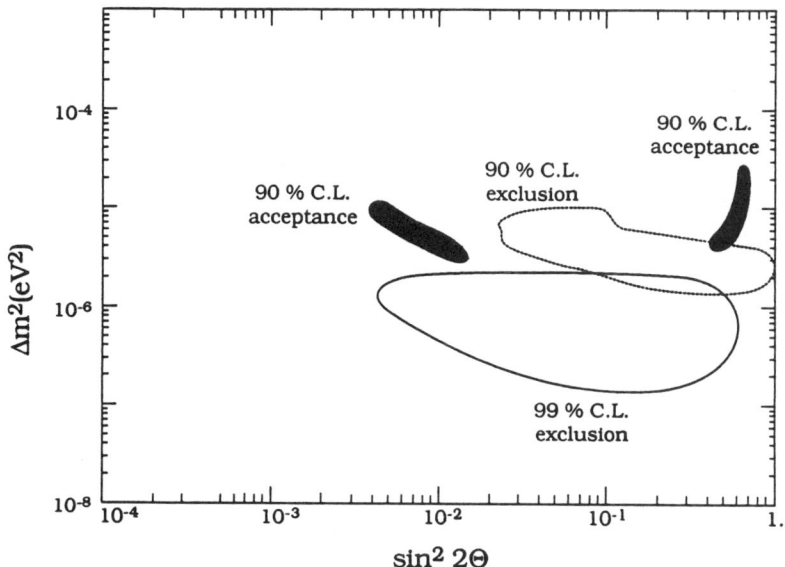

Figure 2. Δm^2 versus $\sin^2 2\theta$ for neutrino oscillation parameters. Within the black areas, the MSW effect successfully (90% C.L.) reconciles the chlorine, Kamiokande and GALLEX experiments with standard solar models. The area inside the dotted line is excluded at 90% C.L. by Kamiokande from a study of day-night effects [15]. The area inside the full line is excluded at 99% C.L. by the GALLEX result.

Chièze et al. model [3] for example give similar results.

5. NEXT STEPS

The GALLEX 1 data period ended on April 29, 1992. Final results for this period will be determined at the end of 1992, after at least 6 months of counting for all runs. The GALLEX 2 data period started in August 1992, after transfer of the gallium solution from the B-tank to the A-tank and removal of the residual ^{68}Ge.

As a test of the whole experimental procedure, the experiment will be calibrated with an artificial neutrino source (> 50 PBq of ^{51}Cr) [16]. This source will be made by irradiating chromium enriched in ^{50}Cr in the reactor Siloé at Grenoble, France. A first irradiation is scheduled for fall 1993.

At least two more years of running are planned after the source exposures. Four years of exposure to solar neutrinos should lead to a statistical error smaller by a factor 2 (about 8 SNU), which should severely constrain the different interpretations.

6. CONCLUSION

The GALLEX experiment has observed, for the first time, the primary pp-neutrinos coming from the core of the Sun. The central value, 83 SNU, is to be compared with the 74 SNU attributed to the pp and pep cycles in the standard models. A "consistent" explanation could be the observation the expected ν_{pp} flux (strongly constrained by the luminosity of the Sun) plus a ν_{Be} and ν_B flux reduced as observed by the chlorine and Kamiokande experiments. The result is also compatible within 2σ with the SSM predictions (122-132 SNU).

To fit this result together with those of the chlorine and Kamiokande experiments requires severe stretching of solar models, but does not rule out such a procedure. The Mikheyev-Smirnov-Wolfenstein (MSW) mechanism provides a good fit, and the GALLEX result fixes the Δm^2 and $\sin^2 2\theta$ parameters in two very confined ranges (around $\Delta m^2 = 6 \times 10^{-6}$ eV^2 and $\sin^2 2\theta = 7 \times 10^{-3}$ and around $\Delta m^2 = 8 \times 10^{-6}$ eV^2 and $\sin^2 2\theta = 0.6$).

Acknowledgements : It is a pleasure to thank M. Cribier, J. Rich and M. Spiro for their fruitful comments.

REFERENCES

1. GALLEX Collaboration : P.Anselmann, W.Hampel, G.Heusser, J.Kiko, T.Kirsten, E.Pernicka, R.Plaga, U.Rönn, M.Sann, C.Schlosser, R.Wink, M.Wójcik (Heidelberg), R.von Ammon, K.H.Ebert, T.Fritsch, K.Hellriegel, E.Henrich, L.Stieglitz, F.Weyrich (Karlsruhe), M.Balata, E.Bellotti, N.Ferrari, H.Lalla, T.Stolarczyk (Gran Sasso), C.Cattadori, O.Cremonesi, E.Fiorini, S.Pezzoni, L.Zanotti (Milano), F.von Feilitzsch, R.Mössbauer, U.Schanda (München), G.Berthomieu, E.Schatzman (Nice), I.Carmi, I.Dostrovsky (Rehovot), C.Bacci, P.Belli, R.Bernabei, S.d'Angelo, L.Paoluzi (Roma), S.Charbit, M.Cribier, G.Dupont, L.Gosset, J.Rich, M.Spiro, C.Tao, D.Vignaud (Saclay), R.L.Hahn, F.X.Hartmann, J.K.Rowley, R.W.Stoenner, J.Weneser (Brookhaven)

2. J.N.Bahcall and R.K.Ulrich, *Rev. Mod. Phys.* 60, pp. 297-372 (1988), J.N.Bahcall and M.H.Pinsonneault, "Standard solar models, with and without helium diffusion and the solar neutrino problem", *Rev. Mod. Phys.*, to appear in Oct. 1992

3. S.Turck-Chièze, S.Cahen, M.Cassé and C.Doom, *Ap. J.* 335, pp. 415-424 (1988) S.Turck-Chièze and I.Lopes, "Towards a unified classical model of the Sun", submitted to Ap. J., Aug. 1992

4. G.Berthomieu, J.Provost, P.Morel and Y.Lebreton, "Standard solar models with CESAM codes", June 1992, to appear in Astron. Astrophys.

5. GALLEX Coll., P. Anselmann et al., *Phys. Lett.* B285, pp. 376-389 (1992).

6. GALLEX Coll., P. Anselmann et al., *Phys. Lett.* B285, pp. 390-397 (1992).

7. B.T.Cleveland, *Nucl. Instr. Methods* 214, pp. 451-458 (1983).

8. A.I.Abazov et al., *Phys. Rev. Lett.* 67, pp. 3332-3335 (1991).

9. V.N.Gavrin, these proceedings.

10. R.Davis, D.S.Harmer and K.C.Hoffmann, *Phys. Rev. Lett.* 20, pp. 1205-1207 (1968) R.Davis, in *Proc. of the 21st ICRC*, ed. R.J.Protheroe, 1990, vol. 12, pp. 143-151.

11. K.S.Hirata et al., *Phys. Rev. Lett.* 65, pp. 1297-1300 (1990), K.S.Hirata et al., *Phys. Rev.* D44, pp. 2241-2260 (1991), T.Kajita, these proceedings

12. S.Bludman et al., *Nucl. Phys.* B374, pp. 373-391 (1992), S.Bludman et al., "Implications of combined solar neutrino observations and their theoretical uncertainties", July 1992.

13. S.P.Mikheyev and A.Yu. Smirnov, *Nuovo Cimento* 9C, pp. 17-28 (1986) L.Wolfenstein, *Phys. Rev.* D17, pp. 2369-2374 (1978) For a review see for example T.K.Kuo and J.Pantaleone, *Rev. Mod. Phys.* 61, pp. 937-979 (1989).

14. S.P.Rosen, these proceedings

15. K.S.Hirata et al., *Phys. Rev. Lett.* 66, pp. 9-12 (1991).

16. M.Cribier et al., *Nucl. Instr. Methods* A275, pp. 574-585 (1988).

LATEST RESULTS FROM THE SOVIET-AMERICAN GALLIUM EXPERIMENT

V.N. Gavrin, O.L. Anosov, E.L. Faizov, A.V. Kalikhov, T.V. Knodel, I.I. Knyshenko,
V.N. Kornoukhov, I.N. Mirmov, A.V. Ostrinsky, A.M. Pshukov, A.A. Shikhin,
P.V. Timofeyev, E.P. Veretenkin, V.M. Vermul, G.T. Zatsepin
Institute for Nuclear Research, Russian Academy of Sciences, Moscow 117312, RUSSIA

T.J. Bowles, S.R. Elliott*, J. S. Nico, H.A. O'Brien, D.L. Wark**, J.F. Wilkerson
Los Alamos National Laboratory, Los Alamos, NM 87545 USA

B.T. Cleveland, R. Davis, Jr., K. Lande
University of Pennsylvania, Philadelphia, PA 19104 USA

M. L. Cherry
Louisiana State University, Baton Rouge, LA 70803 USA

R. T. Kouzes ***
Princeton University, Princeton, NJ 08544 USA

A radiochemical ^{71}Ga-^{71}Ge experiment to determine the primary flux of neutrinos from the Sun began measurements of the solar neutrino flux at the Baksan Neutrino Observatory in 1990. The number of ^{71}Ge atoms extracted from 30 tons of gallium in 1990 and from 57 tons of gallium in 1991 was measured in twelve runs during the period of January 1990 to December 1991. The combined 1990 and 1991 data sets give a value of 58 +17/-24 (stat) ± 14 (syst) SNU. This is to be compared with 132 SNU predicted by the Standard Solar Model.

INTRODUCTION

A fundamental problem during the last two decades has been the large deficit of the solar neutrino flux observed in the radiochemical chlorine experiment[1] compared with the theoretical predictions[2-3] based on the Standard Solar Model (SSM). Recent results of the Kamiokande II water Cherenkov experiment[4] have confirmed this deficit. These results may be explained by deficiencies in the solar model in predicting the ^{8}B neutrino flux or may indicate the possible existence of new properties of the neutrino.[5] The role new neutrino properties may play in the suppression of the high energy solar neutrino flux[5], as possibly indicated in the chlorine and Kamiokande II experiments, can be determined by a radiochemical gallium experiment. An experiment using ^{71}Ga as the capture material[6] provides the only feasible means at present to measure low energy solar neutrinos produced in the proton-proton (p-p) reaction. Exotic hypotheses aside, the rate of the p-p reaction is directly related to the solar luminosity and is insensitive to alterations in the solar models. An observation in a gallium experiment of a strong suppression of the low energy solar neutrino flux requires the invocation of new neutrino properties.

THE BAKSAN GALLIUM EXPERIMENT

In this paper we present the results of the measurements of the solar neutrino flux by the Soviet-American Gallium solar neutrino Experiment (SAGE) during 1990 and 1991 carried out at the Baksan Neutrino Observatory.

Extraction Procedure

The chemical extraction of germanium from liquid metallic gallium was first tested on a small scale in the U.S.[7] and later developed and tested at a 7.5 ton pilot installation in Russia [8]. The experimental layout as well as the chemical and counting procedures have been described previously and are only briefly outlined here.[9]

Each measurement of the solar neutrino flux begins by adding approximately 700 micrograms of natural Ge carrier equally divided among the reactors holding the gallium. After a typical exposure interval of 1 month, the Ge carrier and any ^{71}Ge atoms that have been produced by neutrino capture are chemically extracted from the Ga using the following procedure. A weak HCl solution is mixed with the Ga metal in the presence of H_2O_2 which results in the extraction of Ge into the aqueous phase. The extracted solutions from the reactors are combined and reduced in volume by vacuum evaporation. Additional HCl is then added and an Ar purge is initiated which sweeps the Ge as $GeCl_4$ from the acid solution into 1.2 liters of H_2O. The Ge is then extracted into CCl_4 and back extracted into 0.1 liters of low-tritium H_2O. The counting gas GeH_4 (germane) is then synthesized and purified by gas chromatography. The efficiency of extraction of the germanium carrier is measured at two stages of the extraction procedure by atomic absorption analysis. The final determination of the quantity of germanium is made by measuring the volume of synthesized GeH_4. The overall extraction efficiency is typically 80% with an uncertainty of $\pm$ 6%.

Counting Procedure

The GeH_4 is then mixed with a measured amount of Xe and inserted into a low-background proportional counter. The proportional counter (with a volume of about 0.75 cm^3) is placed in the well of a NaI detector inside a large passive shield and counted for 2-4 months. ^{71}Ge decays by electron capture to the ground state of ^{71}Ga with an 11.4 day half life. The low-energy K- and L-shell Auger electrons and x-rays produced during electron shell relaxation in the ^{71}Ga daughter atom are detected by the proportional counter. Due to considerably higher backgrounds in the L-peak, only the K-peak has been used in the present analysis.

Pulse shape discrimination based on rise time measurements is used to separate the ^{71}Ge decays from background. The energy, amplitude of the differentiated pulse, and any associated NaI signal are recorded for each event in the counter.

The counter is typically calibrated at one month intervals using an external ^{55}Fe source. The K-peak acceptance window is then determined by extrapolation from the ^{55}Fe peak. The extrapolation procedure was verified by filling a counter with $^{71}GeH_4$ together with the standard counter gas.

EXTRACTION HISTORY

The experiment began operation in May, 1988 when purification of 30 tons of Ga commenced. The large quantities of long-lived ^{68}Ge (half life = 271 days) produced by cosmic rays while the Ga was on the surface were removed. New extraction procedures were implemented beginning with the January

1990 extraction which resulted in the elimination of radon contamination in the extractions.

By January 1990, the backgrounds had been reduced to levels sufficiently low to begin measurements of the solar neutrino flux. Monthly extractions were carried out from January through July of 1990. The May run was not used because rise time information was lost due to electronic problems. The resulting high background gave essentially no sensitivity to the solar neutrino flux. But the May run is shown here for completeness. The extraction sample for the June run was lost due to a vacuum accident.

Useful solar neutrino data were not obtained after the July 1990 run due to the Cr engineering test run. Following completion of the test run, a total of about 30 tons of new Ga were purified to remove ^{68}Ge. At the same time, the old gallium used in the previous solar neutrino runs was removed and the chemical reactors were extensively cleaned. The chemical extraction system was carefully cleaned followed completion of the purification of the new gallium. Tests in May of 1991 indicated that the levels of residual ^{68}Ge and radon were well below the signal predicted by the SSM. Separate extractions of the new and old Ga were carried out in June and July. Rise time information was lost for the June 1991 run due to unstable electronics, but the background was still sufficiently low that a measurement of the solar neutrino flux could be made. Beginning in August 1991, combined extractions of the old and new Ga were begun. The run from October 1991 was lost due to a counter failure.

MEASUREMENT OF THE SOLAR NEUTRINO FLUX

Statistical Analysis

Results from measurements carried out in 1990 and 1991 are reported here. Earlier data taken during 1989 are not presented here due to the presence of radon and ^{68}Ge residual contaminations.

The data analysis selects events that have no NaI activity in coincidence within the ^{71}Ge K-peak acceptance window. The K-peak acceptance window in energy is a 2 FWHM wide energy cut centered on the K-peak. The inverse rise time cuts are 95% acceptance, with 1% being cut on fast rise time pulses (i.e., noise) and 4% cut on slow rise time pulses (i.e., background). A maximum likelihood analysis[10] is then carried out on these events by fitting the time distribution to an 11.4-day half-life exponential decay plus a constant rate background. Table 1 shows the results of the maximum likelihood analysis.

The data from each of the twelve extractions are shown in Figure 1, which shows the integral plot of events versus time within the ^{71}Ge K-peak acceptance window. In this figure, the value of the curve is incremented by one count every time an event occurs and thus shows the time distribution of ^{71}Ge-like events. The best fit line to each data set is shown by the dashed line. The Smirnov-Cramer-Von Mises parameter N_w^2 provides a measure of the goodness of fit[11], which is independent of the binning of the data. For this parameter, it is expected that 50% of the fits should have values greater than 0.119, and 50% less than 0.119. (In some sense, one can consider a N_w^2 value of 0.119 as being analogous to a χ^2 value of 1.0.) The probability that a measurement would exceed the value of N_w^2 determined for each of the runs is also given in Table 1.

Systematic Effects

Table 1.
Statistical analysis of runs

Extraction Date	Ga Mass (Tons)	Best Fit(SNU)	N_w^2	68% CL (SNU)	Probability
Jan 24, 90	28.7	0	0.367	60	9%
Feb 28, 90	28.6	39	0.310	83	13%
Mar 29, 90	28.5	90	0.035	175	96%
Apr 20, 90	28.4	0	0.060	94	81%
May 22, 90	28.3	79	0.073	204	73%
Jul 24, 90	21.0	0	0.250	149	19%
Combined 1990 (w/o May)		20	0.223	35	23%
Jun 28, 91	27.4	8	0.142	100	41%
Jul 23, 91	27.4	27	0.079	131	70%
Aug 25, 91	49.3	300	0.050	421	96%
Sep 23, 91	56.6	48	0.064	91	79%
Nov 22, 91	56.3	75	0.088	131	65%
Dec 20, 91	56.2	93	0.037	147	95%
Combined 1991		85	0.142	107	41%
Combined 1990 and 1991		**58**	**0.094**	**80**	**61%**

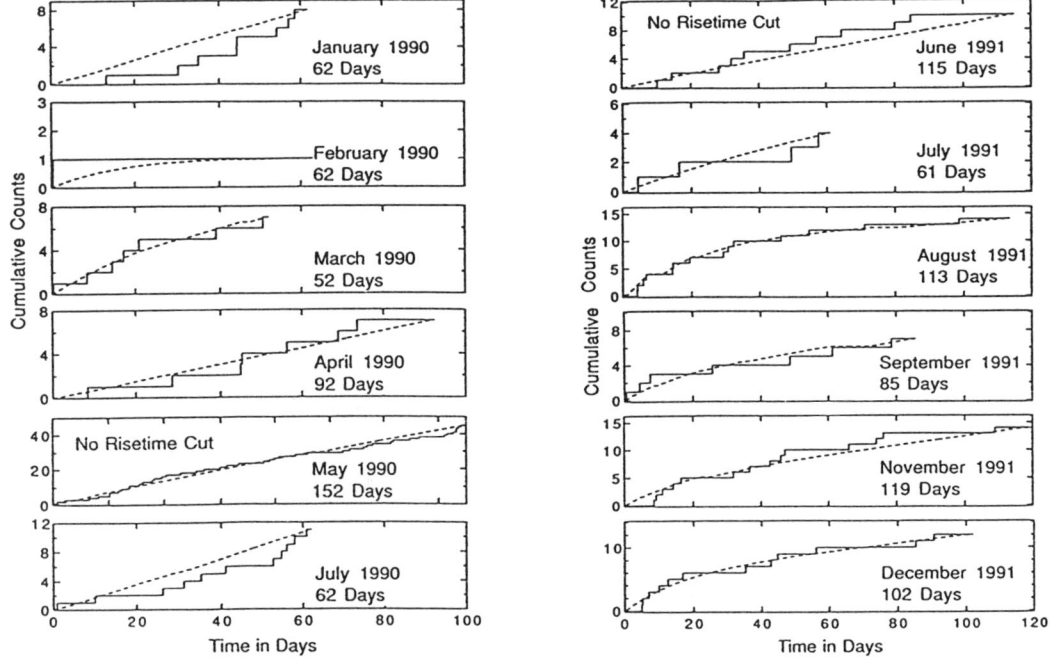

Figure 1. Data from the 1990 runs (final) and 1991 runs (preliminary).

The systematic uncertainties in the chemical extraction and counting efficiencies were typically 6% and 10%, respectively.

The systematic uncertainty in extrapolating the inverse rise-time cuts is estimated using a cut that includes all events not in coincidence with the NaI counter which are within the energy cut of the K-peak acceptance window with no cut made on inverse rise time. This results in an uncertainties of 9 SNU (68% CL) for the combined 1990 and 1991 data.

The uncertainty in background determination under the ^{71}Ge decay curve due to possible time variations of the counter background was checked in a number of ways[12] in the 1990 data set in which there was an apparent increase in the background at late counting times (see figure 1). This apparent change in the background rate at late counting times is reflected in the somewhat poor N_w^2 value of 0.223 (corresponding to a statistical probability of 23%) for the 1990 data. Analyses were made by truncating the data sets in time event by event and determining the resultant change in the value of the best fit and limits. We also fit the time distribution for events in the K-peak window with a first-order polynomial expansion in time of the background rate. We then generated Monte Carlo data sets using the fit values of the polynomial expansion and determined the maximal change. Finally, we binned the data from each of the 1990 extractions (except May which had no rise time cut) in 2-keV wide bins from 1 to 13 keV. We then compared fits to each of these bins assuming a background flat in time and one varying with time. The fit to a constant background rate is very good, giving a N_w^2 value of 0.053, corresponding to obtaining a better fit only 14% of the time. The fit assuming an increasing background rate was very bad with a N_w^2 value of 1.129, corresponding to less than a 0.2% probability. Thus, all tests are consistent with the hypothesis that the apparent increase in background at late times is purely a statistical fluctuation. But such a fluctuation could suppress the solar neutrino signal by causing an overestimation of the background at early times. In order to minimize any assumptions and allow for a possible statistical fluctuation in the background which would give a low apparent signal, we assigned an uncertainty to any possible time variation of the background for the 1990 data to be 30 SNU (68% CL). A possible time variation of the background was checked for in the 1991 data set and none was found. The 1991 data set had lower backgrounds, the ^{71}Ge signal rate was increased by doubling the amount of gallium from 27 tons to 57 tons, and the extractions were counted for longer periods to search for any time variation in the background. No increase in the background rate at late counting times was observed in the 1991 data. The combined 1990 and 1991 data sets show no change in the background rate at late counting times. This is reflected in the good N_w2 value of 0.094 (corresponding to a statistical probability of 61%) for the combined 1990 and 1991 data sets. As there is no evidence for any time variation in the 1990 or the combined 1990 and 1991 data, we assume the background is constant in time and do not assign any systematic uncertainty for a possible time variation in the background to the 1991 and the combined 1990 and 1991 data sets.

The final possible systematic effect is due to possible background reactions which could produce ^{71}Ge and the possible presence of radon, which can mimic a ^{71}Ge signal. The total background production rate in 30 tons of liquid gallium metal of all germanium activities from external neutrons, internal radioactivity, and cosmic ray muons has been calculated to be less than 2.5% of the SSM production rate[9], based upon our studies and measurements of these various background processes. This results in an uncertainty of 3 SNU (68% CL).

The data has been examined to search for a possible presence of radon. Checks included looking at overflow events (due to alpha decays in the radon chain), looking outside of the K-peak acceptance window, looking for delayed coincidences of events (due to subsequent decays in the radon chain), and fitting the data to allow for both ^{71}Ge and radon. No evidence for radon was found and a systematic uncertainty of 8 SNU (68% CL) was assigned to the combined 1990 and 1991 data. Table 2 shows the systematic uncertainties assigned to the combined 1990 and 1991 data sets.

Table 2.
1-σ Systematic uncertainties for the combined 1990 and 1991 data sets.

Efficiencies	7 SNU
Rise time	9 SNU
Backgrounds	3 SNU
Radon	8 SNU
TOTAL	14 SNU

RESULTS

The results[12] of the analysis of the five runs with rise time selection in the 1990 data are, assuming that the extraction efficiency for ^{71}Ge atoms produced by solar neutrinos is the same as that measured using natural Ge carrier:

^{71}Ga Capture Rate =
20 + 15/-20 (stat) ± 32 (syst) SNU.

Upper limits were determined by adding the statistical and systematic errors in quadrature and then adding this linearly to the best fit value. The upper limits are:

^{71}Ga Capture Rate < 55 SNU (68% CL),
< 79 SNU (90% CL).

Including the May 1990 run, which did not have rise time selection, increases the best fit and limits by only 1 SNU.

In terms of the total number of ^{71}Ge atoms observed, these values correspond to a best fit of 2.6 atoms observed in the five runs. The SSM predicts a production rate of 1.2 ^{71}Ge atoms/day in 30 tons of Ga, corresponding to a total 17.0 atoms expected for the runs reported here in 1990 with rise time selection.

For the 1991 data, the rate was determined to be:

^{71}Ga Capture Rate =
85 + 22/-32 (stat) ± 20 (syst) SNU.

For the combined 1990 and 1991 data, the rate was determined to be:

**^{71}Ga Capture Rate =
58 + 17/-24 (stat) ± 14 (syst) SNU.**

This assumes that the extraction efficiency for ^{71}Ge atoms produced by solar neutrinos is the same as that measured using natural Ge carrier. This corresponds to 24 counts assigned to ^{71}Ge decay, compared to the SSM prediction of 55 counts.

Figure 2 shows the histogram of the best fit values and 1-σ uncertainties for each of the runs, together with the best fit values for the 1990, 1991, and combined 1990 and 1991 data. The figure also shows the systematic uncertainty of 30 SNU assigned to a possible time variation of the background for the 1990 data. From this figure, one sees that there is slightly more than a 2-σ statistical difference between the 1990 and 1991 data sets. However, including the systematic uncertainty due to a possible time variation in the background in the 1990 data shows that the 1990 and 1991 data sets are in reasonable agreement.

Table 3.
Extraction efficiency of Ge carrier and ^{71}Ge.

Run	Carrier(μg)	^{71}Ge atoms	Carrier Efficiency	^{71}Ge Efficiency
Amount Added	525 ± 26	6555 ± 359		
1	410 ± 10	5188 ± 195	78 ± 4%	79 ± 5%
2	97 ± 2	1131 ± 107	84 ± 20%	84 ± 26%
3	21 ± 1	<200		
Sum	528 ± 10	6519 + 222/-422	101 ± 5%	99 + 6/-8%

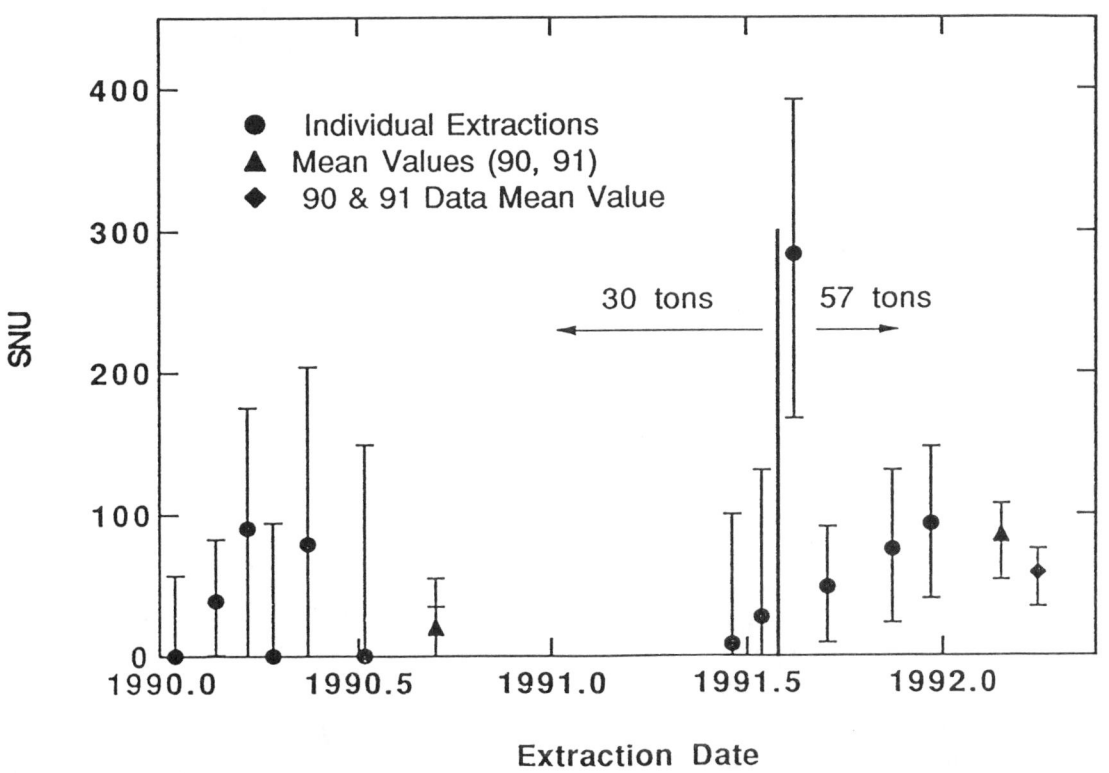

Figure 2. Results from the 1990 and 1991 runs. The first error bar shown for the 1990 mean value is the statistical uncertainty and the second error bar shown is the systematic uncertainty assigned to a possible variation in time of the background.

EXTRACTION EFFICIENCIES

While all available information leads one to expect that the extraction efficiency for ^{71}Ge atoms produced by solar neutrinos should be the same as for the carrier, it is important to test this assumption. A test to search for possible losses in the extraction of ^{71}Ge atoms compared with the natural Ge isotopes was carried out in which the Ge carrier was doped with a known number of ^{71}Ge atoms. The doped carrier was added to one of the reactors holding 7 tons of gallium, three successive extractions were carried out, and the number of ^{71}Ge atoms in each extraction was determined by counting. Table 3 shows the results of this measurement, and indicates that the extraction efficiency of the natural Ge carrier and ^{71}Ge track very closely. The third extraction had a sensitivity of only 200 atoms detected due to electronic problems with one channel of the counting system. The half-life of ^{71}Ge in this extraction test was measured to be 11.0 ± 2.4 days, in good agreement with the known half-life of 11.4 days.

The measurement with the ^{71}Ge doped carrier tests the overall extraction procedures of the experiment. However, it does not test for possible losses which might occur during the formation process. In inverse beta decay, the resultant ^{71}Ge atom may be in an excited state or in some fraction, the ^{71}Ge atom is ionized. It is possible, albeit very unlikely, that these excitations may drive some chemical reaction which may result in the ^{71}Ge atom being tied up in a chemical form which we cannot efficiently extract. That atoms which have been produced in excited states can be extracted from metallic gallium has been demonstrated at some level during the cleanup of the gallium, as we efficiently extracted in excess of 99.9% of the cosmogenic ^{68}Ge.

We are currently carrying out a set of measurements designed to directly test this question. Rather than looking at inverse beta decay, we can look at beta decay in the gallium. In the sudden impulse approximation, atomic excitations of an atom during beta decay should be the same as atomic excitations formed in inverse beta decay. In this experiment, we have taken a few grams of gallium from the reactors and then removed all of the residual Ge carrier. The gallium was then irradiated to form ^{70}Ga and ^{72}Ga from ^{69}Ga and ^{71}Ga respectively by (n,γ) reactions. We expect to produce ^{70}Ga/^{72}Ga in the ratio of 1/1.80. The ^{70}Ge and ^{72}Ge subsequently decay to stable ^{70}Ge and ^{72}Ge with half-lives of 21.1 minutes and 14.1 hours respectively. The gallium metal is kept liquid (in order to simulate conditions in the solar neutrino runs) and allowed to sit for a few weeks so that all of the ^{70}Ga and ^{72}Ga has decayed. The stable ^{70}Ge and ^{72}Ge are then extracted from the irradiated gallium using the same procedure as in the full scale solar neutrino runs. Both the absolute amounts of ^{70}Ge and ^{72}Ge and their ratio are determined by mass spectroscopy. Table 4 shows the preliminary results from our first measurements.

Table 4.
Results from Ga(n, γ).

Ge Isotope	Expected	Measured	Efficiency
^{70}Ge	9.4 μg	8.6 μg	98 ± 10 %
^{72}Ge	18.0 μg	15.5 μg	92 ± 10 %

It appears that ^{70}Ge and ^{72}Ge are formed in the amounts expected. The ratio of the two isotopes agrees quite well with that expected. As the absolute extraction efficiency was not very accurately determined in these first experiments, a more precise set of measurements is now underway.

Finally, an experiment using a neutrino source is planned in order to test the overall extraction efficiency in situ. A suitable neutrino calibration source can be made using ^{51}Cr, which decays with a 27.7 day half-life by electron capture, emitting monoenergetic neutrinos of 751 keV (90.2% BR) and 426 keV (9.8% BR). An engineering test run with a lower-intensity ^{51}Cr source was carried out during the fall of 1990. A full-scale calibration run is scheduled for 1993 using a 1-MCi ^{51}Cr source.

CURRENT STATUS AND FUTURE PLANS

With the combined 1990 and 1991 data sets, SAGE is now observing a signal consistent with ^{71}Ge produced by solar neutrinos. The first results from SAGE, and the data from 1991 appear consistent taking into account the systematic uncertainties. The combined data sets show a good overall fit to a value of 58 SNU. However, these results are still based on limited statistics and assume that the extraction efficiency for ^{71}Ge atoms produced by solar neutrinos is the same as that measured using natural Ge carrier. It is clearly necessary to accumulate more data with higher signal to noise and better efficiencies, as well as to test the extraction efficiency.

Intensive work has been carried out to reduce noise pulsing and backgrounds in the L peak. Preliminary data indicates that the noise and backgrounds in the L-peak have been greatly reduced. Thus, we believe we may be able to count the L peak in the near future, which would almost double our counting efficiency.

Preparations are also underway to fully calibrate the system using an artificial ^{51}Cr source. We expect to be able to carry out this experiment in 1993.

Finally, we are continuing to study possible systematic effects from the data, including additional studies of possible background sources and Monte Carlo simulations.

CONCLUSIONS

Different SSMs predict that the total expected capture rate in ^{71}Ga to be in the range[2,3] of 125 to 132, with the dominant contribution (71 SNU) coming from the p-p neutrinos. The minimum expected rate in a Ga experiment, assuming only that the Sun is presently generating nuclear energy at the rate at which it is radiating energy, is 79 SNU[5]. Observation of significantly less than 79 SNU in a gallium experiment is difficult to explain without invoking new neutrino properties.

The first measurements from a gallium solar neutrino experiment have observed fewer ^{71}Ge atoms than predicted by the SSM. From the 1990 and 1991 data, we observe only 44% of the predicted flux. Assuming the extraction efficiency for ^{71}Ge atoms produced by solar neutrinos is the same as for natural Ge carrier, the first measurements indicate that the flux may be less than that expected from p-p neutrinos alone.

ACKNOWLEDGEMENTS

The SAGE collaboration wishes to thank A.E. Chudakov, G.T. Garvey, M.A. Markov, V.A. Matveev, J.M. Moss, S.P. Rosen, V.A. Rubakov, and A.N. Tavkhelidze. We are also grateful to J.N. Bahcall, Yu. Smirnov, and many members of the GALLEX collaboration for useful discussions. We acknowledge the support of the Russian Academy of Sciences,, the Institute for Nuclear Research, the Ministry of Sciences of the Russian Federation, the Division of Nuclear Physics of the US Department of Energy, the National Science

Foundation, Los Alamos National Laboratory, and the Univ. of Pennsylvania.

*Present Address: L-421, Lawrence Livermore National Laboratory, Livermore, CA 94550

** Present address: Dept. of Particle and Nuclear Physics, Oxford University, Keble Road, Oxford, OX1 3RH, England

***Present address: Batelle Pacific Northwest Laboratories, P.O. Box 999, Richland, WA 99352

REFERENCES

1) R. Davis et al., Proc. 25th Int. Conf. on High Energy Physics, edited by K.K. Phua and Y. Yamaguchi, (World Scientific, Singapore, 1991), p. 667

2) J.N. Bahcall and R. Ulrich, Rev. Mod. Phys. **60**, 297 (1988).

3) S. Turck-Chieze, S. Cahen, M. Casse, and C. Doom, Ap. J. **335**, 415 (1988).

4) K.S. Hirata et al., Phys. Rev. Lett. **65**, 1297 (1990).

5) J.N. Bahcall, Neutrino Astrophysics, Cambridge University Press, 343 (1989).

6) V.A. Kuzmin, Sov. Phys. JETP **22**, 1051 (1966).

7) J.N. Bahcall et al., Phys. Rev. Lett. **40**, 1351 (1978).

8) I.R. Barabanov et al., Proc. COnf. on Solar Neutrinos and Neutrino Astronomy, Homestake, 1984, edited by M.L. Cherry, K. Lande, and W.A. Fowler, AIP COnf. Proc. **126**, 175 (1985).

9) V.N. Gavrin et al., Proc. of "Inside the Sun", Conference, Versailles, edited by G. Berthomieu and M. Cribier (Kluwer Academic, Dordrecht, 1989), p. 201.

10) B.T. Cleveland, Nucl. Instrum. Methods **214**, 451 (1983).

11) A.W. Marshall, Ann. Math. Stat. **29**, 307 (1958).

12) A.I. Abazov et al., Phys. Rev. Lett. **67**, 3332 (1991).

WHAT DO SOLAR MODELS TELL US ABOUT SOLAR NEUTRINO EXPERIMENTS?

John N. Bahcall
Institute for Advanced Study
Olden Lane
Princeton, New Jersey

Abstract

If the published event rates of the chlorine and Kamiokande solar neutrino experiments are correct, then the energy spectrum of neutrinos produced by the decay of ^{8}B in the sun must be different from the energy spectrum determined from laboratory nuclear physics measurements. This change in the energy spectrum requires physics beyond the standard electroweak model. In addition, the GALLEX and SAGE experiments, which currently have large statistical uncertainties, differ from the predictions of the standard solar model by 2σ and 3σ, respectively.

At the conference, I presented a review of recent improvements in the calculations of neutrino fluxes from solar models and then used the most recent results to draw some conclusions about what we have learned by comparing the results of solar neutrino experiments with calculations from solar models of the neutrino fluxes. The analysis of solar models has now been published in detail[1], so there is no need to repeat that material here. My main goal at the conference was, in any event, not to elucidate technical issues in solar model theory, but rather to clarify and make quantitative the conclusions that folllow from the confrontation of the solar model calculations with the four operating experiments. I will therefore take the model results as given and concentrate here on what they teach us about the four solar neutrino experiments.

The first point to recognize is that the individual rates of the four solar neutrino experiments tell us nothing about the possibility of new physics until these rates are compared with solar models. The analogy to an accelerator experiment is clear: we need to know what the beam intensity is, as well as the flavor composition and energy spectrum, in order to know if we are surprised or not by the experimental rates.

The standard solar model[2] predicts the absolute fluxes from each of the important nuclear fusion reactions and furthermore says that all solar neutrinos are ν_e's. What is more, to an accuracy of one part in 10^5, the energy spectrum of the ^{8}B solar neutrinos must have the same shape as the spectrum determined from laboratory nuclear physics experiments[3].

The invariance of the energy spectrum allows us to compute the rate of neutrino capture in the chlorine experiment–independent of any considerations of solar models– provided only that we know from the Kamiokande experiment the flux of the higher energy ($>$ 7.5 MeV) ^{8}B neutrinos. In this process, we ignore the expected contributions to the chlorine experiment, which has a threshold of only 0.8 MeV (an order of magnitude less than the Kamiokande experiment), from ^{7}Be, CNO, and pep neutrinos. Using the empirical result obtained for the Kamiokande experiment[4], one finds that the predicted rate *from ^{8}B neutrinos alone* in the chlorine experiment is 6.20 SNU (from the standard model)$\times 0.48$ (from the Kamiokande measurement), or

$$<\phi\sigma>_{Cl;\ Kamiokande\ only} =$$
$$[3.0 \pm 0.3(1\sigma) \pm 0.4(syst)\ ]\ SNU. \quad (1)$$

This minimum rate, which ignores the contributions of all other neutrino sources to the chlorine experiment, exceeds by 2σ the observed chlorine rate,[5]

$$<\phi\sigma>_{\text{Cl exp}} = (2.2\pm0.2) \text{ SNU}, \quad 1\sigma \text{ error.} \quad (2)$$

Moreover, the lower-energy contributions from ^{7}Be and pep neutrinos –which together amount to about 1.4 SNU– are much more reliably determined by the theoretical calculations than is the contribution from ^{8}B neurinos. If a fraction equal to 0.48 of the less-reliably calculated high energy ^{8}B neutrinos are detected, then presumably more than 0.7 SNU of the expected 1.4 SNU from pep and ^{7}Be neutrinos should be added to the minimum rate of 3.0 SNU calculated above. On the basis of this comparison, Hans Bethe and I concluded[6] that, if the chlorine and Kamiokande experiments are both correct, then physics beyond the standard electroweak model is required to change the ^{8}B neutrino energy spectrum.

More recently, Hans and I have sharpened this argument[7] using a detailed Monte Carlo simulation of how the sun works. The basis for our investigation is a collection of 1000 precise solar models[2] in which each input parameter (the principal nuclear reaction rates, the solar composition, the solar age, and the radiative opacity) for each model was drawn randomly from a normal distribution with the mean and standard deviation appropriate to that variable. The uncertainties in the neutrino cross sections[2] for chlorine and for gallium were included by assuming a normal distribution for each of the absorption cross sections with its estimated mean and error.

We know that Monte Carlo simulations are necessary to understand the results of complicated experiments in nuclear and particle physics. It should therefore seem natural to physicists that Monte Carlo simulations are necessary to interpret the results of solar neutrino experiments; the sun may be as complicated as a terrestrial particle accelerator or detector.

The Monte Carlo study automatically takes account of the nonlinear relations among the different neutrino fluxes that are imposed by the coupled partial differential equations of stellar structure and by matching the stringent boundary conditions of reproducing the observed solar luminosity, the heavy element to hydrogen ratio, and the effective temperature at the present solar age. Attempts to simulate the uncertainties using average scaling laws of the dependence of fluxes upon a single parameter, the central temperature, can lead to serious errors. A full Monte Carlo calculation is required to determine the interrelations and absolute values of the different solar neutrino fluxes. For example, the fact that the ^{8}B flux may be crudely described as $\phi(^8\text{B}) \propto T^{18}_{\text{central}}$ and $\phi(^7\text{Be}) \propto T^{8}_{\text{central}}$ does not specify whether the two fluxes increase and decrease together or whether their changes are out of phase with each other.

Figure 1 shows the number of solar models with different predicted event rates for the chlorine solar neutrino experiment. The solar model with the best input parameters predicts[2] an event rate of about 8 SNU. None of the 1000 calculated solar models yields a capture rate below 5.8 SNU. Therefore, none of the 1000 solar models is within 16σ of the observed rate. The discrepancy that is apparent in Figure 1 was for two decades the entire "solar neutrino problem." We can conclude from Figure 1 that something is wrong with either the standard solar model or the standard electroweak description of the neutrino.

The largest and the most uncertain contribution to the predicted chlorine rate is the ^{8}B neutrino flux. This quantity is completely unimportant for all astronomical purposes since the reaction by which it is produced is extremely rare. Suppose therefore some mistake has been made in calculating the ^{8}B neurino flux and we normalize this flux, as before, by using the empirical determination in the Kamiokande experiment. What do we obtain for the 1000 solar models when we replace–for each model–the calculated flux by a value determined by the Kamiokande experiment?

Figure 2 provides a quantitative expression

of the difficulty in reconciling the Kamiokande and chlorine experiments by changing solar physics, i. e., by arbitrarily changing the ^{8}B neutrino flux. We constructed Figure 2 using the same 1000 solar models as were used in constructing Figure 1, but for Figure 2 we artificially replaced the ^{8}B flux for each standard model by a value drawn randomly for that model from a normal distribution with the mean and the standard deviation measured by Kamiokande. The peak of the resulting distribution is moved to 4.7 SNU (from 8 SNU) and the full width of the peak is decreased by about a factor of three. The peak is displaced because the measured (i.e., Kamiokande) value of the ^{8}B flux is smaller than the calculated value. The width of the distribution is decreased because the error in the Kamiokande measurement is less than the estimated theoretical uncertainty ($\approx 12.5\%$) and because ^{8}B neutrinos constitute a smaller fraction of each displaced rate than of the corresponding standard rate.

Figure 2 was constructed by assuming that something is seriously wrong with the standard solar model, something that is sufficient to cause the ^{8}B flux to be reduced to the value measured in the Kamiokande experiment. Nevertheless, there is no overlap between the distribution of fudged standard model rates and the measured chlorine rate. None of the 1000 fudged models lie within 3σ (chlorine measurement errors) of the experimental result.

The results presented in Figures 1-2 suggest that new physics is required beyond the standard electroweak theory if the existing solar neutrino experiments are correct within their quoted uncertainties. Even if one abuses the solar models by artificially imposing consistency with the Kamiokande experiment, the resulting predictions of all 1000 of the "fudged" solar models are inconsistent with the result of the chlorine experiment(see Figure 2).

Figures 3a-3b show the number of solar models with different predicted event rates for gallium detectors and the recent measurements by the SAGE[8] ($58^{+17}_{-24} \pm 14(\text{syst})$ SNU) and GALLEX[9] ($83 \pm 19(1\sigma) \pm 8(\text{syst})$ SNU) collaborations. Figure 3a compares the gallium experimental results with the "unfudged" histogram of standard solar model calculations and Figure 3b compares the results when the ^{8}B neutrino flux is taken from the Kamiokande mesurement. Unlike the chlorine case (cf. Figures 1 and 2), in which almost 80% of the predicted event rate is from ^{8}B neutrinos, Figures 3a and 3b are not qualitatively different because ^{8}B neutrinos contribute very little (only about 10 %) to the predicted event rate in the gallium experiments.

With the current large statistical errors, the gallium measurements differ from the best-estimate theoretical value of 132 SNU by approximately 2 σ (GALLEX) and 3.5 σ (SAGE). The gallium results provide modest support for the existence of a solar neutrino problem, but by themselves do not constitute a definitive conflict with standard theory.

ACKNOWLEDGMENTS

This work was supported in part by the NSF via grant PHY-91-06210 at I.A.S.

REFERENCES

1. J. N. Bahcall, and M. H. Pinsonneault, *Rev. Mod. Phys.*, **64**, 885, 1992.
2. J. N. Bahcall and R. K. Ulrich, *Rev. Mod. Phys.*, **60**, 297, 1988; J. N. Bahcall, *Neutrino Astrophysics* (Cambridge University Press, Cambridge, England, 1989).
3. J.N. Bahcall, *Phys. Rev. D.*, **44**, 1644, 1991.
4. K. S. Hirata et al., *Phys. Rev. Lett.*, **63**, 16, 1989; **65**, 1297, 1990.
5. R. Davis Jr., in *Proc. of Seventh Workshop on Grand Unification, ICOBAHN'86*, edited by J. Arafune (World Scientific, Singapore, 1987), p. 237; R. Davis Jr., K. Lande, C. K. Lee, P. Wildenhain, A. Weinberger, T. Daily, B. Cleveland, and J. Ullman, in *Proceedings of the 21st International Cosmic Ray Conference*, Adelaide, Australia, 1990, in press; J. K. Rowley, B. T. Cleveland, and R. Davis Jr., in *Solar Neutrinos and Neutrino Astronomy*,

edited by M. L. Cherry, W. A. Fowler, and K. Lande (American Institute of Physics, New York, 1985), Conf. Proceeding No. 126, p. 1.

6. J. N. Bahcall and H. A. Bethe, *Phys. Rev. Lett.*, **65**, 2233, 1990.

7. J. N. Bahcall and H. A. Bethe, *Phys. Rev. D.*, in press, 1992.

8. V. N. Gavrin, et al., *XXVI International Conference on High Energy Physics*, Dallas, Texas, 1992; A. I. Abazov et al., *Phys. Rev. Lett.*, **67**, 332, 1991.

9. P. Anselmann et al., *Phys. Lett. B*, **285**, 376, 1992.

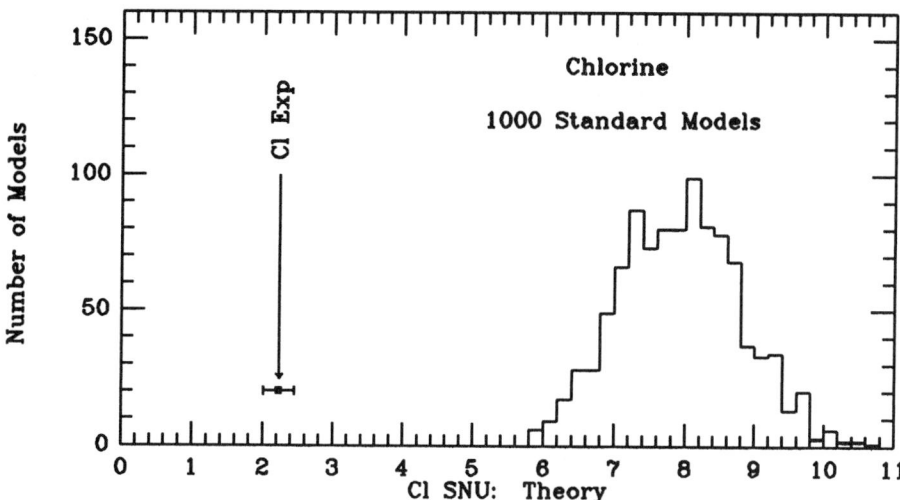

Figure 1. 1000 Solar Models versus Experiments. The number of precisely-calculated solar models that predict different solar neutrino event rates are shown for the chlorine experiment. The solar models from which the fluxes were derived satisfy the equations of stellar evolution including the boundary conditions that the model luminosity, chemical composition, and effective temperature at the current solar age be equal to the observed values[2]. Each input parameter in each solar model was drawn independently from a normal distribution having the mean and the standard deviation appropriate to that parameter. The experimental error bars include only statistical errors (1σ).

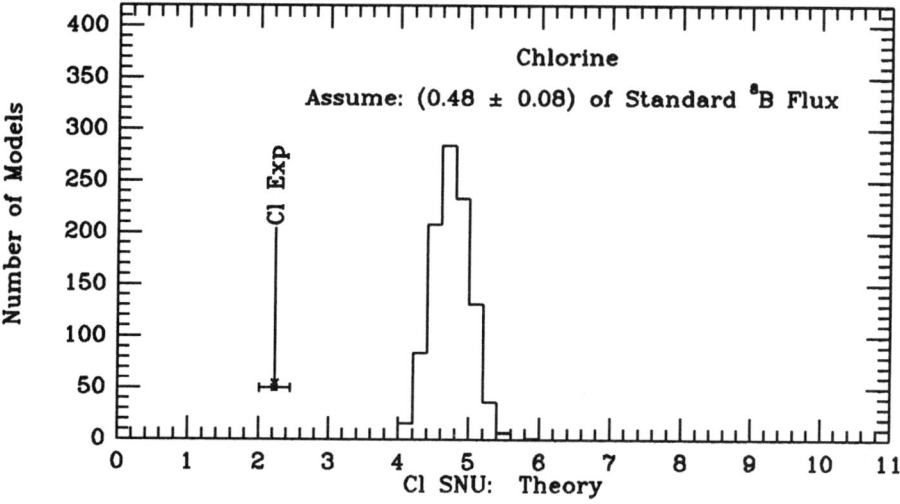

Figure 2. 1000 Artifically Modified Fluxes. The ^{8}B neutrino fluxes computed for the 1000 accurate solar models were replaced in the figure shown by values drawn randomly for each model from a normal distribution with the mean and the standard deviation measured by the Kamiokande experiment[4].

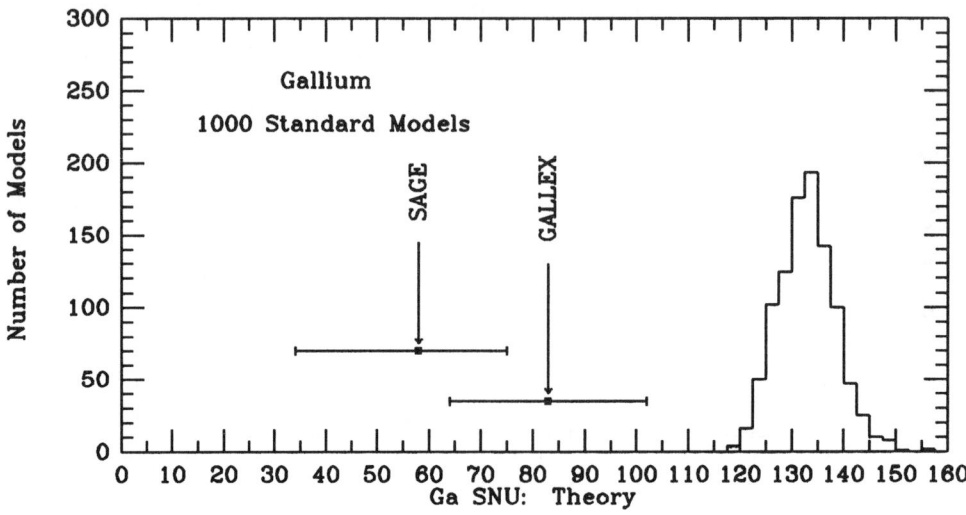

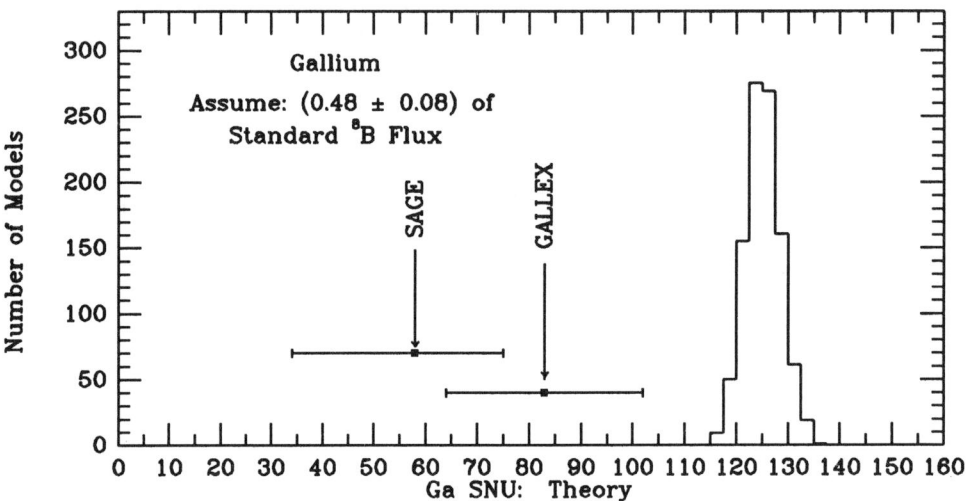

Figure 3. Gallium experiments versus 1000 solar models. Figure 3a compares the gallium experimental results with the "unfudged" solar model calculations and Figure 3b compares the experimental results with the solar model predictions assuming the ^{8}B flux is equal to the measured Kamiokande value.

MSW IMPLICATIONS OF SOLAR NEUTRINO EXPERIMENTS*

S. P. Rosen
College of Science
The University of Texas at Arlington
Arlington, Texas 76019-0047
U. S. A.

Abstract

I discuss the implications for future solar neutrino experiments of the most recent gallium data. To choose between different MSW solutions, we must either measure the ^{7}Be and pp neutrinos separately or determine the spectral shape of ^{8}B neutrinos.

INTRODUCTION

John Bahcall[1] concluded his talk by telling us that the Standard Solar Model (SSM) is in good shape and that the observed deficit of solar neutrinos must be due either to serious flaws in the existing experiments or to new physics. I am going to adopt the second alternative, namely that the experiments are broadly correct and that the deficit is caused by the elegant Mikheyev-Smirnov-Wolfenstein (MSW) matter enhancement of neutrino oscillations. The new SAGE[2] and GALLEX[3] results have a clear interpretation in terms of the MSW effect and definite implications for the next generation of experiments.

Today I shall briefly review the MSW[4] effect, then discuss the interpretation of the latest gallium data and develop the implications for new experiments[5-8] now under construction. It will be useful to distinguish between the 'low energy' end of the solar neutrino spectrum and the 'high energy' end: the low energy end consists of pp neutrinos with energies up to 430 keV and

---

*This work is supported in part by the U. S. Department of Energy grant DE-FG05-92ER40691.

^{7}Be neutrinos with an energy of 860 keV. The high energy part of the spectrum consists of ^{8}B neutrinos with energies all the way up to 14 MeV. Future experiments include BOREXINO, which is designed to look at the ^{7}Be neutrinos, and SNO, Super Kamiokande, and ICARUS which will detect the high energy neutrinos.

BRIEF REVIEW OF MSW

In the standard electroweak model, neutrinos of all flavors can scatter from electrons via Z^0 exchange; the electron-neutrino ν_e can, in addition, scatter through charged $W^\pm$ exchange. This unique diagram gives ν_e a different refractive index, or effective mass, in matter and it also yields a much larger cross-section for neutrino–electron scattering than does the neutral current diagram.

The time development equation for MSW matter oscillations between the electron flavor and another active flavor x can be written as:

$$i\frac{d}{dt}\begin{pmatrix} a_e \\ a_x \end{pmatrix} = \begin{pmatrix} X & Y \\ Y & Z \end{pmatrix} \begin{pmatrix} a_e \\ a_x \end{pmatrix} \quad (1)$$

where a_e, a_x are the probability amplitudes for ν_e to survive, and to oscillate into ν_x respectively. The matter Hamiltonian is given by:

$$X = \frac{m_1^2 c^2 + m_2^2 s^2}{2E} + \sqrt{2} G_F N_e$$
$$Z = \frac{m_1^2 s^2 + m_2^2 c^2}{2E}$$
$$Y = \frac{m_2^2 - m_1^2}{2E} cs = \frac{\Delta m^2}{2E} cs \quad (2)$$

where $c = \cos\theta$, and $s = \sin\theta$. With the right electron density N_e in the matrix element X we can tune the Hamiltonian to maximal mixing, i.e. $X = Z$, and thereby enhance the conversion of ν_e to ν_x.

There are two types of solution to eq. (1): the adiabatic solution in which the instantaneous eigenvectors change slowly as the neutrino travels through the sun, and the nonadiabatic solution, which involves a transition between eigenvectors. The criterion for validity of the adiabatic solution is:

$$\frac{\sin^2 2\theta}{\cos 2\theta} \frac{\Delta m^2}{E} \gg \frac{1}{N}\frac{dN}{dR} \quad (3)$$

At the enhancement point, the scale height $1/N\, dN/dR$ of the solar density comes from the standard solar model,[9] and we find that

$$E \ll 3.5 \times 10^8 \frac{\sin^2 2\theta}{\cos 2\theta} \Delta m^2 \text{ MeV}, \quad (4)$$

where Δm^2 is measured in eV^2. The electron-neutrino survival probability at Earth[4] is

$$P_{ad}(\nu_e \to \nu_e) = \frac{1}{2}\left(1 - \frac{T\cos 2\theta}{\sqrt{T^2 + \sin^2 2\theta}}\right)$$
$$T = 1.53 \times 10^{-5} \frac{E}{\Delta m^2}\frac{N}{N_c} - \cos 2\theta. \quad (5)$$

where N_c is the density at the core of the sun. Key points on the curve of P_{ad} as function of energy are shown in Table 1.

For small mixing angles θ_s, the enhancement energy E_0 and the rapid fall in survival probability between the origin and $2E_0$ are insensitive to θ_s; E_0 is, however, directly proportional to Δm^2.

Table 1. $P_{ad}(\nu_e \to \nu_e)$ as a function of neutrino energy

E_ν	T	$P(\nu_e \to \nu_e)$
0	$-\cos 2\theta$	$(1 - \frac{1}{2}\sin^2 2\theta)$
E_0	0	$\frac{1}{2}$
$2E_0$	$+\cos 2\theta$	$\frac{1}{2}\sin^2 2\theta$
∞	∞	$\sin^2\theta$

The adiabatic probability (eq. (5)) depends on the density at the neutrino birthplace. Since the pp and ^{7}Be neutrinos are produced over broader regions of the solar interior than ^{8}B neutrinos[9], this will contribute to differences in behavior of low and high energy neutrinos. To account for this effect, we integrate P_{ad} over the relevant production regions:

$$\langle P_{ad} \rangle = \int \phi(r) P_{ad}(\nu_e \to \nu_e) dr \quad (6)$$

where $\phi(r)$ is the fraction of neutrinos of a given type produced at radius r inside the sun, with $\int \phi(r) dr = 1$. Its values are tabulated by Bahcall and Ulrich[9].

The nonadiabatic solution is valid when the inequality of eq. (3) is reversed and it yields a very simple form for the electron-neutrino survival probability at Earth[4]:

$$P_{nonad}(\nu_e \to \nu_e) = \exp(-C/E),$$
$$C = \frac{\pi \Delta m^2 \sin^2 2\theta}{4\cos 2\theta}\left(\frac{1}{N}\frac{dN}{dR}\right)^{-1}$$
$$= 2.75 \times 10^8 \Delta m^2 \frac{\sin^2 2\theta}{\cos 2\theta} \text{ MeV}. \quad (7)$$

This expression is essentially independent of the neutrino birthplace. From the fit to the ^{37}Cl data, we find that[10]

$$\Delta m^2 \frac{\sin^2 2\theta}{\cos 2\theta} \approx 3 \times 10^{-8} \text{ eV}^2, \quad (8)$$

and hence that C is approximately 9 MeV.

For small mixing angles, these solutions meet at a point approximately[11] where the

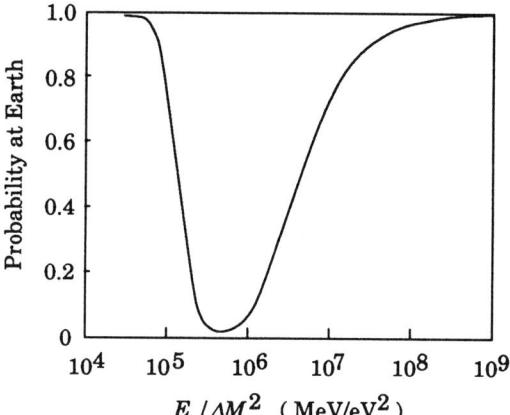

Figure 1. $P(\nu_e \to \nu_e)$ as a function of $E/\Delta m^2$ for $\sin^2 2\theta = 0.01$. The left-hand side of the curve, decreasing as energy increases, is the adiabatic solution and the right-hand side, increasing with energy, is the nonadiabatic solution. The production zone for pp neutrinos is used in the adiabatic component; had the ^{8}B zone been used, the slope would have been steeper.

nonadiabatic probability of eq. (7) equals the asymptotic value $\sin^2 2\theta$ of the adiabatic survival probability (see Fig. 1). The neutrino energy versus mass^2 difference at this point is

$$\left(\frac{E_j}{\Delta m^2}\right) = \left(\frac{-2.75 \times 10^8 \sin^2 2\theta}{\cos 2\theta \ln(\sin^2 \theta)}\right). \quad (9)$$

Imposing the ^{37}Cl nonadiabatic constraint of eq. (8), we obtain

$$E_j = \frac{-8.25}{\ln(\sin^2 \theta)} = 1.3 \text{ MeV} \quad (10)$$

for $\sin^2 2\theta = 0.007$. This formula means that not all solar ν_e will have a nonadiabatic survival probability: those with energies greater than some value near E_j will obey P_{nonad}, but those with energies *less than* E_j will survive according to the *adiabatic* probability. The corresponding behavior of the ν_e survival probability is shown in Figs. 1 and 2.

THE GALLIUM RESULTS

We heard from Vignaud[3] that the GALLEX signal is approximately 80±20 SNU,

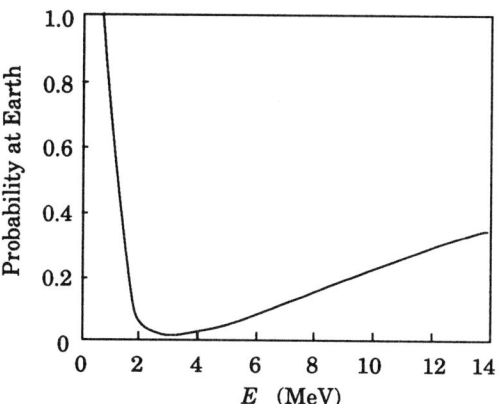

Figure 2. $P(\nu_e \to \nu_e)$ as a function of E for $\sin^2 2\theta = 0.01$ and $\Delta m^2 = 1.7 \times 10^{-5}$.

and from Gavrin[2] that the latest SAGE result is about 60±20 SNU. Both measurements are significantly larger than the original SAGE result which favored the nonadiabatic MSW solution over the adiabatic and large angle ones. As pointed out by GALLEX[3], the present results are still consistent with a nonadiabatic solution, but now admit a large angle solution as an alternative. Thus we need to distinguish between them.

Nonadiabatic, Small Angle Solution

The nonadiabatic, small angle solution given by GALLEX[3] consists of a small region around the point

$$\sin^2 2\theta = 0.007, \quad \Delta m^2 = 6 \times 10^{-6}. \quad (11)$$

Since E_j (eq. (9)) is in the vicinity of 1 MeV, ^{8}B neutrinos survive nonadiabatically and low energy ones adiabatically. For a density of 70 gm/cc, which is typical of the solar production zones for pp and ^{7}Be neutrinos[9], the key points of P_{ad} in Table 1 are given numerically by the second and third columns of Table 2.

Since pp neutrinos are well below 564 keV in energy and ^{7}Be ones well above, the former have a survival probability much greater than 1/2 and the latter much less. The same

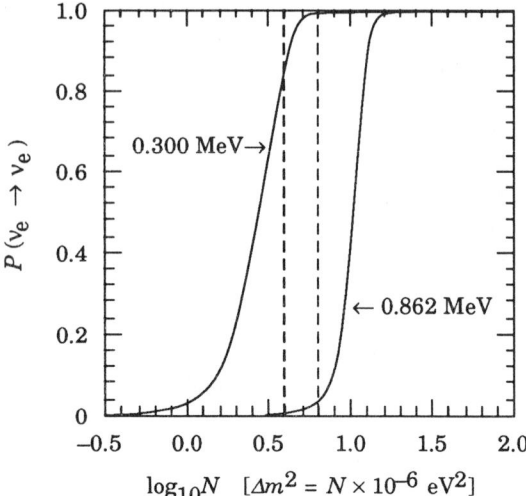

Figure 3. The small angle solution as a function of Δm^2 with $\sin^2 2\theta = 0.007$. The curve on the right is for the ^{7}Be line and that on the left for a typical pp neutrino. The dashed lines represent reduced errors on the GALLEX data.

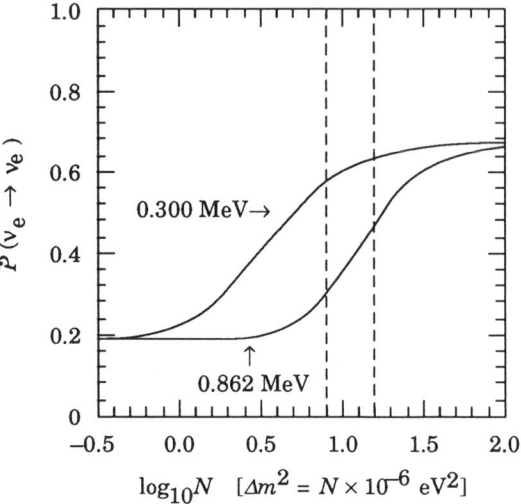

Figure 4. The large angle solution as a function of Δm^2 with $\sin^2 2\theta = 0.007$. The curve on the right is for the ^{7}Be line and that on the left for a typical pp neutrino. The dashed lines represent reduced errors on the GALLEX data.

holds true for most densities encountered in the low energy neutrino production zone and so the gallium signal will be predominantly composed of pp neutrinos. Calculations of $\langle P_{ad} \rangle$ (see eq. (6)) for the 860 keV ^{7}Be neutrino and for a 'typical' pp of 300 keV are plotted as functions of Δm^2 in Fig. 3. For $\Delta m^2 = 6 \times 10^{-6}$, ^{7}Be neutrinos have about a 5% survival probability and pp neutrinos have close to 100%.

Large Angle Solution

The large angle GALLEX[3] solution is centered upon the point

$$\sin^2 2\theta = 0.6, \quad \Delta m^2 = 8 \times 10^{-6}. \quad (12)$$

Again the adiabatic formula is relevant for the low energy neutrinos and, with a density of 70 gm/cc, the key points are shown in the fourth and fifth columns of Table 2.

The pp and ^{7}Be neutrinos are still on opposite sides of the enhancement energy, but the difference in their survival probabilities is not as marked as in the small angle case. The integrated probabilities $\langle P_{ad} \rangle$ (see eq. (6)) for the 860 keV ^{7}Be neutrino and for a 'typical' pp of 300 keV maintain this feature, as can be seen in the plots of Fig. 4. For $\Delta m^2 = 8 \times 10^{-6}$, the ^{7}Be neutrinos have a 30% survival probability and pp neutrinos have a 60% one.

IMPLICATIONS FOR EXPERIMENTS AT THE LOW ENERGY END

New experiments at the low energy end of

Table 2. $P_{ad}(\nu_e \to \nu_e)$ as a function of neutrino energy (in keV).

Point	small angle		large angle	
	E_ν	$P(\nu_e)$	E_ν	$P(\nu_e)$
origin	0	0.997	0	0.7
E_0	564	$\frac{1}{2}$	475	$\frac{1}{2}$
$2E_0$	1,128	0.003	950	0.3
∞	∞	0.002	∞	0.2

the spectrum must attempt to measure the *pp* and ⁷Be neutrinos either by themselves alone, or in some combination distinct from existing experiments. One proposal, the BOREXINO[5] experiment, plans to measure the ⁷Be neutrinos by neutrino–electron scattering and in the SSM it expects to see 47 recoil electrons per day in the energy range of 250 to 663 keV. Contributions from other branches of the solar spectrum are either outside this kinematic range, or are negligible. Thus this experiment would give us a clear shot at the ⁷Be neutrinos.

I assume that ν_e oscillates into another active flavor ν_x. The ν_e itself will scatter from an electron via a coherent combination of charged- and neutral-currents and the ν_x via neutral-currents alone; thus the former will have a cross-section roughly 6 times larger than the latter. If the survival probability for ⁷Be ν_e is P_7, then the number of events predicted per day will be

$$R_7 = (0.79 P_7 + 0.21) \times 47. \quad (13)$$

By the time BOREXINO comes on the air in a few years, the errors on the gallium experiments should be considerably smaller than they are today. I therefore reduce the current errors on $\sin^2 2\theta$ and Δm^2 by a factor of two and assume that the central values will not change significantly.

In the small angle solution, survival probabilities are not sensitive to the mixing angle and our reduced errors indicate that Δm^2 should lie between 4 and 6×10^{-6}. From Fig. 3, it can be seen that P_7 lies between 0 and 5% and that the expected rate in BOREXINO would be:

$$10 \leq R_7 \text{ (small)} \leq 12 \text{ events/day.} \quad (14)$$

In the large angle solution, Δm^2 lies between 8 and 15×10^{-6} and P_7 between 0.3 and 0.5 (see Fig. 4). The expected rate in this case turns out to be much larger,

$$21 \leq R_7 \text{ (large)} \leq 28 \text{ events/day.} \quad (15)$$

Obviously, BOREXINO should be able to make a clear distinction between the two MSW solutions[12].

For *pp* neutrinos, the small angle solution predicts a survival probability of 85% or more, while the large angle one predicts about 65%. Accurate measurements are needed to distinguish the two cases directly.

IMPLICATIONS FOR EXPERIMENTS AT THE HIGH ENERGY END

At the high energy end of the spectrum, the small angle solution yields an exponential survival probability as in eq. (7) with the parameter C between 9 and 11 MeV. The large angle solution yields a constant probability of $\sin^2\theta$ and equal to 0.2 for $\sin^2 2\theta = 0.6$. Our question is whether the next generation of experiments will accumulate sufficient statistics with sufficient accuracy to distinguish between these two survival probabilities.

To make this distinction, it is necessary to measure the spectrum of ν_e arriving at Earth.[10] When the survival probability is flat, the spectral *shape* will be exactly as predicted by the SSM, but the overall normalisation will be reduced; when the probability is a function of neutrino energy, the spectral shape will be distorted. For the exponential formula of eq. (7), the largest distortions come at the lower energies and so it is important for the next generation of experiments to push thresholds to as low an energy as possible.

Some idea of the difficulties involved in measuring spectral shapes can be obtained from a recent analysis of the Kamiokande II data[13]. Differences in the spectral shapes of the neutrinos themselves become muted in the convolutions that lead to the observed recoil electron spectrum, especially when the threshold on electron energy is as high as 7.5 MeV. Were it 5, or better still 3 Mev, then it might be possible to use the observed electron spectrum directly, but at higher thresholds we need clever tricks.

One such trick, originally used by Kamiokande II, is to compare the observed recoil electron spectrum with that predicted by the SSM on a point-by-point basis. This emphasises differences between the effects of different neutrino survival probabilities. In particular, the ratio is flat as a function of energy for the large angle MSW solution, and has a positive slope for the small angle one. Unfortunately, the errors on the present Kamiokande II data are too large to yield any definite conclusion. If the errors were largely statistical, then several thousand events would be needed to reduce them to a useful level. Super Kamiokande[7], collecting events at the rate of 10 per day, could collect a useful sample in one year and, depending upon its size and threshold, so could ICARUS[8].

SNO[6] would take several years for neutrino–electron scattering, but it will have a much more powerful tool in neutrino-induced disintegration of the deuteron. Through the neutral-current process it can establish whether or not oscillations into active flavors are taking place, and through measurements of the electron spectrum in charged-current disintegration, it can measure the spectrum of electron-neutrinos arriving at Earth. Here too, the distinctions between different cases are muted by convolutions of spectra and reaction cross-sections, and it may be necessary to resort to tricks to extract useful information.

CONCLUSION

Matter oscillations and the MSW Effect still provide a viable and attractive solution of the solar neutrino problem, but we must wait until the next generation of experiments before we can know for sure that this is so. As the gallium experiments acquire more data and reduce errors, our confidence in the outcome will increase or decrease according as the central value of the signal decreases or increases. The next five years should be very interesting.

The author is endebted to Dr. Waikwok Kwong for his help in TeXing this article.

REFERENCES

1. J. N. Bahcall, "Solar Neutrino Predictions," in these proceedings, Session PA-11A.
2. A. V. Gavrin, "Summary of the Soviet-American Gallium Experiment," in these proceedings, Session PA-11A.
3. J. Vignaud, "Report on GALLEX," in these proceedings, Session PA-11A.
4. J. N. Bahcall, *Neutrino Astrophysics*. Cambridge: Cambridge University Press, 1989; T. K. Kuo and J. Panteleone, *Rev. Mod. Phys.* **61**, 937 (1989); S. P. Mikheyev and A. Y. Smirnov, *Prog. Part. Nuc. Phys.* **23**, 41 (1989).
5. R. S. Raghavan et al., *Phys. Rev.* **D44**, 3786 (1991).
6. R. G. H. Robertson, "The Sudbury Neutrino Observatory", in these proceedings, Session PA-12A.
7. T. Kajita, "Solar and Atmospheric Neutrinos" in these proceedings, Session PA-12A.
8. D. Cline (private communication).
9. J. N. Bahcall and R. Ulrich, *Rev. Mod. Phys.* **60**, 297 (1988).
10. S. P. Rosen and J. M. Gelb, *Phys. Rev.* **D34**, 969 (1986).
11. S. T. Petcov, *Nucl. Phys. B (Proc. Suppl.)* **13**, 527 (1990).
12. J. M. Gelb, W. Kwong, and S. P. Rosen, *Phys. Rev. Lett.* (to be published), September 28 (1992).
13. W. Kwong and S. P. Rosen, *Phys. Rev. Lett.* **68**, 748 (1992).

A MASSIVE NEUTRINO IN NUCLEAR BETA DECAY ?

Eric B. Norman[1], Yuen-Dat Chan[1], M. T. F. da Cruz[1,2], Alejandro Garcia[1], M. M. Hindi[1,3], K. T. Lesko[1], Ruth-Mary Larimer[1] Robert G. Stokstad[1], Bhaskar Sur[1]*, Fred E. Wietfeldt[1,4], Igor Zlimen[1]

(1) Nuclear Science Division, Lawrence Berkeley Laboratory, Berkeley, CA, U.S.A.
(2) Physics Institute, University of Sao Paulo, Sao Paulo, Brazil
(3) Physics Department, Tennessee Technological University, Cookeville, TN, U.S.A.
(4) Physics Department, University of California, Berkeley, CA, U.S.A.

Abstract

We have continued our studies of the β-spectrum of ^{14}C using a germanium detector doped with ^{14}C. There is a feature in the β-spectrum 17 keV below the endpoint which could be explained by the hypothesis that there is a heavy neutrino emitted in the β-decay of ^{14}C with a mass of 17±1 keV and an emission probability of 1.26±0.25%. However, we also have performed a high statistics measurement of the inner bremsstrahlung spectrum of ^{55}Fe and find no indication of the emission of a 17-keV neutrino. We conclude that the origin of the "kink" that has been observed in some recent beta spectral measurements is not a neutrino.

BETA SPECTRUM OF ^{14}C

Over the past two and a half years, we have studied the beta spectrum of ^{14}C to search for evidence of heavy neutrino emission.[1,2] In this experiment, we have used a detector containing a 1.28-cm thick planar germanium crystal doped with ^{14}C.[3] The collection electrode is divided by a 1-mm wide circular groove into a "center region" 3.2 cm in diameter and an outer "guard ring". By operating the guard ring in anti-coincidence with the center region, we reject events occurring near the boundary which are not fully contained within the center region. The ^{14}C β decay counting rate from the center region of the crystal is 20 s^{-1}. The experiment was conducted at Lawrence Berkeley Laboratory's Low Background Counting Facility. Signals from the center region and the guard ring portions of the ^{14}C crystal were processed through separate amplifiers. Data were acquired and recorded using a PC-based acquisition system. Three separate spectra were accumulated: (1) center region, (2) center region in anti-coincidence with the guard ring, and (3) guard ring. The guard ring veto signal used to generate spectrum (2) required that an event deposit more than 20 keV but less than 183 keV in the guard ring portion of the crystal. Data were collected in 4096 channels of 0.144 keV width. The ^{14}C crystal was counted for a total of 334 days. The background was measured by removing the ^{14}C crystal from the cryostat and replacing it with a similarly shaped ^{14}C-free planar guard-ring germanium crystal. 111 days of background data were accumulated with this crystal.

The experimental data were compared to the theoretically expected spectrum using a least-squares fitting procedure that we have described previously.[1,2] This analysis was performed on data in the energy range 100-160 keV. To illustrate the degree to which the calculated spectra agree with the data, we have divided the type (2) data by the results of the best fit obtained under the assumption of only massless neutrinos. This is illustrated in Fig. 1. For display purposes, the data were compressed into 1-keV wide bins. The horizontal line is the expectation for massless neutrinos. The curve shown is what is obtained by taking a spectrum containing a 1.3% admixture of a 17-keV neutrino (i.e., the best fit to the experimental data) and dividing it by the best fit obtained for $m_v = 0$. The difference in X^2 between these two

---

* Present address: Physics Department, Queen's University, Kingston, ON, Canada

curves is approximately 23 units, thus indicating that we have nearly a 5-σ effect.

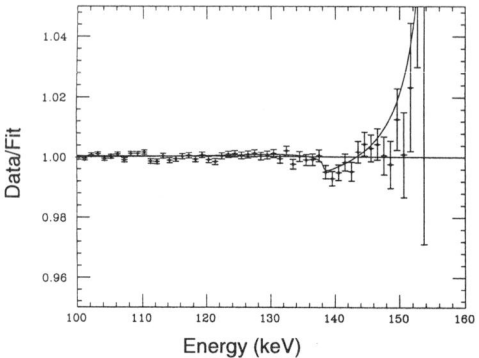

Fig. 1. The ratio of the ^{14}C data to a theoretical fit performed over the region 100-160 keV assuming $m_\nu = 0$. The horizontal line is the shape expected for zero-mass neutrinos. The curve illustrates the shape expected if a 17-keV neutrino were emitted in 1.3% of the ^{14}C beta decays.

In another type of analysis, we fit the region from 141 to 156 keV (i.e., the last 15 keV of the ^{14}C β spectrum). The results of this fit were extrapolated to lower energies and then divided into the experimental data. The results of this analysis are shown in Fig. 2. The horizontal line is the expectation for massless neutrinos. The curve, which shows the expectation for a 1.3%

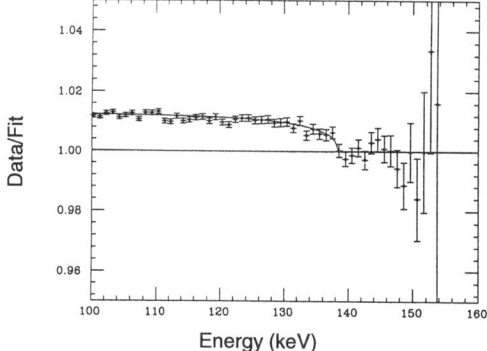

Fig. 2. The ratio of the ^{14}C data to the extrapolation of a theoretical fit performed over the region 141-156 keV assuming $m_\nu = 0$. The horizontal line is the shape expected for zero-mass neutrinos. The curve illustrates the shape expected if a 17-keV neutrino were emitted in 1.3% of the ^{14}C beta decays.

admixture of a 17-keV neutrino, agrees quite well with the data. It should be pointed out, however, that with this normalization, the error bars exaggerate the apparent significance of the result.

We have performed a number of tests to determine if some aspect of the detector response or the electronics could account for the "kink." Using external γ-ray sources, we searched for an anomaly 17 keV below the photopeak and found no such feature.[4] We did observe the Ge x-ray escape peak which occurs 10 keV below the photopeak. For a 166-keV γ ray, this peak is 0.1% as large as the photopeak and therefore cannot account for our result. We have also used a ramped pulse generator to test our ADC for non-linearities and found no evidence for a variation that could cause the effect observed in our data.

INNER BREMSSTRAHLUNG SPECTRUM
OF ^{55}Fe

The electron-capture decay of ^{55}Fe is an allowed ground-state to ground-state transition with a Q_{EC} value of 231.7 keV and the probability of radiative electron capture, or inner bremsstrahlung (IB), is 3.25x10^{-5} (Ref. 5). In our search for small distortions in the spectrum of inner-bremsstrahlung, we used a source of ^{55}Fe purchased from New England Nuclear Co. Because γ-ray counting showed impurities of ^{60}Co, ^{54}Mn, ^{123}Te, ^{127}Te, and ^{59}Fe, we chemically purified the iron using ion-exchange techniques. The strength of the purified iron source was about 25 mCi. The ^{55}Fe source was placed inside a 1-mm thick plastic container and attached to the beryllium window of a 109-cm^3 HPGe detector. An additional absorber, made of copper and aluminum foils and placed between the source and the detector, reduced the intensity of Mn x-rays (from the EC decay of ^{55}Fe) in the spectrum.

As shown in Fig. 3, this entire assembly was then placed inside a NaI anti-coincidence shield consisting of a 30-cm by 30-cm annular detector and a 7.5-cm by 15-cm plug detector. These NaI detectors vetoed both Compton-scattered IB photons as well as external background radiation. The ^{55}Fe IB counting rate in the germanium detector was approximately 8000 s^{-1} Pileup-suppression was done using an Ortec 572 amplifier. Three types of data were recorded simultaneously on three separate ADC's: (1) Ge detector singles, (2) Ge detector singles with pileup rejection, and (3) Ge detector singles with pileup rejection and NaI detector veto. A total of approximately 145 days of ^{55}Fe data taking were recorded in 2-3 day intervals on a PC based acquisition system. The energy scale was internally calibrated using Pb x-rays and the γ rays from the ^{59}Fe impurity contained in the source, and subsequently it was verified with external calibration sources. Background and ^{59}Fe spectra were accumulated and stored between measurements of the ^{55}Fe source. After summing all the ^{55}Fe spectra and subtracting the ^{59}Fe and background contributions, we have approximately 1.1×10^7 counts per keV at an energy of 208 keV (i.e. the expected position of the "kink").

In all previous searches for heavy neutrino emission in beta or inner bremsstrahlung spectra (including our own ^{14}C experiment described above), in order to have the statistical sensitivity to see a 1% distortion, it was necessary to fit a fairly wide energy interval. In the present ^{55}Fe experiment, we have sufficiently high statistics that a true "local" analysis could be performed. It is well known that taking the second derivative of a spectrum can sometimes reveal small peaks that might otherwise be missed.[6] We have found that the second derivative technique is also a powerful way to reveal the distortion in a spectrum produced by the emission of a massive neutrino.

Fig. 4 illustrates the results of numerically taking the second derivative of ^{55}Fe IB spectra that contain 1.1×10^7 counts per keV at an energy of 208 keV. Fig. 4(a) shows the second derivative of Monte Carlo data generated with a 1% admixture of a 17-keV neutrino. The feature observed around 208 keV is the signature of the 17-keV neutrino. Fig. 4(b) shows the results for Monte Carlo data generated with no heavy neutrino. As can be seen, for data of this type there is no structure near 208 keV. Parts (c) and (d) show the results of taking the second derivative of our type (3) experimental data. The winged-shape pattern around 192 keV seen in Fig. 4(c) is caused by an ^{59}Fe contaminant γ ray at this energy. A separate measurement done with a ^{59}Fe source allowed us to subtract this contaminant. The second derivative of the resulting spectrum is shown in Fig. 4(d). There is clearly no hint of a structure near 208 keV. Thus our ^{55}Fe experiment shows no evidence for the emission of a 17 keV neutrino.

In order to quantify the null result of our ^{55}Fe experiment, we have performed a fit to the region of the spectrum from 200-220 keV. Rather than calculating in detail the theoretically expected shape of the ^{55}Fe IB spectrum and then convoluting it with the detector response function, we have taken a much simpler approach. We make use of the fact that if the neutrino is massless, then an IB spectrum is a cubic function of the photon energy.[7] An admixture of a neutrino with mass, m_ν, and mixing angle, θ, modifies this shape in the follwing way.[8]

$$N_\gamma(E) = [a_0 + a_1E + a_2E^2 + a_3E^3] \times [1 + \tan^2\theta (1 - \frac{m_\nu^2}{(E_0-E)^2})^{1/2}]. \quad (1)$$

In the fitting procedure, for each choice of m_ν and $\tan^2\theta$, a_0-a_3 were allowed to vary freely so

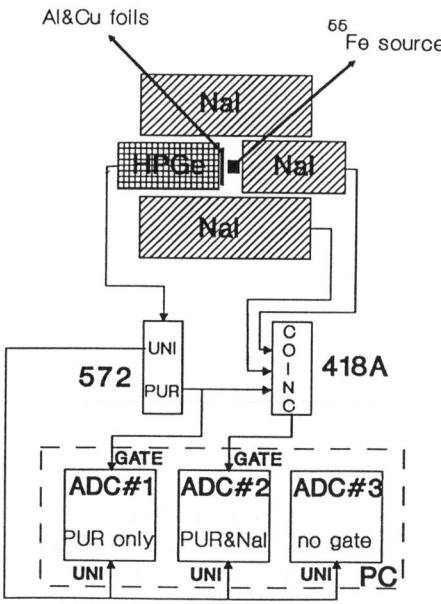

Fig.3. Schematic view of the system used in the ^{55}Fe IB experiment.

as to minimze X^2. The results of this procedure for our type (3) data for $m_\nu = 0$ and for $m_\nu = 17$ keV, $\tan^2\theta = 0.007$ are shown in Fig. 5 (a) and (b), respectively. In part (a), the horizontal line is the expectation for massless neutrinos, and the X^2 for this fit is 157 for 149 degrees of freedom. In part (b) the horizontal line is the expectation for $m_\nu = 17$ keV, $\tan^2\theta = 0.007$, and the X^2 is 203 for 149 degrees of freedom. The results of the fits in which m_ν and $\tan^2\theta$ were varied are shown in Fig. 6. As can be clearly seen, our ^{55}Fe data excludes a 0.7% admixture of a 17-keV neutrino at about the 7σ level.

CONCLUSIONS

We have observed a distortion in the beta spectrum of ^{14}C that looks very much like that which would be produced if a neutrino with a mass of 17 keV were emitted in approximately 1% of all ^{14}C beta decays. However, our much higher statistics ^{55}Fe inner bremsstrahlung spectrum shows no evidence for the emission of a heavy neutrino. Recent magnetic spectrometer experiments on ^{35}S (Ref. 9) and ^{63}Ni (Ref. 10) and a ^{35}S search that used a solid-state silicon detector (Ref. 11) have also failed to see any evidence of heavy neutrino emission. We thus conclude that, whatever causes the "kink" in our ^{14}C spectrum, it is not a neutrino.

ACKNOWLEDGEMENTS

We wish to thank C. P. Cork, F. S. Goulding, E. E. Haller, W. L. Hansen, T. R. Ho, D. A. Landis, P. N. Luke, A. R. Smith, K. L. Wedding, and J. T. Witort for their assistance in various aspects of this work. This work was supported by the U. S. Department of Energy under Contract No. DE-AC03-76SF00098. M. T. F. da Cruz is supported in part by Fundacao de Amparo a Pesquisa do Estado de Sao Paulo, FAPESP, Sao Paulo, Brazil. M. M. Hindi is also supported in part by the U. S. Department of Energy under Contract No. DE-FG05-87ER40314.

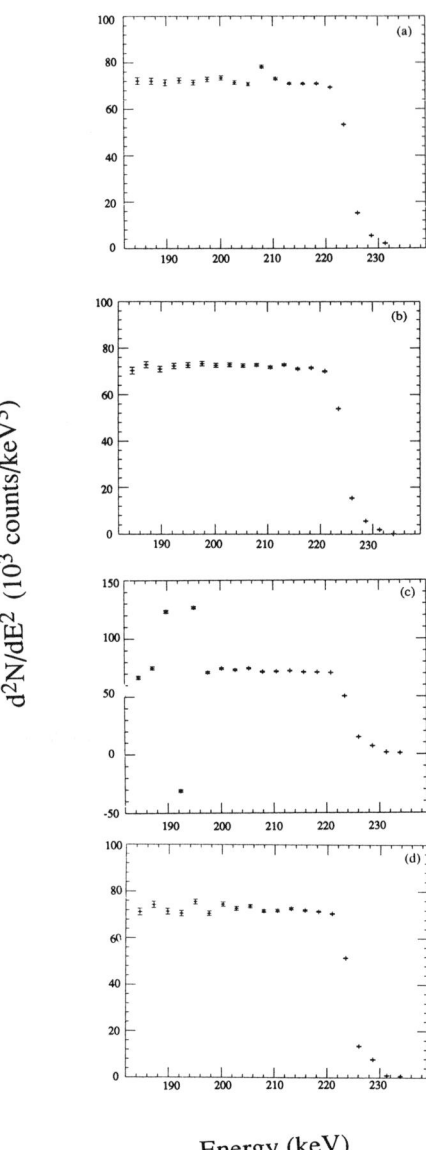

Fig. 4. The second derivative of inner bremsstrahlung spectra of ^{55}Fe. (a) Monte Carlo data generated with a 1% admixture of a 17-keV neutrino, (b) Monte Carlo data generated with $m_\nu = 0$, and (c) experimental data before subtracting ^{59}Fe contaminant, (d) experimental data after subtracting ^{59}Fe.

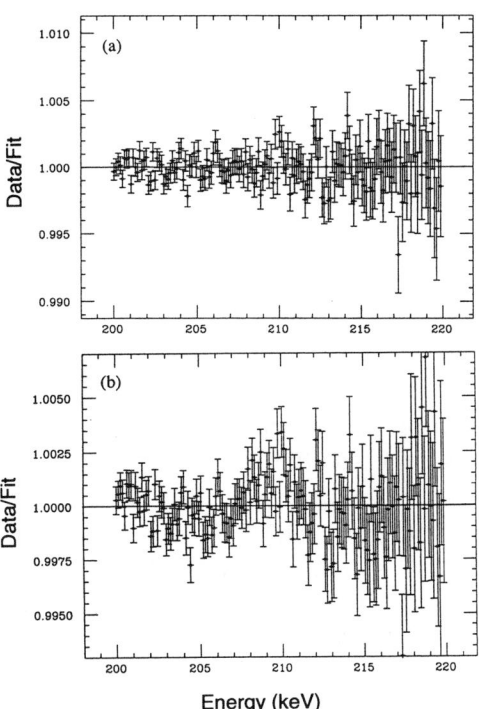

Fig. 5. Ratio of the type (3) ^{55}Fe data to fits using Eqn. 1. (a) Fit performed assuming $m_\nu = 0$. The horizontal line is the expectation for massless neutrinos. (b) Fit performed assuming $m_\nu = 17$ keV and $\tan^2\theta = 0.007$. The horizontal line is the expectation for $m_\nu = 17$ keV and $\tan^2\theta = 0.007$.

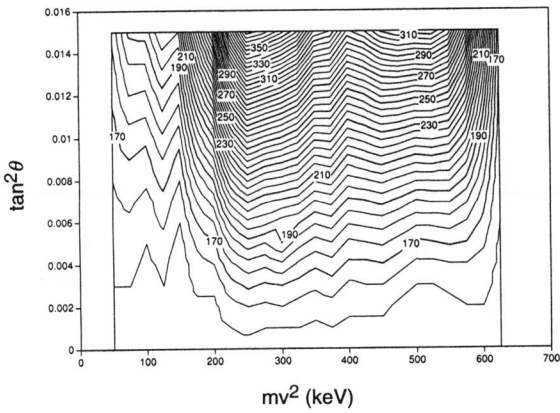

Fig. 6. Contours of X^2 as a function of m_ν and $\tan^2\theta$ for our type (3) ^{55}Fe data.

REFERENCES

1. B. Sur et al., "Evidence for the Emission of a 17-keV Neutrino in the β Decay of ^{14}C," *Phys. Rev. Lett.* 66, 2444-2447 (1991).
2. B. Sur et al., "The Evidence for 17-keV Neutrinos," AIP Conf. Proc. No. 243, edited by W. T. H. Van Oers, New York: Am. Inst. Phys., 1992, p. 1089-1099.
3. E. E. Haller et al., "Carbon in High-Purity Germanium," IEEE Trans. Nucl. Sci. Vol. NS-29, No. 1, 745-750. (1982).
4. B. Sur et al., "Reply to a Comment by N. K. Sherman," Phys. Rev. Lett. 68, 1435 (1992).
5. E. Browne and R. B. Firestone, *Table of Radioactive Isotopes* New York: John Wiley & Sons, 1986.
6. M. A. Mariscotti, "A Method for Automatic Identification of Peaks in the Presence of Background and Its Application to Spectrum Analysis," Nucl. Instr. and Meth. 50, 309-320 (1967).
7. W. Bambyneck et al., "Orbital Electron Capture By the Nucleus," *Rev. Mod. Phys.* 49, No. 1, 77-221 (1977).
8. A. Hime, "Heavy Neutrino Emission in Nuclear Beta Decay Spectra," D. Phil thesis, Oxford Univ., p. 7, 1991.
9. F. Boehm et al., "New Limits on the 17 keV Neutrino," Caltech preprint (1992).
10. T. Oshima et al., "Search for a 17-keV Neutrino," contribution to this conference.
11. S. Freedman et al., "Search for a 17-keV Neutrino," contribution to this conference.

0.073% (95% C.L.) UPPER LIMIT ON 17keV NEUTRINO ADMIXTURE

Takayoshi Ohshima[*]
National Laboratory for High Energy Physics, KEK
Oho 1-1, Tsukuba, Ibaraki 305, Japan

Abstract

A direct search was made for a threshold kink in ^{63}Ni β-ray spectrum due possibly to a sizeable admixture of 17keV neutrino. A fine energy scan was performed using a magnetic spectrometer over the specific energy region with very high statistics and a very high signal-to-background ratio. The resultant mixing strength is $|U|^2 = (-0.011 \pm 0.033(\text{stat.}) \pm 0.030(\text{sys.}))\%$ and its upper limit $|U|^2 < 0.073\%$ (95% C.L.). The result clearly excludes neutrinos with $|U|^2 \geq 0.1\%$ for the mass range from 11 to 24keV.

INTRODUCTION

It was already found 58 years ago in 1934[1], that is a few years after the birth of the Fermi's β-decay theory, that the Kurie plots of measured β-ray spectra were not as straight as theoretically expected. A noticeable excess was seen in a low energy part of the spectrum. This discrepancy was at the time interpreted as an indication of necessity in revising the Fermi's theory, but not as a sign of heavy neutrino production. For instance, Konopinski and Uhlenbeck[2] even proposed a revised theory in 1935 to explain the experiment. Various investigations[3] carried out over about 20 years finally revealled that the low energy excess was the result of energy-losses in a source substance and of backscatterings at a source backing plate.

It is interesting to see that we have been again faced with a controversy on the β-ray spectral shape. This time, however, it is frequently interpreted as arising from a 1% admixture of 17keV neutrino, thus causing excitement among particle physicists.

CONSIDERATIONS

Even though the low-energy excess mentioned above can be reduced by experimental efforts, it never disappears. The first key to successful experiment is therefore how reliably one can control this effect, generally represented by a low-energy tail in the so-called Response Function R(E). The backscattering usually dominates 20-30% of R(E) in an experiment with a β-ray source separated from a detector, and exhibits approximately a flat distribution with the amplitude δ (keV)$^{-1}$ down to zero energy. β-rays of energy E' flow in to lower energies. This contribution can be estimated by approximating the spectrum N(E) to be proportional to the square of the

[*]Author would like to thank the Foundation For High-Energy Accelerator Science, Japan, for its financial support making his attendance to this Conference possible.

energy measured from the endpoint E_0,

$$N(E) \propto (\Delta E)^2, \quad \Delta E \equiv E_0 - E.$$

Then

$$\frac{(Tail\ effect)}{N(E)} \sim \frac{\int_E^{E_0} \delta \cdot N(E')dE'}{N(E)}$$
$$\sim \frac{\delta}{3}\Delta E. \quad (1)$$

This is a cumulative effect over E to E_0, so that, even though the amplitude is only $\delta \sim O(10^{-3}/\text{keV})$, the net effect will be amplified by a factor ΔE of typically a few tens keV, easily resulting in a few % excess. Experimenter's knowledge might be limited to $O(10^{-4}/\text{keV})$ precision, and then one ends up with $O(10^{-3})$ uncertainty in the measured spectrum.

Such effect can not be ignored in a search for a heavy neutrino[4] with a mixing strength $|U|^2$ of 1% level. It has to be taken into account, for instance, as the so called shape correction term in a spectral formula. This term also includes other ambiguities, depending on experimental method. It should never be neglected unless experimentally justified.

Previous experiments[5] measured a spectrum over a wide energy region and made a χ^2 fit to the spectrum looking for a difference in curvature above and below the heavy neutrino production threshold E_{th}. In such data, the lower the β-ray energy E is, the higher the statistical accuracy becomes because of $N(E) \propto (E_0 - E)^2$, but the larger the ambiguity is due to the tail effect. The situation is reversed at higher energies. As a result, the analysis will be strongly biased by the low energy portion of the data where the uncertainty is large.

In a high energy region, especially above E_{th}, an accuracy in background estimate significantly affects the curvature determination unless the signal-to-background ratio is large. For instance, when a solid state detector is used, that ratio is only 6 - 60 even at the 17keV neutrino threshold due to signal pile-up and residual radioactive nuclei. Under such condition, inaccurate knowledge of the background might artificially create or blow out a heavy neutrino effect.

EXPERIMENT

We adopted the strategy to search directly for a kink due to the 17keV neutrino emission by means of a fine scan over narrow energy region in question with equally high statistics both above and below E_{th}. It is to avoid confusion between the kink and the usual shape correction term based on their distinct energy dependences. For the same purpose, R(E) was experimentally determined with high precision using a monochromatic β-source. To be decisive on the mixing strength of 1% level, we aimed to achieve 0.1% sensitivity in $|U|^2$.

The β-ray source was ^{63}Ni (580μCi, 4×20mm^2, E_0=66.7keV, life-time=100years). It provides a high count rate, about 12 times the usual ^{35}S at E_{th}. The monochromatic β-source for calibration was ^{109}Cd (150μCi, 62.5keV K conversion line, life-time=453days). To have the same thickness and the same backing plate as the active Ni source, this calibration source was made by mixing ^{109}Cd with natural Ni atoms. Measured spectrum of the K conversion line accurately represents our response function that combines all the effects such as the spectrometer optics, the energy-loss and the backscattering. Fig.1(a) shows the R(E) thus obtained after subtracting contributions from higher energy conversion lines.

The β-ray was analyzed by the $\pi\sqrt{2}$ type iron-free magnetic spectrometer[6] at INS (Institute for Nuclear Study). The avail-

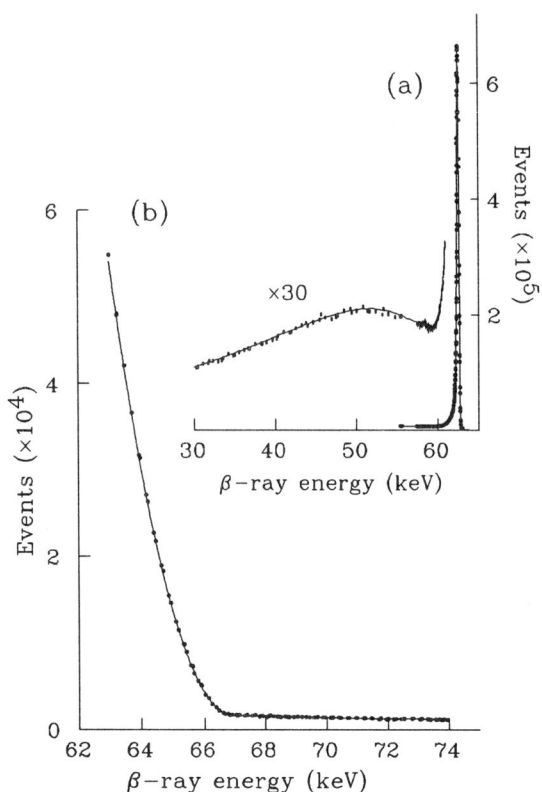

Figure 1. (a): ^{109}Cd K-line spectrum served as the response function R(E). (b): Measured ^{63}Ni β-spectrum near E_0 with the best fit curve.

achieving two orders of magnitude improvement in statistics over the previous measurements. Signal-to-background ratio was as high as 1000 at the E_{th} since problems with pile up or residual radioactive nuclei were a negligible contribution in the present setup.

A measurement was also made near E_0 where the heavy-neutrino effect should not exist and the shape correction term can be ignored. It provided us with one of the important ways to perform consistency checks.

ANALYSIS

The measured spectra $N^{exp}(E)$ were analyzed in terms of a χ^2-fit to the formula,

$$N^{exp}(E) = A_o \int \tilde{N}(E')[1 + \alpha(\Delta E')]$$
$$\times R(E', E) dE' + BG(E), \quad (2)$$
$$\tilde{N}(E') \equiv F(Z, E') p' E'_T [\, (1-|U|^2)(\Delta E')^2$$
$$+ |U|^2 \Delta E' \sqrt{(\Delta E')^2 - m_H^2}\,], \quad (3)$$

where A_o represents a normalization constant, α a shape correction factor, $|U|^2$ a mixing strength, and m_H^2 the heavy neutrino mass squared. E_T and p are total energy and the momentum of the β-ray, respectively. F(Z,E) is the radiatively corrected, relativistic Fermi-function. BG(E) represents the background spectrum as a combination of a constant and a small linear term.

First, the spectrum around E_0 was analyzed by setting both $|U|^2$ and α to zero. The best fit curve is shown in Fig.1(b), where the resultant χ^2/degrees-of-freedom(d.o.f.) was 116.6/102 and the endpoint energy was

$$E_0 = (66945.9 \pm 4.4)\ eV. \quad (4)$$

BG(E) determined here was used in following analyses.

Next, thirty individual spectra measured near E_{th} in individual chamber cells were an-

able momentum resolution, $\Delta p/p$(FWHM) $=2.6\times10^{-3}$, was equivalent to the 250eV energy resolution at 50keV. The β-rays were detected at the spectrometer focal plane by a proportional chamber with 30 isolated cells.

The measurement covered the energy region from 39 to 60keV, where each of three sub-regions was scanned at 20 magnetic field settings in 260eV steps. Three data-points at both ends of the central sub-region always overlapped with those of neighbours. One million or more events were accumulated at every point, resulting in 2.4 billions events distributed over 1800 data-points and thus alyzed with the following five or six free parameters; $|U|^2$ for $m_H=17$keV, α, A_o, and two additional normalization factors between two sub-regions[7], with E_0 either fixed to the value (4) or left free. $R(E)$ was slightly different among 30 spectra because of the spectrometer optics. The resulting reduced χ^2 values and $|U|^2$ values are shown in Fig.2 for the spectra recorded at individual chamber cells. Fig.3 compares the deviations from the best fits with either $|U|^2$ left free or fixed to 1%. They are the results of the 6-parameter fits, but there is no significant difference from those obtained with 5 parameters. It is clear from Fig.2(a) that $|U|^2$ fixed to 1% results in much larger χ^2 values. 30 values of $|U|^2$ shown in Fig.2(b) average to (-0.029 ± 0.038)% and give $\chi^2/$d.o.f. of 1.13. χ^2 vs. $|U|^2$ relation is shown in Fig.4 for

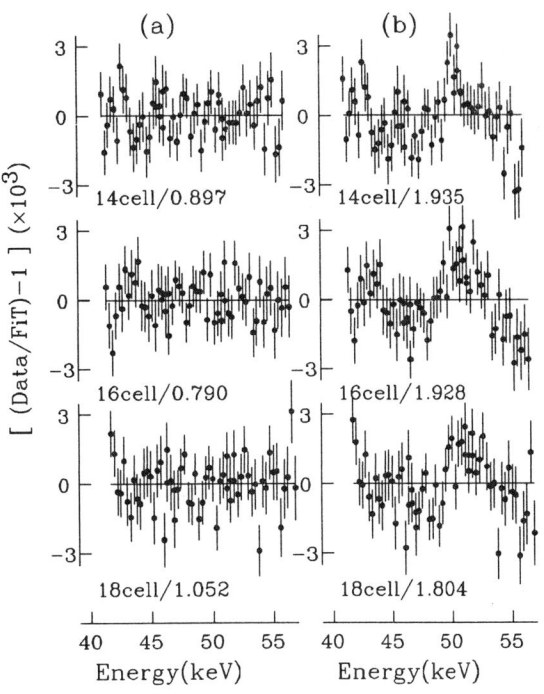

Figure 3. Deviations from the best fit with $|U|^2$ being a free parameter (a) and fixed to 1% (b). $\chi^2/$d.o.f. is indicated with the chamber cell-number.

individual spectra, while Fig.5 shows other parameters resulting from the same fits. The averaged endpoint energy comes out to be

$$<E_0> = (66942.8 \pm 5.5) \ eV, \qquad (5)$$

in good agreement with the separately measured result (4), and the shape correction factors (α's) are the order of $10^{-4}(\text{keV})^{-1}$. On the other hand, when $|U|^2$ is set to 1%, the average endpoint becomes $<E_0>=(66881.3\pm4.6)$eV, in clear contradiction with (4). The relative normalization factors obtained out of the fits are plotted in Fig.5(c) and 5(d). These factors can be independently evaluated as the ratio η of the total counts summed over the 3 overlapped data points between successive sub-regions.

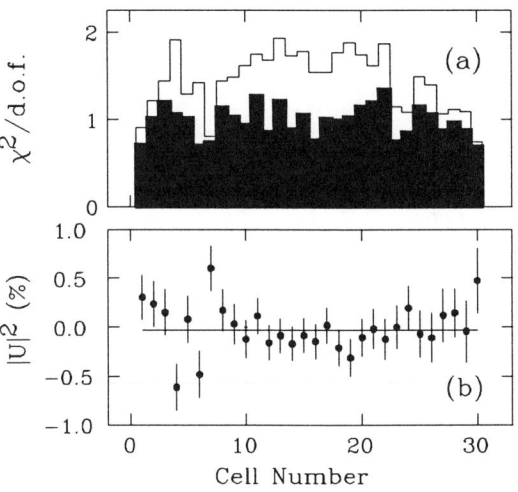

Figure 2. (a): Reduced χ^2 values from the 6-parameter fits with $|U|^2$ left free (closed histogram) and fixed to 1% (open) for 30 individual spectra. (b): Mixing strength $|U|^2$ obtained as the best fit. The average of the 30 values is $<|U|^2>=(-0.029\pm0.038)$% with $\chi^2/$d.o.f. = 1.13.

As seen in the figure, these two methods give completely consistent results only when $|U|^2$ is left free.

Finally global fits were made for all of the thiry spectra by treating E_0 and $|U|^2$ as common parameters. Fig.6 shows the results, where 1800 data points are displayed in 50eV bins for the sake of illustration. Here we find $\chi^2/\text{d.o.f.}=1701/1678=1.01$ and the best fit parameters turn out to be

$$E_0 = (66943.3 \pm 4.1)\ eV, \qquad (6)$$
$$|U|^2 = (-0.011 \pm 0.033)\ \%. \qquad (7)$$

The expected size of 1% heavy-neutrino mixing effect is indicated by the curve in Fig.6(a). Fig.6(b) is the result obtained with $|U|^2=1\%$ fixed, and in this case we find $\chi^2/\text{d.o.f.}=2467/1679=1.47$ and $E_0=(66882.4\pm4.6)$eV. This E_0 value is again in clear contradiction with the separate measurement (4).

Table 1 summarizes the results of various fits. It is clear that the heavy-neutrino mixing

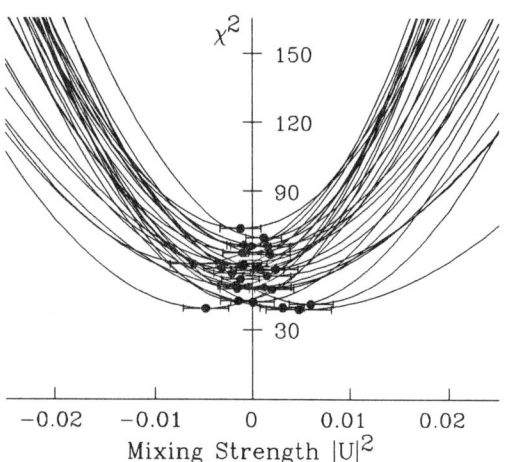

Figure 4. Relation of χ^2 vs. $|U|^2$ for individual 30 spectra with 54 d.o.f. each. Closed circles are the $|U|^2$'s obtained by the best fit.

$|U|^2$ is equivalent to zero in both individual and global fits. χ^2 values are always larger by about 1000 units if 1% mixing is assumed, as seen in Table 1. Systematic errors arise from uncertainties in the β-ray transmission through the detector window and the tail component in R(E). Compared to these, the error from the background estimation is an order of magnitude smaller because of our excellent signal-to-background ratio. The final result with the statistical and systematic

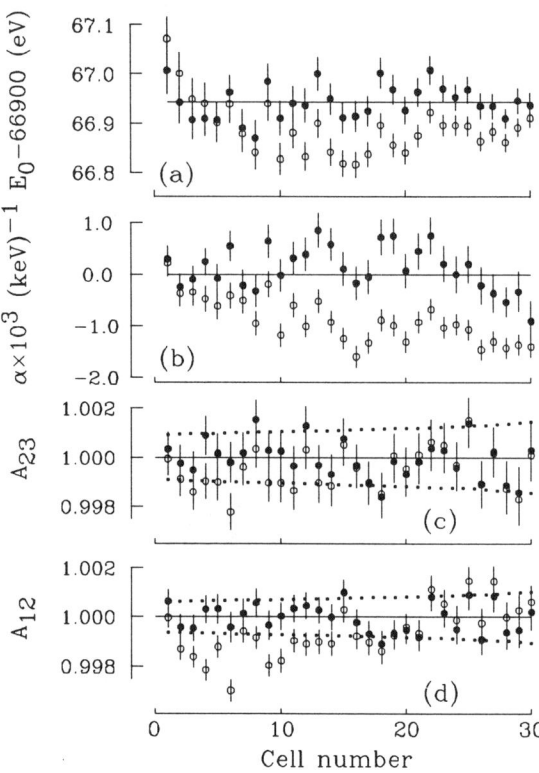

Figure 5. E_0 (a), α (b), and two relative normalization factors (c) and (d) from the individual fits made with $|U|^2$ left free (closed circles) and fixed to 1% (open). The horizontal line in (a) indicates the averaged E_0, Eq.(5), obtained by the fit with $|U|^2$ free. The relative normalization factors are normalized by the η (see text). The dotted curves indicate a range of the statistical uncertainty of the η.

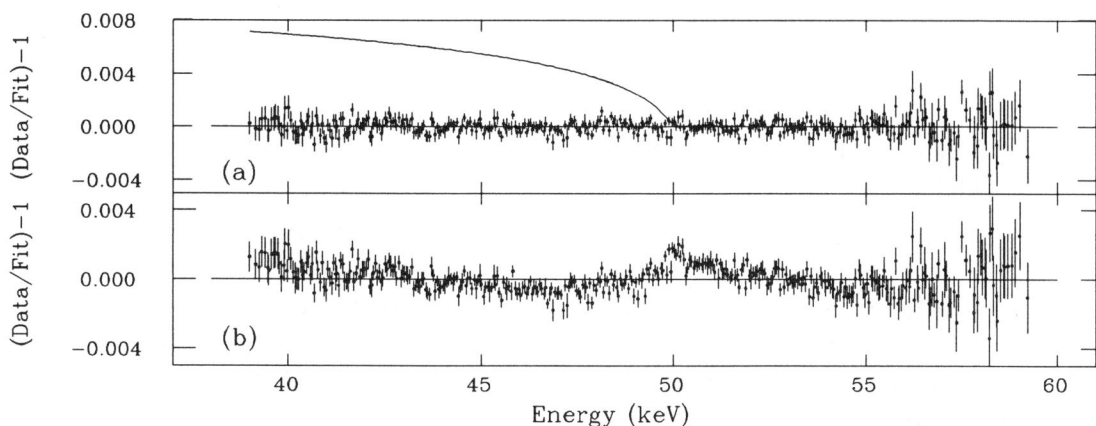

Figure 6. Deviations from the best global fit with $|U|^2$ free (a) and fixed to 1% (b). The curve in (a) indicates the size of 1% mixing effect of the 17keV neutrino. Resultant parameters of these fits are indicated in Table 1.

Table 1. Results of various fits with m_H=17keV. The number in < > is the average of the 30 resultant values obtained by the individual fit. Details are described in text.

	χ^2	d.o.f.	$|U|^2$ (%)	E_0−66900 (eV)
Fit near E_0				
	116.6	104	fixed to 0.0	45.9±4.4(stat.)±3.2(sys.)
Individual Fits				
			fixed to 1.0	fixed to 45.9
			<−0.022±0.033>	fixed to 45.9
			fixed to 1.0	<−18.7±4.6>
			<−0.029±0.038>	< 42.8±5.5>
	$|U|^2$=(−0.029±0.038±0.028)%,		$|U|^2 < 0.077\%$ at 95% C.L.	
	systematic errors; ±0.014% (window trans.), ±0.024% (R-tail)			
Global Fits				
	2744.3	1680	fixed to 1.0	fixed to 45.9
	1701.0	1679	−0.024±0.033	fixed to 45.9
	2466.9	1679	fixed to 1.0	−17.7±4.6
	1701.1	1678	−0.011±0.033	43.3±4.1
	$|U|^2$=(−0.011±0.033±0.030)%,		$|U|^2 < 0.073\%$ at 95% C.L.	
	systematic errors; ±0.013% (window trans.), ±0.027% (R-tail)			
Study of smoothness (see text)			<−0.008>	fixed to 45.9

systemtic errors is consistent with zero,

$$|U|^2 = (-0.011 \pm 0.033 \pm 0.030) \%, \quad (8)$$

and thus one can set its upper limit for the 17keV neutrino to

$$|U|^2 < 0.073\%. \quad (95\% \, C.L.) \quad (9)$$

The endpoint E_0 was obtained from two independent data sets, one measured near E_0 and the other around E_{th}. The two results, Eq.(4) and Eq.(6), agree very well with each other and also with the value of $E_0 = (66946 \pm 20)$eV measured by Hetherington et al.[5].

In order to demonstrate that the shape correction term does not harm our experimental sensitivity on $|U|^2$, an additional analysis was performed in the following way. Each of the 30 spectra was prepared with the relative normalization among three energy regions done based on the number of counts in overlapped data points, the η normalization not by a fit. It was then divided into two parts: One below 50keV (A) that is sensitive to the 17keV neutrino effect, and the other above 50keV (B) that is not. First, a fit was made to the data in B with two free parameters α and A_o fixing E_0 to (4) and $|U|^2$ to zero. Then the resulting fit was extrapolated to the A region, and there we found $\chi^2/d.o.f.=1.52$ as an average over 30 spectra. When the 1% mixing effect was added to the extrapolation, it considerably increased to 21.72. A next fit was made to the data in A with a single parameter $|U|^2$ fixing α and A_o to those obtained in the B region. This resulted in $|U|^2 = -0.008\%$ with $<\chi^2/d.o.f.> = 0.97$. As a whole, the study concludes that the spectra measured above and below E_{th} exhibit smooth continuation, that the shape correction factors determined above E_{th} well reproduce the data below E_{th} with a null mixing, and consequently that there is no structure at all hinting at the existence of the heavy neutrino.

CONCLUSIONS

New measurement, directly sensitive to a possible kink in the β-ray spectrum of ^{63}Ni due to the 17keV neutrino, was carried out by a fine energy scan with very high statistics and a very high signal-to-background ratio. The resulting mixing strength is

$$|U|^2 = (-0.011 \pm 0.033 \pm 0.030) \%, \quad (10)$$

which is 22 σ away from 1%, its upper limit being

$$|U|^2 < 0.073\%. \quad (95\% \, C.L.) \quad (11)$$

The study was extended to different masses of a heavy neutrino and its result is plotted in Fig.7. The curve corresponds to the upper limit at 95% C.L. No neutrino exists with $|U|^2 \geq 0.1\%$ in the mass range from 11 to 24keV.

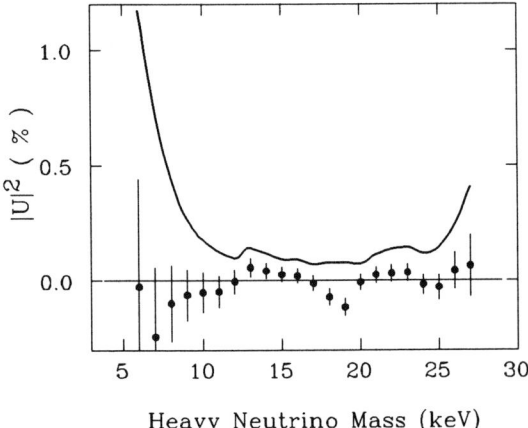

Figure 7. $|U|^2$ for different masses of a hypothetical heavy neutrino. The curve indicates the obtained upper limit at 95% C.L.

ACKNOWLEDGEMENT

I wish to thank Professor R. G. H. Robertson (LANL) for his significant suggestion on the analysis of treating the normalization factors of the spectra as mentioned in ref.7. I also would like thank Professor S. Iwata (KEK) for reading and editing this manuscript. This work has been carried out in collaboration with H. Kawakami, S. Kato, C. Rosenfeld, H. Sakamoto, T. Sato, S. Shibata, J. Shirai, Y. Sugaya, T. Suzuki, K. Takahashi, T. Tsukamoto, K. Ueno, K. Ukai, S. Wilson and Y. Yonezawa.

REFERENCES

1. C. D. Ellis and W. J. Henderson, *Proc. Roy. Soc.* A146, pp. 206–216, (1934).

2. E.J. Konopinski and G.E. Uhlenbeck, *Phys. Rev.* 48, pp. 7–12, (1935).

3. For instance, see C. S. Wu and R. D. Albert, *Phys. Rev.* 75, pp. 315–316, (1948); C. S. Cook, L. M. Langer and H. C. Price Jr., *Phys. Rev.* 74, pp. 548–552, (1948); L. M. Langer and R. J. D. Moffat, *Phys. Rev.* 88, pp. 689–694, (1952); A review at the time on nuclear β-decay can be found in E. J. Konopinski, *Rev. Mod. Phys.* 15, pp. 209–245, (1943).

4. J. J. Simpson, *Phys. Rev. Lett.* 54, pp. 1891–1893, (1985).

5. A. Apalikov et al., *JETP Lett.* 42, 289–293, (1985); J. Markey and F. Boehm, *Phys. Rev.* C32, 2215–2216, (1985); T. Altzitzoglou et al., *Phys. Rev. Lett.* 55, pp. 799–802, (1985); T. Ohi et al., *Phys. Lett.* 160B, 322–324, (1985); *Proc. 12th Int. Conf. on Neutrino Physics and Astrophysics,* Sendai, Japan, June, 1986, pp. 69–76; V. M. Datar et al., *Nature,* 318, pp. 547–548, (1985); D. Wark and F. Boehm, *Proc. Int. Symp. on Nuclear Beta Decays and Neutrinos,* Osaka, Japan, June, 1986, pp. 391–393; M. J. G. Borge et al., *Phys. Scripta,* 34, pp. 591–596, (1986); D. W. Hetherington et al., *Phys. Rev.* C36, pp. 1504–1513, (1987); I. Žlimen et al., *Phys. Scripta,* 38, pp. 539–542, (1988); A. Hime and J. J. Simpson, *Phys. Rev.* D39, pp. 1837–1850, (1989); J. J. Simpson and A. Hime, *Phys. Rev.* D39, pp. 1825–1836, (1989); A. Hime and N. A. Jelly, *Phys. Lett.* B257, pp. 441–449, (1991); I. Žlimen et al., *Phys. Rev. Lett.* 67, pp. 560–563, (1991); B. Sur et al., *Phys. Rev. Lett.* 66, 2444–2447, (1991).

6. H. Kawakami et al., *Phys. Lett.* B256, pp. 105–111, (1991).

7. In our previous analysis reported in H. Kawakami et al., *Phys. Lett.* B287, pp. 45–50, (1992), the spectra of three sub-regions were uniquely normalized to have the same total counts over the three overlapped data-points between the sub-regions. The normalization errors were then statistically added to each data-point. The analysis described in the text should be taken as the revised one, although the difference between present and previous ones does not substantially affect the results.

AN EXPERIMENT TO SEARCH FOR A 17 KEV NEUTRINO*

Robert Shrock
Institute for Theoretical Physics
State University of New York
Stony Brook, NY 11794-3840
USA

Abstract

We discuss an experiment to search for a 17 KeV neutrino in nuclear beta decay via detection of the recoil nucleus a well as the electron, and hence a complete kinematic reconstruction of the final state.[1] A generalization to search for possible heavier admixed neutrinos is noted. From retroactive data analysis we also report the first upper limits on the decays $K^+ \to \pi^- \mu^+ \mu^+$ and $\Xi^- \to p\mu^-\mu^-$.[2,3]

INTRODUCTION

The issue of possible nonzero neutrino masses remains a fundamental and unresolved one. A natural concomitant of such masses is lepton mixing.[4] Strangely, up to 1980, although the awareness of possible lepton mixing led to experiments to search for neutrino oscillations, experimental studies of weak decays yielding neutrinos continued to assume implicitly that the weak and mass eigenstates of neutrinos were the same. In 1980, the author suggested tests for massive neutrinos which could occur, via mixing, in various weak decays including nuclear beta decay, $\pi^+_{\ell 2}$, $K^+_{\ell 2}$, and μ decay and set the first upper limits on the emission of massive, admixed neutrinos in these decays.[5,6] Writing the weak eigenstate ν_e as a linear combination of mass eigenstates, $\nu_e = \sum_i U_{1j}\nu_j$, the limit given in Ref. 5 for nuclear beta decay was $|U_{1j}|^2 < 0.1$ for $m(\nu_j)$ in the range from 0.1 KeV to $\sim$ 3 MeV.

The suggestion to carry out searches for the kinks in Kurie plots due to possible emission of massive, admixed neutrinos led to several experiments directed toward this goal. The current situation is unsettled; some experiments[7] have reported evidence for the emission, via mixing, of an (anti)neutrino with mass $m(\nu_j) = 17$ KeV and mixing strength $|U_{1j}|^2 \simeq 0.01$, while others[8], including the most recent ones, have reported no evidence for such an effect or, indeed, the emission of any massive neutrino in nuclear beta decay. For example, the Tokyo-KEK experiment (see Ref. 8) finds $|U_{1j}|^2 < 1 \times 10^{-3}$ (95 % CL) for $m(\nu_j)$ in the range from 10 to 24 KeV. None of the magnetic spectrometer experiments or the proportional counter experiment has seen any evidence for a 17 KeV neutrino, while several, but not all, of the solid state counter experiments have.

EXPERIMENT

In view of this situation, it would be useful to employ a fundamentally different type of ex-

---

*email address: shrock@max.physics.sunysb.edu; research partially supported by NSF grant NSF-91-08054.

periment to try to address the issue. We have considered an experiment in which one detects and measures not only the kinetic energy and 3-momentum of the electron but also those of the recoil nucleus.[1] This information suffices, in principle, to reconstruct the kinematics of the final state completely and thus to determine, on an event-by-event basis, the mass of the emitted (anti)neutrino.[9] This is to be contrasted with the kink search, which is a statistical, not event-by-event, test for an admixed massive neutrino. The real question is whether an experiment using the recoil method would be feasible in practice and would have sufficient sensitivity to help settle the question of the 17 KeV neutrino. We find that an experiment with ^{3}H in molecular form (T$_2$) looks promising.

The beta decay of one of the ^{3}H nuclei in a T$_2$ molecule dominantly yields the molecular ion (^{3}He ^{3}H)$^+$ together with an electron and the various (kinematically allowed) mass eigenstates in $\bar{\nu}_e$, including the dominantly coupled $\bar{\nu}_1$ with coupling strength $|U_{11}|^2$ close to unity and perhaps another, denoted $\bar{\nu}_j$ here, with $m(\nu_j) = 17$ KeV and $|U_{1j}|^2 \simeq 0.01$. Since one must detect and measure the energy of the recoil ion, it is necessary to have the decaying nucleus in the form of a gas, in order to preserve the initial 3-momentum of the recoil ion. The source is gaseous T$_2$. A jet (molecular beam) is formed via effusion from a small hole in the gas container. The jet is collimated by skimmers and passes through a small region where it is viewed by the detectors. The space between the beam and the detectors is kept at a high vacuum, with pressure $p \sim 10^{-9}$ torr. The beam then impinges on a beam stop, and is periodically or continually recirculated. To minimize the scattering of the recoil ion from the beta decay with the T$_2$ molecules of the beam, one must choose the experimental parameters so that its mean free path is much larger than the radial size of the beam. This is done by keeping the gas in the original container at low temperature T and pressure p, which yield a low-density beam. The low temperature is also important to minimize the uncertainty in the kinematic reconstruction due to the random Maxwellian distribution of velocities of the T$_2$ molecules in the beam. From an optimization study, we choose $T = 10$ K and $p = 10^{-4}$ torr.

Since $Q = 18.6$ KeV for ^{3}H beta decay, the maximum kinetic energy, T_e for an electron emitted in conjunction with a $\bar{\nu}_j$ with $m(\nu_j) = 17$ KeV is 1.6 KeV. This is too low for an accurate measurement with an external solid-state detector without preacceleration, which could introduce large measurement uncertainties. Instead, we would use an electrostatic spectrometer to measure T_e. We conservatively estimate that for our setup this would yield a 1 σ fractional measurement accuracy of $\Delta T_e/T_e = 0.02$.

The recoil ion in the decay has a very small kinetic energy $T_i \lesssim 0.15$ eV for ^{3}H beta decay. While this is too small to measure directly, the fact that the recoil ion is nonrelativistic enables one to perform an accurate indirect measurement of T_i by a time-of-flight method. For our experimental setup, the ion time of flight t_i is of order 100 μsec. Since T_i is so small, we would use an electrostatic grid to preaccelerate it before it is incident upon a detector. In the simplest design, this detector would be a microchannel plate; alternatively, one could use a spectrometer (followed by a detector), which would afford a measurement of the charge/mass ratio for the ions.

From an initial optimization study, we choose the following experimental parameter values: diameter of effusion hole $2r_h = 0.1$ cm; length of decay region viewed by detectors (= length from skimmer to collector): 1 cm; half-angular divergence of jet after skimmers: 10°; and radial distance from center of decay region to electron and ion detectors: $\ell_e = \ell_i = 20$ cm.

For the kinematics, we first consider the decay in the rest frame of the parent nucleus and then incorporate the gas velocity. When the electron is detected and T_e measured, one can reconstruct its velocity and hence the time $t = 0$ when the decay occured. For a particular decay, in order to test the hypothesis of the emission of a $\bar{\nu}_i$ with a given mass $m(\nu_j)$, one can set a window in time for the detection of the recoil nucleus. For our setup we estimate the fractional uncertainty in the ion kinetic energy measurement to be $\Delta T_i/T_i = 0.02$. We have checked that the mean free path of the recoil ions, estimated as 29 cm, is much larger than the radius of the beam, $r_b = 0.3$ cm, as necessary. Further, we estimate the uncertainty in the measurement of the angle θ_{ei} between the electron and the recoil nucleus to be $\Delta\theta_{ei} \simeq 1.9°$

We have carried out (independent) Monte Carlo (MC) simulations which take into account the decay from within the moving molecular beam and from a decay region of finite size. To suppress events from the dominant decay mode and greatly enhance the sensitivity of the search to the possible 17 KeV neutrino events, we impose several cuts. First, we require $T_e < 1.6$ KeV. The recoil ion has substantially lower T_i and velocity v_i for the decay into a hypothetical $\bar{\nu}_j$ with $m(\nu_j) = 17$ KeV than for the dominant decay into $\bar{\nu}_1$ with $m(\nu_1)$ negligibly small (or zero). Our MC simulations show that the time separation which this v_i difference produces is not seriously weakened by inclusion of the random gas velocity effects. From our MC study we thus choose a cut $t_i > 90$ μsec. To suppress accidentals, we also include the cut $t_i < 200$ μsec. Secondly, in a decay to a 17 KeV $\bar{\nu}_j$, the electron and ion are emitted more nearly back-to-back than in the dominant decay to a massless neutrino. For ^3H decay, there is a large difference between θ_{ie} values for events due to the dominant decay mode and hypothetical events

with a 17 KeV $\bar{\nu}_j$, typically 20° − 30°. To restrict to a kinematic region where this angular separation is maximal, we impose a lower bound on T_e so that the full cut is 0.5 KeV $< T_e < 1.6$ KeV. Our MC study shows that this angular separation is also not significantly weakened by the random gas velocity effects. We thus set the cut $|180° - \theta_{ei}| < 12°$. We find that these cuts on T_e, t_i, and θ_{ei} suppress the observed events from the dominant decay mode by a factor of about 10^{-3} while retaining about half of the events from a decay yielding a 17 KeV $\bar{\nu}_j$. We have also studied sources of backgrounds and accidentals. The main source may be adsorption of T_2 molecules on the inner walls of the vacuum chamber.

RESULTS AND DISCUSSION

In Fig. 1 we show a histogram, from the MC simulation, of the reconstructed values of $m(\nu_i)^2$ for T_2 decay, assuming decay into a 17 KeV $\bar{\nu}_j$ with mixing strength $|U_{1j}|^2 = 0.01$ as well as the dominant mode. One sees that the peak at $(17)^2$ KeV^2 is reasonably well sepa-

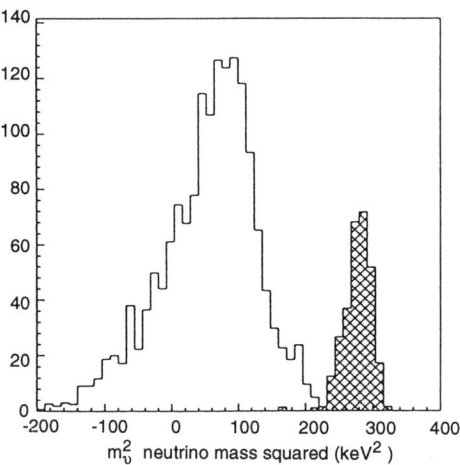

FIG. 1. Histogram from MC simulation of reconstructed values of squared neutrino mass in the T_2 experiment for a sample of 2×10^6 decays (before cuts). For clarity, the events with $m(\nu_j) = 17$ keV are cross-hatched. (Note that the cuts used to isolate the latter events skew the distribution from the dominant decay slightly.)

rated from the peak due to the dominant decay mode. With our parameters, the decay rate folded with the solid angle acceptance and cuts gives about 10 events/day for ^{3}H decay to a 17 KeV $\bar{\nu}_j$ with $|U_{1j}|^2 = 0.01$, and about 90 events/day for the dominant ^{3}H decay. Thus, our study suggests that although certain backgrounds need further study, if these can be controlled, then this type of experiment with ^{3}H using a T_2 molecular beam should be able to detect a hypothetical 17 KeV $\bar{\nu}_j$ emitted in the decay with $|U_{1i}|^2 = 0.01$ if it exists.[10]

OTHER NUCLEI

We have also considered a possible experiment with atomic T. Here one would use an rf discharge to dissociate the T_2 molecules in the initial beam. This has the advantage of a simple final state ion, $(^3\text{He})^+$. However, there would be some small contamination of molecular T_2 (about 10 - 15 % for the rf discharge) and this would give rise to a smooth background of fake massive neutrino events owing to the misinterpretation of the recoil ion as $^3\text{He}^+$ when in fact it was $(^3\text{H}\,^3\text{He})^+$. In view of this, it appears that atomic T is not as promising as T_2.

Finally, we have also studied the feasibility of a recoil experiment with ^{35}S. While a hypothetical decay to a 17 KeV $\bar{\nu}_j$ decay mode is less suppressed by phase-space in this case, this is more than compensated for by the loss of kinematic sensitivity. There is very little change in the recoil ion velocity and hence time-of-flight for the 17 KeV versus the massless neutrino modes, and the typical angular separation of the events corresponding to these modes is much smaller: typically $2° - 3°$ instead of $20° - 30°$ as in the T_2 case. We find that in this case the experiment would not be sufficiently sensitive to detect a 17 KeV neutrino signal at the level of 1 % admixture.

SEARCH FOR GENERAL $m(\nu_j)$

Just as was the case with the original application of the kink search test[5], a recoil experiment can be used to search for and set limits on possible admixed neutrinos over a wide range of masses $m(\nu_j)$. Hence, this method has an interest which is independent of the possible 17 KeV neutrino. For a nuclear beta decay with a given Q value, there are two opposite effects: (i) as $m(\nu_j)$ approaches Q, a decay $(Z,A) \to (Z \pm 1, A) + e^{\mp} + \bar{\nu}_j(\nu_j)$ is more strongly phase-space suppressed; (ii) for small $m(\nu_j)$, the recoil method loses sensitivity. Our results show that a recoil experiment is sensitive to values of $m(\nu_j)$ in a window such that $(Q - m(\nu_j))/Q$ is small. Hence, by using several convenient beta sources with a variety of different Q values ranging up to a few MeV, it should be possible to cover a correspondingly large range of $m(\nu_j)$.

LIMITS ON $|\Delta L| = 2$ DECAYS

Generic neutrino masses lead to the violation of total lepton number L, as well as family lepton number. Indeed, L violation is a common feature of schemes for new physics beyond the standard model. The $|\Delta L| = 2$ decay $K^+ \to \pi^- \mu^+ \mu^+$ had not previously been searched for. Accordingly, we carried out a retroactive analysis of old data and have set the 90 % confidence level upper limit[2]

$$BR(K^+ \to \pi^- \mu^+ \mu^+) < 1.5 \times 10^{-4} \quad (1)$$

This bound can be greatly improved via an analysis of auxiliary data obtained by the BNL rare K decay experiment E787, whose primary goal is to measure $K^+ \to \pi^+ \nu \bar{\nu}$. We have also suggested[2] that it could be worthwhile for neutral K experiments to set limits on $K_L^0 \to \{\pi^- \pi^- \ell^+ \ell'^+ \text{ or } \pi^+ \pi^+ \ell^- \ell'^-\}$ where $\ell\ell' = ee, \mu e$ or $\mu\mu$. This is now being done by E799 at FNAL.[11]

In addition to neutrinoless double beta decay, there are a number of $|\Delta L| = 2$ baryon decays for which no search had been conducted. From another retroactive data analysis, we have set the 90 % CL bound[3]

$$BR(\Xi^- \to p\mu^-\mu^-) < 3.7 \times 10^{-4} \quad (2)$$

It would also be worthwhile to search for analogous decays involving heavy-quark baryons such as $\Xi_c^+ \to \Xi^- \mu^+ \mu^+$.

ACKNOWLEDGEMENTS

I would like to thank G. Finocchiaro and L. Littenberg for collaboration on the work discussed here.

REFERENCES

1. G. Finocchiaro and R. E. Shrock, Phys. Rev. **D46** (Rapid Commun.) R888 (1992) and work in progress.

2. L. Littenberg and R. E. Shrock, Phys. Rev. Lett. **68**, 443 (1992).

3. L. Littenberg and R. E. Shrock, Phys. Rev. **D46** (Rapid Commun.), R892 (1992).

4. Z. Maki, N. Nakagawa, and S. Sakata, Prog. Theor. Phys. **28**, 870 (1962); M. Nakagawa, H. Okonagi, S. Sakata, and A. Toyoda, ibid., **30**, 727 (1963).

5. R. E. Shrock, Phys. Lett. **96B**, 159 (1980).

6. R. E. Shrock, Phys. Rev. **D24**, 1232, 1275 (1981).

7. e.g. J. J. Simpson and A. Hime, Phys. Rev. **D39**, 1825 (1989); A. Hime and J. J. Simpson, Phys. Rev. **D39**, 1837 (1989); A. Hime and N. A. Jelley, Phys. Lett. **257B**, 441 (1991); B. Sur et al., Phys. Rev. Lett. **66**, 2444 (1991).

8. e.g., J. Markey and F. Boehm, Phys. Rev. **C32**, 2215 (1985); D. W. Hetherington et al., Phys. Rev. **C36**, 1504 (1987); M. Bahran and G. R. Kalbfleisch, Oklahoma preprint OKHEP-91-005; H. Kawakami, S. Kato, T. Ohshima, et al., KEK-INS preprint 92-14/INS-Rep-921; M. Chen, D. Imel, T. Radcliffe, H. Henrickson, and F. Boehm, Caltech preprint CALT-63-638; S. Freedman and (for ^{55}Fe IBEC) E. Norman, talks at this conference.

9. For an early beta decay experiment in which the recoil nucleus and electron were both measured (with the source being a solid film of ^{32}P, not a gas jet), see C. Sherwin, Phys. Rev. **75**, 1799 (1949). I thank S. D. Drell for informing me of this work.

10. A brief report of our results was given by R. S. at the Berkeley Workshop on the 17 KeV Neutrino Question (Dec., 1991). A similar study of the tritium experiment by J. Jaros and M. Swartz was reported at this workshop, with results in agreement with ours.

11. Y. Wah, private communication, this conference.

RECENT DOUBLE BETA DECAY RESULTS*

A. Balysh [b], M. Beck [a], S.T. Belyaev [b,1)], F. Bensch [a], J. Bockholt [a], A. Demehin [b],
A. Gurov [b], G. Heusser [a], M. Hirsch [a], H.V. Klapdor-Kleingrothaus [a,1)],
I. Kondratenko [b], V.I. Lebedev [b], B. Maier [a], A. Müller [c], F. Petry [a],
A. Piepke [a], H. Strecker [a], M. Völlinger [a], K. Zuber [a]

[a] Max-Planck-Institut für Kernphysik, W-6900 Heidelberg, Germany
[b] Kurchatov Institute, 123 182 Moscow, Russia
[c] Istituto Nazionale di Fisica Nucleare LNGS, 67010 Assergi, Italy

Abstract

The status and recent results of second generation ββ-experiments using isotopically enriched source materials are described. These experiments are at present the most sensitive tools to distinguish Dirac from Majorana neutrinos. The at present most advanced experimental techniques, namely the use of high-resolution calorimetric detectors and of time projection chambers are compared. New limits on the Majorana neutrino mass as well as for the Majoron-neutrino coupling are presented.

INTRODUCTION

Double beta (ββ) decay is the rarest known decay mode in nature. The interest in the investigation of this process is motivated by its unique sensitivity to the existence of a non-zero Majorana-mass of the neutrinos and right-handed weak currents (RHC).
Non-zero Majorana-masses of the neutrinos are generated in non-standard models (SM) by a breaking of the B-L (Baryon- minus Lepton-number symmetry), giving rise to neutrinoless ββ-decay. On the other hand the predictive power of these theories is very limited, resulting in a giant mass range for the Majorana neutrino mass from few eV to 10^{-11} eV (see e.g. [1)]).
Second generation ββ-experiments, either planned or already in operation, using isotopically enriched source materials, are able to test the very upper range of the above mass scale.

*) Presented by: A. Piepke
1) Speakers of the collaboration

Candidates for the observation of ββ-decay are those even-even nuclides where single beta decay is energetically forbidden but a decay to the next nearest isobar is allowed. Three possible decay modes are discussed:

(ββ2ν) $A(Z,N) \rightarrow A(Z+2, N-2) + 2e^- + 2\bar{\nu}_e$
(ββ0ν) $A(Z,N) \rightarrow A(Z+2, N-2) + 2e^-$
(ββ0νχ) $A(Z,N) \rightarrow A(Z+2, N-2) + 2e^- + \chi$

The ββ2ν-decay is allowed in the SM and can be calculated in second order Fermi theory. Its decay rate is given by:

$$(T_{1/2}^{2\nu})^{-1} = F^{2\nu} \cdot |M^{2\nu}|^2 \qquad (1)$$

$F^{2\nu}$ = leptonic phase space, $M^{2\nu}$ = nuclear matrix-element. The decay rate does not depend on any unknown particle physics parameter.
In the case of the non-SM ββ0ν-decay, the decay rate is, when neglecting contributions from RHC, proportional to the neutrino mass squared:

$$(T_{1/2}^{0\nu})^{-1} = F^{0\nu} \cdot |M^{0\nu}|^2 \cdot \left(\frac{<m_\nu>}{m_e}\right)^2 \quad (2)$$

$F^{0\nu}$ = leptonic phase space, $M^{0\nu}$ = nuclear matrix-element, m_e = electron mass and $<m_\nu>$ = effective Majorana mass of the neutrinos. The effective mass is defined as:

$$<m_\nu> = \sum \omega_i \cdot U_{ei}^2 \cdot m_i \quad (3)$$

ω_i = CP eigenvalue ± 1, U_{ei} = element of the unitary matrix describing the mixing of the mass and flavour eigenstates, m_i = mass of the mass eigenstate i.

In order to extract information on $<m_\nu>$ from a measured decay rate or from a limit of the rate, according to (2) it is necessary to use theoretically calculated nuclear matrix-elements (see ref. [2]). The reliability of such calculations has been verified and discussed recently [3]. Note that $F^{0\nu}/F^{2\nu} \sim 5 \cdot 10^6$ for ^{76}Ge.

If the neutrinoless ββ-decay is the manifestation of the spontaneous breaking of a global B-L symmetry than a massless Goldstone boson the so-called Majoron should exist. The ββ-decay rate would then be proportional to the effective neutrino-Majoron coupling $<g_{\nu\chi}>$ squared:

$$(T_{1/2}^{0\nu\chi})^{-1} = F^{0\nu\chi} \cdot |M^{0\nu}|^2 \cdot <g_{\nu\chi}>^2 \quad (4)$$

The existence of singlet Majorons is not in conflict with the measured Z^0-width [4] and would have consequences in astrophysics, cosmology as well as for possible decay channels of the 17 keV neutrinos.

The different decay modes can be identified through the different sum-energy spectra of the emitted electrons. The ββ2ν- and ββ0νχ-decay are leading to continuous spectra (with different maxima) while the ββ0ν-decay results in a peak at the decay energy. Limits on the neutrino mass obtained from ββ0ν-experiments are complementary to those obtained from the investigation of the shape of the β-spectrum -measuring the masses and admixtures directly- and neutrino oscillation experiments -testing the mass differences squared versus the admixture. Both other methods are not able to distinguish Dirac from Majorana neutrinos if a possible anticorrelation of the solar-neutrino flux to the sunspot number and hence magnetic moments of the neutrinos are neglected.

Neutrinoless ββ-decay is also sensitive to the existence of 17 keV (Majorana) neutrinos (see eq. 3). The fact, that no corresponding ββ0ν-amplitude is observed requires at least a very rigid pattern of neutrino masses and mixtures [5].

EXPERIMENTAL PROCEDURE

The common problem of all ββ-experiments are the extremely large half lives of 10^{20} y or even longer and hence the small counting rates to be identified. Among the direct experiments, detecting the emitted electrons and not the decay products, as geochemical experiments, those using the radioactive ββ-source simultaneously as detector have at present the highest sensitivity. In this way selfabsorption losses in the source can be avoided, but on the other hand the choice of the isotope to be investigated is limited. There are at present more than 30 ββ-experiments underway.

The following discussion will be limited to those two experiments giving the most stringent results, as far as the non-SM decay modes are concerned.

The Heidelberg-Moscow (HD-MO) collaboration uses Ge semiconductor detectors made from isotopically enriched Ge. The ^{76}Ge abundance is 86 % compared to 7.8 % in natural Ge. These detectors are high-resolution ($\Delta E/E_0 \sim 2 \cdot 10^{-3}$) calorimeters.

Since the Q-value of the ^{76}Ge ββ-decay at $E_0 = 2038.56 \pm 0.32$ keV [6] is within the energy range of the natural radioactivity the background reduction is a difficult problem.

The detectors are made from carefully selected low-activity materials. To shield them from external activities the detectors are operated in a common lead shield of 30 cm thickness. The inner 10 cm are made from very clean so-called LC2 grade Pb. All near-detector construction materials have specific activities below 10 µBq/kg !

The experiment is performed in the Gran Sasso underground laboratory of the INFN in Italy in a depth of 3500 m w.e. (water equivalent).

The Caltech-PSI-Neuchâtel (CPN) experiment which studies the decay of ^{136}Xe uses a TPC filled with a Xe+CH$_4$-counting gas. The ^{136}Xe abundance is 62.5 % compared to 8.9 % in natural Xe. The resolution at the endpoint energy of 2.48 MeV is 6.6 %. The detection efficiency for the electrons is ~25 %.

Since two-electron events can be discriminated from the one-electron background lower radiopurities of the used materials can be compensated by the tracking capabilities. The TPC is made from selected materials and shielded by 5 cm Cu and 20-30 cm Pb. The purity of the counting gas can be maintained on the 0.1 ppm level. The experiment is being performed at the Gotthard underground lab in Switzerland with ~3000 m w.e. shielding thickness (see ref. [7]).

RESULTS

Heidelberg-Moscow

The integral counting rates of the used Ge detectors as well as the backgrounds around 2.04 MeV where the hypothetical ββ0ν-peak would be expected are summarized in table 1 for each detector.

Table 1. Measuring times and backgrounds of the different enriched detectors. A fourth enriched detector with a mass of 2.9 kg (~33 mol of ^{76}Ge) is currently under construction and already working in a test cryostat. It will be installed in the Gran Sasso lab in 1993.

Det.	m_a (kg)	t (d)	Background (c/keV·y·kg)	
			B_1	B_2
1	0.92	404.3	9.8	0.56
2	2.76	238.7	6.1	0.29
3	2.32	15.7	7.4	0.20
1+2+3	6.00	-	7.4	0.38

m_a = active mass, t = measuring time, B_1 = 100-2800 keV, B_2 = 2000-2100 keV

ββ2ν-decay

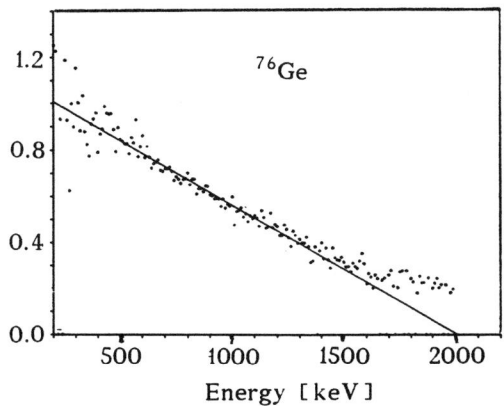

Fig. 1 Kurie plot of the ββ2ν-spectrum. $y=(n(E)/E \cdot [E^4+10 \cdot E^3+40 \cdot E^2+60 \cdot E+30])^{1/5}$ was taken as ordinate, where E=energy in units of the electron mass and n(E)=number of counts. The binwidth is 10 keV/channel.

For the analysis of the ββ2ν-decay, data taken during 223 days with det.#2, before the installation of det.#3 have been used. The exposure is 19.3 mol·y.

Since the continuous ββ2ν-spectrum is superimposed to the Compton-continua of the different background peaks and to continuous background components their contributions have to be unfolded from the data. For this purpose a Monte Carlo background model based on the CERN code GEANT 3 has been developed. The details of this model will be discussed elsewhere [8].

On the basis of this background model the signal-to-background ratio can be estimated to be 1.20 for the energy interval from 500 to 1500 keV, containing 73 % of the ββ2ν-intensity.

A maximum likelihood fit of the background model added to calculated ββ2ν-spectra yields a best fit of $T_{1/2} = (1.50 \pm 0.04 \text{stat}) \cdot 10^{21}$ y. The statistical error has been calculated from the logarithmic likelihood ratio. Since the systematic error has not been estimated yet, the result is still preliminary.

Fig. 1 shows the residual spectrum after subtraction of the background model. The spectrum has been linearized like in a Kurie plot, but the theoretical shape of the ββ2ν-decay from [9] has been used. Linear regression yields for the energy interval 500-1500 keV an endpoint of $E_0 = (2013 \pm 31)$ keV, showing that the continuum present in the data has indeed the shape and endpoint expected for the ββ2ν-decay.

If this is interpreted as evidence for the ββ2ν-decay, the experimental matrix-element shows remarkable agreement to the theoretical of ref. [2] $|M^{2\nu}_{\text{exp}} / M^{2\nu}_{\text{theo}}| = 1.41$.

ββ0ν-decay

Fig. 2 shows the measured spectrum around 2.04 MeV. The measuring time is 33.5 mol·y or 2.92 kg·y. No peak at the correct energy can be identified. Assuming that the ββ0ν-decay is a Poisson-process superimposed to background 7.8 (4.7) ββ-events are excluded with 90(68) % c.l. using the method recommended by the PDG. The corresponding half life limits are:

$T^{0\nu}_{1/2} > 1.8(3.0) \cdot 10^{24}$ y with 90(68) % c.l..

If the Poisson-distribution is approximated by the Gauß-distribution, a squareroot estimate yields $T^{0\nu}_{1/2} > 2.8 \cdot 10^{24}$ y with 90 % c.l.. The obtained half life limit results in new upper limits for $<m_\nu>$, $<\lambda>$ and $<\eta>$ (effective right-right- and left-right-handed admixtures to the weak interaction, respectively). $<m_\nu> < 1.33$ eV, $<\eta> < 1.29 \cdot 10^{-8}$ and $<\lambda> < 2.06 \cdot 10^{-6}$. If RHC are neglected we find: $<m_\nu> < 1.14$ eV.

All limits with 90 % c.l. are derived from the Poisson estimate using the matrix-elements of [2].

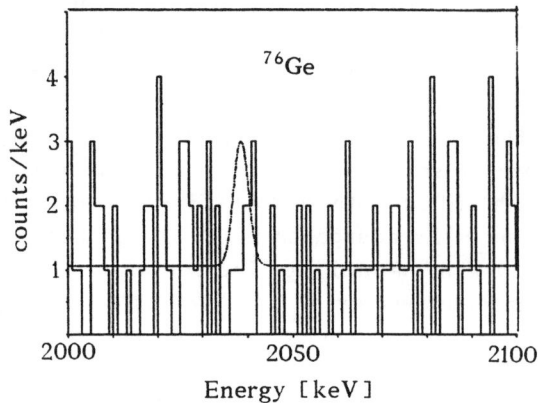

Fig. 2 Measured spectrum around the hypothetical ββ0ν-peak. The dotted curve represents the signal excluded with 90 % c.l..

ββ0νχ-decay

For the analysis of the ββ0νχ-decay the same piece of data as for the ββ2ν-spectrum has been used. If the ββ2ν-decay as well as the discussed background model is subtracted from the data, the ββ0νχ-spectrum should remain.

In the energy interval from 1.1 MeV to 2.05 MeV, corresponding to 71.5 % of the ββ0νχ-intensity, 156 events are contained. This number is on the 1.38·σ level greater than zero. However the analysis of the spectral shape shows, that these extra events can hardly be interpreted as evidence for ββ0νχ-decay. The extra events are also visible in fig. 1 as a deviation of the points above 1.5 MeV from the linear behaviour. From the data a lower limit for the half life of $T^{0\nu\chi}_{1/2} > 3.89(4.08) \cdot 10^{22}$ y with 90(68) % c.l. is estimated.

Using the matrix-element of ref. [2] and the phase space integral of [10] this $T_{1/2}$-value yields a new upper limit for the neutrino-

Majoron coupling of $\langle g_{\nu\chi} \rangle < 1.1 \cdot 10^{-4}$ (90 % c.l.).

Caltech-PSI-Neuchâtel

ββ0ν-decay

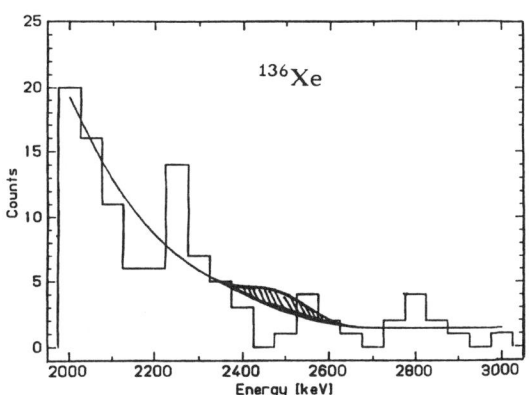

Fig. 3 Spectrum of the two-electron events around the ββ0ν-energy together with the excluded signal with 90 % c.l.. The binwidth is 50 keV/channel.

Fig. 3 depicts the energy spectrum of the two-electron candidates around the ββ-decay energy, measured during 20.7 mol·y. The solid curve, representing the 90 % limit, has been calculated assuming Poisson statistics. Since no peak is present 5.21 (2.76) ββ-events are excluded with 90 (68) % c.l.. A non-zero neutrino mass and RHC are corresponding to different efficiencies resulting in two different half lives beeing derived:

m_ν: $T_{1/2}^{0\nu} > 4.2 (7.8) \cdot 10^{23}$ y with 90(68)% c.l.

RHC: $T_{1/2}^{0\nu} > 2.8 (5.0) \cdot 10^{23}$ y with 90(68)% c.l.

Using once more the matrix-elemets of ref. [2] the following limits (90 % c.l.) can be deduced: $\langle m_\nu \rangle < 2.30$ eV with $\langle \eta \rangle = \langle \lambda \rangle = 0$ and $\langle \eta \rangle < 2.29 \cdot 10^{-8}$, $\langle \lambda \rangle < 4.21 \cdot 10^{-6}$ for RHC Due to the steep rise of the background below 2.2 MeV no results for the ββ2ν-mode are reported [7].

ββ0νχ-decay

Also this experiment at present does not show evidence for a ββ0νχ-decay of ^{136}Xe. The 77 events contained in the energy interval from 2000 to 2550 keV, corresponding to 27 % of the ββ0νχ-intensity, are converted to a half life limit of $7.2 \cdot 10^{21}$ y with 90 % c.l.. Using the same matrix-element and phase space as before a Majoron-neutrino coupling larger than $1.8 \cdot 10^{-4}$ can be excluded.

DISCUSSION

How do the discussed technical parameters and results compare to the other large scale experiments using enriched isotopes ?

Table 2. Technical parameters of large scale ββ-experiments using enriched isotopes.

Exp.		N (mol)	ΔE (keV)	B $\left(\frac{\text{counts}}{\text{keV}\cdot\text{y}\cdot\text{mol}}\right)$	S
CPN[7]	^{136}Xe	26.6	164.	0.003	1.9
Milano[11]	^{136}Xe	20.7	124.	2.2	0.1
ITEP–Y[12]	^{76}Ge	13.3	3.7	0.19	4.4
HD-MO	^{76}Ge	68.7	3.5	0.033	23.0

N=amount of decaying isotope, ΔE=energy resolution at the decay energy, B=background at the decay energy per mol of decaying isotope, S=sensitivity defined as $S = \sqrt{\frac{N}{\Delta E \cdot B}} \cdot \varepsilon$ with ε being the efficiency for the detection of electrons.

It should be mentioned, that the ITEP-Yerevan group was the first one to use a large quantity of an enriched isotope in a ββ-experiment. They also gave the first evidence for the ββ2ν-decay of ^{76}Ge ($T_{1/2}^{2\nu} = 9 \cdot 10^{20}$ y) [13]. The sensitivity of the experiments towards a non-zero Majorana neutrino mass can, in the case of a background limited null experiment, be measured through the achievable

$T^{0\nu}_{1/2}$- limit. It is, independent from the isotope, determined by:

$$T^{0\nu}_{1/2} > (3.17 \cdot 10^{23} \text{mol}^{-1}) \cdot S \cdot \sqrt{t} / f \quad (5)$$

t = measuring time (y), f = factor depending on the c.l. of the limit.

From table 2 it is clear, that the lower background of the TPC is compensated by the much higher efficiency and better energy resolution of the Ge detectors.

Due to the $\sqrt{t}$-dependence it is impossible to balance the lower sensitivity by a longer measuring time. In case of zero background the limit will be linear in time and source strength but it seems to be very difficult to expect zero background during many years of experiment.

Table 3. Exposures and limits for non-SM parameters obtained by the discussed $\beta\beta$-decay experiments.

Exp.	N·t	$T^{0\nu}_{1/2}$	$\langle m_\nu \rangle$	$T^{0\nu\chi}_{1/2}$	$\langle g_{\nu\chi} \rangle$
	(mol·y)	(10^{24}y)	(eV)	(10^{22}y)	(10^{-4})
CPN	20.7	0.42	2.3	0.72	1.8
Milano	14.6	0.02	10.5	-	-
ITEP-Y	14.5	1.0	1.5	1.0	2.2
HD-MO	33.5	1.8	1.2	3.9	1.1

None of these large source strength experiments shows at present evidence for physics beyond the standard model.

When comparing the limits summarized in table 3 to results obtained with other isotopes, one should bear in mind, that the neutrino mass as well as the neutrino-Majoron coupling are entering quadratically into the decay rate. This is especially important for the interpretation of indications for the existence of the $\beta\beta0\nu\chi$-decay, being discussed during the conference.

Using the measured experimental parameters the potential within 5 years of experiment can be estimated to be $\langle m_\nu \rangle < 0.8$ eV for the CPN experiment and $\langle m_\nu \rangle < 0.2$ eV for the HD-MO experiment in its full scale.

The Ge experiments are still hard to beat, at least in their domain - the $\beta\beta0\nu$-decay. Nevertheless there are several very promising new experiments for other $\beta\beta$-emitters in preparation:

The NEMO-collaboration is planning to use 10 kg of Mo enriched to 99 % in ^{100}Mo in multiwire drift tubes operated in the Geiger mode[14]. They hope to achieve a sensitivity of ~0.2 eV.

The other challenge of the semiconductor detectors are the cryogenic detectors of the Milano group (see this volume) which have now reached considerable size and energy resolution, approaching those of the Ge experiments.

The investigation of different isotopes is of big importance in order to avoid systematic errors due to our limited knowledge about the precision of the theoretically calculated nuclear matrix-elements, even if these experiments are not reaching ultimate sensitivities.

Several other interesting applications of the low-background detectors available in $\beta\beta$-experiments as: limits on heavy neutrinos with masses ~ 10^4 GeV, stability of the electron, dark matter searches, limits for the exchange of super-symmetric particles could not be discussed in this article.

$\beta\beta$-decay research is a living field of physics which allows us to look behind the SM without using large accelerators.

The use of isotopically enriched source materials has opened a new order of magnitude for this type of observational physics. The two at present most sophisticated, working experiments can probe for the first time the sub-eV range of the Majorana mass of the neutrinos.

REFERENCES

1. P. Langacker, *Neutrinos*, edited by H.V. Klapdor, Springer Verlag Berlin, Heidelberg, New York, 1988, 71
2. A. Staudt, K. Muto and H.V. Klapdor-Kleingrothaus, Europhys. Lett. 13, 1990, 31
3. M. Hirsch, X.R. Wu, H.V. Klapdor-Kleingrothaus, Ching Cheng-rui and Ho Tso-hsiu, Phys. Rep., in press
4. Z.G. Berezhiani, A.Yu. Smirnov and J.W.F. Valle, Phys. Lett. B 291, 1992, 99
5. D.O. Caldwell and P. Langacker, Phys. Rev. D 44, 1991, 823
   D.O. Caldwell, Phys. Lett. B 289, 1992, 389
6. J.G. Hykawy et al., Phys. Rev. Lett. 67, 1991, 1708
7. H.T. Wong et al., Phys. Rev. Lett. 67, 1991, 1218
   J.-L. Vuilleumier, private communication
8. Heidelberg-Moscow collaboration, in preparation
9. E.J. Konopinski, *The Theory of Beta Radioactivity*, Oxford at the Clarendon Press, 1966, 257
10. M. Doi, T. Kotani and E. Takasugi, Phys. Rev. D 37, 1988, 2575
11. E. Bellotti et al., Phys. Lett. B 266, 1991, 193
12. I.V. Kirpichnikov, Nucl. Phys. B (Proc. Suppl.), 28A, 1992, 210
13. A.A. Vasenko et al., Mod. Phys. Lett. A 5, 1990, 1299
14. R. Arnold et al., Nucl. Phys. B (Proc. Suppl.), 28A, 1992, 223

THE DOUBLE BETA DECAY SPECTRA OF ^{82}Se, ^{100}Mo, AND ^{150}Nd *

M.A. Nelson, M.K. Moe and M.A. Vient
Department of Physics, University of California, Irvine, CA 92717, USA

S.R. Elliott
Lawrence Livermore National Laboratory, Livermore, CA 94550, USA

Abstract

The double beta decay electron energy spectra of ^{82}Se, ^{100}Mo, and ^{150}Nd have been measured with a time projection chamber, and departures from the expected two-neutrino spectral shapes have been observed. Efforts to reduce possible background contamination have been made, and tests are now being done in an effort to determine whether the anomalous signals are real effects, or simply experimental artifacts.

INTRODUCTION

During the past several years, the U.C. Irvine (UCI) group has been studying the double beta decay of several isotopes, and has accumulated enough statistics to produce good spectra of the sum of the energies of the two decay electrons. Standard electroweak theory describes the shape of the sum energy spectrum if double beta decay proceeds according to the standard model, and extensions to this theory predict that measurements of the spectral shape should be very sensitive to non-standard phenomena such as the existence of massive Majorana neutrinos, lepton number non-conservation, right-handed weak couplings, and singlet Majorons. The UCI group has observed significant departures from the spectral shape predicted by the standard model, and although these anomalous signals could be considered evidence for exciting new physics, such claims are not being put forth. Instead, efforts are being directed toward further reducing background levels in the experiment.

Double beta decay theory has been discussed in detail in the literature,[1-3] and will not be covered here. The relevant modes of double beta decay for this discussion are two-neutrino decay ($\beta\beta_{2\nu}$), where two neutrinos accompany the decay electrons; neutrinoless decay ($\beta\beta_{0\nu}$), which has only the two electrons in the final state, and neutrinoless decay with Majoron emission ($\beta\beta_{0\nu,M}$), where the two decay neutrinos couple to a massless boson, leaving a three-body final state. The characteristic spectral shapes of these three modes are distinct (Figure 1), so double beta decay experiments are very sensitive in distinguishing between the signatures of these decay modes.

THE DETECTOR

A time projection chamber (TPC) is be-

---

*Supported by the U.S. Department of Energy, Grant No. DE-FG03-91ER40679.

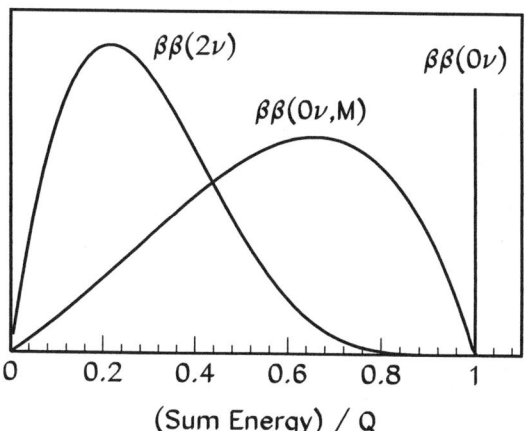

Figure 1. Spectral shapes for electron sum energies in $\beta\beta_{2\nu}$, $\beta\beta_{0\nu}$, and $\beta\beta_{0\nu,M}$ double beta decay.

ing utilized for the study of the emitted electrons in double beta decay. The TPC volume is defined by a Lexan box 80 cm square and 20 cm deep, and is divided into two 10 cm drift regions by a thin aluminized mylar film sandwich, which also serves as a substrate for the isotope deposit. Each drift region of the TPC is instrumented with anode and cathode wire planes, with the anode wires perpendicular to the cathodes. This arrangement provides two orthogonal views of the electrons' ionization tracks as they drift into the wires. The TPC is surrounded by lead shielding and a 4π wire chamber cosmic ray veto. This entire assembly is located between two large Helmholtz coils, which provide the magnetic field required to produce helical electron tracks. All of the materials of construction were chosen for their low radioactivity. A more detailed description of the Irvine TPC may be found elsewhere.[4]

The TPC allows for reconstruction of the emitted beta particle tracks, which provides a simple means for characterization of the events. Two-electron events (such as double beta decay) are qualitatively distinct from other types of events, such as electron-positron pair or an electron passing through the volume of the detector, and this event characterization is the primary technique for background rejection. Events accepted as double beta decays must have two clean electron tracks on opposite sides of the source plane, and must share a common origin on the source. These good candidate event tracks are fitted with helices, which allows for calculation of the energies of the electrons, as well as the opening angle between them.

The three isotopes which have been studied in the Irvine TPC are ^{82}Se, ^{100}Mo, and ^{150}Nd. Two-electron backgrounds caused by the beta decay of ^{214}Bi followed by internal conversion of the ^{214}Po daughter (β-IC) can be identified by the subsequent α decay of ^{214}Po. The detection of this α decay dictated that the source deposits be very thin, and the total mass of the ^{82}Se, ^{100}Mo, and ^{150}Nd sources were 13.4 g, 5.5 g, and 11.2 g, respectively. Although the source materials were highly enriched in the desired isotope (97% elemental ^{82}Se, 97% ^{100}MoO$_3$, and 91% ^{150}Nd$_2$O$_3$), the small source masses and long half-lives demanded very long counting times in order to produce good spectra.

^{82}Se RESULTS

The first isotope studied at Irvine was ^{82}Se. The data were obtained in several distinct stages. The first run had insufficient shielding, and although it's background levels were too high to allow for a good determination of the $\beta\beta_{2\nu}$ half-life, it was possible to include these data in $\beta\beta_{0\nu}$ half-life limit. The shielding was then increased, and, later, the chamber was rebuilt using lower-activity materials. In these subsequent runs, the background was much lower, which permitted the study of $\beta\beta_{2\nu}$ and the search for $\beta\beta_{2\nu,M}$. The

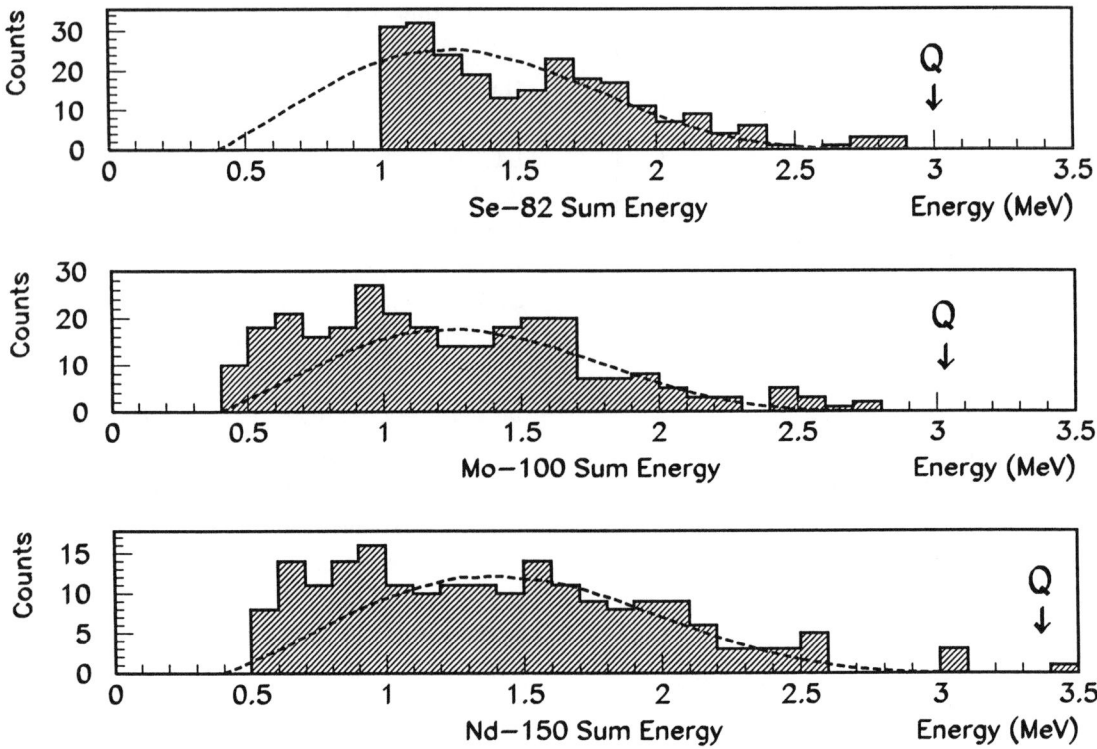

Figure 2. Measured sum energy spectra for two electron events accepted as double beta decays for ^{82}Se (top), ^{100}Mo (center), and ^{150}Nd (bottom). $Q_{\beta\beta}$-values are 3.00, 3.03, and 3.37 MeV, respectively. A 200 keV threshold was imposed on each electron. Energy asymmetry and conversion line energy cuts were imposed to suppress contribution due to Møller scattering and ^{208}Tl β-IC events, eliminating 4 events, all of them in ^{82}Se. The dashed lines are spectra derived from Monte Carlo studies. The low energy excesses are due to Møller scattering and untagged ^{214}Pb β-IC events; in the case of ^{82}Se, this excess was so severe that a 1.0 MeV sum energy cut has been imposed.

final $\beta\beta_{0\nu}$ half-life limit for ^{82}Se is 2.7×10^{22} years (68% CL), and was derived from all of the available data (21 924 hours of livetime), and in the final $\beta\beta_{2\nu}$ half life calculation of $(1.08^{+0.26}_{-0.06}) \times 10^{20}$ years (68% CL), the first high background run was excluded (20 244 hours of livetime).[5]

The $\beta\beta_{2\nu}$ spectrum obtained from these data is shown in Figure 2. In addition to the expected two-neutrino spectrum, there is a significant excess of events in the 2.5 MeV to 3.0 MeV region. Monte Carlo studies predict 1.5 events in this region, while 7 have been observed. The Poisson probability of a statistical fluctuation this large is 7.6×10^{-4}. Since this excess falls in the part of the spectrum which could be populated by $\beta\beta_{0\nu,M}$ decay, these counts, if not due to an unrecognized background process, could be very interesting. Background, however, has always been a problem with double beta decay experiments, so the first reaction was to attempt to identify these anomalous counts as being a background phenomenon.

The most obvious source of potential background is the cosmic ray muon flux. In order to reduce the number of muons passing through the chamber, the experiment was moved from the Irvine campus to an underground site in a tunnel at Hoover Dam in

Arizona. This site provides a minimum of 72 m of rock overburden, and the muon rate in the chamber is down from 40 s^{-1} at the UCI site, to 0.3 s^{-1}. The last 4112 hours of run time were at this location, and 1 of the 7 anomalous events was recorded during this time. The anomalous event rate above and below ground remained constant, within statistics, in spite of two orders of magnitude reduction of cosmic ray flux. This suggests that the anomalous events are probably not due to cosmic rays.

^{100}Mo RESULTS

If the anomalous high energy events are due to a background, their rate should be independent of the $\beta\beta_{2\nu}$ half life. A source of ^{100}Mo, which has a shorter theoretical half life,[6,7] was prepared for the TPC. During 6327 hours of livetime with the new source, a relatively clean spectrum emerged (Figure 2). The measured $\beta\beta_{2\nu}$ half life is $(1.16^{+0.34}_{-0.08}) \times 10^{19}$ years (68% CL),[8] an order of magnitude faster than the ^{82}Se. Once again, there is an excess of events in the 2.5 MeV to 3.0 MeV region; Monte Carlo studies predict two $\beta\beta_{2\nu}$ events in this region, and 12 were observed. The Poisson probability for such a fluctuation is 2×10^{-6}. There were no events observed in the $\beta\beta_{0\nu}$ region at 3.0 MeV.

The energy spectrum of the single electrons (not members of pairs) from the ^{100}Mo source is very similar to that of the ^{82}Se at energies above 1 MeV, which is suggestive that high energy two-electron backgrounds due to trace radioactive contamination should be roughly the same for the two measurements. Since the anomalous event rate seen in the ^{100}Mo corresponds to roughly five times that observed in the ^{82}Se, it is inconsistent with what would be expected from a background process. This result hints that the anomalous event rate might be associated with the $\beta\beta$ process.

^{150}Nd RESULTS

In an attempt to gain insight to the origin of the mysterious high-energy events, a source of ^{150}Nd was prepared, and is currently being studied in the TPC. ^{150}Nd has a $Q_{\beta\beta}$-value higher than either the ^{82}Se or the ^{100}Mo (Figure 2), which helps push the $\beta\beta_{2\nu}$ signature above the energy of ^{214}Pb β-IC, the dominant background below 1 MeV. A preliminary analysis of the ^{150}Nd data indicates a $\beta\beta_{2\nu}$ half-life of $\approx 8 \times 10^{19}$ years. If the high energy anomalous events observed in ^{82}Se and ^{100}Mo are due to some unidentified background process, it should be safe to assume that the anomalous event energies should be independent of the $Q_{\beta\beta}$-value of the isotope being studied. During the first 2534 hours of data from the ^{150}Nd (Figure 2), high energy events were observed at 3 MeV and above, where none had been observed during more than 26 000 hours of running with other isotopes. This energy shift may be further evidence of a link between these high energy events and the $\beta\beta$ decay process.

CONCLUSIONS

If the high energy counts observed by this experiment are to be taken as evidence of $\beta\beta_{0\nu,M}$ decay, then values of Majoron coupling constant can be calculated. For the isotopes ^{82}Se, ^{100}Mo, and ^{150}Nd, this constant would be $(2.4 \pm 0.5) \times 10^{-4}$, $(4.2 \pm 0.9) \times 10^{-4}$, and $(2.1 \pm 0.6) \times 10^{-4}$, respectively. These values are somewhat larger than what would be allowed by results from other experiments with different isotopes.[9] In addition, the location of the anomalous signal is not in good agreement to the $\beta\beta_{0\nu,M}$ spectrum peak location, although this discrepancy could be accounted for by the TPC's poor energy resolution at high energies.

It is possible that the presence of anomalous high energy counts in the spectrum is due to a systematic upward shift of reconstructed energies. Analysis of ^{207}Bi conversion lines shows that the fitter has good linearity, and thus does not support this scenario. In addition, a uniform upward shift of energies would produce a continuous population of the spectrum at the high energies; the lack of counts just below the observed anomalous signal does not fit this model.

In the case of the ^{82}Se and ^{100}Mo, the anomalous signal could be consistent with with $\beta\beta_{0\nu}$ decay to the 2^+ level of the daughter. It is not clear, however, why this decay mode should be favored over the $\beta\beta_{0\nu}$ decay to the ground state, so this explanation is not very satisfying.

In summary, the UCI TPC experiment has observed anomalous signals in the double beta decay spectra of three different isotopes, despite efforts to identify them with some background process. Although these signals could be taken as evidence for new physics beyond the standard model, we prefer to assume they are spurious activity until all efforts to reduce background levels have been completed. In addition, the high-energy resolution of the detector is being improved through the use of a new track fitting algorithm, and a recent increase in magnetic field strength. Continuing data collection with these improvements is expected to shed new light on this persistent puzzle.

ACKNOWLEDGEMENTS

The authors would like to thank Frederick Reines for his continued encouragement and support, and the U. S. Department of the Interior for generously supporting our efforts at Hoover Dam.

REFERENCES

1. M. Doi, T. Kotani, E. Takasugi, "Double Beta Decay and Majorana Neutrino," *Prog. Theor. Phys. Supp.* 83, pp. 1-175, (1985).

2. W.C. Haxton, G.J. Stephenson Jr., "Double Beta Decay," *Prog. Part. Nucl. Phys.* 12, pp. 409-479, (1984).

3. T. Tomoda, "Double beta decay," *Rep. Prog. Phys.* 54, pp. 53-126, (1991).

4. S.R. Elliott, A.A. Hahn, M.K. Moe, "A Time Projection Chamber for Detection of Double Beta Decay," *Nucl. Instr. and Meth. in Phys. Res.* A273, pp. 226-239, (1988).

5. S.R. Elliott, A.A. Hahn, M.K. Moe, M.A. Nelson, M.A. Vient, "Double beta decay of ^{82}Se," *Phys. Rev. C,* 46 (4) (1992) (in print).

6. P. Vogel, "Nuclear Structure and Double Beta Decay," in *Nuclear Beta Decays and Neutrino: Proceedings of the International Symposium, Osaka, Japan,* June 1986, pp. 243-250.

7. A. Staudt, K. Muto and H.V. Klapdor-Kleingrothaus, "Calculation of 2ν and 0ν Double-Beta Decay Rates," *Europhys. Lett.* 13 (1), pp. 31-36, (1990).

8. S.R. Elliott, M.K. Moe, M.A. Nelson, M.A. Vient, "The Double Beta Decay Spectrum of ^{100}Mo as Measured with a TPC," *J. Phys. G: Nucl. Part. Phys.,* 17, pp. S145-S153, (1991).

9. A. Piepke, "Recent $\beta\beta$-Decay Results," presented at the XXVI International Conference on High Energy Physics, Dallas, Texas, August 6-12, 1992.

WHAT NEUTRINOLESS DOUBLE BETA DECAY WOULD TELL US ABOUT NEUTRINO MASS

Boris Kayser
Division of Physics, National Science Foundation
Washington, DC 20550 USA

Abstract

We identify several types of gauge theories, which together comprise a very broad range, in which the observation of neutrinoless double beta decay would imply a significant lower bound on neutrino mass. We explain why these gauge theories have this property.

INTRODUCTION

Neutrinoless double beta decay ($\beta\beta_{0\nu}$) is the process

$$\text{Nuc} \to \text{Nuc}' + 2e^-, \quad (1)$$

in which a nucleus "Nuc" decays to another "Nuc'" by emitting two electrons and nothing else. The observation of this manifestly lepton-number violating decay would imply that neutrinos are Majorana particles (that is, they are their own antiparticles). In addition, it would imply nonzero neutrino mass.[1] To see this, note that at the quark level, $\beta\beta_{0\nu}$ is the process $dd \to uue^-e^-$. If, for whatever reason, the amplitude for this process is nonzero, then, by crossing, so is that for the reaction $e^+\bar{u}d \to e^-u\bar{d}$. Combining the latter reaction with known weak processes, we have the chain $(\overline{\nu_e})_R \to e^+W^- \to e^+\bar{u}d \to e^-u\bar{d} \to e^-W^+ \to \nu_{eL}$. This chain is a "Majorana" mass term for ν_e. Of course, this mass term, being quite high order in the weak interactions, could be very tiny. However, as shall be explained, for a quite broad variety of gauge theories, the observation of $\beta\beta_{0\nu}$ would imply neutrino mass which is not tiny but, rather, is large enough to be observable in experiments beyond $\beta\beta_{0\nu}$.[2] This implied large mass would depend on the observed $\beta\beta_{0\nu}$ lifetime $\tau_{0\nu}$ as $\tau_{0\nu}^{-1/2}$. If, for example, $^{76}\text{Ge} \to {}^{76}\text{Se} + 2e^-$ were seen and found to have a lifetime $\tau_{0\nu}(\text{Ge})$, that would imply that the mass M_{Heaviest} of the heaviest neutrino satisfies the lower bound

$$M_{\text{Heaviest}} \gtrsim 1\,\text{eV}\sqrt{\frac{10^{24}\,\text{yr}}{\tau_{0\nu}(\text{Ge})}}. \quad (2)$$

Since the present limit on $\tau_{0\nu}(\text{Ge})$ is of order 10^{24} yr,[3] we see that observation of $^{76}\text{Ge} \to {}^{76}\text{Se} + 2e^-$ with a lifetime not much longer than the present limit would imply that at least one neutrino has a mass exceeding ~ 1 eV.

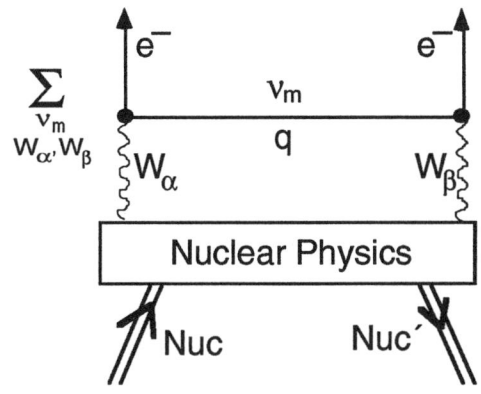

Figure 1. Neutrino exchange mechanism for $\beta\beta_{0\nu}$. The quantity q is the momentum transfer carried by the neutrino.

Now, why does the bound (2) hold, and in what theories does it hold? We assume, as usual, that if $\beta\beta_{0\nu}$ occurs, it results from the neutrino exchange diagram in Fig. 1. There, nuclear processes produce a pair of virtual W bosons, W_α and W_β, either of which may or may not be the W boson already discovered. Then, W_α and W_β exchange a neutrino ν_m to produce the outgoing electrons. It is convenient to work in a basis in which W_α and W_β are gauge eigenstates, while ν_m is a mass eigenstate. As indicated in Fig. 1, the $\beta\beta_{0\nu}$ amplitude $A^{\beta\beta_{0\nu}}$ is a sum over all the ν_m and all the $W_{\alpha,\beta}$ that may exist.

The currents acting at the leptonic vertices in Fig. 1 may both be left-handed (LH), or both right-handed (RH), or one may be LH and the other RH. Accordingly, we decompose $A^{\beta\beta_{0\nu}}$ into three pieces:

$$A^{\beta\beta_{0\nu}} = A_{LL} + A_{RR} + A_{LR} . \quad (3)$$

The couplings occurring at the leptonic vertices are described by the completely general phenomenological charged-current interaction

$$L_{cc} = \sum_\alpha W_\alpha^\mu \left(g_{\alpha L} \sum_{\substack{f=e,\mu,\tau \\ m}} \overline{\ell}_{fL} \gamma_\mu L_{fm}^{(\alpha)} \nu_{mL} \right.$$
$$\left. + g_{\alpha R} \sum_{\substack{f \\ m}} \overline{\ell}_{fR} \gamma_\mu R_{fm}^{(\alpha)} \nu_{mR} \right) + h.c. \quad (4)$$

Here, $g_{\alpha L}$ and $g_{\alpha R}$ are coupling constants, ℓ_e is the electron, ℓ_μ the muon, and so on, and the ν_m are the (Majorana) neutrino mass eigenstates. Finally, $L^{(\alpha)}$ and $R^{(\alpha)}$ are leptonic mixing matrices describing, respectively, the LH and RH currents coupled to the boson W_α.

In the gauge theory underlying the interaction (4), there will be LH gauge eigenstate neutrinos ψ_{iL} and RH ones ψ_{jR}. The lepton-number violation (L) required to produce $\beta\beta_{0\nu}$ will, in most theories, come from Majorana neutrino mass terms. These have the form

$$\overline{(\psi_{iL})^c} \psi_{kL} , \quad (5)$$

where χ^c is the "charge-conjugate" of χ. Since χ^c has handedness opposite to that of χ, a Majorana mass term, like any mass term, flips neutrino handedness. In addition, while a neutrino field absorbs "neutrinos" and creates "antineutrinos," the charge conjugate of one absorbs antineutrinos and creates neutrinos. Thus, from (5) we see that a Majorana mass term turns a neutrino into an antineutrino. (By contrast, the more familiar Dirac mass term, of the form $\overline{\psi_{jR}} \psi_{iL}$, does not.)

Rather than arising from Majorana mass terms, the L required by $\beta\beta_{0\nu}$ can come instead from the charged current interactions through currents of the form

$$\overline{e}_R \gamma_\mu (\psi_{iL})^c . \quad (6)$$

Currents of this type couple an e^- to an antineutrino, rather than to a neutrino.

WHEN $W_{80\,GEV}$ DOMINATES

In the simplest situation, the $\beta\beta_{0\nu}$ diagram of Fig. 1 is dominated by a term in which both W_α and W_β are the 80 GeV W boson already known, and the couplings at the leptonic vertices are Standard Model LH couplings. As is well-known,[4] $A^{\beta\beta_{0\nu}}$ ($\equiv A_{LL}$ in this situation) is then proportional to an effective neutrino mass for double beta decay, M_{eff}, given by

$$M_{eff} = \sum_m \omega_{em} \left| L_{em}^{(80)} \right|^2 M_m . \quad (7)$$

In this expression, M_m is the mass of ν_m, $L^{(80)}$ is the mixing matrix in the LH current to which $W_{80\,GeV}$ couples, and ω_{em} is a phase factor.[5] A determination of the rate for $\beta\beta_{0\nu}$ would yield a measured value for $|M_{eff}|$. Now, in the Standard Model (extended to include neutrino masses), $\sum_m \left| L_{em}^{(80)} \right|^2 = 1$. Thus, from Eq. (7),

$$|M_{eff}| \leq M_{Heaviest} , \quad (8)$$

where, as before, M_{Heaviest} is the mass of the heaviest of the neutrino mass eigenstates.

Suppose that the decay $^{76}\text{Ge} \to {}^{76}\text{Se} + 2e^-$ is observed. Given the calculated nuclear matrix element for this process,[6] the measured $|M_{\text{eff}}|$ determined by the observation is related to the observed lifetime by

$$|M_{\text{eff}}| \simeq 1 \text{ eV} \sqrt{\frac{10^{24} \text{ yr}}{\tau_{0\nu}(\text{Ge})}} \ . \qquad (9)$$

Combining Eqs. (8) and (9), we obtain the lower bound (2) on M_{Heaviest}.

WHEN A_{LR} DOMINATES

Suppose, now, that $A^{\beta\beta_{0\nu}}$ is dominated, not by A_{LL} as in the situation just discussed, but by A_{LR}. This could occur if, for example, the M_{eff} of Eq. (7) vanishes as a result of some symmetry, as it does in several models. If some particular W_α couples to LH currents, and some particular W_β to RH ones, then we see from Fig. 1 and Eq. (4) that the contribution of W_α–W_β exchange to A_{LR}, $A_{LR}(W_\alpha W_\beta)$, is proportional to

$$\sum_m L_{em}^{(\alpha)} \frac{q}{q^2 + M_m^2} R_{em}^{(\beta)} \ . \qquad (10)$$

Here, the factors $L_{em}^{(\alpha)}$ and $R_{em}^{(\beta)}$ come from the leptonic vertices, and we have used the readily-demonstrated fact that in the ν_m propagator, $(\slashed{q} - iM_m) / (q^2 + M_m^2)$, only the $\slashed{q}$ term in the numerator contributes to A_{LR}. Now, it appears naively as if the quantity (10), which contains the neutrino-mass dependence of $A_{LR}(W_\alpha W_\beta)$, remains nonzero even when all the neutrino masses M_m vanish. If this were truly so, then $\beta\beta_{0\nu}$ could be produced by a nonzero A_{LR} even in the absence of neutrino mass, and the observation of $\beta\beta_{0\nu}$ would not imply neutrino mass. However, we shall now show that for a broad variety of gauge theories, A_{LR} actually does *not* remain nonzero in the absence of neutrino mass, and that, even if A_{LR}, rather than A_{LL}, dominates $\beta\beta_{0\nu}$, the neutrino mass bound (2) still holds.

Consider first gauge models in which (a) all of the currents to which a given W_α couples have the same handedness, and (b) there is no doubly-charged gauge boson. Suppose that in some model of this type, the particular boson W_α couples only to LH currents, while the boson W_β couples only to RH ones. Consider the reaction $W_\alpha^- W_\beta^- \to e_L^- e_R^-$. In lowest order, this reaction is produced by the ν_m exchanges in the upper part of Fig. 1 (with fixed W_α and W_β), and has an amplitude proportional to the quantity (10). Now, taken by itself, each ν_m exchange is found to lead to a cross section for $W_\alpha^- W_\beta^- \to e_L^- e_R^-$ which, at center-of-mass energies E_{CM} large compared to the particle masses, is given by

$$\frac{d\sigma}{d\Omega} = (\text{Constant}) E_{CM}^2 \sin^2 \theta \ . \qquad (11)$$

Here, θ is the angle between one of the incoming particles and one of the outgoing ones. Obviously, at sufficiently large E_{CM}, the cross section (11) violates unitarity. Indeed, this cross section has precisely the same form as the unitarity-violating cross section produced by the electron-exchange diagram for $W^+W^- \to \nu_e \bar{\nu}_e$ in the Standard Model. Now, in the Standard Model, the bad high-energy behavior of the electron exchange diagram is cancelled by a direct-channel Z° pole diagram. In principle, a similar cancellation could occur in $W_\alpha^- W_\beta^- \to e_L^- e_R^-$. However, since the initial and final states are now doubly charged, the boson playing the role of the Z° would have to be doubly charged. Thus, if, as we are assuming, there is no doubly-charged gauge boson, this type of cancellation cannot occur. Furthermore, the bad high-energy behavior cannot be cancelled by a pole diagram containing a doubly-charged Higgs particle, even if such a particle should exist. The reason is that Higgs particles are spinless, and the high-energy behavior of Eq. (11), with its non-isotropic angular distribution, can certainly not be cancelled by a J = 0 particle in the intermediate state.

Now, it is a hallmark of gauge theories that the complete lowest-order amplitude for a process such as $W_\alpha^- W_\beta^- \to e_L^- e_R^-$ always has acceptable high-energy behavior. Assuming there is no doubly-charged gauge boson, we see that this acceptable behavior can come about only if the unacceptable contributions of the individual neutrino exchanges cancel *each other* as $E_{CM} \to \infty$. Now, at high energy (and q^2), we see from (10) that the sum of the neutrino-exchange contributions is proportional to

$$\sum_m L_{em}^{(\alpha)} R_{em}^{(\beta)} \equiv Q_{\alpha\beta} \ . \qquad (12)$$

Cancellation of the neutrino-exchange contributions against each other then requires that[7]

$$Q_{\alpha\beta} = 0 \ . \qquad (13)$$

Suppose that all neutrino masses M_m are small compared to 50 MeV, the typical momentum transfer $\langle q \rangle$ in $\beta\beta_{0\nu}$. Then, in the quantity (10), to which the $\beta\beta_{0\nu}$ amplitude $A_{LR}(W_\alpha W_\beta)$ is proportional, we may expand the factor $(q^2+M_m^2)^{-1}$ in powers of M_m^2/q^2. Since $Q_{\alpha\beta} = 0$, the 0th order term in this expansion contributes nothing to the sum on m, and we have

$$A_{LR}(W_\alpha W_\beta) \propto \frac{q}{q^2} \sum_m L_{em}^{(\alpha)} R_{em}^{(\beta)} \frac{M_m^2}{q^2} + O\left(\frac{M_m^4}{q^4}\right). \quad (14)$$

We see that, in gauge models of the type we are considering, A_{LR} is *second order* in neutrino mass. Contrary to the naive expectation, it vanishes when all the M_m do. Furthermore, A_{LL}, which is proportional to the M_{eff} of Eq. (7), is only *first order* in neutrino mass. Since the relevant scale factor is $\langle q \rangle \sim 50$ MeV, this means that when all $M_m \ll 50$ MeV, a given rate for $\beta\beta_{0\nu}$ requires larger neutrino masses when A_{LR} dominates than when A_{LL} does. Thus, if $\beta\beta_{0\nu}$ is observed, and its measured rate implies the lower bound (2) on neutrino mass if we assume A_{LL} dominates, then that bound still holds even if in fact A_{LR} dominates. Indeed, if A_{LR} dominates, some neutrino must be *much* heavier than the bound (2). To see how much heavier, we can relate the nuclear matrix element in A_{LR} to that in the amplitude for the double beta decay of the given initial nuclear state to the given final one, *but with the emission of two antineutrinos*. That these nuclear matrix elements are simply related follows from the fact that A_{LR} is second order in neutrino mass and a clever recent analysis by Haxton.[8] Now, the observed lifetime of approximately 10^{21} yr[9] for $^{76}\text{Ge} \to ^{76}\text{Se} + 2e^- + 2\bar{\nu}_e$ determines the nuclear matrix element for this process. Consequently, it determines the one in the amplitude A_{LR} for $^{76}\text{Ge} \to ^{76}\text{Se} + 2e^-$. Using the latter matrix element, we find that if A_{LR} dominates the latter decay, then the heaviest neutrino has a mass obeying

$$M_{\text{Heaviest}} \gtrsim 20 \text{keV} \left[\frac{10^{24}\text{yr}}{\tau_{0\nu}(\text{Ge})}\right]^{\frac{1}{4}} . \quad (15)$$

THEORIES WHERE $A_{LR} = O[(\text{NEUTRINO MASS})^2]$

As one might expect, A_{LR} is second order in neutrino mass, not only in theories where all the currents coupled to a given W boson have the same handedness and there is no doubly-charged gauge boson, but in other gauge theories as well. Rather than go through the mathematical analysis[2] that identifies the theories in which $A_{LR} = O[(\text{Neutrino Mass})^2]$, we shall just state which theories these are, and try to explain in simple terms why they have this property.

Suppose first that some theory contains no lepton-number violating currents of the form $\overline{e_R}\gamma_\mu(\psi_{iL})^c$ or $\overline{e_L}\gamma_\mu(\psi_{jR})^c$, which couple an electron to an antineutrino. In such a theory, the $\rlap{\,/}L$ required to produce $\beta\beta_{0\nu}$ must occur in the neutrino mass terms. In that case, the diagrams for $\beta\beta_{0\nu}$ had better "know" about the neutrino mass terms. If we draw these diagrams in the neutrino gauge eigenstate basis, this means that they must contain neutrino mass insertions.

In Fig. 2, we show a sample diagram which contributes to A_{LR}. It has some RH current at one leptonic vertex, and some LH current at the other. As illustrated by this diagram, when the currents at the two leptonic vertices are of opposite handedness, the exchanged neutral lepton emitted by the one vertex must be of the same handedness as the one absorbed by the other. Thus, since any mass insertion flips handedness, the number of insertions must be even. Consequently, the minimal nonzero number of insertions is two, so that A_{LR} is second order in neutrino mass.

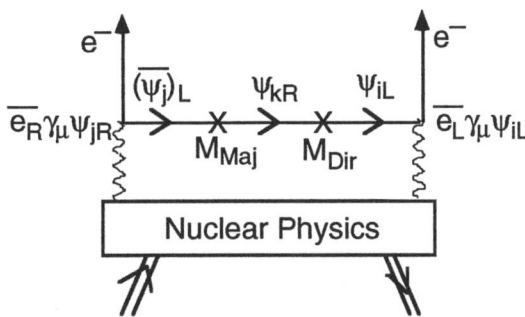

Figure 2. A sample diagram contributing to A_{LR}. At least two mass insertions are required. In this particular diagram, one must be a Majorana mass M_{Maj} and the other a Dirac mass M_{Dir}. The particle ψ_{kR} is some RH neutrino in the theory.

Suppose next that some theory does contain "lepton-number violating" currents of the form $\overline{e_R}\gamma_\mu(\psi_{iL})^c$, but that no neutrino field which occurs charge-conjugated in the charged-current couplings also occurs un-charge-conjugated. Then, by redefinitions of the form $(\psi_{iL})^c \equiv \chi_{iR}$, we can redefine every "antineutrino" coupled to a negatively-charged lepton by the charged currents to be a "neutrino." All $\rlap{\,/}L$ is thereby removed from the charged-current interactions and relocated to the neutrino mass terms. But, if the theory can be cast in a form in which the $\rlap{\,/}L$ required to produce $\beta\beta_{0\nu}$ is in the neutrino mass terms, then, just as in the type of theory considered previously, A_{LR} must be of second order in neutrino mass.

Finally, suppose that in some theory certain neutrino fields do occur both charge-conjugated and un-charge-conjugated in the charged currents, but that these currents never couple both a neutrino field and its own charge conjugate to the electron. An interesting theory of this kind might contain both the currents $\overline{e_R}\gamma_\mu(\psi_{iL})^c$ and $\overline{e_L}\gamma_\mu\psi_{kL}$, but only with $\psi_{iL} \neq \psi_{kL}$. This pair of currents can contribute to A_{LR} via the diagram in Fig. 3, in which the first current emits a ψ_{iL} but the second absorbs a ψ_{kL}. Now, one can prove that in any gauge theory, the quanta of distinct neutrino fields such as ψ_{iL} and ψ_{kL} are orthogonal. Thus, on the neutrino line in Fig. 3, something must intervene to convert ψ_{iL} into

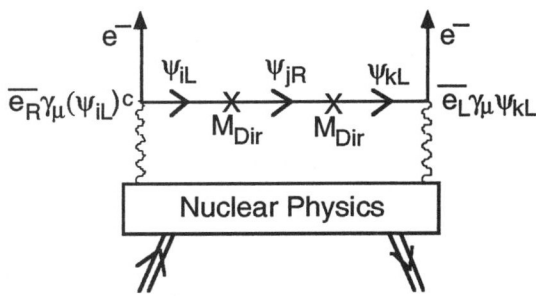

Figure 3. A contribution to A_{LR} arising from the currents $\overline{e_R}\gamma_\mu(\psi_{iL})^c$ and $\overline{e_L}\gamma_\mu\psi_{kL}$. At least two mass insertions, which may either both be Dirac masses as shown or Majorana masses, are required. The particle ψ_{jR} is some RH neutrino in the theory.

ψ_{kL}. That "something" is mass insertions. Since ψ_{iL} and ψ_{kL} have the same handedness, two insertions are required, so that yet again A_{LR} is second order in neutrino mass.

SUMMARY

If $\beta\beta_{0\nu}$ is dominated by the amplitude A_{LL} corresponding to the diagram in which both of the virtual W bosons are the 80 GeV Standard Model W, with its Standard Model LH couplings, then the amplitude is *first* order in neutrino mass and the observation of the decay would imply the lower bound

$$M_{\text{Heaviest}} \gtrsim 1\,\text{eV}\sqrt{\frac{10^{24}\,\text{yr}}{\tau_{0\nu}(\text{Ge})}}\,. \qquad (2)$$

If, instead, $\beta\beta_{0\nu}$ is dominated by A_{LR}, then for a broad variety of gauge theories the amplitude is *second* order in neutrino mass and the observation of the decay would still imply the bound (2). This variety includes all gauge theories in which—

(i) no antineutrinos are coupled to negatively-charged leptons by the charged-current interactions, which means that the Ł required for $\beta\beta_{0\nu}$ must come from the neutrino mass terms,

or—

(ii) any antineutrinos coupled to negatively-charged leptons by the charged currents can be redefined to be neutrinos, so that all Ł is relocated to the neutrino mass terms,

or—

(iii) the charged currents never couple both a neutrino and its own antineutrino to the electron,

or—

(iv) each W_α couples only to currents of one handedness, and there is no doubly-charged gauge boson.

Of course, conditions (i) and (ii) are special cases of condition (iii). However, since, to our knowledge, most lepton-number violating theories satisfy the simple condition (i), this condition and the related one (ii) would seem to deserve special mention.

ACKNOWLEDGMENTS

It is a pleasure to thank N. Christ, P. Pal, P. Ramond, and L. Wolfenstein for very helpful discussions or remarks, and to thank the leaders of the Aspen Center for Physics, where part of this work was carried out, for their hospitality.

REFERENCES

1. J. Schechter and J. Valle, *Phys. Rev.* D **25**, 2951 (1982); E. Takasugi, *Phys. Lett.* **149B**, 372 (1984).
2. B. Kayser, S. Petcov, and S. P. Rosen, in preparation. See also B. Kayser, in *Proceedings of the Seventh Moriond Workshop on New and Exotic Phenomena*, edited by O. Fackler and J. Tran Thanh Van (Editions Frontieres, Gif-sur-Yvette, France, 1987), pp. 349-354; S. Bilenky and S. Petcov, *Rev. Mod. Phys.* **59**, 671 (1987); S. P. Rosen, to be published in the *Proceedings of the Franklin Symposium in Celebration of the Discovery of the Neutrino*.
3. D. Caldwell *et al.*, *Nucl Phys. B (Proc. Suppl.)* **13**, 547 (1990); F. Avignone, talk presented at the April, 1992 Meeting of the American Physical Society, Washington, D.C.; A. Piepke, in these Proceedings.
4. For an explanation, see, for example, B. Kayser, F. Gibrat-Debu and F. Perrier, *The Physics of Massive Neutrinos* (World Scientific, Singapore, 1989).
5. The significance of ω_{em} is discussed in Doi *et al.*, *Phys. Lett.* **102B**, 323 (1981); L. Wolfenstein, *Phys. Lett.* **107B**, 77 (1981); B. Kayser and A. Goldhaber, *Phys. Rev. D* **28**, 2341 (1983); S. Bilenky, N. Nedelcheva, and S. Petcov, *Nucl. Phys.* **B247**, 61 (1984).
6. W. Haxton and G. Stephenson, Jr., *Prog. Part. Nucl. Phys.* **12**, 409 (1984); M. Doi, T. Kotani, and E. Takasugi, *Prog. Theor. Phys. Suppl.* **83**, 1 (1985); K. Grotz and H. Klapdor, *Phys. Lett.* **B153**, 1 (1985); T. Tomoda

et al., *Nucl. Phys.* **A452**, 591 (1986); T. Tomoda and A. Faessler, *Phys. Lett.* **B199**, 475 (1987); J. Engel, P. Vogel, and M. Zirnbauer, *Phys. Rev. C* **37**, 731 (1988).

7. For early discussion of this relation in specific models and conjectures concerning its generality, see T. Kotani, in *Proceedings of the 1984 Moriond Workshop on Massive Neutrinos in Astrophysics and in Particle Physics,* edited by J. Tran Thanh Van (Editions Frontieres, Gif-sur-Yvette, France, 1984), pp. 397-424; M. Doi *et al.*, Ref. 6.

8. W. Haxton, *Phys. Rev. Lett.* **67**, 2431 (1991).

9. F. Avignone and R. Brodzinski, *Prog. Part. Nucl. Phys.* **21**, 99 (1988); A. Vasenko *et al.*, *Mod. Phys. Lett.* A **5**, 1299 (1990); F. Avignone *et al.*, *Phys. Lett.* **B256**, 559 (1991); A. Piepke, in these Proceedings.

SEARCH FOR ISOSINGLET NEUTRAL HEAVY LEPTON WITH THE L3 DETECTOR AT LEP

Sergey Shevchenko
Lauristen Laboratory
California Institute of Technology
Pasadena, CA 91125

Abstract

We have searched for neutral heavy leptons that are isosinglets under the Standard $SU(2)_L$ gauge group. Such neutral heavy leptons are expected in many extensions of the Standard Model. Three types of heavy leptons N_e, N_μ, N_τ associated with the three neutrino types ν_e, ν_μ, ν_τ have been directly searched for and no evidence for a signal has been found. We set the limit $\text{Br}(Z^0 \to \nu_l N_l) < 3 \times 10^{-5}$ at the 95% C.L. for the mass range from 3 GeV up to M_Z.

INTRODUCTION

Isosinglet Neutral Heavy Leptons (INHL) arise in many models that attempt to unify the presently observed interactions into a single gauge scheme such as Grand Unified Theories or Superstring inspired models.[1] Their existence also is predicted in many extended electroweak models such as left-right symmetric and see-saw models.[2]

Constraints on the isosinglet neutral lepton admixture in gauge currents have been placed by several experiments.[3,4] The mass range covered, however, has been below 10 GeV, except for the limit obtained by the OPAL Collaboration which extends from 4 GeV up to M_Z.[4]

In this paper we describe our search for an INHL within the mass range from about 1 GeV up to M_Z. This search is based on the data collected at LEP with the L3 detector[5] during 1990 and 1991 at centre of mass energies between 88.2 and 94.3 GeV. The total integrated luminosity is 17.5 pb^{-1}, corresponding to about 424,000 hadronic Z^0 decays.

PRODUCTION AND DECAYS

In our search we assume that one INHL N_l is associated with each generation of light neutrino. We do not consider mixing of the light neutrinos with higher isodoublet states (sequential leptons) nor the possibility of mixing among light neutrinos.[6]

INHL are singly produced in Z^0 decays:

$$Z^0 \to N_l + \overline{\nu_l}$$

The production cross section is reduced from the light neutrino pair production cross-section by a phase-space factor and by the square of a mixing amplitude U_l,[6,7] which expresses the amount of mixing between the light and hypothetical heavy neutrino generations.

INHL decay via the charged or neutral weak currents:

$$N_l \to Z^* \nu \text{ and } Z^* \to ee, \mu\mu, \tau\tau, \nu\nu, qq$$
$$N_l \to W^* l \text{ and } W^* \to e\nu, \mu\nu, \tau\nu, qq$$

To calculate the branching ratios of the decays into the final states, we use the formulae

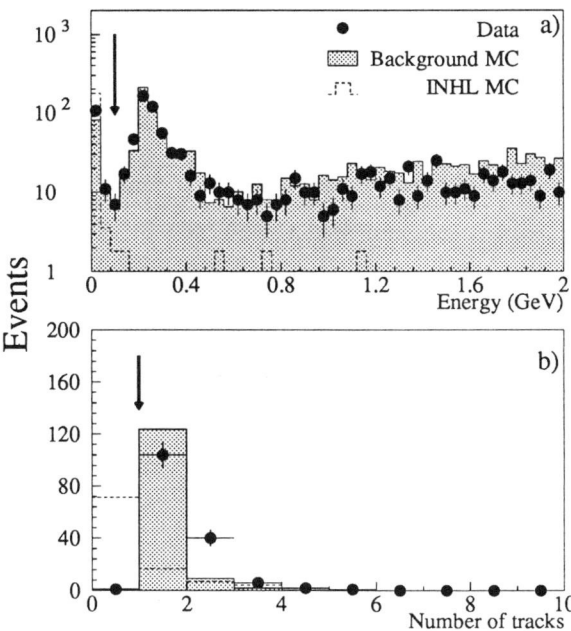

Figure 1. a) Energy in the 30° cone around the opposite direction of the jet and b) Number of tracks in the 90° cone around the opposite direction of the jet. The circles are data, the shaded area is the background MC. The dashed line is the predicted signal $Z^0 \to N_l \nu_l$ for a mass of 10 GeV. The arrows indicate the position of cuts.

in References 6 and 7. The dominant decay mode is via charged currents, with the production of a lepton and two quarks, which is about 50%. For very low masses, the branching ratios change due to kinematical constraints. For the τ family and M_{N_τ} below 3 GeV, for example, the dominant decay mode is via a light neutrino and two quarks.

The mean decay lenght is a function of the coupling constant $|U|^2$ and the INHL mass. It is given by[6]

$$l_N = \beta\,\gamma\,c\,\tau_N \;\propto\; \beta\,|U|^{-2} M_N^\alpha,$$

where $\alpha \approx -6$. This implies that the decay can occur far from the beam vertex if the particle has a low mass or a very small coupling. We will consider in our searches the case where the decay occurs far from the interaction vertex (in the electromagnetic calorimeter or hadron calorimeter), which will allow us to consider mean decay lengths up to around 2 m.

EVENT SIGNATURES AND SELECTION

Because of the boost in the laboratory frame, the decay signature depends on the mass of the INHL. For low mass, we have dominantly monojet events, while for high mass, events with two or more reconstructed jets are dominant. Jets are reconstructed using an algorithm described in Reference 8.

Search for monojets

By searching for monojet events, we cover inclusively all visible decay modes of a low mass INHL ($\leq$ 15 GeV). To select this topology, we select events that have exactly one reconstructed jet and at least two tracks in the vertex detector.

The background to this topology comes from two-photon processes and $e^+e^- \to Z^0 \to \tau^+\tau^-(\gamma)$: when a large fraction of the energy is taken away by neutrinos, the resulting energy deposition can be below the jet reconstruction algorithm threshold of about 2 GeV. In this case, however, some calorimetric energy is expected in the opposite hemisphere and in addition, at least one low momentum track should be present. We therefore require the energy in the 30° cone, which is opposite to the monojet direction, to be less than 0.1 GeV and that no tracks be reconstructed in the 90° cone centered on this direction. The peak in the energy distribution centered at approximately 250 MeV corresponds to the energy deposition of minimum ionizing particles in the BGO calorimeter.

From two-photon processes background is produced at low polar angle and the energy spectra decreases sharply as a function of the energy. We therefore require the energy of the monojet to be greater than 15 GeV and its polar angle to satisfy $20° < \theta < 160°$.

After applying all the cuts, we are left with 2 events from data, while we expect 0.6 ± 0.4 from $Z^0 \to \tau^+\tau^-(\gamma)$ events.

As mentioned above, low masses or small mixing amplitudes $|U_l|^2$ can result in decays far from the interaction point. To find these events, we use cuts which are similar to those used to select monojet events, except we remove the requirement of tracks in vertex chamber. To estimate the acceptance for such events, we generate all visible decay modes with decay lengths from a few cm up to 2 m. After applying the cuts, two events are left in data and 0.6 ± 0.6 in Monte Carlo.

Search for two acoplanar jets

The event topology consists of a pair of acoplanar and acollinear jets with large missing energy and transverse imbalance. This search covers all decay modes containing a neutrino in the final state for the mass region ≥ 15 GeV and the modes containing hadrons and a lepton for the mass region 15 GeV $\leq M_N \leq 50$ GeV.

Background to this topology comes from events where some energy is either unseen or not well measured in the detector.

We select all events which have exactly two reconstructed jets and at least two tracks. Almost all dilepton and hadronic decays of the Z^0 are removed by requiring an acollinearity between the two jets greater than $35°$ and an acoplanarity greater than $20°$. The acollinearity and acoplanarity distributions are shown in Figure 2. Initial state radiation and two-photon events are reduced by the requirement that the polar angle θ of the missing momentum should satisfy $20° < \theta < 160°$. The most energetic jet must have at least 10 GeV

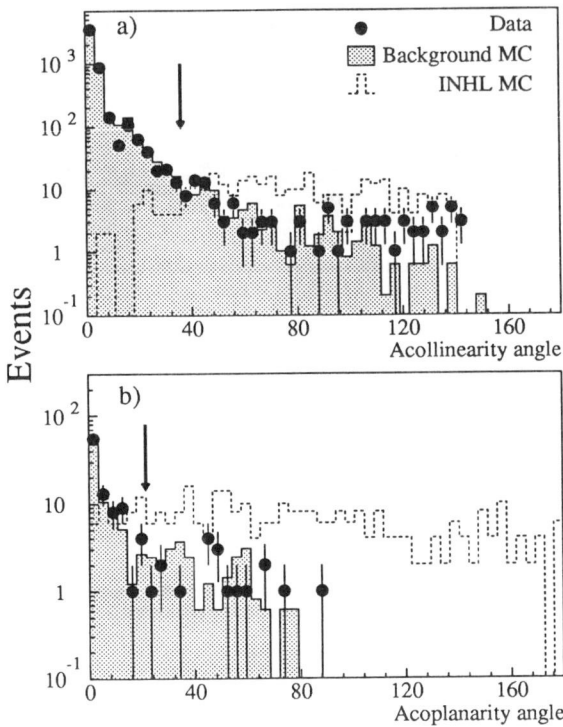

Figure 2. a) Acollinearity and b) Acoplanarity angle of the two jets. The circles are data, the shaded area is the background MC. The dashed line shows the predicted signal $Z^0 \to N_l \nu_l$ for a mass of 50 GeV. The arrows indicate the position of cuts.

and the second jet at least 5 GeV. Remaining background is removed by requiring that the energy in the $30°$ cone around the direction of the missing momentum be less than 0.2 GeV and that the number of tracks in this cone be zero.

After applying all cuts, one event is left in the data while we expect 0.2 ± 0.2 from the $Z^0 \to \tau^+\tau^-(\gamma)$ decay.

Search for isolated lepton in three or more jet events

By selecting hadronic events with an isolated lepton, we deal with the lqq decay modes for the mass region ≥ 50 GeV.

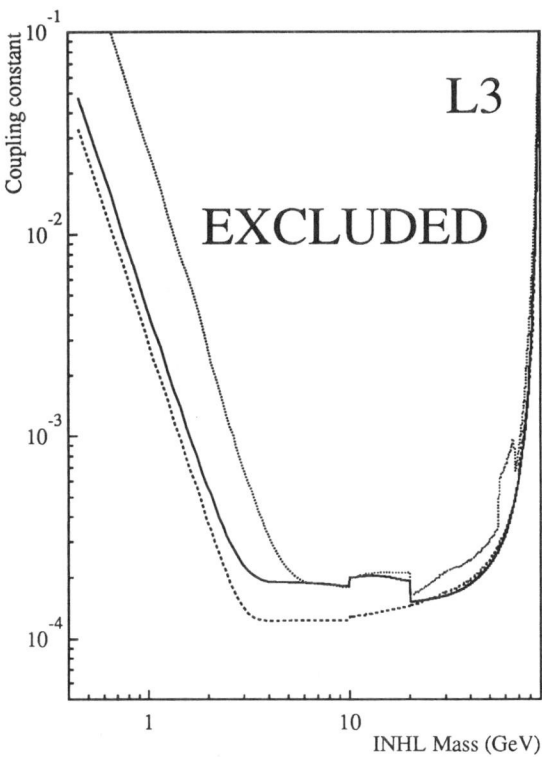

Figure 3. 95% C.L. upper limit on the coupling constant $|U|^2$ vs the mass of the INHL. The solid line is the limit for N_e, the dashed line is the limit for N_μ and dotted line is for N_τ.

Table 1. List of selected events in data and Monte Carlo.

Decay Mode	Data	Monte-Carlo
eqq	6	7 ± 2
μqq	10	7 ± 2
τqq	26	23 ± 3

The main background to this topology comes from the semileptonic decays of heavy quarks. Radiative hadronic decays $Z^0 \to q\bar{q}\gamma$ where a hard photon converts in the beam-pipe can also fake an isolated electron.

We select events with three or more reconstructed jets. The visible energy must be greater than $0.4 \sqrt{s}$. The energy of the third jet must be at least 5 GeV, to remove the QCD background. For the eqq mode, the isolation criteria is that the energy in the 30° cone around the electron candidate is less than 3 GeV. For the μqq mode, the energy in the 30° around the muon must be less than 5 GeV (in this case, we do not subtract the calorimetric loss of the muon). To improve the rejection of hadronic background where a jet is not well measured, we require that the energy in a 20° cone around the missing momentum direction be less than 2 GeV. This cut is applied only when the visible energy is less than $0.9\sqrt{s}$, i.e. when the direction of missing momentum is well defined.

In the τqq mode, the identification of an isolated τ is made more difficult because of the potentially large background from hadronic events. We look for the most isolated track with momentum greater than 2 GeV. There should not be any tracks inside the 20° cone around the track. The energy inside the 10° cone around the track should be more than 3 GeV, and the difference of energies in the 20° and 10° cones around the track should be less than 1 GeV. The acoplanarity between the two most energetic jets has to be greater than 30°. The data events and Monte-Carlo background expectations, after applying our selection cuts, are shown in Table 1.

In these decay modes, due to the presence of only one light neutrino in the final state, the reconstruction of the invariant mass of the isosinglet neutral lepton is possible. Almost all surviving events and are grouped in the mass region $\approx$ 80-90 GeV. These events are compatible with the expected backgrounds.

RESULTS

We calculate the 95% confidence level upper limits on the coupling constant $|U_l|^2$ and the branching ratio Br($Z^0 \to \nu_l N_l$) for each generation, by applying a Poisson statistics.[9]

The results for the square of the mixing amplitude as a function of the mass is shown in Fig.3. The coupling constant $|U_l|^2$ is constrained to be less than 2×10^{-4} for the mass range $4 < M_N < 40$ GeV. The branching ratio Br($Z^0 \to \nu_l N_l$) limit is 3×10^{-5} for masses from 3 GeV up to M_Z.

CONCLUSION

We searched for all visible decay modes of an isosinglet neutral heavy lepton from very low masses up to M_Z. We also searched for displaced vertex decays. No excess was found in the data. We set limits of the order of 10^{-4} on the coupling constant $|U|^2$ as a function of the mass. Branching ratios limits of the order of 10^{-5} were also set.

ACKNOWLEDGMENTS

I would like to thank my colleagues in L3 and specially A.Rubbia and V.Shoutko for their collaboration in this project, and H.Newman for useful discussions and comments.

REFERENCES

1. For a review, see J.W.F. Valle, Nucl. Phys. (Proc. Suppl) 11 (1989) p.118.

2. M. Gell-Mann, P. Ramond, R. Slansky in Supergravity, ed. by D. Freedman *et al.*, (North Holland, 1979);
T. Yanagida, KEK lectures, ed. O. Sawada *et al.* (1979);
R. Mohaparta, G. Senjanovic, Phys. Rev. Lett. 44 (1980) 912, Phys. Rev. D23 (1981) 165.

3. G. J. Feldman *et al.*, Phys. Rev. Lett. 54 (1985) 2289.
A. M. Cooper-Sarkar *et al.*, Phys. Lett. 160B (1985) 207.
J. Dorenbosch *et al.*, Phys. Lett. 166B (1986) 473 and references therein.
S. R. Mishra *et al.*, Phys. Rev. Lett. 59 (1987) 1397.
M. E. Duffy *et al.*, Phys. Rev. D38 (1988) 2032.
W. Bartel *et al.*, Phys. Lett. 123B (1983) 353.
CCFR Collaboration, K. Bachmann *et al.*, Preprint UR-1157.

4. OPAL Collaboration, M. Z. Akrawy et. al., Phys. Lett. 247B (1990) 448.

5. L3 Collaboration, B. Adeva *et al.*, Nucl. Instr. and Meth. A289 (1990) 35.

6. M. Gronau, C. Leung and J. Rosner, Phys. Rev. D29 (1984) 2539.

7. M. Dittmar, M.C. Gonzalez-Garcia, A. Santamaria and J.W.F. Valle, Nucl. Phys. B332 (1990) 1.

8. O. Adriani *et al.*, Nucl. Instr. and Meth. A302 (1991) 53.

9. G. Zech, Nucl. Instr. and Meth. A277 (1989) 608;
O. Helene, Nucl. Instr. and Meth. A212 (1983) 319.

HINTS FOR NEUTRINO MASSES: A THEORETICAL OVERVIEW

J. W. F. Valle

Instituto de Física Corpuscular - IFIC/C.S.I.C.

Departament de Física Teòrica, Universitat de València

46100 Burjassot, València, SPAIN

Abstract

I briefly review the existing limits on neutrino masses and discuss some of the positive hints, with emphasis on solar neutrino observations, neutrinoless double beta decays and the possible implications of COBE results for hot dark matter and neutrino oscillations. I stress that the possible observation of sizeable majoron emission effects in neutrinoless double beta decays need not be in conflict with LEP measurements of the invisible width of the Z boson.

INTRODUCTION

There is no solid reason for why neutrinos are massless. This becomes a bigger puzzle when one considers extensions of the standard $SU(2) \otimes U(1)$ model [1]. In these extensions more often than not neutrinos are likely to pick up a mass. This may occur even in those extensions that in principle are unrelated to neutrino physics, like supersymmetry (SUSY). Indeed, the most general supersymmetric extension of the standard model has no conserved R parity symmetry and, as a result, neutrinos are massive [2].

There is, unfortunately, no solid theoretical guidance as to what sets the scale of neutrino masses. After all, how could one predict neutrino masses when the masses of the electron or muon are not understood? Therefore one must rely on experiment for clues. Despite the tremendous effort in this direction, the existing information is still not enough to pin down the neutrino sector as one would like to. However, there has been progress. In this regard it is interesting to note the role that high energy experiments can play in restricting neutrino properties. A remarkable example is provided by LEP observations [3]. By accurately measuring the number of neutrinos LEP has given us not only yet more confidence on the existence of the ν_τ but also an important constraint on possible neutrino mass generation mechanisms. For example, if neutrino masses arise from the spontaneous violation of lepton number [4] then this breaking should be driven by *isosinglet* vacuum expectation values so that the associated Goldstone boson (majoron) is mostly singlet and as a result the Z does not decay by majoron emission. I will come back to this later.

LIMITS

Observation restricts neutrino masses in many ways. The present bounds from laboratory experiments may be summarized as [5]

$$m_{\nu_e} \lesssim 9\,\text{eV}, \tag{1}$$
$$m_{\nu_\mu} \lesssim 500\,\text{keV},$$
$$m_{\nu_\tau} \lesssim 31\,\text{MeV}$$

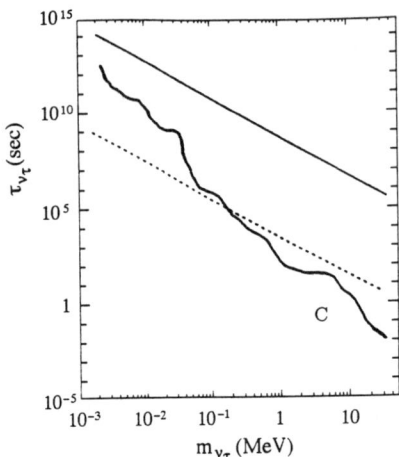

Figure 1. ν_τ lifetime versus observational limits

In addition there is a cosmological limit that follows from considerations related to the abundance of relic neutrinos [6]

$$\sum_i m_{\nu_i} \lesssim 50\, eV \qquad (2)$$

However the limit in eq. (2) only holds if neutrinos are stable. There are many ways to make neutrinos decay by majoron emission [7, 8, 9], e.g.

$$\nu_\tau \to \nu_\mu + J \qquad (3)$$

where J denotes the majoron. The resulting lifetime can be sufficiently short that neutrino mass values as large as eq. (1) are allowed. This remains true when constraints related to primordial nucleosynthesis are taken into account. Moreover, since these decays are *invisible*, they are consistent with all astrophysical observations and the isotropy of the cosmic background radiation.

Examples of seesaw type models where this is possible have been discussed in ref. [8]. A supersymmetric example is provided by the spontaneously broken R parity model [9]. Curve C in Fig. 1 illustrates the ν_τ lifetimes allowed in this model, versus the ν_τ mass. On the other hand the ν_τ decay lifetime required in order to efficiently suppress the relic ν_τ contribution is shown as the solid straight line in Fig. 1. An additional constraint, shown as the dashed line in Fig. 1 [10], may be derived by demanding that the universe has become matter-dominated at a redshift of 1000, so that fluctuations have grown sufficiently by today (this limit is less reliable than the one derived from the critical density). Clearly the decay lifetimes can be shorter than required. It is worth noting in this context, that the controversial hints from recent beta decay experiments in favour of a 17 keV neutrino [11] would require a decay of this type, eq. (3).

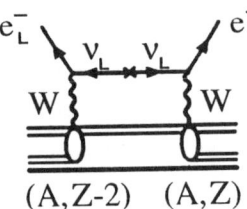

Figure 2. Mass mechanism for $\beta\beta_{0\nu}$ decay.

Limits on neutrino mass and mixing also follow from the nonobservation of neutrino oscillations [5]. Further improvements on oscillation limits involving ν_τ are expected from new experiments such as CHORUS and NOMAD at CERN and the proposed P803 at Fermilab.

The nonobservation of nuclear decay processes $(A, Z-2) \to (A, Z) + 2\,e^-$, or neutrinoless double beta decays, provides complementary limits on neutrino mass and mixing. This process is of great interest because its existence would imply the violation of total lepton number. The standard way to induce this decay is via neutrino exchange, as shown in Fig. 2, and proceeds only if the virtual neutrino is a Majorana particle. The decay amplitude is proportional to

$$\langle m \rangle = \sum_\alpha K_{e\alpha}^2 m_\alpha \qquad (4)$$

where α sums over the light neutrinos. The

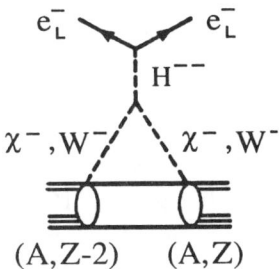

Figure 3. Scalar-induced $\beta\beta_{0\nu}$ decay.

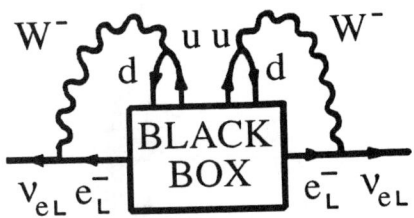

Figure 4. $\beta\beta_{0\nu}$ decay and Majorana neutrinos.

existing data lead to

$$\langle m \rangle \lesssim 1 - 3 \; eV \qquad (5)$$

depending on nuclear matrix elements. A better sensitivity is expected from the enriched germanium experiments. The parameter $\langle m \rangle$ may differ substantially from the neutrino mass inferred from beta decay since in eq. (4) there can be a destructive interference amongst different neutrino types. The simplest way to enforce such cancellation is via the lepton number symmetry characteristic of a Dirac neutrino. Such neutrino can be decomposed as two Majorana neutrinos degenerate in mass so that $\langle m \rangle$ vanishes automatically [12].

In addition to Majorana neutrino exchange gauge theories can engender $\beta\beta_{0\nu}$ decays through the exchange of scalars, as illustrated in Fig. 3. However one can show that, in a gauge theory, whatever the origin of neutrinoless double beta decay is, it requires neutrinos to be Majorana particles, is illustrated in Fig. 4 [13]. The figure shows how any generic "black box" mechanism inducing neutrinoless double beta decay also produces a nonzero Majorana neutrino mass.

Gauge theories may lead to new types of neutrinoless double beta decay involving the *emission of light scalars*, such as the majoron or its real partner ρ

$$(A, Z-2) \rightarrow (A, Z) + 2\, e^- + J\,. \qquad (6)$$

Since such light scalars are very weakly coupled to matter, their emission would only be detected through their effect on the β spectrum. The simplest model leading to sizeable majoron emission in $\beta\beta$ decays involving an isotriplet majoron [14] is no longer phenomenologically viable, since it leads to a new invisible decay mode for the neutral gauge boson by the emission of light scalars,

$$Z \rightarrow \rho + J, \qquad (7)$$

now ruled out by LEP measurements of the invisible Z width [3]. However it has been recently shown that a large majoron-neutrino coupling leading to observable majoron emission in neutrinoless double beta decay can easily be reconciled with the LEP results in models where the majoron is an isosinglet and lepton number is broken at a sufficiently low scale [15]. This is specially interesting in view of the puzzling features recently hinted in double beta decay spectra, suggestive of the emission of very light scalars [16].

POSITIVE HINTS

In addition to limits, observation also provides some hints for neutrino masses. Perhaps the most significant come from solar neutrino data. Indeed, the impressive results of GALLEX [17] can not be used to "eliminate" the solar neutrino puzzle, in view of the persisting deficit of high energy neutrinos seen in the Kamiokande and Homestake experiments. The astrophysical explanation of the

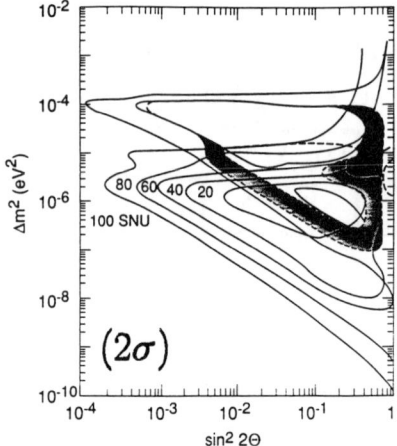

Figure 5. Allowed MSW oscillation parameters

latter would involve too large a drop in the solar core temperature and incorrectly predict the relative degree of suppression observed in these experiments [18].

The most attractive way to interpret the existing solar neutrino data is via the MSW effect. The region of parameters allowed by present experiments is illustrated in Fig. 5 [17]. It will be possible to discriminate between these large and small angle MSW solutions, as discussed by Rosen, in the BOREXINO experiment, by measuring the ^7Be neutrino line with sufficient accuracy.

Here I stress that, amusing as it may sound, in some particle physics models, these two solutions can be discriminated against even in high energy particle physics experiments, such as the LEP experiments [19]. This is at first surprising in view of the small neutrino masses $m_\odot \sim 10^{-3}$ eV required by the MSW effect. To illustrate that such models exist I mention the spontaneously broken R parity models [20]. In these models neutrinos get masses from mixing with the heavy neutral R-odd fermions or neutralinos, i.e. neutral SUSY partners of gauge or Higgs bosons. The basic feature is that the heavier neutrino is ν_τ and its mass scales as $m_{\nu_\tau} \propto v_R^2$ whereas the ν_μ mass scales only as $m_{\nu_\mu} \propto v_L^2$ and ν_e is massless in the tree approximation. Because of the large hierarchy $v_R \gg v_L$ that characterizes this model, the ν_e-ν_μ mass difference is very small on the scale of the ν_τ mass. For typical values $v_R = 1\,TeV$ and $v_L = 100\,MeV$ one gets a ν_μ-ν_τ mass ratio $m_{\nu_\mu}/m_{\nu_\tau} = \mathcal{O}(10^{-8})$ or so. Thus a ν_τ mass in the $10\,KeV - 1\,MeV$ range (consistent with cosmological limits, since the heavy ν_τ decays via majoron emission) corresponds to m_{ν_μ} at a scale $m_\odot \sim 10^{-3\pm1}$ eV, as needed for the MSW effect. Similarly the ν_e-ν_μ mixing angle can easily be chosen in the range where the MSW effect is operative. In such SUSY models there can be sizeable ($\sim 1\%$) R parity violating electroweak neutral (and charged) currents that do not conflict any of the observational constraints from high energy LEP/$Sp\bar{p}S$/Tevatron colliders or from searches for lepton violation (e.g. the non-observation of neutrino oscillations and neutrinoless double β decay, the failure to observe anomalous peaks on the energy distribution of the electrons and muons coming from decays such as $\pi, K \to \ell\nu$ ($\ell = e, \mu$) etc.). One finds that the detectability of new effects, such as

1. *single chargino production in Z decays*

$$Z \to \chi^\pm \tau^\pm, \qquad (8)$$

where $\chi^\pm$ denotes a chargino

2. *μ and τ decays with majoron emission*

$$\mu \to e + J \,,\, \tau \to \mu + J \,,\, \tau \to e + J. \quad (9)$$

is experimentally viable [9, 21, 22]. The first could originate recognizable events at LEP, while the latter could affect the spectra of the decay-produced leptons, leading to bumps at half of the parent muon or tau mass in its rest frame. Such exotic decays can occur with branching ratios well within the sensitivities of LEP and muon or tau factories, respectively. The allowed branching ratios for eq. (8) can

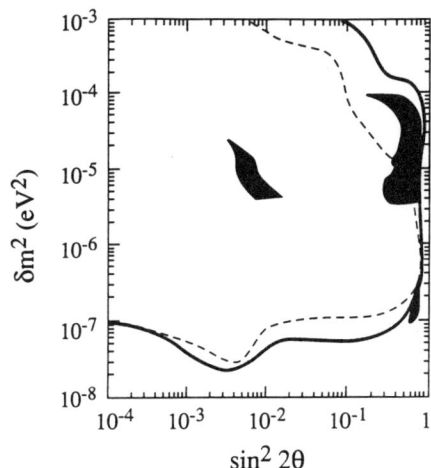

Figure 6. Probing MSW solutions with rare Z decays

exceed $\sim 10^{-5}$ [9, 21] while those of eq. (9) can exceed $\sim 10^{-6}$ and $\sim 10^{-4}$, respectively [22].

These processes may be used to distinguish between the large and small angle MSW solutions left out by the recent GALLEX results. This is illustrated in Fig. 6. Clearly, only in the small mixing solution one can have sizeable branching ratios for *both* $Z \to \chi + \tau$ and $\tau \to \mu + J$ in excess of 10^{-5} [19]. The contours in Fig. 6 correspond to $BR(Z \to \tilde{\chi} + \tau) = 10^{-6}$ (solid) and $BR = 10^{-5}$ (dashed). More explanation on the model parameters corresponding to the figure are given in ref. [19]. This example illustrates that, under certain circumstances, one may be able to probe MSW parameters in high energy accelerator experiments, thus providing an independent check upon the solar neutrino oscillation parameters as determined from solar neutrino data.

Another important hint for neutrino masses arises from recent data on large-scale structure in the Universe suggesting the existence of a hot component in the dark matter [23]. The simplest candidate for such hot dark matter is a massive and stable (on cosmological scales) neutrino with mass in the range $1 \, \text{eV} \lesssim m_{DM} \lesssim 50 \, \text{eV}$. *How can one reconcile the existence of this hot dark matter with the MSW explanation of the solar neutrino data?*

In principle these two scales m_{DM} and $m_\odot$ may be accommodated if the smallness of neutrino masses follows from the exchange of superheavy neutral leptons, *a la seesaw*. Within many grand unified models one expects a hierarchy of neutrino masses of the type $m_{\nu_e} : m_{\nu_\mu} : m_{\nu_\tau} \sim m_u^2 : m_c^2 : m_t^2$. If one chooses a 10 eV ν_τ as the hot dark matter then one gets m_{ν_μ} in the range required by the MSW effect [26]. This suggests that the forthcoming experiments at CERN [24] and Fermilab [25] may be able to observe $\nu_\mu - \nu_\tau$ oscillations.

An alternative scenario is to use radiative corrections associated to new Higgs bosons at the electroweak scale as the origin of the two neutrino mass scales m_{DM} and $m_\odot$ [27]. The simplest model of this type involves the existence of a fourth light neutral lepton ν_S, singlet under $SU(2) \otimes U(1)$ (we call it a sterile lepton). In the model of ref. [27] the dark matter scale arises at the one-loop level while the MSW scale arises only in two-loops. The model allows observable $\nu_e - \nu_\tau$ and $\nu_\mu - \nu_\tau$ oscillation rates. The latter are more strongly constrained by limits from primordial big bang nucleosynthesis, so that $\nu_e - \nu_\tau$ emerges as a theoretically interesting channel to probe (unfortunately an experimentally difficult one!). In contrast with seesaw models, there are other potentially large related phenomena that can probe the model. These include muon number violating processes such as $\mu \to e + \gamma$ and $\mu \to 3e$. Their rates can well lie within the sensitivities of present experiments. Finally, if we ignore nucleosynthesis limits we can have also a common explanation for the atmospheric neutrino deficit via ν_μ oscillations to the sterile neutrino ν_S with maximal mixing and $\delta m^2 \sim 10^{-2} - 10^{-3} \text{eV}^2$.

In summary, there are intriguing hints for massive neutrinos. Taking them for granted opens the way towards a wealth of new phenomena covering a broad range of energies and experimental situations. Apart from providing a natural explanation of the solar neutrino data and accounting for the hot dark matter component suggested by COBE [23], neutrino masses may be seen in neutrino oscillations (with good prospects for ν_e - ν_τ and ν_μ - ν_τ searches), in observable rates for neutrinoless double beta decays (including majoron emission), and (if large enough) they may also show up as distortions in beta spectra. Finally, a variety of related rare decay processes can manifest themselves at muon and tau factories and at LEP.

REFERENCES

1. J. W. F. Valle, *Prog. Part. Nucl. Phys.* **26**, 91 (1991) and references therein.

2. J. W. F. Valle, *Physics at New Accelerators: Looking Beyond the Standard Model*, talk at Neutrino 92, Granada, Spain, June, 1992, CERN-TH.6626/92

3. J. Steinberger, in *Electroweak Physics Beyond the Standard Model*, ed. J. W. F. Valle and J. Velasco (World Scientific, Singapore, 1992), p. 3.

4. Y. Chikashige, R. Mohapatra, R. Peccei, *Phys. Lett.* **98B**, 265 (1981).

5. Particle Data Group, *Phys. Rev.* **D45**, S1 (1992).

6. E. Kolb and M. Turner (Addison Wesley, California, 1990) and references therein.

7. J. W. F. Valle, *Phys. Lett.* **B131**, 87 (1983).

8. G. Gelmini and J. W. F. Valle, *Phys. Lett.* **B142**, 181 (1984). M. C. Gonzalez-Garcia and J. W. F. Valle, *Phys. Lett.* **B216**, 360 (1989). A. Joshipura, S.Rindani, PRL-TH/92-10.

9. P. Nogueira, J. C. Romao, and J. W. F. Valle, *Phys. Lett.* **B251**, 142 (1990).

10. G. Steigman and M. Turner, *Nucl. Phys.* **B253**, 375 (1985).

11. See, e.g. E. Norman, *these proceedings*.

12. See e.g. J. W. F. Valle, *Phys. Rev.* **D27**, 1672 (1983).

13. J. Schechter and J. W. F. Valle, *Phys. Rev.* **D25**, 2951 (1982); talk by Kayser.

14. G. Gelmini and M. Roncadelli, *Phys. Lett.* **B99**, 411 (1981). H. Georgi, S. Glashow, and S. Nussinov, *Nucl. Phys.* **B193**, 297 (1981).

15. Z. Berezhiani, A. Smirnov, and J. W. F. Valle, *Phys. Lett.* **291B**, 99 (1992).

16. See talks by A. Piepke and M. Nelson.

17. See talks by D. Vignaud and P. Rosen.

18. A. Smirnov, *private communication*.

19. J. C. Romao and J. W. F. Valle, *Phys. Lett.* **B272**, 436 (1991); *Nucl. Phys.* **B381**, 87 (1992).

20. A. Masiero and J. W. F. Valle, *Phys. Lett.* **B251**, 273 (1990). J. C. Romao, C. A. Santos, and J. W. F. Valle, *Phys. Lett.* **B288**, 311 (1992).

21. R. Barbieri, etal *Phys. Lett* **B238**, 86 (1990). M. C. Gonzalez-Garcia and J. W. F. Valle, *Nucl. Phys.* **B355**, 330 (1991).

22. J. C. Romao, N. Rius, and J. W. F. Valle, *Nucl. Phys.* **B363**, 369 (1991).

23. G. F. Smoot et al., and E. L. Wright et al., Berkeley preprints (1992).

24. N. Armenise et al., and P. Astier et al., CERN-SPSC/91-21 and CERN-SPSC/90-42.

25. K. Kodama et al., FNAL proposal P803

26. See e.g. S. Bludman and P. Langacker, *Nucl. Phys.* **B374**, 373 (1992).

27. J. Peltoniemi, D. Tommasini, J. W. F. Valle, CERN-TH.6624/92.

BARYOGENESIS AND NEUTRINO MASSES

R. D. Peccei
Department of Physics
Univ. of California at Los Angeles,
Los Angeles, CA. 90024

Abstract

The erasure of any preexisting $B + L$ asymmetry in the universe in its late stages suggests that the B asymmetry observed today either originated at the electroweak scale or it arose from an original L asymmetry. For the latter case to be viable either neutrino masses are much below the eV scale or the L asymmety itself is generated at an intermediate scale. Several features of the generation of a B asymmetry via an L asymmetry are discussed, including the interesting possibility that the present baryon asymmetry in the universe originates as a result of CP violating phases in the neutrino mass matrix.

THE KRS MECHANISM

The sum of baryon plus lepton number, $B + L$, in the standard model is classically conserved but violated at the quantum level [1] by chiral anomalies [2].

$$\partial_\mu J^\mu_{B+L} = 2N_g [\frac{\alpha_2}{8\pi} W^{\mu\nu}_a \tilde{W}_{a\mu\nu} + \frac{\alpha_1}{8\pi} Y^{\mu\nu} \tilde{Y}_{\mu\nu}] \ . \tag{1}$$

Of particular relevance is the $SU(2)$ anomaly since it implies that changes of $B + L$ in any process are necessarily associated with gauge field configurations of non trivial index ν:

$$\Delta(B + L) = 2N_g \nu \tag{2}$$

where N_g is the number of generations and

$$\nu = \frac{\alpha_2}{2\pi} \int d^4 x W^{\mu\nu}_a \tilde{W}_{a\mu\nu} \ . \tag{3}$$

As was shown by 't Hooft [1], this circumstance, at least semiclassically, leads to a strong suppression of $B+L$ violating processes at zero temperature. This is, however, no longer the case at temperatures T of the order of the electroweak phase transition [3], so that standard model $B+L$ violating processes become of cosmological significance.

't Hooft [1] showed that one can estimate the $B+L$ violating amplitudes in the standard model by focusing on the changes in the gauge vacuum needed to have $\nu \neq 0$. In essence, a transition with $\nu \neq 0$ can be viewed as occuring by tunneling between two vacuum states and is thus suppressed by a tunneling factor

$$A(\nu) \sim e^{-\frac{2\pi}{\alpha_2}\nu} \ . \tag{4}$$

This factor is irrelevantly small for the standard model. However, the situation is different at finite temperatures, since the transition between different vacuum states can occur by thermal fluctuations, rather than by tunneling. This important point was realized a few years ago by Kuzmin, Rubakov and Shaposhnikov [3] who suggested that the rate of $B + L$ violation at finite temperature is determined, instead than by a tunneling factor, by a Boltzman factor

$$A(\nu) \sim e^{-\frac{V_0}{T}\nu} , \qquad (5)$$

with V_0 being the height of the barrier separating the vacuum states. Kuzmin, Rubakov and Shaposhnikov estimated V_0 by semiclassical methods, associating V_0 with the energy of a static field configuration with $\nu = 1/2$ of the standard model - known as a sphaleron - discovered earlier by Klinkhamer and Manton [4]. Because the sphaleron energy decreases with temperatures as one approaches the electroweak phase transition, Kuzmin, Rubakov and Shaposhnikov made the seminal observation that near this transition $B + L$ violating processes are sufficiently rapid compared to the expansion of the universe, so that they are in equilibrium. As a consequence, one should then expect that any preexisting asymmetry in $B + L$ in the universe would get washed out at these temperatures.

The early estimates of Kuzmin, Rubakov and Shaposhnikov [3] of the rate for $B + L$ violation in the universe due to standard model processes have been verified by more complete calculations and extended to temperatures T above that of the electroweak phase transition T_c [5]. The transition probability for $\nu = 1$ $B + L$ violating processes, per unit volume per unit time, $\gamma_{B+L \text{ viol}} = \frac{\Gamma_{B+L \text{ viol}}}{V}$ for T both below and above T_c is found to be

$$\gamma_{B+L \text{ viol}} = C\left[\frac{\pi^6 M_W^7(T)}{\alpha_2^3 T^3}\right]e^{-E_{\text{sph}}(t)/T}, T < T_c$$
$$\gamma_{B+L \text{ viol}} = C'(\alpha_2 T)^4 , \quad T > T_c \qquad (6)$$

where C and C' are constants of $0(1)$ and

$$E_{\text{sph}}(T) = \frac{\pi M_W(T)}{\alpha_2}K(\lambda/g_2) \qquad (7)$$

with K, being a function of $0(1)$ calculated by Klinkhamer and Manton [4].

Using the above formulas, it is easy to check that the rate for $B + L$ violation $\Gamma_{B+L \text{ viol}}$ is faster than the expansion rate of the universe

$$H = \frac{5}{3}(g^*)^{1/2}\frac{T^2}{M_P} \equiv \frac{T^2}{M_0} , \qquad (8)$$

with $M_0 \simeq 10^{18} GeV$, for large periods of the Universe's lifetime:

$$\Gamma_{B+L \text{ viol}} > H$$

$$T_{\min} \sim 10^2 GeV < T < T_{\max} \sim 10^{12} GeV .$$

It is easy to see that if the above obtains, so that $B + L$ violating processes are in equilibrium in the universe, then any $B + L$ asymmetry produced before $T_{\max}$ is erased. In equilibrium one can write for the rate of change of the $B + L$ and $\overline{B + L}$ densities (n and $\bar{n}$, respectively) the formula

$$\frac{d}{dt}\left[n - \bar{n}\right] = \gamma_{B+L \text{ viol}} e^{-\mu/T} - \gamma_{B+L \text{ viol}}e^{+\mu/T}$$
$$\simeq -\frac{2\mu}{T}\gamma_{B+L \text{ viol}} , \qquad (9)$$

where μ is the chemical potential. In the high temperture limit, μ is simply related to the densities themselves

$$n - \bar{n} = \frac{4\mu}{\pi^2}T^2 , \qquad (10)$$

so that one is lead to an exponential dilution of any preexisting asymmetry

$$n - \bar{n} = (\Delta n)_0 \; e^{-\frac{\pi}{2}t\Gamma_{B+L \text{ viol}}} . \qquad (11)$$

OPTIONS FOR THE BARYON ASYMMETRY

Given the presence of $B + L$ erasing processes in the temperature range $10^2 \; GeV < T < 10^{12} \; GeV$, two possibilities appear open to explain the present observed B asymmetry of the universe [$N_B \equiv n_B/s$

$$\simeq n_B/7n_\gamma \simeq (0.6 - 1) \times 10^{-10}[6] \;]:$$

i) The observed asymmetry is generated at the electroweak phase transition. Furthermore, this transition is sufficiently strongly first order so that after the transition the rate of $B+L$ violating processes is already so small that the generated asymmetry is not erased [7].

ii) The observed B asymmetry is a result of some primordial $B+L$ asymmetry or an L asymmetry. In either case, since $B-L$ is not affected by the weak interaction anomalies, any asymmetry in this number density survives to present times, so that the observed B asymmetry is simply related to this primordial asymmetry.

If the observed baryon asymmetry is indeed generated at the electroweak scale, one has the exciting possibility that N_B is computable from "low energy" physics, i.e. from the standard model or simple extensions thereof. This is an extremely active research area, which has been reviewed by D. Brahm[8] in this conference. A number of interesting ideas have been suggested which in principle could lead to $N_B \sim 10^{-10}$ but, to my mind, a really convincing scenario is still lacking. For this reason, it appears sensible to concentrate also on the second option above and I shall try to detail here some of its consequences.

If the B asymmetry is not generated at the electroweak scale, the observed baryon asymmetry is related to some primordial $B-L$ violation. It turns out that the final B asymmetry is not just simply $N_B = \frac{1}{2}(N_{B-L})$, as the trivial equation $B = \frac{1}{2}(B+L) + \frac{1}{2}(B-L)$ would suggest, but has a slightly more complicated form [9]:

$$N_B = \left(\frac{8N_g + 4N_H}{22N_g + 13N_H}\right)(N_{B-L})_{\text{prim}} \quad (12)$$

where N_H is the number of Higgs doublets. At any rate, for the above equation to hold one has to assume that in the epoch between $T_{\max} \simeq 10^{12} GeV$ and $T_{\min} \simeq 10^2 GeV$, where $B+L$ violating processes are in equilibrium in the universe, $B-L$ violating processes must be out of equilibrium

$$\Gamma_{B-L \text{ viol}} < H \quad (13)$$

$$T_{\min} \sim 10^2 GeV < T < T_{\max} \sim 10^{12} GeV,$$

otherwise also $(N_{B-L})_{\text{prim}}$ would be erased. In this temperature range, effectively, any purely lepton number violating processes is also $B-L$ violating. Hence, one must also require that $\Gamma_{L \text{ viol}} < H$ in this temperature range. As noted originally by Fukugita and Yanagida [10], and as will be seen in more detail below, this requirement leads to constraints on neutrino masses.

If the rates for $B-L$ violating (or L violating) processes are **faster** than the universe's expansion rate below $T_{\max}$, it is actually still possible to generate the present day B asymmetry, provided that somewhere above $T_{\min}$ one can generate a new $B-L$ (or L) asymmetry of sufficient magnitude. This asymmetry is then transformed as before into a B asymmetry by the KRS mechanism. Although the conditions for producing a significant $B-L$, or L, asymmetry at intermediate scales are somewhat more challenging, this scenario allows eV neutrino masses and relates in an interesting way the universe's baryon asymmetry to CP violating phases in the neutrino mass matrix. I will return to this option shortly.

BOUNDS ON L VIOLATING INTERACTIONS

If the B asymmetry of the universe is due to some primordial $B-L$ asymmetry generated at a temperature $T > T_{\max}$, it is necessary that no $B - L$ violating processes be fast enough so as to erase this primordial asymmetry below $T_{\max} \simeq 10^{12}$ GeV. However, in theories where $B - L$ is violated at very high scales (GUT or Planck), one expects at lower energies the appearance of $B - L$ violating interactions of dimension $d > 4$. These interactions, if they are not weak enough, could bring $B - L$ violating processes into equilibrium below $T_{\max}$. Thus the condition $\Gamma_{B-L \text{ viol}} < H$ for $T < T_{\max}$ imposes, in general, some constraints on the parameters of the theory.

As pointed out by Fukugita and Yanagida [10], a particularly interesting L-violating interaction term, which is generic to theories where $B-L$ is violated at high scales, is characterized by the effective Lagrangian

$$\mathcal{L}_{\Delta L = 2} = \frac{m_\nu}{v^2} L L \Phi \Phi \qquad (14)$$

where v is the vacuum expectation value of the standard model Higgs doublet Φ. Thus m_ν above is the neutrino Majorana mass matrix associated with the left-handed neutrinos in the lepton doublets L. The above interaction can lead to rapid $B - L$ violating processes, like $\nu\nu \to \Phi\Phi$, below $T_{\max}$. If ν_H is the largest eigenvalue of m_ν, requiring that

$$\Gamma_{\Delta L=2} = < n\sigma(\nu_H \nu_H \to \Phi\Phi) > \simeq \frac{m_{\nu_H}^3}{\pi v^4} T^3 \qquad (15)$$

be slow compared to $H = T^2/M_0$ implies a bound on m_{ν_H} [10][11].

$$m_{\nu_H} < \frac{4 eV}{[T_{\max}/10^{10} GeV]^{1/2}} \qquad (16)$$

Thus for $T_{\max} = 10^{12}$ GeV one has that $m_{\nu_H} < 0.4$ eV.

The above bound can be considerably strengthened in models where one can directly compute the $B - L$, or L violating decays of heavy states [12]. For example, consider the decay of a heavy Majorana neutrino N_R which has a standard Yukawa coupling to the doublet Higgs and the lepton doublets L. Since

$$\mathcal{L}_{\text{Yukawa}} = \lambda \bar{N}_R \Phi L + h \cdot c \cdot \qquad (17)$$

can lead to both the decays $N_R \to \nu_L H$ and $N_R \to \bar{\nu}_L H$ it is necessary that the decay rate of N_R at a temperature $T < T_{\max}$ be less than H. Since λ^2 is related to the left-handed neutrino mass matrix m_ν by the see saw mechanism, requiring that $\Gamma_D/H < 1$ for $T \sim M_N$ implies a strong constraint on m_ν. Focusing again on the largest eigenvalue, one has

$$\frac{\Gamma_D}{H}\bigg|_{T \simeq M_H} \simeq \frac{5}{24\sqrt{2}\pi^2} m_{\nu_H} \frac{M_0}{v^2} < 1 \qquad (18)$$

which yields the bound [12]

$$m_{\nu_H} \lesssim 10^{-3} eV \qquad (19)$$

Similar considerations, and quite analogous analyses, can be used to bound a variety of other $B - L$ violating interactions in various extensions of the standard model. A very thorough discussion of the restrictions on all possible B violating and L violating operators 0_i^d of high dimension ($d = 4 + n$)

$$\mathcal{L}_{\text{viol}} = \sum_i \frac{0_i^d}{M^n}, \qquad (20)$$

both in the standard model and its supersymmetric extension, has been carried out recently by Campbell, Davidson, Ellis and Olive [13]. In general, these authors find that the requirement that these operators should not lead to rapid $B - L$ violating processes in the temperature range between $T_{\max}$ and $T_{\min}$ provides stronger limits on the scales M associated with the various operators 0_i^d than can be provided purely by laboratory experiments. In particular, these considerations lead to very strong limits on the strenth of the $d = 4$ R symmetry violating operators in supersymmetric extensions of the standard model [13].

L ASYMMETRY AT INTERMEDIATE SCALES

If the universe's baryon asymmetry originates from a primordial $B-L$ asymmetry, the above discussion makes it difficult to contemplate the possibility that at least one neutrino has a mass in the eV range - a range which is of interest for the dark matter problem. There appears, however, to be three interesting ways to obviate this conclusion. The first way is simply to believe that the B asymmetry is generated at the electroweak scale, rendering our preceeding discussion moot. The second way, is a somewhat wild, but perfectly consistent, recent speculation of Gelmini and Yanagida [14]. They arrive at eV neutrino masses, not from effective interactions originating from a high scale, but as a result of having vacuum expectation values in the KeV range. In this way they avoid altogether the $B-L$ constraints, but the price they pay is a new hierarchy of VEV's. The third way obtains the B asymmetry through a $B-L$ (or L) asymmetry generated at an intermediate scale. By so doing one avoids the direct constraints on neutrino masses of the preceeding section and one is left with other interesting features which are related to neutrinos. This type of scenario was advocated sometime ago by Fukugita and Yanagida [10] and Langacker, Yanagida and I [16], and has been analyzed recently in a much more thorough manner by Luty [17].

The general idea of the above scenario is to generate an L asymmetry at an intermediate scale ($M_N < 10^{12}\ GeV$) from out of equilibrium processes involving a heavy Majorana right handed neutrino N_R. If the light neutrino to which N_R decays has a mass of $0(eV)$, then our earlier computation indicates that at $T \simeq M_N$, $\Gamma_D \gg H$, so that any primordial L (or $B-L$) asymmetry would be erased. However, eventually for $T < M_N$ a new L asymmetry can be established when the inverse decays $\nu H \to N_R$, $\bar{\nu} H \to N_R$ go out of equilibrium. The amount of lepton asymmetry that one can generate in these circumstances is considerably reduced from what one would expect in the standard "delayed decay" scenario, but the reduction may be tolerable. Roughly [6], one expects

$$N_L \simeq 0.3 \left(\frac{H}{\Gamma_D}\right)_{T=M_N} \cdot (N_L)_{\text{standard}} \ , \quad (21)$$

which in our case would amount to a reduction of around a factor of order 10^{-3}.

Apart from the above thermodynamic reduction factor, the asymmetry N_L is diluted by the number of degrees of freedom at $T \simeq M_N$ ($g^* \sim 100$) and is proportional to the amount of microscopic L violation produced by the decays of N_R into νH and $\bar{\nu} H$. That is,

$$\begin{aligned}(N_L)_{\text{standard}} &\simeq \frac{1}{g^*}\left[\frac{\Gamma(N_R \to \nu H)-\Gamma(N_R \to \bar{\nu} H)}{\Gamma(N_R \to \nu H)+\Gamma(N_R \to \bar{\nu} H)}\right]\\ &= \frac{\epsilon_L}{g^*}\end{aligned} \quad (22)$$

The calculation of ϵ_L requires detailed assumptions on the structure of the neutrino mass matrix, both for the Yukawa couplings λ between N_R, L and Φ and for the masses of the heavy Majorana neutrinos. In particular, ϵ_L depends explictly on CP violating phases in the neutrino sector and it would vanish if CP were conserved. Note also that the first non vanishing contribution for ϵ_L is of $0(\lambda^4)$, since the rates for $N_R \to \nu H$ and $N_R \to \bar{\nu} H$ are identical in lowest order.

Fukugita and Yanagida [10], and more recently Luty [17], have estimated ϵ_L by retaining the leading Yukawa coupling and assuming a simple hierarchical structure for the heavy Majorana states. With these approximations one obtain

$$\epsilon_L \simeq \frac{m_{D_3}^2}{\pi v^2} \cdot \frac{M_{N_2}}{M_{N_3}} \cdot \sin\delta \ , \quad (23)$$

where the three terms above represent, respectively, the leading Yukawa coupling contribution, some structure in the Majorana mass matrix and a typical CP violating phase. With this formula in hand, it is not impossible to imagine that one could obtain values for ϵ_L in the range of $10^{-4} - 10^{-6}$. Such values would allow a value of $N_L \sim 10^{-10}$ to be generated at an intermediate temperature $T_{min} < M_N < T_{max}$. By the KRS mechanism such a leptonic asymmetry would then generate a baryonic asymmetry N_B of the desired order of magnitude. These rough estimates are confirmed by the results of the recent detailed calculations of Luty [17], in which he uses the Boltzman equation to study the evolution of N_B and includes the effects of 2 to 2 processes. From his calculations it appears that, without stretching parameters to much, one can produce both a baryon asymmetry $N_B \sim 10^{-10}$ today and a 3rd generation neutrino with an eV mass.

CONCLUDING REMARKS

Let me reiterate the main points discussed. First and foremost, as discovered by Kuzmin, Rubakov and Shaposhnikov, in the standard model there exist rapid $B+L$ violating processes which are in equilibrium in the early universe. This circumstance has important implications for baryogenesis. Because of the KRS mechanism, either one must suppose that the universe's baryon asymmetry is generated at the electroweak scale, or that this asymmetry arises via the transmutation of some $B - L$ or L asymmetry into a B asymmetry. In this latter case, either neutrinos are very light ($m_{\nu_H} < 10^{-3} eV$) allowing some primordial ($B - L$) asymmetry to survive to become the observed B asymmetry, or an L or ($B - L$) asymmetry is generated at some intermediate scale ($M_N \sim 10^{10} GeV$), eventually transmuting itself into the observed B asymmetry. In the last scenario, eV neutrinos are permissible and one has the amusing result that the universe's baryon asymmetry is related to CP phases in the neutrino sector. However, it does not appear that all these CP phases are observable, even in principle, in neutrino oscillations. They should, however, enter in CP violating phenomena involving neutrino - antineutrino oscillations. Unfortunately, neutrino-antineutrino oscillations are suppressed by chirality factors, which renders them only an academic curiosity [18].

ACKNOWLEDGEMENTS

I am grateful to Graciela Gelmini for numerous helpful conversations on this subject matter. I would also like to thank Marcela Carena, Boris Kayser and Sandip Pakvasa for illuminating remarks on CP violating phases in the neutrino sector. I am thankful to V. Zakharov for his hospitality at the Max Planck Institute, where this note was written up. This work was partially supported by the Department of Energy under the contract DE-FG03-91ER40662, TASK C.

REFERENCES

1. G. 't Hooft, *Phy. Rev. Lett*, 37, (1976) 8

2. S. L. Adler, *Phys. Rev.* 177, (1969) 2426; J. S. Bell and R. Jackiw, *Nuovo Cimento* 60A (1969) 49; W. A. Bardeen, *Phys. Rev* 184 (1969) 1841

3. V. A. Kuzmin, V. A. Rubakov and M. E. Shaposhnikov, *Phys. Lett* 166B (1985) 36

4. R. F. Klinkhamer and N. S. Manton, *Phys. Rev.* D30 (1984) 3212

5. P. Arnold and L. McLerran, *Phys.*

Rev. D36 (1987) 581; *Phys. Rev.* D37, (1988) 1020; S. Yu Khlebnikov and M. E. Shaposhnikov, *Nucl. Phys.* B308 (1938) 885; M. Dine, O. Lechtenfeld, B. Sakita, W. Fischler and J. Polchinski, *Nucl. Phys.* B342 (1990) 381

6. See for example, E. W. Kolb and M. S. Turner, *The Early Universe.* (Addison, Wesley, Reading, Mass. 1989)

7. M. E. Shaposhnikov, *Nucl. Phys.* B287 (1987) 757; B299 (1988) 797

8. D. Brahm, these Proceedings

9. S. Yu. Khlebnikov and M. E. Shaposhnikov, *Nucl. Phys* B308 (1988) 835; J. Harvey and M. Turner, *Phys. Rev.* D42 (1990) 3344; A. Nelson and S. Barr, *Phys. Lett.* B246 (1990) 141

10. M. Fukugita and T. Yanagida, *Phys. Rev.* D42 (1990) 1285

11. J. Harvey and M. S. Turner, *Phys. Rev.* D42 (1990) 3344; A. Nelson and S. Barr, *Phys. Lett.* B246 (1990) 141

12. W. Fischler, G. F. Giudice, R. G. Leigh and S. Paban, *Phys. Lett.* B258 (1991) 45

13. B. Campbell, S. Davidson, J. Ellis and K. Olive, *Phys. Lett.* B256 (1991) 457

14. G. Gelmini and T. Yanagida, UCLA preprint UCLA/92/TEP/24

15. M. Fukugita and T. Yanagida, *Phys. Lett.* 174B (1986) 45

16. P. Langacker, R. D. Peccei and T. Yanagida, *Mod. Phys. Lett.* A1 (1986) 541

17. M. Luty, *Phys. Rev.* D45 (1992) 455

18. B. Kayser, in *CP Violation*, ed. by C. Jarlskog (World Scientific, Singapore, 1988)

STRINGY ORIGIN OF NEUTRINO MASSES WITHIN THE MINIMAL SUPERSYMMETRIC STANDARD MODEL

Mirjam Cvetič
Department of Physics
University of Pennsylvania
Philadelphia, PA 19104-6396

Abstract

We present a "gravity-induced" seesaw mechanism, which accommodates neutrino masses ($m_\nu \propto m_u^2/M_I$, with m_u the corresponding quark mass and $M_I \simeq 4 \times 10^{11}$ GeV) compatible with the MSW study of the Solar neutrino deficit within the minimal supersymmetric Standard Model (the grand desert with the gauge coupling unification at $M_U \simeq 2 \times 10^{16}$ GeV). We show that for large radius ($R^2/\alpha' = \mathcal{O}(20)$) Calabi-Yau spaces, threshold corrections ensure $M_U^2 = M_C^2/\mathcal{O}(2R^2/\alpha')$ and the magnitude of the non-renormalizable terms in the superpotential yields $M_I = \mathcal{O}(e^{-R^2/\alpha'})M_C$. Here $M_C = g \times 5.2 \times 10^{17}$ GeV is the scale of the tree level (genus zero) gauge coupling (g) unification.

Precise data from the LEP experiments indicate that the gauge couplings of the Standard Model meet at $M_U \simeq (1-4) \times 10^{16}$ GeV in the minimal supersymmetric extension of the Standard Model.[1] Another set of intriguing data arise from the Solar neutrino experiments. The deficit of Solar neutrinos can most efficiently be explained through the MSW[2] mechanism of matter-enhanced neutrino oscillations. In particular, current data favor[3] the mass splitting of the electron and muon neutrinos to be $\Delta m^2 \equiv |m_{\nu_\mu}^2 - m_{\nu_e}^2|$ in the range $(1-16) \times 10^{-7}$eV^2. In the grand unified gauge theories (GUT's) the seesaw scenario[4] the light neutrino masses are given by: $m_{\nu_{e,\mu}} \simeq c\, m_{u,c}^2/M_I$, where $m_{u,c}$ are the corresponding quark masses and $c \simeq 0.05 - 0.09$ is a factor due to the renormalization down to the low energy scale[3]. This implies that $M_I \simeq (4 \pm 3) \times 10^{11}$ GeV, the central value corresponding to the neutrino mixing angle $\theta_{\nu_\mu \nu_e} \sim \theta_{Cabibbo}$.

Each of the two sets of experimental data has therefore an elegant theoretical explanation. Unfortunately, the two theoretical models are mutually exclusive at first glance. Here, we present a scenario[5] which implements the neutrino masses in the minimal supersymmetric Standard Model in such a way that there is still a grand desert with the gauge coupling unification at $M_U \sim (1-4) \times 10^{16}$GeV, while the effective scale M_I governing the neutrino masses is in the range of $(4 \pm 3) \times 10^{11}$ GeV. We are proposing a "gravity-induced"[6, 7] seesaw mechanism (an extension of a mechanism proposed by Nandi and Sarkar[8] within the context Calabi-Yau vacua with E_6 gauge group[9]), realized through an interplay between the nonrenormalizable and renormalizable terms in the superpotential, as the origin of the neutrino masses.

The essence of the idea is based on a supersymmetric theory with an extended gauge symmetry, which contains an additional sterile neutrino (a Standard Model singlet), and a *restricted* representation of the Higgs fields. Such fields break the extended gauge symmetry at the scale M_U. However, they cannot give the sterile neutrino a large Majorana mass proportional to M_U through the renormalizable (cubic) terms in the superpotential. On the other hand, through the nonrenormalizable (*e.g.*, quartic) terms of the superpotential, which are suppressed by a scale $M_{NR} > M_U$, such Higgs fields can give a Majorana mass of order $M_U^2/M_{NR} < M_U$.

One can demonstrate[5] the gravity-induced seesaw in an explicit (minimal) model with all the essential features, by choosing an enhanced gauge symmetry, $SU(3)_C \times SU(2)_L \times U(1)_Y \times U(1)_{Y'}$, where Y is the ordinary weak hypercharge. The matter consists of the particle content of the minimal Standard Model as well as of the Standard Model singlets, L_i, S_1 and $\bar{S}_1$. L_i supermultiplets with $i = (1,2,3)$ contain a sterile neutrino which accompanies each of the three families. S_1 and $\bar{S}_1$ contain the Higgs fields which break the enhanced gauge symmetry with VEV's of order M_U.

Consistent with the anomaly constraint we choose the following values for the Y' charges: quark $SU(2)_L$ doublets, u_L^c quarks and e_L^c leptons have (-1), lepton doublets and d_L^c quarks have (+3), L_i and S_1 have (-5), $\bar{S}_1$ has (+5), while Higgs doublets $H_{(1,2)}$ have (-2) and (+2), respectively.[10] In the neutrino sector, the only renormalizable terms allowed in the superpotential are of the type $W = L_i \nu_i H_2$ which yields the neutrino mass matrix:

$$\begin{bmatrix} 0 & m \\ m & 0 \end{bmatrix}, \quad (1)$$

where m is proportional to the VEV of the Higgs doublet H_2. Since H_2 gives mass to the quarks as well, m is of the order of the corresponding quark masses. On the other hand, the only allowed nonrenormalizable term in the superpotential with a leading contribution to the neutrino mass matrix is of the type $W_{NR} = L_i L_i \bar{S}_1 \bar{S}_1 / M_{NR}$. This modifies the neutrino mass matrix:

$$\begin{bmatrix} 0 & m \\ m & M_I \end{bmatrix}, \quad (2)$$

where $M_I = M_U^2/M_{NR}$. The quantum numbers prevent the contribution of any nonrenormalizable term to the $\nu\nu$ and νL masses that would be of the order of M_U^K/M_{NR}^{K-1} for any $K > 1$.

While the above scenario is appealing on its own terms, its origin can be motivated from the properties of superstring vacua, which provide a natural framework for the restricted representation of the chiral supermultiplets. This can be exhibited in explicitly[8, 5] in the case of E_6 gauge group and the restricted representation (**27**'s and $\overline{\mathbf{27}}$'s of E_6) of the matter supermultiplets.

However, the requirement of the minimal supersymmetric Standard Model particle content below $M_U \simeq (1-4) \times 10^{16}$ GeV imposes severe constraints on the allowed superstring vacua. (2,2) string vacua, *e.g.*, Calabi-Yau manifolds with gauge and spin connection identified, possess a large number of additional multiplets. In particular, for vacua without Wilson lines, the gauge group is E_6, with **27**'s, $\overline{\mathbf{27}}$'s, and **1**'s of E_6. Some of the particles in these multiplets acquire large masses if there are flat directions in the space of specific string vacua. Finding flat directions allows one to give large VEV's to fields in a particular set of **27**'s and $\overline{\mathbf{27}}$'s, which in turn can give mass to some of the unwanted massless multiplets. At the same time, E_6 is broken down to $SO(10)$ or $SU(5)$. Such directions were found for orbifolds[11, 12] as well as for a class of Calabi-Yau manifolds[13] based on Gepner's[14] construction.

In addition, Wilson lines allow for a breakdown of the simple gauge group (E_6, $SO(10)$, or $SU(5)$) to a direct product of simple groups and $U(1)$'s. It is in general possible[15, 13] to introduce Wilson lines which break the gauge group down to the Standard Model. At the same time, this procedure decouples a large number of unwanted modes.

Thus, a viable scenario is to construct (2,2) string vacua with flat directions as well as Wilson lines. However, in spite of the progress made in the construction of such string vacua there exists no explicit example of a supersymmetric string vacuum which would contain only the minimal Standard Model particle spectrum below the gauge coupling unification scale. On the other hand, string theory can shed light on the scale of the gauge coupling unification M_U, and the magnitude the non-renormalizable terms in the superpotential in a quantitive manner.

While string vacua in general do not possess gauge group grand unification, there is a notion gauge coupling unification. The scale M_C associated with the gauge coupling unification at the tree level of the string theory is determined[16] in the $\overline{DR}$ scheme by the value of the Planck mass M_{Pl} and of the gauge coupling g in the following way:

$$\begin{aligned} M_C &= \frac{e^{(1-\gamma)/2}\sqrt{2}}{3^{3/4}\sqrt{\pi\alpha'}} = g \times \frac{e^{(1-\gamma)/2}}{3^{3/4}4\pi}M_{Pl} \\ &= g \times 0.043 M_{Pl} \\ &= g \times 5.2 \times 10^{17} \text{ GeV}. \end{aligned} \quad (3)$$

where $\gamma = 0.57722$ is the Euler constant, $g^2 = 32\pi/(\alpha' M_{Pl}^2)$, with g defined according to the GUT convention and $M_{Pl} = 1.2 \times 10^{19}$ GeV. For the expected value $g \sim 0.7$ this is one order of magnitude too large compared with $M_U \sim (1-4) \times 10^{16}$ GeV, which is the scale of the gauge coupling unification of the minimal supersymmetric Standard Model. However, threshold effects, i.e., genus one corrections to the gauge couplings, can split the gauge couplings at M_C, thus in principle allowing for an effective unification scale $M_U < M_C$. The part of the threshold corrections which does not depend on the vacuum expectation values (VEV's) of fields is small and at most few percent (i.e. $\mathcal{O}(1)/16\pi^2$) correction to the tree level gauge coupling.[16] This result is to be expected; heavy modes should not affect the evolution of gauge couplings at scales much lower than M_{Pl}.[17] On the other hand, field dependent threshold corrections could be large; namely, massless fields which could acquire VEV's (i.e. of the order of M_U) can drastically affect the threshold corrections.

In (2,2) string vacua, there exist massless chiral superfields, moduli (T_i) which are in one to one correspondence with the matter fields, i.e., **27**'s and $\overline{\mathbf{27}}$'s of E_6. Since moduli do not have a potential to all orders in string loops, their VEV's parametrize a whole class of string vacua. Explicit calculations of the moduli dependence for the threshold corrections for a class of orbifolds are given in Refs. [16, 18]. Extensive study[19] of threshold corrections in orbifolds indicate that $M_U < M_C$ if the massless spectrum possesses certain modular weights which should also be compatible with the target space one-loop modular anomaly. In such examples one would obtain $M_U = \mathcal{O}(e^{(-cR^2/\alpha')})M_C$ when the orbifold radius is large ($R^2/\alpha' \gg 1$). The positive coefficient c depends[19] on the modular weights of the massless states. As we shall see later, the heavy Majorana mass turns out to be $M_I = \mathcal{O}(e^{-c'R^2/\alpha'})M_C$. In order to ensure $M_U \sim 10^{16}$ GeV and $M_I \sim 10^{12}$ GeV , this in turn involves detailed constraints on coefficients c and c'.

In the following, we shall pursue a different approach, i.e., study of smooth Calabi-Yau spaces. Calabi-Yau spaces are related to the underlying solvable (2, 2) conformal field theory (CFT) (e.g. orbifolds,[20] and Gepner's models[14]) by taking the VEV's of *all* the moduli $\langle T_i \rangle \gg 1$. The internal space is then large and smooth everywhere and one can reliably use (e.g., blown-up orbifolds[21]) the $1/\langle T_i \rangle$ expansion of the point field limit approach, thus obtaining qualitatively different results.[22, 23] A large Calabi-Yau space asymptotically approaches the untwisted sector of the orbifolds, i.e. it is a smooth, almost flat space. Thus, the massive modes (Kaluza-Klein states) contributing to the threshold corrections are those of $N = 4$ sector (for each vector supermultiplet there are three chiral superfields) and therefore their contributions go to zero with the inverse powers of the moduli VEV's.[22] The leading threshold corrections therefore arise only from the light fields and they can be calculated using a field theory one-loop calculation.[24] Such corrections are milder, logarithmic in the VEV's of moduli.

The fields theory calculation for the (2,2) vacua without an enhanced gauge symmetry yields[25] the following dependence on the T_i moduli associated with the scaling deformations ((1,1) forms)):

$$\Delta \left(\frac{16\pi^2}{g_{E_6}^2}\right) = \frac{1}{3} b_{E_6} K_1$$
$$+ T(27) \left[2 \log \det \left(\frac{\partial^2 K_1}{\partial T_a \partial T_b^*}\right) - \frac{4}{3} b_{(1,1)} K_1 \right]. \quad (4)$$

Here $b_{E_6} = C_{E_6} - b_{(1,1)} T(27) - b_{(1,2)} T(\overline{27})$. is related to the one-loop $N = 1$ beta function $\beta_{E_6} = b_{E_6} g^3/16\pi$ with C_{E_6} corresponding to the quadratic Casimir operator of E_6 gauge group and $T(27)$, $T(\overline{27})$ are $tr(Q^2_{(27,\overline{27})})$, respectively. Note, that there are $b_{(1,1)}$ ($b_{(1,2)}$) **27**'s (**$\overline{27}$**'s), in one to one correspondence with the number of moduli associated with the scaling deformations (complex structure deformations). The Kähler potential for the moduli $T_i \gg 1$ is of the form $K_1 = \log \sum_{i,j,k=1}^{b_{(1,1)}} (T_i + T_i^*) (T_j + T_j^*) (T_k + T_k^*) h_{ijk}$ where h_{ijk} are intersection numbers of (1,1) forms. Taking $\langle T_i \rangle = \alpha_i T$ with $T \gg 1$ corresponding to an overall size of the manifold and $\alpha_i = \mathcal{O}(1)$ one finds:[22, 23]

$$\Delta \left(\frac{16\pi^2}{g_{E_{6,8}}^2}\right) = -b_{E_{6,8}} [\log(T + T^*) + \mathcal{O}(\log \alpha_i)] \quad (5)$$

which are proportional to the $N = 1$ beta function $\beta_{E_{(6,8)}} = b_{E_{(6,8)}} g^3/16\pi$ of E_6 (E_8) gauge groups. This implies that while the slope of the running gauge couplings is not changed, the effective gauge coupling unification scale is lowered:

$$M_U^2 = \frac{M_C^2}{(T + T^*)} = \mathcal{O}(\frac{M_C^2}{2R^2/\alpha'}) \quad (6)$$

The above results apply only to simply connected Calabi-Yau spaces. The gauge group E_6 can be broken if Wilson lines are introduced. The study of threshold corrections in this case is under way.[26] We proceed under the assumption that the same features persist also in this case.

From eq. (6) one then sees that for $R^2/\alpha' \sim \mathcal{O}(20)$ the gauge unification scale is lowered to $M_U \sim 6 \times 10^{16}$ GeV, which is slightly too large. However, eq. (6) relates M_U^2 to R^2/α' only by orders of magnitude. Thus, an additional factor of 2 in the relation of an overall modulus ReT to R^2/α' enables one to obtain M_U in the preferred range 4×10^{16} GeV.

We turn now address the size of the nonrenormalizable terms. In string theory the magnitude of the coefficient M_{NR} is proportional to M_C. However, one can prove explicitly[28] that for all $(0, 2)$ string vacua the nonrenormalizable terms are suppressed by an additional factor $e^{-R^2/\alpha'}$, i.e., the origin of the nonrenormalizable terms is due only to worldsheet instanton effects. This is a stringy result, proven explicitly on (blown-up) orbifolds[27] as well as in sigma model perturbations[28] of Calabi-Yau manifolds. Therefore:

$$\frac{1}{M_{NR}} = \frac{\mathcal{O}(e^{-R^2/\alpha'})}{M_C}. \quad (7)$$

By choosing vacuum expectation values along the flat direction to be M_C (the only natural scale in the four-dimensional string vacuum) nonrenormalizable terms yield the heavy Majorana mass:

$$M_I = \frac{M_C^2}{M_{NR}} = \mathcal{O}(e^{-R^2/\alpha'})M_C. \qquad (8)$$

It follows from eq. (6) that we need $R^2/\alpha' = \mathcal{O}(20)$ in order to achieve $M_U \sim 10^{16}$ GeV. In this case $M_I \sim 10^{-8} M_C \sim 10^{10}$ GeV. Although these are only order of magnitude statements, it is instructive to set the coefficients in eqs. (6) and (8) equal to unity. In that case the range $M_I \sim (4 \pm 3) \times 10^{11}$ GeV suggested by the Solar neutrino deficit implies $R^2/\alpha' \sim (13-15)$, yielding a slightly too large $M_U \sim 7 \times 10^{16}$ GeV.

To summarize, the desired scale of the gauge coupling unification $M_U \sim 2 \times 10^{16}$ GeV and the scale of Majorana neutrino masses $M_I \sim 4 \times 10^{11}$ GeV, may be achieved in a class of large radius ($R^2/\alpha' = \mathcal{O}(20)$) Calabi-Yau spaces allowing for $M_U^2 = M_C^2/\mathcal{O}(2R^2/\alpha')$ and $M_I = \mathcal{O}(e^{-R^2/\alpha'})M_C$.

A major part of the presented work has been done in collaboration with P. Langacker. This research was supported in part by the U.S. DOE Grant DE-AC02-76-ERO-3071, and by a junior faculty SSC fellowship (M.C.).

REFERENCES

1. P. Langacker and M. Luo, *Phys. Rev.* **D44**, 817 (1991); U. Amaldi, W. de Boer, and H. Furstenau, *Phys. Lett.* **260B**, 447 (1991); J. Ellis, S. Kelley, and D. V. Nanopoulos, *Phys. Lett.* **249B**, 441 (1990).

2. S. P. Mikheyev and A. Yu. Smirnov, *Yad. Fiz.* **42**, 1441 (1985) (*Sov. J. Nucl. Phys.* **42**, 913 (1985)); *Nuo. Cim.* **9C**, 17 (1986); L. Wolfenstein, *Phys. Rev.* **D17**, 2369 (1968); **D20**, 2634 (1979).

3. For a recent discussion, see S. Bludman, D. Kennedy, and P. Langacker, *Phys. Rev.* **D45**, 1810 (1992), and *Nucl. Phys.*, **B374**, 373; S. Bludman, D. Kennedy, Hata and P. Langacker, UPR- 516-92 (July 1992).

4. M. Gell-Mann, P. Ramond, and R. Slansky, in *Supergravity*, ed. F. van Nieuwenhuizen and D. Freedman (North Holland, Amsterdam 1979) p. 315; T. Yanagida, *Prog. Th. Phys.* **B135**, 66 (1978).

5. M. Cvetič and P. Langacker, UPR-505-T (May 1992), *Phys. Rev.* **D** in press.

6. R. Barbieri, J. Ellis and M. K. Gaillard, *Phys. Lett.* **90B**, 249 (1980).

7. For related scenarios, primarily concerned with the 17 KeV neutrino see: E. Akhmedov, Z. Berezhiani and G. Senjanović, ICTP preprint IC/92/79 and D. Grasso, M. Lusignoli, and M. Roncadelli, U. of Rome preprint FNT/T 92/07.

8. S. Nandi and U. Sarkar, *Phys. Rev. Lett.* **56**, 564 (1986).

9. In the context superstring models the neutrino masses were also studied by R. Mohapatra, *Phys. Rev. Lett.* **56**, 561 (1986), and R. Mohapatra and J. Valle, *Phys. Rev.* **34**, 1642 (1986).

10. The choice of the particle spectrum and the $U(1)_{Y'}$ quantum numbers are motivated by $SO(10)$ gauge symmetry.

11. M. Cvetič, *Phys. Rev. Lett.* **59**, 2829 (1987).

12. A. Font, L. Ibañez, H. P. Nilles, and F. Quevedo, *Nucl. Phys.* B **307**, 105 (1988).

13. B. Greene, *Phys. Rev.* **D40**, 1645 (1989).

14. D. Gepner, *Phys. Lett.* **199B**, 380 (1987).

15. A. Font, L. Ibañez, F. Quevedo and A. Sierra, *Nucl. Phys.* **B331**, 421(1990).

16. V. Kaplunovsky, *Nucl. Phys.* **B307**, 145 (1988) and erratum.

17. S. Weinberg, *Phys. Lett.* **B91**, 51 (1980).

18. L. Dixon, V. Kaplunovsky, and J. Louis, *Nucl. Phys.* **B355**, 649 (1991).

19. L. Ibañez, D. Lüst, and G. Ross, *Phys. Lett.* **B272**, 251 (1991); L. Ibañez and D. Lüst, CERN preprint, CERN-TH.6380/92.

20. L. Dixon, J. Harvey, C. Vafa, and E. Witten, Nucl. Phys. B**274**, 285 (1986).

21. M. Cvetič, in Proceedings of "Superstrings, Cosmology, and Composite Structures", College Park, Maryland, March 1987, S. J. Gates and R. Mohapatra eds. (World Scientific 1987).

22. L. Dixon, unpublished.

23. M. Cvetič, UPR–489–T, September 1991 (unpublished).

24. J. Louis, in the Particles, Strings and Cosmology PASCOS91 (P. Nath and S. Reucroft eds.), World Scientific 1991; G. Lopez-Cardoso and B. Ovrut, in Strings and Symmetries 1991, (P. van Nieuwenhuizen et al. eds.), World Scientific 1991, Nucl.Phys. **B369**, 351 (1992) and UPR-502-T (May 1992); J.P. Derendinger, S. Ferrara, C. Kounnas and F. Zwirner, Nucl. Phys. **B372**, 145 (1992).

25. L. Dixon, unpublished; see also S. Ferrara, C. Kounnas, D. Lüst and F. Zwirner, Nucl. Phys. **365**, 431 (1991).

26. M. Cvetič and J. Erler, work in progress.

27. M. Cvetič, *Phys. Rev. Lett.* **59**, 1795 (1987).

28. M. Cvetič, *Phys. Rev.* **D 37**, 2366 (1987); M. Dine and C. Lee, *Phys. Lett.* **203B**, 371 (1988).